SEVENTH EDITION

Microbiology

A HUMAN PERSPECTIVE

Nester
Anderson
Roberts

LearnSmart™

McGraw-Hill LearnSmart™ is ~~~~~~~~~~~~~~~~~~ ~~vered by Connect Microbiology. Learn~~~~~~~ ~~sed on artificial intelligence and constantly assesses a student's knowledge of the course material. Sophisticated diagnostics adapt to each student's individual knowledge base in order to match and improve what they know. Students actively learn the required concepts more easily and efficiently.

Try It Now at **www.mhlearnsmart.com.**

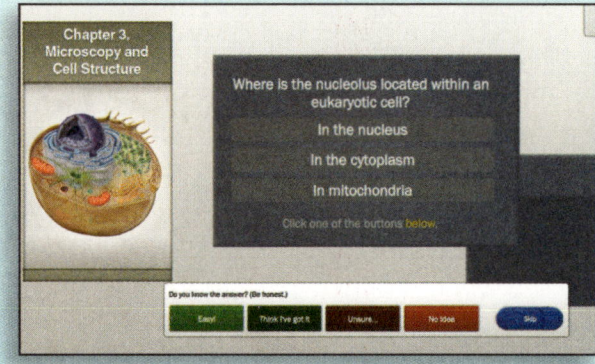

Do More

McGraw-Hill Higher Education and Blackboard® have teamed up! What does this mean for you?

Your life, simplified. Now you and your students can access McGraw-Hill's Connect and Create™ right from within your Blackboard course—all with one single sign-on! Say goodbye to the days of logging in to multiple applications.

Deep integration of content and tools. Not only do you get single sign-on with Connect and Create, you also get deep integration of McGraw-Hill content and content engines right in Blackboard. Whether you're choosing a book for your course or building Connect assignments, all the tools you need are right where you want them—inside of Blackboard.

Seamless gradebooks. Are you tired of keeping multiple gradebooks and manually synchronizing grades into Blackboard? We thought so. When a student completes an integrated Connect assignment, the grade for that assignment automatically (and instantly) feeds your Blackboard grade center.

A solution for everyone. Whether your institution is already using Blackboard or you just want to try Blackboard on your own, we have a solution for you. McGraw-Hill and Blackboard can now offer you easy access to industry leading technology and content, whether your campus hosts it, or we do. Be sure to ask your local McGraw-Hill representative for details.

SEVENTH EDITION

Microbiology
A Human Perspective

Eugene W. Nester
UNIVERSITY OF WASHINGTON

Denise G. Anderson
UNIVERSITY OF WASHINGTON

C. Evans Roberts, Jr.
UNIVERSITY OF WASHINGTON

Martha T. Nester

Mc
Graw
Hill

*Connect
Learn
Succeed*™

MICROBIOLOGY: A HUMAN PERSPECTIVE, SEVENTH EDITION

Published by McGraw-Hill, a business unit of The McGraw-Hill Companies, Inc., 1221 Avenue of the Americas, New York, NY 10020. Copyright © 2012 by The McGraw-Hill Companies, Inc. All rights reserved. Printed in the United States of America. Previous editions © 2009, 2007, and 2004. No part of this publication may be reproduced or distributed in any form or by any means, or stored in a database or retrieval system, without the prior written consent of The McGraw-Hill Companies, Inc., including, but not limited to, in any network or other electronic storage or transmission, or broadcast for distance learning.

Some ancillaries, including electronic and print components, may not be available to customers outside the United States.

This book is printed on acid-free paper.

5 6 7 8 9 0 QVS/QVS 1 0 9 8 7 6 5 4

ISBN 978–0–07–337531–1
MHID 0–07–337531–4

Vice President, Editor-in-Chief: *Marty Lange*
Vice President, EDP: *Kimberly Meriwether David*
Senior Director of Development: *Kristine Tibbetts*
Publisher: *Michael S. Hackett*
Sponsoring Editor: *Lynn M. Breithaupt*
Senior Developmental Editor: *Fran Simon*
Director of Digital Content Development: *Barbekka Hurtt, PhD*
Digital Project Manager: *Amber M. Bettcher*
Marketing Manager: *Amy L. Reed*
Project Manager: *Mary Jane Lampe*
Senior Buyer: *Laura Fuller*
Senior Media Project Manager: *Tammy Juran*
Senior Designer: *David W. Hash*
Cover/Interior Designer: *Elise Lansdon*
Cover Image: *Enterococcus faecalis Bacteria. SEM X3000. © Dennis Kunkel Microscopy, Inc./Visuals Unlimited, Inc.*
Senior Photo Research Coordinator: *John C. Leland*
Photo Research: *David Tietz/Editorial Image, LLC*
Compositor: *Electronic Publishing Services Inc., NYC*
Typeface: *10/12 Times LT Std*
Printer: *Quad/Graphics*

All credits appearing on page or at the end of the book are considered to be an extension of the copyright page.

Library of Congress Cataloging-in-Publication Data

Nester, Eugene W.
  Microbiology : a human perspective / Eugene Nester, Denise Anderson, C. Evans Roberts, Jr. ; contributors, Deborah Allen, Sarah Salm. — 7th ed.
       p. ; cm.
  Includes index.
  ISBN 978–0–07–337531–1 — ISBN 0–07–337531–4 (hard copy : alk. paper) 1. Microbiology. I. Anderson, Denise G. (Denise Gayle). II. Roberts, C. Evans. III. Title.
  [DNLM: 1. Microbiological Techniques. 2. Communicable Diseases—microbiology. QW 4]
  QR41.2.M485 2012
  616.9'041—dc22
                                                             2011010352

www.mhhe.com

Brief Contents

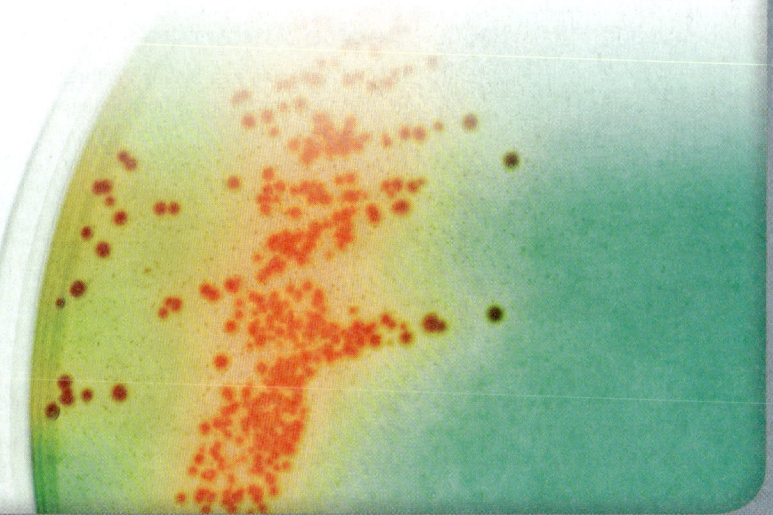

About the Authors

Eugene Nester

Eugene (Gene) Nester performed his undergraduate work at Cornell University and received his Ph.D. in microbiology from Case Western University. He then pursued post-doctoral work in the Department of Genetics at Stanford University with Joshua Lederberg. Since 1962, Gene has been a faculty member in the Department of Microbiology at the University of Washington. Gene's research has focused on gene transfer systems in bacteria. His laboratory demonstrated that *Agrobacterium* transfers DNA into plant cells, the basis for the disease crown gall. He continues to study this unique system of gene transfer, which has become a cornerstone of plant biotechnology.

In 1990, Gene Nester was awarded the inaugural Australia Prize along with an Australian and a German scientist for their work on *Agrobacterium* transformation of plants. In 1991, he was awarded the Cetus Prize in Biotechnology by the American Society of Microbiology. He has been elected to Fellowship in the National Academy of Sciences, the American Academy for the Advancement of Science, the American Academy of Microbiology, and the National Academy of Sciences in India. Throughout his career, Gene has been actively involved with the American Society for Microbiology in several leadership positions.

In addition to his research activities, Gene has taught an introductory microbiology course for students in the allied health sciences for many years. He wrote the original version of the present text, *Microbiology: Molecules, Microbes and Man,* with C. Evans Roberts, Brian McCarthy, and Nancy Pearsall more than 30 years ago because they felt no suitable text was available for this group of students. The original text pioneered the organ system approach to the study of infectious disease.

Gene enjoys traveling, museum hopping, and the study and collecting of Northwest Coast Indian Art. He and his wife, Martha, live on Lake Washington with their labradoodle, Twana, and a well-used kayak. Their two children and four grandchildren live in the Seattle area.

Denise Anderson

Denise Anderson is a Senior Lecturer in the Department of Microbiology at the University of Washington, where she teaches a variety of courses including general microbiology, recombinant DNA techniques, medical bacteriology laboratory, and medical mycology/parasitology laboratory. Equipped with a diverse educational background, including undergraduate work in nutrition and graduate work in food science and in microbiology, she first discovered a passion for teaching when she taught microbiology laboratory courses as part of her graduate training. Her enthusiastic teaching style, fueled by regular doses of Seattle's famous caffeine, receives high reviews by her students.

Outside of academic life, Denise relaxes in the Phinney Ridge neighborhood of Seattle, where she lives with her husband, Richard Moore, and dog, Dudley (neither of whom are well trained). When not planning lectures, grading papers, or writing textbook chapters, she can usually be found chatting with the neighbors, fighting the weeds in her garden, or enjoying a fermented beverage at the local pub.

C. Evans Roberts, Jr.

Evans Roberts was a mathematics student at Haverford College when a chance encounter landed him a summer job at the Marine Biological Laboratory in Woods Hole, Massachusetts. There, interactions with leading scientists awakened an interest in biology and medicine. After finishing his degree at Haverford, he went on to get an M.D. degree at Columbia University College of Physicians and Surgeons, complete an internship at University of Rochester School of Medicine and Dentistry, and a residency in medicine at University of Washington School of Medicine, where he also completed a fellowship in infectious diseases under Dr. William M. M. Kirby, and a traineeship in diagnostic microbiology under Dr. John Sherris.

Subsequently, Dr. Roberts taught microbiology at the University of Washington, University of Oregon, and Chiang Mai University, in Chiagmai, Thailand, returning to the University of Washington thereafter. He has directed diagnostic medical microbiology laboratories, served on hospital infection control committees, and taught infectious diseases to nurse practitioners in a camp for Karen refugees in Northern Thailand. He has had extensive experience in the practice of medicine as it relates to infectious diseases. He is certified both by the American Board of Microbiology and the American Board of Internal Medicine.

Evans Roberts worked with Gene Nester in the early development of *Microbiology: A Human Perspective*. His professional publications concern susceptibility testing as a guide to treatment of infectious diseases, etiology of Whipple's disease, group A streptococcal epidemiology, use of fluorescent antibody in diagnosis, bacteriocin typing, antimicrobial resistance in gonorrhea and tuberculosis, Japanese B encephalitis, and rabies. For relaxation, he enjoys hiking, bird watching, and traveling worldwide.

Martha Nester

Martha Nester received an undergraduate degree in biology from Oberlin College and a Master's degree in education from Stanford University. She has worked in university research laboratories and has taught elementary school. She currently works in an environmental education program at the Seattle Audubon Society. Martha has worked with her husband, Gene, for more than 40 years on microbiology textbook projects, at first informally as an editor and sounding board, and then as one of the authors of *Microbiology: A Human Perspective*. Martha's favorite activities include spending time with their four grandchildren, all of whom live in the Seattle area. She also enjoys playing the cello with a number of musical groups in the Seattle area.

About the Contributors

Deborah Allen is an associate professor at Jefferson College in Missouri, where she teaches microbiology as well as several other courses for students entering allied health careers. Her graduate work was in zoology at the University of Oklahoma and in neurobiology and behavior at Cornell University. She participated in cancer research at the University of Arkansas Medical Center before embarking on a career in publishing, acquiring and developing books in the life sciences. She is now thrilled to be working on the other end of the desk with the Nester team. Away from campus, Deborah reads or listens to her favorite Eve Dallas novels, floats the rivers and listens to folk music in the Ozarks, and fully appreciates the local microbes while visiting Missouri wineries.

Nancy Boury is a Senior Lecturer at Iowa State University, where she also received her Doctorate of Molecular Biology. She received her Master's degree from the University of Wisconsin–Madison in medical microbiology. She currently teaches biology, microbiology, and genetics courses. Nancy is developing a mouse model for oral challenge salmonellosis and its treatment with bacteriophage; intracellular pathogens in general, their mechanism of pathogenicity, and the immune response to pathogen challenge; and molecular and evolutionary immunology.

Sarah Salm is a Professor at the Borough of Manhattan Community College (BMCC) of the City University of New York, where she teaches microbiology, anatomy and physiology, and general biology. She earned her undergraduate and doctoral degrees at the University of the Witwatersrand in Johannesburg, South Africa. She later moved to New York, working first as a postdoctoral fellow and then an Assistant Professor of Research at NYU Langone Medical Center. Her research has covered a range of subjects, from plant virus identification through prostate stem cell characterization. She left her research position at NYU to focus exclusively on her position at BMCC, where she has been enthusiastically teaching since 2004.

Teri Shors is a Professor in the Department of Biology and Microbiology at the University of Wisconsin–Oshkosh. Teri began a tenure track position in the Department of Biology and Microbiology in September 1997. Prior to her arrival at UW–Oshkosh, she served two years as a postdoctoral fellow at the National Institutes of Health, in Bethesda, Maryland, under the direction of Dr. Bernard Moss, Chief of the Laboratory of Viral Diseases, National Institute of Allergies and Infectious Diseases. Her research training/specialty has been in the field of poxviruses. The courses she teaches center around laboratory and lectures in microbiology, virology, molecular and cellular biology, and graduate seminar.

DEDICATION

We dedicate this book to our students; we hope it helps to enrich their lives and to make them better-informed citizens,

to our families whose patience and endurance made completion of this project a reality,

to Anne Nongthanat Panarak Roberts in recognition of her invaluable help, patience, and understanding, and

to our colleagues for continuing encouragement and advice.

Letter from Gene Nester

Instructors and Students:

As this textbook enters its 7th edition, it's time for me to step back from being an active writer to becoming a senior consultant. Denise Anderson, who has been an author since the 2nd edition, will take on increasing authorship responsibilities. I enlisted Denise to become a member of the author team because she had an obvious love for microbiology, had a broad training in all aspects of microbial biology, and was an enthusiastic and dynamic teacher who successfully conveyed her passion for the microbial world to her undergraduate students. Her membership on the team of authors has contributed enormously to the success of the present as well as past editions and I have every confidence will enhance the quality of all future editions. With Denise at the helm, this textbook will remain the very best choice for non-majors and students of the allied health sciences.

—Gene Nester

Customize Your Perfect Foundation

Customize your course materials to your learning outcomes!

Create what you've only imagined.

Introducing McGraw-Hill Create™—a new, self-service website that allows you to create custom course materials—print and eBooks—by drawing upon McGraw-Hill's comprehensive, cross-disciplinary content. Add your own content quickly and easily. Tap into other rights-secured third-party sources as well. Then, arrange the content in a way that makes the most sense for your course. Even personalize your book with your course name and information! Choose the best format for your course: color print, black-and-white print, or eBook. The eBook is now even viewable on an iPad! And, when you are done, you will receive a free PDF review copy in just minutes!

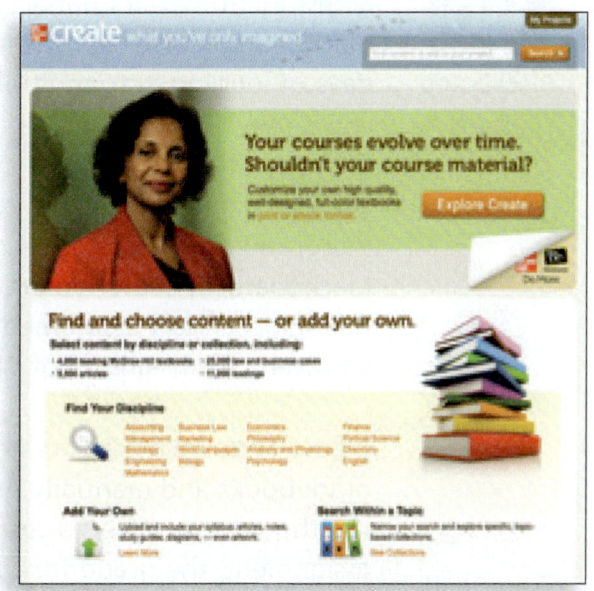

Finally, a way to quickly and easily create the course materials you've always wanted.

Imagine that.

Need a lab manual for your microbiology course? Customize any of these manuals—add your text material—and *Create* your perfect solution!

McGraw-Hill offers several lab manuals for the microbiology course. Contact your McGraw-Hill representative for packaging options with any of our lab manuals.

Kleyn: *Microbiology Experiments,* 7th edition
    ISBN 978-0-07-731554-2

Brown: *Benson's Microbiological Applications: Laboratory Manual in General Microbiology,* 12th edition
    Short Version ISBN 978-0-07-337527-4
    Complete Version ISBN 978-0-07-730213-9

Chess: *Laboratory Applications in Microbiology: A Case Study Approach,* 2nd edition ISBN 978-0-07340237-6

Chess: *Photographic Atlas for Laboratory Applications in Microbiology,* 1st edition ISBN 978-0-07-737159-3

Harley: *Laboratory Exercises in Microbiology,* 8th edition
    ISBN 978-0-07-729281-2

Morello: *Lab Manual and Workbook in Microbiology: Applications to Patient Care,* 10th edition ISBN 978-0-07-352253-1

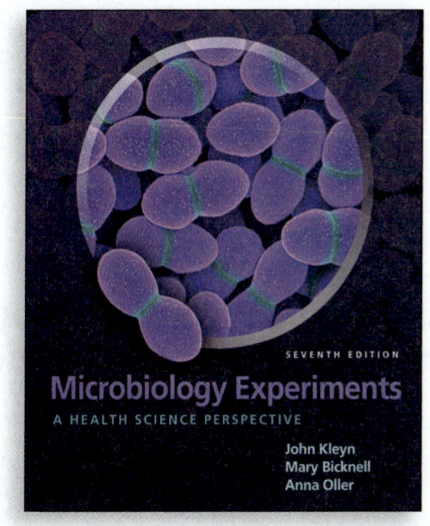

Visit McGraw-Hill Create—www.mcgrawhillcreate.com—today and begin building your perfect book.

Connecting Instructors to Students

McGraw-Hill Higher Education and Blackboard® have teamed up! What does this mean for you?

Do More

Your life, simplified. Now you and your students can access McGraw-Hill Connect® and Create™ right from within your Blackboard course—all with one single sign on! Say goodbye to the days of logging in to multiple applications.

Deep integration of content and tools. Not only do you get single sign on with Connect and Create, you also get deep integration of McGraw-Hill content and content engines right in Blackboard. Whether you're choosing a book for your course or building Connect assignments, all the tools you need are right where you want them—inside of Blackboard.

Seamless gradebooks. Are you tired of keeping multiple gradebooks and manually synchronizing grades into Blackboard? We thought so. When a student completes an integrated Connect assignment, the grade for that assignment automatically (and instantly) feeds your Blackboard grade center.

A solution for everyone. Whether your institution is already using Blackboard or you just want to try Blackboard on your own, we have a solution for you. McGraw-Hill and Blackboard can now offer you easy access to industry-leading technology and content, whether your campus hosts it, or we do. Be sure to ask your local McGraw-Hill representative for details.

Join iTeachMicrobiology.com to connect with other instructors teaching microbiology across the country.

iteach microbiology

INSTRUCTOR COLLABORATION MADE POSSIBLE BY MCGRAW-HILL EDUCATION

and Students to Course Concepts

Introducing McGraw-Hill ConnectPlus™ Microbiology

 McGraw-Hill ConnectPlus™ Microbiology interactive learning platform provides a customizable, assignable eBook, auto-graded assessments, an adaptive diagnostic tool, digital lecture capture, access to instructor resources, and powerful reporting—all in an easy-to-use interface.

Save time with auto-graded assessments and tutorials.

Fully editable, customizable, auto-graded interactive assignments using high-quality art from the textbook, animations, and videos from a variety of sources take you way beyond multiple choice. Assignable content is available for every Learning Outcome in the book. Extremely high-quality content includes case study modules, concept mapping activities, animated tutorials, and more!

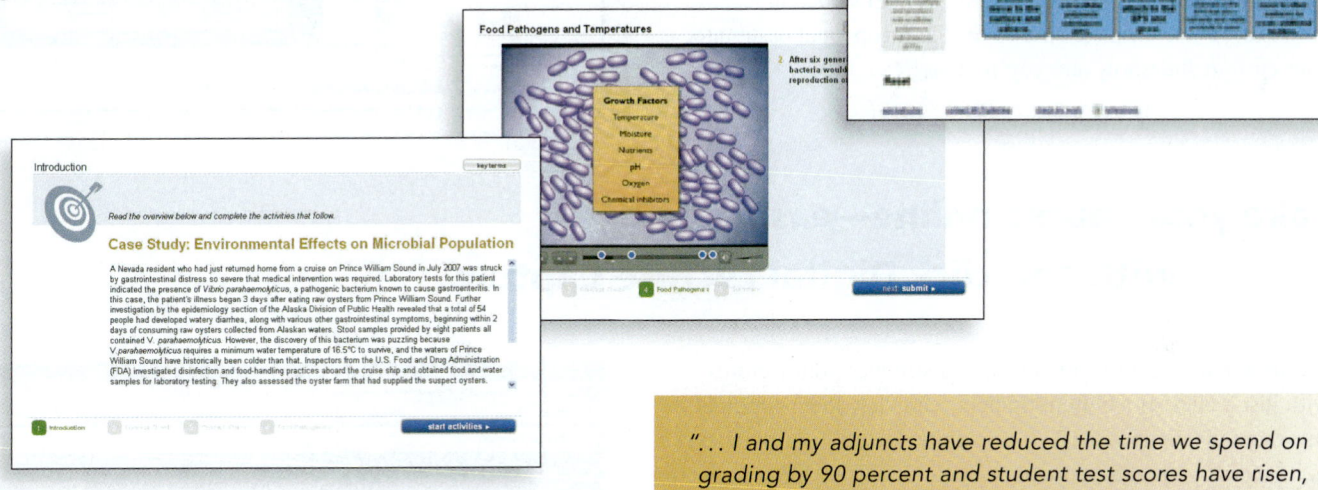

"... I and my adjuncts have reduced the time we spend on grading by 90 percent and student test scores have risen, on average, 10 points since we began using Connect!"

—William Hoover, Bunker Hill Community College

Gather assessment information.

Generate powerful data related to student performance against Learning Outcomes, specific topics, level of difficulty, and more.

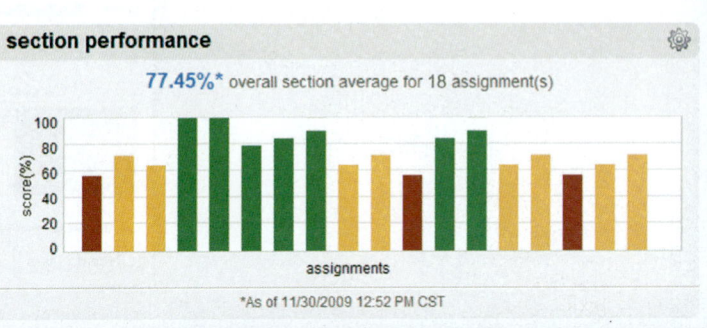

reports

section performance

77.45%* overall section average for 18 assignment(s)

*As of 11/30/2009 12:52 PM CST

Instructors Connect via Customization

Presentation Tools
allow you to customize your lectures.

Enhanced Lecture Presentations contain lecture outlines, Flex Art, art, photos, tables, and animations embedded where appropriate. Fully customizable, but complete and ready to use, these presentations will enable you to spend less time preparing for lecture!

Flex Art   Fully editable (labels and leaders) line art from the text, with key figures that can be manipulated. Take the images apart and put them back together again during lecture so students can understand one step at a time.

Animations   Over 100 animations bringing key concepts to life, available for instructors and students.

Animation PPTs   Animations are truly embedded in PowerPoint® for ultimate ease of use! Just copy and paste into your custom slideshow and you're done!

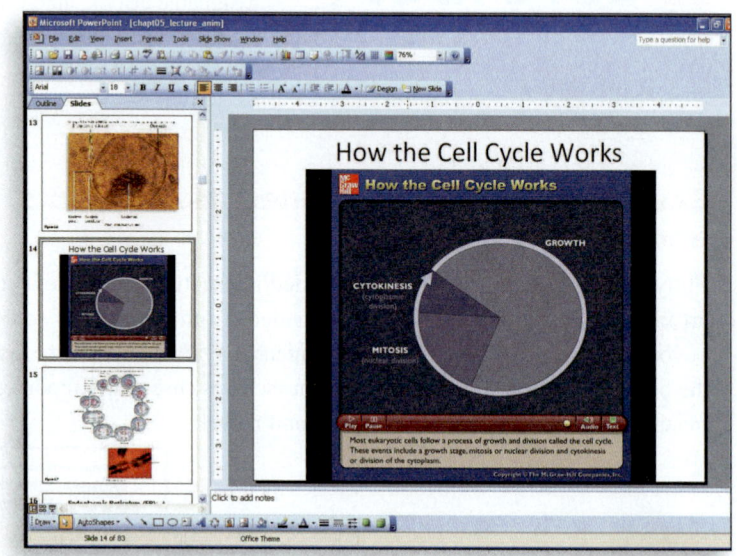

Take your course online—*easily*—
with one-click Digital Lecture Capture.

McGraw-Hill Tegrity™ records and distributes your lectures with just a click of a button. Students can view them anytime/anywhere via computer, iPod, or mobile device. Tegrity indexes as it records your slideshow presentations, and anything shown on your computer, so **students can use key words to find exactly what they want to study.**

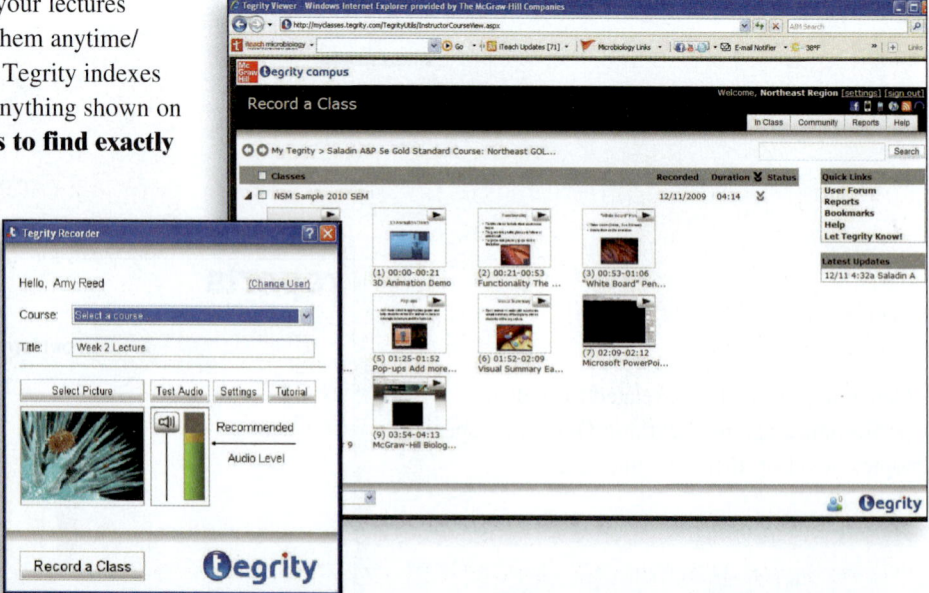

Students Connect 24/7 with Personalized Learning Plans

Access content anywhere, anytime, with a customizable, interactive eBook.

McGraw-Hill ConnectPlus® eBook takes digital texts beyond a simple PDF. With the same content as the printed book, but optimized for the screen, ConnectPlus has embedded media, including animations and videos, which bring concepts to life and provide "just in time" learning for students. Additionally, fully integrated, self-study questions and in-line assessments allow students to interact with the questions in the text and determine if they're gaining mastery of the content.

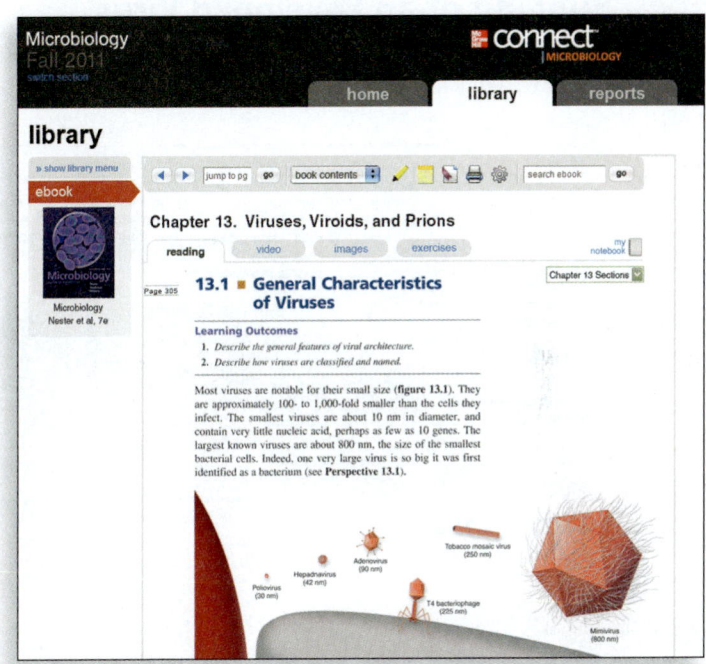

McGraw-Hill LearnSmart™
A Diagnostic, Adaptive Learning System

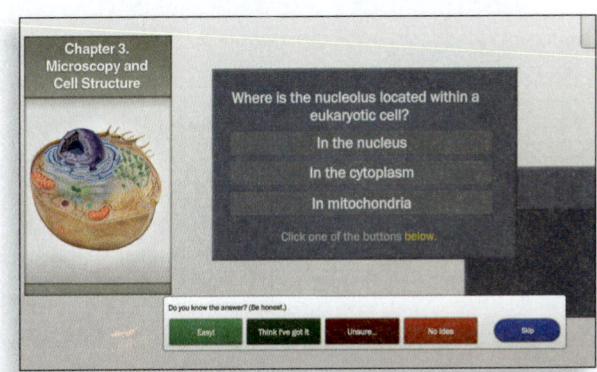

McGraw-Hill LearnSmart™ is an adaptive diagnostic tool, powered by McGraw-Hill Connect® Microbiology, which is based on artificial intelligence and constantly assesses a student's knowledge of the course material.

Sophisticated diagnostics adapt to each student's individual knowledge base in order to match and improve what they know. Students actively learn the required concepts more easily and efficiently.

Self-study resources are also available at www.mhhe.com/nester7

ALL NEW Instructional Art Program

Every piece of art has been revised or replaced to make it more vibrant, three-dimensional, and instructional. The goal of this extensive update was to evaluate every piece, give colors and shapes consistency throughout, and make all art more engaging and useful (for students and instructors). A more modern, instructional art program helps students understand key concepts.

Budding of an Enveloped Virus

As you can see, the seventh edition figure is much more instructive. The student will have a much easier time understanding how this process works!

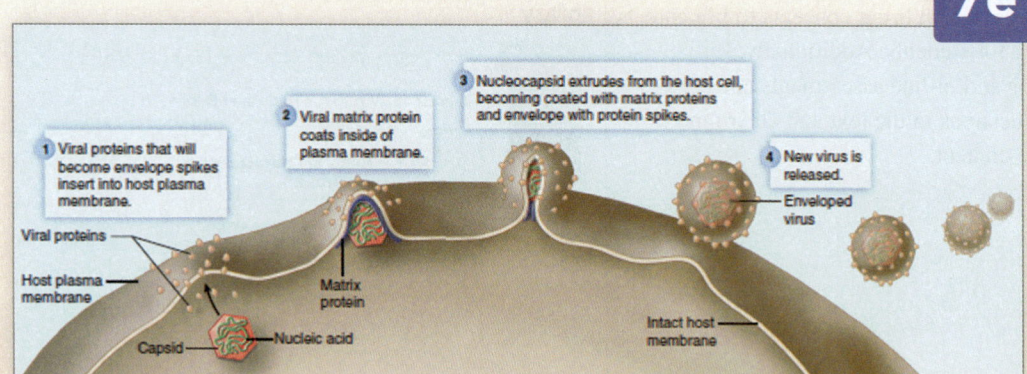

FIGURE 13.15a

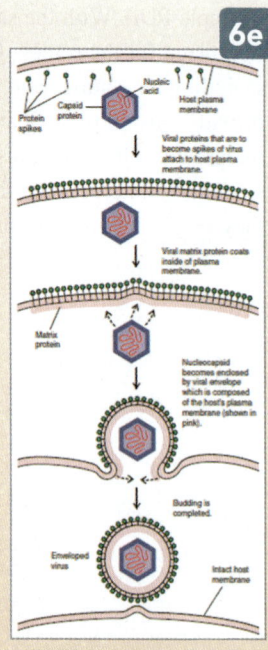

FIGURE 14.6a

Replication/Transcription/Translation

Focus Figures highlight key topics.

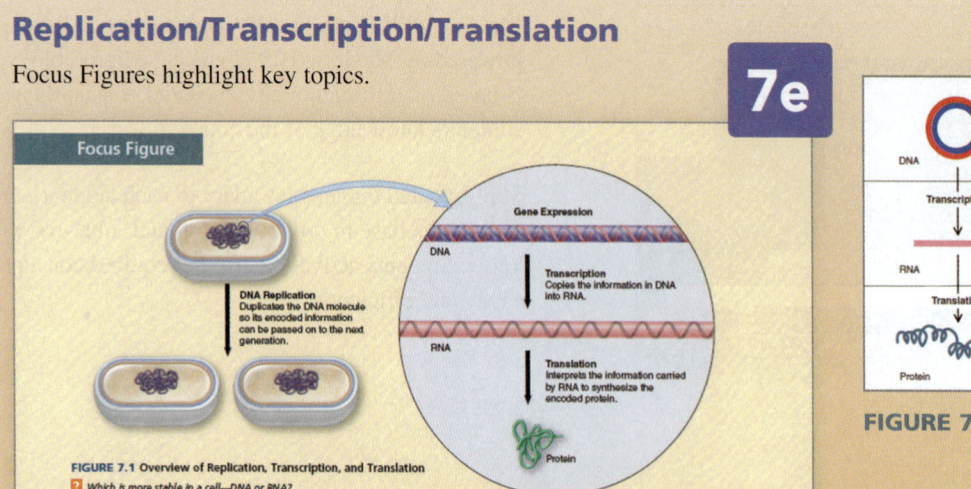

FIGURE 7.1

FIGURE 7.1

System Overviews

The system chapters have anatomically correct figures to help the students throughout the chapter.

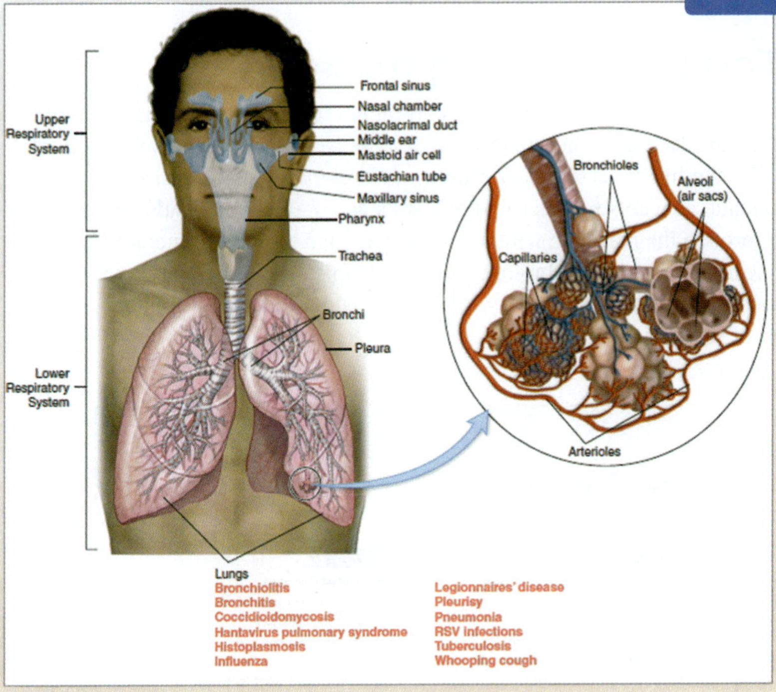

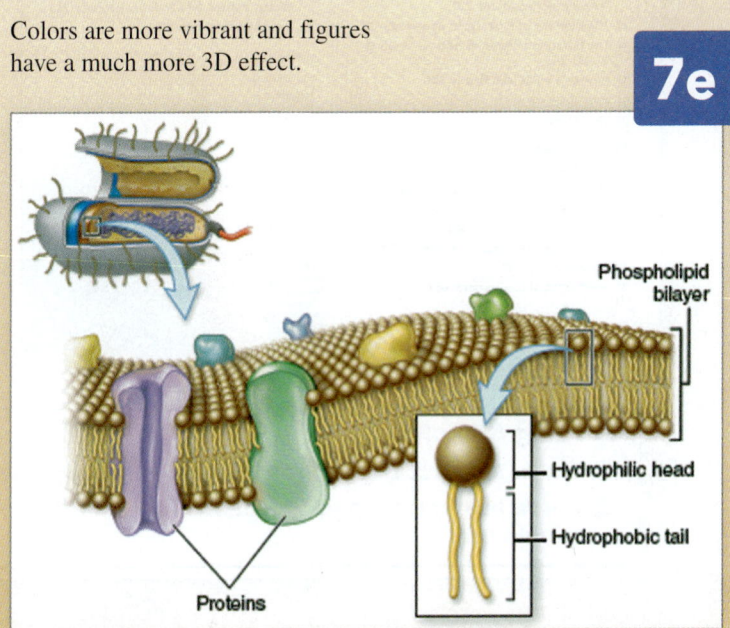

FIGURE 21.1b

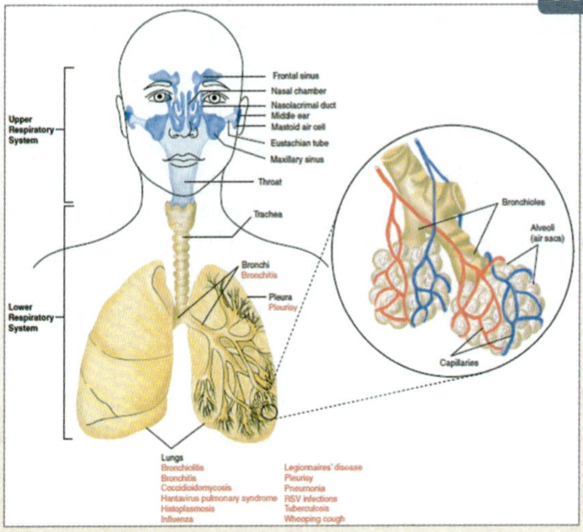

FIGURE 22.2b

Colors are more vibrant and figures have a much more 3D effect.

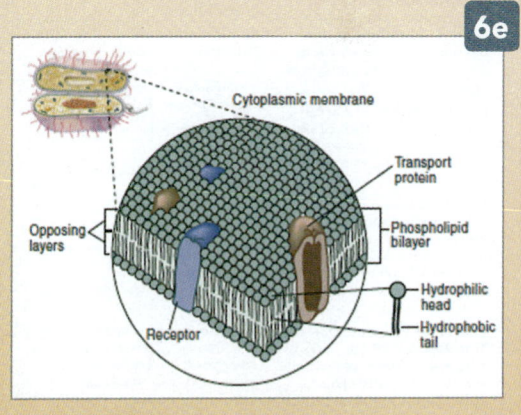

FIGURE 3.24

FIGURE 3.24

Clarified Writing Helps Students Focus on Learning the Science

Microbiology is one of the fastest-moving fields, so it is not enough to teach students only the current hot topics. Students must be armed with a strong foundation to understand the next year's—even the next decade's—hot topics as well. The Nester author team works hard, both in their textbook and in their classrooms, to teach students for the future, as well as for today. *Microbiology: A Human Perspective* remains focused on providing an excellent foundation in fundamental concepts for microbiology students of all backgrounds. With this edition, the authors underwent the tremendous task of combing through the narrative with a critical eye—for example, general vocabulary and reading level were simplified so that students can focus on learning about microbiology and the complex terminology required. The seventh edition of *Microbiology: A Human Perspective* incorporates the changes necessary to ensure it is readable for students—easier to comprehend—yet still maintains an appropriately rigorous level of science.

Table of Contents has been updated as necessary to improve comprehension, reduce redundancy, and help students build a solid foundation in microbiology.

Brief Contents

KEY TERMS

Antigenic Drift Minor changes that occur naturally in influenza virus antigens as a result of mutation.

Antigenic Shift Major changes in the antigenic composition of influenza viruses that result from reassortment of viral nucleic acids during infection of the same host cell by different viral strains.

Croup Acute obstruction of the larynx occurring mainly in infants and young children, often resulting from respiratory syncytial or other viral infection.

Directly Observed Therapy Short-Course (DOTS) Method used to ensure that patients comply with their tuberculosis treatment; the healthcare worker watches while the patient takes each dose of medication.

Extensively Drug-Resistant Tuberculosis (XDR-TB) Strains of *Mycobacterium tuberculosis* resistant to the first-line anti-TB drugs isoniazid and rifampicin, and at least three of the second-line TB drugs.

Mucociliary Escalator Layer of mucus moved by cilia lining the respiratory tract that traps bacteria and other particles and moves them into the throat.

Multidrug-Resistant Tuberculosis (MDR-TB) Strains of *Mycobacterium tuberculosis* resistant to isoniazid and rifampicin—two of the first-line anti-TB drugs.

Otitis Media Infection of the middle ear.

Pharyngitis Inflammation of the throat.

Pneumonia Inflammation of the lungs accompanied by filling of the air sacs with fluids such as pus and blood.

Sputum Pus and other material coughed up from the lungs.

Tubercle Granuloma formed in tuberculosis; granulomas are collections of lymphocytes and macrophages found in a chronic inflammatory response, an attempt by the body to wall off and contain persistent organisms and antigens.

Key Terms begin every chapter to start building a foundation!

Learning Outcomes

1. *Outline the functions of the upper and lower respiratory tract.*
2. *List the parts of the respiratory system that are normally microbe-free.*

NEW!

Learning Outcomes begin each major section. They are numbered to correspond to digital materials.

MicroAssessments

Major sections end with a MicroAssessment that summarizes the major concepts in that section and offers both review questions and critical-thinking questions (marked with a ➕) to assess understanding of the preceding section.

MicroAssessment 21.1

The respiratory system provides a warm, moist environment for microorganisms. It is protected by tonsils and adenoids, and by the mucociliary escalator. The upper respiratory tract contains highly diverse microbiota, including aerobes, anaerobes, facultative anaerobes, and aerotolerant bacteria. Although most of them are of low virulence, these organisms can sometimes cause disease opportunistically. The lower respiratory system is free of a normal microbiota.

1. *What is the normal function of the tonsils and adenoids?*
2. *Describe the normal microbiota of the respiratory system.*
3. *How would paralysis of the cilia impact the respiratory system?* ➕

NEW!

Figure Questions offer another assessment opportunity for students. Most figures have a critical-thinking question associated with them. Animations found online correlate to many images within the text.

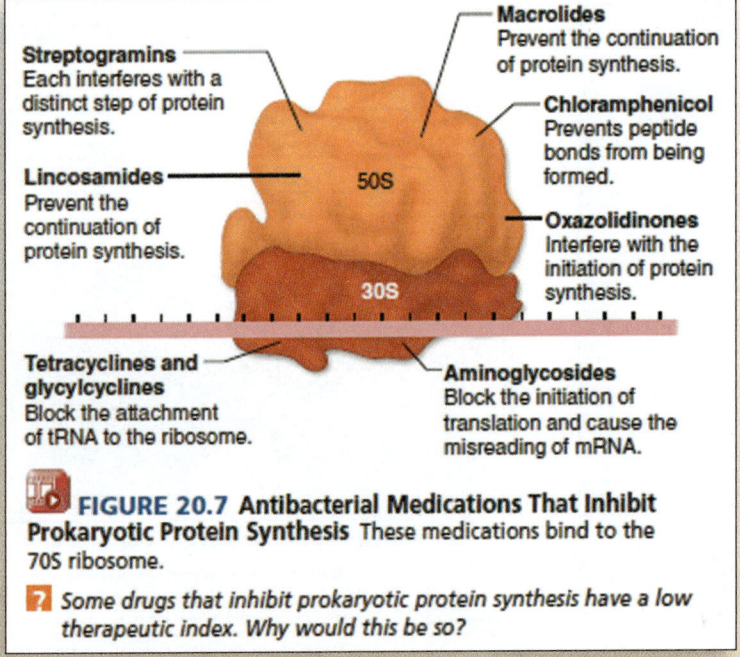

Streptogramins
Each interferes with a distinct step of protein synthesis.

Lincosamides
Prevent the continuation of protein synthesis.

Macrolides
Prevent the continuation of protein synthesis.

Chloramphenicol
Prevents peptide bonds from being formed.

50S

Oxazolidinones
Interfere with the initiation of protein synthesis.

30S

Tetracyclines and glycylcyclines
Block the attachment of tRNA to the ribosome.

Aminoglycosides
Block the initiation of translation and cause the misreading of mRNA.

FIGURE 20.7 Antibacterial Medications That Inhibit Prokaryotic Protein Synthesis These medications bind to the 70S ribosome.

❓ *Some drugs that inhibit prokaryotic protein synthesis have a low therapeutic index. Why would this be so?*

End of Chapter Review

- **Short Answer** questions review major chapter concepts.
- **Multiple Choice** questions allow self-testing; answers are provided in Appendix IV.
- **Applications** provide an opportunity to use knowledge of microbiology to solve real-world problems.
- **Critical Thinking** questions encourage practice in analysis and problem solving that can be used by the student in any subject.

MicroBytes

Found throughout the chapters, these are small "bytes" of interesting information.

MicroByte

Cilia beat at a rate of about 1,000 times a minute, propelling mucus along the mucociliary escalator.

BUILDING UP THE BASE

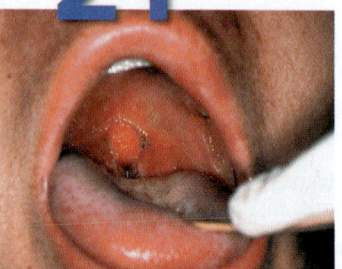

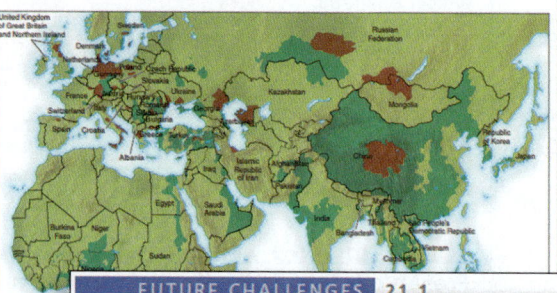

A Glimpse of History

Each chapter opens with an engaging story about the men and women who pioneered the field of microbiology.

Perspective Boxes

Perspective boxes introduce a *human* perspective by showing how microorganisms and their products influence our lives in many different ways.

Future Challenges

Many chapters end with a pending challenge facing current and future microbiologists.

Case Presentations

Each infectious disease chapter includes a case presentation of a realistic clinical situation.

Unmatched Clinical Coverage

Organized by human body systems, the infectious disease chapters (21 to 28) are highlighted with purple shading in the top corner of the page for easy reference.

Incomparable Presentation of Diseases

Each disease is presented systematically and predictably. Individual sections describe the disease's symptoms, causative agents, pathogenesis, epidemiology, and prevention and treatment.

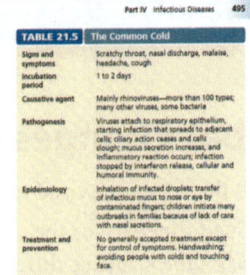

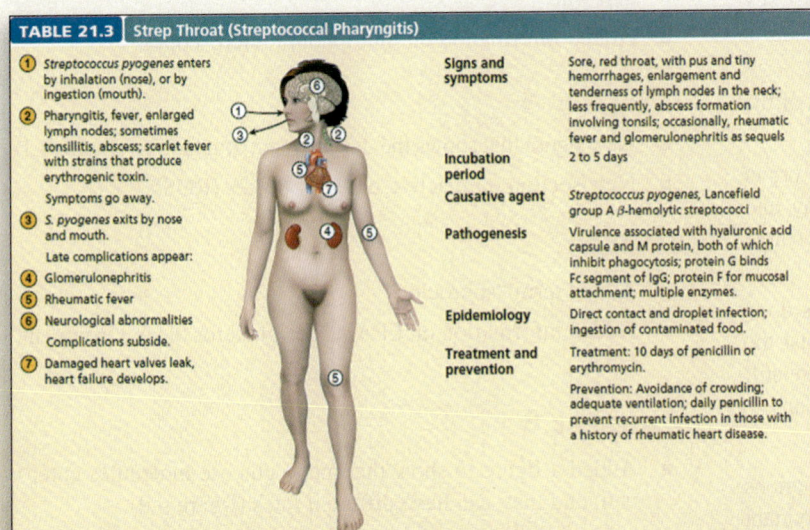

TABLE 21.3 Strep Throat (Streptococcal Pharyngitis)

① *Streptococcus pyogenes* enters by inhalation (nose), or by ingestion (mouth).

② Pharyngitis, fever, enlarged lymph nodes; sometimes tonsillitis, abscess; scarlet fever with strains that produce erythrogenic toxin.

Symptoms go away.

③ *S. pyogenes* exits by nose and mouth.

Late complications appear:

④ Glomerulonephritis

⑤ Rheumatic fever

⑥ Neurological abnormalities

Complications subside.

⑦ Damaged heart valves leak, heart failure develops.

Signs and symptoms	Sore, red throat, with pus and tiny hemorrhages, enlargement and tenderness of lymph nodes in the neck; less frequently, abscess formation involving tonsils; occasionally, rheumatic fever and glomerulonephritis as sequels
Incubation period	2 to 5 days
Causative agent	*Streptococcus pyogenes*, Lancefield group A β-hemolytic streptococci
Pathogenesis	Virulence associated with hyaluronic acid capsule and M protein, both of which inhibit phagocytosis; protein G binds Fc segment of IgG; protein F for mucosal attachment; multiple enzymes.
Epidemiology	Direct contact and droplet infection; ingestion of contaminated food.
Treatment and prevention	Treatment: 10 days of penicillin or erythromycin. Prevention: Avoidance of crowding; adequate ventilation; daily penicillin to prevent recurrent infection in those with a history of rheumatic heart disease.

IMPROVED!
Enhanced Disease Summaries

Major diseases are represented with a summary table that includes an outline of pathogenesis keyed to a human figure showing the entry and exit of the pathogen.

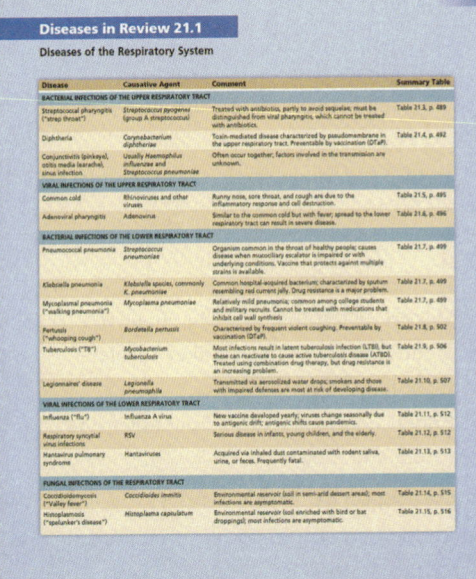

NEW!
Diseases in Review Table

Each infectious disease chapter ends with a table that summarizes the key features of the diseases discussed in that chapter.

UPDATES—Enhancing the Foundation

Global Changes

- All the art was revised to make it more engaging and clear
- Added diversity to the "disease person" in the disease summary tables, using different "people" rather than a generic "person"
- New feature—MicroBytes—short snippets of information, usually 140 characters or less
- New feature—Questions for most figures are meant to engage students in the figure and encourage students to read the accompanying information in the narrative
- New feature—Diseases in Review tables highlight key characteristics of diseases discussed in the chapter, helping students to focus on the "big picture"
- New feature—Focus Figures emphasize the key points of the general topic of the chapter
- In the disease chapters, emphasized the ecology of the organ system by combining the previous sections "Anatomy and Physiology" and "Normal Microbiota," renaming the single section "Anatomy, Physiology, and Ecology"
- Decreased the number of bolded key terms—this was done so that students focus more on the explanations, not just the terms. Note that most of the previously bolded terms are still in the text; students can highlight these terms if they want
- Most bolded terms are in the glossary
- Tightened and simplified the explanations and descriptions throughout the text, with the aim of making the most important information stand out; overall, the information is easier to read

Key Changes in Individual Chapters

Chapter 1

- Combined two sections to create one: "The Dispute Over Spontaneous Generation"
- Added "The Golden Age of Microbiology" to the timeline (figure 1.2)

Chapter 2

- Updated depiction of atoms to match that used in chemistry texts, using only Lewis figures to show the number of electrons in the valence shell (figures 2.1 to 2.4)
- Updated the depiction of polar compounds by using electron density molecules (figures 2.5 and 2.6)
- Introduced the term "electronegativity" (table 2.2)
- Added a table to illustrate biologically important functional groups (table 2.4)
- Added a section on protein domains

Chapter 3

- New figure that shows permeability of the lipid bilayer of the cytoplasmic membrane (figure 3.25a)
- New figure that shows aquaporins (figure 3.25b)
- Combined the figures that show types of transport systems (figure 3.29)
- New figure on secretion (figure 3.30)
- New section on osmosis
- Added information about the gel-like material sandwiched between the cytoplasmic membrane and the Gram-positive cell wall
- Added information on multiphoton microscopes

Chapter 4

- New figure that shows the development of a biofilm (figure 4.3)
- New section on reactive oxygen species (ROS)

Chapter 5

- Added term "sporocide"
- Added information on EPA's "designed for the environment" label

Chapter 6

- Added a figure to show that many glucose molecules enter a cell, and they can have different fates (figure 6.9)
- Added colored icons to the table that shows precursor molecules (table 6.2), and these have been incorporated in other figures in place of the previous gray boxes
- Moved the figure that shows the effect of the energy source versus the terminal electron acceptor forward in the chapter, and used it to give examples (figure 6.7)
- Changed the topic of the "Future Challenges" story to biofuels

Chapter 7

- New figure to show three kinds of RNA involved in protein synthesis (figure 7.3)
- New figure to show amplification with transcription/translation (figure 7.4)
- New figure to show the regions on DNA that direct transcription (figure 7.8)
- New figure to show the regions on mRNA that direct translational (figure 7.12)
- New figure that uses an analogy to show the principles of regulation (figure 7.20)
- Reordered the sections on regulation, so that sensing mechanisms are covered first.

- Added inducer exclusion to the mechanisms of carbon catabolite repression

Chapter 8

- New figure that shows the mutagenic effects of an alkylating agent (figure 8.7)

Chapter 9

- Updated by no longer italicizing restriction enzyme names
- Introduced the term "high throughput"
- Added a MicroByte about the Genetic Information Non-discrimination Act (GINA)
- Added information about the successful replacement of a bacterial chromosome by a machine-made chromosome.

Chapter 10

- Updated taxonomic outline of *Bergey's Manual of Systematic Bacteriology* (table 10.3)
- Moved a figure on serotypes from chapter 11 (now figure 10.9)
- Added a section on nucleic acid amplification tests (NAATs), incorporating the previous information on using PCR for identification
- Added information on multilocus typing to the section on characterizing strain differences using molecular typing (previously titled "Genomic Typing")

Chapter 11

- Updated to use the term "endoflagella" instead of axial filaments
- Added information about the problem associated with *Coxiella* in the placenta of infected animals

Chapter 12

- Rearranged the order of the chapter so that fungi, which more students are probably familiar with, are covered first, followed by algae and protozoa
- Updated classification of fungi
- Updated classification of protozoa
- Updated classification of eukaryotes (figures 12.10 and 12.13)
- New photo of water mold (figure 12.16)
- Separated the section that previously combined the coverage of arthropods and helminths and switched the order
- New photo of elephantiasis (figure 12.17)
- New life cycle figure of ascariasis (figure 12.18)
- Created consistent subheadings within the sections on fungi, algae, and protozoa to cover types, structure, habitats, reproduction, and economic or medical importance
- New box on river blindness (Perspective 12.1)
- Expanded coverage of insect vectors (table 12.6)
- New photo of *Anopheles* mosquito, vector of malaria (figure 12.20)

Chapter 13

Viruses, Viroids, and Prions (Note that chapter titles are included from this point on because of the change in the Table of Contents.)

- Condensed the information previously spread over two chapters into one (to do this, overlapping information, including details that overlapped with coverage in the disease chapters, was eliminated)
- New section on CRISPR, including a figure (figure 13.11)

Chapter 14

The Innate Immune Response

- Added an overview figure (figure 14.1)
- Highlighted the information on pattern recognition receptors by creating a new section (14.5); added information on danger-associated molecular patterns (DAMPs), inflammasomes, and RIG-like receptors
- Added a small section on regulation of the complement system to prepare students for understanding how some pathogens hijack that mechanism to avoid activating the system
- Added a section on the damaging effects of the inflammatory response to prepare students for understanding how the response can both protect against disease and also cause disease symptoms
- Added a section on cell death and the inflammatory response to explain the difference between cell death due to direct damage, apoptosis, and pyroptosis

Chapter 15

The Adaptive Immune Response

- Overview figure has been incorporated as a "mini-figure" into other figures
- New figure showing a Peyer's patch (figure 15.5)
- Modified figures of B-cell activation and T-cell activation to show that lack of accessory signals induces anergy, thereby promoting tolerance (figures 15.11 and 15.20)
- New figure showing natural killer (NK) cells destroying "stressed" cells that lack MHC class I molecules (figure 15.23)

Chapter 16

Host-Microbe Interactions

- Shortened figure that shows mechanisms viruses use to avoid detection by the MHC class I antigen presentation (figure 16.13) to avoid overlap with the new figure in chapter 15

Chapter 17

Immunologic Disorders

- New photo of a goiter (figure 17. 9)

Chapter 18

Applications of Immune Responses

- New figure that shows the host-pathogen "trilogy"—the immune wars, the microbes fight back, and the return of the humans (figure 18.1)

- Added a section on the basic principles of using labeled antibodies, including a figure that highlights the difference between direct and indirect tests (figure 18.9)
- Moved the information on radial immunodiffusion and immunoelectrophoresis to the Web, thereby placing more emphasis on the other techniques covered

Chapter 19

Epidemiology

- New figure on vector transmission, showing the action of a mechanical vector and a biological vector (figure 19.6)
- Split a previous section into two: "Pathogen Factors That Influence the Epidemiology of Infectious Disease" and "Host Factors That Influence the Epidemiology of Infectious Disease"
- Created a separate primary section on "Emerging Infectious Diseases"
- Updated terminology introducing case-fatality rate
- Separated one figure into three separate figures showing (1) reservoirs of infection (figure 19.2), (2) portals of entry (figure 19.3), and (3) mechanisms of transmission (figure 19.4)
- Combined portals of exit and portals of entry into single section

Chapter 20

Antimicrobial Medications

- New section that describes carbapenem-resistant *Enterobacteriaceae* (CRE)
- New section on integrase inhibitors, used to treat HIV infection

Chapter 21

Respiratory System Infections

- New section "Post-Streptococcal Sequelae" added to emphasize the complications that can follow strep throat
- Increased coverage of the signs and symptoms of whooping cough to emphasize the three stages; also added information about the recent outbreak in California
- Updated coverage of tuberculosis, adding the terms "latent tuberculosis infection (LTBI)" and "active tuberculosis disease (ATBD)"; added a new figure to illustrate the pathogenesis of tuberculosis (figure 21.19)
- Updated the coverage of influenza, emphasizing how antigenic drift is responsible for seasonal influenza, and antigenic shift is responsible for pandemic influenza

Chapter 22

Skin Infections

- New photograph of the bull's-eye rash of Lyme disease (figure 22.10)
- New photograph of shingles (figure 22.16)
- New section "Acne Vulgaris" on acne
- Moved the information on *Malassezia furfur* and *Candida albicans,* highlighting it in a new section "Other Fungal Diseases"

Chapter 23

Wound Infections

- Reordered coverage in "Bacterial Infections of Bite Wounds" so that "Human Bites" comes first, reflecting the student population's interest in the topic
- Eliminated some of the overlap in the coverage of *Staphylococcus aureus* in chapters 22 and 23, referring the reader back to chapter 22

Chapter 24

Digestive System Infections

- Added a new figure to highlight the difference between infections of the teeth and infections of the gums (figure 24.3)
- Updated the information on periodontal diseases, emphasizing the polymicrobial nature of the disease and the role of the inflammatory response in the disease process
- Added a new section "General Characteristics of Diarrheal Diseases" to highlight the importance of oral rehydration therapy and other overlapping concepts related to these diseases; a new figure (figure 24.10) summarizes the common mechanisms of pathogenesis of intestinal pathogens
- Moved the information on *Clostridium difficile,* highlighting it in a new section: "*Clostridium difficile*–Associated Disease (CDAD)"; added a photo of pseudomembranous colitis (figure 24.14)
- Updated the information on cholera to include its recent introduction into Haiti
- Updated the information on *E. coli* gastroenteritis to emphasize the different virulence factors and epidemiology of the different pathovars
- Added paratyphoid fever to the section on typhoid fever
- Separated viral diseases of the lower digestive system into two sections: those that cause intestinal diseases and those that cause liver diseases
- Increased the coverage on hepatitis B virus antigens to emphasize their clinical significance

Chapter 25

Genitourinary Tract Infections

- Updated by substituting the term "sexually transmitted infection (STI)" in place of "sexually transmitted disease (STD)" to reflect that many of the infections do not cause noticeable disease
- Added Weil's disease to the coverage of leptospirosis
- Added "Latent Syphilis" to the stages of the disease

Chapter 26

Nervous System Infections

- Moved the information on pneumococcal meningitis, *Haemophilus influenzae* meningitis, and neonatal meningitis into three separate new sections, thereby increasing the emphasis on these diseases

- New figure showing meningitis belt in Africa (figure 26.5)
- New figure that highlights the use of bacteriophage preparations to decrease the risk of listeriosis associated with ready-to-eat foods (figure 26.7)
- New photo showing early signs of leprosy (figure 26.8a)
- New graph showing the incidence of the three types of botulism in the United States (figure 26.11)
- Updated the information on *Cryptococcus gattii;* added map of its spread in the Pacific Northwest (figure 26.18)
- Added a section on "Primary Amebic Meningoencephalitis (PAM)"
- Updated box on rabies survivors

Chapter 27

Blood and Lymphatic Infections
- Updated the coverage of sepsis and septic shock to emphasize the role of the inflammatory response in the disease process
- New section on "Dengue Fever"
- New section on "Chikungunya"
- New section on "Emerging Hemorrhagic Fevers" (Ebola and Marburg)

- Added information about *Plasmodium knowlesi* to the malaria section
- Altered the coverage of plague to emphasize the importance on the route of entry of *Yersinia pestis* in the disease symptoms and outcome

Chapter 28

HIV Disease and Complications of Immunodeficiency
- Incorporated more structure in the coverage of HIV/AIDS by subdividing the information further
- Added information on the role of viral set point in HIV disease progression
- Updated the information on HIV prevention by adding information on male circumcision and new vaccine candidates

Chapter 29

Microbial Ecology
- Revised figures of biogeochemical cycles to highlight the role that the reaction plays in microbial growth (biosynthesis, energy source, terminal electron acceptor)

Unique Interactive Question Types Pre-tagged to ASM's Curriculum Guidelines for Undergraduate Microbiology

1. **Case Study:** Case study questions use tutorial and assessment to help students see the world of microbiology and how it applies to their lives. These static case studies come to life in a learning activity that is interactive, self-grading, and assessable. The integration of the cases with videos and animations enhances the depth of the content covered, and the use of integrated questions forces students to stop, think, and evaluate their understanding. Pre- and post-testing allow instructors and students to assess the overall comprehension of the activity.

2. **System Summary Figures:** System overview figures from the text are made interactive to give students additional opportunity to practice and assess their knowledge. Key components of each disease can be reviewed including signs and symptoms, causative agent, incubation period, epidemiology, and more.

3. **Key Concept Activities:** These learning modules take students deep into the key microbiology concepts they need to know for long-term success in the course. Pre- and post-assessment is used in addition to multimedia tutorials to ensure students gain mastery of the key topics in each chapter.

4. **Disease Pathway Figures:** This question correlates to the Diseases In Review table found at the end of each disease chapter and will allow students to practice on the most important information to remember for each disease covered in the chapter.

5. **Tutorial Animation Learning Modules:** Making use of McGraw-Hill's collection of videos and animations, this new question type presents an interactive, self-grading, and assessable activity. Pre- and post-testing is used to assess student comprehension and the use of integrated questions forces students to stop, think, and evaluate their understanding of the process being presented. These tutorials take a stand-alone, static animation and turn it into an interactive learning experience for your students with real-time remediation.

6. **Labeling:** Using the high-quality art from the textbook, check your students' visual understanding as they practice interpreting figures and learning structures and relationships. Easily edit or remove any label you wish!

7. **Classification:** Ask students to organize concepts or structures into categories by placing them in the correct "bucket."

8. **Sequencing:** Challenge students to place the steps of a complex process in the correct order.

9. **Composition:** Fill in the blanks to practice vocabulary, and then reorder the sentences to form a logical paragraph (these exercises may qualify as "writing across the curriculum" activities!)

All McGraw-Hill ConnectPlus® content is pre-tagged to Learning Outcomes for each chapter as well as topic, section, Bloom's Level, and ASM Curriculum Guidelines (once they're finalized) to assist you in both filtering out unneeded questions for ease of creating assignments and in reporting on your students' performance against these points. This will enhance your ability to effectively assess student learning in your courses by allowing you to align your learning activities to peer-reviewed standards from an international organization.

Acknowledgments

Gene Nester, Evans Roberts, Brian McCarthy, and Nancy Pearsall shared a vision many years ago to write a new breed of microbiology textbook especially for students planning to enter nursing and other health-related careers. Today there are other books of this type, but we were extremely gratified to learn that a majority of the students we surveyed intend to keep their copies of *Microbiology: A Human Perspective* because they feel it will benefit them greatly as they pursue their studies in these fields. Special thanks to the many students who used *Microbiology: A Human Perspective* over the years and who shared their thoughts with us about how to improve the presentation for the students who will use this edition of the text.

We offer our sincere appreciation to the many gracious and expert professionals who helped us with this revision by offering helpful suggestions. In addition to thanking those individuals listed here who carefully reviewed chapters, we also thank those who responded to our information surveys, those who participated in regional focus groups, and those participants who chose not to be identified. All of you have contributed significantly to this work and we thank you.

Reviewers of the Seventh Edition

Ahmmed Ally, *Massachusetts College of Pharmacy and Health Sciences*
Cindy B. Anderson, *Mt. San Antonio College*
James Barbaree, *Auburn University*
Kris Baumgarten, *Delta College*
Suzanne D. Clutter, *West Virginia Northern Community College–Wheeling*
Lorraine A. Cramer, *University of North Carolina at Chapel Hill*
Thomas R. Danford, *West Virginia Northern Community College*
Harry G. Deneer, *University of Saskatchewan*
William Dew, *Nipissing University*
Judy Earl, *Montgomery County Community College*
Kathryn Germain, *Southwest Tennessee Community College*
Robert Gessner, *Valencia Community College*
Julie Gibbs, *College of DuPage*
Judy Gnarpe, *University of Alberta*
Amy D. Goode, *Illinois Central College*
Ellen J. Gower, *Greenville Technical College*
Judy Haber, *California State University Fresno*
John Ireland, *Jackson Community College*
Shirley E. Kangas, *Cuyahoga Community College, Eastern Campus*
Amine Kidane, *Columbus State Community College*
P. Kourtev, *Central Michigan University*
Timothy A. Kral, *University of Arkansas*
Judith Marie Krey, *Waubonsee Community College*
William Lorowitz, *Webster State University*

Jennifer Leigh Myka, *Galen College of Nursing–Cincinnati*
Karen Nakaoka, *Webster State University*
Steven Obenauf, *Broward College*
Erica Perrer, *Delgado Community College*
Madhura Pradhan, *Ohio State University*
Dr. Marceau Ratard, *Delgado Community College*
Pele Eve Rich, *North Hennepin Community College*
Seth Ririe, *Brigham Young University–Idaho*
Eleftherios "Terry" Saropoulos, *Vanier College*
Deemah N. Schirf, *University of Texas at San Antonio*
Arif Sheena, *MacEwan College, Alberta, Canada*
George F. Spiegel, Jr., *College of Southern Maryland*
Ronald J. Stewart, *Humber ITAL, Toronto, Ontario*
Steve Thurlow, *Jackson Community College*
Floyd L. Wormley, Jr., *University of Texas at San Antonio*
Malcolm Zellars, *Georgia State University*

Focus Group Participants

Corrie Andries, *Central New Mexico Community College*
John Bacheller, *Hillsborough Community College*
Michelle Badon, *University of Texas at Arlington*
David Battigelli, *University of North Carolina–Greensboro*
Nancy Boury, *Iowa Sate University*
Lance Bowen, *Truckee Meadows Community College*
William Boyko, *Sinclair Community College*
Dave Brady, *Southwestern Community College*
Antonia Brem, *Wayne County Community College*
Chad Brooks, *Austin Peay State University*
Lisa Burgess, *Broward College*
Liz Carrington, *Tarrant Country College–Southeast Campus*
Joseph Caruso, *Florida Atlantic University*
Erin Christensen, *Middlesex Community College*
Robin Cotter, *Phoenix College*
John Dahl, *Washington State University*
Alison Davis, *East Los Angeles College*
Ana Dowey, *Palomar College*
Elizabeth Emmert, *Salisbury University*
Susan Finazzo, *Georgia Perimeter College*
Clifton Franklund, *Ferris State University*
Jason Furrer, *University of Missouri*
Chris Gan, *Highline Community College*
Edwin Gines-Candalaria, *Miami–Dade*
Zaida Gomez-Kramer, *University of Central Arkansas*
Amy Good, *Illinois Central College*
Todd Gordon, *Kansas City Kansas Community College*
Brinda Govindan, *San Francisco State University*
Julianne Grose, *Brigham Young University*
Gabriel Guzman, *Triton College*
Judy Haber, *California State University–Fresno*

Julie Harless, *Lone Star College*
Zafer Hatahet, *Northwestern State University*
James Herrick, *James Madison University*
Jennifer Herzog, *Herkimer County Community College*
Dena Johnson, *Tarrant County College*
Jim Johnson, *Central Washington University*
Eunice Kamunge, *Essex County College*
Amine Kidane, *Columbus State Community College*
Jeff Kiggins, *Monroe Community College*
Terri Lindsey, *Tarrant County College*
Suzanne Long, *Monroe Community College*
Kimberly Maznicki, *Seminole State College of Florida*
Caroline McNutt, *Schoolcraft College*
Elizabeth McPherson, *University of Tennessee–Knoxville*
A'mee Mehta, *Seminole State College of Florida*
Sharon Miles, *Itawamba Community College*
Tracey Mills, *Ivy Tech Community College–Lawrence*
Rita Moyes, *Texas A & M University*
Valerie Narey, *Santa Monica College*
Ruth Negley, *Harrisburg Area Community College*
Steven Obenauf, *Broward College*
Julie Oliver, *Cosumnes River College*
Marcia Pierce, *Eastern Kentucky University*
Madhura Pradhan, *Ohio State University*
Todd Primm, *Sam Houston State University*
Jean Revie, *South Mountain Community College*
Jackie Reynolds, *Richland College*
Beverly Roe, *Erie Community College*
Silvia Rossbach, *Western Michigan University*
Benjamin Rowley, *University of Central Arkansas*
Donald Rubbelke, *Lakeland Community College*
Mark Schneegurt, *Wichita State University*
Teri Shors, *University of Wisconsin–Oshkosh*
Sasha Showsh, *University of Wisconsin–Eau Claire*
Heidi Smith, *Front Range Community College*
Louise Thai, *University of Missouri–Columbia*
Steve Thurlow, *Jackson Community College*
Sanjay Tiwary, *Hinds Community College*
Stephen Wagner, *Stephen F. Austin State University*
Delon Washo-Krupps, *Arizona State University*
George Wawrzyniak, *Milwaukee Area Technical College*
Janice Webster, *Ivy Tech Community College*
Mary Weintraub, *University of Cincinnati*
John Whitlock, *Hillsborough Community College*
Samia Williams, *Santa Fe Community College*
Fadi Zaher, *Gateway Technical College*

We thank our colleagues in the Department of Microbiology at the University of Washington who have lent their support of this project over many years. Our special thanks go to John Leigh, Mary Bicknell, Mark Chandler, Kendall Gray, Janis Fulton, Jimmie Lara, Sharon Schultz, Michael Lagunoff, Heather Felise, Julie Overbaugh, and James Staley for their general suggestions and encouragement.

We would also like to thank Denise's husband, Richard Moore, who was "forced" to proofread and critique many of the chapters. Although he has no formal scientific education, or perhaps because of that fact, his suggestions have been instrumental in making the text more "reader-friendly." Much to his own surprise, Richard has learned enough about the fundamentals of microbiology to actually become intrigued with the subject.

Special thanks to the reviewers and other instructors who helped guide us in this revision. Deciding what to eliminate, what to add, and what to rearrange is always difficult, so we appreciate your input.

Thanks also to Jason Overby, who provided guidance as we updated our chemistry coverage in this edition, and David Hurley, who helped us navigate the murky waters during a substantial revision of an earlier edition that updated the coverage of innate and adaptive immunity.

A list of acknowledgments is not complete without thanking our fearless leaders from McGraw-Hill. Our sponsoring editor Lynn Breithaupt and developmental editor Fran Simon not only gave inspiration and sound advice, but laughed at our jokes and became true friends. Our project manager Mary Jane Lampe and copyeditor Bea Sussman were instrumental in making sure the correct words actually made it onto paper.

Additionally, we would like to thank John Bacheller, Mira Beins, Chad Brooks, Liz Carrington, Greg Frederick, Renee Krohne, Jonathan Miller, Bethany Morgan, Julie Oliver, Seth Ririe, Ben Rowley, and George Wawrzyniak for producing our digital resources to support us and other instructors who lecture from our text.

We hope very much that this text will be interesting, educational for students, a help to their instructors, and will convey the excitement that we all feel for the subject. We would appreciate any comments and suggestions from our readers.

Eugene Nester
Denise Anderson
C. Evans Roberts, Jr.
Martha Nester

Contents

About the Authors iv ■ Preface vi

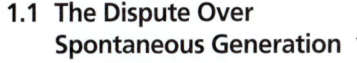

4 Dynamics of Prokaryotic Growth 82

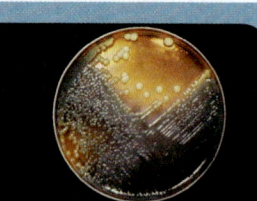

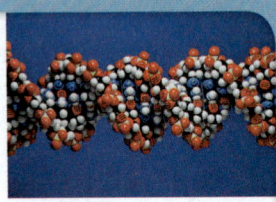
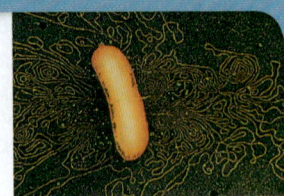

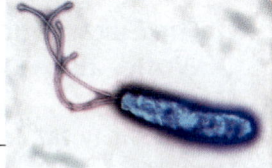

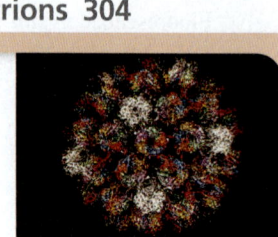

16 Host-Microbe Interactions 380

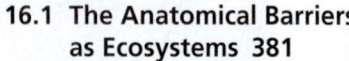

17 Immunologic Disorders 401

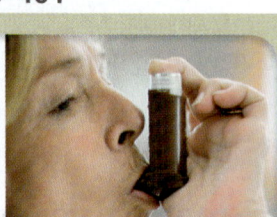

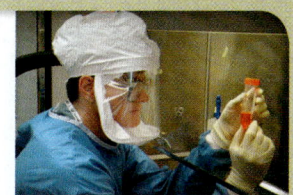

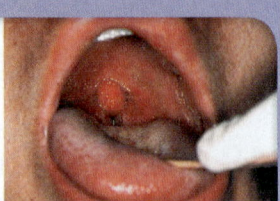

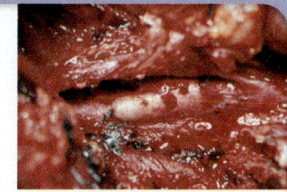
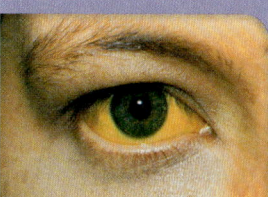

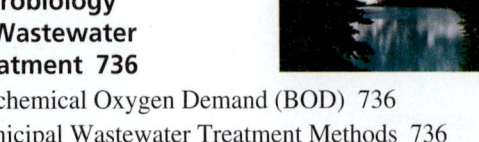

1

Humans and the Microbial World

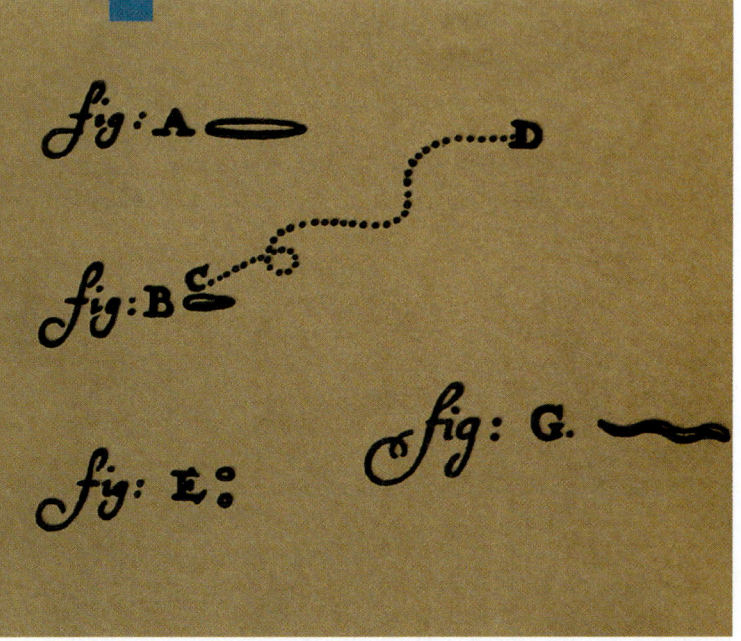

Drawings that van Leeuwenhoek made in 1683 of the shapes of microorganisms he saw through his single lens microscope. He also observed organism B moving from position C to D.

A Glimpse of History

Microbiology as a science was born in 1674 when Antony van Leeuwenhoek (1632–1723), an inquisitive Dutch drapery merchant, peered at a drop of lake water through a glass lens he had carefully made. For several centuries it was known that curved glass would magnify objects, but the skilled hands of a craftsman and his questioning mind uncovered an entirely new part of our world. What he observed through his simple magnifying glass was undoubtedly one of the most startling and amazing sights that humans have ever beheld—the world of microbes. As van Leeuwenhoek wrote in a letter to the Royal Society of London, he saw

> Very many little animalcules, whereof some were roundish, while others a bit bigger consisted of an oval. On these last, I saw two little legs near the head, and two little fins at the hind most end of the body. Others were somewhat longer than an oval, and these were very slow a-moving, and few in number. These animalcules had diverse colours, some being whitish and transparent; others with green and very glittering little scales, others again were green in the middle, and before and behind white; others yet were ashed grey. And the motion of most of these animalcules in the water was so swift, and so various, upwards, downwards, and round about, that 'twas wonderful to see.

Before van Leeuwenhoek observed bacteria, Robert Hooke, an English microscopist saw another kind of microorganism. In 1665, he described what he called a "microscopical mushroom." His drawing was so accurate that his specimen could later be identified as a common bread mold. Hooke also described how to make the kind of microscope that van Leeuwenhoek constructed almost 10 years later. Both men deserve equal credit for revealing the world of microbes—the organisms you are about to study.

Microorganisms are the foundation for all life on earth. They have existed on this planet for about 3.5 billion years, and over this time, plants, animals, and modern microorganisms evolved from them. Microorganisms may be invisible to the naked eye, but our life depends on their activities.

1.1 ■ The Dispute Over Spontaneous Generation

Learning Outcome

1. *Describe the key experiments of scientists who disproved spontaneous generation.*

The discovery of microorganisms raised an intriguing question: "Where did these microscopic forms originate?" Some people believed that worms and other forms of life arise from non-living material in a process referred to as **spontaneous generation.** This was challenged by an Italian biologist and physician, Francesco Redi. In 1668, he used a simple experiment to show that worms found on rotting meat originated from the eggs of flies, not from the decaying meat as proponents of spontaneous generation believed. In his experiment, Redi covered the meat with fine gauze that prevented flies from depositing their eggs. When he did this, no worms appeared.

Despite Redi's work that explained the source of worms on decaying meat, convincing proof that microorganisms did not arise from spontaneous generation took over 200 more years! One reason for the delay was that various experiments in different laboratories gave conflicting results.

Early Experiments

In 1749, John Needham, a scientist and Catholic priest, showed that flasks containing various broths (made by soaking hay, chicken, or other nutrient source in water) gave rise to microorganisms even when the flasks were boiled and sealed with a cork. At that time, brief boiling was thought to kill all organisms, so this suggested that microorganisms did indeed arise spontaneously.

In 1776, the animal physiologist and priest, Father Spallanzani, obtained results that contradicted the experiments of Needham; no bacteria appeared in Spallazani's broths after boiling. His experiments differed from Needham's in two significant ways: Spallanzani boiled the broths for longer periods and he sealed the flasks by melting their glass necks. Using these techniques, he repeatedly demonstrated that broths remained sterile (free of microorganisms). However, if the neck of the flask cracked, the broth rapidly became cloudy due to growth of the organisms. Spallanzani concluded microorganisms had entered the broth with the air, and the corks used by many investigators did not keep them out.

Spallanzani's experiments did not stop the controversy. Some people argued that the heating process destroyed a "vital force" necessary for spontaneous generation. The debate continued and further experiments, with proper controls, were needed to settle the matter.

Experiments of Pasteur

One giant in science who helped disprove spontaneous generation was Louis Pasteur, the French chemist considered by many to be the father of modern microbiology. In 1861, he did a series of clever experiments. First, he demonstrated that air contains microorganisms. He did this by filtering air through a cotton plug, trapping microorganisms. He then examined the trapped microorganisms with a microscope and found that many looked identical to those described by others who had been studying broths. When Pasteur dropped the cotton plug into a sterilized broth, the broth became cloudy.

Most importantly, Pasteur demonstrated that sterile broths in specially constructed swan-necked flasks remained sterile even when left open to air (**figure 1.1**). Microorganisms from the air settled in the bends and sides of these flasks, never reaching the broth. Only when the flasks were tipped would bacteria enter the broth and grow. These simple and elegant experiments ended the arguments that unheated air or the broths themselves contained a "vital force" necessary for spontaneous generation.

Experiments of Tyndall

Although most scientists were convinced by Pasteur's experiments, some remained skeptical because they could not reproduce his results. An English physicist, John Tyndall, finally explained the conflicting data and, in turn, proved Pasteur correct. Tyndall found that various types of broths required different boiling times to be sterilized. Some materials were sterilized by boiling for 5 minutes, whereas others, most notably broths made from hay, still contained living microorganisms even after boiling for 5 hours! Even when hay was merely present in the laboratory, broths that had previously been sterilized by boiling for 5 minutes were sometimes impossible to sterilize. What was going on? Tyndall finally realized that hay contained heat-resistant forms of microbial life. When hay was brought into the laboratory, dust particles must have transferred these heat-resistant forms to the broths.

Tyndall concluded that some microorganisms exist in two forms: a cell readily killed by boiling, and one that is heat resistant. In the same year (1876), a German botanist, Ferdinand Cohn, discovered **endospores,** the heat-resistant forms of bacteria.

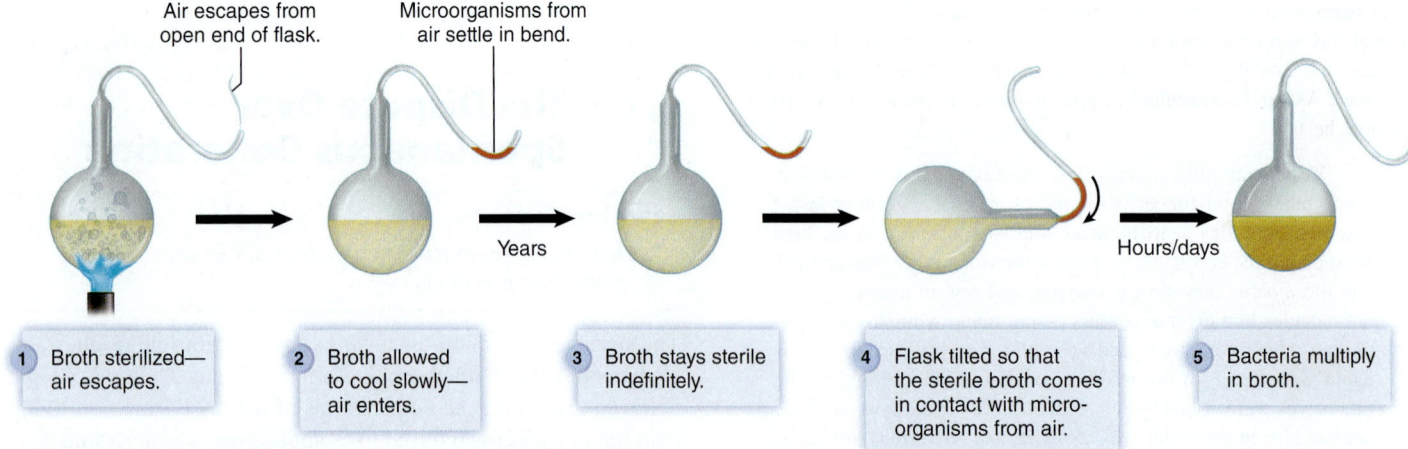

FIGURE 1.1 Pasteur's Experiment with the Swan-Necked Flask If the flask remains upright, no microbial growth occurs. If the flask is tipped, the microorganisms trapped in the neck reach the sterile broth and grow.

If the broth in Pasteur's swan-necked flasks had contained endospores, what results would have been observed?

Robert Koch then demonstrated that anthrax was caused by a spore-forming bacterium. ▸▸ **endospores, p. 67**

The extreme heat resistance of endospores explains the differences between Pasteur's results and those of other investigators. Organisms that produce endospores are commonly found in the soil and most likely were present in broths made from hay. Pasteur used only broths made with sugar or yeast extract, so his experiments probably did not have endospores. At the time, scientists did not appreciate the importance of the source of the broth, but in hindsight, the source was critical. This points out an important lesson for all scientists. In repeating an experiment, it is essential to reproduce all conditions as closely as possible. What may seem like a trivial difference may be extremely important.

It may seem surprising that spontaneous generation was disproved only about a century and a half ago. How far the science of microbiology and all biological sciences have advanced since that time! **Figure 1.2** lists some of the more important advances made over the years in the context of other historical events. It also highlights a period called the Golden Age of Microbiology—a time during which the field of microbiology blossomed. Rather than cover more history now, we will return to many of these milestones in brief stories that open each chapter.

MicroByte

In 2001, someone mailed anthrax endospores to selected individuals in the United States, an act of bioterrorism.

MicroAssessment 1.1

Pasteur and Tyndall finally disproved spontaneous generation about 150 years ago.

1. *Give two reasons why it took so long to disprove spontaneous generation.*
2. *What experiment disproved the notion that a "vital force" in air was responsible for spontaneous generation?*
3. *What conclusions could Tyndall reach on the properties of the agent that entered the broth from hay?* ✚

1.2 ■ Microbiology: A Human Perspective

Learning Outcomes

2. *Explain why life could not exist without microorganisms.*
3. *Describe three applications of microbiology.*
4. *Describe the role of microbes in disease, including the triumphs, present and future challenges, and host-microbe interactions.*

Microorganisms have an enormous impact on all living things. We could not survive without them, and they also make our lives far more comfortable. At the same time, microorganisms and other infectious agents, collectively called **microbes**, have killed far more people than have ever been killed in war. Microbes have even been used as weapons in times of war, and continue today to be used as agents of bioterrorism.

Vital Activities of Microorganisms

The activities of microorganisms are required for the survival of not just humans, but all other organisms on this planet. A few examples readily prove this point. Nitrogen is an essential part of nucleic acids and proteins. Because of this, all life-forms need a continual supply of this element. A plentiful source is N_2—the most common gas in the atmosphere—yet neither plants nor animals can use this form. Instead, we depend on certain microbes that convert N_2 into a form other organisms can use. Without these microbes, life as we know it would not exist.

Humans and other animals all require oxygen gas (O_2) to breathe. The supply of O_2 in the atmosphere, however, would be depleted in about 20 years were it not continually replenished. Plants produce O_2 during photosynthesis, but so do many photosynthetic microorganisms.

Microorganisms are the only organisms that can degrade certain materials. For example, humans and other animals cannot digest cellulose—an important component of plants—but some microorganisms can. Because microbes are able to degrade cellulose, leaves and fallen trees do not pile up in the environment. Cellulose is also degraded by billions of microorganisms in the digestive tracts of cattle, sheep, deer, and other ruminants. The digestion products are used by the ruminants for energy, and without them the animals would starve. Microorganisms also play indispensable roles in degrading a wide variety of materials in sewage and wastewater. ▸▸ **ruminants, p. 732**

Applications of Microbiology

In addition to the crucial roles that microorganisms play in maintaining all life, they also have made life more comfortable for humans over the centuries.

Food Production

Egyptian bakers as early as 2100 B.C. used yeast to make bread. Today, bakeries use essentially the same technology. ▸▸ **breadmaking, p. 755**

The excavation of early tombs in Egypt revealed that by 1500 B.C., Egyptians used a complex procedure for fermenting cereal grains to produce beer. Today, brewers use the same fundamental techniques to make beer and other fermented drinks. ▸▸ **beer, p. 754**

Virtually every population that raised milk-producing animals such as cows and goats also developed procedures to ferment milk. This allowed them to make foods such as yogurt, cheeses, and buttermilk. Today, the bacteria added to some fermented milk products are touted as probiotics, protecting against intestinal infections and bowel cancer. ▸▸ **fermented milk products, p. 751** ▸▸ **probiotics, p. 397**

Biodegradation

Microorganisms degrade environmental pollutants. For example, bacteria can destroy polychlorinated biphenyls (PCBs), dichlorodiphenyltrichloroethane (DDT), trichloroethylene, (a toxic solvent used in dry cleaning), and other chemicals in contaminated soil and water. Bacteria also lessen the damage from oil spills. In some cases, microorganisms are added to pollutants to hasten their decay, a process called **bioremediation.** ▸▸ **bioremediation, p. 744**

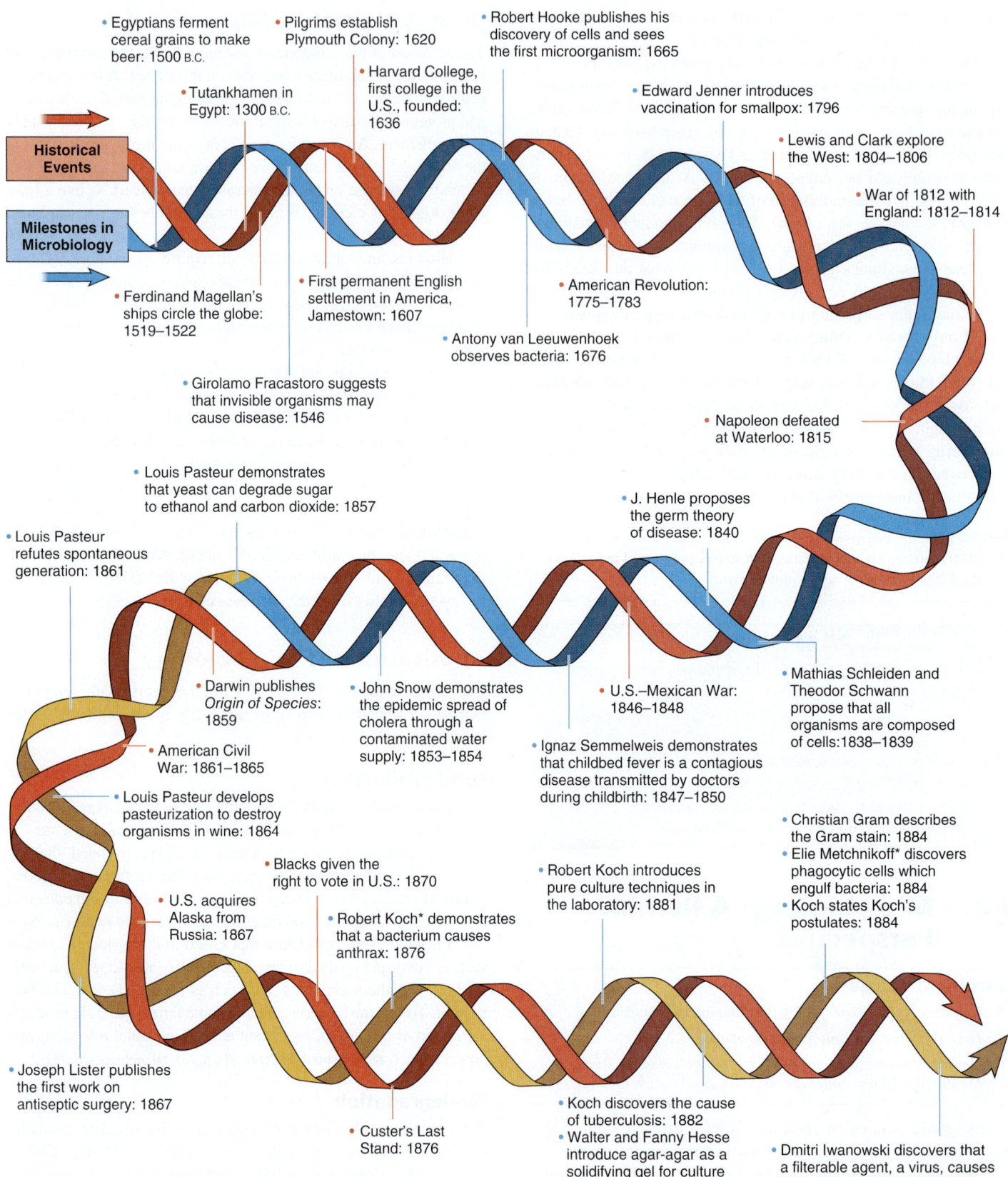

FIGURE 1.2 Some major milestones in microbiology—and their timeline in relation to other historical events. The asterisks indicate scientists who won the Nobel Prize. The gold band indicates the Golden Age of Microbiology.

❓ *What is the Golden Age of Microbiology?*

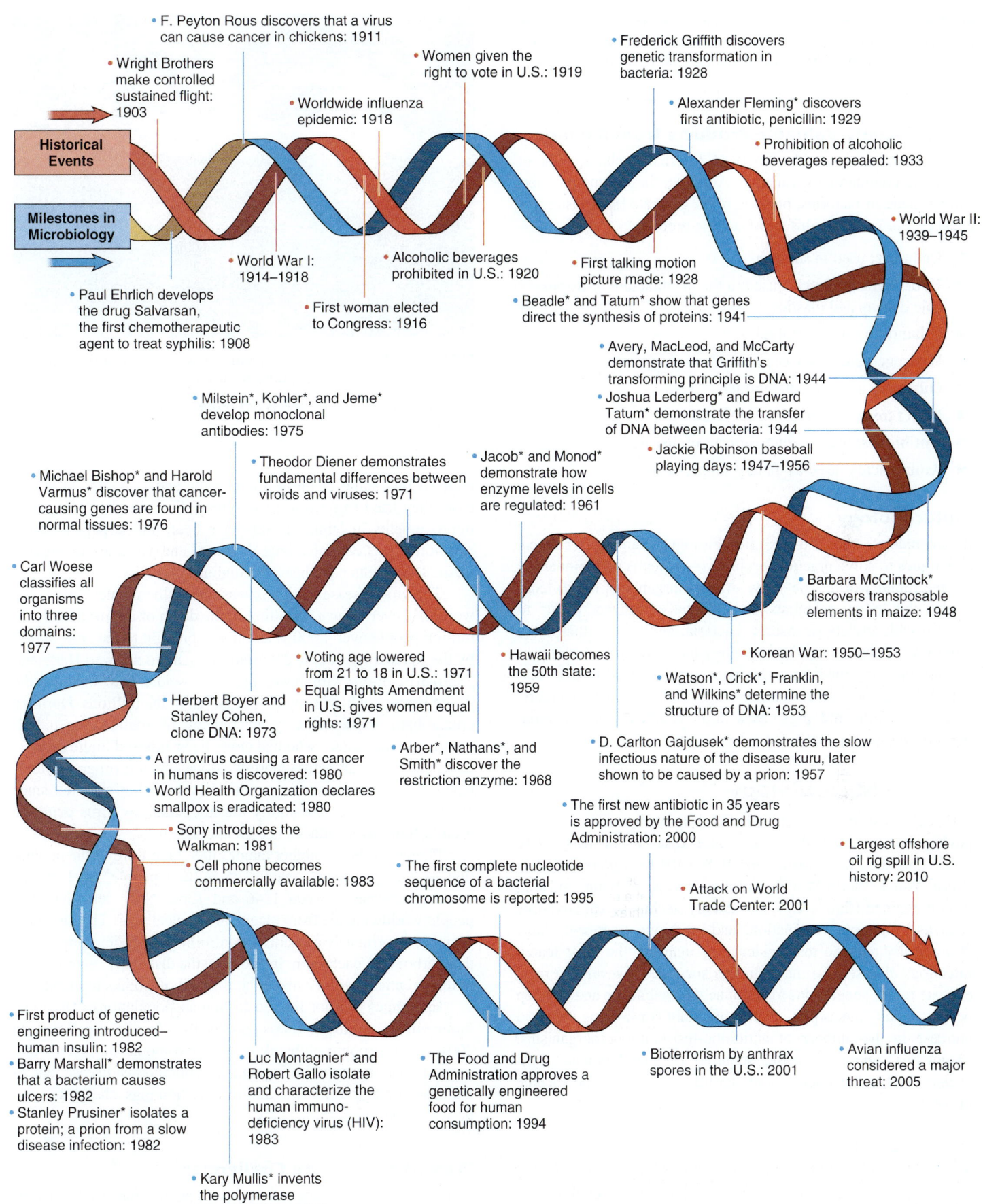

- F. Peyton Rous discovers that a virus can cause cancer in chickens: 1911
- Wright Brothers make controlled sustained flight: 1903
- Worldwide influenza epidemic: 1918
- Women given the right to vote in U.S.: 1919
- Frederick Griffith discovers genetic transformation in bacteria: 1928
- Alexander Fleming* discovers first antibiotic, penicillin: 1929
- Prohibition of alcoholic beverages repealed: 1933

Historical Events

Milestones in Microbiology

- World War I: 1914–1918
- Alcoholic beverages prohibited in U.S.: 1920
- First talking motion picture made: 1928
- World War II: 1939–1945
- Paul Ehrlich develops the drug Salvarsan, the first chemotherapeutic agent to treat syphilis: 1908
- First woman elected to Congress: 1916
- Beadle* and Tatum* show that genes direct the synthesis of proteins: 1941
- Avery, MacLeod, and McCarty demonstrate that Griffith's transforming principle is DNA: 1944
- Milstein*, Kohler*, and Jeme* develop monoclonal antibodies: 1975
- Theodor Diener demonstrates fundamental differences between viroids and viruses: 1971
- Jacob* and Monod* demonstrate how enzyme levels in cells are regulated: 1961
- Joshua Lederberg* and Edward Tatum* demonstrate the transfer of DNA between bacteria: 1944
- Jackie Robinson baseball playing days: 1947–1956
- Michael Bishop* and Harold Varmus* discover that cancer-causing genes are found in normal tissues: 1976
- Carl Woese classifies all organisms into three domains: 1977
- Voting age lowered from 21 to 18 in U.S.: 1971
- Equal Rights Amendment in U.S. gives women equal rights: 1971
- Hawaii becomes the 50th state: 1959
- Barbara McClintock* discovers transposable elements in maize: 1948
- Herbert Boyer and Stanley Cohen, clone DNA: 1973
- Korean War: 1950–1953
- Watson*, Crick*, Franklin, and Wilkins* determine the structure of DNA: 1953
- A retrovirus causing a rare cancer in humans is discovered: 1980
- World Health Organization declares smallpox is eradicated: 1980
- Arber*, Nathans*, and Smith* discover the restriction enzyme: 1968
- D. Carlton Gajdusek* demonstrates the slow infectious nature of the disease kuru, later shown to be caused by a prion: 1957
- Sony introduces the Walkman: 1981
- Cell phone becomes commercially available: 1983
- The first new antibiotic in 35 years is approved by the Food and Drug Administration: 2000
- The first complete nucleotide sequence of a bacterial chromosome is reported: 1995
- Attack on World Trade Center: 2001
- Largest offshore oil rig spill in U.S. history: 2010
- First product of genetic engineering introduced– human insulin: 1982
- Barry Marshall* demonstrates that a bacterium causes ulcers: 1982
- Stanley Prusiner* isolates a protein; a prion from a slow disease infection: 1982
- Luc Montagnier* and Robert Gallo isolate and characterize the human immuno-deficiency virus (HIV): 1983
- The Food and Drug Administration approves a genetically engineered food for human consumption: 1994
- Bioterrorism by anthrax spores in the U.S.: 2001
- Avian influenza considered a major threat: 2005
- Kary Mullis* invents the polymerase chain reaction: 1983

Commercially Valuable Products from Bacteria

Bacteria synthesize a wide variety of different products, some of which are commercially valuable. Although these same products can be made in factories, bacteria often generate them faster and cheaper. For example, different bacteria produce:

- **Cellulose:** used in stereo headsets
- **Hydroxybutyric acid:** used in the manufacture of disposable diapers and plastics
- **Ethanol:** used as a biofuel
- **Hydrogen gas:** used as a possible biofuel
- **Oils:** used as a possible biofuel
- **Insect toxins:** used in insecticides
- **Antibiotics:** used in the treatment of disease
- **Amino acids:** used as dietary supplements

Biotechnology

Biotechnology—the use of microbiological and biochemical techniques to solve practical problems—depends on members of the microbial world. The study of microorganisms led to techniques that allow scientists to genetically engineer plants to resist a wide variety of insects, bacteria, and viruses that could otherwise attack them. Biotechnology has also led to easier production of many medications such as insulin (used to treat diabetes). In the past, insulin needed to be isolated from pancreatic glands of cattle and pigs. Now it is isolated from bacteria.

▶▶| genetic engineering, p. 216

Medical Microbiology

Although most microbes are beneficial or not harmful, some are **pathogens,** meaning they can cause disease. In fact, more Americans died of influenza in 1918–1919 than were killed in World War I, World War II, and the Korean, Vietnam, and Iraq wars combined **(figure 1.3)**. Fortunately, technological advances such as sanitation, vaccination, and antibiotic treatments have dramatically reduced the incidence of many of the most feared infectious diseases. To maintain this success, however, we must educate future generations to continue their vigilance and develop new weapons. This is particularly important considering the rapid increase in the number of antibiotic-resistant microorganisms. Meanwhile, another disease, acquired immunodeficiency syndrome (AIDS), has risen as a modern-day plague that no vaccine prevents.

Past Triumphs

The Golden Age of Microbiology included an important period when scientists were learning a great deal about medically important microbes. Between 1876 and 1918, most pathogenic bacteria were identified, and early work on viruses had begun. Once people

FIGURE 1.3 Students wearing gauze masks to protect themselves against infection with the influenza virus in 1918.

❓ *Why might the gauze masks not protect against the influenza virus?*

realized that some of these microscopic agents could cause disease, they tried to prevent their spread. Over the last 100 years, improvements in human health have been due largely to using antibiotics to treat infectious diseases, and vaccines to prevent them. The results have been astounding!

The viral disease smallpox was one of the greatest killers the world has ever known, resulting in the death of approximately 10 million people over 4,000 years. It was brought to the New World by the Spaniards, and the devastation it caused allowed Hernando (Hernan) Cortez, with fewer than 600 soldiers, to conquer the Aztec Empire, whose subjects numbered in the millions. During a crucial battle in Mexico City, an epidemic of smallpox raged, killing mainly the Aztecs who had never been exposed to the disease. In recent times, an active worldwide vaccination program eliminated the disease in nature, with no cases being reported since 1977. However, the possibility that the smallpox virus could be used in bioterrorist attacks is a great concern.

Plague has been another major killer. One-third of the population of Europe, approximately 25 million people, died of this bacterial disease between 1346 and 1350. Now, less than 100 people worldwide die from plague in a typical year. This dramatic decrease is primarily a result of controlling the rodent population that harbors the bacterium. In addition, the discovery of antibiotics in the twentieth century made the isolated outbreaks treatable.

Epidemics are not limited to human populations. The great famine in Ireland in the 1800s was due to a disease of potatoes. In 2001, a catastrophic outbreak of foot-and-mouth disease of animals ran out of control in England. To contain this disease, one of the most contagious known, almost 4 million pigs, sheep, and cattle were destroyed.

Present and Future Challenges

Although progress has been impressive against bacterial diseases, much more still needs to be done. This is especially true for the treatment of viral infections and diseases associated with poverty. Respiratory infections and diarrheal diseases cause most illness and deaths in the world today. Even in wealthy developed countries with their sophisticated healthcare systems, infectious diseases remain a serious threat. In the United States, about 750 million cases of infectious diseases occur each year, leading to 200,000 deaths and costing tens of billions of dollars.

Emerging Diseases In addition to the long-recognized diseases, emerging diseases continue to arise (**figure 1.4**). These include swine flu, severe acute respiratory syndrome (SARS), multidrug-resistant tuberculosis, Lyme disease, hepatitis C, acquired immunodeficiency syndrome, hemolytic uremic syndrome, hantavirus pulmonary syndrome, mad cow disease (bovine spongiform encephalopathy), and West Nile encephalitis. Most of these diseases are merely newly recognized rather than actually new. Their increased occurrence and wider distribution have brought them to the attention of health workers. Using the latest techniques, scientists have isolated, characterized, and identified the agents causing the diseases. Now, better methods need to be developed to control them.

A number of factors account for the emergence of diseases. Changing lifestyles bring new opportunities for infectious agents to spread. For instance, the suburbs of cities are expanding into rural areas, bringing people into closer contact with animals. Consequently, people become exposed to viruses and infectious organisms that once were far removed from their environment. A good example is the hantavirus. This virus infects rodents, usually without causing disease. The infected animals, however, shed virus in urine, feces, and saliva; from there it can be inhaled by humans as an aerosol. Hantavirus disease is only one of many emerging human diseases associated with small animals.

Some emerging diseases arise because the infectious agents change abruptly and gain the ability to infect new hosts. It appears that HIV (Human Immunodeficiency Virus), the cause of AIDS, arose from a virus that once could infect only chimpanzees. The virus causing SARS is related to viruses found in animals and may have been transmitted from animals to humans. Some bacterial pathogens differ from their normally harmless relatives in that the pathogens contain large pieces of DNA that encode toxins or other characteristics that allow the organism to cause disease. These pieces of DNA may have originated in unrelated organisms.

Are other agents out there that may cause "new" diseases in the future? The answer is undoubtedly yes!

Re-Emerging Diseases Infectious diseases that were under control can increase again. In some cases, the preventive measures become victims of their own success. For example, decades of vaccination have nearly eliminated measles, mumps, and whooping cough in the United States, so many people have no firsthand knowledge of the dangers of the diseases. Couple this with misinformation about vaccines, and people develop an irrational fear of the vaccine more than the disease itself. If parents refuse to have their children vaccinated, the diseases can easily re-emerge.

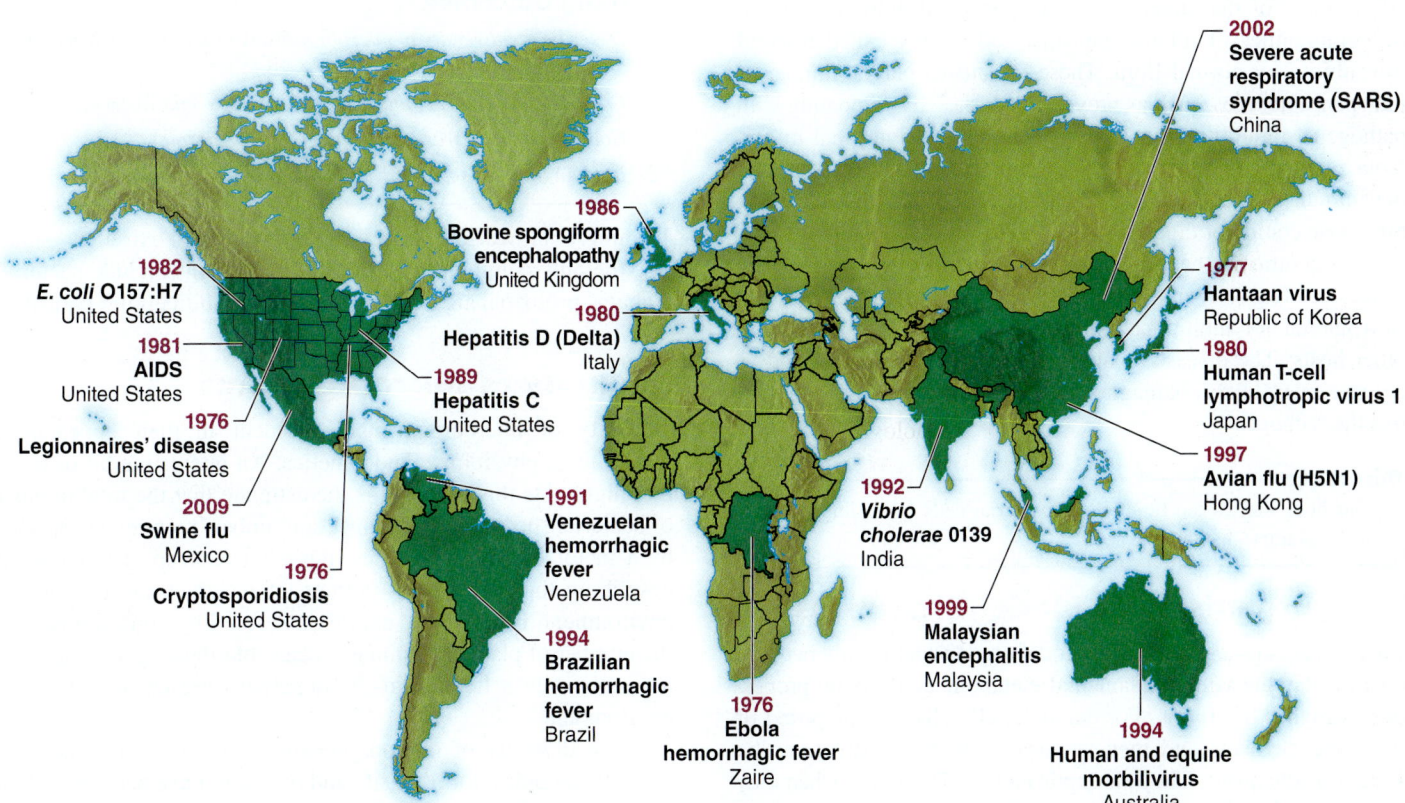

FIGURE 1.4 "New" Infectious Diseases in Humans and Animals Since 1976 Countries where cases first appeared or were identified appear in a darker shade.

Why might so many of the "new" diseases first appear or be identified in the United States and Western European countries?

Diseases also re-emerge when pathogens become resistant to antimicrobial medications. Tuberculosis and malaria are re-emerging, in part because they are often now more difficult to treat.

Travelers and immigrants can contribute to disease re-emergence by inadvertently carrying pathogens around the globe. Diseases such as malaria, cholera, plague, and yellow fever have largely been eliminated from developed countries, but they still exist in many parts of the world. An international traveler incubating a disease could theoretically circle the globe, touch down in several countries, and expose many people before becoming ill.

Changes in the characteristics of a population can also cause re-emergence of diseases. Elderly people typically have weaker immune systems than the young, so aging populations are more susceptible to infectious agents. Individuals with AIDS are especially susceptible to a wide variety of diseases, including tuberculosis.

Chronic Diseases Caused by Bacteria In addition to the diseases long recognized as being caused by pathogens, some illnesses once attributed to other sources may, in fact, be caused by microbes. The best-known example is peptic ulcers. This common affliction—once thought to be due to stress—is caused by a bacterium (*Helicobacter pylori*) and treatable with antibiotics. Chronic indigestion may be caused by the same bacterium. Infectious microbes may play important roles in other chronic diseases as well.

Host-Microbe Interactions

All surfaces of the human body are populated with characteristic communities of microorganisms, collectively called **normal microbiota,** or normal flora. These organisms play a number of indispensable roles, such as preventing disease by competing with pathogens. In addition, members of the normal intestinal microbiota play critical roles in the development of the intestine and the immune system, and they help degrade foods that the body otherwise could not digest.

In contrast to beneficial microbes, pathogens damage body tissues, leading to symptoms of disease. They use the human body as a habitat for multiplication, persistence, and transmission to other hosts. The disease symptoms often result from the body's defense mechanisms damaging the host while attempting to control the pathogen.

MicroByte
Your body carries ten times more bacterial cells (including in the intestinal tract) than human cells!

Microorganisms as Model Organisms

Microorganisms are wonderful model organisms to study because they display the same fundamental metabolic and genetic properties as higher life-forms. For example, all cells are composed of the same elements and they synthesize their cell structures by similar mechanisms. They all duplicate their DNA, and when they degrade foods to harvest energy, they do so via the same metabolic pathways. To paraphrase a Nobel Prize–winning microbiologist, Dr. Jacques Monod—what is true of elephants is also true of bacteria, and bacteria are much easier to study. In addition, bacteria

can be used to obtain results very quickly because they grow rapidly and form billions of cells per milliliter on simple inexpensive growth media. In fact, most major advances made in the last century toward understanding life have come through the study of microbes. The number of Nobel Prizes awarded to microbiologists highlights this point. Such studies constitute basic research, and they continue today.

MicroAssessment 1.2

Microorganisms have an enormous impact on all living things, and their activities are vital to human life. Microorganisms have been used for food production for thousands of years. They are now being used to degrade toxic pollutants and produce a variety of compounds. Enormous progress has been made in preventing and curing infectious diseases, but new ones continue to emerge and re-emerge. Microbes are model organisms, and many principles of biochemistry and genetics have been discovered from their study.

4. *Describe two microbial activities essential to life and three that make our lives more comfortable.*

5. *Describe three reasons why some diseases re-emerge.*

6. *Why would it seem logical, even inevitable, that at least some bacteria would attack the human body and cause disease?*

1.3 ■ Living Members of the Microbial World

Learning Outcomes

5. *Describe the diversity of microorganisms in terms of their numbers and ability to be grown in culture.*

6. *Compare and contrast the* Bacteria, Archaea, *and* Eucarya.

7. *Compare and contrast algae, fungi, and protozoa.*

8. *Explain how the scientific name of an organism is written.*

The microbial world is incredibly diverse, with extremely small size the only common feature. Microorganisms include bacteria, archaea, protozoa, algae, fungi, and some multicellular parasites.

Diversity of Microorganisms

Diversity of microorganisms is evident in their appearance, metabolism, physiology, and genetics. An interesting bit of trivia that highlights this diversity is the estimate that the total number of bacterial species on this planet outnumbers mammalian species by a factor of 10,000! This is important to keep in mind when describing the biodiversity (variety of different species) in any environment. Biodiversity is often measured by considering only the number of plant and animal species, but these species actually represent only a fraction of the organisms present in any given environment.

The diversity of microorganisms results from the fact that they are the oldest forms of life and therefore have had the longest time to evolve. However, they must be even more varied than what we now understand because less than 1% of all microbial species can be grown and studied in the laboratory. Thus, the vast majority of microorganisms have yet to be studied.

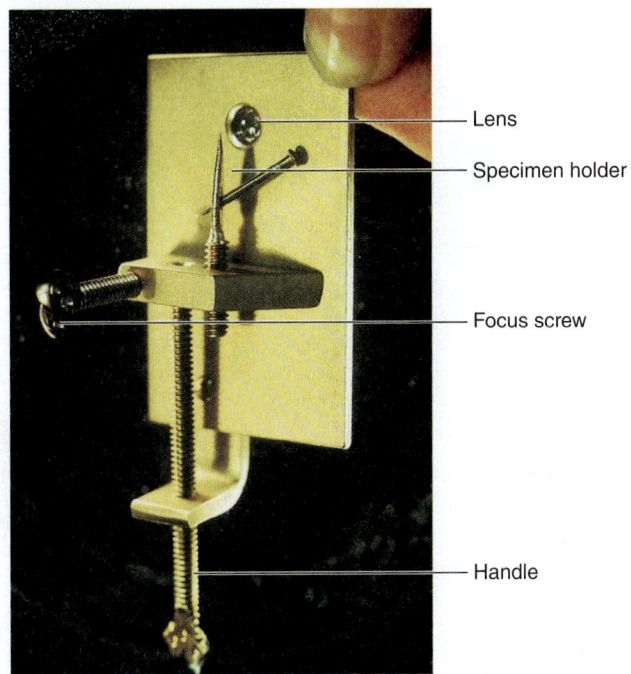

FIGURE 1.5 Model of van Leeuwenhoek's Microscope The original made in 1673 could magnify the object being viewed almost 300 times. The object being viewed is brought into focus with the adjusting screws. Note the small size.

Lens

Specimen holder

Focus screw

Handle

The living microbial world includes the kinds of cells that van Leeuwenhoek observed looking through his simple microscope (**figure 1.5**). Although he could not realize it at the time, members of the microbial world, in fact, all living organisms, can be classified into three distinct **domains**—*Bacteria, Archaea,* and *Eucarya.* Organisms in each domain share properties that distinguish them from members of the other domains. Many characteristics, however, are common among members of different domains. **Table 1.1** compares members of the domains *Bacteria, Archaea,* and *Eucarya.*

Bacteria

Members of the domain *Bacteria* are **prokaryotes,** meaning they are single-celled organisms consisting of a **prokaryotic cell** (meaning "prenucleus"). This cell type typically does not contain a membrane-bound nucleus or any other membrane-bound organelles. Instead, the genetic material resides in a region called the **nucleoid.**

Most *Bacteria* have specific shapes, commonly cylindrical (rod-shaped), spherical (round), or spiral. They typically have rigid cell walls that contain peptidoglycan, a compound unique to bacteria (see figure 3.31). Many of the *Bacteria* can move using flagella (singular: flagellum), appendages that extend from the cell. ▶▶ bacterial shapes, p. 50 ▶▶ flagella, pp. 63, 74

Bacteria typically multiply by binary fission, a process in which one cell enlarges and then divides. This forms two cells, each generally identical to the original. ▶▶ binary fission, p. 82

MicroByte
One group of organisms in the domain *Bacteria* has a membrane surrounding its genetic material as well as other internal membranes.

Archaea

Like members of the *Bacteria,* the *Archaea* have a prokaryotic cell structure. They have similar shapes, sizes, and appearances to the *Bacteria.* In addition, they multiply by binary fission, move primarily by means of flagella, and have rigid cell walls. Considering how much they look like *Bacteria,* scientists initially did not believe they could be so different chemically. We now know, however, that there are major differences between the *Bacteria* and the *Archaea,* and that *Archaea* are more closely related to humans than they are to *Bacteria.* In fact, you are more closely related to a plant than *Archaea* are to *Bacteria!*

Archaea and *Bacteria* have important differences in their chemical composition. For example, the archaeal cell wall does not contain peptidoglycan, whereas the bacterial wall does. One of the most significant distinctions is in their ribosomal RNA sequence, a characteristic discussed in other chapters. ▶▶ ribosomal RNA, p. 246

An interesting feature of many of the *Archaea* is their ability to grow in extreme environments that kill most other organisms. Some, for example, grow in salt concentrations 10 times higher than that of seawater. These organisms grow in such habitats as the Great Salt Lake and the Dead Sea. Others grow best at extremely high temperatures. One member can grow at a temperature of 121°C! (100°C is the temperature at which water boils at sea level). Some *Archaea* can be found in the hot springs at Yellowstone National Park.

While the *Archaea* that grow in extreme environments are the most intensively studied, others are abundant in moderate

TABLE 1.1	Comparison of Traits of Typical *Bacteria, Archaea,* and *Eucarya*		
	Bacteria	*Archaea*	*Eucarya*
Size	0.3–2 μm	0.3–2 μm	5–50 μm
Nuclear membrane	No	No	Yes
Cell wall	Peptidoglycan present	No peptidoglycan	No peptidoglycan
Membrane-bound organelles	No	No	Yes
Where found	In all environments	In all environments	In environments that are not extreme

environments. They are widely distributed in soils, the oceans, marshes, as well as in the intestinal tract. Some have been implicated in severe cases of periodontitis, a destructive inflammation of the gums.

Eucarya

Members of the *Eucarya* are **eukaryotes,** meaning they are composed of one or more **eukaryotic cells** ("true nucleus"). These cells have a membrane-bound nucleus and other organelles, making them far more complex than prokaryotes.

The microbial members of the *Eucarya* include **algae** (singular: alga), **fungi** (singular: fungus), and **protozoa** (singular: protozoan). Algae and protozoa are also referred to as **protists.** Some multicellular parasites are considered in this text because they kill millions of people around the world, especially in developing nations. This group of organisms, **helminths,** include roundworms and tapeworms. The eukaryotic members of the microbial world are compared in **table 1.2.**

FIGURE 1.6 Alga *Micrasterias,* a green alga composed of two symmetrical halves (100×).

❓ *What general features of algae distinguish them from other eukaryotic microorganisms?*

Algae

Algae are a diverse group of photosynthetic eukaryotes. Some are single-celled, whereas others are multicellular, and many different size and shapes are represented (**figure 1.6**). They all contain chloroplasts, many of which have chlorophyll, a green pigment. Some also contain other pigments that give them characteristic colors. The pigments absorb the energy of light, which is used in photosynthesis. Algae are usually found near the surface of either salt or fresh water. Their cell walls are rigid, but their chemical composition is quite distinct from that of the *Bacteria* and *Archaea.* Many algae move by means of flagella, which are structurally far more complex and unrelated to flagella of prokaryotes.

Fungi

Fungi are another diverse group of eukaryotes. Some are single-celled yeasts, but many are large multicellular organisms such as molds and mushrooms (**figure 1.7**). In contrast to algae, fungi gain their energy from degrading organic materials and are found wherever organic materials are present. Unlike algae, which live primarily in water, fungi live mostly on land.

Protozoa

Protozoa are a diverse group of microscopic, single-celled organisms that live in both aquatic and terrestrial environments. Although microscopic, they are very complex organisms and generally much larger than prokaryotes (**figure 1.8**). Unlike algae and

(a)   10 µm   (b)   10 µm

Reproductive structures (spores)

Mycelium

FIGURE 1.7 Fungi (a) Yeast, *Malassezia furfur.* **(b)** *Aspergillus,* a typical mold form whose reproductive structures rise above the mycelium.

❓ *What type of cells make up molds and mushrooms?*

TABLE 1.2	Comparison of Eukaryotic Members of the Microbial World		
	Algae	**Fungi**	**Protozoa**
Cell organization	Single- or multicellular	Single- or multicellular	Single-celled
Source of energy	Sunlight	Organic compounds	Organic compounds
Size	Microscopic or macroscopic	Microscopic or macroscopic	Microscopic

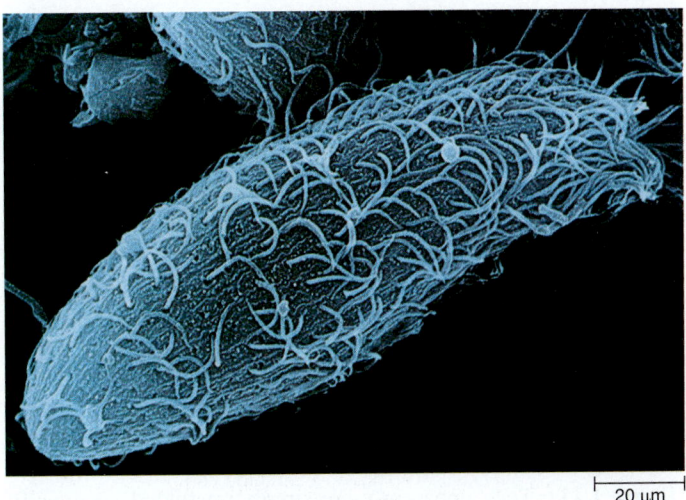

20 μm

FIGURE 1.8 Protozoan A paramecium moves with the aid of hair-like appendages on the cell surface.

❓ *How do protozoa differ from both fungi and algae?*

fungi, protozoa do not have a rigid cell wall. Most protozoa ingest organic compounds as food sources and are motile.

Nomenclature

In biology, the Binomial System of Nomenclature refers to a two-word naming system. The first word in the name indicates the **genus,** with the first letter always capitalized; the second indicates the specific epithet, or **species** name, and is not capitalized. Both words are usually italicized or underlined—for example, *Escherichia coli.* The genus name is commonly abbreviated, with the first letter capitalized—as in *E. coli.* A number of different species are included in the same genus. Members of the same species may vary from one another in minor ways, but not enough to give the organisms different species names. These differences, however, may result in the organism being given different strain designations, for example, *E. coli* strain B or *E. coli* strain K12.

Bacteria are often referred to informally by names resembling genus names but are not italicized. For example, species of *Staphylococcus* are often called staphylococci.

MicroAssessment 1.3

Three domains of life exist: *Bacteria, Archaea,* and *Eucarya.* Members of the *Bacteria* and *Archaea* are prokaryotes, but they are distinctly different from each other in certain features of their chemical composition. Members of the *Eucarya* are eukaryotes; algae, fungi, protozoa, and multicellular parasites belong to this group. Organisms are named according to the Binomial System of Nomenclature.

7. *Name one feature that distinguishes members of the* Bacteria *from the* Archaea.

8. *List two features that distinguish prokaryotes from eukaryotes.*

9. *The binomial system of classification uses both a genus and a species name. Why are two names used?* ➕

1.4 ■ Non-Living Members of the Microbial World

Learning Outcome

9. *Compare and contrast viruses, viroids, and prions.*

Viruses, viroids, and prions are not living; they are acellular infectious agents. By definition, an organism must be composed of one or more cells to be alive. Non-living agents are not microorganisms, so the term microbe is often used as a general term to include any member of the microbial world. ▶▶ **viroids, p. 328**
▶▶ **prions, p. 328**

Viruses consist of nucleic acid packaged within a protein coat, and come in a variety of shapes (**figure 1.9**). To multiply, viruses use the machinery and nutrients of living cells, referred to as **hosts.** Outside the hosts, however, viruses are inactive. Thus, viruses are **obligate intracellular parasites.** All forms of life including members of the *Bacteria, Archaea,* and *Eucarya* can be infected by viruses but of different types. Although viruses frequently kill the cells in which they multiply, some types can remain within the host cell without causing obvious ill effects. As the host cells multiply, they copy the viral genetic information, passing it along to their progeny.

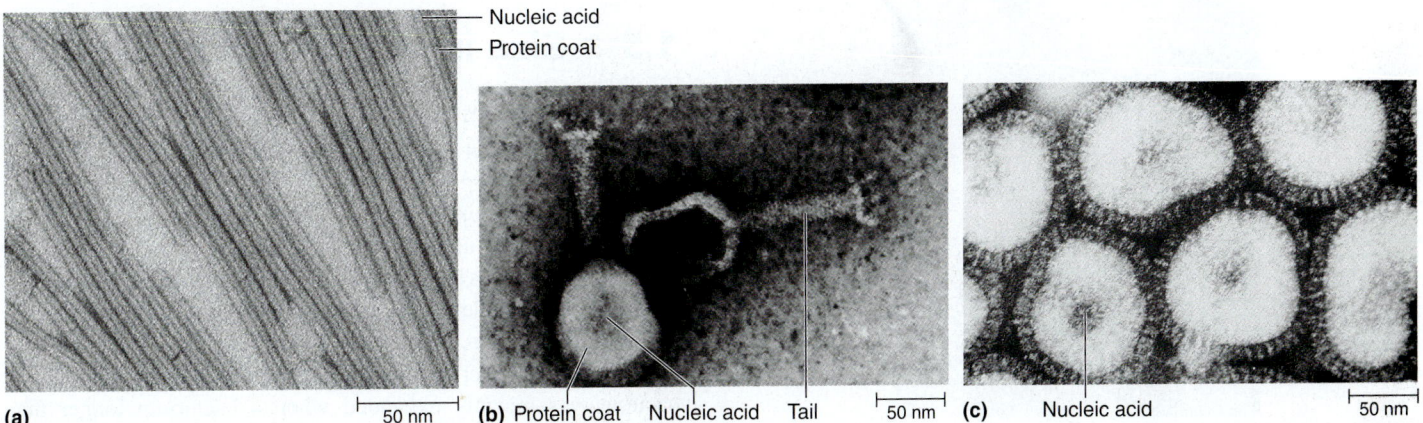

— Nucleic acid
— Protein coat

(a) 50 nm (b) Protein coat Nucleic acid Tail 50 nm (c) Nucleic acid 50 nm

FIGURE 1.9 Viruses That Infect Three Kinds of Organisms **(a)** Tobacco mosaic virus that infects tobacco plants. A long hollow protein coat surrounds a molecule of nucleic acid. **(b)** A bacterial virus (bacteriophage), which infects bacteria. Nucleic acid is surrounded by a protein coat (head) which is attached to a hollow protein tail. **(c)** Influenza virus. This virus infects humans and causes flu.

❓ *Why can viruses be so much smaller than cells and still replicate?*

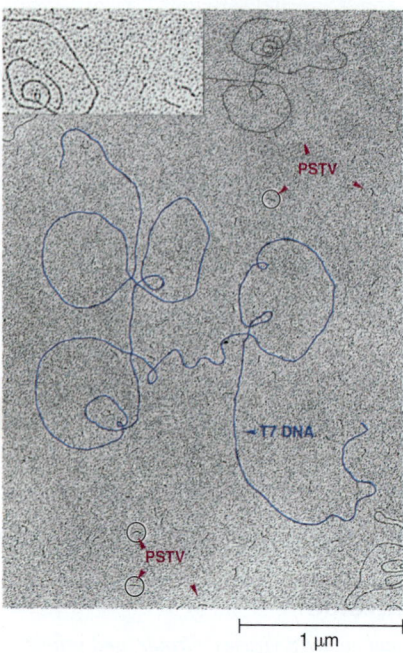

FIGURE 1.10 The Size of a Viroid Compared with a Molecule of DNA from a Virus That Infects Bacteria (T7) The red arrows point to potato spindle tuber viroids (PSTV); the blue arrow points to the bacterial virus T7 DNA.

❓ *How does a viroid differ from a virus?*

Viroids are simpler than viruses, consisting of only a single, short piece of ribonucleic acid (RNA) (**figure 1.10**). Like viruses, they multiply only inside cells. Viroids cause a number of plant diseases, and some scientists speculate that they may cause diseases in humans, although no evidence for this yet exists.

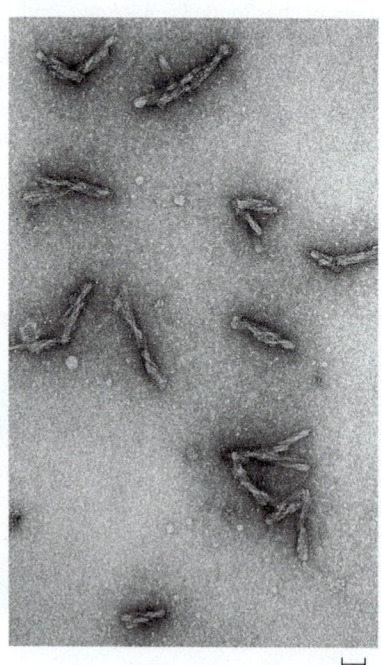

FIGURE 1.11 Prion Prions isolated from the brain of a scrapie-infected hamster. This neurodegenerative disease is caused by a prion.

❓ *Why are prions visible here when normal cellular proteins are not?*

TABLE 1.3	Distinguishing Characteristics of Viruses, Viroids, and Prions	
Viruses	**Viroids**	**Prions**
Obligate intracellular agents	Obligate intracellular agents	Abnormal form of a cellular protein
Consist of either DNA or RNA, surrounded by a protein coat	Consist only of RNA; no protein coat	Consist only of protein; no DNA or RNA

Prions are infectious proteins responsible for several fatal neurodegenerative diseases in humans and other animals (**figure 1.11**). They apparently are simply misfolded versions of normal cellular proteins found in the brain. When the misfolded version contacts the normal cellular protein, it forces the normal protein to misfold, thereby resulting in cells becoming filled with abnormal proteins that bind together to form threadlike structures called fibrils. The fibril-filled cells are not able to function. The disease is acquired by eating prion-containing nervous tissues. Prions are more resistant to degradation by cellular enzymes than are their normal counterparts. Prions are also resistant to the usual sterilization procedures that destroy viruses and bacteria. Some scientists speculate that Alzheimer's and Parkinson's diseases are caused by mechanisms similar to those of prion diseases.

The distinguishing features of the non-living members of the microbial world are listed in **table 1.3.** The relationships of the major groups of the microbial world to one another are presented in **figure 1.12.**

MicroAssessment 1.4

Viruses, viroids, and prions are infectious agents.

10. *Describe the chemical composition of viruses, viroids, and prions.*
11. *Which of the non-living members of the microbial world seems to be the least threat to human health?* ➕

1.5 ■ Size in the Microbial World

Learning Outcome

10. *Compare the size differences among microbes.*

Although all members of the microbial world are small, microbes span a tremendous range of sizes—the largest eukaryotic cells are about a million times bigger than the smallest viruses (**figure 1.13**). Even within a single group, wide variations exist. For example, *Bacillus megaterium* and *Mycoplasma pneumoniae* are both bacteria, but the former are much larger. The variation in size of bacteria was recently expanded when a bacterium longer than 0.5 mm was discovered (see **Perspective 1.1**). It is so big, in fact, that it is visible to the naked eye. More recently, an even larger bacterium, round in shape, was discovered. Its volume is 70 times larger than the previous record holder. Likewise, a eukaryotic cell was recently discovered that is not much larger than a typical

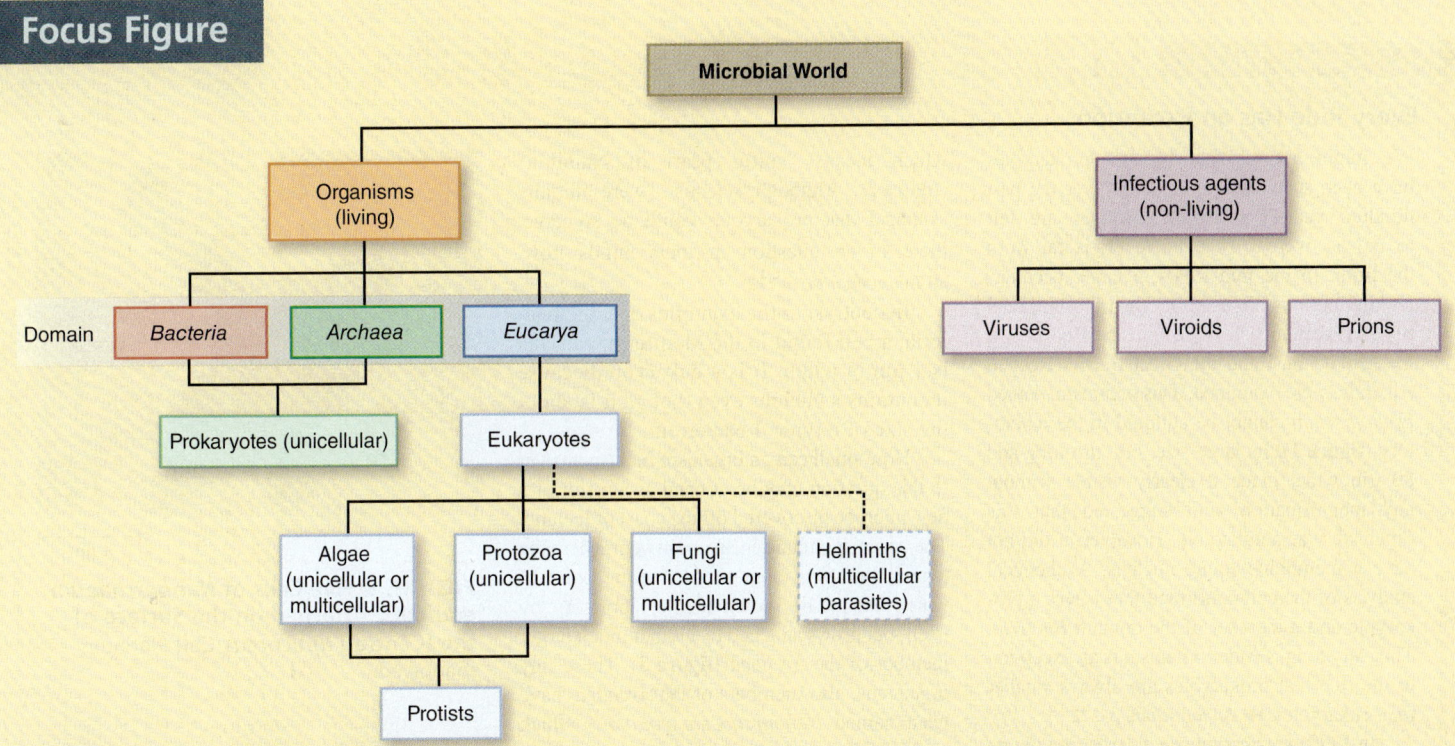

FIGURE 1.12 The Microbial World Although adult helminths are generally not microscopic, some stages in the life cycle of many disease-causing helminths are.

❓ *The members of which two domains cannot be distinguished microscopically?*

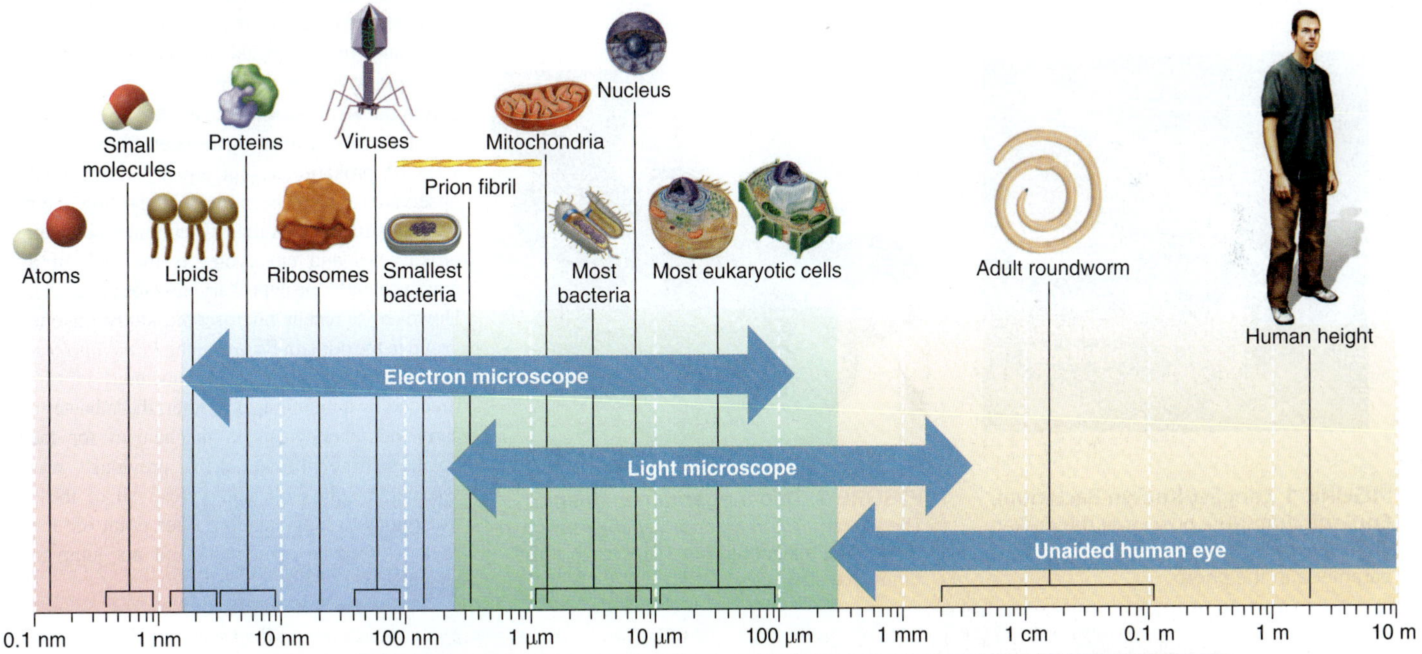

The basic unit of length is the meter (m), and all other units are fractions of a meter.

These units of measurement correspond to units in an older but still widely used convention.

nanometer (nm) = 10^{-9} meter = .000000001 meter
micrometer (μm) = 10^{-6} meter = .000001 meter
millimeter (mm) = 10^{-3} meter = .001 meter
1 meter = 39.4 inches

1 angstrom (Å) = 10^{-10} meter
1 micron (μ) = 10^{-6} meter

FIGURE 1.13 Sizes of Molecules, Non-Living Agents, and Organisms

❓ *Why is a logarithmic scale necessary when comparing sizes of members of the microbial world?*

PERSPECTIVE   1.1

Every Rule Has an Exception

We might assume that because prokaryotes have been so intensively studied over the past hundred years, no major surprises are left to be discovered. This, however, is far from the truth. In the mid-1990s, a large, peculiar-looking organism was seen when the intestinal tracts of certain fish from both the Red Sea in the Middle East and the Great Barrier Reef in Australia were examined. This organism, named *Epulopiscium* cannot be cultured in the laboratory (**figure 1**). Its large size, 600 μm long and 80 μm wide, made it clearly visible without any magnification, and suggested that this organism was a eukaryote. However, it did not have a membrane-bound nucleus. A chemical analysis of the cell confirmed that it was a prokaryote and a member of the domain *Bacteria*. This very long, slender organism is an exception to the rule that prokaryotes are always smaller than eukaryotes. ▶▶ *Epulopiscium*, p. 272

In 1999, a prokaryote even larger in volume was isolated from the muck of the ocean floor off the coast of Namibia in Africa. It is a spherical organism 70 times larger in volume than *Epulopiscium*. Since it grows on sulfur compounds and contains glistening globules of sulfur, it was named *Thiomargarita namibiensis*,

which means "sulfur pearl of Namibia" (**figure 2**). Although scientists were initially skeptical that prokaryotes could be so large, there is no question in their minds now. ▶▶ *Thiomargarita*, p. 273

In contrast to the examples of large bacteria, a cell found in the Mediterranean Sea is 1 μm in width. It is a eukaryote because it contains a nucleus even though it is about the size of a typical bacterium.

How small can an organism be? An answer to this question may be at hand as a result of a new microbe discovered off the coast of Iceland. The organism, found in an ocean vent where the temperature was close to the boiling point of water, cannot be grown in the laboratory by itself, but grows attached to another much larger member of the *Archaea* (**figure 3**). These tiny organisms, also members of the *Archaea*, have been named *Nanoarchaeum equitans*, which means "riding the fire sphere." The organism to which *N. equitans* is attached is *Ignicoccus*, which means "fire ball." *Ignicoccus* grows very well without its rider. *N. equitans* is spherical and only about 400 nanometers in diameter, about a quarter the diameter of *Ignicoccus*. Also, the amount of genetic information (DNA)

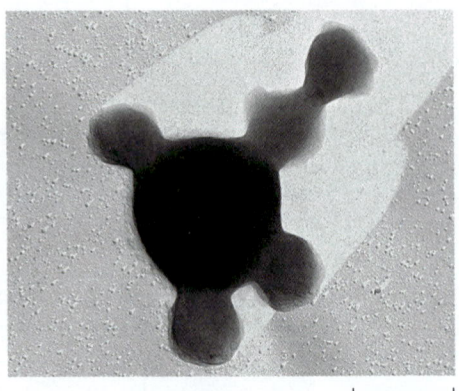

1 μm

FIGURE 3 Five Cells of *Nanoarchaeum equitans*, Attached on the Surface of the (Central) *Ignicoccus* Cell Platinum shadowed.

contained in *N. equitans* is less than in any known organism, and only about one-tenth the amount found in the common gut organism, *Escherichia coli*. Further analysis of these cells suggests that they may resemble the earliest cells and therefore the ancestor of all life. The scientists who discovered *N. equitans* suggest that many more unusual organisms related to *N. equitans* will be discovered. They are probably right! ▶▶ *Nanoarchaeum*, p. 279

Not only have surprises been found in the sizes of prokaryotes, but equally unexpected observations have been made regarding their internal structures. A major distinction between prokaryotes and eukaryotes is the lack of a nuclear membrane in prokaryotes (see table 1.1). However, a recent microscopic study on one group in the domain *Bacteria*, the *Planctomyces*, has clearly shown that the nucleoid is surrounded by a membrane. Further, these bacteria carry out endocytosis, a mechanism for the uptake of foreign material that previously was associated only with eukaryotes. All of these exceptions to long-standing rules point out the need to keep an open mind and not jump to conclusions.

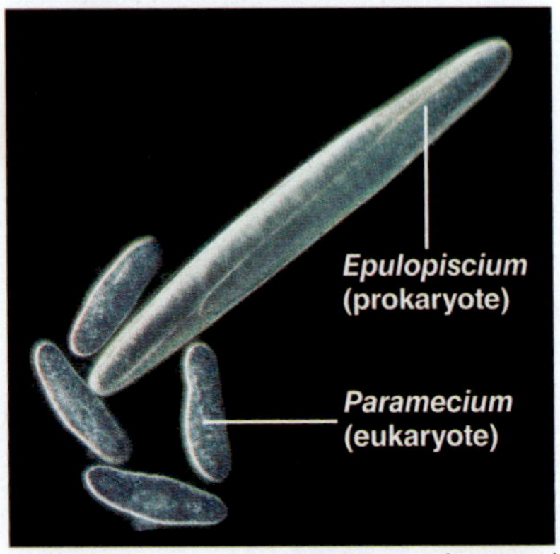

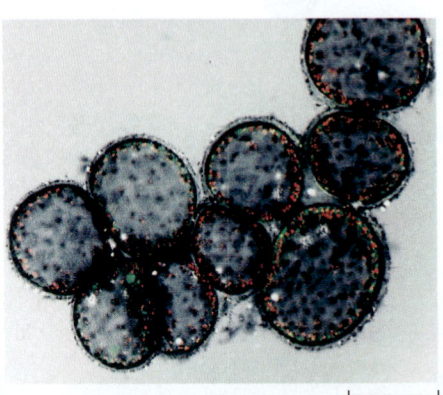

Epulopiscium
(prokaryote)

Paramecium
(eukaryote)

0.1 mm

FIGURE 1 Longest Known Bacterium, *Epulopiscium* Note how large this prokaryote is compared with the four eukaryotic paramecia in the photo.

0.2 mm

FIGURE 2 *Thiomargarita namibiensis* The average *Thiomargarita namibiensis* is two-tenths of a millimeter, but some reach three times that size.

bacterium. These, however, are rare exceptions to the rule that eukaryotes are larger than prokaryotes, which in turn are larger than viruses.

The small size of microbes requires measurements not commonly encountered in everyday life. Logarithms are enormously helpful, especially in designating the sizes of prokaryotes and viruses. A brief discussion of measurements and logarithms is given in Appendix I.

MicroAssessment 1.5

The size range of microbes is tremendous. As a general rule, eukaryotes are the largest and obligate intracellular parasites the smallest.

12. *Place in order with respect to typical size (arrange from smallest to largest) bacteria, eukaryotic cells, and viruses.*

13. *What factor limits the size of free-living cells?* ✛

Entering a New Golden Age

For all the information that has been gathered about the microbial world, it is remarkable how little we know about its prokaryotic members. This is not surprising in view of the fact that less than 1% of the prokaryotes have ever been studied. In large part, this is because only one in a hundred of the prokaryotes in the environment can be cultured in the laboratory. Part of the current revolution in microbiology, however, will allow us to inventory the millions of species yet to be discovered. This is now being done. Using techniques that helped decipher the human genome, scientists have begun to analyze the biological content of the oceans. In a small volume of water from the Sargasso Sea—an area of the ocean that contains few nutrients and therefore presumably few organisms—scientists reported the finding of 1,800 previously unknown bacterial species. This estimate is based on the analysis of the DNA isolated from these microorganisms. The biodiversity of the microbial world is astounding!

Exploring the unknowns in the microbial world is a major challenge and should answer many intriguing questions fundamental to understanding the biological world. What are the extremes of temperature, salt, pH, radioactivity, and pressure in which prokaryotes can live? Are there organisms growing in even more extreme environments? If life can exist on this planet under such extreme conditions, what does this mean about the possibility of finding living organisms on other planets? Although considered highly unlikely, is it possible that living organisms exist whose chemical structure is not based on the carbon atom? Will living organisms be found whose genetic information is coded in a chemical other than deoxyribonucleic acid? What new metabolic pathways remain to be discovered? As extreme environments are mined for their living biological diversity, there seems little doubt that many surprises will be found. In many cases these surprises will be translated into new biotechnology products on this planet, and they will help shape the way we look for life on other planets.

One hundred years ago we were in the Golden Age of Microbiology. With new technologies available, we are entering a second Golden Age of identifying unknown members of the microbial world.

Summary

1.1 ■ The Dispute Over Spontaneous Generation

The belief in **spontaneous generation** was challenged by Francesco Redi in the seventeenth century.

Early Experiments

The experiments of Needham supported the idea of spontaneous generation while those of Spallazani did not.

Experiments of Pasteur

The experiments of Louis Pasteur disproved spontaneous generation (figure 1.1).

Experiments of Tyndall

John Tyndall showed that some microbial forms are not killed by boiling. He and Ferdinand Cohn discovered **endospores,** the heat-resistant forms of bacteria.

1.2 ■ Microbiology: A Human Perspective

Vital Activities of Microorganisms

The activities of microorganisms are vital for the survival of all other organisms, including humans. Bacteria convert the nitrogen gas in air into a form that plants and other organisms can use. Microorganisms replenish O_2 and degrade certain materials that no other organisms can.

Applications of Microbiology

Bread, wine, beer, and cheeses are made today using technology developed 4,000 years ago. Bacteria are used to degrade toxic pollutants and to synthesize a variety of different products, such as cellulose, hydroxybutyric acid, ethanol, antibiotics, and amino acids. Biotechnology depends on members of the microbial world.

Medical Microbiology

Many devastating diseases such as smallpox, plague, and influenza have determined the course of history. Diseases emerge because of new lifestyles and changing infectious agent (figure 1.4). Diseases once under control re-emerge when successful preventive measures become victims of their own success, **pathogens** become resistant to antimicrobial medications, travelers and immigrants carry pathogens around the globe, and populations age or otherwise become more susceptible to diseases. People have become lax in having children vaccinated. The normal microbiota plays a protective role, but pathogens cause disease.

Microorganisms as Model Organisms

Microorganisms are excellent model organisms because they grow rapidly on simple, inexpensive media. Their growth reveals the same genetic, metabolic, and biochemical principles as higher organisms.

1.3 ■ Living Members of the Microbial World (table 1.1)

Diversity of Microorganisms

The microbial world is incredibly diverse (figure 1.12).

Bacteria

Members of the *Bacteria* are single-celled **prokaryotes** that have peptidoglycan in their cell wall.

Archaea

Members of the *Archaea* are identical in appearance to the *Bacteria,* but are very different in their chemical composition. They do not contain peptidoglycan. Many *Archaea* grow in extreme environments.

Eucarya

Members of the *Eucarya* have **eukaryotic cell** structures (table 1.2). **Algae** can be single-celled or multicellular, and use sunlight as an energy source (figure 1.6). **Fungi** include single-celled yeasts and multicellular molds and mushrooms; they use organic compounds as food (figure 1.7). **Protozoa** are typically motile single-celled organisms that use organic compounds as food (figure 1.8).

Nomenclature

Organisms are named according to a binomial system. Each organism has a **genus** and a **species** name, written in italics or underlined.

1.4 ■ Non-Living Members of the Microbial World

The non-living members of the microbial world are not composed of cells (table 1.3). **Viruses** consist of nucleic acid within a protein coat (figure 1.9). **Viroids** consist of a single, short RNA molecule (figure 1.10). **Prions** consist only of protein; apparently, they are mis-folded versions of normal cellular protein (figure 1.11).

1.5 ■ Size in the Microbial World

Although they are all small, sizes of members of the microbial world vary tremendously (figure 1.13).

Review Questions

Short Answer

1. How did Louis Pasteur help disprove spontaneous generation?
2. Give three reasons why life could not exist without the activities of microorganisms.
3. List five beneficial applications of bacteria.
4. State three reasons why there is a resurgence of infectious diseases today.
5. Name the prokaryotic groups in the microbial world.
6. Name one location where you could isolate members of the *Archaea*.
7. How might you distinguish a prokaryotic cell from a eukaryotic cell?
8. In the designation *Escherichia coli* B, what is the genus? What is the species? What is the strain?
9. Why are viruses not microorganisms?
10. Name three non-living groups in the microbial world and describe their major properties.

Multiple Choice

1. The property of endospores that led to confusion in the experiments on spontaneous generation is their
   a) small size.
   b) ability to pass through cork stoppers.
   c) heat resistance.
   d) presence in all infusions.
   e) presence on cotton plugs.
2. The "Golden Age of Microbiology" was the time when
   a) microorganisms were first used to make bread.
   b) microorganisms were first used to make cheese.
   c) most pathogenic bacteria were identified.
   d) a vaccine against influenza was developed.
   e) antibiotics became available.
3. Microorganisms play a role in
   a) disease.   b) biodegradation. c) cheese production.
   d) nitrogen recycling.   e) all of the above.
4. Which disease was once thought to be due to stress but is now known to be caused by a bacterium?
   a) smallpox   b) peptic ulcers   c) AIDS
   d) plague   e) influenza
5. The prokaryotic members of the microbial world include
   1. algae.   2. fungi.   3. prions.   4. bacteria.   5. archaea.
   a) 1, 2   b) 2, 3   c) 3, 4   d) 4, 5   e) 1, 5

6. The *Archaea*
   1. are microscopic.
   2. are commonly found in extreme environments.
   3. contain peptidoglycan.
   4. contain mitochondria.
   5. are most commonly found in the soil.
   a) 1, 2   b) 2, 3   c) 3, 4   d) 4, 5   e) 1, 5
7. Prokaryotes typically do not have
   a) cell walls.   b) flagella.   c) a nuclear membrane.
   d) specific shapes.   e) genetic information.
8. Nucleoids are associated with
   1. genetic information.   2. prokaryotes.
   3. eukaryotes.   4. viruses.   5. prions.
   a) 1, 2   b) 2, 3   c) 3, 4   d) 4, 5   e) 1, 5
9. Viruses
   1. contain both protein and nucleic acid.
   2. infect all domains of life.
   3. can grow in the absence of living cells.
   4. are generally the same size as prokaryotes.
   5. always kill the cells they infect.
   a) 1, 2   b) 2, 3   c) 3, 4   d) 4, 5   e) 1, 5
10. Antony van Leeuwenhoek could not have observed
    a) roundworms.   b) *Escherichia coli.*
    c) yeasts.   d) viruses.

Applications

1. The American Society for Microbiology is preparing a "Microbe-Free" banquet to emphasize the importance of microorganisms in the diet. What foods could not be on the menu?
2. If you were asked to nominate one of the individuals mentioned in this chapter for the Nobel Prize, who would it be? Make a statement supporting your choice.

Critical Thinking ✚

1. A microbiologist obtained two pure biological samples: one of a virus, and the other of a viroid. Unfortunately, the labels had been lost. The microbiologist felt she could distinguish the two by analyzing for the presence or absence of a single molecule. What molecule would she search for and why?
2. Why is the bacterium that causes anthrax such an effective agent of bioterrorism?

2

The Molecules of Life

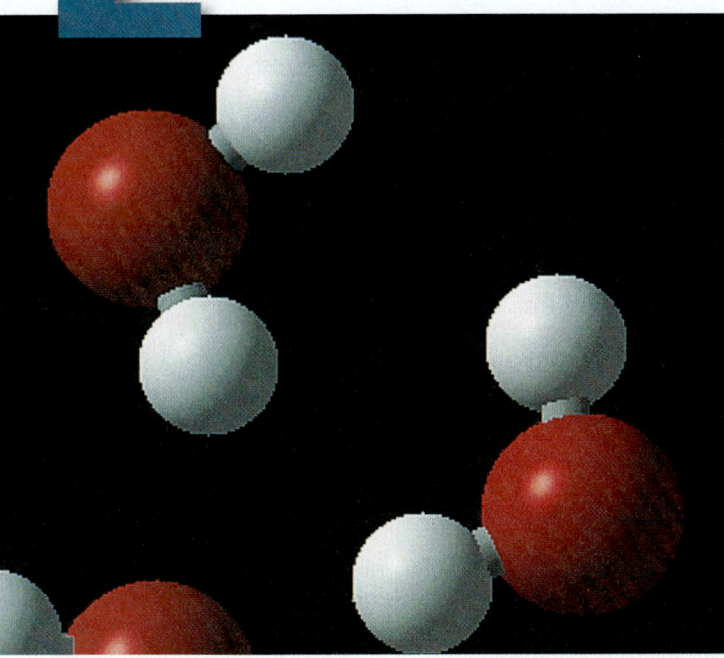

Ball-and-stick model of water molecules.

A Glimpse of History

Louis Pasteur (1822–1895) is often considered the father of bacteriology because of his many contributions to the field. However, he started his scientific career as a chemist, researching crystals for the French wine industry. In his studies, he worked with tartaric and paratartaric acids, which form thick crusts within wine barrels. These two substances form crystals that are identical in their chemical composition, but affect polarized light very differently. When polarized light passes through tartaric acid crystals, the light rotates (twists) to the right. In contrast, paratartaric acid crystals have no effect on the light. Pasteur wanted to learn how the crystals differed.

Pasteur noticed that under a microscope, all tartaric acid crystals looked identical, whereas paratartaric acid crystals had two kinds of structures. Using tweezers, Pasteur carefully separated the two types of paratartaric acid crystals and dissolved them in separate flasks of water. He found that one solution rotated polarized light to the left and the other, to the right. Solutions containing equal numbers of the two crystals did not rotate the light. Pasteur concluded that paratartaric acid must be a mixture of two types of chemical structures, each being the mirror image, or stereoisomer, of the other. The stereoisomers counteracted the light-rotating effects of each other and, as a result, the mixture did not rotate the light.

What Pasteur studied as a simple problem in chemistry has implications far beyond what he ever imagined. In fact, stereoisomers of the same molecule can have very different properties. For example, one form of a key ingredient in the artificial sweetener aspartame is sweet but the other form is bitter. Pasteur's work demonstrates that seemingly unimportant observations sometimes lead to significant discoveries with far-reaching implications.

To understand how cells interact with one another and their environment, you must be familiar with the components that make up all living matter. The principles described in this chapter are fundamental to information throughout the text.

2.1 ■ Atoms and Elements

Learning Outcome

1. *Describe the characteristics of atoms and elements and how atoms can be depicted diagrammatically.*

Atoms, the basic units of all matter, are made up of three major components (**figure 2.1**):

■ **Neutrons:** uncharged particles

■ **Protons:** positively charged particles

■ **Electrons:** negatively charged particles

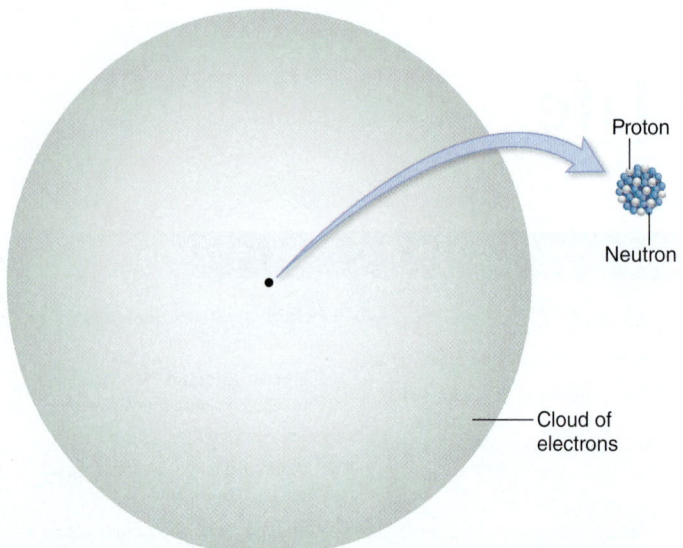

FIGURE 2.1 Atom A proton has a positive charge, a neutron has a neutral charge, and an electron has a negative charge. The electrons move around the nucleus as a cloud, arranged in shells of different energy levels.

? *How does the number of electrons in an atom compare to the number of protons?*

The number of protons normally equals the number of electrons, so the atom as a whole has no charge. The neutrons and protons make up the heaviest part of the atom—the nucleus—and the electrons form an electron "cloud" around the nucleus (figure 2.1).

An atom is distinguished by its atomic number, the number of protons in its nucleus (**table 2.1**). For example, a hydrogen atom has one proton so its atomic number is 1; a carbon atom has 6 protons so its atomic number is 6. Each atom has a mass number, the sum of the number of protons and neutrons (electrons are too light to contribute to the mass). A hydrogen atom that has 1 proton and no neutrons has both an atomic number and mass number of 1. A carbon atom that has 6 protons and 6 neutrons has an atomic number of 6 and a mass number of 12 (**figure 2.2**). The symbols for atoms are sometimes written with the atomic number in subscript to the left, and the mass number in superscript to the left.

An **element** consists of only one kind of atom and cannot be separated into simpler parts by chemical methods. There are 92

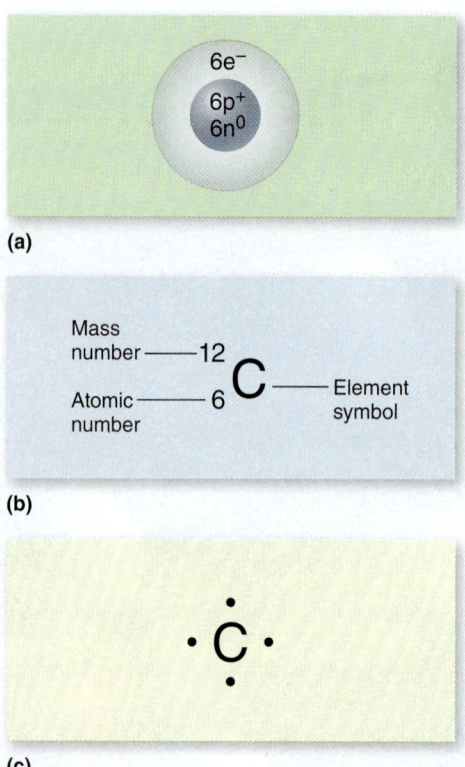

FIGURE 2.2 Depictions of an Atom An atom of carbon (C) is shown in three different ways: **(a)** cloud structure, **(b)** symbol designation, and **(c)** Lewis structure.

? *How would the Lewis structure of hydrogen be different from that of carbon?*

naturally occurring elements, but living material is primarily made up of only four: carbon (the symbol is C), hydrogen (H), oxygen (O), and nitrogen (N). Two other elements, phosphorus (P), and sulfur (S), are also found in all living systems. All atoms of an element have the same atomic number, but they can have different mass numbers. That is, they have the same number of protons but can have different numbers of neutrons. Nearly 99% of naturally occurring carbon atoms have 6 neutrons, but a few have 7 or 8. These various forms—**isotopes**—are useful tools in biological research (see **Perspective 2.1**).

Electrons of an atom are arranged in shells around the nucleus, with each shell having a limit to the number of electrons it can hold.

TABLE 2.1	Characteristics of Atoms Common in Biology				
Atom	Symbol	Atomic Number (Number of Protons)	Mass Number (Protons + Neutrons)	Number of Covalent Bonds Formed	Approximate % in Cells
Hydrogen	H	1	1	1	49
Carbon	C	6	12	4	25
Nitrogen	N	7	14	3	0.5
Oxygen	O	8	16	2	25
Phosphorus	P	15	31	3	0.1
Sulfur	S	16	32	2	0.4

Isotopes: Valuable Tools for the Study of Biological Systems

One important tool in the analysis of living cells is the use of isotopes, variant forms of atoms that have different atomic masses. The nuclei of certain atoms can have greater or fewer neutrons than usual and thereby be heavier or lighter than typical. For example, the most common form of the hydrogen atom contains one proton and no neutrons and has a mass number of 1 (^{1}H). Another form also exists in nature in very low amounts. This isotope, ^{2}H (deuterium), contains one neutron. A third, even heavier isotope, ^{3}H (tritium), is not found in nature but can be made by a nuclear reaction in which stable atoms are bombarded with high-energy particles. This latter isotope is unstable and gives off radiation (decays) in the form of rays or electrons, which can be very sensitively measured by a radioactivity counter. Once the atom has finished disintegrating, it no longer gives off radiation and is stable.

The other properties of radioactive isotopes, called radioisotopes, are very similar to their non-radioactive counterparts. For example, tritium combines with oxygen to form water and with carbon to form hydrocarbons, and both molecules have biological properties similar to those of their non-radioactive counterparts. The only difference is that the molecules containing tritium can be detected by the radiation they emit.

Radioisotopes are used in numerous ways in biological research. They are frequently added to growing cells in order to label particular molecules, thereby making them detectable. For example, tritiated thymidine (thymidine is a component of DNA) added to growing bacteria will label only DNA and no other molecules. Tritiated uridine will label RNA (uridine is a component of RNA). Radioisotopes are also used in medical diagnosis. For example, to evaluate proper functioning of the human thyroid gland—which produces the iodine-containing hormone thyroxin—doctors often administer radioactive iodine. He or she would then scan the gland later to determine if the amount and distribution of iodine is normal (**figure 1**).

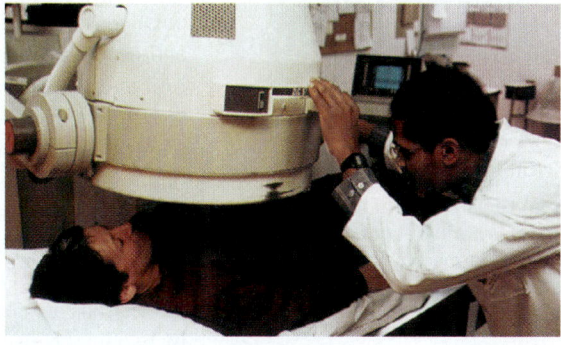
(a)

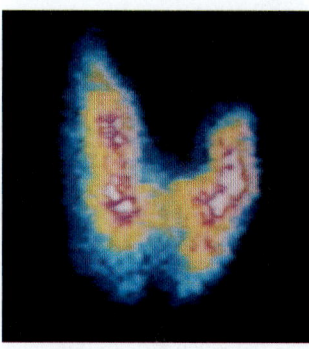
(b)

Figure 1 Radioisotopes (a) Physicians use scintillation counters such as this to detect radioisotopes. **(b)** A scan of the thyroid gland 24 hours after the patient received radioactive iodine.

The electrons are attracted to the protons in the nucleus, so shells closer to the nucleus are filled before electrons occupy other shells. The first shell (closest to the nucleus) holds no more than 2 electrons. These electrons are most highly attracted to the nucleus and have the lowest energy levels in that particular atom. Once that shell is filled, additional electrons occupy the next shell, to a maximum of 8 electrons. Larger atoms have additional shells that hold even more electrons. Most atoms of biological significance follow the "octet rule," meaning they are most stable when their outer shell has 8 electrons. An important exception is hydrogen, with a single shell; recall that the first shell has a limit of 2 electrons.

Valence electrons are those found in an atom's outer shell. They are particularly important in chemical bond formation, which will be discussed next. Lewis structures are often used to show atoms in a way that highlights the number of valence electrons (figure 2.2c). Each of the valence electrons is illustrated as a dot surrounding the atomic symbol; any other electrons are ignored.

MicroByte

If the nucleus of an atom were a marble in the center of a football field, the electrons would occupy the entire space of the stadium.

MicroAssessment 2.1

The basic unit of all matter, the atom, is composed of protons, electrons, and neutrons. The four most important elements in biology are carbon, hydrogen, oxygen, and nitrogen. In diagrams, atoms can be shown as a cloud structure, symbol designation, or Lewis structure.

1. *Why are electrons not considered in determining the mass number of an element?*
2. *What is the "octet rule" and its biologically important exception?*
3. *Why is the energy level of an electron higher the farther it is from the nucleus?* ➕

2.2 ■ Chemical Reactions and Bonds

Learning Outcomes

2. *Describe the importance of valence electrons.*
3. *Compare and contrast ionic bonds, covalent bonds, and hydrogen bonds.*
4. *Explain the concept of a mole and why molarity is important in chemistry.*

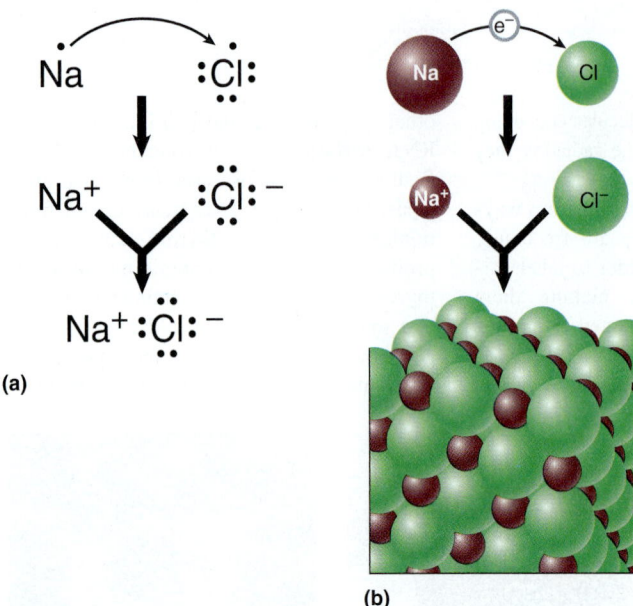

(a)

(b)

FIGURE 2.3 Ions and Ionic Bonds (a) Lewis model of sodium and chloride ions being formed, and an ionic bond between them. **(b)** Space-filling model of a salt crystal being formed by ionic bonding. Note that by convention, each type of atom is depicted using a consistent color. Also, an atom gets smaller when it loses an electron and larger when it gains one.

❓ *Which of the ions in this figure is an anion, and which is a cation?*

Atoms lose, gain, or share the electrons in their outer shell to achieve the most stable state. This is the basis for chemical bond formation. Recall that with the exception of hydrogen, biologically important atoms are most stable when they have eight valence electrons (the octet rule).

Ions and Ionic Bonds

An atom that gains or loses an electron is no longer neutral—it is an **ion (figure 2.3)**. Positively charged ions are called **cations;** negatively charged ions are **anions.** Atoms that gain an electron become negatively charged, whereas those that lose an electron become positively charged. The type and amount of charge are indicated by a superscript to the right of the chemical symbol. For example, Na^+ indicates a sodium atom that has lost one electron and therefore carries a +1 (positive) charge; Mg^{2+} indicates a magnesium atom that has lost two electrons and therefore has a +2 charge.

Ionic bonds form between cations and anions because of strong attractions between positive and negative charges (figure 2.3). The resulting product is called a **salt.** A common type of salt, sodium chloride (table salt), is composed of Na^+ (sodium cations) and Cl^- (chloride anions) and forms a solid crystalline structure. Water often breaks ionic bonds because of reasons described later. In aqueous solution, salts are electrolytes, meaning they conduct electricity.

MicroByte
Electrical charges from the heart are conducted by electrolytes and can be detected on the body surface as an electrocardiogram (ECG).

Covalent Bonds

Atoms often share pairs of valence electrons, the basis for **covalent bond** formation. One pair of shared electrons is a covalent bond. The number of covalent bonds an atom can form—its valence—is the number of electrons the atom must gain or lose to fill its outer shell (table 2.1). Two or more atoms joined together by covalent bonds form a **molecule.** The atoms that make up a molecule may be of the same or different elements. H_2 is a molecule of hydrogen gas formed from two atoms of hydrogen; a water molecule (H_2O) contains two hydrogen atoms bonded to one oxygen atom. Water is a **compound,** a molecule consisting of more than one element.

Carbon (C)—a critical atom in biology—frequently forms covalent bonds. This element has four electrons in its outer shell, and needs four more to fill it. A hydrogen (H) atom has one electron and requires an additional one to fill its shell. If one C atom shares electrons with four H atoms, methane (CH_4) is created (**figure 2.4;** see also table 2.1). Carbon atoms joined by covalent bonds to H atoms form an **organic compound.** All other compounds are **inorganic compounds.**

A covalent bond is indicated by a dash between the two atoms sharing the pair of electrons; for example C—H. Sometimes two pairs of electrons are shared. This forms a double covalent bond indicated by two lines between the atoms; for example O=C=O (CO_2).

Covalent bonds are strong. Double bonds are stronger than single bonds, and triple bonds are stronger still. The stronger the bond, the more difficult it is to break. Consequently, covalent bonds do not break unless exposed to certain chemicals or large amounts of energy, generally as heat. The temperatures required are incompatible with

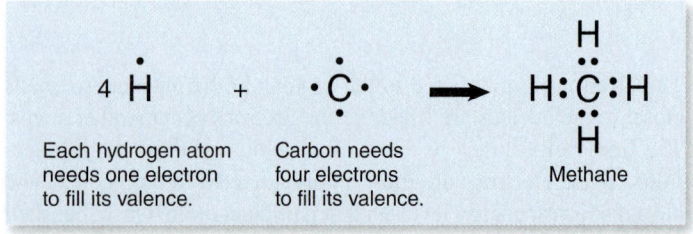

(a)

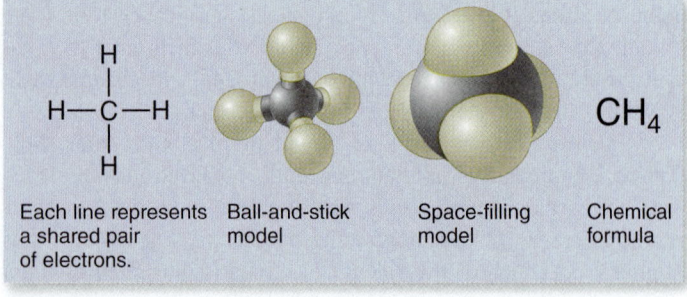

(b)

FIGURE 2.4 Covalent Bonds Covalent bonds are formed when atoms share electrons. **(a)** Methane is formed when a carbon atom fills its outer electron shell by sharing eight electrons—four belong to H atoms and four belong to the carbon atom. **(b)** Different ways of depicting the methane molecule.

❓ *In terms of its bonding properties, why is carbon such an important element in biological systems?*

TABLE 2.2	The Relative Electronegativities of Some Atoms Common in Biology*
Hydrogen	2.1
Carbon	2.5
Nitrogen	3.0
Oxygen	3.5

*Electronegativity is the ability of an atom to attract electrons to itself in a molecule; an atom with a significantly higher electronegativity is better able to attract electrons.

TABLE 2.3	Non-Polar and Polar Covalent Bonds	
Type of Covalent Bond	**Atoms Involved and Charge Distribution**	
Non-polar	C—C C—H H—H	C and H have similar attractions for electrons, so there is a nearly equal charge on each atom.
Polar	O—H N—H O—C N—C	The O and N atoms have a stronger attraction for electrons than do C and H, so the O and N have a slight negative charge; the C and H have a slight positive charge.

life, so cells use protein catalysts called enzymes that help break covalent bonds at lower temperatures. Enzymes and their functions are covered in chapter 6. ▶▶ enzymes, p. 135

Electrons in a covalent bond may or may not be equally shared between the two atoms, depending on the relative electronegativities of the atoms; electronegativity is the ability of an atom to attract electrons to itself in a molecule (**table 2.2**). A **non-polar covalent bond** forms when the electrons are shared equally, such as when identical atoms share electrons. The same can occur between different atoms, if both have a similar attraction for electrons (**table 2.3**). Examples of non-polar covalent bonds are H—H and C—H.

If one atom in a covalent bond is significantly more electronegative than the other, then the electrons are shared unequally,

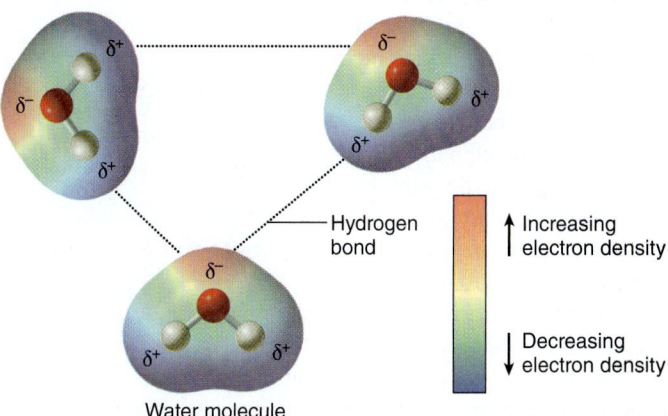

Water molecule

FIGURE 2.6 Hydrogen Bond Formation Hydrogen bonds form between water molecules because the electron-rich oxygen atom attracts electron-poor hydrogen atoms.

❓ *Explain why two identical atoms joined by a covalent bond cannot form a hydrogen bond.*

resulting in a **polar covalent bond.** The slight separation of charge is called a dipole, and is indicated by the Greek symbol delta (δ); the atom that has a slight positive charge is δ^+ and the atom with the slight negative charge is δ^-. An example is the O—H bonds in water; the oxygen atom is more electronegative than the hydrogen atom (**figure 2.5**). Consequently, the oxygen atom pulls the electrons toward it, giving it a slight negative charge and leaving each of the two hydrogen atoms with a slight positive charge. Polar covalent bonds play a key role in biological systems because hydrogen bonds often result.

Hydrogen Bonds

Hydrogen bonds are weak bonds formed when a hydrogen atom in a polar molecule is attracted to an electronegative atom in the same or another polar molecule (**figure 2.6;** see also table 2.2). Compounds that contain electronegative atoms such as oxygen (O) or nitrogen (N) are common in biological systems, creating the possibility for many hydrogen bonds. Like other weak bonds, these are important in molecule-molecule recognition (**figure 2.7**). For

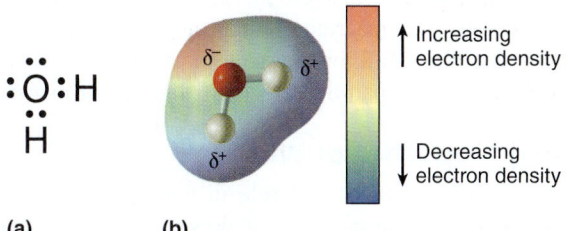

(a)   (b)

FIGURE 2.5 Polar Covalent Bonds Electrons move closer to the more electronegative atom in a compound, creating a polar molecule. **(a)** Lewis diagram of a water molecule. **(b)** Electron density model of a water molecule. The symbol δ indicates a partial charge.

❓ *Why is the oxygen atom in a water molecule more electron-rich than the hydrogen atoms?*

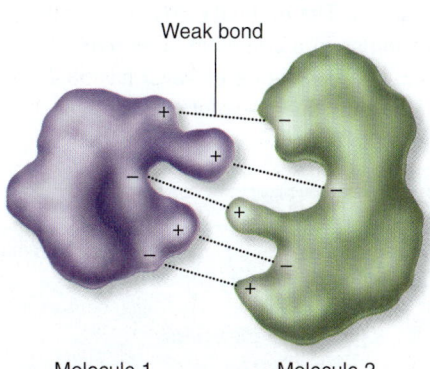

Molecule 1   Molecule 2

FIGURE 2.7 Weak Bonds and Molecular Recognition Weak bonds, such as ionic and hydrogen bonds, are important for molecules to recognize each other. Many weak bonds are required to hold the two molecules together.

❓ *Why would it be important for certain molecules to be held together by hydrogen bonds instead of covalent bonds?*

example, in order for an enzyme to break covalent bonds of a compound (the substrate), the enzyme binds to the substrate through many weak non-covalent bonds, such as hydrogen and ionic bonds.

A single hydrogen bond often exists for only a fraction of a second, and enzymes are not needed to form or break them. The hydrogen bonds between water molecules are constantly being formed and broken at room temperature because the energy produced by the movement of water is enough to break the bonds.

Although a single hydrogen bond is too weak to keep molecules together, a large number can hold them together firmly. An analogy would be the hook and loop fasteners of Velcro. A single hook-and-loop attachment does not provide much strength, but many such attachments result in a strong connection. Like hook and loop attachments, weak bonds can be formed and broken quickly and easily, allowing the molecules to separate. A good example is the double-stranded DNA molecule. The two strands are held together by many hydrogen bonds along the molecule. The two strands will come apart if energy is supplied, usually in the form of heat approaching temperatures of 100°C.

Molarity

A chemical reaction can be likened to a recipe—it uses relative quantities of different substances. But while chefs work with measures such as a dozen, chemists work with moles. One **mole** is 6.022×10^{23} particles. That number is not important from a practical standpoint, but the concept is essential in chemistry—a mole of one compound has the same number of molecules as a mole of any other.

One mole of sodium chloride (NaCl), for example, weighs 58.4 grams, whereas one mole of potassium chloride (KCl) weighs 74.55 grams. The **molarity** (*M*) of a solution is defined as the number of moles of a compound dissolved in 1 liter of solution. Therefore, a 1 *M* solution of NaCl has 58.4 grams of NaCl dissolved in 1 liter of aqueous solution.

MicroAssessment 2.2

Valence electrons are important in chemical bond formation. Ionic bonds form between positively and negatively charged ions. Covalent bonds result from sharing electrons. Hydrogen bonds form between polar molecules or portions of molecules.

4. *Compare the relative strengths of covalent, hydrogen, and ionic bonds.*
5. *Which type of bond requires an enzyme to break it?*
6. *Why does an atom that loses electrons become positively charged? What causes the positive charge?* ✚

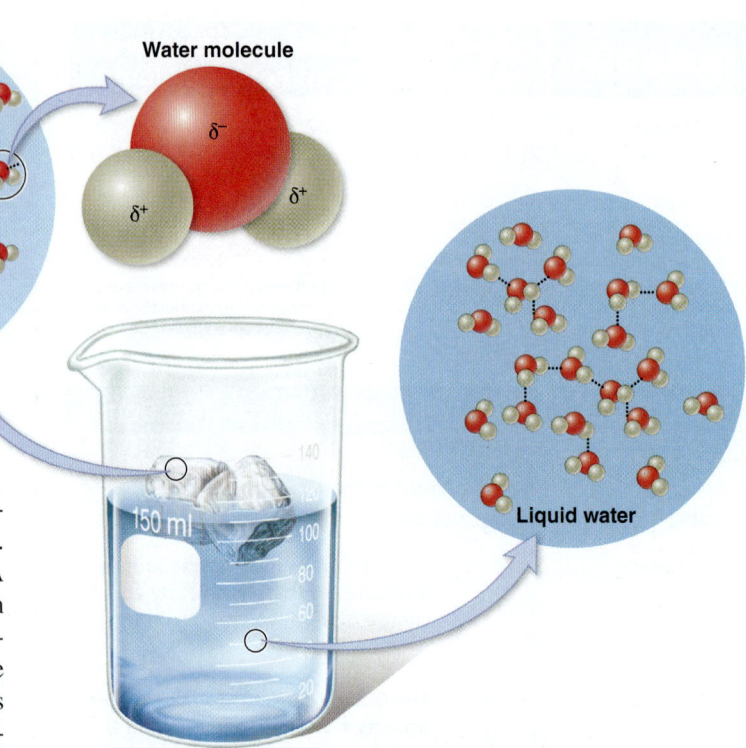

FIGURE 2.8 Water In ice, each H_2O molecule hydrogen bonds to four other H_2O molecules, forming a rigid crystalline structure. In liquid water, the hydrogen bonds continuously break and re-form and the molecules can move closer together.

❓ *Why does water expand as it freezes?*

2.3 ■ Chemical Components of the Cell

Learning Outcomes

5. *Describe the bonding properties of water molecules, and explain why they are important in biology.*
6. *Explain the concept of pH, and how the pH of a solution relates to its acidity.*
7. *Name the four types of macromolecules found in all cells.*

Water

The most important molecule in the cell is water and the life of all organisms depends on it. In addition, cells contain many elements, small molecules, and macromolecules (large molecules). Water makes up over 70% of all living organisms by weight. The importance of water in large part depends on its unusual properties.

Bonding Properties of Water

Hydrogen bonds play a critical role in the properties of water. Because water is a polar molecule, hydrogen bonds form between the δ^- O portions of one molecule and the δ^+ H portions of another. The extent and stability of hydrogen bonding between water molecules depends on the temperature (**figure 2.8**). At freezing temperatures, the water molecules form a lattice-like crystalline structure (ice). At room temperature, however, the hydrogen bonds continually break and reform. Because of the fluctuations in hydrogen bonding at room temperature, the water molecules can move closer together, so liquid water is denser than ice. This is why ice floats.

The polar nature of water molecules also accounts for water's ability to dissolve a large number of compounds. To dissolve in water, compounds must be polar or have a positive or negative charge. The polar water molecules surround these compounds. In the case of salts, water molecules split them into their component ions. For example, NaCl dissolves in water to form Na^+ and Cl^-; the surrounding water molecules prevent them from associating (**figure 2.9**). Salts and polar molecules are **hydrophilic** ("water loving"). In contrast, non-polar molecules are **hydrophobic** ("water-fearing"); they do not dissolve in water because they cannot form hydrogen bonds.

Water containing dissolved substances freezes at a lower temperature than pure water, so most water does not freeze unless the temperature drops below 0°C. Consequently, some microorganisms can multiply below 0°C, because at least some of the water remains liquid.

pH of Aqueous Solutions

An important property of aqueous solutions is their **pH,** a measure of their acidity. It is measured on a logarithmic scale of 0 to 14 in which the lower the number, the more acidic the solution (**figure 2.10**).

pH is a measure of the concentration of H^+ in moles per liter. Water has a slight tendency to split (ionize) into hydrogen ions H^+ (protons), which are acidic, and OH^- ions (hydroxyl), which are

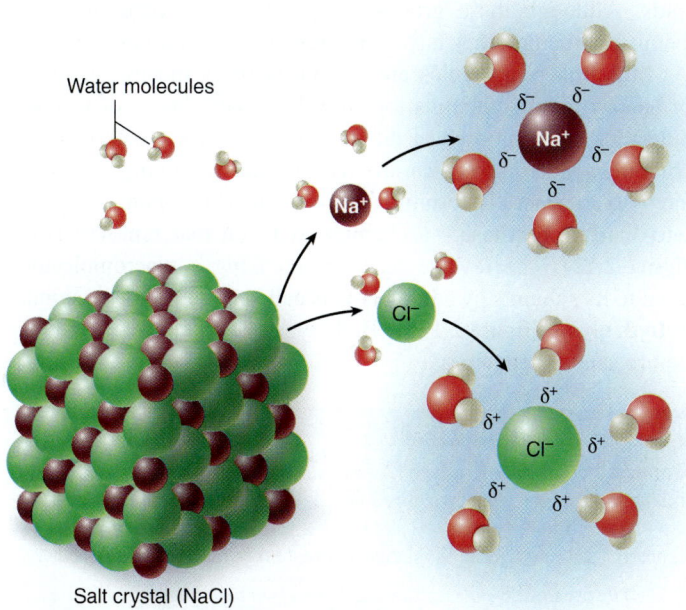

FIGURE 2.9 Salt (NaCl) Crystal Dissolving in Water In water, the Na^+ and Cl^- are separated by H_2O molecules. The Na^+ is attracted to the slightly negatively charged O^- and the Cl^- is attracted to the slightly positively charged H^+ portion of the water molecules. In the absence of water, the salt is highly structured because of ionic bonds between Na^+ and Cl^- ions.

❓ *If water were not polar, would it dissolve sodium chloride? Explain.*

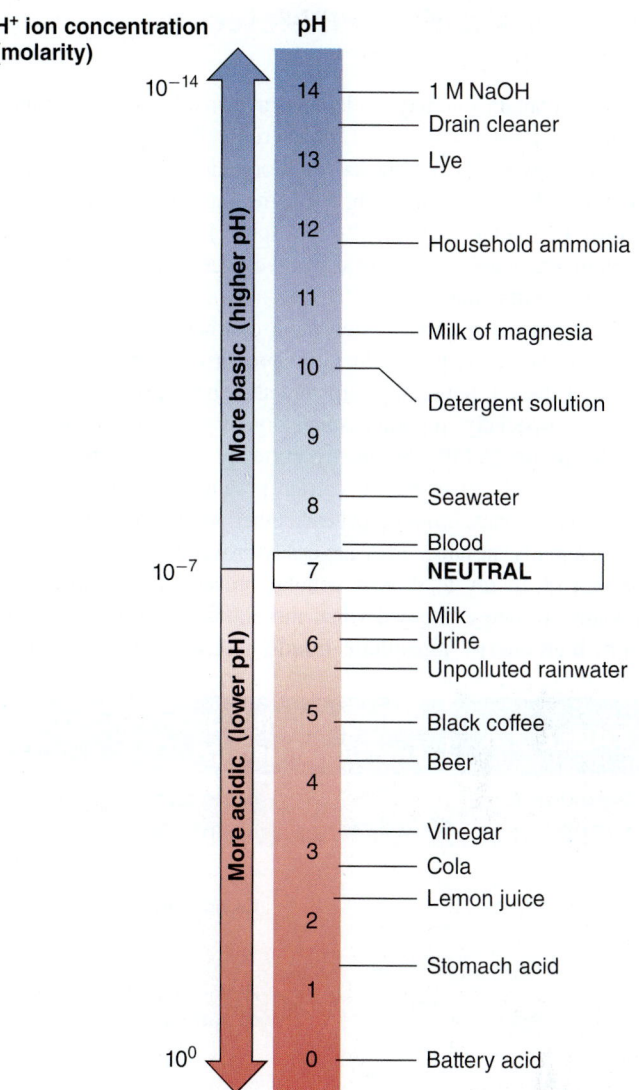

FIGURE 2.10 pH Scale The concentration of H^+ ions varies by a factor of 10 between each pH number since the scale is logarithmic.

❓ *Does the H^+ concentration increase or decrease when the pH drops from 5 to 4? What about the OH^- concentration?*

basic or alkaline. In pure water, the concentration of H^+ and OH^- ions is equal, and the concentration of each is 10^{-7} molar (10^{-7} M). The product of the concentration of H^+ and OH^- is always equal to 10^{-14} M ($10^{-7} \times 10^{-7}$) (exponents are added when numbers are multiplied). Thus, if H^+ ions are added to an aqueous solution, increasing the concentration of H^+ 10-fold (to 10^{-6} M), then the concentration of OH^- must decrease by a factor of 10 (to 10^{-8} M).

The pH scale ranges from 0 to 14 because the concentrations of H^+ and OH^- ions vary within these limits. When the concentrations of H^+ and OH^- are equal (10^{-7}), the pH of the solution is 7.0 and neutral. For every unit on the log scale, however, the concentration of H^+ ions changes by a factor of 10.

Compounds called **buffers** stabilize the pH of solutions. They are frequently added to bacterial growth media to prevent a dramatic rise or fall in pH resulting from metabolic processes. This is important to do because most bacteria can live only within a narrow pH range, usually near neutrality.

Elements and Small Molecules in the Cell

All cells contain a variety of elements and small molecules, many of which are ions. About 1% of a bacterial cell's dry weight (weight excluding water) is composed of inorganic ions, principally Na^+ (sodium), K^+ (potassium), Mg^{2+} (magnesium), Ca^{2+} (calcium), Fe^{2+} (iron), Cl^- (chloride), PO_4^{3-} (phosphate), and SO_4^{2-} (sulfate). Certain enzymes require positively charged ions in very small amounts to function.

Organic compounds often have distinctive chemical groups that contribute to the molecule's properties. Characteristics of some of these functional groups are shown in **table 2.4.**

An especially important small organic molecule is **adenosine triphosphate (ATP),** the energy currency of a cell (**figure 2.11**). ATP is an effective energy carrier because the three negatively charged phosphate groups repel each other, so the bonds joining them are inherently unstable. They are easily broken to release a sufficient amount of energy to drive a cellular process. The relatively high amount of energy released when the bonds are hydrolyzed makes them **high-energy phosphate bonds,** indicated by the symbol ~.

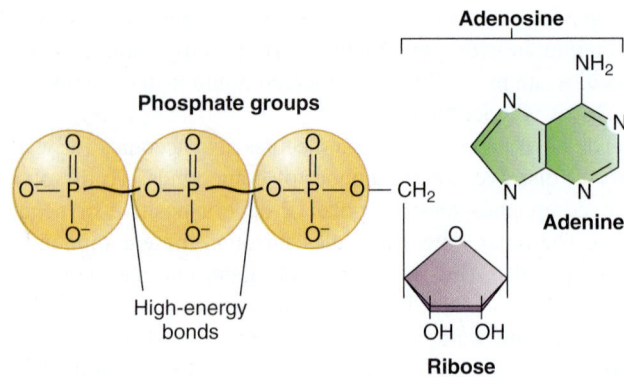

FIGURE 2.11 ATP Adenosine triphosphate (ATP) serves as the energy currency of a cell. When the terminal phosphate bonds break, the energy released can be used to drive cellular reactions.

? *Why are the bonds between the phosphate groups of ATP "high energy"?*

When the terminal phosphate bond of ATP breaks, inorganic phosphate and **adenosine diphosphate (ADP)** are formed. The role of ATP in energy metabolism is covered more fully in chapter 6.

Other organic small molecules in cells are the subunits, or building blocks, of macromolecules that will be described next. The subunits include amino acids, nucleotides, and various sugars.

Macromolecules and Their Component Parts

Macromolecules are large molecules (*macro* means "large"). The four major classes are proteins, carbohydrates, nucleic acids, and lipids. Although these groups differ from each other in their chemical structure, they have some features in common.

Most macromolecules are **polymers** (*poly* means "many"), formed by joining subunits together. Different classes of macromolecules are composed of different subunits, each with distinct structures. Macromolecules are synthesized by joining subunits, one by one, generally forming a long chain. This joining involves **dehydration synthesis**—a chemical reaction that removes H_2O (**figure 2.12**). The reverse reaction, breaking a macromolecule down to its subunits by adding H_2O is called a hydrolytic reaction, or **hydrolysis** (figure 2.12b). Both types of chemical reactions require specific enzymes.

TABLE 2.4	Biologically Important Functional Groups	
Functional Group	**Structure**	**Biological Significance**
Aldehyde	O‖ —C—H	Carbohydrates
Amino	—N⟨H H	Amino acids, the subunit of protein
Carboxyl	—C⟨O OH	Organic acids, including amino acids and fatty acids
Hydroxyl	—OH	Carbohydrates, fatty acids, alcohol, some amino acids
Keto	O‖ —C—	Carbohydrates, polypeptides
Methyl	H| —C—H |H	Some amino acids, attached to DNA
Phosphate	O⁻| —O—P—O⁻ ‖O	Nucleotides (subunit of nucleic acids), ATP, signaling molecules
Sulfhydryl	—S—H	Part of the amino acid cysteine

MicroAssessment 2.3

The weak polar bonds between water molecules are responsible for the many properties of water required for life. The degree of acidity of an aqueous solution is expressed as pH. Macromolecules consist of many repeating subunits, joined together in a reaction that releases water.

7. *Why is water a polar molecule? Give two examples of why this property is important in microbiology.*

8. *Name the four important classes of large molecules in cells.*

9. *In pure water, what must be done to decrease the OH^- concentration? To decrease the H^+ concentration?* ✚

Dehydration Synthesis

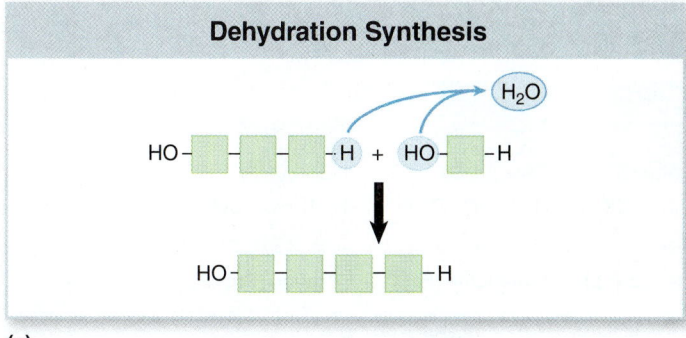

(a)

Hydrolysis

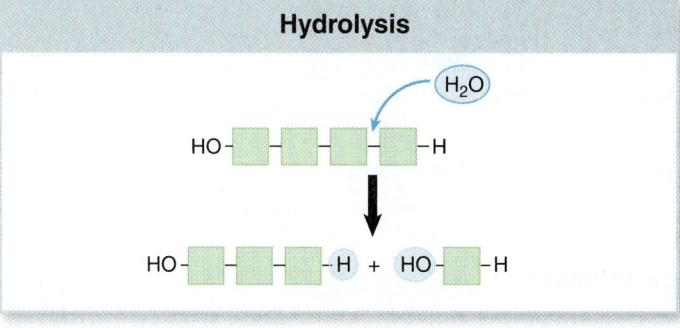

(b)

FIGURE 2.12 Synthesis and Breakdown of Macromolecules
(a) Subunits are joined together (polymerized) by removing water, a dehydration reaction. **(b)** In the reverse reaction, hydrolysis, the addition of water breaks bonds between the subunits.

? *What are the four major classes of macromolecules?*

2.4 ■ Proteins

Learning Outcomes

8. *Describe the important functions of proteins in cells.*

9. *Describe the characteristics of amino acids and the peptide bonds that hold them together.*

10. *Compare and contrast the four levels of protein structure.*

11. *Describe protein domains, substituted proteins, and protein denaturation.*

Proteins make up more than half of the dry weight of cells. Some of their most important roles include:

■ **Catalyzing reactions.** Enzymes are proteins that speed up the various chemical reactions in cells. ▶▶ enzymes, p. 135

■ **Transporting molecules.** Transport proteins move molecules either into or out of cells. ▶▶ transport proteins, p. 55

■ **Motility.** Proteins are essential components of flagella and cilia, structures that move cells. ▶▶ flagella, pp. 63, 74

■ **Cell framework.** Proteins make up the cytoskeleton, the structural framework of many cells. ▶▶ cytoskeleton, p. 73

■ **Sensing and responding to conditions outside the cell.** Proteins on the cell surface recognize conditions in the external environment and relay that information to the cellular machinery. ▶▶ two-component system, p. 178

■ **Regulating gene expression.** Proteins bind to DNA and regulate gene expression. ▶▶ gene expression, pp. 162, 168

MicroByte
The genetic information of a typical bacterial cell encodes up to 4,000 proteins.

Amino Acids

Amino acids are the subunits that make up proteins. Twenty major amino acids can be arranged in a functionally infinite number of combinations in a protein. The characteristics of a protein depend mainly on its shape, which in turn depends on the sequence of amino acids.

All amino acids have a central carbon atom bonded to (1) a carboxyl group, (2) an amino group, and (3) a side chain (R group) (**figure 2.13**). The side chain—the part that distinguishes different amino acids—gives an amino acid its characteristic properties.

Amino acids are subdivided into several different groups based on properties of their side chains (**figure 2.14**). Non-polar amino acids are characterized by side chains that lack polar bonds; an example is the methyl group (—CH_3) of alanine. In contrast, polar amino acids have side chains that contain a polar bond; an example is the hydroxyl group (—OH) of serine. Charged amino acids carry a positive or negative charge because their side chains contain functional groups that can ionize; these include carboxyl groups (—COOH; acidic) and amino groups (—NH_2; basic).

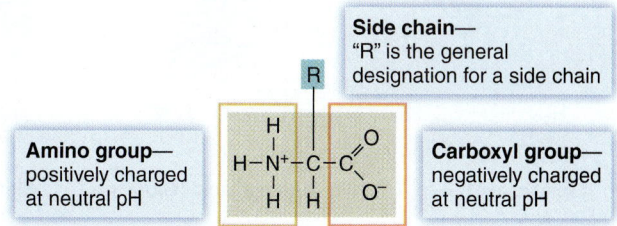

FIGURE 2.13 Generalized Amino Acid This figure illustrates the three groups that all amino acids possess.

? *Which portion of an amino acid is responsible for the unique properties of the molecule?*

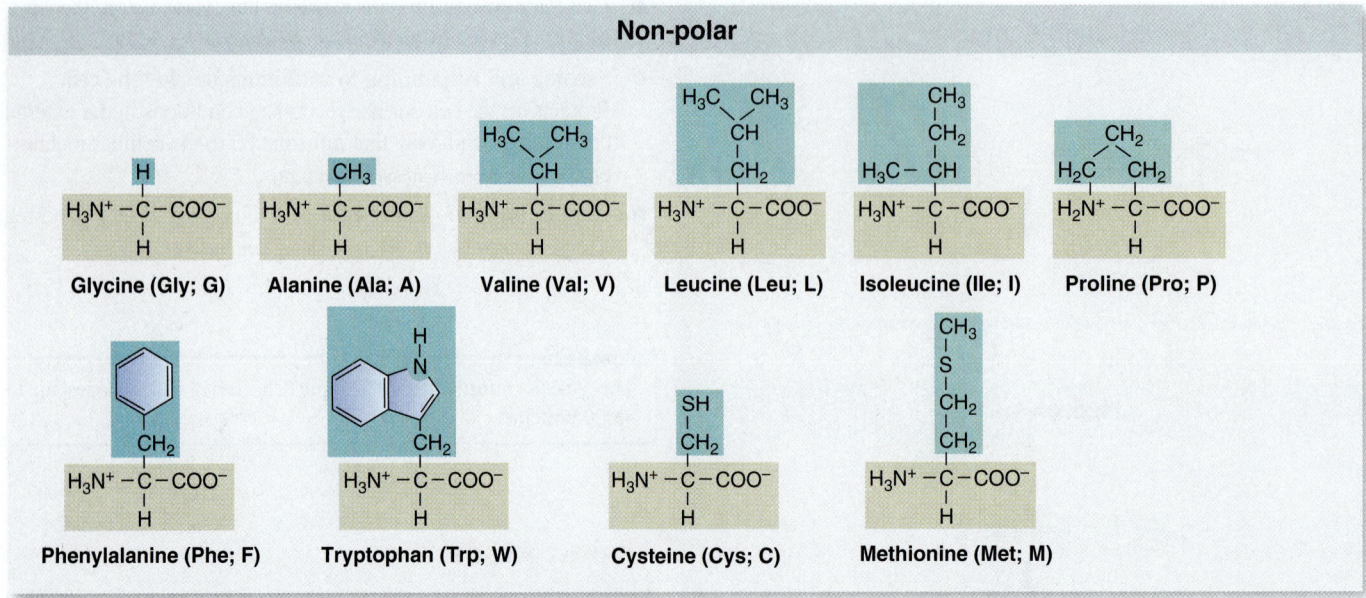

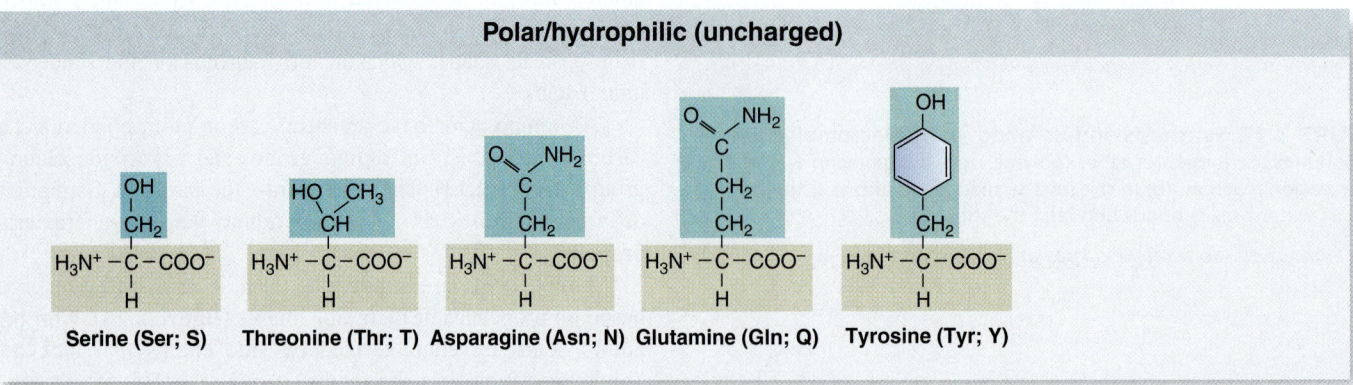

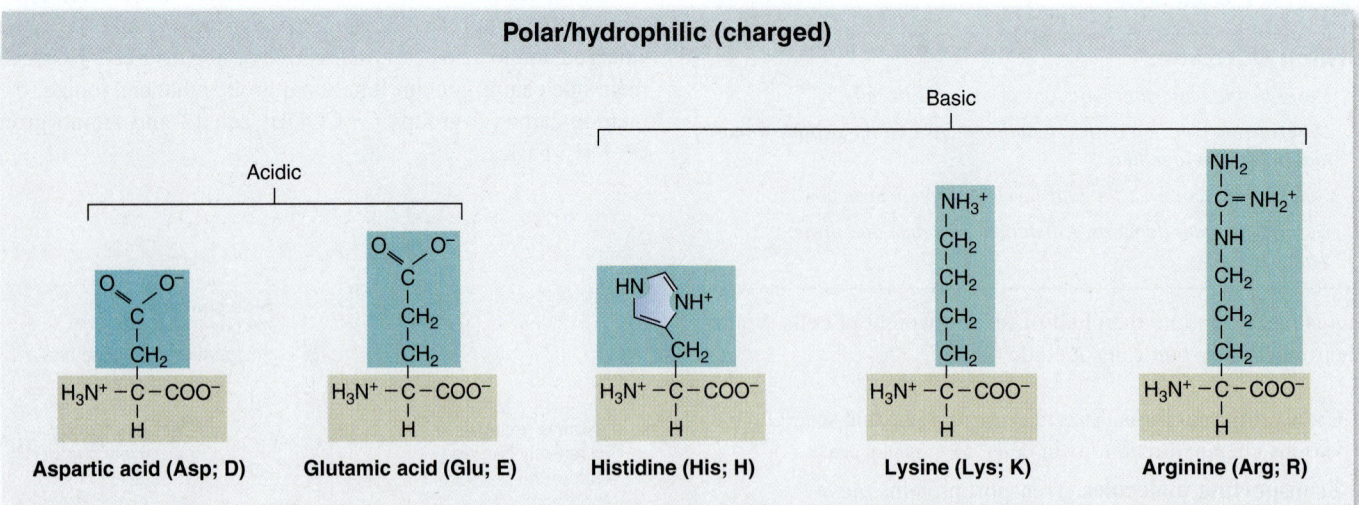

FIGURE 2.14 Common Amino Acids The groupings are based on the polarity and the overall charge of the amino acid. The basic and acidic amino acids have a net positive and negative charge, respectively. For simplicity, tyrosine is shown in only one group but it has both non-polar and polar characteristics. The three-letter and single-letter code names for each amino acid are given.

? *What chemical groups characterize a hydrophobic amino acid? A hydrophilic amino acid?*

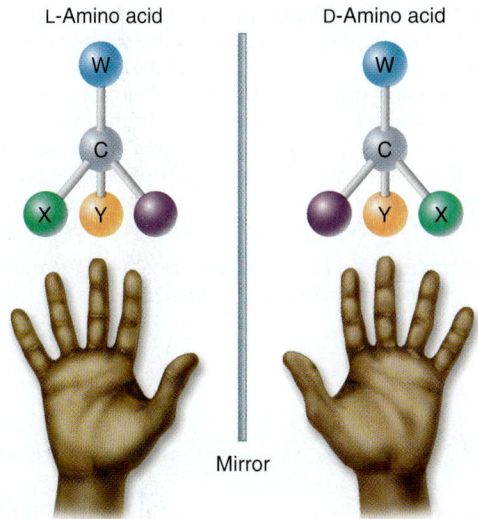

FIGURE 2.15 Mirror Images (Stereoisomers) of an Amino Acid The joining of a carbon atom to four different groups leads to asymmetry in the amino acid. The molecule can exist in either the L or D form, each being the mirror image of the other. The two molecules cannot be rotated in space to give two identical molecules.

❓ *Which form (L or D) is found in proteins?*

All amino acids except glycine can exist in two stereoisomers, or spatial arrangements (**figure 2.15;** see **A Glimpse of History**). These forms—L (left-handed) and D (right-handed)—are mirror images of each other. Proteins only have L-amino acids. The D-amino acids are found in a few structures in bacteria, but otherwise are rare in nature.

Peptide Bonds

Amino acids are held together in an unbranched chain by **peptide bonds.** This type of covalent bond forms when the carboxyl group of one amino acid reacts with the amino group of another, releasing water (dehydration synthesis) (**figure 2.16**). The joining of amino acid subunits by peptide bonds creates a **polypeptide.** One end of the molecule has a free amino group (the N terminal or amino terminal end), and the other has a free carboxyl group (the C terminal or carboxyl terminal end). ▶▶ protein synthesis, p. 170

A **protein** is one or more long polypeptides folded to create a functional molecule. Note, however, that the distinction between a polypeptide and a protein is not always clear, and the terms are often used interchangeably.

Protein Structure

Proteins have four levels of structure: primary, secondary, tertiary, and quaternary. The number and sequence of amino acids in the polypeptide determines its **primary structure** (**figure 2.17**). Proteins vary greatly in size, but an average-size polypeptide

consists of about 250 amino acids. The primary structure in large part determines the final shape of the protein and thus is responsible for its properties.

The **secondary structure** is the three-dimensional shape of localized regions; certain amino acid sequences (the primary structure) lead to characteristic spirals and folds due to weak forces such as hydrogen bonds (figure 2.17b). A spiral or helical structure is an alpha (α) helix, whereas parallel strands make up a beta (β) pleated sheet.

The entire protein folds into its distinctive three-dimensional shape, its **tertiary structure** (figure 2.17c). The tertiary structure is determined primarily by the sequence of amino acids and whether or not they interact with water. Amino acids that have polar side chains are hydrophilic, and are typically located on the outside of the protein molecule, where they can interact with charged polar water molecules. Amino acids that have non-polar side chains are hydrophobic, so these tend to cluster inside the protein molecule, thereby avoiding water molecules. In addition to these interactions, a covalent bond can form between sulfur atoms in different cysteine molecules, creating a disulfide (S—S) bond. The combination of strong and weak bonds between the various amino acids results in the proteins' tertiary structure. Two major shapes exist: globular, which tend to be spherical and water soluble, and fibrous, which are elongated and insoluble.

Proteins sometimes consist of more than one polypeptide, either identical or different from one another, which are held together by many weak bonds. The specific shape that results is the **quaternary structure** (figure 2.17d). Of course, only proteins that consist of more than one polypeptide have a quaternary

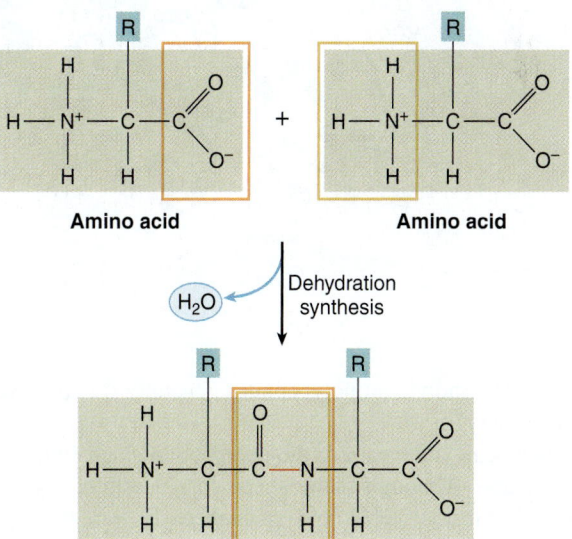

FIGURE 2.16 Peptide Bond Formation Dehydration synthesis forms the peptide bond, shown in red.

❓ *What two chemical groups are involved in the formation of a peptide bond?*

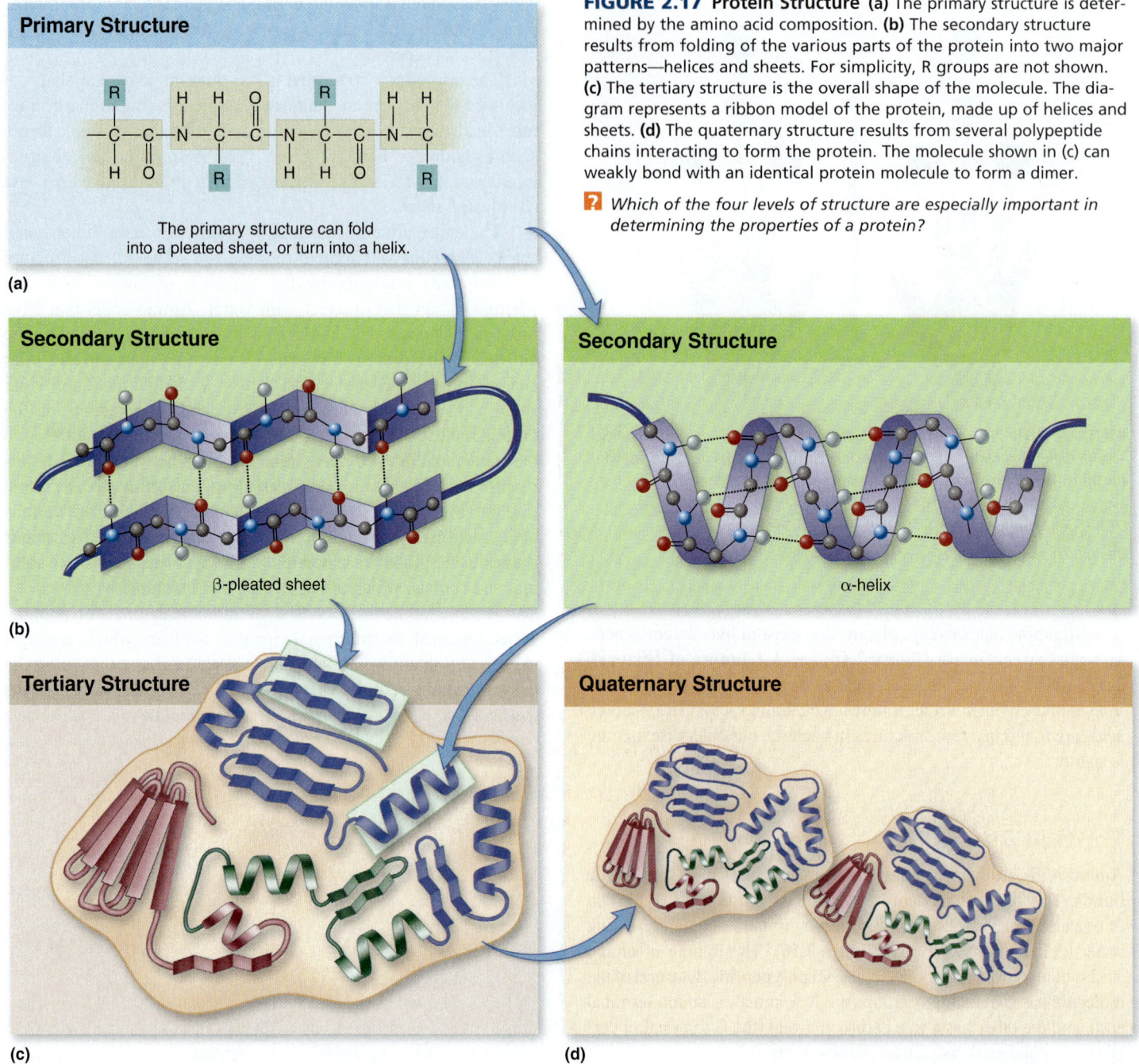

Primary Structure

The primary structure can fold into a pleated sheet, or turn into a helix.

(a)

Secondary Structure

β-pleated sheet

(b)

Secondary Structure

α-helix

Tertiary Structure

(c)

Quaternary Structure

(d)

FIGURE 2.17 Protein Structure **(a)** The primary structure is determined by the amino acid composition. **(b)** The secondary structure results from folding of the various parts of the protein into two major patterns—helices and sheets. For simplicity, R groups are not shown. **(c)** The tertiary structure is the overall shape of the molecule. The diagram represents a ribbon model of the protein, made up of helices and sheets. **(d)** The quaternary structure results from several polypeptide chains interacting to form the protein. The molecule shown in (c) can weakly bond with an identical protein molecule to form a dimer.

❓ *Which of the four levels of structure are especially important in determining the properties of a protein?*

structure. The individual polypeptides generally do not have biological activity.

After being synthesized, the polypeptide chain folds into its correct shape. Although many shapes are possible, only one is functional. Most proteins fold spontaneously into their correct state, but protein **chaperones** are sometimes needed to help with the process. Misfolded proteins are degraded into their amino acid subunits, which are then used to make more proteins.

MicroByte

The computer game "Foldit" allows players to help researchers predict how a specific amino acid sequence will fold to become a functional protein.

Protein Domains

A unit of organization distinct from the four levels just described is the **protein domain** (**figure 2.18**). A domain is a protein substructure consisting of sheets and helices that fold into a stable structure independently of other parts of the molecule. Large proteins sometimes contain many domains, whereas a small protein may contain only one. The domains, usually consisting of 40 to 350 amino acids, are connected to each other by short lengths of polypeptides.

Different domains are associated with specific functions. For example, a certain domain may grasp DNA; another may carry out a catalytic activity. Thus, a single protein may have several different

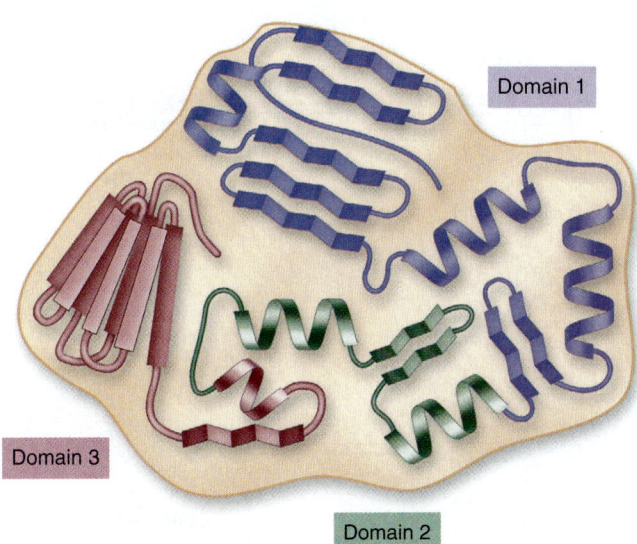

FIGURE 2.18 Domain Structure of a Protein Each domain has different functions.

❓ *Which levels of protein structure determine the properties of domains?*

functions. Once a known function can be attributed to a specific domain, the function of an unknown protein can be inferred if it has that domain.

Substituted Proteins

Substituted proteins have other molecules covalently bonded to the side chains of some of their amino acids. Many proteins found on the surface of cells are substituted. The proteins are named after the molecules covalently joined to the amino acids. If sugar molecules are bonded, the protein is a glycoprotein; if lipids are attached, the protein is a lipoprotein. Sugars and lipids are covered later in this chapter.

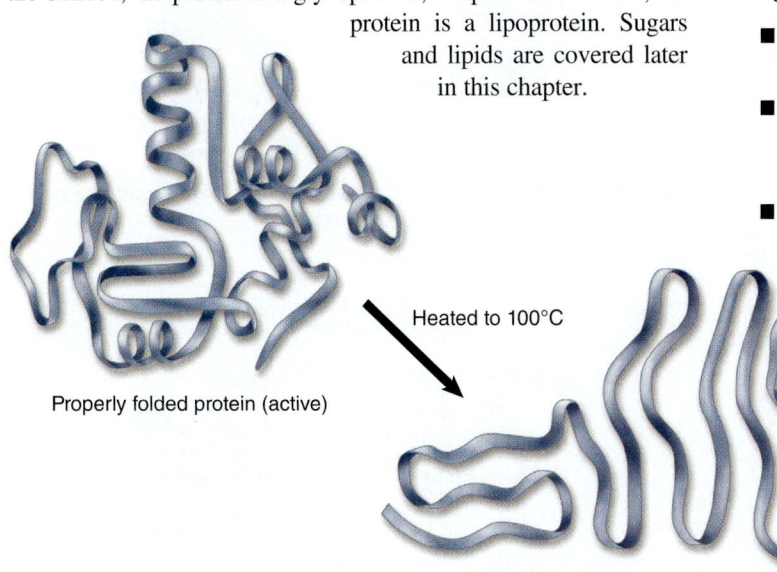

Properly folded protein (active)

Heated to 100°C

Denatured protein (inactive)

▶ **FIGURE 2.19 Denaturation of a Protein** The denatured protein loses its function.

❓ *Describe two environmental conditions that can denature a protein.*

Protein Denaturation

High temperature, extreme pH, and certain solvents can break bonds within a protein, causing its shape to change (**figure 2.19**). The protein becomes **denatured** and no longer functions. This is why most organisms cannot grow at very high temperatures—their enzymes denature. Denaturation may be reversible in some, but not all, cases. If the denaturing agent is a chemical that is then removed, for example, the protein may refold spontaneously into its original shape. However, boiling an egg denatures the egg white protein irreversibly, which is why cooling the egg does not restore it to its original appearance.

MicroAssessment 2.4

The properties of amino acids are determined by their side chains. The sequence of amino acids in a protein determines how the protein folds into a functional three-dimensional shape. Substituted proteins have other molecules attached.

10. *What type of bond joins amino acids to form proteins?*
11. *Describe five roles of proteins.*
12. *What elements must all amino acids contain? What element will only some amino acids contain?* ➕

2.5 ■ Carbohydrates

Learning Outcomes

12. *Describe the important functions of carbohydrates in a cell.*
13. *Compare and contrast monosaccharides, disaccharides, and polysaccharides.*

Carbohydrates are a diverse group of compounds that include sugars and starches. They play several critical roles in biology:

- **Energy source.** Organisms degrade carbohydrates to harvest the energy they contain. ▶▶ **metabolism, p. 126**

- **Energy storage.** Organisms can store excess energy and nutrients for later use by producing certain carbohydrates that function as reserve material. ▶▶ **storage granules, p. 66**

- **Source of carbon for biosynthetic products.** Many microbes can make all of their cell components from a single carbohydrate—glucose. ▶▶ **precursor metabolites, p. 132**

  - **Component of DNA and RNA.** The subunits of DNA and RNA contain sugars. ▶▶ **nucleic acids, p. 32**

    - **Structural components of cells.** Some types of cell walls are composed of sugar-containing material. ▶▶ **cell wall structure, p. 58**

      Carbohydrates all contain carbon, hydrogen, and oxygen atoms in an approximate ratio of 1:2:1. Their general chemical formula, $(CH_2O)_n$, indicates this ratio. The "H_2O" in that formula is reflected in the term "carbohydrate" (meaning "hydrate of carbon"); note, however, that the arrangement of atoms in carbohydrate molecules has little in common with water.

Monosaccharides

Monosaccharides, or simple sugars, have only a single unit (*mono* means "one"). They are classified by the number of carbon atoms they contain, and most common monosaccharides have five or six carbon atoms (**table 2.5**). Each carbon atom is numbered using a characteristic scheme, allowing scientists to describe the position of various functional groups attached to the molecules (**figure 2.20**).

TABLE 2.5	Common Monosaccharides, Disaccharides, and Polysaccharides	
Name	**Components**	**Significance**
Monosaccharides		
(5-carbon)		
Ribose		Component of RNA
Deoxyribose		Component of DNA
(6-carbon)		
Glucose		Common subunit of disaccharides
Galactose		Component of milk sugar (see below)
Fructose		Fruit sugar
Mannose		Found on the surface of some microbes
Disaccharides		
Lactose	Glucose + galactose	Milk sugar
Maltose	Glucose + glucose	Breakdown product of starch
Sucrose	Glucose + fructose	Table sugar from sugar cane and beets
Polysaccharides		
Agar	Polymer of galactose	Gelling agent in bacteriological media; extracted from the cell walls of some algae
Cellulose	Polymer of glucose, in a β 1,4 linkage; no branching	Main structural polysaccharide in plant cell walls
Chitin	Polymer of *N*-acetyl-glucosamine	Major organic component in exoskeleton of insects and crustaceans
Dextran	Polymer of glucose in an α 1,6 linkage; branching	Storage product in some bacterial cells
Glycogen	Polymer of glucose in an α 1,4 linkage; branching	Main storage polysaccharide in animal and bacterial cells
Starch	Polymer of glucose	Main storage product in plants

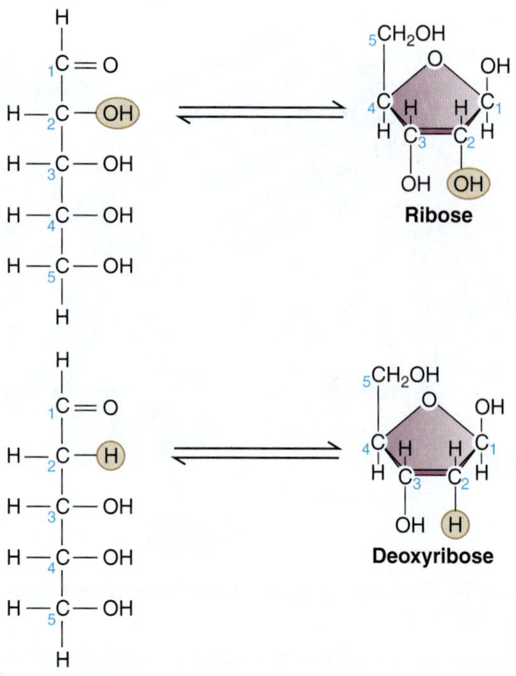

FIGURE 2.20 Ribose and Deoxyribose Linear and ring forms. Although both structures occur in the cell, the ring form predominates. The plane of the ring is perpendicular to the plane of the paper with the thick line on the ring closest to the reader.

? *What is the major chemical difference between ribose and deoxyribose?*

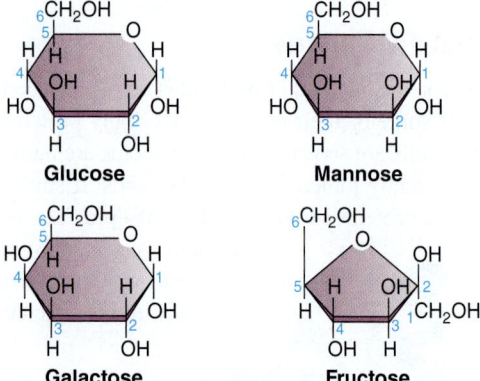

FIGURE 2.21 Common 6-Carbon Sugars These sugars are structural isomers with different properties.

? *What is a structural isomer?*

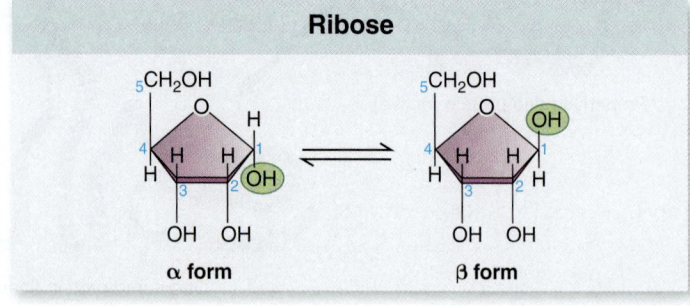

FIGURE 2.22 α and β Forms The α and β forms of ribose are interconvertible and differ only in whether the OH group on carbon 1 is above or below the plane of the ring.

? *When are the α and β forms not interconvertible?*

Sugars occur in two interchangeable forms: linear and ring (figure 2.20). Both naturally occur in the cell, but most are in the ring form. In diagrams, the lower portion of the ring form is thickened to suggest a three-dimensional structure.

Ribose and deoxyribose are 5-carbon sugars found in nucleic acids (figure 2.20). These sugars are identical except deoxyribose has one less molecule of oxygen than does ribose (*de* means "away from"); ribose has a hydroxyl group on the number 2 carbon (also called the 2 prime carbon, written 2′ carbon), whereas deoxyribose has only a hydrogen at that position.

Glucose, galactose, fructose, and mannose are all 6-carbon sugars. These are all structural isomers, meaning that they contain the same atoms but differ in their chemical arrangements (**figure 2.21**). Structural isomers result in distinct sugars with different properties and different names. For example, both glucose and mannose have a sweet taste, but mannose has a bitter aftertaste. Glucose, an important energy source of many cells, will be discussed extensively in the chapter on metabolism. Mannose is found on the surface of some microbes.

Sugars can exist in two different forms—alpha (α) and beta (β)—based on the relative position of the hydroxyl group joined to the number 1 carbon atom (**figure 2.22**). The α and β forms are interconvertible, but once the carbon atom is joined to another sugar molecule, the α or β form is essentially locked in place.

Disaccharides

Disaccharides are two monosaccharides joined together by covalent bonds (table 2.5). The two most common examples are sucrose (table sugar) and lactose (milk sugar). Sucrose, which comes from sugar cane and sugar beets, is composed of glucose and fructose, whereas lactose consists of glucose and galactose. Another disaccharide, maltose, composed of two glucose molecules, is a breakdown product of starch.

To form a disaccharide, two monosaccharides are joined together by a dehydration reaction between a pair of their hydroxyl groups, with the loss of water (**figure 2.23**). Note that this reaction is similar to that used to join two amino acids. The reaction is reversible, so hydrolysis, which adds a water molecule, yields the two original molecules.

Polysaccharides

Polysaccharides are large molecules composed of long chains of monosaccharide subunits or their derivatives (table 2.5). Shorter chains are called **oligosaccharides.**

Polysaccharides often contain only glucose molecules, but are structurally diverse because some polymers are branched. In addition, some types have linkages between α forms of the sugars and others between β forms. The position of the carbon atoms involved in the bonding can also differ (**figure 2.24**).

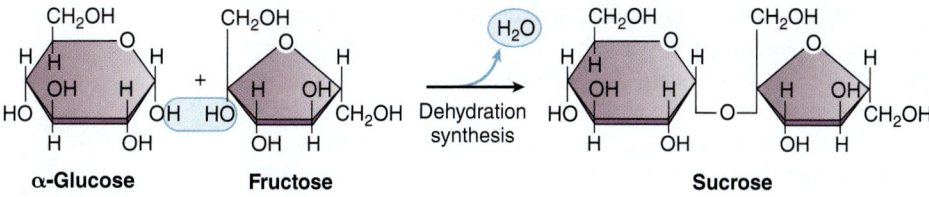

FIGURE 2.23 Formation of a Disaccharide The sucrose molecule is formed by the removal of water.

❓ *What type of reaction would reverse the step shown in this diagram?*

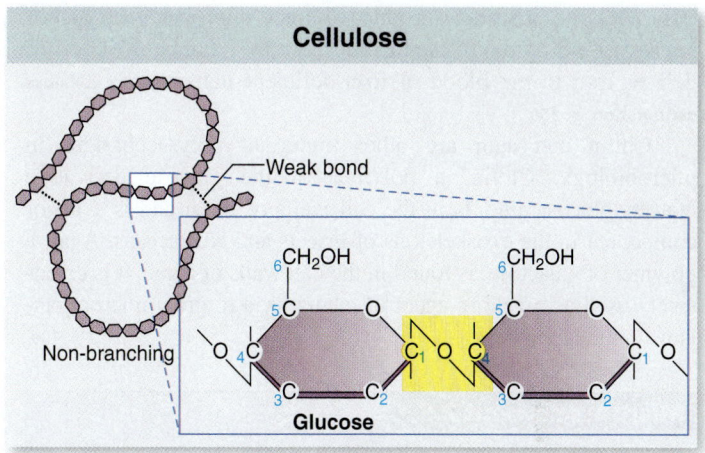

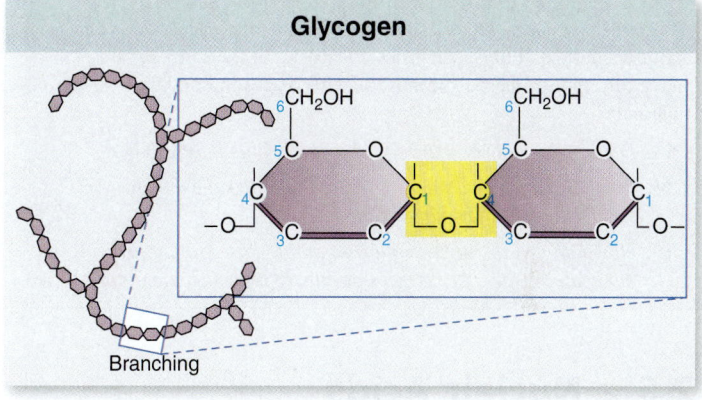

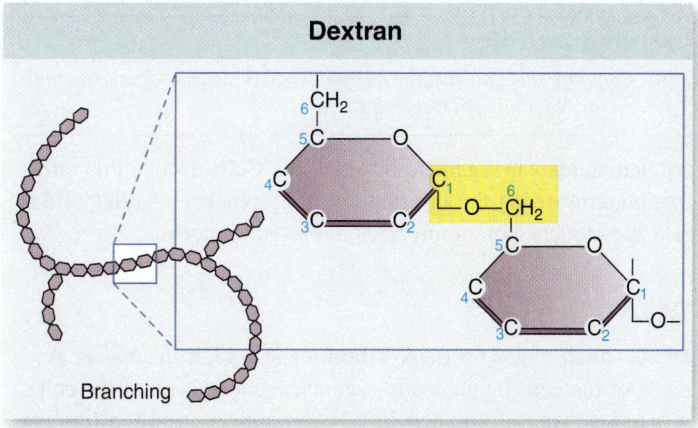

FIGURE 2.24 Structures of Three Important Polysaccharides The molecules shown have the same subunit (glucose) yet they are distinct because of differences in linkage that join the molecules (α and β; 1,4 or 1,6), the degree of branching, and the bonds involved in branching (not shown). Weak bonding forces are also involved.

❓ *Where are the three polysaccharides shown found in nature?*

Cellulose, starch, glycogen, and dextran, all polymers of glucose, are important polysaccharides. Cellulose is the principle component of plant cell walls and the most abundant organic molecule on earth. Most organisms cannot degrade this substance, however, because they lack the enzyme that breaks the bonds joining the subunits. Certain bacteria and fungi have that enzyme, which is why they play such an important role in recycling organic material. Starch, the energy storage form produced by plants, can be used as a food source by many organisms. Glycogen is an energy storage product of animals and some bacteria. Dextran, a storage product of some microbes is a component of some products used to increase the volume of blood or deliver iron to the blood of iron-deficient patients. ▶▎ **cellulose degradation, p. 150**

Chitin and agar are other important polysaccharides in microbiology. Chitin, a polymer of the glucose derivative *N*-acetylglucosamine, is in the cell walls of fungi and is a major component in the exoskeletons of insects and crustaceans. Agar, a polymer of galactose, is found in the cell walls of algae; it is extensively used as a gelling agent in media used to grow microorganisms in the laboratory.

MicroAssessment 2.5

Carbohydrates all have the general formula $(CH_2O)_n$. Sugars can exist in a number of different isomeric forms that have different properties. Monosaccharides are the subunits of disaccharides and polysaccharides. Polysaccharides consisting of the same sugars have different properties because of differences in bonding between subunits.

13. *Distinguish between structural isomers and stereoisomers.*
14. *What is the general name given to a single sugar? How can single sugars differ from another?*
15. *How can you distinguish sucrose and lactose from a protein molecule by identifying the elements in the molecules?* ✚

2.6 ■ Nucleic Acids

Learning Outcome

14. *Compare and contrast the chemical compositions, structures, and major functions of DNA and RNA.*

Nucleic acids carry genetic information. Cells decode this information, converting the information in a sequence of **nucleotides** into the sequence of amino acids in protein molecules.

DNA

DNA, which stands for **deoxyribonucleic acid,** is the master molecule of the cell. Its nucleotide sequence encodes all of the cell's properties. The information in DNA is converted to the form of RNA, which is then translated to make proteins, a process covered in chapter 7.

The nucleotides of DNA have three different parts: a **nucleobase,** deoxyribose, and a phosphate group (**figure 2.25**). The four different nucleobases in DNA can be characterized by their ring structures: purines (adenine and guanine), each consisting

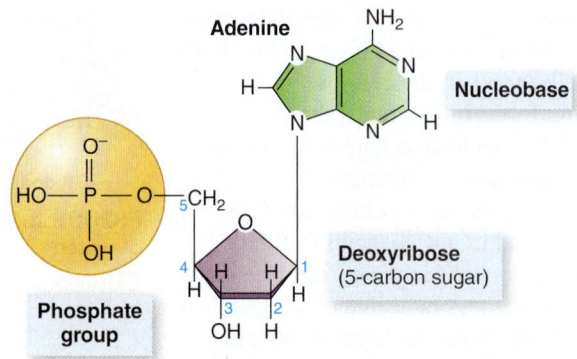

FIGURE 2.25 A Nucleotide This is one subunit of DNA. This subunit is called adenylic acid or deoxyadenosine-5′-phosphate because the nucleobase is adenine. If the nucleobase is thymine, the nucleotide is thymidylic acid; if guanine, guanylic acid; and if cytosine, cytidylic acid. If the nucleotide lacks the phosphate group, it is called a nucleoside—in this case, deoxyadenosine.

❓ *What are the three components of a nucleotide?*

of two fused rings; and pyrimidines (cytosine and thymine), single ring structures (**figure 2.26**).

Nucleic acids consist of linear chains of nucleotide subunits with a covalent bond between the phosphate of one nucleotide and the sugar of the next (**figure 2.27**). Thus, the phosphate is a bridge that joins the number 3 carbon (3′, pronounced "3 prime") of one sugar to the number 5 carbon (5′) of the other. This results in a molecule with a backbone of alternating sugar and phosphate molecules. The 5′ end of the chain has a phosphate attached to the number 5 carbon of sugar; the 3′ end has a hydroxyl group attached to the number 3 carbon (figure 2.27). During DNA synthesis, the chain is elongated by adding more nucleotides to the hydroxyl group at the 3′ end. This topic is covered in chapter 7.

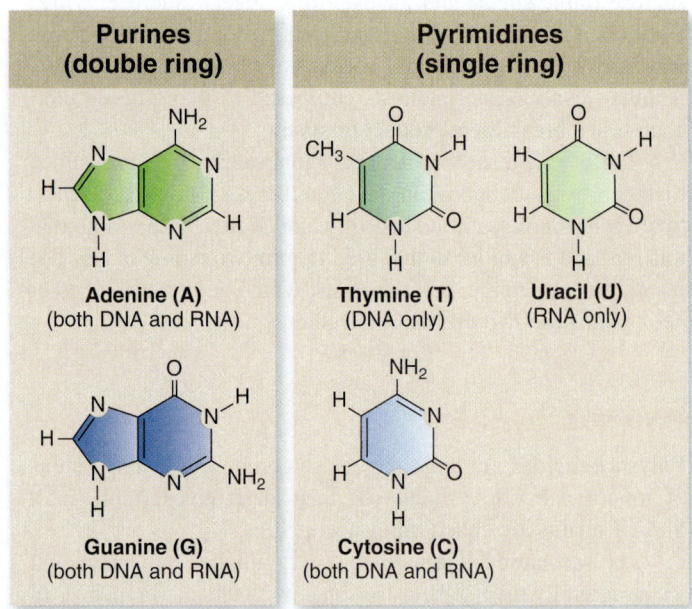

FIGURE 2.26 Formulas of Purines and Pyrimidines

❓ *Which of the nucleobases are found in DNA? In RNA?*

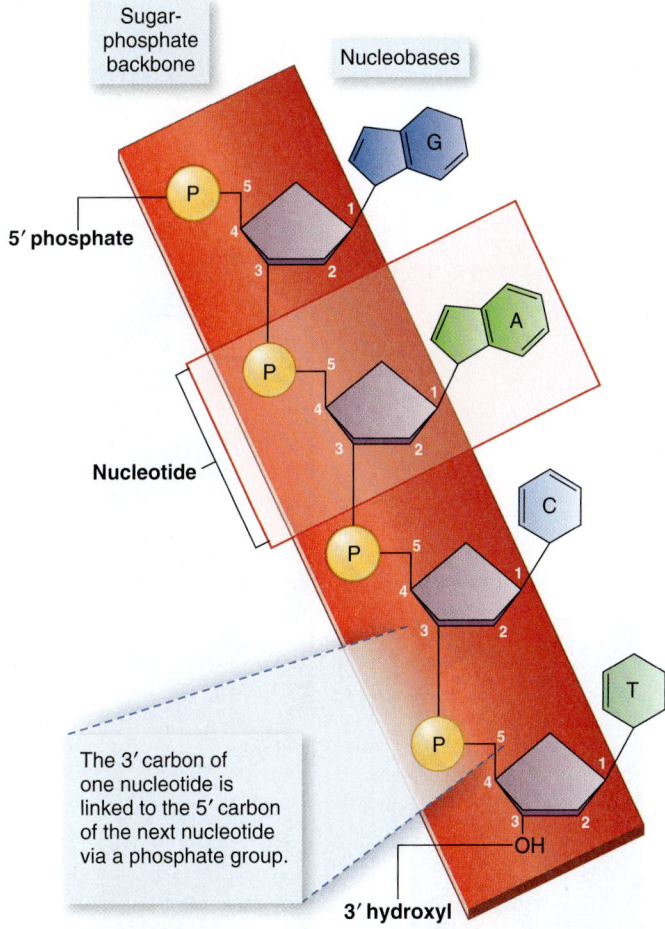

Sugar-
phosphate
backbone

Nucleobases

5′ phosphate

Nucleotide

The 3′ carbon of
one nucleotide is
linked to the 5′ carbon
of the next nucleotide
via a phosphate group.

3′ hydroxyl

FIGURE 2.27 Single Strand of DNA Single chain of nucleotides in DNA showing the differences between 5′ end and 3′ end.

? *What parts of the nucleotides are joined together?*

The DNA of a typical bacterium is a single double-stranded helix, arranged somewhat like a spiral staircase with two railings (**figure 2.28**). The railings represent the sugar-phosphate backbone of the molecule, and the stairs are pairs of nucleobases attached to the railings. The two strands, each about 4 million nucleotides in a typical bacterium, are antiparallel, meaning they are oriented in different directions. One goes in the 3′ to the 5′ direction; the other, 5′ to 3′. Each pair of nucleobases is held together by hydrogen bonds. These bonds are weak, but the large number of them holds the two strands of the DNA molecule firmly together.

The hydrogen bonding between nucleobases is specific in that adenine (A) can only bond to thymine (T), and guanine (G) to cytosine (C). Three hydrogen bonds join each G to C, but only two join A to T. The pair of nucleobases that bond are said to be **complementary** to each other. Thus, G is complementary to C, and A to T. These are referred to as **base-pairing rules.** As a result, one strand of DNA is complementary to the other, and the sequence of one strand determines the sequence of the other.

MicroByte

Scientists once believed that DNA could not play an important role in the cell because it has such a simple structure.

RNA

RNA, which stands for **ribonucleic acid,** is involved in the process that decodes the information in DNA to create a sequence of amino acids in proteins. This complex multistep process will be described in chapter 7.

Although the structure of RNA resembles that of DNA, several differences should be noted. First, the nucleotides in RNA contain the pyrimidine uracil in place of thymine and the sugar ribose in place of deoxyribose (see figures 2.20 and 2.26). Also, whereas DNA is a long, double-stranded helix, RNA is considerably shorter and exists as a single chain of nucleotides. Although single-stranded, RNA may form short, double-stranded stretches as a result of hydrogen bonding between complementary nucleobases in the single strand.

MicroAssessment 2.6

DNA is a double-stranded helical molecule composed of repeating subunits that consist of a nucleobase, a phosphate molecule, and the sugar deoxyribose. RNA is a single-stranded molecule of repeating nucleotides that contain uracil in place of thymine and ribose in place of deoxyribose. DNA carries the genetic code in the sequence of nucleotides. The information in the code is transferred to RNA and then into a sequence of amino acids in proteins.

16. *How do the nucleotides of DNA differ from those of RNA?*
17. *How does the structure of DNA differ from that of RNA?*
18. *If the DNA molecule were placed in boiling water, how would the molecule change?* ✚

2.7 ■ Lipids

Learning Outcome

15. *Compare and contrast the structure and function of simple lipids, compound lipids, and steroids.*

Lipids play an indispensable role in all cells. They are critically important in the structure of membranes, which function as a cell's gatekeepers. Membranes prevent cell contents from leaking out, and also keep many molecules from entering cells. ►► cytoplasmic membrane, p. 52

Lipids are a very diverse group of non-polar, hydrophobic molecules. Their single common feature is that they are only slightly soluble in water but highly soluble in organic solvents such as ether, benzene, and chloroform. Thus, their defining characteristic is a physical, rather than a chemical, property. Unlike other macromolecules, lipids are not composed of similar subunits; rather, they consist of a wide variety of chemically distinct substances.

Simple Lipids

Simple lipids contain only carbon, hydrogen, and oxygen. The most common simple lipids are fats—fatty acids linked to glycerol (**figure 2.29**). Fatty acids are long chains of C atoms bonded to H atoms, with a carboxyl group on one end (figure 2.29a). The length of the chain varies, depending on the fatty acid, which usually has an even number of carbon atoms. Glycerol is a 3-carbon molecule with a hydroxyl group attached to each carbon. Fatty acids can join to glycerol via covalent bonds between a hydroxyl group of glycerol

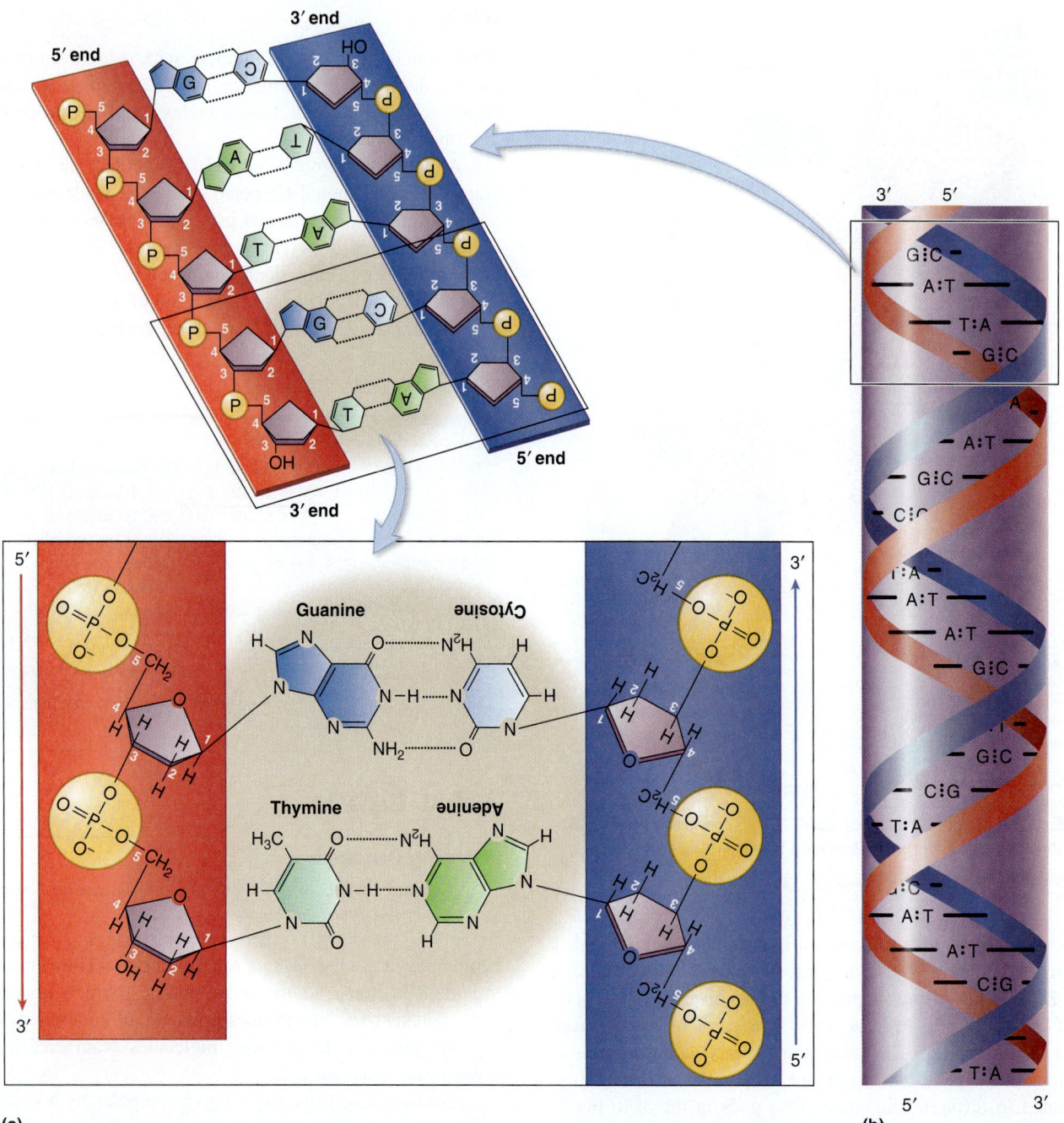

FIGURE 2.28 DNA Double-Stranded Helix (a) The sugar-phosphate backbone and the hydrogen bonding between bases. There are two hydrogen bonds between adenine and thymine and three between guanine and cytosine. **(b)** The "spiral staircase" of the sugar-phosphate backbone, with the nucleobases on the inside.

? *Which would require a higher temperature to denature—a DNA strand composed primarily of A-T base pairs or one that is the same length but composed primarily of G-C base pairs?*

and the carboxyl group of the fatty acid (figure 2.29b). A monoglyceride has only one fatty acid bound to glycerol; a diglyceride has two, and a triglyceride, three. In nature, the most common fats are triglycerides, which are stored in the body as an energy reserve.

Although hundreds of different fatty acids exist, they can be divided into two groups based on the presence of double bonds between carbon atoms. Saturated fatty acids have no double bonds (figure 2.29a). The term "saturated" means they have the maximum number of hydrogen atoms. Unsaturated fatty acids contain one or

more double bonds. Those that have one double bond are monounsaturated, and those with more than one double bond are polyunsaturated. Oleic acid is a common monounsaturated fatty acid (figure 2.29a). Most naturally occurring fatty acids are *cis*, meaning the hydrogen atoms attached to the double-bonded carbon molecules are on the same side of the bond. *Trans* fatty acids have hydrogen atoms on opposite sides of the double bond.

The type of fatty acids in a triglyceride affects the melting point of the fat. Fats that contain only saturated fatty acids are typically solid

Saturated fatty acid (palmitic acid)

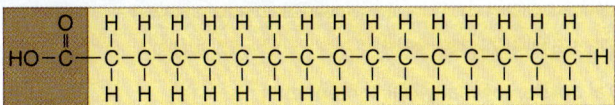

Unsaturated fatty acid (oleic acid)

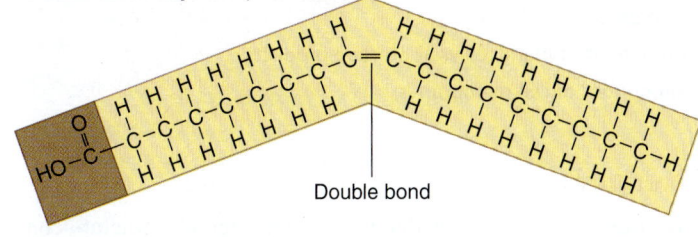

Double bond

(a)

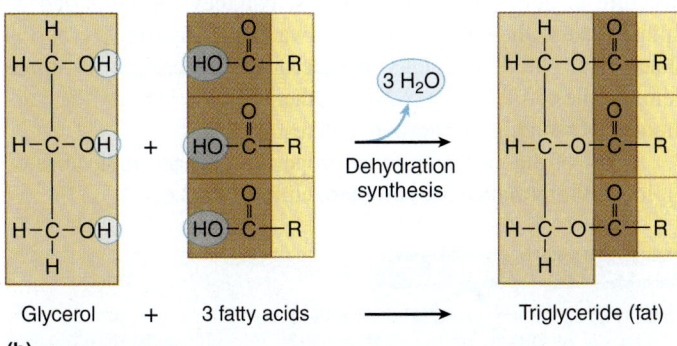

Glycerol   +   3 fatty acids   →   Triglyceride (fat)

(b)

FIGURE 2.29 Fat Formation from Fatty Acids and Glycerol
(a) Two common fatty acids. Most fatty acids contain an even number of carbon atoms (commonly 16 or 18) and may be saturated or unsaturated. The unsaturated fatty acids may be *cis* or *trans*.
(b) Dehydration synthesis in the joining of fatty acids with glycerol; the R groups are carbon-hydrogen chains, such as those shown in (a).

❓ *What characteristic of the fat in this figure makes it a triglyceride?*

at room temperature because the straight, long tails of the fatty acids can pack tightly together. Fats that contain one or more unsaturated fatty acids tend to be liquid at room temperature because these fatty acids have kinks in their long tails that prevent tight packing. Oils are fats that are liquid at room temperature. Manufacturing processes that hydrogenate oils, thereby making them solid at room temperature, sometimes convert *cis* fatty acids to *trans* fatty acids.

Microʙyte

Diets rich in saturated fats and *trans* fats raise blood cholesterol levels, leading to clogged arteries.

Compound Lipids

Compound lipids contain fatty acids and glycerol as well as elements other than carbon, hydrogen, and oxygen, Biologically, some of the most important of these are **phospholipids,** which contain a phosphate group linked to one of a variety of other polar molecules (**figure 2.30**). This phosphate-containing portion is the polar head and is soluble in water (hydrophilic). In contrast, the fatty acid portion is insoluble in water (hydrophobic).

Phospholipids are an essential component of cytoplasmic membranes, the structure that separates the internal contents of a

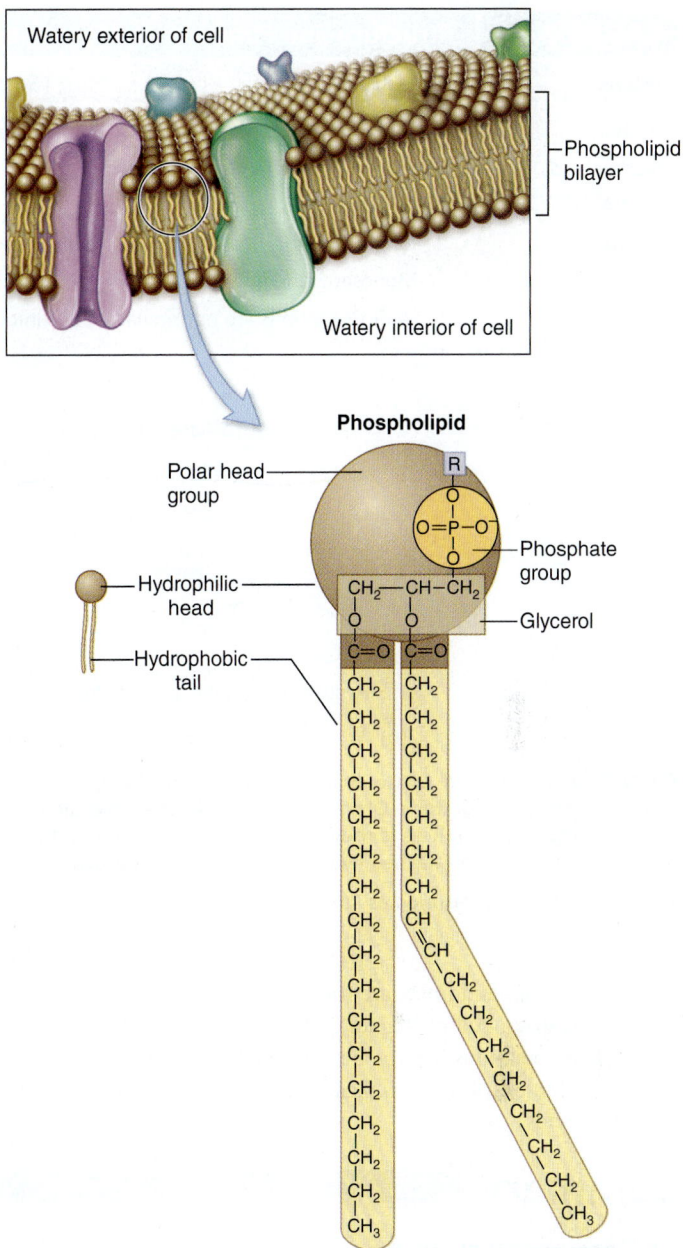

FIGURE 2.30 Phospholipids Are an Essential Component of Cell Membranes In phospholipids, two of the —OH groups of glycerol are linked to fatty acids and the third —OH group is linked to a hydrophilic head group, which contains a phosphate ion and a polar molecule (labeled R).

❓ *What about the structure of a phospholipid makes one portion hydrophilic and the other hydrophobic?*

cell from the outside environment (figure 2.30). The phospholipid molecules orient themselves in the membrane as opposing layers, forming a bilayer. The fatty acids face inward, interacting with the fatty acids of the phospholipid molecules in the opposing layer. The hydrophilic polar heads face outward, toward either the aqueous external environment or the internal (cytoplasmic) environment. Water-soluble substances cannot pass through the hydrophobic portion, so the cell uses special transport mechanisms to move these across the membrane. These will be discussed in chapter 3.

TABLE 2.6	Structure and Function of Macromolecules	
Name	**Subunit**	**Some Functions of Macromolecules**
Protein	Amino acid	Catalysts; structural portion of many cell components
Nucleic acids		
DNA	Deoxyribonucleotide	Carrier of genetic information
RNA	Ribonucleotide	Various roles in protein synthesis
Polysaccharide	Monosaccharide	Structural component of plant cell wall; storage products
Lipids	Varies—subunits are not similar	Important component of cell membranes

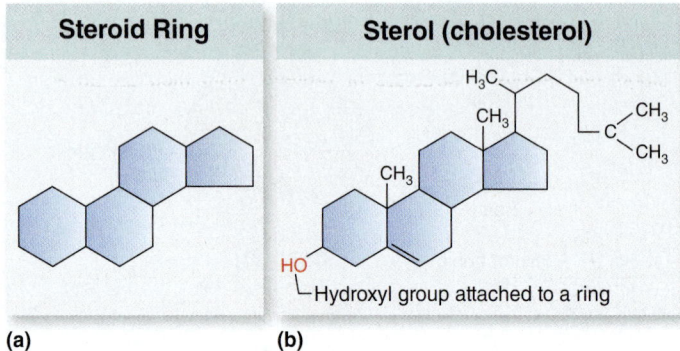

FIGURE 2.31 Steroid (a) General formula showing the four-membered ring and **(b)** the —OH group that makes the molecule a sterol. The sterol shown here is cholesterol. The carbon atoms in the ring structures and the attached hydrogen atoms are not shown.

❓ *Why are steroids classified as lipids?*

Other compound lipids are found in bacterial membranes and will also be discussed in chapter 3. These include the lipoproteins (covalent associations of proteins and lipids) and lipopolysaccharides (molecules of lipid linked with polysaccharides through covalent bonds).

Steroids

Steroids are simple lipids that have a characteristic structure consisting of four connected rings (**figure 2.31**). Their chemical structure is quite different from fats, but they are classified as lipids because they are insoluble in water. If a hydroxyl group is attached to one of the rings, the steroid is a sterol, an example being cholesterol (figure 2.31b). Other steroids include the hormones cortisone, progesterone, and testosterone.

Some of the most important properties of macromolecules of biological importance are summarized in **table 2.6.**

MicroAssessment 2.7

Lipids are a diverse group of hydrophobic molecules. Fats are composed of fatty acids joined to glycerol. Phospholipids, with one end hydrophilic and the other hydrophobic, form a major part of cell membranes where they exist as a bilayer. They limit the entry and exit of molecules into and out of cells.

19. *What are the main functions of lipids in cells?*
20. *What features in the chemical composition of phospholipids make them ideal components of the cytoplasmic membrane?*
21. *How could you determine if a solid compound were a lipid or a carbohydrate based on its solubility properties?* ➕

Summary

2.1 ■ Atoms and Elements

Atoms are composed of **electrons, protons,** and **neutrons** (figures 2.1, 2.2). An element consists of a single type of atom. Valence electrons are particularly important in chemical bond formation.

2.2 ■ Chemical Reactions and Bonds

Ions and Ionic Bonds

Ionic bonds are weak attractions between cations and anions (figure 2.3).

Covalent Bonds

Covalent bonds are formed by atoms sharing electrons (figure 2.4). **Organic compounds** have C—H bonds. When atoms have an equal attraction for electrons, a **non-polar covalent bond** is formed (table 2.3). When one atom is electronegative, a **polar covalent bond** is formed (figure 2.5).

Hydrogen Bonds

Hydrogen bonds are weak bonds that result from the attraction of a hydrogen atom in a polar molecule to an electronegative atom in another polar molecule (figure 2.6). Hydrogen bonds and ionic bonds are important in the weak association of various molecules with one another (figure 2.7).

Molarity

A **mole** of one compound has the same number of molecules as a mole of another compound.

2.3 ■ Chemical Components of the Cell

Water

Water is the most important molecule in the cell, making up over 70% of all living organisms by weight. Hydrogen bonding plays a very important role in the properties of water (figures 2.8, 2.9). **pH** is the degree of acidity of a solution and is measured on a scale of 0 to 14. **Buffers** prevent a dramatic rise or fall of pH (figure 2.10).

Elements and Small Molecules in the Cell

All cells contain a variety of elements and small molecules. Organic molecules often contain distinct chemical groups important in the functioning of the molecule (table 2.4). **ATP** carries energy in two **high-energy phosphate bonds,** which, when broken, release the energy (figure 2.11).

Macromolecules and Their Component Parts

Macromolecules are usually polymers of subunits. Different classes of macromolecules have different subunits. Macromolecules are made through **dehydration synthesis** and degraded by **hydrolysis** (figure 2.12).

FUTURE CHALLENGES 2.1

Fold Properly: Do Not Bend or Mutilate

An organism's properties depend on its proteins, including structural proteins and enzymes. Even though a cell may be able to synthesize a protein, that protein will not function properly unless it is folded correctly and has its correct shape. A major challenge is to understand how proteins fold correctly—a puzzle known as the protein-folding problem. Not only is this important from a purely scientific point of view, but a number of serious neurodegenerative diseases result from protein misfolding. These include Alzheimer's disease and the neurodegenerative diseases caused by prions. If we could understand why proteins fold incorrectly, we might be able to prevent such diseases.

The information that determines how a protein folds into its three-dimensional shape is contained in the sequence of its amino acids. It is not yet possible, however, to predict accurately how a protein will fold from its amino acid sequence. The folding occurs in a matter of seconds after the protein is synthesized. The protein folds rapidly into its secondary structure and then more slowly into its tertiary structure. These slower reactions are still poorly understood, but various attractive and repulsive interactions between the amino acid side chains allow the flexible molecule to "find its way" to the correct tertiary structure. Proteins called chaperones can assist the process by preventing detrimental interactions. Mistakes still occur, but improperly folded proteins can be recognized and degraded by enzymes called proteases. The protein-folding problem has such important implications for medicine, and is so challenging a scientific question, that a supercomputer with a huge memory is now being used to help predict the three-dimensional structure of proteins from their amino acid sequences. ◀◀▶▶ prions, pp. 12, 328

2.4 ■ Proteins

Proteins are made of amino acid subunits.

Amino Acids

Amino acids have a central carbon atom, bonded to a carboxyl group, an amino group, and a side chain. The side chain gives an amino acid its characteristic properties (figure 2.13). Twenty major amino acids function as subunits of proteins (figure 2.14). Most amino acids can exist in two stereoisomers—D and L (figure 2.15).

Peptide Bonds

Peptide bonds join the amino group of one amino acid with the carboxyl group of another (figure 2.16).

Protein Structure (figure 2.17)

The **primary structure** of a protein is its amino acid sequence. The **secondary structure** is the three-dimensional shape of localized regions, characterized by helices and sheets. The **tertiary structure** is the three-dimensional shape, usually either globular or fibrous. The **quaternary structure** is the shape that results from the interaction of multiple polypeptide chains. Some proteins need **chaperones** to help them fold into their functional shape.

Protein Domains

A **protein domain** is a region that folds into a stable structure independently of other parts of the molecule (figure 2.18).

Substituted Proteins

Substituted proteins contain other molecules such as sugars and lipids, bonded to the side chains of amino acids in the protein.

Protein Denaturation

When bonds within a protein break, the protein changes shape and no longer functions; the protein is **denatured** (figure 2.19).

2.5 ■ Carbohydrates

Carbohydrates have carbon, hydrogen, and oxygen atoms in a ratio of approximately $1:2:1$.

Monosaccharides

Monosaccharides include the 5-carbon sugars ribose and deoxyribose (figure 2.20) and the common 6-carbon sugars glucose, galactose, fructose, and mannose (figure 2.21, table 2.5). Sugars can exist in two interchangeable forms: α and β (figure 2.22).

Disaccharides

Disaccharides are two monosaccharides joined together by covalent bonds (figure 2.23). They include lactose, sucrose, and maltose (table 2.5).

Polysaccharides

Polysaccharides are large molecules composed of chains of monosaccharide subunits or their derivatives. They include cellulose, starch, glycogen, dextran, chitin, and agar (figure 2.24, table 2.5).

2.6 ■ Nucleic Acids

Nucleic acids carry genetic information, which is decoded to produce proteins.

DNA

The nucleotide sequence of **DNA (deoxyribonucleic acid)** encodes a cell's properties. The nucleotides have a **nucleobase,** deoxyribose, and a phosphate group (figures 2.25, 2.26, 2.27). DNA is a double-stranded helical molecule with a sugar phosphate backbone; the two strands are antiparallel. The purine and pyrimidine nucleobases extend into the center of the helix (figure 2.28a). The two strands of DNA are **complementary**—meaning there are G-C and A-T base pairs—and are held together by hydrogen bonds between the nucleobases (figure 2.28b).

RNA

RNA (ribonucleic acid) is involved in the process that decodes the information contained in DNA. RNA is a single-stranded molecule and its nucleotides contain uracil in place of thymine and ribose in place of deoxyribose (figures 2.20, 2.26).

2.7 ■ Lipids

Lipids are a diverse group of hydrophobic, non-polar molecules.

Simple Lipids

Simple lipids contain carbon, hydrogen, and oxygen. Fats consist of fatty acids linked to glycerol and may be liquid or solid at room temperature (figure 2.29). Saturated fatty acids have no double bonds, whereas unsaturated fatty acids have one or more double bonds.

Compound Lipids

Compound lipids contain fatty acids and glycerol as well as elements other than carbon, hydrogen, and oxygen. **Phospholipids** are essential components of cytoplasmic membranes (figure 2.30).

Steroids

Steroids are simple lipids that have a characteristic structure consisting of four connected rings (figure 2.31). The sterol cholesterol is an example.

Review Questions

Short Answer

1. Differentiate between an atom, a molecule, and a compound.

2. Why is water a good solvent?

3. Which solution is more acidic, one with a pH of 4 or a pH of 5? What is the concentration of H^+ ions in each? The concentration of OH^- ions?

4. Name the subunits of proteins, polysaccharides, and nucleic acids.

5. Give an example of dehydration synthesis. Give an example of a hydrolysis reaction. How are these reactions related?

6. List four functions of proteins.

7. What are the four levels of protein structure, and what is the distinguishing feature of each?

8. How do the two types of nucleic acids differ from one another in (a) composition, (b) size, and (c) function?

9. What are the two major groups of lipids? Give an example of each group. What feature is common to all lipids?

10. What features do all lipids share?

Multiple Choice

1. Choose the list that goes from the lightest to the heaviest:
   a) proton, atom, molecule, compound, electron.
   b) atom, proton, compound, molecule, electron.
   c) electron, proton, atom, molecule, compound.
   d) atom, electron, proton, molecule, compound.
   e) proton, atom, electron, molecule, compound.

2. The strongest chemical bonds between two atoms in solution are
   a) covalent.   b) ionic.
   c) hydrogen bonds.   d) hydrophobic interactions.

3. Dehydration synthesis is involved in the synthesis of all of the following *except*
   a) DNA.   b) proteins.   c) polysaccharides.
   d) lipids.   e) monosaccharides.

4. The primary structure of a protein relates to its
   a) sequence of amino acids. b) length.   c) shape.
   d) solubility.   e) bonds between amino acids.

5. Pure water has all of the following properties *except*
   a) polarity.   b) ability to dissolve lipids.   c) pH of 7.
   d) covalent joining of its atoms.
   e) ability to form hydrogen bonds.

6. The macromolecules that are composed of carbon, hydrogen, and oxygen in an approximate ratio of 1:2:1 are
   a) proteins.   b) lipids.   c) polysaccharides.
   d) DNA.   e) RNA.

7. In proteins, α helices and β pleated structures are associated with the
   a) primary structure.   b) secondary structure.
   c) tertiary structure.   d) quaternary structure.
   e) multiprotein complexes.

8. Complementarity plays a major role in the structure of
   a) proteins.   b) lipids.   c) polysaccharides.
   d) DNA.   e) RNA.

9. A bilayer is associated with
   a) proteins.   b) DNA.   c) RNA.
   d) complex polysaccharides.   e) phospholipids.

10. Isomers are associated with
    1. carbohydrates.   2. amino acids.   3. nucleotides.
    4. RNA.   5. fatty acids.
    a) 1, 2   b) 2, 3   c) 3, 4   d) 4, 5   e) 1, 5

Applications

1. A group of prokaryotes known as thermophiles thrive at high temperatures that would normally destroy other organisms. Yet these thermophiles cannot survive well at the lower temperatures normally found on the earth. Propose an explanation for this observation.

2. Microorganisms use hydrogen bonds to attach to surfaces. Many of the cells lose hold of the surface because of the weak nature of these bonds. Contrast the benefits and disadvantages of using covalent bonds as a means of attaching to surfaces.

Critical Thinking ✚

1. What properties of the carbon atom make it ideal as the key atom for all molecules in organisms?

2. A biologist determined the amounts of several amino acids in two separate samples of pure protein. The data are shown here:

Amino Acid	Leucine	Alanine	Histidine	Cysteine	Glycine
Protein A	7%	12%	4%	2%	5%
Protein B	7%	12%	4%	2%	5%

The scientist concluded that protein A and protein B were the same protein. Do you agree with this conclusion? Justify your answer.

3. This table indicates the freezing and boiling points of several molecules:

Molecule	Freezing Point (°C)	Boiling Point (°C)
Water	0	100
Carbon tetrachloride (CCl_4)	−23	77
Methane (CH_4)	−182	−164

Carbon tetrachloride and methane are non-polar molecules. How does the polarity and non-polarity of these molecules explain why the freezing and boiling points for methane and carbon tetrachloride are so much lower than those for water?

3 Microscopy and Cell Structure

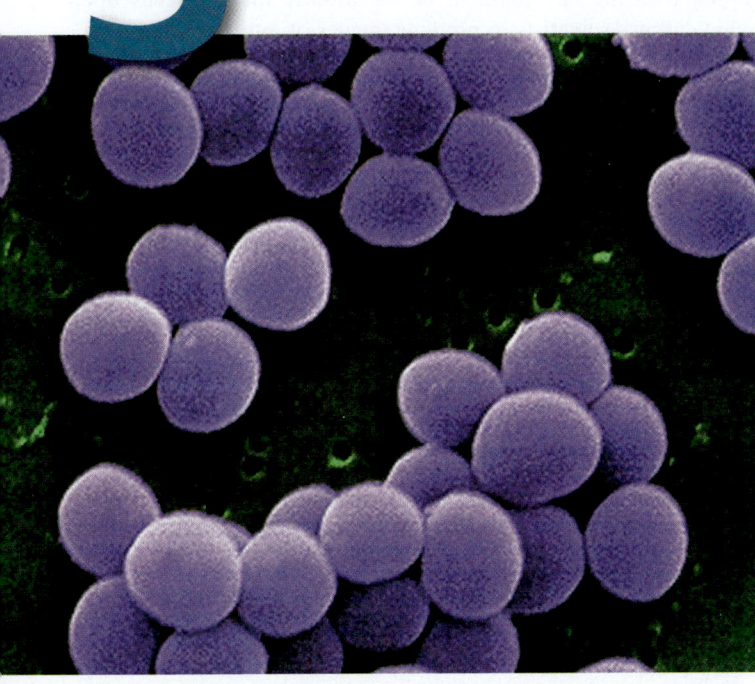

Bacterial cells (color-enhanced scanning electron micrograph).

A Glimpse of History

Hans Christian Joachim Gram (1853–1938) was a Danish physician working in a laboratory at the morgue of the City Hospital in Berlin, microscopically examining the lungs of patients who had died of pneumonia. He was working under the direction of Dr. Carl Friedlander, who was trying to identify the cause of pneumonia by studying patients who had died of it. Gram's task was to stain the infected lung tissue to make the bacteria easier to see under the microscope. Strangely, one of the methods he developed did not stain all bacteria equally; some types retained the first dye applied in this multistep procedure, whereas others did not. Gram's staining method revealed that two different kinds of bacteria were causing pneumonia, and these types retained the dye differently. We now recognize that this important staining method, called the Gram stain, efficiently identifies two large, distinct groups of bacteria: Gram-positive and Gram-negative. The staining outcome reflects a fundamental difference in the structure and chemistry of the cell walls of these two groups, and is a key test in the initial identification of bacterial species.

Imagine the astonishment Antony van Leeuwenhoek must have felt in the 1600s when he first observed microorganisms with his handcrafted microscopes, instruments that could magnify images approximately 300-fold. Even today, observing diverse microbes interacting in a sample of pond water can provide enormous education and entertainment.

Microscopic study of cells has revealed two fundamental types: prokaryotic and eukaryotic. The cells of all members of the domains *Bacteria* and *Archaea* are prokaryotic. In contrast, cells of all animals, plants, protozoa, fungi, and algae are eukaryotic.

The similarities and differences between these two basic cell types are important from a scientific standpoint. They also have significant consequences to human health. For example, chemicals that interfere with processes unique to prokaryotic cells can be used to selectively destroy bacteria without harming humans. ◀◀ prokaryotic cells, p. 9 ◀◀ eukaryotic cells, p. 10

Prokaryotic cells are generally much smaller than most eukaryotic cells—a trait that carries with it certain advantages as well as disadvantages. On one hand, their high surface area relative to low volume makes it easier for these cells to take in nutrients and excrete waste products. Because of this, they can multiply much faster than eukaryotic cells. On the other hand, their small size makes them vulnerable to a variety of threats, including predators, parasites, and competitors. To cope, prokaryotes have evolved many unique features that increase their chances of survival.

Eukaryotic cells are much more complex than prokaryotic cells. Not only are they larger, but many of their cellular processes take place within membrane-bound compartments. Eukaryotic cells are defined by the presence of one of these—the nucleus. Although eukaryotic cells have many of the same characteristics as prokaryotic cells, some structures and processes are fundamentally different.

MICROSCOPY AND CELL MORPHOLOGY

3.1 ■ Microscopic Techniques: The Instruments

Learning Outcomes

1. *Describe the importance and principles of magnification, resolution, and contrast in microscopy.*
2. *Compare and contrast light microscopes, electron microscopes, and atomic force microscopes.*

One of the most important tools for studying microorganisms is the **light microscope,** which uses visible light for observing objects. These instruments can magnify images approximately $1,000\times$ (1,000-fold), making it relatively easy to observe cell size, shape, and motility. The **electron microscope,** introduced in 1931, can magnify images in excess of $100,000\times$, revealing many fine details of cell structure. A major advancement came in the 1980s with the development of the **atomic force microscope,** which allows scientists to produce images of individual atoms on a surface.

Principles of Light Microscopy: The Bright-Field Microscope

In light microscopy, light passes through a specimen and then through a series of magnifying lenses. The most common type of light microscope is the **bright-field microscope,** which evenly illuminates the field of view. Characteristics of this and other microscopes are summarized in **table 3.1.**

TABLE 3.1	A Summary of Microscopic Instruments and Their Characteristics	
Instrument	**Mechanism**	**Comment**
Light Microscopes	Visible light passes through a series of lenses to produce a magnified image.	Relatively easy to use; considerably less expensive than electron and atomic force microscopes.
Bright-field	Illuminates the field of view evenly.	Most common type of microscope.
Dark-field	Light is directed toward the specimen at an angle.	Makes unstained cells easier to see; organisms stand out as bright objects against a dark background.
Phase-contrast	Increases contrast by amplifying differences in refractive index.	Makes unstained cells easier to see.

Instrument	Mechanism	Comment
Differential interference contrast	Two light beams pass through the specimen and then recombine.	The image of the specimen appears three-dimensional.
Fluorescence	Projects ultraviolet light, causing fluorescent molecules in the specimen to emit longer wavelength light.	Used to observe cells stained or tagged with a fluorescent dye.
Scanning laser	Mirrors scan a laser beam across successive regions and planes of a specimen. From that information, a computer constructs an image.	Used to construct a three-dimensional image of a structure; provides detailed sectional views of intact cells.
Electron Microscopes	Electron beams are used in place of visible light to produce the magnified image.	Can clearly magnify images 100,000×.
Transmission	Transmits a beam of electrons through a specimen.	Elaborate specimen preparation is required.
Scanning	A beam of electrons scans back and forth over the surface of a specimen.	Used for observing surface details; produces a three-dimensional effect.
Atomic Force Microscope	A probe moves in response to even the slightest force between it and the sample.	Produces a map showing the bumps and valleys of the atoms on the surface of the sample.

Magnification

The modern light microscope, called a **compound microscope,** has two types of magnifying lenses: an **objective lens** and an **ocular lens** (**figure 3.1**). In combination, these lenses visually enlarge an object by a factor equal to the product of each lens' magnification. For example, a structure is magnified 1,000-fold when viewed through a 10× ocular lens in conjunction with a 100× objective lens. Most compound microscopes have a selection of objective lenses that are of different powers—typically 4×, 10×, 40×, and 100×. This makes a choice of different magnifications possible with the same instrument.

The **condenser lens,** positioned between the light source and the specimen, does not affect the magnification. It focuses the light on the specimen.

Resolution

The usefulness of a microscope depends primarily on its **resolving power,** which determines how much detail can actually be seen (**figure 3.2**). Resolving power is a measure of the ability to distinguish two objects that are very close together. It is defined as the minimum distance between two points at which those points can still be observed as separate objects.

The resolving power of a microscope depends on the quality and type of lens, the wavelength of the light, the magnification, and how the specimen has been prepared. The maximum resolving power of the best light microscope is 0.2 μm. This is sufficient to see the general morphology of a prokaryotic cell but too low to distinguish an object the size of most viruses.

To obtain maximum resolution when using certain high-power objectives such as the 100× lens, immersion oil must be used to displace the air between the lens and the specimen. This prevents the refraction (bending of light rays) that occurs when light passes from glass to air (**figure 3.3**). Light rays bend when they pass from

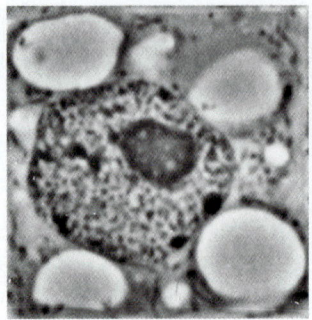
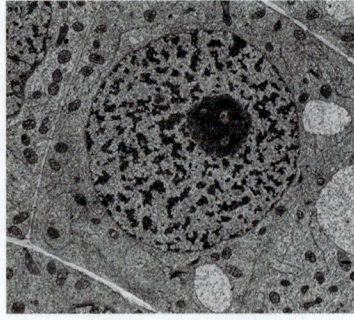

Light microscope (450×)   Electron microscope (450×)

FIGURE 3.2 Resolving Power These images of an onion root tip magnified 450× illustrate the difference in resolving power between a light microscope and an electron microscope. Note the difference in the degree of detail that can be seen at the same magnification.

❓ *Which type of microscope—a light microscope or an electron microscope—has the higher resolving power?*

a medium of one **refractive index** (a measure of the relative speed of light as it passes through the medium) to another. If refraction occurs, some light rays will miss the relatively small openings of higher-power objective lenses. The oil prevents refraction because it has nearly the same refractive index as glass.

Contrast

The amount of **contrast** affects how easily cells can be seen. As an example, bacteria are essentially transparent against a bright colorless background, so the lack of contrast makes them harder

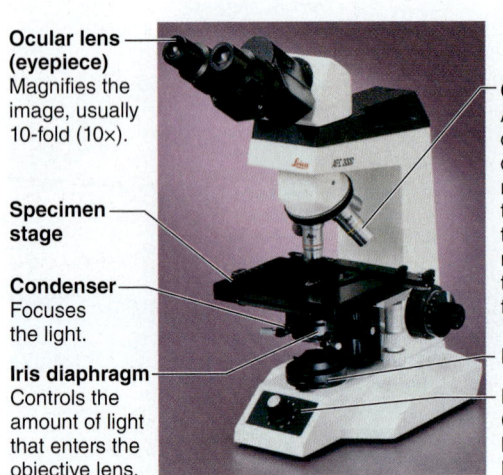

Ocular lens (eyepiece)
Magnifies the image, usually 10-fold (10×).

Specimen stage

Condenser
Focuses the light.

Iris diaphragm
Controls the amount of light that enters the objective lens.

Objective lens
A selection of lens options provide different magnifications. The total magnification is the product of the magnifying power of the ocular lens and the objective lens.

Light source

Rheostat
Controls the brightness of the light.

FIGURE 3.1 A Modern Light Microscope The compound microscope uses a series of magnifying lenses.

❓ *What are the two sets of magnifying lenses called, and how do these relate to total magnification?*

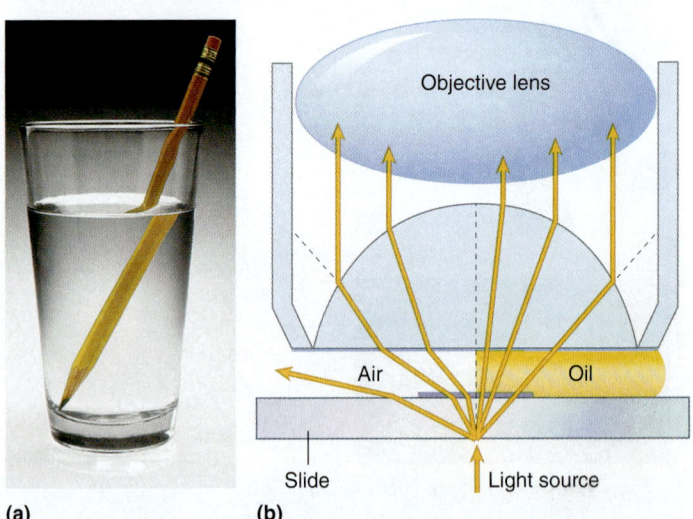

(a)   (b)

FIGURE 3.3 Refraction As light passes from one medium to another, the light rays may bend, depending on the refractive index of the two media. **(a)** The pencil in water appears bent because the refractive index of water is different from that of air. **(b)** Light rays bend as they pass from air to glass because of the different refractive indexes of these media. Immersion oil and glass have the same refractive index, and therefore the light rays are not bent.

❓ *What would the pencil in part (a) look like if oil were in the glass instead of water?*

to see (see figure 11.17). One way to overcome this difficulty is to stain the cells with one of the various dyes that will be discussed shortly. The staining process kills microbes, however, so living cells cannot be observed.

Light Microscopes That Increase Contrast

Special light microscopes that increase the contrast between microorganisms and their surroundings overcome some of the difficulties of observing unstained cells. They are invaluable when examining characteristics of living organisms such as motility. When observing live organisms, the specimen is prepared as a **wet mount**—a drop of liquid on which a coverslip has been placed.

The Dark-Field Microscope

Cells viewed through a **dark-field microscope** stand out as bright objects against a dark background (**figure 3.4**). The microscope works in the same way that a beam of light shining into a dark room makes dust visible. For dark-field microscopy, a special mechanism directs light toward the specimen at an angle, so that only light scattered by the specimen enters the objective lens. A simple "dark-field stop" can be attached to the condenser lens of a bright-field microscope to temporarily convert it to a dark-field microscope.

> **MicroByte**
>
> Dark-field microscopy is used to see *Treponema pallidum*, which causes syphilis; bright-field microscopy will not work because the cells are thin and stain poorly.

The Phase-Contrast Microscope

The **phase-contrast microscope** makes cells and other dense material appear darker (**figure 3.5**). It does this by amplifying the

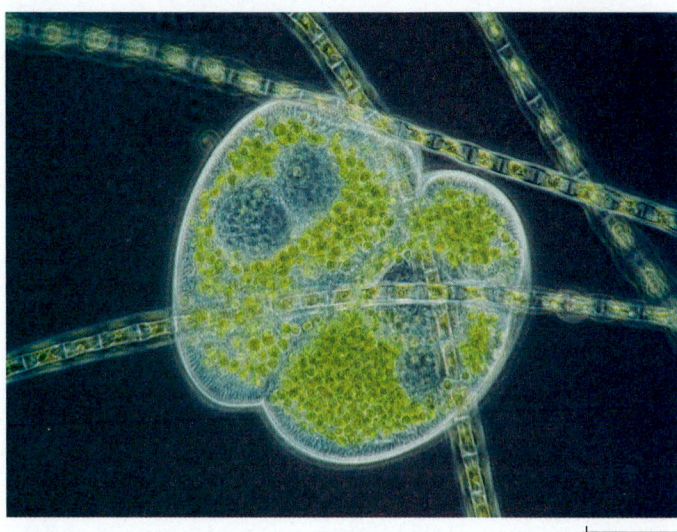

25 µm

FIGURE 3.5 Phase-Contrast Photomicrograph *Paramecium bursaria* containing endosymbiotic *Chlorella* (a green alga).

? *How does a phase-contrast microscope increase contrast?*

slight difference between the refractive index of dense material and that of the surrounding medium. As light passes through material, it is refracted slightly differently than when it passes through its surroundings. Special optical devices increase those differences, thereby enhancing the contrast.

The Differential Interference Contrast Microscope (DIC)

The **differential interference contrast microscope (DIC)** causes the image of the specimen to appear three-dimensional (**figure 3.6**). This microscope, like the phase-contrast microscope, depends on

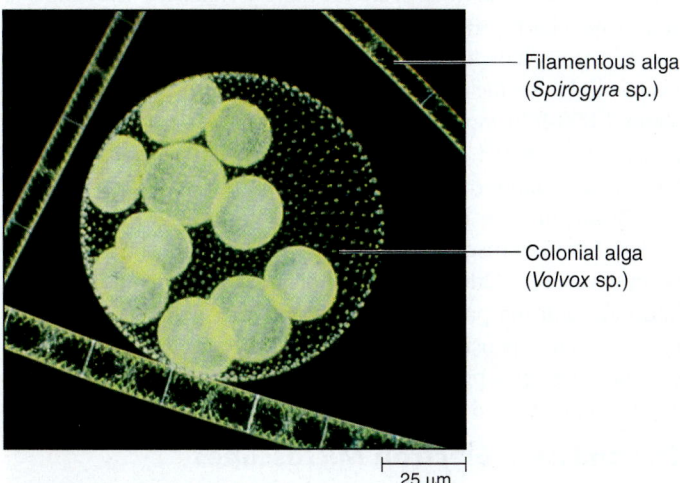

Filamentous alga (*Spirogyra* sp.)

Colonial alga (*Volvox* sp.)

25 µm

FIGURE 3.4 Dark-Field Photomicrograph *Volvox* (sphere) and *Spirogyra* (filaments), both of which are eukaryotes.

? *How does a dark-field microscope increase contrast?*

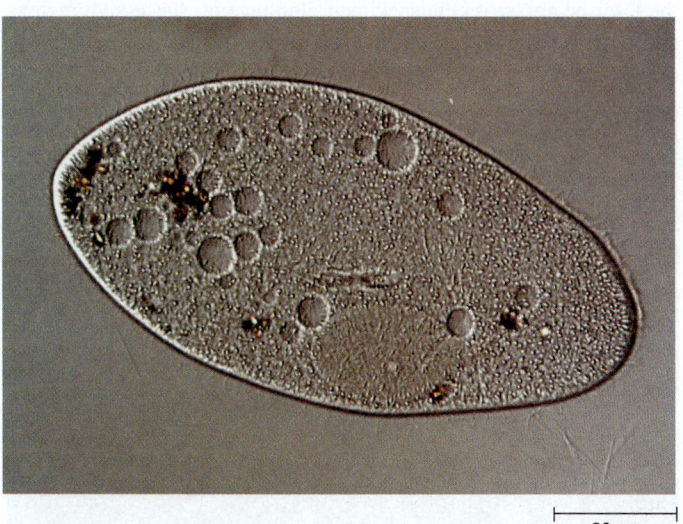

30 µm

FIGURE 3.6 Differential Interference Contrast (DIC) Photomicrograph *Paramecium multimicronuclatum*, a protozoan.

? *How does a DIC microscope increase contrast?*

differences in refractive index as light passes through different materials. It has a device for separating light into two beams that pass through the specimen and then recombine. The light waves are out of phase when they recombine, thereby creating the three-dimensional appearance of the image.

Other Light Microscopes

Various other light microscopes have special characteristics that make them useful in certain situations.

Fluorescence Microscopes

Fluorescence microscopes are used to observe cells or other materials that are either naturally fluorescent or stained or tagged with fluorescent dyes (**figure 3.7**). Fluorescent molecules absorb light at one wavelength (usually ultraviolet light) and then emit light of a longer wavelength. The types and characteristics of fluorescent dyes and tags will be discussed shortly. ▶▶ **fluorescent dyes and tags, p. 49**

Most fluorescence microscopes used today are epifluorescence microscopes, meaning they project ultraviolet light onto the specimen rather than through it. They then capture the light emitted by the fluorescent molecules to form the image, allowing fluorescent cells to stand out as bright objects against a dark background. Because the light does not need to travel through the specimen, cells attached to soil or other opaque particles can be observed.

Scanning Laser Microscopes

A **scanning laser microscope (SLM)** can be used to get detailed interior views of intact cells (**figure 3.8**). Frequently, the specimens are first stained or tagged with a fluorescent dye. By using fluorescent tags that bind only to certain compounds, the precise cellular location of those compounds can be determined. Some scanning laser microscopes can also be used to make three-dimensional images of microbial communities or other thick structures.

In **confocal microscopy,** lenses focus a laser beam to illuminate a given point on one vertical plane of a specimen. Mirrors then scan the laser beam across the specimen, illuminating successive regions and planes until the entire specimen has been scanned. Each plane corresponds to an image of one fine slice of the specimen. A computer

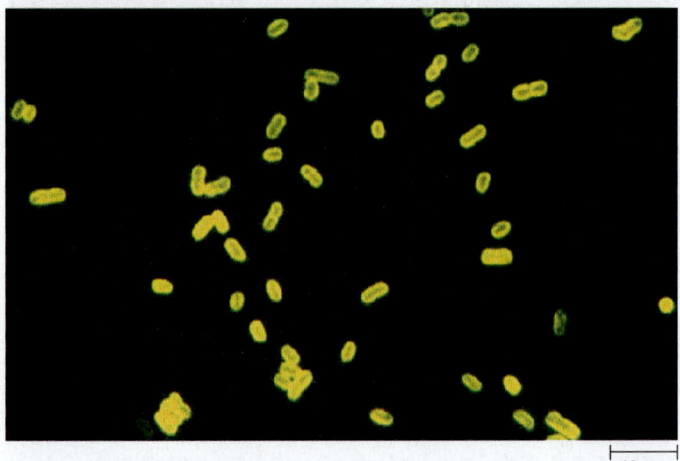

FIGURE 3.7 Fluorescence Photomicrograph A rod-shaped bacterium tagged with fluorescent marker.

❓ *What is an epifluorescence microscope?*

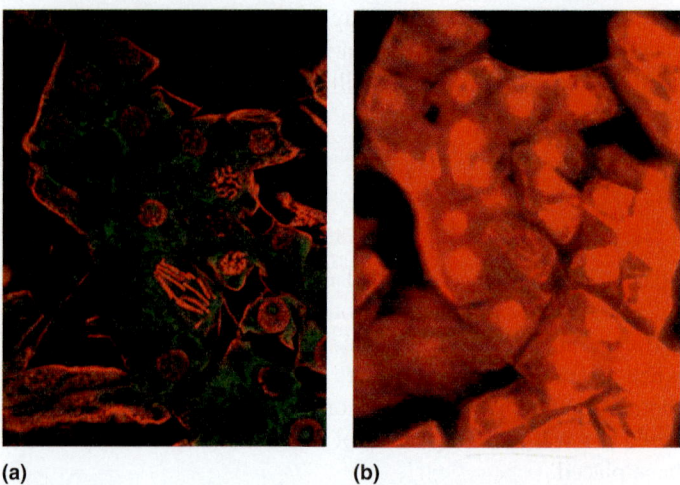

(a)   **(b)**

FIGURE 3.8 Scanning Laser Microscopy (a) Confocal photomicrography of fava bean mitosis. **(b)** Regular photomicrograph.

❓ *How is multiphoton microscopy different from confocal microscopy?*

then assembles the data and constructs a three-dimensional image, which is displayed on a screen. In effect, this microscope is a miniature computerized axial tomography (CAT) scan for cells.

Multiphoton microscopy is similar to confocal microscopy, but lower energy light is used. This light is less damaging to cells, so time-lapse images of live cells can be obtained. In addition, the light penetrates deeper, making it possible to get interior views of relatively thick structures.

Electron Microscopes

Electron microscopy is in some ways comparable to light microscopy. Rather than using glass lenses, visible light, and the eye to observe the specimen, an electron microscope uses electromagnetic lenses, electrons, and a fluorescent screen to produce the magnified image (**figure 3.9**). That image can be captured on photographic film to create an electron micrograph. Sometimes the black-and-white images are artificially enhanced with color to add visual clarity. ◀◀ **electrons, p. 17**

Electrons have a wavelength about 1,000 times shorter than visible light, so the resolving power of electron microscopes is about 1,000-fold more than light microscopes—about 0.3 nanometers (nm) or 0.3×10^{-3} μm (see figure 1.13). Consequently, considerably more detail can be observed. These instruments can clearly magnify an image $100,000\times$. One of the biggest drawbacks of the microscope is that the lenses and specimen must all be in a vacuum. Otherwise, the molecules composing air would interfere with the path of the electrons. The need for a vacuum results in an expensive, bulky unit, and requires substantial and complex specimen preparation. It also means that electron microscopy cannot be used to observe the activities of living cells.

Transmission Electron Microscopes

A **transmission electron microscope (TEM)** is used to observe fine details of cell structure (**figure 3.10**). It works by directing a beam of electrons that will either pass through the specimen or be scattered, depending on the density of the region. The darker areas of the resulting image correspond to the denser portions of the specimen.

Light Microscope   **Transmission Electron Microscope**

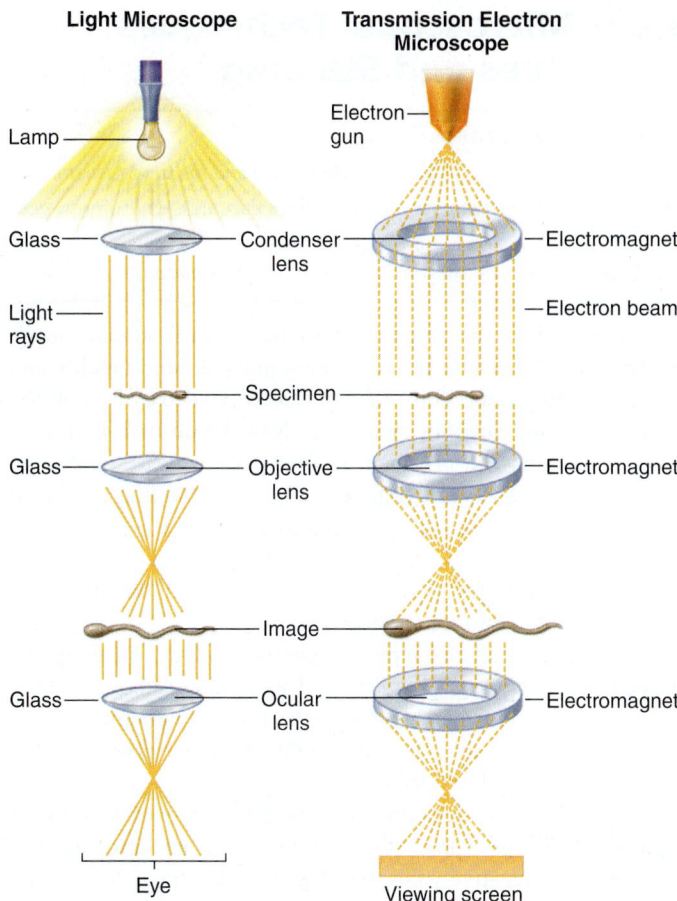

FIGURE 3.9 Comparison of the Principles of Light and Electron Microscopy For the sake of comparison, the light source for the light microscope has been inverted (the light is shown at the top and the ocular lens at the bottom).

❓ *Some electron micrographs are "color enhanced." What does this mean?*

A process called thin-sectioning is used to view details of a specimen's internal structures. First, the sample is treated with a preservative and dehydrated before embedding it in plastic. Once embedded, the sample can be cut into exceptionally thin slices with a diamond or glass knife and then stained with heavy metals. Even a single bacterial cell must be cut into slices to be viewed via TEM. Unfortunately, the procedure can severely distort cells, causing artifacts (artificial structures introduced as a result of the process). Distinguishing actual cell components from artifacts that result from specimen preparation is a major concern.

A method called freeze-fracturing is used to observe the shape of structures within a cell. The specimen is rapidly frozen and then fractured by hitting it with a knife blade. The cells break open, usually along the middle of internal membranes. Next, the surface of the section is coated with a thin layer of carbon to create a replica of the surface. This replica is then examined in the electron microscope. A variation of freeze-fracturing is freeze-etching. In this process, the frozen surface exposed by fracturing is dried slightly under vacuum, which allows underlying regions to be exposed.

A newer method of preparing and observing specimens with an electron microscope is **cryo-electron microscopy (cryo-EM).** This method involves rapidly freezing the specimen at very low

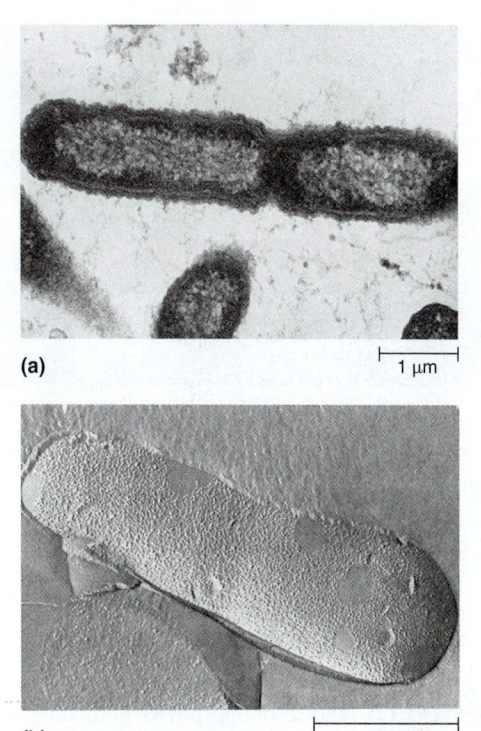

(a)   1 μm

(b)   1 μm

FIGURE 3.10 Transmission Electron Photomicrograph A rod-shaped bacterium prepared by **(a)** thin-sectioning; **(b)** freeze-etching.

❓ *How is thin-sectioning different from freeze-etching?*

temperatures, thereby avoiding some of the sample preparation processes that damage cells. A variation called cryo-electron tomography compiles images taken at different angles to create three-dimensional images of specimens.

Scanning Electron Microscopes

A **scanning electron microscope (SEM)** is used to observe surface details of cells (**figure 3.11**). A beam of electrons scans back and forth over the surface of a specimen coated with a thin film of metal. As those beams move, electrons are released from the specimen and reflected back into the viewing chamber. This reflected radiation is observed with the microscope. Relatively large specimens can be viewed, and a dramatic three-dimensional effect is observed with the SEM.

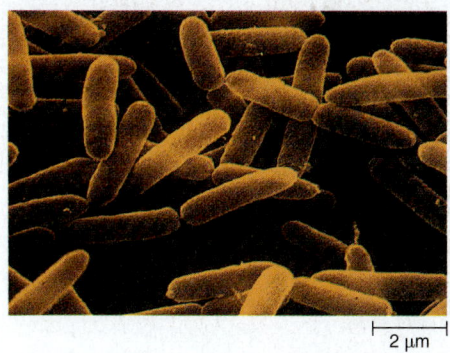

2 μm

FIGURE 3.11 Scanning Electron Photomicrograph A rod-shaped bacterium.

❓ *In what way is scanning electron microscopy different from transmission electron microscopy?*

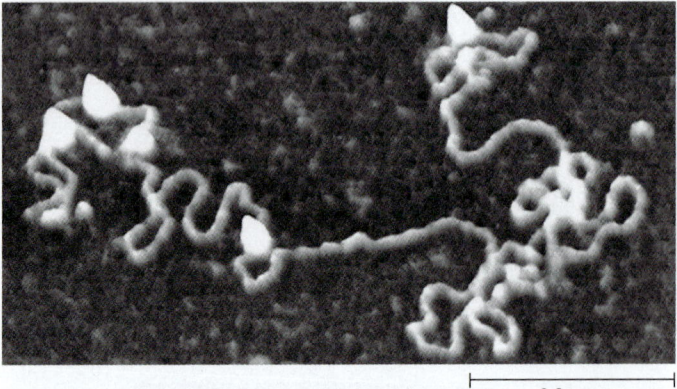

FIGURE 3.12 Atomic Force Photomicrograph A DNA fragment. The bright peaks are enzymes attached to the DNA.

❓ *How does the resolving power of atomic force microscopy compare to that of electron microscopy?*

Atomic Force Microscopes

An **atomic force microscope (AFM)** produces detailed images of surfaces (**figure 3.12**). The resolving power is much greater than that of an electron microscope, and the samples do not need the special preparation required for electron microscopy. In fact, the instrument can inspect samples either in air or submerged in liquid.

The mechanics of AFM can be compared to that of a stylus mounted on the arm of a record player. A very sharp probe (stylus) moves across the sample's surface, "feeling" the bumps and valleys of the atoms. As the probe scans the sample, a laser measures its motion, and a computer produces a surface map of the sample. ◀◀ atom, p. 17

MicroAssessment 3.1

The usefulness of a microscope depends on its resolving power. The most common type of microscope is the bright-field microscope. Dark-field, phase-contrast, and DIC microscopes increase the contrast between a microorganism and its surroundings. The fluorescence microscope is used to observe microbes stained with special dyes. Scanning laser microscopes are used to construct three-dimensional images of thick structures. Electron microscopes can magnify images 100,000×. The atomic force microscope produces detailed images of surfaces.

1. *Why must oil be used to obtain the best resolution with a 100× lens?*
2. *What are some drawbacks of electron microscopes?*
3. *If an object being viewed under the phase-contrast microscope has the same refractive index as the background material, how would it appear?* ✚

3.2 ■ Microscopic Techniques: Dyes and Staining

Learning Outcomes

3. *Describe the principles of the Gram stain and the acid-fast stain.*
4. *Describe the techniques used to observe capsules, endospores, and flagella.*
5. *Describe the benefits of using fluorescent dyes and tags.*

Observing living microorganisms with the bright-field microscope can be difficult because the cells often move around quickly and are nearly transparent. This problem can be avoided by immobilizing the cells and staining them with dyes. To do this, a drop of liquid containing the organism is first placed on a microscope slide and allowed to dry (**figure 3.13**). The resulting specimen forms a film, or smear. Next, the slide is passed over a flame to fix (attach) the cells to the slide. Dye is then applied and the excess washed off with water. If only a single dye is used, the procedure is called **simple staining** (**table 3.2**).

Staining procedures typically use basic dyes, meaning the dyes carry a positive charge. These dyes stain cells because they are attracted to the many negatively charged cellular components. Examples of basic dyes include methylene blue, crystal violet, safranin, and malachite green.

Although acidic dyes do not stain cells, they can be used for **negative staining,** a procedure that colors the background. The cells repel the negatively charged dye, allowing the colorless cell to stand out against the background. An advantage of negative staining is that it can be done as a wet mount. This avoids the heat-fixing step, a process that can distort the shape of the cells.

Differential Staining

Differential staining is used to distinguish different groups of bacteria. The two most frequently used differential staining techniques are the Gram stain and the acid-fast stain.

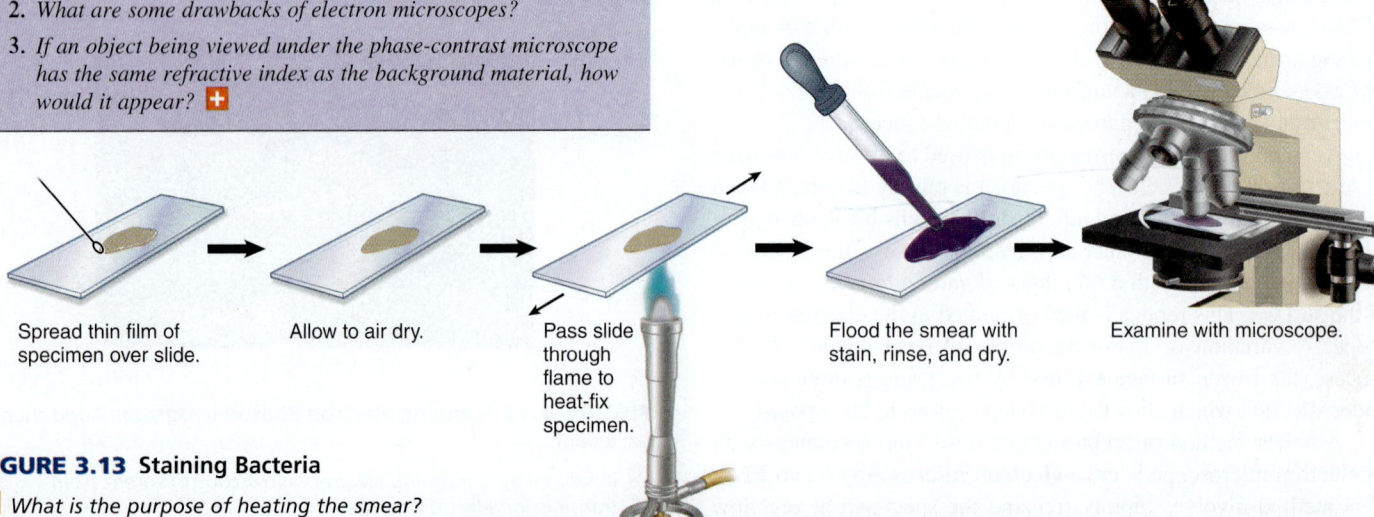

Spread thin film of specimen over slide.

Allow to air dry.

Pass slide through flame to heat-fix specimen.

Flood the smear with stain, rinse, and dry.

Examine with microscope.

FIGURE 3.13 Staining Bacteria

❓ *What is the purpose of heating the smear?*

TABLE 3.2	A Summary of Stains and Their Characteristics
Stain	**Characteristic**
Simple Stains	A basic dye is used to stain cells. Easy way to increase the contrast between otherwise colorless cells and a transparent background.
Differential Stains	A multistep procedure is used to stain cells and distinguish one group of microorganisms from another.
Gram stain	Used to separate bacteria into two major groups, Gram-positive and Gram-negative. The staining characteristics of these groups reflect a fundamental difference in the chemical structure of their cell walls. This is by far the most widely used staining procedure.
Acid-fast stain	Used to detect organisms that do not easily take up stains, such as members of the genus *Mycobacterium*.
Special Stains	Stain specific structures inside or outside of cells.
Capsule stain	These are often negative stains that take advantage of the fact that viscous capsules do not take up certain stains; the capsules stand out against a stained background.
Endospore stain	Stains endospores, a type of dormant cell that does not readily take up stains. Endospores are produced by *Bacillus* and *Clostridium* species.
Flagella stain	The staining agent adheres to and coats the otherwise thin flagella, making them visible with the light microscope.
Fluorescent Dyes and Tags	Fluorescent dyes and tags absorb ultraviolet light and then emit light of a longer wavelength. They are used in conjunction with a fluorescence microscope.
Fluorescent dyes	Some fluorescent dyes bind to compounds found in all cells; others bind to compounds specific to only certain types of cells.
Fluorescent tags	Antibodies to which a fluorescent molecule has been attached are used to tag specific molecules.

Gram Stain

The **Gram stain** is by far the most widely used procedure for staining bacteria. The basis for it was developed over a century ago by Dr. Hans Christian Gram (see **A Glimpse of History**). He showed that bacteria can be separated into two major groups: **Gram-positive bacteria** and **Gram-negative bacteria.** We now know that the difference in the staining properties of these two groups reflects a fundamental difference in the structure of their cell walls.

Gram staining involves four basic steps (**figure 3.14**).

1. The smear is first flooded with the **primary stain,** crystal violet in this case. The primary stain is the first dye applied in differential staining and generally stains all cells.
2. The smear is rinsed to remove excess dye and then flooded with a solution called Gram's iodine. The iodine combines with the crystal violet to form a dye-iodine complex, thereby decreasing the solubility of the dye within the cell.

Steps in Staining	State of Bacteria	Appearance
1 Crystal violet (primary stain)	Cells stain purple.	
2 Iodine (mordant)	Cells remain purple.	
3 Alcohol (decolorizer)	Gram-positive cells remain purple; Gram-negative cells become colorless.	
4 Safranin (counterstain)	Gram-positive cells remain purple; Gram-negative cells appear pink.	

(a)

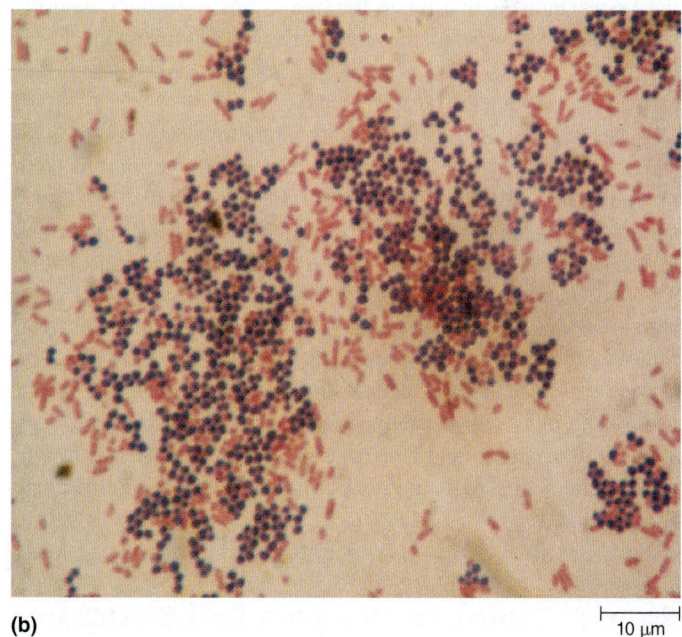

(b)

10 µm

FIGURE 3.14 Gram Stain (a) Steps in the Gram stain procedure. (b) Results of a Gram stain. The Gram-positive cells (purple) are *Staphylococcus aureus;* the Gram-negative cells (reddish-pink) are *Escherichia coli.*

Which step of the Gram stain is most critical with respect to timing?

3. The stained smear is rinsed, and then a **decolorizing agent**—95% alcohol or a mixture of alcohol and acetone—is briefly added. These chemicals quickly remove the dye-iodine complex from Gram-negative, but not Gram-positive, bacteria.

4. A **counterstain** is then applied to give a different color to the now colorless Gram-negative bacteria. For this purpose, the red dye safranin is used. This dye enters Gram-positive cells as well, but because they are already purple, it makes little difference to their color.

To obtain reliable results, the Gram stain must be done properly. One common mistake is to decolorize a smear for the incorrect time period. Even Gram-positive cells can lose the crystal violet-iodine complex during prolonged decolorization. When this happens, over-decolorized Gram-positive cells will appear pink after counterstaining. In contrast, under-decolorizing results in Gram-negative cells appearing purple. Another important consideration is the age of the culture. As bacterial cells age, they lose their ability to retain the crystal violet–iodine dye complex. As a result, cells from old cultures may appear pink. Thus, Gram staining fresh cultures (less than 24 hours old) gives more reliable results.

Acid-Fast Stain

The **acid-fast stain** is a procedure used to detect a small group of organisms that do not readily take up dyes. Among these are members of the genus *Mycobacterium,* including a species that causes tuberculosis and one that causes Hansen's disease (leprosy). The cell wall of these bacteria contains high concentrations of mycolic acid, a waxy fatty acid that prevents uptake of dyes such as those used in Gram staining. Therefore, harsh methods are needed to stain these organisms. Once stained, however, these same cells are very resistant to decolorization. Because mycobacteria are among the few organisms that retain the dye in this procedure, acid-fast staining can be used to presumptively identify them in clinical specimens. **▶▶ tuberculosis, p. 502 ▶▶ Hansen's disease, p. 650 ◀◀ fatty acid, p. 33**

Acid-fast staining, like Gram staining, requires multiple steps. The primary stain is carbolfuchsin, a red dye. In the classic procedure, the dye-flooded smear is heated, which facilitates staining. A current variation does not include heat, but instead uses a concentrated solution of dye for a longer period of time. The slide is then rinsed briefly to remove excess dye before being flooded with acid-alcohol, a strong decolorizing agent. This agent removes the carbolfuchsin from tissue cells and most bacteria; those few species that retain the dye are called **acid-fast.** Methylene blue is then used as a counterstain. As a result of the staining procedure, acid-fast organisms are bright reddish-pink, making them easy to distinguish from the blue non-acid-fast cells (**figure 3.15**).

Special Stains to Observe Cell Structures

Special procedures can also be used to stain specific structures inside or outside the cell. The function of these structures will be discussed in more depth later in the chapter.

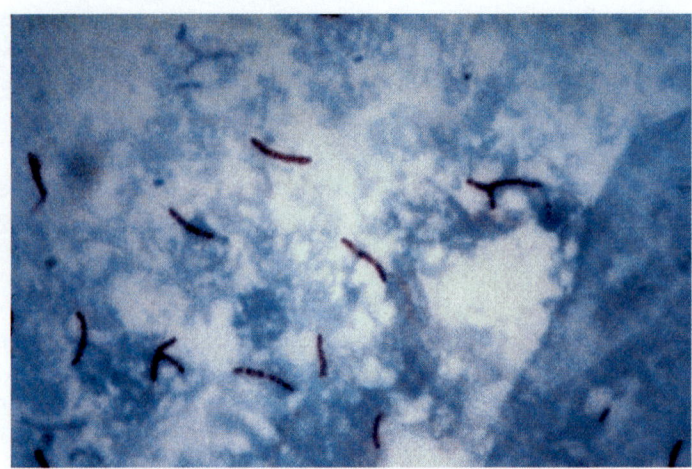

FIGURE 3.15 Acid-Fast Stain *Mycobacterium* species retain the red primary stain, carbolfuchsin. Counterstaining with methylene blue colors the non-acid-fast cells blue.

? *What characteristic of* Mycobacterium *cells makes them acid-fast?*

Capsule Stain

A capsule is a gel-like layer that surrounds certain microbes. This layer has a protective function and often increases an organism's pathogenicity. Capsules stain poorly, so **capsule stains** often use negative staining to make them visible. In one common method, India ink is added to a suspension of cells to make a wet mount. The fine particles of ink darken the background, allowing the capsule to stand out as a clear area surrounding a cell (**figure 3.16**). **▶▶ capsule, p. 62 ◀◀ pathogen, p. 6**

Endospore Stain

Members of certain genera, including *Bacillus* and *Clostridium,* form a special type of resistant, dormant cell called an endospore. These structures do not stain with the Gram stain, but they can often be seen as clear, smooth objects within stained cells. **▶▶ endospore, p. 67**

10 μm

FIGURE 3.16 Capsule Stain *Cryptococcus neoformans,* an encapsulated yeast. The capsules stand out against the India ink–stained background.

? *How is the India ink capsule stain an example of a negative stain?*

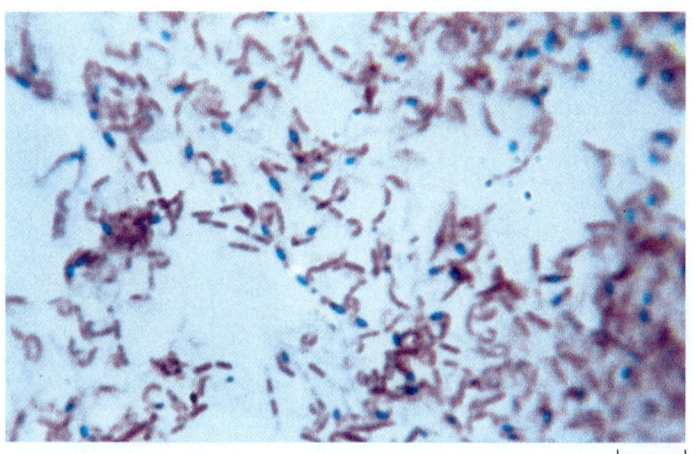

10 µm

FIGURE 3.17 Endospore Stain Endospores retain the primary stain, malachite green. Counterstaining with safranin colors other cells red.

❓ *What color would* Escherichia coli *cells be with the endospore stain shown in the photo?*

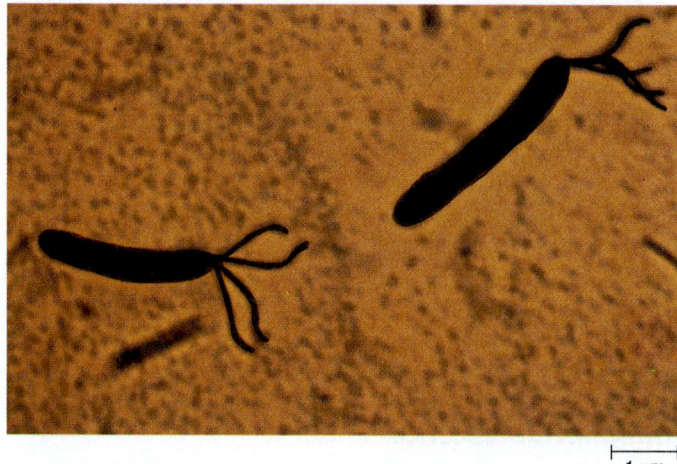

1 µm

FIGURE 3.18 Flagella Stain The staining agent adheres to and coats the flagella. This increases their diameter so they can be seen with the light microscope.

❓ *How can the flagella stain be helpful in identifying bacteria?*

To make endospores easier to see, an **endospore stain** is used. This multistep procedure often uses malachite green as a primary stain, with gentle heating to facilitate uptake of the dye by endospores. The smear is then rinsed with water, which removes the dye from everything but the endospores. The smear is then counterstained, most often with the red dye safranin. Using this method, the endospores will be green while all other cells will be pink (**figure 3.17**).

Flagella Stain

Flagella are appendages that provide the most common mechanism of motility for prokaryotic cells, but they are ordinarily too thin to be seen with the light microscope. The **flagella stain** uses a substance that allows the staining agent to adhere to and coat the thin flagella, making them visible using light microscopy. Not all prokaryotic cells have flagella, but those that do can have them in different arrangements around a cell, so the presence and distribution of these appendages can be used as identifying features (**figure 3.18**). ▶▶ flagella, p. 63

Fluorescent Dyes and Tags

Fluorescence can be used to observe total cells, a subset of cells, or cells with certain proteins on their surface, depending on the procedure employed (**figure 3.19**).

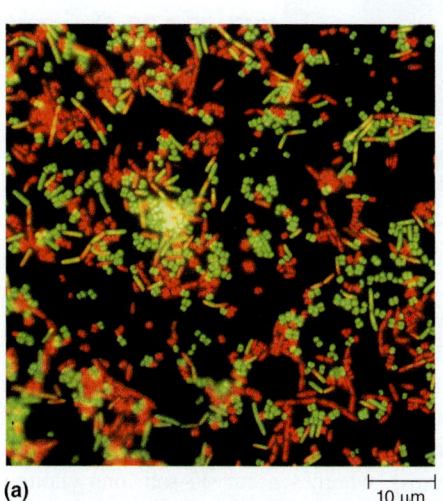

(a)   10 µm

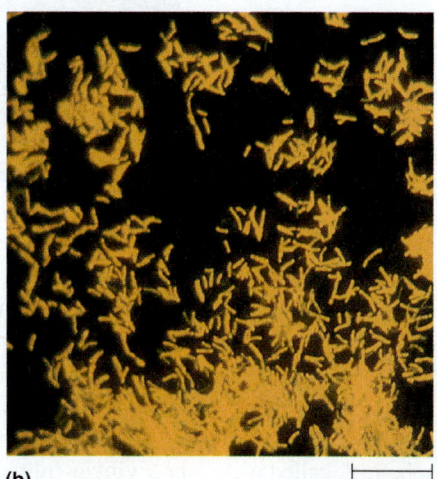

(b)   10 µm

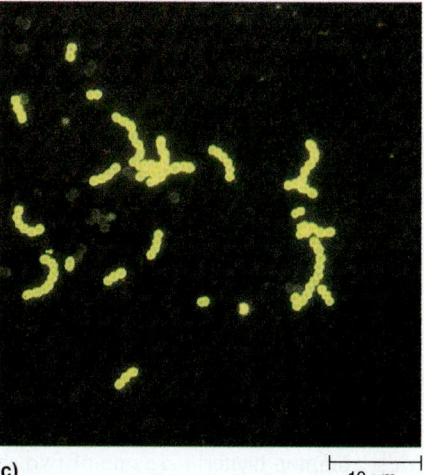

(c)   10 µm

FIGURE 3.19 Fluorescent Dyes and Tags (a) Dyes that cause live cells to fluoresce green and dead ones red. **(b)** Auramine is used to stain *Mycobacterium* species in a modification of the acid-fast technique. **(c)** Fluorescent antibodies tag specific molecules—in this case, the antibody binds to a molecule unique to *Streptococcus pyogenes*.

❓ *How can fluorescent dyes and tags be used to identify bacteria?*

Fluorescent Dyes

Some fluorescent dyes bind to compounds found in all cells whereas others bind only to certain components. As an example, acridine orange binds DNA, making it useful for determining the total number of microbes in a sample. In contrast, calcofluor white binds to a cell wall component of fungi and certain bacteria, causing those organisms to fluoresce bright blue. The fluorescent dyes auramine and rhodamine bind to the mycolic acid in the cell walls of *Mycobacterium* species, making the dyes useful in a staining procedure similar to the acid-fast stain. Other fluorescent dyes are changed by cellular processes, allowing microbiologists to distinguish between living and dead cells. ▶▶ mycolic acid, p. 503

Immunofluorescence

Immunofluorescence is a technique used to tag specific proteins with a fluorescent compound. By tagging a protein unique to a given microbe, immunofluorescence can be used to detect that organism. Immunofluorescence uses an antibody to deliver the fluorescent tag (see figure 18.8). Antibodies, and how they are made, will be described in chapter 15. ▶▶ antibody, pp. 355, 359

MicroAssessment 3.2

Dyes are used to stain cells so they can be seen against an unstained background. The Gram stain is the most commonly used differential stain. The acid-fast stain is used to detect *Mycobacterium* species. Specific dyes and techniques are used to observe cell structures such as capsules, endospores, and flagella. Fluorescent dyes and tags can be used to observe total cells, a subset of cells, or cells that have certain proteins on their surface.

4. *What are the functions of a primary stain and a counterstain?*

5. *Describe one error in the staining procedure that would result in a Gram-positive bacterium appearing pink.*

6. *What color would a Gram-negative bacterium be in an acid-fast stain?* ✚

3.3 ■ Morphology of Prokaryotic Cells

Learning Outcomes

6. *Describe the common bacterial shapes and groupings, and their significance.*

7. *Describe two multicellular associations of bacteria.*

Prokaryotic cells come in a variety of simple shapes and often form characteristic groupings. Some aggregate, living as multicellular associations.

Shapes

Most common bacteria are one of two shapes: spherical, called a **coccus** (plural: cocci), or cylindrical, called a **rod** (**figure 3.20**). A rod-shaped bacterium is sometimes called a **bacillus** (plural: bacilli), and one so short that it can easily be mistaken for a coccus

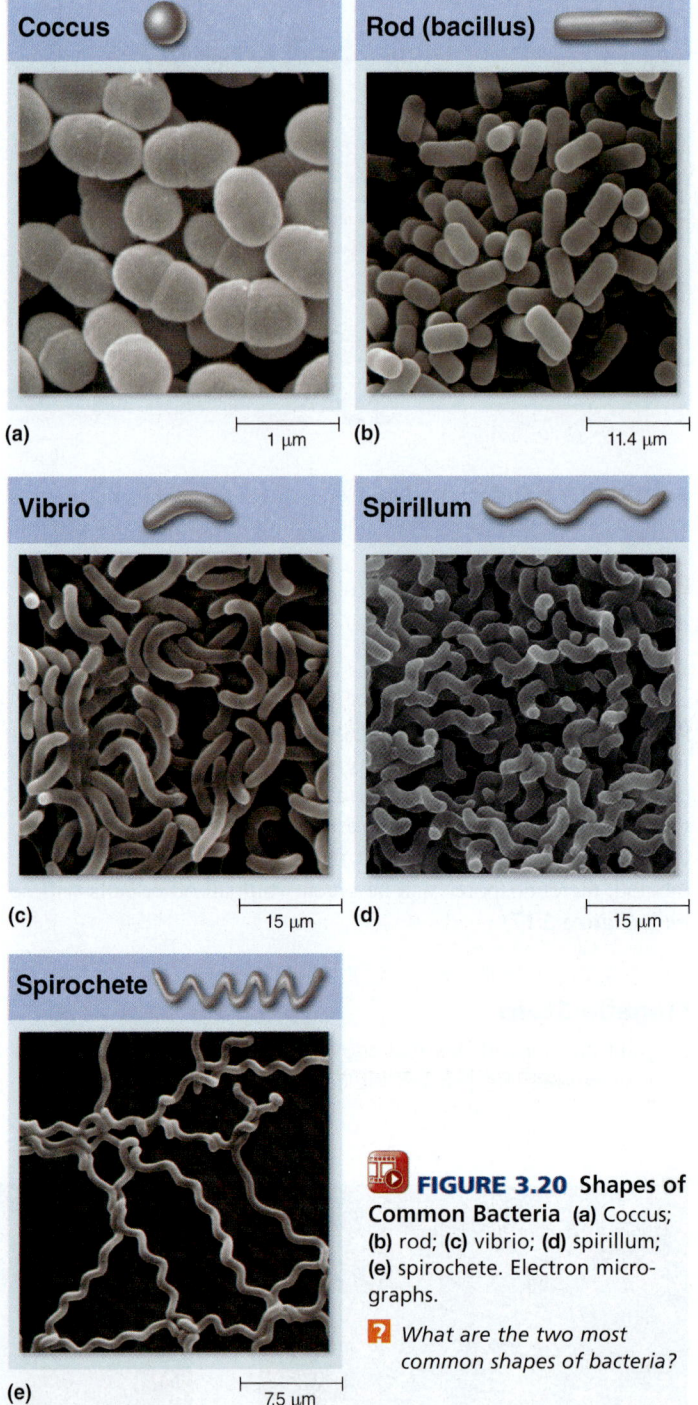

FIGURE 3.20 Shapes of Common Bacteria (a) Coccus; **(b)** rod; **(c)** vibrio; **(d)** spirillum; **(e)** spirochete. Electron micrographs.

❓ *What are the two most common shapes of bacteria?*

is often called a **coccobacillus.** The descriptive term "bacillus" should not be confused with *Bacillus,* the name of a genus. Although members of the genus *Bacillus* are rod-shaped, so are many other bacteria, including *Escherichia coli.*

Cells come in a variety of other shapes. A short, curved rod is a **vibrio** (plural: vibrios), whereas a curved rod long enough to form spirals is a **spirillum** (plural: spirilla). A long, helical cell with a flexible cell wall and a unique mechanism of motility is a **spirochete.** Bacteria that characteristically vary in their shape are **pleomorphic** (*pleo* meaning "many" and *morphic* referring to shape).

The greatest diversity in shapes is found in low-nutrient aquatic environments. Cells there often have a large surface area, which helps them absorb nutrients more easily (**figure 3.21**). For example, some aquatic bacteria have cytoplasmic extensions, giving them a starlike appearance. Square, tilelike archaeal cells have been found in salt ponds.

Groupings

Most prokaryotes divide by binary fission, a process in which one cell divides into two. Cells often stick to each other following division, forming characteristic groupings (**figure 3.22**).

Cells that divide in one plane can form chains of varying length. Cocci that typically occur in pairs are routinely called **diplococci.** An important clue in the identification of *Neisseria gonorrhoeae* is its characteristic diplococcus arrangement. Other cells form long chains, a characteristic typical of some members of the genus *Streptococcus*.

Cocci often divide in more than one plane. Those that divide in perpendicular planes form cubical packets. Members of the genus *Sarcina* form such packets. Cocci that divide in several planes at random may form clusters. *Staphylococcus* species, which typically form grapelike clusters, are an example.

Multicellular Associations

Some prokaryotes characteristically live as multicellular associations. For example, members of a group of bacteria called myxobacteria glide over moist surfaces, forming swarms of cells that move as a pack. The cells release enzymes, and, as a pack, they degrade organic material, including other bacterial cells. When water or nutrients become limiting, the cells come together to form a structure called a fruiting body, which is visible to the naked eye (see figure 11.16). ▶▶ **myxobacteria, p. 268**

In their natural habitat, most bacteria on surfaces live in communities called biofilms. Details about these communities are described in chapter 4. ▶▶ **biofilms, p. 84**

> **MicroAssessment 3.3**
>
> Most common prokaryotes are cocci or rods, but other shapes include vibrios, spirilla, and spirochetes. Cells may form characteristic groupings such as chains or clusters. Some form multicellular associations.
>
> 7. *What shape are* Escherichia coli *cells?*
> 8. *What determines whether a group of dividing cells will form chains or clusters?* ✚

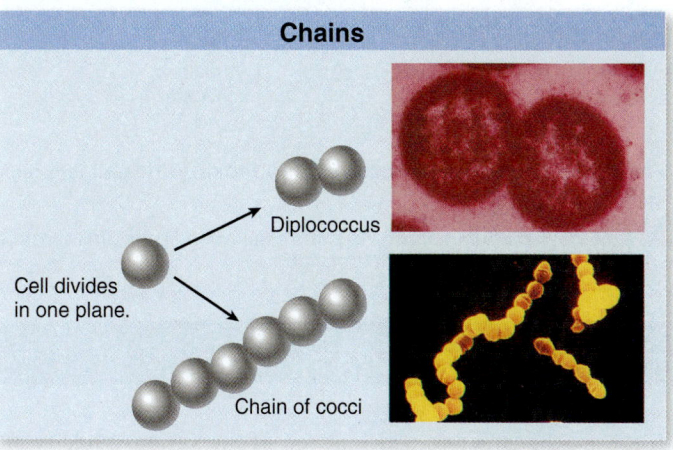

(a)

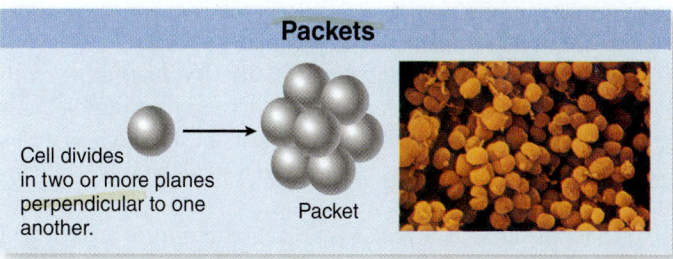

(b)

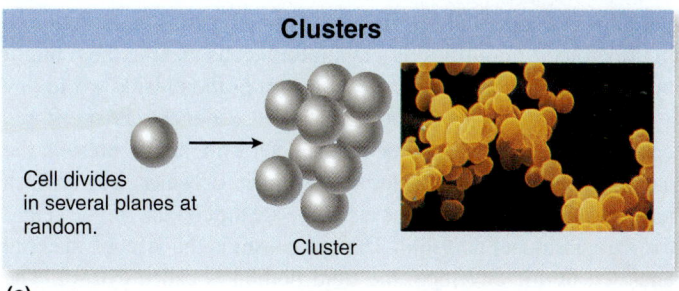

(c)

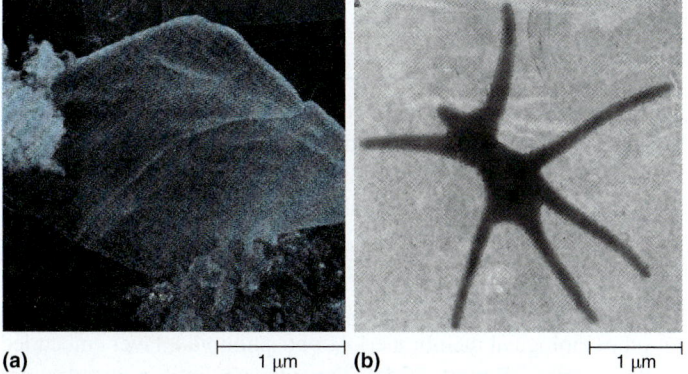

(a)   1 μm   (b)   1 μm

FIGURE 3.21 Diverse Shapes of Aquatic Prokaryotes
(a) Square, tilelike archaeal cell. (b) *Ancalomicrobium adetum,* a bacterium that has a starlike appearance.

❓ *Why would aquatic microbes need maximal surface area?*

FIGURE 3.22 Typical Cell Grouping (a) Chains; (b) packets; (c) clusters.

❓ *What aspect of bacterial cell division determines the characteristic cell arrangements?*

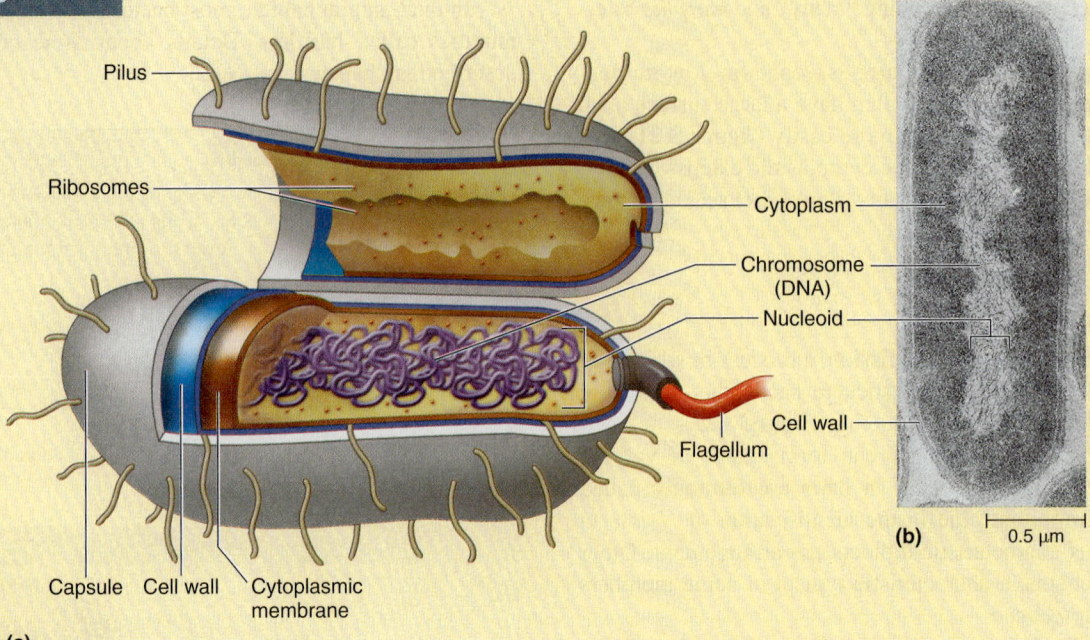

FIGURE 3.23 Typical Structures of a Prokaryotic Cell (a) Diagrammatic representation. Note that archaea rarely have capsules. **(b)** Electron micrograph.

❓ *How does the function of the cytoplasmic membrane differ from that of the cell wall?*

THE PROKARYOTIC CELL

Knowing the structure and function of the various prokaryotic cell components is essential for understanding how microbes survive. The information is particularly relevant from a medical standpoint because characteristics unique to bacteria are potential targets for antibacterial medications. By interfering with these targets, we can selectively kill bacteria or inhibit their growth without harming the human host. In addition, features found in only select groups of prokaryotes can be used to identify different species.

The structures of a prokaryotic cell allow it to survive and multiply in a given environment. Some of the components are essential for growth and therefore found in all prokaryotic cells, but others are optional. While these optional parts are not required in the protected confines of a laboratory, cells lacking them might not survive the competitive environment of the outside world.

The surface layers of a prokaryotic cell, called the **cell envelope,** consist of the cytoplasmic membrane, cell wall, and, if present, the capsule (**figure 3.23**). Bacteria often have capsules, but archaea rarely do. Enclosed within the envelope are the contents of the cell—the cytoplasm and nucleoid. The **cytoplasm** is the viscous material enclosed by the envelope; the fluid portion is called cytosol. The **nucleoid** is the gelatinous region where the genetic material resides. Unlike the nucleus that characterizes eukaryotic cells, the nucleoid is not enclosed by a membrane. The cell may also have appendages that give it useful traits, including motility and the ability to adhere to certain surfaces. The characteristics of typical structures of prokaryotic cells are summarized in **table 3.3.**

MicroByte
Our bodies defenses have "alarm systems" that recognize compounds unique to bacteria, using them as indicators of invading microbes.

3.4 ■ The Cytoplasmic Membrane

Learning Outcomes

8. *Describe the structure and chemistry of the cytoplasmic membrane, focusing on how it relates to membrane permeability.*
9. *Describe how the cytoplasmic membrane is involved with proton motive force.*

The **cytoplasmic membrane** (or plasma membrane) is a thin, delicate structure that surrounds the cytoplasm and defines the boundary of the cell. It serves as the critical permeability barrier between the cell and its external environment (see figure 2.30).

Structure of the Cytoplasmic Membrane

The structure of the prokaryotic cytoplasmic membrane is typical of other biological membranes—a phospholipid bilayer embedded with proteins (**figure 3.24**). The phospholipid molecules are arranged in opposing layers so that their hydrophobic tails face in, toward the other layer, and their hydrophilic heads face outward, interacting freely with aqueous solutions. ◀◀ phospholipids, p. 35

The proteins embedded in the membrane have a variety of different functions. Some act as selective gates, allowing nutrients to

TABLE 3.3	A Summary of Typical Prokaryotic Cell Structures
Structure	**Characteristics**
Extracellular	
Filamentous appendages	Composed of protein subunits that form a helical chain.
Flagella	Provide the most common mechanism of motility.
Pili	Different types of pili have different functions. The common types, often called fimbriae, allow cells to adhere to surfaces. A few types are used for twitching or gliding motility. Sex pili join cells in preparation for DNA transfer.
Capsules and slime layers	Layers outside the cell wall, usually made of polysaccharide.
Capsule	Distinct and gelatinous. Allows bacteria to adhere to specific surfaces; allows some organisms to evade innate defense systems and thus cause disease.
Slime layer	Diffuse and irregular. Allows bacteria to adhere to specific surfaces.
Cell wall	Peptidoglycan provides rigidity to bacterial cell walls, preventing the cells from lysing.
Gram-positive	Thick layer of peptidoglycan that contains teichoic acids and lipoteichoic acids.
Gram-negative	Thin layer of peptidoglycan surrounded by an outer membrane. The outer layer of the outer membrane is lipopolysaccharide.
Cell Boundary	
Cytoplasmic membrane	Phospholipid bilayer embedded with proteins. Surrounds the cytoplasm, separating it from the outside environment. Also transmits information about the external environment to the inside of the cell.
Intracellular	
DNA	Contains the genetic information of the cell.
Chromosomal	Carries the genetic information required by a cell. Typically a single, circular, double-stranded DNA molecule.
Plasmid	Extrachromosomal DNA molecule. Generally carries only genetic information that may be advantageous to a cell in certain situations.
Endospore	A type of dormant cell. Generally extraordinarily resistant to heat, desiccation, ultraviolet light, and toxic chemicals.
Cytoskeleton	Involved in cell division and control of cell shape.
Gas vesicles	Small, rigid structures that provide buoyancy to a cell.
Granules	Accumulations of high-molecular-weight polymers, synthesized from a nutrient available in relative excess.
Ribosomes	Involved in protein synthesis. Two subunits, 30S and 50S, join to form the 70S ribosome.

enter the cell and waste products to exit. Others serve as sensors of environmental conditions, providing the cell with a mechanism to monitor and adjust to its surroundings. So, while the cytoplasmic membrane functions as a permeability barrier, it also transmits information about the external environment to the inside of cell.

◄◄ protein function, p. 25

Membrane proteins are not stationary; rather, they constantly drift in the lipid bilayer. Such movement is necessary for the membrane functions. This structure, with its resulting dynamic nature, is called the **fluid mosaic model.**

Members of the *Bacteria* and *Archaea* have the same general structure of their cytoplasmic membranes, but the lipid compositions are distinctly different. The lipid tails of the archaeal membrane lipids are connected to glycerol by a different type of chemical linkage and are not fatty acids. ◄◄ glycerol, p. 33 ◄◄ fatty acid, p. 33

MicroByte

E. coli's DNA encodes approximately 1,000 different membrane proteins, a large portion of which are involved in transport.

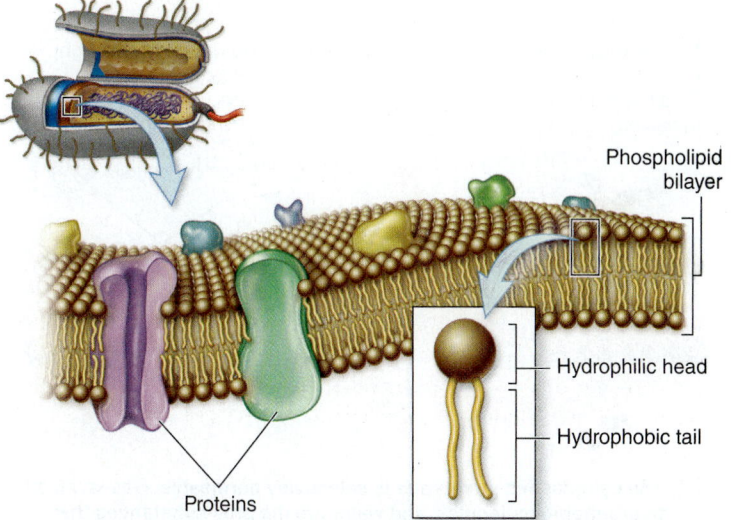

FIGURE 3.24 The Structure of the Cytoplasmic Membrane
Two opposing layers make up the phospholipid bilayer. Various proteins are embedded within the bilayer.

❓ *Which part of the membrane is hydrophobic?*

Permeability of the Lipid Bilayer

The cytoplasmic membrane is **selectively permeable,** meaning that only certain substances can cross. Those that pass through the lipid bilayer freely include gases such as O_2, CO_2, and N_2, small hydrophobic molecules, and water (**figure 3.25**). Some cells facilitate water passage with **aquaporins,** pore-forming membrane proteins that specifically allow water molecules to pass through. Other molecules that enter or exit cells must be moved across the membrane by the transport systems discussed later.

Simple Diffusion

Molecules that can pass through the lipid bilayer move in and out of the cell by **simple diffusion,** in which molecules move from a region of high concentration to one of low concentration, until equilibrium is reached. The speed and direction of movement depend on the relative concentration of molecules on each side of the membrane—the greater the difference in concentration, the higher the rate of diffusion. The molecules continue to pass through at a diminishing rate until their concentration is the same on both sides of the membrane.

Osmosis

Osmosis is the diffusion of water across a selectively permeable membrane. It occurs when the concentrations of solute (dissolved molecules and ions) on two sides of a membrane are unequal. Typical of diffusion, water moves down its concentration gradient from high water concentration (low solute concentration) to low water concentration (high solute concentration).

When describing osmosis, three terms are used to refer to the solutions on opposing sides of a membrane: hypotonic (*hypo* means less; *tonic* refers to solute), hypertonic (*hyper* means more),

Water flows across a membrane toward the hypertonic solution.

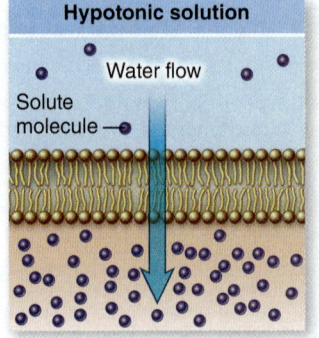
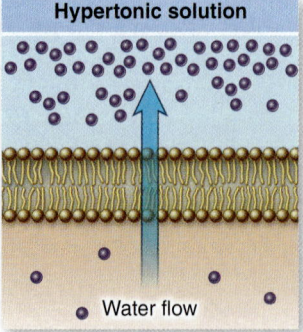

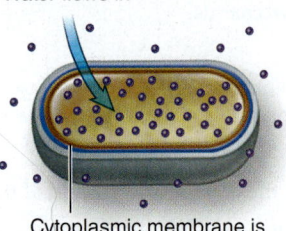
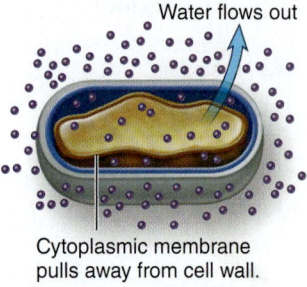

(a)   (b)

FIGURE 3.26 Osmosis (a) The effect of a hypotonic solution on a bacterial cell. **(b)** The effect of a hypertonic solution on a bacterial cell.

❓ *What might happen in part (a) if the cell wall were weakened?*

and isotonic (*iso* means the same). Water flows from the hypotonic solution to the hypertonic one (**figure 3.26**). No net water movement occurs between isotonic solutions.

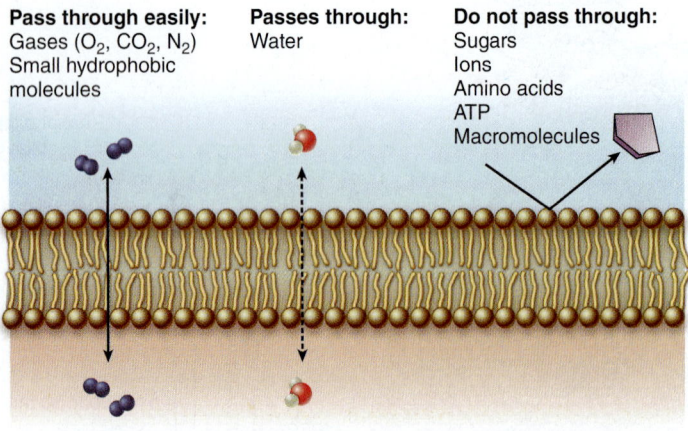
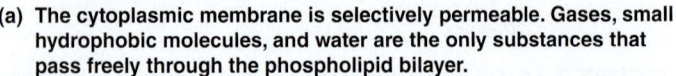
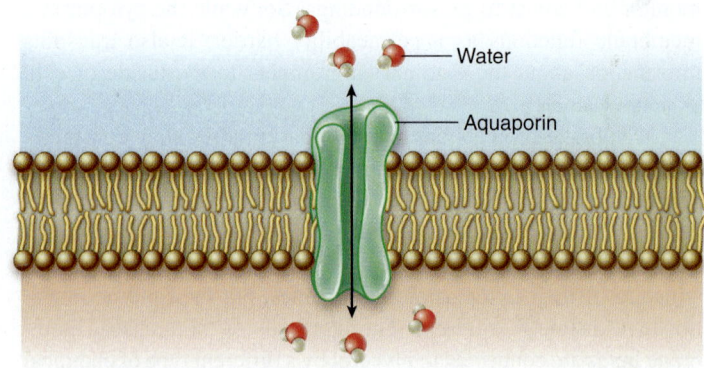

(a) The cytoplasmic membrane is selectively permeable. Gases, small hydrophobic molecules, and water are the only substances that pass freely through the phospholipid bilayer.

(b) Aquaporins allow water to pass through the cytoplasmic membrane more easily.

FIGURE 3.25 Permeability of the Lipid Bilayer (a) Selective permeability. **(b)** The role of aquaporins.

❓ *Why do small charged molecules not pass through the bilayer?*

Osmosis has important biological consequences. The cytoplasm of a cell is a concentrated solution of inorganic salts, sugars, amino acids, and various other molecules. However, the environments in which prokaryotes normally grow are typically very dilute (hypotonic). Water moves toward the high solute concentration, so it flows from the surrounding medium into the cell (figure 3.26b). This inflow of water exerts tremendous osmotic pressure on the cytoplasmic membrane, much more than it generally can resist. However, the strong cell wall surrounding the membrane generally withstands such high pressure. The cytoplasmic membrane is forced against the wall but cannot balloon further. Damage to the cell wall weakens the structure, and consequently, cells may lyse (burst).

The Role of the Cytoplasmic Membrane in Energy Transformation

The cytoplasmic membrane of prokaryotic cells plays a critical role in transforming energy—converting the energy of food or sunlight into ATP, the energy currency of a cell. This is an important distinction between prokaryotic and eukaryotic cells; in eukaryotic cells, this process occurs in membrane-bound organelles, which will be discussed later in this chapter. ◀◀ ATP, p. 24

As part of their energy-transforming processes, most prokaryotes have a series of protein complexes, the **electron transport chain,** embedded in their membrane. These sequentially transfer electrons and, in the process, move protons out of the cell. The details of the process will be explained in chapter 6. The expulsion of protons by the electron transport chain creates a proton gradient across the cell membrane—positively charged protons are concentrated immediately outside the membrane, whereas negatively charged hydroxyl ions remain inside (**figure 3.27**). The charged ions attract each other, so they stay close to the membrane. This separation of protons and hydroxyl ions creates an electrochemical gradient across the membrane; inherent in it is a form of energy called **proton motive force,**

which is analogous to the energy stored in a battery. ▶▶ **electron transport chain, p. 142**

The energy of a proton motive force can be harvested when protons are allowed to move back into the cell. This is used to drive certain cellular processes, including ATP synthesis. It is also used to power one of the transport systems discussed next and some forms of motility.

MicroAssessment 3.4

The cytoplasmic membrane is a phospholipid bilayer embedded with proteins. It serves as a selectively permeable barrier between the cell and the surrounding environment, allowing relatively few substances to pass through freely. Molecules that pass through the lipid bilayer move in and out by simple diffusion. Osmosis is the diffusion of water across the membrane. The electron transport chain within the membrane expels protons, generating a proton motive force that is used to synthesize ATP, and power transport systems and some types of motility.

9. *Explain the fluid mosaic model.*

10. *Name three molecules that pass freely through the lipid bilayer.*

11. *Why do the protons ejected by the electron transport system stay close to the membrane, rather than float away?* ✚

3.5 ■ Directed Movement of Molecules Across the Cytoplasmic Membrane

Learning Outcomes

10. *Explain why transport systems are necessary for a cell.*

11. *Compare and contrast facilitated diffusion, active transport, and group translocation.*

12. *Describe the importance of secretion, and explain how a cell determines which polypeptides are destined for secretion.*

Most molecules that enter or exit a cell must pass via proteins that function as selective gates. This is necessary because the lipid bilayer of the cytoplasmic membrane prevents nearly all substances from passing through. Cells must constantly bring nutrients in through these gates, and expel waste products. They also need to move certain proteins they synthesize to the outside of the cell.

Transport Systems

Mechanisms that allow nutrients and other small molecules to enter the cell are called **transport systems.** These are also used to expel wastes and compounds such as antibiotics and disinfectants that are otherwise damaging to the cell.

Transported molecules enter the cell through transport proteins, sometimes called permeases or carriers. These span the membrane, so that one end projects into the surrounding environment and the other

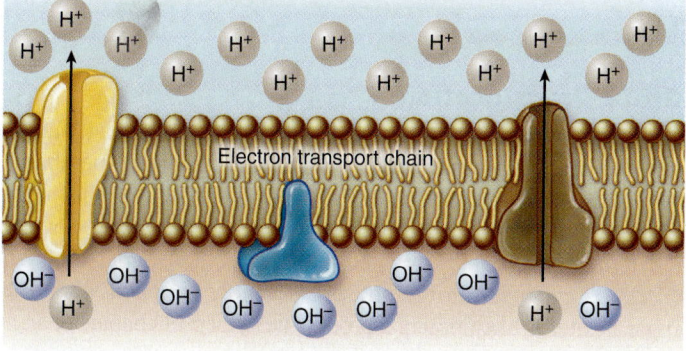

FIGURE 3.27 Proton Motive Force The electron transport chain, a series of protein complexes within the membrane, ejects protons from the cell.

❓ *Why is proton motive force a form of energy?*

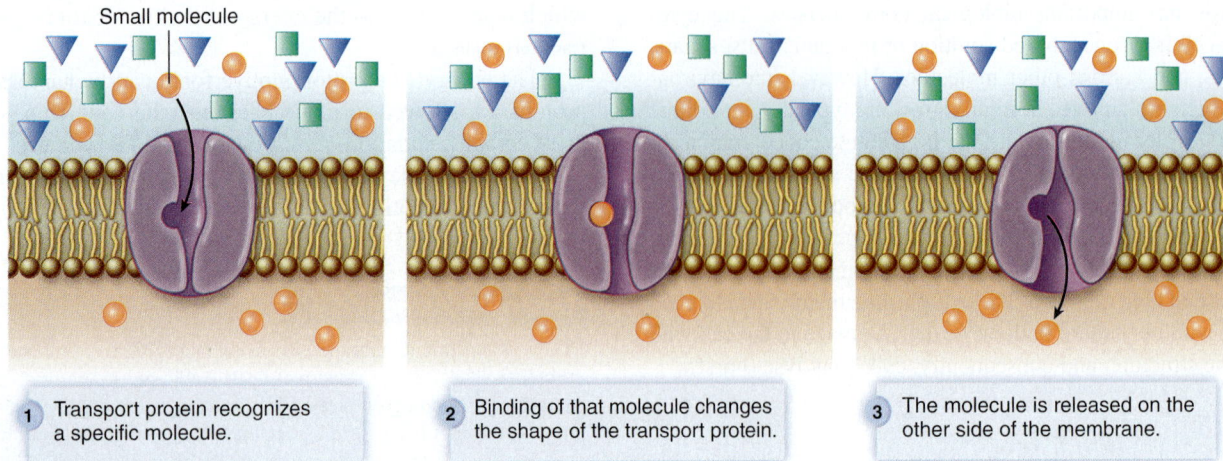

1 Transport protein recognizes a specific molecule.

2 Binding of that molecule changes the shape of the transport protein.

3 The molecule is released on the other side of the membrane.

FIGURE 3.28 Transport Protein

❓ *What types of molecules do prokaryotic cells bring in?*

into the cell (**figure 3.28**). The interaction between the transport protein and the molecule it carries is highly specific. Consequently, a single carrier generally transports only a specific type of molecule. The mechanisms of transport are summarized in **table 3.4.**

Facilitated Diffusion

Facilitated diffusion is a form of passive transport, meaning that it does not require energy. It uses a transport protein to move substances from one side of the membrane to the other, but it can only eliminate a difference in concentration, not create one (**figure 3.29a**). Molecules are transported until their concentration is the same on both sides of the membrane. Because prokaryotes typically grow in relatively nutrient-poor environments, they generally cannot rely on facilitated diffusion to take in nutrients.

TABLE 3.4	A Summary of Transport Mechanisms Used by Prokaryotic Cells
Transport Mechanism	**Characteristics**
Facilitated Diffusion	Rarely used by prokaryotes. Exploits a concentration gradient to move molecules; can only eliminate a gradient, not create one. No energy is expended.
Active Transport	Energy is expended to accumulate molecules against a concentration gradient.
Transporters that use proton motive force	As a proton is allowed into the cell another substance is either brought along or expelled.
ABC transporters	ATP is used as an energy source. Extracellular binding proteins deliver a molecule to the transporter.
Group Translocation	The transported molecule is chemically altered as it passes into the cell.

Active Transport

Active transport moves compounds against a concentration gradient and requires energy. The two main mechanisms use different forms of energy.

Transport Systems That Use Proton Motive Force Many prokaryotic transport systems move small molecules and ions in or out of the cell using the energy of a proton motive force (figure 3.29b). Transporters of this type allow a proton into the cell and simultaneously either bring along or expel another substance. For example, the permease that transports lactose brings the sugar into the cell along with a proton. Expulsion of waste products, on the other hand, relies on transporters that eject the compound as a proton passes in. Efflux pumps, which are used by some bacteria to oust antimicrobial drugs, use this latter mechanism.

Transport Systems That Use ATP Transport mechanisms called ABC transport systems require ATP as an energy source (ABC stands for ATP Binding-Cassette) (figure 3.29b). These systems use specific binding proteins that reside immediately outside of the cytoplasmic membrane to gather and deliver molecules to the respective transport complexes.

Group Translocation

Group translocation is a transport process that chemically alters a molecule during its passage through the cytoplasmic membrane (figure 3.29c). Typically, this is done by adding a phosphate group—a process called phosphorylation. Glucose and several other sugars are phosphorylated during their transport. Although energy is expended in the process, it can be regained when the sugar is later broken down, a process described in chapter 6.

Protein Secretion

Cells actively move certain proteins they synthesize out of the cell—a process called **secretion** (**figure 3.30**). Some of these molecules must be moved to the outside because they will become enzymes that break down macromolecules into their individual subunits. The macromolecules are too large to transport into the

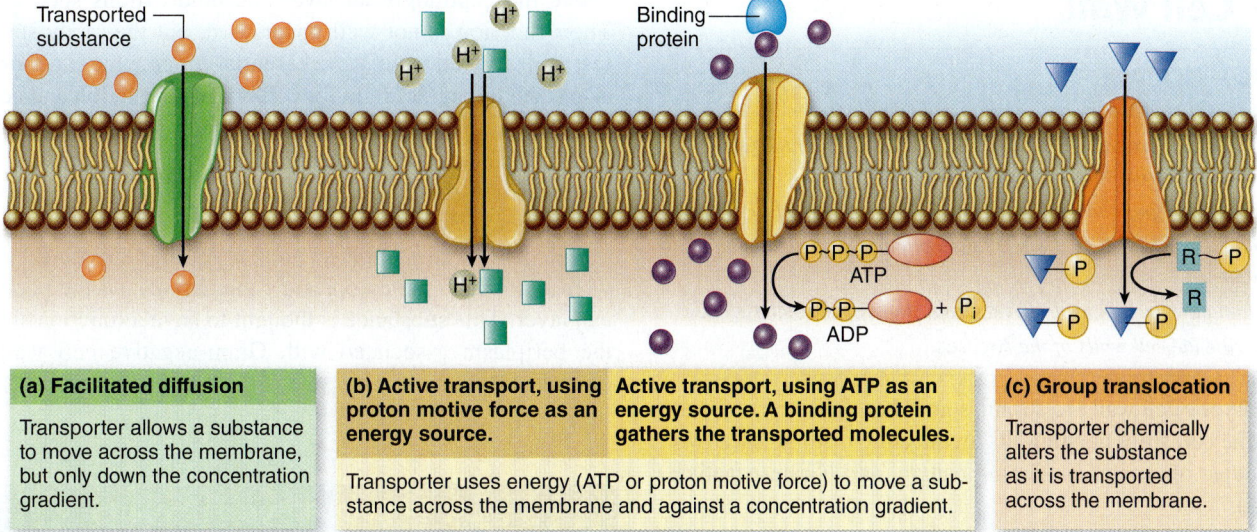

(a) Facilitated diffusion

Transporter allows a substance to move across the membrane, but only down the concentration gradient.

(b) Active transport, using proton motive force as an energy source.

Active transport, using ATP as an energy source. A binding protein gathers the transported molecules.

Transporter uses energy (ATP or proton motive force) to move a substance across the membrane and against a concentration gradient.

(c) Group translocation

Transporter chemically alters the substance as it is transported across the membrane.

FIGURE 3.29 Types of Transport Systems (a) Facilitated diffusion. **(b)** Active transport. **(c)** Group translocation.

? *Why is facilitated diffusion relatively uncommon in prokaryotes?*

cell, but the subunits are not. Others make up external structures such as flagella, the appendages used for motility.

How does the cell machinery know which polypeptides are to be secreted? Those destined for secretion have a characteristic sequence of amino acids, typically at one end of the molecule. This sequence—a **signal sequence**—functions as a tag that directs the secretion machinery to move the preprotein (precursor protein) across the membrane. During the transport process, the signal sequence is removed and the protein ultimately folds into its functional shape. Secretion is a complex process, and a variety of different secretion systems are used by prokaryotes.

MicroAssessment 3.5

Facilitated diffusion does not require energy. Some active transport systems use proton motive force as an energy source and others use ATP. Group translocation chemically modifies a molecule as it enters the cell, usually by phosphorylation. Proteins destined for secretion have a characteristic signal sequence.

12. *Prokaryotes rarely rely on facilitated diffusion. Why is this so?*
13. *Why would a cell need to secrete proteins?*
14. *Can you argue that group translocation is a form of active transport?* ✚

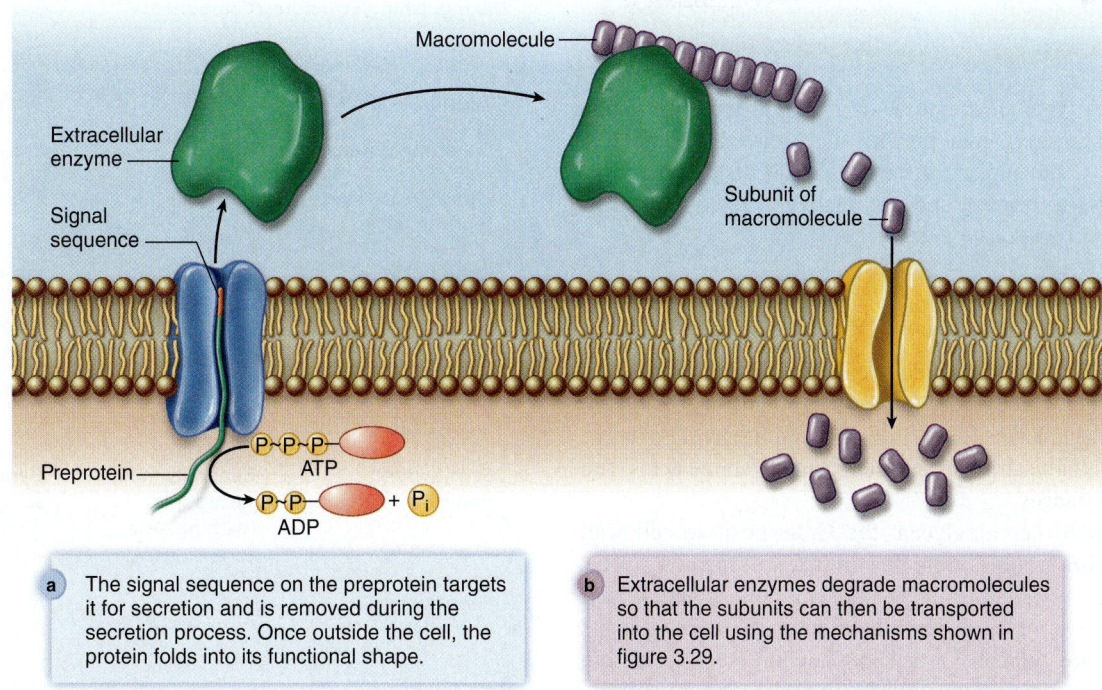

a The signal sequence on the preprotein targets it for secretion and is removed during the secretion process. Once outside the cell, the protein folds into its functional shape.

b Extracellular enzymes degrade macromolecules so that the subunits can then be transported into the cell using the mechanisms shown in figure 3.29.

FIGURE 3.30 Protein Secretion (a) Generalized view of the protein secretion process. **(b)** One function of a secreted protein.

? *Why would a cell secrete enzymes rather than bring intact macromolecules into the cell?*

3.6 ■ Cell Wall

Learning Outcomes

13. *Describe the chemistry and structure of peptidoglycan.*
14. *Compare and contrast the structure and chemistry of the Gram-positive and Gram-negative cell walls.*
15. *Explain the significance of lipid A and the O antigen of LPS.*
16. *Explain how the cell wall affects susceptibility to penicillin and lysozyme.*
17. *Explain how the cell wall affects Gram staining characteristics.*
18. *Describe the cell walls of the* Archaea.

The prokaryotic cell wall is a strong, somewhat rigid structure that prevents the cell from bursting. Its architecture distinguishes two main groups of bacteria—Gram-positive and Gram-negative (**table 3.5**).

Peptidoglycan

The strength of the Gram-positive and Gram-negative bacterial cell walls is due to a layer of **peptidoglycan,** a material found only in bacteria (**figure 3.31**). The basic structure of peptidoglycan is an alternating series of two major subunits: *N*-acetylmuramic acid (NAM) and *N*-acetylglucosamine (NAG). These subunits, which are related to glucose, are covalently joined to one another to form a glycan chain. This linear polymer serves as the backbone of the peptidoglycan molecule.

Attached to each of the NAM molecules is a tetrapeptide chain (a string of four amino acids) that plays an important role in the strength of peptidoglycan. The tetrapeptide chains connect together, linking adjacent glycan chains to form a single, very large three-dimensional molecule, much like a flexible, multi-layered chain-linked fence. In Gram-negative bacteria, tetrapeptides are joined directly. In Gram-positive bacteria, they are usually linked indirectly by a peptide interbridge (a series of amino acids).

Only a few types of amino acids make up the tetrapeptide chains. One of these, diaminopimelic acid (related to lysine), is not found in any other place in nature. Some of the others are D-stereoisomers, a form found in relatively few substances.

◀◀ lysine, p. 26   ◀◀ D-stereoisomer, p. 27

The Gram-Positive Cell Wall

A relatively thick layer of peptidoglycan characterizes the Gram-positive cell wall (**figure 3.32**). As many as 30 layers of interconnected glycan chains make up the polymer. Regardless of its thickness, peptidoglycan is permeable to sugars, amino acids, and many other substances.

In addition to peptidoglycan, the Gram-positive cell wall has **teichoic acids** (from the Greek word *teichos,* meaning wall). These are negatively charged chains of a common subunit (either ribitol-phosphate or glycerol-phosphate) to which various sugars and D-alanine are typically attached. Teichoic acids stick out

above the peptidoglycan layer and bind cations such as Mg^{2+}. Their function is not well understood, but they may serve as a reservoir for cations that are essential for enzyme function. They also seem to be important for cell wall construction and cell division. Some teichoic acids are covalently joined to the peptidoglycan molecule, and these are called wall teichoic acids. Others are linked to the cytoplasmic membrane, and these are called lipoteichoic acids.

Recent studies indicate that a gel-like substance is sandwiched between the cytoplasmic membrane and the peptidoglycan layer. This substance is thought to have a function similar to the periplasm associated with Gram-negative cell walls, discussed next.

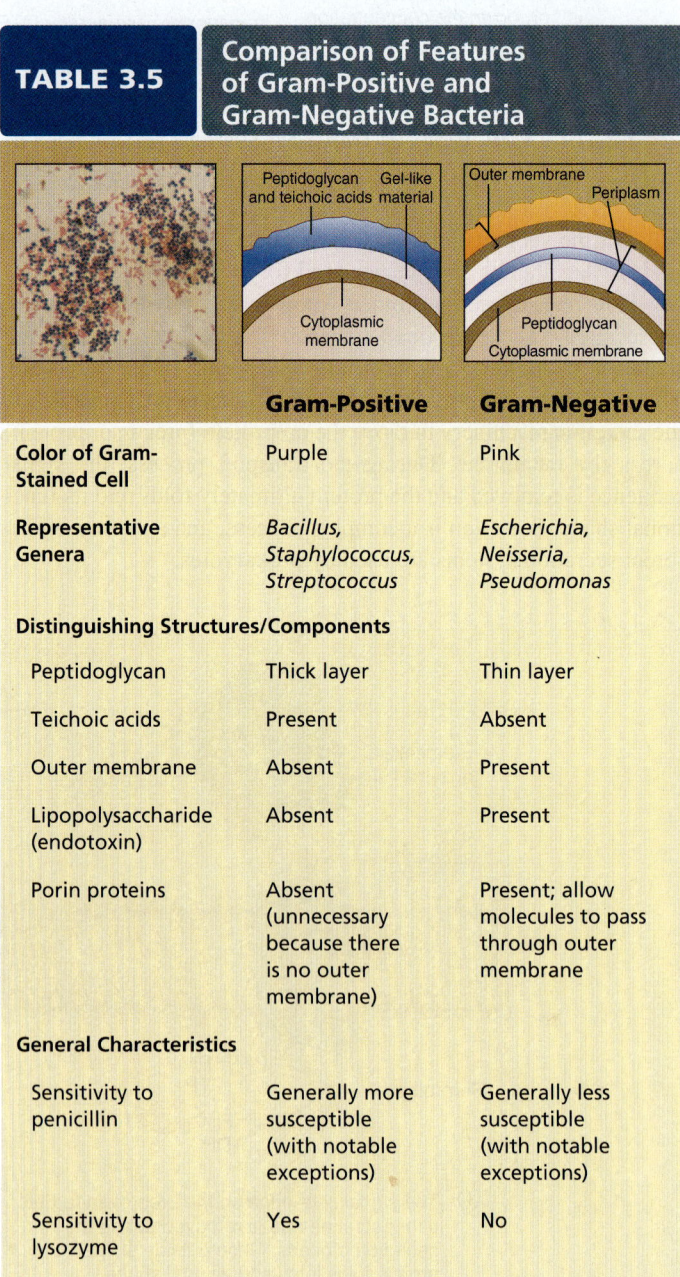

TABLE 3.5	Comparison of Features of Gram-Positive and Gram-Negative Bacteria	
	Gram-Positive	**Gram-Negative**
Color of Gram-Stained Cell	Purple	Pink
Representative Genera	*Bacillus, Staphylococcus, Streptococcus*	*Escherichia, Neisseria, Pseudomonas*
Distinguishing Structures/Components		
Peptidoglycan	Thick layer	Thin layer
Teichoic acids	Present	Absent
Outer membrane	Absent	Present
Lipopolysaccharide (endotoxin)	Absent	Present
Porin proteins	Absent (unnecessary because there is no outer membrane)	Present; allow molecules to pass through outer membrane
General Characteristics		
Sensitivity to penicillin	Generally more susceptible (with notable exceptions)	Generally less susceptible (with notable exceptions)
Sensitivity to lysozyme	Yes	No

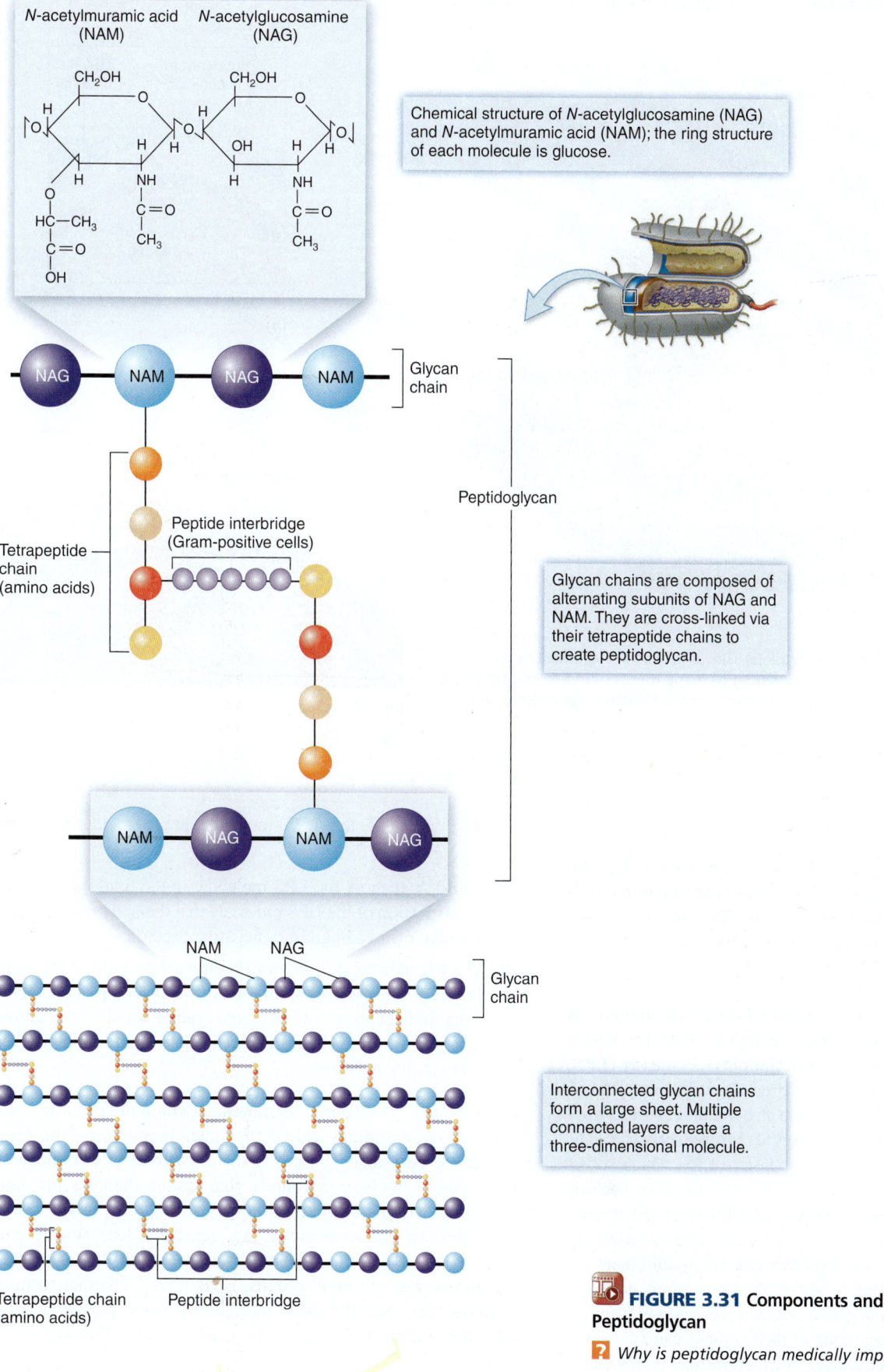

N-acetylmuramic acid (NAM)   *N*-acetylglucosamine (NAG)

Chemical structure of *N*-acetylglucosamine (NAG) and *N*-acetylmuramic acid (NAM); the ring structure of each molecule is glucose.

Glycan chain

Peptidoglycan

Tetrapeptide chain (amino acids)

Peptide interbridge (Gram-positive cells)

Glycan chains are composed of alternating subunits of NAG and NAM. They are cross-linked via their tetrapeptide chains to create peptidoglycan.

NAM   NAG

Glycan chain

Interconnected glycan chains form a large sheet. Multiple connected layers create a three-dimensional molecule.

Tetrapeptide chain (amino acids)   Peptide interbridge

FIGURE 3.31 Components and Structure of Peptidoglycan

? *Why is peptidoglycan medically important?*

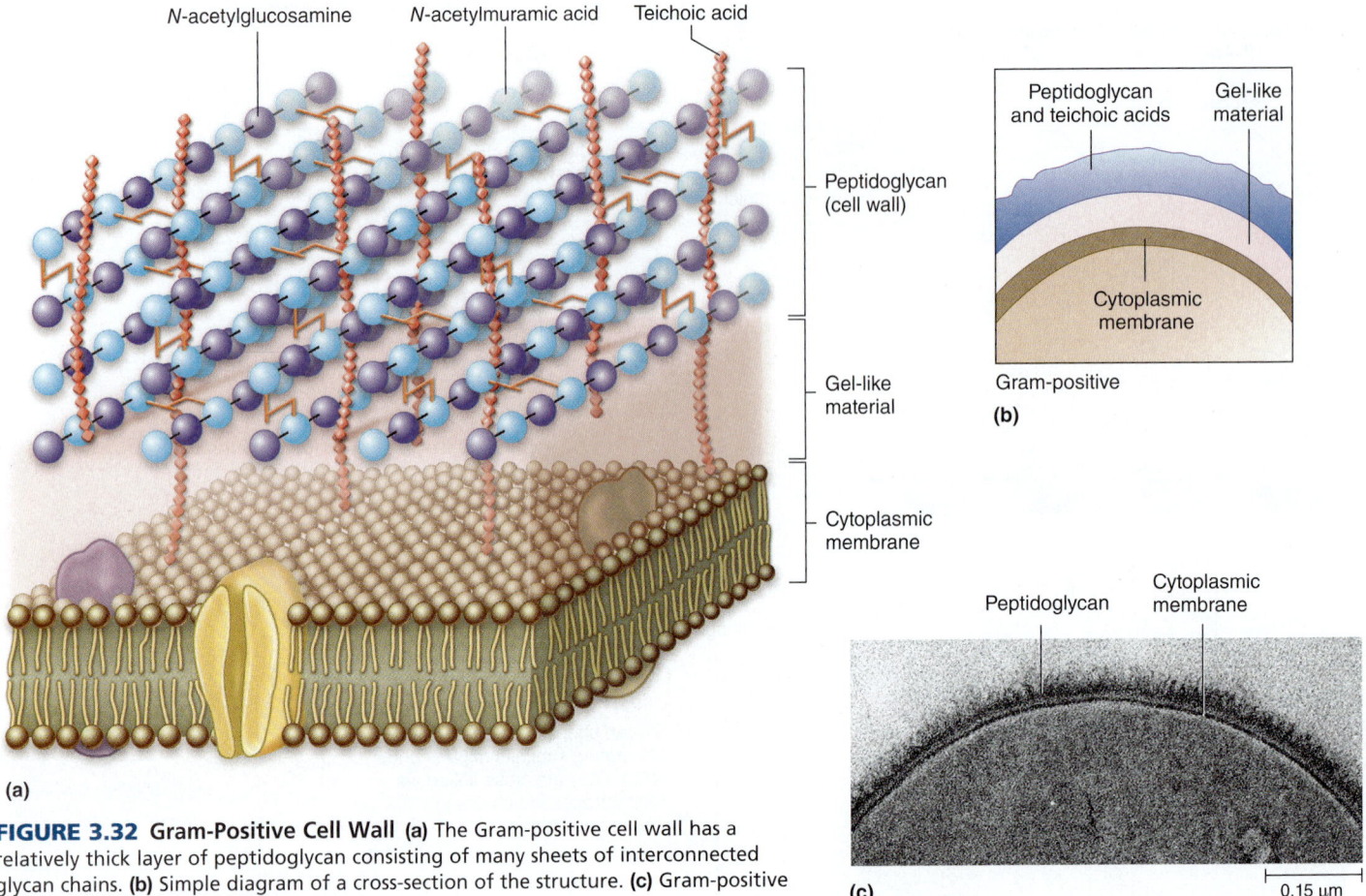

(a)

(b)

Gram-positive

(c)   0.15 μm

FIGURE 3.32 Gram-Positive Cell Wall (a) The Gram-positive cell wall has a relatively thick layer of peptidoglycan consisting of many sheets of interconnected glycan chains. **(b)** Simple diagram of a cross-section of the structure. **(c)** Gram-positive cell wall TEM *(Bacillus subtilis)*.

❓ *What connects the glycan chains in peptidoglycan?*

The Gram-Negative Cell Wall

The Gram-negative cell wall contains only a thin layer of peptidoglycan (**figure 3.33**). Outside of that is the **outer membrane,** a unique lipid bilayer embedded with proteins. The outer membrane is joined to peptidoglycan by lipoproteins. ◄◄ lipoproteins, p. 29

The Outer Membrane

The outer membrane of Gram-negative bacteria is unique. Its bilayer structure is typical of other membranes, but the outside layer is made up of a molecule called **lipopolysaccharide (LPS)** rather than phospholipid. LPS is extremely important from a medical standpoint. When injected into an animal, it causes symptoms characteristic of infections caused by live bacteria. The symptoms are actually the body's response to LPS, which serves as a signal to the defense systems that Gram-negative bacteria have invaded. If very small quantities of LPS enter the tissues, such as when a few bacterial cells contaminate a minor wound, the defense systems respond at a level that can safely eliminate the invader. When large amounts of the molecule accumulate, however, such as when Gram-negative bacteria are actively growing in the bloodstream, the magnitude of the response can be deadly. Because of this lethal effect, LPS is called **endotoxin** (*endo* meaning that it is inside the cell, although it is actually a component of the envelope). ▶▶| endotoxin, p. 394

Two parts of the LPS molecule are particularly notable (figure 3.33b):

- **Lipid A** anchors the LPS molecule in the lipid bilayer. This is the portion of the LPS molecule that the body recognizes as the sign of invading Gram-negative bacteria.

- **O antigen** is the portion of LPS directed away from the membrane, at the end opposite lipid A. It is made up of a chain of sugar molecules, the number and type of which varies among different species. The differences can be used to identify certain species or strains.

Like the cytoplasmic membrane—which in Gram-negative bacteria is sometimes called the inner membrane—the outer membrane serves as a barrier to the passage of most molecules. It excludes many compounds that could damage the cell, including certain antimicrobial medications. This is one reason why Gram-negative bacteria are generally less sensitive to many such medications. Small molecules and ions can cross the membrane through **porins,** specialized channel-forming proteins that span the outer membrane. Some porins are specific for certain molecules; others allow many different molecules to pass.

Gram-negative bacteria have a number of unique secretion systems that move proteins across both the cytoplasmic and outer

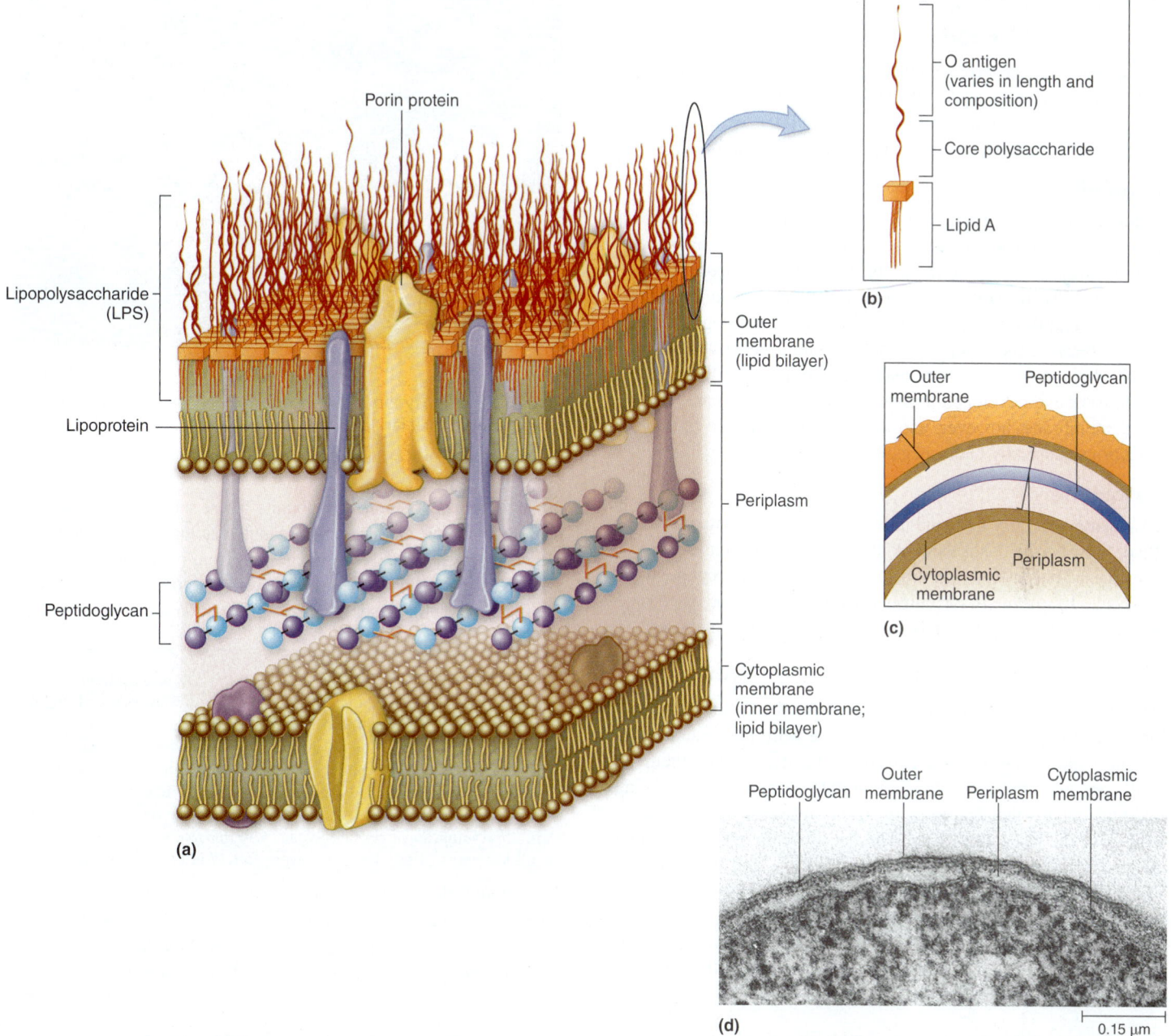

FIGURE 3.33 Gram-Negative Cell Wall **(a)** The Gram-negative cell wall has a thin layer of peptidoglycan made up of only one or two sheets of interconnected glycan chains. The outer membrane is a typical phospholipid bilayer, except the outer layer is lipopolysaccharide. Porins span the membrane to allow specific molecules to pass. Periplasm fills the region between the two membranes. **(b)** Structure of lipopolysaccharide. The lipid A portion is responsible for the symptoms associated with endotoxin. The sugars in the O antigen vary among bacterial species. **(c)** Simple diagram of a cross-section of the structure. **(d)** Gram-negative cell wall TEM *(Pseudomonas aeruginosa)*.

❓ *Why is lipopolysaccharide medically significant?*

membranes. These play a critical role in the disease process of some pathogens, so medical microbiologists are very interested in learning more about how they function. One hope is that medications can be developed to jam these systems. ▶▶| **type III secretion systems, p. 386**

Periplasm

The region between the cytoplasmic membrane and the outer membrane is the periplasmic space, which is filled with a gel-like substance called **periplasm.** All exported proteins accumulate in the periplasm unless specifically moved across the outer membrane as well. Thus, periplasm is filled with proteins involved in a variety of cellular activities, including nutrient degradation and transport. For example, the enzymes that cells export to break down peptides and other molecules are in periplasm. Similarly, the binding proteins of the ABC transport systems are found there.

MicroByte

The "O157" in *E. coli* O157:H7 refers to the characteristic O antigen.

Antibacterial Substances That Target Peptidoglycan

Compounds that interfere with the synthesis of peptidoglycan or alter its structural integrity weaken the molecule to a point where it can no longer prevent the cell from bursting. These substances have no effect on eukaryotic cells because peptidoglycan is unique to bacteria.

Penicillin

Penicillin is the most thoroughly studied of a group of antibiotics that interfere with peptidoglycan synthesis. It functions by preventing the cross-linking of adjacent glycan chains. ▶▶ penicillin, p. 463

Generally, but with notable exceptions, penicillin is far more effective against Gram-positive bacteria than Gram-negative ones. This is because the outer membrane of Gram-negative cells prevents the medication from reaching the peptidoglycan layer. Scientists have developed derivatives of penicillin that cross the outer membrane, however, and these can be effective in treating infections caused by Gram-negative bacteria.

Lysozyme

Lysozyme—an enzyme found in tears, saliva, and many other body fluids—breaks the bonds that link the alternating subunits of the glycan chain. This destroys the structural integrity of the peptidoglycan molecule. Lysozyme is sometimes used in the laboratory to remove the peptidoglycan layer from bacteria for experimental purposes.

Cell Wall Type and the Gram Stain

The type of bacterial cell wall accounts for the Gram stain reaction, but it is the inside of the cell, not the wall, that is stained by crystal violet. Gram-positive cells retain the dye because their cell wall prevents the crystal violet–iodine complex from being washed out by the decolorizing agent, whereas Gram-negative cells lose the dye quite easily. The decolorizing agent is thought to dehydrate the thick layer of peptidoglycan, and in this desiccated state the wall acts as a permeability barrier—the barrier prevents the dye from leaving the cell. In contrast, the solvent action of the decolorizing agent easily damages the outer membrane of Gram-negative bacteria, and the relatively thin layer of peptidoglycan cannot retain the dye complex. ◀◀ Gram stain, p. 47

Bacteria That Lack a Cell Wall

Some bacteria naturally lack a cell wall. *Mycoplasma* species, one of which causes a mild form of pneumonia, have an extremely variable shape because they lack a rigid cell wall (**figure 3.34**). As expected, neither penicillin nor lysozyme affects these organisms. *Mycoplasma* and related bacteria can survive without a cell wall because their cytoplasmic membrane has sterols in it, making it stronger than that of most other bacteria. ◀◀ sterol, p. 36

Cell Walls of the Domain *Archaea*

As a group, members of the *Archaea* have a variety of cell wall types. This is probably due to the fact that they inhabit a wide range of environments, including some that are extreme. The *Archaea* have not been studied as extensively as the *Bacteria*, however, so less is known about the structure of their walls. None of the *Archaea* have peptidoglycan in their cell wall, but some do have a similar molecule called pseudopeptidoglycan. Many have S-layers, which are sheets

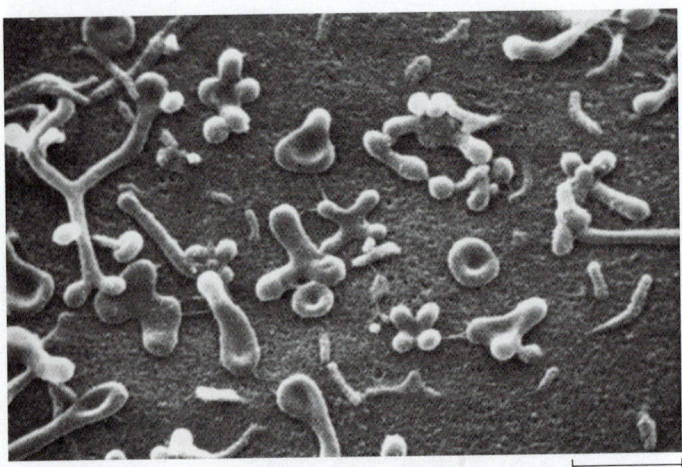

FIGURE 3.34 *Mycoplasma pneumoniae* These cells vary in shape because they lack a cell wall.

❓ *Would lysozyme or penicillin affect M. pneumoniae?*

of flat protein or glycoprotein subunits. These subunits self-assemble, so they may have applications in nanotechnology—a branch of science that seeks to build functional items from molecules and atoms. Some bacterial cells have S-layers as well, but they are in addition to a cell wall, rather than serving as the primary structural component.

MicroAssessment 3.6

Peptidoglycan is a molecule unique to bacteria that provides strength to the cell wall. The Gram-positive cell wall is composed of a relatively thick layer of peptidoglycan as well as teichoic acids. Gram-negative cell walls have a thin layer of peptidoglycan and a lipopolysaccharide-containing outer membrane. The outer membrane excludes many molecules. Penicillin and lysozyme interfere with the structural integrity of peptidoglycan. *Mycoplasma* species lack a cell wall. Members of the *Archaea* have a variety of cell wall types.

15. *What is the significance of lipid A?*
16. *How does the action of penicillin differ from that of lysozyme?*
17. *Explain why penicillin kills only actively multiplying cells, whereas lysozyme kills cells in any stage of growth.* ✚

3.7 ■ Capsules and Slime Layers

Learning Outcome

19. *Compare and contrast the structure and function of capsules and slime layers.*

Many bacteria have a gel-like layer outside the cell wall that either protects the cell or allows it to attach to a surface (**figure 3.35**). If the layer is distinct and gelatinous, it is a **capsule.** If, instead, the layer is diffuse and irregular, it is a **slime layer.** Colonies that form either of these often appear moist and glistening.

Capsules and slime layers vary in their chemical composition, depending on the microbial species. Most are composed of polysaccharides, and are commonly referred to as a **glycocalyx** (*glyco* means "sugar" and *calyx* means "shell"). A few capsules consist of polypeptides made up of repeating subunits of only one or two amino acids. Interestingly, the amino acids are generally of the

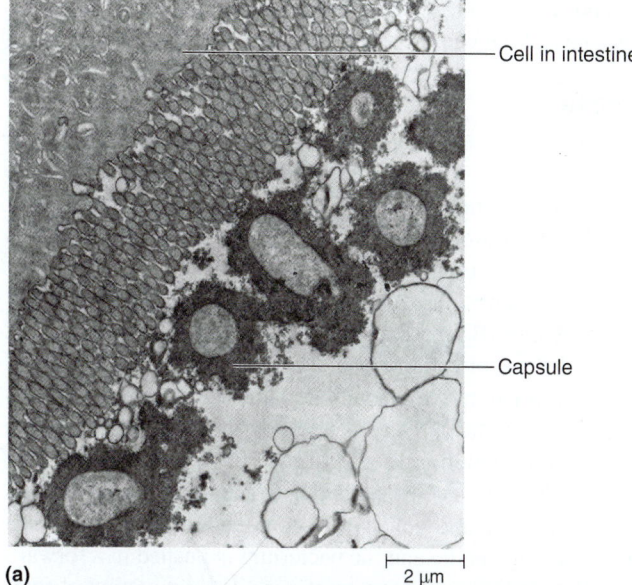

Cell in intestine

Capsule

(a)   2 μm

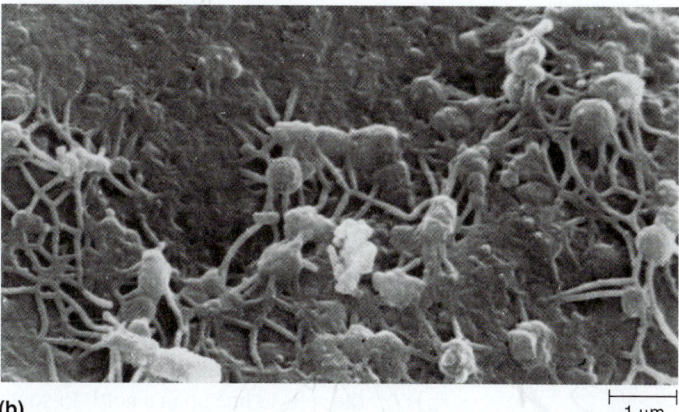

(b)   1 μm

FIGURE 3.35 Capsules and Slime Layers (a) Encapsulated bacteria attaching to intestinal cells (EM). **(b)** Masses of bacteria adhering in a layer of slime (SEM).

❓ *What is the function of capsules and slime layers?*

D-stereoisomeric form, one of the few places these are found in nature. ◀◀ polysaccharide, p. 31 ◀◀ D-amino acid, p. 27

Some capsules and slime layers allow bacterial cells to adhere to specific surfaces, including teeth, rocks, and other bacteria. Once attached, the cells can grow as a biofilm, a polymer-encased community of microbes. One example is dental plaque, a biofilm on teeth. When a person ingests sucrose, *Streptococcus mutans* (a common member of the oral microbiota) can use that to make a glycocalyx. This allows *S. mutans* to attach to other bacterial cells on the teeth. In turn, additional bacteria adhere to the sticky glycocalyx, resulting is an even greater accumulation of plaque. Acids produced by bacteria in dental plaque then damage the tooth surface. ▶▶ biofilm, p. 84 ▶▶ dental plaque, p. 575

Some capsules allow bacteria to avoid host defense systems that otherwise protect against infection. *Streptococcus pneumoniae*, an organism that causes bacterial pneumonia, can cause the disease only if it has a capsule. Unencapsulated cells are quickly engulfed and killed by phagocytes, an important cell of the body's defense system. ▶▶ *Streptococcus pneumoniae*, p. 497 ▶▶ phagocytes, p. 346

MicroAssessment 3.7

Capsules and slime layers allow organisms to adhere to surfaces and sometimes protect bacteria from the host defense systems.

18. *How do capsules differ from slime layers?*
19. *Explain why a sugary diet can lead to tooth decay.* ➕

3.8 ■ Filamentous Protein Appendages

Learning Outcomes

20. *Describe the structure and arrangements of flagella, and explain how they are involved in chemotaxis.*
21. *Compare and contrast the structure and function of fimbriae and sex pili.*

Many prokaryotes have protein appendages anchored in the cytoplasmic membrane that protrude out from the surface. These structures are not essential to the life of microbes, but they do give the cells a competitive advantage.

Flagella

Flagella (singular: **flagellum**) are long protein structures responsible for most types of prokaryotic motility (**figure 3.36**). By spinning like propellers, flagella push a cell through liquid much as a

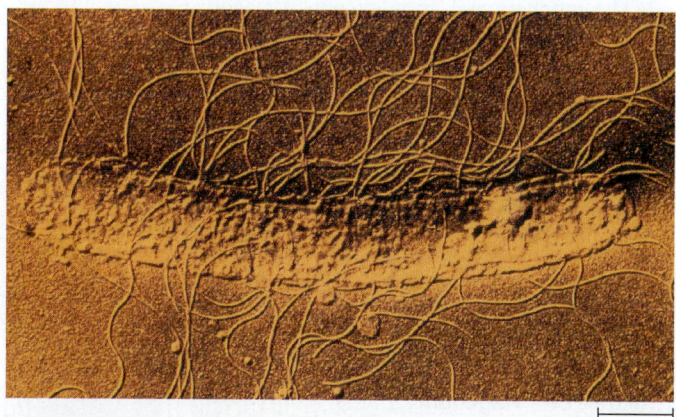

(a)   1 μm

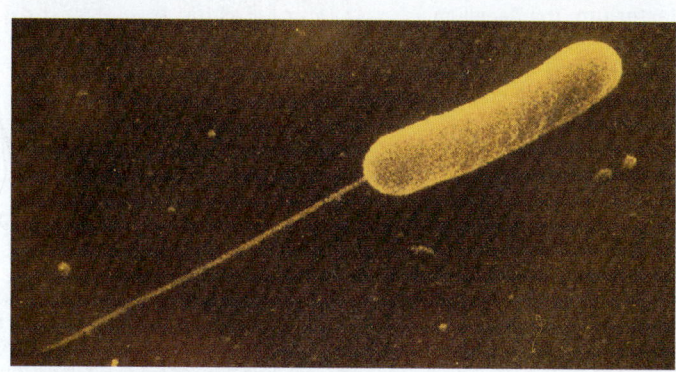

(b)   1 μm

FIGURE 3.36 Flagella (a) Peritrichous flagella (SEM). **(b)** Polar flagellum (SEM).

❓ *How can flagella affect a microbe's ability to cause disease?*

ship is driven through water. Flagella must work very hard to move a cell, because water has the same relative viscosity to prokaryotes as molasses has to humans.

In some cases, flagella are important in disease. For example, *Helicobacter pylori,* the bacterium that causes gastric ulcers, has powerful multiple flagella at one end of the cell. These flagella allow *H. pylori* to penetrate the viscous mucous gel that coats the stomach epithelium. ▶▶ *Helicobacter pylori*, p. 578

Structure and Arrangement of Flagella

A flagellum has three basic parts—a filament, hook, and basal body (**figure 3.37**). The filament is the portion that extends into the external environment. It is made up of identical subunits of a protein called flagellin. These subunits form a chain that twists into a helical structure with a hollow core. The hook, a flexible curved segment, connects the filament to the cell surface. The basal body anchors the structure to the cell wall and cytoplasmic membrane.

While the flagella of bacteria and archaea share the same general structure, they have many differences. For instance, bacterial flagella are powered by proton motive force, whereas archaeal flagella use ATP for energy. The molecules that compose archaeal flagella and the mechanisms of their assembly are also distinct.

The numbers and arrangement of flagella can be used to characterize flagellated bacteria. As an example, *E. coli* have flagella distributed over the entire surface, an arrangement called **peritrichous** (*peri* means "around"). Other common bacteria have a **polar flagellum,** a single flagellum

at one end of the cell. Other arrangements include a tuft of flagella at one or both ends of a cell (see figure 3.18).

Chemotaxis

Motile bacteria sense the presence of chemicals and respond by moving in a certain direction—a phenomenon called **chemotaxis.** If a compound is a nutrient, it may serve as an attractant, causing cells to move toward it. On the other hand, if the compound is toxic, it may act as a repellent, causing cells to move away.

The movement of a bacterial cell toward an attractant is not in a straight line (**figure 3.38**). When an *E. coli* cell travels, it progresses in one direction for a short time, but then stops and tumbles for a fraction of a second. As a consequence, the cell usually finds itself oriented in a completely different direction. It then moves in that direction for a short time, and tumbles again. The seemingly odd pattern of travel is due to the coordinated rotation of the flagella. When flagella rotate counterclockwise, they form a tight propelling bundle and the bacterium is pushed in a forward movement called a run. After a brief period, the direction of rotation of the flagella reverses. This change causes the cell to stop and tumble. When cells sense movement toward an attractant they tumble less frequently, so the runs are longer. Cells also tumble less frequently when they are moving away from a repellent.

In addition to reacting to chemicals, some bacteria respond to the concentration of O_2 (aerotaxis). Organisms that require O_2 for growth will move toward it, whereas bacteria that grow only in its absence tend to be repelled by it. Certain motile bacteria can react to the earth's magnetic field by the process of magnetotaxis. They actually contain a row of magnetic particles that cause the cells to line up in a north-to-south direction much as a compass does (**figure 3.39**). The magnetic forces of

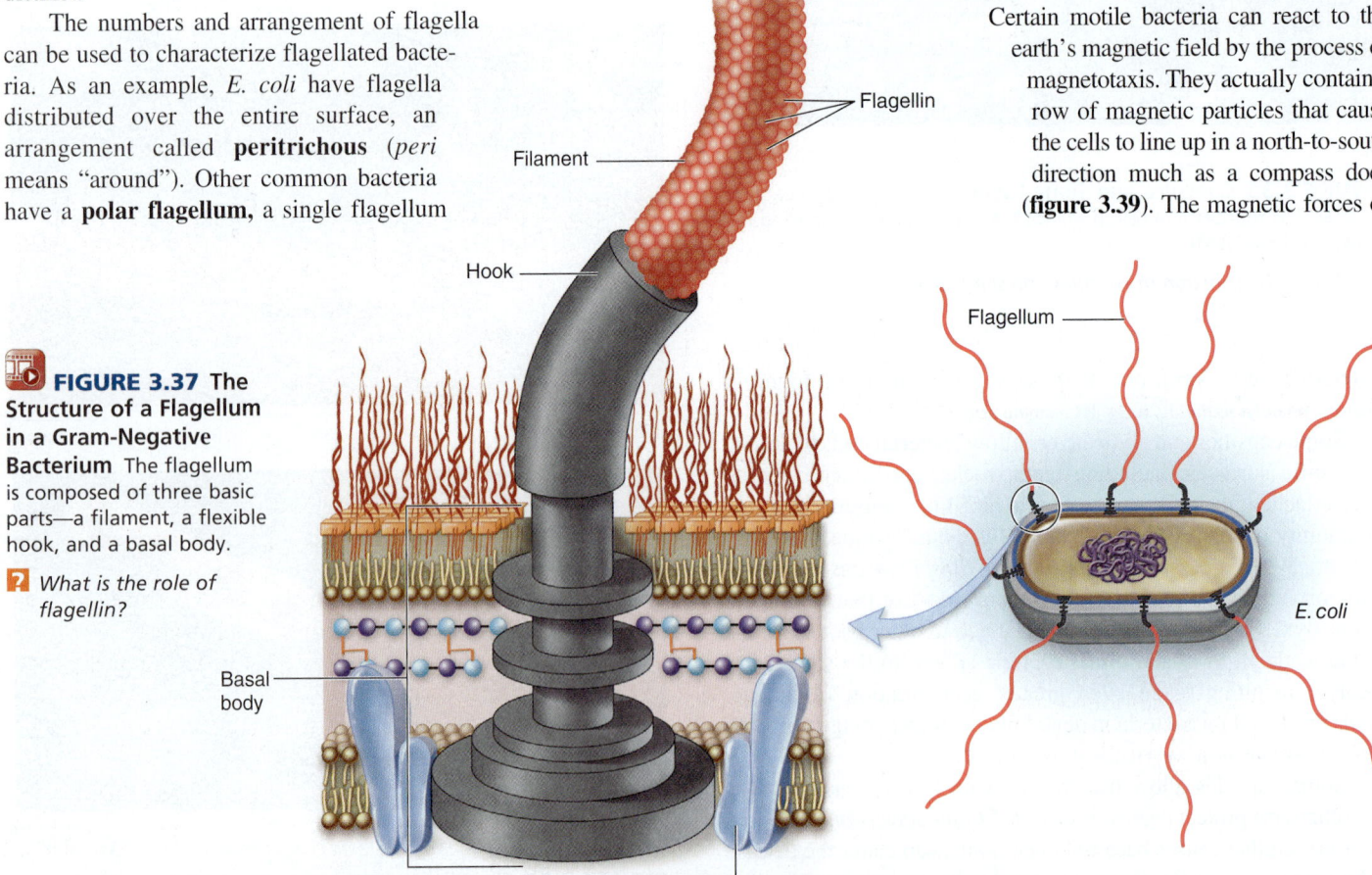

🎬 **FIGURE 3.37 The Structure of a Flagellum in a Gram-Negative Bacterium** The flagellum is composed of three basic parts—a filament, a flexible hook, and a basal body.

❓ *What is the role of flagellin?*

Flagellin

Filament

Hook

Basal body

Harvests the energy of the proton motive force to rotate the flagellum.

Flagellum

E. coli

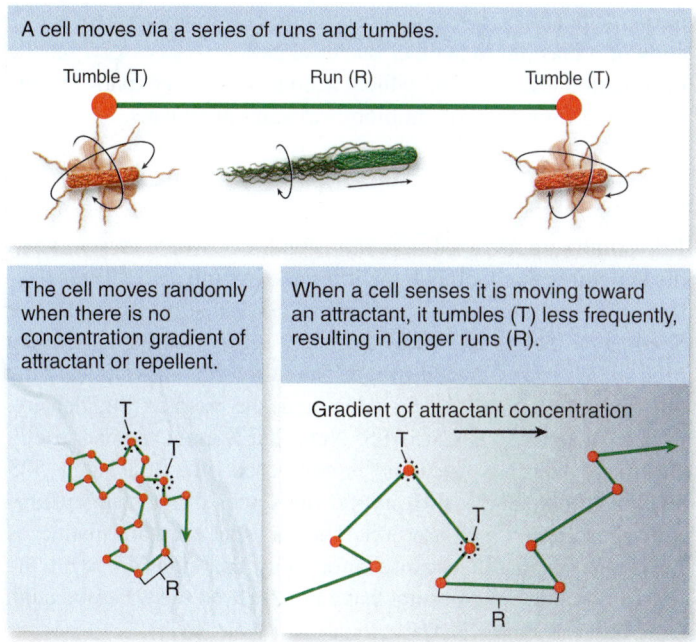

A cell moves via a series of runs and tumbles.

Tumble (T)   Run (R)   Tumble (T)

The cell moves randomly when there is no concentration gradient of attractant or repellent.

When a cell senses it is moving toward an attractant, it tumbles (T) less frequently, resulting in longer runs (R).

Gradient of attractant concentration

FIGURE 3.38 Chemotaxis

? *What mechanism causes a cell to tumble?*

the earth attract the organisms so that they move downward and into sediments where the O_2 concentration is low—the environment best suited for their growth. Some bacteria can move in response to variations in temperature (thermotaxis) or light (phototaxis).

Pili

Pili are considerably shorter and thinner than flagella, and their function is quite different. However, one part of their structure has a theme similar to the filament of a flagellum—a string of protein subunits arranged helically to form a long molecule with a hollow core (**figure 3.40**).

Many types of pili allow cells to attach to specific surfaces; these pili are also called **fimbriae.** Strains of *E. coli* that cause watery diarrhea have pili that allow them to attach to cells that line the small intestine. Without the ability to attach, these cells would

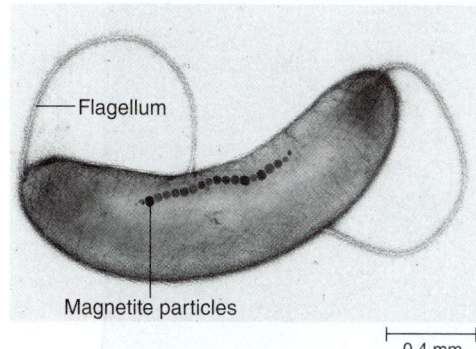

Flagellum

Magnetite particles

0.4 mm

FIGURE 3.39 Magnetotactic Bacterium The chain of magnetic particles (magnetite: Fe_3O_4) within *Magnetospirillum magnetotacticum* helps align the cell along geomagnetic lines (TEM).

? *Why would magnetotaxis benefit a cell?*

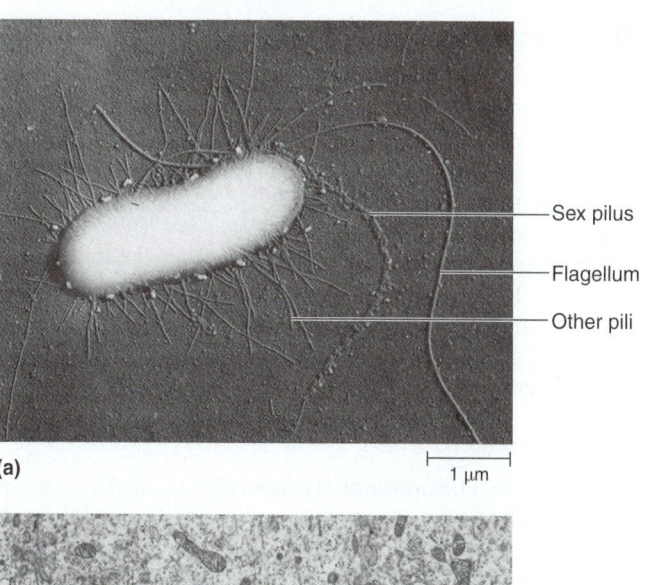

Sex pilus

Flagellum

Other pili

(a)   1 µm

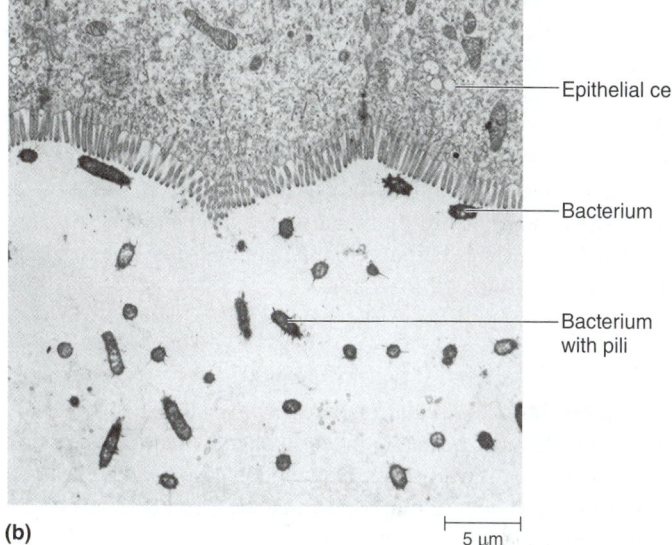

Epithelial cell

Bacterium

Bacterium with pili

(b)   5 µm

FIGURE 3.40 Pili (a) Pili on an *Escherichia coli* cell. The short pili (fimbriae) mediate adherence; the sex pilus is involved in DNA transfer. **(b)** *Escherichia coli* attaching to epithelial cells in the small intestine of a pig.

? *How does the structure and function of pili compare to that of flagella?*

simply be propelled through the intestinal tract along with the other intestinal contents. ▶▶ enterotoxigenic *E. coli*, p. 590

Some types of pili help bacterial cells move on solid media. Twitching motility (characterized by jerking movements) and certain types of smooth, gliding motility involve pili.

Another type of pilus, called a **sex pilus,** is used to join one bacterium to another for a specific type of DNA transfer. This and other mechanisms of DNA transfer will be described in chapter 8. ▶▶ DNA transfer, p. 200

MicroAssessment 3.8

Flagella are the most common mechanism for bacterial motility. Chemotaxis is the directed movement of cells toward an attractant or away from a repellent. Pili provide a mechanism for attachment to specific surfaces.

20. E. coli *cells have peritrichous flagella. What does this mean?*

21. *How are fimbriae different from sex pili in their function?*

3.9 ■ Internal Structures

Learning Outcomes

22. *Describe the structure and function of the chromosome, plasmids, ribosomes, storage granules, gas vesicles, and endospores.*

23. *Describe the significance and processes of sporulation and germination.*

Prokaryotic cells have a variety of structures within the cell. Some, such as the chromosome and ribosomes, are essential for the life of all cells, whereas others give cells certain selective advantages.

The Chromosome

The prokaryotic **chromosome** is typically a single, circular double-stranded DNA molecule that contains all the genetic information required by a cell. It occurs as a mass within the cytoplasm, forming a gel-like region called the **nucleoid.**

Chromosomal DNA is tightly packed into the cell (**figure 3.41**). The compact shape is due partially to nucleoid-associated proteins that bind to DNA, creating a structure that bends and folds. In addition, the DNA is twisted, or supercoiled. This can be visualized by cutting a rubber band and then twisting one end several times before rejoining cut the ends. The resulting circle will twist and coil in response.

Plasmids

Most **plasmids** are circular, supercoiled, double-stranded DNA molecules. They are generally much smaller than the chromosome and carry from a few to several hundred genes. A single cell can have more than one type of plasmid, and these can each be present in multiple copies. ▶▶ **plasmids, p. 208**

A cell generally does not require the genetic information carried by a plasmid. However, the encoded characteristics can be advantageous in certain situations. For example, many plasmids code for the production of enzymes that destroy certain antibiotics, allowing the organism to resist the otherwise lethal effect of these medications. Because a bacterium can sometimes transfer a copy of a plasmid to another bacterial cell, this accessory genetic information can spread, which accounts in large part for the increasing frequency of antibiotic-resistant organisms.

Ribosomes

Ribosomes are involved in protein synthesis, where they serve as the structures that facilitate the joining of amino acids. Their relative size and density is expressed as a distinct unit, S (for Svedberg), that reflects how fast they settle when spun at very high speeds in an ultracentrifuge. The faster they move toward the bottom, the higher the S value and the greater the density. Prokaryotic ribosomes are 70S. Note that S units are not strictly arithmetic; the 70S ribosome is composed of a 30S and a 50S subunit (**figure 3.42**). Prokaryotic ribosomes differ from eukaryotic ribosomes, which are 80S. The fact that they are distinct is important medically because antibiotics that interfere with the function of the 70S ribosome have no effect on the 80S molecule. ▶▶ **function of ribosomes, p. 171**

Cytoskeleton

It was once thought that bacteria lacked a **cytoskeleton,** an interior protein framework. Several bacterial proteins that have similarities to those of the eukaryotic cytoskeleton have now been characterized, and these appear to be involved in cell division and controlling cell shape.

Storage Granules

Storage granules are accumulations of high-molecular-weight polymers synthesized from a nutrient that a cell has in relative excess. If nitrogen and/or phosphorus are lacking, for example, a cell cannot multiply even if a carbon and energy source such as glucose is plentiful. Rather than waste the carbon/energy source, cells use it to produce glycogen, a glucose polymer. Later, when conditions are appropriate, cells degrade and use the glycogen.

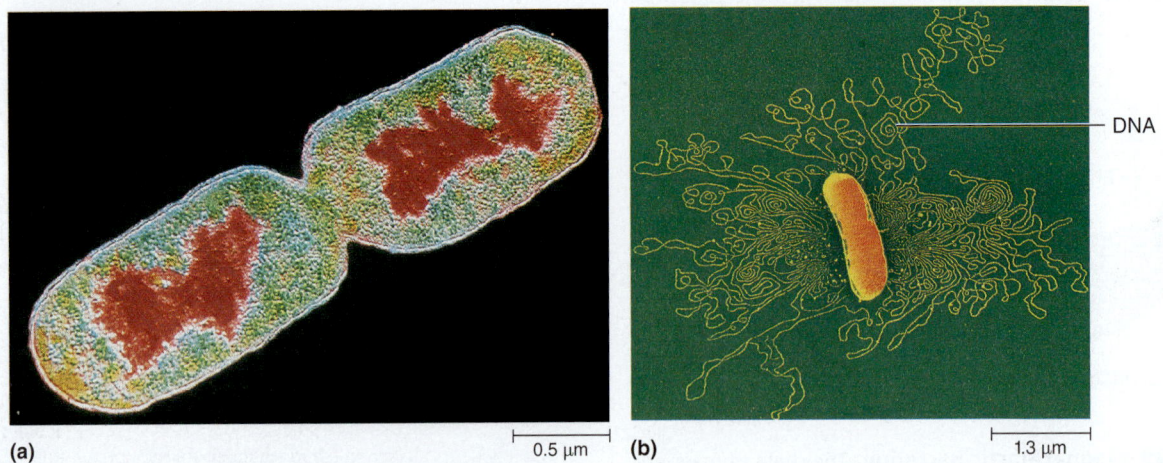

(a)   0.5 µm   (b)   1.3 µm

FIGURE 3.41 The Chromosome (a) Color-enhanced transmission electron micrograph of a thin section of *Escherichia coli*, with the DNA shown in red. **(b)** Chromosome released from a gently lysed *E. coli* cell. Note how tightly packed the DNA must be inside the bacterium.

❓ *What is the gel-like region formed by the chromosome called?*

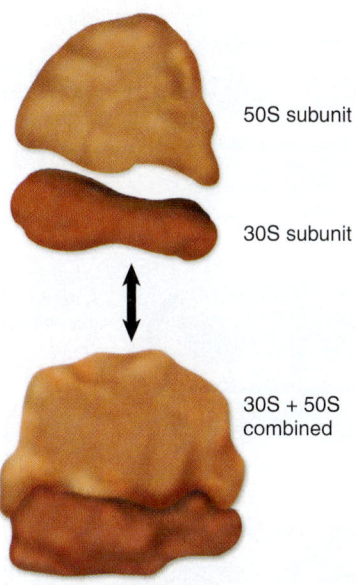

FIGURE 3.42 70S Ribosome The 70S ribosome is composed of 50S and 30S subunits.

❓ *What is the function of ribosomes?*

Other bacterial species store carbon and energy as poly-β-hydroxybutyrate (**figure 3.43**).

Some types of granules can be readily detected by light microscopy. Volutin granules, a storage form of phosphate, stain red with the dye methylene blue, whereas the surrounding cellular material stains blue. Because of this, they are often called metachromatic granules (*meta* means "change" and *chromatic* means "color"). Recent evidence suggests that the role of these granules is more complex than originally thought. The volutin granules of some bacteria are membrane-bound and appear to resemble eukaryotic organelles that are thought to be involved with energy storage and pH balance.

MicroByte ─────────────────────
Poly-β-hydroxybutyrate can be used to make biodegradable plastics. Phosphate-storing cells can be used to remove pollutants from wastewater.

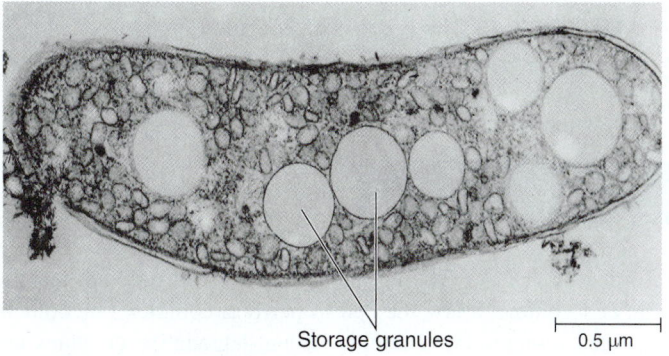

FIGURE 3.43 Storage Granules The large unstained areas in the photosynthetic bacterium *Rhodospirillum rubrum* are granules of *poly-β-hydroxybutyrate.*

❓ *How would storage granules benefit a cell?*

Gas Vesicles

Some aquatic bacteria produce **gas vesicles**—small, rigid, protein-bound compartments that provide buoyancy to the cell. Gases, but not water, flow freely into the vesicles, thereby decreasing the density of the cell. By regulating the number of gas vesicles, a cell can float or sink to its ideal position in the water column. Bacteria that use sunlight as a source of energy use gas vesicles to float closer to the surface, where light is available.

Endospores

An **endospore** is a unique type of dormant cell produced by certain bacterial species such as members of the genera *Bacillus* and *Clostridium* (**figure 3.44**). The structures may remain dormant for perhaps 100 years, or even longer, and are extraordinarily resistant to damaging conditions including heat, desiccation, toxic chemicals, and ultraviolet (UV) light. Immersion in boiling water for hours may not kill them. Endospores that survive these treatments can **germinate,** or exit the dormant stage, to become a typical multiplying cell, called a **vegetative cell.**

Because endospores can survive so long in a variety of conditions, they can be found virtually everywhere. They are common in soil, which can make its way into environments such as laboratories and hospitals and onto products such as medical devices, food, and media used to cultivate microbes. Excluding microbes from these environments and products is very important, so special precautions must be taken to avoid or destroy endospores.

Endospores are sometimes called spores. However, this latter term is also used to refer to structures produced by unrelated microbes such as fungi. Bacterial endospores are much more resistant to environmental conditions than are other types of spores.

MicroByte ─────────────────────
The diseases botulism, tetanus, gas gangrene, and anthrax are all caused by endospore-formers.

Sporulation

Endospore formation, or **sporulation,** is a complex sequence of changes that begin when spore-forming bacteria experience

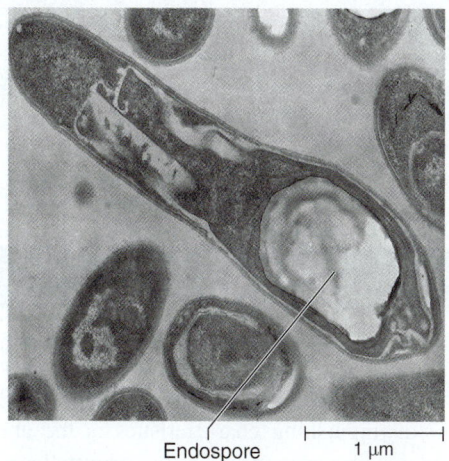

FIGURE 3.44 Endospores Endospore inside a vegetative cell of a *Clostridium* species (TEM).

❓ *What is the function of an endospore?*

limiting amounts of carbon or nitrogen. The cells sense starvation conditions, which triggers them to begin the 8-hour process.

After vegetative growth stops, the DNA is duplicated and a septum forms, dividing the cell asymmetrically (**figure 3.45**). The larger compartment then engulfs the smaller compartment, forming a forespore within a mother cell. These two portions take on different roles in synthesizing the components that will make up the endospore. The forespore will ultimately become the core of the endospore. Peptidoglycan-containing material is laid down between the two membranes that surround the forespore, forming the core wall and the cortex. Meanwhile, the mother cell makes proteins that will form the spore coat. Ultimately, the mother cell is degraded and the endospore released.

The layers of the endospore protect it from damage. The spore coat is thought to function as a sieve, excluding molecules such as lysozyme. The cortex helps maintain the core in a dehydrated state, protecting it from the effects of heat. In addition, the core has small, acid-soluble proteins that bind to DNA, thereby protecting it from damage. The core is rich in an unusual compound called calcium dipicolinate, which seems to also play an important role in spore resistance.

Germination

Germination can be triggered by a brief exposure to heat or certain chemicals. Following such exposure, the endospore takes on water and swells. The spore coat and cortex then crack open, and a vegetative cell grows out. Since one vegetative cell gives rise to one endospore, sporulation is not a means of cell reproduction.

MicroAssessment 3.9

The prokaryotic chromosome is usually a circular, double-stranded DNA molecule that contains all of the genetic information required by a cell. Plasmids generally encode only information that is advantageous to a cell in certain conditions. Ribosomes are the structures that facilitate the joining of amino acids to form a protein. The cytoskeleton is an interior framework involved in cell division and controlling cell shape. Storage granules are polymers synthesized from nutrients a cell has in relative excess. Gas vesicles provide buoyancy to a cell. An endospore is a highly resistant dormant stage produced by certain bacterial species.

22. *Explain how glycogen granules benefit a cell.*

23. *Explain why endospores are an important concern for the canning industry.*

24. *Why are the processes of sporulation and germination not a mechanism of multiplication?* ➕

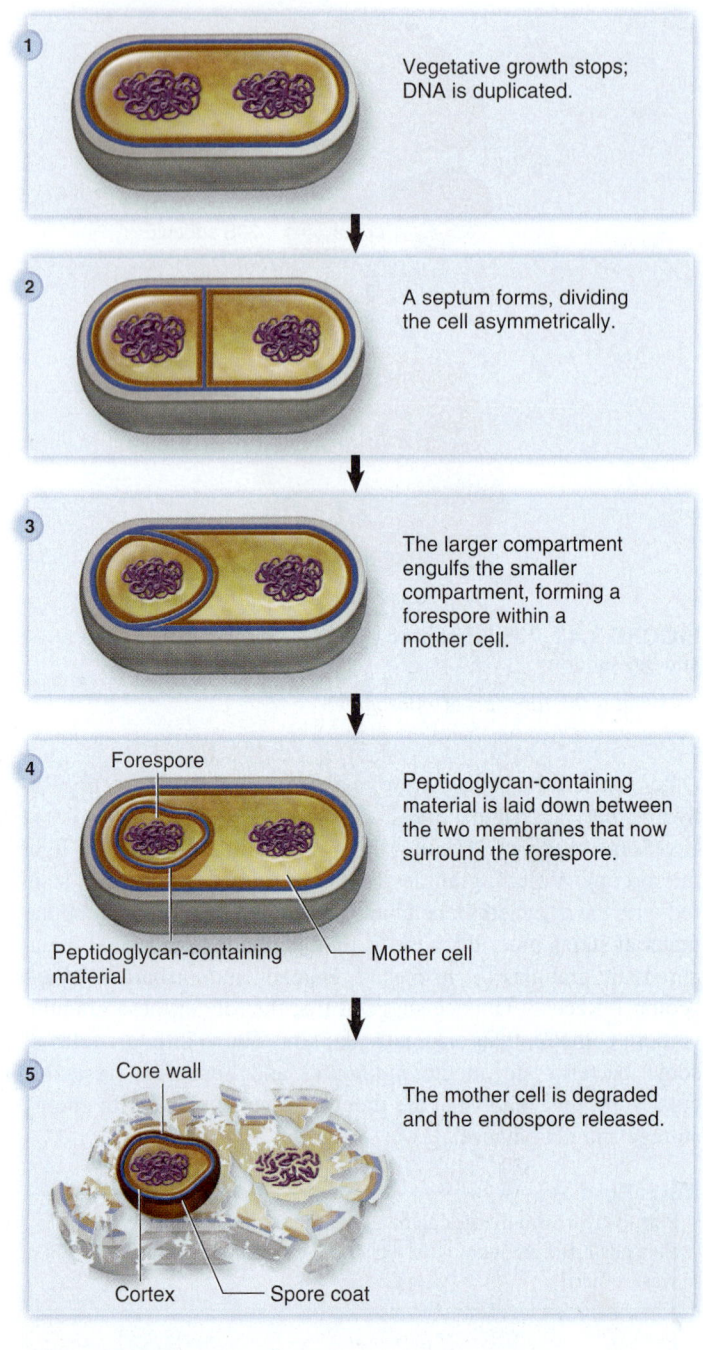

1 — Vegetative growth stops; DNA is duplicated.

2 — A septum forms, dividing the cell asymmetrically.

3 — The larger compartment engulfs the smaller compartment, forming a forespore within a mother cell.

4 — Peptidoglycan-containing material is laid down between the two membranes that now surround the forespore.

Forespore

Peptidoglycan-containing material — Mother cell

5 — The mother cell is degraded and the endospore released.

Core wall

Cortex — Spore coat

FIGURE 3.45 The Process of Sporulation

? *Approximately how long does the sporulation process take?*

THE EUKARYOTIC CELL

Eukaryotic cells are generally much larger than prokaryotic cells, and their internal structures are far more complex (**figure 3.46**). One of their distinguishing characteristics is the abundance of membrane-enclosed compartments or **organelles.** The most important of these is the nucleus, which contains the cell's genetic information (DNA). The organelles, which can take up half the total cell volume, allow the cell to perform complex functions in separated regions. For example, when degradative enzymes are contained within an organelle, they can digest material without posing a threat to the integrity of the cell itself.

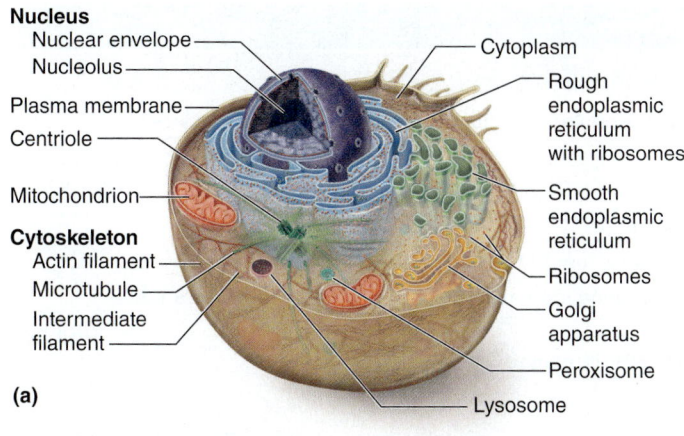

Nucleus
- Nuclear envelope
- Nucleolus

Plasma membrane

Centriole

Mitochondrion

Cytoskeleton
- Actin filament
- Microtubule
- Intermediate filament

Cytoplasm

Rough endoplasmic reticulum with ribosomes

Smooth endoplasmic reticulum

Ribosomes

Golgi apparatus

Peroxisome

Lysosome

(a)

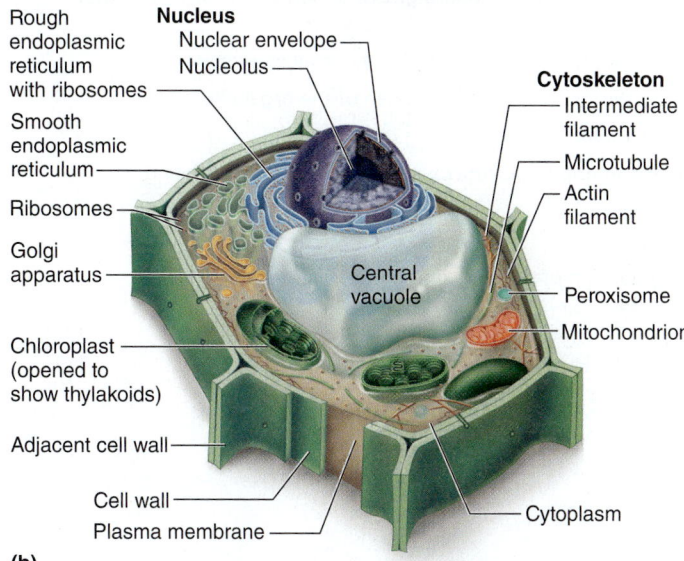

Rough endoplasmic reticulum with ribosomes

Smooth endoplasmic reticulum

Ribosomes

Golgi apparatus

Chloroplast (opened to show thylakoids)

Adjacent cell wall

Cell wall

Plasma membrane

Nucleus
- Nuclear envelope
- Nucleolus

Cytoskeleton
- Intermediate filament
- Microtubule
- Actin filament

Central vacuole

Peroxisome

Mitochondrion

Cytoplasm

(b)

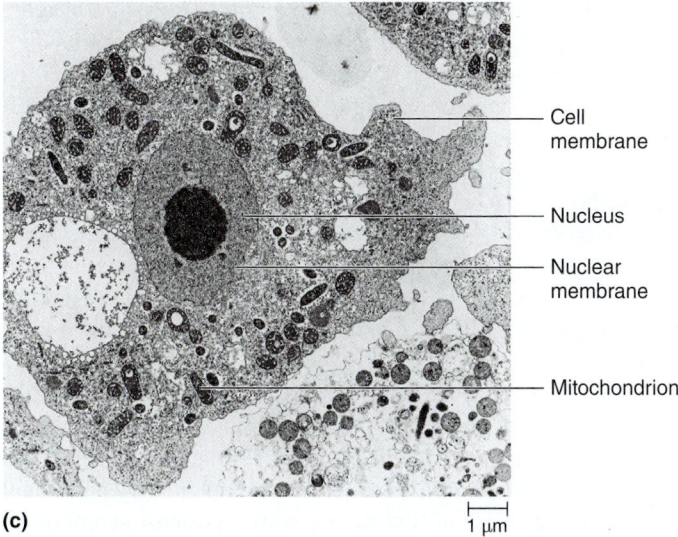

Cell membrane

Nucleus

Nuclear membrane

Mitochondrion

1 μm

(c)

FIGURE 3.46 Eukaryotic Cells (a) Diagrammatic representation of an animal cell. **(b)** Diagrammatic representation of a plant cell. **(c)** Micrograph of an animal cell shows several membrane-bound structures including mitochondria and a nucleus. Note that a prokaryotic cell is about the size of a mitochondrion.

❓ *Which organelle contains the cell's genetic information?*

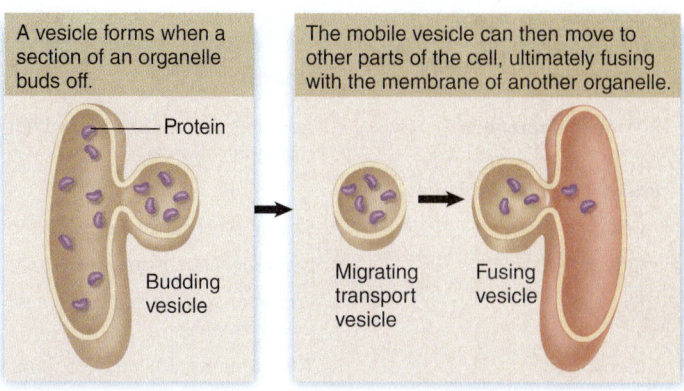

A vesicle forms when a section of an organelle buds off.

Protein

Budding vesicle

The mobile vesicle can then move to other parts of the cell, ultimately fusing with the membrane of another organelle.

Migrating transport vesicle

Fusing vesicle

FIGURE 3.47 Vesicle Formation and Fusion

❓ *The lumen is which part of an organelle?*

Each organelle contains a variety of compounds, many of which are synthesized at other locations. To deliver these to the **lumen** (interior) of another organelle, a section of the organelle buds off, forming a membrane-bound vesicle (**figure 3.47**). This carries a bit of the organelle's contents to other parts of the cell. When the vesicle encounters another organelle, the two membranes fuse to become one unit, similar to two oil droplets merging. By doing so, the vesicle spills its contents into the organelle.

As a group, eukaryotic cells are highly variable. For example, protozoa, which are single-celled organisms, must function exclusively as self-contained units that seek and ingest food. These cells must be mobile and flexible to take in food particles, so they lack cell walls that would otherwise provide rigidity. Animal cells also lack a cell wall, because they too must be flexible to accommodate movement. Fungi, on the other hand, are stationary and benefit from the protection provided by a strong cell wall. Fungal cell walls are composed primarily of polysaccharides including chitin, a polymer of *N*-acetylglucosamine. Plants, which are also stationary, have cell walls composed of cellulose, a polymer of glucose. ◀◀ **polysaccharides, p. 31** ◀◀ **chitin, p. 32** ◀◀ **cellulose, p. 32**

The individual cells of a multicellular organism can be distinctly different from one another. Liver cells, for example, are obviously quite different from bone cells. Different cell types, however, can function in cooperative associations called **tissues.** The tissues in your body include muscle, connective, nerve, epithelial, blood, and lymphoid. Each of these provides a unique function. Various tissues work together to make up **organs,** including skin, heart, and liver. Organs and the tissues that compose them will be covered in more detail in chapters on infectious diseases.

A comprehensive coverage of all aspects of eukaryotic cells is beyond the scope of this textbook. Instead, this section will focus on key characteristics, particularly those that directly affect the interactions of microbe with human hosts. These characteristics

TABLE 3.6	A Summary of Typical Eukaryotic Cell Structures	
Structure	**Characteristics**	
Plasma Membrane	Asymmetric lipid bilayer embedded with proteins. Permeability barrier, transport, and cell-to-cell communication.	
Internal Protein Structures		
Cilia	Beat in synchrony to provide movement. Composed of microtubules in a 9 + 2 arrangement.	
Cytoskeleton	Dynamic filamentous network that provides structure to the cell.	
Flagella	Propel or push the cell with a whiplike or thrashing motion. Composed of microtubules in a 9 + 2 arrangement.	
Ribosomes	Two subunits, 60S and 40S, join to form the 80S ribosome.	
Membrane-Bound Organelles		
Chloroplasts	Site of photosynthesis; the organelle harvests the energy of sunlight to generate ATP, which is then used to convert CO_2 to carbohydrates.	
Endoplasmic reticulum	Site of synthesis of macromolecules destined for other organelles or the external environment.	
Rough	Attached ribosomes thread proteins they are synthesizing into the lumen of the organelle.	
Smooth	Site of lipid synthesis and degradation, and calcium ion storage.	
Golgi apparatus	Site where macromolecules synthesized in the endoplasmic reticulum are modified before being transported in vesicles to other destinations.	
Lysosome	Digestion of macromolecules.	
Mitochondria	Harvest the energy released during the degradation of organic compounds to generate ATP.	
Nucleus	Contains the genetic information (DNA).	
Peroxisome	Site where oxidation of lipids and toxic chemicals occurs.	

are summarized in **table 3.6.** The functions of prokaryotic and eukaryotic cell components are compared in **table 3.7.**

3.10 ■ The Plasma Membrane

Learning Outcome

24. *Describe the structure and function of the eukaryotic plasma membrane, comparing and contrasting it with the prokaryotic cytoplasmic membrane.*

All eukaryotic cells have a cytoplasmic membrane, or **plasma membrane,** which is similar in chemical structure and function to that of prokaryotic cells. It is a typical phospholipid bilayer embedded with proteins. The lipid and protein composition of the layer that faces the cytoplasm, however, differs significantly from that facing the outside of the cell. The same is true for membranes that surround the organelles. The layer facing the inside of the organelle is similar to its plasma membrane counterpart that faces the outside of the cell. This lack of symmetry reflects the important role these membranes play in complex processes within the eukaryotic cell.

The membrane proteins perform a variety of functions. Some are involved in transport and others are attached to internal structures, helping to maintain cell integrity. Those that face the outside often function as **receptors.** A given receptor binds a specific molecule, which is referred to as its **ligand.** These receptor-ligand interactions are extremely important in multicellular organisms because they allow cells to communicate with each other—a process called signaling. For example, when a cell within the human body encounters a compound unique to bacteria, it secretes specific proteins as a call for help. Other cells have receptors for these proteins on their surface, allowing them to recognize the signals, and respond accordingly. Using this cell-to-cell communication, a multicellular organism can function as a cohesive unit.

The membranes of many eukaryotic cells contain sterols, which provide strength to the otherwise fluid structure. The sterol in animal cell membranes is cholesterol, whereas fungal membranes have ergosterol. This difference is exploited by antifungal medications that act by interfering with ergosterol synthesis or function. ▶▶ antifungal medications, p. 476

Within the plasma membrane are cholesterol-rich regions called lipid rafts. The role of these regions is still being studied, but they appear to be important in allowing the cell to detect and respond to signals in the external environment. From a microbiologist's perspective, they are also important because many viruses use these regions when they enter or exit a cell.

TABLE 3.7	Comparison of Typical Prokaryotic and Eukaryotic Cell Structures/Functions	
	Prokaryotic	**Eukaryotic**
General Characteristics		
Size	Generally 0.3–2 μm in diameter.	Generally 5–50 μm in diameter.
Cell division	Chromosome replication followed by binary fission.	Chromosome replication and mitosis followed by division.
Chromosome location	Located in the nucleoid, which is not membrane-bound.	Contained within the membrane-bound nucleus.
Structures		
Cell membrane	Relatively symmetric with respect to the lipid content of the bilayers.	Highly asymmetric; lipid composition of outer layer differs significantly from that of inner layer.
Cell wall	Composed of peptidoglycan (*Bacteria*); Gram-negative bacteria have an outer membrane as well.	Absent in animal cells; composition in other cell types may include chitin, glucans and mannans (fungi), and cellulose (plants).
Chromosome	Single, circular DNA molecule is typical.	Multiple, linear DNA molecules. DNA is wrapped around histones.
Flagella	Composed of protein subunits; attached to the cell envelope.	Made up of a 9 + 2 arrangement of microtubules; covered by an extension of the plasma membrane.
Membrane-bound organelles	Absent.	Present; includes the nucleus, mitochondria, chloroplasts (only in plant cells), endoplasmic reticulum, Golgi apparatus, lysosomes, and peroxisomes.
Nucleus	Absent; DNA resides as an irregular mass forming the nucleoid region.	Present.
Ribosomes	70S ribosomes, which are made up of 50S and 30S subunits.	80S ribosomes, which are made up of 60S and 40S subunits. Mitochondria and chloroplasts have 70S ribosomes.
Functions		
Degradation of extracellular substances	Enzymes are secreted that degrade macromolecules outside of the cell. The resulting small molecules are transported into the cell.	Macromolecules can be brought into the cell by endocytosis. Lysosomes carry digestive enzymes.
Motility	Generally involves flagella, which are composed of protein subunits. Flagella rotate like propellers.	Involves cilia and flagella, which are made up of a 9 + 2 arrangement of microtubules. Cilia move in synchrony; flagella propel a cell with a whiplike motion or thrash back and forth to pull a cell forward.
Protein secretion	Secretion systems transport proteins across the cytoplasmic membrane.	Secreted proteins are moved to the lumen of the rough endoplasmic reticulum as they are being synthesized. From there, they are transported to the Golgi apparatus for processing and packaging.
Strength and rigidity	Peptidoglycan-containing cell wall (*Bacteria*).	Cytoskeleton composed of microtubules, intermediate filaments, and microfilaments. Some have a cell wall; some have sterols in the membrane.
Transport	Primarily active transport. Group translocation.	Facilitated diffusion and active transport. Ion channels.

The plasma membrane plays no role in ATP synthesis; instead, that task is performed by mitochondria (discussed later). Even though the plasma membrane has no electron transport chain to generate a proton motive force, there is still an electrochemical gradient across the membrane. This is maintained by energy-consuming mechanisms that pump either sodium ions or protons to the outside of the cell. ◄◄ electrochemical gradient, p. 55

MicroAssessment 3.10

The plasma membrane is an asymmetric lipid bilayer embedded with proteins. Specific receptors allow cell-to-cell signaling. Sterols provide strength to the fluid membrane.

25. *What is the medical significance of ergosterol in fungal membranes?*
26. *Describe why signaling is important in animal cells.*
27. *How could one argue that the interior of an organelle is "outside" of a cell?* ➕

3.11 ■ Transfer of Molecules Across the Plasma Membrane

Learning Outcomes

25. *Compare and contrast the roles of channels and carriers.*
26. *Describe the processes of endocytosis and exocytosis.*
27. *Describe the role of the endoplasmic reticulum in secretion.*

Various substances—including nutrients, signaling molecules, and waste products—must pass through the plasma membrane. Cells have several mechanisms to accomplish this.

Transport Proteins

The transport proteins of eukaryotic cells function as either carriers or channels. **Carriers** are analogous to proteins in prokaryotic cells that function in facilitated diffusion and active transport. **Channels** form small pores in the membrane that allow only specific ions to diffuse through. To control ion passage, the channel has a gate that can open or close in response to environmental conditions. Eukaryotic cells also have aquaporins. **◀◀ facilitated diffusion, p. 56 ◀◀ active transport, p. 56 ◀◀ aquaporin, p. 54**

Cells of multicellular organisms often take up nutrients by facilitated diffusion. This is possible because systems can regulate the nutrient concentration surrounding the cells. For instance, when the human body maintains a high glucose concentration in the blood, tissue cells can bring the sugar in without using energy.

The active transport mechanisms of eukaryotic cells are structurally similar to those of prokaryotic cells. A medically important example is an ABC transporter used by cancerous cells that ejects therapeutic drugs intended to kill those cells.

Endocytosis and Exocytosis

Endocytosis is the process by which a eukaryotic cell takes up material from the surrounding environment by forming invaginations (inward folds) in its membrane (**figure 3.48**). The type of endocytosis common to most animal cells is **pinocytosis.** In this process, the small invaginations carry along some liquid and any material attached to the membrane. This action ultimately forms a membrane-bound compartment called an **endosome,** which then fuses with digestive organelles called lysosomes to form an **endolysosome**. The characteristics of lysosomes will be discussed shortly. **▶▶ lysosome, p. 77**

Animal cells often take up material by **receptor-mediated endocytosis,** a variation of pinocytosis. It allows cells to take up extracellular ligands that bind to receptors on the cell's surface. When the receptors bind their ligands, the region is internalized to form an endosome that contains the receptors along with their bound ligands. The acidic internal environment of the endosome frees the ligands from the receptors, which are then often recycled. After that, the endosome fuses with lysosomes.

Protozoa and phagocytes, both of which ingest bacteria and large debris, use a specific type of endocytosis called **phagocytosis.** Phagocytes are important cells of the body's defense system against microbial invaders. In phagocytosis, the cells send out armlike extensions called **pseudopods,** which surround and enclose extracellular material, including bacteria. This action brings the material into the cell in an enclosed compartment called a **phagosome.** This fuses with endosomes, and ultimately fuses with lysosomes to form a **phagolysosome. ▶▶ phagocytes, p. 346**

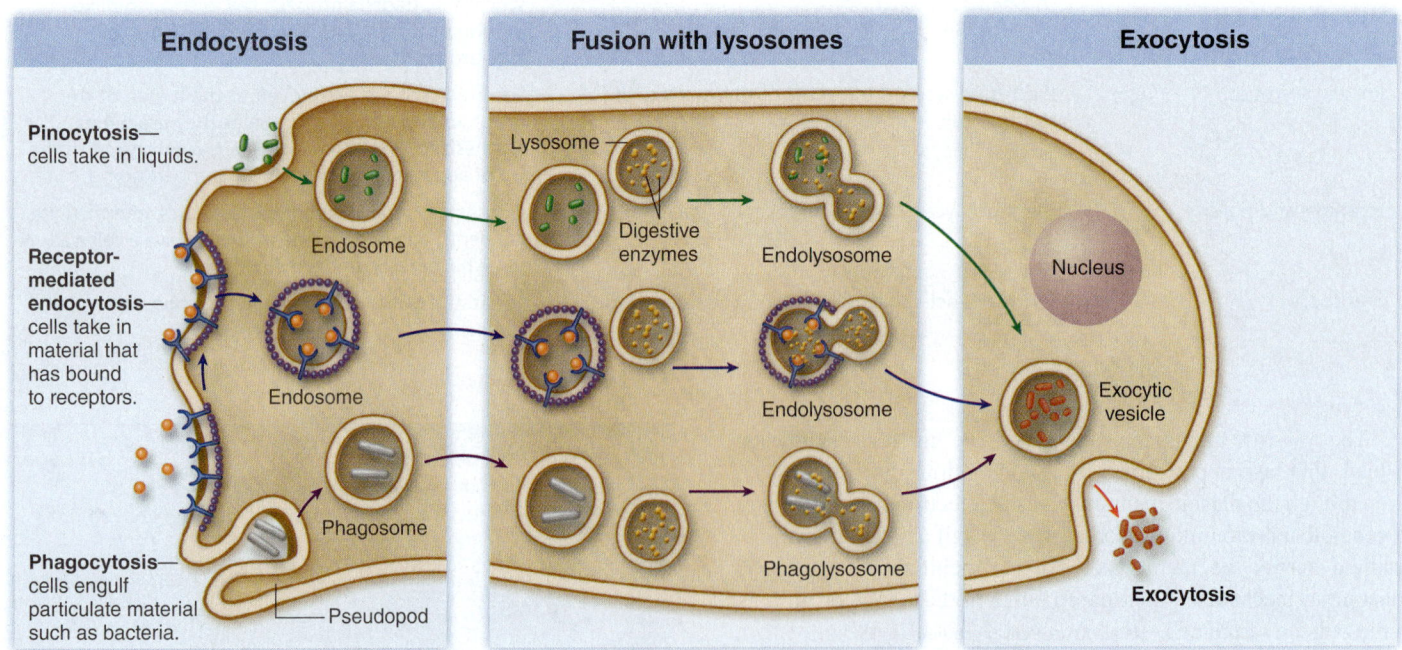

FIGURE 3.48 **Endocytosis and Exocytosis**

❓ *How is pinocytosis different from phagocytosis?*

The process of **exocytosis** is the reverse of endocytosis. Membrane-bound exocytic vesicles inside the cell fuse with the plasma membrane and release their contents to the outside.

MicroByte

Many viruses use receptor-mediated endocytosis to enter animal cells. They bind to a specific receptor, "tricking" the cell into bringing them in.

Secretion

As in prokaryotic cells, proteins destined for secretion have a signal sequence, a characteristic amino acid sequence that functions as a tag. When ribosomes synthesize a protein with such a sequence, they attach to a complex on the membrane of the endoplasmic reticulum (ER)—an organelle that will be described shortly. As the protein is made, it is threaded through the membrane and into the ER. Once inside the ER, the protein can easily be transported by vesicles to the outside of the cell. Proteins destined for various organelles also have specific amino acid tags.
▶▶ endoplasmic reticulum, p. 76

MicroAssessment 3.11

Transport proteins function as either channels or carriers. Pinocytosis allows cells to internalize small molecules. Protozoa and phagocytes internalize bacteria and debris by phagocytosis. Exocytosis is used to expel material. Secreted proteins are threaded through the membrane of the endoplasmic reticulum as they are being made.

28. *How does a cell bring in ligands?*
29. *How is the formation of an endosome different from that of a phagosome?*
30. *How might a bacterium resist the killing effects of a phagolysosome?* ➕

3.12 ■ Protein Structures Within the Cell

Learning Outcome

28. *Describe the structure and function of eukaryotic ribosomes, the cytoskeleton, flagella, and cilia.*

Eukaryotic cells have a variety of important protein structures within the cell. These include ribosomes, the cytoskeleton, flagella, and cilia.

Ribosomes

Ribosomes are the structures involved in protein synthesis. The eukaryotic ribosome is 80S, and is made up of a 60S and a 40S subunit. Recall that the prokaryotic ribosomes are 70S.

Cytoskeleton

The **cytoskeleton** forms the framework of a cell (**figure 3.49**). Its three components—(1) actin filaments (also called microfilaments), (2) microtubules, and (3) intermediate filaments—continually reconstruct to adapt to the cell's constantly changing needs.

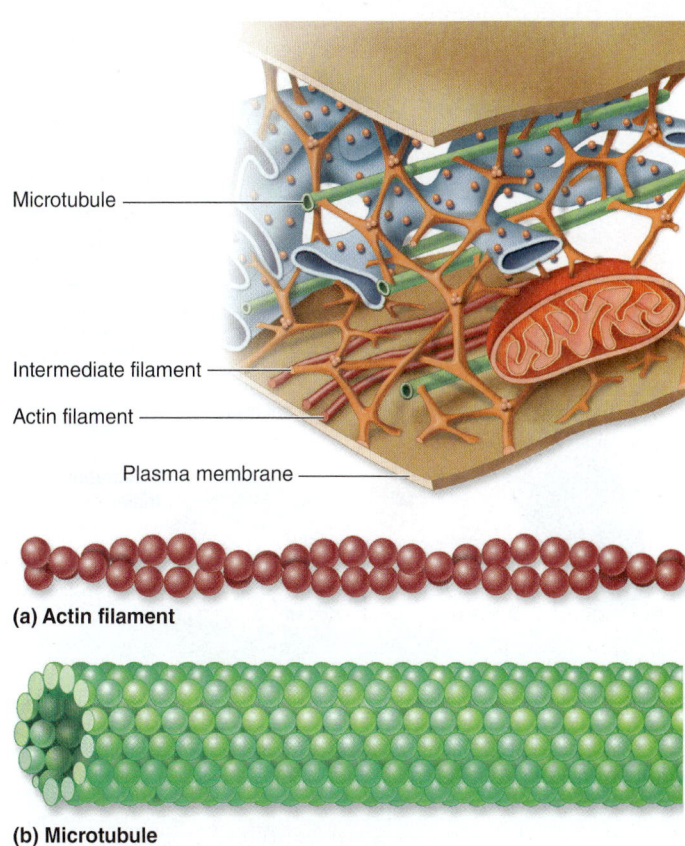

Microtubule

Intermediate filament

Actin filament

Plasma membrane

(a) Actin filament

(b) Microtubule

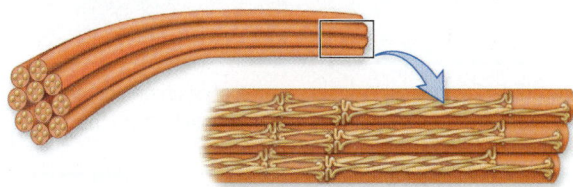

(c) Intermediate filament

FIGURE 3.49 Cytoskeleton Diagrammatic representation of the dynamic filamentous network that provides structure to the cell; the cytoskeleton is composed of three elements: **(a)** actin filaments, **(b)** microtubules, and **(c)** intermediate filaments.

❓ *What is the role of actin filaments?*

Actin filaments allow the cell cytoplasm to move. The filaments are polymers of protein subunits called actin, which rapidly assemble and subsequently disassemble to cause motion. For example, pseudopod formation relies on actin polymerization in one part of the cell and depolymerization in another.

Microtubules, the thickest of the cytoskeleton's components, are long hollow structures made of protein subunits called tubulin. Microtubules form the mitotic spindles, the machinery that separates duplicated chromosomes as cells divide. Without mitotic spindles, cells could not reproduce. In addition, microtubules are the main structures that make up the cilia and flagella, the mechanisms of locomotion in some eukaryotic cells. Microtubules also function as the framework along which organelles and vesicles move within a cell. The antifungal drug griseofulvin interferes with the action of microtubules in some fungi. ▶▶ mitosis, p. 283

Intermediate filaments function like ropes, strengthening the cell mechanically. By providing mechanical support, they allow cells to resist physical stresses.

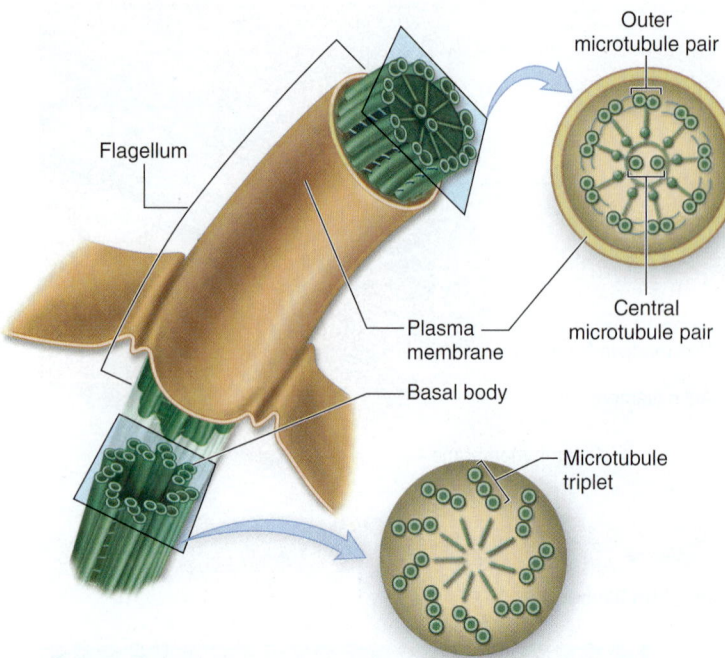

Outer
microtubule pair

Flagellum

Central
microtubule pair

Plasma
membrane

Basal body

Microtubule
triplet

FIGURE 3.50 Flagellum A flexible structure involved in movement.

❓ *How is the structure of a eukaryotic flagellum different from its prokaryotic counterpart?*

MicroByte

Some intracellular pathogens trigger actin polymerization, which propels the pathogens with enough force to push them into an adjacent cell.

Flagella and Cilia

Flagella and **cilia** are flexible structures that appear to project out of a cell yet are covered by extensions of the plasma membrane (**figure 3.50**). They are composed of long microtubules grouped in what is called a 9 + 2 arrangement—nine pairs of microtubules surrounding two individual ones. They originate from basal bodies within the cell; the basal body has a slightly different arrangement of microtubules.

Although eukaryotic flagella function in motility, they are structurally very different from their prokaryotic counterparts. Using ATP as a source of energy, they either propel the cell with a whiplike motion or thrash back and forth to pull the cell forward.

Cilia are shorter than flagella, often covering a cell and moving in synchrony (see figure 1.8). Their motion can move a cell forward in an aqueous solution, or propel surrounding material along a stationary cell. For example, epithelial cells that line the respiratory tract have cilia that beat together in a directed fashion. This moves the mucus film that covers those cells, directing it toward the mouth, where it can be swallowed. Because of this action, called the mucociliary escalator, most microbes that have been inhaled are removed before they can enter the lungs.

MicroAssessment 3.12

The 80S eukaryotic ribosome is composed of 60S and 40S subunits. The cytoskeleton is a dynamic filamentous network that provides structure to the cell; it is composed of actin filaments, microtubules, and intermediate filaments. Flagella function in motility. Cilia either propel a cell or move material along a stationary cell.

31. *Explain how actin filaments are related to phagocytosis.*
32. *Explain what is meant by the 9 + 2 structure of cilia and flagella.*
33. *Why would 60S and 40S ribosomal subunits make an 80S ribosome rather than a 100S ribosome?* ➕

3.13 ■ Membrane-Bound Organelles

Learning Outcome

29. *Describe the function of the nucleus, mitochondria, chloroplasts, endoplasmic reticulum, Golgi apparatus, lysosomes, and peroxisomes.*

Membrane-bound organelles are an important feature of eukaryotic cells that sets them apart from prokaryotic cells.

The Nucleus

An important distinguishing feature of a eukaryotic cell is the **nucleus,** which contains the genetic information. The boundary of this structure is the nuclear envelope, composed of two lipid bilayer membranes—the inner and outer membranes (**figure 3.51**). Complex protein structures span the envelope, forming nuclear pores. These allow large molecules such as ribosomal subunits and proteins to be transported into and out of the nucleus. The nucleolus is a region where ribosomal RNAs are synthesized.

The nucleus contains multiple chromosomes, each one encoding different genetic information. Unlike the situation in most prokaryotic cells, the DNA is linear. It is wound around proteins called histones, a characteristic that adds structure and order to the long molecule.

Events that take place in the nucleus during cell division distinguish eukaryotes from prokaryotes. After DNA is replicated in eukaryotic cells, chromosomes go through a nuclear division process called mitosis, which ensures that the daughter cells receive the same number of chromosomes as the original parent. Because of mitosis, a cell that is diploid (has two copies of each chromosome) will generate two diploid daughter cells. A different process, meiosis, generates haploid daughter cells, which each have a single copy of each chromosome.

Mitochondria

Mitochondria function as ATP-generating powerhouses, and are found in nearly all eukaryotic cells. They are complex structures bounded by two lipid bilayers—the outer and inner membranes (**figure 3.52**). The outer membrane is smooth, but the inner membrane is highly folded, forming invaginations called cristae. The folds increase the membrane's surface area, maximizing the

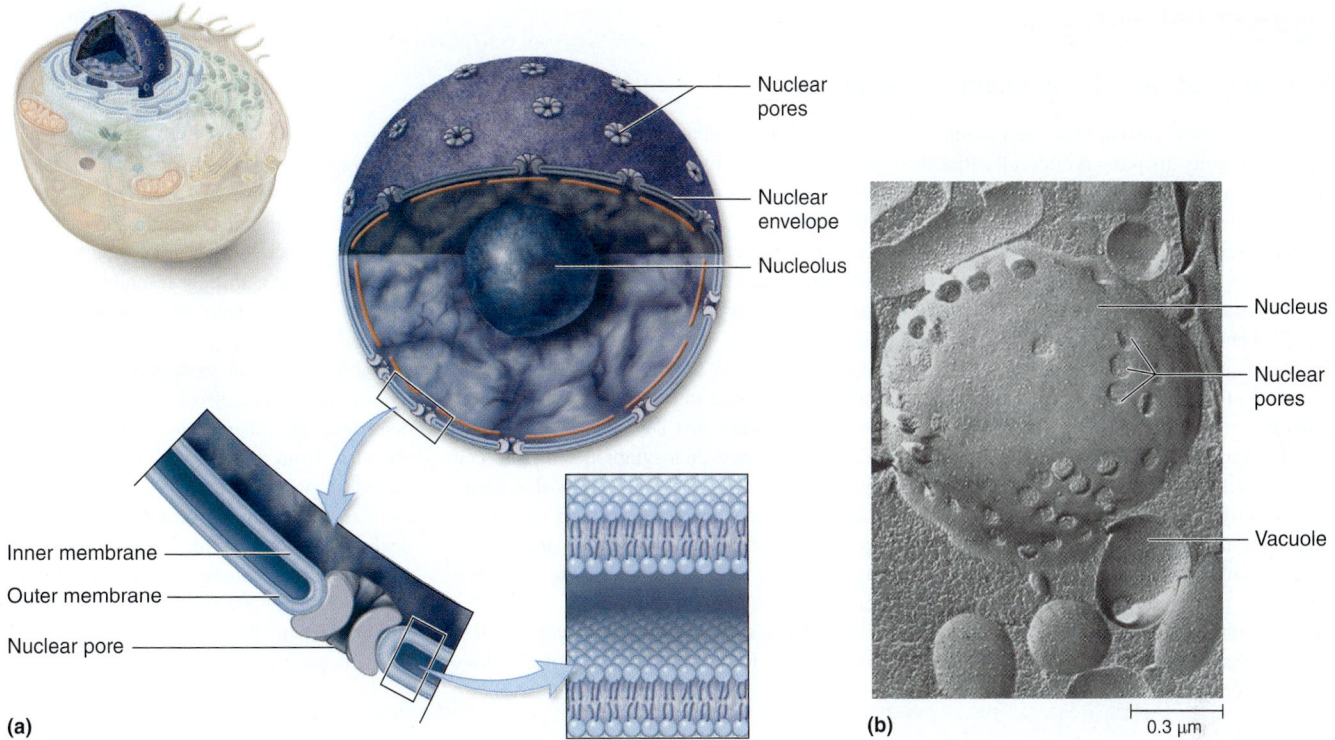

FIGURE 3.51 Nucleus Organelle that contains the cell's genetic information (DNA). **(a)** Diagrammatic representation. **(b)** Electron micrograph of a yeast cell by freeze-fracture technique.

? *What is the function of nuclear pores?*

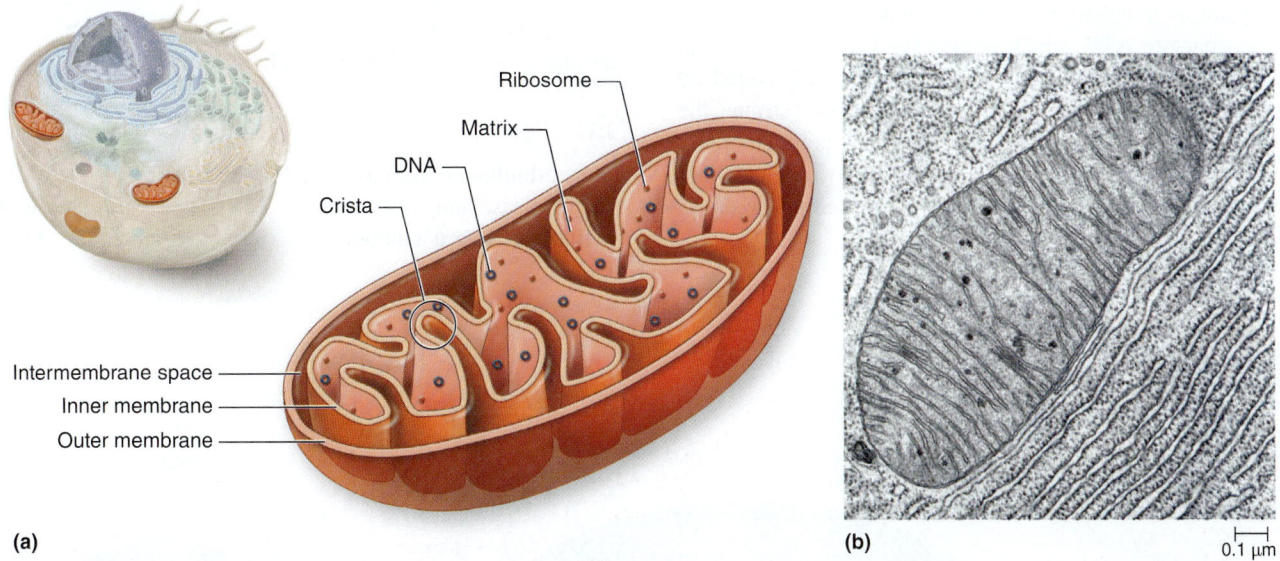

FIGURE 3.52 Mitochondria Organelles that harvest energy released during the degradation of organic compounds to synthesize ATP. **(a)** Diagrammatic representation. **(b)** Electron micrograph.

? *What were the first pieces of evidence that led scientists to conclude that mitochondria evolved from bacterial cells?*

ATP-generating capabilities of the organelle (the processes will be discussed in chapter 6).

The mitochondrial matrix—the gel-like material enclosed by the mitochondrial inner membrane—contains DNA, ribosomes, and other molecules necessary for protein synthesis. Notably, the ribosomes are 70S rather than the 80S ribosome found in the cytoplasm of eukaryotic cells. This observation, and the fact that mitochondria divide in a fashion similar to that of bacteria, were among the first pieces of evidence that led scientists to hypothesize that mitochondria evolved from bacterial cells. This evidence was bolstered by DNA sequencing data, supporting what is now known as the **endosymbiotic theory** (see **Perspective 3.1**).

PERSPECTIVE 3.1

The Origins of Mitochondria and Chloroplasts

Mitochondria and chloroplasts bear such a striking similarity to prokaryotic cells that it is no wonder scientists speculated for many decades that these organelles evolved from bacteria. The **endosymbiont theory** states that the ancestors of mitochondria and chloroplasts were bacteria residing within other cells in a mutually beneficial partnership. The intracellular bacterium in such a partnership is called an **endosymbiont.** As time went on each partner became indispensable to the other, and the endosymbiont eventually lost key features such as a cell wall and the ability to replicate independently.

Several early observations have supported the endosymbiont theory. Mitochondria and chloroplasts, unlike other eukaryotic organelles, both carry some of the genetic information necessary for their function. These include genes for some of the ribosomal proteins and ribosomal RNAs that make up their 70S ribosomes. These ribosomes contrast with the typical 80S ribosomes that characterize eukaryotic cells and, in fact, are equivalent to the prokaryotic 70S ribosomes. Interestingly, nuclear DNA encodes some of the components that make up these ribosomes. Another characteristic that supports the theory that mitochondria and chloroplasts were once intracellular bacteria is the double membrane that surrounds these organelles. Present-day endosymbionts retain their cytoplasmic membranes and live within membrane-bound compartments in their eukaryotic host cell. In addition, mitochondria and chloroplasts multiply by elongating and then dividing (binary fission), just as bacteria do.

Evidence in favor of the endosymbiont theory continues to accumulate. Recent technology allows scientists to easily determine the nucleotide sequence of DNA molecules, making it possible to compare organelle DNA with genomes of different bacteria. This led to the discovery that some mitochondrial DNA sequences bear a striking resemblance to DNA sequences of members of a group of obligate intracellular parasites, the rickettsias. These are probably relatives of modern-day mitochondria.

A tremendous effort is now underway to determine the nucleotide sequence of mitochondria from a wide variety of eukaryotes, including plants, animals, and protists. While the size of mitochondrial DNA varies a great deal among these different eukaryotic organisms, common sequence themes are emerging. Today, researchers are no longer discussing "if" but "when" these organelles evolved from intracellular prokaryotes.

Chloroplasts

Chloroplasts, found exclusively in plants and algae, are the site of photosynthesis in eukaryotic cells. They harvest the energy of sunlight to generate ATP, which is then used to convert CO_2 to sugar and starch. Like mitochondria, chloroplasts are bound by two membranes (**figure 3.53**). Within the chloroplast's stroma, the substance analogous to the mitochondrial matrix, are membrane-bound, disclike structures called thylakoids. Chlorophyll and other pigments that capture radiant energy are embedded in the thylakoid membranes.

As with mitochondria, substantial evidence indicates that chloroplasts evolved from bacterial cells (see Perspective 3.1). Chloroplasts contain DNA and 70S ribosomes, elongate and divide, and have photosynthetic mechanisms and DNA sequences similar to a group of bacteria called cyanobacteria.

Endoplasmic Reticulum (ER)

The **endoplasmic reticulum (ER)** is a complex system of flattened sheets, sacs, and tubes (**figure 3.54**). The **rough endoplasmic reticulum** has a characteristic bumpy appearance due to the

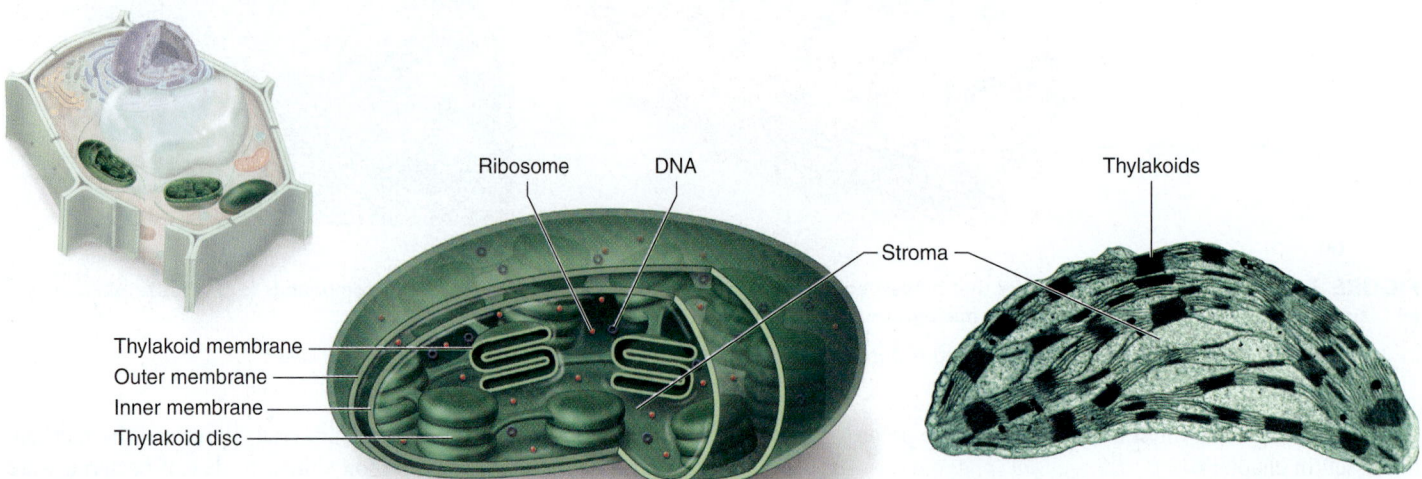

Ribosome   DNA   Thylakoids

Stroma

Thylakoid membrane
Outer membrane
Inner membrane
Thylakoid disc

FIGURE 3.53 Chloroplasts These harvest the energy of sunlight to generate ATP, which is then used to convert CO_2 to an organic form.

? *Chloroplasts evolved from which group of bacteria?*

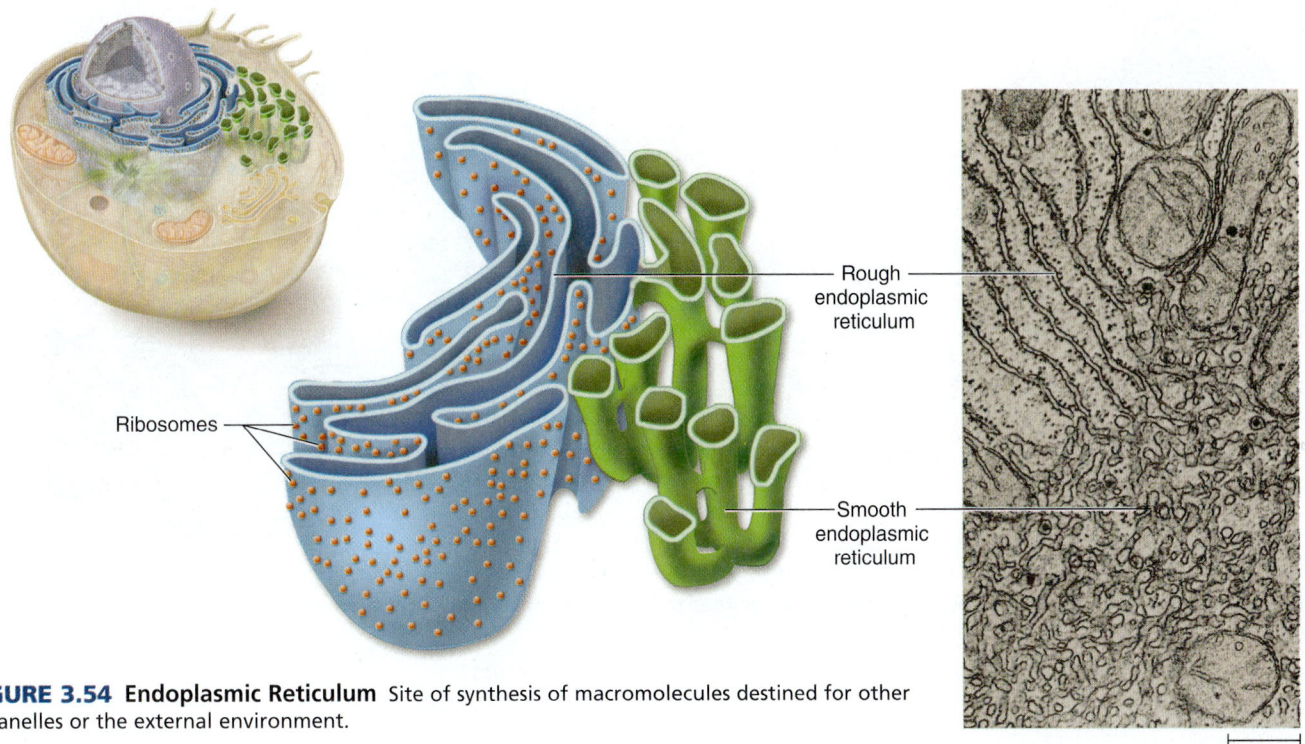

FIGURE 3.54 Endoplasmic Reticulum Site of synthesis of macromolecules destined for other organelles or the external environment.

❓ *What causes the bumpy appearance of the rough endoplasmic reticulum?*

0.08 μm

many ribosomes adhering to the surface. It is the site where proteins not destined for the cytoplasm are synthesized. These include proteins targeted for secretion or transfer to an organelle's lumen. Membrane proteins such as receptors are also synthesized on the rough ER. When ribosomes begin making these proteins, they attach to the ER surface so that the polypeptides they are synthesizing can be threaded through gated pores in the membrane. This delivers the molecules to the ER's lumen, where they then fold to assume their three-dimensional shapes. Vesicles that bud off from the ER then transfer the newly synthesized proteins to the Golgi apparatus for further modification and sorting. ◀◀ polypeptide, p. 27

Some regions of the ER are smooth. This **smooth endoplasmic reticulum** functions in lipid synthesis and degradation, and calcium ion storage. As with material made in the rough ER, vesicles transfer compounds from the smooth ER to the Golgi apparatus.

The Golgi Apparatus

The **Golgi apparatus** consists of a series of membrane-bound flattened compartments (**figure 3.55**). It is the site where macromolecules synthesized in the endoplasmic reticulum are modified before transport to other destinations. These modifications, such as the addition of carbohydrate and phosphate groups, take place in a sequential order in different compartments of the Golgi. Much like an assembly line, the molecules are transferred in vesicles from one compartment to another. The molecules are then sorted and delivered in vesicles to specific cellular compartments or to the outside of the cell.

Lysosomes and Peroxisomes

Lysosomes are organelles that contain a number of powerful degradative enzymes that could destroy the cell if not contained within the organelle. Endosomes and phagosomes fuse with lysosomes, so that material taken up by the cell can be degraded. In a similar manner, old organelles or vesicles formed around unneeded cellular components can fuse with lysosomes. This process, autophagy, digests the cell's own contents.

Peroxisomes are the organelles in which O_2 is used to help break down lipids and detoxify certain chemicals. The organelle's enzymes generate highly reactive molecules such as hydrogen peroxide and superoxide. The peroxisome contains these molecules and ultimately degrades them, protecting the cell from their toxic effects. ▶▶ hydrogen peroxide, p. 90 ▶▶ superoxide, p. 90

MicroAssessment 3.13

The nucleus, which contains the genetic information, is a distinguishing feature of a eukaryotic cell. Mitochondria are ATP-generating powerhouses. Chloroplasts are the site of photosynthesis. The rough endoplasmic reticulum is the site where proteins not destined for the cytoplasm are synthesized. The smooth ER functions in lipid synthesis and degradation, and calcium ion storage. The Golgi apparatus modifies and sorts molecules synthesized in the rough ER. Lysosomes are the organelles within which digestion takes place. In peroxisomes, O_2 is used to break down substances.

34. *Describe the structure of the nucleus.*

35. *How does the function of the rough endoplasmic reticulum differ from that of the smooth endoplasmic reticulum?*

36. *Why is it logical that bacterial cells do not contain chloroplasts?* ➕

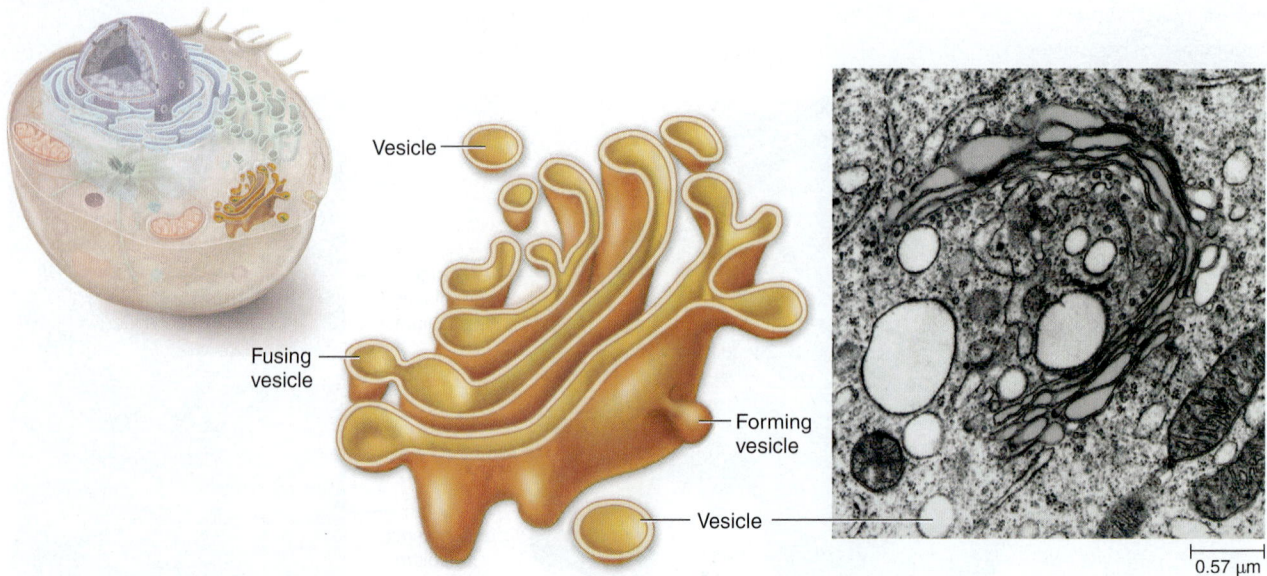

Vesicle

Fusing vesicle

Forming vesicle

Vesicle

0.57 µm

FIGURE 3.55 Golgi Apparatus Site where macromolecules synthesized in the endoplasmic reticulum are modified before being transported to other destinations.

How are the modified macromolecules transported from the Golgi apparatus to other sites?

FUTURE CHALLENGES 3.1

A Case of Breaking and Entering

Unraveling the complex mechanisms that prokaryotic and eukaryotic cells use to transport materials across their membranes can potentially aid in the development of new antimicrobial medications. Armed with a precise model of the structure and function of bacterial transporter proteins, scientists might be able to design new drugs that exploit these systems. One strategy would be to design compounds that irreversibly bind to transporter molecules and jam the mechanism. If the microbes can

be prevented from bringing in nutrients and removing wastes, their growth would cease. Another strategy would be to enhance the uptake or decrease the efflux of a specific compound that interferes with intracellular processes. This is already being done to some extent as new derivatives of current antibiotics are being produced, but more precise understanding of the processes by which bacteria take up or remove compounds could speed drug development.

A more thorough understanding of eukaryotic uptake systems could be used to develop better antiviral drugs. Viruses exploit the process of receptor-mediated endocytosis to gain entry into the cell. Once they are enclosed within the endosome, their protective protein coat is removed, releasing their genetic material. New drugs can potentially be developed that block these steps, preventing the entry or uncoating of infectious viral particles.

Summary

MICROSCOPY AND CELL MORPHOLOGY

3.1 ■ Microscopic Techniques: The Instruments (table 3.1)

Principles of Light Microscopy: The Bright-Field Microscope

The most commonly used type of microscope is the **bright-field microscope** (figure 3.1). The usefulness of a microscope depends largely on its **resolving power** (figure 3.2).

Light Microscopes That Increase Contrast

Cells viewed through a **dark-field microscope** stand out against a dark background (figure 3.4). **Phase-contrast microscopes** increase differences in refraction (figure 3.5). **Differential interference contrast**

(DIC) microscopes cause an image to appear three-dimensional (figure 3.6).

Other Light Microscopes

Fluorescence microscopes are used to observe cells stained with fluorescent dyes (figure 3.7). **Scanning laser microscopes (SLMs)** are used to obtain interior views of intact cells and three-dimensional images of thick structures (figure 3.8).

Electron Microscopes

Transmission electron microscopes (TEMs) transmit electrons through a specimen (figure 3.10). **Scanning electron microscopes (SEMs)** scan a beam of electrons over the surface of a specimen (figure 3.11).

Atomic Force Microscopes

Atomic force microscopes map surfaces on an atomic scale (figure 3.12).

3.2 ■ Microscopic Techniques: Dyes and Staining (table 3.2)

Differential Staining

The **Gram stain** is widely used: **Gram-positive bacteria** stain purple and **Gram-negative bacteria** stain pink (figure 3.14). The **acid-fast stain** is used to stain *Mycobacterium* species; **acid-fast** organisms stain pink and all other organisms stain blue (figure 3.15).

Special Stains to Observe Cell Structures

The **capsule stain** allows the capsule to stand out as a clear zone around a cell (figure 3.16). The **endospore stain** uses heat to facilitate the staining of endospores (figure 3.17). The **flagella stain** uses a substance that allows a stain to adhere to and coat the otherwise thin flagella (figure 3.18).

Fluorescent Dyes and Tags

Some fluorescent dyes bind compounds that characterize all cells; others bind compounds specific to certain cell types (figure 3.19). **Immunofluorescence** is used to tag proteins of interest with fluorescent compounds.

3.3 ■ Morphology of Prokaryotic Cells

Shapes

Most prokaryotes are **cocci** or **rods;** other common shapes are **vibrios, spirilla,** and **spirochetes** (figure 3.20).

Groupings

Cells adhering to one another following division form characteristic arrangements such as chains, packets, and clusters (figure 3.22).

Multicellular Associations

Myxobacteria move as a pack and form fruiting bodies. Most bacteria on surfaces live as biofilms.

THE PROKARYOTIC CELL (figure 3.23; table 3.3)

3.4 ■ The Cytoplasmic Membrane

Structure of the Cytoplasmic Membrane (figure 3.24)

The cytoplasmic membrane is a **phospholipid bilayer** embedded with a variety of different proteins.

Permeability of the Lipid Bilayer

The cytoplasmic membrane is **selectively permeable.** A few compounds pass through by **simple diffusion** (figure 3.26). Some of the membrane proteins function as selective gates.

The Role of the Cytoplasmic Membrane in Energy Transformation

The **electron transport chain** generates an electrochemical gradient, a source of energy called **proton motive force** (figure 3.27).

3.5 ■ Directed Movement of Molecules Across the Cytoplasmic Membrane

Transport Systems (figures 3.28, 3.29; table 3.4)

Facilitated diffusion moves compounds by exploiting a concentration gradient. **Active transport** moves compounds against a concentration gradient and requires energy. **Group translocation** chemically modifies a molecule during its transport.

Protein Secretion

A characteristic signal sequence targets proteins for secretion (figure 3.30).

3.6 ■ Cell Wall

Peptidoglycan (figure 3.31)

Peptidoglycan provides rigidity to the cell wall of bacteria. It is composed of glycan strands connected via tetrapeptide chains.

The Gram-Positive Cell Wall (figure 3.32)

The Gram-positive cell wall contains a relatively thick layer of peptidoglycan. **Teichoic acids** project out of the peptidoglycan. A gel-like substance is sandwiched between the cytoplasmic membrane and peptidoglycan layer.

The Gram-Negative Cell Wall (figure 3.33)

The Gram-negative cell wall has a relatively thin layer of peptidoglycan. An **outer membrane** contains **lipopolysaccharides.** LPS is called **endotoxin. Porins** permit small molecules to pass through the outer membrane. **Periplasm** fills the space between the inner and outer membranes.

Antibacterial Substances That Target Peptidoglycan

Penicillin prevents peptidoglycan synthesis. **Lysozyme** destroys the structural integrity of peptidoglycan.

Cell Wall Type and the Gram Stain

Gram-positive cells retain the crystal violet–iodine dye complex even when subjected to decolorization, but Gram-negative cells do not.

Bacteria That Lack a Cell Wall

Mycoplasma species are variable in shape and not affected by lysozyme or penicillin (figure 3.34).

Cell Walls of the Domain *Archaea*

Archaea have a greater variety of cell wall types than do the *Bacteria,* and they all lack peptidoglycan.

3.7 ■ Capsules and Slime Layers

Capsules and **slime layers** allow bacteria to adhere to surfaces. Some capsules allow pathogens to avoid host defense systems (figure 3.35).

3.8 ■ Filamentous Protein Appendages

Flagella (figure 3.36)

Flagella are used for motility (figure 3.37). **Chemotaxis** is movement toward an attractant or away from a repellent (figure 3.38). Cells use **phototaxis, aerotaxis, magnetotaxis,** and **thermotaxis** to move toward light, O_2, a magnetic field, and temperature, respectively.

Pili (figure 3.40)

Many types of **pili (fimbriae)** allow specific attachment of cells to surfaces. **Sex pili** are involved in a form of DNA transfer.

3.9 ■ Internal Structures

The Chromosome (figure 3.41)

The **chromosome** forms a region called the **nucleoid;** it contains the genetic information required by a cell.

Plasmids

Plasmids encode genetic information that may be helpful to a cell.

Ribosomes (figure 3.42)

Ribosomes are involved in protein synthesis. The 70S bacterial ribosome is composed of a 50S and a 30S subunit.

Cytoskeleton

The **cytoskeleton** is involved in cell division and regulation of shape.

Storage Granules (figure 3.43)

Storage granules are synthesized from nutrients a cell has in excess.

Gas Vesicles

Gas vesicles provide buoyancy to aquatic cells.

Endospores

Endospores are resistant to heat, desiccation, toxic chemicals, and UV light; they can **germinate** to become **vegetative cells** (figures 3.44, 3.45).

THE EUKARYOTIC CELL (figure 3.46; table 3.6)

3.10 ■ The Plasma Membrane

The **plasma membrane** is a phospholipid bilayer embedded with proteins that are involved in transport, structural integrity, and signaling.

3.11 ■ Transfer of Molecules Across the Plasma Membrane

Transport Proteins

Carriers function in facilitated diffusion and active transport. **Channels** form pores in the membrane. These channels are gated.

Endocytosis and Exocytosis (figure 3.48)

Pinocytosis is used to take up liquids. Receptor-mediated endocytosis is used by animal cells to take up material that binds to receptors. Protozoa and phagocytes take up material using **phagocytosis. Exocytosis** expels material.

Secretion

Proteins destined for a secretion are made by ribosomes attached to the **endoplasmic reticulum.**

3.12 ■ Protein Structures Within the Cell

Ribosomes

The **80S** ribosome is composed of **60S** and **40S subunits.**

Cytoskeleton (figure 3.49)

The cytoskeleton is composed of **actin filaments, microtubules,** and **intermediate filaments.**

Flagella and Cilia (figure 3.50)

Flagella propel a cell or pull the cell forward. **Cilia** move in synchrony to either propel a cell or move material along a stationary cell.

3.13 ■ Membrane-Bound Organelles

The Nucleus (figure 3.51)

The **nucleus** contains the cell's genetic information (DNA).

Mitochondria

Mitochondria use the energy released during the degradation of organic compounds to generate ATP (figure 3.52).

Chloroplasts

Chloroplasts capture the energy of sunlight, and then use it to synthesize ATP. This is used to convert CO_2 to an organic form (figure 3.53).

Endoplasmic Reticulum (ER) (figure 3.54)

Proteins not destined for the cytoplasm are synthesized in the **rough endoplasmic reticulum.** The **smooth endoplasmic reticulum** is where lipids are synthesized and degraded, and calcium is stored.

The Golgi Apparatus (figure 3.55)

The **Golgi apparatus** modifies and sorts molecules synthesized in the endoplasmic reticulum.

Lysosomes and Peroxisomes

Lysosomes carry digestive enzymes. **Peroxisomes** are the organelles in which O_2 is used to oxidize certain substances.

Review Questions

Short Answer

1. Explain why resolving power is important in microscopy.
2. Explain why basic dyes are used more frequently than acidic dyes in staining.
3. Describe what happens at each step in the Gram stain.
4. Compare and contrast ABC transport systems with group translocation.
5. Give two reasons why the outer membrane of Gram-negative bacteria is medically significant.
6. Compare and contrast penicillin and lysozyme.
7. Describe how a plasmid can help a cell.
8. How is an organ different from tissue?
9. How is receptor-mediated endocytosis different from phagocytosis?
10. Explain how the Golgi apparatus cooperatively functions with the endoplasmic reticulum.

Multiple Choice

1. Which of the following is most likely to be used in a typical microbiology laboratory?
   a) Bright-field microscope
   b) Confocal scanning microscope
   c) Phase-contrast microscope
   d) Scanning electron microscope
   e) Transmission electron microscope
2. When a medical technologist wants to determine if a clinical specimen contains a *Mycobacterium* species, which should be used?
   a) Acid-fast stain   b) Capsule stain   c) Endospore stain
   d) Gram stain   e) Simple stain
3. Penicillin
   1. is generally effective against Gram-positive bacteria.
   2. is generally effective against Gram-negative bacteria.
   3. functions in the cytoplasm of the cell.
   4. is effective against *Mycoplasma* species.
   5. kills only growing cells.
   a) 1, 2   b) 2, 3   c) 3, 4   d) 4, 5   e) 1, 5
4. Endotoxin is associated with
   a) Gram-positive bacteria.   b) Gram-negative bacteria.
   c) the cytoplasmic membrane.   d) the endospore.

5. The "O157" in the name *E. coli* O157:H7 refers to the type of O antigen. From this information you know that *E. coli*

   a) has a capsule.   b) is a rod.

   c) is a coccus.   d) is Gram-positive.

   e) is Gram-negative.

6. Eliminating which structure is *always* deadly to cells?

   a) Flagella   b) Capsule   c) Cell wall

   d) Cytoplasmic membrane   e) Fimbriae

7. Which of the following do bacterial cells use for attachment?

   1. Capsule   2. Pilus   3. Cytoplasmic membrane.

   4. Periplasm   5. Peptidoglycan

   a) 1, 2   b) 2, 3   c) 3, 4   d) 4, 5   e) 1, 5

8. Endocytosis is associated with

   a) mitochondria.   b) prokaryotic cells.   c) eukaryotic cells.

   d) chloroplasts.   e) ribosomes.

9. Protein synthesis is associated with

   1. lysosomes.   2. the cytoplasmic membrane.

   3. the Golgi apparatus.   4. rough endoplasmic reticulum.

   5. ribosomes.

   a) 1, 2   b) 2, 3   c) 3, 4   d) 4, 5   e) 1, 5

10. If a eukaryotic cell were treated with a chemical that destroys tubulin, all of the following would be directly affected *except*

    a) actin.   b) cilia.   c) eukaryotic flagella.

    d) microtubules.   e) More than one of these.

Applications

1. You are working in a laboratory producing new antibiotics for human and veterinary use. One compound with potential value inhibits the action of prokaryotic ribosomes. The compound, however, was shown to inhibit the growth of animal cells in culture. What is one possible explanation for its effect on animal cells?

2. A research laboratory is investigating environmental factors that inhibit the growth of archaea. They wonder if penicillin would be effective in controlling their growth. Explain the probable results of an experiment in which penicillin is added to a culture of archaea.

Critical Thinking

1. This graph shows facilitated diffusion of a compound across a cytoplasmic membrane and into a cell. As the external concentration of the compound is increased, the rate of uptake increases until it reaches a point where it slows and then begins to plateau. This is not the case with passive diffusion, where the rate of uptake continually increases. Why does the rate of uptake slow and then eventually plateau with facilitated diffusion?

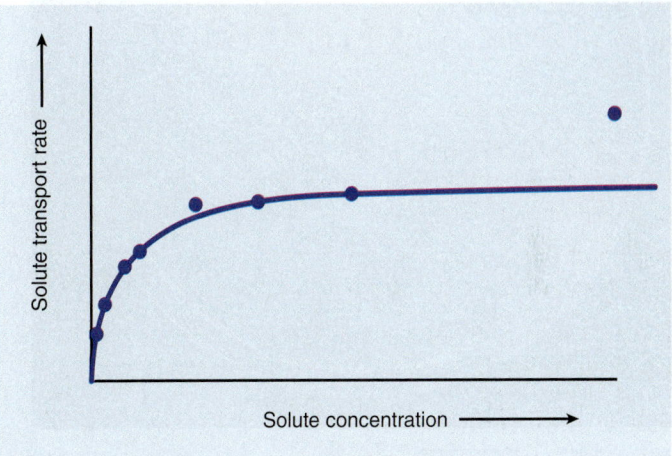

2. Most medically useful antibiotics interfere with either peptidoglycan synthesis or ribosome function. Why would the cytoplasmic membrane be a poor target for antibacterial medications?

4 Dynamics of Prokaryotic Growth

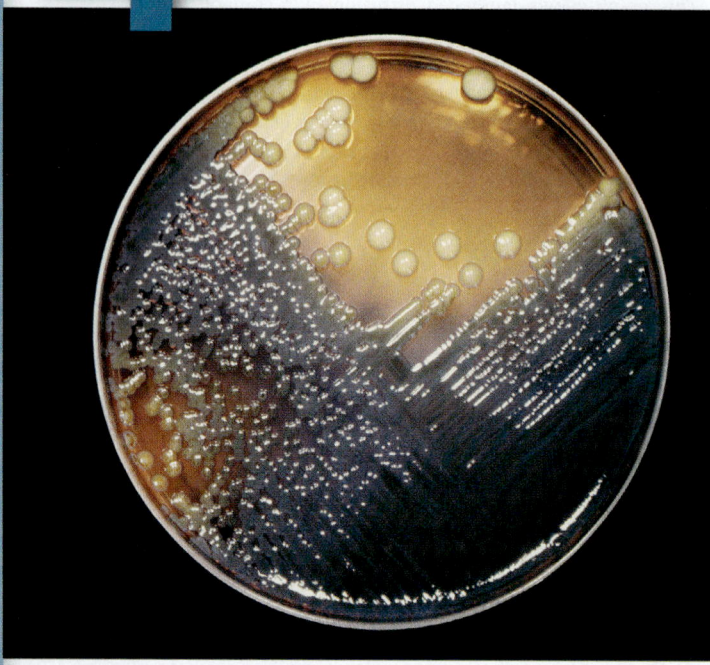
Bacterial colonies on an agar plate.

KEY TERMS

Biofilm Polymer-encased community of microorganisms.

Chemically Defined Medium A culture medium composed of exact quantities of pure chemicals; generally used for specific experiments when nutrients must be precisely controlled.

Complex Medium A culture medium that contains protein digests, extracts, or other ingredients that vary in their chemical composition.

Differential Medium A culture medium with an ingredient that certain microorganisms change in a recognizable way; used to differentiate microbes based on their metabolic traits.

Exponential (Log) Phase Stage in the growth curve during which cells divide at a constant rate; generation time is measured during this period of active multiplication.

Facultative Anaerobe Organism that grows best if O_2 is available, but can also grow without it.

Generation Time The time it takes for a population to double in number.

Obligate Aerobe Organism that requires molecular oxygen (O_2).

Obligate Anaerobe Organism that cannot multiply, and is often killed, in the presence of O_2.

Plate Count Method to measure the concentration of viable cells by determining the number of colonies that develop from a sample added to an agar plate.

Pure Culture A population descended from a single cell.

Selective Medium A culture medium with an ingredient that inhibits the growth of microbes other than the one being sought.

A Glimpse of History

The greatest contributor to methods of cultivating bacteria was Robert Koch (1843–1910), a German physician who combined a medical practice with a productive research career for which he received a Nobel Prize in 1905. Koch was primarily interested in identifying disease-causing bacteria, but to do this, he needed simple methods to isolate and grow these particular species. Koch recognized that a single bacterial cell could multiply on a solid medium in a limited area to form a distinct visible mass of descendants, so he experimented with growing bacteria on the cut surfaces of potatoes. Some species would not grow, however, because the potatoes did not contain enough nutrients, so Koch experimented with methods to solidify any liquid nutrient medium. He used gelatin initially, but there were two major drawbacks—it melts at the temperature preferred by many medically important microbes and some bacteria can digest it. In 1882, Fannie Hess, the wife of an associate of Koch, suggested using agar. This solidifying agent was used to harden jelly at the time and proved to be the perfect answer.

Today, we take pure culture techniques for granted because of their relative ease and simplicity. Their development, however, had a major impact on microbiology. By 1900, the agents causing most of the major bacterial diseases of humans had been isolated and characterized.

Prokaryotes can be found growing even in the harshest climates and most severe conditions. Environments that no unprotected human could survive, such as the ocean depths, volcanic vents, and the polar regions, have thriving prokaryotic species. Indeed many scientists believe that if life exists on other planets, it may resemble these microbes. Each species, however, has a limited set of environmental conditions in which it can grow; even then, it will grow only if specific nutrients are available. Some prokaryotes can grow at temperatures above the boiling point of water but not at room temperature. Others can grow only within an animal host, and then only in specific areas of that host.

Because of the medical significance of some microbes, as well as the nutritional and industrial use of microbial by-products, scientists must be able to grow microorganisms in culture. This is why it is important to understand the basic principles involved in prokaryotic growth while recognizing that much information is yet to be discovered.

4.1 ■ Principles of Prokaryotic Growth

Learning Outcome

1. *Describe binary fission and how it relates to generation time and exponential growth.*

Prokaryotes generally multiply by the process of **binary fission** (**figure 4.1**). After a cell has increased in size and doubled its components, it divides. One cell divides into two, those two divide to become four, those four become eight, and so on. In other words, the increase in cell numbers is exponential. Because it is neither practical, nor particularly meaningful, to determine the relative size of the cells in a given population, microbial growth is defined as an increase in the number of cells in a population. The time it takes for a population to double in number is the **generation time.** This varies greatly from species to species, and is influenced by the conditions in which the cells are grown.

The exponential multiplication of bacteria has important health consequences. For instance, a mere 10 cells of a food-borne pathogen in a potato salad, sitting for 4 hours in the warm sun at a picnic, may multiply to more than 40,000 cells. A simple equation expresses the relationship between the number of cells in a population at a given time (N_t), the original number of cells in the population (N_0), and the number of divisions those cells have undergone during that time (n). If any two values are known, the third can be easily calculated from the equation:

$$N_t = N_0 \times 2^n$$

In this example, assume that 10 cells of the pathogen were initially in the potato salad and that the organism has a generation time of 20 minutes. The first step is to determine the number of cell divisions that will occur in the 4-hour time frame. Because the organism divides every 20 minutes (3 times per hour) we know that in 4 hours it will divide 12 times. Once we know the original number of cells and the number of divisions, we can solve for N_t:

$$10 \times 2^{12} = N_t = 40,960$$

After 4 hours, our potato salad will have 40,960 cells of the pathogen (**table 4.1**). Keep this in mind, and your potato salad in a cooler, the next time you go to a picnic!

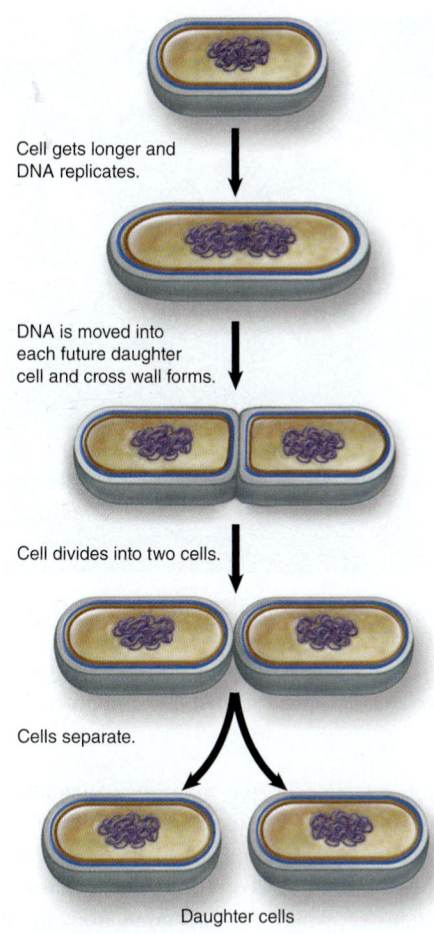

Cell gets longer and DNA replicates.

DNA is moved into each future daughter cell and cross wall forms.

Cell divides into two cells.

Cells separate.

Daughter cells

FIGURE 4.1 Binary Fission

How does the process of binary fission relate to the generation time?

TABLE 4.1	Example of Exponential Growth			
Time in Minutes (t)	**Initial Population (N_0)**	**Number of Generations (n)**	**2^n**	**Population $N_0 \times 2^n$**
0	10	0		10
20	10	1	$2^1 (= 2)$	20
40	10	2	$2^2 (= 2 \times 2)$	40
60 (1 hour)	10	3	$2^3 (= 2 \times 2 \times 2)$	80
80	10	4	$2^4 (= 2 \times 2 \times 2 \times 2)$	160
100	10	5	$2^5 (= 2 \times 2 \times 2 \times 2 \times 2)$	320
120 (2 hours)	10	6	2^6	640
140	10	7	2^7	1,280
160	10	8	2^8	2,560
180 (3 hours)	10	9	2^9	5,120
200	10	10	2^{10}	10,240
220	10	11	2^{11}	20,480
240 (4 hours)	10	12	2^{12}	40,960

MicroByte

Escherichia coli has a generation time of 20 minutes in ideal conditions; *Mycobacterium tuberculosis* needs at least 12 hours to double.

MicroAssessment 4.1

Most prokaryotes multiply by binary fission. The time required for a population to double in number is the generation time.

1. *Explain why microbial growth refers to populations rather than cell size.*
2. *If a bacterium has a generation time of 30 minutes, and you start with 100 cells at time 0, how many cells will you have in 30, 60, 90, and 120 minutes?* ➕

4.2 ■ Prokaryotic Growth in Nature

Learning Outcomes

2. *Describe a biofilm, and give one positive and one negative impact that biofilms have on humans.*
3. *Explain why microbes that grow naturally in mixed communities sometimes cannot be grown in pure culture.*

Historically, microorganisms have been studied by growing them in the laboratory. Scientists now recognize, however, that the dynamic and complex conditions of the natural environment, which differ greatly from the conditions in the laboratory, have profound effects on microbial growth and behavior. In fact, microbial cells are able to adjust to their surroundings by sensing various chemicals and then responding to that input by producing materials appropriate for the situation.

Biofilms

In nature, prokaryotes can live suspended in an aqueous environment, but most attach to surfaces and live in polymer-encased communities called **biofilms** (**figure 4.2**). Biofilms cause the slipperiness of rocks in a stream bed, the slimy "gunk" that coats kitchen drains, the scum that gradually accumulates in toilet bowls, and the dental plaque that forms on teeth. Biofilm formation begins when planktonic (free-floating) prokaryotic cells move to a surface and adhere, where they then multiply and release polysaccharides, DNA, and other hydrophilic polymers to which unrelated cells may attach and grow (**figure 4.3**). The meshlike accumulation of these polymers, referred to as extracellular polymeric substances (EPS), gives a biofilm its characteristic slimy appearance. |◀◀ ▶▶| dental plaque, pp. 63, 573

Surprisingly, biofilms are not random mixtures of microbes in a layer of EPS, but instead have characteristic formations with channels through which nutrients and wastes pass. Cells communicate with one another by synthesizing and responding to chemical signals—an exchange important in establishing structure.

Biofilms are more than just an unsightly annoyance. Dental plaque leads to tooth decay and gum disease. Even troublesome, persistent ear infections and the complications of cystic fibrosis

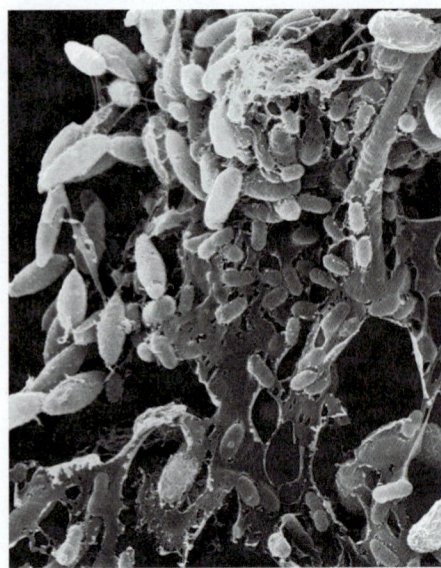

FIGURE 4.2 Biofilm on a Stainless Steel Surface A biofilm is a polymer-encased community of microorganisms.

❓ *Why would microbes in biofilms be more resistant to antibiotics and disinfectants than their planktonic counterparts?*

are due to biofilms. In fact, the majority of bacterial infections seem to involve biofilms. Treatment of these infections is difficult because microbes within the protective EPS often resist the effects of antibiotics as well as the body's defenses. Biofilms are also important in industry, where their accumulations in pipes, drains, and cooling water towers can interfere with processes as well as damage equipment. Again, the structure of the biofilm shields the microbes growing within it, so the bacteria in a biofilm may be hundreds of times more resistant to disinfectants than are their planktonic counterparts. ▶▶| disinfectants, p. 108

Although biofilms can be damaging, they also can be beneficial. Many bioremediation efforts, which use microbes to degrade harmful chemicals, are enhanced by biofilms. So as some industries are exploring ways to destroy biofilms, others, such as wastewater treatment facilities, are looking for ways to foster their development. ▶▶| bioremediation, p. 744 ▶▶| wastewater treatment, p. 736

Interactions of Mixed Microbial Communities

Prokaryotes in the environment regularly grow in close associations with many different species. Sometimes the interactions are cooperative, even fostering the growth of species that otherwise could not survive. For example, organisms that cannot multiply in the presence of O_2 will grow in the mouth if neighboring microbial cells consume that gas; one species creates a microenvironment in which the other can thrive. In a similar manner, the metabolic wastes of one organism can serve as nutrients for another. Often, however, microbes in a community compete for nutrients, and some even resort to a type of biological warfare, synthesizing toxic compounds that inhibit competitors. Understandably, the conditions in these close associations are exceedingly difficult to reproduce in the laboratory. ▶▶| microbial competition and antagonism, p. 720

| Planktonic bacteria move to the surface and adhere. | Bacteria multiply and produce extracellular polymeric substances (EPS). | Other bacteria may attach to the EPS and grow. | Cells communicate and create channels in the EPS that allow nutrients and waste products to pass. | Some cells detach and then move to other surfaces to create additional biofilms. |

FIGURE 4.3 Development of a Biofilm

? *What are extracellular polymeric substances (EPS)?*

MicroAssessment 4.2

Biofilms have a characteristic architecture with channels through which nutrients and wastes can pass. In nature, prokaryotes often grow in associations with many different species.

3. *Water bowls left out for pets sometimes develop a slimy layer if not washed regularly. What causes the slime?*

4. *Describe a situation in which the activities of one species benefit another.*

5. *Why would bacteria in a biofilm be more resistant to harmful chemicals?* ✚

4.3 ■ Obtaining a Pure Culture

Learning Outcome

4. *Describe how the streak-plate method is used to obtain a pure culture, and how the resulting culture can be stored.*

In the laboratory, prokaryotes are generally isolated and grown in pure culture. A **pure culture** is a population descended from a single cell and therefore separated from all other species. Working with pure cultures makes it easier to identify and study the

activities of a particular species. But although the results are much easier to interpret when studying pure cultures, the organisms sometimes behave differently than they do in their natural environment, as discussed earlier. Another complicating issue is that only an estimated 1% of all prokaryotes can currently be grown in culture successfully. This makes it exceedingly difficult to study the vast majority of microorganisms. Fortunately for humanity, most known medically significant bacteria can be grown in pure culture.

To work with a pure culture, all containers, media, and instruments must be sterile, or free of microbes, prior to use. These are then handled using **aseptic techniques,** procedures that minimize the chance of other organisms being accidentally introduced. The medium the cells are grown in, or on, is called a culture medium. It consists of nutrients dissolved in water, and can be a liquid broth or a solid gel. ◀◀ aseptic techniques, p. 85 ▶▶ sterilization, p. 108

Growing Microorganisms on a Solid Medium

The basic requirements for obtaining a pure culture are a solid culture medium, a container to hold and maintain the medium in an aseptic condition, and a method to separate individual microbial

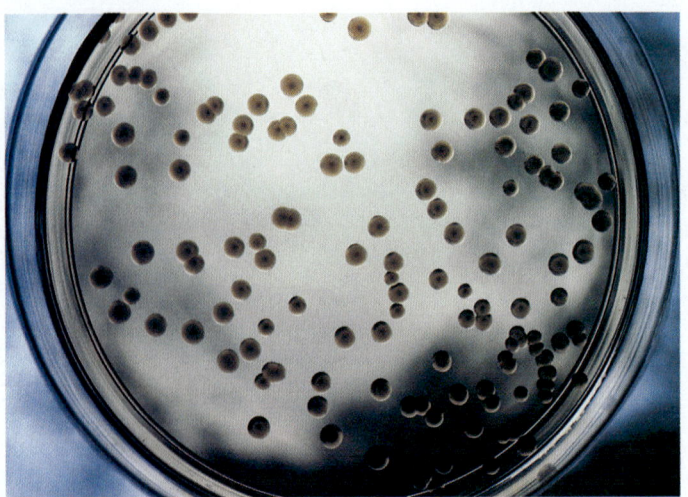

FIGURE 4.4 Colonies Growing on Agar Medium

What is the purpose of agar in the medium?

cells. A single prokaryotic cell, supplied with the right nutrients and conditions, will multiply on the solid medium in a limited area to form a **colony,** a distinct mass of cells (**figure 4.4**). About 1 million cells are required for a colony to be easily visible to the naked eye.

Agar, a polysaccharide extracted from marine algae, is used to solidify culture media. Unlike gelatin and other gelling agents,

very few microbes can degrade agar. It is not destroyed at high temperatures and can therefore be sterilized by heating, a process that also liquefies it. Melted agar will stay liquid until cooled to a temperature below 45°C. Therefore, nutrients that would be destroyed at high temperatures can be added at lower temperatures before the agar hardens. Once solidified, an agar medium will remain so until heated above 95°C. Thus, unlike gelatin—which is liquid at 37°C—agar remains solid over the entire temperature range at which most microbes grow. ◄◄ polysaccharide, p. 31

A solid culture medium is contained in a **Petri dish**—a two-part, covered container made of glass or plastic. While not airtight, the Petri dish does exclude airborne contaminants. A culture medium in a Petri dish is commonly referred to as a plate of that medium—for example, a nutrient agar plate or, more simply, an **agar plate.**

The Streak-Plate Method

The **streak-plate method** is the simplest and most commonly used technique for isolating prokaryotes (**figure 4.5**). A sterile inoculating loop is dipped into a microbe-containing sample and then lightly drawn several times across the surface of an agar plate, creating a set of parallel streaks covering approximately one-third of the agar. The loop is then sterilized and a new series of parallel streaks is made across and at an angle to the previous ones, covering another surface section. This drags some of those cells streaked onto the first portion over to a fresh section, effectively inoculating it with a diluted sample. The loop is sterilized again, and another set of

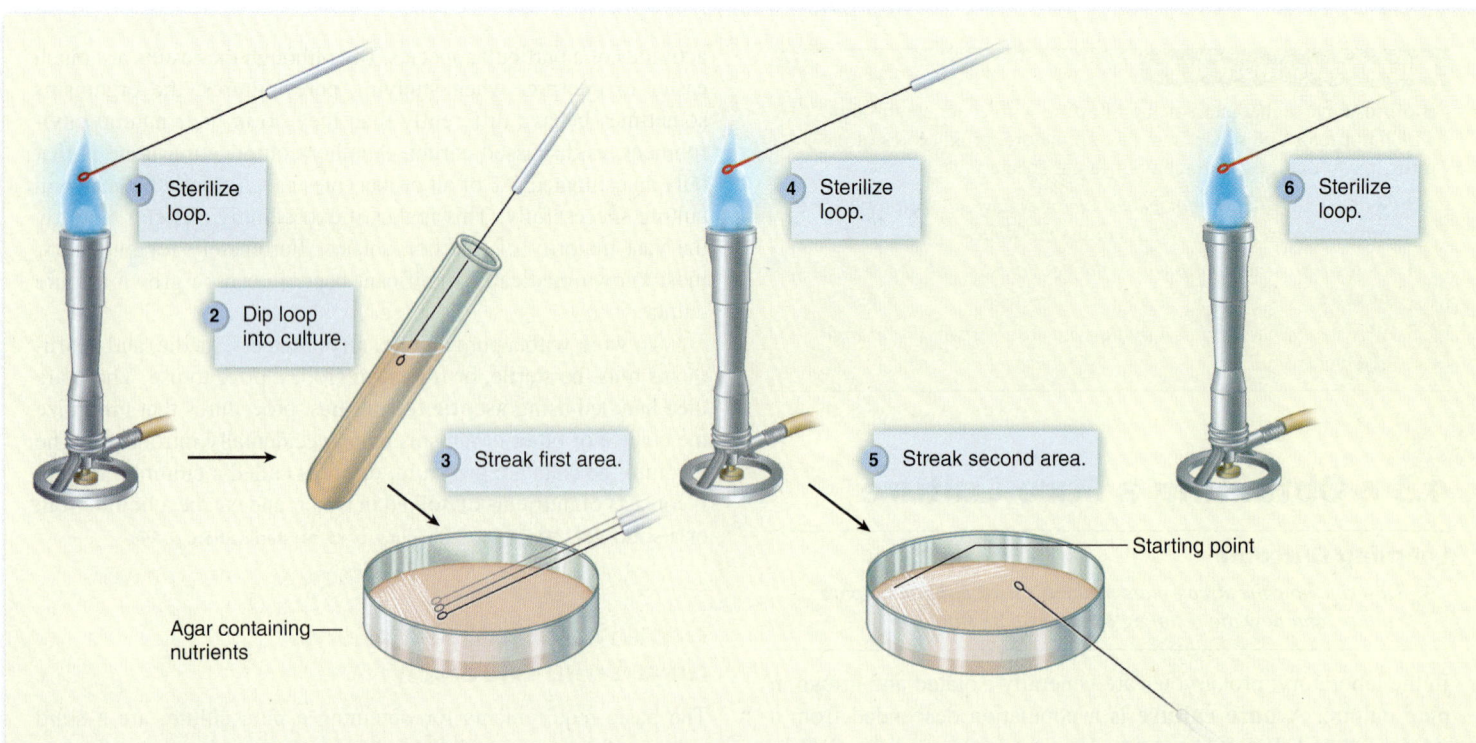

parallel streaks is made, dragging into a third area some of the organisms that had been moved into the second section. The goal is to reduce the number of cells being spread with each successive series of streaks. By the third set of streaks, cells should be separated enough so that distinct, well-isolated colonies will form.

Maintaining Stock Cultures

Once a pure culture has been obtained, it can be maintained as a stock culture, a culture stored for use as an inoculum in later procedures. Often, a stock culture is stored in the refrigerator as growth on the surface of an agar slant (a tube of agar that was held at an angle as it solidified). For long-term storage, stock cultures can be frozen at −70°C in a glycerol-containing solution that prevents ice crystals from forming and damaging cells. Alternatively, cells can be freeze-dried.

MicroAssessment 4.3

Only an estimated 1% of prokaryotes can be grown in culture. Agar is used to solidify nutrient-containing broth. The streak-plate method is used to obtain a pure culture. Stock cultures are stored in the refrigerator or frozen.

6. *What properties of agar make it ideal for use in culture media?*

7. *To identify the causative agent of a given illness, a pure culture is often needed. How is a streak plate used to obtain a pure culture?*

8. *What might be a reason that pathogens can be grown in pure culture more often than environmental organisms?*

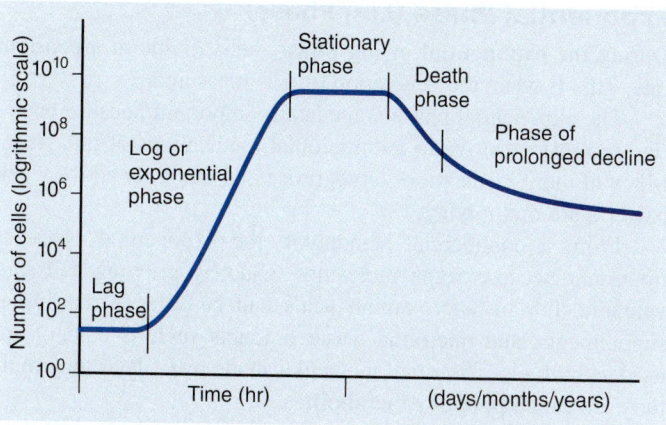

FIGURE 4.6 Growth Curve The growth curve is characterized by five distinct stages: lag phase, exponential or log phase, stationary phase, death phase, and phase of prolonged decline.

❓ *During which phase is generation time measured?*

4.4 ■ Prokaryotic Growth in Laboratory Conditions

Learning Outcome

5. *Describe the stages of a growth curve, and compare this closed system to colony growth and continuous culture.*

In the laboratory, bacteria and archaea are typically grown either on agar plates or in tubes or flasks of broth. These are considered closed systems, or **batch cultures,** because nutrients are not renewed, nor are wastes removed. As the cells grow in this type of system, the population increases in a distinct pattern of stages, and then declines. This characteristic pattern is called a growth curve.

To maintain cells in a state of continuous growth, nutrients must be continuously added and waste products removed. This is an **open system,** or **continuous culture.**

The Growth Curve

A **growth curve** is characterized by five distinct stages—lag phase, exponential or log phase, stationary phase, death phase, and phase of prolonged decline (**figure 4.6**).

Lag Phase

When a dilute culture is transferred into a different medium, the cell number does not immediately increase. During this **lag phase** cells begin synthesizing enzymes required for growth.

The length of the lag phase depends on the conditions in the original culture and the new medium. If cells are transferred into a medium that contains fewer nutrients, the lag phase will be longer because the cells must begin making amino acids or other components not supplied in the new medium. A similar situation occurs when a stock culture is inoculated into fresh medium.

FIGURE 4.5 The Streak-Plate Method The successive streaks dilute the cells. By the third set of streaks, cells should be separated enough so that isolated colonies develop after incubation.

❓ *What is the purpose of obtaining isolated colonies?*

⑦ Streak final area.

⑧ Isolated colonies develop after incubation.

Exponential Phase (Log Phase)

During the **exponential** or **log phase,** cells divide at a constant rate. This is when the generation time is measured.

The exponential phase is medically important because bacteria are most sensitive to antimicrobial medications at this stage. Many of these medications target processes primarily active when bacteria are multiplying.

From a commercial standpoint, the exponential phase is important because some molecules made by growing cells are valuable. For instance, amino acids can be sold as nutritional supplements, and microbial waste products such as ethanol are used as biofuels. The small molecules made by cells as they multiply are called **primary metabolites.**

In the later stages of exponential growth, nutrients gradually become depleted and waste products accumulate. Cells' activities shift as this occurs. If the cells are able to form endospores, they initiate the process of sporulation. If they cannot, they still alter their activities to prepare for starvation conditions. Compounds that begin accumulating at this stage are made for purposes other than growth and are called **secondary metabolites** (**figure 4.7**). Commercially, the most valuable of these are antibiotics.

Stationary Phase

Cells enter the **stationary phase** when the nutrient levels are too low to sustain growth. The total number of viable cells in the population remains relatively constant, but some cells are dying while others are multiplying. How can cells multiply when they have exhausted their supply of nutrients? Dead cells often burst, releasing nutrients that then fuel the growth of other cells.

During the stationary phase, the viable cells continue to synthesize secondary metabolites and maintain the altered properties they demonstrated in late log phase. The length of time cells remain in the stationary phase varies, depending on the species and environmental conditions. Some populations remain in the stationary phase for only a few hours, whereas others remain for days or longer.

Death Phase

The **death phase** is the period when the total number of viable cells in the population decreases as cells die off at a constant rate. Like cell growth, death is exponential, but the rate is usually much slower.

Phase of Prolonged Decline

In many cases, a fraction of the cell population survives the death phase. These cells have adapted to tolerate the worsened conditions and are able to multiply for at least a short time, using the nutrients released from the dead cells. As the conditions continue to deteriorate during this **phase of prolonged decline,** most of these survivors then die. However, the few progeny better equipped for survival can grow. This dynamic process generates successive waves of slightly modified populations, each more fit to survive than the previous ones. Thus, the statement "survival of the fittest" holds true even for closed cultures of microbes.

Colony Growth

Growth of a microbial colony on a solid medium involves many of the same features as growth in liquid, but it is marked by some important differences. After a lag phase, cells multiply exponentially and eventually compete with one another for available nutrients. Unlike the situation in a liquid culture, however, the position of a single cell within a colony determines its environment. Cells multiplying on the edge of the colony face relatively little competition for O_2 and nutrients. In the center of the colony, meanwhile, the high density of cells rapidly depletes available O_2 and nutrients. Wastes such as acids accumulate, and these can be toxic. As a consequence, cells at the edge of the colony may be growing exponentially, while those in the center may be in the death phase. Cells in locations between these two extremes may be in the stationary phase.

Continuous Culture

Microbial cells can be kept in a state of continuous growth by using a **chemostat.** This device continually drips fresh medium into a broth culture contained in a chamber. With each drop that enters, an equivalent volume—containing cells, wastes, and spent medium—leaves through an outlet. By manipulating the nutrient content of the medium and the speed at which it enters the chamber, a constant growth rate and cell density can be maintained. This makes it possible to study a uniform population's response to different nutrient concentrations or environmental conditions.

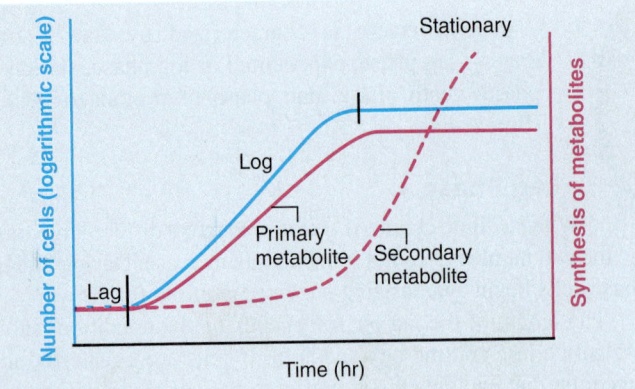

FIGURE 4.7 Primary and Secondary Metabolite Production Primary metabolites are synthesized during the period of active multiplication. Secondary metabolites begin to be synthesized in late log phase.

❓ *What is the most commercially valuable secondary metabolite?*

MicroAssessment 4.4

When grown in a closed system, a prokaryotic population goes through five distinct phases: lag, exponential or log, stationary, death, and prolonged decline. Cells within a colony may be in any one of the phases, depending on their relative location. Chemostats are used to study uniform populations of growing cells.

9. *Explain the difference between the lag and exponential phases.*

10. *Describe how a chemostat keeps cells in a state of continuous growth.*

11. *Why would a compound that prevents bacteria from growing interfere with the function of the antibiotic penicillin?* ➕

4.5 ■ Environmental Factors That Influence Microbial Growth

Learning Outcomes

6. *Describe the importance of a prokaryote's requirements for temperature, O_2, pH, and water availability, and define the terms that express these requirements.*

7. *Explain the significance of reactive oxygen species, and describe the mechanisms cells use to protect against their effects.*

As a group, prokaryotes inhabit nearly every environment on earth. Microbes we associate with disease and rapid food spoilage live in habitats that humans consider quite comfortable. Some, however, live in harsh environments that would kill most other organisms. Most of the examples in this latter group, called **extremophiles** (*phile* means "loving"), are in the domain *Archaea*.

Recognizing the major environmental factors that influence microbial growth—temperature, atmosphere, pH, and water availability—is essential for studying microbes and helps us understand their roles in the complex ecology of the planet. These factors and their associated characteristics are summarized in **table 4.2.**

Temperature Requirements

Each microbial species has a well-defined temperature range in which it grows. Within this range is the optimum growth temperature, the temperature at which the organism multiplies most rapidly. As a general rule, this optimum is close to the upper limit of the organism's temperature range.

Prokaryotes are commonly divided into five groups based on their optimum growth temperatures (**figure 4.8**). Note, however, that this merely represents a convenient organization scheme. In reality, no sharp dividing line exists between each group. Furthermore, not every organism in a group can grow in the entire temperature range typical for its group.

- **Psychrophiles** have an optimum between −5°(These organisms grow in the cold Arctic ar regions and in lakes fed by glaciers.

- **Psychrotrophs** have an optimum between 20°C and 30°C, but grow well at lower temperatures. They are an important cause of spoilage in refrigerated foods. ▸▸ food spoilage, p. 756

- **Mesophiles** have an optimum between 25°C and about 45°C. *E. coli* and most other common bacteria are in this group.

TABLE 4.2	Environmental Factors That Influence Microbial Growth
Environmental Factor/ Descriptive Terms	**Characteristics**
Temperature	Thermostability appears to be due to protein structure.
Psychrophile	Optimum temperature between −5°C and 15°C.
Psychrotroph	Optimum temperature between 20°C and 30°C, but grows well at refrigeration temperatures.
Mesophile	Optimum temperature between 25°C and 45°C.
Thermophile	Optimum temperature between 45°C and 70°C.
Hyperthermophile	Optimum temperature of 70°C or greater.
Oxygen (O_2) Availability	Oxygen (O_2) requirement/ tolerance reflects the organism's energy-harvesting mechanisms and its ability to inactivate reactive oxygen species.
Obligate aerobe	Requires O_2.
Facultative anaerobe	Grows best if O_2 is present, but can also grow without it.
Obligate anaerobe	Cannot grow in the presence of O_2.
Microaerophile	Requires small amounts of O_2, but higher concentrations are inhibitory.
Aerotolerant anaerobe (obligate fermenter)	Indifferent to O_2.
pH	Prokaryotes that live in pH extremes maintain a near-neutral internal pH by pumping protons out of or into the cell.
Neutrophile	Multiplies in the range of pH 5 to 8.
Acidophile	Grows optimally at a pH below 5.5.
Alkalophile	Grows optimally at a pH above 8.5.
Water Availability	Prokaryotes that can grow in high-solute solutions maintain the availability of water in the cell by increasing their internal solute concentration.
Halotolerant	Can grow in relatively high-salt solutions, up to approximately 10% NaCl.
Halophile	Requires high levels of sodium chloride.

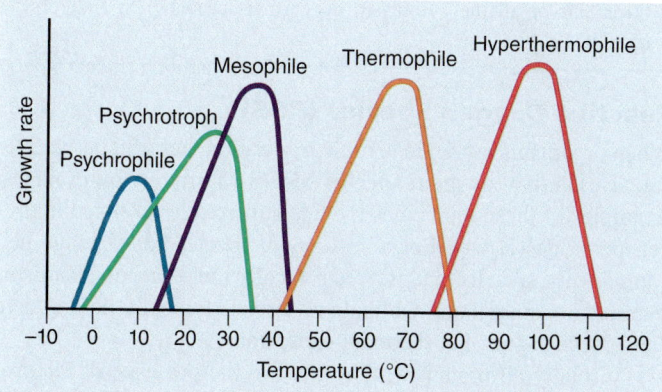

FIGURE 4.8 Temperature Requirements for Growth
Prokaryotes are commonly divided into five groups based on their optimum growth temperatures. This graph depicts a typical example of each group. Note that the optimum temperature, the point at which the growth rate is highest, is near the upper limit of the range.

❓ *Most pathogens fall into which group on this chart?*

Pathogens, adapted to growth in the human body, typically have an optimum between 35°C and 40°C. Mesophiles that inhabit soil, a colder environment, generally have a lower optimum, close to 30°C.

- **Thermophiles** have an optimum between 45°C and 70°C. These organisms commonly live in hot springs and compost heaps. ▶▶| composting, p. 743

- **Hyperthermophiles** have an optimum of 70°C or greater. These are usually members of the *Archaea*. One member, isolated from the wall of a hydrothermal vent deep in the ocean, has a maximum growth temperature of 121°C, the highest yet recorded!

Why can some prokaryotes withstand very high temperatures but most cannot? As a general rule, proteins from thermophiles are not denatured at high temperatures. This thermostability is due to the amino acid sequence of the protein. This controls the number and position of the bonds that form within the protein, which in turn determines its three-dimensional structure. Heat-stable enzymes that degrade fats and other proteins are used in high-temperature detergents. |◀◀ protein denaturation, p. 29

Temperature and Food Preservation

Refrigeration temperatures (approximately 4°C) slow spoilage because they limit the multiplication of otherwise fast-growing mesophiles. Psychrophiles and psychrotrophs can still grow, however, so refrigerated foods will still spoil, but it happens more slowly. ▶▶| low-temperature storage, p. 121 ▶▶| food spoilage, p. 756

Foods and other perishable products that withstand below-freezing temperatures can be frozen for long-term storage. It is important to recognize, however, that freezing is not an effective means of destroying microbes. Recall that freezing is routinely used to preserve stock cultures.

Temperature and Disease

The temperatures of different parts of the human body vary significantly. For example, the heart, brain, and gastrointestinal tract are near 37°C, but the temperature of the extremities is lower. For these reasons, some microbes cause disease more readily in certain body parts. Hansen's disease (leprosy) typically involves the coolest regions (ears, hands, feet, and fingers) because the causative bacterium, *Mycobacterium leprae,* grows best at these lower temperatures. The same situation applies to syphilis, in which lesions appear on the genitalia and then on the lips, tongue, and throat. Indeed, for more than 30 years the major treatment of syphilis was to induce fever by deliberately introducing the agent that causes malaria, which results in very high fevers. ▶▶| Hansen's disease, p. 650 ▶▶| syphilis, p. 626 ▶▶| fever, p. 351

Oxygen (O₂) Requirements

Like humans, some prokaryotes have an absolute requirement for O_2. Others thrive in environments that are **anaerobic,** meaning little or no O_2 is present. One way of determining an organism's O_2 requirement is to grow it in a shake tube. To do this, a tube of nutrient agar is boiled, which both melts the agar and drives off the O_2. The agar is then allowed to cool to just above its solidifying temperature. Next, the test organism is added and dispersed by gentle shaking or swirling. The agar medium is allowed to harden and the tube is incubated. Because the solidified agar slows gas

diffusion, the level of O_2 in the tube is high at the top, whereas the bottom portion is anaerobic. The cells grow in the region that has a suitable O_2 level (**table 4.3**).

Based on their O_2 requirements, prokaryotes can be separated into these groups:

- **Obligate aerobes** have an absolute requirement for oxygen (O_2). They use it in aerobic respiration, an energy-harvesting process. This and other ATP-generating pathways will be discussed in chapter 6. An example of an obligate aerobe is *Micrococcus luteus,* which is common in the environment. ▶▶| aerobic respiration, pp. 134, 144 |◀◀▶▶| ATP, pp. 24, 128

- **Facultative anaerobes** grow better if O_2 is present, but can also grow without it. The term "facultative" means that the organism is flexible, in this case in its requirements for O_2. Facultative anaerobes use aerobic respiration if O_2 is available, but resort to alternative types of metabolism if it is not. Growth is faster when O_2 is present because aerobic respiration yields the most ATP. *E. coli* is one of the most common facultative anaerobes in the large intestine. ▶▶| fermentation, pp. 134, 147 ▶▶| anaerobic respiration, pp. 134, 144

- **Obligate anaerobes** cannot multiply if O_2 is present; in fact, they are often killed by even brief exposure to air. Obligate anaerobes harvest energy using processes other than aerobic respiration; the details of these will be covered in chapter 6. Most inhabitants of the large intestine are obligate anaerobes, as is the bacterium that causes botulism—*Clostridium botulinum.*

- **Microaerophiles** require small amounts of O_2 (2% to 10%) for aerobic respiration; higher concentrations are inhibitory. An example is *Helicobacter pylori*, which causes gastric and duodenal ulcers.

- **Aerotolerant anaerobes** are indifferent to O_2. They can grow in its presence, but do not use it to harvest energy. They are also called obligate fermenters, because fermentation is their only metabolic option. An example is *Streptococcus pyogenes,* which causes strep throat.

MicroByte

Over half of all the cytoplasm on earth is probably in anaerobic microbes!

Reactive Oxygen Species (ROS)

When organisms use O_2 in aerobic respiration, harmful derivatives called **reactive oxygen species** (**ROS**) form as by-products. Examples of these molecules include superoxide (O_2^-) and hydrogen peroxide (H_2O_2). Reactive oxygen species can damage cell components, so cells that grow aerobically must have mechanisms to protect against them. Obligate anaerobes typically do not have these mechanisms, but there are exceptions.

Virtually all organisms that grow in the presence of O_2 produce the enzyme **superoxide dismutase,** which inactivates superoxide by converting it to O_2 and hydrogen peroxide. Nearly all these organisms produce the enzyme **catalase** as well, which converts hydrogen peroxide into O_2 and water. An important exception is the aerotolerant anaerobes. The fact that they do not produce catalase is useful in the laboratory. A simple test for the enzyme can be used to distinguish two groups of medically important

TABLE 4.3	Oxygen (O_2) Requirements of Prokaryotes				
	Obligate aerobe	**Facultative anaerobe**	**Obligate anaerobe**	**Microaerophile**	**Aerotolerant anaerobe**
Growth characteristics	Grows only when O_2 is available.	Grows best when O_2 is available, but also grows without it.	Cannot grow when O_2 is present.	Grows only if small amounts of O_2 are available.	Grows equally well with or without O_2.
Use of O_2 in energy-harvesting processes	Requires O_2 for respiration.	Uses O_2 for respiration, if available.	Does not use O_2.	Requires O_2 for respiration.	Does not use O_2.
Typical mechanisms to protect against reactive oxygen species	Produces superoxide dismutase and catalase.	Produces superoxide dismutase and catalase.	Does not produce superoxide dismutase or catalase.	Produces some superoxide dismutase and catalase.	Produces superoxide dismutase but not catalase.

Gram-positive cocci that grow aerobically—*Staphylococcus* species, which are catalase positive, and *Streptococcus* species, which are catalase negative. ▶▶ catalase test, p. 243

pH

Each prokaryotic species can survive within a range of pH values, and within this range is its pH optimum. Despite the pH of the external environment, however, cells maintain a constant internal pH, typically near neutral. Many that grow in acidic environments quickly pump out protons that enter the cell, whereas those that grow in alkaline conditions bring in protons. ◀◀ pH, p. 23

Most microbes are **neutrophiles**—they live and multiply within the range of pH 5 (acidic) to pH 8 (basic), and have a pH optimum near neutral (pH 7). Food preservation methods such as pickling inhibit bacterial growth by increasing the acidity of the food.

While most neutrophiles cannot withstand highly acidic conditions, one medically important bacterium has found a way. *Helicobacter pylori* grows in the stomach, sometimes causing ulcers. It decreases the acidity of its immediate surroundings by producing urease, an enzyme that splits urea into carbon dioxide and ammonia. The ammonia neutralizes any stomach acid surrounding the cell. ▶▶ ulcers, p. 580

Acidophiles grow optimally at a pH below 5.5. *Picrophilus oshimae*, a member of the *Archaea*, has an optimum pH of less than 1! This prokaryote, which was isolated from the dry, acid soils of a gas-emitting volcanic fissure in Japan, has an unusual cytoplasmic membrane that is unstable at a pH above 4.0.

Alkaliphiles grow optimally at a pH above 8.5. They often live in alkaline lakes and soils.

Water Availability

All microorganisms require water for growth. Even if water is present, however, it may not be available in certain environments. Dissolved substances such as salt (NaCl) and sugars, for example, interact with water molecules, making them unavailable to the cell. In many environments, particularly in certain natural habitats such as salt marshes, prokaryotes are faced with this situation. If the solute concentration is higher in the medium than in the cell, water diffuses out of the cell due to osmosis. This causes the cytoplasm to dehydrate and shrink from the cell wall, a phenomenon called **plasmolysis** (**figure 4.9**). ◀◀ solute p. 54 ◀◀ osmosis, p. 54

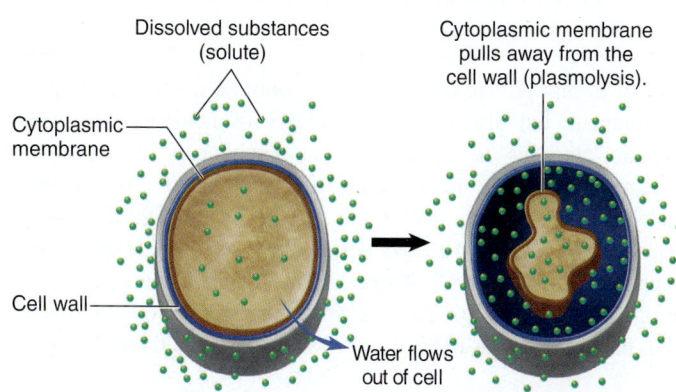

FIGURE 4.9 Effects of Solute Concentration on Cells
The cytoplasmic membrane allows water molecules to pass through freely. If the solute concentration is higher outside of the cell, water moves out.

? *What is plasmolysis?*

The growth-inhibiting effect of high salt and sugar concentrations is used in food preservation. High levels of salt are added to preserve such foods as bacon, salt pork, and anchovies. Many foods with a high sugar content, such as jams, jellies, and honey, are naturally preserved. ▶▶ food preservation, p. 759

Although many microbes are inhibited by high concentrations of salt, some withstand or even require it. Microbes that tolerate high concentrations of salt, up to approximately 10% NaCl, are **halotolerant** (*halo* means "salt"). *Staphylococcus* species, which reside on the dry salty environment of the skin, are an example. **Halophiles** require high levels of sodium chloride. Many marine bacteria are mildly halophilic, requiring concentrations of approximately 3% sodium chloride. Certain members of the *Archaea* are **extreme halophiles,** requiring 9% sodium chloride or more. Extreme halophiles are found in environments such as the salt flats of Utah and the Dead Sea.

TABLE 4.4	Representative Functions of the Major Elements
Chemical	**Function**
Carbon, oxygen, and hydrogen	Component of amino acids, lipids, nucleic acids, and sugars
Nitrogen	Component of amino acids and nucleic acids
Sulfur	Component of some amino acids
Phosphorus	Component of nucleic acids, membrane lipids, and ATP
Potassium, magnesium, and calcium	Required for the functioning of certain enzymes; additional functions as well
Iron	Part of certain enzymes

MicroAssessment 4.5

A prokaryotic species can be categorized according to its optimum growth temperature. A species can also be grouped according to its O_2 requirements. Most microbes grow best at a near-neutral pH, although some prefer acidic conditions, and others grow best in alkaline conditions. Halophiles require high-salt conditions.

12. *Clostridium paradoxum grows optimally at 55ºC, pH 9.3; it will not grow in the presence of O_2. How should this bacterium be categorized with respect to its temperature, pH, and O_2 requirements?*

13. *What is the function of the enzyme catalase?*

14. *Why would hydrogen peroxide be an effective disinfectant?* ➕

4.6 ■ Nutritional Factors That Influence Microbial Growth

Learning Outcomes

8. *List the required elements and examples of common sources.*

9. *Explain the significance of a limiting nutrient.*

10. *Explain why fastidious microbes require growth factors.*

11. *Describe the energy and carbon sources used by photoautotrophs, chemolithoautotrophs, photoheterotrophs, and chemoorganoheterotrophs.*

Growth of any prokaryote depends not only on a suitable physical environment, but also on the availability of nutrients. The cells use the nutrients to synthesize the components discussed in chapter 3—including lipid membranes, cell walls, proteins, and nucleic acids. These are made from subunits such as phospholipids, sugars, amino acids, and nucleotides. In turn, each of these subunits is composed of a variety of chemical elements, including carbon and nitrogen. What sets the prokaryotic world apart from all other forms of life is their remarkable ability to use diverse sources of these elements.

Required Elements

Chemical elements that make up cell constituents are called **major elements;** these include carbon, oxygen, hydrogen, nitrogen, sulfur, phosphorus, potassium, magnesium, calcium, and iron. They are the essential components of proteins, carbohydrates, lipids, and nucleic acids (**table 4.4**). ◀◀ elements, p. 17

The source of carbon distinguishes different groups of prokaryotes. **Heterotrophs** use organic carbon (*hetero* means "different" and *troph* means "nourishment"). Medically important bacteria are typically heterotrophs. **Autotrophs** (*auto* means "self") use inorganic carbon in the form of carbon dioxide (CO_2). They play a critical role in the cycling of carbon in the environment because they can convert inorganic carbon to an organic form, the process of **carbon fixation.** Without carbon fixation, the earth would quickly run out of organic carbon, which is essential to life. ▶▶ carbon cycle, p. 725

Nitrogen is needed to make amino acids and nucleic acids. Some prokaryotes use nitrogen gas (N_2) as a nitrogen source, converting it to ammonia and then incorporating that into cellular material. This process, **nitrogen fixation,** is unique to prokaryotes. Like carbon fixation, it is essential to life because once the nitrogen is incorporated into cellular material such as amino acids, other organisms can easily use it. Many microbes use ammonia as a nitrogen source. Some convert nitrate to ammonia, which is then incorporated into cellular material. All of these nitrogen uses are important steps in the nitrogen cycle ▶▶ nitrogen cycle, p. 726

Sulfur is a component of some amino acids. Many microbes use inorganic sulfur sources such as sulfate, but others require organic sources such as sulfur-containing amino acids.

Phosphorus is a component of nucleic acids, membrane lipids, and ATP. As with sulfur, many organisms can use inorganic phosphorus sources such as phosphate. Some, however, require organic sources, such as phosphorus-containing cell components.

Other elements, including potassium, magnesium, calcium, and iron, are required for some enzyme functions. A variety of inorganic and organic sources may be used by microbes.

Phosphorus and iron are important ecologically because they are often **limiting nutrients**—meaning they are available at the lowest concentration relative to need. To understand this concept, think about using a recipe to make chocolate chip cookies. Just as the quantity of chocolate chips available would determine the number of batches you could make (assuming the other ingredients are on hand), a limiting nutrient dictates the maximum level of microbial growth. This concept is important ecologically, as illustrated when algal blooms in a small Seattle lake resulted in a murky mess. To solve the problem, a chemical treatment was used to remove excess phosphate in the lake. By decreasing the level of phosphorus—the limiting nutrient—algal growth was restricted.

Trace elements are required in such small amounts that most natural environments, including water, have sufficient levels to support microbial growth. Trace elements include cobalt, zinc, copper, molybdenum, and manganese.

Growth Factors

Some microbes cannot synthesize certain organic molecules such as amino acids, vitamins, purines, or pyrimidines. Consequently, these organisms grow only if the molecule they cannot produce is available in the surrounding environment; the molecule is a **growth factor.** ◀◀ purines, p. 32 ◀◀ pyrimidines, p. 32

A microbe's growth factor requirements reflect its biosynthetic capabilities. Most *E. coli* strains, for example, can use glucose as the raw material to synthesize all of their cell components, so they do not need any growth factors. They multiply in a medium containing only glucose and six different inorganic salts. In contrast, *Neisseria* species are less resourceful metabolically, and require numerous growth factors including vitamins and amino acids. Bacteria such as *Neisseria* are **fastidious,** meaning they have complicated nutritional requirements.

Fastidious bacteria are used to measure the quantity of vitamins in food products. To do this, a well-characterized species that requires a specific vitamin is inoculated into a medium that lacks the vitamin but is supplemented with a measured amount of the food product. The extent of growth of the bacterium is related to quantity of the vitamin in the product.

Energy Sources

Organisms harvest energy either from sunlight or chemical compounds, using processes discussed in chapter 6. **Phototrophs** obtain energy from sunlight (*photo* means "light"). They include plants, algae, and photosynthetic bacteria. **Chemotrophs** extract energy from chemical compounds (*chemo* means "chemical"). Mammalian cells, fungi, and many types of prokaryotes use organic chemicals such as sugars, amino acids, and fatty acids as energy sources. Some prokaryotes extract energy from inorganic chemicals such as hydrogen sulfide and hydrogen gas, an ability that distinguishes them from eukaryotes.

Nutritional Diversity

Microbiologists often group prokaryotes according to the energy and carbon sources they utilize (**table 4.5**):

- **Photoautotrophs** use the energy of sunlight along with CO_2 in the atmosphere to make organic compounds. They are primary producers, meaning they support other forms of life by fixing carbon. Cyanobacteria are important primary producers that inhabit soil and aquatic environments. Many fix nitrogen as well, providing another indispensable role in the biosphere. ▶▶ primary producers, p. 719

- **Photoheterotrophs** use the energy of sunlight and derive their carbon from organic compounds. Some are facultative in their nutritional capabilities. For example, some members of a group called the purple nonsulfur bacteria grow anaerobically using light as an energy source and organic compounds as a carbon source (photoheterotrophs). They can also grow aerobically in the dark using organic sources of carbon and energy (chemoheterotrophs). ▶▶ purple nonsulfur bacteria, p. 260

- **Chemolithoautotrophs** (*lith* means "stone"), often referred to simply as chemoautotrophs or chemolithotrophs, use inorganic compounds for energy and derive their carbon from CO_2. These prokaryotes live in seemingly inhospitable places such as sulfur hot springs, which are rich in hydrogen sulfide, and other environments that have reduced inorganic compounds (see **Perspective 4.1**). In some regions of the ocean depths, near hydrothermal vents, chemoautotrophs serve as essential primary producers, supporting rich communities in these habitats that lack sunlight (see figure 29.11). ▶▶ hydrothermal vents, p. 729

- **Chemoorganoheterotrophs,** also referred to as chemoheterotrophs or chemoorganotrophs, use organic compounds for both energy and carbon. They are by far the most common group of microorganism associated with humans and other animals. Individual species of chemoheterotrophs differ in the number of organic compounds they can use. For example, certain members of the genus *Pseudomonas* can derive carbon and/or energy from more than 80 different organic compounds, including such unusual compounds as naphthalene (the ingredient associated with the smell of mothballs). At the other extreme, some organisms can degrade only a few compounds. *Bacillus fastidiosus* can use only urea and certain of its derivatives as a source of both carbon and energy.

TABLE 4.5	Energy and Carbon Sources Used by Different Groups of Prokaryotes	
Type	**Energy Source**	**Carbon Source**
Photoautotroph	Sunlight	CO_2
Photoheterotroph	Sunlight	Organic compounds
Chemolithoautotroph	Inorganic chemicals (H_2, NH_3, NO_2^-, Fe^{2+}, H_2S)	CO_2
Chemoorganoheterotroph	Organic compounds (sugars, amino acids, etc.)	Organic compounds

PERSPECTIVE   4.1

Can Prokaryotes Live on Only Rocks and Water?

Prokaryotes have been isolated from diverse environments that previously were thought to be incapable of sustaining life. For example, members of the *Archaea* have been isolated from environments 10 times more acidic than that of lemon juice. Other *Archaea* have been isolated from oil wells a mile below the surface of the earth at temperatures of 70°C and

pressures of 160 atmospheres (at sea level, the pressure is 1 atmosphere). The discovery of these microbes suggests that thermophiles may be widespread in the earth's crust.

Perhaps the most unusual environment from which prokaryotes have been isolated is volcanic rock 1 mile below the earth's surface near the Columbia River in Washington State.

What do these organisms use for food? They apparently get their energy from the H_2 produced in a reaction between groundwater and the iron-rich minerals in the rock. The groundwater also contains dissolved CO_2, which the organisms use as a source of carbon. These prokaryotes apparently exist on nothing more than rocks and water!

MicroAssessment 4.6

Organisms require a source of major and trace elements. Heterotrophs use an organic carbon source, and autotrophs use CO_2. Phototrophs harvest energy from sunlight, and chemotrophs extract energy from chemicals.

15. *To prevent excess phosphate from entering lakes and streams, certain laws govern the amount of phosphorus allowed in laundry and dishwasher detergents. What can happen if phosphorus levels in a lake increase?*

16. *How would your cells be categorized with respect to their carbon and energy sources?*

17. *Why would human-made materials (such as many plastics) be degraded only slowly or not at all?* ✚

4.7 ■ Cultivating Prokaryotes in the Laboratory

Learning Outcomes

12. *Compare and contrast complex, chemically defined, selective, and differential media.*

13. *Explain how aerobic, microaerophilic, and anaerobic conditions can be provided.*

14. *Describe the purpose of an enrichment culture.*

By knowing the environmental and nutritional factors that influence the growth of specific prokaryotes, it is often possible to provide appropriate conditions for their cultivation. These include a suitable growth medium and the proper atmosphere.

General Categories of Culture Media

Considering the diversity of prokaryotes, it should not be surprising that hundreds of different types of media are available. Even so, some medically important organisms and most environmental ones have not yet been grown on in the laboratory. **Table 4.6** summarizes the characteristics of representative examples of media.

Complex Media

A **complex medium** contains a variety of ingredients such as meat juices and digested proteins, forming what might be viewed as a tasty soup for microbes. This type of medium is easy to make and is used for routine purposes. Although a specific amount of each ingredient is in the medium, the exact chemical compositions of those substances can be highly variable. One common ingredient is peptone, a

mixture of amino acids and short peptides produced by digesting any of a variety of different proteins. Other substances, such as extracts (the water-soluble components of a substance such as lean beef), are often added to provide vitamins and minerals. A common complex medium—nutrient broth—consists of peptone and beef extract in distilled water. If agar is added, then nutrient agar results.

Many medically important bacteria are fastidious, requiring a medium even richer than nutrient agar. Because of this, clinical laboratories often use blood agar. As the name implies, this contains red blood cells, a source of a variety of nutrients including hemin; the medium contains other ingredients as well. A medium used for even more fastidious bacteria is chocolate agar, named for its brownish appearance rather than its ingredients. Chocolate agar contains lysed red blood cells and additional nutrients.

TABLE 4.6	Characteristics of Representative Media Used to Cultivate Bacteria
Medium	**Characteristic**
Blood agar	Complex medium used routinely in clinical labs. Differential because colonies of hemolytic organisms are surrounded by a zone of red blood cell clearing. Not selective.
Chocolate agar	Complex medium used to culture fastidious bacteria, particularly those found in clinical specimens. Not selective or differential.
Glucose-salts	Chemically defined medium. Used in laboratory experiments to study nutritional requirements of bacteria. Not selective or differential.
MacConkey agar	Complex medium used to isolate Gram-negative rods that typically reside in the intestine. Selective because bile salts and dyes inhibit Gram-positive organisms and Gram-negative cocci. Differential because the pH indicator turns pink-red when the sugar in the medium, lactose, is fermented.
Nutrient agar	Complex medium used for routine laboratory work. Supports the growth of a variety of nonfastidious bacteria. Not selective or differential.
Thayer-Martin	Complex medium used to isolate *Neisseria* species, which are fastidious. Selective because it contains antibiotics that inhibit most organisms except *Neisseria* species. Not differential.

TABLE 4.7	Ingredients in Two Types of Media That Support the Growth of *E. coli*
Nutrient Broth (complex medium)	**Glucose-Salts Broth (chemically defined medium)**
Peptone	Glucose
Meat extract	Dipotassium phosphate
Water	Monopotassium phosphate
	Magnesium sulfate
	Ammonium sulfate
	Calcium chloride
	Iron sulfate
	Water

Chemically Defined Media

Chemically defined media are composed of exact amounts of pure chemicals. This type of medium is generally used only for specific research experiments when the type and quantity of nutrients must be precisely controlled. A chemically defined medium called glucose-salts supports the growth of *E. coli,* and contains only those chemicals listed in **table 4.7.** The cells grow more slowly in this medium than in nutrient broth because they must synthesize all of their cell components from glucose. A more elaborate recipe containing as many as 46 different ingredients must be used to make a chemically defined medium that supports the growth of the fastidious bacterium *Neisseria gonorrhoeae* (the cause of gonorrhea).

To maintain the pH near neutral, buffers are often added to culture media. They are especially important in defined media because some bacteria produce so much acid as a by-product of metabolism that they inhibit their own growth. This is usually not a problem in complex media because the amino acids and other natural components provide at least some buffering action. ◀◀ buffer, p. 23

Special Types of Culture Media

To detect or isolate a species from a mixed population, it is often necessary to make that species more prevalent or obvious. For these purposes selective and differential media are used. They can be either complex or chemically defined, depending on the needs of the microbiologist.

Selective Media

Selective media inhibit the growth of certain species, making it easier to isolate the one being sought. For example, Thayer-Martin agar is used to isolate *Neisseria gonorrhoeae* from clinical specimens. Thayer-Martin is chocolate agar to which three or more antimicrobial drugs have been added. The antimicrobials inhibit fungi, Gram-positive bacteria, and Gram-negative rods, but not most *N. gonorrhoeae* strains. Because of this, the pathogen can grow on the medium with little competition.

MacConkey agar is used to isolate Gram-negative rods from various clinical specimens such as urine. In addition to containing peptones and other nutrients, this medium includes two inhibitory compounds—crystal violet (a dye that inhibits Gram-positive bacteria) and bile salts (a compound that inhibits most non-intestinal bacteria).

Differential Media

Differential media contain substances that certain microbes change in a recognizable way. Blood agar is differential because bacteria that produce a hemolysin—a protein that causes red blood cells to burst—are surrounded by a zone of clearing called hemolysis (**figure 4.10**). This readily observable characteristic is

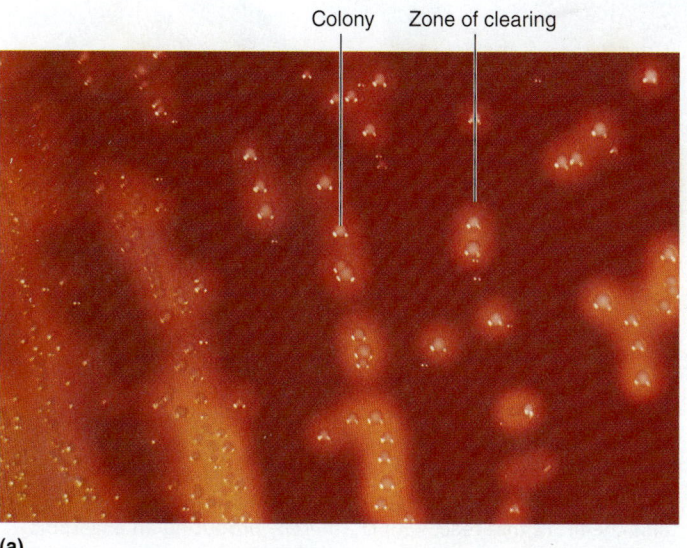
(a)

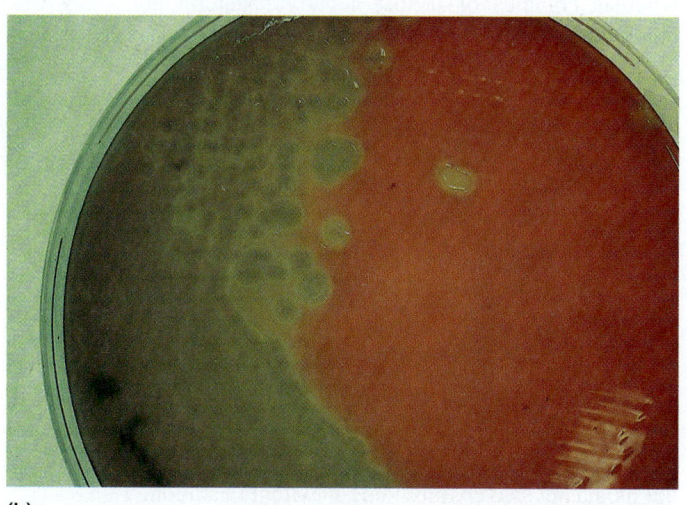
(b)

FIGURE 4.10 Blood Agar This complex medium is differential for hemolysis. **(a)** A zone of complete clearing around a colony growing on blood agar is called beta hemolysis. **(b)** A zone of greenish clearing is called alpha hemolysis.

❓ *Which type of hemolysis characterizes* Streptococcus pyogenes, *the bacterium that causes strep throat?*

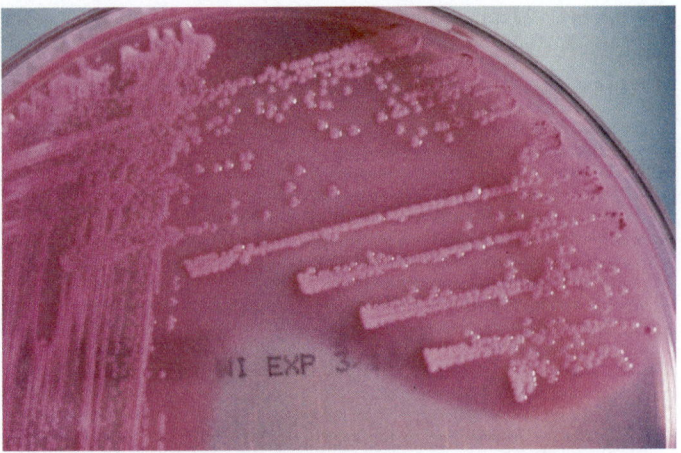

FIGURE 4.11 MacConkey Agar This complex medium is differential for lactose fermentation and selective for Gram-negative rods.

❓ *What specifically causes colonies of lactose fermenters to be dark pink on MacConkey agar?*

important medically because some pathogens can be distinguished by their type of hemolysis. For example, the colonies of *Streptococcus pyogenes* (the cause of strep throat) produce a clear zone of hemolysis called **beta hemolysis**. This makes them stand out from *Streptococcus* species that reside harmlessly in the throat. Many of these show **alpha hemolysis**, meaning their colonies are surrounded by a zone of greenish partial clearing; other streptococci have no effect on red blood cells. ▶▶| *Streptococcus pyogenes*, p. 487

MacConkey agar is differential as well as selective (**figure 4.11**). In addition to its ingredients already mentioned, it contains lactose and a pH indicator. Bacteria that ferment the sugar produce acid, which turns the pH indicator pink. Colonies of lactose-fermenting bacteria are pink on MacConkey agar, whereas colonies of lactose-negative bacteria are colorless. |◀◀ lactose, p. 31

Providing Appropriate Atmospheric Conditions

To grow microorganisms on culture media in the laboratory, appropriate atmospheric conditions must be provided.

Aerobic

When incubating most obligate aerobes and facultative anaerobes on agar media, special atmospheric conditions are not required; they can be incubated in air (approximately 20% O_2). Broth cultures of these organisms grow best when tubes or flasks containing the media are shaken, providing maximum aeration.

Many medically important bacteria, including species of *Neisseria* and *Haemophilus*, grow best in aerobic atmospheres that

have additional CO_2. Some are even capnophiles, meaning they require increased CO_2. One of the simplest ways to build up the level of CO_2 in the incubation atmosphere is to use a candle jar. For this system, the inoculated plates or tubes are added to the jar and then, just before the container is closed, a candle within it is lit. As the candle burns, it consumes some of the O_2 in the air, generating CO_2 and H_2O; the flame soon extinguishes because of insufficient O_2. However, about 17% O_2 remains, enough to support the growth of obligate aerobes and prevent the growth of obligate anaerobes. Special incubators are also available that maintain CO_2 at prescribed levels.

Microaerophilic

Microaerophilic prokaryotes typically require O_2 concentrations less than what is achieved in a candle jar. These microbes are often incubated in a gastight container with a special disposable packet; the packet holds chemicals that react with O_2, reducing its concentration to approximately 5–15%.

Anaerobic

Obligate anaerobes are challenging to grow because of their sensitivity to O_2-containing environments. Those that can tolerate brief exposures to O_2 are incubated in an **anaerobe container** (**figure 4.12**). This is the same type of container used to incubate microaerophiles, but the chemical composition of the disposable packet produces an anaerobic environment. An alternative method is to use a semisolid culture medium that contains a

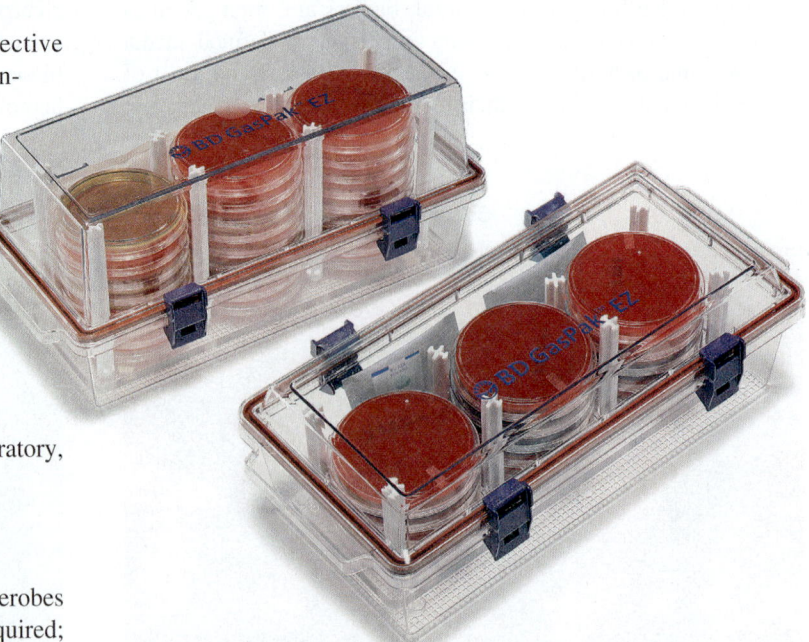

FIGURE 4.12 Anaerobe Containers A disposable packet contains chemicals that react with O_2, thereby producing an anaerobic environment.

❓ *Is the environment in a candle jar anaerobic?*

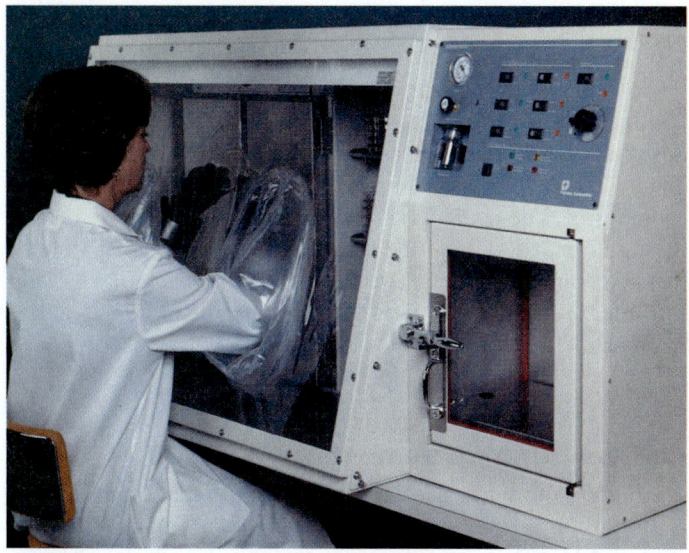

FIGURE 4.13 Anaerobic Chamber The enclosed compartment can be maintained as an anaerobic environment. A special port (visible on the right side of this device) that can be filled with inert gas is used to add or remove items. The airtight gloves allow the researcher to handle items within the chamber.

❓ *Why would the chamber be preferable to an anaerobe container?*

reducing agent such as sodium thioglycolate, which reduces O_2 to water. In many cases, an O_2-indicating dye is included as well. The medium can be heated immediately before use to remove dissolved O_2. ▶▶ **reducing agent, p. 131**

A more stringent method for working with anaerobes is to use an **anaerobic chamber,** an enclosed compartment maintained as an anaerobic environment (**figure 4.13**). A special port that can be

filled with an inert gas is used to add or remove items. Airtight gloves allow researchers to handle items within the chamber.

Enrichment Cultures

An **enrichment culture** is used to isolate an organism present as only a very small fraction of a mixed population. It does this by providing conditions that preferentially enhance the growth of that particular species in a broth (**figure 4.14**).

To enrich for a species, a sample is added to a broth medium designed to favor the growth of the desired organism over others. For example, if the target microbe can grow using atmospheric nitrogen, then that element is left out of the medium. If it can use an unusual carbon source such as phenol, then that compound is added as the only carbon source. A selective agent such as bile salts may also be added. The culture is then incubated in conditions that preferentially promote the growth of the desired organism. As the microbes multiply, the relative concentration of the target organism can increase dramatically. A pure culture can then be obtained by streaking the enrichment onto an appropriate agar medium and selecting a single colony.

MicroAssessment 4.7

Culture media can be complex or chemically defined. Some media contain additional ingredients that make them selective or differential. Appropriate atmospheric conditions must be provided to grow microbes in culture. An enrichment culture increases the relative concentration of an organism growing in a broth.

18. *Distinguish between complex and chemically defined media.*
19. *Describe two methods to create anaerobic conditions.*
20. *Would bacteria that cannot use lactose be able to grow on MacConkey agar?* ➕

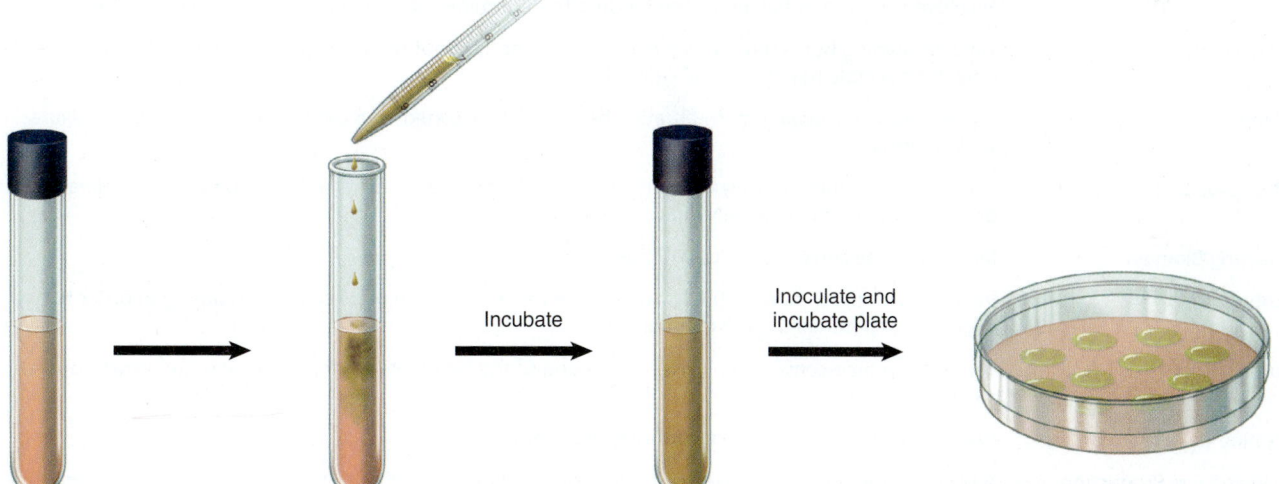

| Medium contains nutrients that few species other than the one of interest can use. | Sample that contains a variety of species, including the one of interest, is added to the medium. | Species of interest multiplies, whereas others cannot. | Enriched sample is plated onto appropriate agar medium. A pure culture is obtained by selecting a single colony of the species of interest. |

FIGURE 4.14 Enrichment Culture Medium and incubation conditions favor the growth of the desired species over other microorganisms in the same sample.

❓ *How would you enrich for an organism that can use phenol as a carbon source?*

4.8 ■ Methods to Detect and Measure Microbial Growth

Learning Outcome

15. *Compare and contrast methods used for direct cell counts, viable cell counts, measuring biomass, and detecting cell products.*

A variety of techniques can be used to monitor microbial growth. Characteristics of common methods are summarized in **table 4.8.**

Direct Cell Counts

Direct cell counts are particularly useful for determining the total numbers of microbes, including those that cannot be grown in culture. Unfortunately, they generally do not distinguish between living and dead cells in the specimen.

Direct Microscopic Count

One of the most rapid methods of determining the cell concentration in a suspension is the direct microscopic count (**figure 4.15**). A liquid specimen is added to special glass slide designed specifically for counting cells. The slide has a thin chamber that holds a known volume of liquid atop a microscopic grid. The contents of the chamber can be viewed under the light microscope, so the number of cells in a given volume can be counted precisely. At least 10 million bacteria (10^7) per milliliter are usually required for enough cells to be seen in the microscope field.

Cell-Counting Instruments

A **Coulter counter** is an electronic instrument that counts cells in a suspension as they pass single file through a narrow channel (**figure 4.16**). The suspending liquid must be an electrically conducting fluid, because the machine counts the brief changes in resistance that occur when non-conducting particles such as bacteria pass.

TABLE 4.8	Methods Used to Measure Microbial Growth
Method	**Characteristics and Limitations**
Direct Cell Counts	Used to determine total number of cells; counts include living and dead cells.
Direct microscopic count	Rapid, but at least 10^7 cells/ml must be present to be effectively counted.
Cell-counting instruments	Coulter counters and flow cytometers count total cells in dilute solutions. Flow cytometers can also be used to count organisms to which fluorescent dyes or tags have been attached.
Viable Cell Counts	Used to determine the number of viable microorganisms in a sample, but that number includes only those that can grow in given conditions. Requires an incubation period of approximately 24 hours or longer. Selective and differential media can be used to determine the number of specific microbial species.
Plate count	Time-consuming but technically simple method that does not require sophisticated equipment. Generally used only if the sample has at least 10^2 cells/ml.
Membrane filtration	Concentrates microorganisms by filtration before they are plated; thus can be used to count cells in dilute environments.
Most probable number	Statistical estimation of likely cell number; it is not a precise measurement. Can be used to estimate numbers of microorganisms in relatively dilute solutions.
Measuring Biomass	Biomass can be correlated to cell number.
Turbidity	Very rapid method; used routinely. A one-time correlation with plate counts is required in order to use turbidity for determining cell number.
Total weight	Tedious and time-consuming; however, it is one of the best methods for measuring the growth of filamentous microorganisms.
Detecting Cell Products	Used to detect growth, but not routinely used for quantitation.
Acid and Gas Production	A pH indicator can be used to detect acid production. Gases can be trapped in an inverted Durham tube in a tube of broth; a more sensitive method uses a fluorescent sensor that detects the slight decrease in pH that accompanies CO_2 production.
ATP	Firefly luciferase catalyzes a light-emitting reaction when ATP is present.

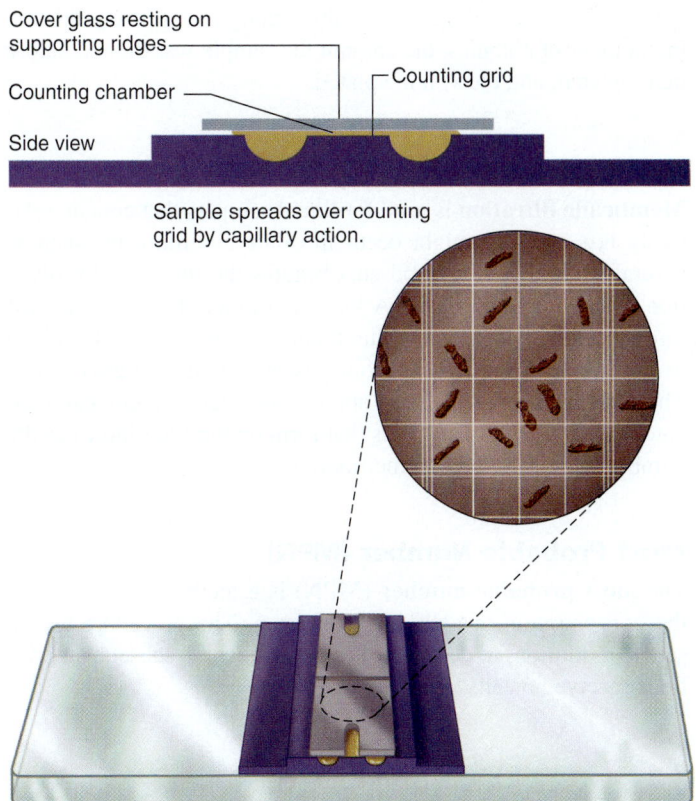

Cover glass resting on supporting ridges

Counting chamber

Counting grid

Side view

Sample spreads over counting grid by capillary action.

Using a microscope, the cells in several large squares like the one shown are counted and the results averaged. To determine the number of cells per ml, that number must be multiplied by 1/volume (in ml) held in the square. For example, if the square holds 1/1,250,000 ml, then the number of cells must be multiplied by 1.25×10^6 ml.

FIGURE 4.15 Direct Microscopic Count The counting chamber holds a known volume of liquid. The number of microbial cells in that volume can be counted precisely.

❓ *What is one disadvantage of doing a direct microscopic count to determine the number of cells in a suspension?*

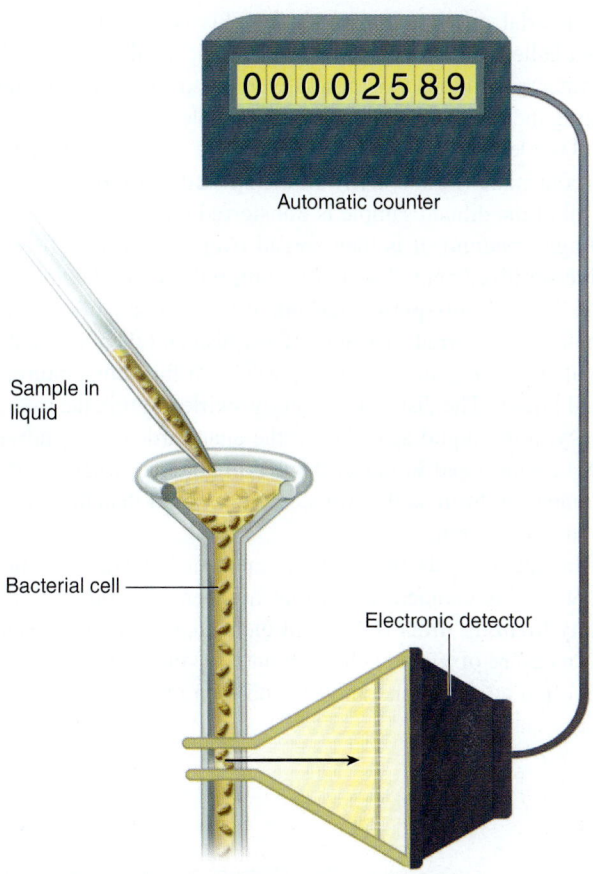

Automatic counter

Sample in liquid

Bacterial cell

Electronic detector

FIGURE 4.16 A Coulter Counter This instrument counts cells as they pass through a narrow channel.

❓ *Why do the microbial cells need to be suspended in an electrically conducting fluid for this method to work?*

A **flow cytometer** is similar in principle to a Coulter counter except it measures light scattered by cells as they pass a laser. The instrument can be used to count either total cells or a specific subset that has been stained with a fluorescent dye or tag. ◀◀ **fluorescent dyes and tags, p. 49**

Viable Cell Counts

Viable cell counts determine the number of cells capable of multiplying. They are particularly valuable when working with samples such as food and water that contain too few microbes for a direct microscopic count. In addition, by using appropriate selective and differential media, these methods can be used to count the cells of a particular microbial species.

Plate Counts

Plate counts measure the number of viable cells in a sample by taking advantage of the fact that an isolated microbial cell on a nutrient agar plate will give rise to one colony. A simple count of the colonies determines how many cells were in the initial sample. Plate counts are generally only done if a sample contains more than 100 organisms/ml. Otherwise, few if any cells will be transferred to the plates. In these situations, alternative methods give more reliable results.

When counting colonies, the ideal number on a plate is between 30 and 300. Numbers outside of that range are more likely to be inaccurate. Samples usually contain many more cells than that, so they generally must be diluted by a stepwise process called serial dilution (**figure 4.17**). This is done using a sterile liquid called the diluent, often physiological saline (0.85% NaCl in water). Dilutions are normally done in 10-fold increments, making the resulting math relatively simple.

Two techniques can be used to plate samples—spread-plate and pour-plate (**figure 4.18**). In the **spread-plate method,** 0.1 to 0.2 ml of the diluted sample is transferred onto a plate of a solidified agar medium. It is then spread over the surface of the agar with a sterilized bent glass rod that resembles a miniature hockey stick. In the **pour-plate method,** 0.1 to 1.0 ml of the diluted sample is transferred to a sterile Petri dish and then overlaid with a melted agar medium cooled to 50°C . At this temperature, agar is still liquid. The dish is then gently swirled to mix the microbial cells with the liquid agar. When the agar hardens, the individual cells become fixed in place; they form colonies when incubated. Colonies that form on the surface will be larger than those embedded in the medium.

In both methods, the plates are incubated and then the number of colonies is counted. From that number, the concentration of **colony-forming units** (**CFUs**) in the sample can be determined. This measure of viable cells accounts for the fact that microbial cells often attach to one another and then grow to form a single colony. When calculating CFUs, three things must be considered: the number of colonies, the amount the sample was diluted before being plated, and the volume plated.

Membrane Filtration

Membrane filtration is used for liquid samples that contain relatively few cells, as might occur in dilute environments such as natural waters. This method concentrates the microbes by filtration before they are plated. A known volume of liquid is passed through a sterile membrane filter that has a pore size small enough to prevent microorganisms from passing through (**figure 4.19**). The filter is then placed on an appropriate agar medium and incubated. The number of colonies that form on the filter indicates the number of cells in the volume filtered.

Most Probable Number (MPN)

The **most probable number** (**MPN**) is a method for estimating the concentration of cells in a specimen. The procedure uses a series of dilutions to determine the point at which subsequent dilutions receive no cells.

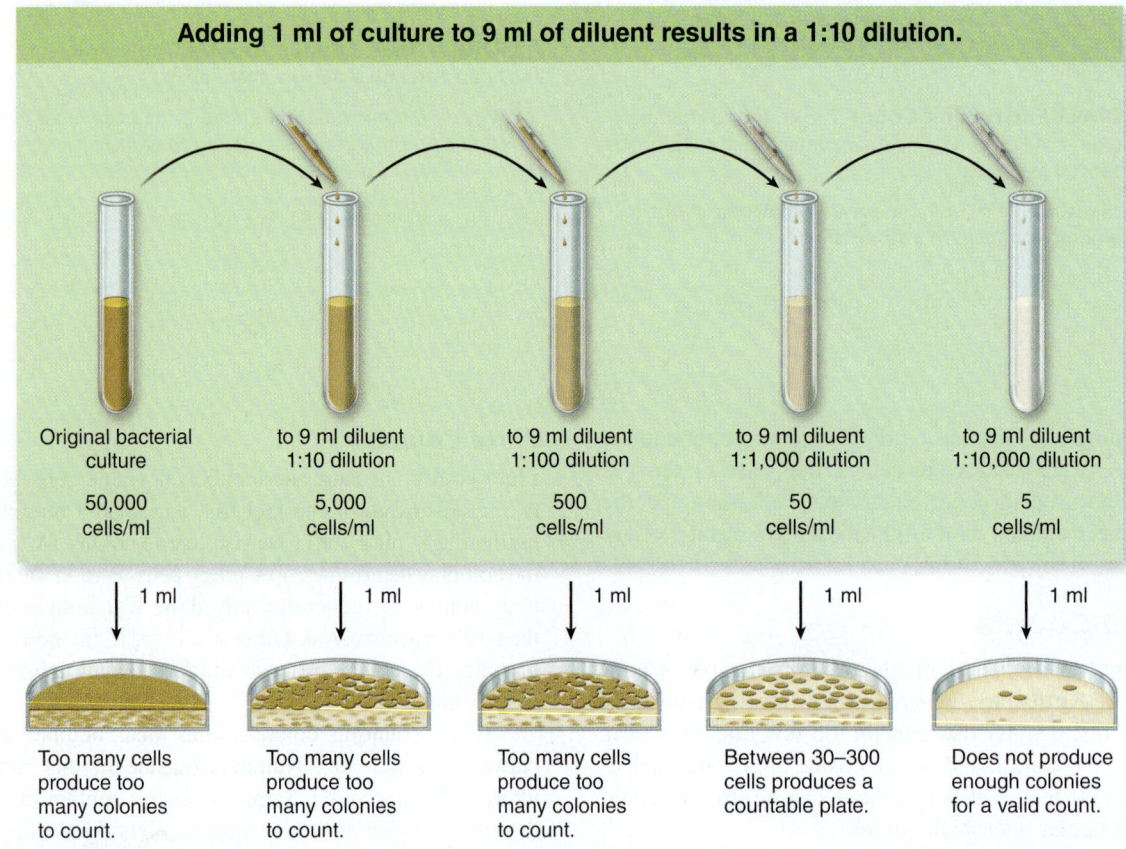

Adding 1 ml of culture to 9 ml of diluent results in a 1:10 dilution.

Original bacterial culture	to 9 ml diluent 1:10 dilution	to 9 ml diluent 1:100 dilution	to 9 ml diluent 1:1,000 dilution	to 9 ml diluent 1:10,000 dilution
50,000 cells/ml	5,000 cells/ml	500 cells/ml	50 cells/ml	5 cells/ml
1 ml	1 ml	1 ml	1 ml	1 ml
Too many cells produce too many colonies to count.	Too many cells produce too many colonies to count.	Too many cells produce too many colonies to count.	Between 30–300 cells produces a countable plate.	Does not produce enough colonies for a valid count.

FIGURE 4.17 Making Serial Dilutions To decrease the concentration of cells in a sample, 10-fold dilutions are often used.

? *If the 1-ml sample had been added to 99 ml of diluent to make the first dilution (instead of 9 ml), how many cells would be in that particular dilution?*

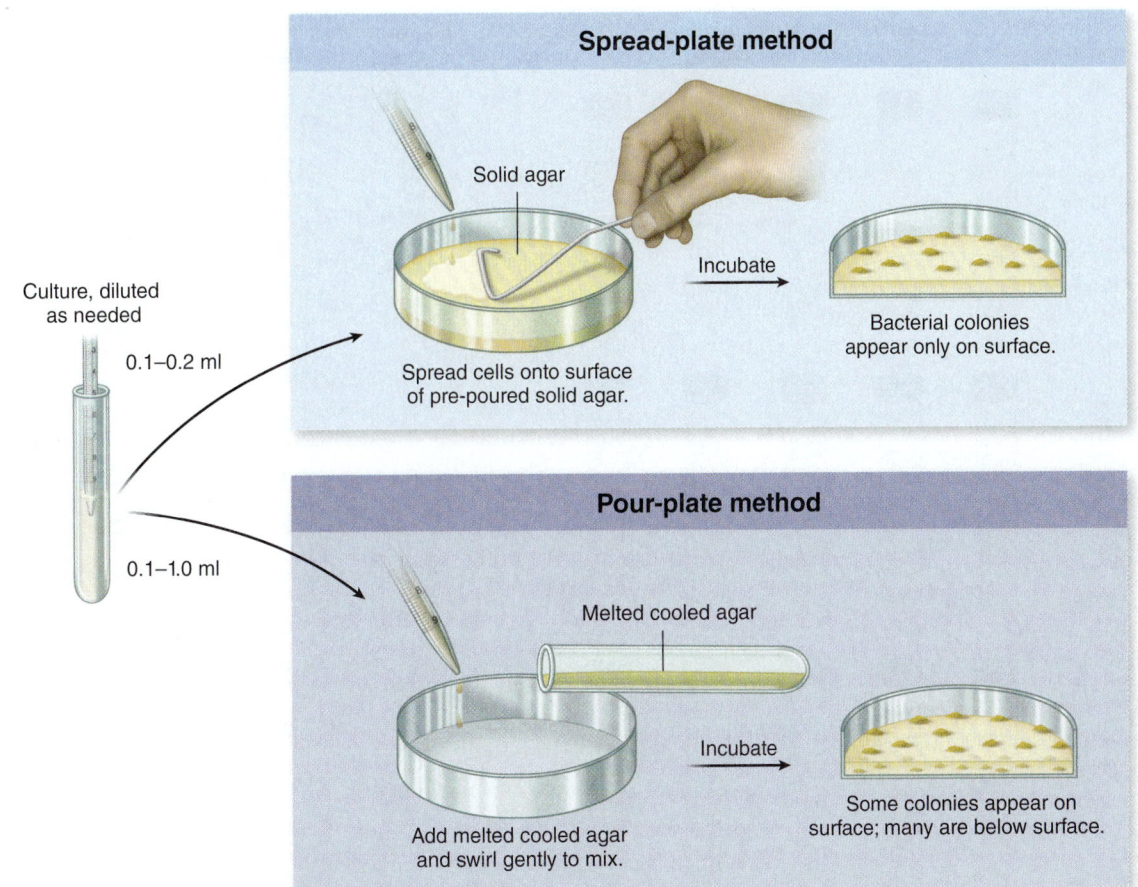

Spread-plate method

Culture, diluted as needed

0.1–0.2 ml

Solid agar

Spread cells onto surface of pre-poured solid agar.

Incubate →

Bacterial colonies appear only on surface.

Pour-plate method

0.1–1.0 ml

Melted cooled agar

Add melted cooled agar and swirl gently to mix.

Incubate →

Some colonies appear on surface; many are below surface.

FIGURE 4.18 Spread Plates and Pour Plates These techniques can be used for plate counts.

? *Why are the results of plate counts expressed as the number of colony-forming units, rather than the number of cells?*

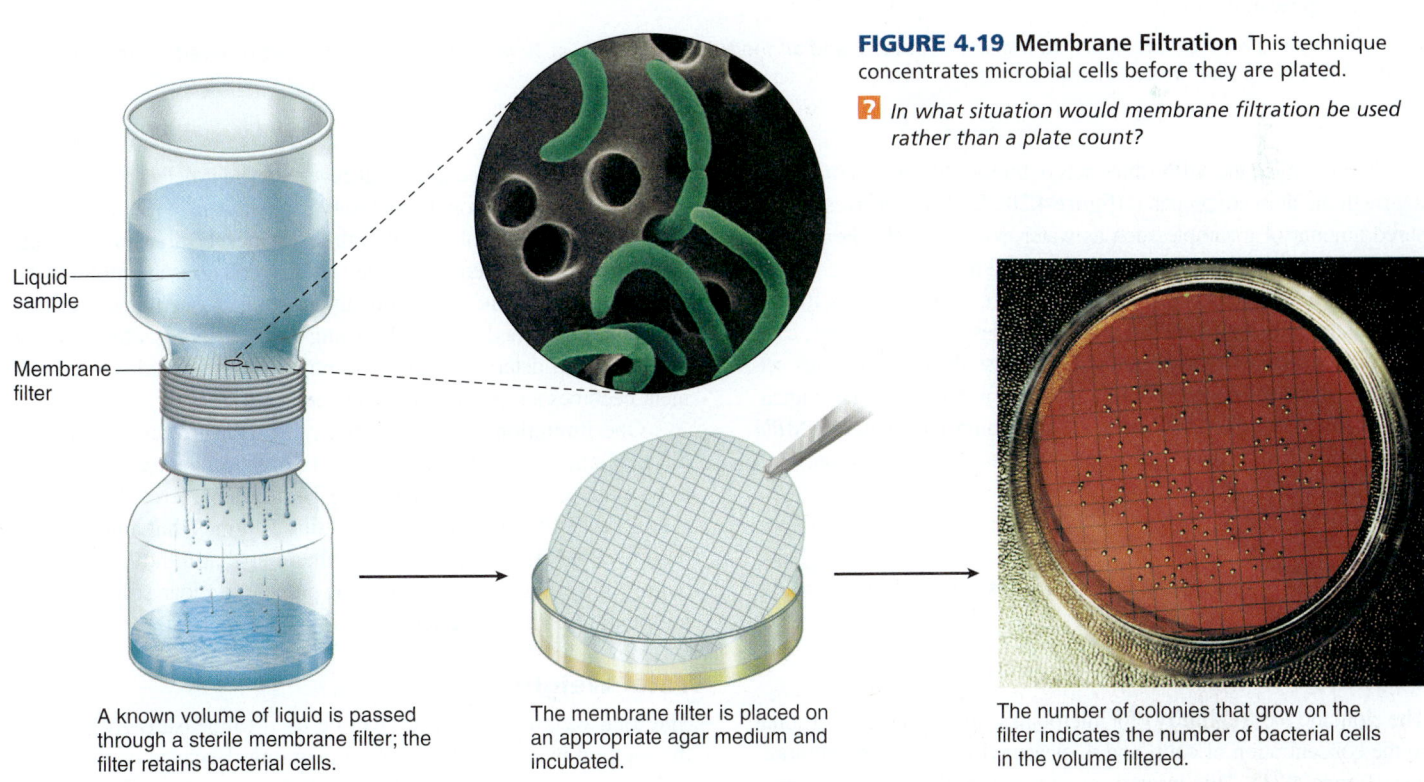

FIGURE 4.19 Membrane Filtration This technique concentrates microbial cells before they are plated.

? *In what situation would membrane filtration be used rather than a plate count?*

Liquid sample

Membrane filter

A known volume of liquid is passed through a sterile membrane filter; the filter retains bacterial cells.

The membrane filter is placed on an appropriate agar medium and incubated.

The number of colonies that grow on the filter indicates the number of bacterial cells in the volume filtered.

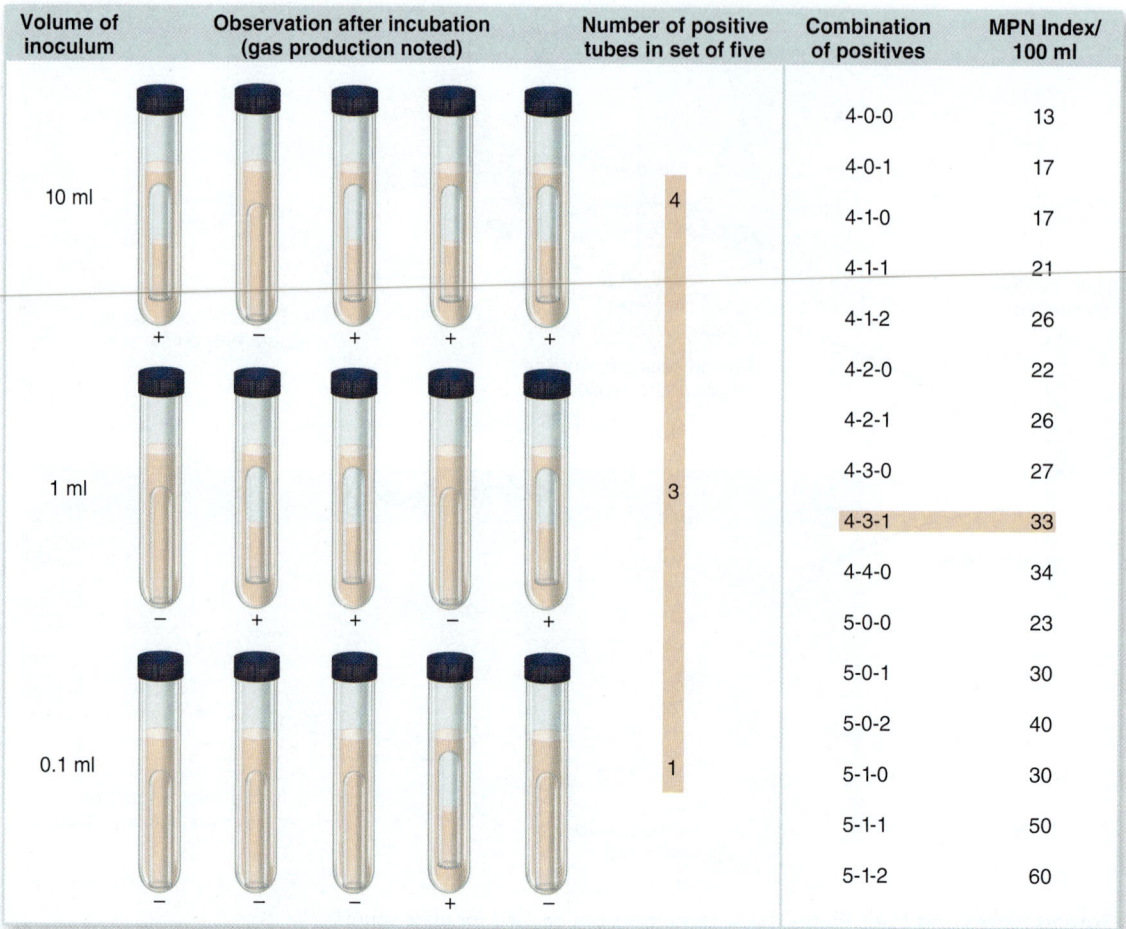

Volume of inoculum	Observation after incubation (gas production noted)					Number of positive tubes in set of five	Combination of positives	MPN Index/ 100 ml
							4-0-0	13
							4-0-1	17
10 ml	+	−	+	+	+	4	4-1-0	17
							4-1-1	21
							4-1-2	26
							4-2-0	22
							4-2-1	26
1 ml	−	+	+	−	+	3	4-3-0	27
							4-3-1	33
							4-4-0	34
							5-0-0	23
							5-0-1	30
							5-0-2	40
0.1 ml	−	−	+	−	−	1	5-1-0	30
							5-1-1	50
							5-1-2	60

FIGURE 4.20 The Most Probable Number (MPN) Method In this example, three sets of five tubes containing the same growth medium were prepared. Each set received the indicated amount of inoculum. After incubation, the presence or absence of gas in each tube was noted. The results were then compared to an MPN table to get a statistical estimate of the concentration in the original sample of gas-producing bacteria that could grow in the medium.

? *If you were using a sample similar to the one illustrated, and all the tubes that received 10 ml had gas, but none that received 1.0 ml or 0.1 ml did, what is the MPN index/100 ml?*

To determine the MPN, three sets of three or five tubes containing a growth medium are prepared (**figure 4.20**). Each set receives a measured amount of a sample such as water, soil, or food. The amount added is determined, in part, by the expected microbial concentration in that sample. What is important is that the second set receives 10-fold less than the first, and the third set 100-fold less. In other words, each set is inoculated with an amount 10-fold less than the previous set. After incubation, the presence or absence of turbidity or other indication of growth is noted; the results are then compared against an MPN table, which gives a statistical estimate of the cell concentration.

Measuring Biomass

Instead of measuring the number of cells, the cell mass can be determined.

Turbidity

The cloudiness or **turbidity** of a microbial suspension is proportional to the concentration of cells, and is measured with a spectrophotometer (**figure 4.21**). This instrument shines light through a specimen and measures the percentage that reaches a light detector. That percentage is inversely proportional to the optical density. To use turbidity to estimate cell numbers, a one-time test must be done to determine the correlation between optical density and cell concentration for the specific organism and conditions under study. Once this correlation has been determined—generally using a direct microscopic count or plate count to determine cell concentration—the turbidity measurement becomes a rapid and relatively accurate assay.

One limitation of using turbidity to measure biomass is that the medium must contain a relatively high concentration of cells to be cloudy. A solution containing 1 million bacteria (10^6) per ml is still perfectly clear, and if it contains 10 times that amount, it is barely turbid. It is important to remember that although a turbid culture indicates that microbes are present, a clear solution does not guarantee their absence.

Total Weight

The total weight of a culture can be used to measure growth, but the method is tedious and time-consuming. Because of this, total weight is usually used to study only filamentous organisms that do

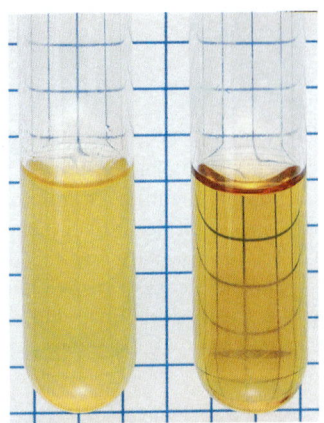

(a) The cloudiness, or turbidity, of the liquid in the tube on the left is proportional to the concentration of cells.

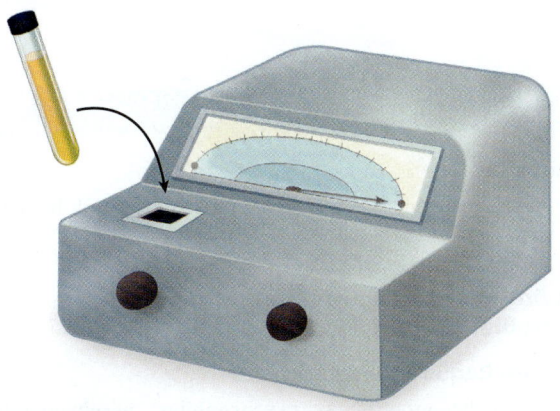

(b) A spectrophotometer is used to measure turbidity.

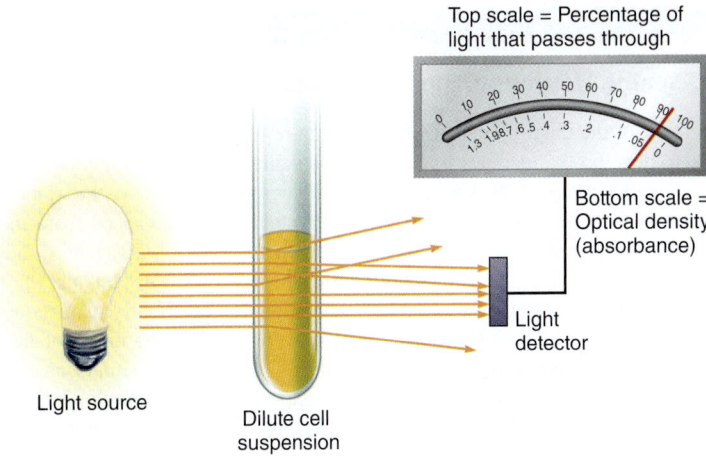

Top scale = Percentage of light that passes through

Bottom scale = Optical density (absorbance)

Light detector

Light source

Dilute cell suspension

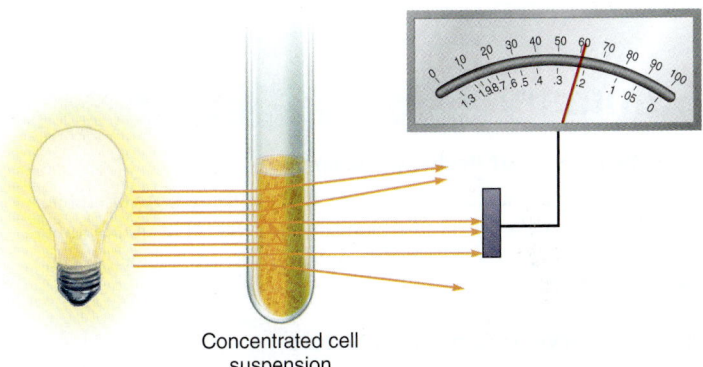

Concentrated cell suspension

(c) The percentage of light that reaches the detector of the spectrophotometer is inversely proportional to the optical density.

FIGURE 4.21 Measuring Turbidity with a Spectrophotometer To use turbidity to estimate cell number, a one-time test must be done to determine the correlation between optical density and cell concentration for the specific organism and conditions under study.

❓ *Approximately how many bacterial cells must be in a suspension for it to be cloudy?*

not readily separate into the individual cells necessary for a valid plate count. To measure the wet weight, cells in liquid culture are centrifuged and the liquid supernate removed. The weight of the resulting packed cell mass is proportional to the number of cells in the culture. The dry weight can be determined by heating the centrifuged cells in an oven before weighing them.

Detecting Cell Products

Products of microbial growth can be used to detect the presence of microbes.

Acid and Gas Production

Microorganisms produce a variety of acids and, sometimes, gases as a result of their metabolism. Acids can be detected by including a pH indicator in the culture medium. Several pH indicators are available, and they differ in the pH value at which their color changes. To detect gas production, inverted tubes (Durham tubes) can be used to trap gas bubbles in broth cultures. Clinical labs use more sensitive methods to detect the slight amounts of CO_2 produced by bacteria growing in patient blood samples. One method uses a fluorescent sensor to detect the slight decrease in pH that accompanies the production of CO_2.

ATP

ATP can be detected by adding the firefly enzyme luciferase. The enzyme catalyzes a chemical reaction that uses ATP as an energy source to produce light. This method is sometimes used to assess the effectiveness of chemical treatments intended to kill bacteria. Light is produced only if living organisms remain.

MicroAssessment 4.8

Direct microscopic counts and cell-counting instruments generally do not distinguish between living and dead cells. Plate counts determine the number of cells capable of multiplying; membrane filtration can be used to concentrate the sample. The most probable number estimates cell concentration. Turbidity of a culture can be correlated to cell number. The total weight of a culture can be correlated to the number of cells present. Microbial growth can be detected by the presence of cell products such as acids, gases, and ATP.

21. *Why is an MPN an estimate rather than an accurate number?*

22. *Water that accumulates in dirty bowls and dishes left in the sink often becomes cloudy over time. Besides suspended food particles, what causes the turbidity?* ➕

Seeing How the Other 99% Lives

One of the biggest challenges for the future is to develop the methods to cultivate and study a wider array of prokaryotes. Without these microbes, humans and other animals would not be able to exist. Yet, considering their importance, we still know very little about most species, including the relative contributions of each to such fundamental processes as O_2 generation and N_2 and CO_2 fixation.

Much of our understanding of prokaryotic processes comes from work with pure cultures. Yet, over 99% of prokaryotes have never been successfully grown in the laboratory. At the same time, when organisms are removed from their natural habitat, and especially when they are separated from other organisms, their environment changes drastically. Consequently, the study of pure cultures may not be the ideal for studying natural situations.

Technological advances such as flow cytometry and fluorescent labeling, along with the DNA sequencing techniques discussed in chapter 9, have made it easier to study uncultivated microorganisms. This may well lead to a better understanding of the diversity and the roles of microbes in our ecosystem. Scientists have learned a great deal since the days of Pasteur, but most of the microbial world is still a mystery.

Summary

4.1 ■ Principles of Prokaryotic Growth

Most prokaryotes multiply by **binary fission** (figure 4.1). Microbial growth is an increase in the number of cells in a population. The time required for a population to double in number is the **generation time** (table 4.1).

4.2 ■ Prokaryotic Growth in Nature

Biofilms (figure 4.2)

Prokaryotes often live in a **biofilm,** a community encased in polysaccharides and other extracellular polymeric substances.

Interactions of Mixed Microbial Communities

Prokaryotes often grow in close associations containing multiple different species. The metabolic activities of one organism often affects the growth of another.

4.3 ■ Obtaining a Pure Culture

Only an estimated 1% of prokaryotes have been grown in the laboratory.

Growing Microorganisms on a Solid Medium

A single microbial cell deposited on a solid medium will multiply to form a visible **colony** (figure 4.4).

The Streak-Plate Method (figure 4.5)

The **streak-plate method** is used to isolate microorganisms in order to obtain a **pure culture.**

Maintaining Stock Cultures

Stock cultures can be stored on agar slants in the refrigerator, frozen, or freeze-dried.

4.4 ■ Prokaryotic Growth in Laboratory Conditions

The Growth Curve (figure 4.6)

When grown in a **closed system,** a population of prokaryotic cells goes through five phases: **lag, exponential** or **log, stationary, death,** and **prolonged decline.**

Colony Growth

The position of a single cell within a colony markedly determines its environment.

Continuous Culture

Microbes can be maintained in a state of continuous growth by using a **chemostat.**

4.5 ■ Environmental Factors That Influence Microbial Growth (table 4.2)

Temperature Requirements (figure 4.8)

Organisms can be grouped as **psychrophiles, psychrotrophs, mesophiles, thermophiles,** or **hyperthermophiles** based on their optimum growth temperatures.

Oxygen (O_2) Requirements (table 4.3)

Organisms can be grouped as **obligate aerobes, facultative anaerobes, obligate anaerobes, microaerophiles,** or **aerotolerant anaerobes** based on their oxygen (O_2) requirements.

pH

Organisms can be grouped as **neutrophiles, acidophiles,** or **alkaliphiles** based on their optimum pH.

Water Availability

Halophiles are adapted to live in high-salt environments.

4.6 ■ Nutritional Factors That Influence Microbial Growth

Required Elements (table 4.4)

The **major elements** make up cell constituents and include carbon, nitrogen, sulfur, and phosphorus. **Trace elements** are required in very minute amounts.

Growth Factors

Microorganisms that cannot synthesize cell constituents such as amino acids and vitamins require these as **growth factors.**

Energy Sources

Organisms harvest energy either from sunlight or from chemical compounds.

Nutritional Diversity (table 4.5)

Photoautotrophs use the energy of sunlight along with the carbon in the atmosphere to make organic compounds. **Chemolithoautotrophs**

use inorganic compounds for energy and derive their carbon from CO_2. **Photoheterotrophs** use the energy of sunlight and obtain their carbon from organic compounds. **Chemoorganoheterotrophs** use organic compounds for energy and as a carbon source.

4.7 ■ Cultivating Prokaryotes in the Laboratory

General Categories of Culture Media (table 4.6)

A **complex medium** contains a variety of ingredients such as peptones and extracts. A **chemically defined medium** is composed of precise mixtures of pure chemicals.

Special Types of Culture Media

A **selective medium** inhibits organisms other than the one being sought. A **differential medium** contains a substance that certain microorganisms change in a recognizable way.

Providing Appropriate Atmospheric Conditions

A candle jar provides increased CO_2, which enhances the growth of many medically important bacteria. Microaerophilic microbes are incubated in a gastight container along with a packet that generates low O_2 conditions. Anaerobes may be incubated in an **anaerobe container** or an **anaerobic chamber** (figures 4.12, 4.13).

Enrichment Cultures (figure 4.14)

An **enrichment culture** provides conditions in a broth that enhance the growth of one particular organism in a mixed population.

4.8 ■ Methods to Detect and Measure Microbial Growth (table 4.8)

Direct Cell Counts

Direct microscopic counts involve counting the number of cells viewed through a microscope (figure 4.15). Both a **Coulter counter** and a **flow cytometer** count cells as they pass through a minute aperture (figure 4.16).

Viable Cell Counts

Plate counts measure the number of viable cells by taking advantage of the fact that an isolated cell will form a single colony (figures 4.17, 4.18). **Membrane filtration** concentrates microbial cells by filtration; the filter is then incubated on an agar plate (figure 4.19). The **most probable number** (MPN) method is used a series of inoculated tubes to statistical estimation of the cell concentration (figure 4.20).

Measuring Biomass

Turbidity of a culture can be correlated with the number of cells; a spectrophotometer is used to measure turbidity (figure 4.21). Wet weight and dry weight are proportional to the number of cells in a culture.

Detecting Cell Products

Products including acid, gas, and ATP can indicate growth.

Review Questions

Short Answer

1. Describe a detrimental and a beneficial effect of biofilms.
2. Define a *pure culture*.
3. Explain what occurs during each of the five phases of growth.
4. Explain how the environment of a colony differs from that of cells growing in a liquid broth.
5. List the five categories of optimum temperature, and describe a corresponding environment in which a representative might thrive.
6. Why would botulism be a concern with canned foods?
7. Explain why O_2-containing atmospheres kill some microbes.
8. Explain why photoautotrophs are primary producers.
9. Distinguish between a selective medium and a differential medium.
10. If the number of microorganisms in lake water were determined using both a direct microscopic count and a plate count, which method would most likely give a higher number? Why?

Multiple Choice

1. If there are 10^3 cells per ml at the middle of log phase, and the generation time of the cells is 30 minutes, how many cells will there be 2 hours later?

   a) 2×10^3 b) 4×10^3 c) 8×10^3

   d) 1.6×10^4 e) 1×10^7

2. Compared with their growth in the laboratory, bacteria in nature generally grow

   a) more slowly.

   b) faster.

   c) at the same rate.

3. Cells are most sensitive to penicillin during which phase of the growth curve?

   a) Lag   b) Exponential   c) Stationary

   d) Death   e) More than one of these.

4. Lactic acid is a primary metabolite. If a company wants to harvest this compound from a bacterial culture, the cells should be in which growth phase?

   a) Lag   b) Exponential   c) Stationary

   d) Death   e) More than one of these.

5. *E. coli,* a facultative anaerobe, is grown for 24 hours on the same solid medium, but under two different conditions: one aerobic, the other anaerobic. The size of the colonies would be

   a) the same under both conditions.

   b) larger when grown under aerobic conditions.

   c) larger when grown under anaerobic conditions.

6. The generation time of a bacterium was measured at two different temperatures. Which results would be expected of a thermophile?

   a) 20 minutes at 10°C; 220 minutes at 37°C

   b) 220 minutes at 10°C; 20 minutes at 37°C

   c) no growth at 10°C; 20 minutes at 37°C

   d) 20 minutes at 45°C; 220 minutes at 65°C

   e) 220 minutes at 37°C; 20 minutes at 65°C

7. Which of the following is *false*?

   a) *E. coli* grows faster in nutrient broth than in glucose-salts medium.

   b) Organisms require nitrogen to make amino acids.

   c) Some eukaryotes can fix N_2.

   d) An organism that grows on ham is osmotolerant.

   e) Blood agar is used to detect hemolysis.

8. If the pH indicator were left out of MacConkey agar, the medium would be

   a) complex.   b) differential.   c) defined.

   d) defined and differential. e) complex and differential.

9. A soil sample is placed in liquid and the number of bacteria in the sample determined in two ways: (1) colony count and (2) direct microscopic count. How would the results compare?

   a) Methods 1 and 2 would give approximately the same results.

   b) Many more bacteria would be estimated by method 1.

   c) Many more bacteria would be estimated by method 2.

   d) Depending on the soil sample, sometimes method 1 would be higher and sometimes method 2 would be higher.

10. The concentration of *E. coli* in a broth is between 10^4 and 10^6 cells per ml. To determine the precise number of living cells in the sample, it would be best to

    a) use a counting chamber.

    b) plate out an appropriate dilution of the sample on nutrient agar.

    c) determine cell number by using a spectrophotometer.

    d) Any of these three methods would be satisfactory.

    e) None of these three methods would be satisfactory.

Applications

1. You are a microbiologist working for a pharmaceutical company and discover a new metabolite that can serve as a medication. You now must oversee its production. What are some factors you must consider if you need to grow 5,000-liter cultures of bacteria?

2. High-performance boat manufacturers know that microbes can collect on a boat, ruining its hydrodynamic properties. A boat-manufacturing facility recently hired you to help with this problem because of your microbiology background. What strategies other than routine cleaning would you pursue to come up with a long-term remedy for the problem?

Critical Thinking

1. This figure shows a growth curve plotted on a non-logarithmic, or linear, scale. Compare this with figure 4.6. In both figures, the number of cells increases dramatically during the log or exponential phase. In this phase, the cell number increases more and more rapidly (this effect is more apparent in the accompanying figure). Why should the increase be speeding up?

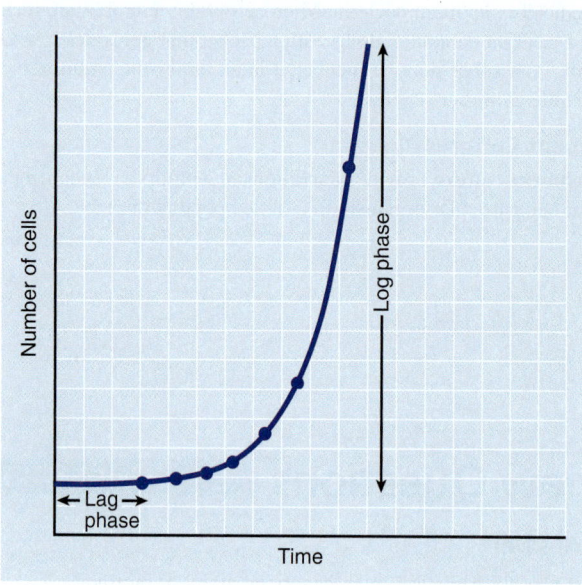

2. In question 1, how would the curve appear if the availability of nutrients were increased?

5 Control of Microbial Growth

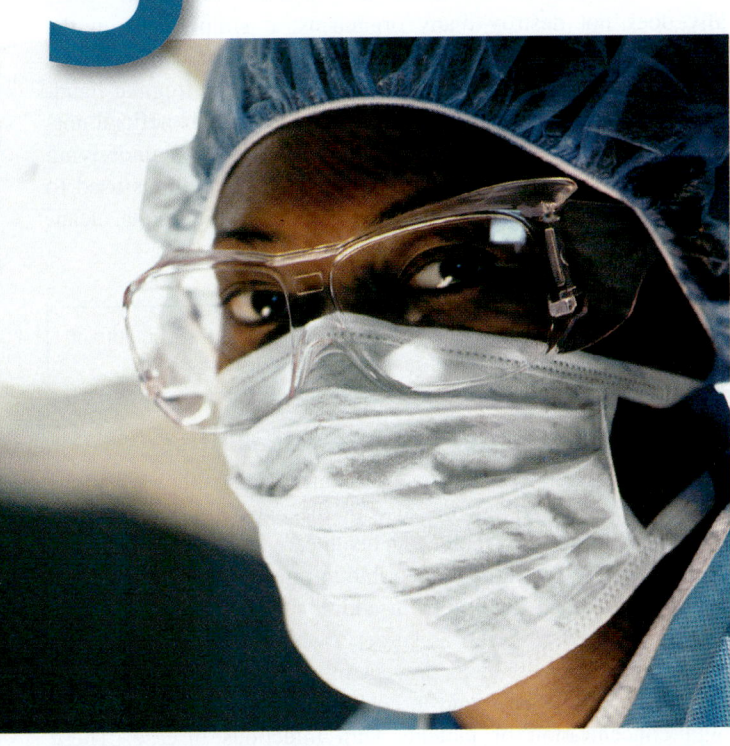

Medical settings warrant a high level of microbial control.

KEY TERMS

Antiseptic A disinfectant non-toxic enough to be used on skin.

Aseptic Technique Procedures that minimize the chance of unwanted microbes being accidentally introduced.

Bactericidal Kills bacteria.

Bacteriostatic Prevents the growth of, but does not kill, bacteria.

Disinfectant A chemical that destroys many microbes.

Germicide Kills microbes.

Pasteurization Brief heat treatment that reduces the number of spoilage organisms and destroys disease-causing microbes.

Preservation Process of inhibiting microbial growth to delay spoilage.

Sterilant A chemical that destroys all microbes.

Sterile Completely free of all viable microbes; an absolute term.

Sterilization The destruction or removal of all microbes through physical or chemical means.

A Glimpse of History

The *British Medical Journal* stated that the British physician Joseph Lister (1827–1912) "saved more lives by the introduction of his system than all the wars of the 19th century together had sacrificed." Lister revolutionized surgery by introducing methods that prevent wounds from becoming infected.

Impressed with Pasteur's work on fermentation (said to be caused by "minute organisms suspended in the air"), Lister wondered if "minute organisms" might be responsible also for the pus that formed in wounds. He then experimented by applying carbolic acid, a toxic compound, directly onto damaged tissues and found it prevented infections. Carbolic acid wound dressings then became standard in his practice, and he took pride in the fact that his patients no longer developed gangrene. His work also provided impressive evidence for the germ theory of disease, even though microorganisms specific for various diseases were not identified for another decade.

Later, Lister improved his methods even further by using surgical procedures that excluded bacteria from wounds. These procedures included sterilizing instruments before use and maintaining a clean environment in the operating room. Doing this was preferable to killing the bacteria after they entered wounds because they avoided the toxic effects of the disinfectant on the wound.

Lister was knighted in 1883 and subsequently became a baron and a member of the House of Lords.

Up until the late nineteenth century, patients undergoing even minor surgeries were at great risk of developing fatal infections due to unsanitary medical practices and hospital conditions. Physicians did not know that their hands could pass diseases from one patient to the next. Nor did they understand that airborne microscopic organisms could infect open wounds. Fortunately, today's modern hospitals use strict procedures to avoid microbial contamination, allowing most surgeries to be performed with relative safety.

Microbial growth affects more than our health. Manufacturers of a wide variety of goods recognize that microorganisms can reduce product quality. Their growth can lead to undesirable changes in the safety, appearance, taste, odor, or function of products ranging from food to lumber.

This chapter covers methods to destroy, remove, and inhibit microbial growth on inanimate objects and some body surfaces. Most of these approaches can damage all forms of life. In contrast, antibiotics and other antimicrobial medications are specifically toxic to microbes, which is why they are so valuable in treating infectious diseases; they will be discussed in chapter 20.

5.1 ■ Approaches to Control

Learning Outcomes

1. *Using the appropriate terminology, describe the principles of sterilization, disinfection, pasteurization, decontamination, sanitization, and preservation.*

2. *Compare and contrast the methods used to control microbial growth in daily life, healthcare settings, microbiology laboratories, food and food production facilities, water treatment facilities, and other industries.*

The processes used to control microorganisms are either physical or chemical, though a combination of both can be used. Physical methods include heat treatment, irradiation, filtration, and mechanical removal (washing). Chemical methods use any of a variety of antimicrobial chemicals. The method chosen depends on the circumstances and level of control required.

Principles of Control

Sterilization is the removal or destruction of all microorganisms and viruses on or in a product. Microbes can be removed by filtration, or destroyed using heat, certain chemicals, or irradiation. Destruction of microorganisms means they cannot be "revived" to multiply even when transferred from the sterilized product to an ideal growth medium. A **sterile** item is one that is free of microbes, including endospores and viruses. It is important to note, however, that the term *sterile* does not consider prions. These infectious protein particles are not destroyed by standard sterilization procedures. ◄◄ endospores, p. 67 ►► prions, p. 328

Disinfection is the elimination of most or all pathogens on or in a material. In practice, the term generally implies the use of antimicrobial chemicals. Unlike sterilization, disinfection suggests that some living microbes may remain. **Disinfectants** are antimicrobial chemicals used for disinfecting inanimate objects. They are toxic to many forms of life, and therefore biocides (*bio* means "life," and *cida* means "to kill"). They are typically used in a manner that targets microorganisms and viruses, however, so they are often called **germicides.** They are also described as **bactericidal,** meaning they kill bacteria. **Antiseptics** are antimicrobial chemicals non-toxic enough to be used on skin or other body tissue. These are routinely used to decrease bacterial numbers on skin before invasive procedures such as surgery. ◄◄ pathogen, p. 6

Pasteurization is a brief heat treatment that reduces the number of spoilage organisms and destroys pathogens. Foods and inanimate objects can be pasteurized.

Decontamination is a process used to reduce the number of pathogens to a level considered safe to handle. The treatment can be as simple as thorough washing, or it may involve the use of heat or disinfectants.

Sanitization generally implies a process that substantially reduces the microbial population to meets accepted health standards. Most people also expect a sanitized object to look clean. Note that this term does not indicate any specific level of control.

Preservation is the process of delaying spoilage of foods or other perishable products. One way to do this is to adjust storage conditions to slow microbial growth. Alternatively, chemical preservatives can be added. These are **bacteriostatic,** meaning they inhibit the growth of bacteria but do not kill them.

Situational Considerations

Methods used to control microbial growth vary greatly depending on the situation and level of control required (**figure 5.1**). Control measures adequate for routine circumstances of daily life might not be sufficient for situations such as hospitals, microbiology laboratories, foods and food production facilities, water treatment facilities, and other industries.

Daily Life

Washing and scrubbing with soaps and detergents is a sufficient level of microbial control in routine situations. Soap itself generally does not destroy many organisms; it simply aids in the mechanical removal of microbes, including most pathogens, as well as dirt, organic material, and some skin cells. Regular handwashing and bathing does not adversely affect the beneficial normal skin microbiota, which reside more deeply on underlying layers of skin cells and in hair follicles. Other methods used to control microorganisms in daily life include cooking foods, cleaning surfaces, and refrigeration.

MicroByte

> Thorough handwashing with soap and water is the most important step in stopping the spread of many infectious diseases.

Hospitals and other Healthcare Facilities

Minimizing the numbers of microorganisms in healthcare settings is particularly important because of the danger of **healthcare-associated infections.** Patients in healthcare facilities, particularly hospitals, are often more susceptible to infectious agents because of their weakened condition. In addition, patients may undergo invasive procedures such as surgery, which cuts the intact skin that would otherwise help prevent infection. Finally, pathogens are more likely to be found in healthcare settings because of the high concentration of patients with infectious disease. These patients can shed pathogens in their feces, urine, respiratory droplets, or other body secretions. Because of this, healthcare facilities must be extremely careful to control microorganisms. Nowhere is this more important than in the operating rooms, where instruments used in invasive procedures must be sterile to avoid introducing even normally harmless microbes into deep body tissue where they could easily cause infection. ►► healthcare-associated infections, p. 449

Prions are a relatively new concern for healthcare facilities. Fortunately, disease caused by these agents is thought to be exceedingly rare, less than 1 case per 1 million persons per year. Healthcare facilities, however, must take special precautions when handling tissue that may be contaminated with prions, because these infectious particles are very difficult to destroy.

Microbiology Laboratories

Microbiology laboratories routinely work with microbial cultures and consequently must use rigorous methods to control microorganisms. To work with pure cultures, all media and instruments that contact the culture must first be sterilized to avoid contaminating the culture with environmental microbes. All materials used to grow microorganisms must again be treated before disposal to avoid contamination of workers and the environment. The use of specific methods to prevent microorganisms from contaminating an environment is called **aseptic technique.** Although all microbiology laboratory personnel must use these measures, those

FIGURE 5.1 Situations That Warrant Different Levels of Microbial Control (a) Daily home life; (b) foods and food production facilities; (c) water treatment facilities; (d) hospitals; (e) other industries.

❓ *Why would a hospital require a more stringent level of microbial control than daily home life?*

who work with known pathogens must be even more careful.
◀◀ aseptic technique, p. 85

The Centers for Disease Control and Prevention (CDC) has established precaution guidelines for laboratories working with

microorganisms. These are known as biosafety levels (BSLs) and range from BSL-1 (for work with microbes not known to cause disease) to BSL-4 (for work with deadly pathogens for which no vaccine or specific treatment exists).

Foods and Food Production Facilities

Foods and other perishable products retain their quality longer when contaminating microbes are destroyed, removed, or inhibited. Heat treatment is the most common and reliable method used to kill microbes, but it can alter the flavor and appearance of the products. Irradiation can be used to destroy microbes, and chemical additives can be used to prevent their growth, but the risk of toxicity must always be a concern. Because of this, the Food and Drug Administration (FDA) regulates these options.

Food-processing facilities need to keep surfaces relatively free of microorganisms to avoid contamination. If machinery used to grind meat is not cleaned properly, for example, it can create an environment in which bacteria multiply, eventually contaminating large quantities of product.

Water Treatment Facilities

Water treatment facilities need to ensure that drinking water is free of pathogenic microbes. Chlorine has traditionally been used to disinfect water, saving hundreds of thousands of lives by preventing the spread of waterborne illnesses such as cholera. Chlorine and other disinfectants, however, can react with naturally occurring chemicals in the water to form **disinfection by-products (DBPs)**. Some of these have been linked to long-term health risks. In addition, certain pathogens, particularly *Cryptosporidium parvum,* a cause of diarrhea, can survive traditional disinfection procedures. To address these problems, water treatment regulations now require facilities to minimize the level of both DBPs and *C. parvum* in treated water. ▶▶ *Cryptosporidium parvum,* **p. 604**

Other Industries

Many diverse industries have specialized concerns regarding microbial growth. Manufacturers of pharmaceuticals, cosmetics, deodorants, or any other product that will be ingested, injected, or applied to the skin must avoid microbial contamination that could affect the product's quality or safety.

MicroAssessment 5.1

The methods used to control microbial growth depend on the situation and the level of control required.

1. *How is sterilization different from disinfection?*
2. *Why would water treatment facilities be concerned about disinfection by-products?*
3. *Why would the term* sterilization *not encompass prions?*

5.2 ■ Selection of an Antimicrobial Procedure

Learning Outcome

3. *Explain why the type and number of microbes, environmental conditions, risk for infection, and composition of the item influence the selection of an antimicrobial procedure.*

Selection of an effective antimicrobial procedure is complicated by the fact that every procedure has disadvantages that limit its use. An ideal, multipurpose, non-toxic method simply does not exist. The ultimate choice depends on many factors including the type and number of microbes, environmental conditions, risk for infection, and the composition of the item.

Type of Microbes

One of the most critical considerations in selecting an antimicrobial procedure is the microbial population. Products contaminated with microbes highly resistant to killing require a more rigorous treatment. Highly resistant microbes include:

- **Bacterial endospores.** The endospores of *Bacillus, Clostridium,* and related genera are the most resistant form of life typically encountered. Only extreme heat or chemical treatment ensures their complete destruction. ◀◀ **endospores, p. 67**
- **Protozoan cysts and oocysts.** Cysts and oocysts are stages in the life cycle of certain intestinal protozoan pathogens such as *Giardia lamblia* and *Cryptosporidium parvum.* These disinfectant-resistant forms are excreted in the feces of infected animals, including humans, and can cause diarrheal disease if ingested. Unlike endospores, they are easily destroyed by boiling. ▶▶ *Cryptosporidium parvum,* **p. 604** ▶▶ *Giardia lamblia,* **p. 602**
- ***Mycobacterium* species.** The waxy cell walls of mycobacteria make them resistant to many chemical treatments. Because of this, stronger, more toxic chemicals must be used to disinfect environments that may contain *Mycobacterium tuberculosis,* the cause of tuberculosis. ▶▶ **tuberculosis, p. 502**
- ***Pseudomonas* species.** These common environmental organisms are not only resistant to some disinfectants, but in some cases actually grow in them. *Pseudomonas* species can cause serious healthcare-associated infections. ▶▶ *Pseudomonas* **infections, p. 554**
- **Naked viruses.** Viruses such as poliovirus that lack a lipid envelope are more resistant to disinfectants. Conversely, enveloped viruses, such as HIV, tend to be very sensitive to these chemicals. ▶▶ **naked viruses, p. 306** ▶▶ **enveloped viruses, p. 306**

Number of Microorganisms

The time it takes for heat or chemicals to kill a microbial population is dictated in part by the number of cells present. It takes more time to kill a large population than it does to kill a small population, because only a fraction of organisms die during a given time interval. For example, if 90% of a bacterial population is killed during the first 3 minutes, then approximately 90% of those remaining will be killed during the next 3 minutes, and so on. Removing organisms by washing or scrubbing can minimize the time necessary to sterilize or disinfect a product.

In the commercial canning industry, the **decimal reduction time,** or **D value,** is the time required for killing 90% of a bacterial population under specific conditions (**figure 5.2**). The temperature

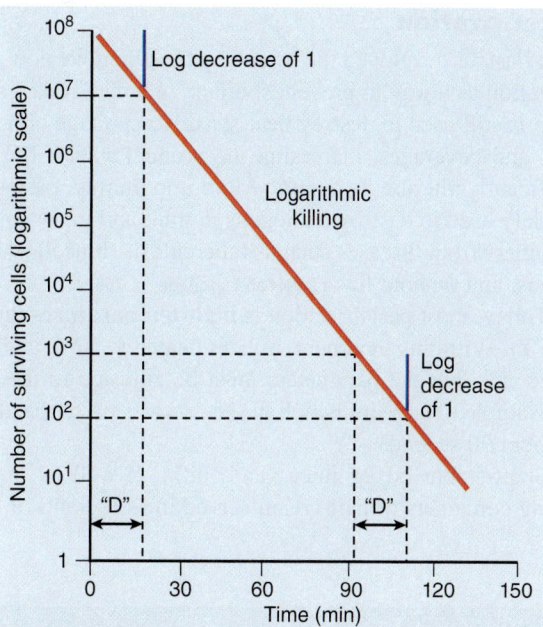

FIGURE 5.2 D Value The D value is the time it takes to reduce the population by 90%.

Based on this graph, what is the approximate D value for the organism?

of the process is indicated by a subscript, for example, D_{121}. A one D process reduces the number of cells by one exponent. Thus, if the D value for an organism is 2 minutes, then it would take 4 minutes (2 D values) to reduce a population of 100 (10^2) cells to only one (10^0) survivor. It would take 20 minutes (10 D values) to reduce a population of 10^{10} cells to only one survivor.

Environmental Conditions

Dirt, grease, and body fluids such as blood can interfere with heat penetration and the action of chemical disinfectants. This is another reason why it is important to thoroughly clean items before disinfection or sterilization.

Temperature and pH influence microbial death rates. A solution of sodium hypochlorite (household bleach), for example, can kill a suspension of *M. tuberculosis* at a temperature of 55°C in half the time it would take if the suspension were held at 50°C. The hypochlorite solution is even more effective at a low pH.

Risk for Infection

To guide the selection of germicidal procedures, medical instruments are categorized according to their risk for transmitting infectious agents. Those that pose greater threats require more rigorous germicidal procedures. The categories are:

- **Critical instruments.** These come into direct contact with body tissues; they include needles and scalpels. Critical instruments must be sterile.

- **Semicritical instruments.** These come into contact with mucous membranes, but do not penetrate body tissue; they include gastrointestinal endoscopes and endotracheal tubes. Semicritical instruments must be free of all viruses and vegetative bacteria. The few endospores that may remain pose little risk for infection because mucous membranes are effective barriers against their entry into deeper tissue.

- **Non-critical instruments and surfaces.** These come into contact only with unbroken skin so they pose little risk for infection. Countertops, stethoscopes, and blood pressure cuffs are examples of non-critical items.

Composition of the Item

Some sterilization and disinfection procedures are inappropriate for certain types of material. For example, heat treatment can damage many types of plastics and other materials. Irradiation provides an alternative to heat, but the process damages some types of plastics. Moist heat (such as boiling water) and liquid chemical disinfectants cannot be used to treat moisture-sensitive material.

MicroAssessment 5.2

The types and numbers of microorganisms initially present, environmental conditions, the potential risks associated with use of the item, and the composition of the item must all be considered when determining which sterilization or disinfection procedure to employ.

4. *Describe three groups of microorganisms that are resistant to certain chemical treatments.*

5. *Why do critical instruments need to be sterile, whereas semicritical instruments need only be free of viruses and vegetative bacteria?*

6. *Would it be safe to say that if all bacterial endospores had been killed, then all other medically important microorganisms had also been killed?*

5.3 ▪ Using Heat to Destroy Microorganisms and Viruses

Learning Outcomes

4. *Compare and contrast pasteurization, sterilization using pressurized steam, and the commercial canning process.*

5. *Explain the drawbacks and benefits of using dry heat rather than moist heat to kill microorganisms.*

Heat treatment is one of the most useful methods of microbial control because it is reliable, safe, relatively fast and inexpensive, and does not introduce potentially toxic substances into materials. Some heat-based methods sterilize the product, whereas others the number of microbes. **Table 5.1** summarizes the characteristics of heat treatment and other physical methods of control.

Moist Heat

Moist heat destroys microbes by irreversibly denaturing their proteins. Examples of moist heat treatment include boiling, pasteurization, and pressurized steam.

Boiling

Boiling (100°C at sea level) easily destroys most microorganisms and viruses. Because of this, drinking water that might be contaminated during floods or other emergency situations should be boiled for at least 5 minutes. Boiling is not a method of sterilization, however, because endospores can survive the process.

Pasteurization

Louis Pasteur developed the brief heat treatment we now call pasteurization as a way to prevent spoilage of wine. Today, pasteurization is still used to destroy heat-sensitive spoilage organisms in foods and beverages, increasing the product's shelf life without significantly altering its quality. More importantly, pasteurization is widely used to destroy pathogens in milk and juices, protecting consumers from diseases such as tuberculosis, brucellosis, salmonellosis, and typhoid fever. ▸▸ food spoilage, p. 756

Today, most pasteurization is **high-temperature–short-time (HTST)**. With this treatment, milk is heated to 72°C and held for 15 seconds, but the parameters must be adjusted to the product. For example, ice cream is rich in fats, so it is pasteurized at 82°C for about 20 seconds.

Shelf-stable boxed juices and milk, as well as the single-serving containers of half-cream served in restaurants, are treated

TABLE 5.1	Physical Methods Used to Destroy Microorganisms and Viruses	
Method	**Characteristic**	**Use**
Moist Heat	Denatures proteins. Relatively fast, reliable, safe, and inexpensive.	Widely used.
Boiling	Boiling for 5 minutes destroys most microorganisms and viruses; a notable exception is endospores.	Boiling for at least 5 minutes can be used to treat drinking water.
Pasteurization	Significantly decreases the numbers of heat-sensitive microorganisms, including spoilage microbes and pathogens (except sporeformers).	Milk is pasteurized by heating it to 72°C for 15 seconds. Juices are also routinely pasteurized.
Pressurized steam (autoclaving)	Typical treatment is 121°C/15 psi for 15 minutes or longer, a process that destroys endospores.	Widely used to sterilize microbiological media, laboratory glassware, surgical instruments, and other items that steam can penetrate. The canning process renders foods commercially sterile.
Dry Heat		
Incineration	Burns cell components to ashes.	Flaming of wire inoculating loops. Also used to destroy medical wastes and contaminated animal carcasses.
Dry heat ovens	Destroys cell components and denatures proteins. Less efficient than moist heat, requiring longer times and higher temperatures.	Laboratory glassware is sterilized by heating at 160°C to 170°C for 2 to 3 hours. Powders, oils, and other dry materials are also sterilized in ovens.
Filtration	Filter retains microbes while letting the suspending fluid or air pass through small holes.	
Filtration of fluids	Various pore sizes are available; 0.2 μm is commonly used to remove bacteria.	Used for beer and wine, and to sterilize some heat-sensitive medications.
Filtration of air	HEPA filters are used to remove microbes that have a diameter of 0.3 μm or greater.	Used in biological safety cabinets, specialized hospital rooms, and airplanes. Also used in some vacuum cleaners and home air purification units.
Radiation	Type of cell damage depends on the wavelength of the radiation.	
Ionizing radiation	Destroys DNA and possibly damages cytoplasmic membranes. Produces reactive molecules that damage other cell components. Items can be sterilized even after packaging.	Used to sterilize heat-sensitive materials including medical equipment, disposable surgical supplies, and drugs such as penicillin. Also used to destroy microbes in spices, herbs, and approved types of produce and meats.
Ultraviolet radiation	Damages DNA. Penetrates poorly.	Used to destroy microbes in the air and drinking water, and to disinfect surfaces.
High Pressure	Treatments of 130,000 psi are thought to denature proteins and alter the permeability of the cell. Products retain color and flavor.	Used to extend the shelf life of certain commercial food products such as guacamole.

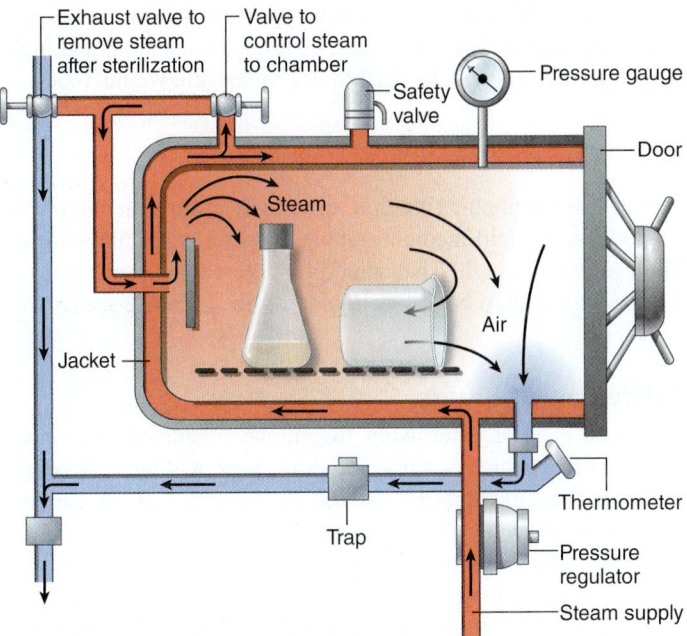

FIGURE 5.3 Autoclave Steam first travels in an enclosed layer, or jacket, surrounding the chamber. It then enters the autoclave, displacing the air downward and out through a port in the bottom of the chamber.

❓ *Why is autoclaving more effective than boiling?*

using **ultra-high-temperature (UHT)** processing. This destroys all microorganisms that can grow under normal storage conditions, so it is referred to as "ultra-pasteurization." The UHT process for milk requires rapidly heating the milk to 140°C, holding it at that temperature for a few seconds, and then quickly cooling it. The product is then aseptically packaged in sterile containers.

MicroByte

Clothes can be pasteurized by regulating the temperature in a washing machine.

Sterilization Using Pressurized Steam

Heat- and moisture-tolerant items such as surgical instruments, most microbiological media, and reusable glassware are sterilized using **autoclaves.** Water in a chamber in the autoclave is heated to form steam, causing the pressure in the chamber to increase (**figure 5.3**). The higher pressure, in turn, increases the temperature at which steam forms. Steam at atmospheric pressure never exceeds 100°C, but steam at an additional 15 psi (pounds per square inch) is 121°C, a temperature that kills even endospores. The pressure itself plays no direct role in the killing.

Typical conditions for sterilization are 15 psi and 121°C for 15 minutes. Longer treatment is necessary for large volumes because it takes more time for heat to completely penetrate the substance. For example, 4 liters of nutrient broth in a flask takes longer to sterilize than the same volume distributed into tubes.

Other autoclave conditions can be used for specific purposes. When rapid processing is important, such as when sterile instruments must always be available in an operating room, flash

sterilization with a higher temperature for a shorter time can be used. To destroy prions, autoclaving at a temperature of 132°C for 1 hour is thought to be effective.

Autoclaving is consistently effective in sterilizing most objects, if done correctly. The temperature and pressure gauges of the autoclave should both be monitored to ensure proper operating conditions. It is also critical that steam enter items to displace air. Because of this, long, thin containers should be placed on their sides, and containers should never be closed tightly.

Tape that contains a heat-sensitive indicator, which turns black at a high temperature, should be attached to items before autoclaving. This provides a visual signal that the items have been processed (**figure 5.4a**). A changed indicator, however, does not mean that the object is sterile, only that it has been heated.

Biological indicators are used to ensure that an autoclave is working properly (figure 5.4b). A tube containing the heat-resistant endospores of *Geobacillus stearothermophilus* can be placed near the center of an item. After that item is autoclaved, the endospores are mixed with a growth medium by manually crushing a container within the tube. If the medium changes color during incubation, microbial growth occurred, indicating an unsuccessful process.

The Commercial Canning Process

Commercial canning uses an industrial-sized autoclave (called a retort). The canning process is designed to ensure that endospores of *Clostridium botulinum* are destroyed. This is critical because surviving spores can germinate in canned foods such as vegetables and meats. The resulting vegetative cells can grow in low-acid anaerobic conditions and produce botulinum toxin, one of the most potent toxins known. ▶▶ botulism, p. 652

In destroying endospores of *C. botulinum,* the canning process also kills all other organisms that grow under normal storage conditions. Canned foods are **commercially sterile,** meaning that the endospores of some thermophiles may survive. These are usually not a concern, however, because they grow only at temperatures well above those of normal storage.

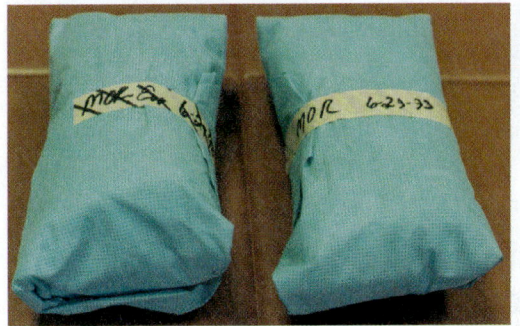

(a)   (b)

FIGURE 5.4 Indicator Used in Autoclaving (a) Chemical indicators. The pack on the left has been autoclaved. Diagonal marks on the tape have turned black, indicating that the object was exposed to heat. **(b)** Biological indicators. Following incubation, a change of color to yellow indicates growth of the endospore-forming organism.

❓ *Why would a biological indicator be better than other indicators to determine if the sterilization procedure was effective?*

Several factors dictate the time and temperature of the canning process. First, as discussed earlier, the higher the temperature, the shorter the time needed to kill all organisms. Second, the higher the concentration of organisms, the longer the heat treatment required to kill them all. To provide a wide safety margin, the commercial canning process is designed to reduce a population of 10^{12} *C. botulinum* endospores to only one spore. In other words, it is a 12 D process. It is virtually impossible for a food to have this many endospores.

Dry Heat

Dry heat is not as efficient as moist heat in killing microbes, and therefore requires longer times and higher temperatures. For example, 200°C for 90 minutes of dry heat has the killing equivalent of 121°C for 15 minutes of moist heat.

Incineration burns the cell components to ashes. In microbiology laboratories, the wire loops continually reused to transfer bacterial cultures are sterilized by flaming—heating them in a flame until they are red hot. Alternatively, they can be heated to the same point in a benchtop incinerator designed for this purpose. Incineration is also used to destroy medical wastes and contaminated animal carcasses.

Temperatures achieved in hot air ovens destroy cell components and irreversibly denature proteins. Glass Petri dishes and glass pipettes are sterilized in ovens with non-circulating air at temperatures of 160°C to 170°C for 2 to 3 hours. Ovens with a fan that circulates the hot air can sterilize in a shorter time because they transfer heat more efficiently. Powders, oils, and other dry material are also sterilized in hot ovens.

MicroAssessment 5.3

Moist heat such as boiling water destroys most microbes. Pasteurization significantly reduces the numbers of heat-sensitive organisms. Autoclaves use pressurized steam to achieve high temperatures that kill microbes, including endospores. The commercial canning process is designed to destroy the endospores of *Clostridium botulinum*. Dry heat takes longer than moist heat to kill microbes.

7. *Why is it important that the commercial canning process destroys the endospores of* Clostridium botulinum?

8. *How does incineration destroy microbes?*

9. *Would pasteurization destroy endospores?* ➕

5.4 ■ Using Other Physical Methods to Remove or Destroy Microbes

Learning Outcomes

6. *Describe how depth filters, membrane filters, and HEPA filters are used to remove microorganisms.*

7. *Compare and contrast the use of gamma radiation, ultraviolet radiation, and microwaves for destroying microorganisms.*

Some materials cannot withstand heat treatment. For these items, other physical methods including filtration, irradiation, and high-pressure treatment can be used to destroy or remove microbes.

Filtration

Recall that membrane filtration, used to determine the number of bacteria in a liquid medium, retains bacteria while allowing the fluid to pass through. That same principle can be used to physically remove microbes from liquids or air. ◀◀ membrane filtration, p. 100

Filtration of Fluids

Filtration is used extensively to remove organisms from heat-sensitive fluids such as sugar solutions, beer, and wine. Specially designed filtration units are also used by backpackers and campers to remove *Giardia* cysts and bacteria from water.

Paper-thin **membrane filters** or **microfilters** have microscopic pores that allow liquid to flow through while trapping particles too large to pass (**figure 5.5**). A vacuum is commonly used to help pull the liquid through the filter; alternatively, pressure may be applied to push the liquid. The filters are made of nitrocellulose or other polymers and come in a variety of different pore sizes, extending below the dimensions of the smallest known viruses. Pore sizes smaller than necessary should be avoided, however, because they slow the flow. Filters with a pore size of 0.2 micrometers (µm) are commonly used to remove bacteria.

Depth filters trap material within thick porous filtration material such as cellulose fibers. They have complex passages that retain microorganisms while letting the suspending fluid pass through the small holes. The diameter of the passages is often much larger than that of microbes, but the electrical charges on the walls help hold the microbial cells. ◀◀ cellulose, p. 32

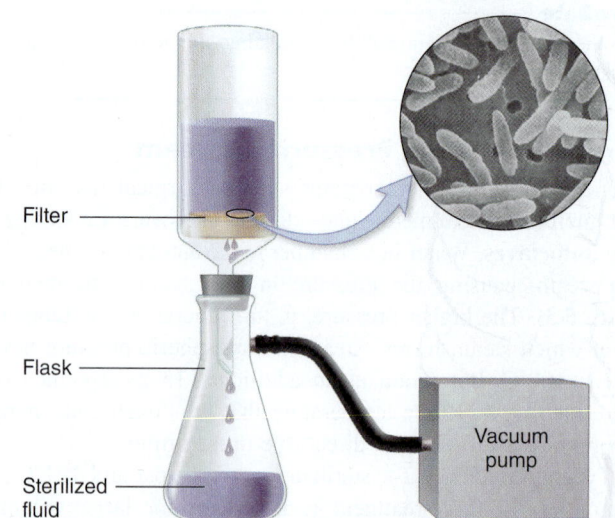

FIGURE 5.5 Sterilization of Fluids Using a Membrane Filter
The liquid to be sterilized flows through the filter on top of the flask in response to a vacuum produced in the flask by means of a pump. Scanning electron micrograph (5,000×) shows a membrane filter retaining cells of *Pseudomonas*.

❓ *Why would a pore size smaller than necessary be a poor choice?*

Filter

Flask

Sterilized fluid

Vacuum pump

Filtration of Air

High-efficiency particulate air (HEPA) filters remove nearly all airborne particles 0.3 μm or larger. These filters are used for keeping microorganisms out of specialized hospital rooms designed for patients extremely susceptible to infection. The filters are also used in biological safety cabinets in which laboratory personnel work with dangerous airborne pathogens such as *Mycobacterium tuberculosis*. A continuous stream of incoming and outgoing air flows through HEPA filters to hold microbes within the cabinet. The cabinets are also used to protect samples from contamination.

> **MicroByte**
>
> HEPA filters are used in passenger aircraft (to remove microbes from the recirculated cabin air) and in vacuum cleaners.

Radiation

Radio waves, microwaves, visible and ultraviolet (UV) light rays, X rays, and gamma rays are all examples of **electromagnetic radiation.** This is a form of energy that travels in waves and has no mass; the amount of energy is related to the wavelength, which is the distance from crest to crest (or trough to trough) of a wave. The wavelength is inversely proportional to the frequency, the number of waves per second. Radiation that has short waves, and therefore high frequency, has more energy than that which has long waves and low frequency. The full range of wavelengths is called the electromagnetic spectrum (**figure 5.6**).

Ionizing Radiation

Ionizing radiation has enough energy to remove electrons from atoms. This harms cells directly by destroying DNA and damaging cytoplasmic membranes. It also causes indirect damage, reacting with O_2 to produce reactive oxygen species (ROS). Bacterial endospores are among the most radiation-resistant microbial forms, whereas Gram-negative bacteria such as *Salmonella* and *Pseudomonas* species are among the most susceptible. Two important forms of ionizing radiation are gamma rays (emitted from decaying radioisotopes such as cobalt-60) and X rays.

◀◀ reactive oxygen species, p. 90 ◀◀ radioisotope, p. 19

Ionizing radiation is used extensively to sterilize heat-sensitive materials including medical equipment, disposable surgical supplies, and drugs such as penicillin. Irradiation can generally be carried out after packaging.

The number of microbes destroyed by irradiating a food product depends on the dose applied. Treatments designed to sterilize food can cause unwanted flavor changes, however, which limits their usefulness. More commonly, food is irradiated as a method of eliminating pathogens and decreasing the numbers of spoilage organisms. For example, it can be used to kill pathogens such as *Salmonella* species in poultry with little or no change in taste of the product.

In the United States, irradiation has been used for many years to control microorganisms on spices and herbs. The FDA has also approved irradiation of fruits, vegetables, and grains (to control insects), pork (to control parasites), and poultry, beef, lamb, and pork (to control bacterial pathogens). Many consumers refuse to accept irradiated products, even though the FDA and officials of the World Health and the United Nations Food and Agriculture Organizations have endorsed the technique. Some people mistakenly believe that irradiated products are radioactive. Others think that irradiation-induced toxins or carcinogens are present in food, even though available scientific evidence indicates that consumption of irradiated food is safe. Another argument raised against irradiation is that it will allow people to ignore other important food-safety practices. Irradiation, however, is intended to complement, not replace, proper food-handling procedures by producers, processors, and consumers.

Ultraviolet Radiation

Ultraviolet light in wavelengths of approximately 220 to 300 nm destroys microbes by damaging their DNA. Actively multiplying organisms are the most easily killed, whereas bacterial endospores are the most UV-resistant.

Ultraviolet light is used extensively to destroy microbes in the air and drinking water and to disinfect surfaces. It penetrates poorly, however, which limits its use. Even a thin film of grease on the UV bulb or material covering microbial cells can make it less effective. Likewise, it cannot be used to destroy microbes in solid substances or turbid liquids. Because most types of glass and plastic screen out ultraviolet radiation, UV light is most effective when used at close range against exposed microorganisms. It must be used carefully because UV rays can also damage the skin and eyes and promote the development of skin cancers.

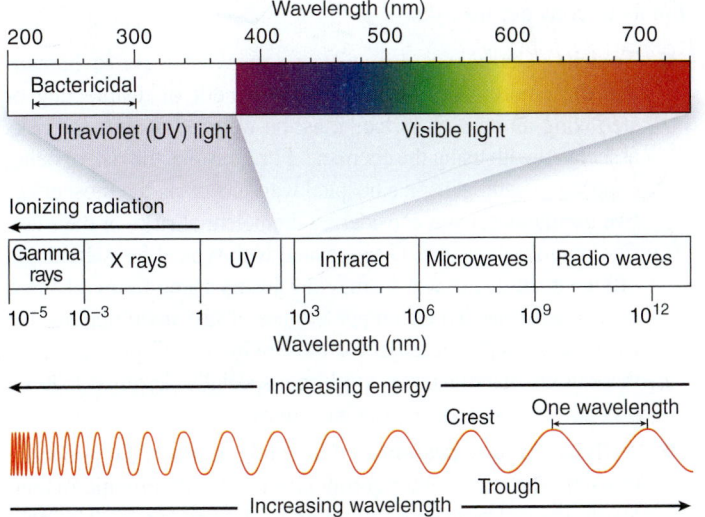

FIGURE 5.6 The Electromagnetic Spectrum Visible wavelengths include the colors of the rainbow.

❓ *Which types of radiation are ionizing?*

Microwaves

Microwaves do not affect microorganisms directly, but the heat they generate can be lethal. It is important to remember, however, that microwave ovens often heat food unevenly, so even heat-sensitive cells can sometimes survive the process.

High Pressure

High-pressure processing is used to decrease the number of microbes in commercial food products, such as guacamole, without using high temperatures. The process, which uses pressures of up to 130,000 psi, is thought to destroy microbes by denaturing proteins and altering cell permeability. High-pressure treated products keep the color and flavor associated with fresh foods.

MicroAssessment 5.4

Filters can be used to remove microorganisms and viruses from liquids and air. Gamma radiation can be used to sterilize products and to decrease the number of microorganisms in foods. Ultraviolet light can be used to disinfect surfaces and air. Microwaves do not kill microbes directly but destroy them by the heat they generate. Extreme pressure can kill microorganisms.

10. *What is the difference between the mechanism of a depth filter and that of a membrane filter?*

11. *How does ultraviolet light kill microorganisms?*

12. *Why could sterilization by gamma irradiation be carried out even after packaging?* ✚

5.5 ■ Using Chemicals to Destroy Microorganisms and Viruses

Learning Outcomes

8. *Describe the difference between sterilants, high-level disinfectants, intermediate-level disinfectants, and low-level disinfectants.*

9. *Describe five important factors to consider when selecting an appropriate germicidal chemical.*

10. *Compare and contrast the characteristics and use of alcohols, aldehydes, biguanides, ethylene oxide gas, halogens, metals, ozone, peroxygens, phenolic compounds, and quaternary ammonium compounds as germicidal chemicals.*

Germicidal chemicals can be used to disinfect and, in some cases, sterilize. Most react irreversibly with proteins, DNA, cytoplasmic membranes or viral envelopes. Their mechanisms of action, however, are often poorly understood. They are generally less reliable than heat but useful for treating large surfaces and heat-sensitive items. Some are sufficiently non-toxic to be used as antiseptics.

Potency of Germicidal Chemical Formulations

Numerous different germicidal chemicals are marketed for medical and industrial use under a variety of trade names. Frequently, they contain more than one antimicrobial chemical as well as other chemicals, such as buffers that can affect their antimicrobial activity. In the United States, the FDA is responsible for regulating chemicals that can be used to process medical devices in order to ensure they work as claimed. Most other chemical disinfectants are considered pesticides and, as such, are regulated by the Environmental Protection Agency (EPA). To be registered with either the FDA or EPA, manufacturers of germicidal chemicals must document the potency of their products using testing

procedures originally defined by the EPA. Germicides are grouped according to their potency:

- **Sterilants.** These can destroy all microorganisms, including endospores and viruses. They are also called sporicides. Destruction of endospores usually requires a 6- to 10-hour treatment. Sterilants are used to treat heat-sensitive critical instruments such as scalpels.

- **High-level disinfectants.** These destroy all viruses and vegetative microorganisms, but they do not reliably kill endospores. Most are simply sterilants used for short time periods, not long enough to ensure endospore destruction. They are used to treat semicritical instruments such as gastrointestinal endoscopes.

- **Intermediate-level disinfectants.** These destroy all vegetative bacteria including mycobacteria, fungi, and most, but not all, viruses. They do not kill endospores even with prolonged exposure. This group of germicides is used to disinfect noncritical instruments such as stethoscopes.

- **Low-level disinfectants.** These destroy fungi, vegetative bacteria except mycobacteria, and enveloped viruses. They do not kill endospores, nor do they always destroy naked viruses. Intermediate-level and low-level disinfectants are also called general-purpose disinfectants. In hospitals, they are used for disinfecting furniture, floors, and walls.

To perform properly, germicides must be used strictly according to the manufacturer's instructions, especially those for dilution, temperature, and the amount of time they must be in contact with the object being treated. It is extremely important to thoroughly clean objects before they are treated to both decrease the number of microbes present and remove organic material that might interfere with the activity of the chemical.

Selecting the Appropriate Germicidal Chemical

Selecting the appropriate germicide is a complex decision. Some points to consider include:

- **Toxicity.** Germicides are at least somewhat toxic to humans and the environment. Therefore, the benefit of disinfecting or sterilizing an item or surface must be weighed against the risks associated with using the chemical. For example, the risk of being exposed to pathogens in a hospital warrants using the most effective germicides, even considering the potential risks of their use. The microbiological risks associated with typical household and office situations, however, may not justify using many of those same chemicals. To encourage the use of less toxic options, the EPA developed a "Design for the Environment" pilot program that allows manufacturers to place a specially designed logo on products considered the least hazardous.

- **Activity in the presence of organic matter.** Hypochlorite (bleach) and many other germicides react with organic matter, losing their effectiveness in doing so. Chemicals such as phenolics, however, tolerate the presence of some organic matter.

- **Compatibility with the material being treated.** Items such as electrical equipment often cannot withstand liquid chemical germicides, and so gaseous alternatives must be employed. Likewise, corrosive germicides such as hypochlorite often damage some metals and rubber.

- **Residue.** Many chemical germicides leave a toxic or corrosive residue. If such a germicide is used to treat an item, the residue must be removed by washing the item.

- **Cost and availability.** Some germicides are less expensive and more readily available than others. Hypochlorite can easily be purchased in the form of household bleach. In contrast, ethylene oxide gas is not only expensive, but requires a special chamber, which affects the cost and practicality.

- **Storage and stability.** Some germicides are sold as concentrated stock solutions, decreasing the required storage space.

The stocks are simply diluted according to the manufacturer's instructions before use. Some have a limited shelf life once prepared.

- **Environmental risk.** Germicides that retain their antimicrobial activity after use can interfere with wastewater treatment systems. The activity of those germicides must be neutralized before disposal. ▶▶ wastewater treatment, p. 736

Classes of Germicidal Chemicals

Germicides are represented in a number of chemical families. Each type has characteristics that make it more or less appropriate for specific uses (**table 5.2**).

TABLE 5.2	Chemicals Used in Sterilization, Disinfection, and Preservation of Non-Food Substances	
Chemical (examples)	**Characteristics**	**Uses**
Alcohols (ethanol and isopropanol)	Easy to obtain and inexpensive. Rapid evaporation limits their contact time.	Aqueous solutions of alcohol are used as antiseptics to clean skin in preparation for procedures that break intact skin, and as disinfectants for treating instruments.
Aldehydes (glutaraldehyde, orthophthalaldehyde, and formaldehyde)	Capable of destroying all microbes. Irritating to the respiratory tract, skin, and eyes.	Glutaraldehyde and orthophthalaldehyde are used to sterilize medical instruments. Formalin is used in vaccine production and to preserve biological specimens.
Biguanides (chlorhexidine)	Relatively low toxicity, destroys a wide range of microbes, adheres to and persists on skin and mucous membranes.	Chlorhexidine is widely used as an antiseptic in soaps and lotions, and is often impregnated into catheters and surgical mesh.
Ethylene Oxide Gas	Easily penetrates hard-to-reach places and fabrics and does not damage moisture-sensitive material. It is toxic, explosive, and may be carcinogenic.	Commonly used to sterilize medical devices.
Halogens (chlorine and iodine)	Chlorine solutions are inexpensive and readily available; however, organic compounds and other impurities neutralize the activity. Some forms of chlorine may react with organic compounds to form toxic chlorinated products. Iodine is more expensive than chlorine and does not reliably kill endospores.	Solutions of chlorine are widely used to disinfect inanimate objects, surfaces, drinking water, and wastewater. Tincture of iodine and iodophores can be used as disinfectants or antiseptics.
Metals (silver)	Most metal compounds are too toxic to be used medically.	Silver sulfadiazine is used in topical dressings to prevent infection of burns. Silver nitrate drops can be used to prevent eye infections caused by *Neisseria gonorrhoeae* in newborns. Some metal compounds are used to prevent microbial growth in industrial processes.
Ozone	This unstable form of oxygen readily breaks down to an ineffective form.	Used to disinfect drinking water and wastewater.
Peroxygens (hydrogen peroxide and peracetic acid)	Readily biodegradable and less toxic than traditional alternatives. The effectiveness of hydrogen peroxide as an antiseptic is limited because the enzyme catalase breaks it down. Peracetic acid is a more potent germicide than hydrogen peroxide.	Hydrogen peroxide is used to sterilize containers for aseptically packaged juices and milk. Peracetic acid is widely used to disinfect and sterilize medical devices.
Phenolic Compounds (triclosan and hexachlorophene)	Wide range of activity, reasonable cost, remains effective in the presence of detergents and organic contaminants, leaves an active antimicrobial residue.	Triclosan is used in a variety of personal care products, including toothpastes, lotions, and deodorant soaps. Hexachlorophene is highly effective against *Staphylococcus aureus*, but its use is limited because it can cause neurological damage.
Quaternary Ammonium Compounds (benzalkonium chloride and cetylpyridinium chloride)	Non-toxic enough to be used on food preparation surfaces. Inactivated by anionic soaps and detergents.	Widely used to disinfect inanimate objects and to preserve non-food substances.

Alcohols

Aqueous solutions of 60% to 80% ethyl or isopropyl alcohol quickly kill vegetative bacteria and fungi. They do not, however, reliably destroy bacterial endospores and some naked viruses. Alcohol denatures essential proteins such as enzymes, and damages lipid membranes. Proteins are more soluble and denature more easily in alcohol mixed with water, which is why the aqueous solutions are more effective than pure alcohol.

Alcohol solutions are commonly used as antiseptics to clean skin before procedures such as injections that break intact skin. In addition, the Centers for Disease Control and Prevention recommends that alcohol-based hand sanitizers be used routinely by healthcare personnel as a means to protect patients.

Alcohol solutions are also used as disinfectants for treating instruments and surfaces. They are relatively non-toxic and inexpensive and do not leave a residue. Unfortunately, they evaporate quickly, which limits their contact time and, consequently, their germicidal effectiveness. In addition, they can damage rubber, some plastics, and other material.

Antimicrobial chemicals such as iodine are sometimes dissolved in alcohol. These alcohol-based solutions, called tinctures, can be more effective than the corresponding aqueous solutions.

Aldehydes

The aldehydes glutaraldehyde, orthophthalaldehyde (OPA), and formaldehyde destroy microorganisms and viruses by inactivating proteins and nucleic acids. A 2% solution of alkaline glutaraldehyde is one of the most widely used liquid chemical sterilants for treating heat-sensitive medical items. Immersion in this solution for 10 to 12 hours destroys all microbes, including endospores and viruses. Vegetative cells can be killed with soaking times as short as 10 minutes. Glutaraldehyde is toxic, however, so treated items must be thoroughly rinsed with sterile water before use.

Orthophthalaldehyde is a relatively new type of disinfectant that provides an alternative to glutaraldehyde. It requires shorter processing times and is less irritating to eyes and nasal passages, but it stains proteins gray, including those of the skin.

Formaldehyde is used as a gas or as formalin, an aqueous 37% solution. It is an extremely effective germicide that kills most microbes quickly. Formalin is used to kill bacteria and inactivate viruses for use as vaccines. It has also been used to preserve biological specimens. Formaldehyde releases irritating vapors and is a suspected carcinogen, limiting its use.

Biguanides

Chlorhexidine, the most effective of a group of chemicals called biguanides, is extensively used in antiseptic products. It stays on skin and mucous membranes, is of relatively low toxicity, and destroys a wide range of microbes, including vegetative bacteria, fungi, and some enveloped viruses. Chlorhexidine is an ingredient in many products including antiseptic skin creams, disinfectants, and mouthwashes. Chlorhexidine-impregnated catheters and implanted surgical mesh are used in medical procedures. Even tiny chips that slowly release chlorohexidine have been developed; these are inserted into periodontal pockets to treat gum

disease. Adverse side effects of chlorhexidine are rare, but severe allergic reactions have been reported.

Ethylene Oxide

Ethylene oxide is an extremely useful gaseous sterilizing agent that destroys all microbes, including endospores and viruses, by reacting with proteins. As a gas, it penetrates well into fabrics, equipment, and implantable devices such as pacemakers and artificial hips. It is particularly useful for sterilizing heat- or moisture-sensitive items such as electrical equipment, pillows, and mattresses. Many disposable laboratory items, including plastic Petri dishes and pipettes, are also sterilized with ethylene oxide.

A special chamber that resembles an autoclave is used to sterilize items with ethylene oxide. This allows careful control of factors such as temperature, relative humidity, and ethylene oxide concentration, all of which influence the effectiveness of the gas. Because ethylene oxide is explosive, it is generally mixed with a non-flammable gas such as carbon dioxide. Under these carefully controlled conditions, objects can be sterilized in 3 to 12 hours. The toxic ethylene oxide must then be eliminated from the treated material using heated forced air for 8 to 12 hours. This is important because the chemical irritates tissues and has a persistent antimicrobial effect, which, in the case of Petri dishes and other items used for growing bacteria, is unacceptable.

Ethylene oxide is mutagenic and may be carcinogenic. Studies have shown a slightly increased risk of malignancies in long-term users of the gas.

Halogens

Chlorine and iodine are common disinfectants that are thought to act by reacting with proteins and other essential cell components.

Chlorine   Chlorine destroys all types of microorganisms and viruses but is too irritating to skin and mucous membranes to be used as an antiseptic. Chlorine-releasing compounds such as sodium hypochlorite can be used to disinfect waste liquids, swimming pool water, instruments, and surfaces, and at much lower concentrations, to disinfect drinking water.

Chlorine solutions are inexpensive, readily available disinfectants (**figure 5.7**). An effective solution can easily be made by diluting liquid household bleach (5.25% sodium hypochlorite) 1:100 in water, resulting in a solution of 500 ppm (parts per million) chlorine. This concentration is several hundred times the amount

FIGURE 5.7 Chlorine in the Form of Household Bleach

[?] *Why should bleach concentrations higher than necessary be avoided?*

Contamination of an Operating Room by a Bacterial Pathogen

A patient with burns infected with *Pseudomonas aeruginosa* was taken to the operating room to have the wounds cleaned and dead tissue removed. After the procedure was completed, samples of various surfaces in the room were cultured to determine the extent of contamination. *P. aeruginosa* was recovered from all parts of the room. **Figure 1** shows how readily and extensively an operating room can become contaminated by an infected patient. Operating rooms and other patient care rooms must be thoroughly cleaned after use, in a process known as terminal cleaning.

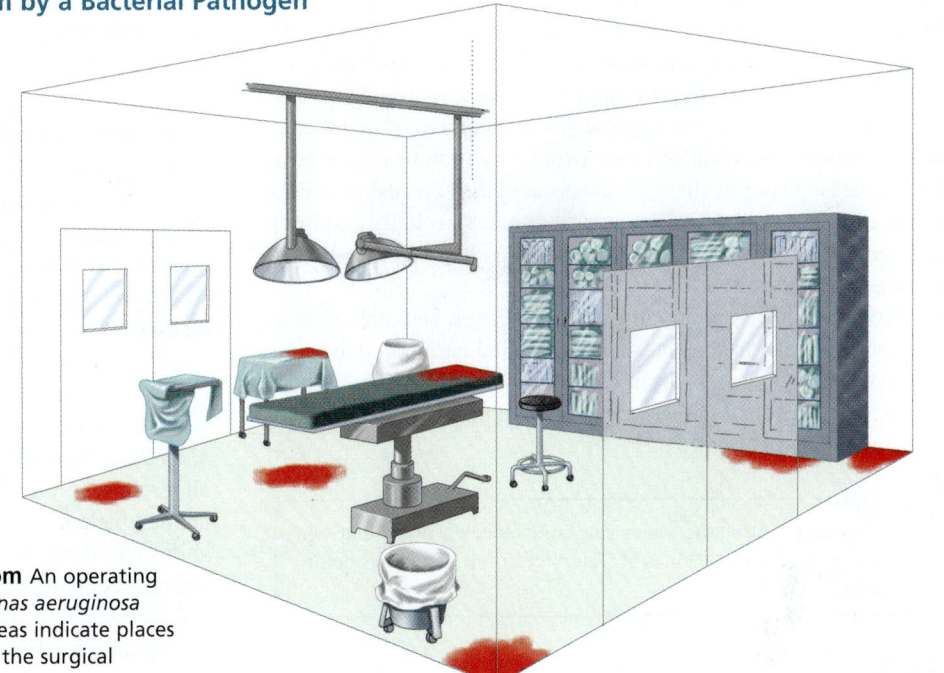

FIGURE 1 Contaminated Operating Room An operating room in which tissue infected with *Pseudomonas aeruginosa* was removed from a burn patient. Reddish areas indicate places where *P. aeruginosa* was recovered following the surgical procedures.

required to kill most pathogens, but usually necessary for fast, reliable effects. In situations when excessive organic material is present, a 1:10 dilution of bleach may be required. This is because chlorine readily reacts with organic compounds and other impurities in water, disrupting its germicidal activity. The use of high concentrations, however, should be avoided when possible, because chlorine is both corrosive and toxic. Diluted bleach deteriorates over time, so fresh solutions should be prepared regularly. More stable forms of chlorine, including sodium dichloroisocyanurate and chloramines, are often used in hospitals.

Chlorine is used at very low levels to disinfect drinking water; the amount needed depends on the amount of organic material in the water. The presence of organic compounds is also a problem because chlorine can react with some organic compounds to form trihalomethanes—disinfection by-products that are potential carcinogens. Although chlorination of drinking water kills many pathogens, it is important to remember that the levels used are not effective against *Cryptosporidium parvum* oocysts and *Giardia lamblia* cysts. ▶▶ *Cryptosporidium parvum*, p. 604 ▶▶ *Giardia lamblia*, p. 602

Chlorine dioxide (ClO_2) is a strong oxidizing agent increasingly being used as a disinfectant and sterilant. It has an advantage over chlorine-releasing compounds in that it does not react with organic compounds to form trihalomethanes or other toxic chlorinated products. Compressed chlorine dioxide gas, however, is explosive and liquid solutions decompose readily, so it must be generated on-site. It is used to treat drinking water, wastewater, and swimming pools. ▶▶ oxidizing agent, p. 131

Iodine Iodine, unlike chlorine, does not reliably kill endospores, but it can be used as a disinfectant. It is used as a tincture, or more commonly as an iodophore, in which the iodine is linked to a carrier molecule that releases free (unbound) iodine slowly. Iodophores are not as irritating to the skin as tincture of iodine nor are they as likely to stain. Iodophores used as disinfectants contain more free iodine (30 to 50 ppm) than do those used as antiseptics (1 to 2 ppm). Stock solutions must be strictly diluted according to the manufacturer's instructions because dilution affects the amount of free iodine available.

Surprisingly, some *Pseudomonas* species survive in the concentrated stock solutions of iodophores. The reasons are unclear, but it could be due to inadequate levels of free iodine in concentrated solutions, because iodine may be released from the carrier only with dilution. *Pseudomonas* species also can form biofilms, which are less permeable to chemicals. Healthcare-associated infections can result if a *Pseudomonas*-contaminated iodophore is unknowingly used to disinfect instruments. ◀◀ biofilm, p. 84

Metal Compounds

Metal compounds kill microorganisms by combining with sulfhydryl groups (—SH) of enzymes and other proteins, thereby interfering with their function. High concentrations of most metals are too toxic to human tissue to be used medically.

Silver is one of the few metals still used as a disinfectant. Creams containing silver sulfadiazine, a combination of silver and

a sulfa drug, are applied topically to prevent infection of second- and third-degree burns. Commercially available bandages with silver-containing pads can be used on minor scalds, cuts, and scrapes. For many years, doctors were required by law to add drops of another silver compound, 1% silver nitrate, into the eyes of newborns. This was to prevent ophthalmia neonatorum, an eye infection caused by *Neisseria gonorrhoeae,* acquired from infected mothers during the birth process. Drops of antibiotics have now largely replaced use of silver nitrate because they are less irritating to the eye and more effective against another genitally acquired pathogen, *Chlamydia trachomatis.* ▶▶ *Neisseria gonorrhoeae,* **p. 622**
▶▶ *Chlamydia trachomatis,* **p. 625**

Compounds of mercury, tin, arsenic, copper, and other metals were once widely used as preservatives in industrial products and to prevent microbial growth in recirculating cooling water. Their extensive use resulted in serious pollution of natural waters, which has prompted strict controls.

MicroByte
In emergencies, drinking water can be disinfected by adding two drops of plain bleach to a liter of water; let sit for 30 minutes before ingesting.

Ozone

Ozone (O_3), an unstable form of oxygen, is a powerful oxidizing agent. It decomposes quickly, however, so it must be generated on-site, usually by passing air or O_2 between two electrodes. Ozone is used as an alternative to chlorine for disinfecting drinking water and wastewater.

Peroxygens

Hydrogen peroxide and peracetic acid are powerful oxidizing agents that can be used as sterilants under controlled conditions. They are readily biodegradable and, in normal concentrations of use, appear to be less toxic than the traditional alternatives (ethylene oxide and glutaraldehyde).

Hydrogen Peroxide   The effectiveness of hydrogen peroxide (H_2O_2) as a germicide depends in part on whether it is used on living tissue or an inanimate object. This is because all cells that use aerobic metabolism, including the body's tissue cells, produce catalase, the enzyme that inactivates hydrogen peroxide by breaking it down to water and O_2. Thus, when a solution of 3% hydrogen peroxide is applied to a wound, our cellular enzymes quickly break it down. When the same solution is used on an inanimate surface, however, it overwhelms the relatively low concentration of catalase produced by microbial cells. ◀◀ catalase, **p. 90**

Hydrogen peroxide is particularly useful as a disinfectant because it leaves no residue and does not damage stainless steel, rubber, plastic, or glass. Hot solutions are commonly used in the food industry to produce commercially sterile containers for aseptically packaged juices and milk. Vapor-phase hydrogen peroxide is more effective than liquid solutions and can be used as a sterilant.

Peracetic Acid   Peracetic acid is an even more potent germicide than hydrogen peroxide. A 0.2% solution of peracetic acid, or a combination of peracetic acid and hydrogen peroxide, can

be used to sterilize items in less than 1 hour. It is effective in the presence of organic compounds, leaves no residue, and can be used on a wide range of materials. It has a sharp, strong odor, however, and like other oxidizing agents, it irritates the skin and eyes.

Phenolic Compounds (Phenolics)

Phenol (carbolic acid) is important historically because it was one of the earliest disinfectants (see **A Glimpse of History**), but it has an unpleasant odor and irritates the skin. Phenolics are derivatives of phenol that have greater germicidal activity. Because they are so effective, more dilute and therefore less irritating solutions can be used. Phenolics are the active ingredients in Lysol.

Phenolics destroy cytoplasmic membranes of microorganisms and denature proteins. They kill most vegetative bacteria and, in high concentrations (from 5% to 19%), many can kill *Mycobacterium* species. They do not, however, reliably inactivate all groups of viruses. The major advantages of phenolic compounds include their wide range of activity, reasonable cost, and ability to remain effective in the presence of detergents and organic contaminants. They also leave an active antimicrobial residue, which in some cases is desirable.

Some phenolics, such as triclosan and hexachlorophene, are sufficiently non-toxic to be used in soaps and lotions. Triclosan is widely used as an ingredient in a variety of personal care products such as deodorant soaps, lotions, and toothpaste. Hexachlorophene has substantial activity against *Staphylococcus aureus,* the leading cause of wound infections, but high levels have been associated with symptoms of neurotoxicity. Although once widely used in over-the-counter products, antiseptic skin cleansers containing hexachlorophene are now available only with a prescription. ▶▶ *Staphylococcus aureus,* **p. 524**

Quaternary Ammonium Compounds (Quats)

Quaternary ammonium compounds (quats) are cationic (positively charged) detergents non-toxic enough to be used to disinfect food preparation surfaces. Like all detergents, quats have both a charged hydrophilic region and an uncharged hydrophobic region. Because of this, they reduce the surface tension of liquids and help wash away dirt and organic material, facilitating the mechanical removal of microbes from surfaces. Unlike most common household soaps and detergents, however, which are anionic (negatively charged) and repelled by the negatively charged microbial cell surface, quats are attracted to the cell surface. They react with membranes, destroying many vegetative bacteria and enveloped viruses. They are not effective, however, against endospores, mycobacteria, or naked viruses.

Quats are economical, effective, and widely used to disinfect clean inanimate objects and preserve non-food substances. The ingredients of many personal care products include quats such as benzalkonium chloride or cetylpyridinium chloride. They also enhance the effectiveness of some other disinfectants. Cationic soaps and organic material such as gauze, however, can neutralize their activity. In addition, *Pseudomonas,* a troublesome cause of healthcare-associated infections, resists the effects of quats and can even grow in solutions preserved with them.

5.6 ■ Preservation of Perishable Products

Learning Outcome

11. *Compare and contrast chemical preservatives, low-temperature storage, and reducing the available water as methods to preserve perishable products.*

Preventing or slowing the growth of microorganisms extends the shelf life of products such as food, medicines, deodorants, cosmetics, and contact lens solutions. Preservatives are often added to these products to prevent or slow the growth of microbes inevitably introduced from the environment. Other common methods of slowing microbial growth include low-temperature storage such as refrigeration or freezing, and reducing available water. These methods are particularly important in preserving foods. ▶▶ **food spoilage, p. 756** ▶▶ **food preservation, p. 759**

Chemical Preservatives

Some of the germicidal chemicals described in the previous section can be used to preserve non-food items. For example, mouthwashes often contain a quaternary ammonium compound, cosmetics may contain thimerosal, and leather belts may be treated with one or more phenol derivatives. Food preservatives, however, must be non-toxic for safe ingestion.

Benzoic, sorbic, and propionic acids are weak organic acids sometimes added to foods such as bread, cheese, and juice to prevent microbial growth. In acidic conditions, these chemicals affect cell membrane functions. Although this action inhibits the growth of many microbes, the low pH itself prevents the growth of most bacteria. Therefore, the primary benefit of adding the preservatives is that they inhibit fungi, which otherwise grow at acidic pH.

Nitrate and its reduced form, nitrite, serve a dual purpose in processed meats. From a microbiological viewpoint, their most important function is to inhibit the germination of endospores and subsequent growth of *Clostridium botulinum*. If low levels of nitrate or nitrite are not added to cured meats such as bologna, ham, bacon, and smoked fish, *C. botulinum* may grow and produce deadly botulinum toxin. At higher concentrations than required for preservation, nitrate and nitrite react with myoglobin in the meat to form a stable pigment that gives a desirable pink color associated with fresh meat. The preservatives pose a potential hazard, however, because they can be converted to nitrosamines—some of which are carcinogens—by the metabolic activities of intestinal bacteria or during cooking.

MicroByte
Swiss cheese naturally contains propionic acid, and cranberries have benzoic acid.

Low-Temperature Storage

Refrigeration inhibits the growth of many pathogens and spoilage microorganisms by slowing or stopping critical enzyme reactions. Psychrotrophic and some psychrophilic organisms, however, can grow at refrigeration temperatures. ◀◀ **psychrophiles, p. 89**

Freezing preserves foods and other products by stopping microbial growth. The ice crystals that form kill some of the microbial cells, but those that survive can grow and spoil foods once thawed.

Reducing the Available Water

Salting and drying decrease the availability of water in food below the limits required for growth of most microbes. The high-solute environment also causes plasmolysis, which damages microbial cells (see figure 4.9). ◀◀ ▶▶ **water availability, pp. 91, 749** ◀◀ **plasmolysis, p. 91**

Adding Sugar or Salt

Sugar and salt draw water out of cells, dehydrating them. High concentrations of these solutes are added to many foods as preservatives (**figure 5.8**). For example, fruit is made into jams and

FIGURE 5.8 Preserved Jams and Jellies
❓ *How does a high concentration of sugar function as a preservative?*

jellies by adding sugar, and fish and meats are cured by soaking them in salty water, or brine. It is important to note, however, that the food-poisoning bacterium *Staphylococcus aureus* can grow in relatively high-salt conditions. ▶▶ *Staphylococcus aureus*, p. 524

Drying Food

Foods that are dried often have added salt, sugar, or chemical preservatives. For example, meat jerkies usually have added salt and sometimes sugar.

Lyophilization (freeze-drying) is widely used for preserving foods such as coffee, milk, meats, and vegetables. In the process of freeze-drying, the food is first frozen and then dried in a vacuum. When water is added to the lyophilized material, it reconstitutes. The quality of the reconstituted product is often much better than that of products dried using ordinary methods. The light weight and stability without refrigeration of freeze-dried foods make them popular with hikers.

Although drying stops microbial growth, it does not reliably kill bacteria and fungi in or on foods. For example, numerous cases of salmonellosis have been traced to dried eggs. Eggshells and even egg yolks may be heavily contaminated with *Salmonella* species from the gastrointestinal tract of the hen. To prevent the transmission of such pathogens, some states have laws requiring dried eggs to be pasteurized before they are sold.

MicroAssessment 5.6

Preservation techniques slow or halt the growth of microorganisms to delay spoilage.

16. *What is the most important function of nitrate in cured meat?*
17. *How do high concentrations of sugar and salt preserve foods?*
18. *Preservation by freezing is sometimes compared to drying. Why would this be so?* ✚

FUTURE CHALLENGES 5.1

Too Much of a Good Thing?

In our complex world, the solution to one challenge may inadvertently lead to the creation of another. Scientists have long been pursuing less toxic alternatives to many traditional biocidal chemicals. For example, glutaraldehyde has now largely replaced the more toxic formaldehyde, chlorhexidine is generally used in place of hexachlorophene, and gaseous alternatives to ethylene oxide are now being sought. Meanwhile, ozone and hydrogen peroxide, which are both readily biodegradable, may eventually replace glutaraldehyde. While these less toxic alternatives are better for human health and the environment, their widespread acceptance and use may be unwittingly contributing to an additional problem—the overuse and misuse of germicidal chemicals. Many products, including soaps, toothbrushes, and even clothing and toys are marketed with the claim of containing antimicrobial ingredients. Already bacterial resistance to some of the chemicals included in these products has been reported.

The issues surrounding the excessive use of antimicrobial chemicals are complicated. On the one hand, there is no question that some microorganisms cause disease. Even those that are not harmful to human health can be troublesome because their metabolic end products can ruin the quality of perishable products. Based on that information, it seems wise to destroy or inhibit the growth of microorganisms whenever possible. The role of microorganisms in our life, however, is not that simple. Our bodies actually harbor a greater number of microbial cells than human cells, and this normal microbiota plays an important role in maintaining our health. Excessive use of antiseptics or other antimicrobials may actually predispose a person to infection by damaging the normal microbiota.

An even more worrisome concern is that overuse of disinfectants and other germicidal chemicals will select for microorganisms that are more resistant to those chemicals, a situation similar to our current problems with

antibiotic resistance. By using antimicrobial chemicals indiscriminately, we may eventually make these useful tools obsolete. Excessive use of disinfectants may even be contributing to the problems of antibiotic resistance. Disinfectant-resistant bacteria sometimes overproduce efflux pumps that expel otherwise damaging chemicals, including antibiotics, from the cell. Thus, by overusing disinfectants, we may be inadvertently increasing antibiotic resistance.

Another concern is over the misguided belief that "non-toxic" or "biodegradable" chemicals cause no harm, and the common notion that "if a little is good, more is even better." For example, concentrated solutions of hydrogen peroxide, though biodegradable, can cause serious damage, even death, when used improperly. Other chemicals, such as chlorhexidine, can elicit severe allergic reactions in some people.

As less toxic germicidal chemicals are developed, people must be educated on the appropriate use of these alternatives.

Summary

5.1 ■ Approaches to Control
The methods used to destroy or remove microbes and viruses can be physical, such as heat treatment, irradiation, and filtration, or chemical.

Principles of Control
A variety of terms are used to describe antimicrobial agents and processes.

Situational Considerations (figure 5.1)
Situations encountered in daily life, hospitals, microbiology laboratories, food production facilities, water treatment facilities, and other industries warrant different degrees of microbial control.

5.2 ■ Selection of an Antimicrobial Procedure

Type of Microbe

One of the most critical considerations in selecting a method of destroying microorganisms and viruses is the type of microbial population thought to be present on or in the product.

Number of Microorganisms

The amount of time it takes for heat or chemicals to kill a population of microorganisms is dictated in part by the number of cells initially present. Microbial death generally occurs at a constant rate. The **D value,** or **decimal reduction time,** is the time it takes to kill 90% of a population of bacteria under specific conditions (figure 5.2).

Environmental Conditions

Factors such as pH and presence of organic materials influence microbial death rates.

Risk for Infection

Medical instruments are categorized as **critical, semicritical,** and **non-critical** according to their risk of transmitting infectious agents.

Composition of the Item

Some sterilization and disinfection procedures are inappropriate for certain types of material.

5.3 ■ Using Heat to Destroy Microorganisms and Viruses

Moist Heat

Moist heat destroys microorganisms by causing irreversible coagulation of their proteins. Pasteurization utilizes a brief heat treatment to destroy spoilage and disease-causing organisms. Pressure cookers and **autoclaves** use pressurized steam to achieve temperatures that can kill endospores (figure 5.3). The most important aspect of the commercial canning process is to ensure that endospores of *Clostridium botulinum* are destroyed.

Dry Heat

Incineration burns cell components to ashes. Temperatures achieved in hot air ovens destroy cell components and irreversibly denature proteins.

5.4 ■ Using Other Physical Methods to Remove or Destroy Microbes

Filtration (figure 5.5)

Membrane filters and **depth filters** retain microorganisms while letting the suspending fluid pass through. **High-efficiency particulate air (HEPA) filters** remove nearly all microorganisms from air.

Radiation (figure 5.6)

Ionizing radiations destroys cells by damaging cell structures and producing reactive oxygen species. Ultraviolet light damages the structure and function of nucleic acids. Microwaves kill microorganisms by generating heat.

High Pressure

High pressure is thought to destroy microorganisms by denaturing proteins and altering the permeability of the cell.

5.5 ■ Using Chemicals to Destroy Microorganisms and Viruses

Potency of Germicidal Chemical Formulations

Germicides are grouped according to their potency as **sterilants, high-level disinfectants, intermediate-level disinfectants,** or **low-level disinfectants.**

Selecting the Appropriate Germicidal Chemical

Factors that must be included in the selection of an appropriate germicidal chemical include toxicity, residue, activity in the presence of organic matter, compatibility with the material being treated, cost and availability, storage and stability, and ease of disposal.

Classes of Germicidal Chemicals (table 5.2)

Solutions of 60% to 80% ethyl or isopropyl alcohol in water rapidly kill vegetative bacteria and fungi by coagulating enzymes and other essential proteins, and by damaging lipid membranes. Glutaraldehyde, orthophthalaldehyde, and formaldehyde destroy microorganisms and viruses by inactivating proteins and nucleic acids. Chlorhexidine is a biguanide extensively used in antiseptic products. Ethylene oxide is a gaseous sterilizing agent that destroys microbes by reacting with proteins. Sodium hypochlorite (liquid bleach) is one of the least expensive and most readily available forms of chlorine. Chlorine dioxide is used as a sterilant and disinfectant. Iodophores are iodine-releasing compounds used as antiseptics. Metals interfere with protein function. Silver-containing compounds are used to prevent wound infections. Ozone is used as an alternative to chlorine in the disinfection of drinking water and wastewater. Peroxide and peracetic acid are both strong oxidizing agents that can be used alone or in combination as sterilants. Phenolics destroy cytoplasmic membranes and denature proteins; triclosan is used in lotions and deodorant soaps. Quaternary ammonium compounds are cationic detergents; they are non-toxic enough to be used to disinfect food preparation surfaces.

5.6 ■ Preservation of Perishable Products

Chemical Preservatives

Benzoic, sorbic, and propionic acids are sometimes added to foods to prevent microbial growth. Nitrate and nitrite are added to some foods to inhibit the germination and subsequent growth of *Clostridium botulinum* endospores.

Low-Temperature Storage

Low temperatures above freezing inhibit microbial growth. Freezing essentially stops all microbial growth.

Reducing the Available Water

Sugar and salt draw water out of cells, preventing the growth of microorganisms. Lyophilization is used for preserving food. The food is first frozen and then dried in a vacuum.

Review Questions

Short Answer

1. How is preservation different from pasteurization?
2. What is the most chemically resistant non-spore-forming bacterial pathogen?
3. Explain why it takes longer to kill a population of 10^9 cells than it does to kill a population of 10^3 cells.
4. What is the primary reason that wine is pasteurized?
5. What is the primary reason that milk is pasteurized?
6. When canning, why are low-acid foods processed at higher temperatures than high-acid foods?
7. How are heat-sensitive liquids sterilized?
8. How does microwaving a food product kill bacteria?
9. How is an iodophore different from a tincture of iodine?
10. Name two products commonly sterilized using ethylene oxide gas.

Multiple Choice

1. Unlike a disinfectant, an antiseptic
   a) sanitizes objects rather than sterilizes them.
   b) destroys all microorganisms.
   c) is non-toxic enough to be used on human skin.
   d) requires heat to be effective.
   e) can be used in food products.
2. The D value is defined as the time it takes to kill
   a) all bacteria in a population.
   b) all pathogens in a population.
   c) 99.9% of bacteria in a population.
   d) 90% of bacteria in a population.
   e) 10% of bacteria in a population.
3. Which of the following is the most resistant to destruction by chemicals and heat?
   a) Bacterial endospores
   b) Fungal spores
   c) *Mycobacterium tuberculosis*
   d) *E. coli*
   e) HIV
4. Ultraviolet light kills bacteria by
   a) generating heat.
   b) damaging DNA.
   c) inhibiting protein synthesis.
   d) damaging cell walls.
   e) damaging cytoplasmic membranes.
5. Which concentration of alcohol is the most effective germicide?
   a) 100%   b) 75%   c) 50%   d) 25%   e) 5%
6. Which of the following can most reliably be used as a sterilant?
   a) Alcohol   b) Phenolic compounds   c) Ethylene oxide gas
   d) Iodine
7. All of the following are routinely used to preserve foods *except*
   a) high concentrations of sugar.
   b) high concentrations of salt.
   c) benzoic acid.
   d) freezing.
   e) ethylene oxide.

8. Aseptically boxed juices and cream containers are processed using which of the following heating methods?
   a) Canning
   b) High-temperature–short-time (HTST) method
   c) Low-temperature–long-time (LTLT) method
   d) Ultra-high-temperature (UHT) method
9. Commercial canning processes are designed to ensure destruction of which of the following?
   a) All vegetative bacteria
   b) All viruses
   c) Endospores of *Clostridium botulinum*
   d) *E. coli*
   e) *Mycobacterium tuberculosis*
10. Which of the following is *false?*
    a) A high-level disinfectant cannot be used as a sterilant.
    b) Critical items must be sterilized before use.
    c) Low numbers of endospores may remain on semicritical items.
    d) Standard sterilization procedures do not destroy prions.
    e) Quaternary ammonium compounds can be used to disinfect food preparation surfaces.

Applications

1. An agriculture extension agent is preparing pamphlets on preventing the spread of disease. In the pamphlet, he must explain the appropriate situations for using disinfectants around the house. What situations should the agent discuss?
2. As a microbiologist representing a food corporation, you have been asked to serve on a health food panel to debate the need for chemical preservatives in foods. Your role is to prepare a statement that compares the benefits of chemical preservatives and the risks. What points must you bring up that indicate the benefits of chemical preservatives?

Critical Thinking ➕

1. This graph shows the time it takes to kill populations of the same microorganism under different conditions. What conditions would explain the differences in lines a, b, and c?

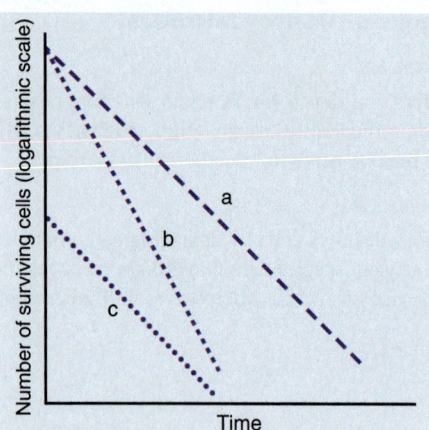

2. This diagram shows the filter paper method used to evaluate the inhibitory effect of chemical agents, heavy metals, and antibiotics on bacterial growth. A culture of a test bacterium is spread

uniformly over the surface of an agar plate. Small filter paper discs containing the material to be tested are then placed on the surface of the medium. A disc that has been soaked in sterile distilled water is sometimes added as a control. After incubation, a lawn (film of growth) will cover the plate, but a clear zone will surround those discs that contain an inhibitory compound. The size of the zone reflects several factors, one of which is the effectiveness of the inhibitory agent. What are two other factors that might affect the size of the zone of inhibition? What is the purpose of the control disc? If a clear area were apparent around the control disc, how would you interpret the observation?

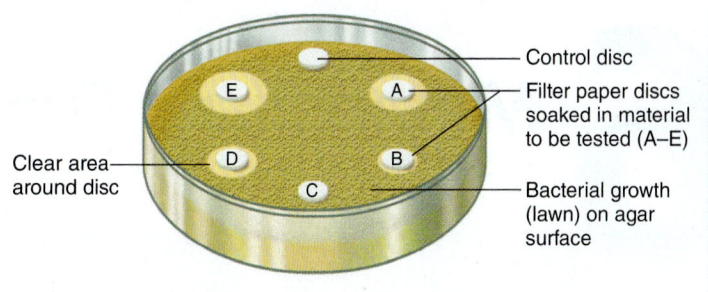

6 Metabolism: Fueling Cell Growth

Wine—a beverage produced using microbial metabolism.

KEY TERMS

Adenosine Triphosphate (ATP) The energy currency of cells. Hydrolysis of the bonds between its phosphate groups can be used to power endergonic (energy-consuming) reactions.

Anabolism Processes that synthesize and assemble the subunits of macromolecules, using energy of ATP; biosynthesis.

Catabolism Processes that harvest energy released during the breakdown of compounds such as glucose, using it to synthesize ATP.

Electron Transport Chain Group of membrane-embedded electron carriers that pass electrons from one to another, and, in the process, create a proton motive force.

Enzyme A protein that functions as a catalyst, speeding up a biological reaction.

Fermentation Metabolic process that stops short of oxidizing glucose or other organic compounds completely, using an organic intermediate as a terminal electron acceptor.

Oxidative Phosphorylation Synthesis of ATP using the energy of a proton motive force created by harvesting chemical energy.

Photophosphorylation Synthesis of ATP using the energy of a proton motive force created by harvesting radiant energy.

Precursor Metabolites Metabolic intermediates that can be either used to make the subunits of macromolecules or oxidized to generate ATP.

Proton Motive Force Form of energy generated as an electron transport chain moves protons across a membrane to create a chemiosmotic gradient.

Respiration Metabolic process that transfers electrons stripped from a chemical energy source to an electron transport chain, generating a proton motive force that is then used to synthesize ATP.

Substrate-Level Phosphorylation Synthesis of ATP using the energy released in an exergonic (energy-releasing) chemical reaction during the breakdown of the energy source.

Terminal Electron Acceptor Chemical that is ultimately reduced as a consequence of fermentation or respiration.

A Glimpse of History

In the 1850s, Louis Pasteur tried to determine how alcohol develops from grape juice. Biologists had already noticed that when grape juice is held in large vats, alcohol and CO_2 are produced and the number of yeast cells increases. They concluded that multiplying cells were converting sugar in the juice to alcohol and CO_2. Pasteur agreed, but could not convince two very powerful and influential German chemists, Justus von Liebig and Friedrich Wöhler, who refused to believe that microorganisms caused the breakdown of sugar. Both men mocked the hypothesis, publishing pictures of yeast cells looking like miniature animals taking in grape juice through one end and releasing CO_2 and alcohol through the other.

Pasteur studied the relationship between yeast and alcohol production using a strategy commonly employed by scientists today—that is, simplifying the experimental system so that relationships can be more easily identified. First, he prepared a clear solution of sugar, ammonia, mineral salts, and trace elements. He then added a few yeast cells. As the yeast grew, the sugar level decreased and the alcohol level increased, indicating that the sugar was being converted to alcohol as the cells multiplied. This strongly suggested that living cells caused the chemical transformation. Liebig, however, still would not believe that the process was occurring inside microorganisms. To convince him, Pasteur tried to extract something from inside the yeast cells that would convert the sugar. He failed, like many others before him.

In 1897, Eduard Buchner, a German chemist, showed that crushed yeast cells could convert sugar to ethanol and CO_2. We now know that the cells' enzymes carried out this transformation. For these pioneering studies, Buchner was awarded a Nobel Prize in 1907. He was the first of many investigators who received Nobel Prizes for studies on the processes by which cells degrade sugars.

All cells need to accomplish two fundamental tasks to grow. They must continually synthesize new parts—including cell walls, membranes, ribosomes, and nucleic acids—that allow the cell to enlarge and eventually divide. In addition, cells need to harvest energy and convert it to a form that can power the various energy-consuming reactions. The sum total of all chemical reactions in a cell is called **metabolism.**

Microbial metabolism is important to humans for a number of reasons. For example, as scientists look for new supplies of energy, some are investigating biofuels—fuels made from renewable biological sources such as plants and organic wastes. Metabolic pathways of microbes are used to produce biofuels, breaking down solid materials such as corn stalks, sugar cane, and wood to make ethanol. As another example, cheese-makers add *Lactococcus* and *Lactobacillus* species to milk because the metabolic wastes of these bacteria contribute to the flavor and texture of various cheeses. Microbial metabolism is also important in the laboratory, because products characteristic of specific microbes can be used as identifying markers. In addition, the metabolic pathways of organisms such as *E. coli* serve as an invaluable model for studying similar pathways in eukaryotic cells, including those of humans. Finally, metabolic processes unique to prokaryotes are potential targets for antimicrobial drugs.

6.1 ■ Principles of Metabolism

Learning Outcomes

1. *Compare and contrast catabolism and anabolism.*
2. *Describe the energy sources used by photosynthetic organisms and chemoorganoheterotrophs.*
3. *Describe the components of metabolic pathways (enzymes, ATP, chemical energy sources, redox reactions, and electron carriers) and the role of precursor metabolites.*
4. *Describe the roles of the three central metabolic pathways.*
5. *Distinguish between respiration and fermentation.*

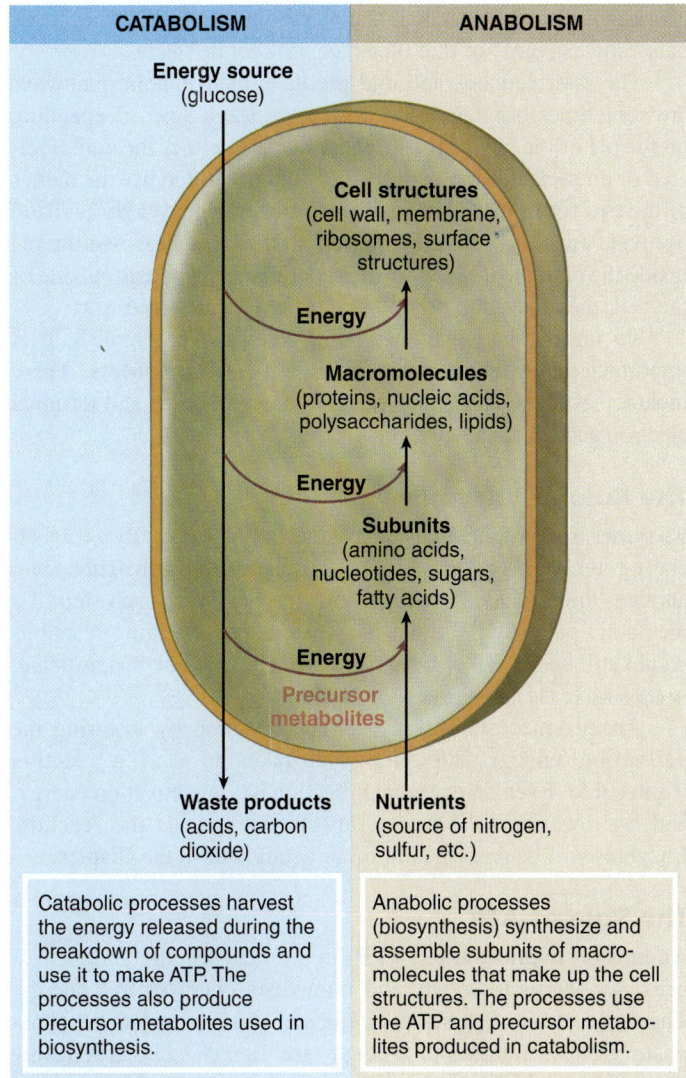

FIGURE 6.1 The Relationship Between Catabolism and Anabolism

❓ *Which subunits make up proteins? Which make up nucleic acid?*

Metabolism can be separated into two components—catabolism and anabolism (**figure 6.1**). **Catabolism** is the set of processes that degrade compounds, releasing their energy. Cells capture that energy and use it to make ATP (see figure 2.11). In contrast, **anabolism,** or **biosynthesis,** is the set of processes that cells use to synthesize and assemble the subunits of macromolecules, using ATP for energy. The subunits of macromolecules include amino acids, nucleotides, and lipids. ◀◀ ATP, p. 24

Although catabolism and anabolism are often discussed separately, they are intimately linked. As mentioned, ATP made during catabolism is used in anabolism. In addition, some of the compounds produced during catabolism serve as precursor metabolites. These are molecules that can be diverted from the catabolic pathways and used to make subunits such as amino acids.

Harvesting Energy

Energy is the capacity to do work. It can exist as potential energy (stored energy) and kinetic energy (energy of motion) (**figure 6.2**). Potential energy can be stored in various forms including chemical bonds, a rock on a hill, or water behind a dam.

Energy in the universe can never be created or destroyed; however, it can be changed from one form to another. In other words, although energy cannot be created, potential energy can be converted to kinetic energy and vice versa, and one form of potential energy can be converted to another. Hydroelectric dams unleash the potential energy of water stored behind a dam, creating the kinetic energy of moving water. This can be captured to generate an electrical current, which can then be used to charge a battery.

Photosynthetic organisms harvest the energy of sunlight, using it to power the synthesis of organic compounds from CO_2. By doing so, they convert the kinetic energy of photons (particles

FIGURE 6.2 Forms of Energy Potential energy is stored energy, such as water held behind a dam. Kinetic energy is the energy of motion, such as movement of water from behind the dam.

❓ *What is energy?*

that travel at the speed of light) to the potential energy of chemical bonds. **Chemoorganotrophs** obtain energy by degrading organic compounds. Thus, most chemoorganotrophs depend on the metabolic activities of photosynthetic organisms (**figure 6.3**).

The amount of energy released by breaking down a compound can be explained by the concept of free energy, the energy available to do work. From a biological perspective, this is the energy released when a chemical bond is broken. In a chemical reaction, some bonds are broken and others are formed. If the starting compounds have more free energy than the products,

energy is released in the reaction. The reaction is said to be **exergonic.** In contrast, if the products have more free energy than the starting compounds, the reaction requires an input of energy and is termed **endergonic.**

The change in free energy for a given reaction is the same regardless of the number of steps involved. For example, converting glucose to CO_2 and water in a single step releases the same amount of energy as degrading it in a series of steps. Cells use multiple steps when degrading compounds, harvesting the energy released in an exergonic reaction by having it power an endergonic one.

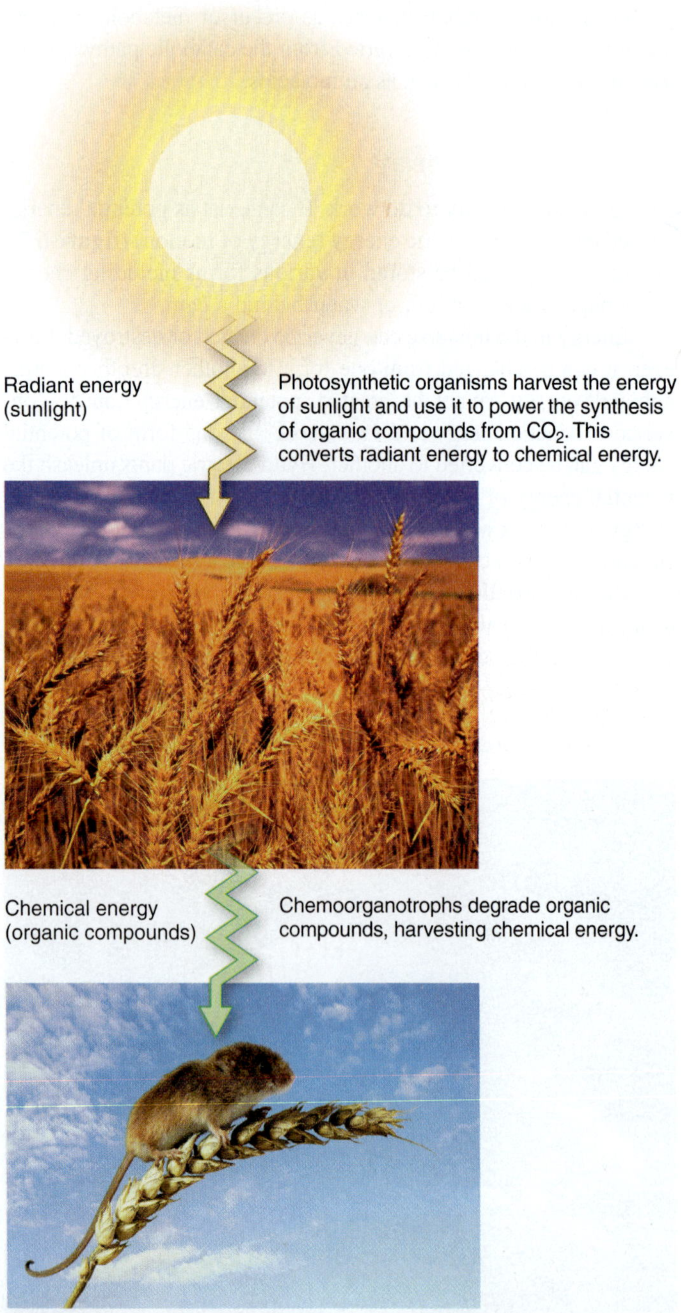

Components of Metabolic Pathways

The series of sequential chemical reactions that converts a starting compound to an end product is called a **metabolic pathway** (**figure 6.4**). Pathways can be linear, branched, or cyclical. Like the flow of rivers controlled by dams, their activities can be adjusted at certain points. This allows a cell to regulate certain processes, ensuring that specific molecules are produced in precise quantities when needed. If a metabolic step is blocked, products "downstream" of that blockage will not be made.

The intermediates and end products of metabolic pathways are sometimes organic acids, which are weak acids. Depending on the pH of the environment, these occur as either the undissociated or dissociated (ionized) form. Biologists often use the names of the two forms interchangeably—for example, pyruvic acid and pyruvate (an ion). Note, however, that at the near-neutral pH inside the cell, the ionized form predominates, whereas outside of the cell, the acid often predominates. ◀◀ ion, p. 20 ◀◀ pH, p. 23

To understand what metabolic pathways accomplish, it is important to be familiar with their critical components. These include enzymes, ATP, the chemical energy source and terminal electron acceptor, and electron carriers.

The Role of Enzymes

Enzymes are proteins that function as biological catalysts, accelerating the conversion of one substance, the **substrate,** into another, the product. A specific enzyme facilitates each step of a metabolic pathway. Without enzymes, energy-yielding reactions would still occur, but at rates so slow they would be insignificant. ▶▶ enzymes, p. 135 ◀◀ proteins, p. 25

An enzyme catalyzes a chemical reaction by lowering the **activation energy**—the energy it takes to start a reaction (**figure 6.5**). Even an exergonic reaction has an activation energy, and by lowering this barrier, an enzyme speeds the reaction. Enzymes will be described in more detail later in the chapter.

The Role of ATP

Adenosine triphosphate (ATP) is the main energy currency of cells, serving as the ready and immediate donor of free energy. The molecule is composed of ribose, adenine, and three phosphate groups arranged in tandem (see figure 2.11). **Adenosine diphosphate (ADP)** can be viewed as an acceptor of free energy.

Cells produce "energy currency" by using energy to add an inorganic phosphate group (P_i) to ADP, forming ATP. That energy

Radiant energy (sunlight)

Photosynthetic organisms harvest the energy of sunlight and use it to power the synthesis of organic compounds from CO_2. This converts radiant energy to chemical energy.

Chemical energy (organic compounds)

Chemoorganotrophs degrade organic compounds, harvesting chemical energy.

FIGURE 6.3 Most Chemoorganotrophs Depend on Photosynthetic Organisms

❓ *How do chemoorganotrophs require the activities of photosynthetic organisms?*

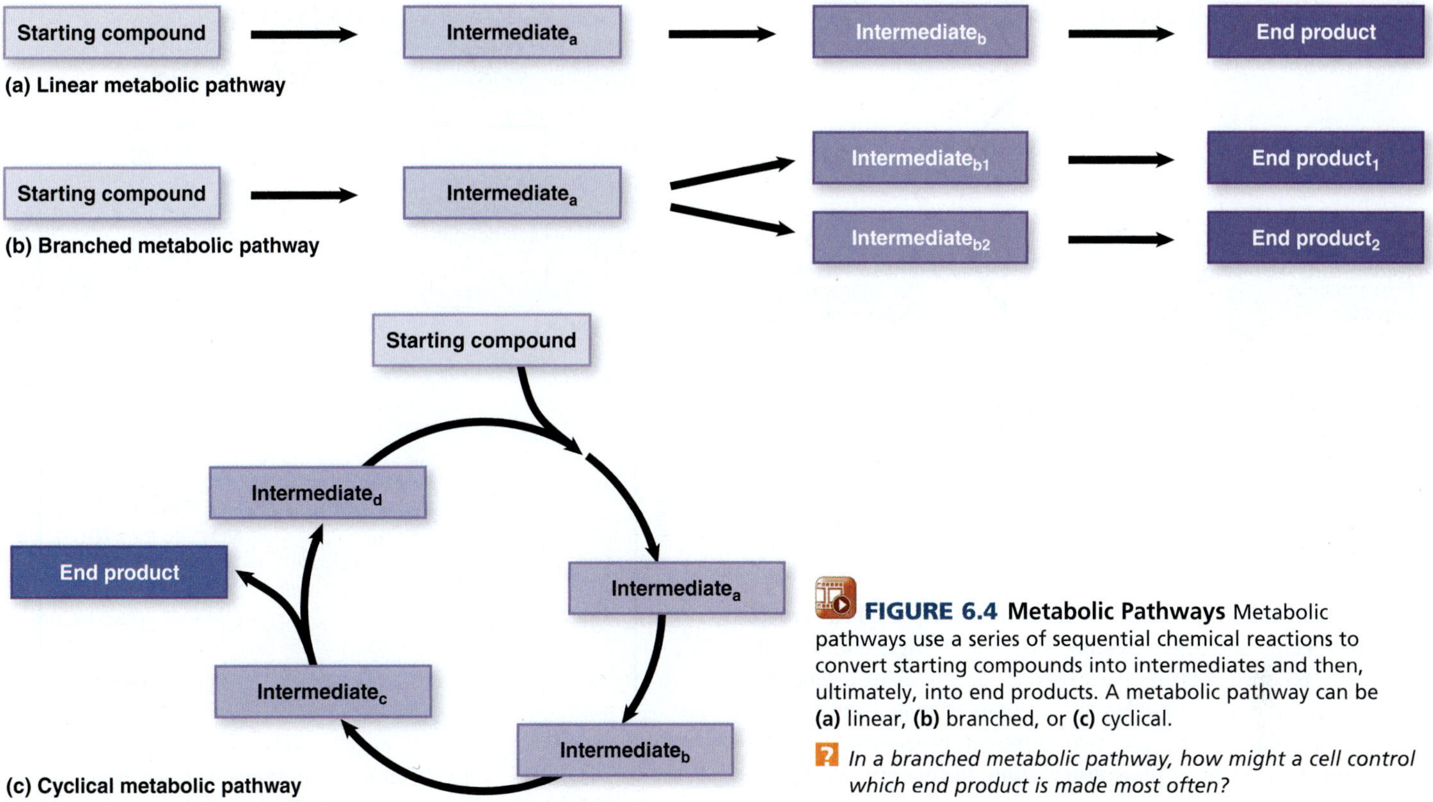

(a) Linear metabolic pathway

Starting compound → Intermediate$_a$ → Intermediate$_b$ → End product

(b) Branched metabolic pathway

Starting compound → Intermediate$_a$ → Intermediate$_{b1}$ → End product$_1$

Intermediate$_{b2}$ → End product$_2$

(c) Cyclical metabolic pathway

Starting compound → Intermediate$_a$ → Intermediate$_b$ → Intermediate$_c$ → Intermediate$_d$ → End product

FIGURE 6.4 Metabolic Pathways Metabolic pathways use a series of sequential chemical reactions to convert starting compounds into intermediates and then, ultimately, into end products. A metabolic pathway can be **(a)** linear, **(b)** branched, or **(c)** cyclical.

? *In a branched metabolic pathway, how might a cell control which end product is made most often?*

FIGURE 6.5 The Role of Enzymes Enzymes function as biological catalysts. **(a)** An enzyme catalyzes a chemical reaction by lowering the activation energy of the reaction. **(b)** A specific enzyme facilitates each step of a metabolic pathway.

? *What is activation energy?*

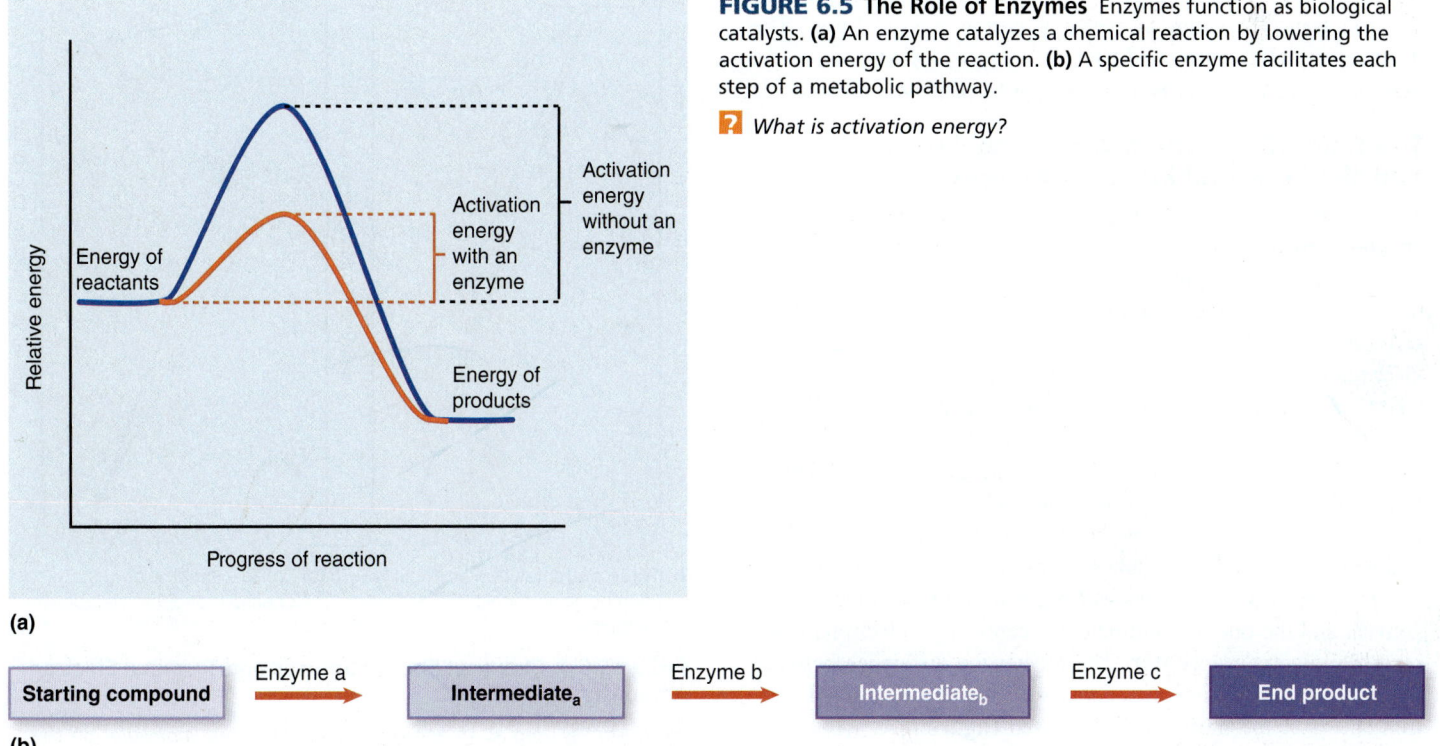

Relative energy

Energy of reactants

Activation energy with an enzyme

Activation energy without an enzyme

Energy of products

Progress of reaction

(a)

Starting compound —Enzyme a→ Intermediate$_a$ —Enzyme b→ Intermediate$_b$ —Enzyme c→ End product

(b)

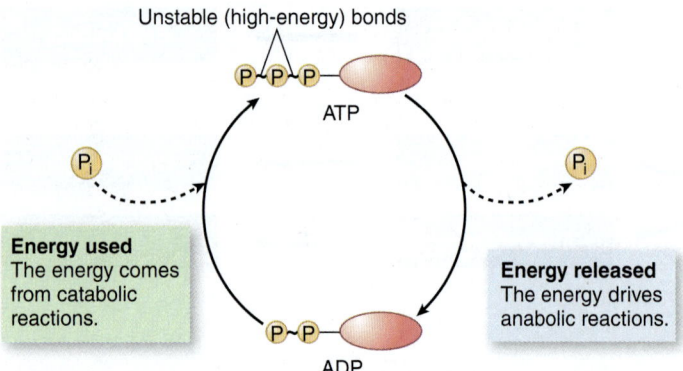

FIGURE 6.6 ATP Cells produce ATP by using energy to add inorganic phosphate (P$_i$) to ADP. Energy is released in the reverse reaction, which breaks the phosphate bond to convert ATP to ADP + P$_i$.

? *What characteristic of the structure of ATP makes its phosphate bonds "high energy"?*

currency can then be "spent" by removing the phosphate group, releasing energy and again yielding ADP and P$_i$ (**figure 6.6**). Cells constantly produce ATP during exergonic reactions of catabolism and then use it to power endergonic reactions of anabolism.

Chemoorganotrophs use two different processes to make ATP. In **substrate-level phosphorylation,** the energy released in an exergonic reaction is used to power the addition of P$_i$ to ADP; in **oxidative phosphorylation** the energy of a proton motive force drives the reaction. Recall from chapter 3 that **proton motive force** is the form of energy that results from the electrochemical gradient established by the electron transport chain (see figure 3.27). Details of this process will be discussed later in the chapter. Photosynthetic organisms can generate ATP by **photophosphorylation.** This uses the sun's radiant energy and an electron transport chain to create a proton motive force. The mechanism will be discussed later. ▸▸ oxidative phosphorylation, p. 142 ◂◂ proton motive force, p. 55 ▸▸ photophosphorylation, p. 152

The Role of the Chemical Energy Source and the Terminal Electron Acceptor

To understand how cells obtain energy, it is helpful to recall from chapter 2 that certain atoms are more electronegative than others, meaning they have a greater affinity (attraction) for electrons (see table 2.2). Likewise, certain molecules have a greater affinity for electrons than other molecules, a characteristic that relates to the electronegativities of the atoms that make up the molecules. When electrons move from a molecule that has a relatively low electron affinity (tends to give up electrons) to one that has a higher electron affinity (tends to accept electrons), energy is released. This is how cells obtain the energy used to make ATP. They remove electrons from glucose or another low electron affinity chemical, and donate them to a molecule such as O$_2$ that has a higher affinity. The chemical that serves as the electron donor is the **energy source,** and the one that ultimately accepts those electrons is the **terminal electron acceptor.** The greater the difference between the electron affinities of the energy source and the terminal electron acceptor, the more energy released (**figure 6.7**). ◂◂ electrons, p. 17 ◂◂ electronegativity, p. 21

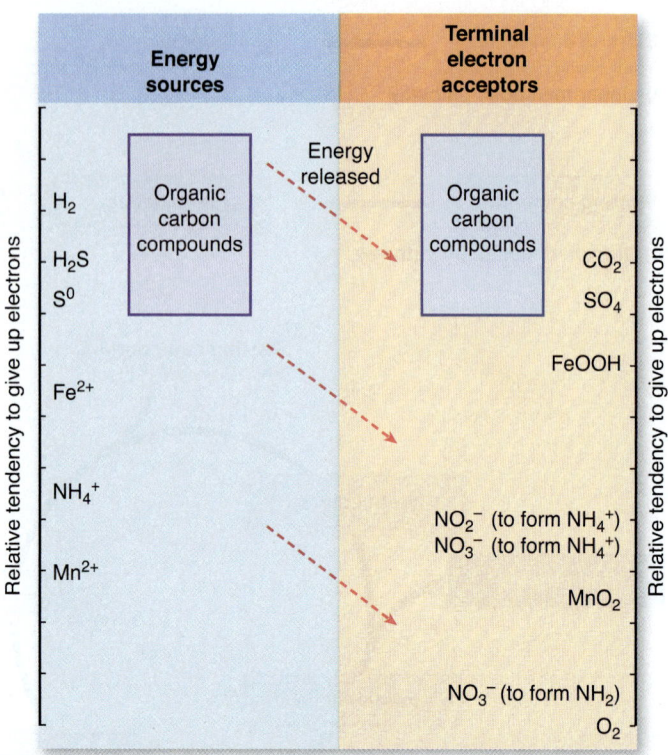

(a) Energy is released when electrons are moved from an energy source with a low affinity for electrons to a terminal electron acceptor with a higher affinity.

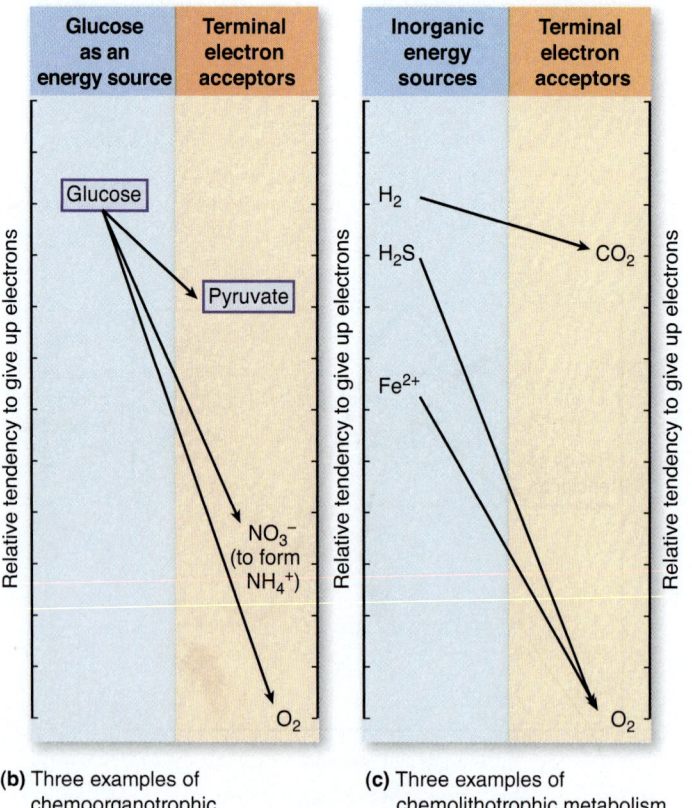

(b) Three examples of chemoorganotrophic metabolism

(c) Three examples of chemolithotrophic metabolism

FIGURE 6.7 Chemical Energy Sources and Terminal Electron Acceptors

? *If Mn^{2+} is used as an energy source, which two molecules on the chart in (a) could serve as terminal electron acceptors?*

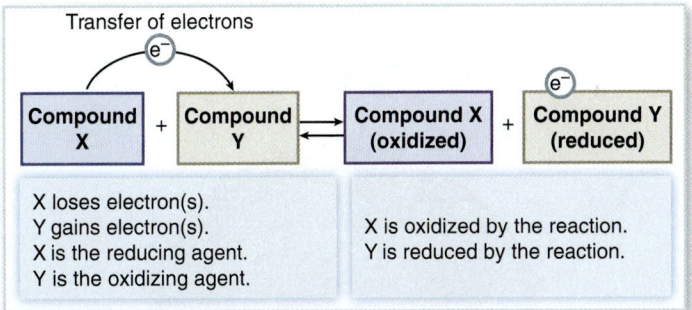

FIGURE 6.8 Oxidation-Reduction Reactions The molecule that loses one or more electrons is oxidized by the reaction; the one that gains those electrons is reduced.

? *Why is hydrogenation a reduction?*

As a group, prokaryotes are remarkably diverse with respect to the energy sources and terminal electron acceptors they can use (see figure 6.7a and b). *E. coli* and other chemoorganotrophs use organic compounds such as glucose for an energy source, whereas chemolithotrophs use hydrogen sulfide or another inorganic chemical. Aerobic organisms can use O_2 as a terminal electron acceptor, whereas anaerobes use only molecules other than O_2.
◀◀ chemoorganotrophs p. 93 ◀◀ chemolithotrophs, p. 93

Cells remove electrons from the energy source through a series of **oxidation-reduction reactions,** or **redox reactions.** The substance that loses electrons is **oxidized** by the reaction; the one that gains those electrons is **reduced** (**figure 6.8**). When electrons are removed from a biological molecule, protons (H^+) often follow. The result is that an electron-proton pair, or hydrogen atom, is removed. Thus, dehydrogenation (the removal of a hydrogen atom) is an oxidation. Correspondingly, hydrogenation (the addition of a hydrogen atom) is a reduction.

The Role of Electron Carriers

When electrons are removed from the energy source, they are not removed all at once, nor are they transferred directly to the terminal electron acceptor. Instead, the energy is harvested in a stepwise process, with the electrons initially transferred to **electron carriers.** Cells have several different types of electron carriers, each with distinct roles (**table 6.1**). The electron carriers are usually referred to by abbreviations that represent their oxidized and reduced forms, respectively:

- **NAD^+/NADH** (nicotinamide adenine dinucleotide)
- **$NADP^+$/NADPH** (nicotinamide adenine dinucleotide phosphate)
- **FAD/FADH$_2$** (flavin adenine dinucleotide)

These electron carriers can also be considered hydrogen carriers because, along with electrons, they carry protons. $FADH_2$ carries two electrons and two protons; NADH and NADPH each carry two electrons and one proton. Note, however, that the location of protons in biological reactions is often ignored because in aqueous solutions protons—unlike electrons—do not require carriers.

Reduced electron carriers represent **reducing power** because they can easily transfer their electrons to another chemical that has a higher affinity for electrons. By doing so, they raise the energy level of the recipient molecule. NADH and $FADH_2$ transfer their electrons to the electron transport chain, which then uses the energy to generate a proton motive force. In turn, this drives the synthesis of ATP in the process of oxidative phosphorylation. Ultimately the electrons are transferred to the terminal electron acceptor. The electrons carried by NADPH have an entirely different fate; they are used to reduce compounds during biosynthetic reactions. Note, however, that many microbial cells can convert reducing power in the form of NADH to NADPH.

TABLE 6.1	Electron Carriers			
Carrier	**Oxidized Form (Accepts Electrons)**		**Reduced Form (Donates Electrons)**	**Typical Fate of Electrons Carried**
Nicotinamide adenine dinucleotide (carries 2 electrons and 1 proton)	$NAD^+ + 2\,e^- + 2\,H^+$	⇌	$NADH + H^+$	Used to generate a proton motive force that can drive ATP synthesis
Flavin adenine dinucleotide (carries 2 electrons and 2 protons; i.e., 2 hydrogen atoms)	$FAD + 2\,e^- + 2\,H^+$	⇌	$FADH_2$	Used to generate a proton motive force that can drive ATP synthesis
Nicotinamide adenine dinucleotide phosphate (carries 2 electrons and 1 proton)	$NADP^+ + 2\,e^- + 2\,H^+$	⇌	$NADPH + H^+$	Biosynthesis

Precursor Metabolites

Certain intermediates of catabolic pathways can be used in anabolic pathways. These intermediates—**precursor metabolites**—serve as carbon skeletons from which subunits of macromolecules can be made (**table 6.2**). For example, the precursor metabolite pyruvate can be converted to any one of three amino acids: alanine, leucine, or valine.

Recall from chapter 4 that *E. coli* can grow in glucose-salts medium, which contains only glucose and a few inorganic salts. This means that the glucose is serving two purposes in the cell: (1) the energy source, and (2) the starting point from which all cell components are made—including proteins, lipids, carbohydrates, and nucleic acids. When *E. coli* cells degrade glucose molecules, they use a series of steps that not only release energy but also form a dozen or so precursor metabolites. Other organisms use the same steps but sometimes lack certain biosynthetic capabilities. Any essential compounds that a cell cannot synthesize must be provided from an external source. ◄◄ glucose-salts medium, p. 95

It is important to realize that the glucose molecules in *E. coli* and other cells can have different fates. In essence, the catabolic pathways are like an extensive highway system for millions of glucose molecules (**figure 6.9**). Glucose molecules that remain on the "highway" are oxidized completely to CO_2, releasing the maximum amount of energy. Some breakdown intermediates, however, can "exit at an off ramp" to be used in biosynthesis. The "off ramps" are located at the steps immediately after a precursor metabolite is made. So once a precursor metabolite is made in catabolism, it can be further oxidized to release energy, or it can be used in biosynthesis.

Overview of Catabolism

Catabolism of glucose, the preferred energy source of many cells, encompasses two key processes: (1) oxidizing glucose molecules to generate ATP, reducing power, and precursor metabolites; and (2) transferring the electrons carried by NADH and $FADH_2$ (reducing power) to the terminal electron acceptor. The central metabolic pathways do the former, and respiration and fermentation do the latter.

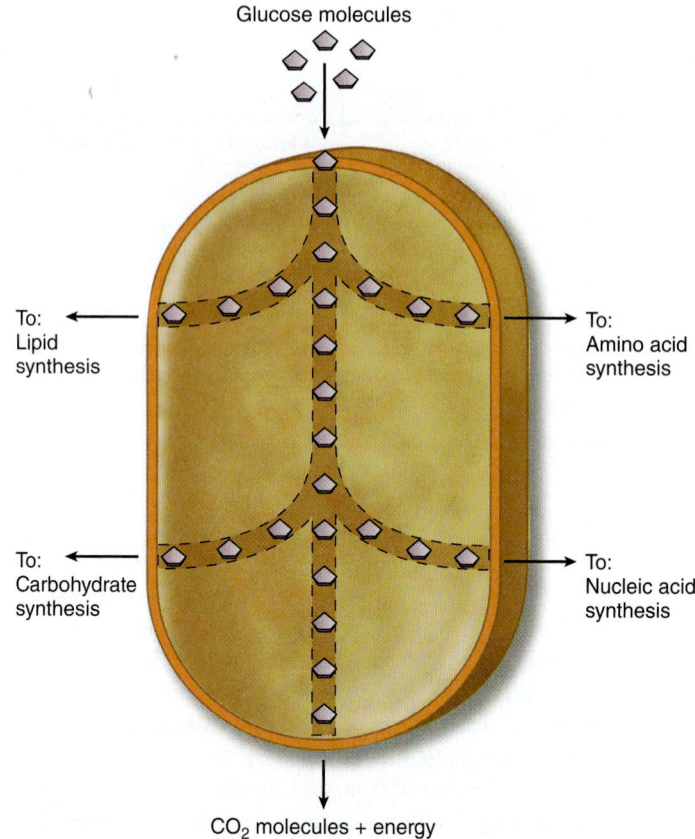

Glucose molecules

To: Lipid synthesis

To: Amino acid synthesis

To: Carbohydrate synthesis

To: Nucleic acid synthesis

CO_2 molecules + energy

FIGURE 6.9 Metabolic Highway The millions of glucose molecules that continually enter a cell can have different fates. Some may be oxidized completely to release the maximum amount of energy, and others will be used in biosynthesis.

❓ *If precursor metabolites were shown on this diagram, where would they be located?*

Central Metabolic Pathways

Three key metabolic pathways—the **central metabolic pathways**—gradually oxidize glucose to CO_2 (**figure 6.10**). The pathways are catabolic, but the precursor metabolites and reducing

TABLE 6.2	Precursor Metabolites	
Precursor Metabolite	**Pathway Generated**	**Biosynthetic Role**
Glucose 6-phosphate	Glycolysis	Lipopolysaccharide
Fructose 6-phosphate	Glycolysis	Peptidoglycan
Dihydroxyacetone phosphate	Glycolysis	Lipids (glycerol component)
3-phosphoglycerate	Glycolysis	Proteins (the amino acids cysteine, glycine, and serine)
Phosphoenolpyruvate	Glycolysis	Proteins (the amino acids phenylalanine, tryptophan, and tyrosine)
Pyruvate	Glycolysis	Proteins (the amino acids alanine, leucine, and valine)
Ribose 5-phosphate	Pentose phosphate cycle	Nucleic acids and proteins (the amino acid histidine)
Erythrose 4-phosphate	Pentose phosphate cycle	Proteins (the amino acids phenylalanine, tryptophan, and tyrosine)
Acetyl-CoA	Transition step	Lipids (fatty acids)
α-ketoglutarate	TCA cycle	Proteins (the amino acids arginine, glutamate, glutamine, and proline)
Oxaloacetate	TCA cycle	Proteins (the amino acids aspartate, asparagine, isoleucine, lysine, methionine, and threonine)

Some organisms use succinyl-CoA as a precursor in heme biosynthesis; *E. coli* uses glutamate.

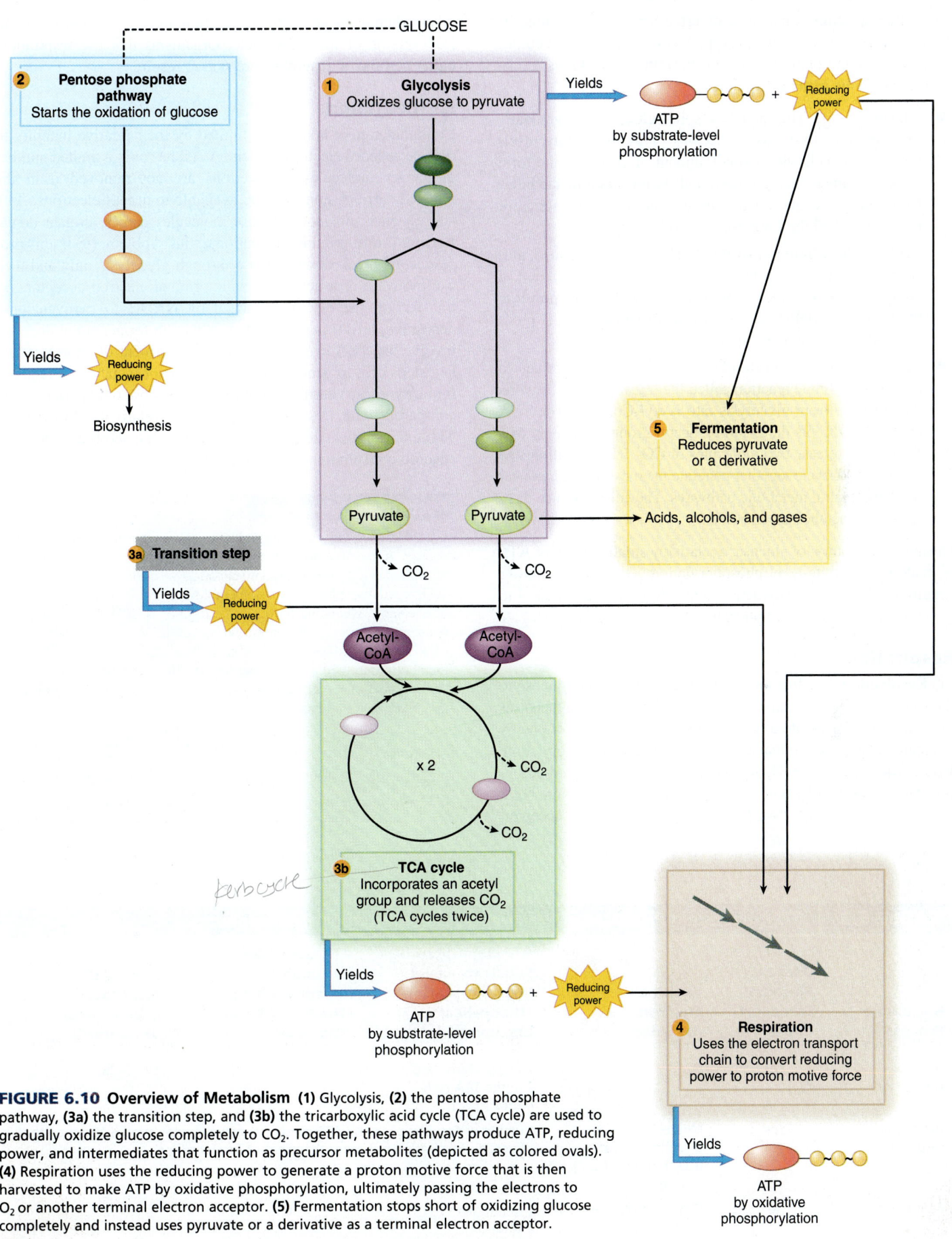

FIGURE 6.10 Overview of Metabolism **(1)** Glycolysis, **(2)** the pentose phosphate pathway, **(3a)** the transition step, and **(3b)** the tricarboxylic acid cycle (TCA cycle) are used to gradually oxidize glucose completely to CO_2. Together, these pathways produce ATP, reducing power, and intermediates that function as precursor metabolites (depicted as colored ovals). **(4)** Respiration uses the reducing power to generate a proton motive force that is then harvested to make ATP by oxidative phosphorylation, ultimately passing the electrons to O_2 or another terminal electron acceptor. **(5)** Fermentation stops short of oxidizing glucose completely and instead uses pyruvate or a derivative as a terminal electron acceptor.

? *What is the terminal electron acceptor in aerobic respiration?*

power they generate can also be diverted for use in biosynthesis. To reflect the dual role of these pathways, they are sometimes called amphibolic pathways (*amphi* meaning "both kinds"). The central metabolic pathways include:

- **Glycolysis. ①** This pathway splits glucose and gradually oxidizes it to form two molecules of pyruvate. Glycolysis provides the cell with a small amount of energy in the form of ATP, some reducing power, and six precursor metabolites. Note that some microbial cells use an alternative series of reactions called the Entner-Doudoroff pathway.

- **Pentose phosphate pathway. ②** This also breaks down glucose, but its primary role in metabolism is the production of compounds used in biosynthesis, including reducing power in the form of NADPH and two precursor metabolites. A product of the pathway feeds into glycolysis.

- **Tricarboxylic acid cycle (TCA cycle). ③a** As a prelude to this cycle, a single reaction called the transition step converts the pyruvate from glycolysis into acetyl-CoA. **③b** The TCA cycle then accepts the 2-carbon acetyl group, ultimately oxidizing it to release two molecules of CO_2. The transition step and the TCA cycle together generate the most reducing power of all the central metabolic pathways. They also produce three precursor metabolites and ATP.

During the oxidation of glucose, a relatively small amount of ATP is made by substrate-level phosphorylation. The reducing power accumulated during the oxidation steps, however, can be used to generate ATP by oxidative phosphorylation.

Respiration

④ Respiration transfers the electrons extracted from glucose to the electron transport chain. The electron transport chain then uses the electrons to generate a proton motive force, a form of energy that can be harvested to make ATP by oxidative phosphorylation. In **aerobic respiration,** O_2 serves as the terminal electron acceptor. **Anaerobic respiration** is similar to aerobic respiration, but it uses a molecule other than O_2 as a terminal electron acceptor.

Also, when using anaerobic respiration, microbes employ a modified version of the TCA cycle. Organisms that use respiration, either aerobic or anaerobic, are said to respire.

Fermentation

Cells that cannot respire are limited by their relative inability to recycle reduced electron carriers. A cell has only a limited number of carrier molecules; if electrons are not removed from the reduced carriers, none will be available to accept electrons. As a consequence, no more glucose molecules can be broken down. **Fermentation** provides a solution to this problem. **⑤** In this process, cells break down glucose through glycolysis only and then use pyruvate or a derivative as a terminal electron acceptor. By transferring the electrons carried by NADH to pyruvate or a derivative, NAD^+ is regenerated. Although fermentation does not involve the TCA cycle, organisms that ferment often use certain key steps of it to generate specific precursor metabolites required for biosynthesis. Fermentation oxidizes glucose only partially and, compared with respiration, produces relatively little ATP. **Table 6.3** summarizes the difference between aerobic respiration, anaerobic respiration, and fermentation.

MicroAssessment 6.1

Catabolic pathways gradually oxidize an energy source to harvest energy. A specific enzyme catalyzes each step. Substrate-level phosphorylation uses exergonic chemical reactions to synthesize ATP; oxidative phosphorylation employs a proton motive force to do the same. Reducing power in the form of NADH and $FADH_2$ is used to generate a proton motive force; the reducing power of NADPH is utilized in biosynthesis. Precursor metabolites are metabolic intermediates that can be used in biosynthesis. The central metabolic pathways generate ATP, reducing power, and precursor metabolites.

1. *How does the fate of electrons carried by NADPH differ from the fate of electrons carried by NADH?*
2. *Why are the central metabolic pathways called amphibolic?*
3. *Why does fermentation supply less energy than respiration?* ➕

TABLE 6.3	ATP-Generating Processes of Prokaryotic Chemoorganoheterotrophs				
Metabolic Process	Uses an Electron Transport Chain	Terminal Electron Acceptor	ATP Generated by Substrate-Level Phosphorylation (Theoretical Maximum)	ATP Generated by Oxidative Phosphorylation (Theoretical Maximum)	Total ATP Generated (Theoretical Maximum)
Aerobic respiration	Yes	O_2	2 in glycolysis (net) 2 in the TCA cycle 4 total	34	38
Anaerobic respiration	Yes	Molecule other than O_2 such as nitrate (NO_3^-), nitrite (NO_2^-), sulfate (SO_4^{2-})	Number varies; however, the ATP yield of anaerobic respiration is less than that of aerobic respiration but more than that of fermentation.		
Fermentation	No	Organic molecule (pyruvate or a derivative)	2 in glycolysis (net) 2 total	0	2

6.2 ■ Enzymes

Learning Outcomes

6. *Describe the active site of an enzyme, and explain how it relates to the enzyme-substrate complex.*

7. *Compare and contrast cofactors and coenzymes.*

8. *List two environmental factors that influence enzyme activity.*

9. *Describe allosteric regulation.*

10. *Compare and contrast competitive enzyme inhibition and non-competitive enzyme inhibition.*

Recall that enzymes are biological catalysts, which increase the rate at which substrates are converted into products (see figure 6.5). They do this with extraordinary specificity and speed, usually acting on only one or a few substrates. They are neither consumed nor permanently changed during a reaction, allowing a single enzyme molecule to be rapidly used again and again. More than a thousand different enzymes exist in a cell. ▶▶| enzymes, p. 135

The name of an enzyme usually reflects its function and ends with the suffix *-ase*. For example, isocitrate dehydrogenase removes a hydrogen atom (a proton-electron pair) from isocitrate. Some groups of enzymes are referred to by their general function—for example, proteases degrade proteins.

MicroByte

In only one second, the fastest enzymes can transform more than 10^4 substrate molecules into products.

Mechanisms and Consequences of Enzyme Action

An enzyme has on its surface an **active site,** typically a relatively small crevice (**figure 6.11**). This is the critical site to which a substrate binds by weak forces. The binding of the substrate to the active site causes the shape of the flexible enzyme to change slightly. This mutual interaction, or induced fit, results in a temporary intermediate called an enzyme-substrate complex. The substrate is held within this complex in a specific orientation so that existing bonds are destabilized and new ones can easily form, lowering the activation energy of the reaction. The products are then released, leaving the enzyme unchanged and free to combine with new substrate molecules. Note that enzymes can also catalyze reactions that join two substrates to create one product. Theoretically, all enzyme-catalyzed reactions are reversible, but the free energy change of certain reactions makes them effectively non-reversible.

The interaction of an enzyme with its substrate is very specific. The substrate fits into the active site like a hand into a glove. Not only must it fit spatially, but appropriate chemical interactions such as hydrogen and ionic bonding need to occur to induce the fit. This requirement for a precise fit and interaction explains why, with few exceptions, a unique enzyme is required to catalyze each reaction in a cell. Very few molecules of any particular enzyme are needed, however, as each can be used repeatedly. |◀◀ hydrogen bonds, p. 21 |◀◀ ionic bonds, p. 20

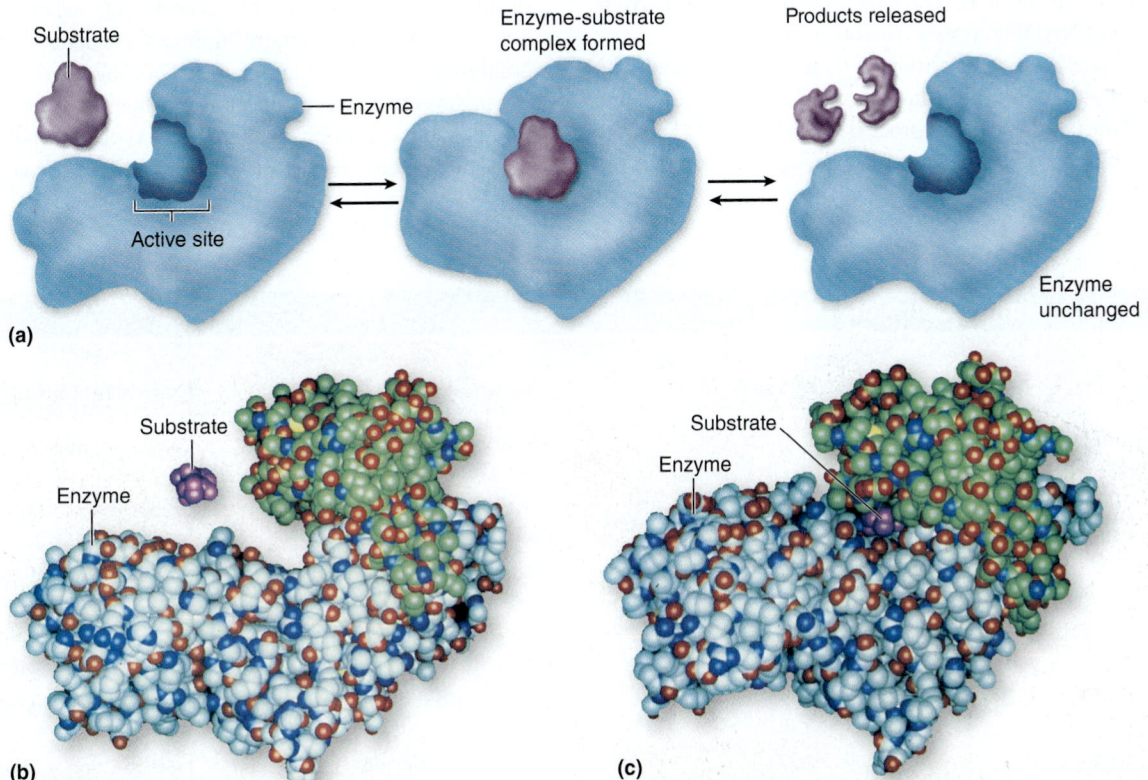

(a)

(b)

(c)

FIGURE 6.11 Mechanism of Enzyme Action (a) The substrate binds to the active site, forming an enzyme-substrate complex. The products are then released, leaving the enzyme unchanged and free to combine with new substrate molecules. (b) A model showing an enzyme and its substrate. (c) The binding of the substrate to the active site causes the shape of the flexible enzyme to change slightly.

▢ *Why are enzymes so specific with respect to the reactions they catalyze?*

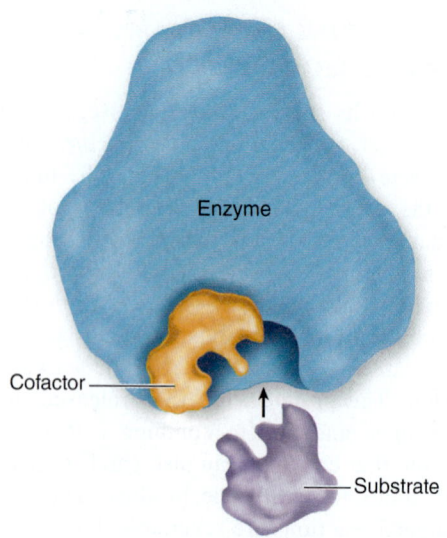

FIGURE 6.12 Some Enzymes Act in Conjunction with a Cofactor Cofactors are non-protein components, either coenzymes or trace elements.

? *What is the function of the coenzyme FAD?*

Cofactors

Some enzymes act with the assistance of a non-protein component called a **cofactor** (**figure 6.12**). Magnesium, zinc, copper, and other trace elements often function as cofactors. **Coenzymes** are organic cofactors that function as loosely bound carriers of molecules or electrons (**table 6.4**). They include the electron carriers FAD, NAD+, and NADP+. ◀◀ trace elements, p. 93

All coenzymes transfer substances from one compound to another, but they function in different ways. Some remain bound to the enzyme during the transfer process, whereas others separate from the enzyme, carrying the substance being transferred along

with them. The same coenzyme can assist different enzymes. Because of this, far fewer different coenzymes are required than enzymes. Like enzymes, coenzymes are recycled as they function and, consequently, are needed only in very small quantities.

Most coenzymes are derived from vitamins (table 6.4). Many prokaryotes can synthesize these vitamins and convert them to the necessary coenzymes. Humans and other animals, however, cannot synthesize most vitamins, so the vitamins usually must be ingested as part of the diet. In some cases, vitamins made by bacteria residing in the intestine can be absorbed. If an animal lacks a vitamin, the functions of all the different enzymes whose activity requires the corresponding coenzyme are impaired. Thus, a single vitamin deficiency has serious consequences.

Environmental Factors That Influence Enzyme Activity

Several environmental factors influence how well enzymes function and in this way determine how rapidly microbes multiply (**figure 6.13**). Each enzyme has a narrow range of conditions—including temperature, pH, and salt concentration—within which it operates best. A 10°C rise in temperature approximately doubles the speed of enzymatic reactions, until optimal activity is reached; this explains why bacteria tend to grow more rapidly at higher temperatures. If the temperature is too high, however, proteins will denature and no longer function. Most enzymes operate best at low salt concentrations and at pH values slightly above 7. Not surprisingly, most microbes grow fastest under these same conditions. Some prokaryotes, however, particularly certain members of the *Archaea,* thrive in environments where conditions are extreme. They may require high salt concentrations, very acidic conditions, or temperatures near boiling. Enzymes from these organisms can be very important commercially because they function in harsh conditions that are often typical of industrial settings. ◀◀ temperature and growth requirements, p. 89 ◀◀ denaturation, p. 29

TABLE 6.4	Some Coenzymes and Their Function		
Coenzyme	**Vitamin from Which It Is Derived**	**Substance Transferred**	**Example of Use**
Nicotinamide adenine dinucleotide (NAD+)	Niacin	Hydride ions (2 electrons and 1 proton)	Carrier of reducing power
Flavin adenine dinucleotide (FAD)	Riboflavin	Hydrogen atoms (2 electrons and 2 protons)	Carrier of reducing power
Coenzyme A	Pantothenic acid	Acyl groups	Carries the acetyl group that enters the TCA cycle
Thiamin pyrophosphate	Thiamine	Aldehydes	Helps remove CO_2 from pyruvate in the transition step
Pyridoxal phosphate	Pyridoxine	Amino groups	Transfers amino groups in amino acid synthesis
Tetrahydrofolate	Folic acid	1-carbon molecules	1-carbon donor in nucleotide synthesis

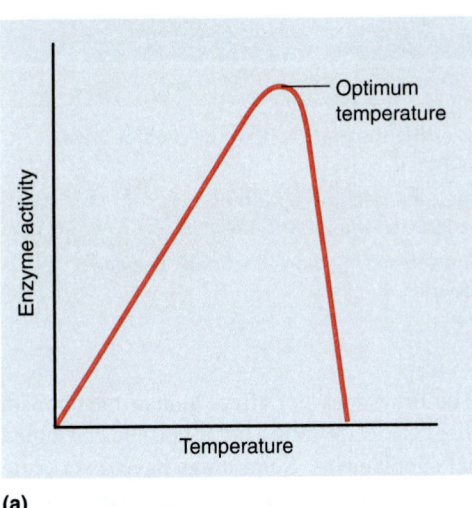

(a)

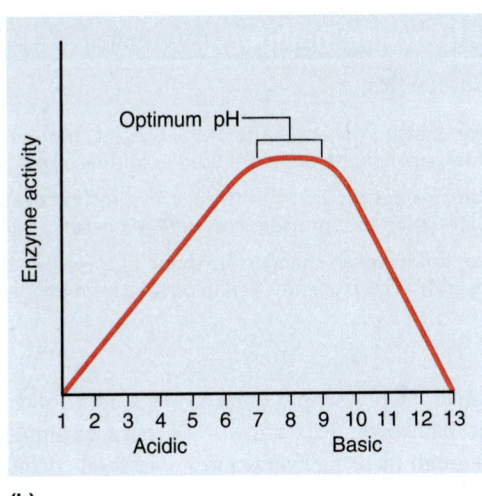

(b)

FIGURE 6.13 **Environmental Factors That Influence Enzyme Activity** (a) A rise in temperature increases the speed of enzymatic activity until the optimum temperature is reached. If the temperature gets too high, the enzyme denatures and no longer functions. (b) Most enzymes function best at pH values slightly above 7.

❓ *Why would an enzyme no longer function once it denatures?*

Allosteric Regulation

Cells can rapidly adjust the activity of certain key enzymes, using other molecules that reversibly bind to and distort them (**figure 6.14**). This is how cells regulate the activity of metabolic pathways. The enzymes that can be controlled are **allosteric** (*allo* means "other" and *stereos* means "shape"). They have an allosteric site as well as an active site. When a regulatory molecule binds to the allosteric site, the shape of the enzyme changes. This distortion alters the relative affinity (chemical attraction) of the enzyme for its substrate. In some cases the binding of the regulatory molecule enhances the affinity for the substrate; in other cases it decreases it.

Allosteric enzymes generally catalyze the first step of a pathway. If the pathway is biosynthetic, the end product generally acts as the allosteric inhibitor—a mechanism called **feedback inhibition** (figure 6.14c). This allows the cell to shut down a pathway when the product begins accumulating. For example, the amino acid isoleucine is an allosteric inhibitor of the first enzyme of the pathway that converts threonine to isoleucine. When the level of isoleucine is relatively high, the pathway is shut down. The binding of the isoleucine is reversible, however, so the enzyme becomes active again if isoleucine levels decrease. Cells can also control the amount of enzyme they synthesize, a topic discussed in chapter 7. ▶▶| bacterial gene regulation, p. 179

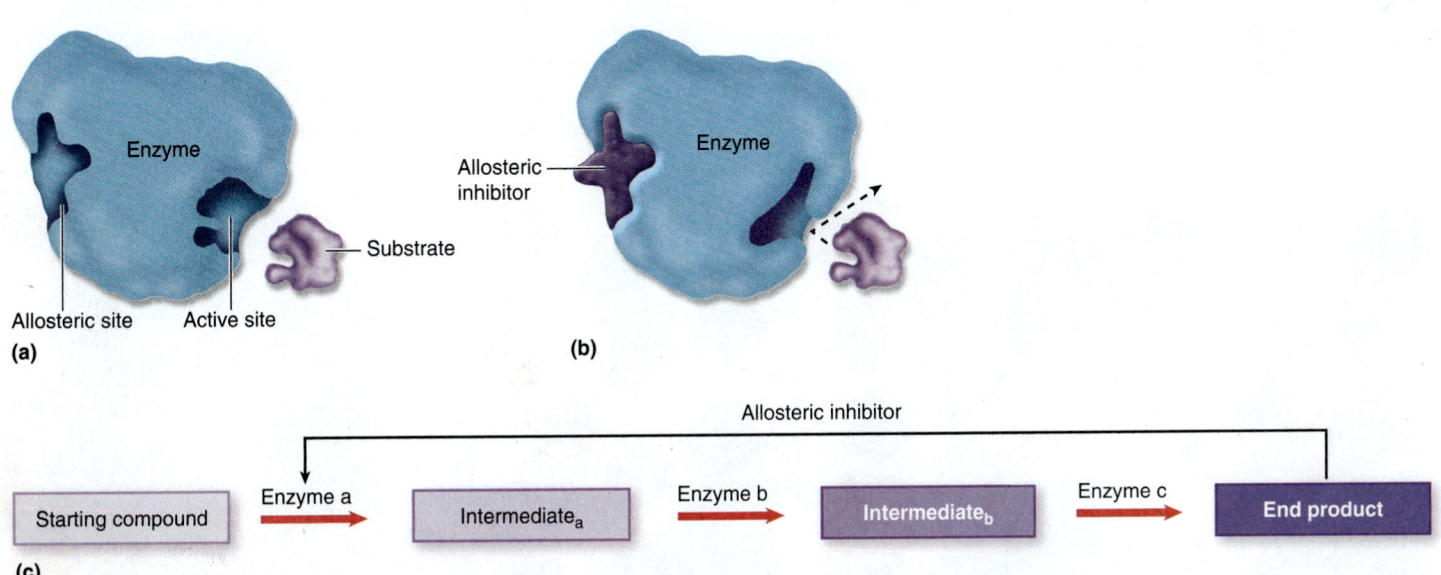

(c)

FIGURE 6.14 Regulation of Allosteric Enzymes (a) Allosteric enzymes have, in addition to the active site, an allosteric site. (b) The binding of a regulatory molecule to the allosteric site causes the shape of the enzyme to change, altering the relative affinity of the enzyme for its substrate. (c) The end product of a given biosynthetic pathway generally acts as an allosteric inhibitor of the first enzyme of that pathway.

❓ *Why would a cell need to regulate enzyme activity?*

TABLE 6.5	Characteristics of Enzyme Inhibitors
Type	**Characteristics**
Competitive inhibition	Inhibitor binds to the active site of the enzyme, blocking access of the substrate to that site. Competitive inhibitors such as sulfa drugs are used as antibacterial medications.
Non-competitive inhibition (by regulatory molecules)	Inhibitor changes the shape of the enzyme, so that the substrate can no longer bind the active site. This is a reversible action that provides cells with a means to control the activity of allosteric enzymes.
Non-competitive inhibition (by enzyme poisons)	Inhibitor permanently changes the shape of the enzyme, making the enzyme non-functional. Enzyme poisons such as mercury are used in certain antimicrobial compounds.

Compounds that reflect a cell's relative energy supply often regulate allosteric enzymes of catabolic pathways. This allows cells to adjust the flow of metabolites through these pathways in response to changing energy needs. High levels of ATP inhibit certain enzymes and, as a consequence, slow down catabolic processes. In contrast, high levels of ADP warn that a cell's energy stores are low and stimulate the activity of some enzymes.

Enzyme Inhibition

Enzymes can be inhibited by a variety of compounds other than the regulatory molecules just described (**table 6.5**). These inhibitory compounds can prevent microbial growth, so they are medically and commercially valuable. The site on the enzyme to which an inhibitor binds determines whether it functions as a competitive or non-competitive inhibitor.

Competitive Inhibition

In **competitive inhibition,** the inhibitor binds to the active site of the enzyme, blocking access of the substrate to that site (**figure 6.15**). Generally the inhibitor has a chemical structure similar to the normal substrate.

A good example of competitive inhibition is the action of the group of antimicrobial medications called sulfa drugs. These inhibit an enzyme in the pathway bacteria use to synthesize the

vitamin folic acid. The drug does not affect human metabolism because humans cannot synthesize folic acid; it must be consumed in foods or nutritional supplements. Sulfa drugs have a structure similar to para-aminobenzoic acid (PABA), an intermediate in the bacterial pathway for folic acid synthesis. Because of this, they fit into the active site of the enzyme that normally uses PABA as a substrate. By doing so, they prevent the enzyme from binding PABA. The greater the number of sulfa molecules relative to PABA molecules, the more likely the active site of the enzyme will be occupied by a sulfa molecule. Once the sulfa is removed, however, the enzyme functions normally with PABA as the substrate. ▶▶ sulfa drugs, p. 467

Non-Competitive Inhibition

Non-competitive inhibition occurs when the inhibitor binds to a site other than the active site. The binding changes the shape of the enzyme so that the substrate can no longer bind the active site. The allosteric inhibitors discussed earlier are non-competitive inhibitors that have a reversible action. The effect of other non-competitive inhibitors is permanent. For example, mercury inhibits growth because it oxidizes the S—H groups of the amino acid cysteine in proteins. This converts cysteine to cystine, which cannot form the important covalent disulfide bond (S—S). As a result, the enzyme shape changes, making it non-functional; the inhibitor "poisons" the enzyme. ◀◀ cysteine, p. 26 ◀◀ disulfide bond, p. 27

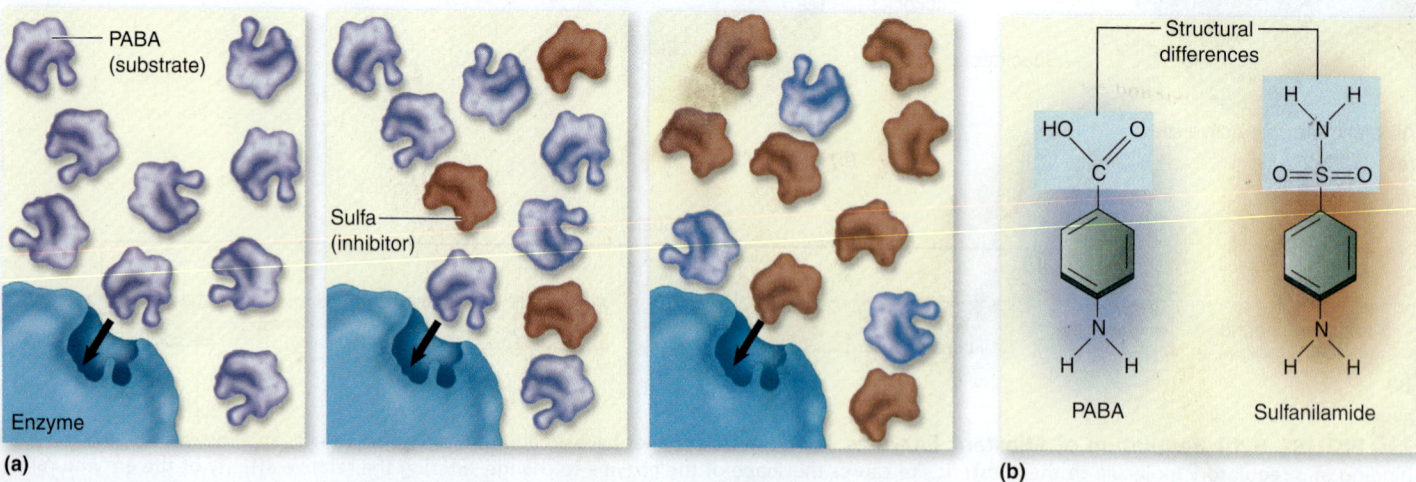

(a)   (b)

FIGURE 6.15 Competitive Inhibition of Enzymes (a) The inhibitor competes with the normal substrate for binding to the active site. The greater the proportion of inhibitor relative to substrate, the more likely the active site of the enzyme will be occupied by an inhibitor. **(b)** A competitive inhibitor generally has a chemical structure similar to the normal substrate.

❓ *Will the enzyme function if sulfa is removed?*

MicroAssessment 6.2

Enzymes facilitate the conversion of a substrate into a product with extraordinary speed and specificity. They are neither consumed nor permanently changed in the reaction. The activity of some enzymes requires a cofactor. Environmental factors influence enzyme activity and, by doing so, determine how rapidly microorganisms multiply. The activity of allosteric enzymes can be regulated. A variety of different compounds and conditions affect enzyme activity.

4. *Explain why sulfa drugs prevent bacterial growth without harming the human host.*
5. *Explain the function of a coenzyme.*
6. *Why is it important for a cell that allosteric inhibition be reversible?*

6.3 ■ The Central Metabolic Pathways

Learning Outcomes

11. *Make a simple diagram that integrates the central metabolic pathways, including their starting and end products.*
12. *Compare and contrast each of the central metabolic pathways with respect to the yield of ATP, reducing power, and number of different precursor metabolites.*

The three central metabolic pathways—glycolysis, the pentose phosphate pathway, and the tricarboxylic acid cycle—modify organic molecules in a step-wise fashion, generating:

■ **ATP** by substrate-level phosphorylation
■ **Reducing power** in the form of NADH, FADH$_2$, and NADPH
■ **Precursor metabolites** (see table 6.2)

This section describes how a molecule of glucose is broken down in the central metabolic pathways. Bear in mind, however, that many millions of molecules of glucose enter a cell, and different molecules can have different fates (see figure 6.9). For example, a cell might oxidize one glucose molecule completely to CO$_2$, thereby producing the maximum amount of ATP. Another glucose molecule might enter glycolysis, or perhaps the pentose phosphate pathway, only to be siphoned off as a precursor metabolite for use in biosynthesis. The step and rate at which the various intermediates are removed for biosynthesis will dramatically affect the overall energy gain of catabolism. This is generally overlooked in descriptions of the ATP-generating functions of these pathways for the sake of simplicity. However, because these pathways serve more than one function, the energy yields are only theoretical.

The pathways of central metabolism are compared in **table 6.6**. The entire pathways with chemical formulas and enzyme names are illustrated in Appendix IV.

Glycolysis

Glycolysis is the primary pathway used by many organisms to convert glucose to pyruvate (**figure 6.16**). For each molecule of glucose that enters, two molecules of pyruvate can be made. This generates a net gain of two molecules of ATP and two molecules of NADH. In addition, the pathway produces six different precursor molecules needed by *E. coli* (see table 6.2).

TABLE 6.6	Comparison of the Central Metabolic Pathways
Pathway	**Characteristics**
Glycolysis	Glycolysis generates: • 2 ATP (net) by substrate-level phosphorylation • 2 NADH + 2 H$^+$ • six different precursor metabolites
Pentose phosphate cycle	The pentose phosphate cycle generates: • NADPH + H$^+$ (amount varies) • two different precursor metabolites
Transition step	The transition step, repeated twice to oxidize two molecules of pyruvate to acetyl-CoA, generates: • 2 NADH + 2 H$^+$ • one precursor metabolite
TCA cycle	The TCA cycle, repeated twice to incorporate two acetyl groups, generates: • 2 ATP by substrate-level phosphorylation (may involve conversion of GTP) • 6 NADH + 6 H$^+$ • 2 FADH$_2$ • two different precursor metabolites

The pathway can be viewed as having two phases (figure 6.16):

■ **Investment** or **preparatory phase.** ① through ⑤ This consumes energy because two different steps each transfer a high-energy phosphate group to the 6-carbon sugar. In eukaryotic cells, both phosphates come from ATP, as shown in figure 6.16. In bacteria, the first one is added as glucose is transported into the cell via group translocation; the next one comes from ATP. The 6-carbon sugar is then split to yield two 3-carbon molecules, each with a phosphate molecule.
◀◀ group translocation, p. 56

■ **Pay-off phase.** ⑥ through ⑩ This oxidizes and rearranges the 3-carbon molecules, generating 1 NADH and 2 ATP in the process, and ultimately forms pyruvate. Note that the steps of this phase occur twice for each molecule of glucose that entered glycolysis. This is because the 6-carbon sugar was split into two 3-carbon molecules in the previous phase. A total of 2 NADH and 4 ATP are made from 1 molecule of glucose.

Yield of Glycolysis

For every glucose molecule degraded, the steps of glycolysis produce:

■ **ATP:** 2 molecules of ATP, net gain (4 ATP molecules are made in the pay-off phase, minus the 2 spent in the investment phase).
■ **Reducing power:** The payoff phase converts 2 NAD$^+$ to 2 NADH + 2 H$^+$.
■ **Precursor metabolites:** Five intermediates of glycolysis as well as the end product, pyruvate, are precursor metabolites used by *E. coli.*

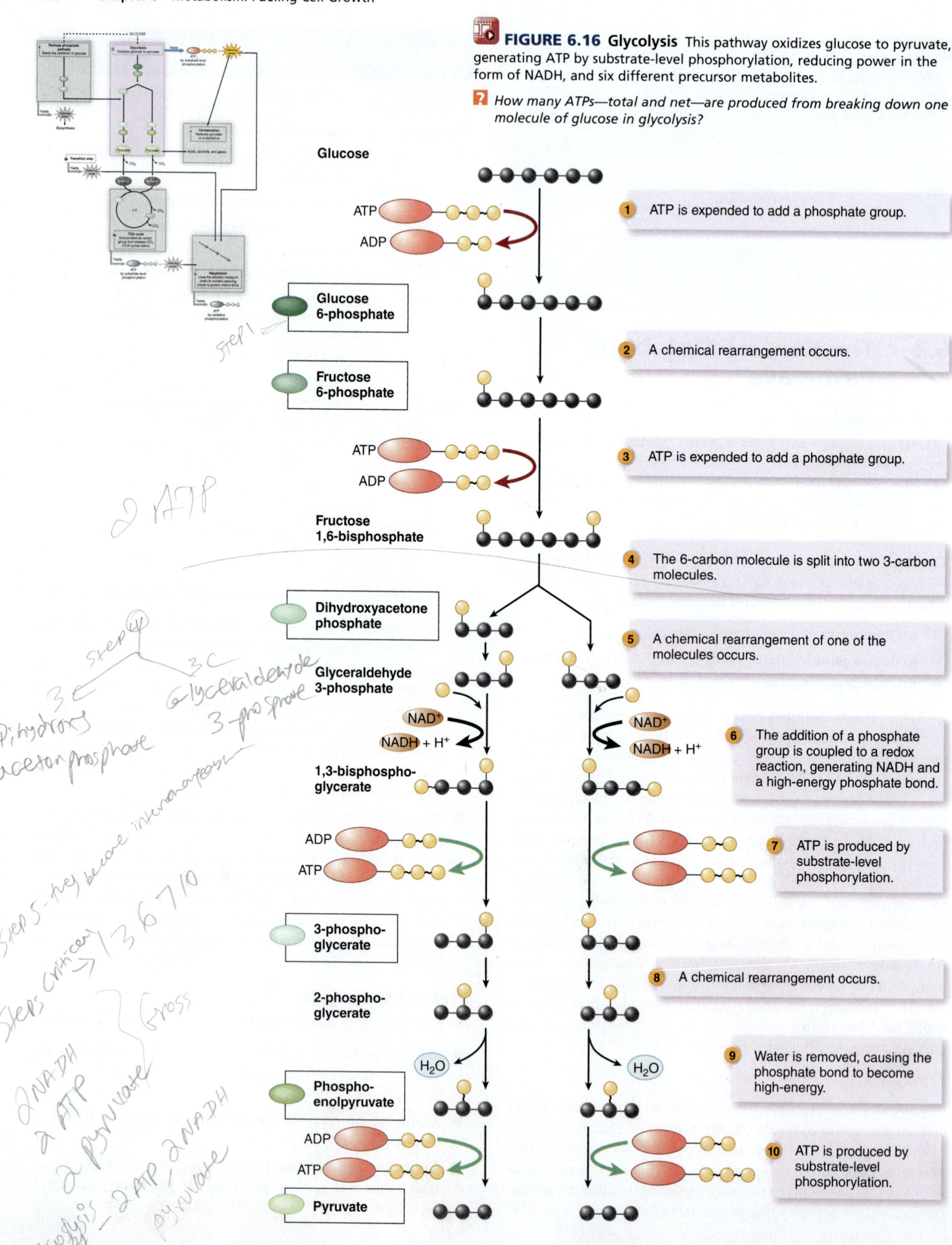

FIGURE 6.16 Glycolysis This pathway oxidizes glucose to pyruvate, generating ATP by substrate-level phosphorylation, reducing power in the form of NADH, and six different precursor metabolites.

? *How many ATPs—total and net—are produced from breaking down one molecule of glucose in glycolysis?*

Glucose

1 ATP is expended to add a phosphate group.

ATP
ADP

Glucose 6-phosphate

2 A chemical rearrangement occurs.

Fructose 6-phosphate

ATP
ADP

3 ATP is expended to add a phosphate group.

Fructose 1,6-bisphosphate

4 The 6-carbon molecule is split into two 3-carbon molecules.

Dihydroxyacetone phosphate

5 A chemical rearrangement of one of the molecules occurs.

Glyceraldehyde 3-phosphate

NAD+
NADH + H+   NAD+
                   NADH + H+

6 The addition of a phosphate group is coupled to a redox reaction, generating NADH and a high-energy phosphate bond.

1,3-bisphospho-glycerate

ADP
ATP

7 ATP is produced by substrate-level phosphorylation.

3-phospho-glycerate

8 A chemical rearrangement occurs.

2-phospho-glycerate

H_2O H_2O

9 Water is removed, causing the phosphate bond to become high-energy.

Phospho-enolpyruvate

ADP
ATP

10 ATP is produced by substrate-level phosphorylation.

Pyruvate

Pentose Phosphate Pathway

The other central metabolic pathway used by cells to break down glucose is the pentose phosphate pathway. This pathway is particularly important because of its contribution to biosynthesis. It generates reducing power in the form of NADPH, and two of its intermediates—ribose 5-phosphate and erythrose 4-phosphate—are important precursor metabolites. A product of this complex pathway is glyceraldehyde 3-phosphate (G3P), which can enter a step in glycolysis for further breakdown.

Yield of the Pentose Phosphate Pathway

The yield of the pentose phosphate pathway varies, depending upon which of several possible alternatives are taken. It can produce:

- **Reducing power:** A variable amount of reducing power in the form of NADPH is produced.
- **Precursor metabolites:** Two intermediates of the pentose phosphate pathway are precursor metabolites.

Transition Step

The transition step, which links the previous pathways to the TCA cycle, involves several reactions catalyzed by a large multi-enzyme complex (**figure 6.17**). CO_2 is first removed

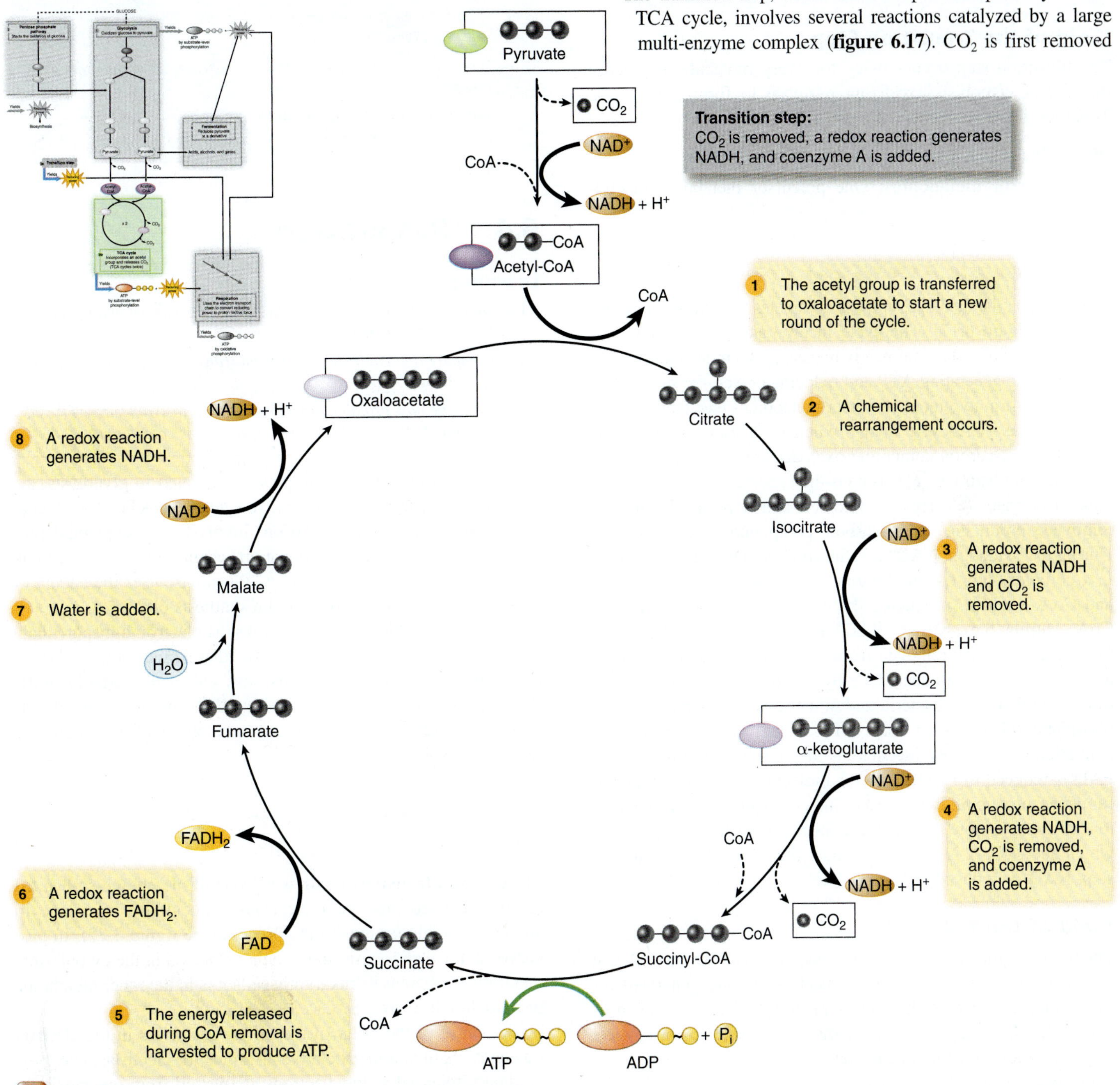

Transition step:
CO_2 is removed, a redox reaction generates NADH, and coenzyme A is added.

1 The acetyl group is transferred to oxaloacetate to start a new round of the cycle.

2 A chemical rearrangement occurs.

3 A redox reaction generates NADH and CO_2 is removed.

4 A redox reaction generates NADH, CO_2 is removed, and coenzyme A is added.

5 The energy released during CoA removal is harvested to produce ATP.

6 A redox reaction generates $FADH_2$.

7 Water is added.

8 A redox reaction generates NADH.

FIGURE 6.17 The Transition Step and the Tricarboxylic Acid Cycle The transition step links glycolysis and the TCA cycle, converting pyruvate to acetyl-CoA; it generates reducing power and one precursor metabolite. The TCA cycle incorporates the acetyl group of acetyl-CoA and, using a series of steps, releases CO_2; it generates ATP, reducing power in the form of both NADH and $FADH_2$, and two different precursor metabolites.

Which generates more reducing power—glycolysis or the TCA cycle?

from pyruvate, a step called decarboxylation. Then, a redox reaction transfers electrons to NAD^+, reducing it to $NADH + H^+$. Finally, the remaining 2-carbon acetyl group is joined to coenzyme A to form acetyl-CoA.

In prokaryotic cells, all the central metabolic pathways occur in the cytoplasm. In eukaryotic cells, however, the enzymes of glycolysis and the pentose phosphate pathways are located in the cytoplasm, whereas those of the transition step and TCA cycle are within the mitochondrial matrix. Because of this, eukaryotic cells must transport pyruvate molecules into mitochondria for the transition step to occur. ◀◀ cytoplasm, p. 52 ◀◀ mitochondria, p. 74

Yield of the Transition Step

The transition step occurs twice for every molecule of glucose that enters glycolysis, oxidizing pyruvate to form acetyl-CoA. Together, these generate:

- **Reducing power:** $2\ NADH + 2\ H^+$.

- **Precursor metabolites:** One precursor metabolite (acetyl-CoA).

Tricarboxylic Acid (TCA) Cycle

The tricarboxylic acid (TCA) cycle completes the oxidation of glucose (see figure 6.17). The cycle incorporates the acetyl groups from the transition step, ultimately releasing two molecules of CO_2. In addition to generating ATP and reducing power, the steps of the TCA cycle form two more precursor metabolites (see table 6.2).
(1) The TCA cycle begins when CoA transfers its acetyl group to the 4-carbon compound oxaloacetate, forming the 6-carbon compound citrate. **(2)** Citrate is then chemically rearranged to make isocitrate. **(3)** This is oxidized and a molecule of CO_2 removed, producing the 5-carbon compound α-ketoglutarate. During the oxidation, NAD^+ is reduced to $NADH + H^+$. **(4)** Then, in a complex step, α-ketoglutarate is oxidized, CO_2 is removed, and CoA is added, producing the 4-carbon compound succinyl-CoA. During the process, NAD^+ is reduced to $NADH + H^+$. **(5)** The CoA is then removed from succinyl-CoA, and the energy released is used to produce ATP by substrate-level phosphorylation. Note that some bacteria and other cells make guanosine triphosphate (GTP) rather than ATP at this step; GTP can be converted to ATP. **(6)** Succinate is then oxidized to fumarate, as FAD is reduced to $FADH_2$. **(7)** A molecule of water is added to fumarate, producing malate. **(8)** This compound is then oxidized to oxaloacetate; note that oxaloacetate is the starting compound to which acetyl-CoA is added to initiate the cycle. During this last step, NAD^+ is reduced to $NADH + H^+$.

Yield of the TCA Cycle

The tricarboxylic acid cycle "turns" once for each acetyl-CoA that enters. Because two molecules of acetyl-CoA are generated for each glucose molecule that enters glycolysis, the breakdown of one molecule of glucose causes the TCA cycle to turn twice. Together, these two turns generate:

- **ATP:** 2 ATP produced in step 5.

- **Reducing power:** Redox reactions at steps 3, 4, 6, and 8 produce a total of $6\ NADH + 6\ H^+$ and $2\ FADH_2$.

- **Precursor metabolites:** Two intermediates of the TCA cycle, formed in steps 3 and 8, are precursor metabolites.

MicroAssessment 6.3

Glycolysis oxidizes glucose to pyruvate, yielding some ATP and NADH and six different precursor metabolites. The pentose phosphate pathway also oxidizes glucose, but more importantly, it produces two different precursor metabolites and NADPH for biosynthesis. The transition step and the TCA cycle, each repeated twice, complete the oxidation of glucose, yielding some ATP, a great deal of reducing power, and three different precursor metabolites.

7. *How does the "investment phase" of glycolysis effect the net yield of ATP in that pathway?*

8. *Which central metabolic pathway generates the most reducing power?*

9. *Which compound contains more free energy—glucose or oxaloacetate? On what did you base your conclusion?* ➕

6.4 ■ Respiration

Learning Outcomes

13. *Describe the components of the electron transport chain and how they generate a proton motive force.*

14. *Compare and contrast the electron transport chains of eukaryotes and prokaryotes.*

15. *Describe how proton motive force is used to synthesize ATP and how the ATP yield of aerobic respiration is calculated.*

Respiration uses the reducing power generated in glycolysis, the transition step, and the TCA cycle to synthesize ATP. The mechanism, **oxidative phosphorylation,** involves two sequential processes. First, the electron transport chain generates a proton motive force. Then, the enzyme ATP synthase uses the energy of the proton motive force to drive the synthesis of ATP.

The remarkable process that links the electron transport chain to ATP synthesis was proposed by the British scientist Peter Mitchell in 1961. His hypothesis, now called the **chemiosmotic theory,** was widely dismissed initially. Only through years of self-funded research was he able to convince others of its validity; he received a Nobel Prize in 1978.

The Electron Transport Chain— Generating Proton Motive Force

The **electron transport chain** is a group of membrane-embedded electron carriers that pass electrons sequentially from one to another, ejecting protons in the process (see figure 3.27). In prokaryotes, the electron transport chain is located in the cytoplasmic membrane, whereas in eukaryotic cells it is in the inner membrane of mitochondria. ◀◀ cytoplasmic membrane, p. 52

Because of the order of the different carriers in the electron transport chain and their relative affinities for electrons, energy is gradually released as the electrons are passed from one carrier to another, much like a ball falling down a flight of stairs (**figure 6.18**). The energy release is coupled to the ejection of protons, moving them from the inside of the cell to the outside or,

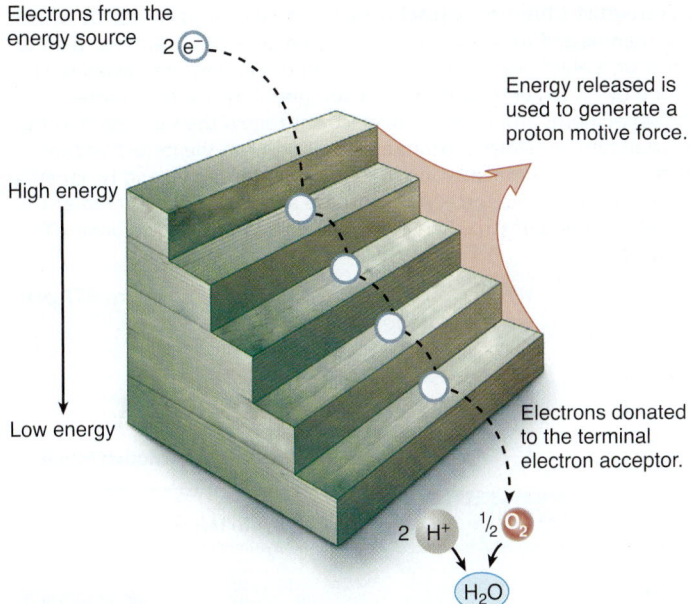

Electrons from the energy source 2 e^-

Energy released is used to generate a proton motive force.

High energy

Low energy

Electrons donated to the terminal electron acceptor.

2 H^+ $\frac{1}{2}$ O_2

H_2O

FIGURE 6.18 Electron Transport As electrons are passed along the electron transport chain, the energy released is used to establish a proton gradient.

? *O_2 is serving as the terminal electron acceptor in this diagram; is it being oxidized, or reduced?*

in the case of mitochondria, from the matrix to the region between the inner and outer membranes. This ejection of protons creates a proton gradient—an electrochemical gradient—across the membrane. Energy of this gradient, proton motive force, is then harvested by the cell to synthesize ATP. Recall from chapter 3 that prokaryotes can also use the energy of proton motive force to transport substances and power the rotation of flagella.

Components of an Electron Transport Chain

Most carriers in the electron transport chain are grouped into several large protein complexes that function as proton pumps. Others shuttle electrons from one complex to the next. Three general groups of electron carriers are notable: quinones, cytochromes, and flavoproteins.

Quinones are lipid-soluble organic molecules that move freely in the membrane and can therefore transfer electrons between different protein complexes in the membrane. Several types of quinones exist, one of the most common being ubiquinone (meaning "ubiquitous quinone"). Menaquinone, a quinone used in the electron transport chain of some prokaryotes, serves as a source of vitamin K for humans and other mammals. This vitamin is required for proper blood coagulation, and mammals obtain much of their requirement by absorbing menaquinone produced by bacteria growing in the intestinal tract.

Cytochromes are proteins that contain heme, a molecule that holds an iron atom in its center. Several different cytochromes exist, each distinguished with a letter, for example, cytochrome *c*. The presence of certain cytochromes can be used as an identifying marker. For instance, the oxidase test—which is used in the steps to identify *Neisseria*, *Pseudomonas*, and *Campylobacter* species—detects the activity of cytochrome *c* oxidase (see table 10.5).

Flavoproteins are proteins to which a flavin is attached. FAD and other flavins are synthesized from the vitamin riboflavin (see table 6.4).

General Mechanisms of Proton Ejection

An important characteristic of the electron carriers is that some accept only hydrogen atoms (proton-electron pairs), whereas others accept only electrons. The spatial arrangement of these two types of carriers in the membrane causes protons to be shuttled from one side of the membrane to the other. This occurs because a hydrogen carrier receiving electrons from an electron carrier must pick up protons, which come from inside the cell (or matrix of the mitochondrion) due to the hydrogen carrier's relative location in the membrane. Conversely, when a hydrogen carrier passes electrons to a carrier that accepts electrons, but not protons, the protons are released to the outside of the cell (or intermembrane space of the mitochondrion). The net effect of these processes is that the components of the electron transport chain pump protons from one side of the membrane to the other, establishing the concentration gradient across the membrane. Note that the gradient could not be established if energy were not released during electron transfer.

The Electron Transport Chain of Mitochondria

The electron transport chain of mitochondria has four different protein complexes, three of which function as proton pumps. In addition, two electron carriers (ubiquinone and cytochrome *c*) shuttle electrons between the complexes (**figure 6.19**):

■ **Complex I** (also called NADH dehydrogenase complex). This accepts electrons from NADH, ultimately transferring them to ubiquinone (also called coenzyme Q); in the process, four protons are moved across the membrane.

■ **Complex II** (also called succinate dehydrogenase complex). This accepts electrons from the TCA cycle, when $FADH_2$ is formed during the oxidation of succinate (see figure 6.17, step 6). Note that the electrons carried by $FADH_2$ enter the electron transport chain "downstream" of those carried by NADH. Because of this, a pair of electrons carried by NADH result in more protons being expelled than does a pair carried by $FADH_2$. Electrons are then transferred from complex II to ubiquinone.

■ **Complex III** (also called cytochrome bc_1 complex). This accepts electrons from ubiquinone, which has carried them from either complex I or II. Complex III pumps four protons across the membrane before transferring the electrons to cytochrome *c*.

■ **Complex IV** (also called cytochrome *c* oxidase complex). This accepts electrons from cytochrome *c* and pumps two protons across the membrane. Complex IV is a terminal oxidoreductase, meaning it transfers the electrons to the terminal electron acceptor, which, in this case, is O_2.

The Electron Transport Chains of Prokaryotes

Considering the versatility and diversity of prokaryotes, it should not be surprising that the types and arrangement of their electron transport components vary tremendously. In fact, a single species can have several alternative carriers, allowing cells to cope with ever-changing growth conditions.

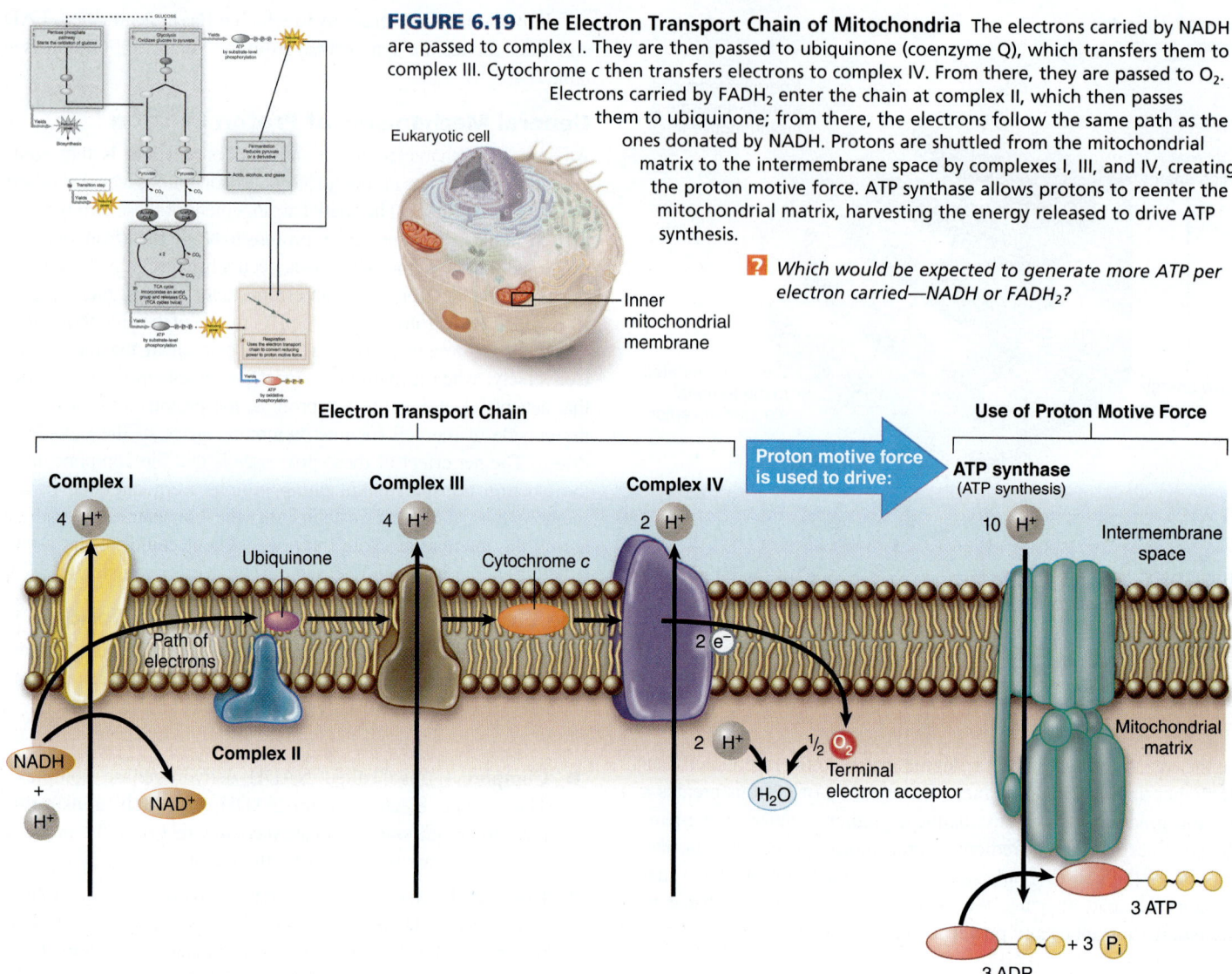

FIGURE 6.19 The Electron Transport Chain of Mitochondria The electrons carried by NADH are passed to complex I. They are then passed to ubiquinone (coenzyme Q), which transfers them to complex III. Cytochrome c then transfers electrons to complex IV. From there, they are passed to O_2. Electrons carried by $FADH_2$ enter the chain at complex II, which then passes them to ubiquinone; from there, the electrons follow the same path as the ones donated by NADH. Protons are shuttled from the mitochondrial matrix to the intermembrane space by complexes I, III, and IV, creating the proton motive force. ATP synthase allows protons to reenter the mitochondrial matrix, harvesting the energy released to drive ATP synthesis.

? *Which would be expected to generate more ATP per electron carried—NADH or $FADH_2$?*

The electron transport chain of *E. coli* provides an excellent example of the versatility of some prokaryotes. This bacterium uses aerobic respiration when O_2 is available, but in the absence of O_2 it can switch to anaerobic respiration if a suitable electron acceptor such as nitrate is present. The *E. coli* electron transport chain serves as a model for both aerobic and anaerobic respiration.

Aerobic Respiration When growing aerobically in a glucose-containing medium, *E. coli* can use two different NADH dehydrogenases; one is a proton pump functionally equivalent to complex I of the mitochondrion (**figure 6.20**). *E. coli* also has a succinate dehydrogenase functionally equivalent to complex II of a mitochondrion. In addition to these protein complexes, *E. coli* can produce several alternatives, allowing the organism to optimally use a variety of different energy sources, including H_2. The bacterium does not have the equivalent of complex III or cytochrome c. Instead, quinones shuttle the electrons directly to one of two variations of a ubiquinol oxidase, which are functionally equivalent to complex IV of a mitochondrion, a terminal oxidoreductase. One variation of the ubiquinol oxidase works optimally in high O_2 conditions and expels

four protons. The other ejects only two protons, but it can more effectively scavenge O_2 and therefore is particularly useful in low O_2 conditions.

Anaerobic Respiration Anaerobic respiration harvests less energy than aerobic respiration, which makes sense considering the lower electron affinities of other electron acceptors (see figure 6.7). Some of the electron transport chain components used during anaerobic respiration are different from those of aerobic respiration. For example, *E. coli* can synthesize a terminal oxidoreductase that uses nitrate as a terminal electron acceptor when O_2 is not available. This produces nitrite, which *E. coli* then converts to ammonia, avoiding the toxic effects of nitrite. Other bacteria can reduce nitrate even further, forming compounds such as nitrous oxide (N_2O), and nitrogen gas (N_2).

A group of obligate anaerobes called the sulfate-reducers use sulfate (SO_4^{2-}) as a terminal electron acceptor, producing hydrogen sulfide as an end product. The diversity and ecology of sulfate-reducing bacteria will be discussed in chapter 11.
▶▶ **sulfate-reducing bacteria, p. 258**

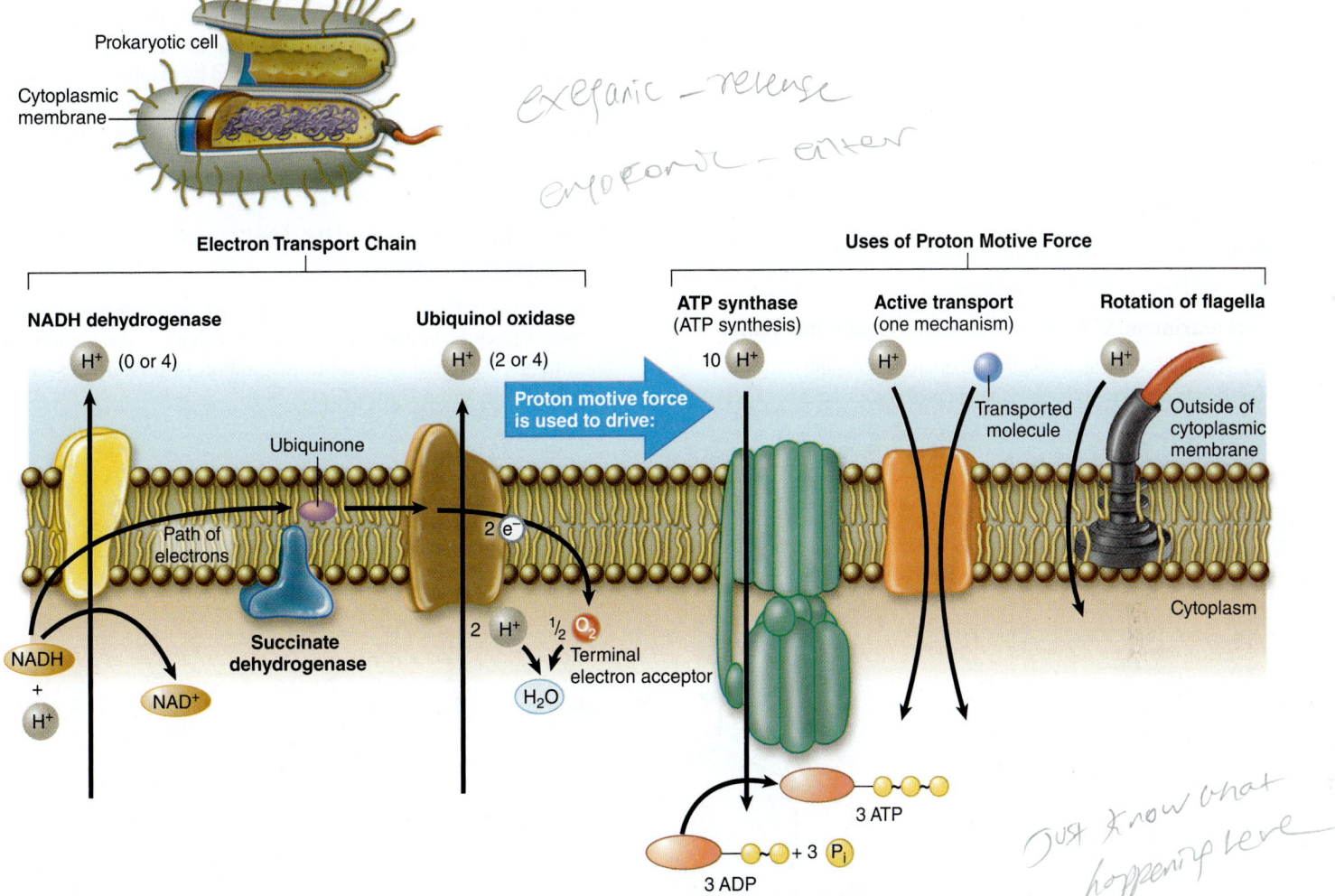

(handwritten annotations: "exeganic — release", "engoganic — enter", "Just know what happening here")

FIGURE 6.20 The Electron Transport Chain of *E. coli* Growing Aerobically in a Glucose-Containing Medium The electrons carried by NADH are passed to one of two different NADH dehydrogenases. They are then passed to ubiquinone, which transfers them to one of two ubiquinol oxidases. From there they are passed to O_2. The electrons carried by $FADH_2$ enter the chain at succinate dehydrogenase, which then transfers them to ubiquinone; from there, the electrons follow the same path as the ones donated by NADH. Protons are ejected by one of the two NADH dehydrogenases and both ubiquinol oxidases, creating the proton motive force. ATP synthase allows protons to reenter the cell, using the energy released to drive ATP synthesis. The proton motive force is also used to drive one form of active transport and to power the rotation of flagella. *E. coli* has other components of the electron transport chain that function under different growth conditions.

? *Succinate dehydrogenase is equivalent to which component of the mitochondrial electron transport chain?*

ATP Synthase—Harvesting the Proton Motive Force to Synthesize ATP

Just as energy is required to establish a concentration gradient, energy is released when the gradient is removed or reduced. The enzyme ATP synthase uses the energy of proton motive force to synthesize ATP. It does this by allowing protons to flow back into the bacterial cell (or matrix of the mitochondrion) in a controlled manner, simultaneously using the energy released to add a phosphate group to ADP. One molecule of ATP is formed from the entry of approximately three protons.

Theoretical ATP Yield of Oxidative Phosphorylation

By calculating the ATP yield of oxidative phosphorylation, the relative energy gains of respiration and fermentation can be

compared. It is not a straightforward comparison, however, because oxidative phosphorylation has so many variables. This is particularly true for prokaryotic cells because they use proton motive force to drive processes other than ATP synthesis. Prokaryotic cells, as a group, use different carriers in their electron transport chain and eject a variable number of protons per pair of electrons passed.

The basis for calculating the ATP yield of oxidative phosphorylation relies on experimental studies using rat mitochondria. These studies indicate that approximately 2.5 ATP are made for each pair of electrons transferred to the electron transport chain by NADH; about 1.5 ATP are made for each pair transferred by $FADH_2$. For simplicity we will use whole numbers (3 ATP/NADH and 2 ATP/$FADH_2$) in calculations. Using these numbers, the maximum theoretical energy yield for oxidative phosphorylation in a prokaryotic cell (assuming the electron transport chain is similar to that of mitochondria) is:

From glycolysis:

- 2 NADH ⟶ 6 ATP (assuming 3 for each NADH)

From the transition step:

- 2 NADH ⟶ 6 ATP (assuming 3 for each NADH)

From the TCA cycle:

- 6 NADH ⟶ 18 ATP (assuming 3 for each NADH)
- 2 $FADH_2$ ⟶ 4 ATP (assuming 2 for each $FADH_2$)

Total maximum ATP yield from oxidative phosphorylation = 34

The ATP gain as a result of oxidative phosphorylation will be slightly less in eukaryotic cells than in prokaryotic cells because of the fate of the reducing power (NADH) generated during glycolysis. Recall that in eukaryotic cells, glycolysis takes place in the cytoplasm, whereas the electron transport chain is located in the mitochondria. Consequently, the electrons carried by cytoplasmic NADH must be moved across the mitochondrial membrane before they can enter the electron transport chain. This requires an expenditure of approximately 1 ATP per NADH generated during glycolysis.

ATP Yield of Aerobic Respiration in Prokaryotes

Now that the ATP-yielding components of the central metabolic pathways have been considered, we can calculate the theoretical maximum ATP yield of aerobic respiration in prokaryotes. This yield is illustrated in **figure 6.21.**

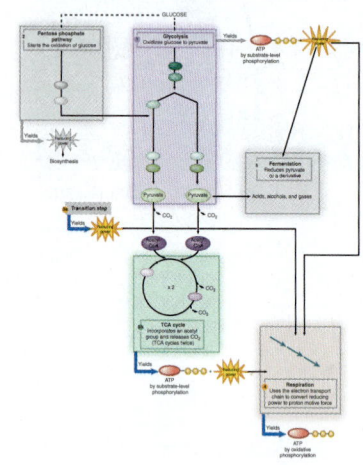

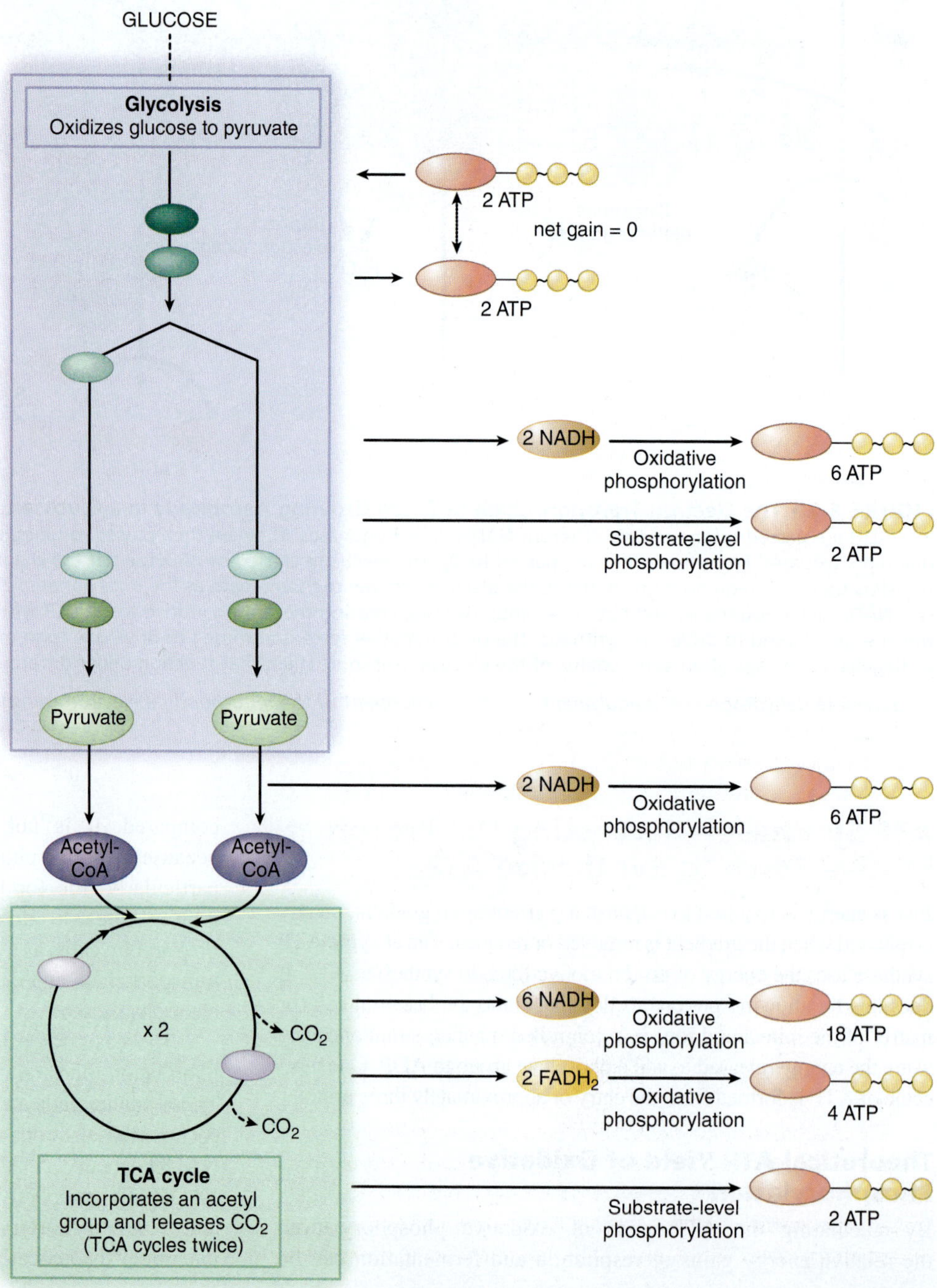

FIGURE 6.21 Maximum Theoretical Energy Yield from Aerobic Respiration in a Prokaryotic Cell This maximum energy yield calculation assumes that for every pair of electrons transferred to the electron transport chain, 3 ATP are synthesized; and for every pair of electrons donated by $FADH_2$, 2 ATP are synthesized.

? *Why is it difficult to calculate the actual maximum ATP yield of respiration in a prokaryotic cell?*

GLUCOSE

Glycolysis
Oxidizes glucose to pyruvate

2 ATP

net gain = 0

2 ATP

2 NADH — Oxidative phosphorylation — 6 ATP

Substrate-level phosphorylation — 2 ATP

Pyruvate   Pyruvate

2 NADH — Oxidative phosphorylation — 6 ATP

Acetyl-CoA   Acetyl-CoA

x 2 CO_2

6 NADH — Oxidative phosphorylation — 18 ATP

2 $FADH_2$ — Oxidative phosphorylation — 4 ATP

CO_2

TCA cycle
Incorporates an acetyl group and releases CO_2
(TCA cycles twice)

Substrate-level phosphorylation — 2 ATP

Substrate-level phosphorylation:

- 2 ATP (from glycolysis; net gain)
- <u>2 ATP</u> (from the TCA cycle)
- 4 total

Oxidative phosphorylation:

- 6 ATP (from the reducing power gained in glycolysis)
- 6 ATP (from the reducing power gained in the transition step)
- <u>22 ATP</u> (from the reducing power gained in the TCA cycle)
- 34 total

Total ATP gain (theoretical maximum) = 38

MicroAssessment 6.4

Respiration uses the NADH and $FADH_2$ generated in glycolysis, the transition step, and the TCA cycle to synthesize ATP. The electron transport chain is used to convert reducing power into proton motive force. ATP synthase then harvests that energy to synthesize ATP. The overall process is called oxidative phosphorylation. In aerobic respiration, O_2 serves as the terminal electron acceptor; anaerobic respiration uses a molecule other than O_2.

10. *In bacteria, what is the role of the molecule that serves as a source of vitamin K for humans?*
11. *Why is the overall ATP yield in aerobic respiration only a theoretical number?*
12. *Why could an oxidase also be called a reductase?* ➕

6.5 ■ Fermentation

Learning Outcome

16. *Describe the role of fermentation and the importance of the common end products.*

Fermentation is used by organisms that cannot respire, either because a suitable inorganic terminal electron acceptor is not available or because they lack an electron transport chain. *E. coli* is a facultative anaerobe able to use any of three ATP-generating options: aerobic respiration, anaerobic respiration, and fermentation. In contrast, the only option for *Streptococcus pneumoniae* is fermentation because it does not have an electron transport chain.

In general, the only ATP-generating reactions of fermentation are those of glycolysis and involve substrate-level phosphorylation. The additional steps simply consume excess reducing power as a way of regenerating NAD^+ (**figure 6.22**). This is a critical function, because NAD^+ is needed to accept electrons in subsequent rounds of glycolysis—without it, glycolysis would stop. To consume reducing power, fermentation uses an organic compound such as pyruvate or a derivative as a terminal electron acceptor.

The end products of fermentation are significant for a number of reasons (**figure 6.23**). For one thing, certain end products help identify bacterial isolates because a given organism uses a characteristic fermentation pathway. In addition, some end products are commercially valuable. Chapter 31 describes how foods and beverages are produced using fermentations. Important end products of fermentation pathways include:

- **Lactic acid.** Lactic acid (the ionized form is lactate) is produced when pyruvate itself serves as the terminal electron acceptor. Lactic acid and other end products of a group of Gram-positive organisms called lactic acid bacteria are instrumental in creating the flavor and texture of cheese, yogurt, pickles, cured sausages, and other foods. Yet lactic acid also causes food spoilage, and contributes to tooth decay when produced by bacteria living on the teeth. Some animal cells use this fermentation pathway temporarily when O_2 is in short supply; the accumulation of lactic acid in muscle tissue causes the pain and fatigue associated with strenuous exercise. ▶▶ lactic acid bacteria, p. 258 ▶▶ cheese, yogurt, and other fermented milk products, p. 751 ▶▶ pickled vegetables, p. 752 ▶▶ fermented meat products, p. 752

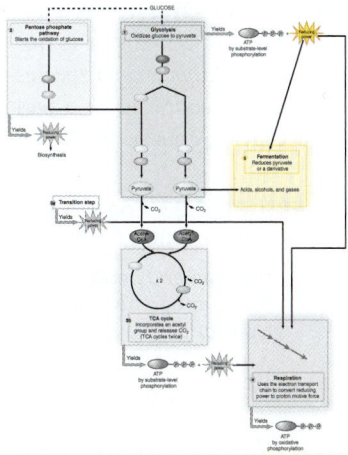

(a) **Lactic acid fermentation**

(b) **Ethanol fermentation**

FIGURE 6.22 Fermentation Pathways Use Pyruvate or a Derivative as a Terminal Electron Acceptor (a) In lactic acid fermentation, the pyruvate generated during glycolysis serves as the terminal electron acceptor, producing lactate. (b) In ethanol fermentation, the pyruvate is first converted to acetaldehyde, which then serves as the terminal electron acceptor, producing ethanol.

❓ *Why is it important for cells to have a mechanism to oxidize NADH?*

- **Ethanol.** Ethanol is produced in a pathway that first removes CO_2 from pyruvate, generating acetaldehyde, which then serves as the terminal electron acceptor. The end products of these sequential reactions—which are used in making wine, beer, spirits, and bread—are ethanol and CO_2, (see figures 31.4, 31.5, and 31.6). Ethanol is also an important biofuel. *Saccharomyces* (yeast) and *Zymomonas* (bacteria) use this pathway. ▶▶| **wine, p. 753** ▶▶| **beer, p. 754** ▶▶| **distilled spirits, p. 755** ▶▶| **bread, p. 755**

- **Butyric acid.** Butyric acid (the ionized form is butyrate) and a variety of other end products are produced in a complex multistep pathway used by *Clostridium* species, which are obligate anaerobes. Under certain conditions, some organisms use a variation of this pathway to produce the organic solvents butanol and acetone.

- **Propionic acid.** Propionic acid (the ionized form is propionate) is generated in a multistep pathway that first adds CO_2 to pyruvate, generating a compound that then serves as a terminal electron acceptor. After NADH reduces this, it is further modified to form propionate. *Propionibacterium* species use this pathway, and their growth is encouraged as a part of Swiss cheese production. The CO_2 they make forms the holes, and propionic acid gives the cheese its unique flavor. ▶▶| **cheese, p. 751**

- **Mixed acids.** These are produced in a multistep branching pathway, generating a variety of different fermentation products including lactic acid, succinic acid (the ionized form is succinate), ethanol, acetic acid (the ionized form is acetate), and gases. This is another pathway used to differentiate members of the family *Enterobacteriaceae;* the methyl-red test detects the low pH resulting from the acidic end products, distinguishing members that use this pathway, such as *E. coli*, from those that do not, such as *Klebsiella* and *Enterobacter* (see table 10.5). ▶▶| **methyl-red test, p. 243**

- **2,3-Butanediol.** 2,3-Butanediol is produced in a multistep pathway that uses two molecules of pyruvate to generate acetoin and two molecules of CO_2. Acetoin is then used as the terminal electron acceptor. The primary significance of this pathway is that it serves to differentiate certain members of the family *Enterobacteriaceae*. The Voges-Proskauer test detects acetoin, distinguishing members that use this pathway (such as *Klebsiella* and *Enterobacter*) from those that do not (such as *E. coli;* see table 10.5). ▶▶| **Voges-Proskauer test, p. 243**

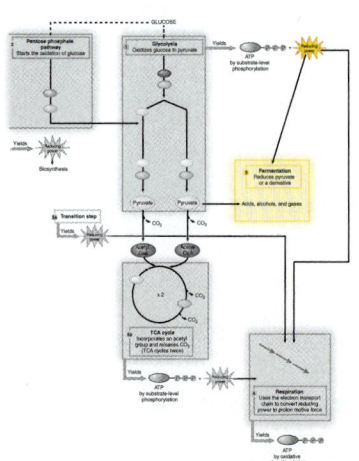

FIGURE 6.23 End Products of Fermentation Pathways

❓ *How can end products of fermentation help identify a bacterium?*

MicroAssessment 6.5

Fermentation stops short of the TCA cycle, using pyruvate or a derivative as a terminal electron acceptor. Many end products of fermentation are commercially valuable.

13. *Why would a cell ferment rather than respire?*

14. *How do the methyl-red and Voges-Proskauer tests differentiate between certain members of the* Enterobacteriaceae?

15. *Fermentation is used as a means of preserving foods. Why would it slow spoilage?* ✚

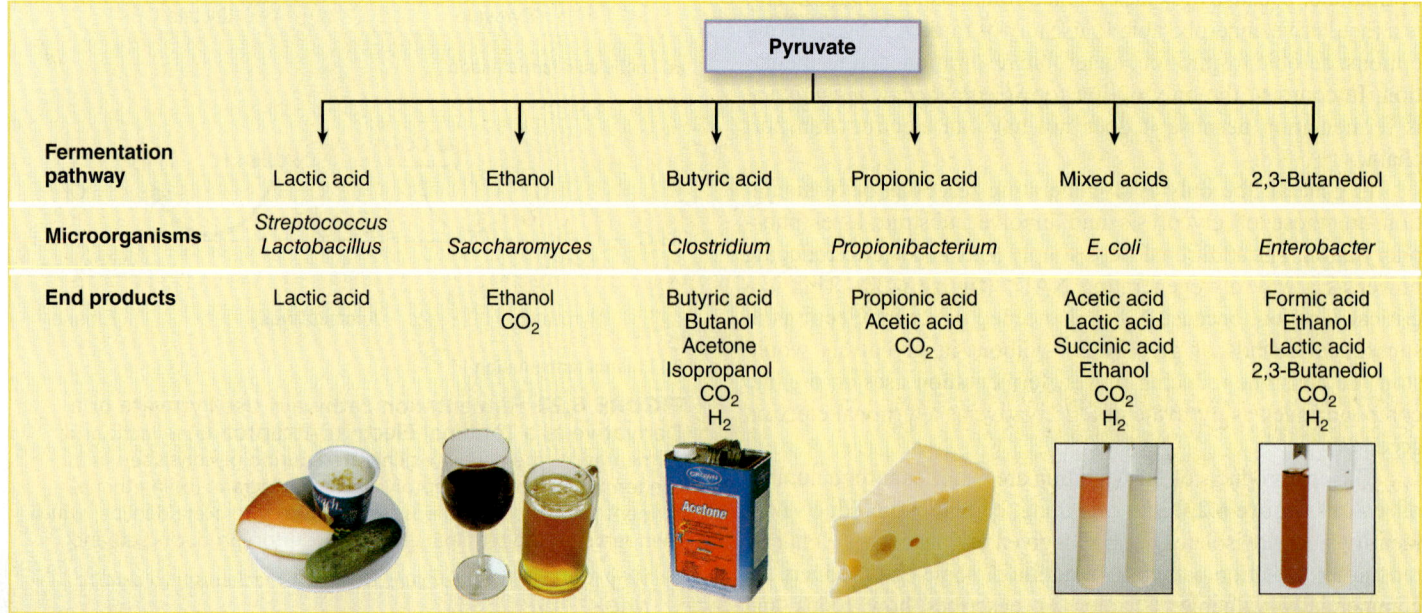

	Pyruvate					
Fermentation pathway	Lactic acid	Ethanol	Butyric acid	Propionic acid	Mixed acids	2,3-Butanediol
Microorganisms	*Streptococcus Lactobacillus*	*Saccharomyces*	*Clostridium*	*Propionibacterium*	*E. coli*	*Enterobacter*
End products	Lactic acid	Ethanol CO_2	Butyric acid Butanol Acetone Isopropanol CO_2 H_2	Propionic acid Acetic acid CO_2	Acetic acid Lactic acid Succinic acid Ethanol CO_2 H_2	Formic acid Ethanol Lactic acid 2,3-Butanediol CO_2 H_2

6.6 ■ Catabolism of Organic Compounds Other Than Glucose

Learning Outcome

17. *Briefly describe how polysaccharides and disaccharides, lipids, and proteins are degraded and utilized by a cell.*

Microbes can use a variety of organic compounds other than glucose as energy sources, including polysaccharides, proteins and lipids. To break these down into their respective sugar, amino acid, and lipid subunits, cells synthesize hydrolytic enzymes, which break bonds by adding water. To use a macromolecule in the surrounding medium, a cell secretes the appropriate hydrolytic enzyme and then transports the resulting subunits into the cell (see figure 3.30). Inside the cell, the subunits are further degraded to form appropriate precursor metabolites (**figure 6.24**). Recall that

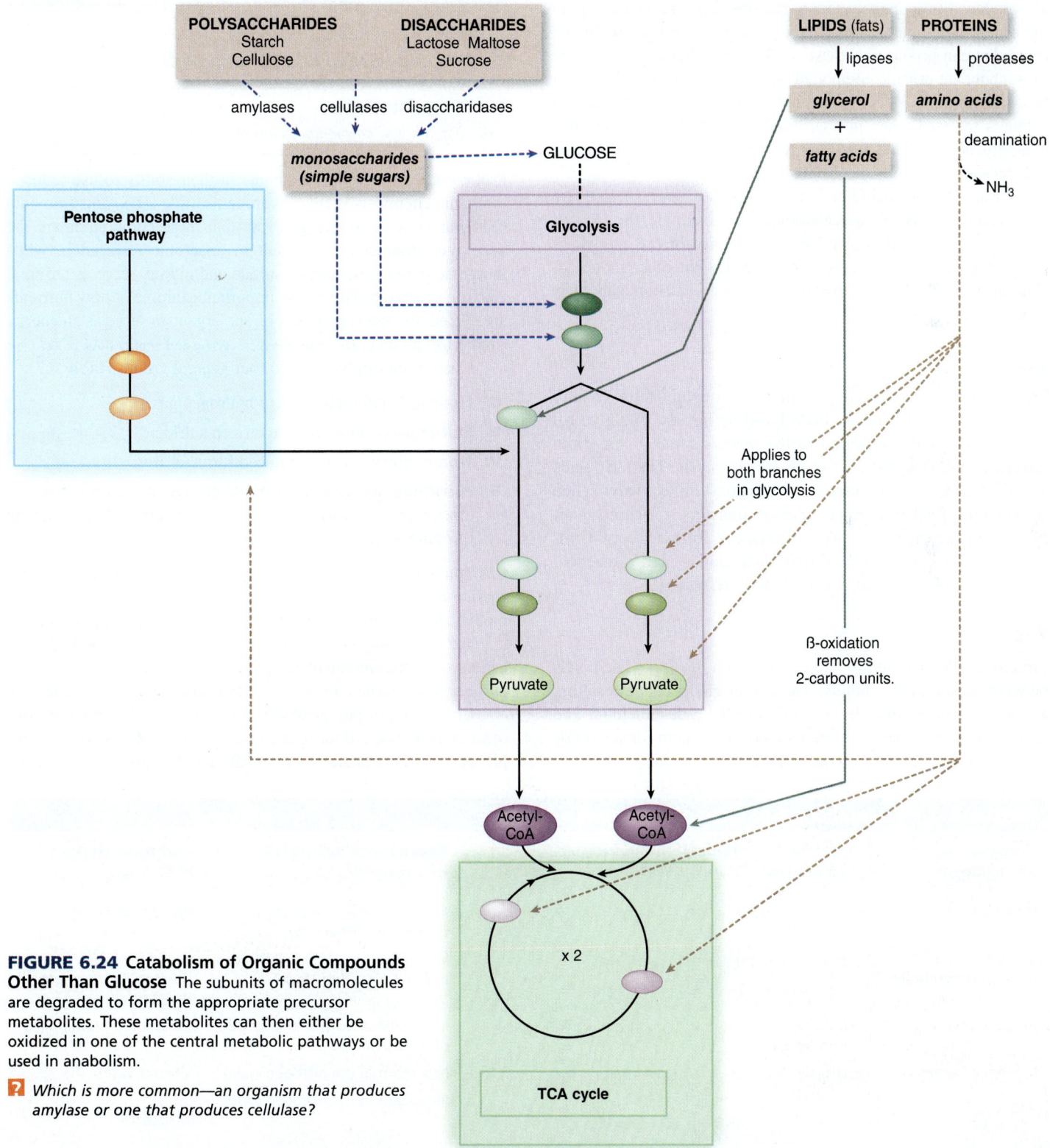

FIGURE 6.24 Catabolism of Organic Compounds Other Than Glucose The subunits of macromolecules are degraded to form the appropriate precursor metabolites. These metabolites can then either be oxidized in one of the central metabolic pathways or be used in anabolism.

❓ *Which is more common—an organism that produces amylase or one that produces cellulase?*

precursor metabolites can be either oxidized in one of the central metabolic pathways or used in biosynthesis. ◀◀ hydrolysis, p. 24

Polysaccharides and Disaccharides

Starch and cellulose are both polymers of glucose, but different types of chemical linkages join their subunits. The nature of this difference profoundly affects their degradation. Enzymes called amylases are made by a wide variety of organisms to digest starches. In contrast, cellulose is broken down by cellulases, which are produced by relatively few organisms. Bacteria that reside in the rumen of animals, as well as many types of fungi, are among those that can produce cellulase. Considering that cellulose is the most abundant organic compound on earth, it is not surprising that fungi are important decomposers in terrestrial habitats. The glucose subunits released when polysaccharides are hydrolyzed can then enter glycolysis to be oxidized to pyruvate. ◀◀ starch, p. 32 ◀◀ cellulose, p. 32 ▶▶ rumen, p. 732

Disaccharides including lactose, maltose, and sucrose are hydrolyzed by specific disaccharidases. For example, the enzyme β-galactosidase breaks down lactose, forming glucose and galactose. Glucose can enter glycolysis directly, but other monosaccharides must first be converted to one of the precursor metabolites. ◀◀ disaccharides, p. 31

Lipids

Fats, the most common simple lipids, are a combination of fatty acids and glycerol. Fats are hydrolyzed by lipases. The glycerol component is then converted to the precursor metabolite dihydroxyacetone phosphate, which enters glycolysis. The fatty acids are degraded using a series of reactions collectively called β-oxidation. Each sequential reaction transfers a 2-carbon unit from the end of the fatty acid to coenzyme A, forming acetyl-CoA, which enters the TCA cycle. Each β-oxidation is a redox reaction, generating 1 NADH + H$^+$ and 1 FADH$_2$. ◀◀ simple lipids, p. 33

Proteins

Proteins are hydrolyzed by proteases, which break peptide bonds between amino acid subunits. The amino group of the resulting amino acids is removed by a reaction called a deamination. The remaining carbon skeletons are then converted into the appropriate precursor metabolites. ◀◀ protein, p. 25

MicroAssessment 6.6

In order for polysaccharides, lipids, and proteins to be used as energy sources, they must be first hydrolyzed to release their respective subunits. These are then converted to the appropriate precursor metabolites so they can enter a central metabolic pathway.

16. *Why do cells secrete hydrolytic enzymes?*
17. *Explain the process used to degrade fatty acids.*
18. *How would cellulose-degrading bacteria in the rumen of a cow benefit the animal?* ➕

6.7 ■ Chemolithotrophs

Learning Outcome

18. *Explain how chemolithotrophs obtain energy.*

Prokaryotes as a group are unique in their ability to use reduced inorganic chemicals such as hydrogen sulfide (H$_2$S) and ammonia (NH$_3$) as sources of energy. Note that these compounds are the very ones produced as a result of anaerobic respiration, when inorganic molecules, such as sulfate and nitrate, serve as terminal electron acceptors. This is one important example of how nutrients are cycled; the waste products of one organism serve as an energy source for another. ▶▶ biogeochemical cycling and energy flow, p. 725

Chemolithotrophs fall into four general groups (**table 6.7**):

■ **Hydrogen bacteria** oxidize hydrogen gas.
■ **Sulfur bacteria** oxidize hydrogen sulfide.
■ **Iron bacteria** oxidize reduced forms of iron.
■ **Nitrifying bacteria** include two groups of bacteria: one oxidizes ammonia (forming nitrite), and the other oxidizes nitrite (producing nitrate).

Chemolithotrophs extract electrons from inorganic energy sources, passing them to an electron transport chain that generates a proton motive force. The energy of this gradient is then harvested to make ATP, using the processes described earlier. As with chemoheterotrophs, the amount of energy gained in metabolism depends on the energy source and the terminal electron acceptor (see figure 6.7).

Chemolithotrophs generally thrive in very specific environments where reduced inorganic compounds are found. For example, certain bacteria are found in sulfur-rich acidic environments;

TABLE 6.7 Metabolism of Chemolithotrophs

Common Name of Organism	Source of Energy	Oxidation Reaction(s) (Energy Yielding)	Important Feature(s) of Group	Common Genera in Group
Hydrogen bacteria	H$_2$	$H_2 + \frac{1}{2} O_2 \longrightarrow H_2O$	Can also use simple organic compounds for energy	*Hydrogenomonas*
Sulfur bacteria (non-photosynthetic)	H$_2$S	$H_2S + \frac{1}{2} O_2 \longrightarrow H_2O + S$ $S + 1\frac{1}{2} O_2 + H_2O \longrightarrow H_2SO_4$	Some members of this group can live at a pH of less than 1.	*Acidithiobacillus, Thiobacillus, Beggiatoa, Thiothrix*
Iron bacteria	Reduced Iron (Fe^{2+})	$2\,Fe^{2+} + \frac{1}{2} O_2 + H_2O \longrightarrow 2\,Fe^{3+} + 2\,OH^-$	Iron oxide present in the sheaths of these bacteria	*Sphaerotilus, Gallionella*
Nitrifying bacteria	NH$_3$	$NH_3 + 1\frac{1}{2} O_2 \longrightarrow HNO_2 + H_2O$	Important in the nitrogen cycle	*Nitrosomonas*
	HNO$_2$	$HNO_2 + \frac{1}{2} O_2 \longrightarrow HNO_3$	Important in the nitrogen cycle	*Nitrobacter*

Mining with Microbes

Microorganisms have been used for thousands of years in the production of bread and wine. Only in the past several decades, however, have microorganisms been used with increasing frequency in another area—the mining industry. The mining process traditionally consists of digging crude ores (mineral-containing rocks) from the earth, crushing them, and then extracting the desired minerals from the contaminants. The extraction process for copper and gold frequently involves harsh conditions, such as smelting, and burning off the contaminants before extracting the metal with cyanide. Such activities are expensive and harmful to the environment. With the development of biomining, some of these problems are being solved.

In the process of biomining copper, the low-grade ore is piled outside the mine and then treated with acid. The acidic conditions encourage the growth of *Acidithiobacillus* species present naturally in the ore. These acidophilic bacteria use CO_2 as a source of carbon and gain energy by oxidizing sulfides of iron first to sulfur and then to sulfuric acid. The sulfuric acid dissolves the insoluble copper and gold from the ore. Currently about 25% of all copper produced in the world comes from the process of biomining. Similar processes are being applied to gold mining.

The current process of biomining uses microbes naturally in the ore. Many improvements should be possible. For example, oxidating the minerals generates heat to the point that the bacteria may be killed, but it may be possible to use thermophiles to overcome this problem. In addition, many ores contain heavy metals, such as mercury, cadmium, and arsenic, which are toxic to the bacteria, but perhaps microorganisms resistant to these metals could be found. Biomining is still in its infancy.

these organisms can be used to enhance the recovery of metals because they oxidize metal sulfides (see **Perspective 6.1**). Thermophilic chemolithotrophs that grow near hydrothermal vents of the deep ocean obtain energy from reduced inorganic compounds that spew from the vents. The diversity and ecology of some chemolithotrophs will be discussed in chapter 11.

Unlike organisms that use organic molecules to fill both their energy and carbon needs, chemolithotrophs incorporate CO_2 into an organic form. This process (carbon fixation) will be described later.

MicroAssessment 6.7

Chemolithotrophs use reduced inorganic compounds as an energy source. They use carbon dioxide as a carbon source.

19. *Describe the roles of hydrogen sulfide and carbon dioxide in chemolithoautotrophic metabolism.*

20. *Which energy source, Fe^{2+} or H_2S, would result in the greatest energy yield when O_2 is used as a terminal electron acceptor (hint: refer to figure 6.7)?*

6.8 ■ Photosynthesis

Learning Outcomes

19. *Describe the role of chlorophylls, bacteriochlorophylls, accessory pigments, reaction-center pigments, and antennae pigments in capturing radiant energy.*

20. *Compare and contrast the tandem photosystems of cyanobacteria and photosynthetic eukaryotes with the single photosystems of purple and green bacteria.*

Plants, algae, and several groups of bacteria harvest the radiant energy of sunlight, and then use it to power the synthesis of organic compounds from CO_2. The capture and subsequent conversion of radiant energy into chemical energy is called **photosynthesis.** The general reaction—with X indicating an element such as oxygen or sulfur—is as follows:

$$6 CO_2 + 12 H_2X \xrightarrow{\text{Light Energy}} C_6H_{12}O_6 + 12 X + 6 H_2O$$

Photosynthetic processes are generally considered in two distinct stages. The **light reactions** (also called the light-dependent reactions)

capture radiant energy and convert it to chemical energy in the form of ATP. An unrelated set of reactions (sometimes called the light-independent reactions or dark reactions) uses the ATP to synthesize organic compounds. This involves carbon fixation, which converts CO_2 into organic compounds. We will describe the steps of carbon fixation in a separate section later in the chapter because chemolithotrophs use the process as well. Characteristics of various photosynthetic mechanisms are summarized in **table 6.8.**

Capturing Radiant Energy

Photosynthetic organisms are highly visible in their natural habitats because they have various colored pigments that capture the energy of light (radiant energy). The colors we observe are due to the wavelengths reflected by the pigments—for example, pigments that absorb only blue and red light are green (see figure 5.6). Multiple pigments are involved in photosynthesis, increasing the range of wavelengths absorbed by a cell. The pigments are located in protein complexes called **photosystems** within photosynthetic membranes (**figure 6.25**). The photosystems specialize in capturing and using the energy of light.

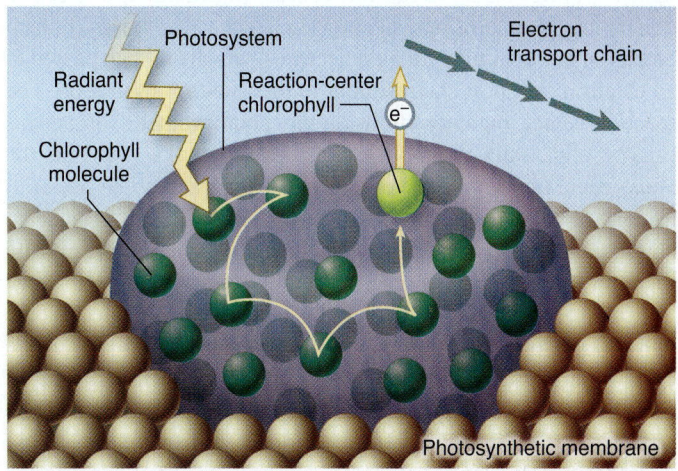

FIGURE 6.25 Photosystem Chlorophyll and other pigments capture the energy of light and then transfer it to reaction-center chlorophyll, which emits an electron that is then passed to an electron transport chain.

❓ *What is the role of an electron transport chain?*

TABLE 6.8   Comparison of the Photosynthetic Mechanisms Used by Different Organisms

	Oxygenic Photosynthesis		Anoxygenic Photosynthesis	
	Plants, Algae	Cyanobacteria	Purple Bacteria	Green Bacteria
Location of the photosystem	In membranes of thylakoids, which are within the stroma of chloroplasts	In membranes of thylakoids, located within the cell	Within the cytoplasmic membrane; extensive invaginations in that membrane increase the surface area.	Primarily within the cytoplasmic membrane; chlorosomes attached to the inner surface of the membrane contain the accessory pigments.
Type of photosystem	Photosystem I and photosystem II		Similar to photosystem II	Similar to photosystem I
Primary light-harvesting pigment	Chlorophyll a	Chlorophyll a	Bacteriochlorophylls	Bacteriochlorophylls
Mechanism for generating reducing power	Non-cyclic photophosphorylation using both photosystems		Reversed electron transport	Non-cyclic use of the photosystem
Source of electrons for reducing power	H_2O	H_2O	Varies among the organisms in the group; may include H_2S, H_2, or organic compounds.	
CO_2 fixation	Calvin cycle	Calvin cycle	Calvin cycle	Reversed TCA cycle
Accessory pigments	Carotenoids	Carotenoids, phycobilins	Carotenoids	Carotenoids

Photosynthetic pigments include chlorophylls, bacteriochlorophylls, and accessory pigments. **Chlorophylls** are found in plants, algae, and cyanobacteria. The various types of chlorophylls are designated with a letter following the term—for example, chlorophyll a. **Bacteriochlorophylls** are found in anoxygenic photosynthetic bacteria ("anoxygenic" means they do not generate O_2). These pigments absorb wavelengths not absorbed by chlorophylls, allowing the bacteria to grow in habitats where other photosynthetic organisms cannot. **Accessory pigments** increase the efficiency of light capture by absorbing wavelengths not absorbed by the other pigments. These pigments include carotenoids—found in a wide variety of photosynthetic prokaryotes and eukaryotes—and phycobilins, which are unique to cyanobacteria and red algae.

Within the photosystems, some pigments function as reaction-center pigments and others function as antennae pigments (figure 6.25). **Reaction-center pigments** are electron donors in the photosynthetic process. When excited by radiant energy, these emit high-energy electrons, which are then passed to an electron transport chain similar to that used in respiration. The reaction-center pigment of oxygenic photosynthetic organisms (plants, algae, and cyanobacteria) is chlorophyll a, whereas the anoxygenic photosynthetic organisms (purple and green bacteria) use one of the bacteriochlorophylls. **Antennae pigments** make up a complex that acts as a funnel, capturing the energy of light and then transferring it to the reaction-center pigment.

The photosystems of cyanobacteria are embedded in the membranes of structures called thylakoids located within the cells. Plants and algae also have thylakoids, in the stroma of the chloroplast (see figure 3.53). The similarity between the structure of chloroplasts and cyanobacteria is not surprising considering that the organelle appears to have descended from an ancestor of a cyanobacterium (see Perspective 3.1).

◀◀ thylakoid, p. 76   ◀◀ stroma, p. 76

The photosystems of the purple and green bacteria are embedded in the cytoplasmic membrane. Purple bacteria have extensive invaginations in the membrane that maximize the surface area. Green bacteria have specialized structures called chlorosomes attached to the inner surface of the cytoplasmic membrane. These structures contain the accessory pigments.

Converting Radiant Energy into Chemical Energy

Photosynthetic organisms use the light-dependent reactions to accomplish two tasks. First, they use radiant energy to fuel ATP synthesis, the process of **photophosphorylation.** They also need to generate reducing power so they can fix CO_2. Depending on the method used to fix CO_2, the type of reducing power required may be either NADPH or NADH.

Light-Dependent Reactions in Cyanobacteria and Photosynthetic Eukaryotic Cells

Cyanobacteria and chloroplasts have two distinct photosystems that work in tandem (**figure 6.26**). The sequential absorption of energy by the two photosystems allows the process to raise the energy of electrons stripped from water to a high enough level to be used to generate a proton motive force as well as produce reducing power. The process is oxygenic—that is, it generates O_2.

First we will consider the simplest situation, which occurs when the cell needs to synthesize ATP, but not reducing power (NADPH). To accomplish this, only photosystem I is used. When radiant energy is absorbed by this photosystem, the reaction-center chlorophylls emit high-energy electrons. The electrons are then passed to an electron carrier, which transports them to a proton pump similar to complex III in the electron transport chain of mitochondria. After being used to move protons across the

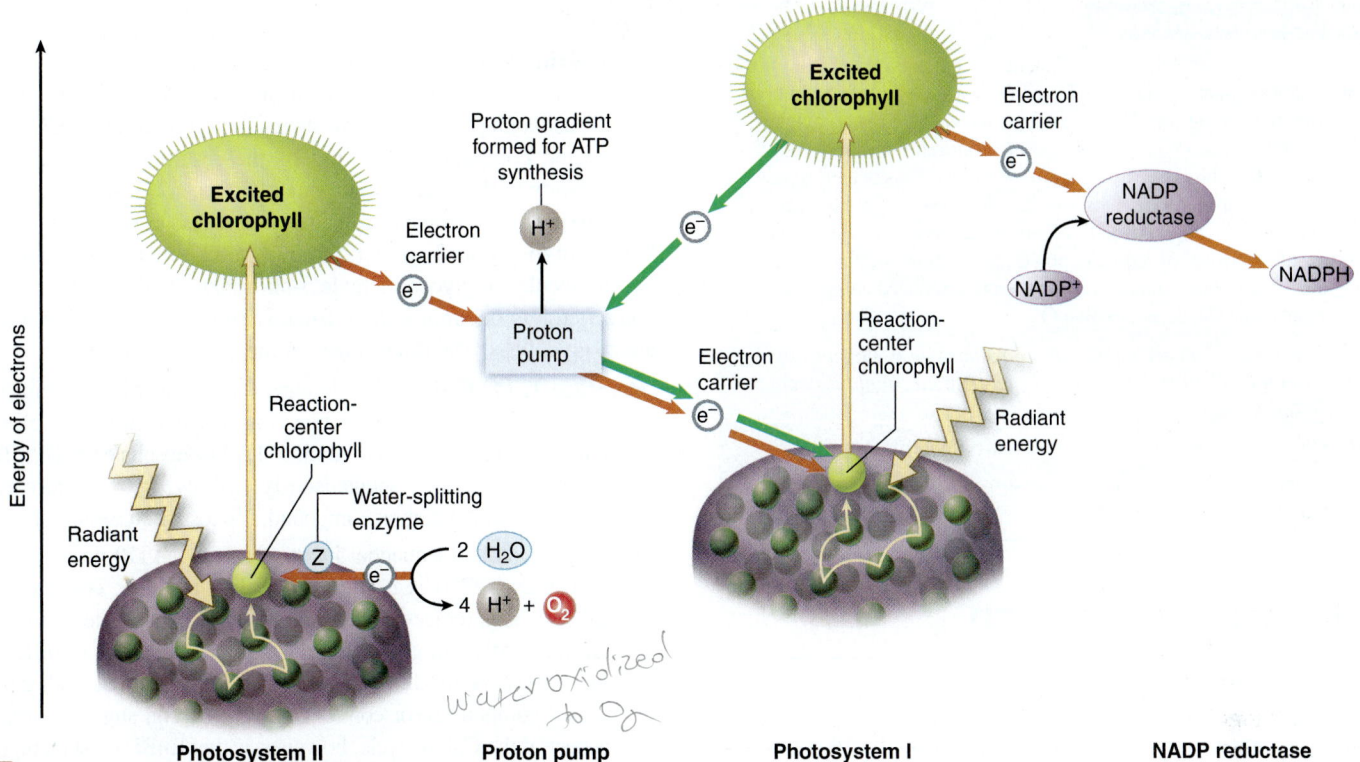

Energy of electrons

Radiant energy

Excited chlorophyll

Reaction-center chlorophyll

Electron carrier

Proton gradient formed for ATP synthesis

H^+

Proton pump

Electron carrier

Excited chlorophyll

Electron carrier

NADP reductase

$NADP^+$

NADPH

Reaction-center chlorophyll

Radiant energy

Water-splitting enzyme

Z

$2\ H_2O$

$4\ H^+ + O_2$

water oxidized to O_2

Photosystem II   **Proton pump**   **Photosystem I**   **NADP reductase**

FIGURE 6.26 The Tandem Photosystems of Cyanobacteria and Chloroplasts Radiant energy captured by photosynthetic pigments excites the reaction-center chlorophyll, causing it to emit a high-energy electron, which is then passed to an electron transport chain. In cyclic photophosphorylation, electrons emitted by photosystem I are returned to that photosystem; the path of the electrons is shown in green arrows. In non-cyclic photophosphorylation, the electrons used to replenish photosystem I are donated by radiant energy–excited photosystem II; the path of these electrons is shown in orange arrows. In turn, photosystem II replenishes its own electrons by stripping them from water, producing O_2.

? *When do the cells need to use non-cyclic photophosphorylation?*

membrane—generating a proton motive force—the electrons are returned to photosystem I. As in oxidative phosphorylation, ATP synthase harvests the energy of the proton motive force to synthesize ATP. This overall process is called **cyclic photophosphorylation** because the electrons have followed a cyclical path—the molecule that serves as the electron donor (reaction-center chlorophyll) is also the terminal electron acceptor.

When photosynthetic cells must produce both ATP and reducing power, **non-cyclic photophosphorylation** is used. In this process, the electrons emitted by photosystem I are not passed to the proton pump but instead reduce $NADP^+$ to NADPH. Although this action provides reducing power, the cell must now use another source to replenish the electrons emitted by reaction-center chlorophyll. In addition, the cell still needs to generate a proton motive force in order to synthesize ATP. Photosystem II plays a critical role in this process. When photosystem II absorbs radiant energy, the reaction-center chlorophylls emit high-energy electrons that can be donated to photosystem I. First, however, the electrons are passed to the proton pump, which uses some of their energy to establish the proton motive force. The electrons emitted from photosystem II are replenished when an enzyme within that complex extracts electrons from water, donating them to the reaction-center chlorophyll. Removal of electrons from two molecules of water generates O_2. In essence, photosystem II captures the energy of light and then uses it to raise the energy level of electrons stripped

from water molecules to a high enough level that they can be used to power photophosphorylation. Photosystem I then accepts those electrons, which still have some energy, and again captures the energy of light to boost the energy of the electrons to an even higher level necessary to reduce NADPH.

Light-Dependent Reactions in Anoxygenic Photosynthetic Bacteria

Anoxygenic photosynthetic bacteria have only a single photosystem and cannot use water as an electron donor for reducing power. This is why they are anoxygenic (do not generate O_2). These bacteria use electron donors such as hydrogen gas (H_2), hydrogen sulfide (H_2S), and organic compounds. Two groups of anoxygenic photosynthetic bacteria are the purple bacteria and green bacteria. ▶▶ purple bacteria, p. 259 ▶▶ green bacteria, p. 260

Purple bacteria synthesize ATP using a photosystem similar to photosystem II of cyanobacteria and eukaryotes. However, this photosystem does not raise the electrons to a high enough energy level to reduce NAD^+ (or $NADP^+$), so the cells must use an alternative mechanism to generate reducing power. They use a process called reversed electron transport, expending ATP to run the electron transport chain in the reverse direction, or "uphill."

Green bacteria have a photosystem similar to photosystem I. The electrons emitted from this can either generate a proton motive force or reduce NAD^+.

6.9 ■ Carbon Fixation

Learning Outcome

21. *Describe the three stages of the Calvin cycle.*

Chemolithoautotrophs and photoautotrophs use carbon dioxide (CO_2) to synthesize organic compounds, the process of **carbon fixation.** In photosynthetic organisms, this process is called the light-independent reactions. Carbon fixation consumes a great deal of ATP and reducing power, which should not be surprising considering that the reverse process (oxidizing those same compounds to CO_2) liberates a great deal of energy. The Calvin cycle is the most common pathway used to fix carbon, but some prokaryotes incorporate CO_2 using other mechanisms. For example, the green bacteria and some members of the *Archaea* use a pathway that reverses the steps of the TCA cycle.

Calvin Cycle

The **Calvin cycle,** or Calvin-Benson cycle, named in honor of the scientists who described much of it, is a complex cycle. It can be viewed as having three essential stages—incorporation of CO_2 into an organic compound, reduction of the resulting molecule, and regeneration of the starting compound (**figure 6.27**). Because of the complexities, it is easiest to consider the process as consisting of six "turns" of the cycle. Together, these six turns generate a net gain of two molecules of glyceraldehyde 3-phosphate (G3P), which can be converted into one molecule of fructose 6-phosphate, an intermediate of glycolysis. The three stages of the Calvin cycle are:

■ **Stage 1.** ① Carbon dioxide enters the cycle when an enzyme commonly called rubisco (ribulose bisphosphate carboxylase)joins it to a 5-carbon compound, ribulose 1,5-bisphosphate (RuBP). The resulting compound spontaneously hydrolyzes to produce two molecules of a 3-carbon compound, 3-phosphoglycerate (3PG).

■ **Stage 2.** ② A sequential input of energy (ATP) and reducing power (NADPH) is used in steps that, together, convert 3PG to G3P. This molecule is also a precursor metabolite formed as an intermediate in glycolysis and can have a variety of different fates. It can be used in biosynthesis, oxidized to make other precursor compounds, or converted to a 6-carbon sugar. A critical aspect of the Calvin cycle, however, is that RuBP must be regenerated from G3P for the cycle to continue. Consequently, in six turns of the cycle, a maximum of 2 G3P can be converted to a 6-carbon sugar; the rest is used to regenerate RuBP.

■ **Stage 3.** ③ Many of the steps used to regenerate RuBP involve reactions of the pentose phosphate cycle.

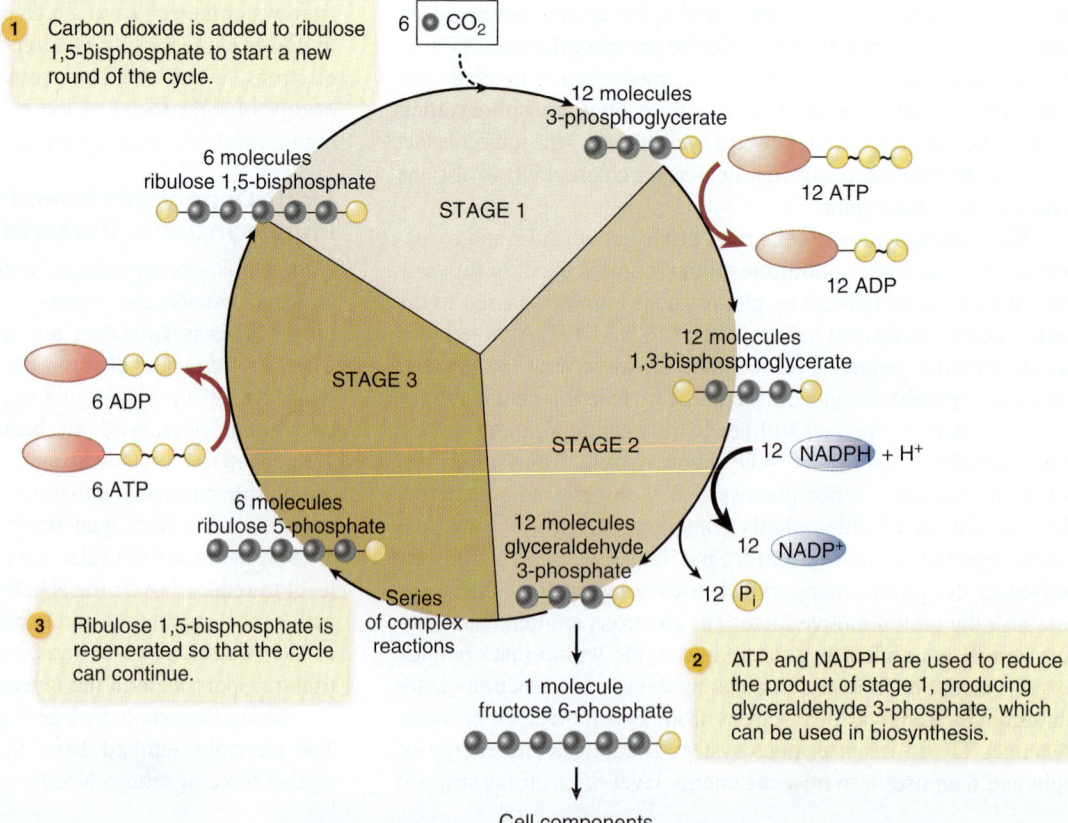

FIGURE 6.27 The Calvin Cycle The Calvin cycle has three essential stages: **(1)** incorporation of CO_2 into an organic compound; **(2)** reduction of the resulting molecule; and **(3)** regeneration of the starting compound.

? *How much ATP and NADPH must be spent to synthesize one molecule of fructose?*

① Carbon dioxide is added to ribulose 1,5-bisphosphate to start a new round of the cycle.

6 molecules ribulose 1,5-bisphosphate

STAGE 1

6 ● CO_2

12 molecules 3-phosphoglycerate

12 ATP

12 ADP

12 molecules 1,3-bisphosphoglycerate

STAGE 2

12 NADPH + H⁺

12 NADP⁺

STAGE 3

6 ADP

6 ATP

6 molecules ribulose 5-phosphate

Series of complex reactions

12 molecules glyceraldehyde 3-phosphate

12 P_i

③ Ribulose 1,5-bisphosphate is regenerated so that the cycle can continue.

② ATP and NADPH are used to reduce the product of stage 1, producing glyceraldehyde 3-phosphate, which can be used in biosynthesis.

1 molecule fructose 6-phosphate

Cell components

Yield of the Calvin Cycle

One molecule of the 6-carbon sugar fructose can be generated for every six turns of the cycle. These six turns consume 18 ATP and 12 NADPH + H$^+$.

MicroByte

Although rubisco is unique to autotrophs, it is probably the most abundant enzyme on earth!

MicroAssessment 6.9

The process of carbon fixation consumes a great deal of ATP and reducing power. The Calvin cycle is the most common pathway used to incorporate inorganic carbon into an organic form.

24. *What is the role of rubisco?*

25. *What would happen if ribulose 1,5-bisphosphate (RuBP) were depleted in a cell?* ✚

6.10 ■ Anabolic Pathways—Synthesizing Subunits from Precursor Molecules

Learning Outcome

22. *Describe the synthesis of lipids, amino acids, and nucleotides.*

Prokaryotes, as a group, are highly diverse with respect to the compounds they use for energy but remarkably similar in their biosynthetic processes. They synthesize the necessary subunits, using specific anabolic pathways that require ATP, reducing power in the form of NADPH, and the precursor metabolites formed in the central metabolic pathways (**figure 6.28**). Organisms lacking one or more enzymes in a given biosynthetic pathway must have the end product provided from an external source. This is why fastidious bacteria, such as lactic acid bacteria, require many different growth factors. Once the subunits are synthesized or transported into the cell, they can be assembled to

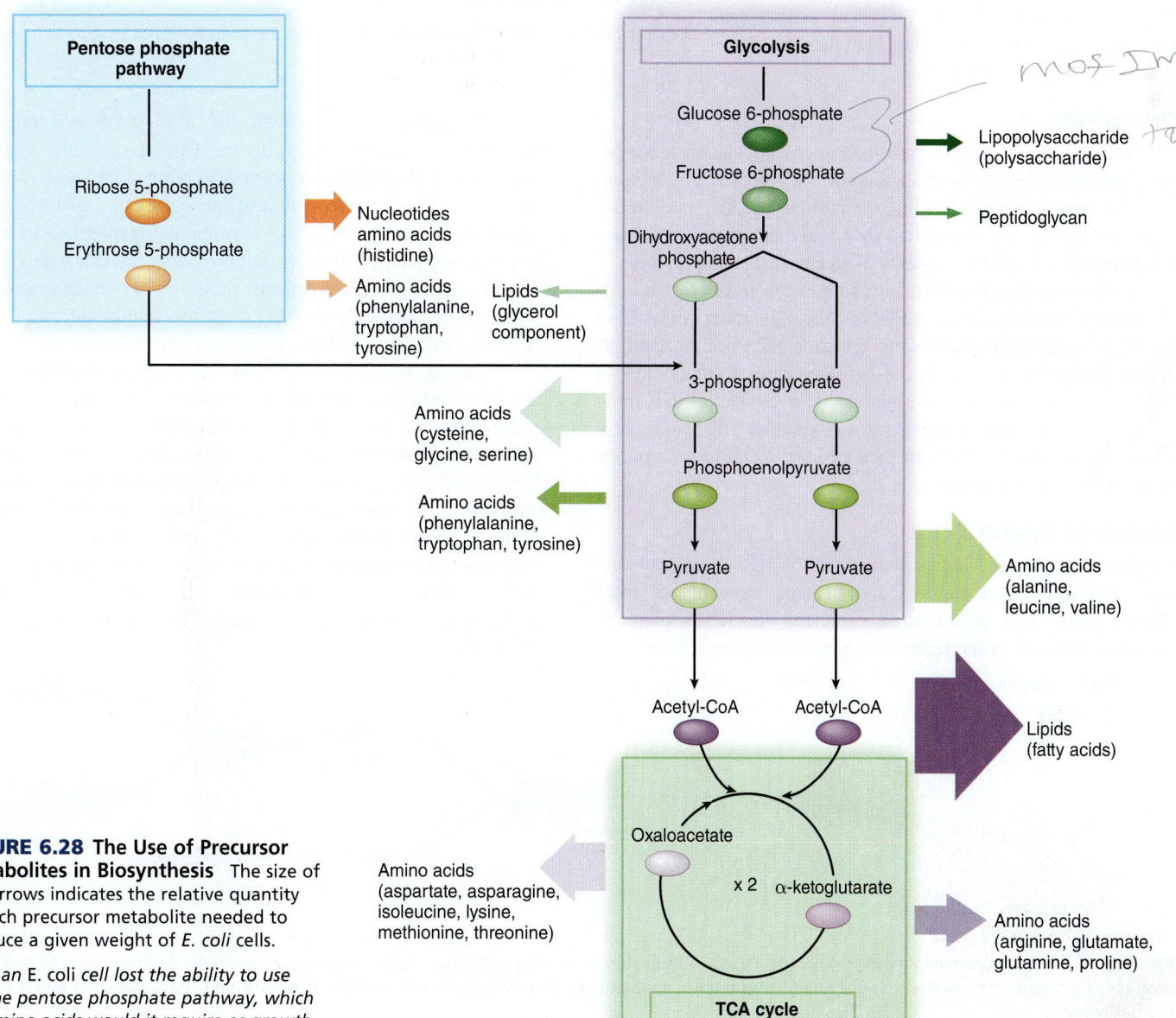

FIGURE 6.28 The Use of Precursor Metabolites in Biosynthesis The size of the arrows indicates the relative quantity of each precursor metabolite needed to produce a given weight of *E. coli* cells.

❓ *If an* E. coli *cell lost the ability to use the pentose phosphate pathway, which amino acids would it require as growth factors?*

make macromolecules. Various different macromolecules can then be joined to form the structures that make up the cell. ◄◄ fastidious, p. 93

Lipid Synthesis

Synthesis of most lipids requires fatty acids and glycerol. To produce fatty acids, the acetyl group of acetyl-CoA (the precursor metabolite produced in the transition step) is transferred to a carrier protein. This carrier holds the developing fatty acid chain as 2-carbon units are progressively added. When the fatty acid reaches its required length, usually 14, 16, or 18 carbon atoms long, it is released. The glycerol component of the fat is synthesized from the precursor metabolite dihydroxyacetone phosphate, which is generated in glycolysis.

Amino Acid Synthesis

Proteins are composed of various combinations of usually 20 different amino acids. These amino acids can be grouped into structurally related families that share common pathways of biosynthesis.

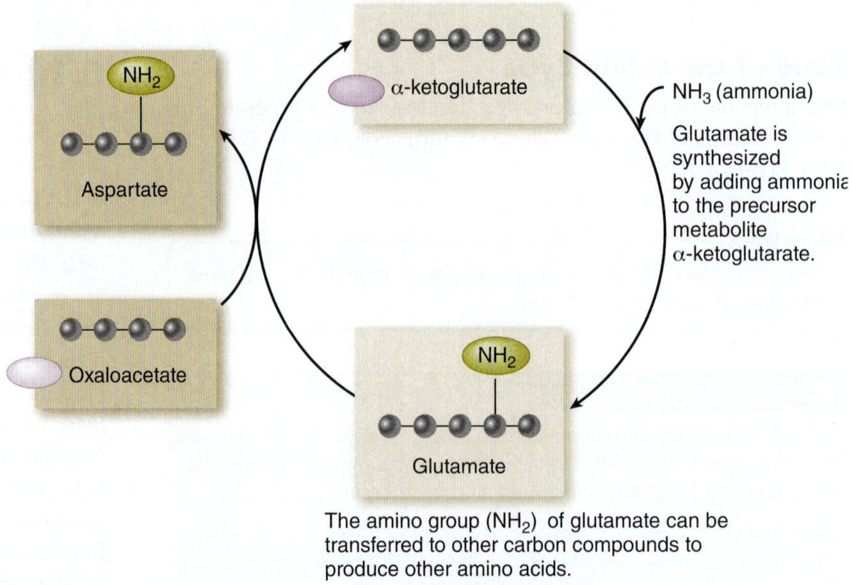

The amino group (NH_2) of glutamate can be transferred to other carbon compounds to produce other amino acids.

FIGURE 6.29 The Role of Glutamate in Amino Acid Synthesis Once glutamate has been synthesized from α-ketoglutarate, its amino group can be transferred to produce other amino acids.

❓ *Alpha-ketoglutarate is produced in which central metabolic pathway?*

Glutamate

Amino acids are necessary for protein synthesis, but glutamate is especially important because its synthesis provides a mechanism for bacteria to incorporate nitrogen into organic material. Recall from chapter 4 that many bacteria use ammonium (NH_4^+) as their source of nitrogen; it is primarily through the synthesis of glutamate that they do this.

Glutamate is synthesized in a single-step reaction that adds ammonia to the precursor metabolite α-ketoglutarate, produced in the TCA cycle (**figure 6.29**). Once glutamate has been formed, its amino group can be transferred to other carbon compounds to produce amino acids such as aspartate. This transfer of the amino group, a transamination, regenerates α-ketoglutarate from glutamate. The α-ketoglutarate can then be used again to incorporate more ammonia. ◄◄ amino group, p. 25

Aromatic Amino Acids

Synthesis of aromatic amino acids such as tyrosine, phenylalanine, and tryptophan requires a multistep, branching pathway (**figure 6.30**). This serves as an excellent illustration of many important features of the regulation of amino acid synthesis.

The pathway begins with the joining of two precursor metabolites—phosphoenolpyruvate (3-carbon) and erythrose 4-phosphate (4-carbon)—to form a 7-carbon compound. The precursors originate in glycolysis and the pentose phosphate pathway, respectively. Then, the 7-carbon compound is modified through a series of steps until a branch point is reached. At this juncture, two options are possible. If synthesis proceeds in one direction, tryptophan is produced. In the other direction, another branch point is reached; from there, either tyrosine or phenylalanine can be made.

When an amino acid is provided to a cell, it would be a waste of carbon, energy, and reducing power for that cell to continue synthesizing it. But with branched pathways, how does the cell stop synthesizing the product of only one branch? In the pathway for aromatic acid biosynthesis, this partly occurs by regulating the enzymes at the branch points. Tryptophan is a feedback inhibitor of the enzyme that directs the branch to its synthesis; this sends the pathway to the steps leading to the synthesis of the other amino acids—tyrosine and phenylalanine. Likewise, these two amino acids each inhibit the first enzyme of the branch leading to their synthesis.

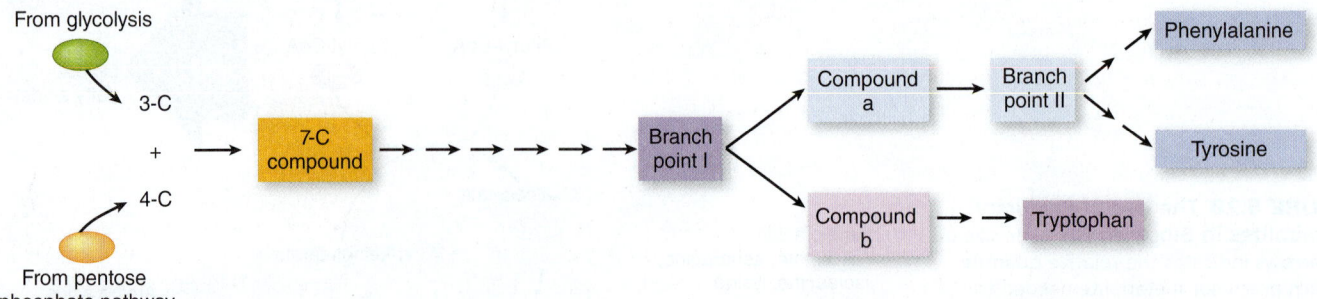

FIGURE 6.30 Synthesis of Aromatic Amino Acids A multistep branching pathway is used to synthesize aromatic amino acids. The end product of a branch inhibits the first enzyme of that branch; that end product also inhibits one of the three enzymes that catalyze the first step of the pathway.

❓ *The synthesis of aromatic amino acids requires precursor metabolites made in which central metabolic pathways?*

In addition to regulating the branchpoints, the three amino acids each control the first step of the pathway, the formation of the 7-carbon compound. In *E. coli,* three different enzymes can catalyze this step; each has the same active site, but they have different allosteric sites. Each aromatic amino acid acts as a feedback inhibitor for one of the enzymes. If all three amino acids are present in the environment, then very little of the 7-carbon compound will be synthesized. If only one or two of those amino acids are present, then proportionally more of the compound will be synthesized. ◀◀ allosteric enzymes, p. 137

Nucleotide Synthesis

Nucleotide subunits of DNA and RNA are composed of three units: a 5-carbon sugar, a phosphate group, and a nucleobase, either a purine or a pyrimidine. They are synthesized as ribonucleotides, but these can then be converted to deoxyribonucleotides by replacing the hydroxyl group on the 2′ carbon of the sugar with a hydrogen atom. ◀◀ nucleotides, p. 32 ◀◀ purine, p. 32 ◀◀ pyrimidine, p. 32

The purine (double-ring structure) and pyrimidine (single ring) nucleotides are synthesized in distinctly different manners. The starting compound of purine synthesis is ribose 5-phosphate, a precursor metabolite generated in the pentose phosphate pathway. Then, in a highly ordered sequence, atoms from the other sources are added to form the purine ring. This can then be converted to a purine nucleotide. To synthesize pyrimidine nucleotides, the pyrimidine ring is made first and then attached to ribose 5-phosphate. After one pyrimidine nucleotide is formed, the nucleobase component can be converted into any of the other pyrimidines.

MicroAssessment 6.10

Biosynthetic processes of different organisms are remarkably similar, using precursor metabolites, NADPH, and ATP to form subunits. Synthesis of the amino acid glutamate provides a mechanism for bacteria to incorporate nitrogen in the form of ammonia into organic material. Synthesis of aromatic amino acids involves branching pathways. The purine nucleotides are synthesized in a very different manner from the pyrimidine nucleotides.

26. *Explain why glutamate synthesis is particularly important for a cell.*
27. *What three general products of the central metabolic pathways does a cell require to carry out biosynthesis?*
28. *With a branched biochemical pathway, why would it be important for a cell to shut down the first step as well as branching steps?*

FUTURE CHALLENGES 6.1

Fueling the Future

Considering that microbes are masters at extracting energy from seemingly unlikely sources, it should not be surprising that they are being enlisted to help us meet our future energy needs. Bacteria, yeast, and algae play starring roles in most of the current research efforts toward developing cost-effective methods to produce **biofuels,** which are fuels made from renewable resources. One goal in producing and using these is to be carbon neutral, meaning that fuel production consumes as much atmospheric carbon as fuel use releases.

First-generation biofuels are made using standard technologies (such as fermentations) and readily available material (including corn and sugar cane). One example is using yeast to ferment the sugars in corn to produce bioethanol. Although this provides an alternative to fossil fuels, it uses an edible resource, contributing to food shortages and increased food prices. In addition, the farming practices used to grow corn on a large-scale basis can adversely effect the environment, and the process for making bioethanol uses fossil fuels so the product is not carbon neutral.

In response to the problems with first-generation biofuels, newer technologies are being developed to produce what are collectively referred to as advanced biofuels. These include:

- **Bioethanol.** Rather than using edible foods, bioethanol produced with advanced technologies is made from cornstalks or other plant waste materials. One problem is that the principle component of plant wastes is cellulose, and relatively few microbes degrade this carbohydrate. Microbes that degrade cellulose generally are not as easy to grow on a large-scale basis as yeast, so technical hurdles must be overcome for this to be a feasible option.
- **Biodiesel.** Composed of fatty acid methyl esters, biodiesel has an advantage over bioethanol in that it has a higher energy content and is more compatible with current fuel storage and distribution systems. Algae and cyanobacteria—both of which are photosynthetic—capture radiant energy to make lipids that can be used to produce biodiesel. This provides a particularly attractive option because it bypasses the need for the two-stage process used to create bioethanol (plants convert sunlight to chemical energy, and then microbes ferment the plant material to produce ethanol). A problem with using photosynthetic microbes to produce biofuels, however, is that areas with enough sunlight to support their abundant growth generally lack sufficient water for the final steps of fuel production. Another option being explored for biodiesel production is to genetically engineer an easy-to-grow bacterium such as *E. coli* to become a biodiesel-producing factory, which a lab has recently done. The strain has also been engineered to produce the enzyme that breaks down cellulose, so it can use waste materials as an energy source. The efficiency is still low, however, so improvements are still necessary. Also, the engineered strain still relies on a second organism—in this case, plants—to convert radiant energy into chemical energy.
- **Biohydrogen.** Some cyanobacteria and anoxygenic phototrophs make hydrogen gas as a by-product of nitrogen fixation and grow using little more than sunlight and water. An advantage of biohydrogen is that it can be easily captured from a culture. As with the other advanced biofuels, however, there are still technical obstacles that must be overcome before the gas can be produced and stored on a large-scale basis.

Will microbes end our reliance on fossil fuels? Only time will tell, but the possibilities are exciting!

Summary

6.1 ▪ Principles of Metabolism

Catabolism is the set of processes that capture and store **energy** by breaking down complex molecules. **Anabolism** includes processes that use energy to make and assemble the building blocks of a cell (figure 6.1).

Harvesting Energy

Photosynthetic organisms harvest the energy of sunlight, using it to power the synthesis of organic compounds. **Chemoorganotrophs** harvest energy contained in organic compounds (figure 6.3). **Exergonic** reactions release energy; **endergonic** reactions use energy.

Components of Metabolic Pathways

A specific **enzyme** facilitates each step of a **metabolic pathway** (figure 6.5). **ATP** is the energy currency of the cell. The **energy source** is **oxidized** to release its energy (figure 6.7). The **redox** reactions **reduce** an **electron carrier** (figure 6.8). **NAD⁺/NADH, NADP⁺/NADPH.** and **FAD/FADH₂** are electron carriers (table 6.1).

Precursor Metabolites

Precursor metabolites are used to make the subunits of macromolecules, and they can also be oxidized to generate energy in the form of ATP (table 6.2).

Overview of Catabolism (figure 6.10)

The **central metabolic pathways** are **glycolysis,** the **pentose phosphate pathway,** and the **tricarboxylic acid cycle (TCA cycle).** **Respiration** uses the reducing power accumulated in the central metabolic pathways to generate ATP by oxidative phosphorylation. **Aerobic respiration** uses O_2 as a terminal electron acceptor; **anaerobic respiration** uses a molecule other than O_2 as a terminal electron acceptor (table 6.3). **Fermentation** uses pyruvate or a derivative as a terminal electron acceptor; this recycles the reduced electron carrier NADH.

6.2 ▪ Enzymes

Enzymes function as biological catalysts; they are neither consumed nor permanently changed during a reaction.

Mechanisms and Consequences of Enzyme Action (figure 6.11)

The substrate binds to the **active site,** forming an enzyme-substrate complex that lowers the activation energy of the reaction.

Cofactors (figure 6.12; table 6.4)

Enzymes sometimes act in conjunction with **cofactors** such as **coenzymes** and trace elements.

Environmental Factors That Influence Enzyme Activity (figure 6.13)

The factors most important in influencing enzyme activities are temperature, pH, and salt concentration.

Allosteric Regulation (figure 6.14)

Cells can fine-tune the activity of an **allosteric** enzyme by using a regulatory molecule that binds to the allosteric site of the enzyme.

Enzyme Inhibition

Non-competitive inhibition occurs when the inhibitor and the substrate act at different sites on the enzyme. **Competitive inhibition** occurs when the inhibitor competes with the normal substrate for the active binding site (figure 6.15).

6.3 ▪ The Central Metabolic Pathways (table 6.6)

Glycolysis (figure 6.16)

Glycolysis converts one molecule of glucose into two molecules of pyruvate; the theoretical net yield of the pathway is 2 ATP, 2 NADH + H⁺. Six different precursor metabolites are produced.

Pentose Phosphate Pathway

The pentose phosphate pathway forms NADPH + 6 H⁺ and two different precursor metabolites.

Transition Step (figure 6.17)

The transition step converts pyruvate to acetyl-CoA. Repeated twice, this produces 2 NADH + 2 H⁺. The end product is a precursor metabolite.

Tricarboxylic Acid (TCA) Cycle (figure 6.17)

The TCA cycle completes the oxidation of glucose; the theoretical yield of two "turns" is 6 NADH + 6 H⁺, and 2 FADH₂, in addition to 2 ATP. Two intermediates are precursor metabolites.

6.4 ▪ Respiration

The Electron Transport Chain—Generating Proton Motive Force

The **electron transport chain** sequentially passes electrons, and, as a result, ejects protons. In the mitochondrial electron transport chain, three different complexes (complexes I, III, and IV) function as proton pumps (figure 6.19). Prokaryotes vary with respect to the types and arrangements of their electron transport components (figure 6.20). Some prokaryotes can use molecules other than O_2 as terminal electron acceptors. This process of anaerobic respiration harvests less energy than aerobic respiration.

ATP Synthase—Harvesting the Proton Motive Force to Synthesize ATP

ATP synthase permits protons to flow back across the membrane, harvesting the energy released to fuel the synthesis of ATP.

ATP Yield of Aerobic Respiration in Prokaryotes (figure 6.21)

The theoretical maximum yield of ATP of aerobic respiration is 38 ATP.

6.5 ▪ Fermentation

In general, the only ATP-yielding reactions of fermentations are those of the glycolytic pathway; the other steps provide a mechanism for recycling NADH (figure 6.22). Some end products of fermentation are commercially valuable (figure 6.23). Certain end products help in bacterial identification.

6.6 ▪ Catabolism of Organic Compounds Other Than Glucose (figure 6.24)

Hydrolytic enzymes break down macromolecules into their respective subunits.

Polysaccharides and Disaccharides

Amylases digest starch, releasing glucose subunits, and are produced by many organisms. Cellulases degrade cellulose. Sugar subunits released when polysaccharides are broken down can then enter glycolysis to be oxidized to pyruvate.

Lipids

Fats are hydrolyzed by lipases, releasing glycerol and fatty acids. Glycerol is converted to dihydroxyacetone phosphate; fatty acids are degraded by β-oxidation, generating reducing power and the precursor metabolite acetyl-CoA.

Proteins

Proteins are hydrolyzed by proteases. Deamination removes the amino group; the remaining carbon skeleton is then converted into the appropriate precursor molecule.

6.7 ■ Chemolithotrophs

Prokaryotes, as a group, are unique in their ability to use reduced inorganic compounds such as hydrogen sulfide (H_2S) and ammonia (NH_3) as a source of energy. Chemolithotrophs are autotrophs.

6.8 ■ Photosynthesis

The **light reactions** capture energy from light and convert it to chemical energy in the form of ATP. That energy is used for carbon fixation.

Capturing Radiant Energy

Various pigments such as **chlorophylls, bacteriochlorophylls,** carotenoids, and phycobilins are used to capture radiant energy. **Reaction center pigments** function as the electron donor in the photosynthetic process; **antennae pigments** funnel radiant energy to the reaction center pigment.

Converting Radiant Energy into Chemical Energy

The high-energy electrons emitted by reaction center chlorophylls are passed to an electron transport chain, which uses them to generate a proton motive force. The energy of proton motive force is harvested by ATP synthase to fuel the synthesis of ATP. Photosystems I and II of cyanobacteria and chloroplasts raise the energy level of electrons stripped from water to a high enough level to be used to generate a proton motive force and produce reducing power; this process is oxygenic (figure 6.26). Purple and green bacteria use only a single photosystem; they must obtain electrons from a reduced compound other than water and therefore do not generate oxygen.

6.9 ■ Carbon Fixation

Calvin Cycle (figure 6.27)

The most common pathway used to incorporate CO_2 into an organic form is the **Calvin cycle.**

6.10 ■ Anabolic Pathways—Synthesizing Subunits from Precursor Molecules (figure 6.28)

Lipid Synthesis

The fatty acid components of fat are synthesized by progressively adding 2-carbon units to an acetyl group. The glycerol component is synthesized from dihydroxyacetone phosphate.

Amino Acid Synthesis

Synthesis of glutamate from α-ketoglutarate and ammonia provides a mechanism for cells to incorporate nitrogen into organic molecules (figure 6.29). Synthesis of aromatic amino acids requires a multistep branching pathway. Allosteric enzymes regulate key steps of the pathway (figure 6.30).

Nucleotide Synthesis

Purines and pyrimidine nucleotides are made in distinctly different manners.

Review Questions

Short Answer

1. Explain the difference between catabolism and anabolism.
2. How does ATP serve as a carrier of free energy?
3. How do enzymes catalyze chemical reactions?
4. Explain how precursor molecules serve as junctions between catabolic and anabolic pathways.
5. How do cells regulate enzyme activity?
6. Why do the electrons carried by $FADH_2$ result in less ATP production than those carried by NADH?
7. Name three food products produced with the aid of microorganisms.
8. In photosynthesis, what is encompassed by the term "light-independent reactions"?
9. Unlike the cyanobacteria, the anoxygenic photosynthetic bacteria do not produce O_2. Why not?
10. What is the role of transamination in amino acid biosynthesis?

Multiple Choice

1. Which of these factors does not affect enzyme activity?
   a) Temperature   b) Inhibitors   c) Coenzymes
   d) Humidity   e) pH

2. Which of the following statements is false? Enzymes
   a) bind to substrates.
   b) lower the energy of activation.
   c) convert coenzymes to products.
   d) speed up biochemical reactions.
   e) can be named after the kinds of reaction they catalyze.

3. Which of these is not a coenzyme?
   a) FAD   b) Coenzyme A   c) NAD^+ d) ATP   e) $NADP^+$

4. What is the end product of glycolysis?
   a) Glucose   b) Citrate   c) Oxaloacetate
   d) α-ketoglutarate   e) Pyruvate

5. The major pathway(s) of central metabolism are
   a) glycolysis and the TCA cycle only.
   b) glycolysis, the TCA cycle, and the pentose phosphate pathway.
   c) glycolysis only.
   d) glycolysis and the pentose phosphate pathway only.
   e) the TCA cycle only.

6. Which of these pathways gives a cell the potential to produce the most ATP?

   a) TCA cycle

   b) Pentose phosphate pathway

   c) Lactic acid fermentation

   d) Glycolysis

7. In fermentation, the terminal electron acceptor is

   a) oxygen (O_2).   b) hydrogen (H_2).

   c) carbon dioxide (CO_2).   d) an organic compound.

8. In the process of oxidative phosphorylation, the energy of proton motive force is used to generate

   a) NADH.   b) ADP.   c) ethanol.   d) ATP.   e) glucose.

9. In the TCA cycle, the carbon atoms contained in acetate are converted into

   a) lactic acid.   b) glucose.   c) glycerol.

   d) CO_2.   e) all of these.

10. Degradation of fats as an energy source involves all of the following *except*

    a) β-oxidation.   b) acetyl-CoA.   c) glycerol.

    d) lipase.   e) transamination.

Applications

1. A worker in a cheese-making facility argues that whey, a nutrient-rich by-product of cheese, should be dumped in a nearby pond where it could serve as fish food. Explain why this proposed action could actually kill the fish by depleting the O_2 in the pond.

2. Scientists working with DNA *in vitro* often store it in solutions that contain EDTA, a chelating agent that binds magnesium (Mg^{2+}). This is done to prevent enzymes called DNases from degrading the DNA. Explain why EDTA would interfere with enzyme activity.

Critical Thinking ✚

1. A student argued that aerobic and anaerobic respiration should produce the same amount of ATP. He reasoned that they both use basically the same process; only the terminal electron acceptor is different. What is the primary error in this student's argument?

2. Chemolithotrophs near hydrothermal vents support a variety of other life-forms there. Explain how their role is analogous to that of photosynthetic organisms in terrestrial environments.

7 The Blueprint of Life, from DNA to Protein

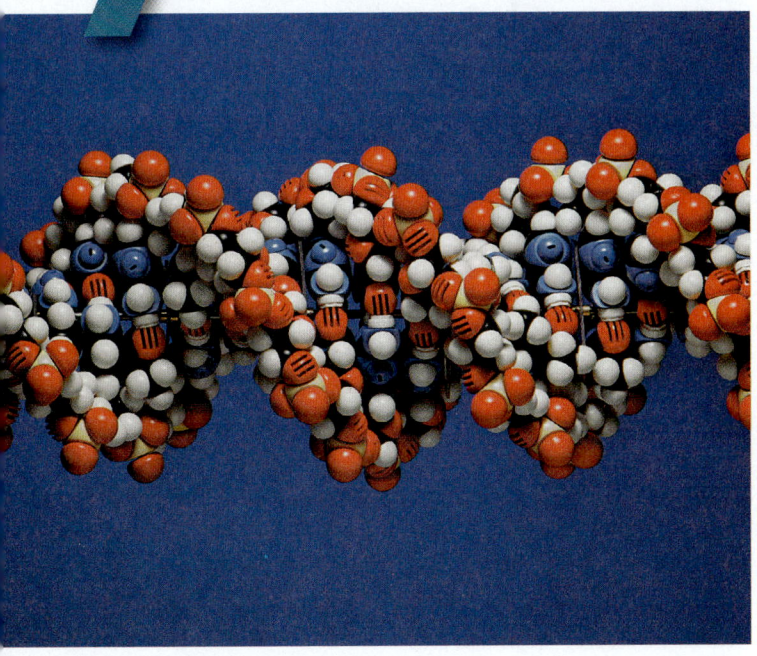

Model of DNA double helix.

A Glimpse of History

In 1866, the Czech-Austrian monk Gregor Mendel determined that traits are inherited as physical units, now called genes. The precise function of genes, however, was not revealed until 1941, when George Beadle and Edward Tatum published a scientific paper reporting that genes direct the production of enzymes.

Beadle and Tatum set out to discover how genes control metabolic reactions by studying common bread molds that have very simple nutritional requirements. These molds, *Neurospora* species, can grow on media containing only sucrose, inorganic salts, and biotin. Beadle and Tatum reasoned that if they created mutant strains which required additional nutrients, they could use them to gain insights into the relationship between genes and enzymes. For example, a mutant that requires a certain amino acid likely has a defect in an enzyme required to synthesize that amino acid.

To generate mutant strains, Beadle and Tatum treated the mold cultures with X rays and then grew the resulting cells on a nutrient-rich medium that supported the growth of both the original strain and any mutants. Next, they screened thousands of the progeny to find the nutrient-requiring mutants. To identify the metabolic defect of each one, they grew them separately in various types of media containing different nutrients.

Eventually, Beadle and Tatum established that the metabolic defect was inherited as a single gene, which ultimately led to their conclusion that a single gene determines the production of one enzyme. That assumption has been modified somewhat, because we now know that some enzymes are made up of more than one polypeptide, and not all proteins are enzymes. In 1958, Beadle and Tatum were awarded a Nobel Prize, largely for these pioneering studies that ushered in the era of modern biology. ◀◀ **polypeptide, p. 27**

Consider for a moment the incredible diversity of life-forms in our world—from the remarkable variety of microorganisms, to the plants and animals consisting of many different specialized cells. Every characteristic of each of these cells, from its shape to its function, is dictated by information within its deoxyribonucleic acid (DNA). DNA encodes the master plan, the blueprint, for all of an organism's features.

DNA itself is a simple structure, a string composed of only four different **nucleotides,** each containing a particular **nucleobase** (also simply called a base): adenine (A), thymine (T), cytosine (C), or guanine (G). A set of three nucleotides encodes a specific amino acid; in turn, a string of amino acids makes up a protein, the structure and function of which is dictated by the order of the amino acid subunits. Some proteins serve as structural components of a cell. Others, such as enzymes, direct cellular activities including biosynthesis and energy conversion. Together, all these proteins control the cell's structure and activities, dictating the overall characteristics of that cell. ◀◀ **deoxyribonucleic acid (DNA), p. 32** ◀◀ **nucleotide, p. 32** ◀◀ **nucleobase, p. 32** ◀◀ **amino acid, p. 25** ◀◀ **enzymes, p. 135**

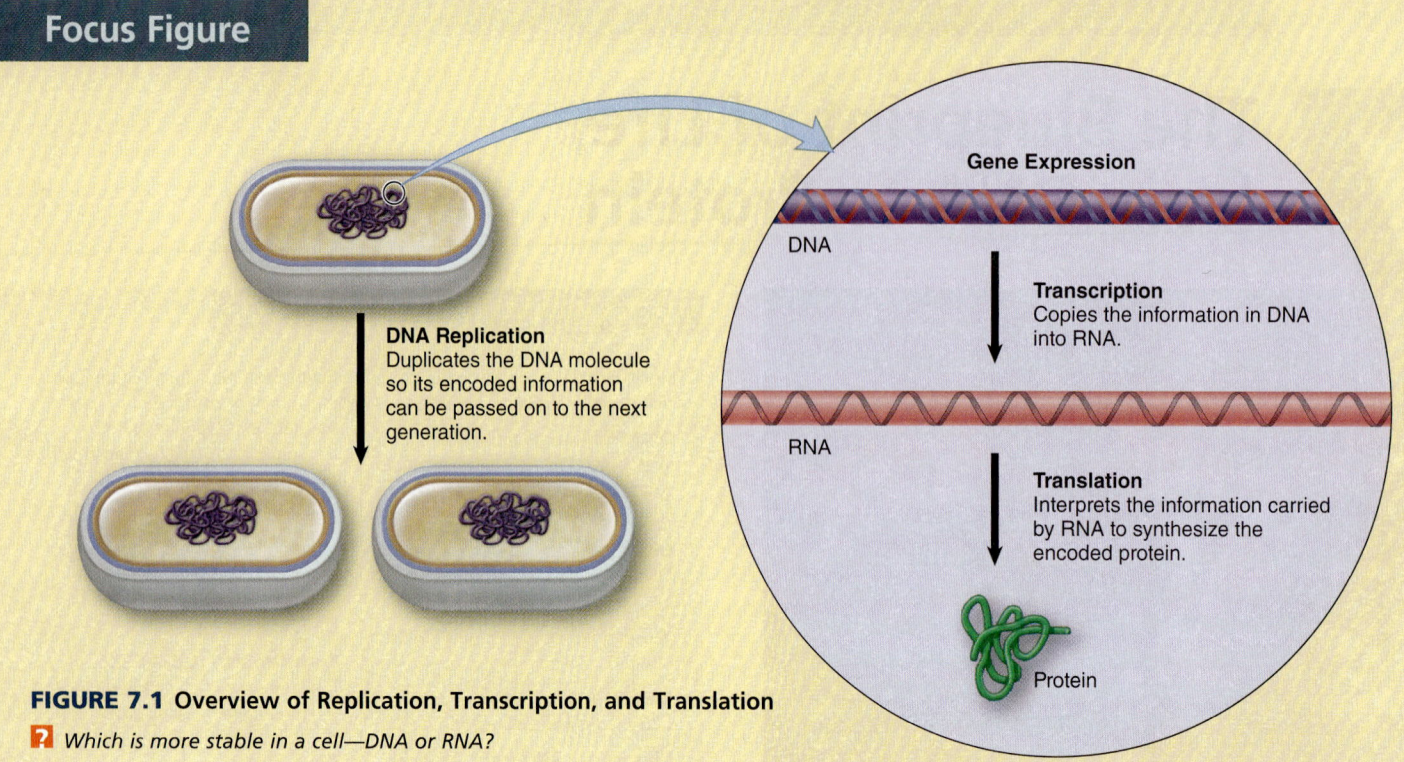

FIGURE 7.1 Overview of Replication, Transcription, and Translation

❓ *Which is more stable in a cell—DNA or RNA?*

While at first it might seem unlikely that the vast array of life-forms could be encoded by a molecule consisting of only four different units (the nucleotides), think about how much information can be transmitted by binary code, the language of all computers. Using only two units, a simple series of ones and zeros, binary can code for each letter of the alphabet. String enough of these together in the right sequence and the letters become words. With longer and longer strings, the words can become complete sentences, chapters, books, or even whole libraries.

This chapter will focus on the processes bacteria use to replicate their DNA and convert the encoded information into proteins. The mechanisms used by eukaryotic cells have many similarities, but are considerably more complicated, and will only be discussed briefly. The processes in archaea are sometimes similar to those of bacteria, but often resemble those of eukaryotic cells.

7.1 ■ Overview

Learning Outcomes

1. *Compare and contrast the characteristics of DNA and RNA.*
2. *Explain why a cell must be able to regulate gene expression.*

The complete set of genetic information of a cell is referred to as its **genome.** Technically, this includes plasmids as well as the chromosome; however, the term "genome" is often used interchangeably with chromosome. The genome of all cells is composed of DNA, but some viruses have an RNA genome. The functional unit of the genome is a **gene.** A gene encodes a product (called the gene product), most commonly a protein. The study and

analysis of the nucleotide sequence of DNA is called **genomics.**
◀◀ chromosome, p. 66 ◀◀ plasmid, p. 66

All cells must accomplish two general tasks in order to multiply. First, the double-stranded DNA must be duplicated before cell division so that its encoded information can be passed on to the next generation (**figure 7.1**). This is the process of **DNA replication.** Second, the information encoded by the DNA must be decoded so that the cell can synthesize the necessary gene products. This process, **gene expression,** involves two interrelated events—transcription and translation. **Transcription** copies the information encoded in DNA into a slightly different molecule, RNA. The RNA serves as a transitional, temporary form of the information and is the one actually deciphered. **Translation** interprets information carried by RNA to synthesize the encoded protein. The flow of information from DNA ⟶ RNA ⟶ protein is often referred to as the central dogma of molecular biology.

Characteristics of DNA

DNA is usually a double-stranded, helical structure (**figure 7.2**). Each strand is composed of a chain of deoxyribonucleotide subunits, more commonly called nucleotides. These subunits are joined together by a covalent bond between the 5′PO₄ (5 prime phosphate) of one nucleotide and the 3′OH (3 prime hydroxyl) of the next. The designations 5′ and 3′ refer to the numbered carbon atoms of the pentose sugar of the nucleotide (see figure 2.20). Joining of the nucleotides in this manner creates a series of alternating sugar and phosphate units, called the sugar-phosphate backbone. Because of the chemical structure of nucleotides and

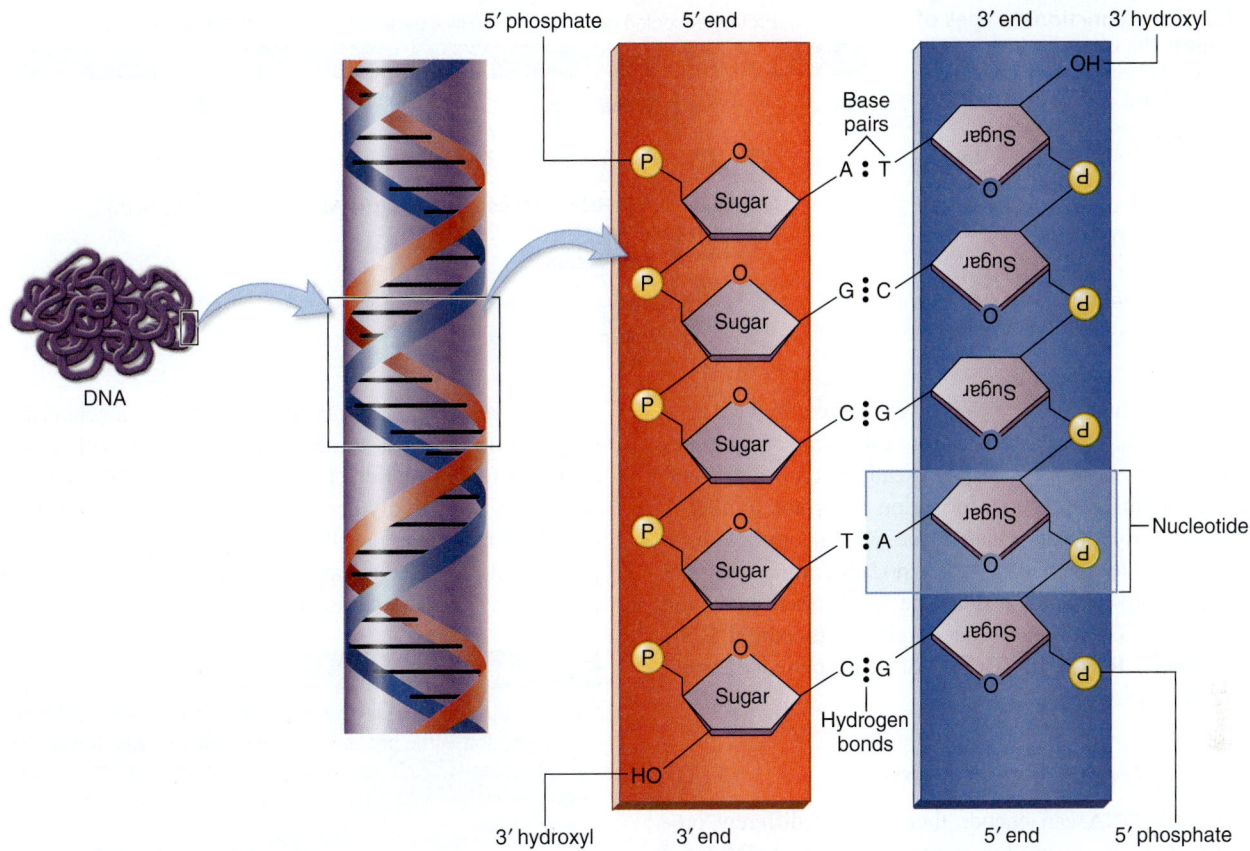

FIGURE 7.2 The Structure of DNA The two strands in the double helix are complementary. Three hydrogen bonds form between a G-C base pair and two between an A-T base pair. The strands are antiparallel; one is oriented in the 5′ to 3′ direction, and its complement is oriented in the 3′ to 5′ direction.

? *If a 100 base-pair double-stranded DNA fragment has 40 cytosines, how many adenines does it contain?*

how they are joined to each other, a single strand of DNA will always have a 5′PO₄ at one end and a 3′OH at the other. These ends are often referred to as the **5′ end** (5 prime end) and the **3′ end** (3 prime end).

The two strands of DNA are **complementary** and are held together by hydrogen bonds between the nucleobases. Wherever an adenine (A) is in one strand, a thymine (T) is in the other; these opposing A-T bases are held together by two hydrogen bonds. Similarly, wherever a guanine (G) is in one strand, a cytosine (C) is in the other. These G-C bases are held together by three hydrogen bonds, a slightly stronger attraction than that of an A-T pair. The characteristic bonding of A to T and G to C is called **base-pairing** and is a fundamental characteristic of DNA. Because of the rules of base-pairing, one strand can always be used as a template for the synthesis of the opposing strand.

Although the two strands of DNA in the double helix are complementary, they are also **antiparallel.** That is, they are oriented in opposite directions. One strand is oriented in the 5′ to 3′ direction and its complement is oriented in the 3′ to 5′ direction.

The duplex structure of double-stranded DNA is generally quite stable because of the numerous hydrogen bonds that occur along its length. Short fragments of DNA have correspondingly fewer hydrogen bonds, so they are easily separated into single

strands. Separating the two strands is called melting, or denaturing. ◄◄ hydrogen bonds, p. 21

Characteristics of RNA

RNA is similar to DNA in many ways, with a few important exceptions. One difference is that RNA is made up of ribonucleotides rather than deoxyribonucleotides, although in both cases these are usually referred to simply as nucleotides. Another distinction is that RNA contains the nucleobase uracil in place of the thymine found in DNA. Like DNA, RNA consists of a chain of nucleotides, but RNA is usually a single-stranded linear molecule much shorter than DNA. ◄◄ ribonucleic acid (RNA), p. 33

RNA is synthesized using a region of one of the two strands of DNA as a template. In making the RNA molecule, or **transcript,** the base-pairing rules apply except that uracil, rather than thymine, pairs with adenine. The interaction of DNA and RNA is only temporary, however, and the transcript quickly separates from the template.

Specific chromosomal regions are used as templates for RNA synthesis, generating numerous distinct transcripts. Either DNA strand may serve as the template, but only one of the two strands is transcribed in a given region.

FIGURE 7.3 Three Functional Types of RNA Molecules The different functional groups of RNA—messenger RNA (mRNA), ribosomal RNA (rRNA), and transfer RNA (tRNA)—are transcribed from different genes. The mRNA is translated, and the tRNA and rRNA fold into characteristic three-dimensional structures that each play a role in protein synthesis.

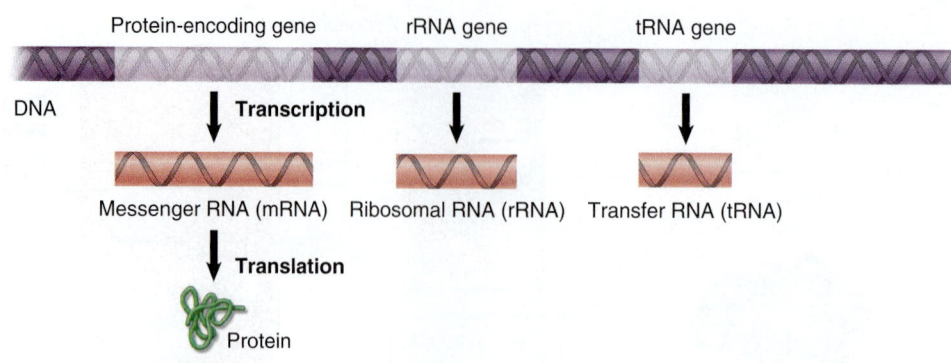

? *Ribosomal RNA is a component of ribosomes. What are ribosomes?*

Three different functional types of RNA are required for gene expression, and these are transcribed from different sets of genes (**figure 7.3**). Most genes encode proteins and are transcribed into **messenger RNA (mRNA)**. The information encrypted in mRNA is deciphered according to the genetic code, which correlates each set of three nucleotides to a particular amino acid. Some genes are never translated into proteins; instead the RNAs themselves are the final products. These genes encode either **ribosomal RNA (rRNA)** or **transfer RNA (tRNA)**, each of which plays a different but critical role in protein synthesis.

Regulating Gene Expression

Although a cell's DNA can encode thousands of different proteins, not all of them are needed at the same time or in equal quantities (**figure 7.4**). Because of this, cells require mechanisms to regulate the expression of certain genes.

A fundamental aspect of gene regulation is the instability of mRNA. Within minutes of being produced, transcripts are degraded by cellular enzymes. Although this might seem wasteful, it actually provides cells with an important regulatory mechanism. If transcription of a gene is turned "on," transcripts will continue to be available for translation. If it is then turned "off," the number of transcripts will rapidly decline. By simply regulating the synthesis of mRNA molecules, a cell can quickly change the levels of protein production. Some cells have mechanisms to adjust the stability of RNA, providing an additional level of control.

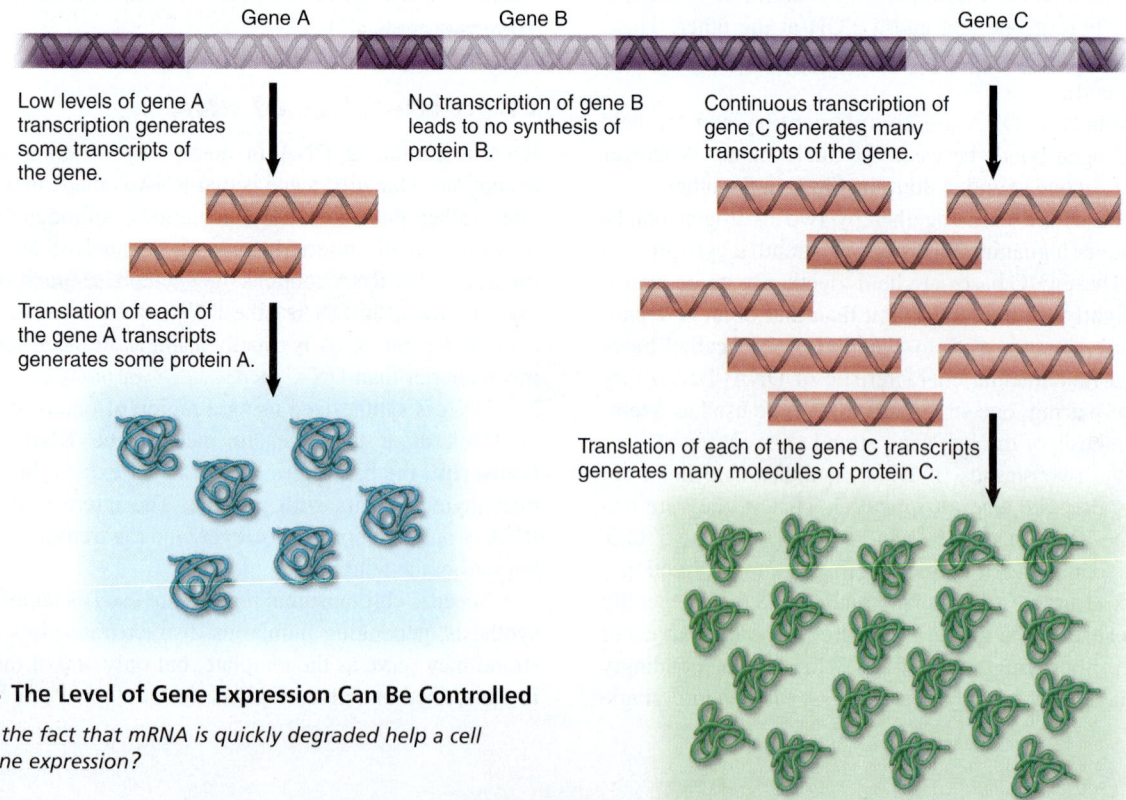

FIGURE 7.4 The Level of Gene Expression Can Be Controlled

? *How does the fact that mRNA is quickly degraded help a cell control gene expression?*

7.2 ■ DNA Replication

Learning Outcome

3. *Describe the DNA replication process, including its initiation and the events that occur at the replication fork.*

DNA is replicated in order to create a duplicate molecule, so that the two cells generated during binary fission can each receive one complete copy. The replication process is generally bidirectional, meaning it proceeds in both directions from the starting point (**figure 7.5**). This allows a chromosome to be replicated in half the time it would take if the process were unidirectional. The progression of bidirectional replication around a circular DNA molecule creates two advancing forks where DNA synthesis is occurring. These regions, called **replication forks,** ultimately meet at a terminating site when the process is complete. ◄◄ binary fission, p. 82

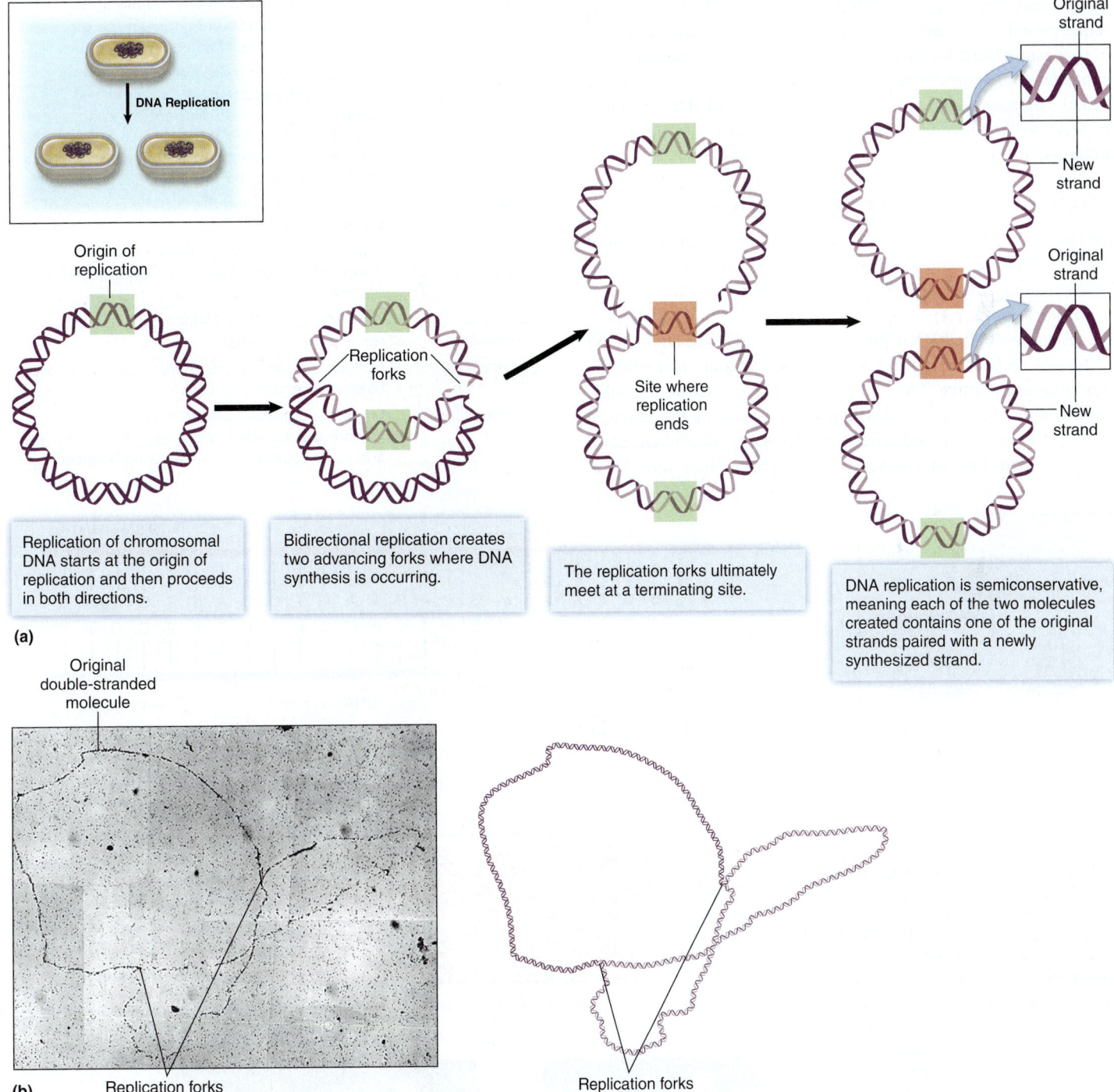

Origin of replication

Replication forks

Site where replication ends

Original strand

New strand

Original strand

New strand

Replication of chromosomal DNA starts at the origin of replication and then proceeds in both directions.

Bidirectional replication creates two advancing forks where DNA synthesis is occurring.

The replication forks ultimately meet at a terminating site.

DNA replication is semiconservative, meaning each of the two molecules created contains one of the original strands paired with a newly synthesized strand.

(a)

Original double-stranded molecule

Replication forks

(b)

Replication forks

FIGURE 7.5 Replication of a Bacterial Chromosome (a) Process of bidirectional replication. (b) Partially replicated chromosome; an electron micrograph and a diagrammatic depiction.

? *What is a replication fork?*

The two DNA molecules created through replication each contain one of the original strands paired with a newly synthesized strand. Because half of the original molecule is conserved in each molecule, replication is said to be **semiconservative.**

Initiation of DNA Replication

To initiate replication of a DNA molecule, specific proteins must recognize and bind to a distinct DNA sequence called an **origin of replication.** Prokaryotic chromosomes and plasmids typically contain only one of these initiating sites. A molecule that lacks this sequence will not be replicated. The proteins that bind to the origin of replication cause localized melting of the double-stranded DNA, exposing single-stranded regions that can act as templates. Enzymes called primases then synthesize short stretches of RNA complementary to the exposed templates. These small fragments, called **primers,** are critical in the next steps of replication.

The Process of DNA Replication

The process of DNA replication requires the coordinated action of many different enzymes and other proteins. The most critical of these exist together in DNA-synthesizing "assembly lines" called replisomes (**table 7.1**).

Enzymes called **DNA polymerases** synthesize DNA in the 5′ to 3′ direction, using one strand as a template to make the complement (**figure 7.6**). To do this, DNA polymerase adds nucleotides onto the 3′ end of a primer, powering the reaction with the energy released when a high-energy phosphate bond of the incoming nucleotide is hydrolyzed. Note that DNA

TABLE 7.1	Components of DNA Replication in Bacteria
Component	**Comment**
Replisome	The complex of enzymes and other proteins that synthesize DNA
DNA gyrase	Enzyme that temporarily breaks the strands of DNA, relieving the tension caused by unwinding the two strands of the DNA helix.
DNA ligase	Enzyme that joins two DNA fragments by forming a covalent bond between the sugar and phosphate residues of adjacent nucleotides.
DNA polymerases	Enzymes that synthesize DNA; they use one strand of DNA as a template to make the complementary strand. Nucleotides can be added only to the 3′ end of an existing fragment—therefore, synthesis always occurs in the 5′ to 3′ direction.
Helicases	Enzymes that unwind the DNA helix ahead of the replication fork.
Okazaki fragment	Nucleic acid fragment produced during discontinuous synthesis of the lagging strand of DNA.
Origin of replication	Distinct region of a DNA molecule at which replication is initiated.
Primase	Enzyme that synthesizes small fragments of RNA to serve as primers for DNA synthesis.
Primer	Fragment of nucleic acid to which DNA polymerase can add nucleotides (the enzyme can add nucleotides only to an existing fragment).

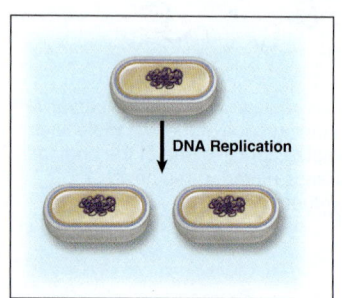

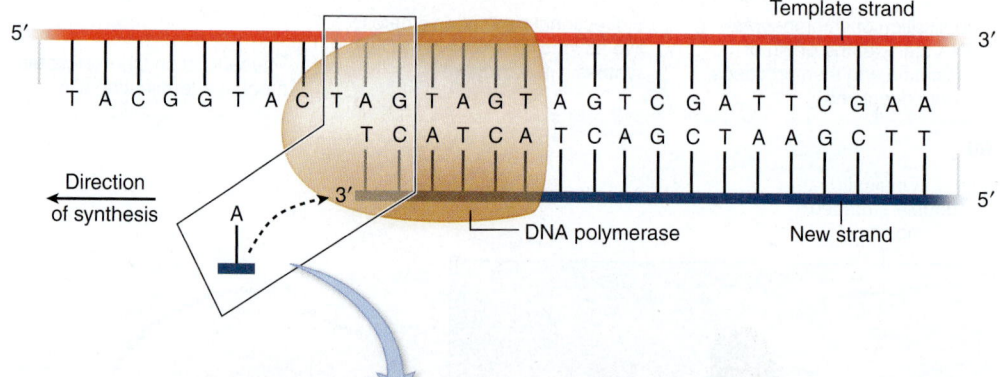

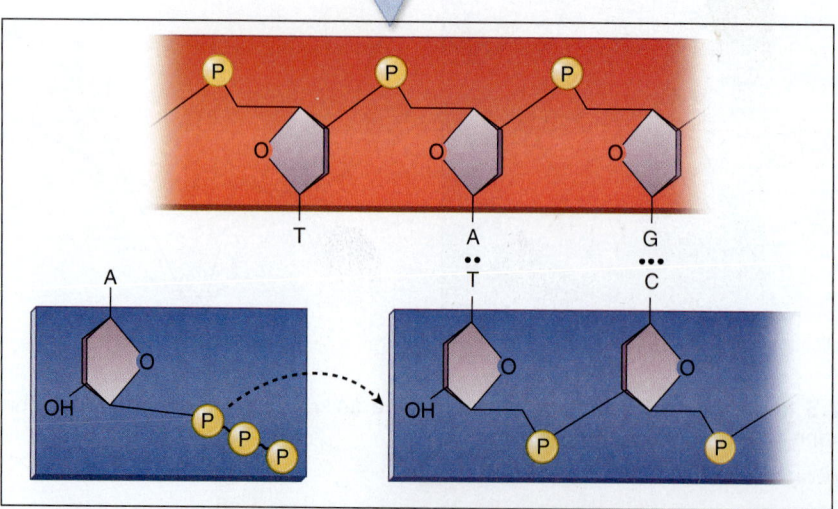

FIGURE 7.6 The Process of DNA Synthesis DNA polymerase synthesizes a new strand by adding one nucleotide at a time to the 3′ end of the elongating strand. The base-pairing rules determine the specific nucleotides that are added.

 Considering that DNA is synthesized in the 5′ to 3′ direction, which direction must DNA polymerase travel along the template strand: 5′ to 3′ or 3′ to 5′?

polymerases add nucleotides only onto an existing DNA strand, so they cannot initiate synthesis. This explains why primers are required at the origin of replication—they provide the DNA polymerase with a molecule to which it can add additional nucleotides.

◄◄ high-energy phosphate bond, p. 24

In order for replication to progress, enzymes called helicases must progressively "unzip" the DNA strands at each replication fork to reveal additional template sequences (**figure 7.7 ①**). **②** Synthesis of one new strand proceeds continuously as fresh template is exposed, because DNA polymerase simply adds nucleotides to the 3′ end. This strand is called the **leading strand.** **③** Synthesis of the other strand, the **lagging strand,** is more complicated. This is because DNA polymerases cannot add nucleotides to the 5′ end, so as additional template is exposed, synthesis must be reinitiated. Each time synthesis is reinitiated, another RNA primer must be made first. The result is a series of small fragments, each of which has a short stretch of RNA at its 5′ end. These fragments are called Okazaki fragments. **④** As DNA polymerase adds nucleotides to the 3′ end of one Okazaki fragment, it eventually reaches the 5′ end of another. A different type of DNA polymerase then removes the RNA primer nucleotides and simultaneously replaces them with deoxynucleotides. **⑤** The enzyme DNA ligase then seals the gaps between fragments by forming a covalent bond between the adjacent nucleotides.

When a circular bacterial chromosome is replicated, the two replication forks eventually meet at a site opposite the origin of replication. Two complete DNA molecules have been produced at this point, and these can be passed on to the two daughter cells.

It takes approximately 40 minutes for the *E. coli* chromosome to be replicated. How, then, can the organism have a generation time of only 20 minutes? This can happen because, under favorable growing conditions, a cell initiates replication before the preceding round of replication is complete. In this way, each of the two daughter cells will get one complete chromosome that has already started another round of replication. ◄◄ generation time, p. 83

MicroByte

Antibacterial medications called fluoroquinolones target the enzyme DNA gyrase, a component of the bacterial replisome.

MicroAssessment 7.2

DNA replication begins at the origin of replication and then proceeds bidirectionally, creating two replication forks. DNA polymerases synthesize DNA in the 5′ to 3′ direction, using one strand as a template to generate the complementary strand.

4. *Why is a primer required for DNA synthesis?*

5. *How does synthesis of the lagging strand differ from that of the leading strand?*

6. *Eukaryotic chromosomes often have multiple origins of replication. Why would this be the case?* ➕

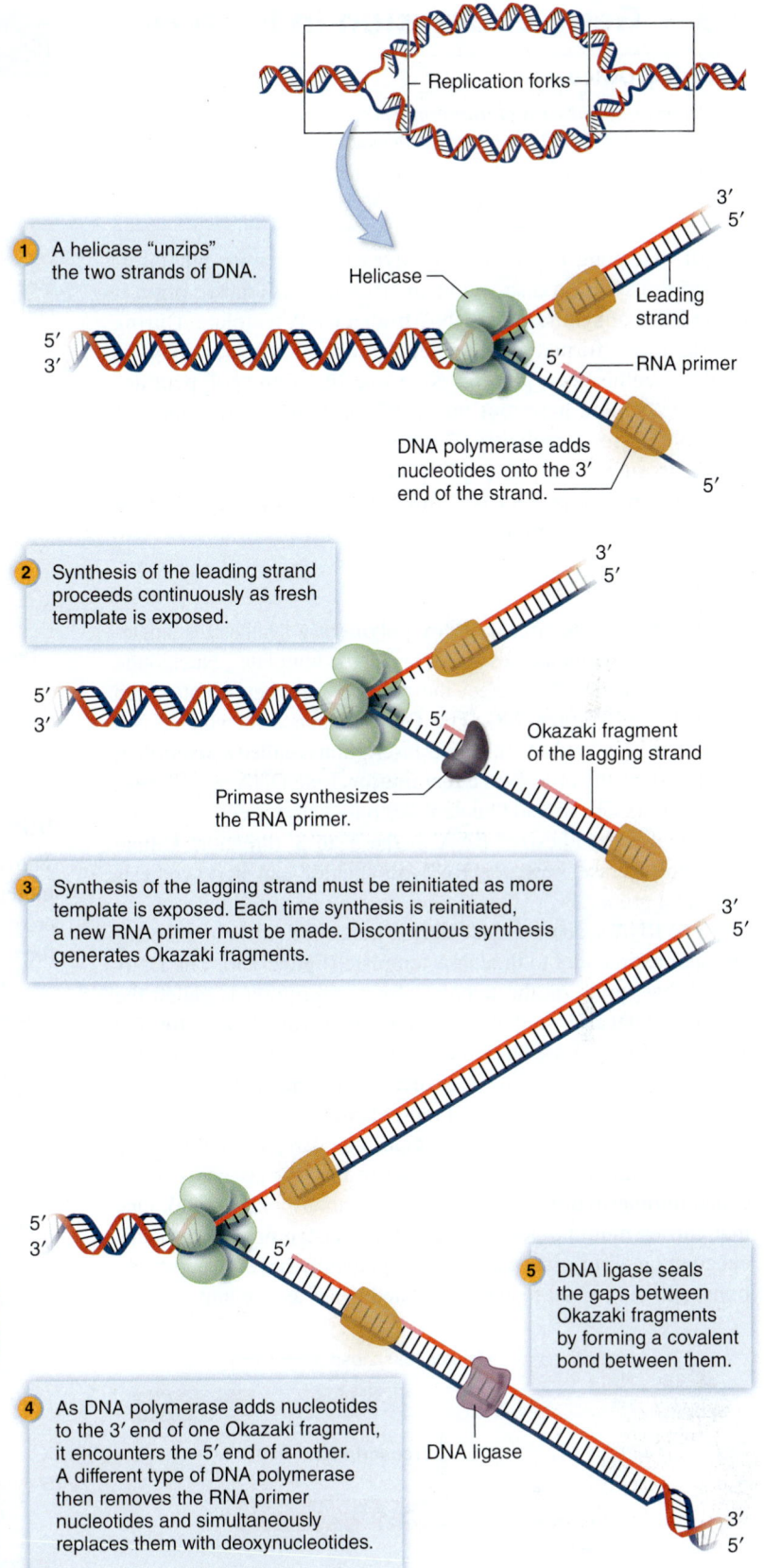

① A helicase "unzips" the two strands of DNA.

② Synthesis of the leading strand proceeds continuously as fresh template is exposed.

③ Synthesis of the lagging strand must be reinitiated as more template is exposed. Each time synthesis is reinitiated, a new RNA primer must be made. Discontinuous synthesis generates Okazaki fragments.

④ As DNA polymerase adds nucleotides to the 3′ end of one Okazaki fragment, it encounters the 5′ end of another. A different type of DNA polymerase then removes the RNA primer nucleotides and simultaneously replaces them with deoxynucleotides.

⑤ DNA ligase seals the gaps between Okazaki fragments by forming a covalent bond between them.

Replication forks

Helicase — Leading strand

RNA primer

DNA polymerase adds nucleotides onto the 3′ end of the strand.

Primase synthesizes the RNA primer.

Okazaki fragment of the lagging strand

DNA ligase

FIGURE 7.7 The Replication Fork This depiction is simplified to highlight the key differences between synthesis of the leading and lagging strands.

❓ *Synthesis of which strand requires the repeated action of DNA ligase?*

7.3 ■ Gene Expression in Bacteria

Learning Outcomes

4. *Describe the process of transcription, focusing on the role of RNA polymerase, sigma (σ) factors, promoters, and terminators.*
5. *Describe the process of translation, focusing on the role of mRNA, ribosomes, ribosome-binding sites, rRNAs, tRNAs, and codons.*

Recall that gene expression involves two separate but inter-related processes, transcription and translation. Transcription is the process of synthesizing RNA from a DNA template. During translation, information encoded by an mRNA transcript is used to synthesize a protein. Note that the term "polypeptide" is more accurate here, but the word "protein" is often used in this context for simplicity. The distinction between these two words is subtle—a polypeptide is simply a chain of amino acids, whereas a protein is a functional molecule made up of one or more polypeptides.

Transcription

In transcription, the enzyme **RNA polymerase** synthesizes single-stranded RNA molecules from a DNA template. Nucleotide sequences in the DNA direct the polymerase where to start and where to end (**figure 7.8**). The DNA sequence to which RNA polymerase can bind and initiate transcription is called a **promoter;** one that stops the process is a **terminator.** Like DNA polymerase, RNA polymerase can add nucleotides only to the 3′ end of a chain and therefore synthesizes RNA in the 5′ to 3′ direction. Unlike DNA polymerase, however, RNA polymerase can start synthesis without a primer.

The RNA sequence made during transcription is complementary and antiparallel to the DNA template (**figure 7.9**). The DNA strand that serves as the template for transcription is called the **minus (–) strand,** and its complement is called the **plus (+) strand** (**table 7.2**). Because the RNA is complementary to the (–) DNA strand, its nucleotide sequence is the same as the (+) DNA strand, except it contains uracil rather than thymine.

In prokaryotes, mRNA molecules can carry the information for one or multiple genes. A transcript that carries one gene is called **monocistronic** (a cistron is synonymous with a gene). One that carries multiple genes is called **polycistronic.** The proteins encoded on a polycistronic message generally have related functions, allowing a cell to express related genes as one unit.

TABLE 7.2	Components of Transcription in Bacteria
Component	**Comment**
(–) strand	Strand of DNA that serves as the template for RNA synthesis; the resulting RNA molecule is complementary to this strand.
(+) strand	Strand of DNA complementary to the one that serves as the template for RNA synthesis; the nucleotide sequence of the RNA molecule is the same as this strand, except it has uracil rather than thymine.
Promoter	Nucleotide sequence to which RNA polymerase binds to initiate transcription.
RNA polymerase	Enzyme that synthesizes RNA using single-stranded DNA as a template; synthesis always occurs in the 5′ to 3′ direction.
Sigma (σ) factor	Component of RNA polymerase that recognizes the promoter regions. A cell can have different types of σ factors that recognize different promoters, allowing the cell to transcribe specialized sets of genes as needed.
Terminator	Nucleotide sequence at which RNA synthesis stops; the RNA polymerase falls off the DNA template and releases the newly synthesized RNA.

Initiation of RNA Synthesis

Transcription is initiated when RNA polymerase binds to a promoter (**figure 7.10**). The binding melts a short stretch of DNA, creating a region of exposed nucleotides that serves as a template for RNA synthesis.

The portion of RNA polymerase that recognizes the promoter is a loosely attached subunit called **sigma (σ) factor.** A cell can produce various types of σ factors, each recognizing different promoters. By controlling which σ factors are made, cells can transcribe specialized sets of genes as needed. The RNA polymerases

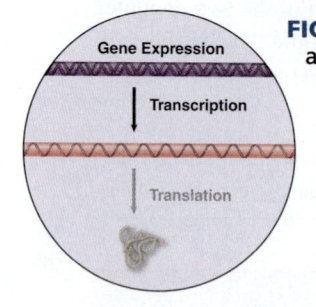

FIGURE 7.9 RNA Is Complementary and Antiparallel to the DNA Template The DNA strand that serves as a template for RNA synthesis is called the (–) strand of DNA. The complement to that is the (+) strand.

❓ *How does the nucleotide sequence of the (+) strand differ from that of the RNA transcript?*

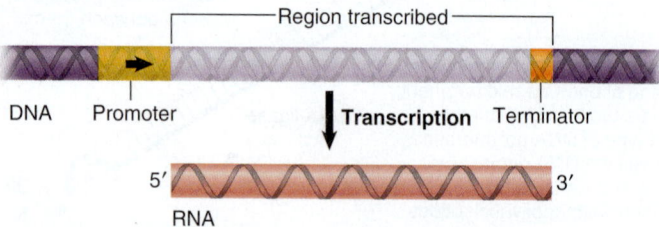

DNA   **Promoter**   **Transcription**   **Terminator**

5′ ━━━━━━━━━━ 3′
RNA

FIGURE 7.8 Nucleotide Sequences in DNA Direct Transcription The promoter is a DNA sequence to which RNA polymerase can bind in order to initiate transcription. The terminator is a sequence at which transcription stops.

❓ *In which direction is RNA synthesized: 5′ to 3′ or 3′ to 5′?*

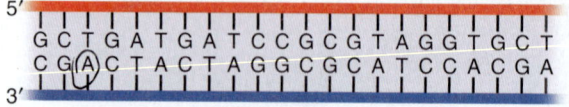

5′ G C T G A T G A T C C G C G T A G G T G C T 3′ Plus (+) strand of DNA
3′ C G A C T A C T A G G C G C A T C C A C G A 5′ Minus (–) strand of DNA

5′ G C U G A U G A U C C G C G U A G G U G C U 3′ RNA

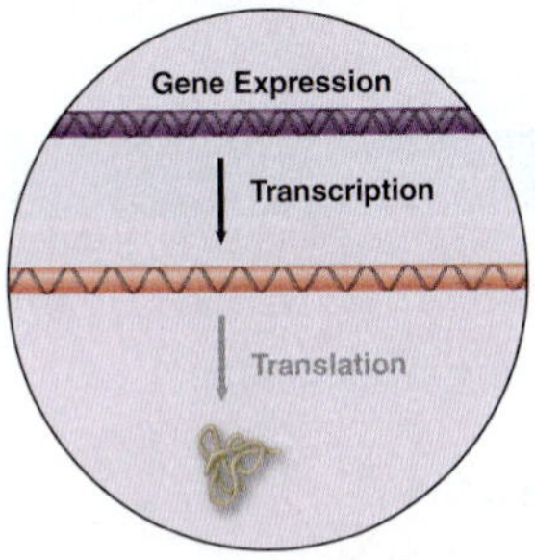

FIGURE 7.10 The Process of RNA Synthesis Bacterial RNA polymerases include a sigma subunit (as illustrated); the RNA polymerases of eukaryotic cells and archaea use transcription factors to recognize promoters.

❓ *Which component of the bacterial RNA polymerase recognizes the promoter?*

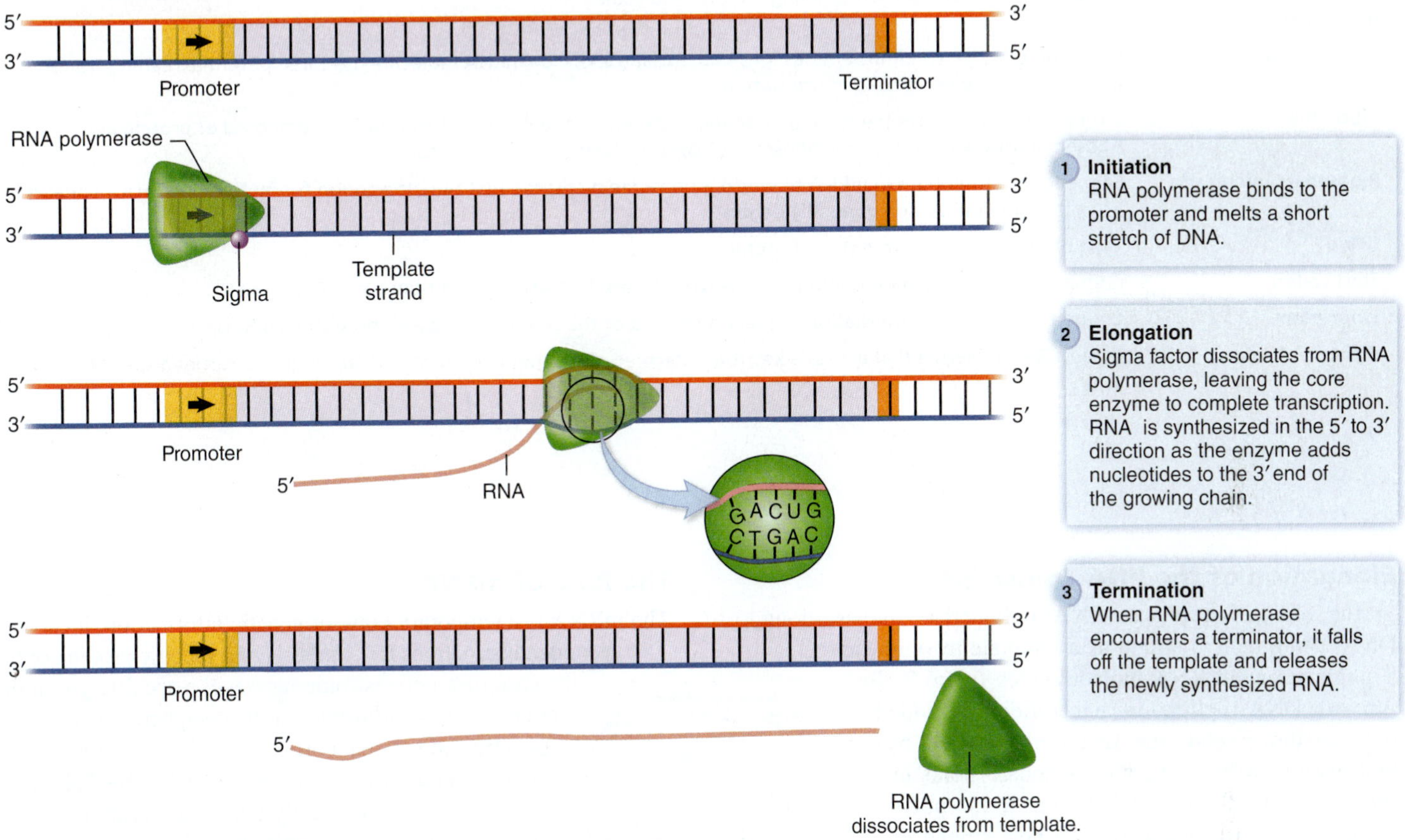

Gene Expression

Transcription

Translation

5′
3′
Promoter   Terminator
3′
5′

RNA polymerase

5′   3′
3′   5′
Sigma   Template strand

1 Initiation
RNA polymerase binds to the promoter and melts a short stretch of DNA.

5′   3′
3′   5′
Promoter
5′   RNA
GACUG
CTGAC

2 Elongation
Sigma factor dissociates from RNA polymerase, leaving the core enzyme to complete transcription. RNA is synthesized in the 5′ to 3′ direction as the enzyme adds nucleotides to the 3′ end of the growing chain.

5′   3′
3′   5′
Promoter
5′

RNA polymerase dissociates from template.

3 Termination
When RNA polymerase encounters a terminator, it falls off the template and releases the newly synthesized RNA.

of eukaryotic cells and archaea use proteins called transcription factors to recognize promoters.

Promoters identify the regions of a DNA molecule that will be transcribed into RNA. In doing so, they also orient the direction of the RNA polymerase on the DNA molecule, thereby dictating which strand will be used as a template (**figure 7.11**). The direction of polymerase movement can be likened to the flow of a river. Because of this, the words "upstream" and "downstream" are used to describe relative positions of other sequences. As an example, promoters are upstream of the genes they control.

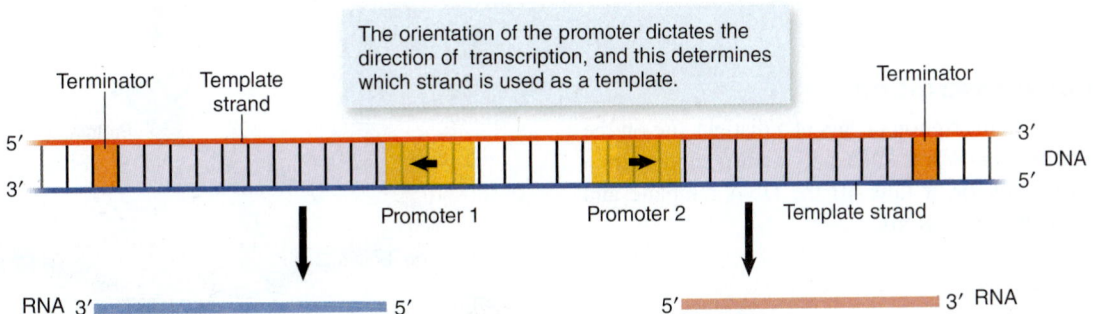

The orientation of the promoter dictates the direction of transcription, and this determines which strand is used as a template.

Terminator Template strand   Terminator
5′   3′ DNA
3′   5′
Promoter 1   Promoter 2   Template strand
RNA 3′ ━━━━━ 5′   5′ ━━━━━ 3′ RNA

FIGURE 7.11 The Promoter Orients RNA Polymerase The orientation of RNA polymerase determines which strand will be used as the template. In this diagram, the color of the RNA molecules indicates which DNA strand was used as the template. The light blue RNA was transcribed from the red DNA strand (and is therefore analogous in sequence to the blue DNA strand), whereas the pink RNA was transcribed from the blue DNA strand (and is therefore analogous in sequence to the red DNA strand).

❓ *The light blue RNA strand is complementary to which DNA strand in this figure? To which DNA strand is the pink RNA strand complementary?*

TABLE 7.3	Components of Translation in Bacteria
Component	**Comment**
Anticodon	Sequence of three nucleotides in a tRNA molecule that is complementary to a particular codon in mRNA. The anticodon allows the tRNA to recognize and bind to the appropriate codon.
mRNA	Type of RNA molecule that contains the genetic information deciphered during translation.
Polyribosome (polysome)	Multiple ribosomes attached to a single mRNA molecule.
Reading frame	Grouping of a stretch of nucleotides into sequential triplets; an mRNA molecule has three potential reading frames, but only one is typically used in translation.
Ribosome	Structure that facilitates the joining of amino acids during the process of translation; composed of protein and ribosomal RNA. The prokaryotic ribosome (70S) consists of a 30S and 50S subunit.
Ribosome-binding site	Sequence of nucleotides in mRNA to which a ribosome binds; the first time the codon for methionine (AUG) appears after that site, translation generally begins.
rRNA	Type of RNA molecule present in ribosomes.
Start codon	Codon at which translation is initiated; it is typically the first AUG after a ribosome-binding site.
Stop codon	Codon that terminates translation, signaling the end of the protein; there are three stop codons.
tRNA	Type of RNA molecule that acts as a key that interprets the genetic code; each tRNA molecule carries a specific amino acid.

Elongation of the RNA Transcript

In the elongation phase, RNA polymerase moves along DNA, using the (–) strand as a template to synthesize a single-stranded RNA molecule (see figure 7.10). As with DNA replication, nucleotides are added only to the 3′ end; the reaction is fueled by hydrolyzing a high-energy phosphate bond of the incoming nucleotide. When RNA polymerase advances, it denatures a new stretch of DNA and allows the previous portion to close. This exposes a new region of the template so elongation can continue.

Once elongation has proceeded far enough for RNA polymerase to clear the promoter, another molecule of the enzyme can bind, initiating a new round of transcription. Thus, a single gene can be transcribed repeatedly very quickly.

Termination of Transcription

Just as an initiation of transcription occurs at a distinct site on the DNA, so does termination. When RNA polymerase encounters a sequence called a terminator, it falls off the DNA template and releases the newly synthesized RNA.

Translation

Translation is the process of decoding the information carried on the mRNA to synthesize the specified protein. The process requires three major structures—mRNA, ribosomes, and tRNAs—in addition to various other components (**table 7.3**).

The Role of mRNA

The mRNA is a temporary copy of genetic information; it carries encoded instructions for synthesis of a specific protein, or in the case of a polycistronic message, a specific group of proteins. That information is deciphered using the **genetic code,** which correlates a series of three nucleotides, a **codon,** with one amino acid (**table 7.4**). The genetic code is practically universal, meaning that it is used in nearly its entirety by all living things.

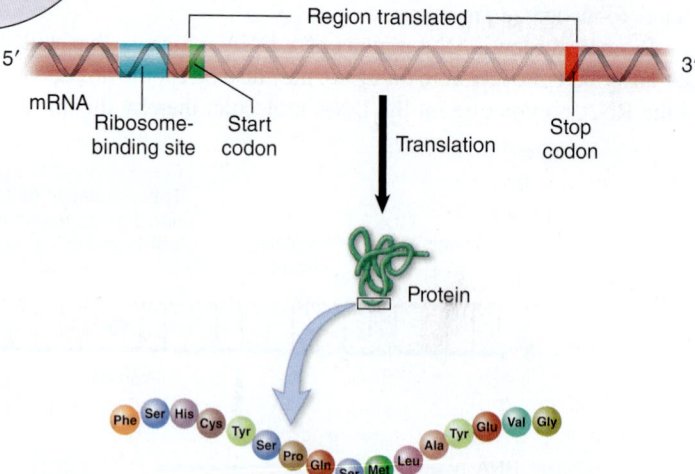

FIGURE 7.12 Nucleotide Sequences in mRNA Direct Translation The ribosome begins to assemble at the ribosome-binding site and starts translating at the start codon. Translation ends at the stop codon.

❓ *In which direction does the ribosome move along RNA?*

TABLE 7.4	The Genetic Code

First Letter	Second Letter				Third Letter
	U	**C**	**A**	**G**	
U	UUU — **Phe** Phenylalanine	UCU	UAU — **Tyr** Tyrosine	UGU — **Cys** Cysteine	U
	UUC	UCC — **Ser** Serine	UAC	UGC	C
	UUA — **Leu** Leucine	UCA	UAA "Stop"	UGA "Stop"	A
	UUG	UCG	UAG "Stop"	UGG — **Trp** Trytophan	G
C	CUU	CCU	CAU — **His** Histidine	CGU	U
	CUC — **Leu** Leucine	CCC — **Pro** Proline	CAC	CGC — **Arg** Arginine	C
	CUA	CCA	CAA — **Gln** Glutamine	CGA	A
	CUG	CCG	CAG	CGG	G
A	AUU	ACU	AAU — **Asn** Asparagine	AGU — **Ser** Serine	U
	AUC — **Ile** Isoleucine	ACC — **Thr** Threonine	AAC	AGC	C
	AUA	ACA	AAA — **Lys** Lysine	AGA — **Arg** Arginine	A
	AUG — **Met** Methionine; "Start"	ACG	AAG	AGG	G
G	GUU	GCU	GAU — **Asp** Aspartate	GGU	U
	GUC — **Val** Valine	GCC — **Ala** Alanine	GAC	GGC — **Gly** Glycine	C
	GUA	GCA	GAA — **Glu** Glutamate	GGA	A
	GUG	GCG	GAG	GGG	G

A codon consists of **three** nucleotides read in the sequence shown. For example, ACU codes for threonine. The first letter, A, is in the First Letter column; the second letter, C, is in the Second Letter column; and the third letter, U, is in the Third Letter column. Many amino acids are specified by more than one codon. For example, threonine is specified by four codons, which differ only in the third nucleotide (ACU, ACC, ACA, and ACG).

Because a codon is a triplet of any combination of the four nucleotides, there are 64 different codons (4^3). Three are stop codons, which will be discussed later. The remaining 61 translate to the 20 different amino acids. This means that more than one codon can code for a specific amino acid. For example, both ACA and ACG encode the amino acid threonine. Because of this redundancy, the genetic code is said to be degenerate.

The nucleotide sequence of mRNA indicates where the coding region begins and ends (**figure 7.12**). The site at which it begins is particularly critical because the translation "machinery" reads the mRNA in groups of three nucleotides. As a consequence, any given sequence has three possible **reading frames,** or ways in which triplets can be grouped (**figure 7.13**). If translation begins in the wrong reading frame, a very different, and generally non-functional, protein would be synthesized.

The Role of Ribosomes

Ribosomes serve as translation "machines," structures that string amino acids together. A ribosome does this by aligning two amino acids so that a ribosomal enzyme can easily create a peptide bond between them. ◀◀ peptide bond, p. 27

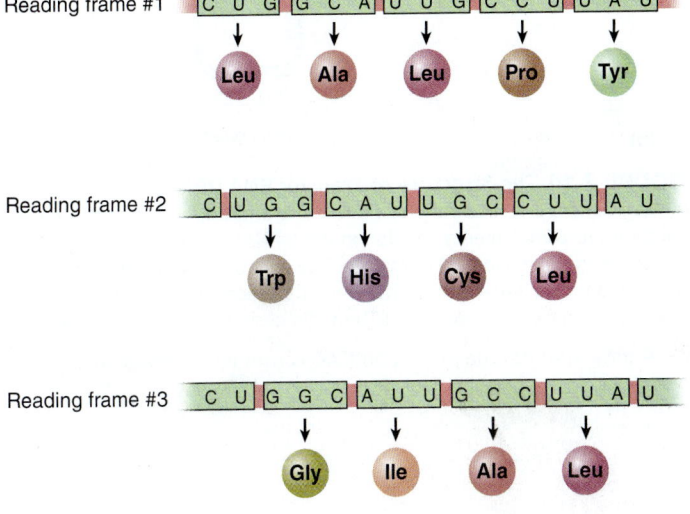

FIGURE 7.13 Reading Frames A nucleotide sequence has three potential reading frames, but only one is typically used for translation.

❓ *Why is it important that the correct reading frame is used?*

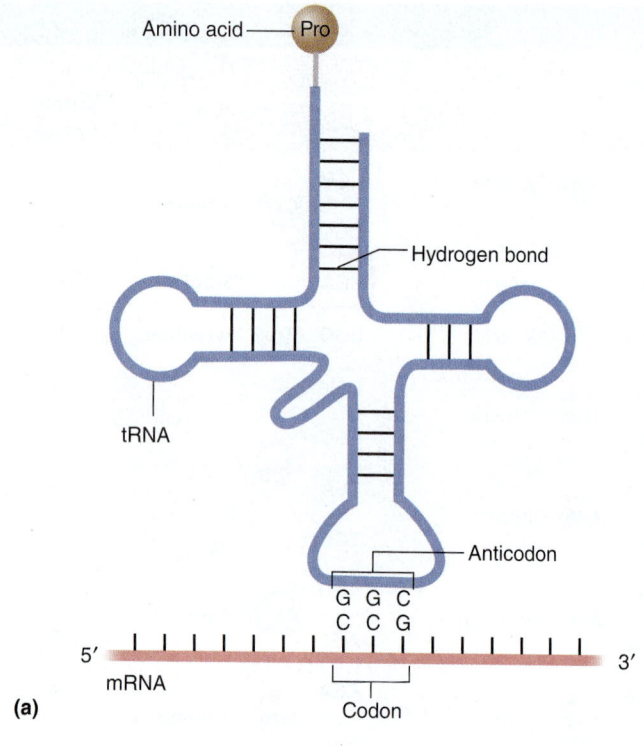

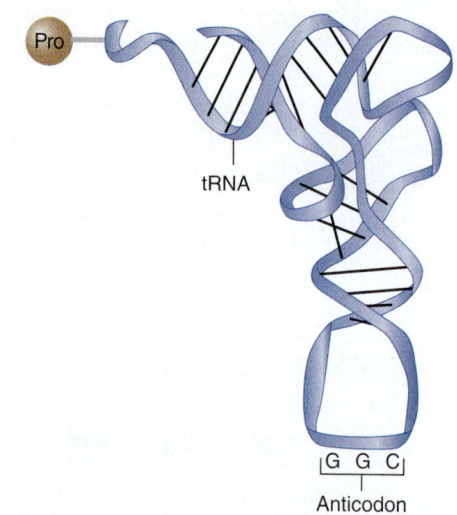

FIGURE 7.14 The Structure of Transfer RNA (tRNA)
(a) Two-dimensional illustration of tRNA. The anticodon of the tRNA base-pairs with its complementary codon in the mRNA; by doing so, it delivers the appropriate amino acid to the site. The amino acid that the tRNA carries is dictated by the genetic code. The tRNA that recognizes the codon CCG carries the amino acid proline. **(b)** Three-dimensional illustration of tRNA.

❓ *A tRNA that has the anticodon GAG carries which amino acid?*

Ribosomes also locate key punctuation sequences on the mRNA molecule, such as the point at which protein synthesis should begin. The ribosome then moves along the mRNA in the 5′ to 3′ direction, "presenting" each codon in a sequential order for deciphering, while maintaining the correct reading frame. ◄◄ ribosomes, p. 66

Prokaryotic ribosomes are composed of a 30S subunit and a 50S subunit, each made up of protein and ribosomal RNA (rRNA) (see figures 3.42 and 10.8); the "S" stands for Svedberg unit, which is a measure of size. Note that Svedberg units are not additive, which is why the 70S ribosome can have 30S and 50S subunits. ►► ribosomal subunits, p. 246

MicroByte

Several types of antibiotics, including tetracycline and azithromycin, interfere with the function of the bacterial 70S ribosome.

The Role of Transfer RNAs

The tRNAs are segments of RNA that act as keys to the genetic code (**figure 7.14**). Each tRNA recognizes and base-pairs with certain codons and, in the process, delivers the appropriate amino acid to that site. This recognition is made possible because each tRNA has an **anticodon**—three nucleotides complementary to a codon in the mRNA. The anticodon of a tRNA molecule dictates which amino acid the molecule carries.

Once a tRNA molecule has donated its amino acid during translation, it can be recycled. An enzyme in the cytoplasm recognizes the tRNA and then attaches the appropriate amino acid.

Initiation of Translation

In prokaryotes, translation begins as the mRNA molecule is still being synthesized (**figure 7.15**). Part of the ribosome binds to a sequence in mRNA called the **ribosome-binding site;** the first AUG after that site usually serves as the **start codon.** The complete ribosome assembles there, joined by an initiating tRNA that carries a chemically altered form of the amino acid methionine (*N*-formylmethionine, or f-Met). The position of the first AUG is critical, as it determines the reading frame used for translation of the remainder of that protein. Note that AUG functions as a start codon only when preceded by a ribosome-binding site; at other sites, it simply encodes methionine.

Elongation of the Polypeptide Chain

The ribosome has two sites to which amino acid–carrying tRNAs can bind—the P-site and the A-site. At the start of translation, the initiating tRNA carrying the f-Met occupies the P-site (**figure 7.16**). A tRNA that recognizes the next codon on the mRNA then fills the unoccupied A-site. Once both sites are filled, an enzyme creates a peptide bond between the two amino acids carried by the tRNAs. This transfers the amino acid from the initiating tRNA to the amino acid carried by the incoming tRNA.

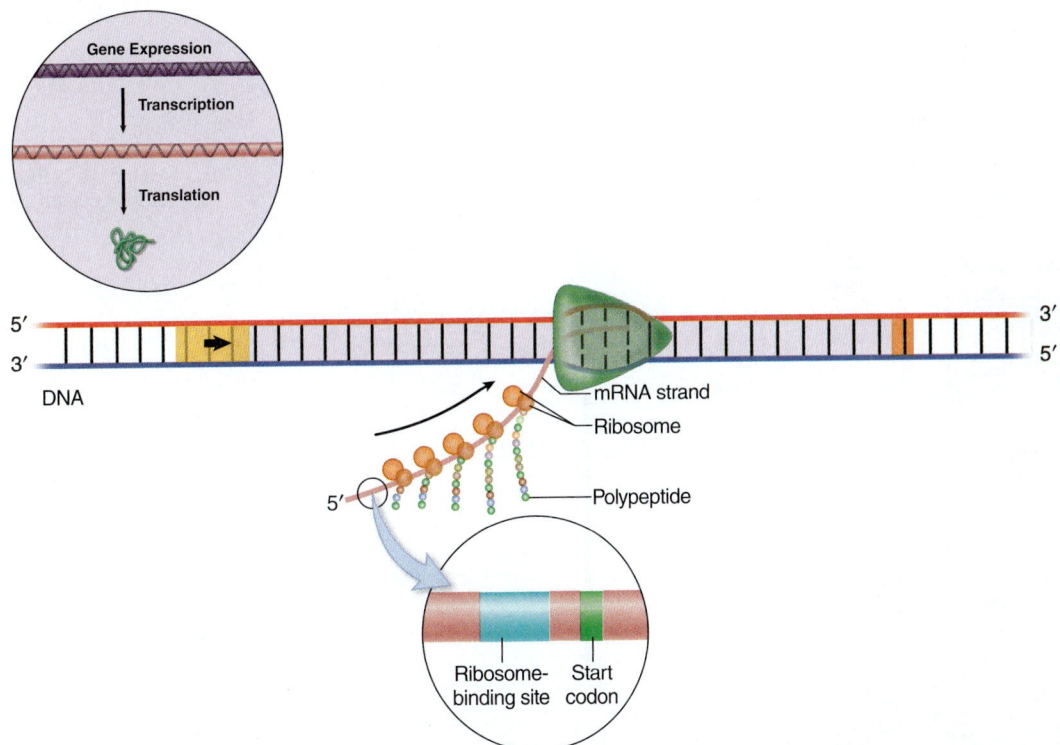

FIGURE 7.15 In Prokaryotes, Translation Begins as the mRNA Molecule Is Still Being Synthesized Ribosomes begin translating the 5' end of mRNA before transcription is complete. More than one ribosome can be translating the same mRNA molecule.

Why is the position of the first AUG after the ribosome-binding site critical?

After the initiating tRNA has donated its amino acid to the tRNA in the A-site, the ribosome advances a distance of one codon, moving along the mRNA in a 5' to 3' direction. As this happens, the initiating tRNA is released through a region called the E-site. The remaining tRNA, which now carries both amino acids, occupies the P-site. The A-site is transiently empty. A tRNA that recognizes the codon in the A-site quickly attaches there, and the process repeats.

Once translation of a gene has progressed far enough for the ribosome to clear the initiating sequences, another ribosome can bind. Thus, at any one time, multiple ribosomes can be translating a single mRNA molecule (see figure 7.15). This allows maximal protein synthesis from a single mRNA template. The assembly of multiple ribosomes attached to a single mRNA molecule is called a polyribosome, or a polysome.

Termination of Translation

Elongation of the polypeptide terminates when the ribosome reaches a **stop codon,** a codon not recognized by a tRNA. At this point, enzymes free the polypeptide by breaking the covalent bond that joins it to the tRNA. The ribosome falls off the mRNA, dissociating into its two component subunits (30S and 50S). The subunits can then be reused to initiate translation at other sites.

Post-Translational Modification

Polypeptides must often be modified after they are synthesized in order to become functional. For example, some must be folded into a specific three-dimensional structure, a process that requires the assistance of proteins called **chaperones.** Polypeptides destined for transport through the cytoplasmic membrane also must be modified. These have a **signal sequence,** a characteristic series of hydrophobic amino acids at their amino terminal end, which "tags" them for transport. The signal sequence must be removed by proteins in the membrane.

MicroAssessment 7.3

Gene expression involves transcription and translation. RNA polymerase synthesizes RNA in the 5' to 3' direction, using one strand of DNA as a template. Ribosomes move along the resulting mRNA in the 5' to 3' direction, synthesizing protein. tRNAs carry specific amino acids, thereby acting as keys to the genetic code.

7. *How does a promoter dictate which DNA strand is used as the template?*

8. *What is the role of tRNA?*

9. *Could two mRNAs have different nucleotide sequences and yet code for the same protein? Explain your answer.*

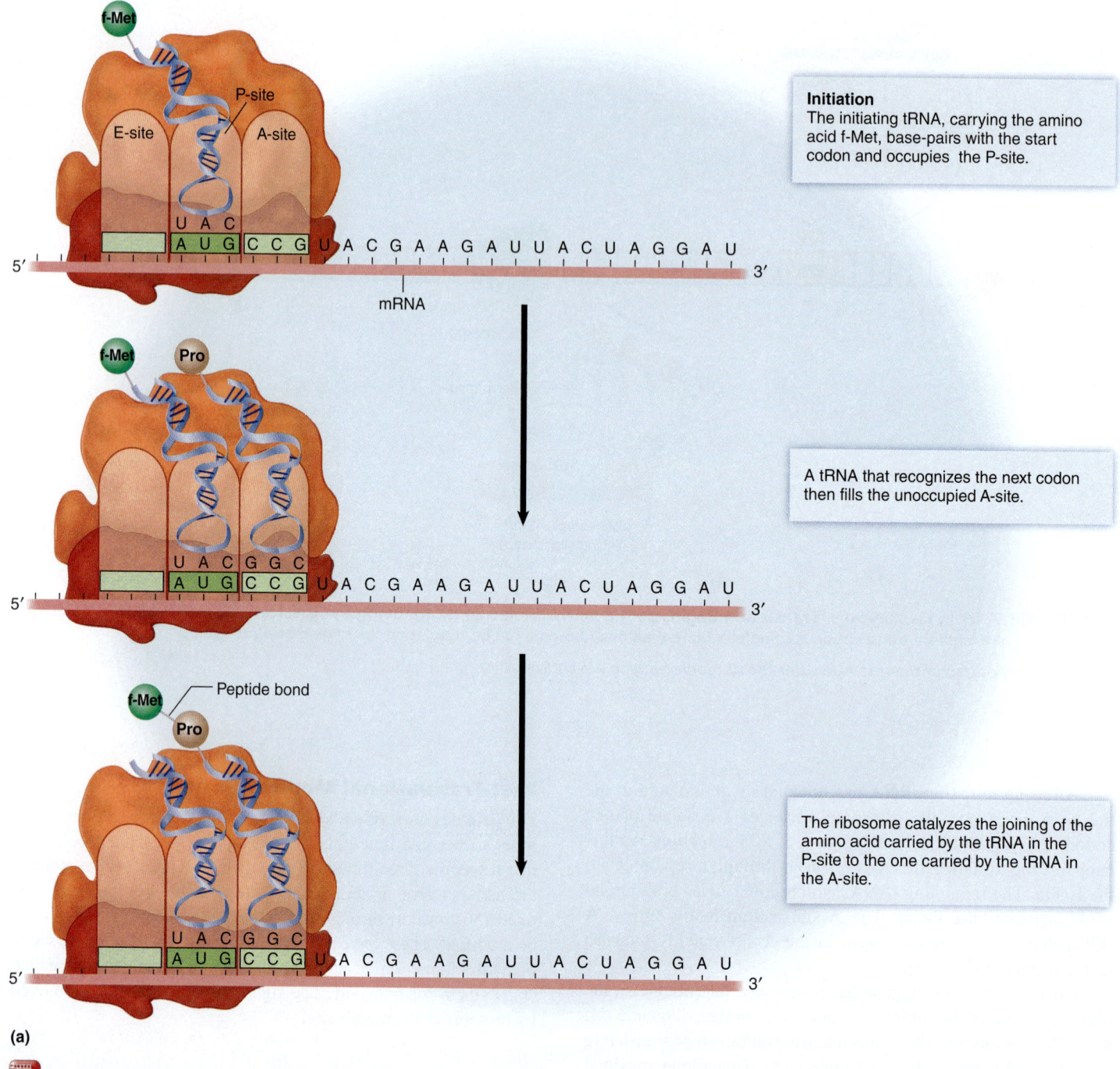

Initiation
The initiating tRNA, carrying the amino acid f-Met, base-pairs with the start codon and occupies the P-site.

A tRNA that recognizes the next codon then fills the unoccupied A-site.

The ribosome catalyzes the joining of the amino acid carried by the tRNA in the P-site to the one carried by the tRNA in the A-site.

(a)

FIGURE 7.16 The Process of Translation (a) Initiation. (b) Elongation. (c) Termination.

What would happen if a tRNA that recognized UAA (the stop codon) were present?

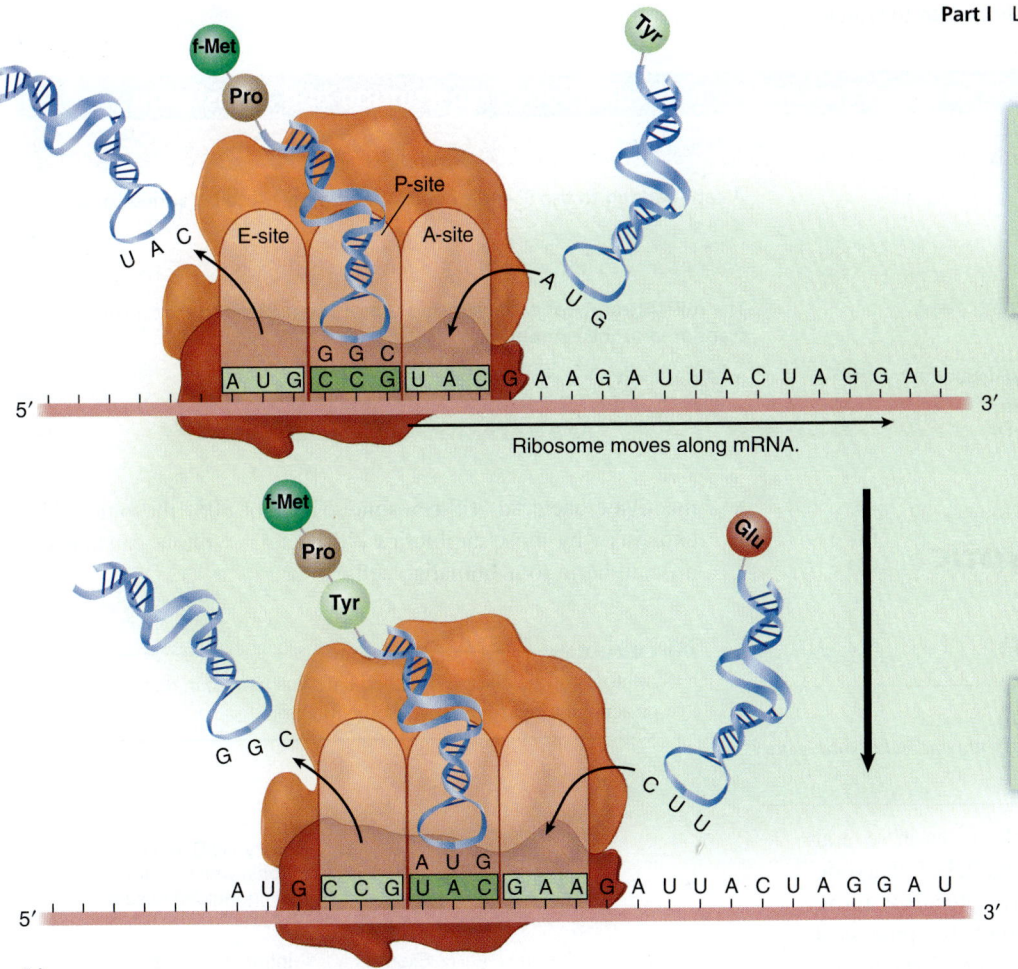

Elongation
The ribosome advances a distance of one codon. The tRNA that occupied the P-site exits through the E-site and the tRNA that was in the A-site occupies the P-site. A tRNA that recognizes the next codon quickly fills the empty A-site.

Ribosome moves along mRNA.

The ribosome continues advancing down the mRNA in the 5′ to 3′ direction, moving one codon at a time.

(b)

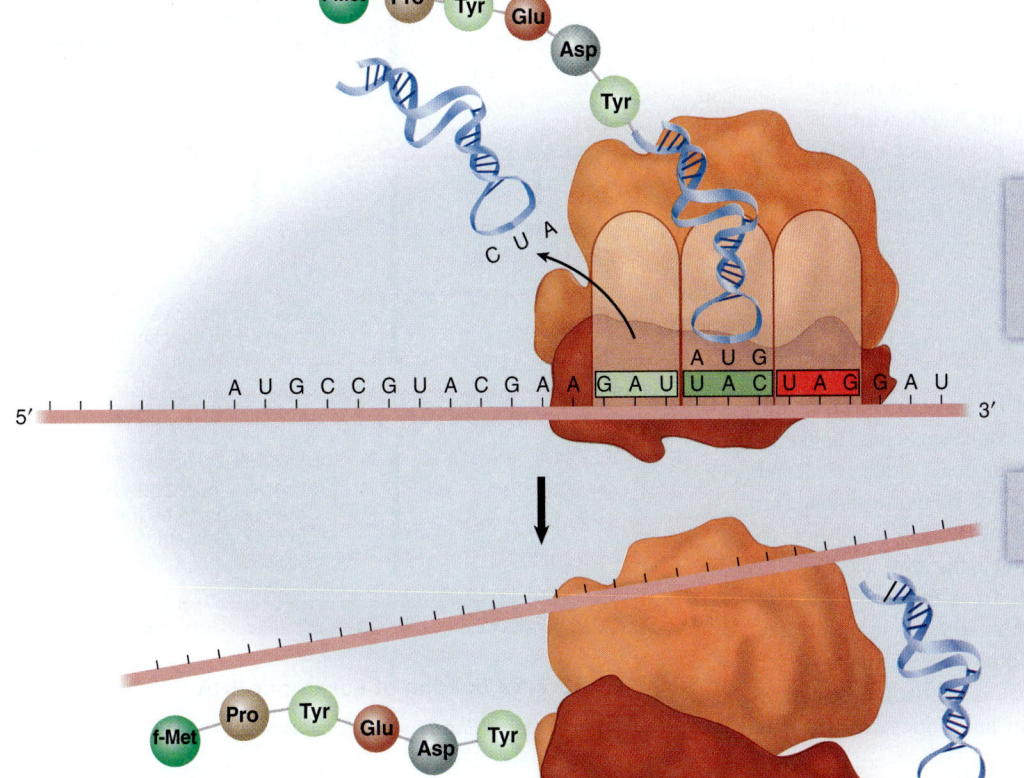

Termination
Translation continues until a stop codon is reached, signaling the end of the process. No tRNA molecules recognize a stop codon.

The components dissemble, releasing the newly formed polypeptide.

(c)

TABLE 7.5	Major Differences Between Prokaryotic and Eukaryotic Transcription and Translation
Prokaryotes	**Eukaryotes**
mRNA is not processed.	A cap is added to the 5′ end of mRNA, and a poly A tail is added to the 3′ end.
mRNA does not contain introns.	mRNA contains introns, which are removed by splicing.
Translation of mRNA begins as it is being transcribed.	The mRNA transcript is transported out of the nucleus so that it can be translated in the cytoplasm.
mRNA is often polycistronic; translation usually begins at the first AUG codon that follows a ribosome-binding site.	mRNA is monocistronic; translation begins at the first AUG.

7.4 ■ Differences Between Eukaryotic and Prokaryotic Gene Expression

Learning Outcome

6. *Describe four differences between prokaryotic and eukaryotic gene expression.*

Eukaryotes differ significantly from prokaryotes in several aspects of transcription and translation (**table 7.5**). Eukaryotic mRNA, for example, is synthesized in a precursor form, called **pre-mRNA.** The pre-mRNA must be processed (altered), both during and after transcription to form mature mRNA. Shortly after transcription begins, the 5′ end of the pre-mRNA is **capped** by adding a methylated guanine derivative. This cap binds specific proteins that stabilize the transcript and enhance translation. The 3′ end of the molecule is also modified, even before transcription has been terminated. This process, **polyadenylation,** cleaves the transcript at a specific sequence and then adds about 200 adenine derivatives to the new 3′ end. This creates a poly A tail, which is thought to stabilize the transcript as well as enhance translation. Another important modification is **splicing,** which removes specific segments of the transcript (**figure 7.17**). Splicing is necessary because eukaryotic genes are often interrupted by non-coding sequences. These intervening sequences, **introns,** are transcribed along with the expressed regions, **exons,** and must be removed from pre-mRNA to create functional mRNA.

The mRNA in eukaryotic cells must be transported out of the nucleus before it can be translated in the cytoplasm. Thus, unlike in prokaryotes, the same mRNA molecule cannot be synthesized and translated at the same time or even in the same cellular location. The mRNA of eukaryotes is generally monocistronic, and translation of the message typically begins at the first AUG in the molecule.

The ribosomes of eukaryotes differ from those of prokaryotes. Whereas the prokaryotic ribosome is 70S, made up of 30S and 50S subunits, the eukaryotic ribosome is 80S, made up of 40S and 60S subunits. The differences in ribosome structure are medically important because certain antibiotics bind to and inactivate bacterial 70S ribosomes, but not 80S ribosomes. This explains why those antibiotics kill bacteria without causing significant harm to mammalian cells.

MicroByte
Symptoms of the disease diphtheria are due to a toxin that inactivates a protein required for 80S ribosome function.

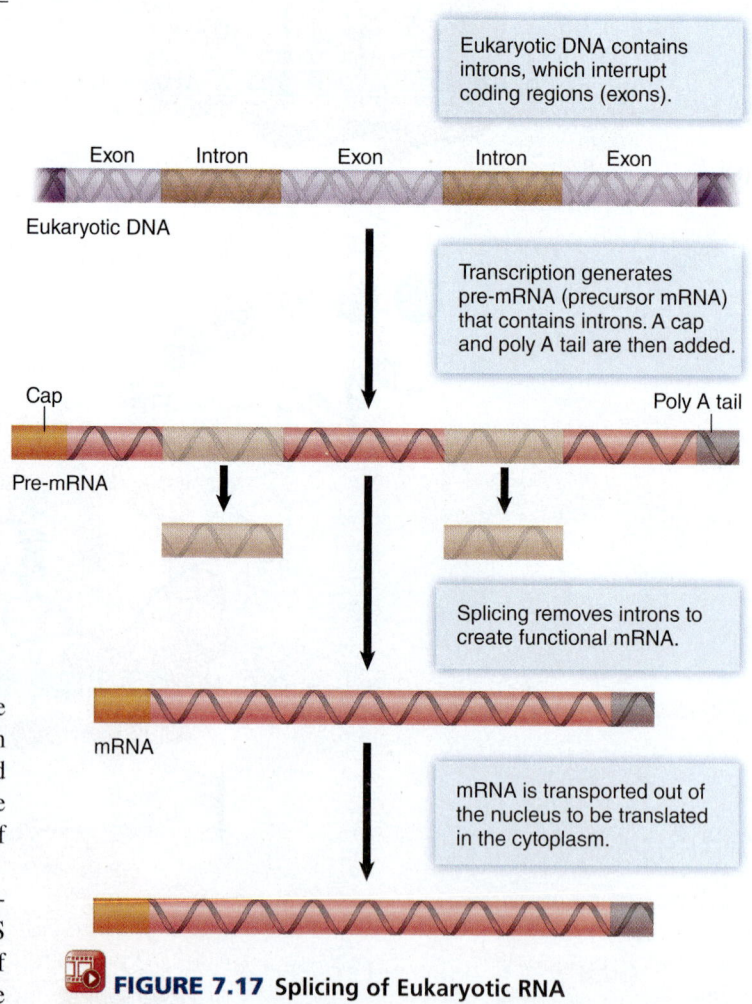

Eukaryotic DNA contains introns, which interrupt coding regions (exons).

Exon   Intron   Exon   Intron   Exon

Eukaryotic DNA

Transcription generates pre-mRNA (precursor mRNA) that contains introns. A cap and poly A tail are then added.

Cap   Poly A tail

Pre-mRNA

Splicing removes introns to create functional mRNA.

mRNA

mRNA is transported out of the nucleus to be translated in the cytoplasm.

FIGURE 7.17 Splicing of Eukaryotic RNA

? *Introns are intervening sequences; what are exons?*

RNA: The First Macromolecule?

The 1989 Nobel Prize in Chemistry was awarded to two Americans, Sidney Altman and Thomas Cech, who independently made the unexpected observation that RNA molecules can act as enzymes. Before their studies, it was believed that only proteins had enzymatic activity.

Cech made a key observation in 1982 when he was trying to understand how introns were removed from ribosomal pre-RNA (pre-rRNA) in a eukaryotic protozoan. Because he was convinced that proteins were responsible for cutting out introns, he added all of the protein in the cells' nuclei to a suspension of pre-rRNA. As expected, the introns were cut out. For a control, Cech used pre-rRNA to which no nuclear

proteins had been added, fully expecting that nothing would happen. Much to his surprise, the introns were removed in the control as well. Based on these results, Cech could only conclude that the RNA acted on itself to cut out pieces of RNA.

The studies of Altman and his colleagues—carried out simultaneously to, and independently of, Cech's—showed that RNA had catalytic properties beyond cutting out introns from pre-rRNA. Altman's group found that RNA could convert a tRNA molecule from a precursor form to its final functional state. Additional studies have shown that enzymatic reactions in which catalytic RNAs, or ribozymes, play a role are very

widespread. Ribozymes have been found in the mitochondria of eukaryotic cells and shown to catalyze other reactions that resemble the polymerization of RNA.

These observations have profound implications for a long-standing question in evolutionary biology: Which came first, proteins or nucleic acids? The answer seems to be nucleic acids, specifically RNA, which acted both as a carrier of genetic information as well as an enzyme. Billions of years ago, before the present universe in which DNA, RNA, and protein are found, the only macromolecule was probably RNA. Once tRNAs became available, they carried amino acids to nucleotide sequences on a strand of RNA.

MicroAssessment 7.4

Eukaryotic pre-mRNA must be processed, which involves capping, polyadenylation, and splicing. In eukaryotic cells, the mRNA must be transported out of the nucleus before it can be translated in the cytoplasm. Eukaryotic mRNA is monocistronic.

10. *What is an intron?*
11. *Would a deletion of two base pairs have a greater consequence if it occurred in an intron or in an exon?* ➕

7.5 ■ Sensing and Responding to Environmental Fluctuations

Learning Outcomes

7. *Describe how quorum sensing and two-component regulatory systems allow cells to adapt to fluctuating conditions.*
8. *Compare and contrast antigenic variation and phase variation.*

Microorganisms are constantly faced with rapidly changing environmental conditions, and must quickly adapt to the fluctuations if they are to survive. Consider the situation of *E. coli* in the intestinal tract of mammals. In this habitat, it must cope with alternating periods of feast and famine. For a limited time after a mammal eats, the bacterial cells prosper, bathed in the mixture of amino acids, vitamins, and other nutrients. The cells actively take up these compounds they would otherwise need to synthesize. Simultaneously, the cells shut down their biosynthetic pathways, channeling the conserved energy into the rapid production of cell components such as DNA and protein. Famine,

however, follows the feast. Between meals—a period of time that can be many days in the case of some mammals—the rich source of nutrients is depleted. Now the bacterial cells' biosynthetic pathways must be activated, using energy and slowing cell growth. Cells dividing several times an hour in a nutrient-rich environment might divide only once every 24 hours in a starved mammalian gut. Later, when the animal defecates, some of the *E. coli* cells are excreted in the feces. Outside of the mammalian host, the bacterial cells must cope with yet a completely different set of conditions to survive.

Signal Transduction

Signal transduction transmits information from outside a cell to the inside. This allows cells to monitor and react to environmental conditions.

Quorum Sensing

Some organisms can "sense" the density of cells within their own population—a phenomenon called **quorum sensing.** This allows cells to activate genes that are only useful when expressed by a critical mass. As an example, the cooperative activities leading to biofilm formation are controlled by quorum sensing. Some pathogens use the mechanism to coordinate expression of genes involved with the infection process. ◀◀ biofilms, p. 84

Quorum sensing involves a process that allows bacteria to "talk" to each other by synthesizing one or more varieties of extracellular signaling molecules. When few cells are present, the concentration of a given signaling molecule is very low. As the cells multiply in a confined area, however, the concentration of that molecule increases proportionally. Only when a signaling

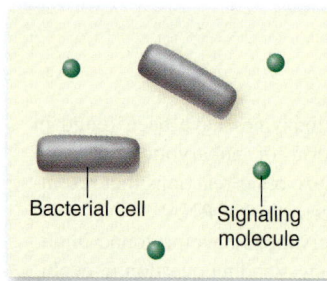
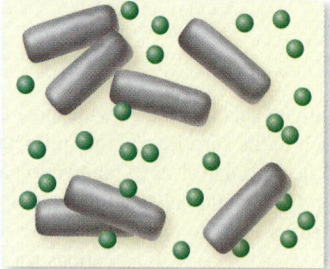

Bacterial cell

Signaling molecule

When few cells are present, the concentration of the signaling molecule is low.

When many cells are present, the signaling molecule reaches a concentration high enough to induce the expression of certain genes.

 FIGURE 7.18 Quorum Sensing

❓ *Why would it be beneficial for cells to wait until a critical population density is present before expressing certain genes?*

molecule reaches a critical level does it induce the expression of specific genes (**figure 7.18**).

Some types of bacteria are able to detect and even interfere with the signaling molecules produced by other species. This allows them to "eavesdrop" and even obstruct "conversations" of other bacteria.

Two-Component Regulatory Systems

An important mechanism that cells use to detect and react to changes in the external environment is a **two-component regulatory system** (**figure 7.19**). *E. coli* uses such a system to control the expression of genes for its alternative types of metabolism. When nitrate is present in anaerobic conditions, the cells activate genes required to use it as the terminal electron acceptor. Some pathogens use two-component regulatory systems to sense and respond to environmental magnesium concentrations. Because the magnesium concentration within certain host cells is generally lower than that of the extracellular environment, these pathogens are able to recognize whether or not they are within a host cell. In turn, they can activate appropriate genes that help them evade host defenses.

◀◀ terminal electron acceptor, p. 130

Two-component regulatory systems consist of two different proteins—a sensor and a response regulator. The sensor spans the cytoplasmic membrane. In response to specific environmental variations, the sensor chemically modifies a region on its internal portion, usually by phosphorylating a specific amino acid. The phosphate group is then transferred to a response regulator. When phosphorylated, the response regulator can turn genes either on or off, depending on the system.

Natural Selection

Natural selection can also play a role in gene expression. The expression of some genes changes randomly in cells, enhancing survival chances of at least a part of a population.

The role of natural selection is readily apparent in bacteria that undergo **antigenic variation,** an alteration in the characteristics of certain surface proteins. Pathogens that do this can stay one step ahead of the body's defenses by altering the very molecules our immune systems must learn to recognize. One of the most

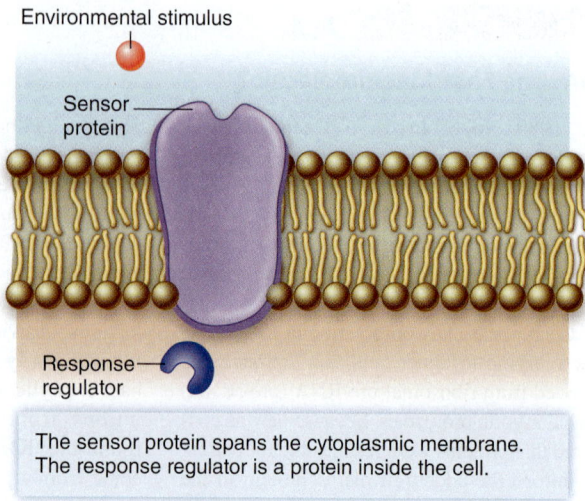

Environmental stimulus

Sensor protein

Response regulator

The sensor protein spans the cytoplasmic membrane. The response regulator is a protein inside the cell.

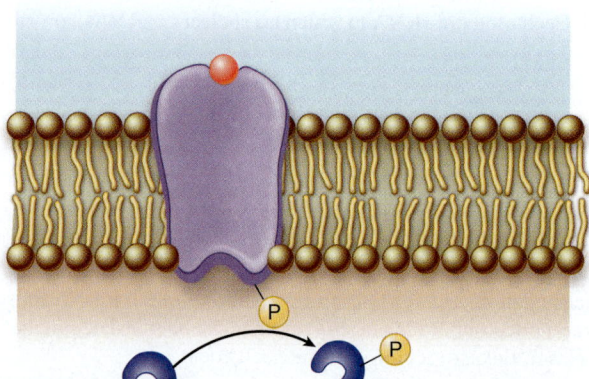

P

P

In response to a specific change in the environment, the sensor phosphorylates a region on its internal portion. The phosphate group is transferred to the response regulator, which can then turn genes on or off, depending on the system.

FIGURE 7.19 Two-Component Regulatory System

❓ *E. coli cells use a two-component system to sense nitrate in the environment. What would happen if they lost the ability to sense that compound?*

well-characterized examples is *Neisseria gonorrhoeae*. This bacterium has many different genes for pilin, the protein subunit that makes up pili, yet most of these genes are silent. The only one expressed is in a particular chromosomal location called an expression locus. *N. gonorrhoeae* cells have a mechanism to shuffle the pilin genes, randomly moving different ones in and out of the expression locus. In a population of 10^4 cells, at least one is expressing a different type of pilin. During an infection, the body's immune system will begin to respond to the dominant pilin type, but the bacterial cells that have "switched" to produce a different pilin type will survive and then multiply. Eventually, the immune system learns to recognize those, but by that time, another subpopulation will have switched its pilin type. ◀◀ pili, p. 65

▶▶ *Neisseria gonorrhoeae*, p. 622

Another mechanism of randomly altering gene expression is **phase variation,** the routine switching on and off of certain genes.

In *E. coli,* for example, certain types of pili required for attachment to epithelial cells undergo phase variation. In an *E. coli* population adhering to an epithelial surface, some bacterial cells will turn off the genes required for pili synthesis, thereby causing those cells to detach from the surface. The process is reversible, so the detached cells will later turn the genes on again, allowing the cells to colonize epithelial cells elsewhere. By altering the expression of genes such as these, at least a part of the population is poised for change.

MicroAssessment 7.5

Quorum sensing and two-component regulatory systems allow a cell to respond to changing environmental conditions. The expression of some genes changes randomly, increasing the chances of survival of at least a subset of a cell population under varying environmental conditions.

12. *Explain how certain bacteria "sense" the density of cells.*

13. *Describe antigenic variation.*

14. *In quorum sensing, why might a bacterium synthesize more than one type of signaling molecule?* ✚

7.6 ■ Bacterial Gene Regulation

Learning Outcomes

9. *Give an example of a constitutive enzyme, an inducible enzyme, and a repressible enzyme.*

10. *Using the* lac *operon as a model, explain the role of inducers, repressors, and inducer exclusion.*

In bacterial cells, many genes are routinely expressed, but others are regulated in response to environmental conditions. Note that scientists describe these regulated genes as capable of being turned on or off, but in reality there are no absolutes. In a population of cells, a gene that is off may still be expressed at very low levels.

A regulatory mechanism sometimes controls the transcription of only a limited number of genes, but in other cases, a wide array of genes is controlled coordinately. A set of regulated genes transcribed as a single polycistronic message is called an **operon.** One of the most well-characterized examples is the *lac* operon, the set of genes required for transporting and hydrolyzing the disaccharide lactose. Separate operons controlled by a single regulatory mechanism constitute a **regulon.** Two-component regulatory systems often control regulons. The simultaneous regulation of numerous genes is called **global control.**
◀◀ disaccharide, p. 31

When describing enzymes, scientists group them according to the type of regulation that governs their synthesis:

■ **Constitutive.** Constitutive enzymes are synthesized constantly; the genes that encode these enzymes are always active. Constitutive enzymes usually play indispensable roles in the central metabolic pathways. For example, the enzymes of glycolysis are constitutive. ◀◀ central metabolic pathways, pp. 132, 139

■ **Inducible.** Inducible enzymes are not routinely produced at significant levels; instead, their synthesis can be turned on when needed (**figure 7.20**). An example is β-galactosidase, the enzyme that hydrolyzes lactose into its component monosaccharides—glucose and galactose. The genes for this enzyme are part of the *lac* operon, which is turned on only when lactose is present. This makes sense because a cell would waste precious resources if it expressed the operon when lactose is not available. Inducible enzymes are often involved in the transport and breakdown of specific energy sources.

■ **Repressible.** Repressible enzymes are produced routinely, but their synthesis can be turned off when they are not required (figure 7.20). Repressible enzymes are generally involved in biosynthetic (anabolic) pathways, such as those that produce amino acids. Cells require a sufficient amount of a given amino acid to multiply; so, if an amino acid is not available in the environment, it needs to be produced by the cell. When the amino acid is available, however, synthesis of the enzymes used in its production would waste energy.

Mechanisms to Control Transcription

The methods a cell uses to prevent or facilitate transcription must be readily reversible, allowing cells to control the relative number of transcripts made. Two of the most common regulatory mechanisms are alternative sigma factors and DNA-binding proteins.

Alternative Sigma Factors

As described earlier, sigma factor is a loose component of RNA polymerase that functions in recognizing specific promoters. Standard sigma factors recognize promoters for genes that need to be expressed during routine growth conditions, but a cell can also produce **alternative sigma factors.** These recognize different sets of promoters, thereby controlling the expression of specific groups of genes. In the endospore-former *Bacillus subtilis,* the sporulation process is controlled by a number of different alternative sigma

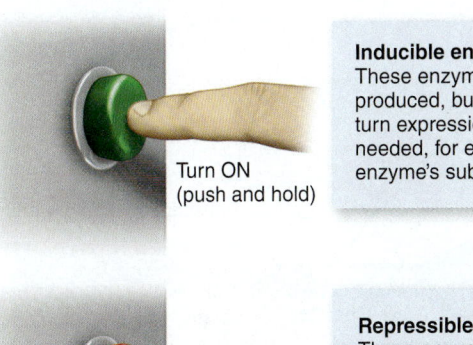

Inducible enzymes
These enzymes are not routinely produced, but mechanisms can turn expression on for as long as needed, for example, when the enzyme's substrate is present.

Turn ON
(push and hold)

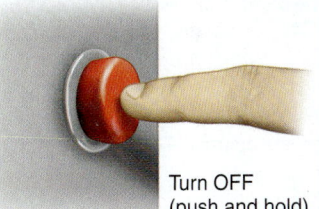

Repressible enzymes
These enzymes are routinely produced, but mechanisms can turn expression off for as long as necessary, for example, when the enzyme's product is present in sufficient quantity.

Turn OFF
(push and hold)

FIGURE 7.20 Principles of Regulation

❓ *Why would the biosynthetic enzymes of a cell be repressible rather than constitutive or inducible?*

factors. One controls the steps at the beginning of sporulation. Others then guide the stages of development in the mother cell and spore. A cell can also express anti-sigma factors, which inhibit the function of specific sigma factors. ◄◄ sporulation, p. 67

DNA-Binding Proteins

Transcription is often controlled by proteins that bind to specific DNA sequences. When a regulatory protein attaches to DNA, it can act either as a repressor, which blocks transcription, or an activator, which facilitates transcription.

Repressors A **repressor** is a regulatory protein that blocks transcription (negative regulation). It does this by binding to an operator, a specific DNA sequence located immediately downstream of a promoter. When a repressor is bound to an operator, RNA polymerase cannot progress past that DNA sequence. Repressors are allosteric proteins, however, meaning that specific molecules can attach to them and change their shape. This can alter the repressor's ability to bind to operator DNA. As shown in **figure 7.21,** there are two general mechanisms by which different repressors can function:

- **Induction.** The repressor is synthesized as a form that binds to the operator, blocking transcription. When a molecule called an **inducer** attaches to the repressor, the shape of the repressor changes so that it can no longer grasp the operator.

With the repressor unable to bind to DNA, RNA polymerase may transcribe the gene.

- **Repression.** The repressor is synthesized as a form that cannot bind to the operator. However, when a molecule termed a **corepressor** attaches to the repressor, the corepressor-repressor complex can then bind to the operator, blocking transcription.

Activators An **activator** is a regulatory protein that facilitates transcription (positive regulation). Genes controlled by an activator have an ineffective promoter preceded by an **activator-binding site.** The binding of the activator to the DNA enhances the ability of RNA polymerase to initiate transcription at that promoter. Like repressors, activators can be changed by the binding of other molecules. When a molecule called an inducer binds to an activator, the shape of the activator is altered so that it can now bind to the activator-binding site (**figure 7.22**). Thus, the term "inducer" applies to a molecule that turns on transcription, either by stimulating the function of an activator or interfering with the function of a repressor.

MicroByte

In *Bacillus subtilis,* the DNA-binding protein that triggers the sporulation process controls a regulon of over 120 genes.

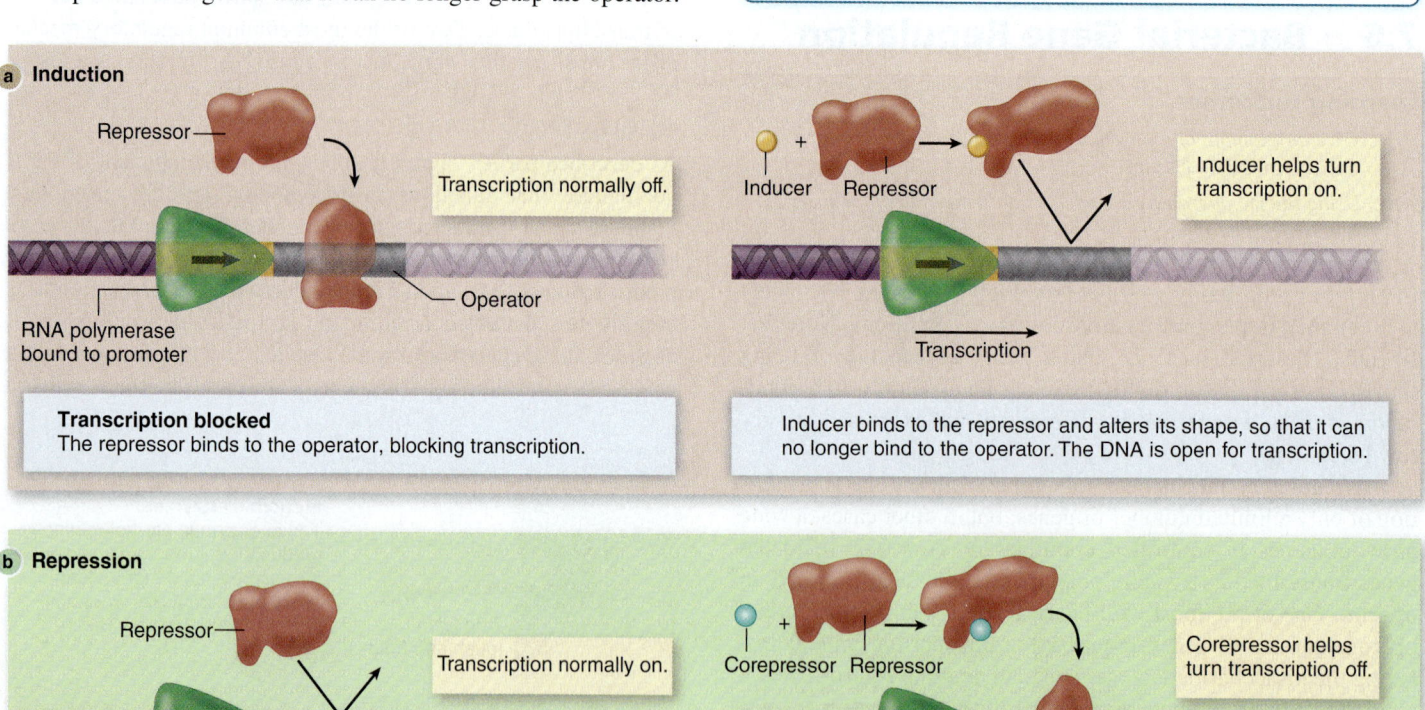

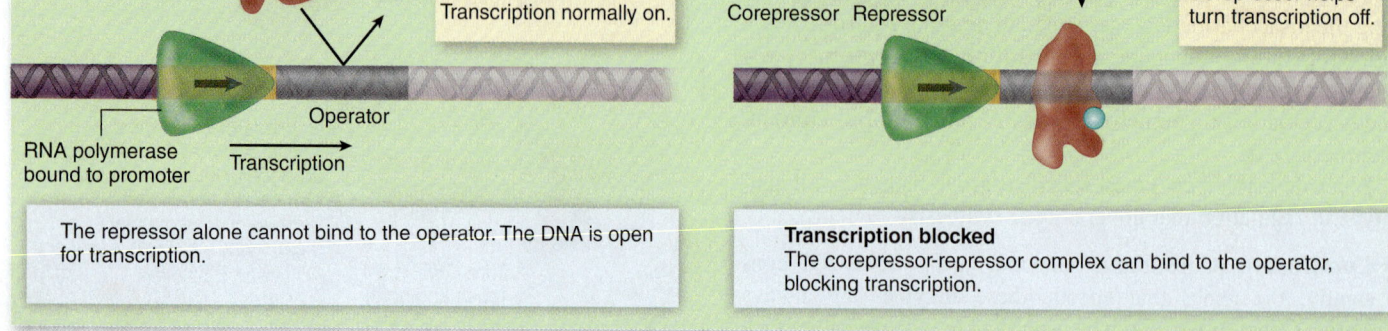

FIGURE 7.21 Transcriptional Regulation by Repressors (a) Induction. **(b)** Repression.

? *How is a corepressor different from an inducer? How is it similar?*

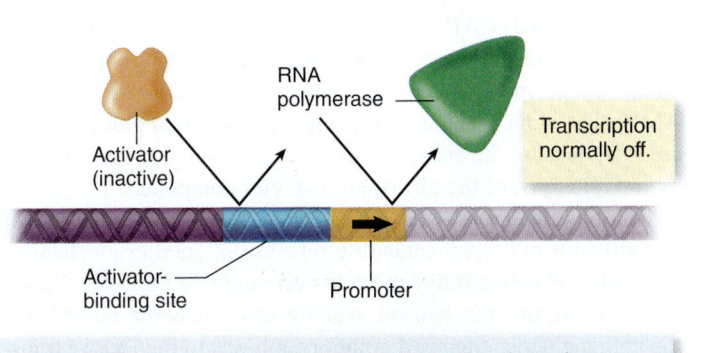

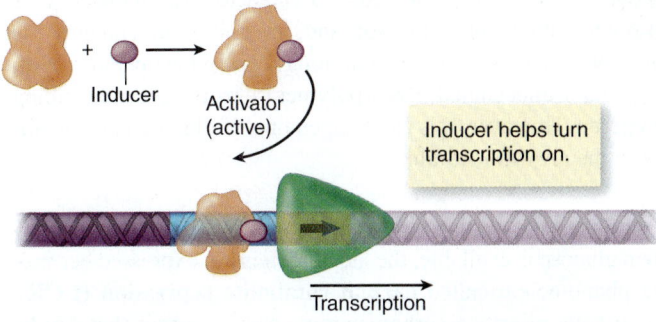

RNA polymerase cannot bind to the promoter unless the activator is bound to the activator-binding site, but the activator is in an inactive form.

Inducer binds to the activator and changes its shape, allowing the activator to bind to the site. RNA polymerase can then bind to the promoter and initiate transcription.

FIGURE 7.22 Transcriptional Regulation by Activators

❓ *How do activators facilitate transcription?*

The *lac* Operon as a Model

Originally described in the early 1960s by Francois Jacob and Jacques Monod, the **lac operon** of *E. coli* has served as an important model for understanding the control of bacterial gene expression. This operon encodes proteins involved with the transport and degradation of lactose, and is only turned on when lactose is in the cell but glucose is not available. The fact that glucose prevents

expression of the genes is significant biologically because it forces cells to use the most efficiently metabolized carbon source first.

Lactose and the *lac* Operon

The *lac* operon uses a repressor that prevents transcription when lactose is not available; the repressor binds the operator, blocking RNA polymerase (**figure 7.23**). When lactose is in the cell,

No lactose in the cell
The repressor binds to operator, blocking transcription.

*lac*Z
(β-galactosidase)

*lac*Y
(permease)

*lac*A
(transacetylase)

DNA

Terminator

RNA polymerase bound to promoter

Repressor bound to operator

Lactose present in the cell
Some lactose is converted to allolactose. This binds to the repressor and alters its shape, so that it can no longer bind to the operator. If glucose is not available, the operon will be transcribed.

Transcription

Transcription

Allolactose

Non-functional repressor

Transcription

Translation

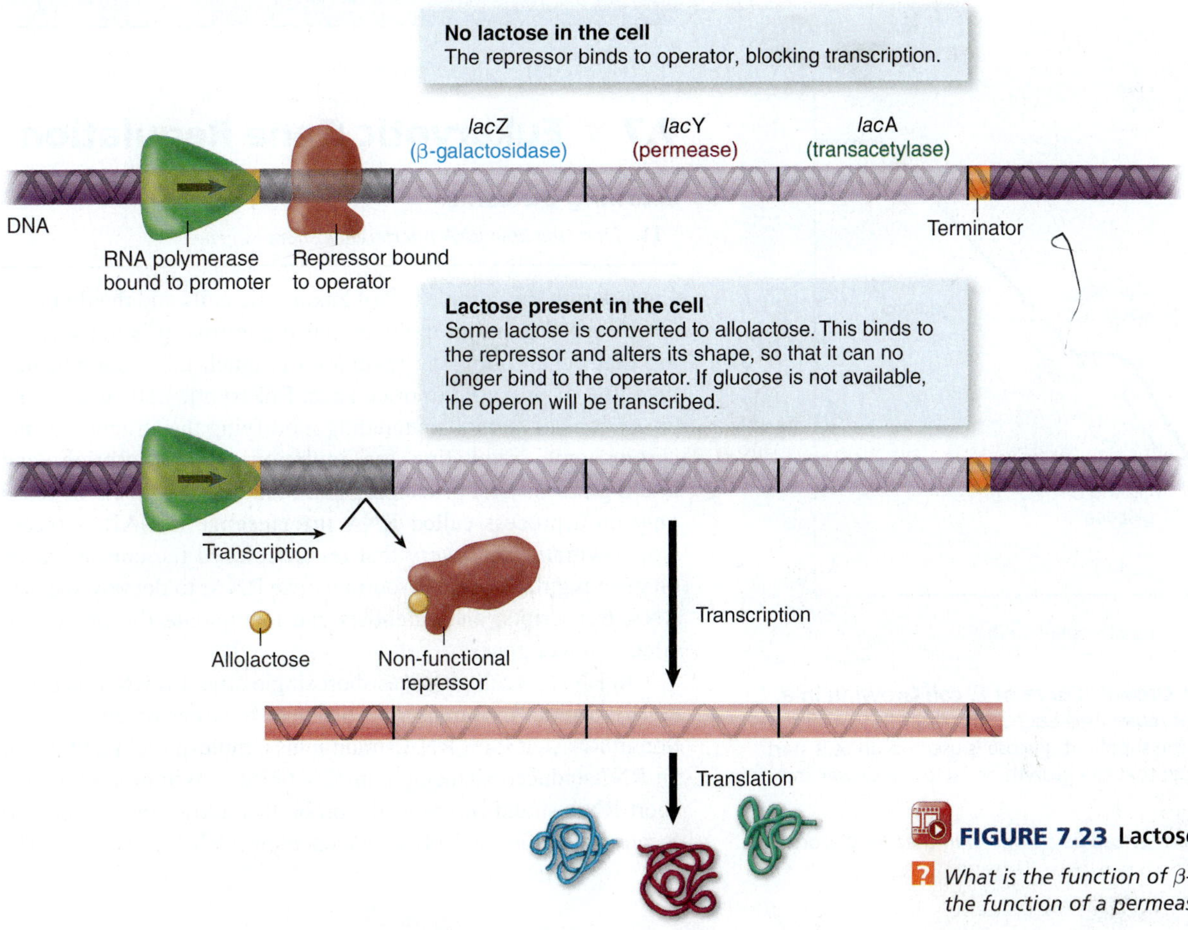

FIGURE 7.23 Lactose and the *lac* Operon

❓ *What is the function of β-galactosidase? What is the function of a permease?*

however, some of it is converted to allolactose, an inducer. This compound binds the repressor and, in doing so, changes the repressor's shape so that it can no longer grasp the operator. With the operator unoccupied, RNA polymerase can begin transcribing the operon. However, this can happen only if glucose is not available in the growth medium.

Glucose and the *lac* Operon

When glucose is available, the lac operon is not expressed because of a phenomenon called **carbon catabolite repression (CCR)**. *E. coli* cells prioritize carbon/energy sources, a trait that can be demonstrated by growing them in a medium containing both glucose and lactose. Initially, the cells multiply using only glucose. Once the supply of that sugar is exhausted, growth stops for a short period as the cells gear up to begin metabolizing lactose. Then, they begin multiplying again, this time using lactose to fuel their growth. This characteristic two-phase growth pattern is called **diauxic growth** (figure 7.24).

Carbon catabolite repression is a global control system that allows glucose to regulate expression of the *lac* operon as well as other sets of genes. Glucose does not act directly in the regulation, however. Instead, the cell's glucose transport system serves as a sensor of glucose availability. When the transport system is moving glucose molecules into the cell, catabolite repression prevents the *lac* operon from being expressed. When the transport system is idle, indicating that glucose is not available, then the *lac* operon can be turned on (**figure 7.25**).

One mechanism of carbon catabolite repression involves an activator called **CAP** (catabolite activator protein), which is required for transcription. To be functional, the activator must be bound by an inducer—an ATP derivative called cAMP (cyclic AMP). The inducer is made only when extracellular glucose levels are low, because the enzyme required for its synthesis is activated by the idle form of the glucose transporter component.

Although a great deal of attention has been paid to the role of the activator in carbon catabolite repression, another mechanism of regulation called **inducer exclusion** might be more significant in *E. coli*. In this mechanism, when glucose is being moved into the cell, a glucose transport component binds to the lactose transporter (permease), locking it in a non-functional position. The locked permease cannot move lactose into the cell, so the *lac* operon will not be induced. Once the glucose supply diminishes, the glucose transporter becomes idle, so lactose can then be brought into the cell.

7.7 ■ Eukaryotic Gene Regulation

Learning Outcome

11. *Describe how RNA interference silences genes.*

Considering the complexity of eukaryotic cells and the diversity of cell types found in multicellular organisms, it is not surprising that eukaryotic gene regulation is much more complicated than that of prokaryotic organisms. Eukaryotic cells use a variety of control methods, including modifying the structure of the chromosome, regulating the initiation of transcription, and altering pre-mRNA processing and modification. We will focus only on a process called **RNA interference (RNAi)**, a recent Nobel-winning discovery that revolutionized the current views on gene regulation. Cells routinely use RNAi to destroy specific RNA transcripts, and scientists can manipulate the process to silence select genes.

In RNAi, a cell produces short single-stranded RNA pieces to locate specific RNA transcripts that need to be destroyed. To function in RNAi, a short RNA strand joins a multi-protein unit called an RNA-induced silencing complex (RISC). Within a RISC, the short RNA strand serves as the probe that allows the complex to locate a specific nucleotide sequence on mRNA molecules. The

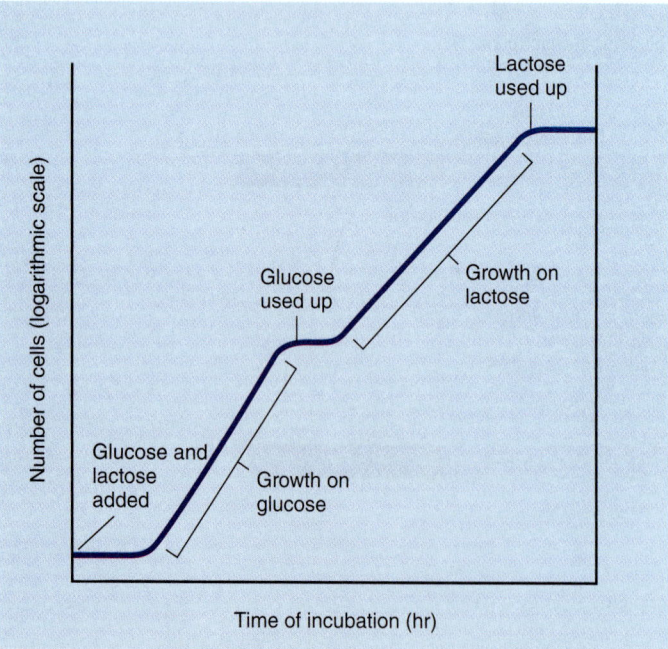

FIGURE 7.24 Diauxic Growth Curve of *E. coli* Growing in a Medium Containing Glucose and Lactose Cells preferentially use glucose. Only when the supply of glucose is used up do cells start metabolizing lactose. Note that the growth on lactose is slower than it is on glucose.

❓ *Why would growth on lactose be slower than that on glucose?*

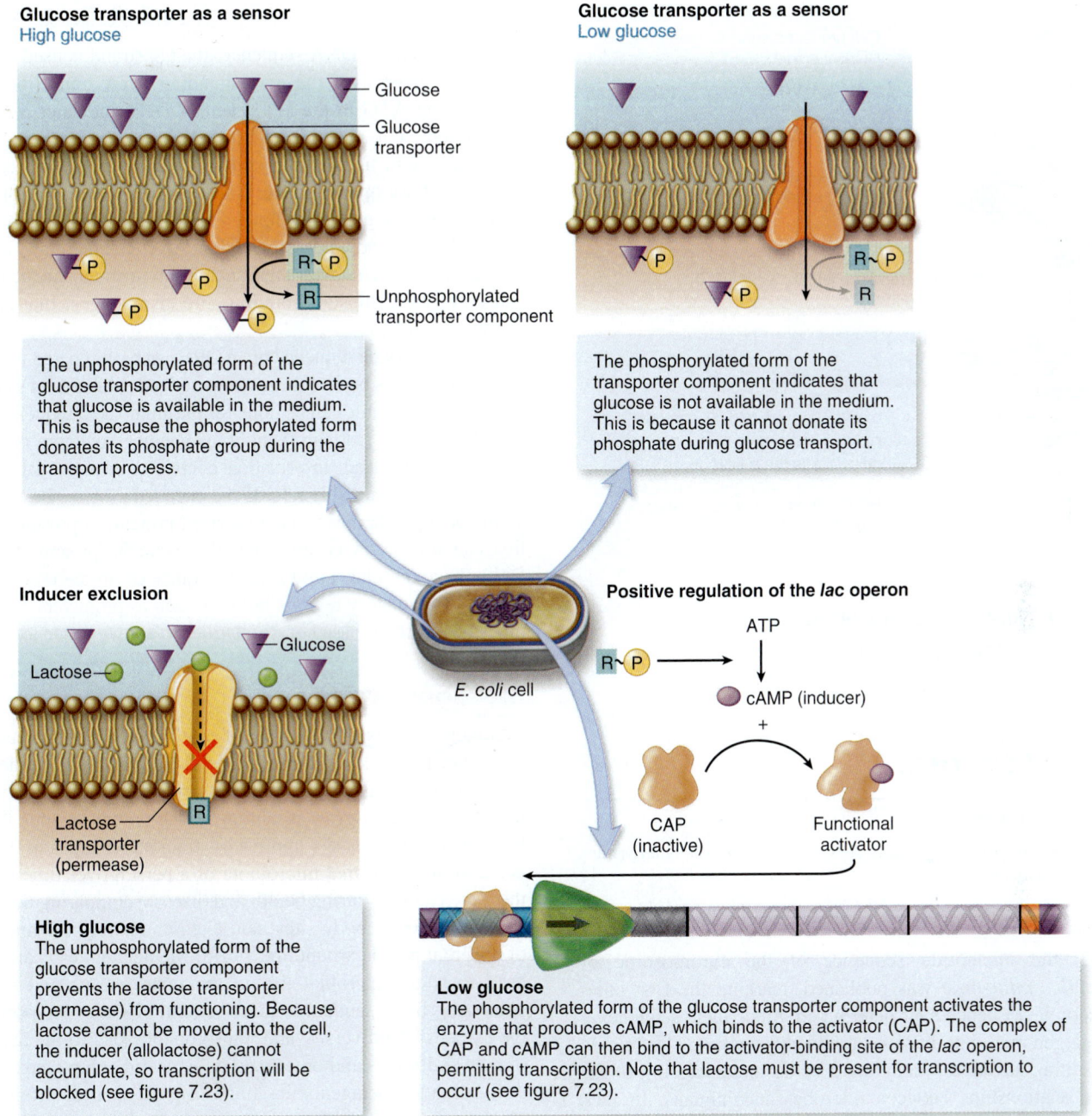

Glucose transporter as a sensor
High glucose

Glucose

Glucose transporter

R~P

R — Unphosphorylated transporter component

The unphosphorylated form of the glucose transporter component indicates that glucose is available in the medium. This is because the phosphorylated form donates its phosphate group during the transport process.

Glucose transporter as a sensor
Low glucose

R~P

R

The phosphorylated form of the transporter component indicates that glucose is not available in the medium. This is because it cannot donate its phosphate during glucose transport.

E. coli cell

Inducer exclusion

Glucose
Lactose

Lactose transporter (permease)

R

High glucose
The unphosphorylated form of the glucose transporter component prevents the lactose transporter (permease) from functioning. Because lactose cannot be moved into the cell, the inducer (allolactose) cannot accumulate, so transcription will be blocked (see figure 7.23).

Positive regulation of the *lac* operon

ATP

R~P →

cAMP (inducer)

+

CAP (inactive)

Functional activator

Low glucose
The phosphorylated form of the glucose transporter component activates the enzyme that produces cAMP, which binds to the activator (CAP). The complex of CAP and cAMP can then bind to the activator-binding site of the *lac* operon, permitting transcription. Note that lactose must be present for transcription to occur (see figure 7.23).

FIGURE 7.25 Glucose and the *lac* Operon

❓ *Why would it be advantageous for a cell to use glucose before lactose?*

short RNA strand does this by binding to complementary sequences on an mRNA molecule, tagging that transcript for destruction by enzymes in the RISC (**figure 7.26**). The components of the RISC are not destroyed in the process, so the complex is catalytic, providing a rapid and effective means of silencing genes that have already been transcribed. Two different types of RNA molecules are used in RNAi—microRNA (miRNA) and short interfering RNA (siRNA). These are each about two dozen nucleotides in length and functionally equivalent, but they differ in how they are produced.

MicroByte

RNAi technology shows promise in treating certain viral infections and tumors.

MicroAssessment 7.7

RNA interference uses short strands of RNA to locate specific RNA transcripts destined for destruction.

18. *What is the role of miRNA and siRNA in regulation of gene expression?*

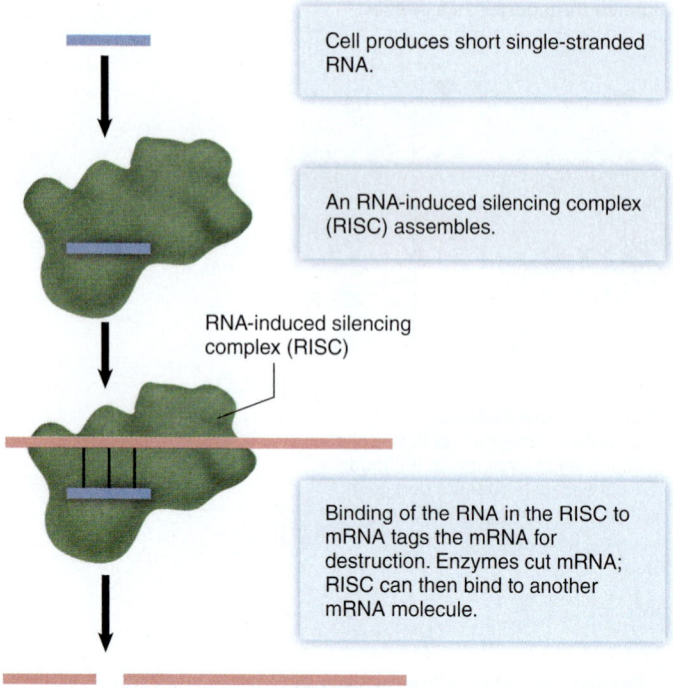

Cell produces short single-stranded RNA.

An RNA-induced silencing complex (RISC) assembles.

RNA-induced silencing complex (RISC)

Binding of the RNA in the RISC to mRNA tags the mRNA for destruction. Enzymes cut mRNA; RISC can then bind to another mRNA molecule.

FIGURE 7.26 RNA Interference (RNAi)

❓ *How could RNAi be used medically?*

7.8 ■ Genomics

Learning Outcomes

12. *Explain how protein-encoding regions are found when analyzing a DNA sequence.*

13. *Describe metagenomics and the information it can provide.*

In 1995, the nucleotide sequence of the chromosome of *Haemophilus influenzae* was published, marking the first complete genomic sequence ever determined. Since then, sequencing microbial genomes has become relatively common, leading to many exciting advances, including a better understanding of the complex relationships between microbes and humans. In fact, many of the recent findings described in this textbook have been discovered through genomics.

Although sequencing methods are becoming more rapid, analyzing the resulting data is far more complex than it might initially seem. Imagine trying to determine the amino acid sequence of a protein encoded by a stretch of DNA, without knowing anything about the orientation of the gene's promoter or the reading frame used for translation. Because either strand of the DNA molecule could potentially be the template strand, two entirely different RNA sequences must be considered as protein-encoding candidates. In turn, each of those two candidates has three reading frames, for a total of six possible reading frames to consider. Yet only one of these actually codes for the protein. Understandably, computers are an invaluable aid and are used extensively in deciphering the meaning of the raw sequence data. As a result, a new field has emerged—**bioinformatics**—which creates the computer technology to store, retrieve, and analyze nucleotide sequence data.

Analyzing a Prokaryotic DNA Sequence

When analyzing a DNA sequence, the (+) strand is used to represent the sequence of the corresponding RNA transcript. As an example, an ATG in the (+) strand of DNA indicates a possible start codon. ◀◀ plus (+) strand, p. 168

Computers help locate protein-encoding regions in DNA. They search for **open reading frames (ORFs)**, stretches of nucleotide sequences generally longer than 300 bp that begin with a start codon and end with a stop codon. An ORF potentially encodes a protein. Other characteristics such as an upstream sequence that can serve as a ribosome-binding site also suggest that an ORF encodes a protein.

The nucleotide sequence of an ORF can be compared with other known sequences by searching computerized databases of published sequences. The same can be done for the amino acid sequence of the encoded protein. Not surprisingly, as genomes of more organisms are being sequenced, information contained in these databases is growing at a remarkable rate. If the encoded protein shows certain amino acid similarities to characterized proteins, a presumed function can sometimes be assigned. For example, proteins that bind DNA have similar amino acid sequences in certain regions. Likewise, regulatory regions in DNA such as promoters can sometimes be identified based on similarities to known sequences.

Metagenomics

Metagenomics is the analysis of total microbial genomes in an environment. By examining the total genomes, researchers can study all microorganisms and viruses in a community, not just the relatively few that grow in culture. Imagine the insights that can be gained from this new approach—tracking changes in the composition of the normal microbiota of a person over time to see if there are changes during health and disease; comparing the microbiota of different body sites; and even comparing the microbiota of different people around the world! Metagenomics is also being used to study microbial life in the open oceans and in soils. Analyzing these sequences gives an entirely new perspective on the extent of biodiversity and will probably lead to the discovery of new antibiotics and other medically useful compounds.

Although metagenomics holds a great deal of promise, the amount of new data presents tremendous challenges. New computing methods are being developed to handle the complications of analyzing such complex information.

MicroByte

So far, over 1,000 prokaryotic genomes have been sequenced!

MicroAssessment 7.8

Sequencing methods are rapid, but analyzing the data and extracting the pertinent information is difficult.

19. *What is an open reading frame?*

20. *Describe two things that you can learn by searching a computerized database for sequences that have similarities to a newly sequenced gene.*

21. *There are characteristic differences in the nucleotide sequences of the leading and lagging strands. Why might this be so?* ✚

Summary

7.1 ■ Overview (figure 7.1)

Characteristics of DNA (figure 7.2)

A single strand of DNA has a **5′ end** and a **3′ end;** the two strands of DNA in the double helix are **complementary** and **antiparallel.**

Characteristics of RNA

A single-stranded RNA molecule is transcribed from one of the two strands of DNA. There are three different functional types of RNA molecules: **messenger RNA (mRNA), ribosomal RNA (rRNA),** and **transfer RNA (tRNA)** (figure 7.3).

Regulating Gene Expression

Protein synthesis is generally controlled by regulating the synthesis of mRNA (figure 7.4).

7.2 ■ DNA Replication

The bidirectional progression of replication around a circular DNA molecule creates two **replication forks** (figure 7.5). DNA replication is **semiconservative.**

Initiation of DNA Replication

DNA replication begins at the **origin of replication.**

The Process of DNA Replication

DNA polymerase synthesizes DNA in the 5′ to 3′ direction, using one strand as a **template** to generate the complementary strand (figures 7.6, 7.7).

7.3 ■ Gene Expression in Bacteria

Transcription

RNA polymerase synthesizes RNA in the 5′ to 3′ direction, producing a single-stranded RNA molecule complementary and antiparallel to the DNA template (figure 7.9). **Transcription** initiates when RNA polymerase recognizes and binds to a **promoter** (figures 7.10, 7.11). When RNA polymerase encounters a **terminator,** it falls off the DNA and releases the newly synthesized RNA.

Translation

The information encoded by mRNA is deciphered using the **genetic code** (table 7.4). The first AUG downstream of a **ribosome-binding site** serves as a **start codon** (figure 7.12). **Ribosomes** are translation "machines." tRNAs carry specific amino acids and act as keys that interpret the genetic code (figure 7.14). The ribosome moves along mRNA in the 5′ to 3′ direction; translation terminates when the ribosome reaches a **stop codon** (figure 7.16).

7.4 ■ Differences Between Eukaryotic and Prokaryotic Gene Expression (table 7.5)

Eukaryotic mRNA is processed; a cap and a poly A tail are added. Eukaryotic genes often contain **introns; splicing** removes these from **pre-mRNA** (figure 7.17). In eukaryotic cells, the mRNA must be transported out of the nucleus before it can be translated in the cytoplasm.

7.5 ■ Sensing and Responding to Environmental Fluctuations

Signal Transduction

Bacteria use **quorum sensing** to activate genes that are useful only when expressed by a critical mass (figure 7.18). **Two-component regulatory systems** use a sensor that recognizes changes outside the cell and then transmits that information to a response regulator (figure 7.19).

Natural Selection

The expression of some genes changes randomly, enhancing survival chances of at least a part of a population. **Antigenic variation** is a routine alteration in the characteristics of certain surface proteins. **Phase variation** is the routine switching on and off of certain genes.

7.6 ■ Bacterial Gene Regulation

Constitutive enzymes are constantly synthesized. The synthesis of **inducible enzymes** can be turned on by certain conditions. The synthesis of **repressible enzymes** can be turned off by certain conditions (figure 7.20).

Mechanisms to Control Transcription

Repressors block transcription (figure 7.21). **Activators** enhance transcription (figure 7.22).

The *lac* Operon as a Model

The *lac* **operon** uses a repressor that prevents transcription of the genes when lactose is not available (figure 7.23). A mechanism called

carbon catabolite repression (CCR) prevents transcription of the *lac* operon when glucose is available (figure 7.25).

7.7 ■ Eukaryotic Gene Regulation

Regulation in eukaryotic cells is much more complicated than that in prokaryotic cells. In **RNA interference (RNAi),** a cell synthesizes short single-stranded RNA pieces to locate specific RNA transcripts destined for destruction (figure 7.26).

7.8 ■ Genomics

Analyzing a Prokaryotic DNA Sequence

When analyzing a DNA sequence, the (+) strand is used to represent the sequence corresponding RNA transcript; computers are used to search for **open reading frames (ORFs).**

Metagenomics

Metagenomics allows researchers to study all microorganisms and viruses in a community, not just the relatively few that grow in culture.

Review Questions

Short Answer

1. Explain what the term *semiconservative* means with respect to DNA replication.

2. What is an origin of replication?

3. Why are primers required in DNA replication but not in transcription?

4. What is polycistronic mRNA?

5. Explain why knowing the orientation of a promoter is critical when determining the amino acid sequence of an encoded protein.

6. What is the function of a sigma factor?

7. What is the fate of a protein that has a signal sequence?

8. Explain how some bacteria sense the density of cells in their own population.

9. Compare and contrast regulation by a repressor and an activator.

10. Explain why it is sometimes difficult to locate genomic regions that encode a protein.

Multiple Choice

1. All of the following are involved in transcription *except*
   a) polymerase.   b) primer.   c) promoter.
   d) sigma factor.   e) uracil.

2. All of the following are involved in DNA replication *except*
   a) polysome.   b) gyrase.   c) polymerase.
   d) primase.   e) primer.

3. All of the following are directly involved in translation *except*
   a) promoter.   b) ribosome.   c) start codon.
   d) stop codon.   e) tRNA.

4. Using the DNA strand shown here as a template, what will be the sequence of the RNA transcript?

   5′ GCGTTAACGTAGGC 3′

   $\xrightarrow{\text{promoter}}$

   3′ CGCAATTGCATCCG 5′
   a) 5′ GCGUUAACGUAGGC 3′
   b) 5′ CGGAUGCAAUUGCG 3′
   c) 5′ CGCAAUUGCAUCCG 3′
   d) 5′ GCCUACGUUAACGC 3′

5. A ribosome binds to the following mRNA at the site indicated by the dark box. At which codon will translation likely begin?

   5′ ■ GCCGGAAUGCUGCUGGC
   a) GCC   b) GGC
   c) AUG   d) AAU

6. Which of the following statements about gene expression is *false*?
   a) More than one RNA polymerase can be transcribing a specific gene at a given time.
   b) More than one ribosome can be translating a specific transcript at a given time.
   c) Translation begins at a site called a promoter.
   d) Transcription stops at a site called a terminator.
   e) Some amino acids are coded for by more than one codon.

7. An enzyme used to synthesize the amino acid tryptophan is most likely
   a) constitutive.   b) inducible.
   c) repressible.   d) a and b.

8. Under which of the following conditions will transcription of the *lac* operon occur?
   a) Lactose present/glucose present
   b) Lactose present/glucose absent
   c) Lactose absent/glucose present
   d) Lactose absent/glucose absent
   e) a and b

9. All of the following are characteristics of eukaryotic gene expression *except*
   a) 5′ cap is added to the mRNA.
   b) a poly A tail is added to the 3′ end of mRNA.
   c) introns must be removed to create the mRNA that is translated.
   d) the mRNA is often polycistronic.
   e) translation begins at the first AUG.

10. Which of the following statements is *false*?
    a) A derivative of lactose serves as an inducer of the *lac* operon.
    b) Signal transduction provides a mechanism for a cell to sense the conditions of its external environment.
    c) Quorum sensing allows bacterial cells to sense the density of like cells.
    d) An example of a two-component regulatory system is the lactose operon, which is controlled by a repressor and an activator.
    e) An ORF is a stretch of DNA that may encode a protein.

Applications

1. A graduate student is trying to identify the gene coding for an enzyme found in a bacterial species that degrades trinitrotoluene (TNT). The student is frustrated to find that the organism does not produce the enzyme when grown in nutrient broth, making it difficult to collect the mRNA needed to help identify the gene. What could the student do to potentially increase the amount of the desired enzyme?

2. A student wants to remove the introns from a segment of DNA coding for protein X. Devise a strategy to do this.

Critical Thinking ➕

1. The study of protein synthesis often uses a cell-free system where cells are ground with an abrasive to release the cell contents and then filtered to remove the abrasive. These materials are added to the system, generating the indicated results:

Materials Added	Results
Radioactive amino acids	Radioactive protein produced
Radioactive amino acids *and* RNase (an RNA-digesting enzyme)	No radioactive protein produced

What is the best interpretation of these observations?

2. In a variation of the experiment in the previous question, the following materials were added to three separate cell-free systems, generating the indicated results:

Materials Added	Results
Radioactive amino acids	Radioactive protein produced
Radioactive amino acids *and* DNase (a DNA-digesting enzyme)	Radioactive protein produced
Several hours after grinding:	
Radioactive amino acids *and* DNase	No radioactive protein produced

What is the best interpretation of these observations?

8 Bacterial Genetics

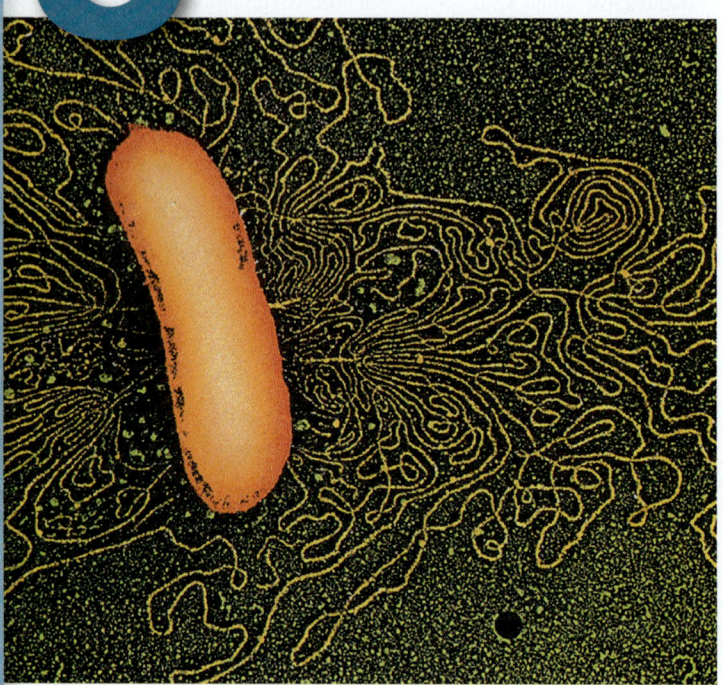

A lysed *E. coli* cell shows the length of its unwound circular chromosome.

KEY TERMS

Auxotroph A microorganism that requires an organic growth factor.

Conjugation Mechanism of horizontal gene transfer in which the donor cell physically contacts the recipient cell.

DNA-Mediated Transformation Mechanism of horizontal gene transfer in which the bacterial DNA is transferred as "naked" DNA.

Genotype The sequence of nucleotides in an organism's DNA.

Homologous Recombination Process by which a cell replaces a stretch of DNA with a segment that has a similar nucleotide sequence.

Horizontal Gene Transfer DNA transfer from one bacterium to another by conjugation, DNA-mediated transformation, or transduction.

Mutation A change in the nucleotide sequence of a cell's DNA that is then passed on to daughter cells.

Non-Homologous Recombination DNA recombination that does not require extensive nucleotide sequence similarity in the stretches that recombine.

Phenotype The observed characteristics of a cell.

Plasmid An extrachromosomal DNA molecule that replicates independently of the chromosome.

Prototroph A microorganism that does not require any organic growth factors.

Transduction Mechanism of horizontal gene transfer in which bacterial DNA is transferred inside a phage coat.

Transposon Segment of DNA that can move from one site to another in a cell's genome.

Wild Type Form of the cell or gene as it typically occurs in nature.

A Glimpse of History

Barbara McClintock (1902–1992) was a remarkable scientist who made several very important discoveries in genetics. She carried out her studies before the age of large interdisciplinary research teams and sophisticated techniques of molecular genetics. Her tools consisted of a clear mind and a curiosity that could make sense of confusing observations. She worked 12-hour days, 6 days a week in a small laboratory at Cold Spring Harbor on Long Island, New York.

McClintock's experimental system consisted of kernels in ears of corn. She noticed that the various kernel colors were not inherited in a predictable manner. In fact, the colors seemed to come and go. Based on extensive data, McClintock concluded that segments of DNA, now called transposons, were moving into and out of genes involved with kernel color. This destroyed the function of the genes, thereby changing kernel color.

At the time that McClintock published her results in 1950, most scientists believed that chromosomal DNA was very stable and changed only through recombination. Consequently, geneticists were skeptical of her conclusions. It was not until the late 1970s that her earlier ideas began to be accepted. By that time, transposons had been discovered in many organisms, including bacteria. Although transposons were first discovered in plants, once they were found in bacteria, the field moved ahead very quickly. In 1983, at age 81, McClintock received a Nobel Prize for her discovery of transposons, popularly called "jumping genes."

*S*taphylococcus aureus, the Gram-positive coccus commonly called Staph, is a frequent cause of skin and wound infections. Since the 1970s, the usual treatment for these infections has been penicillin-like antibiotics, such as methicillin. Today, however, this treatment is likely to fail. In 2004, well over 60% of the *S. aureus* strains from hospitalized patients were resistant to methicillin. In the United States, an estimated 2.3 million healthy people harbor methicillin-resistant *S. aureus* (MRSA) as part of their microbiota. Unfortunately, healthcare-associated MRSA strains (HA-MRSA) are also resistant to a variety of other antibiotics. These are commonly treated with vancomycin, often considered the drug of last resort. The situation became even more worrisome in 2002, when *S. aureus* isolated from foot ulcers on a diabetes patient in Detroit was resistant to vancomycin in addition to other antibiotics.

How do multidrug-resistant strains arise? How are these resistance traits transferred to other bacteria? The answer to these and many other questions important to human health requires a basic understanding of bacterial genetics. This subject encompasses the study of heredity: how genes function (covered in chapter 7), change, and are transferred to other organisms. With this knowledge, you will understand why antibiotics are no longer the miracle drugs they once were against infectious diseases.

8.1 ■ Genetic Change in Bacteria

Learning Outcomes

1. *Distinguish between genotype and phenotype.*
2. *Distinguish between mutation and horizontal gene transfer.*

In the ever-changing conditions that characterize most environments, all organisms need to adapt in order to survive and multiply. If they fail, competing organisms more "fit" to thrive in the new setting will soon predominate. This is the process of **natural selection.** Bacteria have two general means by which they routinely adjust to new circumstances: regulating gene expression (discussed in chapter 7) and genetic change, the focus of this chapter. ◀◀ bacterial gene regulation, p. 179

Bacteria are an excellent experimental system for genetic studies. They grow rapidly in small volumes of simple inexpensive media, accumulating in very large numbers. This makes it easy to study rare events that give rise to strains differing in their genetic makeup. For this reason, it is not surprising that we know more about the genetics of the model organism *E. coli* than any other organism in the world. ◀◀ model organisms, p. 8

A change in an organism's DNA alters the **genotype,** the sequence of nucleotides in the DNA. In bacterial cells, such a change can have a significant impact because bacteria are haploid, meaning they contain only a single set of genes. No "backup copy" of a gene exists in a haploid organism. Because of this, a change in genotype often alters the organism's observable characteristics, or **phenotype.** Note, however, that the phenotype involves more than just the genetic makeup of an organism; it can also be influenced by environmental conditions. For example, colonies of *Serratia marcescens* are red when incubated at 22°C but white when incubated at 37°C. The phenotype, but not the genotype, has changed. However, if the genes responsible for pigment production are removed, the organism's phenotype as well as the genotype changes.

Genetic change in bacteria occurs by two mechanisms—mutation and horizontal gene transfer (**figure 8.1**). **Mutation**

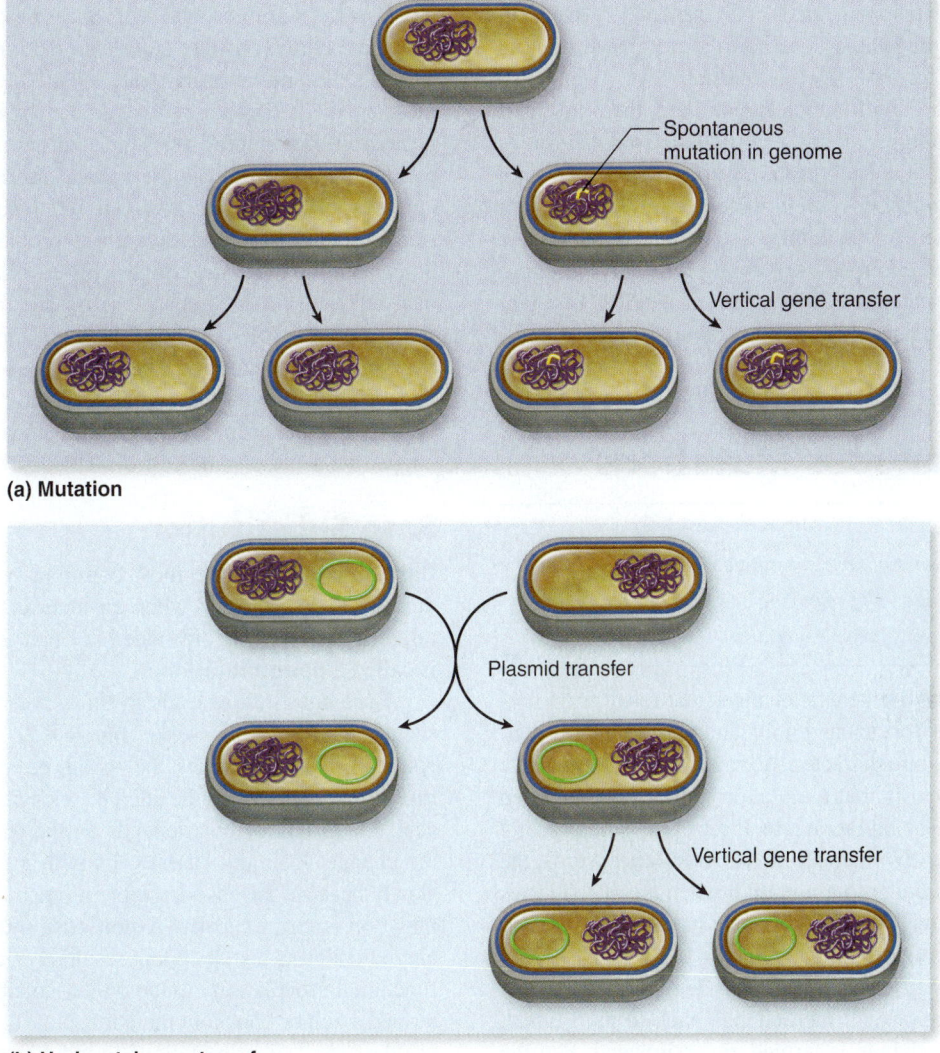

(a) **Mutation**

(b) **Horizontal gene transfer**

FIGURE 8.1 Mechanisms of Genetic Change in Bacteria (a) Mutation. **(b)** Horizontal transfer of plasmid-encoded genes; other DNA can be transferred horizontally as well.

❓ *Which of these mechanisms would have the most pronounced effect on an organism's genotype?*

changes the existing nucleotide sequence of a cell's DNA, which is then passed on to the progeny (daughter cells) through vertical gene transfer. The modified organism and daughter cells are referred to as mutants. **Horizontal gene transfer** is the acquisition of genes from another organism. Like mutations, the changes are then passed on to the progeny.

MicroAssessment 8.1

The properties of bacteria can change either through mutation or horizontal gene transfer.

1. *How is mutation different from horizontal gene transfer?*
2. *Contrast genotype and phenotype.*
3. *Which has a longer-lived effect on a cell—a change in the genotype or a change in the phenotype?* ✚

MUTATION AS A MECHANISM OF GENETIC CHANGE

Mutations can change an organism's phenotype. For example, if a gene required for biosynthesis of the amino acid tryptophan is deleted, then the organism can multiply only if tryptophan is supplied in the growth medium. The same occurs if the gene is disrupted so that the protein it encodes no longer functions properly. A mutant that requires a growth factor is an **auxotroph** (*auxo* means "increase," and *troph* means "nourishment"). This is in contrast to a **prototroph,** which does not require growth factors (*proto* means "earliest form of"). ◀◀ growth factor, p. 93

Geneticists working with mutants compare them to **wild type,** the typical phenotype of strains isolated from nature. A wild-type *E. coli* strain is a prototroph. By convention, a strain's characteristics are designated by three-letter abbreviations, with the first letter capitalized. For example, a strain that cannot make tryptophan is designated Trp⁻. For simplicity, only required growth factors are indicated. Likewise, only if a cell is resistant to an antibiotic is it indicated; streptomycin resistance is indicated as Str^R.

8.2 ■ Spontaneous Mutations

Learning Outcomes

3. *Describe three outcomes of base substitutions.*
4. *Describe the consequences of removing or adding nucleotides.*
5. *Explain how transposons cause mutations.*

Spontaneous mutations are genetic changes that result from normal cell processes. They occur randomly, and genes mutate spontaneously at infrequent but characteristic rates. The mutation rate is defined as the probability that a mutation will occur in a given gene per cell division. The mutation rate of different genes usually varies between 10^{-4} and 10^{-12} per cell division. In other words, the chance that a gene will undergo a mutation when a cell replicates its DNA prior to cell division is between one in 10,000 (10^{-4}) and one in a trillion (10^{-12}). ▶▶ exponents, Appendix 1, p. A-1

Mutations are passed on to a cell's progeny. On rare occasions, however, a mutation will then change back to its original, non-mutant state. This change is termed **reversion** and, like the original mutation, occurs spontaneously at low frequencies.

Because spontaneous mutations occur routinely, every large population contains mutants, so the cells in a colony are not necessarily identical. This gives a population the chance to adapt to changing environments. The environment does not cause the mutations but selects those cells that can grow under its conditions. For example, an organism with a spontaneous mutation to antimicrobial resistance, though rare, will become dominant in an environment where the medication is present. This happens because the antimicrobial kills the sensitive cells, allowing the resistant cells to grow without competition.

A single mutation is a rare event, so two mutations are even more unlikely. Physicians take advantage of this to prevent pathogens from developing resistance to certain antimicrobial medications. In tuberculosis treatment, for example, two or more antimicrobial drugs are given simultaneously. Any mutant bacterium resistant to one medication is probably still sensitive to the other. The chance of a single cell becoming resistant spontaneously to both medications is the product of the mutation rates of the two genes (calculated by taking the sum of the exponents). For example, if the mutation rate to "antibiotic X" resistance is 10^{-6} per cell division and the mutation rate to "antibiotic Y" resistance is 10^{-8}, then the probability that both mutations will spontaneously happen within the same cell is $10^{-6} \times 10^{-8}$, or 10^{-14}.

Base Substitution

Base substitution, the most common type of mutation, occurs during DNA synthesis when an incorrect nucleotide is incorporated (**figure 8.2**). If only one base pair is changed, the mutation is called a **point mutation.**

Base substitution leads to three possible mutation outcomes: silent, missense, or nonsense (**figure 8.3**). A **silent mutation** has a codon that still specifies the wild-type amino acid. A **missense mutation** results when the altered codon specifies a different amino acid. The effect of this depends on the position and the nature of the change. In many cases, cells with a missense mutation grow slowly because the encoded protein functions only partially. Such a mutation is termed leaky. A **nonsense mutation** occurs when the altered codon is a stop codon, resulting in a shorter and often nonfunctional protein. Any mutation that totally inactivates the gene is termed a null or knockout mutation. Note that geneticists sometimes use the term "silent mutation" to indicate a mutation that does not alter the function of the protein. With this broad definition, any base substitution that does not affect the phenotype would be a silent mutation. ◀◀ codon, p. 170 ◀◀ stop codon, p. 173

Base substitutions are more common in aerobic than anaerobic environments. This is because reactive oxygen species (ROS) such as superoxide and hydrogen peroxide are produced from O_2. These chemicals can oxidize the nucleobase guanine, and DNA polymerase often mispairs oxidized guanine with adenine rather than cytosine. ◀◀ reactive oxygen species, p. 90 ◀◀ nucleobase, p. 32

Deletion or Addition of Nucleotides

Deletion or addition of nucleotides during DNA replication also results in spontaneous mutations. The consequence of this depends on how many nucleotides are involved. If three nucleotide pairs are added (or deleted), this adds (or deletes) one codon. When the gene is expressed, one additional (or fewer) amino acid will be in the resulting protein. How serious the effect of this change is depends on its location in the encoded protein.

Adding or subtracting one or two nucleotide pairs causes a **frameshift mutation (figure 8.4)**. This changes the reading frame of the corresponding mRNA molecule so that an entirely different set of codons is translated. Frequently, one of the resulting downstream codons will be a stop codon. As a consequence, a frameshift mutation likely results in a shortened non-functional protein—a knockout mutation. ◀◀ reading frame, p. 171 ◀◀ downstream, p. 169

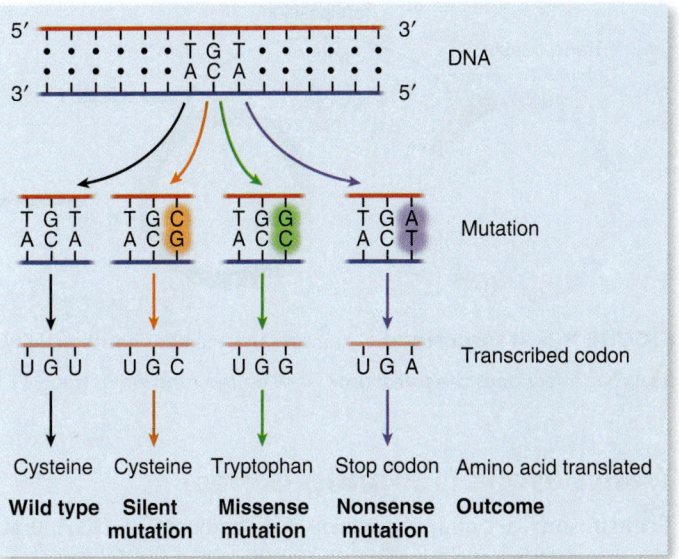

FIGURE 8.3 Potential Outcomes of Base Substitutions

❓ *Which of these outcomes is most likely to result in a leaky mutation?*

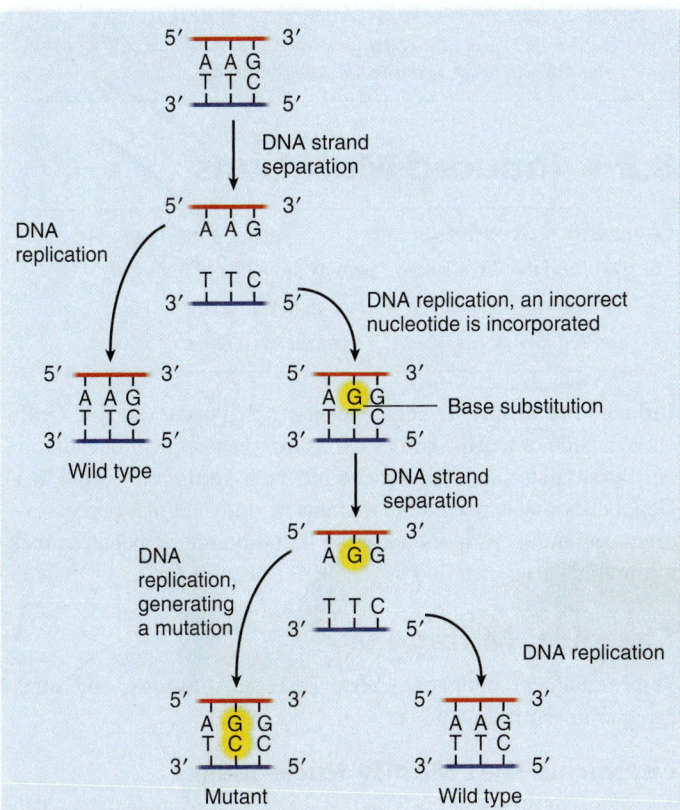

▶ **FIGURE 8.2 Base Substitution** A replication error results in a mismatch between the two DNA strands. Subsequent DNA replication using the altered strand as template results in a point mutation.

❓ *What enzyme incorporated the incorrect nucleotide?*

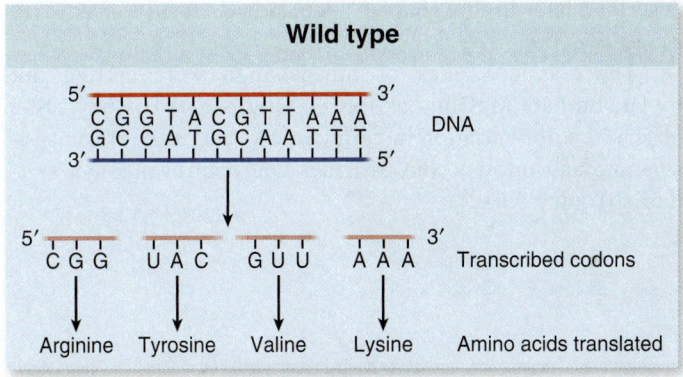

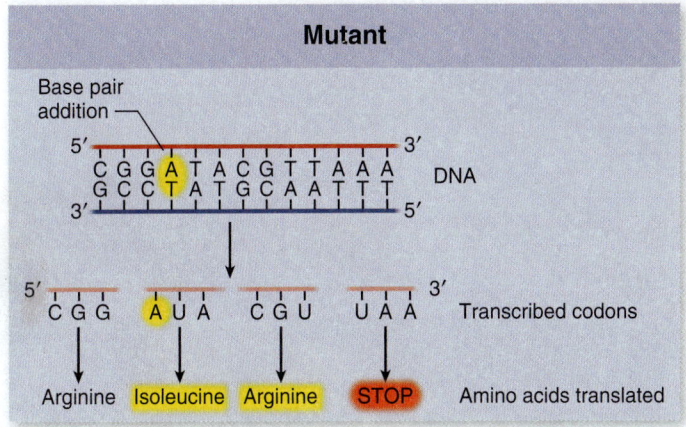

FIGURE 8.4 Frameshift Mutation as a Result of Nucleotide Addition The addition of a nucleotide pair (base pair) to the DNA results in a shift in the reading frame when the sequence is transcribed and translated. Deletion of a single nucleotide pair in the DNA would have a similar effect.

❓ *This figure shows a single nucleotide addition. What would happen if three nucleotides were added?*

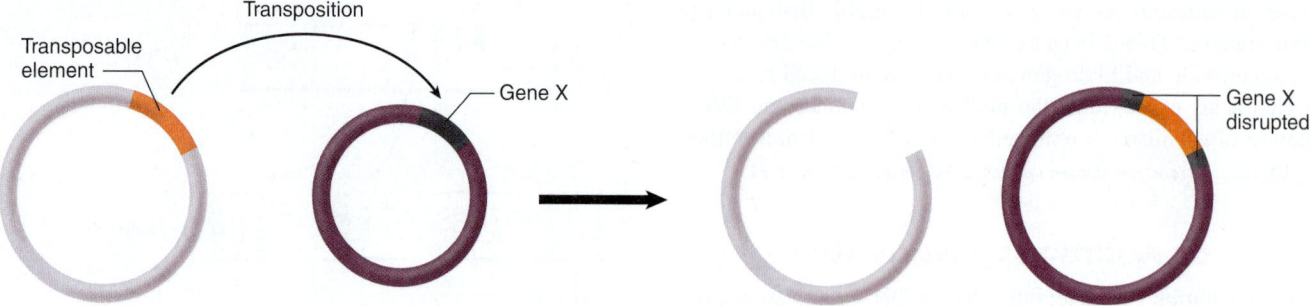

FIGURE 8.5 Transposition A transposable element has the ability to "jump" (transpose) from one piece of DNA into another.

❓ *What effect does the transposon have on the function of gene X?*

Transposons (Jumping Genes)

Transposons, or jumping genes, are segments of DNA that can move from one location to another in a cell's genome, a process called transposition (**figure 8.5**). The gene into which a transposon jumps is insertionally inactivated by the event, meaning that the insertion destroys the function of the gene. Most transposons contain transcriptional terminators, so the expression of downstream genes in the same operon will stop as well. The structure and biology of transposons will be described later in this chapter. ◄◄ transcriptional terminators, p. 170 ◄◄ operon, p. 179

The classic studies of transposition were carried out by Dr. Barbara McClintock (see **A Glimpse of History**). She observed color variation in corn kernels as a result of transposons moving into and out of genes that control pigment synthesis (**figure 8.6**).

FIGURE 8.6 Transposition Detected by Changes in Seed Color Variegation in the color of corn kernels is caused by insertion of transposable elements into genes involved in pigment synthesis.

MicroAssessment 8.2

Spontaneous mutations happen during normal cell processes and can alter the properties of the cell. A leaky mutation results in a partially functional protein; a knockout mutation results in a non-functional protein. Base substitutions that occur during DNA synthesis can lead to silent, missense, and nonsense mutations. Removing or adding nucleotides can cause a frameshift mutation. Transposons can "jump" from one location to another.

4. *Which is generally more severe—a missense mutation or a nonsense mutation?*

5. *What is the likely consequence of a frameshift mutation?*

6. *Is it as effective to take two antibiotics sequentially as it is to take them simultaneously, as long as the total length of time that they are both taken is the same? Explain.* ✚

8.3 ■ Induced Mutations

Learning Outcomes

6. *Describe the three general groups of chemical mutagens.*
7. *Explain why transposons induce mutations.*
8. *Explain how X rays and UV light damage DNA.*

Induced mutations are genetic changes that occur due to an influence outside of a cell, such as exposure to a chemical or radiation. An agent that induces the change is a **mutagen** (**table 8.1**). Geneticists—who depend on mutants to study cellular processes—often use mutagens to increase the mutation rate in bacteria, making mutants easier to find.

Chemical Mutagens

Some chemical mutagens cause base substitutions, and others cause frameshift mutations.

Chemicals That Modify Nucleobases

A number of different chemicals modify the nucleobases in DNA, changing their base-pairing properties. This increases the chance that the incorrect nucleotide will be incorporated during DNA replication. For example, nitrous acid (HNO_2) converts cytosine to

TABLE 8.1	Common Mutagens		
Agent	**Action**		**Result**
Chemical Agent			
Chemicals that modify nucleobases			
Nitrous acid	Converts cytosine to uracil		Nucleotide substitution
Alkylating agents	Adds alkyl groups (CH_3 and others) to nucleobases		Nucleotide substitution
Base analogs	Used in place of normal nucleobases in DNA		Nucleotide substitution
5-Bromouracil			
Intercalating agents	Inserts between base pairs		Addition or subtraction of nucleotides
Ethidium bromide			
Transposons	Randomly insert into DNA		Insertional inactivation
Radiation			
Ultraviolet (UV) light	Causes intrastrand thymine dimer to form		Errors during repair process
X rays	Cause single- and double-strand breaks in DNA		Deletions

uracil, which base-pairs with adenine rather than guanine. It removes amino groups from adenine and guanine as well. A group of chemicals called alkylating agents adds alkyl groups (short chains of carbon atoms) onto nucleobases. An example is nitroso-guanidine, which adds a methyl group to guanine, causing it to base-pair with thymine (**figure 8.7**).

MicroByte

Many chemical mutagens are used in cancer therapy to kill rapidly dividing cancer cells. Unfortunately, they also damage DNA in normal cells.

Base Analogs

Base analogs structurally resemble nucleobases but have different hydrogen-bonding properties. The analogs can be mistakenly used in place of the nucleobases when nucleotides are made, and DNA polymerase then incorporates these into DNA. When a DNA strand has a base analog, the wrong nucleotide can be incorporated when the complementary strand is synthesized. For example, 5-bromouracil resembles thymine, but it often base-pairs with cytosine; 2-amino purine resembles adenine but often pairs with C instead of T (**figure 8.8**). ◄◄ DNA replication, p. 165

FIGURE 8.7 Mutagenic Effects of the Alkylating Agent Nitrosoguanidine (a) Nitrosoguanidine converts guanine bases in DNA to methylguanine, which can base-pair with thymine (T) as well as cytosine (C). (b) The altered base-pairing property of methylguanine can result in point mutations following DNA replication.

? *What are the three possible outcomes of point mutations?*

Hydrogen bonds formed with complementary bases

CH₃ — Added alkyl group

Alkylating agent

deoxyribose   deoxyribose

Guanine (pairs with C)   **Methylguanine** (sometimes pairs with T)

(a)

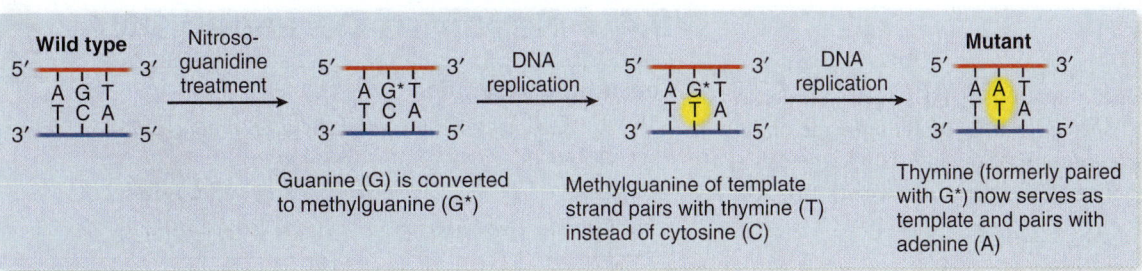

(b)

Wild type   Nitroso-guanidine treatment

5′ A G T 3′ / 3′ T C A 5′ → 5′ A G*T 3′ / 3′ T C A 5′

DNA replication → 5′ A G*T 3′ / 3′ T T A 5′

DNA replication → **Mutant** 5′ A A T 3′ / 3′ T T A 5′

Guanine (G) is converted to methylguanine (G*)

Methylguanine of template strand pairs with thymine (T) instead of cytosine (C)

Thymine (formerly paired with G*) now serves as template and pairs with adenine (A)

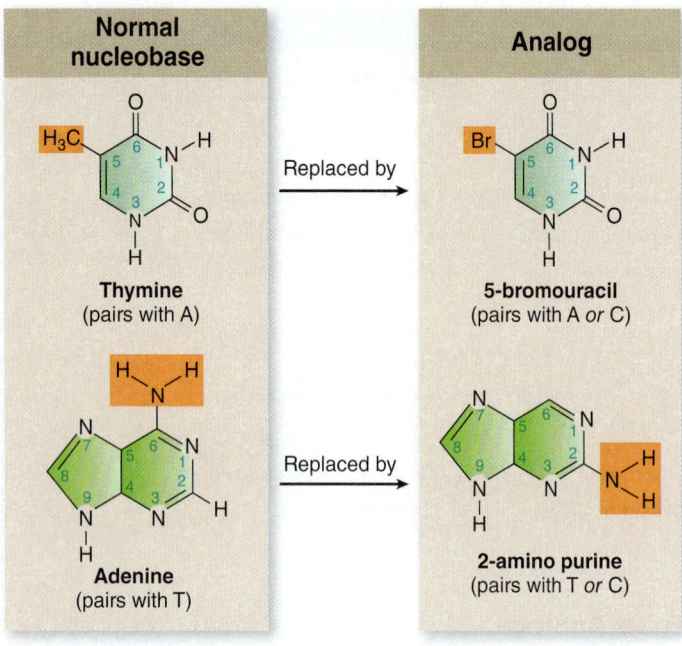

FIGURE 8.8 Two Base Analogs Used in Mutagenesis The altered base-pairing properties of the analogs can lead to point mutations by the same mechanism shown in figure 8.7.

❓ *What base-pair substitution would 5-bromouracil generate?*

Intercalating Agents

Intercalating agents increase the frequency of frameshift mutations. They do this because they are flat molecules that can intercalate (insert) between adjacent base pairs in a strand of DNA. This pushes the nucleotides apart, producing enough space between bases that errors are made during replication. If the intercalating agent inserts into the template strand, a base pair will be added as the new strand is synthesized. If it intercalates into the strand being synthesized, a base pair will be deleted. As in spontaneous frameshift mutants, adding or subtracting a nucleotide in DNA often results in a stop codon being generated prematurely in the mRNA transcript, giving rise to a shortened protein.

Ethidium bromide, a chemical commonly used to stain DNA in the laboratory, is an intercalating agent. The manufacturer now warns users that the chemical should be handled with great care because it likely is a carcinogen (cancer-causing). Another intercalating agent is chloroquine, which has been used for many years to treat malaria.

Transposition

Transposons can be introduced intentionally into a cell in order to generate mutations. The transposon, which cannot replicate on its own because it lacks an origin of replication, inserts into the cell's genome. This generally inactivates the gene into which it inserts (see figure 8.5).

Radiation

Two kinds of radiation are commonly used as mutagens: ultraviolet (UV) light and X rays. ◀◀ wavelengths of radiation, p. 115

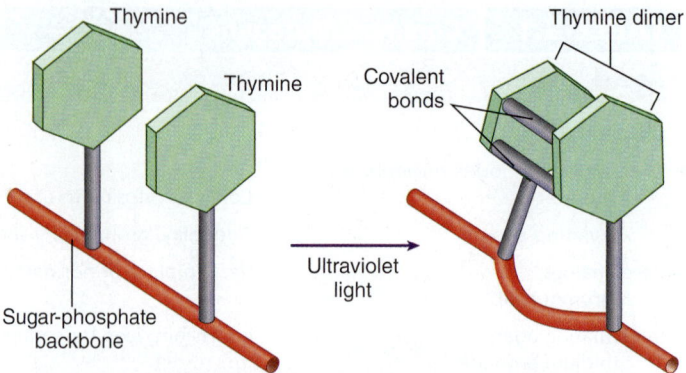

FIGURE 8.9 Thymine Dimer Formation UV light causes covalent bonds to form between adjacent thymine molecules on the same strand of DNA, distorting the shape of the DNA.

❓ *What effect does a thymine dimer have on DNA synthesis?*

Ultraviolet Light

Irradiation of cells with ultraviolet light causes covalent bonds to form between adjacent thymine molecules on a DNA strand, producing **thymine dimers** (**figure 8.9**). The dimer cannot fit properly into the double helix, distorting the DNA molecule. Replication and transcription stall at the distortion, and as a result, the cells will die if the damage is not repaired. How then can UV light be mutagenic? UV mutagenizes cells indirectly. Its major mutagenic action results from the cell's attempting to repair the damage by SOS repair, described in the next section.

X Rays

X rays cause single- and double-strand breaks in DNA, and alterations to the nucleobases. Double-strand breaks often result in deletions that are lethal to the cell.

MicroAssessment 8.3

Mutagens increase the frequency of mutations. Some chemical mutagens cause base substitutions, but intercalating agents cause frameshift mutations. Mutations from UV light are due to SOS repair; X rays cause breaks in DNA strands and alter nucleobases.

7. *How does an intercalating agent cause mutations?*
8. *What mutagen causes thymine dimers, and why does it kill cells?*
9. *Do you think mutations caused by reactive oxygen species should be considered spontaneous or induced? Justify your answer.* ➕

8.4 ■ Repair of Damaged DNA

Learning Outcomes

9. *Explain how errors in nucleotide incorporation can be repaired and the role of methylation in the process.*
10. *Explain how modified nucleobases can be repaired.*
11. *Describe three mechanisms cells use to repair thymine dimers.*

The amount of spontaneous and mutagen-induced damage to DNA is enormous. This damage, if not repaired, can quickly lead to cell death and, in animals, cancer. In humans, two breast cancer susceptibility genes code for enzymes that repair damaged DNA.

Mutations in either gene result in a high probability (80%) of breast cancer.

Mutations are rare because alterations in DNA are generally repaired before they can be passed on to progeny. Over the many millions of years of evolution, cells have developed several different DNA repair mechanisms (**table 8.2**). Eukaryotic and prokaryotic cells share many of the same general mechanisms. Note that cells cannot repair all types of mutations, such as insertional inactivation caused by transposition.

MicroByte

Every 24 hours, the genome of every cell in the human body is damaged more than 10,000 times.

Repair of Errors in Nucleotide Incorporation

DNA polymerase sometimes incorporates the wrong nucleotide as it replicates DNA. The resulting mispairing of nucleobases results in a slight distortion in the DNA helix, which can be recognized by enzymes within the cell that then repair the mistake. By quickly repairing the error before the DNA is replicated, the cell prevents the mutation. Two mechanisms for this are proofreading by DNA polymerase and mismatch repair.

◄◄ DNA polymerase, p.166

Proofreading by DNA Polymerase

DNA polymerases are complex enzymes that not only synthesize DNA, but also verify the accuracy of their actions—a characteristic called **proofreading.** The enzymes can back up and excise (remove) a nucleotide not correctly hydrogen bonded to the opposing nucleobase in the template strand. The DNA polymerase then incorporates the correct nucleotide. Although the proofreading function of DNA polymerases is very efficient, it is not flawless.

Mismatch Repair

Mismatch repair fixes errors missed by the proofreading of DNA polymerase. A specific protein binds to the site of the mismatched nucleobase, directing an enzyme to cut the sugar-phosphate backbone of the strand. Another enzyme then degrades a short region of that DNA strand, thereby eliminating the misincorporated nucleotide. How does the cell know which strand to remove? This is an important question because if the enzyme cuts the template strand and not the new one, then the misincorporated nucleotide would remain. The key to the answer lies in methylation of the DNA nucleobases. Soon after a DNA strand is synthesized, an enzyme adds methyl groups to certain nucleobases. This takes time, however, so the new strand is still unmethylated immediately

TABLE 8.2	Repair of Damaged DNA			
	Type of Defect	**Repair Mechanism**	**Biochemical Mechanism**	**Result**
Spontaneous	Wrong nucleotide incorporated during DNA replication	Proofreading by DNA polymerase	Mispaired nucleotide is removed by DNA polymerase.	Potential mutation eliminated
		Mismatch repair	A protein binds to the site of mismatch and cuts the unmethylated strand. A short stretch of that strand is then degraded and DNA polymerase synthesizes a replacement.	Potential mutation eliminated
	Oxidized guanine in DNA	Action of glycosylase	Glycosylase removes the oxidized guanine. A short stretch of that strand is then degraded, and DNA polymerase synthesizes a replacement.	Potential mutation eliminated
Mutagen-Induced				
Chemical	Wrong nucleotide incorporated during DNA replication	Proofreading and mismatch repair	Same as for spontaneous mutations (see proofreading and mismatch repair)	Potential mutation eliminated
UV light	Thymine dimer formation	Photoreactivation (light repair)	Breaks the covalent bond between the thymine molecules	Original DNA molecule restored
		Excision repair (dark repair)	A short stretch of the strand containing the thymine dimer is removed; DNA polymerase then synthesizes a replacement.	Potential mutation eliminated
		SOS repair	A special DNA polymerase synthesizes DNA even when the template is damaged.	Cell survives but numerous mutations produced

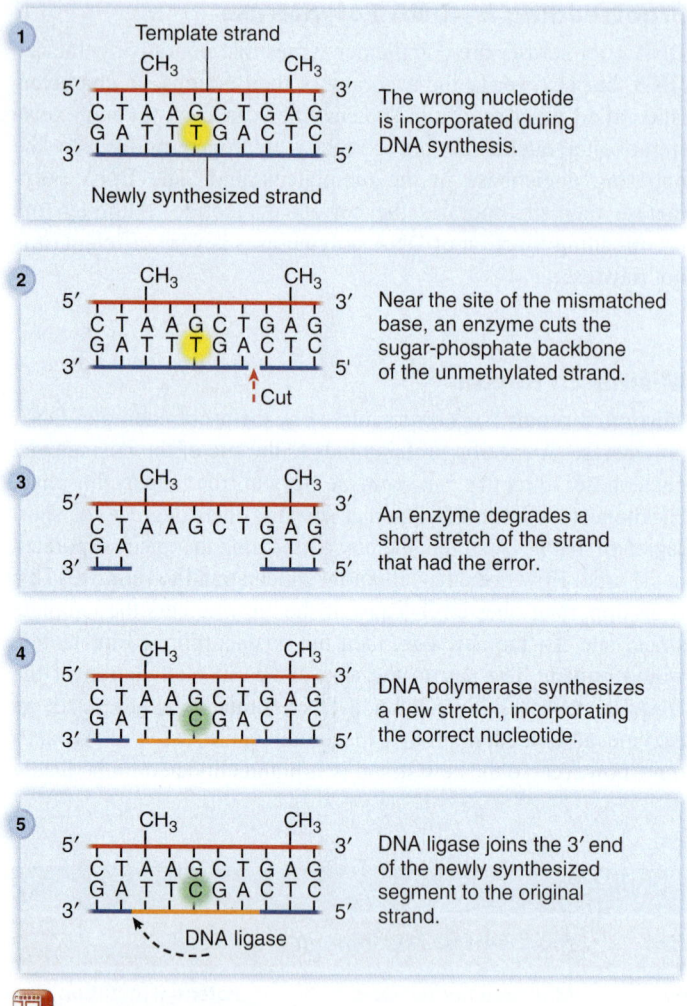

FIGURE 8.10 Mismatch Repair

What role does methylation play in mismatch repair?

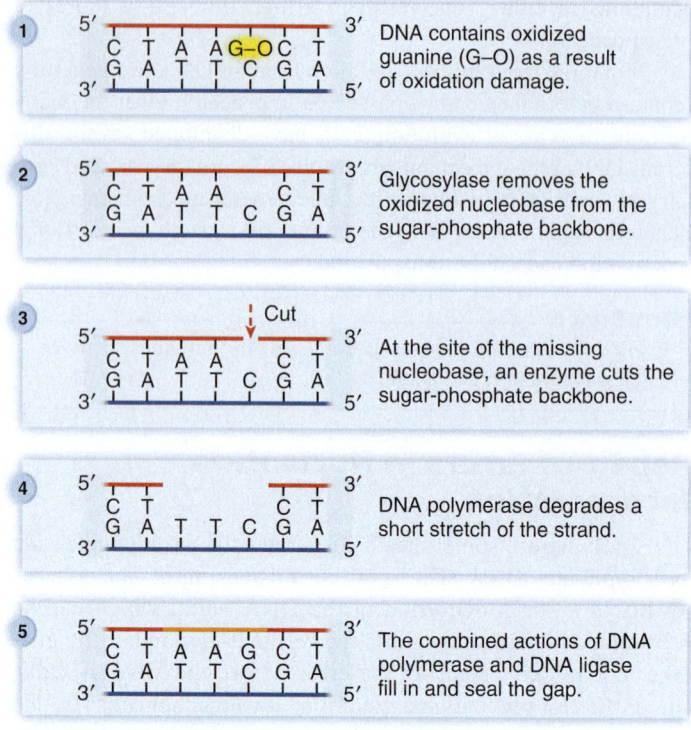

FIGURE 8.11 Repair of Oxidized Guanine

What would be the effect on cells if they did not have the glycosylase enzyme?

MicroByte

In human cells, defects in either mismatch repair or repair of modified nucleobases increases the incidence of colon cancers.

after it is synthesized. Therefore, the template strand is methylated, whereas the new strand is not, allowing the repair enzyme to distinguish between the two (**figure 8.10**). After the nucleotides are removed from the new strand, the combined actions of DNA polymerase and DNA ligase then fill in that section and seal the gap. ◀◀ DNA ligase, p. 167

Repair of Modified Nucleobases in DNA

Modified nucleobases such as oxidized guanine can result in base substitutions if they are not repaired before the DNA is replicated. An important mechanism for repairing this defect uses a glycosylase, an enzyme that removes the oxidized nucleobase from the sugar-phosphate backbone (**figure 8.11**). Another enzyme then recognizes that a nucleobase is missing and cuts the DNA at this site. DNA polymerase degrades a short section of this strand to remove the damage. This same enzyme synthesizes another strand with the proper nucleotides, and DNA ligase seals the gap in the single-stranded DNA.

Repair of Thymine Dimers

Bacteria have several mechanisms to combat the DNA-damaging effects of UV light, a component of sunlight. In one mechanism, an enzyme uses the energy of visible light to break the covalent bonds of the thymine dimer, restoring the DNA to its original state (**figure 8.12**). Because light is required for this mechanism, it is called **photoreactivation**, or light repair. The enzyme used is found only in prokaryotes.

Some bacteria have an enzyme that recognizes the major distortions in DNA that result from thymine dimer formation. In this process, **excision repair**, or dark repair, the enzyme makes single-stranded cuts that flank both sides of the damaged region to remove the strand (figure 8.12). The actions of DNA polymerase and DNA ligase then fill in and seal the gap left by the removal of the segment.

SOS Repair

SOS repair is a last-ditch repair mechanism that copes with extensively damaged DNA. The enzymes that carry out this repair are induced when DNA is so heavily damaged by UV light that photoreactivation and excision repair may not be able to correct all of the damage. DNA and RNA polymerases then stall at sites of unrepaired damage, so the cells cannot replicate or transcribe their DNA. Without SOS repair, the cells would die.

Damaged DNA activates expression of the several dozen genes that encode the SOS system. One component of this system is a DNA polymerase that synthesizes DNA even in extensively damaged regions. Unlike the standard DNA polymerases, however, the SOS DNA polymerase has no proofreading ability. Errors are made as a result, a process called SOS mutagenesis.

MicroAssessment 8.4

DNA polymerases have proofreading ability. Mismatch repair fixes errors missed by the proofreading mechanism; methylation distinguishes the template strand. Specific glycosylases can remove modified nucleobases. Thymine dimers can be repaired through photoreactivation and excision repair; severe damage can be overcome by the SOS repair system.

10. *Distinguish between photoreactivation and excision repair of thymine dimers.*
11. *How does UV light cause mutations?*
12. *To maximize the number of mutations following UV irradiation, should you incubate the irradiated cells in the light or in the dark, or does it make any difference? Explain your answer.*

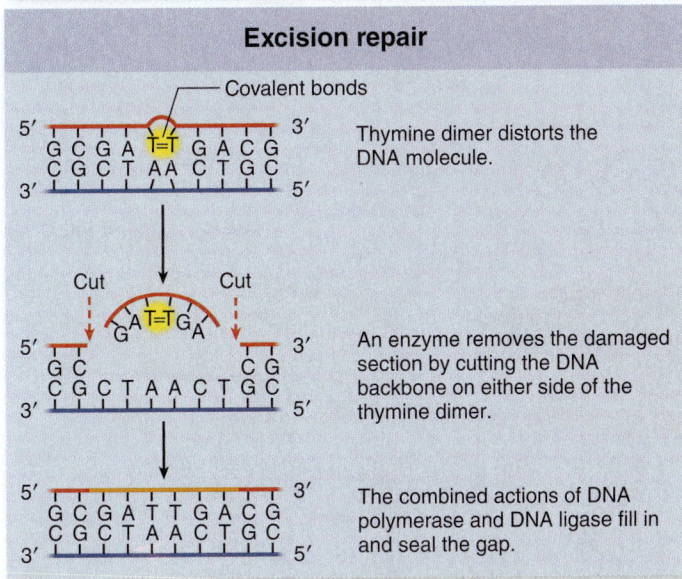

FIGURE 8.12 Repair of Thymine Dimers In photoreactivation, an enzyme uses the energy of light to break the covalent bonds. In excision repair, the section containing the dimer is replaced.

? *How is the mechanism of excision repair similar to that of mismatch repair?*

8.5 ■ Mutant Selection

Learning Outcomes

12. *Compare and contrast direct and indirect selection.*
13. *Describe how direct selection is used to screen for possible carcinogens.*

Mutations are rare events, even when mutagens are used, which presents a challenge to a scientist who wants to isolate a desired mutant. In a culture containing several billion cells, perhaps only one cell has the sought-after mutation. How can that rare cell be isolated and identified? Depending on the type of mutant being sought, either direct or indirect selection can be used.

Direct Selection

Mutants that can grow under conditions in which the parent cells cannot are usually easy to isolate by **direct selection.** In this method, cells are inoculated onto an agar medium that supports the growth of the mutant, but not the parent. For example, antibiotic-resistant mutants can be easily selected directly by inoculating cells onto a medium containing the antibiotic. Only the resistant cells will form colonies (**figure 8.13**).

Indirect Selection

Indirect selection is used to isolate an auxotroph from a prototrophic parent strain. This process is more difficult than direct selection because no medium will allow the growth of the mutant and not the parent. For example, Trp⁻ mutants can grow

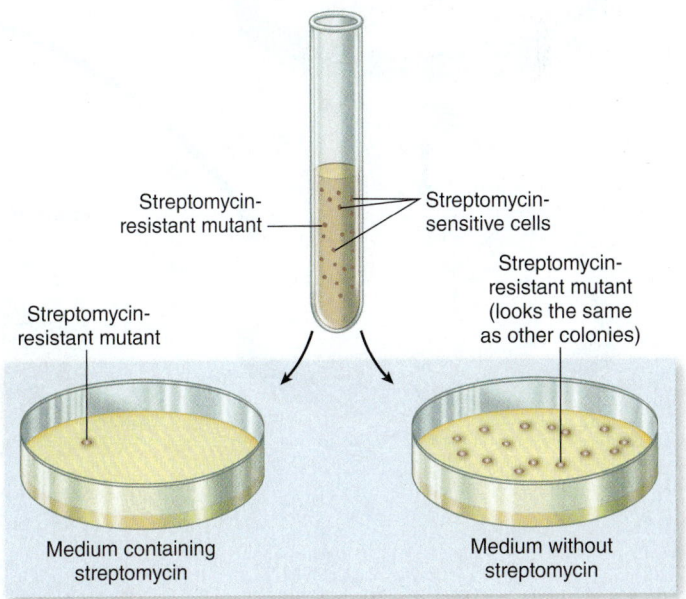

FIGURE 8.13 Direct Selection of Mutants Only cells carrying a mutation that confers resistance to streptomycin can grow on the selective medium used here.

? *Direct selection cannot be used to isolate auxotrophic mutants in a culture of prototrophic parent cells. Why not?*

only on a complex medium such as nutrient agar because this supplies the tryptophan they require, but Trp$^+$ parent cells also grow on this same medium. To overcome this problem, a technique called replica plating is used, sometimes preceded by penicillin enrichment of mutants. ◀◀ **complex medium, p. 94** ◀◀ **nutrient agar, p. 94**

Replica Plating

Replica plating is a clever method for indirect selection of auxotrophic mutants, devised by Joshua and Esther Lederberg in the early 1950s (**figure 8.14**). In this technique, the bacterial culture is first spread onto a nutrient agar plate. Mutant and non-mutant cells will grow on this medium to form colonies, creating what is referred to as the master plate. ① The master plate is then pressed onto sterile velvet, a fabric with tiny threads that stand on end like small bristles. The velvet picks up some cells of every colony. ② Next, two agar plates—one nutrient agar and one glucose-salts agar—are pressed in succession and in the same orientation onto the same velvet. This transfers cells taken from the master plate to both the nutrient agar and the glucose-salts agar, creating replica plates. A line on the plates is used to maintain a consistent orientation. ③ The replica plates are then incubated, allowing cells to grow to form colonies; prototrophs grow on both types of media,

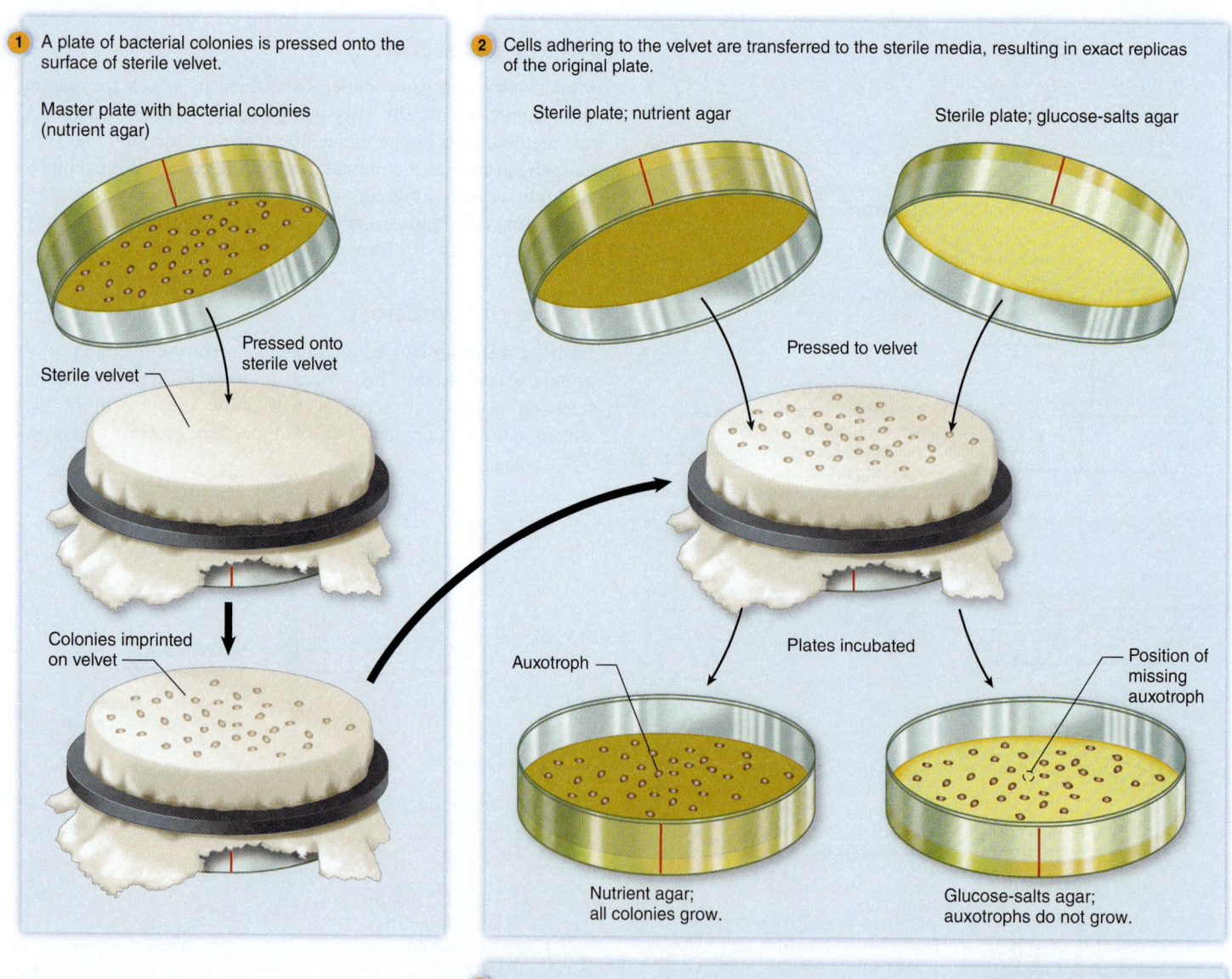

① A plate of bacterial colonies is pressed onto the surface of sterile velvet.

Master plate with bacterial colonies (nutrient agar)

Sterile velvet

Pressed onto sterile velvet

Colonies imprinted on velvet

② Cells adhering to the velvet are transferred to the sterile media, resulting in exact replicas of the original plate.

Sterile plate; nutrient agar

Sterile plate; glucose-salts agar

Pressed to velvet

Plates incubated

Auxotroph

Position of missing auxotroph

Nutrient agar; all colonies grow.

Glucose-salts agar; auxotrophs do not grow.

③ Auxotrophic mutants form colonies on the nutrient agar but not on the glucose-salts agar.

FIGURE 8.14 Indirect Selection of Mutants by Replica Plating The procedure shown was used by the Lederbergs and continues to be used today in many laboratories.

? *Why go to the trouble of creating a master plate (why not simply plate the initial culture on both nutrient agar and glucose-salts agar)?*

but auxotrophs grow only on the nutrient agar; recall that glucose-salts is a chemically defined medium with no added growth factors (see table 4.7). ◀◀ glucose-salts, p. 95

The plates are exact replicas, so colonies on the master plate that cannot grow on glucose-salts can be identified; these are auxotrophs. The particular growth factor required can then be determined by adding nutrients individually to a glucose-salts medium and determining which one promotes growth.

Penicillin Enrichment of Mutants

Penicillin enrichment is sometimes used before replica plating to increase the proportion of auxotrophs in a broth culture. This is helpful because even when mutagenic agents are used, the frequency of mutations in a particular gene is low, sometimes less than one in 100 million cells. By increasing the proportion of mutant cells, it is easier to isolate them.

Penicillin enrichment relies on the fact that penicillin kills only growing cells. The mutagenized cells are incubated in glucose-salts broth containing penicillin. Prototrophs can multiply in this medium so most are killed, whereas the non-multiplying auxotrophs survive (**figure 8.15**). The enzyme

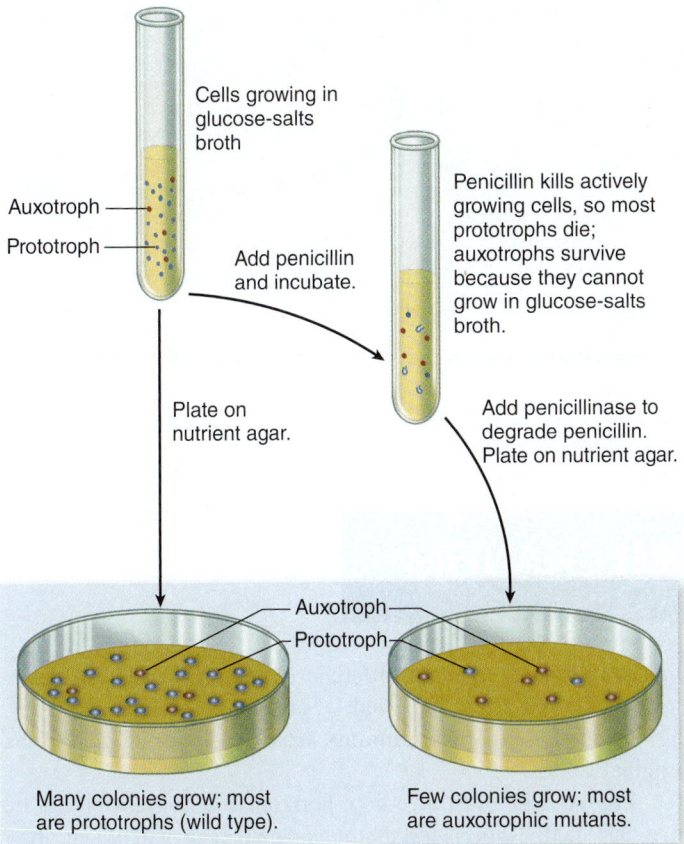

FIGURE 8.15 Penicillin Enrichment of Mutants

❓ *Penicillin causes the growing cells to lyse, releasing all of their contents into the medium. How does this complicate penicillin enrichment?*

penicillinase is then added to destroy the penicillin, and the cells plated on nutrient agar to create the master plate used in replica plating. ▶▶ penicillin, p. 463

Screening for Possible Carcinogens

A substantial proportion of cancers appear to be caused by chemicals that are **carcinogens** (meaning "cancer generating"), and most carcinogens are mutagens. Thousands of chemicals—including pesticides, herbicides, hair dyes, cosmetics, food additives, and the by-products of manufacturing processes—are released into the environment, so how can they all be tested for carcinogenic activity? Testing in animals for tumor formation takes 2 to 3 years and may cost $100,000 or more to test a single compound. To simplify the process, a number of rapid and relatively inexpensive tests have been developed to screen for possible carcinogens. All examine the mutagenic effect of the chemical in a microbiological system. The **Ames test,** devised by Bruce Ames and his colleagues in the 1960s, illustrates the concept of such tests. It relies on the fact that mutagens increase the frequency of spontaneous reversions.

The Ames test uses direct selection to determine the effect of a test chemical on the reversion rate of a histidine-requiring auxotroph of *Salmonella* (**figure 8.16**). If the chemical is mutagenic, the reversion rate will increase relative to that observed when no chemical is added (the control). The test also gives some idea about how hazardous the chemical may be by the number of revertants that arise.

Many chemicals are not carcinogenic themselves but are converted to carcinogens by enzymatic reactions in animals. Bacterial cells may not produce these enzymes, so an extract of ground-up rat liver—which has the enzymes to carry out these conversions—is also added to the suspected mutagen.

Additional testing must be done on any mutagenic chemical to determine if it is carcinogenic. Although data are not available on the percentage of mutagens that are carcinogens, it is clear that the Ames test is useful as a rapid screening test. So far, only compounds that increase the reversion rate have been shown to be carcinogenic.

MicroByte

In 2010, a government advisory panel concluded that environmental pollutants are greatly underestimated as a cause of cancer.

MicroAssessment 8.5

Mutants can be selected using either direct or indirect techniques. Replica plating is used for indirect selection, sometimes preceded by penicillin enrichment. The Ames test is used to screen chemicals to determine which ones are possible carcinogens.

13. *Distinguish between the kinds of mutants that can be isolated by direct and indirect selection.*

14. *How does penicillin enrichment make it easier to isolate mutants?*

15. *How could you demonstrate by replica plating that the environment selects but does not mutate genes in bacteria?* ➕

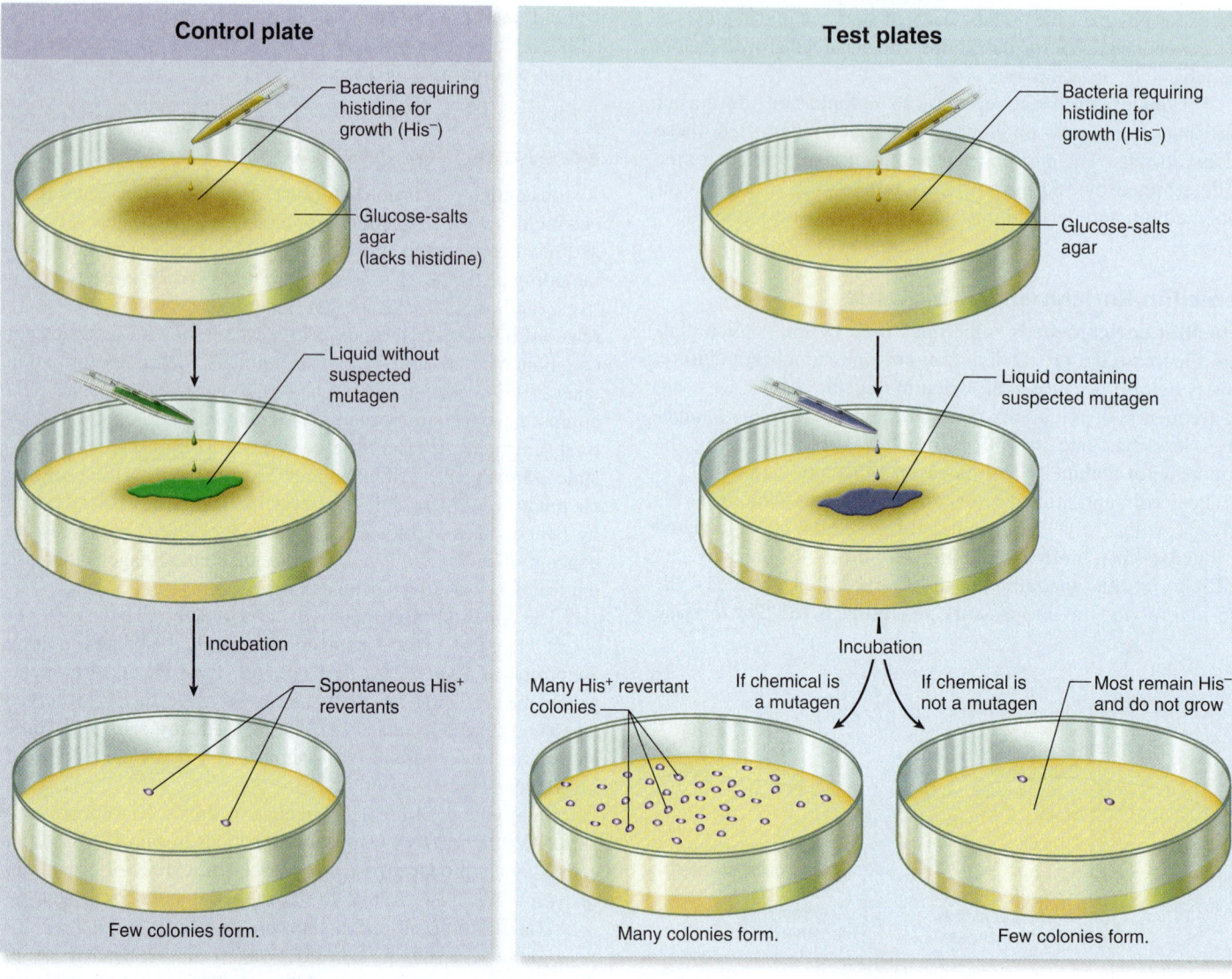

FIGURE 8.16 Ames Test to Screen for Mutagens A control plate determines the spontaneous rate of reversion to His⁺. If the chemical tested is a mutagen, it will increase the rate of His⁺ reversion mutations.

❓ *Why do some cells grow on the control plate in the absence of a mutagen?*

HORIZONTAL GENE TRANSFER AS A MECHANISM OF GENETIC CHANGE

Microorganisms commonly acquire genes from other cells, the process of horizontal gene transfer. The movement of DNA from one cell (the donor) to another (the recipient) is largely responsible for the rapid spread of antibiotic resistance, such as described for *Staphylococcus aureus* earlier in this chapter.

Horizontal gene transfer can be studied only if donor and recipient cells are genetically different, which is one reason why

mutants are useful. These differences make it possible to determine if the recipients have indeed acquired new characteristics. The resulting cells, **recombinants,** have properties of each of the original strains.

Figure 8.17 illustrates how horizontal gene transfer can be demonstrated. Two bacterial strains—neither of which can grow on a glucose-salts medium because of multiple growth factor

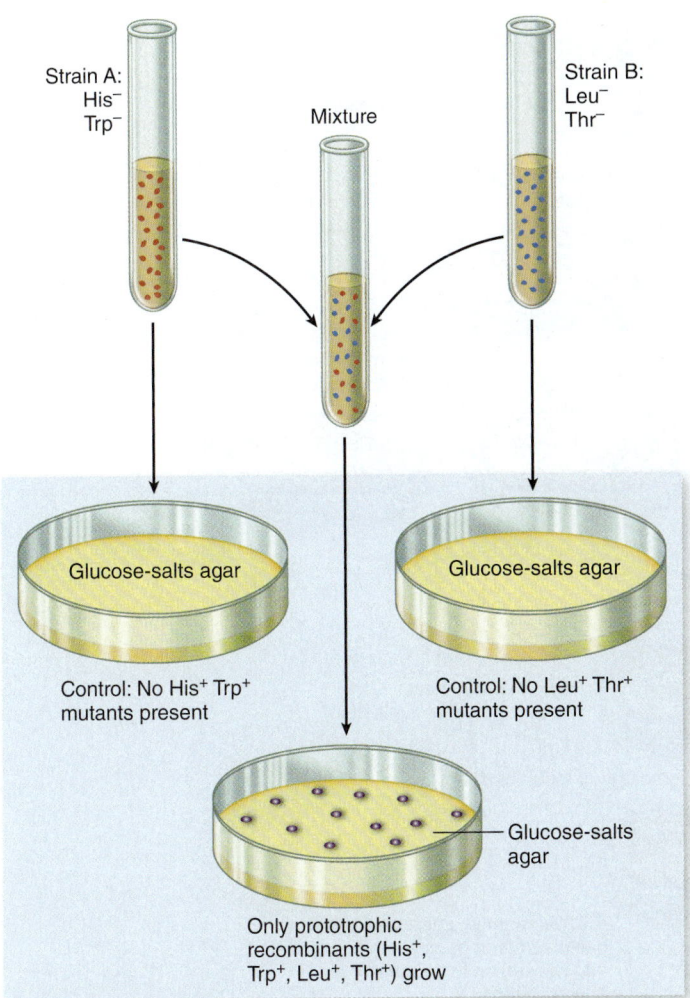

FIGURE 8.17 Experimental Demonstration of Horizontal Gene Transfer in Bacteria Recombinant colonies exhibit genetic traits from both strains present in the mixture. Control plates demonstrate that these colonies are not a result of spontaneous mutation.

❓ *When demonstrating horizontal gene transfer, why is it important to use strains that each require at least two different amino acids?*

requirements—are used. Strain A is His⁻ (requires histidine) and Trp⁻ (requires tryptophan). Strain B is Leu⁻ and Thr⁻, meaning that it requires leucine and threonine. Neither population is likely to give rise to a spontaneous mutant that grows on the glucose-salts agar because two simultaneous mutations in the same cell would be required. The strains are mixed and then spread on a glucose-salts agar plate. If colonies form, it suggests that cells of one strain acquired genes from cells of the other strain.

TABLE 8.3	Mechanisms of DNA Transfer		
Mechanism	**Main Feature**	**Size of DNA Transferred**	**Sensitivity to DNase Addition***
Transformation	Naked DNA transferred	About 20 genes	Yes
Transduction	DNA enclosed in a bacteriophage coat	Small fraction of the chromosome	No
Conjugation			
Plasmid transfer	Cell-to-cell contact required	Entire plasmid	No
Chromosome transfer	Cell-to-cell contact required; only Hfr cells can be donors	Variable fraction of chromosome	No

*DNase is an abbreviation of deoxyribonuclease, an enzyme that degrades DNA.

Genes can be transferred from a donor to a recipient by three different mechanisms (**table 8.3**):

1. **DNA-mediated transformation:** "Naked" DNA is taken up by a cell.
2. **Transduction:** A virus injects bacterial DNA into a cell.
3. **Conjugation:** DNA is transferred during cell-to-cell contact.

Following gene transfer, recipient cells must replicate the DNA to pass it on to daughter cells. This can happen only if the DNA is a **replicon,** meaning it has an origin of replication. Plasmids and chromosomes are replicons, but fragments of chromosomal DNA are not. If a chromosomal fragment is transferred, then it must become part of a replicon to be maintained in a population (**figure 8.18**). This involves a process called **homologous recombination,** which can happen only if the donor DNA is similar in nucleotide sequence to a region in the recipient cell's genome. In homologous recombination, the donor DNA becomes positioned next to the complementary region of the recipient cell's DNA. The donor DNA then replaces a homologous segment of recipient DNA, and the DNA it replaced is degraded. The molecular mechanisms involved in homologous recombination are still not understood, and different mechanisms likely operate in different situations. ◀◀ origin of replication, p. 166

MicroByte

Fungal genes have been found in an aphid's DNA, an extreme example of horizontal gene transfer.

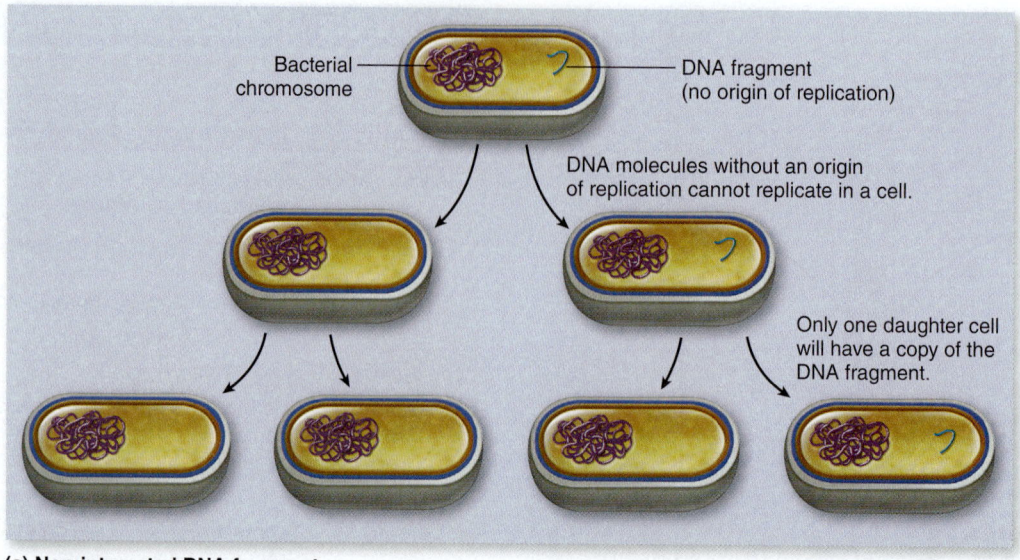

(a) Non-integrated DNA fragment

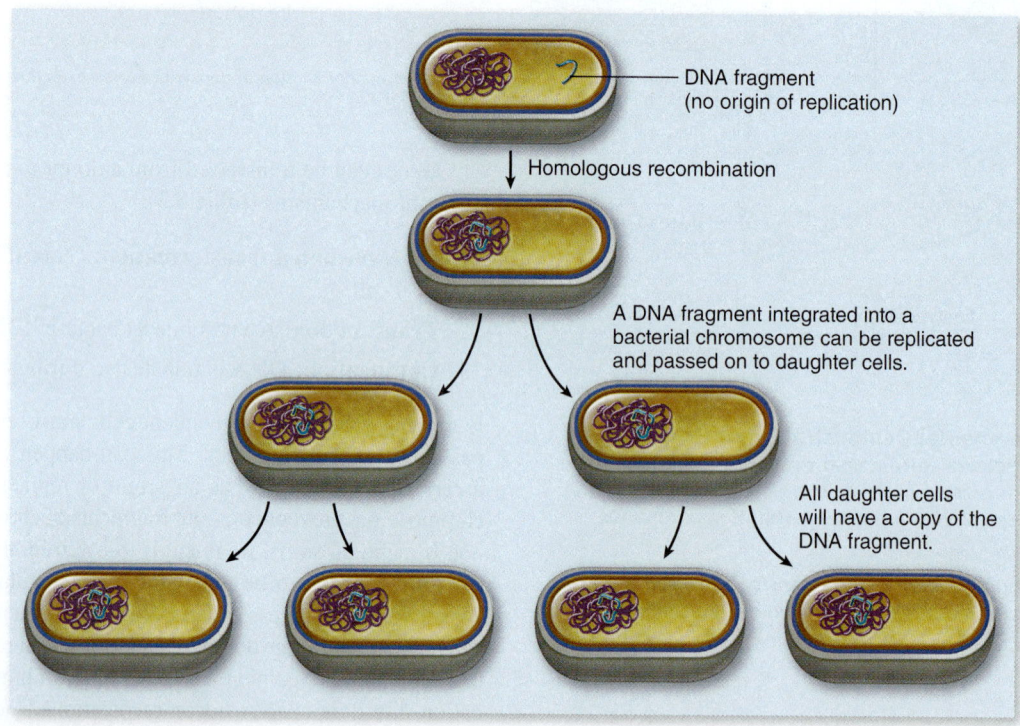

(b) Integrated DNA fragment

FIGURE 8.18 DNA Must Be Part of a Replicon to Be Maintained in a Population (a) DNA without an origin of replication will not be passed on to any new daughter cells during growth of the population. **(b)** If the DNA becomes integrated within a replicon of the cell, it will be inherited by all daughter cells.

❓ *If the DNA fragment encodes penicillin resistance, how could you experimentally distinguish between (a) and (b)?*

8.6 ■ DNA-Mediated Transformation

Learning Outcome

14. *Describe the process of DNA-mediated transformation, including the role of competent cells.*

DNA-mediated transformation, commonly referred to as transformation, involves the uptake of "naked" DNA by recipient cells (see **Perspective 8.1**). Naked DNA is simply free in the cells' surroundings; it is not contained within a cell or a virus. The fact that the DNA is naked can be demonstrated by adding DNase (an enzyme that degrades DNA) to the medium. This prevents transformation, indicating that the process requires naked DNA.

The Biological Function of DNA: A Discovery Ahead of Its Time

In the 1930s, it was well known that DNA was present in all cells, including bacteria. Its function, however, was a mystery. Since it consists of only four repeating subunits, most scientists did not believe it could be a very important molecule. Its critical biological role was discovered through a series of experiments conducted during a 20-year period by scientists in England and the United States.

In the 1920s, Frederick Griffith, an English bacteriologist, was studying pneumococci, the bacteria that cause pneumonia. It was known that pneumococci could cause this disease only if they made a polysaccharide capsule (**figure 1**). In trying to understand the role of this capsule in causing disease, Griffith injected living cells that could not synthesize a capsule and heat-killed encapsulated cells into separate mice. As expected, the mice did not develop pneumonia. He also injected some mice with a mixture of heat-killed encapsulated cells and live non-encapsulated cells. Much to his surprise, these mice developed pneumonia and died. Griffith isolated living encapsulated pneumococci from the dead mice.

Two years after Griffith reported these findings, another investigator, M. H. Dawson, lysed heat-killed encapsulated pneumococci and passed the suspension of ruptured cells through a very fine filter, through which only the cytoplasmic contents of the bacteria could pass. When he mixed the filtrate (the material passing through the filter) with living bacteria unable to make a capsule, some bacteria were able to make a capsule. Moreover, the progeny of these bacteria could also make a capsule. Something in the filtrate was "transforming" the harmless unencapsulated bacteria into ones that could make a capsule.

What was this transforming principle? In 1944, after years of painstaking chemical analysis of lysates capable of transforming pneumococci, three investigators from the Rockefeller Institute, Oswald T. Avery, Colin MacLeod, and Maclyn McCarty, purified the active compound and then wrote one of the most important papers ever published in biology. In it, they reported that the transforming molecule was DNA.

The significance of their discovery was not appreciated at the time. Perhaps the discovery was premature, and scientists were slow to recognize its significance. None of the three investigators received a Nobel Prize, although many scientists believe that they deserved it. Their studies pointed out that DNA is a key molecule in the scheme of life and led to James Watson and Francis Crick's determination of its structure, which they published in 1953. The understanding of the structure and function of DNA revolutionized the study of biology and ushered in the era of molecular biology. Microbial genetics serves as its foundation.

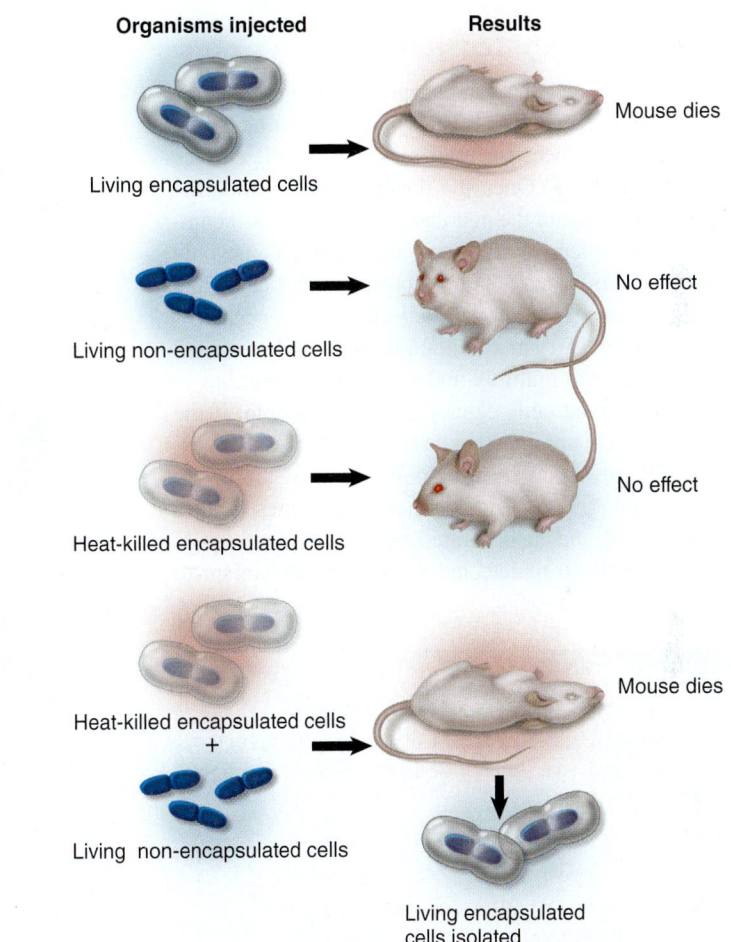

FIGURE 1 Griffith's Demonstration of Genetic Transformation

Naked DNA originates from cells that have either burst or secreted some DNA. When cells burst, the long chromosomes that were tightly jammed into the cells break up into hundreds of pieces. Some bacterial species secrete small pieces of DNA, presumably as a means of promoting transformation.

Competence

In order for transformation to occur, the recipient cell must be **competent**—a specific physiological state that allows the cell to take up DNA. Most competent bacteria take up DNA regardless of its source. Some species accept DNA only from closely related bacteria, however, recognizing it by characteristic nucleotide sequences located throughout the genome.

In the several dozen species that can become competent in nature, the process is tightly controlled. Some species are always competent, whereas others become so only under specific conditions, such as when the population reaches a critical density or under certain nutritional conditions. The fact that some bacterial species become competent only under precise environmental conditions highlights the remarkable ability of these seemingly simple cells to sense their surroundings and adjust their behavior accordingly.

In the case of *Bacillus subtilis,* a two-component regulatory system recognizes when the supply of nitrogen or carbon becomes scarce in the environment and activates a set of genes required for competence. Competence also requires a high concentration of bacteria, a role of quorum sensing. Presumably, this ensures that DNA in the medium will contact the competent bacteria. Even under optimal conditions, however, only a fraction of the population ever becomes competent. This means that seemingly identical cells in a population exposed to the same environment can differ in their physiological properties. ◀◀ **two-component regulatory system, p. 178** ◀◀ **quorum sensing, p. 177**

The Process of Transformation

Double-stranded DNA molecules bind to specific receptors on the surface of competent cells (**figure 8.19 ①**). **②** Only one strand enters the cell, however, because nucleases at the cell surface degrade the other strand. **③** Once the donor DNA is inside the recipient cell, it integrates into the genome by homologous recombination; the strand it replaced will be degraded. **④** Because only a single strand integrates, when the chromosome is replicated and the cell divides, only one daughter cell will inherit donor DNA.

In the laboratory, DNA transformation is most easily detected if the transformed cells can multiply under selective conditions in which the non-transformed cells cannot. **⑤** For example, if the donor cells are Str^R and the recipient cells are Str^S, then cells transformed to Str^R will grow and form colonies on a medium that contains streptomycin. Although many other donor genes besides Str^R will be transferred and incorporated by the recipient strain, these transformants will go undetected without a mechanism to recognize them.

MicroAssessment 8.6

In DNA-mediated transformation, DNA is released from donor cells and taken up by competent recipient cells. Competent cells bind DNA and take up a single strand; that strand then integrates into the genome by homologous recombination.

16. *How does DNAse prevent transformation?*
17. *Describe two ways by which DNA is released from cells.*
18. *How could DNA-mediated transformation be used to test chemicals for their mutagenic activity?* ✚

8.7 ■ Transduction

Learning Outcome

15. *Describe generalized transduction.*

Bacterial viruses, called **bacteriophages,** or simply **phages,** can transfer bacterial genes from a donor to a recipient by **transduction.** There are two types of transduction, generalized and specialized. In this chapter, we will describe only generalized transduction, which transfers any genes of the donor cell. **Specialized transduction,**

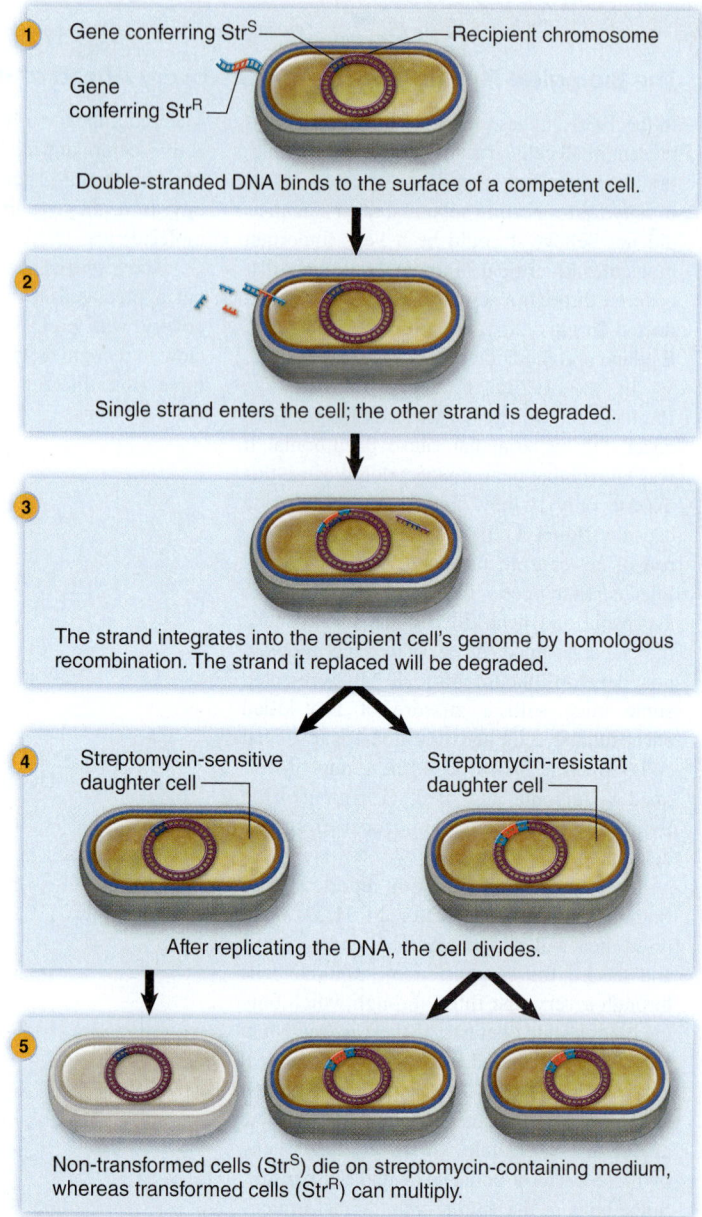

① Gene conferring Str^S ―― ―― Recipient chromosome
Gene conferring Str^R ―
Double-stranded DNA binds to the surface of a competent cell.

② Single strand enters the cell; the other strand is degraded.

③ The strand integrates into the recipient cell's genome by homologous recombination. The strand it replaced will be degraded.

④ Streptomycin-sensitive daughter cell ―― Streptomycin-resistant daughter cell ―
After replicating the DNA, the cell divides.

⑤ Non-transformed cells (Str^S) die on streptomycin-containing medium, whereas transformed cells (Str^R) can multiply.

FIGURE 8.19 DNA-Mediated Transformation The donor DNA in this case contains a gene conferring resistance to streptomycin (Str^R). Genes for resistance and sensitivity to streptomycin may differ by only a single nucleotide.

❓ *The cell at the top of this figure is competent. What does this mean?*

which transfers only a few specific genes, will be described in chapter 13—after the infection cycle of bacteriophages able to carry out this process is discussed.

To understand generalized transduction, you need to know something about phages and how they infect bacterial cells. This subject is covered more fully in chapter 13, so we will cover only the bare essentials here. Phages consist of genetic material, either DNA or RNA (never both), surrounded by a protein coat. A phage infects a bacterium by attaching to the cell and then injecting its nucleic acid. Enzymes encoded by the phage genome then cut the bacterial DNA into small pieces. Next, the cell's enzymes replicate the phage nucleic acid and synthesize proteins that make up the empty phage coat. The phage nucleic acid then enters the phage coat, and the various components assemble to produce complete phage particles. These are then released, usually as a result of host cell lysis. The phage particles then attach to other bacterial cells and begin new cycles of infection.

Generalized transduction results from a rare error that sometimes occurs during the construction of phage particles (**figure 8.20**). A fragment of bacterial DNA—produced when the phage-encoded enzyme cuts the bacterial genome—mistakenly enters the protein coat. The product is called a transducing particle; it carries no phage DNA and therefore is not a phage. Like phage particles, a transducing particle will attach to another bacterial cell

and inject the DNA it contains. The transducing particle, however, injects bacterial DNA. The bacterial DNA may then integrate into the host chromosome by homologous recombination.

MicroAssessment 8.7

Transduction results from an error that occurs during the infection cycle of bacteriophages, leading to the transfer of bacterial genes from one bacterial cell to another.

19. *What error leads to generalized transduction?*
20. *What is a transducing particle?*
21. *Two bacterial genes are transduced simultaneously. What does this suggest about the location of the two genes relative to each other?* ➕

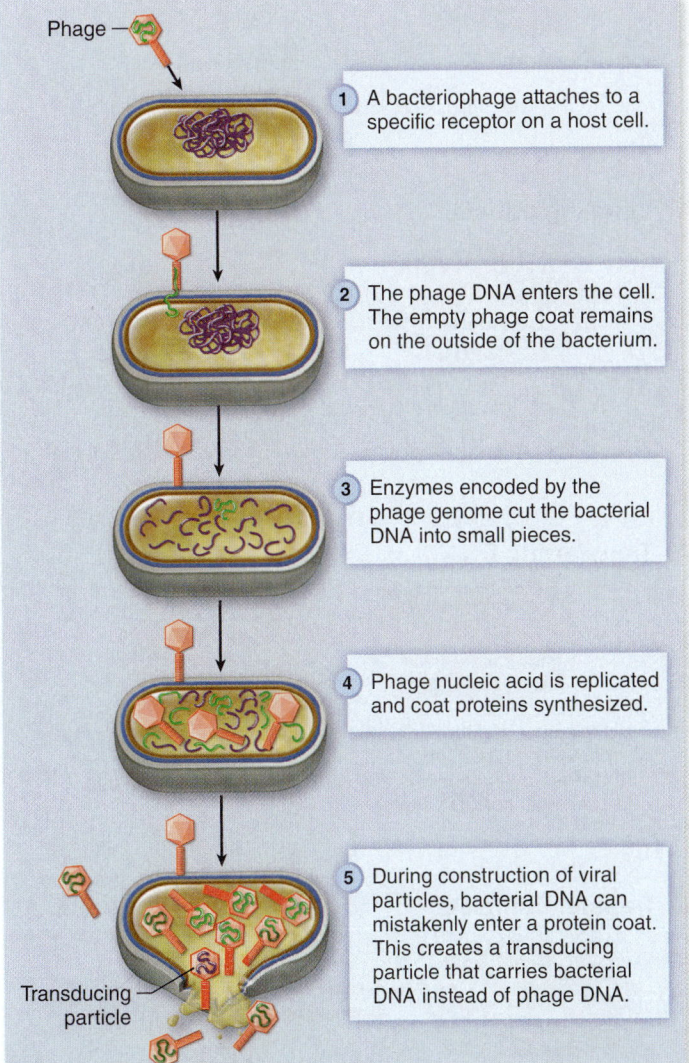

(a) **Formation of a transducing particle**

1. A bacteriophage attaches to a specific receptor on a host cell.
2. The phage DNA enters the cell. The empty phage coat remains on the outside of the bacterium.
3. Enzymes encoded by the phage genome cut the bacterial DNA into small pieces.
4. Phage nucleic acid is replicated and coat proteins synthesized.
5. During construction of viral particles, bacterial DNA can mistakenly enter a protein coat. This creates a transducing particle that carries bacterial DNA instead of phage DNA.

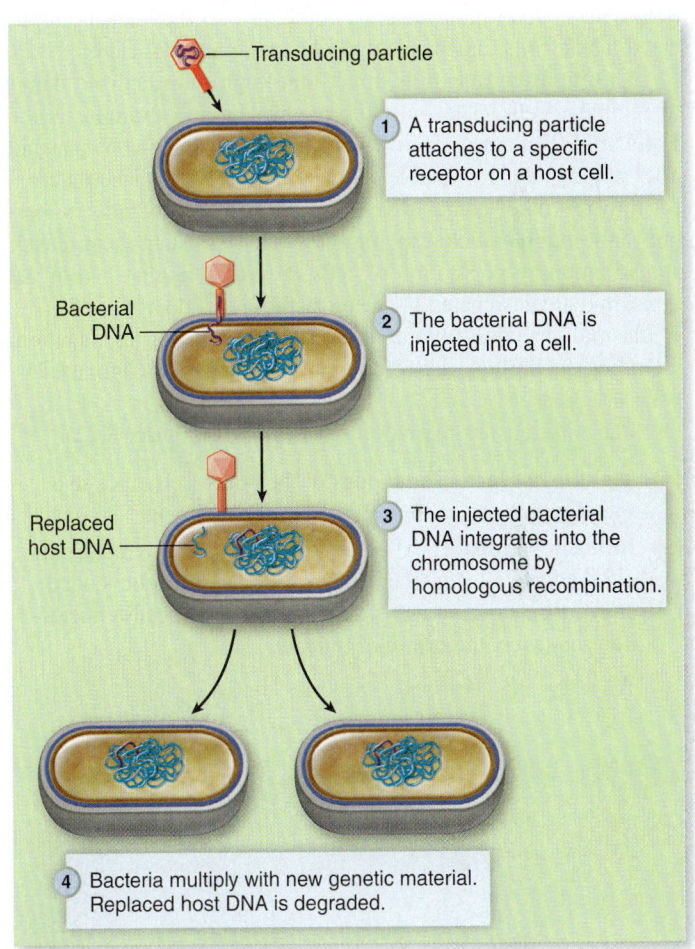

(b) **The process of transduction**

1. A transducing particle attaches to a specific receptor on a host cell.
2. The bacterial DNA is injected into a cell.
3. The injected bacterial DNA integrates into the chromosome by homologous recombination.
4. Bacteria multiply with new genetic material. Replaced host DNA is degraded.

FIGURE 8.20 Generalized Transduction (a) An error during construction of phage particles produces a transducing particle, which contains bacterial DNA instead of phage DNA. (b) The bacterial DNA carried by the transducing particle is injected into a new host, resulting in generalized transduction. Essentially any bacterial gene can be transferred this way.

❓ *After a phage injects its DNA into a bacterial cell, the cell begins making proteins that make up the phage coat. Why does the same thing not happen when a generalized transducing particle injects the DNA it carries?*

8.8 ■ Conjugation

Learning Outcome

16. *Compare and contrast conjugation involving an F$^+$ donor, an Hfr strain, and an F' donor.*

Conjugation is a complex process that requires contact between donor and recipient cells. Gram-positive and Gram-negative bacteria can both transfer DNA this way, but the process is quite different in the two groups. For simplicity, however, we will consider conjugation only in the more intensely studied Gram-negative bacteria.

Plasmid Transfer

Plasmids are most frequently transferred to other cells by conjugation. These DNA molecules are replicons, so they can be replicated inside cells, independent of chromosomal replication.

Conjugative plasmids direct their own transfer from donor to recipient cells. The most thoroughly studied example is the **F plasmid** (F stands for fertility) of *E. coli.* Although this plasmid does not encode any notable characteristics except those required for transfer, other conjugative plasmids encode resistance to certain antibiotics, which explains how such resistance can easily spread among a population of cells. *E. coli* cells that harbor the F plasmid are designated **F$^+$**, whereas those that do not are **F$^-$**. The F plasmid encodes several proteins required for conjugation, including the **F pilus**, also referred to as the sex pilus (**figure 8.21**).

◄◄ sex pilus, p. 65

Plasmid transfer involves a series of steps (**figure 8.22**):

1. **Making contact.** The F pilus of the donor cell binds to a specific receptor on the cell wall of the recipient.

2. **Initiating transfer.** After contact, the F pilus retracts, pulling the two cells together. Meanwhile, a plasmid-encoded enzyme cuts one strand of the plasmid at a specific nucleotide sequence, the origin of transfer.

3. **Transferring DNA.** A single strand of the F plasmid enters the F$^-$ cell, passing through the F pilus. Once inside the recipient cell, that strand serves as a template for synthesis of the complementary strand, generating an F plasmid. Likewise, the strand that remains in the donor serves as a template for DNA synthesis, regenerating the F plasmid. The transfer takes only a few minutes.

4. **Transfer complete.** Both the donor and recipient cells are now F$^+$ so they can act as donors of the F plasmid.

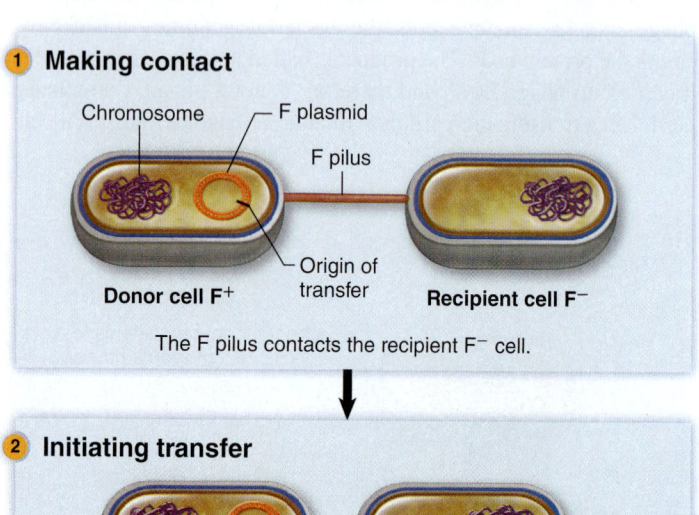

① Making contact

Chromosome ⸺ F plasmid

F pilus

Origin of transfer

Donor cell F$^+$ **Recipient cell F$^-$**

The F pilus contacts the recipient F$^-$ cell.

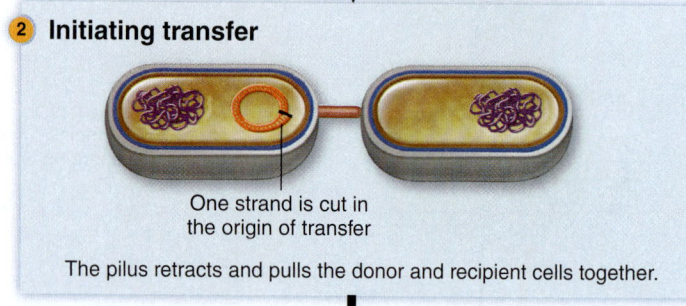

② Initiating transfer

One strand is cut in the origin of transfer

The pilus retracts and pulls the donor and recipient cells together.

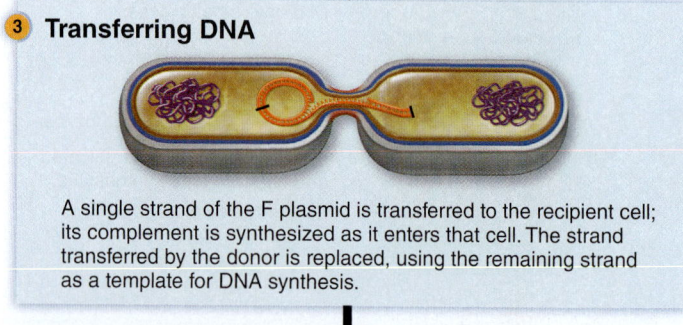

③ Transferring DNA

A single strand of the F plasmid is transferred to the recipient cell; its complement is synthesized as it enters that cell. The strand transferred by the donor is replaced, using the remaining strand as a template for DNA synthesis.

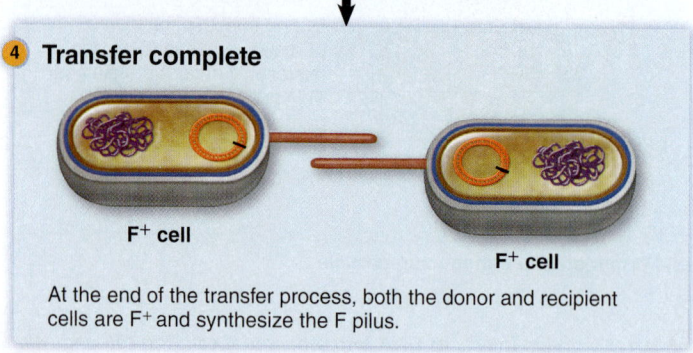

④ Transfer complete

F$^+$ cell

F$^+$ cell

At the end of the transfer process, both the donor and recipient cells are F$^+$ and synthesize the F pilus.

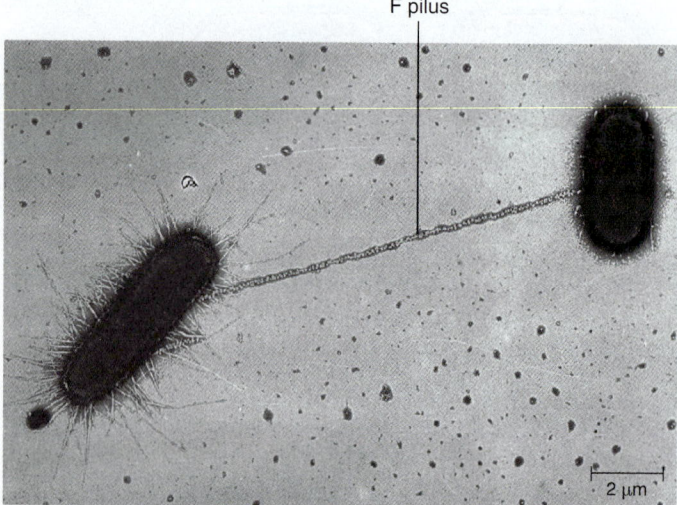

F pilus

2 μm

FIGURE 8.21 F Pilus Joining a Donor and Recipient Cell

❓ *What are the hairlike appendages on the cell on the left?*

FIGURE 8.22 Conjugation—F Plasmid Transfer

❓ *How does the recipient cell change as a result of conjugation?*

Chromosome Transfer

Fplasminchromsome

Chromosomal DNA transfer is less common than plasmid transfer and involves **Hfr cells** (meaning high frequency of recombination). These are strains in which the F plasmid has integrated into the chromosome by homologous recombination, which happens on rare occasions (**figure 8.23**). The transfer of chromosomal DNA involves the same general steps as transfer of the F plasmid. Like F⁺ cells, Hfr cells produce an F pilus, and the F plasmid DNA directs its transfer to the recipient cell. Because the F plasmid DNA is integrated into the chromosome, however, chromosomal DNA is also transferred, beginning with the genes on one side of the origin of transfer (**figure 8.24**). The entire chromosome is generally not transferred because it would

take approximately 100 minutes for this to occur, an unlikely event because the connection between the two cells usually breaks sooner than that. Because the entire integrated F plasmid is not transferred, the recipient remains F⁻. The transferred chromosomal DNA is not a replicon, so it will be maintained only if it integrates into the recipient's chromosome through homologous recombination. Unincorporated DNA will be degraded.

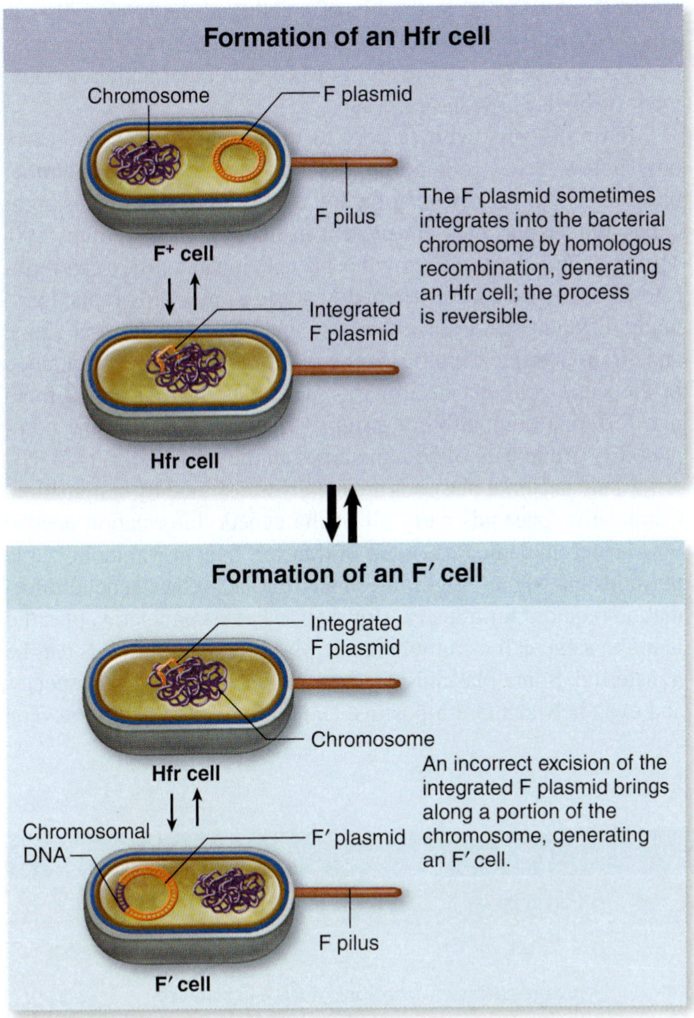

Formation of an Hfr cell

The F plasmid sometimes integrates into the bacterial chromosome by homologous recombination, generating an Hfr cell; the process is reversible.

Formation of an F′ cell

An incorrect excision of the integrated F plasmid brings along a portion of the chromosome, generating an F′ cell.

FIGURE 8.23 Formation of Hfr and F′ Cells An Hfr cell is created when the plasmid integrates into the chromosome as a result of recombination between homologous regions on the F plasmid and chromosome. Note that the process is reversible, as recombination between the same two regions will excise the F plasmid, changing the Hfr cell back to an F⁺ cell. An F′ cell is created when certain recombination events result in an incorrect excision that removes a piece of the chromosome along with the F plasmid. This process is also reversible.

Why does the F plasmid integrate only at specific locations?

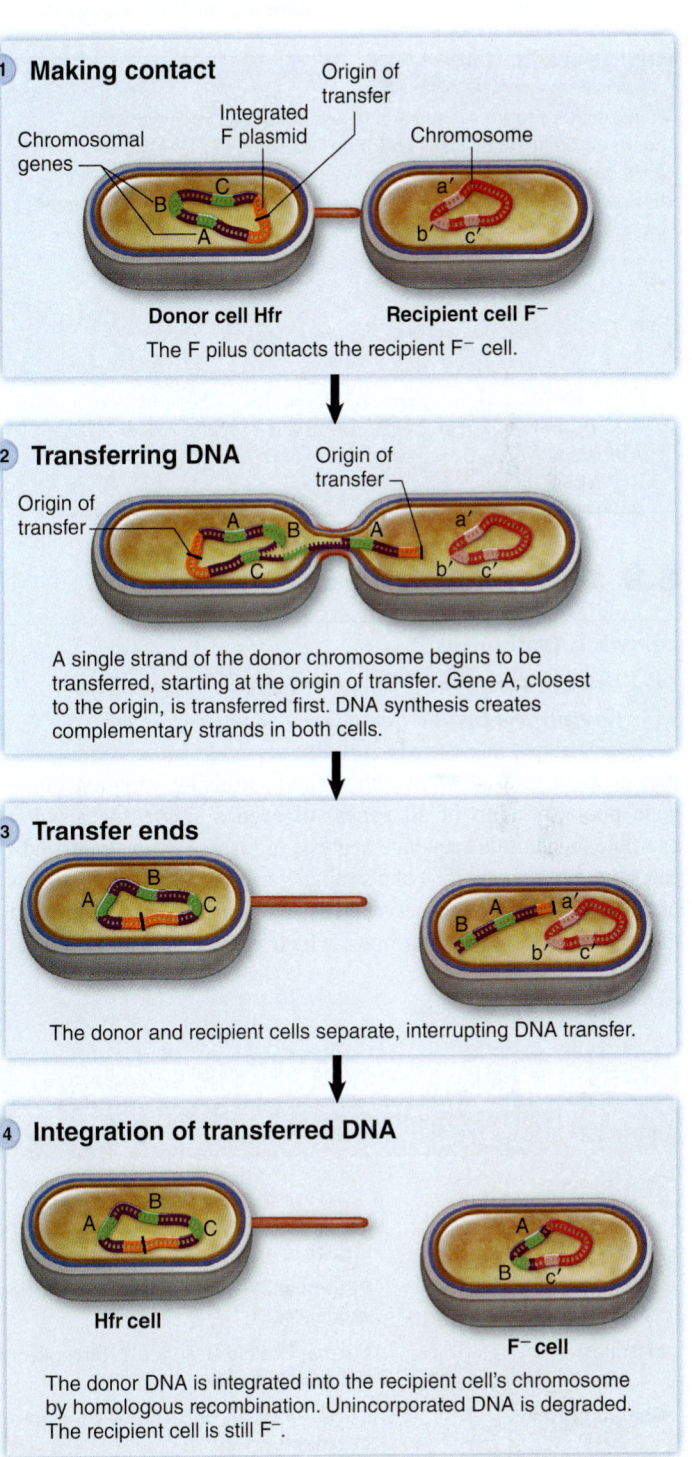

1 Making contact
The F pilus contacts the recipient F⁻ cell.

2 Transferring DNA
A single strand of the donor chromosome begins to be transferred, starting at the origin of transfer. Gene A, closest to the origin, is transferred first. DNA synthesis creates complementary strands in both cells.

3 Transfer ends
The donor and recipient cells separate, interrupting DNA transfer.

4 Integration of transferred DNA
The donor DNA is integrated into the recipient cell's chromosome by homologous recombination. Unincorporated DNA is degraded. The recipient cell is still F⁻.

FIGURE 8.24 Conjugation—Chromosomal DNA Transfer

Why is the entire donor chromosome seldom transferred?

F′ Donors

Hfr strains can revert to F⁺ because the process of F plasmid integration is reversible (see figure 8.23). In some instances, however, an error occurs during excision, and a piece of the bacterial chromosome is removed along with the F plasmid DNA. This action brings a chromosomal fragment into the F plasmid, producing a plasmid called **F′** (F prime). Like the F plasmid, F′ is a replicon that is rapidly and efficiently transferred to F⁻ cells. In the case of F′, however, the chromosomal fragment is transferred as well.

MicroAssessment 8.8

Conjugation requires contact between donor and recipient cells. A donor cell that synthesizes an F pilus transfers the DNA to one that does not. Both plasmid and chromosomal DNA can be transferred. Following transfer, plasmids replicate, but chromosomal DNA must be integrated into a replicon.

22. *The F plasmid encodes which two functions essential for conjugation?*
23. *Describe the outcomes of the three types of matings (F⁺ × F⁻, Hfr × F⁻, and F′ × F⁻).*
24. *Would you expect transfer of chromosomal DNA by conjugation to be more efficient if cells were plated together on solid medium (agar) or mixed together in a liquid in a shaking flask? Explain.* ✚

8.9 ■ The Mobile Gene Pool

Learning Outcomes

17. *Describe how plasmids differ from bacterial chromosomes.*
18. *Compare and contrast transposons and genomic islands.*

Advances in genomics have uncovered surprising variation in the gene pool (the sum of all genes) of even a single species. For example, nucleotide sequence analysis of many *E. coli* strains indicates that only about 75% of a strain's genes are found in all strains of that species. These conserved genes make up the **core genome** of the species. The remaining ones, which vary considerably among different strains, make up the **mobile gene pool** or mobilome. These genes can move from one DNA molecule to another, carried on mobile genetic elements including various plasmids, transposons, regions called genomic islands, and phage DNA (**table 8.4**). Surprisingly, when all the non-conserved genome components of all the various *E. coli* strains are considered as a group, these sequences vastly outnumber sequences in the core genome.

Plasmids

Plasmids are common in the microbial world and are found in most members of the *Bacteria* and *Archaea* and in some *Eucarya*. Like chromosomes, most plasmids are double-stranded DNA molecules. They have an origin of replication and therefore can be replicated by the cell before it divides, but they replicate independently of the chromosome. Plasmids generally do not encode information essential to the life of a cell, and therefore cells can survive their loss (curing). They are important, however, because they provide cells with the ability to cope in a particular environment (**table 8.5**). ◄◄ plasmid, p. 66

Plasmids vary with respect to their properties. Some carry only a few genes, others carry a thousand. Low-copy-number plasmids occur in only one or a few copies per cell, whereas high-copy-number plasmids are present in many copies, perhaps 500. Most plasmids have a narrow host range, meaning they can replicate in only one species. Broad host range plasmids replicate in many different species, sometimes including both Gram-negative and Gram-positive bacteria. Some plasmids cannot be maintained in the same cell. Because of this, scientists have arranged them into different compatibility groups—members of the same compatibility group cannot be maintained in the same cell.

Many bacterial plasmids are readily transferred by conjugation. Conjugative plasmids carry all of the genetic information needed for transfer, including an origin of transfer. In contrast, mobilizable plasmids encode an origin of transfer but lack other genetic information required for transfer. However, when a conjugative plasmid is in the same cell as a mobilizable plasmid, both plasmids can be transferred. Some plasmids can transfer between unrelated species and even between Gram-positive and Gram-negative bacteria. The

TABLE 8.4	The Mobile Gene Pool	
	Composition	**Property**
Transposons		
Insertion sequences (ISs)	Transposase gene flanked by short repeat sequences	Move to different locations in DNA in same cell
Composite transposons	Recognizable gene flanked by insertion sequences	Same as insertion sequences, but encode additional information
Plasmids	Circular double-stranded DNA replicon; smaller than chromosomes	Generally code only for non-essential genetic information
Genomic Islands	Large fragment of DNA in a chromosome or plasmid	Code for genes that allow cell to occupy specific environmental locations
Phage DNA	Phage genome	May encode proteins important to bacteria

TABLE 8.5	Some Plasmid-Coded Traits	
Trait	**Organisms in Which Trait Is Found**
Antibiotic resistance	Many, including *Escherichia coli*, *Salmonella* sp., *Neisseria* sp., *Staphylococcus* sp., and *Shigella* sp.
Pilus synthesis	*E. coli, Pseudomonas* sp.
Tumor formation in plants	*Agrobacterium* sp. (see Perspective 8.2)
Nitrogen fixation	*Rhizobium* sp.
Oil degradation	*Pseudomonas* sp.
Gas vacuole production	*Halobacterium* sp.
Insect toxin synthesis	*Bacillus thuringiensis*
Plant hormone synthesis	*Pseudomonas* sp.
Antibiotic synthesis	*Streptomyces* sp.
Increased virulence	*Yersinia* sp.
Toxin production	*Bacillus anthracis*

genes of an unusual plasmid can even be transferred into plant cells by conjugation (see **Perspective 8.2**).

Resistance Plasmids

Resistance, or R, plasmids encode resistance to many different antimicrobial medications and heavy metals, such as mercury and arsenic, all of which are found in hospital environments. Many of these plasmids are conjugative, composed of two parts: the resistance genes (R genes), which encode the resistance traits, and a resistance transfer factor (RTF), which encodes the properties required for conjugation (**figure 8.25**).

R plasmids can give simultaneous resistance to numerous antimicrobials, and many have a broad host range. As a consequence,

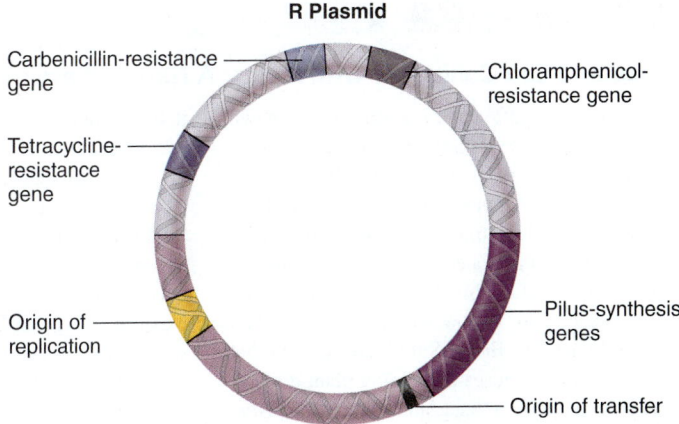

R Plasmid

FIGURE 8.25 Functional Regions of an R Plasmid Genes conferring resistance to antimicrobial compounds tend to be clustered within a specific region (the top half of the illustrated plasmid), whereas genes involved in replication and conjugative transfer are located on a different region (the bottom half of the illustrated plasmid).

? *How is an R plasmid similar to an F plasmid?*

the plasmids can give rise to a wide range of organisms becoming resistant to many different antimicrobials. Members of the normal microbiota can harbor R plasmids, and then transfer them to pathogens. ◀◀ normal microbiota, p. 8

Transposons

In addition to causing mutations, transposons provide a mechanism for transferring various genes. Transposons can move into other replicons in the same cell without any specificity as to where they insert.

Several types of transposons exist, varying in their structural complexity. The simplest, an **insertion sequence (IS)**, encodes only the enzyme responsible for transposition, called transposase (**figure 8.26**). Flanking the gene are inverted repeats—sequences that are identical when read in the 5′ to 3′ direction.

FIGURE 8.26 Transposable Elements The borders of insertion sequences are defined by inverted repeats 15–20 nucleotides in length. The first six nucleotides are shown here in expanded view to demonstrate their inverted orientation. Composite transposons consist of two IS elements and the DNA between them, all of which move as a single unit.

? *Why are some transposons medically important?*

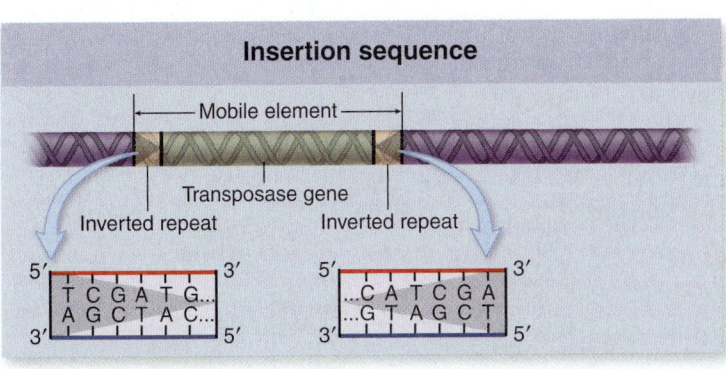

Insertion sequence

Mobile element

Transposase gene

Inverted repeat | Inverted repeat

```
5′          3′        5′          3′
 T C G A T G...        ...C A T C G A
 A G C T A C...        ...G T A G C T
3′          5′        3′          5′
```

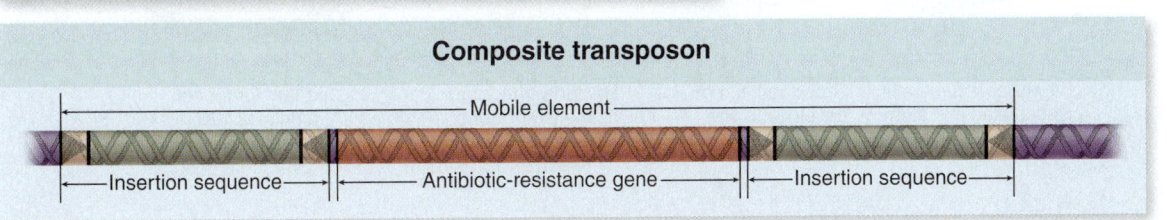

Composite transposon

Mobile element

Insertion sequence —— Antibiotic-resistance gene —— Insertion sequence

PERSPECTIVE   8.2

Bacteria Can Conjugate with Plants: A Natural Case of Genetic Engineering

For more than 50 years, scientists have known that DNA can be transferred between bacteria. Thirty years ago, it was shown that a bacterium can even transfer its genes into cells of plants, including tobacco, carrots, and cedar trees, through a process analogous to conjugation. What led to this discovery started about 100 years ago in the laboratory of a plant pathologist, Dr. Erwin Smith. He showed that the agent that causes a common plant disease, crown gall, is a bacterium—*Agrobacterium tumefaciens*.

Crown gall is characterized by large galls or tumors that occur on the plant at the site of infection, usually near the soil line, the crown of the plant. When other investigators cultured the diseased plant tissue, they found it had properties that differed from normal plant tissue. Whereas normal tissue requires several plant hormones for growth, crown gall tissue grows in the absence of these hormones. In addition, crown gall tissue synthesizes large amounts of a compound, an opine, that neither normal plant tissue nor *A. tumefaciens* synthesizes.

What surprised scientists studying the *Agrobacterium*–crown gall system was the observation that the plant cells maintained their altered growth characteristics and the ability to synthesize opine even after the bacteria were killed by penicillin. Investigators concluded that the crown gall cells are permanently altered. Although *A. tumefaciens* is required to start the infection, it is not needed to maintain the changes to the plant cells.

The explanation of how *Agrobacterium tumefaciens* causes crown gall tumors was established in 1977 following a report that all crown gall–causing strains contain a large plasmid termed the Ti (tumor-inducing) plasmid. A group of microbiologists then showed that a specific piece of the Ti plasmid, called T-DNA (transferred DNA) moves from the bacterial cell to the plant cell, where it becomes incorporated into the plant chromosome (**figure 1**). Because no regions of the plant DNA are similar to those of bacteria, integration occurs through non-homologous recombination. Like conjugation between bacteria, a pilus is required for DNA transfer.

Once incorporated into the plant chromosome, the T-DNA provides the plant cell with additional genetic information, thereby giving it new properties. The promoters in T-DNA resemble those of plants rather than those of bacteria, so the genetic information is expressed in plants but not in *A. tumefaciens*. Scientists learned that T-DNA encodes enzymes for the synthesis of the plant hormones as well as for the opine. The expression of these genes supplies the plant cells with the plant hormones—explaining why the plant cells can grow in the absence of added hormones—and allows them to synthesize the opine. ◀◀ **promoter, p. 168**

Why does *A. tumefaciens* alter plant cells? This bacterium is able to use the opine as a source of carbon and energy, whereas most other bacteria in the soil, as well as plants, cannot. In other words, *A. tumefaciens* alters the metabolism of the plant to produce food that only *Agrobacterium* cells can use. Thus, *Agrobacterium tumefaciens* is a natural genetic engineer of plants.

The *Agrobacterium*–crown gall system is important for several reasons. First, it shows that DNA can be transferred from prokaryotes to eukaryotes. Many people believed that such transfer would be impossible in nature and could occur only in the laboratory. Second, this system has spawned an industry of plant biotechnology dedicated to improving the quality of higher plants. With this technology, it is possible to replace the genes of hormone and opine synthesis in the Ti plasmid with any other genes, which will then be transferred and incorporated into the plant. Examples of beneficial genes that have been transferred into a wide variety of different plants include those conferring resistance to viral pathogens, insects, and different herbicides. Rice has been engineered to synthesize high levels of β-carotene, the precursor of vitamin A.

Genetic engineering of plants became a reality once scientists learned how a common soil bacterium caused a well-recognized and serious plant disease. This system serves as an excellent example of how solving a riddle in basic science can lead to major industrial applications.

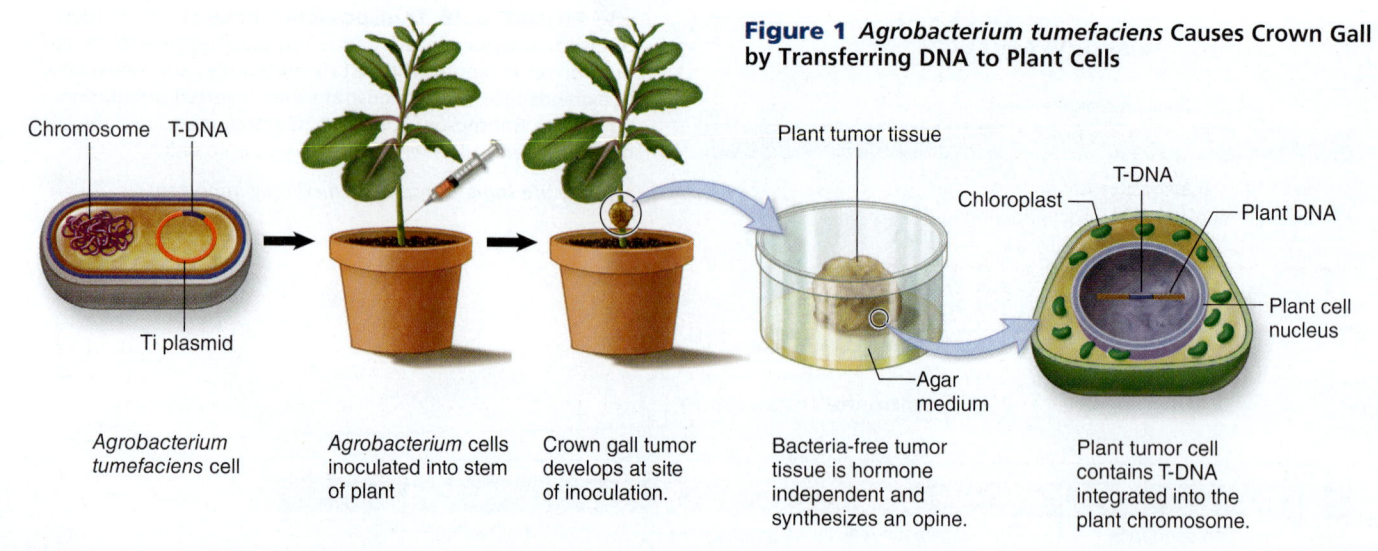

Figure 1 *Agrobacterium tumefaciens* **Causes Crown Gall by Transferring DNA to Plant Cells**

Agrobacterium tumefaciens cell

Agrobacterium cells inoculated into stem of plant

Crown gall tumor develops at site of inoculation.

Bacteria-free tumor tissue is hormone independent and synthesizes an opine.

Plant tumor cell contains T-DNA integrated into the plant chromosome.

Composite transposons consist of one or more genes flanked by ISs (figure 8.26). Like insertion sequences, they can move in the same replicon or from one replicon to another in the cell. Their movement is easily followed if they encode a recognizable gene product such as antibiotic resistance. They integrate into their new location through **non-homologous recombination,** a process that does not require a similar nucleotide sequence in the region of recombination. The transposon simply inserts into a stretch of DNA; it does not replace the existing sequences. If a composite transposon inserts into a conjugative plasmid, it can then be transferred to other cells. Any gene or group of genes can move to another site if flanked by ISs, but transposons that carry genes for antibiotic resistance are particularly important medically.

The introduction in this chapter presented a real-life example of how a transposon allowed a strain of *Staphylococcus aureus* to become resistant to vancomycin. We now know how the resistant strain likely arose (**figure 8.27**). The patient initially was infected with a strain of *S. aureus* susceptible to vancomycin, but resistant to many other antibiotics. After treatment with vancomycin, *S. aureus* cells resistant to that drug were isolated from the patient. The only obvious difference between the sensitive and resistance isolates was that the latter had a gene that encoded resistance to vancomycin, inserted into a plasmid present in both strains. But where did the resistance gene come from? Further analysis suggested that it was acquired from a vancomycin-resistant strain of *Enterococcus faecalis* isolated from the same patient. The resistance gene in that bacterium was identical to the one in the *S. aureus* isolate. It was also part of a transposon integrated into a plasmid, but the plasmid was different from that in *S. aureus*. It appears that *E. faecalis* transferred its transposon-containing plasmid to the sensitive *S. aureus* by conjugation. This entering plasmid was apparently destroyed by enzymes in *S. aureus,* but before that happened, the transposon jumped to the plasmid already in the *S. aureus* cell. Considering the ease with which DNA can move among cells in a bacterial population, is it any wonder that antibiotic resistance is such a serious problem in treating infectious diseases today?

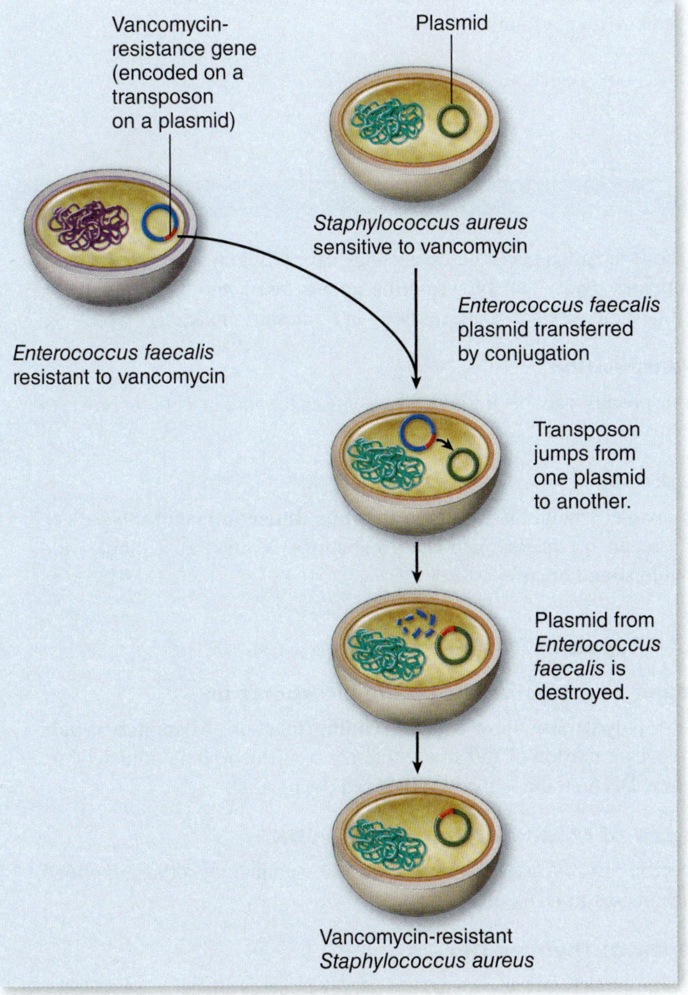

Genomic Islands

Genomic islands are large DNA segments in a cell's genome that originated in other species. This conclusion is based on the fact that their nucleotide composition is quite different from the rest of the cell's genome. In general, each bacterial species has a characteristic proportion of G-C base pairs, so a large segment of DNA that has a very different G-C ratio suggests the segment originated from a foreign source and was transferred to the cell through horizontal gene transfer. ▸▸ GC content, p. 252

The characteristics encoded by genomic islands include utilization of specific energy sources, acid tolerance, development of symbiosis, and ability to cause disease. Genomic islands that encode the latter are called **pathogenicity islands.**

FIGURE 8.27 Transfer of Vancomycin Resistance from *Enterococcus faecalis* to *Staphylococcus aureus* Transfer of resistance involved both a plasmid and a transposon.

❓ *What role did the transposon play in the transfer of resistance?*

MicroByte

Almost 1,400 genes of the pathogen *E. coli* O157:H7 are not in the laboratory strain *E. coli* K-12.

MicroAssessment 8.9

Plasmids vary in size, copy number, host range, genetic composition, compatibility to coexist with other plasmids, and their ability to be transferred to other cells. One type of important plasmid is the R plasmid, which codes for resistance to various antimicrobial medications and heavy metals. Transposons can move from one location to another in the same replicon or to other replicons. Genomic islands are large DNA segments thought to have originated in other species.

25. *What functions must a plasmid encode to be self-transmissible?*
26. *What characteristic of a genomic island suggests that it originated in another species?*
27. *What does the phrase "reservoir for R plasmids" mean when referring to plasmids carried by non-disease-causing bacteria?* ➕

FUTURE CHALLENGES 8.1

Hunting for Magic Bullets

Because of the increasing resistance of microorganisms to current antimicrobial medications, the demand for new antimicrobials that will kill these resistant organisms is rapidly increasing. It is surprising and sobering to realize that only two new classes of antimicrobials have been introduced into clinics in the past 40 years. The great challenge is to develop new antimicrobial agents that strike at targets different from those attacked by current antimicrobials.

With the revolution that has been occurring in biology over the past 15 years, the development of new antimicrobial agents may become a reality. Promising new strategies are based on knowing the genomic sequence of microbial pathogens. Many of the microbial genomes which have now been sequenced are human pathogens. The study and analysis

of the nucleotide sequence of DNA is called genomics. The next step is to identify genes necessary for the survival of the microorganism or required to cause disease. To gain some understanding of the function of any gene, one must compare its DNA sequence with the sequence of all other genes that have been put into a database, called GenBank. If the gene that has been sequenced is similar in nucleotide sequence to any other gene, then it is assumed that the two genes have similar functions. Thus, if the function of a gene that has been sequenced in any organism is known, the function of all genes with a similar sequence is likely to be similar. This is the science of bioinformatics, which involves the analysis of the nucleotide sequence of DNA in order to understand what it codes for. Genes that are required for virulence in the pathogen but are not found

in the host are potential targets. It should be possible to design a protein that inhibits the virulence protein and thereby prevents disease.

To develop an antimicrobial that inhibits an enzyme required for virulence requires a great deal of information about the enzyme, how it folds, what its three-dimensional structure is, and whether or not it interacts with other proteins.

In the past, the search for antimicrobial medications has relied on random screening. Scientists looked for growth inhibition of a pathogen by a large number of organisms isolated from soil samples collected from around the world. Today, new technologies based on microbial genomics should identify new targets and provide a rational approach to developing new antimicrobials.

Summary

8.1 ■ Genetic Change in Bacteria

The **genotype** of bacteria can change either through **mutations** or **horizontal gene transfer** (figure 8.1). Bacteria are haploid, so any changes in DNA can easily alter the **phenotype.**

MUTATION AS A MECHANISM OF GENETIC CHANGE

8.2 ■ Spontaneous Mutations

Spontaneous mutations occur as a result of normal cell processes. They are stable but occasionally revert back to the non-mutant form. The chance that two spontaneous mutations will occur within the same cell is the product of the individual mutation rates.

Base Substitution

Base substitutions occur during DNA synthesis (figure 8.2). They result in silent, missense, and nonsense mutations (figure 8.3).

Deletion or Addition of Nucleotides

Deleting or adding one or two nucleotides causes a **frameshift mutation,** changing the reading frame of the encoded protein (figure 8.4). This often results in a shortened non-functional protein.

Transposons (Jumping Genes)

Transposons can move from one location to another in a cell's genome. The gene into which the transposon jumps is insertionally inactivated by the event (figure 8.5).

8.3 ■ Induced Mutations

Induced mutations are caused by **mutagens** (table 8.1).

Chemical Mutagens

Some chemicals modify nucleobases, altering their hydrogen-bonding properties (figure 8.7). **Base analogs** can be mistakenly incorporated in

place of the usual nucleobases, and they have different hydrogen-bonding properties (figure 8.8). **Intercalating agents** insert into the double helix and push nucleotides apart, resulting in frameshift mutations.

Transposition

Transposons can be introduced intentionally into a cell in order to inactivate genes.

Radiation

Ultraviolet irradiation results in **thymine dimer** formation (figure 8.9). The repair mechanism can cause mutations. X rays cause single- and double-strand breaks.

8.4 ■ Repair of Damaged DNA (table 8.2)

Repair of Errors in Nucleotide Incorporation

DNA polymerases have a **proofreading** function. **Mismatch repair** removes a portion of the strand that has a misincorporated nucleotide. A new DNA strand is then synthesized (figure 8.10).

Repair of Modified Nucleobases in DNA

Specific glycosylases remove oxidized guanine or other modified nucleobases in DNA (figure 8.11).

Repair of Thymine Dimers

In **photoreactivation,** an enzyme uses the energy of light to break the bonds of the thymine dimer (figure 8.12). In **excision repair,** the damaged single-stranded segment is removed and replaced.

SOS Repair

SOS repair is a last-ditch repair mechanism that uses a new DNA polymerase that has no proofreading ability but can bypass the damaged DNA. Consequently, the newly synthesized DNA has many mutations.

8.5 ■ Mutant Selection

Direct Selection

Direct selection is used to obtain mutants that grow under conditions in which the parent cell cannot. These mutants are easy to isolate (figure 8.13).

[handwritten: grow where parent can't]

Indirect Selection

Indirect selection uses **replica plating** to isolate an auxotroph from a prototrophic parent strain (figure 8.14). **Penicillin enrichment** increases the proportion of auxotrophic mutants in a culture (figure 8.15).

[handwritten: isolate]

Screening for Possible Carcinogens

The **Ames test** is used to determine if a chemical is a mutagen and therefore a possible **carcinogen** (figure 8.16).

[handwritten: Ames if a mutagen]

HORIZONTAL GENE TRANSFER AS A MECHANISM OF GENETIC CHANGE (table 8.3)

For newly acquired DNA to replicate in a cell, it must either be a replicon or integrate into the cell's genome through homologous recombination (figure 8.18).

8.6 ■ DNA-Mediated Transformation

DNA-mediated transformation transfers "naked" DNA.

Competence

A cell must be **competent** to take up DNA, and certain species become competent in nature.

The Process of Transformation

Short strands of double-stranded DNA bind to cells, but only one strand enters (figure 8.19).

[handwritten: short double bind one in]

8.7 ■ Transduction

Transduction is the transfer of bacterial DNA by **bacteriophage.** There are two types, **generalized transduction** and **specialized transduction** (figure 8.20).

8.8 ■ Conjugation

Conjugation requires cell-to-cell contact.

Plasmid Transfer

F⁺ cells synthesize an **F pilus,** encoded on an **F plasmid;** the F plasmid is transferred from an F⁺ to an F⁻ cell through the pilus (figure 8.22).

Chromosome Transfer

Hfr strains have the F plasmid integrated into the chromosome (figure 8.23). When the F plasmid is transferred, chromosomal DNA moves into a recipient cell along with it (figure 8.24).

F′ Donors

An **F′** donor carries a modified F plasmid that contains a piece of chromosomal DNA.

8.9 ■ The Mobile Gene Pool (table 8.4)

The mobile gene pool includes plasmids, transposons, genomic islands, as well as phage DNA.

Plasmids (table 8.5)

Plasmids are replicons that code for non-essential information; many are readily transferred by conjugation. **R plasmids** code for antibiotic resistance (figure 8.25).

Transposons

Transposons provide a mechanism for transferring genes. An **insertion sequence (IS)** encodes only transposase (figure 8.26). A **composite transposon** has one or more genes flanked by insertion sequences.

Genomic Islands

Genomic islands are large DNA segments in a cell's genome that originated in other species.

Review Questions

Short Answer

1. How is an auxotroph different from a prototroph?

2. Why is deleting one nucleotide generally more detrimental than deleting three?

3. What type of mutation in an operon is most likely to affect the synthesis of more than one protein?

4. What is meant by "proofreading" with respect to DNA polymerase?

5. Why would a cell use SOS repair, considering that it introduces mutations?

6. Why is replica plating used to isolate an auxotrophic mutant from a prototrophic parent?

7. What is transduction?

8. How is an F⁺ strain different from an Hfr strain?

9. Name four mobile genetic elements.

10. Why are R plasmids important?

Multiple Choice

1. A culture of *E. coli* is irradiated with ultraviolet (UV) light. Answer questions 1 and 2 based on this statement. The UV light specifically
   a) joins the two strands of DNA together by covalent bonds.
   b) joins the two strands of DNA together by hydrogen bonds.
   c) forms covalent bonds between thymine molecules on the same strand of DNA.
   d) forms covalent bonds between guanine and cytosine.
   e) deletes bases.

2. The highest frequency of mutations would be obtained if, after irradiation, the cells were immediately
   a) placed in the dark.
   b) exposed to visible light.
   c) shaken vigorously.
   d) incubated at a temperature below their optimum for growth.
   e) The frequency would be the same no matter what the environmental conditions are after irradiation.

3. Penicillin enrichment of mutants works on the principle that
   a) only Gram-positive cells are killed.
   b) cells are most sensitive to antimicrobial medications during the lag phase of growth.
   c) most Gram-negative cells are resistant to penicillin.
   d) penicillin kills only growing cells.
   e) penicillin inhibits formation of the lipopolysaccharide layer.

4. Repair mechanisms that occur during DNA synthesis are
   1. mismatch repair.
   2. proofreading by DNA polymerase.
   3. light repair.
   4. SOS repair.
   5. excision repair.
   a) 1, 2   b) 2, 3   c) 3, 4   d) 4, 5   e) 1, 5

5. You are trying to isolate a mutant of wild-type *E. coli* that requires histidine for growth. This can best be done using
   1. direct selection.
   2. replica plating.
   3. penicillin enrichment.
   4. a procedure for isolating conditional mutants.
   5. reversion.
   a) 1, 2   b) 2, 3   c) 3, 4   d) 4, 5   e) 1, 5

6. The properties that all plasmids share are that they
   1. all carry genes for antimicrobial resistance.
   2. are self-transmissible to other bacteria.
   3. always occur in multiple copies in the cells.
   4. code for non-essential functions.
   5. replicate in the cells in which they are found.
   a) 1, 2   b) 2, 3   c) 3, 4   d) 4, 5   e) 1, 5

7. The addition of DNase to a mixture of donor and recipient cells will prevent gene transfer via
   a) DNA transformation.
   b) chromosome transfer by conjugation.
   c) plasmid transfer by conjugation.
   d) generalized transduction.

8. An F pilus is essential for
   1. DNA-mediated transformation.
   2. chromosome transfer by conjugation.
   3. plasmid transfer by conjugation.
   4. generalized transduction.
   5. cell movement.
   a) 1, 2   b) 2, 3   c) 3, 4   d) 4, 5   e) 1, 5

9. A plasmid that can replicate in *E. coli* and *Pseudomonas* is most likely a/an
   a) broad host range plasmid.
   b) self-transmissible plasmid.
   c) high-copy-number plasmid.
   d) essential plasmid.
   e) low-copy-number plasmid.

10. The frequency of transfer of an F′ molecule by conjugation is closest to the frequency of transfer of
    a) chromosomal genes by conjugation.
    b) an F plasmid by conjugation.
    c) an F plasmid by transformation.
    d) an F plasmid by transduction.
    e) an R plasmid by DNA transformation.

Applications

1. Some bacteria may have higher mutation rates than others following exposure to UV light. Discuss a reason why this might be the case. What experiments could you do to determine whether this is a likely possibility?

2. A pharmaceutical researcher is disturbed to discover that the major ingredient of a new drug formulation causes frameshift mutations in bacteria. What other information would the researcher want before looking for a substitute chemical?

Critical Thinking ✚

1. You have the choice of different kinds of mutants for use in the Ames test to determine the frequency of reversion by suspected carcinogens. You can choose a deletion, a point mutation, or a frameshift mutation. Would it make any difference which one you chose? Explain.

2. You have isolated a strain of *E. coli* that is resistant to penicillin, streptomycin, chloramphenicol, and tetracycline. You also observe that when you mix this strain with cells of *E. coli* that are sensitive to the four antibiotics, they become resistant to streptomycin, penicillin, and chloramphenicol but remain sensitive to tetracycline. Explain what is going on.

9 Biotechnology and Recombinant DNA

Technicians working at a bench in a DNA laboratory.

KEY TERMS

Colony Blotting Technique used to determine which colonies on an agar plate contain a given nucleotide sequence.

DNA Cloning Procedure in which a fragment of DNA is inserted into a vector and then transferred to another cell, where it then replicates.

DNA Microarray A probe-based technique used to study gene expression patterns.

DNA Probe Single-stranded piece of DNA, tagged with a detectable marker, that is used to detect a complementary sequence.

DNA Sequencing Process of determining the nucleotide sequence of a DNA molecule.

Fluorescence *in situ* Hybridization (FISH) Technique used to detect a given nucleotide sequence within intact cells on a microscope slide.

Gel Electrophoresis A procedure used to separate DNA fragments (or other macromolecules) according to their size.

Genetic Engineering Deliberately altering an organism's genetic information using *in vitro* techniques.

Polymerase Chain Reaction (PCR) *In vitro* technique used to duplicate a specific region of a DNA molecule, increasing the number of copies exponentially.

Recombinant DNA Molecule DNA molecule created by joining DNA fragments from two different sources.

Restriction Enzyme Type of enzyme that recognizes a specific nucleotide sequence and then cuts the DNA within or near that site.

Vector DNA molecule, often a plasmid, that functions as a carrier of cloned DNA.

A Glimpse of History

In 1976, Argentinean newspapers reported a violent shootout between soldiers and the occupants of a house in suburban Buenos Aires, leaving the five extremists inside dead. Conspicuously absent from those reports were the identities of the "extremists"—a young couple and their three children, ages 6 years, 5 years, and 6 months. Over the next 7 years, similar scenarios recurred as the military junta that ruled Argentina killed thousands of citizens it perceived as threats. This "Dirty War," as it came to be known, finally ended in 1983 with the collapse of the military junta and the election of a democratic government. The new leaders opened previously sealed records that confirmed what many had already suspected—more than 200 children survived the bloodshed and had in fact been kidnapped and placed with families in favor with the junta.

Dr. Mary-Claire King was at the University of California at Berkeley when she was enlisted to help in the effort to return the children to the surviving members of their biological families. Dr. King and others recognized that DNA technology could be used for this important humanitarian cause. By analyzing certain DNA sequences, blood and tissue samples from one individual can be distinguished from those of another. These same principles can also be used to show that a particular child is the progeny of a given set of parents. Because a person has two copies of each chromosome—one inherited from each parent—half of a child's DNA will represent maternal sequences and the other half will represent paternal traits. The case of the Argentinean children was complicated, however, because most of the parents were dead or missing. Often, the only surviving relatives were aunts and grandmothers, and it is difficult to use chromosomal DNA to show genetic relatedness between a child and such relatives. Dr. King decided to investigate mitochondrial DNA (mtDNA). This organelle DNA, unlike chromosomal DNA, is inherited only from the mother. A child will have the same nucleotide sequence of mtDNA as his or her siblings, the mother and her siblings, as well as the maternal grandmother.

By comparing the nucleotide sequences of mtDNA in different individuals, Dr. King was able to locate key positions that varied extensively among unrelated people, but were similar in maternal relatives. Dr. King's technique, developed out of a desire to help reunite families victimized by war, has now found many uses. Today her lab, now at the University of Washington, still uses molecular biology techniques for humanitarian efforts, identifying the remains of victims of atrocities around the world.

Biotechnology uses microbiological and biochemical techniques to solve practical problems and produce useful products. In the past, this often meant labor-intensive searches for naturally occurring mutants that had desirable characteristics. Today, **recombinant DNA techniques** have made it possible to genetically alter organisms to give them more useful traits. Researchers can isolate genes from one organism, manipulate the purified DNA *in vitro,* and then transfer the genes into another organism. In fact, biotechnology is now nearly synonymous with **genetic engineering,** the process of deliberately altering an organism's genetic information using *in vitro* techniques.

Since the advent of recombinant DNA techniques, a virtual toolbox of DNA technologies has been developed. The information and innovations these have generated impact society in numerous ways—from agricultural practices and medical diagnoses to evidence used in courtrooms. **Table 9.1** summarizes the applications of the DNA-based technologies described in this chapter.

9.1 ■ Fundamental Tools Used in Biotechnology

Learning Outcome

1. *Describe the role of restriction enzymes and gel electrophoresis in biotechnology.*

Before exploring the application of biotechnology, it is helpful to understand some of the basic components of a molecular biologist's "tool kit." As we describe these, it is important to remember that diagrams focus on only one or a few DNA molecules to illustrate what is happening at a molecular level. In reality, scientists are often working with millions of molecules.

Restriction Enzymes

Restriction enzymes allow scientists to easily cut DNA into fragments in a predictable and controllable manner (**figure 9.1a**). Each enzyme recognizes a specific 4 to 6 base-pair nucleotide sequence (**table 9.2**). The recognition sequences are typically palindromes, meaning they are the same on both strands when read in the 5′ to 3′ direction. The enzyme cuts each strand of DNA within or near that sequence, digesting the DNA to generate **restriction fragments.** ▶▶ restriction enzymes, p. 315

The name of a particular restriction enzyme represents the bacterium from which it was first isolated. The first letter is the

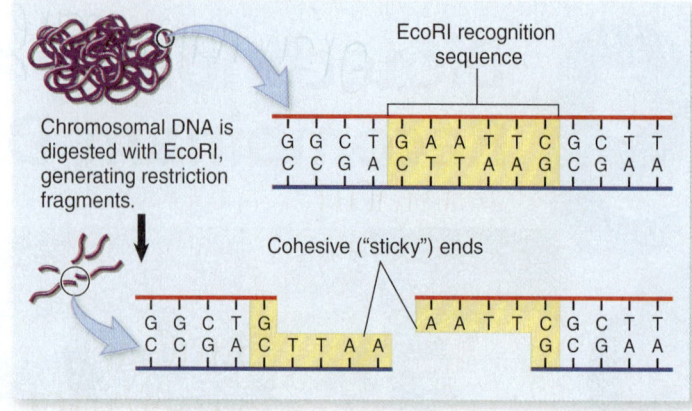

(a)

(b)

▶▶ **FIGURE 9.1 Action of Restriction Enzymes (a)** Digesting DNA with a restriction enzyme generates restriction fragments. **(b)** Fragments that have complementary cohesive ends can anneal, regardless of their original source (N = nucleotide, meaning that it could be any of the four nucleobases as long as base-pairing rules are followed).

❓ *How do restriction enzymes make it easier for scientists to create recombinant DNA molecules?*

TABLE 9.1	Applications of DNA-Based Biotechnologies
Technology	**Applications**
Genetic engineering	Genetically engineered microorganisms are used to produce medically and commercially valuable proteins, to produce specific DNA sequences, and as a tool for studying gene function and regulation.
DNA sequencing	Once the nucleotide sequence of even part of an organism's DNA has been determined, the information can be used to decipher the amino acid sequence of the encoded proteins. It can also be used to determine how similar the nucleotide sequence is to DNA from other organisms, which can give insights into genetic relatedness.
Polymerase chain reaction (PCR)	The presence of a specific segment of DNA can be detected, and the size determined, in only a matter of hours. This can be used in diagnosis of an infectious disease if DNA specific to a pathogen can be amplified. It is also used to "fingerprint" DNA for forensic evidence.
Probe technologies	Colony blots are used to detect colonies that contain a specific DNA sequence; fluorescence *in situ* hybridization (FISH) is used to identify cells directly in a specimen; DNA microarrays are used to study gene expression.

TABLE 9.2	Examples of Common Restriction Enzymes	
Enzyme	**Microbial Source**	**Recognition Sequence (arrows indicate cleavage sites)**
AluI	*Arthrobacter luteus*	↓ 5′ A G C T 3′ 3′ T C G A 5′ ↑
BamHI	*Bacillus amyloliquefaciens* H	↓ 5′ G G A T C C 3′ 3′ C C T A G G 5′ ↑
EcoRI	*Escherichia coli* RY13	↓ 5′ G A A T T C 3′ 3′ C T T A A G 5′ ↑

first letter of the genus name, and the next two are from the species name. In the past, these were italicized, but that rule has been dropped. Other numbers or letters designate the strain and order of discovery. For example, EcoRI is from *E. coli* strain RY13.

Researchers use restriction enzymes not only to cut DNA, but also to create **recombinant DNA molecules,** which are molecules made by joining DNA from two different sources. This is possible because many restriction enzymes produce a staggered cut in the recognition sequence, resulting in ends with short overhangs of usually four nucleotides (figure 9.1b). The overhangs are called sticky ends, or cohesive ends, because they will **anneal** (form base pairs) with one another. Any two complementary cohesive ends can anneal, even those from two different organisms. The enzyme **DNA ligase** is used to form a covalent bond between adjacent nucleotides, joining the two molecules. Thus, if restriction enzymes are viewed as scissors that cut DNA into fragments, then DNA ligase is the glue that pastes the fragments together. ◄◄ DNA ligase, p. 167

Gel Electrophoresis

Gel electrophoresis separates DNA fragments by size (**figure 9.2**). The DNA samples are added to wells in a gelatin-like substance, usually either agarose (a highly purified form of agar) or polyacrylamide. A size standard—a mixture of DNA fragments of known sizes—is routinely put into a well of the same gel. The size standard serves as a basis for comparison, allowing the researcher to determine the sizes of the various DNA fragments in the samples. The gel is then placed in an electrophoresis buffer (an

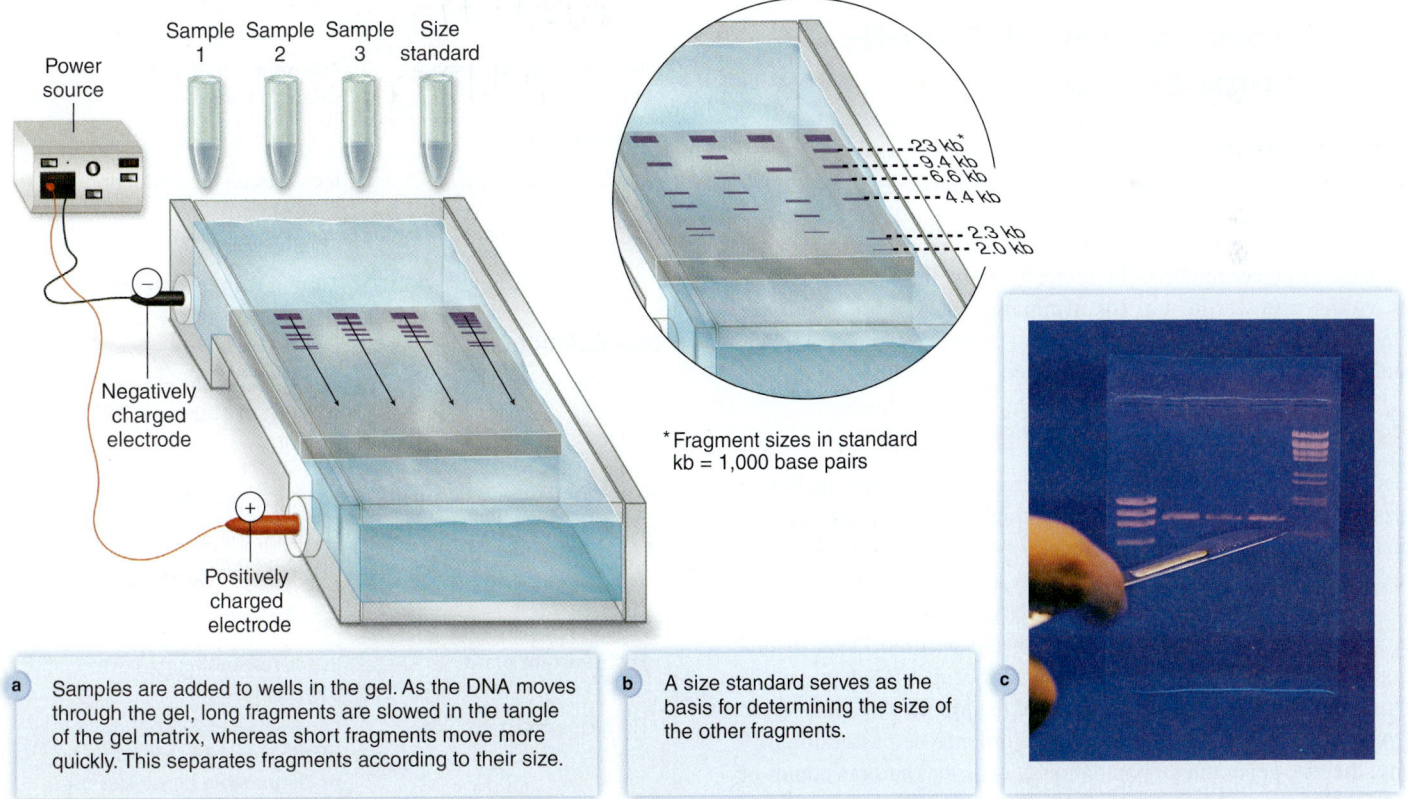

a. Samples are added to wells in the gel. As the DNA moves through the gel, long fragments are slowed in the tangle of the gel matrix, whereas short fragments move more quickly. This separates fragments according to their size.

b. A size standard serves as the basis for determining the size of the other fragments.

c.

*Fragment sizes in standard kb = 1,000 base pairs

FIGURE 9.2 Gel Electrophoresis (a) Gel electrophoresis separates DNA fragments according to size. (b) A size standard serves as a basis for determining the size of the other fragments. (c) DNA on the gel is visible when stained with ethidium bromide and viewed with UV light; each fluorescent band represents millions of molecules of a specific-sized fragment.

? *How does gel electrophoresis separate DNA fragments according to size?*

electrically conductive solution) and subjected to an electrical current. DNA is negatively charged, so the fragments move toward the positively charged electrode. As the DNA moves through the gel, long fragments are slowed in the tangle of the gel matrix, whereas short fragments move more quickly.

The DNA is not visible in the gel unless it is stained. To do this, the gel is immersed in a solution containing ethidium bromide. This dye binds DNA and fluoresces when viewed with UV light (figure 9.2c). Each fluorescent band represents millions of molecules of a specific-sized fragment of DNA.

Gel electrophoresis can also be used to separate other macromolecules, specifically RNA and proteins, according to their size. The basic principles are similar to that just described, but the gel compositions differ.

MicroAssessment 9.1

Restriction enzymes recognize specific nucleotide sequences, and then cut the DNA, generating restriction fragments. Gel electrophoresis is used to separate DNA fragments according to their size.

1. *What is the importance of cohesive ends in genetic engineering?*
2. *How does gel electrophoresis separate different-sized DNA fragments?*
3. *What should a restriction enzyme isolated from* Staphylococcus aureus *strain 3A be called?* ✚

9.2 ■ Applications of Genetic Engineering

Learning Outcome

2. *Describe the applications of genetically engineered bacteria and plants.*

Genetic engineering brought biotechnology into a new era by providing a powerful tool for manipulating microorganisms for medical, industrial, and research uses (**table 9.3**). Plants and animals can now be genetically engineered as well.

Genetically Engineered Bacteria

Genetically engineered bacteria have a variety of uses, including protein production, DNA production, and a tool for research. In fact, much of the information described in this textbook was revealed through research using genetically engineered bacteria.

Genetic engineering relies on **DNA cloning,** a process that involves using a restriction enzyme to cut purified DNA of one organism and then transferring a fragment of that DNA into a different organism (**figure 9.3**). As part of the process, the transferred DNA must replicate in the recipient in order to be passed to progeny, thereby generating a population of cells that harbors copies of the DNA fragment (see figure 8.18). Most DNA fragments, however, will not contain an origin of replication and therefore will not replicate independently in a cell. In addition, they generally have no similarity to chromosomal sequences, so they will not integrate into the chromosome by homologous recombination. To generate a cloned nucleotide sequence that will be replicated in a cell, the DNA fragment can be inserted into a plasmid or other independently

---

TABLE 9.3	Some Applications of Genetic Engineering
Example	**Use**
PROTEIN PRODUCTION	
Pharmaceutical Proteins	
Alpha interferon	Treating cancer and viral infections
Erythropoietin	Treating some types of anemia
Beta interferon	Treating multiple sclerosis
Deoxyribonuclease	Treating cystic fibrosis
Factor VIII	Treating hemophilia
Gamma interferon	Treating cancer
Glucocerebrosidase	Treating Gaucher disease
Growth hormone	Treating dwarfism
Insulin	Treating diabetes
Platelet derived growth factor	Treating foot ulcers in diabetics
Streptokinase	Dissolving blood clots
Tissue plasminogen activator	Dissolving blood clots
Vaccines	
Hepatitis B	Preventing hepatitis
HPV	Preventing cervical cancer
Foot-and-mouth disease	Preventing foot-and-mouth disease in animals
Other Proteins	
Bovine somatotropin	Increasing milk production in cows
Chymosin	Cheese-making
Restriction enzymes	Cutting DNA into fragments
DNA PRODUCTION	
DNA for study	Determining nucleotide sequences; obtaining DNA probes
RESEARCHING GENE FUNCTION AND REGULATION	
Creating gene fusions	Studying the conditions that affect gene activity
TRANSGENIC PLANTS	
Pest-resistant plants	Insect-resistant corn, cotton, and potatoes
Herbicide-resistant plants	Engineered plants (soybean, cotton, corn) are not killed by biodegradable herbicides used to kill weeds.
Plants with improved nutritional value	Rice that produces vitamin A and iron
Plants that function as edible vaccines	Enable researchers to study foods as possible vehicles for edible vaccines

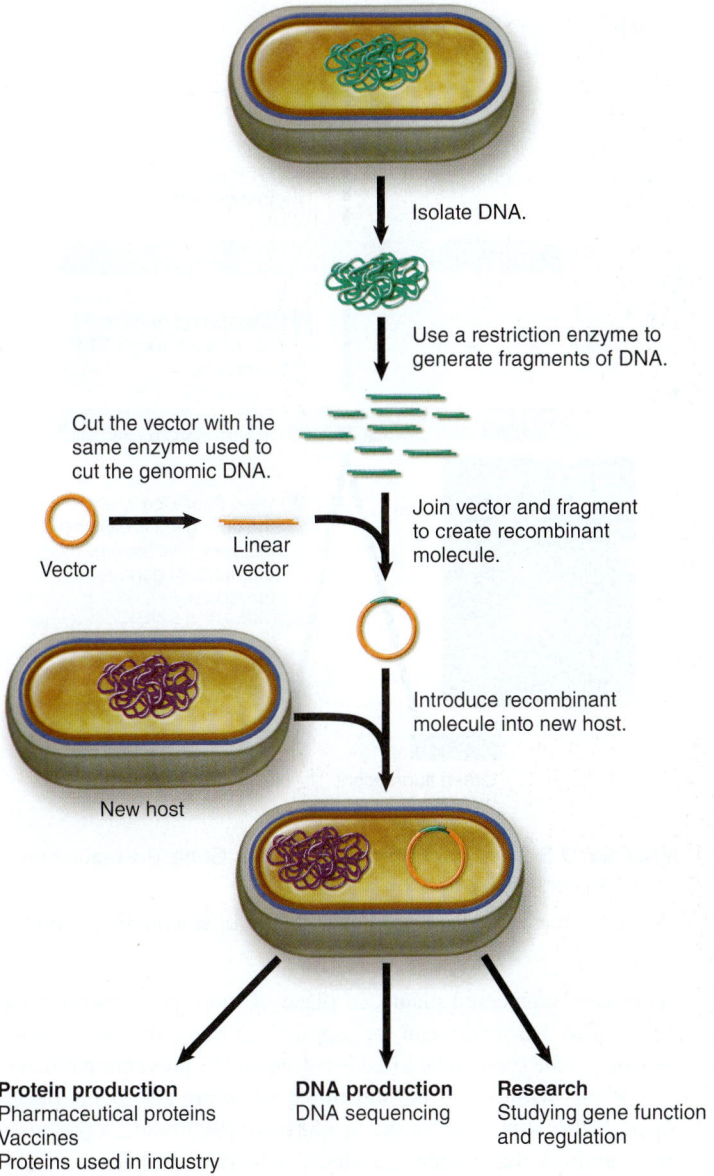

Isolate DNA.

Use a restriction enzyme to generate fragments of DNA.

Cut the vector with the same enzyme used to cut the genomic DNA.

Vector

Linear vector

Join vector and fragment to create recombinant molecule.

Introduce recombinant molecule into new host.

New host

Protein production	DNA production	Research
Pharmaceutical proteins	DNA sequencing	Studying gene function
Vaccines		and regulation
Proteins used in industry		

FIGURE 9.3 Cloning DNA

❓ *Why would it be easiest to use the same restriction enzyme to cut both the vector and the insert DNA?*

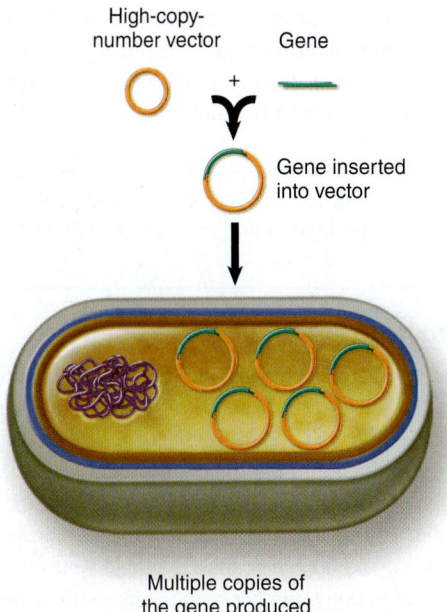

High-copy-number vector   Gene

Gene inserted into vector

Multiple copies of the gene produced

FIGURE 9.4 Cloning into a High-Copy-Number Vector When a gene is inserted into a high-copy-number vector, multiple copies of that gene will be present in a single cell, resulting in the synthesis of many molecules of the encoded protein.

❓ *Why must the gene be inserted into a vector for it to be cloned?*

replicating DNA molecule to form a recombinant molecule. The DNA molecule used as a carrier of the cloned DNA is called a vector and is commonly a plasmid that has been genetically modified. DNA that has been incorporated into the vector is called an insert. ◀◀ origin of replication, p. 166 ◀◀ homologous recombination, p. 201

If a gene coding for a valuable protein is inserted into a high-copy-number vector, a bacterium that harbors the recombinant molecule can make large amounts of that protein. This is because each gene copy can be transcribed and translated (**figure 9.4**). ◀◀ high-copy-number plasmid, p. 208

Protein Production

A number of different pharmaceutical proteins are now produced by genetically engineered microorganisms. In the past, these proteins were extracted from live animal or cadaver tissues, which made them expensive and limited in supply. Human insulin, used

in treating diabetes, was one of the first important pharmaceutical proteins to be produced through genetic engineering. The original commercial product was extracted from pancreatic glands of cattle and pigs and sometimes caused allergic reactions. Once the gene for human insulin was cloned into bacteria, microbes became the major source of insulin. The product is safer and more economical than the version extracted from animal tissues.

Another medically important use of genetically engineered microorganisms is vaccine production. Vaccines protect against disease by harmlessly exposing a person's immune system to killed or weakened forms of the pathogen, or to parts of the pathogen. Although vaccines are generally composed of whole microbes, only specific proteins are necessary to immunize (induce protection) against the disease. The genes coding for these proteins can be cloned into yeast or bacteria, allowing large amounts of the protein to be produced and then purified. This type of vaccine is currently used to prevent hepatitis B and cervical cancer in humans and foot-and-mouth disease of domestic animals. ▶▶ vaccines, p. 421

One of the most widely used proteins made by genetically engineered organisms is chymosin (rennin), an enzyme used in cheese production. It causes milk to coagulate and produces desirable changes in the characteristics of cheeses as they ripen. Traditionally, chymosin was obtained from the stomachs of calves, but genetically engineered bacteria are now the main source. The microbial product is less expensive and more reliably available. Other proteins produced by genetically engineered microbes include various restriction enzymes and bovine somatotropin, a growth hormone used to increase milk production in dairy cows.

Some scientists hope to create microorganisms that can be readily customized and used as miniature factories. With that goal in mind, researchers made a variation of the *Mycoplasma mycoides* chromosome "from scratch," assembling it from machine-made

DNA segments. In 2010, they successfully used it to replace the existing chromosome of a different *Mycoplasma* species. The procedure may pave the way for creating synthetic bacteria that can produce a wide range of commercially valuable substances.

MicroByte

The insulin yield from a 2,000-liter culture of *E. coli* cells with a cloned insulin gene is the same as 1,600 pounds of pancreatic glands!

DNA Production

In many cases, a researcher is interested in obtaining readily available supplies of certain DNA fragments. By cloning a segment of DNA into a well-characterized bacterium such as *E. coli,* that sequence can then be used for study and further manipulation.

Human genes are often cloned into bacteria to make them easier to study. A human cell contains an estimated 25,000 genes, whereas *E. coli* contains only 4,500 genes; thus, a human gene cloned into the bacterium on a high-copy-number vector represents a much higher percentage of the total DNA in the recipient cell than in the original cell. This makes it easier to isolate the DNA as well as the gene product.

Random samples of DNA from any environment can be cloned into *E. coli* and then the nucleotide sequence determined. By doing this, the genomic characteristics of some of the 99% of bacteria that have not been grown in culture can be studied. This "shotgun cloning" is the first step in metagenomics, the study of the total genomes in a sample. ◀◀ metagenomics, p. 184

Research

The function and regulation of genes can be studied with experiments that involve cloning. For example, gene regulation can be studied by creating a gene fusion—joining the gene of interest and a reporter gene (**figure 9.5**). The reporter gene encodes a product that is easy to see, such as green fluorescent protein (GFP). This makes it relatively simple to detect gene expression and, in turn, determine the conditions that affect gene activity.

Genetically Engineered Eukaryotes

Yeasts can be genetically engineered to perform many of the functions described for bacteria. They serve as an important model for gene function and regulation in eukaryotic cells.

Multicellular organisms can also be genetically engineered. A plant or animal that harbors a cloned gene is **transgenic.** Transgenic plants are of particular interest to microbiologists because their development started with basic research studying *Agrobacterium tumefaciens;* the Ti plasmid of this bacterium has now been genetically manipulated so it can be used as a vector to deliver desirable genes to plant cells (see Perspective 8.2).

Corn, cotton, and potatoes have been engineered to produce a biological insecticide called Bt toxin, which is naturally produced by the bacterium *Bacillus thuringiensis* as it forms endospores. Unlike many chemically synthesized toxins, Bt toxin is toxic only to insects, including their larvae. ◀◀ endospore, p. 67

Soybeans, cotton, and corn have been engineered to resist the effects of the herbicide glyphosate (Roundup). This allows growers to apply this biodegradable herbicide, which kills weeds and

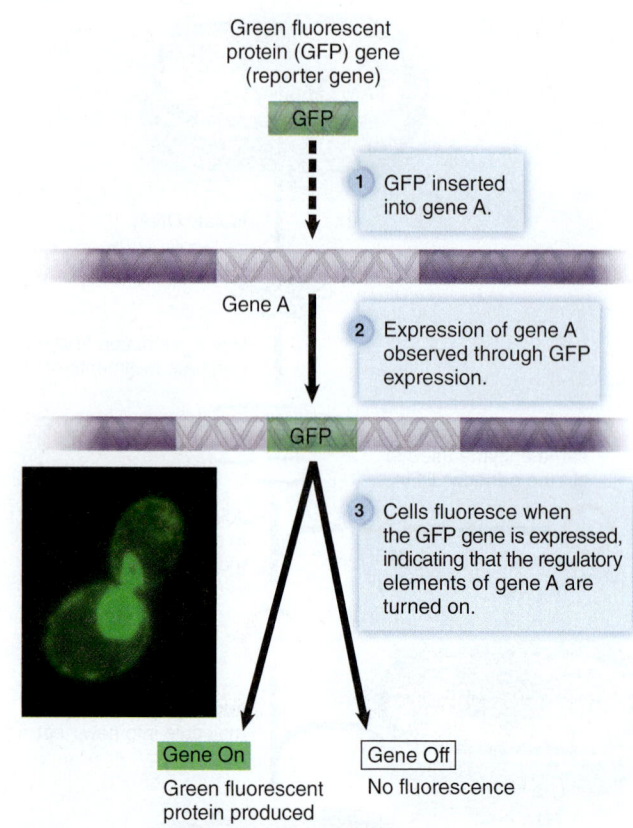

FIGURE 9.5 The Function of a Reporter Gene The regulatory elements are not shown here.

❓ *Why is the green fluorescent protein useful as a reporter gene?*

other non-engineered plants, in place of more persistent alternatives. The herbicide can be applied throughout the growing season, so the soil can be tilled less frequently, preventing erosion.

Plants with improved nutritional value are also being developed. Genes that code for the synthesis of β-carotene, a precursor of vitamin A, have been introduced into rice. The rice was also engineered to provide more dietary iron. The diet of much of the world's population is deficient in these essential nutrients, so advances such as these could profoundly effect global health.

Plants that serve as edible vaccines are also being developed. Researchers have successfully genetically engineered potatoes and rice to produce certain proteins from pathogens and have shown that the immune system responds to these, raising hopes for an edible vaccine. Whether the immune response is strong enough to protect against disease is still unclear.

MicroAssessment 9.2

Genetically engineered bacteria can be used to produce a variety of products, including medically and commercially important proteins, vaccines, and DNA for study. They are also used as research tools. Genetic engineering can be used to develop pest-resistant plants, herbicide-resistant plants, and plants that produce more nutrients.

4. *Name a disease treated with a protein produced by genetically engineered microorganisms.*

5. *Describe the characteristics of two different transgenic plants.*

6. *Why would calves' stomachs have chymosin (rennin)?*

9.3 ■ Techniques Used in Genetic Engineering

Learning Outcomes

3. *Describe how a DNA library is made.*

4. *Explain how introns are removed from eukaryotic genes.*

5. *Describe the characteristics of a typical vector.*

6. *Explain how to obtain cells that harbor recombinant molecules.*

An approach frequently used to clone a specific gene is to make a **DNA library,** a collection of clones that together contain the entire genome (**figure 9.6**). This is made by cutting the DNA of the organism being studied with restriction enzymes and then cloning the entire set of restriction fragments into a population of *E. coli* cells. Although each cell in the resulting population contains only one fragment of the genome, the entire genome is represented in the population as a whole. Once a DNA library has been prepared, colony blots (which will be described later) can be used to determine which cells contain the gene of interest.

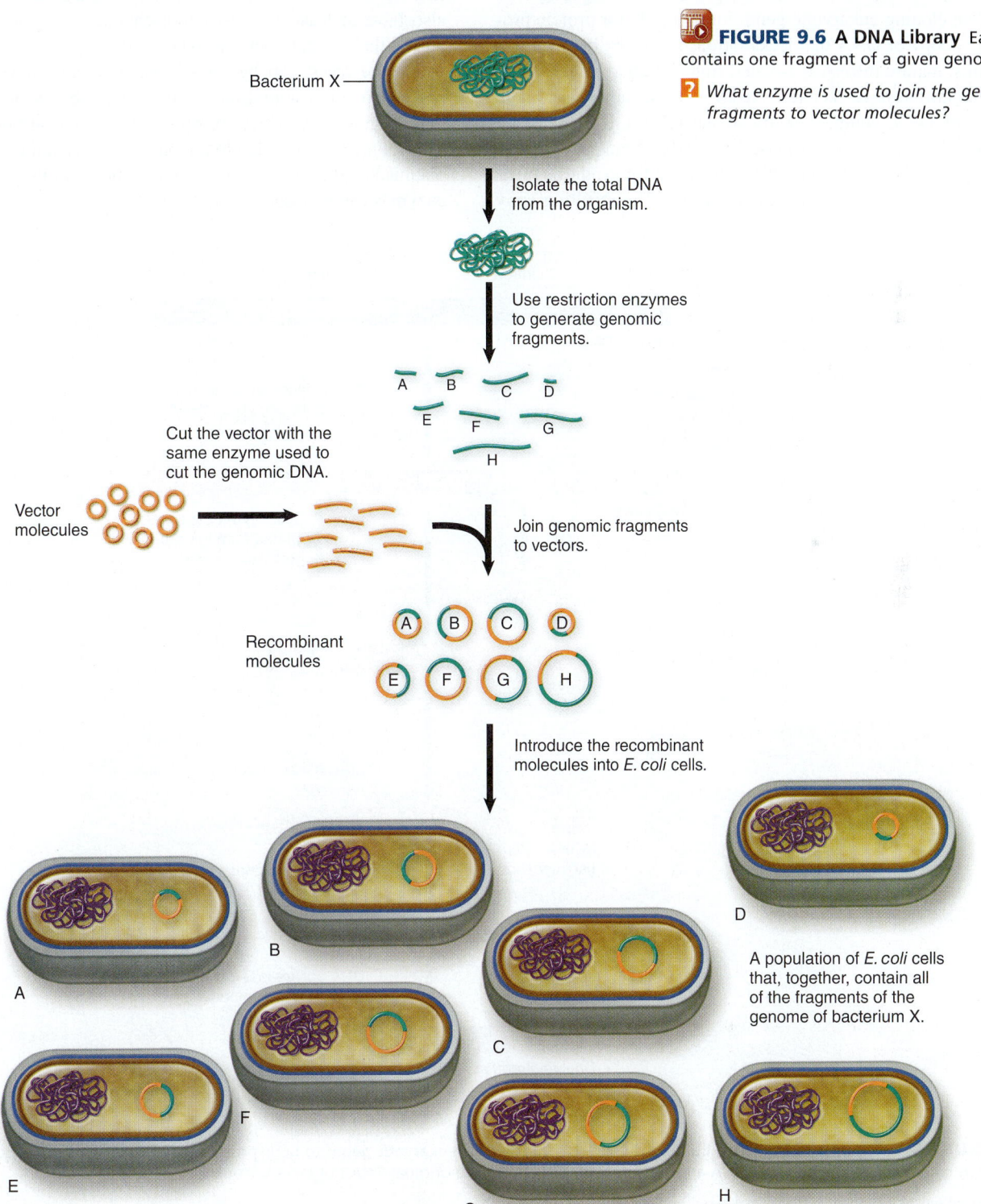

FIGURE 9.6 A DNA Library Each cell contains one fragment of a given genome.

❓ *What enzyme is used to join the genomic fragments to vector molecules?*

Bacterium X

Isolate the total DNA from the organism.

Use restriction enzymes to generate genomic fragments.

A B C D
E F G
H

Cut the vector with the same enzyme used to cut the genomic DNA.

Vector molecules

Join genomic fragments to vectors.

Recombinant molecules

A B C D
E F G H

Introduce the recombinant molecules into *E. coli* cells.

A B C D

A population of *E. coli* cells that, together, contain all of the fragments of the genome of bacterium X.

E F G H

This section will describe methods used to clone DNA in bacterial cells. For information regarding the engineering of eukaryotic organisms, see Perspective 8.2 and visit the text website (**www.mhhe.com/nester7**).

Obtaining DNA

The first step of a cloning experiment is to obtain the DNA that will be cloned. This is done by adding a detergent to cells in a broth culture to lyse them. As the cells burst, the relatively fragile DNA is inevitably sheared into many pieces of varying lengths.

When cloning eukaryotic genes into bacteria for protein production, a copy of DNA that lacks introns must first be obtained. To do this, mature mRNA is isolated from the appropriate tissue; recall that introns have been removed from these molecules. Then, a strand of DNA complementary to the mRNA is synthesized *in vitro* using reverse transcriptase, an enzyme from retroviruses. That strand of DNA is then used as a template for synthesis of its complement, creating double-stranded DNA. The resulting copy

of DNA, or **cDNA,** encodes the same protein as the original DNA but lacks introns (**figure 9.7**). ◄◄ **introns, p. 176** ►► **reverse transcriptase, p. 320**

Generating a Recombinant DNA Molecule

As described earlier, restriction enzymes and DNA ligase are used to create recombinant molecules. The **vector,** usually a modified plasmid or bacteriophage, has an origin of replication and functions as a carrier of the cloned DNA (**figure 9.8**). Vectors must also have at least one restriction enzyme recognition site. This allows the circular vector to be cut, forming a linear molecule to which the insert can be joined. Many vectors have been engineered to contain a multiple-cloning site, a short sequence that has the recognition sequences of several different restriction enzymes. The value of a multiple-cloning site is its versatility—a fragment obtained by digesting with any of a number of different restriction enzymes can be inserted into the site.

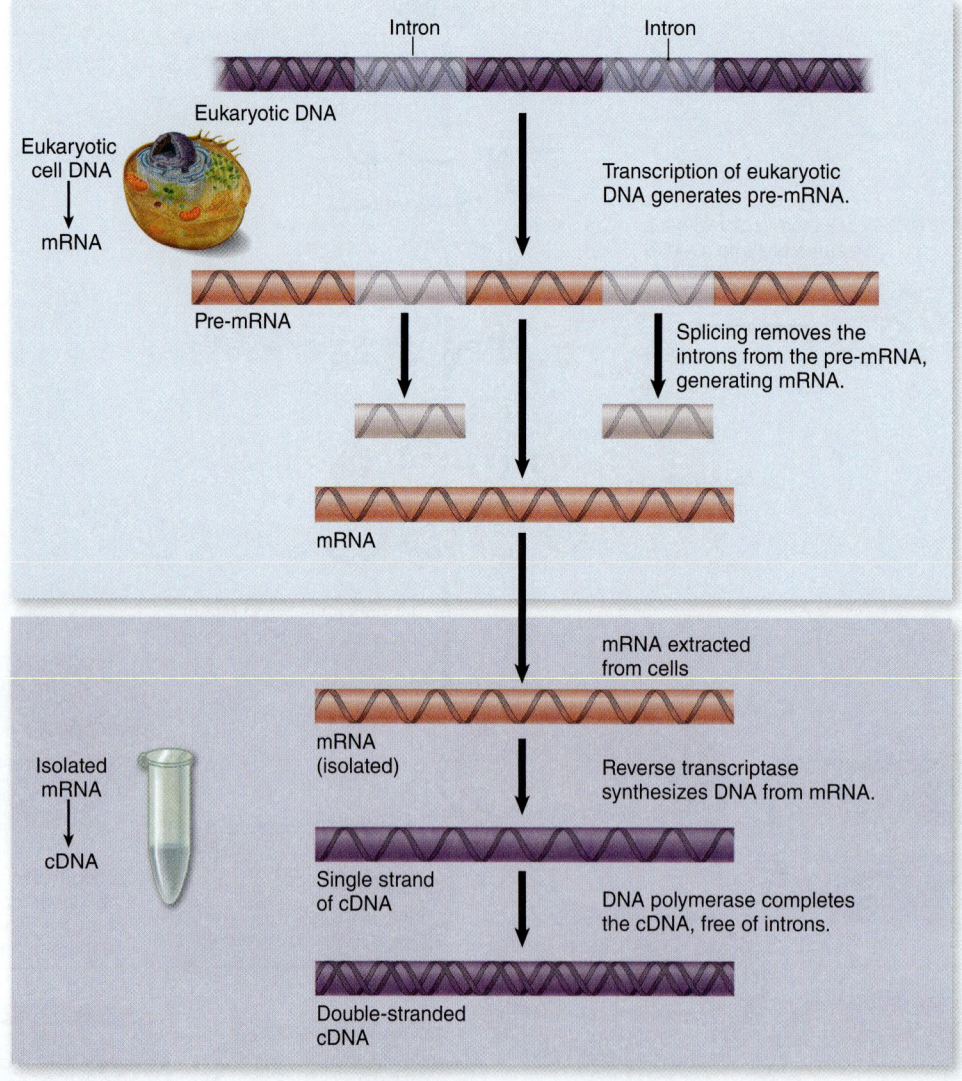

FIGURE 9.7 Making cDNA from Eukaryotic mRNA In order for eukaryotic genes to be expressed by a prokaryotic cell, a copy of DNA without introns must be cloned. The cDNA encodes the same protein as the original DNA but lacks introns.

❓ *What enzyme is used to make cDNA from pre-mRNA?*

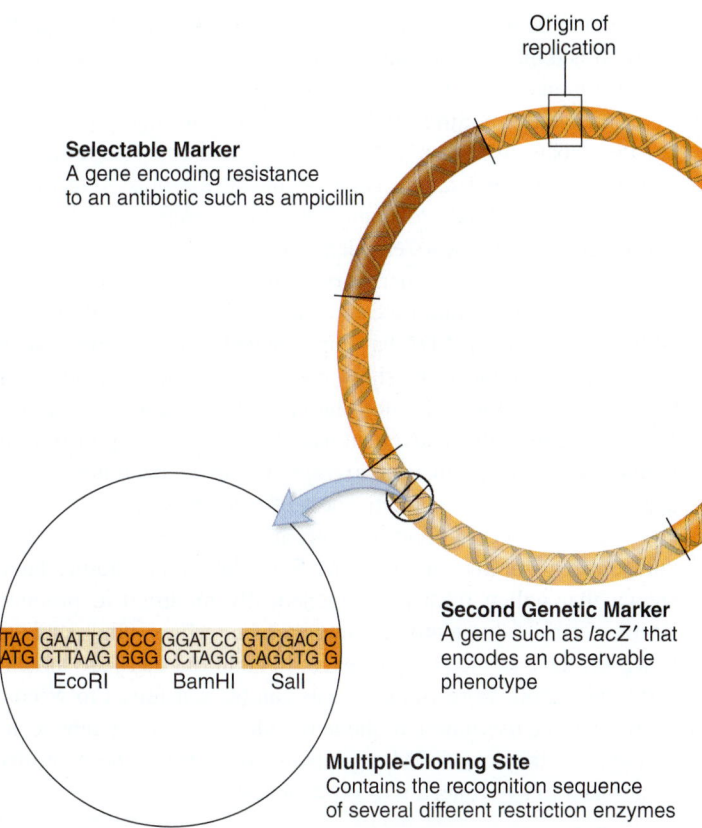

Origin of
replication

Selectable Marker
A gene encoding resistance
to an antibiotic such as ampicillin

```
TAC GAATTC CCC GGATCC GTCGAC C
ATG CTTAAG GGG CCTAGG CAGCTG G
    EcoRI      BamHI    SalI
```

Second Genetic Marker
A gene such as *lacZ'* that
encodes an observable
phenotype

Multiple-Cloning Site
Contains the recognition sequence
of several different restriction enzymes

FIGURE 9.8 Typical Properties of an Ideal Vector Most vectors have an origin of replication, a selectable marker, and a multiple-cloning site. A second genetic marker, used to differentiate cells containing recombinant plasmids from those that contain an intact vector, spans the multiple-cloning site.

? *When a vector has been successfully joined with a vector, will the selectable marker still be functional? Will the second genetic marker still be functional?*

Vectors typically have a **selectable marker,** a gene that allows cells to grow in otherwise inhibitory or lethal conditions. This is important because even under ideal conditions, most cells in a population do not take up DNA. The selectable marker allows the researcher to eliminate those cells that have not taken up either the recombinant molecule or the vector. A common selectable marker is the gene that encodes resistance to the antibiotic ampicillin; cells that have taken up a vector or recombinant molecule are able to grow on a medium containing ampicillin. ▶▶ **ampicillin, p. 464**

Most vectors have a second genetic marker, in addition to the selectable marker, used to distinguish cells that contain recombinant plasmids from those containing an intact vector. This is important because when the vector and insert DNA are both cut with the same restriction enzyme and the fragments mixed, not all will form the desired recombinant molecules. For example, the two ends of the vector can anneal to each other, regenerating the circular vector, which can replicate when introduced into a cell. The second genetic marker is situated so that it will be insertionally inactivated when an insert is in the multiple-cloning site. A good illustration of the utility of the second genetic marker is provided by a vector called pUC18 (**figure 9.9**). The second genetic marker of pUC18 is a gene called *lacZ'*. The product of this gene cleaves a colorless chemical, X-gal,

to form a blue compound. Because the multiple-cloning site of pUC18 is within the *lacZ'* gene, creation of a vector-insert hybrid results in a non-functional gene product. Thus, cells that harbor intact vector have a functional *lacZ'* gene and form blue colonies, whereas those that contain a recombinant molecule form white ones. ◀◀ **insertional inactivation, p. 192**

Various vectors are available for cloning eukaryotic DNA into bacterial cells, and the choice depends largely on the purpose of the procedure. If the goal is to produce the encoded protein, then a vector designed to optimize transcription and translation of the insert DNA is used. To create a DNA library of a human or other eukaryotic genome, a vector that can carry a large insert is generally used. For more information about these vectors, visit the text website (**www.mhhe.com/nester7**).

Obtaining Cells That Harbor Recombinant DNA Molecules

Once recombinant plasmids are generated, they must be transferred into suitable hosts where the molecules can replicate. For routine cloning experiments, one of the many well-characterized laboratory strains of *E. coli* is generally used. These strains are

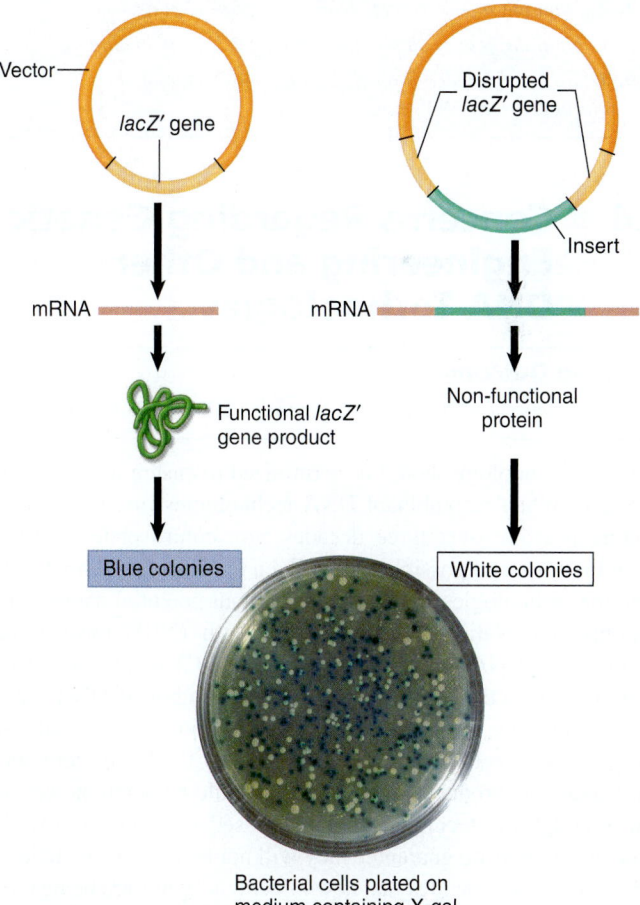

Vector
lacZ' gene

Disrupted
lacZ' gene

Insert

mRNA

mRNA

Functional *lacZ'*
gene product

Non-functional
protein

Blue colonies

White colonies

Bacterial cells plated on
medium containing X-gal

FIGURE 9.9 The Function of the *lacZ'* Gene in a Vector The *lacZ'* gene is used to differentiate cells that contain recombinant plasmid from those that contain vector alone.

? *What color colonies will cells that harbor a recombinant molecule form?*

easy to grow, and they are well characterized with respect to their genetics, biochemistry, and antibiotic sensitivities.

A common method of introducing DNA into a bacterial host is DNA-mediated transformation. *E. coli* cells are not naturally competent and must be specially treated to induce them to take up DNA. An alternative technique is to introduce the DNA by electroporation, a procedure that subjects the cells to an electric current. ◄◄ DNA-mediated transformation, p. 202

After the DNA is introduced into the new host, the transformed bacteria are grown on a medium that both selects for cells containing vector sequences and differentiates those carrying recombinant plasmids (figure 9.9). If the vector's selectable marker encodes ampicillin resistance, for example, then the transformed cells are grown on ampicillin-containing medium. If the second genetic marker is the *lacZ′* gene, then X-gal is added to the medium as well. Cells that form white colonies (rather than blue) should harbor a recombinant molecule; these are then further characterized to determine if they carry the gene of interest.

MicroAssessment 9.3

Introns can be removed from eukaryotic DNA to create cDNA. Vectors typically have an origin of replication, a selectable marker, a multiple-cloning site, and a second genetic marker.

7. *Explain the role of reverse transcriptase in cloning.*
8. *Explain the role of ampicillin and the* lacZ′ *gene in cloning.*
9. *What would happen if a cell took up a molecule of circular insert DNA only?* ✚

9.4 ■ Concerns Regarding Genetic Engineering and Other DNA Technologies

Learning Outcome

7. *Describe some of the concerns regarding DNA technologies.*

Any new technology should be scrutinized to ensure it is safe and effective. When recombinant DNA technologies first made gene cloning possible over three decades ago, controversies swirled about their use and possible abuse. Even the scientists who developed the technologies were concerned about potential dangers. In response, the National Institutes of Health (NIH) formed the Recombinant DNA Advisory Committee (RAC) to develop guidelines for conducting research involving recombinant DNA techniques and gene cloning. Today, we are enjoying the fruits of many of those technologies, as evidenced by the list of commercially available products in table 9.3. It should be noted, however, that although the technologies can be used to make life-saving products, there is no guarantee they will not be used for malicious purposes. Today, the idea that new infectious agents are being created for the purpose of bioterrorism is a disturbing possibility.

Recent advances in genomics have generated new cause for concern, primarily involving ethical issues regarding the appropriateness and confidentiality of information gained by analyzing a person's DNA. For example, will it be in an individual's best interest to be told of a genetic life-terminating disease? Could such

information be used to deny an individual certain rights and privileges? It is important that ongoing discussions about these complex issues continue as the technologies advance.

Genetically modified (GM) organisms hold many promises, but the debate over their use has raised concerns—some logical and others not. For example, some people have expressed fear over the fact that GM foods "contain DNA." Considering that DNA is consumed routinely as we eat plants and animals, this is obviously an irrational concern. Others worry that unanticipated allergens could be introduced into food products, posing a health threat. To address this issue, the FDA has implemented strict guidelines, such as requiring producers to show that GM products intended for human consumption do not unduly elicit allergic reactions. Incidents such as the inadvertent use of GM corn not approved for human consumption in tortilla chips, however, continue to fuel apprehension about the effectiveness of regulatory control.

Another concern about GM products is their possible unintended effects on the environment. Some laboratory studies have shown that pollen from plants genetically modified to produce Bt toxin can inadvertently kill monarch butterflies; other studies, however, have refuted the evidence. In addition, there are indications that herbicide-resistance genes can be transferred to weeds, decreasing the usefulness of the herbicide. As with any new technology, the impact of GM organisms will need to be carefully scrutinized to avoid negative consequences.

MicroByte ─────

The Genetic Information Nondiscrimination Act (GINA) protects Americans against discrimination in healthcare coverage and employment based on their genome.

MicroAssessment 9.4

Concerns about genetic engineering and genomics are varied and include ethical issues. Potentially adverse impacts of genetically modified organisms on human health and the environment are also a concern.

10. *Describe two concerns regarding information that can be gained by analyzing a person's DNA.*
11. *Describe two concerns regarding the use of genetically modified organisms.*

9.5 ■ DNA Sequencing

Learning Outcomes

8. *Describe two applications of DNA sequencing.*
9. *Describe the automated dideoxy chain termination method of DNA sequencing.*

DNA sequencing is the process of determining the nucleotide sequence of a DNA molecule. The **Human Genome Project,** the completed undertaking to sequence the human genome, resulted in highly automated and efficient techniques. These allowed scientists to more readily determine the genomic sequence of other organisms, including both prokaryotes and eukaryotes, fueling the rapidly growing field of genomics. The resulting explosion of data spawned a new field to analyze the information—bioinformatics. ◄◄ genomics, p. 184 ◄◄ bioinformatics, p. 184

By determining the DNA sequence of a genome, the amino acid sequence of the encoded proteins can be established. This makes it possible to compare characteristics of various proteins in different organisms. Non-coding sequences and mobile genetic elements can be compared as well. DNA sequencing also makes it possible to determine the evolutionary relatedness of organisms, a topic discussed in chapter 10. ◄◄ mobile genetic elements, p. 208

Sequencing technologies have become so efficient that a new project has been initiated—the **Human Microbiome Project**—which uses genomics to determine the biological diversity in the normal microbiota of the human body. This project will use both traditional approaches (sequencing the genome of individual microbial strains) and metagenomics. Goals of the Human Microbiome Project include comparing the composition of the microbiota in health and disease, and determining the extent of its variation in different individuals. ◄◄ normal microbiota, p. 8   ◄◄ metagenomics, p. 184

MicroByte —————————
Sequencing the first human genome took 13 years and $440,000,000; a race is on to sequence 100 human genomes in 10 days for about $10,000 each.

Techniques Used in DNA Sequencing

The most widely used technique for sequencing DNA is the **dideoxy chain termination** method. The procedure is now typically done using automated sequencing instruments, making it a fast and efficient process. Several newer sequencing methods have been developed, and these are speeding the process even more. The highly automated rapid methods are high throughput, meaning that large numbers of samples can be processed very quickly.

Dideoxy Chain Termination Method

The fundamental aspect of the dideoxy chain termination method is an *in vitro* DNA synthesis reaction. It requires:

- **Template DNA:** Single-stranded DNA from which complementary copies are synthesized. ◄◄ template, p. 163
- **DNA polymerase:** The enzyme that synthesizes DNA. ◄◄ DNA polymerase, p. 166
- **Primer:** Short DNA molecules that anneal to the complementary sequence in the single-stranded template molecules. The primer allows the researcher to choose where synthesis starts (**figure 9.10**). Recall that DNA polymerase can add nucleotides only to an existing fragment of DNA, thereby extending the length of fragment. ◄◄ primer, p. 166
- **Deoxynucleotides:** Each of the four deoxynucleotides used in DNA synthesis—dATP, dGTP, dCTP, and dTTP.
- **Dideoxynucleotides:** These are identical to their deoxynucleotide counterparts except they lack the 3′OH group.

If only the first four ingredients were in the reaction, full-length molecules complementary to the template DNA would always be synthesized. The dideoxynucleotides, however, function as **chain terminators.** They lack the chemical group to which subsequent nucleotides would be added. Therefore, when a dideoxynucleotide is added to a growing strand of DNA, no additional nucleotides can be incorporated. Elongation of that strand stops (**figure 9.11**).

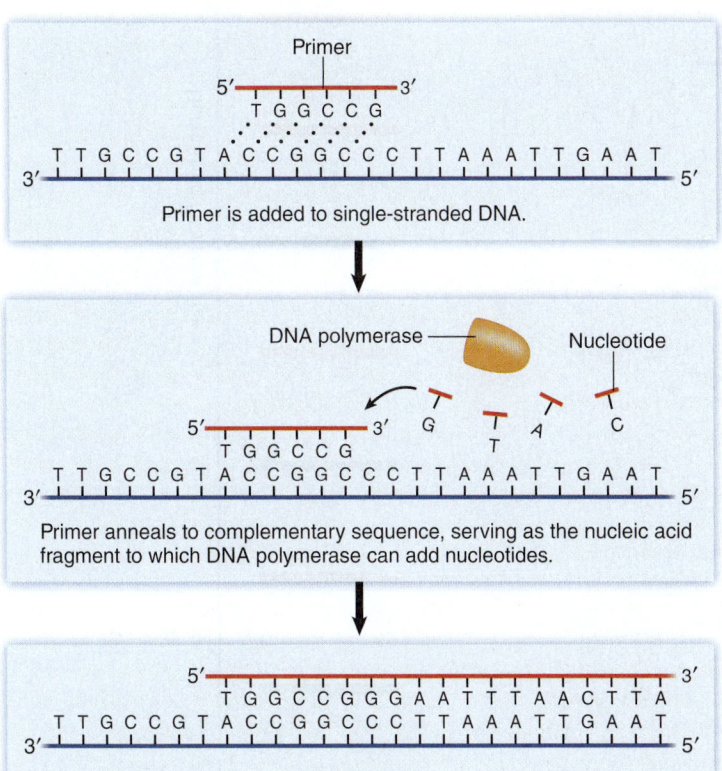

FIGURE 9.10 Primers Through appropriate primer selection, a researcher can choose the site where *in vitro* DNA synthesis will start.

❓ *Why are primers required in DNA sequencing reactions?*

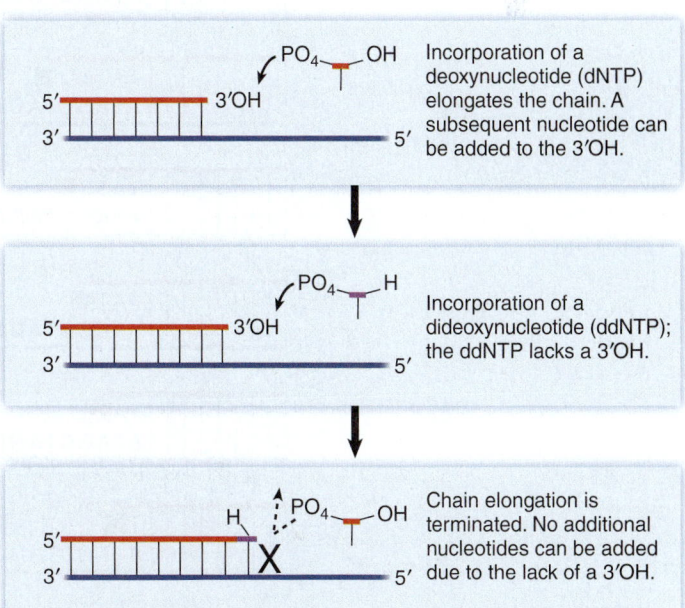

FIGURE 9.11 Chain Termination by a Dideoxynucleotide Once a dideoxynucleotide is incorporated into a growing strand, additional nucleotides cannot be added.

❓ *How is a dideoxynucleotide different from a deoxynucleotide?*

FIGURE 9.12 Dideoxy Chain Termination Method of DNA Sequencing This figure illustrates the principles of the automated method, which uses dideoxynucleotides that carry a fluorescent label.

? *What would happen if the deoxynucleotides were left out of the reaction?*

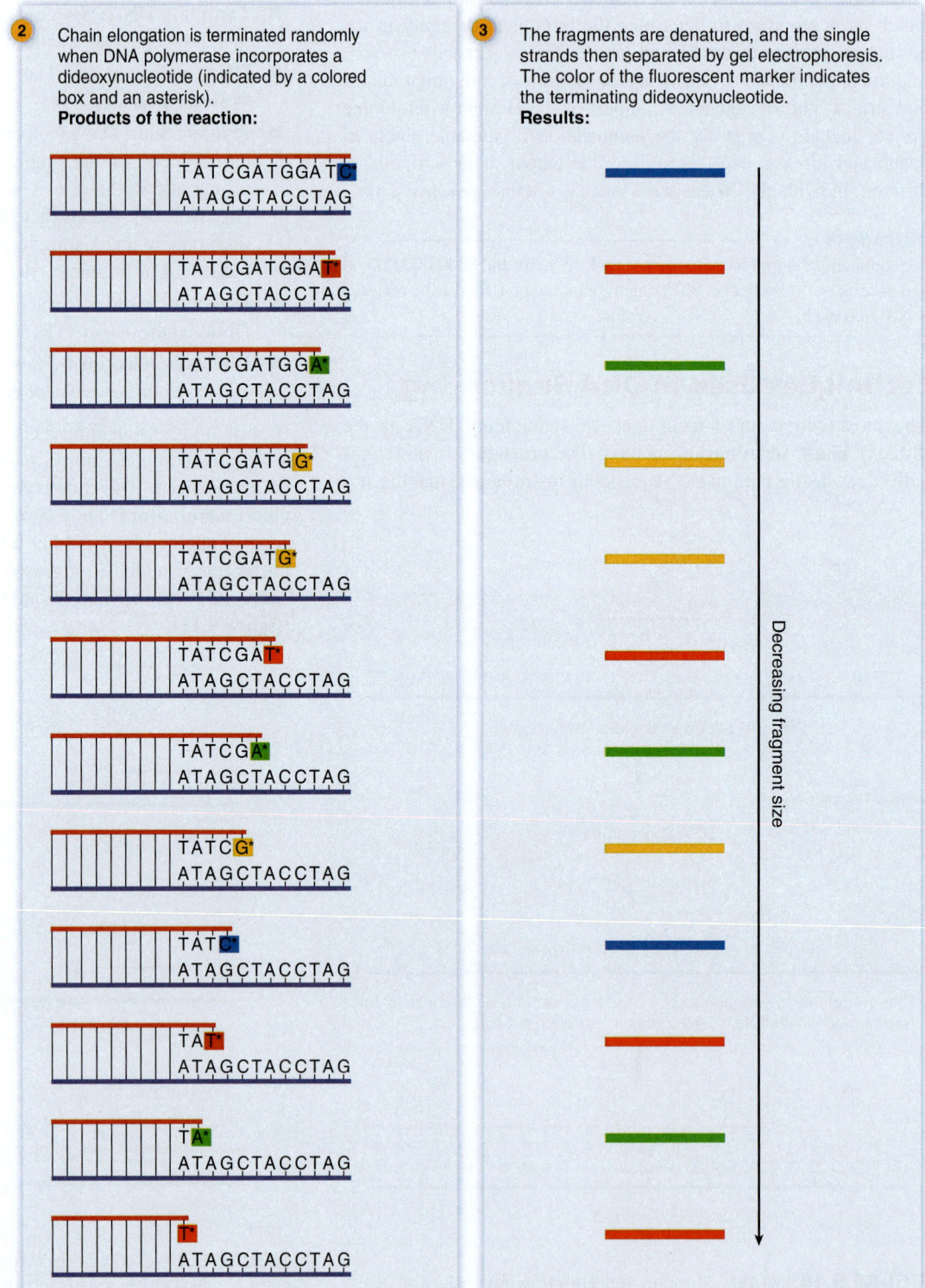

1 **Key ingredients in the reaction:**
• Primer and template (shown annealed)

ATAGCTACCTAG

• DNA polymerase
• Deoxynucleotides (dATP, dTTP, dGTP, dCTP)
• Fluorescently labeled dideoxynucleotides (ddATP, ddTTP, ddGTP, ddCTP; a very small amount of each)

2 Chain elongation is terminated randomly when DNA polymerase incorporates a dideoxynucleotide (indicated by a colored box and an asterisk).
Products of the reaction:

TATCGATGGATC*
ATAGCTACCTAG

TATCGATGGAT*
ATAGCTACCTAG

TATCGATGGA*
ATAGCTACCTAG

TATCGATGG*
ATAGCTACCTAG

TATCGATG*
ATAGCTACCTAG

TATCGAT*
ATAGCTACCTAG

TATCGA*
ATAGCTACCTAG

TATCG*
ATAGCTACCTAG

TATC*
ATAGCTACCTAG

TAT*
ATAGCTACCTAG

TA*
ATAGCTACCTAG

T*
ATAGCTACCTAG

3 The fragments are denatured, and the single strands then separated by gel electrophoresis. The color of the fluorescent marker indicates the terminating dideoxynucleotide.
Results:

Decreasing fragment size

In an automated dideoxy sequencing reaction, DNA polymerase, template DNA, primer, the four deoxynucleotides, and a very small amount of the four dideoxynucleotides are mixed together (**figure 9.12 ①**). The different dideoxynucleotides each carry a distinct fluorescent marker. When the reaction is incubated at an appropriate temperature, the primers anneal to template molecules and DNA synthesis begins. ② Each nucleotide chain is elongated until a chain terminator—a dideoxynucleotide—is incorporated. Termination happens infrequently, however, because of the small amount of ddNTPs relative to dNTPs. The significant aspect is that the color of the fluorescent marker carried by the chain terminator can be used to determine which nucleotide was incorporated at the terminating position.

After the sequencing reaction, the sample is heated, which denatures the DNA. ③ Electrophoresis is then used to separate the DNA fragments and determine their relative size. The conditions (pH, temperature, and gel concentration) keep the DNA single-stranded and allow fragments that differ in length by only one nucleotide to be separated. A laser is used to detect the colors of fluorescent bands as they run past, recording their intensity as a peak. The order of the colored peaks reflects the nucleotide sequence of the DNA (**figure 9.13**).

MicroAssessment 9.5

Efficient DNA sequencing methods have fueled the rapidly growing field of genomics. The automated dideoxy chain termination method is a widely used sequencing technique.

12. *What is the Human Microbiome Project?*
13. *How does a ddNTP terminate DNA synthesis?*
14. *What would happen in the sequencing reaction if the relative concentration of a dideoxynucleotide were increased?* ➕

9.6 ■ Polymerase Chain Reaction (PCR)

Learning Outcomes

10. *Describe how PCR can be used to diagnose diseases.*
11. *Explain how PCR can be used to exponentially amplify a select region of DNA.*

The **polymerase chain reaction (PCR)** makes it possible to create more than a billion copies of a given region of DNA—referred to as target DNA—in a matter of hours. In fact, target DNA can be generated in sufficient concentration to be visible to the unaided eye when fragments in the sample are separated by gel electrophoresis, stained with ethidium bromide, and illuminated with UV light (**figure 9.14**).

One important application of PCR is disease diagnosis. For example, if a woman is suspected of having the sexually transmitted disease gonorrhea, PCR can be used to detect the causative agent (*Neisseria gonorrhoeae*) in a vaginal specimen. This is done by treating the specimen to release the DNA in cells and then amplify target DNA unique to *Neisseria gonorrhoeae*. If that DNA is amplified, then the pathogen must be present. PCR can also be used to detect HIV nucleotide sequences in a sample of blood cells, thereby diagnosing HIV infection.

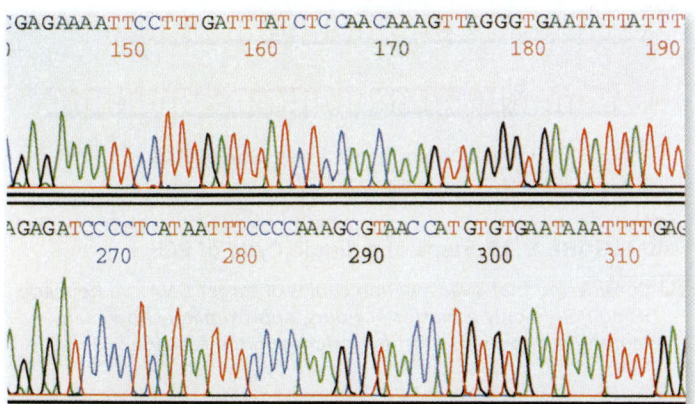

FIGURE 9.13 Results of Automated DNA Sequencing The order of the colored peaks reflects the nucleotide sequence of the DNA.

❓ *What allows the peaks to be different colors?*

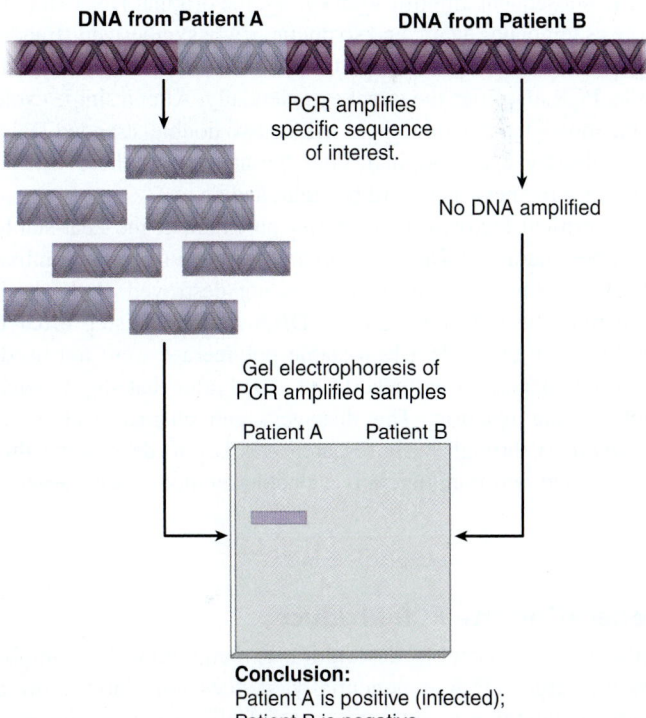

FIGURE 9.14 PCR Amplifies Selected Sequences

❓ *How can PCR be used to diagnose an infectious disease?*

Techniques Used in PCR

The polymerase chain reaction starts with a double-stranded DNA molecule that serves as a template from which more than a billion identical copies of the target DNA can be produced. The process involves DNA synthesis reactions and the following key ingredients:

- **Double-stranded DNA:** This contains the target DNA and serves as a template for DNA synthesis.
- ***Taq* polymerase:** This heat-stable DNA polymerase is from the thermophile *Thermus aquaticus.*
- **Primers:** As with the DNA sequencing reaction, the primers allow the researcher to choose where synthesis starts.
- **Deoxynucleotides:** Each of the four deoxynucleotides used in DNA synthesis—dATP, dGTP, dCTP, and dTTP.

The Three-Step Amplification Cycle

PCR uses a repeating cycle consisting of three steps (**figure 9.15**). **1** In the first step, the sample is heated to near-boiling (about 95°C) in order to denature the DNA. **2** In the second step, the temperature is lowered (to about 50°C); within seconds, the primers anneal to their complementary sequences on the denatured target DNA. **3** In the third step, the temperature is raised to the optimal temperature of *Taq* DNA polymerase (about 70°C), allowing DNA synthesis to occur. After one three-step cycle, the target DNA is duplicated.

In subsequent amplification cycles, the original DNA strands serve as templates again, and so do the newly synthesized strands. Because the number of template molecules increases with each cycle, PCR amplifies the target exponentially. After a single cycle of the three-step reaction, there will be two double-stranded DNA molecules for every original; after the next cycle, there will be four; after the next there will be eight, and so on.

A critical factor in PCR is *Taq* polymerase, the heat-stable DNA polymerase of *Thermus aquaticus.* This polymerase, unlike the DNA polymerase of *E. coli,* is not destroyed at the high temperature used to denature the DNA in the first step of each amplification cycle. If a heat-stable polymerase were not used, fresh polymerase would need to be added after that step in each cycle of the reaction. The discovery and characterization of *T. aquaticus* through basic research was key to developing this widely used and commercially valuable method. ◀◀ thermophile, p. 90

Generating the PCR Product

Although the preceding description explains how PCR amplifies the target DNA exponentially, it does not clarify how a fragment containing only the target DNA becomes the predominant product. This fragment is referred to as the **PCR product.** Generating the PCR product is important, because it allows the technician to use PCR coupled with gel electrophoresis to detect the target DNA in a sample. After PCR, the

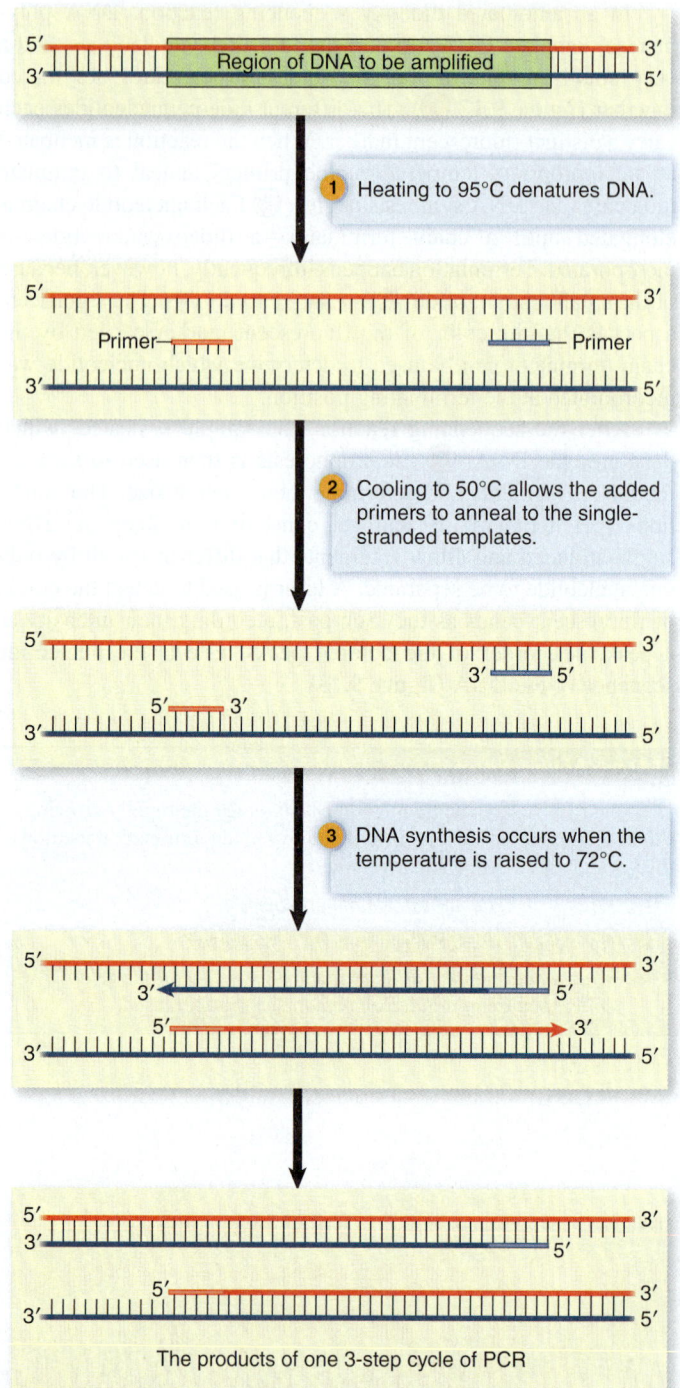

1 Heating to 95°C denatures DNA.

2 Cooling to 50°C allows the added primers to anneal to the single-stranded templates.

3 DNA synthesis occurs when the temperature is raised to 72°C.

The products of one 3-step cycle of PCR

FIGURE 9.15 Steps of a Single Cycle of PCR

? *Considering that over a billion copies of target DNA can be made using PCR in only a matter of hours, approximately how many molecules of primers must be included in the reaction?*

amplified target (PCR product) will appear as a single band on the ethidium bromide–treated gel.

To understand how the PCR product is generated, you must consider the exact sites to which the primers anneal, and visualize

PERSPECTIVE 9.1

Science Takes the Witness Stand

After serving more than 10 years on a rape charge, a wrongfully convicted young man was released from prison when a new DNA typing technique exonerated him. The new technique, based on using the polymerase chain reaction (PCR) to amplify specific sequences, showed that the semen sample taken from the rape victim did not contain the man's DNA. Indeed, a database indicated a DNA match with a man currently in prison for an unrelated rape charge.

Stories abound about the growing power of DNA evidence for obtaining convictions and also clearing the wrongly accused, but how is DNA used in forensics? It is not practical to compare the entire nucleotide sequence of two people; instead, specific regions that vary significantly between individuals are analyzed. In the past, forensics labs have used a method that employs the probe-based technique called Southern blot hybridization to detect differences. This method provides valuable information but cannot be used on small or degraded

samples and is quite time-consuming. Most forensics labs have now switched to using a PCR-based method because results can be obtained in less than 5 hours from a sample as small as a drop of blood the size of a pinhead. In fact, the FBI now catalogs PCR-based DNA profiles from unsolved crimes and convicted violent offenders, making it easier to track or link the crimes of serial offenders. The national database is called CODIS (**Co**mbined **D**NA **I**ndex **S**ystem).

PCR-based DNA typing amplifies certain chromosomal regions that contain short tandem repeats (STRs). These consist of a core sequence of two to six base pairs that repeat a variable number of times in different people. On chromosome 2, for example, in an intron within the thyroid peroxidase gene, the sequence AATG is repeated sequentially between 5 to 14 times. In one individual, there may be 9 of these STRs in one copy of that chromosome and 7 in the other, whereas another individual may have 11 and 5 (**figure 1**). This variation, or polymorphism,

makes tandem repeats a useful genetic marker for distinguishing individuals. With PCR, using primers that bind regions flanking the repeating sequences, the number of repeats can be determined. A fragment that contains 9 repeats, for example, will be longer than one that contains only 7. The PCR-amplified fragments can be quickly separated using a rapid type of gel electrophoresis called capillary electrophoresis. A size standard is incorporated so that the sizes of the PCR-amplified fragments can be determined.

The FBI's CODIS database catalogs the amplification pattern of 13 different STR loci (chromosomal locations). Commercially available kits contain fluorescently labeled primers that allow simultaneous amplification and subsequent recognition of each of the 13 loci. A laser detects the color of each amplified fragment as it moves out of the capillary gel, and computer analysis generates a pattern of peaks that reflect the STR profile of the DNA sample.

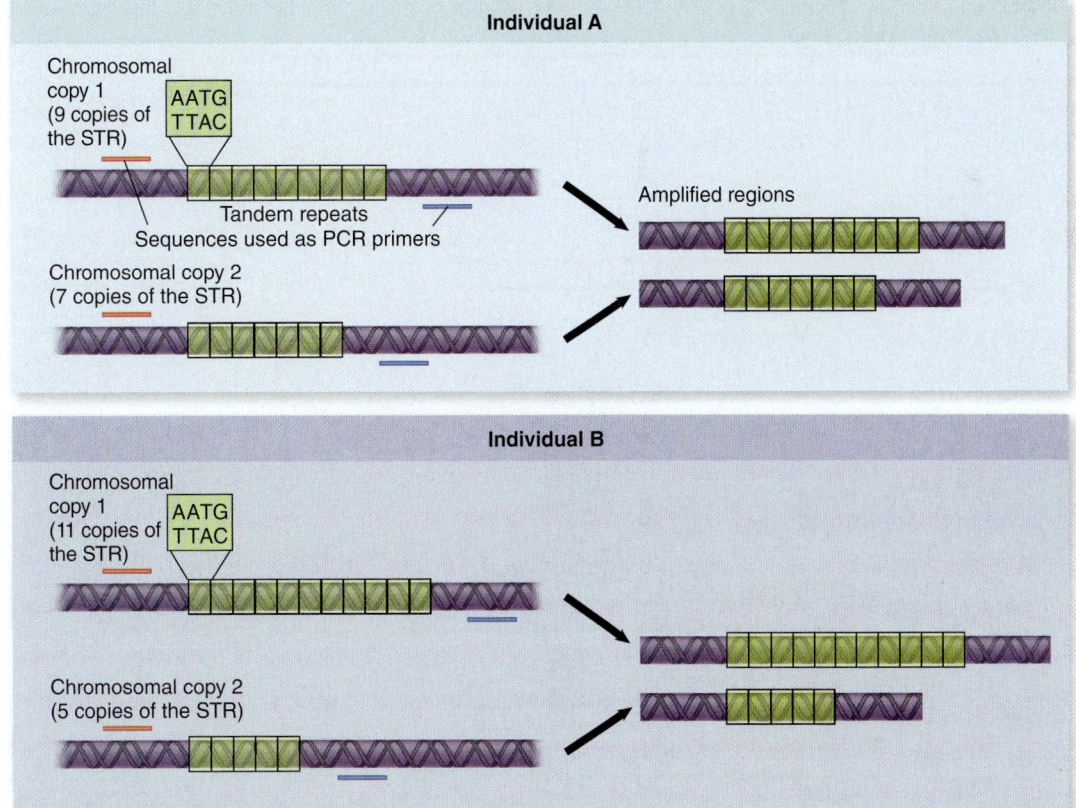

FIGURE 1   Using PCR to Type DNA PCR is used to amplify certain chromosomal regions containing short tandem repeats (STRs). The number of copies of a given STR varies among people, resulting in corresponding differences in the length of the amplified fragments. Typically, at least 13 different STR locations are analyzed.

at least three cycles of PCR (**figure 9.16**). ❶ In the first cycle, two new strands are generated. Note, however, that their 5′ ends are primer DNA; the strands are shorter than the original templates but longer than the target DNA. These mid-length strands will be generated whenever the original full-length molecule is used as a template, which is once per three-step cycle.

In the next cycle, the full-length molecules will again be used as templates, repeating the process just described. ❷ More importantly, the mid-length strands created during the first cycle will be used as templates for DNA synthesis. As before, the primers will anneal to these strands and then nucleotides will be added to the 3′ end. Elongation, however, will stop at the 5′ end of the template molecule, because DNA synthesis requires a template. Recall that the 5′ end of the template is primer DNA. Thus, whenever a mid-length strand is used as a template, a short strand—exactly the length of the target DNA—is generated. The

5′ and 3′ ends of this strand are determined by the sites to which the primers initially annealed.

In the third round of replication, the full-length and the mid-length strands again will be used as templates, repeating the processes just described. ❸ The short strands generated in the preceding round will also be used as templates, however, generating short double-stranded molecules—the PCR product. Continuing to follow the events in further rounds of replication will reveal that this fragment is exponentially amplified (**figure 9.17**).

Selecting Primer Pairs

The nucleotide sequences of the two primers are critical because the primers dictate which portion of the DNA is amplified. Each must be complementary to one end of the target DNA, so that

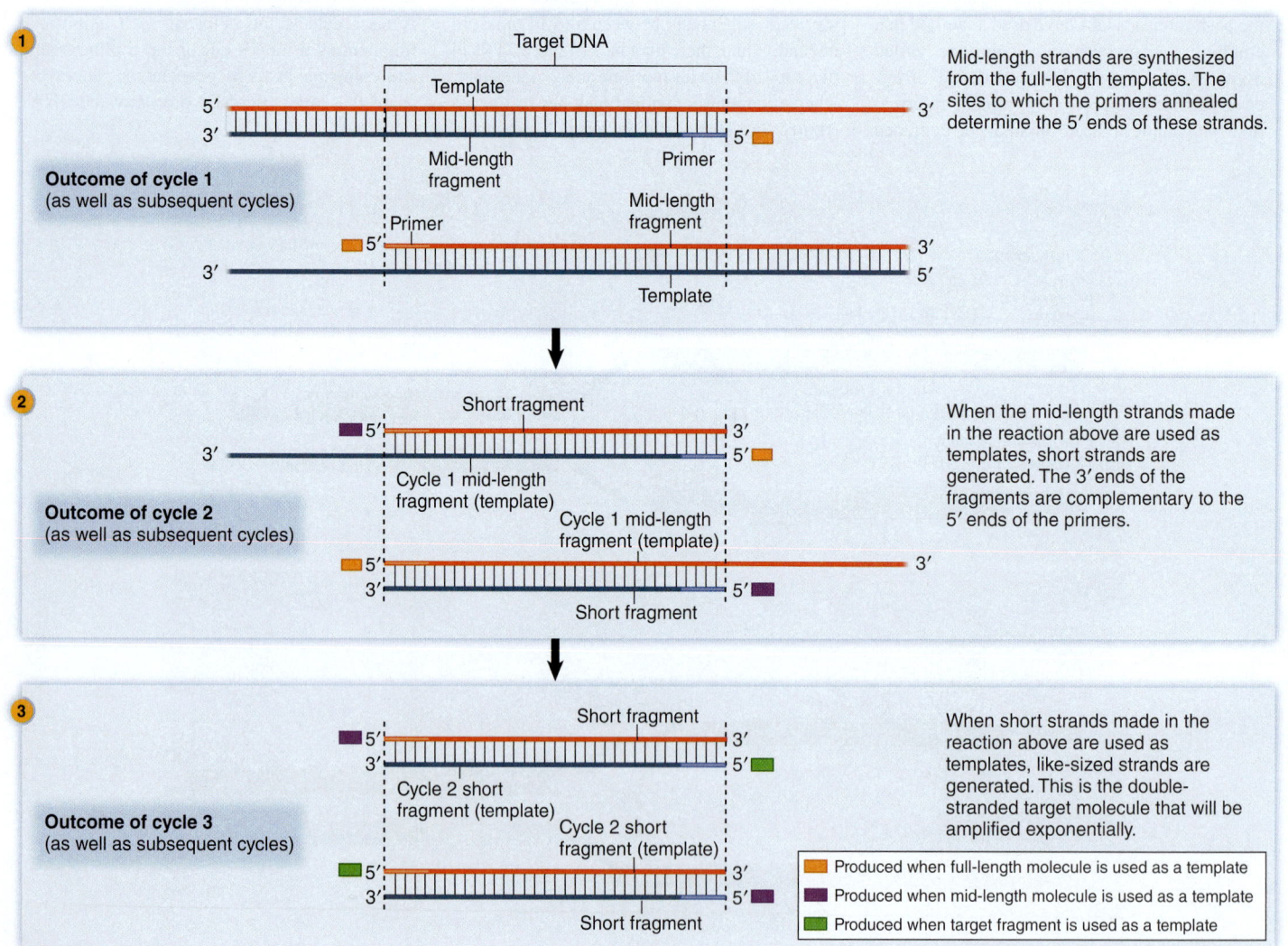

FIGURE 9.16 The PCR Product Is a Fragment of Discrete Size The positions to which the primers anneal to the template dictate the size and sequence of the fragment amplified exponentially.

❓ *If two mid-length strands are present at the end of cycle 1, how many will be present at the end of cycle 3?*

FIGURE 9.17 Exponential Amplification of Target DNA During PCR, mid-length fragments are amplified linearly (arithmetically), whereas the discrete-sized target DNA, referred to as PCR product, is amplified exponentially. After 30 cycles of PCR, more than a billion molecules of PCR product will have been synthesized.

? *How many PCR cycles are required to generate double-stranded target fragments?*

DNA synthesis will extend across that stretch of DNA. If a technician wants to amplify a DNA sequence that encodes a specific protein, he or she must first determine the nucleotide sequences at the ends of the gene. That information can be used to obtain the appropriate pair of primers.

9.7 ■ Probe Technologies

Learning Outcome

12. *Compare and contrast the applications and techniques of colony blotting, FISH, and DNA microarray technologies.*

DNA probes are used to locate specific nucleotide sequences in nucleic acid samples attached to a solid surface. The probe is a single-stranded piece of DNA, complementary to the sequence of interest, that has been labeled with a detectable marker such as a radioactive isotope or a fluorescent dye. The probe will anneal to its complement, a process called **hybridization.** By hybridizing to its complement, the probe "finds" the sequence of interest and makes it detectable (**figure 9.18**).

A variety of technologies use DNA probes to locate specific nucleotide sequences. They include colony blotting, fluorescence *in situ* hybridization (FISH), and DNA microarrays. The technique called Southern blotting also uses probes, but its applications have been largely replaced by PCR. For a description of Southern blotting, visit the text website (**www.mhhe.com/nester7**).

Colony Blotting

Colony blotting uses probes to detect specific DNA sequences in colonies grown on agar plates (**figure 9.19 ①**). This method is commonly used to determine which clones in a DNA library or other collection contain a sequence being studied. ② The term "blot" in the name reflects the fact that the colonies are transferred in place ("blotted") onto a nylon membrane, creating a replica of the colonies on the original plate. The membrane serves as a durable, permanent support for the cells of the colonies and their DNA. ③ After the transfer, the membrane is soaked in an alkaline solution to simultaneously lyse the cells and denature their DNA, generating single-stranded DNA molecules. ④ A solution containing the probe is then added to the

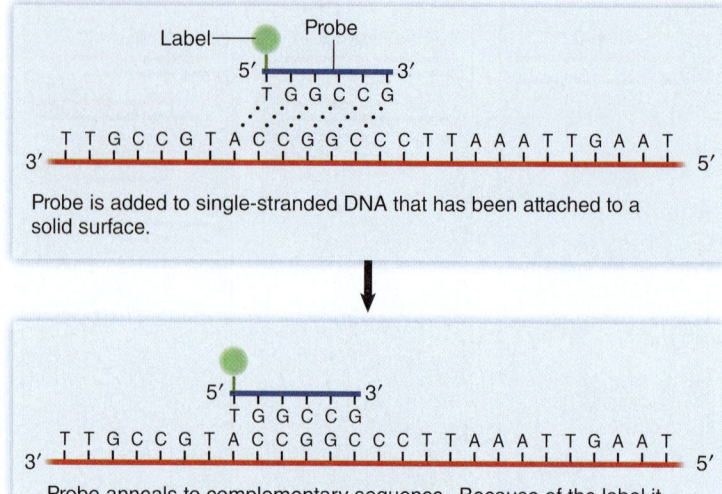

Probe is added to single-stranded DNA that has been attached to a solid surface.

Probe anneals to complementary sequence. Because of the label it carries, its location can easily be determined.

FIGURE 9.18 DNA Probes These single-stranded pieces of DNA tagged (labeled) with a detectable marker are used to detect specific nucleotide sequences in DNA or RNA samples that have been attached to a solid surface.

❓ *Why is it important that the probe be labeled?*

membrane and incubated under conditions that allow the probe to hybridize to complementary sequences on the filter. Any probe that has not bound is then washed off. ⑤ The appropriate method is then used to detect the label carried by the probe, thereby locating the position of the hybridized probe. The positions to which the probe hybridized indicate colonies that have the DNA of interest.

Fluorescence *in situ* Hybridization (FISH)

Fluorescence *in situ* hybridization (FISH) uses a fluorescently labeled probe to detect specific nucleotide sequences within intact cells attached to a microscope slide. Cells containing the hybridized probe can then be observed using a fluorescence microscope. To study prokaryotes, a probe that hybridizes to sequences on ribosomal RNA (rRNA) is generally used. This is because multiplying cells can have thousands of copies of rRNA, increasing the technique's sensitivity. Other characteristics of rRNA that make it useful for identifying prokaryotes are described in chapter 10. ◀◀ fluorescence microscope, p. 44

FISH is revolutionizing microbial ecology research and holds great promise in clinical laboratories. It provides a means to rapidly identify microorganisms directly in a specimen, bypassing the need to grow them in culture. FISH can be used to detect either a group of related organisms or a specific species, depending on the nucleotide sequence of the probe. For example, FISH can be used to determine the relative proportion of two different groups of

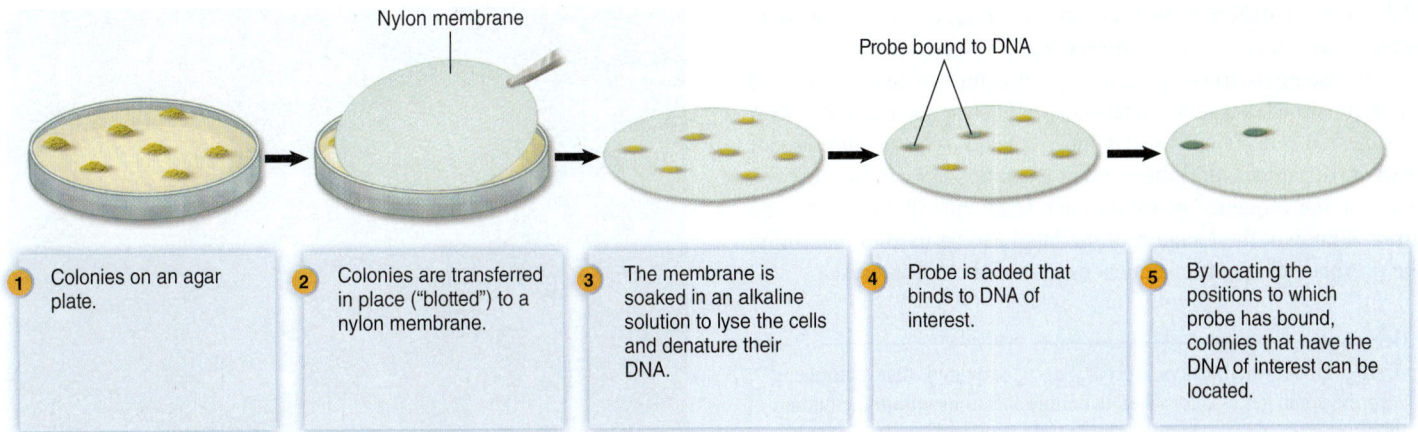

| Nylon membrane | | Probe bound to DNA | |

① Colonies on an agar plate.

② Colonies are transferred in place ("blotted") to a nylon membrane.

③ The membrane is soaked in an alkaline solution to lyse the cells and denature their DNA.

④ Probe is added that binds to DNA of interest.

⑤ By locating the positions to which probe has bound, colonies that have the DNA of interest can be located.

FIGURE 9.19 Colony Blotting This technique is used to determine which colonies on an agar plate contain a given DNA sequence.

? *What is the purpose of soaking the membrane in an alkaline solution?*

prokaryotes in the same specimen by using two separate probes—one specific for each group—labeled with different colored fluorescent markers (**figure 9.20**). It can also be used to identify cells of the bacterium that causes tuberculosis in a sputum specimen.
▶▶ microbial ecology, p. 719 ▶▶ tuberculosis, p. 502

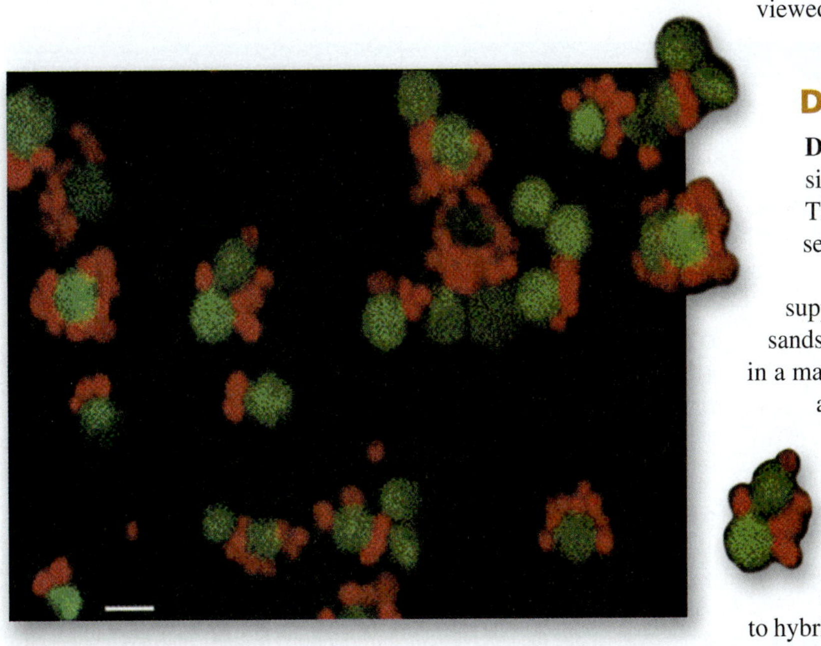

To analyze a sample using FISH, the sample must first be treated with chemicals to preserve the shape of the cells, inactivate enzymes that might otherwise degrade the nucleic acid, and make the cells more permeable so that the labeled probe molecules can easily enter. Once the specimen has been prepared, it is put on a glass slide, bathed with a solution containing the labeled probe, and incubated under conditions that allow hybridization to occur. Unbound probe is then washed off. Finally, the specimen is viewed using a fluorescence microscope.

DNA Microarrays

DNA microarrays are primarily used to study gene expression in organisms whose genomes have been sequenced. They can also be used to detect a specific nucleotide sequence in a DNA sample of interest.

A DNA microarray is a glass slide or other small solid support carrying an arrangement of tens or hundreds of thousands of short DNA fragments. Each DNA fragment functions in a manner analogous to a probe, allowing a researcher to screen a single sample for a vast range of different sequences simultaneously. Unlike typical probes, however, the arrays do not carry a detectable label. Instead, the label is on the nucleic acid of interest.

To study gene expression, mRNA is isolated from an organism and converted to fluorescently labeled, single-stranded cDNA. Those molecules are then allowed to hybridize to a microarray specially created to include sequences specific for each gene of a particular organism. The locations of the labeled cDNA molecules can be detected using a computerized scanner. The genes to which cDNAs hybridize are ones that the organism was expressing. By doing the experiment using cultures grown under different sets of conditions, and labeling their

FIGURE 9.20 Fluorescence *in situ* Hybridization (FISH) Two different probes have been used to stain the cells. The probe that hybridizes to *Ignicoccus* rRNA fluoresces green; the one that hybridizes to *Nanoarchaeum* rRNA fluoresces red. Size bar = 1 μm.

? *Why does FISH typically require a probe that hybridizes to rRNA?*

cDNAs with different fluorescent markers, variations in gene expression can be revealed (**figure 9.21**). ◀◀ **cDNA, p. 222**

To use microarrays to detect specific nucleotide sequences in an organism whose genome has not yet been sequenced, the DNA is digested into small fragments, labeled with a fluorescent marker, denatured, and then added to an appropriate microarray. Because the sequence of each of the fragments that make up the array is known, the location of the label can be used to determine the presence of specific sequences in the DNA of interest.

MicroByte

Using DNA microarrays, researchers discovered that pathogens express some genes only when in certain locations within the human body.

MicroAssessment 9.7

Colony blotting uses probes to identify colonies that contain a given sequence of DNA. Fluorescence *in situ* hybridization is used to observe individual cells that contain a given sequence. DNA microarrays allow researchers to study gene expression and to screen a sample for a vast range of different sequences simultaneously.

18. *What role does colony blotting play in cloning?*
19. *Why is a probe that binds to rRNA used in FISH?*
20. *In FISH, what would happen if unbound probe was not washed off?* ✚

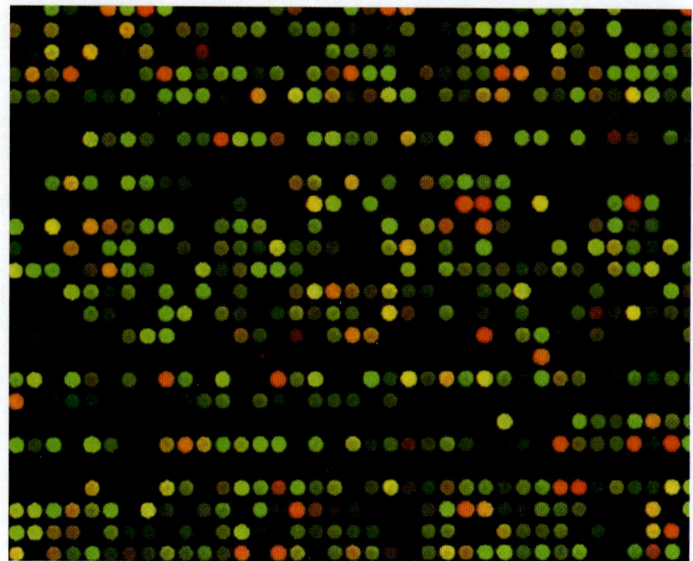

FIGURE 9.21 A DNA Microarray This is an example of an array used to study gene expression. Two different cDNA samples (one labeled with a red fluorescent marker and the other with a green fluorescent marker) were simultaneously hybridized to fragments in the microarray. Red dots indicate the positions to which one sample hybridized, and green dots indicate the positions to which the other hybridized. Yellow dots indicate that both samples hybridized.

❓ *How can DNA microarrays be used to determine which genes of a bacterial pathogen are expressed both inside and outside of a host cell?*

Summary

9.1 ■ Fundamental Tools Used in Biotechnology

Restriction Enzymes (figure 9.1)

Restriction enzymes cut DNA into fragments. Cohesive ends will **anneal** to one another, making it possible to join DNA from two different organisms.

Gel Electrophoresis (figure 9.2)

Gel electrophoresis separates DNA fragments according to their size.

9.2 ■ Applications of Genetic Engineering (table 9.3)

Genetically Engineered Bacteria (figures 9.3, 9.4)

Bacteria can be engineered to produce pharmaceutical proteins, vaccines, and other proteins more efficiently. By **cloning** a segment of DNA into *E. coli*, an easy source of that sequence is available for study and further manipulation.

Genetically Engineered Eukaryotes

Transgenic plants have been engineered to resist pests and herbicides, have improved nutritional value, and function as edible vaccines.

9.3 ■ Techniques Used in Genetic Engineering

A **DNA library** is a collection of clones that together contain the entire genome of an organism (figure 9.6).

Obtaining DNA

To isolate DNA, cells are lysed by adding a detergent. To obtain eukaryotic DNA without introns, reverse transcriptase is used to make a **cDNA** from an mRNA template (figure 9.7).

Generating a Recombinant DNA Molecule

DNA ligase is used to join the **vector** and the insert (figures 9.8, 9.9).

Obtaining Cells That Harbor Recombinant DNA Molecules

The recombinant molecule is introduced into the new host, usually *E. coli*, using transformation or electroporation. The transformed cells are grown on medium that both selects for cells containing vector sequences and differentiates those that carry recombinant molecules.

9.4 ■ Concerns Regarding Genetic Engineering and Other DNA Technologies

Advances in genomics raise ethical issues and concerns about confidentiality. Genetically modified organisms hold many promises, but concerns exist about the inadvertent introduction of allergens into a food product and adverse effects on the environment.

9.5 ■ DNA Sequencing

By sequencing DNA, the encoded information can be compared to that of other organisms.

Techniques Used in DNA Sequencing

A key ingredient in the sequencing reaction is a dideoxynucleotide, a nucleotide that lacks the 3′OH and therefore functions as a chain terminator (figure 9.11). The sizes of fragments in a sequencing reaction indicate the positions of the terminating nucleotide (figures 9.12, 9.13).

9.6 ■ Polymerase Chain Reaction (PCR)

PCR is used to rapidly increase the amount of a specific DNA segment in a sample (figure 9.14).

Techniques Used in PCR

Double-stranded DNA is denatured, primers anneal to their complementary sequences, and then DNA is synthesized, amplifying the target sequence (figure 9.15). The **PCR product** is amplified exponentially (figures 9.16, 9.17). The primers dictate which portion of the DNA is amplified.

9.7 ■ Probe Technologies

DNA probes are used to locate specific nucleotide sequences (figure 9.18).

Colony Blotting

Colony blotting uses a probe to identify colonies that contain a given sequence of DNA (figure 9.19).

Fluorescence *in situ* Hybridization (FISH)

Fluorescence *in situ* hybridization (FISH) uses a fluorescently labeled probe to detect specific nucleotide sequences within intact cells affixed to a microscope slide (figure 9.20).

DNA Microarrays

DNA microarrays contain tens or hundreds of thousands of oligonucleotides that each function in a manner analogous to a probe (figure 9.21).

Review Questions

Short Answer

1. Why are restriction enzymes useful in biotechnology?
2. Describe three general uses of genetically engineered bacteria.
3. Describe the function of a reporter gene.
4. Describe three uses of genetically engineered plants.
5. What is a DNA library?
6. What is cDNA? Why is it used when cloning eukaryotic genes?
7. How many different temperatures are used in each cycle of the polymerase chain reaction?
8. Explain how PCR eventually generates a discrete-sized fragment from a much longer piece of DNA.
9. Describe the function of a probe.
10. How does a DNA microarray function as a set of probes?

Multiple Choice

1. What is the function of a vector?
   a) Destroys cells that do not contain cloned DNA
   b) Allows cells to take up foreign DNA
   c) Carries cloned DNA, allowing it to replicate in cells
   d) Encodes herbicide resistance
   e) Encodes Bt toxin

2. The Ti plasmid of *Agrobacterium tumefaciens* is used to genetically engineer which of the following cell types?
   a) Animals   b) Bacteria   c) Plants
   d) Yeast   e) All of these

3. Which of the following can be used to generate a DNA library?
   a) PCR   b) Sequencing   c) Colony blotting
   d) Microarrays   e) Cloning

4. An ideal vector has all of the following *except*
   a) an origin of replication.
   b) a gene encoding a restriction enzyme.
   c) a gene encoding resistance to an antibiotic.
   d) a multiple-cloning site.
   e) the *lacZ′* gene.

5. Which of the following describes the function of the *lacZ′* gene in a cloning vector?
   a) Means of selecting for cells that contain vector sequences
   b) Means of distinguishing cells that have taken up recombinant molecules
   c) Site required for the vector to replicate
   d) Mechanism by which cells take up the DNA
   e) Gene for a critical nutrient required by transformed cells

6. Which is used for cloning eukaryotic genes but not prokaryotic genes?
   a) Restriction enzymes
   b) DNA ligase
   c) Reverse transcriptase
   d) Vector
   e) Selectable marker

7. Which of the following does a dideoxynucleotide lack?
   a) 5′PO$_4$ b) 3′OH   c) 5′OH
   d) 3′PO$_4$ e) c and d

8. In a sequencing reaction, the dATP was left out of the tube. What would be the result of this error?
   a) No synthesis would occur.
   b) Synthesis would never continue past the first A.
   c) Synthesis would not stop until the end of the template.
   d) Synthesis would terminate randomly, regardless of the nucleotide incorporated.
   e) The error would have no effect.

9. The polymerase chain reaction uses *Taq* polymerase rather than a DNA polymerase from *E. coli*, because *Taq* polymerase
   a) introduces fewer errors during DNA synthesis.
   b) is heat-stable.
   c) can initiate DNA synthesis at a wider variety of sequences.
   d) can denature a double-stranded DNA template.
   e) is easier to obtain.

10. The polymerase chain reaction generates a fragment of a distinct size even when an intact chromosome is used as a template. What determines the boundaries of the amplified fragment?

    a) The concentration of one particular deoxynucleotide in the reaction

    b) The duration of the elongation step in each cycle

    c) The position of a termination sequence, which causes the *Taq* polymerase to fall off the template

    d) The sites to which the primers anneal

    e) The temperature of the elongation step in each cycle

Applications

1. Two students in a microbiology class are arguing about the origins of biotechnology. One student argued that biotechnology started with the advent of genetic engineering. The other student disagreed, saying that biotechnology was as old as ancient civilization. What was the rationale for the argument by the second student?

2. A student wants to clone gene X. On both sides of the gene are the recognition sequences for AluI and BamHI (look at table 9.2). Which enzyme would be easier to use for the cloning experiment and why?

Critical Thinking ➕

1. Discuss some potential issues regarding gene therapy, the use of genetic engineering to correct genetic defects.

2. An effective DNA probe can sometimes be developed by knowing the amino acid sequence of the protein encoded by the gene. A student argued that this is too time-consuming since the complete amino acid sequence must be determined in order to create the probe. Does the student have a valid argument? Why or why not?

Identification and Classification of Prokaryotic Organisms

PulseNet tracks bacterial strains causing foodborne diseases.

A Glimpse of History

In the early 1870s, the German botanist Ferdinand Cohn published several papers on bacterial classification, grouping microorganisms according to shape—spherical, short rods, elongated rods, and spirals. That scheme was not adequate, however, because there were too many different kinds of bacteria that had similar shapes.

The second major attempt at bacterial classification was initiated by Sigurd Orla-Jensen. His early training in Copenhagen was in chemical engineering, but he soon became interested in microbiology. In 1908, he proposed that bacteria be classified according to their physiological properties rather than morphology.

A quarter of a century later, two Dutch microbiologists, Albert Kluyver and C. B. van Niel, proposed classification systems based on evolutionary relationships. They recognized a very serious problem, however: There was no way to distinguish between "resemblance" and "relatedness." The fact that two prokaryotes look alike does not mean they are genetically related.

In 1970, Roger Stanier, a microbiologist at the University of California–Berkeley, pointed out that relationships could be determined by comparing either physical traits, such as proteins and cell walls, or nucleotide sequences. At that time, most microbiologists, including Stanier, assumed that all prokaryotes are basically similar. Chemical analysis of structures from a wide variety of prokaryotes, however, showed that many prokaryotes were fundamentally different from *Escherichia coli,* considered a "typical" bacterium. These "unusual" features included the chemical nature of the cell wall, cytoplasmic membrane, and ribosomal RNA.

In the late 1970s, Carl Woese and his colleagues at the University of Illinois determined the nucleotide sequence of ribosomal RNA in a wide variety of organisms. Based on the data, they recognized that prokaryotes could be divided into two major groups that differ from one another as much as they do eukaryotic cells. This led to a revolutionary system of classification that separates prokaryotes into two domains—the *Archaea* and the *Bacteria.* Each of these is on the same level as the *Eucarya,* which includes the animals, plants, and fungi (all eukaryotes).

Information that is logically organized is easier to use. Newspapers, for instance, do not scatter various subjects throughout the paper; instead, the information is grouped by general topic such as local news, sports, and entertainment. A large library would be extremely difficult to use if the locations of the many books were not organized by subject matter. Likewise, scientists have sorted living organisms into different groups, the better to understand the relationships among the species.

Take a moment and think about how you would group bacteria if you were to create a classification system. Would you group them according to shape? Or would it make more sense to consider their motility? Perhaps you would group them according to their medical significance. But then, how would you classify two apparently identical organisms that differ in their pathogenicity?

10.1 ■ Principles of Taxonomy

Learning Outcome

1. *Describe how prokaryotes are identified, classified, and assigned names.*

Taxonomy is the science that studies organisms in order to arrange them into groups (taxa). Those organisms with similar properties are grouped together and separated from ones that are different. Taxonomy can be viewed as three separate but interrelated areas:

- **Identification:** The process of characterizing an isolate to determine the group (taxon) to which it belongs.
- **Classification:** The process of arranging organisms into similar or related groups, primarily to make it easier to identify and study them.
- **Nomenclature:** The system of assigning names to organisms.

Strategies Used to Identify Prokaryotes

In practical terms, identifying the genus and species of a prokaryote may be more important than understanding its genetic relationship to other microbes. For example, a food manufacturer is most interested in detecting microbial contaminants that can spoil a food product. In a clinical laboratory, identifying a pathogen quickly is critical so that a patient can be treated appropriately.

To identify microorganisms, a wide assortment of procedures may be used, including microscopic examination, culture characteristics, biochemical tests, and nucleic acid analysis. In a clinical laboratory, the patient's disease symptoms play an important role as well. For example, pneumonia in an otherwise healthy adult is typically caused by *Streptococcus pneumoniae,* a bacterium easily differentiated from others using a few specific tests. In contrast, diagnosing the cause of a wound infection is often more difficult, because several different microorganisms could be involved. Often, however, it is only necessary to rule out the presence of organisms known to cause a particular disease, rather than conclusively identify each and every one. For instance, a fecal specimen from a patient complaining of diarrhea and fever would generally be tested only for the presence of organisms that cause those symptoms. The methods used to identify prokaryotes will be discussed later in the chapter. ▶▶ *Streptococcus pneumoniae,* p. 497

Strategies Used to Classify Prokaryotes

Understanding the evolutionary relatedness, or **phylogeny,** of prokaryotes is important in creating a classification scheme that reflects their evolution and biology. Such a scheme is more useful than one that simply groups organisms by arbitrary characteristics, because it is less prone to the bias of human perceptions. It also makes it easier to classify newly recognized species and allows scientists to make predictions, such as which genes are likely to be transferred between organisms.

Unfortunately, determining evolutionary relatedness among prokaryotes is more difficult than it is for plants and animals. Not only do prokaryotes have few differences in size and shape, they do not undergo sexual reproduction. In higher organisms

such as plants and animals, the basic taxonomic unit, a **species,** is generally considered to be a group of morphologically similar organisms capable of interbreeding to produce fertile offspring. Obviously, it is not possible to apply these same criteria to prokaryotes, which makes classification problematic.

Historically, taxonomists have relied heavily on phenotypic characteristics to classify prokaryotes. The development and application of molecular techniques such as nucleotide sequencing, however, is finally making it possible to determine the phylogeny of microorganisms. ◀◀ phenotype, p. 189

Taxonomic Hierarchies

Taxonomic classification categories are arranged in a hierarchical order, with the species being the basic unit. The species designation gives a formal taxonomic status to a group of related isolates, or **strains,** which, in turn, permits their identification. Without classification, scientists and others would not be able to communicate about organisms with any degree of accuracy. Taxonomic categories include:

- **Species:** A group of closely related isolates or strains. Note that members of a species are not all identical; individual strains may vary. The difficulty for the taxonomist is to decide how different two isolates must be in order to be classified as separate species rather than strains of the same species.
- **Genus:** A collection of similar species.
- **Family:** A collection of similar genera. In prokaryotic nomenclature, the name of the family ends in the suffix *-aceae.*
- **Order:** A collection of similar families. In prokaryotic nomenclature, the name of the order ends in the suffix *-ales.*
- **Class:** A collection of similar orders.
- **Phylum** or **Division:** A collection of similar classes.
- **Kingdom:** A collection of similar phyla or divisions.
- **Domain:** A collection of similar kingdoms. The domain is a relatively new taxonomic category that reflects the characteristics of the cells that make up the organism.

Note, however, that microbiologists often group prokaryotes into informal categories based on one or more distinctive characteristics, rather than using the higher taxonomic ranks such as

TABLE 10.1	Taxonomic Ranks of the Bacterium *Escherichia coli*
Formal Rank	**Example**
Domain	*Bacteria*
Phylum	*Proteobacteria*
Class	*Gammaproteobacteria*
Order	*Enterobacteriales*
Family	*Enterobacteriaceae*
Genus	*Escherichia*
Species	*coli*

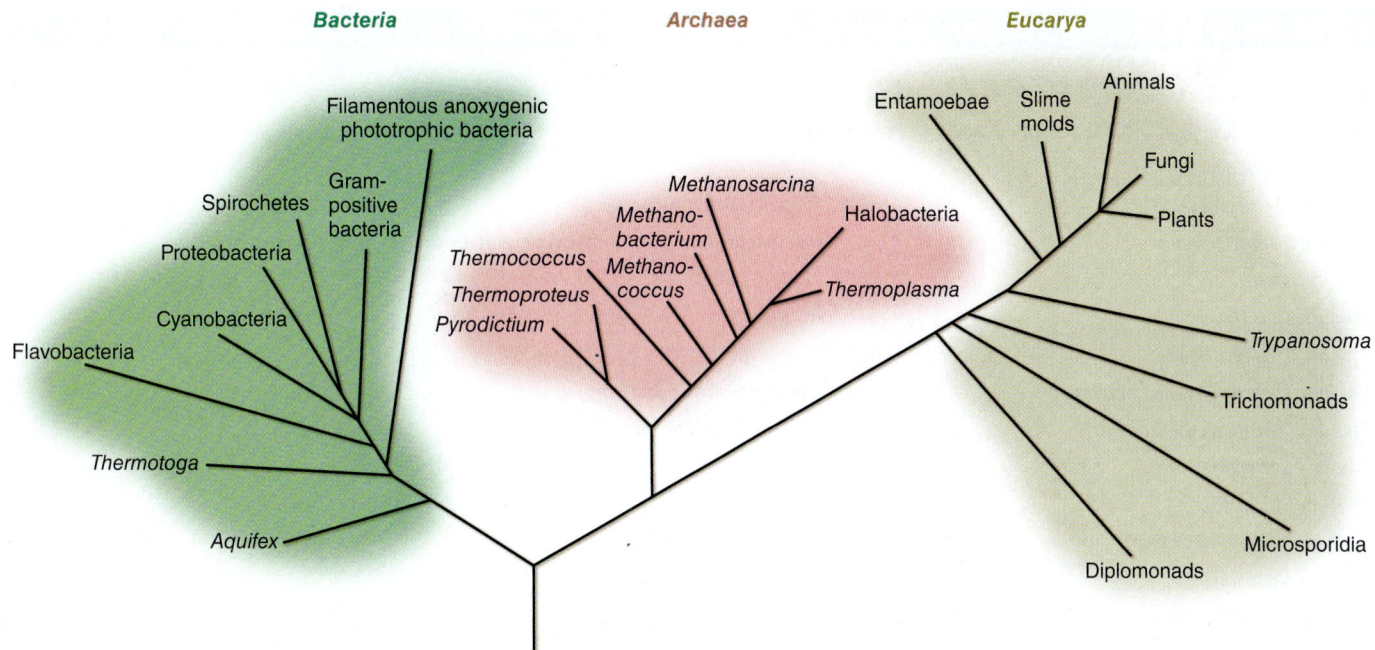

FIGURE 10.1 **The Three-Domain System of Classification** This classification system separates prokaryotic organisms into two domains—*Bacteria* and *Archaea*. The third domain, *Eucarya*, contains all organisms composed of eukaryotic cells.

? *Information about which molecule was used to establish the three-domain system of classification?*

order, class, and phylum. Examples of such informal groupings include the lactic acid bacteria, the anoxygenic phototrophs, the endospore-formers, and the sulfate reducers. Organisms within these groupings share similar phenotypic and physiological characteristics, but may not be genetically related. ▸▸ **lactic acid bacteria, p. 258** ▸▸ **anoxygenic phototrophs, p. 259** ▸▸ **sulfate reducers, p. 258**

An example of how a particular bacterial species is classified is shown in **table 10.1.** The table intentionally omits the taxonomic category of kingdom, because the use of kingdoms within the *Bacteria* and *Archaea* is still in a state of flux.

Classification Systems

Classification systems change over the years as new information is discovered. There is no "official" classification system, and, as new ones are introduced, others become outdated. The classification scheme currently favored by most microbiologists is the three-domain system. This designates all organisms as belonging to one of the three domains—*Bacteria, Archaea,* and *Eucarya* (**figure 10.1**). The system is based on the work of Carl Woese and colleagues, who compared the nucleotide sequences in ribosomal RNA from a wide variety of organisms (see **A Glimpse of History**). The ribosomal·RNA data are consistent with other differences between the *Archaea* and *Bacteria,* including the chemical compositions of their cell walls and cytoplasmic membranes (**table 10.2**). ◂◂ *Bacteria,* **p. 9** ◂◂ *Archaea,* **p. 9** ◂◂ *Eucarya,* **p. 10**

Before the three-domain system was introduced, the most widely accepted scheme was the five-kingdom system, proposed by R. H. Whittaker in 1969. The five kingdoms in this system are Plantae, Animalia, Fungi, Protista (mostly single-celled eukaryotes), and Prokaryotae. Although the five-kingdom system recognizes the obvious morphological differences between plants and animals, it does not reflect the recent genetic insights of the ribosomal RNA data, which indicates that plants and animals are more closely related to each other than *Archaea* are to *Bacteria.*

TABLE 10.2	A Comparison of Typical Properties of the Three Domains—*Archaea, Bacteria,* and *Eucarya*		
Cell Feature	**Archaea**	**Bacteria**	**Eucarya**
Peptidoglycan cell wall	No	Yes	No
Cytoplasmic membrane lipids	Hydrocarbons (not fatty acids) linked to glycerol by ether linkage	Fatty acids linked to glycerol by ester linkage	Fatty acids linked to glycerol by ester linkage
Ribosomes	70S	70S	80S
Presence of introns	Sometimes	No	Yes
Membrane-bound nucleus	No	No	Yes

TABLE 10.3	Taxonomic Outline of *Bergey's Manual of Systematic Bacteriology*, 2nd edition
Representative Genera	

Volume 1: The *Archaea* and the Deeply Branching Phototrophic *Bacteria*

Domain *Archaea*	
Phylum *Crenarchaeota*	*Pyrodictium, Ignicoccus*
Phylum *Euryarchaeota*	*Halobacterium, Methanococcus, Natronococcus, Picrophilus*
Domain *Bacteria*	
Phylum *Aquificae*	*Aquifex*
Phylum *Thermotogae*	*Thermotoga*
Phylum *Thermodesulfobacteria*	*Thermodesulfobacterium*
Phylum *Deinococcus–Thermus*	*Deinococcus, Thermus*
Phylum *Chrysiogenetes*	*Chrysiogenes*
Phylum *Chloroflexi*	*Chloroflexus*
Phylum *Thermomicrobia*	*Thermomicrobium*
Phylum *Nitrospira*	*Nitrospira*
Phylum *Deferribacteres*	*Deferribacter*
Phylum *Cyanobacteria*	*Anabaena, Spirulina, Synechococcus*
Phylum *Chlorobi*	*Chlorobium, Pelodictyon*

Volume 2: The *Proteobacteria*

Phylum *Proteobacteria*	
Class *Alphaproteobacteria*	*Agrobacterium, Caulobacter, Ehrlichia, Nitrobacter, Rhodospirillum, Rickettsia, Rhizobium*
Class *Betaproteobacteria*	*Neisseria, Nitrosomonas, Thiobacillus*
Class *Gammaproteobacteria*	*Azotobacter, Chromatium, Escherichia, Legionella, Nitrosococcus, Pseudomonas, Vibrio*
Class *Deltaproteobacteria*	*Bdellovibrio, Myxococcus*
Class *Epsilonproteobacteria*	*Campylobacter, Helicobacter*

Volume 3: The *Firmicutes*

Phylum *Firmicutes*	
Class I. *Bacilli*	*Bacillus, Streptococcus, Listeria, Staphylococcus*
Class II. *Clostridia*	*Clostridium, Heliobacterium*
Class III. *Erysipelotrichia*	*Erysipelotrichia*

Volume 4: The *Bacteroides, Spirochaetes, Tenericutes, Acidobacteria, Fibrobacteres, Fusobacteria, Dictyoglomi, Gemmatimonadetes, Lentisphaerae, Verrucomicrobia, Chlamydiae,* and *Planctomycetes*

Phylum *Bacteroidetes*	*Bacteroides*
Phylum *Spirochaetes*	*Borrelia, Treponema*
Phylum *Tenericutes*	*Mycoplasma*
Phylum *Acidobacteria*	*Acidobacterium*
Phylum *Fibrobacteres*	*Fibrobacter*
Phylum *Fusobacteria*	*Fusobacterium*
Phylum *Dictyoglyomi*	*Dictyoglomus*
Phylum *Gemmatimonadetes*	*Gemmatinonas*
Phylum *Lentisphaerae*	*Lentisphaera*
Phylum *Verrucomicrobia*	*Verrucomicrobium*
Phylum *Chlamydiae*	*Chlamydia*
Phylum *Planctomycetes*	*Planctomyces*

Volume 5: The *Actinobacteria*

Phylum *Actinobacteria*	*Corynebacterium, Bifidobacterium, Micrococcus, Mycobacterium, Streptomyces*

Bergey's Manual of Systematic Bacteriology Microbiologists generally rely on the classifications listed in the reference text *Bergey's Manual of Systematic Bacteriology.* All known species are described there, including those that have not yet been cultivated. If the properties of a newly isolated organism do not agree with any description in *Bergey's Manual,* then presumably a new organism has been isolated. The newest edition of this comprehensive manual is being published in five volumes and classifies prokaryotes according to the most recent information on their phylogeny (**table 10.3**). In some cases, this classification differs substantially from that of the previous edition, which grouped organisms according to their phenotypic characteristics.

In addition to containing descriptions of organisms, all volumes include information on the ecology, methods of enrichment, culture, and isolation of the organisms as well as methods for their maintenance and preservation. However, the heart of the work is a description of all characterized prokaryotes and their groupings.

Nomenclature

Bacteria are given names according to a set of internationally recognized rules, the *International Code of Nomenclature of Bacteria.* The names may originate from any language, but must include a Latin suffix. In some cases, the name reflects the organism's habitat or other characteristic but often honors a researcher.

Just as classification is always in a state of flux, so is the assignment of names. Although revising names may increase scientific accuracy, it often leads to confusion, particularly when names of medically important bacteria are changed. To ease the transition after a name change, the former name may be included in parentheses. For example, *Lactococcus lactis,* a bacterium that was once included in the genus *Streptococcus,* is sometimes indicated as *Lactococcus (Streptococcus) lactis.*

MicroAssessment 10.1

Taxonomy consists of three interrelated areas: identification, classification, and nomenclature. In clinical laboratories, identifying the genus and species of an isolate is more important than understanding its evolutionary relationship to other organisms.

1. *Why might it be easier to determine the cause of pneumonia than the cause of a wound infection?*
2. *What is* Bergey's Manual?
3. *Some biologists have been reluctant to accept the three-domain system. Why might this be the case?* ✚

10.2 ■ Using Phenotypic Characteristics to Identify Prokaryotes

Learning Outcome

2. *Describe how phenotypic characteristics—including microscopic morphology, culture characteristics, metabolic capabilities, serology, and fatty acid analysis—can be used to identify prokaryotes.*

Phenotypic characteristics can be used to help identify microorganisms, without the need for sophisticated equipment. These and other identification methods are summarized in **table 10.4.**

Microscopic Morphology

An important initial step in identifying a microorganism is to determine its size, shape, and staining characteristics. Microscopic examination gives information very quickly and is sometimes enough to make a presumptive identification.

TABLE 10.4	Methods Used to Identify Prokaryotes
Method	**Comments**
Phenotypic Characteristics	Most of these methods do not require sophisticated equipment and can easily be done anywhere in the world.
Microscopic morphology	Size, shape, and staining characteristics such as Gram stain can give suggestive information as to the identity of the organism. Further testing, however, is needed to confirm the identification.
Culture characteristics	Colony morphology can give initial clues to the identity of an organism.
Metabolic capabilities	A battery of biochemical tests can be used to identify a microorganism.
Serology	Proteins and polysaccharides that make up a prokaryote are sometimes characteristic enough to be considered identifying markers. These can be detected using specific antibodies.
Fatty acid analysis	Cellular fatty acid composition can be used as an identifying marker and is analyzed by gas chromatography.
Genotypic Characteristics	These methods are increasingly being used to identify microorganisms. Even an organism that occurs in very low numbers in a mixed culture can be identified.
Detecting specific nucleotide sequences	Nucleic acid probes and nucleic acid amplification tests can be used to identify prokaryotes grown in culture. In some cases, the method is sensitive enough to detect the organism directly in a specimen.
Sequencing rRNA genes	This requires amplifying and then sequencing rRNA genes, but it can be used to identify organisms that have not yet been grown in culture.

Size and Shape

The size and shape of a microorganism can easily be determined by microscopically examining a wet mount. Based only on the size and shape, the organism can be identified as a prokaryote, fungus, or protozoan. In a clinical lab, this can sometimes be enough to diagnose certain eukaryotic infections. For example, a wet mount of vaginal secretions is routinely used to diagnose yeast infections, and one of stool is examined for the eggs of parasites when certain roundworms are suspected (**figure 10.2**). ◄◄ wet mount, p. 43

Gram Stain

The Gram stain distinguishes between Gram-positive and Gram-negative bacteria (see figure 3.14). This relatively rapid test narrows the list of possible identities of an organism, an essential step in the identification process. ◄◄ Gram stain, p. 47

In a clinical lab, the Gram stain of a specimen is generally not sensitive or specific enough to diagnose the cause of most infections, but it is still an extremely useful tool. The technician can see the Gram reaction, the shape and arrangement of the bacterial cells, and whether the organisms appear to be growing as a pure culture or with other bacteria and/or cells of the host. However, most medically important bacteria cannot be identified by Gram stain alone. For example, *Streptococcus pyogenes,* the bacterium that causes strep throat, cannot be distinguished microscopically from streptococci that are part of the normal throat microbiota. A Gram stain of a stool specimen cannot distinguish *Salmonella* from *E. coli*. These organisms generally must be isolated in pure culture and tested for their biochemical capabilities for accurate identification. ►► strep throat, p. 487

In certain cases, the Gram stain result gives enough information to start appropriate antimicrobial therapy. For example, a Gram stain of sputum showing numerous white blood cells and Gram-positive diplococci is highly suggestive of *Streptococcus pneumoniae,* a bacterium that causes pneumonia (**figure 10.3a**). In some other cases, the Gram stain result is enough for diagnosis. For instance, the presence of Gram-negative diplococci clustered in white blood cells in a sample of a urethral secretion from a man is considered diagnostic for gonorrhea, the sexually transmitted

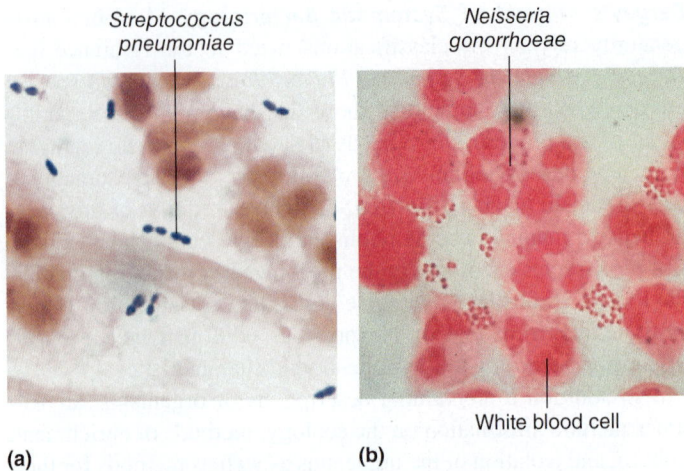

Streptococcus pneumoniae

Neisseria gonorrhoeae

White blood cell

(a)   (b)

FIGURE 10.3 Gram Stains of Clinical Specimens (a) Sputum showing Gram-positive *Streptococcus pneumoniae* and **(b)** male urethra secretions showing Gram-negative *Neisseria gonorrhoeae* inside white blood cells.

❓ *What disease does the patient in (a) have? What disease does the patient in (b) have?*

infection caused by *Neisseria gonorrhoeae* (figure 10.3b). This diagnosis can be made because *N. gonorrhoeae* is the only Gram-negative diplococcus found within white blood cells in the male urethra. ►► pneumonia, p. 486 ►► gonorrhea, p. 622

Special Stains

Certain microorganisms have unique characteristics that can be detected with special staining procedures. As an example, members of the genus *Mycobacterium* are some of the few acid-fast microorganisms (see figure 3.15). If a patient has symptoms of tuberculosis, then an acid-fast stain will be done on a sputum sample to help identify *Mycobacterium tuberculosis*. ◄◄ acid-fast stain, p. 48

Culture Characteristics

Colony morphology can give initial clues to the identity of the organism. For example, colonies of streptococci are generally fairly small relative to many other types of bacteria. Colonies of *Serratia marcescens* are often red when incubated at 22°C due to the production of a pigment. *Pseudomonas aeruginosa* often produces a soluble greenish pigment, which discolors the growth medium (see figure 11.12). In addition, cultures of *P. aeruginosa* have a distinct fruity odor. ◄◄ growth of colonies, p. 88

In a clinical lab, where rapid but accurate diagnosis is essential, specimens are inoculated onto differential media as a preliminary step in the identification process. A specimen taken by swabbing the throat of a patient complaining of a sore throat is inoculated onto blood agar. This makes it possible to detect the characteristic β-hemolytic colonies of *Streptococcus pyogenes,* the cause of strep throat (see figure 4.10). Urine collected from a patient suspected of having a urinary tract infection (UTI) is plated onto MacConkey agar. *E. coli*, the most common cause of UTIs, ferments lactose and therefore forms pink colonies on that agar (see figure 4.11). ◄◄ differential media, p. 95   ◄◄ blood agar, p. 95 ◄◄ MacConkey agar, p. 95

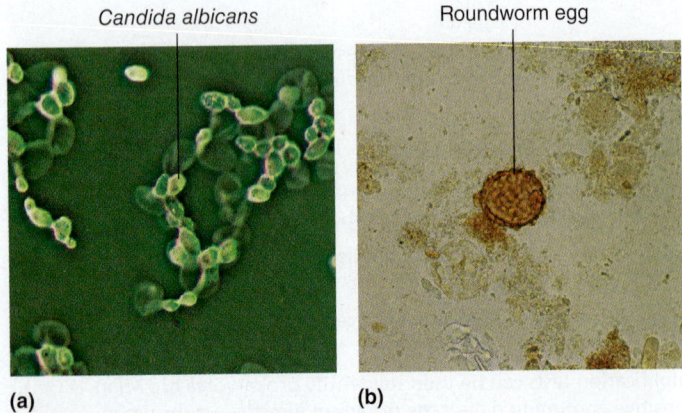

Candida albicans

Roundworm egg

(a)   (b)

FIGURE 10.2 Wet Mounts of Clinical Specimens (a) Vaginal secretions containing yeast (*Candida albicans*, 410×); **(b)** roundworm (*Ascaris*) eggs in a stool (400×).

❓ *Yeast and roundworms belong to which domain?*

  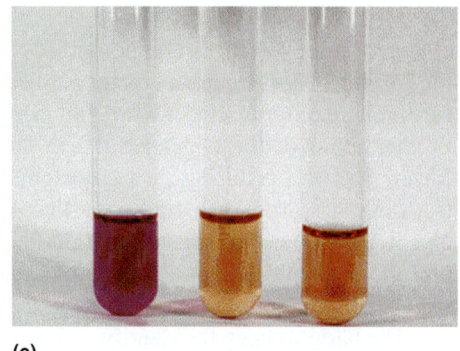

(a)   (b)   (c)

FIGURE 10.4 Biochemical Tests (a) Catalase production (left) and a negative control (right). **(b)** Sugar fermentation (left), negative control (center), and an uninoculated tube (right). **(c)** Urease production (left), negative control (center), and an uninoculated control (right).

❓ *What specifically causes the color to change in the sugar fermentation and urease tests?*

Metabolic Capabilities

The metabolic capabilities such as the types of sugars fermented or the end products made can be used to identify a microorganism.

Biochemical Tests

Colony morphology can give clues as to the identity of an organism, but biochemical tests are generally necessary for a more certain identification (**table 10.5**). One of the simplest tests is an assay for the enzyme catalase (**figure 10.4a**). Most bacteria that grow in the presence of O_2 are catalase-positive. Important exceptions are the lactic acid bacteria, which include members of the genus *Streptococcus*. Thus, if a throat culture yields β-hemolytic colonies but further testing reveals they are all catalase-positive, then *Streptococcus pyogenes* has been ruled out. ◀◀ **catalase, p. 90** ▶▶ *Streptococcus pyogenes*, **p. 487**

TABLE 10.5	Characteristics of Some Important Biochemical Tests	
Biochemical Test	**Principle of the Test**	**Positive Reaction**
Catalase	Detects the activity of the catalase, an enzyme that breaks down hydrogen peroxide to form O_2 and water.	The reagent bubbles.
Citrate	Determines whether or not citrate can be used as a sole carbon source.	Growth, usually accompanied by the color change of a pH indicator.
Gelatinase	Detects enzymatic breakdown of gelatin to polypeptides.	The solid gelatin is converted to liquid.
Hydrogen sulfide production	Detects H_2S released as sulfur-containing amino acids are degraded.	A black precipitate forms due to the reaction of H_2S with iron salts in the medium.
Indole	Detects the enzymatic removal of the amino group from tryptophan.	The product, indole, reacts with an added chemical reagent, turning the reagent a deep red color.
Lysine decarboxylase	Detects the enzymatic removal of the carboxyl group from lysine.	The medium becomes more alkaline, causing a pH indicator to change color.
Methyl red	Detects mixed acids, the characteristic end products of a particular fermentation pathway. ◀◀ **mixed acids, p. 148**	The medium becomes acidic (pH below 4.5); a red color develops upon the addition of a pH indicator.
Oxidase	Detects the activity of cytochrome *c* oxidase, a component of the electron transport chain of specific organisms. ◀◀ **cytochrome *c*, p. 143**	A dark color develops after a specific reagent is added.
Phenylalanine deaminase	Detects the enzymatic removal of the amino group from phenylalanine.	The product of the reaction, phenylpyruvic acid, reacts with ferric chloride to give the medium a green color.
Sugar fermentation	Detects the acidity resulting from fermentation of the sugar incorporated into the medium; also detects gas production.	The medium becomes acidic, causing a pH indicator to change color. An inverted tube traps any gas that is made.
Urease	Detects the enzymatic degradation of urea to carbon dioxide and ammonia.	The medium becomes alkaline, causing a pH indicator to change color.
Voges-Proskauer	Detects acetoin, an intermediate of the fermentation pathway that leads to 2,3-butanediol production. ◀◀ **2,3-butanediol, p. 148**	A red color develops after chemicals that detect acetoin are added.

Most biochemical tests rely on a chemical indicator that changes color when a compound is degraded. To test for the ability of an organism to ferment a given sugar, a broth medium containing that sugar and a pH indicator is used. Fermentation of the sugar results in acid production, which lowers the pH, resulting in a color change; an inverted tube traps any gas produced (figure 10.4b). A medium designed to detect urease, an enzyme that degrades urea to produce carbon dioxide and ammonia, contains urea and a pH indicator (figure 10.4c).

The basic strategy for identifying bacteria based on biochemical tests relies on a **dichotomous key,** a flowchart of tests that give either a positive or negative result (**figure 10.5**). Because each test often requires an incubation period, however, it would be too time-consuming to proceed one step at a time. In addition, relying on a single biochemical test at each step could lead to misidentification. For example, if a strain that normally gives a positive result for a certain test loses the ability to produce a key enzyme, it would instead have a negative result. Therefore, simultaneously inoculating multiple tests identifies the organism faster and more conclusively.

In certain cases, biochemical testing can be done without culturing the organism. *Helicobacter pylori,* the cause of most stomach ulcers, can be detected using the breath test, which assays for the presence of urease. The patient drinks a solution containing urea labeled with an isotope of carbon. If *H. pylori* is present, its urease breaks down the urea, releasing labeled CO_2, which escapes through the airway. Several hours after drinking the solution, the patient exhales into a balloon, and that air is then tested for labeled CO_2. This test is less invasive and, consequently, much cheaper and faster than the stomach biopsy that would be needed to culture the organism.

▶▶ *Helicobacter pylori,* p. 578 ◀◀ isotope, p. 18

Commercial Variations of Traditional Biochemical Tests

Several less labor-intensive commercial variations of traditional biochemical tests are available (**figure 10.6**). The API system uses a strip holding a series of tiny cups that contain dried media. A liquid suspension of the test bacterium is added to

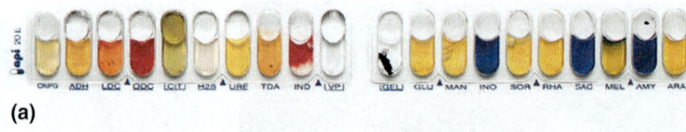

(a)

(b)

FIGURE 10.6 Commercial Modifications of Traditional Biochemical Tests (a) An API test strip. Each small cup contains a dried medium similar in formulation to the traditional tests. A liquid suspension of the isolated test bacterium is added to each compartment; after incubation, the results are read manually. **(b)** An Enterotube II. Each compartment contains a different type of medium. One end of a metal rod that runs through the tube is used to touch a bacterial colony. When the rod is withdrawn, it inoculates each of the compartments. After incubation, the results are read manually.

❓ *What advantage do these commercial methods have over traditional biochemical tests?*

each compartment, inoculating as well as rehydrating the media. The media are similar to those used in traditional tests, giving rise to comparable color changes. After a 16-hour incubation of the inoculated test strip, the results are determined by inspection. The pattern of results is converted to a numerical score, which is then entered into a computer to identify the organism. A similar system is the Enterotube II, a tube with small compartments, each containing a different type of medium. One end of a metal rod that runs through the tube is used to touch a bacterial colony. The other end is used to pull the inoculated

FIGURE 10.5 Dichotomous Key This shows an example of steps that can be used to identify some of the common causes of urinary tract infections. Additional tests need to be done to confirm the identity of the pathogen.

❓ *When identifying organisms, why are certain biochemical tests usually initiated simultaneously, rather than waiting for one result before beginning the next test?*

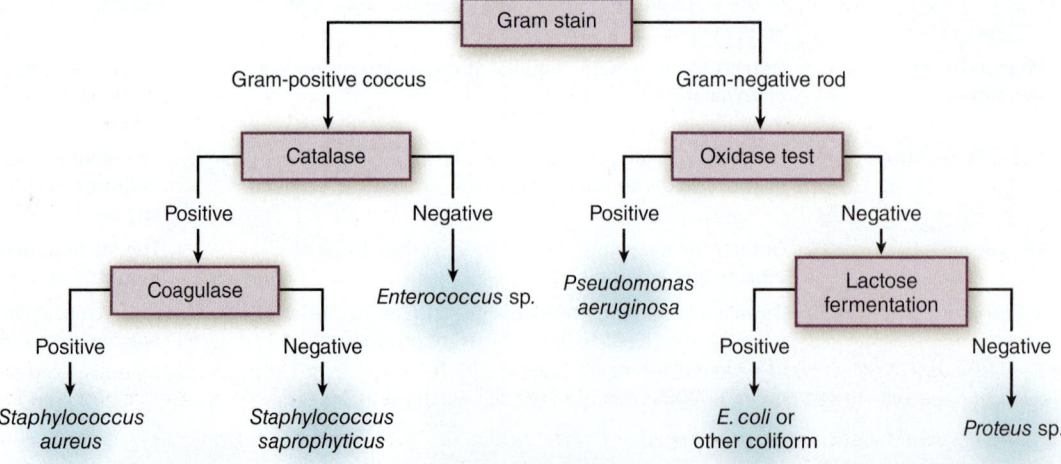

end through the tube, thereby inoculating each of the compartments. A system by Biolog uses a microtiter plate, a small tray containing nearly one hundred wells, to assay simultaneously an organism's ability to use a wide variety of carbon sources. Modifications of these plates allow researchers to characterize the metabolic capabilities of microbial communities, such as those in soil, water, or wastewater.

Highly automated systems also are available. The VITEK 2 system uses a miniature card that contains multiple wells with different types of dried media. After a relatively short incubation period, a computer then reads the growth pattern in the wells.

Serology

The proteins and polysaccharides that make up a prokaryotic cell are sometimes characteristic enough to be identifying markers. The most useful of these are the molecules that make up surface structures including the cell wall, capsule, flagella, and pili. For example, some species of *Streptococcus* contain a unique carbohydrate as part of their cell wall, and this distinguishes them from other species. These carbohydrates, as well as other distinct surface molecules, can be detected using antibodies. Detection methods that use antibodies are called serological tests and will be discussed in more detail in chapter 18. Some serological tests, such as those used to confirm the identity of *S. pyogenes*, are quite simple and rapid. ▶▶ antibodies, p. 359 ▶▶ antigens, p. 359 ▶▶ serology, p. 426

Fatty Acid Analysis (FAME)

Prokaryotic species differ in the type and relative amounts of different fatty acids in their membranes, so fatty acid composition can be used as an identifying marker. To do this, bacterial cells are grown under standardized conditions and then treated with chemicals to release the fatty acids and convert them to their more volatile methyl ester form (FAME stands for fatty acid methyl ester). The resulting FAMEs can then be separated and analyzed using gas chromatography. The chromatogram (pattern of peaks) is then compared to those of known species.

MicroAssessment 10.2

The size, shape, and staining characteristics of a microorganism give important clues to its identity. Conclusive identification generally requires multiple biochemical tests. The proteins and polysaccharides that make up a bacterium are sometimes unique enough to be considered identifying markers. Cellular fatty acid composition can be used as an identifying characteristic.

4. *How does MacConkey agar help identify the cause of a urinary tract infection?*
5. *Describe two methods to test for the enzyme urease.*
6. *A sample must contain many microbial cells for any to be seen by microscopic examination. Why?* ✚

10.3 ■ Using Genotypic Characteristics to Identify Prokaryotes

Learning Outcome

3. *Describe how nucleic acid probes, NAATs, and sequencing 16S rRNA genes can be used to identify prokaryotes.*

Many of the technologies discussed in chapter 9 can be used to identify a microorganism based on it genotype. Some of these methods even make it possible to identify organisms that cannot yet be grown in culture.

Detecting Specific Nucleotide Sequences

Nucleic acid probes and nucleic acid amplification tests (NAATs) can both detect nucleotide sequences unique to a given species or related group. A significant limitation, however, is that each probe or amplification detects only a single possibility. If the organism in question could be one of five different species or related groups, then five distinguishable probes or amplifications would be needed. ◀◀ DNA probe, p. 232

Nucleic Acid Probes

A nucleic acid probe can locate a nucleotide sequence that characterizes a particular species or group (**figure 10.7**). The probe is a

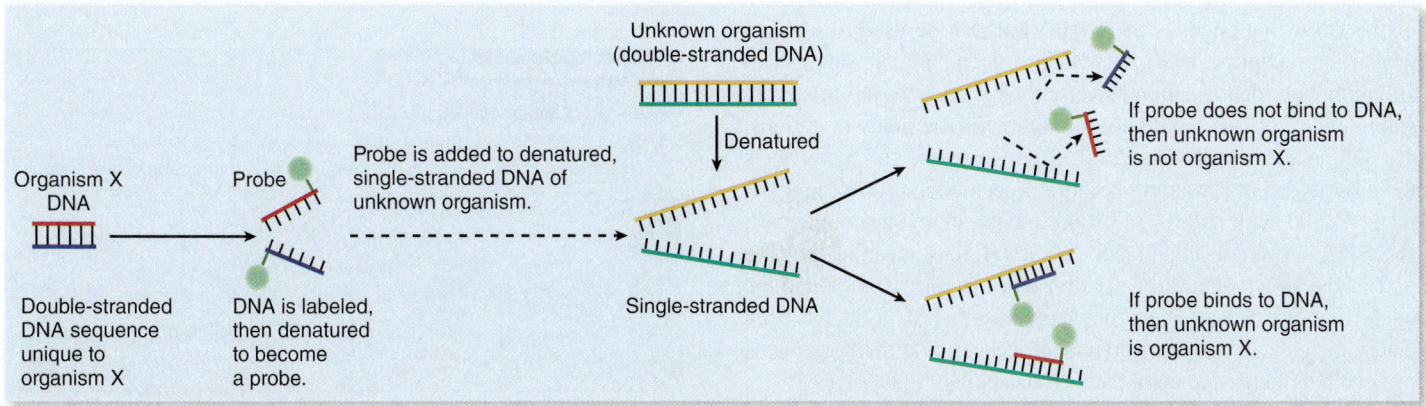

FIGURE 10.7 Nucleic Acid Probes to Detect Specific DNA Sequences The probe, a single-stranded piece of nucleic acid labeled with a detectable marker, is used to locate a unique nucleotide sequence that identifies a particular microbial species.

❓ *What type of label does the probe used in fluorescence* in situ *hybridization (FISH) employ?*

single-stranded piece of nucleic acid, usually DNA, labeled with a detectable tag such as a radioisotope or a fluorescent dye. It is complementary to the sequence of interest.

Most methods that use nucleic acid probes to detect DNA sequences rely on a step that increases the amount of DNA in the sample. This can be done by inoculating the specimen onto an agar medium so that each microbial cell multiplies, forming a colony. Alternatively, a preliminary *in vitro* DNA amplification step can be done.

Fluorescence *in situ* hybridization (FISH) often uses probes that bind 16S ribosomal RNA (rRNA). An amplification step is not needed because numerous copies of rRNA are naturally present in multiplying cells. Various different probes that bind rRNA are available, each specific for a given **signature sequence.** A signature sequence is a sequence in rRNA that characterizes either a certain species or a group of related organisms (see figure 9.20). Applications of FISH, as well as the techniques, were discussed in chapter 9. ◄◄ fluorescence *in situ* hybridization (FISH), p. 232

Nucleic Acid Amplification Tests (NAATs)

Several methods, referred to as **nucleic acid amplification tests (NAATs),** can be used to increase the number of copies of specific DNA sequences. This allows researchers to detect specific sequences in samples such as body fluids, soil, food, and water. These methods can be used to detect organisms present in extremely small numbers as well as those that cannot yet be grown in culture. In most cases, the amplified fragment is visible as a distinct band on an ethidium bromide–stained agarose gel illuminated with UV light. Alternatively, a DNA probe can be used to detect the amplified DNA. ◄◄ ethidium bromide, p. 218 ◄◄ gel electrophoresis, p. 217

One of the most common NAATs is the polymerase chain reaction (PCR), described in chapter 9 (see figure 9.14). To use PCR to detect a microbe of interest, a sample is treated to release and denature the DNA. Specific primers and other ingredients are then added (see figure 9.15). After 30 cycles of PCR, the target DNA will have been amplified approximately a billion-fold (see figure 9.17). ◄◄ PCR, p. 227

Sequencing Ribosomal RNA Genes

The nucleotide sequence of ribosomal RNA molecules (rRNAs), or the DNA that encodes them (rDNAs), can be used to identify prokaryotes (**figure 10.8**). rRNAs are useful in microbial classification and identification because their sequences are relatively stable; the ribosome would not function with too many mutations.

Of the different rRNAs (5S, 16S, and 23S), the 16S molecule is the most useful in taxonomy because of its moderate size (approximately 1,500 nucleotides). 16S RNA and its eukaryotic counterpart, 18S RNA, are small subunit (SS, or SSU) rRNAs, meaning they are part of the small subunit of the ribosomes (figure 10.8). Once the nucleotide sequence of an unknown organism's SSU RNA has been determined, it can be compared with sequences of known organisms by searching extensive computerized databases.

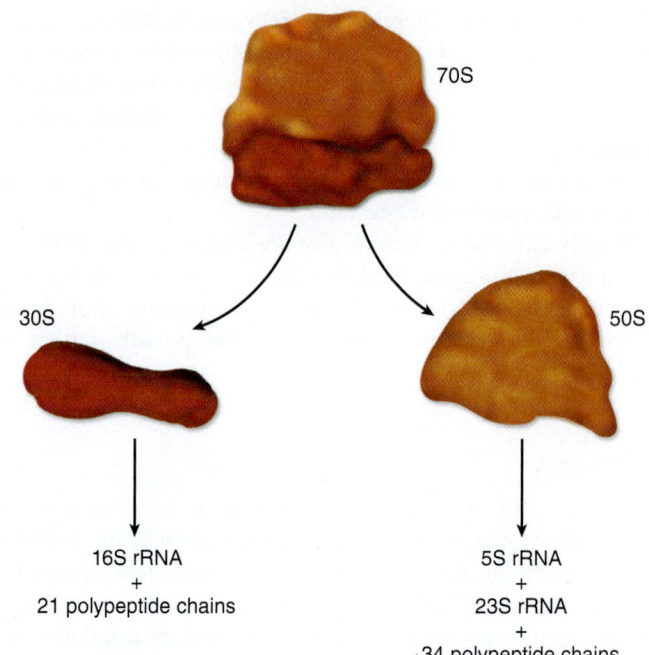

FIGURE 10.8 Ribosomal RNA The 70S ribosome of prokaryotes has three types of rRNA: 5S, 16S, and 23S.

❓ *Which type of ribosomal RNA is most often used in taxonomy?*

Using rDNA to Identify Uncultivated Organisms

The vast majority of microbes cannot yet be grown in culture. However, the DNA from these organisms can be amplified, cloned, and sequenced, making it possible to detect and identify them. The bacterium that causes Whipple's disease, a rare intestinal illness, was identified this way. The organism was given the name *Tropheryma whipplei,* and a specific probe was then developed to detect it in intestinal tissue, well before it could be grown in culture.

MicroByte —
A gram of fertile soil may contain over 4,000 species of prokaryotes, most of which have not been identified.

MicroAssessment 10.3

A microorganism can be identified by using a probe or NAAT to detect a nucleotide sequence unique to that organism. Ribosomal RNA genes can be sequenced to identify organisms, including those that cannot be grown in culture.

7. *When using a probe to detect an organism, why is it necessary to have some idea as to the organism's identity?*
8. *Why is 16S RNA also referred to as SSU RNA?*
9. *Why are molecular methods especially useful when bacteria are difficult to grow?* ➕

10.4 ■ Characterizing Strain Differences

Learning Outcome

4. *Describe five distinct methods to distinguish different strains.*

In some situations, distinguishing different strains of a given species is useful. In 2009, for example, 235 reported salmonellosis cases across 14 states in the United States involved a specific strain of *Salmonella enterica* serotype Saintpaul. Most patients reported consuming raw alfalfa sprouts prior to their illness, leading to a recall of the seeds by the producer. Linking 235 salmonellosis cases amid the thousands that occur nationwide each year would be impossible without methods to distinguish different strains. ▶▶ *Salmonella,* p. 592

Characterizing strain differences is not limited to investigations of foodborne illness. It also plays an important role in forensic investigations of bioterrorism and other biocrimes, and in diagnosing certain diseases. The methods used to characterize different strains are summarized in **table 10.6.**

Biochemical Typing

Biochemical tests are used to identify various species of bacteria, but they can also be used to distinguish strains. A group of strains that have a characteristic biochemical pattern is called a **biovar,** or a **biotype.** A biochemical variant of *Vibrio cholerae* called El Tor caused a worldwide epidemic of cholera beginning in 1961. Because this biovar can be readily distinguished from other strains, its spread can be traced. ▶▶ *Vibrio cholerae,* p. 586

Serological Typing

Proteins and carbohydrates that vary among strains can be used as distinguishing markers. For example, different strains of *E. coli* and related bacteria can be distinguished by the antigenic type of their flagella, capsules, and lipopolysaccharide molecules (**figure 10.9**). The "O157:H7" designation of *E. coli*

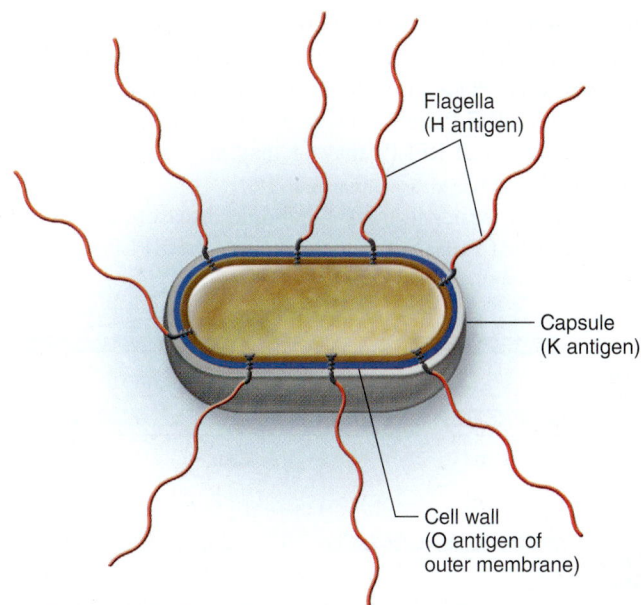

FIGURE 10.9 Serotypes The cell structures used to distinguish different strains of members of the family *Enterobacteriacea* are shown.

❓ *What structures are reflected in the "O157:H7" of* E. coli *O157:H7?*

O157:H7 refers to the antigenic type of its lipopolysaccharide (the O antigen) and flagella (the H antigen). A group of strains that have cell surface antigens different from other strains is called a **serovar,** or a **serotype.** ▶▶ antigen, p. 359 ◀◀ lipopolysaccharide, p. 60 ▶▶ *E. coli* O157:H7, p. 590

Molecular Typing

Subtle differences in DNA sequences can be used to distinguish among phenotypically identical strains, and these genomic variations are important for tracing epidemics of foodborne illness. One method of doing this is to compare the patterns of fragment sizes produced when the same restriction enzyme is used to digest DNA from each isolate; the different patterns are called **restriction fragment length polymorphisms (RFLPs).** Gel electrophoresis

TABLE 10.6	Summary of Methods Used to Characterize Different Strains
Method	**Comment**
Biochemical typing	Biochemical tests are most commonly used to identify various species of bacteria, but in some cases they can be used to distinguish different strains. A group of strains that have a characteristic biochemical pattern is called a biovar or a biotype.
Serological typing	Proteins and carbohydrates that vary among strains can be used to differentiate strains. A group of strains that have a characteristic serological type is called a serovar or a serotype.
Molecular typing	Molecular methods such as pulsed-field gel electrophoresis can be used to detect restriction fragment length polymorphisms (RFLPs). Multilocus sequencing typing compares certain nucleotide sequences.
Phage typing	Strains of a given species sometimes differ in their susceptibility to various types of bacteriophage.
Antibiograms	Antibiotic susceptibility patterns can be used to characterize strains.

is used to separate the fragments so they can be observed; if the fragments are quite large, a special method called pulsed-field gel electrophoresis (PFGE) is used. Isolates of the same species that have different RFLPs are considered different strains, whereas those have identical RFLPs may or may not be the same strain (**figure 10.10**). ▶▶ restriction enzymes, p. 315

To make it easier to track foodborne disease outbreaks, the CDC established **PulseNet,** which catalogs the RFLPs of certain

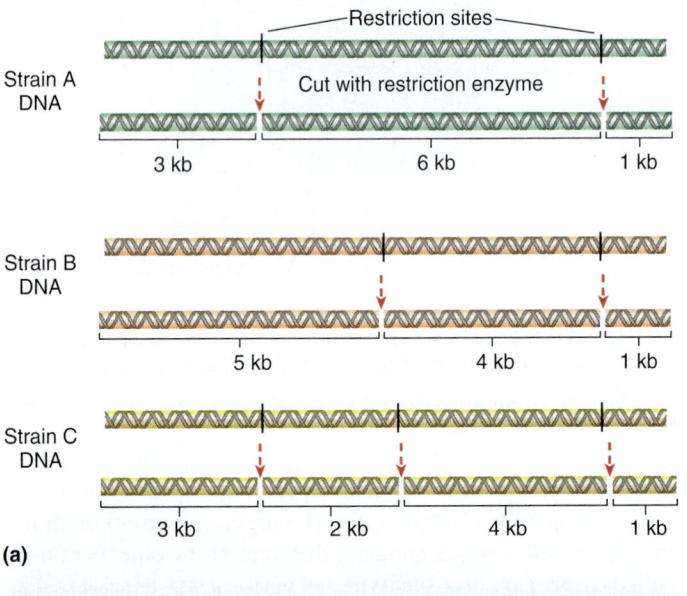

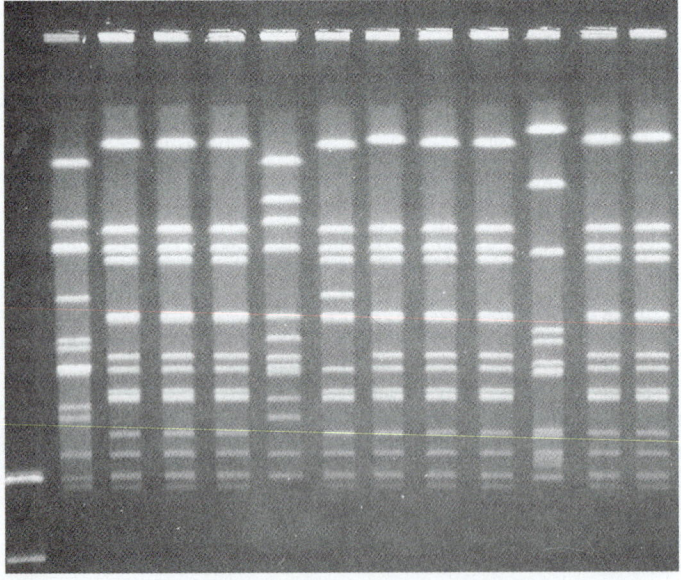

(b)

🎬▶ **FIGURE 10.10 Restriction Fragment Length Polymorphisms (RFLPs) (a)** Different strains of a species may have subtle variations in nucleotide sequences that give rise to a slightly different assortment of restriction fragment sizes (1 kb = 1,000 base pairs). **(b)** In the method shown, genomic DNA is digested with a restriction enzyme that cuts infrequently; the resulting fragments are separated by pulsed-field gel electrophoresis and then stained with ethidium bromide.

❓ *How are RFLPs used to track foodborne disease outbreaks?*

foodborne bacterial pathogens. Laboratories from around the country can submit RFLP patterns to a computer database and quickly receive information about other isolates showing the same pattern. Using this database, multistate foodborne disease outbreaks can more readily be recognized and traced. This is how the salmonellosis cases that led to the alfalfa seeds recall were found to be related (see **Perspective 10.1**). ▶▶ CDC, p. 446

A newer method for distinguishing different strains is **multilocus sequence typing (MLST).** Automated sequencing methods are used to determine the nucleotide sequences of portions of seven or so common genes, and these sequences are then compared.

Phage Typing

Strains of a given species may differ in their susceptibility to bacteriophages. Bacteriophages, or phages, are viruses that infect bacteria, often lysing them; they will be described in more detail in chapter 13. The susceptibility of an organism to a particular type of phage can be easily determined. First, a culture of the test organism is inoculated into melted, cooled nutrient agar and poured onto the surface of an agar plate, creating a uniform layer of cells. Drops of different types of bacteriophage are then placed on the surface of the agar. During incubation, the bacteria multiply, forming a visible haze of cells. If the bacterial strain is susceptible to a specific type of phage, a clear area will form at the spot where bacteriophage was added. The patterns of clearing indicate the susceptibility of the test organism to different phages (**figure 10.11**). Bacteriophage typing has now largely been replaced by molecular methods that detect genomic differences, but it is still a useful tool for laboratories that lack equipment to do molecular typing. ▶▶ bacteriophage, p. 304

Antibiograms

Antibiotic susceptibility patterns, or **antibiograms,** can distinguish different strains. As with phage typing, this method has largely been replaced by molecular techniques. To determine the antibiogram, a culture is uniformly inoculated onto the surface of nutrient agar. Paper discs containing different antibiotics are then placed on the agar. After incubation, clear areas will be visible around discs of antibiotics that inhibit or kill the organism (**figure 10.12**). ▶▶ antibiotic, p. 458

MicroAssessment 10.4

Strains of a given species may differ in phenotypic characteristics such as biochemical capabilities, protein and polysaccharide components, susceptibility to bacteriophages, and sensitivity to antimicrobial drugs. Molecular techniques can be used to detect genomic differences between strains that are phenotypically identical.

10. *Explain the difference between a biotype and a serotype.*

11. *Describe the significance of RFLPs in distinguishing between strains.*

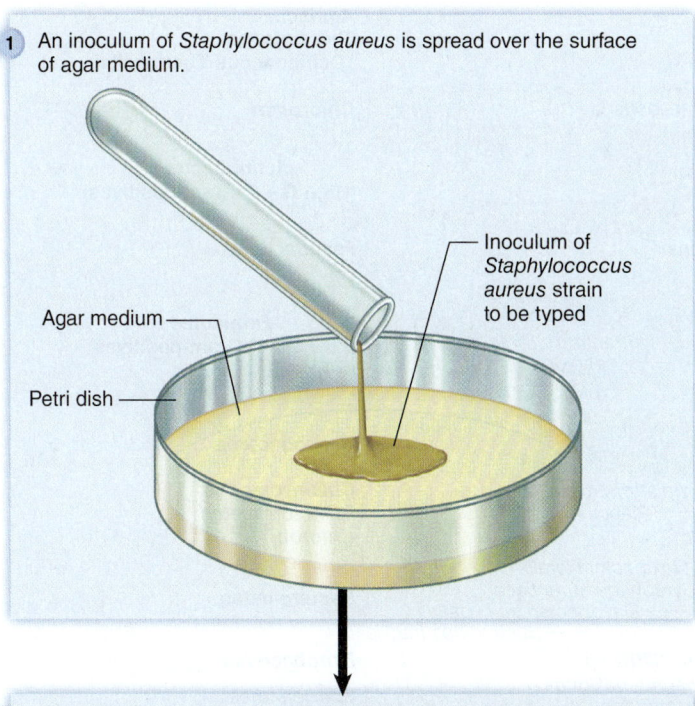

1. An inoculum of *Staphylococcus aureus* is spread over the surface of agar medium.

Inoculum of *Staphylococcus aureus* strain to be typed

Agar medium

Petri dish

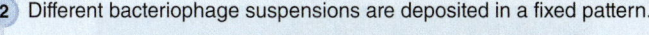

2. Different bacteriophage suspensions are deposited in a fixed pattern.

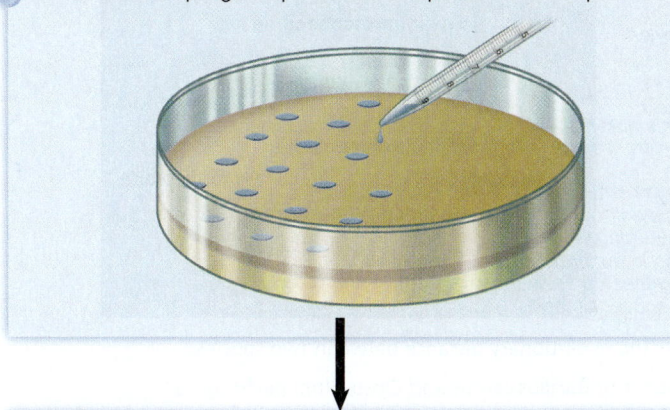

3. After incubation, different patterns of lysis are seen with different strains of *S. aureus*.

Dye marker

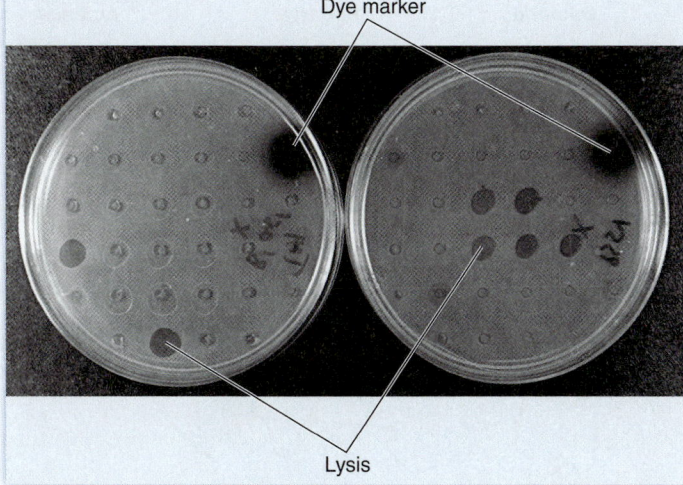

Lysis

FIGURE 10.11 Phage Typing

? *Why would a lab do phage typing rather than molecular typing?*

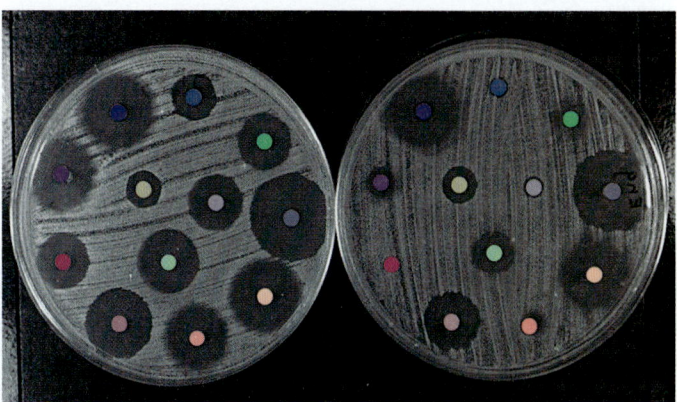

FIGURE 10.12 An Antibiogram In this example, 12 different antimicrobial drugs incorporated in paper discs have been placed on two plates containing different cultures of *Staphylococcus aureus*. Clear areas represent zones of inhibited growth.

? *How can you tell that these* S. aureus *isolates are different strains?*

10.5 ■ Classifying Prokaryotes

Learning Outcome

5. *Describe how 16S rRNA sequences, DNA hybridization, and DNA base ratios are used to classify prokaryotes.*

The goal of phylogenetic classification is to group organisms according to their evolutionary relatedness. Unfortunately, this is difficult when trying to place the diverse types of prokaryotes in their proper positions with respect to evolution. Fossilized stromatolites (coral-like mats of filamentous microorganisms) can be studied, but these remains give few clues to help identify or understand ancient microbes. Because evolutionary relatedness can be difficult to determine, prokaryotic classification schemes historically grouped organisms by their phenotypic traits such as size and shape, staining characteristics, and metabolic capabilities. Although this is easy, there are several significant drawbacks. For instance, phenotypic differences can be due to only a few gene products, and a single mutation resulting in a non-functional enzyme can change an organism's capabilities. In addition, phenotypically similar organisms may be only distantly related; conversely, those that appear dissimilar may be closely related.

Newer molecular techniques such as DNA sequencing avoid some of the problems associated with phenotypic classification while also giving greater insights into the evolutionary relatedness of microorganisms. DNA sequences are viewed as **evolutionary chronometers,** meaning they provide a relative measure of the time elapsed since the organisms diverged from a common ancestor. This is because random mutations cause sequences to change over time. The more time elapsed since two organisms diverged, the greater the differences in the sequences of their DNA.

DNA sequencing makes it possible to more accurately construct a **phylogenetic tree.** These trees are somewhat like a family tree, tracing the evolutionary heritage of organisms. Each branch represents the evolutionary distance between two species, and individual species are represented as nodes (**figure 10.13**).

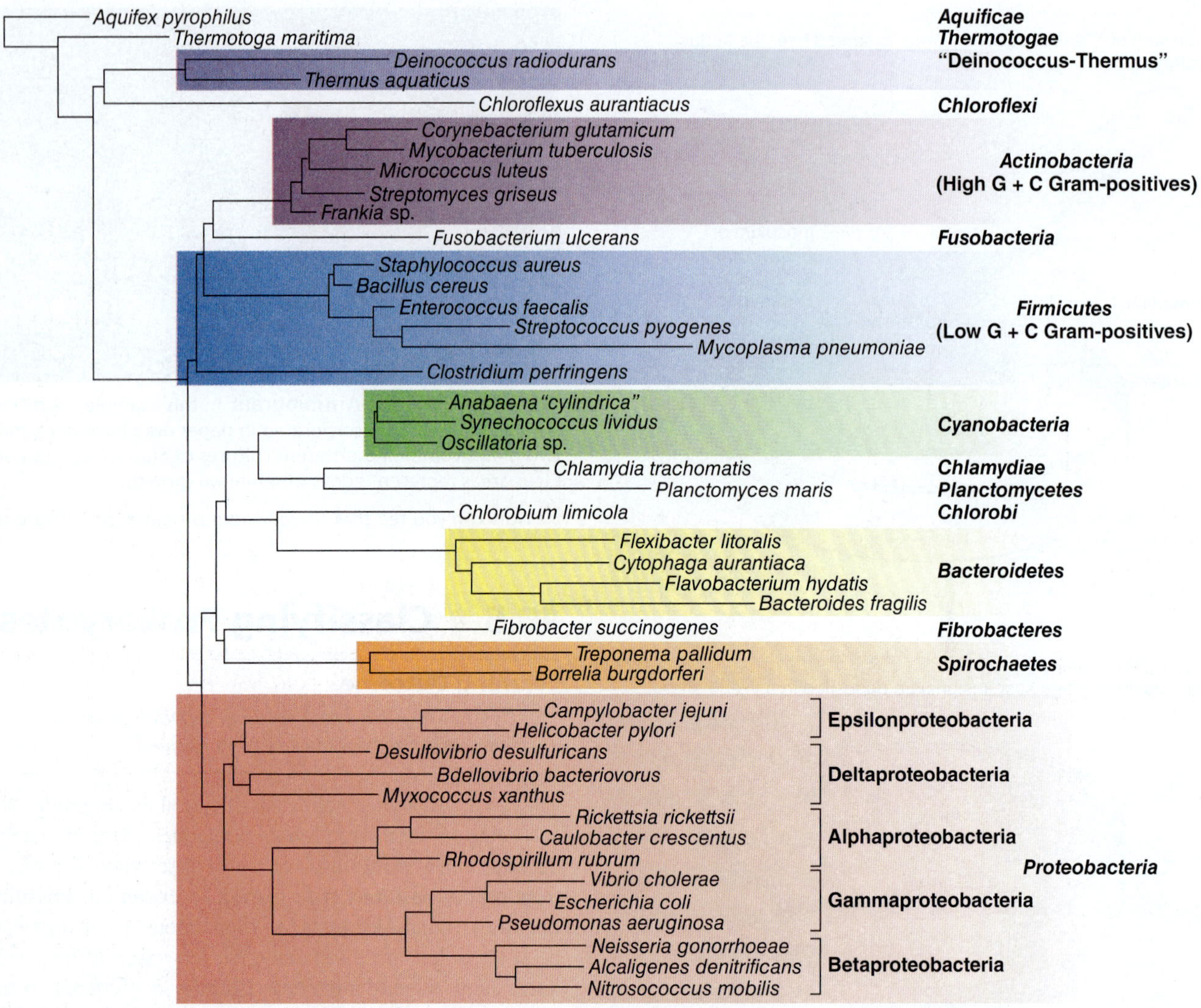

Aquifex pyrophilus	**Aquificae**
Thermotoga maritima	**Thermotogae**
Deinococcus radiodurans	**"Deinococcus-Thermus"**
Thermus aquaticus	
Chloroflexus aurantiacus	**Chloroflexi**
Corynebacterium glutamicum	
Mycobacterium tuberculosis	**Actinobacteria**
Micrococcus luteus	**(High G + C Gram-positives)**
Streptomyces griseus	
Frankia sp.	
Fusobacterium ulcerans	**Fusobacteria**
Staphylococcus aureus	
Bacillus cereus	
Enterococcus faecalis	**Firmicutes**
Streptococcus pyogenes	**(Low G + C Gram-positives)**
Mycoplasma pneumoniae	
Clostridium perfringens	
Anabaena "cylindrica"	
Synechococcus lividus	**Cyanobacteria**
Oscillatoria sp.	
Chlamydia trachomatis	**Chlamydiae**
Planctomyces maris	**Planctomycetes**
Chlorobium limicola	**Chlorobi**
Flexibacter litoralis	
Cytophaga aurantiaca	**Bacteroidetes**
Flavobacterium hydatis	
Bacteroides fragilis	
Fibrobacter succinogenes	**Fibrobacteres**
Treponema pallidum	**Spirochaetes**
Borrelia burgdorferi	
Campylobacter jejuni	**Epsilonproteobacteria**
Helicobacter pylori	
Desulfovibrio desulfuricans	
Bdellovibrio bacteriovorus	**Deltaproteobacteria**
Myxococcus xanthus	
Rickettsia rickettsii	
Caulobacter crescentus	**Alphaproteobacteria**
Rhodospirillum rubrum	**Proteobacteria**
Vibrio cholerae	
Escherichia coli	**Gammaproteobacteria**
Pseudomonas aeruginosa	
Neisseria gonorrhoeae	
Alcaligenes denitrificans	**Betaproteobacteria**
Nitrosococcus mobilis	

FIGURE 10.13 Phylogenetic Tree of the *Bacteria* Each branch represents the evolutionary distance between two species.

❓ *Which is more closely related*—Deinococcus radiodurans *and* Thermus aquaticus *or* Bacillus cereus *and* Clostridium perfringens?

Ancient prokaryotes, which branch early in evolution, are called deeply branching to reflect their position in the phylogenetic tree.

Although analysis of sequence data has solved some difficulties in prokaryotic classification, it has also highlighted an important obstacle. Prokaryotic cells transfer DNA to other species, a process called **horizontal gene transfer**, complicating insights provided by DNA sequence comparison. The bacterium *Thermotoga maritima*, for example, appears to have acquired one-fourth of its genes from a hyperthermophilic archaeal species. Such observations have prompted some scientists to suggest that the tree of life (see figure 10.1) should be depicted as a shrub with interwoven branches (**figure 10.14**). ◀◀ horizontal gene transfer, p. 200

Table 10.7 summarizes some of the methods used to classify prokaryotes.

16S rDNA Sequence Analysis

Analyzing and comparing the nucleotide sequences of 16S ribosomal RNA (rRNA) and the genes that encode rRNA (rDNA) have revolutionized the classification of organisms

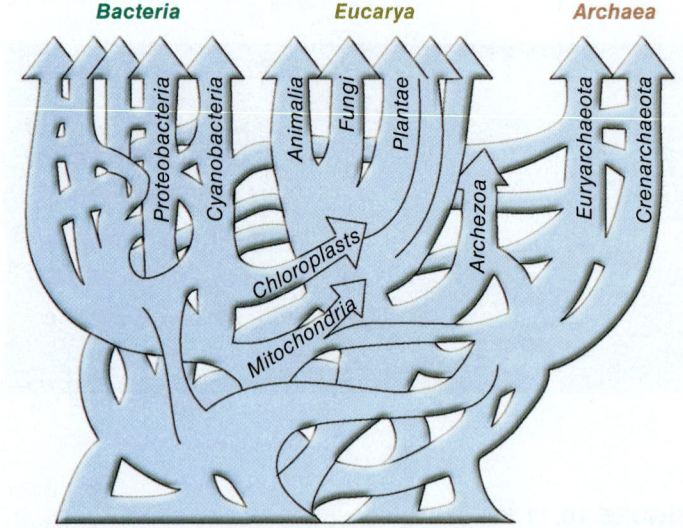

FIGURE 10.14 *"Shrub"* of Life

❓ *Why are chloroplasts and mitochondria depicted as arrows originating from the Bacteria?*

Tracing the Source of an Outbreak of Foodborne Disease

On September 8, 2006, Wisconsin state health authorities alerted the CDC about an *E. coli* O157:H7 outbreak that appeared to involve consumption of fresh spinach. Soon thereafter, Oregon officials reported a similar outbreak. On September 12, the CDC determined that the strains from Wisconsin matched those from Oregon, as well as strains then reported from New Mexico. Within days, the CDC issued a press release advising people not to eat bagged fresh spinach, and a California company that produces several brands of bagged spinach announced a voluntary recall of all fresh spinach products. In the end, the implicated strain was found to have caused 205 cases of illness in 26 states, resulting in three deaths. ▶▶*E. coli* O157:H7, p. 590

Although the sequence of events that led to the recognition of the *E. coli* O157:H7 outbreak may sound quite simple, they are actually very complex. For one thing, most strains of *E. coli* are normal inhabitants of the intestine and can be found in almost every sample of feces. How was this particular strain separated and distinguished from the hundreds of *E. coli* strains that do not cause diarrheal disease? It is also true that sporadic cases of *E. coli* O157:H7 infections occur regularly, originating from unrelated sources. How was it recognized that these cases were connected?

To identify *E. coli* O157:H7 in a stool specimen, the sample is inoculated onto a special agar medium designed to distinguish it from strains that typically inhabit the large intestine. One such medium is sorbitol-MacConkey, a modified version of MacConkey agar in which the lactose is replaced with the carbohydrate sorbitol. On this medium, most *E. coli* O157:H7 isolates are colorless because they do not ferment sorbitol. In contrast, common strains of *E. coli* ferment the carbohydrate, giving rise to pink colonies. Serology is then used to determine if the colorless *E. coli* colonies are serotype O157; those that test positive are then generally tested to confirm they are serotype H7.

The next task is to determine whether or not two isolates of *E. coli* O157:H7 originated from the same source. DNA is extracted and purified from each isolate and is then digested with restriction enzymes. Pulsed-field gel electrophoresis is generally used to compare the resulting restriction fragment length polymorphism (RFLP) patterns of the isolates (see figure 10.10). Those that have identical patterns are presumed to have originated from the same source. Patients from whom those isolates originated can then be questioned to determine where they likely contacted the disease-causing organism. Culture methods are then used to try to isolate the organism from the suspected food source. If that attempt is successful, the RFLP pattern of that isolate is then compared with those of the related cases.

(see **A Glimpse of History**). This is because rRNA is present in all organisms and performs a critical and functionally constant task. The number of mutations that can happen in certain regions without affecting the survival of an organism is limited. Long after organisms have diverged, portions of their 16S rDNA sequences are still similar. Changes in these highly conserved regions occur very slowly over time and therefore are useful for determining distant relationships of diverse organisms. At the same time, certain regions are relatively variable. These sequences can be used to determine more recent divergence. In addition, horizontal gene transfer is unlikely to complicate the analysis because rDNA transfer appears to be rare.

The phylogeny of the many prokaryotes not yet grown in culture can be tentatively determined by 16S rDNA sequence analysis. DNA can be extracted from environmental samples such as soil and water, and the 16S rDNA then amplified, cloned, and sequenced. The sequences can then be compared to databases containing 16S rDNA sequences of known organisms. Using these techniques, a variety of unique prokaryotes have been discovered. In fact, even though most characterized archaeal genera are extremophiles, 16S rDNA sequences indicate that members of this domain are common in non-extreme environments as well. Clearly, the prokaryotic world is much more diverse than previously recognized. ◀◀ using rDNA to identify uncultivated organisms, p. 246

Although 16S rDNA sequence analysis has been very helpful in determining the phylogeny of distantly related organisms, it is often unreliable for distinguishing closely related species. This is because closely related but genetically distinct prokaryotes can have identical 16S rDNA sequences. In these cases, DNA hybridization (discussed next) is a better tool to assess relatedness.

TABLE 10.7	Methods Used to Determine the Relatedness of Different Prokaryotes for Purposes of Classification	
Method	**Comment**	
Genotypic Characteristics	Differences in DNA sequences can be used to determine the point in time at which two organisms diverged from a common ancestor.	
Comparing the sequences of 16S rDNA	This technique has revolutionized classification. Certain regions of the 16S rDNA can be used to determine distant relatedness of diverse organisms; other regions can be used to determine more recent divergence.	
DNA hybridization	The extent of nucleotide sequence similarity between two isolates can be determined by measuring how completely single strands of their DNA hybridize to one another.	
DNA base composition	Determining the G + C content offers a crude comparison of genomes. Organisms with identical G + C contents can be entirely unrelated, however.	
Phenotypic Characteristics	Traditionally, relatedness of different bacteria has been decided by comparing properties such as ability to degrade lactose and the presence of flagella. These characteristics, however, do not necessarily reflect the evolutionary relatedness of organisms.	

DNA Hybridization

The extent of nucleotide sequence similarity between two organisms can be determined by measuring how completely single strands of their DNA hybridize to one another. Just as the complementary strands of DNA from one organism will anneal (base-pair), so will homologous DNA of a different organism. The extent of hybridization reflects the degree of sequence similarity. Two strains that show at least 70% similarity are generally considered to be members of the same species. Surprisingly, DNA hybridization studies have shown that members of the genera *Shigella* and *Escherichia,* which are quite different based on biochemical tests, should actually be grouped in the same species. Note that human and chimpanzee DNA have approximately 99% similarity by DNA hybridization studies. Therefore, by the criteria used to classify prokaryotes, humans and chimpanzees would be members of the same species! ◀◀ DNA hybridization, p. 232

DNA Base Ratio (G + C Content)

One way to roughly compare the genomes of different bacteria is to determine their DNA base ratio, which is the relative portion of adenine (A), thymine (T), guanine (G), and cytosine (C). Because of base-pairing rules, the number of molecules of G in double-stranded DNA always equals the number of molecules of C. Likewise, the number of molecules of T equals that of A. The base ratio of an organism is usually expressed as the percent of guanine plus cytosine, termed the **G + C content,** or more commonly, the GC content. If the GC content of two organisms differs by more than a small percentage, they cannot be closely related. A similarity of base compositions, however, does not necessarily mean that the organisms are related, because the nucleotide sequences and genome sizes could differ greatly.

The GC content is often measured by determining the temperature at which the double-stranded DNA denatures, or melts. DNA that has a high GC content melts at a higher temperature because G-C base pairs are held together by three hydrogen bonds, whereas A-T pairs are held together by only two hydrogen bonds. The temperature at which double-stranded DNA melts can readily be determined by measuring the absorbance of UV light by a solution of DNA as it is heated. The absorbance rapidly increases as the DNA denatures (**figure 10.15**).

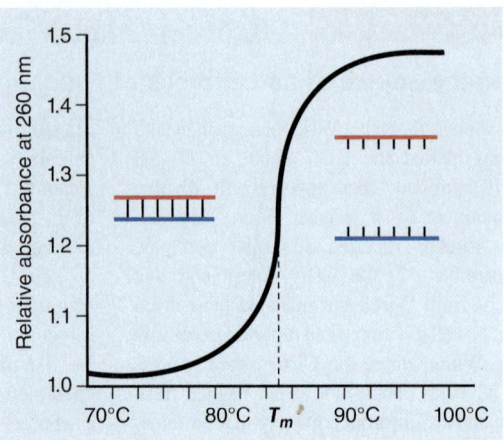

FIGURE 10.15 A DNA Melting Curve The absorbance (relative absorbance at 260 nm) rapidly increases as double-stranded DNA denatures (melts).

❓ *What characteristic of a DNA sequence does the T_m reflect?*

Phenotypic Methods

Classification schemes that group organisms by phenotype have largely been replaced by methods that rely on 16S ribosomal nucleic acid sequence data. Some taxonomists, however, believe classification should be based on more than just genotypic traits. Regardless, phenotypic methods are still important because they provide a foundation for prokaryotic identification.

Numerical taxonomy uses a quantitative approach to phenotypic classification, comparing a battery of characteristics. Information about this method can be obtained by visiting the text website (**www.mhhe.com/nester7**).

MicroAssessment 10.5

Analyzing and comparing the sequences of ribosomal RNA genes has revolutionized the classification of organisms. Other characteristics used to classify prokaryotes include DNA hybridization, GC ratio, and phenotypic properties.

12. *Explain why 16S rDNA is useful for determining phylogeny.*
13. *Explain the role of DNA hybridization in classification.*
14. *Why would it be easier to sequence rDNA than rRNA?* ➕

FUTURE CHALLENGES   10.1

Tangled Branches in the Phylogenetic Tree

Although 16S rDNA sequencing has given remarkable new insight into the evolutionary relatedness of organisms, total genome sequencing is revealing a tremendous amount of additional information. In some cases, this helps clarify the evolutionary picture, providing solid evidence that two organisms are, in fact, closely related. However, genomic sequencing sometimes gives information that conflicts with conclusions based on rDNA

data. This discrepancy is probably due to horizontal gene transfer (HGT), which can result in two distantly related microbes sharing some of the same nucleotide sequences. The outcome, however, is that the phylogenetic tree may be better illustrated as a shrub, with HGT giving rise to intertwining branches (see figure 10.14).

New insights with respect to microbial phylogeny also affect bacterial nomenclature.

As the genetic relatedness among organisms becomes easier to determine, nomenclature is bound to change Already, some species formerly included in the genus *Streptococcus* have been removed to create two relatively new genera, *Enterococcus* and *Lactococcus.* Meanwhile, other microbial species will be moved from one existing genus to another. Any change creates a potential source of confusion for the scientist and layperson alike.

Summary

10.1 ■ Principles of Taxonomy

Taxonomy consists of three interrelated areas: **identification, classification,** and **nomenclature.**

Strategies Used to Identify Prokaryotes

To characterize and identify microorganisms, a wide assortment of technologies is used, including microscopic examination, cultural characteristics, biochemical tests, and nucleic acid analysis.

Strategies Used to Classify Prokaryotes

Taxonomic classification categories are arranged in a hierarchical order, with the species being the basic unit. Taxonomic categories include **species, genus, order, class, phylum** (or **division**), **kingdom,** and **domain.** Individual **strains** within a species vary in minor properties (table 10.1).

Nomenclature

Prokaryotes are assigned names governed by official rules.

10.2 ■ Using Phenotypic Characteristics to Identify Prokaryotes (table 10.4)

Microscopic Morphology

The size, shape, and staining characteristics of a microorganism yield important clues as to its identity (figures 10.2, 10.3).

Culture Characteristics

Selective and differential media used in the isolation process can provide information that helps identify an organism.

Metabolic Capabilities

Most biochemical tests rely on a pH indicator or chemical reaction that shows a color change when a compound is degraded. The basic strategy for identification using biochemical tests relies on the use of a **dichotomous key** (figures 10.4, 10.5; table 10.5).

Serology

Proteins and polysaccharides that make up a prokaryote's surface are sometimes characteristic enough to be identifying markers.

Fatty Acid Analysis (FAME)

Cellular fatty acid composition can be used as an identifying marker.

10.3 ■ Using Genotypic Characteristics to Identify Prokaryotes

Detecting Specific Nucleotide Sequences

A probe complementary to a sequence unique to a given microbe is used to detect that organism (figure 10.7). Nucleic acid amplification tests (NAATs) such as PCR increase the number of copies of a specific nucleotide sequence, thereby determining if a given organism is present.

Sequencing Ribosomal RNA Genes

The nucleotide sequence of ribosomal RNA (rRNA) can be used to identify prokaryotes. Newer techniques simply sequence rDNA, the

DNA that encodes rRNA. Organisms that cannot yet be cultivated can be identified by amplifying, cloning, and then sequencing specific regions of rDNA.

10.4 ■ Characterizing Strain Differences (table 10.6)

Biochemical Typing

A group of strains that has a characteristic biochemical variation is called a **biovar,** or a **biotype.**

Serological Typing

A group of strains that differs serologically from other strains is called a **serovar,** or a **serotype** (figure 10.9).

Molecular Typing

Two isolates that have different **restriction fragment length polymorphisms (RFLPs)** are considered different strains (figure 10.10). **Multilocus sequence typing** determines the nucleotide sequence of representative segments.

Phage Typing

The susceptibility to various types of bacteriophages can be used to demonstrate strain differences (figure 10.11).

Antibiograms

Antibiotic susceptibility patterns can be used to distinguish strains (figure 10.12).

10.5 ■ Classifying Prokaryotes (table 10.7)

DNA sequences can be used to construct a **phylogenetic tree** (figure 10.13). **Horizontal gene transfer** can complicate insights provided by some types of DNA sequence comparison (figure 10.14).

16S rDNA Sequence Analysis

Analyzing and comparing the sequences of rRNA and more recently, rDNA, has revolutionized the classification of organisms.

DNA Hybridization

The extent of nucleotide similarity between two organisms can be determined by measuring how completely single strands of their DNA will anneal to one another.

DNA Base Ratio (G + C Content)

The **G + C content** can be measured by determining the temperature at which double-stranded DNA melts (figure 10.15).

Phenotypic Methods

Classification schemes that group organisms by phenotype have largely been replaced by methods that rely on 16S ribosomal nucleic acid sequence data.

Review Questions

Short Answer

1. Name and describe each of the areas of taxonomy.
2. Compare and contrast the five-kingdom and three-domain systems of classification.
3. Describe how a dichotomous key is used when identifying bacteria.
4. Describe the difference between using a probe and using PCR to detect a specific sequence.

5. Explain how signature sequences are used in bacterial identification.
6. Describe the function of PulseNet.
7. Describe how the GC content of DNA can be measured.
8. Explain why DNA sequences are evolutionary chronometers.
9. What is a phylogenetic tree?
10. Why should a classification scheme reflect the phylogeny of organisms?

Multiple Choice

1. Which of the following is the newest taxonomic unit?
   a) Strain   b) Family   c) Order
   d) Species   e) Domain

2. An acid-fast stain can be used to detect which of the following organisms?
   a) *Cryptococcus neoformans*   b) *Mycobacterium tuberculosis*
   c) *Neisseria gonorrhoeae*   d) *Streptococcus pneumoniae*
   e) *Streptococcus pyogenes*

3. The "breath test" for *Helicobacter pylori* infection determines the presence of which of the following?
   a) Antigens   b) Catalase   c) Hemolysis
   d) Lactose fermentation   e) Urease

4. The "O157:H7" of *E. coli* O157:H7 refers to the
   a) biotype.   b) serotype.   c) phage type.
   d) ribotype.   e) antibiogram.

5. PulseNet catalogs which of the following?
   a) Biotype   b) Serotype   c) Phage type
   d) RFLP   e) Antibiogram

6. Which of the following is an example of an evolutionary chronometer?
   a) Ability to form endospores
   b) 16S ribosomal RNA sequence
   c) Sugar degradation
   d) Motility

7. If the GC content of two organisms is 70%, which of the following is *true*?
   a) The organisms are definitely related.
   b) The organisms are definitely not related.
   c) The AT content is 30%.
   d) The organisms likely have extensive DNA homology.
   e) The organisms likely have many characteristics in common.

8. Which of the molecular methods of assessing similarity gives the crudest approximation of relatedness?
   a) DNA hybridization   b) PCR
   c) 16S rDNA sequencing   d) DNA base composition

9. The sequence of which ribosomal genes are most commonly used for establishing phylogenetic relatedness?
   a) 5S   b) 16S
   c) 23S   d) All of these are commonly used.

10. All of the following statements are correct *except*
   a) *Tropheryma whipplei* could be identified before it had been grown in culture.
   b) the GC content of DNA can be measured by determining the temperature at which double-stranded DNA melts.
   c) sequence differences between organisms can be used to assess their relatedness.
   d) based on DNA homology studies, members of the genus *Shigella* should be in the same species as *Escherichia coli*.
   e) gel electrophoresis is used to determine the serotype of an organism.

Applications

1. Microbiologists debate the use of biochemical similarities and cell features as a way of determining the taxonomic relationships among prokaryotes. Explain why some microbiologists believe these similarities and differences are a powerful taxonomic indicator, whereas others think they are not very useful for that purpose.

2. A researcher interested in investigating the genetic relationship of mitochondria to bacteria must decide on the best method to study this. What advice would you give the researcher?

Critical Thinking ✛

1. In figure 10.15, how would the curve appear if the GC content of the DNA sample were increased? How would the curve appear if the AT content were increased?

2. When DNA probes are used to detect specific sequence similarities in bacterial DNA, the probe is heated and the two strands of DNA are separated. Why must the probe DNA be heated?

The Diversity of Prokaryotic Organisms

Helicobacter pylori (color-enhanced transmission electron micrograph).

A Glimpse of History

Cornelis B. van Niel (1897–1985) earned his Ph.D. from the Technological University in Delft, Holland, the home of an approach to microbiology now commonly referred to as "the Delft School." The outstanding program there was chaired in succession by two well-known microbiologists—Martinus Beijerinck and Albert Kluyver.

As Kluyver's student, van Niel was influenced by his mentor's belief that biochemical processes were fundamentally the same in all cells and that microorganisms could be important research tools, serving as a model to study biochemical process. Thirty years later, Kluyver and van Niel presented lectures that would be published in the book *The Microbe's Contribution to Biology.*

Shortly after completing his dissertation in 1928, van Niel accepted a position at the Hopkins Marine Station in California. There, he continued work he started under Kluyver's direction on the photosynthetic activities of the vividly colored purple bacteria. He demonstrated that these microbes require light for growth, yet, unlike plants and algae, do not evolve O_2. He also showed that to fix CO_2, the purple bacteria oxidize hydrogen sulfide. Furthermore, van Niel noted that the photosynthetic reactions in all photosynthetic organisms are remarkably similar, except the purple bacteria use hydrogen sulfide in place of water and produce oxidized sulfur compounds instead of O_2. This finding raised the possibility that O_2 generated by plants and algae did not come from carbon dioxide, as was believed at the time, but rather from water.

In addition to his scientific contributions, van Niel was an outstanding teacher. During the summers at Hopkins Marine Station, he taught a bacteriology course, inspiring many microbiologists with his enthusiasm for the diversity of microorganisms and their importance in nature. His sharp memory and knowledge of the literature, along with his appreciation for the remarkable abilities of microbes, allowed him to convey the wonders of the microbial world to his students.

Scientists are only beginning to understand the vast diversity of microbial life. Although a million species of prokaryotes are thought to exist, only approximately 6,000 of these, grouped into over 950 genera, have been described and classified. Traditional culture and isolation techniques have not supported the growth and subsequent study of the vast majority. This situation is changing as new molecular techniques make it easier to discover and characterize previously unrecognized species. The sheer volume of information uncovered by modern technologies, however, can be daunting for scientists and students alike.

This chapter covers a variety of prokaryotes, focusing primarily on their extraordinary diversity rather than the phylogenetic relationships discussed in chapter 10. Note, however, that no single chapter could describe all known prokaryotes and, consequently, only a relatively small selection is presented.

METABOLIC DIVERSITY

As a group, prokaryotes use an impressive array of mechanisms to harvest energy in order to produce ATP. This section will highlight this metabolic diversity by describing select prokaryotes. **Table 11.1** summarizes characteristics of the archaea and bacteria covered in this section.

11.1 ■ Anaerobic Chemotrophs

Learning Outcome

1. *Compare and contrast the characteristics and habitats of methanogens, sulfur- and sulfate-reducing bacteria,* Clostridium *species, lactic acid bacteria, and* Propionibacterium *species.*

For approximately the first 1.5 billion years that prokaryotes inhabited earth, the atmosphere was **anoxic,** or devoid of O_2. In that anaerobic environment, some early **chemotrophs** (organisms that harvest energy by oxidizing chemicals) probably used pathways of anaerobic respiration, employing terminal electron acceptors such as carbon dioxide or elemental sulfur, which were plentiful in the environment. Others may have used fermentation, passing the electrons to an organic molecule such as pyruvate. ◀◀ chemotrophs, p. 93 ◀◀ anaerobic respiration, pp. 134, 144 ◀◀ terminal electron acceptor, p. 130 ◀◀ fermentation, pp. 134, 147

Today, anaerobic habitats are still common. Mud and tightly packed soil limit the diffusion of gases, and any O_2 that penetrates is rapidly consumed by aerobically respiring organisms. This creates anaerobic conditions just below the surface. Aquatic environments may also become anaerobic if they contain nutrients that foster the rapid growth of O_2-consuming microbes. This is evident in polluted lakes, where fish may die because of a lack of dissolved O_2. The human body also provides many anaerobic environments, such as in the intestinal tract. Even the skin and the oral cavity, which are routinely exposed to O_2, have anaerobic microenvironments. These are created via the localized depletion of O_2 by aerobes.

MicroByte
Approximately 99% of the prokaryotes that inhabit the intestinal tract are obligate anaerobes.

Anaerobic Chemolithotrophs

Chemolithotrophs oxidize reduced inorganic chemicals such as hydrogen gas (H_2) to obtain energy. Those growing anaerobically obviously cannot use O_2 as a terminal electron acceptor and instead must employ an alternative such as carbon dioxide or sulfur. Relatively few anaerobic chemolithotrophs have been discovered, and most are members of the domain *Archaea*. Some bacterial examples that inhabit aquatic environments will be discussed later. ◀◀ chemolithotrophs, p. 93

The Methanogens

Methanogens are a group of archaea that generate ATP by oxidizing hydrogen gas, using CO_2 as a terminal electron acceptor.

This process generates methane (CH_4), a colorless, odorless, flammable gas:

$$\underset{\text{(energy source)}}{4\,H_2} \quad + \quad \underset{\substack{\text{(terminal electron} \\ \text{acceptor)}}}{CO_2} \quad \longrightarrow \quad CH_4 + 2\,H_2O$$

Many methanogens can also use alternative energy sources such as formate; some can use methanol or acetate as well. Representative genera of methanogens include *Methanospirillum* and *Methanosarcina* (**figure 11.1**).

Methanogens are found in anaerobic environments where H_2 and CO_2 are both available. Because these gases are generated by bacteria that ferment organic material, methanogens often grow in association with these microbes. Methanogens, however, are generally not found in environments containing high levels of sulfate, nitrate, or other inorganic electron acceptors. This is because microorganisms that use these electron acceptors to oxidize H_2 have a competitive advantage; the use of CO_2 as an electron acceptor releases comparatively little energy (see figure 6.7). Environments where methanogens are commonly found include sewage, swamps, marine sediments, rice paddies, and the digestive tracts of humans and other animals. The methane produced can be seen as bubbles rising in swamp waters or the 10 cubic feet of gas discharged from a cow's digestive system each day. As a by-product of sewage treatment plants, methane gas can be collected and used for heating and generating electricity.

Studying methanogens is challenging because they are very sensitive to O_2. Special techniques, including anaerobe chambers, are used to culture them. ◀◀ culturing anaerobes, p. 96

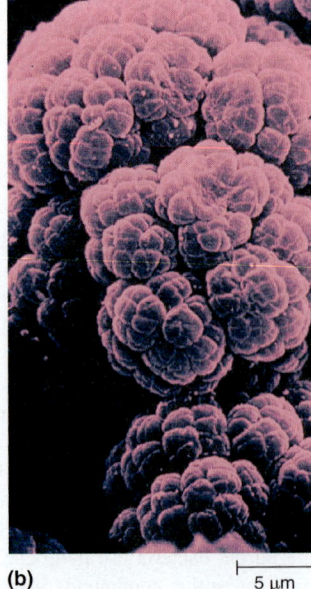

(a)   5 μm   (b)   5 μm

FIGURE 11.1 Methanogens (a) Phase-contrast micrograph of a *Methanospirillum* species. **(b)** Scanning electron micrograph of a *Methanosarcina* species.

❓ *What roles do hydrogen gas and carbon dioxide play in the metabolism of methanogens?*

TABLE 11.1   Metabolic Diversity of Prokaryotes

Group/Genera	Characteristics	Phylum
Anaerobic Chemolithotrophs		
Methanogens—*Methanospirillum, Methanosarcina*	Members of the *Archaea* that oxidize hydrogen gas, using CO_2 as a terminal electron acceptor to generate methane.	*Euryarchaeota*
Anaerobic Chemoorganotrophs—Anaerobic Respiration		
Sulfur- and sulfate-reducing bacteria—*Desulfovibrio*	Use sulfate as a terminal electron acceptor, generating hydrogen sulfide. Found in anaerobic muds rich in organic material. Gram-negative.	*Proteobacteria*
Anaerobic Chemoorganotrophs—Fermentation		
Clostridium	Endospore-forming obligate anaerobes. Inhabitants of soil. Gram-positive.	*Firmicutes*
Lactic acid bacteria—*Streptococcus, Enterococcus, Lactococcus, Lactobacillus, Leuconostoc*	Produce lactic acid as the major end product of their fermentative metabolism. Aerotolerant anaerobes. Several genera are used by the food industry. Gram-positive.	*Firmicutes*
Propionibacterium	Obligate anaerobes that produce propionic acid as their primary fermentation end product. Used in the production of Swiss cheese. Gram-positive.	*Actinobacteria*
Anoxygenic Phototrophs		
Purple sulfur bacteria—*Chromatium, Thiospirillum, Thiodictyon*	Grow in colored masses in sulfur-rich aquatic habitats, using sulfur compounds as a source of electrons when making reducing power. Gram-negative.	*Proteobacteria*
Purple non-sulfur bacteria—*Rhodobacter, Rhodopseudomonas*	Grow in aquatic habitats, preferentially using organic compounds as a source of electrons for reducing power. Many are metabolically versatile. Gram-negative.	*Proteobacteria*
Green sulfur bacteria—*Chlorobium, Pelodictyon*	Found in habitats similar to those preferred by the purple sulfur bacteria. Gram-negative.	*Chlorobi*
Filamentous anoxygenic phototrophic bacteria—*Chloroflexus*	Characterized by their filamentous growth. Gram-negative.	*Chloroflexi*
Others—*Heliobacterium*	Have not been studied extensively.	*Firmicutes*
Oxygenic Phototrophs—Cyanobacteria		
Anabaena, Synechococcus	Important primary producers. Some fix N_2. Gram-negative.	*Cyanobacteria*
Aerobic Chemolithotrophs		
Filamentous sulfur oxidizers—*Beggiatoa, Thiothrix*	Oxidize sulfur compounds as energy sources. Found in sulfur springs and sewage-polluted waters. Gram-negative.	*Proteobacteria*
Unicellular sulfur oxidizers—*Thiobacillus, Acidithiobacillus*	Oxidize sulfur compounds as energy sources. Some species produce enough acid to lower the pH to 1.0. Gram-negative.	*Proteobacteria*
Nitrifiers—*Nitrosomonas, Nitrosococcus, Nitrobacter, Nitrococcus*	Oxidize ammonia or nitrite as energy sources. This converts certain fertilizers to a form easily leached from soils, and depletes O_2 in waters polluted with ammonia-containing wastes. Genera that oxidize nitrite prevent the toxic buildup of nitrite. Gram-negative.	*Proteobacteria*
Hydrogen-oxidizing bacteria—*Aquifex, Hydrogenobacter*	Thermophilic bacteria that oxidize hydrogen gas as an energy source. One of the earliest bacterial forms to exist on earth.	*Aquifacae*
Aerobic Chemoorganotrophs—Obligate Aerobes		
Micrococcus	Widely distributed; common laboratory contaminants. Gram-positive.	*Actinobacteria*
Mycobacterium	Waxy cell wall resists staining; acid-fast.	*Actinobacteria*
Pseudomonas	Common environmental bacteria that, as a group, can degrade a wide variety of compounds. Gram-negative.	*Proteobacteria*
Thermus	*Thermus aquaticus* is the source of *Taq* polymerase (used in PCR). Stains Gram-negative.	*Deinococcus-Thermus*
Deinococcus	Resistant to the damaging effects of gamma radiation. Stains Gram-positive.	*Deinococcus-Thermus*
Aerobic Chemoorganotrophs—Facultative Anaerobes		
Corynebacterium	Widespread in nature. Gram-positive.	*Actinobacteria*
The *Enterobacteriaceae*—*Escherichia, Enterobacter, Klebsiella, Proteus, Salmonella, Shigella, Yersinia*	Most reside in the intestinal tract. Those that ferment lactose are coliforms; their presence in water serves as an indicator of fecal pollution. Gram-negative.	*Proteobacteria*

Anaerobic Chemoorganotrophs— Anaerobic Respiration

Chemoorganotrophs oxidize organic compounds such as glucose to obtain energy. Those that grow anaerobically often use sulfur or sulfate as a terminal electron acceptor. ◀◀ chemoorganotrophs, p. 93

Sulfur- and Sulfate-Reducing Bacteria

When sulfur compounds are used as terminal electron acceptors, they become reduced to form hydrogen sulfide, the compound responsible for the rotten-egg smell of many anaerobic environments. An example of this reaction is:

$$\text{organic compounds} + \text{S} \longrightarrow CO_2 + H_2S$$
$$\text{(energy source)} \quad \text{(terminal electron acceptor)}$$

In addition to its unpleasant odor, H_2S is a problem to industry because it reacts with metals, corroding pipes and other structures. Ecologically, however, prokaryotes that reduce sulfur compounds are an essential component of the sulfur cycle. ▶▶ sulfur cycle, p. 728

Sulfate- and sulfur-reducing bacteria generally live in mud that has organic material and oxidized sulfur compounds. The H_2S they produce causes mud and water to turn black when it reacts with iron molecules. At least a dozen genera are recognized in this group, the most extensively studied of which are species of *Desulfovibrio*. These are Gram-negative curved rods.

Some representatives of the *Archaea* also use sulfur compounds as terminal electron acceptors, but the characterized examples generally do not inhabit the same environments as their bacterial counterparts. Although most of the sulfur-reducing bacteria are either mesophiles or thermophiles, the known sulfur-reducing archaea are hyperthermophiles, inhabiting extreme environments such as hydrothermal vents. They will be discussed later in the chapter. ◀◀ thermophiles, p. 90 ◀◀ hyperthermophiles, p. 90

Anaerobic Chemoorganotrophs— Fermentation

Numerous types of anaerobic bacteria obtain energy by fermenting, producing ATP only by substrate-level phosphorylation. There are many variations of fermentation, using different organic energy sources and producing characteristic end products, but one example is:

$$\text{glucose} \longrightarrow \text{pyruvate} \longrightarrow \text{lactic acid}$$
$$\text{(energy source)} \quad \text{(terminal electron acceptor)}$$

The Genus *Clostridium*

Members of the genus *Clostridium* are Gram-positive rods that can form endospores (see figure 3.45). They are common soil inhabitants, and the vegetative cells live in the anaerobic microenvironments created when aerobic organisms consume available O_2. Their endospores can tolerate O_2 and survive for long periods by withstanding levels of heat, drying, chemicals, and irradiation that would kill vegetative bacteria. When conditions become favorable, these endospores germinate, and the resulting vegetative

bacteria multiply. Vegetative cells that develop from endospores are responsible for a variety of diseases, including tetanus (caused by *C. tetani*), gas gangrene (caused by *C. perfringens*), and botulism (caused by *C. botulinum*). Some species of *Clostridium* are normal inhabitants of the intestinal tract of humans and other animals. ◀◀ endospores, p. 67 ▶▶ tetanus, p. 555 ▶▶ gas gangrene, p. 558 ▶▶ botulism, p. 652

As a group, *Clostridium* species ferment a wide variety of compounds, including sugars and cellulose. Some of the end products are commercially valuable—for example, *C. acetobutylicum* produces acetone and butanol. Some species can ferment amino acids by an unusual process that oxidizes one amino acid, using another as a terminal electron acceptor. This generates a variety of foul-smelling end products associated with rotting flesh.

The Lactic Acid Bacteria

Gram-positive bacteria that produce lactic acid as a major end product of their fermentative metabolism make up a group called the **lactic acid bacteria.** This includes members of the genera *Streptococcus, Enterococcus, Lactococcus, Lactobacillus,* and *Leuconostoc.* Most can grow in aerobic environments, but they only ferment. They can be easily distinguished from other bacteria that grow in the presence of O_2 because they lack the enzyme catalase (see figure 10.4a). ◀◀ obligate fermenters, p. 90 ◀◀ catalase, p. 90

Streptococcus species are cocci that typically grow in chains of varying lengths (**figure 11.2**). They inhabit the oral cavity, generally as part of the normal microbiota. Some, however, are pathogens. One of the most important is *S. pyogenes* (group A strep), which causes pharyngitis (strep throat) and other diseases. Unlike the streptococci that typically inhabit the throat, *S. pyogenes* is β-hemolytic, an important characteristic used to distinguish it from most members of the normal microbiota (see figure 4.10). ▶▶ *Streptococcus pyogenes*, p. 487 ◀◀ hemolysis, p. 96

Species of *Lactococcus* and *Enterococcus* were at one time included in the genus *Streptococcus*. The genus *Lactococcus* now includes species used by the dairy industry to produce fermented milk products such as cheese and yogurt. *Enterococcus* species typically inhabit the intestinal tract of humans and other animals. ▶▶ fermentation of dairy products, p. 751

Members of the genus *Lactobacillus* are rod-shaped bacteria that grow as single cells or loosely associated chains. They are common members of the microbiota in the mouth and the healthy vagina during childbearing years. In the vagina, they break down glycogen that has been deposited in the vaginal lining in response to estrogen (the female sex hormone). The resulting low pH helps prevent vaginal infections. Lactobacilli are also often present in decomposing plant material, milk, and other dairy products. Like the lactococci, they are important in the production of fermented foods (**figure 11.3**).

The Genus *Propionibacterium*

Propionibacterium species are Gram-positive pleomorphic (irregular-shaped) rods that produce propionic acid as their primary fermentation end product. They can also ferment lactic acid, thereby extracting energy from a waste product of other bacteria.

Propionibacterium species are valuable to the dairy industry because their fermentation end products are important in Swiss

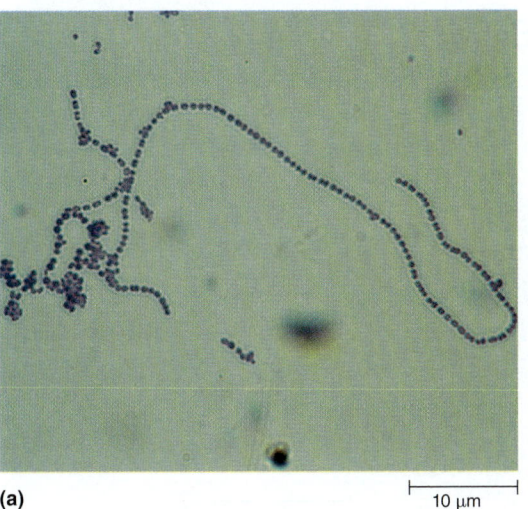

(a)   10 µm

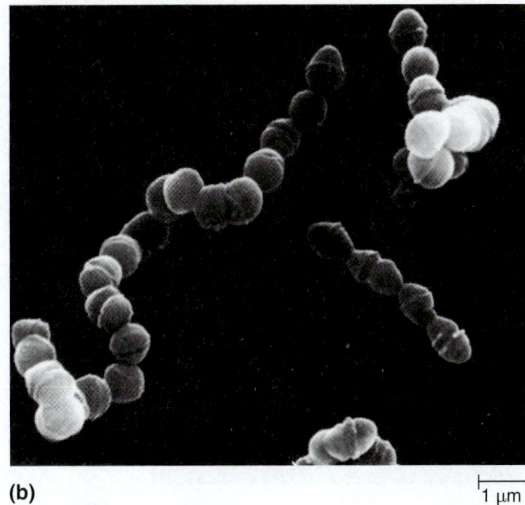

(b)   1 µm

FIGURE 11.2
***Streptococcus* Species**
(a) Gram stain. **(b)** Scanning electron micrograph.

❓ *What is the major metabolic end product of Streptococcus species?*

cheese production. The propionic acid gives the typical nutty flavor of the cheese, and CO_2, also a product of the fermentation, creates the signature holes. *Propionibacterium* species are also found in the intestinal tract and in anaerobic microenvironments on the skin. ▶▶ Swiss cheese production, p. 751

MicroAssessment 11.1

Methanogens are archaea that oxidize hydrogen gas, using CO_2 as a terminal electron acceptor, to generate methane. The sulfur- and sulfate-reducing bacteria oxidize organic compounds, with sulfur or sulfate serving as a terminal electron acceptor, to generate hydrogen sulfide. *Clostridium* species, the lactic acid bacteria, and *Propionibacterium* species oxidize organic compounds, with an organic compound serving as a terminal electron acceptor.

1. *What metabolic process creates the rotten-egg smell characteristic of many anaerobic environments?*

2. *Describe two beneficial contributions of the lactic acid bacteria.*

3. *Relatively little is known about many obligate anaerobes. Why would this be so?* ➕

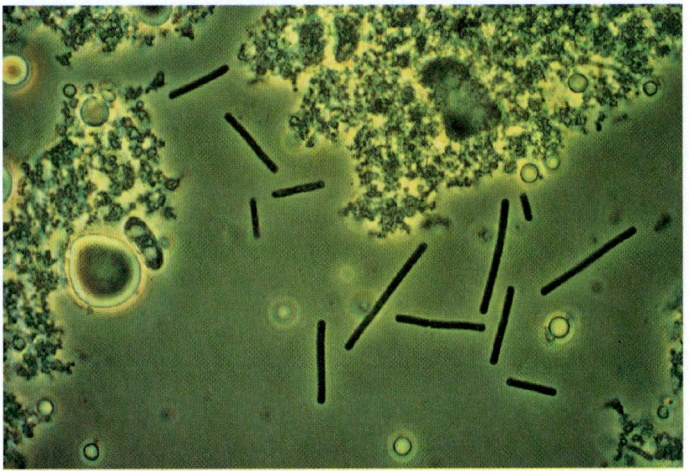

5 µm

FIGURE 11.3 *Lactobacillus* Species from Yogurt

❓ *Lactobacillus species are common members of the normal microbiota of which human body sites?*

11.2 ■ Anoxygenic Phototrophs

Learning Outcome

2. *Compare and contrast the characteristics of the purple bacteria and the green bacteria.*

The earliest photosynthesizing organisms were likely **anoxygenic phototrophs.** Rather than using water as a source of electrons when making reducing power for biosynthesis, these organisms use hydrogen sulfide or organic compounds, and therefore do not generate O_2. For example:

$$6\,CO_2 \;+\; 12\,H_2S \;\longrightarrow\; C_6H_{12}O_6 + 12\,S + 6\,H_2O$$
$$\text{(carbon source)} \quad \text{(electron source)}$$

Modern-day anoxygenic phototrophs are a phylogenetically diverse group of bacteria that live in environments that have little or no O_2 yet light penetrates. Typical habitats include bogs, lakes, and the upper layer of muds. ◀◀ reducing power, p. 131

As discussed in chapter 6, the photosystems of the anoxygenic phototrophs are different from those of plants, algae, and cyanobacteria. They have a unique type of chlorophyll called bacteriochlorophyll. This and their other light-harvesting pigments absorb wavelengths that penetrate deeper than those absorbed by chlorophyll *a*. ◀◀ photosystems, p. 151 ◀◀ chlorophyll *a*, p. 152

The Purple Bacteria

The **purple bacteria** are Gram-negative organisms that appear red, orange, or purple due to their light-harvesting pigments. Unlike other anoxygenic phototrophs, the components of their photosynthetic apparatus are all contained within the cytoplasmic membrane. Folds in this membrane increase the surface area available for the photosynthetic processes.

Purple Sulfur Bacteria

Purple sulfur bacteria can sometimes be seen growing as colored masses in sulfur-rich aquatic habitats such as sulfur springs (**figure 11.4a**). The cells are relatively large, sometimes larger than 5 µm in diameter, and some are motile by flagella. They may also have gas vesicles, allowing them to move up or down to their

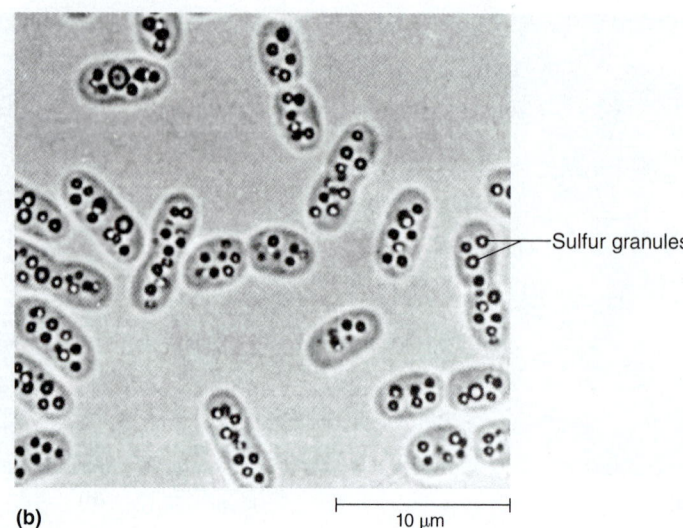

(a)   (b)   10 μm

FIGURE 11.4 Purple Sulfur Bacteria (a) Photograph of bacteria growing in a bog. **(b)** Photomicrograph showing intracellular sulfur granules.

❓ *What is the role of the sulfur granules in the bacterial cells?*

preferred level in the water column. Most store sulfur in intracellular granules (figure 11.4b). ◀◀ **gas vesicles, p. 67**

The purple sulfur bacteria preferentially use hydrogen sulfide to generate reducing power, although some can use other inorganic molecules (such as H_2) or organic compounds (such as pyruvate). Many are strict anaerobes and phototrophs, but some can grow in the absence of light aerobically, oxidizing reduced inorganic or organic compounds as a source of energy. Representative genera include *Chromatium, Thiospirillum,* and *Thiodictyon.*

Purple Non-Sulfur Bacteria

The purple non-sulfur bacteria are found in a wide variety of aquatic habitats, including moist soils, bogs, and paddy fields. One important characteristic that distinguishes them from the purple sulfur bacteria is that they preferentially use a variety of organic molecules rather than hydrogen sulfide as a source of electrons for reducing power. In addition, they lack gas vesicles. If they do store sulfur, the granules form outside the cell.

Purple non-sulfur bacteria are remarkably versatile metabolically. Not only do they grow as phototrophs using organic molecules as a source of electrons, but many can use a metabolism similar to the purple sulfur bacteria, employing hydrogen gas or hydrogen sulfide as an electron source. In addition, most can grow aerobically in the absence of light using chemotrophic metabolism. Representative genera of purple non-sulfur bacteria include *Rhodobacter* and *Rhodopseudomonas.*

The Green Bacteria

The **green bacteria** are Gram-negative organisms that are typically green or brownish in color.

Green Sulfur Bacteria

Green sulfur bacteria are found in habitats similar to those preferred by the purple sulfur bacteria. Like the purple sulfur bacteria, they use hydrogen sulfide as a source of electrons for reducing power and they form sulfur granules. The granules, however, form outside

of the cell (**figure 11.5**). The accessory pigments of the green sulfur bacteria are located in structures called chlorosomes. The bacteria lack flagella, but many have gas vesicles. All are strict anaerobes, and none can use a chemotrophic metabolism. Representative genera include *Chlorobium* and *Pelodictyon.* ◀◀ **accessory pigments, p. 152**

Filamentous Anoxygenic Phototrophic Bacteria

Filamentous anoxygenic phototrophic bacteria form multicellular arrangements and exhibit gliding motility. The most thoroughly studied of this group are members of the genus *Chloroflexus,* particularly the thermophilic strains that grow in hot springs. Many of the filamentous anoxygenic phototrophs have chlorosomes, which initially led scientists to believe they were related to the green sulfur bacteria. Their 16S rDNA sequences indicate otherwise. As a group, filamentous anoxygenic phototrophs are diverse metabolically. Some preferentially use organic compounds to generate reducing power and can also grow in the dark aerobically using chemotrophic metabolism.

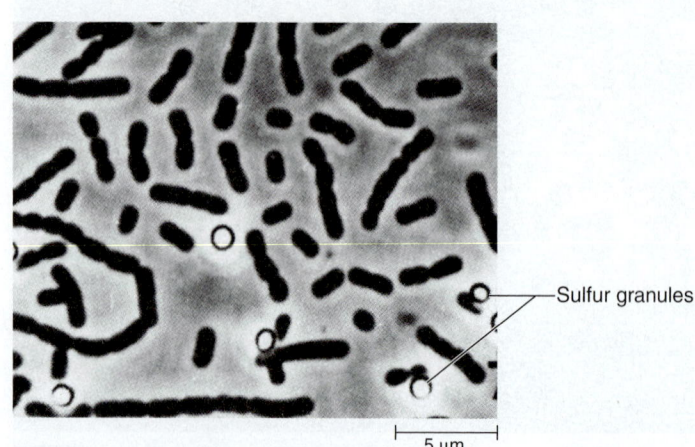

Sulfur granules

5 μm

FIGURE 11.5 Green Sulfur Bacteria Note that the sulfur granules are extracellular.

❓ *What role does sulfur play in the metabolism of green sulfur bacteria?*

Other Anoxygenic Phototrophs

Although the green and purple bacteria have been studied most extensively, other types of anoxygenic phototrophs exist. Among these are members of the genus *Heliobacterium*, Gram-positive endospore-forming rods related to members of the genus *Clostridium*.

MicroAssessment 11.2

Anoxygenic phototrophs harvest the energy of sunlight, but do not generate O_2. The purple sulfur bacteria and green sulfur bacteria use hydrogen sulfide as a source of electrons to generate reducing power; the purple non-sulfur bacteria and many of the filamentous anoxygenic phototrophs preferentially use organic compounds.

4. *Describe a structural characteristic that distinguishes the purple sulfur bacteria from the green sulfur bacteria.*

5. *What is the function of gas vesicles?*

6. *How do anoxygenic phototrophs benefit by having light-harvesting pigments that absorb wavelengths that penetrate deeper than those absorbed by chlorophyll* a? ✚

11.3 ■ Oxygenic Phototrophs

Learning Outcome

3. *Describe the characteristics of cyanobacteria, including how nitrogen-fixing species protect their nitrogenase enzyme from O_2.*

Nearly 3 billion years ago, the earth's atmosphere began changing as O_2 was gradually introduced to the previously anoxic environment. This was probably due to the evolution of the cyanobacteria, thought to be the earliest **oxygenic phototrophs.** These photosynthetic organisms use water as a source of electrons for reducing power, generating O_2:

$$\underset{\text{(carbon source)}}{6\,CO_2} + \underset{\text{(electron source)}}{6\,H_2O} \longrightarrow C_6H_{12}O_6 + 6\,O_2$$

Cyanobacteria still play an essential role in the biosphere. As **primary producers,** they harvest the energy of sunlight, using it to convert CO_2 into organic compounds. They were initially thought to be algae and were called blue-green algae until electron microscopy revealed their prokaryotic structure. ▶▶ primary producers, p. 719

The Cyanobacteria

Cyanobacteria are a diverse group of more than 60 genera of Gram-negative bacteria. They inhabit a wide range of environments, including freshwater and marine habitats, soils, and the surfaces of rocks. In addition to being photosynthetic, many are able to convert nitrogen gas (N_2) to ammonia, which can then be incorporated into cell material. This process, called **nitrogen fixation,** is an exclusive ability of prokaryotes. ▶▶ nitrogen fixation, p. 726

General Characteristics of Cyanobacteria

Cyanobacteria are morphologically diverse. Some are unicellular, with typical prokaryotic shapes such as cocci, rods, and spirals. Others form filamentous multicellular associations called trichomes that may or may not be enclosed within a sheath (a tube

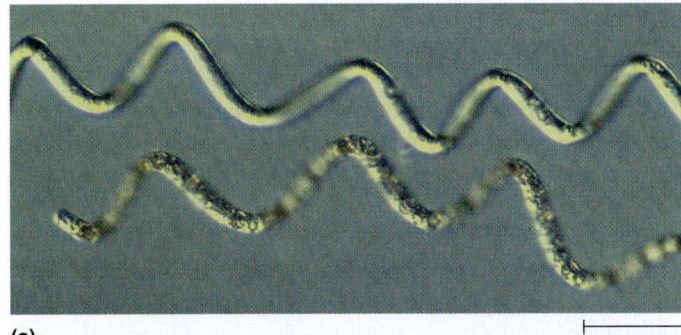

(a)   15 μm

(b)   100 μm

FIGURE 11.6 Cyanobacteria (a) The spiral trichome of *Spirulina* species. **(b)** Differential interference contrast photomicrograph of a species of *Oscillatoria*. Note the arrangement of the individual cells in the trichome.

? *What is a trichome?*

that holds and surrounds a chain of cells) (**figure 11.6**). Motile trichomes glide as a unit. Cyanobacteria that inhabit aquatic environments often have gas vesicles, allowing them to move vertically within the water column. When large numbers of cyanobacteria accumulate in stagnant lakes or other freshwater habitats, it is called a bloom (**figure 11.7**). In the bright, hot conditions of

FIGURE 11.7 Cyanobacterial Bloom Excessive growth of cyanobacteria is evidenced by buoyant masses of cells that have risen to the surface.

? *What allows cyanobacteria cells to rise to the surface?*

summer, the buoyant cells lyse and decay, creating a foul-smelling scum. The ecological effects of these blooms on aquatic habitats are discussed in chapter 29. ▶▶ aquatic habitats, p. 722

The photosystems of the cyanobacteria are like those contained within the chloroplasts of algae and plants. This is not surprising in light of the genetic evidence indicating chloroplasts evolved from a species of cyanobacteria that once resided as an endosymbiont within eukaryotic cells (see Perspective 3.1). In addition to light-harvesting chlorophyll pigments, cyanobacteria have phycobiliproteins. These pigments absorb energy from wavelengths of light not well absorbed by chlorophyll. They contribute to the blue-green, or sometimes reddish, color of the cyanobacteria. ◀◀ photosystem, p. 151

Nitrogen-Fixing Cyanobacteria

Nitrogen-fixing cyanobacteria are critical ecologically. They can incorporate both N_2 and CO_2 into organic material, so they generate a form of these nutrients that can then be used by other organisms. Thus, their activities can ultimately support the growth of a wide range of organisms in environments that would otherwise lack usable nitrogen and carbon. As an example, nitrogen-fixing cyanobacteria in the oceans are essential primary producers that support other sea life. Also, like all cyanobacteria, they help limit atmospheric CO_2 buildup by using the gas as a carbon source.

Nitrogenase, the enzyme complex that catalyzes nitrogen fixation, is destroyed by O_2; therefore, nitrogen-fixing cyanobacteria must protect the enzyme from the O_2 they generate. Species of *Anabaena*, which are filamentous, isolate nitrogenase by confining the process of nitrogen fixation to specialized thick-walled cells called **heterocysts** (figure 11.8). Heterocysts lack photosystem II and consequently do not generate O_2. The heterocysts of some species form at very regular intervals within the filament, reflecting the ability of cells within a trichome to communicate. One species of *Anabaena*, *A. azollae*, forms an intimate relationship with the water fern *Azolla*. The bacterium grows and fixes nitrogen within the protected environment of a special sac in the fern, providing *Azolla* with a source of available

nitrogen. *Synechococcus* species fix nitrogen only in the dark. Consequently, nitrogen fixation and photosynthesis are temporally separated. ◀◀ photosystem II, p. 153

Other Notable Characteristics of Cyanobacteria

Cyanobacteria have various other notable characteristics—some beneficial, others damaging. Filamentous cyanobacteria are responsible for maintaining the structure and productivity of soils in cold desert areas such as the Colorado Plateau. Their sheaths persist in soil, creating a sticky fibrous network that prevents erosion. In addition, these bacteria provide an important source of nitrogen and organic carbon in otherwise nutrient-poor soils. On the negative side, some cyanobacteria produce geosmin, a chemical that has a distinctive "earthy" odor, which makes drinking water taste odd. Some aquatic species such as *Microcystis aeruginosa* produce toxins that can be deadly to an animal when consumed.

MicroAssessment 11.3

The photosystems of cyanobacteria generate O_2 and are similar to those of algae and plants. Many cyanobacteria can fix nitrogen.

7. *What is the function of a heterocyst?*

8. *How do cyanobacteria prevent erosion in cold desert regions?*

9. *How could heavily fertilized lawns foster the development of cyanobacterial blooms?* ➕

11.4 ■ Aerobic Chemolithotrophs

Learning Outcome

4. *Compare and contrast the characteristics of sulfur-oxidizing bacteria, nitrifiers, and hydrogen-oxidizing bacteria.*

Aerobic chemolithotrophs obtain energy by oxidizing reduced inorganic chemicals, using O_2 as a terminal electron acceptor.

The Sulfur-Oxidizing Bacteria

The **sulfur-oxidizing bacteria** are Gram-negative rods or spirals, which sometimes grow in filaments. They obtain energy by oxidizing elemental sulfur and reduced sulfur compounds, including hydrogen sulfide and thiosulfate. O_2 serves as a terminal electron acceptor, generating sulfuric acid. An example of this reaction is:

$$\underset{\text{(energy source)}}{S} + \underset{\substack{\text{(terminal electron} \\ \text{acceptor)}}}{1\tfrac{1}{2}\,O_2} + H_2O \longrightarrow H_2SO_4$$

These bacteria are important in the sulfur cycle. ▶▶ sulfur cycle, p. 728

Filamentous Sulfur Oxidizers

Beggiatoa and *Thiothrix* species are filamentous sulfur oxidizers that live in sulfur springs, in sewage-polluted waters, and on the surface of marine and freshwater sediments. They store sulfur, depositing it as intracellular granules, but differ in the

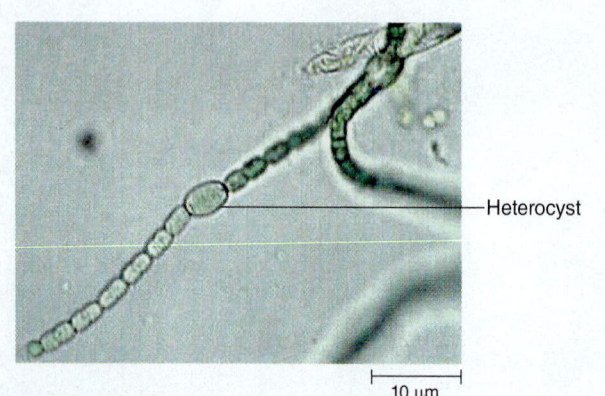

—Heterocyst

10 μm

FIGURE 11.8 Heterocyst of an *Anabaena* Species Nitrogen fixation occurs within these specialized cells.

❓ *Why is it important for heterocysts to not have a functional photosystem II?*

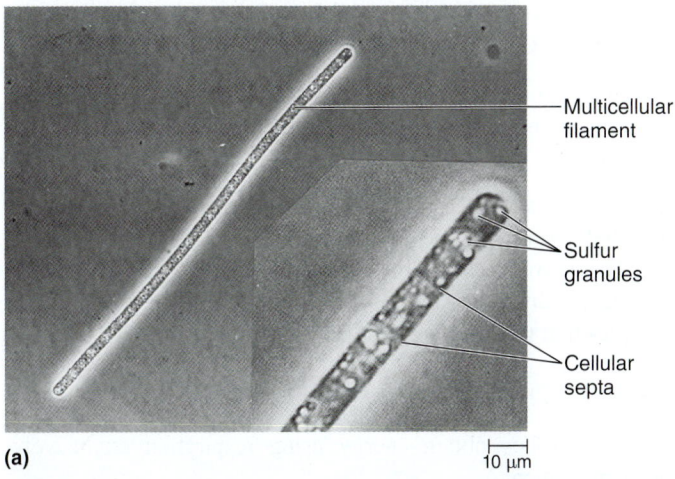

Multicellular filament

Sulfur granules

Cellular septa

(a)

10 μm

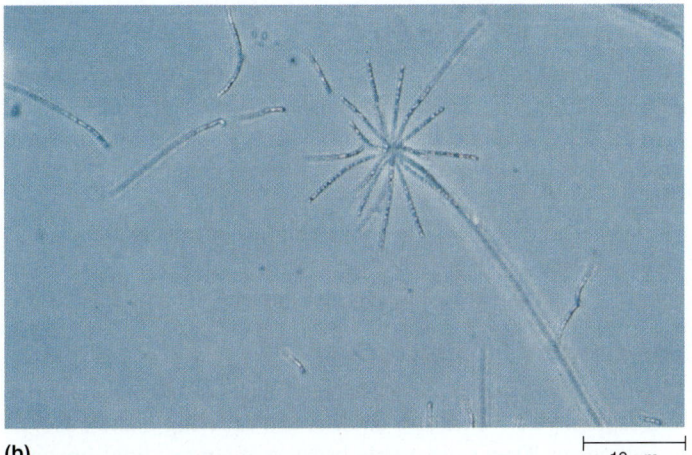

(b)

10 μm

FIGURE 11.9 Filamentous Sulfur-Oxidizing Bacteria Phase-contrast photomicrographs. **(a)** Multicellular filament of a *Beggiatoa* species; the septa separate the cells. **(b)** Multicellular filaments of a *Thiothrix* species, forming a rosette arrangement.

❓ *What is the role of sulfur in the metabolism of sulfur-oxidizing bacteria?*

nature of their filamentous growth (**figure 11.9**). The filaments of *Beggiatoa* species move by gliding motility, a mechanism that does not require flagella. The filaments may flex or twist to form a tuft. In contrast, the filaments of *Thiothrix* species are immobile; they fasten at one end to rocks or other solid surfaces. Often they attach to other cells, forming characteristic rosette arrangements of filaments. Progeny cells detach from the ends of these filaments and use gliding motility to move to new locations, where they form additional filaments. Overgrowth of these filamentous organisms in wastewater at treatment facilities causes a problem called bulking. Because the masses of filamentous organisms do not settle easily, bulking interferes with the separation of the solid sludge from the liquid effluent. ▶▶ **wastewater treatment, p. 736**

Unicellular Sulfur Oxidizers

Acidithiobacillus species are found in both terrestrial and aquatic habitats, where their ability to oxidize metal sulfides can be used for bioleaching, a process used to recover metals (see

Perspective 6.1). The bacteria oxidize insoluble metal sulfides such as gold sulfide, producing sulfuric acid. This lowers the pH, which converts the metal to a soluble form. *Acidithiobacillus* species can also be used to prevent acid rain, a problem that occurs when sulfur-containing coals and oils are burned. To remove the sulfur from the fuels, the bacteria are allowed to oxidize it to sulfate, a form that can then be extracted.

Acidithiobacillus species can also cause severe environmental problems. For example, the strip mining of coal exposes metal sulfides, which the bacteria can then oxidize to produce sulfuric acid; some species produce enough acid to lower the pH to 1.0. The resulting runoff can acidify nearby streams, killing trees, fish, and other wildlife (**figure 11.10**). The runoff may also contain toxic metals made soluble by the bacteria.

The Nitrifiers

Nitrifiers are a diverse group of Gram-negative bacteria that obtain energy by oxidizing inorganic nitrogen compounds such as ammonia or nitrite. These bacteria are a concern to farmers who fertilize their crops with ammonium nitrogen, a form of nitrogen retained by soils because its positive charge causes it to adhere to negatively charged soil particles. The potency and longevity of the fertilizer are affected by nitrifying bacteria converting the ammonia to nitrate. Although plants use the nitrate more easily, it is rapidly leached from soils.

Nitrifying bacteria are also an important consideration when disposing of wastes that have a high ammonia concentration. As nitrifying bacteria oxidize nitrogen compounds, they consume O_2, so waters polluted with nitrogen-containing wastes can quickly become hypoxic (low in dissolved O_2).

The nitrifiers include two metabolically distinct groups that typically grow in close association. Together, they can oxidize

FIGURE 11.10 Acid Drainage from a Mine Sulfur-oxidizing bacteria oxidize exposed metal sulfides, generating sulfuric acid. The yellow-red color is due to insoluble iron oxides.

❓ *How can the metabolic activities that result in this acid drainage be used in a commercially valuable manner?*

ammonia to form nitrate. The ammonia oxidizers, which include the genera *Nitrosomonas* and *Nitrosococcus,* convert ammonia to nitrite in the following reaction:

$$NH_4^+ \quad + \quad 1\tfrac{1}{2} O_2 \quad \longrightarrow \quad NO_2^- + H_2O + 2 H^+$$
(energy source)   (terminal electron acceptor)

The nitrite oxidizers, which include the genera *Nitrobacter* and *Nitrococcus,* then convert nitrite to nitrate as follows:

$$NO_2^- \quad + \quad \tfrac{1}{2} O_2 \quad \longrightarrow \quad NO_3^-$$
(energy source)   (terminal electron acceptor)

The latter group is particularly important in preventing the buildup of nitrite in soils, which is toxic and can leach into groundwater. The oxidation of ammonia to nitrate (nitrification) is an important part of the nitrogen cycle. ▶▶ nitrogen cycle, p. 726

The Hydrogen-Oxidizing Bacteria

Members of the Gram-negative genera *Aquifex* and *Hydrogenobacter* are among the few hydrogen-oxidizing bacteria that are obligate chemolithotrophs. An example of the reaction in their metabolism is:

$$H_2 \quad + \quad \tfrac{1}{2} O_2 \quad \longrightarrow \quad H_2O$$
(energy source)   (terminal electron acceptor)

These related organisms are thermophilic and typically inhabit hot springs. Some *Aquifex* species have a maximum growth temperature of 95°C, the highest of any bacteria. The hydrogen-oxidizing bacteria are deeply branching in the phylogenetic tree, meaning that according to 16S rRNA studies, they were one of the earliest bacterial forms to exist on earth. The fact that they require O_2 seems contradictory to their evolutionary position, but in fact, the low amount they require might have been available early on in certain niches due to photochemical processes that split water.

MicroAssessment 11.4

Sulfur oxidizers use sulfur compounds as energy sources, generating sulfuric acid. The nitrifiers oxidize nitrogen compounds such as ammonium or nitrite. Hydrogen-oxidizing bacteria oxidize H_2.

10. *What is the role of sulfur oxidizers in bioleaching?*
11. *Why would farmers be concerned about nitrifying bacteria?*
12. *Why would sulfur-oxidizing bacteria store sulfur?* ➕

11.5 ■ Aerobic Chemoorganotrophs

Learning Outcomes

5. *Describe the representative obligate aerobes and facultative anaerobes.*
6. *Describe the family* Enterobacteriaceae, *and explain what distinguishes coliforms from other members of this family.*

Aerobic chemoorganotrophs oxidize organic compounds to obtain energy, using O_2 as a terminal electron acceptor:

$$\text{organic compounds} \quad + \quad O_2 \quad \longrightarrow \quad CO_2 + H_2O$$
(energy source)   (terminal electron acceptor)

They include a wide assortment of bacteria, ranging from some that inhabit very specific environments to others that are ubiquitous. This section will profile only representative genera found in a variety of different environments. Later sections will describe examples that thrive in specific habitats.

Obligate Aerobes

Obligate aerobes obtain energy using respiration exclusively; none can ferment.

The Genus *Micrococcus*

Members of the genus *Micrococcus* are Gram-positive cocci found in soil and on dust particles, inanimate objects, and skin. Because they are often airborne, they can easily contaminate bacteriological media. There, they typically form pigmented colonies, a characteristic that helps identify them. The colonies of *M. luteus,* for example, are generally yellow (**figure 11.11**). Like members of the genus *Staphylococcus,* which will be discussed later, they tolerate dry conditions and can grow in salty environments such as 7.5% NaCl.

The Genus *Mycobacterium*

Mycobacterium species are widespread in nature and include harmless saprophytes, which live on dead and decaying matter, as well as pathogens. They stain poorly because of a waxy lipid (mycolic acid) in their unusual cell wall, but special procedures can be used to increase the penetration of certain dyes. Once stained, the cells resist destaining, even with acidic decolorizing solutions. Because of this, *Mycobacterium* species are called **acid-fast,** and the acid-fast staining procedure is an important step in identifying them (see figure 3.15). *Nocardia* species, a related group of bacteria common in soil, are also acid-fast. ◀◀ acid-fast, p. 48

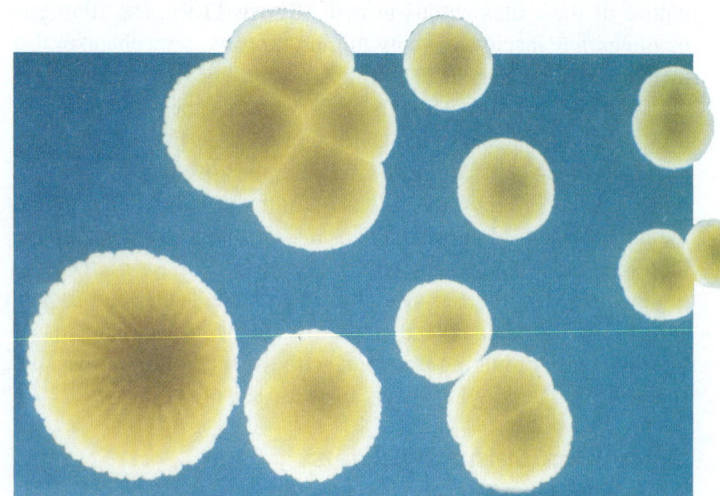

FIGURE 11.11 *Micrococcus luteus* **Colonies**

❓ *Why is* Micrococcus luteus *a common contaminate on bacteriological media?*

Mycobacterium species are generally pleomorphic rods; they often occur in chains that sometimes branch, or bunch together to form cordlike groups. Several species are notable for their effect on human health, including *M. tuberculosis,* which causes tuberculosis, and *M. leprae,* which causes Hansen's disease (leprosy). *Mycobacterium* species are more resistant to disinfectants than most other vegetative bacteria. In addition, they are resistant to many of the most common antimicrobial medications.

The Genus *Pseudomonas*

Pseudomonas species are Gram-negative rods that have polar flagella and often produce pigments (**figure 11.12**). Although most are strict aerobes, some can grow anaerobically if nitrate is available as a terminal electron acceptor. They do not ferment and are oxidase-positive, characteristics that help distinguish them from members of the family *Enterobacteriaceae,* including *E. coli.* ◄ oxidase test, p. 243

As a group, *Pseudomonas* species have extremely diverse biochemical capabilities. Some can metabolize more than 80 different substrates, including unusual sugars, amino acids, and compounds containing aromatic rings. Because of this, *Pseudomonas* species play an important role in the degradation of many synthetic and natural compounds that resist breakdown by most other microorganisms. The ability to carry out some of these degradations is encoded by plasmids.

Pseudomonas species are widespread, typically inhabiting soil and water. Although most are harmless, some cause disease in plants and animals. Medically, the most significant species is *P. aeruginosa.* It is a common opportunistic pathogen, meaning that it primarily infects people who have underlying medical conditions. Unfortunately, it can grow in nutrient-poor environments, such as water used in respirators, and is resistant to many disinfectants and antimicrobial medications. Because of this, hospitals must be very careful to prevent it from infecting patients. ►► *Pseudomonas aeruginosa,* p. 554 ►► opportunistic pathogen, p. 382

The Genera *Thermus* and *Deinococcus*

Thermus and *Deinococcus* are related genera that have scientifically and commercially important characteristics. *Thermus*

species are thermophilic, as their name implies, and this trait is valuable because of their heat-stable enzymes. The bacteria have an unusual cell wall, and stain Gram-negative. ◄ PCR, p. 227

Deinococcus species' unusual cell wall has multiple layers, and they stain Gram-positive. They are unique in their extraordinary resistance to the damaging effects of gamma radiation. *D. radiodurans* can survive a radiation dose several thousand times that lethal to a human being. The dose literally shatters the organism's genome into many fragments, yet enzymes in the cells can repair the extensive damage. Scientists anticipate that through genetic engineering, *Deinococcus* species may eventually help clean up the soil and water contaminated by the radioactive wastes that have accumulated in the United States.

MicroByte
The DNA polymerase of *T. aquaticus* (*Taq* polymerase) is a fundamental part of the polymerase chain reaction (PCR).

Facultative Anaerobes

Facultative anaerobes preferentially use aerobic respiration if O_2 is available. As an alternative, however, they can ferment.

The Genus *Corynebacterium*

Members of the genus *Corynebacterium* are widespread in nature. They are Gram-positive pleomorphic rods, often club-shaped and arranged to form V shapes or palisades (*koryne* is Greek for "club") (**figure 11.13**). Bacteria that exhibit this characteristic morphology are referred to as **coryneforms** or **diphtheroids.** *Corynebacterium* species are generally facultative anaerobes, but some are strict aerobes. Many *Corynebacterium* species reside harmlessly in the throat, but toxin-producing strains of *C. diphtheriae* can cause the disease diphtheria. ►► diphtheria, p. 490

The Family *Enterobacteriaceae*

Members of the family *Enterobacteriaceae,* often referred to as **enterics** or **enterobacteria,** are Gram-negative rods. Their name reflects the fact that most reside in the intestinal tract of

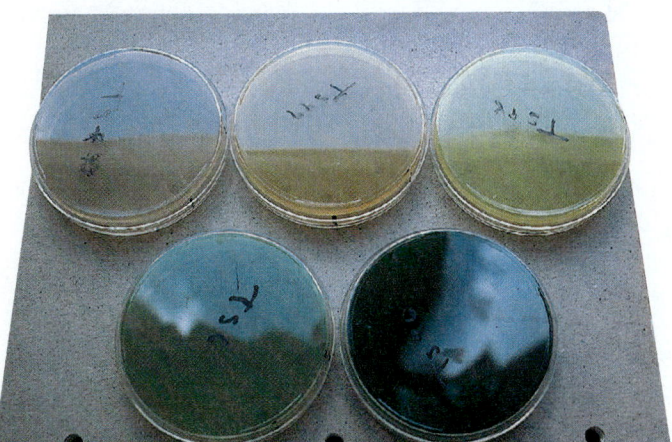

FIGURE 11.12 Pigments of *Pseudomonas* Species Cultures of different strains of *Pseudomonas aeruginosa.* Note the different colors of the water-soluble pigments.

? *Why is the fact that* Pseudomonas *species can grow in nutrient-poor environments medically important?*

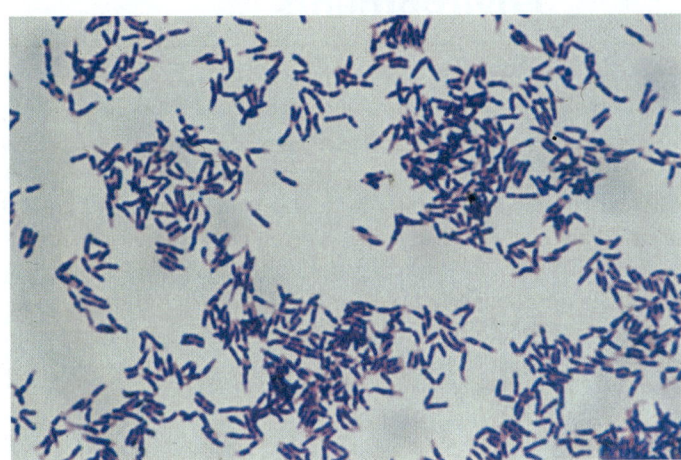

FIGURE 11.13 *Corynebacterium* The Gram-positive pleomorphic rods are often arranged to form V shapes or palisades.

? *How does the name of this genus reflect the shape of the bacteria?*

humans and other animals (in Greek *enteron* means "intestine"), although some thrive in rich soil. Enterics that are part of the normal intestinal microbiota include *Enterobacter, Klebsiella,* and *Proteus* species as well as most strains of *E. coli.* Those that cause diarrheal disease include *Shigella* species, *Salmonella enterica,* and some strains of *E. coli.* Life-threatening systemic diseases include typhoid fever, caused by *Salmonella enterica* serotype Typhi, and both the bubonic and pneumonic forms of plague, caused by *Yersinia pestis.* ▶▶| **diarrheal disease, p. 585** ▶▶| **typhoid fever, p. 592** ▶▶| **plague, p. 678**

Members of the *Enterobacteriaceae* are facultative anaerobes that ferment glucose and, if motile, generally have peritrichous flagella. The family includes over 40 recognized genera that can be distinguished using biochemical tests. Within a given species, many different strains have been described. These are often distinguished using serological tests that detect differences in cell walls, flagella, and capsules (see figure 10.9). |◀◀ **peritrichous flagella, p. 64**

Enteric bacteria that characteristically ferment lactose are included in a group called **coliforms.** This is an informal grouping of certain common intestinal inhabitants such as *E. coli* that are easy to detect in food and water; for years regulatory agencies have considered them to be an indicator of fecal pollution. Their presence indicates a possible health risk because fecal-borne pathogens might also be present. ▶▶| **coliform, p. 742**

MicroAssessment 11.5

Micrococcus, Mycobacterium, Pseudomonas, Thermus, and *Deinococcus* species are obligate aerobes that harvest energy by degrading organic compounds, using O_2 as a terminal electron acceptor. Most *Corynebacterium* species and all members of the family *Enterobacteriaceae* are facultative anaerobes.

13. *What unique characteristic makes members of the genus* Deinococcus *noteworthy?*
14. *What is the significance of finding coliforms in drinking water?*
15. *Why would it be an advantage for a* Pseudomonas *species to encode enzymes for degrading certain compounds on a plasmid rather than the chromosome?* ✚

ECOPHYSIOLOGICAL DIVERSITY

As a group, prokaryotes show remarkable diversity in their physiological adaptations to a wide range of habitats. From the hydrothermal vents of deep oceans to the frozen expanses of Antarctica, prokaryotes have evolved to thrive in virtually all environments, including many that plants and animals would find inhospitable.

This section will highlight the physiological mechanisms prokaryotes use to thrive in terrestrial and aquatic environments; the study of these adaptations is called **ecophysiology.** The section will also describe some examples of bacteria that use animals as habitats. **Table 11.2** summarizes characteristics of the bacteria covered in this section.

11.6 ■ Thriving in Terrestrial Environments

Learning Outcomes

7. *Describe the bacterial groups that form resting stages.*
8. *Compare and contrast* Agrobacterium *species and rhizobia.*

Microorganisms that live in soil must endure a variety of conditions. Daily and seasonally, soil can routinely alternate between wet and dry as well as warm and cold. Nutrient availability can also cycle from abundant to sparse. To thrive in this ever-changing environment, microbes have evolved mechanisms to cope with adverse conditions and to use plants as sources of nutrients.

Bacteria That Form a Resting Stage

Several genera that live in soil can form a resting stage that allows them to survive the dry periods typical in many soils. Of the various types of dormant cells, endospores are by far the most resistant to environmental extremes.

Endospore-Formers

Bacillus and *Clostridium* species are the most common Gram-positive rod-shaped bacteria that form endospores; the position of the spore in the cell can help in identification (**figure 11.14**).

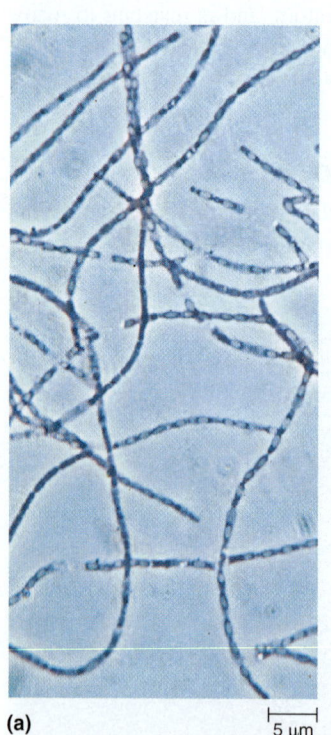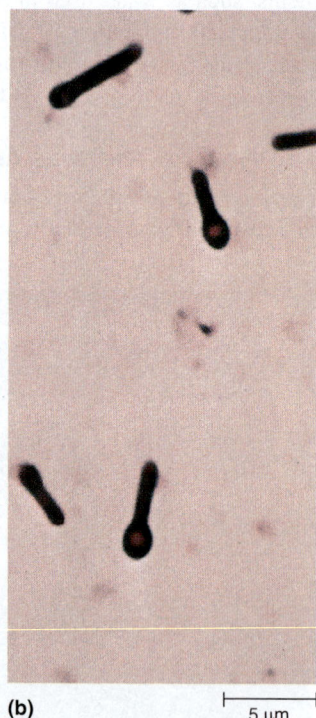

(a) ⊢ 5 μm   (b) ⊢ 5 μm

FIGURE 11.14 Endospore-Formers **(a)** Endospores forming in the mid-portion of the cells of *Bacillus anthracis.* **(b)** Endospores forming at the ends of the cells in *Clostridium tetani.* Both of these species can cause fatal disease, but many other species of endospore-formers are harmless.

❓ *Members of which endospore-forming genus are obligate anaerobes?*

TABLE 11.2 | Ecophysiological Diversity

Group/Genera	Characteristics	Phylum
Thriving in Terrestrial Environments		
Endospore-formers—*Bacillus, Clostridium*	*Bacillus* species include both obligate aerobes and facultative anaerobes; *Clostridium* species are obligate anaerobes. Gram-positive.	*Firmicutes*
Azotobacter	Form cysts. Notable for their ability to fix nitrogen in aerobic conditions. Gram-negative.	*Proteobacteria*
Myxobacteria—*Chondromyces, Myxococcus, Stigmatella*	Congregate to make fruiting bodies; cells within these differentiate to form dormant microcysts. Gram-negative.	*Proteobacteria*
Streptomyces	Resemble fungi in their pattern of growth; produce antibiotics. Gram-positive.	*Actinobacteria*
Agrobacterium	Cause plant tumors. Scientists use their Ti plasmid to move genes into plant cells. Gram-negative.	*Proteobacteria*
Rhizobia—*Rhizobium, Sinorhizobium, Bradyrhizobium, Mesorhizobium, Azorhizobium*	Fix nitrogen; form a symbiotic relationship with legumes. Gram-negative.	*Proteobacteria*
Thriving in Aquatic Environments		
Sheathed bacteria—*Sphaerotilus, Leptothrix*	Form chains of cells enclosed within a protective sheath. Swarmer cells move to new locations. Gram-negative.	*Proteobacteria*
Prosthecate bacteria—*Caulobacter, Hyphomicrobium*	Appendages increase their surface area. Gram-negative.	*Proteobacteria*
Bdellovibrio	Predator of other bacteria. Gram-negative.	*Proteobacteria*
Bioluminescent bacteria—*Photobacterium, Vibrio fischeri*	Some form symbiotic relationships with specific types of squid and fish. Gram-negative.	*Proteobacteria*
Legionella	Often reside within protozoa. Gram-negative.	*Proteobacteria*
Epulopiscium	Very large cigar-shaped bacteria that multiply by releasing several daughter cells; each cell has thousands of copies of the genome. Gram-positive.	*Firmicutes*
Free-living spirochetes—*Spirochaeta, Leptospira* (some species)	Long spiral-shaped bacteria that move by means of endoflagella. Gram-negative.	*Spirochaetes*
Magnetospirillum	Magnetic crystals allow them to move in water and sediments. Gram-negative.	*Proteobacteria*
Spirillum	Spiral-shaped, microaerophilic bacteria. Gram-negative.	*Proteobacteria*
Sulfur-oxidizing, nitrate-reducing marine bacteria—*Thioploca, Thiomargarita*	Use novel mechanisms to compensate for the fact that their energy source (reduced sulfur compounds) and terminal electron acceptor (nitrate) do not coexist.	*Proteobacteria*

Animals as Habitats—See table 11.3

Clostridium species, which are obligate anaerobes, were discussed earlier. *Bacillus* species include both obligate aerobes and facultative anaerobes, and some are medically important. *B. anthracis* causes the disease anthrax, which can be acquired from contacting its endospores in soil or in animal hides or wool. Unfortunately, the spores have also been used as an agent of domestic bioterrorism. ◄◄ endospores, p. 67 ►► anthrax, p. 453

The Genus *Azotobacter*

Azotobacter species are Gram-negative pleomorphic, rod-shaped bacteria that live in soil. They can form a type of resting cell called a cyst (**figure 11.15**). These have negligible metabolic activity and can withstand drying and ultraviolet radiation but are not highly resistant to heat.

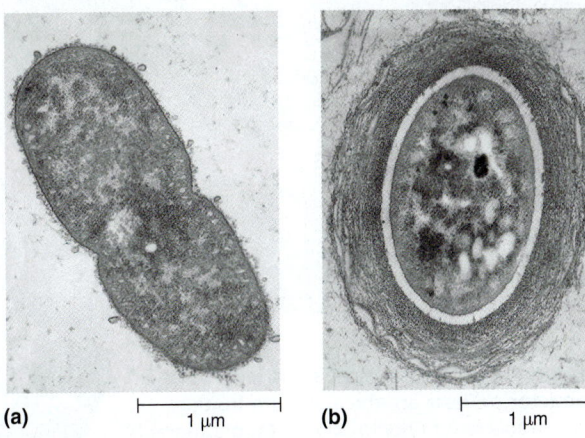

(a)   1 μm   (b)   1 μm

FIGURE 11.15 *Azotobacter* **(a)** Vegetative cells. **(b)** Cyst.

❓ *What types of adverse environmental conditions can Azotobacter cysts withstand?*

Azotobacter species are also notable for their ability to fix nitrogen in aerobic conditions; recall that the enzyme nitrogenase is inactivated by O_2. Apparently, the exceedingly high respiratory rate of *Azotobacter* species consumes O_2 so rapidly that a low O_2 environment is maintained inside the cell. In addition, a protein in the cell binds nitrogenase, thereby protecting it from O_2 damage.

Myxobacteria

The **myxobacteria** are a group of aerobic Gram-negative rods that have a unique developmental cycle as well as a resting stage. When conditions are favorable, cells secrete a slime layer that other cells then follow, creating a swarm of cells. But then, when nutrients are exhausted, the behavior of the group changes. The cells begin to congregate, and then pile up to form a complex structure called a **fruiting body,** which is often brightly colored (**figure 11.16**). In some species, the fruiting body is quite elaborate, consisting of a mass of cells elevated and supported by a stalk made of a hardened slime. The cells within the fruiting body differentiate to become spherical, dormant forms called microcysts. These are considerably more resistant to heat, drying, and radiation than are the vegetative cells of myxobacteria but are much less resistant than bacterial endospores.

Myxobacteria are important in nature as degraders of complex organic substances; they can digest bacteria and certain algae and fungi. Scientifically, these bacteria serve as an important model for studying developmental biology. Included in the myxobacteria are the genera *Chondromyces, Myxococcus,* and *Stigmatella.*

The Genus *Streptomyces*

The genus *Streptomyces* encompasses more than 500 species of aerobic Gram-positive bacteria that resemble fungi in their pattern of growth. Like the fungi, they form a mycelium (a visible mass of branching filaments). The filaments are called hyphae. Chains of characteristic spores called conidia develop at the tips of hyphae (**figure 11.17**). These dormant spores are resistant to drying and are easily spread in air currents. Note that even though this pattern of growth resembles fungi, which are eukaryotes, *Streptomyces* species are much smaller and are prokaryotes.

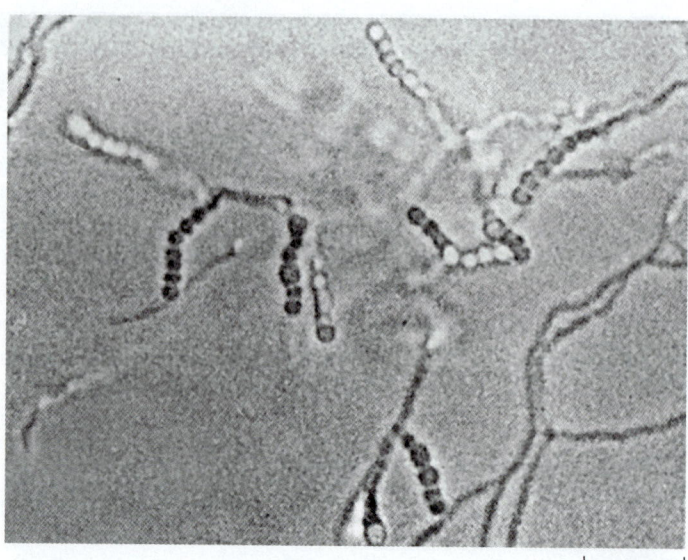

FIGURE 11.17 *Streptomyces* A photomicrograph showing the spherical conidia at the ends of the filamentous hyphae.

❓ *What is a mass of hyphae called?*

Streptomyces species produce a variety of extracellular enzymes that allow them to degrade various organic compounds. They are also responsible for the characteristic "earthy" odor of soil; like the cyanobacteria, they produce geosmin. One species of *Streptomyces, S. somaliensis,* can cause an infection of subcutaneous tissue called an actinomycetoma.

Streptomyces species naturally produce a wide array of medically useful antibiotics, including streptomycin, tetracycline, and erythromycin. The role these compounds play in the life cycle of *Streptomyces* is not entirely understood, but at the low levels produced in soils, they appear to be involved in cell signaling.

Bacteria That Associate with Plants

Members of two related genera use very different means to obtain nutrients from plants. *Agrobacterium* species are plant pathogens that cause tumorlike growths, whereas *Rhizobium* species form a mutually beneficial relationship with certain types of plants.

The Genus *Agrobacterium*

Agrobacterium species are Gram-negative rod-shaped bacteria that have an unusual mechanism of gaining a competitive advantage in soil. They cause plant tumors, the outcome of their ability to genetically alter plants for their own benefit (**figure 11.18**). They do this by attaching to wounded plant tissue, and then transferring a portion of a plasmid to a plant cell; in *A. tumefaciens* the plasmid is called the **Ti plasmid** (for "tumor-inducing"). The transferred DNA encodes the ability to synthesize plant growth hormones, causing uncontrolled growth of the plant tissue and resulting in a tumor. The transferred DNA also encodes enzymes that direct the synthesis of an opine, an unusual amino

(a)   **(b)**

FIGURE 11.16 Fruiting Bodies of Myxobacteria These are the elaborate fruiting bodies of a species of *Chondromyces:* **(a)** photograph; **(b)** scanning electron micrograph.

❓ *How do fruiting bodies help myxobacteria survive adverse conditions?*

FIGURE 11.18 Plant Tumor Caused by *Agrobacterium tumefaciens*

❓ *What is the role of the Ti plasmid in plant tumor formation?*

acid derivative; *Agrobacterium* can then use this compound as a nutrient source (see Perspective 8.2).

MicroByte

Scientists have modified the Ti plasmid, turning it into a commercially valuable tool used to genetically engineer plant cells.

Rhizobia

Rhizobia are a group of Gram-negative rod-shaped bacteria that often fix nitrogen and form intimate relationships with legumes (plants that bear seeds in pods). This group of bacteria includes members of the genera *Rhizobium, Sinorhizobium, Bradyrhizobium, Mesorhizobium,* and *Azorhizobium*. The bacteria live within cells in nodules formed on the roots of the plants (**figure 11.19**). The plants synthesize the protein leghemoglobin, which binds and controls the levels of O_2 (see figure 29.13). Within the resulting microaerobic environment of nodules, the bacteria are able to fix nitrogen. Rhizobia residing within plant cells are examples of endosymbionts, organisms that provide a benefit to the cells in which they reside. ▶▶ symbiotic nitrogen fixers, p. 730

(a)

(b)   5 µm

FIGURE 11.19 Symbiotic Relationship Between Rhizobia and Certain Plants (a) Root nodules. (b) Scanning electron micrograph of bacterial cells within a nodule.

❓ *How do rhizobia benefit plants?*

11.7 ■ Thriving in Aquatic Environments

Learning Outcome

9. *Describe the examples of the four mechanisms aquatic bacteria use to maximize nutrient acquisition and retention.*

Most aquatic environments lack a steady supply of nutrients. To thrive in these habitats, bacteria have evolved various mechanisms to maximize nutrient acquisition and retention.

Sheathed Bacteria

Sheathed bacteria form chains of cells encased within a tube, or sheath (**figure 11.20**). This plays a protective role, helping the bacteria attach to solid objects located in favorable habitats while sheltering them from attack by predators. Masses of filamentous

---

MicroAssessment 11.6

Bacillus and *Clostridium* species make endospores, the most resistant type of dormant cell known. *Azotobacter* species, myxobacteria, and *Streptomyces* species all produce dormant cells that tolerate some adverse conditions but are less resistant than endospores. *Agrobacterium* species and rhizobia obtain nutrients from plants, but the former are plant pathogens and the latter benefit the plant.

16. *Why are myxobacteria important in nature?*

17. *How does* Agrobacterium *benefit from inducing a plant tumor?*

18. *If you wanted to determine the number of endospores in a sample of soil, what could you do before plating it?* ➕

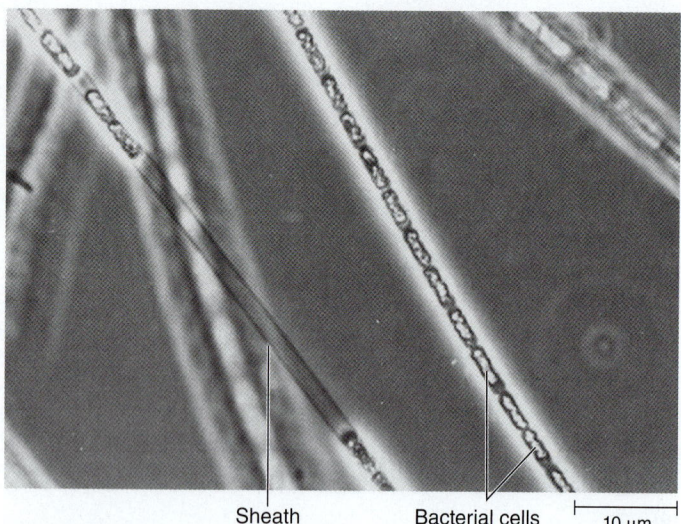

Sheath   Bacterial cells   10 µm

FIGURE 11.20 Sheathed Bacteria Phase-contrast photomicrograph of a *Sphaerotilus* species.

? *What is the role of the sheath?*

sheaths can often be seen streaming from rocks in flowing water polluted by nutrient-rich wastes. They often interfere with sewage treatment and other industrial processes by clogging pipes. Sheathed bacteria include species of *Sphaerotilus* and *Leptothrix,* which are Gram-negative rods.

Sheathed bacteria spread by forming motile cells called swarmer cells that exit through the unattached end of the sheath. These then move to a new solid surface, where they attach. If enough nutrients are present, they can multiply and form a new sheath, which gets longer as the chain of cells grows.

Prosthecate Bacteria

The **prosthecate bacteria** are a diverse group of Gram-negative bacteria that have projections called prosthecae, which are extensions of the cytoplasm and cell wall. These extensions provide increased surface area to facilitate absorption of nutrients. Some prosthecae allow the organisms to attach to solid surfaces.

The Genus *Caulobacter*

Because of their remarkable life cycle, *Caulobacter* species serve as a model for research on cellular differentiation. Entirely different events occur in an orderly fashion at opposite ends of the cell.

Caulobacter cells have a single polar prostheca, commonly called a stalk (**figure 11.21**). At the tip of the stalk is an adhesive holdfast, which provides a mechanism for attachment. To multiply, the cell elongates and divides by binary fission, producing a motile swarmer cell at the end opposite the stalk. This swarmer cell has a flagellum, located at the pole opposite the site of division. The swarmer cell detaches and moves to a new location, where it adheres via a holdfast near the base of its flagellum. It then loses its flagellum, replacing it with a stalk. Only then can the daughter cell replicate its DNA and repeat the process. In favorable conditions, a single cell divides and produces daughter cells many times. With each division, a ring remains at the site of division, allowing a researcher to count the number of progeny.

The Genus *Hyphomicrobium*

Hyphomicrobium species are in many ways similar to *Caulobacter* species, except they have a distinct method of reproduction. The single polar prostheca of the parent cell enlarges at the tip to form a bud (**figure 11.22**). This continues enlarging and develops a flagellum, eventually giving rise to

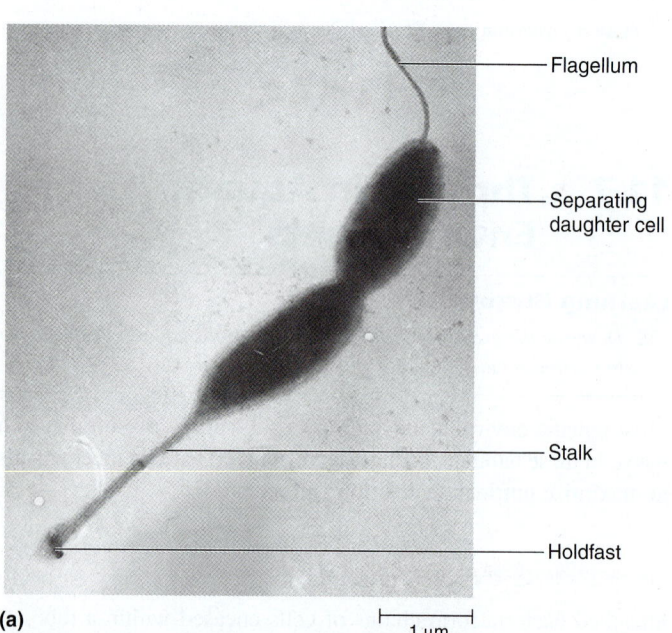

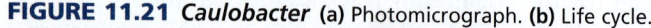

Flagellum

Separating daughter cell

Stalk

Holdfast

(a)
1 µm

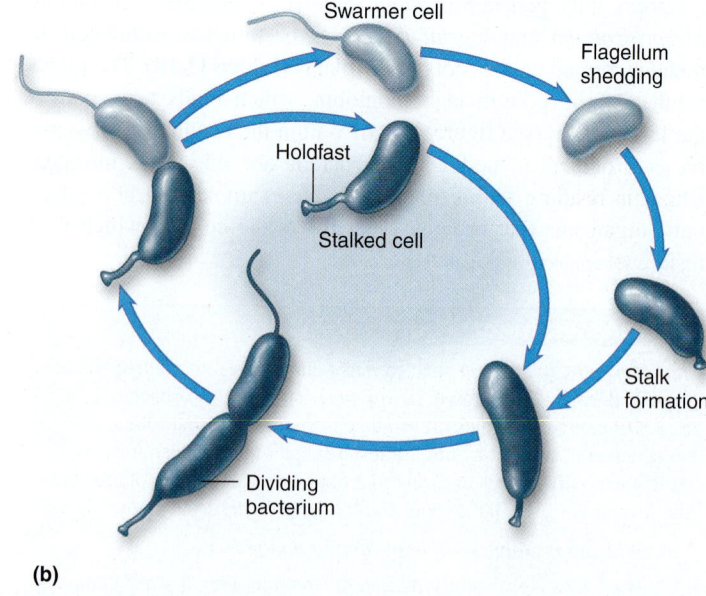

Swarmer cell

Flagellum shedding

Holdfast

Stalked cell

Stalk formation

Dividing bacterium

(b)

FIGURE 11.21 *Caulobacter* (a) Photomicrograph. **(b)** Life cycle.

? *What characteristic of Caulobacter species makes them important research models?*

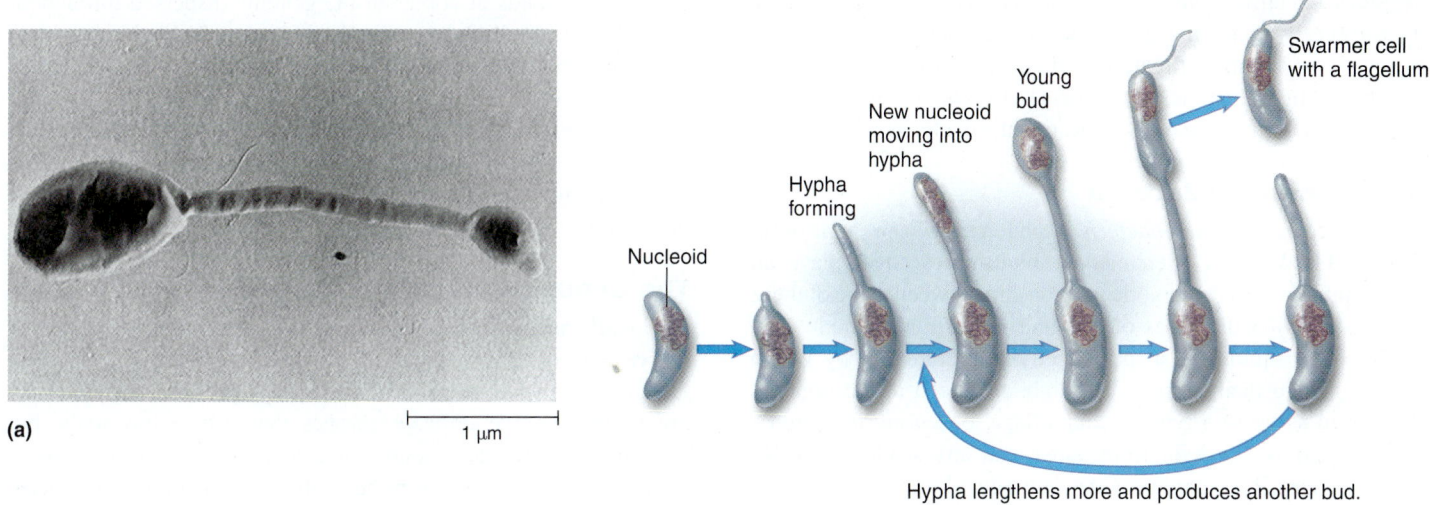

(a)

(b)

Nucleoid

Hypha forming

New nucleoid moving into hypha

Young bud

Swarmer cell with a flagellum

Hypha lengthens more and produces another bud.

FIGURE 11.22 *Hyphomicrobium* **(a)** Photomicrograph. Note the bud forming at the tip of the polar prostheca. **(b)** Life cycle.

❓ *What will happen to the swarmer cell?*

a motile daughter cell. The daughter cell (swarmer cell) then detaches and moves to a new location, eventually losing its flagellum and forming a polar prostheca at the opposite end to repeat the cycle. As with *Caulobacter* species, a single cell can repeatedly produce daughter cells.

Bacteria That Derive Nutrients from Other Organisms

Some bacteria obtain nutrients directly from other organisms. Examples include *Bdellovibrio* species, bioluminescent bacteria, *Epulopiscium* species, and *Legionella* species.

The Genus *Bdellovibrio*

Bdellovibrio species (*bdello,* from the Greek word for "leech") are highly motile Gram-negative curved rods that prey on *E. coli* and other Gram-negative bacteria (**figure 11.23**). When a *Bdellovibrio* cell attacks, it strikes its prey with such force that it propels the prey a short distance. The parasite then attaches to its host and rotates with a spinning motion. At the same time, it makes digestive enzymes that break down lipids and peptidoglycan, eventually forming a hole in the cell wall of the prey. This allows the parasitic bacterium to penetrate the peptidoglycan, lodging in the periplasm. There, over a period of several hours, *Bdellovibrio*

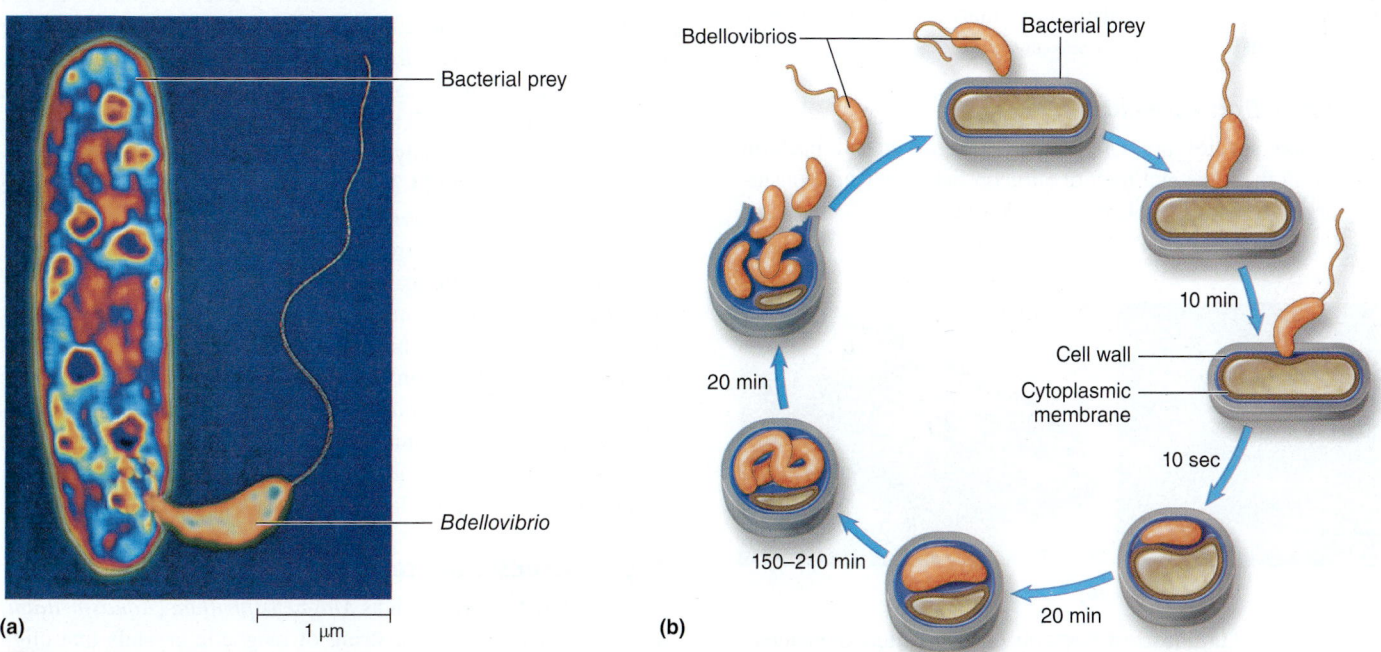

Bacterial prey

Bdellovibrio

(a)

1 µm

Bdellovibrios

Bacterial prey

10 min

Cell wall

Cytoplasmic membrane

10 sec

20 min

20 min

150–210 min

(b)

FIGURE 11.23 *Bdellovibrio* **(a)** Color-enhanced transmission micrograph of a *Bdellovibrio* cell attacking its prey. **(b)** Life cycle of *Bdellovibrio.* Note that the diagram exaggerates the size of the space in which *Bdellovibrio* multiplies.

❓ *How does a* Bdellovibrio *cell penetrate the prey?*

degrades and utilizes the prey's cellular contents. It derives energy by aerobically oxidizing amino acids and acetate. The parasite increases in length, ultimately dividing to form several motile daughter cells. When the host cell lyses, the *Bdellovibrio* progeny are released to find new hosts, repeating the cycle. ◀◀ periplasm, p. 61

Bioluminescent Bacteria

Some species of *Photobacterium* and *Vibrio* can emit light (**figure 11.24**). This phenomenon, **bioluminescence,** plays an important role in the symbiotic relationship between some of these bacteria and specific types of fish and squid. For example, certain types of squid have a specialized organ colonized by *Vibrio fischeri* within their ink sac. The light produced in the organ is thought to serve as a type of camouflage, obscuring the squid's contrast against the light from above and any shadow it might otherwise cast. The squid provides nutrients to the symbiotic bacteria, facilitating their growth. Another example is the flashlight fish, which has a light organ harboring bioluminescent bacteria in a specialized pouch below its eye. By opening and closing a lid that covers the pouch, the fish can control the amount of light released, a tactic believed to confuse predators and prey.

Luminescence is catalyzed by the enzyme luciferase. Studies revealed that the genes encoding it are expressed only when the density of the bacterial population reaches a critical point. This phenomenon of quorum sensing is now recognized as an important mechanism by which a variety of different bacteria regulate the expression of certain genes. ◀◀ quorum sensing, p. 177

Photobacterium and *Vibrio* species are Gram-negative rods (*Vibrio* species are curved rods) with polar flagella. They are facultative anaerobes and typically inhabit aqueous environments; species that require sodium are found in marine environments. Not all are luminescent, and some *Vibrio* species cause human disease. Pathogens include *V. cholerae,* which causes cholera, and *V. parahaemolyticus,* which also causes diarrheal disease; neither of these is bioluminescent. ▶▶ Vibrio cholerae, p. 586

The Genus *Epulopiscium*

Epulopiscium species are Gram-positive cigar-shaped bacteria that reside in the intestinal tract of surgeon fish. They are considerably larger than most prokaryotes (600 μm × 80 μm), and each

cell has thousands of copies of the genome dispersed throughout the cell. The multiple genome copies might help the organism overcome the problem of ensuring that necessary proteins are synthesized even in the far reaches of the large cell.

Epulopiscium species have an unusual life cycle. Rather than undergoing typical binary fission, they enlarge considerably, finally lysing to release up to seven daughter cells. They have not yet been grown in culture.

The Genus *Legionella*

Legionella species are commonly found in aquatic environments, where they often reside within protozoa. They have even been isolated from water in air conditioners and produce misters. They are Gram-negative obligate aerobes that use amino acids, but not carbohydrates, as a source of carbon and energy. *Legionella pneumophila* can cause respiratory disease when inhaled in aerosolized droplets. ▶▶ Legionnaires' disease, p. 506

Bacteria That Move by Unusual Mechanisms

Some bacteria have unique mechanisms of motility that allow them to easily move to desirable locations. These organisms include the spirochetes and the magnetotactic bacteria.

Spirochetes

The **spirochetes** (Greek *spira* for "coil" and *chaete* for "hair") are a group of Gram-negative bacteria with a unique motility mechanism that allows them to move through thick, viscous environments such as mud. Distinguishing characteristics include their spiral shape, flexible cell wall, and motility by means of **endoflagella** (also called an axial filament). Unlike typical flagella, endoflagella are contained within the periplasm. Either a single flagellum or a tuft originates at each end of the cell, and the flagella extend toward each other, overlapping in the mid-region of the cell. Rotation of the endoflagella within the confines of the periplasm causes the cell to move like a corkscrew, sometimes deviating into flexing motions. Many spirochetes are very slender and can be seen only by using special methods such as dark-field microscopy (**figure 11.25**). Many are also difficult or impossible to grow in culture. ◀◀ periplasm, p. 61

Spirochetes include free-living species that inhabit aquatic environments, as well as ones that reside on or in animals. *Spirochaeta* species are anaerobes or facultative anaerobes that thrive in muds and anaerobic waters. *Leptospira* species are aerobic; some are free-living in aquatic environments, whereas others grow within animals. *L. interrogans* causes the disease leptospirosis, which can be transmitted in the urine of infected animals. Spirochetes adapted to reside in body fluids of humans and other animals will be discussed later. ▶▶ leptospirosis, p. 615

Magnetotactic Bacteria

Magnetotactic bacteria such as *Magnetospirillum* (*Aquaspirillum*) *magnetotacticum* contain a string of magnetic crystals that align cells with the earth's magnetism (see figure 3.39). This allows them to move up or down in the water or sediments. It is thought that this unique type of movement allows them to locate the

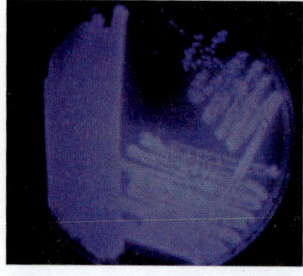

(a)   (b)

FIGURE 11.24 Luminescent Bacteria (a) Plate culture of bioluminescent bacteria. **(b)** Photograph of a flashlight fish; under the eye is a light organ colonized with bioluminescent bacteria.

❓ *What role does quorum sensing play in bioluminescence?*

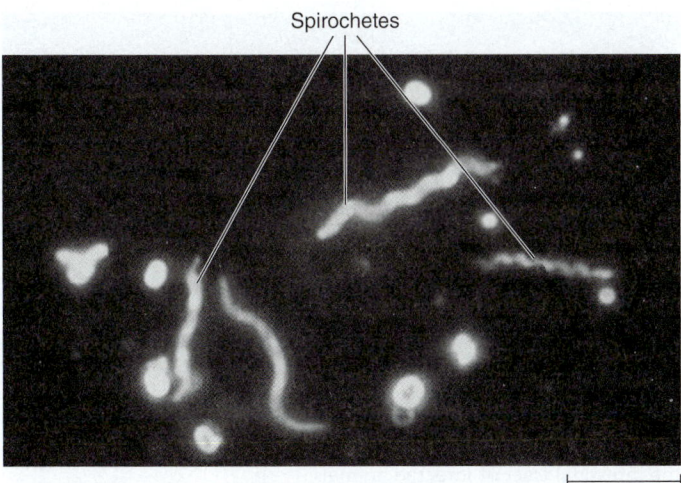

Spirochetes

5 µm

FIGURE 11.25 Spirochetes Dark-field photomicrograph of spirochetes.

❓ *Why are members of this genus difficult to see with bright-field microscopy?*

microaerophilic habitats they require. *Magnetospirillum* species are Gram-negative, spiral-shaped organisms. ◀◀ magnetotaxis, p. 64

Bacteria That Form Storage Granules

A number of aquatic bacteria form granules that store nutrients. Recall that anoxygenic phototrophs often store sulfur granules, which can later be used as a source of electrons for reducing power. Some bacteria store phosphate, and others store compounds that can be used to generate ATP.

The Genus *Spirillum*

Spirillum species are Gram-negative spiral-shaped, microaerophilic bacteria. *Spirillum volutans* forms volutin granules, which are storage forms of phosphate. These are sometimes called metachromatic granules to reflect their characteristic staining with the dye methylene blue. The cells of *S. volutans* are typically large, over 20 µm in length. In wet mounts, *Spirillum* species may be seen moving to a narrow zone near the edge of the coverslip, where O_2 is available in the optimum amount. ◀◀ volutin, p. 67

Sulfur-Oxidizing, Nitrate-Reducing Marine Bacteria

Some marine bacteria store both sulfur (their energy source) and nitrate (their terminal electron acceptor). This provides an advantage to the bacteria, because reduced sulfur compounds are often abundant in anaerobic marine sediments, but these environments lack suitable terminal electron acceptors. In contrast, waters above those sediments lack reduced sulfur compounds but provide a source of nitrate. In other words, the energy source and terminal electron acceptor do not coexist. To cope with this, *Thioploca* species form long sheaths within which cells shuttle between the sulfur-rich sediments and nitrate-rich waters, storing reserves of sulfur and nitrate. The cells of the huge bacterium *Thiomargarita namibiensis* ("sulfur pearl of Namibia") are a pearly white color

due to globules of sulfur in their cytoplasm (see Perspective 1.1). Each cell contains a large nitrate storage vacuole that takes up about 98% of the cell volume. These organisms, which can reach a diameter of 0.75 mm, are not motile and instead rely on storms or other disturbances to bring them into contact with nitrate-rich waters.

MicroByte
> *Thiomargarita namibiensis* can store a 3-month supply of both its energy source (sulfur) and its terminal electron acceptor (nitrate).

MicroAssessment 11.7

Sheathed bacteria attach to solid objects in favorable locations. Prosthecate bacteria produce extensions that maximize the absorptive surface area. *Bdellovibrio* species, bioluminescent bacteria, and *Legionella* species use nutrients from other organisms. Spirochetes and magnetotactic bacteria move by unusual mechanisms. Some organisms form storage granules.

19. *How do squid benefit from having a light organ colonized by luminescent bacteria?*
20. *What is the habitat of* Legionella *species?*
21. *The genomes of free-living spirochetes are larger than those of ones that live within an animal host. Why would this be so?* ➕

11.8 ■ Animals as Habitats

Learning Outcome

10. *Compare and contrast the examples of bacteria that use animals as habitats.*

The bodies of animals, including humans, provide a wide variety of ecological habitats for microbes—from dry, O_2-rich surfaces to moist, anaerobic recesses. **Table 11.3** lists the medically important bacteria covered in this and other sections of the chapter.

Bacteria That Inhabit the Skin

The skin is typically dry and salty, providing an environment inhospitable to many microorganisms. Members of the genus *Staphylococcus,* however, thrive under these conditions. The propionic acid bacteria, which were discussed earlier, inhabit anaerobic microenvironments of the skin. ▶▶ anatomy, physiology, and ecology of the skin, p. 522

The Genus *Staphylococcus*

Staphylococcus species are Gram-positive cocci that are facultative anaerobes. Most, such as *S. epidermidis,* reside harmlessly as part of the normal microbiota of the skin. Like other bacteria that aerobically respire, *Staphylococcus* species are catalase-positive. This distinguishes them from *Streptococcus, Enterococcus,* and *Lactococcus* species, which are also Gram-positive cocci but lack the enzyme catalase. Several species of *Staphylococcus* are notable for their medical significance. *Staphylococcus aureus* causes a variety of diseases, including skin and wound infections, as well as food poisoning. *Staphylococcus saprophyticus* causes urinary tract infections.

TABLE 11.3 | Medically Important Bacteria

Organism	Medical Significance	Phylum
Gram-Negative Rods		
Bacteroides species	Obligate anaerobes that commonly inhabit the mouth, intestinal tract, and genital tract. Cause abscesses and bloodstream infections.	*Bacteroidetes*
Enterobacteriaceae		*Proteobacteria*
Enterobacter species	Normal microbiota of the intestinal tract.	
Escherichia coli	Normal microbiota of the intestinal tract. Some strains cause urinary tract infections; some strains cause specific types of intestinal disease; some cause meningitis in newborns.	
Klebsiella pneumoniae	Normal microbiota of the intestinal tract. Causes pneumonia.	
Proteus species	Normal microbiota of the intestinal tract. Cause urinary tract infections.	
Salmonella enterica serotype Enteritidis	Causes gastroenteritis. Grows in the intestinal tract of infected animals; acquired by consuming contaminated food.	
Salmonella enterica serotype Typhi	Causes typhoid fever. Grows in the intestinal tract of infected humans; transmitted in feces.	
Shigella species	Cause dysentery. Grow in the intestinal tract of infected humans; transmitted in feces.	
Yersinia pestis	Causes bubonic plague, which is transmitted by fleas, and pneumonic plague, which is transmitted in respiratory droplets of infected individuals.	
Haemophilus influenzae	Causes ear infections, respiratory infections, and meningitis in children.	*Proteobacteria*
Haemophilus ducreyi	Causes chancroid, a sexually transmitted disease.	*Proteobacteria*
Legionella pneumophila	Causes Legionnaires' disease, a lung infection. Grows within protozoa; acquired by inhaling contaminated water droplets.	*Proteobacteria*
Pseudomonas aeruginosa	Causes burn, urinary tract, and bloodstream infections. Common in the environment. Grows in nutrient-poor aqueous solutions. Resistant to many disinfectants and antimicrobial medications.	*Proteobacteria*
Gram-Negative Rods—Obligate Intracellular Parasites		
Chlamydophila (Chlamydia) pneumoniae	Causes atypical pneumonia, or "walking pneumonia." Acquired from an infected person.	*Chlamydiae*
Chlamydophila (Chlamydia) psittaci	Causes psittacosis, a form of pneumonia. Transmitted by birds.	*Chlamydiae*
Chlamydia trachomatis	Causes a sexually transmitted disease that mimics the symptoms of gonorrhea. Also causes trachoma, a serious eye infection, and conjunctivitis in newborns.	*Chlamydiae*
Coxiella burnetii	Causes Q fever. Acquired by inhaling organisms shed by infected animals.	*Proteobacteria*
Ehrlichia chaffeensis	Causes human ehrlichiosis. Transmitted by ticks.	*Proteobacteria*
Orientia tsutsugamushi	Causes scrub typhus. Transmitted by mites.	*Proteobacteria*
Rickettsia prowazekii	Causes epidemic typhus. Transmitted by lice.	*Proteobacteria*
Rickettsia rickettsii	Causes Rocky Mountain spotted fever. Transmitted by ticks.	*Proteobacteria*
Wolbachia pipientis	Resides within the filarial worms that cause river blindness and elephantiasis.	*Proteobacteria*
Gram-Negative Curved Rods		
Campylobacter jejuni	Causes gastroenteritis. Grows in the intestinal tract of infected animals; acquired by consuming contaminated food.	*Proteobacteria*
Helicobacter pylori	Causes stomach and duodenal ulcers. Neutralizes stomach acid by producing urease.	*Proteobacteria*
Vibrio cholerae	Causes cholera, a severe diarrheal disease. Grows in the intestinal tract of infected humans; acquired by drinking contaminated water.	*Proteobacteria*
Vibrio parahaemolyticus	Causes gastroenteritis. Acquired by consuming contaminated seafood.	*Proteobacteria*
Gram-Negative Cocci		
Neisseria meningitidis	Causes meningitis.	*Proteobacteria*
Neisseria gonorrhoeae	Causes gonorrhea, a sexually transmitted infection.	*Proteobacteria*

(continued)

TABLE 11.3	Medically Important Bacteria (*Continued*)	
Organism	**Medical Significance**	**Phylum**
Gram-Positive Rods		
Bacillus anthracis	Causes anthrax. Acquired by inhaling endospores in soil, animal hides, and wool. Bioterrorism agent.	*Firmicutes*
Bifidobacterium species	Predominant member of the intestinal tract in breast-fed infants. Thought to play a protective role in the intestinal tract by excluding pathogens.	*Actinobacteria*
Clostridium botulinum	Causes botulism. Disease results from ingesting toxin-contaminated foods, typically canned foods that have been improperly processed.	*Firmicutes*
Clostridium perfringens	Causes gas gangrene. Acquired when soil-borne endospores contaminate a wound.	*Firmicutes*
Clostridium tetani	Causes tetanus. Acquired when soil-borne endospores are inoculated into deep tissue.	*Firmicutes*
Corynebacterium diphtheriae	Toxin-producing strains cause diphtheria, a frequently fatal throat infection.	*Actinobacteria*
Gram-Positive Cocci		
Enterococcus species	Normal microbiota of the intestinal tract. Cause urinary tract infections.	*Firmicutes*
Micrococcus species	Found on skin as well as in a variety of other environments; often contaminate bacteriological media.	*Actinobacteria*
Staphylococcus aureus	Leading cause of wound infections. Causes boils, carbuncles, food poisoning, and toxic shock syndrome.	*Firmicutes*
Staphylococcus epidermidis	Normal microbiota of the skin.	*Firmicutes*
Staphylococcus saprophyticus	Causes urinary tract infections.	*Firmicutes*
Streptococcus pneumoniae	Causes pneumonia and meningitis.	*Firmicutes*
Streptococcus pyogenes	Causes pharyngitis (strep throat), rheumatic fever, wound infections, glomerulonephritis, and streptococcal toxic shock.	*Firmicutes*
Acid-Fast Rods		
Mycobacterium tuberculosis	Causes tuberculosis.	*Actinobacteria*
Mycobacterium leprae	Causes Hansen's disease (leprosy); peripheral nerve invasion is characteristic.	*Actinobacteria*
Spirochetes		
Treponema pallidum	Causes syphilis, a sexually transmitted disease. The organism has never been grown in culture.	*Spirochaetes*
Borrelia burgdorferi	Causes Lyme disease, a tick-borne disease.	*Spirochaetes*
Borrelia recurrentis and *B. hermsii*	Causes relapsing fever. Transmitted by arthropods.	*Spirochaetes*
Leptospira interrogans	Causes leptospirosis, a waterborne disease. Excreted in urine of infected animals.	*Spirochaetes*
Cell Wall-less		
Mycoplasma pneumoniae	Causes atypical pneumonia ("walking pneumonia"). Not susceptible to penicillin because it lacks a cell wall.	*Tenericutes*

Bacteria That Inhabit Mucous Membranes

Mucous membranes of the respiratory, genitourinary, and intestinal tracts provide a habitat for numerous kinds of bacteria, many of which have already been discussed. For example, *Streptococcus* and *Corynebacterium* species reside in the respiratory tract, *Lactobacillus* species inhabit the vagina, and *Clostridium* species and members of the family *Enterobacteriaceae* thrive in the intestinal tract. Some of the other genera are discussed next.

The Genus *Bacteroides*

Bacteroides species are small, strictly anaerobic, Gram-negative rods and coccobacilli. They inhabit the mouth, intestinal tract, and genital tract of humans and other animals. *Bacteroides fragilis* and related species make up about a third of the bacteria in human feces and are often responsible for abscesses and bloodstream infections that follow appendicitis and abdominal surgery. Many are killed by brief exposure to O_2, so they are difficult to study.

The Genus *Bifidobacterium*

Bifidobacterium species are Gram-positive, irregular, rod-shaped anaerobes that reside primarily in the intestinal tract of humans and other animals. They are the predominant members of the intestinal microbiota of breast-fed infants and are thought to provide a protective function by excluding disease-causing bacteria. Formula-fed infants are also colonized with members of this genus, but generally the concentrations are lower.

The Genera *Campylobacter* and *Helicobacter*

Members of the genera *Campylobacter* and *Helicobacter* are curved Gram-negative rods. As microaerophiles, they require specific atmospheric conditions to grow in culture. *Campylobacter jejuni* causes diarrheal disease in humans. It typically lives in the intestinal tract of domestic animals, particularly poultry. *Helicobacter pylori* inhabits the stomach, where it can cause stomach and duodenal ulcers; it has also been linked to stomach cancer. An important factor in its ability to survive in the stomach is its production of the enzyme urease. This breaks down urea to produce ammonia, which neutralizes the acid in the cell's immediate surroundings. ▶▶ stomach ulcers, p. 580

The Genus *Haemophilus*

Haemophilus species are Gram-negative coccobacilli that, as their name reflects, are "blood loving." They require hematin and/or NAD, which are found in blood. Many species are common microbiota of the respiratory tract. *H. influenzae* causes ear infections, respiratory infections, and meningitis, primarily in children. *Haemophilus ducreyi* causes the sexually transmitted disease chancroid. ▶▶ ear infections, p. 492 ▶▶ meningitis, p. 643 ▶▶ chancroid, p. 629

The Genus *Neisseria*

Neisseria species are Gram-negative bacteria, typically kidney-bean-shaped cocci in pairs. They are common microbiota of animals including humans, growing on mucous membranes. *Neisseria* species are typically aerobes, but some can grow anaerobically if a suitable terminal electron acceptor such as nitrite is present. Those noted for their medical significance include *N. gonorrhoeae*, which causes the sexually transmitted disease gonorrhea, and *N. meningitidis*, which causes meningitis; both are nutritionally fastidious. ◀◀ fastidious, p. 93 ▶▶ gonorrhea, p. 622 ▶▶ meningitis, p. 643

The Genus *Mycoplasma*

Members of the genus *Mycoplasma* lack a cell wall, making them flexible and able to pass through the pores of filters that retain other bacteria. Most have sterols in their membrane, providing added strength and rigidity, thereby protecting the cells from osmotic lysis. They are among the smallest forms of life, and their genomes are thought to be the minimum size for encoding the essential functions for a free-living organism.

Medically, the most significant member of this group is *M. pneumoniae*, which, as its name implies, causes a form of pneumonia. This type of pneumonia, often called "walking pneumonia," cannot be treated with penicillin or other antibiotics that interfere with peptidoglycan synthesis, because these organisms lack a cell wall. Colonies of *Mycoplasma* species growing on solid media produce a characteristic "fried egg" appearance (**figure 11.26**).

MicroByte

The genome of *Mycoplasma genitalium* is only 5.8 × 10⁵ base pairs, approximately one-eighth the size of the *E. coli* genome.

FIGURE 11.26 *Mycoplasma pneumoniae* **Colonies** Note the dense central portion of the colony, giving it the typical "fried egg" appearance.

? *Why are members of this genus not affected by penicillin?*

The Genera *Treponema* and *Borrelia*

Members of the genera *Treponema* and *Borrelia* are spirochetes that typically inhabit body fluids and mucous membranes of humans and other animals. Recall that spirochetes are characterized by their corkscrew shape and endoflagella. Although they have a Gram-negative cell wall, they are often too thin to be viewed by conventional microscopy.

Treponema species are obligate anaerobes or microaerophiles that often inhabit the mouth and genital tract. Study of the species that causes syphilis, *T. pallidum*, is difficult because it has never been grown in culture. Its genome has been sequenced, however, providing evidence that it is a microaerophile with a metabolism highly dependent on its host. It lacks critical enzymes of the TCA cycle and a variety of other metabolic pathways.

Three *Borrelia* species are pathogens, transmitted by arthropods such as ticks and lice. *B. recurrentis* and *B. hermsii* both cause relapsing fever; *B. burgdorferi* causes Lyme disease. A striking feature of *Borrelia* species is their genome—a linear chromosome and many linear and circular plasmids.

Obligate Intracellular Parasites

Obligate intracellular parasites cannot reproduce outside a host cell. By living within host cells, the parasites are supplied with a readily available source of compounds they would otherwise need to synthesize for themselves. As a result, most intracellular parasites have lost the ability to make substances needed for extracellular growth. Bacterial examples include members of the genera *Rickettsia, Orientia, Ehrlichia, Coxiella, Chlamydia,* and *Wolbachia*, which are all tiny Gram-negative rods or coccobacilli.

The Genera *Rickettsia, Orientia,* and *Ehrlichia*

Species of *Rickettsia, Orientia,* and *Ehrlichia* are responsible for several serious human diseases spread by blood-sucking arthropods such as ticks and lice. *Rickettsia rickettsii* causes Rocky Mountain spotted fever, *R. prowazekii* causes epidemic typhus, *O. tsutsugamushi* causes scrub typhus, and *E. chaffeensis* causes human ehrlichiosis.

The Genus *Coxiella*

The only characterized species of *Coxiella, C. burnetii,* is an obligate intracellular bacterium that survives well outside the host cell. During its intracellular growth, *C. burnetii* forms spore-like structures called small-cell variants (SCVs) that later allow it to survive in the environment. The structures, however, lack the extreme resistance to heat and disinfectants characteristic of endospores (**figure 11.27**). *Coxiella burnetii* causes Q fever of humans, a disease most often acquired by inhaling bacteria shed from infected animals. This is particularly a problem with pregnant animals because high numbers of the bacteria can be found in the placenta of infected animals.

The Genera *Chlamydia* and *Chlamydophila*

Chlamydia and *Chlamydophila* species are quite different from the other obligate intracellular parasites. They are transmitted directly from person to person, and have a unique growth cycle (**figure 11.28**). Inside the host cell, they initially exist as non-infectious reticulate bodies, which reproduce by binary fission. Later in the infection, the bacteria differentiate into smaller, dense-appearing infectious elementary bodies, which are released when the host cell ruptures. The cell wall of *Chlamydia* and *Chlamydophila* species lacks peptidoglycan, but it has the other features of a Gram-negative type of cell wall. *Chlamydia trachomatis* causes eye infections and a sexually transmitted infection that mimics gonorrhea; *Chlamydophila pneumoniae* causes atypical pneumonia; and *Chlamydophila psittaci* causes psittacosis, a form of pneumonia.

▶▶ *Chlamydia trachomatis,* p. 625

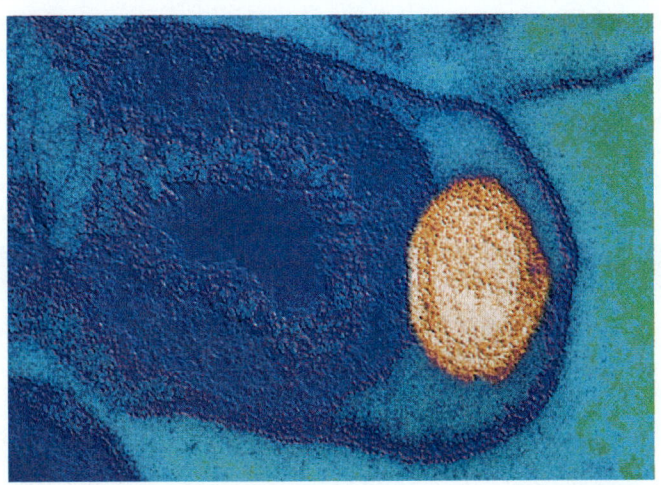

FIGURE 11.27 *Coxiella* Color-enhanced transmission electromicrograph of *C. burnetii.* The oval copper-colored object is the sporelike structure.

❓ *What disease does* C. burnetii *cause?*

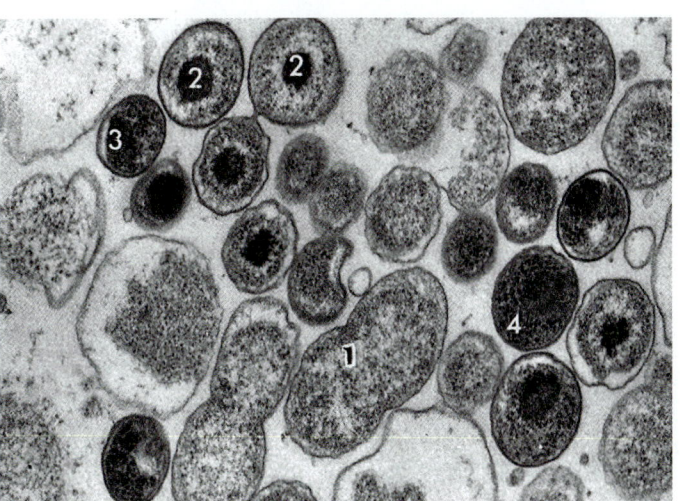

0.2 μm

FIGURE 11.28 *Chlamydia* **Growing in Tissue Cell Culture** The numbers indicate the development from the dividing reticulate body to an infectious elementary body.

❓ *How is the cell wall of* Chlamydia *highly unusual?*

The Genus *Wolbachia*

The only known species of *Wolbachia, W. pipientis,* infects arthropods (including insects, spiders, and mites) and parasitic worms. It is primarily transmitted maternally, via the eggs of infected females to their offspring. In arthropods, the bacterium uses unique strategies to increase the overall population of infected females, including killing male embryos, allowing infected females to reproduce asexually, and causing infected males to gain female traits. In addition, the parasite destroys embryos resulting from the mating of an infected male with either an uninfected female or a female infected with a different strain. *Wolbachia* does not infect mammals, but its medical importance was recognized with the discovery that it resides within filarial worms that cause the diseases river blindness and elephantiasis. The chronic and debilitating inflammation associated with these diseases appears to result from the immune response directed against the bacterial cells in the invading worms. This discovery paves the way for new treatments, because eliminating the bacteria not only lessens the symptoms but also kills the filarial worms.

MicroAssessment 11.8

Staphylococcus species thrive in the dry, salty conditions of the skin. *Bacteroides* and *Bifidobacterium* species reside in the gastrointestinal tract; *Campylobacter* and *Helicobacter* species can cause disease when they reside there. *Neisseria* species, mycoplasmas, and spirochetes inhabit other mucous membranes. Obligate intracellular parasites—including *Rickettsia, Orientia, Ehrlichia, Coxiella, Chlamydia, Chlamydophila,* and *Wolbachia* species—are unable to reproduce outside of a host cell.

22. *How does* Helicobacter pylori *withstand stomach acidity?*

23. *What characteristic of* Mycoplasma *species separates them from other bacteria?*

24. *Why would breast feeding affect the composition of a baby's intestinal microbiota?* ➕

11.9 ■ Archaea That Thrive in Extreme Conditions

Learning Outcome

11. *Compare and contrast the characteristics and habitats of the extreme halophiles and extreme thermophiles.*

Characterized members of the *Archaea* typically thrive in extreme environments (**table 11.4**). These include conditions of high heat, acidity, alkalinity, and salinity. An exception is the methanogens, discussed earlier in the chapter; they inhabit anaerobic niches shared with members of the *Bacteria*. In addition to the characterized archaea, many others have been detected in a variety of non-extreme environments using molecular techniques.

Extreme Halophiles

The extreme halophiles are found in high numbers in salty environments such as salt lakes, soda lakes, and brines used for curing fish. Most grow well in saturated salt solutions (32% NaCl), and they require a minimum of about 9% NaCl. Because they produce pigments, their growth can be seen as red patches on salted fish and pink blooms in concentrated salt water ponds (**figure 11.29**).

Extreme halophiles are aerobic or facultatively anaerobic chemoheterotrophs, but some also can obtain additional energy from light. These organisms have the light-sensitive pigment bacteriorhodopsin, which absorbs energy from sunlight and uses it to expel protons from the cell. This creates a proton gradient that can be used to drive flagella or synthesize ATP.

Extreme halophiles come in a variety of shapes, including rods, cocci, discs, and triangles. They include genera such as *Halobacterium, Halorubrum, Natronobacterium,* and *Natronococcus;* members of these latter two genera are extremely alkaliphilic as well as halophilic.

Extreme Thermophiles

The extreme thermophiles (hyperthermophiles) are found near volcanic vents and fissures that release sulfurous gases and other hot vapors. Because these regions are thought to closely mimic early earth's environment, scientists are interested in studying the prokaryotes that thrive there. Others are found in hydrothermal vents in the deep sea and hot springs. ◄◄ hyperthermophiles, p. 90

FIGURE 11.29 Solar Evaporation Pond

❓ *What causes the pink color in the pond?*

Methane-Generating Hyperthermophiles

In contrast to the mesophilic methanogens discussed earlier, some methanogens are extreme thermophiles. For example, *Methanothermus* species, which can grow in temperatures as high as 97°C, grow optimally at approximately 84°C. Like all methanogens, they oxidize H_2, using CO_2 as a terminal electron acceptor to yield methane.

Sulfur-Reducing Hyperthermophiles

The sulfur-reducing hyperthermophiles are obligate anaerobes that use sulfur as a terminal electron acceptor, generating H_2S. They harvest energy by oxidizing organic compounds and/or H_2. These archaea can be isolated from hot sulfur-containing environments such as sulfur hot springs and hydrothermal vents. They include some of the most thermophilic organisms known, a few even growing above 100°C. One example, *Pyrolobus fumarii,* was isolated from a "black smoker" 3,650 m (about 12,000 feet) deep in the Atlantic Ocean. It grows between 90°C and 113°C. Another hydrothermal vent isolate, *Pyrodictium occultum,* has an optimum temperature of about 105°C, and cannot grow below 82°C. Its disc-shaped cells are connected by hollow tubes, forming a weblike network (**figure 11.30**). The record-holder for the highest maximum growth temperature, dubbed "strain 121" to reflect that it grows at 121°C, appears to be most closely related to *Pyrodictium* species, but it uses iron as an electron acceptor.

TABLE 11.4 | Archaea

Group/Genera	Characteristics	Phylum
Methanogens—*Methanospirillum, Methanosarcina*	Generate methane when they oxidize hydrogen gas as an energy source, using CO_2 as a terminal electron acceptor.	*Euryarchaeota*
Extreme halophiles—*Halobacterium, Halorubrum, Natronobacterium, Natronococcus*	Found in salt lakes, soda lakes, and brines. Most grow well in saturated salt solutions.	*Euryarchaeota*
Extreme thermophiles—*Methanothermus, Pyrodictium, Pyrolobus, Sulfolobus, Thermophilus, Picrophilus, Nanoarchaeum*	Found near hydrothermal vents and in hot springs; some grow at temperatures above 100°C. Includes examples of methane-generating, sulfur-reducing, and sulfur-oxidizing archaea, as well as extreme acidophiles.	*Crenarchaeota, Euryarchaeota,* and *Nanoarchaeota*

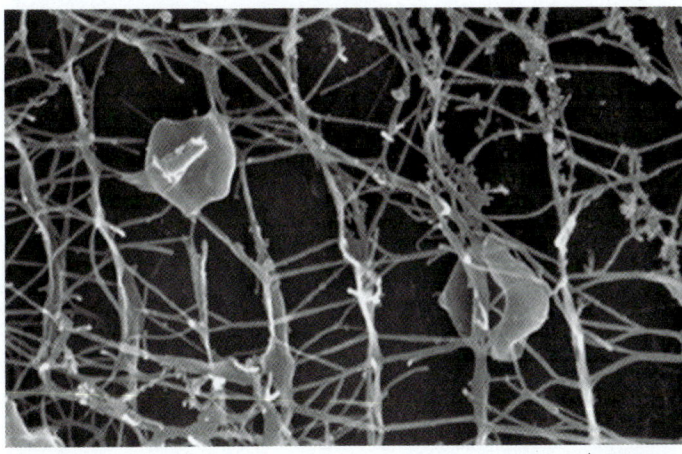

1 μm

FIGURE 11.30 *Pyrodictium* The disc-shaped cells are connected by hollow tubes. Scanning electron micrograph.

🔲 *How would members of this genus be grouped with respect to their temperature preference?*

FIGURE 11.31 Typical Habitat of *Sulfolobus* Sulfur hot spring in Yellowstone National Park.

🔲 *Why must members of this genus be able to tolerate acidic conditions?*

Nanoarchaea

The discovery of an archaeum so unique that it represents an entirely new phylum, *Nanoarcheota* ("dwarf archaea"), was made possible by the earlier discovery of a new genus of sulfur-reducing hyperthermophiles, *Ignicoccus* ("the fire sphere"). *Nanoarchaeum equitans* ("rides the fire sphere") grows as 400-nm spheres attached to the surface of—presumably parasitizing—*Ignicoccus* species (see figure 9.20).

Sulfur Oxidizers

Sulfolobus species are found at the surface of acidic sulfur-containing hot springs such as many of those found in Yellowstone National Park (**figure 11.31**). They are obligate aerobes that oxidize sulfur compounds, using O_2 as a terminal electron acceptor to generate sulfuric acid. In addition, they are thermoacidophilic, only growing above 50°C and at a pH between 1 and 6.

Thermophilic Extreme Acidophiles

Members of two genera, *Thermoplasma* and *Picrophilus,* are notable for growing in extremely acidic, hot environments. *Thermoplasma* species grow optimally at pH 2; in fact, *T. acidophilum* lyses at neutral pH. It was originally isolated from a coal waste pile. *Picrophilus* species tolerate conditions even more acidic, growing optimally at a pH below 1. Two species isolated in Japan inhabited acidic areas in regions that spew sulfurous gases.

MicroAssessment 11.9

Many characterized *Archaea* inhabit extreme environments. These include conditions of high salinity, heat, acidity, and alkalinity.

25. *What is the habitat of* Nanoarchaeum equitans?

26. *At which relative depth in a sulfur hot spring would a sulfur oxidizer likely be found?*

27. *What characteristic of the methanogens makes it logical to discuss them with the* Bacteria *rather than with the* Archaea *described in this section?* ✚

FUTURE CHALLENGES   11.1

Astrobiology: The Search for Life on Other Planets

If life as we know it exists on other planets, it will likely be microbial. The task, then, is to figure out how to find and detect such extraterrestrial microorganisms.

Considering that we still know relatively little about the microbial life on our own planet, coupled with the extreme difficulty of obtaining or testing extraterrestrial samples, this is a daunting challenge with many as yet unanswered questions. What is the most likely source of life on other planets? What is the best way to preserve specimens for study on earth? What will be the culture requirements to grow such organisms? Astrobiology, the study of life in the universe, is a new field

that brings together scientists from a wide range of disciplines, including microbiology, geology, astronomy, biology, and chemistry. The goal is to determine the origin, evolution, distribution, and destiny of life in the universe. Astrobiologists are also developing lightweight, dependable, and meaningful testing devices to be used in future space missions.

Astrobiologists believe that within our solar system, life would most likely be found either on Europa (a moon of Jupiter) or on Mars. This is because Europa and Mars appear to have, or have had, water—crucial for all known forms of life. Europa has an icy crust, beneath which may be liquid water or even

a liquid ocean. Mars is the planet closest to earth, and it has the most similar environment. Images and data from the recent Mars missions indicate that flowing water once existed there.

To prepare for researching life on other planets, microbiologists have turned to some of the most extreme environments here on earth. These include glaciers and ice shelves, hot springs, deserts, volcanoes, deep ocean hydrothermal vents, and subterranean features such as caves. Because select microorganisms can survive in these environs, which are analogous to conditions expected on other planets, they are good testing grounds for the technology to be used on future missions.

Summary

METABOLIC DIVERSITY (table 11.1)

11.1 ■ Anaerobic Chemotrophs

Anaerobic Chemolithotrophs

The **methanogens** are a group of archaea that harvest energy by oxidizing H_2, using CO_2 as a terminal electron acceptor (figure 11.1).

Anaerobic Chemoorganotrophs—Anaerobic Respiration

Desulfovibrio species reduce sulfur compounds to form hydrogen sulfide.

Anaerobic Chemoorganotrophs—Fermentation

Clostridium species form endospores. The **lactic acid bacteria** produce lactic acid as their primary fermentation end product (figures 11.2, 11.3). *Propionibacterium* species produce propionic acid as their primary fermentation end product.

11.2 ■ Anoxygenic Phototrophs

The Purple Bacteria

The **purple bacteria** appear red, orange, or purple; the components of their photosynthetic apparatus are all within the cytoplasmic membrane.

The Green Bacteria

The **green bacteria** are typically green or brownish. Their accessory pigments are often located in chlorosomes.

Other Anoxygenic Phototrophs

Other anoxygenic phototrophs have been discovered, including some that form endospores.

11.3 ■ Oxygenic Phototrophs

The Cyanobacteria (figures 11.6, 11.7)

Genetic evidence indicates that chloroplasts evolved from a species of **cyanobacteria.** Nitrogen-fixing cyanobacteria are critically important ecologically, because they provide an available source of both carbon and nitrogen. Filamentous species maintain the structure and productivity of some soils. Some species of cyanobacteria produce toxins that can be deadly to animals that ingest contaminated water.

11.4 ■ Aerobic Chemolithotrophs

The Sulfur-Oxidizing Bacteria

The filamentous sulfur oxidizers *Beggiatoa* and *Thiothrix* live in sulfur springs, in sewage-polluted waters, and on the surface of marine and freshwater sediments (figure 11.9). *Acidithiobacillus* species are found in both terrestrial and aquatic habitats (figure 11.10).

The Nitrifiers

Ammonia oxidizers convert ammonia to nitrite; they include *Nitrosomonas* and *Nitrosococcus*. Nitrite oxidizers convert nitrite to nitrate; they include *Nitrobacter* and *Nitrococcus*.

The Hydrogen-Oxidizing Bacteria

Aquifex and *Hydrogenobacter* species are thought to be among the earliest bacterial forms to exist on earth.

11.5 ■ Aerobic Chemoorganotrophs

Obligate Aerobes

Micrococcus species are found in soil and on dust particles, inanimate objects, and skin (figure 11.11). *Mycobacterium* species are widespread in nature. They are **acid-fast.** *Pseudomonas* species are widespread in nature and have extremely diverse metabolic capabilities (figure 11.12). *Thermus aquaticus* is the source of *Taq* polymerase, a component of

PCR. *Deinococcus radiodurans* can survive radiation exposure several thousand times that lethal to a human being.

Facultative Anaerobes

Corynebacterium species are widespread in nature (figure 11.13). Members of the family *Enterobacteriaceae* typically inhabit the intestinal tract of animals, although some reside in rich soil. **Coliforms** are used as indicators of fecal pollution.

ECOPHYSIOLOGICAL DIVERSITY (table 11.2)

11.6 ■ Thriving in Terrestrial Environments

Bacteria That Form a Resting Stage

Bacillus and *Clostridium* species form endospores, the most resistant dormant form known (figure 11.14). *Azotobacter* species form cysts. (figure 11.15). **Myxobacteria** aggregate to form fruiting bodies; within a fruiting body, cells form dormant microcysts (figure 11.16). *Streptomyces* species form chains of conidia at the end of hyphae. Many species naturally produce antibiotics (figure 11.17).

Bacteria That Associate with Plants

Agrobacterium species cause plant tumors. They transfer a portion of the **Ti plasmid** to plant cells, genetically engineering the plant cells to produce opines and plant growth hormones. **Rhizobia** reside as endosymbionts in nodules on the roots of legumes, fixing nitrogen (figure 11.18).

11.7 ■ Thriving in Aquatic Environments

Sheathed Bacteria

Sheathed bacteria attach to solid objects in favorable habitats; the sheath shelters them from attack by predators (figure 11.20).

Prosthecate Bacteria

Caulobacter species have a single polar prostheca called a stalk; at the tip of the stalk is a holdfast. The cells divide by binary fission (figure 11.21). *Hyphomicrobium* species divide by forming a bud at the tip of their single polar prostheca (figure 11.22).

Bacteria That Derive Nutrients from Other Organisms

Bdellovibrio species prey on other bacteria (figure 11.23). Certain species of bioluminescent bacteria form symbiotic relationships with specific types of squid and fish (figure 11.24). *Epulopiscium* species reside within the intestinal tract of surgeonfish. *Legionella* species often reside within protozoa; they can cause respiratory disease when inhaled.

Bacteria That Move by Unusual Mechanisms

Spirochetes move by means of endoflagella (figure 11.25). Magnetotactic bacteria contain a string of magnetic crystals that allow them to move up or down in water or sediments to the microaerophilic niches they require.

Bacteria That Form Storage Granules

Spirillum volutans forms polyphosphate granules. *Thioploca* species "commute" to nitrate-rich waters. *Thiomargarita namibiensis*, the largest bacterium known, stores sulfur and has a nitrate-containing vacuole.

11.8 ■ Animals as Habitats (table 11.3)

Bacteria That Inhabit the Skin

Staphylococcus species are facultative anaerobes.

Bacteria That Inhabit Mucous Membranes

Bacteroides species inhabit the mouth, intestinal tract, and genital tract of humans and other animals. *Bifidobacterium* species reside in the intestinal tract of animals, including humans, particularly breast-fed infants. *Campylobacter* and *Helicobacter* species are microaerophilic. *Haemophilus* species require compounds found in blood. *Neisseria* species are nutritionally fastidious aerobes that grow in the oral cavity and genital tract. *Mycoplasma* species lack a cell wall; they often have sterols in their membrane that provide strength and rigidity (figure 11.26). *Treponema* and *Borrelia* species are spirochetes that typically inhabit mucous membranes and body fluids of humans and other animals.

Obligate Intracellular Parasites

Species of *Rickettsia, Orientia,* and *Ehrlichia* are spread when a flea or tick transfers bacteria during a blood meal. *Coxiella burnetii* survives well outside the host due to sporelike structures (figure 11.27). *Chlamydia* and *Chlamydophila* species are transmitted directly from person to person (figure 11.28). *Wolbachia pipientis* alters the reproductive biology of infected arthropods; although it does not infect

mammals, it resides within the filarial worms that cause river blindness and elephantiasis.

11.9 ■ Archaea That Thrive in Extreme Conditions

Extreme Halophiles

Extreme halophiles are found in salt lakes, soda lakes, and brines used for curing fish (figure 11.29).

Extreme Thermophiles

Methanothermus species are hyperthermophiles that make methane. Sulfur-reducing hyperthermophiles are obligate anaerobes that use sulfur as a terminal electron acceptor. Nanoarchaea grow as spheres attached to *Ignicoccus* species. Sulfur-oxidizing hyperthermophiles use O_2 as a terminal electron acceptor, generating sulfuric acid. Thermophilic extreme acidophiles have an optimum pH of 2 or below.

Review Questions

Short Answer

1. What kind of bacteria might compose the subsurface scum of polluted ponds?

2. What kind of bacterium might be responsible for plugging the pipes in a sewage treatment facility?

3. Give three examples of energy sources used by chemolithotrophs.

4. Name two genera of endospore-forming bacteria. How do they differ?

5. How is the life cycle of *Epulopiscium* species unusual?

6. What unique motility structure characterizes the spirochetes?

7. In what way does the metabolism of *Streptococcus* species differ from that of *Staphylococcus* species?

8. How have species of *Streptomyces* contributed to the treatment of infectious diseases?

9. What characteristics of *Azotobacter* species protect their nitrogenase enzyme from inactivation by O_2?

10. Compare and contrast the relationships of *Agrobacterium* and *Rhizobium* species with plants.

Multiple Choice

1. A catalase-negative colony growing on a plate that was incubated aerobically could be which of these genera?

   a) *Bacillus*   b) *Escherichia*   c) *Micrococcus*
   d) *Staphylococcus*   e) *Streptococcus*

2. All of the following genera are spirochetes *except*

   a) *Borrelia.*   b) *Caulobacter.*   c) *Leptospira.*
   d) *Spirochaeta.*   e) *Treponema.*

3. Which of the following genera would you most likely find growing in acidic runoff from a coal mine?

   a) *Clostridium*   b) *Escherichia*   c) Lactic acid bacteria
   d) *Thermus*   e) *Acidithiobacillus*

4. The dormant forms of which of the following genera are the most resistant to environmental extremes?

   1. *Azotobacter*   2. *Bacillus*   3. *Clostridium*
   4. Myxobacteria   5. *Streptomyces*
   a) 1, 2   b) 2, 3   c) 3, 4
   d) 4, 5   e) 1, 5

5. Members of which of the following genera are coliforms?

   a) *Bacteroides*   b) *Bifidobacterium* c) *Clostridium*
   d) *Escherichia*   e) *Streptococcus*

6. Which of the following genera preys on other bacteria?

   a) *Bdellovibrio*   b) *Caulobacter*   c) *Hyphomicrobium*
   d) *Photobacterium*   e) *Sphaerotilus*

7. All of the following genera are obligate intracellular parasites *except*

   a) *Chlamydia.*   b) *Coxiella.*   c) *Ehrlichia.*
   d) *Mycoplasma.*   e) *Rickettsia.*

8. Which of the following genera are known to fix nitrogen?

   1. *Anabaena*   2. *Azotobacter*   3. *Deinococcus*
   4. *Mycoplasma*   5. *Rhizobium*
   a) 1, 3, 4   b) 1, 2, 5   c) 2, 3, 5
   d) 2, 4, 5   e) 3, 4, 5

9. Which of the following archaea would most likely be found coexisting with bacteria?

   a) *Nanoarchaeum*   b) *Halobacterium*   c) *Methanococcus*
   d) *Picrophilus*   e) *Sulfolobus*

10. *Thermoplasma* and *Picrophilus* grow best in which of the following extreme conditions?

    a) Low pH   b) High salt   c) High temperature
    d) a and c   e) b and c

Applications

1. A student argues that it makes no sense to be concerned about coliforms in drinking water because they are harmless members of our normal microbiota. Explain why regulatory agencies are concerned about coliforms.

2. A friend who has lakefront property and cherishes her lush green lawn complains of the green foul-smelling scum on the lake each summer. Explain how her lawn might be contributing to the problem.

Critical Thinking ➕

1. Soil often goes through periods of extreme dryness and extreme wetness. What characteristics of *Clostridium* species make them well suited for these conditions?

2. Some organisms use sulfur as an electron donor (a source of energy), whereas others use sulfur as an electron acceptor. How can this be if there must be a difference between the electron affinity of electron donors and acceptors for an organism to obtain energy?

12

The Eukaryotic Members of the Microbial World

Diatoms, a type of algae, have complex cell walls.

KEY TERMS

Algae Photosynthetic protists.

Arthropod Type of animal such as an insect or arachnid.

Definitive Host Organism in which sexual reproduction or the adult form of a parasite occurs.

Fungus Non-photosynthetic eukaryotic organism with a chitinous cell wall.

Helminth A parasitic worm.

Intermediate Host Organism in which asexual reproduction or

an immature form of a parasite occurs.

Mycosis Disease caused by fungal infection.

Phytoplankton Microscopic free-floating photosynthetic organisms.

Protists Eukaryotes that are not fungi, plants, or animals.

Protozoa Heterotrophic protists.

Yeasts Unicellular fungi that are not chytrids.

A Glimpse of History

Sometimes, a single event can change the course of history. A water mold, *Phytophthora infestans,* which causes a disease called late blight in potatoes, triggered just such an event—the Irish Potato Famine. From 1845 to 1847, the potato crop in Ireland was decimated by the disease, and the entire population of that nation faced starvation. Estimates vary, but it is likely that 1.5 million people died during the famine, while more than 1 million emigrated to other countries—primarily the United States and Canada. The population of Ireland fell by about 25% during this period.

Two hundred years earlier, the Spaniards brought potatoes to Europe from South America. The potato became the main source of nutrition for many Irish people because it was easy to grow and provided a convenient and nearly complete food source that could be stored for months. Irish dependence on this single food source left them open to a major disaster when disease destroyed that food source.

Phytophthora infestans affects every part of the potato plant including the leaves, stems, and tubers (potatoes). The potato turns to a black, mushy mess. Once a field is contaminated, it is very difficult to rid it of the infectious agent. Because the tuber is underground, an infection is often not noticed until it has destroyed the entire plant. The infection is transmitted by spores, which can persist for years in the soil and are spread by wind as well as by those who handle the potatoes.

Plant breeders and scientists still battle potato blight. In 2009, the potato genome was sequenced, and scientists hope to locate resistance genes that can be promoted in crop potatoes. The same year the genome of *Phytophthora infestans* was also sequenced. Long expanses of repetitive genes coding for enzymes that attack the plants were located in the pathogen, providing ready sources of variation that can evade the action of

chemicals and resistance genes. The battle continues as late blight results in global losses of $10 billion per year, causing devastation in developing countries where many still rely on potatoes as their main source of nutrition.

All eukaryotic organisms are in the domain *Eucarya,* a diverse group ranging from microscopic members to plants and animals. Microscopic eukaryotes include fungi, algae, protozoa, slime molds, and water molds. In this chapter, we also include multicellular worms and certain insects. Although usually visible to the naked eye, many are implicated in human disease. Moreover, they are often carried or transmitted in forms that are indeed microscopic. ◄◄ *Eucarya,* p. 10 ◄◄ domains, p. 238

Eukaryotic organisms have traditionally been classified on the basis of gross anatomical characteristics. Such classification has always been problematic, however, because it does not necessarily reflect evolutionary relationships. Modern DNA sequencing makes it possible to examine these organisms at the molecular level, but there are still many problems in developing an accurate classification scheme. Because of this, some microscopic eukaryotes are still discussed in informal groups that do not reflect their evolutionary relatedness. For example, the term **algae** is used to collectively refer to simple autotrophic (photosynthetic) eukaryotes, **fungi** are heterotrophic organisms that have chitin in their cell wall, and **protozoa** are microscopic heterotrophic organisms that are not fungi. **Protists** are eukaryotes that are not fungi, plants, or animals. ◄◄ photosynthesis, p. 151

Eukaryotes are characterized by a membrane-bound nucleus (eukaryote means "true nucleus"). This organelle contains their genetic information, packaged in structures called chromosomes. Because their DNA is located in the nucleus and the ribosomes are in the cytoplasm, the processes of transcription and translation cannot occur simultaneously, as it does in prokaryotic cells. Eukaryotes also have lysosomes, Golgi, and other membrane-bound organelles that allow

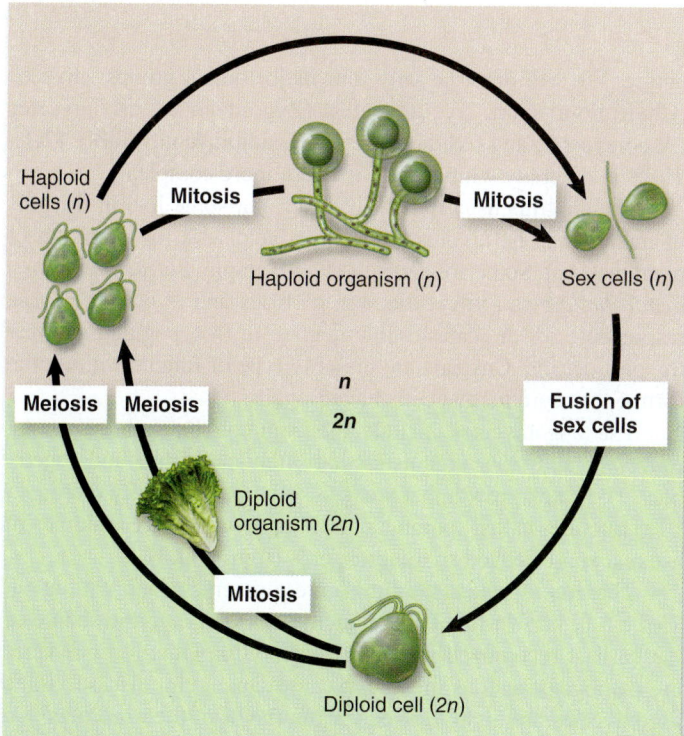

FIGURE 12.1 Cell Division Meiosis gives rise to haploid cells that can grow into a haploid organism by mitosis or can be used as sex cells (gametes) in sexual reproduction. Fusion of gametes forms a diploid cell that can grow into a diploid organism by mitosis or can undergo meiosis to form haploid cells.

❓ *Can a gamete be diploid?*

compartmentalization of cell processes. Energy transformation in eukaryotes primarily occurs in membrane-rich mitochondria and chloroplasts where electron transport chains operate; in prokaryotes, these are associated with the plasma membrane. Many eukaryotes do not have a cell wall, and when they do, it lacks peptidoglycan, the molecule that characterizes the prokaryotic cell wall. Finally, most eukaryotic cells have a well-developed cytoskeleton that is important in movement and maintenance of structure (see figure 3.49) ◄◄ organelles, p. 68 ◄◄ transcription, p. 168

The mode of reproduction in eukaryotes is fundamentally different from that of prokaryotes. Recall that prokaryotes are typically haploid, meaning they carry only one copy of the genome.

They reproduce by binary fission: The chromosome is replicated, and then the cell elongates and divides, with each daughter cell receiving a copy of the chromosome (see figure 4.1). Eukaryotes may be haploid or diploid and often progress through a complex life cycle involving both forms. They may reproduce asexually by a type of nuclear division called **mitosis,** in which each daughter cell receives exactly the same chromosome complement as the parent cell, whether the parent cell is haploid or diploid. They may also reproduce using **meiosis,** in which diploid cells produce haploid cells (often called spores) that can develop into haploid organisms or directly into haploid sex cells called **gametes** that can be used in sexual reproduction (**figure 12.1**). Gametes are often flagellated and highly mobile. Fusion of two gametes forms a diploid cell, allowing recombination of genetic material. This sexual process increases genetic variation that can be used as the raw material for natural selection, the basis of evolution.

MicroByte ──────────
Eukaryotic pathogens can be difficult to target with medication because their cell components are often the same as those of humans.

12.1 ■ Fungi

Learning Outcomes

1. *Describe the structure, habitat, and reproductive strategies of a yeast and a mold.*
2. *Explain the important symbiotic relationships of fungi.*
3. *Name and describe the roles of economically and medically important fungi.*

The study of fungi is known as **mycology.** Fungi include molds, single-celled yeasts, and the familiar mushroom. These terms have nothing to do with the classification of fungi, but instead refer to their morphological forms (**figure 12.2**).

■ **Yeasts** are single-celled fungi.

■ **Molds** are filamentous fungi.

■ **Mushrooms** are simply the reproductive structures of certain fungi, similar to a peach on a peach tree. Like peaches, some large mushrooms are edible.

Mycelium

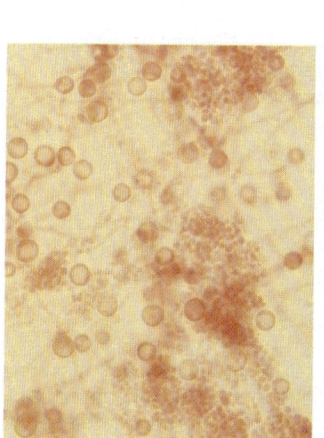

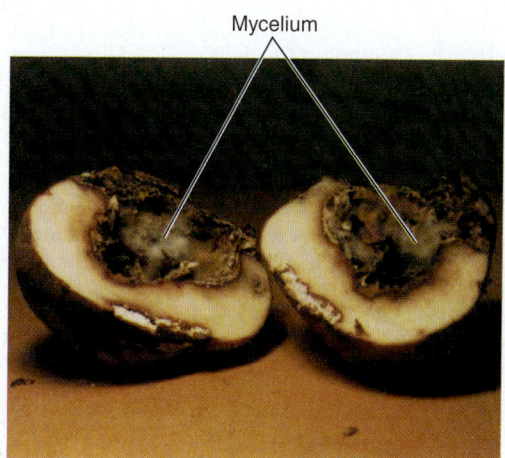

FIGURE 12.2 Morphology of Fungi (a) Microscopic *Candida albicans;* the circles are asexual reproductive spores. (b) The cottony white mass inside the potato is a mold. (c) *Amanita muscaria,* a highly poisonous mushroom.

❓ *What chemical characterizes the cell wall in each of these fungal forms?*

(a)   (b)   (c)

Fungi are characterized by a cell wall that contains chitin—the same molecule found in the exoskeleton of insects. It is somewhat stronger than the cellulose-based cell wall of plants, and chemically distinct from the peptidoglycan cell wall of bacteria. Fungal membranes typically have ergosterol, distinguishing them from animal cell membranes, which have cholesterol. Ergosterol is the target for many antifungal medications.

To obtain nutrients, fungi secrete enzymes into the environment to break large molecules into smaller ones that they can absorb. Along with bacteria, fungi are the principal decomposers of organic compounds because they can degrade both cellulose and lignin, the main components of wood. This decomposition releases carbon dioxide into the atmosphere and nitrogen compounds into the soil, which are then taken up by plants and again converted into organic compounds. Without this recycling of organic material, the earth would quickly be overrun with organic waste. ▶▶ biogeochemical cycles, p. 725

In their key role as decomposers, fungi are **saprophytic,** absorbing nutrients from dead or decaying organic matter. Some fungi, however, can absorb nutrients from living tissue, acting as parasites. Relatively few fungi infect humans—although athlete's foot and vaginal yeast infections are not uncommon—but many plant species suffer devastating fungal infections. More often, fungi partner with other organisms in mutually beneficial relationships. For example, fungi and algae form lichens that can grow on surfaces where neither can survive alone. ▶▶ parasitism, p. 381
▶▶ mutualism, p. 381

Types of Fungi

Classification of fungi is in a state of flux and will continue to be so until the genomes of many more species have been sequenced. About 80,000 fungal species are recognized, but gene studies indicate that there are likely over 1 million species in nature. We will describe only four major fungal groups: chytrids (Chytridiomycota), zygomycetes (Zygomycota), ascomycetes (Ascomycota), and basidiomycetes (Basidiomycota) (**table 12.1**). The real diversity within fungi is much more complex.

The chytrids usually live in water, but some live in the guts of mammalian herbivores where they help with the digestion of plant material. Some are parasitic. For example, *Batrachochytrium dendrobatidis* can infect the skin of frogs and is believed to be responsible for the catastrophic decline in frog populations over the past decade. Chytrids are the only type of fungus with motile forms; the reproductive cells have flagella.

The zygomycetes include the common black bread mold, *Rhizopus,* and other organisms that often spoil fruits and vegetables. The black growth that you see on bread is actually reproductive structures called sporangia that support and house hundreds of asexual spores called sporangiospores (**figure 12.3**).

The ascomycetes (sac fungi) are a diverse group with about 75% of all known fungi. They include *Penicillium,* the source of the earliest recognized antibiotic—penicillin. Other ascomycetes are pathogens such as those that cause Dutch elm disease. Some ascomycetes are highly prized for the flavor of their reproductive structures (morels and truffles). Many ascomycetes form lichens in association with a photosynthetic partner.

The basidiomycetes (club fungi) are commonly recognized by their reproductive structures—mushrooms—that are often collected or grown for food. The group also includes plant parasites like rusts and smuts, which earn their common names because of their appearance on the infected plant. Significant losses in wheat, rye, and corn crops have been caused by these fungi.

Structure of Fungi

Most fungi are multicellular, composed of threadlike filaments called **hyphae** (**figure 12.4**). A visible mass of hyphae is a

TABLE 12.1 Characteristics of Major Groups of Fungi

Group and Representative Member(s)	Appearance	Usual Habitat	Some Distinguishing Characteristics	Asexual Reproduction	Sexual Reproduction
Chytridiomycetes *Batrachochytrium dendrobatidis*		Aquatic, guts of herbivores, parasitic	Unicellular and multicellular. Rhizoids with no true mycelium	Motile zoospores from a sporangium	Flagellated gametes in male and female
Zygomycetes *Rhizopus stolonifer* (black bread mold)		Terrestrial	Multicellular, mycelia of continuous hyphae with many haploid nuclei	Asexual spores develop in sporangia on the tips of aerial hyphae	Sexual spores known as zygospores can remain dormant in adverse environment
Ascomycetes *Neurospora, Saccharomyces cerevisiae* (baker's yeast) *Penicillium, Aspergillus*		Terrestrial, on fruit and other organic materials	Unicellular and multicellular with septated mycelia	Is common by budding; conidiospores	Involves the formation of an ascus (sac) on specialized hyphae
Basidiomycetes *Agaricus campestris* (meadow mushroom) *Cryptococcus neoformans*		Terrestrial	Multicellular, uninucleated mycelia; group includes mushrooms, smuts, rusts that affect the food supply	Commonly absent	Produce basidiospores that are borne on club-shaped structures at the tips of the hyphae

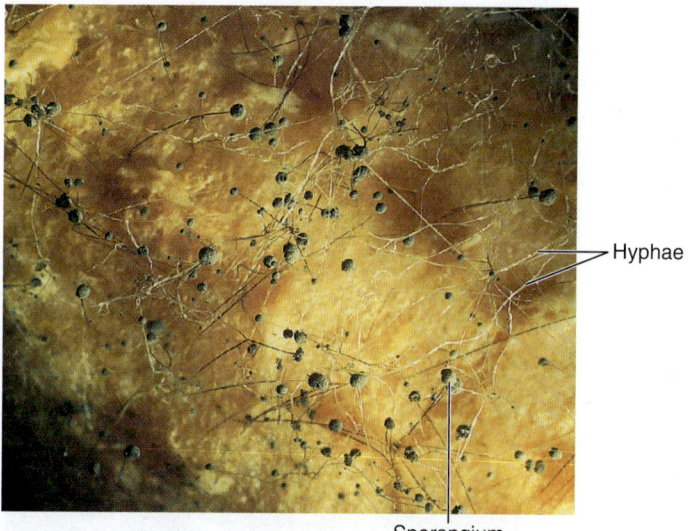

FIGURE 12.3 **Black Bread Mold** *Rhizopus stolonifer,* showing threadlike reproductive hyphae (singular: hypha) and black sporangia housing sporangiospores.

❓ *What is a spore?*

mycelium (plural: mycelia). Fungi are generally not motile, so they cannot move toward a food source. Instead, the tips of the hyphae respond to a nutrient source by growing very quickly in that direction. Cells in the hyphae undergo mitosis, sometimes without accompanying cell division, resulting in a structure with many nuclei. Even when the cytoplasm divides after mitosis, openings remain between cells that allow cytoplasm and organelles to move along the length of the hypha.

The high surface-to-volume ratio of hyphae makes them well adapted to absorb nutrients. The enzymes that they release not only break down nutrient molecules; they also repel the growth of other hyphae. As a result, hyphae branch as they undergo mitosis and spread throughout the food source, ensuring that each hypha will have maximum access to nutrients. When you see mold growing on a food, you typically see only the hyphae involved in reproduction, often bearing reproductive spores. Within the food are the bulk of the hyphae that act in nutrient absorption.

Fungi are most successful in moist environments. Mildew and mold become problems, for example, in a damp basement.

Exposure to such chronic levels of mold or spores can lead to allergies or other health problems.

Parasitic fungi have specialized hyphae called **haustoria** that protrude into host cells to gain nutrients. Haustoria do not penetrate the host cell's plasma membrane, but are surrounded by it, much like a hand in a mitten. Saprophytic fungi sometimes have specialized hyphae called rhizoids, which anchor them to the substrate. Fungal rhizoids do not function in exchange of materials.

Dimorphic fungi can grow as single yeast cells or multicellular mycelia, depending on the environmental conditions. Some of these fungi cause systemic disease in humans. For example, *Histoplasma capsulatum* grows in the soil as a mold, and its reproductive spores easily become airborne. When they are inhaled into the warm, moist environment of the lungs, they develop into the yeast form, which can cause disease. ▶▶| histoplasmosis, p. 515

MicroByte
> The largest organism on earth, called the humongous fungus, has a mycelium that spans over 2,200 acres in Oregon.

Fungal Habitats

Fungi are found in nearly every habitat on earth, including the thermal pools at Yellowstone National Park, volcanic craters, and lakes with very high salt content, such as the Great Salt Lake and the Dead Sea. They are mainly terrestrial organisms. Some fungal species occur only on a particular strain of one genus of plants. Others are widespread because they are extremely versatile in what they can degrade and use as a source of carbon and energy. As a group, fungi can degrade leather, cork, hair, wax, ink, jet fuel, carpet, drywall, and even some synthetic plastics.

Some fungal species can grow in concentrations of salts, sugars, or acids strong enough to kill most bacteria. Because of this, fungi are often responsible for spoiling pickles, fruit preserves, and other foods. Different fungi can grow from pH as low as 2.2 to as high as 9.6, and they usually grow well at a pH of 5.0 or lower. They can grow better than bacteria on acidic fruits and vegetables. ▶▶| food spoilage, p. 756

Most fungi prefer an environment from 20°C to 35°C, but they can easily survive at refrigerator temperatures and below.

FIGURE 12.4 **Formation of Hyphae and Mycelium** The white mass of hyphae form a tangled mass called the mycelium. The mycelium forms when spores of fungi germinate and then elongate to form filaments called hyphae that intertwine.

❓ *Where do fungi digest their food?*

FIGURE 12.5 Lichens (a) A lichen consists of cells of an autotroph, either an alga or a cyanobacterium, entwined within the hyphae of the fungal partner. (b) Lichens on a fallen tree.

❓ *What does the fungus contribute to the lichen?*

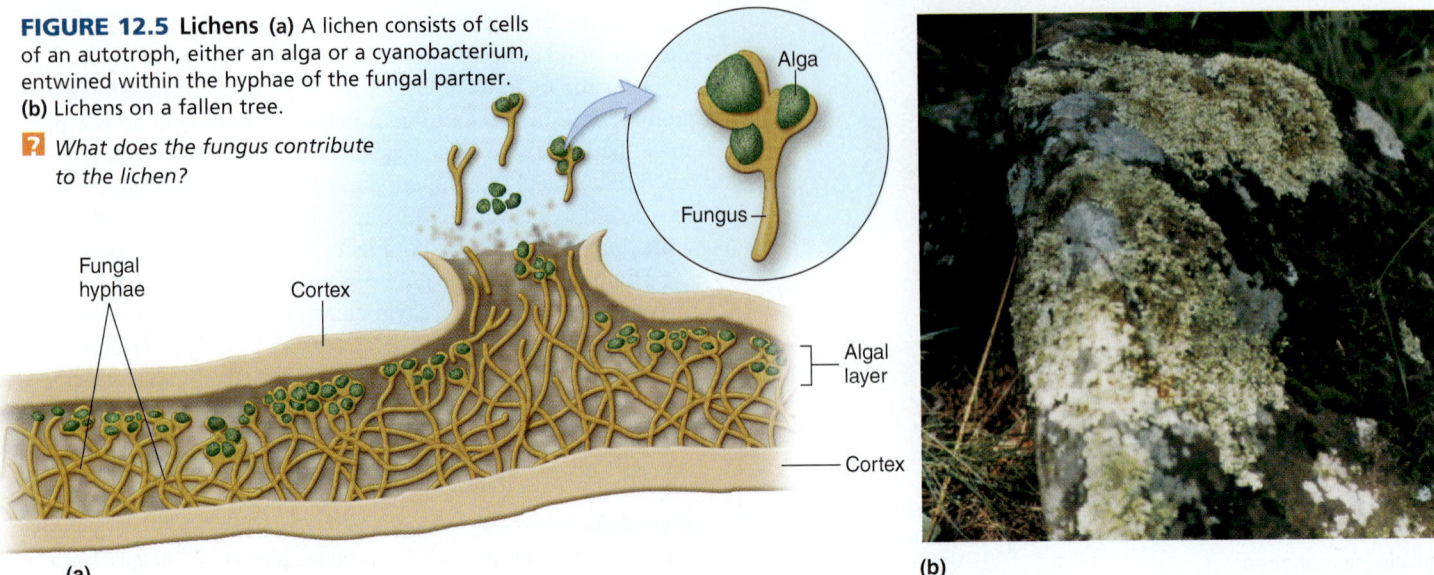

(a)

(b)

Some are resistant to pasteurization, and others can grow at temperatures below the freezing point of water. ◀◀ **pasteurization, p. 112**

Most fungi are aerobic, but some of the yeasts are facultative anaerobes, producing ethanol by fermentation. Some obligate anaerobes live in the rumen of cows, where they are important in the digestion of the plant material. ◀◀ **obligate anaerobe, p. 90**

Symbiotic Relationships of Fungi

Lichens result from the association of a fungus with a photosynthetic organism such as an alga or a cyanobacterium (**figure 12.5**). The fungus provides protection and absorbs water and minerals for the pair. The photosynthetic member supplies the fungus with organic nutrients. Lichens can grow in extreme ecosystems where neither partner could survive alone. For example, they grow in sub-Arctic tundra where they are the primary diet of reindeer. Lichens are often the first organisms to grow on bare rock where they can begin the process of soil formation. In spite of their hardiness, lichens are often a good indicator of air quality. Polluted air is lethal to lichens because they absorb, but are unable to excrete, its common components, including toxic metals, sulfur dioxide, and ozone. You will not find many lichens in industrial cities.

Some fungi grow in a mutually beneficial association with plant roots, forming **mycorrhizas** (**figure 12.6**). The high surface area of fungal hyphae increases the plant's ability to absorb water and minerals. The fungus also supplies the plant with nitrogen and phosphorus from the breakdown of organic material in the soil. The plant, in return, supplies the fungus with organic compounds. It is estimated that 80% of vascular plants have some type of mycorrhizal association. Plants with mycorrhizas grow better than those without. Orchids cannot grow without mycorrhizas that help provide nutrients to the young plant. ▶▶ **mycorrhizas, p. 730**

Certain insects also depend on symbiotic relationships with fungi. For example, leaf-cutting ants farm their own fungal gardens (**figure 12.7**). The ants cannot eat tropical vegetation because the leaves are often poisonous. Instead, the ants chop the plant leaves into bits and add a mycelium. The fungi grow, secreting enzymes that digest the plant material, and eventually produce reproductive structures that the ant then uses as its food source.

Reproduction in Fungi

The reproductive structures of fungi are very important in identification. From a practical standpoint, the most important of these are the forms that develop asexually, because they can be seen in

FIGURE 12.6 Mycorrhizas Fungi form intimate relationships with the roots of most green plants. They supply the plant with nitrogen and phosphorus and increase the plant's ability to absorb water.

❓ *Are mycorrhizas considered parasites?*

FIGURE 12.7 Leaf-Cutter Ants These ants carry food for fungi, whose reproductive structures then serve as a food source for the ants.

❓ *What type of fungal reproduction occurs when the ants place a piece of mycelium on the bits of leaves?*

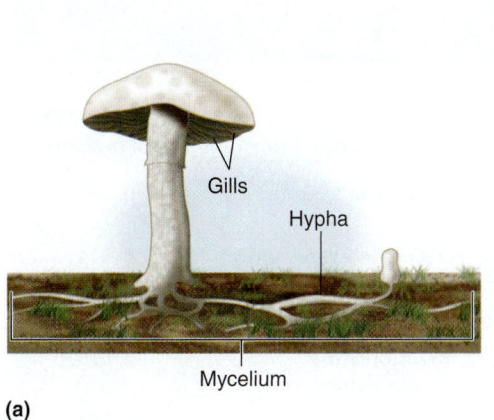

FIGURE 12.8 Mushrooms **(a)** The extensive underground mycelium gives rise to fruiting bodies called mushrooms, composed of dense hyphae protecting the spores produced under the gills. **(b)** *Lepiota rachodes,* showing the fruiting bodies and gills.

❓ *What group of fungi produces mushrooms?*

pure cultures grown in the laboratory. However, the sexual forms play an important role in fungal classification. The sexual forms of some fungi have never been seen, which made it difficult to classify these organisms before DNA sequencing methods were developed. Further complicating matters, many fungi have been inadvertently "discovered" twice—once based on the sexual reproductive forms and once based on the asexual forms—so they are known by two different names!

The reproductive cells of a fungus come in a variety of sizes and shapes. Some mycologists use the term *spore* to refer to reproductive cells that are formed either sexually or asexually. Asexual spores, however, are more typically called **conidia,** or in zygomycetes, sporangiospores. Reproductive cells may be housed in special structures called sporangia in zygomycetes or asci (singular: ascus) in ascomycetes. They may simply be exposed on the tips of reproductive hyphae. Even when exposed, they may be somewhat protected in basidiomycetes beneath the dense hyphae of a mushroom or puffball (**figure 12.8**). Only chytrids produce motile gametes.

Since most fungal reproductive cells are not motile, sexual reproduction results from fusion of hyphae from two different mating types growing toward one another. Often after this fusion, the cells of the fungus will house both haploid nuclei, forming a distinct fungal structure called a dikaryon (meaning two nuclei). Eventually, the nuclei fuse and then undergo meiosis, forming haploid spores. Because the two mating types have no obvious differences, mycologists refer to them as "+" and "−".

Some asexual reproduction does not involve spores. Yeast cells, for example, may simply reproduce by mitosis. Yeasts may also reproduce by budding, in which daughter cells pinch off from a larger parent cell, often leaving a scar on the larger cell, which eventually dies after producing a number of buds (**figure 12.9**). Molds may reproduce asexually by fragmentation. In this type of reproduction, a portion of the parent organism breaks off and grows into a new organism, and the parent organism survives. Hyphae may grow from a piece of a mycelium placed on a suitable nutrient source, as in the farming technique of leaf-cutter ants.

Fungal spores and conidia are small and numerous and easily carried by wind or water. They provide some protection against drying and lack of nutrients and can persist for years until conditions improve. When a fungal spore germinates, it begins to grow

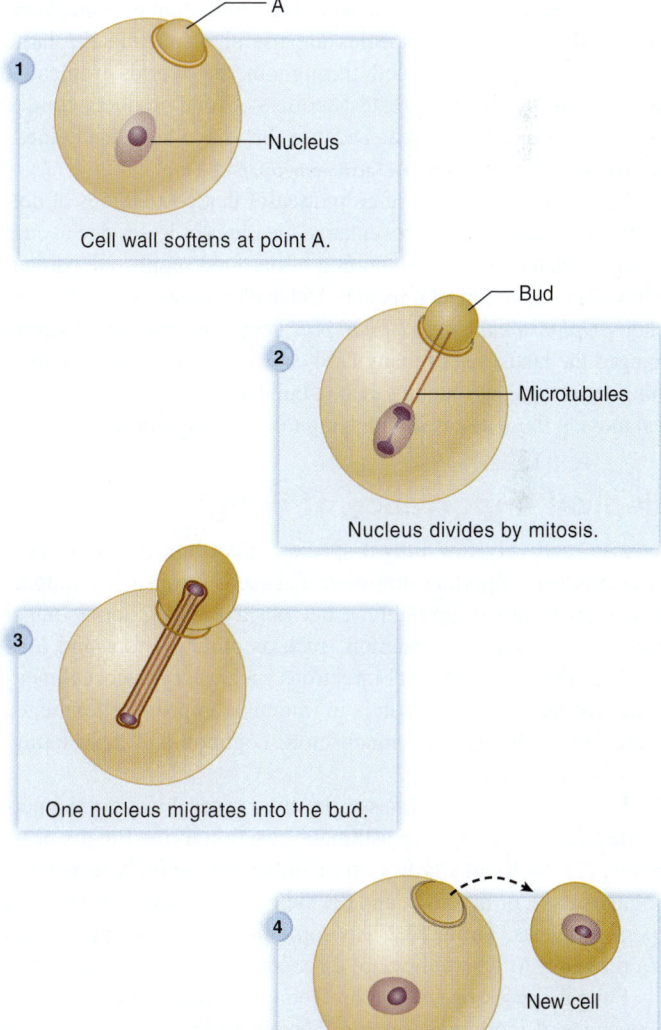

FIGURE 12.9 Budding in Yeast The nucleus divides by mitosis and one nucleus moves into the bud. When the cell wall grows together the bud breaks off, forming a new cell.

❓ *How is budding different from mitosis?*

hyphae in the direction of a food source. When the food source diminishes, spore formation typically increases. Note that fungal spores are quite different from bacterial endospores, which are much more resistant to environmental conditions and are not a means of reproduction. ◄◄ endospore, p. 67

Economic Importance of Fungi

Many fungi are important commercially. They synthesize important antimicrobial medicines such as penicillin and griseofulvin. Fungi are useful tools for studying complex eukaryotic events—such as cancer and aging—within a simple cell. The first eukaryotic genome sequenced was that of the yeast *Saccharomyces cerevisiae,* a model organism used in a variety of genetic and biochemical studies. Yeasts have been genetically engineered to produce numerous important molecules such as human insulin and a vaccine against hepatitis B. ►► penicillin, p. 463 ◄◄ genetic engineering, p. 216

Saccharomyces cerevisiae, also known as brewer's yeast or baker's yeast, has long been used in the production of wine, beer, and bread. Other fungal species are useful in making the large variety of cheeses produced throughout the world. Ironically, fungi are also among the greatest spoilers of food products; tons of food are discarded each year because they have been made inedible by fungal infections. ►► Saccharomyces, p. 753

Fungi cause a number of crop diseases that cost billions of dollars from expenditures on preventative measures or losses due to crop damage. Grain crops, a large portion of our food supply, are particularly vulnerable to fungal infection. Dutch elm disease, caused by the fungus *Ophiostoma ulmi* (*Ulmus* is a genus of elm), dramatically changed the landscape of many U.S. cities when all of the elm trees lining streets and surrounding public buildings were killed. The cost of removing these diseased and dying trees was significant.

Medical Importance of Fungi

Because relatively few fungal species infect humans, and many fungi produce important antimicrobial medications, their impact on human health is probably a net positive. Still, some fungal diseases are relatively common, such as athlete's foot and jock itch. Life-threatening fungal infections such as cryptococcal meningitis are rare. An exception is in immunocompromised patients, where fungal diseases are much more common and devastating. ►► Cryptococcus, p. 660

Fungi cause human illnesses in three general ways: (1) a person develops an allergic or asthmatic reaction to the fungus or its spores; (2) the fungus grows on or in the human body, causing a fungal disease, or **mycosis,** and (3) the fungus produces toxins that a person ingests. **Table 12.2** lists some fungal diseases and serves as a directory to where they are discussed in upcoming chapters.

Fungal spores or conidia are found everywhere on earth up to altitudes of more than 7 miles. The air we breathe may contain more than 10,000 of these cells per cubic meter. People with allergies often monitor published pollen and mold counts and avoid unnecessary outdoor activity when levels are high. Fungal spores can also trigger asthma attacks. ►► asthma, p. 404

The names of mycoses often reflect the causative agent. For example, diseases caused by the yeast *Candida albicans* are called candidiasis and are among the most common mycoses. Infections

TABLE 12.2	Some Medically Important Fungal Diseases	
Disease	**Causative Agent**	**Page for More Information**
Candidial skin infection	*Candida albicans*	p. 545
Coccidioidomycosis	*Coccidioides immitis*	p. 514
Cryptococcal meningoencephalitis	*Cryptococcus neoformans*	p. 660
Histoplasmosis	*Histoplasma capsulatum*	p. 515
Pneumocystis pneumonia	*Pneumocystis jiroveci*	p. 708
Sporotrichosis	*Sporothrix schenckii*	p. 565
Vulvovaginal candidiasis	*Candida albicans*	p. 618
Tinea versicolor	*Malassezia furfur*	p. 544

may also refer to the part of the body affected. Cutaneous mycoses affect the hair, skin, or nails (see figure 22.26). They are caused by skin-invading molds called dermatophytes. Systemic mycoses affect tissues deep within the body and are usually caused by inhalation of spores. ►► dermatophytes, p. 543

Some fungi produce toxins. **Aflatoxins,** produced by *Aspergillus* species, are the most studied of these toxins and are considered carcinogenic. The U.S. Food and Drug Administration (FDA) monitors levels of aflatoxins in foods such as grains and peanuts. The rye mold *Claviceps purpurea,* also known as **ergot,** produces a toxin with hallucinogenic properties. Some believe that some young women in Salem, Massachusetts, in the late 1600s ate contaminated rye, and their resultant strange behavior led to accusations of witchcraft. The active chemical has been purified from *C. purpurea* to yield the drug ergotamine, which decreases blood flow. It is used to control uterine bleeding and relieve migraine headaches. Some fungi, such as *Amanita* species, produce powerful toxins that can cause fatal liver damage (see figure 12.2c).

MicroAssessment 12.1

Fungi have chitin in their cell walls and absorb nutrients after secreting digestive enzymes onto a food source. As saprophytes, they are important recyclers of carbon and other elements. Fungi can also form important relationships with other organisms, such as seen in lichens or mycorrhizas. Fungi may be single-celled yeasts or multicellular molds that grow by extending tubelike hyphae into a food source. They may reproduce sexually by fusion of hyphae of different mating types, or asexually by spores, budding, or fragmentation. Fungi cause significant plant diseases and food spoilage but also are a food source and are used in food production. Relatively few fungi cause human disease.

1. *How can fungi cause disease in humans?*
2. *Describe fungal symbiotic relationships with algae, green plants, and insects, and explain how the relationships benefit each partner.*
3. *Why would spore formation increase when food supplies are diminishing?* ➕

12.2 ■ Algae

Learning Outcomes

4. *Explain how various types of algae differ from one another and from other members of the eukaryotic microbial world.*
5. *Describe how algae can affect human health.*

The term **algae** refers to simple photosynthetic eukaryotes. Algae differ from other eukaryotic photosynthetic organisms such as land plants because they lack an organized vascular system and they have relatively simple reproductive structures. Most algae are aquatic; they may be microscopic or macroscopic. Algae do not directly infect humans, but some produce toxins that may be concentrated in other animals that humans use for food.

Algae contain pigments that absorb radiant energy. That energy is used in the process of photosynthesis, converting CO_2 and H_2O to organic material and oxygen. Algae are essential for maintaining life in aquatic environments because other organisms depend upon the organic molecules for food. ◀◀ photosynthesis, p. 151

Types of Algae

As noted earlier, algae are not a strict classification. They are a diverse group of protists that share some fundamental characteristics, but are not necessarily related. **Figure 12.10** shows a simple taxonomic representation of the protists in this chapter.

The various types of algae are often characterized by the pigments they contain. All algae contain chlorophyll *a*, a pigment also found in green plants and cyanobacteria. Red algae and brown algae also contain other pigments that absorb different wavelengths of light and give them their characteristic appearance. These pigments extend the range of light waves that can be used for photosynthesis. Some of the general characteristics of algal groups are summarized in **table 12.3.**

Structure of Algae

Algae can be either unicellular or multicellular. All algae, however, contain chloroplasts with photosynthetic pigments and have rigid cell walls mostly composed of cellulose. Some multicellular species contain large amounts of other compounds in their cell walls. Cell walls of red algae contain agar used by scientists to solidify growth media; cell walls of brown algae contain alginic acid used to provide the consistency in products like ice cream and cosmetics. ◀◀ chloroplasts, p. 76 ◀◀ agar, p. 86

Microscopic Algae

Microscopic algae can be single-celled organisms floating free or propelled by flagella, or they can grow in long chains or

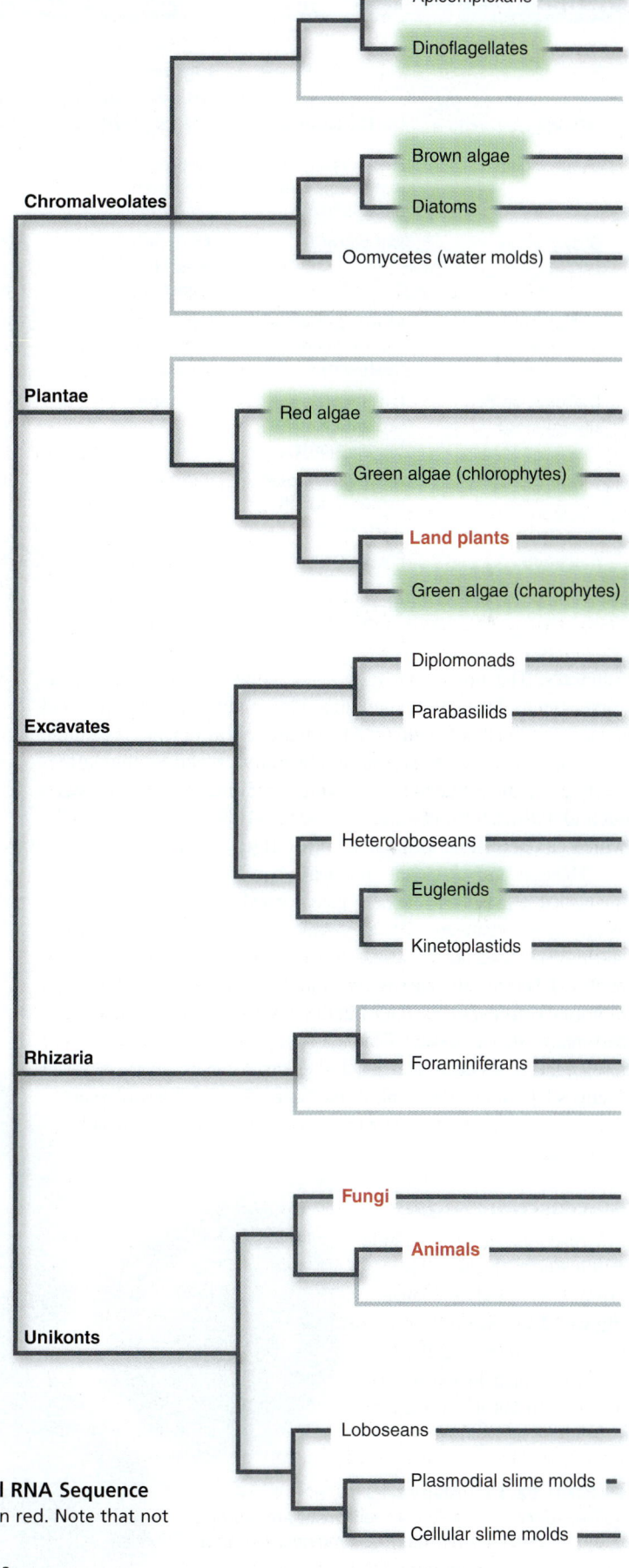

FIGURE 12.10 A Phylogeny of Eukaryotes Based on Ribosomal RNA Sequence Comparisons Algae are highlighted in green. Non-protists are shown in red. Note that not all of the groups shown are at the same taxonomic level.

❓ *Are red algae more closely related to green algae or to brown algae?*

TABLE 12.3	Characteristics of Major Groups of Algae				
Group	Usual Habitat	Principal Pigments (in addition to chlorophyll a)	Storage Product	Cell Wall	Mode of Reproduction
Green algae	Fresh water; salt water; soil; tree bark; lichens	Chlorophyll b; carotenes; xanthophylls	Starch	Cellulose and pectin	Asexual by multiple fission; spores or sexual
Brown algae	Salt water	Xanthophylls, especially fucoxanthin	Starchlike carbohydrates; mannitol; fats	Cellulose and pectin; alginic acid	Asexual, motile zoospores; sexual, motile gametes
Red algae, corallines	Mostly salt water, several genera in fresh water	Phycobilins; carotenes; xanthophylls	Starchlike carbohydrates	Cellulose and pectin; agar; carrageenan	Asexual spores; sexual gametes
Diatoms, golden brown algae	Fresh water; salt water; soil; higher plants	Carotenes	Starchlike carbohydrates	Pectin, often impregnated with silica or calcium	Asexual or sexual
Dinoflagellates	Mostly salt water but common in fresh water	Carotenes; xanthophylls	Starch; oils	Cellulose and pectin	Asexual; rarely sexual
Euglenids	Fresh water	Chlorophyll b; carotenes; xanthophylls	Fats; starchlike carbohydrates	Lacking, but elastic pellicle present	Asexual only by binary fission

filaments. The unicellular algae—including diatoms, as well as some green algae, dinoflagellates and euglenids, and a few red algae—are well adapted to an aquatic environment. As single cells, they have relatively large, absorptive surfaces, thus effectively using the dilute nutrients available. Some microscopic algae such as *Volvox* form colonies of 500 to 60,000 biflagellated cells, which can be visible to the naked eye (**figure 12.11**).

Diatoms are abundant in aquatic environments and are distinguished by silicon dioxide incorporated into their cell walls. When these organisms die, they sink to the bottom of the ocean, and the cell wall does not readily decompose (see chapter-opening photo). Deposits of diatoms are mined for a substance known as **diatomaceous earth,** used for filtering systems, abrasives in polishes, and many other purposes. Diatoms store a type of oil as an energy reserve that helps them float at a desired depth. Chemical and physical changes to diatoms that sank to the ocean floor millions of years ago eventually converted them to a major source of crude oil and natural gas.

Macroscopic Algae

Macroscopic algae include multicellular brown algae, green algae, and red algae (**figure 12.12**). Some of these possess a structure called a holdfast, which looks like a root system but it is not used to obtain water and nutrients for the organism. It functions simply to anchor the organism to a firm substrate. The stalk of a multicellular alga, known as the stipe, usually has leaflike structures or blades attached to it. The blades are the main site of photosynthesis, although some also bear reproductive structures. Many large algae have gas-containing bladders that keep the blades floating on the surface of the water to maximize exposure to sunlight.

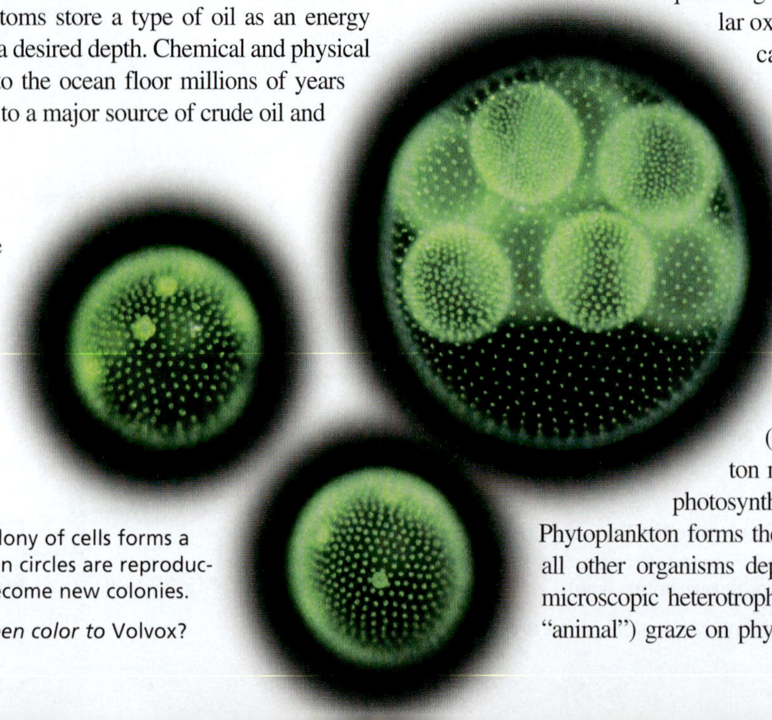

Algal Habitats

Algae are found in both fresh and salt water, as well as in moist soil. Since the oceans cover more than 70% of the earth's surface, aquatic algae are major producers of molecular oxygen as well as important users of carbon dioxide. Algae must live at water levels that allow penetration of light. Algae with different pigments can occupy different levels and thus avoid competition.

Small organisms that float or drift near the surface of aquatic environments are called plankton. Unicellular algae make up a significant part of the **phytoplankton** (phyto means "plant" and plankton means "drifting"), the free-floating, photosynthetic organisms found in plankton. Phytoplankton forms the base of aquatic food chains with all other organisms dependent upon them. For example, microscopic heterotrophs in the **zooplankton** (zoo means "animal") graze on phytoplankton, and then both become

FIGURE 12.11 *Volvox* A colony of cells forms a hollow sphere. The yellow-green circles are reproductive cells that will eventually become new colonies.

❓ *What pigment gives the green color to Volvox?*

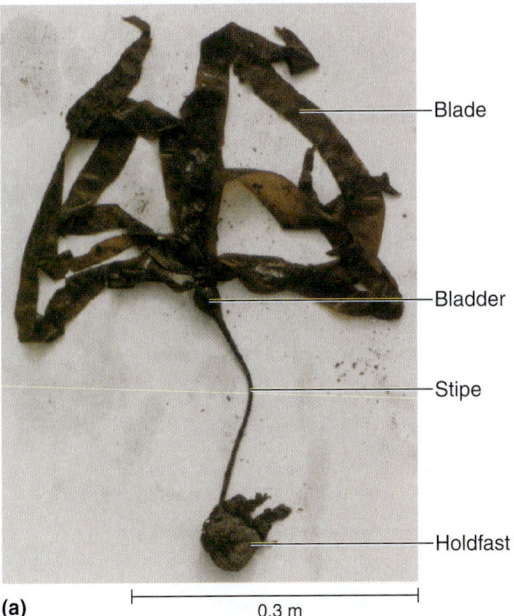

(a)

0.3 m

Blade

Bladder

Stipe

Holdfast

(b)

(c)

FIGURE 12.12 Types of Algae (a) This young brown alga, *Nereocystis luetkeana* (bladder kelp), has a large bladder filled with gas. The blades are the most active sites for photosynthesis. The holdfast anchors the kelp to rocks or other surfaces. In a single season, kelp can grow to lengths of 5 to 15 m. **(b)** Sea lettuce, a green alga. **(c)** *Corallina gracilis,* red coral algae.

❓ *What is the function of the gas bladder in kelp?*

food for other organisms including benthic whales, some of the world's largest mammals. ▶▶| food chain, p. 720

MicroByte

More O_2 is produced by algae in the oceans than by all forests on earth combined.

Algal Reproduction

Some algae, especially multicellular filamentous species, reproduce asexually by fragmentation. Many algae alternate between a haploid generation and a diploid generation. Sometimes, as is the case with *Ulva* (sea lettuce), the generations look physically similar and can be distinguished only by microscopic examination. In other cases, the two forms look quite different.

Medical Importance of Algae

Although algae do not directly cause disease in humans, some do so indirectly. Instances of disease most often occur during algal blooms, during which the numbers of phytoplankton increase dramatically due to changes in water conditions. An upwelling of water due to warmer temperatures often brings more nutrients from the ocean bottom to the surface. When organisms encounter warmer waters and additional nutrients, they multiply rapidly. The runoff of fertilizers along waterways and coastlines, or pollution from untreated sewage, may also cause algae to proliferate. Algal blooms of dinoflagellates are commonly known as **red tides.**

Dinoflagellates of *Gonyaulax* species produce neurotoxins such as saxitoxin and gonyautoxin, some of the most potent non-protein poisons known. Shellfish such as clams, mussels, scallops, and oysters feed on these dinoflagellates without apparent harm and, in the process, accumulate the neurotoxin in their tissues. When humans eat the shellfish, they suffer symptoms of paralytic shellfish

poisoning, including numbness, dizziness, muscle weakness, and impaired respiration. Death can result from respiratory failure. Cooking the shellfish does not destroy these toxins. *Gonyaulax* species are found in both the North Atlantic and the North Pacific and have seriously affected the shellfish industry on both coasts.

MicroAssessment 12.2

Algae are a diverse group of aquatic eukaryotic organisms that contain chlorophyll *a* and carry out photosynthesis. Algae form the base of aquatic food chains and produce much of our atmospheric oxygen. Algae do not cause disease directly, but they can produce toxins that are harmful when ingested by humans.

4. *What primary characteristics distinguish algae from other organisms?*

5. *What harmful effects can algae have on humans?*

6. *Why do algae have a greater variety of photosynthetic pigments than land plants?* ➕

12.3 ■ Protozoa

Learning Outcomes

6. *Describe methods of reproduction in protozoa.*

7. *Describe the major diseases of humans that are caused by protozoa and name the causative agents.*

The term *protozoa* has little meaning today as we learn more about the evolutionary diversity of the eukaryotes through genetic sequencing studies. Organisms classically considered protozoa ("animal-like" unicellular organisms) form a diverse group in which most members bear little relationship to the others. Protozoa are most easily defined by what they are not. In this chapter, protozoa are unicellular heterotrophic organisms that are not fungi, slime molds, or water molds.

Types of Protozoa

Protozoa are protists, along with the algae, slime molds, and water molds. As with algae, protozoa are not a unified group but appear along an evolutionary continuum (**figure 12.13**). Protozoa have traditionally been classified primarily based on their mode of locomotion. Examination of ultrastructure and DNA sequences coding for rRNA, however, show that organisms may not be closely related just because they all use flagella, cilia, or pseudopodia for locomotion. Protozoans are extremely diverse, but we will concentrate on members of this group that cause human disease.

Apicomplexans are parasites with a structure called an apical complex at one end that helps them to penetrate the cell membrane of host cells. Many have complex life cycles that alternate between sexual and asexual forms. This group includes *Plasmodium* species that cause malaria, making them one of the most significant causes of infectious disease in the world. It also includes the human pathogens *Toxoplasma gondii*, *Cryptosporidium parvum*, and *Cyclospora cayetanensis*. ▶▶ malaria, p. 686

Diplomonads and **parabasalids** are groups of flagellated protists that lack mitochondria. Diplomonads typically have two nuclei and reside within hosts where conditions are anaerobic or in stagnant water where O$_2$ levels are low. The group includes *Giardia lamblia,* a common cause of diarrhea in campers who drink water directly from streams or lakes (see figure 24.22). Parabasalids live within a host organism. Some, for example, live within termites and digest cellulose for their hosts. Others, such as *Trichomonas vaginalis,* may cause disease. Parabasalids have a unique structure called a hydrogenosome, which in the absence of true mitochondria, produces some ATP while generating hydrogen. Both groups reproduce asexually; a few parabasalids also reproduce sexually. ▶▶ giardiasis, p. 602 ▶▶ trichomoniasis, p. 635

Kinetoplastids have at least one flagellum and are characterized by a distinctive complex mass of DNA in their large single mitochondrion. This mitochondrial DNA can produce variations in RNA that may account for the rapid changes in cell surface molecules that allow some of these organisms to evade the immune system. This group includes *Leishmania major,* the cause of leishmaniasis, a disease transmitted by sand flies. It also includes *Trypanosoma cruzi,* the cause of Chagas' disease, which kills up to 50,000 people worldwide each year. The most significant species of the group is

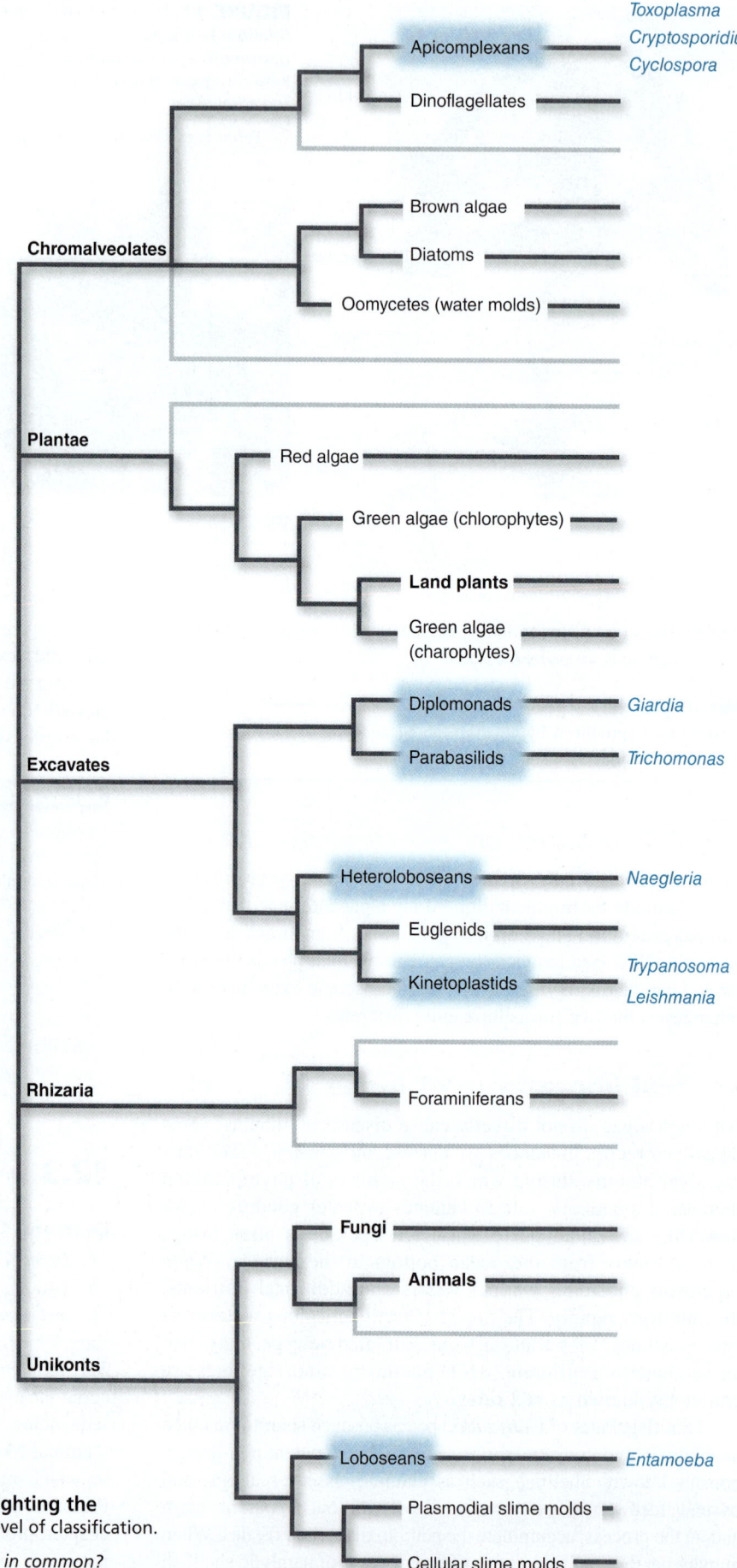

FIGURE 12.13 Classification of Eukaryotes Highlighting the Protozoans Note that not all groups are at the same level of classification.

❓ *What do the diplomonads and the parabasalids have in common?*

Trypanosoma brucei, which causes African sleeping sickness (see figure 26.19). ▶▶┤ sleeping sickness, p. 662

Loboseans and **heteroloboseans** have an ameboid (flexible) body form, but they are only distantly related to one another and to the slime molds discussed later in this chapter. Loboseans change shape as they move by extending and retracting pseudopodia and engulfing food particles by phagocytosis. *Entamoeba histolytica* infects humans, causing diarrhea ranging from mild asymptomatic disease to severe dysentery. Slime molds behave similarly to an ameba, but they aggregate to form much larger complex structures during part of their life cycle. Heteroloboseans exist in an ameboid form during part of their life cycle, but also form flagellated cells. *Naegleria fowleri* can swim through the water in its flagellated form but assumes its ameboid form upon entering the human body, where it literally eats the brain of its host (**figure 12.14**). ▶▶┤ *Entamoeba histolytica,* p. 605 ▶▶┤ *Naegleria fowleri,* p. 663

Structure of Protozoa

By definition, protozoa are not photosynthetic and thus lack chloroplasts. Protozoa also lack the rigid cellulose cell wall found in algae or the chitinous cell wall found in fungi. Foraminifera, however, secrete a hard shell of calcium carbonate. Over millions of years these shells have formed significant limestone deposits, such as the White Cliffs of Dover on the coast of England. Even though protozoa typically lack a cell wall, they often have a specific shape determined by the material beneath the plasma membrane.

Traditionally, protozoa have been grouped by their mode of locomotion. Many have specialized structures for movement such as cilia, flagella, or pseudopodia. As described in chapter 3, eukaryotic flagella and cilia are distinctly different from prokaryotic flagella (see figures 3.37 and 3.50). ┤◀◀ flagella, p. 74

Protozoan Habitats

The majority of protozoa are free-living aquatic organisms. They are essential as decomposers in many ecosystems. Some species, however, are parasitic, living on or in other host organisms. In marine environments, protozoa make up part of the zooplankton. On land, protozoa are abundant in soil as well as in or on plants and animals. Specialized protozoan habitats include the guts of termites, roaches, and ruminants such as cattle.

Protozoa are an important part of the food chain. They eat bacteria and algae and, in turn, serve as food for larger species. The protozoa help maintain ecological balance by devouring vast numbers of bacteria and algae. ▶▶┤ food chain, p. 720

MicroByte
> A single paramecium (a protozoan) can ingest as many as 5 million bacteria in one day.

Protozoan Reproduction

Some protozoa have complex life cycles involving more than one habitat or host. A single protozoan species that can exist as a **trophozoite** (vegetative or feeding form) or **cyst** (resting form) is **polymorphic** (figure 12.14). Environmental conditions—such as the lack of nutrients, moisture, O_2, low temperature, or the presence of toxic chemicals—may trigger the development of a protective cyst. Some protozoans, such as *Cryptosporidium* and *Entamoeba,* develop a protective cell wall during the cyst stage of their life cycle. This protects the organism as it moves from host to host and also helps it withstand stomach acid as it enters a new host. Stomach acid may assist in removal of the cell wall so that the trophozoite can emerge in time to infect the host's intestines. Many parasitic protozoans are transmitted to new hosts during their cyst stage. ▶▶┤ cryptosporidiosis, p. 603 ▶▶┤ amebiasis, p. 605

Both asexual and sexual reproduction are common in protozoa and may alternate during the life cycle of some organisms. Binary fission takes place in many groups of protozoa. Some protozoa divide by multiple fissions, or **schizogony,** in which the nucleus divides a number of times and then the cell produces many single-celled organisms. Multiple fission of the asexual forms of *Plasmodium* in humans results in large numbers of parasites released into the host's circulation at regular intervals, producing the cyclic symptoms of malaria. ▶▶┤ malaria, p. 686

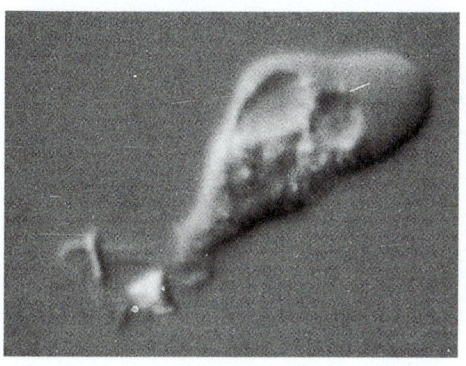
(a)

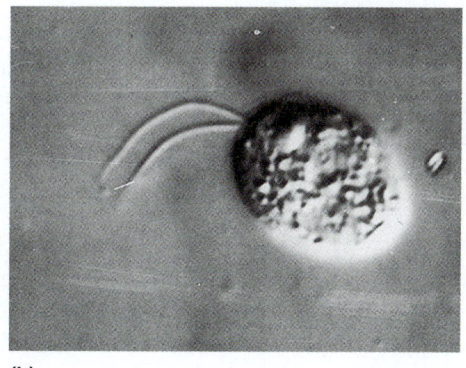
(b)

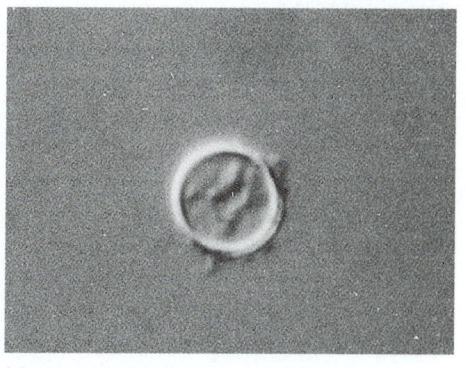
(c)

FIGURE 12.14 Polymorphism in a Protozoan Species of *Naegleria* may infect humans. **(a)** In human tissues, the organism exists in the form of an ameba (10–11 μm at its widest diameter). **(b)** After a few minutes in water, the flagellate form appears. **(c)** Under adverse conditions, a cyst is formed.

❓ *What is the advantage of flagella in water?*

TABLE 12.4	Protozoa of Medical Importance		
rRNA Classification	Genus of Disease-Causing Protozoan	Disease Caused by Protozoan	Page for Additional Information
Apicomplexan	*Plasmodium*	Malaria	p. 686
	Toxoplasma	Toxoplasmosis	p. 709
	Cryptosporidium	Cryptosporidiosis	p. 603
	Cyclospora	Cyclosporiasis	p. 604
Diplomonad	*Giardia*	Giardiasis	p. 602
Parabasalian	*Trichomonas*	Trichomoniasis	p. 635
Kinetoplastid	*Trypanosoma*	African sleeping sickness	p. 662
Heterolobosean	*Naegleria*	Primary amebic meningoencephalitis	p. 663
Lobosean	*Entamoeba*	Amebiasis (diarrhea)	pp. 605, 663

Medical Importance of Protozoa

The majority of protozoa do not cause disease, but those that do have significant impact on world health. Malaria has been one of the greatest killers of humans through the ages. Up to 300 million people in the world still contract malaria each year, and 1 million die of it, most of them in Africa. Amebiasis (amebic dysentery) affects nearly 50 million people worldwide each year, claiming up to 100,000 victims. *Cryptosporidium* and *Giardia* are also among the leading causes of diarrhea worldwide. Trypanosomes that cause sleeping sickness have made some regions of Africa uninhabitable for centuries. **Table 12.4** lists some disease-causing protozoans and directs you to where some of the most significant are discussed in later chapters.

MicroAssessment 12.3

Protozoa are microscopic, single-celled, non-photosynthetic, motile organisms. They occupy a variety of habitats and are a very important part of food chains.

7. *What are some important diseases caused by protozoa?*
8. *What is the role of a cyst in protozoan reproduction?*
9. *Why might a parasitic protozoan lack mitochondria?* ✚

12.4 ■ Slime Molds and Water Molds

Learning Outcomes

8. *Compare and contrast the two types of slime molds with one another and with water molds.*
9. *Explain how fungi and water molds provide an example of convergent evolution.*

The slime molds and water molds are protists that were once considered types of fungi. On the surface they may look like fungi and they may act like fungi, but at a cellular and molecular level, they are completely unrelated to them (see figure 12.13). Fungi and water molds, in particular, are good examples of **convergent evolution,** which occurs when two organisms develop similar characteristics independently because of adaptations to similar environments.

Slime Molds

Slime molds are terrestrial organisms composed of ameboid cells that live on soil, leaf litter, or the surfaces of decaying vegetation, where they ingest organic matter by phagocytosis. They are important links in the terrestrial food chain because they ingest microorganisms and, in turn, serve as food for larger predators.

There are two types of slime molds—cellular and plasmodial:

- **Cellular slime molds** have a vegetative form composed of single, ameba-like cells. When they run out of food, the single cells aggregate into a mass of cells called a slug. Some single cells then form a fruiting body, while others differentiate into spores. These look very much like fungal fruiting bodies and spores (**figure 12.15a**). The model eukaryotic organism, *Dictyostelium discoideum* is a cellular slime mold important for the study of aggregation and multicellular development.

- **Plasmodial slime molds** are large multinucleated "super-amebas" that may easily reach 0.5 m in diameter. They are widespread and readily visible in their natural environment due to their large size and often their bright color (figure 12.15b). Following germination of haploid spores, the cells fuse to form a diploid cell in which the nucleus divides repeatedly, forming a multinucleated stage called a **plasmodium.** The plasmodium oozes over the surface of decaying wood and leaves, ingesting organic debris and microorganisms. When food or water is in short supply, the plasmodium is stimulated to form spore-bearing fruiting bodies, and the process begins again.

MicroByte

Japanese researchers have shown that a slime mold takes the shortest path between two nutrients.

Water Molds

The **water molds,** or oomycetes, were once considered fungi. They form masses of white threads on decaying material (**figure 12.16**). Like fungi, they secrete digestive enzymes onto a substrate and absorb small molecules for nutrients. The cytoplasm in their filaments is continuous with many nuclei. However, water molds have cellulose in their cell walls rather than chitin. They lack chloroplasts and have flagellated reproductive cells.

Oomycetes cause some serious diseases of food crops such as late blight of potato and downy mildew of grapes. The late

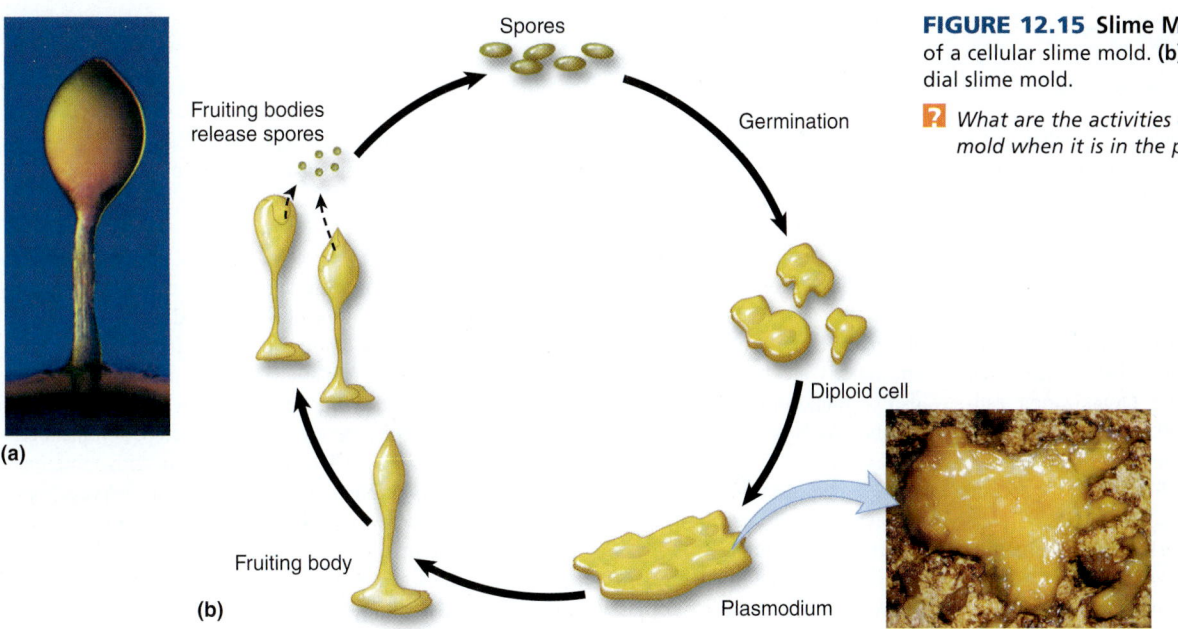

(a)

(b)

Spores

Germination

Fruiting bodies release spores

Diploid cell

Fruiting body

Plasmodium

FIGURE 12.15 Slime Molds (a) Fruiting body of a cellular slime mold. **(b)** Life cycle of a plasmodial slime mold.

❓ *What are the activities of a plasmodial slime mold when it is in the plasmodial form?*

blight of potato was the cause of the potato famine in Ireland in the 1840s that sent waves of immigrants to the United States (see **A Glimpse of History**).

MicroAssessment 12.4

Cellular slime molds may exist as single cells or may form an aggregation called a slug for reproduction. Plasmodial slime molds exist as a multinucleated "super-ameba." Both engulf food by phagocytosis and form fruiting bodies similar to those of fungi. Water molds secrete enzymes and absorb organic nutrients, and they form masses of white threads on organic matter.

10. *How are slime molds and water molds similar to, and different from, fungi?*

11. *What environmental conditions led to the convergent evolution of fungi and water molds?* ➕

FIGURE 12.16 Water Mold The white "threads" secrete digestive enzymes used to break down the organic compounds in this insect.

❓ *How is a water mold like a fungus; how is it different?*

12.5 ■ Multicellular Parasites: Helminths

Learning Outcomes

10. *Explain how disease-causing helminths can be transmitted to humans.*

11. *Describe a disease caused by a roundworm, a tapeworm, and a fluke.*

The **helminths,** which include **nematodes** (roundworms), **cestodes** (tapeworms), and **trematodes** (flukes), cause disease by invading the host's tissues or robbing it of nutrients. These multicellular parasites have been largely controlled in industrialized nations, but they still cause suffering and death of many millions each year in developing parts of the world.

Helminths enter the body in a number of ways. Hookworm larvae, for example, live in the soil and can burrow through human skin. They multiply in the human digestive tract and are eliminated with feces. When sanitation is poor and people are bare-footed, the parasite is easily transmitted. Hookworms can be found in about 740 million individuals, mostly in tropical and subtropical regions.

Some helminths are eaten with food, as when cysts of the nematode *Trichinella spiralis* are ingested in the flesh of animals. Eating undercooked pork is the most common cause of trichinellosis. More often, helminth eggs are ingested on the surface of foods. Children with pinworms (*Enterobius vermicularis*), for example, may pick up eggs by touching their anus and transmit them to a surface that can be touched by another person who then handles food without proper handwashing.

Some helminths are transmitted through insect bites. One example is *Wuchereria bancrofti,* a type of nematode that causes elephantiasis and is transmitted by mosquitoes. These worms lodge in the lymphatic vessels where they block lymphatic drainage. The buildup of fluid can cause massive swelling in various

What Causes River Blindness?

Female black flies, sometimes called gnats or buffalo gnats, swarm around their hosts and bite repeatedly to obtain blood needed for development of their eggs. They require rapidly flowing water for development of their larvae and so are most often found near rivers. These flies are associated with a disease called river blindness that affects at least 18 million individuals, 99% of whom live in Africa. Of those, the World Health Organization estimates that about 270,000 are blind as a result of their infection and another 500,000 have impaired vision. River blindness is the second leading cause of infectious blindness. (The leading cause is trachoma caused by the bacterium *Chlamydia trachomatis*.) But are the black flies the cause of this disease?

When biting, the flies may transmit larvae of a filarial nematode, *Onchocerca volvulus*,

to a human host, leading to river blindness, also called onchocerciasis. Once in a human host, the larvae reside in nodules and mature to adulthood in about a year, at which time adult females produce millions of microfilariae. These can migrate through the skin where they can be picked up by another bite of a black fly to continue the life cycle. When infestations are heavy, the microfilariae can also be found in the blood and in the eye. So are the *Onchocerca* larvae the real cause of the disease?

As microfilariae move throughout the bodies of their hosts, they carry with them a bacterial population of *Wolbachia pipientis* that is necessary for fertility and viability of the worms. When the microfilariae reach the eye and then die, they release the *Wolbachia*. The bacteria cause an inflammatory response

in the host that can damage sensitive tissues and result in vision impairment or blindness. Ultimately, it is the bacteria that cause river blindness.

Efforts to control river blindness have focused on eliminating the black fly vector and providing a medication called ivermectin that targets the worm. Ivermectin reduces the number of microfilariae for a few months, but does not destroy the adults, so repeated medication is needed. A newer more effective approach is to target the *Wolbachia* bacteria with common antimicrobial medications such as doxycycline or tetracycline. This not only reduces effects of the disease, but it sterilizes the worms so that they can no longer produce microfilariae, thus disrupting the life cycle.

parts of the body (**figure 12.17**). Another example is *Onchocerca volvulus*—a nematode that is spread by flies and can cause river blindness (see **Perspective 12.1**).

Some helminths have complex life cycles involving one or more **intermediate hosts** that house a sexually immature stage of the parasite and are necessary for its development. Snails serve as an intermediate host for the fluke *Schistosoma mansoni,* the cause of schistosomiasis. Sexual reproduction of the parasite takes place

in humans, which are its **definitive host.** Humans may become an accidental or **dead-end host** if infected by a parasite that normally completes its life cycle in another host. For example, swimmers are sometimes infected with the larvae of flukes that typically complete their life cycles in fish or water birds. These flukes cannot complete their life cycle in humans, but when they burrow under the skin, they cause local inflammation called "swimmer's itch." **Table 12.5** lists the major diseases caused by helminths.

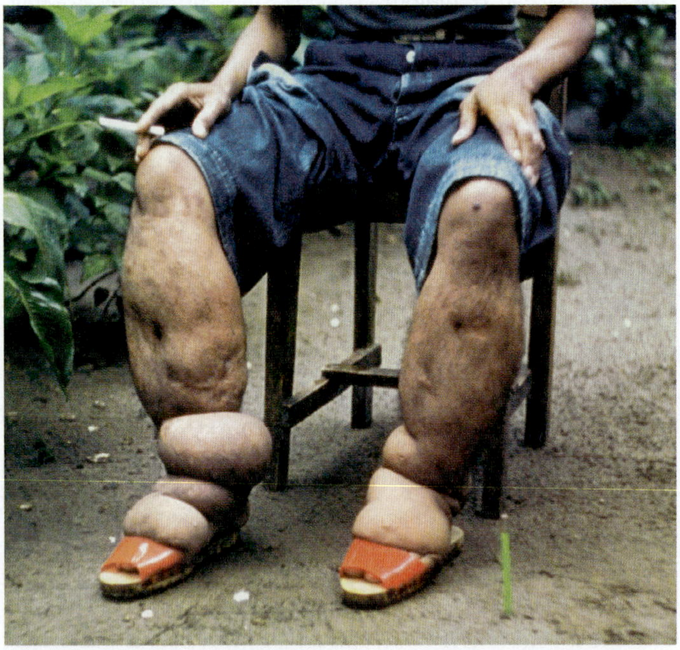

> **MicroByte**
> Over 1 billion people worldwide are believed to carry the roundworm *Ascaris lumbricoides.*

Roundworms

A roundworm, or nematode, has a cylindrical, tapered body with a digestive tract that extends from the mouth to the anus. The nematode *Caenorhabditis elegans* is a model eukaryotic organism that has been the subject of numerous studies in genetics and development because it matures quickly, its genome has been sequenced, and all 959 body cells can be identified. Many nematodes are free-living in soil and water. Others are parasites and produce serious disease. Almost 30,000 nematode species have been identified, but some estimate that there may be over 1 million species.

Ascariasis, caused by *Ascaris lumbricoides*, is the most common human disease caused by roundworms. Females may reach 45 cm long and produce more than 200,000 eggs per day that are eliminated in the feces. Ingested eggs hatch in the digestive tract, releasing immature worms that burrow into the bloodstream (**figure 12.18**). When they reach the lungs, they can be coughed up and swallowed. When the immature worms again reach the intestine, they grow and begin to produce eggs that can again be released with the feces. Although they do not feed on human tissue, they do

FIGURE 12.17 Elephantiasis Buildup of lymphatic fluid has occurred because worms block the vessels leading from the limbs.

❓ *What type of worm causes this condition?*

TABLE 12.5   Nematodes, Cestodes, and Trematodes

Infectious Agents	Disease	Disease Characteristics
Nematodes (roundworms)		
Pinworms (*Enterobius vermicularis*)	Enterobiasis	Anal itching, restlessness, irritability, nervousness, poor sleep
Whipworm (*Trichuris trichiura*)	Trichuriasis	Abdominal pain, bloody stools, weight loss
Hookworm (*Necator americanus* and *Ancylostoma duodenale*)	Hookworm disease	Anemia, weakness, fatigue, physical and intellectual disability in children
Threadworm (*Strongyloides stercoralis*)	Strongyloidiasis	Skin rash at site of penetration, cough, abdominal pains, weight loss
Ascaria (*Ascaris lumbricoides*)	Ascariasis	Abdominal pain, live worms vomited or passed in stools
Trichinella (*Trichinella spiralis*)	Trichinellosis	Fever, swelling of upper eyelids, muscle soreness
Filaria (*Wuchereria bancrofti* and *Brugia malayi*)	Filariasis	Fever, swelling of lymph glands, genitals, and extremities
Cestodes (tapeworms)		
Fish tapeworm (*Diphyllobothrium latum*)	Tapeworm disease	Few or no symptoms, sometimes anemia
Beef tapeworm (*Taenia saginata*)	Tapeworm disease	Few or no symptoms, sometimes anemia
Pork tapeworm (*Taenia solium*)	Cysticercosis	Variable symptoms depending on location and number of eggs that form larval cysts (cysticerci) in the body
Trematodes (flukes)		
Cercaria (*Schistosoma mansoni*)	Schistosomiasis	Liver damage, malnutrition, weakness, and accumulation of fluid in the abdominal cavity
Cercaria of birds and other animals	Swimmer's itch	Inflammation of the skin, itching

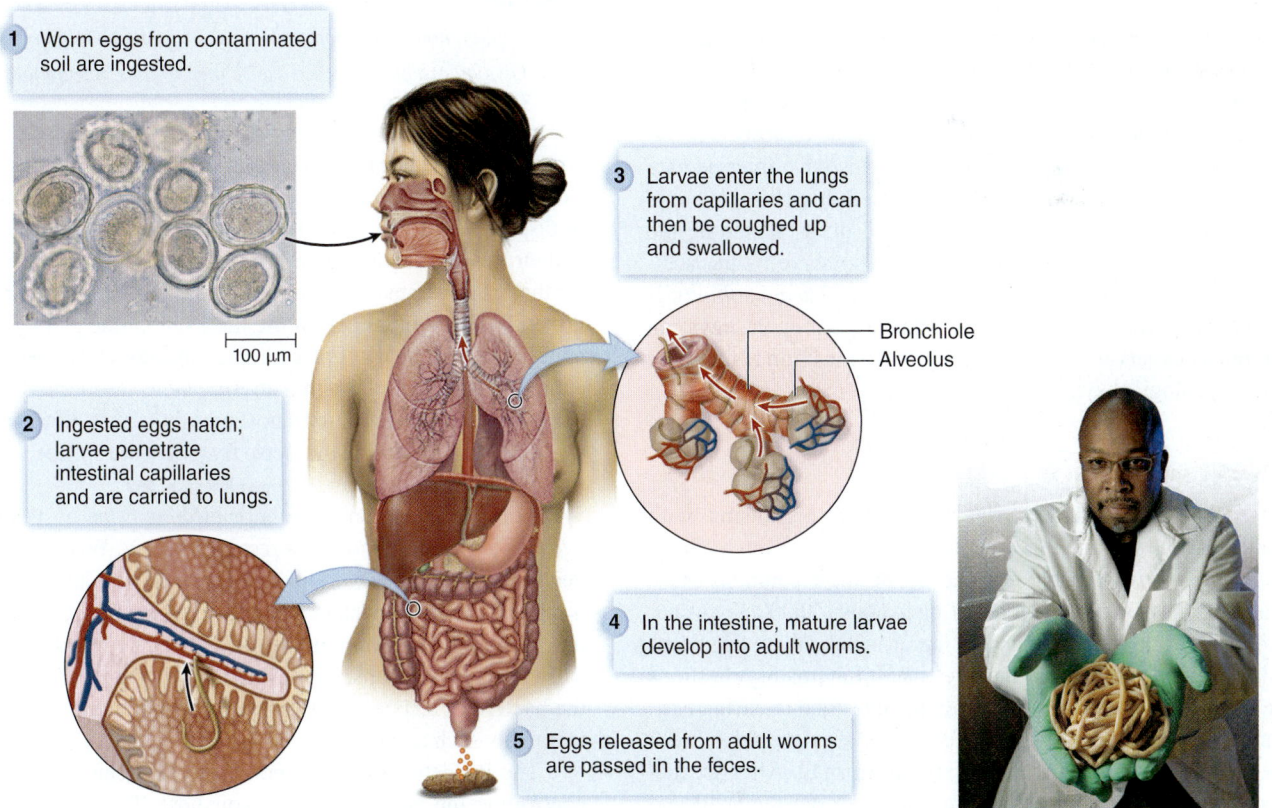

1. Worm eggs from contaminated soil are ingested.

2. Ingested eggs hatch; larvae penetrate intestinal capillaries and are carried to lungs.

3. Larvae enter the lungs from capillaries and can then be coughed up and swallowed.

Bronchiole
Alveolus

100 µm

4. In the intestine, mature larvae develop into adult worms.

5. Eggs released from adult worms are passed in the feces.

FIGURE 12.18 Ascariasis Life cycle of *Ascaris lumbricoides*, the largest roundworm infesting the human intestine, reaching 30–40 cm. Larvae hatching in the intestine migrate through the lungs and back to the intestine before maturing to adulthood. The thick-walled ova are nearly spherical.

❓ *How does* Ascaris *get from the lungs to the intestines?*

FIGURE 12.19

Pork Tapeworm The scolex holds the tapeworm to the intestinal surface. Proglottids contain reproductive structures. They are shed in the feces as the tapeworm elongates by adding new segments.

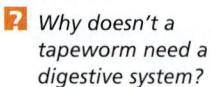 *Why doesn't a tapeworm need a digestive system?*

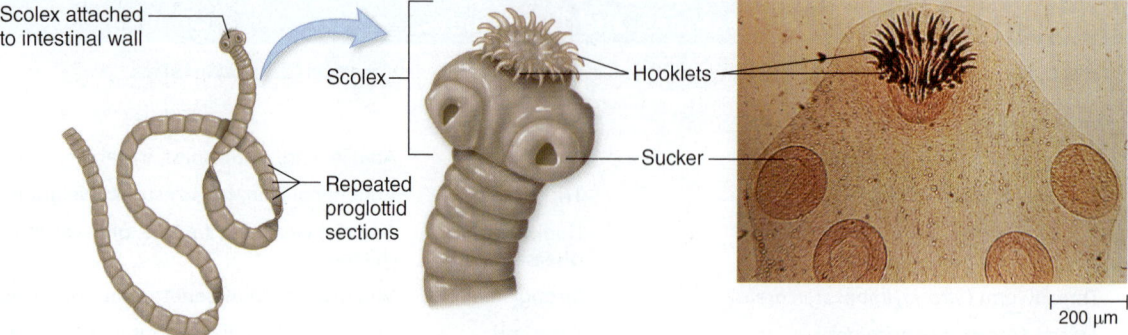

rob the body of nutrients by feeding on material that passes through the digestive tract. They may cause choking and pulmonary symptoms when they enter the respiratory tract.

Tapeworms

Tapeworms, or cestodes, have flat, ribbon-shaped bodies and some types can reach over a meter in length. They have no digestive system, and do not feed directly on the tissues of their host. Rather, adult tapeworms attach within the intestines of the definitive host and absorb predigested nutrients through their body. The head end (scolex) often has suckers and hooks for attachment (**figure 12.19**).

Behind the scolex are a number of segments called proglottids that contain both male and female reproductive structures. The segments farthest from the scolex contain fertilized eggs. As the worm grows, these segments break off and are eliminated in the feces along with tapeworm eggs. When a suitable intermediate host ingests the eggs, the eggs hatch, releasing a larval form that penetrates the intestinal wall and migrates into tissues. These larval forms are infectious when consumed by a definitive host.

The most common tapeworms of humans have intermediate hosts of cattle, pigs, and fish. Humans (the definitive host) become infected when they eat raw or undercooked meat containing the larval forms. Unfortunately, humans can also serve as the accidental intermediate host of the pork tapeworm (*Taenia solium*). If someone inadvertently ingests eggs of this tapeworm, the eggs can hatch, releasing larvae. These may migrate to the brain resulting in serious neurological symptoms.

Flukes

Flukes, or trematodes, are flat leaf-shaped organisms with suckers that hold them in place while sucking fluids from the host. *Schistosoma mansoni*, a blood fluke, is the most common cause of schistosomiasis worldwide, resulting in about 20,000 deaths per year. Female flukes are held by the larger male and lay eggs continually in blood vessels near the intestine. Inflammatory reactions cause the blood vessels to rupture into the intestine, releasing eggs that are carried from the body in feces. Larvae develop in fresh water and are taken up by a certain species of snail. After asexual reproduction, tail-bearing larvae are released from the snail and can penetrate the skin of a human host wading in the water. The larvae enter blood vessels, reproduce sexually in the liver, and begin the cycle again. Schistosomiasis infections do not occur in the United States as there is no appropriate snail intermediate host.

MicroAssessment 12.5

Helminths, including the roundworms, tapeworms, and flukes, cause serious diseases in humans. They may enter a human host through ingestion with food or water, by an insect bite, or by piercing the skin. Many helminths have a complex life cycle with more than one host.

12. *What are the major differences among nematodes, cestodes, and trematodes?*

13. *Differentiate between a definitive host and an intermediate host.*

14. *Why do so many helminth diseases occur in the tropics?* ✚

12.6 ■ Arthropods

Learning Outcomes

12. *Explain how arthropods are related to disease in humans.*

13. *Give an example of a disease related to flies, mosquitoes, fleas, lice, ticks, and mites.*

Arthropods include the insects (such as flies, mosquitoes, lice, and fleas) and the arachnids (such as ticks and mites). Their main role in disease is to serve as **vectors** that can transmit microorganisms and viruses to humans. An arthropod may act as a mechanical vector that simply transfers a pathogen from one surface to another, or it may be an essential part of the life cycle of the pathogen, acting as a biological vector. For example, species of *Plasmodium* that cause malaria multiply within an *Anopheles* mosquito during their life cycle, and species of trypanosomes that cause African sleeping sickness multiply within the tsetse fly (*Glossina* sp.). ▸▸ vector, p. 442
▸▸ malaria, p. 686 ▸▸ African sleeping sickness, p. 662

Animals such as mammals or birds may be hosts in the life cycle of a pathogen and act as reservoirs for it. When an arthropod vector feeds on a reservoir host, it may pick up pathogens and then transfer them to humans in a later bite. Some arthropods bite only one type of host. Certain mosquitoes that carry *Plasmodium* typically bite only humans. However, fleas that carry *Yersinia pestis*, the bacterium responsible for plague, bite both humans and small mammals such as rats. ▸▸ reservoir, p. 438

The incidence of vector-borne diseases can be decreased by controlling the vector or a reservoir host. The risk of mosquito-borne encephalitis, for example, can be minimized by eliminating standing water and using insect repellant. These do not act on the virus that causes encephalitis, but they do reduce the incidence of vector transmission. Plague in the United States has been

TABLE 12.6 Some Arthropods That Transmit Infectious Agents

Arthropod	Infectious Agent	Disease and Characteristic Features	Page for More Information
Insects			
Tsetse fly (*Glossina* species)	Trypanosomes	African sleeping sickness—sleepiness, headache, coma	p. 662
Sand fly (*Phlebotomus* species)	*Leishmania*	Leishmaniasis—ulcers, nosebleeds, diarrhea, fever, cough	p. 521
Black fly (*Simuliidae* species)	*Onchocercus*	Onchocerciasis—rash, itching, visual impairment	p. 296
Mosquito (*Anopheles* species)	*Plasmodium* species	Malaria—chills, bouts of recurring fever	p. 686
Mosquito (*Culex* species)	Togavirus	Equine encephalitis—fever, nausea, convulsions, coma	p. 655
Mosquito (*Aedes aegypti*)	Flavivirus	Yellow fever—fever, vomiting, jaundice, bleeding	p. 682
Mosquito (*Aedes aegypti*)	Flavivirus	Dengue fever—high fever; headache; joint, muscle, and bone pain	p. 683
Flea (*Xenopsylla cheopis*)	*Yersinia pestis*	Plague—fever, headache, confusion, enlarged lymph nodes, skin hemorrhage	p. 678
Louse (*Pediculus humanus*)	*Rickettsia prowazekii*	Typhus—fever, hemorrhage, rash, confusion	p. 521
Arachnids			
Tick (*Dermacentor* species)	*Rickettsia rickettsii*	Rocky Mountain spotted fever—fever, hemorrhagic rash, confusion	p. 529
Tick (*Ixodes* species)	*Borrelia burgdorferi*	Lyme disease—fever, rash, joint pain, nervous system impairment	p. 531

controlled mostly by eliminating rat populations that may infect their fleas with the bacterium *Yersinia pestis*. Cases of plague today occur from occasional transmission by fleas on wild rodents such as rock squirrels or prairie dogs. Examples of some important arthropods, the agents they transmit, and the resulting diseases are shown in **table 12.6**. ▶▶ plague, p. 678

Mosquitoes

Mosquitoes are insects known to transmit diseases such as malaria, yellow fever, dengue fever, and West Nile encephalitis. A female mosquito needs a blood meal for the proper development of her eggs. During a bite, the mosquito forces a sharp, hollow feeding tube through the host's skin to the subcutaneous capillaries (**figure 12.20**). She pumps saliva through the tube, which increases blood flow and prevents clotting of the victim's blood. She can take in as much as twice her body weight in blood, also picking up infectious agents circulating within the host's capillaries. These agents multiply within the mosquito's body and can later be transferred to a new host in a subsequent bite. ▶▶ yellow fever, p. 682 ▶▶ dengue fever, p. 683

MicroByte
The itch of a mosquito bite is caused by an allergic reaction to the mosquito's saliva.

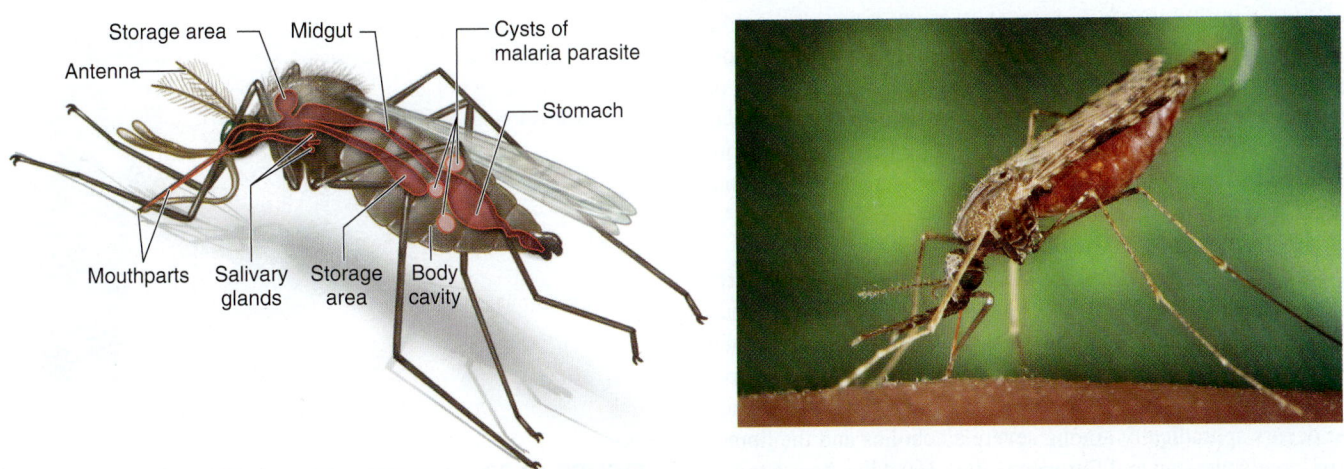

FIGURE 12.20 Internal Anatomy of a Mosquito Storage areas allow ingestion of large amounts of blood, and salivary glands discharge pathogens into the host. The *Anopheles* mosquito adopts a curious head-stand posture when feeding.

? *Is the mosquito an intermediate host or the definitive host for the organism that causes malaria?*

FIGURE 12.21 *Pediculus humanus* A body louse, which is the vector for several human diseases.

❓ *Can body lice act as mechanical vectors?*

Fleas

Fleas are wingless insects, but their appendages are adapted for easily jumping up to 30 cm at a time. They are generally more of a nuisance than a health hazard, but they can transmit a number of pathogens, including the bacterium *Yersinia pestis,* which causes plague. Fleas pick up the pathogen when biting an infected host, and the bacteria multiply within the digestive tract of the flea, causing a blockage. The starving flea bites repeatedly, each time passing bacteria from its blocked digestive tract to a host. Fleas can live in vacant buildings in a dormant stage for many months. When the building becomes inhabited, the fleas quickly mature and hungrily jump to greet the new hosts.

Lice

Like fleas, lice (singular: louse) are small, wingless insects that prey on mammals and birds by piercing their skin and sucking blood. The appendages of lice, however, are adapted for attachment rather than jumping. *Pediculus humanus,* the most notorious of the lice, is 1 to 4 mm long, with a membrane-like lip housing tiny teeth that anchor it firmly to the skin of the host (**figure 12.21**). Lice have a piercing apparatus similar to that of fleas and mosquitoes used for obtaining a blood meal. *Pediculus humanus* has only one host—humans—but easily spreads from one person to another by direct contact or by contact with personal items, especially in areas of crowding and poor sanitation. An infestation of lice is called **pediculosis.**

The two subspecies of *Pediculus humanus* are popularly termed body lice and head lice. Body lice can transmit trench fever, caused by the bacterium *Bartonella quintana;* epidemic typhus, caused by the bacterium *Rickettsia prowazekii;* and relapsing fever caused by the bacterium *Borrelia recurrentis.* Trench fever occurs sporadically among severe alcoholics and the homeless of large American and European cities. Head lice do not transmit disease, but they are easily transmitted from human to human. They are most often discovered when eggs resembling dandruff, or nits, are found clinging to the scalp or hair.

The pubic louse, *Phthirus pubis,* is commonly transmitted during sexual intercourse. It is not a vector of infectious disease, but it can cause an unpleasant itch associated with "crabs."

Ticks

Ticks are arachnids. Arachnids can be distinguished from insects because they lack wings and antennae, their thorax and abdomen are fused, and adults have four pairs of legs rather than three. Ticks generally live in low vegetation where they may contact a suitable host passing by. Once in contact with a host, the tick burrows into skin with its mouthparts (see figure 22.9). When a tick attaches in the scalp or is very small, it may go unnoticed for days, during which time it continually feeds from its host.

Dermacentor andersoni, the wood tick, is the vector for Rocky Mountain spotted fever caused by the bacterium *Rickettsia rickettsii.* Another tick, *Ixodes scapularis,* transmits with its saliva *Borrelia burgdorferi,* the spirochete that causes Lyme disease. A neurotoxin in the saliva of some ticks can cause muscle weakness, a condition called tick paralysis. This occurs especially in children but recovery is rapid following removal of the tick. ▶▶ **Rocky Mountain spotted fever, p. 529** ▶▶ **Lyme disease, p. 531**

Mites

Mites, like ticks, are arachnids. Microscopic mites of the genus *Demodex* typically live in the hair follicles or oil-producing glands without being noticed. Large numbers of dust mites often live indoors and feed on organic material such as shed skin cells. Although they do not transmit infectious disease, inhalation of the mites and their waste products can sometimes trigger asthma. The larvae of some mites are called "chiggers" and may cause intense itching where they attach and feed on fluids within skin cells.

Scabies, a disease caused by a mite, *Sarcoptes scabiei,* is characterized by an itchy rash most prominent between the fingers, under the breasts, and in the genital area. Scabies

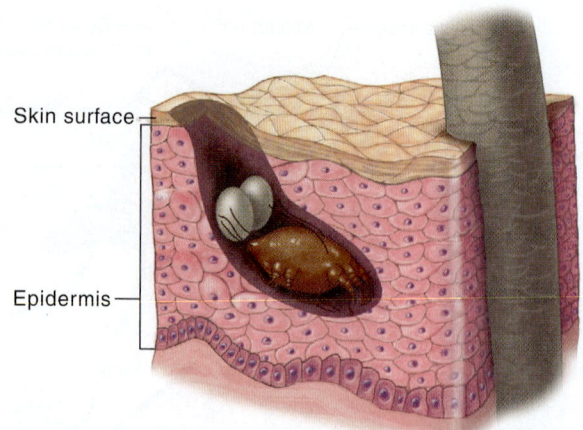

Skin surface
Epidermis

FIGURE 12.22 *Sarcoptes scabiei* (Scabies Mite) The female burrows into outer skin layers to lay her eggs, causing an intensely itchy rash.

❓ *Is a mite an insect or an arachnid?*

is easily transmitted by personal contact, and the disease is commonly acquired during sexual intercourse. Female mites burrow into the outer layers of epidermis (**figure 12.22**), feeding and laying eggs over a lifetime of about 1 month. Allergic reactions are largely responsible for the itchy rash of scabies. Diagnosis requires locating the mites, since scabies mimics other skin diseases. *Sarcoptes scabiei* is not known to transmit infectious agents. Mites of domestic animals and birds can also cause an itchy rash in humans.

MicroAssessment 12.6

Arthropods such as flies, mosquitoes, fleas, lice, ticks, and mites act as vectors for the spread of disease. Some participate in the life cycle of an infectious agent and transmit disease through saliva when biting or burrowing. Others infest the skin and cause itching.

15. *How can arthropods spread disease in humans?*
16. *Name two diseases transmitted by ticks.*
17. *Why are diseases transmitted by insect vectors more common in the summer than in the winter?* ✚

FUTURE CHALLENGES   12.1

The Continued Fight to Eradicate Malaria

Malaria has affected humans for millennia. The symptoms were described in ancient Chinese writings more than 4,500 years ago. Each year about a million people still die from malaria; most are in Africa and most are children under the age of 5 years. With drug-resistant strains spreading rapidly, more people die from malaria today than 30 years ago.

The economic costs to the people living in malaria-infested parts of the world are huge. Drug companies and foundations are continuing their effort to develop new vaccines and drugs to treat malaria. The Bill and Melinda Gates Foundation, among others, has pledged over a billion dollars to aid in the effort. A multifaceted approach includes distribution of insecticide-treated bed nets, spraying interior buildings with insecticide, eliminating breeding grounds for mosquitoes, and providing effective and affordable drugs. These efforts have met with some success. For example, the incidence of malaria decreased by 50% in 26 countries between 2000 and 2006.

In 2010, 39 countries were working to eradicate malaria. This will require billions of dollars along with support of corporations, foundations, and governments worldwide. As the battle continues, mosquitoes will become resistant to pesticides and *Plasmodium* will become increasingly resistant to antimalarial drugs. Researchers will have to work hard to stay ahead of both. To truly eradicate malaria, a vaccine is needed. Clinical trials are underway for a vaccine that could reduce malaria deaths by 50%, but it is not expected to be generally available until 2015. Newer vaccines will attempt to disrupt the life cycle of the parasite by preventing reproduction in the mosquito vector.

Summary

THE EUKARYOTIC MEMBERS OF THE MICROBIAL WORLD

Algae, fungi, and **protozoa** are not accurate classification terms when the rRNA sequences of these organisms are considered. Cell structure in eukaryotic organisms is different from that seen in prokaryotes. Eukaryotes have a membrane-bound nucleus. Reproduction may be asexual using **mitosis** or sexual using **meiosis,** which forms **gametes** (figure 12.1).

12.1 ■ Fungi

Yeast, mold, and **mushroom** are common terms that indicate morphological forms of fungi (figure 12.2). Fungi have chitinous cell walls and are often **saprophytes,** secreting enzymes onto a surface and absorbing nutrients.

Types of Fungi (table 12.1)

Classification of fungi is in flux. Zygomycetes (figure 12.3), ascomycetes, basidiomycetes, and chytridiomycetes are distinctive types of fungi.

Structure of Fungi

Fungal filaments are called **hyphae,** and a group of hyphae is called a **mycelium** (figure 12.4). **Dimorphic fungi** can grow either as a single cell (yeast) or as mycelia.

Fungal Habitats

Fungi inhabit just about every ecological habitat and can spoil a large variety of food materials because they can grow in high concentrations of sugar, salt, and acid. Fungi can be found in moist environments, at a wide range of temperatures, and at pH from 2.2 to 9.6. Fungi can degrade most organic materials.

Symbiotic Relationships of Fungi

Lichens result from an association of a fungus with a photosynthetic organism such as an alga or a cyanobacterium (figure 12.5). **Mycorrhizas** are the result of an intimate association of a fungus and the roots of a plant (figure 12.6). Leaf-cutter ants grow gardens of fungus for food (figure 12.7).

Reproduction in Fungi

Asexual reproduction may occur by a variety of spore-producing structures, by budding, or by fragmentation (figures 12.8, 12.9). Sexual reproduction may involve fusion of hyphae from different mating types.

Economic Importance of Fungi

The yeast *Saccharomyces* is used in the production of beer, wine, and bread. *Penicillium* and other fungi synthesize antibiotics. Fungi spoil many food products and cause diseases of plants such as Dutch elm disease and wheat rust. Fungi have been useful tools in genetic and biochemical studies.

Medical Importance of Fungi (table 12.2)

Relatively few fungi cause human disease, but they cause devastating diseases in plants. Fungi may produce an allergic reaction. Fungi cause **mycoses** such as candidiasis. Fungi can produce toxins that make humans ill. These include **ergot,** those in poisonous mushrooms, and **aflatoxin.**

12.2 ■ Algae

Algae are a diverse group of photosynthetic organisms that contain chlorophyll *a*.

Types of Algae (table 12.3)

Types of algae differ in their major photosynthetic pigments. Organisms are placed on the phylogenetic tree according to rRNA sequences (figure 12.10).

Structure of Algae

Algae may be microscopic (figure 12.11) or macroscopic (figure 12.12). Their cell walls are made of cellulose and other commercially important materials such as agar and alginic acid.

Algal Habitats

Algae are found in fresh and salt water as well as in soil. Unicellular algae make up a significant part of the **phytoplankton.**

Algal Reproduction

Algae reproduce asexually as well as sexually.

Medical Importance of Algae

Algae do not directly cause disease, but produce toxins during algal blooms that are ingested by fish and shellfish. When these fish and shellfish are eaten by humans, they may develop paralytic shellfish poisoning with dizziness, muscle weakness, and even death may result; cooking does not destroy the toxins.

12.3 ■ Protozoa

Protozoa are a diverse group of microscopic, unicellular organisms that lack chlorophyll.

Types of Protozoa (figure 12.13)

In classification schemes based on rRNA, protozoa are not a single group of organisms. **Apicomplexans** (*Plasmodium, Toxoplasma, Cryptosporidium, Cyclospora*) are parasites with an apical complex that helps them to penetrate host cells. **Diplomonads** (*Giardia*) and **parabasalids** (*Trichomonas*) have no mitochondria. **Kinetoplastids** (*Trypanosoma, Leishmania*) have distinct mitochondrial DNA. **Loboseans** (*Entamoeba*) and **heteroloboseans** (*Naegleria*) move by pseudopodia at some stage in their lives (figure 12.14).

Structure of Protozoa

Protozoa lack a cell wall, but most maintain a definite shape using the material lying just beneath the plasma membrane.

Protozoan Habitats

Most protozoa are free-living and are found in marine and fresh water as well as terrestrial environments. They are important decomposers in many ecosystems and are a key part of the food chain.

Protozoan Reproduction

Life cycles are often complex and include more than one habitat or host. Some are **polymorphic** with a vegetative **trophozoite** form and a resting **cyst** form. Reproduction is often by binary fission; some reproduce by multiple fissions or **schizogony.**

Medical Importance of Protozoa (table 12.4)

Protozoa cause diseases such as malaria, African sleeping sickness, toxoplasmosis, and amebic dysentery.

12.4 ■ Slime Molds and Water Molds

Slime molds and water molds were once considered types of fungi.

Slime Molds (figure 12.15)

Cellular slime molds exist as ameba-like single cells, but when food supplies run low, they aggregate into a slug in which some cells differentiate into spores. **Plasmodial slime molds** form a multinucleated plasmodium that oozes over a surface, ingesting organic material. When food runs short, they form fruiting bodies that bear spores.

Water Molds

Oomycetes, also known as **water molds,** cause some serious diseases of plants (figure 12.16). Water molds and fungi are an example of **convergent evolution.**

12.5 ■ Multicellular Parasites: Helminths (table 12.5)

Helminths can be transmitted by burrowing through the skin, being ingested, or being transmitted through insect bites (figure 12.17). Some helminths have complex life cycles with asexual stages occurring in **intermediate hosts** and the sexual or adult stage occurring in the **definitive host.** Humans may be a **dead-end host** in which the organism cannot complete its life cycle.

Roundworms

Most **nematodes** or roundworms are free-living, but they may cause serious disease such as pinworm disease, hookworm disease, and ascariasis (figure 12.18).

Tapeworms

Cestodes are tapeworms with segmented bodies and hooks to attach to the wall of the intestine (figure 12.19). Most tapeworm infections occur in persons who eat uncooked or undercooked meats.

Flukes

Trematodes, or flukes, often have complicated life cycles that necessarily involve more than one host. *Schistosoma mansoni* larvae can penetrate the skin of persons wading in infected waters and cause serious disease.

12.6 ■ Arthropods

Arthropods act as **vectors** for disease (table 12.6).

Mosquitoes

Mosquitoes spread disease by picking up disease-causing organisms when the mosquito bites and later injecting these organisms into subsequent animals that it bites (figure 12.20).

Fleas

Fleas transmit disease such as plague.

Lice

Lice can transmit trench fever, epidemic typhus, and relapsing fever (figure 12.21).

Ticks

Ticks are implicated in Rocky Mountain spotted fever and Lyme disease.

Mites

Mites cause **scabies,** and dust mites are responsible for allergies and asthma (figure 12.22).

Review Questions

Short Answer

1. What are the major differences between a prokaryotic cell and a eukaryotic cell?

2. What are the differences among a yeast, a mold, and a mushroom?

3. How do mycorrhizas improve the growth of a green plant?

4. In what ways are fungi economically important?

5. What is a mycosis? Give an example.

6. What characteristics do all algae have in common?

7. Compare and contrast the organisms that cause malaria and African sleeping sickness and their transmission.

8. Name a disease for which humans are an intermediate host and another for which humans are a definitive host. Give an example of a disease in which humans are a dead-end host.

9. Describe the life cycle of *Schistosoma mansoni*.

10. Explain how a fly might act as a mechanical vector for one disease and a biological vector for another.

Multiple Choice

1. Members of this group have chitinous cell walls.
   a) Algae   b) Protozoa   c) Fungi
   d) Helminths   e) Arthropods

2. Members of this group are photosynthetic.
   a) Algae   b) Protozoa   c) Fungi
   d) Helminths   e) Arthropods

3. This group helps produce many of the foods that we eat.
   a) Algae   b) Protozoa   c) Fungi
   d) Helminths   e) Arthropods

4. Protozoa reproduce asexually by
   a) schizogony.
   b) fragmentation.
   c) meiosis.
   d) polymorphism.

5. Which of the following is mismatched?
   a) *Plasmodium*—malaria
   b) Trypanosomes—dysentery
   c) Dinoflagellates—paralytic shellfish poisoning
   d) Nematode—trichinellosis

6. Which of the following is mismatched?
   a) Trematode—fluke
   b) Tick—arachnid
   c) Baker's yeast—algae
   d) Apicomplexan—protozoa

7. Body lice
   a) can act as a vector to transmit disease.
   b) are not infectious.
   c) have eight legs and sucking mouthparts.
   d) are more closely related to ticks than they are to mosquitoes.

8. All algae have
   a) chlorophyll *a*.
   b) cell walls that contain agar.
   c) holdfasts.
   d) red tides.

9. Which of the following statements regarding protists is *false*?
   a) They include both autotrophic and heterotrophic organisms.
   b) They include both microscopic and macroscopic organisms.
   c) They often act as vectors in disease transmission.
   d) They include algae and protozoa.

10. Which of the following statements regarding tapeworms is *false*?
    a) They absorb nutrients from the host through their body wall.
    b) They complete their life cycles in a single host.
    c) They can form cysts in the tissue of their host.
    d) They cannot be transmitted from human to human.

Applications

1. A molecular biologist working for a government-run fishery in Vietnam is interested in controlling *Pfisteria* in fish farms. *Pfisteria* produces toxins that stun the fish and then causes the skin to slough off, allowing the dinoflagellates to dine on the tissues of the fish. He needs to develop a treatment that kills *Pfisteria* without harming the fish or the beneficial green algae that serve as food for the young fish. What strategy should the biologist consider for developing a selective treatment?

2. Paper recycling companies refuse to collect paper products that are contaminated with food or have been sitting wet for a day. A college sorority member who is running a recycling program on campus wishes to know the reason for this. What reason did the chemist who works for the recycling company probably give her for this policy?

Critical Thinking ✚

1. If you discover a new type of nucleated cell in a lake near your home, how would you determine whether the cell is from a fungus, an alga, a protozoan, or a water mold?

2. Fungi are known for growing and reproducing in a wide range of environmental extremes in temperature, pH, and osmotic pressure. What does this tolerance for extremes indicate about fungal enzymes?

13

Viruses, Viroids, and Prions

SV40 (simian virus 40), a virus of monkeys.

KEY TERMS

Bacteriophage A virus that infects bacteria; often shortened to *phage*.

Latent Infection Viral infection in which the viral genome is present but not active, so viral particles are not being produced.

Lysogen A bacterium that carries phage DNA (the prophage) integrated into its genome.

Lysogenic Conversion A change in the properties of a bacterium conferred by a prophage.

Lytic Infection Viral infection of a host cell with a subsequent production of more virus particles and lysis of the cell.

Plaque Assay Method used to measure the number of viral particles present in a sample.

Prion An infectious agent that causes a neurodegenerative disease; consists of protein similar in amino acid sequence to a normal protein in the body.

Productive Infection Viral infection in which more viral particles are produced.

Viroid An infectious agent of plants that consists only of RNA.

Virion A complete virus in its inert non-replicating form; also referred to as a viral particle.

A Glimpse of History

During the late nineteenth century, many bacteria, fungi, and protozoa were identified as infectious agents. Most of these organisms could be easily seen with the aid of a microscope and grown in the laboratory. In the 1890s, however, D. M. Iwanowsky and Martinus Beijerinck found that mosaic disease—a disease of tobacco plants—was caused by an unusual agent. The agent was too small to be seen with the light microscope, passed through filters that retained most known bacteria, and could be grown only in media that contained living cells. Beijerinck called the agent a filterable virus. About 10 years later, F. W. Twort in England and F. d'Herelle in France discovered a "filterable virus" that destroys bacteria. *Virus* means "poison," a term once applied to all infectious agents. With time, the adjective *filterable* was dropped and only the word *virus* was retained.

Viruses have many features more characteristic of complex chemicals than cells. For example, tobacco mosaic virus (TMV) can be precipitated from a suspension with ethyl alcohol and still remain infective. A similar treatment destroys the infectivity of bacteria. Further, in 1935, Wendell Stanley of the University of California, Berkeley, crystallized TMV. Its physical and chemical properties obviously differed from those of cells, which cannot be crystallized. Surprisingly, the crystallized TMV could still cause disease.

In the simplest of terms, viruses can be viewed as genetic information—either DNA or RNA—contained within a protective coat. They are inert particles, incapable of metabolism, replication, or motility. When a viral genome finds its way into a host cell, however, it can hijack that cell's replication machinery, inducing the cell to produce more virus particles. In essence, viruses straddle the definition of life. Outside of a cell they are inert, but inside a cell they direct activities that have a profound effect on the cell. Most scientists agree, however, that viruses do not fit the definition of life. So although they are infectious agents, they are not organisms.

Viruses can be broadly grouped into two general types based on the category of cells they infect. Some infect prokaryotic cells, and others infect eukaryotic cells. Although both groups are viruses, those that infect bacteria are also referred to as **bacteriophages,** or simply **phages** (phage means "to eat").

The fact that viruses are obligate intracellular parasites makes them very difficult to study. Unlike most bacteria and eukaryotic cells, which can be grown in pure culture, viruses require live organisms as hosts. To add to this difficulty of studying viruses, they are too small to be seen with a light microscope and can be visualized only with an electron microscope.

The first section of this chapter describes the general characteristics of viruses. The primary focus of the remaining sections is on bacteriophages and animal viruses. The latter are obviously important because of the diseases they cause, but the reasons for learning about bacteriophages may not be so apparent. We study them because they are easy to cultivate in the laboratory, and they serve as an important model to understand the molecular biology and relationships of animal viruses with their hosts. What you learn about them will help you understand similar relationships between the medically important viruses and the cells they infect. Bacteriophages are also important because they serve as a vehicle for horizontal gene transfer in bacteria, which was described in chapter 8. In addition, they are important because they kill bacteria, thereby limiting bacterial populations in nature. This is significant ecologically, but also has medical applications. In 2006, the Food and Drug Administration approved a preparation of bacteriophages that infect the food-borne pathogen *Listeria monocytogenes* for use as an antimicrobial agent on ready-to-eat meat and poultry products. *L. monocytogenes* multiplies at refrigeration temperatures and causes a potentially lethal disease, which is why it is a concern on products that are eaten without cooking. The use of bacteriophages as an alternative to antibiotics in treating bacterial infections has also been explored. ◄◄ **horizontal gene transfer, p. 200** ►►| *Listeria monocytogenes*, **p. 648**

13.1 ■ General Characteristics of Viruses

Learning Outcomes
1. *Describe the general features of viral architecture.*
2. *Describe how viruses are classified and named.*

Most viruses are notable for their small size (**figure 13.1**). They are approximately 100- to 1,000-fold smaller than the cells they infect. The smallest viruses are about 10 nm in diameter, and contain very little nucleic acid, perhaps as few as 10 genes. The largest known viruses are about 800 nm, the size of the smallest bacterial cells. Indeed, one very large virus is so big it was first identified as a bacterium (see **Perspective 13.1**).

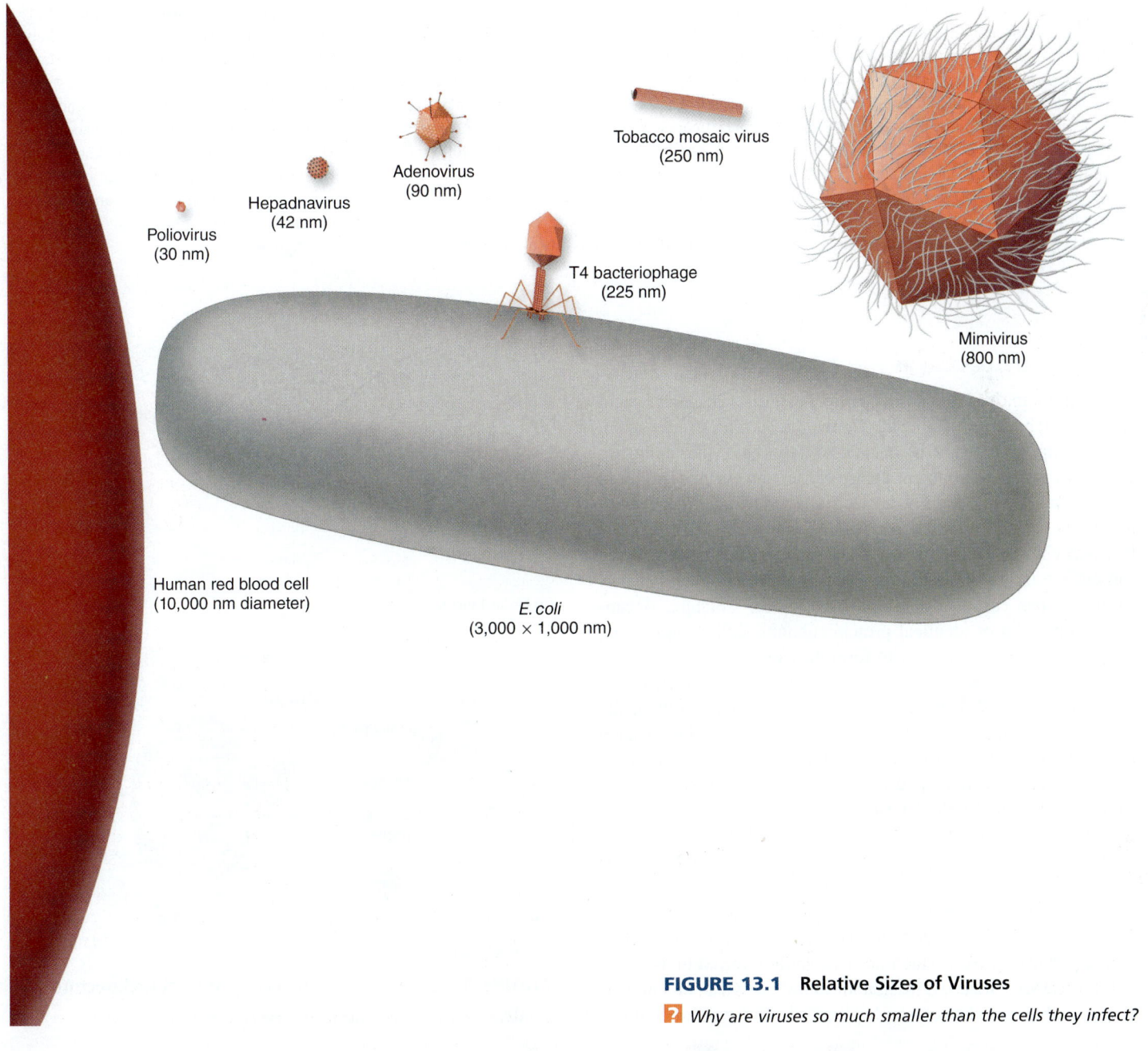

FIGURE 13.1   Relative Sizes of Viruses
❓ *Why are viruses so much smaller than the cells they infect?*

PERSPECTIVE   13.1

Microbe Mimicker

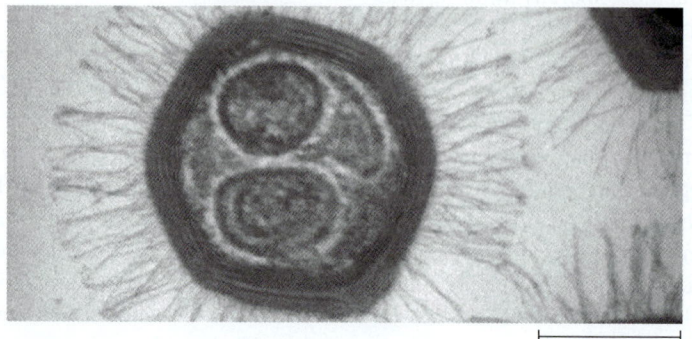

Like bacteria, viruses vary tremendously in size and complexity. This is becoming increasingly apparent as viruses from a variety of animal and bacterial hosts are studied in greater detail. Perhaps the most unusual virus recently characterized came from *Acanthamoeba*, a free-living ameba that may cause certain forms of pneumonia. The virus was first thought to be a Gram-positive bacterium because of its staining characteristics and hairy appearance, with many fibrils sticking out from its capsid (see figure 13.1). Its name, mimivirus, reflects this

FIGURE 1   Mamavirus (Large Mimivirus) with Sputnik Virophages

100 nm

original misconception (**mi**micked a **mi**crobe). Mimivirus is the largest DNA virus ever characterized—the diameter of its capsid is approximately 800 nm. This is twice the size of a small bacterium such as *Mycoplasma*. Its genome size is enormous for a virus—1.2 million base pairs, which is large enough to encode almost 1,000 proteins. These genes encode proteins never before seen in viruses and include enzymes of nucleic acid synthesis, DNA repair, translation, and polysaccharide biosynthesis. However, like all viruses, the mimi virion does not undergo cell division nor does it contain ribosomes. This virus has been placed in a taxonomic group of other large DNA viruses, members of which infect distinctly different organisms, including vertebrates and algae.

Recently, a mimivirus-infected ameba was shown to be infected with an additional virus, a parasite of mimivirus. Its discoverers named the new virus Sputnik and called it a virophage, and the infected virus a mamavirus (**figure 1**). Sputnik can replicate in the ameba only when mimivirus has infected the same cell. When this happens, fewer mimivirus particles are produced and they display unusual morphologies. Thus, Sputnik behaves as a true parasite, the first of its kind but probably not the last. Sputnik appears to contain genes from other viruses, raising an intriguing question—is horizontal gene transfer between viruses a possibility? ◀◀ **horizontal gene transfer, p. 200**

The mimivirus and virophages are unlikely to be the last unusual viruses discovered. Many potential host organisms have not yet been cultured, and thus many likely virus families remain undiscovered. Their discovery and characterization will not only expand our knowledge of viruses but also will challenge our definition of what is living and non-living—and perhaps provide an answer to the question, "Where did viruses originate?"

Viral Architecture

At a minimum, a **virion** (viral particle) consists of nucleic acid surrounded by a protein coat (**figure 13.2**). The protein coat, called a **capsid,** protects the nucleic acid from enzymes and toxic chemicals in the environment. It also carries any enzymes required by the virus for infection of host cells. The capsid together with the nucleic acid it encloses is called the **nucleocapsid.** Given the fact that viral genomes are relatively small in size, and therefore can encode only a limited number of different proteins, it is not surprising that capsids are simple in chemical structure. A capsid is composed of identical protein subunits, called capsomers, arranged in a precise manner to form the capsid.

Some viruses have an envelope, a lipid bilayer outside of the capsid (figure 13.2b). How these **enveloped viruses** obtain the envelope from the host cell will be described later. Sandwiched between the nucleocapsid and envelope is the matrix protein, which is unique to enveloped viruses. Viruses that do not have an envelope are called **naked viruses.** Nearly all phages are naked. In general, enveloped viruses are more susceptible to disinfectants because these chemicals damage the envelope, making the viruses non-infectious.

Viruses contain only a single type of nucleic acid—either RNA or DNA—but never both. This provides a useful method for classifying viruses, which are frequently referred to as either RNA or DNA viruses. The genome may be linear or circular, either double-stranded or single-stranded. The type of genome

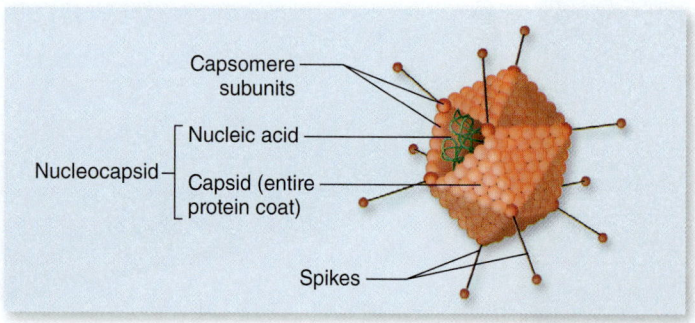

(a) Naked virus

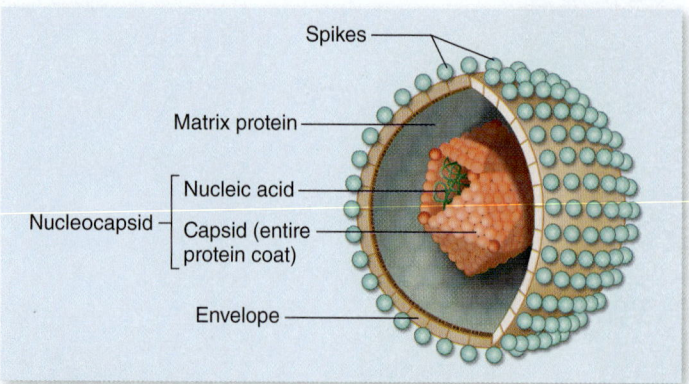

(b) Enveloped virus

FIGURE 13.2   Two Different Types of Viral Architecture

❓ *What components must all viruses have?*

has important implications with respect to the virus's replication strategy, and will be discussed in more detail later.

Viruses have specific protein components that allow the virion to attach (adsorb) to specific receptor sites on host cells. Phages, for example, have tail fibers that attach to host cells, and many animal viruses have protein structures called **spikes** that stick out from either the lipid bilayer of enveloped viruses or the capsid of naked viruses (see figures 13.1 and 13.2).

A virus generally is one of three different shapes—icosahedral, helical, or complex (**figure 13.3**). Icosahedral viruses appear spherical when viewed with the electron microscope, but their surface is actually 20 flat triangles arranged in a manner somewhat similar to a soccer ball. This arrangement is an efficient design for any container that uses identical subunits and requires the least energy to assemble. Helical viruses appear cylindrical when viewed with the electron microscope. Their capsomers are arranged in a helix, somewhat similar to a spiral staircase. Some helical viruses are short and rigid, whereas others are long and filamentous. Complex viruses have more intricate structures. Phages are the most common examples of this, many having an icosahedral nucleocapsid, referred to as the head, with a long helical protein component, the tail.

FIGURE 13.3 Common Shapes of Viruses

❓ *What determines the shape of the virus?*

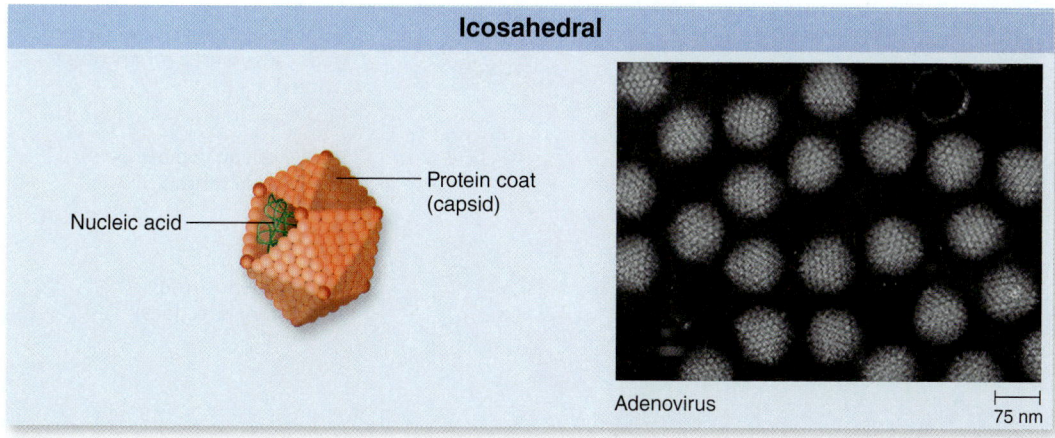

(a)

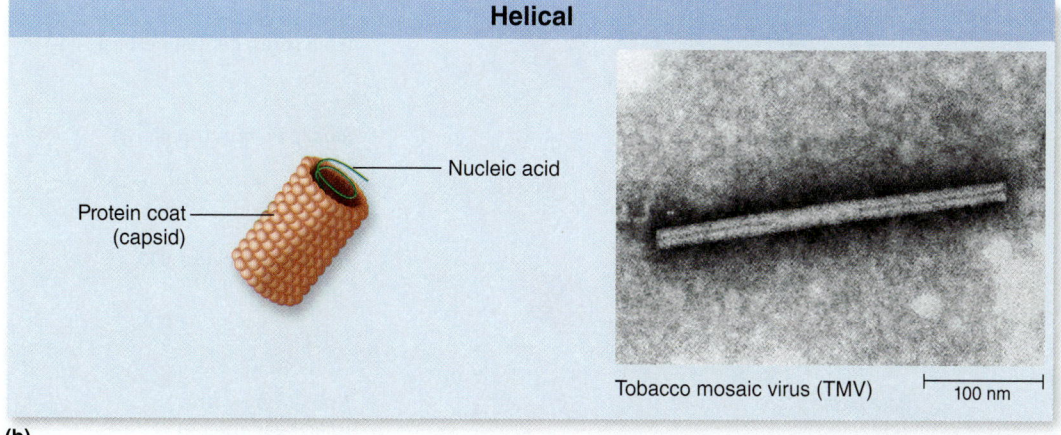

(b)

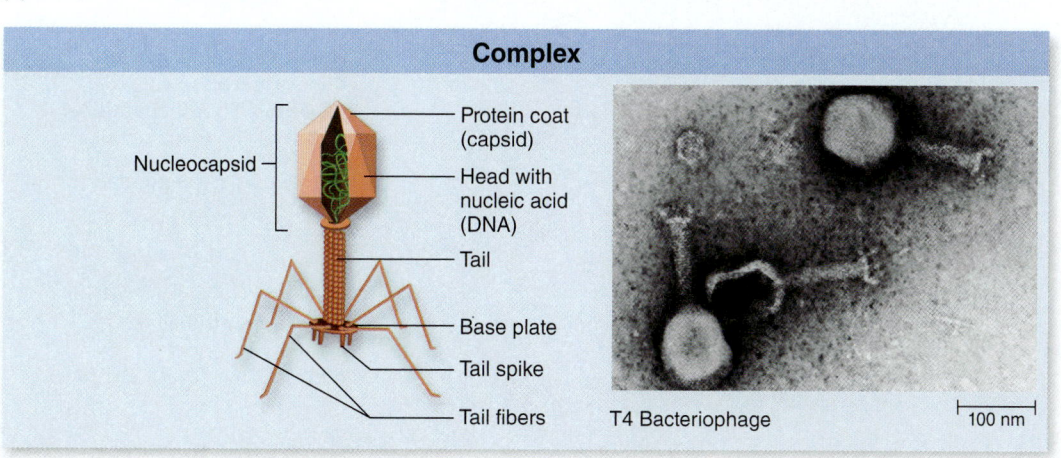

(c)

Viral Taxonomy

Although viruses are not living organisms, they are biological entities that are classified to provide easy identification and study (**table 13.1**). The International Committee on Viral Taxonomy (ICVT) keeps an online database and publishes a report describing the key features, classification, and nomenclature of recognized viruses. The ICVT 2009 report describes over 6,000 viruses belonging to 2,288 species, 348 genera, 87 families, and 6 orders. Key characteristics used in the current classification scheme are the genome structure (type of nucleic acid and strandedness) and the hosts they infect (bacteria, archaea, animals, plants). A variety

TABLE 13.1 | Classification of Some Human Viruses

Nucleic Acid	Outer Covering	Family	Drawing of Virion (not to scale)	Representative Member (and disease caused)
DNA Viruses				
Double-stranded DNA	Naked	*Papillomaviridae*		Human papillomaviruses (some types cause warts; others cause cancers)
		Polyomaviridae		Merkel cell polyomavirus (Merkel skin cancer)
		Adenoviridae		Human adenoviruses (respiratory infections)
	Enveloped	*Herpesviridae*		Herpes zoster virus (chickenpox); herpes simplex viruses (cold sores, genital herpes)
		Poxviridae		Smallpox virus (smallpox)
Single-stranded DNA	Naked	*Parvoviridae*		Human parvovirus B19 (fifth disease)
RNA Viruses				
Double-stranded	Naked	*Reoviridae*		Human rotaviruses (diarrheal disease)
Single-stranded (plus strand)	Naked	*Picornaviridae*		Polioviruses (poliomyelitis); rhinovirus (colds); hepatitis A virus (hepatitis A)
		Caliciviridae		Norovirus (gastroenteritis)
	Enveloped	*Togaviridae*		Rubella virus (rubella); Chikungunya virus (Chikungunya fever)

(continued)

TABLE 13.1	Classification of Some Human Viruses (*Continued*)

Nucleic Acid	Outer Covering	Family	Drawing of Virion (not to scale)	Representative Member (and disease caused)
RNA Viruses, *cont.*				
	Enveloped	*Flaviviridae*		Yellow fever virus (yellow fever); dengue virus (dengue hemorrhagic fever); hepatitis C virus (hepatitis C)
		Coronaviridae		Coronavirus (SARS)
Single-stranded (minus strand)	Enveloped	*Rhabdoviridae*		Rabies virus (rabies)
		Filoviridae		Ebola virus (hemorrhagic fever)
		Paramyxoviridae		Mumps virus (mumps); measles virus (measles)
		Orthomyxoviridae		Influenza virus (influenza)
		Bunyaviridae		Hantavirus (hantavirus pulmonary syndrome)
		Arenaviridae		Lassa virus (lassa fever)
Reverse Transcribing Viruses				
DNA	Enveloped	*Hepadnaviridae*		Hepatitis B virus (hepatitis B)
RNA	Enveloped	*Retroviridae*		Human immunodeficiency virus (AIDS)

TABLE 13.2	Grouping of Human Viruses Based on Route of Transmission	
Virus Group	**Mechanism of Transmission**	**Common Viruses Transmitted**
Enteric	Fecal-oral route	Enteroviruses (polio, coxsackie B); noroviruses; rotaviruses (diarrhea)
Respiratory	Respiratory or salivary route	Influenza; measles; rhinoviruses (colds)
Zoonotic	Vector (such as arthropods)	Sandfly fever; dengue; West Nile encephalitis
	Animal to human directly	Rabies; cowpox
Sexually transmitted	Sexual contact	Herpes simplex virus type 2 (genital herpes); HIV

of other characteristics including viral shape and disease symptoms are also considered.

Our discussion of classification will focus on the viruses that infect animals. The names of virus families are derived from a variety of sources, but they all end in the suffix *-viridae* and are italicized. Their names follow no consistent pattern. In some cases, the name indicates the appearance of the viruses in the family—for example, *Coronaviridae* coming from *corona*, which means "crown." In other cases, the virus family is named for the geographic area from which a member was first isolated. *Bunyaviridae* is derived from Bunyamwera, a locality in Uganda, Africa. Each family contains numerous genera whose names end in *-virus*, making it a single word—for example, *Enterovirus*. The species name is often the name of the disease the virus causes—for example, *poliovirus* causes poliomyelitis. In contrast to bacterial nomenclature, in which an organism is referred to by its genus and species name, viruses are commonly referred to only by their species name. Formal scientific names are italicized, making them easy to recognize.

Although viruses have formal names, virologists most often refer to them by informal names that are not capitalized. In addition, informal terms are often used to refer to groups of animal viruses that are not taxonomically related but share critical characteristics such as the primary route of transmission (**table 13.2**). Viruses transmitted via the fecal-oral route, for instance, are referred to as enteric viruses, and viruses transmitted through the respiratory route are called respiratory viruses. Zoonotic viruses cause zoonoses, which are diseases transmitted from an animal to a human. One group of viruses, the **arboviruses** (meaning **ar**thropod **bo**rne), is so named because they are spread by arthropods such as mosquitoes, ticks, and sandflies. The arthropods are biological vectors; when an infected arthropod takes a blood meal from an animal, it transmits the virus. In many cases, these viruses can infect widely different species. The same arthropod might bite birds, reptiles, and mammals and transfer viruses among those different groups. Arboviruses cause important diseases such as West Nile encephalitis, La Crosse encephalitis, yellow fever, and dengue fever. ▶▶ biological vector, p. 442 ▶▶ West Nile encephalitis, p. 655 ▶▶ La Crosse encephalitis, p. 655 ▶▶ yellow fever, p. 682 ▶▶ dengue fever, p. 683

MicroByte
An estimated 10^{31} bacteriophages exist on earth, the most numerous of all biological entities.

MicroAssessment 13.1

Viruses consist of nucleic acid surrounded by a protein coat and replicate only inside living cells. Viruses are typically icosahedral, helical, or complex. Most phages are naked, whereas animal viruses are either naked or enveloped. Viruses are classified based primarily on the characteristics of their genome, such as type of nucleic acid and strandedness. Viruses are often grouped by their route of transmission.

1. *How are enveloped viruses different from naked viruses?*
2. *List three ways in which viruses can be transmitted from one organism to another.*
3. *Most enteric viruses are naked. Why would this be so?*

13.2 ■ Bacteriophages

Learning Outcomes

3. *Compare and contrast lytic, temperate, and filamentous phage infections.*
4. *Describe two consequences of lysogeny.*

An enormous variety of different phages exist, each with a characteristic shape, size, genome structure, and replication strategy. Although each is interesting, three general types stand out in their different relationships with their hosts (**figure 13.4**).

Lytic Phage Infections: T4 Phage as a Model

By definition, **lytic** or **virulent phages** exit the host at the end of the infection cycle by lysing the cell. These viral infections result in the formation of new virus particles and are called **productive infections.** An intensively studied lytic phage is T4, a double-stranded DNA phage. Its infection cycle illustrates how a phage takes over a cell, directing that cell's activities solely to the synthesis of new phage particles. The infection cycle, which is similar to that of other lytic phages, can be viewed as a five-step process (**figure 13.5**).

Attachment

In liquid, phage particles collide with their host cells by chance. Upon contact, a phage particle attaches by means of a protein on its tail to a receptor on the cell surface or to an appendage such as a pilus. In the case of T4, the receptor is on the bacterial cell wall.

Focus Figure

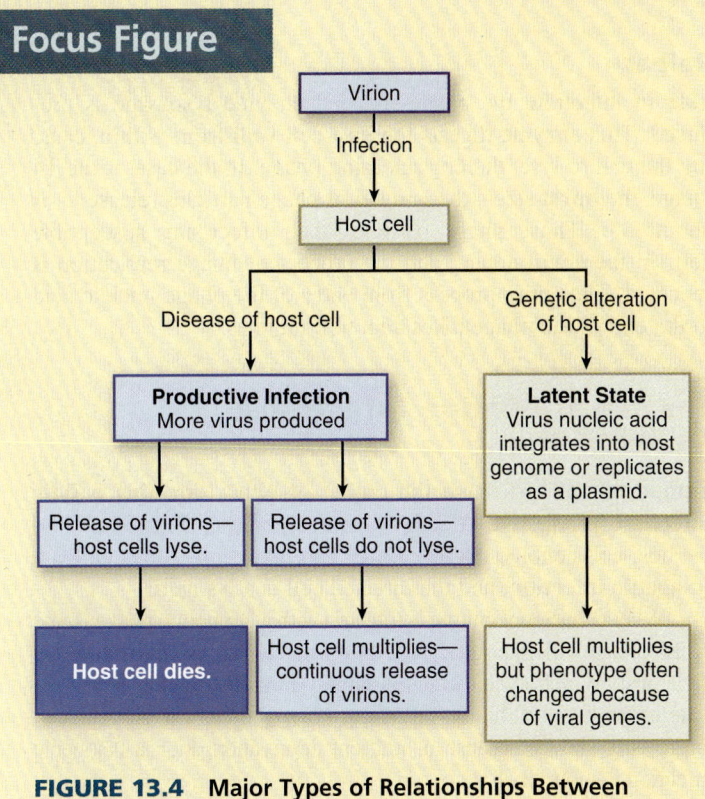

FIGURE 13.4   Major Types of Relationships Between Viruses and the Cells They Infect

❓ *Which type of interaction is least harmful to the host?*

The receptors used by phages normally perform important functions for the cell—the phages merely exploit the molecules for their own use. Cells that lack the receptor used by a given phage are resistant to infection by that specific phage.

Genome Entry

Following attachment, a bacteriophage injects its genome into the cell. T4 does this by degrading a small portion of the bacterial cell wall, using an enzyme located in the tip of its tail. This enzyme, T4 lysozyme, differs from the lysozyme found in tears and eggs, but it serves the same function—to degrade peptidoglycan. The tail contracts so that the phage particle appears to "squat" on the surface of the cell. This action injects the phage DNA through the host's cell wall and membrane, and into the interior of the cell. The capsid remains on the outside of the cell. The separation of the nucleic acid from its protein coat prior to replication is a feature of all viruses. ◀◀ **lysozyme, p. 62**

MicroByte
The nucleic acid is so tightly packed inside the capsid that the internal pressure is 10 times higher than that in a champagne bottle!

Synthesis of Phage Proteins and Genome

Within minutes of entry into a host cell, some T4 genes are transcribed and translated. Thus, the cell begins synthesizing T4 proteins using host ribosomes. Not all phage-encoded proteins are synthesized simultaneously, however. Rather, they are made in a sequential manner during the course of infection. The first viral proteins produced are referred to as early proteins, and are important for the initial steps of phage multiplication. These

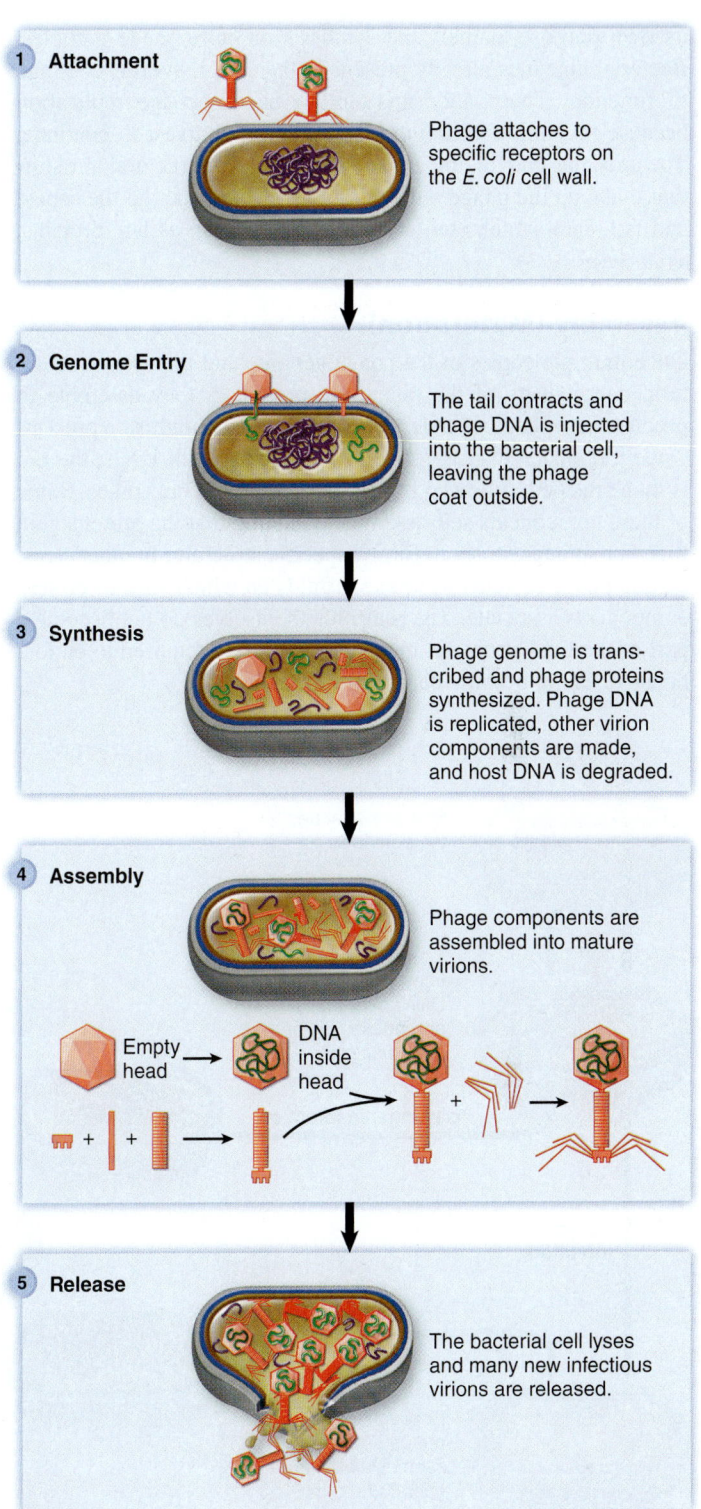

FIGURE 13.5   Steps in the Replication of the Lytic Phage T4 in *E. coli*

❓ *If one phage particle infects one bacterial cell, how many particles are inside the cell 5 minutes after infection? Explain.*

include (1) a nuclease that degrades the host cell's DNA, and (2) proteins that modify a subunit of the host cell's RNA polymerase so that it no longer recognizes bacterial promoters. As a result, soon after infection, no host genes are expressed. In this way, the phage takes over the metabolism of the bacterial cell for its own purpose, namely the synthesis of more phage particles. Bacterial enzymes already present in the cell, however, continue to function. These are important to bacteriophage replication because they allow biosynthesis and energy harvest to continue. Towards the end of the infection cycle, the structural proteins that make up the phage—including those that make up the capsid and tail—are synthesized. These are referred to as late proteins.

◀◀ promoters, p. 168

Assembly (Maturation)

Once multiple copies of the phage genome and the various structural components of the phage are produced, they assemble to produce new phage particles. This is a complex, multistep process. Once the phage head is formed, it is packed with DNA; the tail is then attached, followed by the addition of the tail spikes. Some of these components self-assemble, meaning that the proteins join together spontaneously to form a specific structure. In other steps, certain phage proteins serve as scaffolds on which various protein components associate. The scaffolds themselves do not become a part of the final structure, much as scaffolding required to build a house does not become part of the structure.

Release

Late in infection, the phage-encoded enzyme lysozyme is produced. This enzyme digests the host cell wall from within, causing the cell to lyse, thereby releasing phage. In the case of the T4 phage, the **burst size**—the number of phage particles released—is about 200. These phage particles then infect any susceptible cells in the environment, and the process of phage replication is repeated. The entire process from entry of the phage nucleic acid to the exit of the phage takes about 30 minutes.

Temperate Phage Infections: Lambda Phage as a Model

Temperate phages have the option of either directing a **lytic infection** (productive infection) or incorporating their DNA into the host cell genome (**figure 13.6**). The latter situation is called a **lysogenic infection,** and the infected cell is a **lysogen.**

When a bacterial culture is infected with a temperate phage, some of the phage will enter the lytic cycle, whereas others will lysogenize their host. Which cycle occurs is largely random, but the metabolic state of the host cell has an influence. For example, if a bacterial cell is growing slowly because of nutrient deprivation, then a lysogenic infection is more likely to occur.

The most thoroughly studied temperate phage is lambda (λ). This phage has a linear chromosome, but the two ends have complementary single-stranded overhangs that join together inside of the host cell to form a circular molecule. That molecule can either direct a lytic infection or integrate into the E. coli chromosome; lysogeny occurs when the molecule integrates. The integration process uses a phage-encoded enzyme called an integrase that inserts the phage DNA into the host cell chromosome at specific sites, a process called site-specific recombination. The integrated phage DNA, a **prophage,** replicates along with the host chromosome prior to cell division. Although the prophage can remain integrated indefinitely, it can also be excised from the host chromosome by a phage-encoded enzyme. When this happens a lytic infection begins.

Whether the prophage persists or the lytic cycle begins depends on a complex series of events involving phage-encoded regulatory proteins, the study of which contributed to our current understanding of bacterial gene regulation. One of the proteins, a **repressor,** prevents expression of the gene required for excision, and is therefore essential for maintaining the lysogenic state.

◀◀ bacterial gene regulation, p. 179 ◀◀ repressor, p. 180

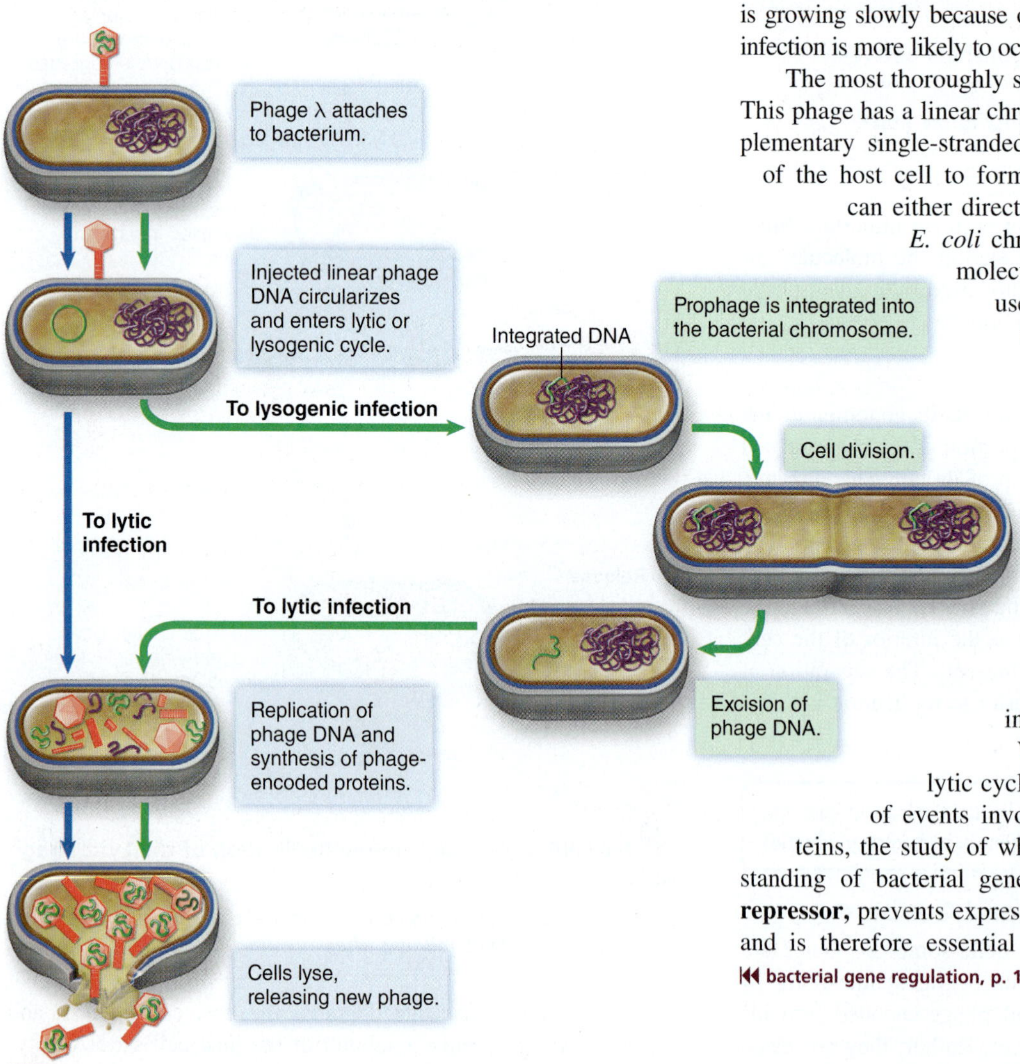

Phage λ attaches to bacterium.

Injected linear phage DNA circularizes and enters lytic or lysogenic cycle.

To lysogenic infection

Integrated DNA

Prophage is integrated into the bacterial chromosome.

To lytic infection

Cell division.

To lytic infection

Excision of phage DNA.

Replication of phage DNA and synthesis of phage-encoded proteins.

Cells lyse, releasing new phage.

FIGURE 13.6 Steps in the Replication of the Temperate Phage λ in *E. coli*

? *When is it advantageous for a temperate phage to excise its DNA from the host chromosome?*

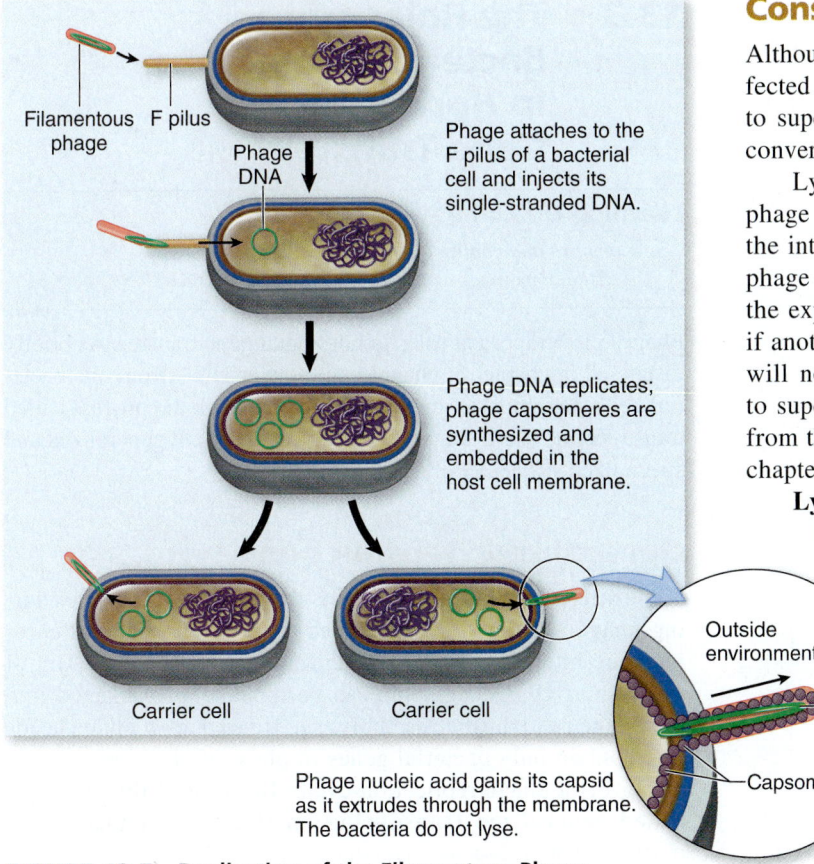

Filamentous phage   F pilus

Phage DNA

Phage attaches to the F pilus of a bacterial cell and injects its single-stranded DNA.

Phage DNA replicates; phage capsomeres are synthesized and embedded in the host cell membrane.

Carrier cell   Carrier cell

Phage nucleic acid gains its capsid as it extrudes through the membrane. The bacteria do not lyse.

Outside environment

Phage DNA

Capsomeres

FIGURE 13.7 Replication of the Filamentous Phage

❓ *Why might infected cells grow more slowly than their uninfected counterparts?*

Under ordinary growth conditions, the phage DNA is excised from the chromosome only about once in 10,000 divisions of the lysogen. If, however, a lysogenic culture is treated with a DNA-damaging agent such as ultraviolet light, the SOS repair system comes into play. This system activates a protease that destroys the repressor protein responsible for maintaining the integration of the prophage. As a consequence, the prophage is excised from the chromosome—allowing the phage to enter the lytic cycle—and a productive infection results. This process, **phage induction,** lets the phage escape from a damaged host, much like rats fleeing a sinking ship. ◀◀ ultraviolet light, p. 115 ◀◀ SOS repair, p. 196 ◀◀ protease, p. 150

Consequences of Lysogeny

Although a lysogen is morphologically identical to an uninfected cell, lysogeny has consequences. These include immunity to superinfection (infection by the same phage) and lysogenic conversion.

Lysogens are protected against infection by the same phage because the repressor that maintains the prophage in the integrated state will also bind to the operator on incoming phage DNA. The operator is a regulatory region that controls the expression of the genes that direct a lytic infection. Thus, if another λ phage injects its DNA into a λ lysogen, that DNA will not be expressed. This phenomenon is called immunity to superinfection. Note that this type of immunity is different from that of the human immune system, which is discussed in chapters 14 and 15. ◀◀ operator, p. 180

Lysogenic conversion is a change in the phenotype of a lysogen as a consequence of the specific prophage it carries. For example, only strains of *Corynebacterium diphtheriae* that are lysogenic for a certain phage (β phage) synthesize the toxin that causes diphtheria. Similarly, lysogenic strains of *Streptococcus pyogenes* and *Clostridium botulinum* produce toxins responsible for scarlet fever and botulism, respectively. If a toxin is encoded exclusively by phage genes, then only bacterial strains that carry the prophage will synthesize the toxin. Some examples of lysogenic conversion are given in **table 13.3.** ▶▶ diphtheria, p. 490

Filamentous Phages: M13 Phage as a Model

Filamentous phages are single-stranded DNA phages that look like long fibers. They cause productive infections, but the process does not kill the host cells. Infected cells, however, grow more slowly than their uninfected counterparts.

M13 is a filamentous phage that initiates infection by attaching to a protein on the F pilus of *E. coli* (**figure 13.7**). The single-stranded DNA phage genome then enters the cytoplasm of the bacterial cell, where a host cell DNA polymerase synthesizes the complementary strand. This double-stranded DNA is referred to as the replicative form (RF). One strand of the RF is then used as a template to make mRNA as well as multiple copies of the phage's single-stranded genome (**figure 13.8**).

TABLE 13.3	Some Properties Encoded by Prophage (Lysogenic Conversion)	
Microorganism	**Medical Importance**	**Property Encoded by Phage**
Corynebacterium diphtheriae	Causes diphtheria	Diphtheria toxin
Clostridium botulinum	Causes botulism	Botulinum toxin
Escherichia coli O157:H7	Causes hemolytic uremic syndrome	Shiga toxin
Streptococcus pyogenes	Causes scarlet fever	Streptococcal pyrogenic exotoxins (SPEs)
Salmonella enterica	Causes food poisoning	Modification of lipopolysaccharide of outer membrane
Vibrio cholerae	Causes cholera	Cholera toxin

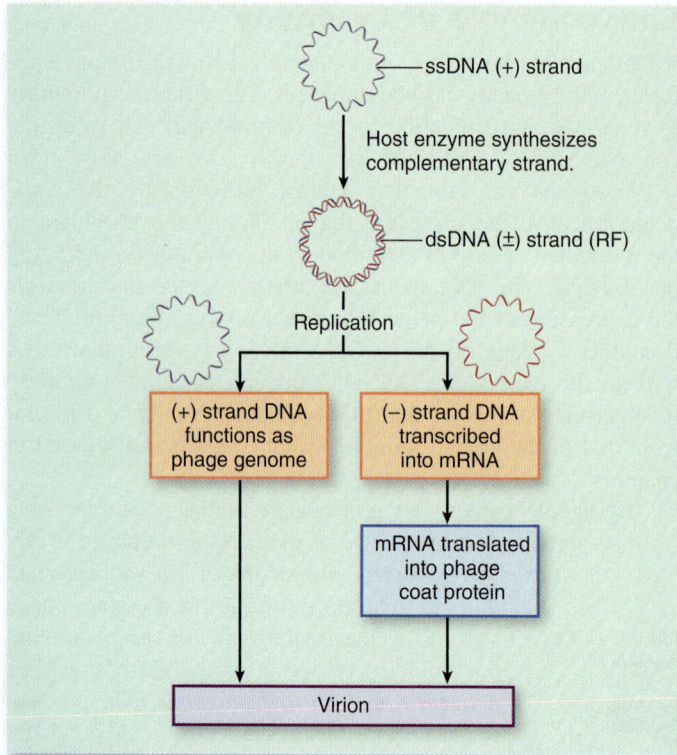

FIGURE 13.8 **Macromolecule Synthesis in Filamentous Phage Replication**

❓ *Which host enzyme synthesizes the complement to the ssDNA to make the RF?*

M13 particles are assembled as the phage DNA is extruded from the cell. In this process, phage coat protein molecules are inserted into the host cell's cytoplasmic membrane. At the same time, other phage-encoded proteins form pores that span the cytoplasmic and outer membrane. Then, as phage DNA is secreted through the pores, the coat protein molecules coat the single-stranded DNA to form the nucleocapsids.

M13 phage is useful in certain recombinant DNA procedures because its replication cycle produces both single- and double-stranded DNA. If a researcher needs a preparation of a specific sequence of single-stranded DNA without its complement, the DNA of interest can be cloned into an RF molecule of M13. When the recombinant DNA is transformed into an *E. coli* cell, it will direct a productive phage infection, generating mature phage particles containing only single-stranded recombinant DNA.

MicroAssessment 13.2

Lytic phages lyse their host cells, whereas temperate phages either lyse their host or integrate their DNA into the host cell's genome. Integrated viral DNA often codes for gene products that confer new properties on the host. Filamentous single-stranded DNA phages are extruded from the host cell without killing the cell.

4. *Describe two ways that phages can replicate without killing their host.*

5. *Give an example of a virulent phage and of a temperate phage.*

6. *Why is lysogenic conversion medically important?* ✚

13.3 ■ The Roles of Bacteriophages in Horizontal Gene Transfer

Learning Outcome

5. *Compare and contrast generalized and specialized transduction.*

Phages play important roles in horizontal gene transfer. As briefly discussed in chapter 8, phages can transfer DNA from one bacterial cell (the donor) to another (the recipient) in the process called transduction. There are two types of transduction: generalized and specialized. ◀◀ horizontal gene transfer, p. 200

Generalized Transduction

Generalized transduction results from a packaging error during phage assembly. Some phages degrade the bacterial chromosome into many fragments during lytic infection. Any of these short DNA fragments can be mistakenly packaged into the phage head during assembly (see figure 8.20). Phage heads that contain only bacterial genes in place of phage genes cannot direct a phage replication cycle. Because of this, they are called **generalized transducing particles** rather than phage particles. ◀◀ generalized transduction, p. 205

Following its release from the phage-infected host, the generalized transducing particle binds to another bacterial cell and injects its DNA. That DNA may then integrate into the recipient cell by homologous recombination, replacing DNA of the host cell. Any gene of the donor cell can be transferred this way, which is why this mechanism is called generalized transduction. ◀◀ homologous recombination, p. 201

Specialized Transduction

Specialized transduction results from an excision mistake made by a temperate phage during its transition from a lysogenic to a lytic cycle. Following induction, the phage DNA is usually excised precisely from the bacterial chromosome. On rare occasions, however, a short piece of bacterial DNA flanking the phage DNA is taken, and a piece of phage DNA remains in the bacterial chromosome (**figure 13.9**).

The excised DNA—containing both bacterial and phage genes—replicates and then becomes incorporated into phage heads during assembly. These phage particles do not carry the entire set of phage genes, so they are defective. The defective phage particles are then released as the host cells lyse. When a defective phage injects its DNA into another bacterial cell, both phage and bacterial DNA enter. The bacterial genes may then integrate into the recipient's genome via homologous recombination. In contrast to generalized transduction, only bacterial genes adjacent to the integrated phage DNA can be transferred, which is why the process is called specialized transduction.

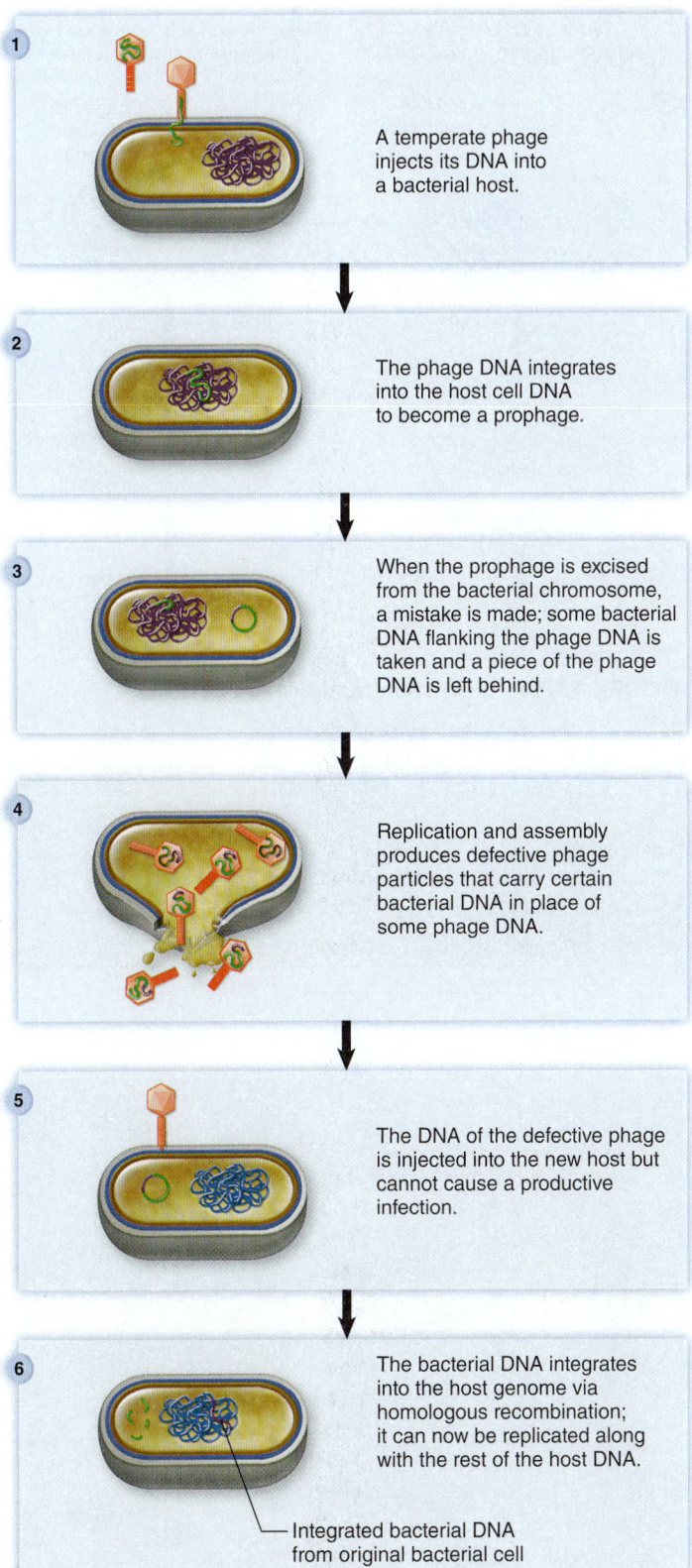

1 A temperate phage injects its DNA into a bacterial host.

2 The phage DNA integrates into the host cell DNA to become a prophage.

3 When the prophage is excised from the bacterial chromosome, a mistake is made; some bacterial DNA flanking the phage DNA is taken and a piece of the phage DNA is left behind.

4 Replication and assembly produces defective phage particles that carry certain bacterial DNA in place of some phage DNA.

5 The DNA of the defective phage is injected into the new host but cannot cause a productive infection.

6 The bacterial DNA integrates into the host genome via homologous recombination; it can now be replicated along with the rest of the host DNA.

Integrated bacterial DNA from original bacterial cell

FIGURE 13.9   Specialized Transduction

❓ *What mistake in the temperate phage replication cycle leads to specialized transduction?*

MicroAssessment 13.3

Transduction is a process by which bacterial DNA is transferred from one bacterium to another by a phage. Generalized transduction results from a DNA packaging error, whereas specialized transduction results from an error in excision of a prophage.

7. *What is meant by a defective phage?*

8. *Which is more likely to be a specialized transducing phage— a lytic or temperate phage?*

9. *Most temperate phages integrate into the host chromosome, whereas some replicate as plasmids. Which kind of relationship would you think would be more likely to maintain the phage in the host cell? Why?* ➕

13.4 ■ Bacterial Defenses Against Phages

Learning Outcome

6. *Describe three mechanisms that confer bacterial resistance to phage infection.*

Bacteria have developed several defense mechanisms to resist invasion by viruses. Without these, bacteria would not be able to survive the ongoing war against phages. Defense mechanisms include altering receptor sites, restriction modification systems, and the more recently discovered CRISPR system.

Preventing Phage Attachment

Recall that the first step in phage infection is attachment to specific receptors on the cell surface. If a bacterium alters or covers a given receptor, that cell becomes resistant to any phage that requires the receptor. In some cases, the alteration has other benefits to the cell as well. For example, *Staphylococcus aureus* produces a protein (called protein A) that covers phage receptors on the cell surface. That same protein also allows *S. aureus* to protect itself against certain human host defenses. Surface polymers bacteria produce, such as some used to form biofilms, may also cover phage receptors.

Restriction-Modification Systems

Restriction-modification systems protect bacteria from phage infection by quickly degrading incoming foreign DNA. They do this through the combined action of two types of enzymes— restriction enzymes and modification enzymes. Different bacteria have different versions of these enzymes, so there are hundreds of varieties, each recognizing different sequences in DNA. The restriction enzyme recognizes specific short nucleotide sequences and then cuts the DNA molecule at these sequences (see figure 9.1). The modification enzyme protects the host cell DNA from the action of the restriction enzyme. It does this by adding methyl groups to the nucleobases recognized by the restriction enzyme. Methylated sequences are invisible to the restriction enzyme. Thus, a restriction enzyme will degrade incoming phage DNA but not the host DNA. Occasionally, the modification enzyme will methylate the incoming phage DNA before the restriction enzyme has acted. When this happens, the phage DNA will not be degraded, allowing it to replicate and lyse the host cell. However, the DNA of the phage progeny will likely be degraded by a restriction-modification system when it enters a different bacterial strain (**figure 13.10**). ◀◀ restriction enzyme, p. 216

Restriction enzymes not only explain why some bacteria can degrade foreign DNA, but they also played an important role in the biotechnology revolution. The enzymes gave scientists a tool to remove genes from one DNA molecule, and join them to another, the basis for recombinant DNA technology. ◀◀ biotechnology, p. 215

CRISPR System

The **CRISPR system** was more recently discovered, and details about how it works are still being investigated. Bacterial cells that survive some phage infections retain small segments of phage DNA, incorporating them into the bacterial genome. The segments of phage DNA—called spacer DNA—are inserted into a chromosomal region called CRISPR, named for the characteristic **c**lusters of **r**egularly **i**nterspersed **s**hort **p**alindromic **r**epeats (**figure 13.11**). The spacer DNA provides a historical record of

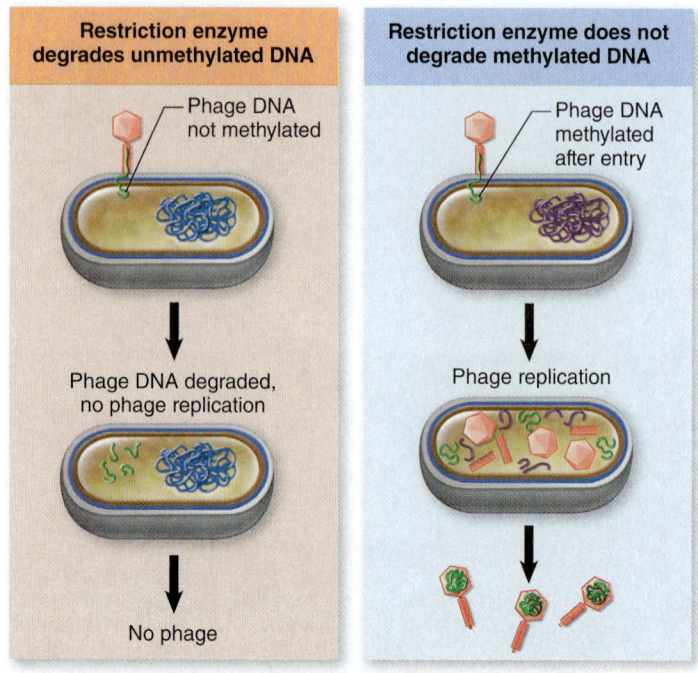

FIGURE 13.10 **Restriction-Modification System**

❓ *How is the DNA modified by this system?*

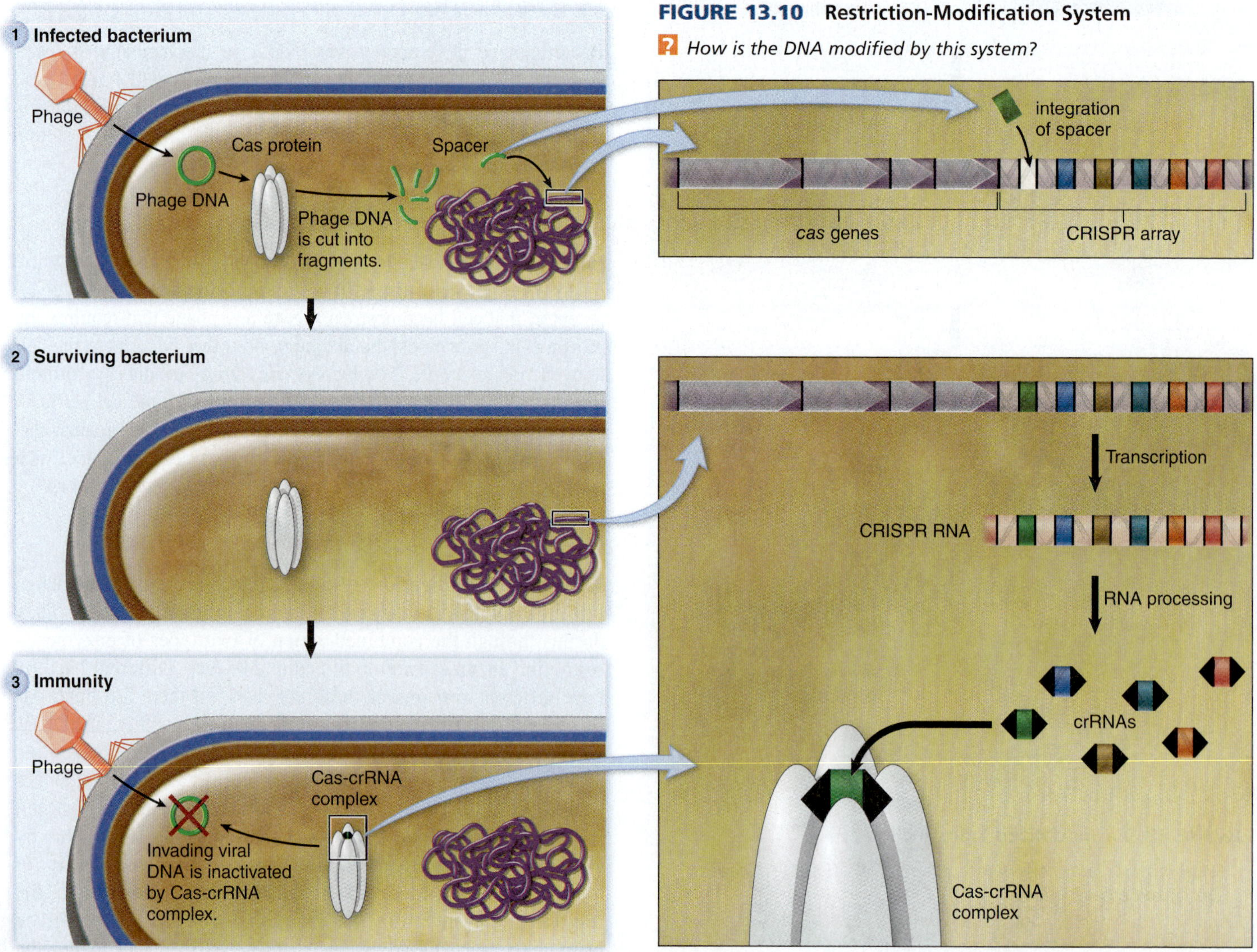

1 **Infected bacterium**

Phage
Phage DNA
Cas protein
Phage DNA is cut into fragments.
Spacer

integration of spacer
cas genes
CRISPR array

2 **Surviving bacterium**

Transcription
CRISPR RNA
RNA processing
crRNAs

3 **Immunity**

Phage
Cas-crRNA complex
Invading viral DNA is inactivated by Cas-crRNA complex.

Cas-crRNA complex

FIGURE 13.11 **CRISPR Defense System Against Phage Infection**

❓ *How does the CRISPR system target nucleic acid of an invading phage for destruction?*

past phage infections, allowing the bacterial cell as well as its progeny to recognize and then block subsequent infections by the same types of phages. How this works is not yet clear, but the CRISPR system may function by a type of RNA interference. The CRISPR array, including the spacer sequences, is transcribed and then cut into small RNAs called crRNAs. These bind to protein complexes called Cas (CRISPR-associated sequences). When the spacer RNA in a Cas base-pairs with nucleic acid of an invading phage, that nucleic acid is targeted for destruction. In addition to defending against phage, the CRISPR system can recognize and destroy other entering foreign DNA that a bacterial strain has encountered in the past. ◀◀ **RNA interference, p. 182**

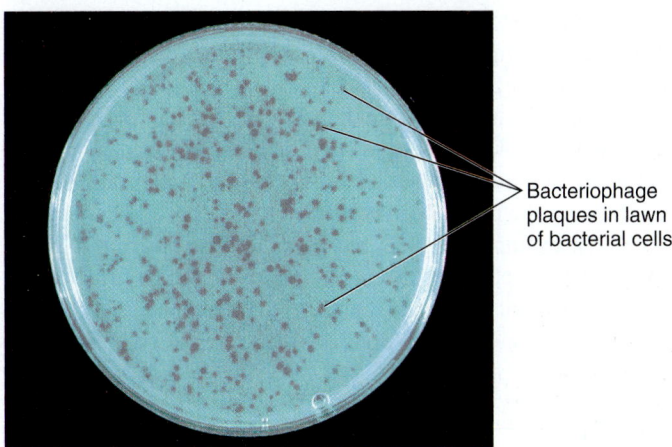

FIGURE 13.12 Phage Plaques

❓ *Why are plaques limited in size?*

MicroAssessment 13.4

Bacteria use several mechanisms to defend against viral infections. These include masking or altering phage receptor sites, and restriction-modification and CRISPR systems.

10. *How do modification enzymes protect host cell DNA from restriction enzymes?*

11. *How does a bacterial cell acquire a historical record of phage infections?*

12. *Which mechanisms that defend against phage infection would also limit conjugation and DNA-mediated transformation?* ➕

MicroAssessment 13.5

Bacteriophages require bacterial host cells. Plaque assays are used to quantitate phage particles.

13. *Explain what it is meant by plaque-forming units?*

14. *Is it important to have fewer phages than bacterial host cells when doing a quantitative plaque assay? Explain.* ➕

13.5 ■ Methods Used to Study Bacteriophages

Learning Outcome

7. *Describe how a plaque assay is used to detect and estimate the numbers of phage particles.*

Viruses can multiply only inside living actively metabolizing cells, so studying viruses in the laboratory requires cultivating appropriate host cells. It is considerably easier to grow bacterial cells than animal cells, which is one reason why the study of bacteriophages advanced much more rapidly than investigations on animal viruses.

Plaque assays are routinely used to quantitate phage particles in samples such as sewage, seawater, and soil. In this assay, a double layer of agar is used; the top layer, called soft agar, is inoculated with both a bacterial host and the phage-containing specimen. It is then poured over the surface of an agar-filled Petri dish. The bacteria present in the soft agar multiply rapidly, producing a turbid layer or dense lawn of bacterial growth. Meanwhile, any phages in the specimen adsorb to susceptible bacteria and lyse them, releasing progeny phage that diffuse and infect neighboring bacteria. **Plaques,** circular zones of clearing, form in the lawn due to cell lysis caused by the phage (**figure 13.12**). Each plaque represents a plaque-forming unit (PFU) initiated by a single phage particle infecting a cell.

The plaque assay is done using different dilutions of the phage suspension to ensure that the number of plaques on one plate can be accurately counted. The number is used to determine the **titer,** the concentration of infectious phage particles in the original phage suspension.

13.6 ■ Animal Virus Replication

Learning Outcomes

8. *Describe the steps of a generalized infection cycle of animal viruses.*

9. *Compare and contrast the replication strategies of DNA viruses, RNA viruses, and reverse-transcribing viruses.*

Understanding the infection cycle of animal viruses is particularly important from a medical standpoint because virus replication often depends on virally encoded enzymes, which offer potential targets of antiviral drugs. By interfering with the activities of the enzymes required for viral replication, antiviral medications can slow the progression of a viral infection, often giving host defense systems enough time to clear the virus before symptoms appear.

As we discuss the infection cycles of animal viruses, it is helpful to recall those described for bacteriophages, because they have features in common. A generalized infection cycle of animal viruses can be viewed as a five-step process.

Attachment

The process of attachment (adsorption) is basically the same in all virus-cell interactions. Animal virus surfaces are studded with attachment proteins or spikes (see figure 13.2). The receptors to which these proteins bind are usually glycoproteins located on the host cell plasma membrane, and often more than one receptor is required for effective attachment. HIV, for example, must bind to two key molecules on the cell surface before it can enter the cell.

As with the receptors used by phages, the normal function of these molecules is completely unrelated to their role in virus attachment. ◄◄ glycoproteins, p. 29 ▶▶ HIV replication cycle, p. 700

Because a virion must bind to specific receptors, a particular virus may be able to infect only a single or a limited number of cell types, and most viruses can infect only a single species. This accounts for the resistance that some animals have to certain diseases. For example, dogs do not contract measles from humans, nor do humans contract distemper from cats. Some viruses, such as the rabies virus, however, can infect unrelated animals such as horses and humans with serious consequences in both. ▶▶ measles, p. 537 ▶▶ rabies, p. 658

MicroByte

Tens to hundreds of thousands of receptors are present on each host cell, allowing ample opportunity for a virion to encounter a receptor on a cell.

Penetration and Uncoating

The mechanism an animal virus uses to enter its host cell depends in part on whether the virion is enveloped or naked. In all cases,

the entire virion is taken into the cell. This differs from most phages, where only nucleic acid enters and the capsid stays outside the bacterium.

Enveloped viruses enter the host cell by one of two mechanisms: fusion with the host membrane or endocytosis (**figure 13.13**). In the case of fusion, the lipid envelope of the virion fuses with the plasma membrane of the host after the virion attaches to the host cell receptor, much as drops of oil in an aqueous medium can fuse together. As a result of fusion, the nucleocapsid is released directly into the cytoplasm. Viruses that enter by endocytosis exploit the process of receptor-mediated endocytosis, a normal mechanism by which cells bring certain extracellular material into the cell (see figure 3.48). The viral particles bind to the receptors that normally trigger and facilitate the process, causing the cell to take them up. As an example, the interaction of the HIV envelope glycoproteins with host receptors triggers entry of HIV into the host cell by endocytosis. After a virion is taken into the cell, the viral envelope fuses with the membrane of the endosome, releasing the nucleocapsid into the cytoplasm. ◄◄ receptor-mediated endocytosis, p. 72

Naked viruses, which have no lipid envelope, cannot fuse to host membranes to enter cells. Therefore, the virions enter only

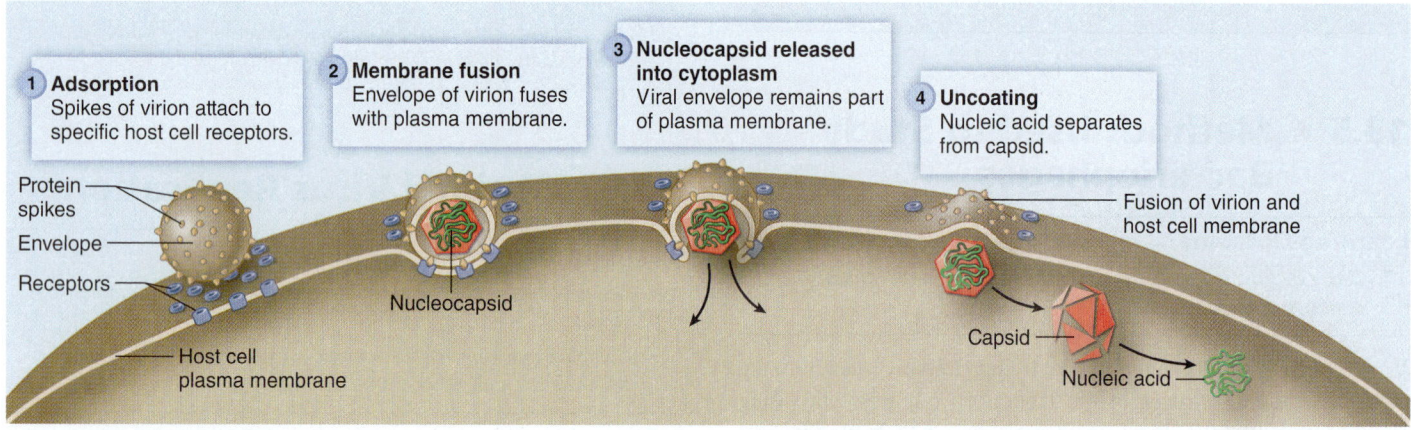

(a) Entry by membrane fusion

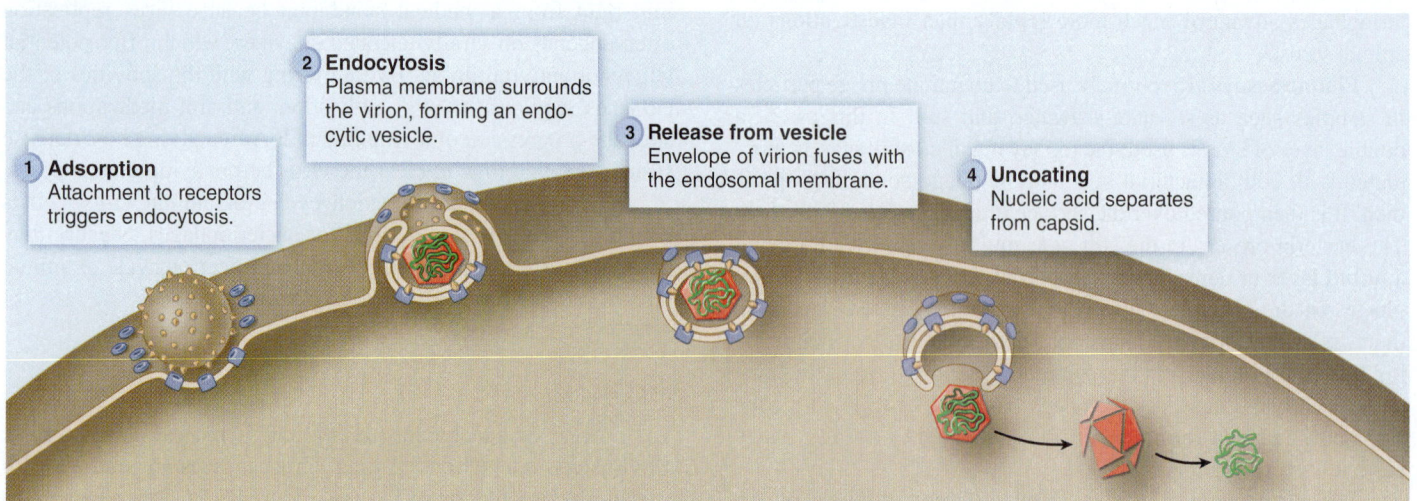

(b) Entry by endocytosis

FIGURE 13.13   Entry of Enveloped Animal Viruses into Host Cells (a) Entry following membrane fusion. (b) Entry by endocytosis.

❓ *Could phages enter bacteria by the process of membrane fusion? Why or why not.*

via endocytosis. Once in the endosome, the virus damages the vesicle membrane, resulting in the nucleocapsid being released into the cytoplasm.

Following penetration, the virion is transferred to its site of replication. Most DNA viruses multiply in the nucleus. They enter the nucleus through nuclear pores using viral proteins that have nuclear localization signals. In all viruses, the nucleic acid separates from its protein coat prior to the start of replication, the process of **uncoating.** The protective coats are disassembled by virus-specific mechanisms to release the viral genome and any accompanying proteins. This may occur simultaneously with entry of the virion into the cell, or after the final intracellular destination for viral replication has been reached.

Synthesis of Viral Proteins and Replication of the Genome

As in the replication of phage, production of viral particles in an infected cell requires two distinct but interrelated events: (1) expression of viral genes to produce structural and catalytic proteins, such as capsid proteins and any enzymes required for replication; and (2) synthesis of multiple copies of the viral genome. The fact that the relatively small viral genome is able to direct production of more viral particles given the constraints posed by eukaryotic cells' strict control of DNA replication and gene expression is a remarkable feat, and an area of intensive research.

The replication strategy of viruses can be divided into three general categories: those used by (1) DNA viruses, (2) RNA viruses, and (3) reverse transcribing viruses (**figure 13.14**). As you will see in the following discussion, the type of genome has a significant influence on the viral replication strategy, a primary reason why virus classification takes genome structure into account (see table 13.1).

Replication of DNA Viruses

DNA viruses usually replicate in the nucleus of the host cell and use the host cell machinery for DNA synthesis as well as gene expression. These viruses often encode their own DNA polymerase, however, thereby freeing them from dependency on certain host cell constraints, such as replicating only when the host is actively synthesizing its chromosome in preparation for division. Poxviruses are an exception to the rule that DNA viruses replicate in the nucleus. These large viruses replicate in the cytoplasm, and encode all of the enzymes and factors necessary for DNA and RNA synthesis.

Replication of double-stranded DNA viruses is fairly simple because they follow the central dogma of molecular biology (see figure 7.1). The viral DNA can be replicated and expressed using the general mechanisms discussed in chapter 7. ◀◀ central dogma of molecular biology, p. 162

Replication of single-stranded DNA viruses is quite similar to that of their double-stranded counterparts, except that the complement to the single-stranded molecule must first be synthesized. To understand this, recall that during DNA synthesis, a (+) strand is used as a template to make the (−) strand, and, likewise, the (−) strand serves as a template to make the

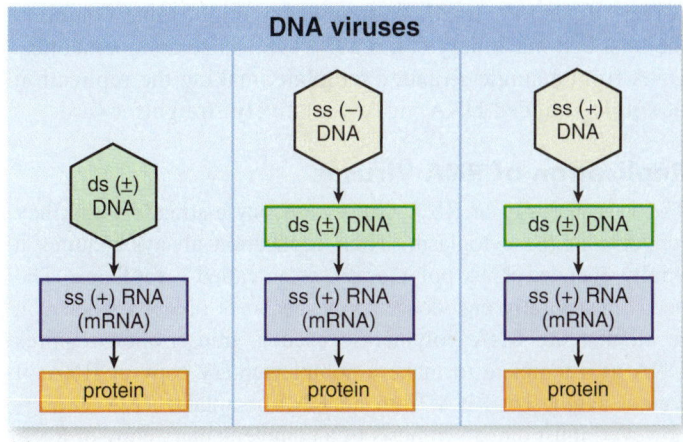

(a)

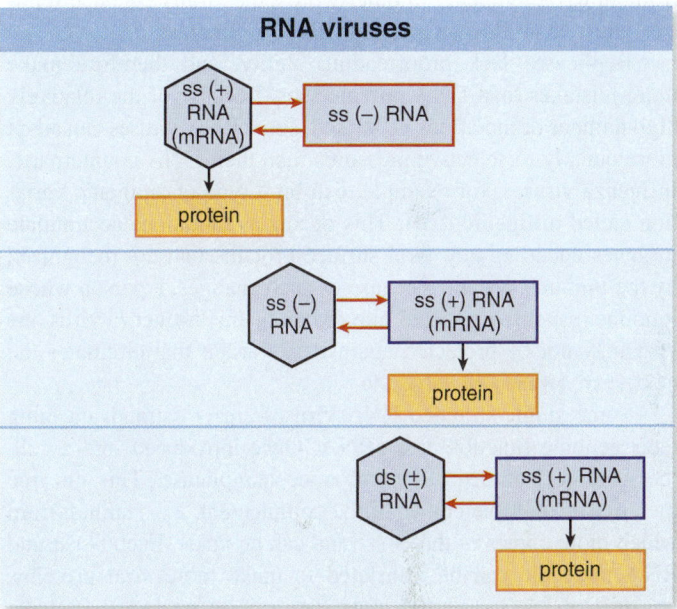

(b)

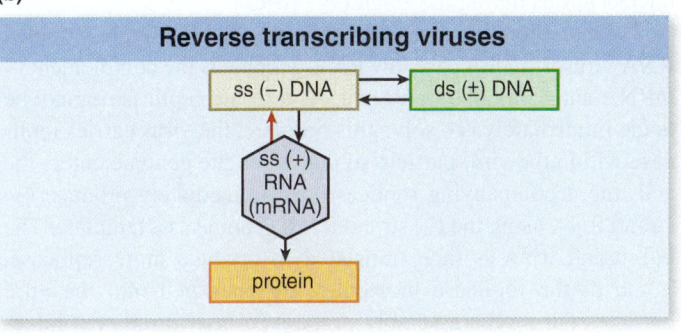

(c)

FIGURE 13.14 Viral Replication Strategies Viral genomes are shown in black hexagons. The steps indicated by black arrows use host enzymes, while those indicated by red arrows use virally encoded enzymes.

❓ *Why do RNA viruses need to encode an enzyme in order to replicate their genome, while DNA viruses do not?*

(+) strand. Therefore, if the genome of a single-stranded DNA virus is the (+) strand, the only way to produce identical copies is to first use the genome as a template to make a (−) strand. Then that (−) strand can serve as a template to make multiple copies of the (+) strand. Although this might seem complex, the host cell machinery can readily synthesize complementary DNA from a single-stranded template, making the replication of single-stranded DNA viruses relatively straightforward.

Replication of RNA Viruses

The vast majority of RNA viruses are single-stranded, and they replicate in the cytoplasm. Their replication always requires a virally encoded RNA polymerase, often called a **replicase.** The need for a virally encoded enzyme becomes obvious when you recall that the RNA polymerase used in transcription requires DNA as a template to make a complementary copy of RNA; it cannot synthesize RNA from an RNA template. RNA viruses, however, must direct synthesis of a complementary copy of RNA from an RNA template. In other words, the virally encoded replicase must be an RNA-dependent RNA polymerase.

Replicases lack proofreading ability, and therefore make more mistakes than DNA polymerases. Because of the relatively high number of mutations generated, some RNA viruses can adapt more quickly to selective pressures than their DNA counterparts. Influenza viruses, for example, exhibit a type of antigenic variation called **antigenic drift.** This occurs as mutations accumulate in genes encoding key viral surface proteins that are recognized by the immune system. Because of such changes, a person whose immune response protected him or her against influenza virus one year may not be protected against the variant that circulates the next year. ▶▶ antigenic drift p. 510

Some single-stranded RNA viruses are (+) strand, meaning their genome functions as mRNA. Once introduced into a cell, the genome is translated to produce a replicase. This enzyme then makes multiple copies of the complement, a (–) strand, from which more copies of the (+) strand can be made. Each (+) strand RNA molecule can be translated to make more viral proteins. In addition, the (+) strand molecules can be packaged to make nucleocapsids during the assembly process.

A more complicated situation arises for (–) single-stranded RNA viruses. In this case, the RNA genome is the complement of mRNA and cannot be translated. As a result, replicase cannot be made immediately. To solve this problem, the virus carries replicase within the viral particle so that when the genome enters the cell, the accompanying replicase can immediately produce (+) strand RNA using the (–) strand RNA genome as a template. The (+) strand RNA is then translated to produce more replicase. Some of the replicase molecules are packaged into the viral particles during the assembly process for use in the next infection cycle.

Double-stranded RNA viruses, which are relatively uncommon, must also carry their own replicase because the host cell translation machinery is unable to translate double-stranded RNA. The replicase immediately uses the (–) strand of the double-stranded molecule as a template to make (+) strand RNA. This molecule is then translated to make more replicase, and the infection cycle can continue.

Some RNA viruses have segmented genomes, with different segments containing different genes. All of the segments must be replicated for new virions to be produced. If two different strains of a segmented virus infect the same cell, reassortment can occur, meaning the viral particles that exit the cell can have various combinations of segments from the initial infecting strains. For example, if a virus has seven segments, numbered one through seven, then the progeny must also have segments numbered one through seven, but the segments can originate from either of the two parent strains. Influenza A virus is an example of a segmented virus that undergoes reassortment, resulting in a phenomenon called **antigenic shift.** Antigenic shift and drift will be discussed in detail in chapter 21. ▶▶ antigenic shift, p. 511

Replication of Reverse-Transcribing Viruses

Reverse-transcribing viruses encode the enzyme **reverse transcriptase,** which synthesizes DNA from an RNA template. This runs counter to the central dogma of molecular biology in that an RNA molecule is used as a template to make DNA. ◀◀ central dogma of molecular biology, p. 162

Retroviruses, which include the human immunodeficiency virus (HIV), have a (+) strand RNA genome, and carry reverse transcriptase within the virion. After entering the host cell, the reverse transcriptase uses the RNA genome as a template to make one strand of DNA. The complement to that strand is then synthesized to make double-stranded DNA, which integrates into the host cell chromosome. Once integrated, the genome can direct a productive infection, or remain in a latent state. This latent state is similar to what occurs with phage λ. Once the DNA copy is made, however, it cannot be eliminated from the cell. ▶▶ replication of HIV, p. 699

Hepatitis B virus (HBV) has a complicated infection cycle that involves a temporary RNA copy of its double-stranded DNA genome. When the HBV genome enters a cell, it is transcribed by the cell's machinery to produce RNA. That molecule codes for viral proteins, and also serves as a temporary genome copy called pregenomic RNA (pgRNA). During the assembly process, the pgRNA along with reverse transcriptase is packaged into the maturing viral particles. There, the reverse transcriptase functions to make DNA using the RNA as a template. DNA synthesis stops once the virion buds from the membrane, resulting in a DNA molecule that is not fully double-stranded. ▶▶ replication of hepatitis B virus, p. 600

Assembly

Viral assembly involves the same general principles described for phages. The protein capsid must be formed, and then the genome and any necessary enzymes packaged within it. In animal viruses, this assembly process takes place in the nucleus or in the cytoplasm, depending on the virus. Many viruses assemble near the plasma membrane close to their site of release. ◀◀ Golgi apparatus p. 77 ◀◀ microtubules, p. 73

Release

Just as the entry mechanism into a host cell depends on whether the virion is enveloped or naked, so does its release. Most enveloped viruses are released by **budding,** a process

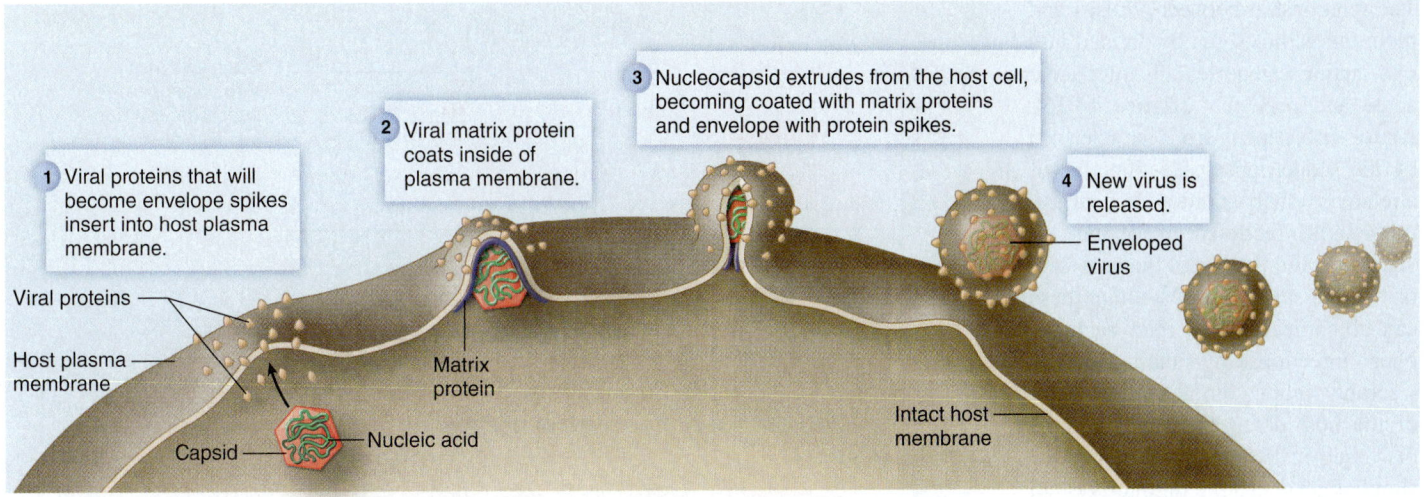

1 Viral proteins that will become envelope spikes insert into host plasma membrane.

2 Viral matrix protein coats inside of plasma membrane.

3 Nucleocapsid extrudes from the host cell, becoming coated with matrix proteins and envelope with protein spikes.

4 New virus is released.

Enveloped virus

Viral proteins

Host plasma membrane

Matrix protein

Capsid

Nucleic acid

Intact host membrane

(a)

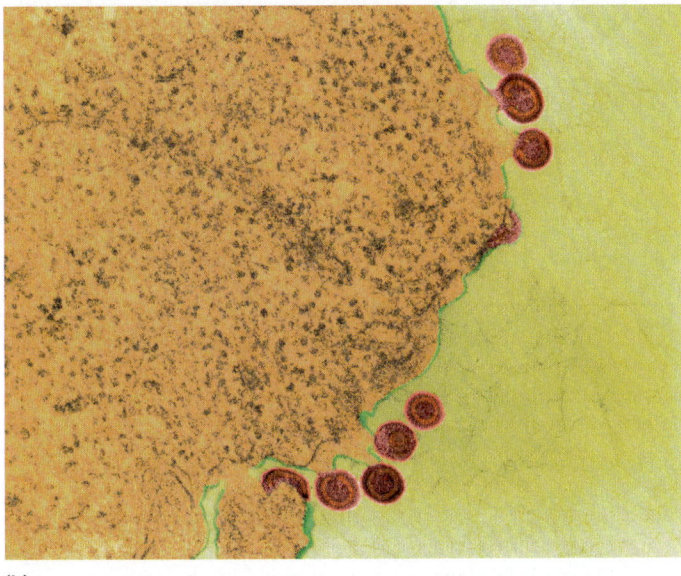

(b)

FIGURE 13.15   Mechanisms for Releasing Enveloped Virions (a) Process of budding. (b) Electron micrograph of virions budding from the surface of a human cell.

? *What component of the virion is gained in the process of budding?*

process called **apoptosis,** or programmed cell death, prior to the release of the virus particles. The immune response of the animal, which is directed toward eliminating the virus, can also lead to the same process. The virions released from the dead cells may then invade any healthy cells in the vicinity. ▶▶ apoptosis, p. 350

To be maintained in nature, infectious virions must leave one animal host and be transmitted to another. Viral particles may be shed in feces, urine, genital secretions, blood, or mucus and saliva released from the respiratory tract during coughing or sneezing. From there, the virus enters the next host to begin another round of infection.

whereby the virus acquires its envelope (**figure 13.15**). Before budding occurs, virally encoded protein spikes insert into specific regions of the host cell's membrane. Matrix protein accumulates on the inside surface of those same regions. Assembled nucleocapsids are then extruded from the cell at these regions, becoming covered with a layer of matrix protein and lipid envelope in the process. Not all enveloped viruses have plasma membrane–derived envelopes, however. Some obtain their envelope from the membrane of an organelle such as the Golgi apparatus or the rough endoplasmic reticulum. They do this by budding into the organelle. From there, they are transported in vesicles to the outside of the cell. Budding may not destroy the cell because the membranes can be repaired after the viral particles exit. ◀◀ rough endoplasmic reticulum, p. 76

Naked viruses are released when the host cell dies. How this occurs, however, is sometimes quite different from how bacteriophages kill their hosts. Many viruses trigger a normal cellular

MicroAssessment 13.6

Animal viruses and bacteriophages share similar features in their infection cycle. Because many animal viruses are enveloped and phages are not, differences exist in their entry and exit. The genomes of animal viruses are diverse, influencing viral replication strategies.

15. *Why are virally encoded enzymes medically important?*

16. *How do enveloped viruses exit a cell?*

17. *Why can viruses not replicate independently of living cells?* ➕

13.7 ■ Categories of Animal Virus Infections

Learning Outcome

10. *Compare and contrast acute infections and the two types of persistent infections caused by animal viruses.*

The relationship between viruses and their animal hosts can be divided into two major categories of infections: acute and persistent (**figure 13.16**). **Acute infections** are characterized by the sudden onset of symptoms of a relatively short duration. In contrast, **persistent infections** can continue for years, or even the life of the host with or without symptoms. Although these can be compared to certain bacteriophage infections, the situation is considerably more complicated because of the host defenses, which will be discussed in later chapters. For now, simply realize that the immune system is constantly working to detect and destroy virus-infected cells. Viruses, meanwhile, have mechanisms that evade some of these defenses.

Although we often categorize viral infections as either acute or persistent, they do not always fall neatly into a single category. When a person is first infected with HIV, for instance, the virus replicates to high levels, causing acute symptoms including fever, fatigue, swollen lymph nodes, and headache. The immune system soon eliminates most virions, and the symptoms subside. However, the DNA copy of the viral genome integrates into the host cell chromosome,

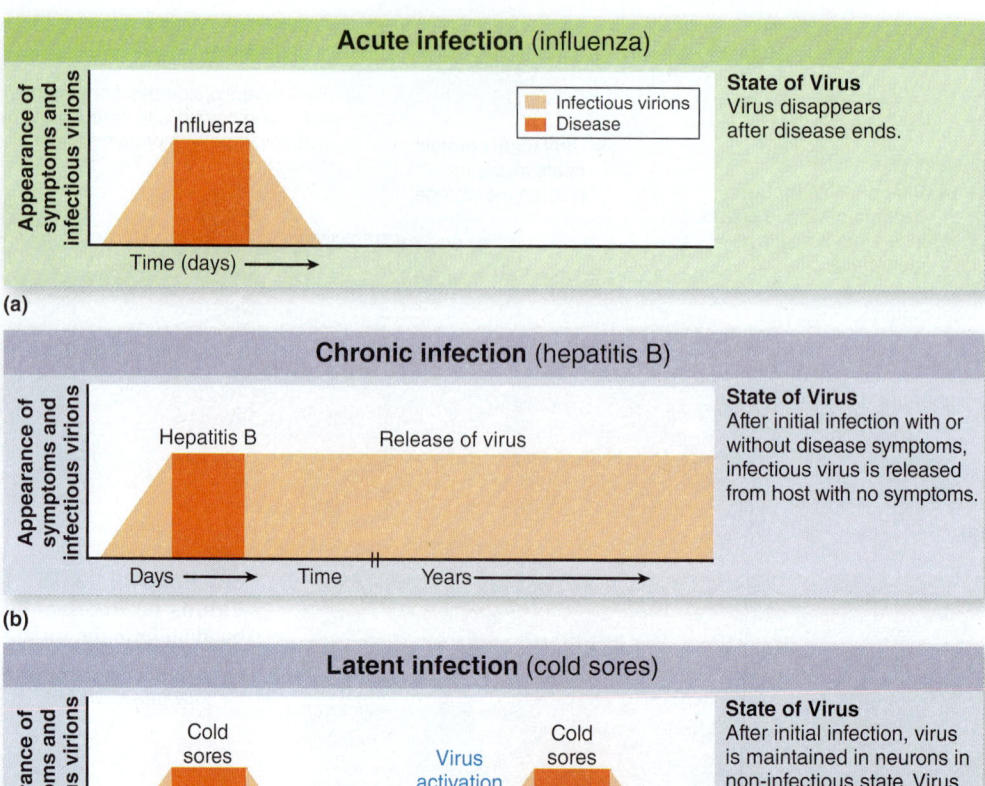

FIGURE 13.16 **Types of Infection by Animal Viruses** (a) Acute infection. (b) Persistent infection—chronic. (c) Persistent infection—latent.

❓ *Which type of infection is most likely to lead to immunity?*

resulting in a persistent infection with subsequent symptoms associated with acquired immunodeficiency syndrome (AIDS) developing after many years. Thus, HIV infection has features of acute and persistent infections.

Acute Infections

On a cellular level, acute infections can be compared to productive lytic infections by bacteriophages, resulting in a burst of virions being released from infected host cells. A critical difference, however, is that while the virus-infected cells often die, the host may survive. This is because the immune system of the animal host may gradually eliminate the virus over a period of days to months. Examples of diseases due to acute viral infections include influenza, mumps, and poliomyelitis. Symptoms of the diseases result from localized or widespread tissue damage following cell death as well as damage caused by the immune response itself.

▶▶ influenza, p. 508 ▶▶ mumps, p. 583 ▶▶ polio, p. 656

Persistent Infections

Persistent infections remain for years, or even the life of the host—sometimes without any symptoms. In laboratory studies, two general types of persistent infections can be identified: chronic and latent (**table 13.4**). These categories may overlap in

an actual infection, however, with different cells experiencing different types of infection.

Chronic infections are characterized by the continuous production of low levels of viral particles. In some cases, this is similar to what occurs in filamentous bacteriophage infections with each infected cell surviving and slowly releasing viral particles. In other cases, the infected cell lyses, but only a small proportion of cells is infected at any given time, resulting in a low number of viral particles being continuously released—for example, hepatitis B virus infection. From a practical standpoint, the important aspect of either type of persistent infection is the continuous production of infectious viral particles, often in the absence of disease symptoms. Consequently, a person can transmit the virus to others even in the absence of symptoms. For example, some people infected with hepatitis B virus develop a chronic infection. They become carriers of the virus, able to pass it to other people through blood and body fluids. ◀◀ filamentous phage infection, p. 313 ▶▶ hepatitis B virus, p. 599

Latent infections are analogous to lysogeny by bacteriophages. The viral genome remains silent within a host cell, yet can reactivate to direct a productive infection. Some viruses do not integrate into the host cell chromosome; rather, they replicate independently of the host genome, much like a plasmid. The silent viral genome is called a **provirus.** The fact that a provirus cannot be eliminated from the body means that the disease can recur even after an extended period without symptoms. This explains why cold

TABLE 13.4	Examples of Persistent Infections		
Virus	**Type of Infection**	**Cells Involved**	**Disease**
Hepatitis B virus	Chronic	Hepatocytes (liver cells)	Hepatitis, cirrhosis, hepatocellular carcinoma
Hepatitis C virus	Chronic	Hepatocytes (liver cells)	Hepatitis, cirrhosis, hepatocellular carcinoma
Herpes simplex virus type 1	Latent	Neurons of sensory ganglia	Primary oral herpes and recurrent herpes simplex (cold sores)
Herpes simplex virus type 2	Latent	Neurons of sensory ganglia	Genital herpes and recurrent genital herpes
Varicella zoster (*Herpesviridae* family)	Latent	Satellite cells of sensory ganglia	Chickenpox and herpes zoster (shingles)
Cytomegalovirus (CMV; *Herpesviridae* family)	Latent	Salivary glands, kidney epithelium, leukocytes	CMV pneumonia, eye infections, mononucleosis, congenital CMV infection
Epstein-Barr virus	Latent	B cells, which are involved in antibody production	Burkitt's lymphoma
Human immunodeficiency virus (HIV)	Chronic	Activated helper T cells, macrophages	AIDS
	Latent	Memory helper T cells	

sores, which are caused by herpes simplex type 1 virus (HSV-1), can recur. HSV-1 causes an acute infection in mucosal epithelial cells, leading to the typical symptoms of cold sores. From there, the virus can spread to sensory nerve cells where it remains latent. Later, the latent virus can reactivate to cause another episode of cold sores (**figure 13.17**). What reactivates the virus is not clear, but certain physiological or immunological changes in the host are often involved. ▶▶ herpes simplex type 1 virus, p. 582

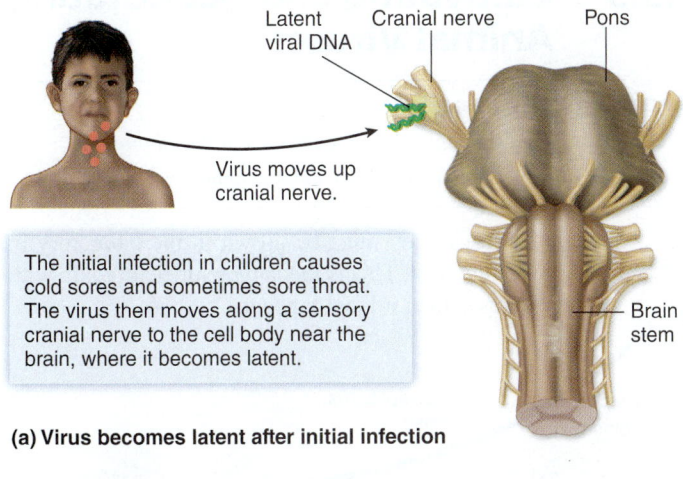

Latent viral DNA   Cranial nerve   Pons

Virus moves up cranial nerve.

The initial infection in children causes cold sores and sometimes sore throat. The virus then moves along a sensory cranial nerve to the cell body near the brain, where it becomes latent.

Brain stem

(a) Virus becomes latent after initial infection

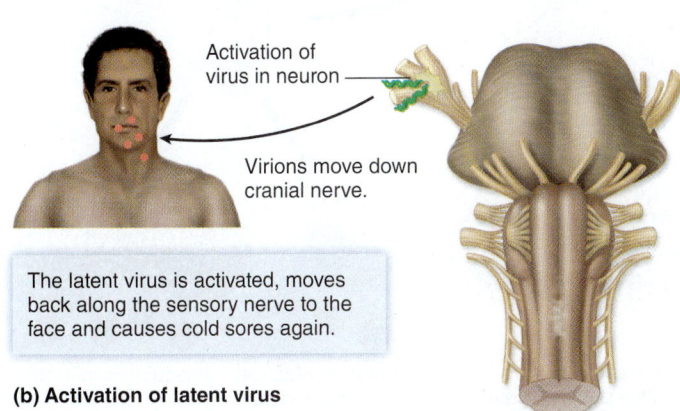

Activation of virus in neuron

Virions move down cranial nerve.

The latent virus is activated, moves back along the sensory nerve to the face and causes cold sores again.

(b) Activation of latent virus

FIGURE 13.17 Infection Cycle of Herpes Simplex Virus, HSV-1

❓ *What term is used to describe a virus in its latent state?*

MicroAssessment 13.7

Many viruses cause acute infections in which viruses multiply and spread rapidly in the host. Some viruses cause persistent infections; these long-term infections can be chronic or latent.

18. *Distinguish between acute and persistent viral infections at the cellular level.*

19. *How are latent viral infections different from chronic ones?*

20. *Could the same type of virus cause both an acute and a persistent infection? Explain.* ➕

13.8 ■ Viruses and Human Tumors

Learning Outcome

11. *Describe the roles of proto-oncogenes and tumor suppressor genes in controlling cell growth, and how some viruses can circumvent this control.*

A tumor is an abnormal growth of tissue resulting from a malfunction in the normally highly regulated process of cell growth. Some tumors are benign, meaning they do not metastasize (spread) or invade nearby normal tissue. Others are cancerous or malignant, meaning they have the potential to metastasize.

Control of cell growth involves genes that stimulate cell growth, called proto-oncogenes, and ones that inhibit cell growth, termed tumor suppressor genes. These genes work together to regulate growth and cell division. Mutations that result in inappropriate timing or level of expression of proto-oncogenes or repression of tumor suppressor genes are the major cause of abnormal and/or uncontrolled growth. A single change in the DNA sequence of these regulatory genes is probably not enough to cause a tumor; rather, multiple changes at different sites are required. The changes can result from different factors, including viruses. Some viruses, for example, carry genes called **oncogenes,** which are very similar in DNA sequence to proto-oncogenes. Consequently, their entry into cells can interfere with the cell's own control mechanisms, leading to tumor formation. However, the majority of tumors are not caused by viruses but by mutations

TABLE 13.5	Viruses Associated with Cancers in Humans	
Virus	**Type of Nucleic Acid**	**Kind of Tumor**
Human papillomaviruses (HPVs)	DNA	Different kinds of tumors, caused by different HPV types
Hepatitis B	DNA	Hepatocellular carcinoma
Epstein-Barr	DNA	Burkitt's lymphoma; nasopharyngeal carcinoma; B-cell lymphoma
Hepatitis C	RNA	Hepatocellular carcinoma
Human herpesvirus type 8	DNA	Kaposi's sarcoma
HTLV-1	RNA (retrovirus)	Adult T-cell leukemia (rare)

in host genes that regulate cell growth. Most virus-induced tumors are caused by certain DNA viruses (**table 13.5**).

Fifteen human papillomaviruses (HPVs) are associated with the development of cancers. Although the mechanism by which the viruses cause cancer is not entirely clear, several HPV-encoded proteins appear to interfere with the function of an important tumor suppressor gene product. The FDA approved a vaccine that contains two types of HPV that cause approximately 70% of cervical cancers, along with two types of HPV that cause genital warts. Other viruses implicated in cancer include human herpesvirus type 8 (HHV-8), which can cause Kaposi's sarcoma; Epstein-Barr virus (EBV), which can cause Burkitt's lymphoma; hepatitis B and C viruses, which can cause hepatocellular carcinoma (liver cancer); and T-cell lymphotrophic virus type 1 (HTLV-1), which can cause adult T-cell leukemia (table 13.5). In all of these cases, only a small percentage of infected individuals develop the cancer, indicating that other important factors must be involved.

▶▶ Epstein-Barr virus, p. 680 ▶▶ Kaposi's sarcoma, p. 707 ▶▶ hepatitis B virus, p. 599 ◀◀ papillomaviruses, p. 308

The various effects that animal viruses can have on their host cells are shown in **figure 13.18**.

MicroByte ───

In 2010, a government advisory panel concluded that environmental pollutants are grossly underestimated as a cause of cancer.

MicroAssessment 13.8

The majority of tumors are not caused by viruses but by mutations in certain host genes that regulate cell growth. The most common viral cause of tumors are the DNA tumor viruses.

21. *Name three viruses that cause tumors in humans.*
22. *Explain why a vaccine can prevent cervical cancer.*
23. *Why would it be advantageous to a virus to interfere with the function of proto-oncogenes or tumor suppressor genes?* ➕

13.9 ■ Cultivating and Quantitating Animal Viruses

Learning Outcomes

12. *Describe how animal cells are cultivated in the laboratory.*
13. *Describe the methods used to quantitate animal viruses.*

To study viruses, they must be grown in the laboratory in an appropriate host. This is much more difficult to do with animal viruses than it is with phages.

FIGURE 13.18   Various Effects of Animal Viruses on the Cells They Infect

❓ *What relationship is the most destructive to the infected cell?*

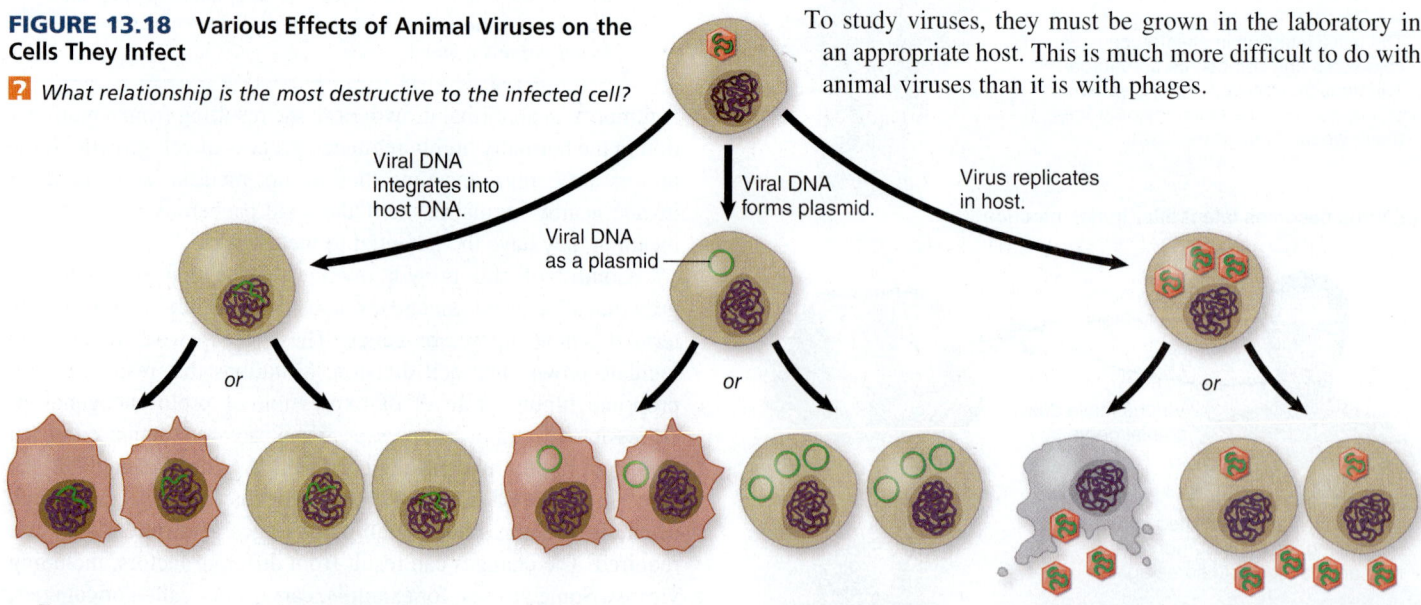

Viral DNA integrates into host DNA.	Viral DNA forms plasmid.	Virus replicates in host.

Viral DNA as a plasmid

Tumor
Normal cells are transformed into tumor cells.

Latent infection
Viral DNA replicates as part of the chromosome without harming the host.

Tumor
Normal cells are transformed into tumor cells.

Latent infection
Viral DNA replicates as a plasmid without harming the host.

Productive infection
New virions released when host cell lyses.

Productive infection
New virions released by budding.

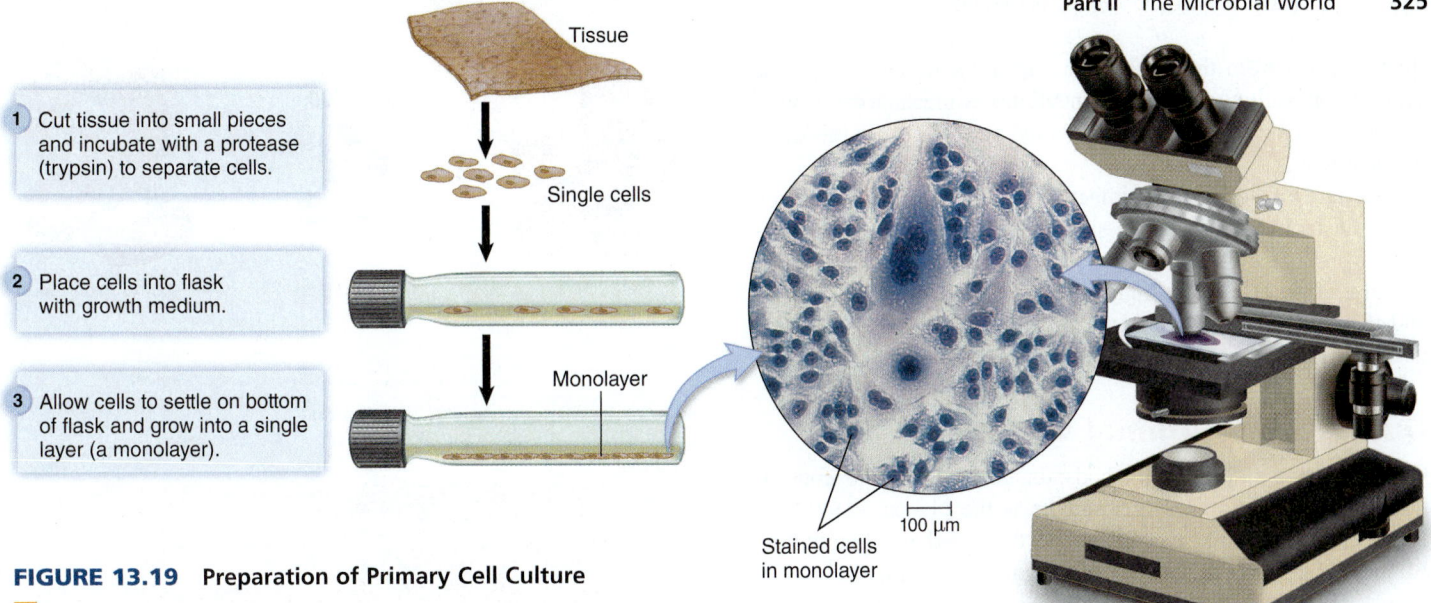

1 Cut tissue into small pieces and incubate with a protease (trypsin) to separate cells.

Tissue

Single cells

2 Place cells into flask with growth medium.

3 Allow cells to settle on bottom of flask and grow into a single layer (a monolayer).

Monolayer

Stained cells in monolayer

100 µm

FIGURE 13.19 **Preparation of Primary Cell Culture**

❓ *Why must primary cell cultures be restarted every so often?*

Cultivating Animal Viruses

Initially, the only way to study animal viruses was to inoculate live animals with a suspension of virus particles. Later, embryonated (fertilized) chicken eggs were used to grow viruses, which made research much easier. Although improved techniques have largely replaced these methods, they are still used to cultivate certain viruses. For example, influenza viruses are grown in embryonated chicken eggs to manufacture vaccines against influenza.
▶▶ vaccine, p. 421 ▶▶ influenza virus, p. 508

Today, **cell culture** or **tissue culture** is commonly used to cultivate most animal viruses. Animal cells—grown in a liquid medium contained in special screw-capped flasks—are used as host cells for the virus culture. Much like bacteria, animal cells bathed in the proper nutrients can divide repeatedly. Animal cells grow much more slowly than most bacteria, however, dividing no faster than once every 24 hours (*E. coli* can divide every 20 minutes). Animal cell types that make up solid tissues typically grow as a monolayer, a single sheet of cells adhering to the bottom of the flask. Cells from blood grow as single cells in suspension. Although cell culture is generally easier than growing viruses in living animals, not all viruses can be grown in this way.

Obtaining and Maintaining Cell Cultures

One way to obtain animal cells for culture is to remove tissue from an animal, and then process it to get individual cells. These cells can then be grown in a Petri dish with a liquid nutrient medium. Cells acquired in this way form **primary cultures (figure 13.19)**. One problem with this approach is that normal cells can divide only a limited number of times, even when diluted into a fresh medium, and so new primary cultures must regularly be made. To avoid this problem, tumor cells are often used in cell culture. These multiply indefinitely *in vitro*, resulting in what is called an established cell line.

MicroByte
The commonly used HeLa cell line was derived from cervical cancer cells taken from a patient named **He**nrietta **La**cks, who eventually died from the cancer.

Effects of Viral Replication on Cell Cultures

Many viruses can be detected by their effect on cells in cell cultures. A virus propagated in tissue culture often causes distinct morphological alterations in infected cells, called a **cytopathic effect (figure 13.20)**. The host cells, for example, may change

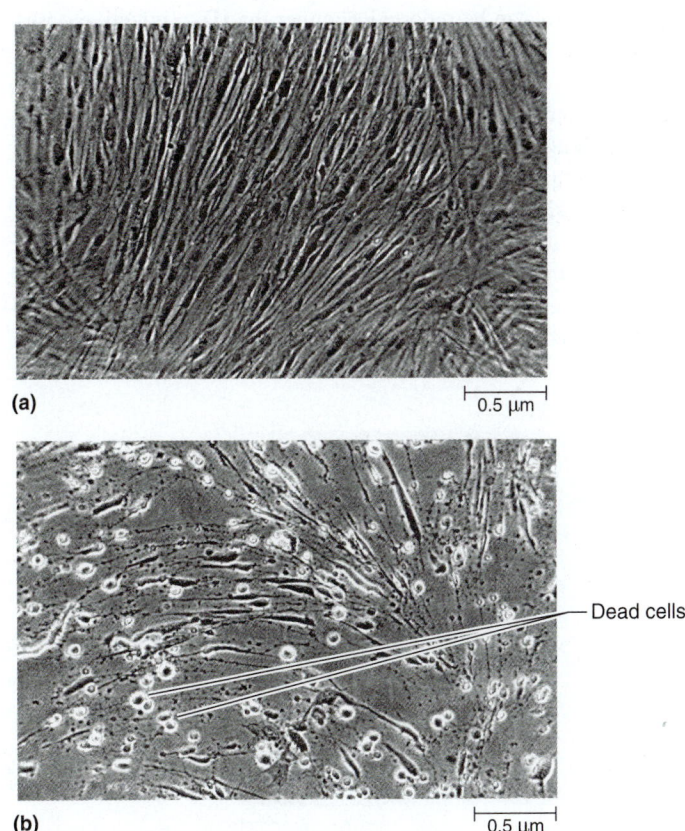

(a)   0.5 µm

(b)   0.5 µm

Dead cells

FIGURE 13.20 **Cytopathic Effects of Virus Infection on Tissue Culture** **(a)** Fetal tonsil diploid fibroblasts growing as a monolayer, uninfected. **(b)** Same cells infected with herpes simplex virus.

shape, detach from the surface, or lyse. Infected cells may fuse into a giant multinuclear cell (syncytium), a mechanism of viral spread. Several viruses, such as HIV and measles, cause this cytopathic effect.

Certain viruses cause an infected cell to form a distinct region called an **inclusion body,** the site of viral replication. The position of an inclusion body in a cell depends on the type of virus. Because cytopathic effects indicate that a particular cell is infected and they are often characteristic for a particular virus, they are useful to scientists studying and identifying viruses.

Quantitating Animal Viruses

One of the most precise methods for determining the concentration of animal viruses in a sample is the plaque assay. This is similar in principle to the method described for quantitating bacteriophages, but in the case of animal viruses, a monolayer of tissue culture cells is the host. Again, clear zones surrounded by uninfected cells are counted to determine the viral titer.

If a sample contains a high enough concentration of viruses to be seen with an electron microscope, direct counts can be used to determine the number of viral particles in a suspension (**figure 13.21**).

A viral titer can be estimated using a **quantal assay.** In this method, several dilutions of the virus preparation are administered to a number of animals, cells, or chick embryos, depending on the host specificity of the virus. The titer of the virus, or the endpoint, is the dilution at which 50% of the inoculated hosts are infected or killed. This can be reported as either the ID_{50} (infective dose), or the LD_{50}, (lethal dose).

Certain viruses cause red blood cells to agglutinate (clump). This phenomenon, **hemagglutination,** occurs when individual viral particles attach to surface molecules of multiple red blood cells simultaneously, connecting the cells to form an aggregate (**figure 13.22**). Hemagglutination is visible only at high concentrations of viruses, and can be used to determine only the relative concentration of viral particles. Hemagglutination is measured by mixing serial dilutions of the viral suspension with a standard amount of red blood cells. The highest dilution showing maximum

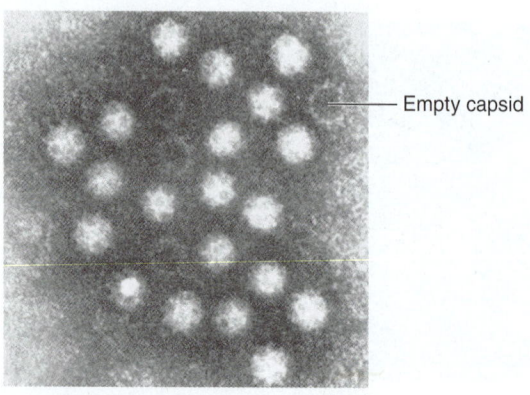

FIGURE 13.21   Electron Micrograph of Calicivirus Note that empty capsids are readily distinguishable.

❓ *Is a virion with an empty capsid infectious?*

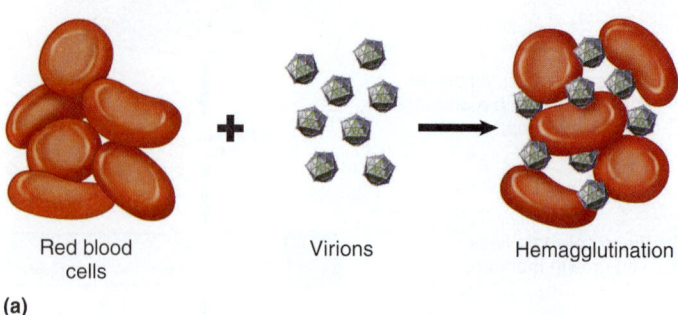

Red blood cells   Virions   Hemagglutination

(a)

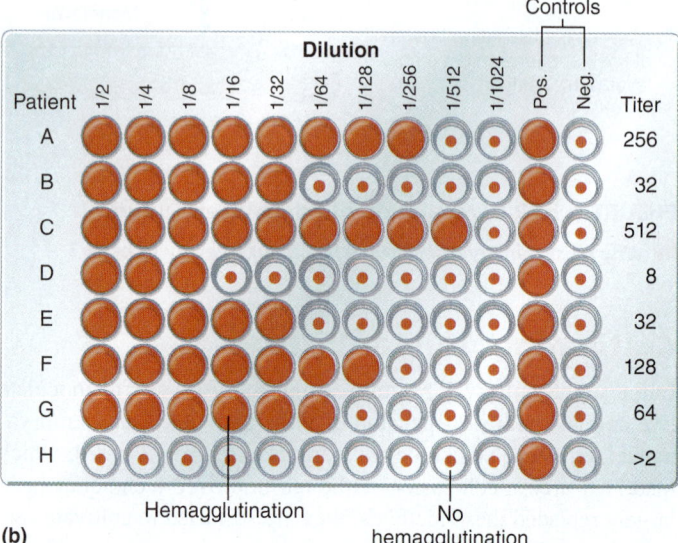

(b)

FIGURE 13.22   Hemagglutination (a) Diagram showing virions combining with red blood cells, resulting in hemagglutination. **(b)** Assay of viral titer. Red blood cells normally form a pellet when they sink to the bottom of the wells, whereas agglutinated ones do not.

❓ *Which patient has the highest titer of antibodies?*

agglutination is the titer of the virus. One group of animal viruses that can agglutinate red blood cells is the orthomyxoviruses, of which the influenza virus is a member.

MicroAssessment 13.9

Various hosts are required to grow different viruses. These include whole animals, embryonated chicken eggs, and cell cultures. Viruses that lyse their host cells can be assayed by counting plaques. Other methods include direct counts, quantal assays, and hemagglutination.

24. *In tissue culture, why is it advantageous to use tumor cells rather than normal cells?*
25. *Discuss two methods used to quantitate animal viruses.*
26. *How might syncytia formation of infected cells benefit viruses?* ➕

13.10 ■ Plant Viruses

Learning Outcomes

14. *Compare and contrast the mechanisms by which plant and animal viruses enter host cells.*
15. *Discuss the ways that plant viruses can be transmitted to their hosts.*

(a)

(b)

(c)

FIGURE 13.23 **Signs of Viral Diseases of Plants** (a) A healthy wheat leaf can be seen in the center. The yellowed leaves on either side are infected with wheat mosaic virus. (b) Typical ring lesions on a tobacco leaf resulting from infection by tobacco mosaic virus. (c) Stunted growth (right) in a wheat plant caused by wheat mosaic virus.

? *How do most plant viruses enter plants?*

Viral diseases of plants are economically important, particularly when they occur in crop plants such as corn, wheat, rice, soybeans and sugar beets. Over half of the crop yields can be lost during a serious virus infection. Viral diseases are especially common among perennial plants (those that live for many seasons), such as tulips and potatoes, and those propagated vegetatively (not by seeds), such as potatoes.

Viral infection of plants can be recognized through various outward signs, including yellowing of foliage with irregular lines appearing on the leaves and fruits (**figure 13.23**). Individual cells or specialized organs of the plant may die, and tumors may appear. Usually, infected plants become stunted in their growth, although in a few cases growth is stimulated, leading to deformed structures. Plants generally do not recover from viral infections, because unlike animals, plants are not capable of developing specific immunity to rid themselves of invading viruses. In severely infected plants, virions may accumulate in enormous quantities. For example, as much as 10% of the dry weight of a tobacco mosaic virus–infected plant may consist of virus. The virus, which causes a serious disease of tobacco, retains its infectivity for at least 50 years, which explains why it is usually difficult to eradicate the virions from a contaminated area. Other plant viruses are also extraordinarily stable in the environment.

In a few instances, plants have been purposely maintained in a virus-infected state. The most well-known example involves tulips, in which a virus transmitted through the bulbs can cause a desirable color variegation of the flowers (**figure 13.24**). The infecting virus was transmitted through bulbs for a long time before the cause of the variegation was even suspected.

In contrast to phages and animal viruses, when plant viruses infect a cell they do not attach to specific receptors. Instead, they enter through wound sites in the cell wall, which is otherwise very tough and rigid. Infection in the plant can then spread from cell to cell through openings (the plasmodesmata) that interconnect cells.

Plant viruses can be transmitted through soil contaminated by prior growth of infected plants, and by growers themselves.

FIGURE 13.24 **Tulips with Symptoms Resulting from Virus Infection**

? *What other symptoms can plant viruses cause?*

A small percentage of the known plant viruses are transmitted through contaminated seeds, tubers, or pollen. Virus infections can also spread through grafting of healthy plant tissue onto diseased plants. Tobacco mosaic virus is transmitted to healthy seedlings on the hands of workers who have been in contact with the virus from infected plants. The most important transmitters of plant viruses are probably insects; thus, insect control is a critical tool for preventing the spread of plant viruses.

MicroByte

Tobacco in cigarettes, chewing tobacco, and cigars may carry the tobacco mosaic virus. Smokers should wash their hands thoroughly before handling plants.

MicroAssessment 13.10

Plant viruses cause many plant diseases and are of major economic importance. They invade through wound sites in the cell wall and are spread through soil, insects, and growers.

27. How does a plant virus penetrate the tough outer coat of the plant cell?
28. How are plant viruses transmitted?
29. Why is it especially important for plant viruses to be stable outside the plant?

13.11 ■ Other Infectious Agents: Viroids and Prions

Learning Outcomes

16. Describe the chemical structure of viroids and prions.
17. Compare hosts of viroids and prions.
18. Describe the process by which prions accumulate in tissues.

Although viruses are composed of only nucleic acid surrounded by a protective protein coat, other infectious agents are even simpler in structure. These are the viroids and prions.

Viroids

Viroids consist solely of a single-stranded RNA molecule that varies in size from 246 to 375 nucleotides and forms a closed ring (**figure 13.25**). This is about one-tenth the size of the smallest infectious viral RNA genome known. A great deal of hydrogen bonding exists between complementary bases in the viroid RNA so that the single-stranded molecule appears to be a double-stranded structure.

All viroids that have been identified infect only plants, where they cause serious diseases including potato spindle tuber, chrysanthemum stunt, citrus exocortis, cucumber pale fruit, hopstunt, and cadang-cadang. Like plant viruses, they enter plants through wound sites rather than binding to specific receptors. A great deal is known about the structure of viroid RNA, but many questions remain. How do viroids replicate? How do they cause disease? How did they originate? Do they have counterparts in animals, or are they restricted to plants? The answers to these questions will provide insights into new and fascinating features of this unusual member of the microbial world.

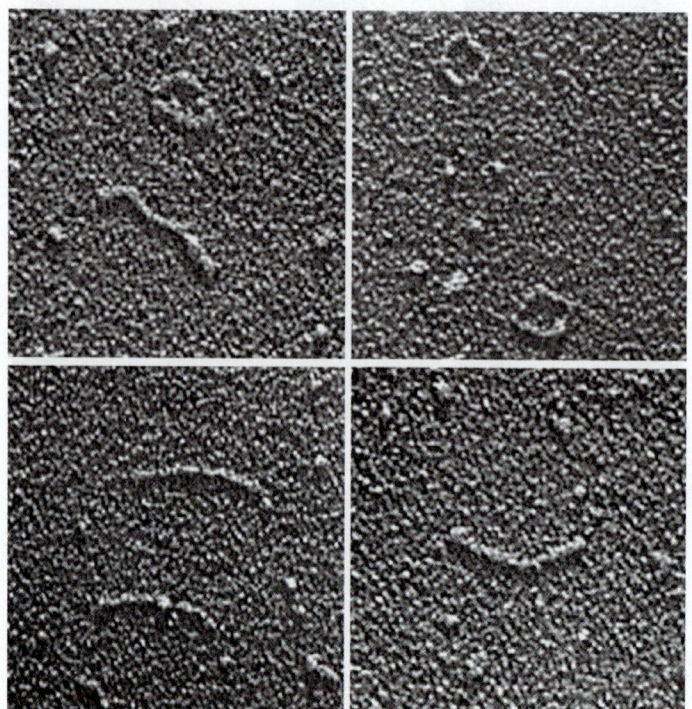

FIGURE 13.25 Electron Micrographs of Viroids

❓ Will heating affect the structure of a viroid?

Prions

Prions are composed solely of protein, which is reflected in the name (derived from **pro**teinaceous **in**fectious agent). These agents have been linked to a number of slow, always fatal, human diseases including Creutzfeldt-Jakob disease and kuru, as well as to animal diseases such as scrapie (sheep and goats), bovine spongiform encephalopathy (cattle), and chronic wasting disease (deer and elk) (**table 13.6**). In all these diseases, prion proteins

TABLE 13.6	Prion Diseases
Disease	**Host**
Scrapie	Sheep and goats
Bovine spongiform encephalopathy	Cattle
Chronic wasting disease	Deer and elk
Transmissible mink encephalopathy	Ranched mink
Exotic ungulate encephalopathy	Antelope in South Africa
Feline spongiform encephalopathy	Cats
Kuru	Humans (caused by cannibalism)
Variant Creutzfeldt-Jakob disease	Humans (caused by consumption of prion-contaminated beef)
Creutzfeldt-Jakob disease	Humans (inherited)
Gerstmann–Straussler Scheinker syndrome	Humans (inherited)
Fatal familial insomnia	Humans (inherited)

accumulate in neural tissue. For unknown reasons, neurons die and brain function deteriorates as the tissues develop characteristic holes (**figure 13.26**). The characteristic spongelike appearance of the brain tissues gave rise to the general term **transmissible spongiform encephalopathies,** which refers to all prion diseases.

▶▶ prions, p. 664

Considering that prions lack any nucleic acid, how they accumulate in tissue and cause disease has long been an intriguing question. An answer began to emerge when it was discovered that uninfected animals synthesize a protein in neurons, especially in the brain, that has an identical or nearly identical amino acid sequence to infectious prions. Prion proteins from different animals are similar but not identical. A key difference between the normal cellular prion protein and the infectious form is the shape of the protein, which influences its stability. The normal cellular form, referred to or **PrP^C** (for **pr**ion **p**rotein, **c**ellular) is readily destroyed by host cell proteases, as a normal turnover process, with older molecules being destroyed as new ones are synthesized. In contrast, infectious prion proteins, referred to as **PrP^{SC}** (for **pr**ion **p**rotein, **sc**rapie) are less susceptible to degradation by proteases and become insoluble, leading to aggregation.

Prions are unusually resistant to heat and chemical treatments that are commonly used to inactivate infectious agents.

Prions violate the central dogma of replication that requires nucleic acid act as a template for replication of macromolecules. Prions are not made up of nucleic acid, making their accumulation inside cells difficult to explain. It is hypothesized that PrP^{SC} interacts directly with PrP^C and converts its folding properties from PrP^C to PrP^{SC} (**figure 13.27**). This conversion may require another unknown factor. As a result, cells that continue to produce PrP^C accumulate PrP^{SC}. How prions cause neurons to die is not known.

In most cases, the disease is transmitted only to members of the same species. However, the barrier to prion transmission between species also depends on the strain of prion. It is now clear that the prion that caused mad cow disease in England killed more than 170 people by causing a disease very similar

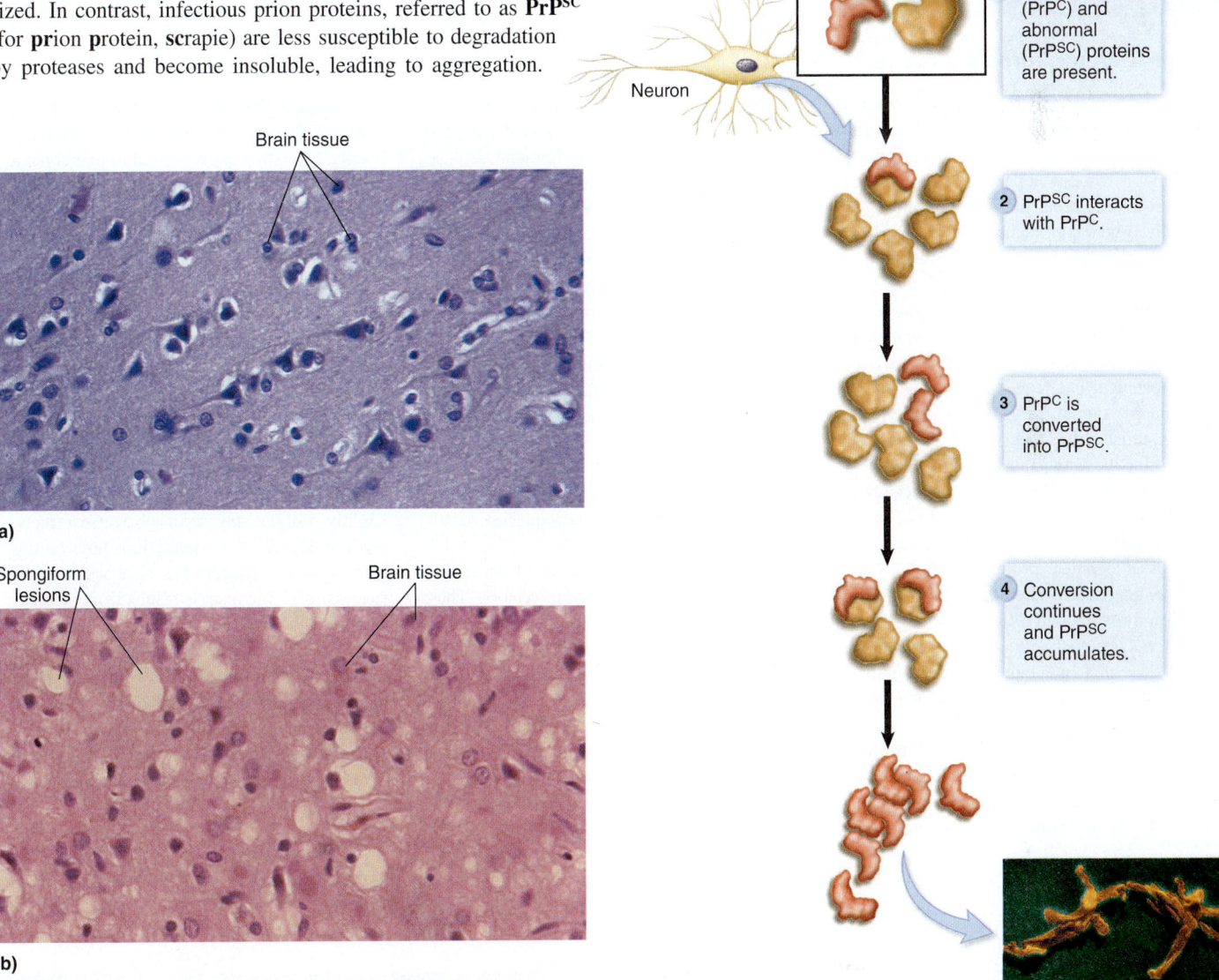

1 Both normal (PrP^C) and abnormal (PrP^{SC}) proteins are present.

2 PrP^{SC} interacts with PrP^C.

3 PrP^C is converted into PrP^{SC}.

4 Conversion continues and PrP^{SC} accumulates.

(a)

Brain tissue

(b)

Spongiform lesions

Brain tissue

FIGURE 13.26 Appearance of a Brain with Spongiform Encephalopathy (a) Normal brain section. **(b)** Brain section of patient with spongiform encephalopathy.

❓ *From these photos, why has this disease been given the name it has?*

FIGURE 13.27 Proposed Mechanism by Which Prions Propagate

❓ *Why would PrP^{SC} accumulate when PrP^C does not?*

to the Creutzfeldt-Jakob disease. Presumably these people consumed tissue of infected animals. Thus far, no human deaths have been attributed to eating sheep infected with the scrapie agent or deer and elk infected with the prion causing chronic wasting disease (CWD).

A good example of how a prion disease spreads in the United States is CWD. Deer at a research facility in Colorado were the first animals to show signs and symptoms of CWD in 1967. Within 40 years, the disease spread from Colorado to wild deer populations through their natural migration and transport of captive farmed animals located in more than a dozen states and parts of Canada. The disease is spread through direct animal-to-animal (nose-to-nose) contact and as a result of indirect exposure to prions in the environment. Feed, water sources and soil may be contaminated with prions shed in saliva, urine and feces, or decomposing carcasses. The stability of prions makes them difficult, if not impossible, to eliminate from the environment. Additional information about the role of prions in human disease is covered in chapter 26.

MicroAssessment 13.11

Two infectious agents that are structurally simpler than viruses are viroids and prions. Viroids contain only single-stranded RNA and no protein; prions contain protein and no nucleic acid.

30. *Distinguish between a viroid and a prion in terms of structure and hosts.*
31. *Discuss how prion proteins accumulate in nervous tissue.*
32. *Must all prion diseases result from eating infected food? Explain.* ✚

FUTURE CHALLENGES 13.1

Gene Therapy: Turning a Corner?

Gene therapy is the treatment of hereditary disorders by introducing new genetic information into cells of the body to correct the defect. The goal is to achieve persistent expression of the introduced normal DNA so as to modify the defective phenotype without unintended harmful consequences. Gene therapy was launched in 1990, and numerous trials treating different conditions are now underway. However, no treatments have gained final approval for use in the United States. It has been a rocky road with technical challenges and the death of a volunteer in one of the trials.

Scientists have focused on viruses as the most effective way to introduce desired genes into cells. This is because viral genomes are small and readily manipulated, so foreign genes can be easily cloned into them. In addition, the attachment proteins of the viral coat can be genetically altered, thereby allowing the virion to attach to different tissues. A variety of well-studied animal viruses are known, some of which integrate their DNA into the host's genome, ensuring their maintenance in the cells.

For all its promises, gene therapy has a number of problems that must be solved. For one thing, the new genetic information must not only be delivered, but also maintained inside cells in the body to allow synthesis of the gene products for an extended period of time. This requires that the immune defenses of the body be circumvented. The genes must also be transferred to enough cells in the proper tissues, and the genes expressed in large enough amounts, to change the phenotype. Not all viruses can transfer genes to nondividing cells, so the properties of the target tissues must be known. If cells are dividing, it may be necessary to use a retrovirus whose DNA becomes part of the host's genome. If this is done, however, the potential exists for the DNA to integrate into a vital gene or into genes concerned with the control of growth. Thus, another major consideration is how to prevent any unintended consequences resulting from gene therapy.

Three groups of viruses have been studied most extensively as potential vectors. These are the double-stranded DNA adenoviruses, the single-stranded DNA adeno-associated viruses, and the retroviruses. In all cases, most of the viral genome has been deleted and replaced with the genes intended to cure the inherited disease. Such deletions eliminate the pathogenicity of the virus and also reduce its ability to provoke an immune response that would otherwise eliminate it.

Has gene therapy achieved any success in humans? The answer is yes, with encouraging results. Gene therapy was used to overcome a fatal inherited disease caused by a deficiency of an enzyme of purine metabolism that results in major defects in the body's immune system. Ten young children were treated with a retrovirus containing the normal gene. All ten patients are alive after up to 10 years and most are leading normal lives without any adverse effects from the treatment. Gene therapy has also been used on three individuals suffering from retinal degeneration, which results in total blindness. An adenovirus containing the normal gene was injected into the retina of the eye. The vision of each patient improved slightly without any serious adverse effects. A brain disorder that usually kills boys before they become teenagers has also been treated successfully. The disease results from a defective gene involved in maintaining the myelin sheath around nerves. A normal gene was incorporated into blood cells of two 7-year-old boys suffering from the condition. Some of the cells made the functional protein and also apparently migrated to the boys' brains. After 2 years, the progressive damage has stopped.

Summary

13.1 ■ General Characteristics of Viruses

Most viruses are approximately 100- to 1,000-fold smaller than the cells they infect (figure 13.1).

Viral Architecture

At a minimum, a **virion** (viral particle) consists of nucleic acid surrounded by a **capsid.** Capsids are composed of capsomers. Some viruses have an envelope surrounding the **nucleocapsid;** other viruses are **naked** (figure 13.2). They contain either RNA or DNA, but never both. The shape of a virus is generally icosahedral, helical, or complex (figure 13.3).

Viral Taxonomy

Viruses are classified primarily on the basis of their genome structure and hosts they infect (table 13.1). The names of virus families end in *-viridae.* Viruses are also given informal names and are sometimes grouped on their routes of transmission (table 13.2).

13.2 ■ Bacteriophages

Lytic Phage Infections: T4 Phage as a Model

Lytic or **virulent phages** exit the host at the end of the cycle by lysing the host, resulting in a **productive infection.** The infection proceeds through five steps: attachment, genome entry, synthesis of phage proteins and genome, assembly (maturation) and release (figure 13.5).

Temperate Phage Infections: Lambda Phage as a Model

Temperate phages have the option of either directing a productive infection or initiating a **lysogenic infection** (figure 13.6); the infected cell is a **lysogen.** A **repressor** maintains the **prophage** in an integrated state, but the prophage can be excised to initiate a lytic infection.

Consequences of Lysogeny

Lysogens are immune to superinfection. **Lysogenic conversion** occurs if a prophage carries genes that change the phenotype of the host cell (table 13.3).

Filamentous Phages: M13 Phage as a Model

Filamentous phages cause productive infections, but the viral particles are continually extruded from the host in the assembly process and the cells are not killed (figure 13.7).

13.3 ■ The Roles of Bacteriophages in Horizontal Gene Transfer

Generalized Transduction

Generalized transduction results from a phage packaging error during assembly. Generalized transducing particles can transfer any gene of a donor cell.

Specialized Transduction

Specialized transduction results from an excision mistake made by a temperate phage during its transition from a lysogenic to a lytic cycle (figure 13.9). Only genes located near the site at which the temperate phage integrates are transduced.

13.4 ■ Bacterial Defenses Against Phages

Preventing Phage Attachment

If a bacterium alters or covers a given receptor, that cell becomes resistant to any phage that requires the receptor.

Restriction-Modification Systems

Restriction-modification systems protect bacteria from phage infection by quickly degrading foreign DNA (figure 13.10). Restriction enzymes recognize and cut specific DNA sequences; modification enzymes protect the hosts' DNA from the action of the restriction enzyme by adding methyl groups to certain nucleobases.

CRISPR System

Bacterial cells that survive some phage infections appear to retain small segments of phage DNA, termed spacer, and incorporate them into a region of DNA called **CRISPR** (figure 13.11). That region is transcribed, and the spacer segments join with Cas proteins, functioning as a type of RNA interference to target the phage DNA for destruction.

13.5 ■ Methods Used to Study Bacteriophages

Plaque assays are used to quantitate the phages particles in samples (figure 13.12). Each plaque represents a single phage particle infecting a cell.

13.6 ■ Animal Virus Replication

The generalized infection cycle of animal viruses can be viewed as a five-step process.

Attachment

Attachment proteins or spikes on the virus particle attach to specific receptors on the cell surface.

Penetration and Uncoating

In the case of animal viruses, the entire virion enters the cell. Enveloped viruses either fuse with the host membrane or are taken in by receptor-mediated endocytosis (figure 13.13). Naked virions enter by receptor-mediated endocytosis. **Uncoating** releases the nucleic acid from the protein coat.

Synthesis of Viral Proteins and Replication of the Genome (figure 13.14)

DNA viruses generally replicate in the nucleus and use the host cell machinery for DNA synthesis as well as gene expression, although they often encode their own DNA polymerase. Replication of ssDNA viruses is similar to that of double-stranded DNA viruses, but the complementary strand must be synthesized first. RNA viruses usually replicate in the cytoplasm. Replication requires a virally encoded **replicase** to synthesize complementary RNA strand. This enzyme lacks proofreading ability and makes more mistakes in replication than DNA polymerase. Reverse-transcribing viruses encode **reverse transcriptase,** which synthesizes DNA from an RNA template. As with replicases, these enzymes are error-prone.

Assembly

Capsids are formed, and then the genome and any necessary proteins are packaged within it. The process may take place in the cytoplasm, nucleus, or in a variety of organelles.

Release

Enveloped virions most often exit by **budding** (figure 13.15). Naked virions are released when the host cell dies.

13.7 ■ Categories of Animal Virus Infections (figure 13.16)

Acute infections are characterized by sudden onset of symptoms of relatively short duration. **Persistent infections** can continue for the life of the host, with or without symptoms.

Acute Infections

On a cellular level, acute infections can be compared to productive lytic infections by bacteriophages, but even though the cells often die, the host may survive because of the immune response.

Persistent Infections

Chronic infections are characterized by the continuous production of low levels of viral particles; **latent infections** are analogous to lysogeny in bacteriophages.

13.8 ■ Viruses and Human Tumors

Some viruses carry **oncogenes** that interfere with the ability of the cell to control growth. Most virus-induced tumors are caused by certain

DNA viruses (table 13.5). A vaccine against human papillomaviruses (HPVs) prevents many cervical cancers.

13.9 ■ Cultivating and Quantitating Animal Viruses

Cultivating Animal Viruses

Cell culture or **tissue culture** is commonly used to cultivate most viruses (figure 13.19). This can be done using **primary cultures** or established cell lines. Many viruses can be detected by their effect on cells in culture, called a **cytopathic effect** (figure 13.20). Certain viruses cause an infected cell to form an **inclusion body.**

Quantitating Animal Viruses

The plaque assay is one of the most precise methods for determining the concentration of animal viruses in a sample. In some cases, virions can be counted with an electron microscope (figure 13.21). **Quantal assays** estimate the titer by determining the ID_{50} or LD_{50}. The concentration of viruses able to cause **hemagglutination** can be measured by determining the highest dilution that clumps red blood cells (figure 13.22).

13.10 ■ Plant Viruses

Viral infection of plants can be recognized by outward signs such as yellowing foliage, stunted growth, and tumor formation; plants usually do not recover from the infection (figure 13.23). Virions do not bind to receptor sites on plant cells but enter through wound sites. Insects are probably the most important transmitter of the viruses.

13.11 ■ Other Infectious Agents: Viroids and Prions

Viroids

Viroids are plant pathogens that consist of small circular, single-stranded RNA molecules (figure 13.25).

Prions

Prions are composed solely of protein, and cause a number of **transmissible spongiform encephalopathies** (table 13.6). Prions accumulate by converting PrP^C to PrP^{SC}, proteins that are less susceptible to proteases and form aggregates (figure 13.27).

Review Questions

Short Answer

1. Why are naked viruses generally more resistant to disinfectants than are enveloped viruses?

2. How is the replication cycle of lambda phage different from that of T4?

3. What is lysogenic conversion?

4. How is specialized transduction different from generalized transduction?

5. How does the CRISPR system protect bacteria from phage infection?

6. Why must (–) strand but not (+) strand RNA viruses bring their own replicase into a cell?

7. Why are RNA viruses and retroviruses more error-prone in their replication than DNA viruses?

8. What is the role of a prophage in persistent infections?

9. How do oncogenes differ from proto-oncogenes?

10. Describe how prions propagate.

Multiple Choice

1. Capsids are composed of
   a) DNA.   b) RNA.   c) protein.
   d) lipids.   e) polysaccharides.

2. The tail fibers on phages are associated with
   a) attachment.
   b) penetration.
   c) transcription of phage DNA.
   d) assembly of virus.
   e) lysis of host.

3. Classification of viruses is based on all of the following *except*
   a) type of nucleic acid.
   b) shape of virus.
   c) size of virus.
   d) host infected.
   e) strandedness of nucleic acid.

4. Temperate phages can do all of the following *except*
   a) lyse their host cells.
   b) change properties of their hosts.
   c) integrate their DNA into the host DNA.
   d) bud from their host cells.
   e) become prophages.

5. All phages must have the ability to
   1. have their nucleic acid enter the host cell.
   2. kill the host cell.
   3. multiply in the absence of living bacteria.
   4. lyse the host cell.
   5. have their nucleic acid replicate in the host cell.
   a) 1, 2   b) 2, 3   c) 3, 4   d) 4, 5   e) 1, 5

6. Filamentous phages
   a) infect animal and bacterial cells.
   b) cause their host cells to grow more quickly.
   c) are extruded from the host cell.
   d) undergo assembly in the cytoplasm.
   e) degrade the host cells' DNA.

7. Influenza vaccines must be changed yearly because the amino acid sequence of the viral proteins change gradually over time. Based on this information, which is the most logical conclusion? The influenza virus

   a) is enveloped.

   b) is naked.

   c) has a DNA genome.

   c) has an RNA genome.

   e) causes a persistent infection.

8. Acute infections of animals

   1. are a result of productive infection.

   2. generally lead to long-lasting immunity.

   3. result from integration of viral nucleic acid into the host.

   4. are usually followed by chronic infections.

   5. often lead to tumor formation.

   a) 1, 2   b) 2, 3   c) 3, 4   d) 4, 5   e) 1, 5

9. Quantitating viral titers of both phage and animal viruses frequently involves

   a) plaque formation.

   b) quantal assays.

   c) hemagglutination.

   d) determining the ID_{50}.

   e) counting of virions by microscopy.

10. Prions

   a) contain only nucleic acid without a protein coat.

   b) replicate like HIV.

   c) integrate their nucleic acid into the host genome.

   d) cause diseases of humans.

   e) cause diseases of plants.

Applications

1. A public health physician isolated large numbers of phages from rivers used as a source of drinking water in western Africa. The physician is very concerned about humans becoming ill from drinking this water, although she knows that phages specifically attack bacteria. Why is she concerned?

2. Researchers debate the evolutionary value to the virus of its ability to cause disease. Many argue that viruses accidentally cause disease and only in animals that are not the natural host. They state that this strategy may eventually prove fatal to the virus's future in that host. It is reasoned that the animals will eventually develop immune mechanisms to combat the virus and prevent its spread. Another group of researchers supports the view that disease is a way to enhance the survival of the virus. What position would you take, and what arguments would you give to support your view?

Critical Thinking

1. A filter capable of preventing bacteria from passing is placed at the bottom of a U tube to separate the two sides. Streptomycin-resistant cells of a bacterial strain are placed on one side of the filter and streptomycin-sensitive cells are placed on the other side. After incubation, the side of the tube that originally contained only streptomycin-sensitive cells now contains some streptomycin-resistant cells. Give three possible reasons for this observation. What further experiments would you do to determine the correct explanation?

2. Why is it virtually impossible to eradicate (eliminate) a disease caused by a zoonotic virus?

14

The Innate Immune Response

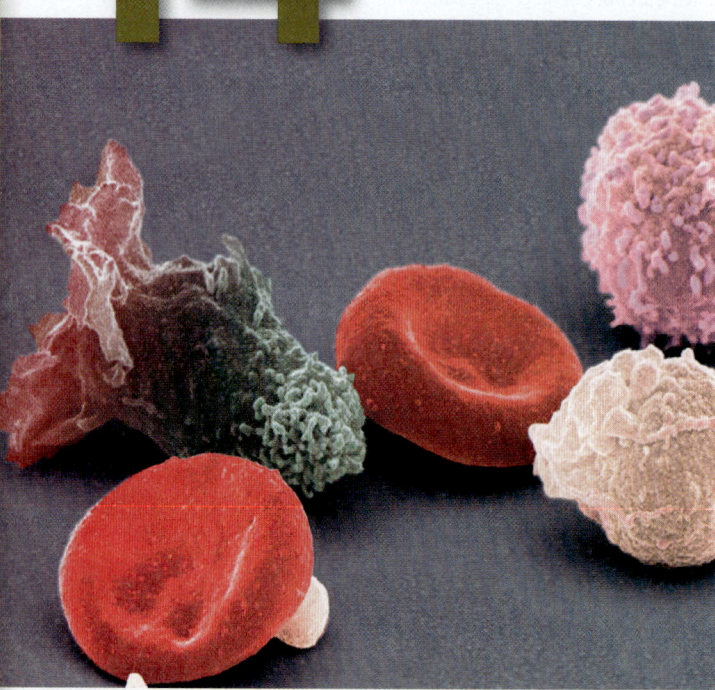

Human blood cells (color-enhanced scanning electron micrograph).

A Glimpse of History

As soon as microorganisms were shown to cause disease, scientists worked to explain how the body defends itself against their invasion. Elie Metchnikoff, a Russian-born scientist, hypothesized that specialized cells in the body destroy invading organisms. His ideas arose while he was studying the larval form of starfish. As he looked at the larvae under the microscope, he could see ameba-like cells within the bodies. He described his observations:

> . . . I was observing the activity of the motile cells of a transparent larva, when a new thought suddenly dawned on me. It occurred to me that similar cells must function to protect the organism against harmful intruders. . . . I thought that if my guess was correct a splinter introduced into the larva of a starfish should soon be surrounded by motile cells much as can be observed in a man with a splinter in his finger. No sooner said than done. In the small garden of our home . . . I took several rose thorns that I immediately introduced under the skin of some beautiful starfish larvae which were as transparent as water. Very nervous, I did not sleep during the night, as I was waiting for the results of my experiment. The next morning, very early, I found with joy that it had been successful.

Metchnikoff reasoned that certain cells in animals are able to ingest and destroy foreign material. He called these cells phagocytes ("cells that eat") and proposed they were primarily responsible for the body's ability to destroy invading microbes. He then studied the process by watching phagocytes ingest and destroy invading yeast cells in transparent water fleas. In 1884, Metchnikoff published a paper supporting his belief

that phagocytic cells were primarily responsible for destroying disease-causing organisms. He spent the rest of his life studying this process and other biological phenomena. Metchnikoff was awarded a Nobel Prize in 1908 for his studies of immunity.

From a microorganism's standpoint, the tissues and fluids of the human body are much like a culture flask filled with a warm nutrient-rich solution. Considering this, it may be surprising that the interior of the body—including blood, muscles, bones, and organs—is generally sterile. If this were not the case, microbes would simply degrade our tissues, just as they readily break down the bodies of dead animals.

How do body tissues remain sterile in this world full of microbes? Like other multicellular organisms, humans have several mechanisms of defense. First, we are covered with skin and mucous membranes that prevent most microbes from entering the tissues. If microbes get past those barriers, sensor systems in the tissues detect them and then direct and assist other defenses that destroy the invaders. **Innate immunity** is the routine protection provided by these mechanisms.

Innate immunity had been called a non-specific defense, but recent discoveries show that most of its components detect molecules associated with invading microbes or tissue damage,

Focus Figure

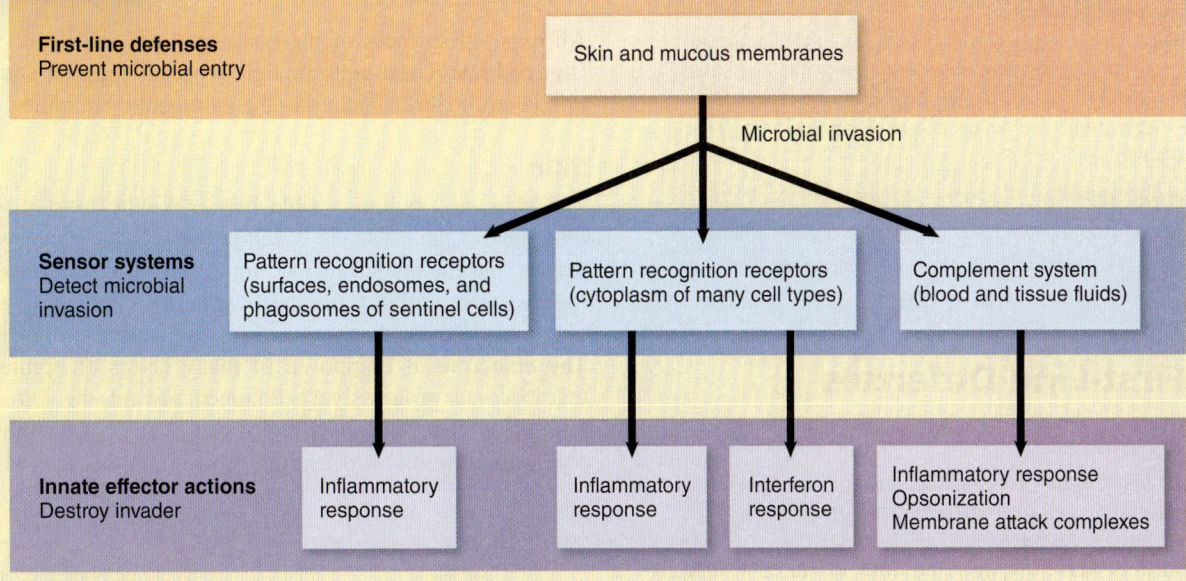

First-line defenses
Prevent microbial entry

Skin and mucous membranes

Microbial invasion

Sensor systems
Detect microbial
invasion

Pattern recognition receptors
(surfaces, endosomes, and
phagosomes of sentinel cells)

Pattern recognition receptors
(cytoplasm of many cell types)

Complement system
(blood and tissue fluids)

Innate effector actions
Destroy invader

Inflammatory
response

Inflammatory
response

Interferon
response

Inflammatory response
Opsonization
Membrane attack complexes

FIGURE 14.1 Overview of the Innate Defenses ❓ *What is the role of the sensor systems in innate immunity?*

---

a feature called **pattern recognition.** The molecules recognized include parts of bacterial cell walls and other compounds unique to microbes, as well as substances associated with damaged host cells. ◀◀ bacterial cell walls, p. 58

In addition to the innate response, vertebrates have evolved a more specialized defense system, providing protection called **adaptive immunity.** This develops throughout life and substantially increases the host's ability to defend itself. Each time the body is exposed to a microbe, or certain other types of foreign material, the adaptive defense system first "learns" and then "remembers" the most effective response to that specific material; it then reacts accordingly if the material is encountered again. The substance that causes an immune response is called an **antigen.** An important action of the adaptive immune response is the production of Y-shaped proteins called **antibodies.** These bind specifically to antigens, thereby targeting them for destruction or removal by other host defenses. The adaptive immune response can also destroy the body's own cells—referred to as **host cells** or **"self" cells**—if they harbor a virus or other invader.

To simplify the description of the immune system, it is helpful to consider it as a series of individual parts. This chapter will focus almost exclusively on innate immunity. Remember, however, that although the various parts are discussed separately, their actions are connected and coordinated. In fact, as you will see in chapter 15, certain components of the innate defenses educate the adaptive defenses, helping them recognize that a particular antigen represents a microbial invader.

14.1 ■ Overview of the Innate Defenses

Learning Outcome

1. *Outline the essential components of the innate defenses.*

First-line defenses are the barriers that separate and shield the interior of the body from the surrounding environment; they are the initial obstacles microbes must overcome to invade the tissues. The anatomical barriers, which include the skin and mucous membranes, not only provide physical separation but are often bathed in secretions that have antimicrobial properties.

Sensor systems within the body recognize when the first-line barriers have been breached and then relay that information to other components of the host defenses. Certain cells serve as sentinels (lookouts or guards). They have **pattern recognition receptors (PRRs)** on their surface and within their endosomes or phagosomes. These receptors recognize groups of compounds unique to microbes, allowing the sentinel cells to detect invaders. When invasion is detected, the cells send chemical signals to alert other components of the host defenses, triggering a protective response (an effector action). Many cell types have a different set of PRRs in their cytoplasm, allowing them to recognize when they have been invaded by a microbe. As with the sentinel cells, they respond by sending out a call for help. Another type of sensor is a series of proteins always present in blood and tissue fluids; these proteins are collectively called the **complement system** because they can "complement" (act in combination with) the adaptive immune defenses. The complement system becomes activated in response to certain stimuli, setting off a chain of events that results in removal and destruction of invading microbes. ◀◀ endosome, p. 72 ◀◀ phagosome, p. 72

When invading microorganisms or tissue damage is detected, an **inflammatory response** occurs, involving many components of the innate defenses. During this response, cells that line local blood vessels undergo changes that allow complement system components and other proteins to leak out into tissues. **Phagocytes** (cells that specialize in engulfing and digesting microbes and cell debris) also leave the bloodstream and accumulate in the tissues. Some types of phagocytes play a dual role, destroying invaders while communicating with cells of the adaptive immune system, enlisting their far more powerful effects. **Figure 14.1** provides a general overview of innate immunity.

MicroAssessment 14.1

First-line defenses are the initial barriers that microbes must pass to invade tissues. Sensor systems within the body recognize invading microbes. Inflammation is a coordinated response to invasion or tissue damage that recruits phagocytes and other protective mechanisms to an area.

1. *List two sensor systems of the innate defenses.*
2. *Describe the dual roles played by some types of phagocytes.*
3. *What molecules might pattern recognition receptors recognize?* ➕

14.2 ■ First-Line Defenses

Learning Outcome

2. *Describe the first-line defenses, including the physical barriers, antimicrobial substances, and normal microbiota.*

The body's borders serve as the first line of defense against invading microbes (**figure 14.2**). Some of these borders are thought of as being "inside" the body, but they directly contact the external environment. For example, the digestive tract, which begins at the mouth and ends at the anus, is simply a hollow tube that runs through the body, allowing intestinal cells to absorb nutrients from food that passes through (see figure 24.1); the respiratory tract is a cavity that allows O_2 and CO_2 to be exchanged (see figure 21.1).

In this section, we will describe the general physical and chemical aspects of the anatomical barriers, as well as the protective contributions of the normal microbiota. These are described in more detail in the chapters dealing with each body system.

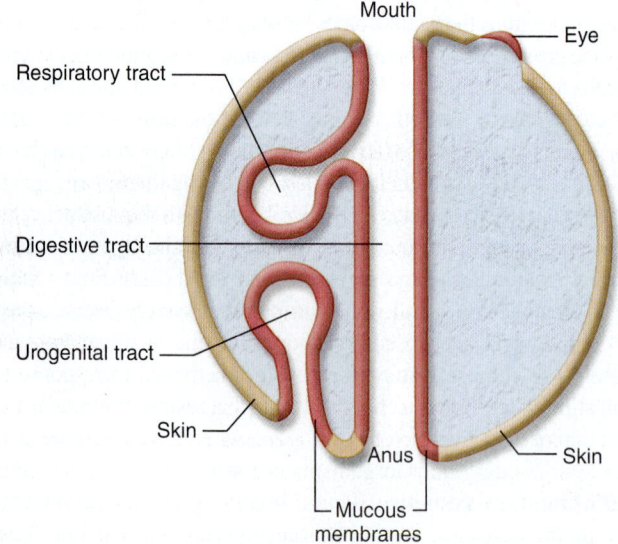

FIGURE 14.2 The Body's Borders These borders separate the interior of the body from the surrounding environment; they are the initial obstacles microorganisms must overcome to invade tissues. The skin is shown in tan, and mucous membranes in pink.

❓ *Why are the contents of the digestive tract considered to be outside the body?*

Physical Barriers

All exposed surfaces of the body are lined with epithelial cells (**figure 14.3**). These cells are tightly packed together and rest on a thin layer of fibrous material, the basement membrane.

Skin

The skin, an obvious visible barrier, is the most difficult for microbes to penetrate. It is composed of two main layers—the dermis and the epidermis (see figure 22.1). The dermis contains tightly woven fibrous connective tissue, making it extremely tough and durable (the dermis of cattle is used to make leather). The epidermis is composed of many layers of epithelial cells that become progressively flattened toward the exterior. The outermost sheets are made up of dead cells filled with a water-repelling protein called keratin, resulting in the skin being a dry

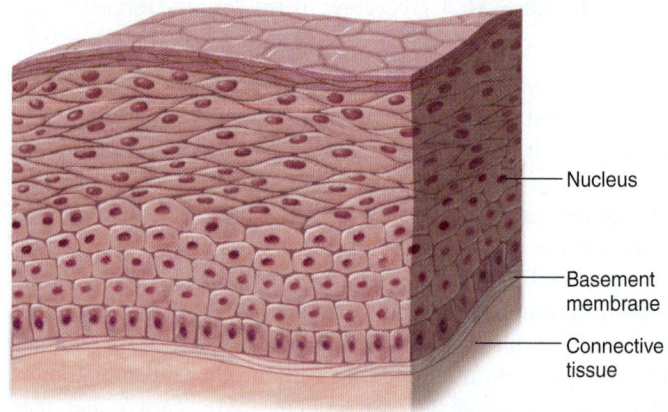

Stratified epithelium
• Skin (the outer cell layers are embedded with keratin)
• Lining of the mouth, vagina, urethra, and anus

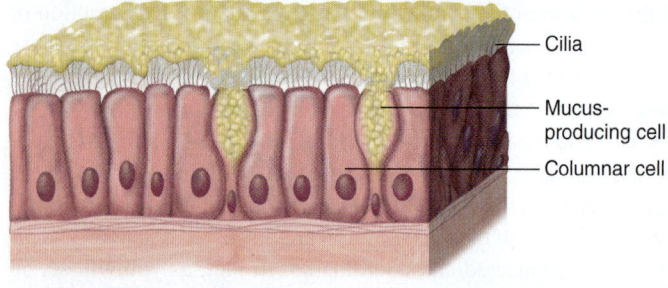

Columnar epithelium
• Passages of respiratory system
• Various tubes of the reproductive systems

FIGURE 14.3 Epithelial Barriers Cells of these barriers are tightly packed together and rest on a layer of thin fibrous material, the basement membrane.

❓ *What is the purpose of the cilia on the respiratory epithelium?*

environment. The cells continually slough off, taking with them any microbes that might be adhering. ▶▶ anatomy, physiology, and ecology of the skin, p. 522

Mucous Membranes

Mucous membranes line the digestive tract, respiratory tract, and genitourinary tract. They are constantly bathed with mucus or other secretions that help wash microbes from the surface. Most mucous membranes have mechanisms that move microbes toward areas where they can be eliminated. For example, **peristalsis**—the contractions of the intestinal tract—propels food and liquid and also helps remove microbes. The respiratory tract is lined with ciliated cells; the hairlike cilia constantly beat in an upward motion, moving materials away from the lungs to the throat where they can then be swallowed. This movement out of the respiratory tract is referred to as the **mucociliary escalator.** ◀◀ cilia, p. 74

Antimicrobial Substances

Skin and mucous membranes are protected by a variety of antimicrobial substances that inhibit or kill microorganisms (**figure 14.4**). For example, the salty residue that accumulates on skin as perspiration evaporates inhibits all but salt-tolerant microbes.

Lysozyme, the enzyme that degrades peptidoglycan, is in tears, saliva, and mucus. It is also found within the body, in phagocytic cells, blood, and the fluid that bathes tissues. ◀◀ lysozyme, p. 62

Peroxidase enzymes, which break down hydrogen peroxide to produce reactive oxygen species, are in saliva and milk; they are also found within body tissues and inside phagocytes. Bacteria that produce the enzyme catalase can break down hydrogen peroxide, potentially destroying it before peroxidase reacts with it. Catalase-negative organisms are more readily killed by peroxidase. ◀◀ reactive oxygen species, p. 90 ◀◀ catalase, p. 90

Lactoferrin is an iron-binding protein in saliva, mucus, and milk; it is also found in some types of phagocytes. A similar compound, **transferrin,** is in blood and tissue fluids. Iron is one of the major elements, so withholding it prevents microbial growth. Some microorganisms can capture iron from the host, however, thwarting this defense. ◀◀ major elements, p. 92

Defensins are short antimicrobial peptides produced by neutrophils and epithelial cells. They insert into bacterial membranes, forming pores that damage cells.

Normal Microbiota (Flora)

The **normal microbiota (flora)** is the population of microorganisms that routinely grow on the body surfaces of healthy humans (see figure 16.1). Although these organisms are not technically part of the immune system, they provide considerable protection.

One protective effect of the normal microbiota is the competitive exclusion of pathogens. For example, the normal microbiota

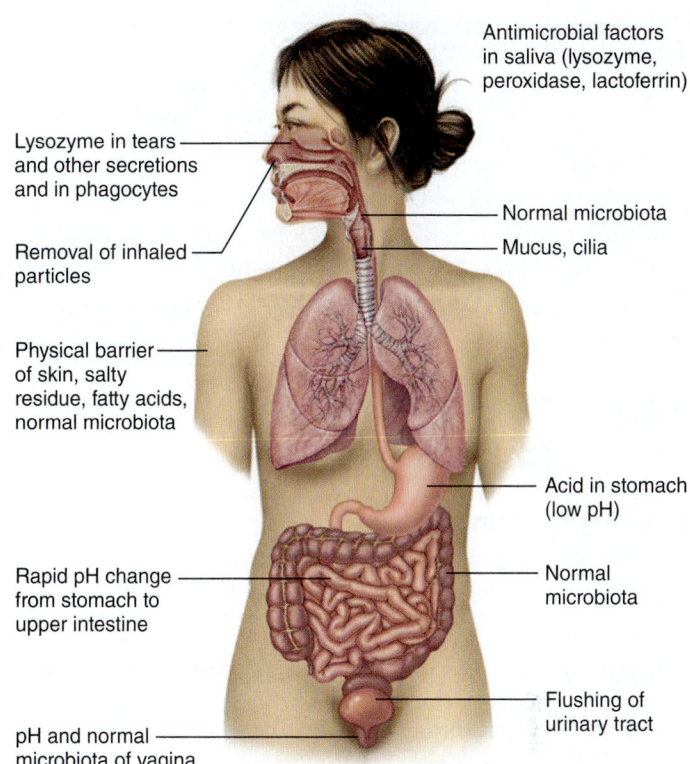

Antimicrobial factors in saliva (lysozyme, peroxidase, lactoferrin)

Lysozyme in tears and other secretions and in phagocytes

Normal microbiota

Removal of inhaled particles

Mucus, cilia

Physical barrier of skin, salty residue, fatty acids, normal microbiota

Acid in stomach (low pH)

Rapid pH change from stomach to upper intestine

Normal microbiota

pH and normal microbiota of vagina

Flushing of urinary tract

FIGURE 14.4 Antimicrobial Substances and the Normal Microbiota These play important roles in protecting the body's borders.

❓ *How is lysozyme antibacterial?*

prevents pathogens from adhering to host cells by covering binding sites that might otherwise be used for attachment. The population also consumes available nutrients that could otherwise support the growth of less desirable organisms.

Some members of the normal microbiota produce compounds toxic to other bacteria. In the hair follicles of the skin, for instance, *Propionibacterium* species degrade lipids, releasing fatty acids that inhibit the growth of many pathogens. In the gastrointestinal tract, some strains of *E. coli* synthesize colicins, a group of proteins toxic to certain bacteria. *Lactobacillus* species growing in the vagina produce lactic acid as a fermentation end product, resulting in an acidic pH that inhibits the growth of some pathogens.

Disruption of the normal microbiota, which occurs when antibiotics are used, can predispose a person to various infections. Examples include antibiotic-associated diarrhea and pseudomembranous colitis, caused by the growth of toxin-producing strains of *Clostridium difficile* in the intestine, and vulvovaginitis, caused by excessive growth of *Candida albicans* in the vagina. ▶▶ *Clostridium difficile*–associated disease, p. 594 ▶▶ vulvovaginal candidiasis, p. 618

The normal microbiota is also essential to the development of the immune system. As certain microbes are encountered, the system learns to distinguish harmless ones from pathogens. Other effects of the population will be discussed in chapter 16.

14.3 ■ The Cells of the Immune System

Learning Outcome

3. *Describe the characteristics and roles of granulocytes, mononuclear phagocytes, dendritic cells, and lymphocytes.*

The cells of the immune system can move from one part of the body to another, traveling through the body's circulatory systems like vehicles on an extensive interstate highway system. They are always found in normal blood, but their numbers usually increase during infections, recruited from reserves of immature cells in the bone marrow. Some of the cell types are found primarily in the blood, but others leave the blood circulatory system and take up residence in various tissues. Certain cell types play dual functions, having crucial roles in both innate and adaptive immunity.

The formation and development of blood cells is called **hematopoiesis** (Greek for "blood" and "to make"). All blood cells, including those important in the body's defenses, originate from the same cell type, the **hematopoietic stem cell,** found in the bone marrow (**figure 14.5**). As with other types of stem cells, hematopoietic cells are capable of long-term self-renewal, meaning they can divide repeatedly. Hematopoietic stem cells are induced to develop into the various types of blood cells by a group of proteins called colony-stimulating factors (CSFs).

FIGURE 14.5 Blood Cells and Their Derivatives All of these descend from hematopoietic stem cells found in the bone marrow. Multiple steps occur between the hematopoietic stem cell and the final cells produced.

❓ *How are the roles of neutrophils and macrophages similar?*

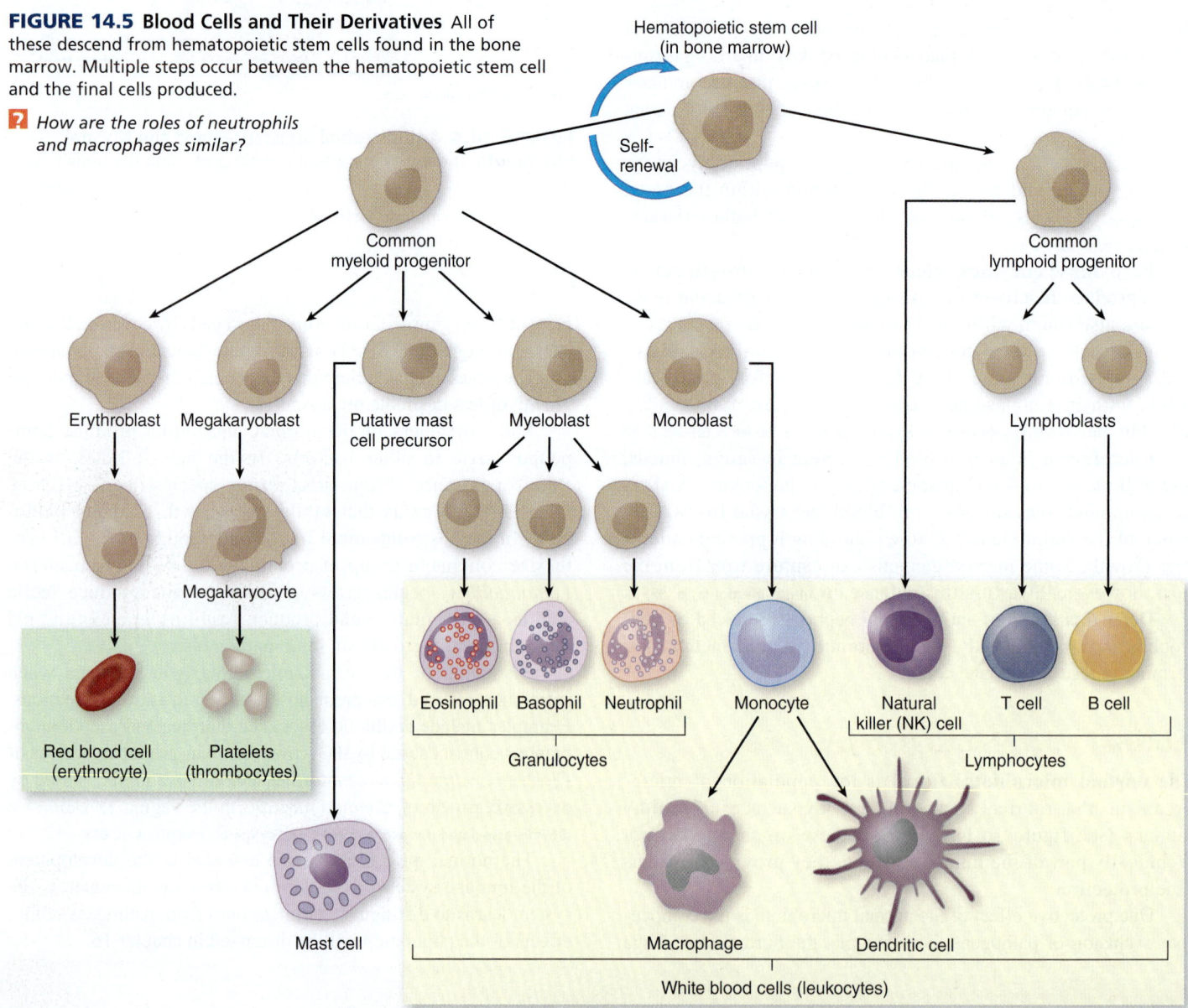

The general categories of blood cells and their derivatives include red blood cells, platelets, and white blood cells. Red blood cells, or **erythrocytes,** carry O_2 in the blood. Platelets, which are actually fragments arising from large cells called megakaryocytes, are important for blood clotting. White blood cells, or **leukocytes,** are important in all host defenses. Leukocytes can be divided into four broad groups—granulocytes, mononuclear phagocytes, dendritic cells, and lymphocytes (**table 14.1**).

Granulocytes

Granulocytes contain cytoplasmic granules filled with biologically active chemicals. There are three types of granulocytes—neutrophils, basophils, and eosinophils; their names reflect the staining properties of their granules.

Neutrophils efficiently engulf and destroy bacteria and other material. Their granules, which stain poorly, contain many enzymes and antimicrobial substances that help destroy the engulfed materials. Neutrophils are the most numerous and important granulocytes of the innate responses. They are also called polymorphonuclear neutrophilic leukocytes, polys, or PMNs, names that reflect the appearance of multiple lobes of their single nucleus. They normally account for over half of the circulating white blood cells, and their numbers increase during most bacterial infections. Few neutrophils are generally found in tissues, except during inflammation. Because of their importance in innate immunity, they will be described in more detail later in the chapter. ▶▶ specialized attributes of neutrophils, p. 348

Basophils are involved in allergic reactions and inflammation. Their granules, which stain dark purplish-blue with

TABLE 14.1	Leukocytes		
Cell Type (% of blood leukocytes)		**Location in Body**	**Function**
Granulocytes			
Neutrophils (polymorphonuclear neutrophilic leukocytes or PMNs, often called polys; 55–65%)		Most common type of circulating leukocyte; few in tissues except during inflammation	Phagocytize and digest engulfed materials
Eosinophils (2–4%)		Few in tissues except in certain types of inflammation and allergies	Participate in inflammatory reaction and immunity to some parasites
Basophils (0–1%), mast cells		Basophils in circulation; mast cells present in most tissues	Release histamine and other inflammation-inducing chemicals from the granules
Mononuclear Phagocytes			
Monocytes (3–8%)		In circulation; they differentiate into either macrophages or dendritic cells when they migrate into tissue	Phagocytize and digest engulfed materials
Macrophages		Present in virtually all tissues; known by various names based on the tissue in which they are found	Phagocytize and digest engulfed materials
Dendritic cells		Initially in tissues, but they migrate to secondary lymphoid organs (e.g., lymph nodes, spleen, appendix, tonsils)	Collect antigen from the tissues and then bring it to lymphocytes that gather in the secondary lymphoid organs
Lymphocytes			
Several types (25–35%)		In lymphoid organs (e.g., lymph nodes, spleen, thymus, appendix, tonsils); also in circulation	Participate in adaptive immune responses

the basic dye methylene blue, contain histamine and other chemicals that increase capillary permeability during inflammation. **Mast cells** are similar in appearance and function to basophils but are found in tissues rather than blood. They do not come from the same precursor cells as basophils. Mast cells are important in the inflammatory response and are responsible for many allergic reactions.

Eosinophils are thought to be primarily important in ridding the body of parasitic worms. They are involved in allergic reactions, causing some of the symptoms associated with allergies, but reducing others. The granules of eosinophils, which stain red with the acidic dye eosin, contain antimicrobial substances and also histaminase, an enzyme that breaks down histamine.

Mononuclear Phagocytes

Mononuclear phagocytes make up the mononuclear phagocyte system (MPS) (**figure 14.6**). This grouping includes **monocytes**—which circulate in the blood—and the cell types that develop from them as they leave the bloodstream and migrate into tissues.

Macrophages are a differentiated form of monocytes, meaning they have gained specialized properties. They are an important sentinel cell, present in nearly all tissues, and particularly abundant in the liver, spleen, lymph nodes, lungs, and the peritoneal (abdominal) cavity. When residing in tissues, they are given different names based on their location (figure 14.6). Macrophages will be discussed in more detail later in the chapter. ▶▶ specialized attributes of macrophages, p. 348

Dendritic cells also develop from monocytes. Their function goes beyond engulfment and destruction of an invader, so they will be considered separately.

Dendritic Cells

Dendritic cells are sentinel cells that function as "scouts." They engulf material in the tissues and then bring it to the cells of the adaptive immune system for "inspection." Most dendritic cells develop from monocytes, but some descend from other cell types. Details regarding the interactions of dendritic cells with the cells of the adaptive immune response will be discussed in chapter 15.

Lymphocytes

Lymphocytes are responsible for adaptive immunity, the focus of chapter 15. In contrast to the generic pattern recognition that characterizes the innate defenses, cells of the two major groups of lymphocytes, **B cells** and **T cells,** are remarkably specific in their recognition of antigen. These cell types generally reside in lymph nodes and other lymphatic tissues. **Natural killer (NK) cells** are another type of lymphocyte; unlike B cells and T cells, however, they lack specificity in their mechanisms of antigen recognition. ▶▶ lymphatic tissues, p. 357

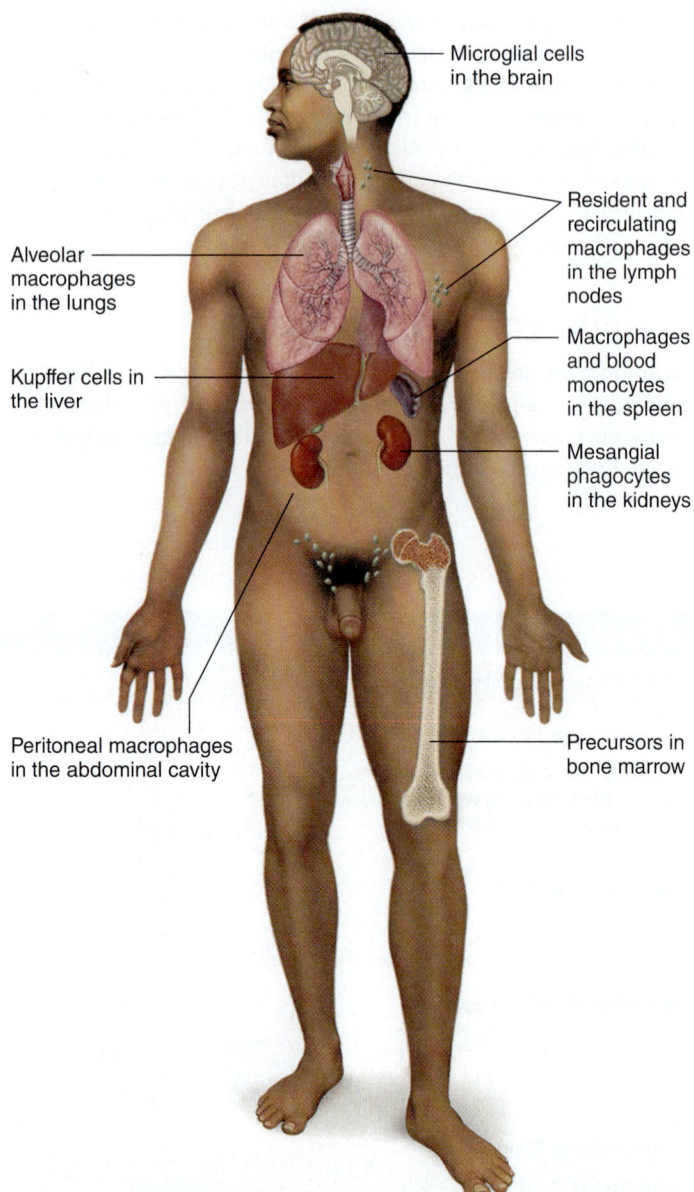

FIGURE 14.6 Mononuclear Phagocyte System Cells in this system sometimes have special names depending on their location—for example, Kupffer cells (in the liver) and alveolar macrophages (in the lung).

❓ *Macrophages descend from which type of blood cell?*

Microglial cells in the brain

Resident and recirculating macrophages in the lymph nodes

Alveolar macrophages in the lungs

Macrophages and blood monocytes in the spleen

Kupffer cells in the liver

Mesangial phagocytes in the kidneys

Peritoneal macrophages in the abdominal cavity

Precursors in bone marrow

MicroAssessment 14.3

Granulocytes include neutrophils, basophils, and eosinophils. Mononuclear phagocytes include monocytes and macrophages. Dendritic cells function as scouts for the adaptive immune system. Lymphocytes are responsible for adaptive immunity.

7. *Which type of granulocyte is the most abundant?*

8. *How are macrophages related to monocytes?*

9. *Why can bone marrow transplants be used to replace defective lymphocytes?* ➕

14.4 ■ Cell Communication

Learning Outcome

4. *Describe the characteristics and roles of surface receptors, cytokines, and adhesion molecules in innate immunity.*

In order for immune system cells to mount a coordinated response to invasion, they must communicate with each other. They do this through surface receptors, cytokines, and adhesion molecules.

Surface Receptors

Surface receptors can be viewed as the "eyes" and "ears" of a cell. They are proteins that generally span the plasma membrane, connecting the outside of the cell with the inside, allowing the inner workings of the cell to sense and respond to external signals. Each receptor is specific with respect to the compound or compounds it will bind; a molecule that can bind to a given receptor is called a **ligand** for that receptor. When a ligand binds to its surface receptor, the internal portion of the receptor becomes modified in some manner. This change then triggers some type of response by the cell, such as chemotaxis. Cells can alter the types and numbers of surface molecules they make, allowing them to respond to signals relevant to their immediate situation. ◀◀ chemotaxis, p. 64

Cytokines

Cytokines can be viewed as the "voices" of a cell. A cytokine produced by one cell diffuses to another and binds to the appropriate **cytokine receptor** of that cell. Binding of a cytokine to its receptor induces a change in the cell such as growth, differentiation, movement, or cell death. Cytokines act at extremely low concentrations, having local, regional, or systemic effects. They often act together or in sequence, in a complex fashion. The source and effects of some cytokines are listed in **table 14.2.**

Chemokines are cytokines important in chemotaxis of immune cells. Certain types of cells have receptors for chemokines, allowing the cells to sense the location where they are needed, such as an area of inflammation.

Colony-stimulating factors (CSFs) are important in the multiplication and differentiation of leukocytes (see figure 14.5). When more leukocytes are needed during an immune response, a variety of different colony-stimulating factors direct immature cells into the appropriate maturation pathways.

Interferons (IFNs) are important in the control of viral infections. In addition to being antiviral, IFN-gamma helps regulate the function of cells involved in the inflammatory response and adjusts certain actions of adaptive immunity. The role of interferons during viral infections will be described later in the chapter.

Interleukins (ILs) are produced by leukocytes and have diverse, often overlapping, functions. As a group, they are important in both innate and adaptive immunity.

Tumor necrosis factor (TNF) was discovered because of its role in killing tumor cells, a characteristic reflected by the name, but it has multiple roles. It helps initiate the inflammatory response and triggers the process of "cell suicide," a programmed cell death called apoptosis. ▶▶ apoptosis, p. 350

Groups of cytokines often act together to generate a response. For example, certain cytokines referred to as **pro-inflammatory cytokines** (TNF, IL-1, IL-6, and others) contribute to inflammation. Others are involved in promoting antibody responses (IL-4 and others). A different group stimulates certain types of T cells (IL-2, IFN-gamma, and others).

MicroByte

HIV takes advantage of two chemokine receptors, CCR5 and CXCR4, using them as attachment sites for infection.

TABLE 14.2	Some Important Cytokines	
Cytokine	**Source**	**Effect**
Chemokines	Various cells	Chemotaxis
Colony-Stimulating Factors (CSFs)	Various cells	Stimulate growth and differentiation of different kinds of leukocytes
Interferons		
Interferon alpha	Leukocytes	Antiviral
Interferon beta	Fibroblasts	Antiviral
Interferon gamma	T cells, NK cells	Macrophage activation; promotes certain adaptive immune responses
Interleukins (ILs)		
IL-1	Macrophages, epithelial cells	T-cell activation; macrophage activation; induces fever
IL-2	T cells	T-cell proliferation
IL-4	T cells, mast cells	Promotes antibody responses
IL-6	T lymphocytes, macrophages	T- and B-cell growth; inflammatory response; fever
Tumor Necrosis Factor (TNF)	Macrophages, T cells, NK cells	Promotes inflammation; cytotoxic for some tumor cells; regulates certain immune functions

Adhesion Molecules

Adhesion molecules on the surface of cells allow those cells to "grab" other cells. For example, when phagocytic cells in the blood are needed in tissues, the endothelial cells that line the blood vessels synthesize adhesion molecules that bind to passing phagocytic cells. This slows the rapidly moving phagocytes, allowing them to then leave the bloodstream. Other types of adhesion molecules are used by cells to form connections so that one cell can deliver cytokines or other molecules directly to another cell.

MicroAssessment 14.4

Surface receptors allow a cell to detect molecules present outside of that cell. Cytokines provide cells with a mechanism of communication. Adhesion molecules allow one cell to adhere to another.

10. *What is a ligand?*

11. *What is the function of a colony-stimulating factor?*

12. *How could colony-stimulating factors be used as a therapy?* ➕

14.5 ■ Pattern Recognition Receptors (PRRs)

Learning Outcome

5. Describe the role of TLRs, NLRs, RLRs, and the interferon response in the host defenses.

Pattern recognition receptors (PRRs) allow the body's cells to "see" signs of microbial invasion. They detect generic microbe-associated patterns, called **pathogen-associated molecular patterns (PAMPs)**. These substances are common on all microbes, not just pathogens, so they are sometimes referred to as microbe-associated molecular patterns (MAMPs). You'll recognize many of the PAMPs from earlier reading in the textbook—various cell wall components (lipopolysaccharide, peptidoglycan, lipoteichoic acid, and lipoproteins), flagellin subunits, and RNA molecules that characterize viruses. Some PRRs recognize **danger-associated molecular patterns (DAMPs)**, molecules that indicate host cell damage. Although PRRs were discovered relatively recently, and much is still being learned about them, they have stimulated tremendous interest in innate immunity.

Toll-Like Receptors (TLRs)

Toll-like receptors (TLRs) are anchored in membranes of sentinel cells such as macrophages, dendritic cells, and cells that line sterile body sites. A number of different TLRs have been described (at least 10 in humans), and each recognizes a distinct compound or group of compounds associated with microbes. TLRs are related to Toll receptors, first identified in *Drosophila* species (fruit flies). Their name reflects the reaction of a researcher to the sight of a fruit-fly with a mutant form of the gene—she exclaimed "toll!"— a German word meaning "awesome."

Some TLRs are anchored in plasma membranes, allowing cells to "see" PAMPs in the extracellular environment

(**figure 14.7**). Others are in phagosomal or endosomal membranes, facing the lumen of the organelle. These let macrophages and dendritic cells characterize the material they ingested. ◀◀ endosome, p. 72 ◀◀ phagosome, p. 72 ◀◀ lumen, p. 69

When a TLR detects a compound, a signal is transmitted to the cell's nucleus, causing certain genes to be expressed. The actual response is dictated by the cell type and the array of TLRs triggered, so it can be tailored to the situation and category of pathogen. For example, when macrophages "see" bacterial products, they begin producing pro-inflammatory cytokines, leading to an inflammatory response. If a TLR detects viral nucleic acid, the cell synthesizes products that promote an antiviral response.

NOD-Like Receptors (NLRs)

NOD-like receptors (NLRs) are cytoplasmic proteins that detect bacterial components, allowing the cell to recognize when its own borders have been breached; some also detect signs of cell damage (**figure 14.8**). At least 23 NLRs have been described, but details about their roles are still being uncovered.

When NLRs detect PAMPs or DAMPs, they unleash a series of events that lead to outcomes that protect the host, sometimes at the expense of the cell itself. In macrophages, some NRLs can join with other proteins in the cytoplasm to form an **inflammasome**.

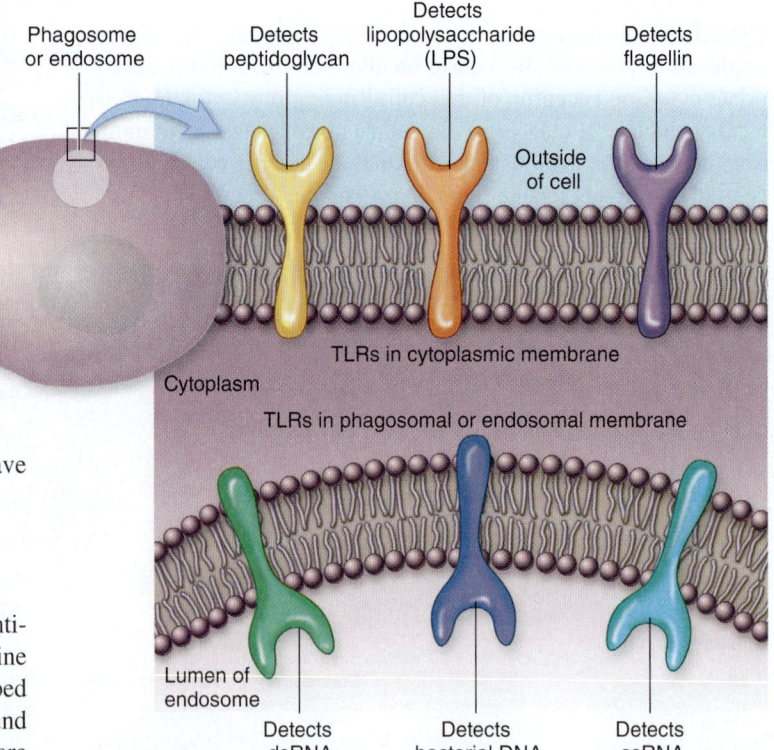

FIGURE 14.7 Toll-Like Receptors (TLRs) These pattern recognition receptors are anchored in membranes of sentinel cells, allowing these cells to "see" microbial compounds that originated outside of the cell. Not all of the compounds recognized by TLRs are shown in this figure.

❓ *Why would cells that line the blood vessels have TLRs?*

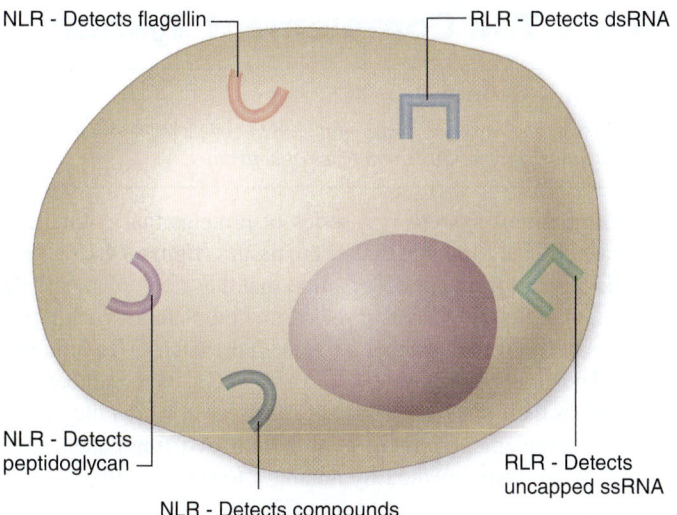

NLR - Detects flagellin

RLR - Detects dsRNA

NLR - Detects peptidoglycan

RLR - Detects uncapped ssRNA

NLR - Detects compounds that indicate cell damage

FIGURE 14.8 NOD-Like Receptors (NLRs) and RIG-Like Receptors (RLRs) These pattern recognition receptors are found within cells and detect microbial components that indicate cell invasion. Not all NLRs and RLRs are shown.

? *How do NLRs and RLRs differ from TLRs with respect to the source of the microbial compounds detected?*

This complex then activates a potent pro-inflammatory cytokine, thereby initiating an inflammatory response. Mutations in the genes encoding NLRs seem to be a predisposing factor in certain inflammatory diseases such as Crohn's disease (an inflammatory bowel disease). Scientists hope that by studying NLRs, they will better understand a wide range of inflammatory disorders.

RIG-Like Receptors (RLRs)

RIG-like receptors (RLRs) are cytoplasmic proteins that detect viral RNA (figure 14.8). As with pattern recognition by NLRs, this provides a mechanism for an infected cell to detect the invader. RLRs can distinguish viral RNA from normal cellular RNA because at least two characteristics differ. First, viral RNA often has three phosphates at the 5′ end; recall that cellular RNA is processed after transcription, and one of those steps (capping) modifies the 5′ end, hiding the phosphates. Second, viral RNA is often double-stranded. RNA viruses other than retroviruses routinely generate long dsRNA as a result of viral replication. Even DNA viruses often generate long dsRNA because both strands are sometimes used as templates for transcription, leading to production of complementary RNA molecules. Cellular RNA is typically not double-stranded because only one DNA strand in a gene is used as a template for mRNA synthesis.

◀◀ capping, p. 176

The Interferon Response

When a cell's RLRs detect viral RNA, the cell responds by synthesizing and secreting interferons (**figure 14.9**). These proteins then attach to specific receptors on both the infected cell and neighboring cells, causing the cells to express what can be viewed as inactive "suicide enzymes" (protein kinase R, RNase L, and others). For convenience, we will refer to these collectively as inactive antiviral proteins (iAVPs). The activated forms of the antiviral proteins (AVPs) degrade mRNA and stop protein synthesis, leading to apoptosis (a programmed cell death). A key feature of this

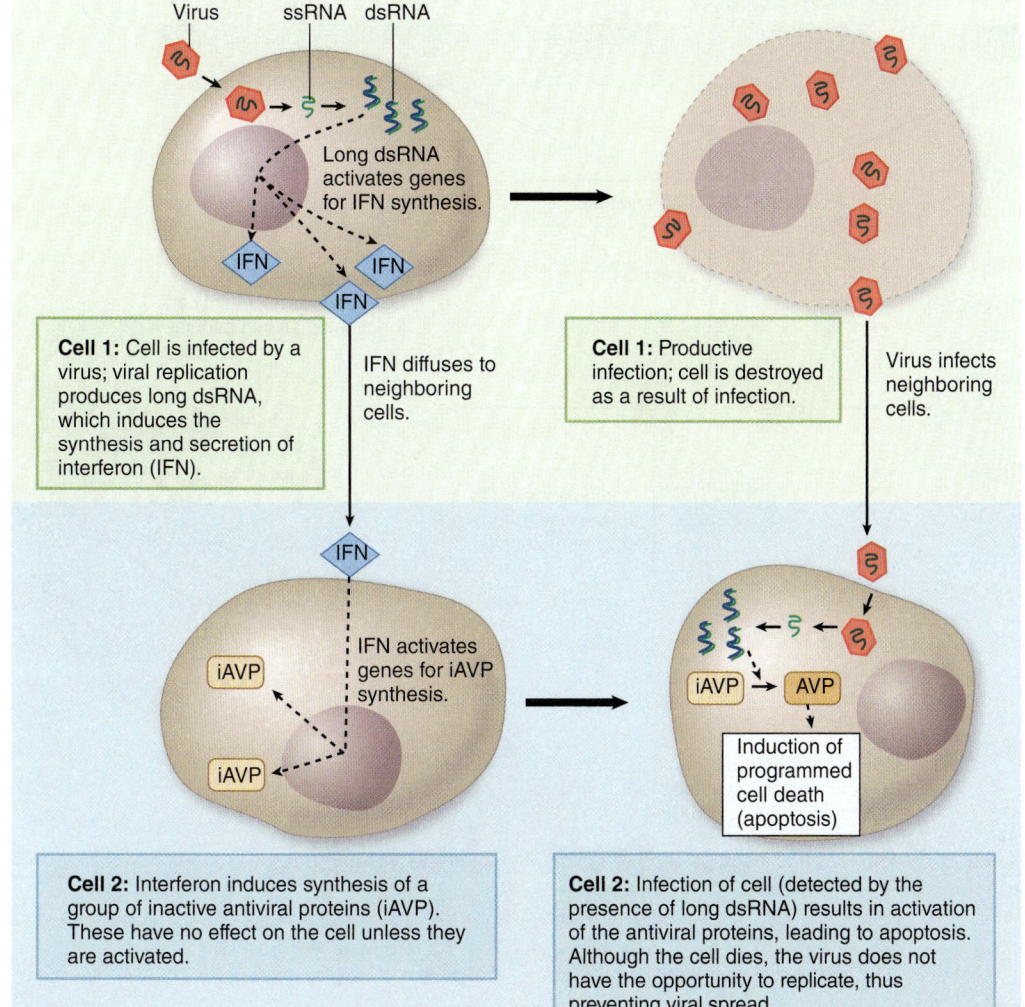

Virus   ssRNA   dsRNA

Long dsRNA activates genes for IFN synthesis.

IFN   IFN

IFN

Cell 1: Cell is infected by a virus; viral replication produces long dsRNA, which induces the synthesis and secretion of interferon (IFN).

IFN diffuses to neighboring cells.

Cell 1: Productive infection; cell is destroyed as a result of infection.

Virus infects neighboring cells.

IFN

IFN activates genes for iAVP synthesis.

iAVP

iAVP

iAVP → AVP

Induction of programmed cell death (apoptosis)

Cell 2: Interferon induces synthesis of a group of inactive antiviral proteins (iAVP). These have no effect on the cell unless they are activated.

Cell 2: Infection of cell (detected by the presence of long dsRNA) results in activation of the antiviral proteins, leading to apoptosis. Although the cell dies, the virus does not have the opportunity to replicate, thus preventing viral spread.

FIGURE 14.9 Antiviral Effects of Interferon

? *Why would it be beneficial to a host for a virally infected cell to undergo apoptosis?*

response is that the iAVPs are activated by dsRNA. Thus, when cells bind interferon, only the infected ones are sacrificed. Their uninfected counterparts remain functional but are prepared to undergo apoptosis should they become infected. ▶▶ apoptosis, p. 350

MicroAssessment 14.5

Cells use TLRs to detect microbial compounds that originated outside of the cell. NLRs and RLRs detect microbial compounds within an infected cell. Viral RNA triggers an interferon response.

13. *Why would macrophages have TLRs in their endosomal membranes?*

14. *If a cell produces antiviral proteins, what happens to that cell when those proteins encounter long dsRNA?*

15. *Why would the discovery of TLRs alter the view that innate immunity is non-specific?* ✚

14.6 ■ The Complement System

Learning Outcome

6. *Describe the three pathways that lead to complement system activation and the three outcomes of activation.*

The **complement system** is a series of proteins that circulate in the blood and the fluid that bathes the tissues (**figure 14.10**). Their name is derived from the observation that they "complement" the function of antibodies. Each of the major complement system proteins has been given a number along with the letter C (for complement). The nine major proteins, C1 through C9, were numbered in the order of their discovery and not the order in which they react. When a complement protein is split into two fragments, those fragments are distinguished by adding a lowercase letter to each name. For example, C3 is split into C3a and C3b.

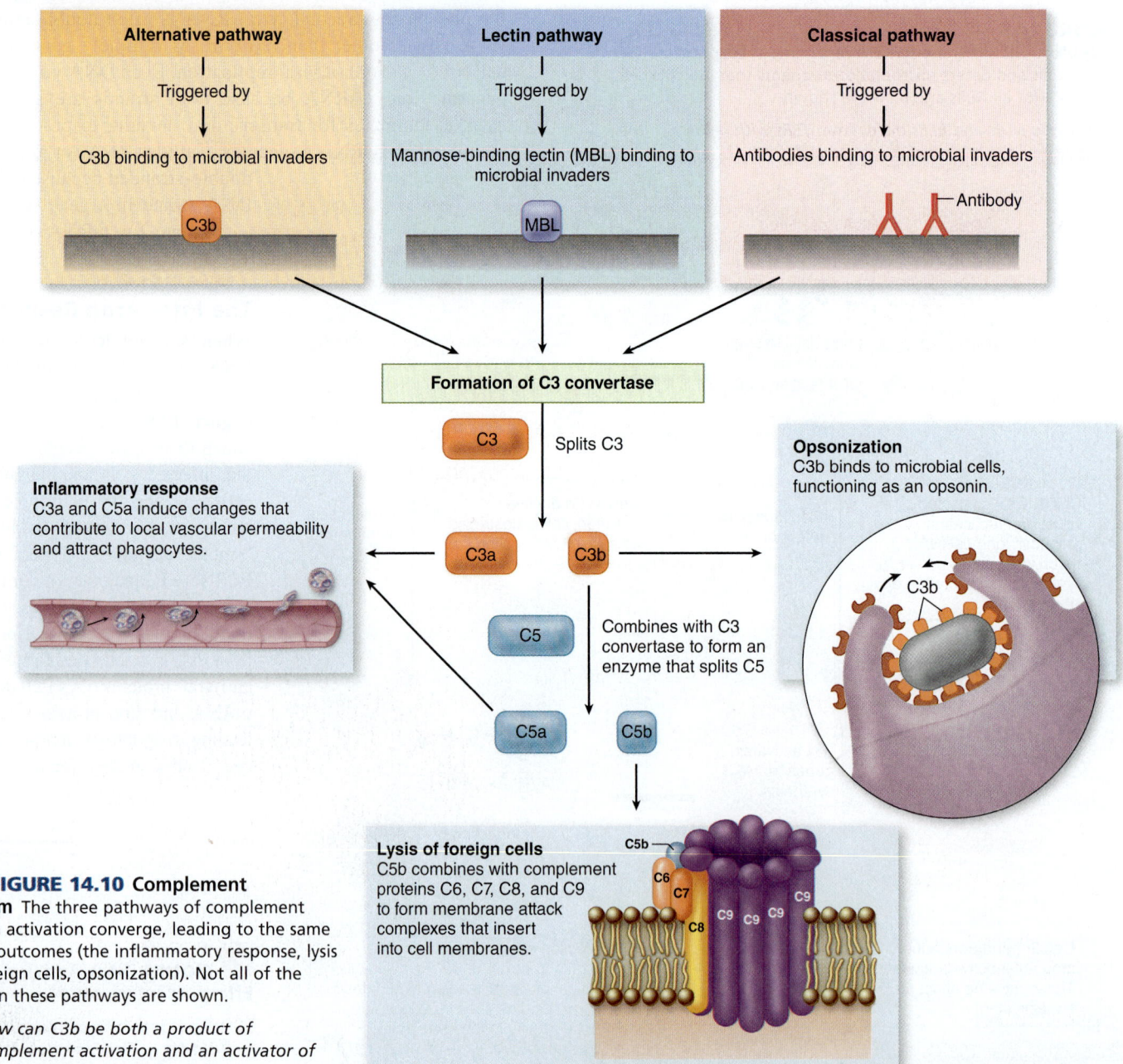

FIGURE 14.10 Complement System The three pathways of complement system activation converge, leading to the same three outcomes (the inflammatory response, lysis of foreign cells, opsonization). Not all of the steps in these pathways are shown.

❓ *How can C3b be both a product of complement activation and an activator of the complement system?*

Complement proteins routinely circulate in an inactive form, but in response to certain signals indicating the presence of microbial invaders, a cascade of reactions occurs. This results in the rapid activation of the system. The activated forms of the complement proteins have specialized functions that cooperate with other host defenses to quickly remove and destroy the invader.

Complement System Activation

The complement system can be activated by three different pathways that converge when a complex called C3 convertase is formed (figure 14.10). C3 convertase then splits C3, leading to additional steps of the activation cascade.

Alternative Pathway

The name of the **alternative pathway** may seem to imply that the pathway is "second choice," but it actually reflects the fact that the pathway was not discovered first. It is quickly and easily triggered, providing vital early warning that an invader is present.

The alternative pathway is triggered when C3b binds to foreign cell surfaces. The binding of C3b allows other complement proteins to then attach, eventually forming the C3 convertase. What might seem confusing is the fact that C3b is a product of complement activation, yet it also triggers the alternative pathway. How can it be both a product and a trigger? This can occur because C3 is somewhat unstable, and spontaneously splits to C3a and C3b at a low rate even when the complement system has not been activated. The C3a and C3b formed this way are rapidly inactivated by regulatory proteins, but some C3b is always present to trigger the alternative pathway when needed.

Lectin Pathway

Activation of the complement system via the **lectin pathway** involves pattern recognition molecules called mannose-binding lectins (MBLs). These bind to certain arrangements of multiple mannose molecules that characterize microbial cells. Once an MBL attaches to a surface, it can interact with other complement system components to form a C3 convertase. ◄◄ mannose, p. 31

Classical Pathway

Complement system activation by the **classical pathway** requires antibodies, a component of adaptive immunity. When antibodies bind to an antigen (forming an antigen-antibody complex, also called an immune complex), they interact with the same complement system component involved with the lectin pathway to form a C3 convertase.

Effector Functions of the Complement System

Activation of the complement system eventually leads to three major protective outcomes: opsonization, an inflammatory response, and lysis of foreign cells (figure 14.10).

Opsonization

The C3b concentration increases substantially when the complement system is activated, and these molecules bind to bacterial cells or other foreign particles. This has two effects: (1) continued complement activation via the alternative pathway, and (2) **opsonization.** Material that has been opsonized (meaning "prepared for eating") is easier for phagocytes to bind to and engulf because phagocytes have receptors that attach specifically to the opsonins (in this case, C3b).

MicroByte —
Think of opsonization as coating a microbe with one layer of Velcro, with phagocytes having the opposing layer on their surface.

Inflammatory Response

The complement component C5a is a potent chemoattractant, drawing phagocytes to the area where the complement system has been activated. In addition, C3a and C5a induce changes in the endothelial cells that line the blood vessels, contributing to the vascular permeability associated with inflammation. They also cause mast cells to release various pro-inflammatory cytokines.

Lysis of Foreign Cells

Complexes of complement system proteins (C5b, C6, C7, C8, and multiple C9 molecules) spontaneously assemble in cell membranes, forming doughnut-shaped structures called **membrane attack complexes (MACs) (figure 14.11)**. This creates pores in the membrane, disrupting the integrity of the cell. MACs have little effect on Gram-positive bacteria because the peptidoglycan layer of these cells prevents the complement system components from reaching their cytoplasmic membranes. The outer membranes of Gram-negative bacteria, however, make them susceptible.

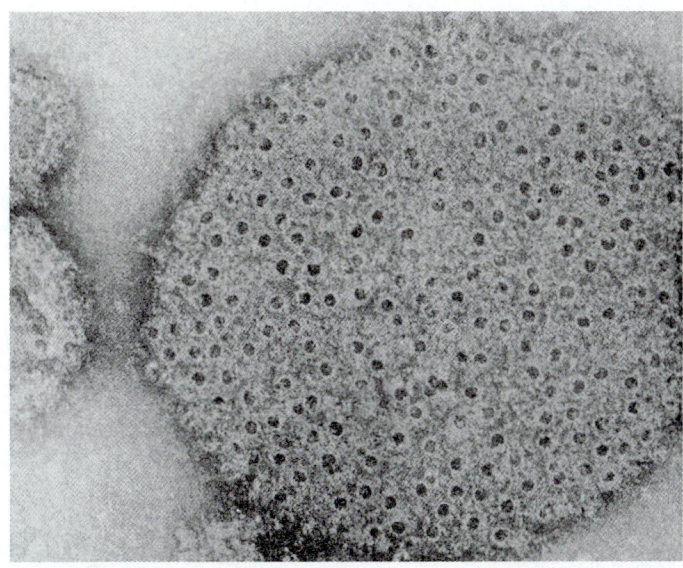

FIGURE 14.11 Membrane Attack Complexes (MACs) in a Cell Membrane Each dark dot in this electron micrograph is a MAC.

? *How do MACs cause cells to lyse?*

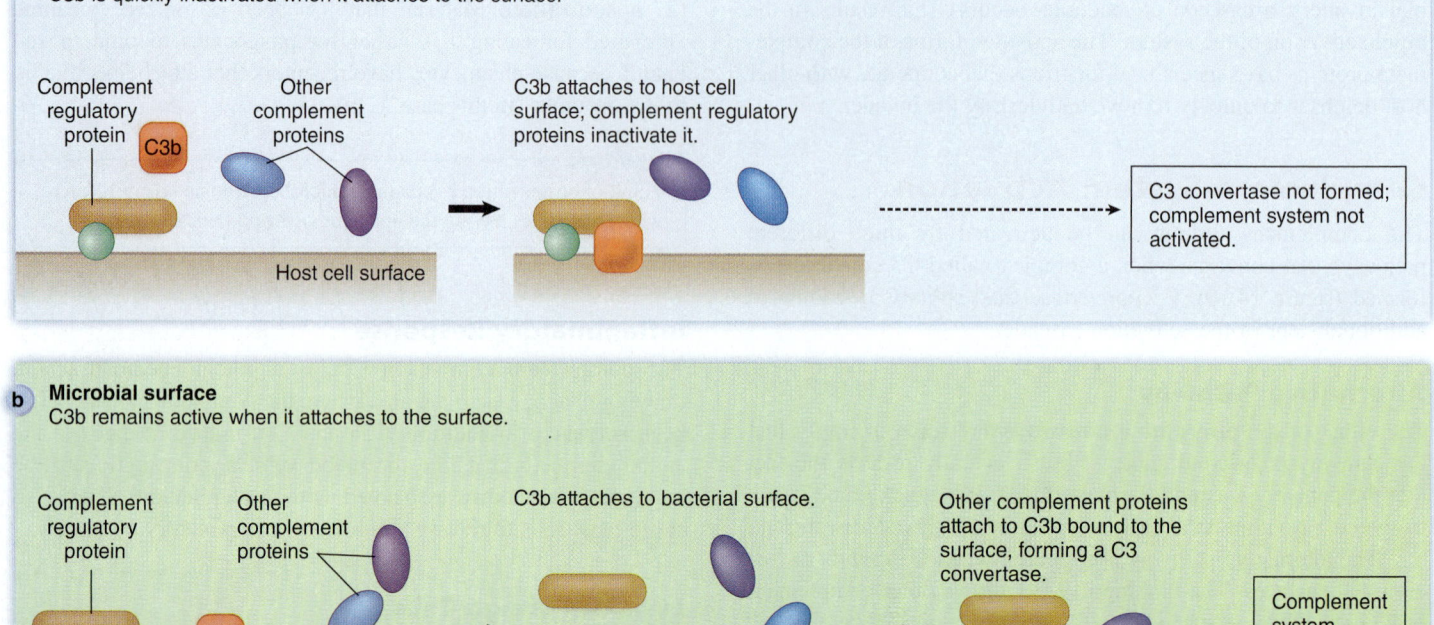

FIGURE 14.12 Complement Regulatory Proteins Inactivate C3b (a) Molecules in host cell membranes bind regulatory proteins that quickly inactivate C3b. **(b)** Most microbial cell surfaces do not bind the regulatory proteins, so they trigger the alternative pathway of complement activation.

❓ *Some pathogens attract complement regulatory proteins to their surfaces. How would this help the pathogens avoid destruction?*

Regulation of the Complement System

A number of different control mechanisms prevent host cells from activating the complement system and also protect them from the effector functions of the complement proteins. For example, molecules in host cell membranes bind regulatory proteins that quickly inactivate C3b. This prevents host cell surfaces from triggering the alternative pathway of complement regulation (**figure 14.12**). It also prevents host cells from being opsonized. Most microbial cell surfaces do not bind the regulatory proteins, but as we will discuss in chapter 16, some pathogens have mechanisms to hijack the host's protective mechanisms.

MicroAssessment 14.6

The complement system can be activated by three different pathways that each lead to opsonization, an inflammatory response, and lysis of foreign cells.

16. *What role does C3b play in opsonization and complement system activation?*
17. *The body's own cells do not trigger the alternative pathway of complement pathway activation. How come?*
18. *From a pathogen's standpoint, why would it be beneficial to bind regulatory proteins that inactivate C3b?* ➕

14.7 ■ Phagocytosis

Learning Outcomes

7. *Outline the steps of phagocytosis.*
8. *Compare and contrast the roles of macrophages and neutrophils.*

Phagocytes routinely engulf and digest material, including invading organisms. In routine situations, such as when microbes enter through a minor skin wound, resident macrophages in the tissues destroy the relatively few invaders that enter. If the microbes are not rapidly cleared, macrophages produce cytokines to recruit additional phagocytes—particularly neutrophils—for extra help.

The Process of Phagocytosis

Phagocytosis involves a series of steps (**figure 14.13**). These are particularly important medically, because most pathogens have evolved the ability to evade one or more of them, a topic explored in chapter 16.

Chemotaxis ①

Phagocytic cells are recruited to the site of infection or tissue damage by chemicals that act as chemoattractants. These include

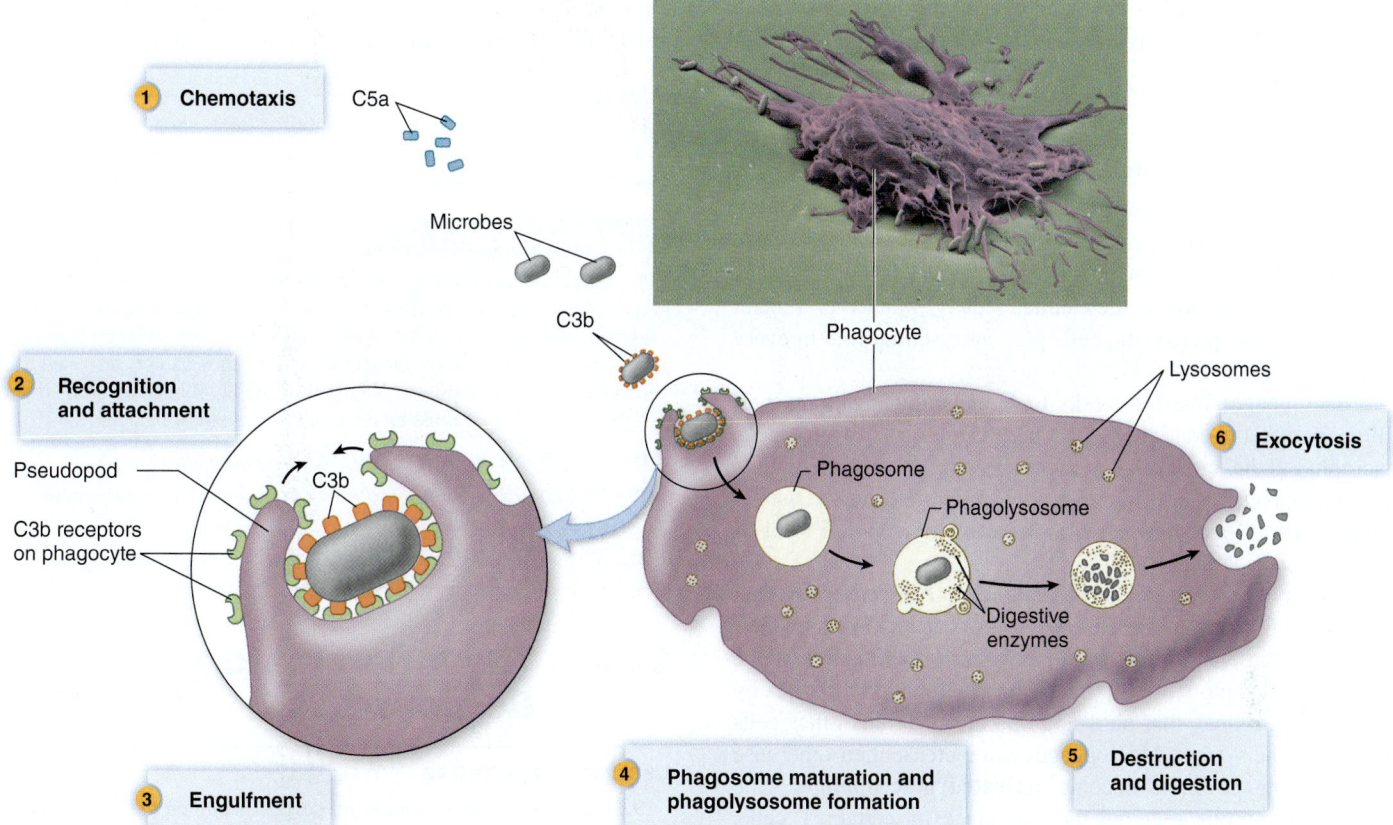

FIGURE 14.13 Phagocytosis This diagram shows a microbe that has been opsonized by the complement protein C3b; certain classes of antibodies can also function as opsonins.

❓ *What would happen if a bacterium could prevent the phagosome from fusing with lysosomes?*

products of microorganisms, phospholipids released by injured host cells, chemokines, and the complement system component C5a.

Recognition and Attachment ②

Phagocytic cells use various receptors to bind invading microbes either directly or indirectly. For example, direct binding occurs when a phagocyte's receptors bind mannose, a sugar found on some bacteria and yeasts. Indirect binding happens when a particle has first been opsonized. **Opsonins** are secreted proteins that tag particles for phagocytosis and include the complement component C3b and certain classes of antibody molecules. Phagocytes have specific receptors for opsonins, making it easier for the cells to attach to and subsequently engulf the material.

Engulfment ③

Once the phagocyte has attached to a particle, it sends out pseudopods that surround and engulf the material. This action brings the material into the cell, enclosed in a **phagosome.** If a phagocyte encounters something too large to engulf, it releases its toxic contents as a means of destroying it. ◀◀ **pseudopod, p. 72** ◀◀ **phagosome, p. 72**

Phagosome Maturation and Phagolysosome Formation ④

Initially, a phagosome has no antimicrobial capabilities, but it matures to develop these. As part of this process, it fuses with

various endosomes, allowing it to gain properties that characterize those endosomes. For example, the pH becomes progressively more acidic. Finally, the phagosome fuses with enzyme-filled lysosomes, forming a **phagolysosome.** ◀◀ **lysosome, p. 77**

The phagosome maturation stages are highly regulated, and appear to depend on the type of material ingested. For instance, a phagosome that contains host cell material has a different fate than one that carries microbial components.

Destruction and Digestion ⑤

A number of factors within the phagolysosome work together to destroy an engulfed invader. O_2 consumption increases dramatically—a phenomenon called respiratory burst—allowing an enzyme to produce reactive oxygen species (ROS), which are toxic. Another enzyme makes nitric oxide, which reacts with ROS to produce additional toxic compounds. Special pumps move protons into the phagolysosome, lowering the pH. The various enzymes contributed by the lysosomes degrade peptidoglycan and other components. Defensins damage membranes of the invader, and lactoferrin ties up iron. ◀◀ **reactive oxygen species, p. 90**

Exocytosis ⑥

Following digestion, the vesicle fuses with the plasma membrane, expelling the remains. In the case of macrophages, some of the ingested material is also put on the cell's surface as a way of displaying bits of invaders to certain cells of the adaptive immune system. Details of this process will be discussed in chapter 15.

◀◀ exocytosis, p. 72

Specialized Attributes of Macrophages

Macrophages can be viewed as the scavengers and sentries—routinely phagocytizing dead cells and debris, but ready to destroy invaders and call in reinforcements when needed. They are always present in tissues, where they either slowly wander or remain stationary. These phagocytic cells play an essential role in every major tissue in the body.

Macrophages live for weeks to months, and maintain their killing power by continually regenerating their lysosomes. As macrophages die, circulating monocytes—which can differentiate into macrophages—leave the blood and migrate to the tissues to replace them. Monocyte migration increases in response to invasion and tissue damage.

Macrophages have several important characteristics that allow them to carry out their diverse tasks. Various TLRs on their surfaces and in phagosomes help them sense microbial invaders. When these receptors are triggered, the macrophage responds by producing cytokines that alert and stimulate various other cells of the immune system. Macrophages can increase their otherwise limited killing power to become **activated macrophages.** One mechanism for this requires the assistance of certain T cells, an example of the cooperation between the innate and adaptive defenses. The compounds produced by activated macrophages also damage tissues when released, so it would be potentially harmful for macrophages to routinely be in an activated state. Details of the activation process, including the roles of T cells, will be discussed in chapter 15. ▶▶ macrophage activation, p. 373

If activated macrophages fail to destroy microbes, the phagocytes can fuse together to form **giant cells.** Macrophages, giant cells, and T cells form concentrated groups called **granulomas** that wall off and retain organisms or other material that cannot be destroyed; again, this is an example of the cooperation between defense systems. Granulomas, which are part of the disease process in tuberculosis and several other illnesses, prevent the microbes from escaping to infect other cells (see figure 21.20). Unfortunately, they also harm the host because they interfere with normal tissue function. ▶▶ tuberculosis, p. 502

Specialized Attributes of Neutrophils

Neutrophils can be viewed as the rapid response team—quick to move into an area of trouble and ready to eliminate the invaders. They play a critical role during the early stages of inflammation, being the first cell type recruited to the site of damage from the bloodstream. They have more killing power than macrophages. The cost for their effectiveness, however, is a relatively short life span of only 1 to 2 days in the tissues; once they have used their granules, they die. Fortunately, many more neutrophils are in reserve.

Neutrophils not only kill microbes through phagocytosis, they also release the contents of their granules along with DNA to form neutrophil extracellular traps (NETs). The DNA strands in the

NET ensnare microbes, allowing the granule contents (enzymes and peptides) that accumulate within the NET to destroy them.

MicroByte

For every neutrophil in the circulatory system, about 100 more are waiting in the bone marrow, ready to be mobilized when needed.

MicroAssessment 14.7

The process of phagocytosis includes chemotaxis, recognition and attachment, engulfment, phagosome maturation and phagolysosome formation, destruction and digestion, and exocytosis. Macrophages are long-lived and always present in tissues; they can be activated to enhance their killing power. Neutrophils are highly active, short-lived phagocytic cells that must be recruited to the site of damage.

19. *How does a phagolysosome differ from a phagosome?*
20. *Tuberculosis is characterized by granulomas called tubercles. What is a granuloma?*
21. *What could a microorganism do to avoid engulfment?* ➕

14.8 ■ The Inflammatory Response

Learning Outcomes

9. *Describe the inflammatory process, focusing on the factors that initiate the response and the outcomes of inflammation.*
10. *Compare and contrast apoptosis and pyroptosis.*

When tissues have been damaged, such as when an object penetrates the skin or microbes are introduced, **inflammation** occurs. The purpose of this is to contain a site of damage, localize the response, eliminate the invader, and restore tissue function. Everyone has experienced the signs of inflammation; in fact, the Roman physician Celsus described the four cardinal signs in the first century A.D.—swelling, redness, heat, and pain. A fifth sign, loss of function, is sometimes present.

Factors That Trigger an Inflammatory Response

A number of factors trigger an inflammatory response, but they often involve the pattern recognition receptors described in section 14.5—particularly TLRs and NLRs. Recall that these receptors detect pathogen-associated molecular patterns (PAMPs) and damage-associated molecular patterns (DAMPs). The triggers of inflammation cause host cells to release **inflammatory mediators,** a collective term for various pro-inflammatory cytokines and chemicals such as histamine and bradykinin. Inducers of the inflammatory process include:

■ **Microbes.** When TLRs on sentinel cells such as macrophages detect PAMPs, the cells produce inflammatory mediators. One of these, tumor necrosis factor (TNF), induces the liver to synthesize acute-phase proteins, a group of proteins that facilitate phagocytosis and complement activation. When NLRs within cells detect PAMPs, additional inflammatory mediators are released. Meanwhile, microbial surfaces trigger complement activation, also leading to an inflammatory response.

- **Tissue damage.** The sensors of DAMPs are not well understood, but seem to involve NLRs. As with detection of PAMPs, these cause cells to release inflammatory mediators. If blood vessels are injured, two enzymatic cascades are activated. One is the coagulation cascade, which results in blood clotting, and the other produces several molecules such as bradykinin, which increases blood vessel permeability.

The Inflammatory Process

The inflammatory process involves a cascade of events that result in dilation of small blood vessels, leakage of fluids from those vessels, and the migration of leukocytes out of the bloodstream and into the tissues (**figure 14.14**).

The diameter of local blood vessels increases during inflammation due to the action of inflammatory mediators. This results in greater blood flow to the area, causing the heat and redness associated with inflammation. It also slows the blood flow in the capillaries. Because of the dilation, normally tight junctions between endothelial cells are disrupted, allowing more fluid to leak from the vessels and into the tissue. This fluid contains various substances such as transferrin, complement system proteins, and antibodies that help counteract invading microbes. The increase of fluids in the tissues causes the swelling and pain associated with inflammation. The direct effects of chemicals on sensory nerve endings also cause pain (see **Perspective 14.1**).

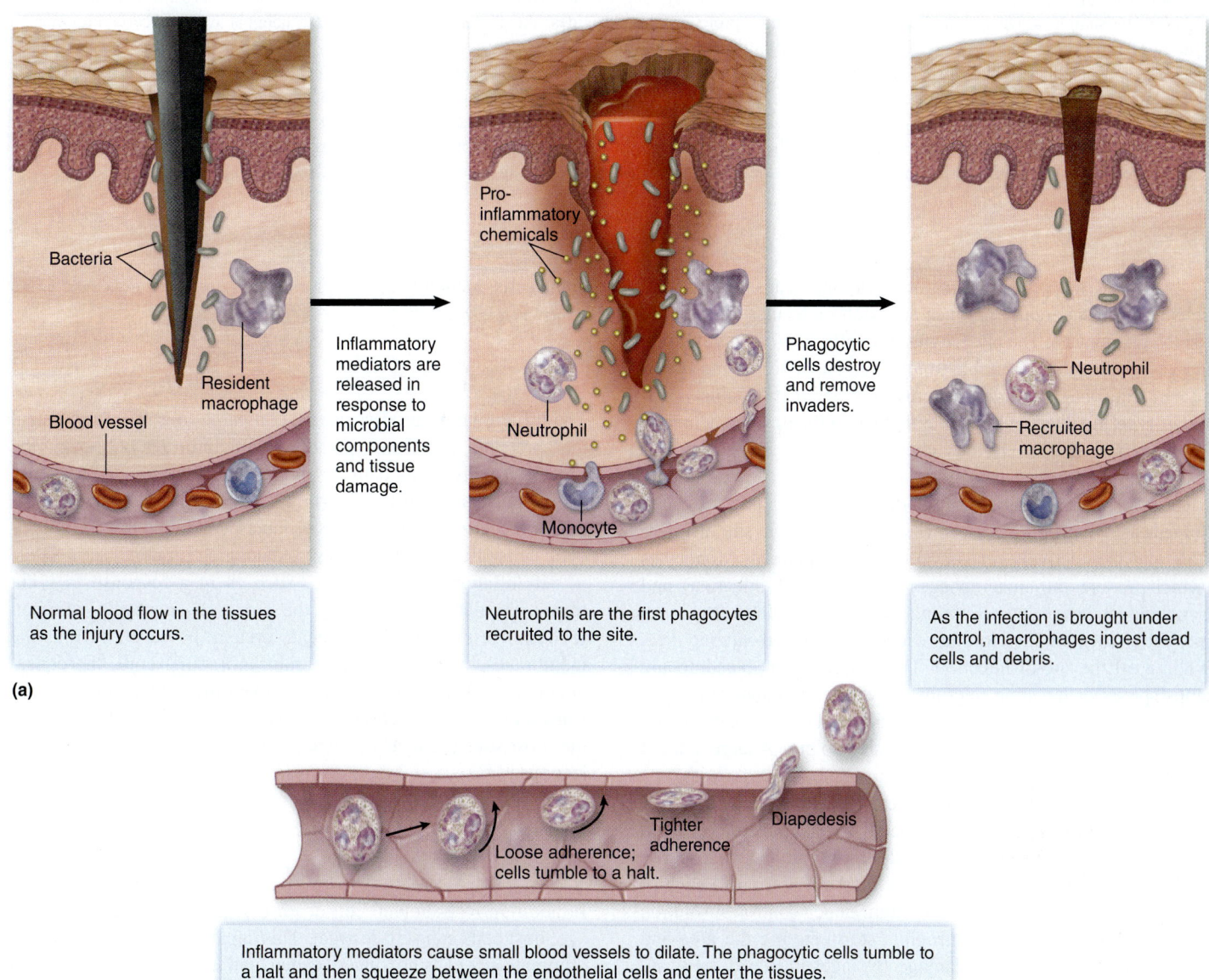

(a)

Normal blood flow in the tissues as the injury occurs.

Inflammatory mediators are released in response to microbial components and tissue damage.

Neutrophils are the first phagocytes recruited to the site.

Phagocytic cells destroy and remove invaders.

As the infection is brought under control, macrophages ingest dead cells and debris.

Pro-inflammatory chemicals

Bacteria

Resident macrophage

Blood vessel

Neutrophil

Monocyte

Neutrophil

Recruited macrophage

(b)

Loose adherence; cells tumble to a halt.

Tighter adherence

Diapedesis

Inflammatory mediators cause small blood vessels to dilate. The phagocytic cells tumble to a halt and then squeeze between the endothelial cells and enter the tissues.

FIGURE 14.14 The Inflammatory Response (a) The process of inflammation. **(b)** Phagocytes leave the blood vessels and move to the site of infection.

Which type of phagocyte is the first to be recruited to a site of inflammation?

PERSPECTIVE 14.1

For *Schistosoma,* the Inflammatory Response Delivers

The parasitic flatworms that cause schistosomiasis do not shy from the immune response when it comes to reproducing; instead they use it to move their ova to an environment where they might hatch. Adult females of *Schistosoma* species, which live in the bloodstream of infected hosts, lay their ova in veins near the intestine or bladder. They then rely on the inflammatory response there to expel the ova, completing one portion of a complex life cycle. The ova released in feces or urine can hatch—liberating a larval stage called a miracidium—if untreated sewage reaches water. The miracidium then infects a specific freshwater snail host, where it multiplies asexually. The infected snail then releases large numbers of another larval form, cercariae, which swim about in search of a human host.

The parasite is acquired when a person wades or swims in contaminated water. The cercariae penetrate the skin by burrowing through with the aid of digestive enzymes; schistosomes are rare among pathogens because they can penetrate intact skin. The larvae then enter the bloodstream where the parasite can live for over a quarter of a century.

Schistosoma species have separate sexes and, remarkably, the male and female worms locate one another in the bloodstream. The male's body has a deep longitudinal groove in which he clasps his female partner to live in copulatory embrace (*schistosoma* means "split-body," referring to the long slit). The adult worms effectively mask themselves from the immune system by adsorbing various blood proteins; this provides them with a primitive stealth "cloaking device."

The worms migrate to the veins of either the intestine or bladder (depending on the schistosome species) to lay hundreds of ova per day. The innate immune system responds vigorously to the highly antigenic eggs, ejecting them in a manner similar to what is experienced as a sliver in the skin works its way to the surface. Only half of the ova are expelled, however, and the remaining ones are often swept by the bloodstream to the liver. The inflammatory process and granuloma formation there gradually destroy liver cells, replacing the cells with scar tissue. Malfunction of the liver results in malnutrition and a buildup of pressure in intestinal and esophageal veins. Fluid accumulates in the abdominal cavity and hemorrhage occurs if the esophageal veins rupture.

Despite their complex life cycle, *Schistosoma* species are highly successful parasites. Not only are they adept at avoiding certain immune responses that would otherwise lead to their destruction, they have learned to exploit the inflammatory response for their own spread. Over 200 million people worldwide are infected with these parasites, resulting in over 500,000 deaths each year.

Some of the pro-inflammatory cytokines cause endothelial cells in the local area to produce adhesion molecules that loosely "grab" phagocytes. The phagocytes normally flow rapidly through the vessels but slowly tumble to a halt as the adhesion molecules attach. The phagocytic cells themselves begin producing a different type of adhesion molecule that strengthens the attachment. Then, in response to various chemoattractants, the phagocytes leave the blood vessels and move into the area. They do this by squeezing between the cells of the dilated vessel, a process called diapedesis. Neutrophils are the first to arrive at the site of infection, and they actively phagocytize foreign material. Monocytes (which mature into macrophages at the site of infection) and lymphocytes arrive later. Clotting factors in the fluid that leaks into the tissues initiate clotting reactions in the surrounding area, walling off the site of infection. This helps prevent bleeding and stops spread of invading microbes. As the inflammatory process continues, large quantities of dead neutrophils accumulate. Those dead cells, along with tissue debris, make up pus. A large amount of pus constitutes an abscess (see figure 22.2). ▶▶| **abscess, p. 550**

The extent of inflammation varies, depending on the nature of the injury, but the response is localized, begins immediately upon injury, and increases rapidly. A short-term inflammatory response is called **acute inflammation** and is marked by a prevalence of neutrophils. As the infection is brought under control, resolution of inflammation begins. Neutrophils stop entering the area and macrophages clean up the damage by ingesting dead cells and debris. As the area heals, new capillaries grow, destroyed tissues are replaced, and scar tissue forms.

If acute inflammation cannot limit the infection, **chronic inflammation** occurs. This is a long-term inflammatory process that can last for years. In chronic inflammation, macrophages and giant cells accumulate, and granulomas form. ▶▶| **giant cells, p. 373** ▶▶| **granulomas, p. 373**

Damaging Effects of Inflammation

The inflammatory process can be compared to a sprinkler system that prevents fire from spreading in a building. Although the process usually limits damage and restores function, the response itself can cause significant harm. One undesirable consequence is that some enzymes and toxic products contained within phagocytic cells are inevitably released, damaging tissues.

If inflammation is limited, such as in a response to a cut finger, the damage caused by the process is normally minimal. If the process occurs in a delicate system, however, such as the membranes that surround the brain and spinal cord, the consequences can be much more severe, even life-threatening. As you learn more about infectious diseases, you will notice that many of the most severe effects of infection result from the inflammatory response.

Cell Death and the Inflammatory Process

In addition to traumatic cell death that results from tissue damage, host cells can self-destruct. This capability allows the host to eliminate any cells no longer needed, and it serves as a mechanism for sacrificing "self" cells that might otherwise spread an infection. One type of programmed cell death avoids an inflammatory response, whereas another type promotes one.

Apoptosis (*apo* means "off"; *ptosis* means "falling") is a programmed cell death that does not trigger an inflammatory response. During apoptosis, the dying cells undergo certain changes. For example, the shape of the cell changes, enzymes cut the DNA, and portions of the cell bud off, effectively shrinking the cell. Some changes appear to signal macrophages that the remains of the cell are to be engulfed without the events associated with inflammation.

If the pattern recognition receptors in a macrophage's cytoplasm are triggered, that cell might initiate **pyroptosis** (*pyro* means "fire"). This programmed self-destruction triggers an inflammatory response, recruiting various components of the immune system to the region.

MicroAssessment 14.8

The inflammatory response is initiated when microbes invade or tissues are damaged. The outcome is dilation of small blood vessels, leakage of fluids from those vessels, and migration of leukocytes out of the bloodstream and into the tissue. Inflammation helps contain an infection, but the response itself can be damaging. Apoptosis destroys "self" cells without initiating inflammation; pyroptosis triggers an inflammatory response.

22. *Describe two general events that can initiate inflammation.*
23. *Describe two changes in cells undergoing apoptosis.*
24. *Infection of the fallopian tubes can lead to infertility. Why would this be so?* ✚

14.9 ■ Fever

Learning Outcome

11. *Describe the induction and outcomes of fever.*

Fever is an important host defense mechanism, and a strong indication of infectious disease, especially of bacterial origin. The temperature within the human body is normally kept around 37°C by a temperature-regulation center in the brain. During an infection, the regulating center "sets" the body's thermostat at a higher level. An oral temperature above 37.8°C is regarded as fever.

A higher temperature setting occurs as a result of certain pro-inflammatory cytokines released by macrophages when they detect microbial products. The cytokines are carried in the bloodstream to the brain, where they act as messages that microbes have invaded the body. These cytokines and other fever-inducing substances are **pyrogens.** Fever-inducing cytokines are called endogenous pyrogens, indicating the body makes them, whereas microbial products are called exogenous pyrogens, indicating they are introduced from external sources. The temperature-regulating center responds to pyrogens by raising body temperature.

The adverse effect of fever on pathogens partly involves ideal growth temperature. Bacteria that grow best at 37°C are less likely to cause disease in people with fever because bacterial growth rates often decline sharply above their optimum growth temperature. A slower growth rate allows more time for other defenses to destroy invaders. ◄◄ temperature requirements, p. 89

A moderate rise in temperature increases the rate of enzymatic reactions. Fever has been shown to enhance the inflammatory response, phagocytic killing by leukocytes, multiplication of lymphocytes, release of substances that attract neutrophils, and production of interferons and antibodies. Release of leukocytes into the blood from the bone marrow also increases.

MicroAssessment 14.9

Fever results when macrophages release pro-inflammatory cytokines; this occurs when the TLRs on the macrophages are engaged by microbial products.

25. *What are endogenous and exogenous pyrogens?*
26. *How does fever inhibit the growth of pathogens?*
27. *Syphilis was once treated by infecting the patient with the parasite that causes malaria, a disease characterized by repeated bouts of fever, shaking, and chills. Why would this treatment cure syphilis?* ✚

Summary

14.1 ■ Overview of the Innate Defenses

First-line defenses prevent entry, sensor systems detect invasion, and effector mechanisms destroy and remove the invader (figures 14.1). Cells have **pattern recognition receptors (PRRs)** that recognize invaders. The **complement system** is activated in response to certain invaders. The **inflammatory response** involves many components of innate immunity, including **phagocytes.**

14.2 ■ First-Line Defenses (figure 14.2)

Physical Barriers

The skin is composed of two main layers—the dermis and the epidermis. Mucous membranes are constantly bathed with mucus and other secretions that help wash microbes from the surfaces (figure 14.3).

Antimicrobial Substances

Lysozyme, peroxidase enzymes, lactoferrin, and **defensins** inhibit or kill microorganisms (figure 14.4).

Normal Microbiota (Flora)

Members of the **normal microbiota** competitively exclude pathogens and stimulate the host defenses.

14.3 ■ The Cells of the Immune System (figure 14.5, table 14.1)

Granulocytes

Granulocytes include **neutrophils, basophils,** and **eosinophils.**

Mononuclear Phagocytes

Monocytes circulate in blood; **macrophages** are in tissues (figure 14.6).

Dendritic Cells

Dendritic cells develop from monocytes; some have other origins.

Lymphocytes

Lymphocytes, which include **B cells, T cells,** and **natural killer (NK) cells,** are involved in adaptive immunity.

14.4 ■ Cell Communication

Surface Receptors

Surface receptors bind **ligands,** allowing the cell to detect substances.

Cytokines (table 14.2)

Cytokines include **chemokines, colony-stimulating factors (CSFs), interferons (IFNs), interleukins (ILs),** and **tumor necrosis factor (TNF).**

Adhesion Molecules

Adhesion molecules allow cells to adhere to other cells.

14.5 ■ Pattern Recognition Receptors (PRRs)

Toll-Like Receptors (TLRs)

TLRs allow sentinel cells to detect extracellular molecules that signify the presence of microbes **(figure 14.7).**

NOD-Like Receptors (NLRs)

NLRs allow cells to detect internal invaders or cell damage **(figure 14.8).**

RIG-Like Receptors (RLRs)

RLRs allow cells to detect viral nucleic acid. Virally infected cells respond by making interferons, causing nearby cells to prepare to undergo apoptosis if they become infected with a virus **(figure 14.9).**

14.6 ■ The Complement System

Complement System Activation

The complement system detects microbial cells and antibodies bound to antigens, and is activated in response **(figure 14.10).**

Effector Functions of the Complement System

The major protective outcomes of complement system activation include **opsonization,** an inflammatory response, and lysis of foreign cells **(figure 14.11).**

Regulation of the Complement System

Complement regulatory proteins prevent host cell surfaces from activating the complement system via the alternative pathway **(figure 14.12).**

14.7 ■ Phagocytosis

The Process of Phagocytosis (figure 14.13)

The steps of **phagocytosis** include chemotaxis, recognition and attachment, engulfment, **phagosome** maturation and **phagolysosome** formation, destruction and digestion, and exocytosis.

Specialized Attributes of Macrophages

Macrophages are always present in tissues to some extent but can call in reinforcements when needed. A macrophage can become an **activated macrophage.** Macrophages, **giant cells,** and T cells form **granulomas** that wall off and retain material that cannot be destroyed.

Specialized Attributes of Neutrophils

Neutrophils are the first cell type recruited from the bloodstream to the site of damage.

14.8 ■ The Inflammatory Response

Swelling, redness, heat, and pain are the signs of **inflammation,** the body's attempt to contain a site of damage, localize the response, eliminate the invader, and restore tissue function.

Factors That Trigger an Inflammatory Response

Inflammation is initiated when microbes are detected by TLRs, NLRs, or the complement system, or when tissue damage occurs.

The Inflammatory Process

The inflammatory process results in dilation of small blood vessels, leakage of fluids from those vessels, and movement of leukocytes from the bloodstream into the tissues **(figure 14.14). Acute inflammation** is marked by the prevalence of neutrophils; **chronic inflammation** is characterized by macrophage and giant cell accumulation, and granuloma formation.

Damaging Effects of Inflammation

The inflammatory process can be damaging to the host, and in some cases this is life-threatening.

Cell Death and the Inflammatory Process

Apoptosis is a mechanism of eliminating "self" cells without triggering an inflammatory response; **pyroptosis** triggers an inflammatory response.

14.9 ■ Fever

Fever results when macrophages release certain pro-inflammatory cytokines. It inhibits the growth of many pathogens and increases the rate of various body defenses.

Review Questions

Short Answer

1. Describe how the skin protects against infection.
2. What factors in saliva aid in protection against microbes?
3. Why is iron metabolism important in body defenses?
4. Name two categories of cytokines and give their effects.
5. What is the function of a TLR?
6. Contrast the pathways of complement activation.
7. How do complement proteins cause foreign cell lysis?
8. How do phagocytes enter tissues during an inflammatory response?
9. How is acute inflammation different from chronic inflammation?
10. Describe the function of apoptosis.

Multiple Choice

1. Lysozyme does which of the following?
   a) Disrupts cell membranes
   b) Hydrolyzes peptidoglycan
   c) Waterproofs skin
   d) Propels gastrointestinal contents
   e) Propels the cilia of the respiratory tract
2. The hematopoietic stem cells in the bone marrow can become which of the following cell types?

1. Red blood cell	2. T cell	3. B cell
4. Monocyte	5. Macrophage	

   a) 2, 3   b) 2, 4   c) 2, 3, 4, 5
   d) 1, 4, 5   e) 1, 2, 3, 4, 5

3. All of the following refer to the same type of cell *except*

   a) macrophage.　　b) neutrophil.

   c) poly.　　　　　d) PMN.

4. TLRs are triggered by all of the following compounds *except*

   a) peptidoglycan.　　b) glycolysis enzymes.

   c) lipopolysaccharide.　d) flagellin.

   e) certain nucleotide sequences.

5. The direct/immediate action of interferon on a cell is to

   a) interfere with the replication of the virus.

   b) prevent the virus from entering the cell.

   c) stimulate synthesis of inactive "suicide enzymes."

   d) stimulate the immune response.

   e) stop the cell from dividing.

6. A pathogen that can avoid the complement component C3b would directly protect itself from

   a) opsonization.　　b) triggering inflammation.

   c) lysis.　　　　　d) inducing interferon.　　e) antibodies.

7. Which of the following statements about phagocytosis is *false*?

   a) Phagocytes move toward an area of infection by chemotaxis.

   b) Digestion of invaders occurs within a phagolysosome.

   c) Phagocytes have receptors that recognize C3b bound to bacteria.

   d) Phagocytes have receptors that recognize antibodies bound to bacteria.

   e) Macrophages die after phagocytizing bacteria, but neutrophils regenerate their lysosomes and survive.

8. All of the following cell types are found in a granuloma *except*

   a) neutrophils.　　b) macrophages.

   c) giant cells.　　d) T cells.

9. All of the following trigger an inflammatory response *except*

   a) engagement of TLRs.

   b) complement system activation.

   c) interferon induction of antiviral protein synthesis.

   d) tissue damage.

10. Which of the following statements about inflammation is *false*?

    a) Vasodilation results in leakage of blood components.

    b) The process can damage host tissue.

    c) Neutrophils are the first to migrate to a site of inflammation.

    d) Apoptosis induces inflammation.

    e) The signs of inflammation are redness, swelling, heat, and pain.

Applications

1. Paraplegic patients often have recurrent urinary tract infections. Why would the condition keep coming back in spite of repeated treatment?

2. A cattle farmer sees a sore on the leg of one of his cows. The farmer feels the sore and notices that the area just around the sore is warm to the touch. A veterinarian examines the wound and explains that the warmth may be due to inflammation. The farmer wants an explanation of the difference between the localized warmth and fever. What would be the vet's explanation to the farmer?

Critical Thinking ➕

1. A student argues that phagocytosis is a wasteful process because after engulfed organisms are digested and destroyed, the remaining material is excreted from the cell (see figure 14.13). A more efficient process would be to release the digested material *inside* the cell. This way, the material and enzymes could be reused by the cell. Does the student have a valid argument? Why or why not?

2. According to figure 14.9, *any* cell infected by viruses may die due to the action of interferons. This strategy, however, seems counterproductive. The same result would occur without interferon—any cell infected by a virus might die directly from the virus. Is there any apparent benefit from the interferon action?

15 The Adaptive Immune Response

Blood clot, with erythrocytes, fibrin filaments, and a lymphocyte (color-enhanced scanning electron micrograph).

A Glimpse of History

Near the end of the nineteenth century, diphtheria was a terrifying disease that killed many infants and small children. The first symptom was a sore throat, often followed by the development of a gray membrane that could come loose and block the airway. Frederick Loeffler, working in Robert Koch's laboratory in Berlin, found club-shaped bacteria growing in the throats of people with the disease but not elsewhere in their bodies. He hypothesized that the organisms were making a poison that spread through the bloodstream. In Paris, at the Pasteur Institute, Emile Roux and Alexandre Yersin followed up by growing the bacteria and extracting the poison, or toxin, from culture fluids. When the toxin was injected into guinea pigs, it generally killed them. ▶▶| Alexandre Yersin, p. 670

Back in Berlin, Emil von Behring injected diphtheria toxin into guinea pigs that had recovered from lab-induced diphtheria. These animals did not become ill, suggesting that something in their blood protected them; von Behring called it *antitoxin.* To test this idea, he mixed toxin with serum (the liquid portion of blood) from a guinea pig that previously had diphtheria, and injected it into one that had not had the disease. The animal remained well. In further experiments, he cured animals with diphtheria by giving them antitoxin.

The effectiveness of antitoxin was put to the test in late 1891, when a diphtheria epidemic broke out in Berlin. On Christmas night, antitoxin was first given to an infected child, who then recovered from the dreaded disease. The substances with antitoxin properties were then given the name *antibodies,* and materials that induced antibody production were called *antigens.*

Emil von Behring received the first Nobel Prize in Medicine in 1901 for his work on antibody therapy. It took many more decades of investigation to reveal the biochemical nature of antibodies. In 1972, Rodney Porter and Gerald Edelman were awarded a Nobel Prize for their part in determining the structure of antibodies.

In contrast to the innate immune response, which is always ready to respond to patterns that signify invasion or damage, the adaptive immune response matures throughout life, developing from the immune system arsenal the most effective means to eliminate a specific invader when it is encountered. The protection provided by the response is called **adaptive immunity.** On first exposure to a given microbe, adaptive immunity takes a week or more to build. During this delay, the host depends on innate immunity for protection, which may not be sufficient to prevent disease. In some cases, a person may not survive long enough for the adaptive immune response to reach an effective level.

An important characteristic of adaptive immunity is memory, a stronger response to re-exposure. Individuals who survive diseases such as measles, mumps, or diphtheria generally never develop the same disease again. Vaccination now prevents these diseases by exposing a person's immune system to harmless forms of the pathogen or its products. It is true that diseases such as influenza can be contracted repeatedly, but that is generally due to the pathogen's ability to continually evade the host defenses, a topic discussed in chapter 16.

▶▶ vaccination, p. 421

The adaptive immune response has molecular specificity. A response that protects an individual from developing measles does not prevent a different disease—for example, chickenpox.

Another important aspect of adaptive immunity is tolerance, the ability to ignore any given molecule. Most significantly, the immune system can distinguish normal host cells from invading microbes, an ability referred to as self versus non-self recognition. A more accurate description might be "healthy self" versus "dangerous," the latter including invading microbes as well as cancerous or other "corrupt" cells. Regardless, the ability to develop self-tolerance is critical because without it the immune system would routinely turn against the body's own cells, attacking them just as it does invading microbes. Self-tolerance is not a fail-safe system, however, which is why autoimmune diseases can occur.

▶▶ autoimmune disease, p. 412

The adaptive immune system is extraordinarily complex, involving an intricate network of cells, cytokines, and other compounds. The need for such complexity is logical given the task of the system, but it makes the system difficult to explain and comprehend! Although it might seem that any description should start at the beginning and then continue in a linear progression, that approach can be bewildering. As an analogy, imagine explaining how to build a car to someone who has never even seen an automobile, much less driven one; the person would have no idea why certain parts were being put together. With that in mind, we have chosen to start the chapter with a general overview of the adaptive immune response, focusing on the essential elements of the process and how the response eliminates invading microbes. Armed with this foundation, it should then be easier for you to follow the more in-depth descriptions of the various parts and how they fit together. At the end of the chapter, we will focus on the development of the system, concentrating on how the **lymphocytes**—the primary participants in the adaptive immune responses—gain the specificity required to respond to an incredibly diverse and ever-changing assortment of microbes.

◀◀ lymphocytes, p. 340

15.1 ■ Strategy of the Adaptive Immune Response

Learning Outcome

1. *Compare and contrast the general aspects of humoral immunity and cell-mediated immunity.*

This first response to a particular antigen is called the **primary response.** As a result of this initial encounter, the adaptive immune system "remembers" the mechanism that proved effective against that specific antigen. As a result, when the same antigen is encountered later in life, there is a stronger antigen-specific immune response called the **secondary response.** ◀◀ antigen, p. 335

The adaptive immune response uses two basic strategies for countering foreign material. One response, **humoral immunity,** works to eliminate extracellular antigens—for example, bacteria, toxins, or viruses in the bloodstream or in tissue fluids. The other, called **cell-mediated immunity (CMI),** or **cellular immunity,** deals with antigens residing within a host cell, such as a virus infecting a cell. Humoral and cell-mediated immunity are both powerful and, if misdirected, can damage the body's own tissues. Because of this, the adaptive immune response is tightly regulated; before a lymphocyte can unleash its power, it generally requires confirmation from a different cell type that the antigen does indeed represent a threat.

Overview of Humoral Immunity

B lymphocytes, or **B cells,** are responsible for humoral immunity (**figure 15.1**). Their name reflects the fact that they develop in an organ in birds called the bursa. In humans, however, B cells develop in the bone marrow.

In response to extracellular antigens, B cells may be triggered to proliferate and then differentiate into **plasma cells,** which function as factories that produce Y-shaped proteins called **antibodies.** These molecules bind to antigens, providing protection by mechanisms described shortly. A high degree of specificity is involved in the binding, so many different antibody molecules are needed to bind to the wide array of antigens encountered throughout life. Some of the B cells form memory B cells, long-lived cells that respond more quickly if the antigen is encountered again.

Antibody molecules have two functional regions—the two identical arms and the stem of the molecule. The arms bind specific antigen molecules; the amino acid sequence of the end of the arms varies among antibodies, providing the basis for their specificity. The stem functions as a "red flag," tagging antigen bound by antibodies and enlisting other immune system components to eliminate the molecule.

Antibodies protect the host by both direct and indirect mechanisms. The direct effect is due to their ability to bind antigens. Simply by binding, they can coat an antigen, thereby preventing it from attaching to a host cell. For example, an antibody-coated viral particle cannot attach to its receptor, and therefore cannot enter the cell. The indirect protective effect is due to the "red flag" region tagging the antigen for elimination. ◀◀ attachment of viruses, p. 317

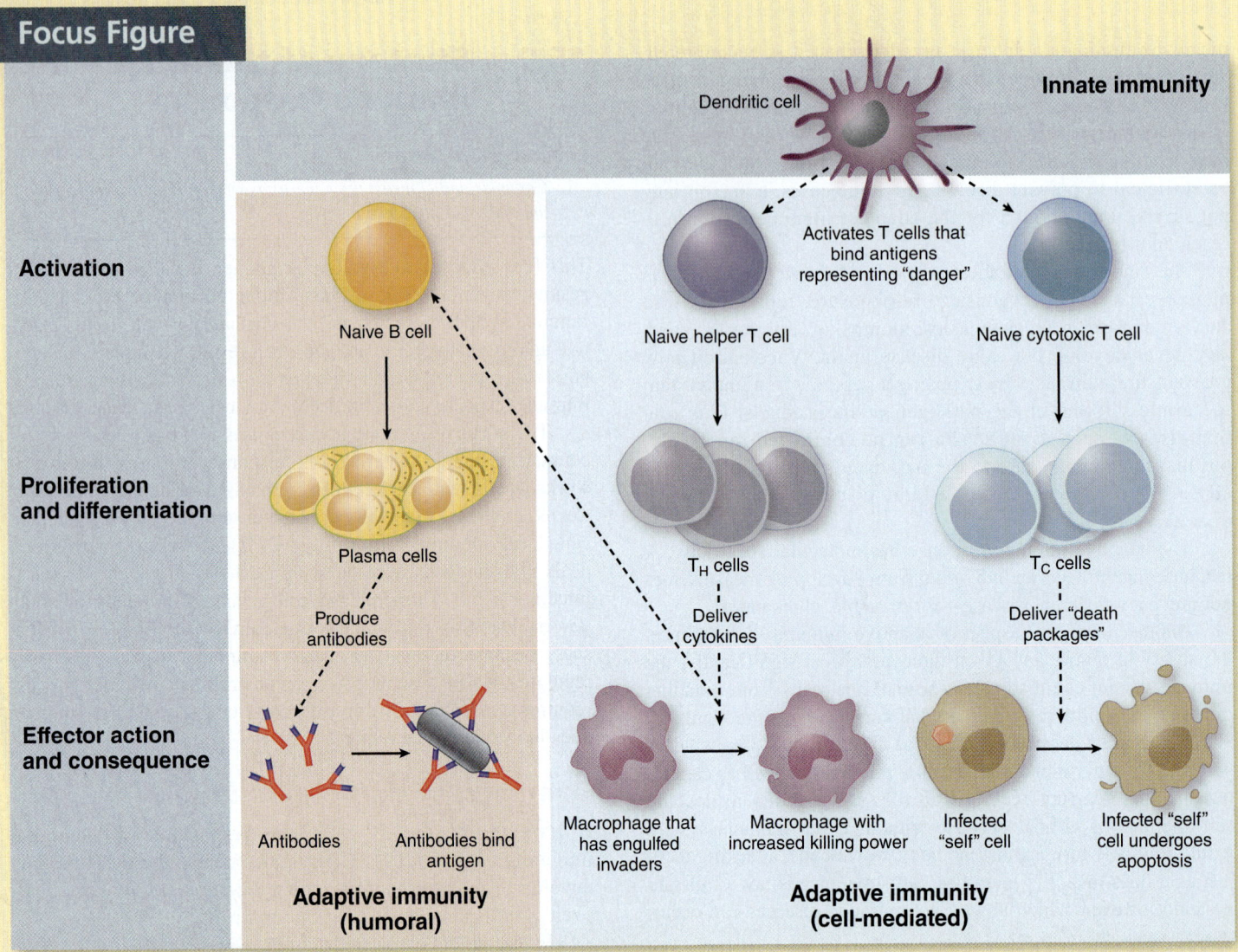

FIGURE 15.1 Overview of Humoral and Cell-mediated Immunity Humoral immunity protects against antigens in blood and tissue fluid (extracellular antigens); cell-mediated immunity protects against antigens within host cells (intracellular antigens). In this diagram, solid arrows represent the path of a cell or molecule; dashed arrows represent a cell's interactions and effector functions; antigen receptors and memory cells are not shown.

How do antibodies protect against infection?

How does a naive B cell (one that has never "seen" antigen before) know when to respond? The B-cell surface has numerous copies of a **B-cell receptor (BCR),** a membrane-bound version of the specific antibody the B cell is programmed to make (**figure 15.2**). If a naive B cell encounters an antigen that its BCR binds, the cell is triggered to multiply. Some of the resulting clones (cell copies) eventually differentiate, becoming plasma cells that secrete huge numbers of antibody molecules. Before the naive B cell can multiply, however, it generally needs to be activated by another type of lymphocyte—a T_H cell. The T_H cell must also recognize the antigen, confirming that it is indeed dangerous.

Overview of Cell-Mediated Immunity

Cell-mediated immunity involves **T lymphocytes,** or **T cells** (see figure 15.1). Their name reflects the fact that they mature in the thymus. Two subsets of T cells help eliminate antigen—**cytotoxic**

T cells and **helper T cells.** Both of these have multiple copies of a surface molecule called a **T-cell receptor (TCR),** which is functionally analogous to a BCR; it allows the cell to bind a specific antigen (figure 15.2). Unlike a BCR, however, a TCR does not recognize free antigen. Instead, the antigen must be presented by one of the body's own cells.

A third T-cell subset, **regulatory T cells** (formerly T suppressor cells), has recently been described and is currently the focus of a great deal of research. Regulatory T cells are similar to the other T cells in that they have a TCR, but their role is entirely different. Instead of fostering a response, they help prevent the immune system from mounting a response against "self" molecules; failure to do this results in autoimmune diseases—a topic discussed in chapter 17. The description of T cells in this chapter focuses exclusively on the cytotoxic and helper subsets.

Like the B cells, naive helper T cells and naive cytotoxic T cells that bind antigen must be activated before they can multiply. Again, this process helps confirm that the antigen signi-

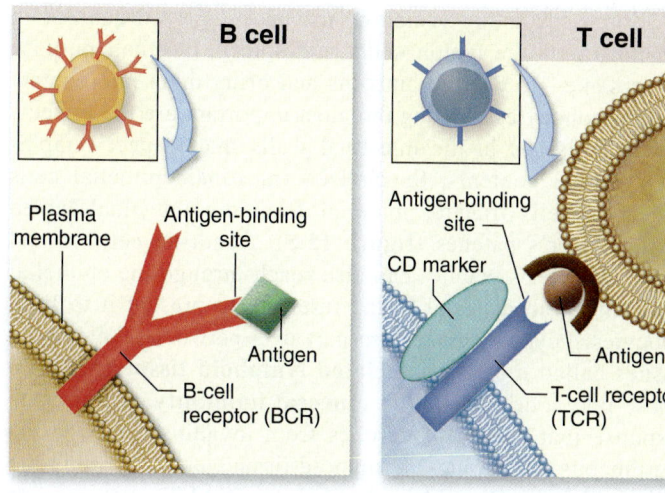

FIGURE 15.2 Antigen Receptors on Lymphocytes B cells and T cells have surface receptors that allow them to recognize specific antigens. The CD marker identifies the type of T cell.

❓ *How is a B-cell receptor similar to an antibody?*

fies danger. Dendritic cells are responsible for T-cell activation; recall that these cells are a part of innate immunity. Once activated, a T cell proliferates and then differentiates to form effector T cells, armed to perform distinct protective roles. For simplicity and clarity, we will refer to effector helper T cells as T_H **cells,** and effector cytotoxic cells as T_C **cells.** Like B cells, both types of T cells can form memory cells that react quickly if the same antigen is encountered later. ◀◀ **dendritic cell, p. 340**

T_C cells respond to intracellular antigens (antigens within a host cell). When they find such an antigen, they induce the "self" cell that harbors it to undergo apoptosis. Virally infected cells provide a good example of the effectiveness of this strategy. By inducing these cells to undergo apoptosis, the immune system destroys cells that would otherwise produce more viral particles. Sacrificing the cells also releases unassembled viral components, provoking a humoral response as well. ◀◀ **apoptosis, p. 350**

T_H cells help orchestrate the various responses of humoral and cell-mediated immunity. They activate B cells and macrophages, and produce cytokines that direct and support T cells.

> **MicroAssessment 15.1**
>
> B cells are programmed to eliminate extracellular antigens. Cytotoxic T cells are programmed to destroy infected "self" cells as a means of eliminating intracellular antigens. Helper T cells are programmed to orchestrate the adaptive immune response.
>
> 1. *How is the strategy of the cell-mediated response different from that of the humoral response?*
> 2. *Why is it important that B cells and T cells be activated before they can mount a response against antigen?*
> 3. *How would you expect a T_C cell to respond if it encountered a T_H cell that was infected with a virus?* ➕

15.2 ■ Anatomy of the Lymphatic System

Learning Outcome

2. *Compare and contrast the roles of lymphatic vessels, the various secondary lymphoid organs, and primary lymphoid organs.*

The **lymphatic system** is a collection of tissues and organs that bring the population of B cells and T cells into contact with antigens (**figure 15.3**). This is important because each lymphocyte recognizes only one or a few different antigens. In order for the body to mount an effective response, the appropriate lymphocyte must encounter the given antigen. ◀◀ **tissues, p. 69** ◀◀ **organs, p. 69**

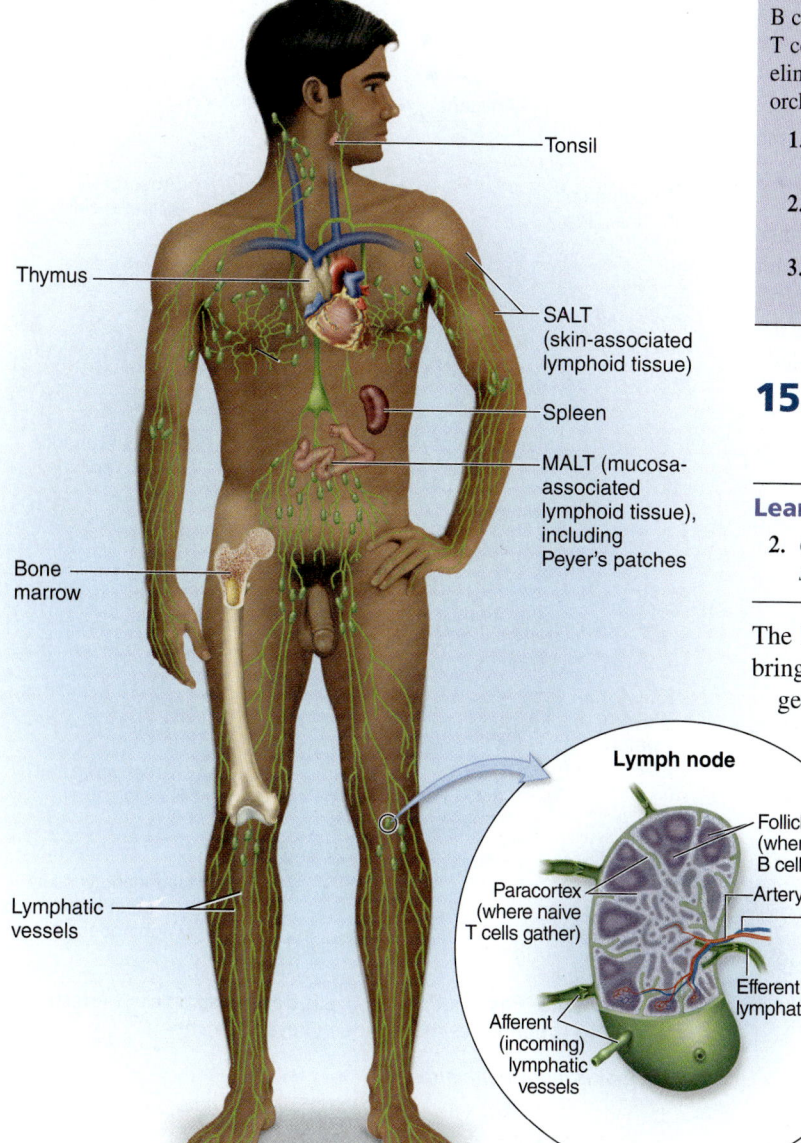

FIGURE 15.3 Anatomy of the Lymphatic System Lymph flows through a system of lymphatic vessels, passing through lymph nodes and lymphoid tissues.

❓ *What is the function of the lymph nodes?*

Lymphatic Vessels

Flow within the lymphatic system occurs via the **lymphatic vessels,** or **lymphatics.** These vessels carry **lymph,** a fluid that forms as a result of the body's circulatory system (see figure 27.1).

As blood travels from the heart and lungs through the capillaries in the tissues, much of the fluid portion filters out, supplying tissues with the O_2 and nutrients carried by the blood (**figure 15.4**). Most of that fluid then reenters the capillaries as they return to the heart and lungs, but some enters the lymphatics instead. The lymph—which also contains white blood cells and antigens that have entered the tissues—travels via the lymphatics to the lymph nodes, where proteins, cells, and other materials are removed. The lymph then empties back into the blood circulatory system at a large vein behind the left collarbone. Note that the inflammatory response causes more fluid to enter the tissues at the site of inflammation; this causes a corresponding increase in the antigen-containing fluids that enter lymphatic vessels.

Secondary Lymphoid Organs

Secondary lymphoid organs are the sites where lymphocytes gather to contact the various antigens. Examples include the lymph nodes, spleen, tonsils, adenoids, and appendix (see figure 15.3). They are situated at strategic positions in the body so that immune responses can be initiated almost anywhere. For instance, lymph nodes capture materials from the lymphatics, and the spleen collects materials from blood.

The secondary lymphoid organs are like busy, highly organized lymphoid coffee shops where many cellular meetings take place. The anatomy of these organs provides a structured center for various cells of the immune system to interact and transfer cytokines. No other places in the body do this, so these organs are the only sites where adaptive immune responses can be initiated.

Some secondary lymphoid organs are less organized in structure than the lymph nodes and spleen, but their purpose is the same—to capture antigens and bring them into contact with lymphocytes. Among the most important are the **Peyer's patches,** tissues in the intestinal walls that inspect samples of intestinal contents. Specialized intestinal epithelial cells called M cells transfer material from the intestinal lumen to the Peyer's patches (**figure 15.5**). Dendritic cells in and near the Peyer's patches can also reach through the epithelial layer and grab material in the intestine to present it to lymphocytes. Peyer's patches are part of a network of lymphoid tissues called **mucosa-associated lymphoid tissue (MALT).** These play a critical role in **mucosal immunity,** the immune response that prevents microbes from invading the body via the mucous membranes. Lymphoid tissues under the skin are called **skin-associated lymphoid tissue (SALT).** ◀◀ lumen p. 69

Primary Lymphoid Organs

Primary lymphoid organs include the bone marrow and thymus. The bone marrow is where hematopoietic stem cells reside; recall that these give rise to all blood cells, including lymphocytes (see

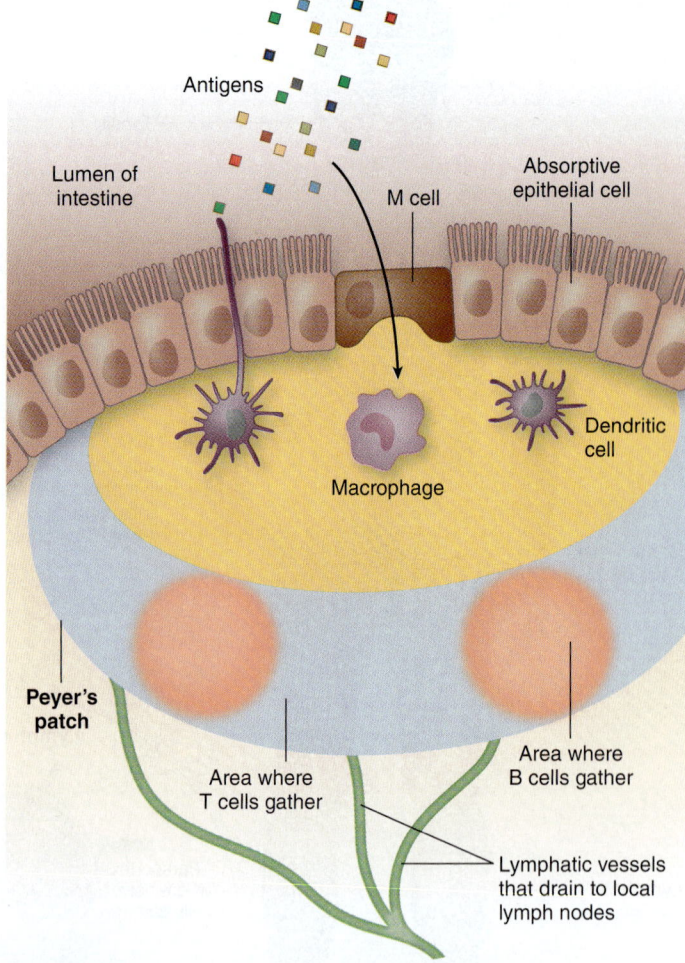

FIGURE 15.5 Peyer's Patch M cells transfer samples of intestinal contents to lymphocytes that reside in Peyer's patches.

❓ *What is mucosal immunity?*

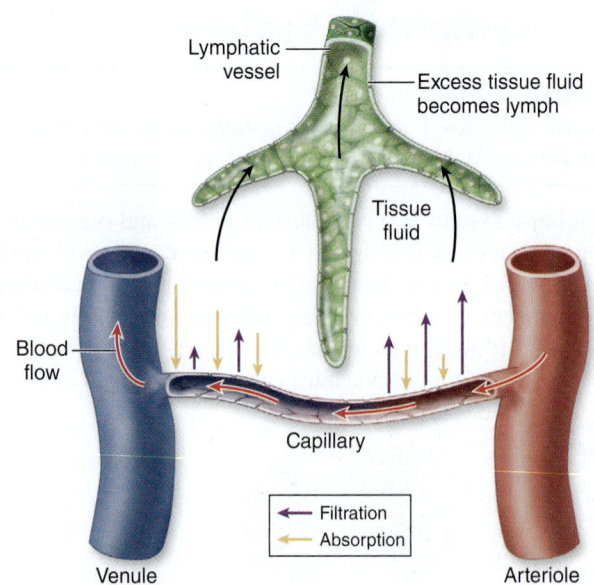

FIGURE 15.4 Formation of Lymph

❓ *What material is typically contained in lymph?*

figure 14.5). B cells and T cells all originate in the bone marrow but only B cells mature there; immature T cells migrate to the thymus and mature there. Once mature, the lymphocytes gather in the secondary lymphoid organs just described, waiting to encounter antigen. ◄◄ hematopoietic stem cells, p. 338

MicroAssessment 15.2

Lymphatic vessels carry fluid collected from tissues to the lymph nodes. Secondary lymphoid organs are where lymphocytes gather to encounter antigens. Hematopoietic stem cells destined to become B cells and T cells mature in the primary lymphoid organs.

4. *How is lymph formed?*
5. *What are Peyer's patches?*
6. *How would mucosal immunity prevent diarrheal disease?* ➕

15.3 ■ The Nature of Antigens

Learning Outcome

3. *Define the terms* antigen, immunogen, T-dependent antigen, T-independent antigen, antigenic, *and* epitope.

The term **antigen** was initially coined in reference to compounds that induce antibody production; it is derived from the descriptive expression <u>anti</u>body <u>gen</u>erator. Today, the term is used more broadly to describe any molecule that reacts specifically with an antibody, a B-cell receptor, or a T-cell receptor; it does not necessarily imply that the molecule can induce an immune response. When referring specifically to an antigen that elicits an immune response in a given situation, the more restrictive term **immunogen** may be used. The distinction between the terms antigen and immunogen helps clarify discussions in which a normal protein from host A elicits an immune response when transplanted into host B; the protein is an antigen because it can react with an antibody or lymphocyte, but it is an immunogen only for host B, not for host A.

Antigens include an enormous variety of materials, from invading microbes and their products to plant pollens, but they fall into two general categories. Most antigens are **T-dependent antigens,** meaning that the responding B cell requires assistance from a T_H cell in order to become activated. T-dependent antigens characteristically have a protein component. In contrast, **T-independent antigens** can activate B cells without T_H cell help. They include lipopolysaccharide (LPS) and molecules with repeating subunits, such as some carbohydrates.

Various antigens differ in their effectiveness in stimulating an immune response. Proteins generally induce a strong response, whereas lipids often do not. The terms **antigenic** and immunogenic are used interchangeably to describe the ability of an antigen to elicit an immune response. Small molecules are usually not antigenic.

Although antigens are generally large molecules, the adaptive immune system recognizes discrete regions of the molecule

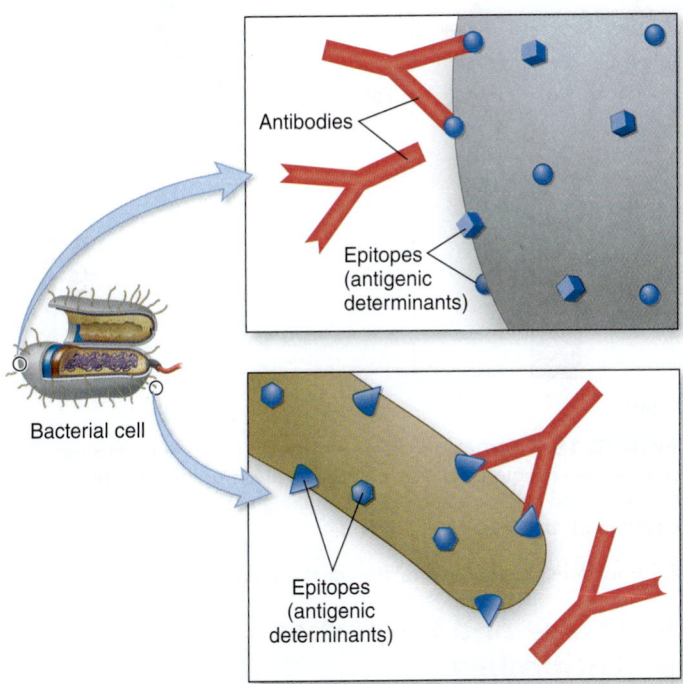

🎬 **FIGURE 15.6 Antibodies Binding to Epitopes**

❓ *Explain why a pathogen would never have only a single epitope.*

known as **epitopes,** or antigenic determinants (**figure 15.6**). Some epitopes are stretches of 10 or so amino acids, whereas others are three-dimensional shapes such as a region that sticks out in a molecule. A bacterial cell usually has a diverse assortment of macromolecules on its surface, each with a number of distinct epitopes, so the entire cell has a multitude of different epitopes.

MicroAssessment 15.3

The adaptive immune response is directed against epitopes on antigens.

7. *How is an epitope different from an antigen?*
8. *Would a denatured antigen be expected to have the same epitopes as its native (undenatured) counterpart?* ➕

15.4 ■ The Nature of Antibodies

Learning Outcomes

4. *Diagram an antibody, labeling the various functional regions.*
5. *Describe six protective outcomes of antibody-antigen binding.*
6. *Compare and contrast the five classes of immunoglobulins.*

The structure and characteristics of antibody molecules are critical for their functional properties. Recognizing these features will help you understand their essential roles in the host defenses.

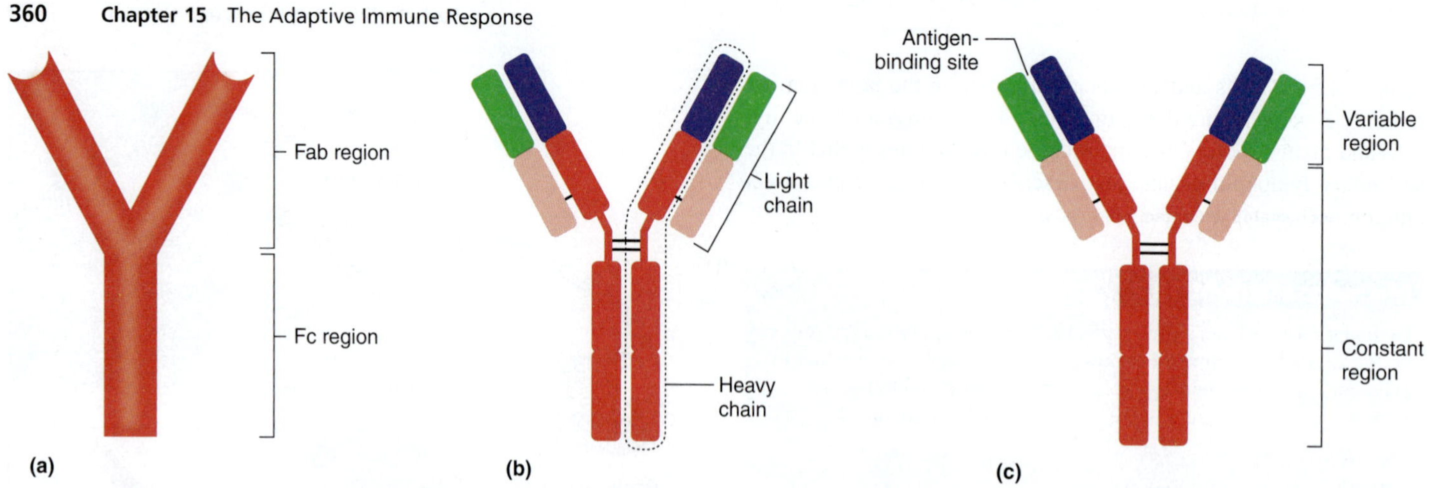

FIGURE 15.7 Structure of an Antibody Molecule (a) Simplified Y-shaped structure; the arms of the Y make up the Fab regions, and the stem is the Fc region. **(b)** The molecule is made up of two identical heavy chains and two identical light chains. Disulfide bonds link the chains. **(c)** The amino acid sequence of the variable region accounts for the antigen-binding specificity; the amino acid sequence of the constant region determines the class of the molecule.

? *How does the function of the constant region differ from that of the variable region?*

Structure and Properties of Antibodies

Antibodies, also called immunoglobulins, are Y-shaped proteins that have two general parts—the arms and the stem (**figure 15.7a**). The two identical arms, called the **Fab regions,** bind antigen. The stem is the **Fc region.** These names were assigned following early studies that showed that enzymatic digestion of antibodies yielded two types of fragments—fragments that were antigen-binding (Fab) and fragments that could be crystallized (Fc).

All antibodies have the same basic Y-shaped structure, called an antibody monomer. It consists of two copies of a high-molecular-weight polypeptide chain, called the **heavy chain,** and two copies of a lower-molecular-weight polypeptide chain, called the **light chain** (figure 15.7b). The amino acids in the chains fold into characteristic domains, referred to as immunoglobulin domains; the light chains have two domains each, and most heavy chains have four. Each light chain is linked to a heavy chain by a disulfide bond. The fork of the Y is a flexible stretch called the hinge region, and one or more disulfide bonds link the two heavy chains there. ◄◄ **domain, p. 28**
◄◄ **disulfide bond, p. 27**

When the amino acid sequences of antibody molecules that bind to different epitopes are compared, there is tremendous variation in the portion referred to as the variable region. The other portion is known as the constant region.

Variable Region

The **variable region** is the portion at the ends of the Fab regions; it accounts for the antigen-binding specificity (figure 15.7c). Part of this region is the **antigen-binding site,** the portion that attaches to a specific epitope. The fit needs to be very precise, because the interaction depends on numerous non-covalent bonds to keep the molecules together. Nevertheless, the antigen-antibody interaction is reversible, and the molecules can separate, leaving both antigen and antibody unchanged.

Constant Region

The **constant region** includes the entire Fc region, as well as part of the two Fab regions (figure 15.7c). The consistent nature of this region allows other components of the immune system to recognize the otherwise diverse antibody molecules.

There are five general types of constant regions, and these correspond to the major classes of immunoglobulin (Ig) molecules—IgM, IgG, IgA, IgD, and IgE. Each class has distinct functions and properties that will be described later, after we consider some general properties of antibodies.

Protective Outcomes of Antibody-Antigen Binding

The protective outcomes of antibody-antigen binding (**figure 15.8**) depend partly on the antibody class, and include:

- **Neutralization.** Toxins and viruses must bind specific molecules on a cell surface before they can damage that cell. A toxin or virus coated with antibodies cannot attach to cells and is said to be neutralized.

- **Opsonization.** Phagocytic cells have receptors for the Fc region of IgG molecules, making it easier for the phagocyte to engulf antibody-coated antigens. Recall from chapter 14 that the complement protein C3b opsonizes antigens; IgG molecules have a similar effect. ◄◄ **opsonization by C3b, p. 345**

- **Complement system activation.** Antigen-antibody complexes (commonly called immune complexes) can trigger the classical pathway of complement system activation. When multiple molecules of certain antibody classes are bound to a cell surface, a specific complement system protein attaches to their Fc regions, initiating the cascade. Recall that activation of the complement system results in production of the opsonin C3b, initiation of an inflammatory response, and formation of membrane attack complexes. ◄◄ **complement system, p. 344**

- **Immobilization and prevention of adherence.** Binding of antibodies to flagella interferes with a microbe's ability to move; binding to pili prevents it from attaching to surfaces. These capabilities are often necessary for a pathogen to infect a host, so antibodies that bind to flagella or pili prevent infection.

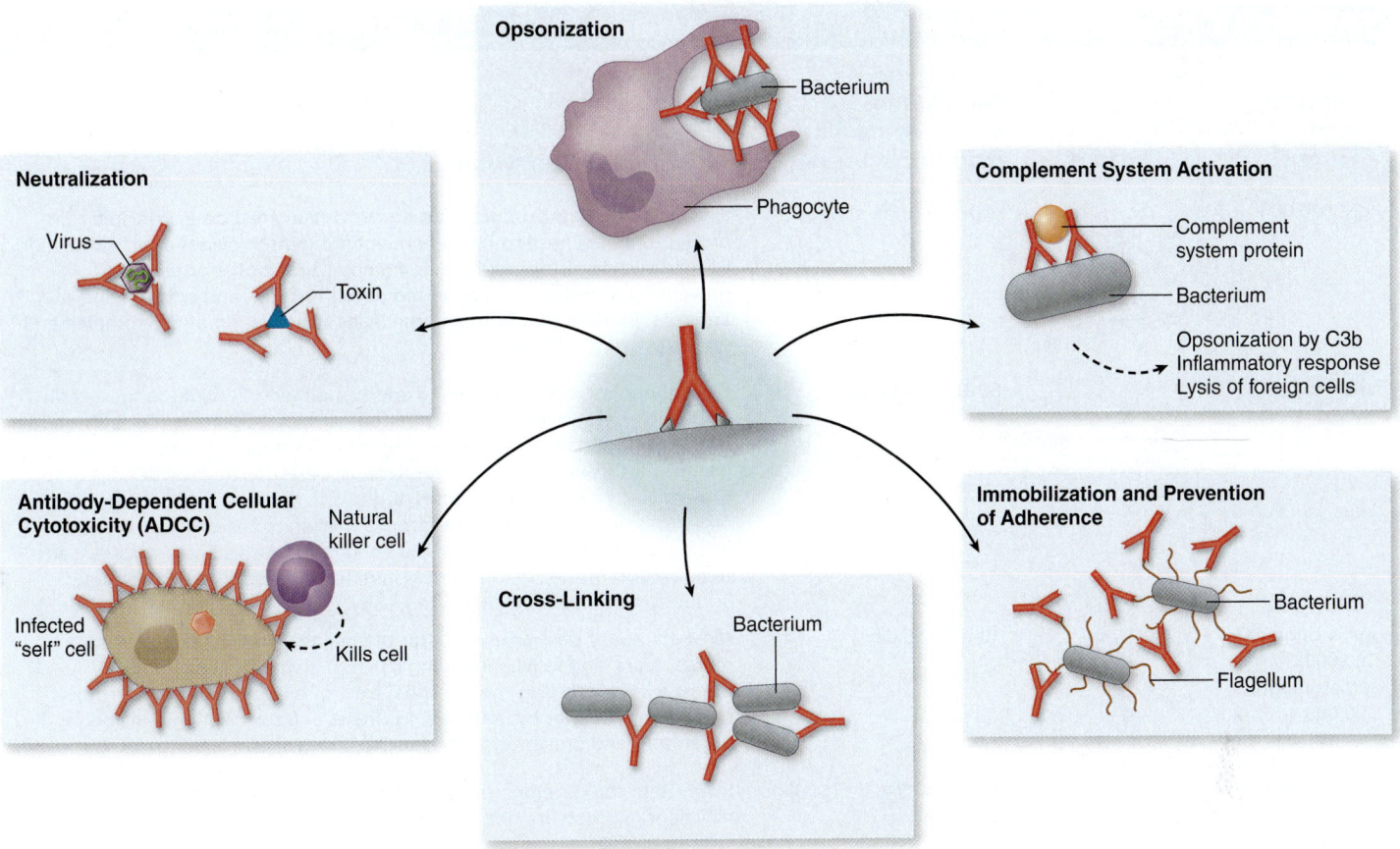

FIGURE 15.8 Protective Outcomes of Antibody-Antigen Binding

❓ *Which pathway of complement activation is depicted in this figure?*

■ **Cross-linking.** The two arms of an antibody can bind separate but identical antigen molecules, linking them. The overall effect is that large antigen-antibody complexes form, creating big "mouthfuls" of antigens for phagocytic cells to engulf.

■ **Antibody-dependent cellular cytotoxicity (ADCC).** When multiple IgG molecules bind to a virally infected cell or a tumor cell, that cell becomes a target for destruction by **natural killer (NK) cells.** The NK cell attaches to the Fc regions of IgG and once attached, kills the target cell by delivering compounds directly to it. ◄◄ ►►| natural killer cells, pp. 340, 374

Immunoglobulin Classes

All five major classes of immunoglobulins have the same basic monomeric structure, but each class has a different constant portion of the heavy chain. Some of the immunoglobulins form multimers of the basic monomeric structure. Characteristics of the various immunoglobulin classes are summarized in **table 15.1.**

IgM

IgM accounts for 5% to 13% of the circulating antibodies and is the first class produced during the primary response to an antigen. It is the principal class produced in response to some T-independent antigens.

IgM is a pentamer. Its large size prevents it from crossing from the bloodstream into tissues, so its primary role is to control bloodstream infections. The five monomeric subunits give IgM a total of 10 antigen-binding sites, so it cross-links antigens very effectively. It is the most efficient class in triggering the classical pathway of complement system activation.

A fetus is normally sterile until the birth membrane is ruptured, so IgM generally begins being made about the time of birth. However, a fetus infected *in utero* can make IgM antibodies.

IgG

IgG accounts for about 80% to 85% of the total serum immunoglobulin (serum is the liquid portion of blood). It circulates in the blood but exits the vessels to enter the tissues as well.

IgG provides the longest-term protection of any antibody class; its half-life is 21 days, meaning that a given number of IgG molecules will be reduced by about 50% after 21 days. In addition, IgG is generally the first and most abundant circulating class produced during the secondary response. The basis for this will be discussed later in the chapter. IgG provides protection by neutralization, opsonization, complement activation, immobilization and prevention of adherence, cross-linking, and ADCC.

An important characteristic of IgG is that it is transported across the placenta into the fetus's bloodstream, so it protects the developing fetus against infections. Women who are not already

TABLE 15.1   Characteristics of the Main Classes of Antibodies

Class and Molecular Weight (daltons)	Structure	Percent of Total Serum Immunoglobulin (half-life in serum)	Properties and Functions
IgM 970,000		5–13% (10 days)	First antibody class produced during the primary response. Principle class produced in response to T-independent antigens. Provides direct protection by neutralizing viruses and toxins, immobilizing motile organisms, preventing microbes from adhering to cell surfaces, and cross-linking antigens. Binding of IgM to antigen leads to activation of the complement system (classical pathway).
IgG 146,000		80–85% (21 days)	Most abundant class in the blood and tissue fluids. Provides longest-term protection because of its long half-life. Transported across the placenta, providing protection to a developing fetus; long half-life extends the protection through the first several months after birth. Provides direct protection by neutralizing viruses and toxins, immobilizing motile organisms, preventing microbes from adhering to cell surfaces, and cross-linking antigens. Binding of IgG to antigen facilitates phagocytosis, leads to activation of the complement system (classical pathway), and allows antibody-dependent cellular cytotoxicity.
IgA monomer 160,000; secretory IgA 390,000		10–13% (6 days)	Most abundant class produced, but the majority of it is secreted into mucus, tears, and saliva, providing mucosal immunity. Also found in breast milk, protecting the intestinal tract of breast-fed infants. Protects mucous membranes by neutralizing viruses and toxins, immobilizing motile organisms, and preventing attachment of microbes to cell surfaces.
IgD 184,000		<1% (3 days)	Involved in the development and maturation of the antibody response. Its functions in blood are not well understood.
IgE 188,000		<0.01% (2 days)	Binds via the Fc region to mast cells and basophils. This bound IgE allows those cells to detect parasites and other antigens and respond by releasing their granule contents. Involved in many allergic reactions.

immune to a given pathogen lack IgG against that microbe, so they are warned to take extra precautions during pregnancy to avoid pathogens that can infect and damage a fetus. For example, pregnant women are advised not to eat raw meat or become first-time cat owners to avoid a primary infection by *Toxoplasma gondii,* a parasite found in steak tartare and other raw meat as well as the feces of infected cats. ▶▶ *Toxoplasma gondii,* p. 709

Maternal IgG not only protects the developing fetus, but also the newborn. The maternal antibodies present at birth gradually degrade over a period of about 6 months, but during this time the infant begins producing protective antibodies (**figure 15.9**).

IgG is also in colostrum, the first breast milk produced after giving birth. The newborn's intestinal tract absorbs this antibody.

IgA

The monomeric form of IgA accounts for about 10% to 13% of antibodies in the serum. Most IgA, however, is the secreted form, a dimer called **secretory IgA** (**sIgA**). In fact, IgA is the most abundant immunoglobulin class produced. The secreted form is important in mucosal immunity and is found on the mucous membranes that line the gastrointestinal, genitourinary, and respiratory tracts. It is also in secretions such as saliva, tears, and breast milk. Secretory IgA in breast milk protects breast-fed infants against intestinal pathogens. ◀◀ mucosal immunity, p. 358

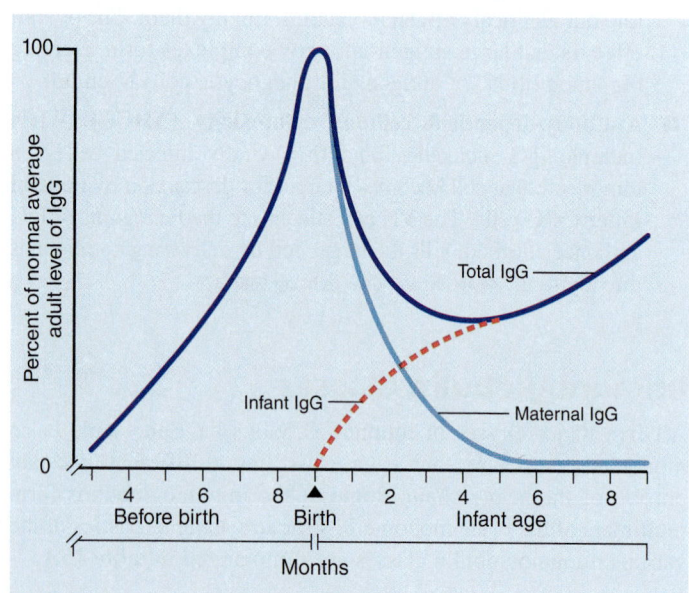

FIGURE 15.9 Immunoglobulin G Levels in the Fetus and Infant As the fetus develops, maternal IgG is transported across the placenta to provide protection. After birth, the infant begins producing antibodies.

❓ *Why does the level of maternal antibodies decrease over time?*

Protection by secretory IgA is primarily due to the direct effect of its binding. These include neutralizing toxins and viruses and interfering with the attachment of microbes to host cells.

IgA is produced by the plasma cells that reside in the mucosa-associated lymphoid tissues (MALT). Recall that plasma cells are the antibody-secreting form of B cells. As IgA is transported across the mucosa (mucous membranes), a polypeptide called the secretory component is added. This attaches the antibody to the layer of mucus that coats the mucosal surface and protects it from destruction by enzymes there. ◄◄ MALT, p. 358

IgD

IgD accounts for less than 1% of all serum immunoglobulins. It is involved with the development and maturation of the antibody response, but its functions in blood have not been clearly defined.

IgE

IgE is barely detectable in blood, because most is tightly bound via the Fc region to basophils and mast cells, rather than being free in the circulation. The bound IgE molecules allow these cells to detect and respond to antigens. For example, when antigen binds to two adjacent IgE molecules carried by a mast cell, the cell releases histamine and other inflammatory mediators. IgE-mediated responses seem to be important in eliminating parasites, particularly helminths. ◄◄ inflammatory mediators, p. 348 ◄◄ helminths, p. 295

Unfortunately for allergy sufferers, basophils and mast cells also release their chemicals when IgE binds to normally harmless materials such as foods, dusts, and pollens, leading to immediate reactions such as coughing, sneezing, and tissue swelling. In some cases these allergic, or hypersensitivity, reactions can be life-threatening. ►► hypersensitivity reactions, p. 401

MicroAssessment 15.4

Antibody-antigen binding results in neutralization, immobilization, and prevention of adherence, cross-linking, opsonization, complement activation, and antibody-dependent cytotoxicity. Immunoglobulin classes include IgM, IgG, IgA, IgD, and IgE.

9. Why is IgM particularly effective at cross-linking antigens?
10. Which maternal antibody classes protect a breast-fed newborn?
11. In opsonization with IgG, why is it important that phagocytes recognize the antibodies only after they bind antigen? ✚

15.5 ■ Clonal Selection and Expansion of Lymphocytes

Learning Outcome

7. Outline the process of clonal selection and expansion.

Early on, immunologists recognized that the immune system is capable of making a seemingly infinite array of antibody specificities. The clonal selection theory describes how this occurs (**figure 15.10**). Each B cell is programmed to make only a single specificity of antibody. When antigen is introduced, only the B cells capable of making the appropriate antibody can bind to the

antigen, and only those cells begin multiplying. This generates a population of clones—copies of the specific B cells capable of making the appropriate antibodies.

Clonal selection is a critical theme in the adaptive immune response, pertaining to both B cells and T cells. As lymphocytes mature in the primary lymphoid organs, a population of cells able to recognize a functionally limitless variety of antigens is generated; each individual cell, however, recognizes and responds to only one epitope. Thus, if a person's immune system can make antibodies to billions of different epitopes, that person must have billions of different B cells, each interacting with a single epitope. The process of generating the diversity in antigen recognition is random and does not require previous exposure to an epitope; the mechanisms will be described later.

Each lymphocyte residing in the secondary lymphoid organs is waiting for the "antigen of its dreams"—an antigen that has an epitope to which that particular lymphocyte is programmed to respond. When an antigen enters a lymphoid organ, only those rare lymphocytes that recognize it can respond; the specificity of the antigen receptor they carry on their surface (B-cell receptor or T-cell receptor) determines this recognition. Lymphocytes that do not recognize the antigen remain inactive. Recall that in most cases, lymphocytes require additional signals—confirmation from another cell type—in order to multiply. This helps prevent the immune system from mounting a response against "self" molecules by mistake.

Some progeny of the lymphocytes that encountered their "dream antigen" leave the secondary lymphoid organs and migrate to the tissues where they continue responding for as long as the antigen is present. Without continuous stimulation by antigen, these cells will undergo apoptosis. ◄◄ apoptosis, p. 350

The activities of individual lymphocytes change as they encounter antigen. As a means of clarifying discussions of lymphocyte characteristics, descriptive terms are sometimes used:

■ **Immature lymphocytes.** These have not fully developed their antigen-specific receptors.

■ **Naive lymphocytes.** These have antigen receptors, but have not yet encountered the antigen to which they are programmed to respond.

■ **Activated lymphocytes.** These are able to proliferate; they have bound antigen via their antigen receptor and have received the required accessory signals from another cell, confirming that the antigen merits a response.

■ **Effector lymphocytes.** These are descendants of activated lymphocytes, armed with the ability to produce specific cytokines or other protective substances. Plasma cells are effector B cells, T_C cells are effector cytotoxic T cells, and T_H cells are effector helper T cells.

■ **Memory lymphocytes** are long-lived descendants of activated lymphocytes; they can quickly become activated when an antigen is encountered again. Memory lymphocytes are responsible for the speed and effectiveness of the secondary response.

MicroByte

The body is thought to have about 1 billion B cells, and only one or a few will recognize a given epitope.

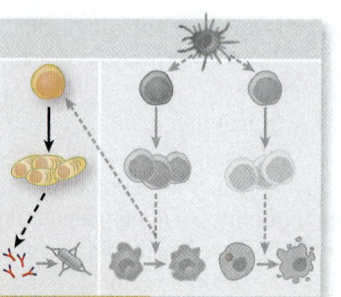

FIGURE 15.10 Clonal Selection and Expansion During the Antibody Response

? *What is meant by clonal selection?*

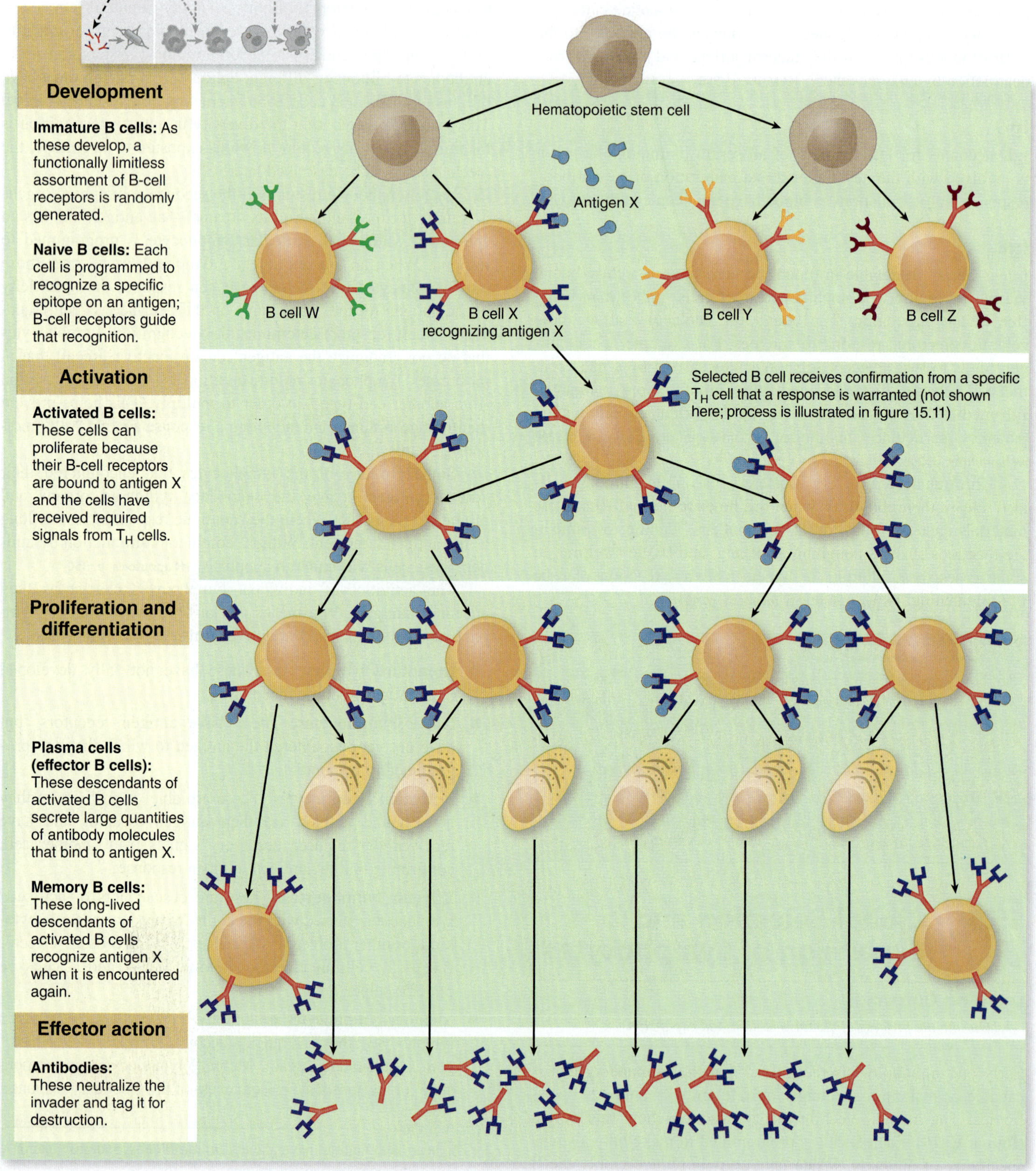

Hematopoietic stem cell

Antigen X

B cell W

B cell X
recognizing antigen X

B cell Y

B cell Z

Selected B cell receives confirmation from a specific T$_H$ cell that a response is warranted (not shown here; process is illustrated in figure 15.11)

Development

Immature B cells: As these develop, a functionally limitless assortment of B-cell receptors is randomly generated.

Naive B cells: Each cell is programmed to recognize a specific epitope on an antigen; B-cell receptors guide that recognition.

Activation

Activated B cells: These cells can proliferate because their B-cell receptors are bound to antigen X and the cells have received required signals from T$_H$ cells.

Proliferation and differentiation

Plasma cells (effector B cells): These descendants of activated B cells secrete large quantities of antibody molecules that bind to antigen X.

Memory B cells: These long-lived descendants of activated B cells recognize antigen X when it is encountered again.

Effector action

Antibodies: These neutralize the invader and tag it for destruction.

15.6 ■ B Lymphocytes and the Antibody Response

Learning Outcomes

8. *Describe the role of T_H cells in B-cell activation.*

9. *Compare and contrast the primary and the secondary responses.*

10. *Compare and contrast the response to T-dependent antigens and T-independent antigens.*

Recall that most antigens are **T-dependent antigens,** meaning the B cells that recognize them require help from T_H cells; the response to these antigens will be the primary focus of this section. The B-cell response to T-independent antigens will be covered at the end of this section. Later in the chapter, we will explain how naive helper T cells become activated to develop their effector functions.

B-Cell Activation

Naive B cells gather in the secondary lymphoid organs to encounter antigens. When a B cell's antigen receptor (B-cell receptor) binds to a T-dependent antigen, the B cell takes the antigen in by endocytosis, enclosing it within an endosome. There, the antigen is degraded into peptide fragments that are then delivered to protein structures called **MHC class II molecules.** These move to the B-cell surface, where they "present" pieces of the antigen for inspection by T_H cells—a process called **antigen presentation** (**figure 15.11**). ◀◀ endocytosis, p. 72

T_H cells, which also gather in the secondary lymphoid organs, scan the naive B cells there to determine if any have encountered an antigen they recognize. If a T_H cell's antigen receptor (T-cell receptor) binds one of the peptide fragments being presented by a B cell, then that T cell activates the B cell. It does this by delivering cytokines to the B cell, initiating the process of clonal expansion of that particular B cell.

If no T_H cells recognize the peptides presented by a B cell, that B cell may become anergic (unresponsive to future exposure to the antigen). This results in tolerance to that antigen, a mechanism the adaptive immune system uses to avoid responses against "self" and other harmless antigens. ◀◀ tolerance, p. 355

Characteristics of the Primary Response

In the first (primary) exposure to an antigen, it takes about 10 to 14 days for a substantial amount of antibodies to accumulate

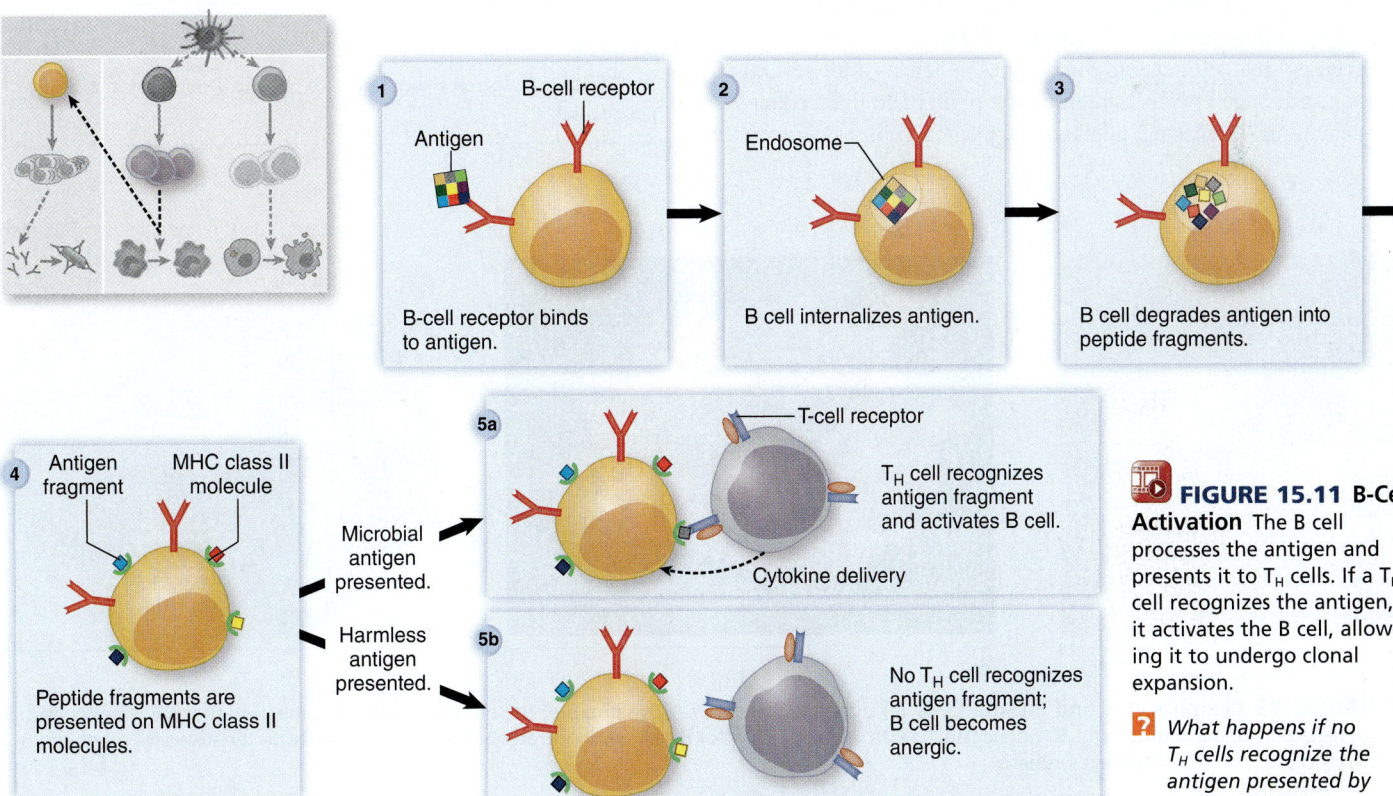

FIGURE 15.11 B-Cell Activation The B cell processes the antigen and presents it to T_H cells. If a T_H cell recognizes the antigen, it activates the B cell, allowing it to undergo clonal expansion.

❓ *What happens if no T_H cells recognize the antigen presented by the B cell?*

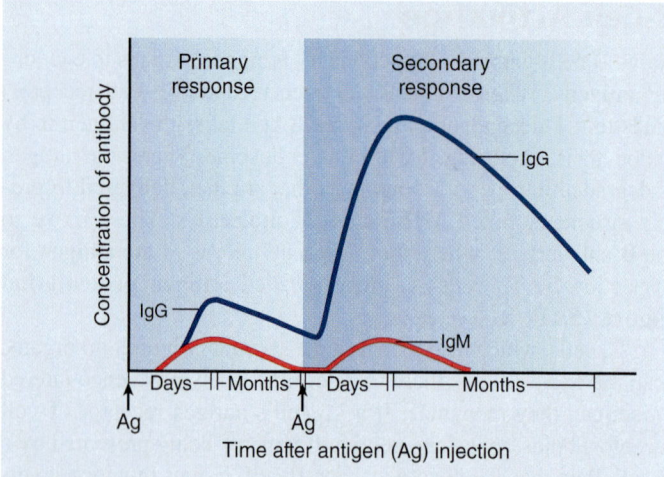

FIGURE 15.12 The Primary and Secondary Responses to Antigen The first exposure to antigen elicits relatively low amounts of first IgM, followed by IgG in the blood. The second exposure, which characterizes the memory of the adaptive immune system, elicits rapid production of relatively large quantities of IgG.

How would this graph be different if it illustrated antibody levels on a mucosal surface?

(**figure 15.12**). During this delay, the person might experience signs and symptoms of an infection, which could be life-threatening. The immune system, however, is actively responding. Naive B cells that bind the antigen present the peptide fragments to T_H cells. Once activated, those B cells multiply, generating a population of cells that recognize the antigen. As some of the activated B cells continue dividing, others differentiate to form antibody-secreting plasma cells (**figure 15.13**).

Each plasma cell generally undergoes apoptosis after several days, but activated B cells continue multiplying and differentiating, generating increasing numbers of plasma cells as long as

antigen is present. The result is a slow but steady increase in the titer (concentration) of antibody molecules.

Over time, the proliferating B cells undergo changes that improve the immune response. These include:

- **Affinity maturation.** This is a form of natural selection among proliferating B cells (**figure 15.14**). As activated B cells multiply, spontaneous mutations commonly occur in certain regions of the antibody genes. Some of these result in slight changes in the antigen-binding site of the antibody (and therefore the B-cell receptor). B cells that bind antigen for the longest duration are most likely to proliferate.

- **Class switching.** All B cells are initially programmed to differentiate into plasma cells that secrete IgM. Cytokines produced by T_H cells, however, induce some activated B cells to switch that genetic program, causing them to differentiate into plasma cells that secrete other antibody classes. B cells in the lymph nodes most commonly switch to IgG production (**figure 15.15**). B cells in the mucosa-associated lymphoid tissues generally switch to IgA production, providing mucosal immunity.

After class switching, some B cells become memory B cells. These persist in the body for years and are present in numbers sufficient to give a prompt secondary response if the same antigen is encountered again later.

The antibody response begins to wane as the accumulating antibodies clear the antigen. Progressively fewer molecules of antigen remain to stimulate the lymphocytes, and, as a result, the activated lymphocytes undergo apoptosis. Memory B cells, however, are long-lived even in the absence of antigen.

MicroByte

A plasma cell secretes thousands of identical antibody molecules per second.

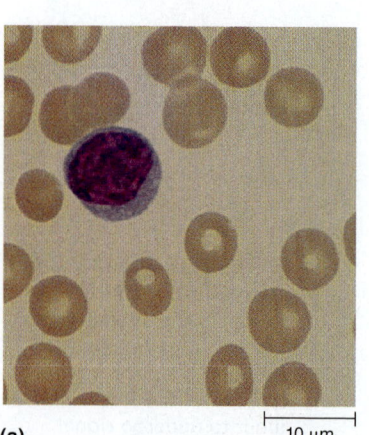 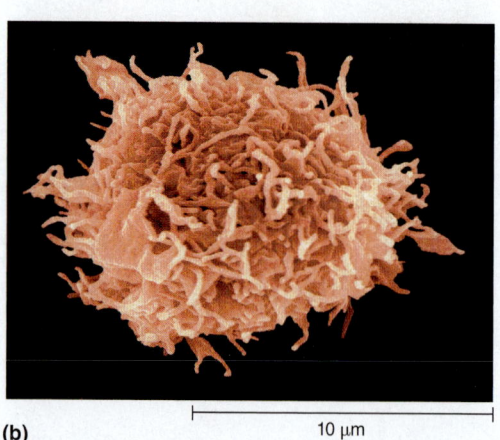 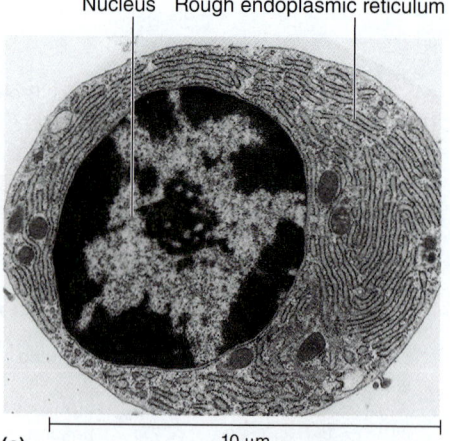

FIGURE 15.13 Lymphocytes and Plasma Cells (a) Light micrograph of a T lymphocyte. The morphology is the same as that of a B lymphocyte. (b) Scanning electron micrograph of a B lymphocyte. (c) Plasma cell, an effector B cell, which secretes antibody molecules. Note the extensive rough endoplasmic reticulum, the site of protein synthesis.

Would a single plasma cell produce antibody molecules of the same specificity or different specificities? Explain your answer.

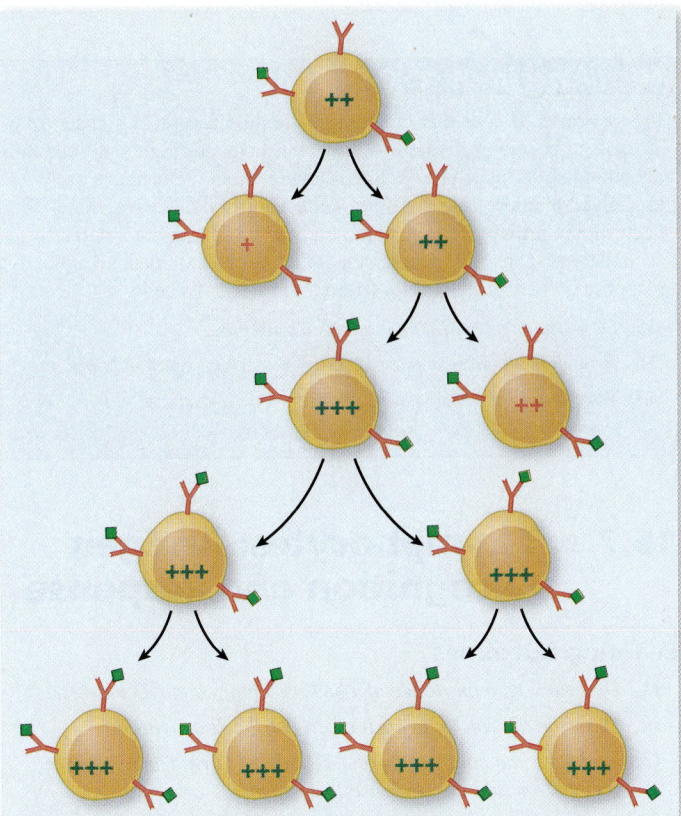

FIGURE 15.14 Affinity Maturation B cells that bind antigen for the longest duration are the most likely to proliferate. The plus signs indicate the relative quality of binding of the antibody to the antigen; those in green indicate the most "fit" to continue proliferating.

❓ *What accounts for the change in a B cell's ability to bind antigen?*

Characteristics of the Secondary Response

The secondary response is significantly faster and more effective than the primary response. In fact, repeat invaders are generally eliminated before they cause noticeable harm. This is why a person who has recovered from a particular disease generally has long-lasting immunity to that disease. Vaccination takes advantage of this naturally occurring phenomenon. ▶▶ **vaccination, p. 421**

Memory B cells are responsible for the efficiency of the secondary response. For one thing, there are more cells—memory B cells as well as memory helper T cells—that can respond to a specific antigen. In addition, the memory B cells are able to scavenge even low concentrations of antigen because their receptors have been fine-tuned through affinity maturation. Likewise, the antibodies coded for by these cells bind antigen more effectively.

When memory B cells become activated, some quickly differentiate to form plasma cells, resulting in the rapid production of antibodies. Because of class switching, these antibodies are often IgG or IgA. Other activated cells begin proliferating, once again undergoing affinity maturation to generate even more effective antibodies. Subsequent exposures to antigen lead to an even stronger response.

The Response to T-Independent Antigens

T-independent antigens can activate B cells without the aid of T$_H$ cells. Relatively few antigens are T-independent, but they can be very important medically.

FIGURE 15.15 Class Switching Naive B cells are programmed to produce IgM antibodies. Activated B cells undergo class switching. Class switching does not alter the variable region and therefore has no effect on the antibody specificity. The plasma cells descended from B cells in lymph nodes most commonly produce IgG after class switching.

❓ *How would this illustration change if it showed class switching by B cells residing in Peyer's patches?*

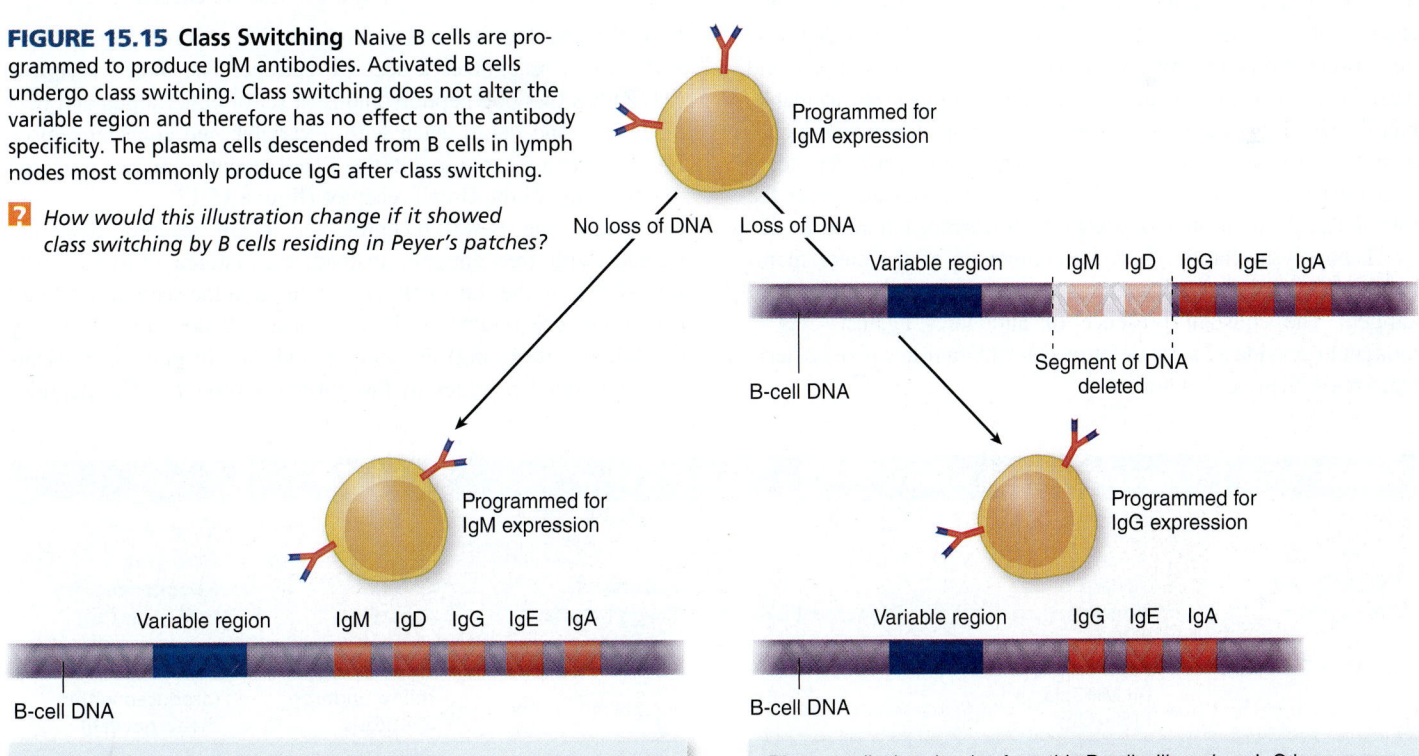

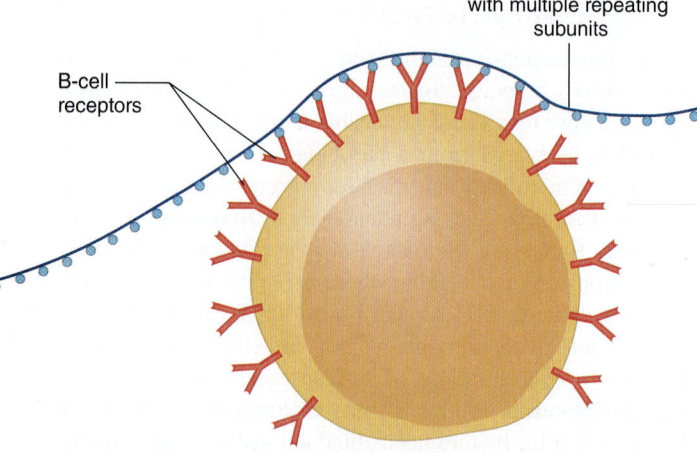

Polysaccharide antigen with multiple repeating subunits

B-cell receptors

FIGURE 15.16 T-Independent Antigens Antigens such as some polysaccharides have multiple repeating epitopes. Because of the arrangement of epitopes, clusters of B-cell receptors bind to the antigen simultaneously, leading to B-cell activation without the involvement of T_H cells.

❓ *Why is the response to T-independent antigens important medically?*

Polysaccharides and other molecules that have numerous identical evenly spaced epitopes are one type of T-independent antigen. Because of the arrangement of epitopes, clusters of B-cell receptors bind the antigen simultaneously, leading to activation of that B cell without the involvement of helper T cells (**figure 15.16**). These T-independent antigens are particularly significant because they are not very immunogenic in young children. This is why children less than 2 years of age are more susceptible to pathogens such as *Streptococcus pneumoniae* and *Haemophilus influenzae,* which cloak themselves in polysaccharide capsules. Antibodies that bind the capsules would be protective if the child's immune system could make them. Vaccines made from purified capsules are available, but, likewise, they do not elicit an immune response in young children. Fortunately, newer vaccines designed to induce a T-dependent response have been developed. They will be discussed later, when the role of T_H cells in the antibody response is described in more detail.

Lipopolysaccharide (LPS), a component of the outer membrane of Gram-negative bacteria, is another type of T-independent antigen. The constant presence of antibodies against LPS is thought to provide an early defense against Gram-negative bacteria that breach the body's barriers.

MicroAssessment 15.6

In most cases, B cells that bind antigen require accessory signals from T_H cells to become activated. Activated B cells proliferate, ultimately becoming either plasma cells that secrete antibody molecules or long-lived memory cells. Affinity maturation and class switching occur in the primary response; these allow a swift and more effective secondary response. T-independent antigens can stimulate an antibody response by activating B cells without the aid of T_H cells.

15. *Describe the significance of class switching.*

16. *How does the ability to bind antigen increase as B cells multiply?*

17. *Why should B cells residing in the mucosa-associated lymphoid tissues produce IgA?* ➕

15.7 ■ T Lymphocytes: Antigen Recognition and Response

Learning Outcomes

11. *Describe the importance of T-cell receptors and CD markers.*

12. *Describe the role of dendritic cells in T-cell activation.*

13. *Compare and contrast T_H and T_C cells with respect to antigen recognition and the response to antigen.*

The role of T cells is very different from that of B cells (see figure 15.1). For one thing, T cells never produce antibodies. Instead, effector T cells directly interact with other cells—target cells—to cause distinct changes in those cells (**table 15.2**).

General Characteristics of T Cells

Like B cells, T cells have multiple copies of a receptor on their surface that recognizes a specific epitope. The **T-cell receptor (TCR)** has two polypeptide chains (a set of either alpha and beta or gamma and delta), each with a variable and constant region. From a structural standpoint, the T-cell receptor can be compared to one "arm" of the B-cell receptor (**figure 15.17**).

Unlike the B-cell receptor, the T-cell receptor does not interact with free antigen. Instead, the antigen must be "presented" by another host cell, as described in the section on B-cell activation (see figure 15.11). The host cell does this by partly degrading (processing) the antigen and then displaying (presenting) individual peptides of the antigen's proteins. The peptides

TABLE 15.2	Characteristics of T Cells				
T Cell Type/ CD Marker	Antigen Recognition	Effector Form	Potential Target Cells	Effector Function	Source of Antigen Recognized by Effector Cell
Cytotoxic T cell/CD8	Peptides presented on MHC class I molecules	T_C cell	All nucleated cells	Induces target cell to undergo apoptosis	Endogenous (produced within the target cell)
Helper T cell/CD4	Peptides presented on MHC class II molecules	T_H cell	B cells, macrophages	Activates target cell	Exogenous (produced outside of the target cell)

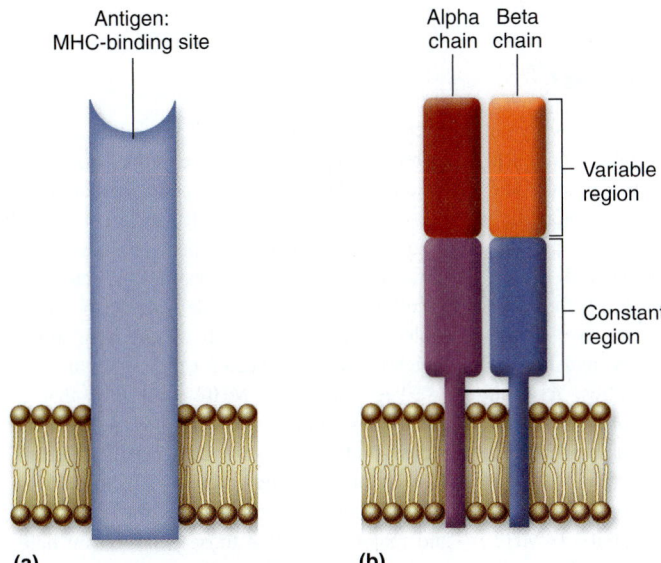

(a)   **(b)**

FIGURE 15.17 Structure of a T-Cell Receptor **(a)** Simplified structure, showing the antigen:MHC-binding site. **(b)** The molecule is made of two different chains, each with a constant region and a variable region, linked by a disulfide bond.

❓ *A T-cell receptor is similar to which part of a B-cell receptor?*

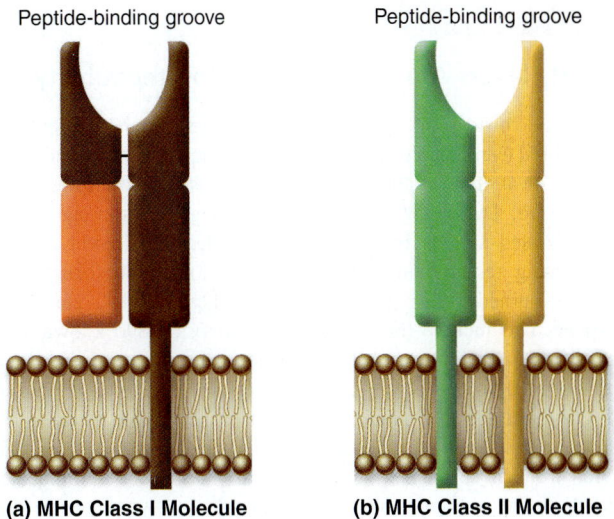

(a) MHC Class I Molecule   **(b) MHC Class II Molecule**

FIGURE 15.18 MHC Molecules **(a)** MHC class I molecule; cytoplasmic proteins (endogenous antigens) are presented in the groove of these molecules. **(b)** MHC class II molecule; proteins taken in by the cell (exogenous antigens) are presented in the groove of these molecules.

❓ *Which cell type recognizes peptides presented in MHC class I molecules? Which cell type recognizes peptides presented in MHC class II molecules?*

from the antigen are cradled in the groove of proteins—**major histocompatibility complex (MHC) molecules**—on the surface of the presenting cell.

Two types of MHC molecules present antigen—**MHC class I** and **MHC class II** (**figure 15.18**). Both are shaped somewhat like an elongated bun and hold the peptide lengthwise, like a bun holds a hot dog. When a T cell recognizes an antigen, it is actually recognizing both the peptide and MHC molecule simultaneously. In other words, the T-cell receptor recognizes the "whole sandwich"—the peptide:MHC complex.

MHC class I molecules present endogenous antigens (antigens made within the cell). MHC class II molecules present exogenous antigens (antigens taken up by a cell). All nucleated cells produce MHC class I molecules, but only specialized cell types (dendritic cells, B cells, and macrophages)—collectively referred to as **antigen-presenting cells (APCs)**—make MHC class II molecules (see **Perspective 15.1**).

Recall from the introductory section that two functionally distinct T-cell populations are involved in eliminating antigen—cytotoxic T cells and helper T cells. These differ in their roles, and also how they recognize antigen: Cytotoxic T cells recognize antigen presented on MHC class I molecules, whereas helper T cells recognize antigen presented on MHC class II molecules. Because of these recognition characteristics, effector cytotoxic T (T_C) cells respond to endogenous antigens, whereas effector helper T (T_H) cells respond to exogenous antigens (**figure 15.19**).

Cytotoxic and helper T cells are identical microscopically, so scientists distinguish them based on the presence of surface proteins called **cluster of differentiation (CD) markers**. Most cytotoxic T cells have the CD8 marker and are frequently referred to as CD8 T cells; most helper T cells carry the CD4 marker and are often called CD4 T cells. Note that CD4 is also a receptor for HIV, which explains why the virus infects helper T cells.

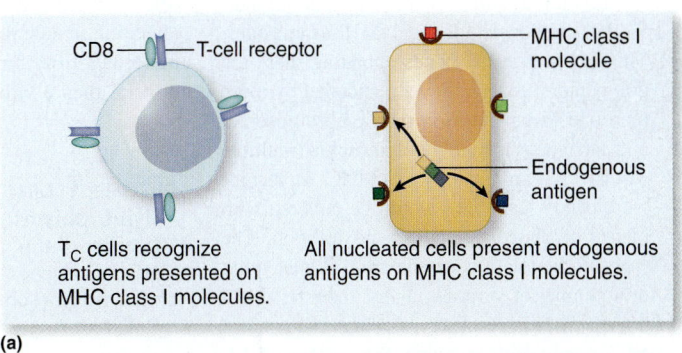

(a)

T_C cells recognize antigens presented on MHC class I molecules.   All nucleated cells present endogenous antigens on MHC class I molecules.

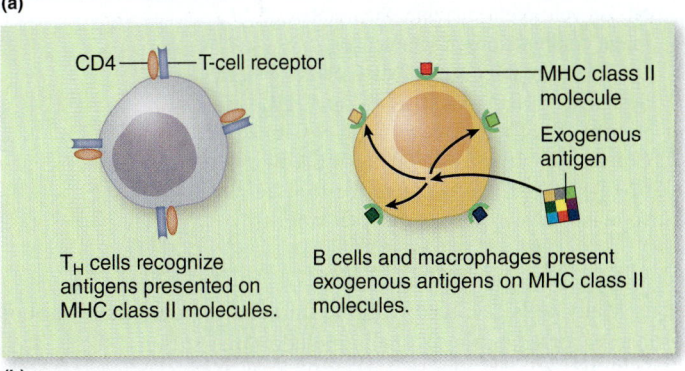

(b)

T_H cells recognize antigens presented on MHC class II molecules.   B cells and macrophages present exogenous antigens on MHC class II molecules.

FIGURE 15.19 Antigen Recognition by Effector T Cells **(a)** Cytotoxic T (T_c) cell. **(b)** Helper T (T_H) cell.

❓ *What is the fate of a cell that presents antigen recognized by a T_C cell? What is the fate of a cell that presents antigen recognized by a T_H cell?*

PERSPECTIVE   **15.1**

What Flavors Are Your Major Histocompatibility Complex Molecules?

The major histocompatibility molecules were discovered over half a century ago, long before their critical role in adaptive immunity was recognized. During World War II, bombing raids caused serious burns in many people, stimulating research into skin transplants to replace burned tissue. That research quickly expanded to include transplants of a variety of other tissues and organs. Unfortunately, the transplanted material was generally rejected by the recipient's immune system because certain molecules on the donor cells differed from those on the recipient's cells; the donor tissue was perceived as an "invader" by the recipient's immune system. To overcome this problem, researchers tried to more closely match donor and recipient tissues, relying on newly developed tissue-typing tests. The typing tests looked for leukocyte surface molecules called **human leukocyte antigens (HLAs),** which serve as markers for tissue compatibility. Later, researchers determined that HLAs were encoded by a cluster of genes, now called the major histocompatibility complex. Unfortunately, the terminology can be confusing because the molecules that transplant biologists refer to as HLAs are called MHC molecules by immunologists.

It is highly unlikely that two random individuals will have identical MHC molecules. This is because the genes encoding them are polygenic, meaning they are encoded by more than one locus (position on the chromosome) and each locus is highly polymorphic (multiple forms). As an analogy, if MHC molecules were candy, each cell would be covered with pieces of chocolate, taffy, and lollipop. The type of chocolate could be dark, white, milk, or a number of various flavors; the taffy could be peppermint, raspberry, or cinnamon, and so on. As you might imagine, the number of different combinations are immense.

There are three loci of MHC class I genes, designated HLA-A, HLA-B, and HLA-C (**figure 1**). There are at least 890 alleles (forms) for HLA-A, 1,400 for HLA-B, and 620 for HLA-C. In addition, the loci are all co-dominantly expressed. In other words, you inherited one set of the three genes from your mother and one set from your father; both sets are expressed. Putting this all together, and assuming that you inherited two completely different sets of alleles from your parents, your cells express six different varieties of MHC class I molecules—two of the over 890 known HLA-A possibilities, two of the over 1,400 HLA-B possibilities, and two of the over 620 HLA-C possibilities. As you can imagine, the likelihood that anyone you encounter in a day will have those same MHC class I molecules is extremely unlikely, unless you have an identical twin.

Why is there so much diversity in MHC molecules? The answer lies in the complex demands of antigen presentation. MHC class I molecules bind peptides that are only 8 to 10 amino acids in length; MHC class II molecules bind peptides that are only 13 to 25 amino acids in length. Somehow, within that constraint, the MHC molecules must bind as many different peptides as possible in order to ensure that a representative selection from the proteins within a cell can be presented to T cells. The ability to bind a wide variety of peptides is particularly important considering how readily microbes evolve in response to selective pressure. For example, if a single alteration in a viral protein prevented all MHC molecules from presenting peptides from that protein, then a virus with that mutation could overwhelm the defenses. The ability to bind different peptides is not enough, however, because, ideally, a given peptide should be presented in several slightly different orientations so that distinct aspects of the three-dimensional structures can be inspected by T-cell receptors. No single variety of MHC molecule can accomplish all of these aims, which explains the need for their diversity.

The variety of MHC molecules that a person has on his or her cells affects that individual's adaptive response to certain antigens. This is not surprising because MHC molecules differ in the array of peptides they can bind and manner in which those peptides are held in the molecule. Thus, they affect what the T cells actually "see." In fact, the severity of certain diseases has been shown to correlate with the MHC type of the infected individual. For example, rheumatic fever, which can occur as a consequence of *Streptococcus pyogenes* infection, develops more frequently in individuals with certain MHC types. The most serious manifestations of schistosomiasis have also been shown to correlate with certain MHC types. Epidemics of life-threatening diseases such as plague and smallpox have dramatically altered the relative proportion of MHC types in certain populations, killing those whose MHC types ineffectively present peptides from the causative agent. ▶▶ **rheumatic fever, p. 489** ◀◀ **schistosomiasis, p. 350**

FIGURE 1 MHC Polymorphisms The order of the MHC class I genes on the chromosome is B, C, and A.

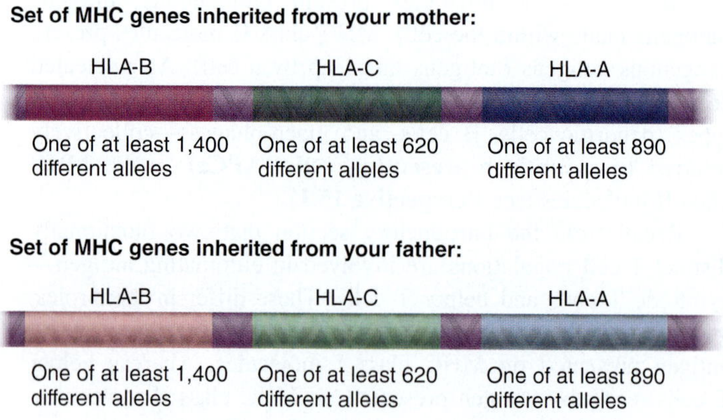

Set of MHC genes inherited from your mother:

HLA-B	HLA-C	HLA-A
One of at least 1,400 different alleles	One of at least 620 different alleles	One of at least 890 different alleles

Set of MHC genes inherited from your father:

HLA-B	HLA-C	HLA-A
One of at least 1,400 different alleles	One of at least 620 different alleles	One of at least 890 different alleles

---

Many cells involved with the immune response have various subsets (for example, T_H cells include three subsets—T_H1, T_H2, and T_H17). For simplicity, however, we will focus on the general characteristics of the cell types, rather than the subsets.

MicroByte

Of the approximately 10^{10} T cells in the body, only a few will recognize a given epitope the first time it is encountered.

Activation of T Cells

Dendritic cells, the scouts of innate immunity, play a crucial role in T-cell activation (**figure 15.20**). Immature dendritic cells reside in peripheral tissues, particularly under the skin and mucosa, gathering various materials from those areas. The cells use both phagocytosis and pinocytosis to take up particulate and soluble material that could contain foreign proteins. Dendritic

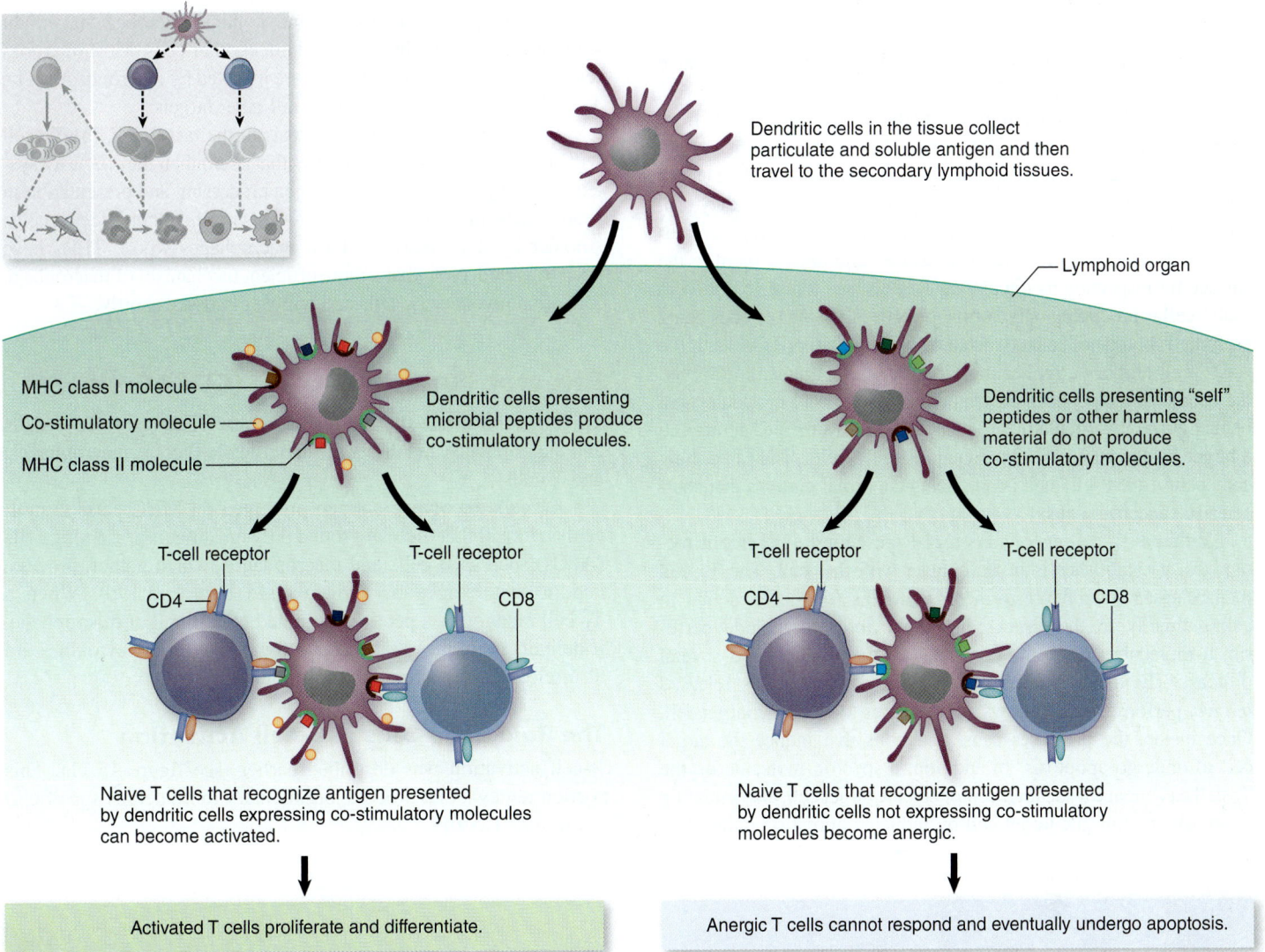

Dendritic cells in the tissue collect particulate and soluble antigen and then travel to the secondary lymphoid tissues.

Lymphoid organ

MHC class I molecule

Co-stimulatory molecule

MHC class II molecule

Dendritic cells presenting microbial peptides produce co-stimulatory molecules.

Dendritic cells presenting "self" peptides or other harmless material do not produce co-stimulatory molecules.

T-cell receptor

T-cell receptor

CD4

CD8

T-cell receptor

T-cell receptor

CD4

CD8

Naive T cells that recognize antigen presented by dendritic cells expressing co-stimulatory molecules can become activated.

Naive T cells that recognize antigen presented by dendritic cells not expressing co-stimulatory molecules become anergic.

Activated T cells proliferate and differentiate.

Anergic T cells cannot respond and eventually undergo apoptosis.

FIGURE 15.20 T-Cell Activation

What causes a dendritic cell to produce co-stimulatory molecules?

cells located just below the mucosal barriers are even able to send tentacle-like extensions between the epithelial cells of the barriers. Using this action, the dendritic cells gather material from the respiratory tract and the lumen of the intestine.

Dendritic cells have TLRs (toll-like receptors) and other receptors that allow them to recognize pathogens. If pathogens are detected, the dendritic cell takes up even more material, and then travels to the secondary lymphoid organs where it will encounter naive T cells. Dendritic cells near the end of their life span do this as well. ◀◀ TLRs, p. 342

En route to the secondary lymphoid organs, the dendritic cells mature into a form that presents antigen to naive T cells. Dendritic cells that detected pathogens produce surface proteins called **co-stimulatory molecules.** These molecules function as "emergency lights" that interact with the T cell, communicating that the material being presented indicates "danger." Naive T cells that recognize antigen presented by dendritic cells displaying co-stimulatory molecules can become activated. In contrast, naive T cells that recognize antigen presented by a dendritic cell not displaying co-stimulatory molecules become anergic and eventually

undergo apoptosis. Recall that inducing anergy is one mechanism the adaptive immune response uses to eliminate lymphocytes that recognize "self" proteins, a critical aspect of self-tolerance.

Dendritic cells are able to "cross present" antigens, meaning they can present peptides on both types of MHC molecules—class I and class II—regardless of the origin. This allows them to present antigen to, and therefore activate, cytotoxic T cells as well as helper T cells.

Once a T cell is activated, it undergoes clonal selection and expansion as described previously, eventually forming effector cells and memory cells. These can leave the secondary lymphoid organs and circulate in the bloodstream. They can also enter tissues, particularly at sites of infection.

Macrophages and B cells also present antigens they have gathered on MHC class II molecules and can produce co-stimulatory molecules. Based on this information, it seems they should also be able to activate naive T cells, but studies suggest they do not contact naive T cells *in vivo*. They do encounter memory T cells, however, so they probably activate these cells during the secondary response.

Effector Functions of T$_C$ (CD8) Cells

T$_C$ cells induce apoptosis in infected "self" cells. They also destroy cancerous "self" cells.

How do T$_C$ cells distinguish infected or cancerous cells from their normal counterparts? The answer lies in the significance of antigen presentation on MHC class I molecules (**figure 15.21**). All nucleated cells routinely degrade a portion of the proteins they produce (endogenous proteins). They then load the resulting peptides into the groove of MHC class I molecules and deliver them to the surface for inspection by circulating T$_C$ cells (see figure 15.19a). If a "self" cell is producing only normal proteins, then the peptides being presented should not be recognized by any circulating T$_C$ cells. If the "self" cell harbors a replicating virus or microorganism, however, then some of the peptides presented on MHC class I molecules will be recognized by circulating T$_C$ cells. This makes the presenting cell a target for the lethal effector functions of T$_C$ cells. The same thing can occur if the "self" cell is producing abnormal proteins that characterize cancerous cells.

When a T$_C$ cell encounters a cell presenting a peptide it recognizes, it establishes intimate contact with that cell. The T$_C$ cell then releases several pre-formed cytotoxins (molecules lethal to cells) directly to the target cell. The cytotoxins include perforin, a molecule that forms pores in cell membranes, and several proteases. Evidence indicates that at the concentrations released *in vivo*, perforin simply allows the proteases to enter the target cell. Once inside, the proteases cause reactions that induce the target cell to undergo apoptosis. In addition, a specific molecule on the T$_C$ cell can engage a "death receptor" on the target cell, also initiating apoptosis. Killing the target cell by inducing apoptosis rather than

lysis minimizes the number of intracellular microbes that might spill into the surrounding area and infect other cells. Most microbes remain in cell remnants until they are ingested by macrophages. The T$_C$ cell survives and can go on to kill other targets.

In addition to inducing apoptosis in the target cell, the T$_C$ cell will also produce various cytokines that strengthen the "security system." One cytokine increases antigen processing and presentation in nearby cells, making it easier for T$_C$ cells to find other infected cells. Another cytokine activates local macrophages whose TLRs have been triggered. Note that a more efficient mechanism of macrophage activation involves T$_H$ cells and will be discussed shortly.

Effector Functions of T$_H$ (CD4) Cells

T$_H$ cells orchestrate the immune response. They activate B cells and macrophages and direct the activities of B cells, macrophages, and T cells.

T$_H$ cells recognize antigen presented on MHC class II molecules. Recall that these are found only on antigen-presenting cells (APCs) such as B cells and macrophages, which gather, process, and present exogenous antigens (see figure 15.19b). When a T$_H$ cell recognizes a peptide presented by a B cell or macrophage, it delivers cytokines that activate the cell. Various cytokines are also released, depending on the subset of T$_H$ cell.

The Role of T$_H$ Cells in B-Cell Activation

B-cell activation was described earlier (see figure 15.11). This section reviews that process, but includes more details specific to the role of T$_H$ cells.

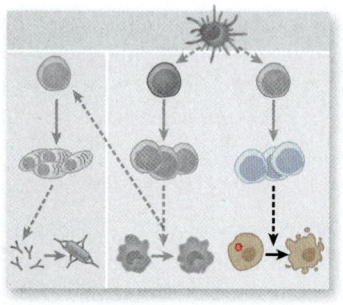

FIGURE 15.21 Functions of T$_C$ Cells **(a)** T$_C$ cells ignore healthy "self" cells. **(b)** T$_C$ cells induce apoptosis in virally infected "self" cells.

❓ *Could a T$_C$ cell induce apoptosis in a "self" cell that lost the ability to produce MHC class I molecules?*

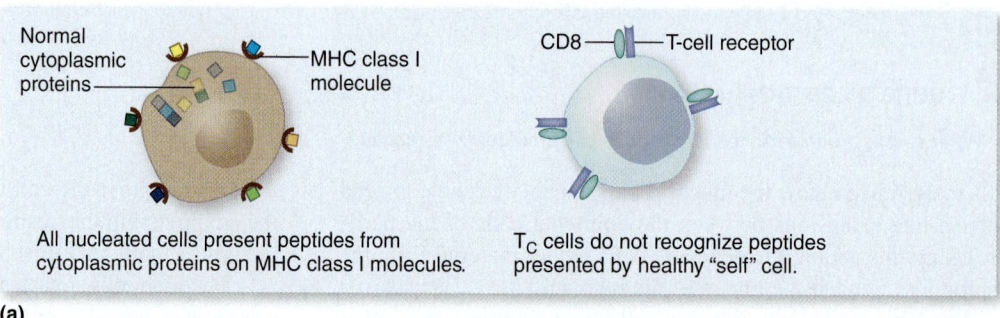

Normal cytoplasmic proteins — MHC class I molecule

All nucleated cells present peptides from cytoplasmic proteins on MHC class I molecules.

CD8 — T-cell receptor

T$_C$ cells do not recognize peptides presented by healthy "self" cell.

(a)

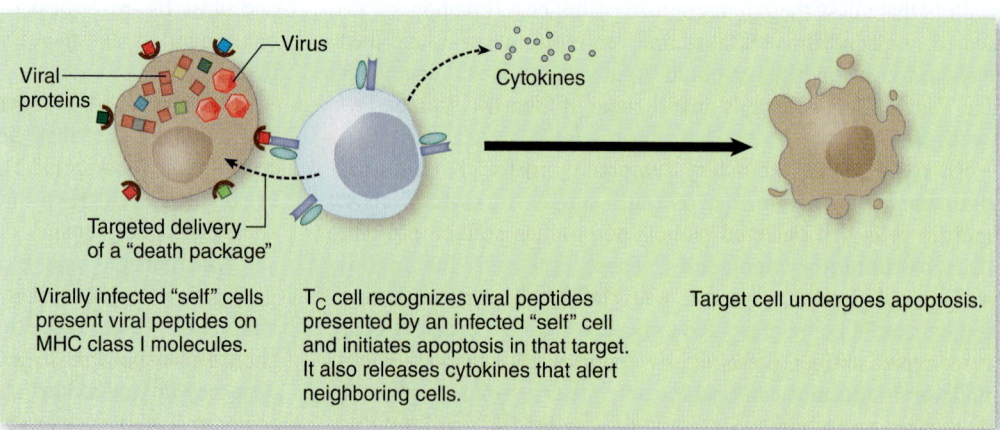

Virus

Viral proteins

Cytokines

Targeted delivery of a "death package"

Virally infected "self" cells present viral peptides on MHC class I molecules.

T$_C$ cell recognizes viral peptides presented by an infected "self" cell and initiates apoptosis in that target. It also releases cytokines that alert neighboring cells.

Target cell undergoes apoptosis.

(b)

When a naive B cell binds antigen via its B-cell receptor, the cell takes the antigen in by endocytosis. Proteins within the endosome are then degraded to produce short peptides that can be loaded into the groove of MHC class II molecules. If a T_H cell encounters a B cell presenting a peptide it recognizes, it delivers cytokines to that cell. These activate the B cell, allowing it to proliferate and undergo class switching. The cytokines also drive the formation of memory B cells.

A T_H cell does not need to recognize the same epitope as the B cell it activates. This is because the B cell presents numerous peptides from the antigen, and the T-cell receptor could bind any of those. In fact, a responding B cell probably recognized an epitope on a pathogen's surface, whereas a T_H cell could very well recognize a peptide from within the pathogen.

Understanding the role of antigen processing and presentation led to a vaccine that now prevents infection by what was once the most common cause of meningitis in children—*Haemophilus influenzae*. Recall that young children are particularly susceptible to meningitis caused by this organism because it produces a polysaccharide capsule, an example of a T-independent antigen to which this age group responds poorly. A polysaccharide antigen can be converted to a T-dependent antigen by covalently attaching it to a large protein molecule, making a **conjugate vaccine.** The polysaccharide component of the vaccine binds to a B-cell receptor and the entire molecule is taken in. The protein component will then be processed and presented to T_H cells. Although the B cell recognizes the polysaccharide component of the vaccine, the T cells recognize peptides from the protein component. This recognition leads to B-cell activation and subsequent production of antibodies that bind the capsule. Conjugate vaccines against other pathogens have also been developed. ▶▶ conjugate vaccines, p. 424

Antigen processing and presentation also explains how some people develop allergies to penicillin. This medication is a **hapten,** a molecule that binds a B-cell receptor yet does not normally elicit antibody production. In the body, penicillin can react with proteins, forming a penicillin-protein conjugate. This functions in a manner analogous to the *H. influenzae* conjugate vaccine, resulting in antibodies that bind penicillin. If IgE antibodies are formed, their binding to penicillin can result in allergic reactions, ruling out the further use of the antimicrobial medication in these individuals. ▶▶ allergy, p. 401 ◀◀ ▶▶ penicillin, pp. 62, 463

The Role of T_H Cells in Macrophage Activation

As discussed in chapter 14, macrophages routinely engulf and degrade invading microbes, clearing most organisms even before an adaptive response is mounted. If this alone is not sufficient to control the invader, macrophages can be activated by T_H cells, allowing the phagocytes to produce more potent destructive mechanisms. ◀◀ macrophages, pp. 340, 348

The steps of macrophage activation are very similar to those described for B-cell activation. When macrophages engulf material, they enclose it within a membrane-bound phagosome (**figure 15.22**). The proteins within the phagosome are then degraded, and the resulting peptides presented on MHC class II molecules. If a T_H cell recognizes one of the peptides, it delivers cytokines that activate the macrophage. ◀◀ phagosome, p. 72

When a macrophage is activated, it gets larger, the plasma membrane becomes ruffled and irregular, and the cell increases its metabolism so that the lysosomes—which contain antimicrobial substances—increase in number. The activated macrophage also begins producing nitric oxide, a potent antimicrobial chemical, along with various compounds that can be released to destroy extracellular microorganisms.

If the response is still not sufficient to control the infection, activated macrophages fuse together, forming **giant cells.** These, along with other macrophages and T cells, can form granulomas that wall off the offending agent, preventing infectious microbes from escaping to infect other cells. Activated macrophages are an important aspect of the immune response against diseases such as tuberculosis that are caused by organisms capable of surviving within macrophages. ◀◀ giant cell, p. 348 ◀◀ granuloma, p. 348

Subsets of Dendritic Cells and T Cells

Various subsets of dendritic cells (DC1 and DC2) and effector helper T cells (T_H1, T_H2, T_H17) steer the immune system toward

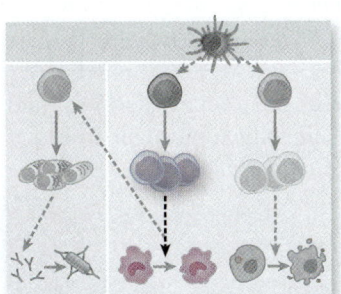

FIGURE 15.22 The Role of T_H Cells in Macrophage Activation

? *How does macrophage activation help prevent disease?*

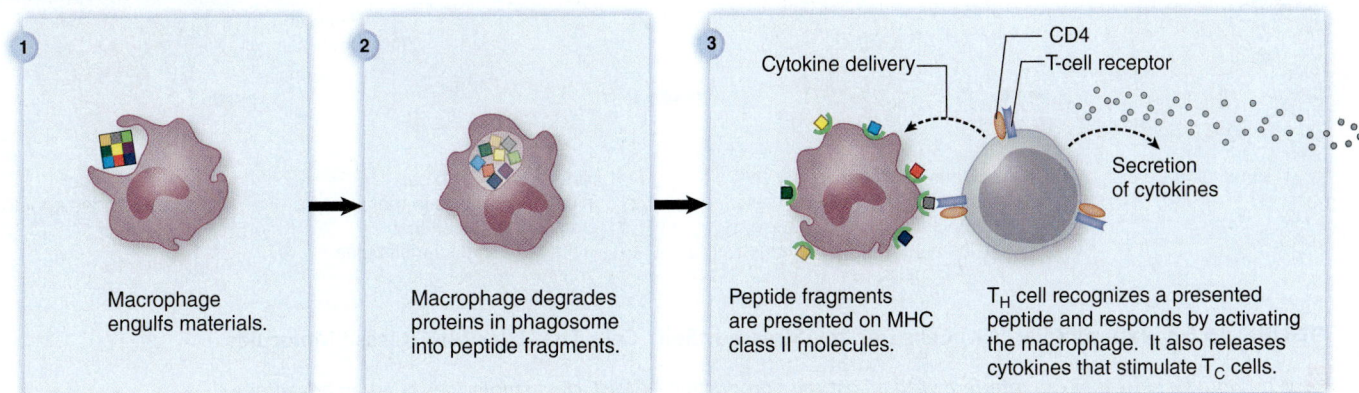

1. Macrophage engulfs materials.
2. Macrophage degrades proteins in phagosome into peptide fragments.
3. Peptide fragments are presented on MHC class II molecules. — Cytokine delivery — CD4 — T-cell receptor — Secretion of cytokines. T_H cell recognizes a presented peptide and responds by activating the macrophage. It also releases cytokines that stimulate T_C cells.

an appropriate response. The roles of these subsets are still being clarified, but the different types of dendritic cells produce specific cytokines that cause activated helper T cells to differentiate into specialized effector subsets. For example, T_H1 cells activate macrophages, thereby promoting a response against intracellular pathogens; T_H2 cells direct a response against multicellular pathogens by recruiting eosinophils and basophils; and T_H17 cells recruit neutrophils, thereby directing a response against extracellular pathogens. The outcome of some conditions, such as Hansen's disease (leprosy), appears to correlate with the type of helper T-cell response.

MicroAssessment 15.7

Dendritic cells expressing co-stimulatory molecules activate T cells that recognize the presented antigen. T_C (CD8) cells recognize antigen presented on MHC class I molecules; they induce apoptosis in target cells and produce cytokines that increase the level of surveillance. T_H (CD4) cells recognize antigen presented on MHC class II molecules (found on B cells and macrophages); they activate the target cells and secrete various cytokines that orchestrate the immune response.

18. *Name three types of antigen-presenting cells.*

19. *If an effector CD8 cell recognizes antigen presented on an MHC class I molecule, how should it respond?*

20. *Why would a person who has AIDS be more susceptible to the bacterium that causes tuberculosis?* ➕

15.8 ■ Natural Killer (NK) Cells

Learning Outcome

14. *Describe two distinct protective roles of NK cells.*

Natural killer (NK) cells descend from lymphoid progenitor cells, but lack the antigen-specific receptors that characterize B cells and T cells. Their activities, however, assist the adaptive immune responses. ◀◀ lymphoid progenitor cells, p. 338

NK cells induce apoptosis in antibody-bound "self" cells. This process, antibody-dependent cellular cytotoxicity (ADCC) allows them to destroy host cells that have viral or other foreign proteins inserted into their membrane (see figure 15.8). NK cells can do this because they have Fc receptors for IgG molecules on

their surface; recall that Fc receptors bind the "red flag" portion of antibody molecules. The NK cell attaches to the antibodies and then delivers perforin- and protease-containing granules directly to the cell, initiating apoptosis. ◀◀ ADCC, p. 361

NK cells also recognize and destroy stressed host cells that do not have MHC class I molecules on their surface (**figure 15.23**). This is important because some viruses have evolved mechanisms to dodge the action of cytotoxic T cells by interfering with the process of antigen presentation; cells infected with such a virus will essentially be bare of MHC class I molecules and thus cannot be a target of cytotoxic T cells. The NK cells recognize the lack of MHC class I molecules on those cells, along with certain molecules that indicate the cells are under stress, and induce the infected cells to undergo apoptosis.

Recent evidence indicates that NK cells are more than killing machines. For example, they produce cytokines that help regulate and direct certain immune responses. Unfortunately, studying these actions is difficult because there are different subsets of NK cells, and the activities of the various subsets are influenced by cues in their local environment.

MicroAssessment 15.8

Natural killer (NK) cells can kill antibody-bound cells by antibody-dependent cellular cytotoxicity (ADCC). NK cells also kill cells not bearing MHC class I molecules on their surface.

21. *What mechanism do NK cells use to kill "self" cells?*

22. *Why would a "self" cell not display MHC class I molecules?*

23. *Why might a virus encode its own version of an MCH class I molecule?* ➕

15.9 ■ Lymphocyte Development

Learning Outcomes

15. *Describe the roles of gene rearrangement, imprecise joining, and combinatorial associations in the generation of diversity.*

16. *Describe positive and negative selection of lymphocytes.*

As descendants of hematopoietic stem cells develop into B cells and T cells, they acquire their ability to recognize distinct epitopes. B cells undergo the developmental stages in the bone

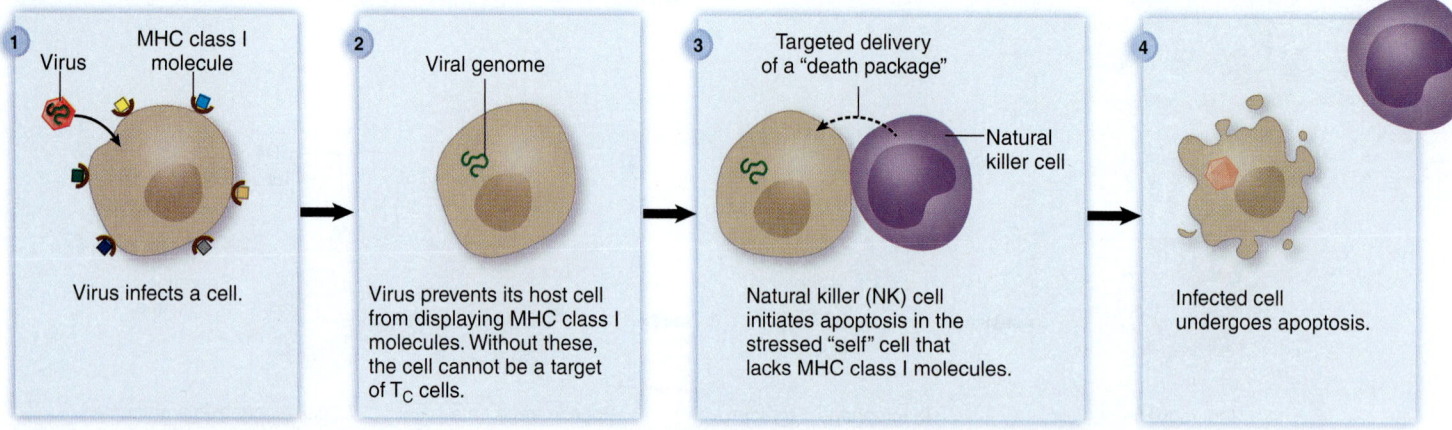

FIGURE 15.23 Natural Killer (NK) Cells Destroy Stressed "Self" Cells That Lack MHC Class I Molecules

❓ *Why would a virus that can interfere with a host cell's production of MHC class I molecules be at an advantage?*

marrow; T cells go through the maturation processes described in this section in the thymus. ◄◄ hematopoietic stem cells, p. 338

The events involved in the adaptive immune response, from the maturation of lymphocytes to the development of their effector functions, are summarized in **figure 15.24.**

Generation of Diversity

The mechanisms lymphocytes use to produce a seemingly limitless assortment of antibodies and antigen-specific receptors were first discovered in studies using B cells. Because the processes in

FIGURE 15.24 Summary of the Adaptive Immune Response

❓ *How would the adaptive immune response be affected if memory cells could not be produced?*

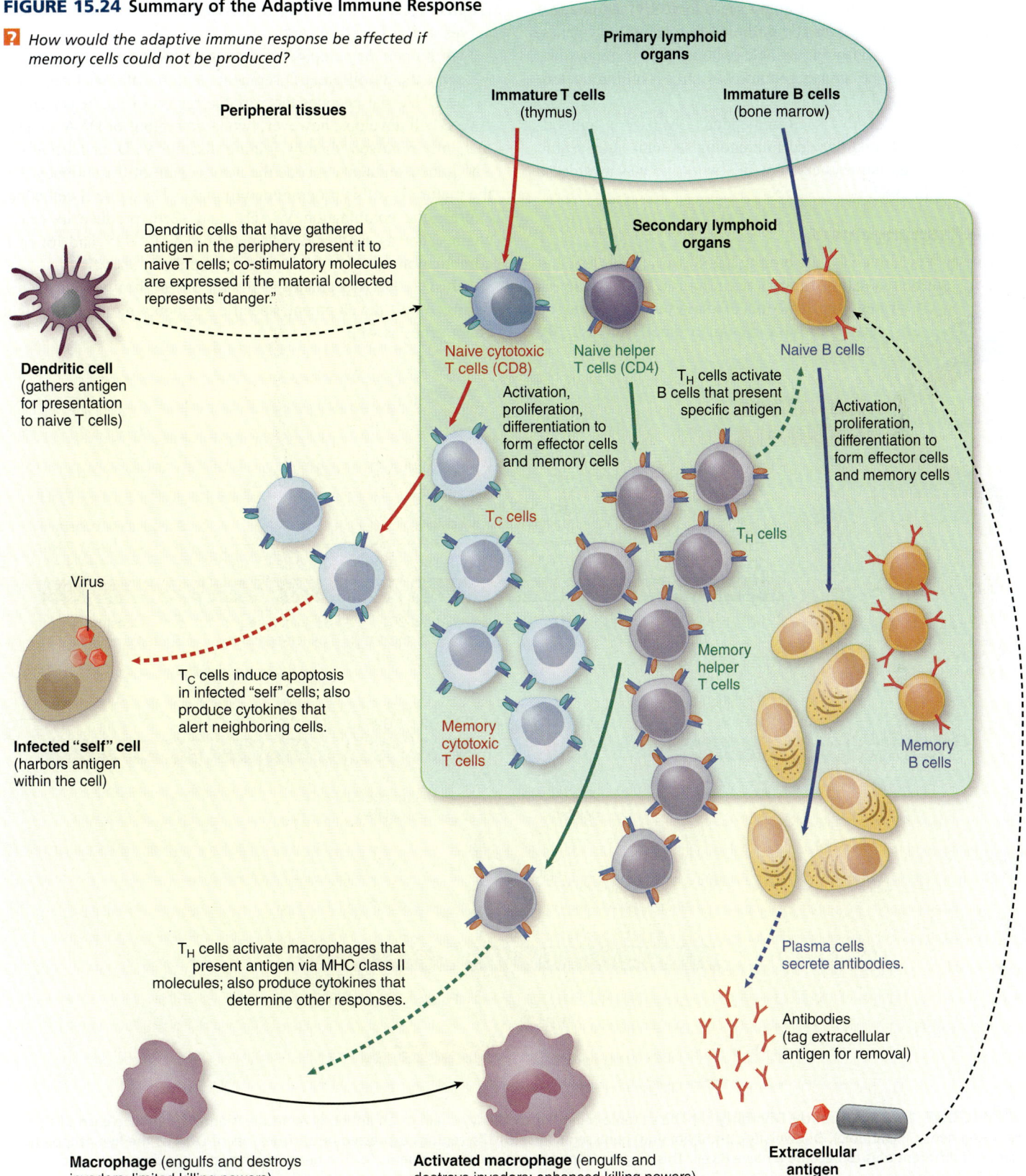

Primary lymphoid organs

Immature T cells (thymus)   Immature B cells (bone marrow)

Peripheral tissues

Secondary lymphoid organs

Dendritic cells that have gathered antigen in the periphery present it to naive T cells; co-stimulatory molecules are expressed if the material collected represents "danger."

Dendritic cell (gathers antigen for presentation to naive T cells)

Naive cytotoxic T cells (CD8)   Naive helper T cells (CD4)   Naive B cells

T_H cells activate B cells that present specific antigen

Activation, proliferation, differentiation to form effector cells and memory cells

Activation, proliferation, differentiation to form effector cells and memory cells

T_C cells T_H cells

Virus

T_C cells induce apoptosis in infected "self" cells; also produce cytokines that alert neighboring cells.

Memory helper T cells

Memory B cells

Infected "self" cell (harbors antigen within the cell)

Memory cytotoxic T cells

T_H cells activate macrophages that present antigen via MHC class II molecules; also produce cytokines that determine other responses.

Plasma cells secrete antibodies.

Antibodies (tag extracellular antigen for removal)

Macrophage (engulfs and destroys invaders; limited killing powers)   **Activated macrophage** (engulfs and destroys invaders; enhanced killing powers)   **Extracellular antigen**

B cells are very similar to those for T cells, we will use B cells as a general model to describe the generation of diversity with respect to antigen recognition.

Each B cell responds to only one epitope, yet the population of B cells within the body appears able respond to more than 100 million different epitopes. Based on this number and the information presented in chapter 7, it might seem logical to assume that the human genome has over 100 million different antibody genes, each encoding specificity for a single epitope. This is impossible, however, because the human genome has only 3 billion nucleotides and contains about 25,000 genes.

The question of how such tremendous diversity in antibodies could be generated puzzled immunologists until Dr. Susumu Tonegawa solved the mystery. For this work, he was awarded a Nobel Prize in 1987.

Gene Rearrangement

A primary mechanism for generating a wide variety of different antibodies using a limited-size region of DNA employs a strategy similar to that of a savvy and well-dressed traveler living out of a small suitcase. By

mixing and matching different shirts, pants, and shoes, the traveler can create a wide variety of unique outfits from a limited number of components. Likewise, a B cell expresses three gene segments, one each from DNA regions called V (variable), D (diversity), and J (joining), to form an ensemble that encodes a nearly unique variable region of the heavy chain of an antibody (**figure 15.25**). Recall that a derivative of this molecule also serves as the B-cell receptor.

A human hematopoietic stem cell has about 40 different V segments, 25 different D segments, and 6 different J segments in the DNA that encodes the variable region of the heavy chain. As a B cell develops, however, two large regions of DNA are permanently removed, thereby joining discrete V, D, and J regions. The joined segments encode the heavy chain of the antibody that the mature B cell is programmed to make. Thus, one B cell could express the combination V3, D1, and J2 to produce its heavy chain, whereas another B cell might use V19, D25, and J6; each combination would result in a unique antibody specificity.

Just as DNA segments rearrange in the heavy chain genes during B-cell development, specific segments move in the light chain genes as well.

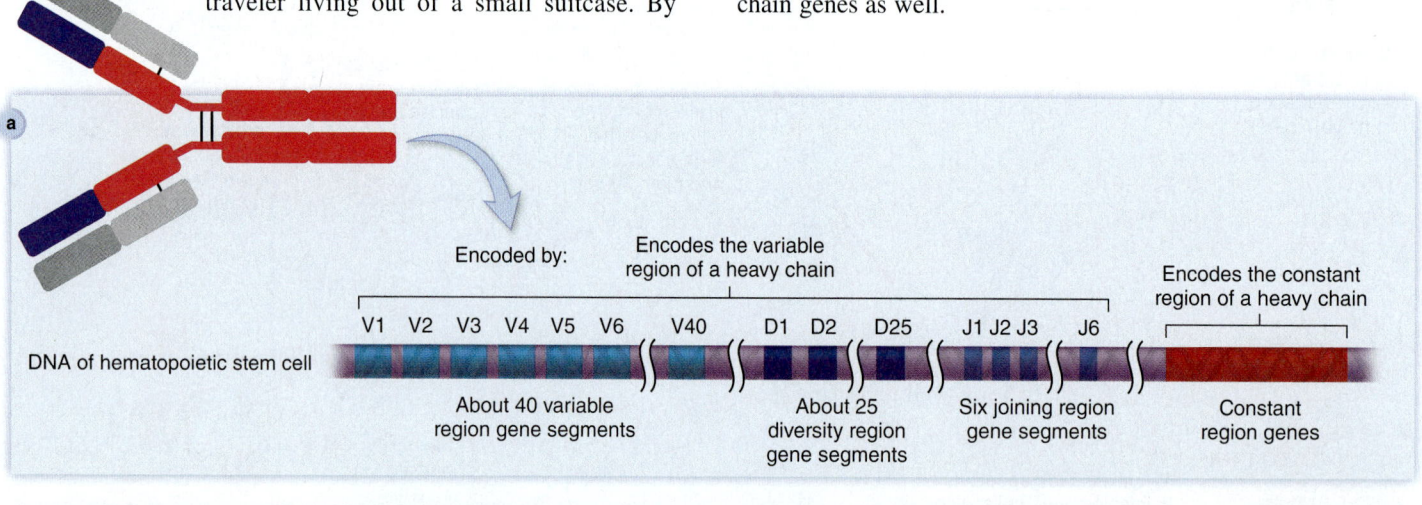

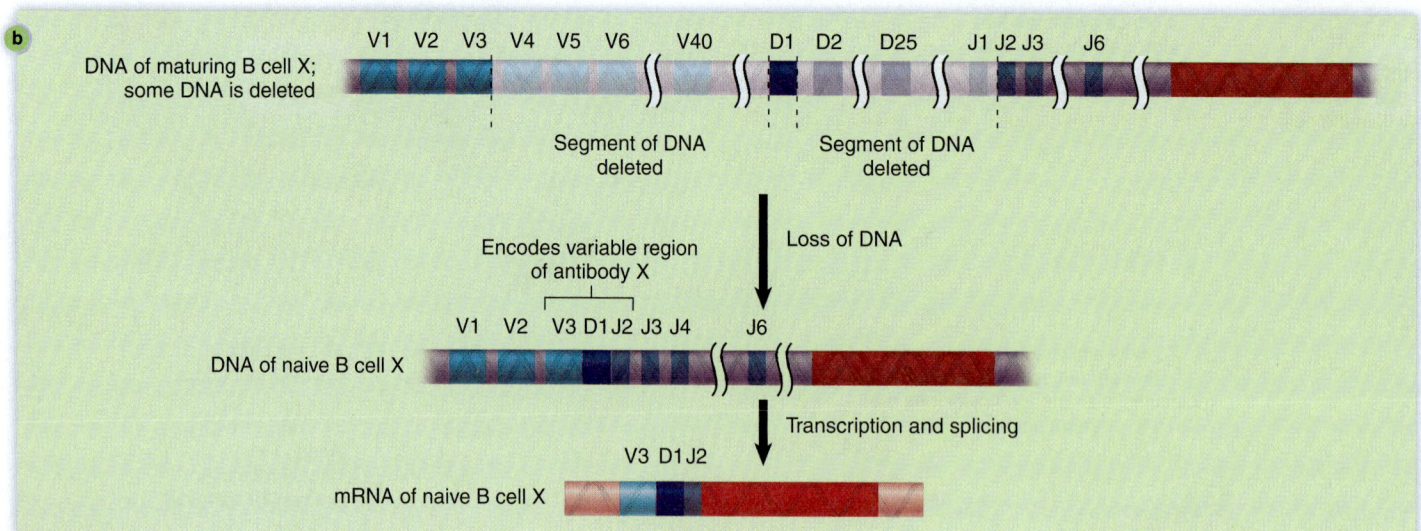

FIGURE 15.25 Antibody Diversity (a) The variable region of the heavy chain of an antibody molecule is encoded by one each of three gene segments: V (variable), D (diversity), and J (joining) gene segments. (b) The regions expressed result from loss of DNA as the hematopoietic stem cell differentiates to become a naive B cell. This diagram shows only the gene segments for the heavy chain and is not drawn to scale.

Which three aspects of this process in developing B cells contribute to antibody diversity?

Imprecise Joining

As the various segments are joined during gene rearrangement, nucleotides are often deleted or added between the sections. This imprecise joining changes the reading frame of the encoded protein so that two B cells that have the same V, D, and J segments for their heavy chain could potentially give rise to antibodies with very different specificities. Likewise, the segments of the light chain often join imprecisely.

Combinatorial Associations

Combinatorial association refers to the specific groupings of light chains and heavy chains that make up the antibody molecule. Both types of chains independently acquire diversity through gene rearrangement and imprecise joining. Additional diversity is then introduced when these two molecules join, because the combination of the two chains creates the antigen-binding site (see figure 15.7b, c).

Negative Selection of Self-Reactive B Cells

Once a B cell has developed its antigen receptor (B-cell receptor), it passes through rigorous checkpoints in the bone marrow. One of the most important is negative selection, which eliminates any B cell that binds "self." Most developing B cells fail negative selection and, as a consequence, are induced to undergo apoptosis. If these cells are not eliminated, then the immune system may attack "self" substances by mistake.

Positive and Negative Selection of Self-Reactive T Cells

The fate of developing T cells rests on two phases of trials—positive and negative selection. Positive selection permits only those T cells that recognize MHC to develop further. Recall that the T-cell receptor, unlike the B-cell receptor, recognizes a peptide: MHC complex. The T-cell receptor, therefore, must show at least some recognition of an MHC molecule regardless of the peptide it is carrying. T cells that show insufficient recognition fail positive selection and, as a consequence, are eliminated. Each T cell that passes positive selection is also subjected to negative selection. T cells that recognize "self" peptides presented on MHC molecules are eliminated. Positive and negative selection processes are so strict that over 95% of developing T cells undergo apoptosis in the thymus.

MicroAssessment 15.9

Mechanisms lymphocytes use to generate diversity of antigen specificity include rearrangement of gene segments, imprecise joining of those segments, and combinatorial associations of heavy chains and light chains. Negative selection eliminates B cells and T cells that recognize normal "self" molecules. Positive selection permits only those T cells that recognize the MHC molecules to develop further.

24. *What three gene segments encode the variable region of the heavy chain of an antibody molecule?*
25. *Why is negative selection important?*
26. *How is imprecise joining similar to a frameshift mutation?*

Summary

15.1 ■ Strategy of the Adaptive Immune Response

As a result of the **primary response** to an antigen, the **secondary response** is more effective. **Humoral immunity** works to eliminate extracellular antigens; **cell-mediated immunity (CMI)** deals with antigens residing within a host cell.

Overview of Humoral Immunity

Humoral immunity involves **B cells** (figure 15.1); in response to extracellular antigens, B cells proliferate and then differentiate into **plasma cells** that function as antibody-producing factories. Memory B cells are also formed.

Overview of Cell-Mediated Immunity

Cell-mediated immunity involves **T cells**; in response to intracellular antigens, **cytotoxic T cells** proliferate and then differentiate into T_C cells that induce apoptosis in "self" cells harboring the intruder. Memory cytotoxic T cells are also formed. **Helper T cells** proliferate and then differentiate to form T_H cells that help orchestrate the various responses of humoral and cell-mediated immunity. Memory helper T cells are also formed.

15.2 ■ Anatomy of the Lymphatic System (figure 15.3)

Lymphatic Vessels

Lymph, which contains antigens that have entered tissues, flows in the **lymphatic vessels** to the lymph nodes (figure 15.4).

Secondary Lymphoid Organs

Secondary lymphoid organs are the sites at which lymphocytes gather to contact antigens.

Primary Lymphoid Organs

Primary lymphoid organs are the sites where B cells and T cells mature.

15.3 ■ The Nature of Antigens

Antigens are molecules that react specifically with an antibody or lymphocyte; **immunogen** refers specifically to an antigen that elicits an immune response. The immune response is directed to **epitopes** (antigenic determinants) on the antigen (figure 15.6).

15.4 ■ The Nature of Antibodies

Structure and Properties of Antibodies (figure 15.7)

Antibodies have a Y shape with an antigen-binding site at the end of each arm. The tail of the Y is the **Fc region.** The antibody monomer is composed of two identical **heavy chains** and two identical **light chains.** The **variable region** contains the antigen-binding site; the **constant region** encompasses the entire Fc region as well as part of the Fab regions.

Protective Outcomes of Antibody-Antigen Binding (figure 15.8)

Antibody-antigen binding results in neutralization, **opsonization,** complement activation, immobilization and prevention of adherence, cross-linking, and **antibody-dependent cellular cytotoxicity (ADCC).**

Immunoglobulin Classes (table 15.1)

The five major antibody classes—IgM, IgG, IgA, IgD, and IgE—each have distinct functions.

15.5 ■ Clonal Selection and Expansion of Lymphocytes

When an antigen enters a secondary lymphoid organ, only the lymphocytes that specifically recognize that antigen will respond; the antigen receptor they carry on their surface governs this recognition (figure 15.10). Lymphocytes may be immature, naive, activated, effector, or memory cells.

15.6 ■ B Lymphocytes and the Antibody Response

Most antigens are **T-dependent antigens,** meaning the B cells that recognize them require help from T_H cells.

B-Cell Activation

B cells present peptides from T-dependent antigens to T_H cells for inspection. If a T_H cell recognizes a peptide, it delivers cytokines to the B cell, initiating the process of clonal expansion, which ultimately gives rise to plasma cells that produce antibodies (figure 15.11).

Characteristics of the Primary Response

In the primary response, the expanding B-cell population undergoes **affinity maturation.** Under the direction of T_H cells, **class switching** and memory cell formation also occurs (figures 15.14, 15.15).

Characteristics of the Secondary Response

Memory cells are responsible for the swift and effective secondary response, eliminating invaders before they cause noticeable harm (figure 15.12).

The Response to T-Independent Antigens

T-independent antigens include polysaccharides that have multiple identical evenly spaced epitopes, and LPS (figure 15.16).

15.7 ■ T Lymphocytes: Antigen Recognition and Response

General Characteristics of T Cells (table 15.2, figure 15.19)

Cytotoxic T cells (CD8) recognize antigen presented on **major histocompatibility complex (MHC) class I molecules.** Helper T cells (CD4) recognize antigen presented on **major histocompatibility complex (MHC) class II molecules.**

Activation of T Cells (figure 15.20)

Dendritic cells sample material in tissues and then travel to secondary lymphoid organs to present antigens to naive T cells. The dendritic cells that detect molecules associated with danger produce **co-stimulatory molecules** and are able to activate both subsets of T cells.

Effector Functions of T_C (CD8) Cells

T_C cells induce apoptosis in cells that present peptides they recognize on MHC class I molecules; they also produce cytokines that allow neighboring cells to become more vigilant against intracellular invaders (figure 15.21). All nucleated cells present peptides from endogenous proteins in the groove of MHC class I molecules.

Effector Functions of T_H (CD4) Cells (figures 15.11, 15.22)

T_H cells activate cells that present peptides they recognize on MHC class II; various cytokines are released, depending on subset of the responding T_H cell. Macrophages and B cells present peptides from exogenous proteins in the groove of MHC class II molecules.

Subsets of Dendritic Cells and T Cells

Subsets of dendritic cells and T_H cells direct the immune system to an appropriate response.

15.8 ■ Natural Killer (NK) Cells

NK cells mediate antibody-dependent cellular cytotoxicity (ADCC). NK cells also induce apoptosis in host cells that are not bearing MHC class I molecules on their surface (figure 15.23).

15.9 ■ Lymphocyte Development

Generation of Diversity

Mechanisms used to generate the diversity of antigen specificity in lymphocytes include rearrangement of gene segments, imprecise joining of those segments, and combinatorial associations of heavy and light chains (figure 15.25).

Negative Selection of Self-Reactive B Cells

Negative selection occurs as B cells develop in the bone marrow; cells to which material binds to their B-cell receptor are induced to undergo apoptosis.

Positive and Negative Selection of Self-Reactive T Cells

Positive selection permits only those T cells that show moderate recognition of the MHC molecules to develop further. Negative selection also occurs.

Review Questions

Short Answer

1. What is a secondary lymphoid organ?
2. Diagram an IgG molecule and label (a) the Fc region and (b) the areas that combine with antigen.
3. What are the protective outcomes of antibodies binding to antigen?
4. Which antibody class is the first produced during the primary response?
5. Which antibody class neutralizes viruses in the intestinal tract?
6. Describe clonal selection and expansion in the immune response.
7. How do T-independent antigens differ from T-dependent antigens?
8. What are antigen-presenting cells (APCs)?
9. Describe the role of dendritic cells in T-cell activation.
10. How does the role of natural killer cells differ from cytotoxic T cells?

Multiple Choice

1. The variable regions of antibodies are located in the
   1. Fc region.   2. Fab region.   3. light chain.
   4. heavy chain.   5. light chain *and* heavy chain.
   a) 1, 3   b) 1, 5   c) 2, 3   d) 2, 4   e) 2, 5
2. Which of the following statements about antibodies is *false*?
   a) If you removed the Fc portion, antibodies would no longer be capable of opsonization.
   b) If you removed the Fc portion, antibodies would no longer be capable of activating the complement system.
   c) If you removed the Fab portion, an antibody would no longer be capable of cross-linking antigen.
   d) If IgG were a pentamer, it would bind antigens more efficiently.
   e) If IgE had longer half-life, it would protect newborn infants.

3. Which class of antibody can cross the placenta?
   a) IgA   b) IgD   c) IgE   d) IgG   e) IgM
4. A person who has been vaccinated against a disease should have primarily which of these types of serum antibodies against that agent 2 years later?
   a) IgA   b) IgD   c) IgE   d) IgG   e) IgM
5. Which of the following statements about B cells/antibody production is *false*?
   a) B cells of a given specificity initially have the potential to make more than one class of antibody.
   b) In response to antigen, all B cells located close to the antigen begin dividing.
   c) Each B cell is programmed to make a single specificity of antibody.
   d) The B-cell receptor allows B cells to detect antigen.
   e) The cell type that makes and secretes antibody is called a plasma cell.
6. Which term describes the loss of specific heavy chain genes?
   a) Affinity maturation
   b) Apoptosis
   c) Clonal selection
   d) Class switching
7. Which of the following specifically refers to an effector lymphocyte?
   a) B cell   b) Cytotoxic T cell
   c) Helper T cell   d) Plasma cell
8. Which markers are found on all nucleated cells?
   a) MHC class I molecules
   b) MHC class II molecules
   c) CD4
   d) CD8
9. Which of the following are examples of an antigen-presenting cell (APC)?
   1. Macrophage   2. Neutrophil   3. B cell
   4. T cell   5. Plasma cell
   a) 1, 2   b) 1, 3   c) 2, 4   d) 3, 5   e) 1, 2, 3
10. What is the appropriate response when antigen is presented on MHC class II molecules?
    a) An effector CD8 cell should kill the presenting cell.
    b) An effector CD4 cell should kill the presenting cell.
    c) An effector CD8 cell should activate the presenting cell.
    d) An effector CD4 cell should activate the presenting cell.

Applications

1. Many dairy operations keep cow's milk for sale and use formula and feed to raise any calves. One farmer noticed that calves raised on the formula and feed needed to be treated for diarrhea more frequently than calves left with their mothers to nurse. He had some tests run on the diets and discovered no differences in the calories or nutritional content. The farmer called a veterinarian and asked him to explain the observations. What was the vet's response?
2. What kinds of diseases would be expected to occur as a result of lack of T or B lymphocytes?

Critical Thinking ✚

1. The development of primary and secondary immune responses to an antigen differ significantly. The primary response may take a week or more to develop fully and establish memory. The secondary response is rapid and relies on the activation of clones of memory cells. Would it not be better if clones of reactive cells were maintained regardless of prior exposure? In this way, the body could always respond rapidly to *any* antigen exposure. Would there be any disadvantages to this approach? Why?
2. Early investigators proposed two hypotheses to explain the specificity of antibodies. The clonal selection hypothesis states that each lymphocyte can produce only one specificity of antibody. When an antigen binds to that B-cell receptor, the lymphocyte is selected to give rise to a clone of plasma cells producing the antibody. The template hypothesis states that any antigen can interact with any lymphocyte and act as a template, causing newly forming antibodies to be specific for that antigen. In one experiment to test these hypotheses, an animal was immunized with two different antigens. After several days, lymphocytes were removed from the animal and individual cells placed in separate small containers. Then, the original two antigens were placed in the containers with each cell. What result would support the clonal selection hypothesis? The template hypothesis?

16 Host-Microbe Interactions

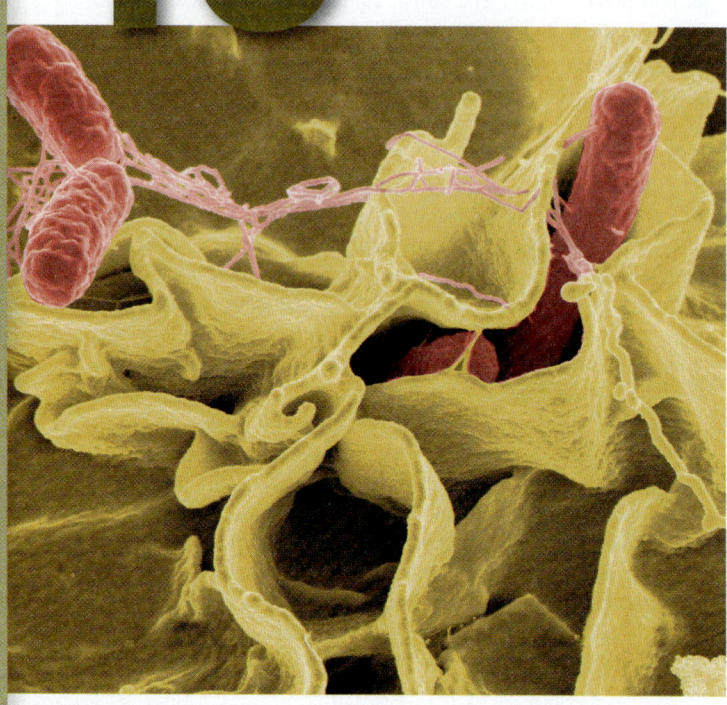

Salmonella enterica serotype Typhimurium invading cultured human cells (color-enhanced scanning electron micrograph).

A Glimpse of History

The ancients thought epidemics and diseases were divine punishment of the people for their sins. By the time of Moses, however, the Egyptians and Hebrews had come to believe that leprosy could be transmitted by contact with lepers. In Europe, around 430 B.C. Thucydides had concluded that some plagues were contagious. By the Middle Ages, many accepted this, and fled cities to escape the diseases. Fracastorius, in 1546, first proposed that communicable diseases were caused by living agents passed from one person or animal to another. He had no way to test this idea, however.

With Leeuwenhoek's discovery of microorganisms in the late seventeenth century, people began to suspect that microorganisms might cause disease, but the techniques of the times could not prove this. It was not until 1876 that Robert Koch offered convincing proof of what is now known as the germ theory of disease. He showed that *Bacillus anthracis* causes anthrax, an often fatal disease of humans, sheep, and other animals. With his microscope, he observed *B. anthracis* cells in the blood and spleen of dead sheep. He then inoculated mice with the infected sheep blood, and recovered *B. anthracis* from the blood of those mice. In addition, he grew the bacteria in pure culture and showed that they caused anthrax when injected into healthy mice. From these experiments and later work with *Mycobacterium tuberculosis,* Koch formalized a group of criteria for establishing the cause of an infectious disease, known as Koch's postulates.

Every day we contact an enormous number and variety of microorganisms. Some enter our respiratory system as we breathe; other are ingested with each bite of food or sip of drink; and still more adhere to our skin whenever we touch an object or surface. It is important to recognize, however, that the vast majority of these microbes generate no ill effects whatsoever. Some may colonize the body surfaces, taking up residence with the variety of other harmless microbes that live there; others are sloughed off with dead epithelial cells. Most of those swallowed are either killed in the stomach or eliminated in feces.

Relatively few microbes cause noticeable damage to the human body, such as invading tissues or producing toxic substances. Those that can are called **pathogens.** They have distinct characteristics that allow them to avoid at least some of the body's defenses. Research into how these microbes evade our innate and adaptive defenses is unraveling an impressive array of ploys. The knowledge we are gaining in areas such as genomics and immunology have given new insights into the pathogenic strategies, fueling hope that therapies targeted to specific pathogens can be developed.

This chapter will explore some of the ways in which microbes colonize the human host, living either as members of the normal microbiota or causing disease. It will also describe how pathogens evade or overcome the immune responses and damage the host.

MICROBES, HEALTH, AND DISEASE

Many people think of microorganisms as "germs" that should routinely be killed or avoided. Most microbes are harmless, however, and many are beneficial. The organisms that routinely reside on the body's surfaces are the normal microbiota, or normal flora. This relationship is a delicate balancing act, though, because some members of the normal microbiota, as well as microbes that make incidental contact with humans, can cause disease if the opportunity arises. Weaknesses or defects in the innate or adaptive defenses can leave people vulnerable to invasion; these individuals are said to be **immunocompromised.** Factors that can lead to an individual becoming immunocompromised include malnutrition, cancer, AIDS or other diseases, surgery, wounds, genetic defects, alcohol or drug abuse, and immunosuppressive therapy that accompanies procedures such as organ transplants.

16.1 ■ The Anatomical Barriers as Ecosystems

Learning Outcome

1. *Compare and contrast mutualism, commensalism, and parasitism.*

The skin and mucous membranes are barriers against invading microorganisms, but they also host a complex ecosystem—an interacting biological community. The intimate relationships between the microorganisms and the human body are an example of **symbiosis,** meaning "living together."

Microorganisms can have a variety of symbiotic relationships with each other and with the human host. These relationships may take on different characteristics depending on the closeness of the association and the relative advantages to each partner. Symbiotic associations can be one of several forms, and these may change, depending on the state of the host and the traits of the microbes:

- **Mutualism** is an association in which both partners benefit. In the large intestine, for example, some bacteria synthesize vitamin K and certain B vitamins. These nutrients are then available for the host to absorb. The bacteria residing in the intestine benefit as well, supplied with warmth and a variety of different energy sources.

- **Commensalism** is an association in which one partner benefits but the other remains unharmed. Many microbes living on the skin are neither harmful nor helpful to the human host, but they obtain food and other necessities from the host.

- **Parasitism** is an association in which one organism, the parasite, benefits at the expense of the other. All pathogens are parasites, but medical microbiologists often reserve the word *parasite* for eukaryotic pathogens such as protozoa and helminths.

MicroAssessment 16.1

Depending on the relative benefit to each partner such as a human host and a microbe, the relationship can be described as mutualism, commensalism, or parasitism.

1. *How is mutualism different from commensalism?*

16.2 ■ The Normal Microbiota

Learning Outcomes

2. *Describe three protective roles of the normal microbiota.*
3. *Describe how the composition of the normal microbiota can change over time.*

The **normal microbiota** is the population of microorganisms routinely found growing on the body of healthy individuals (**figure 16.1**). Microbes that typically inhabit body sites for extended periods are resident microbiota, whereas temporary occupants are transient microbiota. Considering how important this population is to human health, relatively little is known about its members. With that in mind, the Human Microbiome Project is aimed at studying this diverse population. ◄◄ **Human Microbiome Project, p. 225**

MicroByte

There are more bacteria in just one person's mouth than there are people in the world!

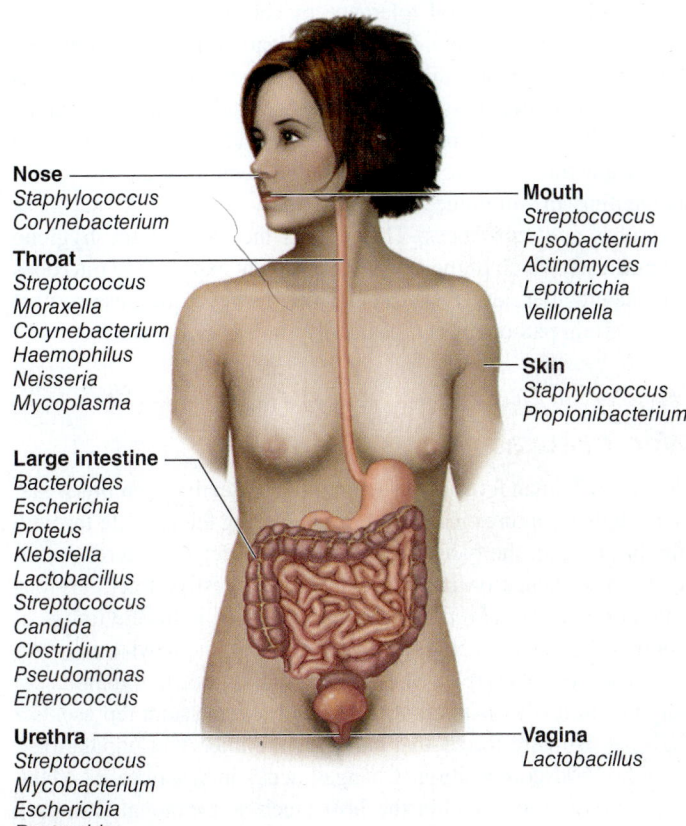

Nose
Staphylococcus
Corynebacterium

Throat
Streptococcus
Moraxella
Corynebacterium
Haemophilus
Neisseria
Mycoplasma

Large intestine
Bacteroides
Escherichia
Proteus
Klebsiella
Lactobacillus
Streptococcus
Candida
Clostridium
Pseudomonas
Enterococcus

Urethra
Streptococcus
Mycobacterium
Escherichia
Bacteroides

Mouth
Streptococcus
Fusobacterium
Actinomyces
Leptotrichia
Veillonella

Skin
Staphylococcus
Propionibacterium

Vagina
Lactobacillus

FIGURE 16.1 Normal Microbiota This shows only some of the common genera; many others may also be present.

❓ *What can happen if members of the normal microbiota in the intestinal tract are killed or their growth suppressed?*

The Protective Role of the Normal Microbiota

One of the most significant contributions of the normal microbiota to health is protection against pathogens. As discussed in chapter 14, the normal microbiota excludes pathogens by (1) covering binding sites that might otherwise be used for attachment, (2) consuming available nutrients, and (3) producing compounds toxic to other bacteria. When members of the normal microbiota are killed or their growth suppressed, as can happen during antibiotic treatment, pathogens may colonize and cause disease. For instance, certain antibiotics inhibit the *Lactobacillus* species that normally predominate in the vagina of mature females. These bacteria normally suppress the growth of the yeast *Candida albicans,* and without their protective action, the yeast cells can multiply to high numbers, resulting in vulvovaginal candidiasis. Oral antibiotics can also inhibit members of the normal intestinal microbiota, allowing the overgrowth of toxin-producing strains of *Clostridium difficile* that cause antibiotic-associated diarrhea and colitis. ▶▶ **vulvovaginal candidiasis, p. 618** ▶▶ *Clostridium difficile–associated disease,* **p. 594**

Another critical role of the normal microbiota is to stimulate the adaptive immune system. The importance of this can be shown in mice reared in a microbe-free environment. These "germ-free" animals have greatly underdeveloped mucosa-associated lymphoid tissue (MALT). In addition, antibodies produced against members of the normal microbiota bind to pathogen surfaces as well. ◀◀ **MALT, p. 358**

The normal microbiota appears to play an important role in the development of oral tolerance by the immune system. In a complex series of events, our defenses learn to lessen the immune response to the many microbes that routinely inhabit the gut, as well as foods that pass through. Recent studies into the actions of regulatory T cells indicate that early and consistent exposure to certain microbes in the gut stimulates these T cells, thereby preventing the immune system from overreacting to harmless microbes and substances. This idea is the basis of the **hygiene hypothesis,** which proposes that insufficient exposure to microbes can lead to allergies. It is a fine balance, however, because contact with certain pathogens can be deadly. ◀◀ **tolerance, p. 355**

The Dynamic Nature of the Normal Microbiota

A healthy human fetus is sterile until the protective membrane that surrounds it ruptures just before birth. During the passage through the birth canal, the baby is exposed to a variety of microbes that take up residence on its skin and in the digestive tract. Various microorganisms in food, on other humans, and in the environment soon also become established as residents on the newborn.

The composition of the normal microbiota is dynamic. At any one time, the makeup of this complex ecosystem represents a balance of many forces that can alter the microbial population's quantity and composition. Changes occur in response to physiological variations within the host (such as hormonal changes), and as a direct result of the activities of the human host (such as consuming food). An intriguing example of the dynamic nature of the microbiota was the discovery that the intestinal microbiota of obese and lean people differs. Obese people have more members

of the *Firmicutes,* a phylum that includes *Clostridium* and *Bacillus* species, whereas thin individuals typically have more members of the *Bacteroidetes,* a phylum that includes *Bacteroides* species. As obese people lost weight, their intestinal microbiota changed to resemble that of typically lean people. A variety of studies to track changes in the normal microbiota in both health and disease are now underway as part of the Human Microbiome Project.

MicroAssessment 16.2

The normal microbiota provides protection against potentially harmful organisms and stimulates the immune system.

2. *What factor favors abundant growth of* Clostridium difficile *in the intestine?*
3. *Why would the immune response to members of the normal microbiota cross-react with pathogens?* ✚

16.3 ■ Principles of Infectious Disease

Learning Outcomes

4. *Define the terms* primary pathogen, opportunist, *and* virulence.
5. *Describe the characteristics of infectious diseases, including the course of disease, duration of symptoms, and distribution of the pathogen.*

The term **colonization** refers to a microbe establishing itself and multiplying on a body surface. If the microbe has a parasitic relationship with the host, then the term **infection** can be used. That is, a member of the normal microbiota is said to have colonized the host, but a pathogen is described as having either colonized or infected the host. Infection does not always lead to illness. It can be **subclinical,** meaning that symptoms either do not appear or are mild enough to go unnoticed.

An infection that results in disease (a noticeable impairment of body function) is called an **infectious disease.** Diseases are characterized by symptoms and signs; **symptoms** are the subjective effects of the disease experienced by the patient, such as pain and nausea, whereas **signs** are the objective evidence, such as rash, pus formation, and swelling.

Effects of one disease may leave a person predisposed to developing another. For example, a respiratory illness that damages the mucociliary escalator makes a person more likely to develop pneumonia. The initial infection is a **primary infection;** an additional infection that occurs as a result of the primary infection is a **secondary infection.** ◀◀ **mucociliary escalator, p. 337**

Pathogenicity

A **primary pathogen,** or more simply, a pathogen, is a microbe or virus that causes disease in otherwise healthy individuals. Diseases such as plague, malaria, measles, influenza, diphtheria, tetanus, and tuberculosis are caused by primary pathogens.

An **opportunistic pathogen,** or opportunist, causes disease only when the body's innate or adaptive defenses are compromised, or when introduced into an unusual location. Opportunists can be members of the normal microbiota or they can be common in the environment. For instance, *Pseudomonas* species are environmental bacteria that routinely come into contact with healthy individuals without harmful effect, yet they can cause fatal infections in individuals who have the genetic disease cystic fibrosis and also in burn patients (see figure 23.6). Ironically, as our healthcare systems improve, extending the life span of patients through surgery and immunosuppressive drugs, diseases caused by opportunists are becoming more common. Also, many organisms not previously known to cause disease have now been shown to do so in severely immunocompromised patients.

The term **virulence** refers to the degree of pathogenicity of an organism. An organism described as highly virulent is more likely to cause disease, particularly severe disease, than might otherwise be expected. *Streptococcus pyogenes* causes strep throat, for example, but certain strains are particularly virulent, causing diseases such as necrotizing fasciitis ("flesh-eating disease"). **Virulence factors** are the traits of a microorganism that specifically allow it to cause disease. The genes encoding these traits can sometimes be transferred horizontally. ▶▶ necrotizing fasciitis, p. 553 ◀◀ horizontal gene transfer, p. 200

Characteristics of Infectious Disease

Infectious diseases that spread from one host to another are called **communicable, or contagious, diseases.** Some contagious diseases, such as colds and measles, are easily transmitted. The ease with which a contagious disease spreads partly reflects the **infectious dose**—the number of microbes necessary to establish an infection. For example, the intestinal disease shigellosis is quite contagious in humans because only 10 to 100 cells of a *Shigella* species need be ingested to establish an infection; in contrast, salmonellosis does not spread as readily because as many as 10^6 cells of *Salmonella enterica* serotype Enteritidis must be ingested to cause illness. The difference in these infectious doses reflects, in part, the pathogen's ability to survive the acidic conditions encountered as the cells pass through the stomach. Generally, the infectious dose is expressed as the ID_{50}, an experimentally derived figure that indicates the number of microbial cells administered that resulted in disease in 50% of the test population. ▶▶ *Shigella*, p. 588 ▶▶ *Salmonella enterica* serotype Enteritidis, p. 592

Course of Infectious Disease
The course of an infectious disease includes several stages (**figure 16.2**). The time between introduction of a microbe to a susceptible host and the onset of illness is the **incubation period.** This varies considerably, from only a few days for the common cold, to several weeks for hepatitis A, to many months for rabies, and even years for Hansen's disease (leprosy). The length of the incubation period depends on a variety of factors, including the growth rate of the pathogen, the host's condition, and the number of infectious cells or virions encountered. ▶▶ Hansen's disease, p. 650

A phase of **illness** follows the incubation period. During this period, a person will experience the signs and symptoms of the disease. In some cases, onset of illness is heralded by a **prodromal** phase—the early,

vague symptoms such as malaise and headache. After the illness subsides, there is a period of **convalescence,** the stage of recuperation and recovery from the disease. Even though there is no indication of infection during the incubation and convalescent periods, many infectious agents can still be spread during these stages. Some individuals, called **carriers,** harbor an infectious agent for months or years and continue to spread the pathogen, even though they show no signs or symptoms of the disease. The impact of carriers on spread of disease will be discussed in chapter 19. ▶▶ carriers, p. 439

Following recovery from infection, or after immunization, the host normally has accumulated protective antibodies and memory lymphocytes that prevent reinfection with the same microbe. In most cases, the host is no longer susceptible to infection with that particular infectious agent.

Duration of Symptoms
Infections and the associated diseases are often described according to the timing and duration of the symptoms (figure 16.2):

- **Acute infections** are characterized by symptoms that develop quickly but last only a short time; an example is strep throat.
- **Chronic infections** develop slowly and last for months or years; an example is tuberculosis.
- **Latent infections** are never completely eliminated; the microbe continues to exist in host tissues, often within host cells, without causing any symptoms. If there is a decrease in immunity, the latent infection may become reactivated and symptomatic. Note that the symptomatic phase of the disease may be either acute or chronic. For example, the infection caused by the varicella-zoster virus results in the characteristic symptoms of chickenpox, an acute illness. The illness is stopped by an effective immune response, leaving the host immune to reinfection. The virus, however, is not completely eliminated. It takes refuge in sensory nerves, held in check by the immune system. Later in life, infectious viral particles may be produced again, causing the skin disease shingles (herpes zoster). In tuberculosis, the mycobacteria are often

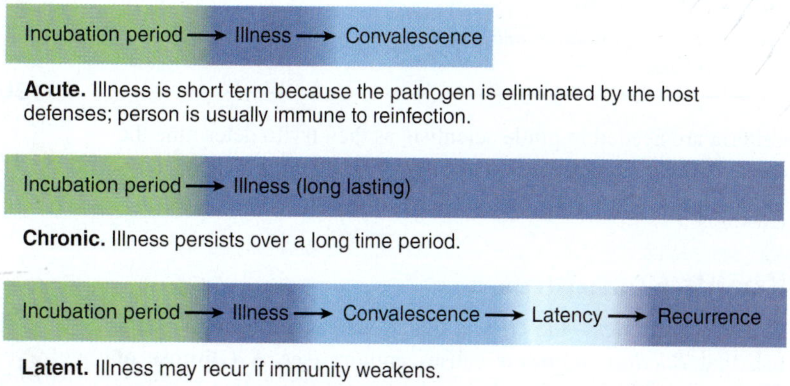

Incubation period → Illness → Convalescence

Acute. Illness is short term because the pathogen is eliminated by the host defenses; person is usually immune to reinfection.

Incubation period → Illness (long lasting)

Chronic. Illness persists over a long time period.

Incubation period → Illness → Convalescence → Latency → Recurrence

Latent. Illness may recur if immunity weakens.

FIGURE 16.2 The Course of Infectious Diseases Infections can be acute, chronic, or latent.

❓ *Some diseases include a prodromal phase. Where would this phase fit in the figure?*

initially confined within a small area by host defenses, causing no symptoms; much later, the bacteria may begin multiplying again, resulting in a chronic illness. Other diseases in which the causative agent becomes latent include cold sores and genital herpes. ▶▶ chickenpox, p. 534 ▶▶ tuberculosis, p. 502

Distribution of the Pathogen

Infections are often described according to the distribution of the causative agent in the body. In a **localized infection,** the microbe is limited to a small area; an example is a boil caused by *Staphylococcus aureus*. In a **systemic infection,** the infectious agent is disseminated (spread) throughout the body; an example is measles.

The suffix *-emia* means "in the blood." Thus, **bacteremia** indicates that bacteria are circulating in the bloodstream. Note that this term does not necessarily imply a disease state. A person can become transiently bacteremic after vigorous tooth brushing. **Toxemia** indicates that toxins are circulating in the bloodstream. The organism that causes tetanus, for instance, produces a localized infection yet its toxins circulate in the bloodstream. The term **viremia** indicates that viral particles are circulating in the bloodstream.

MicroAssessment 16.3

A primary pathogen can cause disease in an otherwise healthy individual; an opportunist causes disease in an immunocompromised host. The course of infectious disease includes an incubation period, illness, and a period of convalescence. Infections can be acute or chronic, latent, localized, or systemic.

  4. *Why are diseases caused by opportunists becoming more frequent?*
  5. *Give an example of a microbe that causes a latent infection.*
  6. *What factors might contribute to a long incubation period?* ➕

16.4 ■ Establishing the Cause of Infectious Disease

Learning Outcome

  6. *List Koch's postulates, and compare them to the Molecular Koch's postulates.*

Criteria are needed to guide scientists as they try to determine the cause of infectious diseases. They can also be helpful when studying the disease process.

Koch's Postulates

Koch's postulates—the criteria that Robert Koch used to establish that *Bacillus anthracis* causes anthrax (see **A Glimpse of History**)—provide a foundation for establishing that a given microbe causes a specific infectious disease (**figure 16.3**):

  ① The microorganism must be present in every case of the disease.
  ② The organism must be grown in pure culture from diseased hosts.

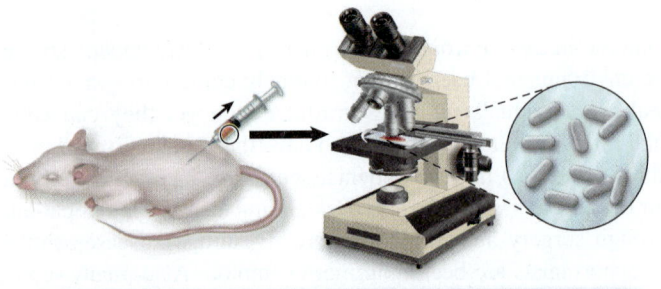

① The microorganism must be present in every case of the disease, but not in healthy hosts.

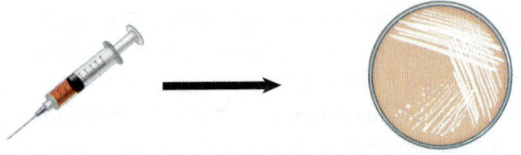

② The microorganism must be grown in pure culture from diseased hosts.

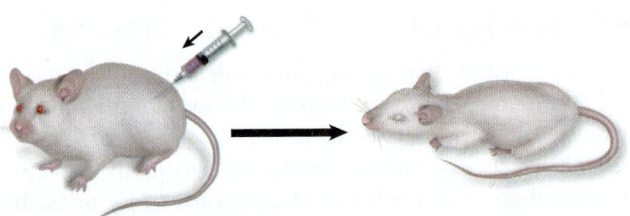

③ The same disease must be produced when a pure culture of the microorganism is introduced into susceptible hosts.

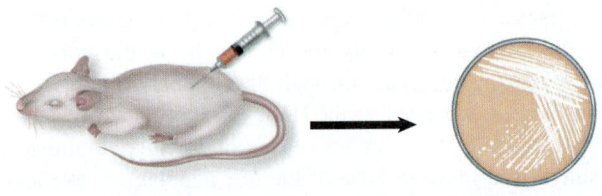

④ The same microorganism must be recovered from the experimentally infected hosts.

FIGURE 16.3 Koch's Postulates These criteria provide a foundation for establishing that a given microbe causes a specific disease.

❓ *Why cannot Koch's postulates be used to show that* Treponema pallidum *causes syphilis?*

  ③ The same disease must be produced when a pure culture of the organism is introduced into susceptible hosts.
  ④ The organism must be recovered from the experimentally infected hosts.

When Koch studied anthrax, he grew *B. anthracis* from all cases examined; he introduced pure cultures of the organisms into healthy susceptible mice, again causing the disease anthrax.

Finally, he recovered the organism from the experimentally infected mice.

It is important to note that there are many situations in which Koch's postulates cannot be carried out. For example, the second postulate cannot be fulfilled for organisms that cannot be grown in laboratory medium, such as *Treponema pallidum* (causes syphilis). In other cases, the third postulate does not always hold true. There are many examples, including cholera and polio, in which some infected people do not have symptoms of disease. In addition, some diseases are polymicrobial, meaning that multiple species act together to cause the illness; an example is chronic periodontal disease. Also, suitable experimental animal hosts are not available for some diseases and it would not be ethical to test the postulates on humans because of safety concerns. Nevertheless, despite the limitations of the postulates, they have provided scientists with a logical framework for determining the causes of infectious diseases. ▶▶ syphilis, p. 626 ▶▶ cholera, p. 586 ▶▶ polio, p. 656 ▶▶ periodontal disease, p. 577

Molecular Koch's Postulates

Molecular Koch's postulates are similar in principle to Koch's postulates, but they rely on molecular techniques to study a microbe's virulence factors. They are particularly relevant in the study of pathogens such as *E. coli* and *Streptococcus pyogenes,* which can cause several different diseases depending on the virulence factors of a given strain. Molecular Koch's postulates are as follows:

1. The virulence factor gene or its product should be found in pathogenic strains of the organism.
2. Mutating the virulence gene to disrupt its function should reduce the virulence of the pathogen.
3. Reversion of the mutated virulence gene or replacement with a wild-type version should restore virulence to the strain.

As with the traditional Koch's postulates, it is not always possible to apply all of these criteria, but they provide an approach to studying how infectious agents cause disease.

> **MicroAssessment 16.4**
>
> Koch's postulates can be used to establish that a given microbe causes a specific infectious disease. Molecular Koch's postulates are used to identify the virulence factors responsible for disease.
>
> 7. *How were Koch's postulates used to prove the cause of anthrax?*
> 8. *Why can Koch's postulates not be used to identify the causes of diseases due to polymicrobial infections?* ✚

MECHANISMS OF PATHOGENESIS

From a microbe's perspective, the interior of the human body is a rich source of nutrients guarded by the innate and adaptive defenses. The ability to get past these defenses and cause damage is what distinguishes pathogens from other microbes. Understanding how they do this helps illustrate why only certain microbes can cause disease in a healthy host. Pathogenic mechanisms generally follow one of several patterns:

- **Production of toxins that are then ingested.** The microbe does not grow on or in the host, so this is not an infection but rather a foodborne intoxication, a form of food poisoning. The only virulence determinant is toxin production. Relatively few bacteria cause foodborne intoxication; these include *Clostridium botulinum* (causes botulism), and toxin-producing strains of *Staphylococcus aureus* (cause staphylococcal food poisoning). ▶▶ botulism, p. 652 ▶▶ *Staphylococcus aureus* foodborne intoxication, p. 757

- **Colonization of mucous membranes of the host, followed by toxin production.** The microbe adheres to a mucous membrane such as the lining of the intestinal or upper respiratory tracts and multiplies to high numbers. There, it produces a toxin that interferes with cell function. Examples of bacteria that do this include *Vibrio cholerae* (causes cholera), *E. coli* O157:H7 (causes bloody diarrhea), and *Corynebacterium diphtheriae* (causes diphtheria). ▶▶ cholera, p. 586 ▶▶ *E. coli* O157:H7 diarrhea, p. 590 ▶▶ diphtheria, p. 490

- **Invasion of host tissues.** The microbe penetrates the first-line defenses and then multiplies within the tissues. Organisms that do this generally have mechanisms to avoid destruction by macrophages; some also have mechanisms to avoid antibodies. There are numerous examples of bacteria that invade, including *Mycobacterium tuberculosis* (causes tuberculosis), *Yersinia pestis* (causes plague) and *Salmonella enterica* (most strains cause gastroenteritis and one causes typhoid fever). ▶▶ *Mycobacterium tuberculosis*, p. 503 ▶▶ *Yersinia pestis*, p. 678

- **Invasion of host tissues, followed by toxin production.** These microbes are similar to those in the previous category, but in addition to invading, they also make toxins. Examples include *Shigella dysenteriae* (causes diarrhea) and *Clostridium tetani* (causes tetanus). ▶▶ *Clostridium tetani*, p. 555

A successful pathogen needs only to overcome the host defenses long enough to multiply and then exit the host. In fact, a pathogen that completely overwhelms the host defenses is actually at a disadvantage because it will likely kill the host. If the host dies, the pathogen loses an exclusive source of nutrients and perhaps the opportunity to be transmitted.

Pathogens and their hosts generally evolve over time to a state of **balanced pathogenicity.** The pathogen becomes less virulent while the host becomes less susceptible. This was demonstrated when the myxoma virus was intentionally introduced into Australia in the early 1950s to kill the rapidly increasing rabbit population. As expected, the rabbit population dropped dramatically after the virus was introduced. Eventually, however, the numbers of rabbits again began rising. Viruses isolated from these rabbits were shown to be less virulent than the original strain, and the rabbits were more resistant to the original virus strain.

The next sections will describe how pathogens adhere to and colonize host tissue, avoid innate defenses, avoid adaptive defenses, and cause the damage associated with disease. We will focus on mechanisms of bacterial pathogenesis because these are by far the most thoroughly characterized; later in the chapter, we will discuss pathogenesis of viruses and eukaryotic organisms. As we describe various virulence factors, recognize that their roles are not mutually exclusive—a single structure can serve more than one purpose. Also note that one microbe can have more than one virulence factor, and various strains of the same species can have different virulence factors.

16.5 ■ Establishing Infection

Learning Outcomes

7. *Describe the requirements for adherence and colonization.*
8. *Explain the role of type III secretion systems in infection.*

To cause disease, most pathogens must first adhere to a body surface and then multiply. In some cases, they deliver molecules to epithelial cells, causing changes in those cells.

Adherence

The first-line defenses are very effective in sweeping microbes away, so pathogens must adhere to host cells to initiate infection. Microbes that attach to cells, however, do not necessarily cause disease. For example, members of the normal microbiota often adhere to epithelial cells with no ill effect whatsoever. Other factors such as toxin production or invasion generally must come into play before disease results. |◀◀ first-line defenses, p. 336

Bacteria use **adhesins** to attach to host cells (**figure 16.4**). These are often located at the tips of pili (pili used for attachment are often called fimbriae; see figure 3.40). Adhesins can also be a component of other surface structures such as capsules or various cell wall proteins (see figure 3.35). |◀◀ pili, p. 65 |◀◀ capsule, p. 62

The molecule to which an adhesin attaches is called the **receptor.** Note that receptors have distinct roles for the host cells; the microbes merely exploit the molecules for their own use. For example, the normal role of the receptor used by *Neisseria gonorrhoeae* is to help protect host cells from damage by the complement system. Receptors are typically glycoproteins or glycolipids, and the adhesin binds to the sugar portion. |◀◀ complement system, p. 344 |◀◀ glycoprotein, p. 29

Adhesin-receptor binding is highly specific, dictating the type of cells to which the bacterium can attach. For instance, the adhesin of common *E. coli* strains allows them to adhere to cells that line the large intestine, where the strains multiply as part of the normal microbiota. Pathogenic *E. coli* strains have additional adhesins, broadening the range of tissues to which they can attach.

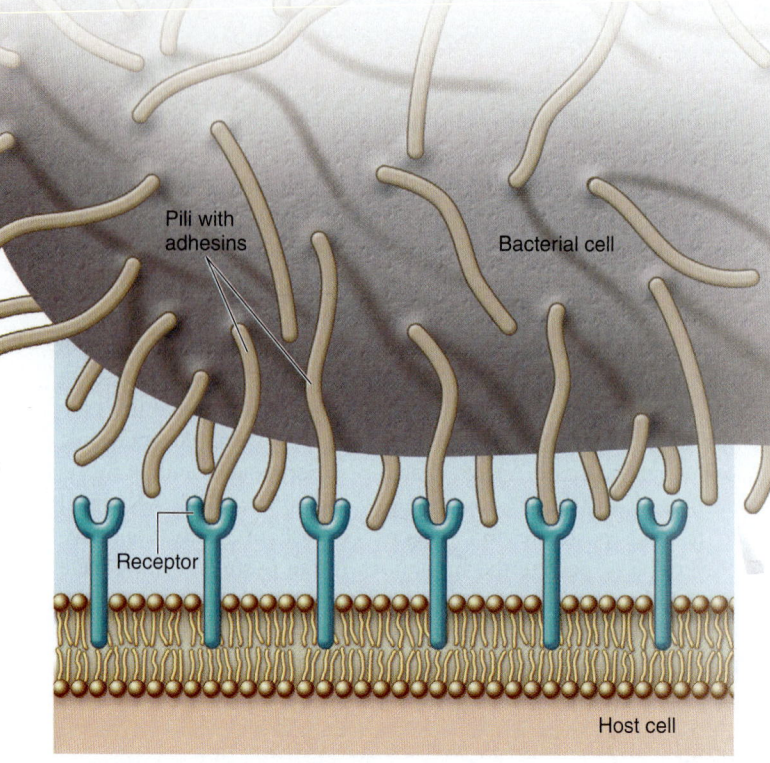

FIGURE 16.4 Pili Attachment to Host Cell An adhesin at the tip of a pilus attaches to a molecule on the host cell surface.

❓ *What would happen if a pathogen lost the ability to produce adhesins?*

Strains that cause urinary tract infections have pili that attach to the bladder, and strains that cause watery diarrhea have pili that adhere to cells of the small intestine.

Colonization

A microorganism must multiply in order to colonize the host. In many cases, pathogens grow in biofilms. |◀◀ biofilms, p. 84

To colonize a mucosal surface, the pathogen must deal with the host's defenses that protect those surfaces. Recall, for example, that the body uses lactoferrin and transferrin to bind iron, thereby limiting the growth of microbes. Some pathogens respond by producing their own iron-binding molecules, **siderophores;** others can use the iron bound to the host proteins. |◀◀ lactoferrin and transferrin, p. 337

Secretory IgA also protects mucosal surfaces. Pathogens, however, have evolved mechanisms to avoid those antibodies. Mechanisms include rapid turnover of pili (to shed any bound antibody), antigenic variation, and **IgA proteases** (enzymes that cleave IgA antibodies). |◀◀ IgA, p. 362 |◀◀ antigenic variation, p. 178

If the body site has normal microbiota, the new arrival must compete for space and nutrients. It must also tolerate any toxic products such as fatty acids produced by the competitors.

Delivering Effector Proteins to Host Cells

Some Gram-negative pathogens deliver proteins directly into host cells using secretion systems. For example, a **type III secretion system,** or injectisome, is a syringelike structure that injects proteins into eukaryotic cells (**figure 16.5**). The injected proteins,

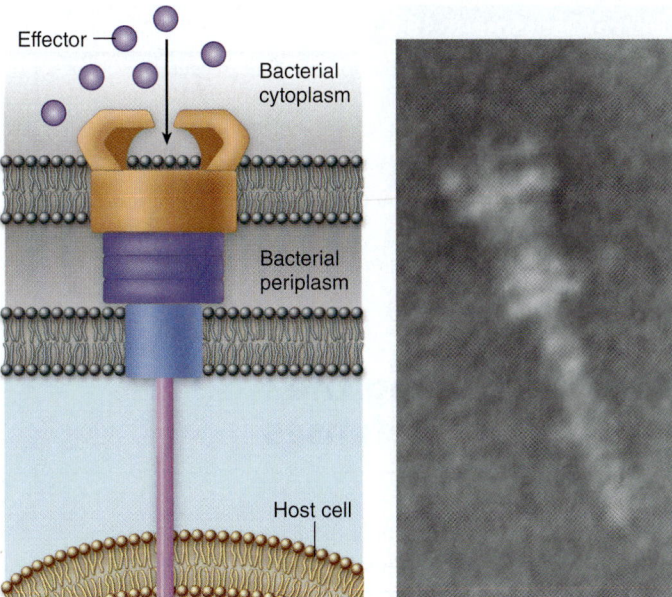

FIGURE 16.5 Type III Secretion Systems Gram-negative bacteria use type III secretion systems to deliver certain molecules directly to host cells, inducing changes in those cells. Peptidoglycan is not shown in this figure.

❓ *What bacterial structure do type III secretion systems resemble?*

referred to as effector proteins, induce changes such as altering the cell's cytoskeleton structure. Some effector proteins direct the host cell to engulf the bacterial cell, a process discussed in the next section. Several types of secretion systems have been discovered, and some can inject molecules other than proteins. ◀◀ secretion, p. 56
◀◀ cytoskeleton, p. 73

MicroAssessment 16.5

Pathogens use adhesins, often on pili, to bind to a body surface. To colonize a surface, the pathogen must often compete with the normal microbiota, prevent binding of secretory IgA, and obtain iron. Some bacteria deliver effector proteins to epithelial cells, inducing a specific change in those cells.

9. *What are siderophores?*
10. *What is a type III secretion system?*
11. *Why is it a good strategy for a microbe to adhere to a receptor that plays a critical function for a host cell?* ➕

16.6 ■ Invasion—Breaching the Anatomical Barriers

Learning Outcome

9. *Describe the mechanisms pathogens use to penetrate the skin and mucous membranes.*

Some bacterial pathogens cause disease while remaining on the mucosal surfaces, but many others penetrate the anatomical barriers. By crossing the epithelial barrier, invading microbes can multiply in the nutrient-rich tissues without competition.

Penetrating the Skin

Skin is the most difficult anatomical barrier for microbes to penetrate. Bacterial pathogens that invade via this route rely on skin-damaging injury. *Staphylococcus aureus* enters tissues via a cut or other wound. *Yersinia pestis* is injected by infected fleas.
▶▶ plague, p. 678

Penetrating Mucous Membranes

Mucous membranes are the entry points for most pathogens, but the invasive processes are complex and difficult to study. It appears, however, that there are at least two mechanisms used for invasion: directed uptake by cells and exploiting antigen-sampling processes.

Directed Uptake by Cells

Some pathogens induce non-phagocytic cells to engulf them. The pathogen first attaches to a cell, then triggers the process of endocytosis. ◀◀ endocytosis, p. 72

Gram-negative bacteria often inject effector proteins that induce engulfment by host cells. *Salmonella* species, for example, use a type III secretion system to deliver specific proteins to intestinal epithelial cells. These cause actin molecules in the host cell cytoplasm to rearrange, resulting in characteristic **membrane ruffling** on the cell's surface (**figure 16.6**). The ruffles enclose the bacterial cells, bringing them into the intestinal cell.

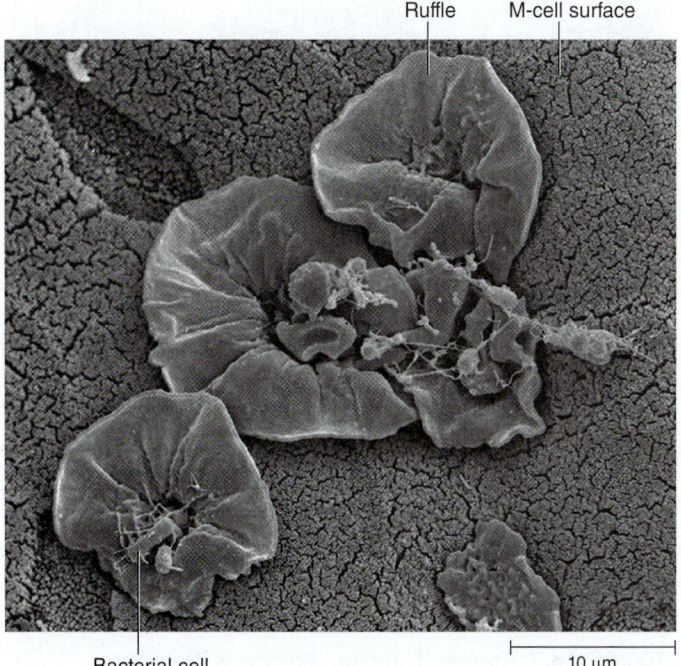

FIGURE 16.6 Ruffling *Salmonella enterica* serotype Typhimurium inducing ruffles on an M cell (a specialized epithelial cell), leading to uptake of the bacterial cells.

❓ *How do* Salmonella *cells induce ruffling?*

Exploiting Antigen-Sampling Processes

Recall that mucosa-associated lymphoid tissue (MALT) samples material from the mucosal surface. Some pathogens use this process to cross the membranes. **◀◀ MALT, p. 358**

Several pathogens use M cells to cross the intestinal barrier. Recall that M cells transport material from the lumen of the intestine to the Peyer's patches (see figure 15.5). Most microbes delivered this way are destroyed by the macrophages in the Peyer's patches, but pathogens have mechanisms to avoid this fate. When *Shigella* cells are transferred to the macrophages, for instance, the bacteria survive, and eventually induce the phagocyte to undergo apoptosis (**figure 16.7**). The freed bacterial cells then bind to the base of the mucosal epithelial cells and cause these nonphagocytic cells to engulf them, using a mechanism similar to that of *Salmonella*. **◀◀ M cell, p. 358 ◀◀ Peyer's patches, p. 358**

Some pathogens invade by means of alveolar macrophages, which engulf material that enters the lungs. *Mycobacterium tuberculosis* produces surface proteins that direct their uptake by macrophages. Although this might seem to be a disadvantage to the bacteria, it actually allows them to avoid a process that could otherwise lead to macrophage activation. *Mycobacterium* cells survive within macrophages that have not been activated.

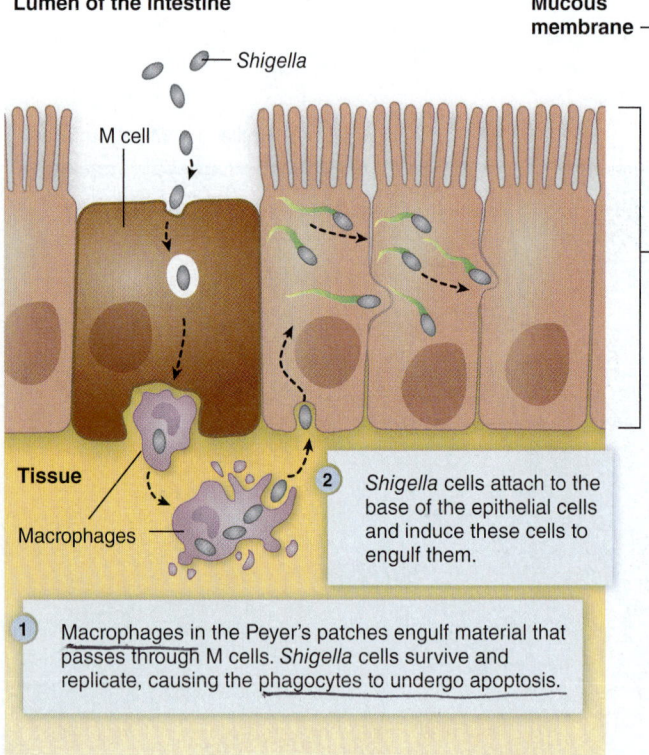

3 Within an epithelial cell, *Shigella* cells cause the host actin to polymerize. This propels the bacterial cell, sometimes with enough force to push it into the next cell.

Lumen of the intestine

— *Shigella*

M cell

Mucous membrane

Tissue

Macrophages

2 *Shigella* cells attach to the base of the epithelial cells and induce these cells to engulf them.

1 Macrophages in the Peyer's patches engulf material that passes through M cells. *Shigella* cells survive and replicate, causing the phagocytes to undergo apoptosis.

FIGURE 16.7 Antigen-Sampling Processes Provide a Mechanism for Invasion *Shigella* species use M cells to move across the epithelial barrier. Once the bacterial cells are on the other side, macrophages ingest them, but the bacteria are able to escape and then infect other cells.

? *What is the normal function of M cells?*

MicroAssessment 16.6

Skin is the most difficult barrier for microbes to penetrate. Some pathogens induce mucosal epithelial cells to engulf the bacterial cells. Some take advantage of antigen-sampling processes.

12. *How do* Shigella *species enter intestinal epithelial cells?*

13. *Why does* Mycobacterium tuberculosis *direct macrophages to engulf them?*

14. *Why would it be difficult to study invasion of mucous membranes?* ➕

16.7 ■ Avoiding the Host Defenses

Learning Outcome

10. *Describe mechanisms that bacteria use to avoid complement system proteins, antibodies, and destruction by phagocytes.*

Inside the body, invading microorganisms soon encounter the innate and adaptive immune defenses. Pathogens as a group have evolved a variety of mechanisms to avoid the otherwise lethal effects of these defenses.

Hiding Within a Host Cell

Some pathogens enter host cells, where they hide from complement proteins, phagocytes, and antibodies. Once a *Shigella* cell is within an intestinal epithelial cell, it directs its own transfer to adjacent cells (figure 16.7). It does this by causing the host cell actin to polymerize at one end of the bacterial cell. This forms an "actin tail" that propels the bacterium within the cell. The force of the propulsion is so great that the bacterial cells are often driven into neighboring cells. *Listeria monocytogenes* (causes meningitis) does the same thing (**figure 16.8**). **◀◀ actin, p. 73 ▶▶ listeriosis, p. 648**

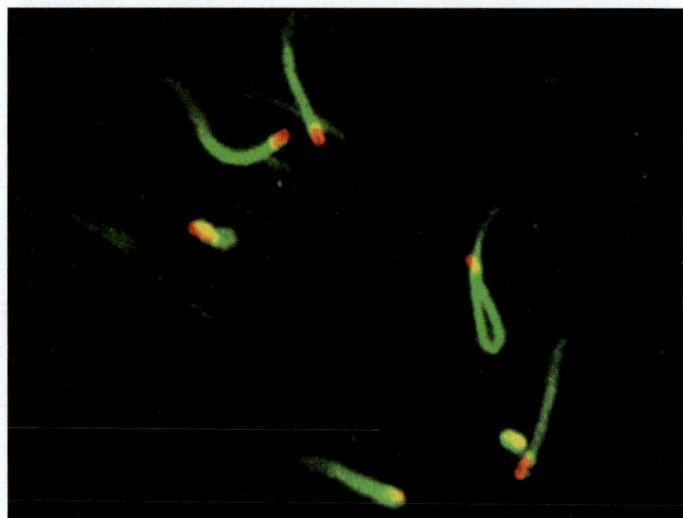

FIGURE 16.8 Actin Tail of Intracellular *Listeria monocytogenes* Rapid polymerization of host cell actin (green) at one end of the bacterial cell (orange) propels the bacterium within the cell.

? *How does an actin tail benefit a bacterial cell?*

Avoiding Killing by Complement System Proteins

As described in chapter 14, activation of the complement system leads to three primary outcomes—lysis of foreign cells by membrane attack complexes (MACs), opsonization, and inflammation (see figure 14.10). Because the latter two outcomes are associated with phagocytosis, mechanisms that bacteria use to avoid them will be discussed in the next section. Here, we will focus on how bacteria avoid the lethal effects of MACs. Recall that Gram-negative bacteria are susceptible to MACs because the outer membrane serves as a target; MACs have little effect on Gram-positive bacteria. ◀◀ complement system, p. 344

Bacteria that use mechanisms to avoid killing by the complement proteins are said to be **serum resistant;** strains of *Neisseria gonorrhoeae* that cause disseminated gonococcal infection are an example. These strains hijack the mechanism that host cells use to prevent their own surfaces from activating the complement system (see figure 14.12). By binding to the host's complement regulatory proteins, they avoid complement activation by the alternative pathway, thereby postponing MAC formation. ▶▶ disseminated gonococcal infection, p. 622

Avoiding Destruction by Phagocytes

Phagocytosis involves multiple steps—including chemotaxis, recognition and attachment, engulfment, and fusion of the phagosome with lysosomes—that lead to the destruction of invading microbes (see figure 14.10). Pathogens have evolved several mechanisms to avoid these destructive effects (**figure 16.9**).

① Preventing Encounters with Phagocytes

Some pathogens prevent phagocytosis by avoiding macrophages and neutrophils altogether. The mechanisms include:

- **C5a peptidase.** This enzyme degrades the complement system component C5a, a chemoattractant that recruits phagocytic cells. *Streptococcus pyogenes* (causes strep throat) makes C5a peptidase. ◀◀ C5a, p. 344 ▶▶ *Streptococcus pyogenes*, p. 487

- **Membrane-damaging toxins.** These kill phagocytes and other cells, often by forming pores in their membranes. *S. pyogenes* makes a membrane-damaging toxin called streptolysin O. ▶▶ membrane-damaging toxins, p. 392

② Avoiding Recognition and Attachment

Some pathogens avoid being recognized by phagocytes. Recall that phagocytes recognize and attach to foreign material more efficiently if opsonins such as C3b or antibodies coat it. Mechanisms that bacteria use to avoid opsonization include: ◀◀ opsonins, p. 347

- **Capsules.** These have long been recognized for their ability to prevent phagocytosis. Scientists now know that some capsules do this by interfering with opsonization. In some cases, they bind the host's complement regulatory proteins that inactivate C3b—a mechanism identical to that described earlier for serum-resistant bacteria. Rapid inactivation of C3b prevents the molecule from being an effective opsonin, and it also avoids activation of the complement system by the alternative pathway. *Streptococcus pneumoniae* (causes pneumonia) produces a capsule that does this.

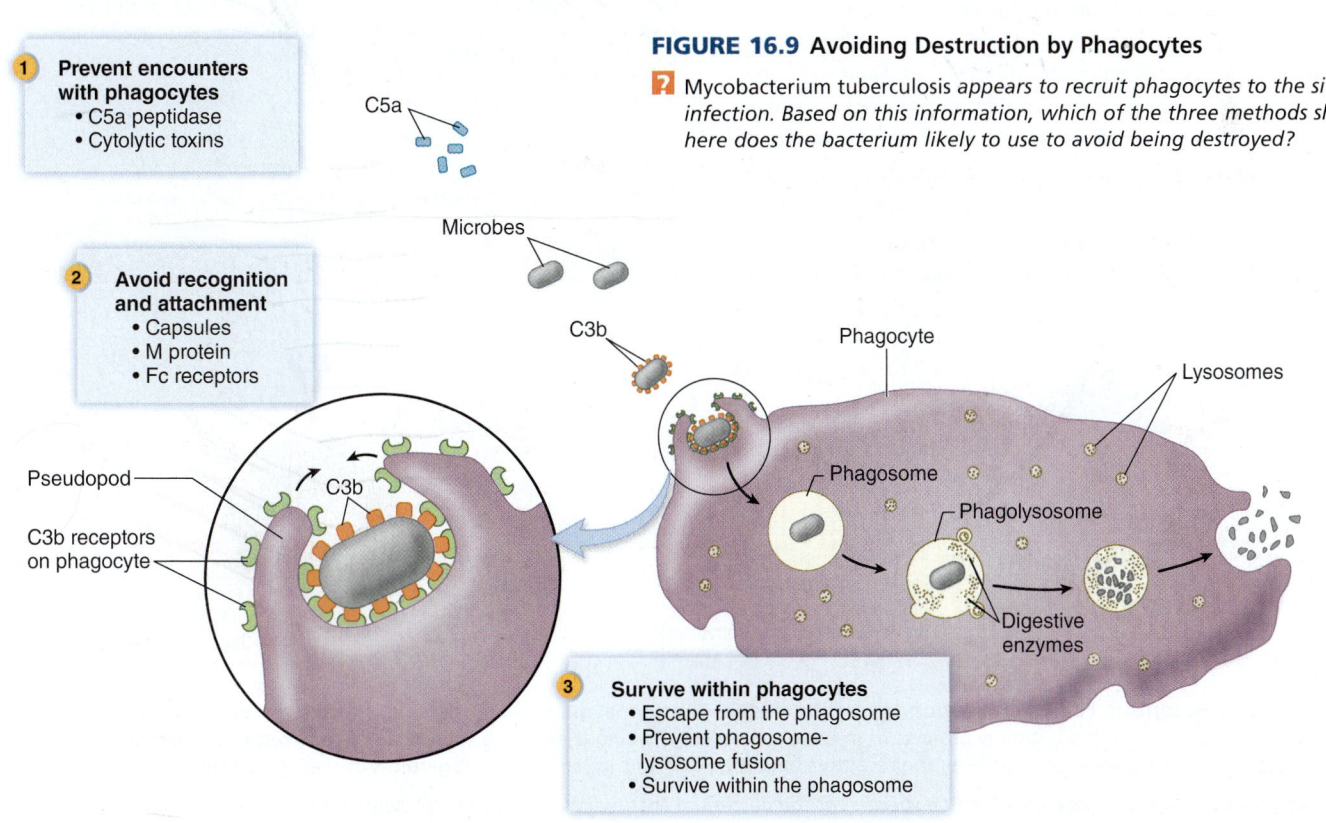

FIGURE 16.9 Avoiding Destruction by Phagocytes

❓ *Mycobacterium tuberculosis appears to recruit phagocytes to the site of infection. Based on this information, which of the three methods shown here does the bacterium likely to use to avoid being destroyed?*

1 Prevent encounters with phagocytes
- C5a peptidase
- Cytolytic toxins

C5a

Microbes

2 Avoid recognition and attachment
- Capsules
- M protein
- Fc receptors

C3b

Phagocyte

Lysosomes

Pseudopod

C3b

C3b receptors on phagocyte

Phagosome

Phagolysosome

Digestive enzymes

3 Survive within phagocytes
- Escape from the phagosome
- Prevent phagosome-lysosome fusion
- Survive within the phagosome

- **M protein.** This component of the cell wall of *Streptococcus pyogenes* functions in a manner similar to that described for capsules. It binds a complement regulatory protein that inactivates C3b, thereby preventing it from being an effective opsonin and avoiding the alternative pathway of complement system activation.

- **Fc receptors.** These proteins bind the Fc region of antibodies, interfering with their function as opsonins (**figure 16.10**). Recall that antibodies have two parts—the Fab region, which binds specifically to antigens, and the Fc region, which functions as a "red flag" (see figure 15.7). Bacterial cells that have Fc receptors coat themselves with antibody molecules with the Fab region projecting outward. The phagocytic cell has no mechanism for recognizing the Fab regions, so this masks the bacterial cell from phagocytes. *Staphylococcus aureus* and *Streptococcus pyogenes* make Fc receptors (protein A and protein G, respectively). ◀◀ Fc and Fab regions, p. 360

③ Surviving Within Phagocytes

Some bacteria make no attempt to avoid engulfment by phagocytes, instead using it as an opportunity. It allows them to hide from antibodies, control some aspects of the immune response, and be transported to other locations in the body. Mechanisms used to survive within phagocytes include:

- **Escape from the phagosome.** Some pathogens escape from the phagosome before it fuses with lysosomes. The bacteria then multiply within the cytoplasm of the phagocyte, protected from other host defenses. *Listeria monocytogenes* produces a molecule that forms pores in the phagosomal membrane, allowing the bacterial cells to escape. *Shigella* species lyse the phagosome before it fuses with lysosomes.

- **Preventing phagosome-lysosome fusion.** Bacteria that prevent phagosome-lysosome fusion avoid the otherwise inevitable exposure to the destructive components of lysosomes. *Salmonella* species can sense they have been ingested by a macrophage, and then respond by producing a protein that blocks the fusion process.

- **Surviving within the phagolysosome.** Relatively few microbes can survive the destructive environment within the phagolysosome. *Coxiella burnetii* (causes Q fever), however, is able to withstand the conditions. It appears that once the organism has been ingested by a macrophage, it delays fusion of the phagosome with the lysosome, allowing additional time for the microbe to equip itself for growth within the phagolysosome.

Avoiding Antibodies

Pathogens that survive the innate defenses soon encounter an additional obstacle, the adaptive defenses. For bacteria, the most important of these are antibodies. Mechanisms for avoiding them include:

- **IgA protease.** This enzyme cleaves IgA, the class of antibody found in mucus and other secretions. *Neisseria gonorrhoeae* and a variety of other pathogens produce IgA protease. This enzyme may also have other roles.

- **Antigenic variation.** Some pathogens routinely alter the structure of their surface antigens. This allows them to stay ahead of antibody production by altering the very molecules antibodies would otherwise recognize. *Neisseria gonorrhoeae* is able to vary the antigenic structure of its pili; antibodies produced by the infected host in response to one variation of the pili cannot bind effectively to another. ◀◀ antigenic variation, p. 178

- **Mimicking host molecules.** Pathogens sometimes cover themselves with molecules similar to those normally found in the host. This molecular mimicry takes advantage of the fact that the immune system typically does not mount an attack against "self" molecules. Certain strains of *Streptococcus pyogenes* have a capsule composed of hyaluronic acid, a polysaccharide found in human tissues.

MicroAssessment 16.7

Serum-resistant bacteria avoid the killing effects of complement system proteins. Mechanisms bacteria use to avoid destruction by phagocytes include preventing encounters with phagocytes, avoiding recognition and attachment, and surviving within the phagocyte. Mechanisms for avoiding antibodies include IgA protease, antigenic variation, and mimicking host molecules.

15. *Describe how Fc receptors prevent phagocytosis.*

16. *Describe three mechanisms pathogens may use to survive within phagocytic cells.*

17. *Encapsulated organisms can be phagocytized once antibodies against the capsule have been produced. Why would this be so?* ✚

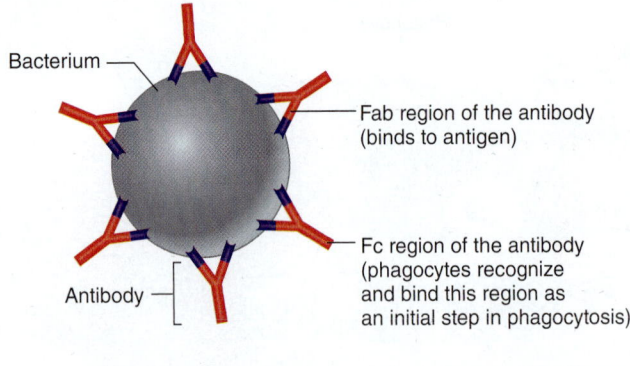

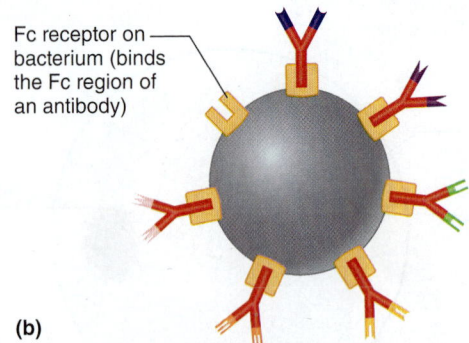

Bacterium

Fab region of the antibody (binds to antigen)

Fc region of the antibody (phagocytes recognize and bind this region as an initial step in phagocytosis)

Antibody

(a)

Fc receptor on bacterium (binds the Fc region of an antibody)

(b)

FIGURE 16.10 Fc Receptors Foil Opsonization by Antibodies (a) The normal orientation of antibody molecules on the surface of a bacterium; note that the Fc region of the antibody projects from the bacterial cell, making it available for a phagocyte to recognize and bind. **(b)** The effect of Fc receptors on a bacterial cell's surface; the receptors bind the Fc portion of antibodies, regardless of their specificity.

❓ *Would antibodies that the Fc receptors bind be specific for the bacterium that makes the receptors? Why or why not?*

16.8 ◼ Damage to the Host

Learning Outcomes

11. *Describe the difference between exotoxins and endotoxins.*
12. *Compare and contrast neurotoxins, enterotoxins, and cytotoxins, giving two examples of each.*
13. *Explain how inflammation and antibodies can cause damage.*

Damage due to infection can be the result of direct effects of the pathogen, such as toxins produced, or indirect effects, such as the immune response. In many cases, the damage helps the organism exit the host, allowing it to spread to others. For example, *Vibrio cholerae* (causes cholera) induces watery diarrhea—up to 20 liters of microbe-containing fluid in one day! In areas of the world without adequate sewage treatment, this can lead to contaminated water supplies and widespread outbreaks. *Bordetella pertussis* (causes whooping cough) triggers severe bursts of coughing, propelling the respiratory pathogens into the air.

Exotoxins

A number of Gram-positive and Gram-negative pathogens produce **exotoxins**—proteins that have very specific damaging effects (**table 16.1**). Exotoxins are often a major cause of damage to an infected host.

Exotoxins are either secreted by the bacterium or leak into the surrounding fluid following lysis of the bacterial cell. In most cases, the pathogen must colonize a body surface or tissue to produce enough toxin to cause damage. With foodborne intoxication, however, the bacterial cells multiply in a food product where they produce toxin that is then consumed. In the case of botulism, ingestion of even tiny amounts of botulinum toxin is sufficient to cause paralysis. Like most other exotoxins, botulinum toxin can be destroyed by heating. ▶▶ botulism, p. 652

Exotoxins can act locally, or they may be carried in the bloodstream throughout the body, causing systemic effects. *Corynebacterium diphtheriae* (causes diphtheria) grows and releases its exotoxin in the throat. There, the toxin destroys local cells, leading

TABLE 16.1	Exotoxins Produced by Various Primary Pathogens

Example	Name of Disease; Name of Toxin	Characteristics of the Disease	Mechanism	Page Reference
A-B TOXINS—Composed of two subunits, A and B. The A subunit is the toxic, or active, part; the B subunit binds to the target cell.				
Neurotoxins				
Clostridium botulinum	Botulism; botulinum toxin	Flaccid paralysis	Blocks transmission of nerve signals to the muscles by preventing the release of acetylcholine.	p. 652
Clostridium tetani	Tetanus; tetanospasmin	Spastic paralysis	Blocks the action of inhibitory neurons by preventing the release of neurotransmitters.	p. 555
Enterotoxins				
Enterotoxigenic *E. coli*	Traveler's diarrhea; heat-labile enterotoxin (cholera-like toxin)	Severe watery diarrhea	Modifies a regulatory protein in intestinal cells, causing those cells to continuously secrete electrolytes and water.	p. 590
Vibrio cholerae	Cholera; cholera toxin	Severe watery diarrhea	Modifies a regulatory protein in intestinal cells, causing those cells to continuously secrete electrolytes and water.	p. 586
Cytotoxins				
Bacillus anthracis	Anthrax; edema factor, lethal factor	Inhaled form—septic shock; cutaneous form—skin lesions	Edema factor modifies a regulatory protein in cells, causing accumulation of fluids. Lethal factor inactivates proteins involved in cell signaling functions.	p. 497
Bordetella pertussis	Pertussis (whooping cough); pertussis toxin	Sudden bouts of violent coughing	Modifies a regulatory protein in respiratory cells, causing accumulation of respiratory secretions and mucus. Other factors also contribute to the symptoms.	p. 501
Corynebacterium diphtheriae	Diphtheria; diphtheria toxin	Pseudomembrane in the throat; heart, nervous system, kidney damage	Inhibits protein synthesis by inactivating an elongation factor of eukaryotic cells. Kills local cells (in the throat) and is carried in the bloodstream to various organs.	p. 490
E. coli O157:H7	Bloody diarrhea, hemolytic uremic syndrome; shiga toxin	Diarrhea that may be bloody; kidney damage	Inactivates the 60S subunit of eukaryotic ribosomes, halting protein synthesis.	p. 590
Shigella dysenteriae	Dysentery, hemolytic uremic syndrome; shiga toxin	Diarrhea that contains blood, pus, and mucus; kidney damage	Inactivates the 60S subunit of eukaryotic ribosomes, halting protein synthesis.	p. 588

(continued)

TABLE 16.1	**Exotoxins Produced by Various Primary Pathogens (Continued)**			
Toxins	Name of Disease; Name of Toxin	Characteristics of the Disease	Mechanism	Page Reference
MEMBRANE-DAMAGING TOXINS (cytotoxins)—Disrupt plasma membranes.				
Clostridium perfringens	Gas gangrene; α-toxin	Extensive tissue damage	Removes the polar head group on the phospholipids in the membrane, destroying membrane integrity.	p. 558
Staphylococcus aureus	Wound and other infections; leukocidin	Accumulation of pus	Inserts into membranes, forming pores that allow fluids to enter the cells.	p. 551
Streptococcus pyogenes	Pharyngitis and other infections; streptolysin O	Accumulation of pus	Inserts into membranes, forming pores that allow fluids to enter the cells.	p. 487
SUPERANTIGENS—Override the specificity of the T-cell response.				
Staphylococcus aureus (certain strains)	Foodborne intoxication; staphylococcal enterotoxins	Nausea and vomiting	Not well understood with respect to how the ingested toxins lead to the characteristic symptoms of foodborne intoxication.	p. 757
Staphylococcus aureus (certain strains)	Staphylococcal toxic shock; toxic shock syndrome toxin (TSST)	Fever, vomiting, diarrhea, muscle aches, rash, low blood pressure	Systemic toxic effects due to the resulting massive release of cytokines.	p. 619
Streptococcus pyogenes (certain strains)	Streptococcal toxic shock; streptococcal pyrogenic exotoxins (SPE)	Fever, vomiting, diarrhea, muscle aches, rash, low blood pressure	Systemic toxic effects due to the resulting massive release of cytokines.	p. 553
OTHER TOXIC PROTEINS				
Staphylococcus aureus	Scalded-skin syndrome; exfoliatin	Separation of the outer layer of skin	Thought to break ester bonds that hold the layers of skin together.	p. 527
Various organisms	Various diseases; proteases, lipases, and other hydrolases	Tissue damage	Degrades proteins, lipids, and other compounds that make up tissues.	

to the accumulation of dead host cells, pus, and blood. This forms a membrane that sometimes dislodges and blocks the airway. The toxin can be absorbed and carried to the heart, nervous system, and other organs, causing additional damage. ▶▶diphtheria, p. 490

Because exotoxins are proteins, the immune system can generally produce neutralizing antibodies (see figure 15.8). Unfortunately, many exotoxins are so powerful that fatal damage occurs before an adequate immune response is mounted. This is why vaccination is so important. It prevents otherwise common and often fatal diseases such as tetanus and diphtheria (see table 18.1). The vaccines against tetanus and diphtheria are **toxoids,** which are inactivated toxins. A vaccine against botulinum toxin is also available, but it is not part of routine vaccination because the risks of developing the disease are extremely low if sensible food preparation procedures are followed. If a person develops symptoms of a toxin-mediated disease, he or she can be treated with **antitoxin,** a suspension of neutralizing antibodies. ▶▶vaccines, p. 421 ▶▶antitoxin, p. 420

Many exotoxins can be grouped into functional categories according to the tissues they affect (table 16.1). **Neurotoxins** damage the nervous system, causing symptoms such as paralysis. **Enterotoxins** cause symptoms associated with intestinal disturbance, such as diarrhea and vomiting. **Cytotoxins** damage a variety of different cell types, either by interfering with essential cellular mechanisms or by lysing cells. Some exotoxins do not fall into any of these groups, instead causing symptoms associated with excessive stimulation of the immune response. Most exotoxins fall into three general categories that reflect their structure and general mechanism of action: A-B toxins, membrane-damaging toxins, and superantigens.

MicroByte

Botox is a dilute suspension of botulinum toxin (see Perspective 31.1).

A-B Toxins

A-B toxins consist of two parts: the A subunit is the toxic (active) portion and the B subunit binds to a specific surface molecule on cells (**figure 16.11**). In other words, the A subunit, usually an enzyme, is responsible for the effects of the toxin on a cell, whereas the B subunit dictates the type of cell to which the toxin is delivered.

The structure of A-B toxins offers novel approaches for the development of vaccines and therapies. For example, a fusion protein that contains diphtheria toxin is used to treat a type of cancer called cutaneous T-cell lymphoma. By fusing the toxin to a cytokine that binds T cells, the toxin is delivered to those cells, including cancerous ones. In some countries, the B subunit of cholera toxin is used as an orally administered vaccine against cholera. Antibodies that bind the B subunit prevent cholera toxin from attaching to intestinal cells, thus protecting the vaccine recipient. Researchers are now experimenting with joining medically useful compounds to B subunits, allowing medications to be delivered specifically to the cell type targeted by the B subunit.

Membrane-Damaging Toxins

Membrane-damaging toxins are cytotoxins that disrupt plasma membranes, causing the cell to lyse. Many lyse red blood cells, causing hemolysis that can be observed when the organisms are grown on blood agar; because of this, many of these toxins are referred to as **hemolysins.** ◀◀hemolysis, p. 96 ◀◀blood agar, p. 95

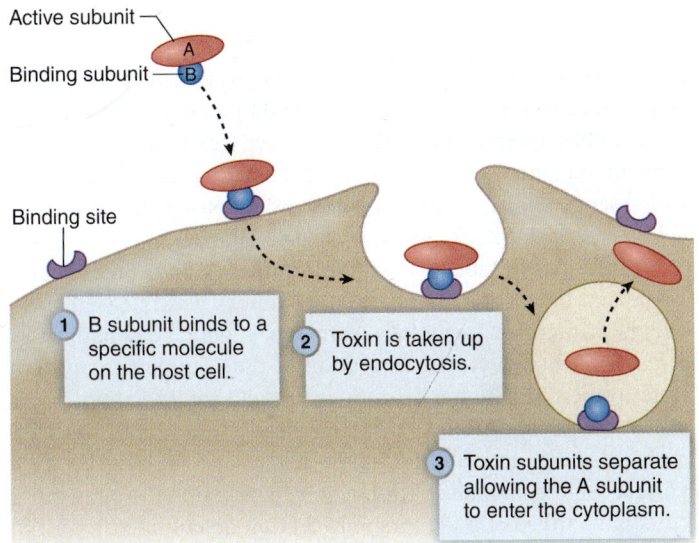

FIGURE 16.11 The Action of A-B Toxins The A subunit is toxic and the B subunit binds to specific receptors on cells. After being take up by endocytosis, the A and B subunits separate in the acidic endosome. The A subunit then enters the cell's cytoplasm and exerts its toxic effect.

? *Would an antibody response against the B subunit protect against the effects of the toxin?*

Some membrane-damaging toxins insert themselves into membranes, forming pores that allow fluids to enter the cell. One pore-forming toxin is streptolysin O, the compound responsible for the characteristic β-hemolysis of *Streptococcus pyogenes* grown anaerobically on blood agar (see figure 4.10). Recall that streptolysin O also helps *S. pyogenes* avoid phagocytosis.

Phospholipases hydrolyze phospholipids in the plasma membrane. The α-toxin of *Clostridium perfringens* (causes gas gangrene) is an example. ◀◀ **phospholipid, p. 35** ▶▶ **gas gangrene, p. 558**

Superantigens

Superantigens override the specificity of the helper T-cell response, causing toxic effects due to the massive release of cytokines by T_H cells (effector helper T cells). Superantigens include toxic shock syndrome toxin (TSST) as well as several other toxins produced by *Staphylococcus aureus* and *Streptococcus pyogenes*. ◀◀ **helper T cells, p. 360**

Superantigens short-circuit the normal specificity of antigen recognition by a helper T cell. They do this by binding simultaneously to the outer portion of the major histocompatibility (MHC) class II molecule on antigen-presenting cells and the T-cell receptor (**figure 16.12**). The T-cell machinery interprets the binding to mean that the T-cell receptor recognized the antigen presented on the MHC molecule, when it probably did not. Whereas an antigen usually stimulates about one in 10,000 helper T cells, superantigens stimulate as many as one in five. This results in a massive release of cytokines, leading to fever, nausea, vomiting, and diarrhea. Shock may occur, with organ failure, circulatory collapse, and even death. In addition, the immune response is suppressed because many T cells undergo apoptosis following the stimulation. Superantigens are also suspected of contributing to autoimmune diseases. By overriding the normal control mechanisms of adaptive immunity, superantigens may promote proliferation of T cells that respond to healthy "self." ◀◀ **MHC class II molecules, p. 369** ◀◀ **apoptosis, p. 350**

The exotoxins produced by *Staphylococcus aureus* strains that cause foodborne intoxication are superantigens. Although they cause nausea and vomiting and are therefore referred to as enterotoxins, their structure and action are very different from the enterotoxins of *Vibrio cholerae* and *E. coli* strains. The mechanism by which they induce vomiting is poorly understood. Unlike most other exotoxins, the enterotoxins produced by *Staphylococcus aureus* are heat-stable. Even thorough cooking of foods contaminated with these toxins will not prevent illness. ▶▶ *Staphylococcus aureus* **food poisoning, p. 757**

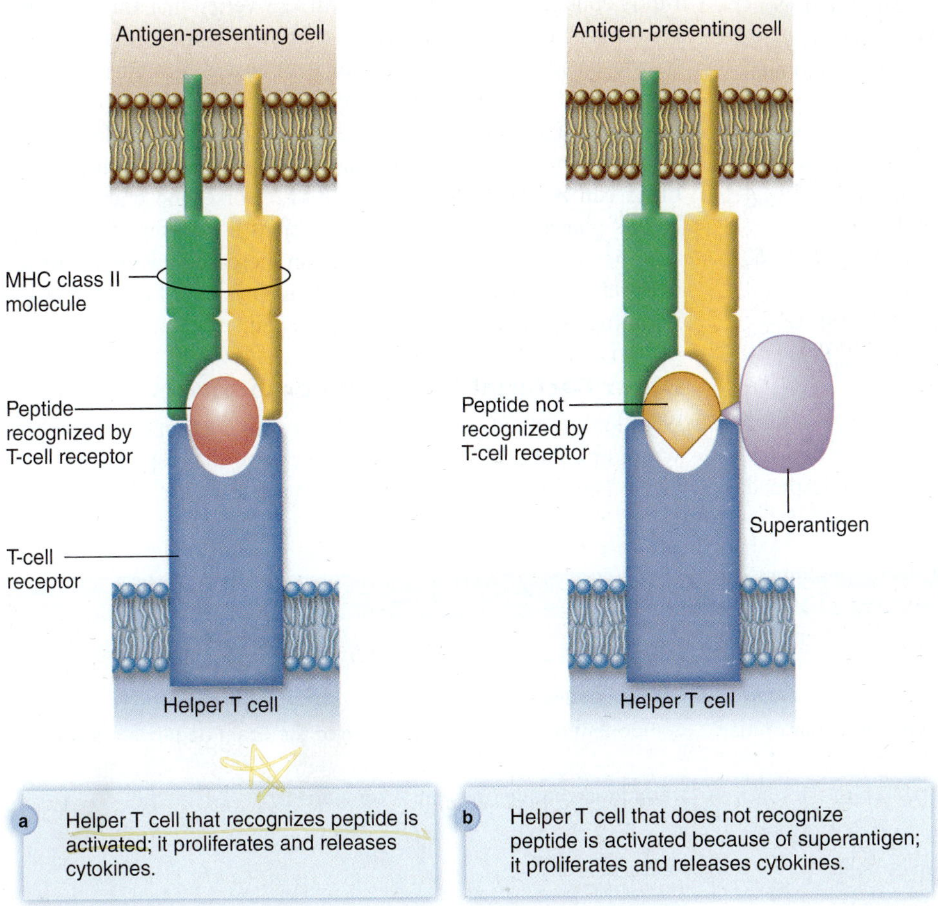

a Helper T cell that recognizes peptide is activated; it proliferates and releases cytokines.

b Helper T cell that does not recognize peptide is activated because of superantigen; it proliferates and releases cytokines.

FIGURE 16.12 Superantigens (a) Normal antigen presentation. (b) Superantigens short-circuit the normal specificity of the helper T-cell response by binding simultaneously to the outer portion of the MHC class II molecule and the T-cell receptor.

? *What specifically causes the toxic effects of superantigens?*

Other Toxic Proteins

Various proteins that are not A-B toxins, superantigens, or membrane-damaging toxins can have damaging effects. An important example is the toxin produced by strains of *Staphylococcus aureus* that cause scalded skin syndrome. This toxin, **exfoliatin,** destroys material that binds together the layers of skin, causing the outer layer to separate (see figure 22.4). The bacteria might be growing in a small, localized lesion but the toxin spreads systemically.

Various hydrolytic enzymes including proteases, lipases, and collagenases break down tissue components. Along with destroying tissues, some of these enzymes help the bacteria spread.

Endotoxin and Other Bacterial Cell Wall Components

The host defenses are primed to respond to various bacterial cell wall components, including lipopolysaccharide and peptidoglycan. A strong and widespread immune response to these compounds, however, can have toxic effects.

Endotoxin $\text{Endotoxin} = (LPS)$

Endotoxin is lipopolysaccharide (LPS), the molecule that makes up the outer layer of the outer membrane of the Gram-negative cell wall (see figure 3.33). The name is somewhat unfortunate, because it implies that endotoxin is "inside the cell," and, conversely, that exotoxins are "outside the cell." This is misleading, because endotoxin is an integral part of the outer membrane, whereas exotoxins are proteins that may or may not be secreted by the bacterial cell. Unlike most exotoxins, endotoxin cannot be converted to an effective toxoid for immunization. **Table 16.2** summarizes some of the other differences between exotoxins and endotoxin. ◄◄ **lipopolysaccharide, p. 60**

Recall from chapter 3 that the lipopolysaccharide molecule contains lipid A. This component triggers an inflammatory response and is responsible for the adverse effects of LPS. When lipid A is present in a localized region, the response helps clear an infection. It is a different situation entirely, however, when the infection is systemic, as in a bloodstream infection. Imagine a state in which inflammation occurs throughout the body—extensive leakage of fluids from permeable blood vessels and widespread

activation of the coagulation cascade. The overwhelming systemic response causes a dramatic drop in blood pressure, disseminated intravascular coagulation (DIC), and fever (see figure 27.3). This array of symptoms associated with systemic bacterial infection is called **septic shock;** when it is caused by endotoxin, it may also be called endotoxic shock. ▶▶ **DIC, p. 674** ▶▶ **septic shock, p. 674**

Lipid A is embedded in the Gram-negative outer membrane and does not cause a response unless it is released. This occurs primarily when a bacterium lyses, which can happen as a result of phagocytosis, activation of the complement system (due to membrane attack complexes), and treatment with certain types of antibiotics. Once released from a cell, the LPS molecules can activate the innate and adaptive defenses by a variety of mechanisms. Monocytes, macrophages, and other cells have toll-like receptors (TLRs) that detect liberated LPS, inducing the cells to produce pro-inflammatory cytokines. LPS also functions as a T-independent antigen; at high concentrations it activates a variety of different B cells, regardless of the specificity of their B-cell receptor. ◄◄ **TLRs, p. 342** ◄◄ **T-independent antigens, p. 367**

Endotoxin is heat-stable; it is not destroyed by autoclaving. Consequently, solutions intended for intravenous (IV) administration must not only be sterile, but free of endotoxin as well. Disastrous results including death have resulted from IV fluids contaminated with endotoxin. To verify that fluids are not contaminated, a very sensitive test known as the Limulus amoebocyte lysate (LAL) assay is done. This uses proteins extracted from blood of the horseshoe crab (*Limulus polyphemus*) that form a gel-like clot when exposed to endotoxin; as little as 10 to 20 picograms (1 picogram = 10^{-12} grams) of endotoxin per milliliter can be detected using the LAL. Horseshoe crabs are one of this planet's more unique and ancient life-forms. The critical role they play in this test has led to an increased awareness of the importance of their habitat. A non-lethal system of capture, blood sampling, and release has been developed.

Other Bacterial Cell Wall Components

Peptidoglycan and other bacterial cell wall components can cause symptoms similar to those that characterize the response to endotoxin. The systemic response leads to septic shock.

TABLE 16.2	Comparison of Exotoxins and Endotoxin	
Property	**Exotoxins**	**Endotoxin**
Bacterial source	Gram-positive and Gram-negative species	Gram-negative species only
Location in the bacterium	Synthesized in the cytoplasm; may or may not be secreted	Component of the outer membrane of the Gram-negative cell wall
Chemical nature	Protein	Lipopolysaccharide (the lipid A component)
Ability to form a toxoid	Generally	No
Heat stability	Generally inactivated by heat	Heat-stable
Mechanism	A distinct toxic mechanism for each	Innate immune response; a systemic response leads to fever, a dramatic drop in blood pressure, and disseminated intravascular coagulation
Toxicity	Generally very potent; some are among the most potent toxins known.	Small amounts in a localized area lead to an appropriate immune response that helps clear an infection, but systemic distribution can be deadly

Damaging Effects of the Immune Response

Although the immune response eliminates invading microbes, it can inadvertently damage host tissues as well. The reactions to endotoxin and other cell wall components are examples of damaging effects of the immune response, but are typically considered a toxic effect of the bacterium because the reactions can be immediate and overwhelming. The damaging responses discussed next become apparent more slowly.

Damage Associated with Inflammation

The inflammatory response itself can destroy tissue because phagocytic cells recruited to the area release some of the enzymes and toxic products they contain. The life-threatening aspects of bacterial meningitis, for example, are due to the inflammatory response itself. Complications of certain sexually transmitted infections are also due to the damage associated with inflammation. For example, if *Neisseria gonorrhoeae* or *Chlamydia trachomatis* infections involve the fallopian tubes, the inflammatory response can lead to scarring that obstructs the tubes, either preventing fertilization or predisposing a woman to an ectopic pregnancy (meaning the fertilized egg implants outside of the uterus).

Damage Associated with Adaptive Immunity

The adaptive immune response can also lead to damaging effects. Mechanisms include:

- **Immune complexes.** When antibodies bind to antigens, the complexes can settle in the kidneys and joints, where they activate the complement system, causing destructive inflammation. An example is acute glomerulonephritis, a complication that can follow skin and throat infections caused by *Streptococcus pyogenes;* the immune complexes that form trigger a response that damages kidney structures called glomeruli. ▶▶ acute glomerulonephritis, p. 490

- **Cross-reactive antibodies.** Certain antibodies produced in response to an infection bind to the body's own tissues, promoting an autoimmune response. Acute rheumatic fever—a complication that can follow strep throat—is thought to be due to antibodies against *S. pyogenes* binding to normal tissue proteins. This occurs most frequently in people with certain MHC types (see Perspective 15.1). ▶▶ acute rheumatic fever, p. 489

MicroAssessment 16.8

Damage can be due to exotoxins, including A-B toxins, superantigens, membrane-damaging toxins, and other toxic proteins. Endotoxin and other cell wall components in the bloodstream can result in septic shock. The inflammatory response can cause tissue damage that leads to scarring. Immune complexes can trigger damaging inflammation in the kidneys and joints; cross-reactive antibodies can lead to an autoimmune response.

18. *What is an A-B toxin?*

19. *How is an enterotoxin different from endotoxin?*

20. *Home-canned foods should be boiled before consumption to prevent botulism. Considering that this treatment does not destroy endospores, why would it prevent the disease?* ✚

16.9 ■ Mechanisms of Viral Pathogenesis

Learning Outcomes

14. *Describe how viruses bind to host cells, and how they spread to other cells.*

15. *Describe how viruses avoid interferon, regulate apoptosis, and avoid antibodies.*

To infect a host, a virus must enter an appropriate cell, use the cell's machinery for replication, keep the host from recognizing and destroying the infected cell, and then move to new cells or hosts. In carrying out these functions, many viruses damage host cells and induce inflammatory responses, causing disease.

Binding to Host Cells and Invasion

As discussed in chapter 13, viruses attach to target cells via specific receptors. Only cells that bear the receptor can be infected, so this influences the host range and tissue specificity of a particular virus. For example, HIV's receptor is CD4, a molecule found on helper T cells and macrophages. Picornaviruses and reoviruses bind to receptors on M cells in the Peyer's patches of the intestinal tract. Once attached, viruses then enter the cells using either receptor-mediated endocytosis or fusion with the plasma membrane (see figure 13.13). ◀◀ CD4, p. 369

Virions released from a cell can either infect neighboring cells or disseminate via the bloodstream or lymphatic system to other tissues. Polioviruses, for example, initially infect cells in the throat and intestinal tract. Upon release from these cells, the virus enters the bloodstream and may then spread to infect motor nerve cells of the brain and spinal cord, causing paralysis. ▶▶ polio, p. 656

Avoiding Immune Responses

Viruses must avoid the host defenses that detect and eliminate invaders. These include interferon, apoptosis, and antibodies. ◀◀ interferon, p. 341   ◀◀ apoptosis, p. 350

Avoiding the Antiviral Effects of Interferons

Early in viral infection, interferons play an important role in limiting viral spread (see figure 14.9). They cause cells to produce enzymes that, when activated, prevent viral replication. To avoid this, some viruses encode proteins that shut down expression of host genes. Others avoid interferon's effects by interfering with activation of the enzymes.

Regulating Host Cell Death

Many viruses regulate host cell death. Some kill the host cell once it has produced virions, and others prevent the host cell from dying prematurely.

Recall that the interferon response induces apoptosis in virally infected cells. Some viruses prevent this from happening, thereby giving the virus more time to replicate. They do this by controlling a protein called p53 that regulates apoptosis in cells. When this protein is inhibited, tumors sometimes develop. Papillomaviruses, a group of viruses that cause various types of warts, interfere with the normal function of p53. ◀◀ **papillomaviruses, p. 308**

To avoid being detected by the cellular immune response, many types of viruses interfere with antigen presentation by MHC class I molecules. Herpesviruses, for example, block the movement of the molecules to the surface of the infected cell, so that T$_C$ cells (effector cytotoxic T cells) cannot inspect the proteins being made. The immune system, however, is prepared for such a strategy. Natural killer (NK) cells recognize stressed cells that lack MHC class I molecules and destroy them (see figure 15.23). Perhaps not surprisingly, some viruses have methods to trick NK cells. Cytomegalovirus (CMV), which causes disease in immunocompromised people, encodes production of "fake" MHC class I molecules. These decoy versions are displayed on the surface of the host cell, tricking the immune system into believing that all is well in the cell (**figure 16.13**). ◀◀ **antigen presentation by MHC class I, p. 369** ◀◀ **natural killer cells, p. 374** ▶▶ **cytomegaloviruses, p. 711**

Avoiding Antibodies

Antibodies generally control the spread of viruses by neutralizing extracellular virions (see figure 15.8). To avoid antibodies, some viruses move directly from one cell to its immediate neighbors. Other viruses remain intracellular by forcing cellular neighbors to fuse, forming a syncytium. HIV induces syncytia formation. ◀◀ **syncytium, p. 326**

The surface antigens of some viruses change rapidly, outpacing the body's capacity to produce effective neutralizing antibodies. This rapid evolution occurs because the replicases of RNA viruses and the reverse transcriptase of HIV have no proofreading ability, so mutations occur fairly often. Thus, as the viruses replicate, they give rise to pools of genetically altered virions. Eventually, an immune response against essential proteins that cannot tolerate extensive mutations may eventually clear the viruses. This is not the case for HIV, however. An HIV-infected person produces many antibodies against HIV, but none are effective in clearing the virus. ◀◀ **replicase, p. 320** ◀◀ **reverse transcriptase, p. 320**

Some viruses actually take advantage of antibodies, using them to help enter macrophages. Opsonized viral particles are engulfed by phagocytes, which they then infect. Dengue hemorrhagic fever, a severe illness caused by dengue viruses, is associated with antibody-dependent enhancement of infection. ▶▶ **dengue hemorrhagic fever, p. 683**

MicroAssessment 16.9

Proteins on the surface of virus particles function as adherence factors. Viruses, as a group, have several mechanisms for avoiding the effects of interferon, regulating host cell death, and avoiding antibodies.

21. *From a virus's perspective, why would it be beneficial to prevent apoptosis?*
22. *How do cytomegaloviruses avoid the cellular immune response?*
23. *Why would various unrelated viral infections often include a similar set of symptoms (fever, headache, fatigue, and runny nose)?* ➕

16.10 ■ Mechanisms of Eukaryotic Pathogenesis

Learning Outcomes

16. *Describe the mechanisms of pathogenesis of dermatophytes,* Candida albicans, *and the dimorphic fungi.*
17. *Compare and contrast the mechanisms of pathogenesis of protozoa and helminths.*

Pathogenesis of eukaryotic cells, including fungi and protozoa, involves the same basic scheme as that of bacterial pathogens—colonization, evasion of host defenses, and damage to the host. The mechanisms, however, are generally not as well understood.

Fungi

Most fungi, such as yeasts and molds, are saprophytes, meaning they acquire nutrients from dead and dying material; those that can cause disease are generally opportunists, although notable exceptions exist.

Members of a group of fungi referred to as dermatophytes cause superficial infections of hair, skin, and nails, but do not

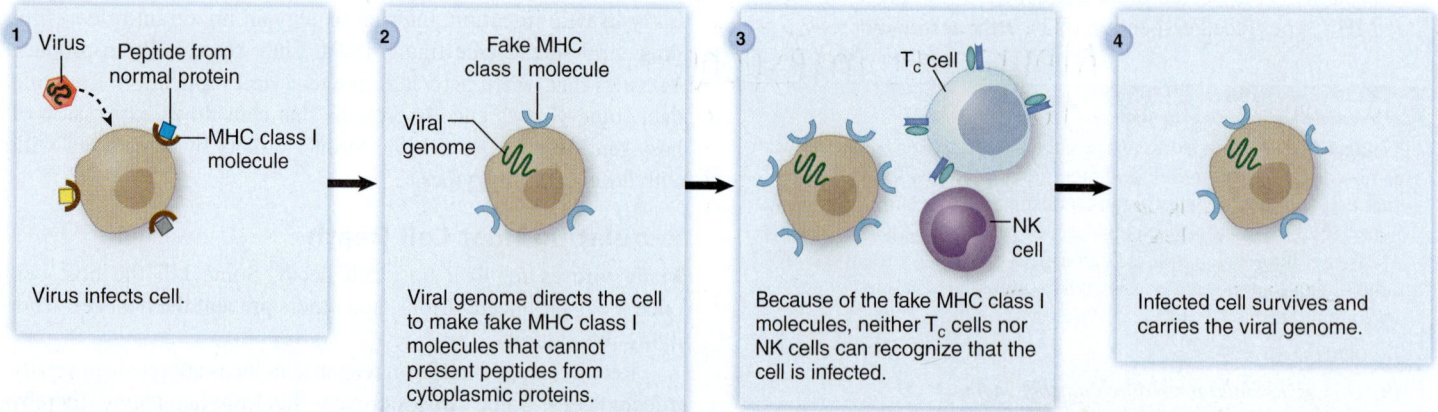

① **Virus** Peptide from normal protein — MHC class I molecule

Virus infects cell.

② Fake MHC class I molecule — Viral genome

Viral genome directs the cell to make fake MHC class I molecules that cannot present peptides from cytoplasmic proteins.

③ T$_C$ cell — NK cell

Because of the fake MHC class I molecules, neither T$_C$ cells nor NK cells can recognize that the cell is infected.

④ Infected cell survives and carries the viral genome.

FIGURE 16.13 The Role of MHC Class I Decoys in Viral Pathogenesis

❓ *What is the role of MHC class I molecules in adaptive immunity?*

invade deeper tissues. These fungi have keratinase enzymes that break down the keratin in superficial tissues, allowing the fungi to use this protein as a source of nutrition. Dermatophytes cause diseases such as ringworm and athlete's foot.

Fungi in the normal microbiota, especially the yeast *Candida albicans,* can cause disease in immunocompromised hosts. Factors that can lead to excessive growth of *C. albicans* include AIDS, uncontrolled diabetes, severe burns, and inhibition of normal microbiota due to hormonal influences or antibiotic treatment. *C. albicans* generally infects the mucous membranes, causing thrush (an infection of the throat and mouth) or vulvovaginitis.

The most serious fungal infections are caused by dimorphic fungi. These occur as molds in the environment, and when the small airborne conidia are inhaled, they lodge deep within the lungs. There, they develop into other forms, usually yeasts. The immune system generally controls the infection, so most cases are asymptomatic. The situation is entirely different if the person is immunocompromised, however, because the infections can become life-threatening. ◀◀ dimorphic fungi, p. 285 ◀◀ conidia, p. 287

Some fungi produce toxins, collectively referred to as mycotoxins. *Aspergillus flavus,* for example, a fungus that grows on certain grains and nuts, produces aflatoxin. Ingestion of this toxin can damage the liver and increases the risk of liver cancer. Fungal spores and other products can cause hypersensitivities in some people. ▶▶ hypersensitivities, p. 402

Protozoa and Helminths

Most protozoa and helminths either live within the intestinal tract or enter the body via the bite of an arthropod. *Schistosoma* species, however, can enter the skin directly (see Perspective 14.1).

Like bacteria and viruses, eukaryotic parasites attach to host cells via specific receptors. *Plasmodium vivax,* one of the two most common causes of malaria, attaches to the Duffy blood group antigen on red blood cells. Most people of West African ancestry lack this antigen and are therefore resistant to infection

by this species. *Giardia lamblia* uses a disc that functions as a suction cup to attach to the intestinal surface; it also has an adhesin associated with the disc that helps the initial attachment.

Protozoa and helminths use a variety of mechanisms to avoid antibodies. Some hide within cells, thus avoiding exposure to antibodies as well as certain other defenses. Malarial parasites, for example, produce enzymes that allow them to penetrate red blood cells. This cell type does not present antigen to T_C cells, so the parasite escapes antibodies as well as the cell-mediated immune defenses. *Leishmania* species survive and multiply within macrophages when phagocytized. Parasites such as the African trypanosomes, the cause of sleeping sickness, prevent antibodies from recognizing them by routinely varying their surface antigens through antigenic variation. *Schistosoma* species coat themselves with host proteins, thereby disguising themselves. Some parasitic worms appear to suppress immune responses in general.

The extent and type of damage caused by parasites varies tremendously. In some cases, the parasites compete for nutrients in the intestinal tract, contributing to malnutrition of the host. Helminths may accumulate in high enough numbers or grow long enough to block the intestines or other organs. Some parasites produce enzymes that digest host tissue, causing direct damage. In other cases, damage is due to the immune response; examples include the high fevers that characterize malaria, and the granulomatous response to *Schistosoma* eggs.

MicroAssessment 16.10

Pathogenic mechanisms of fungi, protozoa, and helminths are not as well understood as those of bacteria; however, they involve the same basic scheme—colonization, evasion of host defenses, and damage to the host.

24. *What is the importance of keratinase?*
25. *How do the African trypanosomes avoid the effects of antibodies?*
26. *Why would relatively few people of West African ancestry have the Duffy blood group antigen?* ➕

FUTURE CHALLENGES 16.1

The Potential of Probiotics

Considering the important protective roles the normal microbiota play in human health, it is not surprising that administering live beneficial microbes, referred to as **probiotics,** has been suggested as therapy for diarrheal and other diseases. *In vitro* studies certainly support their use, indicating that some strains of bacteria interfere with the growth or toxin production of certain pathogens. Animal studies using probiotics have also shown promising results. For example, chicks fed a mixture of *Lactobacillus* species that normally inhabit poultry intestines were less likely to be subsequently colonized by the pathogen *Salmonella enterica.* But which probiotics will be successful in treating humans?

Marketing claims touting probiotics abound, but there are still many unanswered questions. For example, will a microbe that has beneficial effects *in vitro* be stable in a food or supplement so that an adequate dose can be consumed? And even if it is stable in the package, will it survive passage through the stomach so that viable cells enter the intestinal tract? If it does access the intestines, can it colonize there, or will it simply pass through as part of feces?

Beneficial effects of probiotics observed in humans so far are species and strain specific. One species may be protective, whereas another closely related organism is not. For example, several small studies indicate that consumption of *Lactobacillus rhamnosus* GG

can shorten the duration of certain diarrheal illnesses. In contrast, studies that used a mixture containing *L. bulgaricus* and *L. acidophilus* generally showed no helpful effect. Because of this, products that list ingredients such as "live active culture" or "lactic acid bacteria" do not necessarily contain a beneficial strain.

Complicating matters is that probiotics are considered "dietary supplements," so there is little regulatory control over health claims. Early studies such as those using *L. rhamnosus* GG certainly highlight their potential, but larger and well-controlled scientific studies of probiotics are needed to provide more data as to their effectiveness in curing and preventing various medical conditions.

Summary

MICROBES, HEALTH, AND DISEASE

16.1 ■ The Anatomical Barriers as Ecosystems

In **mutualism**, both partners benefit, in **commensalism**, one partner benefits while the other is unaffected, and in **parasitism**, the parasite benefits at the expense of the host.

16.2 ■ The Normal Microbiota (figure 16.1)

The Protective Role of the Normal Microbiota

The **normal microbiota** excludes pathogens by covering binding sites that might otherwise be used for attachment, consuming available nutrients, and producing compounds toxic to other bacteria. The normal flora primes the adaptive immune system.

The Dynamic Nature of the Normal Microbiota

The composition of the normal microbiota continually changes in response to host factors including hormonal changes, and type and quantity of food consumed.

16.3 ■ Principles of Infectious Disease

Infectious diseases have characteristic **signs** and **symptoms**. A **secondary infection** can occur as the result of a **primary infection**.

Pathogenicity

A **primary pathogen** causes disease in otherwise healthy individuals; an **opportunistic pathogen** causes disease only when the body's innate or adaptive defenses are compromised. **Virulence** refers to the degree of pathogenicity of an organism.

Characteristics of Infectious Disease

Communicable or **contagious diseases** spread from one host to another; ease of spread partly reflects the **infectious dose**. Stages of infectious disease include the **incubation period, illness,** and **convalescence**; during the illness a person experiences signs and symptoms of the disease (figure 16.2). **Infections** can be described as **acute, chronic,** or **latent,** depending on the timing and duration of symptoms. Infections can be **localized** or **systemic**.

16.4 ■ Establishing the Cause of Infectious Disease

Koch's Postulates

Koch's postulates are used to establish the cause of infectious disease (figure 16.3).

Molecular Koch's Postulates

Molecular Koch's postulates are used to identify virulence factors that contribute to disease.

MECHANISMS OF PATHOGENESIS

16.5 ■ Establishing Infection

Adherence

Bacteria use **adhesins** to bind to host cells (figure 16.4).

Colonization

Rapid turnover of pili, antigenic variation, and **IgA proteases** allow bacteria to avoid the effects of secretory IgA. **Siderophores** enable microbes to scavenge iron.

Delivering Effector Proteins to Host Cells

Type III secretion systems of Gram-negative bacteria allow them to deliver proteins directly to host cells (figure 16.5).

16.6 ■ Invasion—Breaching the Anatomical Barriers

Penetrating the Skin

Bacteria take advantage of trauma that destroys the integrity of the skin or rely on arthropods to inject them.

Penetrating Mucous Membranes

Some pathogens induce mucosal epithelial cells to engulf bacterial cells; others exploit antigen-sampling processes (figures 16.6, 16.7).

16.7 ■ Avoiding the Host Defenses

Hiding Within a Host Cell

Some bacteria can evade the innate defenses, as well as some aspects of the adaptive defenses, by remaining inside host cells.

Avoiding Killing by Complement System Proteins

Some **serum-resistant** bacteria postpone the formation of membrane attack complexes by interfering with activation of the complement system via the alternative pathway.

Avoiding Destruction by Phagocytes (figure 16.9)

Mechanisms to prevent encounters with phagocytes include C5a peptidase and membrane-damaging toxins. Mechanisms to avoid recognition and attachment by phagocytes include **capsules, M protein,** and **Fc receptors** (figure 16.10). Mechanisms to survive within the phagocyte include escape from the phagosome, preventing phagosome-lysosome fusion, and surviving within the phagolysosome.

Avoiding Antibodies

Mechanisms to avoid antibodies include **IgA protease, antigenic variation,** and mimicking "self."

16.8 ■ Damage to the Host

Exotoxins (table 16.1)

Exotoxins are proteins that have very specific damaging effects; they may act locally or cause dramatic systemic effects. Many can be grouped into categories such as **neurotoxins, enterotoxins,** or **cytotoxins**. The toxic activity of **A-B toxins** is mediated by the A subunit; binding to specific cells is mediated by the B subunit (figure 16.11). **Membrane-damaging toxins** disrupt cell membranes either by forming pores or by removing the polar head group on phospholipids in the membrane. **Superantigens** override the specificity of the T-cell response, causing systemic effects due to the massive release of cytokines (figure 16.12).

Endotoxin and Other Bacterial Cell Wall Components

The symptoms associated with **endotoxin** are due to a vigorous host response. Lipid A of lipopolysaccharide is responsible for its toxic properties. Peptidoglycan and certain other components induce various cells to produce pro-inflammatory cytokines.

Damaging Effects of the Immune Response

The release of enzymes and toxic products from phagocytic cells can damage tissues. **Immune complexes** can cause kidney and joint damage; cross-reactive antibodies can result in an autoimmune response.

16.9 ■ Mechanisms of Viral Pathogenesis

Binding to Host Cells and Invasion

Viruses attach to specific receptors on the target cell.

Avoiding Immune Responses

Some viruses can avoid the effects of interferon; some can regulate apoptosis of the host cell (figure 16.13). To subvert the role of antibodies, some viruses transfer directly from cell to cell; the surface antigens of some viruses change quickly, outpacing the production of antibodies.

16.10 ■ Mechanisms of Eukaryotic Pathogenesis

Fungi

Saprophytes are generally opportunists; dermatophytes cause superficial infections of skin, hair, and nails. The most serious fungal infections are caused by dimorphic fungi.

Protozoa and Helminths

Eukaryotic parasites attach to host cells via specific receptors. They use a variety of mechanisms to avoid antibodies; the extent and type of damage they cause varies tremendously.

Review Questions

Short Answer

1. Describe three types of symbiotic relationships.
2. Describe two situations that can lead to changes in the composition of the normal microbiota.
3. How are acute, chronic, and latent infections different?
4. Why are Koch's postulates not sufficient to establish the cause of all infectious diseases?
5. Describe the four general mechanisms by which microorganisms cause disease.
6. Describe two mechanisms that bacteria use to invade via mucous membranes.
7. Explain how a capsule can allow an organism to be serum resistant and avoid phagocytosis.
8. Give an example of a neurotoxin, an enterotoxin, and a cytotoxin.
9. Describe two mechanisms a virus might use to prevent the induction of apoptosis in an infected cell.
10. How do *Schistosoma* species avoid antibodies?

Multiple Choice

1. Opportunistic pathogens are *least* likely to affect which of the following groups?
   a) AIDS patients
   b) Cancer patients
   c) College students
   d) Drug addicts
   e) Transplant recipients

2. Capsules and M protein are thought to interfere with which of the following?
   a) Opsonization by complement proteins
   b) Opsonization by antibodies
   c) Recognition by T cells
   d) Recognition by B cells
   e) Phagosome-lysosome fusion

3. The C5a peptidase enzyme of *Streptococcus pyogenes* breaks down C5a, resulting in
   a) lysis of the *Streptococcus* cells.
   b) lack of opsonization of *Streptococcus* cells.
   c) killing of phagocytes.
   d) decreased accumulation of phagocytes.
   e) inhibition of membrane attack complexes.

4. All of the following are known mechanisms of avoiding the effects of antibodies *except*
   a) antigenic variation.
   b) mimicking "self."
   c) synthesis of an Fc receptor.
   d) synthesis of IgG protease.
   e) remaining intracellular.

5. Which of the following statements about diphtheria toxin is *false*? It
   a) is an example of an endotoxin.
   b) is produced by a species of *Corynebacterium*.
   c) inhibits protein synthesis.
   d) can cause local damage to the throat.
   e) can cause systemic damage (that is, to organs such as the heart).

6. Which of the following statements about botulism is *true*?
   a) It is caused by *Bacillus botulinum*, an obligate aerobe.
   b) The toxin is heat-resistant, withstanding temperatures of 100°C.
   c) The organism that causes botulism can cause disease without avoiding the immune response.
   d) Vaccinations are routinely given to prevent botulism.
   e) Symptoms of botulism include uncontrolled contraction of muscles.

7. Superantigens
   a) are exceptionally large antigen molecules.
   b) cause a very large antibody response.
   c) elicit a response from a large number of T cells.
   d) attach non-specifically to B-cell receptors.
   e) assist in a protective immune response.

8. Which of the following statements about endotoxin is *true*? It
   a) is an example of an A-B toxin.
   b) is a component Gram-positive bacteria.
   c) can be converted to a toxoid.
   d) is heat-stable.
   e) causes T cells to release cytokines.

9. The tissue damage caused by *Neisseria gonorrhoeae* is primarily due to
   a) cross-reactive antibodies.
   b) exotoxins.
   c) hydrolytic enzymes.
   d) the inflammatory response.
   e) all of the these.

10. Which of the following statements about viruses is *false*? They may

    a) colonize the skin.

    b) enter host cells by endocytosis.

    c) enter host cells by fusion of the viral envelope with the cell membrane.

    d) induce apoptosis in infected host cells.

    e) suppress expression of MHC class I molecules on host cells.

Applications

1. A group of smokers suffering from *Staphylococcus aureus* infections are suing the cigarette companies. They claim that the disease was aggravated by smoking. The group is citing studies indicating that phagocytes are inhibited in their action by compounds in cigarette smoke. A statement prepared by their lawyers states that the *S. aureus* would not have caused such a severe disease if the phagocytes were functioning properly. During the proceedings, a microbiologist was called in as a professional witness for the court. What were her conclusions about the validity of the claim?

2. A microbiologist put forth a grant proposal to study the molecules bacteria use to communicate. Her principal rationale was that the damaging effects of many pathogenic microorganisms could be prevented by inactivating the molecules these bacteria use to communicate. Is this a reasonable proposal? Why or why not?

Critical Thinking ✚

1. A student argued that no distinction should be made between commensalism and parasitism. Even in commensalism, the microorganisms are gaining some benefit (such as nutrients) from the host, and this represents a loss to the host. In this sense, the host is being damaged. Does the student have a valid argument? Why or why not?

2. A microbiologist argued that there is no such thing as "normal" microbiota in the human body, since the population is dynamic and is constantly changing, depending on diet and external environment. What would be an argument against this microbiologist's view?

17 Immunologic Disorders

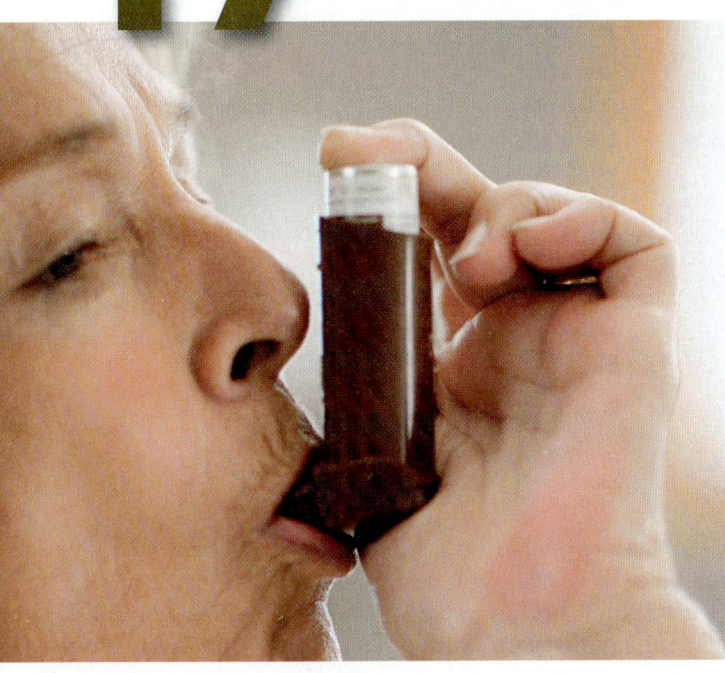

Asthmatic individual using an inhaler.

A Glimpse of History

Pasteur is widely quoted as saying, "Chance favors the prepared mind." This was certainly the case when physiologist Charles Richet discovered hypersensitivities, commonly called allergies. He and his colleague, Paul Portier, were cruising in the South Seas on Prince Albert of Monaco's yacht when they hypothesized that the Portuguese man-of-war jellyfish must produce a toxin that causes the swollen painful reactions to its stings. Prince Albert encouraged them to study the reactions, so they made an extract of the jellyfish tentacles and showed that it was indeed toxic to rabbits and ducks.

Upon returning to France, Richet and Portier continued their work by studying the effects of the toxin from a readily available sea anemone. When they tested it on dogs, some survived the first exposure to the potent toxin, but when given a small second dose at least 3 weeks later, the dogs died within minutes. Richet and Portier had worked with toxins and antisera earlier so they had prepared minds. They recognized that the dogs' reactions were probably caused by an immune response. Unlike protective immune responses they had worked with before, however, this one was destructive. They called the response *anaphylaxis*, indicating it was the opposite of *phylaxis*, the Greek term for protection. Richet received a Nobel Prize in 1913 for his work on anaphylaxis.

The immune system does a superb job protecting the body from infection, but, as Richet showed, these same protective mechanisms can be harmful if uncontrolled. Just as a building's sprinkler system that protects against fire can also cause significant water damage, particularly noticeable if set off by accident, immune responses can also be destructive. An immune response that injures tissue is a **hypersensitivity** and can be categorized into one of four groups according to the mechanisms and timing of the response (**table 17.1**).

Various terms are helpful to describe hypersensitivities and other immune disorders. An **allergy,** or allergic reaction, is a hypersensitivity reaction to an **allergen,** a normally harmless environmental substance. Although these terms can refer to any type of hypersensitivity, they are most often used to describe those that involve an IgE response. An **autoimmune disease** is an inappropriate immune response that targets body tissues. An **immunodeficiency** is a disorder that occurs when the immune system is too weak to prevent infection.

TABLE 17.1	Some Characteristics of the Major Types of Hypersensitivities			
Characteristic	Type I Hypersensitivity Immediate IgE-Mediated	Type II Hypersensitivity Cytotoxic	Type III Hypersensitivity Immune Complex-Mediated	Type IV Hypersensitivity Delayed-Type Cell-Mediated
Effector	B cells	B cells	B cells	T cells
Type of antigen	Soluble	Cell-bound	Soluble	Soluble or cell-bound
Type of antibody	IgE	IgG, IgM	IgG, IgM	None
Other immune cells involved	Basophils, mast cells	NK cells	Phagocytes	Dendritic cells
Mediators	Histamine, leukotrienes, prostaglandins	Complement, ADCC	Complement, neutrophil proteases, cytokines	Cytokines
Time of reaction after challenge with antigen	Immediate, up to 30 minutes	Hours to days	Hours to days	Peaks at 48 to 72 hours
Skin reaction	Wheal and flare	None	Arthus reaction	Induration, inflammation
Examples	Hives, hay fever, asthma, anaphylactic shock	Transfusion reaction, hemolytic disease of the newborn	Serum sickness, farmer's lung, malarial kidney damage, disseminated intravascular coagulation	Tuberculin reaction, contact dermatitis, tissue transplant rejection

17.1 ■ Type I Hypersensitivities: Immediate IgE-Mediated

Learning Outcomes

1. *Describe the immunologic reactions involved in type I hypersensitivities.*
2. *Compare and contrast localized allergic reactions and systemic anaphylaxis.*
3. *Explain the treatments for allergies.*

Type I hypersensitivities involve IgE, the class of antibody that binds by its Fc portion to mast cells or basophils. When bound to the cells, the antibodies function as captured antigen receptors, allowing mast cells and basophils to detect invaders. If the attached IgE molecules bind an invader, the cell is triggered to release inflammatory mediators, a response that is particularly effective against parasitic worms. Unfortunately, when IgE binds to allergens, the effect is the same but, like a sprinkler system set off by accident, the response has no benefits, only drawbacks. A wide variety of common substances can cause allergic reactions, including inhaled substances such as pollen, pet dander, and mold; ingested substances such as peanuts, milk, and seafood; and substances injected into the bloodstream such as insect venom or certain drugs. ◄◄ mast cell, p. 340 ◄◄ inflammatory mediators, p. 348 ◄◄ IgE, p.363

One of the questions immunologists are investigating is why some people are much more likely to develop allergies than others. The tendency to have type I allergic reactions is inherited, but the specific allergen to which a person reacts is due to environmental exposures rather than genetic traits. Food allergies affect about one out of every five children, but these often go away by the time the children reach school age.

Allergic reactions occur only in sensitized individuals who have had prior exposure to that specific antigen. **Sensitization** begins when contact with the allergen induces an antibody response. Recall that all B cells are initially programmed to produce IgM molecules, but they can then switch to produce other classes. B cells that reside in tissues under the mucous membranes often switch to IgA production. The B cells can also switch to IgE production, which happens more often in people who are prone to allergies. As IgE accumulates in this area, which is rich in mast cells, the IgE molecules attach by their Fc portion to receptors on the mast cells and also on circulating basophils (**figure 17.1**). Once attached, the IgE molecules are stable for weeks with their antigen-binding sites available to interact with allergens. A typical sequence of events occurs when pollen grains contact the mucous membrane in an individual's respiratory tract, triggering IgE production, which then sensitizes that individual to pollen. ◄◄ class switching, p. 366 ◄◄ basophil, p. 339

People with type I hypersensitivities have many IgE antibodies attached to mast cells throughout their bodies. Unless exposed to the allergen recognized by these antibodies, the mast cells are harmless. However, when at least two cell-bound IgE molecules react with specific antigen, the IgE molecules come together in a reaction called cross-linking (figure 17.1). Within seconds, the cross-linking causes the mast cell to release histamine and other inflammatory mediators in a process called **degranulation.** These chemicals cause capillary dilation, smooth muscle contraction, and increased mucus secretion. They are the direct cause of hives (urticaria), hay fever (allergic rhinitis), asthma, anaphylactic shock, and other allergic manifestations. Symptoms begin within minutes of exposure to the allergen.

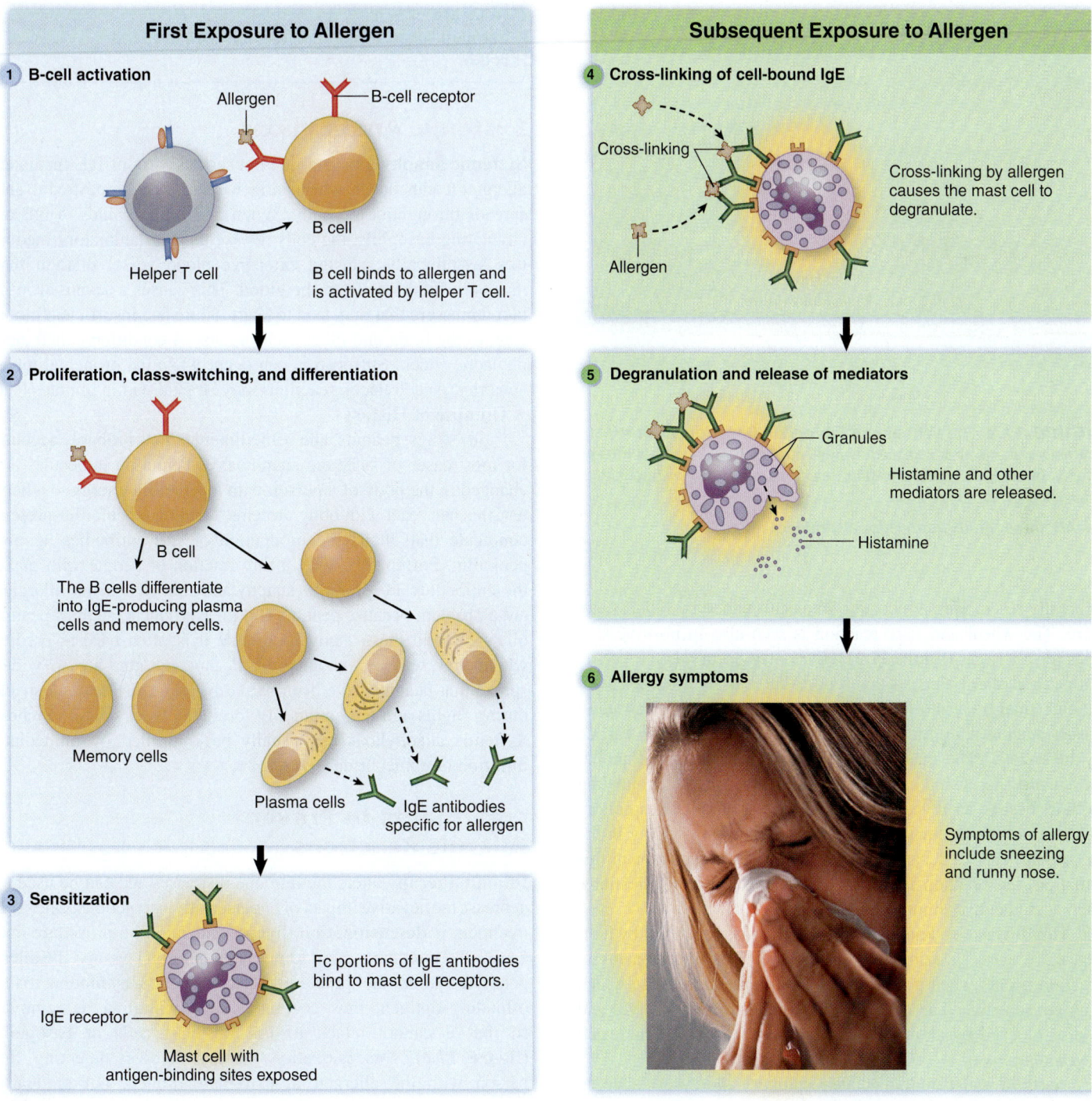

First Exposure to Allergen

1 **B-cell activation**

Allergen

B-cell receptor

B cell

Helper T cell

B cell binds to allergen and is activated by helper T cell.

2 **Proliferation, class-switching, and differentiation**

B cell

The B cells differentiate into IgE-producing plasma cells and memory cells.

Memory cells

Plasma cells

IgE antibodies specific for allergen

3 **Sensitization**

Fc portions of IgE antibodies bind to mast cell receptors.

IgE receptor

Mast cell with antigen-binding sites exposed

Subsequent Exposure to Allergen

4 **Cross-linking of cell-bound IgE**

Cross-linking

Cross-linking by allergen causes the mast cell to degranulate.

Allergen

5 **Degranulation and release of mediators**

Granules

Histamine and other mediators are released.

Histamine

6 **Allergy symptoms**

Symptoms of allergy include sneezing and runny nose.

FIGURE 17.1 Type I Hypersensitivity: Immediate IgE-Mediated First exposure to an allergen induces an IgE antibody response, leading to sensitization. With subsequent exposures to the allergen, cross-linking of IgE molecules on mast cells results in release of histamine and other substances that can cause itching, swelling, and pain, and conditions such as asthma, hay fever, and anaphylactic shock.

Why are antihistamines sometimes used to combat allergy symptoms?

Localized Allergic Reactions

Localized allergic reactions often occur in the skin or in the lungs. Typical responses in the skin include swelling due to fluid leaking from capillaries that dilate as a result of histamine release from

mast cells. Responses in the lung may also involve constriction of smooth muscle that can interfere with gas exchange across the respiratory surface.

Hives (urticaria) is an allergic skin condition characterized by the formation of a wheal and flare; the wheal is an itchy swelling

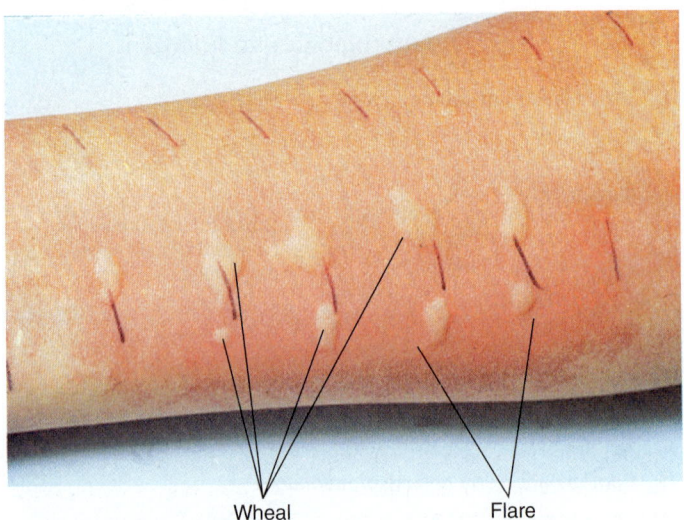

Wheal   Flare

FIGURE 17.2 The Wheal and Flare Skin Reaction A variety of antigens are injected or placed in small cuts in the skin to test for sensitivity. Immediate wheal and flare reactions occur with antigens to which the person is sensitive.

? *Are these results positive for all antigens tested?*

generally resembling a mosquito bite, surrounded by redness, the flare. The wheal and flare reaction is seen also in positive skin tests for allergens (**figure 17.2**). Hives may occur when a person with a food allergy ingests that food. Allergens are absorbed from the intestinal tract into the bloodstream where they are carried to tissues such as skin. In the skin, the allergens bind to IgE carried by mast cells. The mast cells degranulate, releasing inflammatory mediators, which causes swelling due to fluid accumulating in skin tissues as it leaks from dilated capillaries. Respiratory obstruction may occur if there is extensive tissue swelling in the throat and larynx. Because histamine is a major mediator in this situation, the reaction is blocked by antihistamine medications, such as diphenhydramine (Benadryl).

Hay fever, also known as allergic rhinitis, is marked by itching, teary eyes, sneezing, and runny nose. This hypersensitivity reaction occurs when people inhale an airborne antigen to which they are sensitive. The mechanism is similar to that of hives and is also blocked by antihistamines or drugs that inhibit mast cell degranulation.

Asthma is also an immediate respiratory allergy. An allergen reacting with IgE-sensitized mast cells causes their degranulation and release of inflammatory mediators into the lower respiratory system. The mediators cause spasms of the bronchial tubes, which interfere with breathing. Mediators other than histamine, mainly lipids such as leukotrienes and prostaglandins, are responsible for bronchospasm and increased mucus production. Eosinophils also contribute to an inflammatory response. Antihistamines are not effective in treating asthma, but several other drugs are available to block the reaction, including bronchodilating drugs that relax constricted muscles and relieve the bronchospasm. These are often self-delivered at the onset of an attack through an inhaler (see the chapter-opening photo). Cortisone-like steroids are often used to decrease the inflammatory reaction over a longer period of time. An anti-IgE therapy (described later) may also be effective.

MicroByte
The most common airborne allergen in the United States is ragweed pollen.

Systemic Anaphylaxis

Systemic anaphylaxis is a rare but serious form of IgE-mediated allergy. It can occur when antigen enters the bloodstream and spreads throughout the body. When this antigen binds to IgE on circulating basophils, the cells release their inflammatory mediators systemically, causing extensive blood vessel dilation that results in fluid loss from the blood. This causes a severe drop in blood pressure that may lead to heart failure and insufficient blood flow to the brain and other vital organs—a condition called anaphylactic shock. Suffocation can occur when the bronchial tubes constrict. Anaphylactic reactions may be fatal within minutes (see **A Glimpse of History**).

Bee stings, peanuts, and penicillin injections probably account for most cases of systemic anaphylaxis. Penicillin molecules are changed in the body of a person with a penicillin allergy to a hapten; this can react with body proteins, forming a penicillin-protein conjugate that elicits the production of IgE antibodies against penicillin. Fortunately, only a tiny fraction of people who make the antibodies are prone to anaphylactic shock. Peanut allergies are a problem because peanuts and their products (such as peanut oil) are found in so many foods that it is often hard to predict where they will be encountered. Peanut allergies are so widespread that peanuts have been barred by some airlines as a snack during flights and are commonly forbidden from school lunches. Systemic anaphylaxis can usually be controlled by immediate injection of epinephrine. ◀◀ **hapten, p. 373**

Treatments to Prevent Allergic Reactions

Immunotherapy alters the immune responses and can be used to decrease the negative impact of hypersensitivity reactions. One such treatment is **desensitization** (hyposensitization), a procedure that causes the immune system to produce more IgG against the allergen. These antibodies probably protect the patient by binding to the offending antigen, thus coating it and facilitating its removal so that it cannot attach to IgE on mast cells or basophils (**figure 17.3**). Desensitization therapy involves injecting the person with the allergen in extremely diluted, but gradually increasing concentrations over a period of several months. The antigen must be diluted enough to avoid an anaphylactic reaction. As the concentration in the injections increases during the course of treatment, the individual becomes less and less sensitive and may even lose the hypersensitivity entirely. Concurrently, antibodies may allow IgG to outcompete IgE antibodies for antigen binding sites, neutralizing the antigen before it contacts sensitized mast cells. Activated regulatory T cells may also play a role through release of cytokines that suppress the IgE response. ◀◀ **IgG, p. 361** ◀◀ **regulatory T cells, p. 356**

Another allergy treatment is to prevent IgE antibodies from binding to mast cells and basophils. A medication called omalizumab is a genetically engineered form of an IgG molecule that binds specifically to the Fc portion of IgE molecules, thereby

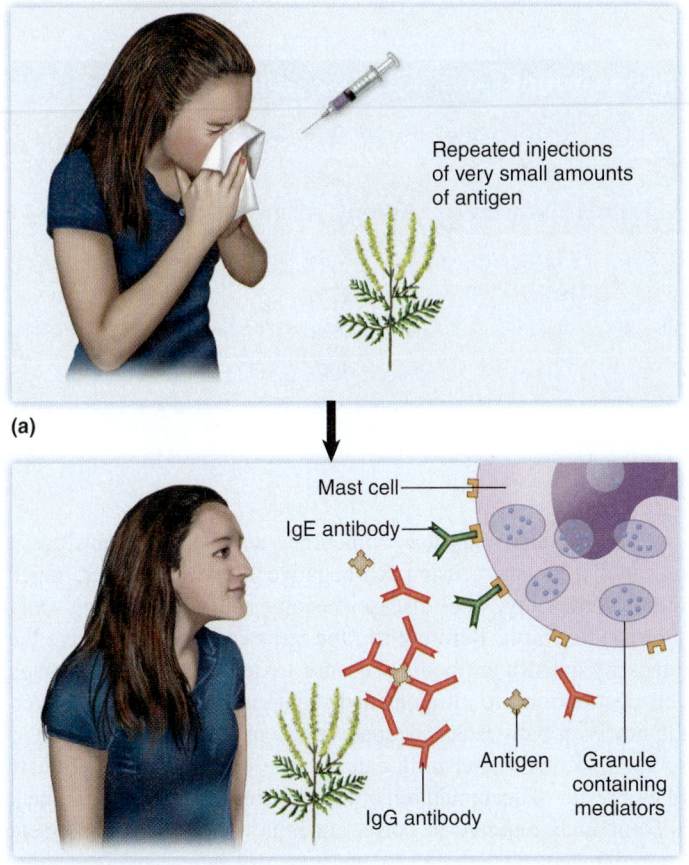

(a)

(b)

FIGURE 17.3 Immunotherapy for IgE Allergies (a) Repeated injections of very small amounts of antigen are given over several months. **(b)** This regimen leads to the formation of specific IgG antibodies. The IgG reacts with antigen before it can bind to IgE, and therefore it blocks the IgE reaction.

❓ *Why are only very small amounts of antigen injected?*

blocking the site that would otherwise attach to mast cells and basophils. Injections of omalizumab are effective in treating most asthma cases, reducing the need for more toxic medications. However, the medication is expensive and can cause anaphylactic shock on rare occasions, so it is generally used only for the most serious asthma cases. It is an example of a **rhuMab** (**r**ecombinant **hu**manized **M**onoclonal **a**nti**b**ody) used in treating various other conditions.

MicroAssessment 17.1

Hypersensitivity reactions are immunological reactions that cause tissue damage. Type I hypersensitivity reactions mediated by cell-bound IgE antibodies occur immediately after exposure to antigen. The reactions are caused by the release of histamine and other inflammatory mediators from cell granules. Localized allergic reactions include hives, hay fever, and asthma; systemic reactions can lead to anaphylactic shock. Immunotherapy is used to induce IgG rather than IgE-mediated responses.

1. *How do localized allergic and systemic anaphylactic reactions differ?*
2. *Define* allergen, *and give five common examples of substances that act as allergens.*
3. *Why does leaking of plasma from blood vessels during an allergic reaction cause a wheal?* ✚

17.2 ■ Type II Hypersensitivities: Cytotoxic

Learning Outcomes

4. *Explain the mechanisms of lysis of red blood cells in transfusion reactions.*
5. *Explain the cause of hemolytic disease of the newborn and how it can be prevented.*

In type II hypersensitivity reactions, antibodies react with molecules on the surface of a cell and trigger its destruction by the complement system or by antibody-dependent cellular cytotoxicity (ADCC). Recall that antibodies bound to an antigen trigger the classical pathway of complement activation, and one outcome of activation is cell lysis due to membrane-attack complexes (MACs) assembling in cell membranes. In ADCC, the Fc regions of antibodies bound to a cell mark the cell for destruction. Natural killer (NK) cells bind to the Fc regions and then deliver chemicals that destroy the marked cell. Because the responses destroy cells, type II reactions are called cytotoxic hypersensitivities. Examples include transfusion reactions and hemolytic disease of the newborn. Some autoimmune diseases also involve type II reactions. ◄◄ complement system, p. 344 ◄◄ antibody-dependent cellular cytotoxicity (ADCC), p. 361

Transfusion Reactions

Erythrocytes (red blood cells) have various antigenic determinants on their surface that can differ from one person to another. A major blood group system is reflected in ABO blood types. People have blood type A, B, AB, or O, depending on which specific carbohydrate antigens are present on their red blood cells; they have antibodies to the antigens not present on their own red blood cells. For example, people who have type A antigens are blood type A; they have anti-B antibodies. Those who have type B antigens are blood type B and have anti-A antibodies. People with blood type O lack A and B antigens, but have antibodies to both. People with blood type AB have both A and B antigens, and have antibodies to neither.

Anti-A and anti-B antibodies are called natural antibodies because they occur without obvious pre-exposure. They probably arise due to multiple exposures to substances similar to blood group antigens that occur in environmental materials such as bacteria, dust, and food. Anti-A and anti-B antibodies are not present at birth but generally appear within 6 months. They are mostly of the class IgM, and therefore cannot cross the placenta. ◄◄ IgM, p. 361

When an individual receives a transfusion of red blood cells with different antigens from his or her own, a **transfusion reaction** occurs when natural antibodies bind to those cells, resulting in agglutination (clumping). The transfused cells are rapidly destroyed by membrane attack complexes or NK cells. Release of hemoglobin into the bloodstream and the presence of damaged membranes can block blood vessels and initiate clotting reactions, damaging kidneys and producing fever along with respiratory and digestive problems. Because of the potentially life-threatening

TABLE 17.2	Antigens and Antibodies in Human ABO Blood Groups				
			Incidence of Blood Type in United States		
Blood Type	Antigen Present on Erythrocyte Membranes	Antibody in Plasma	Among Whites	Among Asians	Among Blacks
A	A	Anti-B	41%	28%	27%
B	B	Anti-A	10%	27%	20%
AB	A and B	Neither anti-A nor anti-B	4%	5%	7%
O	Neither	Anti-A and anti-B	45%	40%	46%

consequences of giving the wrong blood to a patient, donor and recipient blood types are carefully cross-matched in clinical settings. The cells and serum of the donor are mixed with the cells and serum of the recipient. Any clumping indicates an incompatibility between the two blood types, either in the ABO group **(table 17.2)** or in another of the many antigens on the surface of the red blood cell. That blood would not be used in the transfusion.

Hemolytic Disease of the Newborn

Another important antigen on red blood cells is the Rh (rhesus) antigen. People with the Rh antigen are Rh-positive; those without are Rh-negative. In contrast to the ABO system, Rh-negative individuals do not have natural antibodies to the Rh antigen. These antibodies develop only when the individual is exposed to Rh-positive cells. Just as donor and recipient bloods are matched for the ABO system, they are also matched for Rh system antigens. If Rh-positive blood is given in error, an Rh-negative recipient will then develop antibodies to the Rh-positive red blood cells. If the person is exposed to Rh-positive blood again, his or her immune system will destroy the transfused cells. A more common event that exposes an Rh-negative person to Rh-positive cells occurs during pregnancy.

An Rh-negative woman is likely to become sensitized to the Rh antigen during childbirth if she carries an Rh-positive baby. Entry of Rh-positive cells into the mother's bloodstream will cause her to develop antibodies to the Rh antigen. Sensitization usually occurs as the mother's blood is exposed to the baby's blood when the placenta ruptures during childbirth, but it can also occur during pregnancy, or after induced or spontaneous abortion. Antibodies produced by the woman will not affect her red blood cells because her cells lack the antigen. They will not affect her first baby because IgM, which is produced upon first exposure, cannot cross the placenta. If the woman carries a second Rh-positive baby, however, her anti-Rh IgG antibodies can cross the placenta and damage her developing fetus, causing **hemolytic disease of the newborn** **(figure 17.4)**. This is also called erythroblastosis fetalis since the affected fetus responds by producing immature red blood cells called erythroblasts.

To save the fetus it is sometimes necessary to transfuse it repeatedly *in utero,* using Rh-negative blood. However, much more commonly, the disease becomes life-threatening only shortly after birth. Before birth, the fetus can usually survive the attack by anti-Rh antibodies because toxic products of red blood cell destruction are eliminated by maternal enzymes. However, the newborn baby lacks adequate levels of these enzymes, and as soon as 36 hours after birth can become jaundiced and seriously ill as a result of accumulation of these toxins. Immediate treatment is sometimes required to correct anemia and prevent permanent brain damage.

Hemolytic disease of the newborn is rare today because of a medication called RhoGAM, which contains anti-Rh antibodies. Rh-negative women who are carrying an Rh-positive fetus are injected with anti-Rh antibodies during pregnancy and again shortly after delivery. The anti-Rh antibodies bind to any Rh-positive erythrocytes that may have entered the mother's circulation from the baby. This prevents these red blood cells from stimulating a primary immune response with the development of memory cells. Unfortunately, injecting anti-Rh antibody is not effective if the Rh-negative mother is already sensitized and has formed memory cells. ◄◄ **primary immune response, p. 365**

MicroByte

Phototherapy converts toxic erythrocyte breakdown products to a form that is more readily removed by the kidneys and excreted.

MicroAssessment 17.2

Type II cytotoxic hypersensitivity reactions are mediated by antibodies, either by lysis of cells via the complement system or by antibody-dependent cellular cytotoxicity. Blood transfusion reactions and hemolytic disease of the newborn are examples.

4. *Describe the mechanism of cell damage in a blood transfusion where ABO antigens are mismatched.*

5. *Why do Rh-negative but not Rh-positive mothers sometimes have babies with hemolytic disease of the newborn?*

6. *Why is it surprising that people who lack the A antigen on their red blood cells have antibodies to that antigen?* ✛

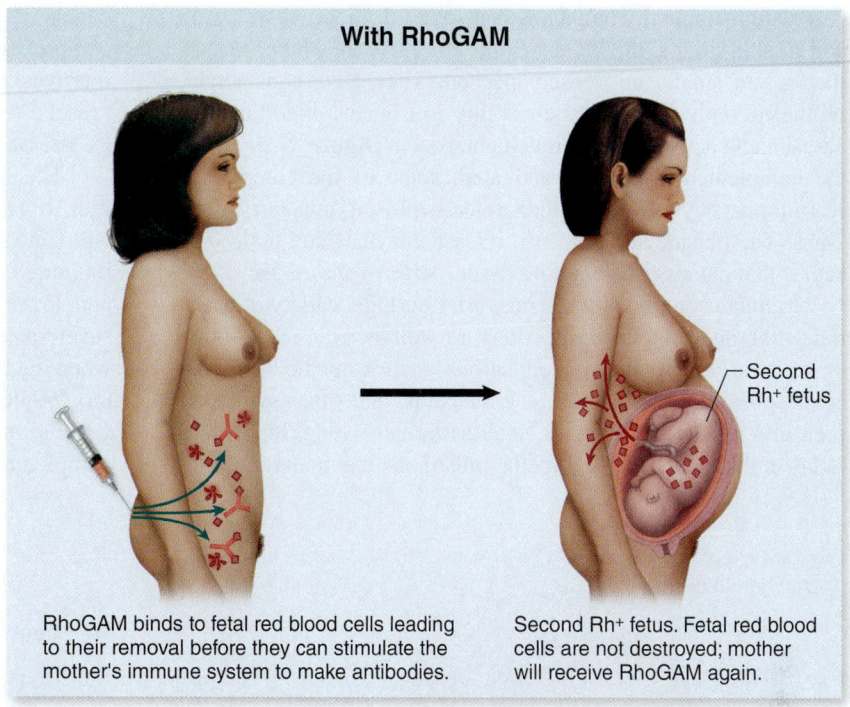

With RhoGAM

RhoGAM binds to fetal red blood cells leading to their removal before they can stimulate the mother's immune system to make antibodies.

Second Rh+ fetus. Fetal red blood cells are not destroyed; mother will receive RhoGAM again.

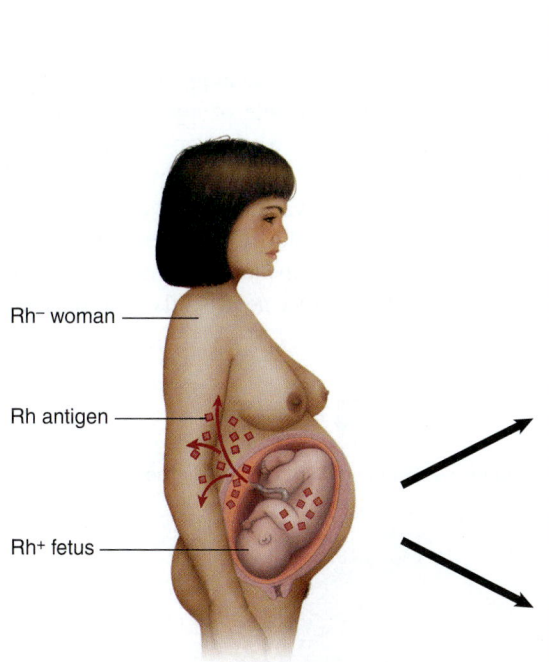

Rh⁻ woman

Rh antigen

Rh⁺ fetus

First Rh+ fetus. Baby's red blood cells with Rh+ antigen enter mother's circulation during birth.

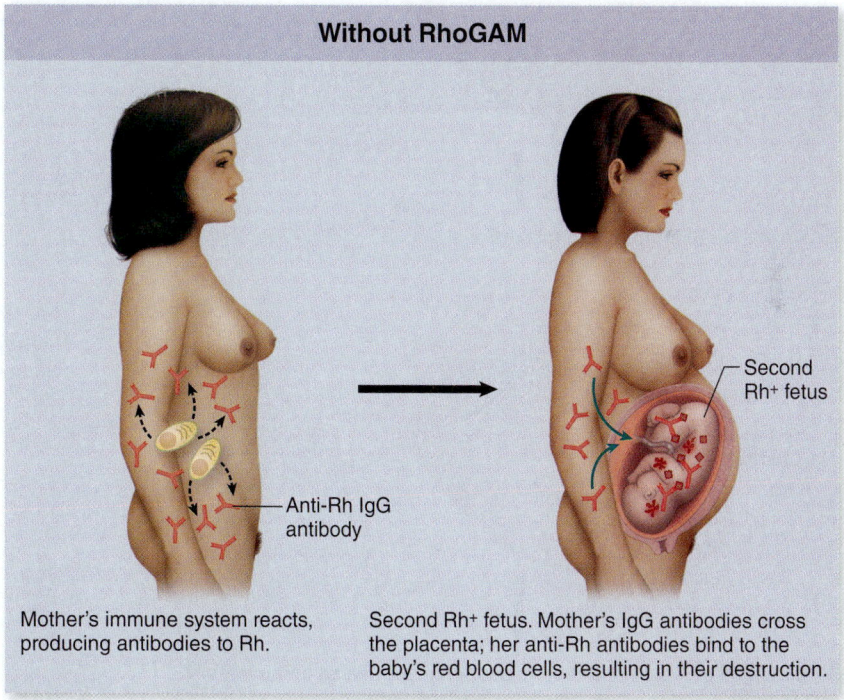

Without RhoGAM

Anti-Rh IgG antibody

Mother's immune system reacts, producing antibodies to Rh.

Second Rh+ fetus. Mother's IgG antibodies cross the placenta; her anti-Rh antibodies bind to the baby's red blood cells, resulting in their destruction.

FIGURE 17.4 Hemolytic Disease of the Newborn RhoGAM contains antibodies to Rh antigens and removes them from the mother's blood before her immune system can respond to them.

❓ *Would a mother need RhoGAM again with a third Rh⁺ fetus?*

17.3 ■ Type III Hypersensitivities: Immune Complex-Mediated

Learning Outcomes

6. *Describe the role of immune complexes in type III hypersensitivity reactions.*
7. *Compare and contrast the Arthus reaction with serum sickness.*

Type III hypersensitivities are caused by the formation of **immune complexes** consisting of IgG or IgM antibodies bound to soluble antigen. Generally, these complexes are rapidly removed from the bloodstream, particularly if they are large. This occurs because the Fc receptors on phagocytes bind the Fc regions of the antibodies, allowing the phagocytes to engulf and destroy the complexes. If there is slightly more antigen than antibody, however, smaller complexes form. These are not quickly destroyed but remain in circulation or at their sites of formation in tissue.

Circulating immune complexes can lodge in blood vessel walls and in skin, joints, the kidneys, and other tissues where the capillaries are small and densely packed. They have considerable biological activity due to their ability to trigger a blood-clotting cascade and activate the complement system (**figure 17.5**). When the complement system is activated, some of the components recruit phagocytes, which then release pro-inflammatory cytokines. The phagocytes may also release enzymes and toxic molecules that damage surrounding tissue. ◄◄ **Fc receptors, p. 390**

Immune complex disease may arise during a variety of bacterial, viral, and protozoan infections, as well as from inhaled dust or bacteria, and injected medications such as penicillin. They are responsible for the rashes, joint pains, and other symptoms seen in a number of diseases, such as farmer's lung, lupus, bacterial endocarditis, early rubella infection, and malaria. When deposited in the kidneys, they cause glomerulonephritis. Immune complexes can also trigger a devastating condition, **disseminated intravascular coagulation,** in which clots form in small blood vessels, leading to failure of vitalorgans. ◄◄ ►► **glomerulonephritis, pp. 395, 490** ►► **disseminated intravascular coagulation, p. 674**

The **Arthus reaction** is a localized immune complex reaction. If antigen is injected into a previously immunized person who already has high levels of circulating specific antibody, immune complexes can form at the site of injection where antigen levels are high. The result is a reaction characterized by severe pain, swelling, and redness. This can happen, for example, when tetanus-diphtheria booster vaccine is given to a person too frequently. The immune complexes form outside the blood vessels, in the skin tissues, and activate complement, producing complement components that attract neutrophils. The release of

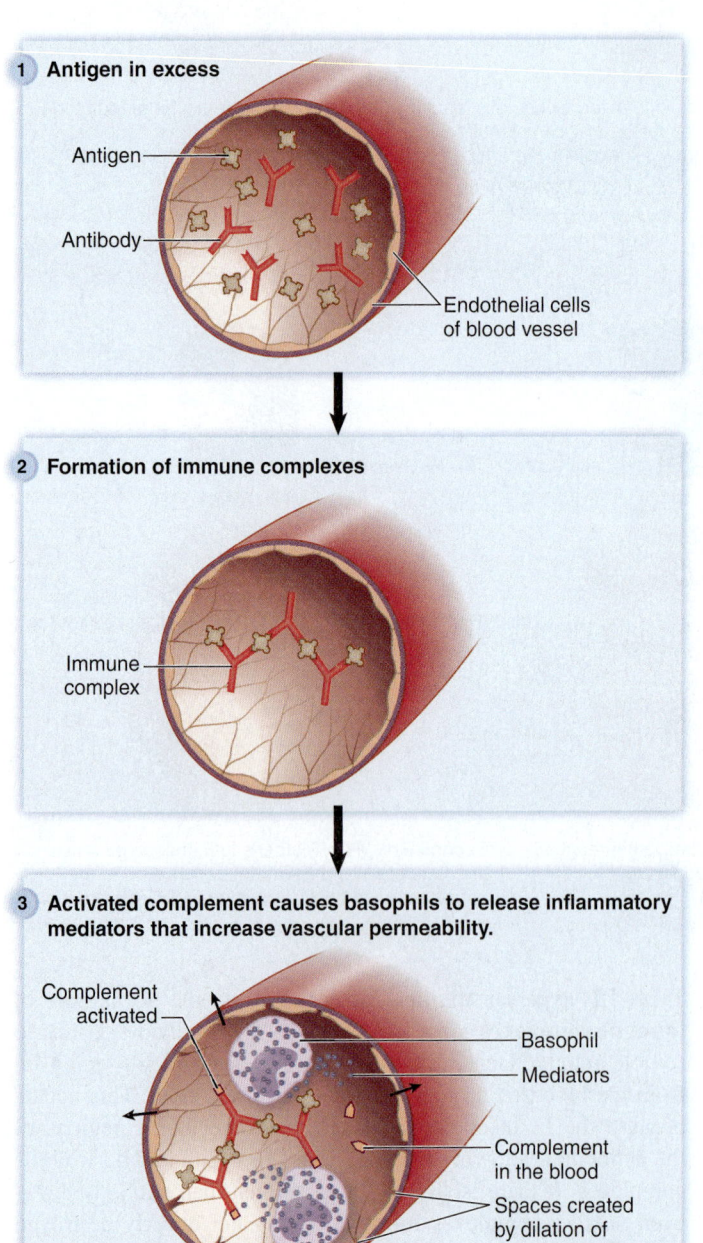

1. **Antigen in excess**

   Antigen

   Antibody

   Endothelial cells of blood vessel

2. **Formation of immune complexes**

   Immune complex

3. **Activated complement causes basophils to release inflammatory mediators that increase vascular permeability.**

   Complement activated

   Basophil

   Mediators

   Complement in the blood

   Spaces created by dilation of blood vessel

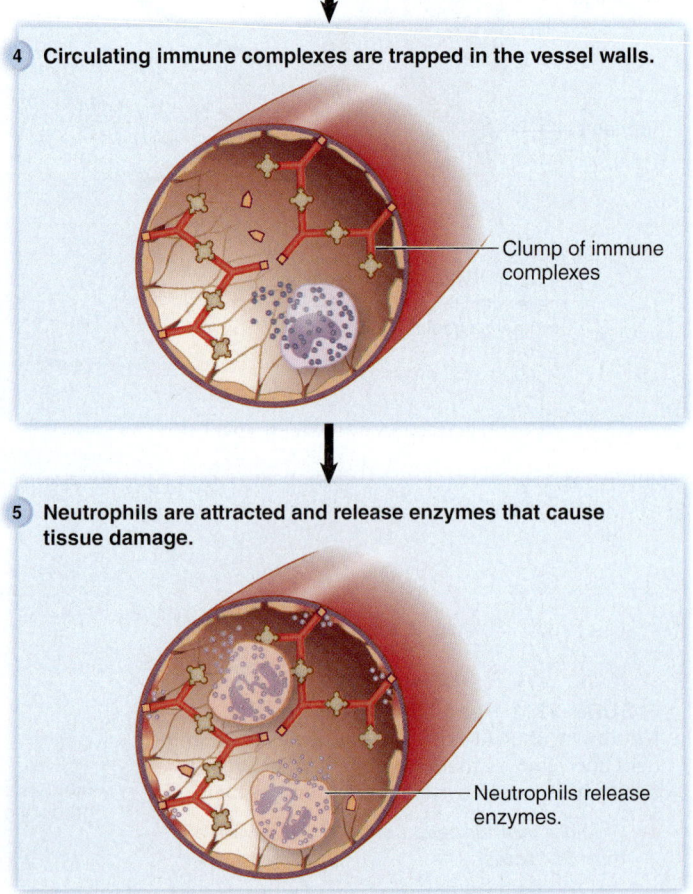

4. **Circulating immune complexes are trapped in the vessel walls.**

   Clump of immune complexes

5. **Neutrophils are attracted and release enzymes that cause tissue damage.**

   Neutrophils release enzymes.

FIGURE 17.5 Type III Hypersensitivity: Immune Complex-Mediated When antigen is in slight excess **(1)**, immune complexes form and activate complement **(2)**, resulting in increased vascular permeability **(3)**. Immune complexes are trapped in the blood vessels **(4)** and complement attracts neutrophils that release enzymes, damaging the tissue **(5)**.

❓ *Why do immune complexes form only when an antigen is soluble?*

neutrophil enzymes contributes to a local inflammatory response that peaks in 6 to 12 hours.

Serum sickness is a systemic immune complex disease caused by passive immunization (antibodies provided as a source of protection). When an antibody-containing serum from a horse or other animal is injected into humans to prevent or treat a disease, the recipient may mount an immune response to antigens in the foreign serum. After 7 to 10 days, enough immune complexes form to cause signs such as fever, inflammation of blood vessels, arthritis, and kidney damage. People with serum sickness usually recover as the antigens of the animal serum are cleared from the body. Because of the potential to cause serum sickness, horse serum is no longer used for passive immunization against diphtheria and tetanus; instead, hyperimmune human globulin is used. The serum sickness form of hypersensitivity is rare now, but can occur following treatment of heart attack patients with the bacterial enzyme streptokinase to dissolve clots, or after injection of horse-derived anti-venom to treat snakebites. ▶▶ **passive immunity, p. 420** ▶▶ **hyperimmune globulin, p. 420**

MicroByte ─────────────
> Like farmer's lung that develops from inhaling moldy hay, librarian's lung can develop in response to inhaled mold and dust from books.

MicroAssessment 17.3

Type III immune complex–mediated hypersensitivities arise when small complexes of antigen-antibody persist in the tissues and activate the complement system, triggering an inflammatory response.

7. *How are immune complexes normally cleared from the body?*
8. *How can immune complexes cause tissue damage?*
9. *Why are the kidneys particularly vulnerable to immune complex damage?* ✚

17.4 ■ Type IV Hypersensitivities: Delayed-Type Cell-Mediated

Learning Outcomes

8. *Describe the key immunologic reactions involved in type IV hypersensitivities.*
9. *Outline the role of type IV hypersensitivity reactions in contact dermatitis or contact allergies.*

Delayed-type hypersensitivity reactions (type IV hypersensitivities) are due to antigen-specific T-cell responses and can occur almost anywhere in the body. Whereas immediate hypersensitivities involve antibodies and occur within minutes, these cell-mediated reactions peak 2 to 3 days following antigen exposure. ◀◀ **cell-mediated immunity, p. 355**

Tuberculin Skin Test

The tuberculin skin test, which can be used to detect latent *Mycobacterium tuberculosis* infections, relies on a delayed-type hypersensitivity reaction. When a very small quantity of

mycobacterial protein antigens are injected into the skin, people who are infected with the organism will exhibit a delayed-type hypersensitivity reaction. The site of injection reddens and gradually becomes indurated (thickened) within 6 to 24 hours, reaching a peak in 2 to 3 days (**figure 17.6**). No wheals (as seen in IgE-mediated reactions) form. The reaction results mainly from effector helper T cells that recognize the antigens and then release pro-inflammatory cytokines. The inflammatory response that characterizes a positive skin test also occurs in response to the infecting organisms, often resulting in the formation of tubercles in the lung.

Delayed-Type Hypersensitivity in Infectious Diseases

Recall from chapter 15 that cell-mediated immunity plays a central role in combating intracellular microbial infections. Effector cytotoxic T cells are particularly important because they destroy the infected host cells. Although this prevents the infection from spreading, it also damages the tissues. With chronic infections, delayed-type hypersensitivity causes extensive host cell destruction and progressive loss of tissue function, as seen in the damaged sensory nerves of leprosy. The immune response is indeed a two-edged sword, with a delicate balance between protective immunity and harmful hypersensitivity. ◀◀ **cell-mediated immunity, p. 355** ▶▶ **leprosy, p. 650**

Contact Hypersensitivities

Contact hypersensitivity, also called **contact dermatitis** or contact allergy, is due to effector T cells responding to small molecules that penetrate intact skin. The mechanisms of tissue damage are similar to those of other types of delayed-type hypersensitivities,

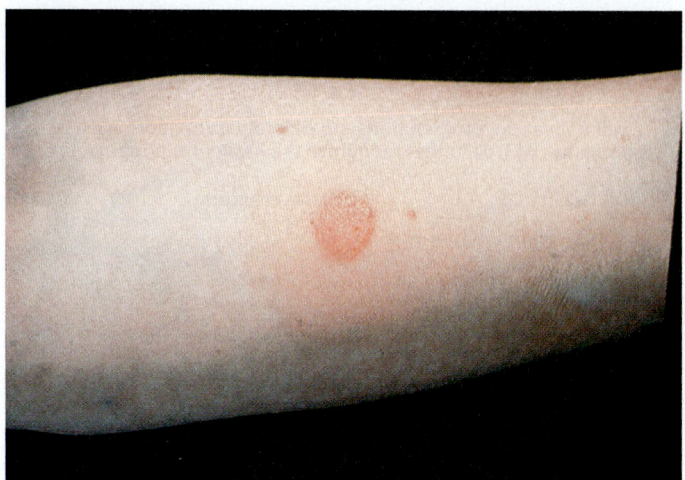

FIGURE 17.6 Positive Skin Test for Delayed-Type Hypersensitivity Reaction Injection of tuberculin protein into the skin of a person sensitized to *Mycobacterium tuberculosis* causes a firm red raised area to form by 48 to 72 hours. A reaction greater than 10 mm in diameter is considered positive.

❓ *How many days after a tuberculin skin test should results be evaluated?*

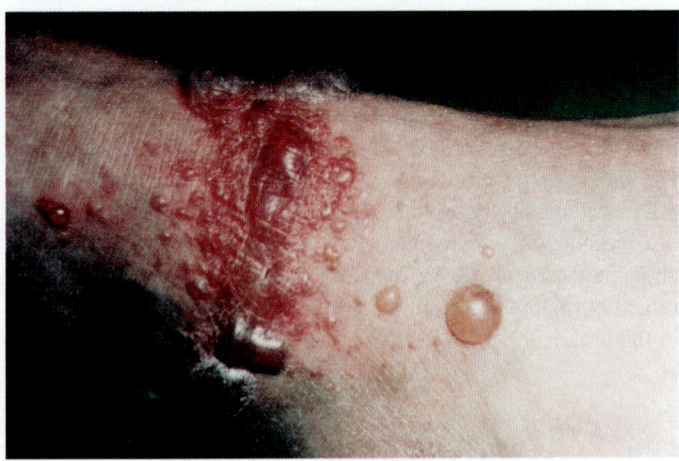

FIGURE 17.7 Severe Contact Hypersensitivity Note the redness and blisters.

❓ *What type of immune cells cause this reaction?*

▶ **FIGURE 17.8 Poison Oak Dermatitis**
This skin reaction results from delayed-type hypersensitivity.

❓ *Why does a sensitized person not experience a skin reaction immediately after exposure to poison oak?*

but damage to skin cells causes an irritating rash and sometimes blisters (**figure 17.7**).

Familiar substances like the nickel of metal jewelry, the chromium salts in certain leather products, or components of some cosmetics are common culprits in contact dermatitis. An oily product of poison ivy and poison oak is another example of a substance that causes the reaction (**figure 17.8**). All of these materials shed small chemicals that act as haptens and combine chemically with proteins in the skin. Dendritic cells take up and process these hapten:protein complexes and present the resulting hapten:peptides to T cells in nearby lymph nodes. T cells that become activated form memory cells, sensitizing the individual to the substance. Once sensitized, a second exposure causes a damaging cell-mediated response.
◀◀ hapten, p. 373

Latex products are a frequent cause of hypersensitivity reactions. Latex, a product of the rubber tree, contains a plant protein that readily induces sensitization.

A hapten from poison oak combines with skin proteins.

First exposure

Subsequent exposures activate memory cells

① **Dendritic cells present the hapten:peptide complex to T cells.**

Hapten:peptide

Dendritic cell   Naive helper   Naive cytotoxic
                          T cell   T cell

② **T-cell activation, proliferation, and differentiation increases the population of T cells that recognize the hapten:peptide complex.**

Naive helper   Naive cytotoxic
T cell   T cell

Effector
T cells

Memory
cells

③a **Helper T cells activate macrophages presenting hapten:peptide complex.**

Macrophage

Cytokines   Helper T cell

③b **Cytotoxic T cells destroy skin cells presenting hapten:peptide complex.**

Skin cell

Cytotoxic T cell

④ **Activated macrophages release inflammatory mediators.**

Macrophage

Inflammatory mediators

⑤ **A rash appears after 24 hours, reaching a peak at 48–72 hours after exposure.**

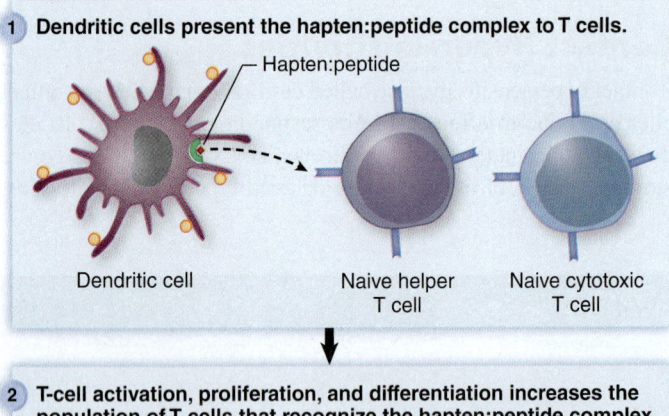

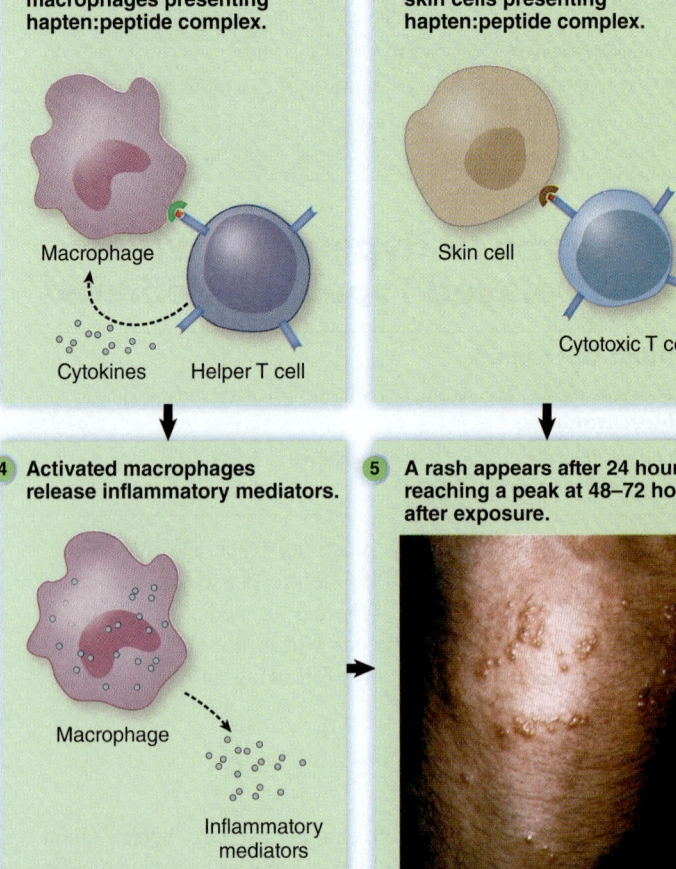

Many products contain latex, such as fabrics, elastics, toys, and contraceptive condoms, but latex gloves probably account for most latex sensitization. Gloves are used extensively by healthcare and laboratory workers, food preparers, and many others. Typically, a person will notice redness, itching, and a rash on the hands after wearing gloves. To prevent the reaction, latex gloves should be replaced by vinyl or other synthetic gloves. Topical cortisone-like medications that inhibit T-cell responses are effective treatments.

The causative substance in contact hypersensitivity is commonly detected by patch tests, in which suspect substances are applied to the skin under an adhesive bandage. Positive reactions consisting of redness, itching, and blisters reach their peak in about 3 days.

MicroByte

Approximately half of the people with latex allergies are allergic to bananas as well.

MicroAssessment 17.4

Type IV delayed-type hypersensitivity depends on the action of sensitized T cells. The reaction peaks 2 to 3 days after exposure to antigen. Examples are contact dermatitis, damage in a variety of infectious diseases, rejection of tissue grafts, and some autoimmune diseases.

10. *Explain the events that occur in the skin during a positive delayed-type hypersensitivity skin test.*

11. *Describe a patch test for contact hypersensitivity.*

12. *In the tuberculin skin test, why does a positive result indicate prior exposure to* M. tuberculosis?

17.5 ■ Rejection of Transplanted Tissues

Learning Outcomes

10. *Explain the process of rejection of transplanted tissues.*

11. *Describe the mode of action of medications used to prevent rejection of transplanted tissue.*

A special case of delayed-type cell-mediated hypersensitivity occurs with the rejection of transplanted organs or tissues. Most human transplants involve **allografts** in which the tissues of the donor and recipient are not genetically identical. Antigenic differences, particularly in the major histocompatibility complex (MHC) molecules, lead to rejection of the graft (see Perspective 15.1). Larger differences result in stronger immune reactions. Certain types of transplants, including autografts (tissue transplanted from elsewhere on the recipient's body) and isografts (grafts donated by an identical sibling), avoid these problems. Xenografts, involving tissues from pigs or non-human animals, typically evoke vigorous immune responses, but may become more widely used as we gain the ability to genetically engineer animals that express human MHC molecules. **Perspective 17.1** discusses a very special transplantation situation: the fetus as an allograft. MHC molecules, p. 369

Transplant rejection is predominantly a cell-mediated immunological reaction, involving a complex set of events centered around the actions of effector cytotoxic T cells and natural killer (NK) cells. To avoid rejection, antigen incompatibilities are minimized by using MHC tissue typing and ABO blood typing to match donor and recipient tissues. Even in well-matched tissue transplants, however, recipients must use immunosuppressive drugs indefinitely to prevent graft rejection. These drugs are needed to prevent immune reaction to the many minor antigens present on allograft cells. cytotoxic T cells, p. 357 NK cells, p. 374

Immunosuppressive drugs minimize the rejections, but they make the recipient more susceptible to infections and cancer. Cyclosporin A (produced by a fungus) and tacrolimus (produced from a species of *Streptomyces*) are both effective immunosuppressants. These drugs interfere with cellular signaling, thereby inhibiting clonal selection and expansion of activated T lymphocytes. They suppress only T-cell proliferation, so the patient still has some level of immunity mediated by other types of immune cells. clonal selection and expansion, p. 363

Many thousands of bone marrow transplants have saved the lives of individuals with cancers such as multiple myeloma and leukemia, and other serious conditions. Generally, the patient's bone marrow and immune system are intentionally destroyed by intense radiation and chemotherapy before transplantation of the new bone marrow. Donor bone marrow is given to the recipient by infusing it through the large subclavian vein under the left collar bone. The marrow hematopoietic stem cells circulate until they reach the recipient's bone marrow, where they leave the bloodstream and begin producing hematopoietic cells. Close monitoring is required for months after a transplant because the recipient is prone to infection and bleeding while the grafted marrow restores the supply of erythrocytes, leukocytes, and platelets. Recovery may be complicated by a graft-versus-host reaction in which the graft-derived immune system recognizes the patient's tissues as an invader and attacks them. Therefore, many of these patients also require immunosuppressive treatment. hematopoietic stem cells, p. 338

MicroByte

Hematopoietic stem cells can sometimes be filtered from donor blood without the use of general anesthetic or punctures into the marrow.

MicroAssessment 17.5

Successful organ and tissue transplantation depend on matching major histocompatibility antigens and using immunosuppressive agents such as cyclosporin and tacrolimus to minimize the immune response to other antigens of the graft. Rejection of transplants is complex, but type IV cellular immune responses are the major mechanism of rejection for allografts.

13. *What differences make an allograft more likely to be rejected than an autograft?*

14. *How is graft-versus-host disease different from a recipient's rejection of donor tissue?*

15. *Why is the recipient of a transplanted kidney unlikely to suffer from a graft-versus-host reaction?*

The Fetus as an Allograft

One allograft that does not face the usual threat of rejection is a very special one—the mammalian fetus. Half of the fetal antigens are of paternal origin and are likely to differ from those of the mother. In spite of immunological differences, the fetus develops in the uterus for 9 months and is not rejected. The mechanisms for survival of the fetal allograft have long been the subject of research, but they are not yet fully understood.

Do paternal antigens from the fetus reach the mother's immune system where they might cause a response? Clearly, they do. Mothers make antibodies to paternal antigens, such as the Rh antigen on red blood cells. Antibodies used for typing major histocompatibility antigens can be obtained from women who have had several children with the same father—they have made antibodies to his MHC molecules. Furthermore, small numbers of fetal cells are found in the maternal circulation during pregnancy. The placenta, however, prevents most fetal cells from entering the mother and most maternal T cells from reaching the fetus.

The fetus is protected in the uterus, an immunologically privileged site. The outer layer of the placenta, the trophoblast, acts as a buffer zone between the fetus and the mother. The trophoblast does not express MHC class I or II molecules and is not subject to T-cell attack. It also avoids destruction by natural killer cells. A few other areas in the body—the brain, the eyes, and the testes—are also immunologically privileged sites. Antigens leaving these sites do not drain through lymphatic vessels and reach lymphoid tissues where antigen-presenting cells are abundant. Also, antigen leaves these privileged sites accompanied by immunosuppressive cytokines that direct the immune response toward tolerance rather than destruction. It is well recognized that maternal immune responses are suppressed to some extent during pregnancy, though the reasons for this are not clear.

17.6 ■ Autoimmune Disease

Learning Outcomes

12. *Define* autoimmunity, *and explain some possible causes.*
13. *Give five examples of autoimmune diseases and the mechanism of tissue injury in each.*

The immune system generally eliminates or silences developing lymphocytes that would respond to and attack normal body tissues, or autoantigens ("self" antigens). **Autoimmune disease** results when the system fails to do this, resulting in attacks against autoantigens as if they were invaders.

The causes of autoimmune disease are not clear and probably involve many things. A deficiency in the action or control of regulatory T cells (T_{reg}) that normally inhibit responses to "self" may be at fault. A genetic component, perhaps related to major histocompatibility molecules, may be a predisposing factor, but disease occurrence is also influenced by environmental factors, including infection. In some cases, a pathogen's ability to evade the immune system by mimicking "self" likely plays a role. As a result of this molecular mimicry, the immune response mounted against the pathogen attacks healthy tissues as well, causing inflammation and damage even in the absence of the pathogen. Autoimmune responses may also occur after injury in which self antigens are released from privileged (isolated) or damaged organs and stimulate production of antibodies to them (autoantibodies). ◄◄ regulatory T cells, p. 356

The Spectrum of Autoimmune Diseases

Autoimmune diseases can be organ-specific or systemic, depending on whether the autoantigens are localized to a specific organ or are found throughout the body (**table 17.3**). In both situations, however, the damage may be caused by antibodies, cell-mediated immune mechanisms, or both. Examples include the following:

■ **Type 1 diabetes mellitus.** Also known as insulin-dependent or juvenile diabetes, this is an organ-specific autoimmune disease caused when cytotoxic T cells destroy a group of pancreatic cells called β cells. These cells produce insulin, a protein that allows various cell types in the body to take up glucose from the bloodstream. The lack of insulin results in increased blood levels of glucose, causing water to be drawn from the cells. In turn, this leads to symptoms of increased thirst and urination. Because the cells do not take in glucose

TABLE 17.3	Characteristics of Some Autoimmune Diseases	
Disease	**Organ Specificity**	**Major Mechanism of Tissue Damage**
Type 1 diabetes mellitus	Pancreas	T-cell destruction of pancreatic β cells
Graves' disease	Thyroid	Autoantibodies bind thyroid-stimulating hormone receptor, causing overstimulation of thyroid
Systemic lupus erythematosus	Systemic	Autoantibodies to DNA and other nuclear components form immune complexes in small blood vessels
Myasthenia gravis	Muscle	Autoantibodies bind to acetylcholine receptor on muscle, preventing muscle contraction
Rheumatoid arthritis	Systemic, especially joints	Lymphocyte destruction of joint tissues; immune complexes of IgG and anti-IgG

that would otherwise be used as their energy source, symptoms also include extreme hunger, fatigue, and weight loss. Persons with diabetes are vulnerable to high blood pressure, stroke, blindness, limb amputation, and kidney disease. Insulin injections are used to treat the disease.

- **Graves' disease.** This is an organ-specific autoimmune disease that affects the thyroid gland. In most cases, antibodies are directed at receptors on the thyroid gland for thyroid-stimulating hormone (TSH). Attachment of the antibody to the receptor activates the receptor inappropriately, leading to increased thyroid hormone production and enlargement of the gland. An enlarged thyroid is sometimes evident as a goiter (**figure 17.9**). Metabolic rate increases in affected individuals, resulting in symptoms that include weight loss, fatigue, irritability, heat intolerance, rapid heartbeat, and bulging eyes.

- **Systemic lupus erythematosus (SLE).** As the name implies, this is a systemic disease. Antibodies made against molecules found in the nuclei of cells provoke an autoimmune response throughout the body. Symptoms vary considerably, but typically include joint pain, swelling in the joints, and rashes. Severe cases involve damage to the kidneys or other organs. Curiously, the disease affects mainly women.

FIGURE 17.9 Goiter A goiter may result from overstimulation of the thyroid gland.

❓ *What receptors in the thyroid gland are stimulated in Graves' disease?*

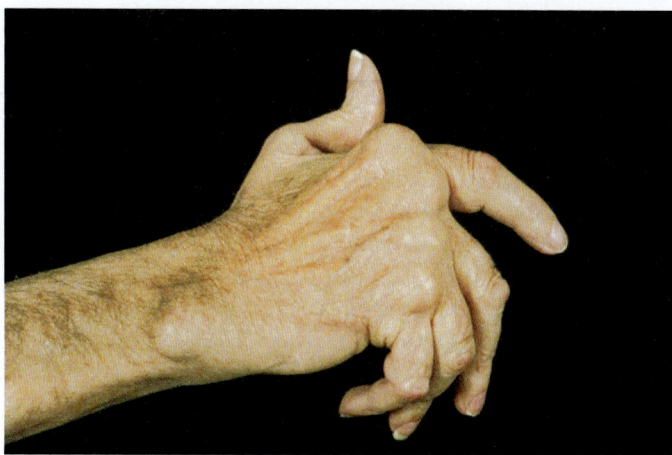

FIGURE 17.10 Rheumatoid Arthritis Autoimmune reactions cause chronic inflammation and destruction of the joints.

❓ *What type of hypersensitivity has damaged the joints in this hand?*

- **Myasthenia gravis.** The name means "grave muscle weakness" and reflects the fact that this systemic disease is caused by a disruption in nerve transmission to muscles. Antibodies bind to acetylcholine receptors at the neuromuscular junction, thereby blocking transmission of impulses that normally cause muscle contraction. Moreover, binding of the antibodies may activate the complement system, which then damages the acetylcholine receptors. Drugs that inhibit the enzyme cholinesterase (the enzyme that degrades acetylcholine) allow acetylcholine to accumulate at the neuromuscular junction and may lessen the effects of the disease. Immunosuppressive medications and removal of the thymus gland are helpful in many cases. The role of the thymus in this disease is not well understood.

- **Rheumatoid arthritis.** This is a systemic disease in which antibodies and cellular mechanisms target collagen in connective tissues, most often within joints (**figure 17.10**). T cells infiltrate the joints and, when stimulated by specific antigens, they release cytokines that cause inflammation. Antibodies to collagen form immune complexes characteristic of type III hypersensitivities that further damage the joint tissue. This crippling inflammatory condition is one of the most common autoimmune diseases, and unlike arthritis that arises from wear and tear of joints, it may occur at any age. Rheumatoid arthritis is most common in women ages 30 to 50. ◀◀ **cytokines, p. 341** ◀◀ **immune complexes, p. 395**

MicroByte
Babies born to mothers with myasthenia gravis experience temporary muscle weakness, since IgG antibodies cross the placenta.

Treatment of Autoimmune Diseases

Autoimmune diseases are usually treated with anti-inflammatory medications or immunosuppressant drugs that interfere with T-cell signaling or kill dividing cells, thereby limiting the immune response. Replacement therapy may be necessary to restore damaged tissues or tissue products, such as when patients with type 1 diabetes receive insulin. Because the risk of infection and cancer increase when the immune system is suppressed, the options described next offer more promising approaches to the treatment of autoimmune disease.

A strategy being explored to treat autoimmune diseases is to induce **tolerance,** a process that decreases the reactivity of the immune system to a specific antigen. One promising approach is to introduce the antigen by the oral route, so that the immune system "learns" to tolerate it, just as it does the many antigens ingested in food. This antigen-specific immunotherapy causes fewer side effects than drugs that generally suppress the immune system. Trials designed to induce oral tolerance in humans have successfully relieved the symptoms of several autoimmune diseases, but there is much still to learn about the immunological mechanisms, antigen preparations, doses, and duration of treatment. ◀◀ immunological tolerance, p. 355

Attempts have been made to cure type 1 diabetes by replacing the tissues destroyed by immune cells. Transplantation of the pancreas or insulin-producing cells of the pancreas has been successful in many cases, but dangerous immunosuppressive agents must be given to prevent rejection. Because of this, typically the only patients who are given pancreas transplants are those with advanced diabetes who also require a kidney transplant, so must take the immunosuppressive drugs anyway. Research efforts are directed toward developing better methods of transplantation, such as injecting pancreas cells harvested from cadavers into the vein that carries blood to the liver. The cells establish themselves in the liver and produce insulin, often eliminating or decreasing the need for injected insulin. However, the cells usually die in months or a few years, and the immunosuppressant side effects are often severe. Research on manipulating stem cells to produce insulin-secreting β cells holds greater promise for a cure of type 1 diabetes.

MicroAssessment 17.6

Autoimmune disease can result when the immune system attacks autoantigens. Damage from autoimmune reactions may be caused by antibody action or by cell-mediated immune functions, or both. Some autoimmune diseases are organ-specific, but others are systemic. Typical treatment involves drugs that suppress immune responses, but these have dangerous side effects.

16. *What might make an autoimmune reaction general rather than tissue-specific?*

17. *What are the negative side effects of taking immunosuppressive drugs?*

18. *How could viruses or bacteria potentially cause autoimmune disease?* ✚

17.7 ■ Immunodeficiency Disorders

Learning Outcomes

14. *Contrast the two main categories of immunodeficiency disorders.*

15. *Explain how immunodeficiency can lead to multiple and unusual infections.*

In contrast to hypersensitivities and autoimmune diseases that arise from an overblown or inappropriate immune response, **immunodeficiencies** arise when the body cannot make or sustain an immune response. Immunodeficiency disorders may be primary (congenital) or secondary (acquired). Primary immunodeficiency can result from a genetic defect or from environmental factors that cause developmental abnormalities. Secondary immunodeficiency can be acquired as the result of infection or other stresses on the immune system such as malnutrition. People with either type of immunodeficiency are subject to repeated infections. The types of these infections will often depend on which part of the immune system is affected. Some of the more important immunodeficiency diseases are listed in **table 17.4.**

Primary Immunodeficiencies

Primary immunodeficiencies are generally rare. They may affect B cells, T cells, natural killer (NK) cells, phagocytes, or complement components. Many of the gene defects that cause primary immunological disorders are known and work is underway to correct them. Deficiencies and defects can occur at any point in the complex steps that lead to an effective immune response. A few categories of primary immunodeficiencies include the following:

- **Antibody deficiencies.** One of the most common primary immunodeficiencies known is selective IgA deficiency, in which very little or no IgA is produced. Studies indicate that it occurs as often as one per 333 to 700 people. Although people

TABLE 17.4 | **Immunodeficiency Diseases**

Disease	Part of the Immune System Involved
Primary Immunodeficiencies	
Severe combined immunodeficiency (SCID)	Bone marrow stem cells (defect)
Selective IgA deficiency	B cells making IgA (deficiency)
Congenital agammaglobulinemia	B cells (deficiency)
Infantile X-linked agammaglobulinemia	Early B cells (deficiency)
DiGeorge syndrome	T cells (deficiency)
Chediak-Higashi disease	Phagocytes (defect)
Chronic granulomatous disease (CDG)	Phagocytes (defect)
Hereditary angioneurotic edema	Complement regulator (deficiency)
Secondary Immunodeficiencies	
Acquired immunodeficiency syndrome (AIDS)	T cells (destroyed by viral infection)
Monoclonal gammopathy	B cells (multiply out of control)

with this disorder may appear healthy, many have repeated bacterial infections of the respiratory, gastrointestinal, and genitourinary tracts, where secretory IgA normally protects against colonization or invasion by pathogens. Another antibody deficiency is agammaglobulinemia, a disease in which few or no antibodies are produced.

- **Lymphocyte deficiencies.** A disease called severe combined immunodeficiency (SCID) results when hematopoietic stem cells in the bone marrow produce neither T nor B lymphocytes. Children with SCID generally die of infectious disease at an early age unless they are successfully treated with a bone marrow transplant to reconstitute the bone marrow with healthy cells. A variety of gene defects can cause SCID. One is a mutation in an enzyme necessary to form B- and T-cell receptors required for antigen recognition. Another is in a gene for the interleukin-2 receptor on lymphocytes, preventing the cells from receiving the signal to proliferate. Other individuals with SCID lack adenosine deaminase, an enzyme important in the proliferation of B and T cells. A few children with SCID have responded well temporarily to receiving their own defective T cells, with an inserted adenosine deaminase gene linked to a retrovirus vector. Unfortunately, the genetically altered cells do not live long, and the treatments must be repeated. Still, these results are promising, and the possibility of treating other severe disorders with gene therapy is exciting.

  In children with DiGeorge syndrome, lymphocyte deficiency results when the thymus fails to develop in the embryo. As a result, T cells do not differentiate and are absent in circulation. Affected individuals have other developmental defects as well, such as heart and blood vessel abnormalities, and a characteristic appearance with low-set deformed ears, small mouth, and wide-set eyes. As expected from a lack of T cells, affected people are susceptible to infections by eukaryotic pathogens, such as *Pneumocystis jiroveci* and other fungi, as well as viruses and obligate intracellular bacteria.
  ▶▶ *Pneumocystis jiroveci*, p. 708

- **Defects in phagocytic cells.** Chronic granulomatous disease (CGD) is caused by a defect that results in failure of lymphocytes to produce hydrogen peroxide and certain other active products of oxygen metabolism. These phagocytes are unable to kill some organisms, especially the catalase-positive *Staphylococcus aureus*. In Chediak-Higashi disease, lysosomes in the phagocyte lack certain enzymes and cannot destroy phagocytized bacteria. People with this condition suffer from recurring pyogenic (pus-forming) bacterial infections. In leukocyte adhesion deficiency, white blood cells fail to leave the circulation to concentrate at sites of infection.
  ◀◀ catalase test, p. 243

- **Defects in complement system components.** People with deficiencies in the early components of the complement system (such as C1 and C2) may develop immune complex diseases, because these components normally help clear immune complexes from the circulation. People who lack

late components (C5, C6, C7, C8) have recurrent *Neisseria* infections, because the membrane attack complexes typically destroy these bacteria. People who lack one of the important control proteins of the sequence, C1-inhibitor, experience uncontrolled complement activation. This causes fluid accumulation and potentially fatal tissue swelling, a condition called hereditary angioneurotic edema. ◀◀ complement system, p. 344

MicroByte
SCID is sometimes called "bubble boy disease" after a patient who survived for 12 years in a sterile inflated plastic bubble.

Secondary Immunodeficiencies

Secondary, or acquired, immunodeficiency diseases may result from malignancies, advanced age, pregnancy, certain infections (especially viral infections), immunosuppressive drugs, or malnutrition. Some viral infections will deplete certain cells of the immune system. The measles virus, for example, replicates in various cells of the immune system, killing many of them and leaving the body temporarily open to other infections. Syphilis, leprosy, and malaria affect the T-cell population and also macrophage function, causing defects in cell-mediated immunity.

One of the most serious and widespread secondary immunodeficiencies is AIDS (acquired immunodeficiency syndrome), caused by human immunodeficiency virus (HIV). This retrovirus infects and destroys helper T cells, leaving the affected person highly susceptible to infections. Opportunistic infections become more common, caused by agents that are usually unable to establish infection in a healthy individual. AIDS and opportunistic infections are covered in chapter 28. ◀◀ retrovirus, p. 320

Cancers involving the lymphatic system often decrease effective antibody-mediated immunity. For example, multiple myeloma is a malignancy arising from a single plasma cell that proliferates out of control and in most cases produces large quantities of immunoglobulin. This overproduction of a single kind of molecule results in the body using its resources to produce immunoglobulin of a single specificity at the expense of others needed to fight infection. The result is an overall immunodeficiency. Other lymphoid disorders include macroglobulinemia (overproduction of IgM) and some forms of leukemia.

MicroAssessment 17.7
Primary immunodeficiencies may be caused by genetic or developmental defects in components of the immune response. Secondary immune deficiencies result from infection or environmental influences.

19. *Compare and contrast severe combined immunodeficiency disease (SCID) and chronic granulomatous disease (CGD).*
20. *How can a person with multiple myeloma be immunodeficient?*
21. *How could you determine whether an immunodeficiency disease affects T lymphocytes, B lymphocytes, or macrophages?*

New Approaches to Correcting Immunologic Disorders

In recent years, many of the genes responsible for immunodeficiency diseases have been identified. It has been possible to correct some of these gene defects in cells in the laboratory and, rarely, in patients. In the near future, research will be directed toward developing the existing technology for gene transfer to make it more effective in correcting these gene defects in human patients. It is important also to continue the search for other defective genes; with increasing knowledge of the human genome it is likely that more will be found soon. A continuing challenge is finding ways to overcome graft rejection to make bone marrow and other transplants more acceptable. The challenge of treating cancer and of preventing rejection of essential transplants may be met, at least in part, by the development of gene transfer technology and by better understanding of the mechanisms of cell-mediated immunological mechanisms.

One interesting line of research stems from the observation that parasitic worm infestations appear to protect people from allergies and autoimmune diseases. Experiments with mice support this idea and also show that feeding mice parasitic worms effectively treats experimental autoimmune disease. The effect appears to be due at least in part to down-regulation of the immune response. Research into understanding the regulation of immune responses will help in controlling autoimmune and immunodeficiency diseases, and it will permit development of improved vaccines.

A promising and challenging area is the development of human stem cell research. Stem cells have an almost unlimited capacity to divide, and some of them can differentiate into most of the tissues in the body. They could be used to generate cells for transplantation and to replace defective or injured tissues such as nerve tissue. They might also be used to test the effects of drugs on human cells, without the danger of testing on human beings. A major stumbling block is the fact that the stem cells come from fetal material, either from early stage embryos obtained from fertility treatments or from aborted fetuses. The legal and ethical guidelines for the use of these cells are matters of considerable debate.

Summary

17.1 ■ Type I Hypersensitivities: Immediate IgE-Mediated (table 17.1)

IgE attached to mast cells or basophils reacts with specific antigen, resulting in the release of powerful mediators of the allergic reaction (figure 17.1).

Localized Allergic Reactions

Localized allergic (type I) reactions include **hives** (urticaria), **hay fever** (allergic rhinitis), and **asthma** (figure 17.2).

Systemic Anaphylaxis

Systemic anaphylaxis is a rare but serious reaction that can lead to shock and death.

Treatments to Prevent Allergic Reactions

Desensitization, or **immunotherapy,** is often effective in decreasing the type I hypersensitivity state (figure 17.3). A new treatment, using an engineered anti-IgE, promises to be effective in treating asthma.

17.2 ■ Type II Hypersensitivities: Cytotoxic (table 17.1)

Type II hypersensitivity reactions, or cytotoxic reactions, are caused by antibodies that can destroy normal cells by complement lysis or by antibody-dependent cellular cytotoxicity (ADCC).

Transfusion Reactions (table 17.2)

The ABO blood group antigens have been the major cause of **transfusion reactions.**

Hemolytic Disease of the Newborn (figure 17.4)

The Rhesus blood group antigens are usually responsible for this potentially fatal disease. Injected anti-Rh antibody helps prevent Rh sensitization of Rh-negative mothers.

17.3 ■ Type III Hypersensitivities: Immune Complex-Mediated (table 17.1)

Type III hypersensitivity reactions are mediated by small antigen-antibody complexes that activate complement and other inflammatory systems, attract neutrophils, and contribute to inflammation. The **immune complexes** are often deposited in small blood vessels in organs, where they cause inflammatory disease—for example, glomerulonephritis in the kidney or arthritis in the joints (figure 17.5).

17.4 ■ Type IV Hypersensitivities: Delayed-Type Cell-Mediated (table 17.1)

Delayed-type hypersensitivity reactions depend on the actions of sensitized T lymphocytes.

Tuberculin Skin Test (figure 17.6)

A positive reaction to tuberculin protein introduced under the skin peaks 2 to 3 days after exposure to antigen.

Delayed-Type Hypersensitivity in Infectious Diseases

Delayed-type hypersensitivity is important in responses to many chronic, long-lasting infectious diseases.

Contact Hypersensitivities

Contact allergy, or **contact dermatitis,** occurs frequently in response to substances such as poison ivy, nickel in jewelry, and chromium salts in leather products (figures 17.7, 17.8).

17.5 ■ Rejection of Transplanted Tissues

Transplantation rejection of **allografts** is caused largely by type IV cellular reactions.

17.6 ■ Autoimmune Disease

Responses against autoantigens can lead to **autoimmune diseases.**

The Spectrum of Autoimmune Diseases

Autoimmune diseases can be organ-specific (table 17.3, figure 17.9) or widespread (figure 17.10). Some autoimmune diseases are caused by antibodies produced to body components, and others by cell-mediated reactions or a combination of antibodies and immune cells.

Treatment of Autoimmune Diseases

Autoimmune diseases are usually treated with drugs that suppress the immune and/or inflammatory responses.

17.7 ■ Immunodeficiency Disorders (table 17.4)

Immunodeficiencies can be primary genetic or developmental defects in any components of the immune response, or they can be secondary and acquired.

Primary Immunodeficiencies

B-cell immunodeficiencies result in diseases involving a lack of antibody production, such as agammaglobulinemias and selective IgA deficiency. T-cell deficiencies result in diseases such as DiGeorge syndrome. Lack of both T- and B-cell functions results in combined immunodeficiencies, which are generally severe. Defective phagocytes are found in chronic granulomatous disease and Chediak-Higashi disease.

Secondary Immunodeficiencies

Acquired immunodeficiencies can result from malnutrition, immunosuppressive agents, infections (such as AIDS), and malignancies such as multiple myeloma.

Review Questions

Short Answer

1. Why are antihistamines useful for treating many IgE-mediated allergic reactions but not effective in treating asthma?

2. Penicillin is a very small molecule, yet it can cause any of the types of hypersensitivity reactions, especially type I. How can this occur?

3. What are some major differences between an IgE-mediated skin reaction, such as hives, and a delayed-type hypersensitivity reaction, such as a positive tuberculin skin test?

4. What causes insulin-dependent diabetes mellitus?

5. Compare and contrast the autoimmune processes causing myasthenia gravis and Graves' disease.

6. Give an example of an organ-specific autoimmune disease and one that is widespread, involving a variety of tissues and organs.

7. Compare and contrast the Arthus reaction and serum sickness.

8. Why might malnutrition and starvation lead to immunodeficiencies?

9. What is the most common primary immunodeficiency disorder?

10. How can genetic abnormalities leading to immunodeficiency disorders be corrected? Give an example.

Multiple Choice

1. An IgE-mediated allergic reaction
   a) reaches a peak within minutes after exposure to antigen.
   b) occurs only to polysaccharide antigens.
   c) requires complement activation.
   d) requires considerable macrophage participation.
   e) is characterized by induration.

2. Which of the following statements is true of the ABO blood group system in humans?
   a) A antigen is present on type O red cells.
   b) B antigen is the most common antigen in the population of the United States.
   c) Natural anti-A and anti-B antibodies are of the class IgG.
   d) People with blood group O do not have natural antibodies against A and B antigens.
   e) In blood transfusions, incompatibilities cause complement lysis of red blood cells.

3. All of the following are true of immune complexes *except*
   a) the most common complexes consist of antigen and IgE.
   b) an immune complex consists of soluble antigen attached to antibody.
   c) complement components are activated by antigen-antibody complexes.
   d) immune complexes cause strong inflammatory reactions.
   e) immune complexes deposit in kidneys, joints, and skin.

4. Delayed-type hypersensitivity reactions in the skin
   a) are characterized by a wheal and flare reaction.
   b) peak at 4 to 6 hours after exposure to antigen.
   c) require complement activation.
   d) show induration because of the influx of sensitized T cells and macrophages.
   e) depend on activities of the Fc portion of antibodies.

5. Organ transplants, such as of kidneys
   a) are experimental at present.
   b) can be successful only if there are exact matches between donor and recipient.
   c) survive best if radiation is used for immunosuppression.
   d) survive best if B cells are suppressed.
   e) are rejected by a complex process in which cellular mechanisms predominate.

6. All of the following are true of autoimmune disease *except*
   a) some show association with particular major histocompatibility types.
   b) induction of tolerance may alleviate symptoms.
   c) damage to organs occurs due to long-term exaggerated production of IgE.
   d) disease may result from reaction to viral antigens that are similar to autoantigens.
   e) some are organ-specific and some are widespread in the body.

7. Autoantibody-induced autoimmune diseases
   a) can sometimes be passively transferred from mother to fetus.
   b) include diabetes mellitus.
   c) are always organ-specific.
   d) are never organ-specific.
   e) cannot be treated.

8. All of the following approaches are used to treat autoimmune diseases *except*

   a) immunosuppressant drugs.

   b) induction of tolerance.

   c) antibiotics.

   d) anti-inflammatory medications.

   e) replacement therapy, as with insulin in diabetes.

9. Patients with primary immunodeficiencies in the complement system

   a) who lack late-acting components (C5, C6, C7, C8) show increased susceptibility to *Neisseria* infections.

   b) who lack C3 are prone to develop tuberculosis.

   c) generally have no symptoms.

   d) only show defects in the major components C1 through C9.

   e) usually handle infections normally.

10. One of the most serious of the secondary immunodeficiencies is

    a) acquired immunodeficiency syndrome, caused by the human immunodeficiency virus.

    b) severe combined immunodeficiency.

    c) DiGeorge syndrome.

    d) chronic granulomatous disease.

    e) Chediak-Higashi disease.

Applications

1. Jack and Jill were badly burned in an accident at the well and both were taken to the burn unit of the local hospital. The burns covered only a small area of skin so grafts were prepared for both patients from the skin of Jack's thigh. Jack's graft was successful and his burn healed completely. Jill, however, rejected the grafted skin. Explain the immune responses of both patients to these grafts. What treatments could have helped Jill to avoid rejection of her graft?

2. Horse serum containing specific antibody to snake venom has been a successful approach to treating snakebite in humans. How do you think this anti-venom could be generated? What are some advantages of using horses to produce the antibody instead of humans? Why might it be unsafe to administer the anti-venom more than once?

Critical Thinking ✚

1. Hypersensitivity reactions, by definition, lead to tissue damage. Can they also be beneficial? Explain.

2. Explain why people with B-cell deficiencies are more prone to bacterial infections, but people with T-cell deficiencies are more prone to viral infections.

18 Applications of Immune Responses

An immunoassay.

A Glimpse of History

Long before people knew that microbes caused disease, they recognized that individuals who recovered from a disease such as smallpox rarely got it a second time. Old Chinese writings dating from the Sung dynasty (A.D. 960–1280) describe a procedure known as variolation, in which small amounts of the powdered scabs from smallpox lesions were inhaled or placed into a scratch made in the skin. The resulting disease was usually mild, and the person was then immune to smallpox. Occasionally, however, severe disease developed, often resulting in death. In addition, the person became contagious, so the disease could spread.

Although variolation was practiced in China and the Mideast a thousand years ago, it was not widely used in Europe until after 1719. At that time, Lady Mary Wortley Montagu, wife of the British ambassador to Turkey, had their children immunized against smallpox in this way. Variolation then became popular in Europe. Because of the dangers, however, and the fact that the procedure was expensive, many people remained unprotected.

As an apprentice physician, Edward Jenner noted that milkmaids who had recovered from cowpox (a disease of cows that caused few or no symptoms in humans) rarely got smallpox. Then, in 1796, long before viruses had been discovered, he conducted a classic experiment in which he deliberately transferred material from a cowpox lesion on the hand of a milkmaid, Sarah Nelmes, to a scratch on the arm of a young boy named James Phipps. Six weeks later, when exposed to pus from a smallpox victim, Phipps did not develop the disease. The boy had been made immune to smallpox when he was inoculated with pus from the cowpox lesion.

Using the less dangerous cowpox material in place of the scabs from smallpox cases, Jenner and others worked to spread the practice of variolation. Later, Pasteur used the word vaccination (from the Latin *vacca* for "cow") to describe any type of protective inoculation. By the twentieth century, most of the industrialized world was generally free of smallpox as the result of routine vaccination.

In 1967, the World Health Organization (WHO) started a program of intensive smallpox vaccination. Because there were no animal hosts and no non-immune humans to whom it could be spread, the disease died out. The last naturally contracted case occurred in Somalia, Africa, in 1977, and two years later WHO declared the world free of smallpox. Nevertheless, a few laboratories around the world still have the virus. In this age of bioterrorism concerns, some see smallpox as a major threat should the deadly virus ever be released into the largely unprotected populations of the world. Because of this, vaccine stores in the United States have been increased.

FIGURE 18.1
The Host-Pathogen Trilogy

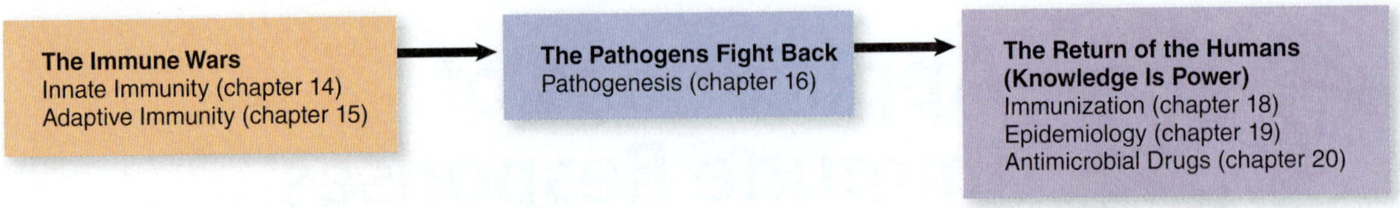

? *How does immunization prevent disease?*

| The Immune Wars
Innate Immunity (chapter 14)
Adaptive Immunity (chapter 15) | → | The Pathogens Fight Back
Pathogenesis (chapter 16) | → | The Return of the Humans
(Knowledge Is Power)
Immunization (chapter 18)
Epidemiology (chapter 19)
Antimicrobial Drugs (chapter 20) |

Chapters 14 and 15 discussed the innate and adaptive defense systems, describing antibodies and lympho-cytes. This chapter will consider how **immunization,** the process of inducing immunity, can be used to protect against disease. In fact, immunization has probably had the greatest impact on human health of any medical procedure, and it is just one example of how knowledge is power with respect to fighting disease (**figure 18.1**). We will also explore some useful applications of immunological reactions in diag-nostic tests.

IMMUNIZATION

18.1 ■ Principles of Immunization

Learning Outcome

1. *Compare and contrast naturally acquired active immunity, artificially acquired active immunity, naturally acquired passive immunity, and artificially acquired passive immunity.*

Naturally acquired immunity is the acquisition of adaptive immunity through normal events, such as exposure to an infec-tious agent. Immunization mimics those same events, protecting against disease by inducing **artificially acquired immunity** (**figure 18.2**). The protection provided by immunization can be either active or passive.

Active Immunity

Active immunity is the result of an immune response in an individual upon exposure to antigen. Specific B and T cells are activated and they then proliferate, giving the individual lasting protection due to immunological memory. Active immunity can develop either naturally from an actual infection or artificially from vaccination. ◀◀ memory, p. 355

Passive Immunity

Passive immunity occurs naturally during pregnancy; the moth-er's IgG antibodies cross the placenta and protect the fetus. These antibodies remain active in the newborn during the first few months of life, when his or her own immune responses are still developing. This is why a number of infectious diseases typically do not develop until a baby is 3 to 6 months of age, after the maternal antibodies have been degraded. Passive immunity also occurs as a result of breast feeding; the secretory IgA in breast milk protects the digestive tract of the child. Note that passive immunity provides no memory; once the transferred antibodies are degraded, the protection is lost.

Artificially acquired passive immunity involves injecting a person with antibodies produced by other people or animals. This can be used to (1) prevent disease immediately before or after likely exposure to a pathogen, (2) limit the duration of certain dis-eases, and (3) block the action of microbial toxins. A preparation of serum (the fluid portion of blood that remains after blood clots) containing the protective antibodies is referred to as **antiserum.** One that protects against a given toxin is called an **antitoxin.**

Two kinds of antisera (or antitoxins) are used. **Hyperimmune globulin**—prepared from the sera of donors with high amounts of antibodies to certain disease agents—is used to prevent or treat specific diseases. Examples include tetanus immune globulin (TIG), rabies immune globulin (RIG), and hepatitis B immune globulin (HBIG). If these preparations are given during the incubation period, they can often prevent severe diseases from developing. **Immune globulin,** the IgG fraction of pooled blood plasma from many donors, has a variety of antibodies due to typical infections and vaccines experienced by the donors. It is used to protect unvaccinated people who have been recently exposed to the measles virus and immunosuppressed people who have low levels of antibodies. ◀◀ incubation period, p. 383

MicroAssessment 18.1

Immunity is natural or artificial, active or passive. Active immunity occurs naturally in response to infections, and artificially in response to vaccination. Passive immunity occurs naturally from maternal antibodies transferred during pregnancy and breast feeding, and artificially through administration of immune serum globulin or hyperimmune globulin.

1. *How is naturally acquired active immunity different from artificially acquired active immunity?*
2. *What is antitoxin?*
3. *What would be a primary advantage of passive immunity with diseases such as tetanus?* ✚

Active Immunity	Passive Immunity

Natural Active Immunity

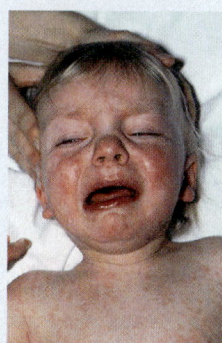

Immunity that results from an immune response in an individual after exposure to an infectious agent.

Natural Passive Immunity

Immunity that results when antibodies from a woman are transferred to her developing fetus during pregnancy or to an infant during breast feeding.

Artificial Active Immunity

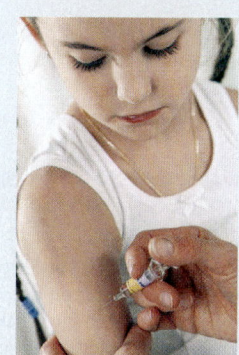

Immunity that results from an immune response in an individual after vaccination.

Artificial Passive Immunity

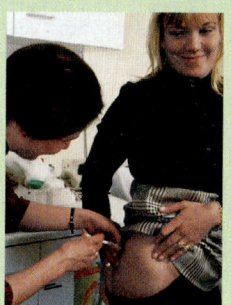

Immunity that results when antibodies contained in the serum of other people or animals are injected into an individual.

FIGURE 18.2 Acquired Immunity Acquired immunity can be natural or artificial, active or passive.

? *Why does active immunity last longer than passive immunity?*

18.2 ■ Vaccines and Immunization Procedures

Learning Outcomes

2. *Compare and contrast the characteristics of attenuated and inactivated vaccines.*

3. *List six diseases that routine childhood immunizations have reduced in occurrence by at least 95%.*

A **vaccine** is a preparation of a pathogen or its products used to induce active immunity. Vaccines not only protect an individual against disease, they can also prevent diseases from spreading in a population. This is because a phenomenon called **herd immunity** develops when a critical portion of a population is immune to a disease, either through natural immunity or vaccination. The infectious agent is unable to spread because there are not enough susceptible hosts. Herd immunity is responsible for dramatic declines in childhood diseases, both in the United States and in developing countries. Unfortunately, we periodically see some of these diseases reappear and spread as a direct consequence of parents failing to have their children vaccinated. **Table 18.1** lists a number of human diseases for which vaccines are available. As the table indicates, some are routinely given, whereas others are used only in special circumstances.

Effective vaccines should be safe, with few side effects, while giving lasting protection against the illness. They should induce protective antibodies or immune cells, or both, as appropriate. For example, polio vaccine should induce antibodies that neutralize the virus, thereby preventing it from reaching and attaching to nerve cells to cause the paralysis of severe poliomyelitis. In contrast, an effective vaccine against tuberculosis would induce cell-mediated immunity that can limit growth of the intracellular bacterium. Of course, vaccines ideally should be low in cost, stable with a long shelf life, and easy to administer. ◄◄ cell-mediated immunity, p. 356

Vaccines fall into two general categories—attenuated and inactivated, based on whether or not the immunizing agent can

	Type of Vaccine	Persons Who Should Receive the Vaccine
	Subunit	Adults in occupations that put them at risk of exposure, such as military personnel
	Toxoid	Children (the "D" in the DTaP vaccine given to children); adolescents and adults receive a booster every 10 years (the "d" in the Td and Tdap booster vaccines)
Haemophilus influenzae type b infections	Polysaccharide-protein conjugate	Children
Hepatitis A	Inactivated virus	Children; adolescents who live in selected areas; adults with indications that put them at increased risk (such as traveling to certain countries; men who have sex with men); close contacts of internationally adopted children.
Hepatitis B	Subunit	Newborns and children; also adults with indications that put them at increased risk (such as healthcare workers who might be exposed to blood, people who have multiple sexual partners, and contacts of infected people)
Human papillomavirus (HPV) infection	VLP (two or four serotypes)	Girls/women ages 11–26
Influenza	Three age-specific forms—inactivated virus (TIV), hemagglutinin antigen (Fluzone High Dose), and attenuated virus (LIAV); the first two are given by injection and the latter as a nasal mist	All people 6 months of age or older; given yearly, as the antigens of the virus change frequently. When the vaccine supply is limited, then vaccination should focus on high-risk groups
Measles	Attenuated virus	Children (an "M" in the MMRV or MMR vaccine given to children); booster(s) for adults born after 1956 who do not have evidence of immunity and do not have a medical contraindication
Meningococcal disease	Two age-specific forms active against four serotypes—meningococcal conjugate vaccine (MCV4 and MenACWY-CRM) and meningococcal polysaccharide vaccine (MPSV4)	Adolescents; also children and adults with certain medical conditions that put them at greater risk (for example, those without a spleen or who have certain complement system defects); also adults in certain high-risk groups (such as college students living in dormitories and people traveling to sub-Saharan Africa)
Mumps	Attenuated virus	Children (an "M" in the MMRV or MMR vaccine given to children); booster(s) for adults born after 1956 who do not have evidence of immunity and do not have a medical contraindication
Pertussis (whooping cough)	Subunit (acellular vaccine)	Children (the "aP" in the DTaP vaccine given to children); adolescents should receive a booster (the "ap" in the Tdap booster vaccine); adults younger than age 65 may receive a booster
Pneumococcal infection	Two age-specific forms—purified polysaccharide (PPV) and polysaccharide protein conjugate (PCV)	Children; adults age 65 and over, people with certain chronic infections, and others in high-risk groups
Polio	Two forms—inactivated virus (Salk vaccine) and attenuated virus (Sabin vaccine)	Children; attenuated virus is used for global control
Rabies	Inactivated virus	People exposed to the virus, people at high risk for exposure, such as veterinarians and other animal handlers
Rotavirus infection	Attenuated virus	Children
Rubella (German measles)	Attenuated virus	Children (the "R" in the MMR or MMRV vaccine given to children); women who do not have evidence of immunity and do not have a medical contraindication
Shingles	Attenuated virus	Adults age 60 and over
Tetanus	Toxoid	Children (the "T" in the DTaP vaccine given to children); adults receive a booster every 10 years (the "T" in the Td and Tdap vaccines)
Tuberculosis	Attenuated bacterium (BCG strain)	Used only in special circumstances in the United States; widely used in other countries
Typhoid fever	Two forms—attenuated bacterium (Ty21a strain; taken orally) and purified polysaccharide (ViCPS)	People traveling to certain parts of the world
Varicella-zoster (chickenpox)	Attenuated virus	Children ("V" in the MMRV vaccine given to children); also adults without evidence of immunity
Yellow fever	Attenuated virus	Travelers to endemic areas

TABLE 18.2	A Comparison of Characteristics of Attenuated and Inactivated Vaccines	
Characteristic	**Attenuated Vaccines**	**Inactivated Vaccines**
Antibody response (memory)	IgG; secretory IgA if administered orally or nasally	IgG
Cell-mediated immune response	Good	Poor
Duration of protection	Long-term	Short-term
Need for adjuvant	No	Yes
Number of doses	Usually single	Multiple
Risk of mutation to virulence	Very low	Absent
Risk to immunocompromised recipient	Can be significant	Absent
Route of administration	Injection, oral, or nasal	Injection
Stability in warm temperatures	Poor	Good
Types	Attenuated viruses, attenuated bacteria	Inactivated whole agents, toxoids, subunit vaccines, VLPs, polysaccharide vaccines

replicate. Each type has characteristic advantages and disadvantages (**table 18.2**).

Attenuated Vaccines

An **attenuated vaccine** is a weakened form of the pathogen that generally cannot cause disease. The attenuated strain replicates in the vaccine recipient, causing an infection with undetectable or mild disease that typically results in long-lasting immunity. Because infection with the attenuated strain mimics that of the wild-type strain, it evokes the type of immune response appropriate for controlling the infection. For instance, attenuated vaccines given orally induce mucosal immunity (a secretory IgA response), protecting against pathogens that infect via the gastrointestinal tract. Some attenuated vaccines are able to stimulate cytotoxic T cells, inducing cell-mediated immunity. ◀◀ mucosal immunity, p. 358

Attenuated strains are often produced by growing a microbe under conditions that cause mutations to accumulate, making the microbe less pathogenic. Viruses of humans can sometimes be attenuated by growing them in cells of a different animal species; the mutations that allow them to multiply in the other animal cells often cause them to grow poorly in human cells. Genetic manipulation is now being used to produce strains of pathogens with low virulence. Specific genes are mutated and used to replace wild-type genes. The inserted mutant genes are engineered so they cannot revert to the wild type.

Attenuated vaccines have several advantages compared to their inactivated counterparts. For one thing, a single dose of an attenuated agent is often enough to induce long-lasting immunity. This is because the microbe multiplies in the body, causing the immune system to be exposed to the antigen for a longer period and in greater amounts than with inactivated agents. In addition, the vaccine strain has the added potential of being spread from an individual being immunized to other non-immune people, inadvertently immunizing the contacts of the vaccine recipient.

The disadvantage of attenuated agents is they sometimes cause disease in immunosuppressed people, and can occasionally mutate to become pathogenic again. Attenuated vaccines are generally not advised for pregnant women because of the possibility that the vaccine strain may cross the placenta and damage the developing fetus. Another disadvantage of attenuated vaccines, especially in

developing countries where they are desperately needed, is that they usually require refrigeration to keep them active.

Attenuated vaccines currently in widespread use include those against measles, mumps, rubella, chickenpox, rotavirus, and yellow fever. The Sabin vaccine against polio is also an attenuated vaccine.

Inactivated Vaccines

An **inactivated vaccine** is unable to replicate, but retains the immunogenicity of the pathogen or toxin. The advantage of inactivated vaccines is that they cannot cause infections or revert to pathogenic forms. Because they do not replicate, however, there is no amplification of the dose *in vivo*, so the magnitude of the immune response is limited. To compensate for the relatively low effective dose, several booster doses are usually needed to induce sufficient immunity to be protective. Some inactivated vaccines include the whole infectious agent, and others include only fractions of the agent. Examples include:

■ **Inactivated whole agent vaccines.** These contain killed microorganisms or inactivated viruses. The vaccines are made by treating the pathogen with formalin or another chemical that does not significantly change the surface epitopes. The treatment leaves the agent immunogenic even though it cannot reproduce. Vaccines in this category include those against influenza, rabies, and polio (Salk vaccine). ◀◀ formalin, p. 118

■ **Toxoids.** These are inactivated toxins used to protect against diseases caused by bacterial toxins. They are prepared by treating the toxins to destroy the toxic part of the molecules while retaining the antigenic epitopes. Diphtheria and tetanus vaccines are toxoids.

■ **Subunit vaccines.** These consist of key protein antigens or antigenic fragments of a pathogen. Obviously, they can be developed only after research has revealed which of the microbe's components are most important in triggering a protective immune response. Their advantage is that cell parts that may cause undesirable side effects are not included. The vaccine currently used to prevent whooping cough (pertussis) is a subunit vaccine, referred to as the acellular pertussis (aP) vaccine. It does not cause the side effects that sometimes occurred with the killed whole-cell vaccine used previously.

- **Recombinant vaccines.** These are subunit vaccines produced by genetically engineered microorganisms. An example is the vaccine against the hepatitis B virus; it is produced by yeast cells engineered to produce part of the viral protein coat.

- **VLP (virus-like particle) vaccines.** These are empty capsids. Laboratory organisms are genetically engineered to produce the major capsid proteins of a virus, which then self-assemble. The human papillomavirus (HPV) vaccines are VLPs.

- **Polysaccharide vaccines.** These contain the polysaccharides that make up the capsules of certain organisms. They are not effective in young children because polysaccharides are T-independent antigens; recall that these antigens generally elicit a poor response in this age group. The pneumococcus vaccines given to adults are polysaccharide vaccines.

- **Conjugate vaccines.** These are polysaccharides linked to proteins, a modification that converts the polysaccharides into T-dependent antigens. The first conjugate vaccine developed was against *Haemophilus influenzae* type b (Hib) and has nearly eliminated Hib meningitis in children. The conjugate vaccine developed against certain *Streptococcus pneumoniae* strains promises to do the same for a variety of infections caused by those strains. ▶▶| *Haemophilus influenzae* type b, p. 647 ▶▶| *Streptococcus pneumoniae*, p. 497

Many inactivated vaccines contain an **adjuvant,** a substance that enhances the immune response to antigens (*adjuvare* means "to help"). These are necessary additives because purified antigens such as toxoids and subunit vaccines are often poorly immunogenic by themselves because they lack the "danger" signals—the patterns associated with tissue damage or invading microbes. These patterns cause dendritic cells to produce co-stimulatory molecules, allowing them to activate helper T cells, which, in turn, activate B cells (see figure 15.20). Adjuvants are thought to function by providing the danger signals to dendritic cells. Some adjuvants appear to adsorb the antigen, releasing it at a slow but constant rate to the tissues and surrounding blood vessels. Unfortunately, many effective adjuvants trigger an intense inflammatory response, making them unsuitable for use in vaccines for humans. Alum (aluminum hydroxide and aluminum phosphate) is the most common adjuvant used, but others, including one that uses a derivative of lipid A, have recently been developed. |◀◀ pattern recognition, p. 342 |◀◀ dendritic cells, pp. 340, 370 |◀◀ lipid A, p. 60

An Example of Vaccination Strategy—The Campaign to Eliminate Poliomyelitis

Vaccines against poliomyelitis provide an excellent illustration of the complexity of vaccination strategies. The virus that causes this disease enters the body orally, infects the throat and intestinal tract, and then invades the bloodstream. From there, it can invade nerve cells and cause the disease poliomyelitis (see figure 26.13). There are three types of poliovirus, any of which can cause poliomyelitis. The Salk vaccine, developed in the mid-1950s, consists of inactivated viruses of all three types. It successfully lowered the rate of the disease dramatically but had the disadvantage of requiring a series of injections for maximum protection. In 1961, the Sabin vaccine became available, with the advantage of cheaper oral administration. The attenuated vaccine strains replicate in the

intestine, but the vaccine still has to be given in a series of three doses because of interactions among the viruses.

Attenuated and inactivated polio vaccines both cause the immune system to produce antibodies that protect against viral invasion of the central nervous system and consequent paralytic poliomyelitis. The Sabin vaccine, however, has a distinct advantage over the Salk vaccine in that it induces better mucosal immunity (secretory IgA response), and therefore provides better herd immunity. A disadvantage of the Sabin vaccine is that the attenuated viruses can mutate to become virulent. Approximately one case of poliomyelitis arises for every 2.4 million doses of Sabin vaccine administered. ▶▶| poliomyelitis, p. 656

An obvious way to avoid vaccine-related poliomyelitis is to abandon the Sabin vaccine in favor of the Salk. As usual, however, the situation is not as simple as it might seem. The Sabin vaccine, unlike the Salk vaccine, provides better protection against transmission of the wild-type virus. If only the inactivated vaccine is used, the virus can still replicate in the gastrointestinal tract and be transmitted to others, rapidly spreading in a population.

A campaign to eliminate polio using the Sabin vaccine was so successful that by 1980 the United States was free of wild-type polioviruses (see figure 26.14). By 1991, the viruses had been eliminated from the Western Hemisphere. Because of the continued risk of vaccine-associated paralytic polio, a vaccine strategy that attempted to capture the best of both vaccines was adopted in the United States. Children first received doses of the Salk vaccine, protecting them from poliomyelitis. Following these doses, the Sabin vaccine was given, providing mucosal protection while also boosting immunity. The routine use of the Sabin vaccine was then discontinued altogether in the United States.

The original goal of global eradication of polio by 2000 was not achieved, but efforts continue. For these eradication programs, the Sabin vaccine must be used because it prevents transmission of the virus.

The Importance of Childhood Immunizations

Before vaccination was available for common childhood diseases, thousands died or were permanently disabled from these diseases (**table 18.3**). Unfortunately, many people still become ill or even die from diseases easily prevented by vaccines.

One reason some children are not protected is that their parents have refused to have them vaccinated, fearing that vaccination might be harmful. In situations such as this, vaccines have become victims of their own success. They have been so effective at preventing diseases that people have been lulled into a false sense of security. Reports of adverse effects of vaccination have led some people to falsely believe that the risk of vaccination is greater than the risk of diseases.

There will always be at least some risk associated with almost any medical procedure, but there is no question that the benefits of routine vaccinations greatly outweigh the very slight risks. Data show that a child with measles has a 1:2,000 chance of developing serious brain inflammation, compared with a 1:1,000,000 chance from the measles vaccine. Between 1989 and 1991 measles immunization rates dropped 10% and an outbreak of 55,000 cases occurred, with 120 deaths. Now that the vaccination rates have increased again, measles outbreaks are rarely seen.

TABLE 18.3	The Effectiveness of Universal Immunization in the United States	
Disease	**Cases per Year Before Immunization**	**Decrease After Immunization**
Smallpox	48,164 (1900–1904)	100%
Diphtheria	175,885 (1920–1922)	Nearly 100%
Pertussis (whooping cough)	147,271 (1922–1925)	93.4%
Tetanus	1,314 (1922–1926)	98.1%
Paralytic poliomyelitis	16,316 (1951–1954)	100%
Measles	503,282 (1958–1962)	Nearly 100%
Mumps	152,209 (1968)	99.8%
Rubella (German measles)	50,230 (1966–1969)	98%
Haemophilus influenzae type b invasive disease in children	20,000 (estimated)	99.8%

Routine immunization against pertussis (whooping cough) caused a significant decrease in its incidence in the United States and saved many lives. Because of some adverse reactions to the killed whole-cell vaccine being used at the time, however, many parents refused to allow their babies to get the vaccine. By 1990, this resulted in the highest incidence of pertussis in 20 years and the deaths of some children, mostly those under one year of age. An acellular subunit pertussis vaccine, which is more effective and has fewer side effects than the whole-cell vaccine, is now used.

The suggestion that vaccines are associated with autism has again threatened the acceptance of immunization. It is important to note, however, that scientific studies show no evidence of a link between the two.

The U.S. Centers for Disease Control and Prevention (CDC) regularly publishes recommended immunization schedules for children, adolescents, and adults. These are updated regularly as vaccines are developed and modified. Because of the complexity of the schedules and the frequency at which they have been updated recently, it is important to know how to access the most current schedule. A link can be found in the chapter 18 readings at the text website (**www.mhhe.com/nester7**).

TABLE 18.4	Some Diseases for Which New or Improved Vaccines Are Sought	
Disease	**Estimated Impact**	
HIV/AIDS	40 million infected worldwide, with approximately 14,000 new infections daily	
Malaria	300–500 million cases/yr and up to 3 million deaths/yr worldwide	
Influenza	30–50 million cases/yr worldwide; 10,000–40,000 deaths/yr in the United States	
Strep throat	20 million cases/yr in the United States	
Genital herpes	45 million infected and 500,000 new infections/yr in the United States	
Hepatitis C	170 million infected worldwide	
Cancer	1 in 3 in the United States may get cancer, resulting in 560,000 deaths/yr	

Vaccines in the CDC recommended immunization schedule are generally covered by the National Vaccine Injury Compensation Program. This no-fault alternative for resolving vaccine injury claims was established to stabilize the U.S. vaccine market. An excise tax on every vaccine dose purchased funds the program.

Current Progress in Immunization

Advances in understanding the immune system allow researchers to make safer and more effective vaccines. An excellent example of this is the introduction of conjugate vaccines designed to enlist T-cell help. Another is the new adjuvants being developed based on insights gained from the discovery of toll-like receptors (TLRs). The immune response can also be improved by administering certain cytokines with vaccines to guide that response. ◀◀ TLRs, p. 342

Novel types of vaccines being studied include peptide vaccines, edible vaccines, and DNA-based vaccines. None of these rely on whole cells, so they eliminate the possibility of infection with the immunizing agent, but some are only weakly immunogenic. Peptide vaccines are composed of key antigenic peptides from disease-causing organisms. They are stable to heat and do not contain extra materials that might cause unwanted reactions or side effects. Edible vaccines are created by transferring genes encoding key antigens from infectious agents into plants. If appropriate plants can be genetically engineered to function as vaccines, they could potentially be grown throughout the world, eliminating difficulties involving transport and storage. DNA-based vaccines are segments of naked DNA from infectious organisms that can be introduced directly into muscle tissue. The host tissue actually expresses the DNA for a short period of time, producing the encoded microbial antigens, which induces an immune response.

There are several serious and widespread diseases for which new or more effective vaccines are currently being sought (**table 18.4**). Many of the pathogens involved are quite good at avoiding the host defenses, complicating the development of long-lasting effective vaccines. In addition to seeking vaccines that protect against infectious diseases, other uses of vaccines are also being studied, including vaccines to treat cancer.

MicroAssessment 18.2

An attenuated vaccine is a weakened form of the pathogen. An inactivated vaccine is unable to replicate but retains the immunogenicity of the pathogen or toxin; examples include killed microorganisms, inactivated viruses, and fractions of the agents, including toxoids. Routine childhood immunizations have prevented millions of cases of disease and many deaths during the past decades. Many experimental vaccines are under study or in clinical trials.

4. *What is the difference between an attenuated and inactivated vaccine?*

5. *Childhood diseases such as measles and mumps are rare now, so why is it important for children to be immunized against them?*

6. *What would be a primary advantage of using an attenuated agent for a vaccine rather than just an antigen from that agent?*

IMMUNOLOGICAL TESTING

Immunoassays take advantage of the specificity of antibody-antigen interactions, using them for diagnosis. For example, if a person is suspected of having syphilis, antibodies that bind specifically to the causative bacterium, *Treponema pallidum,* can be added to fluid collected from a genital lesion. If the antibodies bind to a bacterium in the specimen, they identify the organism as *T. pallidum,* indicating that the patient does indeed have syphilis (**figure 18.3**). In addition to identifying microbes, immunoassays can be used to detect specific antibodies. If a patient's blood has antibodies to a given microbe, then the patient's immune system must have responded to the microbe at some point, indicating either previous or current infection.

One of the earliest examples of immunological testing is the tuberculin skin test (also called the Mantoux test), which is still used for diagnosing tuberculosis. People infected with *Mycobacterium tuberculosis* develop a strong cell-mediated response to the bacterium and its products, which is the basis of the test. When a purified protein derivative (PPD) from *M. tuberculosis* is injected into the skin of someone infected with the organism, redness and a firm swelling usually develop at the site (see figure 17.6). In contrast, people who have not been infected show little, if any, response.

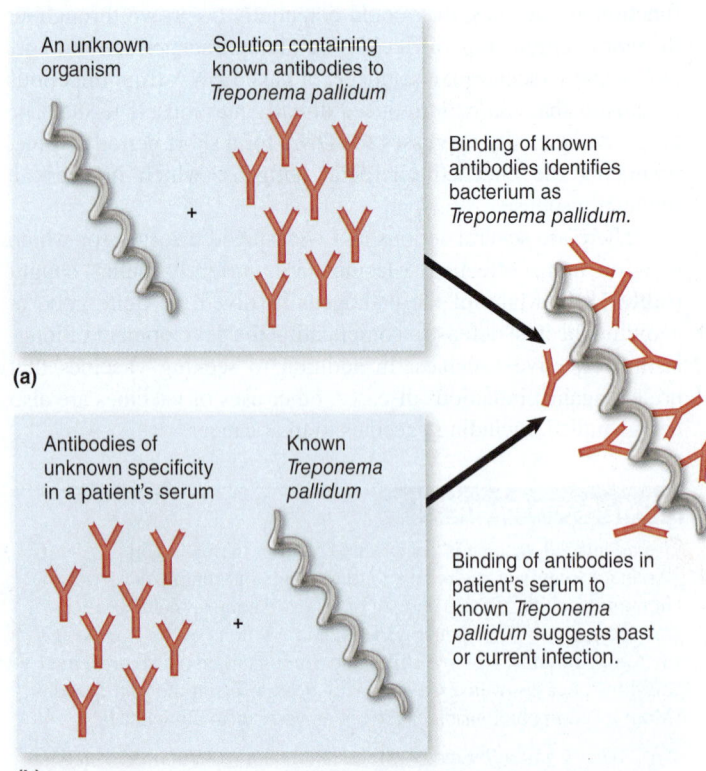

(a)

(b)

FIGURE 18.3 Principles of Immunoassays These assays can be used to **(a)** identify unknown bacteria (or other antigens); **(b)** detect specific antibodies.

❓ *What three occurrences could account for a person having antibodies to a specific infectious agent?*

Just as the field of immunology has advanced significantly in the last few decades, so has immunological testing. New tests are continually being developed, supplementing or gradually replacing many of the older methods. This section will focus primarily on tests commonly used today; information about other tests—including radial immunodiffusion, immunoelectrophoresis, complement fixation, radioimmunoassay and hemagglutination inhibition—which are declining in use but illustrate key immunological principles, can be found in the chapter 18 readings at the text website (**www.mhhe.com/nester7**).

18.3 ■ Principles of Immunological Testing

Learning Outcomes

4. *Describe the difference between polyclonal and monoclonal antibodies.*
5. *Describe how the antibody titer is determined.*

A person who has not been exposed to a given pathogen typically lacks specific antibodies against the microbe in their serum, and is referred to as seronegative. Once infected, that person will begin producing specific antibodies about a week to 10 days later, becoming seropositive. This change from seronegative to seropositive is referred to as **seroconversion.** As the infection progresses, increasing amounts of specific antibodies are produced. A rise in the concentration (titer) of specific antibodies is characteristic of an active infection. In contrast, small but steady levels of specific antibodies indicate a previous infection or vaccination.

To determine if a person has specific antibodies in the blood, then either the serum or plasma is tested. **Serum** is the fluid portion of blood that remains after blood clots; **plasma** is the fluid portion of blood treated with an anticoagulant to prevent clotting. Because serum is so often used as a source of antibodies, the study of *in vitro* antibody-antigen interactions is referred to as **serology.** Most frequently it implies testing a patient for specific antibodies to diagnose a disease. Cerebrospinal fluid, tissues, and other clinical specimens can also be tested for antibodies.

Obtaining Antibodies

Several different methods are used to obtain antibodies. The approach chosen depends largely on the intended use of the product.

Polyclonal Antibodies

Laboratory animals are used to produce antibodies known to bind a certain infectious agent. The animals are immunized with either the whole agent or part of the agent, and the resulting antibodies are then collected by harvesting the animal's serum. The antibody preparation will be polyclonal, meaning that multiple naive B cells responded to the immunization, giving rise to a mixture of antibodies that together recognize a variety of epitopes on the antigen. The more complex the antigen, the greater the number of different epitopes recognized by the **polyclonal antibodies.** For instance, injecting whole bacteria will result in a wider array of antibody specificities than injecting purified toxin.

One problem with polyclonal antibodies is that some may bind to closely related organisms, resulting in a false positive reaction. As an example, *Shigella* species have outer membrane proteins in common with *E. coli,* so an animal immunized with whole *Shigella* cells would produce some antibodies that also bind *E. coli* cells. If those antibodies were used in a diagnostic test for *Shigella,* a specimen containing *E. coli* but not *Shigella* would yield a false positive result.

Certain serological tests discussed in this chapter use antibodies that bind to the constant region of human IgG molecules. These are referred to as **anti-human IgG antibodies.** They are produced by animals that have been immunized with IgG from human serum. Anti-human IgG antibodies are available commercially as are antibodies that bind to the other immunoglobulin classes.

Monoclonal Antibodies

Monoclonal antibodies recognize only a single epitope. They are difficult and expensive to develop, so a given specificity is generally available only if it has commercial value due to widespread use.

Monoclonal antibodies are obtained through a complicated process that involves taking B cells from an immunized animal, and then fusing those short-lived B cells with other cells that will divide repeatedly in culture (see **Perspective 18.1**). Because each B cell is programmed to produce antibody molecules that recognize only a single epitope, antibody preparations produced by the descendants (clones) of a single B cell are all identical.

A monoclonal antibody derived from a mouse or other laboratory animal can be "humanized." To do this, recombinant DNA techniques are used to replace most of the antibody molecule with the human equivalents. This gives the molecule a longer half-life in humans because the human immune system is less apt to destroy it. Some humanized monoclonal antibodies are given as a form of passive immunity to treat certain types of cancers. Others are tagged with a drug or toxic substance and then used to deliver that tag to a specific cell type *in vivo.* These tagged monoclonal antibodies have been used to treat non-Hodgkin's lymphoma that has not responded to traditional treatment.

Quantifying Antigen-Antibody Reactions

The concentration of antibody molecules in a specimen such as serum is usually determined by making serial dilutions similar to those done to count bacterial cells (see figure 4.17). Sequential two-fold or ten-fold dilutions are used to dilute the specimen, and then antigen is added to each dilution. The **titer** (concentration) is expressed as the reciprocal of the last dilution that gives a detectable antigen-antibody reaction. Thus, if a positive reaction is observed in the dilution 1:256 but not in 1:512, then the antibody titer is 256.

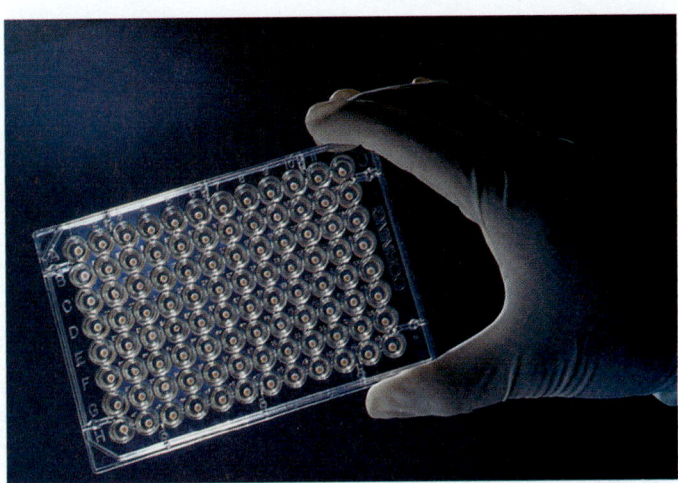

FIGURE 18.4 Microtiter Plate Serological tests can be done in the wells of these small plates.

❓ *What is the advantage of doing serological tests in a microtiter plate?*

Serology tests can be done in test tubes, but this requires many tubes and large amounts of reagents. Therefore, the tests are usually done using plastic microtiter plates, which have 96, 384, or 768 wells (**figure 18.4**). The volumes used in each well are a mere fraction of those needed for even a small test tube, so tests can be done on very small samples. Special equipment can be used to mix the reagents and read the results.

MicroAssessment 18.3

Antibodies used in serological tests can be either polyclonal or monoclonal. Serial dilution of serum or plasma permits quantification of antibodies in the specimen.

7. *What is the significance of a rise in titer of specific antibodies in serum samples taken at different times?*

8. *How are polyclonal antibodies different from monoclonal antibodies?*

9. *Would antibodies produced by a patient in response to infection be monoclonal or polyclonal?* ✚

18.4 ■ Observing Antigen-Antibody Aggregates

Learning Outcome

6. *Compare and contrast precipitation reactions and agglutination reactions.*

Recall from chapter 15 that antibodies can cross-link antigens, creating large "mouthfuls" for phagocytic cells (see figure 15.8). This type of antigen-antibody interaction can be observed in precipitation and agglutination reactions.

Precipitation Reactions

When antibodies bind to soluble antigens, the molecules sometimes cross-link to form latticelike insoluble complexes

PERSPECTIVE   **18.1**

Monoclonal Antibodies

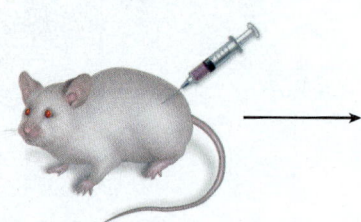

Immunize a mouse with antigen X to activate and induce proliferation of specific B cells.

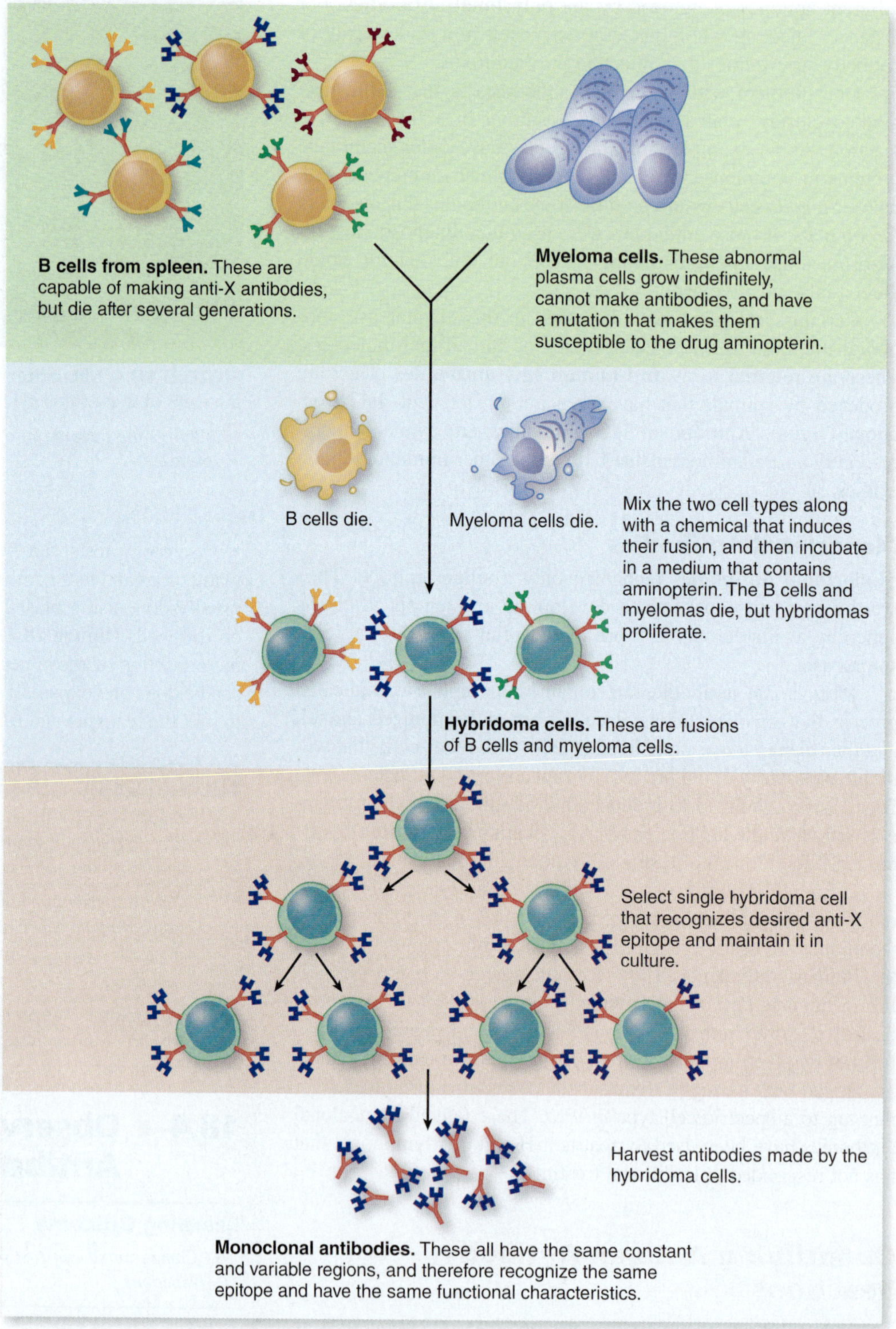

FIGURE 1 Production of Monoclonal Antibodies

When an animal is injected with an antigen, its immune system responds by making antibodies directed against the antigen's different epitopes. Even though there is a single antigen, a variety of different B cells respond, resulting in the production of polyclonal antibodies. Unfortunately, this makes it difficult to standardize experimental results, because the antibody composition is different each time the antiserum is made.

In 1975, Georges Köhler and Cesar Milstein overcame the problem of variable antiserum preparations by developing a technique to make monoclonal antibodies. These are antibodies produced by a single B clone, so all molecules in a preparation will have the same constant and variable regions and, thus, the same functional characteristics and specificity. With such consistency, tests can be standardized more easily and with greater reliability.

To make monoclonal antibodies, a laboratory animal is immunized with the agent being studied, and B lymphocytes are then isolated. These are then fused with myeloma cells, which are malignant (cancerous) plasma cells. Unlike normal plasma cells, these myeloma cells can divide repeatedly in culture, do not make antibodies, and have lost the capacity to produce a critical enzyme, so they cannot grow in a medium that contains the drug aminopterin. When the B cells and myeloma cells are mixed and grown in a medium that contains aminopterin, only the fusion products, called hybridomas, can proliferate. The hybridoma cells retains critical

B cells from spleen. These are capable of making anti-X antibodies, but die after several generations.

Myeloma cells. These abnormal plasma cells grow indefinitely, cannot make antibodies, and have a mutation that makes them susceptible to the drug aminopterin.

B cells die.   Myeloma cells die.

Mix the two cell types along with a chemical that induces their fusion, and then incubate in a medium that contains aminopterin. The B cells and myelomas die, but hybridomas proliferate.

Hybridoma cells. These are fusions of B cells and myeloma cells.

Select single hybridoma cell that recognizes desired anti-X epitope and maintain it in culture.

Harvest antibodies made by the hybridoma cells.

Monoclonal antibodies. These all have the same constant and variable regions, and therefore recognize the same epitope and have the same functional characteristics.

traits from the fused cells: The B cell supplies the genes for the specific antibody production, and the myeloma cell supplies the cellular machinery for producing the antibodies and multiplying indefinitely.

In the laboratory, monoclonal antibodies are the basis of a number of diagnostic tests.

For example, monoclonal antibodies against a hormone can detect pregnancy only 10 days after conception. Specific monoclonal antibodies are used for rapid diagnosis of hepatitis, influenza, herpes simplex, and *Chlamydia* infections. Köhler and Milstein won the Nobel Prize in 1984 for their work.

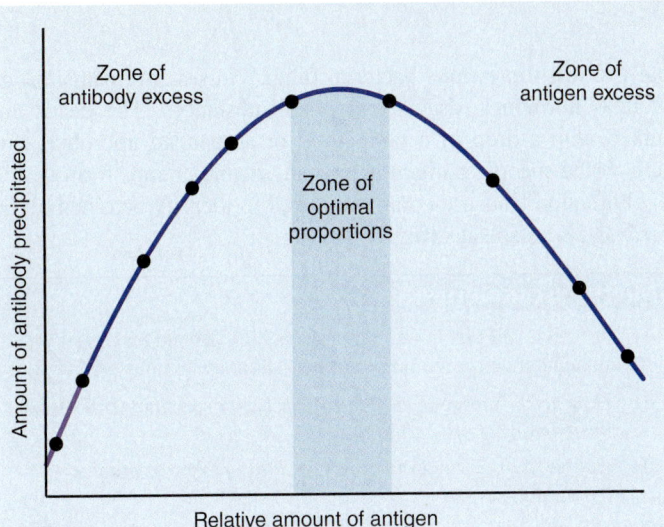

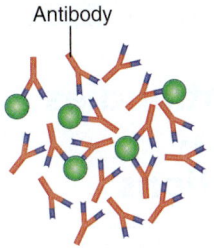 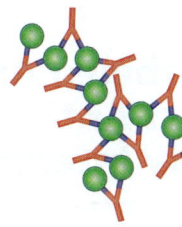 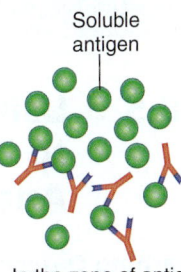

In the zone of antibody excess, little or no cross-linking occurs; no visible precipitate forms.

In the zone of optimal proportions, extensive cross-linking occurs; a visible precipitate forms.

In the zone of antigen excess, little or no cross-linking occurs; no visible precipitate forms.

FIGURE 18.5 Antigen-Antibody Precipitation Reactions
The maximum amount of precipitate forms in the zone of optimal proportion.

❓ *It takes fewer molecules of IgM than IgG to cause precipitation. Why would this be so?*

that then precipitate out of solution (**figure 18.5**). This is the basis of **precipitation reactions,** which can be used to detect specific antibodies or antigens. The complexes (aggregates) can take several hours to form, and develop only at certain relative concentrations of antibody and antigen molecules. If there is a great excess of either, the aggregates do not form, and consequently, no precipitate will be seen. The easiest way to get the proper concentration is to place the antigen and antibody suspensions near each other in a gel, and let the molecules diffuse toward each other. A precipitate will form in a distinct region called the zone of optimal proportions.

The use of precipitation reactions in diagnosis has largely been replaced by methods that will be discussed shortly. However, the principle of the reactions is nicely demonstrated by the Ouchterlony technique, which can be done in a Petri dish (**figure 18.6**). Antigen and antibody solutions are placed into separate wells cut in the gel contained in the dish. These solutions will gradually diffuse outward, meeting between the wells. If the antibody molecules recognize the antigen, a line of precipitation will form at the zone of optimal proportions. Because there are often multiple antigens present in the sample, as well as different specificities of antibodies in the serum, more than one line can form, each in its area of optimal proportions. The Ouchterlony test can be used to detect autoantibodies associated with certain connective tissue disorders. Other immunodiffusion tests (precipitation tests carried out in a gel) are described on the text website (**www.mhhe.com/nester7**). ◀◀ autoantibody, p. 412

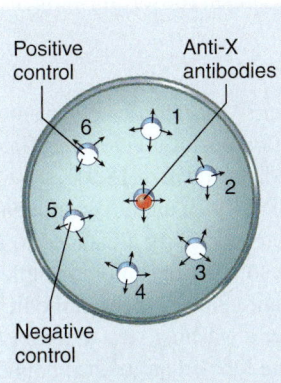 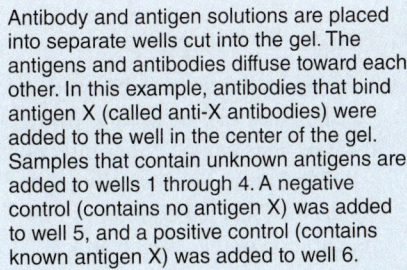 

Antibody and antigen solutions are placed into separate wells cut into the gel. The antigens and antibodies diffuse toward each other. In this example, antibodies that bind antigen X (called anti-X antibodies) were added to the well in the center of the gel. Samples that contain unknown antigens are added to wells 1 through 4. A negative control (contains no antigen X) was added to well 5, and a positive control (contains known antigen X) was added to well 6.

When antibody molecules that recognize the antigen meet at the zone of optimal proportions, antigen-antibody complexes precipitate out of solution, forming a visible line. In this example, a line has formed between the center well and well 6 (the positive control). A line has also formed between the center well and the sample in well 4, indicating that the sample contains antigen X. The other samples do not contain detectable amounts of antigen X.

(a)

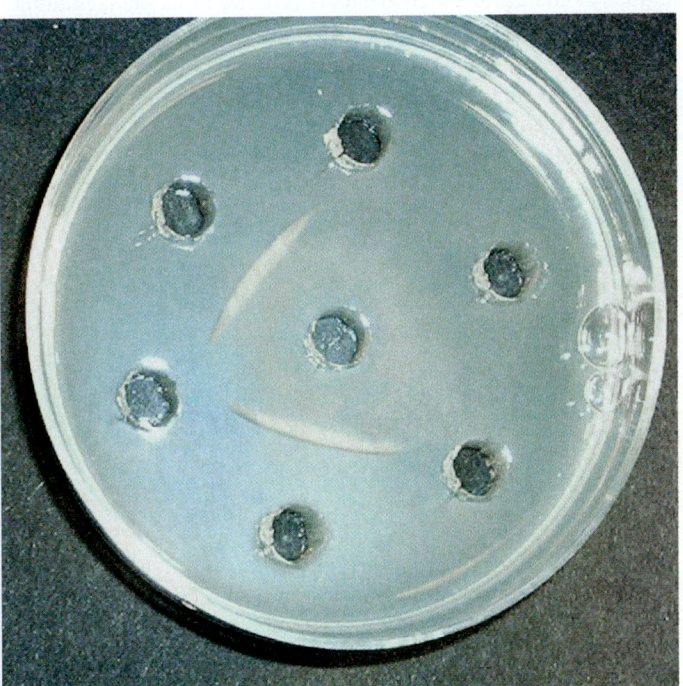

(b)

FIGURE 18.6 Ouchterlony Technique (a) Method. (b) Photograph of results.

❓ *What is the purpose of including positive and negative controls?*

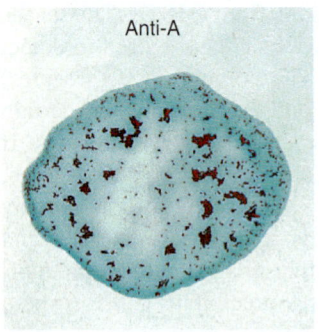

Anti-A

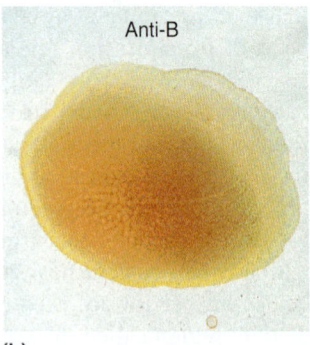

Anti-B

(a)   **(b)**

FIGURE 18.7 ABO Blood Typing This method tests for two different antigens (A and B) on red blood cells. **(a)** The red blood cells agglutinated when mixed with anti-A antibodies. **(b)** The red blood cells did not agglutinate when mixed with anti-B antibodies. Together, the results in panels (a) and (b) indicate that the blood group is type A.

? *What term is used to describe the agglutination of red blood cells?*

Agglutination Reactions

Agglutination and precipitation reactions are similar in principle—both depend on cross-linking and lattice formation. In **agglutination reactions,** however, relatively large insoluble particles are involved rather than soluble molecules. Because of this, bulkier aggregates form, which are much easier to see.

In direct agglutination tests, an antibody suspension is mixed with the insoluble antigen, such as red blood cells, bacteria, or fungi. If the antibodies bind to the antigens, visible clumping will occur—a positive test. The agglutination of red blood cells by antibody binding or other means is referred to as hemagglutination, and is used in blood typing (**figure 18.7**).

Passive agglutination tests use either antibodies or antigens attached to particles such as latex beads to make the aggregates larger and therefore easier to see. Latex beads to which specific antibodies have been attached are produced commercially and

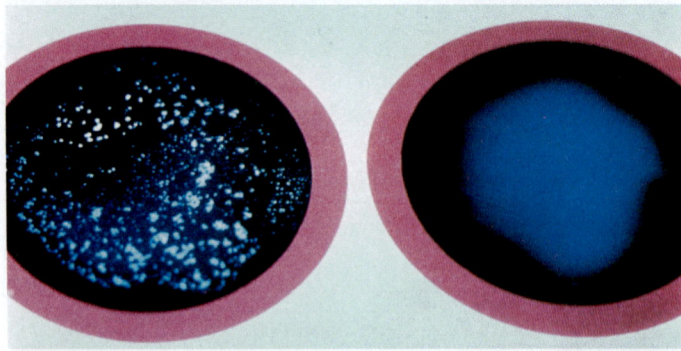

FIGURE 18.8 Latex Agglutination In this example, the latex beads are coated with antibodies that bind specifically to cell wall antigens of *Streptococcus pyogenes*. Visible clumping (shown on the left), confirms that the organism is *S. pyogenes*. Negative test results are shown on the right.

? *Why coat latex beads with antibodies rather than simply adding the antibodies to a bacterial suspension?*

used to test for various bacteria, fungi, viruses, and parasites, as well as hormones, drugs, and other substances. The beads are mixed with a drop of a body fluid or suspended microbial culture. If the specific antigen is present, visible clumps form. Latex agglutination tests are commonly used to identify beta-hemolytic *Streptococcus* species (**figure 18.8**).

MicroAssessment 18.4

Agglutination and precipitation reactions both depend on cross-linking and lattice formation of antigen-antibody complexes.

10. *How do the antigens used in precipitation reactions differ from those in direct agglutination tests?*
11. *How is a direct agglutination test different from a passive agglutination test?*
12. *In precipitation reactions, why can cross-linked lattices not form when there is an excess of antibody?* ✚

18.5 ■ Using Labeled Antibodies to Detect Antigen-Antibody Interactions

Learning Outcomes

7. *Explain how labeled antibodies are used in direct and indirect tests.*
8. *Compare and contrast fluorescent antibody tests, ELISAs, and Western blots.*
9. *Describe how the fluorescence-activated cell sorter is used in immunoassays.*

Detectable markers such as enzymes, fluorescent dyes, and radioactive tags can be attached to antibodies, which are then used to detect certain antigens. Marking antigens with fluorescently labeled antibodies also provides a mechanism for sorting antigens.

Basic Principles

Labeled antibodies can be used to identify bacteria or other antigens, or they can be used to detect antibodies of a given specificity. The tests can be either direct or indirect (**figure 18.9**).

Direct tests are typically used to identify an unknown antigen. To do this, labeled antibodies of known specificity are added to a preparation of the antigen attached to a solid surface. For example, if the antigen is suspected of being "antigen X," then antibodies specific to that antigen are added. After a washing step to remove unbound antibodies, the presence of the labeled antibodies bound to the antigen identifies the antigen. A variety of antibodies that bind specifically to common pathogens are commercially available, and these can be purchased with different detectable labels.

Indirect tests are often used to detect antibodies of a given specificity in a patient's serum. The tests are called indirect because they require a labeled secondary antibody to detect the first (or primary) antibody. To determine if a serum sample contains antibodies of a given specificity, the serum is added to a known antigen attached to a solid surface. Any unbound serum

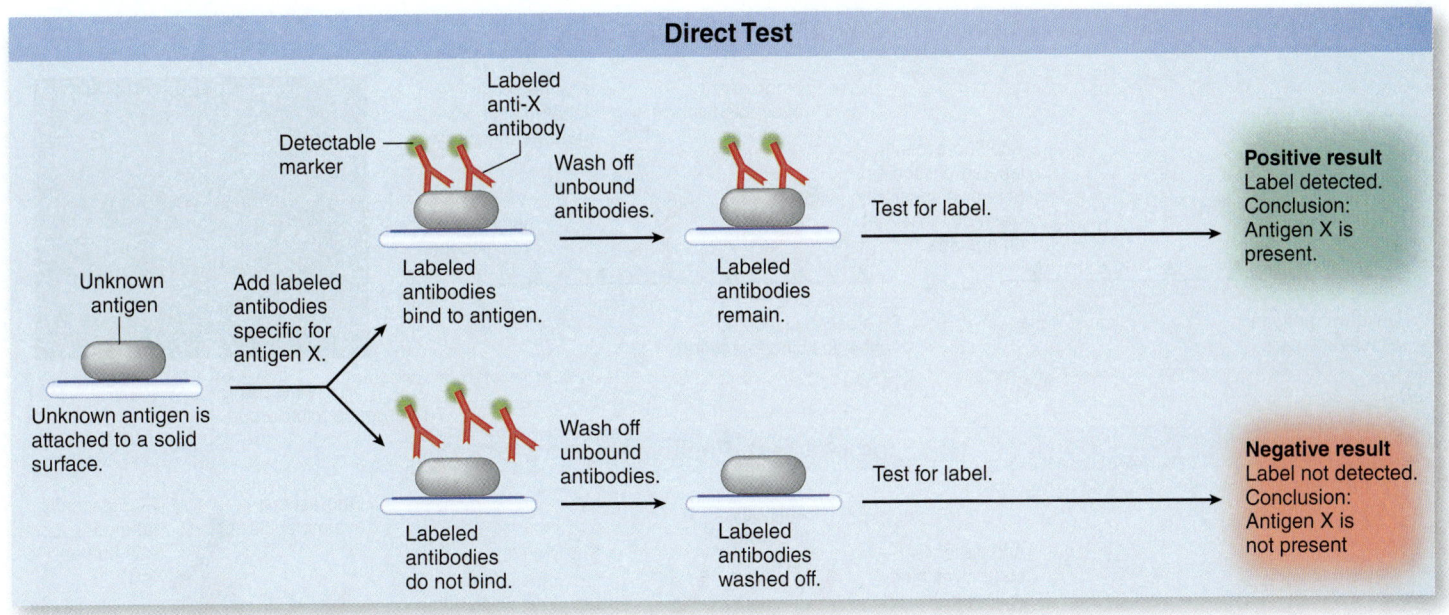

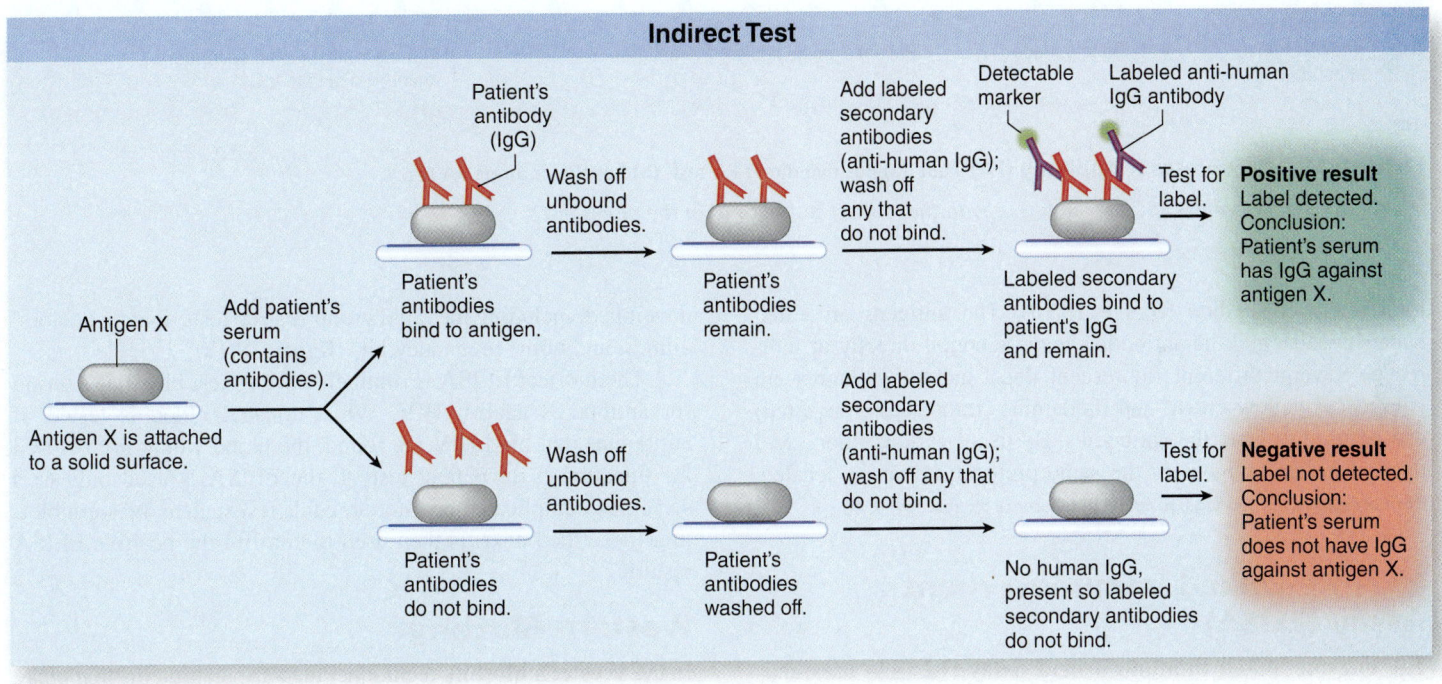

FIGURE 18.9 Basic Principles of Tests That Use Labeled Antibodies to Detect Antigen-Antibody Interactions (a) Positive and negative results of a direct test. (b) Positive and negative results of an indirect test.

? *In (b), why would it be more efficient to use labeled anti-human IgG rather than label the patient's antibodies?*

antibodies are then washed off. The secondary antibody is then added, which in this case would be labeled anti-human IgG antibodies. These then bind to any IgG molecules bound to the antigen. After another washing step, the presence of the labeled secondary antibody identifies the primary antibody bound to the antigen. Note that the usefulness of the labeled secondary antibodies is they can bind to any human IgG molecules, regardless of the source or specificity. Thus, they can be used to detect specific antibodies in a variety of different patients. The alternative, labeling the antibodies from each patient tested, would be cost-prohibitive. ◄◄ anti-human IgG antibodies, p. 427

Fluorescent Antibody (FA) Test

The **fluorescent antibody (FA) test** use fluorescence microscopy to locate fluorescently labeled antibodies bound to antigens fixed

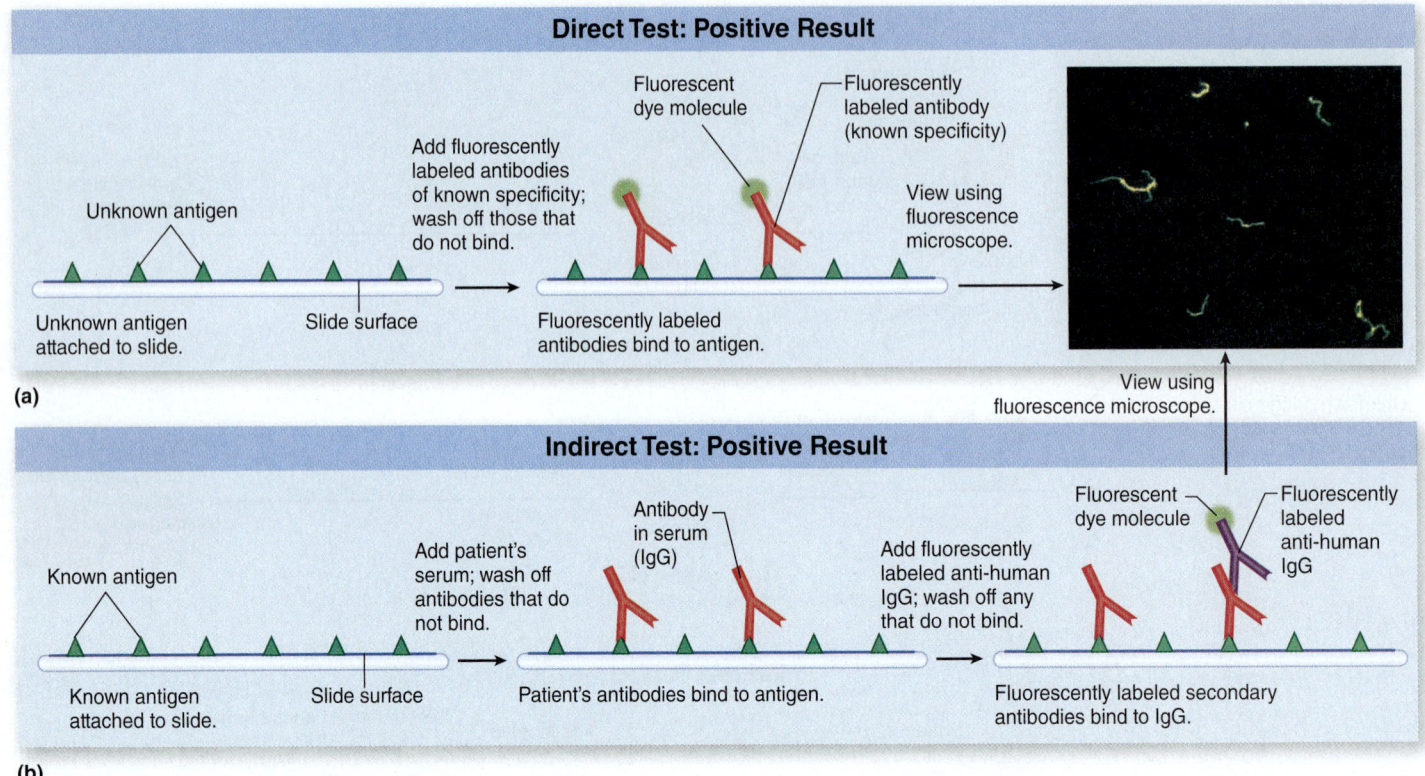

FIGURE 18.10 Fluorescent Antibody (FA) Test **(a)** Positive direct FA test. **(b)** Positive indirect FA test.

❓ *In this diagram, why is anti-human IgG used in the indirect test but not in the direct test?*

to a microscope slide (**figure 18.10**). The antigens are often bacterial cells, and the antibodies may be bound directly or indirectly. Several different fluorescent dyes, including fluorescein (fluoresces yellow-green) and rhodamine (fluoresces orange-red) can be used to label the antibodies. By using various fluorescent dyes, different antigens in the same preparation can be located.

◀◀ fluorescent dyes and tags, p. 49 ◀◀ fluorescence microscopy, p. 44

Enzyme-Linked Immunosorbent Assay (ELISA)

The **enzyme-linked immunosorbent assay (ELISA)** uses antibodies labeled with an enzyme such as peroxidase from the horseradish plant. The enzyme can be detected using a colorimetric assay, which measures the enzymatic conversion of a colorless substrate into a colored product (**figure 18.11**). The ELISA is often done in microtiter plates, making it useful for screening large numbers of specimens. The method is widely used, and a variety of prepared ELISA plates are commercially available.

In direct ELISAs, the antigen is sometimes "captured" by antibodies that have been attached to the surface. For example, the wells of ELISA plates used to diagnose giardiasis are coated with antibodies that bind antigens of the causative agent, *Giardia lamblia*. When a stool sample is added to the well, the antibodies capture the antigens. An enzyme-labeled antibody is then used to detect, and therefore identify, the captured antigens. In addition to standard ELISAs, commercial modifications of the test have been

developed, including the rapid group A strep tests done in doctors' clinics, and home pregnancy kits (**figure 18.12**).

The indirect ELISA is routinely used to test blood and serum for antibodies against HIV. When donated blood is tested, if antibodies that bind HIV are found, the blood would not be used for transfusion. In patient testing, the ELISA is used only as a screen. A complicated but more reliable test such as Western blotting (described next) is then used to confirm the positive ELISA results.

Western Blotting

In the **Western blotting** technique, the various proteins that make up an antigen are separated by size before reacting them with antibodies. This makes it possible to determine exactly which proteins the antibodies are recognizing, an essential part of accurate HIV testing (**figure 18.13**).

To separate the proteins, a special type of gel electrophoresis called SDS-PAGE is used. This separates proteins of different sizes into a series of bands, as smaller proteins move faster than the larger ones. The separated proteins are then transferred to a nylon membrane to immobilize them before a solution of antibody molecules is added. The steps after that are very similar to those of an ELISA. To determine if a patient's serum has antibodies specific for the separated proteins, a sample is added to the blot, and then unbound antibodies are washed off. Enzyme- or radioactively labeled anti-human IgG antibodies are then added, which attach to any antibodies remaining from the serum. Unbound labeled

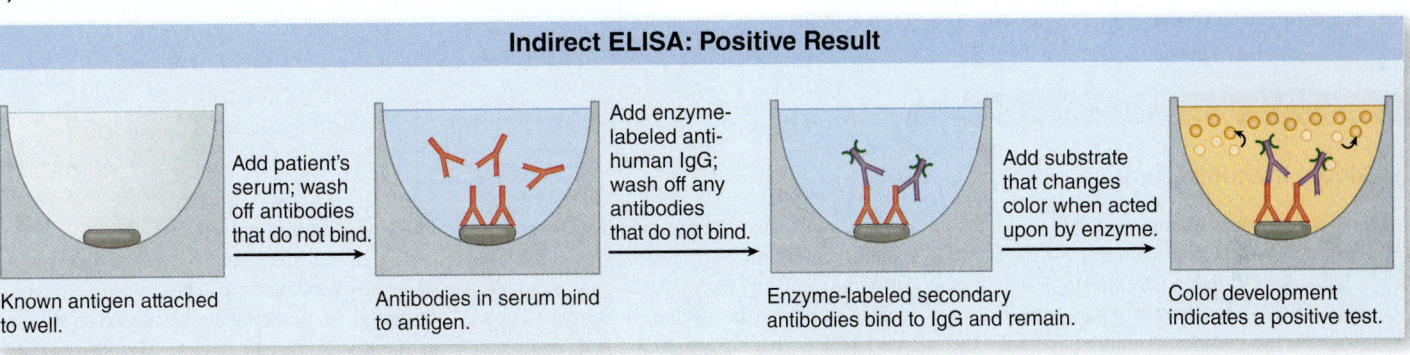

Direct ELISA (Sandwich Method): Positive Result

Antibodies to known antigen attached to well.

Add specimen.

Antibodies in the well "capture" antigen in specimen.

Add enzyme-labeled antibodies of known specificity; wash off any that do not bind.

Enzyme

Enzyme-labeled secondary antibody

Enzyme-labeled antibodies of known specificity bind antigen and remain.

Add substrate that changes color when acted upon by enzyme.

Color development indicates a positive result.

(a)

Indirect ELISA: Positive Result

Known antigen attached to well.

Add patient's serum; wash off antibodies that do not bind.

Antibodies in serum bind to antigen.

Add enzyme-labeled anti-human IgG; wash off any antibodies that do not bind.

Enzyme-labeled secondary antibodies bind to IgG and remain.

Add substrate that changes color when acted upon by enzyme.

Color development indicates a positive test.

(b)

FIGURE 18.11 Enzyme-Linked Immunosorbent Assay (ELISA) (a) Positive direct ELISA. (b) Positive indirect ELISA.

? *In the first panel of part (a), what is the purpose of the antibodies attached to the well?*

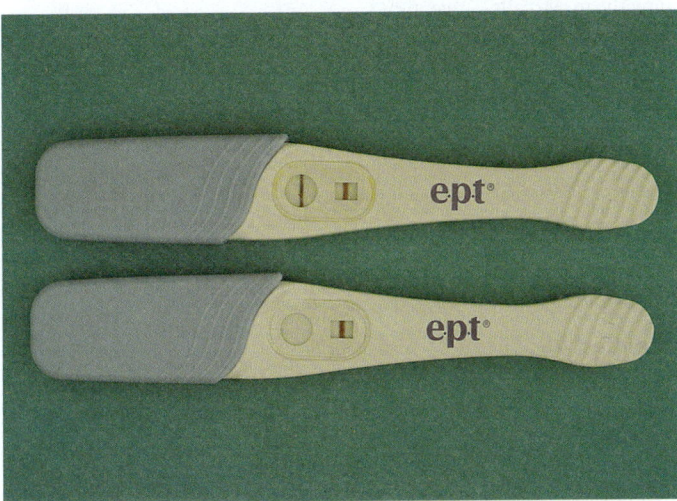

FIGURE 18.12 ELISA Test for Pregnancy The test detects human chorionic gonadotropin (HCG), an antigen present only in pregnant women. A urine sample is applied on the left. Two pink lines indicate reaction of HCG with antibodies, a positive test. A single line indicates absence of HCG in the urine, a negative test.

? *What captures the antigen on the strip?*

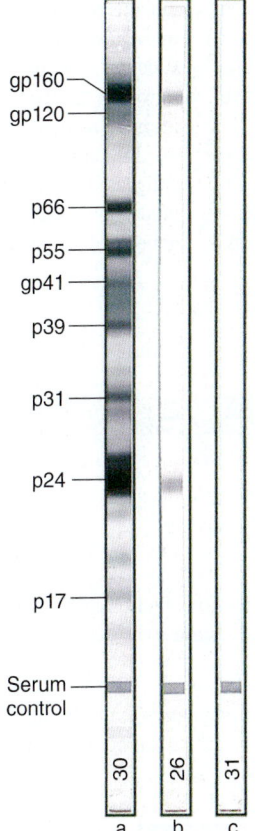

gp160
gp120
p66
p55
gp41
p39
p31
p24
p17
Serum control

30 26 31
a   b   c

(a) Strong reactive control
(b) Weak reactive control
(c) Non-reactive control

FIGURE 18.13 Western Blotting Results Each strip in this photograph contains HIV proteins that have been separated by gel electrophoresis and then transferred in-place to the strips, along with a serum control. The dark bands indicate the locations of bound antibodies. The results for the strong reactive control (lane a) show that antibodies have bound all HIV proteins on the strip. The results for the weak reactive control (lane b) show binding of antibodies to only certain critical HIV proteins (gp160/120 and p24). The results for the non-reactive control (lane c) show no binding to HIV proteins.

? *Why are the results of the Western blot more reliable than those of the ELISA?*

antibodies are then washed off. Finally, the label is detected, generating a bar code–like pattern that indicates which proteins were recognized by the patient's antibodies.

Fluorescence-Activated Cell Sorter (FACS)

As described in chapter 4, flow cytometry counts cells or other particles by measuring the light scattered as they pass single-file by a laser. A specialized version called the **fluorescence-activated cell sorter (FACS)** can be used to count and separate cells labeled with fluorescent antibodies. For example, subsets of T cells (CD4$^+$ and CD8$^+$) can be counted and even separated by first mixing the cells with two preparations of monoclonal antibodies (one binds CD4 markers and the other CD8 markers), each labeled with a different fluorescent dye. ◀◀ flow cytometry, p. 99 ◀◀ CD markers, p. 369

MicroAssessment 18.5

Antibodies labeled with a detectable marker can be used to identify an antigen (direct test) or to detect a patient's antibodies bound to a known antigen (indirect test). Examples of methods that use labeled antibodies include fluorescence antibody tests, ELISA, and Western blotting. Fluorescence-activated cell sorters can separate and count cells labeled with fluorescent antibodies.

13. *What equipment is required for a fluorescent antibody test that is not required for an ELISA?*

14. *In HIV testing, why is the Western blot used to confirm ELISA results?*

15. *Why is a false positive more significant in HIV testing of patients than in screening donated blood for transfusions?* ✚

FUTURE CHALLENGES 18.1

Global Immunization

In addition to developing new safe and effective vaccines, a major challenge for the future is delivering available vaccines to populations worldwide. When this was done in the case of smallpox, the disease was eliminated from the world; and poliomyelitis is almost eradicated now. The World Health Organization and the governments that support it deserve much of the credit for these achievements. Much remains to be done, however, and gifts totaling well over a billion dollars from Bill Gates, one of the founders of Microsoft, and his wife have stimulated further progress in this area.

To immunize universally, it is necessary to have vaccines that are easily administered, inexpensive, stable under a variety of environmental conditions, and preferably painless.

Instead of expensive needles and syringes and painful injections, vaccines may be delivered in a number of easy ways. For example, naked DNA can be coated onto microscopic gold pellets, which are shot from a gunlike apparatus directly through the skin into the muscle, painlessly. Skin patches deliver antigens slowly through the skin. Vaccines against mucous membrane pathogens can be delivered by a nasal spray, as in some new influenza vaccines, or by mouth. Time-release pills introduce antigen steadily to give a sustained immune response.

In addition to sprays and pills that deliver antigens to mucous membranes in order to induce mucosal immunity, techniques are being developed to get antigens directly to M cells

in the gastrointestinal mucosa. These are the cells that transfer antigens across the membrane to the Peyer's patches, where immune responses occur. Antigens are incorporated into substances known to bind to M cells, thereby making it easier for them to enter M cells and Peyer's patches. ◀◀ M cells, p. 358

One of the major remaining challenges is development of an effective vaccine against HIV that can be administered universally. Although anti-HIV vaccines are being studied, prospects are dim for a truly effective immunization program for HIV disease. It is also likely that even if HIV disease is brought under control, a new and currently unforeseen challenge will arise during the twenty-first century.

Summary

IMMUNIZATION

18.1 ■ Principles of Immunization (figure 18.2)

Active Immunity

Active immunity occurs naturally in response to infections or other natural exposure to antigens, and artificially in response to vaccination.

Passive Immunity

Passive immunity occurs naturally during pregnancy and through breast feeding, and artificially by transfer of preformed antibodies, as in **immune globulin** and **hyperimmune globulin.**

18.2 ■ Vaccines and Immunization Procedures

A **vaccine** is a preparation of a pathogen or its products used to induce active immunity. It protects an individual against disease, and can also provide **herd immunity** (table 18.1).

Attenuated Vaccines

An **attenuated vaccine** is a weakened form of the pathogen that can replicate but is generally unable to cause disease.

Inactivated Vaccines

Inactivated vaccines are unable to replicate but they retain the immunogenicity of the infectious agent or toxin. They include inactivated whole agents, toxoids, subunits, VLPs, polysaccharides, and conjugates. **Adjuvants** increase the intensity of the immune response to the antigen in a vaccine.

An Example of Vaccination Strategy—The Campaign to Eliminate Poliomyelitis

Both the Sabin and the Salk polio vaccines protect against paralytic poliomyelitis; the Sabin vaccine induces mucosal immunity.

The Importance of Childhood Immunizations (table 18.3)

Routine childhood immunizations have prevented millions of cases of disease and many deaths.

Current Progress in Immunization

Progress in vaccination includes enhancing the immune response to vaccines and developing new or improved vaccines against certain diseases (table 18.4).

IMMUNOLOGICAL TESTING

18.3 ■ Principles of Immunological Testing

A seronegative individual becomes seropositive after initial infection with an agent; this **seroconversion** takes about a week to 10 days. To determine if a patient has antibodies in the blood against a specific infectious agent, the patient's **serum** or **plasma** is tested.

Obtaining Antibodies

Polyclonal antibodies recognize multiple epitopes, whereas **monoclonal antibodies** recognize only a single epitope. **Anti-human IgG antibodies** can detect IgG molecules in a patient specimen. Recombinant DNA techniques have been used to produce humanized monoclonal antibodies.

Quantifying Antigen-Antibody Reactions

The concentration of antibody molecules in a specimen is usually determined by making serial dilutions; the last dilution that gives a detectable antigen-antibody reaction reflects the **titer.**

18.4 ■ Observing Antigen-Antibody Aggregates

Precipitation Reactions

Antibodies bound to soluble antigens may form complexes that precipitate out of solution (figure 18.5). An example of a test that involves a **precipitation reaction** is the **Ouchterlony** technique (figures 18.6).

Agglutination Reactions

Agglutination reactions are similar in principle to precipitation reactions, but insoluble particles are involved, causing obvious aggregates to form; examples include direct agglutination tests and passive agglutination tests (figures 18.7, 18.8).

18.5 ■ Using Labeled Antibodies to Detect Antigen-Antibody Interactions

Basic Principles

Tests that use labeled antibodies can be either direct or indirect (figure 18.9).

Fluorescent Antibody (FA) Test

The **fluorescent antibody (FA) test** relies on fluorescence microscopy to locate fluorescently labeled antibodies bound to antigens fixed to a microscope slide (figure 18.10).

Enzyme-Linked Immunosorbent Assay (ELISA)

The **enzyme-linked immunosorbent assay (ELISA)** uses antibodies labeled with a detectable enzyme (figure 18.11).

Western Blotting

In the **Western blotting** technique, the various proteins that make up an antigen are separated by size before reacting them with antibodies (figure 18.13).

Fluorescence-Activated Cell Sorter (FACS)

The **fluorescence-activated cell sorter (FACS)** can be used to count and separate antigens labeled with fluorescent antibodies.

Review Questions

Short Answer

1. How is immune globulin different from hyperimmune globulin?
2. Describe two advantages of an attenuated vaccine over an inactivated one.
3. Describe two advantages of an inactivated vaccine over an attenuated one.
4. What is herd immunity?
5. Describe how both active and passive immunization can be used to combat tetanus.
6. Why are humanized monoclonal antibodies better for therapy than the original versions?
7. In a precipitation reaction, what is meant by "optimal proportions"?
8. Is blood typing an example of a precipitation reaction or an agglutination reaction?
9. An ELISA test is used to screen patient specimens for HIV. A positive ELISA test is confirmed by a Western blot test. Why not the other way around, with the ELISA second?
10. What is the purpose of anti-human IgG antibodies in immunological testing?

Multiple Choice

1. Which is an example of immunization that elicits active immunity?
   a) Giving antibodies against diphtheria
   b) Immune globulin injections to prevent hepatitis
   c) Sabin polio immunization
   d) Rabies immune globulin
   e) Tetanus immune globulin
2. Breast feeding provides which of the following to an infant?
   a) Artificial active immunity
   b) Artificial passive immunity
   c) Natural active immunity
   d) Natural passive immunity
3. Vaccines ideally should be all of the following, *except*
   a) effective in protecting against the disease.
   b) inexpensive.
   c) stable.
   d) living.
   e) easily administered.

4. Severely immunosuppressed people should not receive the measles vaccine. Based on this information, the vaccine is
   a) an inactivated whole agent.
   b) a toxoid.
   c) a subunit vaccine.
   d) a genetically engineered vaccine against hepatitis B.
   e) an attenuated vaccine.

5. All of the following are attenuated vaccines *except*
   a) chickenpox.   b) mumps.   c) rubella.
   d) Salk polio.   e) yellow fever.

6. An important subunit vaccine that is widely used is the
   a) pertussis vaccine.   b) Sabin vaccine.   c) Salk vaccine.
   d) measles vaccine.   e) mumps vaccine.

7. In quantifying antibodies in a patient's serum
   a) total protein in the serum is measured.
   b) the antibody is usually measured in grams per ml.
   c) the serum is serially diluted.
   d) both antigen and antibody are diluted.
   e) the titer refers to the amount of antigen added.

8. Which of the following about immunological testing is *false*?
   a) Polyclonal antibody preparations recognize multiple epitopes.
   b) Monoclonal antibodies recognize a single epitope.
   c) Serum and plasma can both be tested for antibodies.
   d) The direct ELISA uses anti-human IgG antibodies.
   e) A rise in specific antibody titer indicates an active infection.

9. All of the following are matching pairs *except*
   a) ELISA—radioactive label.
   b) fluorescence-activated cell sorter—flow cytometry.
   c) fluorescent antibody test—microscopy.
   d) Western blot—gel electrophoresis.

10. Which of the following would be most useful for screening thousands of specimens for antibodies that indicate a certain disease?
    a) Western blot   b) Fluorescent antibody   c) ELISA
    d) All of the above   e) None of the above

Applications

1. A new parent asks you which vaccines the CDC recommends for a 2-month-old infant. What is your answer? The chapter 18 readings at the text website (**www.mhhe.com/nester7**) provide a link to the CDC's recommended immunization schedules.

2. There has been debate about keeping smallpox virus stored, since the disease has been eradicated. What would be an argument for keeping the virus? What should be done to protect against use of the virus in biological warfare?

Critical Thinking ✚

1. In figure 18.5, how would the curve change if the concentration of antibody in the original sample were increased? (Would the shape of the curve change? Would the curve be shifted left, right, up, or down?) Briefly explain your answer.

2. *Staphylococcus aureus* makes a protein called protein A, which binds to the Fc region of antibody molecules from a wide variety of species. How could protein A be exploited in immunoassays?

19 Epidemiology

CDC microbiologist studies influenza virus.

A Glimpse of History

Puerperal fever, a bacterial infection of the uterus following childbirth, rose to epidemic proportions in the eighteenth century when more women chose to deliver their babies in hospitals. By the middle of the nineteenth century in the hospitals of Vienna, the major medical center of the world at that time, about one of every eight women died of puerperal fever following childbirth.

In 1841, Ignaz Semmelweis, a Hungarian, traveled to Vienna to study medicine. After medical school, he worked for Professor Johann Klein, who managed one section of the Lying-In Hospital with the medical students. Midwives and midwifery students served a second section. Semmelweis noticed that the incidence of puerperal fever in Dr. Klein's section was as high as 18%, four times that in the section served by the midwives. During this period, a friend of Semmelweis incurred a scalpel wound while doing an autopsy and died of symptoms very similar to those of puerperal fever. Semmelweiss reasoned that the "poison" that killed his friend probably also contaminated the hands of the medical students who did autopsies. Perhaps these students were transferring the "poison" from the cadavers to the women in childbirth. Midwives, after all, did not perform autopsies. This was before Pasteur and Koch established the germ theory of disease, so Semmelweis had no way of knowing that the "poison" being transferred was probably *Streptococcus pyogenes,* a common cause of many infections, including puerperal fever.

To test his hypothesis that physicians and students were transferring "poison" to patients, Semmelweis had them wash their hands with a strong disinfectant before attending patients. The incidence of puerperal fever dropped to one-third its previous level. Instead of appreciating these findings, Semmelweis's colleagues refused to admit responsibility for the deaths of so many patients. His work was so fiercely attacked that he was forced to leave Vienna and return to his native Hungary. Although his disinfection techniques achieved a remarkable reduction in the number of deaths from puerperal fever there as well, Semmelweis became increasingly outspoken and bitter. He was finally confined to a mental institution, where he died one month later of a generalized infection similar to the kind that had killed his friend and the many women who had contracted puerperal fever following childbirth.

Epidemiology is the study of disease patterns in populations. Epidemiologists—the "disease detectives"—collect and compile data about sources of disease and associated risk factors to design infection control strategies and to prevent or predict the spread of disease. They approach a disease outbreak much as a criminal detective describes the scene of a crime using expertise in diverse disciplines including ecology, microbiology, sociology, statistics, and psychology. Many of our daily habits, from handwashing to waste disposal, are based on the work of epidemiologists.

19.1 ■ Principles of Epidemiology

Learning Outcomes

1. *Explain why epidemiologists are most concerned with the rate of disease rather than the number of cases.*

2. *Describe movement of a pathogen from its reservoir to a host, and back to its reservoir or to another host.*

3. *Explain how characteristics of a pathogen or of a host can influence the epidemiology of a disease.*

Diseases that can be transmitted from one host to another, such as measles, colds, and influenza, are **contagious, or communicable, diseases.** Disease transmission is determined by interactions between the environment, the pathogen, and the host. Control of any of these factors may break the cycle of infection, hindering or preventing the disease. For example, improved sanitation prevents the pathogen from accessing the host; antimicrobial medications can kill or inhibit infecting organisms; and vaccination increases resistance of the host.

Diseases that do not spread from one host to another are called **non-communicable diseases.** Microorganisms that cause these diseases most often arise from an individual's normal microbiota or from the environment. *Clostridium tetani,* the bacterium that causes tetanus, can enter a host from the environment but is not then transmitted to another host.

Rate of Disease in a Population

Epidemiologists generally are less concerned with the absolute number of cases of a disease than they are with the rate. For example, 100 people in a city of 1,000,000 developing genital herpes in a given time period is not as alarming as 100 people in a town of 5,000 developing the same disease over the same time period. The much higher rate or proportion of the population infected is of greater concern to the epidemiologist.

The **attack rate** describes the percentage of people who become ill in a population after exposure to an infectious agent. For example, if 100 people at a party eat chicken contaminated with *Salmonella,* and 10 people develop symptoms of salmonellosis, then the attack rate is 10%. The attack rate reflects many factors, including the infectious dose of the organism and the immune status of the population.

The **incidence** of a disease is the number of new cases in a specific time period in a given population; the **prevalence** of a disease is the total number of cases at any time or for a specific period in a given population. Usually these are expressed as a rate, the number of cases per 100,000 people. Incidence provides a useful measure of the risk of an individual contracting the disease. Prevalence reflects the overall impact of a disease on society because it includes both old and new cases, taking the duration of the disease into account.

Morbidity refers to the incidence of disease in a population at risk. Contagious diseases such as influenza often have a high morbidity rate because each infected individual may transmit the infection to several others. **Mortality** refers to the overall death rate in a population. In developed countries, mortality is most often associated with non-communicable diseases such as cancer or heart attack. In developing countries, however, communicable diseases are still a major cause of death. **Case-fatality rate** is the percentage of a population that dies from a specific disease. Diseases such as plague and Ebola hemorrhagic fever are feared because of their very high case-fatality rate. As the case-fatality rate for AIDS has decreased due to improved treatment, the prevalence of AIDS has increased because more people with the disease survive within the population.

Endemic diseases are constantly present in a given population. For example, the common cold is endemic in the United States. When cases occur only from time to time, they are **sporadic.** An unusually large number of cases in a population constitutes an **epidemic.** Epidemics may be caused by diseases not normally present in a population, such as cholera being reintroduced to the Western Hemisphere, or by fluctuations in the incidence of endemic diseases such as influenza and pneumonia (**figure 19.1**). An **outbreak** describes a cluster of cases occurring during a brief time interval and affecting a specific population; an outbreak may herald the onset of an epidemic. When an epidemic spreads worldwide, as AIDS has, it is called a **pandemic.**

Reservoirs of Infection

A pathogen must have a suitable environment in which to live. That natural habitat, the **reservoir of infection,** may be on or in an animal, including humans, or in an environment such as soil or water (**figure 19.2**). The reservoir of infection affects the extent

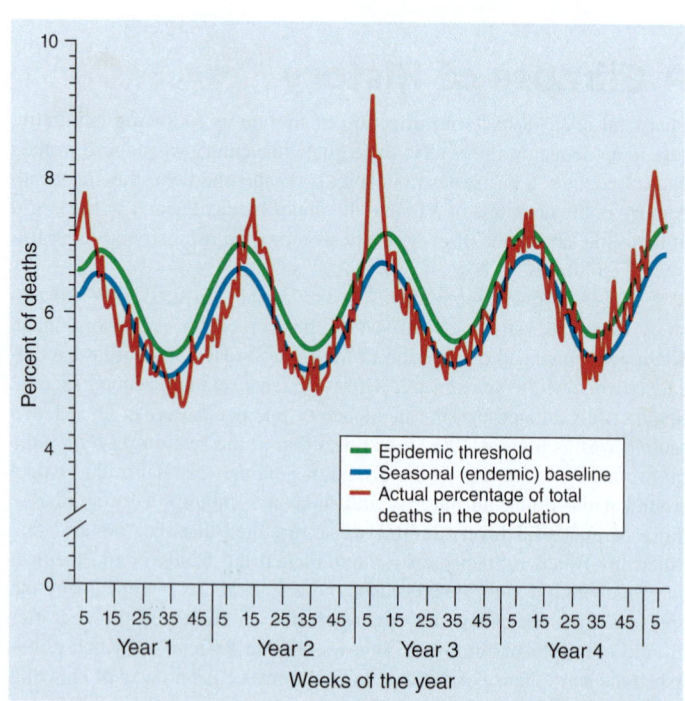

FIGURE 19.1 Endemic Disease Can Be Epidemic Example of yearly fluctuation of pneumonia and influenza deaths (expressed as a percentage of all deaths).

❓ *During what season(s) of the year is a pneumonia epidemic most likely?*

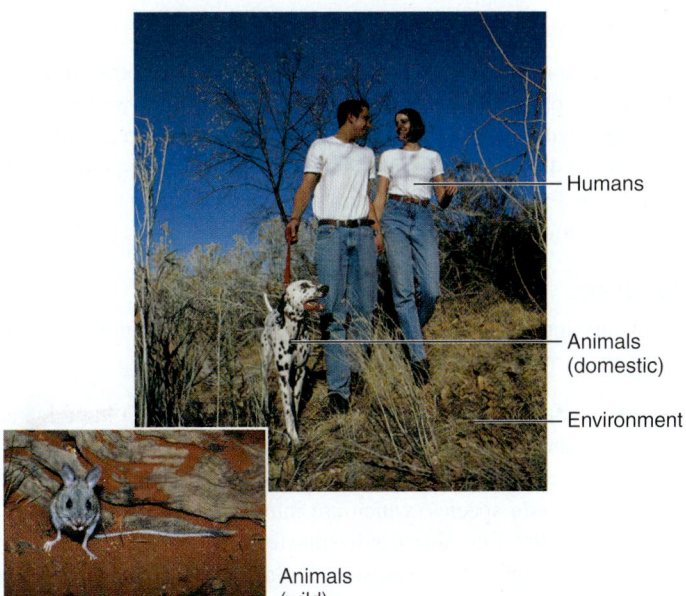

- Humans
- Animals (domestic)
- Environment

Animals (wild)

FIGURE 19.2 Reservoirs of Infection

❓ *How might you be a reservoir of infection?*

and distribution of a disease. Once the reservoir is identified, people can be prevented from coming into contact with the disease source. The United States does not have epidemics of plague, in part because populations of wild rats, mice, and prairie dogs are controlled. These are the natural reservoirs of *Yersinia pestis,* the bacterium that causes plague. ▶▶ *Yersinia pestis,* **p. 678**

Human Reservoirs

Infected humans are a significant reservoir of most communicable diseases. In some cases, humans are the only reservoir. In other instances, the pathogen can also exist in other animals and, occasionally, in the environment as well. When infected humans are the only reservoir, the disease can be easier to control because it is more feasible to set up prevention and control programs in humans than in wild animals. The eradication of smallpox is an excellent example. The combined effects of widespread vaccination programs (which resulted in fewer susceptible people) and isolation of those who became infected eliminated the smallpox virus from nature. The virus no longer had a reservoir in which to multiply.

Symptomatic Infections People with symptomatic illnesses are an obvious source of infectious agents. Ideally they understand the importance of taking precautions to avoid transmitting their illness to others. Staying home and resting while ill both helps the body recover and protects others from exposure to the disease-causing agent. Even conscientious people, however, can unintentionally be a source of infection. For example, people who are in the incubation period of mumps shed virus before symptoms appear. ◀◀ incubation period, **p. 383** ▶▶ mumps, **p. 583**

Asymptomatic Carriers A person can harbor a pathogen with no ill effects, acting as a carrier of the disease agent. These people may shed the organism intermittently or constantly for months, years, or even a lifetime.

Some carriers have an asymptomatic infection; their immune system is actively responding to the invading microorganism, but they have no obvious clinical symptoms. Because these people often have no reason to consider themselves a reservoir, they move freely about, spreading the pathogen. People with asymptomatic infections are a significant complicating factor in the control of sexually transmitted infections (STIs) such as gonorrhea. Up to 50% of women infected with *Neisseria gonorrhoeae* have no symptoms, and may unknowingly transmit the disease to their sexual partners. In contrast, infected men are more often symptomatic and seek medical treatment. Gonorrhea infections can be treated with antibacterial drugs, but locating sexual contacts of infected people is often difficult and costly. ▶▶ *Neisseria gonorrhoeae,* **p. 622**

Some potentially pathogenic microbes can colonize the skin or mucosal surfaces, establishing themselves as part of a person's microbiota. For instance, many people carry *Staphylococcus aureus* as a part of their nasal or skin microbiota. Carriers of *S. aureus* may never have any illness or disease as a result of the organism, but they remain a potential source of infection to themselves and others. Unfortunately, ridding a colonized carrier of the infectious organism is often difficult, even with the use of antimicrobial drugs. ▶▶ *Staphylococcus aureus,* **p. 524**

Non-Human Animal Reservoirs

Non-human animal reservoirs are the source of many pathogens. Poultry are a reservoir of gastrointestinal pathogens such as species of *Campylobacter* and *Salmonella.* In the United States, raccoons, skunks, and bats are reservoirs of the rabies virus. Occasional transmission of plague to humans is still reported in the southwestern states where *Y. pestis* is endemic in prairie dogs and other rodents. Rodents, particularly the deer mouse, are the reservoir for hantavirus. ▶▶ *Campylobacter,* **p. 593** ▶▶ rabies, **p. 658**

Diseases such as plague and rabies that can be transmitted to humans but primarily exist in other animals are called zoonotic diseases, or **zoonoses.** Zoonotic diseases are often more severe in humans than in the typical animal host because the infection in humans is accidental; there has been no evolution toward the balanced pathogenicity that normally exists between host and parasite. ◀◀ balanced pathogenicity, **p. 385**

Environmental Reservoirs

Some pathogens have environmental reservoirs. *Clostridium botulinum,* which causes botulism, and *Clostridium tetani,* which causes tetanus, are both widespread in soils. Pathogens that have environmental reservoirs are difficult or impossible to eliminate. ▶▶ *Clostridium botulinum,* **p. 652** ▶▶ *Clostridium tetani,* **p. 555**

Portals of Exit and Portals of Entry

A pathogen must leave its reservoir to be transmitted to a susceptible host. If the reservoir is an animal, the body orifice or surface from which a microbe is shed is called the **portal of exit.** To complete the cycle of infection, the pathogen is transmitted to the next

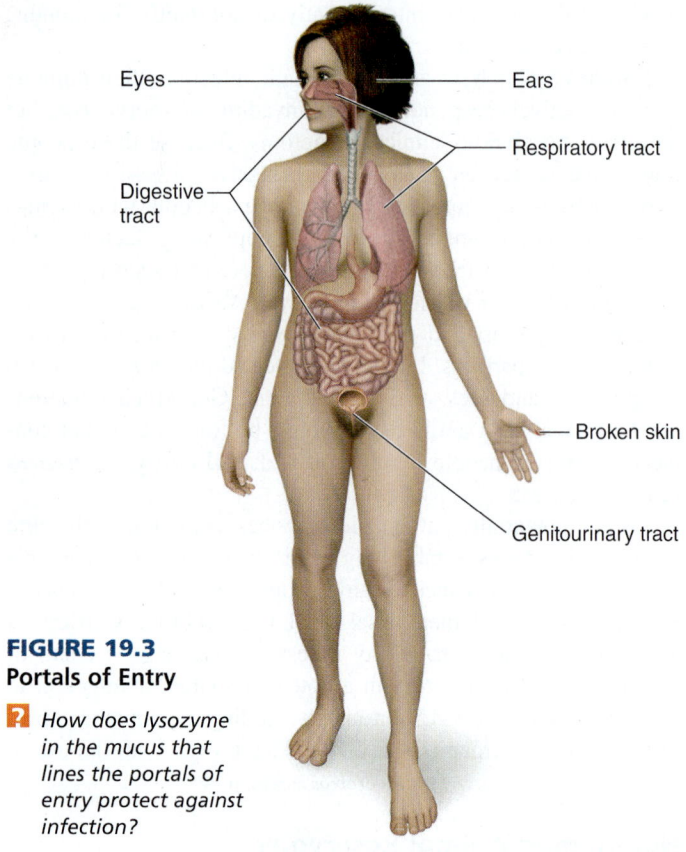

Eyes
Ears
Digestive tract
Respiratory tract
Broken skin
Genitourinary tract

FIGURE 19.3
Portals of Entry

? *How does lysozyme in the mucus that lines the portals of entry protect against infection?*

host and enters through a body surface or orifice called the **portal of entry** (**figure 19.3**).

Microorganisms that inhabit the intestinal tract are shed in the feces. Pathogens such as *Vibrio cholerae,* which cause massive volumes of watery diarrhea, can contaminate drinking water and food. Pathogens such as *Mycobacterium tuberculosis* and various respiratory viruses exit the body in droplets of saliva and mucus when people talk, laugh, sing, sneeze, or cough. Organisms that inhabit the skin are constantly shed on skin cells. Even as you read this text you are shedding skin cells, some of which may have *Staphylococcus aureus* on their surface. Genital pathogens such as *Neisseria gonorrhoeae* can be carried in semen and vaginal secretions. Hantavirus is found in the saliva, urine, and droppings of deer mice. ▶▶| *Vibrio cholerae,* p. 586 ▶▶| *Mycobacterium tuberculosis,* p. 502 ▶▶| hantavirus, p. 512

To cause disease, not only must a pathogen be transmitted to a new host, it must also colonize a surface of or enter that new host. Cells of a *Shigella* species can be transferred by a handshake, for example, but they will only cause disease if they are then transferred to the mouth or to food that is then eaten. In this case, the mouth is the portal of entry. Respiratory pathogens released into the air during a cough generally cause disease only when someone inhales them. For these pathogens, the nose is the portal of entry. Many organisms can cause disease if they enter one body site, but are harmless if they enter another. For instance, *Enterococcus faecalis* may cause a bladder infection if it enters the urinary tract, but it is harmless when it is ingested and then colonizes the large intestine, where it may reside as part of the normal microbiota.

Disease Transmission

Transmission of a pathogen from one person to another through the air, by physical contact, by ingestion of food or water, or via a living agent such as an insect is called **horizontal transmission** (**figure 19.4**). This contrasts with **vertical transmission,** which is the transfer of a pathogen from a pregnant woman to her fetus, or from a mother to her infant during childbirth or breast feeding.

Contact

Transmission of a pathogen from one host to another often involves direct or indirect contact.

Direct Contact **Direct contact** can be as simple as a handshake, or as intimate as sexual intercourse. Sometimes the transfer of very few microbes can initiate an infection. For example, the infectious dose of *Shigella* species, which are intestinal pathogens, is about 10 to 100 cells, a number easily transferred when shaking hands. Once on the hands, the organisms can inadvertently be ingested—resulting in **fecal-oral transmission.** Handwashing, a simple routine that physically removes microbes, is important in preventing this type of spread of disease. Even washing in plain water reduces the numbers of potential pathogens on the hands, which decreases the possibility of transferring or ingesting enough cells to establish an infection. In fact, routine handwashing is considered to be the single most important measure for preventing the spread of infectious disease. When handwashing is not possible, use of alcohol gels may provide some protection, but they are not effective against all pathogens. ▶▶| *Shigella* sp., p. 588

Pathogens that cannot survive for extended periods in the environment must generally be transmitted through direct contact. *Treponema pallidum,* which causes syphilis, and *Neisseria gonorrhoeae,* which causes gonorrhea, both die quickly when exposed to a relatively cold, dry environment and thus require intimate sexual contact for their transmission. ▶▶| *Treponema pallidum,* p. 627 ▶▶| *Neisseria gonorrhoeae,* p. 622

Indirect Contact **Indirect contact** involves transfer of pathogens via inanimate objects, or **fomites,** such as clothing, table-tops, doorknobs, and drinking glasses. As an example, carriers of *Staphylococcus aureus* may inoculate their fingers with the organism when touching a skin lesion or colonized nostril. Organisms on the fingers can then easily be transferred to a fomite. Another person can acquire the microbes when handling that object. Again, handwashing is an important control measure.

Droplet Transmission Large microbe-laden respiratory droplets generally fall to the ground no farther than a meter (approximately 3 feet) from release. People in close proximity can inhale those droplets, resulting in the spread of respiratory disease via **droplet transmission.** Although physical contact is not necessary, droplet transmission is considered contact transmission because of the close range involved. Droplet transmission is particularly important as a source of contamination in densely populated buildings such as schools and military barracks. Desks or beds in such locations should be spaced more than 1.5 meters and preferably about 3 meters apart to minimize the transfer of infectious agents. The spread of respiratory diseases is minimized if people cover their mouths when they cough or sneeze.

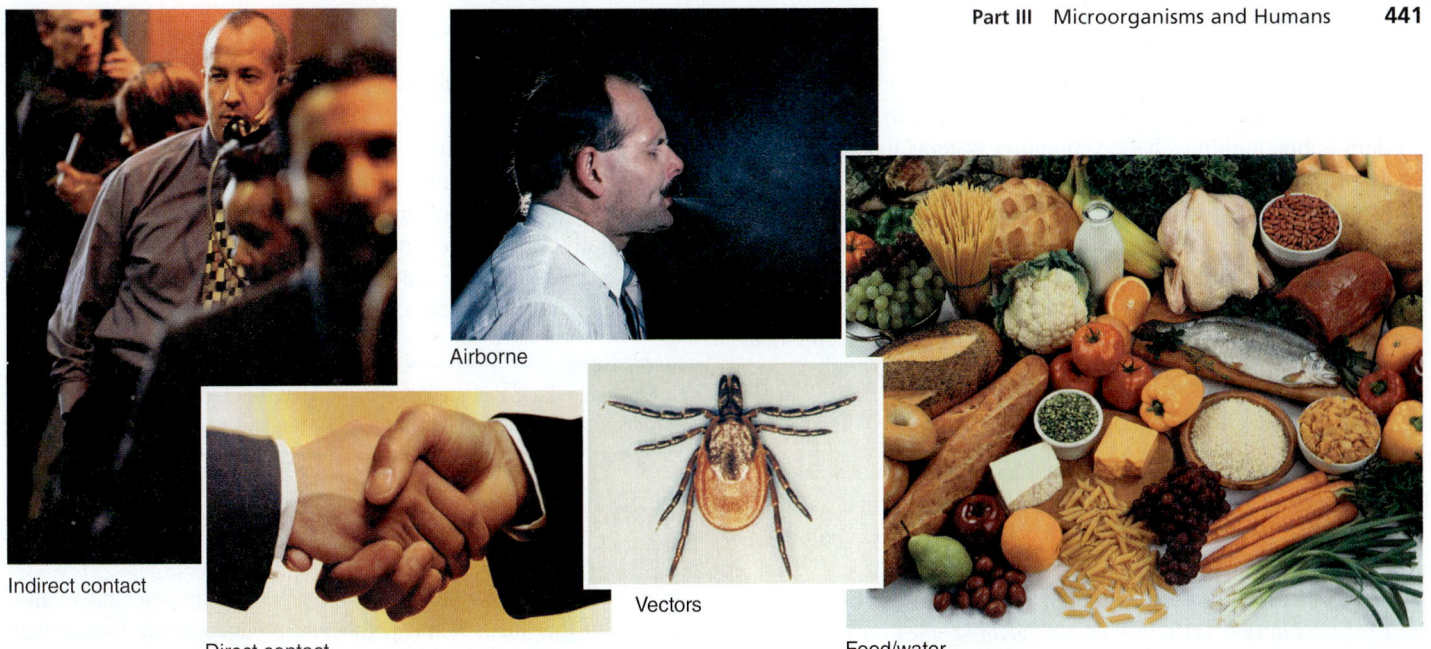

Indirect contact

Direct contact

Airborne

Vectors

Food/water

FIGURE 19.4 Transmission of Pathogens Microbes may be spread by indirect contact with fomites, by direct contact, through the air, by vectors, or by food or water.

? *List three potential fomites that you contacted today.*

Food and Water

Pathogens, particularly those that infect the digestive tract, can be transmitted through contaminated food or water. Foods can become contaminated in a number of ways. Animal products such as meat and eggs can have pathogens that originated from the animal's intestinal tract. This is the case with poultry contaminated with species of *Salmonella* or *Campylobacter* and hamburger contaminated with *E. coli* O157:H7. Pathogens can also be inadvertently added during food preparation. *Staphylococcus aureus* carriers who do not wash their hands prior to preparing food can easily contaminate the food. **Cross-contamination** results when pathogens from one food are transferred to another. A cutting board used first to carve raw chicken and then to cut cooked potatoes can transfer *Salmonella* cells from the chicken onto the potatoes. Because many foods are a rich nutrient source, microorganisms can multiply to high numbers if the contaminated food is not refrigerated. Sound food-handling methods, including sanitary preparation as well as thorough cooking and proper storage, can prevent foodborne diseases. ▶▶ **foodborne disease, p. 756**

Waterborne disease outbreaks can involve large numbers of people because municipal water systems distribute water to widespread areas. The 1993 waterborne outbreak of *Cryptosporidium parvum*, an intestinal parasite, in Milwaukee, Wisconsin, was estimated to have involved over 400,000 people. Prevention of waterborne diseases requires disinfection and filtration of drinking water and proper disposal and treatment of sewage. ▶▶ *Cryptosporidium parvum*, **p. 604** ▶▶ **drinking water treatment, p. 742** ▶▶ **sewage treatment, p. 736**

Air

Respiratory diseases are often transmitted through the air. When particles larger than 10 μm are inhaled, they are usually trapped in the mucus lining of the nose and throat and eventually swallowed. Smaller particles, however, can enter the lungs, where any pathogens they carry can potentially cause disease.

As mentioned earlier, when people talk, laugh, sing, sneeze, or cough they continually discharge microorganisms in liquid droplets. Large droplets quickly fall to the ground, but smaller droplets dry, creating **droplet nuclei** composed of microbes attached to a thin coat of the dried material. These can remain suspended indefinitely in the presence of even slight air currents. Other airborne particles, including dead skin cells, household dust, and soil disturbed by the wind, may also carry respiratory pathogens. An air-conditioning system can distribute contaminated air.

The number of viable organisms in air can be estimated by using a machine that pumps a measured volume of air, including any suspended dust and particles, against the surface of a nutrient-rich medium in a Petri dish. This technique has shown that the number of bacteria in the sampled air rises in proportion to the number of people in a room (**figure 19.5**).

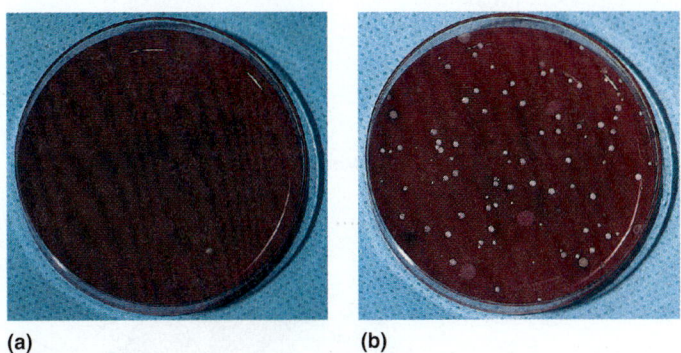

(a)   (b)

FIGURE 19.5 Air Sample Cultures (a) Air from a clean, empty hospital room. **(b)** Air from a small room containing 12 people. In both situations, 5 cubic feet of air was sampled.

? *Name an organism that you might expect to find in the plate shown in (b).*

Understandably, airborne transmission of pathogens is very difficult to control. To prevent the buildup of airborne pathogens, modern public buildings have ventilation systems that constantly change the air. The air pressure in hospital microbiology laboratories can be lowered so that air flows in from the corridors, preventing microorganisms and viruses from being swept out of the lab to other parts of the building. Air in specialized hospital rooms, jetliners, and some laboratories is circulated through high-efficiency particulate air (HEPA) filters to remove airborne organisms that may be present. ◀◀ HEPA filters, p. 115

Vectors

A **vector** is any living organism that can carry a disease-causing microbe, but most commonly they are arthropods such as mosquitoes, flies, fleas, lice, and ticks. A vector can carry a pathogen externally or internally.

Flies that land on feces can pick up intestinal pathogens such as *Escherichia coli* O157:H7 and *Shigella* species on their legs. If the fly then lands on food and transfers the microorganisms, it serves as a **mechanical vector,** carrying the microbe on its body from one place to another (**figure 19.6a**).

Diseases such as malaria, plague, and Lyme disease are transmitted via arthropods that harbor the pathogen internally. The vector either injects the infectious agent while taking a blood meal, or deposits the pathogen when it defecates on skin where it can be inoculated when the individual scratches the bite. For example, infected fleas inject *Yersinia pestis* while attempting to take a blood meal. In the case of malaria, caused by species of the eukaryotic pathogen *Plasmodium,* the mosquito not only transmits the parasite but also plays an essential role in its reproductive life cycle. Such vectors are called **biological vectors** (figure 19.6b). The pathogen multiplies to high numbers within this vector. ◀◀ *Plasmodium*, p. 292

Prevention of vector-borne disease relies largely on control of arthropods. Malaria, once endemic in the continental United States, was successfully eliminated from the nation through a combination of mosquito elimination and prompt treatment of infected patients. Unfortunately, worldwide eradication efforts that initially showed great promise ultimately failed, in part due to the decreased vigilance that accompanied the dramatic but short-lived decline of the disease.

MicroByte
An outbreak of dengue fever in Key West, Florida, in 2009 was transmitted by *Aedes* mosquitoes that bite during the day.

Pathogen Factors That Influence the Epidemiology of Disease

An infectious agent transmitted to a new host can potentially cause disease in that host. Some pathogens produce more disease than others. For example, 95% of people infected with the measles virus will develop symptoms of measles. On the other hand, less than 1% of those infected with poliovirus will develop poliomyelitis. The outcome of transmission is affected by many factors, including virulence of the pathogen, the dose, and the incubation period.

Virulence

Successful pathogens have virulence factors that increase their ability to cause disease, as discussed in chapter 16. These may allow pathogens to adhere to a host or penetrate host cells. *Neisseria gonorrhoeae* binds to mucosal epithelial cells; *Shigella* species adhere to intestinal epithelial cells and promote their own uptake. Other virulence factors allow pathogens to thwart immune defenses of the host. Capsules such as those of *Streptococcus pneumoniae* often protect microbial cells from phagocytosis.

FIGURE 19.6 Vector Transmission (a) A mechanical vector moves microbes from one place to another. **(b)** A biological vector participates in the life cycle of the pathogen and provides a place for it to multiply.

❓ *Name one disease spread by a mechanical vector and one disease involving a biological vector.*

(a) Mechanical vector

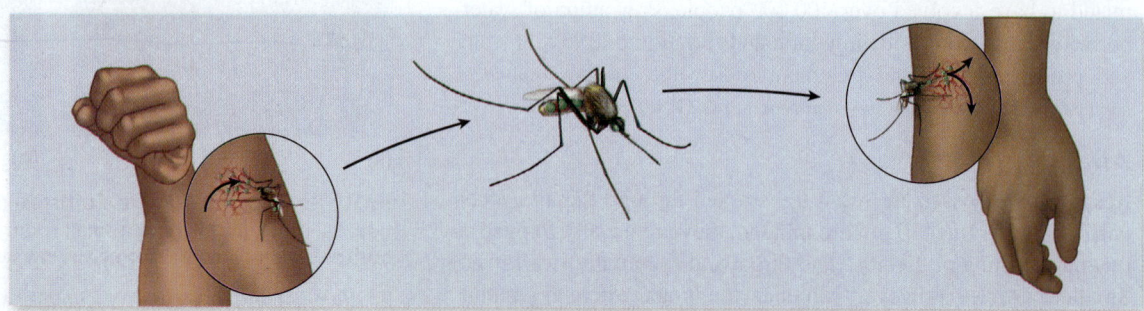

(b) Biological vector

Mycobacterium tuberculosis can survive within activated macrophages. Finally, pathogens damage the host, perhaps by production of toxins or destructive enzymes, or perhaps by stimulating a strong inflammatory response that damages host tissue. *E. coli* O157:H7 produces a cytotoxin that destroys cells that line blood vessels; *Clostridium perfringens,* the cause of gas gangrene, destroys host cell membranes with a phospholipase. Organisms that possess a variety of virulence factors are more likely to cause severe disease. ◀◀ virulence factors, p. 383 ▶▶ *S. pneumoniae*, p. 497 ▶▶ *C. perfringens*, p. 558

The Dose

The probability of infection and disease is generally lower when an individual is exposed to small numbers of a pathogen. This is because a certain minimum number of cells of the pathogen must enter the body and produce enough damage to cause disease. If 30 cells enter the body, yet 1,000,000 cells are required to produce symptoms, then it will take some time for the bacterial population to increase to that number. Because host defenses are mobilized at the same time and are racing to eliminate the invader, small doses often result in asymptomatic infections—the immune system eliminates the organism before symptoms appear. On the other hand, there are few if any infections for which immunity is absolute. An unusually large exposure to a pathogen, such as can occur in a laboratory accident, may produce serious disease in a person who is ordinarily immune to that microbe. Therefore, even immunized persons should take precautions to minimize exposure to infectious agents. This principle is especially important for medical workers taking care of patients with infectious diseases.

The Incubation Period

The extent of the spread of an infectious agent is influenced by its incubation period. Diseases with typically long incubation periods can spread extensively before the first symptomatic cases appear. An excellent example was the spread of typhoid fever from a ski resort in Switzerland in 1963. As many as 10,000 people were exposed to drinking water containing small numbers of *Salmonella enterica* serotype Typhi, the bacterium that causes the disease. The long incubation period of the disease, 10 to 14 days, allowed the organism to be spread by the skiers, who flew home to various parts of the world before they became ill. As a result, more than 430 cases of typhoid fever appeared in at least six countries. ▶▶ typhoid fever, p. 592 ◀◀ incubation period, p. 383

Host Factors That Influence the Epidemiology of Disease

Some populations more than others are likely to be affected by a given disease-causing agent. Many population characteristics influence the occurrence of disease.

Immunity to the Pathogen

Previous exposure or immunization of a population to a disease agent or an antigenically related agent decreases incidence of the disease. A disease is unlikely to spread in a population in which 90% of the individuals are immune to the disease agent. **Herd immunity** protects non-immune individuals when an infectious

agent cannot spread in a population because most potential hosts are immune. Unfortunately, some infectious agents can undergo antigenic variation and overcome herd immunity. Currently, humans have no herd immunity to the H5N1 (avian) influenza strain. ◀◀ antigenic variation, pp. 178, 390

General Health

Malnutrition, overcrowding, and fatigue increase people's susceptibility to infectious disease, enhancing its spread. Infectious diseases are more of a problem in developing areas of the world where individuals are crowded together without proper food or sanitation. Factors that promote good general health result in increased resistance to diseases such as tuberculosis. When infection does occur in a healthy individual, it is more likely to be asymptomatic or result in mild disease.

Age

The very young and the elderly are generally more susceptible to infectious agents. The immune system of young children is not fully developed and, consequently, they are vulnerable to certain diseases. For example, young children are particularly susceptible to meningitis caused by *Haemophilus influenzae.* The introduction of Hib vaccine in the late 1980s has dramatically decreased the incidence of meningitis in this age group. The elderly are more prone to disease because immunity wanes over time. Influenza outbreaks in nursing homes often have high case-fatality rates. Older adults are also less likely to update their immunizations, making them more susceptible to diseases such as tetanus (**figure 19.7**). To prevent tetanus, a booster vaccine is needed once every 10 years. ▶▶ Hib vaccine, p. 647 ▶▶ influenza, p. 508 ▶▶ tetanus, p. 555

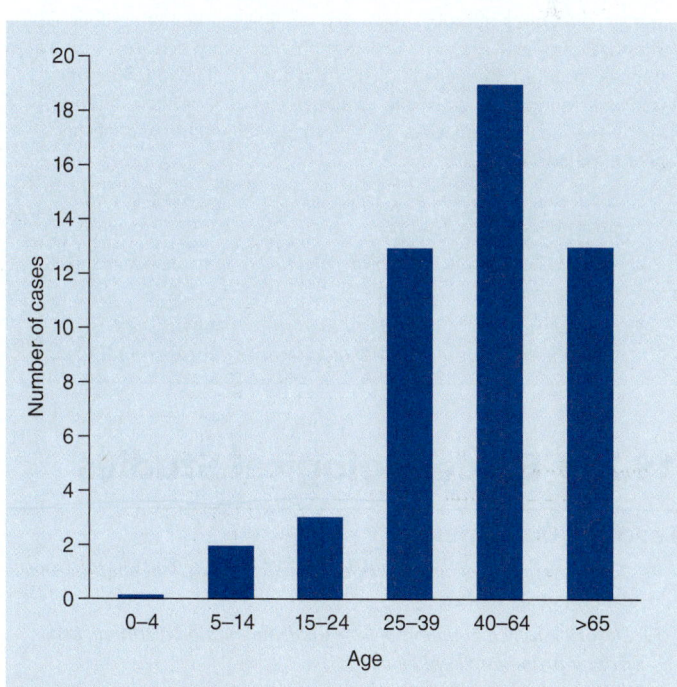

FIGURE 19.7 New Cases of Tetanus by Age Group

❓ *Why do you think individuals from 40–64 contract tetanus more often than younger people?*

Gender

Gender may influence disease distribution. Women are more likely to develop urinary tract infections because their urethra (the tube that connects the urinary bladder to the external environment) is relatively short. Microbes can ascend the urethra into the bladder. Pregnant women are more susceptible to listeriosis, caused by *Listeria monocytogenes.* ▶▶ listeriosis, p. 648

Religious and Cultural Practices

The distribution of disease is influenced by religious and cultural practices. An infant who is breast fed is less likely to have infectious diarrhea because of the protective antibodies in the mother's milk. Groups who eat traditional dishes made from raw freshwater fish are more likely to acquire the tapeworm, *Diphyllobothrium latum,* a parasite normally killed by cooking.

Genetic Background

Natural immunity can vary with genetic background, but it is usually difficult to determine the relative importance of genetic, cultural, and environmental factors. In a few instances, however, the genetic basis for resistance to infectious disease is known. For example, many people of black African ancestry are not susceptible to malaria caused by *Plasmodium vivax* because they lack a specific red blood cell receptor used by the infecting organism. Some populations of Northern European ancestry are less susceptible to HIV infection because they lack a certain receptor on their white blood cells.

MicroAssessment 19.1

Rates of disease within a population are a concern of epidemiologists who study disease patterns. Disease may be prevented by controlling the environment, the pathogen, or the host. The reservoir of a disease agent can be infected people, other animals, or the environment. To spread, infectious microbes must exit one host and enter another. Handwashing and vector control can prevent many diseases; airborne transmission of pathogens is difficult to control. The outcome of transmission is affected by the virulence and dose of the infecting agent, the incubation period, and by various characteristics of the host population.

1. *Explain the difference between incidence of a disease and prevalence of the disease.*
2. *Explain how a low dose of an infecting organism can result in an asymptomatic infection.*
3. *Considering that circulating blood is not normally released from the body, describe how blood-borne microbes might exit.* ✚

19.2 ■ Epidemiological Studies

Learning Outcomes

4. *Compare and contrast descriptive studies, analytical studies, and experimental studies.*
5. *Describe how a common-source epidemic can be distinguished from a propagated epidemic.*

British physician John Snow illustrated the power of a well-designed epidemiological study over 150 years ago. Long before the relationship between microbes and disease was accepted, he documented that the cholera epidemics plaguing England from 1849 to 1854 were due to contaminated water supplies. He did this by carefully comparing the conditions of households that were affected by cholera to those that were not, eventually determining that the primary difference was their water supply. His analysis saved countless lives. ▶▶ cholera, p. 586

Descriptive Studies

When a disease outbreak occurs, epidemiologists conduct a **descriptive study** by collecting data that characterize the occurrence, from the time and place of the outbreak to the individuals affected. That information is used to compile a list of possible risk factors involved in the spread of disease.

The Person

Determining the profile of those who become ill is critical to defining the population at risk. Variables such as age, gender, ethnicity, occupation, personal habits, previous illnesses, socioeconomic class, and marital status may all yield clues about risk factors for developing the disease. For example, in the Swiss ski resort epidemic of typhoid fever mentioned earlier, cases occurred only in tourists because the local people rarely drank water, preferring wine instead.

The Place

The geographic location of disease acquisition identifies the general site of contact between the person and the infectious agent (see figure 21.28). This helps pinpoint the exact source. The location may also give clues about potential reservoirs, vectors, or geographical boundaries that might affect disease transmission. For example, malaria can be transmitted only in regions that have the appropriate mosquito vector.

Microßyte

The first case of H1N1 influenza in the pandemic of 2009 was traced to a 5-year-old boy in Veracruz, Mexico.

The Time

The timing of an outbreak may also yield clues. A rapid rise in the numbers of people who become ill suggests that they were all exposed to a single, common source of the infectious agent, such as contaminated chicken at a picnic. This is called a **common-source epidemic.** In contrast, if the numbers of ill people rise gradually, then the disease is likely contagious, with one person transmitting it to several others, who each then transmit it to several more, and so on. This is called a **propagated epidemic (figure 19.8).** The first case in such an outbreak is called the **index case.** The time between the onset of symptoms in this case and the next cases reflects the incubation period of the disease.

The season in which the epidemic occurs may also be significant. Respiratory diseases including influenza, respiratory syncytial virus infections, and the common cold are more easily transmitted in crowded indoor conditions during the winter. Conversely, vector and foodborne diseases are more often

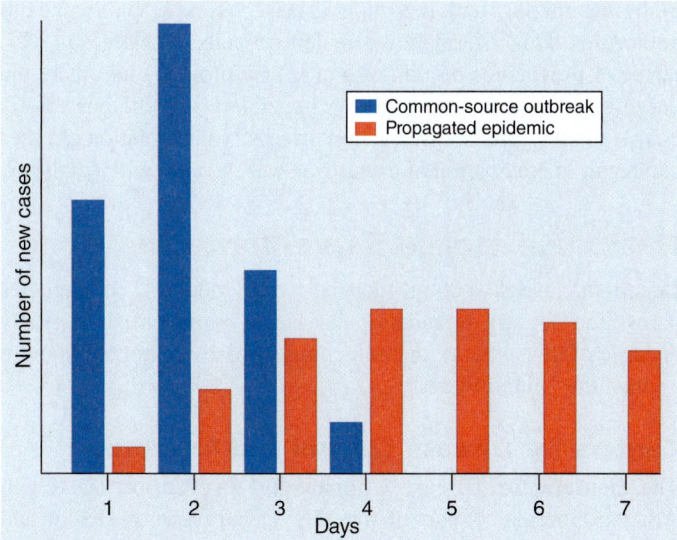

FIGURE 19.8 Comparison of Common-Source Versus Propagated Epidemics The graph depicts the number of new cases that develop over a period of days.

❓ *How does comparing day 1 and day 7 help you to determine if an epidemic is common-source or propagated?*

transmitted in warm weather when people are more likely to be exposed to mosquitoes and ticks, or eat picnic food that has not been stored properly (**figure 19.9**).

Analytical Studies

Analytical studies are designed to determine which of the potential risk factors identified by the descriptive studies are actually relevant in the spread of the disease.

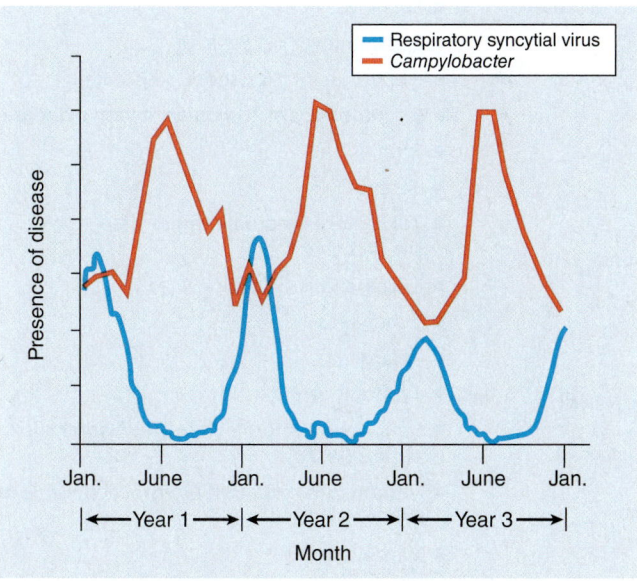

FIGURE 19.9 Seasonal Occurrence of Respiratory and Gastrointestinal Infections

❓ *Why does the level of foodborne disease increase in the summer?*

Cross-Sectional Studies

A **cross-sectional study** surveys a range of people to determine the prevalence of such characteristics as infection, presence of risk factors associated with disease, or previous exposure to a disease-causing agent. This population assessment may suggest associations between risk factors and disease. A cross-sectional survey does not attempt to establish cause of disease.

Retrospective Studies

A **retrospective study** is done following a disease outbreak. The actions and events surrounding clinical cases (individuals who developed the disease) are compared to those surrounding appropriate controls (those who remained healthy). A **case-control study** starts with the disease and attempts to identify the causative chain of events leading to it. If an activity or event is common among the cases, but not among the controls, it may have been a factor in the development of the disease. It is important to select controls that match the cases with respect to variables not thought to be associated with disease. Matching variables such as age, gender, and socioeconomic status, ensures that all controls had equal probability of coming in contact with the disease agent.

Prospective Studies

A **prospective study** looks ahead to see if the risk factors identified by the retrospective study predict a tendency to develop the disease. **Cohort groups,** which are study groups that have a known exposure to the risk factor, are selected and then followed over time. The incidence of disease in those who were exposed to the risk factor and those who were not is then compared. By following cohort groups, the study attempts to determine if the suspected cause does indeed correlate with the expected effect. This type of study is less prone to the bias of inaccurate recall than a retrospective study. It is generally more time-consuming and expensive, however, particularly when examining a disease with a long incubation period.

Experimental Studies

An **experimental study** is used to judge the cause-and-effect relationship of the risk factors or, more commonly, the preventive factors, and the development of disease. Experimental studies are done most frequently to assess the value of a particular intervention or treatment, such as antimicrobial drug therapy. The effectiveness of the treatment is compared with one of known value or with a **placebo.** A placebo is a mock drug—it looks and tastes like the experimental drug but has no medicinal value. To assess the value of the experimental drug, a group of patients is divided into two subgroups, one of which is given the treatment and the other an alternative or a placebo. To avoid bias, the study should ideally be **double-blind,** where neither the researchers nor the patients know who is receiving the experimental treatment. Ethical issues sometimes necessitate the use of animals rather than human patients in experimental studies.

19.3 ■ Infectious Disease Surveillance

Learning Outcomes

6. *Compare and contrast the roles of the Centers for Disease Control and Prevention, state public health departments, and the World Health Organization.*

7. *Explain the value of maintaining a current list of notifiable diseases.*

8. *Describe the conditions that may allow eradication of a disease.*

Infectious disease surveillance, nationally and worldwide, is one of the most important aspects of disease prevention. It involves both recognizing and reporting disease cases to public health authorities. This information can alert officials to changes in incidence or prevalence of a disease or it may provide clues as to the cause of disease. In some cases, cooperative surveillance efforts coupled with global immunization programs and isolation of cases can result in eradication of disease, as was the case with smallpox.

National Disease Surveillance Network

Infectious disease control depends on a network of agencies across the country to monitor disease development. It is partly because of this network that infectious diseases do not claim more lives in the United States.

Centers for Disease Control and Prevention

The **Centers for Disease Control and Prevention (CDC)** in Atlanta, Georgia, is part of the U.S. Department of Health and Human Services. It provides support for infectious disease laboratories in the United States and abroad and collects data on diseases that impact public health. Each week, the CDC publishes the *Morbidity and Mortality Weekly Report (MMWR)*, which summarizes the status of a number of diseases. The *MMWR* is available online (http://www.cdc.gov/mmwr/), making it readily accessible to anyone in the world.

The number of new cases of over 50 notifiable diseases is reported to the CDC by individual states (**table 19.1**). The list of

TABLE 19.1	Notifiable Infectious Diseases, 2011

Individual states and territories require physicians to report cases of these notifiable diseases. In turn, the number of cases is reported to the CDC, where they are collated and published in the *MMWR*.

■ Anthrax	■ HIV infection	■ Shiga toxin–producing *Escherichia coli* (STEC)
■ Arboviral neuroinvasive and non-neuroinvasive diseases	■ Influenza-associated pediatric mortality	■ Shigellosis
■ Babesiosis	■ Legionellosis (Legionnaires' disease)	■ Smallpox
■ Botulism	■ Listeriosis	■ Spotted fever rickettsiosis
■ Brucellosis	■ Lyme disease	■ Streptococcal toxic shock syndrome
■ Chancroid	■ Malaria	■ *Streptococcus pneumoniae*, invasive disease
■ *Chlamydia trachomatis*, genital infections	■ Measles	■ Syphilis
■ Cholera	■ Meningococcal disease	■ Tetanus
■ Coccidioidomycosis	■ Mumps	■ Toxic shock syndrome (other than streptococcal)
■ Cryptosporidiosis	■ Novel influenza A infections	■ Trichinellosis (Trichinosis)
■ Cyclosporiasis	■ Pertussis	■ Tuberculosis
■ Dengue	■ Plague	■ Tularemia
■ Diphtheria	■ Poliomyelitis, paralytic	■ Typhoid fever
■ Ehrlichiosis/Anaplasmosis	■ Poliovirus infection, nonparalytic	■ Vancomycin-intermediate *Staphylococcus aureus* (VISA)
■ Giardiasis	■ Psittacosis	■ Vancomycin-resistant *Staphylococcus aureus* (VRSA)
■ Gonorrhea	■ Q fever	
■ *Haemophilus influenzae*, invasive disease	■ Rabies, animal and human	■ Varicella
■ Hansen's disease (leprosy)	■ Rubella	■ Vibriosis
■ Hantavirus pulmonary syndrome	■ Rubella, congenital syndrome	■ Viral hemorrhagic fevers
■ Hemolytic uremic syndrome, post-diarrheal	■ Salmonellosis	■ Yellow fever
■ Hepatitis, viral, acute and chronic	■ Severe acute respiratory syndrome–associated coronavirus (SARS-CoV) disease	

diseases considered notifiable is determined through collaborative efforts of the CDC and state health departments. Typically the diseases are of relatively high incidence or pose potential danger to public health. The data collected by the CDC are published in the *MMWR* along with historical numbers to reflect any trends. Potentially significant case reports, such as the 1981 report of a cluster of opportunistic infections in young homosexual men that heralded the AIDS epidemic, are also included in the *MMWR*. This publication is an invaluable aid to physicians, public health agencies, teachers, students, and others concerned about infectious disease or public health. In fact, many of the epidemiological charts and stories in this textbook are from the *MMWR*.

The CDC also conducts research relating to infectious diseases and can dispatch teams worldwide to assist with identifying and controlling epidemics. After all, an epidemic anywhere in the world is a potential threat. In addition, the CDC provides refresher courses for laboratory and infection control personnel.

MicroByte

Chlamydia, an STI, is the most commonly reported infectious disease in the United States with over 1.2 million reported cases in 2009.

Public Health Departments

Each state has a public health laboratory responsible for infection surveillance and control as well as other health-related activities. Individual states have the authority to mandate which diseases must be reported by physicians to the state laboratory. The prompt response of health authorities in Washington State that helped stop an outbreak of *Escherichia coli* O157:H7 in 1993 caused by contaminated hamburger patties was partly because Washington then was one of the few states with surveillance and reporting measures for the organism. The epidemic had actually started in other states but had gone unrecognized. Likewise, the fungus *Cryptococcus gattii* appeared in Washington State in 2007, causing lung infections and meningitis in healthy persons, and is now closely watched by health authorities in Oregon and California. ▶▶ *Cryptococcus gattii*, p. 660

Other Components of the Public Health Network

The public health network also includes public schools, which report absentee rates, and hospital laboratories, which report on the isolation of pathogens with epidemiological significance. In conjunction with these local activities, the news media alert the general public to the presence of infectious disease.

Worldwide Disease Surveillance

The **World Health Organization (WHO),** an agency of the United Nations with 193 member states, is devoted to achieving the highest possible level of health around the globe. It has four main functions: (1) provide worldwide guidance in the field of health; (2) set global standards for health; (3) cooperatively strengthen national health programs; and (4) develop and transfer appropriate health technology. To accomplish its goals, the WHO provides education and technical assistance to member countries.

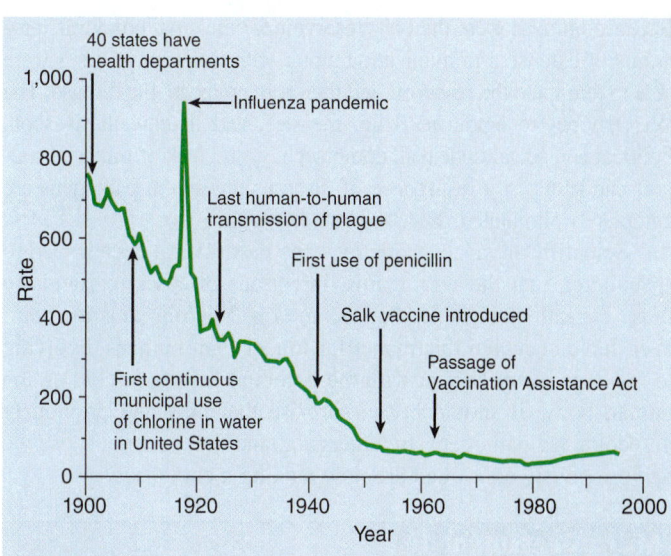

FIGURE 19.10 Crude Mortality Rate for Infectious Diseases, United States, per 100,000 Population per Year

❓ *What factors might cause an increase in the mortality rate over the next two decades?*

The WHO disseminates information through a series of periodicals and books. For example, the *Weekly Epidemiological Record* reports timely information about epidemics of public health importance, particularly those of global concern.

Reduction and Eradication of Disease

Humans have been enormously successful in eliminating or reducing the occurrence of certain diseases through efforts in improved sanitation, reservoir and vector control, vaccination, and antibiotic treatment (**figure 19.10**). In the United States, many diseases that were once common and claimed many lives are now relatively rare. Successful vaccination programs have led to dramatic decreases in the number of deaths caused by *Haemophilus influenzae*, *Corynebacterium diphtheriae*, *Clostridium tetani*, and *Bordetella pertussis*. Meanwhile, recognizing and controlling the source of diseases such as malaria, plague, and cholera have been effective in limiting their spread.

One human disease—smallpox—has been globally eradicated, eliminating the natural occurrence of a disease that had a 25% case-fatality rate and disfigured many who survived (**figure 19.11**).

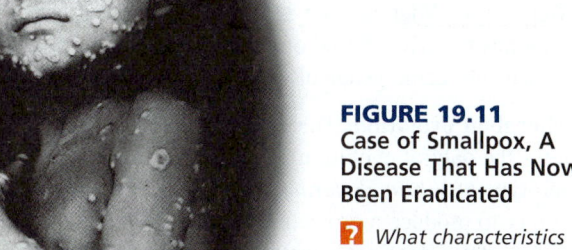

FIGURE 19.11
Case of Smallpox, A Disease That Has Now Been Eradicated

❓ *What characteristics make a disease a good candidate for eradication?*

Because humans were the only reservoir for the smallpox virus, programs of extensive immunization along with isolation of cases were able to eliminate the reservoir and therefore eradicate the disease. The WHO hopes to eradicate polio, measles, and dracunculiasis soon. Political and social upheaval, complacency, and lack of financial support can result in a resurgence of diseases unless the pathogens are completely eliminated. ▶▶ polio, p. 656 ▶▶ measles, p. 537

Scientific advances made over the past several decades led to speculation that the war against infectious diseases, particularly those caused by bacteria, had been won. Microorganisms, however, have occupied this planet far longer than humans, evolving to reside in every habitat with the potential for life, including the human body. It should be no surprise that new and previously unrecognized pathogens are emerging, and that some of those we had previously controlled are now making a comeback.

MicroAssessment 19.3

Across the United States, a network of agencies, including the Centers for Disease Control and Prevention (CDC) and state and local public health departments, monitors disease development. The World Health Organization (WHO) is devoted to achieving the highest possible level of health for people around the globe. Humans have been successful in reducing or eliminating certain diseases.

7. *What is the MMWR?*
8. *Explain why smallpox was successfully eradicated, but rabies probably never will be.*
9. *Explain why we have relatively accurate data on the number of cases of measles that occur in the United States but not on the number of cases of the common cold.* ➕

19.4 ■ Emerging Infectious Diseases

Learning Outcomes

9. *Explain two ways in which microbial evolution can lead to emergence of disease.*
10. *Describe how human behavior can contribute to the emergence and reemergence of disease.*
11. *Explain how climate can influence disease emergence.*

Emerging diseases are those that are novel or have recently increased in incidence. These include new or newly recognized diseases such as severe acute respiratory syndrome (SARS), mad cow disease, and avian flu, as well as familiar ones such as malaria and tuberculosis that are reemerging after years of decline (see figure 1.4). Microbes are adept at taking advantage of new opportunities to thrive. Some of the factors that contribute to the emergence and reemergence of diseases include:

■ **Microbial evolution.** The emergence of some diseases follows the natural evolution of microbes. For example, a new serotype of *Vibrio cholerae,* designated O139, has gained the ability to produce a protective capsule. Immunity against the earlier strain does not guarantee immunity to the newer one.

In the case of avian influenza, scientists are concerned that the virus will evolve to spread from person to person. Resistance to antimicrobial drugs is contributing to the reemergence of many diseases, including malaria. ▶▶ avian flu, p. 509 ▶▶ malaria, p. 686

■ **Complacency and public health efforts.** As an infectious disease is controlled and therefore of less concern, complacency can develop, paving the way for resurgence of the disease. The early success of the plan to eliminate tuberculosis in the United States by the year 2000 caused news reports, education, and research money to be diverted to more common diseases. Also, some social welfare programs were curtailed as the AIDS crisis demanded more funding. As a result, poor health and living conditions put more people at risk of developing active tuberculosis and the disease reemerged as an increasing threat. Fortunately, new public health measures have brought the disease back under control in the United States (see figure 21.17). ▶▶ tuberculosis, p. 502

■ **Changes in human society.** Daycare centers, where diapered infants mingle, oblivious to sanitation and hygiene, are a relatively new component of American society. For obvious reasons, these centers can be hotbeds of contagious diseases. This is particularly true for the intestinal pathogens *Giardia* and *Shigella,* which have a low infectious dose, because infants often explore through taste and touch and are thus likely to ingest fecal organisms (**figure 19.12**). Moreover, many young children have not yet acquired immunity to communicable diseases such as colds and diarrhea that are readily transmitted among this susceptible population. ▶▶ *Giardia*, p. 602 ▶▶ *Shigella*, p. 588

■ **Advances in technology.** Technology can make life easier, but can inadvertently create new habitats for microorganisms. The advent of contact lenses gave microorganisms the opportunity to grow in a new location—the lenses and storage solutions of

FIGURE 19.12 Children in a Daycare Center

❓ *Explain why* Shigella *and* Giardia *species are readily transmitted in daycare centers.*

users who did not employ proper disinfection techniques. In turn, this resulted in new types of eye infections.

- **Population expansion.** In a growing population humans can expand into areas where they come into contact with new reservoirs of disease (such as occurred with the Ebola virus), or have increased contact with established reservoirs. As more homes are built along the borders of forests, for example, humans are exposed to more deer that carry ticks, the host for *Borrelia burgdorferi,* the cause of Lyme disease.
▶▶| Lyme disease, p. 531

MicroByte ————————————————
About 75% of emerging infectious diseases affecting humans are of animal origin.

- **Development.** Dams, which provide important sources of power necessary for economic development, have inadvertently extended the range of certain diseases. Transmission of the disease schistosomiasis relies on the presence of an aquatic snail that serves as a host for the *Schistosoma* parasite. Construction of dams such as the Aswan Dam on the Nile River has increased the habitat for the snail, thus extending the distribution of the disease. |◀◀ schistosomiasis, p. 350

- **Mass production, widespread distribution, and importation of food.** Foodborne illness has always existed, but the ease with which we can now transport items worldwide from a single production plant can create new problems. Widespread distribution of foods contaminated with pathogens can result in a similarly broad outbreak of disease. Contaminated hotdogs shipped from a single food manufacturing plant in Canada in 2008 led to a country-wide epidemic of listeriosis that killed 21 people.

- **War and civil unrest.** Wars and civil unrest can disrupt the infrastructure on which disease prevention relies. Refugee camps that crowd people into substandard living quarters lacking toilet facilities and safe drinking water are hotbeds of infectious diseases such as cholera and dysentery. Unfortunately, war also disrupts disease eradication efforts.

- **Climate changes.** Warm temperatures favor the reproduction and survival of some arthropods, which can serve as vectors for diseases such as malaria and West Nile encephalitis. The heavy rainfall and flooding caused by the effects of El Niño may be related to surges of cholera cases in Africa.

MicroAssessment 19.4

Microbes are adept at responding to change. Evolution of the pathogen, changes in human behavior, or climate change can provide new opportunities for pathogens to thrive.

10. *Why might a person immunized against cholera contract the disease anyway?*
11. *How would you expect a trend toward warmer climates to affect the spread of vector-borne disease?*
12. *What political and societal factors might lead to a decrease in childhood immunizations?* ✚

19.5 ■ Healthcare-Associated Infections

Learning Outcomes

12. *Describe four reservoirs of infectious agents in healthcare settings and three mechanisms by which the agents can be transferred to patients.*
13. *Describe the role of Infection Control Committees in preventing nosocomial infections.*

Healthcare-associated infections (HAIs), which are infections individuals acquire while receiving treatment in a healthcare setting, are one of the top 10 causes of death in the United States. Perhaps this should not be surprising—hospitals, in particular, can be viewed as densely populated communities of unusually susceptible people where the most antimicrobial resistant and virulent pathogens can potentially circulate. In fact, hospital-acquired infections, or **nosocomial infections,** have been a problem since hospitals began (*nosocomial* is derived from the Greek word for hospital). Modern medical practices, however, including the extensive use of antimicrobial drugs and invasive therapeutic procedures, have increased the incidence of these infections. In the United States alone, an estimated 5% to 10% of patients admitted to a hospital develop a nosocomial infection, adding billions to the price of healthcare. Many of these infections originate from the patient's own normal microbiota, but approximately one-third are potentially preventable. **Figure 19.13** shows the relative frequency of different types of nosocomial infections.

The severity of healthcare-associated infections can range from very mild to fatal. Some infections arise during a hospital stay, but others are not discovered until after the patient has been discharged. Many factors influence the development of these infections such as the length of exposure, the manner in which a patient is exposed, the virulence and number of organisms involved, and the immune state of the host. Characteristics of some of the microorganisms and viruses commonly implicated in healthcare-associated infections are summarized in **table 19.2**.

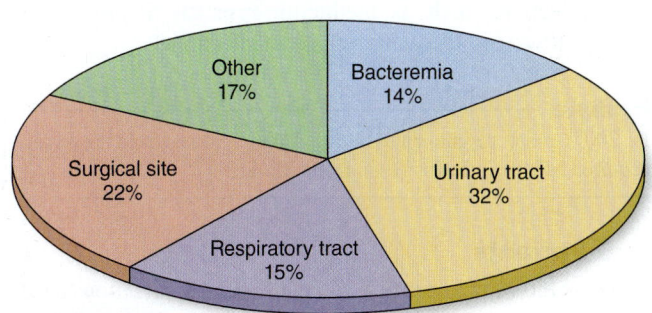

FIGURE 19.13 Relative Frequency of Different Types of Nosocomial Infections

❓ *What is bacteremia?*

 Part III Microorganisms and Humans **449**

TABLE 19.2	Common Causes of Healthcare-Associated Infections
Infectious Agent	**Comments**
Acinetobacter baumannii	This environmental bacterium can be found on skin of healthy people and is resistant to a number of antimicrobial medications. It causes a variety of healthcare-associated infections including bloodstream and surgical site infections and pneumonia.
Candida species	These yeasts, part of the normal microbiota, are a common cause of healthcare-associated bloodstream infections. Some are resistant to a number of antifungal medications. ▶▶ *Candida albicans*, p. 545
Clostridium difficile	Toxin-producing strains of this bacterium can cause diarrhea and colitis in people taking antibiotics. Because the bacterium produces endospores, which are not killed by disinfectants, thorough handwashing is an important means of preventing transmission. ▶▶ *Clostridium difficile-associated disease*, p. 594
Enterococcus species	These bacteria, part of the normal intestinal microbiota, are a common cause of nosocomial urinary tract infections as well as wound and bloodstream infections. Some strains are resistant to all conventional antimicrobial drugs. ◀◀ enterococci, p. 258
Escherichia coli and other members of the Enterobacteriaceae	These members of the normal intestinal microbiota commonly cause healthcare-associated urinary tract and other infections. ◀◀ *Enterobacteriaceae*, p. 265
Norovirus	This virus infects the gastrointestinal tract, causing vomiting and diarrhea. It has a very low infectious dose, so scrupulous cleaning and handwashing are necessary to prevent its transmission. ▶▶ norovirus, p. 596
Pseudomonas species	These bacteria grow readily in many moist, nutrient-poor environments such as the water in the humidifier of a mechanical ventilator. *Pseudomonas* species are resistant to many disinfectants and antimicrobial drugs. They are a common cause of healthcare-associated pneumonia, and infections of the urinary tract and burn wounds. ▶▶ *Pseudomonas*, p. 554
Respiratory syncytial virus (RSV)	This virus is easily transmitted, and causes outbreaks of healthcare-associated lower respiratory tract infections, particularly at times when the virus is circulating in non-healthcare settings. ▶▶ respiratory syncytial virus, p. 511
Staphylococcus aureus	Many people, including healthcare personnel, are carriers of this bacterium. Because it survives for prolonged periods in the environment, it is readily transmissible on fomites. It is a common cause of healthcare-associated pneumonia and surgical site infections. Hospital-acquired strains are often resistant to a variety of antimicrobial drugs. ▶▶ *Staphylococcus aureus*, p. 524
Staphylococcus species other than *S. aureus*	These members of the normal skin microbiota can colonize the tips of intravenous catheters (small plastic tubes inserted into the veins). The resulting biofilms continuously seed organisms into the bloodstream and increase the likelihood of a systemic infection. ◀◀ biofilm, p. 84

Reservoirs of Infectious Agents in Healthcare Settings

The organisms that cause healthcare-associated infections can originate from a number of different sources, including other patients, the healthcare environment, healthcare workers, and the patient's own microbiota. Because of the widespread use of antimicrobial drugs in hospitals, many organisms that cause nosocomial infections such as methicillin-resistant *Staphylococcus aureus* (MRSA) are resistant to these medications.

MicroByte
In 2007, more people died in the United States from hospital-acquired MRSA infections than from AIDS.

Other Patients

Because of the very nature of healthcare settings, infectious agents are always present. In fact, patients are often hospitalized because they have a severe infectious disease. The pathogens that these patients harbor can be discharged into the environment via skin cells, respiratory droplets, and other body secretions and excretions. Scrupulous cleaning and the use of disinfectants minimize the spread of these pathogens.

Healthcare Environment

Some Gram-negative rods, particularly the common opportunistic pathogen *Pseudomonas aeruginosa*, can thrive in healthcare environments such as sinks, respirators, and toilets. Not only is *P. aeruginosa* resistant to many disinfectants and antimicrobial drugs, it requires few nutrients, enabling it to multiply in environments containing little other than water. Many nosocomial infections have been traced to soaps, disinfectants, and other aqueous solutions that were contaminated with the organism. ▶▶ *Pseudomonas aeruginosa*, p. 554

Healthcare Workers

Outbreaks of healthcare-associated infections are sometimes traced to infected personnel. Clearly, those who report to work with even a mild case of influenza can expose patients to an infectious agent that has serious or fatal consequences to those in poor health. A more troublesome source of infection is a healthcare worker who is a carrier of a pathogen such as *Staphylococcus aureus* or *Streptococcus pyogenes*. These personnel may not recognize they pose a risk to patients until they are implicated in an outbreak. Carriers who are members of a surgical team pose a particular threat because inoculation of a pathogen directly into a surgical site can result in a systemic infection.

Patient Microbiota

Many healthcare-associated infections originate from the patient's own microbiota. Nearly any invasive procedure can transmit these organisms to otherwise sterile body sites. When intravenous fluids are administered, for example, *Staphylococcus epidermidis,* a common member of the skin microbiota, can potentially gain access to the bloodstream causing bacteremia. Although the immune system can usually eliminate these normally benign organisms, the underlying illness of many hospitalized patients compromises their immunity and they can develop a bloodstream infection. Patients who undergo intestinal surgery are prone to surgical site infection by their normal bowel microbiota. Similarly, patients who are on certain medications or have impaired cough reflexes can inadvertently inhale their normal oral microbiota, resulting in healthcare-associated pneumonia. ◄◄ bacteremia, p. 384

Severely immunocompromised patients, such as people who have undergone cancer chemotherapy or are on immunosuppressive drugs, are prone to activation of latent infections their immune system could previously control. For example, latent infections of *Toxoplasma gondii,* a protozoan parasite commonly acquired during childhood, can become activated and cause a life-threatening disease. ◄◄ latent infection, p. 322

Transmission of Infectious Agents in Healthcare Settings

Diagnostic and therapeutic procedures can potentially transmit infectious agents to patients. This is particularly true in intensive care units (ICUs), where patients generally have indwelling catheters used to deliver intravenous fluids or monitor the patient's condition (**figure 19.14**).

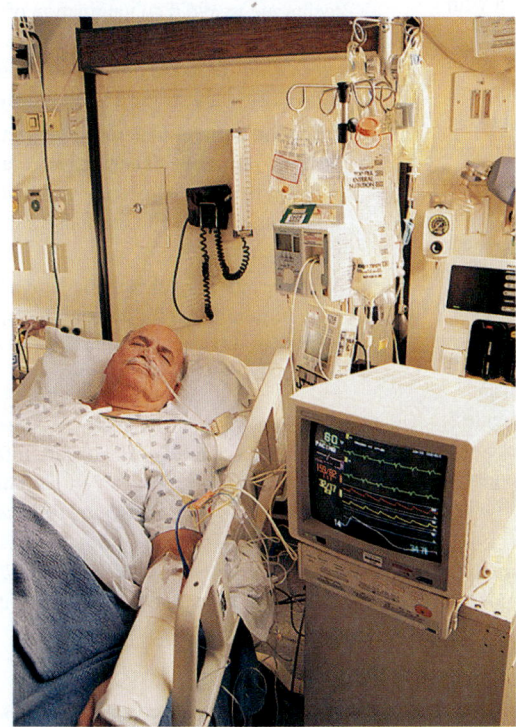

FIGURE 19.14 Patient in an Intensive Care Unit

❓ *To what sources of nosocomial infections is this patient exposed?*

Fomite Transmission—Medical Devices

Healthcare-associated infections most often result from medical devices that breach the first-line barriers of the normal host defense. Catheterization of the urinary tract can readily introduce microorganisms into the normally sterile bladder. Because urine is an excellent growth medium, the urinary tract often becomes infected. Urinary tract infections are the most common type of nosocomial infection (see figure 19.13).

Just as urinary catheters can introduce bacteria into the bladder, intravenous (IV) catheters can introduce microorganisms into the bloodstream. This can happen when normal skin microbiota colonize the tip of an indwelling catheter or when environmental organisms contaminate IV fluids or the lines that deliver them. Even normally benign skin microbiota can cause life-threatening bacteremia if they gain access to the bloodstream.

Mechanical respirators that assist a patient's breathing by pumping air directly into the trachea can potentially deliver microorganisms. For example, if a nutritionally versatile organism such as a *Pseudomonas* species gains access to water droplets in the machine, it can multiply and be pumped into the lungs.

Inadequately sterilized instruments that are used in invasive procedures such as surgery or biopsy can also transmit infectious agents. Endoscopes and other heat-sensitive instruments are often treated with chemical sterilants to render them microbe-free. Improper use of these chemicals, however, can result in the survival of some organisms. ◄◄ sterilants, p. 116

Direct Transmission—Healthcare Personnel

Healthcare personnel must be extremely vigilant to avoid transmitting infectious disease agents, particularly from patient to patient. What Ignaz Semmelweis found to be true in the 1800s is equally true today—handwashing between contact with individual patients helps prevent the spread of disease (see **A Glimpse of History**). Unfortunately, this relatively simple procedure is too often overlooked.

Healthcare personnel should routinely wash or disinfect their hands after touching one patient before going to the next. They should perform a more thorough hand scrubbing lasting 10 minutes with a strong disinfectant before participating in a surgical operation, or when working in a nursery, or an intensive care or isolation unit. Healthcare workers are instructed to wear gloves when they have contact with blood, mucous membranes, broken skin, or body fluids.

Airborne Transmission

Most hospitals are designed to minimize the airborne spread of microorganisms. Airflow to operating rooms is usually regulated so that it is supplied under slight pressure, thereby preventing contaminated air in the corridors from flowing into the room. Floors are washed with a damp mop or floor washer rather than swept in order to avoid resuspending microbes into the air. To exclude airborne microorganisms and viruses from rooms in which extremely susceptible patients reside, such as those who have recently undergone a bone marrow transplant, high-efficiency particulate air (HEPA) filters are used. These filter out most airborne particles, including microorganisms. ◄◄ HEPA filters, p. 115

PERSPECTIVE 19.1

Standard Precautions—Protecting Patients and Healthcare Personnel

One of the biggest challenges for a hospital has always been to prevent spread of disease within that confined setting. Patients with infectious diseases were once in separate hospitals; those with similar diseases were sometimes housed in clusters on the same floor. In 1910, a cubicle system of isolation was introduced in which patients were placed in multiple-bed wards. Aseptic nursing procedures were aimed at preventing the transmission of infectious agents to other patients and personnel. These scrupulous measures were so successful that general hospitals were able to accommodate infectious disease patients, ultimately resulting in the closure of many infectious disease hospitals beginning in the 1950s. To help general hospitals with isolation procedures, in 1970 the CDC began publishing a manual that recommended a category system of seven isolation procedures for patients based on their diagnosis.

In the early 1980s, healthcare workers began to acquire hepatitis B and, later, HIV from contact with the blood or other body fluids of patients. Existing guidelines were primarily aimed at preventing patient-to-patient transmission of disease, rather than patient-to-personnel. In response, the CDC recommended an additional set of guidelines, the *Universal (Blood and Body Fluid) Precautions,* to be followed when working with any patient, regardless of the diagnosis. These defined the situations in which gloves, gowns, masks, and eye protection were required to prevent contact with blood. Many hospitals then broadened this concept, requiring the use of gloves to isolate all moist and potentially infectious body substances. This approach, Body Substance Isolation, made the traditional diagnosis-dependent isolation procedures largely unnecessary.

The advent of Universal Precautions and Body Substance Isolation, in addition to the previous diagnosis-dependent isolation procedures, resulted in such a mix of recommendations that it generated a great deal of confusion. No existing, single set of guidelines was sufficient, and it was not clear when each should be used. In response, the CDC and the Hospital Infection Control Practices Advisory Committee (HICPAC) established a new set of guidelines in 1996 that incorporated the strengths of each of the alternatives. The

guidelines were updated and expanded in 2007 to include recommendations that can be applied to all healthcare facilities. The advisory committee has also been renamed, substituting the word "healthcare" for "hospital," to reflect the expanded scope of the infection control recommendations.

The current guidelines have two tiers of isolation procedures: The fundamental measures are the Standard Precautions, designed for the care of all patients. The Transmission-Based Precautions are supplementary measures to be used in addition to the Standard Precautions if a patient is, or might be, infected with a highly transmissible or epidemiologically important pathogen. These are separated into three sets—Airborne Precautions, Droplet Precautions, and Contact Precautions—that are used singly or in combination as appropriate. The Standard Precautions can be summarized as:

- **Hand hygiene.** During the delivery of healthcare, avoid unnecessary touching of surfaces in close proximity to the patient. When hands are visibly dirty or contaminated, wash them with soap and water; if non-antimicrobial soap is used, decontaminate hands with an alcohol-based hand rub. If hands are not visibly soiled, decontaminate hands before direct contact with patients; after contact with blood, body fluids, secretions, excretions, contaminated items; immediately after removing gloves; and between patient contacts. If contact with spores is likely to have occurred, wash with soap and water.

- **Personal protective equipment.** Gloves are worn when touching blood, body fluids, secretions, excretions, and contaminated items; and for touching mucous membranes and non-intact skin. A gown is worn during procedures and patient-care activities when contact of clothing/exposed skin with blood/body fluids, secretions, and excretions is anticipated. Mask, goggles, or a face shield is worn during procedures and patient-care activities likely to generate splashes or sprays of blood, body fluids, or secretions.

- **Respiratory hygiene/cough etiquette.** Instruct symptomatic persons to cover

mouth/nose when sneezing/coughing; use tissues and dispose in a no-touch receptacle; observe hand hygiene after soiling of hands with respiratory secretions; wear surgical mask if tolerated or maintain spatial separation of more than 3 feet if possible.

- **Patient placement.** Prioritize for single-patient room if patient is at increased risk of transmission, is likely to contaminate the environment, does not maintain appropriate hygiene, or is at increased risk of acquiring infection or developing adverse outcome following infection.

- **Patient-care equipment and instruments/devices.** If the equipment is soiled, handle in a manner that prevents transfer of microorganisms to others and to the environment; wear gloves if visibly contaminated; perform hand hygiene.

- **Care of the environment.** Develop procedures for routine care, cleaning, and disinfection of environmental surfaces, especially frequently touched surfaces in patient-care areas.

- **Textiles and laundry.** Handle in a manner that prevents transfer of microorganisms to others and to the environment.

- **Safe injection practices.** Use aseptic technique to avoid contamination of sterile injection equipment. Specific precautions describe how medications and IV solutions are stored and administered.

- **Infection control practices for special lumbar puncture procedures.** Wear a surgical mask when placing a catheter or injection material into the spinal canal or subdural space.

- **Worker safety.** Adhere to federal and state requirements for protection of healthcare personnel from exposure to bloodborne pathogens.

From Siegel, J.D., Rhinehart, E., Jackson, M., Chiarello, L., and the Healthcare Infection Control Practices Advisory Committee, *2007 Guidelines for Isolation Precautions: Preventing Transmission of Infectious Agents in Healthcare Settings,* June 2007. http://www.cdc.gov/ncidod/dhqp/pdf/isolation2007.pdf

Preventing Healthcare-Associated Infections

The most important steps in preventing healthcare-associated infections are to first detect their occurrence and then establish

policies to prevent their development. To accomplish this, nearly every hospital has an Infection Control Committee, composed of representatives of the various professionals in the hospital, including nurses, physicians, dietitians, housekeeping staff, and microbiology laboratory personnel. On this committee, and sometimes chairing it, is often a hospital epidemiologist, a professional specially trained in hospital infection control. Hospitals may employ an infection control practitioner (ICP), whose role is to perform active surveillance of the types and numbers of infections that arise in the hospital. The Infection Control Committee, in conjunction with the ICP, drafts and implements preventive policies following the guidelines suggested by the Standard Precautions and the Transmission-Based Precautions (see **Perspective 19.1**).

The CDC also takes an active role in preventing healthcare-associated infections, and has established the Healthcare Infection Control Practices Advisory Committee (HICPAC). The role of this national committee is to provide advice to hospitals and recommend guidelines for surveillance, prevention, and control of healthcare-associated infections.

MicroAssessment 19.5

Healthcare-associated infections may originate from other patients, the healthcare environment, healthcare workers, or the patient's own normal microbiota. Diagnostic and therapeutic procedures can potentially transmit infectious agents. The most important steps in preventing healthcare-associated infections are to first detect their occurrence and then establish policies to prevent their development.

13. *Explain why an IV catheter poses a risk to a patient.*

14. *Describe two ways in which infectious agents can be transmitted to a patient.*

15. *The rate of nosocomial infections is often relatively high in emergency room settings. Explain why this might be so.* ✚

FUTURE CHALLENGES   19.1

Maintaining Vigilance Against Bioterrorism

Today, an unfortunate challenge in epidemiology is to maintain vigilance against **bioterrorism**—the deliberate release of infectious agents or their toxins as a means to cause harm. Even as we work to control diseases, we must be aware that microbes pose a threat as agents of bioterrorism. Hopefully, future attacks will never occur, but it is crucial to be prepared for the possibility. Prompt recognition of such an event, followed by rapid and appropriate isolation and treatment procedures, can help to minimize the consequences. The CDC, in cooperation with the Association for Professionals in Infection Control and Epidemiology (APIC), has prepared a bioterrorism readiness plan to be used as a template by healthcare facilities. Many of the recommendations are based on the Standard Precautions already used by hospitals to prevent the spread of infectious agents (see Perspective 19.1).

The CDC separates bioterrorism agents into three categories based on the ease of spread and severity of disease. **Category A agents** pose the highest risk because they are easily spread or transmitted from person to person and result in high mortality. These agents include:

■ *Bacillus anthracis.* Endospores of this bacterium were used in the bioterrorism events of 2001. The most severe outcome, inhalational anthrax, results when an individual breathes in the airborne spores. It can lead to a rapidly fatal systemic illness. Cutaneous anthrax, which occurs when the organism enters the skin, manifests as a blister that develops into a skin ulcer with a black center. Although this usually heals without treatment, it can also progress to a fatal bloodstream

infection. Gastrointestinal anthrax results from consuming contaminated food, leading to vomiting of blood and severe diarrhea; it is not common but has a high case-fatality rate. Anthrax can be prevented by vaccination, but that option is not widely available. Prophylaxis with antimicrobial medications is possible for those who might have been exposed, but this requires prompt recognition of exposure. Fortunately, person-to-person transmission of the agent is not likely.

■ **Botulism.** Botulism is caused naturally by the ingestion of botulinum toxin, produced by *Clostridium botulinum.* Any mucous membrane can absorb the toxin, so aerosolized toxin could be used as a weapon. Botulism can be prevented by vaccination, but that option is not widely available. An antitoxin is also available in limited supplies. Botulism is not contagious.

■ *Yersinia pestis.* Pneumonic plague, caused by inhalation of *Yersinia pestis,* is the most likely form of plague to result from a biological weapon. Although no effective vaccine is available, post-exposure prophylaxis with antimicrobial medications is possible. Special isolation precautions must be used for patients who have pneumonic plague because the disease is easily transmitted by respiratory droplets.

■ **Smallpox.** Although a vaccine is available to prevent infection with this virus, routine immunization was stopped over 30 years ago because the natural disease has been eradicated. As is the case with nearly all infections caused

by viruses, effective drug therapy is not available. Special isolation precautions must be used for smallpox patients because the virus can be acquired through droplet, airborne, or contact transmission.

■ *Francisella tularensis.* This bacterium, naturally found in animals such as rodents and rabbits, causes the disease tularemia. Inhalation of the bacterium results in severe pneumonia, which is incapacitating but would probably have a lower case-fatality rate than inhalational anthrax or plague. A vaccine is not available, but post-exposure prophylaxis with antimicrobial medications is possible. Fortunately, person-to-person transmission of the agent is not likely.

■ **Viruses that cause hemorrhagic fevers.** These include various viruses such as Ebola and Marburg. Symptoms vary depending on the virus, but severe cases show signs of bleeding from many sites. There are no vaccines against these viruses, and generally no treatment. Some, but not all, of these viruses can be transmitted from person to person, so patient isolation in these cases is important.

Category B agents pose moderate risk because they are relatively easy to spread and cause moderate morbidity. These agents include organisms that cause food- and waterborne illness, various biological toxins, *Brucella* species, *Burkholderia mallei* and *pseudomallei, Coxiella burnetii,* and *Chlamydophila (Chlamydia) psittaci.* **Category C agents** are emerging pathogens that could be engineered for easy dissemination. These include Nipah virus, which was first recognized in 1999, and hantavirus, first recognized in 1993.

Summary

19.1 ■ Principles of Epidemiology

Epidemiologists study the frequency and distribution of disease in order to identify its cause, source, and route of transmission.

Rate of Disease in a Population

Epidemiologists focus on the rate of disease. Diseases that are constantly present in a population are **endemic;** an unusually large number of cases in a population constitutes an **epidemic** (figure 19.1).

Reservoirs of Infection (figure 19.2)

Preventing susceptible people from coming in contact with a **reservoir of infection** can prevent infectious disease. People who have asymptomatic infections or are colonized with a pathogen are **carriers** of the infectious agent. **Zoonoses** such as plague and rabies can be transmitted to humans but exist primarily in other animals. Pathogens with environmental reservoirs are probably impossible to eliminate.

Portals of Exit and Portals of Entry

Pathogens may be shed in feces, in respiratory droplets, on skin cells, in genital secretions, and in urine. Many organisms can cause disease if they enter one body site, but are harmless if they enter another (figure 19.3).

Disease Transmission (figure 19.4)

Handwashing is a key control measure in preventing diseases that are spread through direct or indirect contact, as well as those that spread via contaminated food. **Direct contact** occurs when one person physically touches another. **Indirect contact** involves transfer of pathogens via **fomites. Droplet transmission** of respiratory pathogens is considered contact because of the close proximity involved. Foodborne pathogens can originate from the animal reservoir or from contamination during food preparation. Waterborne pathogens often originate from sewage contamination. Airborne transmission of pathogens is the most difficult to control (figure 19.5). A **mechanical vector** carries a microbe on its body from one place to another. Pathogens can multiply to high numbers in a **biological vector** (figure 19.6b). Prevention of vector-borne disease relies on vector control.

Pathogen Factors That Influence the Epidemiology of Disease

Pathogens with a variety of virulence factors are more likely to cause severe disease. The probability of infection and disease is generally lower if an individual is exposed to small numbers of pathogens. Diseases with a long incubation period can spread before the first cases appear.

Host Factors That Influence the Epidemiology of Disease

A disease is unlikely to spread very widely in a population showing **herd immunity** in which most people are immune to the disease agent. Malnutrition, overcrowding, and fatigue increase the susceptibility of people to infectious diseases. The very young and the elderly are generally more susceptible to infectious agents (figure 19.7). Natural immunity can vary with genetic background, but it is difficult to determine the relative importance of genetic, cultural, and environmental factors.

19.2 ■ Epidemiological Studies

Descriptive Studies

Descriptive studies attempt to identify potential risk factors that correlate with the development of disease. Determining the time that the illness occurred helps distinguish a **common-source epidemic**

from a **propagated epidemic** (figure 19.8). Some epidemics are seasonal (figure 19.9).

Analytical Studies

Analytical studies try to determine which risk factors are actually relevant to disease development. A **retrospective study** compares the activities of cases with controls to determine the cause of the epidemic. A **prospective study** compares **cohort groups,** to determine if the identified risk factors predict a tendency to develop disease.

Experimental Studies

Experimental studies are generally used to evaluate the effectiveness of a treatment or intervention in preventing disease.

19.3 ■ Infectious Disease Surveillance

National Disease Surveillance Network

The **Centers for Disease Control and Prevention (CDC)** collects data on diseases of public health importance and summarizes their status in the *Morbidity and Mortality Weekly Report (MMWR);* other activities of the CDC include research, assistance in controlling epidemics, and support for infectious disease laboratories. State public health departments are involved in infection surveillance and control (table 19.1).

Worldwide Disease Surveillance

The **World Health Organization (WHO)** is devoted to achieving the highest possible level of health for people around the globe.

Reduction and Eradication of Disease (figure 19.10)

Smallpox has been eradicated (figure 19.11). The WHO hopes to soon eliminate polio, measles, and dracunculiasis.

19.4 ■ Emerging Infectious Diseases

Emerging diseases are new or newly recognized, or are reemerging after years of decline. Factors that contribute to the emergence and reemergence of diseases include microbial evolution, the breakdown of public health infrastructure, changes in human behavior, advances in technology, population expansion, economic development, mass distribution and importation of food, war, and climate changes (figure 19.12).

19.5 ■ Healthcare-Associated Infections

Healthcare-associated infections (HAIs) are acquired by individuals in a healthcare setting. **Nosocomial infections** are acquired in a hospital (figure 19.13).

Reservoirs of Infectious Agents in Healthcare Settings

The organisms that cause healthcare-associated infections may originate from other patients, the healthcare environment, healthcare workers, or the patient's own microbiota.

Transmission of Infectious Agents in Healthcare Settings

Healthcare-associated infections can result from medical devices that breach the first-line barriers of the normal host defense (figure 19.14). Healthcare personnel should routinely wash or disinfect their hands after touching one patient before going to the next.

Preventing Healthcare-Associated Infections

The most important steps in preventing healthcare-associated infections are to first recognize their occurrence and then establish policies to prevent both their development and spread.

Review Questions

Short Answer

1. Describe the impact on a society of high incidence and high prevalence of an endemic debilitating disease.
2. What is the epidemiological significance of people who have asymptomatic infections?
3. Explain why zoonotic diseases are often severe in humans.
4. List the main portals of exit from the human body.
5. Name the most important control measure for preventing person-to-person transmission of a disease.
6. Describe the factors within a population that may make it more susceptible to infectious disease.
7. Draw a representative graph (time versus number of people ill) depicting both a propagated and a common-source epidemic.
8. Describe the differences between a retrospective (case-control) study and a prospective (cohort) study.
9. What information is available in the *Weekly Epidemiological Record?*
10. Explain how smallpox was eradicated.
11. Describe the factors that contribute to the emergence or re-emergence of disease.
12. What are the main reservoirs of nosocomial infections?

Multiple Choice

1. Which of the following is an example of a fomite?
   a) Table   b) Flea   c) *Staphylococcus aureus* carrier
   d) Water   e) Air
2. Which of the following would be the easiest to eradicate?
   a) A pathogen that is common in wild animals but sometimes infects humans
   b) A disease that occurs exclusively in humans, always resulting in obvious symptoms
   c) A mild disease of humans that often results in no obvious symptoms
   d) A pathogen found in marine sediments
   e) A pathogen that readily infects both wild animals and humans
3. Which of the following methods of disease transmission is the most difficult to control?
   a) Airborne   b) Foodborne   c) Waterborne
   d) Vector-borne   e) Direct person-to-person
4. Which of the following statements is *false?*
   a) A botulism epidemic that results from improperly canned green beans is an example of a common-source outbreak.
   b) Droplet nuclei fall quickly to the ground.
   c) Congenital syphilis is an example of a disease acquired through vertical transmission.
   d) Plague is endemic in the prairie dog population in parts of the United States.
   e) The first case in an outbreak is called the index case.

5. Which of the following statements is *false?*
   a) A disease with a long incubation period might spread extensively before an epidemic is recognized.
   b) A person exposed to a low dose of a pathogen might not develop disease.
   c) The young and the aged are more likely to develop certain diseases.
   d) Malnourished populations are more likely to develop certain diseases.
   e) Herd immunity occurs when a population does not engage in a given behavior, such as eating raw fish, that would otherwise increase their risk of disease.
6. The purpose of an analytical study is to
   a) identify the person, place, and time of an outbreak.
   b) identify risk factors that result in high frequencies of disease.
   c) assess the effectiveness of preventive measures.
   d) determine the effectiveness of a placebo.
   e) None of the above
7. Which of the following causes of emerging diseases is thought to be a new pathogen?
   a) *Giardia*   b) *Vibrio cholerae* O139
   c) *Mycobacterium tuberculosis* d) *Shigella dysenteriae*
   e) *Schistosoma*
8. All of the following are thought to contribute to the emergence of disease *except*
   a) advances in technology.
   b) breakdown of public health infrastructure.
   c) construction of dams.
   d) mass distribution and importation of food.
   e) widespread vaccination programs.
9. Which of the following common causes of healthcare-associated infections is an environmental organism that grows readily in nutrient-poor solutions?
   a) *Enterococcus*   b) *Escherichia coli*
   c) *Pseudomonas aeruginosa*   d) *Staphylococcus aureus*
10. What is the most common type of nosocomial infection?
   a) Bloodstream infection   b) Gastrointestinal infection
   c) Pneumonia   d) Surgical wound infection
   e) Urinary tract infection

Applications

1. A news station reported about a potentially fatal epidemic disease occurring in a small Laotian village. An epidemiologist from the CDC was interviewed to discuss the disease and was very distressed that it was not being contained. Why did the epidemiologist feel the disease was a concern for people in North America?
2. An international team was gathered to discuss how funding should be spent to eliminate human infectious disease. There is only enough

funding to eliminate one disease. How would the scientists go about choosing the next disease to be eliminated from the planet?

Critical Thinking ✚

1. *Yersinia pestis* and hantavirus are both found in wild rodents in the southwestern United States. What is the risk of trying to stop a hantavirus epidemic by destroying rodents in that region?

2. A student disagreed with the presentation of the examples in figure 19.8. She claimed that the number of cases from a common-source outbreak could remain high over a much longer period of time in some cases and not decrease to zero. Is the student's claim reasonable? Why or why not?

20 Antimicrobial Medications

KEY TERMS

Acquired Resistance Resistance that develops through mutation or acquisition of new genes.

Antibiotic A compound naturally produced by molds or bacteria that inhibits the growth of or kills other microorganisms.

Antimicrobial Drug A chemical that inhibits the growth of or kills microorganisms; the term encompasses antibiotics and chemically synthesized drugs.

Antiviral Drug A drug that interferes with the infection cycle of a virus.

Bactericidal Kills bacteria.

Bacteriostatic Inhibits the growth of bacteria.

Broad-Spectrum Antimicrobial An antimicrobial drug that is effective against a wide range of microorganisms, often including both Gram-positive and Gram-negative bacteria.

Chemotherapeutic Agent A chemical used to treat disease.

Intrinsic (Innate) Resistance Resistance due to inherent characteristics of the organism.

Narrow-Spectrum Antimicrobial An antimicrobial drug that is effective against a limited range of microorganisms.

R Plasmid A plasmid that encodes resistance to one or more antimicrobial drugs.

Antibiotic susceptibility testing.

A Glimpse of History

Paul Ehrlich (1854–1915), a German physician and bacteriologist, was born into a wealthy family. He was the only son after many daughters, so family and servants indulged his interests, even though his large collection of frogs and snakes occasionally entered the laundry room. As an adult, he was rarely without a good cigar and habitually scribbled notes on his shirt cuffs. After receiving a degree in medicine, he became intrigued with the way various types of body cells differ in their ability to take up dyes and other substances. When he observed that certain dyes stain bacterial cells but not animal cells, indicating that the two cell types are somehow fundamentally different, it occurred to him that it might be possible to find a chemical that selectively harms bacteria without affecting human cells.

Ehrlich began searching for a "magic bullet," a term he used to describe a drug that would kill a microbial pathogen without harming the human host. He began by looking for a chemical that would cure the sexually transmitted disease syphilis, which is caused by the spirochete *Treponema pallidum*. Much of the mental illness during this time resulted from tertiary syphilis, a late stage of the disease. Ehrlich knew that an arsenic compound had shown some success in treating a protozoan disease of animals, and so he and his colleagues began synthesizing hundreds of different arsenic compounds in search for a cure for syphilis. In 1910, the 606th compound tested, arsphenamine, proved to be highly effective in treating the disease in laboratory animals. Although the drug itself was potentially lethal for patients, it did cure infections previously considered hopeless. The drug was given the name Salvarsan, a term derived from the words salvation and arsenic. Ehrlich's discovery proved that some chemicals could indeed selectively kill microbes.

Think back to the last time you were prescribed an antimicrobial medication. Could you have recovered from the infection without the drug? The prognosis for people with common diseases such as bacterial pneumonia and severe staphylococcal infections was grim before the discovery and widespread availability of penicillin in the 1940s. Physicians were able to identify the cause of the disease, but the only treatment option was usually bedrest. Today, however, antimicrobials are routinely prescribed, and this simple cure is often taken for granted. Unfortunately, the misuse of these life-saving medications, coupled with the amazing ability of microorganisms to adapt, has led to an increase in the number of drug-resistant organisms. Some people even speculate that we are in danger of seeing an end to the era of antimicrobial medications. In response, scientists are scrambling to develop new drugs.

20.1 ■ History and Development of Antimicrobial Drugs

Learning Outcomes

1. *Describe the discovery of antimicrobial drugs and antibiotics.*
2. *Explain how new generations of antimicrobial drugs are developed.*

To appreciate the unique antibiotic era in which we now live, it is important to understand the history and development of these life-saving remedies.

Discovery of Antimicrobial Drugs

The development of Salvarsan by Paul Ehrlich was the first documented example of a chemical used successfully as an antimicrobial medication (see **A Glimpse of History**). The next breakthrough came almost 25 years later, when the German chemist Gerhard Domagk discovered that a red dye called Prontosil could be used to treat streptococcal infections in animals. Surprisingly, Prontosil had no effect on streptococci growing in test tubes. It was later discovered that enzymes in the blood of the animal split the Prontosil molecule, producing a smaller molecule called sulfanilamide; this breakdown product acted against the infecting streptococci. Thus, the discovery of sulfanilamide, the first sulfa drug, was based on luck as well as scientific effort. If Prontosil had been screened only against bacteria in test tubes and not given to infected animals, its effectiveness might never have been discovered.

Salvarsan and Prontosil are **chemotherapeutic agents,** meaning chemicals used to treat disease. Because they are used to treat microbial infections, they can also be called **antimicrobial drugs** or, more simply, antimicrobials.

Discovery of Antibiotics

In 1928, Alexander Fleming, a British scientist, was working with cultures of *Staphylococcus aureus* when he noticed that colonies growing near a contaminating mold looked as if they were dissolving. Recognizing that the mold might be secreting a substance that killed bacteria, he proceeded to study it more carefully. He identified the mold as a species of *Penicillium* and found it was indeed producing a bacteria-killing substance; he called this penicillin. Even though Fleming was unable to purify penicillin, he showed that it was remarkably effective in killing many different bacterial species and did not cause adverse effects when injected into rabbits and mice. Fleming recognized the potential medical significance of his discovery, but became discouraged with his inability to purify the compound and eventually abandoned his study of it.

About 10 years after Fleming's discovery of penicillin, two other scientists in Britain, Ernst Chain and Howard Florey, successfully purified the compound. In 1941, the drug was tested for the first time on a police officer with a life-threatening *Staphylococcus aureus* infection. The patient improved so dramatically that within 24 hours his illness seemed under control. Unfortunately, the supply of purified penicillin ran out, and the man eventually died of the infection. Later, with greater supplies of the drug, two deathly ill patients were successfully cured.

World War II spurred cooperation of British and American scientists to determine the chemical structure of penicillin and to develop the means for its large-scale production. Several different penicillins were found in the *Penicillium* cultures, and were designated alphabetically. Penicillin G (or benzyl penicillin) was found to be the most suitable for treating infections. This was the first of what we now call **antibiotics**—antimicrobial drugs naturally produced by microorganisms.

Soon after the discovery of penicillin, Selman Waksman isolated a bacterium from soil, *Streptomyces griseus,* that produced an antibiotic he called streptomycin. The realization that bacteria as well as molds could produce antibiotics prompted researchers to begin screening hundreds of thousands of different microbial strains for antibiotic production. Even today, pharmaceutical companies examine soil samples from around the world for microbes that produce novel antibiotics.

Development of New Generations of Drugs

In the 1960s, scientists discovered they could alter the chemical structure of certain antimicrobial drugs, giving them new properties. For example, penicillin G, which is active mainly against Gram-positive bacteria, can be altered to produce ampicillin, a drug that kills a variety of Gram-negative species as well. Other changes to penicillin created methicillin, which is less susceptible to enzymes used by some bacteria to destroy penicillin. Today a variety of penicillin-like medications exist, making up what is referred to as the family of penicillins (**figure 20.1**). Other unrelated antimicrobial drugs have also been altered to give them new characteristics.

MicroAssessment 20.1

Antimicrobials are chemotherapeutic agents that are effective in treating microbial infections. Antibiotics are antimicrobial chemicals naturally produced by particular microorganisms.

1. *How is the microbe that makes penicillin different from the one that makes streptomycin?*
2. *Define and contrast the terms* chemotherapeutic agent, antimicrobial drug, *and* antibiotic.
3. *How might* Streptomyces griseus *cells protect themselves from the effects of streptomycin?*

20.2 ■ Features of Antimicrobial Drugs

Learning Outcome

3. *Describe selective toxicity; antimicrobial action; spectrum of activity; tissue distribution/metabolism/excretion; effects of combinations; adverse effects; and resistance to antimicrobials.*

Most modern antibiotics come from microorganisms that normally reside in the soil, including species of *Streptomyces* and *Bacillus* (bacteria), and *Penicillium* and *Cephalosporium* (fungi). To commercially produce an antibiotic, a carefully selected strain of the appropriate species is inoculated into a broth medium and incubated in a huge vat. As soon as the maximum antibiotic concentration is reached, the drug is extracted from the medium and

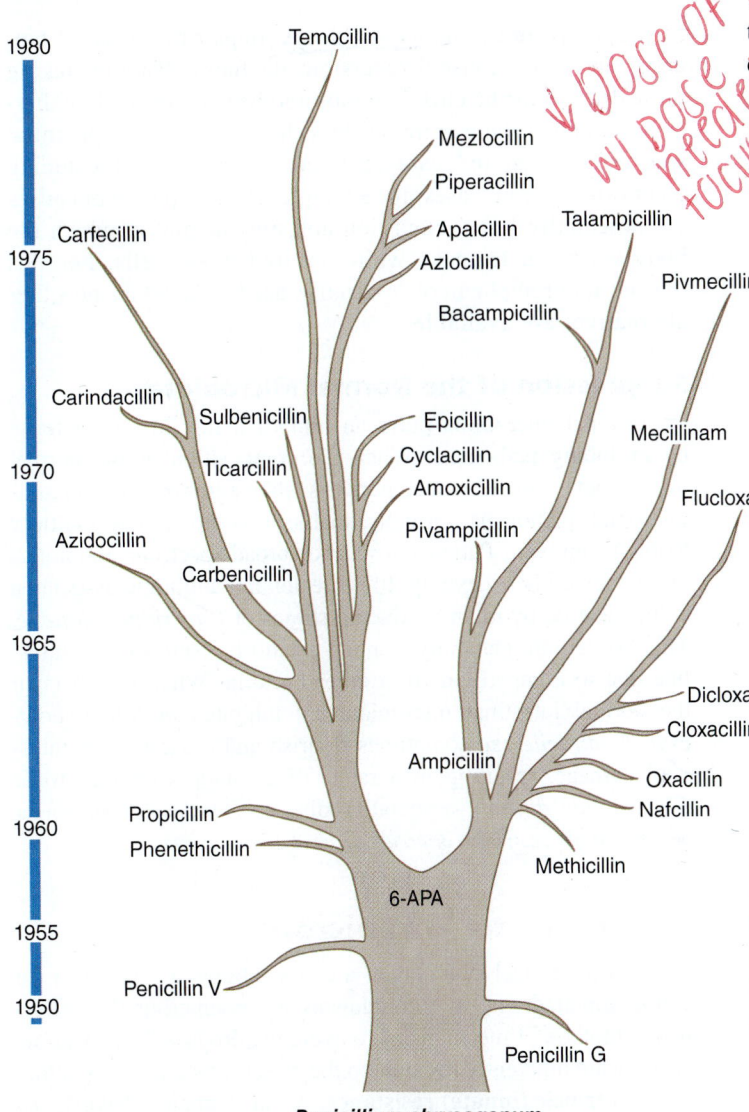

*↓ dose of toxic
w/ dose
needed
to cure*

FIGURE 20.1 Family Tree of Penicillins All of the derivatives contain 6-aminopenicillanic acid (6-APA), the core portion of penicillin G.

❓ *Why is it necessary to develop new generations of antimicrobial medications?*

extensively purified. In many cases, the antibiotic is chemically altered after purification to give it new characteristics such as increased stability. These chemically modified compounds are called semisynthetic. In some cases, the entire drug can be synthesized in the laboratory. By convention, these partially or totally synthetic chemicals are still called antibiotics because microorganisms can produce the core structure naturally. Numerous different antimicrobial drugs are now available, each with characteristics that make it more or less suitable for a given clinical situation. Hundreds of tons and many millions of dollars' worth of antibiotics are now produced each year.

Selective Toxicity

Medically useful antimicrobial drugs exhibit **selective toxicity,** causing greater harm to microbes than to the human host. They do this by interfering with essential structures or biochemical processes that are common in microbes but not human cells.

Although the ideal antimicrobial drug is non-toxic to humans, most can be harmful at high concentrations. In other words, selective toxicity is a relative term. The toxicity of a given drug is expressed as the **therapeutic index,** which is the lowest dose toxic to the patient divided by the dose typically used for therapy. Antimicrobials that have a high therapeutic index are less toxic, often because the drug acts against a vital biochemical process of microorganisms that does not exist in human cells. For example, penicillin G, which interferes with bacterial cell wall synthesis, has a very high therapeutic index.

When an antimicrobial that has a low therapeutic index is used, the concentration in the patient's blood must be carefully monitored to make sure it does not reach a toxic level. Drugs too toxic for systemic use can sometimes be used for topical applications, such as first-aid antibiotic skin ointments.

Antimicrobial Action

Some antimicrobial drugs usually kill microbes, whereas others only inhibit their growth. Both actions are medically important.

Bacteriostatic drugs inhibit bacterial growth. A patient taking a bacteriostatic drug must rely on his or her body's defense systems to kill or eliminate the pathogen after its growth has been stopped. Sulfa drugs, for example, are frequently prescribed for urinary tract infections. They prevent bacteria in the bladder from growing, so that urination can more effectively eliminate them.

Bactericidal drugs kill bacteria. These are particularly useful when the host defenses cannot be relied on to eliminate pathogens. Bactericidal drugs are sometimes only inhibitory, depending on the drug concentrations and the stage of bacterial growth. Surprisingly, the lethal effects of different bactericidal drugs seem to involve the same mechanism, regardless of the drug's target. It appears that the antimicrobial-induced cell damage overwhelms the bacterium's ability to detoxify reactive oxygen species, leading to extensive oxidative damage.

◀◀ reactive oxygen species, p. 90

Spectrum of Activity

Antimicrobial drugs vary with respect to the range of microorganisms they kill or inhibit. Some kill or inhibit a narrow range of microorganisms, such as only Gram-positive bacteria, whereas others affect a wide range, generally including both Gram-positive and Gram-negative organisms.

Broad-spectrum antimicrobials affect a wide range of bacteria. These drugs are important for treating acute life-threatening diseases when immediate antimicrobial therapy is essential and there is no time to culture and identify the pathogen. The disadvantage of broad-spectrum antimicrobials is that by affecting a wide range of microbes they disrupt the normal microbiota that play an important role in excluding pathogens. This in turn can leave the patient predisposed to other infections.

Narrow-spectrum antimicrobials affect a limited range of bacteria. Their use requires that the pathogen be identified and its antimicrobial susceptibility tested, but they cause less disruption to the normal microbiota. ◀◀ normal microbiota, pp. 337, 381

Effects of Combinations

Combinations of antimicrobials are sometimes used to treat infections, but these must be chosen carefully because some drugs counteract the effects of others. Bacteriostatic drugs, for example, interfere with the action of drugs that kill only actively dividing cells. Counteracting combinations such as this are antagonistic. In contrast, combinations in which the activity of one drug enhances the activity of the other are synergistic. Combinations that are neither synergistic nor antagonistic are additive.

Tissue Distribution, Metabolism, and Excretion of the Drug

Antimicrobials differ not only in their action and activity, but also in how they are distributed, metabolized, and excreted by the body. Only some drugs cross from the blood into the cerebrospinal fluid, an important factor in treating meningitis. Drugs that are unstable at low pH are destroyed by stomach acid when swallowed, so these drugs must instead be administered through intravenous or intramuscular injection. ▶▶ meningitis, p. 643

Another important characteristic of an antimicrobial drug is its rate of elimination, expressed as the half-life. The half-life of a drug is the time it takes for the serum concentration to decrease by 50%. This dictates the frequency of doses required to maintain an effective level in the body. Penicillin V, which has a very short half-life, needs to be taken four times a day, whereas azithromycin, which has a half-life of over 24 hours, is taken only once a day or less. Patients who have kidney or liver dysfunction often excrete or metabolize drugs more slowly, and so their drug dosages must be adjusted accordingly to avoid toxic levels.

Adverse Effects

As with any medication, several concerns and dangers are associated with antimicrobial drugs. It is important to remember, however, that antimicrobials are extremely valuable drugs that save countless lives when properly prescribed and used.

Allergic Reactions

Some people develop allergies to antimicrobial drugs. An allergy to penicillin or related drug usually results in a fever or rash but can abruptly cause life-threatening anaphylactic shock. For this reason, people who have allergic reactions to a given antimicrobial must alert their physicians and pharmacists so an alternative drug can be prescribed. They should also wear a bracelet or necklace that records that information in case of emergency. ◀◀ anaphylactic shock, p. 404

Toxic Effects

Several antimicrobials are toxic at high concentrations or occasionally cause adverse reactions. Aminoglycosides such as streptomycin can damage kidneys, impair the sense of balance, and even cause irreversible deafness. Patients taking these drugs must be closely monitored because of the low therapeutic index. Some antimicrobials have such severe potential side effects that they are reserved for only life-threatening conditions. In rare cases, for example, chloramphenicol causes the potentially lethal condition aplastic anemia, in which the body is unable to make white and red blood cells. For this reason, chloramphenicol is usually used only when no other alternatives are available.

Suppression of the Normal Microbiota

The normal microbiota plays an important role in host defense by excluding pathogens. When the composition of the normal microbiota is altered, which happens when a person takes an antimicrobial, pathogens normally unable to compete may multiply to high numbers. Patients who take broad-spectrum antibiotics orally sometimes develop life-threatening antibiotic-associated colitis, caused by toxin-producing strains of *Clostridium difficile*. This bacterium generally cannot establish itself in the intestine due to competition from other bacteria. When members of the normal intestinal microbiota are inhibited or killed, however, *C. difficile* can sometimes flourish and cause serious intestinal damage, resulting in a range of symptoms referred to as *Clostridium difficile*–associated disease. ◀◀ normal microbiota, pp. 337, 381 ▶▶ *Clostridium difficile*–associated disease, p. 594

Resistance to Antimicrobials

As pharmaceutical companies are assembling a vast array of antimicrobial drugs, microorganisms are countering the efforts with a toolbox of mechanisms to avoid the drugs' effects. Certain bacteria are inherently resistant to the effects of some drugs, a trait called **intrinsic (innate) resistance.** As an example, *Mycoplasma* species lack a cell wall, so they are resistant to penicillin and any other drug that interferes with peptidoglycan synthesis. Many Gram-negative bacteria are intrinsically resistant to certain drugs because the lipid bilayer of their outer membrane prevents the drug from entering (see figure 3.33). In other cases, previously sensitive organisms develop **acquired resistance** through spontaneous mutation or horizontal gene transfer. The mechanisms involved will be discussed later. ◀◀ the genus *Mycoplasma*, p. 276 ◀◀ Gram-negative cell wall, p. 60

MicroAssessment 20.2

When choosing an antimicrobial to prescribe, a variety of factors must be considered including the therapeutic index, antimicrobial action, spectrum of activity, effects of combinations, tissue distribution, half-life, adverse effects, and resistance of the microbe.

4. *Which would be safest: an antimicrobial that has a low therapeutic index or one that has a high therapeutic index? Why?*

5. *In what clinical situation is it most appropriate to use a broad-spectrum antimicrobial?*

6. *Why would antimicrobials that have toxic effects be used at all?* ✚

Focus Figure

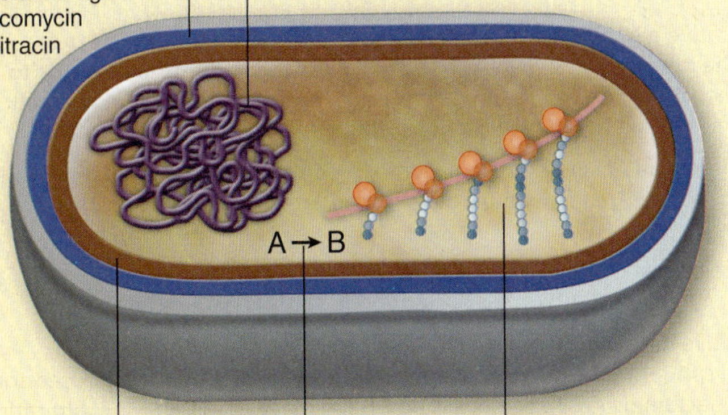

Cell wall
(peptidoglycan)
synthesis
β-lactam drugs
Vancomycin
Bacitracin

Nucleic acid synthesis
Fluoroquinolones
Rifamycins

A→B

**Cell membrane
integrity**
Polymyxin B
Daptomycin

**Metabolic pathways
(folate biosynthesis)**
Sulfonamides
Trimethoprim

Protein synthesis
Aminoglycosides
Tetracyclines
Macrolides
Chloramphenicol
Lincosamides
Oxazolidinones
Streptogramins

FIGURE 20.2 Targets of Antibacterial Medications

❓ *Why would* Mycoplasma pneumoniae *be intrinsically resistant to β-lactam drugs?*

20.3 ■ Mechanisms of Action of Antibacterial Drugs

Learning Outcomes

4. *Describe the β-lactam drugs and other antimicrobials that inhibit cell wall synthesis.*
5. *Describe the antimicrobial drugs that inhibit protein synthesis.*
6. *Describe the antimicrobial drugs that inhibit nucleic acid synthesis.*
7. *Describe the antimicrobial drugs that inhibit metabolic pathways.*
8. *Describe the antimicrobial drugs that interfere with cell membrane integrity.*
9. *Describe the antibacterial medications used to treat* Mycobacterium tuberculosis *infections.*

This section describes the mechanisms of action of various antibacterial drugs, highlighting the bacterial processes and structures targeted (**figure 20.2**). A group of antibiotics called β-lactam drugs will be covered in the greatest detail, because they serve as excellent examples of some of the important features of antimicrobials in general. **Table 20.1** summarizes the antimicrobials discussed in this section.

TABLE 20.1	Characteristics of Antibacterial Drugs
Target/Drug	**Comments/Characteristics**
Cell Wall Synthesis	
β-lactam drugs	Bactericidal against a variety of bacteria; inhibit penicillin-binding proteins. Resistance is due to synthesis of β-lactamases, decreased affinity of penicillin-binding proteins, or decreased uptake.
Penicillins	A family of antibacterial medications; different groups vary in their spectrum of activity and their susceptibility to β-lactamases.
Natural penicillins: penicillin G, penicillin V	Active against Gram-positive and a few Gram-negative bacteria. Penicillin G is destroyed by stomach acid, and so it usually must be administered by injection. Penicillin V can be taken orally.
Penicillinase-resistant: methicillin, dicloxacillin	Similar to natural penicillins, but resistant to inactivation by the penicillinase of staphylococci.
Broad-spectrum: ampicillin, amoxicillin	Similar to the natural penicillins, but more active against Gram-negative organisms.
Extended-spectrum: ticarcillin, piperacillin	Increased activity against Gram-negative rods, including *Pseudomonas* species.
Cephalosporins Cephalexin, cephradine, cefaclor, cefprozil, cefixime, ceftibuten, cefepime	A family of antibacterial medications. The later generations are generally more effective against Gram-negative bacteria and less susceptible to destruction by β-lactamases but susceptible to extended-spectrum β-lactamases.
Carbapenems Imipenem, meropenem, ertapenem, and doripenem	Resistant to inactivation by β-lactamases, but susceptible to carbapenemases. Imipenem must be given in combination with a drug that inhibits certain kidney enzymes in order to avoid its inactivation.
Monobactams Aztreonam	Resistant to β-lactamases but susceptible to extended-spectrum β-lactamases; can be given to patients who are allergic to penicillin. Primarily active against members of the family *Enterobacteriaceae*.
Vancomycin	Bactericidal against Gram-positive bacteria; binds to the peptide side chain of *N*-acetylmuramic acid. Used to treat serious systemic infections and severe *Clostridium difficile*–associated disease. In enterococci, resistance is due to a plasmid-encoded altered target.
Bacitracin	Bactericidal against Gram-positive bacteria; interferes with the transport of peptidoglycan precursors. Common ingredient in non-prescription antibiotic ointments.

(continued)

TABLE 20.1 Characteristics of Antibacterial Drugs (*Continued*)

Target/Drug	Comments/Characteristics
Protein Synthesis	
Aminoglycosides Streptomycin, gentamicin, tobramycin, amikacin, neomycin	Bactericidal against aerobic and facultative bacteria; bind to the 30S ribosomal subunit, blocking the initiation of translation and causing the misreading of mRNA. Toxicity limits the use. Resistance is due to a plasmid-encoded inactivating enzyme, alteration of the target molecule, or decreased uptake by a cell. Neomycin is commonly used in non-prescription topical antibiotic ointments.
Tetracyclines Tetracycline, doxycycline, glycylcyclines	Bacteriostatic against some Gram-positive and Gram-negative bacteria; bind to the 30S ribosomal subunit, blocking the attachment of tRNA. Resistance is generally due to decreased accumulation, either through decreased uptake or increased efflux.
Glycylcyclines Tigecycline	Bacteriostatic against many Gram-positive and Gram-negative bacteria; bind to the 30S ribosomal subunit, blocking the attachment of tRNA.
Macrolides Erythromycin, clarithromycin, azithromycin	Bacteriostatic against many Gram-positive bacteria as well as the most common causes of atypical pneumonia; bind to the 50S ribosomal subunit, preventing the continuation of protein synthesis. Used for treating patients who are allergic to β-lactam drugs. Resistance is due to an inactivating enzyme, alteration of the target molecule, or decreased uptake by a cell.
Chloramphenicol	Bacteriostatic and broad-spectrum; binds to the 50S ribosomal subunit, preventing peptide bonds from being formed. Generally used only as a last resort for life-threatening infections. Resistance is often due to a plasmid-encoded inactivating enzyme.
Lincosamides Lincomycin, clindamycin	Bacteriostatic against a variety of Gram-positive and Gram-negative bacteria, including the anaerobe *Bacteroides fragilis.* Bind to the 50S ribosomal subunit, preventing the continuation of protein synthesis. Associated with an even greater risk of developing *Clostridium difficile*–associated disease.
Oxazolidinones Linezolid	Bacteriostatic against a variety of Gram-positive bacteria. Bind to the 50S ribosomal subunit, interfering with the initiation of protein synthesis.
Streptogramins Quinupristin, dalfopristin	A synergistic combination of two drugs that bind to two different sites on the 50S ribosomal subunit, inhibiting distinct steps of protein synthesis. Individually each drug is bacteriostatic, but together they are bactericidal. Effective against a variety of Gram-positive bacteria.
Nucleic Acid Synthesis	
Fluoroquinolones Ciprofloxacin, moxifloxacin	Bactericidal against a wide variety of Gram-positive and Gram-negative bacteria; inhibit topoisomerases. Resistance is most often due to structural alterations in the topoisomerase target.
Rifamycins Rifampin	Bactericidal against Gram-positive and some Gram-negative bacteria. Bind RNA polymerase, blocking the initiation of RNA synthesis. Primarily used to treat infections caused by *Mycobacterium tuberculosis* and as prophylaxis for patients who have been exposed to *Neisseria meningitidis.*
Metronidazole	Bactericidal against anaerobes. Activated by anaerobic metabolism and then binds DNA, interfering with synthesis and causing damaging breaks. Used to treat bacterial vaginosis and *Clostridium difficile*–associated disease.
Folate Biosynthesis	
Sulfonamides	Bacteriostatic against a variety of Gram-positive and Gram-negative bacteria. Structurally similar to para-aminobenzoic acid (PABA) and therefore inhibit the enzyme for which PABA is a substrate. Resistance is most commonly due to a plasmid-encoded alternative enzyme.
Trimethoprim	Often used in combination with a sulfa drug for a synergistic effect; inhibits the enzyme that catalyzes a step following the one inhibited by the sulfonamides. Resistance is commonly due to a plasmid-encoded alternative enzyme; the genes that encode resistance to sulfa drugs are often carried on the same plasmid.
Cell Membrane Integrity	
Daptomycin	Bactericidal against Gram-positive bacteria by damaging the cytoplasmic membrane.
Polymyxin B	Bactericidal against Gram-negative bacteria by damaging cell membranes. Toxicity limits its use primarily to topical applications, but it is a common ingredient in non-prescription antibiotic ointments.
Mycobacterium tuberculosis	
Ethambutol	Inhibits the synthesis of a component of the mycobacterial cell wall.
Isoniazid	Inhibits synthesis of mycolic acid, a major component of the mycobacterial cell wall.
Pyrazinamide	Mechanism unknown.

Antibacterial Medications That Inhibit Cell Wall Synthesis

Bacterial cell walls are unique in that they contain peptidoglycan (see figures 3.32 and 3.33). Because of this, antimicrobial medications that interfere solely with synthesis of this cell wall component do not affect eukaryotic cells, and often have a very high therapeutic index (**figure 20.3**).

Penicillins, Cephalosporins, and Other β-Lactam Drugs

Penicillins and cephalosporins are members of a group of antimicrobial medications referred to as **β-lactam drugs.** This group, which also includes the monobactams and carbapenems, all have a shared chemical structure called a β-lactam ring (**figure 20.4**).

The β-lactam drugs competitively inhibit a group of enzymes that catalyze the formation of peptide bridges between adjacent glycan strands, an essential step in the final stages of peptidoglycan synthesis (see figure 3.31). These enzymes are commonly called **penicillin-binding proteins (PBPs),** reflecting the fact

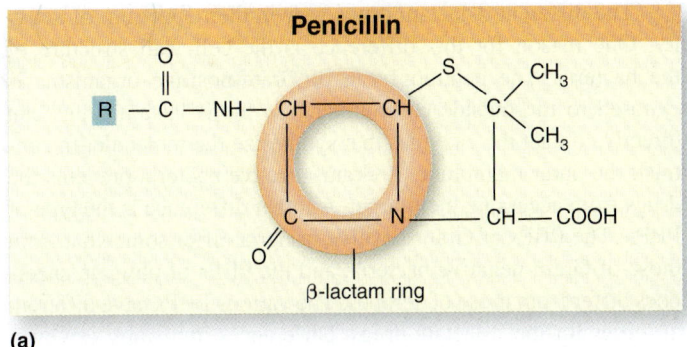

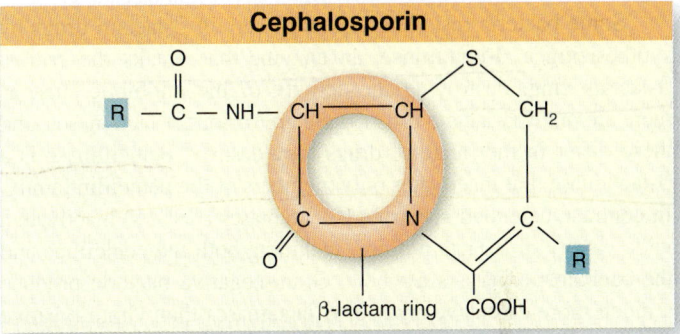

(a)

(b)

FIGURE 20.4 The β-Lactam Ring of Penicillins and Cephalosporins The core chemical structure of **(a)** a penicillin; **(b)** a cephalosporin. The β-lactam rings are marked by an orange circle. The R groups vary among different penicillins and cephalosporins.

❓ *Why is it not surprising that penicillin and cephalosporin both have a high therapeutic index?*

that they bind penicillin and were initially discovered during experiments to study the effects of the medication. The disruption in cell wall synthesis weakens the wall, causing the cell to lyse (**figure 20.5**). The β-lactam drugs are bactericidal only against growing bacteria, because these cells continuously synthesize peptidoglycan. ◄◄ **competitive enzyme inhibition, p. 138**

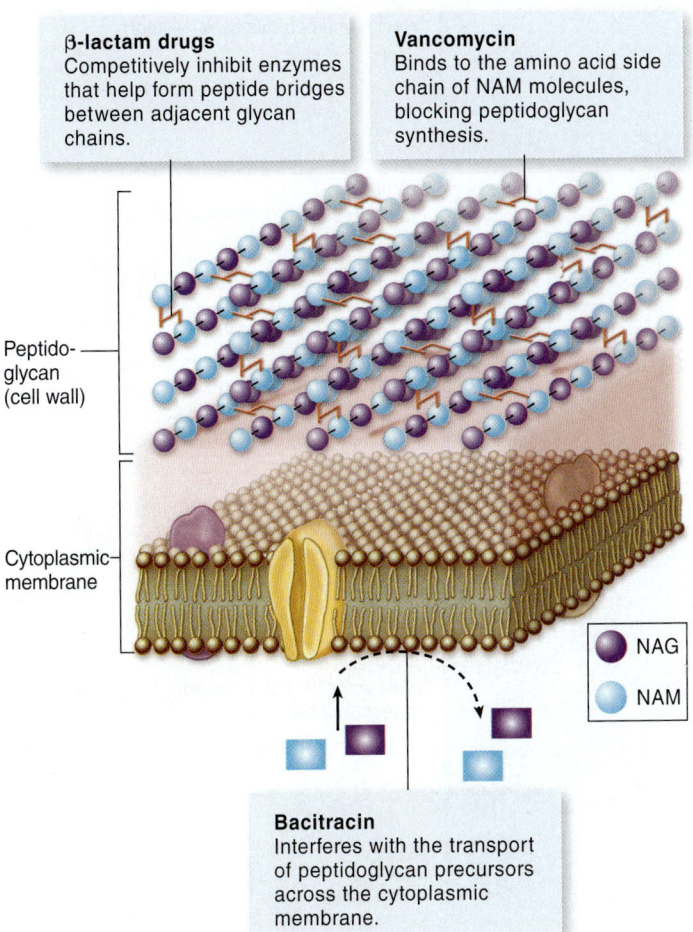

β-lactam drugs
Competitively inhibit enzymes that help form peptide bridges between adjacent glycan chains.

Vancomycin
Binds to the amino acid side chain of NAM molecules, blocking peptidoglycan synthesis.

Peptido-glycan (cell wall)

Cytoplasmic membrane

● NAG
○ NAM

Bacitracin
Interferes with the transport of peptidoglycan precursors across the cytoplasmic membrane.

FIGURE 20.3 Antibacterial Medications That Interfere with Cell Wall Synthesis

❓ *What is the function of penicillin-binding proteins?*

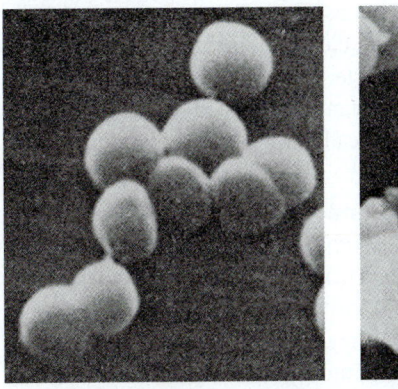
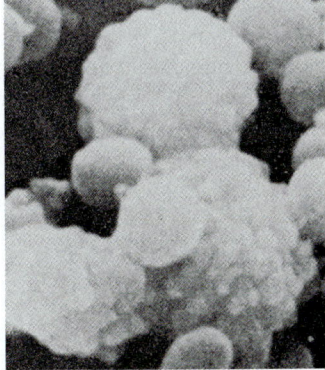

FIGURE 20.5 Effect of a β-Lactam Drug on a Cell The drug disrupts cell wall synthesis, leading to weakening of the cell wall, ultimately causing the cells to lyse.

❓ *Why would a bacteriostatic drug be an antagonistic combination with penicillin?*

The different β-lactam drugs vary in their spectrum of activity. One reason for this difference is the cell wall structure of the bacteria. The peptidoglycan of Gram-positive organisms is exposed to the outside environment, so the β-lactam drugs can directly contact the enzymes that synthesize the molecule. In contrast, the outer membrane of Gram-negative bacteria prevents the drugs from accessing their target. Another difference is the type of PBPs. The PBPs of Gram-positive bacteria differ somewhat from those of Gram-negative bacteria, and the PBPs of obligate anaerobes differ from those of aerobes. The various PBPs have different affinities for the β-lactam drugs. Differences in affinity can even exist among related organisms such as Gram-positive cocci.

Some bacteria resist the effects of certain β-lactam drugs by synthesizing a **β-lactamase,** an enzyme that breaks the critical β-lactam ring, destroying the activity of the antibiotic. Just as there are many β-lactam drugs, there are various β-lactamases, and these differ in the range of drugs they destroy. Penicillinase is a β-lactamase that inactivates only members of the penicillin family. In contrast, extended-spectrum β-lactamases (ESBLs) inactivate a wide variety of β-lactam drugs, including both the penicillins and the cephalosporins. As a whole, Gram-negative bacteria produce a much more extensive array of β-lactamases than Gram-positive organisms can.

The Penicillins All members of the penicillin family share a common basic structure. This structure's side chain has been chemically modified to create penicillin derivatives, each with unique characteristics (**figure 20.6**). Penicillins can be loosely grouped into several categories:

- **Natural penicillins.** These are the original penicillins produced naturally by the mold *Penicillium chrysogenum.* Natural penicillins are narrow-spectrum antibiotics, effective against Gram-positive and a few Gram-negative bacteria. Penicillin V is more stable in acid and, therefore, better absorbed than penicillin G when taken orally. Bacteria that produce penicillinase are resistant to the natural penicillins.

- **Penicillinase-resistant penicillins.** Scientists developed these in a response to the problem of penicillinase-producing *Staphylococcus aureus* strains. Penicillinase-resistant penicillins include methicillin and dicloxacillin. Unfortunately, some penicillinase-producing bacteria acquired the ability to make altered PBPs to which β-lactam drugs no longer bind. *S. aureus* strains that can do this are called MRSA (methicillin-resistant *S. aureus*).

- **Broad-spectrum penicillins.** The modified side chains of these drugs give them a broad spectrum of activity. They retain their activity against penicillin-sensitive Gram-positive bacteria, yet they are also active against Gram-negative bacteria. Unfortunately, they can be inactivated by many β-lactamases. Broad-spectrum penicillins include ampicillin and amoxicillin.

- **Extended-spectrum penicillins.** These have greater activity against *Pseudomonas* species—Gram-negative bacteria that are unaffected by many antimicrobial drugs. However, this group of penicillins has less activity against Gram-positive bacteria. Like the other broad-spectrum penicillins, they are

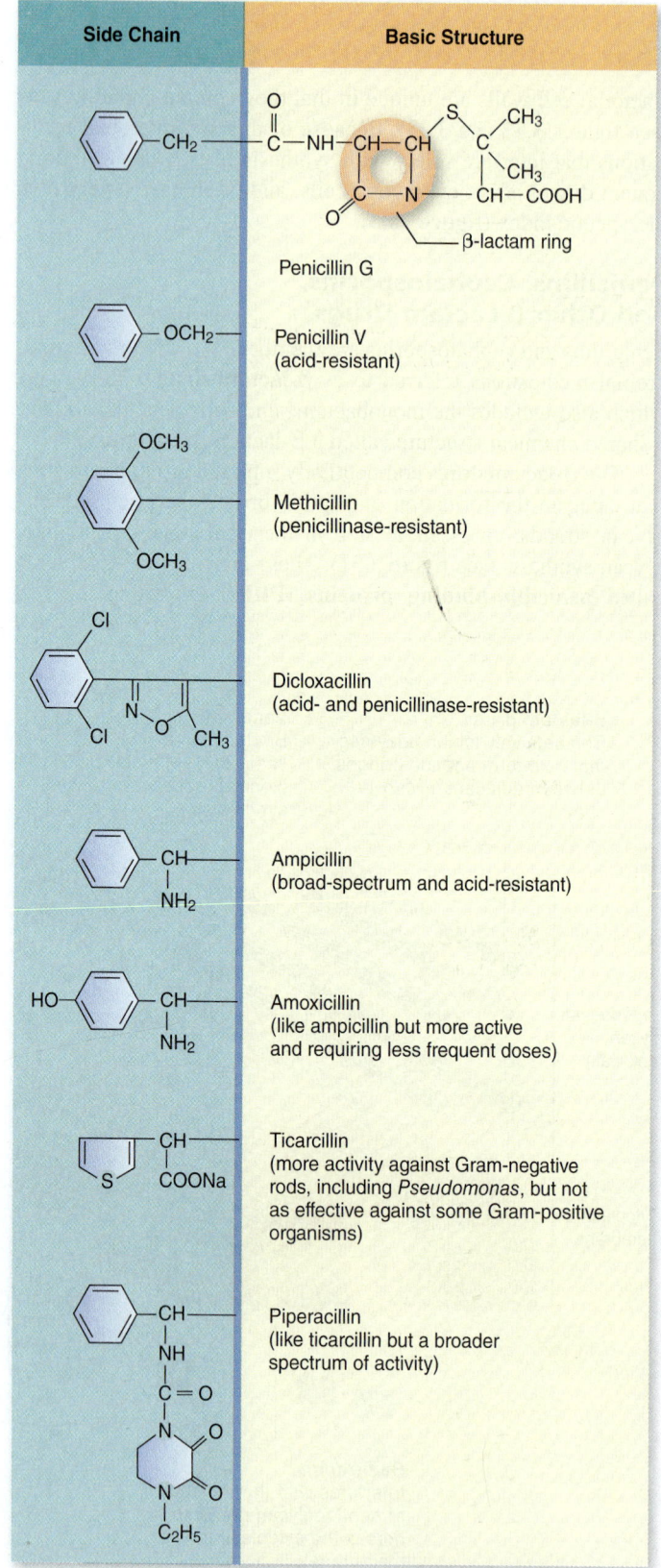

Side Chain	Basic Structure

Penicillin G

Penicillin V (acid-resistant)

Methicillin (penicillinase-resistant)

Dicloxacillin (acid- and penicillinase-resistant)

Ampicillin (broad-spectrum and acid-resistant)

Amoxicillin (like ampicillin but more active and requiring less frequent doses)

Ticarcillin (more activity against Gram-negative rods, including *Pseudomonas*, but not as effective against some Gram-positive organisms)

Piperacillin (like ticarcillin but a broader spectrum of activity)

FIGURE 20.6 Chemical Structures and Properties of Representative Members of the Penicillin Family The entire structure of penicillin G and the side chains of other penicillins are shown.

? *Could penicillin be used to treat MRSA (methicillin-resistant Staphylococcus aureus) infections? Why or why not?*

destroyed by many β-lactamases. Extended-spectrum penicillins include ticarcillin and piperacillin.

- **Penicillins + β-lactamase inhibitor.** Rather than a new drug, this is a combination of agents. The β-lactamase inhibitor interferes with the activity of some types of β-lactamases, thereby protecting the penicillin against enzymatic destruction. An example is Augmentin, a combination of amoxicillin and clavulanic acid.

The Cephalosporins These are derived from an antibiotic produced by the fungus *Acremonium cephalosporium*. Generally included in the cephalosporin family is a closely related group of antibiotics made by filamentous bacteria related to *Streptomyces*.

The chemical structure of the cephalosporins makes them resistant to inactivation by certain β-lactamases. Some are not very effective against Gram-positive bacteria, however, because the drugs have a low affinity for the PBPs of these organisms.

The cephalosporins have been chemically modified, giving rise to first-, second-, third-, and fourth-generation versions. These include cephalexin and cephradine (first generation), cefaclor and cefprozil (second generation), cefixime and ceftibuten (third generation), and cefepime (fourth generation). The later generations are generally more effective against Gram-negative bacteria and less susceptible to destruction by β-lactamases.

Other β-Lactam Antibiotics Two other groups of β-lactam drugs, carbapenems and monobactams, are resistant to most β-lactamases. The carbapenems are effective against a wide range of Gram-negative and Gram-positive bacteria. Four types are available—imipenem, meropenem, ertapenem, and doripenem. Imipenem is rapidly destroyed by a kidney enzyme and is therefore given in combination with a drug that inhibits that enzyme. Bacteria that produce carbapenemases are resistant. The only monobactam used therapeutically, aztreonam, is primarily effective against members of the family *Enterobacteriaceae*, which are Gram-negative rods. Aztreonam has a slightly different structure than other β-lactam drugs, so it can be given to patients who are allergic to penicillin. Bacteria that produce ESBLs are resistant to aztreonam. ◀◀ *Enterobacteriaceae, p. 265*

Vancomycin

Vancomycin blocks peptidoglycan synthesis by binding to the peptide side chain of NAM molecules being assembled to form glycan chains. This weakens the cell wall, causing cell lysis. The drug is poorly absorbed from the intestinal tract, so it must be administered intravenously except when used to treat intestinal infections. ◀◀ NAM, p. 58

Vancomycin is a very important medication for treating infections caused by Gram-positive bacteria resistant to β-lactam drugs. It is also used to treat severe cases of *Clostridium difficile*–associated disease that do not respond to metronidazole. In fact, vancomycin is sometimes referred to as the antibiotic of last resort because it is generally reserved for organisms resistant to all other options. Fortunately, new alternatives have been developed, but vancomycin is still the drug of choice for treating most infections caused by Gram-positive bacteria resistant to other drugs.

▶▶ metronidazole, p. 478

Vancomycin does not cross the outer membrane of Gram-negative bacteria, so these organisms are intrinsically resistant. Acquired resistance to vancomycin—an increasing problem—is most often due to a change in the peptide side chain of the NAM molecule that prevents vancomycin from binding.

Bacitracin

Bacitracin inhibits cell wall biosynthesis by interfering with the transport of peptidoglycan precursors across the cytoplasmic membrane. Its toxicity limits its use to topical applications (used only on the surface of the skin); however, it is a common ingredient in non-prescription first-aid ointments.

Antibacterial Medications That Inhibit Protein Synthesis

Several types of antibacterial drugs inhibit prokaryotic protein synthesis (**figure 20.7**). Although all cells synthesize proteins, the structure of the prokaryotic 70S ribosome—composed of a 30S and a 50S subunit—is different enough from the eukaryotic 80S ribosome to make it a suitable target for selective toxicity. The mitochondria of eukaryotic cells also have 70S ribosomes, however, which may partially account for the toxicity of some of these drugs. ◀◀ ribosome structure, p. 66

The Aminoglycosides

The aminoglycosides are bactericidal drugs that irreversibly bind to the 30S ribosomal subunit, causing it to distort and malfunction. This blocks the initiation of translation and causes misreading of mRNA by ribosomes that have already passed the initiation step.

Aminoglycosides are generally not effective against anaerobes, enterococci, and streptococci. This is because they enter bacterial cells by an active transport process that requires respiratory metabolism. To extend their spectrum of activity, the aminoglycosides are sometimes used in a synergistic combination

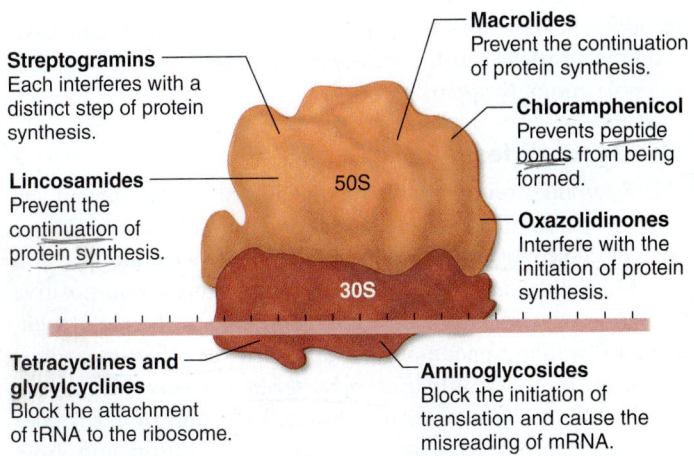

FIGURE 20.7 Antibacterial Medications That Inhibit Prokaryotic Protein Synthesis These medications bind to the 70S ribosome.

❓ *Some drugs that inhibit prokaryotic protein synthesis have a low therapeutic index. Why would this be so?*

with a β-lactam drug. The β-lactam drug interferes with cell wall synthesis, which, in turn, allows the aminoglycoside to enter cells that would otherwise be resistant. ◀◀ active transport, p. 56

Examples of aminoglycosides include streptomycin, gentamicin, tobramycin, and amikacin. Unfortunately, all of these can cause severe side effects including hearing loss and kidney damage; consequently, they are generally used only when other alternatives are not available. A form of tobramycin that can be administered through inhalation rather than injection makes treatment of lung infections in cystic fibrosis patients caused by *Pseudomonas aeruginosa* safer and more effective. Another aminoglycoside, neomycin, is too toxic for systemic use; however, it is a common ingredient in non-prescription topical ointments.

The Tetracyclines

The tetracyclines reversibly bind to the 30S ribosomal subunit, blocking the attachment of tRNA and preventing translation from continuing. These bacteriostatic drugs are actively transported into prokaryotic but not animal cells, which effectively concentrates them inside bacteria. This, in part, accounts for their selective toxicity.

The tetracyclines are effective against certain Gram-positive and Gram-negative bacteria. Derivatives such as doxycycline have a longer half-life, allowing less-frequent doses. Resistance to the tetracyclines is primarily due to a decrease in their accumulation by the bacterial cell, either by decreased uptake or increased excretion.

MicroByte
Tetracyclines can cause discoloration in teeth when used by young children.

The Glycylcyclines

The glycylcyclines are a new class of antibacterial drugs related to the tetracyclines. They have the same mechanism of action and effect as the tetracyclines, but a wider spectrum of activity. In addition, they are effective against many bacteria that have acquired resistance to the tetracyclines. Tigecycline is the only example currently approved.

The Macrolides

The macrolides reversibly bind to the 50S ribosomal subunit and prevent the continuation of translation. They often serve as the drug of choice for patients who are allergic to penicillin.

Macrolides are bacteriostatic against many Gram-positive bacteria as well as the most common causes of atypical pneumonia ("walking pneumonia"). They are not effective against members of the family *Enterobacteriaceae* because they do not pass through the outer membrane. Examples of macrolides include erythromycin, clarithromycin, and azithromycin. Both clarithromycin and azithromycin have a longer half-life than erythromycin, so that they can be taken less frequently. Resistance can occur through modification of the ribosomal RNA target, production of an enzyme that chemically modifies the drug, and alterations that result in decreased drug uptake. ▶▶ walking pneumonia, p. 500

Chloramphenicol

Chloramphenicol binds to the 50S ribosomal subunit, preventing peptide bonds from being formed and, consequently, blocking translation. Its action is bacteriostatic.

Chloramphenicol is active against a wide range of bacteria, but it is generally only used as a last resort for life-threatening infections in order to avoid a rare but lethal side effect. This complication, aplastic anemia, can occur in response to even a small amount of chloramphenicol and is characterized by the inability of the body to form white and red blood cells.

The Lincosamides

The lincosamides bind to the 50S ribosomal subunit and prevent the continuation of translation. Their action is bacteriostatic.

Lincosamides inhibit a variety of Gram-negative and Gram-positive bacteria. They are particularly useful for treating infections resulting from intestinal perforation because they inhibit *Bacteroides fragilis,* a member of the normal intestinal microbiota that is frequently resistant to other antimicrobials. Unfortunately, the risk of developing *Clostridium difficile*–associated disease is greater for people taking lincosamides than some other antimicrobials because most *C. difficile* strains are resistant to the lincosamides. The most commonly used lincosamide is clindamycin.

The Oxazolidinones

The oxazolidinones bind to the 50S ribosomal subunit, interfering with the initiation of translation. They are bacteriostatic against a variety of Gram-positive bacteria and are useful in treating infections caused by bacteria that are resistant to β-lactam drugs and vancomycin. Linezolid is an example.

The Streptogramins

Two streptogramins (quinupristin and dalfopristin) are administered together in a medication called Synercid. These act as a synergistic combination, binding to two different sites on the 50S ribosomal subunit and inhibiting distinct steps of translation. Individually, each drug is bacteriostatic but together they are bactericidal. Synercid is effective against a variety of Gram-positive bacteria, including some that are resistant to β-lactam drugs and vancomycin.

Antibacterial Medications That Inhibit Nucleic Acid Synthesis

Enzymes required for nucleic acid synthesis are the targets of some groups of antimicrobial drugs.

The Fluoroquinolones

The fluoroquinolones are synthetic drugs that inhibit one or more of a group of enzymes called topoisomerases, which maintain the supercoiling of DNA within the bacterial cell. One type of topoisomerase, DNA gyrase, breaks and rejoins strands to relieve the strain caused by the localized unwinding of DNA during replication and transcription. Consequently, inhibition of this enzyme prevents these essential cell processes. ◀◀ supercoiled DNA, p. 66

The fluoroquinolones are bactericidal against a wide variety of bacteria, including both Gram-positive and Gram-negative organisms. Examples of fluoroquinolones include ciprofloxacin and moxifloxacin. Acquired resistance is most commonly due to an alteration in the DNA gyrase target.

The Rifamycins

The rifamycins block prokaryotic RNA polymerase from initiating transcription. Rifampin, the most widely used rifamycin, is bactericidal against many Gram-positive and some Gram-negative bacteria as well as members of the genus *Mycobacterium*.

Rifampin is primarily used to treat tuberculosis and Hansen's disease (leprosy) and to prevent meningitis in people who have been exposed to *Neisseria meningitidis*. In some patients, a reddish-orange pigment appears in urine and tears. Resistance to the drug develops rapidly and is due to a mutation in the gene that encodes RNA polymerase.

Metronidazole

Metronidazole (Flagyl) interferes with DNA synthesis and function, but only in anaerobic microorganisms. The selective toxicity is due to the fact that anaerobic metabolism is required to convert the medication to its active form. The active form then binds DNA, interfering with synthesis and causing damaging breaks. Metronidazole is used to treat bacterial vaginosis and *Clostridium difficile*–associated disease. ▶▶ bacterial vaginosis, p. 617

Antibacterial Medications That Interfere with Metabolic Pathways

Relatively few antibacterial medications interfere with metabolic pathways. Among the most useful are the folate inhibitors—sulfonamides and trimethoprim. These each inhibit different steps in the pathway that leads initially to the synthesis of folate and ultimately to the synthesis of a coenzyme required for nucleotide biosynthesis (**figure 20.8**). Animal cells lack the enzymes in the folate synthesis portion of the pathway, which is why folic acid is a dietary requirement. ◀◀ coenzyme, p. 136

The Sulfonamides

Sulfonamides and related compounds, collectively referred to as **sulfa drugs,** inhibit the growth of many Gram-positive and Gram-negative bacteria. They are structurally similar to para-aminobenzoic acid (PABA), a substrate in the pathway for folate biosynthesis. Because of this similarity, the enzyme that normally binds PABA binds sulfa drugs instead, an example of competitive inhibition (see figure 6.15). Human cells lack this enzyme, providing the basis for the selective toxicity of the sulfonamides. Resistance to the sulfonamides is often due to the acquisition of a plasmid-encoded enzyme that has a lower affinity for the drug. ◀◀ competitive inhibition, p. 138

Trimethoprim

Trimethoprim inhibits the bacterial enzyme that catalyzes a metabolic step following the one inhibited by sulfonamides. Fortunately, the drug has little effect on the enzyme's counterpart in human cells. The combination of trimethoprim and a sulfonamide has a synergistic effect, and they are often used together to treat urinary tract infections. The most common mechanism

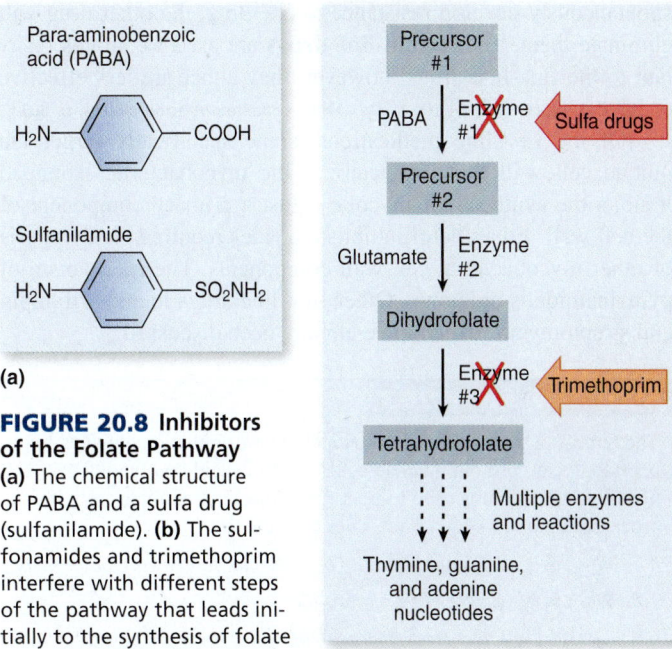

FIGURE 20.8 Inhibitors of the Folate Pathway (a) The chemical structure of PABA and a sulfa drug (sulfanilamide). (b) The sulfonamides and trimethoprim interfere with different steps of the pathway that leads initially to the synthesis of folate and ultimately to the synthesis of a coenzyme required for nucleotide biosynthesis.

❓ *Sulfa drugs have a high therapeutic index. Why would this be so?*

of resistance is a plasmid-encoded alternative enzyme that has a lower affinity for the drug. Unfortunately, the genes encoding resistance to trimethoprim and sulfonamide are often carried on the same plasmid.

Antibacterial Medications That Interfere with Cell Membrane Integrity

A few antimicrobial drugs damage bacterial membranes. They cause the cells to leak, leading to cell death.

Daptomycin inserts into bacterial cytoplasmic membranes, and is used to treat certain infections caused by Gram-positive bacteria resistant to other drugs. It is not effective against Gram-negative bacteria because it cannot penetrate the outer membrane.

Polymyxin B, a common ingredient in first-aid skin ointments, binds to the membranes of Gram-negative cells. Unfortunately, these drugs also bind to eukaryotic cells, though to a lesser extent, which generally limits their use to topical applications.

Antibacterial Medications Effective Against *Mycobacterium* Species

Relatively few antimicrobials are effective against *Mycobacterium tuberculosis* and related species. This is due to several factors, including the organism's waxy cell wall (which prevents the entry of many drugs) and slow growth. A group of five medications—the **first-line drugs**—are preferred because they are the most effective as well as the least toxic. These are generally given in combination of two or more to patients who have active tuberculosis disease. This combination therapy decreases the chance that resistant mutants will develop; if some cells in the infecting population spontaneously develop resistance to one drug, the other drug will eliminate them. The **second-line drugs** are used for strains resistant to the first-line drugs; however, they either are less effective or have greater risk of toxicity. ►► *Mycobacterium tuberculosis*, p. 503

Of the first-line medications, some specifically target the unique cell wall that characterizes the mycobacteria. Isoniazid inhibits the synthesis of mycolic acids, a primary component of the cell wall. Ethambutol inhibits enzymes required for synthesis of other mycobacterial cell wall components. The mechanism of pyrazinamide is unknown. Other first-line drugs include rifampin and streptomycin, which have already been discussed.

MicroAssessment 20.3

The targets of antimicrobial drugs include biosynthetic pathways for peptidoglycan, protein, nucleic acid, and folic acid and the integrity of membranes. Drugs used to treat tuberculosis often interfere with processes unique to *Mycobacterium tuberculosis*.

7. *Why are β-lactam drugs only bactericidal to growing bacteria?*

8. *What is the target of the macrolides?*

9. *Considering that all β-lactam drugs have the same target, why do they vary in their spectrum of activity?* ✚

20.4 ■ Determining the Susceptibility of a Bacterial Strain to an Antimicrobial Drug

Learning Outcomes

10. *Describe how the minimum inhibitory concentration (MIC) and the minimum bactericidal concentration (MBC) are determined.*

11. *Compare and contrast the Kirby-Bauer disc diffusion test with commercial modifications of antimicrobial susceptibility testing.*

Susceptibility of a pathogen to a specific antimicrobial drug is often unpredictable. In these cases, laboratory tests are used to determine the susceptibility of the organism to various drugs and then choose the one that acts against the pathogen but as few other bacteria as possible.

Minimum Inhibitory and Bactericidal Concentrations (MIC and MBC)

The **minimum inhibitory concentration (MIC)** is the lowest concentration of a specific antimicrobial drug needed to prevent the growth of a given bacterial strain *in vitro*. This is determined by growing the test strain in broth cultures containing different concentrations of the antimicrobial (**figure 20.9**). To do this, serial dilutions are used to generate decreasing concentrations of the drug in tubes containing a suitable growth medium. Then, a fixed concentration of bacterial cells is added to each tube. The tubes are incubated for at least 16 hours and then examined for turbidity, which indicates growth. The lowest concentration of the drug that prevents growth of the microorganism is the MIC.

The fact that a strain is inhibited by a given concentration of drug does not necessarily mean that an infection caused by that strain can be successfully treated with the drug. For example, an organism with an MIC of 16 μg/ml would be considered resistant to a drug if the level that can be achieved in a person's blood is less than that. Microbes that have an MIC on the borderline between susceptible (treatable) and resistant (untreatable) are called intermediate.

The **minimum bactericidal concentration (MBC)** is the lowest concentration of a specific antimicrobial drug that kills 99.9% of cells of a given bacterial strain *in vitro*. The MBC is determined by finding out how many live organisms remain in tubes from the MIC test that showed no growth. A small sample from each of those tubes is transferred to a plate containing an antibiotic-free agar medium. If a sample gives rise to no colonies, then no cells survived that particular drug concentration, indicating it was bactericidal. If colonies form, they can be counted to determine the number of cells that survived.

Determining the MIC and MBC using these methods gives precise information regarding an organism's susceptibility. The techniques, however, are labor-intensive and therefore expensive. In addition, individual sets of tubes must be inoculated to determine susceptibility to each drug tested (see **Perspective 20.1**).

Conventional Disc Diffusion Method

The **Kirby-Bauer disc diffusion test** is routinely used to determine the susceptibility of a given bacterial strain to a battery of antimicrobial drugs. A standard concentration of the strain is first uniformly spread on the surface of an agar plate. Then 12 or so discs, each containing a known amount of a different drug, are placed on the surface of the medium (**figure 20.10**). During incubation, the various drugs diffuse outward from the discs, forming a concentration gradient around each disc. Meanwhile the bacterial cells multiply, eventually forming a film of growth on the plate, except in regions around the discs where the bacteria were killed or their growth inhibited.

A clear zone of inhibition around an antimicrobial disc reflects, in part, the degree of susceptibility of the organism to the drug. The zone size is also influenced by characteristics of the drug, including its molecular weight and stability, as well as the amount in the disc.

Special charts have been prepared correlating the size of the zone of inhibition to susceptibility of bacteria to the drug. Based on the size of the zone, organisms can be described as susceptible, intermediate, or resistant to the drug.

Commercial Modifications of Antimicrobial Susceptibility Testing

Commercial modifications of the conventional methods offer certain advantages. They are less labor-intensive, and the results can be obtained in as little as 4 hours. One system uses a small card with tiny wells containing specific antimicrobial concentrations. The highly automated system inoculates and incubates the cards, determines the growth rate by reading the turbidity, and uses mathematical formulas to interpret the results and determine the MICs in 6 to 15 hours (**figure 20.11**).

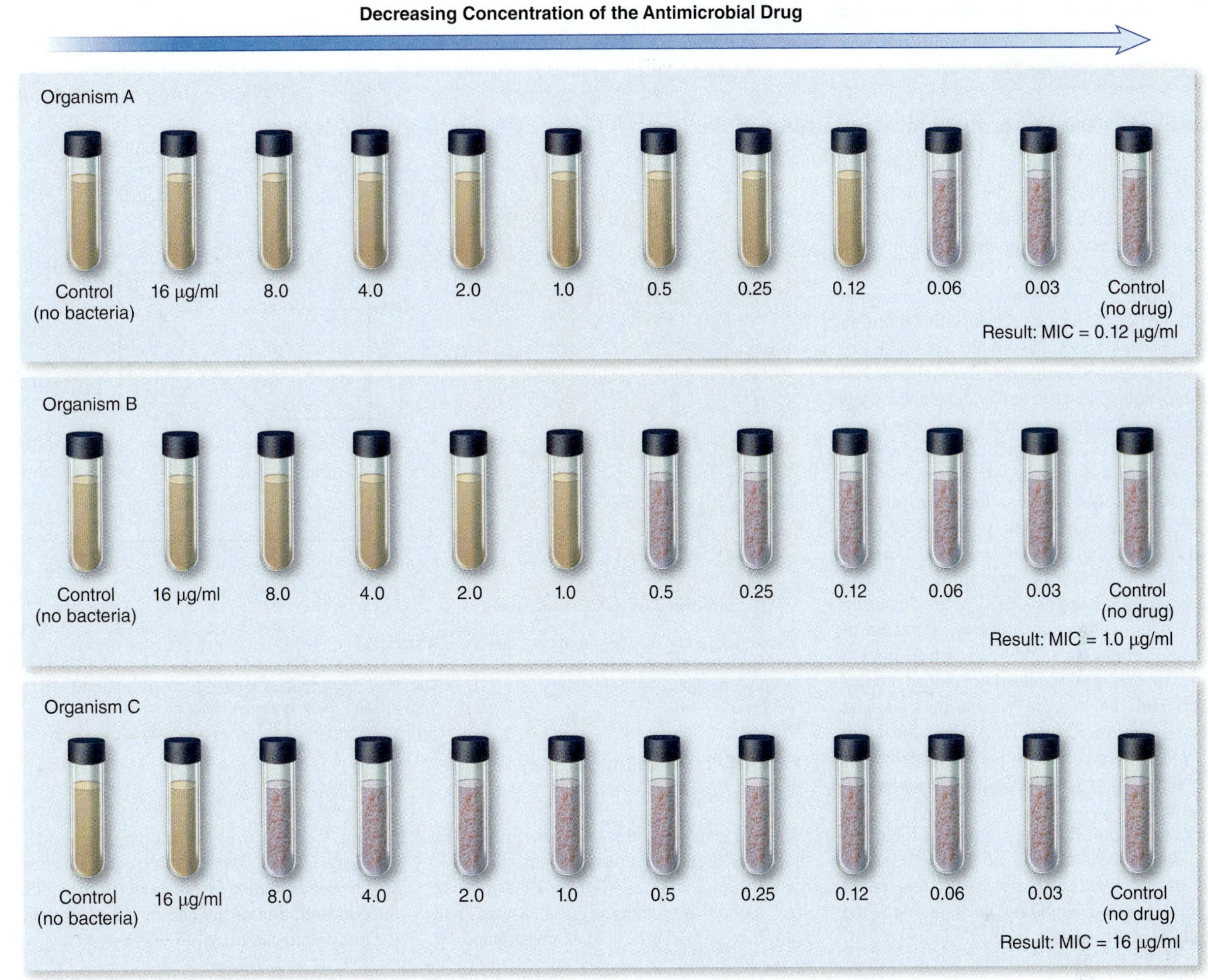

Decreasing Concentration of the Antimicrobial Drug
Organism A
Control (no bacteria)
16 µg/ml
8.0
4.0
2.0
1.0
0.5
0.25
0.12
0.06
0.03
Control (no drug)
Result: MIC = 0.12 µg/ml
Organism B
Control (no bacteria)
16 µg/ml
8.0
4.0
2.0
1.0
0.5
0.25
0.12
0.06
0.03
Control (no drug)
Result: MIC = 1.0 µg/ml
Organism C
Control (no bacteria)
16 µg/ml
8.0
4.0
2.0
1.0
0.5
0.25
0.12
0.06
0.03
Control (no drug)
Result: MIC = 16 µg/ml

FIGURE 20.9 Determining the Minimum Inhibitory Concentration (MIC) of an Antimicrobial Drug The lowest concentration of drug that prevents growth of the culture is the MIC.

? *Considering that organism C is susceptible to 16 µg/ml of the drug, why might it still be considered resistant to the drug?*

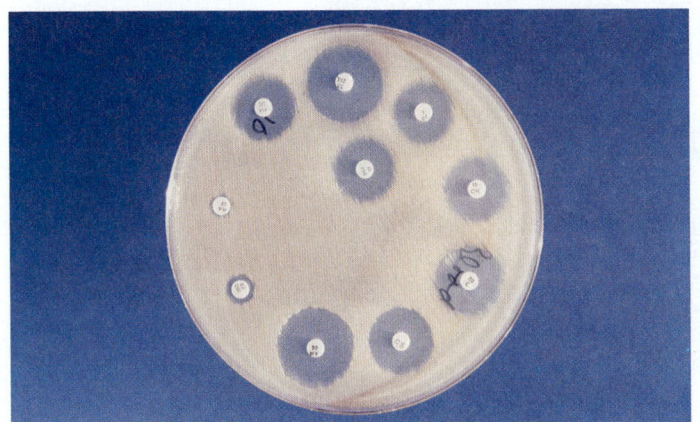

FIGURE 20.10 Kirby-Bauer Method for Determining Drug Susceptibility The size of the zone of inhibition surrounding the disc reflects, in part, the sensitivity of the bacterial strain to the drug.

? *When using the Kirby-Bauer test to determine drug susceptibility, a chart must be consulted. Why not simply choose the drug that gives the largest zone size?*

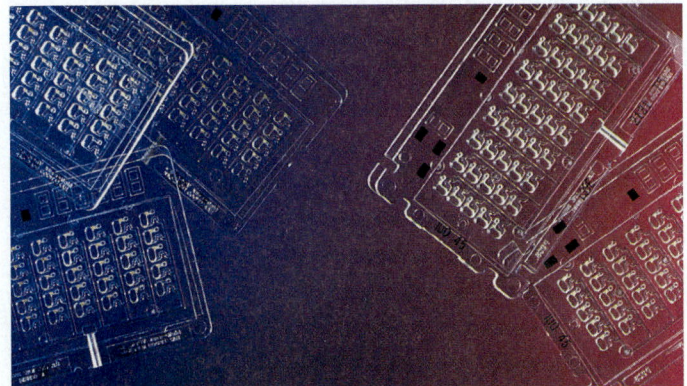

FIGURE 20.11 Automated Tests Used to Determine Antimicrobial Susceptibility The miniature wells in the card contain specific concentrations of an antimicrobial drug. An automated system inoculates and incubates the cards, determines the growth rate by reading turbidity, and uses mathematical formulas to interpret the results and derive the MICs.

? *What are two advantages of automated tests used to determine antimicrobial susceptibility?*

PERSPECTIVE 20.1

Measuring the Concentration of an Antimicrobial Drug in Blood or Other Body Fluids

Sometimes it is necessary to determine the concentration of an antimicrobial drug in a patient's blood or other body fluid. For example, patients who are being administered an aminoglycoside must often be carefully monitored to ensure that the concentration of the drug in their blood does not reach an unsafe level, particularly if they have kidney or liver dysfunction that interferes with normal elimination. Likewise, newly developed drugs must be tested to determine achievable levels in the blood, urine, or other body fluids.

A diffusion bioassay is used to measure the concentration of an antimicrobial drug in a fluid specimen. The test relies on the same principle as the Kirby-Bauer test, except in this case it is the concentration of drug being determined, not the sensitivity of the organism.

To do a diffusion bioassay, a culture of a stock organism highly susceptible to the drug in question is added to melted cooled agar, and the mixture poured into an agar plate and allowed to solidify. This results in a solid medium uniformly inoculated with the sensitive organism. Cylindrical holes are then punched out of the agar, creating wells. Standards (known concentrations of the drug) are then added to some of the wells, while others are filled with the body fluid being tested. Zones of inhibition develop around the wells during overnight incubation, the sizes of

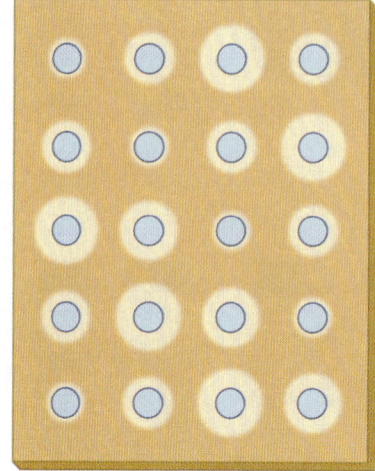

Standards and patient's serum are added to agar that has been seeded with susceptible strain of bacteria.

(a)

A standard curve that correlates the zone diameter with the antimicrobial drug concentration is constructed. The drug concentration in the serum can be read from the line relating zone size to concentration.

(b)

FIGURE 1 Diffusion Bioassay

which correspond to the concentrations of the drug—the higher the concentration, the larger the zone of inhibition (**figure 1**). The zone sizes around the standards are measured, and from this a standard curve is constructed by

plotting the zone sizes against the corresponding drug concentration. A line relating zone size to concentration is obtained, from which the concentration of the antimicrobial drug in the body fluid can be read.

The E test, a modification of the disc diffusion test, uses a strip containing a gradient of concentrations of an antimicrobial drug. Multiple strips, each one containing a different drug, are placed on the surface of an agar medium that has been uniformly inoculated with the test organism. During incubation, the test organism will grow, and a zone of inhibition will form around the strip. Because of the gradient of drug concentrations, the zone of inhibition will be shaped somewhat like a teardrop that intersects the strip at some point (**figure 20.12**). The MIC is determined by reading the printed number at the point where the bacterial growth intersects the strip.

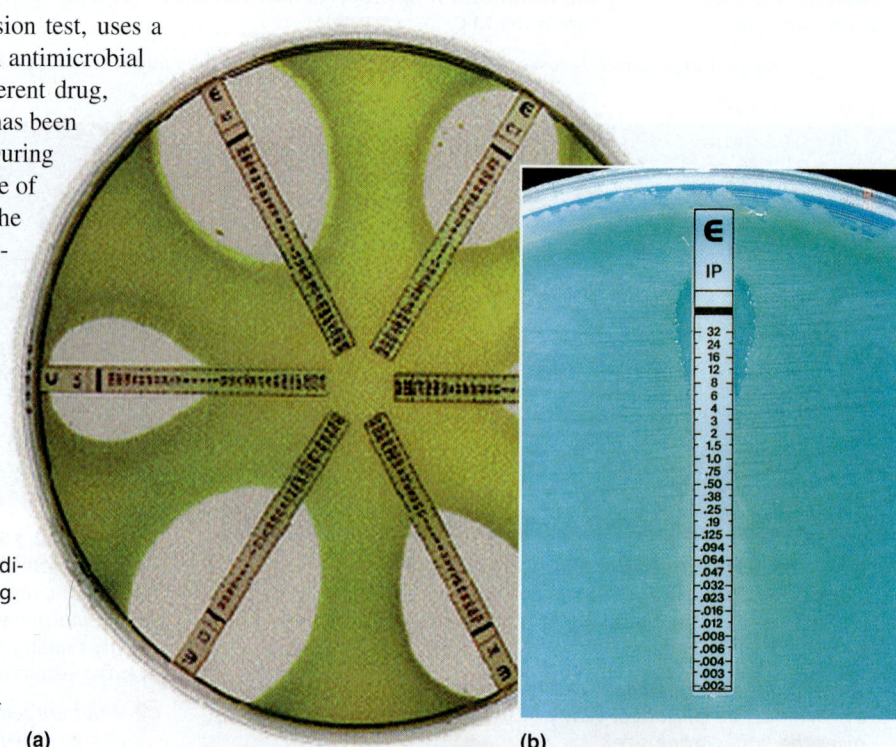

FIGURE 20.12 The E Test (a) Each strip has a gradient of concentrations of a different antimicrobial drug. (b) The MIC is determined by reading the number on the strip at the point at which growth intersects the strip.

❓ *Why is the zone of inhibition larger at one end of the strip?*

(a)

(b)

MicroAssessment 20.4

The MIC and MBC are quantitative measures of a bacterial strain's susceptibility to an antimicrobial drug. Disc diffusion tests can determine whether an organism is susceptible, intermediate, or resistant to a battery of different drugs. Commercial tests for determining antimicrobial sensitivity are less labor-intensive and often more rapid.

10. *Explain the difference between the MIC and the MBC.*

11. *List two factors other than a strain's sensitivity that influence the size of the zone of inhibition around an antimicrobial disc.*

12. *Why would it be important for the Kirby-Bauer disc diffusion test to use a standard concentration of the bacterial strain being tested?* ✚

20.5 ■ Resistance to Antimicrobial Drugs

Learning Outcomes

12. *Describe four general mechanisms of antimicrobial resistance.*

13. *Describe how antimicrobial resistance can be acquired.*

14. *List five examples of emerging antimicrobial resistance.*

15. *Describe how the emergence and spread of antimicrobial resistance can be slowed.*

After sulfa drugs and penicillin were introduced, people hoped that such drugs would eliminate most bacterial diseases. We now recognize, however, that drug resistance limits the usefulness of all known antimicrobials.

As antimicrobial drugs are increasingly used and misused, resistant bacterial strains have a selective advantage over their sensitive counterparts (**figure 20.13**). For example, when penicillin G was first introduced, less than 3% of *Staphylococcus aureus* strains were resistant to its effects. Heavy use of the drug, measured in hundreds of tons per year, eliminated sensitive strains, so that 90% or more are now resistant.

Antimicrobial resistance is alarming because of the impact on the cost, complications, and outcomes of treatment. Dealing with the problem requires an understanding of the mechanisms of resistance and how they are spread. A better understanding of drug resistance may also allow scientists to develop new drugs that foil common resistance mechanisms.

Mechanisms of Acquired Resistance

Figure 20.14 depicts the most common mechanisms of acquired resistance to antimicrobial drugs.

Drug-Inactivating Enzymes

Some bacteria produce enzymes that chemically modify a specific drug, interfering with its function. One example is penicillinase, which destroys penicillin. Another is the enzyme chloramphenicol acetyltransferase, which chemically alters the antibiotic chloramphenicol, making it ineffective.

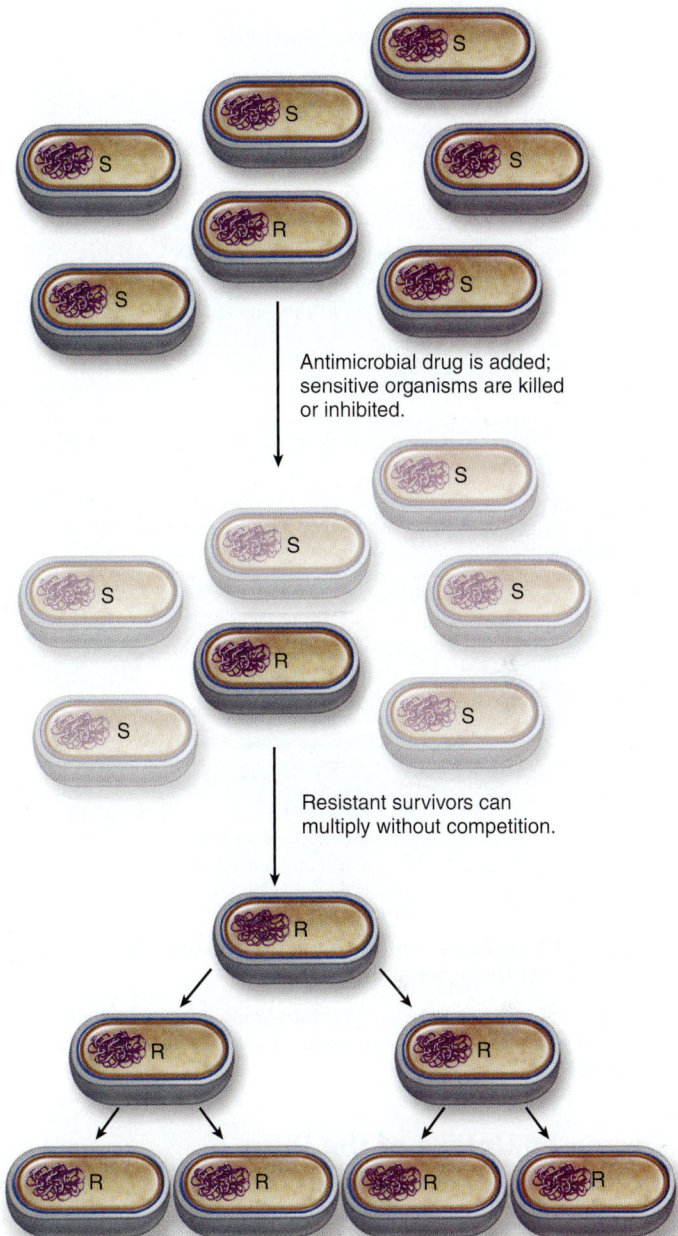

Antimicrobial drug is added; sensitive organisms are killed or inhibited.

Resistant survivors can multiply without competition.

FIGURE 20.13 The Selective Advantage of Drug Resistance When antimicrobial drugs are used, bacterial strains that are resistant (R) to their effects have a selective advantage over their sensitive (S) counterparts.

❓ *How does overuse of antibiotics contribute to increasing numbers of drug-resistant bacteria?*

Alteration in the Target Molecule

Antimicrobial drugs generally act by recognizing and binding to specific target molecules in a bacterium, interfering with their function. Minor structural changes in the target can prevent the drug from binding. Modifications in the penicillin-binding proteins (PBPs) prevent β-lactam drugs from binding to them. Similarly, a change in ribosomal RNA, the target for the macrolides, prevents those drugs from interfering with ribosome function.

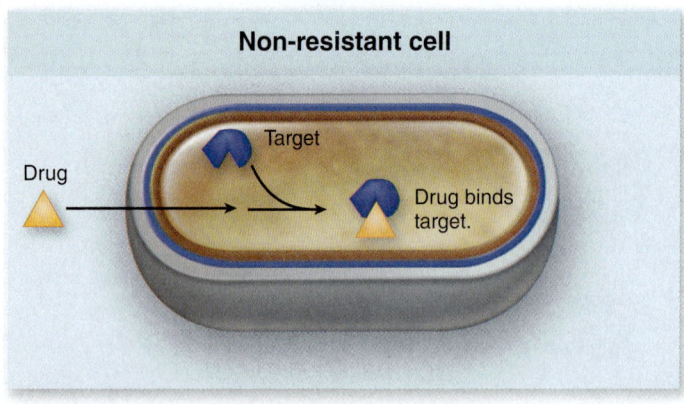

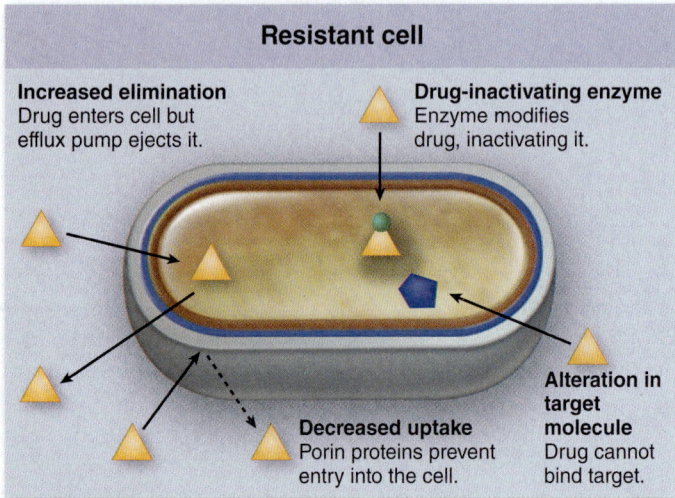

FIGURE 20.14 Common Mechanisms of Acquired Antimicrobial Drug Resistance

❓ *Strains of MRSA (methicillin-resistant* Staphylococcus aureus*) use which two methods to avoid the bactericidal effects of β-lactam drugs?*

Decreased Uptake of the Drug

The porin proteins in the outer membrane of Gram-negative bacteria selectively permit small hydrophobic molecules to enter a cell. Changes in these proteins can therefore prevent certain drugs from entering the cell. By stopping entry of a drug, an organism avoids its effects. ◄◄ porins, p. 60

Increased Elimination of the Drug

The systems that bacteria use to transport damaging compounds out of a cell are called **efflux pumps.** When a cell makes more of these pumps, it can eject the drug faster. In addition, structural changes in the pumps can influence the range of drugs that can be pumped out. Resistance that develops by this mechanism is particularly worrisome because it might allow an organism to become resistant to several drugs simultaneously. ◄◄ efflux pumps, p. 56

Acquisition of Resistance

Antimicrobial resistance can be due to either spontaneous mutation, which alters existing genes, or acquisition of new genes (see figure 8.1).

Spontaneous Mutation

As cells replicate, spontaneous mutations happen at a relatively low rate. Those few mutations that occur, however, can have a significant effect on the resistance of a bacterial population to an antimicrobial drug.

Acquired resistance to streptomycin (an aminoglycoside) is a good example of the consequence of spontaneous mutation. A single base-pair change in the gene encoding a ribosomal protein alters the target enough to make the cell streptomycin-resistant. When a streptomycin-sensitive strain is grown in streptomycin-free medium to a population of 10^9 cells, at least one cell in the population probably has that particular mutation. If streptomycin is then added to the medium, only that cell and its progeny will be able to replicate, giving rise to a streptomycin-resistant population. ◄◄ spontaneous mutation, p. 190

When an antimicrobial drug has several different targets or has multiple binding sites on a single target, resistance through spontaneous mutation is less likely to occur. This is because several different mutations are required to prevent binding of the drug. The newer aminoglycosides bind to several sites on the ribosome, so resistance due to spontaneous mutation is unlikely.

Drugs such as streptomycin for which a single point mutation causes resistance are sometimes used in combination with one or more other drugs. If any cell spontaneously develops resistance to one drug, another drug will still kill it. **Combination therapy** is effective because the chance of a cell simultaneously developing mutational resistance to multiple drugs is extremely low.

Gene Transfer

Genes encoding resistance to antimicrobial drugs can spread to different strains, species, and even genera, most commonly through conjugative transfer of **R plasmids.** These plasmids often carry several different resistance genes, each one encoding resistance to a specific antimicrobial drug. Thus, when an organism acquires an R plasmid, it becomes resistant to several different medications simultaneously. ◄◄ R plasmid, p. 209 ◄◄ conjugation, p. 206

In some cases, resistance genes originate through spontaneous mutation of common bacterial genes, such as one encoding the target of the drug. In other cases, the genes may have originated from the soil microbes that naturally produce that antibiotic. For example, a gene coding for an enzyme that chemically modifies an aminoglycoside likely originated from the *Streptomyces* species that produces the drug.

Examples of Emerging Antimicrobial Resistance

Some of the problems associated with the increasing antimicrobial resistance are highlighted by the following examples.

Enterococci

One of the most dramatic examples of antimicrobial resistance is the enterococci, a group of bacteria that is part of the normal intestinal microbiota and a common cause of healthcare-associated infections. Enterococci are intrinsically less susceptible to many common antimicrobials. For example, their penicillin-binding proteins have low affinity for certain β-lactam antibiotics. In addition, many enterococci have R plasmids. Some strains, called

vancomycin-resistant enterococci (VRE), are even resistant to vancomycin. Recall that this drug is usually reserved as a last resort for treating life-threatening infections caused by Gram-positive organisms resistant to all β-lactam drugs. Because vancomycin resistance in these strains is encoded on a plasmid, the resistance is transferable to other organisms. ◀◀ healthcare-associated infections, p. 449

Staphylococcus aureus

Staphylococcus aureus, another common cause of healthcare-associated infections, is becoming increasingly resistant to antimicrobials. Although nearly all strains were susceptible to penicillin when it was first introduced 70 years ago, most are now resistant due to their acquisition of a gene encoding penicillinase. Until recently, infections by these strains could be treated with methicillin or other penicillinase-resistant penicillins. New strains have emerged, however, that not only produce penicillinase, but also have penicillin-binding proteins with low affinity for all β-lactam drugs. These strains, called **methicillin-resistant Staphylococcus aureus (MRSA),** are resistant to methicillin as well as all other β-lactam drugs.

There are two categories of MRSA strains—healthcare-associated (HA-MRSA) and community acquired (CA-MRSA). HA-MRSA strains are generally resistant to a wide range of antimicrobial medications, so they are generally treated with vancomycin. A few hospitals, however, have reported isolates that are no longer susceptible to normal levels of vancomycin. So far, strict hospital guidelines designed to immediately halt the spread of these **vancomycin-intermediate S. aureus (VISA)** and **vancomycin-resistant S. aureus (VRSA)** strains have been successful. Fortunately, most CA-MRSA strains are currently susceptible to antibiotics other than β-lactam drugs. ▶▶ *Staphylococcus aureus,* p. 524

Streptococcus pneumoniae

Until recently, *Streptococcus pneumoniae,* the leading cause of pneumonia in adults, has remained very sensitive to penicillin. Some isolates, however, are now resistant to the drug. This acquired resistance is due not to the production of a β-lactamase, but rather to changes in the chromosomal genes coding for the targets of penicillin—the penicillin-binding proteins. The modified targets have lower affinities for the drug. The nucleotide changes do not appear to have come from point mutations as one might expect; instead, they are due to the acquisition of chromosomal DNA from other species of *Streptococcus.* As you may recall from earlier reading, *S. pneumoniae* can acquire DNA through DNA-mediated transformation. ▶▶ *Streptococcus pneumoniae,* p. 497
◀◀ DNA-mediated transformation, p. 202

Mycobacterium tuberculosis

Tuberculosis treatment has always been a long and complicated process, requiring a combination of two or more different drugs taken for a period of 6 months or more. Unfortunately, *Mycobacterium tuberculosis* can easily become resistant to the first-line drugs (the preferred medications) through spontaneous mutation. Large numbers of bacterial cells are found in an active infection, so it is likely that at least one cell has developed spontaneous resistance to a drug, which is why combination therapy is required. The length of treatment is due to the very slow growth of *M. tuberculosis.* ▶▶ *Mycobacterium tuberculosis,* p. 503

Many tuberculosis patients do not comply with the complex course of combination therapy, skipping doses or stopping treatment too soon. As a consequence, strains of *M. tuberculosis* develop resistance to the first-line drugs. This results in even longer, more expensive treatments that are also less effective.

Strains resistant to two of the favored drugs for tuberculosis treatment—isoniazid and rifampin—are called **multidrug-resistant M. tuberculosis (MDR-TB).** To prevent the emergence of these strains, some cities are using directly observed therapy; healthcare workers routinely visit patients and watch them take their drugs to make sure they comply with their prescribed antimicrobial treatment. Strains of **extensively drug-resistant M. tuberculosis (XDR-TB)** are an even greater concern. These are defined as *M. tuberculosis* strains resistant to isoniazid and rifampin, plus three or more of the second-line drugs. ▶▶ directly observed therapy, p. 505

Enterobacteriaceae

Members of the family *Enterobacteriaceae* are intrinsically resistant to many antimicrobial medications because their outer membrane prevents the drugs from entering cells. The situation became more complicated when some enterics developed the ability to produce a β-lactamase, allowing the strains to resist the effects of ampicillin and other penicillins. Some strains then developed the ability to produce extended-spectrum β-lactamases (ESBLs), making them resistant to most cephalosporins and aztreonam, as well as penicillins. More recently, strains referred to as **carbapenem-resistant Enterobacteriaceae (CRE)** have been discovered. These strains, which are resistant to nearly all available antimicrobial drugs, produce an enzyme that inactivates carbapenems, a group of drugs considered a last resort for treating infections caused by ESBL-producing bacteria.

Slowing the Emergence and Spread of Antimicrobial Resistance

To reverse the alarming trend of increasing antimicrobial drug resistance, everyone must cooperate. On an individual level, physicians as well as the general public must take more responsibility for the appropriate use of these life-saving medications. On a global scale, countries around the world need to make important policy decisions about what is, and what is not, an appropriate use of these drugs.

The Responsibilities of Physicians and Other Healthcare Workers

Physicians and other healthcare workers need to increase their efforts to identify the cause of a given infection and, only if appropriate, prescribe suitable antimicrobials. They must also educate patients about the proper use of prescribed drugs in order to increase compliance. Although these efforts may be more expensive in the short term, they will ultimately save both lives and money.

The Responsibilities of Patients

Patients need to carefully follow the instructions that accompany their prescriptions, even if those instructions seem inconvenient. It is essential to maintain adequate blood levels of the antimicrobial for a specific time period. When a patient skips a scheduled dose of a drug, the blood level of the drug may not remain high enough to inhibit the growth of the least-sensitive members of the population. If these less-sensitive organisms then have a chance to grow, they will give rise to a population that is not as sensitive as the original. Likewise, failure to complete the prescribed course of treatment may not kill the least-sensitive organisms, allowing their subsequent multiplication. Misusing antimicrobials by skipping doses or failing to complete the prescribed treatment increases the likelihood that resistant mutants will develop.

The Importance of an Educated Public

A greater effort must also be made to educate people about the role and limitations of antibiotics in order to make sure they are used wisely. First and foremost, people need to understand that antibiotics are not effective against viruses. Taking antibiotics will not cure the common cold or any other viral illness. A few antiviral drugs are available, but they are effective against only certain viruses. Unfortunately, surveys indicate that far too many people mistakenly believe that antibiotics are effective against viruses, and often seek prescriptions to "cure" viral infections. This misuse only selects for antibiotic-resistant bacteria in the normal microbiota. Even though these organisms typically do not cause disease, they can serve as a reservoir for R plasmids, eventually transferring their resistance genes to an infecting pathogen.

Global Impacts of the Use of Antimicrobial Drugs

The overuse of antimicrobial drugs is a worldwide concern. Countries may vary in their laws and customs, but antimicrobial resistance recognizes no political boundaries. An organism that develops resistance in one country can quickly be transported globally.

In many parts of the world, particularly in developing countries, antimicrobial drugs are available on a non-prescription basis. Because of the consequences of inappropriate use, many people believe that over-the-counter availability of these drugs should be restricted or eliminated.

Another worldwide concern is the use of antimicrobial drugs in animal feeds. Low levels of these drugs in feeds enhance the growth of animals, a seemingly attractive option. This use, like any other, however, selects for drug-resistant organisms, which has caused some scientists to question its ultimate wisdom. In fact, infections caused by drug-resistant *Salmonella* strains have been linked to animals whose feed was supplemented with those drugs. In response to these concerns, there is growing pressure worldwide to ban the use of antimicrobial drugs in animal feeds.

MicroAssessment 20.5

Mutations and transfer of genetic information allow microorganisms to become resistant to antimicrobial medications. Drug resistance affects the outcomes of medical treatment. Slowing the emergence and spread of resistant microbes involves the cooperation of healthcare personnel, educators, and the general public.

13. *Explain how using a combination of two antimicrobial drugs helps prevent the development of spontaneously resistant mutants.*

14. *Explain the significance of a member of the normal flora that harbors an R plasmid.*

15. *A student argued that "spontaneous mutation" meant that a drug could cause mutations. Is the student correct? Why or why not?* ➕

20.6 ■ Mechanisms of Action of Antiviral Drugs

Learning Outcomes

16. *Describe the antiviral drugs that prevent viral entry or uncoating.*

17. *Compare and contrast the antiviral drugs that interfere with nucleic acid synthesis or genome integration.*

18. *Describe the antiviral drugs that interfere with the assembly and release of viral particles.*

Viruses rely almost exclusively on the host cell's metabolic machinery for their replication, making it difficult to find a target for selective toxicity. They have no cell wall, ribosomes, or any other structure targeted by antibiotics. Because of this, viruses are completely unaffected by antibiotics. Many encode their own polymerases, however, and these are potential targets of antiviral drugs (**figure 20.15**). Relatively few other targets have been discovered.

Many scientists are trying to develop more effective **antiviral drugs,** medications that interfere with viral replication. The relatively few options available are generally effective against only a specific type of virus, and none can eliminate latent viral infections (**table 20.2**). ◄◄ latent viral infections, p. 322

Prevent Viral Entry

A new group of drugs effective against HIV are the entry inhibitors, which prevent the virus from entering host cells. Enfuvirtide does this by binding to an HIV protein that promotes fusion of the viral envelope with the cell membrane. Maraviroc blocks the HIV co-receptor CCR5. ►► HIV attachment and entry, p. 699

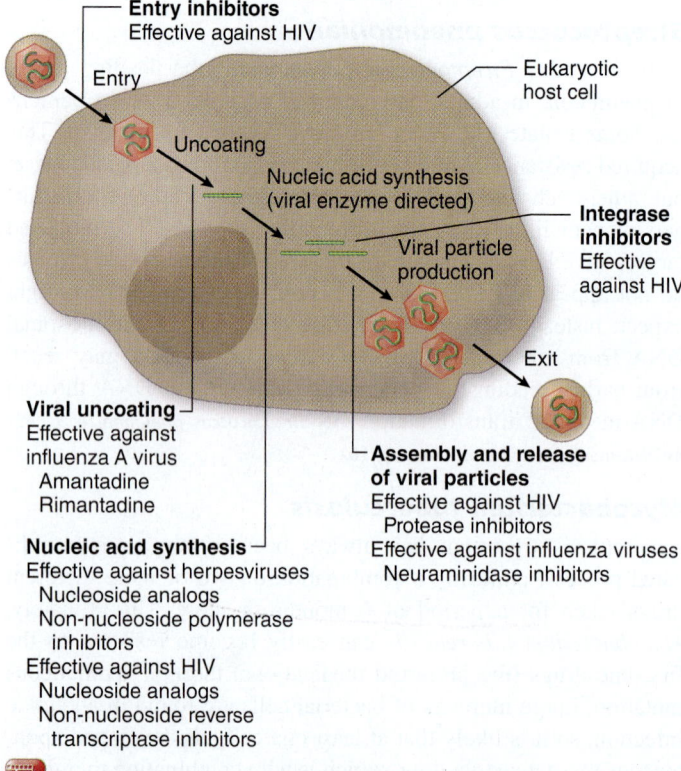

FIGURE 20.15 Targets of Antiviral Drugs

❓ *Why are there relatively few options for antiviral drugs?*

TABLE 20.2	Characteristics of Antiviral Drugs

Target/Drug Examples	Comments/Characteristics
Viral Entry	
Enfuvirtide, Maraviroc	Used to treat HIV infections.
Viral Uncoating	
Amantadine and rimantadine	Reduce severity and duration of influenza A infections, but resistance limits their use.
Nucleic Acid Synthesis	
Nucleoside analogs Acyclovir, ganciclovir, ribavirin, zidovudine (AZT), didanosine (ddI), lamivudine (3TC)	Primarily used to treat infections caused by herpesviruses and HIV; they do not cure latent infections. The drugs are converted within eukaryotic cells to a nucleotide analog; virally encoded enzymes are prone to incorporate these, resulting in premature termination of synthesis or improper base-pairing of the viral nucleic acid. Acyclovir is used to treat herpes simplex virus (HSV) and varicella-zoster virus (VZV) infections. Ganciclovir is used to treat cytomegalovirus infections in immunocompromised patients. Ribavirin is used to treat respiratory syncytial virus (RSV) infections in newborns. Combinations of nucleoside analogs such as zidovudine (AZT), didanosine (ddI), and lamivudine (3TC) are used to treat HIV infections.
Non-nucleoside polymerase inhibitors Foscarnet	Primarily used to treat infections caused by herpesviruses. They inhibit the activity of viral polymerases by binding to a site other than the nucleotide-binding site. Foscarnet is used to treat ganciclovir-resistant cytomegalovirus (CMV) and acyclovir-resistant herpes simplex virus (HSV).
Non-nucleoside reverse transcriptase inhibitors Nevirapine, delavirdine, efavirenz	Used to treat HIV infections. They inhibit the activity of reverse transcriptase by binding to a site other than the nucleotide-binding site and are often used in combination with nucleoside analogs.
Genome Integration	
Raltegravir	Used to treat HIV infections.
Assembly and Release of Viral Particles	
Protease inhibitors Indinavir, ritonavir, saquinavir, nelfinavir	Used to treat HIV infections. They inhibit protease, an essential enzyme of HIV, by binding to its active site.
Neuraminidase inhibitors Zanamivir, oseltamivir	Used to treat influenza virus infections.

Interfere with Viral Uncoating

After a virus enters a host cell, the nucleic acid must separate from the protein coat for replication to occur. Because of this, drugs that interfere with the uncoating step prevent viral replication. Only two drugs target this step—amantadine and rimantadine—and they block influenza A viruses. They both prevent or reduce the severity of the disease, but viral strains develop resistance easily, limiting their usefulness. ▶ influenza A virus, p. 508

Interfere with Nucleic Acid Synthesis

Many of the most effective antiviral drugs take advantage of the error-prone, virally encoded enzymes that replicate viral nucleic acid. With few exceptions, however, these drugs are generally limited to treating infections caused by herpesviruses or HIV.

Nucleoside Analogs

A number of antiviral drugs are nucleoside analogs, compounds similar in structure to a nucleoside. These analogs can be phosphorylated in vivo by a virally encoded or normal cellular enzyme to form a nucleotide analog, a chemical structurally similar to the nucleotides of DNA and RNA. When a nucleotide analog is incorporated into a growing nucleotide chain, it sometimes prevents additional nucleotides from being added. In other cases, the analog results in a defective strand with altered base-pairing properties. ◀ nucleotide, p. 32

The basis for the selective toxicity of most nucleoside analogs is the fact that virally encoded enzymes are more likely than the host cell polymerases to incorporate a nucleotide analog. Therefore, more damage is done to the rapidly replicating viral genome than to the host cell genome. The analogs, however, are only effective against replicating viruses. Viruses such as herpesvirus and HIV can remain latent in cells, so the drugs do not cure these infections; they simply shorten the active infection. Latent virus can still reactivate, causing symptoms to recur.

Most nucleoside analogs are reserved for severe infections because of their significant side effects. An important exception is acyclovir, a drug used to treat herpesvirus infections. This drug causes little harm to uninfected cells because normal cellular enzymes do not convert the drug into a nucleotide analog. Instead, a virally encoded enzyme does this. The enzyme is only present in cells infected by herpesviruses such as herpes simplex virus (HSV) and varicella-zoster virus (VZV). Other nucleoside analogs include ganciclovir, which is used to treat life- or sight-threatening cytomegalovirus (CMV) infections in immunocompromised patients, and ribavirin, which is used to treat respiratory syncytial virus infections (RSV) in newborns. Nucleoside analogs used to treat HIV infection interfere with the activity of reverse transcriptase and are called nucleoside reverse transcriptase inhibitors (NRTIs). Unfortunately, the virus rapidly develops mutational resistance to these drugs, which is why they are often used in combination with other anti-HIV drugs. NRTIs include zidovudine (AZT), didanosine (ddI), and lamivudine (3TC). Two of these are often used in combination for HIV therapy. ▶▶ herpesviruses, pp. 535, 582

Non-Nucleoside Polymerase Inhibitors

Non-nucleoside polymerase inhibitors are compounds that inhibit the activity of viral polymerases by binding to a site other than the nucleotide-binding site. One example, foscarnet, is used to treat infections caused by ganciclovir-resistant CMV and acyclovir-resistant HSV.

Non-Nucleoside Reverse Transcriptase Inhibitors (NNRTIs)

Non-nucleoside reverse transcriptase inhibitors (NNRTIs) inhibit the activity of reverse transcriptase by binding to a site other than the nucleotide-binding site. They are often used in combination with nucleoside analogs to treat HIV infections. The medications include nevirapine, delavirdine, and efavirenz.

Prevent Genome Integration

Integrase inhibitors offer a new option for treating HIV infections. They inhibit the HIV-encoded enzyme integrase, thereby preventing the virus from inserting the DNA copy of its genome into that of the host cell. Raltegravir is the first approved drug of this class. ▶▶ HIV integrase, p. 698

Prevent Assembly and Release of Viral Particles

Virally encoded enzymes required for the assembly and release of viral particles are the targets of medications used to treat certain viral infections.

Protease Inhibitors

Protease inhibitors are used to treat HIV infections. These medications inhibit the HIV-encoded enzyme protease, which plays an essential role in the production of infectious viral particles. When HIV replicates, several of its proteins are translated as a polyprotein, a single amino acid chain that must be cleaved by the protease to release the individual proteins. The various protease inhibitors—including indinavir, ritonavir, saquinavir, and nelfinavir—differ in dosage and side effects. ▶▶ HIV protease, p. 698

Neuraminidase Inhibitors

Neuraminidase inhibitors inhibit neuraminidase, an enzyme encoded by influenza viruses. This enzyme is important for the release of viral particles from infected cells. Two neuraminidase inhibitors are currently available—zanamivir, which is administered by inhalation, and oseltamivir, which is taken orally. Both shorten the infection when taken within two days of the onset of symptoms.

Viral replication generally uses host cell machinery; because of this, there are few targets for selectively toxic antiviral drugs. Available antiviral drugs are virus-specific; targets include viral entry, viral uncoating, nucleic acid synthesis, integrase, and the assembly and release of viral particles.

16. *Explain why acyclovir has fewer side effects than do other nucleoside analogs.*
17. *How do protease inhibitors interfere with the production of infectious viral particles?*
18. *Why are nucleoside analogs active only against replicating viruses?* ➕

20.7 ■ Mechanisms of Action of Antifungal Drugs

Learning Outcomes

19. *Describe the antifungal drugs that interfere with plasma membrane synthesis and function.*
20. *Compare and contrast echinocandins, griseofulvin, and flucytosine.*

Eukaryotic pathogens such as fungi more closely resemble human cells than do bacteria. This is why relatively few drugs are available for systemic use against fungal pathogens (**table 20.3**). The targets of antifungal drugs are illustrated in **figure 20.16**.

Interfere with Plasma Membrane Synthesis and Function

The target of most antifungal drugs is ergosterol, a sterol found in the plasma membrane of fungal but not human cells. ◀◀ sterol, p. 36

Polyenes

The polyenes, a group of antibiotics produced by certain species of *Streptomyces,* bind to ergosterol. This disrupts the fungal

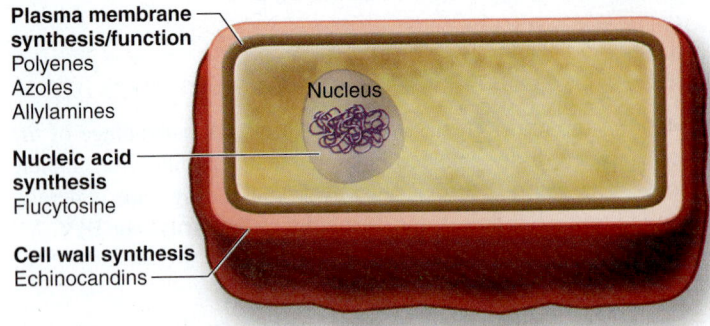

FIGURE 20.16 Targets of Antifungal Drugs

❓ *What compound in the plasma membrane is the target of most antifungal drugs?*

TABLE 20.3	Characteristics of Antifungal Drugs
Drug Target/Drug	**Comments**
Plasma Membrane	
Polyenes Amphotericin B, nystatin	Bind to ergosterol, disrupting the plasma membrane and allowing the cytoplasm to leak out. Amphotericin B is very toxic but the most effective drug for treating life-threatening infections; newer lipid-based emulsions are less toxic but very expensive. Nystatin is too toxic for systemic use, but can be used topically.

Azoles
  Imidazoles: ketoconazole, miconazole, clotrimazole
 	
  Triazoles: fluconazole, voriconazole
	Interfere with ergosterol synthesis, leading to defective cell membranes; active against a wide variety of fungi. Used to treat a variety of systemic and localized fungal infections. Triazoles are less toxic than imidazoles.

Allylamines Naftifine, terbinafine	Inhibit an enzyme in the pathway of ergosterol synthesis. Administered topically to treat dermatophyte infections. Terbinafine can be taken orally.
Cell Wall Synthesis	
Echinocandins Caspofungin	Interfere with β-1,3 glucan synthesis. Used to treat *Candida* infections as well as invasive aspergillosis that resists other treatments.
Cell Division	
Griseofulvin	Used to treat skin and nail infections. Taken orally for months; concentrates in the dead keratinized layers of the skin; taken up by fungi invading those cells and inhibits their division. Active only against fungi that invade keratinized cells.
Nucleic Acid Synthesis	
Flucytosine	Used to treat systemic yeast infections; enzymes within yeast cells convert the drug to 5-fluorouracil, which inhibits an enzyme required for nucleic acid synthesis; not effective against most molds; resistant mutants are common.

membrane, killing the cell by allowing its cytoplasmic contents to leak out. Unfortunately, the polyenes are quite toxic to humans, which limits their systemic use to life-threatening infections. Amphotericin B causes severe side effects, but it is the most effective drug for treating certain systemic infections. Newer lipid-based emulsions are less toxic but more expensive. Nystatin is too toxic to be given systemically, but is used topically.

Azoles

The azoles are a family of chemically synthesized drugs, some of which have antifungal activity. They include two classes—the imidazoles and the newer triazoles; the latter are generally less toxic. Both classes inhibit ergosterol synthesis, resulting in defective fungal membranes that leak cytoplasmic contents. Fluconazole and voriconazole, which are triazoles, are increasingly being used to treat systemic fungal infections. Ketoconazole, an imidazole, is also used systemically, but has more severe side effects. Other imidazoles, including miconazole and clotrimazole, are commonly used in non-prescription creams, ointments, and suppositories to treat vaginal yeast infections. They are also applied to the skin to treat dermatophyte infections. ▶◀ dermatophyte, p. 543

Allylamines

The allylamines inhibit an enzyme in the pathway of ergosterol synthesis. Naftifine and terbinafine can be applied to the skin to treat dermatophyte infections. Terbinafine can also be taken orally.

Interfere with Cell Wall Synthesis

Fungal cell walls contain some components not produced by animal cells. Drugs in a family called echinocandins interfere with synthesis of the fungal cell wall component β-1,3 glucan, causing fungal cells to burst. Caspofungin, the first member to be approved, is used to treat *Candida* infections as well as invasive aspergillosis that resists other treatments.

Interfere with Cell Division

The target of one antifungal drug, griseofulvin, is cell division. Griseofulvin interferes with the action of tubulin, a structure required for nuclear division. Because tubulin is a part of all eukaryotic cells, the selective toxicity of the drug may be due to its greater uptake by fungal cells. When taken orally for months, it is absorbed and eventually concentrated in the dead keratinized layers of the skin. Fungi that then invade keratin-containing structures such as skin and nails take up the drug, which prevents them from multiplying. Griseofulvin is only active against fungi that invade keratinized cells, and is used to treat skin and nail infections. ◀◀ tubulin, p. 73 ◀◀ keratin, p. 336

Interfere with Nucleic Acid Synthesis

Nucleic acid synthesis is a common feature of all eukaryotic cells, generally making it a poor target for antifungal drugs. The drug flucytosine, however, is taken up by yeast cells and then converted by yeast enzymes to an active, inhibitory form.

Flucytosine

Flucytosine is a synthetic derivative of cytosine, one of the nucleobases. Enzymes within infecting yeast cells convert flucytosine to 5-fluorouracil, which inhibits an enzyme required for nucleic acid synthesis. Unfortunately, resistant mutants are common, and therefore, flucytosine is used mostly in combination with amphotericin B or as an alternative drug for patients with systemic yeast infections who are unable to tolerate amphotericin B. Flucytosine is not effective against molds. ◀◀ nucleobase, p. 32

MicroAssessment 20.7

Because fungi are eukaryotic cells, there are relatively few targets for selectively toxic antifungal drugs. Most antifungal drugs interfere with ergosterol function or synthesis. Other targets include cell wall synthesis, cell division, and nucleic acid synthesis.

19. *Why is amphotericin B used only for treating life-threatening infections?*

20. *Why is flucytosine generally used only in combination with other drugs?*

20.8 ■ Mechanisms of Action of Antiprotozoan and Antihelminthic Drugs

Learning Outcome

21. *Describe the targets of most antiparasitic drugs.*

Most antiparasitic drugs probably interfere with biosynthetic pathways of protozoan parasites or the neuromuscular function of worms. Unfortunately, compared with antibacterial, antifungal, and antiviral drugs, little research and development goes into these drugs, because most parasitic diseases are concentrated in the poorer areas of the world where people simply cannot afford to spend money on expensive medications.

Some of the most important antiparasitic drugs and their characteristics are summarized in **table 20.4.**

TABLE 20.4 Characteristics of Some Antiprotozoan and Antihelminthic Drugs

Causative Agent/Drug	Comments
Intestinal protozoa	
Iodoquinol	Mechanism unknown; poorly absorbed but taken orally to eliminate amebic cysts in the intestine.
Nitazoxanide	New drug used to treat cryptosporidiosis and giardiasis.
Nitroimidazoles Metronidazole	Activated by the metabolism of anaerobic organisms. Interfere with electron transfer and alter DNA. Do not reliably eliminate the cyst stage. Metronidazole is also used to treat infections caused by anaerobic bacteria.
Quinacrine	Mechanism of action is unknown, but may be due to interference with nucleic acid synthesis.
Plasmodium (Malaria) and Toxoplasma	
Folate antagonists Pyrimethamine, sulfonamide	Interfere with folate metabolism; used to treat toxoplasmosis and malaria.
Malarone	A synergistic combination of atovaquone and proguanil hydrochloride used to treat malaria. Atovaquone interferes with mitochondrial electron transport while proguanil disrupts folate synthesis. The combination is active against both the blood stage and early liver stage of *Plasmodium* species.
Quinolones Chloroquine, mefloquine, primaquine, tafenoquine	The mechanism of action is not completely clear. Chloroquine is concentrated in infected red blood cells and is the drug of choice for preventing or treating the red blood cell stage of the malarial parasite. Its effects may be due to inhibition of an enzyme that protects the parasite from the toxic by-products of hemoglobin degradation. Primaquine and tafenoquine destroy the liver stage of the parasite and are used to treat relapsing forms of malaria. Mefloquine is used to treat infection caused by chloroquine-resistant strains of the malarial parasite.
Trypanosomes and Leishmania	
Eflornithine	Used to treat infections caused by some types of *Trypanosoma;* inhibits the enzyme ornithine decarboxylase.
Heavy metals Melarsoprol, sodium stibogluconate, meglumine antimonate	These inactivate sulfhydryl groups of parasitic enzymes, but are very toxic to host cells as well. Melarsoprol is used to treat trypanosomiasis, but the treatment can be lethal. Sodium stibogluconate and meglumine antimonate are used to treat leishmaniasis.
Nitrofurtimox	Widely used to treat acute Chagas' disease; forms reactive oxygen radicals that are toxic to the parasite as well as the host.

(continued)

TABLE 20.4	Characteristics of Some Antiprotozoan and Antihelminthic Drugs (*Continued*)
Causative Agent/Drug	**Comments**
Intestinal and Tissue Helminths	
Avermectins Ivermectin	Ivermectin causes neuromuscular paralysis in parasites; used to treat infections caused by *Strongyloides* and tissue nematodes.
Benzimidazoles Mebendazole, thiabendazole, albendazole	Mebendazole binds to tubulin of helminths, blocking microtubule assembly and inhibiting glucose uptake. It is poorly absorbed in the intestine, making it effective for treating intestinal, but not tissue, helminths. Thiabendazole may have a similar mechanism, but it is well absorbed and has many toxic side effects. Albendazole is used to treat tissue infections caused by *Echinococcus* and *Taenia solium*.
Phenols Niclosamide	Absorbed by cestodes in the intestinal tract, but not by the human host.
Piperazines Piperazine, diethylcarbamazine	Piperazine causes a flaccid paralysis in worms and can be used to treat *Ascaris* infections. Diethylcarbamazine immobilizes filarial worms and alters their surface, which enhances killing by the immune system. The resulting inflammatory response, however, causes tissue damage.
Pyrazinoisoquinolines Praziquantel	A single dose of praziquantel is effective in eliminating a wide variety of trematodes and cestodes. It is taken up but not metabolized by the worm, ultimately causing sustained contractions of the worm.
Tetrahydropyrimidines Pyrantel pamoate, oxantel	Pyrantel pamoate interferes with neuromuscular activity of worms, causing a type of paralysis. It is not readily absorbed from the gastrointestinal tract and is active against intestinal worms including pinworm, hookworm, and *Ascaris*. Oxantel can be used to treat *Trichuris* infections.

FUTURE CHALLENGES   20.1

War with the Superbugs

With respect to antimicrobial drugs, the future challenge is already upon us. The challenge is to maintain the effectiveness of antimicrobials by (1) preventing the continued spread of resistance, and (2) developing new drugs that have even more desirable properties. New ways must be developed to fight infections caused by the ever greater numbers of bacterial strains that are resistant to the effects of conventional antimicrobial drugs. Several strategies are being used to develop potential weapons against these "superbugs."

One way to combat resistance is to continue developing modifications of existing antimicrobial drugs. Researchers constantly work to modify drugs chemically, trying to keep at least one step ahead of bacterial resistance. Another method to foil drug resistance is to interfere with the resistance mechanisms, as is done currently in using β-lactamase inhibitors in combination with β-lactam drugs to protect the antimicrobial drug from enzymatic destruction. Alternatively, perhaps chemicals can be used to inactivate or interfere with bacterial efflux systems.

Some researchers are focusing on developing medications entirely unrelated to conventional antimicrobials. One example is a class of compounds called defensins, which are short peptides, approximately 29 to 35 amino acids in length, produced naturally by a variety of eukaryotic cells to fight infections. Various defensins and related compounds are being intensively studied as promising antimicrobials.

Knowing the nucleotide sequences of pathogens is allowing researchers to identify genes associated with pathogenicity. This information, in turn, might help researchers design antimicrobials that interfere directly with those processes. The "master switch" that controls expression of virulence determinants is one such potential target. Nucleotide sequence information and determination of the three-dimensional conformation of proteins may uncover new targets for antimicrobial drug therapy.

Summary

20.1 ■ History and Development of Antimicrobial Drugs

Discovery of Antimicrobial Drugs

Salvarsan, developed by Paul Ehrlich, was the first documented example of an antimicrobial medication.

Discovery of Antibiotics

Alexander Fleming discovered that a species of the fungus *Penicillium* produces a drug he called penicillin.

Development of New Generations of Drugs

Antimicrobial drugs can be chemically modified to give them new properties (figure 20.1).

20.2 ■ Features of Antimicrobial Drugs

Most modern antibiotics come from species of *Streptomyces* and *Bacillus* (bacteria) and *Penicillium* and *Cephalosporium* (fungi).

Selective Toxicity

Medically useful antimicrobials are **selectively toxic;** the relative toxicity of a drug is expressed as the **therapeutic index.**

Antimicrobial Action

Bacteriostatic drugs inhibit the growth of bacteria; drugs that kill bacteria are **bactericidal.**

Spectrum of Activity

Broad-spectrum antimicrobials affect a wide range of bacteria; those that affect a narrow range are called **narrow-spectrum.**

Effects of Combinations

Combinations of antimicrobial drugs can be synergistic, antagonistic, or additive.

Tissue Distribution, Metabolism, and Excretion of the Drug

Some antimicrobials cross the blood-brain barrier into the cerebrospinal fluid; these can be used to treat meningitis. Drugs that are unstable in acid must be administered through injection. Drugs that have a long half-life need to be administered less frequently.

Adverse Effects

Some people develop allergies to certain antimicrobials. Some antimicrobials can have potentially damaging side effects such as kidney damage. When the composition of the normal microbiota is altered, which happens when a person takes antimicrobials, pathogens normally unable to compete may grow to high numbers.

Resistance to Antimicrobials

Certain types of bacteria have **intrinsic (innate) resistance** to a particular drug. **Acquired resistance** is due to spontaneous mutation or the acquisition of new genetic information.

20.3 ■ Mechanisms of Action of Antibacterial Drugs
(figure 20.2, table 20.1)

Antibacterial Medications That Inhibit Cell Wall Synthesis (figure 20.3)

The **β-lactam drugs,** which include the penicillins, cephalosporins, carbapenems, and monobactams, irreversibly inhibit **penicillin-binding proteins (PBPs),** ultimately leading to cell lysis (figure 20.4). Vancomycin binds to the peptide side chain of NAM, blocking peptidoglycan synthesis. Bacitracin interferes with the transport of peptidoglycan precursors across the cytoplasmic membrane; it is a common ingredient in non-prescription ointments.

Antibacterial Medications That Inhibit Protein Synthesis (figure 20.7)

Antibiotics that inhibit protein synthesis by binding to the 70S ribosome include the aminoglycosides, the tetracyclines, the macrolides, chloramphenicol, the lincosamides, the oxazolidinones, and the streptogramins.

Antibacterial Medications That Inhibit Nucleic Acid Synthesis

The fluoroquinolones interfere with DNA replication and transcription by inhibiting one or more topoisomerases. The rifamycins block prokaryotic RNA polymerase from initiating transcription. Metronidazole is activated by anaerobic metabolism, and the activated form then binds DNA.

Antibacterial Medications That Interfere with Metabolic Pathways (figure 20.8)

Sulfa drugs competitively inhibit an enzyme in the metabolic pathway that leads to folic acid synthesis. Trimethoprim inhibits the enzyme that catalyzes a metabolic step following the one inhibited by sulfonamides.

Antibacterial Medications That Interfere with Cell Membrane Integrity

Daptomycin damages the cytoplasmic membrane. Polymyxin B damages membranes of Gram-negative bacteria.

Antibacterial Medications Effective Against *Mycobacterium* Species

First-line drugs that specifically target *Mycobacterium* species include isoniazid, ethambutol, and pyrazinamide. **Second-line drugs** are used for strains resistant to the first-line drugs.

20.4 ■ Determining the Susceptibility of a Bacterial Strain to an Antimicrobial Drug

Minimum Inhibitory and Bactericidal Concentrations (MIC and MBC) (figure 20.9)

The **minimum inhibitory concentration (MIC)** is the lowest concentration of a specific antimicrobial drug needed to prevent the growth of a bacterial strain *in vitro*. The **minimum bactericidal concentration (MBC)** is the lowest concentration of a specific antimicrobial drug that kills 99.9% of cells of a given strain of bacterial *in vitro*.

Conventional Disc Diffusion Method (figure 20.10)

The **Kirby-Bauer disc diffusion test** determines the susceptibility of a bacterial strain to a battery of antimicrobial drugs.

Commercial Modifications of Antimicrobial Susceptibility Testing

Automated methods can determine antimicrobial susceptibility in as little as 4 hours (figure 20.11).

20.5 ■ Resistance to Antimicrobial Drugs (figure 20.13)

Mechanisms of Acquired Resistance (figure 20.14)

Enzymes that chemically modify a drug render it ineffective. Structural changes in the target can prevent the drug from binding. Altered porin proteins prevent drugs from entering cells. **Efflux pumps** actively pump drugs out of cells.

Acquisition of Resistance

Resistance can be acquired through spontaneous mutation or horizontal gene transfer. The most common mechanism of transfer of antibiotic resistance genes is through the conjugative transfer of **R plasmids.**

Examples of Emerging Antimicrobial Resistance

Vancomycin-resistant enterococci (VRE), methicillin-resistant *Staphylococcus aureus* (MRSA), vancomycin-intermediate *S. aureus* (VISA), vancomycin-resistant *S. aureus* (VRSA), penicillin-resistant *Streptococcus pneumoniae,* **multidrug-resistant *Mycobacterium tuberculosis* (MDR-TB), extensively drug-resistant *Mycobacterium tuberculosis* (XDR-TB),** and **carbapenem-resistant *Enterobacteriaceae* (CRE)** are all examples of emerging antimicrobial resistance.

Slowing the Emergence and Spread of Antimicrobial Resistance

Physicians should prescribe antimicrobial medications only when appropriate. The public must be educated about the appropriateness and limitations of antimicrobial therapy. Patients need to carefully follow prescribed instructions when taking antimicrobials.

20.6 ■ Mechanisms of Action of Antiviral Drugs
(figure 20.15, table 20.2)

Prevent Viral Entry
Enfuvirtide and maraviroc prevents HIV from entering cells.

Interfere with Viral Uncoating
Amantadine and rimantadine block the uncoating of influenza A virus after it enters a cell.

Interfere with Nucleic Acid Synthesis
Most antiviral drugs take advantage of the error-prone virally encoded enzymes used to replicate viral nucleic acid. Nucleoside analogs are phosphorylated *in vivo* to form nucleotide analogs; when these are incorporated into viral DNA they interfere with replication.

Prevent Genome Integration
Raltegravir inhibits HIV integrase.

Prevent Assembly and Release of Viral Particles
Protease inhibitors bind to and inhibit protease, the enzyme required for the production of infectious HIV particles. Neuraminidase inhibitors interfere with the release of influenza virus particles from a cell.

20.7 ■ Mechanisms of Action of Antifungal Drugs
(figure 20.16, table 20.3)

Interfere with Plasma Membrane Synthesis and Function
The polyenes disrupt fungal cell membranes by binding to ergosterol. The azoles inhibit the synthesis of ergosterol. The allylamines inhibit an enzyme in the pathway of ergosterol synthesis.

Interfere with Cell Wall Synthesis
Echinocandins interfere with the synthesis of β-1,3 glucan.

Interfere with Cell Division
Griseofulvin is concentrated in keratinized skin cells, where it inhibits fungal cell division.

Interfere with Nucleic Acid Synthesis
Flucytosine is taken up by yeast cells and converted by yeast enzymes to an active form.

20.8 ■ Mechanisms of Action of Antiprotozoan and Antihelminthic Drugs (table 20.4)
Most antiparasitic drugs are thought to interfere with biosynthetic pathways of protozoan parasites or the neuromuscular function of worms.

Review Questions

Short Answer

1. Describe the difference between the terms *antibiotic* and *antimicrobial*.
2. Define *therapeutic index* and explain its importance.
3. Explain the role of penicillin-binding proteins in drug susceptibility.
4. Name three classes of antimicrobial drugs that target ribosomes.
5. Explain the roles of the first-line drugs versus the second-line drugs in the treatment of tuberculosis.
6. Compare and contrast the method for determining the minimum inhibitory concentration (MIC) with the Kirby-Bauer disc diffusion test.
7. Name three targets that can be altered sufficiently via spontaneous mutation to result in resistance to an antimicrobial drug.
8. What is MRSA? Why is it significant?
9. Why is it difficult to develop antiviral drugs?
10. Explain the difference between the mechanism of action of an azole and that of a polyene.

Multiple Choice

1. Which of the following targets would you expect to be the most selective with respect to toxicity?
   a) Cytoplasmic membrane function
   b) DNA synthesis
   c) Glycolysis
   d) Peptidoglycan synthesis
   e) 70S ribosome

2. Penicillin has been modified to make derivatives that differ in all of the following *except*
   a) spectrum of activity.
   b) resistance to β-lactamases.
   c) potential for allergic reactions.
   d) a and c.

3. Which of the following is the target of β-lactam antibiotics?
   a) Peptidoglycan synthesis   b) DNA synthesis
   c) RNA synthesis   d) Protein synthesis
   e) Folic acid synthesis

4. Which of the following statements is *false*?
   a) A bacteriostatic drug stops the growth of a microorganism.
   b) The lower the therapeutic index, the less toxic the drug.
   c) Broad-spectrum antibiotics are associated with the development of *Clostridium difficile*–associated disease.
   d) Azithromycin has a longer half-life than does penicillin V.
   e) Chloramphenicol can cause a life-threatening type of anemia.

5. All of the following interfere with the function of the ribosome *except*
   a) fluoroquinolones.   b) lincosamides.   c) macrolides.
   d) streptogramins.   e) tetracyclines.

6. The target of the sulfonamides is
   a) cytoplasmic membrane proteins.
   b) folate synthesis.
   c) gyrase.
   d) peptidoglycan biosynthesis.
   e) RNA polymerase.

7. Routine antimicrobial therapy to treat tuberculosis involves taking

    a) one drug for 10 days.

    b) two or more drugs for 10 days.

    c) one drug for at least 6 months.

    d) two or more drugs for at least 6 months.

    e) five drugs for 2 years.

8. *Staphylococcus aureus* strains referred to as HA-MRSA are sensitive to

    a) methicillin.   b) penicillin.   c) cephalosporin.

    d) vancomycin.   e) none of the above.

9. Acyclovir is a

    a) nucleoside analog.   b) non-nucleoside polymerase inhibitor.

    c) protease inhibitor.   d) none of the above.

10. The antifungal drug griseofulvin is used to treat

    a) vaginal infections.   b) systemic infections.

    c) nail infections.   d) eye infections.

Applications

1. A physician was treating one young woman and one elderly patient for urinary tract infections caused by the same type of bacterium. Although the patients had similar body dimensions and weight, the physician gave a smaller dose of drug to the older patient. What was the physician's rationale for this decision?

2. An advocacy group in Washington, D.C., is petitioning the U.S. Department of Agriculture (USDA) to stop the use of low-dosage antimicrobial agents used to enhance the growth of cattle and chickens. Why is the group against this practice? Why does the USDA permit it?

Critical Thinking ✚

1. Figure 20.12 shows the E-test procedure for determining an MIC value. How would the zone of inhibition appear if the drug concentrations in the strip were decreased slightly?

2. Why is acyclovir converted to a nucleotide analog only in cells infected with herpes simplex virus?

21 Respiratory System Infections

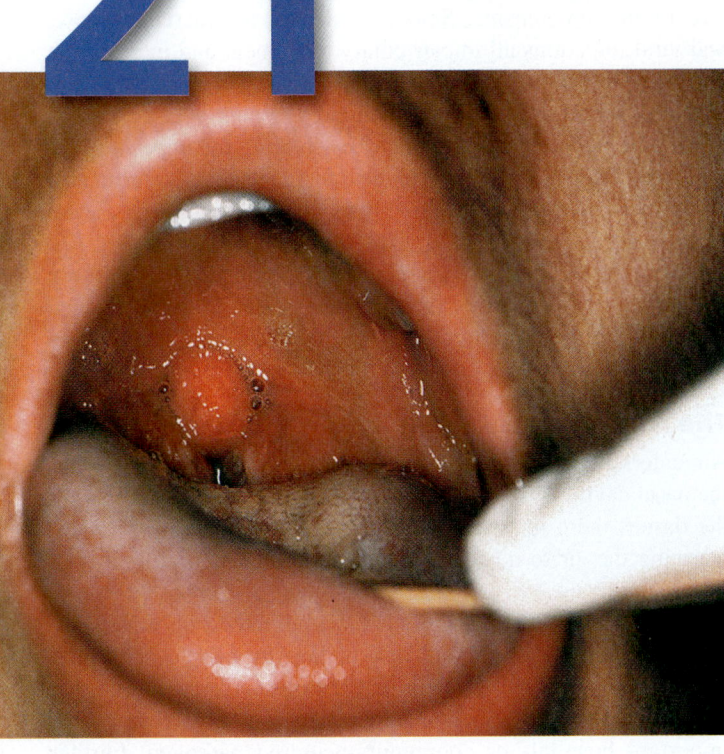

Individual with streptococcal pharyngitis.

A Glimpse of History

Rebecca Lancefield's mother was a direct descendent of Lady Mary Wortley Montagu, who promoted smallpox variolation in England more than 75 years before Edward Jenner. Lancefield attended Wellesley College in 1912, and became interested in biology. She was awarded a scholarship to Columbia University, where she studied under the famous microbiologist Hans Zinsser, a pioneer in the science of immunology. She received a master's degree, but her studies were interrupted when her husband was drafted into the armed services in World War I. Fortunately he was assigned to the Rockefeller Institute, where Rebecca got a job as a laboratory technician for the distinguished microbiologists O. T. Avery and A. R. Dochez. These scientists were known for their research on pneumococci, and were studying other streptococci from personnel at army camps. At that time, classification of streptococci was largely based on whether their colonies produced colorless clearing (β-hemolysis) or a greenish partial clearing (α-hemolysis) when grown on blood agar. Avery and Dochez showed that streptococci could be more effectively divided into groups based on their surface antigens.

◀◀ Mary Wortley Montagu, p. 419

The Lancefields returned to Columbia University, where Rebecca received her Ph.D. in 1925 for her studies of α-hemolytic streptococci. Thereafter, she went back to the Rockefeller Institute where she spent the rest of her scientific career studying streptococci. She showed that β-hemolytic streptococci could be classified according to cell wall carbohydrates, a system now known as "Lancefield grouping." Lancefield demonstrated that almost all strains of β-hemolytic streptococci isolated from human infections had the same cell wall carbohydrate—"A"—whereas streptococci from other sources had different wall carbohydrates: "B" from cattle infections; "C" from cattle, horses, and guinea pigs; "D" from cheese and human normal microbiota; and so on. The "Lancefield grouping" proved to be a much better predictor of pathogenic potential than hemolysis on blood agar. Lancefield became the first woman president of the American Association of Immunologists and in 1970 was elected to the prestigious National Academy of Sciences. She died in 1981 at the age of 86.

Respiratory infections include an enormous variety of illnesses ranging from the trivial to the fatal. They can be divided into infections of the upper respiratory system (in the head and neck) and infections of the lower respiratory system (in the chest). Upper respiratory infections such as colds are extremely common and uncomfortable, but they are not life-threatening and go away without treatment in about a week. Diseases of the lower respiratory system such as pneumonia and tuberculosis are often serious, however, and may be fatal.

21.1 ■ Anatomy, Physiology, and Ecology

Learning Outcomes

1. *Outline the functions of the upper and lower respiratory tracts.*
2. *List the parts of the respiratory system that are normally microbe-free.*

A critical function of the respiratory system is gas exchange. Each breath brings in O_2, which replenishes the supply in the blood, and then releases CO_2, the waste product of cellular metabolism. In addition, movement of the vocal cords makes sound, which allows us to speak, and sensors in the nose detect odors. Although the mouth can also be used for breathing, it is considered part of the digestive tract, so it is covered in chapter 24.

The respiratory system has two general parts—the upper respiratory tract and the lower respiratory tract. We also include the eyes and ears in our discussion. Even though these are not considered part of the respiratory tract, the eyes and the nose are important from a microbiological standpoint because they serve as two main portals of entry to the body. Pathogens often first establish themselves there and then spread to other parts of the body. ◀◀ portals of entry, p. 439

The Upper Respiratory Tract

The upper respiratory tract includes the nose and nasal cavity, pharynx (throat), and epiglottis (**figure 21.1**). Mucous membranes line the respiratory tract. These are coated with mucus, a slimy glycoprotein material that traps air-borne dust and other particles, including microorganisms. Mucus is produced by specialized cells called **goblet cells** that are scattered among the cells of the membrane. Their name reflects their shape, which is narrow at the base and wide at the surface, like a wine glass or goblet. ◀◀ mucous membrane, p. 337

Ciliated epithelium lines most of the mucous membranes of the respiratory system. These cells have cilia—tiny, hairlike

projections—along their exposed free border. The cilia beat synchronously, continually propelling the mucus film out of the respiratory tract. The mucus, along with any entrapped microbes, is then swallowed and digested. This mechanism, the **mucociliary escalator,** normally keeps the lower respiratory tract completely free of microorganisms. Smoking, alcohol and narcotic abuse, and viral infections all impair ciliary movement and increase the chance of infection. ◀◀ cilia, p. 74

The tonsils are secondary lymphoid organs, located so that they come into contact with microbes entering the upper respiratory tract. They are important in the immune response but can also be the sites of infection, resulting in tonsillitis. ◀◀ secondary lymphoid organs, p. 358

Some of the bacterial genera that inhabit the upper respiratory system are listed in **table 21.1.** Although generally harmless, they are opportunists that can cause disease when host defenses are impaired. ◀◀ opportunists, p. 382

The Nose and Nasal Cavity

Air enters the respiratory system at the nostrils and flows into the nasal cavity. When cold air enters the nose, the blood flow to the tissues there immediately increases due to nervous reflexes, warming the air to near body temperature and saturating it with water vapor.

The nasal entrance usually contains diphtheroids and staphylococci. In addition, about 20% of healthy people constantly carry *Staphylococcus aureus* in their noses; an even higher percentage of hospital personnel are likely to be carriers of this important hospital- or community-acquired pathogen. Further inside the nasal passages, the normal microbiota are similar to that of the throat. Infection of the nasal passages, usually by viruses, results in rhinitis; this inflammation of the nasal tissues causes the familiar "runny nose."

Pharynx (Throat) and Epiglottis

Once past the nose, the air moves down to the throat (pharynx). Inflammation of the throat, **pharyngitis,** is commonly the result of viral infection. In addition to being part of the respiratory system,

TABLE 21.1	Normal Microbiota of the Upper Respiratory System	
Genus	**Characteristics**	**Comments**
Staphylococcus	Gram-positive cocci in clusters	Facultative anaerobes. Commonly includes the potential pathogen *Staphylococcus aureus,* inhabiting the nostrils.
Corynebacterium	Pleomorphic, Gram-positive rods; non-motile; non-spore-forming	Aerobic or facultatively anaerobic. Diphtheroids include anaerobic and aerotolerant organisms.
Moraxella	Gram-negative diplococci and diplobacilli	Aerobic. Some microscopically resemble pathogenic *Neisseria* species such as *N. meningitidis.*
Haemophilus	Small, Gram-negative rods	Facultative anaerobes. Commonly include the potential pathogen *H. influenzae.*
Bacteroides	Small, pleomorphic, Gram-negative rods	Obligate anaerobes.
Streptococcus	Gram-positive cocci in chains	Aerotolerant (obligate fermenters). α (especially viridans, meaning green hemolysis), β (clear hemolysis), and γ (non-hemolytic) types; the potential pathogen, *S. pneumoniae* is often present.

FIGURE 21.1 Anatomy and Infections of the Respiratory System (a) Lateral view of upper respiratory tract. **(b)** Frontal view of the upper and lower respiratory tracts.

? *Why would inflammation in the lungs be life-threatening?*

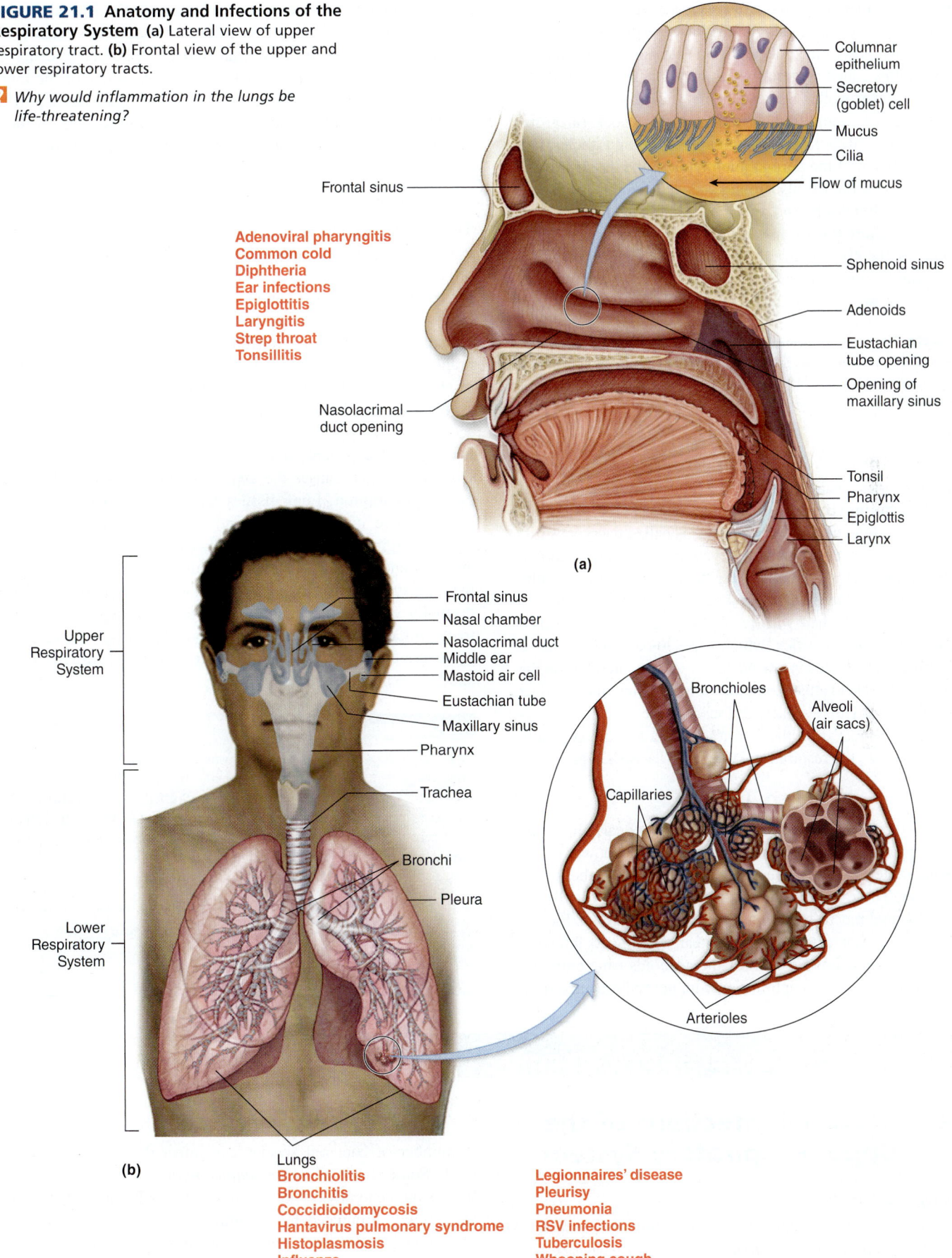

Columnar epithelium
Secretory (goblet) cell
Mucus
Cilia
Flow of mucus

Frontal sinus

**Adenoviral pharyngitis
Common cold
Diphtheria
Ear infections
Epiglottitis
Laryngitis
Strep throat
Tonsillitis**

Nasolacrimal duct opening

Sphenoid sinus
Adenoids
Eustachian tube opening
Opening of maxillary sinus
Tonsil
Pharynx
Epiglottis
Larynx

(a)

Upper Respiratory System

Frontal sinus
Nasal chamber
Nasolacrimal duct
Middle ear
Mastoid air cell
Eustachian tube
Maxillary sinus
Pharynx
Trachea

Bronchioles
Alveoli (air sacs)
Capillaries
Arterioles

Bronchi
Pleura

Lower Respiratory System

(b)

Lungs
**Bronchiolitis
Bronchitis
Coccidioidomycosis
Hantavirus pulmonary syndrome
Histoplasmosis
Influenza**

**Legionnaires' disease
Pleurisy
Pneumonia
RSV infections
Tuberculosis
Whooping cough**

the throat is essential to the digestive system. A small muscular flap, the epiglottis, covers the opening to the lower respiratory tract during swallowing, preventing material from entering it. Inflammation of the epiglottis, called epiglottitis, can be a life-threatening emergency because the swollen flap can block the airway.

Streptococci, including viridans streptococci (a group of α-hemolytic streptococci) and non-hemolytic species, are common members of the normal microbiota of the throat. A variety of other bacteria are also found there, including *Moraxella catarrhalis,* diphtheroids, and anaerobic Gram-negative bacteria, such as *Bacteroides* species. In addition, a number of opportunistic pathogens frequently colonize the throat, including *Streptococcus pneumoniae, Haemophilus influenzae,* and *Neisseria meningitidis.*
◀◀ α-hemolysis, p. 96

MicroByte

> Cilia beat at a rate of about 1,000 times a minute, propelling mucus along the mucociliary escalator.

Eyes

The surface of the eyes and lining of the eyelids are covered by mucous membranes called conjunctiva. Infection of the conjunctiva is called **conjunctivitis.** The tear ducts connect the eyes to the nasal chamber. Infection of the tear ducts is called dacryocystitis.

Surprisingly, even though the eyes are constantly exposed to large numbers of microorganisms, the conjunctiva of healthy people commonly have very few bacteria. Presumably, this is because the eye is bathed with lysozyme-rich tears and cleaned by the eyelid's blinking reflex, which wipes the eye surface like a car windshield wiper cleans a windshield. Unless microbes on the conjunctiva are able to attach to it, they are swept into the tear duct and nasopharynx. The few bacteria found on the normal conjunctiva usually originate from the skin microbiota and are generally unable to colonize the respiratory system. ◀◀ lysozyme, p. 62

Ears

The ear has three parts: external, middle, and inner ear. The external ear is protected from microbes by cerumen (ear wax). The middle ear is sterile. It is connected by the eustachian tubes to the nasopharynx. These tubes equalize the pressure in the middle ear and drain normal mucus secretions. They are also a route through which microbes can enter the middle ear, leading to infection called **otitis media.** Enlargement of the adenoids can contribute to middle ear infections by interfering with normal drainage from

the eustachian tubes. The inner ear, which is fluid-filled, is also microbe-free. Occasionally, however, viruses or bacteria enter the inner ear, often following an upper respiratory tract infection, causing labyrinthitis (named for the chambers of the inner ear which are called labyrinths). The skull also has air-filled chambers—the sinuses and mastoid air cells. Infections of these chambers are called sinusitis and mastoiditis, respectively.

The Lower Respiratory Tract

The lower respiratory tract includes the larynx (voice box), trachea, bronchi, and lungs (figure 21.1b). Inflammation of the larynx is called laryngitis, and is manifest as hoarseness. The trachea (windpipe) is a continuation of the larynx and branches into two bronchi. Inflammation of the bronchi is called bronchitis, commonly the result of viral infection or smoking. The bronchi branch repeatedly, becoming bronchioles, the site of an important viral infection called bronchiolitis. The smallest branches of the bronchioles end in the alveoli, which are tiny, thin-walled air sacs that make up the bulk of lung tissue. Inflammation of the lungs is called pneumonitis, often the result of viral infections. Pneumonitis that causes the alveoli to fill with pus and fluid is called **pneumonia.** Lung tissues have many macrophages that readily move into the alveoli and airways to engulf infectious agents, helping to prevent pneumonia from developing. The lungs are surrounded by two membranes called pleura: One adheres to the lung and the other to the chest wall and diaphragm. The pleura normally slide against each other as the lung expands and contracts. Inflammation of the pleura is called pleurisy, characterized by severe chest pain. The lower respiratory tract is usually sterile and has no normal microbiota.

MicroAssessment 21.1

The respiratory system provides a warm, moist environment for microorganisms. It is protected by tonsils and adenoids, and by the mucociliary escalator. The upper respiratory tract contains highly diverse microbiota, including aerobes, anaerobes, facultative anaerobes, and aerotolerant bacteria. Although most of them are of low virulence, these organisms can sometimes cause disease opportunistically. The lower respiratory system is free of a normal microbiota.

1. *What is the normal function of the tonsils and adenoids?*
2. *Describe the normal microbiota of the respiratory system.*
3. *How would paralysis of the cilia impact the respiratory system?* ✚

INFECTIONS OF THE UPPER RESPIRATORY SYSTEM

21.2 ■ Bacterial Infections of the Upper Respiratory System

Learning Outcomes

3. *Compare the distinctive characteristics of strep throat and diphtheria.*
4. *List the parts of the upper respiratory system commonly infected by* Streptococcus pneumoniae *and* Haemophilus influenzae.

A number of bacterial species can infect the upper respiratory tract. Some, such as *Haemophilus influenzae,* can cause sore throats but generally do not require treatment because the bacteria are quickly eliminated by the immune system. Others, such as *Streptococcus pyogenes,* cause infections that require treatment because they are not as easily eliminated and can cause serious complications.

Streptococcal Pharyngitis ("Strep Throat")

Sore throat is one of the most common reasons that people in the United States seek medical care. Many of these visits are due to justifiable concerns about streptococcal pharyngitis, commonly known as strep throat. One concern about streptococcal infections is the risk of post-streptococcal sequelae. These are complications that develop after the initial infection and will be discussed in the next section.

Signs and Symptoms

Strep throat is characterized by a sore throat, difficulty swallowing, and fever. The throat is red, with patches of pus and scattered tiny hemorrhages. The lymph nodes in the neck are enlarged and tender. Abdominal pain or headache may occur in older children and young adults. Patients do not usually have a cough, weepy eyes, or runny nose. Most patients with strep throat recover spontaneously after about a week. In fact, many infected people have only mild symptoms or no symptoms at all.

Causative Agent

Strep throat is caused by *Streptococcus pyogenes,* a Gram-positive coccus that grows in chains (**figure 21.2**). The organism can be differentiated from other streptococci that normally inhabit the throat by its colony morphology on blood agar—*S. pyogenes* colonies are surrounded by a characteristic clear zone of β-hemolysis (**figure 21.3**). In contrast, most species of *Streptococcus* that are typically part of the normal throat microbiota are either α-hemolytic, producing a zone of greenish partial clearing around colonies on blood agar, or non-hemolytic. A few other streptococci and some other bacteria are also β-hemolytic, so further tests are needed to identify *S. pyogenes.* ◀◀ hemolysis, p. 96 ◀◀ blood agar, p. 95

Streptococcus pyogenes is commonly referred to as the group A streptococcus (GAS), reflecting its Lancefield grouping (see **A Glimpse of History** in this chapter). This grouping system uses antibodies to distinguish the cell wall carbohydrates in streptococcal

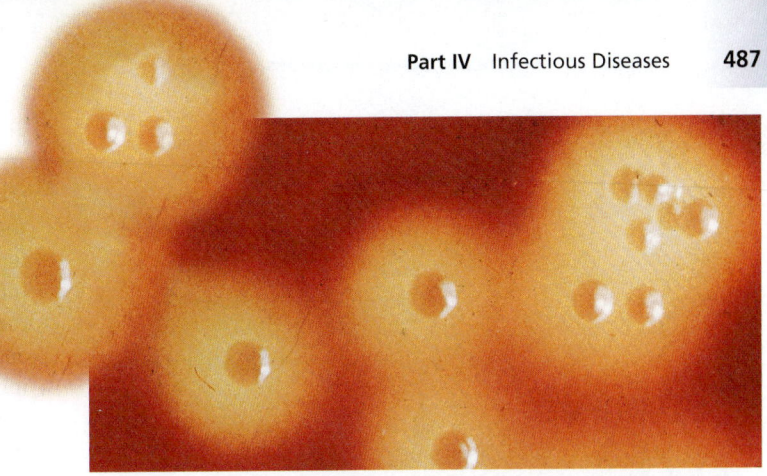

FIGURE 21.3 *Streptococcus pyogenes* **Colonies on Blood Agar** The colonies are surrounded by a wide zone of β-hemolysis.

❓ *What causes β-hemolysis?*

species. *S. pyogenes* is characterized by the "A" carbohydrate in its cell wall. Antibodies that bind to the group A carbohydrate are the basis for many of the rapid diagnostic tests done on throat specimens in a physician's office (see figure 18.9).

Different *S. pyogenes* strains within GAS are distinguished by variations of a surface antigen called **M protein,** an important virulence factor. Strains with certain types of M protein are the cause of strep throat, whereas others with different M protein are more likely to cause skin infections.

Pathogenesis

Streptococcus pyogenes has many virulence factors (**table 21.2**). Some of these disease-causing mechanisms are structural components of the cell wall that allow the bacterium to avoid host

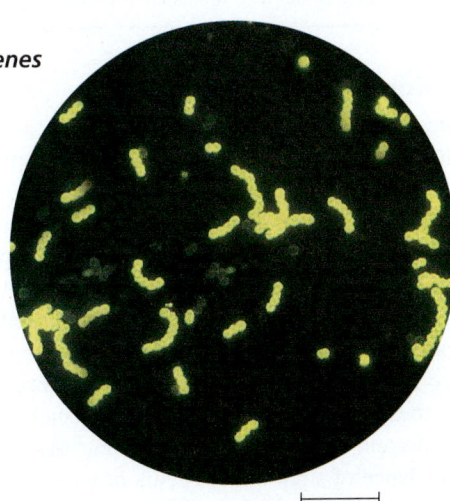

FIGURE 21.2
Streptococcus pyogenes
Chain formation of
S. pyogenes, revealed
by fluorescence
microscopy.

❓ *What are two sequelae that may occur after strep throat?*

10 μm

TABLE 21.2	Virulence Factors of *Streptococcus pyogenes*
Product	**Effect**
C5a peptidase	Inhibits recruitment of phagocytes by destroying complement C5a
Hyaluronic acid capsule	Inhibits phagocytosis
M protein	Interferes with phagocytosis by causing breakdown of complement C3b, an opsonin
Protein F	Responsible for attachment to host cells
Protein G	Binds to Fc portion of antibody, thereby interfering with opsonization
Streptococcal pyrogenic exotoxins (SPEs)	Superantigens responsible for scarlet fever, toxic shock, "flesh-eating" fasciitis
Streptolysins O and S	Lyse leukocytes and erythrocytes
Tissue-degrading enzymes	Enhance spread of bacteria by breaking down DNA, proteins, blood clots, tissue, hyaluronic acid

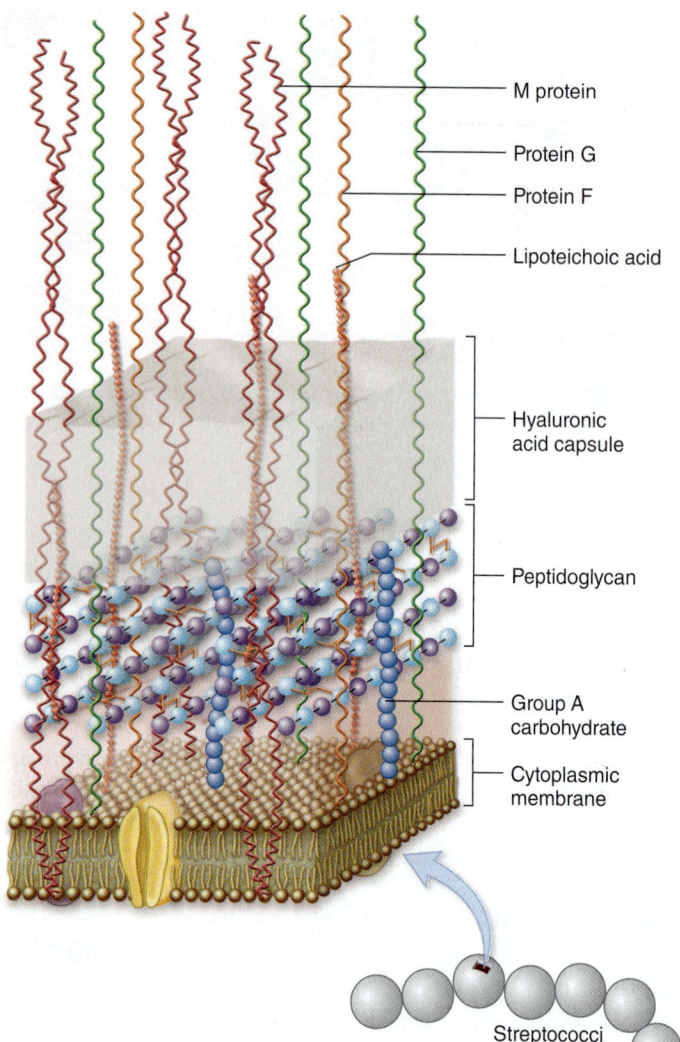

FIGURE 21.4 Virulence Factors on the Cell Envelope of *Streptococcus pyogenes*

❓ *What is the role of M protein in pathogenesis?*

phagocytosis) from being deposited on the bacterial cell wall. Another cell wall protein—protein G—is an Fc receptor that binds the Fc portion of IgG, preventing opsonization by antibodies (see figure 16.11). The organism releases C5a peptidase, an enzyme that destroys complement component C5a, normally responsible for attracting phagocytes to the site of a bacterial infection, thus further inhibiting phagocytosis. It also produces streptolysins O and S—enzymes that destroy erythrocytes and leukocytes by making holes in their cell membranes. Leukocyte destruction inhibits the immune response, whereas destruction of erythrocytes causes the β-hemolysis exhibited by *S. pyogenes*. Some *S. pyogenes* strains have a hyaluronic acid capsule that further prevents phagocytosis. Hyaluronic acid is a normal component of human tissue, so the capsule is thought to be a cloaking disguise, making it difficult for the host immune system to detect the pathogen. ◄◄ complement system, p. 344 ◄◄ C5a peptidase, p. 389 ◄◄ membrane-damaging toxins, p. 392 ◄◄ opsonins, p. 347

A few strains of *S. pyogenes* produce **streptococcal pyrogenic exotoxins (SPEs),** a family of exotoxins that cause severe streptococcal diseases characterized by high fever (*pyro* means "fire"). SPEs are encoded by bacteriophages, an example of lysogenic conversion. These toxins are superantigens, causing massive activation of T cells. The resulting uncontrolled release of cytokines is probably responsible for the seriousness of these infections. A person with strep throat caused by an SPE-producing strain of *S. pyogenes* may develop scarlet fever, characterized by high fever, roughening of the skin (texture like sandpaper) and a pink-red rash. Although the toxin that causes these symptoms is an SPE, it is traditionally referred to as erythrogenic toxin (*erythro* means "red"). The rash of scarlet fever is found on the head, neck, chest and thighs, but usually not around the mouth. The toxin also causes the tongue to look like a ripe strawberry—red and spotted. Both the skin and the tongue may peel. Some SPE-producing strains of *S. pyogenes* have additional virulence factors that allow them to cause severe invasive diseases such as streptococcal toxic shock syndrome, and "flesh-eating" necrotizing fasciitis. ◄◄ lysogenic conversion, p. 313 ◄◄ superantigens, p. 393 ◄◄ cytokines, p. 341 ►► necrotizing fasciitis, p. 553 ►► streptococcal toxic shock syndrome, p. 553

Microᛒyte

Streptococcal streptokinase is used as a clot-dissolving medication in treating some heart attacks and pulmonary embolisms.

defenses (**figure 21.4**). Others are destructive enzymes and toxins released by the bacterial cell that damage or kill host cells.

Proteins in the cell wall of *S. pyogenes* allow the bacteria to attach to host cells. M protein is an important adhesin involved in attachment—antibodies that bind to it prevent infection. There are more than 80 antigenic types of M protein among the many strains of *S. pyogenes,* however, and antibodies to one type do not prevent infection by a strain that has a different kind. Another protein, protein F, mediates attachment of *S. pyogenes* to cells of the throat by adhering to fibrin, a protein found on epithelial cells.

Once *S. pyogenes* colonizes host tissues, it produces enzymes such as DNase, hyaluronidase, and proteases that break down intracellular connections and allow the organism to spread rapidly to other cells. This spread is assisted by **s**treptokinase, which causes the breakdown of blood clots.

S. pyogenes has many mechanisms for avoiding the host immune system. M protein, in addition to aiding attachment to the host cell, also interferes with phagocytosis. It prevents complement component C3b (an opsonin that would usually increase

Epidemiology

Streptococcus pyogenes naturally infects only humans. The strains that cause strep throat spread easily by respiratory droplets generated by shouting, coughing, and sneezing. Epidemics have also originated from food contaminated with *S. pyogenes*. Nasal carriers of *S. pyogenes* are more likely than pharyngeal carriers to spread the organisms. Anal carriers are not common but can be a dangerous source of healthcare-associated infections. A person may be an asymptomatic carrier of *S. pyogenes* for weeks, if they are not given treatment. Some people become long-term carriers; in these cases, the infecting strain usually becomes deficient in M protein and is not a threat to the carrier or to others. The peak incidence of strep throat occurs in winter or spring and is highest in grade school children. ◄◄ healthcare-associated infections, p. 449

Treatment and Prevention

People with fever and sore throat should be taken to a physician so that a throat swab can be taken for a rapid diagnostic test and throat culture. Confirmed strep throat is treated with a full 10 days of penicillin or erythromycin, which eliminates the organism in about 90% of the cases. Treatment given as late as 9 days after the onset prevents post-streptococcal sequelae (covered next).

Adequate ventilation and avoiding crowded situations help to control the spread of streptococcal infections. No vaccine is available. However, genome sequencing has revealed some new possibilities for vaccine targets. **Table 21.3** summarizes some important facts about strep throat.

Post-Streptococcal Sequelae

Post-streptococcal sequelae—complications that can develop after strep throat or other streptococcal infections—are thought to be a result of immune responses to *Streptococcus pyogenes*. They are uncommon in Western industrialized nations because of quick identification and treatment of *S. pyogenes* infections. Globally, however, they occur much more frequently and can be serious.

Acute Rheumatic Fever

Acute **rheumatic fever** usually begins about 3 weeks after recovery from strep throat. Signs and symptoms include fever, joint pains, chest pains, rash, and nodules under the skin. Uncontrollable body movements (chorea, commonly called St. Vitus' dance) can occur and, when present, are major criteria for diagnosis. Carditis, the most serious complication, develops in about a third to a half of patients. This can lead to chronic rheumatic heart disease in which one or more of the heart valves are damaged, causing them to leak and resulting in heart failure later in life. The damaged valves are also prone to infection, usually by bacteria from the normal skin or mouth microbiota, resulting in subacute bacterial endocarditis. ▶▶ **subacute bacterial endocarditis, p. 673**

Rheumatic fever only develops in people who are genetically predisposed to the disease—some MHC class II alleles are involved in susceptibility (see **Perspective 15.1**). The pathogenesis of rheumatic fever is not well understood but is thought to be an autoimmune response involving both humoral and cell-mediated immunity—in some individuals, antibodies to *S. pyogenes* cross react with host tissue antigens that are similar to the pathogen antigens, a case of molecular mimicry. In chronic rheumatic heart disease, for example, molecular mimicry between cardiac myosin and epitopes of streptococcal M protein may result in production of antibodies that bind to both *S. pyogenes* M protein and host cardiac myosin. If such antibodies bind to cardiac tissue, the tissue is then targeted for attack by the host's own effector T helper (T_H) cells. Several other shared antigens have also been identified. This recognition of self-antigens starts an autoimmune

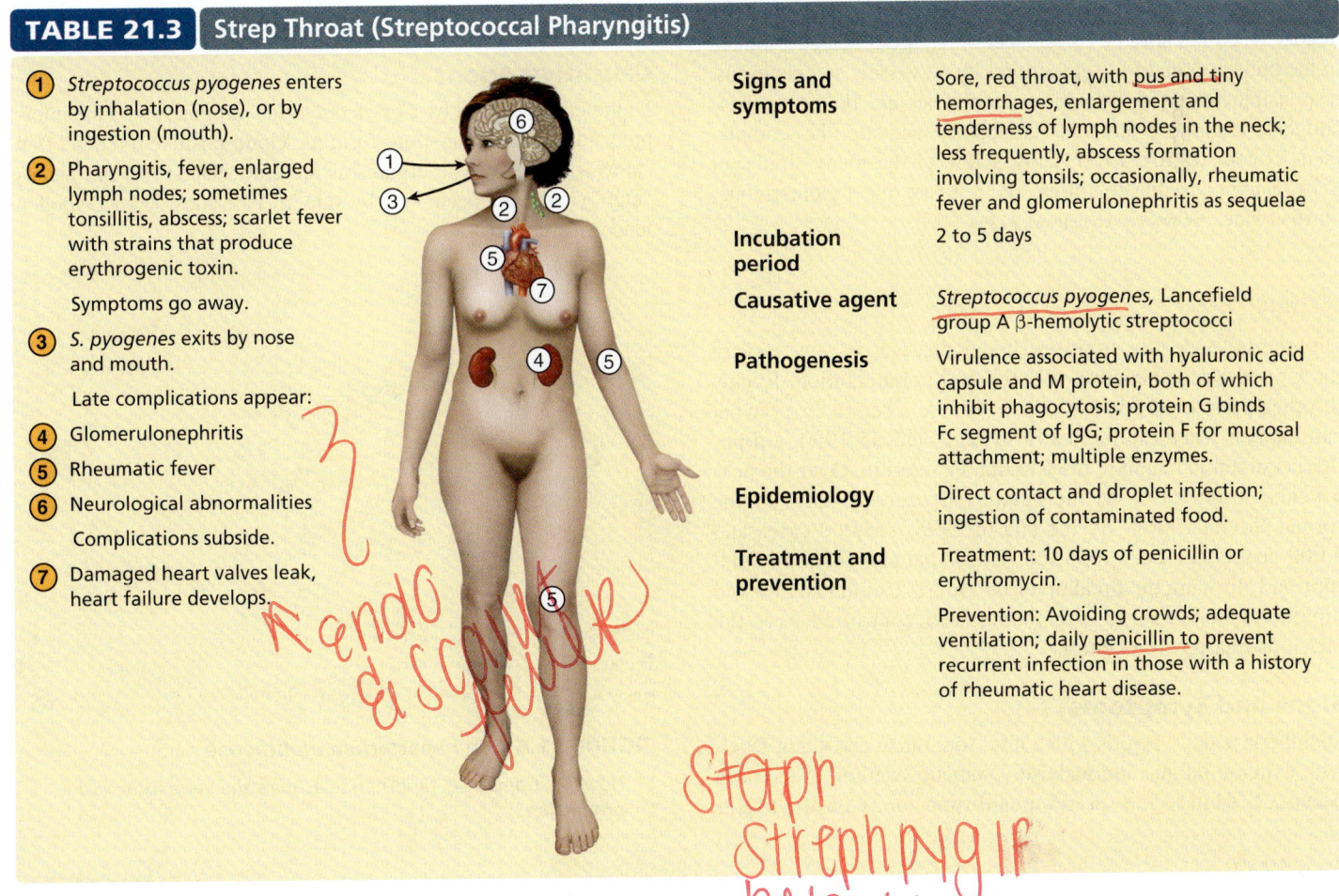

TABLE 21.3   Strep Throat (Streptococcal Pharyngitis)

1. *Streptococcus pyogenes* enters by inhalation (nose), or by ingestion (mouth).

2. Pharyngitis, fever, enlarged lymph nodes; sometimes tonsillitis, abscess; scarlet fever with strains that produce erythrogenic toxin.

   Symptoms go away.

3. *S. pyogenes* exits by nose and mouth.

   Late complications appear:

4. Glomerulonephritis
5. Rheumatic fever
6. Neurological abnormalities

   Complications subside.

7. Damaged heart valves leak, heart failure develops.

Signs and symptoms	Sore, red throat, with pus and tiny hemorrhages, enlargement and tenderness of lymph nodes in the neck; less frequently, abscess formation involving tonsils; occasionally, rheumatic fever and glomerulonephritis as sequelae
Incubation period	2 to 5 days
Causative agent	*Streptococcus pyogenes*, Lancefield group A β-hemolytic streptococci
Pathogenesis	Virulence associated with hyaluronic acid capsule and M protein, both of which inhibit phagocytosis; protein G binds Fc segment of IgG; protein F for mucosal attachment; multiple enzymes.
Epidemiology	Direct contact and droplet infection; ingestion of contaminated food.
Treatment and prevention	Treatment: 10 days of penicillin or erythromycin. Prevention: Avoiding crowds; adequate ventilation; daily penicillin to prevent recurrent infection in those with a history of rheumatic heart disease.

response—the T cells release proinflammatory cytokines, causing inflammation that leads to permanent damage to the local tissues, particularly the heart valves. ◀◀ **MHC, p. 369** ◀◀ **inflammation, p. 348** ◀◀ **APC, p. 369** ◀◀ **helper T cells, p. 356** ◀◀ **molecular mimicry, p. 412**

Although outbreaks of rheumatic fever still occur in the United States, incidence has generally declined. This is probably a result of quick treatment of strep throat with antibiotics, and decreased prevalence of the strains associated with the disease. Overall, the risk of developing acute rheumatic fever after severe untreated strep throat is 3% or less, and those with untreated mild pharyngitis have an even lower risk. Nonetheless, 10 to 20 million new cases of rheumatic fever occur globally each year, mostly in economically disadvantaged countries. Signs and symptoms of rheumatic fever usually subside with rest and anti-inflammatory medicines such as aspirin. Individuals with cardiac damage due to acute rheumatic fever take penicillin daily for years to prevent recurrence of the disease.

Acute Post-Streptococcal Glomerulonephritis

Acute post-streptococcal glomerulonephritis occasionally follows strep throat but is more often an aftermath of a strep skin infection. It begins 7–21 days after the initial infection, usually about 10 days in the case of strep throat. The signs and symptoms include fever, fluid retention, high blood pressure, and blood and protein in the urine (making it look like tea or coca cola). There are no streptococci in the urine or the diseased kidney tissues—*S. pyogenes* has generally been eliminated from the throat by the body's immune response by the time the symptoms of glomerulonephritis appear. The damage to the kidneys is due to an inflammatory reaction caused by streptococcal antigens that accumulate in the kidney glomeruli (tufts of tiny blood vessels and nephrons, responsible for urine formation); antibodies are bound to these antigens and these immune complexes activate the complement system (**figure 21.5**). Only a few of the many strains of *S. pyogenes* cause the condition, and it is rare to get glomerulonephritis twice. ◀◀ **immune complexes, p. 395**

Diphtheria

Diphtheria, a deadly toxin-mediated disease, is now rare in the United States because of childhood immunization. Events in other parts of the world, however, are a reminder of what can happen when public health is neglected. In 1990, a diphtheria epidemic began in the Russian Federation. Over the next 5 years, it spread to all the newly independent states of the former Soviet Union. By the end of 1995, 125,000 cases and 4,000 deaths had been reported. The social and economic disruption following the breakup of the Soviet Union had allowed diphtheria to reemerge after being well controlled over the previous quarter of a century.

Signs and Symptoms

Diphtheria usually begins with a mild sore throat and slight fever, with extreme fatigue and malaise. Dramatic swelling of the neck occurs. A whitish-gray **pseudomembrane** forms on the tonsils

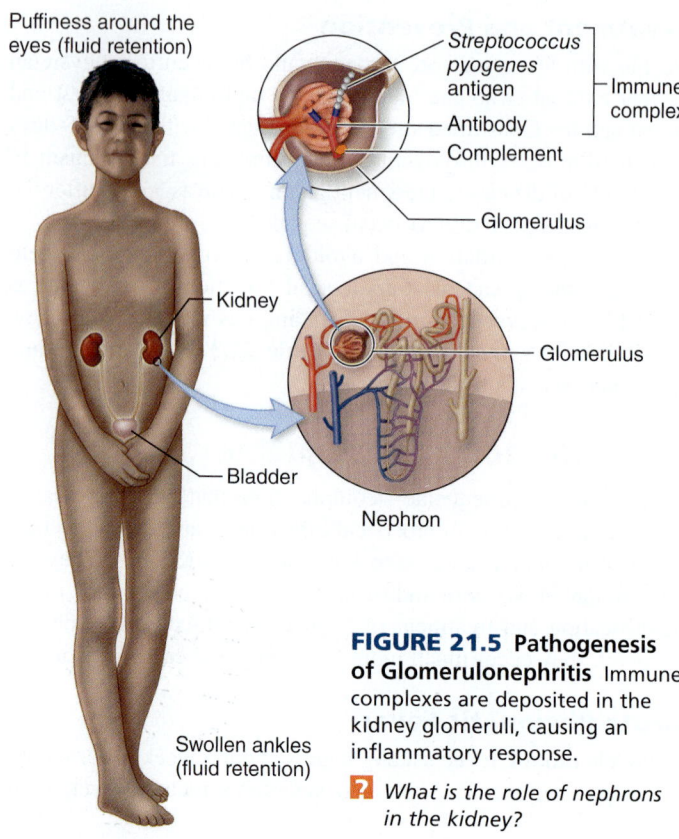

FIGURE 21.5 Pathogenesis of Glomerulonephritis Immune complexes are deposited in the kidney glomeruli, causing an inflammatory response.

❓ *What is the role of nephrons in the kidney?*

and throat or in the nasal cavity. Heart and kidney failure and paralysis may occur later.

Causative Agent

Diphtheria is caused by *Corynebacterium diphtheriae,* a pleomorphic, non-motile, non-spore-forming, Gram-positive rod that often stains irregularly. The organisms are club-shaped (*koryne* means "club") and often occur side by side in "palisades" (like a wooden fence) (**figure 21.6**).

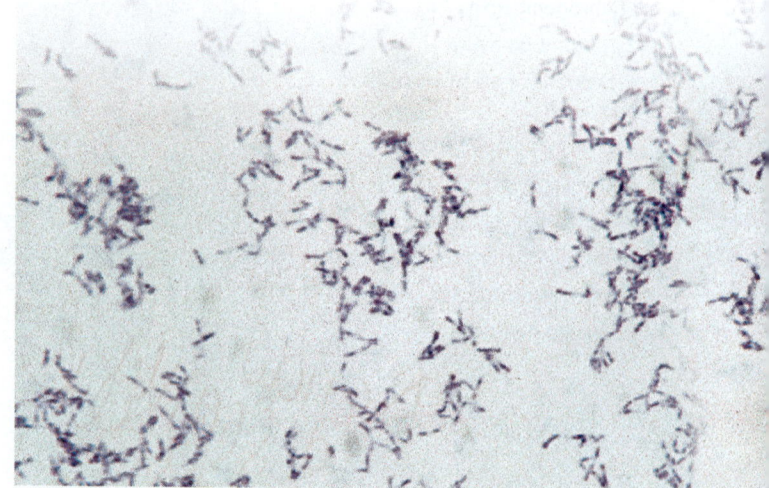

FIGURE 21.6 *Corynebacterium diphtheriae*

❓ *How do strains of* C. diphtheriae *acquire the gene for toxin production?*

Most strains of *C. diphtheriae* release diphtheria toxin, a powerful exotoxin responsible for the serious symptoms of the disease. The gene for this toxin is carried by a specific lysogenic bacteriophage, another example of lysogenic conversion.

To isolate *C. diphtheriae* from throat material, a special medium that contains potassium-tellurite is used. This chemical inhibits most of the normal microbiota of the upper respiratory tract and causes *C. diphtheriae* to form black colonies, aiding in identification. The organism can also be grown on Loeffler's, a medium that enhances volutin granule formation at the poles of the cells, which can be seen with methylene blue stains. **◀◀ lysogenic conversion, p. 313 ◀◀ volutin granules, p. 67**

Pathogenesis

Corynebacterium diphtheriae has little invasive ability, and rarely enters the blood or tissues. The disease is caused by the diphtheria exotoxin released by the bacteria growing in the throat. The pseudomembrane that forms is made up of dead epithelial cells (killed by the exotoxin) and clotted blood, along with fibrin and the leukocytes that accumulate during inflammation. This membrane may come loose and obstruct the airways, causing the patient to suffocate. The toxin may be absorbed into the bloodstream, allowing it to access and damage heart, nerve, and kidney tissues.

Diphtheria toxin is an A-B toxin (**figure 21.7**). The B subunit attaches to specific receptors on a host cell membrane, and the entire toxin molecule is taken into the cell by endocytosis. Cells lacking these receptors do not take up the toxin and are unaffected by it—this receptor specificity explains why some tissues of the body are not affected in diphtheria, while others are severely damaged. Once the toxin molecule is inside the cell, the A subunit separates from the B subunit. It becomes an active enzyme and catalyses a chemical reaction that inactivates elongation factor 2 (EF-2), which is required for movement of the eukaryotic ribosome on mRNA. This stops protein synthesis, and the cell dies. Since the toxin A subunit is an enzyme, it is not used up in the process, so one or two molecules of it can inactivate nearly all the cell's EF-2. This explains the extreme potency of diphtheria toxin. **◀◀ A-B toxins, p. 392 ◀◀ endocytosis, p. 72**

Epidemiology

Humans are the primary reservoir for *Corynebacterium diphtheriae*. The organisms are typically spread by air from infected people. They are then acquired either by inhalation or from fomites. *C. diphtheriae* can sometimes cause cutaneous diphtheria and the chronic skin ulcers that develop may become a source of infection to others. This problem primarily occurs in homeless populations. **◀◀ fomites, p. 440**

Treatment and Prevention

Diphtheria is treated by injecting the patient with antiserum against diphtheria toxin as soon as possible. Delaying treatment while waiting for culture results can be fatal, so the antiserum must be given immediately once the disease is suspected. The bacteria are sensitive to antibiotics such as erythromycin and penicillin, but such treatment only stops transmission of the disease; it has no effect on toxin that has already been absorbed. Even with treatment, about 1 of 10 diphtheria patients dies. **◀◀ antiserum, p. 420**

Because diphtheria results from toxin production rather than microbial invasion, it can be effectively prevented by immunization with toxoid. The toxoid, prepared by formalin treatment of diphtheria toxin, causes the body to produce antibodies that specifically neutralize the toxin. The childhood vaccination DTaP consists of diphtheria and tetanus toxoids along with components of *Bordetella pertussis*. Unfortunately, these immunizations are often neglected, and serious epidemics of the diseases have occurred periodically. Since the 1980s, there has been an active campaign in most of the United States to ensure that children who are entering school are immunized against diphtheria. As a result, the incidence of the disease has been reduced to only a few cases a year. Immunity decreases after childhood, however, so booster

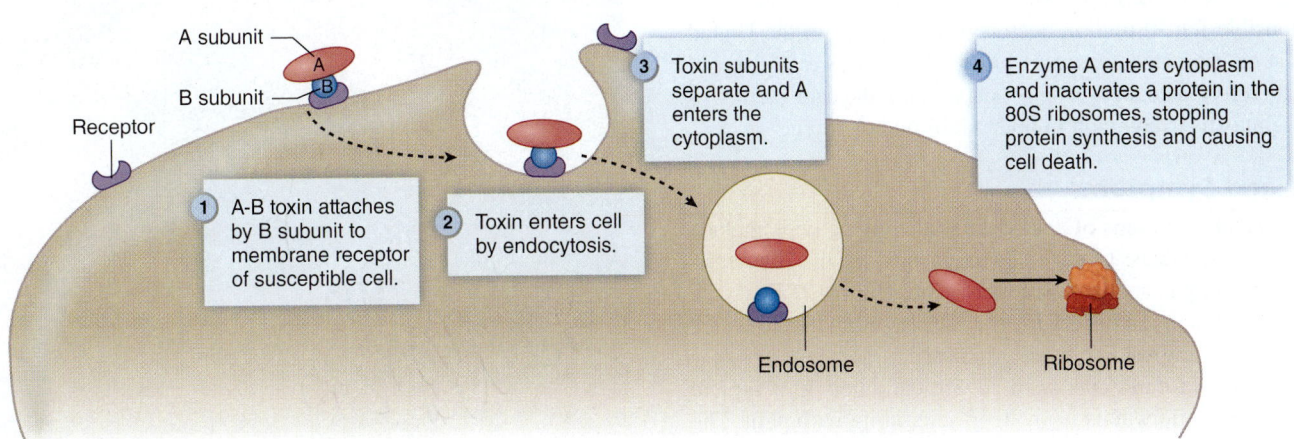

FIGURE 21.7 **Mode of Action of Diphtheria Toxin**

❓ *Why is diphtheria toxin so potent?*

TABLE 21.4 | Diphtheria

① *Corynebacterium diphtheriae* enters by inhalation.

② Infection established in nasal cavity and/or throat.

③ Toxin released, pseudomembrane forms.

④ Toxin causes paralysis, damages heart muscle, kidneys, nerves.

⑤ Membrane may come loose and obstruct breathing.

⑥ Exit from body by respiratory secretions.

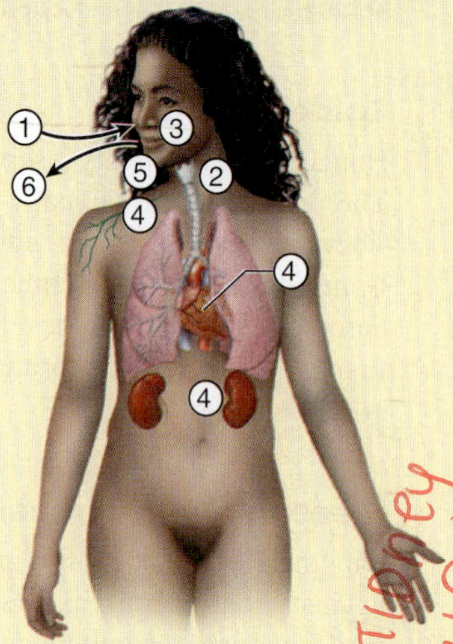

(handwritten note: toxin - ♡ & kidney suffocate)

(handwritten note: erymro & Anti w/ DTAP)

Signs and symptoms	Sore throat, fever, fatigue, and malaise; pseudomembrane forms on tonsils and throat or in nose; paralysis, heart and kidney failure
Incubation period	2 to 6 days
Causative agent	*Corynebacterium diphtheriae*, an A-B toxin–producing, non-spore-forming Gram-positive rod
Pathogenesis	Infection in upper respiratory tract; exotoxin is released and absorbed by bloodstream; toxin kills cells by interfering with protein synthesis; effect is on cells that have receptors for the toxin—mainly heart, kidney, and nerve tissue.
Epidemiology	Inhalation of infectious droplets; direct contact with patient or carrier; indirect contact with contaminated articles.
Treatment and prevention	Treatment: antitoxin; erythromycin to prevent transmission. Prevention by immunization with diphtheria toxoid; given to children at 6 weeks, 4 months, 6 months, 18 months, and 4 to 6 years in DTaP vaccine; boosters every 10 years.

injections must be given every 10 years to maintain protection. **Table 21.4** summarizes some important facts about diphtheria.

◀◀ **toxoid, p. 423**

Pinkeye, Earache, and Sinus Infections

Bacterial infections of the eye surface (conjunctivitis or "pinkeye"), middle ear (otitis media), and sinuses (sinusitis) are very common, often occur together, and often have the same causative agent. Otitis media is especially common in children, and is responsible for 30 million doctor visits per year in the United States, at an estimated cost of $1 billion. Pinkeye is easily spread and is therefore a concern when it occurs in a daycare facility or school. Sinusitis is common in both adults and children.

Signs and Symptoms

The signs and symptoms of acute bacterial conjunctivitis include increased tears, redness of the conjunctiva, swollen eyelids, sensitivity to bright light, and large amounts of pus (**figure 21.8**). Acute bacterial conjunctivitis differs from viral conjunctivitis, in which eyelid swelling and pus are usually minimal.

Otitis media manifests with severe earache—in a typical case, a young child wakes from sleep screaming with pain. The intense pain often causes vomiting. Fever is generally mild or absent.

In sinusitis, facial pain and a pressure sensation characteristically occur in the region of the involved sinus. Headache and

severe malaise also occur. A thick green nasal discharge that may contain pus and blood sometimes develops as well.

Causative Agents

Pinkeye, earache, and sinus infections are often caused by two common bacterial pathogens: (1) *Haemophilus influenzae*, a tiny

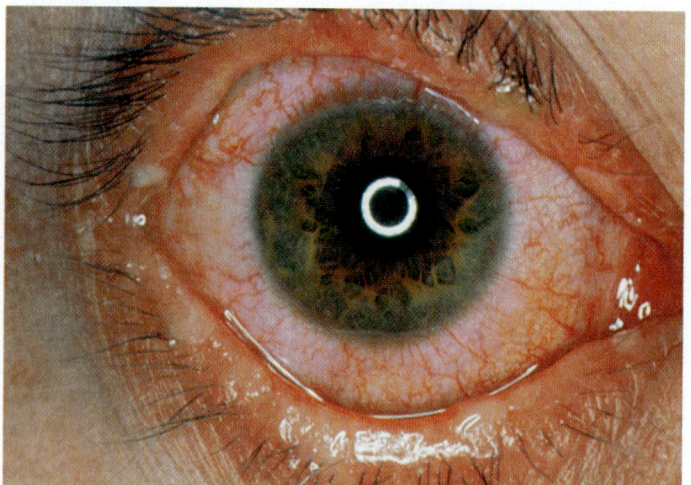

FIGURE 21.8 Bacterial Conjunctivitis Redness, swollen eyelids, and pus, characteristic of bacterial conjunctivitis.

❓ *What symptom shown here is usually minimal in viral conjunctivitis?*

Gram-negative rod; and (2) *Streptococcus pneumoniae*, the Gram-positive encapsulated diplococcus known as the pneumococcus. Strains that infect the conjunctiva have adhesins that allow them to attach firmly to the epithelium. ▶▶ *Haemophilus influenzae*, **p. 646** ▶▶ *Streptococcus pneumoniae*, **p. 497** ◀◀ **adhesins, p. 386**

Conjunctivitis can also be caused by a variety of other organisms, including *Moraxella lacunata*, enterobacteria and *Neisseria gonorrhoeae*. Although not very common, some eye infections are caused by environmental microbes that contaminate eye medications and contact lens solutions. When these infections occur they are usually serious and can lead to permanent eye damage. People who wear contact lenses should follow the instructions on use of cleansing solutions exactly; cloudy or outdated solutions and eye medications should be thrown away.

Otitis media and sinusitis can also be caused by *Mycoplasma pneumoniae*, *Streptococcus pyogenes*, *Moraxella catarrhalis*, and *Staphylococcus aureus*. About one-third of the cases are caused by respiratory viruses, explaining why some infections do not respond to antibiotics, which have no effect on viruses.

Pathogenesis

Few details are known about the pathogenesis of acute bacterial conjunctivitis. The organisms are probably inoculated directly onto the conjunctiva from airborne respiratory droplets or rubbed in from contaminated hands. Like most bacterial pathogens, they resist destruction by lysozyme. Attachment is aided in some cases by degradation of mucin, a protective component of epithelial surface mucus. Following attachment, the bacteria release proteases, collagenases, and coagulases, combined with toxins in some cases, that further damage the tissue and allow the entry of the organisms. ◀◀ **lysozyme, p. 62**

Otitis media and sinusitis are usually preceded by infection of the nasal chamber and nasopharynx that probably spreads upward through the eustachian tube (**figure 21.9**). The infection damages the ciliated cells, resulting in inflammation and swelling. Because the damaged eustachian tube cannot move secretions from the middle ear, fluid and pus collect behind the eardrum. This leads to a buildup of pressure, causing the ear to ache. The eardrum may perforate (burst), discharging blood or pus from the ear and giving immediate relief from pain. With treatment, holes in the eardrum usually heal quickly. The pressure in the middle ear can force infected material into the mastoid air cells, resulting in mastoiditis. The fluid behind the eardrum may also impair hearing ability, resulting in some young children showing a delay in speech development. Both middle ear and sinus infections sometimes spread to the brain coverings, causing meningitis. ▶▶ **meningitis, p. 643**

Epidemiology

The ecological factors involved in the appearance and spread of the eye, ear, and sinus infections caused by *H. influenzae* and *S. pneumoniae* are largely unknown; carrier rates can sometimes reach 80% in the absence of disease. The virulence of the bacteria, crowding, and presence of respiratory viruses are probably all important factors in these epidemics.

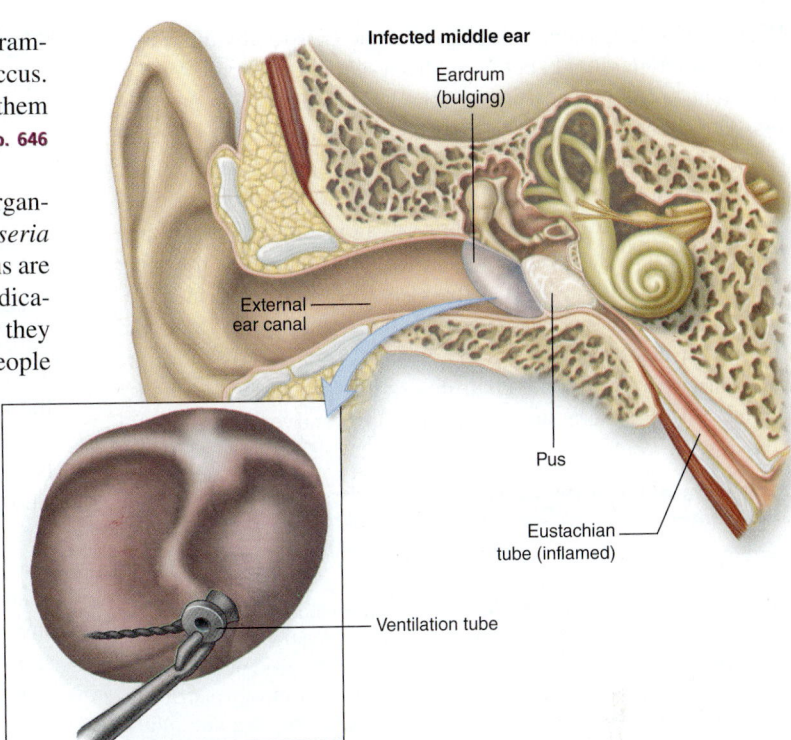

FIGURE 21.9 Otitis Media Inset shows a ventilation tube placed in the eardrum to equalize middle ear pressure in individuals with chronically malfunctioning eustachian tubes.

? *Why is otitis media painful?*

A preceding or simultaneous viral illness is common in otitis media and sinusitis; the virus probably damages the mucociliary mechanism that would normally protect against bacterial infection. Otitis media is rare in the first month of life, but it becomes very common in early childhood. Children who use pacifiers beyond the age of 2 years have an increased risk of developing otitis media. Conditions that cause inflammation of the nasal mucosa—including viral infections, nasal allergies, exposure to air pollution and cigarette smoke—play a role in some cases. Older children develop immunity to *H. influenzae*, and the bacterium rarely causes otitis media in children beyond age five. Sinusitis tends to affect adults and older children in whom the sinuses are more fully developed.

Treatment and Prevention

Bacterial conjunctivitis is effectively treated with eyedrops or ointments containing an antibacterial medication to which the infecting strain is sensitive. Antibacterial therapy with amoxicillin is generally effective against both otitis media and sinusitis; alternative medications are available for communities where antibiotic-resistant strains of *H. influenzae* and *S. pneumoniae* are common. In general, antibiotics have been used indiscriminately in treating otitis media. However, when properly used, they decrease the risk of serious complications such as mastoiditis and meningitis. Decongestants and antihistamines generally are ineffective and can be harmful because they reduce the immune response.

General preventive measures for conjunctivitis include handwashing, and protecting the eyes from contamination by avoiding rubbing or touching them, particularly with shared towels.

Bacterial conjunctivitis is highly contagious, so individuals suspected of having it are kept home from school or daycare settings for diagnosis and start of treatment. Otitis media during the "flu" season can be substantially decreased by giving influenza vaccine to infants in daycare facilities. Ampicillin or sulfasoxazole given continuously over the winter and spring are useful preventives in people who have three or more cases of otitis media within a 6-month period. Surgical removal of enlarged adenoids improves drainage from the eustachian tubes and can help prevent recurrences in certain patients. In those with chronically malfunctioning eustachian tubes and hearing loss, tiny plastic tubes are often inserted through the eardrums so that pressure can equalize (figure 21.9). There are no proven preventive measures for sinusitis.

MicroAssessment 21.2

Streptococcus pyogenes, a group A streptococcus with many virulence factors, causes a sore throat commonly known as strep throat. Untreated *S. pyogenes* infections can sometimes cause serious diseases (sequelae) in other parts of the body long after the initial infections have resolved. In diphtheria, toxin is absorbed into the bloodstream and circulates throughout the body, selectively damaging certain tissues such as heart, kidneys, and nerves. Conjunctivitis, otitis media, and sinusitis are common infections that often occur together and are caused by the same pathogens.

4. *Name two post-streptococcal sequelae.*
5. *How does diphtheria toxin kill cells?*
6. *How would adequate ventilation help to prevent the spread of streptococcal infections?* ✚

21.3 ■ Viral Infections of the Upper Respiratory System

Learning Outcomes

5. *List the strategies helpful in avoiding common colds.*
6. *Give the distinctive characteristics of adenoviral pharyngitis.*

The average person in the United States gets two to five viral upper respiratory infections each year. Although hundreds of different viruses cause these infections, the range of symptoms they produce is similar. These infections generally resolve without any treatment and rarely cause permanent damage. However, they impair respiratory tract defenses and allow for more serious secondary bacterial infections.

The Common Cold

The common cold is the most frequent infectious disease in humans, and accounts for more than half of the upper respiratory tract infections that people get every year. Colds are the leading cause of absences from school, and result in people missing 150 million workdays per year.

MicroByte

The average adult gets between 2 and 4 common colds a year, whereas children can get as many as 8 colds in this time.

Signs and Symptoms

Colds begin with malaise, followed by a scratchy or mildly sore throat, runny nose, cough, and hoarseness. The nasal secretions are initially profuse and watery, then thicken in a day or two, finally becoming cloudy and greenish. There is no fever unless secondary bacterial infection occurs. Symptoms are mostly gone within a week, but a mild cough sometimes continues for longer.

Causative Agents

Viruses that cause the common cold are often referred to simply as "cold viruses." Between 30% and 50% of colds are caused by the 100 or more types of human rhinoviruses (*rhino* means "nose," as in rhinoceros, or "horny nose") (**figure 21.10**). These are members of the picornavirus family (*pico* means "small" and *rna* is ribonucleic acid; thus, "small RNA viruses"), a group of naked viruses that have a single-stranded RNA genome. Rhinoviruses can usually be grown in cell cultures under temperature and pH conditions that mimic the upper respiratory tract (33°C and at a slightly acid pH). They are inactivated if the pH drops below 5.3, and therefore are usually destroyed in the stomach. Many other viruses and some bacterial species can also produce the signs and symptoms of the common cold.

Pathogenesis

Rhinoviruses attach to specific receptors on respiratory epithelial cells and then infect these cells. The replication cycle produces large numbers of virions and these are released to infect other cells. Ciliary motion in the infected cells stops and the cells may die and slough off. The damage causes the release of pro-inflammatory cytokines and stimulates nervous reflexes, resulting in increased nasal secretions, tissue swelling that partially or completely obstructs the airways, and sneezing. Later in the inflammatory response, blood vessels dilate, allowing plasma to ooze out and leukocytes to migrate to the infected area. Secretions from the area may then contain pus and blood. The infection is eventually stopped by the innate and adaptive responses, but it can spread into the ears, sinuses, or even the lower respiratory tract

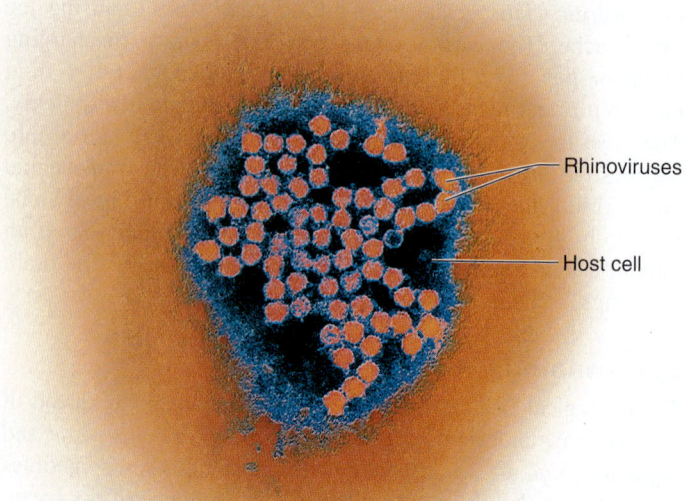

— Rhinoviruses

— Host cell

FIGURE 21.10 Rhinovirus-Infected Cell (TEM)
❓ *What kinds of cells do rhinoviruses infect?*

before this occurs. Rhinoviruses can even cause life-threatening pneumonia in individuals with AIDS.

Epidemiology

Humans are the only source of cold viruses, which are spread by close contact with an infected person. In adults the disease is usually contracted when airborne virus-containing droplets are inhaled. Transmission can also occur when secretions from infected people are accidently rubbed into the eyes or nose by contaminated hands. Viruses introduced into the eye quickly enter the nasal passage via the nasolacrimal duct. A person with severe symptoms early in the course of a cold is much more likely to transmit the virus than is someone with mild symptoms or in the late stage of the disease. This is because infected people have very high concentrations of virus in their nasal secretions and on their hands during the first 2 or 3 days of a cold. By the fourth or fifth day, virus levels are often undetectable, but low levels can be present for 2 weeks. Although a few virions are sufficient to infect the nasal mucosa, colds are actually not highly contagious if reasonable preventive measures such as handwashing are taken. In a study in which non-immune adults were exposed to infected individuals, less than half contracted colds. Young children, however, transmit cold and other respiratory viruses very effectively because they are often careless with their respiratory secretions. Experimental and epidemiological studies show that there is no relationship between exposure to low temperatures and development of colds, contrary to popular belief. Emotional stress, however, can almost double the risk of catching a cold.

Treatment and Prevention

There are no proven treatments for the common cold. Like all viruses, rhinoviruses are not affected by antibiotics or other antibacterial medications. Analgesics (painkillers) and antipyretics (fever-reducers) such as aspirin and ibuprofen can help reduce symptoms. However, experimental evidence suggests that these drugs somewhat prolong symptoms and duration of virus excretion, and delay antibody production and thus recovery.

Mechanisms to prevent the spread of rhinoviruses include handwashing (even in plain water) to physically remove the viruses, keeping hands away from the face, and avoiding crowds and crowded places like subways when respiratory diseases are prevalent. It is especially important to avoid people with colds during the first few days of their symptoms, when they are shedding high numbers of viral particles. It has not been possible to develop a vaccine because such a large number of immunologically different viruses cause colds. However, all known rhinovirus strain genomes have been sequenced, revealing new possibilities for vaccines. **Table 21.5** summarizes some facts about the common cold.

Adenoviral Respiratory Tract Infections

Adenoviruses are widespread and can cause a variety of different types of infections, depending on the viral serotype. For example, some cause a sore throat, whereas others cause eye infection. They are representative of the many viruses that cause febrile (meaning with fever) upper respiratory tract infections.

TABLE 21.5	The Common Cold
Signs and symptoms	Scratchy throat, nasal discharge, malaise, headache, cough
Incubation period	1 to 2 days
Causative agent	Mainly rhinoviruses—more than 100 types; many other viruses, some bacteria
Pathogenesis	Viruses attach to respiratory epithelium, starting infection that spreads to adjacent cells; ciliary action ceases and cells slough; mucus secretion increases, and inflammatory reaction occurs; infection stopped by interferon release, cell-mediated and humoral immunity.
Epidemiology	Inhalation of infected droplets; transfer of infectious mucus to nose or eye by contaminated fingers; children initiate many outbreaks in families because of lack of care with nasal secretions.
Treatment and prevention	No generally accepted treatment except for control of symptoms. Handwashing; avoiding people with colds and touching face.

(handwritten annotation: Attach to respiratory epimel)

Signs and Symptoms

Adenoviruses that infect the upper respiratory tract generally cause a runny nose, but unlike the common cold, fever is typically present. The throat is usually sore, with regions of gray-white pus on the pharynx and tonsils, signs and symptoms that may be confused with those of strep throat. The lymph nodes of the neck become large and tender and a mild cough is common. In some epidemics, conjunctivitis is a prominent symptom. Diarrhea may occur. Patients sometimes develop a severe cough with chest pain (symptoms that can be confused with pneumonia), whooping cough, or pleurisy. Recovery usually takes about 1 to 3 weeks.

◀◀ pleurisy, p. 486

Causative Agent

More than 50 antigenic types of adenoviruses infect humans. The viruses are non-enveloped, with double-stranded DNA. Like many other naked viruses, they can remain infectious in the environment for long periods of time and are resistant to destruction by detergents and alcohol solutions. They are easily inactivated, however, by heat (56°C), adequate levels of chlorine, and various other disinfectants.

Pathogenesis

Adenoviruses infect epithelial cells by attaching to receptors near the basement membrane. Once inside the cell, the genome is transported to the host cell nucleus, where the virus multiplies. The virus has many mechanisms for avoiding host defenses, including delaying apoptosis, blocking interferon function, and interfering with antigen presentation by MHC class I molecules. Once replication is complete, a virally encoded "death protein" is produced that causes host cell lysis. In severe infections,

TABLE 21.6	Adenoviral Pharyngitis
Signs and symptoms	Fever, very sore throat, severe cough, swollen lymph nodes of neck, pus on tonsils and throat, sometimes conjunctivitis; less frequently, pneumonia
Incubation period	5 to 10 days
Causative agent	Adenoviruses—more than 45 types
Pathogenesis	Virus multiplies in host cells; cell destruction and inflammation occur; different types produce different symptoms.
Epidemiology	Inhalation of infected droplets; possible spread from gastrointestinal tract.
Treatment and prevention	No treatment except for relief of symptoms. No vaccine. Avoided by handwashing, avoiding people with symptoms.

[handwritten in margin: fever gray/white pus]

extensive cell destruction and inflammation occur. Different types of adenoviruses affect different tissues. ◀◀ **basement membrane, p. 336** ◀◀ **apoptosis, p. 350** ◀◀ **interferon, p. 341** ◀◀ **MHC class I molecule, p. 369**

Epidemiology

Humans are the only reservoir of adenoviruses. Because these viruses can survive in the environment, healthcare personnel must take precautions to prevent transferring them from one patient to another on medical instruments. Adenoviruses commonly infect schoolchildren, usually resulting in sporadic cases, but occasionally causing outbreaks in winter and spring. Summer epidemics can occur when viruses are transmitted in inadequately chlorinated swimming pools. Adenoviral disease is spread by respiratory droplets so it can be a problem in military recruits and groups living together in crowded conditions. Asymptomatic infections are common, which fosters epidemic spread. The viruses are shed from the respiratory tract during the acute illness and continue to be eliminated in the feces for months thereafter.

Treatment and Prevention

As with colds, there is no specific treatment for adenoviral disease, and most patients recover on their own. Secondary bacterial infections may occur, however, and these require treatment with an antibacterial medication.

An attenuated vaccine administered orally was helpful for preventing acute respiratory disease in military recruits, but the vaccine program was abandoned in 1996 because of cost. Now, with more than 2,500 adenoviral illnesses monthly and some deaths, reinstitution of adenoviral vaccination of recruits is planned. Vaccines for two strains are currently under evaluation. **Table 21.6** summarizes some important facts about adenoviral pharyngitis.

MicroAssessment 21.3

Many different kinds of infectious agents cause upper respiratory tract diseases, often with the same signs and symptoms. Emotional stress significantly increases the risk of contracting the common cold, but exposure to cold temperatures probably does not. People with a cold are most likely to transmit it if symptoms are severe, and during the first few days of illness. Adenovirus infections resemble colds, but fever is present.

7. *Why are there no vaccines for the common cold?*
8. *How is an adenovirus infection treated?*
9. *People who work at polar ice stations often do not develop colds. Is this an expected observation? Why or why not?* ➕

INFECTIONS OF THE LOWER RESPIRATORY SYSTEM

21.4 ■ Bacterial Infections of the Lower Respiratory System

Learning Outcomes

7. *Compare the distinctive features of pneumococcal,* Klebsiella, *and mycoplasmal pneumonia.*
8. *Outline the pathogenesis of pertussis and tuberculosis.*

Bacterial infections of the lower respiratory system are less common than those of the upper respiratory system, mostly because they are stopped by the immune defenses at the portal of entry. However, they are generally much more serious. An earache or sore throat is unlikely to be life-threatening, but the causative organism can cause serious illnesses if it infects the lungs. Pneumonias are inflammatory diseases of the lung in which fluid fills the alveoli. They top the list of fatal community-acquired infections (meaning in the general population) in the United States, and are common healthcare-associated infections. Whooping cough, tuberculosis, and Legionnaires' disease are other serious infections of the lungs.

Pneumococcal Pneumonia

Pneumococci are an important cause of community-acquired pneumonia, accounting for about 60% of the adult pneumonia patients requiring hospitalization (see **Perspective 21.1**).

Signs and Symptoms

The typical signs and symptoms of pneumococcal pneumonia are cough, fever, chest pain, and **sputum** production (sputum is pus and other material coughed up from the lungs). These are usually preceded by a day or two of runny nose and upper respiratory congestion, ending with a sudden rise in temperature and a single, intense chill. The sputum quickly becomes pinkish or rust colored because it contains blood from the lungs. The severe chest pain is aggravated by each breath or cough, causing shallow, rapid

Terror by Mail: Inhalation Anthrax

During October and November 2001, 22 human cases of anthrax were reported in the United States, half due to inhalation of *Bacillus anthracis* spores (inhalation anthrax) and half due to skin infections (cutaneous anthrax). Five deaths occurred, all among the inhalation cases. More than 30,000 individuals potentially exposed to the spores were given antibiotic treatment as a preventive measure. Most of the anthrax cases could reasonably be linked to four mailing envelopes containing purified *B. anthracis* spores, which contaminated people, air, and surfaces during their journey through the postal system to their destinations.

Sometimes called "anthrax pneumonia," inhalation anthrax victims generally fail to show the typical symptoms of pneumonia. Inhaled *B. anthracis* spores are quickly taken up by lung phagocytes and carried to the regional lymph nodes, where they germinate, kill the phagocytes, and invade the bloodstream. Although the organisms are susceptible to various antibacterial medications, treatment must be given as soon as possible after infection, because the bacteria produce a powerful toxin that usually kills the victim within a day or two after bloodstream invasion occurs.

Bacillus anthracis endospores have long been considered for use in biological warfare. The organism is readily available—anthrax is endemic in livestock in most areas of the world including the United States and Canada.

The spores are easy to produce and remain viable for years. When anthrax is acquired by inhalation, the fatality rate is very high, yet the disease is easily confined to the attack area because it does not spread person-to-person. *Bacillus anthracis* spores were used as a weapon as early as World War I, although ineffectively, in an attack on livestock used for food. Just before World War II, Japan, the United States, USSR, Germany, and Great Britain secretly began to develop and perfect anthrax weapons, but they were not used during the war. During the "cold war" that followed, both the United States and the USSR developed massive biological warfare programs that employed many thousands of people, perfecting techniques for preparing the spores and delivering them to enemy targets. An executive order by President Nixon ended the U.S. program in 1969, and the stockpiled weapons were ordered destroyed, but other countries continued weapon development. In 1979, an accidental release of a tiny amount of *B. anthracis* spores from a biological warfare plant in the USSR caused more than 90 deaths downwind of the facility. The 1990 war with Iraq revealed their large anthrax weapon program, and several times during the 1990s a Japanese terrorist group tried ineffectively to attack Tokyo institutions with anthrax spores.

During this long history of anthrax weapon development, remarkably little was accomplished that would help defend against an attack, presumably because the best defense was considered an opponent's fear of counter-attack. As a result of the 2001 anthrax-by-mail incident, a number of questions came into focus: How do you diagnose anthrax quickly, and what is the best way to teach and organize medical practitioners to meet an anthrax attack? What tests are available to identify *B. anthracis* rapidly and reliably, and what is the best way to sample people and the environment? How can decontamination of buildings and other objects be accomplished? Is there sufficient vaccine available, can better vaccines be developed, and what is their role before or after exposure to *B. anthracis*? What is the best antimicrobial treatment, is there enough of it, is it readily accessible, and how long should it be given to exposed individuals? Can other potential treatments such as antitoxins and designer medications to block the lethal effect of *B. anthracis* toxin be developed?

The mail attack has stimulated remarkable progress toward answering these questions, especially in the area of using the latest technology for early detection, and teaching medical personnel about symptoms, signs, and X-ray findings. Antibacterial treatment proves highly effective if given soon after exposure. Approaches to combatting the toxin are under development, using designer drugs and antibodies. More information about anthrax is available at **www.mhhe.com/nester7**.

breathing; the patient becomes short of breath and develops a dusky color because of poor oxygenation. People who survive without treatment show profuse sweating and a rapid fall in temperature to normal after 7 to 10 days.

Causative Agent

Pneumococcal pneumonia is caused by *Streptococcus pneumoniae*, a Gram-positive diplococcus known as pneumococcus (**figure 21.11**). The most striking characteristic of *S. pneumoniae* is its thick polysaccharide capsule, which is responsible for the organism's virulence. There are 90 different serotypes of *S. pneumoniae*, each with different capsular antigens. Certain serotypes are more commonly associated with pneumonia and other invasive diseases (invasive means they infect normally sterile body sites). Strains of the organism that lack a capsule do not cause invasive disease. ◀◀ capsules, p. 62

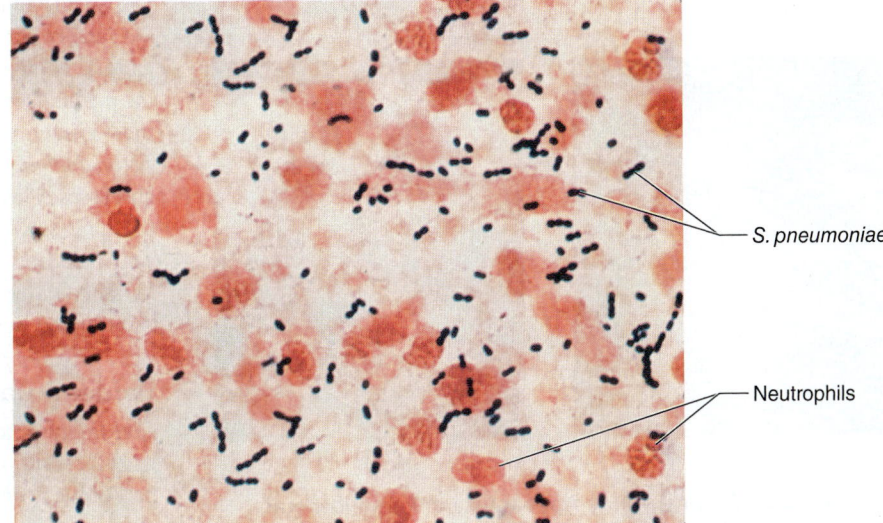

FIGURE 21.11 *Streptococcus pneumoniae* Gram stain of sputum from a person with pneumococcal pneumonia.

❓ *Are the bacteria pictured here Gram-positive or Gram-negative?*

Pathogenesis

When encapsulated pneumococci are inhaled into the alveoli, they can multiply rapidly and cause an inflammatory response. This response often affects nerve endings in the pleura, causing pain, a condition called pleurisy. The bacteria are resistant to phagocytosis because their capsule interferes with the action of the complement system component C3b, an important opsonin. Pneumococcal surface protein (PspA) also interferes with the action of C3b. The bacteria produce pneumolysin, a membrane-damaging toxin that destroys ciliated epithelium. The inflammatory response leads to an accumulation of serum and phagocytic cells in the lung alveoli, causing breathing difficulty. This fluid can be seen as abnormal shadows on chest X-ray films of patients (**figure 21.12**). Sputum coughed from the lungs increases in amount and contains pus, blood, and many pneumococci. ◀◀ **C3b, p. 344** ◀◀ **pleurisy, p. 486** ◀◀ **membrane-damaging toxin, p. 392**

Pneumococci may enter the bloodstream from the inflamed lungs causing three complications that are often fatal: sepsis (a symptomatic infection of the bloodstream); endocarditis (an infection of the heart valves); and meningitis (an infection of the membranes covering the brain and spinal cord). People who do not develop complications usually produce enough specific anticapsular antibodies within about a week to allow phagocytosis and destruction of the pneumococci. Complete recovery usually results. Most pneumococcal strains do not destroy lung tissue. ▶▶ **sepsis, p. 674** ▶▶ **meningitis, p. 643** ▶▶ **endocarditis, p. 672**

Epidemiology

Up to 30% of healthy people carry encapsulated pneumococci in their throat. These bacteria seldom reach the lung because the mucociliary escalator effectively removes them. The risk of pneumococcal pneumonia rises dramatically, however, when this defense mechanism is impaired, as it is with alcohol and narcotic use, and with viral respiratory infections such as influenza. There is also an increased risk of the disease with underlying heart or lung disease, diabetes, and cancer, and with age over 50.

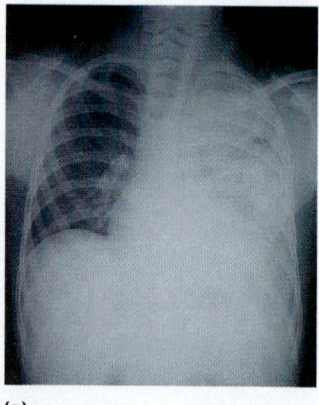
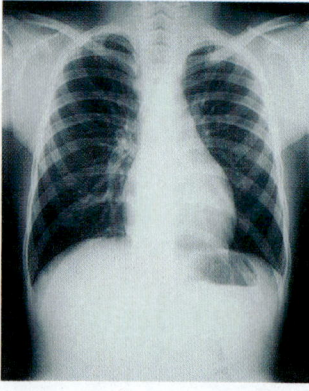

(a)   (b)

FIGURE 21.12 Chest X-Ray Appearance in Pneumococcal Pneumonia (a) Pneumonia. The left lung (right side of figure) appears white because the alveoli are filled with fluid. **(b)** Normal X-ray film after recovery.

❓ *Why does the sputum of a pneumonia patient quickly become pinkish or rust-colored?*

Treatment and Prevention

Most pneumococcal infections can be cured with penicillin or erythromycin if they are given early in the illness. However, strains of pneumococci resistant to one or more antibiotics are becoming increasingly common. ◀◀ *S. pneumoniae* **antibiotic resistance, p. 473**

A vaccine is available that gives immunity to 23 pneumococcus strains that cause over 90% of serious pneumococcal disease. A conjugate vaccine against 13 strains is available for infants and children. **Table 21.7** describes some features of pneumococcal pneumonia. ◀◀ **conjugate vaccine, p. 424**

Klebsiella Pneumonia

Enterobacteria such as *Klebsiella* sp. and other Gram-negative rods can cause pneumonia, especially if host defenses are impaired. *Klebsiella* sp. are common hospital-acquired pathogens and cause most of the deaths from healthcare-associated infections. ◀◀ **enterobacteria, p. 265** ◀◀ **healthcare-associated infections, p. 449**

Signs and Symptoms

The general signs and symptoms of *Klebsiella* pneumonia—cough, fever, and chest pain—are indistinguishable from those of pneumococcal pneumonia. *Klebsiella* pneumonia patients typically have repeated chills, however, and they produce thick, bloody gelatinous sputum that resembles red current jelly.

Causative Agent

Several species of *Klebsiella* cause pneumonia, but *Klebsiella pneumoniae* is the best known. It is a Gram-negative rod with a large capsule that produces big, noticeably mucoid colonies when grown on agar (**figure 21.13**).

Pathogenesis

Klebsiella pneumoniae is contracted through inhalation, by person-to-person contact, or from medical equipment such as ventilators. Organisms first colonize the throat and gain access to the lung via inhaled air or mucus. Specific adhesins aid colonization. The capsule is an essential virulence factor, probably functioning like the pneumococcal capsule in interfering with the action of complement system component C3b. Unlike *Streptococcus pneumoniae*, *K. pneumoniae* causes tissue death and rapid formation of lung abscesses. Therefore, even with effective antibacterial medication, the lung can be permanently damaged and the patient may die. The infection often enters the bloodstream, causing abscesses in other tissues such as the liver and brain, and septic shock. ▶▶ **septic shock, p. 674** ▶▶ **abscess, p. 550** ◀◀ **complement C3b, p. 344**

Epidemiology

Klebsiella sp. are widespread in nature. In humans, they can colonize the skin, throat. and gastrointestinal tract, where they form part of the normal microbiota in some people. Typically, individuals who contract *Klebsiella* pneumonia are very old, very young, or have a compromised immune system (such as alcoholics, or those in a hospital or other institutional setting). The strains that circulate in hospitals and nursing homes are frequently resistant to antimicrobial medications and are increasingly multidrug-resistant.

TABLE 21.7	Pneumococcal, *Klebsiella*, and Mycoplasmal Pneumonias Compared		
	Pneumococcal Pneumonia	***Klebsiella* Pneumonia**	**Mycoplasmal Pneumonia**
Signs and symptoms	Cough, fever, single shaking chill, rust-colored sputum from degraded blood, shortness of breath, chest pain	Chills, fever, cough, chest pain, and grossly bloody, mucoid sputum	Gradual onset of cough, fever, sputum production, headache, fatigue, and muscle aches
Incubation period	1 to 3 days	1 to 3 days	2 to 3 weeks
Causative agent	The pneumococcus, *Streptococcus pneumoniae*; encapsulated strains	*Klebsiella pneumoniae*, an enterobacterium	*Mycoplasma pneumoniae*; lacks cell wall
Pathogenesis	Inhalation of encapsulated pneumococci; colonization of the alveoli incites inflammatory response; plasma, blood, and inflammatory cells fill the alveoli; pain results from involvement of nerve endings.	Aspiration of colonized mucus droplets from the throat. Destruction of lung tissue and abscess formation common; infection spreads via blood to other body tissues.	Cells attach to specific receptors on the respiratory epithelium; inhibition of ciliary motion and destruction of cells follow.
Epidemiology	High carrier rates for *S. pneumoniae*. Risk of pneumonia increased with conditions such as alcoholism, narcotic use, chronic lung disease, and viral infections that impair the mucociliary escalator. Other predisposing factors are chronic heart disease, diabetes, and cancer.	Often resistant to antibiotics, and colonize individuals who are taking them. *Klebsiella* sp. and other Gram-negative rods are common causes of fatal healthcare-associated pneumonias.	Inhalation of infected droplets; mild infections common and foster spread of the disease.
Treatment and prevention	Treatment with penicillin, erythromycin, and others. Capsular vaccine available contains 23 capsular antigens; conjugate vaccine for infants.	Treated with a cephalosporin with an aminoglycoside. No vaccine available.	Treated with tetracycline or erythromycin. No vaccine available; avoiding of crowding in schools and military facilities advisable.

Treatment and Prevention

Klebsiella pneumonia is treated with antibiotics. Drug sensitivity tests must be carried out to determine which medications should be used. In seriously ill patients, immediate combination antibiotic therapy is given, often followed by a cephalosporin. Surgery may be required to drain abscesses.

Treatment of these infections is becoming increasingly challenging. *Klebsiella* species commonly produce a plasmid-encoded β-lactamase, an enzyme that makes the bacteria resistant to all β-lactam medications (such as the penicillins). Many strains now also produce an extended-spectrum lactamase (ESBL), which makes them resistant to many of the

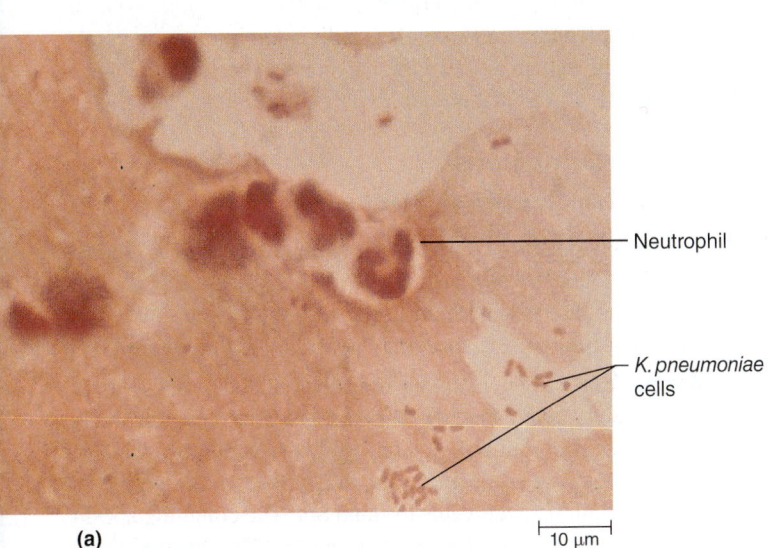

Neutrophil

K. pneumoniae cells

(a)
10 μm

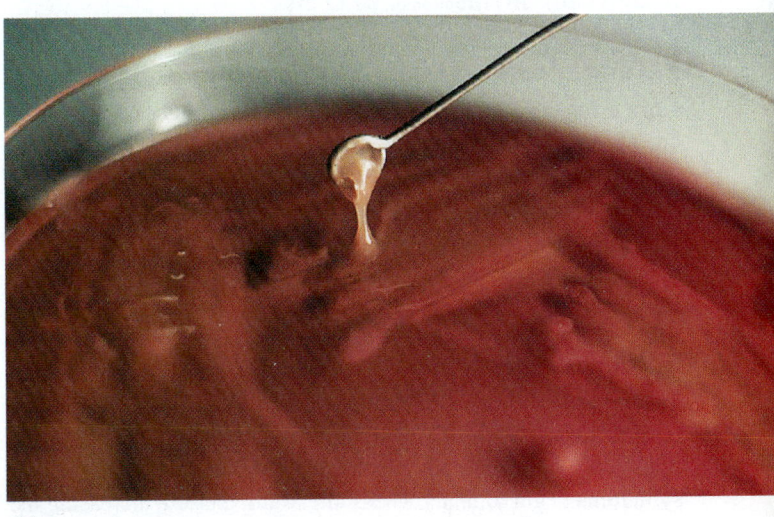

(b)

FIGURE 21.13 *Klebsiella pneumoniae* (a) In sputum from a pneumonia patient. (b) *Klebsiella* colonies. Notice the mucoid nature of the colonies.

? Why is *Klebsiella* pneumonia often fatal?

cephalosporins as well. There are very few antibiotic treatment choices available in these cases. Medications such as chloramphenicol and gentamicin are used to treat these infections with some success, but they can have serious adverse side effects. The case fatality rate even with treatment is as high as 50%, and patients tend to die more quickly than other pneumonia patients. ◄◄ **plasmid, p. 208** ◄◄ **β-lactamase, p. 464**

There are no specific preventive measures such as vaccination for *Klebsiella* pneumonia. To prevent spread between patients, healthcare workers must follow infection control measures such as wearing gloves and gowns when in a room with a *Klebsiella* patient, and washing hands. Disinfection of the environment, use of sterile respiratory equipment, and use of antimicrobial medications only when necessary help control the development of resistance of organisms in hospitals. See table 21.7 for a description of the main features of this disease.

Mycoplasmal Pneumonia ("Walking Pneumonia")

Mycoplasmal pneumonia is the leading kind of pneumonia in college students and is also common among military recruits. The disease is generally mild (as reflected by its popular name "walking pneumonia") and seldom requires hospitalization.

Signs and Symptoms

The onset of mycoplasmal pneumonia is typically gradual. The first symptoms are fever, headache, muscle pain, and fatigue. After several days, a dry cough begins, but mucoid sputum may be produced later. About 15% of cases also have otitis media.

Causative Agent

Mycoplasmal pneumonia is caused by *Mycoplasma pneumoniae*, a small, easily deformed bacterium that has no cell wall (see figure 3.34). It grows slowly and is aerobic. Like other mycoplasmas, *M. pneumoniae* colonies on agar look like fried eggs (see figure 11.26). ◄◄ **mycoplasmas, pp. 62, 276**

Pathogenesis

Only a few inhaled *M. pneumoniae* cells are necessary to start an infection. The organisms attach to specific receptors on the respiratory epithelium, interfere with ciliary action, and cause the ciliated cells to slough off (**figure 21.14**). An inflammatory response characterized by accumulation of lymphocytes and macrophages causes the walls of the bronchial tubes and alveoli to thicken.

Epidemiology

Mycoplasma pneumoniae is spread by aerosolized droplets of respiratory secretions. The organisms are shed in these secretions for a long time period, ranging from about 1 week before symptoms begin to many weeks afterward, thereby increasing the likelihood of transmission. Mycoplasmal pneumonia accounts for about one-fifth of bacterial pneumonias, and has a peak incidence in young people. Immunity after recovery is not permanent, and repeat attacks have occurred within 5 years.

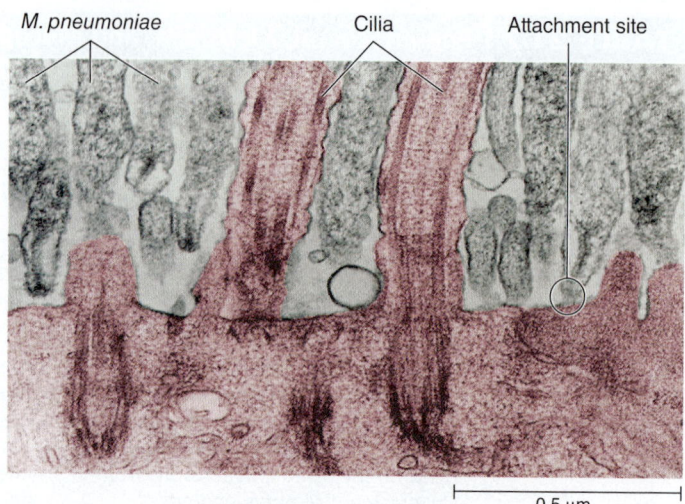

FIGURE 21.14 *Mycoplasma pneumoniae* **Infecting Respiratory Epithelium (TEM)** The tips of the *Mycoplasma* cells adjacent to the host epithelium have a distinctive appearance—they are probably specialized for attachment.

❓ *Why is penicillin ineffective in treating mycoplasmal pneumonia?*

Treatment and Prevention

Mycoplasma pneumoniae lacks a cell wall, so antibiotics that act against bacterial cell wall synthesis are not effective. Tetracycline and erythromycin shorten the illness if given early, but these medications are only bacteriostatic. ◄◄ **bacteriostatic, p. 108**

No practical preventive measures exist for mycoplasmal pneumonia, except avoiding crowding in schools and military facilities. See table 21.7 for a description of the main features of this disease.

Pertussis ("Whooping Cough")

Whooping cough is the common name for pertussis, a vaccination-preventable disease. It is still endemic in many countries, including industrialized nations such as the United States. Worldwide, the disease causes up to half a million deaths yearly.

Signs and Symptoms

Pertussis is characterized with three stages:

- **Catarrhal stage** (meaning inflammation of the mucous membranes). This typically begins with 1 to 2 weeks of signs and symptoms that resemble an upper respiratory tract infection including runny nose, sneezing, low fever, and mild cough.

- **Paroxysmal stage** (meaning repeated sudden attacks). This is characterized by frequent bursts of violent, uncontrollable coughing. The cough is dry but is severe enough to rupture small blood vessels in the eyes, and to cause the tongue to protrude and the neck veins to stand out. The coughing spasm is followed by forceful attempts to inhale. The inspired air is gasped in, causing the characteristic "whoop" of this disease. Vomiting and seizures can occur during this stage and the patient may also become cyanotic (blue from lack of O_2).

■ **Convalescent stage.** During this stage the person is no longer contagious. The coughing attacks gradually become less frequent, and the person slowly recovers.

Causative Agent

Whooping cough is caused by *Bordetella pertussis,* a tiny, encapsulated, strictly aerobic, Gram-negative rod (**figure 21.15**). These organisms are sensitive to drying and sunlight, and die quickly outside the host.

Pathogenesis

When *Bordetella pertussis* is inhaled, it attaches specifically to ciliated cells of the respiratory epithelium. Attachment is aided by two colonization factors: (1) filamentous hemagglutinin (FHA), a pilus that extends from the bacterial surface; and (2) pertussis toxin (PTx), a protein that functions as an adhesin, as well as having toxic effects (described later). Both factors are needed for colonization. The areas colonized by *B. pertussis* include the nasopharynx, trachea, bronchi, and bronchioles. The organisms grow in dense masses on the epithelial surface, but generally do not invade the cells. Instead, they release three toxins (described next) that play critical roles in the disease process.

Pertussis toxin (PTx) is an A-B exotoxin (**figure 21.16**). The B subunit attaches specifically to receptors on the host cell surface, allowing the A subunit to move through the cytoplasmic membrane of the host cell. As it does so, it activates a membrane-bound regulatory protein that controls the production of cyclic adenosine monophosphate (cAMP), leading to increased production of this molecule. High levels of cAMP interfere with cell signaling pathways, resulting in marked increase in mucus output, decreased killing ability of phagocytes, massive release of

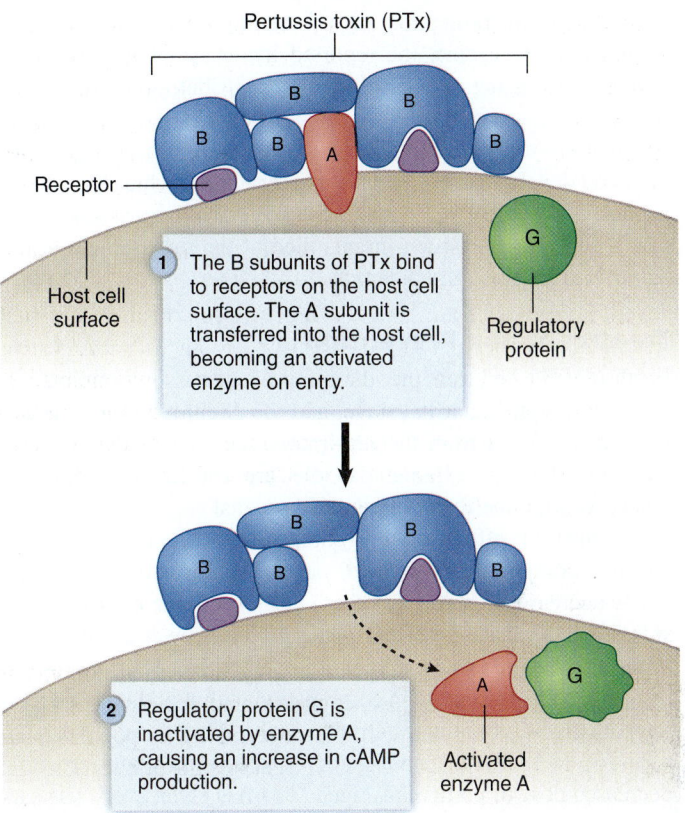

Pertussis toxin (PTx)

Receptor

Host cell surface

Regulatory protein

1 The B subunits of PTx bind to receptors on the host cell surface. The A subunit is transferred into the host cell, becoming an activated enzyme on entry.

G

2 Regulatory protein G is inactivated by enzyme A, causing an increase in cAMP production.

Activated enzyme A

FIGURE 21.16 Mode of Action of Pertussis Toxin (PTx)

? *What is the result of increased cAMP production?*

lymphocytes into the bloodstream, ineffectiveness of natural killer cells, and low blood sugar. ◀◀ **A-B toxins, p. 392** ▶▶ **cAMP, p. 587**

Adenylate cyclase is both a membrane-damaging toxin and an enzyme. This toxin reduces phagocytosis by causing lysis of accumulating leukocytes. Inside the cell, it also catalyzes the reaction that converts ATP to cAMP.

Tracheal cytotoxin is a fragment of peptidoglycan that *B. pertussis* releases during growth, and it causes host cells to release a fever-inducing cytokine (interleukin-1; IL-1). It is also toxic to ciliated epithelial cells, causing them to die and slough off, resulting in a rapid decline in ciliary action. The combination of increased mucus production and decreased ciliary action results in the cough of pertussis because only the cough reflex remains for clearing chest secretions. Some of the bronchioles become completely obstructed by mucus, resulting in small areas of collapsed lung. Spasms or partial mucus plugging in other bronchioles let air enter but not escape, causing hyperinflation. Pneumonia due to *B. pertussis* or, more commonly, secondary bacterial infection is the main cause of death. ◀◀ **interleukin-1, p. 341**

Epidemiology

Pertussis is highly contagious, spread via respiratory secretions suspended in air. Patients are most infectious during the catarrhal stage. Once the paroxysmal period begins, the numbers of expelled organisms decreases substantially. Pertussis is classically a disease of infants. However, it occurs in a milder form in older children and adults, and may be overlooked as a persistent

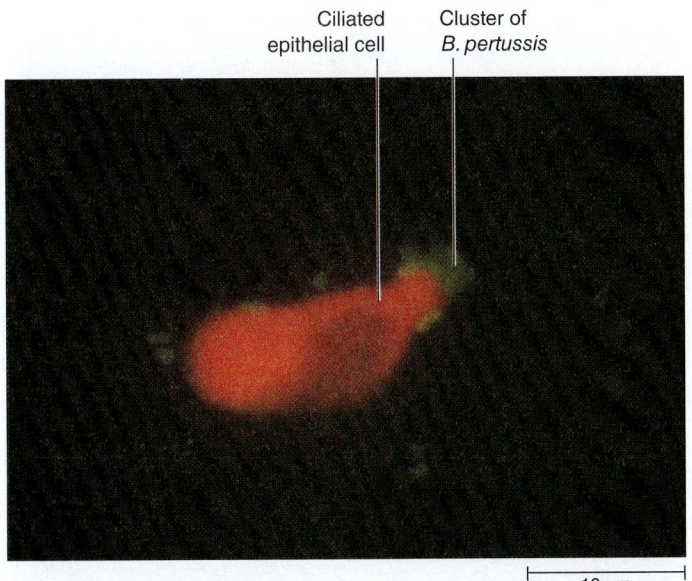

Ciliated epithelial cell   Cluster of *B. pertussis*

10 μm

FIGURE 21.15 *Bordetella pertussis* Fluorescent antibody stain of respiratory secretions from a person with whooping cough.

? *Why are the* B. pertussis *cells clustered at one end of the ciliated cell?*

cold, asthma, or bronchitis, thus fostering transmission. Despite the successful vaccine, the reported incidence of the disease is steadily increasing. Possible factors contributing to this trend include better surveillance and diagnostic techniques, suboptimal vaccination (especially in developing nations), decreasing immunity, and development of *B. pertussis* strains for which the current vaccine is not effective. In 2010, an outbreak of pertussis occurred in California. At least ten infants died of the disease, and health authorities reported an epidemic.

Treatment and Prevention

Erythromycin reduces the duration of pertussis symptoms if given during the catarrhal stage, and the antibiotic usually eliminates *B. pertussis* from the respiratory secretions. Azithromycin and trimethoprim-sulfamethoxazole are effective alternatives. Antibiotics are ineffective in the paroxysmal stage of pertussis.

Pertussis is effectively prevented with a vaccine. The original vaccine, composed of whole *B. pertussis* cells, produced a dramatic decrease in pertussis cases, but sometimes caused severe side effects. It has now been replaced with a newer acellular pertussis vaccine (aP), that includes only part of the bacterium instead of whole cells. It is given in combination with diphtheria and tetanus toxoids—a grouping referred to as the DTaP vaccine—at 2, 4, 6, and 15 to 18 months, and again at 4 years old. A booster with a decreased dose of pertussis antigen (Tdap) is given every 10 years after that. The main features of pertussis are shown in **table 21.8**.

◀◀ acellular vaccines, p. 423

Tuberculosis ("TB")

Tuberculosis (TB) was once a very common disease but the number of cases gradually declined in industrialized countries as living standards improved. In 1985, however, the incidence of TB began to rise again, the trend associated with the expanding AIDS

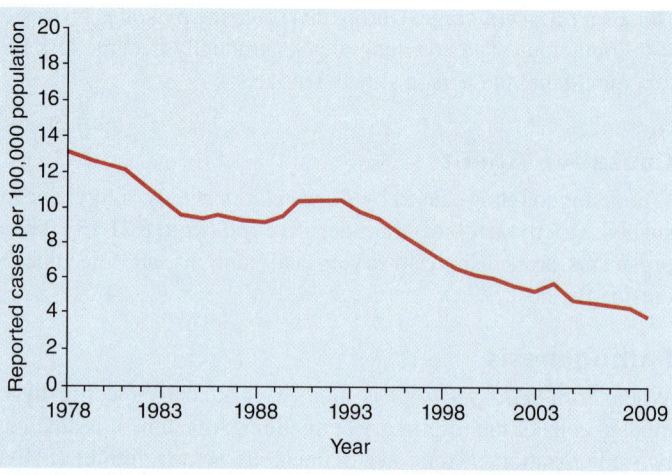

FIGURE 21.17 **Incidence of Tuberculosis, United States, 1978–2009**

❓ *What caused the rise in TB incidence between 1988 and 1994?*

epidemic and increasing prevalence of drug-resistant strains of the causative agent (**figure 21.17**). To respond to the problem, the CDC developed a Strategic Plan for the Elimination of Tuberculosis in the United States, published in 1989. The plan depends largely on increased efforts in identifying and treating cases among the high-risk groups, particularly poor people, people with AIDS, prisoners, and immigrants from countries with high rates of TB. By 1993, the incidence began to decrease again and by 2009, only 3.8 cases per 100,000 population were recorded.

MicroByte

It is estimated that one-third of the global population is infected with *Mycobacterium tuberculosis* and almost 2 million die of tuberculosis annually.

TABLE 21.8	Pertussis
Signs and symptoms	Runny nose followed after a number of days by spasms of violent coughing; vomiting and possible convulsions
Incubation period	7 to 21 days
Causative agent	*Bordetella pertussis*, a tiny Gram-negative rod
Pathogenesis	Colonization of the surfaces of the upper respiratory tract and tracheobronchial system; ciliary action slowed; toxins released by *B. pertussis* cause death of epithelial cells and increased cAMP; fever, excessive mucus output, and a rise in the number of lymphocytes in the bloodstream result.
Epidemiology	Inhalation of infected droplets; older children and adults have mild symptoms.
Treatment and prevention	Erythromycin, somewhat effective if given before coughing spasms start, eliminates *B. pertussis*. Acellular vaccine (DTaP), for immunization of infants and children.

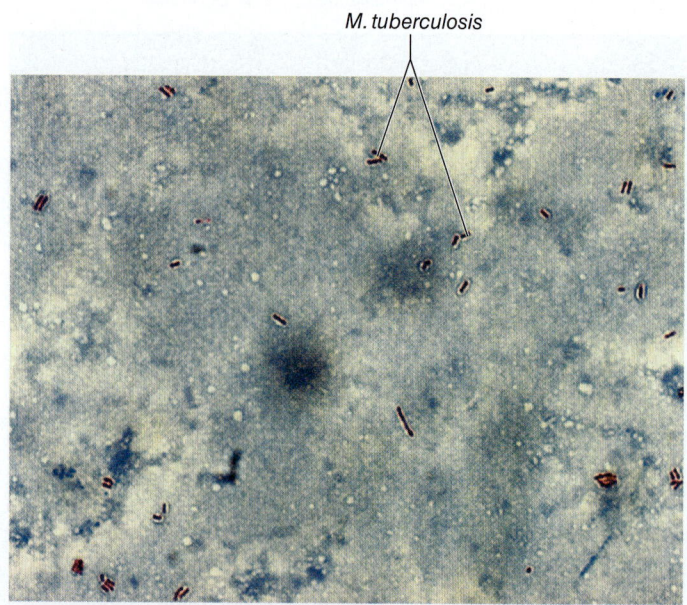

FIGURE 21.18 **Acid-fast Stain of *Mycobacterium tuberculosis***

❓ *Why is* M. tuberculosis *so resistant to drying?*

Signs and Symptoms

The initial infection with *Mycobacterium tuberculosis* typically results in an asymptomatic lung infection. The immune response generally controls this primary infection but is not able to eliminate it entirely—the person is left healthy but with a latent infection, known as **latent tuberculosis infection (LTBI).** Much later in life, the person may develop **active tuberculosis disease (ATBD),** a chronic illness characterized by slight fever, progressive weight loss, night sweating, and persistent cough, often producing blood-streaked sputum. Some people, especially children or those with compromised immune systems, may develop ATBD on primary infection. ◀◀ **latent infection, p. 383**

Causative Agent

Tuberculosis is caused by *Mycobacterium tuberculosis,* commonly called the tubercle bacillus. The organism is a slender, acid-fast, rod-shaped bacterium (**figure 21.18**). It is a strict aerobe that grows very slowly, with a generation time of over 16 hours. This slow growth makes it difficult to diagnose TB quickly. The organism has an unusual cell wall that contains a large amount of complex glycolipids called mycolic acids—these make it unusually resistant to drying, disinfectants, and strong acids and alkali, although it is easily killed by pasteurization. The mycolic acids are also largely responsible for its acid-fast staining. *M. tuberculosis* primarily infects the lung but can also cause disease in many other tissues, including bones, kidneys, joints, and the central nervous system. ◀◀ **pasteurization, p. 112** ◀◀ **lipids, p. 33** ◀◀ **acid-fast staining, p. 48**

Pathogenesis

When airborne *M. tuberculosis* cells from a person who has active TB disease are inhaled, they may enter the lungs. Alveolar macrophages quickly engulf the bacteria but are unable to destroy them because the mycolic acids in the bacterial cell wall prevent fusion of the phagosome with lysosomes. The bacteria leave the phagosome

and multiply within the cytoplasm of the macrophages (**figure 21.19**).

The multiplying bacilli trigger an inflammatory response, recruiting more macrophages to the site, thereby providing *M. tuberculosis* with additional host cells in which to multiply. Some of the macrophages fuse together to form giant multinucleated cells. Others are induced by the bacteria to accumulate large numbers of oil droplets, becoming foamy macrophages. The lipids in foamy macrophages are thought to help the bacteria

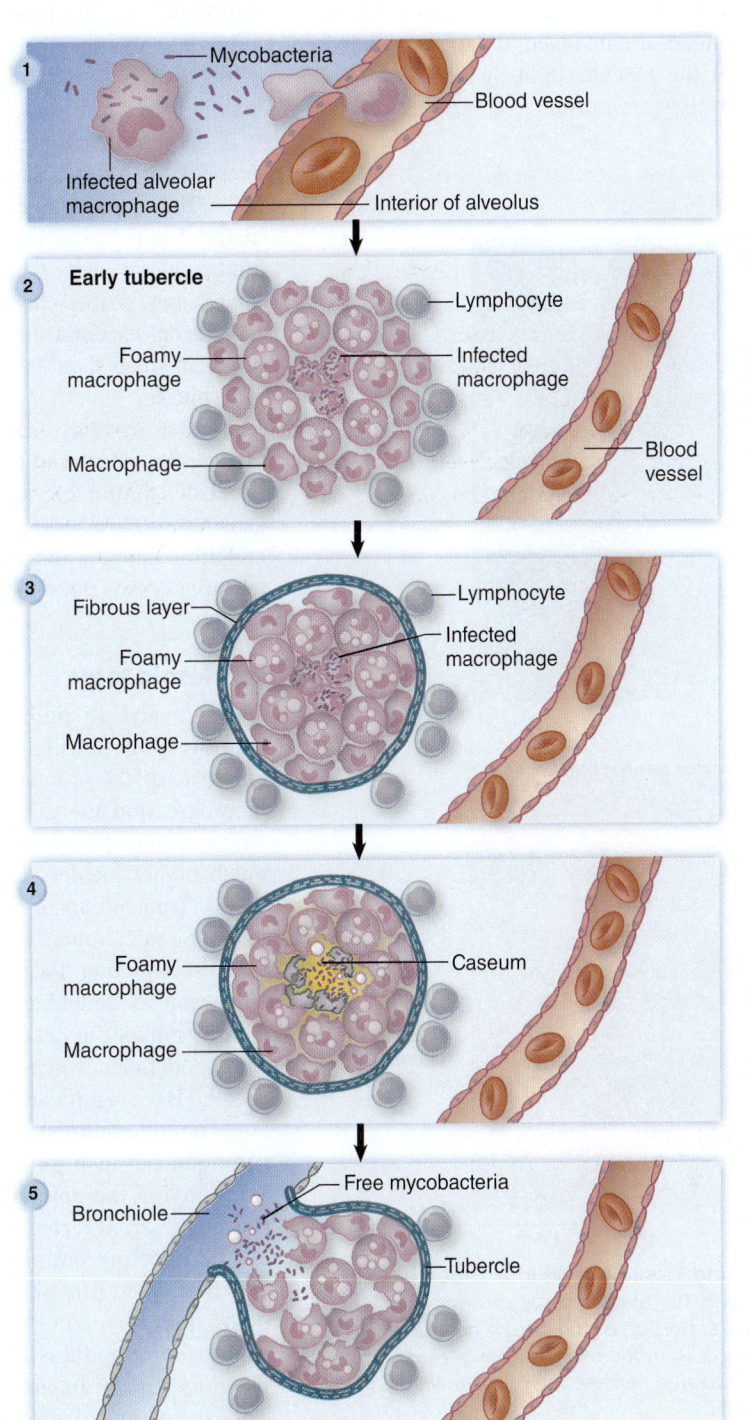

FIGURE 21.19 Pathogenesis of Tuberculosis

❓ *Why is prolonged treatment necessary for tuberculosis?*

survive within the cells. Lymphocytes collect around the macrophages, walling off the infected area from the surrounding tissue. This localized collection of inflammatory cells—a granuloma—is the body's characteristic response to microorganisms and other foreign substances that resist destruction and removal by phagocytosis. The granulomas of tuberculosis are called **tubercles** (**figure 21.20**; see also figure 21.19). Within the granuloma, effector helper T cells release cytokines that activate macrophages to destroy the bacteria infecting them. At this time, a fibrous layer forms around the macrophages, keeping the lymphocytes outside of the tubercle. This fibrous tissue can be seen on X rays as Ghon foci. If the adjacent lymph nodes are involved, the focus is called a Ghon complex. Some of the mycobacteria in the Ghon foci and complexes survive, but they are prevented from multiplying

by conditions in the tubercle, including low pH and low available O_2. The bacteria remain in this state for many years, causing a latent TB infection (LTBI). Individuals with LTBI are asymptomatic and non-infectious. In many cases, the infection resolves.

◀◀ granuloma, p. 348

Active TB disease results if the inflammatory response cannot contain or destroy the mycobacteria. This can occur during primary infection but can also happen in a person with LTBI—the infection reactivates (reactivation TB) if the person's immunity becomes impaired by stress, advanced age, or disease such as AIDS. Within the tubercle, macrophages containing mycobacteria die, releasing bacteria, enzymes, and cytokines. An area of necrosis is formed in the center of the tubercle—this has the texture of soft white cheese and is referred to as caseous necrosis. It is thought that foamy macrophages play an important role in necrosis formation and that the caseum contains lipids from these cells. The tubercle then ruptures, releasing the bacteria and dead material into the airways. This causes a large lung defect called a tuberculous cavity that spreads the bacteria to other parts of the lung. Lung cavities characteristically persist, slowly enlarging for months or years and shedding bacterial cells into the bronchi. The organisms can then be transmitted to other people by coughing and spitting.

Other systems that can be affected by reactivation TB include the pleura and pericardium, lymph nodes, kidneys, bones (typically the lumbar and thoracic vertebrae which break down, causing Pott's disease), joints, and central nervous system. In a condition called miliary tuberculosis, tiny tubercles are found in multiple organs throughout the body.

Epidemiology

An estimated 15 million Americans have LTBI, but the vast majority of these will never develop active TB disease (ATBD)—only 5% to 10% of latent infections will later reactivate, resulting in progression to ATBD. LTBI rates are highest among non-whites and elderly poor people. Foreign-born U.S. residents have much higher incidence of LTBI than those born in the United States. Transmission of *M. tuberculosis* occurs almost entirely by the respiratory route; 10 or fewer inhaled organisms are enough to cause infection. Factors important in transmission include the frequency of coughing, the adequacy of ventilation (transmission is unlikely to occur outdoors), and the degree of crowding. Immunodeficiency increases activation of *M. tuberculosis* in those with LTBI, a significant problem in AIDS patients.

The tuberculin skin test (TST), also known as the Mantoux (pronounced man-too) test, is an extremely important tool for studying the epidemiology of the disease and in detecting those who are infected with *M. tuberculosis*. The test is carried out by injecting into the skin a small amount of a sterile fluid called purified protein derivative (PPD), derived from cultures of *M. tuberculosis*. People who are infected with the bacterium develop redness and a firm swelling at the injection site, reaching a peak intensity after 48 to 72 hours (**figure 21.21**). This reaction is due to the accumulation of macrophages and T lymphocytes at the injection site, a manifestation of delayed

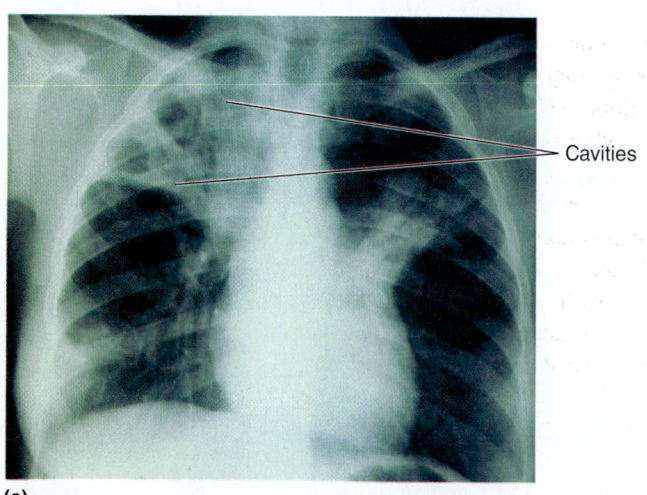

(a)

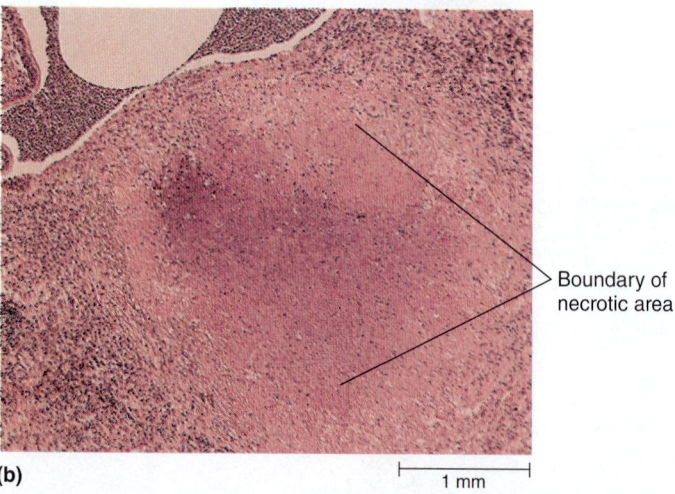

(b) ⊢————1 mm————⊣

FIGURE 21.20 Stained Lung Tissue Showing a Tubercle
(a) Chest X ray of a person with TB. (b) Lung tissue showing tubercle. The dark dots around the outer portion of the picture are nuclei of lung tissue and inflammatory cells. In the center of the photograph, most of the nuclei have disappeared because the cells are dead and the tissue has begun to liquefy.

▤ *What is a tubercle?*

Cavities — (labels in figure a)

Boundary of necrotic area — (labels in figure b)

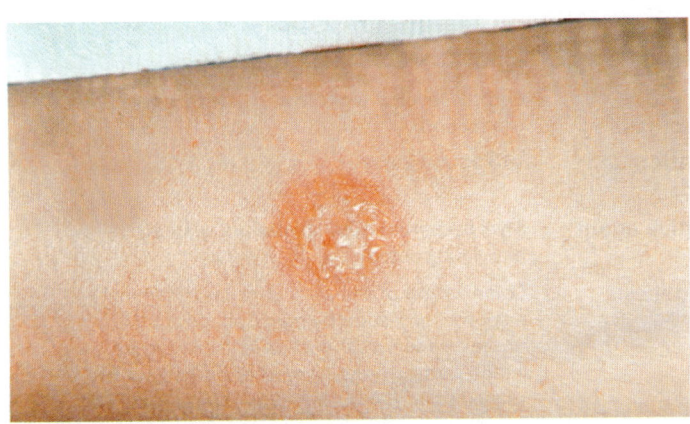

FIGURE 21.21 Tuberculin Skin Test This positive test is caused by delayed hypersensitivity to *Myobacterium tuberculosis* antigens injected into the skin.

❓ *What does a positive tuberculin skin test indicate?*

hypersensitivity to the tubercle bacillus. A positive reaction to the test (a skin reaction more than 10 mm in diameter) does not mean that the person has ATBD, only that the person has been infected by *M. tuberculosis* at some time in the past; they may have either LTBI or ATBD. They may also have received a TB immunization. ◀◀ **delayed hypersensitivity, p. 409**

Beside the TST, blood tests are available for diagnosing *M. tuberculosis* infection. Interferon gamma–release assays (IGRAs) detect interferon gamma released in response to *M. tuberculosis*. IGRAs generally give results similar to TST but more quickly and with greater specificity. Nucleic acid amplification methods are fast, sensitive, and accurate, and are used on clinical specimens. ◀◀ **IFN-gamma, p. 341** ◀◀ **nucleic acid amplification tests, p. 246**

Treatment and Prevention

Treatment of tuberculosis requires multiple medications for many months. A combination of drugs must be given because people with ATBD have high numbers of *M. tuberculosis* cells in the body, so there is a strong likelihood that some of the cells will have acquired resistance to a given drug by spontaneous mutation. A cell is much less likely to develop resistance to two drugs given simultaneously. Treatment must be continued for months to cure the disease because of the long generation time of *M. tuberculosis* and its resistance to destruction by body defenses. ◀◀ **antibiotic resistance, p. 471** ◀◀ **generation time, p. 83**

The most effective and least toxic first-line medications for treating ATBD include rifampin (RIF), isoniazid (INH), pyrazinamide (PZA), and ethambutol (EMB), which are bactericidal against actively growing organisms. Some kill metabolically inactive intracellular organisms as well. In the initial treatment phase of ATBD, all four first-line drugs are given for 2 months. After that, INH and RIF are given in combination for another 4 to 7 months, depending on the number of doses scheduled each week.

Strains resistant to one or more first-line medications often evolve when people fail to comply with the complex treatment regime. What commonly happens is that once a person feels better and the symptoms of active TB disease disappear, he or she becomes careless about continuing to take the prescribed medications. To combat this problem, **DOTS (directly observed therapy short-course)** may be used—healthcare workers that supply the medications can watch the patients swallow the prescribed tablets.

Despite the success of DOTS, the problem of drug-resistant *M. tuberculosis* strains has nevertheless reached alarming proportions in some areas. During the 1990s, **multidrug-resistant TB (MDR-TB)** became an increasing problem. These strains resist rifampin and isoniazid—the two most effective first-line drugs available—and therefore must be treated with less effective, more toxic (causing liver damage), and more expensive second-line medications. By the end of the decade, **extensively drug-resistant TB (XDR-TB)** strains that resist both first-line and many of the second-line drugs evolved. These strains now threaten tuberculosis control efforts around the world. Fortunately, promising new anti-tuberculosis medications representing at least five different chemical families are being developed. One entirely new and highly active medication—a diarylquinoline (DARQ)—is among those under evaluation. ◀◀ **MDR-TB, p. 473** ◀◀ **XDR-TB, p. 473**

In the United States, TB prevention involves identifying unsuspected cases using skin tests and lung X rays. Individuals with active disease are then treated, thereby interrupting the spread of the causative agent. People who have LTBI are also treated, reducing the risk of developing ATBD later in life. The preferred treatment for LTBI is isoniazid for 9 months. Treatment is particularly recommended for high-risk individuals such as those with underlying conditions (including those with HIV infection, drug abuse, and diabetes), the very young, the elderly, those recently exposed to ATBD, those recently converted to a positive tuberculin skin test, and those entering the country from regions with high prevalence of the disease. The National Tuberculosis Indicators Project (NTIP) collects data on individual TB cases to measure the performance of prevention and control measures in the United States. The CDC has also initiated the Tuberculosis Genotyping Information Management System (TB GIMS) to disseminate data on suspected TB outbreaks so that they can be managed more efficiently.

Prevention and control of TB is a global challenge and many countries lack the resources to track and treat both ATBD and LTBI. Vaccination against TB has been widely used in many countries. The vaccine used, BCG (Bacille Calmette-Guérin), is a live attenuated vaccine derived from *M. bovis*, a cattle-infecting species that has little virulence in humans. Although this vaccine prevents childhood TB disease, it appears ineffective in preventing LTBI, which can later reactivate. Use of the vaccine is discouraged in the United States, because people who receive it usually develop a positive tuberculin skin test. By causing a positive test, this vaccination eliminates an important way of diagnosing TB early in the disease when it can most easily be treated. BCG is also not safe to use in severely immunocompromised patients. Several new genetically engineered

TABLE 21.9 | Tuberculosis

1. Airborne *Mycobacterium tuberculosis* cells are inhaled and lodge in the lungs.

2. The bacteria are phagocytized by lung macrophages and multiply within them, protected by lipid-containing cell walls and other mechanisms.

3. Infected macrophages are carried to various parts of the body such as the kidneys, brain, lungs, and lymph nodes; release of *M. tuberculosis* occurs.

4. Delayed hypersensitivity develops; wherever infected *M. tuberculosis* has lodged, an intense inflammatory reaction develops.

5. The bacteria are surrounded by macrophages and lymphocytes; growth of the bacteria ceases.

6. Intense inflammatory reaction and release of enzymes can cause caseation necrosis and cavity formation.

7. With uncontrolled or reactive infection, *M. tuberculosis* exits the body through the mouth with coughing.

Signs and symptoms	Chronic fever, weight loss, cough, sputum production
Incubation period	2 to 10 weeks
Causative agent	*Mycobacterium tuberculosis*; unusual cell wall with high lipid content
Pathogenesis	Colonization of the alveoli incites inflammatory response; ingestion by macrophages follows; organisms survive ingestion and are carried to lymph nodes, lungs, and other body tissues; tubercle bacilli multiply; granulomas form.
Epidemiology	Inhalation of airborne organisms; latent infections can reactivate.
Treatment and prevention	Treatment: two or more antitubercular medications given simultaneously long term, such as isoniazid (INH) and rifampin; DOTS; BCG vaccination preventive but not used in the United States; tuberculin (Mantoux) skin test for detection of infection, allows early therapy of cases; treatment of all high-risk cases including young people with positive tests and individuals whose skin test converts from negative to positive.

[handwritten annotations: weightloss cold, chills fever, sputum isonaized rif; weightloss cold, chills, fever sputum]

vaccines are being developed, many of which are currently being tested in TB-endemic regions. The main features of tuberculosis are shown in **table 21.9**.

Legionnaires' Disease

Legionnaires' disease was unknown until 1976, when a number of people attending an American Legion Convention in Philadelphia developed a mysterious pneumonia that was fatal in many cases. Months of scientific investigation eventually paid off when the cause was discovered to be a previously unknown bacterium commonly present in the natural environment.

Signs and Symptoms

Legionnaires' disease typically begins with headache, muscle aches, high fever, confusion, and shaking chills. A dry cough develops that later produces small amounts of sputum, sometimes containing blood. Pleurisy can also occur. About one-fourth of the cases also have some digestive tract symptoms such as diarrhea, abdominal pain, and vomiting. Shortness of breath is common, and O_2 therapy is often needed. Recovery is slow, and weakness and fatigue last for weeks.

Causative Agent

Legionnaires' disease is caused by *Legionella pneumophila*, a Gram-negative rod that requires a special medium for laboratory

culture (**figure 21.22**). Its fastidious growth requirements and the fact that the cells stain poorly in tissue partly explain why this organism was undetected for so long. *L. pneumophila* is a facultative intracellular parasite and survives well in freshwater

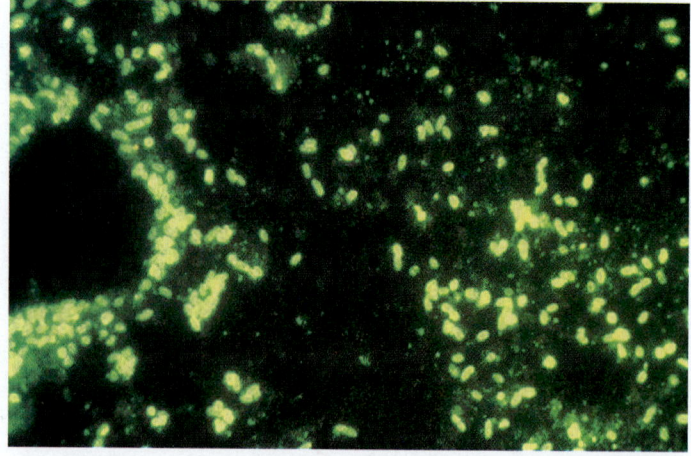

FIGURE 21.22 *Legionella pneumophila* in Lung Tissue **Stained with Fluorescent Antibody** The bacterium does not stain with most of the usual microbiological stains in tissue or sputum.

? *How is* L. pneumophila *acquired?*

amebas such as *Acanthamoeba*. These protozoa form cysts during adverse environmental conditions, allowing the bacteria within them to survive. *Legionella* also persists in biofilms—if the biofilm is disturbed, huge numbers of *Legionella* are released into the water.

Pathogenesis

Legionella pneumophila is acquired by breathing aerosolized water contaminated with the organism. Healthy people are quite resistant to infection, but smokers and those with impaired host defenses are susceptible.

The organisms lodge in and near the alveoli of the lung. Rather than avoiding phagocytosis by alveolar macrophages, they promote it. One of their surface proteins, macrophage invasion potentiator (Mip), aids entry into the macrophages. The bacterial cells also bind complement component C3b, an opsonin. Once ingested by the macrophages, the bacteria survive by preventing phagosome-lysosome fusion. They also manipulate other events within the phagocyte, creating an environment in which they can multiply. The host cell eventually dies, releasing bacterial cells that can then infect other tissues. Necrosis (tissue death) of alveolar cells and an inflammatory response result, causing multiple small abscesses, pneumonia, and pleurisy. Bacteremia is often present. Fatal respiratory failure (meaning the lungs can no longer adequately oxygenate the blood or expel CO_2) occurs in about 15% of hospitalized cases. Curiously, *L. pneumophila* infections remain confined to the lung in most cases.

Epidemiology

Legionella pneumophila is widespread in warm natural waters containing other microorganisms such as amebas, in which the bacteria live and multiply. The organism also survives well in the water systems of buildings, particularly in hot water systems, where chlorine levels are generally low. Increased chlorine levels sometimes fail to decontaminate water systems, probably because the *L. pneumophila* cells are protected inside amebas. Legionnaires' disease cases have originated from contaminated aerosols from large central air-conditioning systems, nebulizers, sprays to freshen produce, and from showers and water faucets. People have even contracted this disease from car windshield sprays that do not contain detergent. Direct person-to-person spread, however, does not occur.

Treatment and Prevention

Legionnaires' disease is treated with high doses of erythromycin, sometimes concurrently with rifampin. *L. pneumophila* produces a β-lactamase, which makes it resistant to many penicillins and some cephalosporins. Also, since the bacteria multiply inside the alveolar macrophages, the medication used must be able to accumulate within these cells, which β-lactam drugs do poorly.

◀◀ β-lactamase, p. 464

Most efforts at control of Legionnaires' disease have focused on designing equipment to minimize the risk of infectious aerosols and on disinfecting procedures. Environmental surveillance is not practical because of the lack of a simple method for detecting virulent strains. The main features of Legionnaires' disease are given in **table 21.10.**

confusion, muscle ache, fever

TABLE 21.10	Legionnaires' Disease
Signs and symptoms	Muscle aches, headache, fever, cough, shortness of breath, chest and abdominal pain, diarrhea
Incubation period	2 to 10 days
Causative agent	*Legionella pneumophila,* a Gram-negative bacterium that stains poorly in clinical specimens
Pathogenesis	Organism multiplies within phagocytes; released with death of the cell; necrosis of cells lining the alveoli; inflammation and formation of microabscesses
Epidemiology	Originates mainly from warm water contaminated with other microorganisms, such as found in air-conditioning systems.
Treatment and prevention	Treatment: erythromycin and rifampin. Avoidance of contaminated water aerosols; regular cleaning and disinfection of humidifying devices.

contam H_2O

MicroAssessment 21.4

Pneumococcal pneumonia is typically acquired in the community and leads the list of pneumonias in adults requiring hospitalization. Pneumonia due to *Klebsiella* sp. and other Gram-negative rods is mainly hospital-acquired and leads the causes of death from healthcare-associated infections. Mycoplasmal pneumonia usually does not require hospitalization. Whooping cough (pertussis) is mainly a threat to infants but is commonly spread by adults; childhood immunization against the disease protects them, but immunity often does not persist to adulthood. Tuberculosis is a chronic disease spread from one person to another by aerosol drops. Most *Mycobacterium tuberculosis* infections become latent, posing the risk of reactivation throughout life. *Legionella pneumophila,* the bacterium that causes Legionnaires' disease, originates from water containing other microorganisms, where it can grow within protozoa. In the human lung, the bacterium readily multiplies within macrophages.

10. *What structural feature of the* S. pneumoniae *cell is responsible for its virulence?*

11. *Outline the pathogenesis of tuberculosis.*

12. *Why is pertussis toxin not eliminated by the mucociliary escalator?* ✚

21.5 ■ Viral Infections of the Lower Respiratory System

Learning Outcomes

9. *Describe antigenic drift and antigenic shift and discuss how they affect the epidemiology of influenza.*

10. *Compare the distinctive characteristics of respiratory syncytial virus infection and hantavirus pulmonary syndrome.*

DNA viruses such as the adenoviruses sometimes cause serious pneumonias, but RNA viruses are of greater overall importance because of the large number of people they infect and their potential for serious outcomes. The following section covers some diseases caused by RNA viruses.

Influenza ("Flu")

Influenza is a good example of the constantly changing interaction between people and infectious agents. Antigenic changes in the influenza viruses are responsible for serious annual epidemics of the disease—almost 20% of the world population gets infected with flu virus every year. There are three major influenza virus types, named A, B, and C, based on differences in their protein coat. Type A, considered here, causes the most serious disease, and its epidemics are widespread. Outbreaks due to type B strains occur each year, but they are less extensive and the disease is not as severe. Type C strains are of relatively little importance. Influenza viruses do not cause "stomach flu."

Signs and Symptoms

After a short incubation period averaging 2 days, influenza typically begins with headache, fever, sore throat, and muscle pain, peaking in 6 to 12 hours. A dry cough develops and worsens over a few days. These acute symptoms usually go away within a week, leaving the patient with a lingering cough, fatigue, and generalized weakness for additional days or weeks. Infection by influenza viruses is occasionally associated with Reye's syndrome, a complication linked to certain other viral infections as well. ▸▸ Reye's syndrome, p. 535

Causative Agents

Influenza A virus belongs to the orthomyxovirus family. It has eight segments of single-stranded RNA, which are enclosed in a protein capsid. The capsid is surrounded by a lipid envelope derived from the host cell membrane (**figure 21.23**). Embedded in

- Lipid envelope
- Nucleoprotein
- Hemagglutinin (HA)
- Neuraminidase (NA)
- RNA segments
- Matrix protein

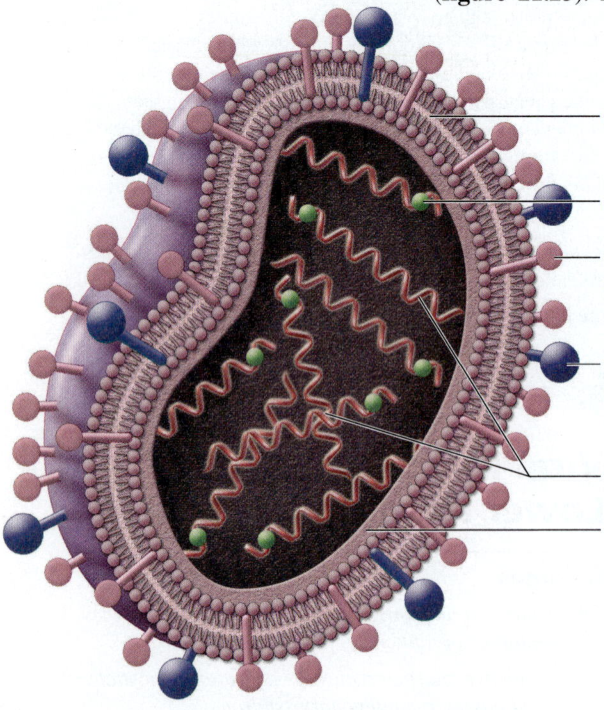

FIGURE 21.23 Structure of Influenza Virus

❓ *What is the role of the HA and NA spikes?*

the envelope are two kinds of glycoprotein spikes—hemagglutinin antigen (HA) and neuraminidase antigen (NA)—which have a role in viral pathogenesis. The HA spikes allow the virus to recognize and attach to specific receptors on ciliated host epithelial cells, initiating infection. Viruses with HA spikes cause red blood cells to stick together (hemagglutination), a useful characteristic for virus identification. NA is an enzyme that plays a critical role in the release of newly formed virions from host cells. As new virions are made, they bud out of the host cell but remain bound to surface receptors in the host membrane. NA destroys these receptors, allowing the virions to leave the infected host cell, and aiding the spread of the virus to uninfected host cells.

There are different subtypes of influenza A viruses, characterized by antigenically distinct HA and NA spikes. The subtypes are given numbers according to these variations—H1, H2, N1, N2 and so on. For example, the "avian flu" epidemic of 1997 was caused by influenza virus H5N1, whereas the "swine flu" epidemic of 2009 was caused by influenza virus H1N1. There are 16 HA and 9 NA subtypes, but only H1, 2, and 3, and N1 and 2 spread among humans (see **Perspective 21.2**).

Pathogenesis

Individuals acquire influenza by inhaling aerosolized respiratory secretions from a person who has the disease. They can also contract the virus from fomites such as doorknobs, banknotes, and household items, accidentally transferring it into the eyes or nasal passages by touch. The virions attach by their HA spikes to specific receptors on ciliated respiratory epithelial cells, and enter the cell by endocytosis. Viral RNA is released into the host cell cytoplasm following fusion with the endosomal membrane. Host cell protein and nucleic acid synthesis stop, and rapid synthesis of viral RNA and proteins begins. Regions of the host cell membrane become embedded with virally encoded HA and NA glycoproteins. Within 6 hours, mature virions bud from the host cell, acquiring host cell-derived membrane containing these HA and NA spikes as they do so. The virus spreads rapidly to nearby cells, including mucus-secreting cells and cells of the alveoli. Infected cells die and slough off, thus destroying the mucociliary escalator. The damage to this important first line of defense makes the person susceptible to secondary respiratory infections. The immune response quickly controls the influenza virus infection in most cases, although complete recovery of the respiratory epithelium may take 2 months or more.

Epidemiology

Usually, only a small percentage of people with influenza die, but many people fall ill during an epidemic so the total number of deaths is high. Although influenza virus infection alone can kill otherwise healthy people, most deaths are due to bacterial secondary infections such as pneumonia. People are predisposed to these infections because of the virally induced damage to the respiratory epithelium. In fact, *H. influenzae* got its name because it was often found in the lungs of people who died after having influenza, leading to the incorrect conclusion that it was the cause of the disease.

What to Do About Bird Flu

The known antigenic types of influenza A are 16 different hemagglutinins, designated H 1–16, and nine neuraminidases, N types 1–9. All of these are represented among the influenza A viruses infecting wild waterfowl. Generally, only H types 1–3, and N types 1 and 2 viruses have been important in human disease. Occasionally, however, avian viruses infect humans and cause illness. For example, there were 123 known human cases caused by avian viruses between 1996 and 2004, caused by five avian strains: H7N7, H5N1, H9N2, H7N2, and H7N3. There was a high mortality among the 29 cases caused by the H5N1 strains (seven died), whereas only one death occurred among the remaining 103 cases due to the other four avian viruses. The high virulence of the H5N1 strain has been apparent since the first H5N1 outbreak occurred in 1997. Moreover, the virus has spread rapidly from Asia to the Middle East, Europe, and Africa (**figure 1**), devastating flocks of domestic fowl and causing several hundreds of deaths among humans. Fortunately, so far there have been only a few instances that suggest person-to-person spread of the disease.

The H5N1 avian virus concerns us because it has two of the three characteristics required of a pandemic influenza virus: (1) It is infectious for humans who have no herd immunity; and (2) it causes severe disease in humans. The virus does not yet meet the third requirement—easy spread from person to person, but it might acquire this ability through mutation or antigenic shift. ◀◀ **herd immunity, p. 421**

The H5N1 virus may never become a pandemic strain, but even if it does not, it is highly likely that other avian strains will do so sometime in the future. We have a historical example in the 1918–1919 "Spanish flu" that killed 40 to 100 million people. This pandemic was caused by an avian virus.

Today, we are using some important tools to defend ourselves that were not available in 1918. Rapid viral diagnosis is now possible, and global communication can inform us promptly of outbreaks anywhere in the world so we can take defensive measures. We can also make and stockpile influenza vaccines, as well as vaccines against bacterial secondary invaders such as *Streptococcus pneumoniae* and *Haemophilus influenzae.* We can also place antiviral and antibacterial medications at strategic locations. A great deal depends on international cooperation.

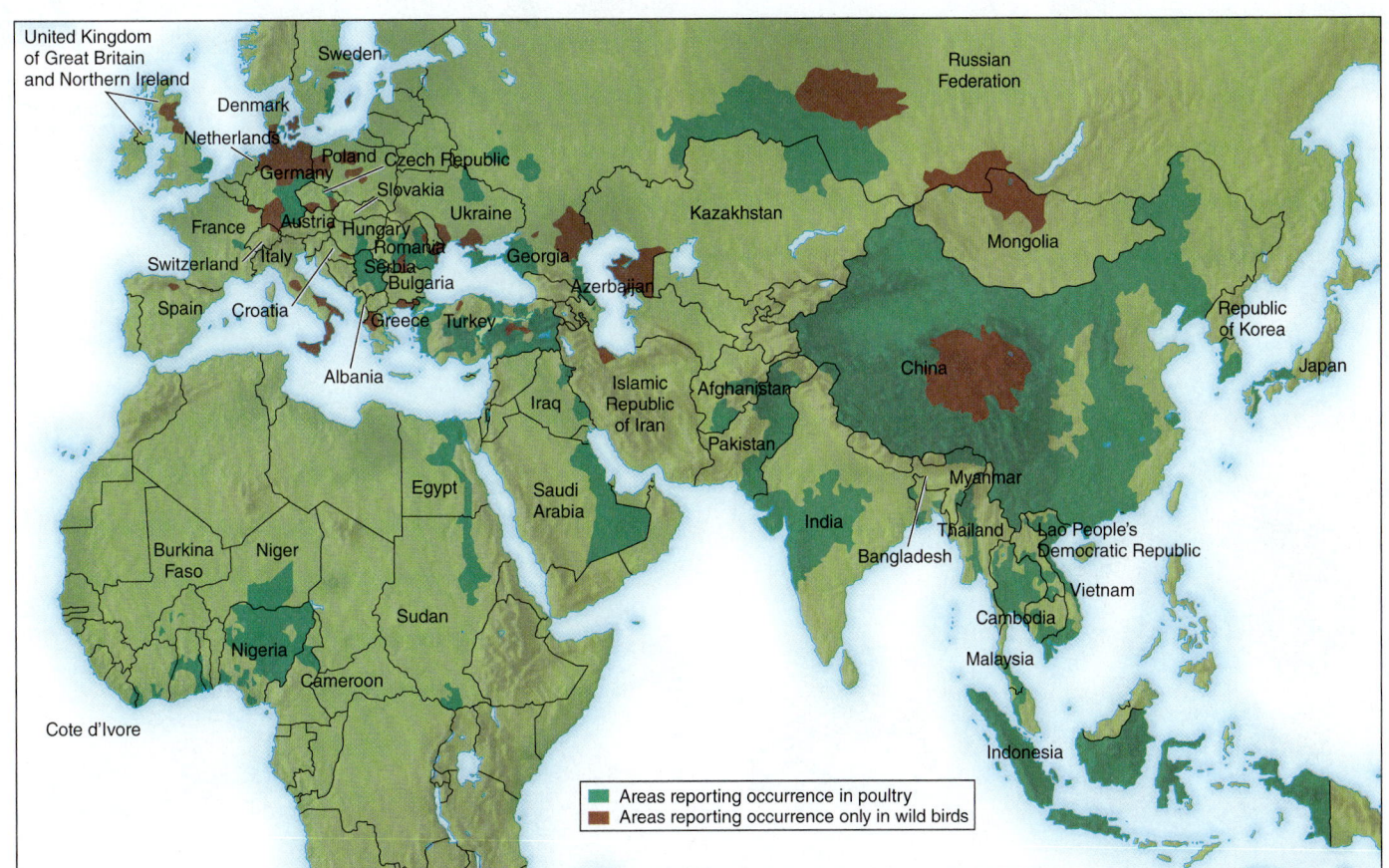

Areas reporting occurrence in poultry

Areas reporting occurrence only in wild birds

FIGURE 1 Areas Reporting H5N1 Avian Influenza in Poultry and Wild Birds Confirmed reports since 2003, last updated September 2007.

It actually causes pneumonia, among other diseases. ◂◂**secondary infection, p. 382**

Influenza epidemics occur every year. Pandemics occur periodically over the years, marked by rapid spread of the viruses around the globe and higher than normal morbidity. Several factors are involved in the spread of influenza viruses, but major attention has focused on their antigenic changeability.

Two types of variation occur: antigenic drift and antigenic shift (**figure 21.24**):

■ **Antigenic drift.** This is caused by minor mutations in the genes that code for the HA and NA antigens and is responsible for the yearly occurrence of influenza outbreaks, called **seasonal influenza.** The mutations happen during normal viral

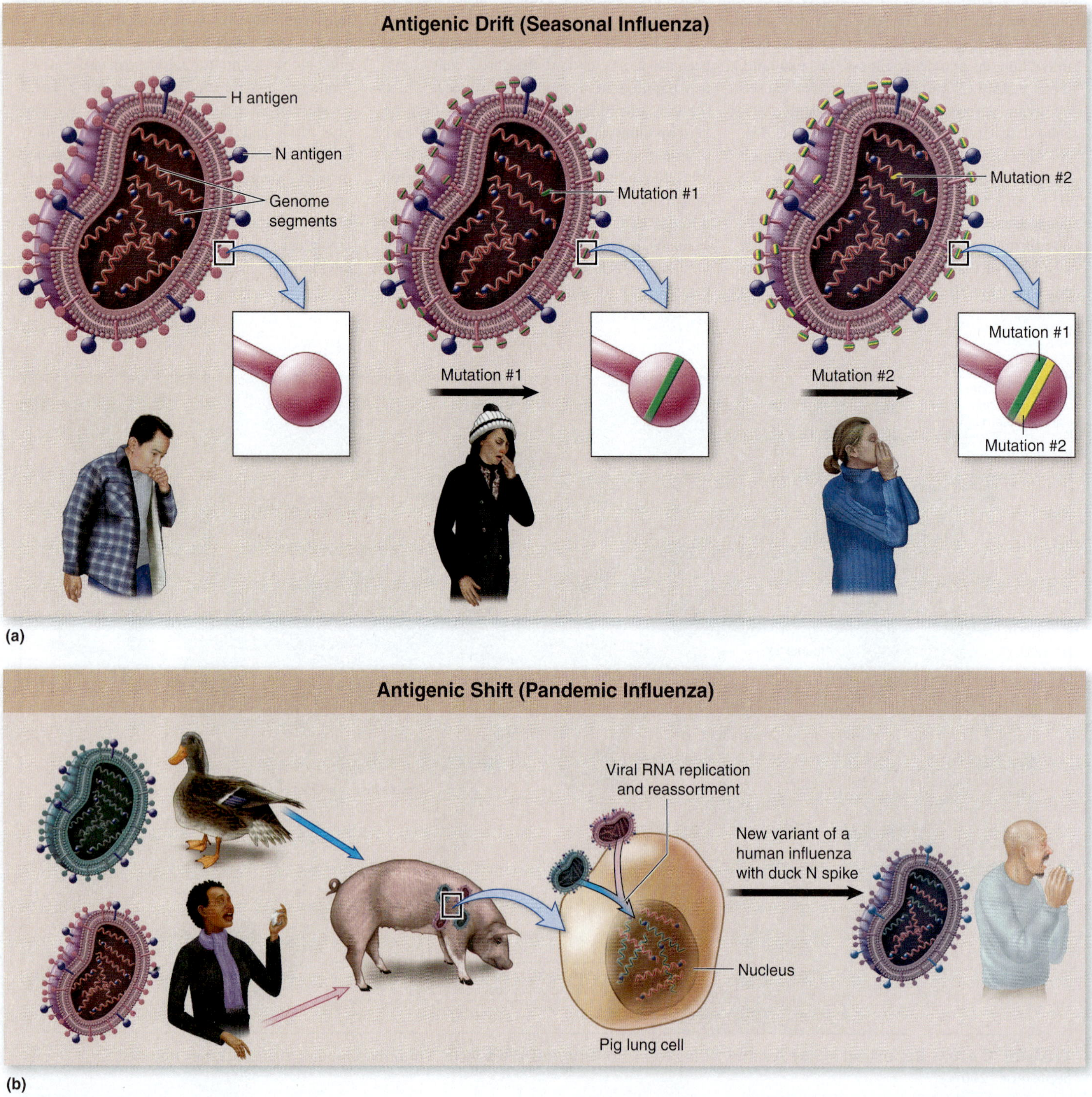

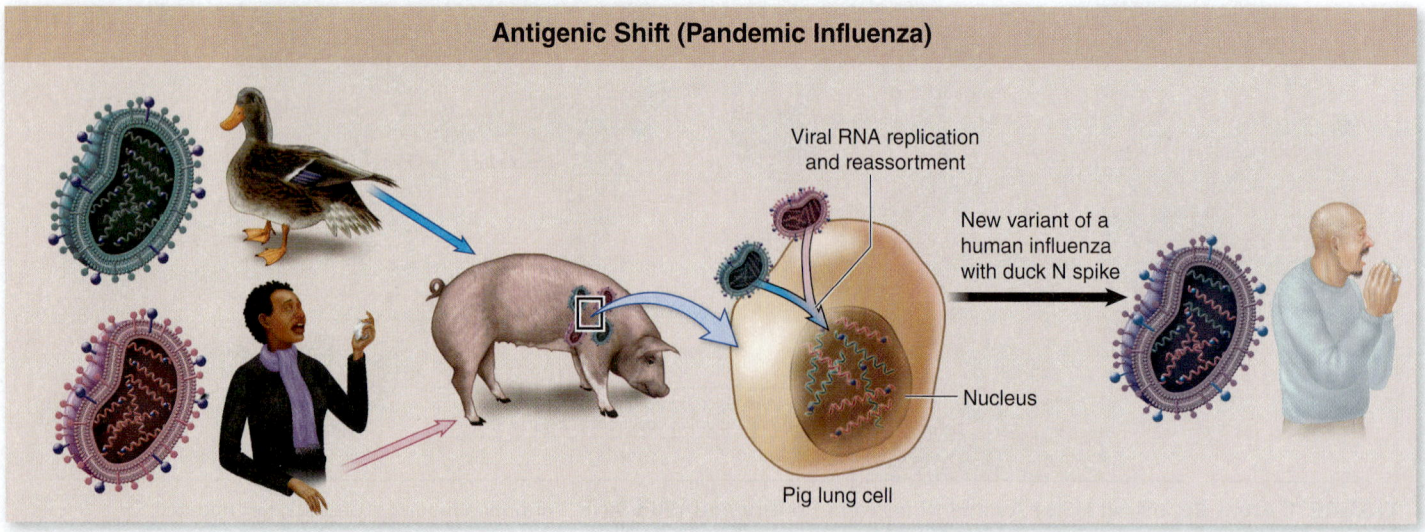

FIGURE 21.24 Influenza Virus: Antigenic Drift and Antigenic Shift With drift, repeated mutations cause a gradual change in the HA and/or NA spikes, so that antibody against the original virus becomes progressively less effective. With shift, there is a sudden major change in the spikes because the virus acquires a new genome segment.

❓ *Why can antigenic shifts cause pandemics?*

replication and often cause a change in only a single amino acid in the HA or NA spikes. They occur frequently, however, and are enough to make immunity developed to virus strains of previous years less effective, ensuring a continual supply of susceptible hosts within which the virus can multiply. Strains that arise because of antigenic drift are given names indicating the year and location they were isolated. For example, a strain referred to as A/Texas/77 (H3N2) is an influenza A strain isolated in 1977 in Texas, whereas A/Bangkok/79 (H3N2) occurred in Bangkok in 1979. The Bankok/79 strain has minor HA spike mutations that distinguish it from the Texas/77 strain. The antibody produced by people who have recovered from A/Texas/77 (H3N2) is only partially effective against the A/Bangkok/79 (H3N2). Thus, the newer Bangkok strain might be able to spread and cause a minor epidemic in people previously exposed to the Texas strain.

■ **Antigenic shift.** This is an uncommon but more dramatic change that occurs as a result of viral genome reassortment and is the cause of **pandemic influenza.** Recall that the influenza virus genome is segmented, meaning that viral proteins are encoded on eight different RNA segments rather than being encoded on one long molecule. Because of this characteristic, when two different influenza viruses infect a cell at the same time, the progeny produced can have RNA segments from either of the viruses. From an infectious disease standpoint, this is particularly a problem when a genome segment from a virus that normally infects only a non-human host is acquired by a strain that infects humans. For example, if a pig is simultaneously infected with two virus strains—one that normally infects birds and pigs and another that usually infects only pigs and humans—the viruses that emerge will still have eight genome segments, but the segments may originate from either of the initial infecting strains. This can result in a human-infecting strain that has novel HA and/or NA antigens for which populations have no immunity (figure 21.24). In 2009, a new strain of H1N1 virus appeared that resulted from reassortment between bird, swine (pig), and human viruses. This new strain spread quickly, causing a pandemic (swine flu). ◄◄ **genetic reassortment, p. 320**

Another situation arises when a strain that does not typically infect humans gains the ability to do so. Ecological studies show that all the known influenza A virus types exist in aquatic birds, generally causing chronic intestinal infections. Bird influenza viruses (avian flu) readily infect domestic fowl, and from them can infect other domestic animals and humans. The 1997 "bird flu" epidemic in Hong Kong involved an H5N1 virus from chickens that spread to humans, sometimes causing fatal infections. Different H5N1 strains have been categorized as low pathogenic avian influenza (LPAI) and highly pathogenic avian influenza (HPAI), depending on whether they cause fatal disease or not. Fortunately, the HPAI strain that started the Hong Kong outbreak did not spread easily from person to person and so the number of human cases remained relatively low. Subsequent outbreaks show that HPAI H5N1 viruses are now endemic in a number of Asian communities. LPAI viruses have been isolated in the United States. A constant concern is that an HPAI strain will evolve through antigenic drift or antigenic shift to spread person to person.

Treatment and Prevention

Like all viral infections, antibiotic treatment is not effective for influenza. Medications such as amantadine (Symmetrel) and rimantadine (Flumadine) have been used for short-term prevention of influenza A disease when vaccination is not an option or while waiting for the protection of vaccination to take effect. However, they are not currently recommended because many circulating viruses are resistant to them. Zanamivir (Relenza) and oseltamivir (Tamiflu) are neuraminidase inhibitors, active against both A and B viruses. These medicines are generally used together with vaccination to protect exposed individuals until they can develop immunity. ◄◄ **neuraminidase inhibitors, p. 476**

Prevention is by vaccine, but there is no vaccine that produces lifelong immunity to influenza, because of antigenic drift—every year a slightly new strain of virus emerges, against which immunity must be generated. Inactivated vaccines are developed annually against the three most important influenza strains in circulation at that time. It takes 6 to 9 months from the appearance of a new influenza strain before adequate amounts of vaccine can be manufactured. However, these multivalent vaccines can be 80% to 90% effective in preventing influenza. Because influenza vaccines are inactivated, they are relatively safe even for the immunocompromised. An attenuated vaccine that is given as a nasal spray to children has been developed and is safe and effective. The main features of influenza are summarized in **table 21.11.**

Respiratory Syncytial Virus Infections

Respiratory syncytial virus (RSV) infection is the leading cause of serious lower respiratory tract infections in infants and young children, resulting in an estimated 90,000 hospitalizations and 4,500 deaths in the United States each year. It is also responsible for serious disease in elderly people, and for healthcare-associated epidemics.

Signs and Symptoms

Signs and symptoms of RSV infection begin after an incubation period of 1 to 4 days with runny nose followed by cough, wheezing, and difficulty breathing. Fever may or may not be present. Patients often develop a dusky color, indicating that they are not getting enough O_2. Healthy older children and adults with RSV generally show symptoms of a bad cold. RSV is one of the causes of **croup,** which manifests as a loud high-pitched cough and noisy inhalation due to airway obstruction.

Hospitalized infants seldom die from RSV, but it is sometimes fatal for elderly patients with underlying diseases such as heart and lung disease, cancer, and immunodeficiency.

Causative Agent

RSV is a single-stranded, enveloped RNA virus of the paramyxovirus group. It causes cells in cell cultures to fuse together; the clumps of fused cells are known as **syncytia,** thus the name of the virus. ◄◄ **paramyxovirus family, p. 309**

TABLE 21.11 | Influenza

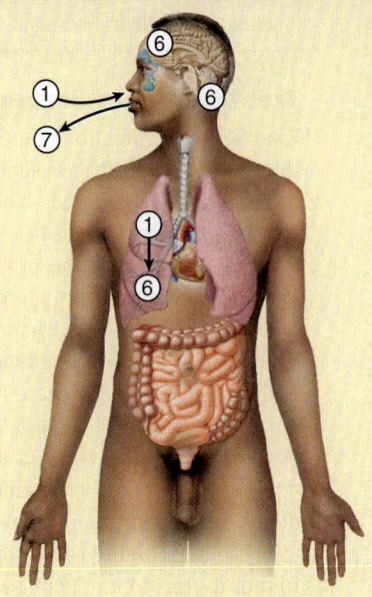

1. Influenza virus is inhaled and carried to the lungs.
2. Viral hemagglutinin attaches to specific receptors on ciliated epithelial cells, the viral envelope fuses with the epithelial cell, and the virus enters the cell by endocytosis.
3. Host cell synthesis is diverted to synthesizing new virus.
4. Newly formed virions bud from infected cells; they are released by viral neuraminidase and infect ciliated epithelium, mucus-secreting, and alveolar cells.
5. Infected cells ultimately die and slough off; recovery of the mucociliary escalator may take weeks.
6. Secondary bacterial infection of the lungs, ears, and sinuses is common.
7. The virus exits with coughing.

Signs and symptoms	Fever, muscle aches, lack of energy, headache, sore throat, nasal congestion, cough
Incubation period	1 to 2 days
Causative agent	Influenza virus, an orthomyxovirus
Pathogenesis	Infection of respiratory epithelium; cells destroyed and virus released to infect other cells. Secondary bacterial infection results from damaged mucociliary escalator.
Epidemiology	Antigenic drift and antigenic shift prevent immunity.
Treatment and prevention	Amantadine and rimantadine are sometimes effective for preventing type A but not type B virus disease; neuraminidase inhibitors effective against both A and B viruses. These medications somewhat effective for treatment when given early in the disease. Vaccines usually 80% to 90% effective.

Pathogenesis

The virus enters the body by inhalation and infects the respiratory tract epithelium, causing cells to die and slough off. Bronchiolitis is a common feature of the disease; the inflamed bronchioles become partially plugged by sloughed cells, mucus, and clotted plasma that has oozed from the walls of the bronchi. The initial obstruction causes wheezing when air rushes through the narrowed passageways, sometimes causing the condition to be confused with asthma. The obstruction often acts like a one-way valve, allowing air to enter the lungs, but not leave them. In many cases the inflammatory process extends into the alveoli, causing pneumonia. There is a high risk of secondary infection because of the damaged mucociliary escalator. ◄◄ **mucociliary escalator, p. 337**

Epidemiology

RSV outbreaks are common from late fall to late spring, peaking in mid-winter. Recovery from infection produces only weak and short-lived immunity, so that infections can recur throughout life. Healthy children and adults usually have mild illness and readily spread the virus to others.

Treatment and Prevention

There are no effective antiviral medications for treating RSV.

Preventing healthcare-associated RSV illness requires strict isolation techniques. People with underlying illnesses can be protected from the disease by giving them passive immunity via monthly injections of immune globulin or a monoclonal antibody called palivizumab. No vaccines are available, although there are at least six programs underway to develop a vaccine. The main features of RSV infections are summarized in **table 21.12**. ◄◄ **passive immunity, p. 420**

Hantavirus Pulmonary Syndrome

In the spring of 1993, a newly emerging disease made a dramatic appearance in the "Four Corners" region of the American Southwest, an area where the states of Arizona, Colorado, New Mexico, and Utah come together. It was a small outbreak but quite alarming because most victims were vigorous young adults who developed influenza-like symptoms, and then died within days. Scientists from the CDC rushed to join local epidemiologists and

TABLE 21.12 | Respiratory Syncytial Virus (RSV) Infections

Signs and symptoms	Runny nose, cough, fever, wheezing, difficulty breathing, dusky color
Incubation period	1 to 4 days
Causative agent	RSV, a paramyxovirus that produces syncytia
Pathogenesis	Sloughing of respiratory epithelium and inflammatory response plug bronchioles, cause bronchiolitis; pneumonia results from bronchiolar and alveolar inflammation, or secondary infection.
Epidemiology	Yearly epidemics during the cool months; readily spread by otherwise healthy older children and adults who often have mild symptoms; no lasting immunity.
Treatment and prevention	No satisfactory antiviral treatment. No vaccine. Preventable by injections of immune serum globulin or a monoclonal antibody.

health officials investigating the outbreak. Their studies quickly established that the disease was associated with exposure to mice. They soon learned that the patients had been infected with a hantavirus closely related to one that plagued American troops during the 1950s Korean War.

Signs and Symptoms

Hantavirus pulmonary syndrome usually begins with fever, muscle aches (especially in the lower back), nausea, vomiting, and diarrhea. Unproductive cough and increasingly severe shortness of breath appear within a few days, followed by shock and death.

Causative Agents

Certain hantaviruses, enveloped viruses of the bunyavirus family, cause hantavirus pulmonary syndrome. Their genome consists of three segments of single-stranded RNA. In nature, these viruses primarily infect rodents, causing lifetime infections, without any apparent harm to the animals. Each type of hantavirus generally infects a particular rodent species. ◄◄ bunyavirus family, p. 309

Pathogenesis

The virus enters the body by inhalation of airborne dust contaminated with the urine, feces, or saliva of infected rodents. The virus then enters the circulation and is carried throughout the body, infecting the cells that line tissue capillaries. Massive amounts of the viral antigen appear in lung capillaries, but the antigen is also found in capillaries of the heart and other organs. The inflammatory response to the viral antigen causes the capillaries to leak large amounts of plasma into the lungs, suffocating the patient and causing the blood pressure to fall. Shock and death occur in more than 40% of the cases. Luckily, despite the large amount of viral antigen in the lung capillaries, few mature infectious virions enter the air passages of the lung, so person-to-person transmission occurs rarely, if ever.

Epidemiology

Hantavirus pulmonary syndrome is a zoonosis. It is considered an emerging disease because of its recent discovery and apparent increase in frequency but has probably existed for centuries. Since the description of the syndrome, cases have been identified from Canada to Argentina, including several hundred from the United States (**figure 21.25**). Most of the cases have occurred west of the Mississippi River and were due to a type of hantavirus carried by deer mice, but hantavirus carried by other rodents can cause disease as well. Epidemics have been associated with increases in mouse populations in poor communities with substandard housing. Thirty percent or more of the mice can become carriers of the disease. Complex ecological factors such as numbers of foxes, owls, snakes, and other predators, and weather (which affect food supply) all play a role in mouse population levels. The emergence of hantavirus pulmonary syndrome is a convincing example of how environmental change can result in infectious human disease. ◄◄ zoonoses, p. 439 ◄◄ emerging diseases, p. 448

Treatment and Prevention

There is no proven antiviral treatment for hantavirus pulmonary syndrome, a highly fatal disease.

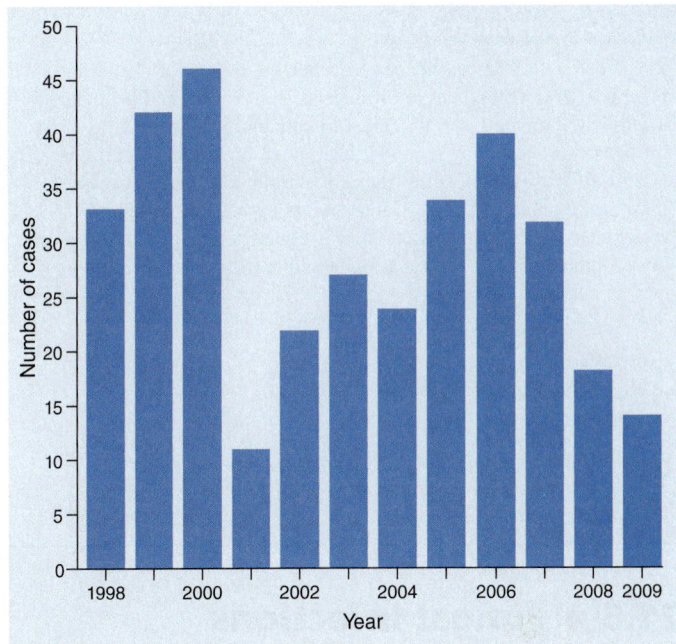

FIGURE 21.25 Total Hantavirus Pulmonary Syndrome Cases, United States, 1996–2009

❓ *Which animals most commonly carry hantaviruses?*

Prevention of the syndrome is based on minimizing exposure to rodents and dusts contaminated by their urine, saliva, and feces. Rodent populations should be controlled by keeping foods in containers and making buildings as mouse-proof as possible. When cleaning an area where rodents are found, maximal ventilation should be ensured and the area should be mopped with a disinfectant solution rather than being swept with brooms and vacuum cleaners that can stir up dust. Lethal traps and poisons may be necessary to decrease the rodent population in the area. The main features of hantavirus pulmonary syndrome are summarized in **table 21.13**.

TABLE 21.13	Hantavirus Pulmonary Syndrome
Signs and symptoms	Fever, muscle aches, vomiting, diarrhea, cough, shortness of breath, shock
Incubation period	3 days to 6 weeks
Causative agent	Sin Nombre and related hantaviruses of the bunyavirus family
Pathogenesis	Viral antigen localizes in capillary walls in the lungs; inflammation.
Epidemiology	Zoonosis likely to involve humans in proximity to increasing mouse populations; generally no person-to-person spread.
Treatment and prevention	Avoid contact with rodents; seal access to houses, food supplies; good ventilation, avoid dust, use disinfectants in cleaning rodent-contaminated areas. No proven antiviral treatment.

MicroAssessment 21.5

The speed with which influenza travels around the world and the potential for development of virulent influenza virus strains make the disease an extremely serious threat to humankind. Most deaths from influenza are caused by secondary bacterial infections. Respiratory syncytial virus is the leading cause of serious respiratory disease in infants and young children. Hantavirus pulmonary syndrome, first recognized in 1993, is often fatal. It is contracted from inhalation of dust contaminated by urine, feces, or saliva from mice infected with certain hantaviruses.

13. *Why are there so many deaths from influenza when it is generally a mild disease?*

14. *What is the source of the virus that causes hantavirus pulmonary syndrome?*

15. *Why might you expect an influenza epidemic to be more severe following an antigenic shift in the virus than after antigenic drift?* ✚

21.6 ■ Fungal Infections of the Lung

Learning Outcomes

11. *Describe the pathogenesis of histoplasmosis.*

12. *Outline the epidemiology of coccidioidomycosis*

Serious lung diseases caused by fungi are quite unusual in healthy, immunocompetent individuals. Symptomatic and asymptomatic infections that subside without treatment, however, are common. Coccidioidomycosis and histoplasmosis are two examples of widespread mycoses of the respiratory tract.

MicroByte

Coccidioidomycosis and histoplasmosis can both manifest with symptoms that resemble tuberculosis.

Coccidioidomycosis ("Valley Fever")

In the United States, coccidioidomycosis occurs mainly in California, Arizona, Nevada, New Mexico, Utah, and West Texas. People who are exposed to dust and soil, such as farm workers, are most likely to become infected, but only 40% develop symptoms.

Signs and Symptoms

"Flulike" signs and symptoms such as fever, cough, chest pain, and loss of appetite and weight are common manifestations of coccidioidomycosis. The majority of people with the disease recover spontaneously within a month. A small percentage of patients, however, develop chronic disease.

Causative Agent

Coccidioidomycosis is caused by *Coccidioides immitis,* a dimorphic fungus that grows in soil as a mold. Its hyphae give rise to numerous barrel-shaped, highly infectious structures called arthroconidia (**figure 21.26a**). These become airborne and can be inhaled. In infected tissues, the arthroconidia develop into

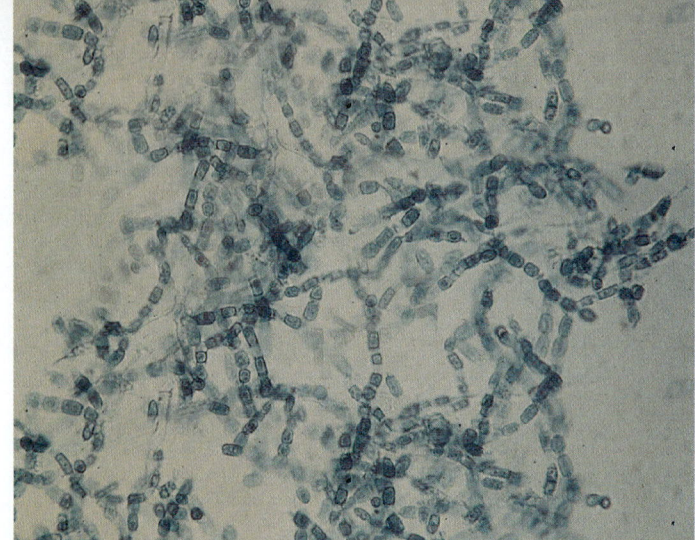

(a)   20 µm

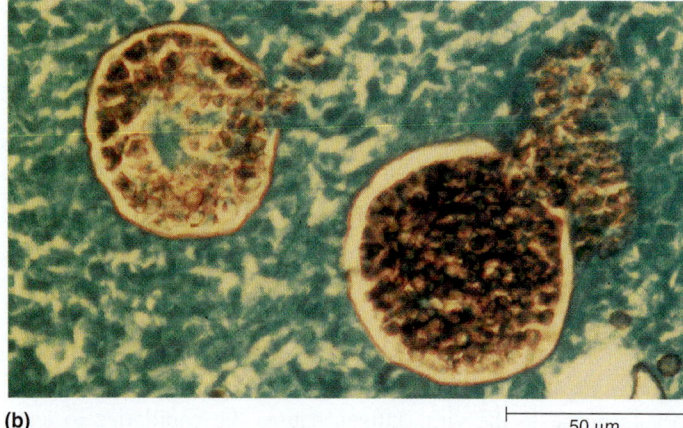

(b)   50 µm

FIGURE 21.26 *Coccidioides immitis* **(a)** Mold-phase hyphae fragmenting into arthroconidia. **(b)** Spherules containing fungal endospores.

❓ *Which form of this fungus is found in host tissues?*

thick-walled spherules that may contain several hundred small cells called endospores, not to be confused with bacterial endospores (figure 21.26b). ◀◀ dimorphic fungi, p. 285

Pathogenesis

Arthroconidia enter the lung with inhaled air and develop into spherules. Endospores develop in the spherules, which mature and rupture, releasing them. The endospores develop into more endospore-containing spherules and the process is repeated. Each time this cycle occurs, an inflammatory response is provoked. Tissue injury and symptoms are caused mainly by the host's immune response to coccidioidal antigens.

The organisms are usually eliminated by body defenses, but caseous necrosis (dead tissue that has cheeselike consistency) occurs in a small percentage of individuals, resulting in lung cavities similar to those seen in tuberculosis. Occasionally, organisms are carried throughout the body by the bloodstream and infect the skin, mucous membranes, brain, and other organs. This disseminated form of the disease occurs more often in people with AIDS or other immunodeficiencies and is fatal without treatment.

◀◀ caseous necrosis, p. 504

Epidemiology

Coccidioides immitis grows only in semi-arid desert areas of the Western Hemisphere. In these areas, infections occur only during the hot, dry, dusty seasons when airborne arthroconidia are easily dispersed from the soil. Dust stirred up by earthquakes can result in epidemics. Rainfall encourages growth of the fungus, which then produces increased numbers of spores when dry conditions return. People can contract coccidioidomycosis by simply traveling through the endemic area. Infectious spores have unknowingly been transported to other areas, but the organism apparently is unable to establish itself in moist climates.

Treatment and Prevention

Medications approved for treatment of serious cases of coccidioidomycosis include amphotericin B and fluconazole or itraconazole. They must be given for long periods of time, and cause troublesome side effects. Even with treatment, disseminated disease can reactivate months or years later.

Preventive measures include avoiding dust in the endemic areas, and watering and planting vegetation to aid in dust control. **Table 21.14** describes the main features of coccidioidomycosis.

Histoplasmosis ("Spelunker's Disease")

Histoplasmosis, like coccidioidomycosis, is usually benign but occasionally mimics tuberculosis. Rare, serious forms of the disease suggest that the patient has an underlying immunodeficiency such as AIDS. The distribution is more widespread than that of coccidioidomycosis and is associated with different soil types and climate.

Signs and Symptoms

Most infections are asymptomatic. Fever, cough, and chest pain are the most common symptoms, sometimes with shortness of breath. Mouth sores may develop, especially in children.

Causative Agent

Histoplasmosis is caused by the dimorphic fungus *Histoplasma capsulatum,* although the name is misleading because the fungus does not have a capsule. This organism grows in soils contaminated by bat or bird droppings, but it is not pathogenic for these animals. In pus or tissue from people with active disease, *H. capsulatum* is a tiny oval yeast that grows within host macrophages (**figure 21.27a**). The mold form of the organism characteristically produces two kinds of conidia: macroconidia, which often have numerous projecting knobs (figure 21.27b), and tiny pear-shaped or spherical microconidia.

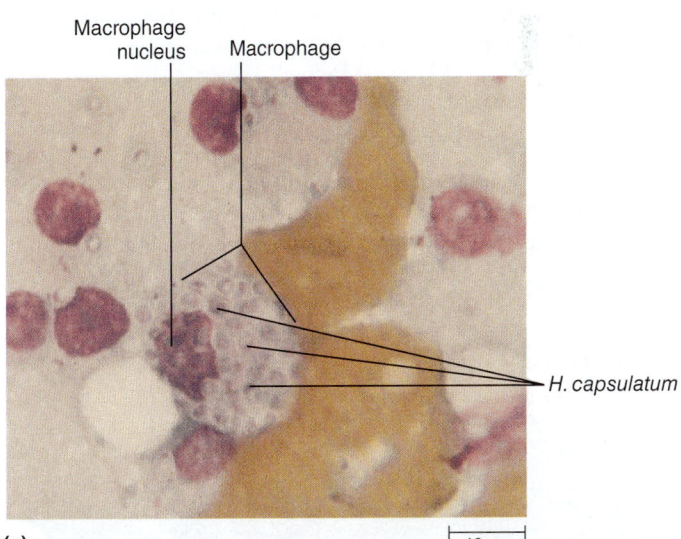

(a)

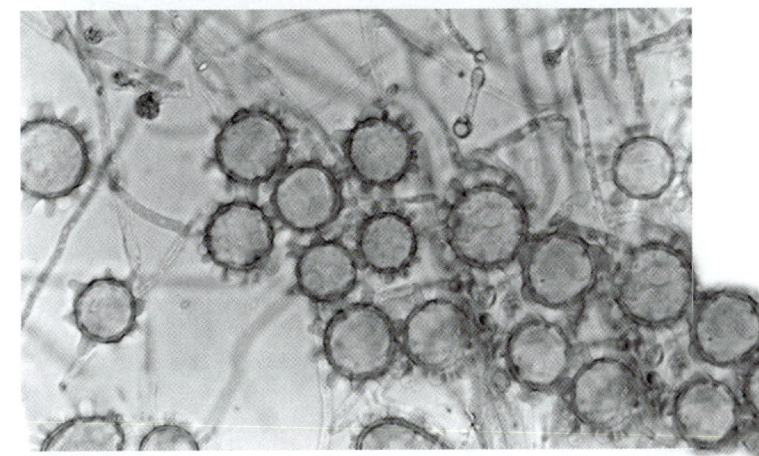

(b)

FIGURE 21.27 *Histoplasma capsulatum* (a) Yeast-phase organisms in the cytoplasm of a macrophage. (b) Mold phase, showing macroconidia.

❓ *Is this fungus encapsulated?*

TABLE 21.14	Coccidioidomycosis (Valley Fever)
Signs and symptoms	Fever, cough, chest pain, loss of appetite and weight; less frequently, painful nodules on extremities, pain in joints; skin, mucous membranes, brain, and internal organs sometimes involved
Incubation period	2 days to 3 weeks
Causative agent	*Coccidioides immitis,* a dimorphic fungus
Pathogenesis	After lodging in lung, arthrospores develop into spherules that mature and discharge endospores, each of which then develops into another spherule; inflammatory response damages tissue; hypersensitivity to fungal antigens causes painful nodules and joint pain.
Epidemiology	Inhalation of airborne *C. immitis* spores with dust from soil growing the organism. Occurs only in certain semi-arid regions of the Western Hemisphere.
Treatment and prevention	Treatment: amphotericin B and fluconazole or itraconazole. Prevention by dust control methods such as grass planting and watering.

Pathogenesis

Histoplasma capsulatum conidia inhaled into the lungs are taken up by resident macrophages. The fungus then develops into the yeast form, which multiplies within the phagocytes. Granulomas develop in infected areas, closely resembling those seen in tuberculosis, sometimes even showing caseous necrosis. Eventually, the lesions are replaced with scar tissue, and many calcify, becoming visible on X rays. In rare cases, particularly in those who are immunodeficient, the disease spreads throughout the body.

Epidemiology

The distribution of histoplasmosis is quite different from that of coccidioidomycosis. **Figure 21.28** shows the distribution of histoplasmosis in the United States, but the disease also occurs in tropical and temperate zones scattered around the world. Cave explorers (spelunkers) are at risk for histoplasmosis because many caves contain soil contaminated with bat droppings. Most cases in the United States have occurred in the Mississippi and Ohio River drainage area and in South Atlantic states. Skin tests reveal that millions of people living in these areas have been infected.

Treatment and Prevention

Treatment of histoplasmosis is similar to that of coccidioidomycosis. Amphotericin B and itraconazole are used for treating severe disease, but both medications have potentially serious side effects.

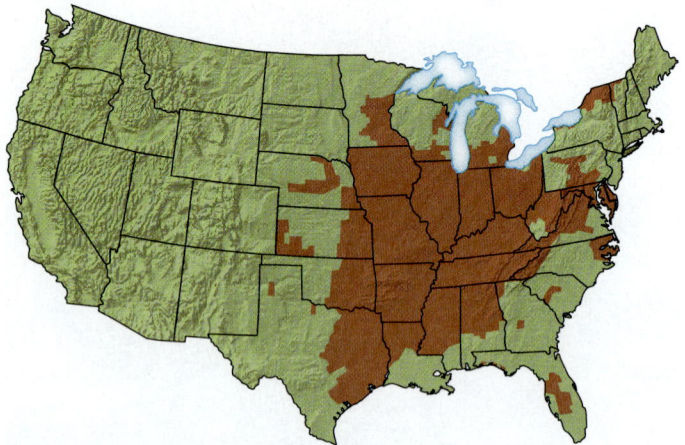

FIGURE 21.28 Geographical Distribution of *Histoplasma capsulatum* in the United States

❓ *Why is histoplasmosis also called "spelunker's disease"?*

TABLE 21.15	Histoplasmosis (Spelunker's Disease)
Signs and symptoms	Mild respiratory symptoms; less frequently, fever, chest pain, cough, chronic sores
Incubation period	5 to 8 days
Causative agent	*Histoplasma capsulatum,* a dimorphic fungus
Pathogenesis	Spores inhaled, change to yeast phase, multiply in macrophages; granulomas form; disease spreads in individuals with AIDS or other immunodeficiencies.
Epidemiology	The fungus prefers to grow in soil contaminated by bird or bat droppings, especially in Ohio and Mississippi River valleys, and in the U.S. Southeast. Spotty distribution in many other countries around the world. Spelunkers are at risk of infection.
Treatment and prevention	Treatment: amphotericin B and itraconazole for serious infections. Prevented by avoiding soils contaminated with chicken, bird, or bat droppings.

No proven preventive measures are known other than to avoid areas where soil is heavily enriched with bat and bird droppings, especially if the soil has been left undisturbed for a long period. Some researchers have recommended placing several inches of clay soil over soils containing large quantities of old droppings. The main features of histoplasmosis are described in **table 21.15.**

The key features of the diseases covered in this chapter are highlighted in the **Diseases in Review 21.1** table.

MicroAssessment 21.6

Coccidioidomycosis and histoplasmosis are two diseases caused by fungi that live in the soil. The body responds to these infections by forming granulomas, mimicking tuberculosis. Each fungus has its own ecological niche: *Coccidioides immitis* in semi-arid regions of the Western Hemisphere, and *Histoplasma capsulatum* in moist soils enriched with bird or bat droppings around the world.

16. *Why should an immunodeficient person avoid traveling through hot, dry, dusty areas of the Southwest?*

17. *Why might cave exploration increase the risk of histoplasmosis?*

18. *Several students staying in a hotel next to a bulldozing operation developed histoplasmosis. How might the bulldozing explain the outbreak?* ➕

Diseases in Review 21.1

Diseases of the Respiratory System

Disease	Causative Agent	Comment	Summary Table
BACTERIAL INFECTIONS OF THE UPPER RESPIRATORY TRACT			
Streptococcal pharyngitis ("strep throat")	*Streptococcus pyogenes* (group A streptococcus)	Treated with antibiotics, partly to avoid sequelae; must be distinguished from viral pharyngitis, which cannot be treated with antibiotics.	Table 21.3, p. 489
Diphtheria	*Corynebacterium diphtheriae*	Toxin-mediated disease characterized by pseudomembrane in the upper respiratory tract. Preventable by vaccination (DTaP).	Table 21.4, p. 492
Conjunctivitis (pinkeye), otitis media (earache), sinus infection	Usually *Haemophilus influenzae* and *Streptococcus pneumoniae*	Often occur together; factors involved in the transmission are unknown.	
VIRAL INFECTIONS OF THE UPPER RESPIRATORY TRACT			
Common cold	Rhinoviruses and other viruses	Runny nose, sore throat, and cough are due to the inflammatory response and cell destruction.	Table 21.5, p. 495
Adenoviral pharyngitis	Adenovirus	Similar to the common cold but with fever; spread to the lower respiratory tract can result in severe disease.	Table 21.6, p. 496
BACTERIAL INFECTIONS OF THE LOWER RESPIRATORY TRACT			
Pneumococcal pneumonia	*Streptococcus pneumoniae*	Organism common in the throat of healthy people; causes disease when mucociliary escalator is impaired or with underlying conditions. Vaccine that protects against multiple strains is available.	Table 21.7, p. 499
Klebsiella pneumonia	*Klebsiella* species, commonly *K. pneumoniae*	Common hospital-acquired bacterium; characterized by sputum resembling red current jelly. Drug resistance is a major problem.	Table 21.7, p. 499
Mycoplasmal pneumonia ("walking pneumonia")	*Mycoplasma pneumoniae*	Relatively mild pneumonia; common among college students and military recruits. Cannot be treated with medications that inhibit cell wall synthesis.	Table 21.7, p. 499
Pertussis ("whooping cough")	*Bordetella pertussis*	Characterized by frequent violent coughing. Preventable by vaccination (DTaP).	Table 21.8, p. 502
Tuberculosis ("TB")	*Mycobacterium tuberculosis*	Most infections result in latent tuberculosis infection (LTBI), but these can reactivate to cause active tuberculosis disease (ATBD). Treated using combination drug therapy, but drug resistance is an increasing problem.	Table 21.9, p. 506
Legionnaires' disease	*Legionella pneumophila*	Transmitted via aerosolized water drops; smokers and those with impaired defenses are most at risk of developing disease.	Table 21.10, p. 507
VIRAL INFECTIONS OF THE LOWER RESPIRATORY TRACT			
Influenza ("flu")	Influenza A virus	New vaccine developed yearly; viruses change seasonally due to antigenic drift; antigenic shifts cause pandemics.	Table 21.11, p. 512
Respiratory syncytial virus infections	RSV	Serious disease in infants, young children, and the elderly.	Table 21.12, p. 512
Hantavirus pulmonary syndrome	Hantaviruses	Acquired via inhaled dust contaminated with rodent saliva, urine, or feces. Frequently fatal.	Table 21.13, p. 513
FUNGAL INFECTIONS OF THE RESPIRATORY TRACT			
Coccidioidomycosis ("Valley fever")	*Coccidioides immitis*	Environmental reservoir (soil in semi-arid desert areas); most infections are asymptomatic.	Table 21.14, p. 515
Histoplasmosis ("spelunker's disease")	*Histoplasma capsulatum*	Environmental reservoir (soil enriched with bird or bat droppings); most infections are asymptomatic.	Table 21.15, p. 516

Global Preparedness Versus Emerging Respiratory Viruses

In November 2002, a frightening new respiratory disease emerged in Guangdong province, China. Symptoms included cough and fever, and X rays showed characteristics of viral pneumonia. The disease, labeled "SARS" (severe acute respiratory syndrome), quickly spread around the world, involving 25 countries and causing more than 700 deaths. Many of its victims were medical personnel. It had two characteristics that quickly led to its control: a long incubation of 6 days, and dramatic symptoms that were easily recognized. This allowed healthcare personnel time to identify contacts and institute quarantines to stop the spread of the disease. The causative agent proved to be a previously unknown coronavirus, and its source was never identified, although it was probably a wild animal sold for meat in the markets. On three occasions, SARS escaped from laboratories where the virus was being studied, and the disease may arise again from its natural source. However, its main importance is that it caused many countries to learn to work together to help forge a more effective global response to emerging pandemic diseases. More details on SARS can be found at www.sarsreference.com.

The problem of avian influenza is more difficult to solve than SARS because the avian viruses are carried in the intestines of wild waterfowl that readily infect domestic flocks, which then infect humans. Epidemic influenza in humans generally has a short incubation period, and many victims have mild, coldlike symptoms not easily recognized as part of an influenza epidemic. Therefore, quarantines are not helpful in disease control. Many experts think we have been lucky that no virus capable of causing a pandemic has appeared. However, thanks to the battles against the H5N1 avian virus and the SARS outbreaks, the world is much closer to being prepared for the next pandemic. The challenge is to sustain and improve advances in cooperative surveillance, and action planning to avoid a global catastrophe.

Summary

21.1 ■ Anatomy, Physiology, and Ecology

The Upper Respiratory Tract

The upper respiratory tract includes the nose and nasal cavity, pharynx (throat), and epiglottis (figure 21.1). Ciliated cells line much of the respiratory tract and remove microorganisms by constantly propelling mucus out of the respiratory system. A wide variety of microorganisms colonize parts of the system (table 21.1).

The Lower Respiratory Tract

The lower respiratory system includes the larynx, trachea, bronchi, and lungs. Pleural membranes surround the lungs. Viruses and microorganisms are normally absent from the lower respiratory system.

INFECTIONS OF THE UPPER RESPIRATORY SYSTEM

21.2 ■ Bacterial Infections of the Upper Respiratory System

Streptococcal Pharyngitis ("Strep Throat") (tables 21.2, 21.3)

Streptococcus pyogenes is a β-hemolytic Gram-positive coccus that causes strep throat (figures 21.2, 21.3). Strains that produce SPEs (**streptococcal pyrogenic exotoxins**) can cause scarlet fever. Other diseases are associated with SPEs as well.

Post-Streptococcal Sequelae

Post-streptococcal sequelae, including rheumatic fever and glomerulonephritis, may follow strep throat and are due to the immune response and molecular mimicry (figure 21.5).

Diphtheria (table 21.4)

Diphtheria, caused by *Corynebacterium diphtheriae*, is a toxin-mediated disease that can be prevented by immunization (figures 21.6, 21.7).

Pinkeye, Earache, and Sinus Infections

Conjunctivitis (pinkeye) is usually caused by *Haemophilus influenzae* or *Streptococcus pneumoniae* (pneumococcus) (figure 21.8). Viral agents, including adenoviruses and rhinoviruses, usually result in a milder illness. **Otitis media** and sinusitis develop when infection spreads from the nasopharynx (figure 21.9).

21.3 ■ Viral Infections of the Upper Respiratory System

The Common Cold (table 21.5)

The common cold can be caused by many different viruses, rhinoviruses being the most common (figure 21.10).

Adenoviral Respiratory Tract Infections (table 21.6)

Adenoviruses cause illnesses that can resemble a common cold or strep throat, with symptoms varying from mild to severe.

INFECTIONS OF THE LOWER RESPIRATORY SYSTEM

21.4 ■ Bacterial Infections of the Lower Respiratory System

Pneumococcal Pneumonia (table 21.7; figures 21.11, 21.12)

Streptococcus pneumoniae, one of the most common causes of pneumonia, is virulent because of its capsule.

Klebsiella Pneumonia (table 21.7)

Klebsiella pneumonia is representative of many healthcare-associated pneumonias that cause permanent damage to the lung (figure 21.13). Serious complications such as lung abscesses and bloodstream infection are more common than with many other bacterial pneumonias.

Mycoplasmal Pneumonia ("Walking Pneumonia")
(table 21.7, figure 21.14)

Mycoplasmal pneumonia is often called walking pneumonia; serious complications are rare. Penicillins and cephalosporins are not useful in treatment because the causative agent, *M. pneumoniae*, lacks a cell wall.

Pertussis ("Whooping Cough") (table 21.8; figures 21.15, 21.16)

Whooping cough is characterized by violent spasms of coughing and gasping. Childhood immunization against the causative agent, *Bordetella pertussis*, prevents the disease. Mild disease is common in adults.

Tuberculosis ("TB") (table 21.9; figures 21.17–21.21)

Tuberculosis, caused by the acid-fast rod *Mycobacterium tuberculosis,* is generally slowly progressive or heals and remains latent, presenting the risk of later reactivation.

Legionnaires' Disease (table 21.10, figure 21.22)

Legionnaires' disease occurs when there is a high infecting dose of the causative microorganisms or an underlying lung disease. The cause, *Legionella pneumophila,* is a rod-shaped bacterium common in the environment.

21.5 ▪ Viral Infections of the Lower Respiratory System

Influenza ("Flu") (table 21.11; figures 21.23, 21.24)

Widespread epidemics are characteristic of influenza A viruses. Antigenic shifts and drifts are responsible. Deaths are usually but not always caused by secondary infection. Reye's syndrome may rarely occur during recovery from influenza and other viral infections but is probably not caused by the virus itself.

Respiratory Syncytial Virus Infections (table 21.12)

RSV is the leading cause of serious respiratory disease in infants and young children.

Hantavirus Pulmonary Syndrome (table 21.13, figure 21.25)

Hantavirus pulmonary syndrome is an often fatal disease contracted when airborne dust contaminated by urine from mice infected with a hantavirus is inhaled.

21.6 ▪ Fungal Infections of the Lung

Coccidioidomycosis ("Valley Fever") (table 21.14)

Coccidioidomycosis occurs in hot, dry areas of the Western Hemisphere and is initiated by airborne spores of the dimorphic soil fungus *Coccidioides immitis* (figure 21.26).

Histoplasmosis ("Spelunker's Disease") (table 21.15)

Histoplasmosis is similar to coccidioidomycosis but occurs in tropical and temperate zones around the world (figure 21.28). The causative fungus, *Histoplasma capsulatum,* is dimorphic and is found in soils contaminated by bat or bird droppings (figure 21.27).

Review Questions

Short Answer

1. How does contamination of the eye lead to upper respiratory infection?
2. After you recover from strep throat, can you get it again? Explain.
3. Where is the gene for diphtheria toxin production located?
4. Describe two ways to decrease the chance of contracting a cold.
5. What kinds of diseases are caused by adenoviruses?
6. How do alcoholism and cigarette smoking predispose a person to pneumonia?
7. Give a mechanism by which *Klebsiella* sp. become antibiotic-resistant.
8. Why does the incidence of whooping cough rise promptly when pertussis immunizations are stopped?
9. Why are two or more antitubercular medications used together to treat tuberculosis?
10. Why did it take so long to discover the cause of Legionnaires' disease?

Multiple Choice

1. The following are all complications of streptococcal pharyngitis *except*
   a) glomerulonephritis.
   b) scarlet fever.
   c) subacute bacterial endocarditis.
   d) acute rheumatic fever.
   e) Reye's syndrome.
2. All of the following are true of diphtheria *except*
   a) a membrane that forms in the throat can cause suffocation.
   b) a toxin is produced that interferes with ribosome function.
   c) the causative organism typically invades the bloodstream.
   d) immunization with a toxoid prevents the disease.
   e) nerve injury with paralysis is common.

3. Adenoviral infections generally differ from the common cold in all the following ways, except adenoviral infections are
   a) not caused by picornaviruses.
   b) often associated with fever.
   c) associated with severe sore throat.
   d) much more likely to cause pneumonia.
   e) avoided by handwashing.
4. All are true of mycoplasmal pneumonia *except*
   a) it is a mycosis.
   b) it usually does not require hospitalization.
   c) penicillin is ineffective for treatment.
   d) it is the leading cause of bacterial pneumonia in college students.
   e) the infectious dose of the causative organism is low.
5. All of the following are true of Legionnaires' disease *except*
   a) the causative organism can grow inside amebas.
   b) it spreads readily from person to person.
   c) it is more likely to occur in long-term cigarette smokers than in nonsmokers.
   d) it is often associated with diarrhea or other intestinal symptoms.
   e) it can be contracted from household water supplies.
6. Which of the following infectious agents is most likely to cause a pandemic?
   a) Influenza A virus
   b) *Streptococcus pyogenes*
   c) *Histoplasma capsulatum*
   d) Sin Nombre virus
   e) *Coccidioides immitis*

7. Respiratory syncytial virus

   a) is a leading cause of bronchiolitis in infants.

   b) is an enveloped DNA virus of the adenovirus family.

   c) attaches to host cell membranes by means of neuraminidase.

   d) poses no threat to elderly people.

   e) mainly causes disease in the summer months.

8. In the United States, hantaviruses

   a) are limited to southwestern states.

   b) are carried only by deer mice.

   c) infect human beings with a fatality rate above 40%.

   d) were first identified in the early 1970s.

   e) are contracted mainly in bat caves.

9. All of the following are true of coccidioidomycosis *except*

   a) it is contracted by inhaling arthrospores.

   b) it is caused by a dimorphic fungus.

   c) endospores are produced within a spherule.

   d) it is more common in Maryland than in California.

   e) it is often associated with painful nodules on the legs.

10. The disease histoplasmosis

    a) is caused by an encapsulated bacterium.

    b) is contracted by inhaling arthrospores.

    c) occurs mostly in hot, dry, and dusty areas of the American Southwest.

    d) is a threat to AIDS patients living in areas bordering the Mississippi River.

    e) is commonly fatal for pigeons and bats.

Applications

1. A physician is advising the family on the condition of a diphtheria patient. How would the physician explain why the disease affects some tissues and not others?

2. How should a physician respond to a mother who asks if her daughter can get pneumococcal pneumonia again?

Critical Thinking

1. If all transmission of *Mycobacterium tuberculosis* from one person to another was stopped, how long would it take for the world to be rid of the disease?

2. Medications that prevent and treat influenza by binding to neuraminidase on the viral surface act against all the kinds of influenza viruses that infect humans. What does this imply about the nature of the interaction between the medications and the neuraminidase molecules?

Skin Infections

A dividing *Staphylococcus epidermidis* cell.

A Glimpse of History

Howard T. Ricketts was born in Ohio in 1871. He studied medicine in Chicago, and then specialized in pathology, the study of the nature of disease and its causes. In 1902, he was appointed to the faculty of the University of Chicago, where his research interests turned to Rocky Mountain spotted fever (RMSF).

While patients with RMSF have a dramatic rash and often die, the disease was poorly understood. Ricketts noticed that people and laboratory animals with it had tiny rod-shaped bacteria in their blood. He was sure these tiny invaders caused the disease, because he could transmit the disease by injecting laboratory animals with blood from an infected person. However, he was never able to grow the bacteria on laboratory media.

Based on observations of patients with RMSF, scientists suspected that the disease was contracted from tick bites. Ricketts went on to prove that certain species of ticks could transmit the disease from one animal to another. The infected ticks remained healthy but were capable of transmitting the disease for long periods of time. Surprisingly, the offspring of infected ticks were often born infected as well. Ricketts showed that the eggs of infected ticks frequently contained large numbers of the same bacteria found in the patients' blood. When these eggs were fertilized they developed into infected ticks. Ricketts was unable to cultivate these bacteria for further studies so he declined to give them a scientific name and went off to Mexico to study a very similar disease—louse-borne typhus. Unfortunately, he contracted typhus and died at the age of 39. Stanislaus Prowazek, a European scientist studying the same disease, met the same fate at almost the same age.

The martyrdom of the two young scientists struggling to understand infectious diseases is memorialized in the name of the louse-borne typhus agent, *Rickettsia prowazekii.* Both the genus and species names of the RMSF agent, *Rickettsia rickettsii,* recognize Howard Ricketts. We now know that these bacteria are obligate intracellular parasites. Since most culture media lack live cells, the bacteria did not grow in the culture media used by Ricketts and Prowazek.

Much of the human body's contact with the outside world occurs at the skin surface. This tough, flexible outer covering provides a remarkable barrier to invasion and infection. Its exposed state, however, leaves it vulnerable to a variety of injuries. Cuts, punctures, burns, chemical injury, and insect or tick bites can break this barrier and provide a way for pathogens to enter underlying tissues. For example, *Staphylococcus aureus* can enter a surgical wound and then invade the bloodstream, and sandfly bites can introduce *Leishmania* species, the cause of leishmaniasis—more can be learned about this disease at the text website (**www.mhhe.com/nester7**). Skin infections also occur when microorganisms or viruses that enter the body from another site (such as the respiratory or gastrointestinal systems) are carried by the bloodstream to the skin.

22.1 ■ Anatomy, Physiology, and Ecology

Learning Outcomes

1. *Describe the function of skin in health and disease.*
2. *Explain the role of normal skin microbiota in health and disease.*

The skin is far more than an inert wrapping for the body. It prevents the entry of microbes, regulates body temperature, and restricts the loss of fluid from body tissues. The multiple sensory receptors found in the skin provide the central nervous system with information about the environment. The skin also plays an essential role in the function of the immune system. If the barrier of the skin is breached, resident macrophages and dendritic cells produce cytokines that aid the development of an immune response. Collections of lymphocytes found in skin-associated lymphoid tissues (SALT) are ready to proliferate when needed. ◄◄ cytokines, p. 341 ◄◄ skin-associated lymphoid tissue, p. 358

The skin is composed of two layers—the **epidermis** and the **dermis** (**figure 22.1**). The subcutaneous layer lies beneath the dermis and supports the skin.

■ **Epidermis.** This is the surface layer, made up of multiple layers of flat cells. The outermost cells are dead and filled with keratin, a tough, water-resistant protein also found in hair and nails. These skin cells regularly flake off, exposing their replacements immediately below. Cells at the boundary between the dermis and epidermis are actively dividing. These newly formed cells migrate to the skin's surface, become flattened and die as keratin fills their cytoplasm. Microbes living on the outermost cells are shed with the flaking skin cells. This process is a valuable defense against infection and results in the complete regeneration of the epidermis about once a month. Dandruff represents excessive shedding of skin cells.

■ **Dermis.** The dermis is the layer of skin cells directly below the epidermis. This layer contains many tiny nerves, glands, blood vessels, and lymphatic vessels. The dermis has connective tissue that binds irregularly to the fat and other cells that make up the subcutaneous layer.

■ **Subcutaneous tissue.** While not part of the skin itself, the subcutaneous layer is made up of fat and other cells that support the layers of the skin.

The outermost layers of skin are typically bathed in secretions. Glands in the dermis produce sweat and an oily secretion called sebum. Sweat is delivered via fine tubules from the sweat glands to the surface of the skin where it evaporates to leave a salty residue that inhibits many microbes. The sebaceous glands open into hair follicles, so that sebum flows up through the follicles and out over the skin surface. Sebum keeps the hair and skin soft, pliable, and water-repellent.

In addition to being a physical barrier, skin represents a distinct ecological habitat. Compared to the moist warm conditions in the respiratory tract, however, the skin is a cool desert. Members of the normal skin microbiota are uniquely suited to thrive in this environment. They use the substances in sebum and sweat as nutrients to fuel their growth, and in doing so, inhibit other microbes. For example, resident bacteria degrade lipids in sebum, producing fatty acids that are toxic to many bacteria. In fact, the skin surface is an unfriendly habitat for most pathogens, being too dry, salty, acidic, and toxic for their survival. Those that tolerate the conditions are often shed with the dead skin cells. ◄◄ normal microbiota, pp. 8, 381

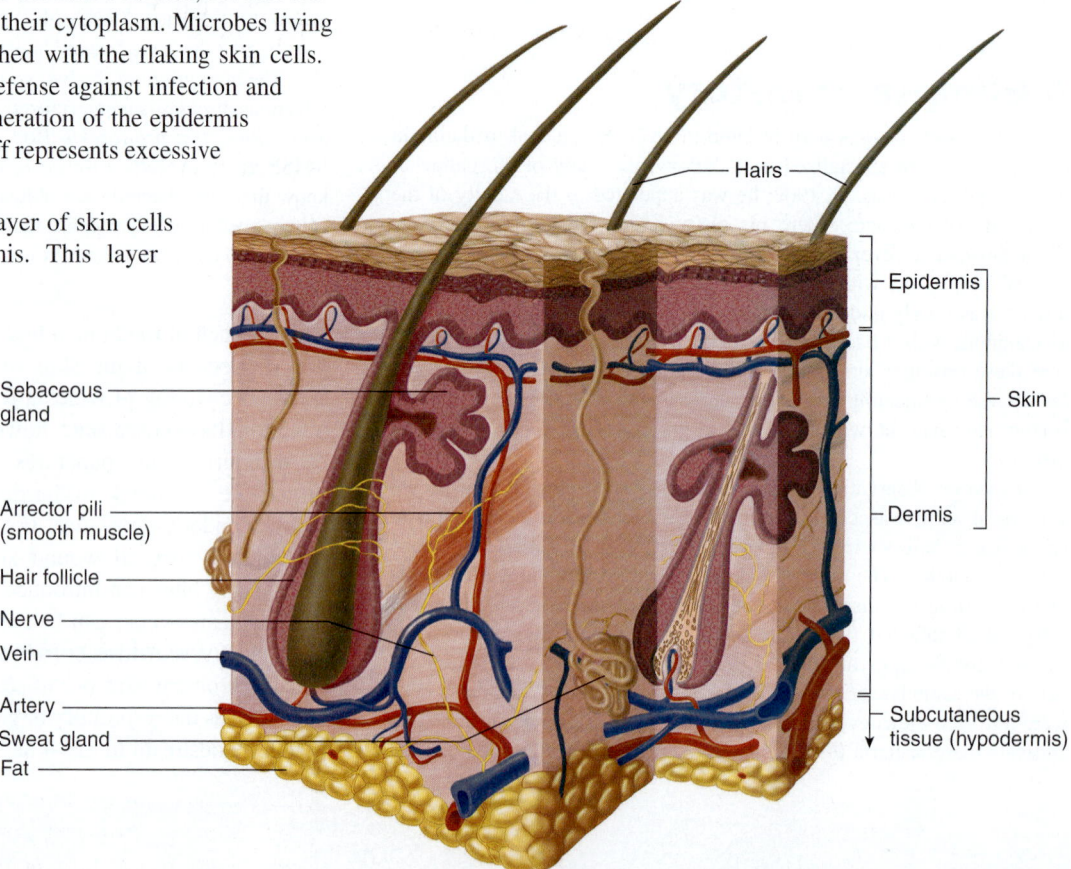

FIGURE 22.1 Microscopic Anatomy of the Skin Notice that the sebaceous unit, composed of the hair follicle and the attached sebaceous gland, almost reaches the subcutaneous tissue.

❓ *How can anaerobic bacteria, such as Propionibacteria species, grow on human skin?*

Labels on figure: Hairs · Epidermis · Skin · Dermis · Subcutaneous tissue (hypodermis) · Sebaceous gland · Arrector pili (smooth muscle) · Hair follicle · Nerve · Vein · Artery · Sweat gland · Fat

Although members of the normal microbiota play an essential protective role on the skin, they can also be troublesome. For one thing, their metabolic products are responsible for body odor. Sweat is odorless when first secreted, but bacteria degrade some of its components, producing foul-smelling compounds. Most antiperspirants prevent body odor by decreasing the number and metabolic activity of bacteria at the site of application. Some members of the normal skin microbiota are opportunistic pathogens, causing disease in people with impaired body defenses.
◀◀ opportunistic pathogens, p. 382

Different regions of the skin can be compared to unique neighborhoods, made up of distinct numbers and types of inhabitants. Depending on the body location and amount of moisture, the number of bacteria on the skin surface may range from only about 1,000 organisms per square centimeter on the back to more than 10 million in the groin and armpit, where there is plenty of moisture. Most of the microbial skin inhabitants can be categorized in three groups: diphtheroids, staphylococci, and fungi (**table 22.1**).

In oily regions like the forehead, upper chest, and back, diphtheroids are common. The name of this informal grouping refers to the fact that these Gram-positive bacteria physically resemble *Corynebacterium diphtheriae*, the pathogen that causes diphtheria. Among the most common diphtheroids on the skin are *Propionibacterium* species. These obligate anaerobes grow within hair follicles, where O_2 is limited. ◀◀ *Corynebacterium*, p. 265
◀◀ diphtheroids, p. 265   ◀◀ anaerobe, p. 90

Staphylococci such as *Staphylococcus epidermidis* are common on the skin surface. These salt-tolerant Gram-positive cocci grow well on the dry environment of the skin surface. Staphylococci use available nutrients on the skin, preventing pathogen colonization. They also produce antimicrobial substances highly active against other Gram-positive bacteria, helping to maintain a balance among the microbial inhabitants of the skin ecosystem.
◀◀ *Staphylococcus*, p. 273

Malassezia species are tiny lipid-dependent yeasts (fungi) present on human skin from late childhood onward. These yeasts can be grown on laboratory media containing fatty substances such as olive oil.

MicroByte

The average person sheds about 40,000 skin cells daily, or 1,500 during a typical microbiology lecture.

TABLE 22.1	Principal Members of the Normal Skin Microbiota
Name	**Characteristics**
Diphtheroids	Variably shaped, non-motile, Gram-positive rods of the *Corynebacterium* and *Propionibacterium* genera
Staphylococci	Gram-positive cocci arranged in packets or clusters; coagulase negative; facultatively anaerobic
Fungi	Small yeasts of the genus *Malassezia* that require oily substances for growth

MicroAssessment 22.1

The skin resists colonization by most microbial pathogens, provides a physical barrier to infection, transmits sensory input from the environment, and assists the body's regulation of temperature and fluid balance. Members of the normal skin microbiota help protect the skin from colonization by pathogens. Resident microbes are responsible for body odor and can cause disease when the host's defenses are impaired.

1. *Describe four characteristics of skin that help it resist infection.*
2. *Name and describe three groups of microorganisms generally present on normal skin.*
3. *Would you expect a person living in the tropics or in the desert to have larger numbers of bacteria living on the surface of his or her skin?* ✚

22.2 ■ Bacterial Skin Diseases

Learning Outcomes

3. *Compare and contrast acne and hair follicle infections.*
4. *Describe the characteristics of staphylococcal scalded skin syndrome and impetigo.*
5. *Compare and contrast Lyme disease and Rocky Mountain spotted fever.*

Few bacterial species invade intact skin directly. However, strands of hair provide a route of invasion by extending past the epidermis to hair follicles that penetrate into the dermis. Acne and hair follicle infections both begin with the hair shaft and are two of the most common infections due to direct invasion.

Acne Vulgaris

Acne in its most common form begins at puberty in association with a rise in sex hormones. Although the incidence of acne drops dramatically after puberty, some individuals have acne well into their 30s and 40s.

Signs and Symptoms

Acne is characterized by enlarged sebaceous glands and increased secretion of sebum. The hair follicle epithelium thickens and sloughs off in clumps, gradually blocking the flow of sebum to the skin surface. Continued sebum production by the infected gland can force a plug of material to the surface, where it is visible as a blackhead. With complete obstruction, the follicle fills with sebum, causing the epidermis to bulge outward. This produces a whitish lesion called a whitehead.

Causative Agent

The bacterium typically associated with acne is *Propionibacterium acnes*. The cells are Gram-positive rods that grow anaerobically in and around hair follicles. They have lipases that break down the oily sebum in the sebaceous glands and use the resulting fatty acids and glycerol as a food source. Sebum accumulates when a gland is blocked, providing more food for *P. acnes*. These bacteria then multiply to enormous numbers in the trapped sebum.

Pathogenesis

The metabolic products of the dividing bacteria cause an inflammatory response, attracting neutrophils whose enzymes damage the wall of the enlarged follicle. This can cause the follicle to burst, releasing its contents into the surrounding tissue. The result is an **abscess**—a collection of pus surrounded by inflamed tissue. The pus is made up of living and dead neutrophils, bacteria, and tissue debris. Most abscesses will eventually heal, but may leave a scar. ◄◄ inflammatory response, p. 348 ◄◄ neutrophils, p. 339

Epidemiology

Most people have *P. acnes* on their skin throughout their lives. The increased incidence of acne during puberty is probably due to excess sebum production caused by increased hormone levels, particularly testosterone. Infants may also have acne caused by maternal hormones stimulating sebum production.

Treatment and Prevention

Acne is usually a relatively mild infection that will run its course. The length of infection can be limited by medications such as antibiotics and benzoyl peroxide that inhibit the growth of *P. acnes*. Other medications such as isotretinoin (Accutane) act by reducing sebum production. Accutane is generally reserved for the most serious cases of acne because it has potentially serious side effects. Squeezing acne lesions is ill-advised, because it may cause the inflamed follicles to rupture, leading to more acne scars. It can also introduce bacteria into the bloodstream, which could spread the infection to multiple locations throughout the body.

Hair Follicle Infections

Hair follicle infections are generally mild and commonly clear up without treatment. In some instances, however, they progress into severe or even life-threatening disease.

Signs and Symptoms

There are three outcomes of hair follicle infections, listed by increasing severity: folliculitis, furuncles, and carbuncles. **Folliculitis** is inflammation of one or more hair follicles, causing small red bumps, or pimples. The hair can be pulled from its follicle, releasing a small amount of pus. The infection often goes away without further treatment. If the infection extends from the follicle to adjacent tissues, it will cause a localized redness, swelling, severe tenderness, and pain. This lesion is called a **furuncle** or boil. Pus may drain from the boil along with a plug of inflammatory cells and dead tissue. A furuncle may worsen to form a **carbuncle,** a large area of redness, swelling, and pain with several sites of draining pus. Carbuncles usually develop in areas of the body where the skin is thick, such as the back of the neck. Fever is often present, along with other signs of a serious infection. ◄◄ fever, p. 351

Causative Agent

Most furuncles and carbuncles, as well as many cases of folliculitis, are caused by *Staphylococcus aureus*. The name derives from the Greek root *staphyle* or "a bunch of grapes," referring

TABLE 22.2	Some Diseases Often Caused by *Staphylococcus aureus*
Disease	**Page for More Information**
Carbuncles	p. 524
Endocarditis	p. 672
Folliculitis	p. 524
Food poisoning	p. 757
Furuncles	p. 524
Impetigo	p. 528
Scalded skin syndrome	p. 527
Toxic shock syndrome	p. 619
Wound infections	p. 551

to the arrangement of the bacteria as seen in stained smears. The species name, *aureus,* or "golden," refers to the typical creamy color of the *S. aureus* colonies. This bacterium is an extremely important potential pathogen and is mentioned frequently throughout this text as the cause of a number of medical conditions (**table 22.2**).

One of the most useful identifying characteristics of *S. aureus* that distinguishes it from most other staphylococci is that it produces **coagulase** and **clumping factor.** Despite the -*ase* ending, coagulase is not an enzyme, but a secreted protein. Clumping factor is a functionally similar, but genetically distinct, protein. It is easy to test for in the laboratory and it is often called "slide coagulase." Both coagulase and clumping factor are important virulence factors for *S. aureus*. Other staphylococcal species, such as *Staphylococcus epidermidis,* cause disease infrequently and lack the genes and the proteins for coagulase and clumping factor. ◄◄ virulence factor, p. 383

Pathogenesis

Infection begins when *Staphylococcus aureus* attaches to the cells of a hair follicle, multiplies, and spreads to involve the sebaceous glands. The infection induces an inflammatory response with swelling and redness, followed by attraction and accumulation of neutrophils. If the infection continues, the follicle becomes a plug of inflammatory cells and dead tissue overlying a small abscess (**figure 22.2**). The infectious process sometimes spreads deeper, reaching the subcutaneous tissue where a large abscess forms. This subcutaneous abscess is responsible for the painful localized swelling called a boil. Without effective treatment, pressure within the abscess may increase, causing it to expand to other hair follicles and form a carbuncle. If organisms enter the bloodstream, the infection can spread to other parts of the body, such as the heart, bones, or brain.

S. aureus strains can have many different virulence factors, although not all pathogenic strains make the same factors (**table 22.3**). Nearly all strains have **protein A,** a cell wall component that interferes with phagocytosis (see figure 16.11).

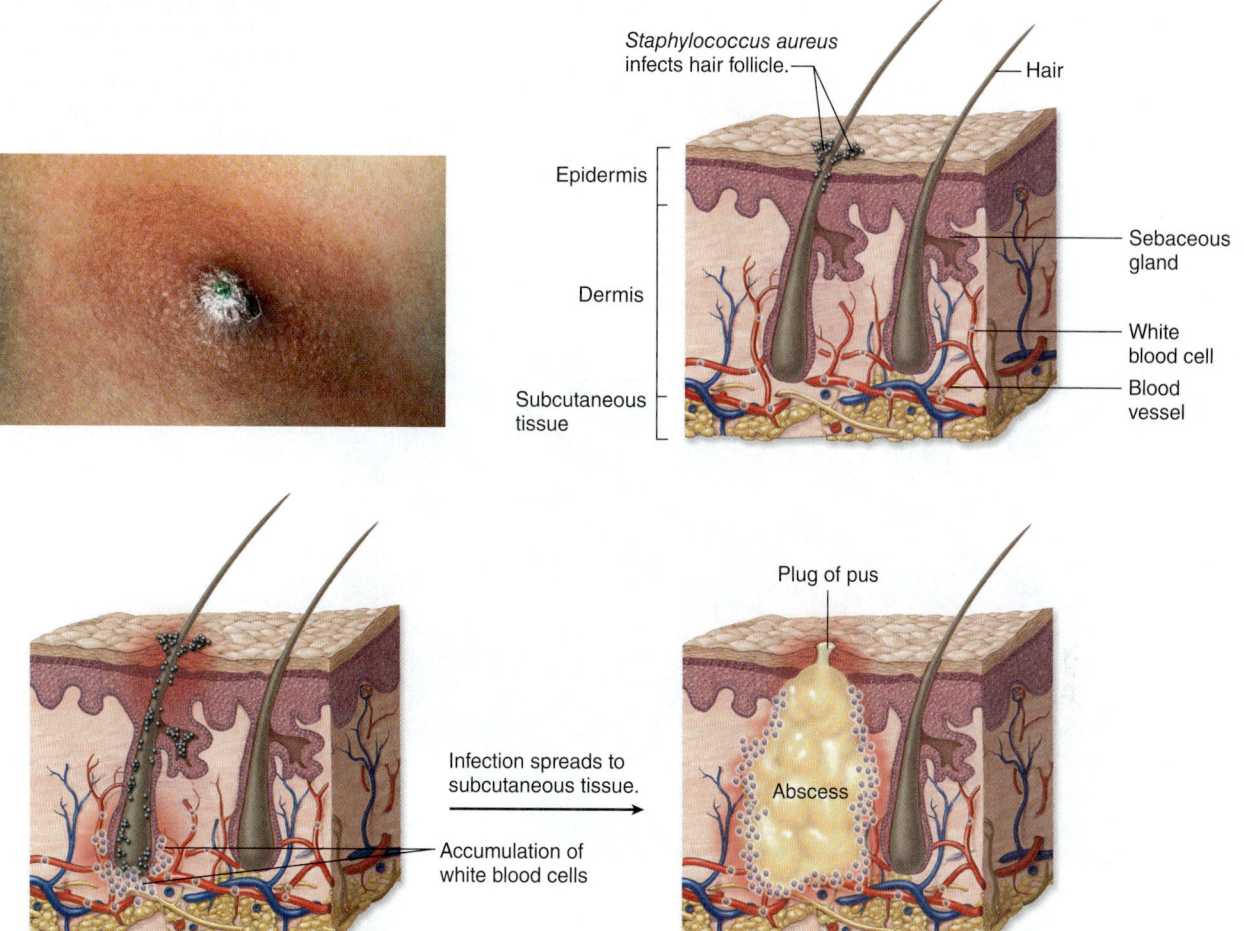

FIGURE 22.2 Pathogenesis of a Boil (Furuncle) *Staphylococcus aureus* infects a hair follicle through its opening on the skin surface. The infection produces a plug of necrotic material, a small abscess in the dermis, and finally, a larger abscess in the subcutaneous tissue.

Why is it a bad idea to squeeze a pimple?

TABLE 22.3	Properties of *Staphylococcus aureus* Implicated in Its Virulence
Product	**Effect**
Capsule	Inhibits phagocytosis
Clumping factor	Attaches the bacterium to fibrin, fibrinogen, and plastic devices
Coagulase	May slow progress of leukocytes into infected area by producing clots in the surrounding capillaries
Enterotoxins	Superantigens cause food poisoning if ingested, cause toxic shock if systemic
Exfoliatin	Separates layers of epidermis, causing scalded skin syndrome
Fibronectin-binding protein	Attaches bacterium to acellular tissue substances, endothelium, epithelium, clots, indwelling plastic devices
Hyaluronidase	Breaks down hyaluronic acid component of tissue, thereby allowing infection to spread
Leukocidin	Kills neutrophils or causes them to release their enzymes
Lipase	Breaks down fats by hydrolyzing the bond between glycerol and fatty acids
Proteases	Degrade collagen and other tissue proteins
Protein A	Binds to Fc portion of antibody, thereby interfering with opsonization
Toxic shock syndrome toxin	Causes rash, diarrhea, and shock
α-toxin	Makes holes in host cell membranes

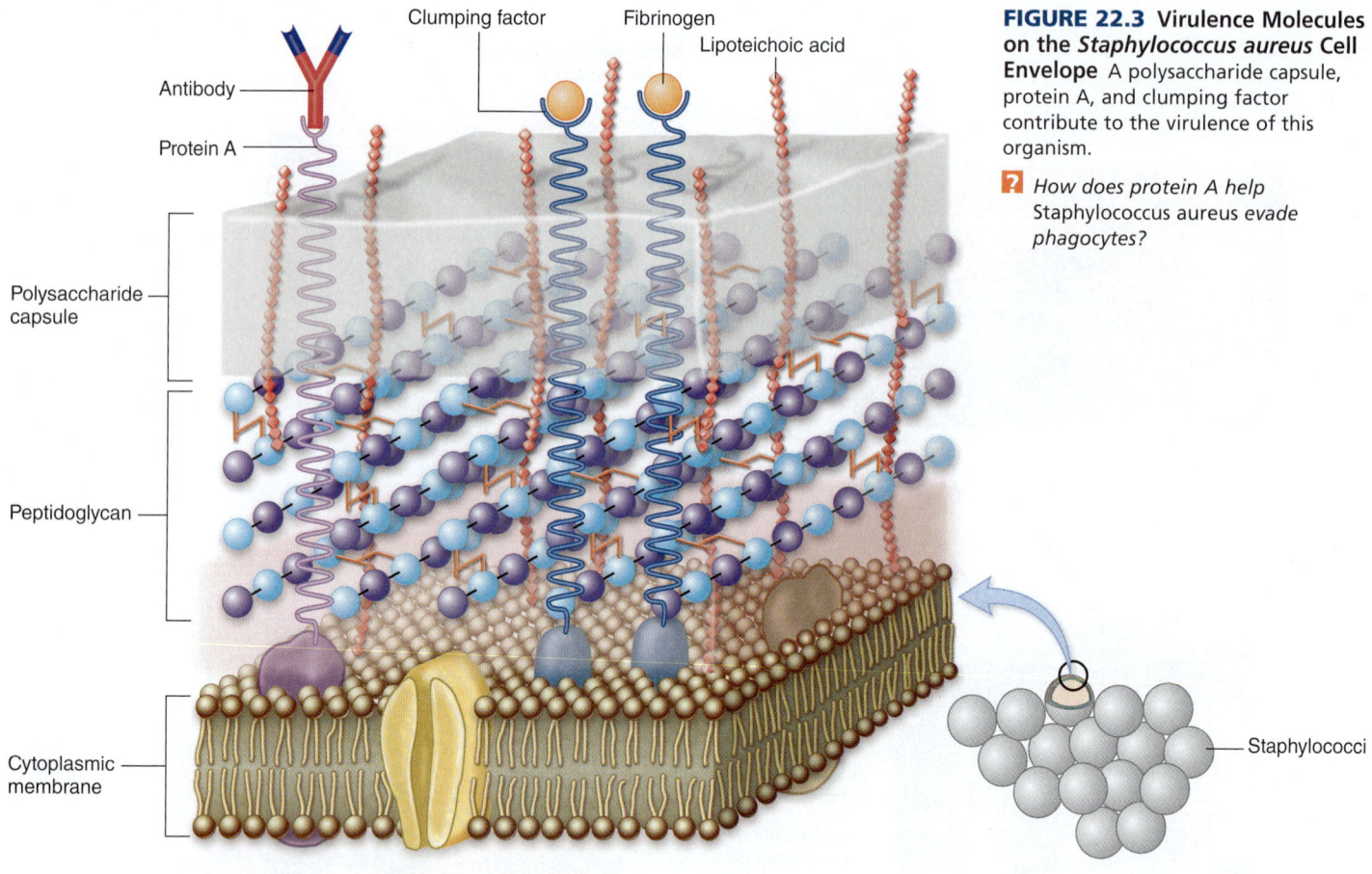

Antibody

Protein A

Clumping factor

Fibrinogen

Lipoteichoic acid

Polysaccharide capsule

Peptidoglycan

Cytoplasmic membrane

Staphylococci

FIGURE 22.3 Virulence Molecules on the *Staphylococcus aureus* Cell Envelope A polysaccharide capsule, protein A, and clumping factor contribute to the virulence of this organism.

❓ *How does protein A help* Staphylococcus aureus *evade phagocytes?*

Many strains of *S. aureus* also synthesize a polysaccharide capsule that inhibits phagocytosis (**figure 22.3**). ◀◀ protein A, p. 390

Coagulase and clumping factor allow *S. aureus* to colonize soft tissues while avoiding the immune response. Coagulase is an important virulence factor because it activates a blood protein called prothrombin to form thrombin. Thrombin then converts fibrinogen to fibrin, the fibrous protein that forms the weblike foundation of blood clots. While most coagulase is secreted, some is tightly bound to the bacterial cells and coats their surface with fibrin when they come into contact with blood. This helps disguise the bacteria, hiding them from the immune system. Fibrin-coated staphylococci resist phagocytosis. Clumping factor is a virulence factor for *S. aureus* because it attaches to fibrinogen and fibrin present in damaged tissues, and helps the bacteria colonize wound surfaces. Because plastic devices such as intravenous catheters and heart valves quickly become coated with fibrinogen after insertion into the bloodstream, they also become targets for colonization.

S. aureus secretes many enzymes that can damage host tissues. Hyaluronidase degrades hyaluronic acid, a component of host tissue that helps hold the cells together. Proteases degrade various host proteins including collagen, the white fibrous protein found in skin, tendons, and connective tissue. Lipases, which degrade lipids, provide a food source when *S. aureus* colonizes oily hair follicles.

Most strains of *S. aureus* also produce one or more toxins that damage or kill host cells. The membrane-damaging α-toxin

kills host cells by making holes in their membranes. Leukocidins kill white blood cells. A relatively small percentage of *S. aureus* strains produce one or more additional disease-specific toxins. One of these, exfoliatin, causes staphylococcal scalded skin syndrome, the next disease described in this chapter. ◀◀ membrane-damaging toxin, p. 392

Epidemiology

Staphylococcus aureus can be found in the nostrils of virtually everyone at one time or another. About 20% of healthy adults have continually positive nasal cultures for a year or more, whereas over 60% will be colonized at some time during a given year. The organisms are mainly spread to other parts of the body and to the environment by the hands. Although the nostrils seem to be the preferred habitat of *S. aureus*, moist areas of skin are also frequently colonized. People with boils and other staphylococcal infections shed large numbers of *S. aureus* and should not work with food, or near patients with surgical wounds or chronic illnesses. Staphylococci survive well in the environment, so they are easily transferred from one host to another via fomites. ◀◀ fomites, p. 440

Because *S. aureus* is so common, epidemics of staphylococcal disease can generally be traced to their sources only by identification of the epidemic strain. To precisely identify an epidemic strain, techniques such as molecular typing, phage typing, and antibiogram determination must be used (see section 10.4). ◀◀ characterizing strain differences, p. 247

Treatment and Prevention

Treatment of boils and carbuncles may require minor surgery to drain the pus from the lesion. Afterward, patients are usually given a course of oral antibiotics.

Treating *S. aureus* infections can be complicated because many strains are resistant to multiple antibacterial medications. As described in chapter 20, the organism is an important example of emerging antibiotic resistance. When penicillin was first introduced, nearly all strains were susceptible. However, strains that produce a penicillinase (a type of β-lactamase) spread quickly, so that most isolates are now resistant to this medication. Strains producing modified versions of penicillin-binding proteins also appeared. These strains, referred to as MRSA (methicillin-resistant *Staphylococcus aureus*) are resistant to all β-lactam medications, including penicillin derivatives and cephalosporins. Many MRSA strains are resistant to a number of other medications as well. Before 1997, even the most resistant examples were reliably treated with vancomycin. Since then, however, the first vancomycin-intermediate *S. aureus* (VISA) and vancomycin-resistant *S. aureus* (VRSA) strains were identified. Several medications active against VRSA have been marketed, including linezolid, daptomycin, and tigecycline. Linezolid represents a new class of antibacterials, the oxazolidinones. Hopefully, development of new medications will keep pace with the growth of bacterial resistance, but this will depend on humans avoiding overuse of these valuable substances. ◀◀ β-lactamase, p. 464 ◀◀ VISA and VRSA, p. 473

When they first appeared, most MRSA strains could be traced to hospitals and clinics. More recently however, completely different strains have become widespread among healthy carriers in the community. These new strains are called CA-MRSA (community-acquired MRSA) to distinguish them from hospital-acquired strains (HA-MRSA). The most common CA-MRSA strain in the United States is USA-300. All current isolates are thought to be the progeny of one original population. Generally, these CA-MRSA strains are resistant only to methicillin and other β-lactam antibiotics, and macrolides. They are highly virulent and possess a group of genes that codes for a leukocyte-destroying leukocidin. There is some debate as to whether this leukocidin contributes to higher morbidity and mortality or if it is simply a trait shared by most strains of CA-MRSA. ◀◀ penicillins and other β-lactam antibiotics, p. 463 ◀◀ macrolides, p. 466

In an effort to control the spread of MRSA, many hospitals screen patients at admission. The patients who carry a MRSA strain are isolated and given appropriate antibacterial treatment to limit spread of the bacteria to other patients or staff. Some hospitals also screen patients for MRSA when they are discharged from the hospital to make sure they do not take a MRSA strain home with them. ◀◀ carriers, p. 439

Prevention of staphylococcal skin disease is very difficult because so many individuals carry the organism. Applying an antibacterial cream to the nostrils and using soaps containing an antibacterial agent such as hexachlorophene to wash the skin reduce the bacterial burden and may eliminate the carrier state. Because *S. aureus* is most commonly transmitted by the hands, handwashing and regular use of hand sanitizers are both effective means of limiting the spread of these bacteria.

MicroByte

Individuals colonized with *S. aureus* may have as many as 10^8 cells of the bacterium per nostril.

Staphylococcal Scalded Skin Syndrome

Staphylococcal scalded skin syndrome (SSSS) is a potentially fatal, toxin-mediated disease. This disease occurs mainly in infants.

Signs and Symptoms

As the name suggests, staphylococcal scalded skin syndrome causes a patient's skin to appear burned (**figure 22.4**). It begins as a generalized redness of the skin, sometimes affecting the entire body. Other signs and symptoms include malaise—a vague feeling of discomfort and uneasiness—irritability, and fever. The nose, mouth, and genitalia may be painful for one or more days before the typical features of the disease become apparent. Within 48 hours after the redness appears, the skin becomes wrinkled, and large blisters filled with clear fluid develop. The skin is tender to the touch, peels easily, and feels like sandpaper.

Causative Agent

SSSS is caused by *Staphylococcus aureus* strains that produce a toxin called **exfoliatin.** Only about 5% of *S. aureus* strains produce this toxin, which destroys material that binds together the outer layers of epidermis. At least two kinds of exfoliatins exist: One is coded for by a plasmid gene, and the other by a chromosomal gene. ◀◀ plasmids, p. 208

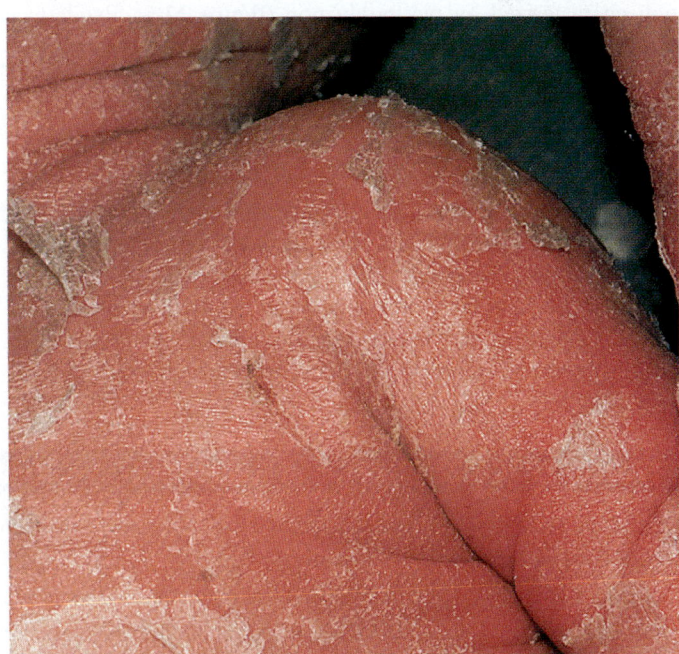

FIGURE 22.4 Staphylococcal Scalded Skin Syndrome (SSSS) A toxin called exfoliatin, produced by certain strains of *Staphylococcus aureus,* causes the outer layer of skin to separate.

❓ *Why is SSSS sometimes fatal?*

Pathogenesis

Exfoliatin causes the epidermis to split just below the dead keratinized outer layer. The *S. aureus* strains that cause SSSS generally grow in a relatively small area. However, the exfoliatin produced by the colonizing bacteria is carried through the bloodstream to damage large areas of skin. Because the outer layers of skin are lost (as happens with a severe burn), there is significant loss of body fluid and danger of secondary infection with Gram-negative bacteria such as *Pseudomonas* sp. ◀◀ **secondary infection, p. 382** ◀◀ *Pseudomonas* **sp., p. 265**

Epidemiology

Staphylococcal scalded skin syndrome can appear in any age group but occurs most frequently in newborn infants. Elderly and immunocompromised individuals are also at increased risk. Transmission is generally from person to person. The disease usually appears in isolated cases, although small epidemics in nurseries sometimes occur.

Treatment and Prevention

Initial therapy for staphylococcal scalded skin syndrome includes an antibiotic such as methicillin. As with severe burns, dead skin is removed to help prevent secondary infection. Although the disease can be fatal, prompt therapy usually leads to full recovery. There are no preventive measures except to place patients suspected of having SSSS in protective isolation. This limits the spread of the pathogen to others and helps prevent secondary infections in the isolated patient. **Table 22.4** describes the main features of this disease.

Streptococcal Impetigo

Impetigo is the most common type of **pyoderma,** a skin infection characterized by pus production (**figure 22.5**). Pyodermas can result from infection of an insect bite, burn, scrape, or other wound. Sometimes, the injury is so slight it is not apparent.

Signs and Symptoms

Impetigo is a superficial skin infection, causing inflammation in patches of epidermis just beneath the dead, scaly outer layer. Thin-walled blisters develop, then break, and are replaced by oozing, yellowish crusts of drying plasma. Usually, patients experience little fever or pain, but lymph nodes near the involved areas often enlarge, indicating that bacterial products have entered the lymphatic system and an immune response is occurring. ◀◀ **plasma, p. 426**

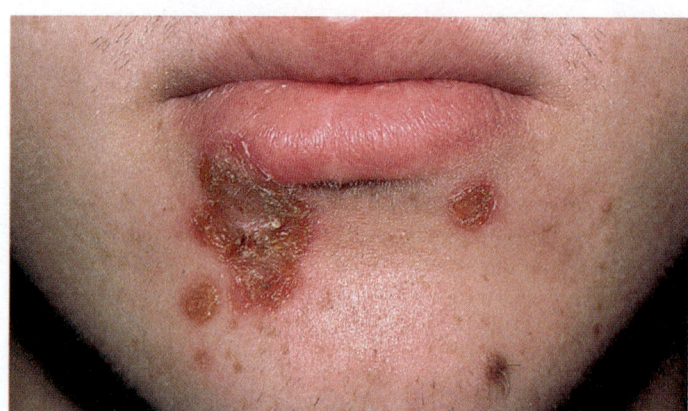

FIGURE 22.5 Impetigo This type of pyoderma is often caused by *Streptococcus pyogenes* and may result in glomerulonephritis.

❓ *How is impetigo spread in a population?*

Causative Agent

Many cases of impetigo, even epidemics, are due to *Streptococcus pyogenes.* These Gram-positive, chain-forming organisms are group A streptococci, bacteria characterized by the presence of a polysaccharide called group A carbohydrate in their cell walls. Like *Staphylococcus aureus, S. pyogenes* causes a variety of different diseases and is mentioned in numerous places throughout this text. The most detailed description is in chapter 21. *S. aureus* also causes impetigo. **Table 22.5** compares *S. aureus* and *S. pyogenes.* ◀◀ **hemolysis, p. 96** ◀◀ *Streptococcus pyogenes,* **p. 487**

Pathogenesis

Many different strains of *S. pyogenes* exist, some of which can colonize the skin. Minor injuries introduce the bacteria into the deeper layer of epidermis, often leading to an infection. Scratching an infected area may spread the bacteria to other areas of the skin.

The surface components of *S. pyogenes,* notably a hyaluronic acid capsule and a cell wall component called M protein, interfere with phagocytosis (see figure 21.4). A number of extracellular products contribute to the virulence of *S. pyogenes,* similar to those described for *S. aureus.* These include enzymes such as proteases, nucleases, and hyaluronidase. None of them appears to be essential, however, because antibodies against them fail to protect experimental animals. A more detailed description of these virulence factors is in chapter 21. ◀◀ **virulence factors of** *S. pyogenes,* **p. 488** ◀◀ **M protein, p. 390**

TABLE 22.4	Staphylococcal Scalded Skin Syndrome
Signs and symptoms	Sensitive red rash with sandpaper texture, malaise, irritability, fever, large blisters, peeling of skin
Incubation period	Variable, usually days
Causative agent	Strains of *Staphylococcus aureus* that produce exfoliatin toxin
Pathogenesis	Exfoliatin is produced by organisms at an infection site, usually of the skin, and carried by the bloodstream to the epidermis, where it causes a split in a cellular layer; loss of body fluid and secondary infections contribute to mortality.
Epidemiology	Person-to-person transmission; seen mainly in infants, but can occur at any age
Treatment and prevention	Treatment with penicillinase-resistant penicillins; removal of dead tissue. Isolate the patient to prevent secondary infection and to limit the spread of the disease to others.

TABLE 22.5	*Streptococcus pyogenes* Versus *Staphylococcus aureus*	
	Streptococcus pyogenes	**Staphylococcus aureus**
Morphology	Chains of Gram-positive cocci	Clusters of Gram-positive cocci
Growth characteristics	Beta-hemolytic colonies; catalase negative, coagulase negative, obligate fermenter	Golden, hemolytic colonies, catalase-positive, coagulase-positive, facultative anaerobe
Virulence factors	Cell wall contains group A polysaccharide, an Fc receptor (protein G), and M protein. Bacterium produces hemolysins (streptolysins O and S), streptokinase, DNase, hyaluronidase, and others	Cell wall contains an Fc receptor (protein A). Bacterium produces superantigens, hemolysins, leukocidin, hyaluronidase, nuclease, protease, and others
Diseases	Impetigo, strep throat, wound infections, scarlet fever, puerperal fever, toxic shock, and necrotizing fasciitis. Complications: glomerulonephritis and rheumatic fever	Boils, staphylococcal scalded skin syndrome, wound infections, abscesses, bone infections, impetigo, food poisoning, and staphylococcal toxic shock syndrome

Even though impetigo is limited to the epidermis, streptococcal products are absorbed into the circulation. The immune response to these proteins is thought to cause post-streptococcal sequelae, particularly acute glomerulonephritis. ◀◀ post-streptococcal sequelae, p. 489 ◀◀ glomerulonephritis, p. 490

Epidemiology

Impetigo is most common among poor young children living in hot, humid areas. Person-to-person contact spreads the disease, as do insects, and fomites such as toys and towels. Impetigo patients often become throat and nasal carriers of *S. pyogenes.* ◀◀ fomites, p. 440

Treatment and Prevention

So far, *S. pyogenes* strains remain susceptible to penicillin. For patients allergic to penicillin, erythromycin can be substituted.

People with impetigo should avoid contact with others to prevent spreading the bacteria. The transmission of impetigo can be limited by keeping skin clean and avoiding sharing personal items such as towels and washrags. Antiseptics and wound cleansing also decrease the chance of infection. **Table 22.6** summarizes the main features of impetigo.

Rocky Mountain Spotted Fever

Rocky Mountain spotted fever (RMSF) was first recognized in the Rocky Mountain area of the United States—thus its name. The disease is caused by an obligate intracellular pathogen called rickettsia and is transmitted by certain species of ticks, mites, and lice. ◀◀ the genus *Rickettsia*, p. 277

Signs and Symptoms

Rocky Mountain spotted fever begins suddenly with a headache, pains in the muscles and joints, and fever. Within a few days of infection, a rash characterized by faint pink spots appears on the palms, wrists, ankles, and soles. This rash, caused by blood leaking from damaged vessels, spreads up the arms and legs to the rest of the body (**figure 22.6**). Bleeding may occur at other sites, such

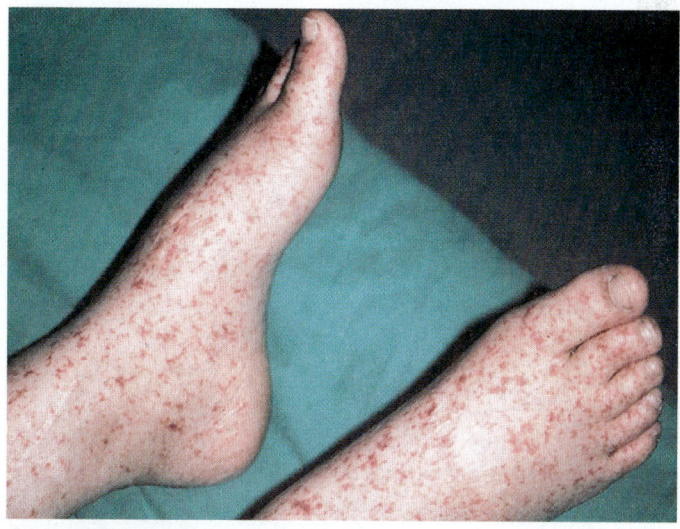

FIGURE 22.6 Rash Caused by Rocky Mountain Spotted Fever Characteristically, the rash begins on the arms and legs, spreads centrally, and as shown in this photo, becomes hemorrhagic.

? *What processes cause this rash?*

TABLE 22.6	Impetigo
Signs and symptoms	Blisters that break, releasing plasma and pus; formation of golden-colored crusts; lymph node enlargement
Incubation period	2 to 5 days
Causative organisms	*Streptococcus pyogenes, Staphylococcus aureus*
Pathogenesis	Organisms entering the skin through minor breaks; certain strains of *S. pyogenes* that cause impetigo can also cause glomerulonephritis
Epidemiology	Spread by direct contact with carriers or patients with impetigo, insects, and fomites
Treatment and prevention	An oral penicillin if cause is known to be *S. pyogenes;* otherwise, an anti-staphylococcal antibiotic orally or topically. Cleanliness; care of skin injuries.

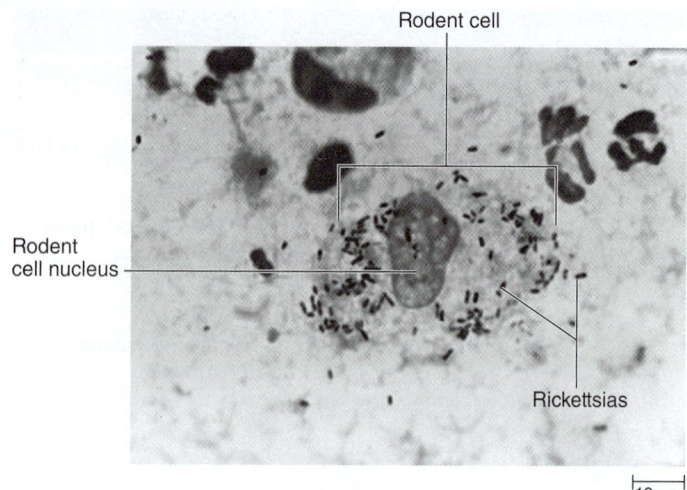

FIGURE 22.7 The Obligate Intracellular Bacterium *Rickettsia rickettsii* Growing in a Rodent Cell

❓ *Do you think* Rickettsia rickettsii *can grow on blood agar?*

as the nose and mouth. Damage to the heart, kidneys, and other body tissues can cause a drop in blood pressure, shock, and death unless treatment is given promptly.

Causative Agent

Rocky Mountain spotted fever is caused by *Rickettsia rickettsii* (**figure 22.7**). The organisms are tiny, Gram-negative, non-motile coccobacilli that require eukaryotic cells to grow. This obligate intracellular growth makes the bacteria very difficult to grow in culture. *R. rickettsii* can sometimes be identified early in an infection by visualizing the organisms in skin lesion biopsies (pieces of tissue surgically removed from the lesion site). Also, their DNA can be amplified by the polymerase chain reaction (PCR) and identified with a probe. ◀◀ **PCR, p. 227** ◀◀ **probe, p. 232**

Pathogenesis

The bite of a tick infected with *R. rickettsii* transmits Rocky Mountain spotted fever. The tick remains attached for hours while it feeds, but the process is usually painless and goes unnoticed. The pathogen is not transmitted until a tick has fed from 4 to 10 hours. The bacteria are released into capillary blood with the tick saliva and are taken up by the endothelial cells lining the small blood vessels. There is evidence that *R. rickettsii* force the endothelial cells to engulf

them, because endothelial cells are not normally capable of phagocytosis. Once inside the host cell, the bacteria leave the phagosome and multiply in both the cytoplasm and nucleus. They coat themselves in host cell actin and use this protein to move into adjacent host cells. Eventually, the host cell membrane is so damaged by this process that the cell ruptures, releasing the rickettsias. These travel through the bloodstream, infecting even more cells.

Infection can also extend into the walls of the small blood vessels, causing an inflammatory reaction. This leads to clotting in the blood vessels, and small areas of necrosis (tissue death). This process causes the hemorrhagic skin rash. More ominously, clotting occurs throughout the body, resulting in damage to vital organs such as the kidneys and heart. Potentially even more serious is the release of endotoxin (lipopolysaccharide) into the bloodstream from the rickettsial cell walls, causing shock and generalized bleeding because of **disseminated intravascular coagulation.**
◀◀ **lipopolysaccharide, p. 60** ▶▶ **disseminated intravascular coagulation, p. 674**

Epidemiology

Rocky Mountain spotted fever occurs in a sporadic distribution throughout the Americas. Despite the name of the disease, the highest incidence in the United States has generally been in the south Atlantic and south-central states (**figure 22.8**).

RMSF is a zoonosis, maintained in nature in various species of ticks and mammals. Rickettsias generally do not cause disease in their natural hosts—the host and pathogen co-evolved to reach a state of balanced pathogenicity. Humans, being an accidental host, often develop severe disease. ◀◀ **balanced pathogenicity, p. 385**

Several species of ticks transmit RMSF to humans. The main vector in the western United States is the wood tick, *Dermacentor andersoni* (**figure 22.9**), while in the East it is the dog tick, *D. variabilis*. Once infected, ticks remain so for life, transmitting *R. rickettsii* from one generation to the next through their eggs. Ticks are most active from April to September. Not surprisingly, most cases of RMSF occur during this time of year.
◀◀ **zoonoses, p. 439**

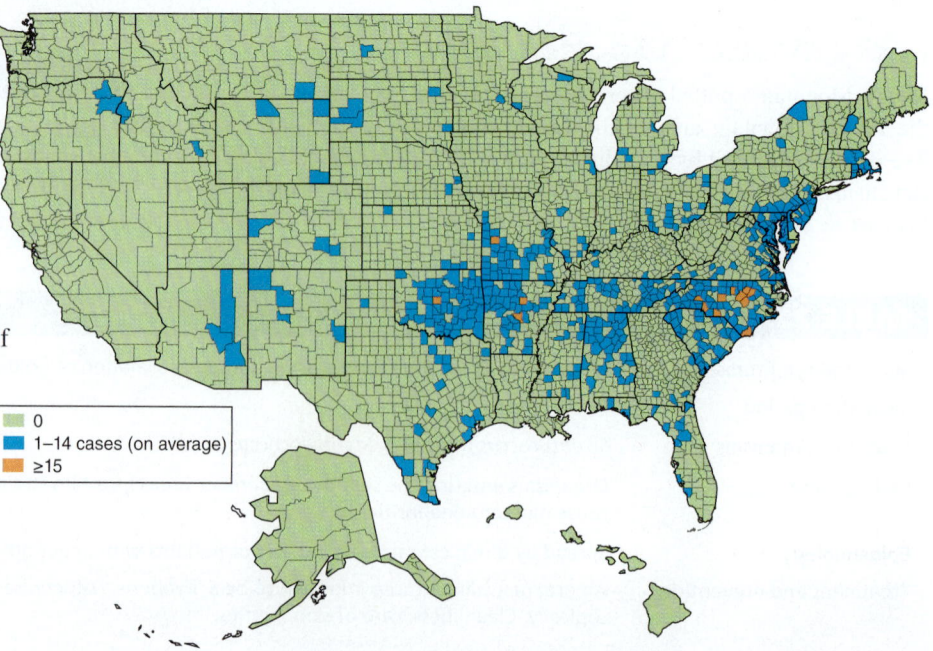

⬜	0
🟦	1–14 cases (on average)
🟧	≥15

FIGURE 22.8 Rocky Mountain Spotted Fever Number of reported cases, by county—United States, 2005.

❓ *Why do most cases of RMSF occur in the summer months?*

FIGURE 22.9 *Dermacentor andersoni,* **the Wood Tick**
The wood tick is the main vector of Rocky Mountain spotted fever in the western United States.

❓ *What can be done to limit exposure to these ticks?*

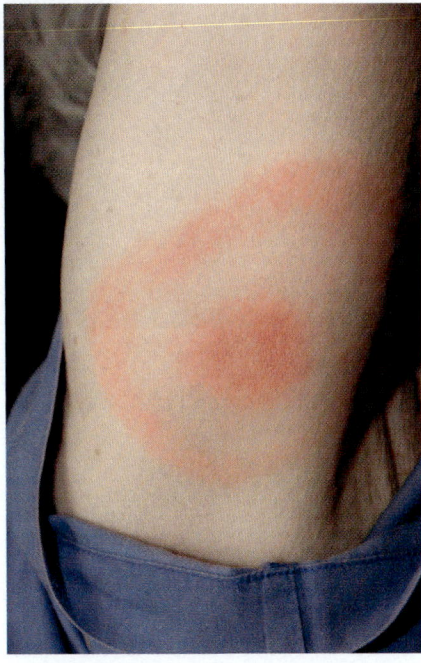

FIGURE 22.10
Erythema Migrans, the Characteristic Rash of Lyme Disease The rash usually has a targetlike or bull's-eye appearance. Although highly suggestive of Lyme disease, many people with the disease do not develop the rash.

❓ *What causes the bull's-eye pattern of this rash?*

Treatment and Prevention

The antibiotics doxycycline and chloramphenicol are very effective in treating Rocky Mountain spotted fever if given early in the disease, before irreversible damage to vital organs has occurred. Without treatment, the overall case-fatality rate from the disease is about 20%, but it can be much higher in elderly patients. With early diagnosis and treatment, the case-fatality rate is reduced to less than 5%.

No vaccine against RMSF is currently available to the public. The disease can be prevented if people take precautions to (1) avoid tick-infested areas when possible; (2) wear protective clothing; (3) use tick repellents such as DEET (N,N-diethyl-meta-toluamide); (4) carefully inspect their bodies (especially the scalp, armpits, and groin) for ticks several times daily; and (5) remove attached ticks carefully to avoid crushing them and thereby contaminating the bite wound with their infected tissue fluids. To remove ticks, blunt tweezers can be used to grasp the tick by its mouthparts and then gently pull it away from the skin. The site of the bite should be treated with an antiseptic. Touching the tick with a hot object, gasoline, or whiskey is ineffective. The main features of Rocky Mountain spotted fever are summarized in **table 22.7.**

Lyme Disease

Lyme disease was first recognized as a distinct entity in the mid-1970s, when a cluster of cases occurred in Lyme, Connecticut. It was not until 1982 that the causative agent was first identified in ticks by Dr. Willy Burgdorfer at the Rocky Mountain Laboratories in Montana. We now know that Lyme disease was widespread long before its recognition at Lyme. The ecology of the disease is complex and still incompletely understood, but we are beginning to get some answers as to why the disease has increased and extended its range. About 25,000 new cases occur each year, making it the most common vector-borne disease in the United States.

Signs and Symptoms

Signs and symptoms of Lyme disease can be divided roughly into three stages, although individual patients may be asymptomatic in one or more of these:

■ **Early localized infection.** This stage typically begins a few days to several weeks after a bite by an infected tick. It is characterized by enlargement of nearby lymph nodes and a circular skin rash called erythema migrans (**figure 22.10**). The rash begins as a red spot or bump at the site of the tick

TABLE 22.7	Rocky Mountain Spotted Fever
Signs and symptoms	Headache, pains in muscles and joints, and fever, followed by a hemorrhagic rash that begins on the extremities
Incubation period	4 to 8 days
Causative organism	*Rickettsia rickettsii,* an obligate intracellular bacterium
Pathogenesis	Organisms multiply at site of tick bite; invade the bloodstream and then infect endothelial cells; vascular lesions and immune response to endotoxin account for pathologic changes.
Epidemiology	A zoonosis transmitted by bite of infected tick, usually *Dermacentor* sp.
Treatment and prevention	Treated with: doxycycline or chloramphenicol. Prevented by avoiding tick-infested areas, using tick repellent, removal of ticks within 4 hours.

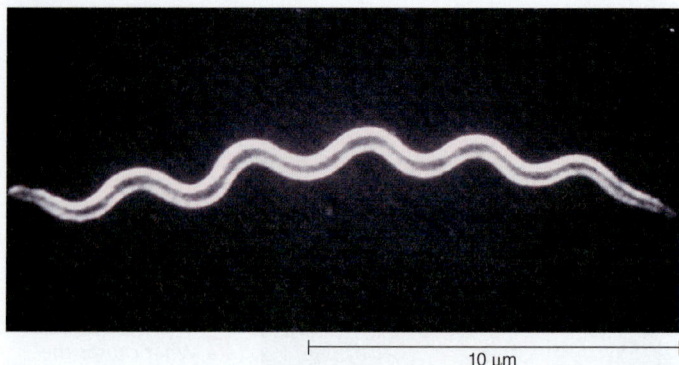

10 µm

FIGURE 22.11 Scanning Electron Micrograph of *Borrelia burgdorferi*, the Cause of Lyme Disease

? *How would you describe the shape of this organism?*

bite and slowly enlarges to eventually reach a diameter of about 15 cm (6 inches). The advancing edge is bright red, leaving behind an area of fading redness as the lesion enlarges. This characteristic bull's-eye rash is the hallmark of Lyme disease but is not present in all cases. Most of the other signs and symptoms that occur during this stage are flulike—malaise, chills, fever, headache, stiff neck, joint and muscle pains, and backache.

■ **Early disseminated infection.** This generally begins 2 to 8 weeks after the first signs and symptoms appear, and affects the nervous system. Electrical conduction within the heart is impaired, leading to dizzy spells or fainting. A temporary pacemaker may be required to maintain a normal heartbeat. Nervous system effects also include one or more of the following: paralysis of the face, severe headache, pain when moving the eyes, difficulty concentrating, emotional instability, fatigue, and impairment of the nerves of the legs or arms.

■ **Late persistent infection.** This begins around 6 months after the skin rash. This stage is characterized by joint pain and swelling, usually of large joints such as the knee. These signs and symptoms slowly disappear over subsequent years. Chronic nervous system impairments such as localized pain, paralysis, and depression can occur.

Causative Agent

Lyme disease is caused by *Borrelia burgdorferi*, a large, Gram-negative, microaerophilic spirochete (**figure 22.11**). The *Borrelia* chromosome is unusual in that it is linear and present in multiple copies. *B. burgdorferi* also contains numerous circular and linear plasmids that contain genes usually found on bacterial chromosomes. Studying this unusual bacterium may lead to an understanding of how it can infect such widely differing species as mice, lizards, and ticks. ◄◄ spirochetes, p. 272 ◄◄ microaerophilic, p. 90

Pathogenesis

The spirochetes are introduced into the skin by the bite of an infected tick. They multiply and migrate outward in a circular fashion. The lipopolysaccharide layer of these Gram-negative bacteria causes an inflammatory reaction in the skin, producing the expanding rash. The host's immune response is initially suppressed, allowing continued multiplication of the spirochete at the site. The organisms then enter the bloodstream and spread to all parts of the body. This dissemination accounts for the flulike signs and symptoms of the first stage. At this point, an intense immune response occurs, but after the first few weeks, it becomes very difficult to recover *B. burgdorferi* from blood or body tissues. The immune response against the bacterial antigens is probably responsible for the signs and symptoms of the second and third stages. The third stage of Lyme disease is characterized by arthritis, and the affected joints have high concentrations of reactive immune cells and immune complexes. Evidence suggests a role for autoimmune response against host tissues in some cases. ◄◄ autoimmunity, p. 412 ◄◄ lipopolysaccharide, p. 60

Epidemiology

Lyme disease is a zoonosis with humans being an accidental host. It is widespread in the United States, and its incidence depends on complex ecological factors (**figure 22.12**). Several species of ticks have been implicated as vectors, but the most important in the eastern United States is the black-legged (deer) tick, *Ixodes scapularis* (**figure 22.13**). In some areas of the East Coast, 80% of these ticks

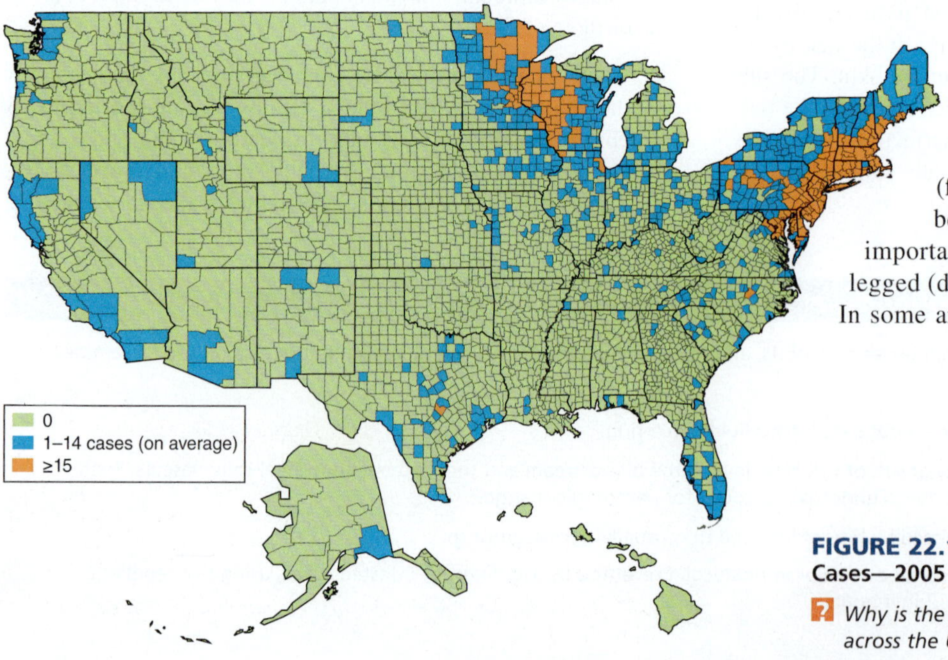

■ 0
■ 1–14 cases (on average)
■ ≥15

FIGURE 22.12 Number of Reported Lyme Disease Cases—2005

? *Why is the distribution of Lyme disease cases not uniform across the United States?*

FIGURE 22.13 The Black-Legged (Deer) Tick, *Ixodes scapularis*, Adult and Nymph This tick is the most important vector of Lyme disease in the eastern and north-central United States.

❓ *Why are people not always aware that they have been bitten by this tick?*

are infected with *B. burgdorferi*. Because of the tick's small size, they often feed and drop off their host without being detected—two-thirds of Lyme disease patients are unable to recall a tick bite.

The ticks mature during a 2-year cycle (**figure 22.14**). A larval form emerges from the egg. After growing, it molts, shedding its outer covering to become a nymph. After another molt, the nymph becomes the sexually mature adult form. The nymph actively seeks blood meals and is therefore mainly responsible for transmitting Lyme disease. *B. burgdorferi* is easily transferred from the tick to its preferred host, the white-footed mouse. Once infected, the mouse develops a sustained bacteremia and becomes a source of infection for other ticks. The spirochete is rarely passed from adult tick to its offspring. ◀◀ **bacteremia, p. 384**

Infected ticks and mice are the main reservoirs of *B. burgdorferi*, but deer play an important role as well. They are the preferred host of the adult ticks and the site where tick-mating occurs. Moreover, deer can quickly spread the disease over a wide area. Tick nymphs are most active from May to September, corresponding to the peak occurrence of Lyme disease. Adult ticks sometimes bite humans late in the season and transmit the disease. Infectious ticks can be present in well-mowed lawns as well as in wooded areas. ◀◀ **reservoirs, p. 438**

Treatment and Prevention

Several antibiotics are effective in patients with early stages of Lyme disease. In late disease, antibiotic treatment is less effective, most likely because the spirochetes are not actively multiplying and antibacterial medications usually target dividing bacteria. Nevertheless, prolonged treatment with intravenous ampicillin or ceftriaxone has cured many cases.

General preventive measures for Lyme disease are the same as those for Rocky Mountain spotted fever and include avoiding

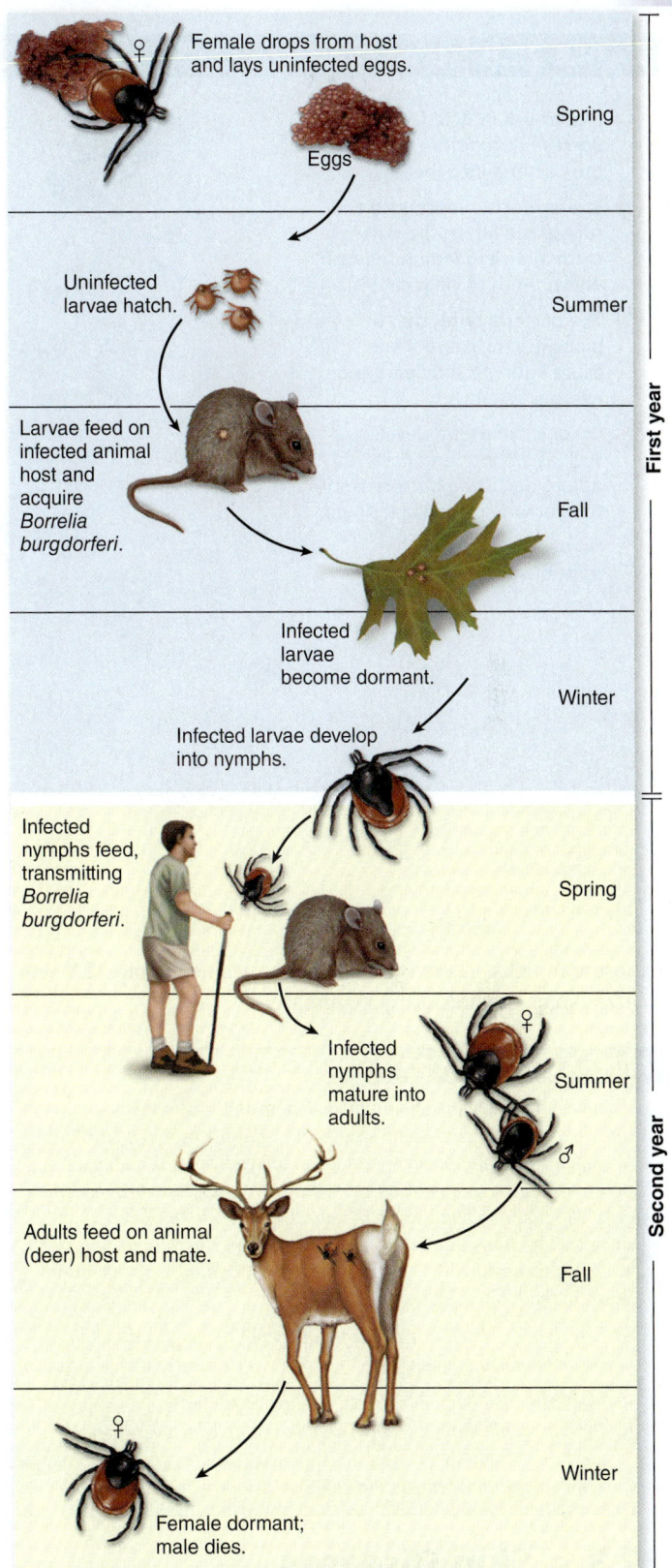

FIGURE 22.14 Life Cycle of the Black-Legged (Deer) Tick, *Ixodes scapularis* The life cycle covers 2 years, during which the tick obtains three blood meals. The males die soon after mating, the females after depositing their eggs in the following spring. Variations in the life cycle occur, probably dependent on climate and food availability for the natural hosts.

❓ *What time of the year has the highest number of new cases of Lyme disease?*

TABLE 22.8   Lyme Disease

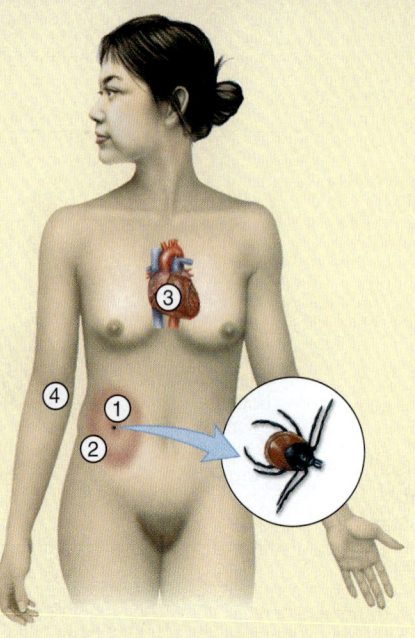

① Bite of tick infected with *Borrelia burgdorferi* introduces the bacteria into the skin.

② *B. burgdorferi* reproduce and spread radially in the skin, causing an expanding red rash which tends to clear centrally.

③ The bacteria enter the bloodstream, cause fever, acute injury to the heart and nervous system.

④ Chronic symptoms develop, such as arthritis and paralysis due to persisting bacteria and the immune response to them.

No person-to-person transmission.

Signs and symptoms	*Early localized infection:* Enlarging rash at the site of the bite; lymph node enlargement near bite, flulike symptoms. *Early disseminated infection:* Heart and nervous system involvement. *Late persistent infection:* Chronic arthritis and nervous system impairment.
Incubation period	Approximately 1 week
Causative agent	*Borrelia burgdorferi*, a spirochete
Pathogenesis	Spirochetes injected into the skin by an infected tick multiply and spread radially; the spirochetes enter the bloodstream and are carried throughout the body; the immune reaction to bacterial antigen causes tissue damage.
Epidemiology	Spread by the bite of ticks, *Ixodes* sp., usually found in association with animals such as white-footed mice and white-tailed deer living in wooded areas.
Treatment and prevention	Early treatment with doxycycline and others; prolonged antibiotic therapy in chronic cases. Protective clothing; tick repellents.

exposure to ticks. There is no available vaccine. **Table 22.8** summarizes some features of Lyme disease.

MicroAssessment 22.2

Folliculitis, furuncles, and carbuncles are usually caused by *Staphylococcus aureus.* Some strains of *S. aureus* produce exfoliatin and can cause staphylococcal scalded skin syndrome. Impetigo is a kind of pyoderma that is often caused by *Streptococcus pyogenes.* Rocky Mountain spotted fever is caused by *Rickettsia rickettsii* and is transmitted by ticks. Lyme disease is caused by *Borrelia burgdorferi* and is also transmitted by ticks. It is the most common vector-borne disease in the United States.

4. *List four extracellular products of* Staphylococcus aureus *that contribute to its virulence.*

5. *Describe the characteristic rash of Lyme disease.*

6. *Patients with staphylococcal scalded skin syndrome often do not have* Staphylococcus aureus *growing in the affected areas. Why do you think this is the case?* ➕

22.3 ■ Skin Diseases Caused by Viruses

Learning Outcomes

6. *Compare and contrast chickenpox, measles, and rubella.*

7. *Compare and contrast fifth disease and roseola.*

8. *Describe the process that leads to warts.*

Several different viruses initially infect the upper respiratory tract, but are carried in the blood to the skin where they cause distinctive skin rashes. These diseases are usually diagnosed based on the appearance of the rash and other clinical findings. When the disease is not typical, however, immunological testing can be used to identify antibodies against each virus. Other viruses enter through skin lesions (see **Perspective 22.1**). ◀◀ immunological testing, p. 426

Varicella (Chickenpox)

Chickenpox is the popular name for varicella, a rash that was common in childhood before the varicella vaccine was introduced in 1995. The causative agent is a member of the herpesvirus family. All herpesviruses produce latent infections that can reactivate long after recovery from the initial illness. ◀◀ latent infections, p. 322

Signs and Symptoms

Most cases of childhood chickenpox are mild, and sometimes go unnoticed. The typical case begins as small, red spots called macules, progressing to little bumps called papules, and then to small blisters called vesicles. The lesions appear at different times, so that any time during the rash, they are at various stages of development from macule to papule or vesicle (**figure 22.15**). They can occur anywhere on the body, but usually first appear on the back of the head, then the face, mouth, main body, and arms and legs. Patients can have only a few lesions or many hundreds. The lesions are pruritic (itchy) and scratching them can lead to serious, even fatal, secondary infection by *S. pyogenes* or *S. aureus*.

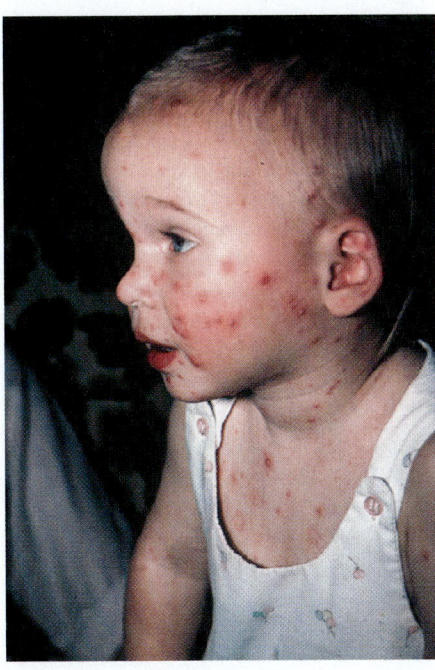

FIGURE 22.15
A Child with Chickenpox (Varicella) Characteristically, lesions in various stages of evolution—macules, papules, vesicles, and pustules—are present.

❓ *Is chickenpox caused by a poxvirus?*

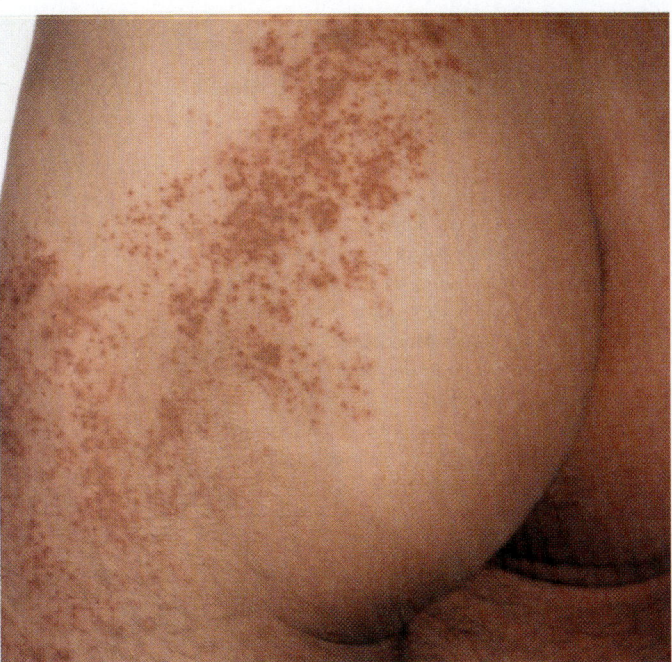

FIGURE 22.16 **Shingles (Herpes Zoster)** The rash mimics that of chickenpox, except that it is limited to a sensory nerve distribution on one side of the body.

❓ *Can you catch chickenpox or shingles from a person with shingles?*

MicroByte

Non-immune people can contract chickenpox from someone with shingles.

The signs and symptoms of chickenpox tend to be more severe in older children and adults. About 20% of adults develop pneumonia. This may lead to rapid breathing, cough, shortness of breath, and a dusky skin color from lack of O₂. The pneumonia subsides with the rash, but respiratory signs and symptoms often persist for weeks. Chickenpox is a threat to immunocompromised patients of any age—the virus damages the lungs, heart, liver, kidneys, and brain, resulting in death in about 20% of the cases.

Chickenpox is a major threat to babies. It has a case-fatality rate as high as 30% in newborn infants. Babies born to mothers who had chickenpox early in pregnancy occasionally develop congenital varicella syndrome. These babies may be born with underdeveloped head and limbs, and cataracts.

Reactivation of the chickenpox virus results in **shingles.** This disease can occur at any age but becomes increasingly common later in life. It begins with pain in the area supplied by a sensory nerve, often on the chest or abdomen but sometimes on the face or an arm or leg. After a few days to 2 weeks, a rash characteristic of chickenpox appears, but unlike chickenpox, the rash is usually restricted to the area supplied by the branches of the involved sensory nerve (**figure 22.16**). The rash generally subsides within a week, but pain may persist for weeks, months, or longer. In people with AIDS or other serious immunodeficiency, the rash often spreads to involve the entire body, as in a severe case of chickenpox. Shingles does not spread from person to person, since it is caused by a reactivation of latent virus.

A curious, rare affliction known as **Reye's syndrome** occasionally occurs in association with chickenpox and a number of other viral infections, usually within 2 to 12 days of the onset of the infection. The patient begins vomiting and then slips into a coma. The syndrome occurs predominantly in children between 5 and 15 years old and is characterized by liver and brain damage and a death rate as high as 30%. Epidemiologic evidence suggests that aspirin therapy increases the risk of Reye's syndrome and should be avoided in children with fever.

Causative Agent

Chickenpox is caused by varicella-zoster virus (VZV), a member of the herpesvirus family. It is an enveloped, double-stranded DNA virus, indistinguishable from other herpesviruses in appearance (**figure 22.17**).

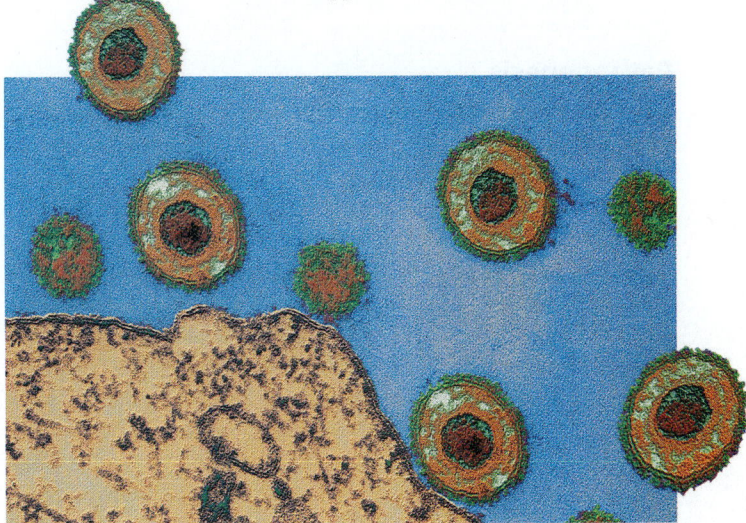

100 nm

FIGURE 22.17 **Varicella-Zoster Virus** Electron micrograph of the viral cause of chickenpox and shingles.

❓ *How does this virus cause both shingles and chickenpox?*

The Ghost of Smallpox: An Evil Shade

Historically, smallpox epidemics have been devastating to the Americas. In the 1500s smallpox virus introduced into Central and South America caused horrendous loss of life and may have contributed to the downfall of the Inca and Aztec nations. An epidemic that swept the Massachusetts coast in the 1600s killed so many Native Americans that in some communities there were not enough survivors to bury the dead. Even in the latter 1700s, during the Revolutionary War, a smallpox epidemic raged through the American colonies. General George Washington suspected that the virus had been deliberately introduced by the British. So many of his men were ill after his defeat at Quebec in 1777 that he ordered the mass variolation of remaining troops. ◀◀ smallpox, pp. 419, 447 ◀◀ variolation, p. 419

Why does the ghost of smallpox concern us now, when the last case, acquired through a laboratory accident, occurred in 1978? The answer is that the smallpox virus still exists, locked in high-security laboratories in the United States and the Russian Federation, and perhaps held in secret locations by countries or individuals that could use it to harm others. A number of factors are notable regarding the possibility of the virus being used as an agent of bioterror.

Factors That Might Encourage Its Use

- It spreads easily from person to person, mainly through close contact with respiratory secretions, but also by airborne virus from the respiratory tract, skin lesions, and contaminated bedding or other objects.

- It can be highly lethal, with mortality rates generally above 25%. After the virus establishes infection of the respiratory system, it enters the lymphatics and bloodstream, finally causing lesions of the skin and throughout the body.

- The virus is relatively stable, probably remaining infective for hours in the air of a building; viable smallpox has been demonstrated in dried crusts from skin lesions after storage for 10 years at room temperature in ordinary envelopes.

- Large numbers of people are highly susceptible to the virus. Routine vaccination against smallpox was discontinued in the United States several decades ago.

- The relatively large genome of the smallpox virus probably permits genetic modifications that could enhance its virulence.

Factors That Discourage Its Use

- Propagating the smallpox virus is dangerous and requires advanced knowledge and laboratory facilities.

- A proven highly effective vaccine (vaccinia virus) is available. A protective antibody response occurs rapidly and prevents fatalities for 10 years or more. Smallpox is prevented even when the vaccine is administered up to 4 days after exposure to the virus.

- Infected people do not spread the disease during the long incubation period, generally 12 to 14 days; they become infectious only with the onset of fever.

- The disease can usually be diagnosed rapidly, by the characteristic appearance of skin lesions that predominate on the face and hands, and by laboratory examination of material from skin lesions.

- There is already widespread experience on how to watch for and contain the disease.

In a simulated attack on an American city in the summer of 2001, only 24 primary cases of smallpox increased in 2 months to 3 million, with 1 million deaths. This study probably greatly exaggerated the risk. Nevertheless, even though unlikely, the potential danger from a smallpox introduction has caused the United States to begin preparations for this possibility, including markedly expanding its stockpile of smallpox vaccine. More information on smallpox is available at the text website (**www.mhhe.com/nester7**).

Pathogenesis

The VZV enters the body through the respiratory tract, establishes an infection, replicates, and travels to the skin via the bloodstream. After the dermis and epidermis are infected, the virus spreads to adjacent cells, and the characteristic skin lesions appear.

Stained preparations of infected cells show **intranuclear inclusion bodies** (pink bodies in the nucleus) where the virus reproduces. Some infected cells fuse together, forming multinucleated giant cells. The infected cells swell and eventually lyse. When an area of skin infection involves a sensory nerve ending, the virus enters the nerve. From there, it travels to ganglia (singular: ganglion), which are small bulges in sensory nerves located near the spine. Conditions inside the ganglia prevent full expression of the VZV genome. However, viral DNA is present and fully capable of coding for mature infectious virus. The mechanism of suppression of viral replication within the nerve cell is not known but is probably under the control of immune cells.

Shingles is most likely to occur when cell-mediated immunity declines. With a weakened cellular immune system, infectious viral particles are produced in the nuclei of the nerve cells. The virus particles are then carried to the skin by the normal circulation of cytoplasm within the nerve cell. With the appearance of the skin lesions, a prompt, intense secondary (memory) response of adaptive immunity begins. As immune cells accumulate in the ganglion, the signs and symptoms of shingles quickly disappear. Sometimes, however, the inflammatory response leads to scars and chronic pain. ◀◀ secondary response, p. 367 ◀◀ cell-mediated immunity, p. 356

Epidemiology

Chickenpox is a notifiable disease, requiring that confirmed cases are reported to the CDC. However, many cases are so mild that they go unreported. In the early 1990s, before a vaccine was available, there were approximately 4 million cases of chickenpox each year in the United States, resulting in 10,000 hospitalizations and approximately 100 to 150 deaths. A decade after the release of the vaccine, chickenpox caused fewer than half a million cases and 8 deaths each year.

VZV can be transmitted by both respiratory secretions and skin lesions. As with many diseases transmitted by the respiratory route, most cases occur in the winter and spring months. Humans are the only reservoir, and the disease is highly contagious. The incubation period averages about 2 weeks, with a range of 10 to 21 days. Individuals are infective from 1 to 2 days before the rash appears until all the lesions have crusted (usually 4 days after the onset). ◀◀ reservoir, p. 438

The ability of the chickenpox virus to form a latent infection allows it to persist within a population. By contrast, a virus such as measles spreads quickly through a population and infected individuals become immune or die. When this happens, the measles virus eventually runs out of available hosts and disappears from the community. The chickenpox virus can reappear from cases of shingles if enough susceptible individuals are present. Shingles occurs in about one out of three individuals in their lifetime, and as many as 25% of these patients develop eye involvement.

Treatment and Prevention

The antiviral medications acyclovir and famciclovir, among others, are helpful in preventing and treating VZV infections.

A safe attenuated chickenpox vaccine has been used in the United States since 1995. All healthy children and adults without a history of chickenpox are advised to receive the vaccine. HIV-infected children and adults should receive the vaccine if their immune system is still intact. The vaccine should not be given to people with immunodeficiencies although healthy, non-immune contacts of such people should be vaccinated. ◀◀ attenuated vaccines, p. 423

By preventing chickenpox, the vaccine reduces the chance of developing shingles. Another vaccine is available to prevent shingles in individuals 60 years of age or older. This vaccine is composed of the same attenuated virus used in the chickenpox vaccine, but in a much higher dose. It is given in a single dose and reduces the risk of developing shingles by 50%.

People with impaired immunity are at risk of severe disseminated VZV infections. These individuals include newborns and people with cancer, AIDS, or organ transplants. Immunocompromised individuals can be partially protected from severe disease by being passively immunized with VZIG, a hyperimmune globulin with high concentrations of antibody to VZV. The main features of chickenpox are summarized in **table 22.9**. ◀◀ passive immunity, p. 420 ◀◀ hyperimmune globulin, p. 420

Rubeola (Measles)

Measles, "hard measles," and "red measles" are common names for rubeola. One of the great success stories of the last half of the twentieth century was the dramatic reduction in measles cases as a result of immunizing children with an attenuated vaccine against the disease. The number of cases has increased in recent years, due in large part to some parents' refusal to vaccinate their children.

Signs and Symptoms

Measles begins with fever, runny nose, cough, and swollen, red, weepy eyes. Within a few days, a fine red rash appears on the forehead and spreads outward over the rest of the body

TABLE 22.9	Varicella (Chickenpox)	
① Varicella-zoster virus is inhaled; infects nose and throat.		
② The virus infects nearby lymph nodes, reproduces, and enters the bloodstream.	**Signs and symptoms**	Itchy bumps and blisters in various stages of development, fever; latent infections can reactivate, resulting in shingles (herpes zoster), years later.
③ Infection of other body cells occurs, resulting in showers of virions into the bloodstream.	**Incubation period**	10 to 21 days
④ These virions cause successive skin lesions, which evolve into blisters and crusts.	**Causative agent**	Varicella-zoster virus; enveloped double-stranded DNA virus of the herpesvirus family.
⑤ Immune system eliminates the infection except for some virions inside the nerve cells.	**Pathogenesis**	Multiplication in the upper respiratory tract followed by dissemination via bloodstream to the skin; cytopathic effect of virus includes giant cell formation.
⑥ If immunity decreases with age or other reason, the virus persisting in the nerve ganglia can infect the skin, causing shingles (herpes zoster).	**Epidemiology**	Highly infectious. Acquired by the respiratory route; people infected with either chickenpox or shingles the only source; spread via skin lesions and respiratory secretions.
⑦ Transmission to others occurs from respiratory secretions and skin.	**Treatment and prevention**	Acyclovir or similar antiviral medication for prevention and treatment. Attenuated vaccine. Passive immunization with zoster immune globulin (VZIG) for immunocompromised individuals.

FIGURE 22.18 A Child with Measles (Rubeola) The rash is usually accompanied by fever, runny nose, and a bad cough.

❓ *Why is it important to continue immunizing children in the United States against measles, even though it is a rare disease here?*

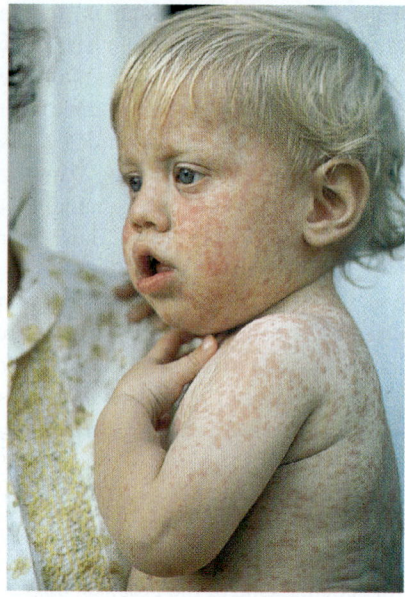

(**figure 22.18**). Unless complications occur, signs and symptoms generally disappear in about a week. Unfortunately, many cases are complicated by secondary infections caused mainly by *Staphylococcus aureus, Streptococcus pneumoniae, Streptococcus pyogenes,* and *Haemophilus influenzae.* These bacteria easily invade the body because the measles virus damages the normal body defenses. Secondary infections most commonly cause earaches and bacterial pneumonia.

In about 5% of cases, the measles virus causes viral pneumonia, with rapid breathing, shortness of breath, and dusky skin color from lack of adequate O_2 exchange in the lungs. Encephalitis (inflammatory disease of the brain) is another serious complication, marked by fever, headache, confusion, and seizures. This complication occurs in about one out of every 1,000 measles cases, and sometimes results in permanent brain damage, with mental disability, deafness, and epilepsy.

Very rarely, measles is followed 2 to 10 years later by a disease called subacute sclerosing panencephalitis (SSPE). This disease is marked by slowly progressive degeneration of the brain, generally resulting in death within 2 years. Defective measles virus particles that cannot complete replication can be detected in the brains of these patients. High levels of measles antibodies are present in their blood. Successful immunization programs in the United States have decreased SSPE, but there is concern that the recent increase in measles cases may lead to more SSPE in the future.

Measles that occurs during pregnancy increases the risk of miscarriage, premature labor, and low birth weight babies. Birth defects, however, generally do not occur.

Causative Agent

Measles is caused by rubeola virus, an enveloped, single-stranded RNA virus of the paramyxovirus family. Two biologically important proteins are found on the viral envelope—hemagglutinin and fusion protein. The virus uses hemagglutinin to attach to host cells and the fusion protein to fuse the viral envelope with the host cell's cytoplasmic membrane. The fusion protein also causes adjacent infected host cells to join together, producing multinucleated giant cells.

Pathogenesis

Rubeola virus is acquired by the respiratory route. It replicates in the upper respiratory epithelium, spreads to lymphatic tissues, and eventually spreads to all parts of the body. The rash is caused by a cell-mediated immune response against viruses multiplying in the skin.

Mucous membrane involvement is responsible for an important diagnostic sign, **Koplik spots** (**figure 22.19**). These look like grains of salt lying on red and rough oral mucosa, resembling red sandpaper, and are usually best seen in the back of the mouth, opposite the molars.

The measles virus temporarily suppresses cell-mediated immunity, causing cold sores to appear and latent tuberculosis disease to reactivate in some individuals. It also damages the respiratory mucous membranes, thereby increasing the susceptibility of measles patients to secondary bacterial infections, especially infection of the middle ear and lung. Damage to the intestinal epithelium may explain the diarrhea that sometimes occurs in measles and contributes to high measles death rates in impoverished countries. In the United States, deaths from measles occur in about one to two of every 1,000 cases, mainly from pneumonia and encephalitis. ▶▶ cold sores, p. 581 ◀◀ tuberculosis, p. 502

Epidemiology

Humans are the only natural host of rubeola virus. Before vaccination became widespread in the 1960s, probably less than 1% of the global population escaped infection with this highly contagious virus. Continued use of the measles vaccine resulted in a progressive decline in the number of cases, so that measles is no longer endemic in the Western Hemisphere. Small outbreaks can occur, however, when the virus enters the area with people travelling to and from other countries. In these situations, the virus spreads among non-immune populations including (1) children too young to be vaccinated; (2) children and adults inadequately vaccinated; and (3) unvaccinated people.

Worldwide, about 750,000 children still die from measles every year. This ranks among the leading causes of death and disability among the impoverished. The case-fatality rate may reach 15% and the rate of secondary infections may reach 85% when patients are poorly nourished. Measles vaccination is a high priority in areas struck by natural disasters.

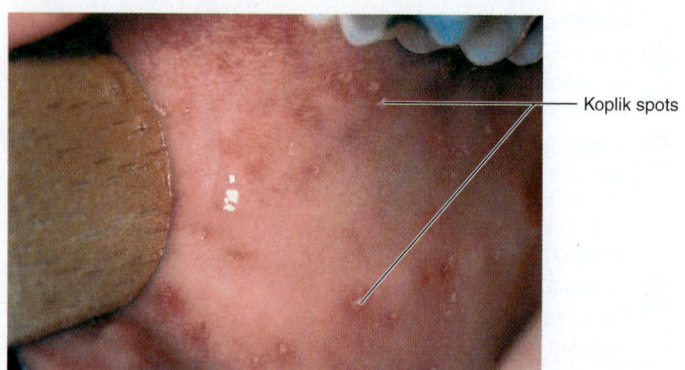

Koplik spots

FIGURE 22.19 Koplik Spots, Characteristic of Measles (Rubeola) These spots resemble grains of salt on a red base.

❓ *Can someone have measles without having Koplik spots?*

TABLE 22.10 | Rubeola (Measles)

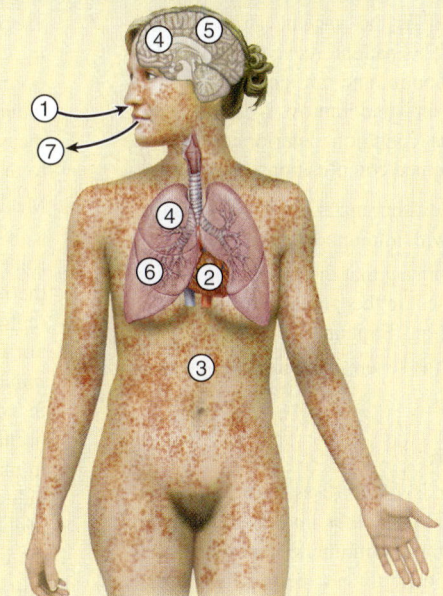

1. Airborne rubeola virus infects eyes and upper respiratory tract, then the lymph nodes in the region.
2. Virus enters the bloodstream and is carried to all parts of the body including the brain, lungs, and skin.
3. Skin cells infected with the rubeola virus are attacked by cytotoxic T cells, causing a generalized rash.
4. Virus replicating in the lungs can cause pneumonia; the brain can also be infected.
5. In rare cases, virus persisting in the brain causes subacute sclerosing panencephalitis, months or years after the acute infection.
6. Secondary infection of the ears and lungs is common.
7. Transmission is by respiratory secretions.

Signs and symptoms	Rash, fever, weepy eyes, cough, and nasal discharge
Incubation period	10 to 12 days
Causative agent	Rubeola virus, a single-stranded RNA virus of the paramyxovirus family
Pathogenesis	Virus multiplies in respiratory tract; spreads to lymphatic tissues, then to all parts of body, notably skin, lungs, and brain; damage to respiratory tract epithelium leads to secondary infection of ears and lungs.
Epidemiology	Acquired by respiratory route; highly contagious; humans only source.
Treatment and prevention	No antiviral treatment currently available. Attenuated vaccine after age 12 months; second dose on entering elementary school or at adolescence.

Treatment and Prevention

No antiviral treatment currently exists for measles, but an attenuated vaccine can prevent the disease. In 1980, the worldwide incidence of measles was estimated to be 100 million with 5.8 million deaths, but vaccination programs have lowered the number of cases dramatically since then. Unfortunately, the incidence of measles in the United States is on the rise. This is largely caused by importation and spread of the disease by travelers from countries where it is still endemic.

The measles vaccine is usually given together with mumps, rubella, and varicella vaccines (MMRV). The first injection of vaccine is given near an infant's first birthday. Since 1989, a second injection of vaccine is given when children enter elementary school. The two-dose regimen has resulted in at least 99% of the recipients becoming immune. In an epidemic, vaccine is given to babies as young as 6 months, who are then reimmunized before their second birthday. Students entering high school or college are often required to get a second dose of vaccine if they have not had one earlier. Those at high risk of acquiring measles, such as medical personnel, should be immunized unless they definitely have had the disease or have laboratory proof of immunity. Some features of measles are summarized in **table 22.10.**

Rubella (German Measles)

German measles and "three-day measles" are common names for rubella. The term *German measles* arose because the disease was first described in Germany. In contrast to chickenpox and measles, rubella is typically a mild, often unrecognized disease that is difficult to diagnose. It is of concern because infection of pregnant women can have tragic consequences.

Signs and Symptoms

Characteristic signs and symptoms of German measles are slight fever, mild cold symptoms, and enlarged lymph nodes behind the ears and on the back of the neck. After about a day, a faint rash consisting of many pink spots appears over the face, chest, and abdomen (**figure 22.20**). Unlike measles, there are no diagnostic mouth lesions. Adults commonly develop painful joints, with pain generally lasting 3 weeks or less. Other signs and symptoms typically last only a few days. The significance of German measles, however, lies with its threat to a developing fetus.

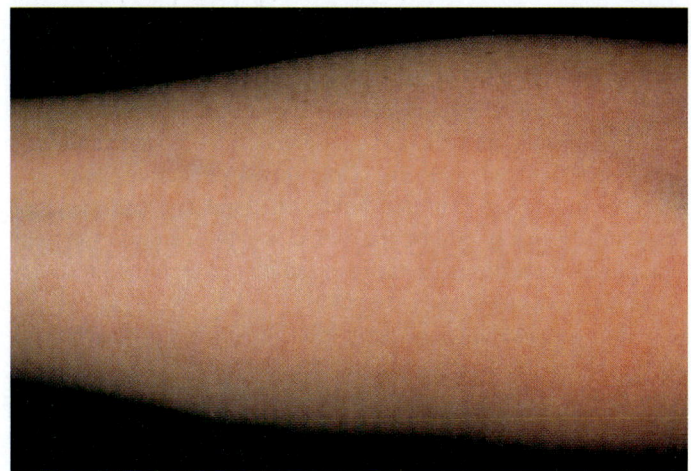

FIGURE 22.20 Adult with German Measles (Rubella) Symptoms are often very mild, but the effects on a fetus can be devastating.

❓ *Is congenital rubella syndrome more likely to occur if a pregnant woman is infected early or late in the pregnancy?*

The patient was a 20-year-old asymptomatic man who was immunized against measles as a requirement for starting college. He had received his first dose of measles vaccine at approximately 1 year of age. Past medical history revealed that he was a hemophiliac and had contracted the human immunodeficiency virus (HIV) from clotting factor (a blood product given to control bleeding) contaminated with the virus. Laboratory tests showed that he had a very low CD4+ lymphocyte count, indicating a severely damaged immune system.

About a month after his precollege immunization, he developed *Pneumocystis* pneumonia, a lung infection characteristic of AIDS, was hospitalized, had a good response to treatment, and was discharged. Ten months later, he was again hospitalized for symptoms of a severe lung infection. He had no rash. Multiple laboratory tests to determine the cause of his infection were negative. Finally, a lung biopsy was performed and revealed "giant cells"—very large cells with multiple nuclei. Cytoplasmic and intranuclear inclusion bodies were also present. This picture was highly suggestive of measles pneumonia, and measles virus subsequently was recovered from cell cultures of the biopsy material. Other studies

showed it to be the vaccine strain of measles virus. The patient received intravenous immune globulin and an antiviral medication—ribavirin—and improved. Subsequently, however, his condition deteriorated, and he died of presumed complications of AIDS.

1. Is measles immunization a good idea for people with immunodeficiency?

2. Is it surprising that the vaccine virus was still present in this patient 11 months after vaccination? Explain.

3. Despite the severe infection, there was no rash. Why?

Discussion

1. Measles is often disastrous for persons with AIDS or other immunodeficiencies. They should be immunized as soon as possible in their illness, before the immune system becomes so weakened it cannot respond effectively to the vaccine. Also, as this and other cases have shown, the vaccine virus can itself cause disease when immunodeficiency is severe. With the worldwide effort to eliminate measles, the risk of exposure to the wild-type measles virus, as opposed to the laboratory-derived

vaccine virus, is declining, but outbreaks in colleges and other institutions still occur. A severely immunodeficient individual can be passively immunized against measles with immune globulin if exposed to the wild-type virus.

2. Measles is often given as an example of a persistent viral infection, meaning that following infection the virus can persist in the body for months or years in a slowly replicating form. It has been suggested but not proven that this explains the lifelong immunity conferred by measles infection in normal people. Rarely in presumably normal individuals and, more commonly, in malnourished or immunodeficient individuals, persistent infection leads to damage to the brain, lung, liver, and possibly, the intestine.

3. Following acute infection, the measles virus floods the bloodstream and is carried to various tissues of the body, including the skin. The rash that characterizes measles is caused by T cells attacking measles virus antigen lodged in the skin capillaries. In the absence of functional T cells, the rash does not occur.

Causative Agent

German measles is caused by the rubella virus, a member of the togavirus family. The virus is an enveloped, single-stranded RNA virus that can easily be cultivated in cell cultures. Proteins on the viral surface cause *in vitro* hemagglutination, which can be inhibited by specific antibody. This allows serological identification of the virus.

Pathogenesis

The rubella virus enters the body via the respiratory tract. It multiplies in the nasopharynx and enters the bloodstream, causing a sustained viremia. The virus travels to various body tissues, including the skin and joints. Humoral and cell-mediated immunity develop against the virus, and the resulting immune complexes probably cause the rash and joint symptoms. ◄◄ viremia, p. 384 ◄◄ immune complexes, p. 395

If a woman is infected early in the pregnancy, virus particles in the bloodstream can cross the placenta and infect the fetus. This is less likely to happen later in pregnancy. Virtually all types of fetal cells are susceptible to infection; some cells are killed, whereas others develop a persistent infection in which cell division is impaired and chromosomes are damaged. The result is a characteristic pattern of fetal abnormalities referred to as the **congenital rubella syndrome.** The abnormalities include cataracts and other eye defects, brain damage, deafness, heart defects, and low birth weight despite normal gestation. Babies may be stillborn. Those

that live continue to excrete rubella virus in throat secretions and urine for many months. The likelihood of the syndrome varies according to the age of the fetus when infection occurs. Infections within the first 6 weeks of pregnancy result in most of the fetuses having a detectable injury, commonly deafness. Even infants who are apparently normal, however, excrete rubella virus for extended periods and thus can infect others.

Epidemiology

Humans are the only natural host for rubella virus. The disease is highly contagious although less so than measles; it is estimated that in the prevaccine era, only 10% to 15% of people reached adulthood without being infected. Complicating the epidemiology of rubella is the fact that over 40% of infected individuals fail to develop symptoms, but can still spread the virus. People who develop typical rubella can be infectious for as long as 7 days before the rash appears until 7 days afterward. Before widespread use of the vaccine began in 1969, periodic major epidemics arose. One epidemic in 1964 resulted in about 30,000 cases of congenital rubella syndrome. This caused a "rubella bulge" of hearing-impaired children that affected the educational system in the United States for decades.

Treatment and Prevention

No specific antiviral therapy for rubella is available. The disease can be prevented by an attenuated rubella virus given to babies at 12 to 16 months old with a second dose at age 4 to 6 years. The vaccine produces long-lasting immunity in nearly all recipients.

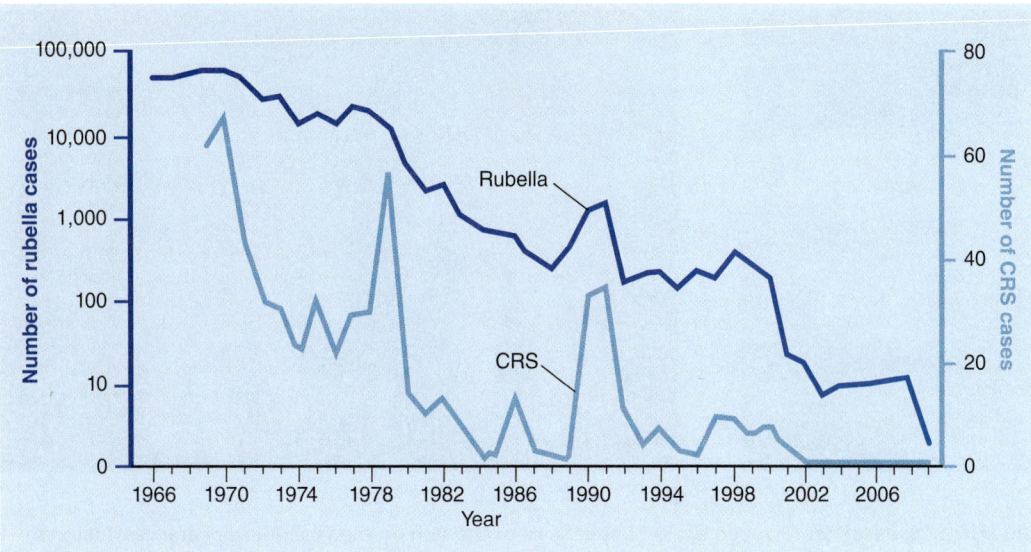

FIGURE 22.21 Reported Cases of German Measles (Rubella) and Congenital Rubella Syndrome (CRS), United States, 1966–2009

❓ *Why are there fewer cases of CRS than of German measles at every time tested?*

The vaccine is not given to pregnant women for fear it might result in congenital defects. As an added precaution, women are advised not to become pregnant for 28 days after receiving the vaccine.

Use of the vaccine has markedly reduced the incidence of rubella in the United States to generally less than 50 cases per year (**figure 22.21**). There were less than 1,000 confirmed cases in the entire Western Hemisphere in 2004. Some features of German measles are summarized in **table 22.11.**

Other Viral Rashes of Childhood

The kinds of viruses that can cause childhood rashes probably number in the hundreds. One group alone, the enteroviruses, has about 50 members associated with skin lesions. In the early 1900s the causes of the common childhood rashes were largely unknown, and it was the practice to number them 1 to 6 as follows: (1) rubeola; (2) scarlet fever; (3) rubella; (4) Duke's disease—a mild disease with fever and bright red generalized rash, now thought to have been due to an enterovirus; (5) erythema infectiosum; and (6) exanthem subitum. The causes of erythema infectiosum (fifth disease) and exanthem subitum (roseola) have only been established in recent years.

Fifth disease (erythema infectiosum) occurs in both children and young adults. The illness begins with fever, malaise, and head and muscle aches. A diffuse redness appears on the cheeks, giving the appearance of the face as if it were slapped

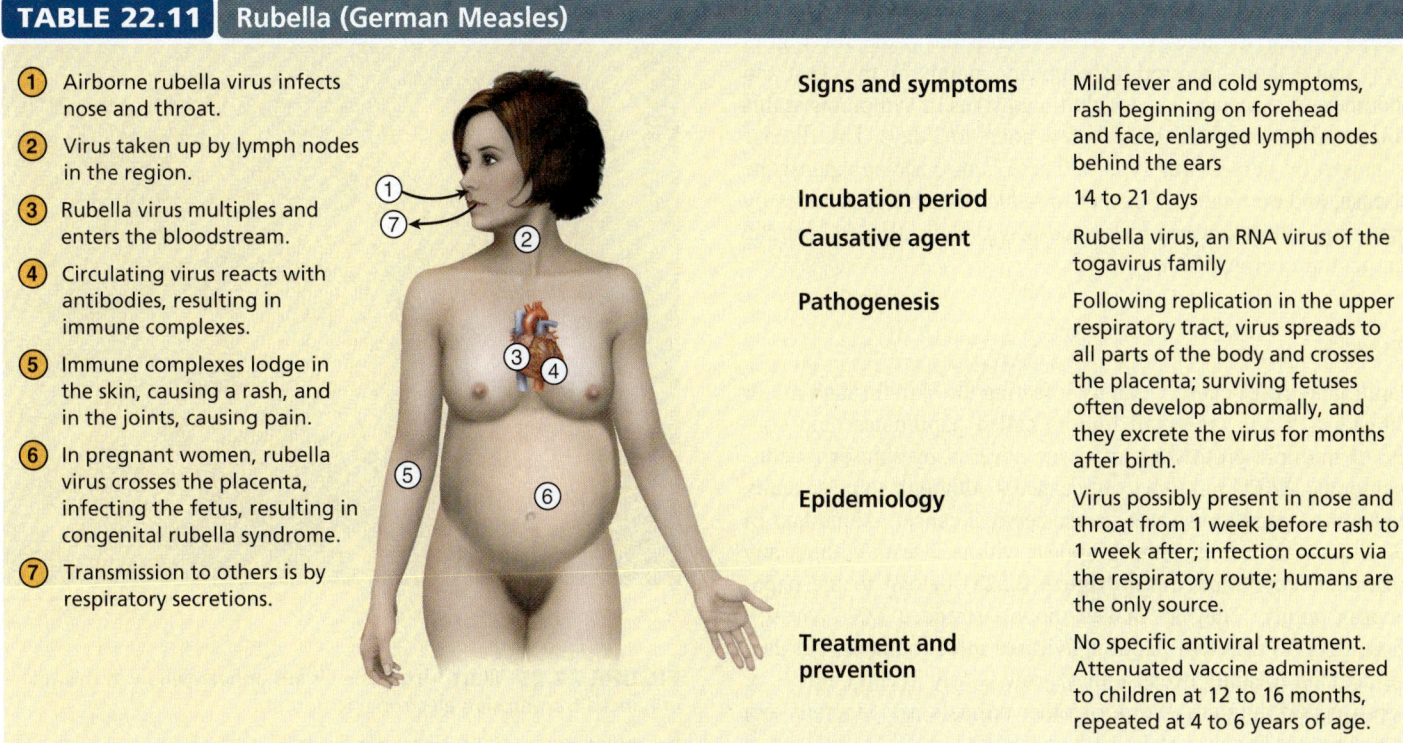

TABLE 22.11	Rubella (German Measles)

① Airborne rubella virus infects nose and throat.

② Virus taken up by lymph nodes in the region.

③ Rubella virus multiples and enters the bloodstream.

④ Circulating virus reacts with antibodies, resulting in immune complexes.

⑤ Immune complexes lodge in the skin, causing a rash, and in the joints, causing pain.

⑥ In pregnant women, rubella virus crosses the placenta, infecting the fetus, resulting in congenital rubella syndrome.

⑦ Transmission to others is by respiratory secretions.

Signs and symptoms Mild fever and cold symptoms, rash beginning on forehead and face, enlarged lymph nodes behind the ears

Incubation period 14 to 21 days

Causative agent Rubella virus, an RNA virus of the togavirus family

Pathogenesis Following replication in the upper respiratory tract, virus spreads to all parts of the body and crosses the placenta; surviving fetuses often develop abnormally, and they excrete the virus for months after birth.

Epidemiology Virus possibly present in nose and throat from 1 week before rash to 1 week after; infection occurs via the respiratory route; humans are the only source.

Treatment and prevention No specific antiviral treatment. Attenuated vaccine administered to children at 12 to 16 months, repeated at 4 to 6 years of age.

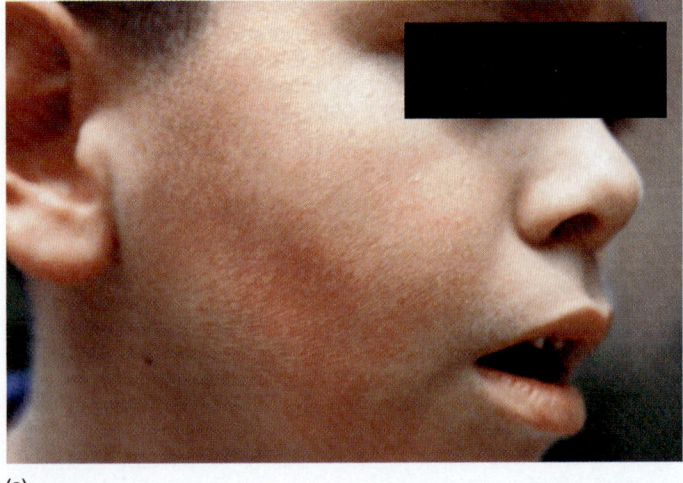

(a)

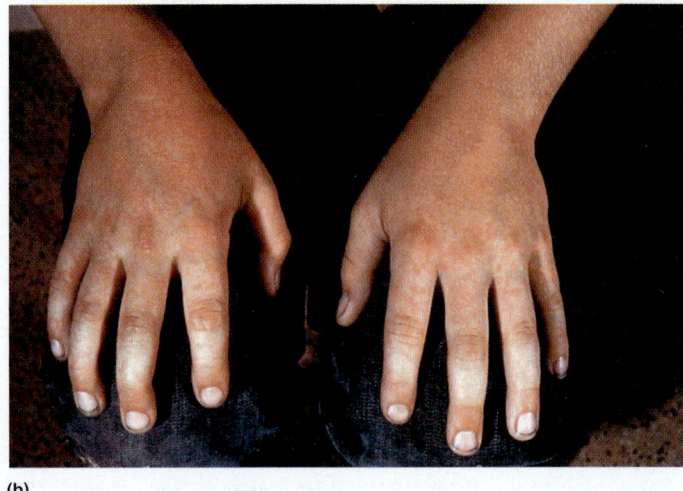

(b)

FIGURE 22.22 Erythema Infectiosum (Fifth Disease) (a) "Slapped cheek" appearance of the rash on the face. **(b)** Appearance of the rash on the extremities.

❓ *Why is this disease a major threat to individuals with sickle cell anemia?*

(**figure 22.22a**). The rash commonly spreads in a lacy pattern to involve other parts of the body, especially the extremities (figure 22.22b). The rash may come and go for 2 weeks or more before recovery. Adults with fifth disease often have joint pains. The disease is caused by parvovirus B-19, a naked, single-stranded DNA virus. The virus preferentially infects certain bone marrow cells and is a major threat to people with sickle cell and other anemias. The infected marrow sometimes stops producing blood cells, a condition known as aplastic crisis. Also, about 10% of women infected with the virus during pregnancy suffer spontaneous abortion.

Roseola (exanthem subitum, roseola infantum), is a common disease in children 6 months to 3 years old. It causes a great deal of parental anxiety because it begins abruptly with a fever that may reach 105°F and cause convulsions. The children generally do not appear ill, however. After several days, the fever goes away and a short-lived red rash appears, mainly on the chest and abdomen. The patient has no additional signs or symptoms at this point, and the rash vanishes in a few hours to 2 days. This disease is caused by herpesvirus type 6. There is no vaccine against the disease, and no treatment except to reduce the risk of seizures by sponging with lukewarm water and using medication to keep the temperature below 102°F.

Warts

Papillomaviruses cause warts by infecting the skin through minor abrasions. Warts are small tumors called papillomas, and consist of multiple protrusions of tissue covered by skin or mucous membrane. Warts rarely become cancers, although some sexually transmitted papillomaviruses cause cervical cancer. About half of the time, warts on the skin disappear within 2 years without any treatment. Papillomaviruses (**figure 22.23**) belong to the papovavirus family. They are naked, double-stranded DNA viruses. More than 50 different papillomaviruses infect humans, but they are difficult to study because they grow poorly in cell cultures or experimental animals. Warts of other animals are generally not infectious for humans. ▶▶ cervical cancer, p. 631

Papillomaviruses can survive on inanimate objects such as wrestling mats, towels, and shower floors, and infection can be acquired from such contaminated objects. The viruses infect the deeper cells of the epidermis and reproduce in the nuclei. Some of the infected cells grow abnormally, forming the wart. The incubation period ranges from 2 to 18 months. Infectious virus is present in the wart and can contaminate fingers or objects that pick or rub the lesions.

Like other tumors, warts can be treated effectively only by killing or removing all of the abnormal cells. This can usually be done by freezing the wart with liquid nitrogen, by cauterization (burning the tissue usually with an electrically heated needle), or by surgically

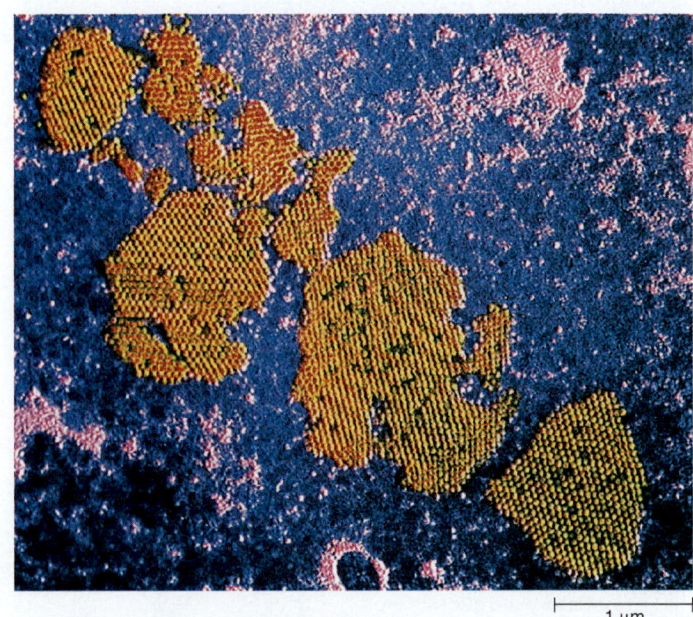

1 μm

FIGURE 22.23 Wart Virus The virions appear yellow in this color-enhanced transmission electron micrograph.

❓ *Which group of viruses causes warts?*

removing the wart. Virus generally remains in the adjacent normal-appearing skin and may cause additional warts.

Warts that grow on the soles of the feet are called plantar warts (often mistakenly called planter's warts; *plantar* is a word meaning "referring to the sole of the foot"). These warts are very difficult to treat because the pressure of standing on them causes them to grow wide and deep.

MicroAssessment 22.3

Chickenpox, measles, and rubella can be controlled by vaccines. Viral diseases that may only inconvenience a pregnant woman can be disastrous to her fetus. A viral infection acquired in childhood can remain latent for years only to reactivate in a different form. One group of viruses causes benign skin tumors.

7. *What important diagnostic sign is often present in the mouth of measles (rubeola) patients?*

8. *Does exposure to a person with shingles transmit shingles or chickenpox?*

9. *Why is it a good idea to immunize both boys and girls against rubella?* ➕

22.4 ■ Skin Diseases Caused by Fungi

Learning Outcomes

9. *Describe the characteristics of superficial cutaneous mycoses, including the role of dermatophytes.*

10. *Describe the condition tinea versicolor and its causative agent, Malassezia furfur.*

Diseases caused by fungi are called mycoses. Several fungi are responsible for mild to serious infections of the skin. The condition of the host's defenses against infection determines the severity of most fungal infections.

Superficial Cutaneous Mycoses

A group of molds called **dermatophytes** can invade hair, nails, and the keratinized portion of the skin. The resulting mycoses have colorful names such as jock itch, athlete's foot, and ringworm, as well as Latinized names that describe their location: tinea capitis (scalp), tinea barbae (beard), tinea axillaris (armpit), tinea corporis (body), tinea cruris (groin), and tinea pedis (feet), to list a few. Tinea simply means "worm," which probably reflects early incorrect ideas about the cause.

Signs and Symptoms

Most people colonized by dermatophytes have no signs or symptoms at all. Others complain of itching, a bad odor, or a rash. In ringworm, a rash occurs at the site of the infection and consists of a scaly area surrounded by redness at the outer margin, producing irregular rings or a lacy pattern on the skin. On the scalp, patchy areas of hair loss can occur, with a fine stubble of short hair left behind. Infected nails become thickened and brittle and may separate from the nailbed. Sometimes, a rash consisting of fine papules

and vesicles develops distant from the infected area. This rash is referred to as a dermatophytid, or "id" reaction, a reflection of allergy to products of the infecting fungus.

Causative Agents

Dermatophytes are a group of skin-invading molds including members of the genera *Epidermophyton, Microsporum,* and *Trichophyton* (**figure 22.24**). They can be grown on media especially designed for molds and are usually identified by their colonial and microscopic appearance. In some cases their nutritional requirements and biochemical tests are important as well.

Pathogenesis

The normal skin is generally resistant to invasion by dermatophytes. Some species, however, are relatively virulent and can even cause epidemics, especially among children. In moist areas of the skin, dermatophytes can invade keratin-containing cells and structures. They produce an enzyme called keratinase that

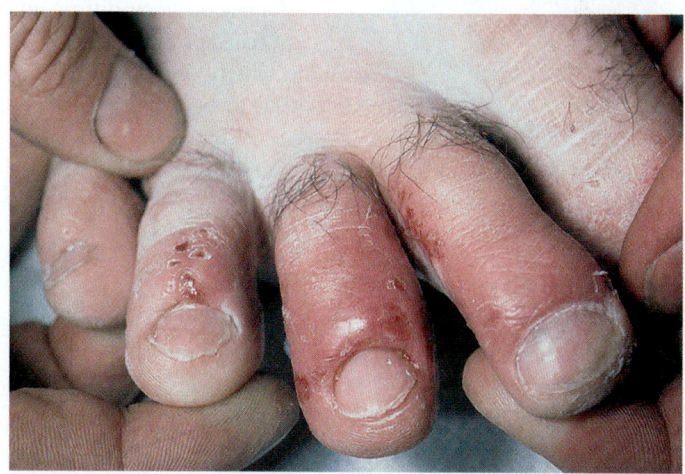

(a)

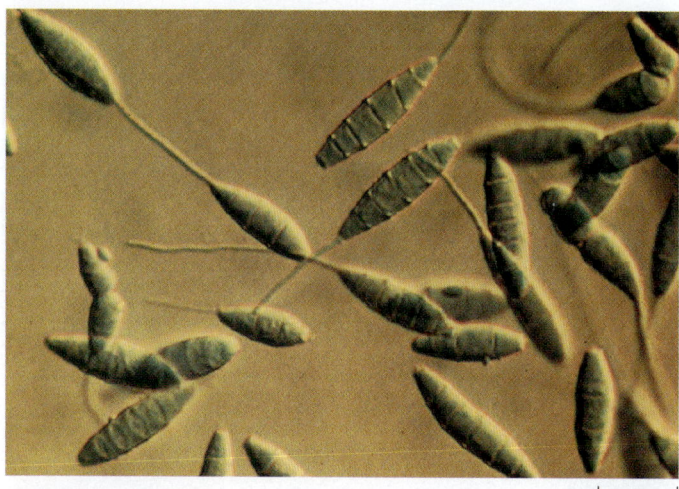

(b)   20 μm

FIGURE 22.24 Dermatophytosis (a) Tinea pedis, usually caused by species of *Trichophyton*. **(b)** Large boat-shaped conidia of *Microsporum gypseum*, a cause of scalp ringworm in children.

❓ *What is the common name for tinea pedis?*

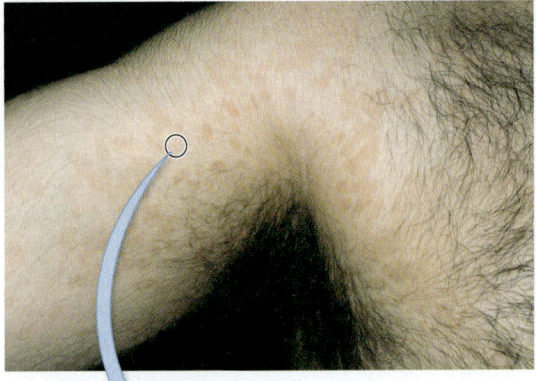

(a)

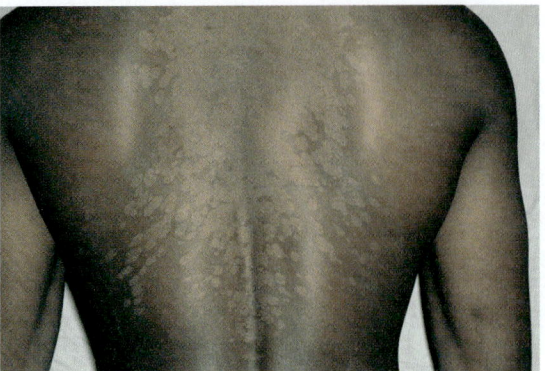

(b)

FIGURE 22.25 Tinea Versicolor Appearance in **(a)** a fair-skin individual and **(b)** a dark-skin individual. **(c)** Microscopic appearance of skin scraping showing *Malassezia furfur* yeast and filamentous forms.

❓ *What other conditions can be caused by this fungus?*

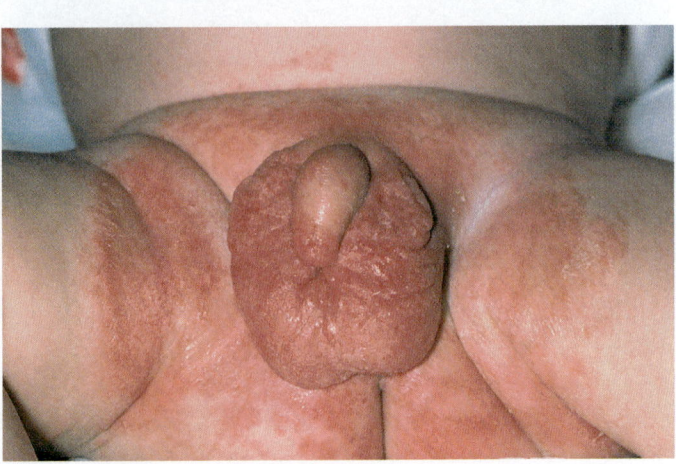

(c)

breaks down the protein, allowing them to use it as a nutrient. Dermatophytes can invade the epidermis down to the level of the keratin-producing cells. Hair is invaded at the follicle, which is relatively moist.

Fungal products diffuse into the dermis and provoke an immune reaction, which probably explains why adults tend to be more resistant to infection than children. Children are more likely to have hypersensitivities, like asthma and eczema. Diffusion of fungal products also explains why some people develop the allergic "id" reactions.

Epidemiology

Patient age, virulence of the infecting strain of mold, and moisture availability are important factors in determining the course of infection. Common causes of excessive moisture include obesity where folds of skin lie together, tight clothing, and plastic or rubber footwear. Potentially pathogenic molds may be present in soil and on pets such as young cats and dogs. Fungi acquired through soil or animal contact tend to cause more noticeable signs and symptoms in humans.

Treatment and Prevention

Numerous prescription and over-the-counter medications can be used to treat superficial skin infections. However, nail infections are often much more difficult to cure, requiring taking medication by mouth for months, and sometimes surgical removal of the nail.

Attention to cleanliness and maintenance of normal dryness of the skin and nails effectively prevent most dermatophyte infections. Powders, open shoes, changing of socks, and applying rubbing alcohol after bathing may help prevent toenail infections.

Other Fungal Diseases

Although *Malassezia furfur* is generally harmless and commonly found on the skin, in some people, it causes skin conditions such as a scaly face rash, dandruff, or tinea versicolor (**figure 22.25**). The latter is a common skin disease characterized by patchy scaliness and increased pigment in light-skin people, or a decrease in pigment in dark-skin people. Scrapings of the affected skin show large numbers of *M. furfur,* both in its yeast form and as short filaments called hyphae. Unknown host factors are important in these diseases because most people carry the organism on their skin without any disease. AIDS patients often have a severe rash with pus-filled pimples caused by *Malassezia* yeasts, and the organisms may even infect internal organs in patients receiving lipid-containing intravenous feedings. ◀◀ yeasts, p. 283 ◀◀ hyphae, p. 284

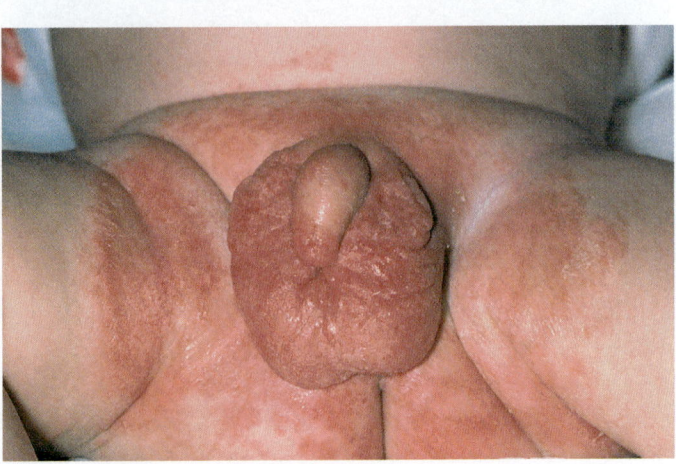

(a)

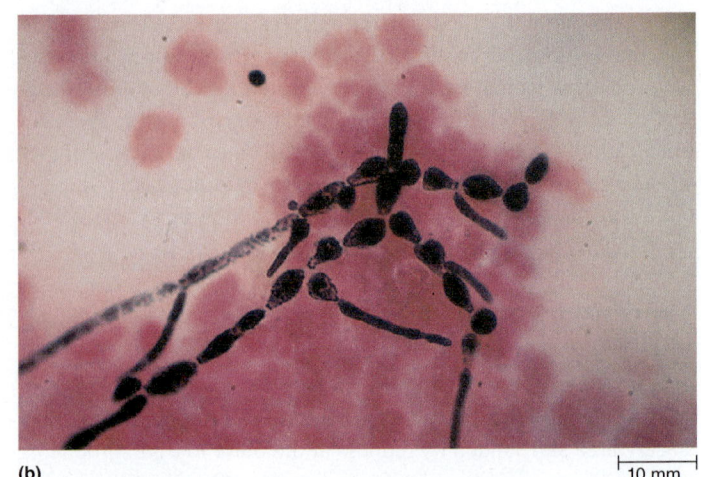

(b)

10 mm

FIGURE 22.26 *Candida albicans* **(a)** Causing a diaper rash; **(b)** Gram stain of pus showing *C. albicans* yeast forms and filamentous forms called pseudohyphae.

❓ *Would an antibiotic ointment containing penicillin be effective against this candidal infection?*

Diseases in Review 22.1

Common Bacterial, Viral, and Fungal Skin Diseases

Disease	Causative Agent	Comment	Summary Table
BACTERIAL SKIN DISEASES			
Acne	*Propionibacterium acnes* commonly associated	Most common during puberty, probably due to excess sebum secretion in response to increased hormone levels.	
Hair follicle infections	*Staphylococcus aureus*	Causative agent commonly colonizes the nostrils and moist skin areas; skin infections include folliculitis, furuncles, and carbuncles. Organism is often resistant to multiple antibiotics.	Table 22.2, p. 524
Staphylococcal scalded skin syndrome	Exfoliatin-producing strains of *S. aureus*	Characterized by peeling of the outer layer of skin; occurs in newborns and infants, as well as the elderly and immunocompromised.	Table 22.4, p. 528
Impetigo	Usually *Streptococcus pyogenes*; sometimes *Staphylococcus aureus*	Characterized by blisters, which break and are replaced by oozing yellow crusts; some people later develop glomerulonephritis.	Table 22.6, p. 529
Rocky Mountain spotted fever (RMSF)	*Rickettsia rickettsii*	Spread by ticks; characterized by rash that spreads and then becomes hemorrhagic.	Table 22.7, p. 531
Lyme disease	*Borrelia burgdorferi*	Spread by ticks; characterized by "bull's-eye rash"; later symptoms include injury to heart and nervous system, and arthritis.	Table 22.8, p. 534
VIRAL SKIN DISEASES			
Varicella (chickenpox)	Varicella-zoster virus (VZV)	Virus enters via the respiratory tract; infection characterized by itchy skin lesions; VZV becomes latent and can later reactivate to cause shingles. Preventable by vaccination.	Table 22.9, p. 537
Rubeola (measles)	Rubeola virus	Virus enters via the respiratory tract. Disease manifests with respiratory symptoms and a spreading rash; Koplik spots occur in mouth. Preventable by vaccination.	Table 22.10, p. 539
Rubella (German measles)	Rubella virus	Virus enters via the respiratory tract, causes mild respiratory symptoms, joint pain, and fine rash; can damage developing fetus (congenital rubella syndrome). Preventable by vaccination.	Table 22.11, p. 541
Warts (dermal warts)	Papillomaviruses	Warts, which are benign skin tumors, can be removed by freezing, burning, surgery, or topical medication.	
FUNGAL SKIN DISEASES			
Superficial cutaneous mycoses	Usually *Epidermophyton, Microsporum,* or *Trichophyton* species	Fungi invade keratinized skin, causing what is commonly known as athlete's foot, jock itch, and ringworm.	
Other fungal diseases	*Malassezia furfur, Candida albicans*	Both organisms are usually harmless on the skin but *M. furfur* sometimes causes skin conditions such as scaly face rash, dandruff, or tinea versicolor. *C. albicans* sometimes invades deeper layers and subcutaneous tissues.	

The yeast *Candida albicans* may live harmlessly among the normal microbiota of the skin, but in some people it invades the deep layers of the skin and subcutaneous tissues (**figure 22.26**). In many people with candidal skin infections, no precise cause for the invasion can be determined. Certain molds also cause cutaneous mycoses, but they are not as likely as *C. albicans* to invade the deep skin layers.

The key features of the diseases covered in this chapter are highlighted in the **Diseases in Review 22.1** table.

MicroAssessment 22.4

The fungi that cause skin mycoses can commonly colonize skin without causing signs or symptoms. The best protection against fungal skin infections is to maintain normal skin dryness.

10. *What is a mycosis?*

11. *What kinds of structures are invaded by dermatophytes?*

12. *Why do you think it is so more difficult to treat nail infections than other superficial dermatophytoses?*

The Ecology of Lyme Disease

Lyme disease is often referred to as one of the emerging diseases. Unrecognized in the United States before 1975, it is now the most commonly reported vector-borne disease. Because of the seeming explosion in the numbers of Lyme disease cases, and its apparent extension to new geographical areas, the ecology of Lyme disease is under intense study. In the northeastern United States, large increases in white-footed mouse populations occur in oak forests during years in which there is a heavy acorn crop, with a corresponding increase in *Ixodes scapularis* ticks. Both deer and mice feed on the acorns and subsequently spread the disease to adjacent areas. Variations in weather conditions, and their effect on food supply for these animals, might therefore be an important ecological factor, although it is not clear that weather cycles completely explain the emerging nature of the disease. The presence of animals other than white-footed mice for the ticks to feed on is another factor. Alternative tick hosts usually do not have a sustained *Borrelia* *burgdorferi* bacteremia following infection from a tick, and the blood of a common lizard host along the West Coast even kills the spirochetes. The role of snakes, foxes, and birds of prey that control mouse populations and that of birds, spiders, and wasps that feed on ticks are also under study. The challenge is to define more completely the ecology of Lyme and other tick-borne diseases in order to predict their emergence and find new ways for their prevention.

Summary

22.1 ■ Anatomy, Physiology, and Ecology
The skin prevents the entry of microbes, regulates body temperature, restricts the loss of fluid from body tissues, and plays an essential role in the function of the immune system. It is composed of the **epidermis** and the **dermis** (figure 22.1). Common members of the skin microbiota include diphtheroids, staphylococci, and fungi (table 22.1).

22.2 ■ Bacterial Skin Diseases

Acne Vulgaris
Acne is characterized by enlarged sebaceous glands. Sebum accumulation within these glands allows *Propionibacterium acnes* to grow to high numbers. Their metabolic products cause an inflammatory response.

Hair Follicle Infections
Folliculitis, boils and **carbuncles** are caused by *Staphylococcus aureus* which is **coagulase**-positive and often resists penicillin and other antibiotics (figures 22.2 and 22.3, table 22.3). There are many different strains that vary in virulence. A carbuncle is more serious because the infection is more likely to be carried to the heart, brain, or bones.

Staphylococcal Scalded Skin Syndrome (table 22.4)
Staphylococcal scalded skin syndrome results from **exfoliatin** produced by certain strains of *Staphylococcus aureus* (figure 22.4).

Streptococcal Impetigo (table 22.6)
Impetigo is a superficial skin infection caused by *Streptococcus pyogenes* and *Staphylococcus aureus* (figure 22.5).

Rocky Mountain Spotted Fever (table 22.7; figure 22.6)
Rocky Mountain spotted fever, caused by the obligate intracellular bacterium *Rickettsia rickettsii,* is an often fatal disease transmitted to humans by the bite of an infected tick (figure 22.8).

Lyme Disease (table 22.8)
Lyme disease is characterized by stages of disease and is caused by a spirochete, *Borrelia burgdorferi* (figure 22.11). A bull's-eye rash is the hallmark of the early stage disease (figure 22.10). *B. burgdorferi* is transmitted by certain ticks (figures 22.13, 22.14).

22.3 ■ Skin Diseases Caused by Viruses

Varicella (Chickenpox) (table 22.9)
Chickenpox, once a common disease of childhood, is caused by the varicella-zoster virus (figure 22.15). **Shingles** can occur months or years after chickenpox and is due to reactivation of the virus (figure 22.16). Shingles cases can be sources of chickenpox epidemics.

Rubeola (Measles) (table 22.10)
Rubeola (measles) is a potentially dangerous viral disease that can lead to serious secondary bacterial infections, and fatal lung or brain damage. Measles can be controlled by vaccinating with an attenuated vaccine (figures 22.18, 22.19).

Rubella (German Measles) (table 22.11)
German measles (**rubella**), if contracted by a woman early in pregnancy, often results in birth defects, making up the **congenital rubella syndrome.** Immunization with an attenuated virus protects against this disease (figure 22.21).

Other Viral Rashes of Childhood
Numerous viruses can cause rashes. Fifth disease (erythema infectiosum), caused by parvovirus B-19, is characterized by a "slapped cheek" rash (figure 22.22). Roseola (exanthem subitum), caused by herpesvirus type 6, is marked by a high fever and a rash that appears as the temperature returns to normal.

Warts
Warts are skin tumors caused by a number of papillomaviruses (figure 22.23). They rarely become cancers, but some sexually transmitted papillomaviruses cause cervical cancer.

22.4 ■ Skin Diseases Caused by Fungi

Superficial Cutaneous Mycoses
Dermatophytes cause athlete's foot, ringworm, and invasions of the hair and nails (figure 22.24).

Other Fungal Diseases
Malassezia sp. can cause tinea versicolor and dandruff, as well as serious skin disease in AIDS patients (figure 22.25). *Candida albicans* may live harmlessly among the normal flora, but it can invade deeper layers of the skin and subcutaneous tissues (figure 22.26).

Review Questions

Short Answer

1. What is the difference between a furuncle and carbuncle?
2. Why do only certain strains of *Staphylococcus aureus* cause scalded skin syndrome?
3. How is impetigo spread?
4. How does the fact that Rocky Mountain spotted fever is a zoonosis relate to the relative severity of the disease symptoms?
5. Describe the causative agent of Lyme disease.
6. What is characteristic about the rash of varicella?
7. What is the relationship between chickenpox (varicella) and shingles (herpes zoster)?
8. Why are many cases of measles complicated by secondary infections?
9. What is the significance of rubella viremia during pregnancy?
10. How does a person contract warts?

Multiple Choice

1. Which of the following conditions is important in the ecology of the skin?
   a) Temperature   b) Salt concentration   c) Lipids
   d) pH   e) All of the above
2. *Staphylococcus aureus* can be responsible for which of these following conditions?
   a) Impetigo   b) Food poisoning
   c) Toxic shock syndrome   d) Scalded skin syndrome
   e) All of the above
3. The main effect of staphylococcal protein A is to
   a) interfere with phagocytosis.
   b) enhance the attachment of the Fc portion of antibody to phagocytes.
   c) coagulate plasma.
   d) kill white blood cells.
   e) degrade collagen.
4. Which of the following is essential for the virulence of *Streptococcus pyogenes*?
   a) Protease   b) Hyaluronidase   c) DNase
   d) All of the above   e) None of the above
5. Which of the following statements is true of streptococcal impetigo?
   a) It is caused by a Gram-negative rod.
   b) It cannot be transmitted from one person to another.
   c) Pathogenic streptococci all produce coagulase.
   d) All of the above.
   e) None of the above.
6. All of the following are true of Rocky Mountain spotted fever *except*
   a) the disease is most prevalent in the western United States.
   b) it is caused by an obligate intracellular bacterium.
   c) it is a zoonosis transmitted to human beings by ticks.
   d) those with the disease characteristically develop a hemorrhagic rash.
   e) antibiotic therapy is usually curative if given early in the disease.

7. All of the following are true of Lyme disease *except*
   a) it is caused by a spirochete.
   b) it is transmitted by certain species of ticks.
   c) it occurs only in the region around Lyme, Connecticut.
   d) most cases get a rash that looks like a target.
   e) it can cause heart and nervous system damage.
8. Which of the following statements is more likely to be true of measles (rubeola) than German measles (rubella)?
   a) Koplik spots are present.
   b) It causes birth defects.
   c) It causes only a mild illness.
   d) Human beings are the only natural host.
   e) Attenuated virus vaccine is available for prevention.
9. All of the following must be cultivated in cell cultures instead of cell-free media *except*
   a) *Rickettsia rickettsii*.   b) rubella virus.
   c) varicella-zoster virus.   d) *Borrelia burgdorferi*.
   e) rubeola virus.
10. All of the following might contribute to development of ringworm or other superficial cutaneous mycoses *except*
    a) obesity.   b) playing with kittens.
    c) rubber boots.   d) using skin powder.
    e) dermatophyte virulence.

Applications

1. A school administrator in a small Iowa community prohibited a child with chickenpox from attending school. He said this was the first case of chickenpox in the school in 6 years and he did not want to have an outbreak. Several parents argued to the school board that an outbreak would benefit the school in the long term. Discuss the pros and cons of allowing this child to attend school.
2. A public health official was asked to speak about immunization during a civic group luncheon. One parent asked if rubella was still a problem. In answering the question, the official cautioned women planning to have another child to have their present children immunized against rubella. Why did the official suggest this?

Critical Thinking ➕

1. When Lyme disease was first being investigated, the observation that frequently only one person in a household was infected was a clue leading to the discovery that the disease was spread by arthropod bites. Why was this so?
2. Why might it be more difficult to eliminate a disease like Lyme disease or Rocky Mountain spotted fever from the earth than rubeola or rubella?

23 Wound Infections

Shotgun wound of the torso.

A Glimpse of History

Endospores of *Clostridium tetani*, the bacterium that causes tetanus (lockjaw), are found in soil and dust—virtually everywhere. This disease used to be common before its cause and pathogenesis were understood, and it often ended in an agonizingly painful death. Dr. Shibasaburo Kitasato (1856–1931), working on tetanus in Robert Koch's laboratory in Germany, was the first to discover that *C. tetani* is an obligate anaerobe. This critical information helped him develop a method to grow the bacterium in pure culture, an essential step toward characterizing a pathogen and learning how it causes disease.

Kitasato showed that laboratory animals injected with *C. tetani* developed tetanus. He was puzzled, however, by a surprising finding: Although the animals died of generalized disease, there were no *C. tetani* cells anywhere other than the injection site. By doing experiments in which he injected the tails of mice and then removed the inoculated tissue at hourly intervals, he showed that the animals developed tetanus only if the bacteria remained in the animals for more than an hour. He also showed that the organisms stayed at the site of inoculation; at no time were they found in the rest of the body. Kitasato reasoned that something other than bacterial invasion was causing the disease.

While Kitasato was working with tetanus, another scientist, Emil von Behring, was busy investigating how *Corynebacterium diphtheriae* caused the disease diphtheria. Together, Kitasato and von Behring showed that toxins produced by the bacteria caused both diseases. The concept that a bacterial toxin could cause disease was an extremely important advance in the understanding and control of infectious diseases.

Kitasato published his studies in 1890 and 2 years later returned to Japan. The Japanese government did not support basic research at that time, so Kitasato established his own institute for infectious diseases where he worked and trained Japanese scientists for the rest of his life. In 1908, Koch visited Kitasato in Japan and a Shinto shrine was built in Koch's honor. During Kitasato's later years, he played an important role in establishing laws regulating health practices in Japan. When he died at age 75, a shrine in his honor was erected next to Koch's.

Most people occasionally suffer wounds that cause breaks in the skin or mucous membranes. Microorganisms originating from the environment or the object causing the wound almost always contaminate these injuries. Infection may result, depending on several factors, including (1) virulence of the microbes; (2) number of microbial cells in the wound; (3) status of the host's immune system; and (4) the type of wound, especially whether the tissues are crushed or contain foreign matter. Wounds that contain foreign materials such as dirt, leaves, bits of rubber, or cloth usually become infected and do not heal until this matter is removed. Such wounds often provide places for microorganisms to multiply out of the reach of phagocytes and other immune defenses. In some cases, foreign materials create surfaces for biofilm development. They may also reduce available O_2, thereby inhibiting phagocytic function and allowing the growth of anaerobic pathogens. Clean wounds often heal without treatment despite microbial colonization, but sometimes even a minor wound can result in a severe, or possibly fatal, infection.

MicroByte

Costs for treating postoperative wound infections are almost $1.5 billion per year in the United States.

23.1 ■ Anatomy, Physiology, and Ecology

Learning Outcomes

1. *Name three tissue components exposed by wounds to which pathogens specifically attach.*
2. *Describe the beneficial and harmful aspects of abscess formation.*

Wounds vary in their characteristics, severity, and associated risks. The general categories of wounds include the following:

■ **Incisions:** Produced by a knife or other sharp object.

■ **Punctures:** Result from penetration by a small sharp object, such as a needle or a nail.

■ **Lacerations:** Occur when the tissue is torn.

■ **Contusions:** Produced by a blow that crushes tissue.

■ **Abrasions:** Occur when the epidermis is scraped off.

■ **Gunshot wounds:** Caused by bullets or other projectiles.

■ **Burns:** Caused by heat (thermal burns), electricity, chemicals, radiation, or friction.

Wounds expose tissue components normally protected by skin or mucous membranes, providing surfaces to which pathogens can attach and then colonize. These tissue components include collagen, fibronectin, fibrinogen, and fibrin. **Collagen** is a fibrous material—the main supportive protein of skin, tendons, scars, and other body structures. **Fibronectin** is a fibrous glycoprotein that occurs both as a circulating form and as a component of tissue, where it binds cells and other tissue substances together. **Fibrinogen** is a blood protein; when a wound occurs, this protein is converted to **fibrin,** which forms clots in the damaged vessels. The clots stop the flow of blood as the first step in the wound repair process.

Wound healing begins with the outgrowth of connective tissue cells (fibroblasts), and capillaries from the surfaces of the wound, producing a red, translucent fibrous material called **granulation tissue.** In clean wounds, granulation tissue fills the space created by the wound. This tissue shrinks and is converted to collagen, a component of scar tissue that is eventually covered by skin or mucous membrane (**figure 23.1**).

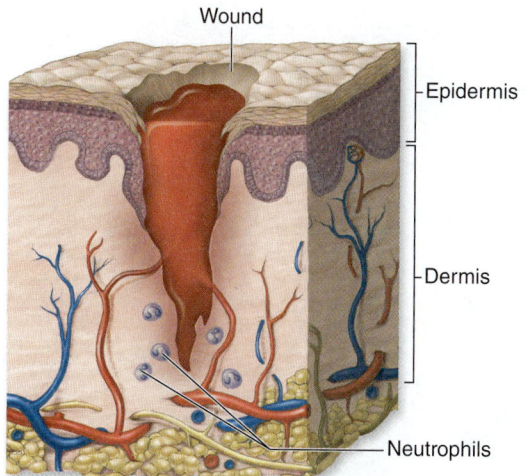

Wound — Epidermis — Dermis — Neutrophils

1 Cut blood vessels bleed into the wound. Fibrin forms clots in severed capillaries.

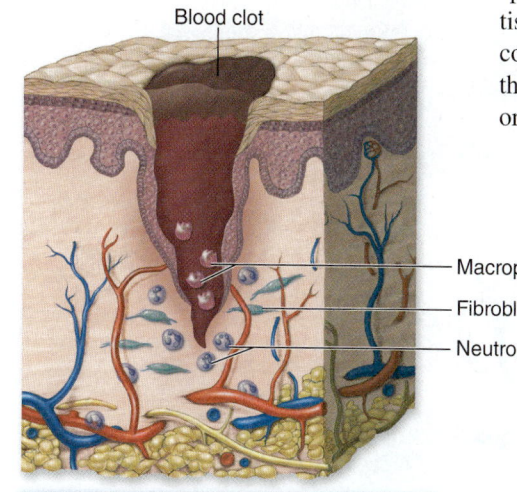

Blood clot — Macrophages — Fibroblast — Neutrophils

2 Blood clot forms in wound, and phagocytes destroy microbes.

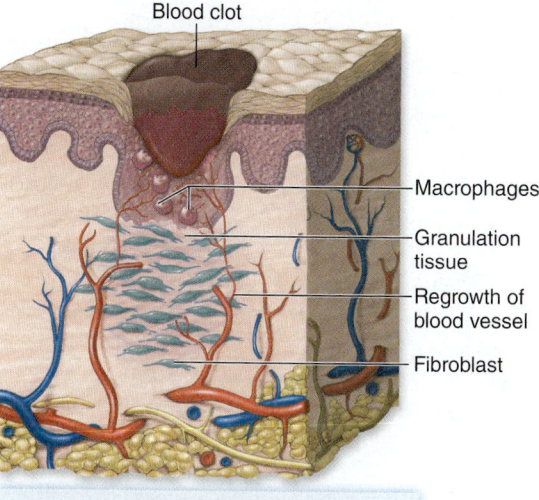

Blood clot — Macrophages — Granulation tissue — Regrowth of blood vessel — Fibroblast

3 Wound fills with granulation tissue and blood vessels regrow.

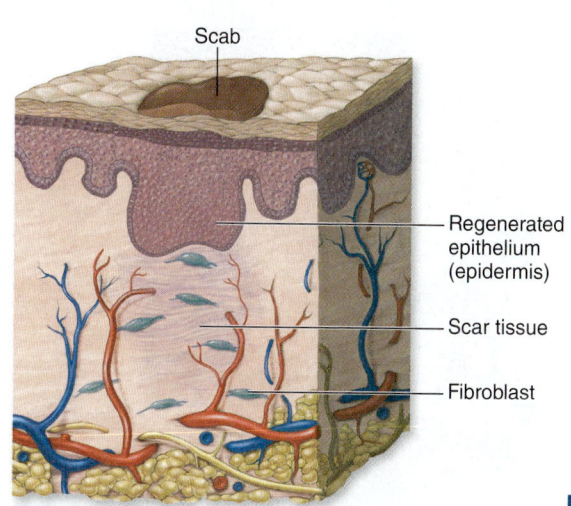

Scab — Regenerated epithelium (epidermis) — Scar tissue — Fibroblast

4 Fibroblasts secrete collagen, forming scar tissue. Collagen contracts and epithelium regenerates.

FIGURE 23.1 The Process of Wound Repair

❓ *What is the function of granulation tissue?*

Wound Abscesses

An **abscess** (**figure 23.2**) is a localized collection of **pus** surrounded by inflamed body tissue. The pus—a thick yellowish fluid—is composed of living and dead leukocytes, tissue debris, and proteins. Abscesses form as a result of the body's immune defenses and usually indicate an infection. Although an abscess helps localize the infection and prevents its spread, it also indicates a potentially serious situation. If some cells of the pathogen escape the abscess, they can enter the blood or lymph, leading to infection in other parts of the body.

The very nature of abscesses makes them difficult to treat. They have no blood vessels, because the developing pus pocket destroys or pushes them aside, and adjacent blood vessels are often blocked by clots. This lack of blood circulation makes it difficult for antimicrobial drugs to reach the infected site. Even if the drugs do enter the site, the chemical nature of pus interferes with the action of some antibiotics. In addition, antimicrobial drugs are often ineffective against microorganisms in the abscess because the microbes stop multiplying, and most antimicrobial medications work against actively dividing cells. Abscesses usually must therefore burst to a body surface or be drained surgically to be cured. ◀◀ **antimicrobial medications, p. 457**

Anaerobic Wounds

An important feature of many wounds is that they are relatively anaerobic, which allows the growth of obligate anaerobes such as *Clostridium tetani*. Wounds that are likely to be anaerobic are those that have extensive tissue damage, are contaminated with dirt or are small in diameter but deep, such as punctures. Punctures may have foreign material and microorganisms forced deep into the tissues. Anaerobic conditions are also often created when different microbial species grow in a wound (polymicrobial infections), because facultative anaerobes use up the available O_2, converting it to water as they respire.

FIGURE 23.2 Abscess Formation

❓ *What is the composition of pus?*

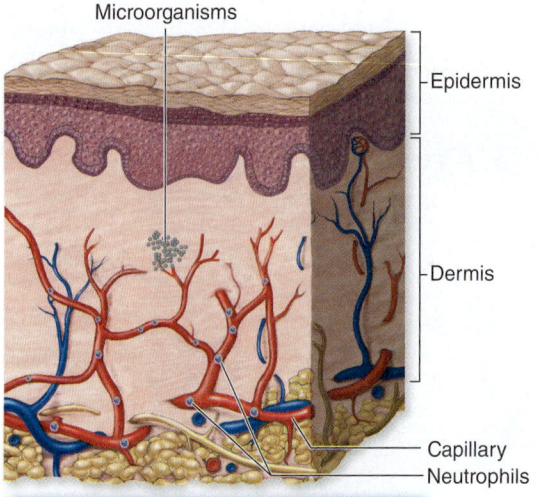

1) Microorganisms enter the tissue from a wound or from the bloodstream.

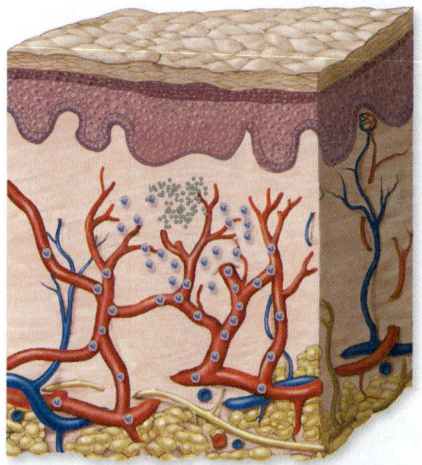

2) Blood vessels dilate, and leukocytes migrate to the area of the developing infection.

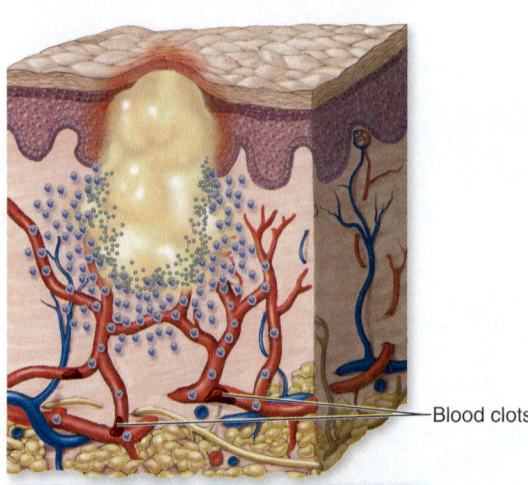

3) Pus forms and an abscess develops; clotting occurs in adjacent blood vessels.

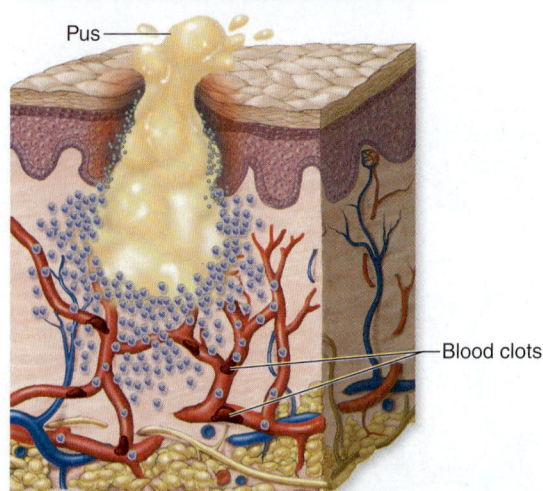

4) Buildup of pressure causes the abscess to expand in the direction of least resistance; if it reaches a body surface, it may rupture and discharge its contents.

Wounds can be classified as incisions, punctures, lacerations, contusions, abrasions, gunshots, or burns. Wounds expose tissue components to which pathogens can attach. Healing involves the outgrowth of fibroblasts and capillaries from the sides of the wound to produce granulation tissue that fills the wound. Abscess formation provides a way of isolating infections and preventing their spread. Anaerobic conditions in wounds are created by the presence of dead tissue and foreign material.

1. *Name and describe two substances in wounds to which pathogens attach.*

2. *Give two reasons why an abscess might not respond to antibiotic treatment.*

3. *Why is it important that the granulation tissue shrinks after it is formed?* ✚

23.2 ■ Common Bacterial Infections of Wounds

Learning Outcomes

3. *Give distinctive characteristics of three common wound infections caused by bacteria that grow aerobically.*

4. *Discuss the significance of fibronectin binding by* S. epidermidis.

If a wound becomes infected, several serious consequences are possible. These include (1) delayed healing, (2) formation of abscesses, and (3) spread of the bacteria or their toxins to other areas of the body. Infected surgical wounds often split open as swelling causes the stitches to pull through tissues weakened by the infection. The infection can spread to devices such as an artificial hip or knee, which may then have to be removed before the infection can be eliminated. **Table 23.1** summarizes the characteristics of the leading causes of wound infections.

Staphylococcal Wound Infections

Staphylococcus species, common inhabitants of the nostrils and skin, are the leading causes of wound infections (**figure 23.3**). Of the 30 or more recognized species, only two cause most human wound infections—*Staphylococcus aureus* and *Staphylococcus epidermidis*. The more important of these two, *S. aureus,* was covered extensively in chapter 22. ◄◄ the genus *Staphylococcus*, p. 273 ◄◄ bacterial skin diseases, p. 523

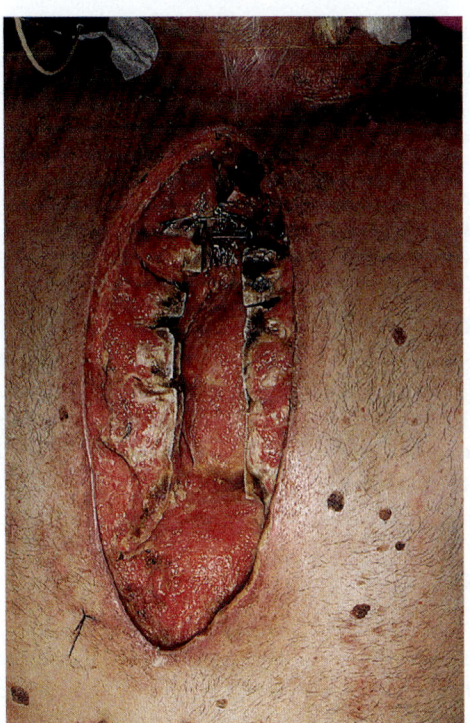

FIGURE 23.3 Surgical Wound Infection Due to *Staphylococcus aureus*

❓ *Why do infected surgical wounds sometimes split open?*

Signs and Symptoms

Staphylococcus species are **pyogenic,** meaning that they cause the production of pus (*pyo* means "pus" and *genic* means "generating"). Staph infections are usually characterized by an inflammatory reaction, with swelling, redness, and pain. Fever occurs if the infected area is large or if the infection has spread to the blood or lymph.

Toxic shock syndrome can occur if the wound is infected with a toxin-producing staphylococcal strain. Signs and symptoms of this include high fever, muscle aches, and a life-threatening drop in blood pressure and shock. Sometimes the infected person will also have a rash and diarrhea. ▶▶ staphylococcal toxic shock syndrome, p. 619

TABLE 23.1	Leading Causes of Wound Infections	
Causative Organism	**Characteristics**	**Consequences**
Staphylococcus aureus	Gram-positive cocci in clusters, coagulase-positive	Delayed healing; abscess formation; extension into tissues, artificial devices, or bloodstream; some strains can cause toxic shock syndrome
Streptococcus pyogenes	Gram-positive cocci in chains; Lancefield group A	Same as above, except some strains can cause "flesh-eating" necrotizing fasciitis
Pseudomonas aeruginosa	Gram-negative rod, green pigment	Delayed healing; abscess formation; extension into tissues, artificial devices, or bloodstream; septic shock

Causative Agents

Staphylococcus aureus and *S. epidermidis* are Gram-positive cocci that grow in clusters (**figure 23.4**). These facultative anaerobes are quite hardy, which is not surprising because they have evolved to thrive on skin, which is dry and salty. They survive well in the environment and are easily transferred from person to person. ◀◀ **facultative anaerobe, p. 90**

The coagulase test is used to distinguish *S. aureus* from other staphylococci. *S. aureus* is often referred to as "coag-positive staph" because it makes coagulase, whereas the other staphylococcal species are collectively referred to as "coag-negative staph." Of the coag-negative staphylococci, *S. epidermidis* is the most common cause of healthcare-associated infections, including those of surgical wounds. However, *S. aureus* causes serious wound infections much more commonly than *S. epidermidis*. ◀◀ **Healthcare-associated infections, p. 449** ◀◀ **coagulase, p. 526**

Pathogenesis

Staphylococcus aureus   *S. aureus* produces multiple virulence factors that act together in the disease process (see table 22.3). These virulence factors are covered extensively in chapter 22 (skin infections), but it is important to recognize that those same factors play a critical role in wound infections as well. For example, clumping factor and other proteins allow the cells to attach to clots and tissue components, an initial step in colonization. Lipases, proteases, and hyaluronidases together cause tissue damage. Capsules, coagulase, and protein A protect the cells from attack by the complement system, phagocytes, and antibodies. ◀◀ **S. aureus pathogenesis, p. 524**

Immunity to staphylococcal infection is generally weak or non-existent, probably because the organism so effectively evades the immune defenses. However, some protein A is released from the bacterial surface, and it reacts with circulating antibodies. The resulting immune complexes can activate the complement system, probably contributing to the intense inflammatory response and accumulation of pus that characterize *S. aureus* infections. ◀◀ **complement system, p. 344**

Staphylococcus aureus infections may cause systemic complications. Bacteria growing in a wound can spread, leading to abscesses in the heart, bones, or other tissues. Some *S. aureus* strains produce **superantigens** that can enter the circulation. These exotoxins react with helper T cells, activating them and causing them to release large amounts of cytokines that lead to toxic shock. ◀◀ **superantigens, p. 393** ◀◀ **cytokines, p. 341**

Staphylococcus epidermidis   *S. epidermidis* is not particularly virulent and cannot invade healthy tissues. However, the bacterium often causes minor abscesses around the stitches used in surgery. It also adheres to and then colonizes medical devices, including indwelling catheters and artificial joints. It can do this because it binds to fibronectin, the blood protein that quickly coats surgical implants in the body. The bacterial cells may then produce a slime layer or glycocalyx, a critical step in biofilm formation.

Biofilms are a serious problem for several reasons. Diffusion of antibacterial medications into them is slow and inefficient. Even if the medication does enter the biofilm, it may be ineffective because bacteria within the biofilm are often metabolically inactive. In addition, bacteria in biofilms can come loose, and are then carried by the bloodstream to the heart and other tissues. In people with a compromised immune system, such as those with cancer, AIDS, or diabetes mellitus, this can result in subacute bacterial endocarditis or multiple tissue abscesses, which require surgical treatment. In healthy people, wound infections by *S. epidermidis* are usually cleared by host defenses without additional treatment. ◀◀ **glycocalyx, p. 62** ◀◀ **biofilm, p. 84** ▶▶ **bacterial endocarditis, p. 672**

MicroByte

It takes more than 100,000 *S. aureus* cells injected into skin to cause an abscess, whereas only 100 injected into a suture site to do the same.

Epidemiology

S. aureus carriers are at an increased risk for surgical wound infections caused by this species. Other factors that increase a person's risk include advanced age, poor general health, immunosuppression, prolonged preoperative hospital stay, and an infection at another site. Additional information about the epidemiology of *S. aureus* is discussed in chapter 22. ◀◀ **S. aureus epidemiology, p. 526**

S. epidermidis is found on the skin and mucous membranes of most people, residing as part of their normal microbiota. It is an opportunist that can cause disease in individuals with a compromised immune system. ◀◀ **opportunist, p. 382**

Treatment and Prevention

Treating staphylococcal infections can often be difficult because of widespread antibiotic resistance. **Methicillin-resistant *S. aureus* (MRSA)** is a serious problem in wound infections. This organism is extremely difficult to treat, being resistant to multiple β-lactam antibiotics, including methicillin, penicillin, nafcillin, and oxacillin. MRSA infections may be subcategorized into healthcare-associated (HA-MRSA) infections and community-acquired (CA-MRSA) infections. Both HA-MRSA and CA-MRSA are resistant to multiple antibiotics, although CA-MRSA is more susceptible than HA-MRSA and can be treated successfully with sulfa drugs, tetracyclines, and clindamycin in about 75% of cases. HA-MRSA is resistant even to these and is often susceptible only

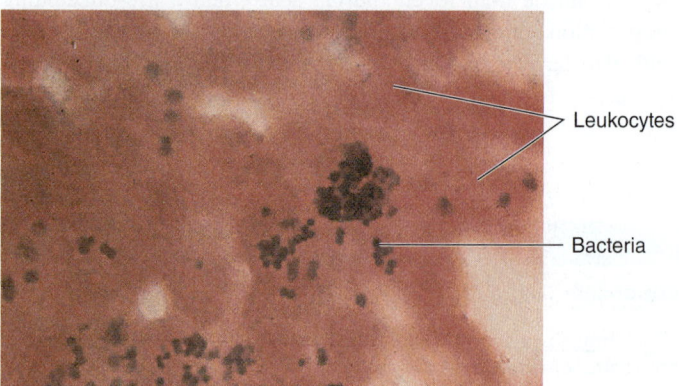

Leukocytes

Bacteria

FIGURE 23.4 *Staphylococcus aureus* in Pus

❓ *What does the name* Staphylococcus *indicate about the typical growth arrangement of the bacterial cells?*

to vancomycin. **Vancomycin-intermediate** *S. aureus* (**VISA**) and **vancomycin-resistant** *S. aureus* (**VRSA**) have emerged, making treatment of these infections extremely difficult. New medications belonging to the oxazolidinones class may be effective in treating them. ◄◄ **antibacterial resistance, p. 471** ◄◄ **MRSAs, p. 473** ◄◄ **VISA and VRSA, p. 473**

To reduce the chance of infection, wounds should be thoroughly cleaned, removing any dirt or dead tissue. Clean, deep wounds and surgical wounds should be quickly closed by sutures to help avoid infection. Surgical wound infections can be reduced by half if the patient is given an effective anti-staphylococcal medication immediately before surgery. For unknown reasons, the infection rate is actually increased if the medication is given more than 3 hours before or 2 hours after the surgical incision.

Group A Streptococcal "Flesh-Eating Disease"

Streptococcus pyogenes is another common cause of wound infections. *S. pyogenes* infections have generally been easy to treat because all known strains of the organism are still susceptible to penicillin. Occasionally, however, the infections can progress rapidly, even leading to death despite antimicrobial treatment. These more severe infections are called "invasive" because they spread into tissues and organs, causing pneumonia, meningitis, puerperal fever, **fasciitis**—inflammation of the fascia that surround muscles and body organs—and streptococcal toxic shock. This section will focus on **necrotizing fasciitis** ("flesh-eating disease"), a rare but serious complication of *S. pyogenes* infection (**figure 23.5**). ◄◄ *Streptococcus pyogenes*, **p. 487** ►► **streptococcal toxic shock, p. 553**

Signs and Symptoms

Signs and symptoms of necrotizing fasciitis appear suddenly and are very serious. Severe pain develops at the site of the wound, which sometimes can be so minor that no break in the skin is even seen. Within a short time, swelling occurs, and the injured

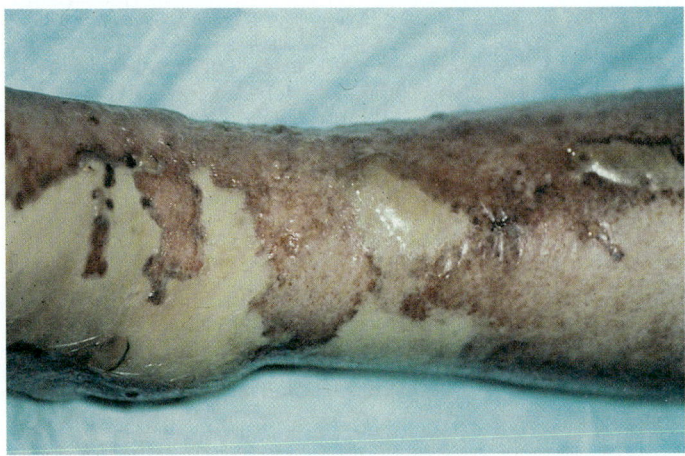

FIGURE 23.5 Individual with *Streptococcus pyogenes* "Flesh-Eating Disease" (Necrotizing Fasciitis)

❓ *Why does necrotizing fasciitis require immediate surgery?*

person develops fever and confusion. The overlying skin becomes stretched and discolored because of the swelling. Unless treatment is started quickly, shock and death usually follow in a short time.

Causative Agent

Streptococcus pyogenes is a β-hemolytic, Gram-positive, chain-forming, aerotolerant organism, with Lancefield group A cell wall polysaccharide and a nonantigenic capsule of hyaluronic acid. Strains of *S. pyogenes* that cause invasive disease are more virulent because they produce various enzymes and toxins that cause severe tissue damage (covered next). ◄◄ **aerotolerance, p. 90** ◄◄ **β-hemolysis, p. 96** ◄◄ **pyrogens, p. 351**

Pathogenesis

The pathogenesis of *Streptococcus pyogenes* has been covered in detail in chapter 21. Like *Staphylococcus aureus*, *S. pyogenes* has a fibronectin-binding protein (F protein) that helps colonization of wounds (see table 22.5). In necrotizing fasciitis, *S. pyogenes* destroys the subcutaneous fatty tissue and fascia, the bands of fibrous tissue that underlie the skin and surround muscle and body organs. In some cases, the organism destroys the muscle tissue itself. It does this damage by producing a variety of destructive enzymes such as streptokinases (break down blood clots), hyaluronidases (break down connections between cells), deoxyribonucleases (break down DNA), and streptolysins (break down red blood cells and neutrophils). Strains that cause necrotizing fasciitis also produce exotoxin A, a superantigen that causes toxic shock, and exotoxin B, a protease that destroys tissue. As the tissues break down, osmotic pressure increases, and fluid moves into the area, causing intense swelling. The organisms multiply in the dead tissue, using the breakdown products as nutrients.

As the bacteria grow, they shed M protein, a surface protein which binds to fibrinogen. The M protein–fibrinogen complexes bind to neutrophils, causing them to release strong inflammatory molecules called heparin-binding proteins that increase the vascular permeability in the host. The blood vessels leak fluid, resulting in a massive, life-threatening drop in blood pressure and shock. M protein also helps the organism avoid phagocytosis by breaking down complement component C3b.

In many cases, *S. pyogenes* also releases a variety of streptococcal pyrogenic exotoxins (SPEs), which are superantigens. These enter the bloodstream, activating T cells and causing them to release large amounts of cytokines. The high levels of cytokines causes symptoms associated with streptococcal toxic shock syndrome (TSS) including fever, nausea, vomiting, rash and sometimes shock and death. Table 22.5 compares and contrasts the pathogenesis of *S. pyogenes* and *S. aureus*.

Epidemiology

Cases of flesh-eating disease in the United States are generally sporadic. Of the deaths caused by invasive *S. pyogenes* infections in the United States yearly, less than 2% are due to necrotizing fasciitis. Underlying conditions that increase the risk of necrotizing fasciitis and other invasive *S. pyogenes* infections include diabetes, cancer, alcoholism, AIDS, recent surgery, abortion, childbirth, chickenpox, and injected-drug abuse. Invasive infections seldom occur in healthy individuals with minor injuries.

Treatment and Prevention

The toxins produced by *S. pyogenes* spread with such speed that immediate surgery is often essential to reduce the pressure of the swollen tissue and to remove dead tissue. Amputation is sometimes necessary to quickly remove the source of toxins. Penicillin is effective for treating early infection, but it has little or no effect on streptococci in necrotic tissue and no effect on toxins. Therefore surgery remains the best treatment.

There are no proven preventive measures for *S. pyogenes* fasciitis, although M protein vaccines are being tested.

Pseudomonas aeruginosa Infections

Pseudomonas aeruginosa, an opportunistic pathogen, is a major cause of healthcare-associated infections. In hospitals, *P. aeruginosa* is an important cause of lung infections and a common cause of wound infections, especially of thermal burns. Burns have large exposed areas of dead tissue that are not protected by normal body defenses and are therefore ideal sites for infection by bacteria. Almost any opportunistic pathogen can infect burns, but *P. aeruginosa* is among the most common and most difficult to treat. ◀◀ *Pseudomonas, p. 265*

P. aeruginosa also occasionally causes community-acquired infections. Such infections include skin rashes and external ear canal infections from contaminated swimming pools and hot tubs, and eye infections from contaminated contact lens solutions. Other infections caused by this organism include those of foot bones from stepping on sharp objects, heart valve infections in injected-drug abusers, scarring infections from ear piercing, and biofilms in the lungs of individuals with the inherited disease cystic fibrosis.

Signs and Symptoms

Signs and symptoms of *Pseudomonas aeruginosa* infection are serious and include chills, fever, skin lesions, and shock, which are caused by bloodstream invasion by this organism. A very noticeable symptom of *P. aeruginosa* infection of burns and other wounds is a characteristic green color, caused by water-soluble pigments produced by the organisms (**figure 23.6a**; see also figure 11.12). These pigments include fluorescent yellow pyoverdin and blue pyocyanin, which together produce the green color.

Causative Agent

Pseudomonas aeruginosa is a motile Gram-negative rod with a single polar flagellum (figure 23.6c). It is found in a wide variety of environments such as soil and water, where it grows easily and fast. The bacterium is classified as an aerobe. However, it also respires anaerobically in the absence of O_2 if nitrate is present. The ability of *P. aeruginosa* to respire anaerobically is critical in the development of biofilms by this organism. ◀◀ **anaerobic respiration, p. 134** ◀◀ **the genus** *Pseudomonas*, **p. 265**

Pathogenesis

Pseudomonas aeruginosa infection of burns and other wounds causes additional tissue damage, delays healing, and increases the risk of septic shock. Virulence of the organism depends mainly on

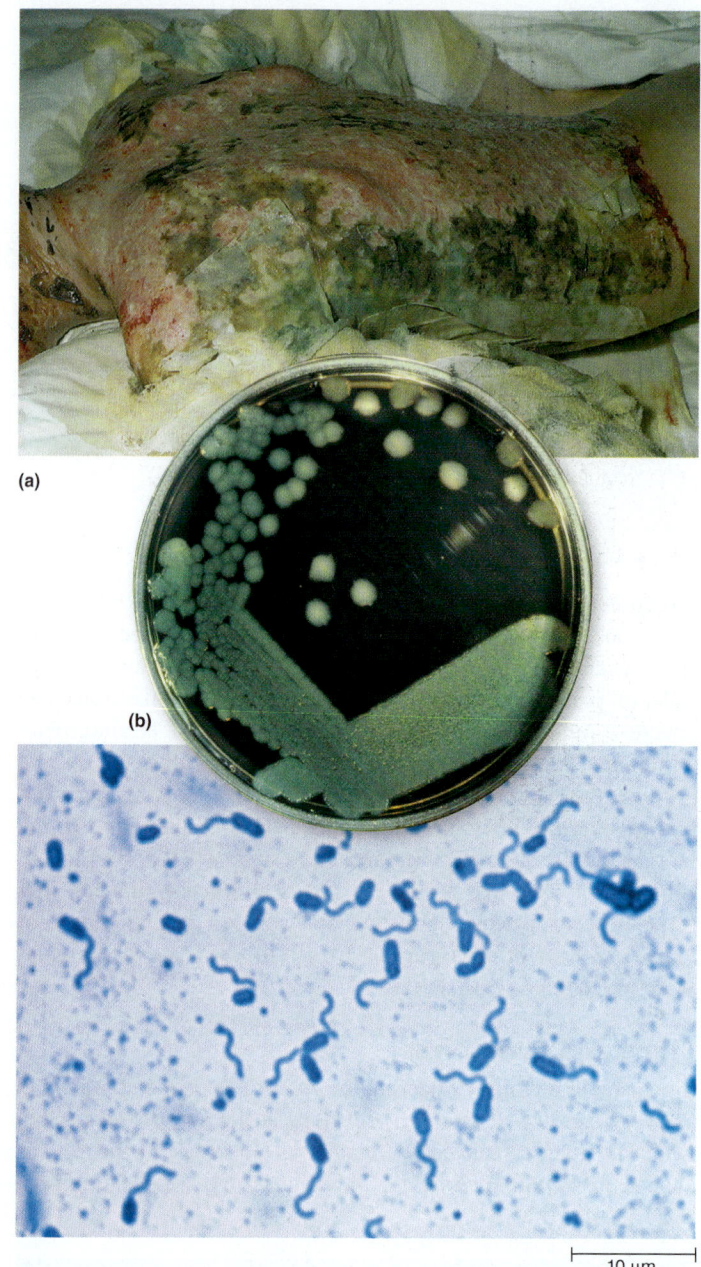

(a)

(b)

(c)

|—— 10 μm ——|

FIGURE 23.6 *Pseudomonas aeruginosa* **(a)** Extensive burn infected with *P. aeruginosa*. **(b)** Culture. **(c)** *P. aeruginosa* has a single polar flagellum.

❓ *What causes the green discoloration of the wound and the culture medium?*

production of two extracellular proteins—exotoxin A and exoenzyme S. Exotoxin A stops host cell protein synthesis. Exoenzyme S is a phospholipase that acts synergistically with a protease called lecithinase to hydrolyze lecithin, an important lipid component of cell membranes. Destruction of lecithin causes membrane disruption and cell death.

The pigment pyocyanin produced by *P. aeruginosa* may further add to pathogenicity. A derivative of this pigment acts as a siderophore, helping the pathogen acquire iron from the environment, thereby inhibiting competing bacteria by reducing their available iron. This pigment also impairs the function of human nasal cilia and disrupts respiratory epithelium. No role in virulence is known for the pigment pyoverdin. ◄◄ **siderophores, p. 386**

Epidemiology

Pseudomonas aeruginosa is widespread in nature. The organism can grow in most places where there is moisture, including soaps, ointments, eyedrops, contact lens solutions, cosmetics, disinfectants, swimming pools, hot tubs, and even distilled water. *P. aeruginosa* is introduced into hospitals on ornamental plants, flowers, and produce. For this reason, visitors are not allowed to take flowers or fruit into hospital burn wards or intensive care units. It can also be found on many kinds of hospital equipment, the inner soles of shoes, and in illegal injectable drugs, all of which have been sources of serious infections.

Treatment and Prevention

Pseudomonas aeruginosa infections are treated with antimicrobial medications. Established infections, however, are very difficult to treat because *P. aeruginosa* is resistant to a wide range of antibiotics. Only a few antibiotics are effective against this pathogen, including fluoroquinolones, gentamicin, and imipenem—and even these drugs may not be effective against all strains. Antibiotic sensitivity tests are done to determine the most appropriate medication for treatment. For systemic infection, antibacterial medications must usually be given intravenously in high doses.

P. aeruginosa infections can be prevented by eliminating possible sources of the bacterium and prompt care of wounds. Removing dead tissue from burn wounds, followed by application of an antibacterial cream, also helps prevention of infection by this organism.

MicroAssessment 23.2

Staphylococcus aureus is the most important cause of wound infections because it is commonly carried by humans, it transfers easily from one person to another, and it has multiple virulence factors. *Staphylococcus epidermidis* forms biofilms on foreign materials, protecting the bacteria from body defenses and antibacterial medications. Flesh-eating strains of *Streptococcus pyogenes* are uncommon but can cause life-threatening disease. *Pseudomonas aeruginosa*, a pigment-producing organism, is widespread in the environment and is a major cause of healthcare-associated infections.

4. *Why are antibacterial medications not effective for treating flesh-eating disease (necrotizing fasciitis)?*

5. *Why do wound infections caused by* Pseudomonas aeruginosa *sometimes produce green pus?*

6. *Why is it not surprising that staphylococci are the most common cause of wound infections?* ➕

23.3 ■ Diseases Due to Anaerobic Bacterial Wound Infections

Learning Outcomes

5. *Describe the conditions that lead to the development of anaerobic wound infections.*

6. *Discuss why it is difficult to treat wounds infected with toxin-producing bacteria.*

Wounds are often anaerobic and may be colonized by certain strictly anaerobic species of bacteria. This section describes three distinctive diseases that result from anaerobic bacterial wound infections: lockjaw, gas gangrene, and lumpy jaw.

Tetanus ("Lockjaw")

Tetanus is often fatal. Fortunately, it is uncommon in economically advanced countries. Exposure to the causative organism cannot be avoided because it produces endospores that are widespread in dust and dirt, frequently contaminating clothing, skin, and wounds. Even a minor wound in a non-immunized person can result in tetanus if the wound provides conditions that allow germination of the spores. ◄◄ **endospores, p. 67**

Signs and Symptoms

Tetanus is characterized by continuous, painful, and uncontrollable cramplike muscle spasms that are usually generalized but may be limited to one area of the body (**figure 23.7**). The spasms often begin with the jaw muscles, giving the disease the common name "lockjaw." The early symptoms include restlessness, irritability, difficulty swallowing, contraction of the jaw muscles, and sometimes seizures. As more muscles go into sustained contraction (called tetany), breathing becomes difficult, abnormal heart rhythms may occur, and in some cases bones can fracture. After a period of almost unbearable pain, the infected person often dies of pneumonia or from lung damage caused by regurgitation of stomach contents into the lung. In addition, because people being treated for tetanus remain in the hospital for long periods of time, healthcare-associated infections can develop.

Causative Agent

Tetanus is caused by *Clostridium tetani,* an anaerobic, spore-forming, Gram-positive, rod-shaped bacterium (**figure 23.8**).

FIGURE 23.7 Infant with Neonatal Tetanus

❓ *How does a newborn infant contract tetanus?*

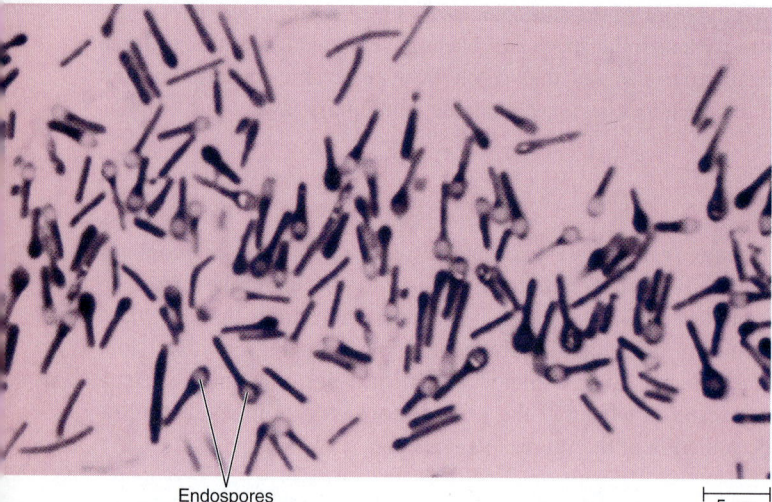

Endospores

5 µm

FIGURE 23.8 *Clostridium tetani* Terminal endospores are characteristic of this species.

❓ *How are the endospores shown here different from endospores of other clostridial species?*

It has two characteristic features: (1) a spherical endospore that forms at the end of the cell, and (2) swarming growth that quickly spreads over the surface of solid media. Identification of *C. tetani* depends on characterizing its plasmid-encoded toxin, although tetanus is typically diagnosed by signs and symptoms. ◄◄ exotoxins, p. 391 ◄◄ *Clostridium*, p. 258 ◄◄ plasmids, p. 208

Pathogenesis

Clostridium tetani is not invasive, and colonization is generally localized to a wound. Its pathologic effects are caused by **tetanospasmin,** an exotoxin released from multiplying vegetative cells. Tetanospasmin is an A-B toxin. The B portion attaches to receptors on motor neurons, which then take up the A portion by endocytosis. The toxin is carried to the neuron cell body in the spinal cord. There, the motor neuron contacts other neurons that normally control its action by means of chemicals called neurotransmitters. Some of these other neurons stimulate the motor nerve cell, causing muscle contraction, while others make the motor neuron resistant to stimulation, inhibiting muscle contraction. Tetanospasmin prevents the release of the neurotransmitter

from inhibitory neurons, blocking their action and causing the muscles to contract without control (**figure 23.9**). ◄◄ A-B toxins, p. 392

The toxin generally spreads across the spinal cord to the side opposite the wound and then downward, typically causing spastic muscles to first appear on the side of the wound, then on the opposite side, and then downward, depending on the amount of toxin. Tetanospasmin released from the infected wound often enters the bloodstream, which carries it to the central nervous system. In these cases, inhibitory neurons of the brain are first affected, and the muscles of the jaw are among the first to become spastic.

Epidemiology

Clostridium tetani occurs not only in dirt and dust, but also in the gastrointestinal tract of humans and other animals that have eaten foods contaminated with its spores. Therefore, fecal contamination is a potential source of infection. Many cases of tetanus result from puncture wounds, which occur in stepping on a nail, body piercing, tattooing, animal bites, splinters, injected-drug abuse, and insect stings. Cases can also occur following surgery. Surface abrasions and burns can result in tetanus if the wound is anaerobic because of dirt or dead tissue.

In economically developed countries, people who get tetanus have either never been immunized or are no longer immune because they received their last vaccine booster shot more than ten years ago. In less economically advanced countries, tetanus occurs more frequently because vaccination is not available, and because of improper wound care. In some parts of the world, babies commonly die of neonatal tetanus as a result of their umbilical cord being cut with instruments contaminated with *C. tetani* (see figure 23.7).

Treatment and Prevention

Tetanus is treated by injecting the affected person with human tetanus immune globulin (TIG), a preparation of antibodies to the tetanus toxin. The antibodies bind to toxin molecules not yet attached to nerve cells, thereby neutralizing their effects and providing passive immunity. TIG, however, cannot neutralize tetanospasmin that is already attached to nerve tissue and it does not repair nerve damage that has already occurred. In this case, the affected person is given muscle relaxants and supportive care,

FIGURE 23.9
Tetanus and Inhibitory Neuron Function

❓ *How does tetanospasmin cause muscle contraction?*

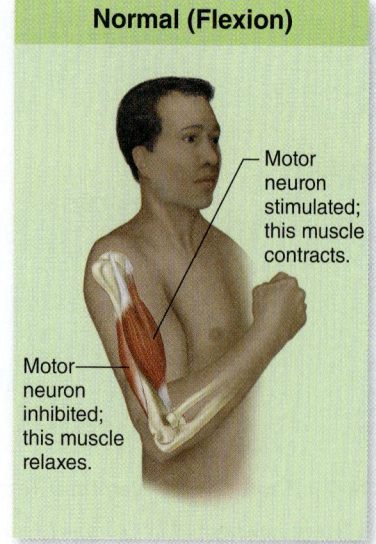

Normal (Flexion)

Motor neuron stimulated; this muscle contracts.

Motor neuron inhibited; this muscle relaxes.

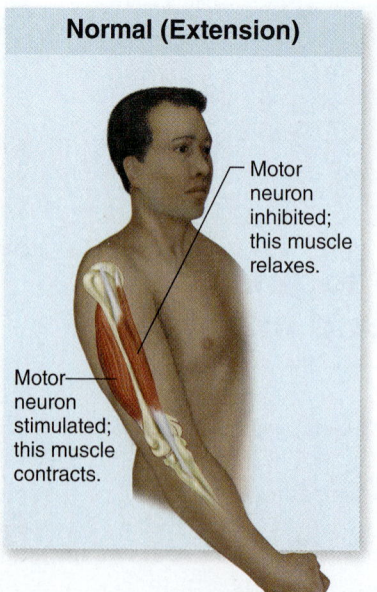

Normal (Extension)

Motor neuron inhibited; this muscle relaxes.

Motor neuron stimulated; this muscle contracts.

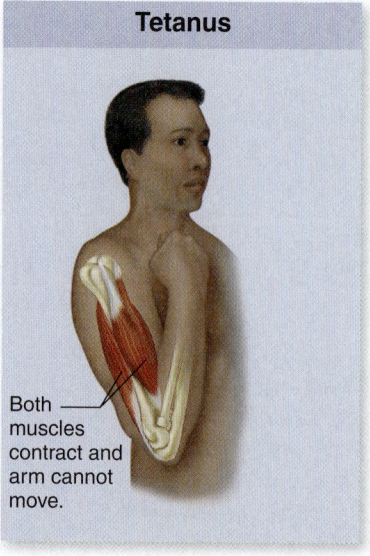

Tetanus

Both muscles contract and arm cannot move.

TABLE 23.2 | **Tetanus Prevention in the Management of Wounds**

Immunization History	CLEAN MINOR WOUNDS		ALL OTHER WOUNDS	
	Toxoid	TIG	Toxoid	TIG
Unknown, or fewer than three injections	Yes	No	Yes	Yes
Fully vaccinated (three or more injections of toxoid)				
■ 5 years or less since last dose	No	No	No	No
■ 5 to 10 years since last dose	No	No	Yes	No
■ More than 10 years since last dose	Yes	No	Yes	No

including being placed on a ventilator if needed. Over time, the affected nerves repair themselves. ◄◄ TIG, p. 420

In addition to TIG treatment, the wound is thoroughly cleaned of all dead tissue and foreign material that could cause anaerobic conditions. An antibacterial medication such as metronidazole is given to kill any actively multiplying clostridia, preventing the production of more tetanospasmin. The person is also given tetanus vaccine, at a different site from the TIG injection.

Tetanus can easily be avoided by vaccination with tetanus toxoid, which is inactivated tetanospasmin. Immunization with tetanus toxoid in combination with diphtheria toxoid and acellular pertussis vaccine, called DTaP, is recommended in children under the age of 7. DTaP is given at 2, 4, 6, and 18 months of age, with a booster dose when children enter school. Additional booster doses of tetanus toxoid are given to people at 10-year intervals to maintain an adequate level of immunity. Currently DTaP is recommended for booster doses in people ages 10 to 64 years. Updating of booster doses of toxoid more frequently than every 10 years can be dangerous because there is risk of an allergic reaction in individuals who already have large amounts of antibodies against it. Individuals who have recovered from tetanus are not immune to the disease and must be immunized. People who have been injured or burned or are planning surgery should receive a booster shot of the tetanus vaccine. **Table 23.2** summarizes prevention of tetanus by wound management. ◄◄ toxoid, p. 423

Neonatal tetanus can be prevented by vaccinating pregnant women. Their infants are then protected by maternal antibodies crossing the placenta. **Table 23.3** describes the main features of tetanus.

TABLE 23.3 | **Tetanus ("Lockjaw")**

1. *Clostridium tetani* spores from dust or dirt enter a wound.
2. In anaerobic wounds, the spores germinate, and vegetative bacteria release an exotoxin called tetanospasmin.
3. Tetanospasmin is carried to the central nervous system by motor nerve axons or by the bloodstream.
4. The toxin prevents any inhibitory neurons it reaches from functioning.
5. The corresponding neurons, which cause muscles to contract, act unopposed by inhibitory neurons.
6. The result is a sustained, painful cramplike muscle spasm.

Symptoms	Restlessness, irritability, difficulty swallowing; muscle pain and spasm in jaw, abdomen, back, or entire body
Incubation period	3 days to 3 weeks; average 8 days
Causative agent	*Clostridium tetani,* an anaerobic, spore-forming, Gram-positive rod
Pathogenesis	Tetanus results from tetanospasmin, an exotoxin produced by the bacterium. The toxin is carried to the brain and spinal cord by motor nerve axons or circulating blood; toxin acts against nerve cells that normally inhibit muscle contraction. Other nerves that normally cause muscle contraction then act unopposed, causing muscle spasms.
Epidemiology	Organisms common in soil; spores contaminate wounds, germinate in those having anaerobic conditions, particularly dirty or puncture wounds.
Treatment and prevention	Treated with metronidazole, tetanus antitoxin. Immunization of children at ages 2 months, 4 months, 6 months, 18 months; booster dose at time of entering school and at 10-year intervals after that; tetanus immune globulin (TIG), cleaning wound.

Clostridial Myonecrosis ("Gas Gangrene")

Before antibiotics were discovered, clostridial myonecrosis (commonly called "gas gangrene") killed many soldiers wounded in wars. Endospores of *Clostridium perfringens,* the bacterium that can cause this disease, are found in soil and dust everywhere, and frequently can be isolated from wounds. However, infection with *C. perfringens* rarely leads to gas gangrene, because like *C. tetani,* the organism grows only in anaerobic conditions. Gas gangrene occurs mainly in neglected wounds with fragments of bone, foreign material, and serious tissue damage. It also occurs occasionally in surgical sites, especially in people with underlying diseases.

Signs and Symptoms

Gas gangrene signs and symptoms appear suddenly and dramatically. They include severe pain that quickly increases in the infected wound, followed by swelling in the area, and a thin, bloody or brownish fluid that leaks from the wound. This fluid may look frothy because of gas bubbles released by the organism. The overlying skin becomes stretched tight and mottled with black (**figure 23.10**). Although seriously ill, the victim remains quite alert until late in the illness when, near death, he or she becomes delirious and goes into a coma.

Causative Agent

Several species of *Clostridium* can cause gas gangrene when they invade injured muscle, but by far the most common is *C. perfringens,* an encapsulated Gram-positive anaerobic toxin-producing rod.

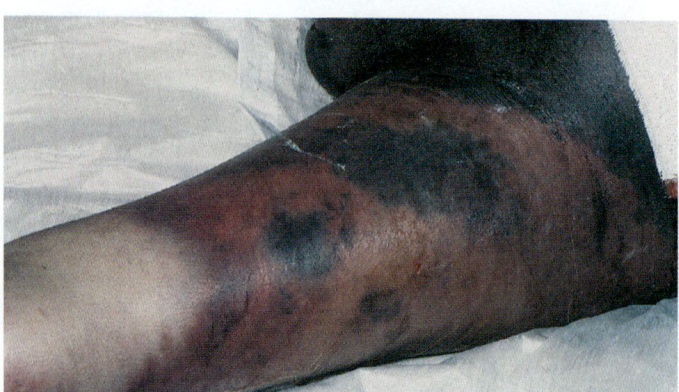

FIGURE 23.10 Individual with Gas Gangrene (Clostridial Myonecrosis) Fluid leaking from the involved area typically shows bits of muscle digested by *Clostridium perfringens.* Leukocytes are absent.

❓ *How is gas gangrene treated?*

Although *C. perfringens* is a spore-forming organism, it usually does not produce spores in wounds or cultures.

Pathogenesis

Two main factors lead to the development of gas gangrene: (1) the presence of dirt and dead tissue in the wound, and (2) long delays before the wound is treated. *Clostridium perfringens* is unable to infect healthy tissue but grows easily in poorly oxygenated and dead tissues that provide it with a source of growth factors and amino acids. It releases α-toxin, an enzyme that destroys lecithin in host cell membranes, leading to cell lysis. The toxin diffuses from the area of infection, killing leukocytes and tissue cells. Several enzymes produced by the pathogen, including collagenase and hyaluronidase, break down the host tissues. The organisms multiply, using the tissue breakdown products, and release hydrogen and carbon dioxide. These gases accumulate in the tissue and cause a rise in pressure, which leads to further spread of the infection. Without quick surgical treatment, massive amounts of α-toxin diffuse into the bloodstream and destroy red blood cells, tissue capillaries, and other structures throughout the body. *C. perfringens* infections are generally not dangerous until they invade muscle, but if they do, severe toxicity occurs quickly for unknown reasons. It is not reversed by treatment with antibody to α-toxin. ◀◀ **growth factors, p. 93**

Epidemiology

Clostridium perfringens is widespread in soil and is common in the feces of many animals and humans. It is sometimes found in the vagina of healthy women. Besides neglected trauma wounds and occasional surgical wounds, gas gangrene of the uterus is fairly common after self-induced abortions, and occasionally it can occur after miscarriages and childbirth. In other cases, diseases such as arteriosclerosis or diabetes—which lead to poor oxygenation of tissue from restricted blood flow—are predisposing factors. Cancer patients also have increased susceptibility to gas gangrene.

Treatment and Prevention

Treatment of gas gangrene includes prompt surgical removal of all dead and infected tissues, and may require amputations. Antibiotics such as penicillin are given to help stop bacterial growth and toxin production. However, because the drugs do not diffuse well into large areas of dead tissue and do not inactivate toxin, they are of little use in treating the disease. Hyperbaric oxygen treatment is sometimes used, in which the patient is placed in a special chamber and breathes pure O_2 under increased pressure. The pressure treatment inhibits growth of the clostridia, and so stops the release of toxin. The high levels of O_2 also improve oxygenation of injured tissues, lessening anaerobic conditions. Hyperbaric treatment is not always successful and may lead to other complications.

There is no available vaccine for gas gangrene. The disease can be prevented by prompt cleaning and removal of dead tissue from wounds (debridement). **Table 23.4** describes the main features of this disease.

TABLE 23.4 | Clostridial Myonecrosis ("Gas Gangrene")

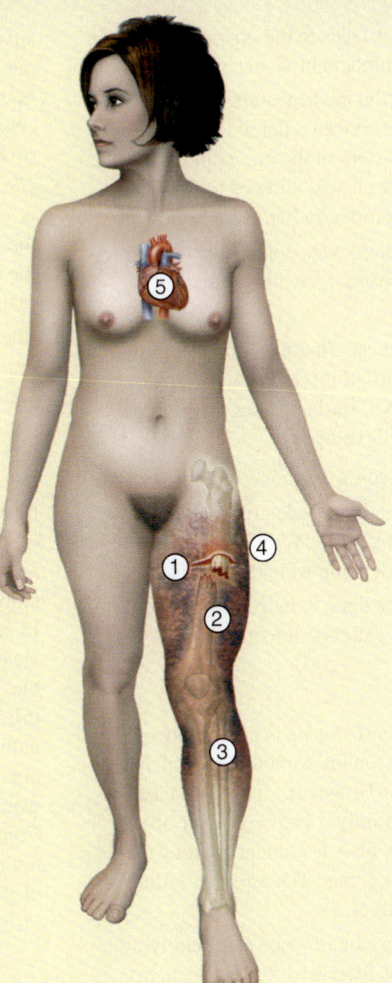

① *Clostridium perfringens* spores enter a wound with two essential characteristics: dead tissue and anaerobic conditions.

② The spores germinate, and the vegetative bacteria multiply in dead tissue, producing α-toxin.

③ α-toxin diffuses into normal tissue and kills it. The infection spreads into the dead tissue, the bacteria using amino acids and growth factors released from the tissue by bacterial enzymes.

④ Swelling and gas produced by fermenting amino acids and muscle glycogen aid rapid progress of the infection.

⑤ Massive amounts of α-toxin are produced and diffuse into the bloodstream, destroying blood cells and other cells throughout the body.

Signs and Symptoms	Severe pain; gas and fluid seep from wound, blackening of overlying skin; shock and death commonly follow
Incubation period	Usually 1 to 5 days
Causative agent	Usually *Clostridium perfringens;* other clostridia less frequently
Pathogenesis	Organism grows in dead and poorly oxygenated tissue and releases α-toxin; toxin kills leukocytes and normal tissue cells by degrading lecithin in their cell membranes; involvement of muscle causes shock by unknown mechanism.
Epidemiology	Wounds of war; dirt contamination of wounds, tissue death, impaired circulation to tissue as in people with diabetes or arteriosclerosis; self-induced abortions.
Treatment and prevention	Treated by surgical removal of dirt and dead tissues of primary importance; hyperbaric oxygen of possible value; antibiotics to kill vegetative *C. perfringens* of marginal value. Prevented by prompt cleaning and debridement of wounds; no vaccine available.

Actinomycosis ("Lumpy Jaw")

Many wound infections are caused by anaerobes other than clostridia, frequently by members of the normal microbiota of the mouth and intestine. The disease commonly called "lumpy jaw" was originally thought to be a fungal disease, and was therefore given the name actinomycosis. We now know that lumpy jaw is not a mycosis, but its original name is still used.

Signs and Symptoms

Actinomycosis is characterized by slowly progressing, sometimes painful, swellings under the skin that eventually burst and drain pus. The openings usually heal, only to reappear at the same or nearby areas days or weeks later. Most cases involve the area of the jaw and neck and cause swellings and scars, leading to the popular name lumpy jaw (**figure 23.11**). In some cases, the swellings and drainage develop on the chest or abdominal wall, or in the genital tract of women.

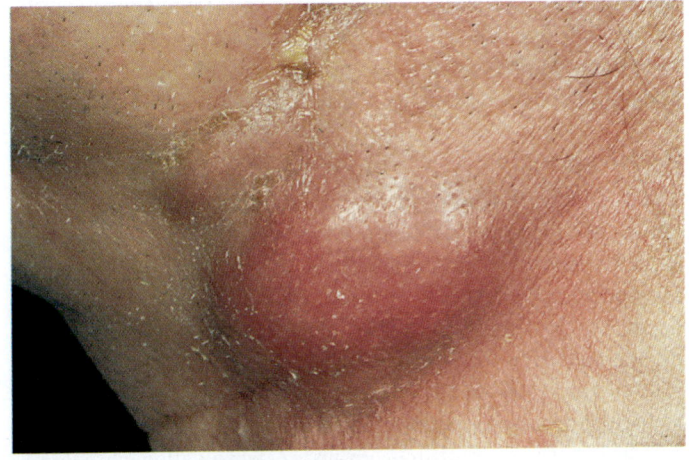

FIGURE 23.11 Actinomycosis ("Lumpy Jaw")

❓ *Is actinomycosis caused by a fungus?*

CASE PRESENTATION

A 63-year-old woman, healthy except for mild diabetes, had her gallbladder surgically removed. The surgery went well, but within 72 hours the surgical incision became swollen and pale. Within hours, the swollen area widened and developed a bluish discoloration. The woman's surgeon suspected gangrene. Antibiotic therapy was started, and she was rushed back to the operating room where the entire swollen area, including the repaired operative incision, was surgically removed. After that, the wound healed normally, although she required a skin graft to close the large wound. Large numbers of *Clostridium perfringens* grew from the wound culture.

Six days later, a 58-year-old woman had surgery in the same operating room for a malignant tumor of the colon. The surgery was performed without difficulty, but 48 hours later she developed rapidly advancing swelling and bluish discoloration of her surgical wound. As with the first case, gangrene was suspected and she was treated with antibiotics and surgical removal of the affected tissue. She also needed a skin graft. Her wound culture also showed *C. perfringens*.

Because the surgery department had never had any of its patients develop surgical wound infections with *C. perfringens* before this, the hospital epidemiologist was asked to do an investigation. Among the findings of the investigation:

1. Cultures of surfaces in the operating room grew large numbers of *C. perfringens*.
2. Unknown to the medical staff, a workman had recently serviced a fan in the ventilation system of the operating room, and for a time air was allowed to flow into the operating room, rather than out of it.
3. Heavy machinery was doing road repair outside the hospital, creating clouds of dust.

As a result of these findings, the operating room and its ventilating system were cleaned and upgraded. No further cases of surgical wound gangrene developed.

1. Was the surgeon's diagnosis correct?
2. Many other patients had surgery in the same operating room. Why did only these two patients develop wound gangrene?
3. What could be done to help identify the source of the patients' infections?

Discussion

1. *Clostridium perfringens* is commonly cultivated from wounds without any evidence of infection. However, in these cases, there was not only a heavy growth of the organism but also a clinical picture that suggested gangrene. The surgeon's diagnosis was correct.
2. In both cases, there was an underlying condition that increased the two patients' susceptibility to infection—cancer in one, and diabetes in the other. Moreover, both had a recognized source for the organism. Cultures of as many as 20% of diseased gallbladders are positive for *Clostridium perfringens,* and the organism is commonly found in large numbers in the human intestine—a potential source in the case involving removal of the bowel malignancy.
3. The surgeon thought that the infecting organism might have come from the patients themselves because such strains are often much more virulent than strains that live and sporulate in the soil. On the other hand, the gross contamination of the operating room, as revealed by the cultures of its surfaces, could indicate a very large infecting dose at the operative site, possibly compensating for lesser virulence. In addition, no further cases occurred after cleaning the operating room and fixing the ventilation system. Unfortunately, in this case, no cultures of the removed gallbladder or bowel tumor were done, nor were the strains isolated from the wounds and the environment compared. Comparing the antibiotic susceptibility, toxin production, and other characteristics of the different isolates could have helped identify the source of the infections.

Causative Agent

Most cases of actinomycosis are caused by *Actinomyces israelii,* a Gram-positive, filamentous, branching, slow-growing anaerobic bacterium. A number of similar species can cause the disease in animals and humans. Despite its name, this organism is not a fungus.

Pathogenesis

The detailed mechanisms by which *Actinomyces israelii* causes lumpy jaw have not yet been established. *A. israelii* cannot penetrate the normal mucosal surface, but it can establish an infection in association with other members of the normal microbiota if introduced into tissue by wounds. The infectious process is characterized by alternate cycles of abscess formation and healing. The disease usually progresses to the skin, where pus is discharged, but occasionally it penetrates into bone or into the central nervous system.

In the tissue, *A. israelii* grows as dense yellowish colonies called sulfur granules because they are the color and size of particles of sulfur (**figure 23.12**). Finding these sulfur granules in

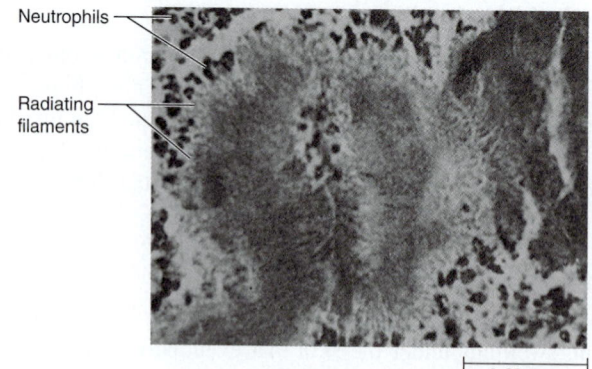

Neutrophils

Radiating filaments

0.25 mm

FIGURE 23.12 *Actinomyces israelii* **Sulfur Granule**
Microscopic view of a stained smear of pus from an infected person. Filaments of the bacteria can be seen radiating from the edge of the colony.

❓ *What are the sulfur granules in the pus?*

the wound drainage helps significantly with diagnosis. Typically, diagnosis depends on culturing the organism from the pus discharge, but because *A. israelii* is slow growing, other more rapidly growing bacteria present in the pus may outcompete it in culture, confusing the diagnosis. The presence of sulfur granules in the wound discharge confirms *A. israelii* infection.

Epidemiology

Actinomyces israelii can be part of the normal microbiota of the mucosal surfaces of the mouth, upper respiratory tract, intestine, and sometimes the vagina. It is common in the gingival crevice, particularly in people with poor dental care. Pelvic actinomycosis can complicate use of intrauterine contraceptive devices (IUDs). The species of *Actinomyces* responsible for actinomycosis of dogs, cattle, sheep, pigs, and other animals generally do not cause human disease. The disease is sporadic and is not transmitted from person to person. ▶▶| gingival crevice, p. 573

Treatment and Prevention

Actinomycosis can be treated with a number of antibacterial medications, including penicillin and tetracycline. To be effective, however, treatment must be given over weeks or months because the organisms grow very slowly, and in dense colonies, into which medications do not easily diffuse.

There are no proven preventive measures for lumpy jaw. The main features of actinomycosis are presented in **table 23.5**.

TABLE 23.5	Actinomycosis ("Lumpy Jaw")
Signs and symptoms	Chronic disease; recurrent, sometimes painful swellings open and drain pus, heal with scarring; usually involves face and neck; chest, abdomen, and pelvis are other common sites
Causative agent	*Actinomyces israelii*, a filamentous, branching, Gram-positive, slow-growing, anaerobic bacterium
Incubation period	Months (usually indeterminate)
Pathogenesis	Usually begins in a mouth wound, extends to the face, neck, or upper chest; sometimes begins in the lung, intestine, or female pelvis. *A. israelii* always accompanied by normal microbiota. In tissue, grows as dense yellowish colonies called sulfur granules.
Epidemiology	No person-to-person spread. *A. israelii* commonly part of normal mouth, upper respiratory, intestine, and vagina microbiota. Dental procedures, intestinal surgery, insertion of IUDs can initiate infections.
Treatment and prevention	Because of its slow growth, *A. israelii* infections require prolonged treatment; a number of antibacterial medications are effective. No proven preventive measures.

23.4 ■ Bacterial Infections of Bite Wounds

Learning Outcomes

7. *Explain why human mouth microbiota can cause serious bite wound infections.*

8. *Describe two zoonotic wound infections.*

Each year, more than 3 million animal bites occur in the United States, the most feared result being the viral disease, rabies. Human bites are also common. The infection risk following a bite depends partly on the type of injury (crushing, lacerated, or puncture), and partly on the kinds of pathogens introduced into the bite wound. ▶▶| rabies, p. 658

Human Bites

Wounds caused by human teeth (such as bites that break the skin, hitting the teeth of another person, or being injured with objects that have been in a person's mouth) are common and can result in very serious infections. These infections are caused by normal mouth microbiota. Occasionally, diseases such as syphilis, hepatitis B, and hepatitis C are transmitted this way, because the causative agents of these diseases are found in the blood or saliva (see **Perspective 23.1**).

Signs and Symptoms

The wound may appear minor at first but then becomes painful and swells massively. Pus that smells very bad often leaks out of the wound.

Causative Agents

The most frequent causes of infection in human bite wounds are both aerobic and anaerobic members of the normal mouth microbiota, including streptococci, fusiforms, spirochetes, and *Bacteroides* species, often in association with *Staphylococcus aureus*.

Infection Caused by a Human "Bite"

A 26-year-old man injured a finger of his right hand during a bar fight when he punched someone in the mouth. The man did not seek medical help until more than 36 hours after the fight. At that time, his entire hand was very swollen, red, and tender, and the swelling was spreading to his arm. The surgeon cut open the infected tissues, allowing pus to discharge. He removed the damaged tissue and washed the wound with sterile fluid. Smears and cultures of the infected material showed aerobic and anaerobic bacteria characteristic of mouth normal microbiota, including species of *Bacteroides* and *Streptococcus*. The patient was given antibiotics to fight infection, but the wound did not heal well and continued to drain pus. Several weeks later, X rays revealed that infection had spread to the bone at the base of the finger. To cure the infection, the finger had to be amputated.

Pathogenesis

Bite wounds are usually infected with both aerobic and anaerobic species. Crushed tissue and the presence of facultative anaerobes, which reduce available O_2, create anaerobic conditions that allow anaerobic bacteria to establish infection at the site. Although most mouth microbiota are harmless alone, together they produce large numbers of toxins and enzymes that damage the host tissues and cells involved in the immune response. These include leukocidin, collagenase, hyaluronidase, ribonuclease, various proteinases, neuraminidase, and enzymes that destroy antibodies and proteins of the complement system. The result of all these factors is a **synergistic infection,** meaning that the sum effect of all the organisms acting together is greater than the sum of their individual effects. Irreversible destruction of tissues such as tendons and permanent loss of function can be the result.

◀◀ facultative anaerobes, p. 90

Epidemiology

Most serious human bite infections occur because of fights related to drunkenness, or during forcible restraint, as found in law enforcement and in mental institutions. The risk is greatly increased when the biting individual has poor mouth care and extensive dental disease. Bites by young children are usually not serious, mostly because children seldom break the skin or cause crushing wounds when they bite.

Treatment and Prevention

Bite wounds should be treated immediately to avoid serious infections. This involves opening the infected area with a scalpel, washing the wound thoroughly with sterile fluid, and removing dirt, foreign matter (such as broken teeth), and dead tissue. Antibacterial medications should also be given. Usually more than one drug is used because many oral microbiota involved in bite wound infections are resistant to various drugs such as penicillin. One drug effective against anaerobes should be included. If bite wounds are not treated quickly, serious complications may occur, including infection of muscles and tendons, which may require surgery.

Prevention of bite wound infections involves avoiding situations that lead to biting and hitting. If a bite wound is sustained, prompt cleaning of the wound followed by application of an antiseptic can help prevent infection. The main features of human bite wound infections are presented in **table 23.6.**

Pasteurella multocida Bite Wound Infections

Infections caused by bacteria that live in the mouth of a biting animal are very common. Surprisingly, bite infections from a number of kinds of animals are generally caused by a single bacterial species, *Pasteurella multocida*.

Signs and Symptoms

Early signs of infection in wounds caused by animal bites include spreading redness, tenderness and swelling of tissues next to the wound, followed by pus discharge. Abscesses commonly form. Without quick treatment, infection can lead to bloodstream invasion or permanent loss of function.

Causative Agent

Pasteurella multocida is a Gram-negative, facultatively anaerobic coccobacillus. Most isolates have antiphagocytic capsules. There are a number of different antigenic types.

Pathogenesis

The details of pathogenesis are not yet known. Some strains produce a toxin that kills host cells by affecting protein synthesis and cell cycle regulation. The organisms have antiphagocytic capsules

TABLE 23.6	Human Bite Wound Infections
Signs and symptoms	Rapid onset, pain, massive swelling, drainage of foul-smelling pus
Incubation period	Usually 6 to 24 hours
Causative agent	Mixed mouth microbiota: anaerobic streptococci, fusiforms, spirochetes, anaerobic Gram-negative rods; sometimes *Staphylococcus aureus*
Pathogenesis	Various mouth bacteria act synergistically to destroy tissue.
Epidemiology	Alcohol-related violence; forcible restraint; poor mouth care and extensive dental disease.
Treatment and prevention	Treatment is usually surgical. No proven preventive measures except to avoid fights. Prompt cleaning of wound and application of antiseptic is advised.

but when antibodies bind to them, phagocytes are able to ingest and kill the opsonized bacteria. If untreated, *P. multocida* can cause bacteremia, leading to endocarditis or meningitis.

Epidemiology

Humans—and many other healthy animals such as cats, dogs, and monkeys—carry *P. multocida* as normal oral and upper respiratory microbiota. Both diseased and healthy animals are reservoirs for human infections. Cats are more likely to carry *P. multocida* than dogs, and so cat bites more frequently cause the infection. *P. multocida* also causes diseases in a number of other animal species, sometimes causing epidemics of fatal pneumonia and bloodstream infection in them.

Treatment and Prevention

Unlike many Gram-negative pathogens, *P. multocida* is susceptible to penicillin, so amoxicillin clavulanate (Augmentin) is given immediately, even before the diagnosis has been confirmed by culturing the bacterium. This drug is a combination of a penicillin derivative (amoxicillin) and a β-lactamase inhibitor (clavulanate). The clavulanate is helpful because many bite wounds often also contain strains of β-lactamase-producing *Staphylococcus aureus*. This and other antibacterial medications are effective if given early in the infection.

No vaccines for *P. multocida* are available for use in humans. Immediate cleaning of bite wounds and quick medical attention usually prevent the development of serious infection. **Table 23.7** gives the main features of *Pasteurella multocida* bite wound infections.

TABLE 23.7	*Pasteurella multocida* Bite Wound Infections
Signs and symptoms	Spreading redness, tenderness, swelling, discharge of pus
Incubation period	24 hours or less
Causative agent	*Pasteurella multocida*, a Gram-negative, facultatively anaerobic, encapsulated coccobacillus
Pathogenesis	Introduced by bite, *P. multocida* attaches to tissue, resisting phagocytes because of its capsule; probable cell-destroying toxin. Extensive swelling, abscess formation. Opsonins develop, allow phagocytic killing, limit spread.
Epidemiology	Carried by many animals in their mouth or upper respiratory tract.
Treatment and prevention	Treatment with penicillin, other antibacterials, effective if given promptly. No vaccines to use in humans. Prompt wound care is preventive.

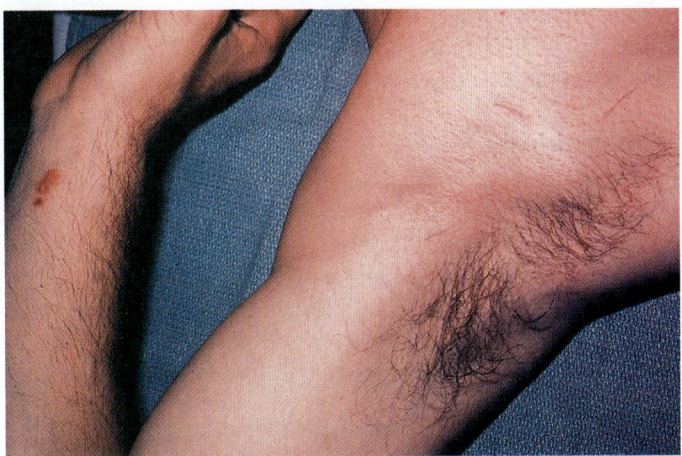

FIGURE 23.13 Bartonellosis ("Cat Scratch Disease") Note the wrist lesion and enlarged lymph nodes

❓ *Is this disease only transmitted by cat scratches?*

Bartonellosis ("Cat Scratch Disease")

In the United States, bartonellosis (commonly known as "cat scratch disease") is the most common cause of chronic lymph node enlargement at one body site in young children. Despite its name, this disease can also be transmitted by bites and possibly by other means such as fleabites or saliva on mucous membranes. Typically the infection is mild and localized, lasting from several weeks to a few months.

Signs and Symptoms

Cat scratch disease begins within a week of a scratch or bite with the appearance of a pus-filled pimple at the site of injury. Painful enlargement of the lymph nodes of the area develops in 1 to 7 weeks (**figure 23.13**). Patients may develop fever, and in some cases the lymph nodes become pus-filled and soft. The disease generally disappears without treatment in 2 to 4 months. A small number of patients develop irritation of an eye with local lymph node enlargement, epileptic seizures and coma due to encephalitis, or acute or chronic fever associated with bloodstream or heart valve infection.

Causative Agent

Cat scratch disease is caused by *Bartonella henselae*, a curved, Gram-negative rod.

Pathogenesis

The virulence factors of *B. henselae* and the process by which it causes disease are not yet known. The organisms can enter the body by a cat bite or scratch, or when cat saliva contaminates a mucous membrane. The bacteria are carried to the lymph nodes, and the disease is resolved by the immune system in most cases. Spreading by the bloodstream, however, occurs in some individuals, causing serious complications. Peliosis hepatis and bacillary angiomatosis are two complicating conditions seen mostly in people with AIDS. In peliosis hepatis, blood-filled cysts form in the liver. In bacillary angiomatosis, nodules composed of proliferating blood vessels develop in the skin and other parts of the body.

Epidemiology

Cat scratch disease occurs worldwide, often in children and young adults. The disease is a zoonosis, transmitted to humans mainly by bites or scratches from cats. Person-to-person spread does not occur. Asymptomatic bacteremia is common in cats, however, and transmission from cat to cat occurs by cat fleas. Fleas probably are responsible for transmission in cases where people develop the disease after they handle cats but are not scratched or bitten.

Treatment and Prevention

Any cat scratch or bite should be promptly cleaned with soap and water and then treated with an antiseptic. If signs of infection develop, or if the cat's immunization status against rabies is uncertain, prompt medical evaluation is needed. Severe *B. henselae* infections can usually be treated with antibacterial medications such as ampicillin, but some strains are resistant to this drug.

There are no proven preventive measures for cat scratch disease, other than to avoid handling stray cats, especially those with fleas. The main features of cat scratch disease are presented in **table 23.8.**

TABLE 23.8	Bartonellosis ("Cat Scratch Disease")
Signs and symptoms	Pimple appears at the bite or scratch site, followed by local lymph node enlargement; nodes may soften and drain pus; prolonged fever and convulsions indicate spread to other body parts
Incubation period	Usually less than 1 week
Causative agent	*Bartonella henselae*, a Gram-negative rod
Pathogenesis	Bacteria enter with cat bite or scratch and reach lymph nodes, where the disease is usually stopped. May spread by bloodstream, cause infections of the heart, brain, or other organs. Peliosis hepatis or bacillary angiomatosis may occur, mostly in people with AIDS.
Epidemiology	A zoonosis, spread among cats by fleas, which are biological vectors and have *B. henselae* in their feces. Infected cats usually asymptomatic, but often bacteremic. Humans are accidental hosts. No person-to-person spread.
Treatment and prevention	Promptly wash skin breaks, apply antiseptic. Most *B. henselae* infections respond to antibacterial treatment. Avoiding rough play with cats. Flea control in cats.

Streptobacillary Rat Bite Fever

Rat bites are fairly common among poor people in large cities and among workers who handle laboratory rats.

Signs and Symptoms

The bite wound usually heals quickly without any problems. Two to 10 days later, however, the person develops chills, fever, head and muscle aches, and vomiting. The fever comes and goes. A rash usually appears after a few days, followed by joint pain.

Causative Agent

The cause of streptobacillary rat bite fever is *Streptobacillus moniliformis,* a facultatively anaerobic, Gram-negative rod that varies in shape (pleomorphic). The organism is unique in that it spontaneously develops **L-forms.** L-forms are variants that lack a cell wall, first identified at the famous Lister Institute (hence, the *L* in *L-form*). As might be expected, L-form colonies resemble those of mycoplasmas—bacteria that lack a cell wall. ◀◀ mycoplasmas, p. 276

Pathogenesis

Streptobacillus moniliformis enters the body through a bite or scratch, and sometimes by ingestion. There is typically little enlargement of the local lymph nodes, and the bacteria quickly enter the bloodstream and spread throughout the body. The majority of people recover without treatment in about 2 weeks, but some cases are rapidly fatal, and others develop serious complications such as brain abscesses or infection of the heart valves. The main features of streptobacillary rat bite fever are presented in **table 23.9.**

TABLE 23.9	Streptobacillary Rat Bite Fever
Signs and symptoms	Chills, fever, muscle aches, headache, and vomiting; later, rash and pain in one or more of the large joints
Incubation period	Usually 2 to 10 days (range, 1 to 22) after a rat bite
Causative agent	*Streptobacillus moniliformis*, a highly pleomorphic, Gram-negative bacterium that spontaneously produces L-forms
Pathogenesis	Bite wound heals without treatment; *S. moniliformis* quickly invades bloodstream. Fevers come and go irregularly. Most victims recover without treatment; in others, infection established in various body organs, results in death if treatment is not given.
Epidemiology	Wild and laboratory rats can carry *S. moniliformis*. Bites of other rodents and animals that prey on them can transmit the disease to humans. Food or drink contaminated with rodent feces can also transmit the infection.
Treatment and prevention	Effectively treated with penicillin, other antibacterial medications. Control wild rats and mice; care in handling laboratory animals.

Epidemiology

Many healthy laboratory rats, mice, and other rodents carry *S. moniliformis* in their nose and throat. Laboratory and pet store workers are at increased risk of contracting the disease. Rat bites and scratches are the usual source of human infections. Epidemics of the disease have occurred when people have eaten or drunk anything contaminated with *S. moniliformis* from rodent feces. Cases of rat bite fever have also been associated with exposure to animals that eat rodents, including cats and dogs.

Treatment and Prevention

Penicillin, given intravenously, is used to treat cases of rat bite fever. Since this drug inhibits cell wall synthesis, the fact that it can be used to treat rat bite fever shows that either *S. moniliformis* L-forms do not occur *in vivo*, or if they do, they are avirulent. Prevention of rat bite fever includes control of wild rat populations and care in handling laboratory rats and their feces.

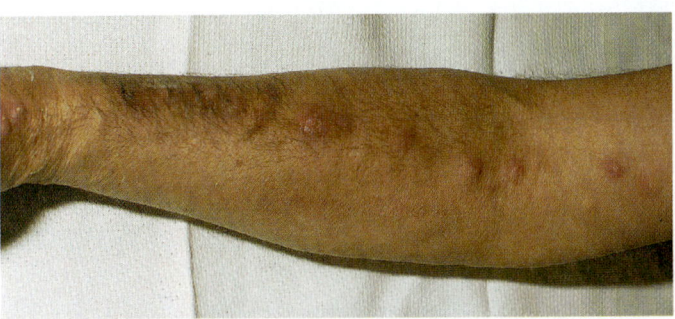

FIGURE 23.14 Individual with Sporotrichosis ("Rose Gardener's Disease")

❓ *Why do the abscesses occur along a relatively straight line?*

MicroAssessment 23.4

Human bite infections can be dangerous because certain members of the mouth microbiota, which have little invasive ability when growing alone, can invade and destroy tissue when growing synergistically. *Pasteurella multocida* can infect bite wounds caused by a number of different animals, especially cats. Cat bites and scratches can also transmit *Bartonella henselae*, the cause of cat scratch disease, which is characterized typically by local lymph node enlargement, although it may involve other parts of the body. Streptobacillary rat bite fever, caused by *Streptobacillus moniliformis*, is acquired from bites of rats and mice, and from animals that eat them. It is characterized by fever that comes and goes and a rash. *S. moniliformis* spontaneously develops L-forms.

10. *What Gram-negative organism commonly infects wounds caused by animal bites?*
11. *What is the most common cause of chronic localized lymph node enlargement in young children?*
12. *Why are normal mouth microbiota a cause of serious infection in human bites?* ➕

23.5 ■ Fungal Wound Infections

Learning Outcome

9. *Give the distinctive features of rose gardener's disease.*

Fungal infections of wounds are more serious than dermal mycoses. Typically, they are caused by soil fungi that enter through injuries that allow them to invade subcutaneous tissue. Sporotrichosis is a common fungal infection that occurs worldwide. It is typically a mild disease but may become life-threatening under certain circumstances.

Sporotrichosis ("Rose Gardener's Disease")

Sporotrichosis, also known as "rose gardener's disease," occurs around the world and is associated with activities that lead to puncture wounds from plant material. Although many cases are sporadic, the disease can occur in groups of people working in the same occupation, such as farmers, gardeners, and agricultural workers. Veterinarians are also at risk from animals such as cats that have contracted the disease.

Signs and Symptoms

The site of infection in most cases is a hand or arm but the body, legs, and face can also be involved. Typically, a chronic ulcer forms at the wound site, followed by a slowly progressing series of ulcerating nodules that develop sequentially toward the center of the body (**figure 23.14**). Lymph nodes in the region of the wound enlarge, but patients generally do not become ill. If they have AIDS or other immunodeficiency, however, the disease can spread throughout the body, infecting joints and the central nervous system, and becoming life-threatening.

Causative Agent

Sporotrichosis is caused by the dimorphic fungus *Sporothrix schenckii*, which lives in soil and on vegetation (**figure 23.15**).
◀◀ dimorphic fungus, p. 285

Pathogenesis

Sporothrix schenckii spores enter the body through an injury caused by plant material. After an incubation period that usually ranges from 1 to 3 weeks but can be longer, the multiplying fungi cause a small pimple to form at the site of the injury. This slowly enlarges and ulcerates, producing a painless, red lesion that bleeds easily. There is little or no pus unless the ulcer becomes secondarily infected with bacteria. After a week or longer, the process repeats itself—progression of the disease usually follows the flow of a lymphatic vessel. In healthy individuals, the process does not proceed beyond the lymph node. Sometimes, however, satellite

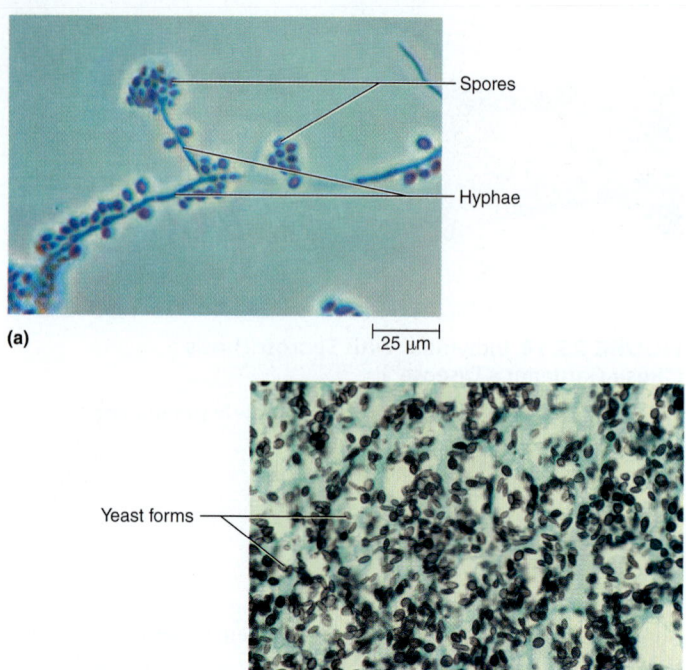

(a) |——| 25 µm

(b) |——| 30 µm

Yeast forms

FIGURE 23.15 *Sporothrix schenckii* **(a)** Mold form. **(b)** Yeast form, as seen in infected tissue.

❓ *Sporothrix schenckii is dimorphic; what does this mean?*

lesions appear irrespective of the lymphatic vessels. Without treatment, the disease can become chronic and go on for years.

◀◀ lymphatic vessel, p. 358

Epidemiology

Sporotrichosis is distributed worldwide, mostly in the warmer regions but extending into temperate climates. It is an occupational disease of farmers, carpenters, gardeners, greenhouse workers, and others who work with plant materials. Risk factors for the disease besides occupation include diabetes, immunosuppression, and alcoholism. Individuals with chronic lung disease can contract *S. schenckii* lung infections from inhaling dust from hay or cattle feed. Deaths from sporotrichosis are rare.

Treatment and Prevention

Surprisingly, unlike other fungal infections, sporotrichosis can usually be cured by oral treatment with the simple chemical compound potassium iodide (KI). KI somehow enhances the body's

TABLE 23.10	**Sporotrichosis ("Rose Gardener's Disease")**
Signs and symptoms	Painless, ulcerating nodules appearing in sequence in a linear pattern
Incubation period	Usually 1 to 3 weeks
Causative agent	*Sporothrix schenckii*, a dimorphic fungus
Pathogenesis	Spores multiply at site of introduction by thorn, splinter, or other plant material, causing a small nodule that ulcerates. Spores carried by lymph flow repeat the process along the lymphatic vessel. May spread beneath the skin irrespective of lymphatic vessels.
Epidemiology	Distributed worldwide in tropical and temperate climates. Occupations requiring contact with sharp plant materials at particular risk.
Treatment and prevention	Most cases effectively treated with potassium iodide. Generalized infections require amphotericin B or itraconazole. Protective gloves and clothing.

ability to fight the fungus. Itraconazole or the antifungal medication amphotericin B is used in rare cases when the disease spreads throughout the body. Sporotrichosis is often misdiagnosed, leading to delayed and inappropriate treatment.

Sporotrichosis can be prevented by wearing protective gloves and a long-sleeved shirt when working with soil and vegetation. **Table 23.10** presents some of the main features of sporotrichosis.

The key features of the diseases covered in this chapter are highlighted in the **Diseases in Review 23.1** table.

MicroAssessment 23.5

Rose gardener's disease (sporotrichosis), caused by the dimorphic fungus *Sporothrix schenckii*, is widely distributed around the world, affecting mainly those associated with occupations that expose them to splinters and sharp vegetation. Unlike most invasive fungal infections, sporotrichosis is usually easy to treat.

13. *List some of the risk factors associated with contracting rose gardener's disease.*
14. *What is unique about the treatment of sporotrichosis?*
15. *Why is sporotrichosis often misdiagnosed?* ➕

Diseases in Review 23.1

Wound Infections

Disease	Causative Agent	Comment	Summary Table
BACTERIAL INFECTIONS			
Staphylococcal infections	*Staphylococcus aureus* and *S. epidermidis*	*S. aureus* (described in chapter 22) is the most common cause of wound infections, but *S. epidermidis,* part of the normal skin microbiota, can form biofilms on medical devices.	Table 23.1, p. 551
Necrotizing fasciitis ("flesh-eating disease")	SPE-producing strains of *Streptococcus pyogenes*	SPEs of *S. pyogenes* (described in chapter 21) are superantigens, which cause widespread and inappropriate activation of helper T cells.	Table 23.1, p. 551
***Pseudomonas aeruginosa* infections**	*Pseudomonas aeruginosa*	Widespread environmental organism that commonly infects burned tissues; produces a green pigment; resistant to a wide variety of antimicrobial medications.	Table 23.1, p. 551
DISEASES DUE TO ANAEROBIC BACTERIA			
Tetanus ("lockjaw")	*Clostridium tetani*	Endospore-forming anaerobe grows in wound-damaged tissues; tetanus toxin blocks the action of inhibitory neurons, leading to continuous painful muscle contractions; typically fatal without treatment. Preventable by vaccination (DTaP).	Tables 23.2 and 23.3, p. 557
Clostridial myonecrosis ("gas gangrene")	Usually *Clostridium perfringens*	Endospore-forming anaerobe that can germinate and grow in necrotic and poorly oxygenated tissue; gases produced accumulate in tissues; typically fatal without treatment.	Table 23.4, p. 559
Actinomycosis ("lumpy jaw")	Usually *Actinomyces israelii*	Characterized by recurring and slowly progressing swellings, usually near the neck and jaw; these burst, draining pus and sulfur granules.	Table 23.5, p. 561
BITE WOUND INFECTIONS			
Human bites	Mixed aerobic and anaerobic species, often members of the normal mouth microbiota	Crushing nature of wound causes tissue damage and anaerobic conditions; synergistic infection adds to that tissue damage.	Table 23.6, p. 562
***Pasteurella multocida* infections**	*Pasteurella multocida*	Characterized by abscess at the site of the wound; causative agent is a common member of the normal mouth microbiota of animals; results most often from cat bites.	Table 23.7, p. 563
Bartonellosis ("cat scratch disease")	*Bartonella henselae*	Most common cause of lymph node enlargement at one site in children; acquired via a cat bite or scratch.	Table 23.8, p. 564
Streptobacillary rat bite fever	*Streptobacillus moniliformis*	Characterized by relapsing fevers, rash, and joint pain. Acquired via a rat bite or scratch, or ingestion of food or water contaminated with rodent feces.	Table 23.9, p. 564
FUNGAL WOUND INFECTIONS			
***Sporotrichosis* ("rose gardener's disease")**	*Sporothrix schenckii*	Characterized by painless ulcerating nodules that develop sequentially along a lymphatic vessel; dimorphic fungus lives on soil and vegetation and enters the body through an injury.	Table 23.10, p. 566

Staying Ahead in the Race with *Staphylococcus aureus*

Over the last 50 years, scientists have generally kept us at least one step ahead in the race between *Staphylococcus aureus* and human health. Today, "staph" seems again ready to cause problems, but new knowledge from molecular biology promises to help us fight this pathogen.

In the past, we have relied on developing antibacterial medications to kill or inhibit the growth of *S. aureus,* but there is continuing investigation of alternative approaches. This search is aided by improvement in understanding staphylococcal pathogenesis. We now understand the mechanism by which *S. aureus* obtains iron from hemoglobin. We also know that the bacterium's protein A can attach to tumor necrosis factor (TNF) receptors on host cells, causing release of cytokines that recruit large numbers of neutrophils. There have also been advances in vaccine development. An experimental vaccine composed of staphylococcal capsule material conjugated with a non-toxic form of *Pseudomonas* exotoxin A has been shown to significantly reduce staphylococcal infections in kidney dialysis patients.

Other options now seem to be within reach. For example, it should be possible to determine the structure of the active part of toxin molecules and to design medications that would bind to that site and inactivate it. Also, it is now possible to determine which staphylococcal genes code for virulence factors. The products of these genes may prove to be unexpected vaccine candidates.

The race continues!

Summary

23.1 ■ Anatomy, Physiology, and Ecology

Wounds expose tissue components to which pathogens specifically attach. Wounds heal by forming **granulation tissue** which, in the absence of dirt and infection, fills the defect, and subsequently contracts to minimize scar tissue (figure 23.1). Thermal burns often result in large areas of dead tissue lacking competing organisms and body defenses, ideal conditions for microbial growth.

Wound Abscesses

An **abscess** is composed of **pus,** which contains leukocytes, tissue breakdown products, and infecting organisms. Abscess formation localizes an infection within tissue, preventing its spread. Inflammatory cells and clotted blood vessels separate the abscesses from normal tissue (figure 23.2).

Anaerobic Wounds

Anaerobic conditions are likely to occur in wounds containing dead tissue or foreign material, and those with limited opening to the air.

23.2 ■ Common Bacterial Infections of Wounds (table 23.1)

Possible consequences of wound infections include delayed healing, abscess formation, and spread of bacteria or toxins into nearby tissue or the bloodstream. Infections can cause surgical wounds to open, and they can spread to create biofilms on artificial devices.

Staphylococcal Wound Infections (figures 23.3, 23.4)

Staphylococci are the leading cause of wound infections, both surgical and accidental; *Staphylococcus aureus* and *S. epidermidis* are the most common wound-infecting species. *Staphylococcus aureus* possesses many virulence factors; some strains release a toxin that causes toxic shock syndrome (table 23.1). *Staphylococcus epidermidis* is less virulent but can form biofilms on catheters and other devices.

Group A Streptococcal "Flesh-Eating Disease"

Streptococcus pyogenes (group A, β-hemolytic streptococcus) causes strep throat, scarlet fever, wound infections, and other conditions. **Necrotizing fasciitis**–causing strains of *S. pyogenes* produce exotoxin B, a protease thought to be responsible for the tissue destruction (figure 23.5).

Pseudomonas aeruginosa Infections

Pseudomonas aeruginosa, an aerobic, Gram-negative rod with a single polar flagellum, is an opportunistic pathogen widespread in the environment, and a cause of both healthcare-associated infections and those acquired outside the hospital. Production of two pigments by the bacterium often colors infected wounds green (figure 23.6).

23.3 ■ Diseases Due to Anaerobic Bacterial Wound Infections

Tetanus ("Lockjaw") (table 23.3)

The characteristic symptom of tetanus is sustained, painful, cramp-like spasms of one or more muscles (figure 23.7). The disease is often fatal. Tetanus is caused by an exotoxin, **tetanospasmin,** produced by *Clostridium tetani,* a non-invasive, anaerobic, Gram-positive rod (figure 23.8). The toxin prevents inhibitory neurons from releasing their neurotransmitter, causing muscles to contract without control (figure 23.9). The spores of *C. tetani* are widespread in dust and dirt. Tetanus can be prevented by vaccination with toxoid (inactivated tetanospasmin), and maintaining immunity throughout life with regular booster injections. People sustaining any wound, including surgeries, no matter how minor, should make sure that their immunization is up to date (table 23.2).

Clostridial Myonecrosis ("Gas Gangrene") (table 23.4)

Usually caused by the α-toxin-producing anaerobe *Clostridium perfringens*. Signs and symptoms begin suddenly with pain and swelling followed by discharge of a thin, brown, bubbly fluid and dark discoloration of the overlying skin (figure 23.10). The toxin causes tissue necrosis; hydrogen and carbon dioxide are produced from fermentation of amino acids and glycogen in the dead tissue. There is no vaccine or toxoid. Prevention depends on prompt medical care of dirty wounds containing dead tissue. Treatment depends on urgent surgical removal of dead and infected tissue, and it may require amputation.

Actinomycosis ("Lumpy Jaw") (table 23.5)

Actinomycosis is a chronic, slowly progressive disease characterized by repeated swellings, discharge of pus, and scarring, usually of the face and neck (figure 23.11). The causative agent is *Actinomyces israelii*, a member of the normal microbiota of the mouth, intestine, and vagina that enters tissues with wounds such as those with dental and intestinal surgery (figure 23.12). Despite its name, it is not a fungus. The organism is slow growing; treatment must be continued for weeks or months.

23.4 ■ Bacterial Infections of Bite Wounds

The kind of bite wound infection depends on the kinds of infectious agents in the mouth of the biting animal, and the nature of the wound—whether punctured, crushed, or torn.

Human Bites   (table 23.6)

Wounds caused by human teeth are common and can result in very serious infections, with pain, massive swelling, and foul-smelling pus. The infections are usually caused by synergistic activity among members of the normal mouth microbiota, including anaerobic streptococci, fusiforms, spirochetes, and *Bacteroides* sp., often with *Staphylococcus aureus*. The crushing nature of bite wounds causes death of tissue and conditions suitable for growth of anaerobes; prompt cleansing and use of an antiseptic are advised.

Pasteurella multocida Bite Wound Infections   (table 23.7)

Pasteurella multocida, a Gram-negative rod, can infect bite wounds inflicted by a number of animal species. *P. multocida* causes disease in animals, but many animals are asymptomatic carriers.

Bartonellosis ("Cat Scratch Disease")   (table 23.8, figure 23.13)

Cat scratch disease is the most common cause of chronic, localized lymph node enlargement in children. Caused by *Bartonella henselae,* it begins with a pimple at the site of a bite or scratch, followed by enlargement of local lymph nodes, which often become pus-filled. Most individuals with cat scratch disease recover without treatment.

Streptobacillary Rat Bite Fever   (table 23.9)

Streptobacillary rat bite fever is characterized by relapsing fevers, head and muscle aches, and vomiting. A rash and joint pains often develop. It is usually caused by *Streptobacillus moniliformis,* a pleomorphic, Gram-negative rod that characteristically produces cell wall–deficient variants called **L-forms.** People who live in rat-infested areas and those who handle laboratory rats are at greatest risk of contracting the disease.

23.5 ■ Fungal Wound Infections

Fungal wound infections are usually more serious than dermal mycoses. They are typically caused by fungi from soil.

Sporotrichosis ("Rose Gardener's Disease")   (table 23.10)

Rose gardener's disease, also called sporotrichosis, is a chronic fungal disease mainly of people who work with soil or vegetation. The usual case is characterized by painless, ulcerating nodules that develop one after the other along the path of a lymphatic vessel (figure 23.14). The causative organism is the dimorphic fungus, *Sporothrix schenckii* (figure 23.15), usually introduced into wounds caused by thorns or splinters. The fungus is distributed worldwide in tropical and temperate climates; people at risk include farmers, carpenters, gardeners, and greenhouse workers.

Review Questions

Short Answer

1. What property of *Staphylococcus epidermidis* help it to colonize plastic materials used in medical procedures?
2. What is the relationship between the superantigens of *S. aureus* and the organism's production of toxic shock?
3. Name two underlying conditions that predispose a person to *Streptococcus pyogenes* flesh-eating disease.
4. Give two sources of *Pseudomonas aeruginosa.*
5. Outline the pathogenesis of tetanus.
6. Explain why *C. tetani* can be cultivated from wounds in the absence of tetanus.
7. What characteristics of bite wounds lead to anaerobic infections?
8. What is the causative agent of cat scratch disease? Why is it a threat to patients with AIDS?
9. What is a synergistic infection? How might one be acquired?
10. Why is sporotrichosis sometimes called rose gardener's disease?

Multiple Choice

1. Which of the following about *Staphylococcus aureus* is *false*?
   a) It is generally coagulase-positive.
   b) Its infectious dose is increased in the presence of foreign material.
   c) Some strains infecting wounds can cause toxic shock.
   d) Nasal carriers have an increased the risk of surgical wound infection.
   e) It is pyogenic.

2. Which of these statements about *Streptococcus pyogenes* is *false*?
   a) It is a Gram-positive coccus occurring in chains.
   b) Some strains that infect wounds can cause toxic shock.
   c) Some strains that infect wounds can cause necrotizing fasciitis.
   d) It can cause puerperal sepsis.
   e) A vaccine is available for preventing *S. pyogenes* infections.

3. Choose the one *false* statement about *Pseudomonas aeruginosa.*
   a) It is widespread in nature.
   b) Some strains can grow in distilled water.
   c) It is a Gram-positive rod.
   d) It produces a hemolytic toxin.
   e) Under certain circumstances, it can grow anaerobically.

4. Which of these statements about tetanus is *true*?
   a) It can start from a bee sting.
   b) Immunization is carried out using tiny doses of killed *C. tetani.*
   c) Those who recover from the disease are immune for life.
   d) Tetanus immune globulin does not prevent the disease.
   e) It is easy to avoid exposure to spores of the causative organism.

5. Choose the one *true* statement about gas gangrene.
   a) There are few or no leukocytes in the wound drainage.
   b) It is best to rely on antibacterial medications and avoid disfiguring surgery.
   c) A toxoid is generally used to protect against the disease.
   d) Only one antitoxin is used for treating all cases of the disease.
   e) It is easy to avoid spores of the causative agent.

6. Which of the following statements about actinomycosis is *false*?
   a) It can occur in cattle.
   b) It is caused by a branching filamentous bacterium.
   c) It always appears on the jaw.
   d) It can arise from intestinal surgery.
   e) Its abscesses can penetrate bone.

7. Which of the following statements about *Pasteurella multocida* is *false*?
   a) Infections generally respond to a penicillin.
   b) It can cause epidemics of fatal disease in domestic animals.
   c) It is commonly found in the mouths of biting animals, including humans.
   d) A vaccine is used to prevent *P. multocida* disease in people.
   e) Cat bites are more likely to result in *P. multocida* infections than dog bites.

8. Which of these statements about cat scratch disease is *false*?
   a) It is a common cause of chronic lymph node enlargement in children.
   b) It is a serious threat to individuals with AIDS.
   c) Cat scratches are the only mode of transmission to humans.
   d) It is a zoonosis of cats transmitted by fleas.
   e) It can affect the brain or heart valves in a small percentage of cases.

9. The following statements about *Streptobacillus moniliformis* are all *true* except
   a) it can be transmitted by food.
   b) its colonies can resemble those of mycoplasmas.
   c) it can be transmitted by the bites of animals other than rats.
   d) human infection is characterized by irregular fevers, rash, and joint pain.
   e) it is a Gram-positive spore-forming rod.

10. Which statement concerning sporotrichosis is *false*?
    a) It is characterized by ulcerating lesions along the path of a lymphatic vessel.
    b) Person-to-person transmission is common.
    c) It can occur in epidemics.
    d) It can persist for years if not treated.
    e) The causative organism is a dimorphic fungus.

Applications

1. Clinicians become concerned when the laboratory reports that organisms capable of digesting collagen and fibronectin are present in a wound culture. What is the basis of their concern?

2. An army field nurse working at a mobile surgical hospital asks this question of all the ambulance drivers: "Was the soldier wounded while in a field with cows?" Why does the nurse ask this question?

Critical Thinking ✚

1. In what way would the incidence of tetanus at various ages in a developing country differ from age incidence in developed countries?

2. Could colonization of a wound by a non-invasive bacterium cause disease? Explain your answer.

24 Digestive System Infections

Yellow color of eye and skin as seen in jaundice.

A Glimpse of History

An army surgeon wrote a vivid description of cholera in 1832 when the disease first appeared in the United States:

> The face was sunken . . . perfectly angular, and rendered peculiarly ghastly by the complete removal of all the soft solids, in their places supplied by dark lead-colored lines. The hands and feet were bluish white, wrinkled as when long macerated in cold water; the eyes had fallen to the bottom of their orbes, and evinced a glaring vitality, but without mobility, and the surface of the body was cold.

Cholera is a very old disease, thought to have originated in the Far East thousands of years ago. Sanskrit writings indicate it existed in India many centuries before Christianity. With the increased shipping of goods and the mobility of people during the nineteenth century, cholera spread from Asia to Europe and then to North America. The disease caused major epidemics in the nineteenth century.

John Snow, a London physician, demonstrated that cholera was transmitted by contaminated water. He observed that almost all people who contracted the disease during an outbreak in 1854 got their water from a well on Broad Street, whereas neighbors who got their water elsewhere were unaffected. Even though the germ theory of disease had not yet been described, Snow was able to convince local authorities to remove the pump handle so that people were forced to get their water elsewhere. The number of new cases decreased. The outbreak had already begun subsiding before the handle was removed, however, so Snow's explanation that cholera was a waterborne disease was not accepted by most doctors and government officials. By 1866, however, it was evident that cholera did indeed appear in areas that lacked sanitation, just as Snow had proposed. Public health agencies then played a major role in preventing epidemic cholera. In 1883 Robert Koch, who proved the germ theory of disease, isolated *Vibrio cholerae,* the bacterium that causes cholera.

In the fall of 2008, the first reports of a cholera outbreak in Zimbabwe began trickling in to the World Health Organization. Volunteers from the medical relief organization Medecins Sans Frontieres (Doctors Without Borders) rushed to set up medical facilities, while people from the World Health Organization, Red Cross, and local health agencies worked to provide filtered and chlorinated drinking water, construct latrines, and educate people about sanitary measures. Patients were treated with oral and intravenous rehydration fluids. The lack of public health resources, political instability in Zimbabwe, and global economic crisis combined to amplify this outbreak. The resulting epidemic sickened almost 100,000 people and killed nearly 2,300 individuals in Zimbabwe and surrounding countries by its end in July 2009.

24.1 ■ Anatomy, Physiology, and Ecology

Learning Outcomes

1. *Describe the functions of the main components of the upper and lower digestive tract.*
2. *Identify the role of the liver and accessory organs in health and disease.*
3. *Describe the significance of the normal intestinal microbiota.*

The main purpose of the digestive system is to convert the food we eat into a form that the body's cells can use as a source of energy and raw materials for growth. It does this by first breaking down food into small particles, processing those even further, and then absorbing the available nutrients. The waste material that remains is discharged as feces.

The digestive system encompasses two general components—the digestive tract and the accessory organs (**figure 24.1**). The digestive tract is a hollow tube that extends from the mouth to the anus. When referring only to the stomach and the intestines, the term gastrointestinal tract is often used. The accessory organs, which include the salivary glands, liver, and pancreas, support the process of digestion by producing vital enzymes and other substances that help break down the food.

Like the respiratory system and skin, the digestive tract is one of the body's major boundaries with the environment. It is distinct, however, in that the foods being digested by the host

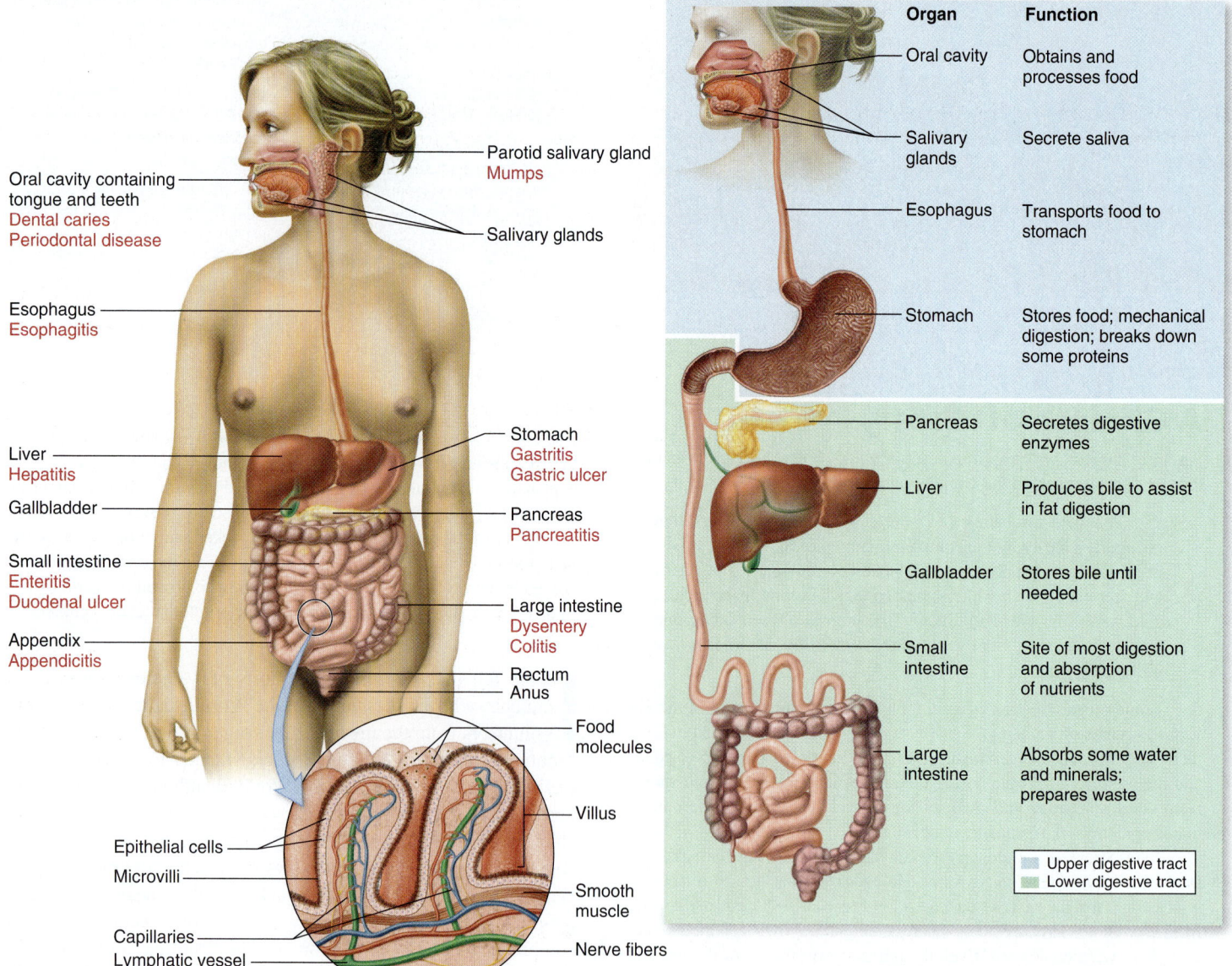

Organ	Function
Oral cavity	Obtains and processes food
Salivary glands	Secrete saliva
Esophagus	Transports food to stomach
Stomach	Stores food; mechanical digestion; breaks down some proteins
Pancreas	Secretes digestive enzymes
Liver	Produces bile to assist in fat digestion
Gallbladder	Stores bile until needed
Small intestine	Site of most digestion and absorption of nutrients
Large intestine	Absorbs some water and minerals; prepares waste

Upper digestive tract
Lower digestive tract

Parotid salivary gland
Mumps

Oral cavity containing tongue and teeth
Dental caries
Periodontal disease

Salivary glands

Esophagus
Esophagitis

Liver
Hepatitis

Stomach
Gastritis
Gastric ulcer

Gallbladder

Pancreas
Pancreatitis

Small intestine
Enteritis
Duodenal ulcer

Large intestine
Dysentery
Colitis

Appendix
Appendicitis

Rectum
Anus

Food molecules

Villus

Epithelial cells
Microvilli

Smooth muscle

Capillaries
Lymphatic vessel

Nerve fibers

FIGURE 24.1 The Digestive System Some of the disease conditions that can affect the system are shown in color. For example, mumps is a viral disease affecting a parotid gland.

? *What is the role of villi and microvilli in the small intestine?*

provide a ready source of carbon and energy for microbial growth. Every mouthful of food you ingest feeds not only yourself but the microbial population in your digestive tract as well.

Most microbes that inhabit the digestive tract live in harmony with the host, but the balance in this complex ecosystem is delicate. A mucous membrane only a single cell layer thick separates the microbial population from some underlying tissues, so damage to this layer allows resident microbes to penetrate. Ingested pathogens often have mechanisms to breach intact barriers.

The Upper Digestive System

The upper digestive system includes the mouth, salivary glands, esophagus, and stomach.

The Mouth and Salivary Glands

The mouth is mainly a grinding apparatus that begins the physical and chemical digestion of food. The act of chewing allows the teeth to grind the food into smaller pieces, while the saliva moistens the food and provides amylase, an enzyme that helps start the breakdown of starches.

The outer portion of teeth is made up of a hard protective substance called enamel (**figure 24.2**). Proteinaceous material from the saliva adheres to the enamel, creating a thin film or pellicle. Various types of bacteria can attach to receptors on the pellicle, colonizing the tooth surface to create a biofilm called **dental plaque.** Over time, mineral salts deposit in plaque, creating a crusty substance called dental calculus or tartar. Routine brushing and flossing can remove plaque, but tartar removal requires professional cleaning.

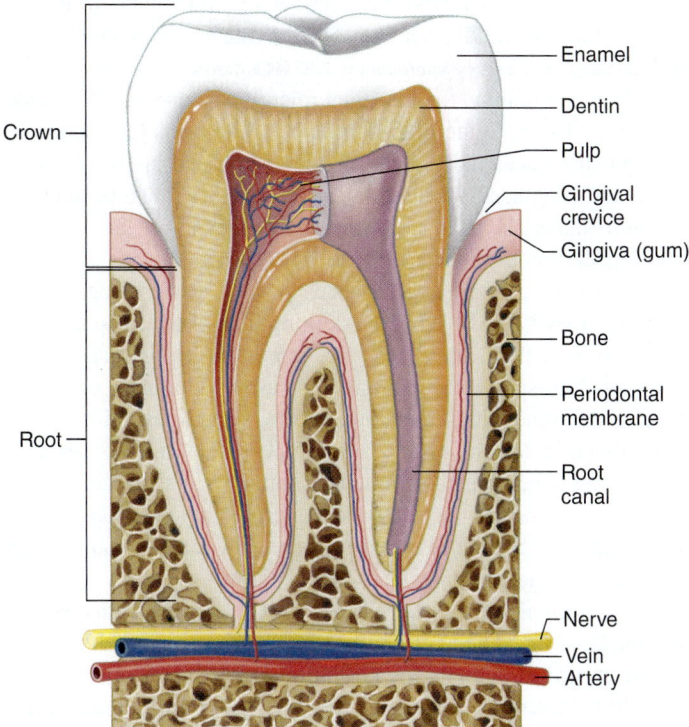

FIGURE 24.2 Structure of a Tooth and Its Surrounding Tissues

❓ *What is the difference between dental plaque and tartar?*

When the tooth enamel deteriorates or is otherwise damaged, microorganisms can enter the substance of the tooth. Tooth decay, or **dental caries,** can usually be repaired with dental fillings. If bacteria reach the root canal, a region of the tooth filled with pulp (tissue in the tooth that contains blood vessels and nerves), the tooth must either be removed or the pulp replaced with an inert substance. This replacement procedure—like the region that is filled—is called a root canal.

Another frequent site of microbial accumulation and infection is the gingival crevice, the space between the gum and a tooth. In response to accumulated plaque, the gums become inflamed, a condition called **gingivitis**. The irritated gums may recede from the tooth root, allowing bacteria to reach that region as well, which can ultimately result in tooth loss.

Saliva is produced by various salivary glands that empty into the mouth. Normally, 1,500 ml of saliva are secreted each day, equivalent to approximately four cans of soda pop. The largest of these glands, the parotid glands, are notable because the mumps virus infects them. In addition to saliva's function in digestion, it plays an important role in protecting the oral cavity against microbes. Saliva is rich in secretory IgA (the antibodies that provide mucosal immunity), and it contains antibacterial compounds such as lysozyme and lactoferrin. People with poor saliva production are subject to severe tooth decay. ◀◀ lactoferrin, p. 337 ◀◀ secretory IgA, p. 362 ◀◀ lysozyme, pp. 62, 337

Of all the microorganisms entering the mouth, relatively few can colonize the surfaces there. Those that do are able to bind specifically to molecules on host tissues or attached microbes, allowing them to resist the scrubbing action of food and the tongue and the flushing effect of salivary flow. They are also able to withstand the antibacterial effects of saliva. The fact that the superficial layers of cells of the mucous membranes are constantly shed also limits the numbers of microbes in the mouth.

Although over 600 types of bacteria can be found in the mouth, members of the genus *Streptococcus* are among the most common. The different streptococcal species inhabit different microenvironments. Some preferentially colonize the upper part of the tongue, others the teeth, and still others the mucosa of the cheek. This difference in distribution is due to the fact that the epithelial cells lining various parts of the mouth have different receptors. ◀◀ receptors, p. 386

In addition to *Streptococcus* species—which are obligate fermenters—obligate anaerobes are also surprisingly common in the mouth. The anaerobes thrive in complex communities, protected because aerobic species consume available O_2. The foul-smelling end products of anaerobic metabolism are associated with halitosis, also known as bad breath.

Microβyte

There can be up to 100 billion bacteria per gram of plaque.

The Esophagus

The esophagus is a collapsible tube about 10 inches long, connecting the mouth to the stomach. Peristalsis, the rhythmic contractions of the digestive tract, pushes food to the stomach. The esophagus has a relatively sparse microbial population, consisting mostly of bacteria from the mouth and upper respiratory tract.

One reason for the relative lack of microbes is the secretory IgA–containing mucus and saliva that bathe the lining of the tube. Microbes trapped in those secretions are moved down the esophagus, along with food and liquids, propelled by peristalsis. The esophagus rarely becomes infected except in individuals with AIDS or other immunodeficiencies. ▶▶ **AIDS, pp. 633, 695**

The Stomach

The stomach is an elastic, saclike structure with a muscular wall. Its primary function is to break down and store food particles as they await controlled entry into the small intestine. The gastric juices are highly acidic, which denatures the proteins in food particles. The acidic environment also activates pepsinogen produced in the stomach to form pepsin, a protein-splitting enzyme. The cells that line the stomach protect themselves from the acid and enzymes by secreting a thick alkaline mucus. Most bacterial cells cannot survive the hostile environment, however, so a normal empty stomach has few microorganisms.

The Lower Digestive System

The lower digestive system includes the small and large intestines, as well as the pancreas and liver.

The Small Intestine

As the stomach contents enter the small intestine, digestive fluids from the pancreas and liver are mixed in. These alkaline fluids neutralize the stomach acid and also contain **bile,** an emulsifying agent that helps break up fats. Substances in bile called bile salts help the intestine absorb oils, fats, and fat-soluble vitamins. Many types of bacteria are killed by bile, but those adapted to live in the intestinal tract resist its bactericidal effects. This has practical applications in the laboratory, because bile salts are used in selective media designed to isolate intestinal bacteria. ◀◀ **bactericidal, p. 108** ◀◀ **selective media, p. 95**

Additional enzymes are produced by intestinal cells. Some are secreted into the intestinal lumen (the region inside the tube), but many are permanently attached to the cell membranes. Microbial infections that damage these cells interfere with digestion.

If the intestinal tract were a simple pipe, it would have a surface area of less than a bath towel. Although only about 6 meters in length, the small intestine has an enormous surface area, approximately 250 square meters, or the size of a tennis court. The inside surface of the intestine is covered with many small, fingerlike projections called **villi.** Each of these is lined with cells that have cytoplasmic projections called **microvilli** (see figure 24.1). Bundles of actin filaments make up the structural core of these projections. This microvilli-coated surface is sometimes referred to as the brush border. At the base of the villi are pits called crypts. These are small glands that continuously secrete large amounts of enzyme-containing fluids and mucus into the lumen. Certain cells in the crypts give rise to new intestinal epithelial cells. Intestinal epithelial cells are completely replaced every 9 days—one of the fastest turnover rates in the body. ◀◀ **actin, p. 73**

A major role of the small intestine is nutrient and fluid absorption. Cells that line the villi take in amino acids and monosaccharides using active transport mechanisms, bringing in sodium ions (Na^+) simultaneously. Fatty acids, vitamins, and minerals such as

iron are absorbed as well. In addition, the small intestine absorbs fluids. Taking into account the volume of digestive juices, as well as saliva and the fluids ingested as food and drink, this amounts to approximately 9 liters per day! Understandably, disruption in this fluid balance can result in diarrhea. ◀◀ **active transport, p. 56**

Few microbes reside in the upper small intestine, because the flushing action of rapidly passing digestive juices limits their ability to colonize the surface. As the involuntary muscle action of peristalsis propels the intestinal contents toward the large intestine, the movement slows and the bacterial population increases. The immune system monitors this population with numerous dendritic cells sampling the environment and M cells delivering small amounts of the lumen's contents to Peyer's patches. ◀◀ **dendritic cells, p. 340** ◀◀ **M cells, p. 358** ◀◀ **Peyer's patches, p. 358**

The Large Intestine

The main function of the large intestine is to absorb water as well as vitamins produced by the resident microbiota. Because the small intestine absorbs so much water, only 300 to 1,000 ml of fluid normally reach the large intestine per day, and most of this is absorbed. The semisolid feces, composed of indigestible material and bacteria, remain. Infection of the large intestine can interfere with absorption and stimulate the painful contractions known as "stomach cramps."

Microbes flourish in the large intestine, supported by the abundance of nutrients in undigested and indigestible food material. In fact, bacteria make up about one-third of the fecal weight, reaching concentrations of approximately 10^{11} cells per gram! Anaerobic bacteria, particularly members of the genus *Bacteroides,* make up about 99% of the microbial population. The remaining microbes are facultative anaerobes, primarily *Escherichia coli* and other members of the *Enterobacteriaceae.* ◀◀ **members of the genus *Bacteroides*, p. 275** ◀◀ ***Enterobacteriaceae*, p. 265**

As a community, the intestinal microbiota can degrade a wide variety of foods. Some high-fiber foods, like beans and broccoli, contain substances indigestible by the gastric and intestinal juices but readily degraded by intestinal organisms. Microbial breakdown of these materials often produces large amounts of gas, causing abdominal discomfort and intestinal gas (flatus). Bacterial enzymes can also convert various substances in food to carcinogens and therefore may be involved in the development of intestinal cancers.

The intestinal microbes are important to human health because they synthesize a number of useful vitamins, including niacin, thiamine, riboflavin, pyridoxine, vitamin B_{12}, folic acid, pantothenic acid, biotin, and vitamin K. They are also essential for normal development of mucosal immunity. At the same time, many members of the intestinal microbiota are opportunistic pathogens that can cause disease if they gain access to other body sites such as the urinary tract.

The normal microbiota helps prevent pathogens from colonizing the large intestine. Antibiotic treatment, especially with broad-spectrum drugs, disrupts the normal microbiota and often results in mild to severe **antibiotic-associated diarrhea.** In some cases, this is caused by toxin-producing strains of *Clostridium difficile,* which readily colonizes the intestine of people whose normal intestinal microbiota has been reduced by antimicrobial chemotherapy. Suppression of intestinal microbiota with antibacterial medications can also increase susceptibility to other pathogens such as *Salmonella enterica*. ▶▎ *Clostridium difficile*–associated disease, p. 594 ▶▎ *Salmonella enterica*, p. 592

The Pancreas

The pancreas is a large organ located behind the stomach. Some pancreatic cells produce hormones, and others make digestive enzymes. About 2 liters of pancreatic digestive juices discharge directly into the upper portion of the small intestine each day. These fluids, as well as digestive juices of the small intestine itself, are alkaline and neutralize the stomach acid as it passes into the small intestine.

The Liver

The liver is a large, dark-red organ located in the upper right portion of the abdomen. One of its roles is to produce bile, which is then concentrated and stored in a saclike structure called the gallbladder. The bile then flows through a system of tubes into the upper small intestine. Severe liver disease or obstruction of the bile ducts can cause **jaundice,** a yellow color of the skin and eyes caused by buildup of a bile component (bilirubin) in the blood.

The liver also inactivates toxic substances that enter the bloodstream, generally from the digestive tract. For example, ammonia produced by intestinal bacteria and then absorbed into the bloodstream could reach poisonous levels if not detoxified by the liver. These same processes can also chemically alter and remove many medications. Therefore, if the liver is damaged, as it might be by a viral infection, lower medication doses might be needed to prevent an accidental overdose.

MicroAssessment 24.1

The digestive tract is a complex ecosystem and a major route for pathogens to enter the body. The distribution of bacterial microbiota in the mouth is governed by different receptors on the epithelium. Bacteria colonizing the tooth surface can create a biofilm called plaque. People with poor saliva production risk severe tooth decay. Infections of the esophagus are so unusual that their occurrence suggests immunodeficiency. The stomach is responsible for acid digestion of food and destruction of most microbes before they reach the intestines. The small intestine carries out the majority of the digestion and absorption of nutrients. Undigested material becomes feces in the large intestine. Metabolic activity of intestinal microbiota produces vitamins, but it causes gas and can contribute to cancer. Disruption of the normal microbiota through antibiotic treatment can result in antibiotic-associated diarrhea.

1. *What causes halitosis (bad breath)?*
2. *What makes the surface area of the small intestine so large, when it is only about 6 meters long?*
3. *Two patients recently had spinal surgery, but one is jaundiced. Which one would likely need a lower dose of acetaminophen?* ➕

UPPER DIGESTIVE SYSTEM INFECTIONS

24.2 ■ Bacterial Diseases of the Upper Digestive System

Learning Outcomes

4. *Compare and contrast dental caries, periodontal disease, and acute necrotizing ulcerative gingivitis.*
5. *Describe Helicobacter pylori gastritis and how it relates to gastric ulcers and stomach cancer.*

Bacterial diseases of the upper digestive system often involve the teeth and gums (**figure 24.3**). Some of these local infections—which often go unnoticed for years—may have consequences for the rest of the body as well. For example, some studies suggest that chronic gum infections contribute to hardening of the arteries, arthritis, and premature birth. In addition, members of the normal oral microbiota can enter the bloodstream during dental procedures and cause subacute bacterial endocarditis. Another site of upper digestive infections is the stomach. ▶▎subacute bacterial endocarditis, p. 673

Dental Caries (Tooth Decay)

Dental caries (tooth decay) is the most common infectious disease and the main reason for tooth loss. In the United States, about 60% of all teenagers have tooth decay, making it four times more common than asthma in this group.

Signs and Symptoms

Dental caries is usually far advanced before any symptoms develop. Sometimes there is noticeable discoloration, roughness, or defect, and a tooth can break during chewing. However, the severe, throbbing pain of a toothache is often the first symptom.

Causative Agent

Streptococcus mutans and related species are **cariogenic** (meaning "caries generating"). These Gram-positive cocci live only on teeth and therefore do not colonize the mouth in the absence of teeth. Unlike many other bacteria, they thrive in acidic environments (below pH 5), which allows them to survive the conditions that result from the lactic acid they produce during fermentation. They convert sucrose (table sugar) into extracellular insoluble polysaccharides called glucans. Glucans are essential for the development of dental caries on smooth tooth surfaces. ▎◀ polysaccharides, p. 31 ▎◀ fermentation, pp. 134, 147

Pathogenesis

Formation of dental caries is a complex process. First, oral streptococci adhere to the pellicle on the tooth to create dental plaque, but this alone does not lead to caries (figure 24.3). If dietary sucrose is present, extracellular enzymes produced by *S. mutans* split the sucrose into its two monosaccharide components—glucose and fructose. The glucose is polymerized to make glucans, and the fructose is fermented, producing lactic acid. The glucans help

Tooth infection

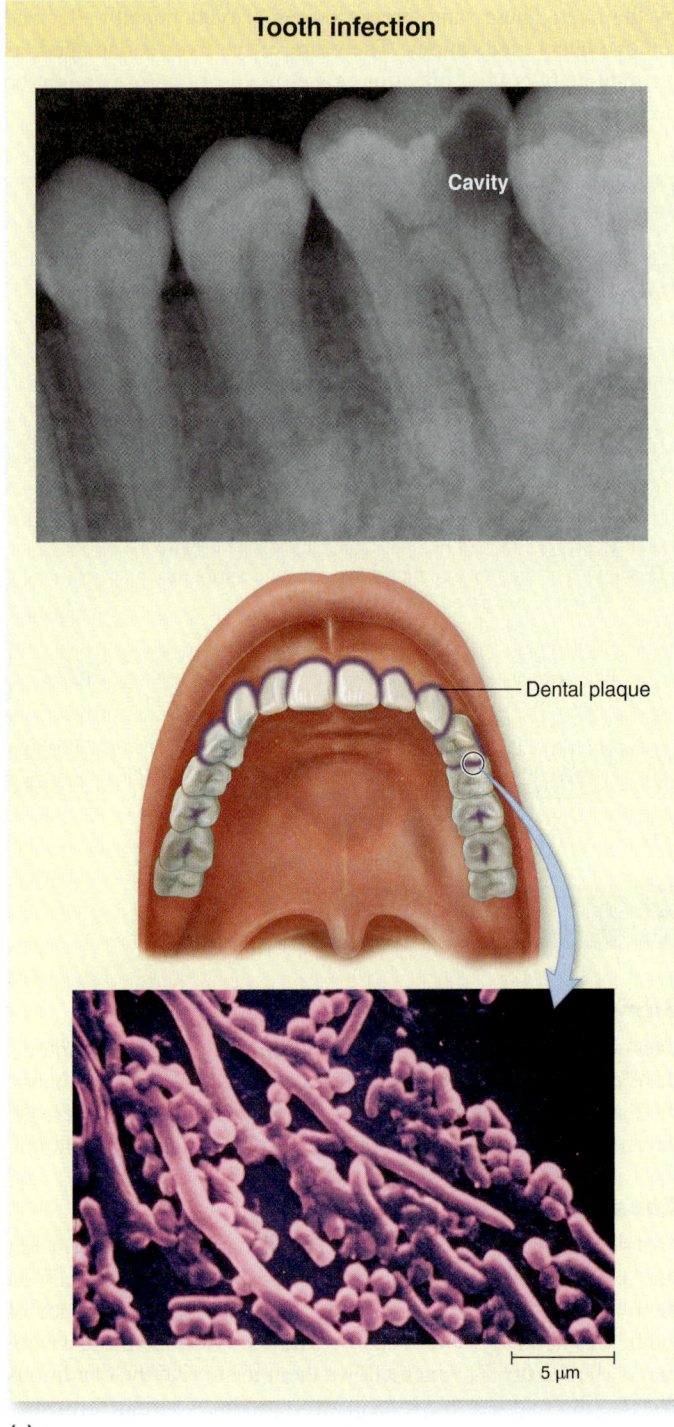

(a)

Gum infection

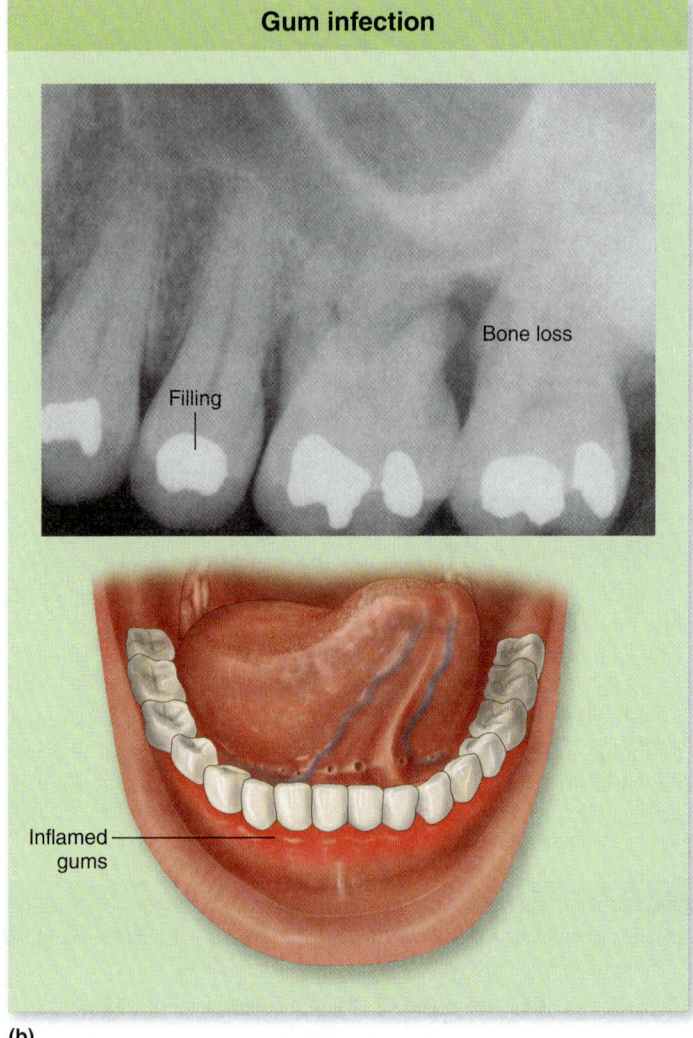

(b)

FIGURE 24.3 Infections of the Teeth and Gums Note the different types of bacteria in the community forming dental plaque (shown as a purple color). If plaque hardens between teeth and gingiva, it can increase the likelihood of periodontal disease, which may result in bone loss and loosening or loss of teeth.

❓ *How does brushing away plaque decrease the chance of developing cavities?*

create an even thicker biofilm by binding a mixed population of microbes together and to the tooth, making the plaque impenetrable to saliva. ◄◄ dental plaque, p. 573

Plaque that contains cariogenic microbes can act as a tiny acid-soaked sponge closely applied to the tooth. In fact, when sugar enters the mouth, the pH of the plaque drops from its normal value of about 7 to below 5 within minutes (**figure 24.4**). This 100-fold increase in acidity begins dissolving the calcium phosphate of the teeth. After food leaves the mouth, the pH of the plaque

rises slowly to neutrality. The delay is due to the storage granules made by cariogenic bacteria—polysaccharides in these granules can be fermented, generating acid even after the mouth is empty. ◄◄ storage granules, p. 66

Both *S. mutans* and a sucrose-rich diet are required to produce dental caries on smooth surfaces of the teeth. In deep fissures or pits, plaque can accumulate in the absence of *S. mutans,* and tooth decay can occur if lactic acid–producing bacteria and fermentable substances are present.

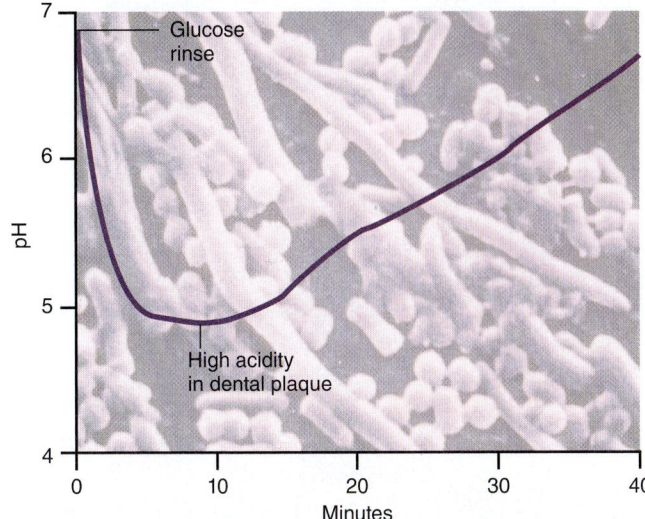

FIGURE 24.4 Acidity of Dental Plaque Plaque becomes more acidic after rinsing the mouth with a glucose solution. Tooth enamel begins to dissolve at about pH 5.5.

❓ *Would you expect this dip in pH if* Streptococcus mutans *metabolized sugar via aerobic respiration rather than fermentation?*

Epidemiology

Dental caries is worldwide in distribution, but the incidence varies markedly depending mainly on intake of dietary sucrose and access to preventive dental care. Genetics also plays an important role because some people inherit resistance to the disease. The incidence of dental caries peaks during teen years and falls off with age. This is probably because the pits and fissures on the tooth—ideal sites for bacterial growth—wear down with time.

Treatment and Prevention

Treatment of dental caries requires drilling out the cavity, filling the defect with material such as amalgam (an alloy of mercury and some other metal), and restoring the contour of the tooth.

The most important method for controlling dental caries is restricting dietary sucrose and other refined carbohydrates. This reduces *Streptococcus mutans* colonization as well as acid production by cariogenic plaque. However, the quantity of sugar in the diet is not the most important factor leading to dental caries—the frequency of eating sugary foods and the length of time the food stays on the teeth are more critical. In fact, chewing paraffin or sorbitol-sweetened gum reduces dental caries, probably because it increases the flow of saliva.

Trace amounts of fluoride prevent dental caries by making tooth enamel harder and more resistant to dissolving in acid. In the United States, more than half the population is supplied with fluoridated public drinking water, resulting in a 60% reduction in dental caries. This saves over $4.5 billion in dental costs each year. In areas where fluoridated drinking water is not available, fluoride tablets or solutions can be used. To have optimum effect, children should begin receiving fluoride before their permanent teeth erupt. Fluoride applied to tooth surfaces in the form of mouthwashes, gels, or toothpaste is generally less effective.

Toothbrushing and flossing remove plaque, reducing the incidence of dental caries by about half. Unfortunately, toothbrush bristles cannot remove plaque from the deep narrow pits and fissures normally present in children's teeth, which is where most childhood tooth decay starts. Caries can be prevented by using a sealant—a kind of epoxy glue that seals the fissures, kills the bacteria in plaque, and prevents bacterial recolonization. Older people have less fissure-related caries, but are prone to receding gums that expose root surfaces, which become sites for dental caries.

MicroByte ────────────────
Eliminating sucrose-containing sweets reduces dental caries by 90%.

Periodontal Disease

Periodontal disease often results from plaque accumulation near the gum margin (periodontal means "around the tooth"). Bacterial products in plaque trigger an inflammatory response in the gums and the local tissues, leading to at least some degree of tissue damage. Periodontal diseases include **gingivitis** (swelling and redness of the gums) and **chronic periodontitis,** a destructive response that progressively damages the structures that support the teeth (see figure 24.3). Chronic periodontitis is an important cause of tooth loss from middle age onward.

Signs and Symptoms

Gingivitis is marked by gums that are tender and bleed easily. This can occur within days of plaque accumulation and typically resolves once plaque is removed.

Chronic periodontitis is a long-term response to plaque and is characterized by bad breath, red shiny gums that bleed easily, and loosening of the teeth. The base of the teeth becomes discolored—ranging from yellowish to black—and the gums recede. Exposed roots of the teeth are susceptible to dental caries.

Causative Agent

Periodontal disease is associated with dental plaque that forms at the point where the gum joins the tooth. The plaques are typically quite different from those associated with dental caries in that most members of the microbial community are Gram-negative anaerobes. An association of three types bacteria—*Porphyromonas gingivalis, Treponema denticola,* and *Tannerella (Bacteroides) forsythia*—seem to play an important role. However, hundreds of kinds of bacteria have been identified in plaques associated with periodontal disease, many of which have not been cultivated, so other microbes are likely important as well.

Pathogenesis

As plaque (and tartar) accumulates at the gum margin—especially in hard-to-clean areas between the teeth—it gradually extends into the gingival crevice. Bacterial products incite an inflammatory response, leading to the characteristic symptoms of gingivitis. If the level of plaque remains small, the host response limits the process. ◀◀ tartar, p. 573

In some cases, the inflammatory response does not control the microbial population, leading to the tissue destruction associated with chronic periodontitis. The process appears to involve many factors, including the microbial population and host factors. Tissue-degrading enzymes released by the plaque microbes weaken the gingival tissue and cause the gingival crevice to widen and deepen. This allows the plaque to spread further toward the root of the tooth. The toll-like receptors of surrounding tissues detect the lipopolysaccharide of the Gram-negative outer membrane (endotoxin), causing the release of pro-inflammatory cytokines. Some of these evoke such a strong inflammatory response that nearby tissues are damaged. With progressive tissue damage, the membrane that attaches the tooth root to the bone weakens, and the bone gradually softens. The tooth becomes loose and may fall out. ◀◀ **toll-like receptors, p. 342**

Epidemiology

Periodontal disease such as gingivitis can begin in childhood. After age 65, almost 90% of individuals have some degree of chronic periodontitis. Smokers and those with underlying defects in acquired or innate immunity (including those with AIDS) often have severe periodontitis that leads to tooth loss.

Treatment and Prevention

Periodontal disease can be treated in its early stages by cleaning out the inflamed gingival crevice and removing plaque and tartar. In advanced cases, minor surgery is usually required to expose and clean the roots of the teeth. Careful flossing and toothbrushing can prevent periodontal disease, especially when combined with twice-yearly polishing and removal of tartar at a dental office.

Acute Necrotizing Ulcerative Gingivitis

Acute necrotizing ulcerative gingivitis (ANUG) is a severe, acute condition distinct from other forms of periodontitis (**figure 24.5**). The disease was first called trench mouth because it was common among soldiers living in trenches during World War I, unable to care for their teeth and gums. Since 2005, dentists have noticed a dramatic increase in ANUG cases associated with methamphetamine use, coining the term "meth mouth" as a form of ANUG. Meth mouth is further complicated by rampant dental caries.

Signs and Symptoms

ANUG is characterized by bleeding painful gums, abscessed and broken teeth, and extremely foul breath.

Causative Agent

The suspected causative agent of ANUG is an oral, not yet cultivated spirochete of the *Treponema* genus. The organism probably acts synergistically with other anaerobic species, with which it is always found. The disease is associated with heavy growth of anaerobes at the gum line. Only about 50% of oral microbiota members have been grown in culture, so their role in ANUG is poorly understood. ◀◀ **the genus** *Treponema*, **p. 276**

Pathogenesis

The spirochetes and the other anaerobes are presumed to act together to destroy tissues in the oral cavity, but the precise mechanisms are unknown. Plaque is always present, but its bacterial composition shows much larger numbers of spirochetes and other anaerobes than are present in chronic periodontal disease. The spirochetes invade the tissue, causing necrosis and ulceration, mainly of the gums between the teeth.

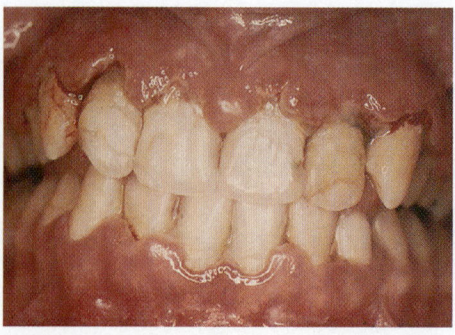

(a)

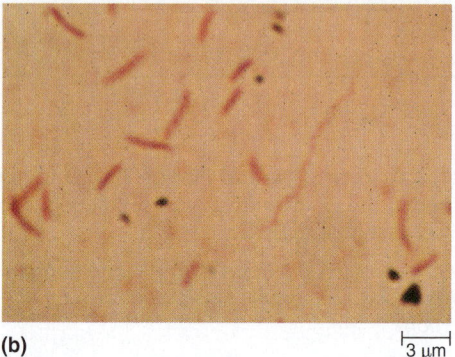

(b)   3 μm

FIGURE 24.5 Acute Necrotizing Ulcerative Gingivitis (ANUG) (a) Red, swollen gums with loss of tissue, especially between the teeth. **(b)** Gram stain of exudate showing a spirochete and rod-shaped bacteria.

? *What aspects of methamphetamine abuse increase the risk of developing ANUG?*

Methamphetamine use often lowers saliva production, leading to a dry mouth and increased risk of biofilm formation, tooth decay, and ANUG. Tooth grinding, also associated with methamphetamine use, increases the risk of fracturing teeth weakened by severe decay and gingivitis.

Epidemiology

ANUG can occur at any age in association with poor oral hygiene, particularly when combined with poor nutrition, high sugar diet, chronic stress, and immunodeficiency. Methamphetamine use appears to intensify the problems.

Treatment and Prevention

Most bacteria at the gingival crevice are anaerobes, so hydrogen peroxide treatment rapidly relieves ANUG symptoms. Long-term cures include removing plaque and tartar. Prevention begins with daily brushing and flossing, and twice-yearly professional cleaning. The main features of teeth and gum infections are presented in **table 24.1**.

Helicobacter pylori Gastritis

Many people have gastritis, meaning inflammation of the stomach, without even knowing it. The association of the causative agent to ulcers was demonstrated in the 1980s when Barry Marshall, one of the scientists who discovered the organism, intentionally drank a culture of *Helicobacter pylori*.

TABLE 24.1	Important Infections of the Teeth and Gums		
	Dental Caries	**Periodontal Diseases**	**Acute Necrotizing Ulcerative Gingivitis**
Signs and symptoms	None until advanced disease. Late: discoloration, roughness, broken tooth, throbbing pain	Gingivitis characterized by tender bleeding gums; chronic periodontitis characterized by bad breath, red shiny gums that bleed easily, loose teeth, and exposed discolored tooth roots	Bleeding painful gums, abscessed and broken teeth, and extremely foul breath
Incubation period	1 to 24 months before cavity is detectable	Gingivitis—days; periodontitis—months or years	Undetermined
Causative agent	Dental plaque, particularly when populated with *Streptococcus mutans*	Microbes in dental plaque; an association of *Porphyromonas gingivalis*, *Treponema denticola*, and *Tannerella forsythia* seem to play an important role in periodontitis.	Probably a spirochete of the genus *Treponema* acting with other anaerobes
Pathogenesis	Bacteria in plaque produce acid from dietary sugars; slowly dissolves the calcium phosphate composing the tooth; sucrose is critical for cariogenic plaque formation.	Plaque formed at the gum margins causes the inflammatory response associated with gingivitis. If the plaque extends into the gingival crevices, the strong inflammatory response damages tissues that support the teeth, causing chronic periodontitis.	The spirochetes and certain other anaerobes act synergistically to destroy tissues. The spirochetes invade tissue, causing necrosis and ulceration.
Epidemiology	Worldwide distribution, incidence depending on dietary sucrose, natural or supplemental fluoride. The young are more susceptible than the old.	Gingivitis can begin in childhood; periodontitis primarily affects older people, particularly smokers and immunodeficient individuals.	All ages are susceptible in association with poor oral hygiene, malnutrition, immunodeficiency, or methamphetamine use.
Treatment and prevention	Restriction of dietary sucrose, supplemental fluoride, mechanical removal of plaque, sealing pits and fissures in children's teeth.	Avoid buildup of plaque. Surgical treatment in severe cases of periodontitis to clean tooth roots.	Avoid buildup of plaque. Antibiotic treatment, followed by removal of plaque and tartar.

Signs and Symptoms

Most *Helicobacter pylori* infections are asymptomatic. Early in infection, however, the signs and symptoms may range from belching to vomiting. When the infection results in peptic ulcers of the stomach and duodenum (*peptic* meaning "caused by digestive juices"), the patient may experience localized abdominal pain, tenderness, and bleeding. Stomach cancer can also develop.

Causative Agent

Helicobacter pylori is a short, curved, Gram-negative, microaerophilic bacterium. It has multiple polar flagella that are unusual because they are covered by sheaths. ◀◀ the genus *Helicobacter*, p. 276

Pathogenesis

Helicobacter pylori cells survive the acidic environment of the stomach by (1) producing urease, an enzyme that converts urea to ammonia, thereby creating an alkaline microenvironment, and (2) burrowing within the mucus layer that coats the stomach lining (**figure 24.6**). Urea is normally found in gastric juices because it is released as proteins are degraded. *H. pylori* cells use their flagella to move through the mucus, following the pH gradient from the acidic gastric lumen to the nearly neutral underlying epithelial cells. The bacterial cells then attach to the mucus-secreting epithelium or multiply adjacent to it. The ability to both move away from the acidic lumen and produce ammonia from urea are required for *H. pylori* to colonize the stomach. Non-motile strains are removed with mucus turnover, and urease-deficient ones can cause disease only if injected directly into the mucus layer.

Virulent *H. pylori* strains produce CagA, a protein they inject into host cells that changes the shape and surface characteristics of the cells. Evidence suggests that these changes represent a prelude to cancer. Another bacterial product, VacA, acts on mucosal cells to promote flow of urea into the stomach. This protein also might damage the mucus-producing epithelial cells.

Bacterial products from *H. pylori* cells growing near the epithelium incite an inflammatory response in the wall of the stomach. As the epithelial cells are damaged by a combination of products released from the responding neutrophils and toxins produced by *H. pylori*, mucus production decreases. The thinning of the protective mucus layer and damage to local cells probably accounts for the development of peptic ulcers.

H. pylori infections persist for years, often for life. The outcome of the infection is quite variable. Only about one in six infected persons develops ulcers. A small percentage of chronically infected people develops stomach cancer, but more than 90% of those with stomach cancer are infected.

Epidemiology

Overall, about one in five adults in the United States is infected with *H. pylori*, but the incidence increases with age, reaching almost 80% for those over age 75. Infections tend to cluster in families.

FIGURE 24.6 *Helicobacter pylori* **Pathogenesis**
Neutrophils are early responders to invasion,
followed by other cells such as lymphocytes.

❓ *How does urease allow*
H. pylori to live in the stomach?

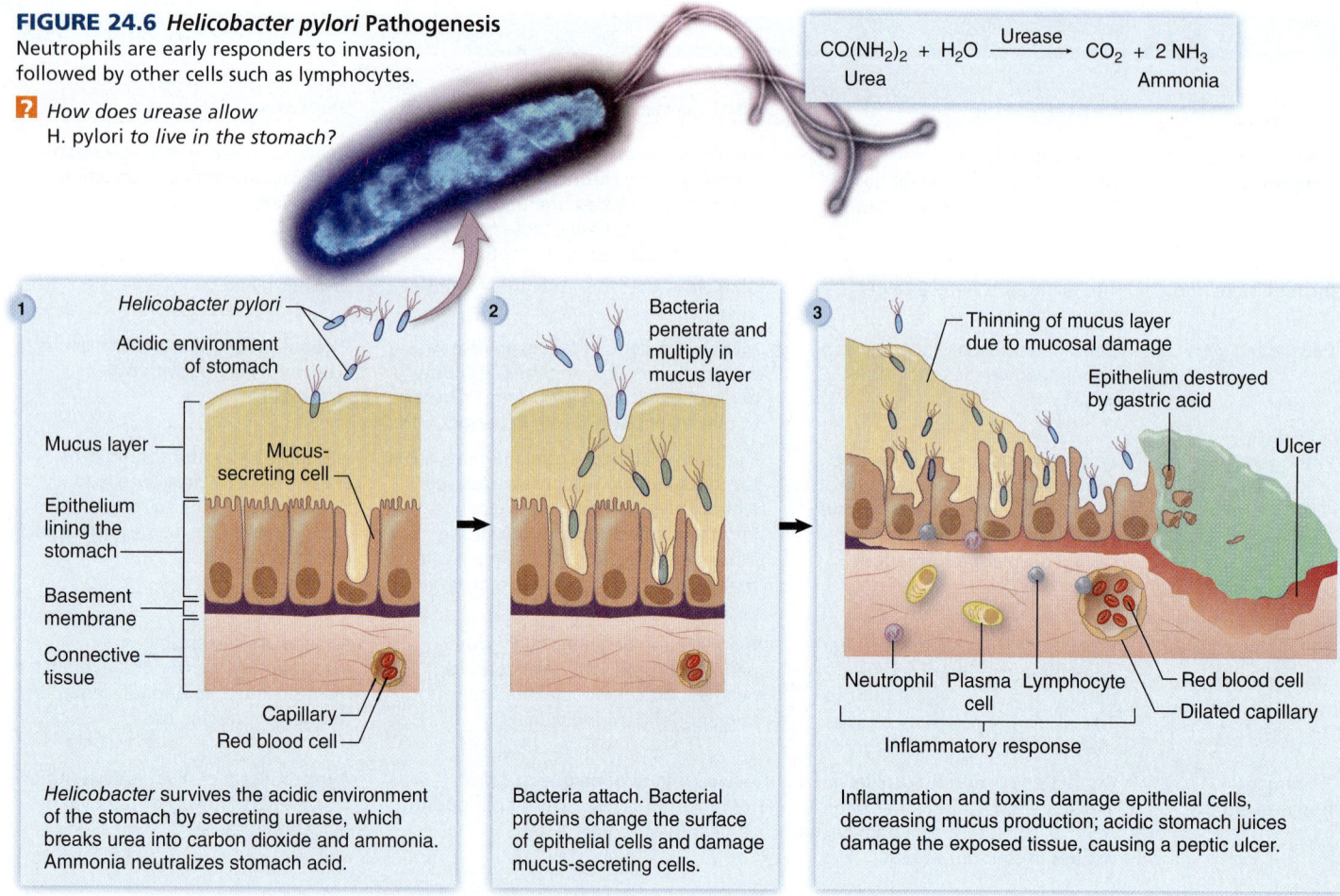

$$CO(NH_2)_2 + H_2O \xrightarrow{\text{Urease}} CO_2 + 2 NH_3$$
Urea   Ammonia

1 *Helicobacter pylori*
Acidic environment
of stomach
Mucus layer — Mucus-secreting cell
Epithelium lining the stomach
Basement membrane
Connective tissue
Capillary
Red blood cell

Helicobacter survives the acidic environment
of the stomach by secreting urease, which
breaks urea into carbon dioxide and ammonia.
Ammonia neutralizes stomach acid.

2 Bacteria penetrate and multiply in mucus layer

Bacteria attach. Bacterial
proteins change the surface
of epithelial cells and damage
mucus-secreting cells.

3 Thinning of mucus layer due to mucosal damage
Epithelium destroyed by gastric acid
Ulcer
Neutrophil Plasma cell Lymphocyte Red blood cell
Dilated capillary
Inflammatory response

Inflammation and toxins damage epithelial cells,
decreasing mucus production; acidic stomach juices
damage the exposed tissue, causing a peptic ulcer.

TABLE 24.2	*Helicobacter pylori* Gastritis
Signs and symptoms	Infections are often asymptomatic, but peptic ulcers can cause abdominal pain, tenderness, and bleeding. Stomach cancer can develop.
Incubation period	Usually undetermined
Causative agent	*Helicobacter pylori*, a curved, Gram-negative, microaerophilic bacterium, with multiple polar flagella covered by sheaths
Pathogenesis	*H. pylori* cells survive stomach acidity by producing urease and burrowing within the stomach's mucus coating. The bacterial cells attach to and transfer CagA to epithelial cells, causing changes that may be a prelude to cancer. The VacA they produce increases flow of urea into the stomach. The inflammatory response to infection and bacterial toxins may account for peptic ulcers. Cancer rarely develops, but most people with stomach cancer are infected with *H. pylori*.
Epidemiology	Probably fecal-oral transmission. Incidence increases with age.
Treatment and prevention	No proven preventive. Treated with a combination of two antibiotics and a medication that inhibits stomach acid production.

Transmission of *H. pylori* probably occurs by the fecal-oral route, and the bacteria have been found in well water. Flies that have landed on feces are also capable of transmitting the organism. Infection rates are highest in low-socioeconomic groups.

Treatment and Prevention

Helicobacter pylori infections can be treated with a combination of two antibiotics and a medication that inhibits stomach acid production. This usually results in clearing of the gastritis and healing of any ulcers. There are no proven preventive measures. The main features of *H. pylori* gastritis are shown in **table 24.2.**

MicroAssessment 24.2

The pathogenesis of tooth decay depends on both dietary sucrose and acid-forming bacteria in biofilms on teeth. Periodontal disease is an inflammatory disease associated with plaque at the gum margin and can lead to loosening of teeth. ANUG destroys gingival tissue. Chronic infection by *Helicobacter pylori* is a key factor in the development of stomach and duodenal ulcers and stomach cancer.

4. *Why are new cavities less common, but loss of teeth more common, in people over age 65?*

5. *What enzyme produced by* Helicobacter pylori *helps it survive in stomach acid?*

6. *Do you agree or disagree with the statement "*Helicobacter pylori *causes stomach cancer." Explain.* ➕

CASE PRESENTATION

The patient was a 35-year-old man who consulted his physician because of upper abdominal pain. The pain was described as a steady burning or gnawing sensation, like a severe hunger pain. Usually it came on $1\frac{1}{2}$ to 3 hours after eating, sometimes waking him from sleep. Generally, it was relieved in a few minutes by food or antacid medicines.

On examination, the patient appeared well, without evidence of weight loss. The only positive finding was tenderness slightly to the right of the midline in the upper part of the abdomen. A test of the patient's feces was positive for blood. The remaining laboratory tests were normal.

Endoscopy, a procedure that uses a long, flexible fiber-optic device passed through the mouth, showed a patchy redness of parts of the stomach lining. A biopsy was taken. The endoscopy tube was passed through the pylorus and into the duodenum. About 2 cm into the duodenum, there was a lesion 8 mm in diameter that lacked a mucous membrane and appeared to be "punched out." The base of the lesion was red and showed an adherent blood clot. After the endoscopy, a portion of material obtained by biopsy was placed on urea-containing medium. Within a few minutes, the medium began to turn color, indicating a developing alkaline pH.

1. What is the patient's diagnosis?
2. What would you expect microscopic examination and culture of the gastric mucosa biopsy to show?
3. Outline the pathogenesis of this patient's disease.

4. It took a long time for doctors to accept that this condition had an infectious etiology. Why?

Discussion

1. This patient had a duodenal ulcer. The ulcer had penetrated deeply beyond the mucosa, involving small blood vessels and causing bleeding. This was apparent from the clot seen at endoscopy and the positive test for blood in the stool.

2. Microscopic examination of the biopsy showed curved bacteria, confirmed by culture to be *Helicobacter pylori*.

3. *Helicobacter pylori* cells enter the gastrointestinal tract by the fecal-oral route. In the stomach, they escape the lethal effect of gastric acid because they produce urease. Highly motile, they enter the gastric mucus and follow a gradient of acidity ranging from pH 2 in the gastric juices to pH 7.4 at the epithelial surface. Mutant strains that lack the ability to produce urease are infectious only if they are introduced directly into the mucus layer. Multiplication occurs just above the epithelial surface, but some of the bacteria attach to the epithelial cells and cause a loss of microvilli and thickening at the site of attachment. An inflammatory reaction develops beneath the affected mucosa. Two genes, *cagA* and *vacA*, correlate with virulence. The bacteria inject the protein CagA into host cells, causing the cells to elongate and spread out. CagA

also provokes a strong immune response. The protein VacA promotes the flow of urea into the stomach. Once established, *H. pylori* infections persist for years and often for a lifetime. It is not known why some people develop gastric or duodenal ulcers and others do not. Both host and bacterial factors are almost certainly involved. For example, strains of *H. pylori* isolated from peptic ulcer patients tend to be more virulent than those from patients who just have gastritis; patients with blood group O have more receptors for the bacterium and a higher incidence of peptic ulcers than do other people. Stomach acid and peptic enzymes probably play a role in ulcer formation by acting on damaged epithelium unprotected by normal mucus.

4. Claude Bernard, a scientist of Pasteur's time, put it this way: "It is that which we do know which is the greatest hindrance to our learning that which we do not know." In 1983, when Dr. Barry J. Marshall proclaimed before an international gathering of infectious disease experts that a bacterium caused stomach and duodenal ulcers, everyone "knew" it could not be true because no organism was thought to exist that could survive stomach acidity and enzymes. Indeed, almost everyone already "knew" the cause of ulcers to be psychosomatic. There is much still to be learned about the cause of ulcers, however, and Bernard's statement remains relevant.

24.3 ■ Viral Diseases of the Upper Digestive System

Learning Outcome

6. *Compare and contrast herpes simplex (cold sores) and mumps.*

In this section, we focus on two viral diseases that have dramatic signs and symptoms involving the upper digestive system. Herpes simplex (cold sores) is characterized by painful oral ulcers. Mumps causes enlarged and painful parotid glands.

Some viral diseases involve the upper digestive system but produce more dramatic symptoms elsewhere in the body. For example, measles causes an obvious skin rash and respiratory symptoms, and it causes Koplik spots in the mouth. Chickenpox causes a skin rash and oral blisters. Infectious mononucleosis causes impressively enlarged lymph nodes and spleen and may give rise to oral ulcers and bleeding gums. ◀◀ chickenpox, p. 534

▶▶ infectious mononucleosis, p. 680 ◀◀ measles, p. 537

Herpes Simplex (Cold Sores)

Herpes simplex (cold sores) is an extremely widespread disease with many symptoms. In its most common form, it begins in the mouth and throat where it causes cold sores (also called fever blisters). Involvement of the esophagus suggests AIDS or other immunodeficiency. Although the disease is usually insignificant, it can have tragic consequences in newborn infants or people with immunodeficiency.

Signs and Symptoms

The initial signs and symptoms include fever and small blisters in the mouth. The blisters break within a day or two, producing superficial ulcers that can be so painful they make it difficult to eat or drink. Although the lesions heal without treatment within about 10 days, the virus persists as a latent infection; the affected person may suffer the recurrent cold sores, also known as herpes simplex labialis (the word *labialis* indicates location of the lesions

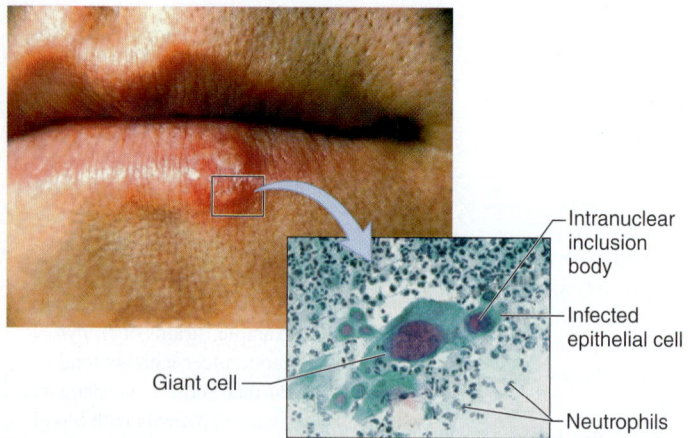

FIGURE 24.7 Herpes Simplex Labialis (Cold Sores or Fever Blisters) The photomicrograph is of stained material from a herpes simplex lesion. It shows a multinucleated giant cell and intranuclear inclusion bodies. The pink areas within the epithelial cell nuclei are inclusion bodies, the sites of viral replication.

❓ *How do the signs and symptoms of recurrent herpes simplex differ from those of the primary infection?*

on the lips) (**figure 24.7**). The signs and symptoms of recurrences are usually less severe than those of initial infection. They include tingling, itching, and burning on the lips, followed by blisters and painful ulcerations, which usually heal within 7 to 10 days.

Causative Agent

Cold sores are caused by herpes simplex viruses (HSV), enveloped viruses containing double-stranded, linear DNA. There are two types of the virus—HSV-1 and HSV-2. Most oral infections are due to HSV-1; HSV-2 usually causes genital infections. Both forms persist throughout life as latent viruses that reactivate periodically. ▶▶ **genital herpes, p. 630** ◀◀ **latent virus, p. 322**

Pathogenesis

HSV multiplies in the epithelium of the mouth or throat, and destroys the cells. Some epithelial cells fuse together, producing large, multinucleated giant cells. The nucleus of an infected cell typically contains a deeply staining area called an intranuclear inclusion body—the site of viral replication (figure 24.7). The characteristic blisters contain many infectious virions. Although an immune response develops and quickly limits the infection, some virions enter the sensory nerves in the area.

Viral DNA persists in nerve cells in a latent form—non-infectious and non-replicating. This latent virus can occasionally reactivate. The reactivated virus is carried by the nerves to skin or mucous membranes where it produces recurrent disease. Stresses that can precipitate recurrences include menstruation, sunburn, and any illness associated with fever.

Epidemiology

The initial infection with HSV typically occurs during childhood. The virus is extremely widespread and infects up to 90%

of some U.S. populations, usually resulting in mild symptoms. Approximately one in three Americans suffers from cold sores. The virus is transmitted primarily by close physical contact, but can survive for several hours on plastic and cloth, so transmission by fomites is also possible. The greatest risk of infection is from contact with lesions or saliva from patients within a few days of disease onset, because at this time large numbers of virions are present. Even the saliva of asymptomatic people can be infectious, however, posing a risk to dentists and other healthcare workers. ◀◀ **fomite, p. 440**

HSV can infect almost any body tissue. For example, herpetic whitlow, a painful finger infection, is not uncommon among nurses. Wrestlers can develop infections at almost any skin site because saliva containing HSV can contaminate wrestling mats and get rubbed into abrasions. Blindness can result if the virus is rubbed into the eye. Although uncommon, HSV is the most frequently identified cause of sporadic viral encephalitis, a serious brain disease.

Treatment and Prevention

Medications such as acyclovir and penciclovir target HSV DNA polymerase and are useful for treating severe cases and for preventing disabling recurrences. They do not affect the latent virus, however, and therefore cannot cure the infection. Sunlight exposure can trigger disease recurrence, so sunscreens are sometimes a helpful preventive. Some of the main features of herpes simplex are shown in **table 24.3**. ◀◀ **DNA polymerase, p. 166** ◀◀ **acyclovir, p. 475**

TABLE 24.3	**Herpes Simplex**
Signs and symptoms	Initial infection: fever, severe throat pain, ulcerations of the mouth and throat. Recurrences: itching, tingling, or pain usually localized to the lip, followed by blisters that break leaving a painful sore, which usually heals in 7 to 10 days.
Incubation period	2 to 20 days
Causative agent	Herpes simplex virus (HSV), usually type 1
Pathogenesis	The virus multiplies in the epithelium and destroys the cells. Blisters contain large numbers of infectious virions. An immune response limits the infection, but non-infectious HSV DNA persists in sensory nerves. This DNA becomes the source of infectious virions that are carried to the skin or mucous membranes, usually of the lip, causing recurrent sores.
Epidemiology	Widespread virus, transmitted by close physical contact. The saliva of asymptomatic individuals can be infectious.
Treatment and prevention	Acyclovir, penciclovir, and similar medications that target HSV DNA polymerase can shorten the duration of the symptoms or prevent recurrences. Sunscreens can prevent recurrences due to ultraviolet exposure.

Mumps

Mumps is an acute viral illness that often affects glands such as the parotids. Formerly common in the United States, the disease is now relatively rare because of routine childhood immunization. Outbreaks do occur, however, mostly due to waning immunity in college students and other young adults. A large outbreak in 2006 resulted in over 6,000 reported cases.

Signs and Symptoms

The onset of mumps is marked by fever, loss of appetite, and headache. These symptoms are typically followed by painful swelling of one or both parotid glands (**figure 24.8**). Spasm of the underlying muscle makes it difficult to chew or talk, perhaps giving rise to the name of the disease (*mump* means "to mumble" or "whisper"). Symptoms usually disappear in about a week.

Painful parotid swelling is characteristic of mumps, but symptoms can arise elsewhere in the body with or without parotid swelling. For example, headache and stiff neck indicate that the virus is causing meningitis—infection of the coverings of the brain. Pregnant women with mumps often miscarry. Rare but serious consequences of mumps, such as death from brain infection, are most likely to occur in older people. Sudden-onset deafness due to asymptomatic infection has also been reported. ▶▶ **meningitis, p. 643**

Mumps symptoms are generally much more severe in individuals past the onset of puberty. About one-quarter of cases in

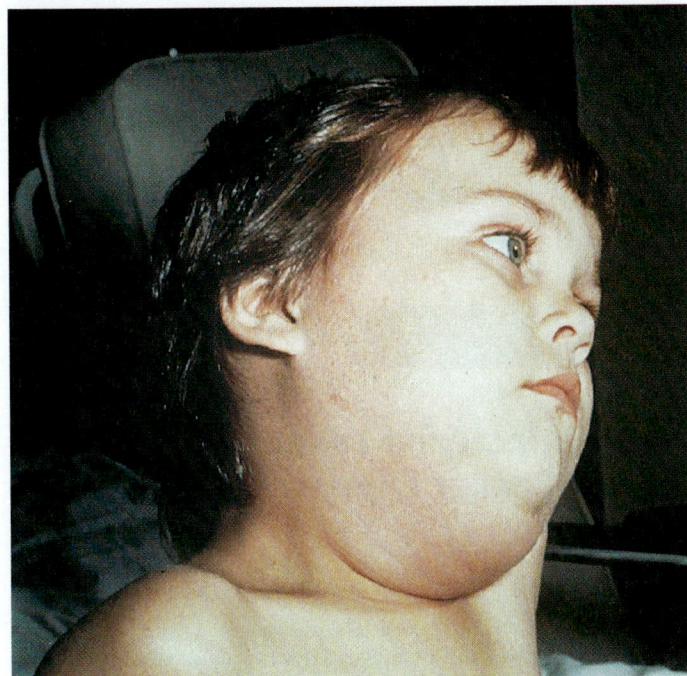

FIGURE 24.8 A Child with Mumps The swelling directly below the earlobe is due to enlargement of the parotid gland.

❓ *When might mumps result in sterility?*

post-pubertal boys and men are complicated by orchitis—a rapid, intensely painful swelling of one or both testicles to three to four times their normal size. Atrophy (shrinkage) of the involved testicles commonly develops after recovery from the illness, and in rare cases causes sterility. In women and post-pubertal girls, ovarian involvement occurs in about one of 20 cases and is manifested by pelvic pain.

Causative Agent

Mumps virus is an enveloped single-stranded RNA virus. It is a member of the paramyxovirus family, a group that includes the rubeola and respiratory syncytial viruses. Only one antigenic type of the mumps virus is known.

Pathogenesis

The mumps virus enters the body when virus-laden droplets of saliva are inhaled. It reproduces first in the upper respiratory tract, then spreads throughout the body in the bloodstream. Symptoms begin only after tissues such as the parotid glands, meninges, pancreas, ovaries, or testicles are infected. Because of this, the incubation period is relatively long, generally 15 to 21 days.

In the parotid salivary glands, the virus multiplies in the epithelium of ducts that convey saliva to the mouth. This destroys these cells, releasing enormous quantities of virus into the saliva and eliciting a strong inflammatory response, which causes the severe swelling and pain characteristic of the disease. A similar sequence of events occurs in the testicles, where the virus infects the system of tubules that convey the sperm. The swelling and pressure often impair the blood supply, which can lead to hemorrhage and death of testicular tissue. Kidney tubules are also infected, and the virus can be cultivated from the urine for 10 or more days following the onset of illness.

Epidemiology

Humans are the only natural host of mumps virus. There is only one antigenic type, so infection confers lifelong immunity. Individuals sometimes claim to have had mumps more than once, probably because other infectious and non-infectious diseases can cause parotid swelling.

The mumps virus is often spread by individuals who have asymptomatic infections but secrete the virus in their saliva and continue to mingle with other people. In symptomatic patients, the virus can be present in saliva from almost a week before symptoms appear to 2 weeks afterward. Peak infectivity however, is from 1 to 2 days before parotid swelling until the gland begins to return to normal size.

Treatment and Prevention

There is no effective treatment for mumps. However, an effective attenuated vaccine has been available in the United States since 1967. It is part of the measles, mumps, rubella, and varicella vaccine (MMRV). **Figure 24.9** shows how the incidence of mumps has generally declined, although it increased in the 1980s because of cuts to funding for vaccinations. Mumps is a

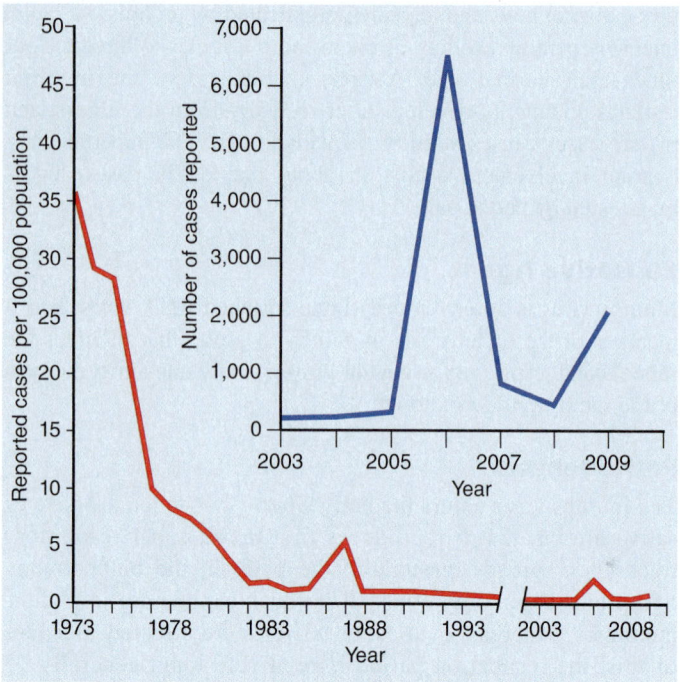

FIGURE 24.9 Reported Cases of Mumps per 100,000 Population, United States, 1973–2009 Mumps vaccine was licensed in 1967. Two doses of MMR vaccine were recommended in 1989 following an outbreak of over 20,000 cases in 1986–1987. An epidemic among college students resulted in more than 6,000 cases of mumps in 2006. Another outbreak of about 2,000 cases occurred in 2009–2010, prompting one New York county to begin offering a third dose of MMR vaccine.

❓ What events or factors might have contributed to the 2006 mumps epidemic among college students?

good candidate for eradication because it infects only humans, latent infections do not occur, and there is only one serotype. The main features of mumps are presented in **table 24.4.**

◄◄ eradication, p. 447

TABLE 24.4	Mumps
Signs and symptoms	Fever, headache, loss of appetite, followed by painful swelling of one or both parotid glands. Meningitis can occur. Painful enlargement of the testicles in men, pelvic pain in women.
Incubation period	Generally 15 to 21 days
Causative agent	Mumps virus, a single-stranded RNA virus of the paramyxovirus family
Pathogenesis	The virus initially replicates in the upper respiratory tract, then spreads throughout the body in the bloodstream. In the parotid salivary glands, the virus multiplies in the cells that line the ducts. The inflammatory response to cell destruction causes the swelling and pain.
Epidemiology	Humans are the only source of the virus. Infections are often asymptomatic, so the disease can be spread unknowingly.
Treatment and prevention	No antiviral therapy is available. An effective attenuated vaccine is available.

MicroAssessment 24.3

Herpes simplex (cold sores) is characterized by acute infection followed by lifelong latency and the possibility of recurrent disease. Infectious virus is often present in saliva in the absence of symptoms. Mumps virus infections characteristically cause enlargement of the parotid glands, but they can involve the brain, testicles, and ovaries, and cause miscarriages.

   7. *In what type of cell does HSV-1 persist?*

   8. *Why is mumps virus a good candidate for worldwide eradication?*

   9. *Why would acyclovir not cure HSV infections even though it prevents recurrent cold sores?* ➕

LOWER DIGESTIVE SYSTEM INFECTIONS

One out of five children in developing countries dies of diarrhea before the age of 5. Most of the 5 million individuals who die of diarrhea each year are infants, but no age group is spared. Fatal cases are less common in the United States, but millions of diarrhea cases still occur each year.

Microbes infecting the digestive tract can cause intestinal symptoms, and so can foodborne microbial toxins (foodborne intoxication). This chapter focuses on intestinal infections, and chapter 31 gives examples of foodborne intoxications. Tapeworm and roundworm infections are discussed in chapter 12. Information about other helminth infections can be found at the text website (**www.mhhe.com/nester7**). ▶▶ foodborne intoxication, p. 756

◄◄ helminth infections, p. 295

24.4 ■ Bacterial Diseases of the Lower Digestive System

Learning Outcomes

   7. *Describe the general characteristics of bacterial intestinal diseases.*

   8. *Compare and contrast cholera, shigellosis, the various types of* E. coli *gastroenteritis, campylobacteriosis, and* Clostridium difficile–*associated disease.*

Diarrheal illness is a common result of bacterial infection of the intestinal tract. Bacteria can also use the intestines as an entry to the rest of the body, thereby causing other types of illness.

General Characteristics

Although many different bacteria can infect the intestinal tract, the diseases they cause share some general characteristics.

Signs and Symptoms

The all-too-familiar signs and symptoms of intestinal diseases include diarrhea, loss of appetite, nausea and vomiting, and sometimes fever. The incubation period is typically a day or two, but varies according to the dose consumed. Some physicians often refer to diarrheal disease as **gastroenteritis** (*gastro-*, "stomach," *entero-*, "intestine," *-itis*, "inflammation"), whereas others prefer the term "stomach flu." (Note "stomach flu" has no relationship to "the flu," or influenza.)

Diarrhea can be copious and watery ("the runs") when infection involves the small intestine. Large intestine invasion causes smaller amounts of diarrhea ("the squirts") that contains mucus, pus, and sometimes blood. The name **dysentery** is given to diarrheal illnesses when pus and blood are present in the feces. A few bacterial pathogens first establish an intestinal infection and then spread systemically. This causes **enteric fever,** characterized by systemic signs such as shock. ◄◄ shock, p. 394

Causative Agent

Members of the family *Enterobacteriaceae* are among the most common causes of bacterial diarrhea. These include species of *Shigella* and *Salmonella,* and some strains of *E. coli.* Other causes are *Vibrio cholerae* and *Campylobacter jejuni.* In individuals with predisposing conditions, *Clostridium difficile* causes diarrhea.

Pathogenesis

The pathogenesis of bacterial intestinal disease involves a number of different mechanisms. Various strains within the same species can differ in the way they cause disease, and the same strain can use more than one mechanism. Moreover, many of the responsible genes are on plasmids, phages, or other mobile genetic elements, which can be transferred from one species to another by horizontal gene transfer. ◄◄ mobile genetic elements, p. 208

Attachment to the intestinal surface is often a prerequisite for infection, and this typically involves adhesins on pili. Additional common mechanisms of pathogenesis include the following (**figure 24.10**): ◄◄ adhesins, p. 386

- **Toxin production.** Toxins involved in intestinal infections fall into two groups: **enterotoxins,** which cause water and electrolytes to flow from intestinal cells; and **cytotoxins,** which cause cell death. Some types of cytotoxins produced in the intestine can be absorbed into the bloodstream, resulting in systemic effects. ◄◄ exotoxins, p. 391

- **Alterations in intestinal epithelial cells.** Some pathogens alter the characteristics of intestinal epithelial cells, using type III secretion systems to deliver effector proteins to those cells (see figure 16.5). For example, certain strains of *E. coli* inject protein molecules that assemble in the host plasma membrane. The bacteria then use these molecules as receptors for closer attachment. After that, the bacterial cells inject proteins that rearrange actin filaments, resulting in replacement

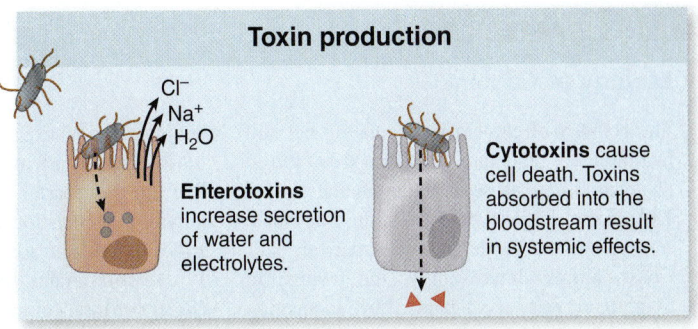

Toxin production

Enterotoxins increase secretion of water and electrolytes.

Cl^-
Na^+
H_2O

Cytotoxins cause cell death. Toxins absorbed into the bloodstream result in systemic effects.

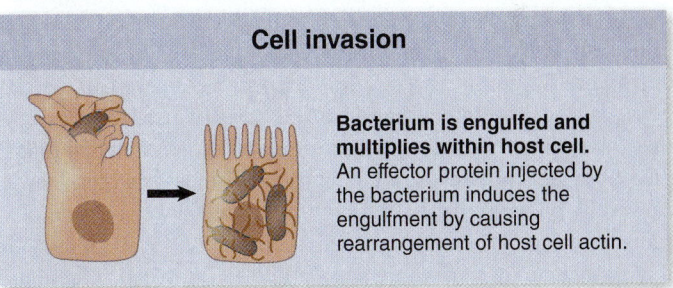

Alterations in the host cells

Pedestal

Inject effector proteins

Attachment and effacing (A/E) lesions formed after bacterium injects various effector proteins. One protein functions as a receptor for the bacterium. Another induces rearrangement of actin filaments, resulting in the formation of a pedestal under the bacterium.

Cell invasion

Bacterium is engulfed and multiplies within host cell. An effector protein injected by the bacterium induces the engulfment by causing rearrangement of host cell actin.

FIGURE 24.10 Common Mechanisms of Pathogenesis of Intestinal Pathogens Invasion or cell damage elicits a strong inflammatory response that also contributes to intestinal pathology.

❓ *Why is adhesion generally a prerequisite for pathogenicity?*

of microvilli on the intestinal cell surface with a thick structure—a "pedestal"—under the bacterium. The characteristic damage caused by these bacteria is called **attaching and effacing (A/E) lesions.** ◄◄ type III secretion systems, p. 386 ◄◄ effector proteins, p. 387

- **Cell invasion.** Some pathogens use type III secretion systems to deliver effector molecules that induce the intestinal epithelial cells to engulf the bacteria (see figure 16.6). Once inside, they can multiply.

Severe watery diarrhea occurs as a result of rapid loss of water and electrolytes from intestinal cells. This causes severe dehydration, as the intestinal cells draw fluid from the bloodstream. Because of the reduced blood volume, the flow may not be sufficient to keep vital organs working properly, a potentially fatal situation.

Invasion or cell damage elicits a strong inflammatory response. This primarily occurs in the large intestine and results in the pus and blood in the feces that characterize dysentery. Proteins injected by type III secretion systems can also elicit an inflammatory response.

PERSPECTIVE 24.1

Ecology of Cholera

The ecology of cholera is fascinating but still incompletely understood. The O139 *Vibrio cholerae* strain arose in India in the fall of 1992 from a serotype O1 strain, the same serotype as the pandemic and American Gulf Coast strains. Interestingly, the change in O antigen was accompanied by acquisition of a capsule—these changes probably resulting from genes acquired by transduction or conjugation. Various strains of *V. cholerae*, most of which are not pathogenic for humans, live in the coastal seas around the world, largely in association with zooplankton. These zooplankton can sometimes also harbor the *V. cholerae* strains responsible for pandemic cholera. Moreover, the zooplankton feed on phytoplankton and therefore increase markedly in number when warm seas and abundant nutrients cause explosive growth of phytoplankton. These findings could explain (1) how new pandemic strains could arise through genetic interchange, (2) an association between cholera and climatic changes such as El Niño, and (3) the onset and rapid spread of epidemic cholera by ocean currents along coastal areas. ◀◀ **zooplankton, p. 290**

Epidemiology

The epidemiology of intestinal diseases involves transmission by the fecal-oral route. A common way this occurs is by ingesting food or drinking water contaminated with animal or human feces. Sexual practices that lead to oral-anal contact can also transmit intestinal pathogens.

Intestinal pathogens sensitive to acid generally have a high infectious dose, because most of the ingested microbes are destroyed by stomach acid. In contrast, acid-resistant pathogens have a low infectious dose. Diseases caused by bacteria that have a low infectious dose are easily transmitted through direct contact as well as in contaminated foods or water. Those caused by bacteria that have a high infectious dose are generally transmitted only in contaminated foods or water. People with low gastric acidity are more susceptible to intestinal infections. ◀◀ **infectious dose, p. 383**

Treatment and Prevention

Oral rehydration therapy (ORT)—giving appropriate fluids by mouth—can be used to counteract the loss of fluid and electrolytes (salt) from diarrhea. If the patient drinks water alone, however, the intestinal tract cannot absorb enough to keep pace with the amount lost. Fortunately, researchers discovered that glucose increases the absorptive capacity of the intestine. This breakthrough led to the development of what is called oral rehydration salts (ORS) solution, a highly effective lifesaver in severe diarrheas regardless of cause. ORS is a mixture of glucose and various salts (sodium chloride, potassium chloride, and trisodium citrate) that is commercially available as pre-measured packets to be dissolved in clean water. The World Health Organization has also developed a list of recommended home fluids (RHF) that can be used to prevent dehydration. If oral rehydration cannot be tolerated—for example, if the patient is vomiting—then intravenous hydration may be required.

Antibacterial medications are not helpful in most intestinal infections, and often prolong the illness because they suppress the normal microbiota. On the other hand, these drugs can be lifesaving in cases where the bacterium invades beyond the intestine. For children in developing countries, zinc supplements decrease the length and severity of diarrheal disease.

Sewage treatment, handwashing, and chlorinating drinking water are important measures to control diarrheal diseases. In the United States, surveillance using **PulseNet**—a DNA subtyping resource—helps track illness caused by specific intestinal pathogens. This helps public health agencies detect foodborne outbreaks so that intervention strategies can be implemented. ◀◀ **PulseNet, p. 248**

Only a few vaccines are available to prevent diarrheal diseases, and these are pathogen-specific. Most are not very effective because they fail to elicit a strong enough secretory IgA response on the intestinal mucosa to reliably limit pathogen colonization. ◀◀ **secretory IgA, p. 362**

Cholera

Cholera causes potentially fatal diarrhea. Seven cholera pandemics have occurred since the early 1800s, with the seventh one beginning in 1961 in Indonesia and spreading to South Asia, the Middle East, and parts of Europe and Africa. South America had remained cholera-free for 100 years until January 1991, when the disease abruptly appeared in Peru (see **Perspective 24.1**). The source of the bacterium was likely a freighter that discharged bilgewater into a Lima harbor. Lima's water supply was not chlorinated and quickly became contaminated. The disease then spread rapidly, so that in 2 years more than 700,000 cases and 6,323 deaths had been reported in South and Central America. The introduction of cholera into Haiti in 2010, the first cholera epidemic there in over a century, poses a new threat. By mid-2011 there were over 300,000 cases and 5,000 deaths reported. ◀◀ **pandemic disease, p. 438**

Signs and Symptoms

Cholera is a classic example of severe watery diarrheal disease. The diarrheal fluid can amount to 20 liters a day and is described as "rice water stool" because of its appearance. Vomiting also occurs in most people at the onset of the disease, and many people suffer muscle cramps caused by loss of fluid and electrolytes. Severe dehydration can lead to multiple organ failure and death.

Causative Agent

The causative agent of cholera is *Vibrio cholerae*, a curved, Gram-negative rod. There are several different serotypes, grouped according to differences in their O antigen; O1 is the serotype currently circulating. *V. cholerae* is halotolerant and can grow in alkaline conditions, two characteristics used to design appropriate selective media. ◀◀ O antigen, pp. 60, 247 ◀◀ selective media, p. 95

Pathogenesis

Vibrio cholerae cells are killed by acid, so large numbers must be ingested before enough survive passage through the stomach to establish infection. Once in the small intestine, surviving bacteria adhere to the small intestinal epithelium using pili and other surface proteins. The bacteria multiply on the epithelial cells but do not visibly damage them. However, the bacteria produce cholera toxin, an enterotoxin (**figure 24.11**). This toxin activates ion transport channels in the epithelial cell membrane, causing chloride and other electrolytes to exit the cell. Water follows the electrolytes, resulting in an outpouring of fluid and salts into the intestinal lumen. Although the toxin does not affect the large intestine, the volume of fluid is too much to be absorbed, causing diarrhea that results in severe dehydration.

Cholera toxin is an A-B toxin. The B (binding) portion attaches irreversibly to specific receptors on the microvilli of the epithelial cells, allowing the A (active) portion to enter the cells. It causes severe diarrhea by activating a G protein, which normally functions as a sophisticated on/off switch that controls activities inside the cell in response to external signals. Cholera toxin chemically modifies the G protein, locking it in the "on" position. This, in turn, results in nonstop activity of an enzyme called adenylate cyclase, which converts ATP to cAMP. The net effect is high levels of cAMP at the cell membrane, a condition that causes the cell to continually secrete chloride. The normal turnover of intestinal cells eventually removes the toxin. ◀◀ A-B toxins, p. 392

Cholera toxin is encoded by a bacteriophage that infects *V. cholerae,* an example of lysogenic conversion. Synthesis of both the toxin and pili is regulated by the same bacterial gene. This means that the two factors required for disease production are synthesized at the same time. ◀◀ lysogenic conversion, p. 313

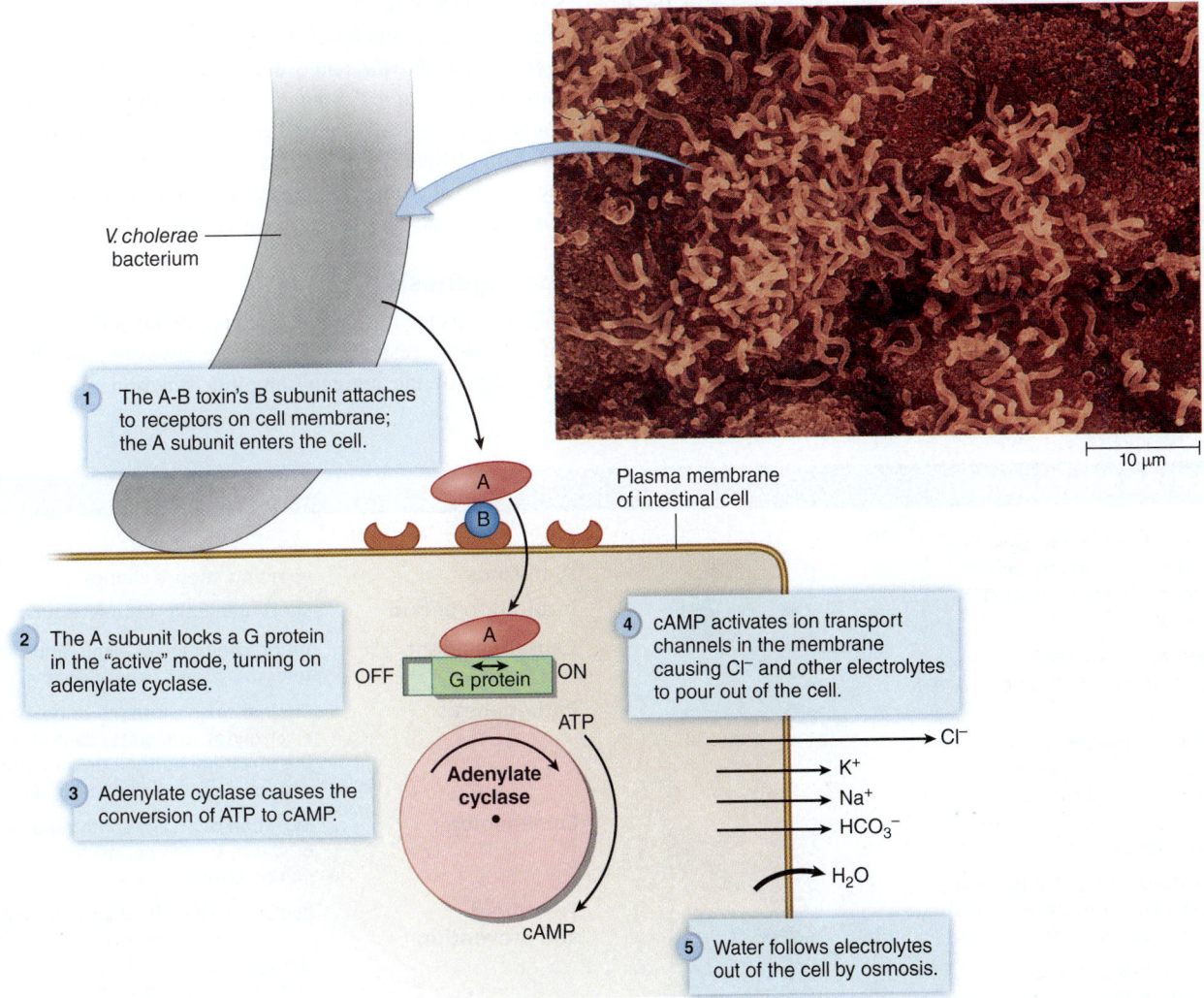

1. The A-B toxin's B subunit attaches to receptors on cell membrane; the A subunit enters the cell.

V. cholerae bacterium

Plasma membrane of intestinal cell

10 μm

2. The A subunit locks a G protein in the "active" mode, turning on adenylate cyclase.

OFF G protein ON

3. Adenylate cyclase causes the conversion of ATP to cAMP.

Adenylate cyclase

ATP

cAMP

4. cAMP activates ion transport channels in the membrane causing Cl⁻ and other electrolytes to pour out of the cell.

Cl⁻
K⁺
Na⁺
HCO₃⁻
H₂O

5. Water follows electrolytes out of the cell by osmosis.

FIGURE 24.11 *Vibrio cholerae* **Pathogenesis** *Vibrio cholerae* attaches to the small intestine using pili and begins to produce an A-B toxin.

❓ *How would vaccination against the B subunit of cholera toxin help prevent disease?*

Epidemiology

Fecally contaminated water is the most common source of cholera infection, although foods such as contaminated crab, oysters, and vegetables have also been implicated in outbreaks. A person with cholera can discharge a million or more *V. cholerae* cells in each milliliter of feces.

Although cholera is relatively common worldwide, the pandemic *V. cholerae* strain has caused relatively few cases in the United States since the early 1900s. The occasional cases along the Gulf of Mexico—traced to eating coastal marsh crabs and oysters—involved a different strain.

Ominously, a new *V. cholerae* strain appeared in India in 1992 and then spread rapidly across South Asia. It belongs to serogroup group O139 and infects even those people with immunity to the seventh pandemic strain. This new strain and its rapid spread suggested it could initiate another pandemic, but fortunately it declined in incidence and remained endemic, causing only resurgent localized epidemics.

Treatment and Prevention

Treatment of cholera depends on the rapid replacement of fluid and electrolytes before irreversible damage to vital organs can occur. If patients are vomiting or are too weak to drink oral rehydration solutions, intravenous lines are started. The prompt administration of intravenous or oral rehydration therapy decreases the mortality of cholera from over 30% to less than 1%.

Cholera control depends largely on adequate sanitation and safe, clean water supplies. Travelers to areas where cholera is occurring are advised to cook food immediately before eating it. Crabs should be cooked for no less than 10 minutes. No fruit should be eaten unless peeled personally by the traveler, and ice should be avoided unless known to be made from boiled water. Orally administered vaccines are available in a number of countries other than the United States. One vaccine consists of a live genetically altered *V. cholerae* strain, and the other is killed *V. cholerae* in combination with the purified recombinant B subunit of cholera toxoid. The main features of cholera are summarized in **table 24.5**.

Shigellosis

Shigellosis is found all over the world, most commonly in areas lacking adequate sewage treatment. Reported cases in the United States average about 21,500 per year, but the true prevalence is much likely higher, because diarrhea often goes unreported.

Signs and Symptoms

Shigellosis classically involves dysentery, but some *Shigella* species cause watery diarrhea. Other signs and symptoms include headache, vomiting, fever, stiff neck, convulsions, and joint pain. The disease is often fatal for infants in developing countries.
◀◀ dysentery, p. 585

Causative Agent

Shigellosis is caused by the four species of *Shigella*—*S. dysenteriae*, *S. flexneri*, *S. boydii*, and *S. sonnei*. Of these, *S. dysenteriae* is the most virulent, and *S. sonnei* the least. Shigellosis in developing countries is typically caused by *S. dysenteriae* and *S. flexneri*. In the United States, however, *S. sonnei* causes over two-thirds of the cases. Like other members of the family *Enterobacteriaceae*, *Shigella* species are Gram-negative rods.

Pathogenesis

Shigella species invade intestinal epithelial cells, causing a strong inflammatory response. To initiate invasion, the bacteria take advantage of the antigen sampling function of M cells, which

---

TABLE 24.5 | Cholera

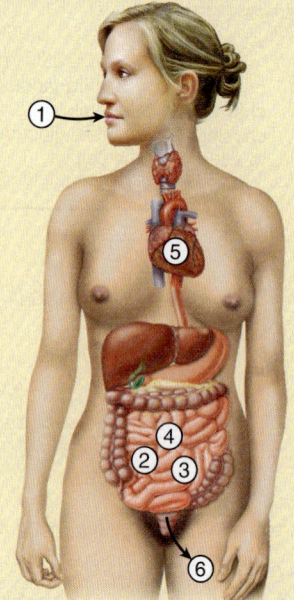

1. *Vibrio cholerae*, the causative bacterium, enters the mouth with fecally contaminated food or drink.

2. The bacteria attach to epithelial cells of the small intestine.

3. Cholera toxin enters the cells and causes them to continuously secrete chloride ions. Other electrolytes and water follow.

4. The outpouring of water and electrolytes into the intestinal lumen causes watery diarrhea.

5. Fluid loss causes severe dehydration, resulting in shock and death unless the fluid can be replaced.

6. The bacteria exit the body with feces.

Signs and symptoms	Abrupt onset of massive diarrhea, vomiting, muscle cramps
Incubation period	Short, generally 12 to 48 hours
Causative agent	*Vibrio cholerae*, an alkali- and salt-tolerant, curved Gram-negative rod
Pathogenesis	Cholera toxin causes chloride, other electrolytes, and water to pour out of the intestinal epithelium and into the lumen; leads to dehydration and shock.
Epidemiology	Ingestion of fecally contaminated food or water; sometimes associated with marine crustaceans
Treatment and prevention	Replacement of fluid and electrolytes using intravenous or oral rehydration therapy. Prevented by purifying water and handwashing; vaccination is available in some countries.

normally transfer microbes to macrophages in Peyer's patches (**figure 24.12**). Once *Shigella* cells enter the macrophages, they escape from the phagosome and multiply in the macrophages' cytoplasm. These hidden bacteria are released when the infected macrophages die. ◀◀ **M cells, p. 358**

With access to the bases of the intestinal epithelial cells, *Shigella* cells attach to specific receptors and induce the epithelial cells to take them in (figure 24.12). Once inside, the *Shigella* cells escape again and multiply in the cytoplasm of the epithelial cells.

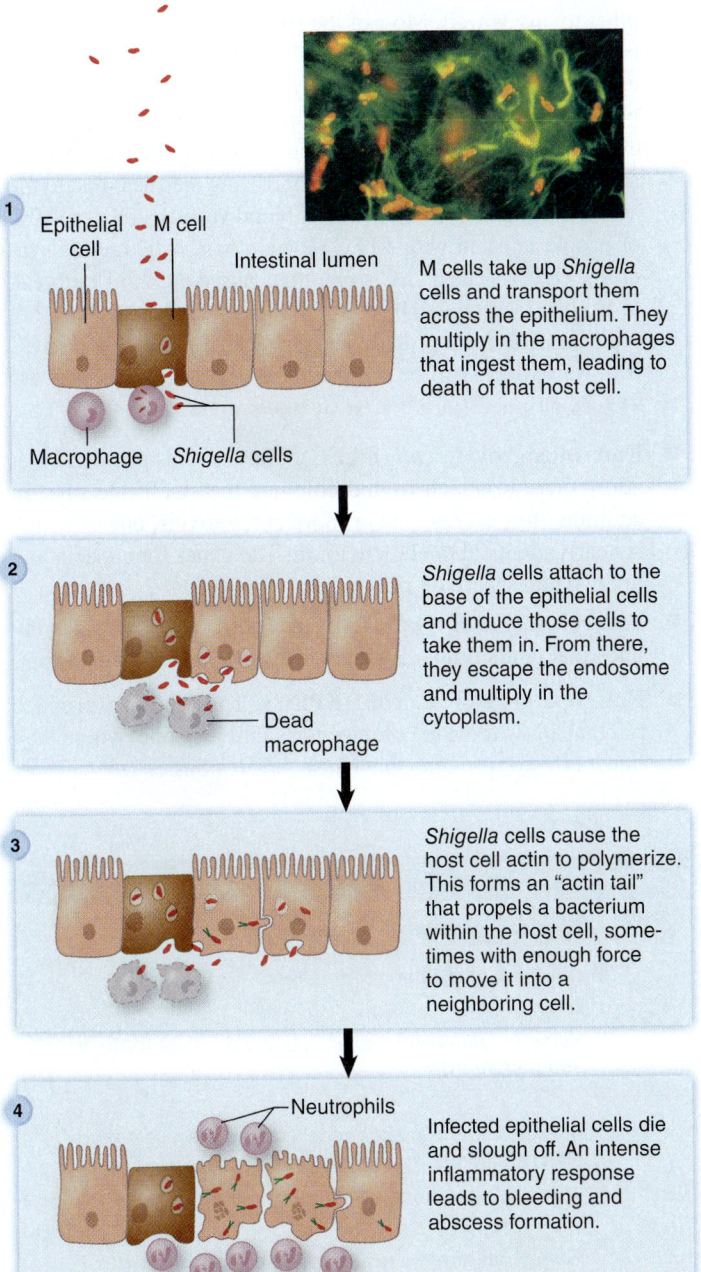

1 Epithelial cell — M cell
Intestinal lumen
M cells take up *Shigella* cells and transport them across the epithelium. They multiply in the macrophages that ingest them, leading to death of that host cell.
Macrophage   *Shigella* cells

2 *Shigella* cells attach to the base of the epithelial cells and induce those cells to take them in. From there, they escape the endosome and multiply in the cytoplasm.
Dead macrophage

3 *Shigella* cells cause the host cell actin to polymerize. This forms an "actin tail" that propels a bacterium within the host cell, sometimes with enough force to move it into a neighboring cell.

4 Neutrophils
Infected epithelial cells die and slough off. An intense inflammatory response leads to bleeding and abscess formation.

FIGURE 24.12 *Shigella* Pathogenesis The photomicrograph shows the actin tails (green) that form on intracellular *Shigella* cells (orange) and rapidly push these non-motile bacteria from cell to cell.

❓ *Why does shigellosis cause dysentery, whereas cholera causes watery diarrhea?*

Although nonmotile, the bacterial cells produce a protein that polymerizes host cell actin. This "actin tail" propels the bacterium within the cell, sometimes with enough force to move it into a neighboring cell. The overall result of invasion and spread is the death and sloughing of patches of epithelium. The bare areas become intensely inflamed, covered with pus and blood, which accounts for the signs and symptoms of dysentery.

Some strains of *Shigella dysenteriae* produce a potent cytotoxin known as Shiga toxin, a chromosomally encoded A-B toxin. The toxin enters the bloodstream and then the B portion binds to endothelial cells that line the small blood vessels, particularly in the kidneys. This allows the A subunit to then enter the cells, where it reacts with ribosomes, halting protein synthesis, which leads to cell death. Shiga toxin is important because it is responsible for **hemolytic uremic syndrome (HUS),** an often fatal condition that can follow *Shigella dysenteriae* dysentery. In HUS, red blood cells break up in the tiny blood vessels, resulting in anemia and kidney failure. Symptoms of HUS sometimes include paralysis or other signs of nervous system injury. Shiga toxin is also produced by strains of *Escherichia coli* that cause HUS.

Epidemiology

Shigellosis is almost exclusively a disease of humans, transmitted by the fecal-oral route. Unlike *V. cholera*, *Shigella* cells are not easily killed by stomach acid, so the infectious dose is small. As few as 10 cells can start an infection. Shigellosis spreads readily in overcrowded populations with poor sanitation, such as in refugee camps and daycare centers. People who engage in anal intercourse are also prone to contracting the disease. Fecally contaminated food and water have caused numerous shigellosis outbreaks.

Treatment and Prevention

Antimicrobial medications such as ampicillin and co-trimoxazole (a combination of trimethoprim and a sulfonamide) are useful against susceptible *Shigella* strains because they shorten the duration of symptoms and the time during which the pathogens are discharged in the feces. Many strains, however, are resistant to these common medications. Unfortunately, these strains often have R plasmids that encode resistance to several antibacterial drugs. ◀◀ **R plasmids, p. 209**

The spread of *Shigella* is controlled by using sanitary measures to avoid fecal contamination of food and water supplies. Two specimens of feces, collected at least 48 hours after stopping antimicrobial medicines, must be negative for *Shigella* before a person is allowed to return to a daycare center or food-handling job. Shigellosis cases can be tracked through PulseNet, making it easier for public health agencies to detect outbreaks. **Table 24.6** describes the main features of shigellosis. ◀◀ **PulseNet, p. 248**

Escherichia coli Gastroenteritis

Escherichia coli strains are almost universal residents of the intestinal tracts of humans and a number of other animals. Although most strains are harmless, certain ones produce specific virulence factors that allow them to cause intestinal disease. Other strains—with different virulence factors—cause urinary tract infections, septicemia, and meningitis. ◀◀ **virulence factors, p. 383**

TABLE 24.6 Shigellosis

Signs and symptoms	Fever, dysentery, vomiting, headache, stiff neck, convulsions, and joint pain
Incubation period	3 to 4 days
Causative agent	Four species of *Shigella*, Gram-negative, non-motile members of the *Enterobacteriaceae*
Pathogenesis	Pass through intestinal barrier in M cells; induce uptake from the base of epithelial cells; induce "actin tails" to spread. Invasion and spread leads to death and sloughing of epithelium; inflammation and ulcerations of intestinal lining. Some strains make Shiga toxin.
Epidemiology	Fecal-oral transmission; low infectious dose; humans generally the only source
Treatment and prevention	Medications such as ampicillin and co-trimoxazole shorten duration of symptoms and pathogen excretion; many strains have R plasmids. Spread is controlled by sanitary measures.

Signs and Symptoms

The signs, symptoms, and severity of *E. coli* gastroenteritis depend on the infecting strain. Some strains cause watery diarrhea and others cause dysentery. One group can cause hemolytic uremic syndrome (HUS), marked by anemia due to lysis of red blood cells and kidney failure. ◄◄ dysentery, p. 585 ◄◄ HUS, p. 589

Causative Agent

Escherichia coli is a Gram-negative rod, closely related to *Shigella* species. Most strains ferment lactose, an easily observable trait that distinguishes them from *Shigella* species.

Pathogenesis

Escherichia coli strains that cause intestinal disease can be grouped into six pathovars (pathogenic varieties), based on their array of virulence factors (table 24.7):

- **Shiga toxin–producing *E. coli* (STEC).** These strains, also referred to as EHEC or enterohemorrhagic *E. coli*, produce Shiga toxins, a family of functionally identical toxins that includes the toxin of *Shigella dysenteriae*. In STEC strains, the genes for Shiga toxins are encoded by various related prophages, an example of lysogenic conversion. Some STEC strains produce multiple types of Shiga toxins, and even make other toxins as well. Most of the strains identified in outbreaks belong to a single serotype, O157:H7, but there are important exceptions. STEC strains typically colonize the large intestine, where they inject effector proteins that cause attaching and effacing (A/E) lesions. The intestinal damage results in diarrhea, which becomes bloody (hemorrhagic diarrhea) due to the action of Shiga toxins on the local blood vessels. About 5–10% of people infected with STEC develop hemolytic uremic syndrome (HUS). The STEC strain that caused the 2011 outbreak in Europe (serotype O104:H4) is unusual because it does not produce A/E lesions and has characteristics of the EAEC pathovar, which will be discussed shortly. ◄◄ Shiga toxin, p. 589
◄◄ lysogenic conversion, p. 313 ◄◄ A/E lesions, p. 585

- **Enterotoxigenic *E. coli* (ETEC).** These strains make pili that allow them to attach to and colonize the small intestine. In addition, they secrete one or more enterotoxins, one of which is nearly identical to cholera toxin. The genes for adhesin and toxin synthesis are on plasmids.

- **Enteroinvasive *E. coli* (EIEC).** These strains invade the intestinal epithelium, causing a disease similar to shigellosis.

- **Enteropathogenic *E. coli* (EPEC).** These strains produce pili that allow them to colonize the small intestine, where they inject effector proteins that cause A/E lesions.

TABLE 24.7 Characteristics of Diarrhea-Causing *Escherichia coli*

Designation	Characteristic Features	Clinical Picture
Diffusely adhering *E. coli* (DAEC)	Grows as a diffuse layer in a thick mucus-associated biofilm on the intestinal epithelium; produces toxins	Diarrhea, particularly in children
Enteroaggregative *E. coli* (EAEC)	Grows in bricklike aggregations in a thick mucus-associated biofilm on the intestinal epithelium; produces toxins	Variable symptoms; nausea, watery diarrhea (both acute and persistent)
Enterotoxigenic *E. coli* (ETEC)	Colonizes the small intestine and produces toxins, one nearly identical to cholera toxin	Nausea, vomiting, abdominal cramps, massive watery diarrhea leading to dehydration
Enteroinvasive *E. coli* (EIEC)	Invades the intestinal epithelium, causing a disease very similar to shigellosis	Fever, cramps, blood and pus in the feces
Enteropathogenic *E. coli* (EPEC)	Colonizes the small intestine; induces changes in actin filaments, causing the microvilli to be replaced by pedestals under the bacterial cells; A/E lesions	Fever, vomiting, watery diarrhea containing mucus
Shiga toxin–producing *E. coli* (STEC)	Same as above, except it colonizes the large intestine rather than the small intestine and releases Shiga toxin	Fever, abdominal cramps, bloody diarrhea without pus; some patients develop hemolytic uremic syndrome; most strains identified in outbreaks are serotype O157:H7

- **Enteroaggregative *E. coli* (EAEC).** These strains produce pili that allow them to adhere to the intestinal epithelium. There, they grow in characteristic aggregations ("bricklike") in a thick mucus-associated biofilm. In addition, they produce enterotoxins and cytotoxins, damaging the intestinal cells and evoking an inflammatory response.

- **Diffusely adhering *E. coli* (DAEC).** These strains are similar to EAEC, but rather than forming aggregations, they grow as a diffuse layer.

Epidemiology

Just as the symptoms and pathogenesis of *E. coli* gastroenteritis depend on the infecting strain, so does the epidemiology.

STEC strains can be foodborne, and epidemics have involved ground beef, unpasteurized milk, apple juice, bean sprouts, and green leafy vegetables. The initial source of infection is often untreated cow manure, reflecting the fact that cattle are an important reservoir. The infectious dose of STEC strains is typically very low, so in addition to foodborne transmission, the bacteria are easily spread by direct contact. ◀◀ reservoir of infection, p. 438

ETEC strains commonly cause diarrhea in infants in developing countries as well as travelers visiting those regions. Their infectious dose is relatively high, so they typically do not spread by direct contact. ETEC are species-specific, and those that infect animals are responsible for significant mortality in young livestock.

EIEC strains primarily cause disease in young children in developing countries. Relative to some *E. coli* strains, they are not easily spread by direct contact, indicating that the infectious dose is not very low. Humans appear to be the only source of infection.

EPEC strains are an important cause of chronic diarrhea in infants. The infection is uncommon in breast-fed infants, probably because of protective antibodies in breast milk. Outbreaks have occurred in hospital nurseries, so the infectious dose for infants is thought to be very low. A variety of animals have been shown to harbor EPEC, but their importance as a source of infection is not known.

EAEC strains cause diarrhea in children, travelers, and AIDS patients. A variety of animals harbor EAEC strains, but it is not clear that the animal strains cause human disease.

DEAC strains have caused diarrhea outbreaks in children, but relatively little is known about their epidemiology.

Treatment and Prevention

Treatment of *E. coli* gastroenteritis varies according to the symptoms and infecting strain, but as with any disease, fluid lost from vomiting and diarrhea should be replaced. Antibacterial medications are no longer routinely used because most cases are self-limiting. Patients with STEC infections who were treated with antibiotics showed an overall worse outcome.

Measures to prevent *E. coli* gastroenteritis include handwashing, pasteurization of drinks, and thorough cooking of food. PulseNet helps track infections caused by *E. coli* O157:H7. Traveler's diarrhea can usually be prevented with bismuth preparations (such as Pepto-Bismol). Unfortunately, the widespread use of antibiotics to prevent diarrhea has promoted development of resistant strains. Some features of *E. coli* gastroenteritis are summarized in **table 24.8.**

TABLE 24.8 | *Escherichia coli* **Gastroenteritis**

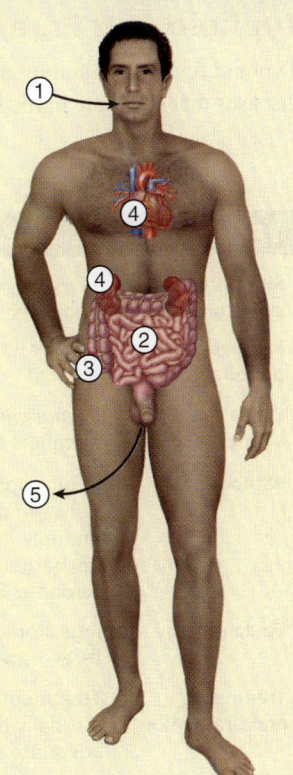

1. Pathogenic strain of *E. coli* enters by the fecal-oral route, either directly from an infected person or with contaminated food or beverage.

2. Most strains colonize the small intestine and produce watery diarrhea.

3. Others invade the large intestine and cause dysentery.

4. Some strains produce Shiga toxin, which is absorbed by the bloodstream and causes hemolytic-uremic syndrome with damage to red blood cells and the kidneys.

5. The bacteria exit the body with feces.

Signs and symptoms	Vomiting and diarrhea; sometimes dysentery
Incubation period	2 hours to 6 days
Causative agent	*Escherichia coli*, certain strains only
Pathogenesis	Various mechanisms; attachment to small intestinal cells allows colonization; some strains produce one or more enterotoxins; some strains invade large intestinal epithelium; others alter actin polymerization to cause attachment and effacing lesions, and may produce Shiga toxin
Epidemiology	Common in travelers; can be foodborne or waterborne; fecal-oral route transmission; some strains have an animal source.
Treatment and prevention	Replacement of fluids and electrolytes. Prevention methods include handwashing; pasteurization of drinks, thorough cooking of meats. Bismuth compounds help prevent traveler's diarrhea.

Salmonella Gastroenteritis

Salmonella gastroenteritis is caused by *Salmonella enterica* and can be acquired from many animal sources. An estimated 1.4 million cases occur in the United States each year, with 400 deaths. Large outbreaks are usually due to commercially distributed foods contaminated by animal feces.

Signs and Symptoms

Signs and symptoms of *Salmonella* gastroenteritis include diarrhea (sometimes bloody), abdominal cramps, nausea, vomiting, headache, and fever. The disease is often short-lived and mild, but that varies depending on the virulence of the infecting strain and the number of cells ingested.

Causative Agent

Salmonella enterica is a Gram-negative rod and, like *Shigella* and *E. coli,* a member of the *Enterobacteriaceae*. The various *Salmonella* strains are subdivided into more than 2,400 serotypes based on differences in their somatic (O), flagellar (H), and capsular (K) antigens (see figure 10.9). Each serotype was once considered a separate species and given a distinct name. However, DNA sequence data indicates there are only two species—*S. enterica* and *S. bongori,* the latter only rarely isolated from humans. ◀◀ serotypes, p. 247

The serotype of a *Salmonella* strain is significant from both an epidemiological and a disease standpoint. For instance, certain serotypes cause enteric fever (discussed next) instead of gastroenteritis. Because of the significance, the serotype is often included with the name—for example, *Salmonella enterica* serotype Dublin. For convenience, the name is often shortened, as in *Salmonella* Dublin. Note that the serotype is not italicized and the first letter is capitalized, which distinguishes it from a species name. *Salmonella* serotype Typhimurium and *Salmonella* serotype Enteritidis are most commonly isolated in the United States.

Pathogenesis

Most *Salmonella* serotypes are sensitive to acid, so millions of cells must generally be ingested for enough to survive passage through the stomach to colonize the intestines. Upon reaching the distal small intestine (the region where the small intestine ends), the bacterial cells attach to specific receptors on the surface of the epithelial cells. Contact activates a type III secretion system that transfers bacterial effector proteins into the epithelial cell. Within minutes, the epithelial cell takes in the bacterial cells by endocytosis (see figure 16.6). The bacteria multiply within a phagosome and are discharged from the base of the cell by exocytosis. Some bacterial cells escape the phagosome and multiply in the cytoplasm. Macrophages and neutrophils take up any bacteria that are released, but the macrophages are often destroyed as a result. The infection, however, remains localized. The inflammatory response increases epithelial cell fluid secretion, causing diarrhea. ◀◀ effector proteins, p. 387

Epidemiology

Most cases of *Salmonella* gastroenteritis originate from a non-human animal source. The bacteria sometimes survive for months in soil and water, so they can be spread in untreated manure. Poultry often carry *S. enterica,* and eggs can be contaminated as

well. Other contaminated products—including tomatoes, brewer's yeast, alfalfa sprouts, protein supplements, raw eggs, dry milk, and even a red dye used to diagnose intestinal disease—have started outbreaks. Children are commonly infected by colonized but seemingly healthy pets—particularly lizards, snakes, and turtles—that shed the bacteria in their feces. The disease has been on the rise, largely due to mass production and distribution of foods.

Treatment and Prevention

Most people with *Salmonella* gastroenteritis recover without antimicrobial treatment, which is fortunate because *S. enterica* strains have shown increasing plasmid-mediated resistance to antimicrobial medications. Many isolates of *Salmonella* serotype Typhimurium are resistant to five or more drugs. Antibiotic-resistant strains are often associated with more severe illness and longer hospital stays. Resistance is likely due to the widespread use of subtherapeutic levels of antibiotics in animal feeds, which is why several countries have now banned this practice. Antibiotics are not advised for treating humans except in cases involving tissue or bloodstream invasion.

Control of *Salmonella* gastroenteritis depends on sanitary handling of animal carcasses, pasteurizing or irradiating animal products, and testing products for contamination. Surveillance and tracing contaminated sources using PulseNet are also important. Adequate cooking kills *Salmonella* cells, but the center of cooked food does not always reach the 160° Fahrenheit recommended to kill foodborne bacteria. This is especially important when cooking frozen poultry—the outside may appear "well done," but the center remains relatively cool, allowing bacteria to survive and infect the unwary diner. **Table 24.9** summarizes the characteristics of *Salmonella* gastroenteritis.

Typhoid and Paratyphoid Fevers

Typhoid fever and paratyphoid fever are enteric fevers—systemic diseases that originate in the intestine. They are caused by specific

TABLE 24.9	*Salmonella* Gastroenteritis
Signs and symptoms	Diarrhea, vomiting, headache, abdominal pain, and fever
Incubation period	Usually 6 to 72 hours
Causative agent	*Salmonella enterica*, motile, Gram-negative, members of the *Enterobacteriaceae*
Pathogenesis	Induce uptake by epithelial cells in the region between the small and large intestines; bacteria multiply in the phagosome and then are discharged at the base of the cell; inflammatory response increases fluid secretion
Epidemiology	Ingestion of food contaminated by animal feces, especially poultry
Treatment and prevention	Treatment with antimicrobial medication is usually not advised. Prevention relies on adequate cooking and proper handling of food.

serotypes of *Salmonella enterica* and spread from person to person through fecal-oral transmission. Although rare in the United States, millions of cases occur in developing countries.

Signs and Symptoms

Patients suffering from enteric fevers often have a progressively increasing fever over a number of days, severe headache, constipation, and abdominal pain. In severe cases, this is followed by intestinal rupture, internal bleeding, shock, and death.

Causative Agent

Typhoid fever is caused by *Salmonella* serotype Typhi, whereas paratyphoid fever is caused by *Salmonella* serotype Paratyphi. These are systemic diseases, so cases are confirmed by blood culture rather than stool culture.

Pathogenesis

Bacteria that cause enteric fever colonize the intestines, cross the mucous membrane via M cells, and then resist killing by macrophages. After multiplying within macrophages, the bacteria are carried in the bloodstream to locations throughout the body. The systemic infection causes fever, abscesses, sepsis, and shock, often with little or no diarrhea. The Peyer's patches are sometimes destroyed, leading to rupture of the intestine, hemorrhage, and death. ▶▶ sepsis, p. 674 ◀◀ Peyer's patches, p. 358

Epidemiology

Humans are the only known host for *Salmonella* serotypes Typhi and Paratyphi, so the bacteria are spread from person to person. Often, this occurs via contaminated food or water.

Some patients surviving typhoid or paratyphoid fever remain colonized with the causative agents. The bacteria actually reside in the gallbladder, where they multiply free of competition because most other bacteria are killed or inhibited by concentrated bile. Carriers can shed high numbers of the bacteria for years. Mary Mallone—"Typhoid Mary"—a young Irish cook living in New York State in the early 1900s, was a notorious *Salmonella* serotype Typhi carrier. She was responsible for at least 53 cases of typhoid fever over a 15-year period. In her day, about 350,000 cases occurred in the United States each year. The incidence is now low due to improved sanitation, pasteurization, and public health surveillance measures.

Treatment and Prevention

Antibiotics are used to treat enteric fever, but some *Salmonella* serotype Typhi strains are multidrug-resistant, so susceptibility testing must be done. Most strains in the United States are susceptible to fluoroquinolones. Surgical removal of the gallbladder and months of antibiotic therapy are often necessary to rid *Salmonella* serotype Typhi carriers of their infection.

An attenuated live oral vaccine for preventing typhoid fever is about 50% to 75% effective. An equally effective injectable vaccine composed of *Salmonella* serotype Typhi capsular polysaccharide is also available. There is no vaccine against *Salmonella* serotype Paratyphi. **Table 24.10** summarizes some of the features of enteric fever. ◀◀ attenuated vaccine, p. 423

TABLE 24.10	Typhoid and Paratyphoid Fevers
Signs and symptoms	Fever, severe headache, constipation, and abdominal pain; sometimes followed by intestinal rupture, internal bleeding, shock, and death
Incubation period	1 to 3 weeks
Causative agent	*Salmonella* serotype Typhi (typhoid fever) and *Salmonella* serotype Paratyphi (paratyphoid fever), Gram-negative, members of the *Enterobacteriaceae*
Pathogenesis	Cross the intestinal epithelium via M cells; taken up by macrophages and then multiply within them; bacteria are carried in the bloodstream to locations throughout the body. The Peyer's patches are sometimes destroyed, leading to rupture of the intestine and hemorrhage.
Epidemiology	Humans are the only source of infection. Person-to-person fecal-oral transmission.
Treatment and prevention	Antibiotics are used for treatment. Attenuated and inactivated vaccines for typhoid fever; no vaccine for paratyphoid fever.

Campylobacteriosis

Campylobacter jejuni was first isolated from a diarrheal stool in 1972, but it then took 5 years to develop a suitable culture protocol. Once this was widely used, *C. jejuni* infection was found to be a common cause of diarrhea.

Signs and Symptoms

Campylobacteriosis typically leads to fever, vomiting, diarrhea, and abdominal cramps. Dysentery occurs in about half the cases.

Causative Agent

Campylobacter jejuni is a curved, Gram-negative rod (**figure 24.13**). It can be cultivated from feces under microaerophilic

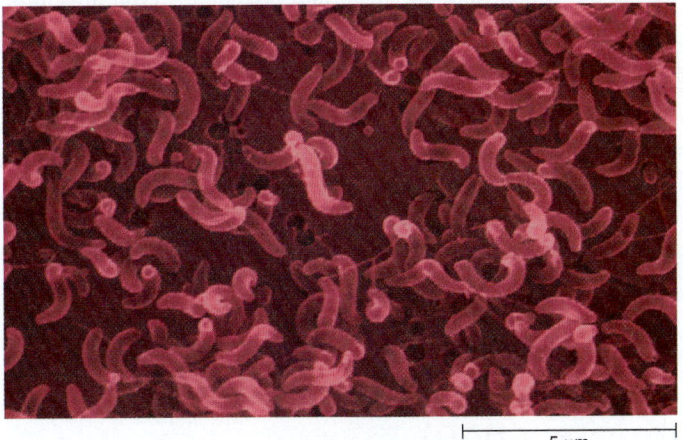

5 μm

FIGURE 24.13 *Campylobacter jejuni* *C. jejuni* is a common bacterial cause of diarrhea in the United States. Color-enhanced scanning electron micrograph.

❓ *Why was the incidence of this common disease largely unknown for decades?*

conditions using a selective medium to suppress the other intestinal organisms. ◄◄ **selective media, p. 95**

Pathogenesis

Once ingested, *Campylobacter jejuni* passes through the stomach and penetrates the epithelial cells of the small and large intestines. The bacteria multiply within and beneath the epithelium and cause a localized inflammatory reaction. Penetration into the bloodstream is uncommon.

A mysterious consequence of *C. jejuni* infection, Guillain-Barré syndrome, occurs in about 0.1% of cases. Up to 40% of all Guillain-Barré cases are preceded by campylobacteriosis, and autoimmunity is likely responsible in these cases. The syndrome begins within about 10 days of the onset of diarrhea, with tingling of the feet followed by progressive paralysis of the legs, arms, and rest of the body. Most patients require hospitalization, but recover completely; about 5% of patients die despite treatment.

Epidemiology

Campylobacteriosis is a leading bacterial diarrheal illness in the United States, with an estimated 2.4 million cases and 125 deaths each year. Most of the deaths occur in the elderly and those with AIDS or other immunodeficiencies. Numerous foodborne and waterborne outbreaks of *C. jejuni* have been reported, involving as many as 3,000 people. Most cases, however, are sporadic.

Like *Salmonella*, *C. jejuni* lives in the intestines of a variety of domestic animals, including pets and migratory birds. Poultry is a common source of infection, and up to 90% of raw poultry products harbor the organism. The infectious dose of *C. jejuni* is as low as 500 organisms, so one drop of juice from raw chicken meat can easily contain an infectious dose. Unpasteurized milk and non-chlorinated surface water have also started epidemics. Despite a low infectious dose, person-to-person spread of *C. jejuni* is rare.

Treatment and Prevention

Campylobacter jejuni gastroenteritis is typically self-limiting and leaves the patient immune to further infection. Erythromycin or azithromycin is used to treat severe cases.

Campylobacteriosis can be prevented by cooking and handling raw poultry properly to avoid cross-contamination. Chicken should be cooked until no longer pink (160° Fahrenheit). Pet owners should wash their hands after contact with animal feces, and outdoor play areas should be kept free of bird droppings. Chlorinating drinking water and pasteurizing beverages are also important control measures. PulseNet helps track *Campylobacter* isolates. The main features of campylobacteriosis are listed in **table 24.11.** ◄◄ **cross-contamination, p. 441**

Clostridium difficile–Associated Disease (CDAD)

Clostridium difficile has been recognized as a cause of antibiotic-associated diarrhea for many years, but the severity of the symptoms and the number of outbreaks has been increasing. The

TABLE 24.11	**Campylobacteriosis**
Signs and symptoms	Diarrhea, fever, abdominal cramps, nausea, vomiting, bloody stools
Incubation period	Usually 3 days (range, 1 to 5 days)
Causative agent	*Campylobacter jejuni*, a curved Gram-negative, microaerophilic rod
Pathogenesis	Low infectious dose. The bacteria multiply within and beneath the epithelial cells, causing an inflammatory response. Bloodstream invasion is uncommon. Complicated by Guillain-Barré syndrome on rare occasions.
Epidemiology	Large foodborne and waterborne outbreaks originating from poultry and other animals; person-to-person spread rare
Treatment and prevention	Treatment with antibacterial medications is usually not required. Prevention relies on adequate cooking and proper handling of food, particularly poultry. Water chlorination and pasteurization of beverages are effective control measures.

disease is particularly a problem in healthcare settings, where *C. difficile* is the most common cause of diarrhea.

Signs and Symptoms

Clostridium difficile–associated disease (CDAD) ranges widely in severity. Some patients experience only mild diarrhea, often accompanied by fever and abdominal pain. In other cases, the disease progresses to colitis (inflammation of the colon), sometimes with patchy areas on the colon called pseudomembranes, a characteristic of **pseudomembranous colitis** (**figure 24.14**).

Causative Agent

Clostridium difficile is a Gram-positive, rod-shaped, endospore-forming obligate anaerobe. The endospores are highly resistant to common disinfectants and environmental conditions, making the disease control difficult. *C. difficile* strains that cause CDAD produce one or more characteristic cytotoxins, and assays that detect the toxins are used to diagnose the disease. Some strains appear to be hypervirulent, and the characteristics of these are still under investigation.

Pathogenesis

Two toxins (A and B) appear to be important in CDAD. Both toxins disrupt the host cell actin, causing lethal effects to the intestinal epithelium. In some cases, they cause pseudomembranes—composed of dead epithelium, inflammatory cells, and clotted blood—to form on the intestine.

Epidemiology

CDAD primarily occurs in hospitalized patients on antibiotic therapy, probably because *C. difficile* is able to grow to high numbers

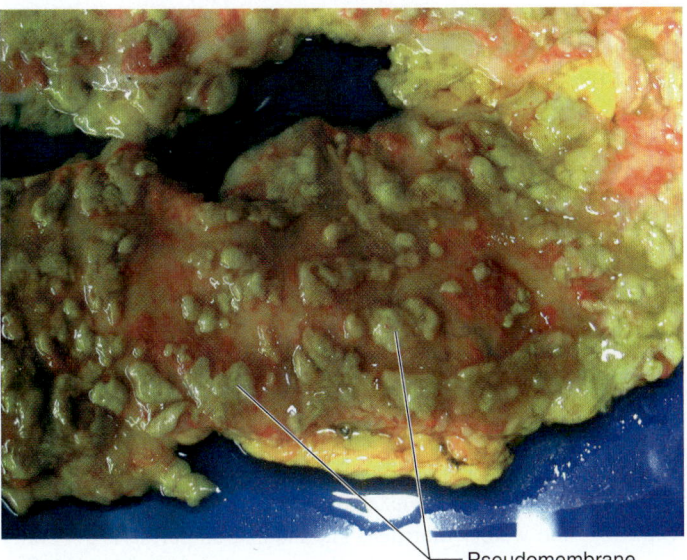

— Pseudomembrane

FIGURE 24.14 Pseudomembranous Colitis

❓ *How does antibiotic therapy predispose a person to this condition?*

only when the normal intestinal microbiota has been disrupted. Some patients acquire the microorganism while in the hospital, but others carry it as a minor component of their normal microbiota. Infectious endospores are shed in the feces. Some cases of CDAD are community acquired, and these patients generally have an underlying chronic intestinal condition.

Treatment and Prevention

When feasible, the antibiotics that predisposed the patient to CDAD are stopped. This alone often causes symptoms to disappear. If this is not an option or it fails, then vancomycin, metronidazole, or fidaxomicin (Dificid), a new drug developed to treat CDAD, is

TABLE 24.12	*Clostridium difficile*–Associated Disease
Signs and symptoms	Variable; mild diarrhea to pseudomembranous colitis, a potentially fatal inflammation of the colon
Incubation period	Not known; probably a week or less
Causative agent	*Clostridium difficile*, a Gram-positive, rod-shaped, endospore-forming anaerobe
Pathogenesis	Toxins disrupt host cell actin, causing lethal effects to the intestinal epithelium.
Epidemiology	Primarily occurs in hospitalized patients on antibiotic therapy
Treatment and prevention	When feasible, stop antibiotics; otherwise, vancomycin, metronidazole, or Dificid.

used. The use of probiotics in treatment has shown promising results. ◀◀ probiotics, p. 397

Measures to prevent CDAD include minimizing the use of antibiotics, and avoiding transmission of *C. difficile* by handwashing, wearing gloves, and keeping surfaces disinfected. Characteristics of *Clostridium difficile*–associated disease are summarized in **table 24.12.**

MicroByte

Fecal transplants—which allow another person's microbiota to repopulate the colon—have been successfully used to treat *Clostridium difficile*–associated disease.

MicroAssessment 24.4

Shigella, Salmonella, various *E. coli* strains, *Vibrio cholerae,* and *Campylobacter jejuni* account for most intestinal bacterial infections. Pathogenic mechanisms of these bacteria include attachment, enterotoxin or cytotoxin production, cell invasion, and destruction of microvilli. Dehydration can be treated with oral rehydration therapy. *Clostridium difficile* causes antibiotic-associated diarrhea.

10. *Explain how* Vibrio cholerae *causes cholera.*
11. *How does the epidemiology of* Salmonella *gastroenteritis differ from that of typhoid fever?*
12. *How would you devise a selective medium for* Vibrio cholerae? ➕

24.5 ■ Viral Diseases of the Lower Digestive System— Intestinal Tract

Learning Outcome

9. *Compare and contrast the diseases caused by rotaviruses and noroviruses.*

Viral infections of the lower digestive system are common in all age groups, resulting in millions of cases each year in the United States alone.

Rotaviral Gastroenteritis

Most cases of viral gastroenteritis in infants and children are caused by rotaviruses. Before a vaccine was available, rotavirus infections in the United States resulted in half a million emergency room or clinic visits, and over 55,000 hospital admissions each year. Worldwide, more than 500,000 children still die from rotavirus infection each year.

Signs and Symptoms

Rotaviral gastroenteritis begins abruptly with vomiting and slight fever, followed in a short time by profuse, watery diarrhea. Signs and symptoms are generally gone in about a week, but fatal dehydration can occur if fluids are not replaced.

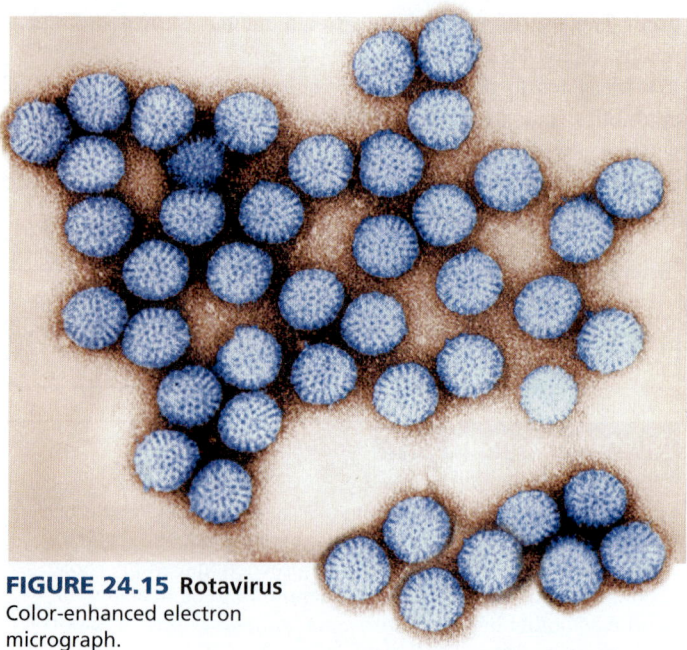

FIGURE 24.15 Rotavirus
Color-enhanced electron
micrograph.

❓ *What age group is most commonly infected with rotavirus?*

Causative Agent

Rotaviruses are naked viruses with a double-walled capsid, and a double-stranded, segmented, RNA genome (**figure 24.15**). The viruses represent a major subgroup of the reovirus family. ◄◄ reovirus family, p. 308

Pathogenesis

Rotaviruses mainly infect the epithelial cells that line the upper part of the small intestine. Infection causes epithelial cell death and decreased production of digestive enzymes. The damaged lining fails to absorb fluids, leading to watery diarrhea. In addition, one of the viral proteins appears to function as an enterotoxin, causing fluid secretion in a manner somewhat similar to cholera toxin.

Epidemiology

Rotaviruses are transmitted by the fecal-oral route. Childhood epidemics generally occur in winter in temperate climates, probably because children are often indoors in groups where the viruses can spread easily. In regions where vaccination is not common, most children are infected before age 5. The infection results in some immunity, so the second and third rotaviral infections cause much milder diarrhea than the first. Rotaviruses also cause about 25% of traveler's diarrhea cases.

Rotaviruses that infect a wide variety of young wild and domestic animals do not cause human disease. Experimentally, however, reassortment of genetic segments can occur with dual infections. It is not known whether new human pathogenic rotaviruses arise by this mechanism.

Treatment and Prevention

Although there is no direct treatment of rotavirus infection, some infants and small children are hospitalized and given intravenous fluids to prevent dehydration.

Handwashing, disinfectant use, and other sanitary measures help limit the spread of rotaviruses. Attenuated vaccines approved since 2006, administered to infants, are resulting in a substantial decline in this disease.

Norovirus Gastroenteritis

Noroviruses are the most common cause of viral gastroenteritis in the United States, responsible for an estimated 23 million cases annually. They were originally called "Norwalk viruses," from Norwalk, Ohio, the place where they were first implicated in an epidemic of gastroenteritis. The viruses are designated a category B bioterrorism agent because they spread easily, with the potential of causing large demoralizing outbreaks. ◄◄ bioterrorism agents, p. 453

Signs and Symptoms

Norovirus gastroenteritis usually causes abrupt onset of nausea, vomiting, and watery diarrhea. The incubation period is generally 1 or 2 days. Vomiting is typically most severe in older children and adults, and generally resolves within the first day or two. Other symptoms take several days to subside.

Causative Agent

Noroviruses are naked, single-stranded RNA viruses (**figure 24.16**). These viruses represent a group of gastroenteritis-producing viruses within the calicivirus family, none of which has been cultivated in the laboratory. ◄◄ calicivirus family, p. 308

Pathogenesis

Noroviruses infect the epithelium of the upper small intestine, causing epithelial cell death and decreased production of digestive enzymes. The epithelium generally recovers fully within about 2 weeks. For reasons that are not understood, immunity to noroviruses is only short term, lasting just several months. Because of this, individuals can have repeated infections.

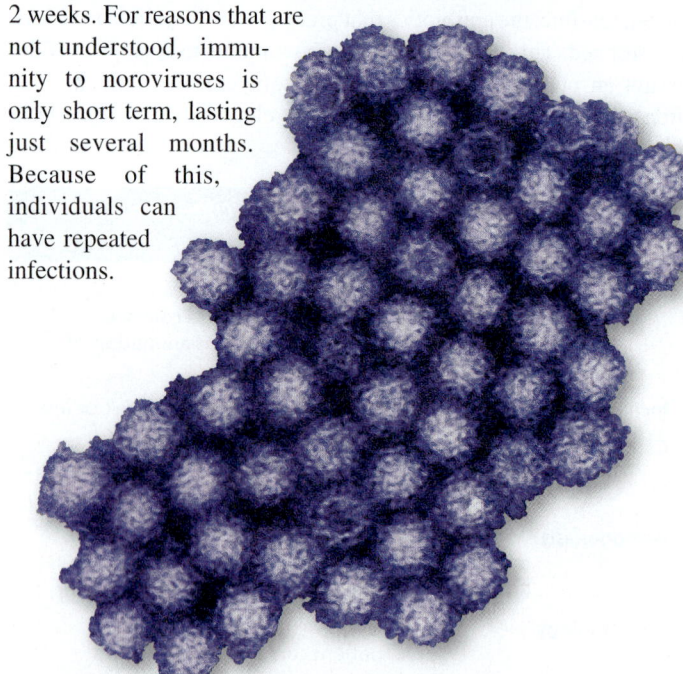

FIGURE 24.16 Norovirus These viruses cause relatively brief bouts of gastroenteritis which sometimes go by the name "stomach flu." Color-enhanced electron micrograph.

❓ *Why are noroviruses so common in crowded settings such as cruises and college dormitories?*

TABLE 24.13	Rotavirus and Norovirus Compared	
Causative Agent	**Rotavirus**	**Norovirus**
Characteristics	Double-walled capsid; double-stranded segmented RNA genome	Single-stranded RNA genome
Incubation period	24 to 48 hours	12 to 48 hours
Typical signs and symptoms	Vomiting, abdominal cramps, diarrhea, lasting 5 to 8 days	Vomiting, abdominal cramps, diarrhea, lasting 12 to 60 hours
Prevention	Attenuated vaccine	No vaccine

24.6 ■ Viral Diseases of the Lower Digestive System—Liver

Learning Outcome

10. *Compare and contrast hepatitis A, B, and C.*

At least five different viruses can cause **hepatitis,** an inflammation of the liver, but three types—A, B, and C—account for most cases (**figure 24.17**). The viruses are unrelated to each other, but all damage the liver and cause similar signs and symptoms during an acute infection (**table 24.14**). The most noticeable sign is jaundice—yellowing of the skin and the whites of the eyes. Patients with any form of hepatitis should avoid alcohol, acetaminophen, and other chemicals known to damage the liver. Note that some other diseases, including yellow fever and infectious mononucleosis, can also damage the liver.

Hepatitis A

Hepatitis A, formerly called infectious hepatitis, is found around the world. Introduction of an effective vaccine in 1995 significantly lowered the incidence of the disease (**figure 24.18**).

Signs and Symptoms

Hepatitis A is an acute illness with no known chronic form or carrier state. Older children and adults with the disease usually develop jaundice, fever, fatigue, clay-colored feces, and vomiting after an incubation period of about 1 month. Most young children (less than 6 years old) and many older children (ages 6 to 14) are asymptomatic. About one in five infected adults requires hospitalization. Patients generally recover within 2 months, but some take up to 6 months.

Epidemiology

Transmission of noroviruses is primarily by the fecal-oral route. Vomit contains infectious viral particles, however, so transmission through aerosols and contaminated surfaces can also occur. The disease is highly contagious due to the low infectious dose of the virus—about 10 viral particles. The virions are relatively resistant to destruction, and stable in the environment.

Norovirus epidemics are common on cruise ships and college dormitories. Their easy spread also makes them a concern for healthcare facilities. More than 50% of foodborne disease outbreaks are due to norovirus, so the CDC has started CaliciNet, a national surveillance network to trace norovirus strains.

Treatment and Prevention

There are no proven anti-noroviral medications. There is no vaccine, and natural immunity to the viruses is short-lived. Thorough handwashing using soap and water is an important preventative measure. Disinfectants and other sanitary measures also minimize viral transmission. Infected foodworkers should be restricted from working for 72 hours after their symptoms subside. **Table 24.13** compares noroviruses and rotaviruses.

MicroAssessment 24.5

Rotaviruses are the leading cause of viral gastroenteritis in infants and children, and they also are a common cause of traveler's diarrhea. Noroviruses are the most common cause of viral gastroenteritis. They are highly infectious and easily spread.

13. *Why are rotaviral infections common in young children but not adults?*

14. *Why is handwashing an important means to control the spread of norovirus?*

15. *Why might it be more difficult to develop an effective vaccine against noroviruses than against rotaviruses?* ✚

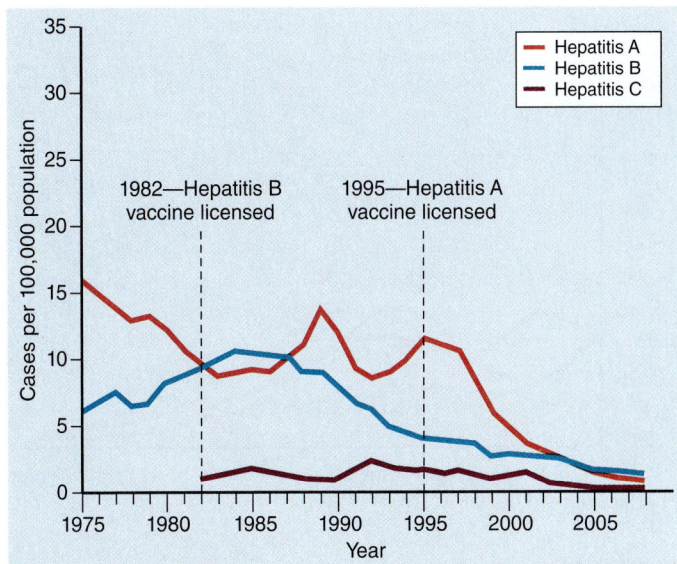

FIGURE 24.17 Incidence of Viral Hepatitis in the United States, 1975–2008

❓ *What organ is affected by hepatitis?*

TABLE 24.14 | Viral Hepatitis

	Hepatitis A	Hepatitis B	Hepatitis C
Causative agent	Naked, single-stranded RNA picornavirus, HAV	Enveloped, double-stranded DNA hepadnavirus, HBV	Enveloped, single-stranded RNA flavivirus, HCV
Transmission	Fecal-oral	Blood, semen	Blood, possibly semen
Incubation period	3 to 5 weeks (range, 2 to 7 weeks)	10 to 15 weeks (range, 6 to 23 weeks)	6 to 7 weeks (range, 2 to 24 weeks)
Prevention	Inactivated vaccine; immune globulin	Subunit vaccine; immune globulin	No vaccine
Comments	Usually mild symptoms, but often prolonged; full recovery; no long-term carriers; combined hepatitis A and B vaccine available	Acute symptoms often more severe than hepatitis A; chronic disease can lead to cirrhosis and cancer; chronic carriers; can cross the placenta; combined hepatitis A and B vaccine available	Usually few or no symptoms; progressive liver damage can lead to cirrhosis and cancer; chronic carriers

	Hepatitis D	Hepatitis E	
Causative agent	Defective single-stranded RNA virus, HDV	Naked, single-stranded RNA calicivirus, HEV	
Transmission	Blood, semen	Fecal-oral	
Incubation period	2 to 12 weeks	2 to 6 weeks	
Prevention	No vaccine	No vaccine	
Comments	Prior or concurrent HBV infection necessary; can cause worsening of hepatitis B; can cross the placenta	Similar to hepatitis A, except severe disease in pregnant women, same or related virus in rats	

Causative Agent

Hepatitis A is caused by the hepatitis A virus (HAV), a naked single-stranded RNA virus of the picornavirus family. There is only one serotype of HAV. ◄◄ picornaviruses, p. 308

Pathogenesis

Following ingestion, the virus reaches the liver by an unknown route. The liver is the main site of replication and the only tissue known to be damaged by the infection. The virus is released into the bile and eliminated with the feces.

Epidemiology

Hepatitis A virus spreads by the fecal-oral route, principally via fecal-contaminated hands, food, or water. Many outbreaks have been traced to restaurants where infected food-handlers failed to wash their hands. Raw shellfish are also a frequent source of infection because they concentrate the virus from fecally polluted seawater. Groups at high risk of contracting the disease include children in day care centers, residents in nursing homes, international travelers, and individuals having sexual contact with an infected person. Because the incubation period averages 30 days, HAV can spread widely through a population before being detected. Infected infants and children can shed the virus in their feces for several months.

Treatment and Prevention

No antiviral treatment for HAV is available. However, immune globulin can be given by injection after exposure. This passive immunization gives short-term protection against the disease if administered within 2 weeks. ◄◄ immune globulin, p. 420

An effective vaccine, composed of inactivated HAV, has been available since 1995. The CDC recommends vaccination for all children 1 year of age, and several high-risk groups such as people traveling to areas of high incidence or in occupations that put them at high risk of exposure. Since the introduction of vaccination, the number of reported cases of hepatitis A has dropped to historic lows. ◄◄ inactivated vaccine, p. 423

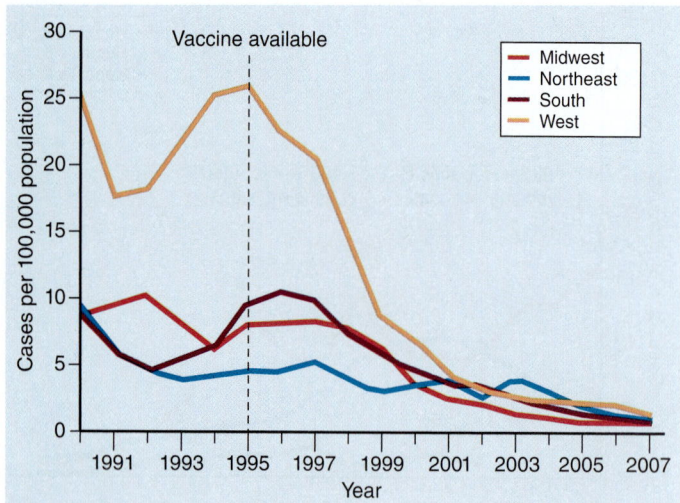

FIGURE 24.18 Acute Hepatitis A Reported Cases, different regions of the United States, 1991–2007.

❓ *What occupations might put a person at risk for hepatitis A?*

Hepatitis B

Hepatitis B, formerly known as serum hepatitis, represents about half of the cases of viral hepatitis in the United States. Unlike HAV, which is passed through the fecal-oral route, the hepatitis B virus (HBV) is transmitted through contact with body fluids.

Signs and Symptoms

Signs and symptoms of acute hepatitis B are similar to those of other forms of hepatitis, ranging from asymptomatic to severe. The incubation period varies considerably—from about 2 to 5 months—depending on the dose received. The acute disease is rarely fatal, and the virus is usually cleared within weeks to months of initial symptoms. It can become chronic, particularly in infants and children. One in five people with a chronic infection develop **cirrhosis** (scarring of the liver), liver failure, liver cancer, or other chronic liver disease. Between 2,000 and 4,000 people die from HBV-related diseases each year.

Causative Agent

Hepatitis B is caused by hepatitis B virus (HBV), an enveloped virus that has a mostly double-stranded DNA genome (a portion is single-stranded) (**figure 24.19a**). Unlike most enveloped viruses, HBV is remarkably resistant to environmental conditions—the

virions can still be infectious after a week outside the body. HBV is a member of the hepadnavirus family (*hepa-,* "liver," and *-dna-* "DNA").

Three components of HBV are notable because they serve as useful markers of infection:

- **Hepatitis B surface antigen (HBsAg).** This envelope protein is produced during viral replication in amounts far in excess of that needed for virion production. The antigen appears in the bloodstream days or weeks after infection, often long before signs of liver damage are evident. Antibodies to HBsAg confer immunity to HBV.

- **Hepatitis B core antigen (HBcAg).** IgM antibodies against this antigen indicate active viral replication.

- **Hepatitis B e antigen (HBeAg).** High levels of this soluble component of the viral core correlate with increased risk of liver damage and increased risk of spreading the disease.

The complete hepatitis B virion is also referred to as a Dane particle. This term provides a contrast to another common viral product of infection—viral envelopes with HBsAg, but lacking DNA (figure 24.19b). These empty envelopes can be 1,000 times more common in the bloodstream than Dane particles, which is why HBsAg is a useful marker of infection.

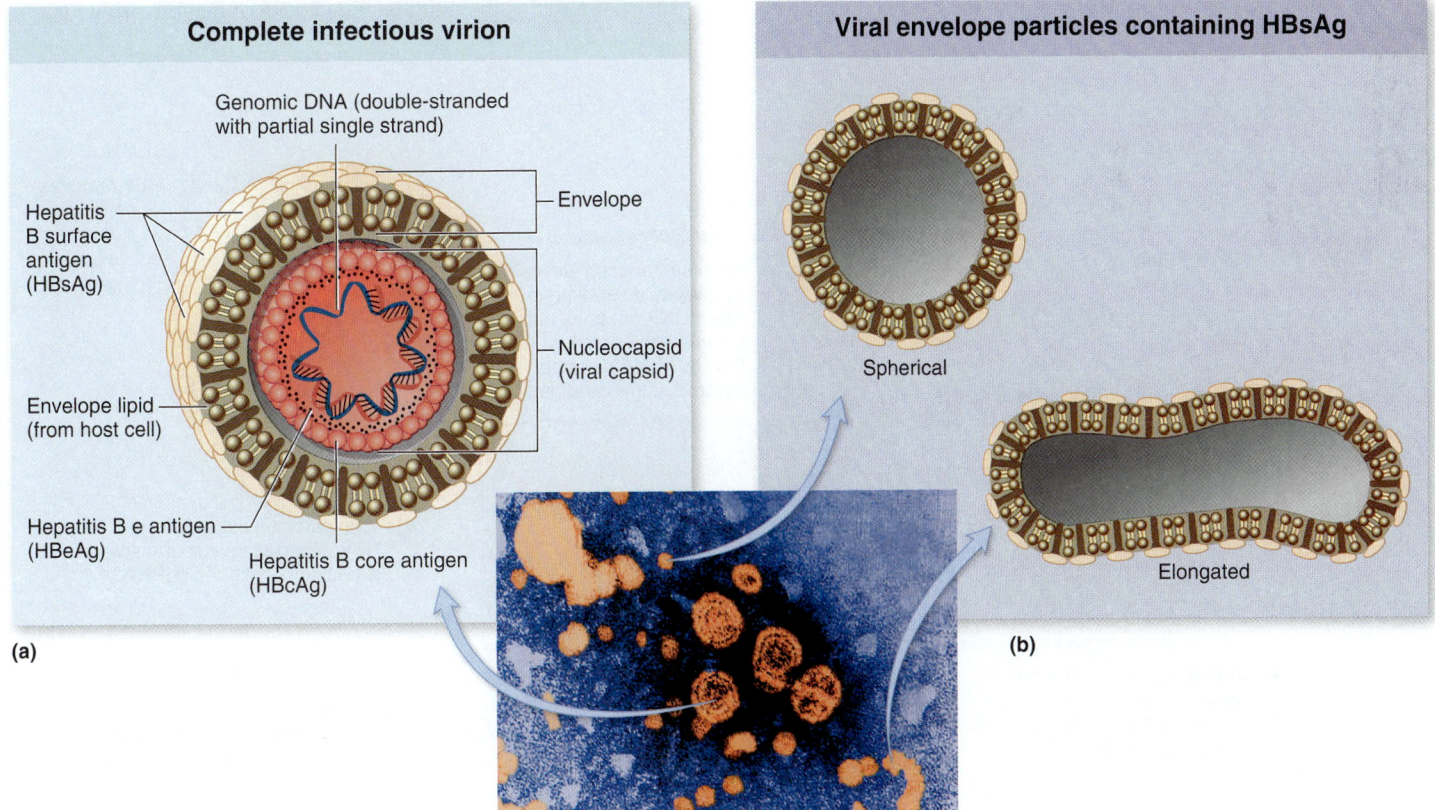

FIGURE 24.19 Hepatitis B Virus Components Found in the Blood of Infected Individuals (a) Complete infectious virion also known as a Dane particle. (b) Smaller spherical and elongated envelope particles lacking DNA.

What does circulating HBeAg indicate about individuals with chronic HBV infection?

Pathogenesis

Following its entry into the body, HBV is carried to the liver by the bloodstream. Surface antigen (HBsAg) allows the virus to attach to and enter host cells.

The next steps of HBV replication are complex, but they help explain why certain antiviral medications can decrease disease symptoms. The replication process begins when the viral genome is transported to the host cell nucleus, and the single-stranded gap filled in (**figure 24.20**). At this point, the genome is a double-stranded covalently closed circular DNA molecule (cccDNA). A host cell RNA polymerase uses the cccDNA as a template to produce mRNA molecules that are then translated to make the various viral proteins. The significant and highly unusual aspect of HBV is that one of the RNA molecules serves as a template for DNA synthesis; in other words, HBV is a reverse transcribing virus. The RNA molecule—referred to as pregenomic RNA (pgRNA)—is packaged into the nucleocapsid along with an HBV-encoded **reverse transcriptase.** This enzyme uses the pgRNA as a template to make one strand of DNA; it then uses that DNA molecule as a template to synthesize the complementary strand. The process is not finished inside the virion, however, leaving the HBV genome only partly double-stranded. ◀◀ reverse transcribing virus, p. 320 ◀◀ reverse transcriptase, p. 320

Liver damage from HBV infection is likely due to the cell-mediated immune response, as effector cytotoxic T cells attack infected liver cells. The liver cell destruction leads to cirrhosis. The role of the virus in the development of liver cancer is not well understood, but in most cases of the cancer, the viral DNA has integrated into the host cell genome.

Epidemiology

Hepatitis B can be transmitted in body fluids, such as saliva, blood, blood products, and semen. Activities that mix these fluids from two individuals are considered risk factors—examples include sharing needles, toothbrushes, razors, or towels, and using unsterile tattooing or ear-piercing instruments. Unprotected sex is a particularly risky behavior, with nearly half of new hepatitis B cases in the United States acquired through sexual intercourse.

When HBV infection becomes chronic, the virus continues replicating—circulating in the blood for many years—even if the patient is asymptomatic. These infected individuals are extremely important in the spread of hepatitis B because they are often unaware of their infection. The failure to clear the infection is age related. More than 90% of infants who contract the virus from an infected mother at or shortly after birth develop a chronic infection. In contrast, 25% to 50% of children infected between the ages of 1 and 5, and 6% to 10% of those infected as an older child or adult, will develop a chronic infection.

A progressive rise in hepatitis B cases was reported in the United States from 1965 to the mid 1980s. Since that time, the HBV vaccine has lowered the incidence of the disease (see figure 24.17). Today, approximately 45,000 new HBV infections occur each year, and about 1 million people in the United States are chronically infected.

Approximately 2 billion people worldwide have been infected by HBV, with more than 350 million chronic infections. This "silent disease" is the ninth leading cause of death worldwide.

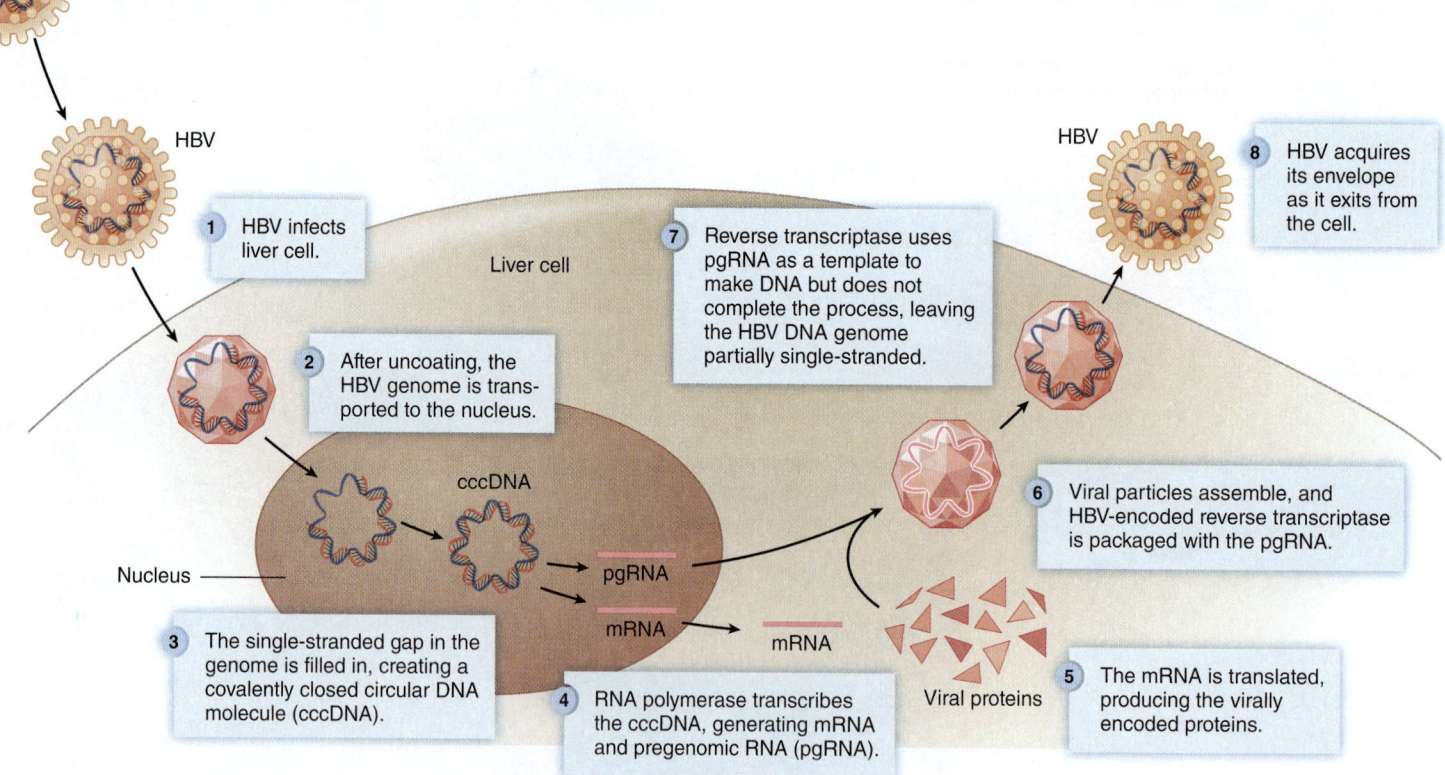

FIGURE 24.20 Replication of Hepatitis B Virus

❓ *With respect to the replication strategy, what unusual feature does HBV have in common with HIV?*

Treatment and Prevention

There is currently no curative antiviral treatment for hepatitis B. However, some chronically infected patients show remarkable improvement when given reverse transcriptase inhibitors such as lamivudine or adefovir dipivoxil along with injections of genetically engineered interferon. ◀◀ **reverse transcriptase inhibitors, p. 476** ◀◀ **interferon, p. 341**

The first vaccine against hepatitis B was approved in the early 1980s. It consisted of HBsAg obtained from the blood of chronic carriers; 1 ml of their blood often had enough antigen to immunize eight people! Since 1986, however, a subunit vaccine produced in genetically engineered yeast has been available. It is administered to all infants before they leave the hospital. A combined vaccine against both hepatitis A and B can be used to complete the vaccination schedule. Hepatitis B vaccination prevents the disease, and may also prevent many of the 500,000 to 1 million new liver cancers that occur worldwide each year.

Educating groups at high risk of contracting hepatitis B helps prevent the disease. Individuals likely to be exposed to blood—like healthcare workers—are taught to consider all blood infectious and use universal precautions. These precautions include wearing gloves and handling potentially contaminated sharp objects with care (see **Perspective 19.1**). Teaching the importance and proper use of condoms also helps limit spread of the infection.

Hepatitis C

Hepatitis C is the most common chronic, blood-borne infection in the United States. It was discovered when post-transfusion hepatitis continued to occur regularly even after HBV was excluded from transfusions. In 1989, scientists were able to clone parts of the genome of a transfusion-associated virus, now known as hepatitis C virus (HCV). One of the gene products of these first isolates was HCV antigen, which was then used to detect anti-HCV antibody in the blood of prospective donors. By eliminating donated blood that contained antibodies against HCV, the incidence of post-transfusion hepatitis fell dramatically. The antibody test has revealed that more than 3 million Americans are infected with HCV, and approximately 20,000 new cases occur each year.

Signs and Symptoms

The symptoms of hepatitis C are similar to those of hepatitis A and B except they are generally milder. About 65% of infected individuals have no symptoms relating to the acute infection, whereas only about 25% have jaundice. In many cases, the infections become chronic.

Causative Agent

HCV is an enveloped, single-stranded RNA virus of the flavivirus family, first cultivated *in vitro* in 2005. There is considerable genetic variability, and the variants are classified into types and subtypes that differ in pathogenicity as well as response to treatment. ◀◀ **flavivirus family, p. 309**

Pathogenesis

Few details are known about the pathogenesis of HCV. Infection generally occurs from exposure to contaminated blood. The incubation period averages about 6 weeks (range, 2 weeks to 6 months). Although most people lack symptoms with acute infection, more than 80% develop chronic infections. The virus infects the liver and triggers inflammatory and immune responses. After that, the disease process starts and stops—at times the liver seems to return to normal, then weeks or months later shows marked inflammation. After up to 20 years, cirrhosis and liver cancer develop in 10% to 20% of patients.

Epidemiology

Although HCV is transmitted in blood, the mechanism of exposure is not always obvious. Approximately 50% of infections in the United States are due to sharing of syringes. Tattoos and body piercing with unclean instruments have also transmitted the disease. Other items that can become contaminated with blood and are therefore possible sources of infection include toothbrushes, razors, and towels. Sexual intercourse is an uncommon means of transmission, but individuals with multiple partners and other sexually transmitted infections are at increased risk for the disease. The risk of contracting the disease from blood transfusion is now extremely low because of effective screening of donated blood.

Treatment and Prevention

A combination treatment of interferon and ribavirin sometimes prevents chronic disease. In addition, two drugs that inhibit HCV protease have recently been approved for use with the standard combination.

No vaccine is available for preventing hepatitis C. However, physicians often recommend vaccination against hepatitis A and B to help prevent a combination of infections that might severely damage the liver.

MicroAssessment 24.6

Hepatitis viruses are a diverse, unrelated group that cause liver inflammation. Most cases of hepatitis are caused by hepatitis viruses A, B, or C. Hepatitis A is transmitted by the fecal-oral route and is preventable by immune globulin and an inactivated vaccine. Hepatitis B, transmitted by exposure to blood and by sexual intercourse, is preventable by a subunit vaccine. A combination hepatitis A and hepatitis B vaccine is available. Hepatitis C is transmitted by blood and occasionally by sexual intercourse. Chronic hepatitis often results in cirrhosis and liver cancer.

16. *What two serious complications can occur late in the course of both chronic hepatitis B and C?*
17. *At what stage in the replication of hepatitis B does a reverse transcriptase inhibitor act?*
18. *Why is it possible to immunize multiple people from the blood of one hepatitis B patient?*

24.7 ■ Protozoan Diseases of the Lower Digestive System

Learning Outcome

11. *Compare and contrast giardiasis, cryptosporidiosis, cyclosporiasis, and amebiasis.*

Protozoa—single-celled eukaryotes—are important causes of human intestinal disease. Intestinal protozoan pathogens are all transmitted by the fecal-oral route. ◄◄ protozoa, p. 291

Giardiasis

Giardiasis is the most commonly identified waterborne illness in the United States. It can be contracted from clear mountain streams, chlorinated city water that has not been filtered, and person-to-person contact. The disease has worldwide distribution and is responsible for many cases of traveler's diarrhea.

Signs and Symptoms

In epidemics of giardiasis, about two-thirds of exposed individuals develop symptoms. The incubation period is generally 6 to 20 days. Symptoms can range from mild (indigestion, "gas," and nausea) to severe (vomiting, explosive diarrhea, abdominal cramps, fatigue, and weight loss). The symptoms usually end without treatment in 1 to 4 weeks, but some cases become chronic. Both symptomatic and asymptomatic persons can become long-term carriers, unknowingly excreting infectious cysts with their feces.

Causative Agent

Giardia lamblia is a flagellated protozoan shaped like a pear cut lengthwise. It has two side-by-side nuclei that resemble eyes and an adhesive disc on its undersurface that together give the organism a distinctive appearance (**figure 24.21**). It can exist in two forms: a growing, feeding **trophozoite** and a dormant **cyst.** The cysts have thick walls composed of a tough, flexible, chitinlike polysaccharide that protects the organism from harsh environmental conditions. ◄◄ chitin, p. 32

Because *G. lamblia* lacks mitochondria, it was thought to have evolved before eukaryotes acquired these organelles. More recent evidence appears to refute this idea because the protozoan contains atypical energy-metabolizing structures called mitosomes that likely evolved from mitochondria.

Pathogenesis

The cysts of *G. lamblia* are infectious because, unlike the trophozoites, they are resistant to stomach acid. From each cyst that reaches the upper part of the small intestine, two trophozoites emerge. Some of these attach to the epithelium by their adhesive disc, whereas others use their flagella to move freely in the intestinal mucus. Some may even migrate up the bile duct to the gallbladder and cause cramping or jaundice.

The trophozoites interfere with the intestine's ability to absorb nutrients and secrete digestive enzymes. The result is bulky feces containing fat, excessive intestinal gas from bacterial digestion of unabsorbed food material, and malnutrition. Some of the

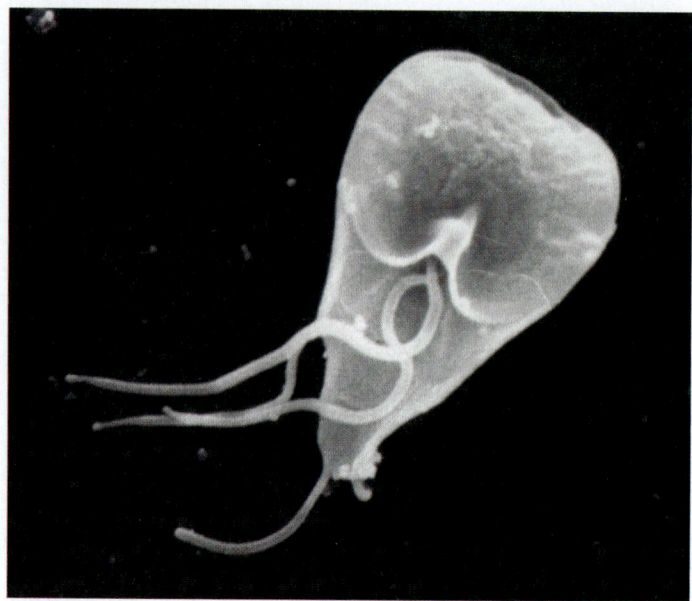

FIGURE 24.21 *Giardia lamblia*

❓ *Does this image show a cyst or a trophozoite?*

intestinal impairment is probably due to the host immune system attacking the parasites.

When trophozoites detach from the epithelium, they are carried by intestinal contents toward the large intestine. If the transit time is long enough, they develop into cysts. Thus, a person who has formed stools is more likely to excrete cysts, whereas a person with diarrhea is apt to excrete trophozoites (**figure 24.22**).

Epidemiology

Giardia lamblia is easily spread by the fecal-oral route, because only 10 cysts are required to establish infection. Water contaminated with human feces is a common source of infection, but feces from animals such as beavers, raccoons, muskrats, dogs, and cats can also be implicated. The cysts are infectious and can remain viable in cold water for more than 2 months. Hikers who drink from streams, even in remote areas thought to contain safe water, are at risk of contracting giardiasis.

Although waterborne outbreaks are the most common, person-to-person contact can also transmit the disease. This is especially likely in daycare centers where workers' hands become contaminated while changing diapers. Sexual practices that lead to oral-anal contact can also transmit giardiasis. Transmission by fecally contaminated food has also been reported. Good personal hygiene, especially handwashing, decreases the chance of passing on the infection.

MicroByte
A single human stool can contain 300 million *G. lamblia* cysts.

Treatment and Prevention

Several medicines can be used to treat giardiasis, including tinidazole, metronidazole (Flagyl), and the newest option—nitazoxanide.

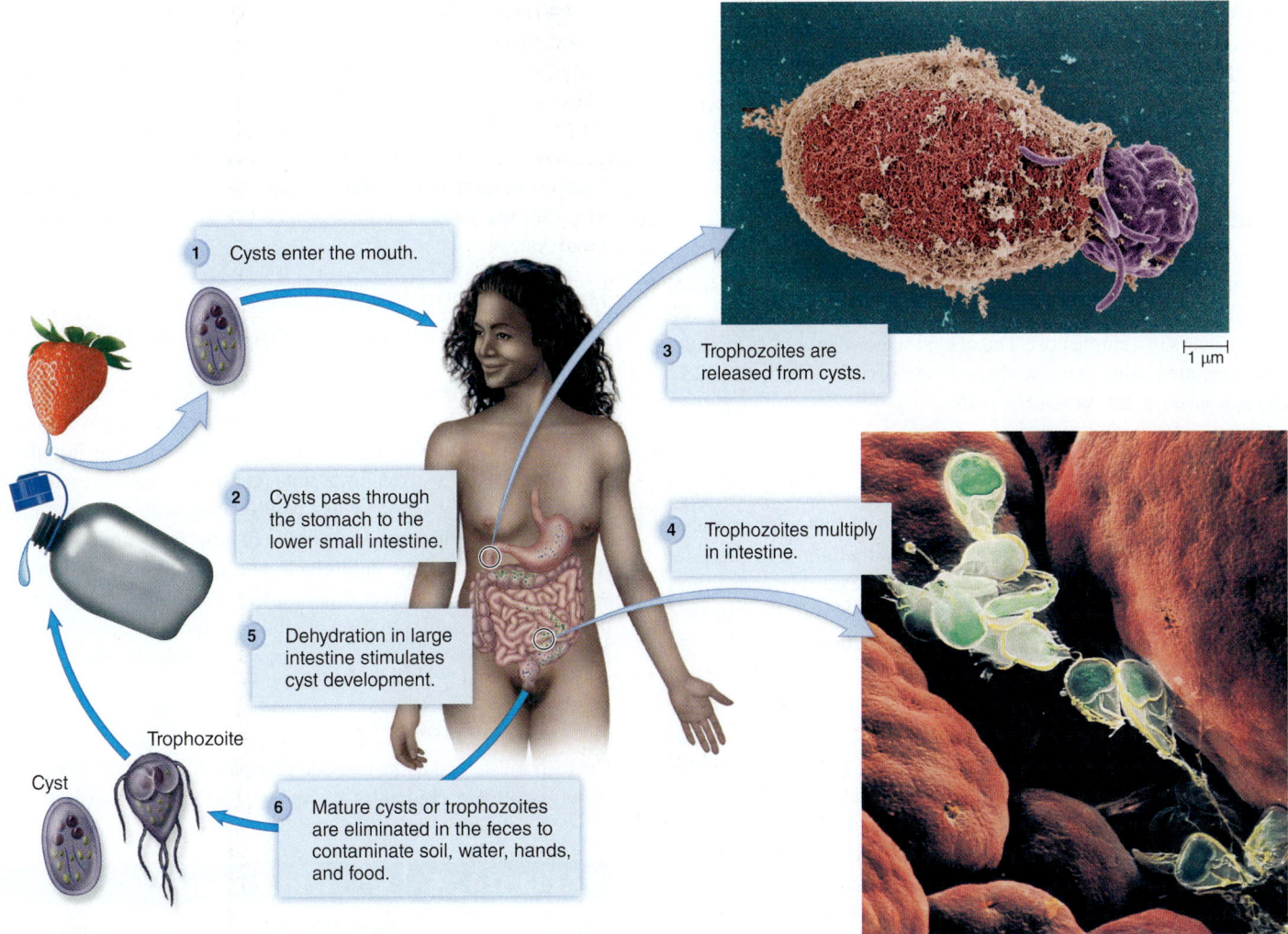

FIGURE 24.22 Life Cycle of *Giardia lamblia*

❓ *Why is an infected person with diarrhea less likely to transmit giardiasis than one who has formed stools?*

The level of chlorine used to treat municipal water supplies does not destroy *Giardia* cysts, so the water is generally filtered to remove them. For hikers, the best way to make drinking water safe from giardiasis is to boil it for 1 minute or use a portable filter of 1 μm pore size or smaller. Other methods, including adding commercial water-purifying tablets, tincture of iodine, or household bleach are time-consuming and much less reliable. As in all chemical microbial control procedures, time and temperature are important. Only an hour may be necessary to treat warm water, but many hours are required to kill cysts in cold water. **Table 24.15** describes the main features of giardiasis.

Cryptosporidiosis ("Crypto")

Cryptosporidiosis, commonly called "crypto," was first identified as a threat to humans when the AIDS epidemic struck in the early 1980s. We now know that the disease is a hazard not only to those with immunodeficiency but to the public at large. A 1993 waterborne outbreak in Milwaukee affected more than 400,000 people!

TABLE 24.15	Giardiasis
Signs and symptoms	Mild: indigestion, flatulence (intestinal gas), nausea; severe: vomiting, diarrhea, abdominal cramps, weight loss
Incubation period	6 to 20 days
Causative agent	*Giardia lamblia*, a flagellated pear-shaped protozoan with two nuclei
Pathogenesis	Ingested cysts survive stomach passage; trophozoites emerge from the cysts in the small intestine, where some attach to epithelium and others move freely; mucosal function is impaired by adherent protozoa and host immune response.
Epidemiology	Ingestion of fecally contaminated water; low infectious dose; person-to-person spread
Treatment and prevention	Several treatment options including tinidazole, metronidazole (Flagyl), and nitazoxanide. Prevention: boiling, filtering, or disinfecting drinking water.

Signs and Symptoms

Signs and symptoms of cryptosporidiosis include fever, loss of appetite, nausea, abdominal cramps, and profuse watery diarrhea. These begin after an incubation period of 4 to 12 days. The symptoms generally last 10 to 14 days, but in people with immunodeficiency diseases, they can last for months and be life-threatening.

Causative Agent

Cryptosporidiosis is caused by *Cryptosporidium parvum,* an apicomplexan. The organism multiplies intracellularly in the small intestinal epithelium. Unlike other Apicomplexa, its entire life cycle occurs in a single host. The oocyst stage is an acid-fast sphere that contains four banana-shaped sporozoites (**figure 24.23**).

◀◀ Apicomplexa, p. 292 ◀◀ acid-fast, p. 48

Pathogenesis

The digestive fluids of the small intestine release sporozoites from ingested oocysts. The sporozoites then invade the epithelial cells of the small intestine, altering the epithelium and intestinal villi and causing inflammation. Water and electrolyte secretion increases, and nutrient absorption decreases. Cell-meditated immunity is important in controlling the infection.

Epidemiology

Oocysts of *Cryptosporidium parvum* passed in feces are immediately infectious. The infectious dose is as few as 10, so person-to-person spread readily occurs with poor sanitation. The organism is responsible for many cases of traveler's diarrhea.

Infected individuals often have prolonged diarrhea and can continue to eliminate infectious oocysts for 2 weeks or more after the diarrhea ceases. These oocysts can survive for up to 6 months in food and water, and are even more resistant to chlorine than

Giardia cysts. They are too small to be removed from drinking water by some filtration methods.

C. parvum is particularly difficult to control because it has a wide host range, infecting domestic animals such as dogs, pigs, and cattle. Feces from these animals, as well as from humans, can contaminate food and drinking water. Epidemics have arisen from drinking water, swimming pools, a water slide, a zoo fountain, daycare centers, unpasteurized apple juice, and other food and drink.

Treatment and Prevention

A new medication has recently been approved to treat cryptosporidiosis—nitazoxanide.

Effective measures for preventing the disease include sanitary disposal of human and animal feces and treating water using filtration, ultraviolet radiation, or ozone. Pasteurizing liquids for consumption is also an important control measures. All food-handlers should wash their hands with soap and water regularly, and those with diarrhea should not handle food until symptom-free. Immunodeficient individuals are advised to avoid contact with animals, boil or filter drinking water using a 1 μm or smaller pore size filter, and avoid recreational water activities. The main features of cryptosporidiosis are shown in **table 24.16.**

Cyclosporiasis

Cyclosporiasis first came to medical attention in the late 1980s, with widely scattered epidemics of severe diarrhea. For several years,

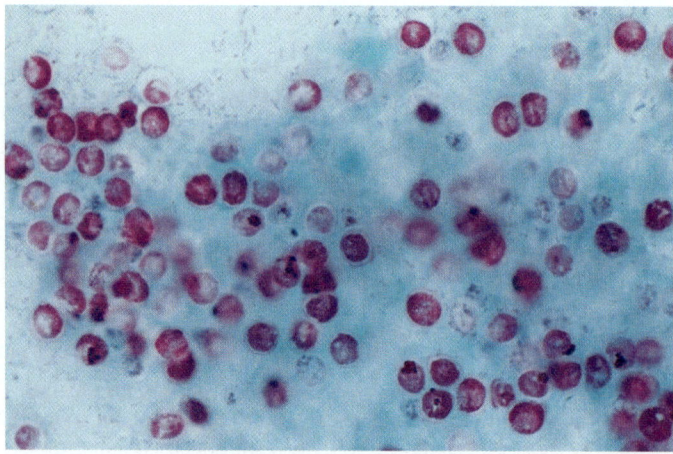

FIGURE 24.23 Oocysts of *Cryptosporidium parvum* Acid-fast stain of feces. Person-to-person and waterborne transmission are common.

❓ *Why is person-to-person transmission more likely with* Cryptosporidium *infection than with* Cyclospora cayetanensis?

TABLE 24.16	Cryptosporidiosis
Signs and symptoms	Fever, loss of appetite, nausea, crampy abdominal pain, watery diarrhea
Incubation period	Usually about 6 days (range, 4 to 12 days)
Causative agent	*Cryptosporidium parvum,* an apicomplexan. Its life cycle takes place entirely within the epithelial cells of the small intestine.
Pathogenesis	Following ingestion of oocysts, sporozoites are released in the intestine. The sporozoites invade the epithelial cells, causing an inflammatory response. Secretion of water and electrolytes increases, and absorption of nutrients decreases.
Epidemiology	Oocysts of *C. parvum* are immediately infectious. The infectious dose is low, so person-to-person spread occurs easily. Infected individuals can discharge the oocysts for weeks after symptoms subside. The oocysts resist chlorination, and pass through many municipal water filtration systems. Wide host range.
Treatment and prevention	Treated with nitazoxanide. Prevented by pasteurization of beverages, boiling or filtering drinking water. Sanitary disposal of human and animal feces.

the causative organism was known only as a "cyanobacterium-like body." Later, the bodies were shown to be the oocysts of a protozoan, *Cyclospora cayetanensis*. ◄◄ **cyanobacteria, p. 261**

Signs and Symptoms

Cyclosporiasis begins after an incubation period of about 1 week, with fatigue, loss of appetite, slight fever, vomiting, and watery diarrhea, followed by weight loss. The diarrhea usually subsides in 3 to 4 days, but relapses occur for up to 4 weeks.

Causative Agent

Cyclospora cayetanensis is an apicomplexan, and its oocysts are similar to those of *Cryptosporidium parvum* but larger. However, *C. cayetanensis* oocysts are not yet infectious when passed in the feces. With favorable conditions outside the body, sporocysts containing sporozoites develop within the infectious oocyst.

Pathogenesis

Little is known about the pathogenesis of *Cyclospora cayetanensis* because there is no good animal model for the disease. Analysis of small intestinal biopsies of infected patients confirms that sexual and asexual forms of the protozoan are both present in the intestinal epithelium.

Epidemiology

The oocysts of *Cyclospora cayetanensis* are immature and not infectious when eliminated in the stool, so person-to-person spread does not occur. In temperate regions, most infections happen in spring and summer months and among travelers to tropical areas. This is likely because warm, moist conditions favor maturation of the oocysts. Fresh produce (raspberries) has been implicated in a number of epidemics. In most instances, the produce was imported from a tropical region, but the source of the contaminating organisms is unknown. *C. cayetanensis* has been found in natural waters, but the source could not be determined.

Treatment and Prevention

Co-trimoxazole (trimethoprim plus sulfamethoxazole) effectively treats most cases of cyclosporiasis.

No specific preventive measures are available. People should boil or filter drinking water and thoroughly wash produce such as berries and leafy vegetables during an outbreak. The main features of cyclosporiasis are presented in **table 24.17.**

Amebiasis

Amebiasis is caused by *Entamoeba histolytica*. Although usually a mild disease, it causes about 30,000 deaths per year worldwide, mostly in developing countries. Life-threatening disease occurs in these areas because virulent *E. histolytica* strains spread easily in crowded, unsanitary living conditions.

Signs and Symptoms

Patients with *Entamoeba histolytica* infections are commonly asymptomatic. Some patients suffer from chronic, mild diarrhea,

TABLE 24.17	Cyclosporiasis
Signs and symptoms	Fatigue, loss of appetite, vomiting, watery diarrhea, and weight loss. Symptoms improve in 3 to 4 days, but relapses can occur for up to a month.
Incubation period	Usually about 1 week (range, 1 to 12 days)
Causative agent	*Cyclospora cayetanensis*, an apicomplexan
Pathogenesis	Little is known. Biopsies show sexual and asexual stages in the intestinal epithelium.
Epidemiology	The oocysts of *C. cayetanensis* are not infectious when discharged in the feces, and so person-to-person spread does not occur. Travelers to tropical countries are at risk of infection. Produce, especially raspberries, imported from tropical Central America, has been implicated in most North American outbreaks.
Treatment and prevention	Treated with co-trimoxazole (trade names Bactrim, Septra). Prevention: Use of boiled or filtered drinking water is advised in the tropics. Thorough washing of imported berries and leafy vegetables.

however, lasting months or years. Acute dysentery occurs in the most severe cases, which can be fatal.

Causative Agent

Entamoeba histolytica forms cysts that have a chitin-containing wall. Mature cysts—the infectious form for the next host—have four nuclei (**figure 24.24**).

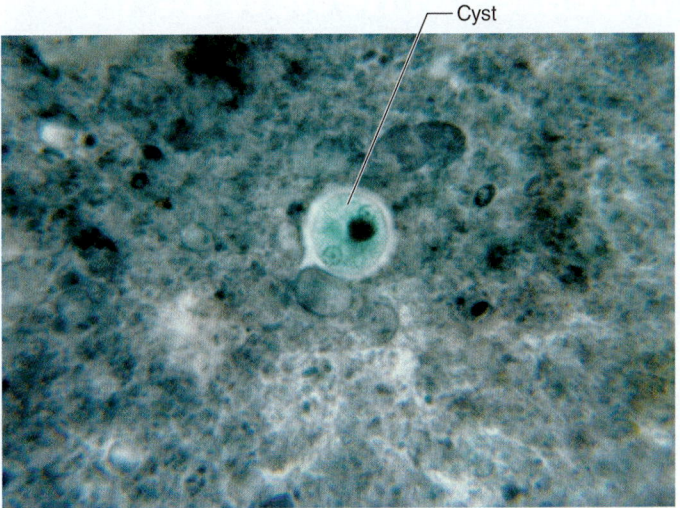

Cyst

FIGURE 24.24 Cysts of *Entamoeba histolytica*

❓ *Why is the organism transmitted in the cyst form and not as a trophozoite?*

TABLE 24.18 | Amebiasis

Signs and symptoms	Diarrhea, abdominal pain, blood in feces
Incubation period	2 days to several months
Causative agent	*Entamoeba histolytica*
Pathogenesis	Ingested cysts excyst, releasing trophozoites; these feed on mucus and bacteria in the large intestine; cytotoxic enzyme kills intestinal cells. The trophozoites may penetrate the intestinal wall, and are sometimes carried to the liver and other organs, resulting in abscesses.
Epidemiology	Ingestion of fecally contaminated food or water; disease associated with poverty, migrant workers, and men who have sex with men
Treatment and prevention	Metronidazole, paromomycin. Prevented by good sanitation and personal hygiene.

Pathogenesis

Ingested cysts of *E. histolytica* survive passage through the stomach. The cysts then excyst, releasing trophozoites. Upon reaching the large intestine, these trophozoites begin feeding on mucus and intestinal bacteria. The irritating effect of the trophozoites on the cells causes intestinal cramps and diarrhea. Many strains produce a cytotoxic enzyme that kills intestinal epithelium on contact. The organisms may penetrate the lining cells and enter deeper tissues of the intestinal wall. Sometimes, they penetrate into blood vessels and are carried to the liver or other body organs. Multiplication of the organisms and tissue destruction in the intestine and in other body tissues can result in amebic abscesses. Due to intestinal ulceration, the diarrhea is often bloody, and the condition is referred to as **amebic dysentery.**

Epidemiology

Entamoeba histolytica has a low infectious dose, spreads by the fecal-oral route, and is distributed worldwide. The disease is more common in tropical areas where sanitation is poor. In the United States, cases occur mainly in poverty-stricken areas, and among migrant farm workers and men who have sex with men. Humans are the only important reservoir.

Treatment and Prevention

Medications such as metronidazole and paromomycin are available for treatment. Prevention of amebiasis depends on sanitary measures and avoiding fecal contamination of foods and drinking water. **Table 24.18** gives the main features of amebiasis.

The key features of the diseases covered in this chapter are highlighted in the **Diseases in Review 24.1** table.

MicroAssessment 24.7

Giardiasis and cryptosporidiosis are common waterborne diseases that can also be transmitted person to person. Cyclosporiasis is a similar disease, but is not transmissible person to person. Amebiasis is caused by an ameba that ulcerates the large intestinal epithelium, resulting in dysentery.

19. *What causes the excessive intestinal gas that characterizes giardiasis?*
20. *Why is cryptosporidiosis difficult to control?*
21. *How could the fact that* Cryptosporidium parvum *is acid-fast be used in diagnosis?* ✚

Diseases in Review 24.1

Digestive System Diseases

Disease	Causative Agent	Comment	Summary Table
BACTERIAL INFECTIONS OF THE UPPER DIGESTIVE SYSTEM			
Dental caries (tooth decay)	*Streptococcus mutans*	Cariogenic organisms adhere to teeth and produce acids that damage tooth surfaces; incidence correlates with sugar consumption and lack of preventive dental care.	Table 24.1, p. 579
Periodontal disease	Certain groups of Gram-negative anaerobes	Gingivitis is the result of the inflammatory response to plaque and tartar at the gum line; chronic periodontitis damages the structures that support the teeth, leading to tooth loss.	Table 24.1, p. 579
Acute necrotizing ulcerative gingivitis (ANUG)	*Treponema* sp. with other anaerobes	Characterized by painful, bleeding gums, abscessed and broken teeth, and extremely foul breath; associated with poor oral hygiene, particularly in combination with other stresses.	Table 24.1, p. 579
Helicobacter gastritis	*Helicobacter pylori*	Causes peptic ulcers; associated with gastric cancers; *H. pylori* withstands stomach acid by producing urease.	Table 24.2, p. 580

(continued)

Digestive System Diseases

Disease	Causative Agent	Comment	Summary Table
VIRAL INFECTIONS OF THE UPPER DIGESTIVE SYSTEM			
Herpes simplex (cold sores)	Herpes simplex virus	Initial infection typically occurs during childhood; latent virus reactivates, causing recurrent cold sores.	Table 24.3, p. 582
Mumps	Mumps virus	Characterized by painful swelling of the parotid glands; virus enters via the respiratory tract and then spreads to glands; can cause sterility in males; preventable by vaccination.	Table 24.4, p. 584
BACTERIAL INFECTIONS OF THE LOWER DIGESTIVE SYSTEM			
Cholera	*Vibrio cholerae*	Classic example of watery diarrhea; *V. cholerae* colonizes small intestine and produces cholera toxin, causing secretion of water and electrolytes; rehydration is critical.	Table 24.5, p. 588
Shigellosis	*Shigella* species	Classic example of bacterial dysentery; *Shigella* species invade cells of large intestine; low infectious dose; some species produce Shiga toxin.	Table 24.6, p. 590
***Escherichia coli* gastroenteritis**	Certain strains of *E. coli*	Various symptoms depending on the pathovar; STEC strains cause bloody diarrhea and HUS; symptoms of ETEC are similar to cholera; symptoms of EIEC are similar to shigellosis.	Table 24.7, p. 590; Table 24.8, p. 591
***Salmonella* gastroenteritis**	*Salmonella enterica* (not serotypes Typhi or Paratyphi)	Foodborne, particularly poultry	Table 24.9, p. 592
Typhoid and paratyphoid fevers	*Salmonella enterica* serotypes Typhi and Paratyphi	*Salmonella* serotypes Typhi and Paratyphi cross intestinal mucosa and invade bloodstream, causing life-threatening enteric fever; vaccine for typhoid fever is available.	Table 24.10, p. 593
Campylobacteriosis	*Campylobacter jejuni*	Foodborne, particularly poultry	Table 24.11, p. 594
***Clostridium difficile*–associated disease**	*Clostridium difficile*	Typically occurs only in patients on antibiotic therapy; most common cause of diarrhea in healthcare settings.	Table 24.12, p. 595
VIRAL INFECTIONS OF THE LOWER DIGESTIVE SYSTEM—INTESTINAL TRACT			
Rotaviral gastroenteritis	Rotaviruses	Most common cause of viral gastroenteritis in infants and children in United States; preventable by vaccine approved in 2006.	Table 24.13, p. 597
Norovirus gastroenteritis	Noroviruses	Most common cause of viral gastroenteritis in the United States; highly infectious; epidemics are common on cruise ships and college dormitories.	Table 24.13, p. 597
VIRAL INFECTIONS OF THE LOWER DIGESTIVE SYSTEM—LIVER			
Hepatitis A	Hepatitis A virus (HAV)	Spreads via fecal-oral route; usually mild symptoms, but often prolonged; preventable by vaccination.	Table 24.14, p. 598
Hepatitis B	Hepatitis B virus (HBV)	Spreads in blood and semen; reverse transcribing virus; chronic infections can result in cirrhosis and liver cancer; chronic carriers; preventable by vaccination.	Table 24.14, p. 598
Hepatitis C	Hepatitis C virus (HCV)	Spreads in blood, possibly semen; most common chronic blood-borne infection in the United States; chronic infections can result in cirrhosis and liver cancer; chronic carriers.	Table 24.14, p. 598
PROTOZOAN INFECTIONS OF THE LOWER DIGESTIVE SYSTEM			
Giardiasis	*Giardia lamblia*	Most commonly identified waterborne illness in the United States; low infectious dose; cysts are chlorine-resistant.	Table 24.15, p. 603
Cryptosporidiosis	*Cryptosporidium parvum*	Wide host range; oocysts are chlorine-resistant; low infectious dose.	Table 24.16, p. 604
Cyclosporiasis	*Cyclospora cayetanensis*	No person-to-person spread because oocysts must mature in the environment to become infectious; imported produce has been implicated in outbreaks.	Table 24.17, p. 605
Amebiasis	*Entamoeba histolytica*	Most common in tropical regions where sanitation is poor; low infectious dose; amebic dysentery results from intestinal damage.	Table 24.18, p. 606

Defeating Digestive Tract Diseases

Development of better preventive and treatment techniques for digestive tract diseases has an urgency arising from massive food and beverage production and distribution methods. For example, in 1994, an estimated 224,000 people became ill because a tanker truck used to transport ice cream mix had previously carried liquid eggs. One day's production from a ground beef factory can yield hundreds of thousands of pounds of hamburgers, which are soon sent to many parts of this or other countries. The challenge is to better educate the producers and transporters, to develop guidelines to help them avoid contamination, to develop fast and accurate ways of identifying pathogens in food, and to utilize newly approved methods such as meat irradiation.

Other challenges include:

- **Developing additional effective vaccines against intestinal pathogens.** There is special need for vaccines that can be administered by mouth and evoke long-lasting mucosal immunity.
- **Exploring the influence of global warming on digestive tract diseases.** A study of Peruvian children by Johns Hopkins University scientists found an 8% increase in clinic visits for diarrhea with each 1°C increase in temperature from the normal.
- **Exploring new prevention and treatment options.** Scientists at the University of Florida have developed a genetically engineered *Streptococcus mutans* that does not produce lactic acid but readily displaces wild strains of the dental decay-causing bacterium. Researchers at the University of Alberta have custom-designed a molecule that binds circulating Shiga toxin, potentially preventing hemolytic uremic syndrome.

Summary

24.1 ■ Anatomy, Physiology, and Ecology

The digestive system encompasses the digestive tract and accessory organs. (figure 24.1)

The Upper Digestive System

The upper digestive system includes the mouth and salivary glands, the esophagus, and the stomach. Bacteria grow on teeth, attached to the pellicle that adheres to the enamel (figure 24.2), forming a biofilm called **dental plaque.** Peristalsis in the esophagus propels microbes to the stomach. The acidity and enzymes in the stomach destroy most bacterial cells.

The Lower Digestive System

The lower digestive system includes the small and large intestines, as well as the pancreas and liver. As the stomach contents enter the small intestine, digestive fluids from the pancreas and liver are mixed in. **Villi** and **microvilli** increase the surface area of the intestinal lining. The large intestine absorbs remaining nutrients and water from the intestinal contents and feces are excreted. Microbes make up about one-third of the weight of feces. The liver produces **bile,** which is stored in the gallbladder.

UPPER DIGESTIVE SYSTEM INFECTIONS

24.2 ■ Bacterial Diseases of the Upper Digestive System

Dental Caries (Tooth Decay) (table 24.1)

Plaque that contains *Streptococcus mutans* or other cariogenic microbes can act as a tiny acid-soaked sponge applied closely to the tooth (figures 24.3, 24.4). Control of dental caries depends mainly on restricting dietary sucrose, supplying fluoride, flossing, and brushing teeth.

Periodontal Disease (table 24.1)

Periodontal disease is caused by an inflammatory response to the plaque bacteria at the gum line; it is an important cause of tooth loss in older people (figure 24.3).

Acute Necrotizing Ulcerative Gingivitis (table 24.1)

Acute necrotizing ulcerative gingivitis (ANUG) can occur at any age in association with poor mouth care and drug abuse (figure 24.5).

Helicobacter pylori Gastritis (table 24.2)

Helicobacter pylori predisposes the stomach and the uppermost part of the duodenum to peptic ulcers (figures 24.6). Treatment with antimicrobial medications can cure the infection and prevent peptic ulcer recurrence.

24.3 ■ Viral Diseases of the Upper Digestive System

Herpes Simplex (Cold Sores) (table 24.3; figure 24.7)

Herpes simplex is caused by an enveloped DNA virus. HSV persists as a latent infection inside sensory nerves; active disease occurs when the body is stressed.

Mumps (table 24.4; figure 24.8)

Mumps is caused by an enveloped RNA virus that infects the parotid glands and a variety of other body tissues. Mumps virus generally causes more severe disease in persons beyond the age of puberty; it can be prevented using an attenuated vaccine (figure 24.9).

LOWER DIGESTIVE SYSTEM INFECTIONS

24.4 ■ Bacterial Diseases of the Lower Digestive System

General Characteristics (figure 24.10)

Bacterial infections of the small intestine typically cause watery diarrhea, whereas invasion of the large intestine causes **dysentery.** The bacteria are transmitted via the fecal-oral route, often through contaminated foods and water. Intestinal pathogens that have a low infectious dose can be transmitted by direct contact as well. Dehydration can be treated with **oral rehydration therapy (ORT),** using oral rehydration salts (ORS).

Cholera (table 24.5)

Cholera is a severe form of watery diarrhea caused by a toxin of *Vibrio cholerae* that acts on the small intestinal epithelium (figure 24.11).

Shigellosis (table 24.6)

Shigella species invade the epithelium of the large intestine, causing dysentery (figure 24.12). *Shigella dysenteriae* produces Shiga toxin, which causes **hemolytic uremic syndrome (HUS).**

Escherichia coli Gastroenteritis (tables 24.7, 24.8)

Escherichia coli strains that cause intestinal disease can be grouped into six pathovars: STEC (or EHEC), ETEC, EIEC, EPEC, EAEC, and DAEC. STEC strains cause hemolytic uremic syndrome.

Salmonella Gastroenteritis (table 24.9)

Most cases of *Salmonella* gastroenteritis originate from animals. The organisms are often foodborne, commonly in poultry and eggs.

Typhoid and Paratyphoid Fevers (table 24.10)

Typhoid fever is caused by *Salmonella* serotype Typhi, and paratyphoid fever is caused by *Salmonella* serotype Paratyphi. Both of these infect only humans. The diseases are characterized by high fever, headache, and abdominal pain. Untreated, they can be fatal. Vaccination helps prevent typhoid fever.

Campylobacteriosis (table 24.11)

Campylobacter jejuni is the most common bacterial cause of diarrhea in the United States; it usually originates from domestic animals (figure 24.13).

Clostridium *difficile*–Associated Disease (CDAD) (table 24.12)

Clostridium difficile causes diarrhea and **pseudomembranous colitis,** primarily in patients on antibiotic therapy (figure 24.14).

24.5 ■ Viral Diseases of the Lower Digestive System—Intestinal Tract

Rotaviral Gastroenteritis (table 24.13)

Rotaviral gastroenteritis is the main diarrheal illness of infants and young children but also can involve adults, as in traveler's diarrhea. Rotaviruses are segmented RNA viruses of the reovirus family (figure 24.15).

Norovirus Gastroenteritis (table 24.13)

Norovirus is the most common cause of viral gastroenteritis in the United States. The viruses are RNA viruses of the calicivirus family (figure 24.16).

24.6 ■ Viral Diseases of the Lower Digestive System—Liver

Hepatitis A (table 24.14; figures 24.17, 24.18)

Hepatitis A virus (HAV) is a picornavirus spread by fecal contamination of hands, food, or water. Infection is often asymptomatic in young children; disease in older children and adults is prolonged.

Hepatitis B (table 24.14)

Hepatitis B virus (HBV) is a hepadnavirus spread by blood, blood products, semen, and from mother to baby (figures 24.19, 24.20). Chronic infection can lead to **cirrhosis** and liver cancer.

Hepatitis C (table 24.14)

Hepatitis C virus (HCV) is a flavivirus transmitted mainly by blood. Acute infection is often asymptomatic, and many infections become chronic, leading to cirrhosis and liver cancer.

24.7 ■ Protozoan Diseases of the Lower Digestive System

Giardiasis (table 24.15)

Transmission of *Giardia lamblia* is usually via drinking water contaminated by feces. It is a common cause of traveler's diarrhea (figures 24.21, 24.22). The **cysts** survive in chlorinated water but can be removed by filtration.

Cryptosporidiosis ("Crypto") (table 24.16)

The life cycle of *Cryptosporidium parvum* takes place in the small intestinal epithelium. Its oocysts are infectious, resist chlorination, and are too small to be removed by many filters (figure 24.23). It is a cause of many water- and foodborne epidemics, and traveler's diarrhea.

Cyclosporiasis (table 24.17)

Transmission of *Cyclospora cayetanensis* is fecal-oral, via water or produce such as berries; it causes traveler's diarrhea. Oocysts are not infectious when passed in feces, so there is no person-to-person spread; no hosts other than humans are known.

Amebiasis (table 24.18)

Entamoeba histolytica is an important cause of dysentery; infection can spread to the liver and other organs (figure 24.24). Cysts are infectious.

Review Questions

Short Answer

1. Describe two characteristics of *Streptococcus mutans* that contribute to its ability to cause dental caries.
2. Describe the process of periodontal disease.
3. How does *Helicobacter pylori* cause stomach ulcers?
4. When would a case of mumps likely be complicated by swelling of the testicles?
5. What characterizes the solutions used for oral rehydration therapy?
6. How do *Shigella* cells move from one host cell to another even though they are non-motile?
7. Name four different pathogenic groups of *Escherichia coli*.
8. What predisposes someone to a *Clostridium difficile* infection?
9. Name two kinds of hepatitis that can be prevented by vaccines.
10. Contrast the cause and epidemiology of giardiasis and amebiasis.

Multiple Choice

1. Which of the following about intestinal bacteria is *false*?
   a) They produce vitamins.
   b) They can produce carcinogens.
   c) They are mostly aerobes.
   d) They produce gas from indigestible substances in foods.
   e) They include potential pathogens.

2. All of the following attributes of *Streptococcus mutans* are important in tooth decay *except*
   a) it produces endotoxin, which triggers an inflammatory response.
   b) it can grow at pH below 5.
   c) it produces lactic acid.
   d) it synthesizes glucan.
   e) it stores fermentable polysaccharide.

3. *Helicobacter pylori* has all of the following characteristics *except*
   a) it is a helical bacterium with sheathed flagella.
   b) it has not been cultivated *in vitro.*
   c) it produces a powerful urease.
   d) it causes long-term infections, lasting for years.
   e) it can cause stomach ulcers.

4. *Vibrio cholerae* pathogenesis involves all of the following *except*
   a) attachment to the small intestinal epithelium.
   b) production of cholera toxin.
   c) lysogenic conversion.
   d) acid resistance.

5. Which of the following statements concerning *Salmonella enterica* serotype Typhi is *false?*
   a) It is commonly acquired from domestic animals.
   b) It can colonize the gallbladder for years.
   c) It is highly resistant to killing by bile.
   d) It can destroy Peyer's patches.
   e) It causes typhoid fever.

6. Which statement about rotaviral gastroenteritis is *false?*
   a) A vaccine is available to prevent the disease.
   b) On a worldwide basis, most of the deaths are due to dehydration.
   c) Most cases of the disease occur in infants and children.
   d) The causative agent infects mainly the stomach.
   e) The disease is transmitted by the fecal-oral route.

7. Which of the following statements about noroviruses is *false?*
   a) They are the most common cause of viral gastroenteritis in the United States.
   b) They have a low infectious dose.
   c) They generally cause vomiting lasting 1 to 2 weeks.
   d) Immunity does not last long.
   e) They are a category B bioterrorism agent.

8. Which of the following statements about hepatitis is *false?*
   a) Both RNA and DNA viruses can cause hepatitis.
   b) Some kinds of hepatitis can be prevented by vaccines.
   c) HCV infections are often associated with injected-drug abuse.
   d) Lifelong carriers of hepatitis A are common.
   e) Hepatitis A spreads by the fecal-oral route.

9. Which of the following statements about hepatitis B virus is *false?*
   a) Replication involves reverse transcriptase.
   b) Infected persons may have large numbers of non-infectious viral particles circulating in their bloodstream.
   c) In the United States, infection rates have been steadily increasing over the last few years.
   d) Asymptomatic infections can last for years.
   e) Infection can result in cirrhosis.

10. Choose the most accurate statement about cryptosporidiosis.
    a) Waterborne transmission is unlikely.
    b) The host range of the causative agent is narrow.
    c) It is prevented by chlorination of drinking water.
    d) Person-to-person spread does not occur.
    e) The life cycle of the causative agent occurs within small intestinal epithelial cells.

Applications

1. One reason given by Peruvian officials for not chlorinating their water supply is that chlorine can react with substances in water or in the intestine to produce carcinogens. How do you assess the relative risks of chlorinating or not chlorinating drinking water?

2. A medical scientist is designing a research program to determine the effectiveness of hepatitis B vaccine in preventing liver cell cancer. Because liver cell cancer probably has multiple causes, how would you measure the success of an anticancer vaccination program?

Critical Thinking ➕

1. Why does the lack of a brown color in feces indicate hepatitis?

2. Mutant strains of *Helicobacter pylori* that lack the ability to produce urease fail to cause infection when they are swallowed. Infection occurs, however, if a tube is used to introduce them directly into the layer of mucus that overlies the stomach epithelium. What does this imply about the role of urease in the bacterium's pathogenicity?

25 Genitourinary Tract Infections

Color-enhanced electron micrograph of a human papillomavirus.

KEY TERMS

Catheter A flexible plastic or rubber tube inserted into the bladder or other body space, in order to drain it or deliver medication.

Chancre Sore resulting from an ulcerating infection.

Cystitis Inflammation of the bladder.

Papilloma A kind of tumor characterized by rough projections of tissue; warts are an example.

Pelvic Inflammatory Disease (PID) Infection of the fallopian tubes, uterus, or ovaries.

Puerperal Fever Infection of the uterus following childbirth, commonly caused by *Streptococcus pyogenes*.

Pyelonephritis Infection of the kidneys.

Toxic Shock Syndrome Collapse of the blood pressure due to a circulating bacterial toxin.

Urinary Tract Infection (UTI) Bacterial infection affecting any part of the urinary system.

difficult to see. Using dark-field microscopy, he saw that the organisms looked like a corkscrew without a handle—bacteria that are now known as spirochetes. The spirochetes were found in specimens from other cases of syphilis, and Schaudinn felt certain he had discovered the cause of syphilis, even though the organism could not be cultured on laboratory media. Schaudinn named the organisms *Spirochaeta pallida,* "the pale spirochete." This organism is now called *Treponema pallidum,* from the Greek words *trep* and *nema,* which mean "turning thread."

A Glimpse of History

Syphilis is an ancient disease that was once very common. History and literature describe many cases of famous people who went insane or were seriously disabled because of this disease. A few examples are Henry VIII (King of England, 1509–1547), Merriwether Lewis (Lewis and Clark Expedition, 1803–1808), Oscar Wilde (writer, 1854–1900), and Al Capone (gangster, 1899–1947).

Syphilis was first named the "French pox" or the "Neapolitan disease" because people thought it came from France or Italy. In 1530, the Italian physician Girolamo Fracastoro wrote a poem about a shepherd named Syphilis, who had ulcerating sores all over his body. The description matched the signs and symptoms of syphilis, and from then on, the disease was known by the shepherd's name.

Syphilis was a very severe disease in the early years of the epidemic, and people often died from it, because the available treatments of mercury and natural plant products were not very effective. With time, the disease became less serious because mutations and natural selection led to more resistant hosts and a less virulent microbe.

At first it was not known how syphilis was contracted, but gradually people realized the disease was sexually transmitted. In 1905, the German biologist Fritz Schaudinn examined fluid from a syphilitic sore. He observed some motile organisms that were very pale, slender, and

Infections of the reproductive and urinary tracts are very common. They are often uncomfortable and can have serious consequences, sometimes without advance symptoms. **Urinary tract infections (UTIs)** are the most frequent healthcare-associated infections and are the main origin of fatal, bacterial bloodstream invasions. Healthcare-associated uterine infections such as **puerperal fever,** once a common cause of maternal deaths from childbirth, still require strict medical attention to be prevented (see **A Glimpse of History,** chapter 19). Sexually transmitted infections (STIs) are widespread and may cause life-threatening diseases. Although most STIs are transmitted by sexual behavior, some can also be transmitted by drug needles, or through childbirth or breast feeding. ◀◀ puerperal fever, p. 437

The United States has the highest reported incidence of STIs among economically developed countries. About 85% of students have had sexual intercourse, and almost a third of these have had six or more sexual partners. Approximately 70% of sexually active students report that they or their partners rarely or never use a condom, suggesting a lack of either knowledge or concern about STI transmission and risks.

25.1 ■ Anatomy, Physiology, and Ecology

Learning Outcome

1. *Describe the anatomy, physiology, and ecology of the urinary and genital systems.*

The reproductive and urinary systems are referred to as the *genitourinary system* because they are positioned close to each other. Both systems are often affected by the same pathogens.

The Urinary System

The urinary system (or tract) consists of the kidneys, ureters, bladder, and urethra (**figure 25.1**). The kidneys act as a filtering system, removing waste materials from the blood and selectively reabsorbing substances that can be reused. Each kidney is connected to the urinary bladder by a ureter. The bladder acts as a holding tank for urine. Once full, it empties through the urethra. Waste materials are excreted in the urine.

The urinary system is protected from infection by a number of mechanisms. Sphincter muscles near the urethra keep the system closed most of the time and help prevent infections by stopping bacteria from ascending to the bladder. The downward flow of urine also helps clean the system by washing away microorganisms before they have a chance to multiply and cause infection. In addition, normal urine contains organic acids that may make it acidic, and antimicrobial substances such as small quantities of antibodies. The length and position of the urethra also play a role in preventing infection. Women have a relatively short urethra (4 cm) emptying close to the anus, and because of this they get UTIs far more frequently than men, who have a longer urethra (20 cm) more distant from the anus.

If an infection occurs in the urinary tract, an inflammatory response recruits phagocytes to the bladder, where they engulf and destroy the invading microorganisms. Lymphocytes in the infected kidneys or bladder wall also respond, secreting large amounts of antibodies into the urine. ◀◀ phagocytes, p. 346

MicroByte
If urine is alkaline, the person probably has an infection with a urease-producing bacterium that converts the urea in urine to ammonia.

The Genital System

The female genital system is composed of two ovaries, two fallopian tubes, a uterus, the vagina and the external genitalia (vulva), including the labia and clitoris (**figure 25.2a**). In women of childbearing age, an egg (ovum) is released from one of the two ovaries each month during ovulation and is swept into the adjacent fallopian tube. When fertilization occurs, it is normally in the fallopian tube. Ciliated epithelium of the tube then moves the fertilized ovum to the uterus, where it implants in the epithelial lining, developing into an embryo and then a fetus. If fertilization does not occur, the epithelial lining of the uterus sloughs off, producing a menstrual period. The lower end of the uterus is the cervix, which is filled with antimicrobial mucus, except during menstruation. The cervix opens to the vagina, which leads to the external genitalia.

Infections of the female genital system can occur in several places. The vagina is a portal of entry for a number of infectious organisms that can move to the uterus and fallopian tubes during menstruation. The cervix is a common site of STIs, and is a place where cancer can develop. Infection of the fallopian tubes can cause scarring and destruction of the ciliated epithelium, so that ova are not moved efficiently to the uterus, leading to infertility. The fallopian tubes are open at both ends, which allows organisms entering the fallopian tubes from the uterus to move into the abdominal cavity, where they may infect the liver and other organs.

In men, the reproductive organs include the testes (testicles), a variety of tubes, ducts and glands, and the penis (figure 25.2b). The testes are found outside the abdominal cavity in the scrotum. Sperm cells from each testis collect in a tightly coiled tubule called the epididymis and are carried to the prostate gland by a long tube called the vas deferens. The sperm and secretions of the prostate gland make up the semen. Prostate secretions have antimicrobial properties. The urinary and reproductive systems join at the prostate gland, which can be infected by urinary or

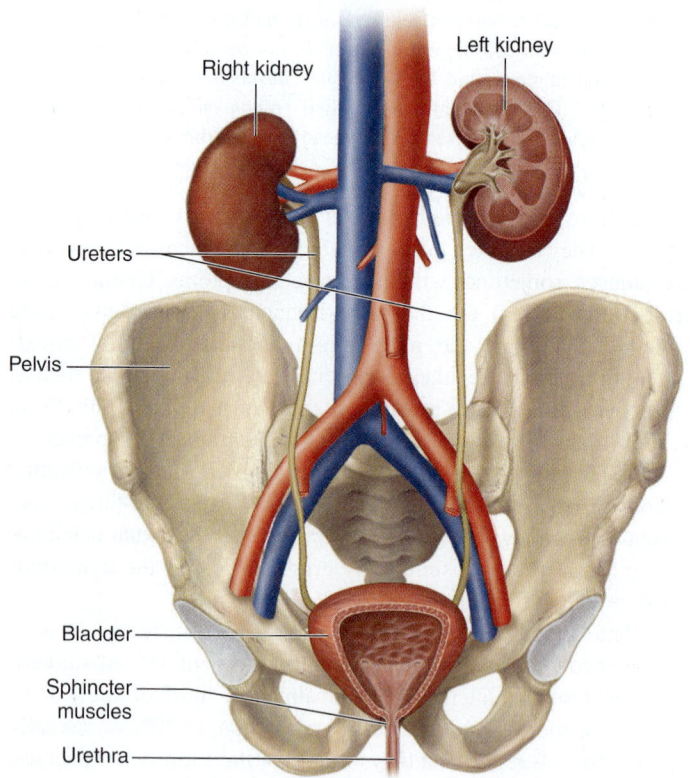

FIGURE 25.1 Anatomy of the Urinary System Urine flows from the kidneys, down the ureters, and into the bladder, which empties through the urethra.

? *How does the length and position of the urethra affect the tendency to get urinary tract infections?*

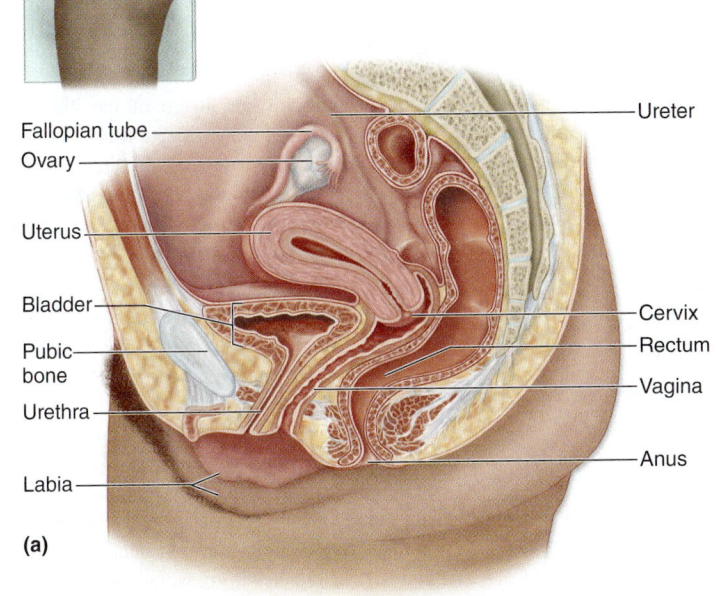

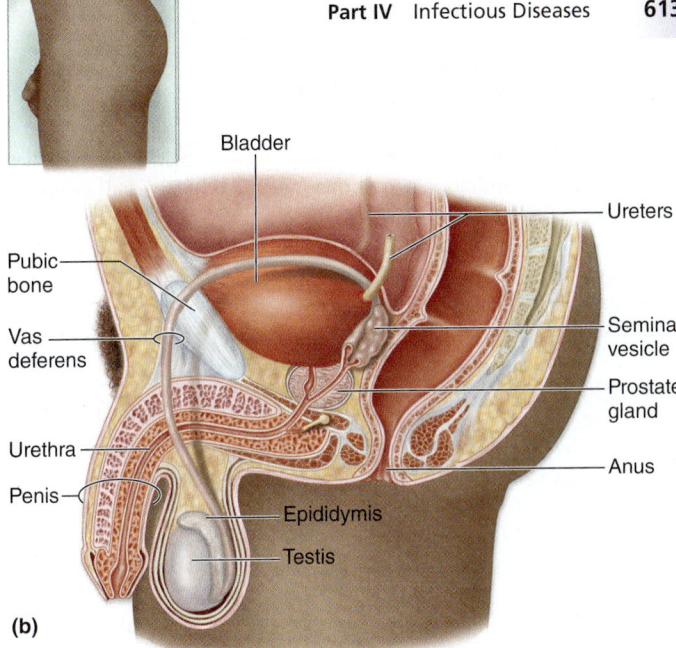

Fallopian tube
Ovary
Uterus
Bladder
Pubic bone
Urethra
Labia

Ureter
Cervix
Rectum
Vagina
Anus

(a)

Bladder
Pubic bone
Vas deferens
Urethra
Penis

Ureters
Seminal vesicle
Prostate gland
Anus
Epididymis
Testis

(b)

FIGURE 25.2 Anatomy of the Genital System (a) Female. **(b)** Male.

❓ *Why can infection of the fallopian tubes result in infertility?*

sexually transmitted pathogens. In older men, the prostate often enlarges and slows the flow of urine, making it more likely that a UTI will develop.

The urine and urinary tract above the urethra are usually free of microorganisms in both men and women. The lower urethra, however, has a normal resident microbiota that includes species of *Lactobacillus, Staphylococcus, Corynebacterium, Haemophilus, Streptococcus,* and *Bacteroides.* The normal microbiota of the genital tract of women is affected by the action of estrogen hormones on the epithelial cells of the vaginal mucosa. When estrogens are present, glycogen is deposited in these cells. The glycogen is converted to lactic acid by lactobacilli, resulting in an acidic pH that inhibits the growth of many potential pathogens. Lactobacilli may also release hydrogen peroxide, an inhibitor of some anaerobic bacteria. Thus, the normal microbiota and resistance to infection of the female genital tract vary considerably with the person's hormonal status. ◀◀ lactobacilli, p. 258

MicroAssessment 25.1

The genitourinary system is one of the portals of entry for pathogens. The fallopian tubes can provide a passageway for pathogens to enter the abdominal cavity. The cervix is a common site of infection by sexually transmitted pathogens. Prostate enlargement predisposes men to UTIs. Urine and the urinary tract above the urethra are normally free of microorganisms, but the lower urethra has several resident genera of bacteria. A woman's hormones affect vaginal resistance to infection.

1. *Describe how the bladder is protected from pathogens.*
2. *Which parts of the genitourinary system are normally sterile?*
3. *What changes might occur in the vagina if lactobacilli were eliminated?* ✚

25.2 ■ Urinary Tract Infections

Learning Outcomes

2. *Describe the features of bacterial cystitis.*
3. *Outline the characteristics of leptospirosis.*

Urinary tract infections (UTIs) can involve the urethra, bladder, or kidneys, alone or in combination. They account for about 7 million visits to the doctor's office each year in the United States. Any situation that causes inhibition of urination increases the risk of developing UTIs. Anesthesia and major surgery, for example, temporarily stop the reflex ability to urinate, and urine accumulates in the bladder. Even being too busy to empty the bladder may predispose a person to infection. Most cases of UTI occur in otherwise healthy young women with normal urinary flow.

Bacterial Cystitis ("Bladder Infection")

Cystitis (inflammation of the bladder) is the most common type of UTI. Bacterial cystitis is common among otherwise healthy women, and is also a frequent healthcare-associated infection.

Signs and Symptoms

Bacterial cystitis is sometimes asymptomatic, especially among children and the elderly. When symptoms do occur, they typically start suddenly and include a burning pain during urination, an urgent need to urinate, and frequent release of small amounts of urine. The urine is cloudy due to accumulation of leukocytes and may be a pale red color due to blood. It also often has a bad smell. The area above the pubic bone may be painful because of the underlying inflamed bladder.

Sometimes a more serious condition called **pyelonephritis** occurs. Symptoms of pyelonephritis include fever, chills, vomiting,

A 32-year-old married woman complained of 1 week of burning pain on urination, and frequent release of small amounts of bloody urine. About 8 days earlier, she had completed 3 days of trimethoprim-sulfamethoxazole therapy for similar symptoms. Tests at that time showed that her urine was infected with *Escherichia coli,* resistant only to amoxicillin. When the symptoms returned, she began drinking 12 ounces of cranberry juice three times daily but had only partial relief. She did not have other symptoms such as chills, fever, back pain, nausea, or vomiting. She was approximately 12 weeks pregnant.

Her medical history showed that she had suffered two or three similar episodes of urinary symptoms every year for a number of years. Sometimes the symptoms would go away when she forced herself to drink more, but at other times the symptoms would persist and she would obtain medical evaluation and treatment with an antibacterial medication. On one occasion several years before the present illness, she had chills, fever, back pain, nausea, and vomiting with her other urinary symptoms, and she was hospitalized for a "kidney infection." There was no history suggesting any underlying disease such as diabetes, cancer, or immunodeficiency.

The patient was examined and appeared well, with no obvious discomfort. Her temperature was normal, as was the rest of her physical examination.

The results of her laboratory tests included normal leukocyte count and kidney function tests. Microscopic examination of her urine showed many red and white blood cells. A Gram-stained smear of the urine showed

numerous rod-shaped, Gram-negative bacteria and neutrophils. Culture of the urine revealed more than 100,000 colony-forming units (CFUs) of *Escherichia coli* per milliliter. The bacterium was resistant to amoxicillin, but sensitive to the other antibacterials useful for treating urinary infections. ◀◀ neutrophils, p. 339

1. What is the diagnosis?
2. What is the treatment?
3. What is the prognosis?
4. What future preventive measures would be advisable?

Discussion

1. This woman's signs and symptoms clearly lead to a diagnosis of bacterial cystitis, but there are a number of clues in the presentation of this case that point to a serious complication. First, most patients with uncomplicated bacterial cystitis are cured by 3 days of an antibacterial medication to which the causative bacterium is susceptible. Her signs and symptoms recurred only 1 day after completing her medication. Second, she had signs and symptoms for a full week before she went for medical evaluation. Third, she was pregnant. Fourth, she gave a past history of being hospitalized for pyelonephritis.

   These clues make it highly likely that she has a condition called subclinical pyelonephritis, in which her bladder infection has spread to her kidneys but has not yet produced the symptoms of pyelonephritis. As many as 30% of patients with cystitis have subclinical pyelonephritis, depending on risk factors such as the clues mentioned in

this case. Kidney infections can be detected with a scintigram, an image of the kidneys produced following injection into the bloodstream of a tiny amount of radioactive material, which is removed from the blood by the kidneys and excreted. However, a scintigram would be dangerous for this patient because of her pregnancy.

2. The patient can be treated with trimethoprim-sulfamethoxazole, the same medication that was given before. Because infection in the kidneys takes much longer to cure than bladder infections, however, the treatment must be continued for 2 weeks or longer. Although this drug is safe to use early in pregnancy, it cannot be used late in pregnancy because it can make jaundice in the newborn worse. A urine culture was done 1 week after completion of treatment to be sure that the infection was truly gone.

3. The outlook is good for a full recovery without any permanent damage to the kidneys. The patient was advised, however, that repeated future infections of the kidneys could lead to kidney failure if not treated quickly.

4. For the future, this patient should use all the usual methods for preventing UTIs, including drinking enough to ensure urinating at least four or five times daily, avoiding delays in emptying the bladder, not using a diaphragm for contraception, urinating promptly after intercourse, and taking a preventive antimicrobial medication. No vaccine for this infection has been approved for use in the United States.

back pain, and tenderness overlying the kidneys. Repeated episodes of pyelonephritis lead to scarring and shrinkage of the kidneys and can cause kidney failure.

Causative Agents

Bladder infections usually originate from the normal intestinal microbiota. More than 80% of cases are caused by specific uropathogenic strains of *Escherichia coli.* The remaining infections in young women are caused by other *Enterobacteriaceae* members such as Gram-negative *Klebsiella* and *Proteus* species, or by Gram-positive *Staphylococcus saprophyticus.* Hospitalized patients, and people with long-standing bladder **catheters** (tubes inserted into the bladder), are often chronically infected with multiple species of bacteria, such as Gram-negative *Serratia marcescens* and *Pseudomonas aeruginosa* and the Gram-positive *Enterococcus faecalis.* Many of these species are resistant to antibiotics and are difficult to treat.

Pathogenesis

The causative agents of cystitis generally reach the bladder by moving up the urethra, a process helped by motility of the organisms. Uropathogenic *E. coli* (UPEC) strains that infect the urinary system have fimbriae that attach specifically to receptors on bladder epithelial cells. Bacterial attachment is followed by the death and sloughing of this superficial layer of cells. The bacteria then enter the underlying epithelium by endocytosis and multiply rapidly to create intracellular bacterial communities (IBCs), which are biofilm-like in nature. Bacteria later detach from the outer surface of the IBC and move into the bladder lumen to attach to surrounding naive epithelium, creating new IBCs. The UPEC may eventually establish a dormant intracellular reservoir that resists antibiotics and is undetected by the immune system, often leading to chronic or recurrent infections. Filamentous forms of UPEC are observed during IBC development. These forms help the causative organisms avoid the innate immune response by preventing formation of neutrophil-recruiting cytokines and possibly by killing neutrophils already present. Bacteria that move up the ureters to the kidneys can cause pyelonephritis. ◀◀ fimbriae, p. 65 ◀◀ endocytosis, p. 72

Epidemiology

About 30% of women develop cystitis at some time during their life. Factors that predispose women to UTIs include:

- **Short urethra.** The length and position of the urethra put it at risk for fecal contamination and colonization with potentially pathogenic intestinal bacteria. From the urethra, the bacteria need only travel a few centimeters to the bladder.

- **Sexual intercourse.** About one-third of UTIs in sexually active women are associated with sexual intercourse. The massaging effect of sexual intercourse on the urethra moves bacteria from the urethra into the urinary bladder. Many women develop bladder infections after having sex for the first time.

- **Birth control devices.** Diaphragms compress the urethra and slow the flow of urine, increasing the risk of UTI.

Other factors involved in development of UTIs include:

- **Enlarged prostate.** UTIs are unusual in men until about age 50, when enlargement of the prostate gland compresses the urethra and makes it difficult to completely empty the bladder.

- **Catheterization.** Medical conditions may require a bladder catheter for periods ranging from several days to months. This allows bacteria to reach the bladder and establish UTIs. Pathogens may create a biofilm either within or on the catheter, making it difficult or impossible to kill them with antibacterial medications

- **Paraplegia.** Paraplegics (individuals with paralysis of the lower half of the body) often have recurrent UTIs. This is because they cannot urinate normally due to lack of bladder control and require a catheter indefinitely to transfer their urine to a container.

MicroByte

In the United States, about 500,000 hospitalized patients develop bladder infections each year, mostly after catheterization.

Treatment and Prevention

Cystitis is usually easily treated with a few days of an antimicrobial medication effective against the causative bacterium. Pyelonephritis, a more serious condition, usually requires hospitalization and intravenous antibiotic treatment.

General ways to prevent UTIs include drinking enough to ensure urinating at least four or five times daily, urinating immediately after sexual intercourse, and wiping from front to back after defecation to minimize fecal contamination of the urethra. Many antimicrobial medications accumulate in the urine, reaching concentrations higher than in the blood, so taking a low dose of an antibiotic daily can be used to prevent recurrent infections.

TABLE 25.1	Bacterial Cystitis
Signs and symptoms	Sudden onset, burning pain on urination, urgency, frequency, foul smell, red-colored urine; fever, chills, back pain, and vomiting with pyelonephritis
Incubation period	Usually 1 to 3 days
Causative agents	Most due to *Escherichia coli*; other *Enterobacteriaceae* members, *Staphylococcus saprophyticus* cause some cases; healthcare-associated infections with antibiotic-resistant strains of *Pseudomonas*, *Serratia*, and *Enterococcus* sp.
Pathogenesis	Usually, bacteria ascend the urethra, enter the bladder, and attach by pili to receptors on urinary tract epithelium. Sloughing of cells and an inflammatory response follow. Spread to the kidneys can occur via the ureters, causing pyelonephritis and possible kidney failure.
Epidemiology	Bacterial cystitis is common in women, promoted by a relatively short urethra, use of a diaphragm, and sexual intercourse. Middle-aged men are at risk of infection because enlargement of the prostate gland partially obstructs their urethra. Catheterization often results in infection.
Treatment and prevention	Treatment: short-term antimicrobial therapy usually sufficient. Longer treatment for pyelonephritis. Prevention: taking enough liquid to urinate at least four to five times daily, wiping from front to back. Low daily dose of antibiotic may help prevent recurrent bacterial cystitis in women.

Cranberry juice and green tea are often used to prevent bacterial cystitis, but the scientific evidence on their effectiveness is inconclusive. The main features of bacterial cystitis are presented in **table 25.1**.

Leptospirosis

Leptospirosis is mainly a disease of animals, but it can be passed to humans. The causative bacterium enters the body through a mucous membrane or wound and is then carried to the urinary system by the bloodstream. The disease is probably one of the most common of all the zoonoses, but many cases are mild, resolve without treatment, and remain undiagnosed. More severe cases require treatment and can sometimes be fatal. ◀◀ zoonoses, p. 439

Signs and Symptoms

Leptospirosis infections are often asymptomatic. When signs and symptoms do occur, they begin after an incubation period averaging 10 days. In mild cases, which are most common, symptoms are flulike and include the sudden development of a headache, spiking fever, chills, and muscle pain. A dry cough may occur. A characteristic feature of this first phase (septicemic) is photophobia (sensitivity to light) and red eyes, caused by dilation of small blood vessels. These symptoms usually fade within a week, and all signs and symptoms of illness are gone in a month or less.

In severe cases of the disease, symptoms recur after 1 to 3 days of feeling well. Symptoms of this second (immune) phase include those of the septicemic phase as well as bleeding from various sites, vomiting, rash, and confusion.

In rare cases, people with leptospirosis develop Weil's disease, which affects the liver and kidneys. Signs and symptoms of Weil's disease include jaundice (yellowing of skin and eyes), liver and kidney failure, hemorrhage in many organs, and meningitis. People who suffer biphasic leptospirosis or Weil's disease need medical attention, since these can be fatal.

Causative Agent

Leptospirosis is caused by *Leptospira interrogans,* a slender aerobic Gram-negative spirochete with hooked ends (**figure 25.3**). There are more than 250 antigenic types of this species, some of which were given different names in the past. Most strains can be cultivated using cell-free media. ◄◄ spirochetes, p. 272

Pathogenesis

Leptospira interrogans enters the body through mucous membranes, eyes and breaks in the skin. No lesion develops at the site of entry, but the organisms multiply and spread throughout the body by way of the bloodstream, penetrating all tissues including the eyes, and the brain. Severe pain is characteristic of this first (septicemic) phase, and victims sometimes are given unnecessary surgery because it is assumed the pain is from appendicitis or gallbladder infection. There are no inflammatory changes or tissue damage at this stage. Within a week, the immune response destroys the organisms present in most tissues, although they continue to multiply in the kidneys. Many people recover at this stage, with no further development of symptoms.

If the disease progresses, the infected person then usually seems healthy for a few days before the second phase of signs and symptoms occurs. The cause of the recurrent symptoms is thought to be an immune response, and this second phase of the illness is therefore called the "immune phase." This phase is characterized by injury to the cells that line the lumen of tiny blood vessels, causing clotting and impaired blood flow in tissues throughout the body, and leading to most of the serious effects of the illness, including kidney failure, the main cause of fatalities.

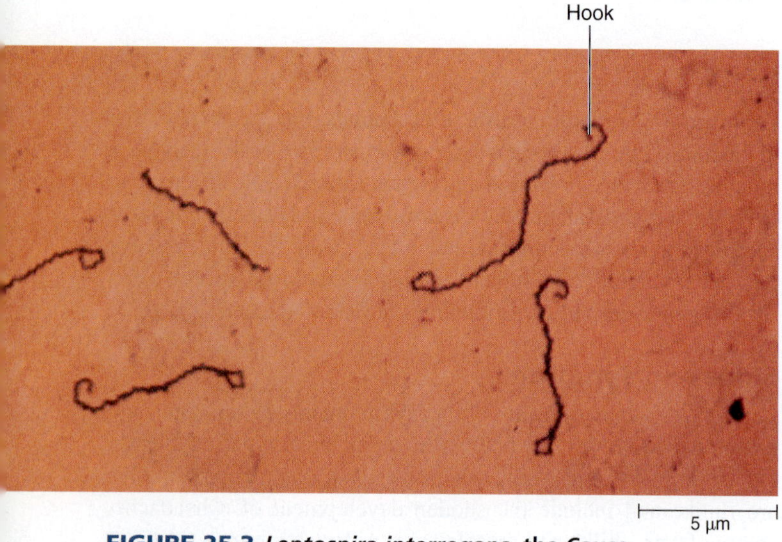

Epidemiology

Leptospirosis occurs around the world in all types of climates, although it is more common in the tropics. *Leptospira interrogans* infects many species of wild and domestic animals, usually causing little or no apparent illness, but sometimes causing fatal, epidemic disease. The organisms are excreted in the animal's urine, and urine spots on the ground remain infectious for as long as they are still moist. Contaminated urine is the main mode of transmission to other hosts. *L. interrogans* can also survive in mud or water for several weeks, and there is a correlation between the amount of rainfall and leptospirosis incidence. Humans contract leptospirosis from water, soil, or food contaminated with infected animal urine. Many cases are caused by swimming in urine-contaminated fresh water. Ingested organisms enter through the mucosa of the upper digestive system. Infected humans usually excrete the organisms for a few weeks to several months, but person-to-person transmission does not seem to occur.

Treatment and Prevention

Mild cases of leptospirosis resolve without treatment, although antibiotics can be given. In biphasic leptospirosis, a number of different antibiotics are effective for treating the disease, but only if they are started during the first 4 days of the initial illness. Once treatment begins, the person often shows a temporary worsening of symptoms due to massive release of antigens from organisms lysed by the antimicrobial medication.

There are few effective measures for preventing leptospirosis other than avoiding animal urine. Multivalent vaccines (that contain a number of different serotypes of *L. interrogans*) are available for preventing the disease in domestic animals, but they do not always prevent the animal from becoming a carrier. Small doses of a tetracycline antibiotic can prevent the disease in high-risk individuals, such as veterinarians, slaughterhouse workers, farmers, sewer workers, and water sport enthusiasts. The main features of leptospirosis are shown in **table 25.2.**

Hook

5 µm

FIGURE 25.3 *Leptospira interrogans,* the Cause of Leptospirosis

❓ *What is the main mode of transmission of this organism?*

MicroAssessment 25.2

Situations that affect normal urine flow predispose a person to UTIs. Bladder infections are common, especially in women, and are usually caused by normal intestinal bacteria ascending from the urethra. Pyelonephritis is a serious complication that may cause death due to kidney failure. In leptospirosis, a widespread zoonosis spread by urine, the causative organisms enter the urinary system from the bloodstream. Although usually mild, the disease can be a severe biphasic illness characterized by tissue invasion and pain in the first phase and tissue destruction and kidney damage in the second.

4. *What organism causes the majority of bladder infections in otherwise healthy women? From where does this organism come?*

5. *How do people become infected with* Leptospira interrogans?

6. *How can catheterization lead to bladder infection?*

TABLE 25.2 Leptospirosis

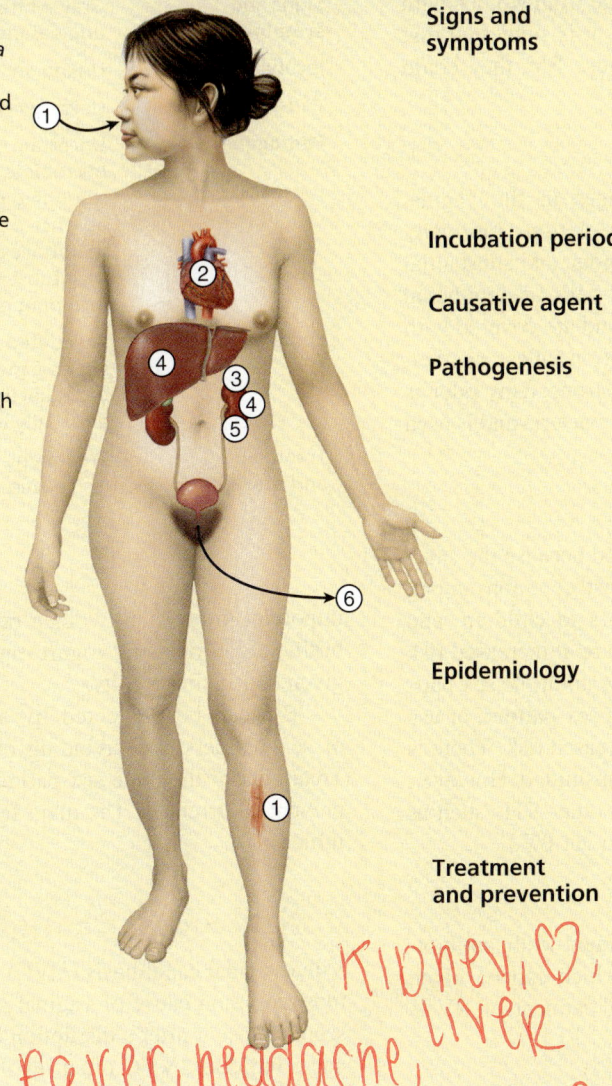

1. Water or animal urine contaminated with *Leptospira interrogans* splashes onto mucous membrane or abraded skin.

2. The bacteria infect the bloodstream and are carried throughout all the body tissue causing fever, intense pain.

3. Signs and symptoms resolve and bacteria disappear from blood and tissues, except kidneys.

4. Biphasic illness associated with severe damage to liver and kidneys.

5. Complete recovery occurs if kidney failure can be effectively treated.

6. Excretion of *L. interrogans* continues in urine.

Signs and symptoms	Many mild and asymptomatic cases. Others have a biphasic illness: spiking fever, headache, muscle pain, bloodshot eyes in the first (septicemic) phase, then 1 to 3 days of improvement; heart, brain, liver, and kidney damage in the second (immune) phase
Incubation period	Usually about 10 days (range, 2 to 30 days)
Causative agent	*Leptospira interrogans*, a spirochete with many serotypes
Pathogenesis	The bacteria penetrate mucous membranes or breaks in the skin, multiply in the bloodstream, and are carried to all parts of the body. Septicemic phase: severe pain with penetration of body tissues, but little or no tissue damage. Immune phase: damage to cells that line small blood vessels and clotting of blood. Causes severe damage to the liver, kidneys, heart, brain, and other organs.
Epidemiology	Worldwide distribution. Wide range of animal hosts chronically excrete the bacteria in their urine, causing contamination of natural waters and soils. Organisms remain infectious under warm, moist, neutral or alkaline conditions for long periods of time.
Treatment and prevention	Treatment: various antibacterial medications useful in treatment of leptospirosis but only if given early in the disease. Prevention: avoiding contact with animal urine. Vaccines prevent disease in domestic animals, may not prevent urinary carriage. Tetracycline antibiotics preventive in epidemics.

[Handwritten annotations: "kidney, ♡, liver" and "fever, headache, muscle ache, scrap in wound"]

25.3 ■ Genital System Diseases

Learning Outcome

4. *Compare and contrast bacterial vaginosis (BV), vulvovaginal candidiasis (VVC), and staphylococcal toxic shock syndrome.*

The genital tract is the portal of entry for many infectious diseases, both non-sexually and sexually transmitted. This section discusses some genital system diseases that are not generally transmitted sexually.

Bacterial Vaginosis (BV)

In the United States, bacterial vaginosis (BV) is the most common vaginal disease of women in their childbearing years. It is termed *vaginosis* rather than *vaginitis* because there are no inflammatory changes. In the United States, BV is common in pregnant women, and puts them at risk for premature deliveries.

Signs and Symptoms

BV is characterized by a thin, grayish-white, slightly bubbly vaginal discharge that has a characteristic strong fishlike smell. The bacteria associated with BV may spread to the uterus or fallopian tubes, causing **pelvic inflammatory disease (PID),** which can lead to sterility. About half of BV cases are asymptomatic.

Causative Agent *[Handwritten: "Prevotella Mobiluncus myco"]*

The cause or causes of BV are unknown. Because most cases show a significant decrease in vaginal lactobacilli, conditions that suppress lactobacilli or promote the growth of other microbiota are thought to play a causative role. The discharge that characterizes BV contains large numbers of bacteria, including aerotolerant *Gardnerella vaginalis*, anaerobic species of the genera *Mobiluncus* and *Prevotella*, *Mycoplasma* sp., and anaerobic streptococci. Although these species are generally present in women with BV, women voluntarily inoculated with cultures of these organisms do not always develop the disease. Also, each of these species can occur in vaginal secretions of healthy women, although typically in much smaller numbers than in women showing symptoms of BV. However, since discharge from women with BV can cause the disease in healthy women, it is suggested that although these species do not necessarily cause BV, they could play a role in developing the disease.

Pathogenesis

Women with BV have characteristic changes in the vagina, including a loss of acidity of the vaginal secretions (normally pH 3.8–4.2), disruption of the normal microbiota, and substantial increase in the numbers of clue cells. Clue cells are epithelial cells that have sloughed off the vaginal wall and are covered with bacteria (**figure 25.4**). There is no inflammation unless another, concurrent vaginal infection is present. The strong fishy odor is caused by metabolic products of the anaerobic bacteria and is used for diagnosis in the whiff test.

Epidemiology

The epidemiology of BV is not well understood because the causative agent is not known. The disease is most common among sexually active women and sometimes occurs in children who have been sexually abused. Pregnant women are at increased risk of BV. Women who wear thongs, douche, have multiple sex partners, have sex with other women, have a new sex partner, or use an intrauterine device (IUD) also have an increased risk. There is no proof that bacterial vaginosis is sexually transmitted. However, women with BV are at higher risk of getting other STIs such as gonorrhea, HIV, or chlamydia. Virgins seldom get BV.

Treatment and Prevention

Most cases of BV respond quickly to treatment with antibiotics such as metronidazole or clindamycin, which can be given orally or vaginally. The disease can recur. Treatment of BV is

TABLE 25.3	Bacterial Vaginosis
Signs and symptoms	Gray-white vaginal discharge and unpleasant fishy odor
Incubation period	Unknown
Causative agent	Unknown
Pathogenesis	Uncertain. Marked change in normal microbiota composition. Increased sloughing of vaginal epithelium in the absence of inflammation. Odor due to metabolic products of anaerobic bacteria. May cause complications of pregnancy, including premature births.
Epidemiology	Associated with many sexual partners or a new partner, but can occur in the absence of sexual intercourse. No evidence that it is a sexually transmitted infection.
Treatment and prevention	Treatment with metronidazole is effective. No proven preventive measures.

important in pregnant women because BV may cause premature birth. Studies on using yogurt vaginally to restore lactobacilli have given conflicting results.

BV can be prevented by abstinence, limiting the number of sex partners, and avoiding douching and the use of thongs. Treatment of the male sex partners of patients with BV does not prevent recurrences. The main features of BV are summarized in **table 25.3**.

Vulvovaginal Candidiasis (VVC)

Vulvovaginal candidiasis (VVC), a fungal infection, is the second most common cause of vaginal symptoms after BV. Like BV, it seems to occur after a disruption of the normal microbiota. As the name indicates, VVC often involves not only the vagina, but the vulva as well.

Signs and Symptoms

The most common signs and symptoms of VVC are constant, intense itching and burning of the vagina or vulva. Typically, there is a large amount of thick, clumpy whitish or whitish-gray vaginal discharge. The vaginal mucosa is usually red and somewhat swollen, and may have patches of cottage cheese–appearing clumps attached to it.

Causative Agent

VVC is caused by *Candida albicans,* a yeast that is part of the normal microbiota of the vagina in about a third of all women (**figure 25.5**). Because *C. albicans* is a fungus, it has a eukaryotic cell structure. ◀◀ yeast, p. 283

Pathogenesis

Normally, vaginal colonization by *Candida albicans* causes no symptoms. The growth of the organism is usually limited by the immune system and the normal vaginal lactobacilli that occupy the same niche and compete for nutrients. When the normal

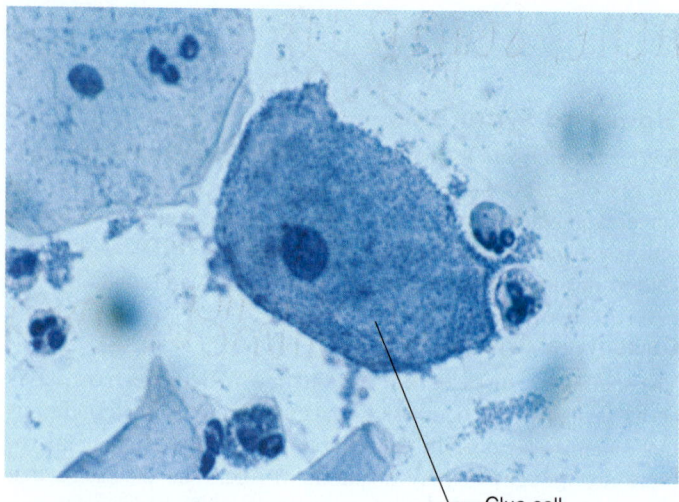

Clue cell

FIGURE 25.4 Clue Cell in an Individual with Bacterial Vaginosis The cells in the photograph are stained epithelial cells that have sloughed from the vaginal wall, one of which, the clue cell, is a dark color.

? *Why does the clue cell appear darker in color than other cells?*

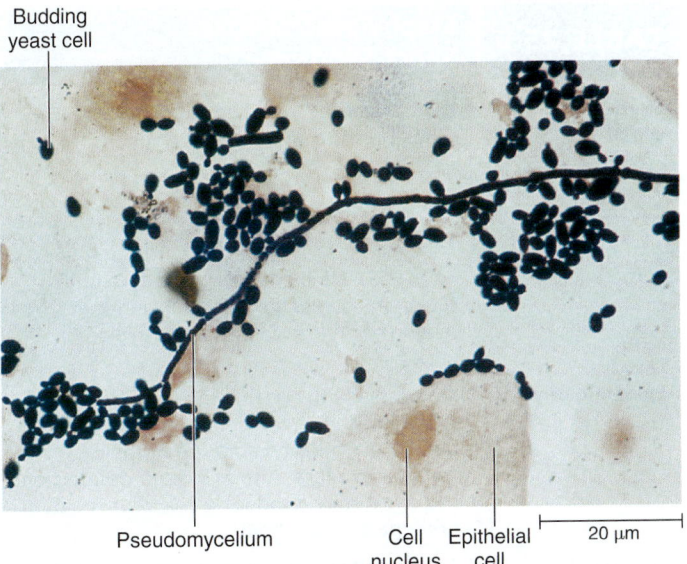

FIGURE 25.5 *Candida albicans* Vaginal discharge of a woman with vulvovaginal candidiasis, showing *C. albicans*.

❓ *What usually prevents VVC from occurring?*

TABLE 25.4	Vulvovaginal Candidiasis
Signs and symptoms	Itching, burning, thick, white vaginal discharge, redness and swelling
Incubation period	Usually unknown. Generally 3 to 10 days when associated with antibacterial medications
Causative agent	*Candida albicans*, a yeast
Pathogenesis	Inflammatory response to overgrowth of the yeast, which is often present among the normal microbiota
Epidemiology	Not contagious. Usually not sexually transmitted. Associated with antibacterial therapy, use of oral contraceptives, pregnancy, and uncontrolled diabetes, but most cases have no identifiable predisposing factor.
Treatment and prevention	Intravaginal antifungal medications such as clotrimazole usually effective. No proven preventive measures.

microbiota balance is disturbed—as occurs during menstruation or pregnancy, or when using oral contraceptives or antibiotics—*C. albicans* can multiply freely, causing an inflammatory response. The signs and symptoms of VVC occur within about 10 days.

Epidemiology

Factors that predispose a woman to *Candida* infection are late pregnancy, poorly controlled diabetes, and the use of oral contraceptives or antibiotics. Hormone replacement therapy may also increase the risk of VVC. However, most patients with VVC have no known predisposing factors. The disease does not spread from person to person.

Treatment and Prevention

Intravaginal treatment of *C. albicans* infections with antifungal medicines such as nystatin, clotrimazole, or fluconazole is usually effective. Fluconazole (Diflucan) taken by mouth is generally safe and effective, although it may cause side effects such as headache and nausea, and it can interact with other medications, causing rare, but serious reactions. Self-diagnosis and treatment with over-the-counter medications can lead to development of drug-resistant organisms.

Prevention of VVC depends on minimizing the use and duration of antibacterial medications and on effective treatment of underlying conditions such as diabetes. The main features of VVC are presented in **table 25.4**.

Staphylococcal Toxic Shock Syndrome

Toxic shock syndrome was described in the late 1970s in several children with staphylococcal infections. In 1980, it became epidemic in young, healthy, menstruating women who were using a brand of high-absorbency tampon that has since

been removed from the market (**figure 25.6**). The term *toxic shock syndrome* was used to describe the signs and symptoms of the illness. Now that we know its cause, it is called *staphylococcal toxic shock syndrome*. This is not a new disease, but one that emerged in a new form and became much more common as a result of changes in technology and human behavior. ◀◀ emerging diseases, p. 448

Signs and Symptoms

Staphylococcal toxic shock syndrome is characterized by the sudden development of high temperature, headache, muscle aches, bloodshot eyes, vomiting, diarrhea, a sunburnlike rash, and confusion. Typically, the skin peels about a week after the development

antifungal medicines - clotrimazole

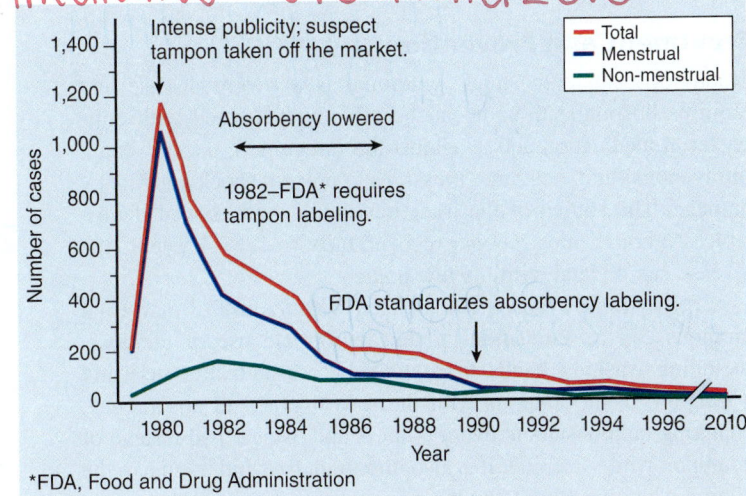

FIGURE 25.6 **Staphylococcal Toxic Shock Syndrome, United States** A sharp drop in cases occurred when a brand of high-absorbency tampon was taken off the market. Since 1996, only six or fewer total cases have generally been reported per 100,000 population each year.

❓ *How does tampon use increase the risk of toxic shock syndrome?*

of the disease. Without treatment, the blood pressure can drop, leading to multiorgan failure, coma, and sometimes death.

Causative Agent

Staphylococcal toxic shock syndrome is caused by strains of *Staphylococcus aureus* that produce toxic shock syndrome toxin-1 (TSST-1) or other related exotoxins. ◄◄ *Staphylococcus aureus*, p. 524 ◄◄ exotoxins, p. 391

Pathogenesis

Tampon-associated toxic shock syndrome usually begins 2 to 3 days after the start of menstruation when tampons are used. The staphylococci grow in the blood-soaked tampon. The bacteria rarely spread throughout the body, but as they multiply they produce TSST-1 or other exotoxins. Staphylococcal toxic shock syndrome results from absorption of these toxins into the bloodstream. The toxins are superantigens that cause activation of large numbers of helper T cells, leading a massive release of cytokines (cytokine storm). This in turn causes a drop in blood pressure and multiorgan failure—the most dangerous aspect of the potentially fatal illness. TSST-1 causes 75% of all toxic shock syndrome cases. ◄◄ superantigen, p. 393

Epidemiology

Staphylococcal toxic shock syndrome can occur after infection with any strain of *Staphylococcus aureus* that produces one of the responsible exotoxins. The syndrome can occur after infection of surgical wounds, infections associated with childbirth, and other types of staphylococcal infections. It does not spread from person to person. Using tampons increases the risk of staphylococcal toxic shock, and the higher-absorbency tampons may pose a greater risk. Use of intravaginal contraceptive sponges also increases risk. People who have recovered from the disease are not always immune to it. Since 1990, there has been a slow, steady decline in the incidence of staphylococcal toxic shock syndrome, now estimated to be six or fewer total cases per 100,000 people per year (figure 25.6).

Treatment and Prevention

Staphylococcal toxic shock syndrome is a severe disease and requires hospitalization. It can be effectively treated with antibacterial medication active against the infecting *S. aureus* strain, intravenous fluid, and other measures to prevent shock and kidney damage. The source of the infection should be removed if possible. Although most people recover fully in 2 to 3 weeks, the disease can be fatal within a few hours.

Toxic shock syndrome associated with the use of menstrual tampons can be prevented by the appropriate use of tampons, including washing hands thoroughly before and after inserting a tampon, using tampons with the lowest practical absorbency, changing tampons about every 6 hours and using a pad instead of a tampon while sleeping. It is also important to avoid trauma to the vagina when inserting tampons, to recognize the signs and symptoms of staphylococcal toxic shock syndrome, and to remove any tampon immediately if symptoms occur. Women who have had staphylococcal toxic shock syndrome previously should not use tampons. **Table 25.5** describes the main features of staphylococcal toxic shock syndrome.

TABLE 25.5	Staphylococcal Toxic Shock Syndrome
Signs and symptoms	Fever, vomiting, diarrhea, muscle aches, low blood pressure, and a rash that peels
Incubation period	3 to 7 days
Causative agent	Certain toxin-producing strains of *Staphylococcus aureus*
Pathogenesis	Toxin (TSST-1 and others) produced by certain strains of *S. aureus*; toxins are superantigens, causing cytokine release and drop in blood pressure.
Epidemiology	Associated with certain high-absorbency tampons, leaving tampons in place for long periods of time, and abrasion of the vagina from tampon use. Also as a result of infection by certain toxin-producing *S. aureus* strains in other parts of the body.
Treatment and prevention	Antimicrobial medication effective against the causative *S. aureus* strain; intravenous fluids. Awareness of symptoms. Prompt treatment of *S. aureus* infections; frequent change of tampons by menstruating women.

MicroAssessment 25.3

Bacterial vaginosis (BV), the most common vaginal disease of women in their childbearing years, is characterized by a significant change in the composition of the normal vaginal microbiota. Vulvovaginal candidiasis (VVC) often occurs as a result of antibacterial therapy suppressing normal vaginal microbiota, but many other cases arise for unknown reasons. Staphylococcal toxic shock syndrome is caused by certain strains of *Staphylococcus aureus* that produce exotoxins, which are absorbed into the bloodstream, causing the massive release of cytokines responsible for shock.

7. *What is the causative agent of bacterial vaginosis (BV)?*

8. *Why was the incidence of staphylococcal toxic shock syndrome higher in menstruating women than in other people during the early 1980s?*

9. *Why would using antibiotics predispose a woman to vulvovaginal candidiasis (VVC)?* ✚

25.4 ■ Sexually Transmitted Infections: Scope of the Problem

Learning Outcomes

5. *Discuss behaviors that increase the risk of acquiring STIs.*

6. *List three ways that STIs can be prevented.*

MicroByte

Despite spending billions each year for STI control, an estimated 19 million Americans are infected annually, almost half in individuals 15 to 24 years old.

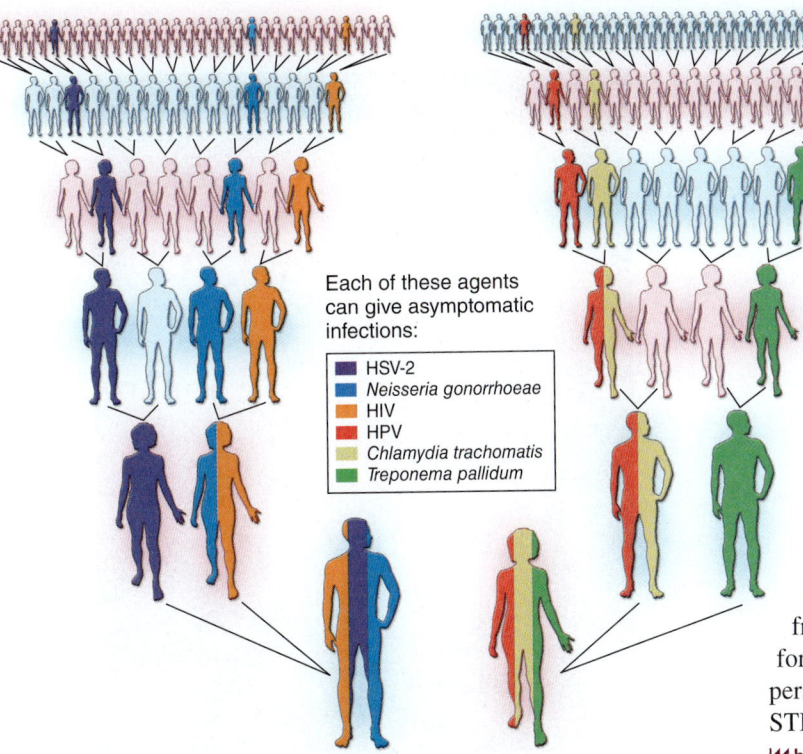

Each of these agents can give asymptomatic infections:

- ■ HSV-2
- ■ *Neisseria gonorrhoeae*
- ■ HIV
- ■ HPV
- ■ *Chlamydia trachomatis*
- ■ *Treponema pallidum*

FIGURE 25.7 The Possible Risk of Acquiring STIs in People Having Unprotected Sex Each partner had two previous sexual partners, and each of these partners had two previous partners, and so on. This risk of contracting an STI rises with the number of sexual partners. (HSV-2—herpes simplex virus type 2, HIV—human immunodeficiency virus, HPV—human papillomavirus)

? *What measures can be taken to avoid getting an STI?*

acquiring an STI, although does not absolutely guarantee protection.

The list of diseases that can be transmitted sexually is very long and includes some that have non-sexual modes of transmission as well—for example, shigellosis, giardiasis, scabies, and viral hepatitis. The epidemiology can be obscure, such as when a drug abuser contracts hepatitis B from sharing a contaminated needle, remains symptom-free for years, and then unknowingly transmits the virus to another person during sexual intercourse. **Table 25.7** lists some common STIs discussed in this chapter. ◀◀ shigellosis, p. 588 ◀◀ giardiasis, p. 602 ◀◀ hepatitis, pp. 599–601

A person who gets one STI may have acquired others without knowing it, so he or she needs to be tested for that possibility. The risk of acquiring an STI increases sharply with the number of partners with whom an individual has unprotected sex, and with the numbers of sexual partners of those partners (**figure 25.7**).

Table 25.6 lists some often-overlooked signs and symptoms that can indicate the possibility of an STI. These symptoms require clinical evaluation even if they go away without treatment, especially if they appear within a few weeks of sexual intercourse or other intimate sexual contact with a new partner. A number of STIs can cause few or no signs or symptoms, and yet they can have serious effects and are transmissible to other people.

Simple measures are highly effective in preventing STIs. These include abstaining from sexual intercourse or having a monogamous relationship with a non-infected person. Using latex or polyurethane condoms also significantly reduces the risk of

TABLE 25.6	**Signs and Symptoms That Suggest an STI**

1. Abnormal discharge from the vagina or penis
2. Pain or burning sensation with urination
3. Sore or blister (painful or painless) on the genitals or nearby; swellings in the groin
4. Abnormal vaginal bleeding or unusually severe menstrual cramps
5. Itching in the vaginal or rectal area
6. Pain in the lower abdomen in women; pain during sexual intercourse
7. Skin rash or mouth lesions

TABLE 25.7	**Common Sexually Transmitted Infections**	
Disease	**Cause**	**Comment**
Bacterial		
Gonorrhea	*Neisseria gonorrhoeae*	Average reported cases per year—340,000. True incidence much higher.
Chlamydial infections	*Chlamydia trachomatis*	Average reported cases per year—800,000. True incidence much higher.
Syphilis	*Treponema pallidum*	Average reported cases (primary and secondary) per year—6,600.
Chancroid	*Haemophilus ducreyi*	Average reported cases per year—50. True incidence much higher.
Viral		
Genital herpes	Herpes simplex virus (HSV)	Not reportable. Estimated 45 million Americans infected; about 85% HSV, type 2.
Papillomavirus infections	Human papillomavirus (HPV)	Not reportable. Estimated 40 million Americans infected.
AIDS	Human immunodeficiency virus (HIV)	Average reported cases per year—40,000.
Protozoal		
Trichomoniasis ("trich")	*Trichomonas vaginalis*	Not reportable. Estimated 5 million Americans infected per year.

25.5 ■ Bacterial STIs

Learning Outcomes

7. *Compare and contrast gonorrhea and chlamydial genital system infections.*

8. *Compare and contrast syphilis and chancroid.*

Most of the bacteria that cause STIs survive poorly in the environment. Because of this, transmission from one person to another usually requires intimate physical contact and is highly unlikely to occur by a handshake or from a toilet seat. Most STI pathogens are adept at avoiding both innate and adaptive immunity.

Gonorrhea

Gonorrhea is the second most commonly reported STI in the United States. The name originates from the Greek words *gonos* meaning "seed" and *rhoia* meaning "to flow," probably because the pus produced during infection looks like semen.

Gonorrhea is not a new problem and was originally believed to be a symptom of syphilis, another common STI. In the mid–nineteenth century, gonorrhea was recognized as a specific disease. Today, it is still common, and its increasing resistance to antibacterial treatment is a cause for concern.

Signs and Symptoms

The incubation period of gonorrhea is usually only 2 to 5 days. Infections are often asymptomatic in both men and women. If genital signs and symptoms occur in men, they include urethritis, pain during urination, and a thick, pus-containing discharge from the penis (**figure 25.8**). These symptoms are noticeable and unpleasant, and men who have them usually go immediately for treatment. Gonorrhea follows a different course in women. Genital signs and symptoms if present include painful urination and vaginal discharge. The causative agent grows well in the cervix and fallopian tubes, in glands in the vaginal wall, and in other areas of the genital tract, causing inflammation, pain, and possible scarring.

Untreated gonorrhea in either men or women can lead to complications. In men, an inflammatory reaction to the infection can cause scar tissue formation that partially obstructs the urethra, slowing urination and predisposing the man to UTIs. The infection may spread to the prostate gland and testes, producing prostatic abscesses and orchitis (inflammation of the testicles). If scar

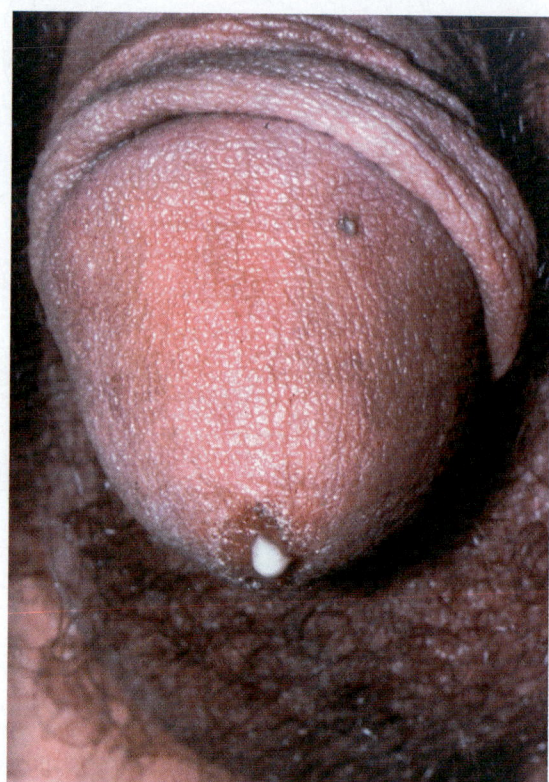

FIGURE 25.8 Urethral Discharge in a Man with Gonorrhea While symptoms of gonorrhea in men are noticeable when present, the disease is often asymptomatic.

❓ *What is the risk of asymptomatic infection?*

tissue blocks the tubes that carry the sperm, or if testicular tissue is destroyed by the infection, infertility can result.

If gonorrhea is untreated in women, the infection can progress upward through the uterus into the fallopian tubes, causing pelvic inflammatory disease (PID) and sometimes infertility. Scarring of a fallopian tube can also lead to **ectopic pregnancy**, in which the embryo develops in the fallopian tube or even in the abdominal cavity outside the uterus. Ectopic pregnancy can lead to life-threatening internal hemorrhaging. Occasionally, the infection spreads from the fallopian tubes into the abdominal cavity, where it can affect the liver or other abdominal organs. It is unclear how *N. gonorrhoeae* (which is non-motile) can move through the uterus to reach the fallopian tubes, but it is possible that the organisms are carried on sperm, to which the bacteria attach.

Occasionally, the causative agent of gonorrhea can produce **disseminated gonococcal infection (DGI).** DGIs are characterized by fever, rash, and arthritis caused by growth of the pathogen within the joint spaces. They can also cause endocarditis and meningitis. DGIs are not usually preceded by urogenital symptoms and are caused by strains of the organism that are serum-resistant (see Causative Agent). ▶▶ endocarditis, p. 672 ▶▶ meningitis, p. 643

Ophthalmia neonatorum is a destructive *N. gonorrhoeae* infection of the eyes of newborn babies whose mothers have symptomatic or asymptomatic gonorrhea. The bacteria are transmitted during the infant's passage through the infected birth canal. Signs and symptoms include irritation of the conjunctiva and production of thick pus.

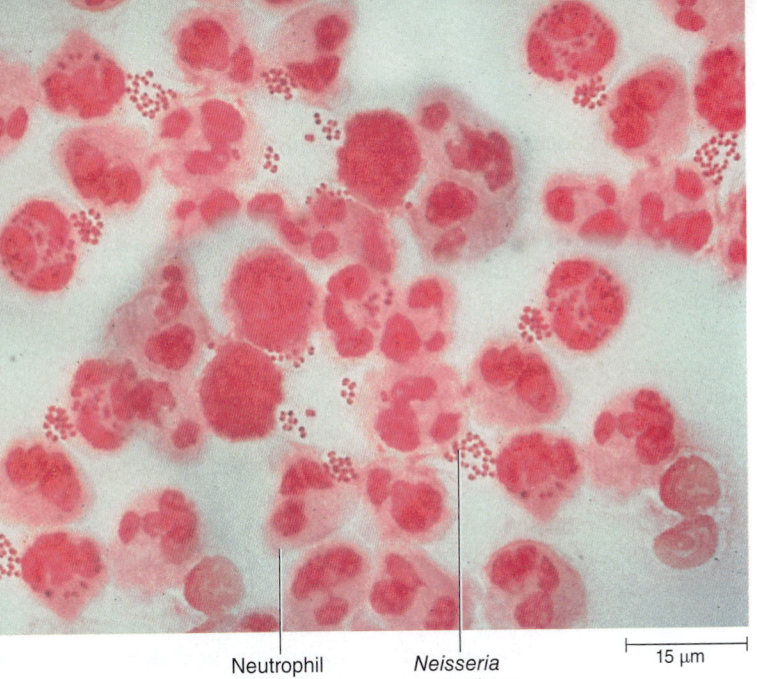

FIGURE 25.9 *Neisseria gonorrhoeae* Stained smear of pus from the urethra showing *N. gonorrhoeae*.

[?] *What is the morphology of* N. gonorrhoeae?

Causative Agent

Gonorrhea is caused by the gonococcus (GC) *Neisseria gonorrhoeae*—a fastidious Gram-negative diplococcus that requires a rich medium such as chocolate agar for cultivation (**figure 25.9**). Some strains of the organism produce pili (fimbriae) for attachment to host cells (**figure 25.10**). The outer membrane of GC has the general structure of a typical Gram-negative LPS membrane. However, the lipid that makes up the membrane has a highly branched oligosaccharide structure, and is referred to as lipooligosaccharide (LOS). LOS plays a role in GC virulence and pathogenicity. *N. gonorrhoeae* is very efficient at obtaining iron (needed for bacterial invasion) from the host during growth. It does this by producing several membrane proteins that take iron from transferrin and lactoferrin under low iron conditions.

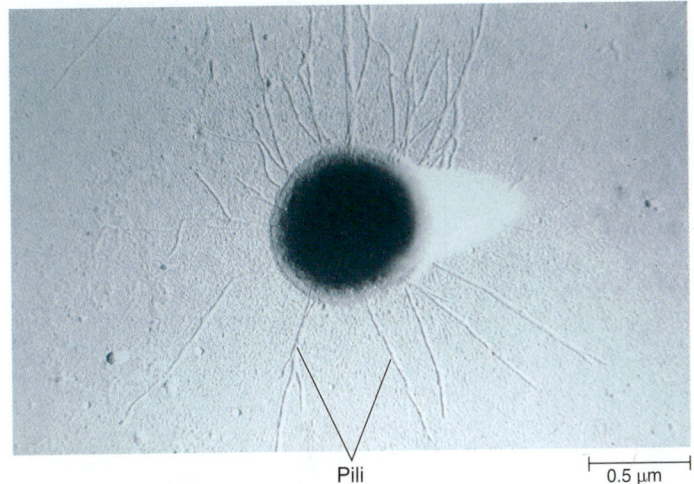

FIGURE 25.10 Pili on Single Coccus of *Neisseria gonorrhoeae* (EM)

[?] *What is the function of the pili?*

Pathogenesis

Neisseria gonorrhoeae has many mechanisms that allow it to avoid host defenses. The organism is a human-specific pathogen that survives poorly in the environment but is well adapted for colonization and proliferation in the host. Infection begins when the organism selectively attaches by means of pili (fimbriae) and Opa (outer membrane) proteins to receptors on certain non-ciliated epithelial cells of the body, including those of the urethra, uterine cervix, mouth, pharynx, anus, and conjunctiva. GC cannot colonize squamous epithelium, such as that lining the adult vagina, or ciliated cells. The bacteria enter the epithelial cells by endocytosis, mediated by a porin protein in the LOS membrane. LOS and peptidoglycan released by autolysis of cells cause complement activation, triggering an inflammatory response that leads to most of the symptoms of gonorrhea. They also stimulate the production of pro-inflammatory cytokines such as tumor necrosis factor (TNF) that cause the release of enzymes such as proteases and phospholipases, leading to cell and tissue damage. ◄◄ **porin, p. 60**

GC is very good at avoiding both innate and adaptive immune responses. For example, it (1) produces proteases that destroy IgA found on mucosal surfaces; (2) uses host sialic acid to make a capsulelike structure that allows the organism to resist phagocytosis; and (3) produces an outer membrane porin protein that allows it to survive within phagocytes by preventing phagolysosome formation. Furthermore, LOS is also antigenically similar to red blood cells, and this similarity to "self" cells prevents an effective immune response.

N. gonorrhoeae avoids adaptive immunity by antigenic and phase variation. It can express several antigenic types of LOS and Opa proteins (antigenic variation), or it may not produce any at all (phase variation). The organism can change the type of LOS they express by an unknown mechanism. A single strain of GC can also express many different kinds of pili, brought about by chromosomal rearrangements within the pili genes. Furthermore, the organisms may produce pili or not. Finally, GC Opa proteins allow the organism to attach specifically to receptors on T helper cells, preventing their activation and proliferation. ◄◄ **antigenic variation, p. 178** ◄◄ **phase variation, p. 178**

Epidemiology

Gonorrhea is among the most common of the STIs. In the United States, its incidence is the highest of any reportable bacterial disease other than *Chlamydia* infection. The number of cases has been declining since the emergence of HIV/AIDS, however, probably due to increased condom use.

N. gonorrhoeae infects humans only, living mainly on the mucous membranes of the host. Most strains are susceptible to UV light, cold and desiccation and so do not survive well outside the host. For this reason, gonorrhea is transmitted almost exclusively by direct contact. Because the bacteria mainly live in the genital tract, this contact is almost always sexual. However, gonorrhea can be transmitted by vaginal, oral, or anal sex. It can also be passed from mother to baby during vaginal childbirth. Factors that influence the incidence of gonorrhea include:

■ **Birth control pills.** Oral contraceptives without use of a condom offer no protection against STIs and may increase susceptibility to them. Use of oral contraceptives leads to migration of gonorrhea-susceptible epithelial cells from the cervical lumen onto more exposed areas of the outer cervix. Oral contraceptives also tend to increase both the pH and the moisture content of the vagina, favoring infection with GC and other agents of STIs. Besides being more susceptible to gonorrhea, women taking oral contraceptives are also more likely to develop serious complications from the disease.

- **Asymptomatic infection.** Male and females who have asymptomatic gonorrhea can unknowingly transmit the infection over months or even years. Up to 20% of men and 60% of women with gonorrhea are asymptomatic.

- **Lack of immunity.** There is little or no immunity following recovery from the disease. Individuals can contract gonorrhea repeatedly.

Treatment and Prevention

N. gonorrhoeae is now frequently resistant to many antibacterial medications, including penicillins, tetracyclines, sulfonamides, and fluoroquinolones. A single class of antibiotics, the cephalosporins, is currently recommended for treatment of gonorrhea. Ciprofloxacin is commonly used.

Prevention of gonorrhea depends on abstinence, monogamous relationships, and consistent, correct use of condoms. Rapid identification and treatment of sexual contacts also prevents gonorrhea by decreasing the overall number of infected individuals. There is no vaccine for preventing gonorrhea because the antigenic variation of the causative organism and the lack of an animal model for the disease have made it impossible to develop one.

Ophthalmia neonatorum is prevented by putting an antibiotic ointment such as erythromycin directly into the eyes of all newborn infants within 1 hour of birth. Some mothers who feel certain that they do not have gonorrhea have challenged this practice, but because gonorrhea is frequently asymptomatic and the disease can cause blindness in infants, this prophylactic treatment of a baby's eyes is required by law. As a result, ophthalmia neonatorum is now unusual in the United States. **Table 25.8** describes the main features of the disease.

TABLE 25.8 | Gonorrhea

1. *Neisseria gonorrhoeae* enters the body through mucous membranes of the genitalia, mouth, or anus. It can also enter the eyes; serious infections leading to loss of vision are likely in newborns.
2. The cervix is the usual site of primary infection in women.
3. The outer covering of the liver is infected when gonococci enter the abdominal cavity from infected fallopian tubes.
4. Prostatic gonococcal abscesses may be difficult to eliminate.
5. Infection of the fallopian tubes results in scarring, which can cause infertility or ectopic pregnancy.
6. Organisms carried by the bloodstream infect the heart valves and joints.
7. Urethral scarring from gonococcal infection can predispose to urinary infections by other organisms.
8. Scarring of testicular tubules can cause sterility.

Signs and symptoms	Men: pain on urination, discharge; complications include impaired urinary flow, sterility, or arthritis. Women: pain on urination, discharge, fever, pelvic pain; sterility, ectopic pregnancy, arthritis can occur. Men and women may be asymptomatic.
Incubation period	2 to 5 days
Causative agent	*Neisseria gonorrhoeae (GC)*, a Gram-negative diplococcus
Pathogenesis	Organisms attach to certain non-ciliated epithelial cells by pili; phase and antigenic variation in surface proteins allows attachment to different host cells and escape from immune mechanisms. Inflammation, scarring; can spread by bloodstream.
Epidemiology	Transmitted by sexual contact. Asymptomatic carriers. No immunity.
Treatment and prevention	Treatment: ceftriaxone. Prevention: abstinence, monogamous relationships, condoms, early treatment of sexual contacts.

Chlamydial Infections

Chlamydial infections are among the most common STIs worldwide. These infections resemble gonorrhea because they can cause urethritis and may lead to testicle and fallopian tube damage that result in infertility. Worldwide, they are the leading cause of infertility. There are several different antigenic types of the causative agent that lead to distinct diseases. ◀◀ *Chlamydia*, p. 277

Signs and Symptoms

The signs and symptoms of *Chlamydia trachomatis* infection generally appear 7 to 14 days after exposure. In men, the main symptom is a thin, gray-white discharge from the penis, sometimes with painful testes. Some men also develop pain on urination, and fever. Women with genital chlamydia most commonly develop an increased vaginal discharge, sometimes accompanied by painful urination, abnormal vaginal bleeding, upper or lower abdominal pain, and pain during sexual intercourse. Many infections of men and women are asymptomatic—nearly 75% of women and 50% of men with chlamydia do not develop signs or symptoms and can have the disease for months or years before being diagnosed.

Some strains of *C. trachomatis* can cause lymphogranuloma venereum, a rare disease in which the lymph nodes of the groin swell and drain pus. Enlarged nodes, called buboes, are commonly painful. Repeated draining and healing of buboes eventually leads to fibrosis, which can obstruct lymphatic vessels, resulting in gross swelling of the genitalia after several years. The long-term effect of this symptom may be genital elephantiasis.

MicroByte

A type of chlamydial infection of the eyes, called trachoma, can cause blindness if left untreated.

Causative Agent

Chlamydia trachomatis is a spherical, obligate intracellular Gram-negative bacterium. During replication, the organism causes inclusion bodies containing a glycogen-like material to form in the cytoplasm of host cells. These inclusions stain with iodine and can provide a fast way of identifying *C. trachomatis* infections.

Different antigenic types of *C. trachomatis* cause distinct diseases. Approximately eight types are responsible for most *C. trachomatis* STIs. Three other types cause lymphogranuloma venereum and four others cause trachoma, a serious eye infection.

◀◀ glycogen, p. 32

Pathogenesis

The infectious form of *C. trachomatis,* called an **elementary body,** attaches specifically to receptors on the surface of the host epithelial cell, inducing the cell to take in the bacterium by endocytosis. Once inside the endosome, the elementary body interacts with glycogen inclusions, enlarges, and becomes a noninfectious **reticulate body.** The reticulate body divides repeatedly by binary fission, resulting in many new elementary bodies. These are released when the host cell lyses, and infect nearby cells. The infected cells produce cytokines that cause an intense inflammatory reaction. Much of the tissue damage results from the

cell-mediated immune response. The cell wall of *C. trachomatis* lacks peptidoglycan but instead has cysteine-rich proteins. This wall inhibits the formation of a phagolysosome, allowing the organism to survive within phagocytes. ◀◀ endocytosis, p. 72

Chlamydial infection usually involves the urethra in both men and women. In men, it spreads to the epididymis, causing acute pain and swelling. It may lead to infertility. In women, the infection commonly involves the cervix, making it bleed easily, especially with sexual intercourse. The uterus also often becomes infected, leading to pain and bleeding. From the uterus, the infection is carried by sperm to the fallopian tubes, where it can lead to PID (pelvic inflammatory disease), causing the risk of ectopic pregnancy or infertility (**figure 25.11**). Like *N. gonorrhoeae, C. trachomatis* can move from the fallopian tubes to the abdominal cavity and infect the surface of the liver.

Epidemiology

Chlamydial genital infections are the most common of all notifiable bacterial infectious diseases. The number of reported cases has been rising, probably because of increased awareness of the disease and better diagnostic tests. Non-sexual transmission of this agent also occurs, for example, in non-chlorinated swimming pools. Newborn babies of infected mothers often develop chlamydial ophthalmia neonatorum and pneumonia from *C. trachomatis* infection contracted during passage through the birth canal.

Treatment and Prevention

Several antibiotics can be used to treat chlamydial infections. Azithromycin can be given as a single dose, whereas tetracyclines and erythromycin require more doses but are less expensive alternatives. Sexual partners are also treated. Chlamydial ophthalmia neonatorum is treated with oral erythromycin.

Chlamydial infections can be prevented by abstinence, monogamous relationships, and condoms used correctly. All sexually active people are advised to get tested for *Chlamydia* each

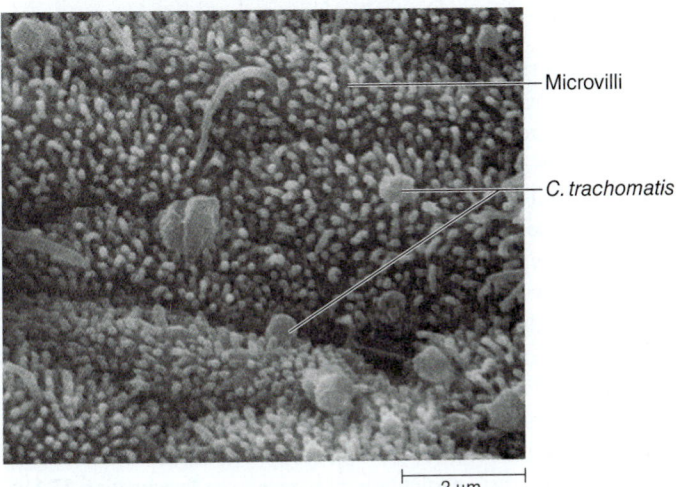

FIGURE 25.11 *Chlamydia trachomatis* Scanning electron micrograph showing *C. trachomatis* attached to fallopian tube mucosa.

❓ *Is* Chlamydia *limited to infections of the genital mucosa?*

TABLE 25.9	Chlamydial Genital System Infections
Signs and symptoms	Men: thin, gray-white penile discharge, painful testes. Women: vaginal discharge, vaginal bleeding, lower or upper abdominal pain
Incubation period	Usually 7 to 14 days
Causative agent	*Chlamydia trachomatis,* an obligate intracellular bacterium, certain serotypes
Pathogenesis	Elementary body attaches to specific receptors on the epithelial cell, causing endocytosis; becomes reticulate body in the endosome; repeated replication by binary fission and differentiation into elementary bodies; host cell bursts, releasing elementary bodies to infect adjacent cells; release of cytokines results in inflammatory response; cell-mediated immune response against infection causes extensive damage; scar tissue forms, causing ectopic pregnancy and infertility.
Epidemiology	The leading reportable bacterial infection in the United States. Large numbers of asymptomatic men and women carriers. Non-sexual transmission can occur in non-chlorinated swimming pools.
Treatment and prevention	Treatment: azithromycin and other antibacterial medications. Prevention: abstinence, monogamous relationship, condom use. Test sexually active people at least once yearly to rule out asymptomatic infection.

(handwritten margin note: women-bleed men-discharge)

year, or twice yearly if they have multiple partners or if their partner has multiple partners. The main features of chlamydial genital infections are summarized in **table 25.9.**

Syphilis

During the first half of the twentieth century, syphilis was a major cause of mental illness and blindness, and a significant cause of heart disease and stroke. The disease was almost eradicated by the mid-1950s through a successful program aimed at locating and treating all people with syphilis and their sexual contacts. Unfortunately, the number of cases then increased in the 1990s due to factors including inner-city poverty, prostitution, and drug use. This trend was reversed through renewed efforts in education, case finding, and treatment that caused a dramatic drop in the number of new cases. The AIDS epidemic added urgency to syphilis control efforts, because syphilis, like other STIs that cause genital sores, promotes the spread of AIDS by increasing the risk of HIV infection. By the end of 1998, the syphilis rate was at the lowest level since it became a notifiable disease. Rates have risen since then, particularly in women (see **Perspective 25.1**).

Signs and Symptoms

Syphilis causes so many different signs and symptoms that it is easily confused with other diseases and is often called "the great imitator." Its symptoms occur in defined clinical stages.

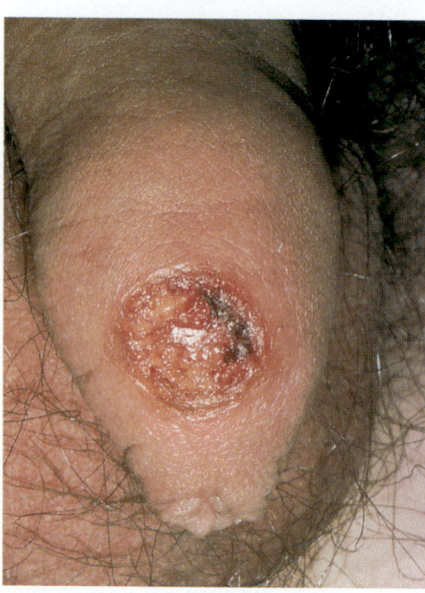

FIGURE 25.12
Syphilitic Chancre
A chancre develops at the site where *Treponema pallidum* entered the body, as on the foreskin of this uncircumcised penis.

❓ *What is the risk associated with an open ulcer, as shown here?*

- **Primary syphilis.** The stage is characterized by a painless, red ulcer called a hard **chancre** (pronounced "shanker") that appears at the site of infection about 3 weeks after exposure (**figure 25.12**). The chancre usually develops on the genitalia but may occur anywhere on the body. The local lymph nodes enlarge. Primary syphilis chancres often go unnoticed in women and men who have sex with men because they are hidden from view in the vagina or anus and are painless. Sometimes no chancre develops, only a pimple small enough to go unnoticed.

- **Secondary syphilis.** The symptoms of secondary syphilis usually appear 2 to 10 weeks (or longer) after the primary stage. The infection has spread systemically by this time, and causes many diverse signs and symptoms, the most common of which is a rash that includes the palms and soles, and white patches on the mucous membranes (**figure 25.13**). Other signs and symptoms include runny nose and watery eyes, aches and pains, sore throat, fever, malaise, weight loss, and headache.

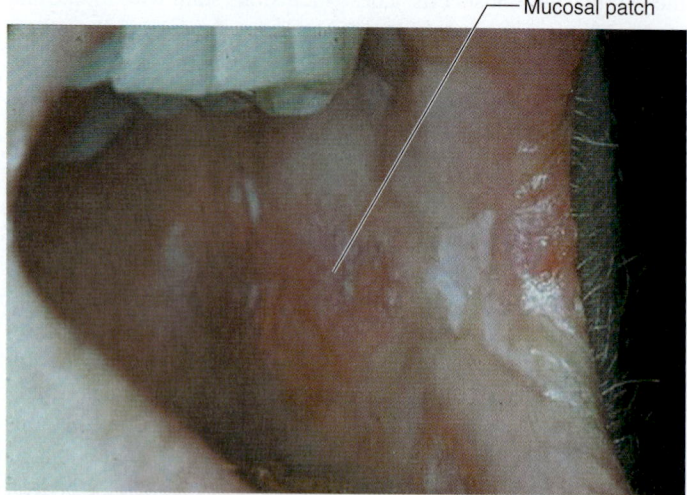

Mucosal patch

FIGURE 25.13 Secondary Syphilis Secondary syphilis lesions such as these oral mucous patches contain numerous *Treponema pallidum* organisms and are highly infectious.

❓ *What causes the manifestations of secondary syphilis?*

Occasionally, hepatitis (liver disease) and renal disease occur. The person is most infectious at this stage.

- **Latent syphilis.** During this stage, the person does not have any signs and symptoms of syphilis, although they have antibodies to the causative agent. A quarter of people in this stage recover; another quarter remain in this stage. The rest progress to the next stage of the disease.

- **Tertiary syphilis.** After a latent period that can last for many years, signs and symptoms of tertiary syphilis sometimes occur. The symptoms of this stage can be grouped into three categories. In gummatous syphilis, localized areas of tissue damage develop as a result of prolonged inflammatory responses. These lesions, called **gummas,** are granulomas similar to the tubercles seen in tuberculosis. Gummas are chronic and may occur anywhere in the body (**figure 25.14**). Cardiovascular syphilis is characterized by aneurysms that form in the ascending aorta, caused by chronic inflammation of the arterioles supplying this vessel. Neurosyphilis occurs in a small number of cases and is characterized by a pattern of symptoms including personality change, emotional instability, delusions, hallucinations, memory loss, impaired judgment, abnormalities of the pupils of the eye, and speech defects. Other signs and symptoms of this stage include blindness, vertigo, insomnia, stroke, and meningitis. Although neurosyphilis symptoms are usually seen during the tertiary stage, they can also be experienced during the first and secondary stages of the disease. ◀◀ tubercles, p. 504

During pregnancy, *T. pallidum* easily crosses the placenta and infects the fetus, causing **congenital syphilis.** Fetal infections can occur in the absence of any signs or symptoms of syphilis in the mother and at any stage of pregnancy, but damage to the fetus does not generally happen until the fourth month. Common outcomes of fetal infection include spontaneous abortion, stillbirth, and neonatal death. Almost all infected infants that survive birth develop characteristic signs and symptoms. Early signs and symptoms resemble those of severe adult secondary syphilis and

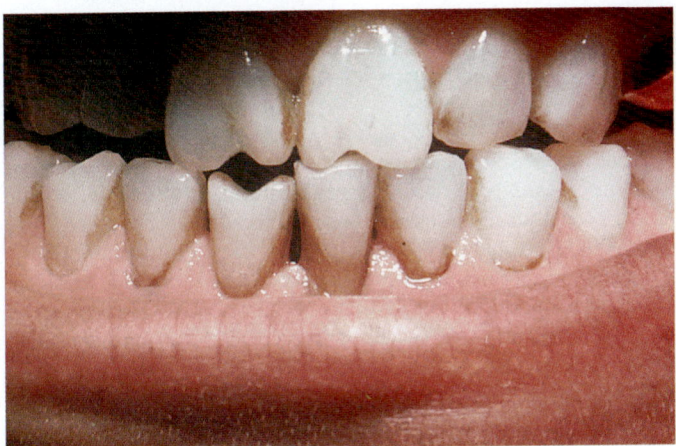

FIGURE 25.15 Hutchinson Teeth Notice the notched, deformed incisors, a late manifestation of congenital syphilis.

? *What are some other manifestations of congenital syphilis?*

include runny nose, rash, and liver and spleen enlargement. Late symptoms occur after 2 years. The bones, cartilage, and teeth can become deformed as the child develops, causing cleft palate, sabre shins (bent tibias), saddle nose (sunken nasal bridge), mulberry molars (many cusps), and Hutchinson teeth (notched incisors) (**figure 25.15**). Eye inflammation and deafness often occur.

MicroByte

Evidence suggests that *T. pallidum* is directly involved in aiding HIV infection and progression. Stopping syphilis would decrease the number of new HIV cases.

Causative Agent

Syphilis is caused by *Treponema pallidum,* a very slender, motile spirochete that is difficult to see unless dark-field microscopy is used (**figure 25.16**). All spirochetes are highly motile, and their endoflagella—which propel them in a corkscrewlike motion—allow them to penetrate a wide variety of tissues and organs. The

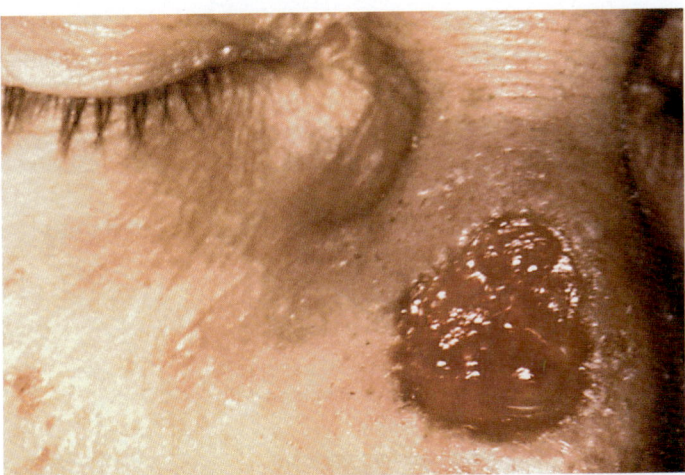

FIGURE 25.14 A Gumma of Tertiary Syphilis Gummas are formed by an inflammatory mass which can perforate tissue, as in this example in the nose.

? *Where in the body do gummas occur?*

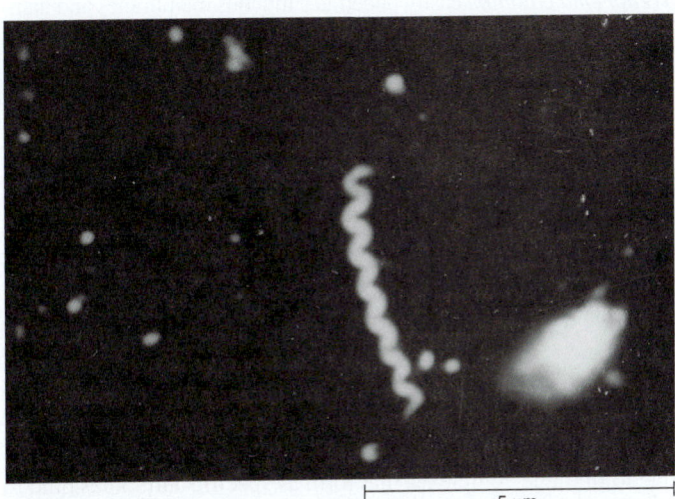

5 µm

FIGURE 25.16 *Treponema pallidum* This organism must be viewed with dark-field microscopy.

? *Why is dark-field microscopy needed to observe* T. pallidum?

The Death of Syphilis?

For 40 years (1932 to 1972), the U.S. Public Health Service conducted a study (now known as the Tuskegee Syphilis Experiment) of 399 poor Alabama black men with advanced syphilis. The study was done to assess the natural progression of the disease in black men. The men were not told their diagnosis nor educated about the nature of their condition and its transmission. For the first half of the study, there were no effective drugs against syphilis, and so the men remained untreated. However, even after effective therapy became available, the study was allowed to continue without giving treatment, in a blatant mix of bad science and racism. Twenty-eight of the men died of the disease, and another hundred died from its complications.

Forty wives became infected, and there were 19 children born with congenital syphilis. Nothing of scientific value was learned, and many people to this day remain bitter and distrustful of the U.S. Public Health Service. In 1997, President Clinton, speaking for the government, formally apologized to the surviving eight subjects.

Fortunately, in more recent years, major governmental efforts to conquer the disease have met with some success. In 2009, only 13,997 cases of primary and secondary syphilis were reported, down from 50,223 in 1990. The Centers for Disease Control and Prevention (CDC) has even considered the possibility that transmission of the disease could be eliminated in the United States.

Syphilis is a good target for eradication because it has no animal reservoir, infectious cases can generally be treated with a single injection of penicillin, and the incubation period is long enough that sexual contacts of an infected person can be found and treated before they spread the disease any further. Moreover, instead of being widely endemic, most syphilis cases are now concentrated in a relatively few areas and certain populations. Also, new techniques are now available for rapid diagnosis, treatment, and epidemiological tracing of the disease. Finally, control of syphilis is closely linked to AIDS control efforts because the risk of HIV transmission is six times as great if either sex partner has syphilis.

---

organism is fragile and easily killed by heat, cold, desiccation, and soap. It is therefore transmitted almost exclusively by sexual or oral contact. In nature, *T. pallidum* infects only humans. ◀◀ dark-field microscopy, p. 43

T. pallidum must be grown in the testicles of laboratory rabbits to study its cell division and pathogenesis, because it does not replicate *in vitro*. The organism lacks metabolic abilities, and does not have enzymes needed for the TCA cycle or electron transport chain, so cannot generate much ATP. It is therefore thought to get most of its essential macromolecules from the host, instead of synthesizing them itself. The sequence of nucleotides in its genome is now known, which should give scientists a better understanding of its virulence, why it cannot be cultivated *in vitro,* and how to make an effective vaccine. ◀◀ TCA cycle, p. 142

Pathogenesis

Treponema pallidum easily penetrates mucous membranes and damaged skin. The infectious dose is very low—less than 100 organisms. In primary syphilis, *T. pallidum* multiplies in a localized area of the genitalia, spreading from there to the lymph nodes and bloodstream. The hard chancre is caused by an intense inflammatory response to the bacteria, which are present in high numbers in the lesion. The chancre disappears within 2 to 6 weeks, even without treatment, and patients may mistakenly believe that they have recovered from the disease. However, the organism avoids destruction by the body's defenses and progression of the disease can continue for years. Few virulence factors have been discovered for *T. pallidum,* and the outer membrane of the cells lack LPS, the molecule that normally triggers an immune response. There are also few proteins in the outer membrane and therefore few targets for opsonizing antibodies to bind.

Many of the signs and symptoms of secondary syphilis are due to immune complexes that form as specific antibodies bind to circulating *T. pallidum.* By this time, the spirochetes have become systemic, and infectious lesions occur on the skin and mucous membranes in various locations, especially in the mouth. The secondary stage lasts for weeks to months, sometimes as long as a

year, and then gradually subsides. About 50% of untreated cases never progress past the secondary stage. After a latent period of from 5 to 20 years or even longer, however, some people with the disease develop tertiary syphilis. ◀◀ immune complexes, p. 395

Tertiary syphilis signs and symptoms occur from hypersensitivity reactions to small numbers of *T. pallidum* that grow and persist in the tissues. In this stage, the patient is no longer infectious. The organisms may be present in almost any part of the body, and the signs and symptoms of tertiary syphilis depend on where the hypersensitivity reactions occur. If they occur in the skin, bones, or other areas not vital to existence, the disease is not life-threatening. If, however, they occur within the walls of a major blood vessel such as the aorta, the vessel may become weakened and even rupture, resulting in death. The main characteristics of the stages of syphilis are summarized in **table 25.10.**

Epidemiology

Syphilis is usually transmitted by sexual intercourse. However, infection can occur from kissing a person with secondary syphilis,

TABLE 25.10	**Stages of Syphilis**	
Stage of Disease	**Main Characteristics**	**Infectious?**
Primary	Firm, painless ulcer (hard chancre) at site of infection; lymph node enlargement	Yes
Secondary	Rash, aches, and pains; rash and mucous membrane lesions	Yes
Latent	No signs or symptoms; serologically positive	Early—yes Late—no
Tertiary	Gummas; damage to large blood vessels, eyes, nervous system; insanity	No

or by contact with a primary ulcer infected with *T. pallidum.* Because there is no animal reservoir, eradicating this disease is possible. To do this, however, cases must be identified and treated, particularly in high-risk groups such as prostitutes, men who have sex with men, and jail inmates. Infected individuals can be detected using a simple blood test.

Treatment and Prevention

Primary and secondary syphilis are easily treated with an antibiotic such as penicillin—the dose and drug type depend on the stage of the disease. Treatment is more difficult for tertiary syphilis, because most of the organisms are not actively multiplying and penicillin acts against dividing cells. Congenital syphilis can be prevented by diagnosing and treating the mother's syphilis before the fourth month of pregnancy, when *T. pallidum* starts to affect the fetus.

There is no vaccine for syphilis. Abstinence, monogamous relationships, effective and correct use of condoms, and other safer sex practices decrease the risk of contracting the disease. Quick identification and treatment of sexual contacts are important in limiting spread of the disease. **Table 25.11** summarizes the main features of syphilis.

Chancroid

Chancroid is another bacterial STI that, like syphilis, causes genital sores, thereby increasing the risk of contracting HIV. The disease is common in developing countries, usually associated with commercial sex workers. The reported incidence of chancroid in the United States is low, but it often goes unreported or misdiagnosed, and may therefore be more common than recorded.

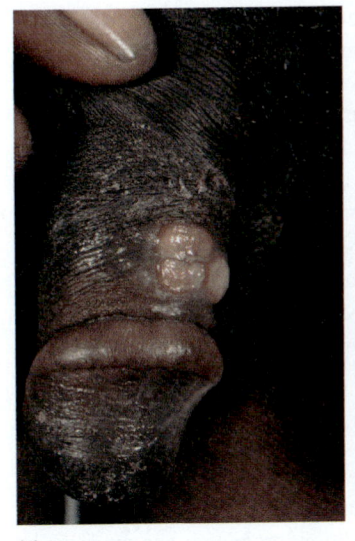
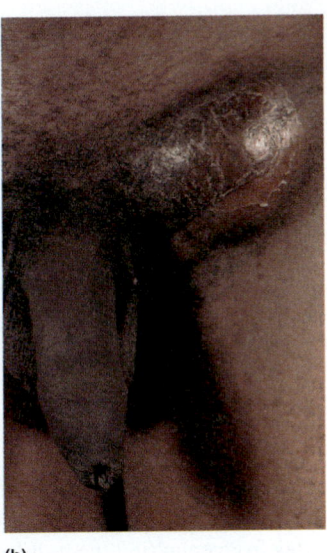

(a)   (b)

FIGURE 25.17 Chancroid (a) Lesions of the penis. These ulcerations are soft and painful. **(b)** Swollen groin lymph nodes. The nodes are tender, often pus-filled, and may break open and drain.

❓ *How are chancroid ulcers different from chancres?*

Signs and Symptoms

Chancroid is characterized by one or more painful genital sores called soft chancres (**figure 25.17a**). Typically, these begin as a small pimple at the site of bacterial entry through the skin, but

TABLE 25.11	Syphilis		
① *Treponema pallidum* enters the body through a microscopic abrasion or mucous membranes, usually genitalia, mouth, or rectum.		**Signs and symptoms**	Chancre, fever, rash, stroke, nervous system deterioration; can imitate many other diseases
② A chancre develops at site of entry.		**Incubation period**	10 to 90 days
③ Organisms multiply locally and spread throughout the body by the bloodstream.		**Causative agent**	*Treponema pallidum,* a non-culturable spirochete
④ Infectious mucous patches and skin rashes of secondary syphilis appear. A fetus may become infected, resulting in miscarriage or a live-born infant with congenital syphilis.		**Pathogenesis**	Primary lesion, or chancre, appears at site of inoculation, heals after 2 to 6 weeks; *T. pallidum* invades the blood vessel system and is carried throughout the body, causing fever, rash, mucous membrane lesions; damage to brain, arteries, and peripheral nerves appears years later.
⑤ An asymptomatic latent period occurs. *T. pallidum* disappears from blood, skin, and mucous membranes.		**Epidemiology**	Sexual contact with infected partner; kissing; transplacental passage
⑥ After months or years, symptoms of tertiary syphilis appear: heart and blood vessel defects, gummas, strokes, eye abnormalities, general neurological symptoms, insanity.		**Treatment and prevention**	Treatment: penicillin. Prevention: abstinence, monogamous relationships, use of condoms, safer sex practices, treatment of sexual contacts, reporting cases.

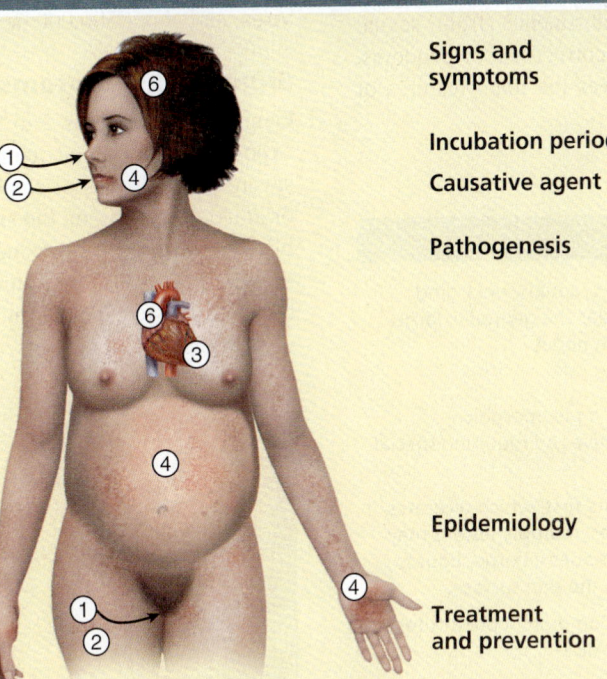

they then ulcerate and quickly get bigger. The lymph nodes in the groin also enlarge and become tender (figure 25.17b). Sometimes the nodes are pus-filled and may rupture, discharging the pus. In women, chancres typically occur on the labia or thighs, causing pain during urination and sexual intercourse.

Causative Agent

Chancroid is caused by *Haemophilus ducreyi*, a fastidious, pleomorphic, Gram-negative coccobacillus that, like *N. gonorrhoeae*, can only be cultivated on a rich medium such as chocolate agar.
◀◀ *Haemophilus*, p. 276

Pathogenesis

Haemophilus ducreyi infection causes an intense inflammatory response, with accumulation of neutrophils and macrophages in the area. The bacteria survive by producing proteins that prevent phagocytes from engulfing them. In addition, proteins in the outer membrane protect the cells from activated complement proteins. The mechanism of these protective proteins is still being investigated.

Epidemiology

Epidemics in American cities associated with prostitution sometimes occur. In some tropical countries, chancroid is second only to gonorrhea in prevalence of the STIs.

Treatment and Prevention

Chancroid can usually be treated with a variety of antibiotic options, including erythromycin, azithromycin, or ceftriaxone. Some strains, however, are resistant to multiple antibacterial medications. Effectiveness of therapy is also significantly reduced if the patient has AIDS.

Chancroid can be prevented by abstinence from sexual intercourse, monogamous relationships, correct use of condoms, and safer sex practices. **Table 25.12** gives the main features of chancroid.

TABLE 25.12	Chancroid
Signs and symptoms	One or more painful, gradually enlarging, soft chancres on or near the genitalia; large, tender regional lymph nodes
Incubation period	3 to 10 days
Causative agent	*Haemophilus ducreyi*, a pleomorphic, fastidious Gram-negative rod requiring special growth media
Pathogenesis	A small pimple appears first, which ulcerates and gradually enlarges; multiple lesions may join together; lymph nodes enlarge, liquefy, and may discharge to the skin surface.
Epidemiology	Sexual transmission. Common in prostitutes; fosters the spread of AIDS.
Treatment and prevention	Treatment: several antibacterial medications effective. Resistance can be a problem. Prevention: abstinence, monogamous relationships; avoiding sexually promiscuous partners; correct use of condoms.

MicroAssessment 25.5

Transmission of bacterial STIs usually requires direct person-to-person contact. Unsuspected STIs (such as gonorrhea and syphilis) of pregnant women are a serious risk to fetuses and newborn babies. Even asymptomatic infections can cause genital tract damage (including PID) and can be spread to other people.

13. *Can a baby have congenital syphilis without its mother ever having had signs and symptoms of syphilis? Explain.*
14. *What, if any, is the relationship between chancroid and development of AIDS?*
15. *Why does scarring of a fallopian tube increase the risk of an ectopic pregnancy?* ➕

25.6 ■ Viral STIs

Learning Outcomes

9. *Compare the infections caused by genital herpes simplex virus and human papillomavirus.*
10. *Outline the relationship between HIV infection and AIDS.*

Viral STIs are probably as common or more common than the bacterial diseases, and they are incurable. Their effects can be severe and long-lasting. Genital herpes can cause recurrent signs and symptoms for years, papillomavirus infections can lead to cancer, and untreated HIV infections usually result in AIDS.

Genital Herpes

Genital herpes is among the top three or four most common STIs. An estimated 45 million Americans are infected with the causative virus, and about 500,000 new cases occur each year.

Signs and Symptoms

Genital herpes begins 2 to 20 days (usually about a week) after exposure, with genital itching and burning, and in some cases, severe pain. Groups of small, red bumps appear on the genitalia or anus, depending on the site of infection. These bumps become blisters surrounded by redness (**figure 25.18**). The blisters break in 3 to 5 days, leaving an ulcerated area that slowly dries and becomes crusted. This eventually heals without a scar. Less common signs and symptoms include discharge from the penis or

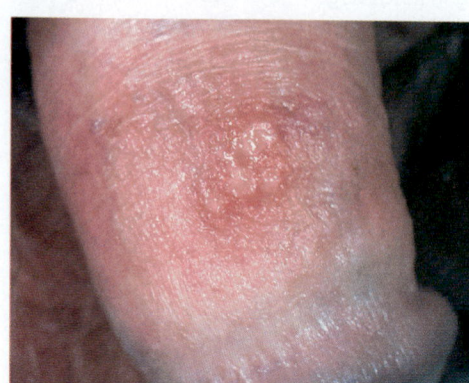

FIGURE 25.18 Genital Herpes on the Shaft of the Penis

❓ *Can genital herpes be transmitted when there are no visible signs or symptoms of the infection?*

vagina, headache, fever, muscle pain, and in the case of women, painful urination. Signs and symptoms are the worst in the first 1 to 2 weeks and disappear within 3 weeks. Genital herpes signs and symptoms often recur, however, because the virus becomes latent. The recurrent signs and symptoms are usually not as severe as those of the first episode, and recurrences usually occur less often with time. Some individuals have recurrences throughout their life, whereas others never have recurrence. ◀◀ latent infections, p. 383

Causative Agents

Genital herpes is usually caused by herpes simplex virus type 2 (HSV-2), an enveloped, double-stranded DNA virus. Herpes simplex virus type 1, the cause of "cold sores" ("fever blisters") can also cause genital herpes. The two virus types look the same, and their genomes are about 50% homologous. Either virus can infect the mouth and the genitalia, but HSV-2 causes more severe genital lesions with a greater frequency of recurrence. ◀◀ HSV-1, p. 582

Pathogenesis

Lesions begin with infection of a group of epithelial cells that lyse following viral replication, creating small, fluid-filled blisters (vesicles) containing large numbers of infectious virions. The vesicles burst, producing painful ulcers. During initial infection and for as long as 1 month after, the virus is found in genital secretions. In recurrence, the virus is usually present in large numbers for less than a week.

Latency of the disease is not completely understood but probably depends on cell-mediated immunity as well as viral products. Most of the time, the viral DNA exists within nerve cells in a circular, non-infectious form, causing no symptoms. In this state, only a single gene is expressed, and it encodes for a small segment of RNA called a microRNA (miRNA), that probably causes latency of the virus. At times, however, the entire viral chromosome can be transcribed and complete infectious virions produced. These reinfect the area supplied by the nerve and cause symptom recurrence.

Epidemiology

HSV-2 can survive for short periods on fomites or in water, but non-sexual transmission is rare. Sexual transmission of HSV is most likely to occur during the first few days of symptomatic disease, but it can happen in the absence of signs or symptoms of the disease. Once infected, an individual is forever at risk of transmitting the virus to another person. HSV, like other ulcerating genital diseases, promotes the spread of AIDS by increasing the risk of HIV infection. Neither HSV-1 nor HSV-2 has animal reservoirs.

Treatment and Prevention

There is no cure for genital herpes, although anti-HSV medications such as acyclovir and famciclovir can decrease the severity of the first attack and the incidence of recurrences.

Genital herpes can be prevented by abstaining from sexual intercourse or having a monogamous relationship with a genital herpes-free person. In addition, avoiding sexual intercourse during active signs or symptoms and using condoms with a spermicide that inactivates HSVs help prevent contracting the disease. However, using a condom does not always give full protection because herpes lesions can occur on parts of the genitalia that a condom does not cover.

Genital herpes can be a serious risk to newborn babies. If the mother has a primary infection near the time of delivery, the baby has about a 30% risk of acquiring the infection. The baby often dies from the infection or is permanently disabled by it. To prevent contact with the infected birth canal, these babies must be delivered by cesarean section. If the mother has recurrent disease, the risk to the baby is very low, presumably because transplacental anti-HSV antibody from the mother protects the baby, but doctors will often do a cesarean section to minimize the risk. The main features of genital herpes are presented in **table 25.13.**

Papillomavirus STIs: Genital Warts and Cervical Cancer

Sexually transmitted human papillomavirus (HPV) strains are among the most common of the STI agents, infecting an estimated 40 million Americans. Some HPV strains cause **papillomas**—warty growths of the external and internal genitalia.

TABLE 25.13	Genital Herpes
Signs and symptoms	Itching, burning pain at the site of infection, painful urination, tiny blisters with underlying redness. The blisters break, leaving a painful superficial ulcer, which heals without scarring. Recurrences are common.
Incubation period	Usually 1 week (range, 2 to 20 days)
Causative agent	Usually herpes simplex virus type 2. The cold sore virus, herpes simplex type 1, can also be responsible. Herpesviruses are enveloped and contain double-stranded DNA.
Pathogenesis	Lysis of infected epithelial cells results in fluid-filled blisters containing infectious virions. Vesicles burst, causing a painful ulcer. The acute infection is controlled by immune defenses; genome persists within nerve cells in a non-infectious form beyond the reach of immune defenses. Replication of infectious virions can occur and cause recurrent symptoms in the area supplied by the nerve. Newborn babies can contract fatal generalized herpes infection if their mother has a primary infection at the time of delivery.
Epidemiology	No animal reservoirs. Transmission by sexual intercourse, oral-genital contact. Transmission risk greatest first few days of active disease. Transmission can occur in the absence of symptoms. Herpes simplex increases the risk of contracting HIV.
Treatment and prevention	No cure. Medications help prevent recurrences, shorten duration of symptoms. Prevention: abstinence, monogamy, and correct use of condoms help prevent transmission.

Other strains cause non-warty lesions of mucosal surfaces such as the uterine cervix, and these are a major factor in the development of cervical cancer.

MicroByte

Between 75% and 80% of sexually active Americans will have an HPV infection at some point during their life.

Signs and Symptoms

Most people who have HPV infections clear them without ever developing signs or symptoms. In fact, they often do not know that they are infected. Some types of HPV do cause symptoms, however, and warts are the most easily recognized of these. Warts usually appear about 3 months after infection (range, 3 weeks to 8 months), developing on the head or shaft of the penis (**figure 25.19**), at the vaginal opening, or around the anus. Different kinds of lesions develop depending on the virus type and the location of the infection. Lesions can be flat or raised, cauliflowerlike, or hidden within the epithelium. Occasionally, warts become inflamed and bleed. After warts are removed, HPV persists in surrounding normal-looking epithelium and can cause additional warts. Warts sometimes partly block the urethra or, if very large, the birth canal. Newborn infants can become infected with HPV at birth and develop warts that block their respiratory tract, a serious condition that occurs occasionally (less than one out of every 100,000 births). Certain HPV strains can cause cervical cancer. Precancerous lesions are generally asymptomatic and can be detected only by examining the cervical tissues. The same strains can also cause other cancers including oral, penile, vaginal, and anal cancers. Accumulating evidence indicates that oral sex is a risk factor for oral cancers. A CDC study implicated HPV as the cause of up to 60% of mouth and throat cancers.

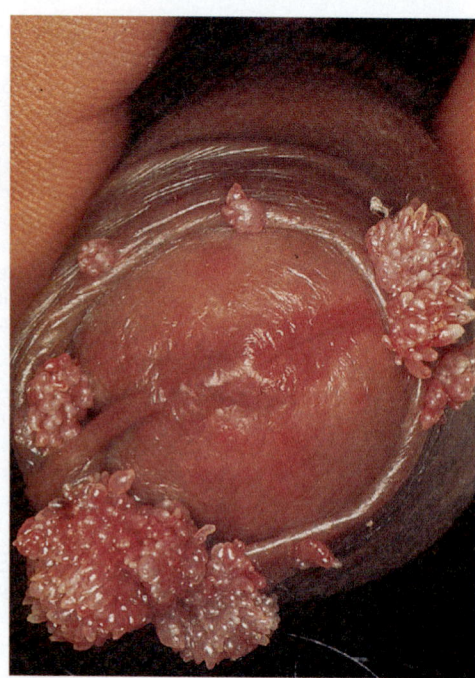

FIGURE 25.19
Genital Warts
Penis with warts, a manifestation of human papillomavirus infection.

[?] *Would condom use always prevent the transmission of HPV?*

Causative Agents

Human papillomaviruses are naked, double-stranded DNA viruses of the papovavirus family. There are more than 100 types of HPVs, and the different types are tissue- and species-specific. For example, HPVs that cause the common warts of the hands and feet generally do not infect the genitalia. Similarly, the HPVs that cause genital warts may infect the cervix but do not cause cancer.

At least 40 HPV types are transmitted by sexual contact. Approximately 15 of these are strongly associated with cancer of the cervix, penis, anus, vagina and throat. These high-risk HPVs include HPVs 16, 18, 31, and 45.

Pathogenesis

HPVs are thought to enter and infect the deeper layers of epithelium through microscopic abrasions. The genome of low-risk (wart-causing) HPV types exists in infected cells as extrachromosomal, closed DNA circles, but the mechanism by which warts arise is unknown. In contrast, the genome of high-risk (cancer-associated) HPV types can integrate into the chromosome of the host cell; they are oncogenic because they code for a protein that allows excessive cell growth. Oncogenic types of HPV tend to cause infections that persist longer than other HPV types, and to cause precancerous lesions. It takes on average 20 years for the infection to cause invasive cancer. A cancer-associated HPV is present in almost 100% of cervical cancers, but only a small percentage of infections by these viruses result in cancer. This indicates that other unknown factors must be present for cancer to develop. ◀◀ oncogenes, p. 323

Epidemiology

HPV strains are easily spread by sexual intercourse. Asymptomatic individuals infected with HPV can transfer the virus to others. HPV infection is the most common reason for an abnormal Papanicolaou (Pap) test in teenage women (see next section). A history of having multiple sex partners is the most important risk factor for acquiring genital HPV infection.

Treatment and Prevention

Warts often regress over time, but several treatments are available to remove them or hasten their clearance. Unfortunately, none of these methods cure the infection, so the warts may recur. Dermal warts can be removed by laser treatment, freezing with liquid nitrogen, or surgical excision. External genital warts can be treated with creams and ointments, including (1) imiquimod (Aldara Cream), an immune modifier that induces cytokine production; (2) podofilox, which prevents cell division of infected cells; and (3) sinecatechin-containing ointments, which cause wart regression by an unknown mechanism.

HPV infections can be prevented by abstinence, having monogamous relationships, or using condoms appropriately. As with HSV, condoms do not provide complete protection against HPV because the virus can be transmitted by exposure to areas not covered by the condom. Women should have a Pap smear every 12 months. This test involves removing cells from the cervix, staining them on a microscope slide, and examining them for

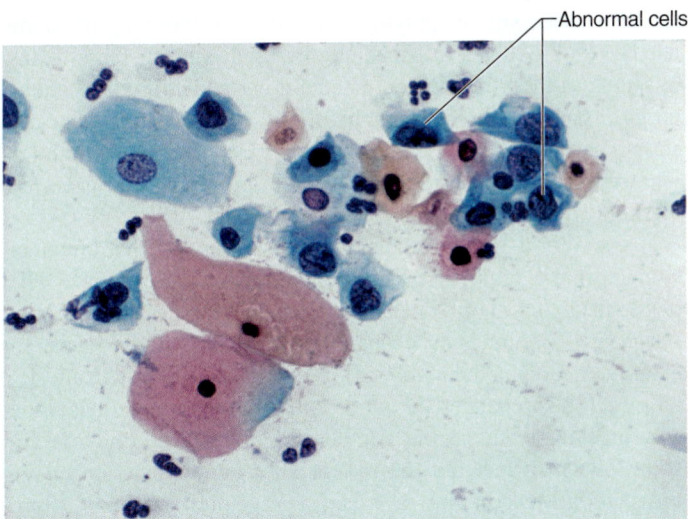

Abnormal cells

FIGURE 25.20 Abnormal Papanicolaou (Pap) Smear The pink and blue objects are squamous epithelial cells; abnormalities include doubling of the nuclei and a clear area around them.

❓ *What is the most frequent cause of an abnormal Pap smear in young women?*

TABLE 25.14	Papillomavirus STIs
Signs and symptoms	Often asymptomatic. Warts of the external and internal genitalia the most common symptom. Symptoms are strain-specific.
Incubation period	Usually 3 months (range, 3 weeks to 8 months)
Causative agents	Human papillomaviruses, many types, naked DNA viruses of the papovavirus family. Different types infect different tissues and produce different lesions.
Pathogenesis	Virus enters epithelium through abrasions, infects deep layer of epithelium; establishes latency; cycles of replication occur when host cell begins maturation; cancer-associated viral types can integrate into the host cell chromosome and can cause precancerous lesions.
Epidemiology	Asymptomatic individuals can transmit the disease; multiple sex partners the greatest risk factor; warts can be transmitted to the mouth with oral sex, and to newborn babies at birth.
Treatment and prevention	Treatment: wart removal by multiple techniques, but does not cure the infection. Prevention: latex condoms advised to minimize transmission and avoiding sexual contact with people having multiple sex partners. Pap tests at least yearly for sexually active women. Vaccines protect against some cancer-causing strains.

any abnormal cells (**figure 25.20**). Cervical cancer is generally preceded by precancerous lesions—areas of abnormal cell growth that can be detected by this test. The lesion can be removed, thereby preventing development of a cancer.

Two vaccines—Gardasil and Cervarix—are currently available that protect against infection by HPV types 16 and 18, which are responsible for about 70% of cancers. Gardasil also protects against HSV types 6 and 11, which cause more than 90% of genital warts. Either vaccine is given in a three-dose series, and is recommended for girls and young women ages 11 to 26. Although not included in the recommended vaccine schedule, Gardasil is available for girls as young as 9, and for boys and men ages 9 through 26. Cervarix is given to females only. **Table 25.14** summarizes some important features of papillomavirus STIs.

HIV/AIDS

Acquired immunodeficiency syndrome (AIDS) is unquestionably the most important sexually transmitted infection of the past century. It is covered briefly here and more fully in chapter 28. The AIDS epidemic was first recognized in the United States in 1981. Over the first 15 years, 500,000 Americans developed AIDS, and 300,000 died of the disease. With major advances in prevention and treatment of the disease, and expenditures of huge amounts of money, the epidemic shows signs of leveling off. Nevertheless, in 2009, an estimated 33.3 million people around the world were infected with human immunodeficiency virus (HIV), the causative agent of AIDS, and more than 2 million died of AIDS. An estimated 1.7 million new infections were in sub-Saharan Africa alone. The advances in prevention and treatment of AIDS are not uniformly available to many of those around the world who suffer because of the disease. ▶▶ AIDS, p. 695

Signs and Symptoms

Six days to 6 weeks after contracting HIV, some individuals develop flulike signs and symptoms—fever, head and muscle aches, and enlarged lymph nodes. These signs and symptoms are often mild, and go away by themselves. Many individuals have no signs or symptoms at all. HIV infection then persists unnoticed for almost 10 years before immunodeficiency develops, with certain malignancies and unusual microbial infections that often attack the lungs, intestines, skin, eyes, or the central nervous system.

Causative Agents

Most AIDS cases are caused by human immunodeficiency virus type 1 (HIV-1), an enveloped, single-stranded RNA virus of the retrovirus family. Each virion contains the enzyme reverse transcriptase and two copies of the viral genome. In parts of western Africa, another retrovirus, HIV-2, is a common cause of AIDS; it is rarely a cause in the United States. ◀◀ retroviruses, p. 320

Pathogenesis

HIV can attack a variety of cell types in the human body, but most critically affects the helper T (T_H) cells. These cells have the CD4 surface protein to which the virus attaches, and they are referred to as CD4$^+$ cells. In order for the virus to enter the cell, it must also attach to certain cytokine receptors, such as CCR5 or CXCR4 on the host cell surface. After the virus enters the CD4$^+$ cell, it makes a DNA copy of its RNA genome using its reverse transcriptase. The DNA copy then integrates into the cell's genome using the virally encoded enzyme integrase. Once integrated, the virus is inaccessible to antiviral chemotherapy.

If the CD4$^+$ cell becomes activated, the viral genes are transcribed, creating RNA copies. These serve dual roles, functioning both as mRNA molecules that can be translated and RNA genomes for new viral particles. The new viral particles bud out of the host cell, ultimately leading to the death of that cell. Despite the body's ability to replace hundreds of billions of CD4$^+$ lymphocytes, the number of these lymphocytes slowly declines over many months or years. Infection of macrophages, which are also CD4$^+$ cells, adds to the decline in cell-mediated immunity. Although macrophages are generally not killed by HIV, their function is impaired both by the viral infection and the lack of interaction with T$_H$ lymphocytes. Eventually, the immune system becomes so impaired it can no longer respond to infections or cancers. ◀◀ helper T cells, p. 356   ◀◀ macrophages, p. 340

Epidemiology

HIV is present in blood, semen, and vaginal secretions in symptomatic and asymptomatic infections. The virus spreads horizontally mainly by sexual intercourse and sharing hypodermic needles. It is not highly contagious, and most transmissions of this kind can be prevented by changes in human behavior. HIV can also be transmitted vertically—from mother to newborn—during pregnancy, during passage through the birth canal, and through breast-feeding. ◀◀ horizontal and vertical transmission, p. 440

Treatment and Prevention

Treatment of HIV disease is designed to block replication of the virus, but it does not affect viral nucleic acid already integrated in the genome of host cells. Four groups of antiretroviral medications are currently available—those that (1) block reverse transcriptase activity, (2) act against viral protease, (3) block viral integrase, and (4) interfere with viral attachment to host cells. The first three block enzymes essential at different stages of viral replication. The fourth type is represented by a drug that interferes with HIV attachment to the host cell cytokine receptor CCR5. Viral mutants resistant to a single medication, however, quickly arise. To help prevent the selection of resistant variants, a "cocktail" of several medications, each acting in a unique way, is given. When combinations of antiviral drugs are given, the treatment is referred to as **highly active antiretroviral therapy (HAART).** In many cases, this therapy can clear the patient's blood of detectable viral nucleic acid, stop the progress of the disease, and even allow partial recovery of immune function. Therapy is expensive and the medications often have serious side effects, two features that preclude use of the regimen in many of the world's HIV disease sufferers. Although not curative, chemotherapy can significantly prolong and improve the quality of life. ▶▶ HIV replication, p. 699   ▶▶ HAART, p. 703

Prevention of HIV disease is aimed at education and safer sex practices. One of the most important developments in AIDS prevention was the finding that vertical transmission from mother to newborn can be stopped in about two-thirds of the cases by using antiretroviral medications (ARVs). Needle-exchange programs

are also successful. Supplying injected-drug abusers with sterile needles and syringes in exchange for used ones decreases transmissions of the virus without encouraging addiction. Educational programs aimed at minimizing high-risk sexual practices have also shown short-term success in decreasing HIV spread. In addition, identifying and treating other STIs and promoting consistent use of latex condoms decreases the spread of HIV.

The World Health Organization has reported that circumcision (surgical removal of the penis foreskin) can reduce HIV incidence by up to 60% and should also be included in prevention strategies. Circumcision changes the anatomy of the penis, for example, reducing the number of Langerhans cells, which are HIV targets. It also changes the normal microbiota populations under the foreskin, reducing the number of certain anaerobic species, some of which increase the risk of HIV infection by enhancing virus stability. Another promising prevention strategy is pre-exposure prophylaxis, in which ARVs are given before sexual intercourse in an effort to reduce transmission (see chapter 28).

About 25% of the estimated 900,000 people in the United States infected with HIV are unaware of their infection and more than half of new HIV infections in the country are caused by those unaware of their HIV status. Individuals who, since the 1980s, have engaged in injected drug use or risky sexual behavior, and anyone having had unprotected sexual intercourse with them, should get a blood or saliva test to rule out HIV infection. People who learn that their HIV test is positive can receive optimum treatment sooner and also prevent transmission of the virus to others. **Table 25.15** presents the main features of HIV disease and AIDS.

TABLE 25.15	HIV Disease and AIDS
Signs and symptoms	No symptoms, or flulike symptoms early in the illness; an asymptomatic period typically lasting years; symptoms of lung, intestine, skin, eyes, brain, and other infections, and certain cancers
Incubation period	About 6 days to 6 weeks for flulike symptoms; many months or years for cancers and unusual infections
Causative agents	Generally human immunodeficiency virus type 1 (HIV-1)
Pathogenesis	The virus infects CD4$^+$ lymphocytes and macrophages, thereby slowly destroying the ability of the immune system to fight infections and cancers.
Epidemiology	HIV present in blood, semen, and vaginal secretions in symptomatic and asymptomatic infections; spread usually by sexual intercourse, sharing of needles by injected-drug abusers, and from mother to infant at childbirth. Other STIs foster transmission.
Treatment and prevention	Treatment: reverse transcriptase and protease inhibitors in combination (HAART). Prevention: abstinence from sexual intercourse and drug abuse; monogamy; consistent use of latex condoms; avoidance of sexual contact with injected-drug abusers, those with multiple partners or history of STIs. Anti-HIV medication for pregnant women and their newborn infants. Circumcision and pre-exposure prophylaxis.

25.7 ■ Protozoal STIs

Learning Outcome

11. *List the distinctive characteristics of the organism that causes trichomoniasis.*

A number of protozoan diseases can be transmitted sexually. Most, however, are intestinal infections that can be transmitted through oral-anal contact during sexual activity. Trichomoniasis is a sexually transmitted protozoan disease that involves the genital system. ◀◀ protozoa, p. 291

Trichomoniasis ("Trich")

Trichomoniasis ranks third after bacterial vaginosis and vulvovaginal candidiasis among the diseases that commonly cause vaginal signs and symptoms. Men can also be infected. An estimated 7.4 million Americans contract this STI each year.

Signs and Symptoms

Trichomoniasis is often asymptomatic, especially in men. Symptomatic infections in women are characterized by itching of the vulva and inner thighs, itching and burning of the vagina, and a frothy, sometimes smelly, yellowish-green vaginal discharge. In some cases, there is burning pain with urination. The vulva and vaginal wall are red and slightly swollen, and may develop pinpoint hemorrhages. Symptomatic men have penile discharge, burning pain with urination, painful testes, or a tender prostate gland.

Causative Agent

Trichomoniasis is caused by *Trichomonas vaginalis,* a motile protozoan that has four anterior flagella and a posterior flagellum attached to an undulating membrane (**figure 25.21**). It also has a slender, posteriorly protruding, rigid rod called an axostyle that is used for attachment. *T. vaginalis* can be identified by its unmistakable, jerky motility on microscopic examination of the vaginal drainage. Unlike most other pathogenic protozoa, *T. vaginalis* lacks a cyst form to aid its survival in the environment away from the host's body. The organism is a eukaryote, but it does not have any mitochondria. Instead, it has interesting cytoplasmic organelles called hydrogenosomes for glucose metabolism and respiration. Enzymes within these double membrane-bound organelles remove the carboxyl group (COOH) from pyruvate and transfer electrons to hydrogen ions, producing hydrogen gas. ◀◀ mitochondria, p. 74

Pathogenesis

The pathogenesis of trichomoniasis is not fully understood. *T. vaginalis* produces a variety of adherence factors (including adhesins, microtubules, and microfilaments) specific for vaginal epithelium, allowing it to colonize the vulva and vagina walls. The signs and symptoms of trichomoniasis are thought to be due to mechanical trauma by the axostyle of the moving protozoa, but toxins or enzymes such as cysteine proteinases might also be involved. The frothy discharge is probably due to the gas produced by the organisms.

Epidemiology

Trichomonas vaginalis is distributed worldwide as a human parasite and has no other reservoirs. It is easily killed by drying, so transmission is usually by sexual contact. The organism can survive for a time on moist objects such as towels and bathtubs, however, so it can occasionally be transmitted non-sexually. Nevertheless, *T. vaginalis* infections in children should at least raise the question of sexual abuse and possible exposure to other

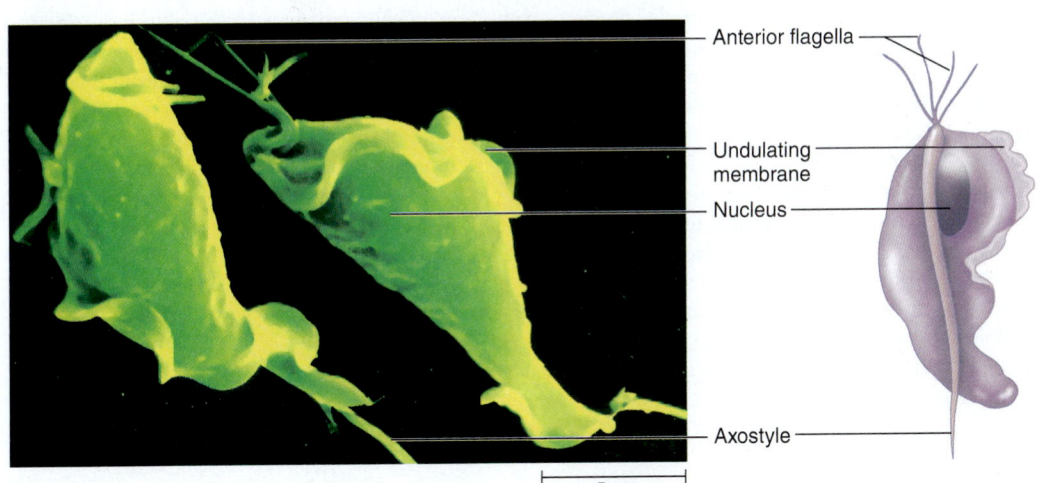

Anterior flagella

Undulating membrane

Nucleus

Axostyle

5 μm

FIGURE 25.21 *Trichomonas vaginalis T. vaginalis* is a common cause of vaginitis.

❓ *What is unusual about the metabolism of this organism?*

STIs. Newborn infants can contract the infection from infected mothers at birth. The high percentage of asymptomatic infections, especially in men, helps transmission of the disease. Infection rates are highest in men and women with multiple sex partners. Infection with *T. vaginalis* causes tissue damage, so people with this disease are at higher risk of contracting HIV.

MicroByte ───────────

An African study indicated that trichomoniasis can cause a two- to threefold increase in HIV transmission.

Treatment and Prevention

Most strains of *T. vaginalis* respond quickly to metronidazole treatment, but a few are resistant. Transmission is prevented by abstinence, monogamy, and the use of condoms. **Table 25.16** gives the main features of trichomoniasis.

The key features of the diseases covered in this chapter are highlighted in the **Diseases in Review 25.1** table.

MicroAssessment 25.7

Trichomoniasis is a protozoan STI that involves the genital tract. The causative agent is a flagellated protozoan that does not form cysts and lacks mitochondria. Organelles called hydrogenosomes produce hydrogen gas from pyruvate.

19. *What significance does the lack of a cyst form of* Trichomonas vaginalis *have on the epidemiology of trichomoniasis?*
20. *What is the relationship between being "easily killed by drying" and "transmission usually by sexual contact"?* ✚

TABLE 25.16	Trichomoniasis
Signs and symptoms	Women: itching, burning, swelling, vaginal redness; frothy, sometimes malodorous, yellow-green discharge, and burning on urination. Men: discharge from penis, burning on urination, painful testes, tender prostate. Many women, most men asymptomatic
Incubation period	4 to 20 days
Causative agent	*Trichomonas vaginalis,* a protozoan with four anterior flagella, and a posterior flagellum attached to an undulating membrane; a rigid rodlike structure called an axostyle protrudes posteriorly; unmistakable jerky motility; no mitochondria; hydrogenosomes are present
Pathogenesis	Unexplained. Inflammatory changes and pinpoint hemorrhages suggest mechanical trauma from the motile organisms.
Epidemiology	Worldwide distribution; asymptomatic carriers foster spread; easily killed by drying due to lack of cyst form, transmission by intimate contact; high rate of infection with multiple sex partners. Newborn infants of infected mothers can acquire the infection at birth.
Treatment and prevention	Treatment: metronidazole. Prevention: abstinence, monogamy, and consistent use of condoms.

FUTURE CHALLENGES 25.1

Getting Control of Sexually Transmitted Infections

Few problems are as complicated as getting control of STIs because of the psychological, cultural, religious, and economic factors that are involved—factors that vary from one population and culture to another. Gaining control means focusing on diagnosis, interruption of transmission, education, and treatment. The challenge is to develop innovative approaches in each of these areas and to apply them worldwide on a sustained basis.

In the area of diagnosis, the challenge is to use molecular biological techniques to identify and produce specific antigens and antibodies that can be used to quickly identify causative agents and determine the extent of their spread. The development of nucleic acid probes to identify genes coding for resistance to therapeutic agents helps in choosing prompt treatment.

Using condoms is an effective method for interruption of STI transmission, but in practice there are a number of problems with using them. When infected college students are asked why they did not use a condom, they often state that "it takes away the romance," "it decreases pleasurable sensation," or "that's for sissies." Also, many individuals are allergic to latex, and the alternatives, except for polyurethane, are unreliable for preventing disease transmission. Worldwide, as many as one out of three condoms is defective, and people in the poorest countries cannot even afford them. The challenge is to develop cheaper, more reliable, and more acceptable barrier methods, perhaps using new polymers or antimicrobial vaginal gels.

Effective vaccines could block the spread of STIs, but their development has been extremely problematic and remains a long-term goal. Sequencing the genomes of the causative agents is helping to identify appropriate antigens to include in vaccines, but using them to stimulate a protective immune response is a continuing challenge.

Educational efforts focusing on groups at high risk for STIs have had mixed success. Education can be a useful tool for gaining control of STIs, and the challenge is to better understand the reasons for its failures.

Diseases in Review 25.1

Genitourinary Infections

Disease	Causative Agent	Comment	Summary Table
UROGENITAL INFECTIONS			
Cystitis (bladder infection)	Usually uropathogenic *Escherichia coli*	Most common in women because of their relatively short urethra; characterized by painful urination; can progress to kidney infection.	Table 25.1, p. 615
Leptospirosis	*Leptospira interrogans*	Spread by water contaminated with the urine of infected animals; enters via mucous membranes or breaks in skin, and then spreads to all tissues, including eyes.	Table 25.2, p. 617
Bacterial vaginosis (BV)	Unknown, possibly bacterial	Characterized by thin, gray-white discharge and fishy odor. Most common in sexually active women and pregnant women.	Table 25.3, p. 618
Staphylococcal toxic shock	Toxin-producing strains of *Staphylococcus aureus*	Toxin is a superantigen and causes cytokine release, leading to drop in blood pressure; an epidemic occurred as a result of high-absorbency tampon use.	Table 25.5, p. 620
BACTERIAL STIs			
Gonorrhea	*Neisseria gonorrhoeae*	Infections are often asymptomatic, particularly in women. Signs and symptoms include painful urination and pus (men) or vaginal discharge; can progress to PID.	Table 25.8, p. 624
Chlamydial infections	*Chlamydia trachomatis*	Most common STI in the United States; infections are often asymptomatic; can progress to PID.	Table 25.9, p. 626
Syphilis	*Treponema pallidum*	Primary syphilis is characterized by a painless chancre, secondary by a widespread rash, and tertiary by cardiovascular and nervous system damage. Congenital infections also occur.	Table 25.10, p. 628 Table 25.11, p. 629
Chancroid	*Haemophilus ducreyi*	Characterized by painful genital sores and enlarged groin lymph nodes.	Table 25.12, p. 630
VIRAL STIs			
Genital herpes	Herpes simplex virus, usually type 2 (HSV-2)	Characterized by itching, burning, painful blisters. Virus becomes latent, so infections are lifelong.	Table 25.13, p. 631
Papillomavirus infections	Human papillomaviruses (HPVs)	Some strains cause genital warts, and others cause cancers. Warts removed using lasers, liquid nitrogen, or immune modifier creams. Vaccines protect against some cancer-causing strains.	Table 25.14, p. 633
HIV/ AIDS	Human immunodeficiency virus, usually type 1 (HIV-1)	HIV disease usually progresses to AIDS. Virus infects helper T cells, resulting in immunodeficiency. HAART slows disease progression.	Table 25.15, p. 634
FUNGAL DISEASE			
Vulvovaginal candidiasis (VVC)	*Candida albicans*	Characterized by intense itching and thick, white discharge. Associated with antibiotics or other conditions that disrupt the normal vaginal microbiota, allowing the overgrowth of *C. albicans*.	Table 25.4, p. 619
PROTOZOAL STI			
Trichomoniasis ("trich")	*Trichomonas vaginalis*	Sexually transmitted; characterized by itching, discharge, and urinary discomfort.	Table 25.16, p. 636

Summary

25.1 ■ Anatomy, Physiology, and Ecology

The Urinary System (figure 25.1)

The urinary system is composed of the kidneys, ureters, bladder, and urethra. Frequent urination with complete bladder emptying is an important defense mechanism against UTI. Infections occur more frequently in women than in men because of the shortness of the female urethra and its closeness to the opening of the intestinal tract.

The Genital System (figure 25.2)

The fallopian tubes, which are open on both ends, provide a passageway for infection to enter the abdominal cavity. The uterine cervix is a frequent site of infection and a place where cancer can develop. The vagina is a portal of entry for a number of infections. In men, the prostate can enlarge and partially obstruct urinary flow.

The lower urethra is inhabited by various microorganisms, sometimes including potential pathogens. The normal vaginal microbiota is affected by estrogen hormones, which cause deposition of glycogen in the cells lining the vagina. Vaginal lactobacilli metabolize the glycogen and release lactic acid and hydrogen peroxide, thereby protecting the vagina from colonization by pathogens.

25.2 ■ Urinary Tract Infections

Any condition that prevents a person from urinating normally increases the risk of infection. The urinary system usually becomes infected by organisms ascending from the urethra, but it can also be infected from the bloodstream.

Bacterial Cystitis ("Bladder Infection") (table 25.1)

Most UTIs in healthy people are caused by *Escherichia coli* or other *Enterobacteriaceae* members from the person's own normal intestinal microbiota. Healthcare-associated UTIs are common and are caused by *Pseudomonas aeruginosa, Serratia marcescens,* and *Enterococcus faecalis,* which are often resistant to many antibiotics. Kidney infection (**pyelonephritis**) may complicate untreated bladder infection when pathogens ascend through the ureters to the kidneys.

Leptospirosis (table 25.2)

In leptospirosis, the urinary system is infected by *Leptospira interrogans* from the bloodstream (figure 25.3). This biphasic illness causes fever, bloodshot eyes, and pain in the septicemic phase. Improvement occurs, then new signs and symptoms emerge, with damage to multiple organs during the immune phase. Many species of animals are chronically infected with *L. interrogans* and excrete the organism in their urine.

25.3 ■ Genital System Diseases

Bacterial Vaginosis (BV) (table 25.3, figure 25.4)

Bacterial vaginosis is the most common cause of vaginal signs and symptoms that include a gray-white discharge from the vagina and a strong fishy odor; there is no inflammation. The causative agent or agents are unknown, but BV occurs with a disruption of the normal microbiota.

Vulvovaginal Candidiasis (VVC) (table 25.4)

Vulvovaginal candidiasis is second among causes of vaginal disorders. Signs and symptoms include itching, burning, vulvar redness and swelling, and a thick, white discharge. The causative agent is a yeast, *Candida albicans,* that is commonly part of the normal vaginal microbiota (figure 25.5). Antibacterial treatment, uncontrolled diabetes, and oral contraceptives are predisposing factors, but in most cases no such factor can be identified.

Staphylococcal Toxic Shock Syndrome (table 25.5)

Staphylococcal toxic shock syndrome became widely known with a 1980 epidemic in menstruating women who used a certain kind of vaginal tampon that has since been removed from the market (figure 25.6). Signs and symptoms include sudden fever, headache, muscle aches, bloodshot eyes, vomiting, diarrhea, a sunburnlike rash that later peels, and confusion. The blood pressure drops, and without treatment, kidney failure and death may occur.

25.4 ■ Sexually Transmitted Infections: Scope of the Problem (tables 25.6, 25.7)

In the United States, 19 million new STIs occur each year, almost half in individuals 15 to 24 years old. Simple measures exist for controlling STIs: abstinence from sexual intercourse, a monogamous relationship with an uninfected person, and consistent and correct use of latex or polyurethane condoms (figure 25.7).

25.5 ■ Bacterial STIs

Most of the bacteria that cause STIs survive poorly in the environment and require intimate contact for transmission.

Gonorrhea (table 25.8)

Gonorrhea, caused by *Neisseria gonorrhoeae* (figures 25.9, 25.10), is one of the most commonly reported bacterial diseases. Men usually develop painful urination and thick pus draining from the urethra (figure 25.8). Women may have similar signs and symptoms, but these tend to be milder and are often unnoticed. Inflammatory reaction to the infection causes scarring, which can partially obstruct the urethra or cause infertility in men and women. Expression of different surface antigens allows attachment of *N. gonorrhoeae* to different types of cells. Antigenic variation has prevented development of a vaccine.

Chlamydial Infections (table 25.9, figure 25.11)

Chlamydial genital infections, caused by *Chlamydia trachomatis,* are reported more often than any other bacterial disease. Signs and symptoms and complications of chlamydial infections are very similar to those of gonorrhea, but are milder. Asymptomatic infections are common and easily transmitted. Sexually active people should be tested at least once a year so that transmission can be stopped and complications prevented.

Syphilis (tables 25.10, 25.11)

Syphilis, caused by *Treponema pallidum,* (figure 25.16), is called "the great imitator" because its many signs and symptoms can resemble other diseases. Primary syphilis is characterized by a painless, firm ulceration called a hard **chancre;** the organisms multiply and spread throughout the body (figure 25.12). In secondary syphilis, skin and mucous membranes show lesions that have large numbers of the organisms (figure 25.13); a latent period of months or years occurs between the secondary and tertiary phases of the disease. Tertiary syphilis is not contagious. It causes damage to the eyes and the cardiovascular and central nervous systems. An inflammatory, necrotizing mass called a **gumma** can involve any part of the body (figure 25.14). Syphilis in pregnant women can spread across the placenta to involve the fetus, resulting in congenital syphilis (figure 25.15).

Chancroid (table 25.12)

Chancroid is another widespread bacterial STI, but it is not commonly reported because of difficulties in making a bacterial diagnosis. Epidemics in the United States have been associated with prostitution. Chancroid is caused by *Haemophilus ducreyi,* and is characterized by single or multiple soft, tender genital ulcers, and enlarged, painful groin lymph nodes (figure 25.17). Some strains of *H. ducreyi* are resistant to various antimicrobial drugs, making treatment more difficult.

25.6 ■ Viral STIs

Viral STIs are at least as common as bacterial STIs, but they are not yet curable.

Genital Herpes (table 25.13)

Genital herpes is a very common disease, important because of the discomfort and emotional trauma it causes, its potential for causing death in newborn infants, and the increased risk it poses of transmitting HIV. Signs and symptoms may include a group of vesicles (figure 25.18) with itching, burning, or pain; these break, leaving painful ulcers. Local lymph nodes enlarge. Many infections have few or no signs or symptoms; some have painful recurrences. The virus establishes a latent infection in sensory nerves and cannot be cured. It can be transmitted in the absence of signs or symptoms, but the risk is greatest when lesions are present.

Papillomavirus STIs: Genital Warts and Cervical Cancer (table 25.14)

Papillomavirus STIs are probably more common than any other kind of STI. The causative agents, human papillomaviruses (HPVs), are small DNA viruses. Genital infection symptoms are warts on or near the genitalia (figure 25.19). Some HPV types cause precancerous lesions that are asymptomatic and can be detected only by medical examination. HPVs are the main cause of abnormal Pap smears in young women

(figure 25.20). Several types of HPV, including HPV 16 and 18, are associated with cancer of the cervix, vagina, penis, anus, or throat.

HIV/AIDS (table 25.15)

AIDS is the end stage of disease caused by human immunodeficiency virus (HIV). Worldwide, millions of new HIV infections and deaths from AIDS occur each year. HIV disease is usually first manifested as a flulike illness that develops about 6 days to 6 weeks after an individual contracts the virus. An asymptomatic interval follows that typically lasts almost 10 years, during which the immune system is slowly and progressively destroyed. Unusual cancers and infectious diseases indicate the start of AIDS. No vaccine or medical cure is yet available, but HIV transmission can be slowed significantly by consistent use of condoms and use of sterile needles by injected-drug abusers. A significant reduction in mother-to-newborn transmission can be achieved with medication. Pre-exposure prophylaxis with antiretroviral medication also shows promise in reducing infection rates.

25.7 ■ Protozoal STIs

Trichomoniasis ("Trich") (table 25.16)

Signs and symptoms include itching, burning, swelling, and redness of the vagina; with frothy, sometimes smelly, yellow-green discharge; and burning on urination. Men may have discharge from the penis, burning on urination, sometimes accompanied by painful testes and a tender prostate gland. Many women and most men are asymptomatic. The causative agent is *Trichomonas vaginalis,* a protozoan with four anterior flagella, and a posterior flagellum attached to an undulating membrane (figure 25.21).

Review Questions

Short Answer

1. Name two substances released by lactobacilli that help protect the vagina from potential pathogens.
2. List four things that predispose to the development of infection of the urinary bladder.
3. Name two genera of bacteria that infect the kidneys from the bloodstream.
4. What possible danger can be found in a spot on the ground where an animal has urinated 1 week earlier?
5. What is a clue cell?
6. What is ophthalmia neonatorum?
7. List three diseases caused by different antigenic types of *Chlamydia trachomatis*.
8. Why is dark-field microscopy used to view *Treponema pallidum*?
9. Give two ways in which the chancre of chancroid differs from the chancre of syphilis.
10. What is the relationship between AIDS and HIV disease?

Multiple Choice

1. Which of the following about bacterial cystitis is *false*?
   a) About one-third of all women will have it at some time during their life.
   b) Catheterization of the bladder markedly increases the risk of contracting the disease.
   c) Individuals who have a bladder catheter in place indefinitely risk bladder infections with multiple species of intestinal bacteria at the same time.
   d) Bladder infections occur as often in men as they do in women.
   e) Bladder infections can be asymptomatic.

2. Choose the one correct statement about leptospirosis.
   a) Humans are the only reservoir.
   b) Most infections produce severe symptoms.
   c) Transmission is by the fecal-oral route.
   d) It can lead to unnecessary abdominal surgery.
   e) Effective vaccine is generally available for preventing human disease.

3. Which one of the following statements about bacterial vaginosis is *false*?
   a) It is the most common vaginal disease in women of childbearing age.
   b) In pregnant women, it is associated with a sevenfold increased risk of obstetrical complications.
   c) Inflammation of the vagina is a constant feature of the disease.
   d) The vaginal microbiota shows a significant decrease in lactobacilli and a marked increase in anaerobic bacteria.
   e) The cause is unknown.

4. Pick the one *false* statement about vulvovaginal candidiasis.
   a) It often involves the external genitalia.
   b) It is readily transmitted by sexual intercourse.
   c) It is caused by a yeast present among the normal vaginal microbiota in about one-third of healthy women.
   d) It is associated with prolonged antibiotic use.
   e) It involves increased risk late in pregnancy.

5. All of the following statements about staphylococcal toxic shock are true *except*
   a) It can quickly lead to kidney failure.
   b) The causative organism usually does not enter the bloodstream.
   c) It occurs only in vaginal tampon users.
   d) Almost one-third of victims of the disease will suffer a recurrence sometime after recovery.
   e) Person-to-person spread does not occur.

6. Which of the following statements about gonorrhea is *false*?
   a) The incubation period is only a few days.
   b) Disseminated gonococcal infection (DGI) is almost invariably preceded by prominent urogenital symptoms.
   c) DGI can result in arthritis of the knee.
   d) Phase variation helps the causative organism evade the immune response.
   e) Pelvic inflammatory disease (PID) is common in untreated women.

7. Which one of these statements about chlamydial genital infections is *false*?
   a) The incubation period is usually shorter than in gonorrhea.
   b) Infected cells develop inclusion bodies.
   c) Pelvic inflammatory disease (PID) can be complicated by infection of the surface of the liver.
   d) Tissue damage largely results from cell-mediated immunity.
   e) Fallopian tube damage can occur in the absence of symptoms.

8. Which symptom is least likely to occur as a result of tertiary syphilis?
   a) Gummas   b) White patches on mucous membranes
   c) Emotional instability   d) Stroke
   e) Blindness

9. During the first 15 years of the AIDS epidemic, approximately how many Americans died of the disease?
   a) 10,000   b) 50,000
   c) 300,000   d) 5 million
   e) 50 million

10. All of the following are true of "trich" (trichomoniasis) *except*
    a) It can cause burning pain on urination and painful testes in men.
    b) It occurs worldwide.
    c) Asymptomatic carriers are rare.
    d) Transmission can be prevented by proper use of condoms.
    e) Individuals with multiple sex partners are at high risk of contracting the disease.

Applications

1. Religious restrictions of a small North African community are preventing a World Health Organization project from reducing the incidence of gonorrhea. The community will not permit the testing of females for the disease. They can be treated, however, if they show outward evidence of the disease. Only males are allowed to participate fully in the project, with testing for the disease and treatment. The village elders argue that eradicating the disease from males would eventually remove it from the population. What would be the impact of these restrictions on the success of the project?

2. Former President Ronald Reagan once commented at a press conference that the best way to combat the spread of AIDS in the United States was to prohibit everyone from having sexual contact for 5 years. What would be the success of such a program if it were possible to carry it out?

Critical Thinking ✚

1. The middle curve of figure 25.6 shows the occurrence of staphylococcal toxic shock syndrome in menstruating women from 1979 to 2010. What aspect of these data argues that high-absorbency tampons were not the only cause of staphylococcal toxic shock syndrome associated with menstruation?

2. In early attempts to identify and isolate the cause of syphilis, various bacteria in the discharge from syphilitic lesions in experimental animals were isolated in pure culture. None of them, however, would cause the disease when used in attempts to infect healthy animals. Why was it considered a critical step to have the cultivated bacteria reproduce the disease in the healthy animals?

26 Nervous System Infections

TEM of West Nile virus, a virus introduced into the United States in 1999.

A Glimpse of History

Today it is hard to appreciate the fear and loathing once attached to leprosy. The Bible refers to several disfiguring skin diseases, including leprosy (*lepros,* meaning "scaly"), and people suffering from the diseases are portrayed as filthy, outcast, or condemned by God for sin. Moses called lepers "unclean" and proclaimed they must live away from others. In the Middle Ages, lepers attended their own symbolic burial before being sent away.

Gerhard Henrik Armauer Hansen (1841–1912) was a burly Norwegian physician with many interests, ranging from science to religion to polar exploration. When he was 32 years old, Hansen went into medical research, and was named assistant to Dr. Daniel C. Danielson, a leading authority on leprosy. Danielson believed that leprosy was a hereditary disease of the blood, and considered the idea that the disease was contagious as a "peasant superstition." Hansen, however, disproved Danielson's hypothesis in careful studies conducted over a number of years. He found a unique bacterium associated with the disease in every leprosy patient he studied. His 1873 report of the findings marked the first time that a specific bacterium was linked to a disease—almost a decade before Koch's proof of the cause of tuberculosis.

In the United States, even during the first half of the twentieth century, persons diagnosed with leprosy risked having their houses burned to destroy the source of infection. Their names were changed to avoid embarrassing their family, and they were whisked to a leprosarium such as the one at Carville, Louisiana, surrounded by a 12-foot fence topped with barbed wire. Sufferers were separated from spouses and children and denied the right to marry or vote. Those who attempted to escape were captured and brought back in handcuffs. The Carville leprosarium was finally closed and converted to a military-style academy for high school dropouts in 1999.

Because the word *leprosy* carries centuries of dark overtones, many people prefer to use the term *Hansen's disease,* a name that honors the discoverer of the causative bacterium.

Nervous system infections are frightening. They threaten a person's ability to move, feel, or even think. Consider poliomyelitis, which can result in a paralyzed limb or the inability to breathe without mechanical assistance. Hansen's disease (leprosy) can result in loss of fingers or toes or deformity of the face. Infections of the brain or its covering membranes can render a child deaf or intellectually disabled. Before the discovery of antibiotics, bacterial infections of the nervous system were often fatal. Fortunately, CNS infections are uncommon.

26.1 ■ Anatomy, Physiology, and Ecology

Learning Outcomes

1. *Describe how information flows through and between neurons.*
2. *Differentiate between the central nervous system and the peripheral nervous system.*
3. *Explain how bone, cerebrospinal fluid, meninges, and the blood-brain barrier protect the central nervous system.*

Nerve cells work together, transmitting electrical impulses throughout the body like a highly sophisticated circuit board. Each nerve cell, or **neuron,** has different regions with distinct functions (**figure 26.1a**). Branching projections called dendrites receive information. They convey that information to the cell body, the command center that contains the cell nucleus. A long, thin extension called an axon then transmits the information from the cell body to another cell.

Most neurons communicate using **neurotransmitters,** chemicals produced in the cell body and stored in vesicles at the end of the axon. When neurotransmitters are released, the molecules diffuse across the short distance to a neighboring cell. That cell has receptors for the neurotransmitter, allowing it to receive and respond to the signal. The region where transmission of information takes place between the two cells is called a synapse. Although movement of neurotransmitter molecules is normally away from the cell body to the synapse, some viruses and toxins can move through the axon toward the cell body in a process called retrograde transport. The rabies virus, for example, moves from a bite on the skin toward the brain or spinal cord where it multiplies, ultimately killing the infected person. ▶▶ rabies, p. 658

The brain and spinal cord (figure 26.1b) make up the **central nervous system (CNS).** The brain is a very complex structure; distinct parts have different functions. If an individual has a brain abscess, physicians can sometimes determine its location by evaluating the resulting loss of function. Generalized inflammation or infection of the brain is called **encephalitis.**

The **peripheral nervous system (PNS)** is made up of nerves composed of bundles of axons that carry information to and from the CNS. Axons of motor neurons carry messages from the CNS to different parts of the body and cause them to respond; axons of sensory neurons transmit sensations like touch, heat, light, and sound to the central nervous system. Most nerves are mixed, carrying both sensory and motor information throughout the body. The cell bodies of sensory neurons from the skin are located in clusters called ganglia (singular: ganglion) near the vertebral column; the cell bodies of motor neurons are located within the CNS.

The central nervous system is much better protected than the peripheral nervous system. Both the brain and spinal cord are enclosed by bone—the brain by the skull and the spinal cord by the vertebral column. Nerves can be damaged if the bones are infected at sites where the PNS penetrates this protective covering. Only rarely do infections in the bone surrounding the CNS spread to the brain or spinal cord. Sometimes, however, they extend through bone from the sinuses, mastoid air cells, or middle ear. Skull fractures commonly predispose a person to recurring infection of the CNS. ◀◀ mastoid air cells, p. 486 ◀◀ middle ear, p. 492

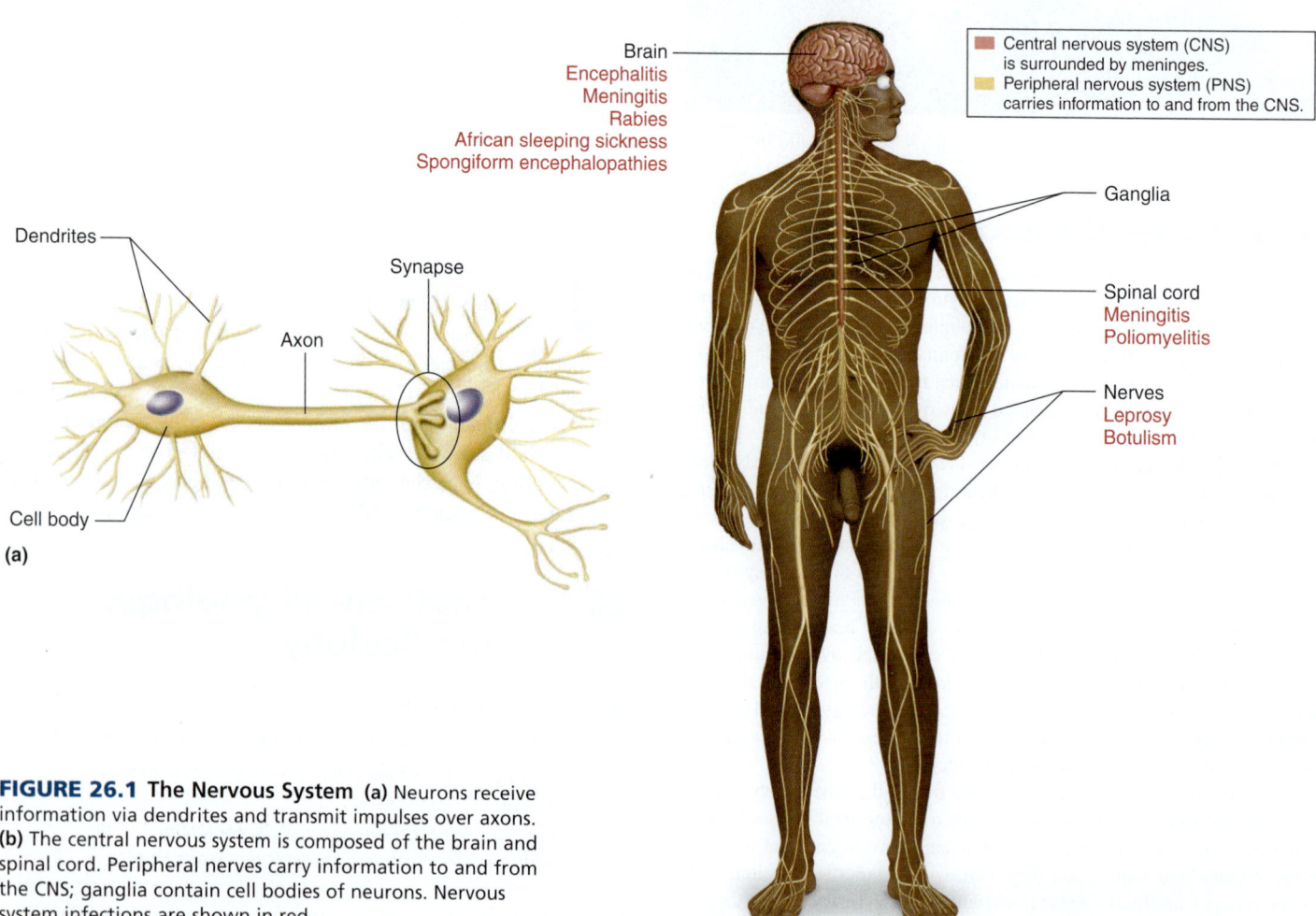

Brain
Encephalitis
Meningitis
Rabies
African sleeping sickness
Spongiform encephalopathies

Central nervous system (CNS) is surrounded by meninges.
Peripheral nervous system (PNS) carries information to and from the CNS.

Ganglia

Spinal cord
Meningitis
Poliomyelitis

Nerves
Leprosy
Botulism

Dendrites

Synapse

Axon

Cell body

(a)

(b)

FIGURE 26.1 The Nervous System (a) Neurons receive information via dendrites and transmit impulses over axons. **(b)** The central nervous system is composed of the brain and spinal cord. Peripheral nerves carry information to and from the CNS; ganglia contain cell bodies of neurons. Nervous system infections are shown in red.

❓ *What part of a neuron is found in the peripheral nerves?*

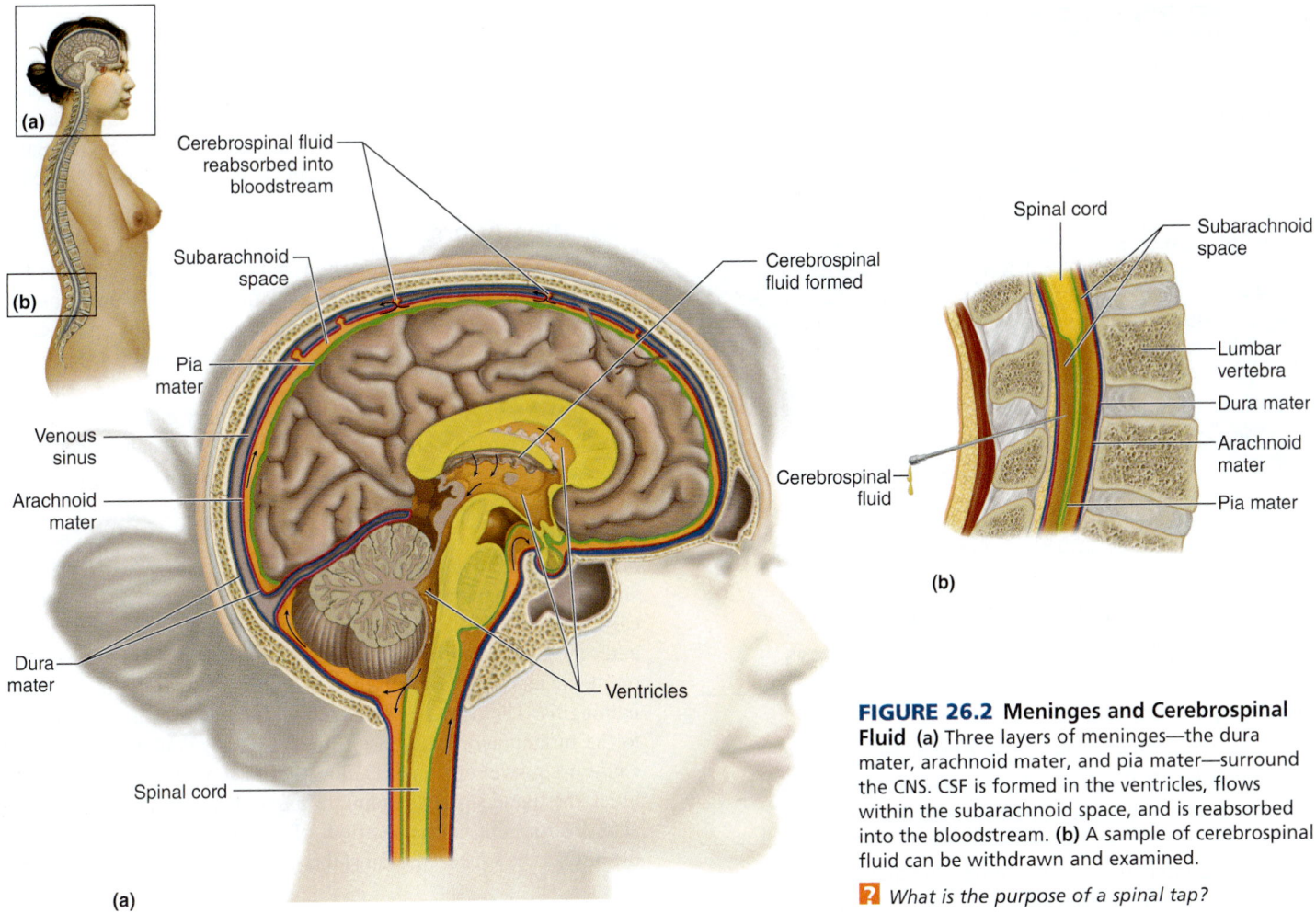

FIGURE 26.2 Meninges and Cerebrospinal Fluid (a) Three layers of meninges—the dura mater, arachnoid mater, and pia mater—surround the CNS. CSF is formed in the ventricles, flows within the subarachnoid space, and is reabsorbed into the bloodstream. (b) A sample of cerebrospinal fluid can be withdrawn and examined.

❓ *What is the purpose of a spinal tap?*

Deep inside the brain are four fluid-filled cavities called ventricles. The fluid within them, **cerebrospinal fluid (CSF),** is continually produced within the walls of ventricles, and spreads over the surface of the brain and spinal cord. CSF is reabsorbed into the bloodstream at specialized sites within a venous sinus (**figure 26.2a**). CSF provides cushion and support for the brain and also transports nutrients and other materials throughout the CNS.

Three layers of membranes called **meninges** cover the surface of the brain and spinal cord. The outer dura mater is tough and provides a barrier to the spread of infection from bones surrounding the central nervous system. It adheres closely to the skull and vertebrae, and in some parts of the brain encloses a venous sinus. The two inner membranes, the arachnoid mater and the pia mater, are separated by the subarachnoid space within which the cerebrospinal fluid flows (figure 26.2a). The innermost pia mater adheres to the nervous tissue. Inflammation or infection of these membranes is called **meningitis.** When both the meninges and the brain are infected, the condition is meningoencephalitis.

The nervous system lies entirely within body tissues and has no normal microbiota. CSF is generally sterile, so the presence of microbes indicates an infection. To diagnose meningitis, a sample of CSF is obtained using a procedure called a spinal tap or lumbar puncture. A needle is inserted into the subarachnoid space between lumbar vertebrae where the spinal cord has tapered to a threadlike

structure, and a sample of CSF is withdrawn (figure 26.2b). The agent causing the infection can then be examined and cultivated for identification. Other characteristics of the CSF are also examined, including glucose levels, protein levels, and white blood cell counts. If bacteria are growing in CSF, glucose levels decrease and protein levels increase. Neutrophils also accumulate. Viral meningitis is characterized by increased numbers of lymphocytes.

The bloodstream is the chief source of CNS infections. However, it is difficult for infectious agents to cross from the bloodstream to the brain because of the **blood-brain barrier.** This barrier depends on special cells lining capillaries in the CNS. These cells, only a single layer thick, are so close to each other that except for essential nutrients actively taken up, most molecules cannot pass through them. The blood-brain barrier generally prevents pathogens from entering nervous tissue except when a rare high concentration of the infectious agent circulates for a long time in the bloodstream. Unfortunately, the barrier also prevents many medications, including penicillin, from crossing into the CNS unless their concentrations in the blood are very high.

MicroByte

Infections on the face can move through bone to veins on the brain surface. A good rule is not to pop pimples on the upper part of the face.

Information within a neuron typically flows from dendrites to the cell body to the axon. Neurotransmitter molecules are released from the axon at a synapse where they bind to receptors on a neighboring cell, initiating a response. The brain and spinal cord make up the central nervous system (CNS). Nerves in the peripheral nervous system (PNS) convey sensory and motor information to and from the CNS. The CNS is protected by bone, by circulating cerebrospinal fluid (CSF), and by membranes called meninges. Inflammation of the meninges is termed meningitis. Infection can reach the CNS through the bloodstream, by axonal transport, or by contact with infected bone. The blood-brain barrier prevents passage of most pathogens from the bloodstream into the CNS, but it also can prevent access to the CNS by antimicrobial medications.

1. *What vital molecules are carried from the cell body to the synapse for release?*
2. *List the components of the CNS and of the PNS.*
3. *Why can encephalitis usually be detected in CSF obtained from the lower back?* ➕

26.2 ■ Bacterial Diseases of the Nervous System

Learning Outcomes

4. *Compare and contrast the various types of bacterial meningitis.*
5. *Explain how Hansen's disease can cause peripheral nerve damage.*
6. *Explain how botulism affects the nervous system.*

Bacteria can infect peripheral neurons, but they most often infect the meninges surrounding the central nervous system—causing meningitis. Early symptoms are like a mild cold, but they can quickly escalate. Bacterial meningitis can kill within hours, so antibiotics are often given while the cause of the inflammation is still being determined. The case-fatality rate for untreated bacterial meningitis can approach 100%. Even with proper treatment, the rate is still relatively high (about 10–20%). Survivors sometimes suffer permanent disabilities including deafness, blindness, paralysis, or mental impairment.

The three most common species that cause meningitis—*Streptococcus pneumoniae, Neisseria meningitidis,* and *Haemophilus influenzae*—are often part of the normal microbiota of the upper respiratory tract. They are transmitted through respiratory droplets and routinely cause common diseases like sinusitis or conjunctivitis. In a small percentage of infected people, however, the organisms invade the bloodstream and then spread to the CNS. People with lowered immunity such as newborns and the elderly are at most risk for meningitis. Young adults living in crowded, confined environments such as military barracks or college dormitories are also at increased risk. ◀◀ *Streptococcus pneumoniae*, p. 497 ◀◀ conjunctivitis, p. 492

Pneumococcal Meningitis

Pneumococcal meningitis is caused by *Streptococcus pneumoniae*, part of the normal microbiota in the throat and nasopharynx of many healthy individuals. Although best known as a cause of pneumonia, it is the leading cause of meningitis in adults.

Signs and Symptoms

Bacterial meningitis usually begins like a mild cold. This is followed by the sudden onset of a severe, throbbing headache; fever; pain and stiffness of the neck and back; nausea; and vomiting. Deafness, confusion, loss of consciousness, and coma may develop. Death may occur rapidly due to inflammation and shock.

Causative Agent

Streptococcus pneumoniae, often called pneumococcus, is a Gram-positive lancet-shaped coccus, often seen in pairs (**figure 26.3**). Many strains are protected from phagocytosis by a polysaccharide capsule. Antigenic differences in the capsules produce over 90 recognized serotypes. When the pathogen is part of the normal respiratory microbiota, the mucociliary escalator usually prevents it from entering the lungs, where it could cause pneumonia or enter the bloodstream. ◀◀ mucociliary escalator, p. 337

Pathogenesis

Streptococcus pneumoniae is a common cause of otitis media, sinusitis, and pneumonia, any of which can precede pneumococcal meningitis. Encapsulated strains resist phagocytosis and may enter the bloodstream of an infected individual, bind to receptors on meningeal cells, and pass into the cerebrospinal fluid, causing meningitis. The damage associated with meningitis is largely due to the inflammatory response. Fluids that accumulate in the area cause brain swelling, and the clots that form in the capillaries can block the blood supply, leading to tissue death. Inflammation can also obstruct the normal outflow of cerebrospinal fluid, causing the brain to be squeezed against the skull by the buildup of internal

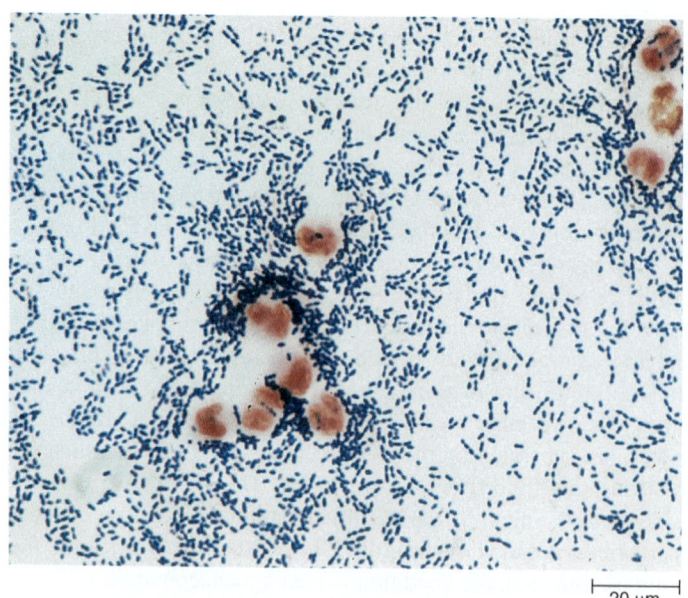

⊢ 20 µm ⊣

FIGURE 26.3 *Streptococcus pneumoniae* **(Pneumococci)** This Gram-stained smear shows large numbers of the Gram-positive diplococci in the CSF of a person with pneumococcal meningitis; the large red objects are polymorphonuclear leukocytes that have entered the spinal fluid in response to the infection.

❓ *How would a Gram stain distinguish this organism from Neisseria meningitidis?*

pressure. Mortality and neurological damage are more common with this type of meningitis than with others. ◀◀ otitis media, p. 492 ◀◀ pneumococcal pneumonia, p. 496

Epidemiology

When the mucociliary escalator is impaired, perhaps by respiratory infections such as influenza, *Streptococcus pneumoniae* cells can enter the lungs from the throat or nasopharynx, where they are part of the normal microbiota. Bacteremia is seen in most cases of pneumococcal meningitis, but trauma that allows contact between the CSF and the nasopharynx can also cause infection. The pathogen is spread by respiratory droplets, but very few who are infected will develop meningitis. The case-fatality rate is about 20% in adults and much higher in the elderly. ◀◀ bacteremia, p. 384

Treatment and Prevention

Penicillin remains the drug of choice for susceptible strains of *Streptococcus pneumoniae,* but some strains show antibiotic resistance. For those cases of pneumococcal meningitis, a cephalosporin such as ceftriaxone or cefotaxime is administered, with the possible addition of vancomycin.

Vaccines against *S. pneumoniae* offer protection from pneumonia and otitis media as well as meningitis, but are not effective against all of the different capsular antigens. A vaccine of the 23 most common capsular antigens has been developed for use in vulnerable adults. This vaccine is not effective in children because polysaccharide antigens are T-cell-independent and children under two respond poorly to them. A different vaccine was specifically developed for this age group. It consists of seven of the capsular antigens attached to a bacterial protein, forming an effective T-cell-dependent conjugate vaccine. Since the vaccine was licensed in 2000, childhood cases of pneumococcal meningitis caused by these seven strains have dropped significantly. However, incidence of cases caused by other strains has increased. A new 13-valent pneumococcal conjugate vaccine was licensed in 2010; it will provide protection against infections caused by six additional pneumococcal serotypes. ◀◀ T-independent antigen, p. 367 ◀◀ conjugate vaccines, p. 424

Meningococcal Meningitis

Neisseria meningitidis, often called meningococcus, is frequently responsible for epidemics of meningitis. Although humans are the only host, meningococcal meningitis is difficult to eliminate because *N. meningitidis* is commonly part of the normal respiratory microbiota of healthy individuals.

Signs and Symptoms

Symptoms of meningococcal meningitis are similar to those of pneumococcal meningitis but often include purplish spots on the skin called **petechiae** (figure 26.4). Septic (endotoxic) shock can lead to rapid death. ◀◀▶▶ septic shock, pp. 394, 671

Causative Agent

Meningococcal meningitis is caused by *Neisseria meningitidis,* a Gram-negative encapsulated diplococcus. The most serious infections are due to serotypes A, B, C, Y, and W135. Groups B and C are most common in the United States and Europe; type A is

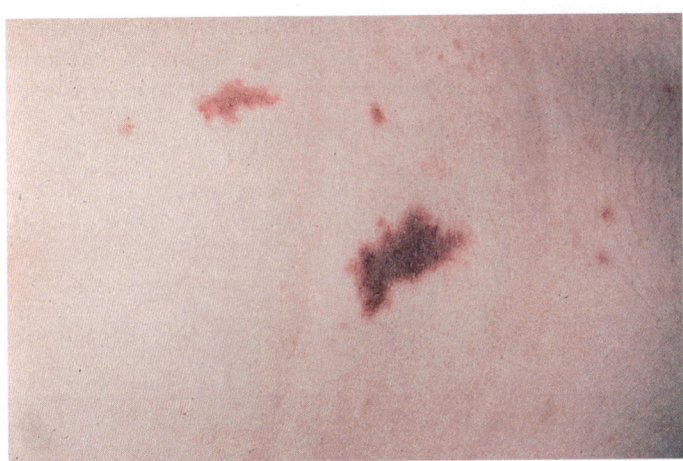

FIGURE 26.4 Petechiae of Meningococcal Disease These skin lesions are characteristic of meningitis caused by *Neisseria meningitidis.*

? *What causes petechiae?*

most common in Africa and Asia. Like *Neisseria gonorrhoeae, N. meningitidis* can vary some of its antigens and easily acquires DNA through horizontal gene transfer. ◀◀ serotype, p. 247 ◀◀ antigenic variation, p. 390 ◀◀ horizontal gene transfer, p. 200

Pathogenesis

Meningococci inhaled in airborne droplets attach by pili to mucous membranes and multiply. Specific proteins in their outer membranes allow the bacterial cells to pass through epithelial cells lining the respiratory tract and into the bloodstream to the meninges, although this is rare. Circulating *Neisseria meningitidis* release blebs of their outer membranes, and the endotoxin causes vasodilation and capillary leakage, leading to a drop in blood pressure. Endotoxic or septic shock results when blood pressure becomes so low that circulation cannot adequately supply O_2 to vital body tissues. Capillary damage causes the petechiae associated with meningococcal meningitis. ◀◀ pili, p. 65 ◀◀ endotoxin, p. 394

The capsule helps the cells avoid phagocytosis and protects them from the lethal actions of the complement system. Large numbers of neutrophils enter the CSF in response to the infection, but the bacteria multiply faster than they can be destroyed. As in pneumococcal meningitis, tissue damage and cerebral edema often result from inflammation. ◀◀ complement system, p. 344

Epidemiology

Transmission of *Neisseria meningitidis* can occur when someone is exposed to a person with the disease or to an asymptomatic carrier. Although most cases are sporadic, meningococcal meningitis causes epidemics if it spreads via respiratory droplets through crowded and stressed populations. *Neisseria meningitidis* can cause meningitis at any age. As the incidence of cases in babies has decreased due to immunization, a higher percentage of cases is reported in young adults. In the United States, it is a relatively rare disease (fewer than 3 cases per 100,000 people annually). The highest incidence worldwide occurs in the "meningitis belt" extending between Senegal and Ethiopia in sub-Saharan Africa, where the population is about 300 million (**figure 26.5**). During the dry season between December and June, incidence may reach 500 cases per 100,000 people.

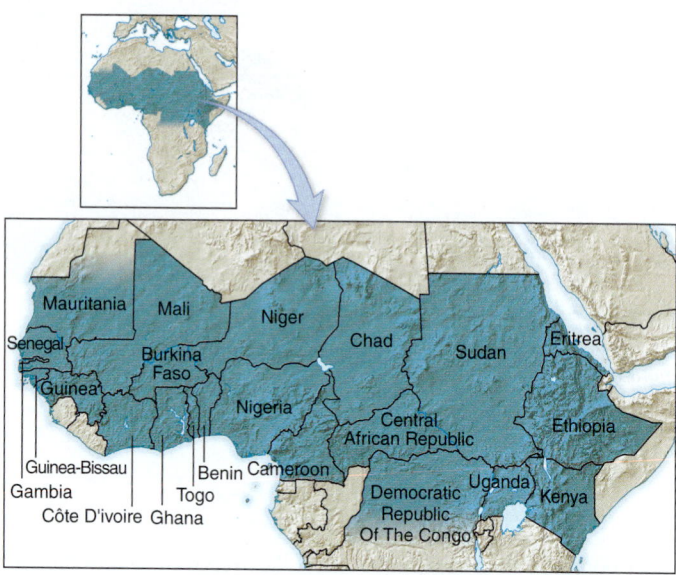

FIGURE 26.5 **The Meningitis Belt** Meningitis is most prevalent in Mali, Burkina Faso, Niger, Chad, Sudan, Ethiopia, and northern Nigeria.

❓ *Why would the incidence of meningitis be seasonal in this region?*

Treatment and Prevention

If treated in time, meningococcal meningitis can usually be cured with penicillin or ceftriaxone. Most patients recover without permanent nervous system damage. The case-fatality rate is less than 10% in treated cases.

A vaccine composed of purified A, C, Y, and W135 capsular polysaccharides has been available for years to control epidemics and to immunize people at high risk. A conjugate vaccine was approved in 2005 for the same four types, and an additional option was licensed in 2010. A conjugate vaccine is recommended for all 11- to 18-year-olds, and for those 2 to 55 years of age who are at increased risk, including college freshmen who will live in dormitories. Group B strains possess a poorly immunogenic polysaccharide, and no vaccine is yet available for their control. Antibacterial medications, however, can be useful in controlling epidemics in confined groups such as people in jails, nursing homes, and schools. People intimately exposed to cases of meningococcal disease are routinely given prophylactic treatment with the antibiotic rifampin. **Table 26.1** gives the main features of this disease.

Haemophilus influenzae Meningitis

Haemophilus influenzae once caused meningitis in about 1 out of 200 children under 5 years of age and was the leading cause of meningitis in that age group. Hib vaccine, introduced in the late 1980s, is a great success story as it has decreased incidence of this type of meningitis in children by over 99%.

TABLE 26.1	Meningococcal Meningitis		
① *Neisseria meningitidis* inhaled, infects upper airways.		**Signs and symptoms**	Mild cold followed by headache, fever, pain, stiff neck and back, vomiting, petechiae
② Bacteria enter the bloodstream and are circulated throughout the body.		**Incubation period**	1 to 7 days
③ The bacteria damage skin capillaries and cause petechiae.		**Causative agent**	*Neisseria meningitidis*, the meningococcus; a Gram-negative diplococcus
④ Bacteria infect the meninges causing meningitis.		**Pathogenesis**	Meningococci adhere by pili, colonize upper respiratory tract, enter bloodstream; carried to meninges and spinal fluid; inflammatory response obstructs normal outflow of fluid; increased pressure caused by obstructed flow impairs brain function; damage to motor nerves produces paralysis; endotoxin release causes shock.
⑤ Lysing bacteria in the circulation release endotoxin, producing shock.			
⑥ Inflammatory response in meninges can damage nerves of hearing, causing deafness, and obstruct the flow of cerebrospinal fluid, causing increased pressure inside the brain.		**Epidemiology**	Close contact with a case or carrier; inhalation of infectious droplets; crowding and fatigue predispose to the disease.
⑦ Bacteria exit with respiratory secretions.		**Treatment and prevention**	Rifampin given to those exposed. Penicillin, ceftriaxone for treatment. Conjugate vaccine against serotypes A, C, Y, and W135 used to immunize ages 2–55 years.

The patient was a 31-month-old girl admitted to the hospital because of fever, headache, drowsiness, and vomiting. She had been previously well until 12 hours before admission, when she developed a runny nose, malaise, and loss of appetite.

Her birth and development were normal. There was no history of head trauma. Her routine immunizations had been neglected.

On examination, her temperature was 40°C (104°F), her neck was stiff, and she did not respond to verbal commands.

Her white blood cell count was elevated and showed a marked increase in the percentage of polymorphonuclear neutrophils (PMNs). Her blood sugar was in the normal range.

A spinal tap was performed, yielding cloudy cerebrospinal fluid under increased pressure. The fluid contained 18,000 white blood cells per microliter (normally, there are few or none), a markedly elevated protein level, and a markedly low glucose level. Gram stain of the fluid showed many tiny, Gram-negative coccobacilli, most of which were outside the white blood cells.

1. What is the diagnosis, and what is the causative agent?
2. What is the prognosis in this case?

3. What age group is most susceptible to this illness?
4. Compare the pathogenesis of this disease in children and adults.

Discussion

1. The patient had bacterial meningitis caused by *Haemophilus influenzae* serotype b. The fact that she had been healthy prior to her illness makes it unlikely that other pathogens could be responsible.

2. With treatment, the fatality rate is approximately 5%. Formerly, ampicillin was effective in most cases, but beginning in 1974, an increasing number of strains possessed a plasmid coding for β-lactamase. So far, however, these strains have been susceptible to the newer cephalosporin-type antibiotics. Unfortunately, about one-third of those who are treated and recover from the infection are left with permanent damage to the nervous system, such as deafness or paralysis of facial nerves. Prompt diagnosis and correct choice of antibacterial treatment minimize the chance of permanent damage.

3. The peak incidence of this disease is in the age range of 6 to 18 months, corresponding to the time when maternal

antibodies are lacking and adaptive immunity has not yet fully developed. Since 1987, vaccines consisting of type b capsular antigen conjugated with a protein, such as the outer membrane protein of *Neisseria meningitidis* or diphtheria toxin, have been used to immunize infants and thus eliminate the immunity gap. As a result, meningitis caused by *H. influenzae* type b is now rare.

4. *Haemophilus influenzae* type b strains are referred to as "invasive" strains because they establish infection of the upper respiratory tract, pass the epithelium, and enter the lymphatic vessels and bloodstream. In this way, they gain access to the general circulation and are carried to the central nervous system. Most strains of *H. influenzae* are non-invasive. Although they can cause infections of respiratory epithelium, they usually do not enter the circulation. Most adults are immune to type b strains, but some adults develop meningitis from non-invasive *H. influenzae* strains that gain access to the nervous system because of a skull fracture or by direct extension to the meninges from an infected sinus or middle ear.

Signs and Symptoms

As with other types of meningitis, symptoms usually begin with a mild cold and progress to severe headache, fever, and vomiting. Older children may report a stiff neck. Infants may show a bulging fontanelle (the gap between the bones of an infant's skull). The disease may progress rapidly to coma and death.

Causative Agent

Haemophilus influenzae, a Gram-negative non-motile rod, was named because it was thought to have caused an influenza epidemic in the late 1890s. Although it did not cause the epidemic, it was and is often isolated from influenza patients. Encapsulated strains labeled a through f cause most disease in children; bacteria with capsular antigens of type b, abbreviated Hib, cause the most serious disease. Unencapsulated strains are part of the normal microbiota of the nasopharynx in many healthy individuals, and some people also carry Hib. In 1991, *H. influenzae* earned the distinction of being the first free-living organism to have its small genome completely sequenced.

Pathogenesis

Encapsulated *Haemophilus influenzae* in the upper respiratory tract can bind to and penetrate the epithelium with the aid of pili

and enter capillaries. The capsule helps them resist phagocytosis. Bacteremia can result in meningitis or other infections such as epiglottitis or cellulitis. Unencapsulated strains generally do not enter the bloodstream but can cause local infections such as sinusitis and otitis media.

Epidemiology

Unlike meningococcal meningitis, *Haemophilus influenzae* usually does not cause epidemics. Healthy adult carriers are often the source of sporadic infection by *H. influenzae* type b. Children rarely carry Hib due to the high rate of immunization. The pathogen is spread by respiratory droplets, but it rarely causes disease except in unimmunized infants. Case-fatality rate when untreated is about 90%. Even with treatment, about 5% die and 10–30% of children have lasting neurological damage.

Treatment and Prevention

Treatment with cefotaxime or ceftriaxone is usually successful against *Haemophilus influenzae*. People who have been in close contact with an infected individual may be given rifampin to help prevent the disease.

The conjugate Hib vaccine contains the type b polysaccharide antigen attached to a protein such as diphtheria toxoid. The protein

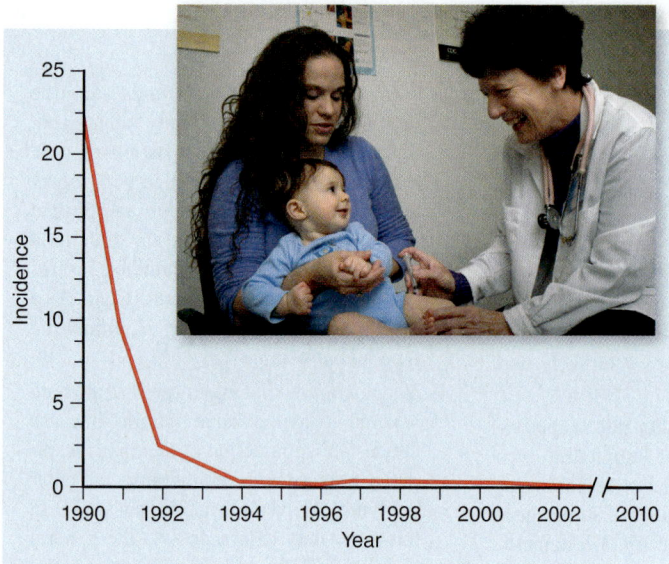

FIGURE 26.6 Rate of Serious *Haemophilus influenzae* Disease per 100,000 Children Less than Age 5, United States, 1990–2010 Before the availability of conjugate vaccines in late 1987, *H. influenzae* type b was the most common cause of bacterial meningitis in preschool children.

❓ *What are the components of the conjugate Hib vaccine?*

component makes the vaccine effective in children under age two, who otherwise respond poorly to polysaccharide antigens. Infants in the United States are routinely vaccinated against Hib beginning at 2 months of age (**figure 26.6**). This vaccine is relatively expensive and must be administered more than once, making it less practical in developing countries, where Hib infections are still responsible for most deaths due to bacterial meningitis. ◀◀ conjugate vaccines, p. 424

Neonatal Meningitis

Most cases of neonatal meningitis (during the first month of life) are caused by bacteria that colonize the mother's birth canal. The common causes of meningitis in adults (*Neisseria meningitidis, Streptococcus pneumoniae,* and *Haemophilus influenzae*) seldom cause meningitis in newborns because most mothers have antibodies against them. These antibodies cross the placenta and protect the baby until it is about 6 months old.

Signs and Symptoms

Meningitis symptoms in an infant may be vague, as it is difficult to tell if a baby has a headache or a stiff neck. Signs of meningitis in a newborn may include fever and vomiting, irritability, poor feeding, lethargy, and bulging fontanels.

Causative Agents

The most common cause of meningitis in newborn infants is *Streptococcus agalactiae,* a Lancefield group B streptococcus that colonizes the vagina of many women. Certain Gram-negative rods, such as encapsulated strains of *E. coli,* originate from the mother's intestinal tract and also can cause neonatal meningitis.

Other cases are caused by *Listeria monocytogenes* from the bloodstream of an infected mother. ◀◀ Lancefield grouping, p. 487

Pathogenesis

Infection of the meninges is usually preceded by bacteremia in the newborn. Inflammation increases intracranial pressure that can block CSF flow, causing hydrocephalus. Disruption of blood flow may damage nervous tissue. Infection may also lead to brain abscess.

Epidemiology

Neonates usually acquire these infections from the mother's genital tract shortly before or during birth. Premature or low birth weight babies are at most risk for meningitis. Neonatal meningitis causes death in 6–20% of affected newborns. Those who survive often have long-lasting consequences such as hearing loss or mental disability.

Treatment and Prevention

Treatment of neonatal meningitis includes intravenous dosage with a mixture of antibacterial drugs such as ampicillin and gentamicin effective against both group B streptococci and *E. coli.* Other drugs may be added after the causative agent is identified.

The Centers for Disease Control and Prevention (CDC) recommends that the vagina and rectum of pregnant women be tested for group B streptococci late in pregnancy. Women with positive cultures can then be treated with an appropriate antibacterial medication shortly before or during labor. Screening and subsequent treatment can decrease the incidence of serious group B streptococcal disease by more than 75%.

Listeriosis

Meningitis is the most common result of listeriosis, a foodborne disease caused by *Listeria monocytogenes.* This organism generally causes only a small percentage of meningitis cases in the United States, but epidemics may occur.

Signs and Symptoms

Listeria monocytogenes infections are generally asymptomatic or mild in most healthy people. Symptomatic listeriosis is usually characterized by fever and muscle aches, and sometimes nausea or diarrhea. Most of the cases requiring medical attention have meningitis with fever, headache, stiff neck, and vomiting. Pregnant women who become infected often miscarry or deliver terminally ill premature or full-term infants.

Causative Agent

Listeria monocytogenes is a motile, non-spore-forming, facultatively anaerobic, Gram-positive rod that can grow at 4°C. The organism can grow in refrigerated foods even if vacuum-packaged.

Pathogenesis

The mode of entry in sporadic cases of listeriosis is usually unclear, but during epidemics it is generally via the gastrointestinal tract. Gastrointestinal symptoms may or may not occur, but the bacteria promptly penetrate the intestinal mucosa—through the M cells and into the Peyer's patches—and then enter the bloodstream. The resulting bacteremia is the source of meningeal infection. In pregnant women, *L. monocytogenes* crosses the placenta and produces widespread abscesses in tissues of the fetus. Babies infected at the time of birth usually develop meningitis after an incubation period of 1 to 4 weeks. ◄◄ M cells, p. 358 ◄◄ Peyer's patches, p. 358

Epidemiology

Listeria monocytogenes is widespread in natural waters and vegetation, and can be carried in the intestines of asymptomatic humans and other animals. Pregnant women, the elderly, and those with underlying illnesses such as immunodeficiency, diabetes, cancer, and liver disease are especially susceptible to listeriosis. Epidemics have resulted from *L. monocytogenes* contaminating foods including coleslaw, non-pasteurized milk, pork tongue in jelly, some soft cheeses, and hot dogs. Because the organisms can grow in commercially prepared food stored at refrigeration temperatures, thousands of infections have originated from a single food-processing plant.

Treatment and Prevention

Most strains of *L. monocytogenes* remain susceptible to antibacterial medications such as penicillin. Even though the disease is often mild in pregnant women, prompt diagnosis and treatment are important to protect the fetus.

Listeria monocytogenes can be killed by thoroughly cooking poultry, pork, beef, and other meats. To reduce the risk of cross-contamination, uncooked meats should not be kept with other foods; countertops and utensils should be cleaned after food preparation; and raw vegetables should be thoroughly washed before eating. Pregnant women and others at high risk are advised to avoid soft cheeses, refrigerated meat spreads, and smoked seafood. They should also heat cold cuts and hot dogs before eating them and avoid the fluids that may be in the packaging. In 2006, the U.S. Food and Drug Administration approved a food additive consisting of a mixture of bacteriophage strains that lyse *L. monocytogenes* (**figure 26.7**). The additive can be sprayed on a variety of meats during the production process, and

FIGURE 26.7 Bacteriophage in Food Safety When sprayed onto ready-to-eat foods, bacteriophage preparations may reduce the risk of listeriosis.

❓ *What is the source of most listeriosis infections?*

its presence is recorded on the food label. Use of this bacteriophage additive has increased the safety of these foods. The main causes of acute bacterial meningitis are compared in **table 26.2**. Some of the main features of listeriosis are presented in **table 26.3**. ◄◄ bacteriophages, p. 304

TABLE 26.2	Main Causes of Acute Bacterial Meningitis Compared			
Agent	**Characteristics**	**Source**	**Vaccine?**	**Age Group Usually Affected**
S. pneumoniae	Gram-positive diplococci	Respiratory system	Yes, multiple types	Mostly late teens and adults
N. meningitidis	Gram-negative diplococci	Respiratory system	Yes, four types	Mostly ages 2–20 years; can cause epidemics
H. influenzae	Gram-negative coccobacilli	Respiratory system	Yes, type b	Young children
S. agalactiae	Gram-positive cocci in chains	Colon, vagina	No	Mostly neonates; others with underlying diseases
E. coli	Gram-negative rods; usually a specific encapsulated type	Colon	No	Mostly neonates
L. monocytogenes	Gram-positive rods; multiply at refrigerator temperatures	Environment; contaminated cheeses, cold cuts, other foods	No	Pregnant women, neonates; elderly

TABLE 26.3 | **Listeriosis**

① Causative organism *Listeria monocytogenes* is ingested with food such as soft cheeses, non-pasteurized milk, hot dogs, or smoked fish.

② The bacteria rapidly penetrate the intestinal epithelium and establish bacteremia, especially in pregnant women, the elderly, and the immunodeficient.

③ In pregnant women, circulating *L. monocytogenes* cross the placenta and fatally infect the fetus; bacteria transmitted to the baby at birth cause meningitis in 1 to 4 weeks. The mother usually does not have a serious illness.

④ In older people and those with underlying diseases, *L. monocytogenes* attacks brain and meninges, causing meningitis, brain abscesses.

Signs and symptoms	Fever and muscle aches, with or without gastrointestinal symptoms; headache and stiff neck mark the onset of meningitis
Incubation period	A few days to 2 to 3 months; in newborn babies, 1 to 4 weeks
Causative agent	*Listeria monocytogenes,* a non-spore-forming Gram-positive rod able to grow at 4°C
Pathogenesis	Ingested *L. monocytogenes* cells penetrate the intestinal epithelium and enter the bloodstream; the resulting bacteremia spreads to the meninges, causing meningitis.
Epidemiology	Epidemics from contaminated soft cheeses, non-pasteurized milk, coleslaw, hot dogs. Pregnant women get bacteremia, fetal infection, miscarriage. Infants contract infection at birth, develop meningitis in the first month of life. Elderly and those with immunodeficiency, cancer, and diabetes also at high risk.
Treatment and prevention	Antibacterial medications such as penicillin, given promptly, are effective treatment. Care in handling, cooking of raw meats; thorough washing of vegetables; reheating of cold cuts, hot dogs, and refrigerated leftovers.

Hansen's Disease (Leprosy)

Hansen's disease, also known as leprosy, is an ancient disease appearing throughout written history from about 600 B.C. Disfigurement, loss of limbs, and blindness can result from skin and peripheral nerve involvement. The number of new cases of Hansen's disease has decreased dramatically since the introduction of multidrug therapy; today it is most often seen in tropical or developing countries, such as India, Mozambique, and Brazil. Still, millions of people continue to suffer from the residual effects of their disease and the stigma associated with it. In the United States, about 200 cases are reported annually. This number has increased in recent years, mostly in states receiving immigrants from countries where the disease is endemic.

Signs and Symptoms

Hansen's disease begins gradually, usually with the onset of pigmentation changes and increased or decreased sensation in certain areas of skin. These areas may thicken—losing hair, sweat glands, and all sensation. The nerves of the arms and legs may become visibly enlarged with accompanying pain, later changing to numbness. Loss of nerve activity can lead to muscle wasting, ulceration, and finally loss of fingers or toes due to unnoticed or untreated injury (**figure 26.8**). In more severe cases, changes are most obvious in the face, with thickening of the nose and ears and deep wrinkling of the facial skin. Collapse of the supporting structure of the nose leads to congestion and bleeding.

Causative Agent

Mycobacterium leprae is aerobic, rod-shaped, and acid-fast (**figure 26.9**). It grows very slowly with a generation time of about 12 days, and prefers the slightly cooler temperatures of the body's extremities. Hansen's disease is usually diagnosed based on clinical findings, but skin biopsies that show acid-fast rods can provide an early indication of nerve invasion. Despite many attempts, *M. leprae* has not been grown in the absence of living cells. It can, however, grow in the footpads of mice, in armadillos, and in mangabey monkeys. Moreover, a gene library of *M. leprae* has been made and expressed in *E. coli*, providing large quantities of the organism's antigens for study. ◀◀ **acid-fast stain, p. 48** ◀◀ *Mycobacterium,* **p. 264** ◀◀ **DNA library, p. 221**

Pathogenesis

Mycobacterium leprae is the only known human pathogen that preferentially infects peripheral nerves. From there, the course of the infection depends on the immune response of the host. In most cases, cell-mediated immunity develops against the invading bacteria, and activated macrophages limit their spread. The chronically infected nerve cells, however, are progressively damaged by attacking immune cells, leading to disabling deformities, resorption of bone, and skin ulcerations that characterize the disease. The disease often spontaneously stops progressing, and the nerve damage, although permanent, does not worsen. This limited type of Hansen's disease, in which cell-mediated immunity

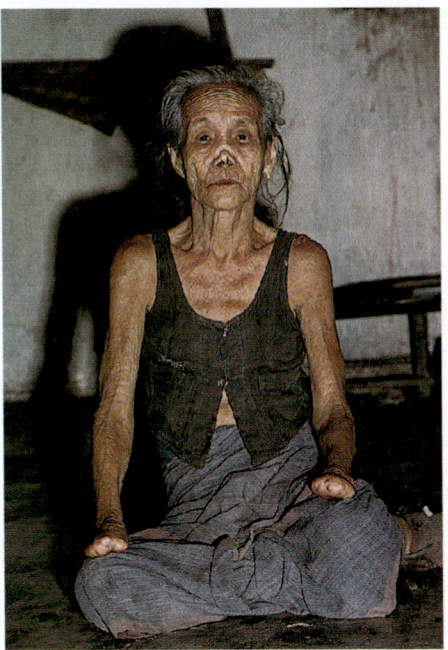

FIGURE 26.8 Effects of Leprosy (a) An early symptom of leprosy is a change in skin pigmentation. **(b)** Notice the absence of fingers and sunken nose as a result of severe disease.

❓ *Why are the fingers, toes, and nose often affected by leprosy?*

(a)   (b)

successfully stops the proliferating bacteria, is called **tuberculoid leprosy** (also called paucibacillary Hansen's disease). People with tuberculoid leprosy rarely, if ever, transmit the disease to others. ◀◀ cell-mediated immunity, p. 356 ◀◀ macrophage, pp. 340, 348

When cell-mediated immunity to *M. leprae* fails to develop or is suppressed, unrestricted growth of *M. leprae* occurs, leading to a form of Hansen's disease called **lepromatous leprosy** (also called multibacillary Hansen's disease). The bacteria first multiply in the cooler tissues of the body, notably in skin macrophages and peripheral nerves, and later throughout the body. The tissues and mucous membranes contain billions of *M. leprae,* but there is almost no inflammatory response to them. Mucus of the nose and throat contains high numbers of the pathogen, which can easily be transmitted to others.

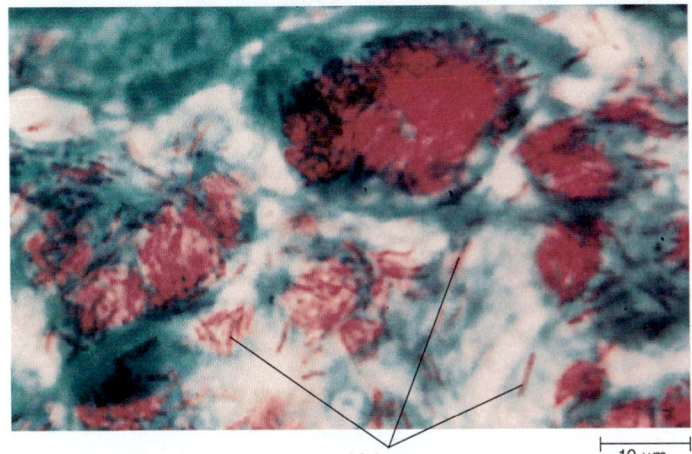

M. leprae   ⊢ 10 μm ⊣

FIGURE 26.9 *Mycobacterium leprae* in a Biopsy Specimen
The red areas are dense masses composed of millions of the bacteria, a typical finding in lepromatous leprosy.

❓ *In what human cells would you expect to find these bacteria?*

Epidemiology

Transmission of *Mycobacterium leprae* is by direct human-to-human contact. The source of the organisms is mainly nasal secretions of lepromatous cases, which can transport *M. leprae* to mucous membranes or skin abrasions of other individuals. Even then, the disease develops in only a tiny minority, being controlled by immune defenses in the rest. Natural infections with *M. leprae* also occur in wild nine-banded armadillos. Although armadillos have not been considered an important source of human leprosy, a recent study suggests that contact with the animals may indeed transmit the disease.

Eradication of Hansen's disease, defined as less than 1 case per 10,000 individuals, has been achieved in most countries where the disease was endemic. It is difficult to completely eradicate, however, due to the long generation time of *M. leprae*. The long generation time results in an incubation period of about 3 years (range, 3 months to 20 years), during which time the disease can remain undetected. ◀◀ generation time, p. 83

MicroByte
During much of history, Hansen's disease was so feared that "lepers" were forced to carry a bell or horn to warn others of their presence.

Treatment and Prevention

Early treatment can keep Hansen's disease from progressing. Since 1995, the World Health Organization has provided free multidrug therapy to patients, radically reducing the number of cases. Tuberculoid leprosy can be successfully treated by a combination of dapsone and rifampin administered for 6 months. Lepromatous leprosy is generally treated for a minimum of 2 years, with addition of a third drug, clofazimine, to the treatment regimen. Multiple drug therapy is required to prevent drug-resistant strains

TABLE 26.4	Hansen's Disease (Leprosy)
Signs and symptoms	Skin lesions that lack sensation, deformed face, loss of fingers or toes
Incubation period	3 months to 20 years; usually 3 years
Causative agent	*Mycobacterium leprae,* an acid-fast rod that has not been grown in the absence of living cells
Pathogenesis	Invasion of small nerves of skin; multiplication in macrophages; course of disease depends on immune response of host; activated macrophages limit growth of bacterium; attack of immune cells against infected nerve cells produces nerve damage, leading to deformity; in lepromatous leprosy, lymphocytes fail to react to the bacteria, allowing unrestrained growth of *M. leprae.*
Epidemiology	Direct contact with *M. leprae* from mucous membrane secretions
Treatment and prevention	Treatment: dapsone plus rifampin for months or years; clofazimine added for lepromatous disease. No vaccine.

from developing. No proven vaccine to control Hansen's disease is yet available. Some features of Hansen's disease are summarized in **table 26.4.**

Botulism

Botulism is not a nervous system infection, but a severe form of intoxication, usually foodborne. The name "botulism" comes from *botulus,* meaning "sausages," chosen because some of the earliest recognized cases occurred in people who ate contaminated sausages. Like other clostridia, *Clostridium botulinum* produces heat-resistant endospores. The endospores themselves do not cause disease, but they germinate in nutrient-rich anaerobic environments and produce a powerful exotoxin that causes the characteristic symptoms. Botulism is considered in this chapter because its key symptom is paralysis.

On a global basis, foodborne botulism is the most common type of this disease. It typically occurs when inadequate canning practices fail to destroy all endospores, and those surviving germinate and grow in the food. Some botulism cases occur when endospores are ingested and then germinate and colonize the intestine (intestinal botulism). Others result from endospore contamination of a wound where they then germinate and multiply (wound botulism). ▶▶| foodborne botulism, p. 757

Signs and Symptoms

Symptoms of foodborne botulism usually begin 12 to 36 hours after ingesting toxin-contaminated food. Most cases begin with dizziness, dry mouth, and blurred or double vision, indicating eye muscle weakness. Abdominal symptoms, including pain, nausea, vomiting, and diarrhea or constipation can also occur. Progressive paralysis generally involves all voluntary muscles; respiratory paralysis is the most common cause of death.

Causative Agent

Botulism is caused by the strictly anaerobic, Gram-positive, spore-forming, rod-shaped bacterium *Clostridium botulinum* (**figure 26.10**). The endospores generally survive boiling and can then germinate to form vegetative cells if the environment is favorable (nutrient-rich, anaerobic conditions, a pH above 4.5, and a temperature above 4°C). Several types of a neurotoxin called botulinum toxin are produced by different strains of *C. botulinum,* and all cause paralysis. Types A, B, and E are responsible for most human cases. ◀◀| neurotoxin, p. 392

Pathogenesis

Vegetative cells of *Clostridium botulinum* growing in food release botulinum toxin, one of the most powerful poisons known. A few milligrams could kill the entire population of a large city. Indeed, cases of botulism have resulted from eating a single, contaminated string bean, or from licking a finger when tasting a food contaminated with botulinum toxin. When a person eats the contaminated food, the toxin passes through the stomach and into the small intestine. Once absorbed into the bloodstream, the toxin can circulate for 3 weeks or more.

Circulating botulinum toxin attaches to motor neurons, blocking transmission of signals to the muscles, producing paralysis. Botulinum toxin is an A-B toxin. The B portion binds to specific receptors on motor nerve endings, and the A portion enters the nerve cell, where it inactivates proteins that regulate the release of the neurotransmitter. Unlike tetanus toxin, which blocks inhibitory neurons and results in spastic muscle contraction, botulinum toxin blocks muscle contraction resulting in **flaccid paralysis.** Because it blocks nerve transmission to muscles, botulinum toxin can be used to treat people with certain chronic, spastic conditions such as crossed eyes. Commercially available botulinum toxin, or Botox, is used in a cosmetic treatment that relaxes facial muscles,

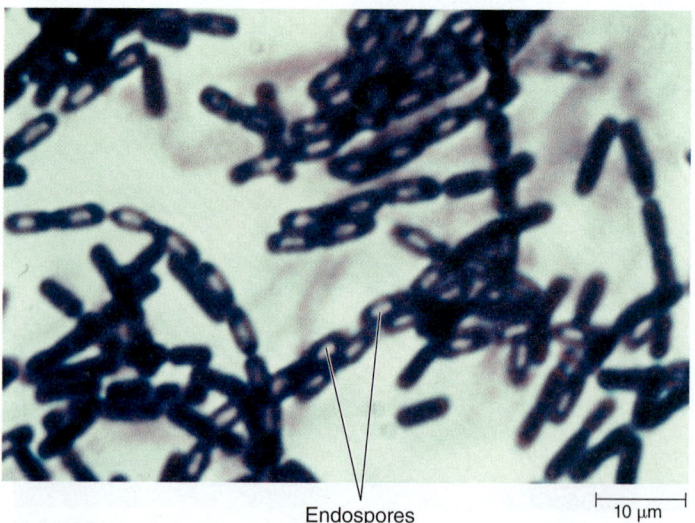

Endospores   10 µm

FIGURE 26.10 *Clostridium botulinum* Notice the lighter-colored endospores in this Gram stain. These are not reliably killed by the temperature of boiling water.

❓ *What is an endospore?*

resulting in fewer wrinkles (see **Perspective 31.1**). ◀◀ A-B toxin, p. 392 ◀◀ neurotransmitter, p. 642 ◀◀ tetanus, p. 555

Intestinal botulism occurs when *C. botulinum* colonizes the intestine. This primarily happens in infants, which is why it is also known as infant botulism. When *C. botulinum* endospores are ingested, they can germinate, and the resulting vegetative cells may grow in the intestine, where they produce toxin. This rarely happens in adults, presumably because the normal intestinal microbiota successfully compete with the germinating *C. botulinum* cells. Exceptions occur, however, particularly in immunodeficient patients whose normal microbiota has been suppressed by antibiotic treatment. Intestinal botulism is characterized by constipation followed by generalized paralysis that can range from mild lethargy to respiratory insufficiency.

Clostridium botulinum can also colonize dirty wounds, especially those containing dead tissue. The bacteria do not invade, but they can multiply in dead tissue. Botulinum toxin then diffuses into the bloodstream.

Epidemiology

Clostridium botulinum endospores are widely distributed in soils and aquatic sediments worldwide. Still, fewer than 30 cases of foodborne botulism per year are typically seen in the United States. Most of these are from eating preserved fish or improperly home-canned foods. Strict controls on commercial canners have drastically reduced the number of botulism outbreaks. Today, in the United States, intestinal (infant) botulism is more common than foodborne botulism, but there are generally fewer than 100 cases per year. Ingestion of honey has been implicated in infant botulism cases, leading to the recommendation that honey—a source of *C. botulinum* spores—not be given to children under 2 years old. Most cases of wound botulism in the United States are due to wounds caused by abuse of injected drugs (**figure 26.11**).

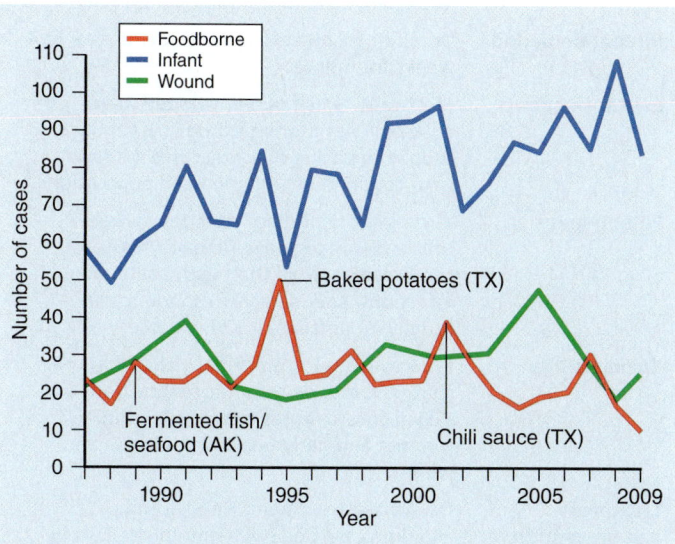

FIGURE 26.11 Botulism in the United States, 1987–2009

❓ *What is the most common form of botulism in the United States?*

Treatment and Prevention

Foodborne botulism is treated by administering antitoxin intravenously as soon as possible after the diagnosis. The antitoxin, however, only neutralizes toxin circulating in the bloodstream. The nerves already affected recover slowly, over weeks or months. Artificial respiration with a mechanical ventilator may be required for prolonged periods until affected nerve endings can regenerate. Most intestinal botulism patients recover without receiving antitoxin treatment, although respiratory support and tube feeding may be needed until the pathogens are replaced by normal intestinal microbiota. For wound botulism, surgical removal of dirt and dead tissue helps to eliminate any organisms and unabsorbed toxin. Enemas and gastric washing are used to do the same for intestinal botulism.

Prevention of foodborne botulism depends on proper sterilization and sealing of food at the time of canning. Contamination does not always result in a spoiled smell, taste, or appearance, but fortunately, the toxin is heat-labile (destroyed by heat). Heating home-canned, low-acid foods (pH over 4.5) to 100°C for 15 minutes just prior to serving should ensure that they are safe. Immunity does not develop in response to the low levels of toxin, so a person may get botulism more than once. The main features of botulism are presented in **table 26.5**.

TABLE 26.5	Botulism
Signs and symptoms	Blurred or double vision, weakness, nausea, vomiting, diarrhea; generalized paralysis and respiratory insufficiency
Incubation period	Usually 12 to 36 hours
Causative agent	*Clostridium botulinum*, an anaerobic, Gram-positive, spore-forming, rod-shaped bacterium
Pathogenesis	*Clostridium botulinum* endospores germinate in food and release neurotoxin. Toxin is ingested, absorbed, and is carried by the bloodstream to motor nerves; toxin acts by blocking the transmission of nerve signals to the muscles, producing paralysis; *C. botulinum* can also colonize intestines or wounds and cause generalized weakness or paralysis.
Epidemiology	Ingestion of toxin produced by vegetative *Clostridium botulinum* cells growing in a food, often an improperly processed home-canned, low-acid food. Endospores widespread in soil, aquatic sediments, and dust. Organisms can colonize the intestines of adults and infants with deficiencies in normal microbiota, and wounds containing dirt and dead tissue, including those caused by injected-drug abuse.
Treatment and prevention	Treatment: enemas and stomach washing to remove toxin, cleaning infected wounds of dirt and dead tissue, intravenous administration of antitoxin, and artificial respiration. Education in proper home-canning methods; heating food to boiling for 15 minutes just prior to serving.

26.3 ■ Viral Diseases of the Nervous System

Learning Outcomes

7. *Compare and contrast viral meningitis with viral encephalitis.*

8. *Explain the history of poliomyelitis and why it has been targeted for global elimination.*

9. *Outline the steps in the pathogenesis of rabies in wild animals and explain how it can be prevented in humans.*

A wide variety of viruses can infect the nervous system causing meningitis or encephalitis. Poliomyelitis, now targeted for global elimination, and rabies, a disease normally found in animals, result from viral infection of neurons. Many viruses that typically affect other body systems can occasionally infect the CNS. These include the Epstein-Barr virus; the mumps, rubeola, varicella-zoster, and herpes simplex viruses; and, most commonly, human enteroviruses.

Viral Meningitis

Viral meningitis is more common and much milder than bacterial meningitis. It usually does not require specific treatment, and individuals generally recover in 7 to 10 days. Viruses are responsible for most cases of "aseptic meningitis," meaning that a lumbar puncture reveals no microorganisms in the cerebrospinal fluid.

Signs and Symptoms

The onset of viral meningitis is typically abrupt, with fever and severe headache above or behind the eyes, sensitivity to light, and a stiff neck. Nausea and vomiting are common. In addition, depending on the causative agent, there may be a sore throat, chest pain, swollen parotid salivary glands, or a skin rash.

Causative Agents

Non-enveloped RNA viruses—members of the enterovirus subgroup of picornaviruses—are responsible for at least half of the cases of viral meningitis. Of these, the most common offenders are coxsackie viruses, which can cause throat or chest pain, and echoviruses, which can cause a rash. ◀◀ picornaviruses, p. 308

Pathogenesis

Enteroviruses characteristically infect the throat, intestinal epithelium, and lymphoid tissue, and then spread to the bloodstream causing viremia. This can result in meningeal infection. The inflammatory response differs from bacterial meningitis in that fewer cells usually enter the cerebrospinal fluid, and more are monocytes rather than neutrophils. Viral meningitis is typically less severe than bacterial meningitis and causes little lasting neurological damage. ◀◀ viremia, p. 384

Epidemiology

Enteroviruses are relatively stable in the environment, sometimes even persisting in chlorinated swimming pools. Enteroviral meningitis is transmitted by the fecal-oral route, unlike bacterial meningitis that is usually transmitted by respiratory droplets. The feces of infected individuals often contain viruses for weeks.

Treatment and Prevention

No specific treatment is available. Handwashing and avoidance of crowded swimming pools are reasonable preventive measures when cases of aseptic meningitis are present. There are no vaccines against coxsackie and echoviruses. The main features of viral meningitis are presented in **table 26.6**.

TABLE 26.6	Viral Meningitis
Signs and symptoms	Abrupt onset, fever, severe headache, stiff neck, often vomiting; sometimes sore throat, large parotid glands, rash, or chest pain
Incubation period	Usually 1 to 2 weeks for enteroviruses, 2 to 4 weeks for mumps
Causative agents	Most cases: small non-enveloped RNA enteroviruses of the picornavirus family, usually coxsackie or echoviruses. Mumps virus common in unimmunized populations.
Pathogenesis	Viremia from primary infection spreads to the meninges. Fewer leukocytes enter cerebrospinal fluid than with bacterial infections, and many are mononuclear, usually no decrease in CSF glucose.
Epidemiology	Enteroviruses transmitted by the fecal-oral route, mumps by respiratory secretions and saliva. Enteroviruses transmission mainly summer and early fall; mumps in fall and winter.
Treatment and prevention	No specific treatment. Handwashing, avoiding crowded swimming pools during enterovirus epidemics; mumps vaccine for mumps prevention.

Viral Encephalitis

Although viral meningitis is usually mild, viral encephalitis is more likely to cause death or permanent disability. Viral encephalitis can be sporadic, resulting in a few widely scattered cases occurring routinely, or it can be epidemic. Sporadic encephalitis is usually due to activation of latent herpes simplex viruses. Most people recover from the disease but are left with permanent damage such as epilepsy, paralysis, deafness, or mental impairment.

◀◀ herpes simplex viruses, p. 582

Signs and Symptoms

The onset of viral encephalitis is usually abrupt, with fever, headache, and vomiting, as well as possible disorientation, localized paralysis, deafness, seizures, or coma.

Causative Agents

Epidemic viral encephalitis is usually caused by **arboviruses** (arthropod-borne viruses), a group of enveloped, single-stranded RNA viruses transmitted by insects, mites, or ticks. The leading causes of epidemic encephalitis in the United States are all transmitted by mosquitoes. These include LaCrosse encephalitis virus, St. Louis and West Nile encephalitis viruses, and eastern and western equine encephalitis viruses (**table 26.7**). ◀◀ arthropods, p. 298

Pathogenesis

Viruses multiply at the site of a mosquito bite and in local lymph nodes, producing mild and brief viremia. Relatively few infected individuals develop encephalitis. Some develop viral meningitis, or mild fever and headache. If viruses infect cells of the blood-brain barrier, they can enter the brain and replicate in neurons, causing destruction of brain tissue. Neutralizing antibodies stop disease progression. Case-fatality rates range from about 2% with LaCrosse encephalitis to 35–50% with eastern equine encephalitis. Emotional instability, epilepsy, blindness, or paralysis may occur in 5–50% of those who recover. The likelihood of disability depends largely on the kind of virus and the age of the patient. As with many diseases, the very young and the elderly are most affected.

Epidemiology

The common types of epidemic viral encephalitis are all zoonoses with a natural reservoir in birds or other small animals. Natural hosts can develop such high levels of virus in the bloodstream that mosquitoes can transmit the virus from one host to another. Humans are an accidental dead-end host—they can acquire the virus from a mosquito bite but do not develop sufficient viremia to transmit it back to the arthropod vector.

The LaCrosse virus infects *Aedes* mosquitoes, which pass it directly from one mosquito to another in semen. They feed on and infect squirrels and chipmunks, where the virus is amplified and spread to uninfected mosquitoes feeding on the rodents' blood.

The West Nile virus, introduced into New York from the Middle East in the summer of 1999, has a bird reservoir. Migrating birds quickly spread the virus across the country (**figure 26.12**).

Some arboviruses have a broad host range and cause disease in horses as well as humans. This gives rise to the names eastern and western equine encephalitis.

Treatment and Prevention

There is no proven antiviral therapy for epidemic viral encephalitis. The blood of sentinel chickens in cages with free access to mosquitoes is tested periodically for evidence of arbovirus infection. A positive test would trigger an encephalitis alert, like this St. Louis encephalitis alert issued in Florida: "Avoid outdoor activity during evening and night, the peak hours of biting for the *Culex* mosquito vector. If outdoors, wear long sleeves and pants. Make sure windows and porches are properly screened. Use insect repellents and insecticides." Equine encephalitis viruses generally infect horses 1 or 2 weeks before the first human cases appear, so these cases provide a warning to increase protection against mosquitoes. A vaccine against eastern equine encephalitis is approved for use in horses

TABLE 26.7	Arboviruses That Cause Encephalitis in the United States
Virus	**Family**
LaCrosse encephalitis virus	Bunyavirus
St. Louis encephalitis virus	Flavivirus
West Nile encephalitis virus	Flavivirus
Eastern equine encephalitis virus	Togavirus
Western equine encephalitis virus	Togavirus

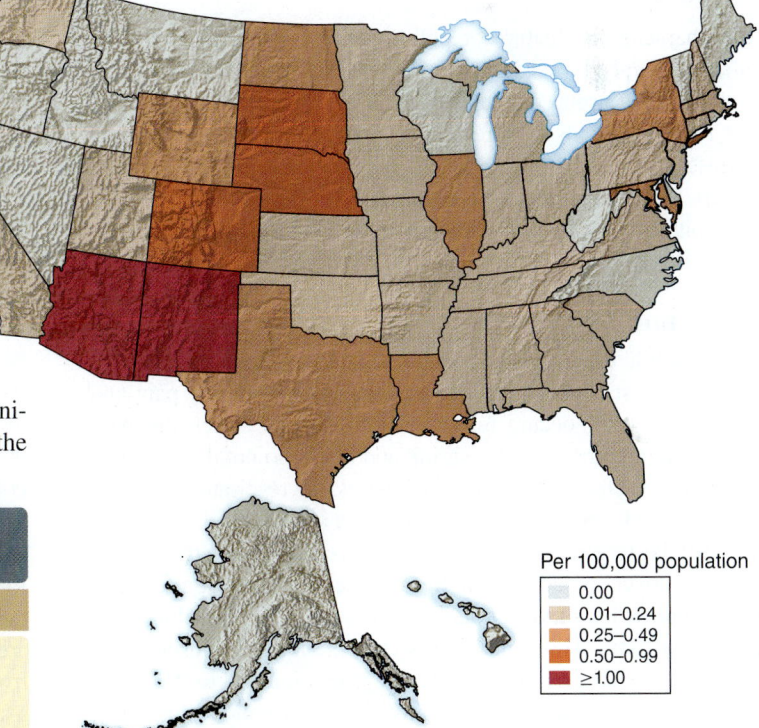

Per 100,000 population
- 0.00
- 0.01–0.24
- 0.25–0.49
- 0.50–0.99
- ≥1.00

FIGURE 26.12 Areas Reporting West Nile Virus Infection, United States, 2010

❓ *What is an arbovirus?*

TABLE 26.8	Epidemic Viral Encephalitis	

① Infected mosquito introduces encephalitis virus.

② Virus multiplies locally, establishes brief low-level viremia.

③ Virus crosses blood-brain barrier and preferentially attacks the brain.

④ Destruction of brain tissue causes death or permanent disabilities such as emotional instability, mental disability, paralysis of face, arm, leg.

⑤ Due to brief viremia, there is no exit for the virus, thus humans are the final host.

Signs and symptoms	Abrupt onset, fever, headache, vomiting, disorientation, paralysis, seizures, deafness, coma
Incubation period	First symptoms within a few days; encephalitic symptoms often within the first week
Causative agent	One of five arboviruses, LaCrosse, St. Louis, West Nile, western equine, or eastern equine
Pathogenesis	Replication of virus at the site of the mosquito bite, replication in lymph nodes, then viremia invasion of brain tissue. Nerve cells in the brain destroyed. Process halted by neutralizing antibody.
Epidemiology	Viruses transmitted to humans from birds or rodents by mosquitoes.
Treatment and prevention	No accepted treatment for arboviral encephalitis. Chicken sentinels to warn of arbovirus epidemics. Insecticides and other anti-mosquito preventive measures.

and has also been used to protect the emu—a large, domesticated, meat-producing bird susceptible to this virus. The main features of epidemic viral encephalitis are presented in **table 26.8.**

Poliomyelitis

The characteristic feature of poliomyelitis (also called polio, or infantile paralysis) is destruction of motor neurons resulting in paralysis of a group of muscles, such as those of an arm or leg. The two individuals most responsible for control of this terrifying disease will not see its imminent elimination from the world— Albert Sabin died in 1992; Jonas Salk in 1995. The two men were bitter rivals, both of whom expected, but did not receive, the Nobel Prize.

Signs and Symptoms

Poliomyelitis usually begins with symptoms of meningitis: head-ache, fever, stiff neck, and nausea. In addition, muscle pain and spasm generally occur, followed by paralysis. Over the next weeks and months, muscles shrink and bones do not develop normally in the affected area. In severe cases, the respiratory muscles are paralyzed, and air must be pumped in and out of the lungs by an artificial respirator. Those who survive this acute stage of the illness recover some function. Sensory neurons are not affected.

People who survive poliomyelitis sometimes develop post-polio syndrome. This condition is characterized by muscle pain, increased weakness, and muscle degeneration 15 to 50 years after recovering from acute poliomyelitis. Post-polio syndrome is not due to a resurgence of polioviruses and sometimes involves muscles not obviously affected by the original illness. Instead it is thought to be a secondary effect of the initial damage. During recovery from acute poliomyelitis, surviving nerve cells branch out to take over the functions of the killed nerve cells. Post-polio

syndrome is probably due to the death of these nerve cells after doing double duty for many years.

Causative Agent

Poliomyelitis is caused by three types of polioviruses— designated 1, 2, and 3—distinguished using different antisera. These non-enveloped, single-stranded RNA viruses are members of the enterovirus subgroup of the picornavirus family. They can be grown *in vitro* in cell cultures, where they cause cell destruction to form plaques. ◀◀ plaque assay, p. 317

Pathogenesis

Polioviruses enter the body orally, infect the throat and intestinal tract, and then invade the bloodstream. Usually, symptoms are mild or absent, the immune system conquers the infection, and recovery is complete. Only rarely does the virus bind to specific receptors on motor neurons, replicate, and destroy the cells when the mature virus is released. Since most people infected with poliovirus do not develop nervous involvement, a single case of poliomyelitis means that the virus is rampant.

Epidemiology

In areas where sanitation is poor, the polioviruses are endemic, transmitted by the fecal-oral route. Newborns in these nations are partially protected for 2 to 3 months until antibodies received from their mothers begin to wane. During this time, the infants are usually exposed to poliovirus because of crowding and unsanitary conditions. They develop mild infections of the throat and intestine, but the maternal antibodies generally prevent the virus from spreading to the motor neurons. Thus, infants develop lifelong immunity to the virus during the time when the maternal antibodies protect them from paralysis.

Poliomyelitis has been most devastating in countries with good sanitation. In these situations, polioviruses generally disappear from the community because they cannot spread to susceptible people. Then, when poliovirus is reintroduced, a high incidence of paralysis results because people lack immunity. This occurred in the United States in the 1950s, resulting in many cases of respiratory paralysis and death (**figure 26.13**). With most people now routinely immunized against the disease, however, and the likelihood of imported disease rapidly waning, this scenario will hopefully no longer occur.

Treatment and Prevention

There is no treatment for polio, but individuals receive supportive care such as rest, intravenous fluids, and pain medications. If the respiratory muscles are paralyzed, use of a ventilator or "iron lung" is required. Exercise therapy may be helpful to those with post-polio syndrome.

Like other enteroviruses, polioviruses are quite stable under natural conditions, and can often be found in swimming pools. They can be inactivated by pasteurization and by chlorinating drinking water. Control of poliomyelitis using vaccines represents one of the greatest success stories in the battle against infectious diseases (**figure 26.14**). Ironically, all cases of paralytic polio acquired in the United States since 1980 were caused by Sabin's oral, attenuated polio vaccine, which was introduced in 1961. These cases arise because in rare instances the vaccine strain mutates and becomes virulent.

The small risk of developing paralytic poliomyelitis from Sabin's vaccine virus (approximately one case per 2.4 million doses given) led the United States to return to the routine use of the inactivated Salk vaccine in 1999. However, the Sabin oral vaccine remains key to controlling polio in areas of the world

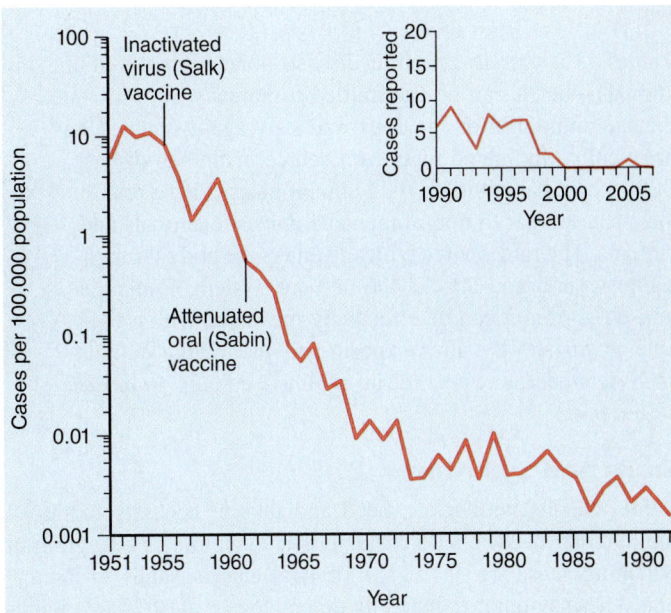

FIGURE 26.14 Incidence of Poliomyelitis in the United States, 1951–2007 Ironically, since 1980, all cases acquired in the United States have been caused by the oral attenuated (Sabin) vaccine. Endemic poliomyelitis was eliminated from the United States by 1980, and the entire Western Hemisphere by 1991.

❓ *Why is the Sabin vaccine against poliomyelitis the preferred vaccine to use in countries where polio is still prevalent?*

where transmission of the wild virus still occurs. It produces better mucosal immunity in the throat and intestine, can spread from person to person, does not require an injection, provides herd immunity, and is less expensive than the Salk vaccine. In the years following 1988, when the World Health Organization resolved to eradicate poliomyelitis, the number of countries with endemic disease has been reduced from 125 to only four (Afghanistan, India, Nigeria, and Pakistan). Poliomyelitis is summarized in **table 26.9**.

◀◀ herd immunity, p. 421 ◀◀ campaign to eliminate poliomyelitis, p. 424

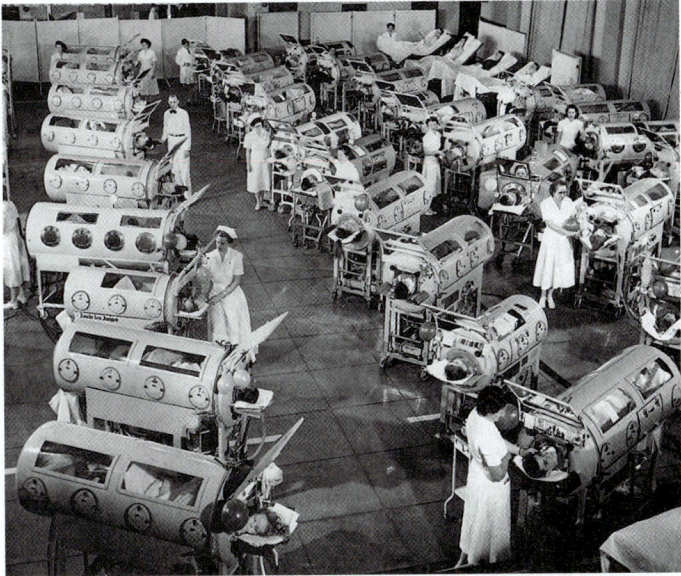

FIGURE 26.13 The Horror of Poliomyelitis These tanklike respirators ("iron lungs") were used during the 1950s epidemics of poliomyelitis to keep alive people whose respiratory muscles were paralyzed by the disease. Now we can hope to see polio forever banished from the earth!

❓ *How are polioviruses transmitted?*

TABLE 26.9	Poliomyelitis
Signs and symptoms	Headache, fever, stiff neck, nausea, pain, muscle spasm, followed by paralysis
Incubation period	7 to 14 days
Causative agents	Polioviruses 1, 2, 3, members of the picornavirus family
Pathogenesis	Virus infects the throat and intestine, circulates via the bloodstream, and enters some motor nerve cells of the brain or spinal cord; infected nerve cells lyse upon release of mature virus.
Epidemiology	Spreads by the fecal-oral route; asymptomatic and nonparalytic cases common
Treatment and prevention	Treatment: artificial ventilation for respiratory paralysis; physical therapy and rehabilitation. Prevented by injecting Salk's inactivated vaccine or by Sabin's orally administered attenuated vaccine in areas of epidemic or endemic disease.

Rabies

Rabies is a classic zoonotic disease—one normally found in animals—but it can be transmitted to humans. In the United States, immunization of dogs and cats against rabies has practically eliminated them as a source of human disease. Still, rabies has multiple wild animal hosts. These remain a constant threat to non-immunized domestic animals and humans. The rabies virus typically enters the body through a bite wound and attacks the nervous system. Rabies is unusual in that it can be effectively prevented with a vaccine given shortly after exposure to the virus. Without such treatment, however, death is almost certain. ◀◀ **zoonotic disease, p. 439**

Signs and Symptoms

Rabies begins with fever, head and muscle aches, sore throat, fatigue, and nausea. The characteristic symptom is a tingling or twitching sensation at the site of viral entry, usually an animal bite. These symptoms generally do not appear until 1 to 2 months after infection, but they progress rapidly to agitation, confusion, hallucinations, seizures, increased sensitivity, and encephalitis. A few days later, the individual typically goes into a coma and dies of respiratory failure or cardiac arrest.

The later stages of rabies are often characterized by increased salivation and difficulty in swallowing. This causes the "frothing at the mouth" classically associated with rabid animals and "mad dogs." Swallowing, or even the sight of fluids, often leads to severe spasms of the throat and respiratory muscles. The common name for rabies is "hydrophobia" or fear of water.

Causative Agent

The cause of rabies is the rabies virus (**figure 26.15a**), a member of the rhabdovirus family. This virus has a striking bullet shape, is enveloped, and contains single-stranded, negative-sense RNA.

Pathogenesis

After the rabies virus is introduced into the body, it typically multiplies in cells at the site of infection for several weeks before entering a sensory neuron. It then travels by retrograde transport up the axon to the spinal cord and eventually to the brain. The length of time before symptoms occur depends on the location of the bite, the amount of virus introduced, and the condition of the host. Individuals with head wounds, for example, tend to show symptoms sooner than those with leg wounds.

Once in brain tissue, the virus multiplies extensively, causing the symptoms of encephalitis. Characteristic inclusion bodies called **Negri bodies,** made up of viral nucleocapsids, are found in most rabies cases (figure 26.15b). From the brain, the virus spreads outward via the nerves to various body tissues, notably the salivary glands, eyes, and fatty tissue under the skin, as well as to the heart and other vital organs. Rabies can be diagnosed before death by identifying the virus in stained smears collected from the surface of the eyes.

MicroByte

> The only documented cases of human-to-human transmission of rabies have been in patients receiving transplants from infected donors.

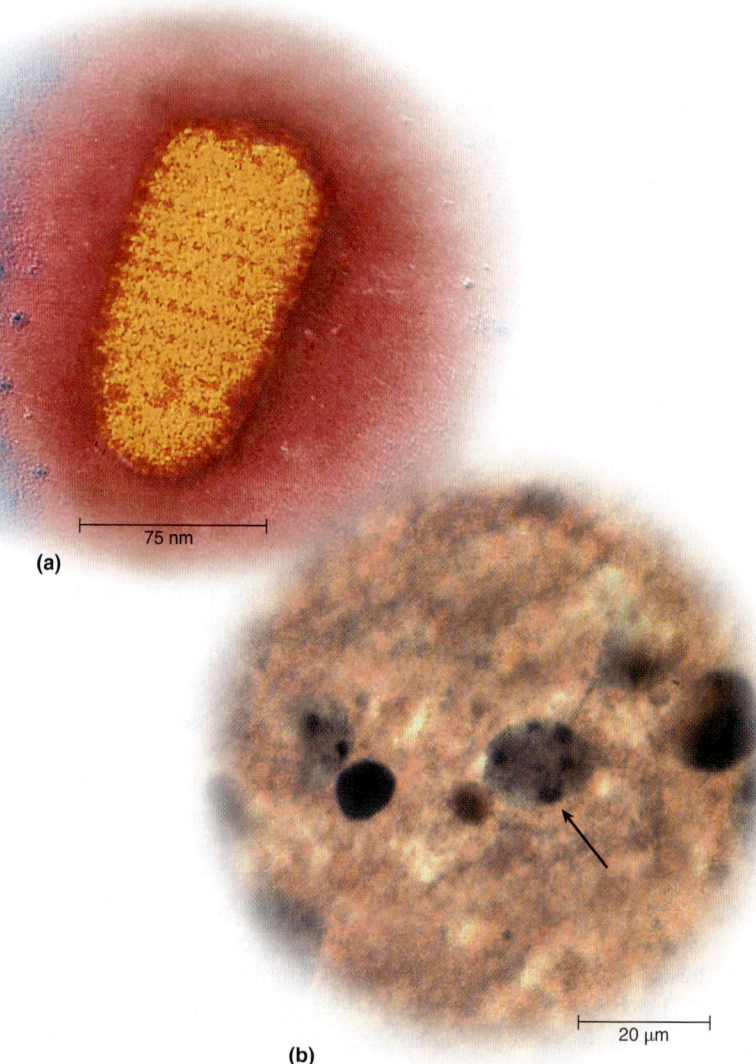

(a)

75 nm

(b)

20 µm

FIGURE 26.15 Rabies Virus (a) Color-enhanced transmission electron micrograph of rabies virus. Notice the bullet shape. **(b)** Stained smear of brain tissue from a rabid dog. A Negri body (see arrow) represents a site of rabies virus replication.

❓ *Why can an animal that appears well still transmit the rabies virus?*

Epidemiology

The primary mode of transmission of rabies to humans is via the saliva of a rabid animal introduced into bite wounds of the skin. It has been reported, but not documented, that individuals can also contract rabies by inhaling aerosols containing the virus, such as from bat feces. Because of the extensive rabies vaccination program for dogs in the United States, the main reservoir for rabies is wild animals such as raccoons, bats, skunks, and foxes (**figure 26.16**). Over 5,000 wild animal cases are reported in the United States each year, representing an enormous reservoir from which infection can be transmitted to domestic animals and humans. Raccoons lead the list of wildlife cases, but almost all human cases are due to contact with infected bats. Human rabies is rare in the United States with only 1 to 3 cases reported per year.

Although rabies is rare in the United States, about 40,000-70,000 people worldwide die of rabies. Most of these are due to dog bites in areas where dogs are not routinely vaccinated. A person bitten by a dog with rabies virus in its saliva has about a 30%

PERSPECTIVE   26.1

Rabies Survivors!

In October 2004, a 15-year-old girl was admitted to the hospital with typical signs and symptoms of rabies. A month earlier, while in church, she picked up a bat that had fallen to the floor. The bat bit her left index finger, and she carried it outside and released it. The small bite wound was cleaned with hydrogen peroxide and healed uneventfully. In the hospital, a tube was placed in the girl's trachea and hooked to a respirator because most rabies deaths result from respiratory failure. She was given medication to put her into a coma to rest her nervous system, and she was given the antiviral medication ribavirin, according to an experimental protocol. After 7 days, the coma-inducing medication was reduced and the girl was allowed to wake up. A little more than a month after her illness began, her breathing tube was removed, she regained speech, solved mathematics problems, and walked with assistance. Slow, steady improvement continued after she returned home. So far, attempts to cure other patients using the treatment used in this case have been unsuccessful.

In February 2009, a 17-year old girl reported to the emergency room with a severe headache that had lasted for 2 weeks, vomiting, neck pain, and other symptoms of nervous system infection. On March 6, she was admitted to the hospital with suspected infectious encephalitis but did not respond to treatment. Two months earlier she had entered a cave in Texas and several flying bats had hit her body, although she did not notice any scratches or bite wounds at the time. On March 11, she tested positive for rabies and was given hyperimmune human rabies globulin and one dose of anti-rabies vaccine. She remained in the hospital with supportive care, but never entered intensive care and was discharged on March 22. After an emergency room visit on April 3 for headache and vomiting, she did not return for follow-up.

Survival after the onset of rabies symptoms is known to have occurred in only five individuals prior to these cases, and all five had received anti-rabies vaccine either before they were exposed to rabies or before they showed signs or symptoms of the disease. All but one suffered persisting neurological damage, unlike these patients, who made complete recoveries. Over the years, many treatments have been tried without benefit, including human rabies immune globulin, anti-rabies vaccines, and interferon. What is to be learned from the survival of these patients? Perhaps their survival had nothing to do with their treatment—but was due to their own immune systems fighting a small infectious dose or a rabies viral strain of low virulence. Is there some clue in their survival that will perhaps lead to the first effective rabies treatment in the 3,000 years the disease has been recognized?

---

risk of developing rabies. Most rabies cases are in Asia, followed by Africa. Rabies does not exist in Australia or New Zealand.

Most rabid dogs excrete the virus in their saliva, sometimes even a few days before they get sick. Therefore, if an unvaccinated dog bites a person, the animal should be confined for 10 days to see if it develops symptoms of rabies. Some dogs become irritable and hyperactive with the onset of rabies, produce excessive saliva, and attack people, animals, and inanimate objects. Perhaps more common is the "dumb" form of rabies, in which an infected dog simply stops eating, becomes inactive, and suffers paralysis of throat and leg muscles. Obviously, one should not try to remove a foreign body from the throat of a sick, choking, unvaccinated dog!

Treatment and Prevention

A person who has been bitten by an animal should immediately wash the wound thoroughly with soap and water, and then apply an antiseptic to avoid any kind of infection. If there is a possibility that the animal is rabid, human rabies immune globulin (anti-rabies antibody) is injected at the wound site and intramuscularly to provide passive immunity. The individual should then receive four injections of inactivated vaccine, the first as soon as possible after exposure, and at 3, 7, and 14 days after the first. The vaccine provokes an immune response that neutralizes free virus and kills infected cells during the long incubation period before the virus enters the neurons.  passive immunity, p. 420

There is no effective treatment for rabies once symptoms appear. Only two people are known to have recovered from the disease without receiving the vaccine (**Perspective 26.1**). Active immunization after a bite can usually generate antibody in time to protect against the disease. In the United States, about 30,000 people annually receive inactivated rabies vaccine after a bite from a suspected rabid animal. Individuals at high risk for rabies, such as veterinarians or animal control personnel, should be immunized before exposure.

It is impossible to eradicate a zoonotic disease without clearing the animal hosts. Routine vaccination of domestic dogs and cats in the United States has dramatically reduced the number of cases in domestic pets. A DNA vaccine has been developed that may make it more practical to vaccinate dogs in developing countries with high incidence of rabies. Vaccination of wild animals is more difficult since it is impossible to catch, immunize, and

FIGURE 26.16
Rabies Carrier?
Skunks are a main reservoir for rabies.

? *What other animals most commonly carry the rabies virus?*

TABLE 26.10	Rabies
Signs and symptoms	Fever, headache, nausea, vomiting, sore throat, cough at onset; later, spasms of the muscles of mouth and throat, coma, and death
Incubation period	Usually 30 to 60 days; sometimes many months or years
Causative agent	Rabies virus, single-stranded RNA, rhabdovirus family; has an unusual bullet shape
Pathogenesis	During incubation period, virus multiplies at site of bite, then travels via nerves to the central nervous system; it multiplies and spreads outward via multiple nerves to infect heart and other organs.
Epidemiology	Bite of rabid animal, usually a bat. Inhalation is another possible mode.
Treatment and prevention	Effective post-exposure measures: immediately wash wound with soap and water and apply antiseptic; inject rabies vaccine and human rabies antiserum as soon as possible. No effective treatment once symptoms begin. Avoid suspect animals; immunize pets.

release all of them. Programs are underway to administer an oral vaccine to widespread wild populations by placing it into a bait food, such as peanut butter or fish meal. Many eastern states have participated in air-drop programs to vaccinate raccoons; Texas has a similar program targeting foxes and coyotes. Some features of rabies are summarized in **table 26.10**.

MicroAssessment 26.3

Many different kinds of viruses can attack the nervous system, but they generally do so in only a small percentage of infected people. At least half of viral meningitis cases are caused by the enterovirus subgroup of picornaviruses, which are generally spread by the fecal-oral route. Viral meningitis is usually mild, but viral encephalitis often causes permanent disability. Arboviruses maintained in nature in a mosquito-bird or mosquito-rodent cycle are a leading cause of epidemic viral encephalitis. Sporadic viral encephalitis is usually due to herpes simplex virus. Poliomyelitis, characterized by paralysis of one or more muscle groups, is caused by three other enteroviruses. Rabies is a widespread zoonosis, almost uniformly fatal for humans, usually transmitted by animal bites.

7. *Explain why epidemic viral encephalitis is called a zoonosis.*

8. *What are the advantages and disadvantages of the Sabin polio vaccine?*

9. *Why is rabies now rare in humans when it is still so common in wildlife?* +

26.4 ■ Fungal Diseases of the Nervous System

Learning Outcome

10. *Compare and contrast the development of cryptococcal meningoencephalitis in AIDS patients and in healthy individuals.*

Inhalation of fungal cells found in soil and bird droppings seldom causes serious lung disease, but phagocytic cells that contain these cells may carry them from the lungs via the bloodstream to the brain. This can lead to inflammation of the brain and meninges.

Cryptococcal Meningoencephalitis

Cryptococcal meningoencephalitis was uncommon until the onset of the AIDS epidemic. Now the disease is among the most important HIV-related opportunistic infections. Because it is difficult to clear without T-cell involvement, AIDS patients with this infection are sometimes maintained indefinitely on antifungal medications. More recently, the incidence of cryptococcal meningoencephalitis has increased among healthy individuals in the western United States and Canada due to an emerging pathogen, *Cryptococcus gattii*. ◀◀ opportunistic infection, p. 382

Signs and Symptoms

In apparently healthy people, symptoms of cryptococcal meningoencephalitis develop gradually and generally consist of difficulty in thinking, dizziness, intermittent headache, and possibly slight fever. After weeks or months, vomiting, weight loss, paralysis, seizures, and coma may appear. In people with immunodeficiency, the disease generally progresses much faster; without treatment, death can occur in as little as 2 weeks.

Causative Agent

Cryptococcal meningoencephalitis is an infection of the meninges and brain by either *Cryptococcus neoformans* or *Cryptococcus gattii*. The organisms are small, spherical yeasts generally 3 to 20 μm in diameter surrounded by a thick capsule that resists the immune response (**figure 26.17**). *Cryptococcus neoformans* is an opportunistic pathogen that usually causes disease only in the immunocompromised. *C. gattii*, on the other hand, causes disease in healthy individuals.

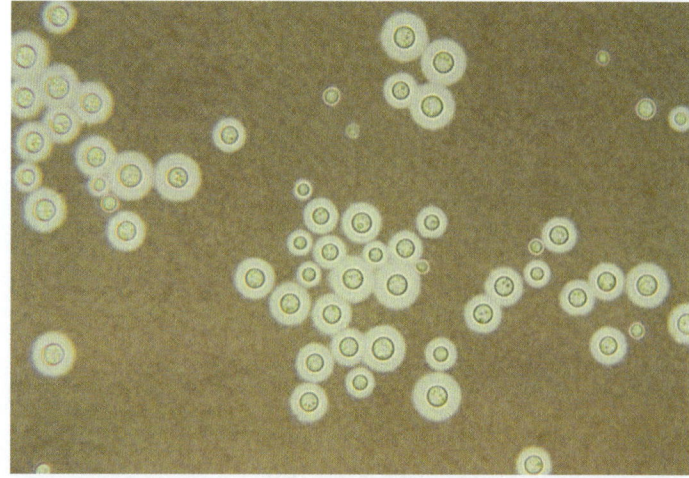

20 μm

FIGURE 26.17 *Cryptococcus neoformans* Note the capsule around the organism.

❓ *How might a capsule benefit a yeast cell?*

Pathogenesis

Infection is first established in the lung, usually producing mild or no symptoms. Immune defenses of healthy people usually eliminate the infection, but phagocytic killing is slow and inefficient. In some cases, particularly in immunocompromised individuals, the organisms multiply, enter the bloodstream, and are distributed throughout the body. Meningoencephalitis is the most common infection outside of the lung, but organisms in the bloodstream can also infect skin, bones, and other body tissues. The capsule inhibits the antibody response and phagocytosis of the organisms. Capsular material can be detected in spinal fluid and urine, aiding diagnosis. In meningoencephalitis, the organisms typically cause thickening of the meninges, sometimes hindering the flow of cerebrospinal fluid and increasing pressure within the brain. They also invade brain tissue, producing multiple abscesses.

Epidemiology

Cryptococcus neoformans is distributed worldwide in soil and vegetation contaminated with bird droppings. Symptomatic infection is sometimes the first indication of AIDS. *Cryptococcus gattii* was once associated with eucalyptus trees in tropical or subtropical regions of the world, but emerged in 1999 in British Columbia, Canada. Pathogenic strains have since spread in the U.S. Pacific Northwest and Canada (**figure 26.18**). Appearance of more virulent strains has increased the case-fatality rate from 5% to about 25%.

The organisms enter the body by inhalation of spores; infection can occur in humans as well as cats, dogs, and other animals. Subsequent onset of disease is rare. Person-to-person transmission of the disease does not occur.

> **MicroByte**
> Several Dall's porpoises have been found washed up on the shores in British Columbia since 2000; all were positive for *C. gattii* infection.

Treatment and Prevention

Treatment with the antibiotic amphotericin B is often effective, particularly if given with flucytosine (5-fluorocytosine) followed by the oral medicine, fluconazole. Amphotericin B must be given intravenously and the dose carefully regulated to minimize its

TABLE 26.11	Cryptococcal Meningoencephalitis
Signs and symptoms	Headache, vomiting, confusion, and weight loss; slight or no fever; symptoms may progress to seizures, paralysis, coma, and death
Incubation period	Widely variable, few to many weeks
Causative agent	*Cryptococcus neoformans, Cryptococcus gattii*—encapsulated yeasts
Pathogenesis	Infection starts in lung; encapsulated organisms multiply, enter bloodstream, and are carried to various parts of the body; phagocytosis inhibited; meninges and adjacent brain tissue become infected.
Epidemiology	Inhalation of material contaminated with the fungus; other sources; most people resistant to the disease
Treatment and prevention	Treatment: amphotericin B with flucytosine followed by fluconazole. No preventive measures.

toxic effects. Because amphotericin B does not reliably cross the blood-brain barrier, the drug may be administered through a plastic tube inserted through the skull into a lateral ventricle of the brain. Treatment is successful in about 70% of cases except in AIDS patients who respond poorly to treatment, most likely because they lack T-cell-dependent killing that normally assists the action of the antifungal medications. Unless their T-cell function can be restored by treatment, AIDS patients are rarely cured of their infection. ◀◀ antifungal medicines, p. 476

There is no vaccine or other preventive measure available. The main features of cryptococcal meningoencephalitis are summarized in **table 26.11.**

> **MicroAssessment 26.4**
> Cryptococcal meningoencephalitis occurs opportunistically in immunocompromised individuals such as AIDS patients. The causative organism is a small yeast with a large capsule, often found in soil contaminated with pigeon droppings. The disease has recently emerged among healthy individuals in the Pacific Northwest.
>
> 10. *Why is it difficult to cure AIDS patients of a fungal infection?*
> 11. *Why are AIDS patients so vulnerable to cryptococcal meningoencephalitis?*
> 12. *What might cause the spread of new strains of* C. gattii *in the northwest United States?* ✚

26.5 ■ Protozoan Diseases of the Nervous System

Learning Outcome

11. *Compare and contrast African sleeping sickness and primary amebic meningoencephalitis.*

Infection of the nervous system by protozoans is quite rare, but usually results in fatal disease when it occurs. Protozoan diseases are difficult to treat because medications that affect eukaryotic cells may also affect fragile neurons. ◀◀ protozoa, p. 291

FIGURE 26.18 Spread of *Cryptococcus gattii* in the Pacific Northwest in 2008

❓ *How does* Cryptococcus gattii *enter the body?*

African Sleeping Sickness

African sleeping sickness—also known as African trypanosomiasis—is transmitted by the tsetse fly, its biological vector. Sickness may persist for years, producing the symptoms for which it is named. The disease is important because it can be contracted by residents and visitors in a wide area across the middle of the African continent. ◀◀ **biological vector, p. 442**

Signs and Symptoms

African trypanosomiasis can be chronic or acute. A tender nodule develops at the site of the bite within a week after a person is bitten by an infected tsetse fly. Regional lymph nodes might enlarge, but symptoms can disappear spontaneously. In the chronic form of the disease, recurrent fevers develop that can continue for months or years. Involvement of the central nervous system is marked by gradual loss of interest in everything, decreased activity, and indifference to food. The eyelids droop, the individual falls asleep even while eating or standing, and speech becomes slurred. Eventually the person becomes comatose and dies. In acute cases, symptoms develop over weeks or months, and neurological symptoms develop much more rapidly.

Causative Agent

African sleeping sickness is caused by the flagellated protozoan, *Trypanosoma brucei*. These organisms are slender and have a wavy, undulating membrane and an anteriorly protruding flagellum (**figure 26.19**). Two subspecies are morphologically identical—*T. brucei rhodesiense* and *T. brucei gambiense*. The *rhodesiense* subspecies causes acute disease and occurs mainly in the cattle-raising areas of East Africa; the *gambiense* subspecies causes chronic disease and occurs mainly in forested areas of Central and West Africa. Both are transmitted by tsetse flies, biting insects of the genus *Glossina*.

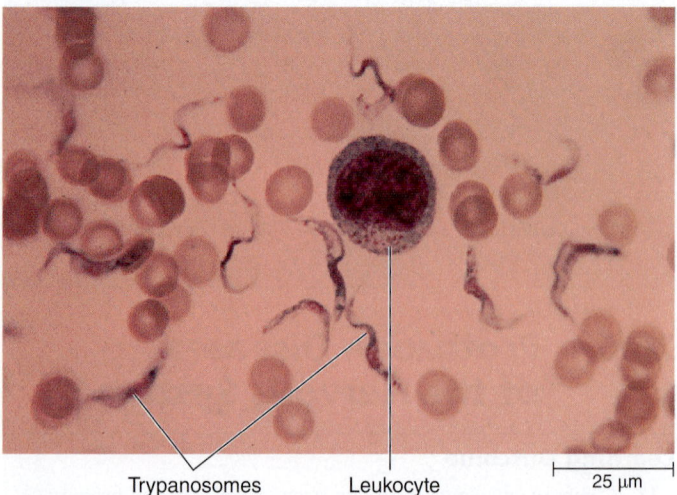

Trypanosomes   Leukocyte   25 µm

FIGURE 26.19 *Trypanosoma brucei* **in a Blood Smear** Notice the slender protozoa among the blood cells of an individual with African sleeping sickness (African trypanosomiasis).

❓ *Name the biological vector of* Trypanosoma brucei.

Pathogenesis

The protozoan enters the bite wound in the saliva of an infected tsetse fly. The parasite multiplies at the skin site and eventually enters the lymphatics and blood circulation. Immune responses of fever and antibody production lessen early symptoms. Within about a week and at roughly weekly intervals thereafter, however, the number of parasites in the blood increases. Each burst coincides with the appearance of a new glycoprotein on the surface of the trypanosomes. The parasite is a master at avoiding the immune response, with more than a thousand different genes coding for variations of that surface glycoprotein, but only one is expressed at a time. Each time a new gene is expressed, a new immune response is needed to produce the appropriate antibody. The recurrent cycles of parasitemia and antibody production continue until the patient is treated or dies.

In *T. brucei rhodesiense* infections, disease can progress rapidly. The parasite enters the heart and brain within 6 weeks of infection. Irritability, personality changes, and mental dullness result from brain involvement, but the patient usually dies from heart failure within 6 months. With *T. brucei gambiense*, progression of the chronic infection is much slower. Years may pass before death occurs, often from secondary infection. Much of the damage to the host is due to immune complexes formed when high levels of protozoan antigen are bound by antibodies that then react with complement system proteins. ◀◀ **immune complex, p. 395**

Epidemiology

African sleeping sickness occurs on the African continent within about 15° of the equator, with 10,000 to 20,000 new cases each year, some in tourists. The acute Rhodesian form of the disease is a zoonosis, and the main reservoirs are wild animals. Humans are the main reservoir for the chronic Gambian form, and human-to-human transmission is more common. The presence of animal hosts makes the Rhodesian form more difficult to control. Less than 5% of the tsetse fly vectors are infected. ◀◀ **reservoir, p. 438**

Treatment and Prevention

As with other eukaryotic pathogens, treatment is problematic because of toxic side effects of the available medications. Nevertheless, treatment of infected people helps reduce the protozoan's reservoir. A single, intramuscular injection of the medication pentamidine prevents the Gambian form of the illness for a number of months, although the infection could progress later. Suramin can be used if the disease has not progressed to involve the central nervous system; melarsoprol and eflornithine cross the blood-brain barrier and can be used when the central nervous system is involved. Early diagnosis and treatment of the Rhodesian form is important to prevent rapid damage to the nervous system. ◀◀ **antiprotozoan medications, p. 478**

Actions directed against tsetse fly vectors include use of insect repellents and protective clothing to prevent bites, use of traps containing bait and insecticides to reduce the vector population, and clearing of brush to reduce breeding habitats for the flies. The main characteristics of African sleeping sickness are presented in **table 26.12.**

TABLE 26.12	African Sleeping Sickness
Signs and symptoms	Tender nodule at site of tsetse fly bite; fever, enlargement of lymph nodes; later, involvement of the central nervous system, uncontrollable sleepiness, headache, poor concentration, unsteadiness, coma, death
Incubation period	Weeks to several years
Causative agent	*Trypanosoma brucei*, a flagellated protozoan
Pathogenesis	The protozoa multiply at site of a tsetse fly bite, then enter blood and lymphatic circulation; as new cycles of parasites are released, their surface protein changes and the body is required to respond with a new antibody.
Epidemiology	Bites of infected tsetse flies transmit the trypanosomes through fly saliva; wild animal reservoir for *T. brucei rhodesiense*.
Treatment and prevention	Treatment: suramin; when central nervous system is involved, melarsoprol or eflornithine. Protective clothing, insecticides, clearing of brush where flies breed, pentamidine.

FIGURE 26.20 *Naegleria fowleri* Notice the suckerlike hooks used in phagocytosis. Keep Your Head Above Water sign: Courtesy of Jay Stahl-Herz, MD

❓ *By what route does* Naegleria fowleri *enter the brain?*

Primary Amebic Meningoencephalitis (PAM)

Naegleria fowleri, the cause of primary amebic meningoencephalitis (PAM), is commonly found in warm fresh water and soils. Nevertheless, fewer than 200 cases of PAM have been reported worldwide. About three people per year become infected after swimming or diving in natural waters in the United States, making it a rare event, but one that is almost certainly fatal.

Signs and Symptoms

Symptoms of PAM are similar to those of bacterial meningitis. Early symptoms include headache, fever, stiff neck, and vomiting. Once neurological symptoms appear—such as confusion or seizures—death quickly follows.

Causative Agent

Naegleria fowleri is one of only a few free-living protozoa pathogenic for humans. This frightening-looking organism with sucker-like structures used in phagocytosis can literally "eat your brain" (**figure 26.20**). The ameboid trophozoite gives rise to flagellated forms and spherical cysts (see figure 12.14). It prefers warm temperatures.

Pathogenesis

Naegleria fowleri penetrates the skull along the olfactory nerves serving the nasal mucosa. It multiplies and migrates to the brain, where it destroys nervous tissue, especially in the frontal lobes. Hemorrhage, coma, and death occur within a week.

Epidemiology

Naegleria fowleri may be carried in the normal microbiota, but it rarely causes disease. For every case of *Naegleria* meningoencephalitis, many millions of people are exposed to the organism without harm. PAM is usually acquired when individuals swim or dive in warm natural fresh water. It is not found in seawater, and chlorination in swimming pools will kill the organism. Contaminated drinking water cannot spread the pathogen, and it is not transmitted from person to person.

MicroByte
A person is over 10,000 times more likely to die from drowning in natural waters than from PAM.

Treatment and Prevention

The antifungal drug amphotericin B has been used in these infections, but with little success. There is no vaccine. The main characteristics of primary amebic meningoencephalitis are presented in **table 26.13**.

TABLE 26.13	Primary Amebic Meningoencephalitis (PAM)
Signs and symptoms	Headache, fever, stiff neck, and vomiting, confusion, seizures
Incubation period	Less than a week
Causative agent	*Naegleria fowleri*
Pathogenesis	Destroys brain tissue
Epidemiology	Rare incidence after swimming or diving in warm, natural waters
Treatment and prevention	Unsuccessful treatment with amphotericin B; no vaccine

Residents and visitors to a wide swath of tropical Africa are at risk of contracting African sleeping sickness, caused by the flagellated protozoan parasite *Trypanosoma brucei,* and transmitted by a biting insect, the tsetse fly. These protozoa can circulate in the bloodstream for extended periods by changing their surface antigens to escape the host's antibodies. Eventually they penetrate the CNS, causing indifference, sleepiness, coma, and death. Only on rare occasions can free-living amebas such as *Naegleria fowleri* cause meningoencephalitis.

13. *How likely is it that a person who swims in warm fresh water will contract primary amebic meningoencephalitis?*

14. *How can one explain repeated, abrupt increases in* T. brucei *in the blood of African sleeping sickness victims?* ➕

26.6 ■ Diseases Caused by Prions

Learning Outcomes

12. *Explain how prions differ from other infectious agents.*

13. *Describe how prions can cause disease.*

A rare and mysterious group of chronic, degenerative brain diseases caused by prions has been seen in wild animals (mink, elk, and deer), domestic animals (sheep, goats, and cattle), and humans. Affected brain tissue has a spongy appearance, which is why these diseases are referred to as spongiform encephalopathies (**figure 26.21**).

Scrapie, a disease of sheep and goats, was named because affected animals had difficulty standing and "scraped" along fences for support. Cattle, presumably fed meat and bone meal from infected sheep, developed bovine spongiform encephalopathy, or "mad cow disease." These diseases assumed new prominence in the 1990s when an outbreak of a spongiform encephalopathy in humans in the United Kingdom was related to an earlier outbreak of mad cow disease. Brain and other tissues from affected animals can transmit the disease to normal animals, even of different species. Although it has not been documented, there is some concern that wild animals with chronic wasting disease caused by prions may also transmit

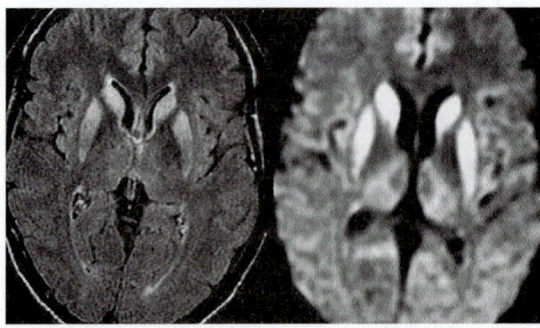

FIGURE 26.21 Appearance of Brain with Spongiform Encephalopathy Notice the spongelike appearance of the infected brain on the right compared to the normal brain on the left.

❓ *What animals are affected by spongiform encephalopathies?*

spongiform encephalopathy to humans who eat them. ◀◀ **chronic wasting disease, p. 328**

Transmissible Spongiform Encephalopathy in Humans

Transmissible spongiform encephalopathy (TSE) is rare in humans, occurring in only 0.5 to 1 case per million people. Most cases occur as Creutzfeldt-Jakob disease (CJD), which affects individuals over 45 years of age and sometimes runs in families. The disease acquired by eating affected animals is distinctively different—but has the same result and is identified as a variant of CJD, termed vCJD. Cases of CJD and vCJD are invariably fatal.

MicroByte ⸻
Another human TSE, kuru, is associated with cannibalism in New Guinea, where natives ate deceased relatives as a sign of respect.

Signs and Symptoms

Early symptoms of TSE include vague behavioral changes, anxiety, insomnia, and fatigue. Symptoms progress to characteristic muscle jerks, lack of coordination, memory loss, and dementia. Although the incubation period may last for years, once symptoms appear, death generally occurs within a year.

Causative Agent

The causative agents of spongiform encephalopathies are called proteinaceous infectious particles, or **prions.** Much smaller than a virus, prions (PrP) appear to be a misfolded form of a normal cellular protein (PrP^c). PrP is encoded by a normal human gene and modified after transcription. The misfolding results in a protein that is protease-resistant, whereas the normal protein is protease-sensitive. ◀◀ **prions, p. 328**

Pathogenesis

Prions increase in quantity during the incubation period of the disease as the misfolded protein (PrP) acts as a template that promotes misfolding of the normal cellular protein (PrP^c) on the surface of neurons (see figure 13.27). Prions aggregate in insoluble masses called plaques in the brain, causing tissue damage. They may be taken up by neurons or phagocytic cells, but cannot be degraded by cellular proteases. Death of neurons produces the spongy appearance of affected brain tissue. Not all prions, however, act the same. They differ in host range, incubation period, and the areas of the nervous system attacked. Transmissible spongiform encephalopathies can be differentiated from encephalitis because they typically do not evoke an immune response. This may be because the plaques consist of host protein.

Epidemiology

Creutzfeldt-Jakob disease (CJD) generally occurs in individuals older than 45 years. It has been transmitted from human to human through corneal transplants, contaminated surgical instruments, and injections of human hormone replacements. It can be transmitted experimentally to chimpanzees.

TABLE 26.14	Transmissible Spongiform Encephalopathy
Signs and symptoms	Behavioral changes, anxiety, insomnia, fatigue, progressing over weeks or months to muscle jerks, lack of coordination, dementia
Incubation period	Usually many years
Causative agents	Proteinaceous infectious particles known as prions; lack nucleic acids; identical amino acid sequence to a normal protein, but folded differently, and relatively resistant to proteases; resistant to heat, radiation, and disinfectants
Pathogenesis	Prions increase in quantity by converting normal protein to more prions; transmission to the brain; aggregation into masses outside the nerve cells; cell malfunction and death.
Epidemiology	Human-to-human transmission by corneal transplantation and by contaminated surgical instruments; probable transmission of cattle prions to humans by eating contaminated beef; sporadic Creutzfeldt-Jakob disease in those over age 45 years; median age of variant Creutzfeldt-Jakob cases only 28 years.
Treatment and prevention	No treatment, invariably fatal. Prions are inactivated by autoclaving in concentrated sodium hydroxide.

Spongiform encephalopathy of sheep (scrapie) has been known for more than two centuries, without any evidence that it is directly transmissible to humans. Current evidence indicates that cattle prions can be transmitted to humans and cause a variant Creutzfeldt-Jakob disease (vCJD) marked by differences in symptoms, brain pathology, and age of onset. The median age of individuals with vCJD is only 28 years. When mad cow disease is suspected, radical measures may minimize its transmission to humans. In 2005, for example, the United States banned importation of cattle from Canada when the disease was identified in several animals from there.

Treatment and Prevention

There is no treatment for the spongiform encephalopathies, and they are always fatal. It is important to avoid eating any animals that show neurological symptoms. Prions are highly resistant to disinfectants, including formaldehyde. They also are resistant to heat and to ultraviolet and ionizing radiation. They can be inactivated by extended autoclaving in 1M (molar) sodium hydroxide. The main features of transmissible spongiform encephalopathies are presented in **table 26.14.**

The key features of the diseases covered in this chapter are highlighted in the **Diseases in Review 26.1** table.

MicroAssessment 26.6

Transmissible spongiform encephalopathies are degenerative nervous system diseases that occur in a variety of wild and domestic animals. They are rare in humans, but disease can be transmitted from animals via ingested tissues. These diseases appear to be caused by prions—infectious agents consisting only of protein and highly resistant to inactivation by heat, radiation, and disinfectants. The diseases are always fatal.

15. *What is a prion?*
16. *What is the best way to prevent transmissible spongiform encephalopathies?*
17. *If you were an eye surgeon, would you rather the donor for a cornea transplant be under 35 or over 45 years of age?*

Diseases in Review 26.1

Nervous System Diseases

Disease	Causative Agent	Comment	Summary Table
BACTERIAL NERVOUS SYSTEM DISEASES			
Pneumococcal meningitis	*Streptococcus pneumoniae* (pneumococcus)	Leading cause of meningitis in adults; 90 serotypes with vaccine to 23 of the most common.	Table 26.2, p. 649
Meningococcal meningitis	*Neisseria meningitidis* (meningococcus)	May lead to endotoxic shock; formation of petechiae due to capillary damage; associated with meningitis epidemics; vaccine against the common serotypes is available, but occurs in unvaccinated children and young adults.	Table 26.1, p. 646
Haemophilus influenzae meningitis	*Haemophilus influenzae*	Once the most common cause of infant meningitis, now largely controlled by a conjugate vaccine.	Table 26.2, p. 649

(continued)

Nervous System Diseases

Disease	Causative Agent	Comment	Summary Table
BACTERIAL NERVOUS SYSTEM DISEASES (Continued)			
Neonatal meningitis	*Streptococcus agalactiae*	Newborns can acquire causative agent from the mother's birth canal; U.S. pregnant women are routinely screened for the organism before delivery and if positive treated with antibiotics.	Table 26.2, p. 649
Listeriosis	*Listeria monocytogenes*	Contaminated foods have caused epidemics; multiplies at refrigeration temperatures; infected pregnant women may miscarry; may cause abscesses in the fetus.	Table 26.3, p. 650
Hansen's disease (leprosy)	*Mycobacterium leprae*	Infects peripheral nerves, causing immune system to attack them; course of disease determined by cell-mediated immune response; long incubation period; loss of limbs, blindness.	Table 26.4, p. 652
Botulism	*Clostridium botulinum*	Botulinum toxin causes flaccid paralysis; foodborne intoxication is most common form worldwide; also occurs when organism grows in wounds or in intestines of infants.	Table 26.5, p. 653
VIRAL NERVOUS SYSTEM DISEASES			
Viral meningitis	Usually enteroviruses	Aseptic meningitis; more common and much milder than bacterial meningitis; often transmitted by fecal-oral route.	Table 26.6, p. 654
Viral encephalitis	Arboviruses	Transmitted in the United States by mosquitoes; more likely to cause death or disability than viral meningitis; case-fatality rate varies from 2–50%, depending upon type of virus.	Table 26.8, p. 656
Poliomyelitis	Polioviruses	Destruction of motor neurons leads to paralysis; fecal-oral transmission, but infection rarely leads to disease; vaccination has eliminated the disease in most parts of the world, and the disease is now targeted for eradication.	Table 26.9, p. 657
Rabies	Rabies virus	Zoonotic disease; virus travels from bite wound through sensory neurons to CNS, causing a generally fatal infection; vaccination shortly after exposure prevents disease.	Table 26.10, p. 660
FUNGAL NERVOUS SYSTEM DISEASES			
Cryptococcal meningoencephalitis	*Cryptococcus neoformans* and *C. gattii*	Caused by inhalation of fungal cells carried by phagocytic cells to the brain; can be an early sign of AIDS, but now also seen in healthy individuals where *C. gattii* is found.	Table 26.11, p. 661
PROTOZOAN NERVOUS SYSTEM DISEASES			
African sleeping sickness	*Trypanosoma brucei*	Chronic (Gambian) or acute (Rhodesian); two subspecies of protozoa transmitted by tsetse fly; loss of interest, drooping eyes, coma, death; early treatment is key in acute form.	Table 26.12, p. 663
Primary amebic meningoencephalitis (PAM)	*Naegleria fowleri*	Swimming or diving in water containing the organism transmits this rare but deadly disease; travels to brain via olfactory neurons where it damages tissue; no vaccine or effective treatment.	Table 26.13, p. 663
PRION DISEASES			
Transmissible spongiform encephalopathy	Prions	Caused by an accumulation of destruction-resistant misfolded proteins, resulting in death of neurons; may come from contaminated animal tissues; no treatment; always fatal.	Table 26.14, p. 665

Eradicate Polio: Then What?

It is hoped that poliomyelitis will be among the next ancient scourges to follow smallpox down the road to eradication. However, for a variety of political and scientific reasons, the causative virus of smallpox still exists many years after the last naturally acquired case of the disease. Will it be any easier to rid the world of the polioviruses after the last case of poliomyelitis occurs?

Dramatic progress toward polio eradication has mainly been accomplished using "immunization days," when the entire population in a given area is immunized using attenuated (Sabin) vaccine, often with one or more follow-up days

to immunize individuals missed earlier. The vaccine viruses then disappear from the population within a few months. Although a small percentage of the population suffers paralytic illness from the vaccine, the population is protected from virulent polioviruses until new generations are born. If there were no polioviruses left in the world, there would be no need to immunize the new generations.

Potential sources of virulent polioviruses include wild viruses circulating in remote populations, individuals with mild or atypical illness, and long-term carriers (especially those with immunodeficiency, who excrete the

viruses for months or years). Also laboratory freezers around the world hold stocks of the viruses, as well as fecal specimens that could harbor them. It is even possible to synthesize a poliovirus in the laboratory. Lastly, vaccine viruses can change genetically and acquire full virulence, as occurred in the Hispaniola polio epidemic of 2000–2001.

The challenge is to have a continuous, reliable, global polio surveillance system, and maintain immunizations and strategically located stockpiles of vaccine during what is likely to be a long time after the last case of paralytic polio occurs.

Summary

26.1 ■ Anatomy, Physiology, and Ecology (figure 26.1)
The brain and spinal cord make up the **central nervous system (CNS);** the **peripheral nervous system (PNS)** is composed of nerves and ganglia. **Cerebrospinal fluid (CSF)** is produced in ventricles in the brain and flows out over the brain and spinal cord (figure 26.2). **Meninges** are the membranes that cover the brain and spinal cord. Infectious agents most often reach the CNS through the bloodstream when they penetrate the **blood-brain barrier.**

26.2 ■ Bacterial Diseases of the Nervous System
Bacteria most often infect the membranes surrounding the brain, causing **meningitis.** Bacterial meningitis is uncommon; immunization has greatly decreased the incidence among children. In most cases, the causative bacterium is part of the normal respiratory microbiota.

Pneumococcal Meningitis (table 26.2)
Streptococcus pneumoniae, or pneumococcus, is the most common cause of meningitis in adults (figure 26.3). Symptoms are similar to other forms of meningitis: cold symptoms followed by abrupt onset of fever, severe headache, pain and stiffness of the neck and back, nausea, and vomiting.

Meningococcal Meningitis (tables 26.1, 26.2)
Meningococcal meningitis caused by *Neisseria meningitidis* is associated with meningitis epidemics. Small hemorrhages in the skin (figure 26.4), deafness, and coma can occur. Shock results from the release of endotoxin into the bloodstream.

Haemophilus influenzae Meningitis (table 26.2)
Haemophilus influenzae, once the leading cause of childhood bacterial meningitis, is largely sporadic and mostly controlled by a vaccine (figure 26.6).

Neonatal Meningitis (table 26.2)
Newborns most often acquire meningitis-causing bacteria from the mother's genital tract shortly before or during birth. Infants who survive often face long-lasting consequences of their infection.

Listeriosis (tables 26.2, 26.3)
Listeriosis is caused by *Listeria monocytogenes,* a non-spore-forming, Gram-positive rod usually associated with foodborne illness. The bacterium is widespread, commonly contaminates foods such as nonpasteurized milk, cold cuts, and soft cheeses, and can grow in refrigerated foods (figure 26.7). The bacteria readily penetrate the gastrointestinal mucus membranes, enter the bloodstream, and infect the meninges.

Hansen's Disease (Leprosy) (figure 26.8; table 26.4)
Hansen's disease is characterized by invasion of peripheral nerves by the acid-fast rod *Mycobacterium leprae,* which has not been cultivated *in vitro* (figure 26.9). The disease occurs in two main forms—tuberculoid and lepromatous, depending on the immune status of the individual.

Botulism (table 26.5)
Botulism is not a nervous system infection, but an intoxication, usually foodborne, that causes severe generalized paralysis. The causative bacterium, *Clostridium botulinum,* is an anaerobic, Gram-positive rod that forms heat-resistant endospores (figure 26.10). Endospores that survive canning or other heat treatment of foods germinate, and the bacteria multiply, releasing a powerful toxin into the food. Because of strict controls on food processing, intestinal or infant botulism is now the most common form of the disease in the United States (figure 26.11). Wound botulism, caused when *C. botulinum* colonizes dirty wounds containing dead tissue, is rare.

26.3 ■ Viral Diseases of the Nervous System
Most viral nervous system infections are caused by human enteroviruses or by the viruses of certain zoonoses. Many common viruses of humans can occasionally infect the nervous system, including those that cause infectious mononucleosis, mumps, measles, chickenpox, and herpes simplex ("cold sores," genital herpes).

Viral Meningitis (table 26.6)
Viral meningitis is much more common than bacterial meningitis. It is generally a mild disease for which there is no specific treatment.

Viral Encephalitis (table 26.8)

Viral encephalitis has a high fatality rate and often leaves survivors with permanent disabilities. Herpes simplex virus is the most important cause of sporadic encephalitis; epidemic encephalitis is usually caused by **arboviruses** (figure 26.12; table 26.7).

Poliomyelitis (figures 26.13, 26.14; table 26.9)

Destruction of motor nerve cells of the brain and spinal cord leads to paralysis, muscle wasting, and failure of normal bone development. Post-polio syndrome occurs years after poliomyelitis, and it is probably caused by the death of nerve cells that had taken over for those killed by the poliomyelitis virus.

Rabies (figure 26.15; table 26.10)

Rabies is a widespread zoonosis transmitted to humans mainly through the bite of an infected animal (figure 26.16). Once symptoms appear in an infected person, the disease is almost always fatal. Because of the long incubation period, prompt immunization with inactivated vaccine after a rabid animal bite is effective in preventing the disease. Passive immunization given at the same time increases protection.

26.4 ■ Fungal Diseases of the Nervous System

Fungi are usually opportunistic but can cause disease in healthy people.

Cryptococcal Meningoencephalitis (table 26.11)

Infection begins in the lung after a person inhales spores of *Cryptococcus neoformans* or *Cryptococcus gattii,* encapsulated yeasts that resist phagocytosis (figure 26.17). *C. neoformans* infects immunocompromised individuals, but *C. gattii* can cause disease in healthy people (figure 26.18). Treatment of these diseases is usually difficult.

26.5 ■ Protozoan Diseases of the Nervous System

Only a few protozoa infect the human nervous system.

African Sleeping Sickness (table 26.12)

African sleeping sickness is a major health problem in a wide area across equatorial Africa. In its late stages, it is marked by indifference, sleepiness, coma, and death. The disease is caused by *Trypanosoma brucei* (figure 26.19), a flagellated protozoan transmitted by its biological vector, the tsetse fly. During infection, the organism shows bursts of growth, each appearing with different surface proteins.

Primary Amebic Meningoencephalitis (PAM) (table 26.13)

PAM is rare but fatal. Infection with *Naegleria fowleri* (figure 26.20) can occur after swimming in warm, fresh water.

26.6 ■ Diseases Caused by Prions

Prions—abnormal proteins that are resistant to heat, radiation, and disinfectants—cause the spongiform encephalopathies, rare diseases characterized by a spongelike appearance of brain tissue caused by loss of nerve cells (figure 26.21). They afflict a variety of wild and domestic animals as well as humans.

Transmissible Spongiform Encephalopathies in Humans (table 26.14)

Examples of **transmissible spongiform encephalopathies (TSEs)** include "mad cow disease," Creutzfeldt-Jakob disease (CJD), and variant Creutzfeldt-Jakob disease (vCJD), which can be transmitted to humans from infected meat. There is no treatment for these diseases, and they are invariably fatal.

Review Questions

Short Answer

1. What sign would differentiate meningococcal meningitis from pneumococcal meningitis?

2. Name and describe the organism that is the leading cause of bacterial meningitis in adults.

3. What measures can be undertaken to prevent neonatal meningitis?

4. Why is listeriosis so important to pregnant women even though it usually causes them few symptoms?

5. Can botulism be spread from person to person?

6. Give two ways in which viral meningitis usually differs from bacterial meningitis.

7. What is the difference between sporadic encephalitis and epidemic encephalitis? Name one cause of each.

8. Explain why the biggest impact of poliomyelitis in the 1950s occurred in countries with good sanitation.

9. Why is it possible to prevent rabies with vaccine given after exposure?

10. If you contract African sleeping sickness on a visit to central Africa, what type do you most likely have?

Multiple Choice

1. Which is the best way to prevent meningococcal meningitis in individuals intimately exposed to the disease?

   a) Vaccinate them against *Neisseria meningitidis.*

   b) Treat them with the antibiotic rifampin.

   c) Culture their throat and hospitalize them for observation.

   d) Withdraw a sample of spinal fluid and begin antibacterial treatment if the cell count is high and the glucose level is low.

   e) Have them return to their usual activities, but seek medical evaluation if symptoms of meningitis occur.

2. Which of these statements concerning the causative agent of listeriosis is *false?*

   a) It can cause meningitis during the first month of life.

   b) It is a Gram-positive rod that can grow in refrigerated food.

   c) It is usually transmitted by the respiratory route.

   d) Infection commonly results in bacteremia.

   e) It is widespread in natural waters and vegetation.

3. Which of these statements concerning Hansen's disease is *false*?

   a) It was once common in the United States.

   b) An early symptom is loss of sensation, sweating, and hair in a localized patch of skin.

   c) The incubation period is usually less than 1 month.

   d) Treatment should include more than one antimicrobial medication given at the same time.

   e) The form the disease takes depends on the individual's immune status.

4. Which of these statements concerning foodborne botulism is *false*?

   a) It is not a central nervous system infection.

   b) Only some strains of the causative agent cause disease in humans.

   c) Food can taste normal but still cause botulism.

   d) Treatment is based on choosing the correct antibiotic.

   e) Control of the disease depends largely on proper food-canning techniques.

5. Which of the following statements about viral meningitis is *true*?

   a) Vaccines are generally available to protect against the disease.

   b) The main symptom is muscle paralysis.

   c) Transmission is often by the fecal-oral route.

   d) The causative agents do not survive well in the environment.

   e) Recovery is rarely complete.

6. Which of these statements concerning arboviral encephalitis is *false*?

   a) It is likely to occur in epidemics.

   b) Mosquitoes can be an important vector.

   c) Epilepsy, paralysis, and thinking difficulties are among the possible sequels to the disease.

   d) Use of sentinel chickens helps warn about the disease.

   e) In the United States, the disease is primarily a zoonosis involving cattle.

7. Which of these statements concerning poliomyelitis is *false*?

   a) The sensory nerves are usually involved.

   b) It can be caused by any of three specific enteroviruses.

   c) Only a small fraction of those infected will develop the disease.

   d) The disease is transmitted via the fecal-oral route.

   e) A post-polio syndrome can develop years after recovery from the original illness.

8. Which of these statements concerning cryptococcal meningoencephalitis is *true*?

   a) It is caused by a yeast with a large capsule.

   b) It is a disease of trees transmissible to humans.

   c) Typically it attacks the meninges but spares the brain.

   d) Person-to-person transmission commonly occurs.

   e) It is seen only in persons who are immunocompromised.

9. Which of these statements concerning African sleeping sickness is *true*?

   a) It is transmitted by a species of biting mosquito.

   b) It is a threat to visitors to tropical Africa.

   c) The onset of sleepiness is usually within 2 weeks of contracting the disease.

   d) It is caused by free-living protozoa.

   e) Distribution of the disease is determined mainly by the distribution of standing water.

10. Which of these statements concerning Creutzfeldt-Jakob disease (CJD) and vCJD is *true*?

   a) CJD occurs in children; vCJD occurs in adults over 45.

   b) CJD and vCJD are sometimes fatal.

   c) CJD is caused by prions; vCJD is a viral infection.

   d) Only humans suffer from diseases like CJD and vCJD.

   e) Both CJD and vCJD produce a spongy appearance in affected brain tissue.

Applications

1. An outbreak of viral meningitis in a small eastern city was linked epidemiologically to a group who swam a non-chlorinated pool in an abandoned quarry outside of town. What might public health officials surmise about the probable cause of the outbreak?

2. Two microbiologists are writing a textbook, but they cannot agree where to place the discussion of botulism. One favored the chapter on nervous system infections, whereas the other insisted on the chapter covering digestive system infections. Where do you think the discussion should be placed, and why?

Critical Thinking ✚

1. A pathologist stated that it was much easier to determine the causative agent of meningitis than of an infection of the skin or intestine. Is her statement valid? Why or why not?

2. Why is it important to learn about rabies when only a few cases occur in the entire United States each year?

Blood and Lymphatic Infections

Bubo An enlarged, tender lymph node characteristic of plague and some sexually transmitted infections.

Disseminated Intravascular Coagulation (DIC) Condition in which clots form in small blood vessels throughout the body, causing organ failure.

Endocarditis Inflammation of the heart valves or lining of the heart chambers.

Lymphangitis Inflammation of lymphatic vessels.

Petechiae Small, purple spots on the skin and mucous membranes caused by hemorrhage from small blood vessels.

Pneumonic Referring to the lung.

Sepsis Acute illness caused by pathogens or their products circulating in the bloodstream.

Septic Shock An array of effects that results from infection of the bloodstream or circulating endotoxin; includes fever, drop in blood pressure, and disseminated intravascular coagulation.

SEM of the tip of the mouthparts of a female mosquito.

A Glimpse of History

Alexandre Emile John Yersin (1863–1943) is one of the most interesting, though relatively unknown, contributors to the understanding of infectious diseases. He developed an interest in science when, as a young boy, he found a microscope and dissecting instruments that belonged to his dead father. A local physician befriended Yersin and influenced him to study medicine. Once he had become a physician, Yersin volunteered at the Pasteur Institute in Paris, where he was hired by Emile Roux, a coworker of Louis Pasteur. Yersin became well recognized for his work at the Institute, but was bored with research, and did not want to practice medicine because he felt it was wrong for physicians to make money from other people's sickness.

Yersin left the Pasteur Institute and was employed as a physician on a ship sailing from France to Vietnam. He became enchanted with Vietnam and the people who lived there. He made it his home for the rest of his life, even establishing a medical school there.

Yersin studied the diseases that affected the Vietnamese people, including plague. The cause and transmission of this serious disease were completely unknown at the time. It had killed millions in Europe in medieval times and had affected France as recently as 1720. In 1894, Yersin went to study an outbreak of plague in Hong Kong. Unfortunately for Yersin, Shibasaburo Kitasato, the famous colleague of Robert Koch (see A Glimpse of History, chapter 23), had arrived 3 days earlier with a large team of Japanese scientists. British authorities had given Kitasato and his colleagues access to patients and laboratory facilities. Yersin did not speak English, so it was difficult for him to communicate his requirements to the authorities. He had to set up a laboratory in a bamboo shack and was forced to do his research on samples from dead plague victims that he got from British soldiers, whose job it was to bury them.

◀◀ Koch, p. 548 ◀◀ Kitasato, p. 548

A week after he arrived in Hong Kong, Yersin reported his discovery of a bacillus that was always present in the swollen lymph nodes of plague victims. The bacterium could be cultivated, and it caused plague-like disease when injected into rats. The disease was transmitted from one rat to another.

Kitasato, working with blood from the hearts of plague victims, also announced soon after his arrival in Hong Kong that he had isolated the bacterium that caused plague. It was later proved to be only a laboratory contaminant, but Kitasato was named co-discoverer of the plague bacillus because of his great prestige.

Yersin's plague bacillus was named *Yersinia pestis*. It was used to make a vaccine, and later, antiserum prepared against the organism was used to cure a patient with plague—the first successful treatment of a plague victim. Not long after, another Pasteur Institute scientist proved that *Y. pestis* is transmitted by rat fleas.

The circulation of blood and lymph fluid supplies nutrients and O_2 to cells and carries away their waste products. The circulatory system also heats and cools body tissues to maintain an optimum temperature. Infection of the system can be serious because infectious agents become **systemic,** meaning they can be carried to all parts of the body, producing disease in one or more vital organs, or causing the circulatory system itself to stop functioning. When a substance is circulating in the bloodstream, the condition is given a name that specifies the nature of the substance and ends in *-emia* as in **bacteremia, viremia,** and **fungemia.** These terms do not imply a disease state—in many cases, the circulating substance does not cause any symptoms. For example, a person is often transiently bacteremic after brushing his or her teeth, when mouth bacteria enter small abrasions caused by the brushing. If illness results from a circulating agent or its toxins,

PERSPECTIVE 27.1

Arteriosclerosis: The Infection Hypothesis

Arteriosclerosis, the main cause of heart attacks and strokes, is characterized by lipid-rich deposits that develop in arteries and can slow or stop the flow of blood.

In 1988, researchers in Finland reported that patients with coronary artery disease—meaning arteriosclerosis of the arteries that supply the heart muscle—often had antibodies against *Chlamydophila pneumoniae* (previously *Chlamydia pneumoniae*). This organism is a tiny, obligate, intracellular bacterium that causes a variety of common respiratory illnesses including sinusitis and pneumonia. *C. pneumoniae* is widespread, and most people have been infected

with it by the time they reach adulthood. What was interesting about the antibody studies was that the higher the titer of antibody, the greater the risk of coronary disease. Then, in 1996, live *C. pneumoniae* was shown to be present in many arteriosclerotic lesions, and other studies established that there was an association between the bacterium and arteriosclerotic lesions of both heart and brain arteries.

Does *C. pneumoniae* cause arteriosclerosis, does it worsen the effects of arteriosclerosis, or does it merely exist harmlessly in the lesions? How can the hypothesis that *C. pneumoniae* contributes to heart attacks

and strokes be tested? Some studies of coronary patients treated with antibiotics effective against the bacterium have suggested a beneficial effect, and others have not. It now appears that many inflammatory conditions, not just *C. pneumoniae* infections, increase the risk of heart attacks and strokes. The risk correlates with the level of acute-phase proteins arising from the release of pro-inflammatory cytokines. Undoubtedly there will be more to come in determining the relationship, if any, between arteriosclerosis and acute-phase proteins. ◀◀ **acute-phase proteins, p. 348**

---

the condition is referred to as **sepsis,** or blood poisoning. Sepsis can cause the blood pressure to fall to such low levels that blood flow to vital organs is insufficient to maintain their functioning, a condition called **septic shock.**

27.1 ■ Anatomy, Physiology, and Ecology

Learning Outcome

1. *Describe the characteristics and functions of the heart, blood vessels, lymphatics, and spleen.*

The cardiovascular system is composed of the heart, blood vessels, and blood. The lymphatic system consists of lymph, lymph vessels, lymph nodes, and lymphoid organs, including the tonsils, appendix, and spleen (**figure 27.1**). Both systems are normally sterile. As blood flows around the body, it passes alternately through the lungs and tissue capillaries. During each circuit, some of the blood passes through the lymphoid organs, which contain phagocytic cells that remove infectious agents and other foreign material. ◀◀ **lymphatic system, p. 357**

The Heart

The heart—a muscular pump enclosed in a fibrous sac called the pericardium—moves the blood around the body. It is divided into right and left sides by a septum, a wall of tissue through which blood normally does not pass after birth (figure 27.1). The right and left sides of the heart are each divided into two chambers—the atria, which receive blood, and the ventricles, which discharge it. Blood from the right ventricle flows through the lungs and into the left atrium. From there, the blood passes into the left ventricle and is then pumped through the aorta to the arteries and capillaries that supply the tissues of the body. The blood is then returned to the right atrium by the veins. Valves at the entrance and exit of each ventricle prevent the blood flow from reversing. Although not common, infections of the heart valves, heart muscle, and pericardium can be serious because they affect blood circulation.

Blood Vessels

Blood vessels include arteries, veins, and capillaries. Arteries have thick muscular walls to withstand the high pressure of the arterial system. Arterial blood is bright red, the color of oxygenated

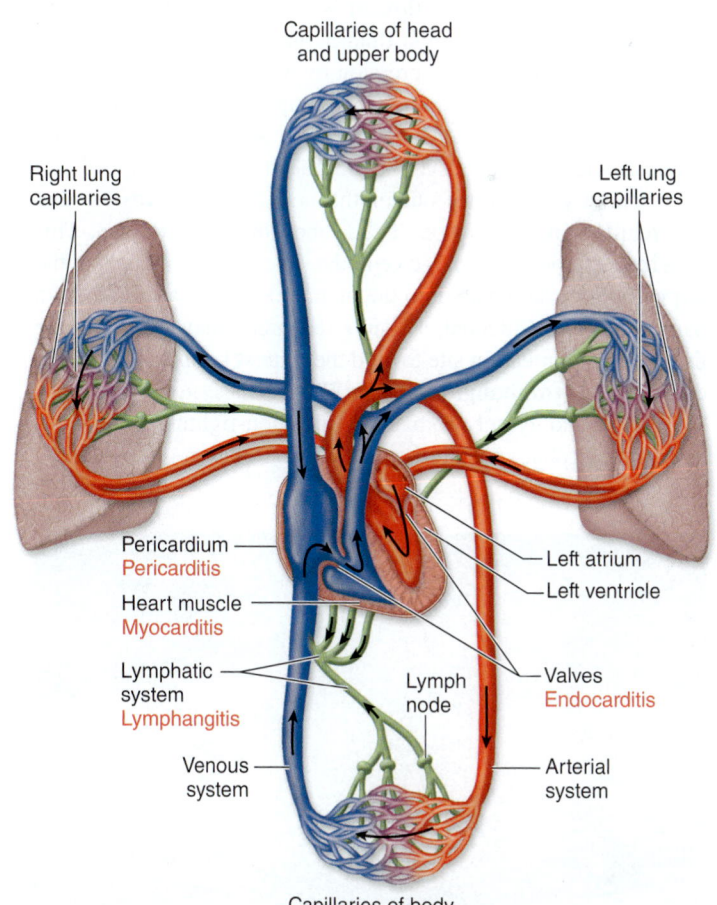

FIGURE 27.1 **The Blood and Lymphatic Systems** Disease conditions are indicated in red type. For simplicity, the spleen is not shown.

❓ *Which chambers of the heart receive blood, and which discharge it?*

hemoglobin. Infection of the arteries is unusual, although the aorta can be dangerously weakened in people with syphilis. Arteries may become affected by arteriosclerosis (lipid-rich deposits that block the vessels), putting people at risk for heart attacks and strokes. Some studies suggest that bacteria and viruses play a role in arteriosclerosis (see **Perspective 27.1**). ◀◀ syphilis, p. 626

Venous blood is dark red, because it becomes depleted of O_2 in the capillaries. Valves in veins play an important role in keeping the blood flowing in one direction, because the venous pressure is too low to do this. Veins are easily compressed, so the action of muscles also aids the flow of venous blood.

Lymphatics (Lymphatic Vessels)

The lymphatic system consists of lymphatic vessels that resemble blood capillaries but are larger (see figure 15.3). The vessels carry an almost colorless fluid called lymph, which comes from plasma, the non-cellular portion of the blood. Lymph fluid seeps through the walls of the blood capillaries to become the interstitial fluid that surrounds tissue cells. This fluid bathes and feeds the tissue cells and then enters the lymphatics (see figure 15.4). Unlike the blood capillaries, lymphatic vessels are easily permeable and take up foreign material such as invading microbes and their products, including toxins and other antigens. Many one-way valves in the lymphatic vessels keep the flow of lymph moving away from the lymphatic capillaries. Lymph fluid is moved by both contraction of the vessel walls and compression by the body's muscles.

Lymphatic vessels drain into multiple lymph nodes. These nodes, which contain phagocytic cells and lymphocytes, are where foreign materials such as microbial cells are trapped and destroyed. Lymph flows out of the nodes through vessels that eventually combine into one large tube—the thoracic duct. This duct then empties into a large vein (the subclavian vein) behind the left collarbone, and back into the main blood circulation. When a hand or a foot is infected, a visible red streak may spread up the limb from the infection site toward the nearest lymph node, a condition called **lymphangitis** (**figure 27.2**). ◀◀ phagocytic cells, p. 346

Blood and lymph both carry infection-fighting leukocytes and antimicrobial proteins including antibodies, complement,

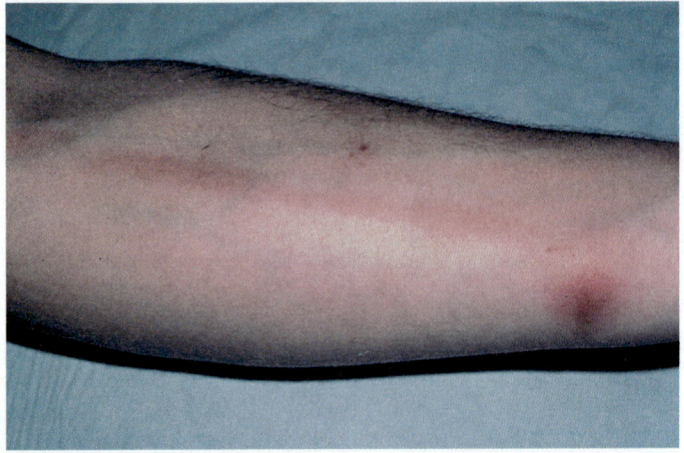

FIGURE 27.2 Lymphangitis Notice the red streak extending up the arm.

❓ *What causes lymphangitis?*

lysozyme, and interferon. An inflammatory response may cause lymph and blood to clot in vessels close to areas of infection, preventing microbes from spreading. ◀◀ leukocytes, p. 339 ◀◀ antimicrobial substances, p. 337 ◀◀ inflammatory response, p. 348

Spleen

The spleen is a fist-sized organ found on the left side of the abdominal cavity, behind the stomach. It contains two kinds of tissue—red pulp (multiple blood-filled passageways) and white pulp (lymphoid tissue). The spleen has several functions. The red pulp cleans the blood by filtration, in the same way that the lymph nodes clean the lymph. Large numbers of phagocytes in red pulp remove aging or damaged red blood cells (RBCs), bacteria, and other foreign materials from the blood. It also produces new blood cells in rare situations where the bone marrow is unable to make enough, and has a reserve of monocytes. The white pulp, which contains both B and T lymphocytes, provides an active immune response to microbial invaders.

> ### MicroAssessment 27.1
>
> The heart has four chambers that work together to move the blood around the body through arteries and veins. The lymphatic system cleans interstitial fluid and returns it to the bloodstream. Systemic infections may disrupt the transport of O_2 and nutrients to body tissues and removal of waste products. Infectious agents and their toxins can be spread throughout the body by the circulatory system.
>
> 1. *What are the functions of the spleen?*
> 2. *What is the function of valves in the lymphatic and circulatory systems?*
> 3. *Which side of the heart—right or left—do bacteria in infected lymph reach first? Where do they go from there?* ➕

27.2 ■ Bacterial Diseases of the Blood Vascular System

Learning Outcomes

2. *Outline the events that lead to subacute bacterial endocarditis.*
3. *Explain the role of the inflammatory response in sepsis and septic shock.*

Bacterial infections of the vascular system can kill a person quickly, or they can progress slowly for months, causing a gradual decline in health. They are not common, but they are always dangerous. Usually the bacteria access the bloodstream as they are carried by lymph flowing from an infected area in the tissues. Some pathogens multiply in the bloodstream, and they may colonize and form biofilms on structures such as the heart valves. ◀◀ biofilm, p. 84

Endocarditis is the term used for infections of the heart valves or the inner surfaces of the heart. **Acute bacterial endocarditis** starts suddenly with fever and is usually caused by virulent species such as *Staphylococcus aureus* and *Streptococcus pneumoniae*, which can infect both normal and abnormal heart valves. These bacteria quickly destroy the valves and form abscesses in the heart muscle, causing heart failure. By contrast, **subacute bacterial endocarditis** is usually caused by organisms with little virulence, progresses much slower, and is less likely to be fatal. ◄◄ *Staphylococcus aureus*, p. 524 ◄◄ *Streptococcus pneumoniae*, p. 497

Sepsis is caused by both Gram-negative and Gram-positive bacteria, as well as other infectious agents.

Subacute Bacterial Endocarditis (SBE)

Subacute bacterial endocarditis (SBE) is usually localized to one of the valves on the left side of the heart. It commonly occurs on valves that are deformed because of a birth defect or a disease such as rheumatic fever. ◄◄ rheumatic fever, p. 489

Signs and Symptoms

People with SBE usually have noticeable fatigue and slight fever. They typically become ill gradually and slowly lose energy over a period of weeks or months. They may suddenly develop a stroke.

Causative Agent

SBE is usually caused by normal bacterial microbiota of the mouth or skin, such as viridans streptococci and *Staphylococcus epidermidis*. The infecting organisms are usually shed from the infected heart valve into the blood and can be identified by cultivating samples taken from an arm vein. However, it is not always possible to culture the causative agent. For example, blood that has passed through the lung before reaching the arm vein is cleared of microorganisms by lung phagocytes. Also, some pathogenic bacteria that cause endocarditis, such as *Coxiella burnetii*, cannot be cultured in cell-free media. In these cases, the causative agent is identified by PCR and nucleic acid probes. ◄◄ *Coxiella burnetii*, p. 277 ◄◄ PCR, p. 227 ◄◄ nucleic acid probes, p. 232

Pathogenesis

The bacteria that cause SBE enter the blood during dental procedures, tooth brushing, or trauma. These microbes can get trapped in the thin blood clots that often form around deformed heart valves or other areas with disturbed blood flow. They multiply there, creating a biofilm that protects them from phagocytosis and antimicrobial medications. The clot grows larger around the multiplying organisms, gradually building up a fragile mass. Bacteria continually wash off the mass into the circulation, and pieces of infected clot (septic emboli) can break off. These can block important blood vessels, leading to death of the tissue supplied by the vessel. They can also cause a vessel to weaken and balloon out, forming an aneurysm.

People with SBE often have high levels of antibodies against the bacteria. These may be harmful because they lead to the formation of immune complexes, which may lodge in the skin, eyes, and other body structures, triggering an inflammatory response. In the kidney, the complexes cause glomerulonephritis. ◄◄ immune complexes, p. 395 ◄◄ glomerulonephritis, p. 490

Even though the bacteria that cause SBE normally have little invasive ability, large numbers of them growing in the heart are sometimes able to penetrate into heart tissue, producing abscesses or damaging valve tissue, and resulting in a leaky valve.

Epidemiology

In recent years, viridans streptococci have caused fewer cases of SBE cases than previously. Cases that do occur, however, are more serious because the causative organisms are becoming increasingly antibiotic-resistant. This may be an unintended result of giving

prophylactic antibiotic treatment before dental procedures to patients with serious heart murmurs. A heart murmur is an abnormal sound the physician hears when listening to the heart, often indicative of a deformed valve or other structural abnormality.

SBE also occurs in injected-drug abusers, hospitalized patients who have plastic intravenous catheters for long periods, and those with artificial heart valves. These people are usually infected with *S. epidermidis*, *S. aureus*, enterococci, or a wide variety of species other than viridans streptococci.

Treatment and Prevention

SBE is treated with antibacterial medications chosen according to the susceptibility of the causative organism. Only bactericidal medications such as penicillin and gentamicin are effective, and usually two or more antimicrobials are used together. Prolonged treatment over one or more months is usually required. Infection of an artificial heart valve or other foreign material is almost always associated with biofilm formation, and the object must often be replaced to cure the person. Sometimes, surgery is needed to remove an infected clot or to drain abscesses.

There are no proven methods for preventing SBE. However, people with known or suspected heart valve abnormalities are commonly given an antibacterial medication shortly before dental or other bacteremia-causing procedures. The medication is chosen according to the expected bacterial species and its likely susceptibility to antimicrobials. Healthcare-associated SBE can be prevented by strict use of sterile technique when inserting plastic intravenous catheters, moving the catheters to a new site every few days, and removing them as soon as possible. Such precautions help prevent bacterial colonization of the catheters and consequent bacteremia that could lead to heart valve infection. The main characteristics of SBE are presented in **table 27.1**. ◄◄ healthcare-associated infections, p. 449

TABLE 27.1	Subacute Bacterial Endocarditis
Signs and symptoms	Fever, loss of energy over a period of weeks or months; sometimes, a stroke
Incubation period	Poorly defined, usually weeks
Causative agents	Usually oral α-hemolytic viridans streptococci or *Staphylococcus epidermidis*
Pathogenesis	Normal microbiota enter bloodstream through dental procedures, other trauma; in an abnormal heart, turbulent blood flow causes formation of a thin clot that traps circulating organisms; a biofilm forms, protecting the organisms from phagocytic killing; pieces of clot break off, block important blood vessels, leading to tissue death.
Epidemiology	People at risk are mainly those with congenital heart defects or with hearts damaged by disease such as rheumatic fever; may develop after dental procedures or other situations that cause bacteremia.
Treatment and prevention	Treatment: bactericidal antibiotics given together, such as penicillin and gentamicin. Prevention: giving an antibiotic immediately before anticipated bacteremia, such as before dental work.

Sepsis and Septic Shock

Sepsis, an infection-induced systemic inflammatory response, is a common healthcare-associated illness. The disease is caused by the release of bacterial products—particularly endotoxins from Gram-negative bacteria—that change the normally beneficial inflammatory response into an excessive damaging response that can lead to life-threatening illness. If uncontrolled, sepsis can progress to septic shock, a dramatic drop in blood pressure. The discovery that sepsis involves pro-inflammatory cytokines and not only invading microorganisms has led to advances in understanding the pathogenesis of this disease. ◀◀ **endotoxic shock, p. 394** ◀◀ **pro-inflammatory cytokines, p. 341**

MicroByte —
An estimated 400,000 cases of Gram-negative sepsis occur in the United States each year.

Signs and Symptoms

The signs and symptoms of severe sepsis include violent shaking, chills, and fever, often with rapid breathing and feelings of anxiety. If septic shock develops, urine output drops, respiration and pulse speed up, and the arms and legs become cool and dusky-colored. Septic patients are often referred to as "looking ill."

Causative Agents

Systemic infection by any microorganism can cause sepsis, but most fatal cases involve Gram-negative bacteria. Recall that the outer membrane of Gram-negative bacteria contains lipopolysaccharide (LPS), also referred to as endotoxin (see figure 3.33). Examples of common causes of fatal sepsis include members of the normal microbiota of the large intestine—particularly facultative anaerobes such as *Escherichia coli* and other *Enterobacteriaceae* members—and anaerobes such as *Bacteroides* species. Environmental bacteria such as *Pseudomonas aeruginosa* are also common causes. ◀◀ **lipopolysaccharide, p. 60** ◀◀ *Enterobacteriaceae*, **p. 265** ◀◀ *Pseudomonas*, **p. 265**

Pathogenesis

Sepsis almost always starts from an infection somewhere in the body other than the bloodstream—a kidney infection, for example. When normal body defenses are compromised by medical treatments such as surgery, catheters, and some medications, microorganisms that normally have little invasive ability are able to infect the blood.

Sepsis progresses in stages and is initially due to an overstimulation of the inflammatory response. When the toll-like receptors (TLRs) on macrophages and neutrophils detect endotoxin or other pathogen-associated molecular patterns (PAMPs), the phagocytes respond by releasing pro-inflammatory cytokines. When this occurs systemically, an uncontrolled release of cytokines—a cytokine storm—results. The complement pathway is also activated, further amplifying the inflammatory response. ◀◀ **toll-like receptors, p. 342** ◀◀ **PAMPs, p. 342** ◀◀ **cytokines, p. 341** ◀◀ **complement system, p. 344**

The systemic release of pro-inflammatory cytokines, along with activation of complement, results in a widespread, self-stimulating inflammatory response. Various pro-inflammatory

compounds recruit additional phagocytes from the bone marrow and cause the phagocytic cells to produce more TLRs. This increases the system's sensitivity to PAMPs, leading to an even greater release of pro-inflammatory mediators. The phagocytic cells and dying host cells then release compounds that function as damage-associated molecular patterns (DAMPs). The host responses to PAMPs and DAMPs leads to multiple devastating outcomes, including the inhibition of systems that normally control inflammation, suppression of the adaptive immune response, and activation of the coagulation cascade (**figure 27.3**). ◀◀ **DAMPs, p. 342**

The net result of this dysregulated inflammatory response is complex and varied. Activation of the coagulation cascade causes small clots to form in the capillaries. This blocks them, cutting off the blood supply to tissues, which leads to tissue hypoxia (low levels of O_2) and subsequent tissue necrosis (death). The widespread clotting—**disseminated intravascular coagulation (DIC)**—is often accompanied by hemorrhage, because the endothelium is damaged. DIC may lead to multiorgan failure due to circulation disruptions. At the same time, phagocytes release tissue-damaging lysosomal enzymes. This is particularly important in the lungs, which are seriously and irreversibly damaged by these enzymes. Lung damage often results in death even if the individual's infection has been cured. The harmful effects of the inflammatory response are made worse by the hypotension (low blood pressure) usually present. The hypotension is caused by decreased muscular tone of the heart and blood vessel walls

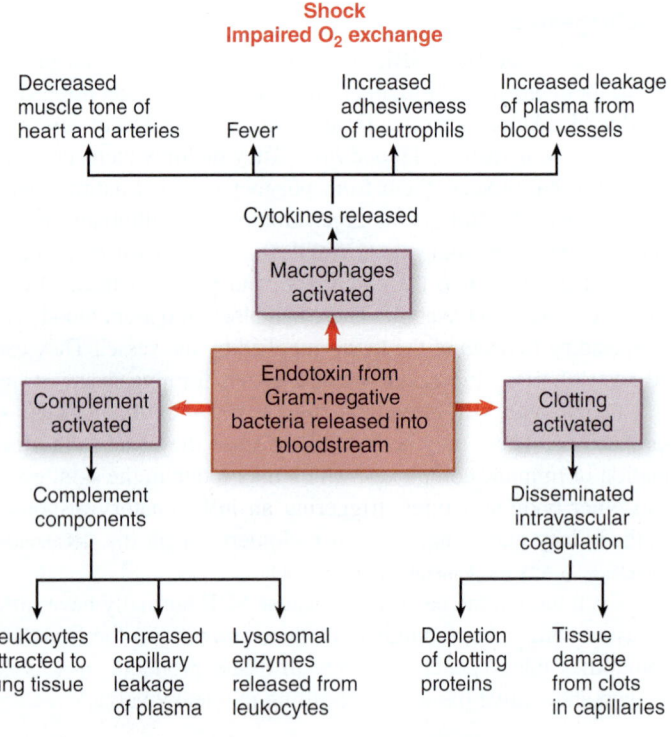

FIGURE 27.3 Events in Gram-Negative Sepsis
❓ *What causes organ failure in sepsis?*

and the low blood volume that results from fluid leakage out of the blood vessels. Current evidence indicates that the pro-inflammatory cytokines lead to septic cardiomyopathy (disorder of the heart muscles) and heart failure. Shock results when the blood pressure falls so low that vital organs are no longer supplied with adequate amounts of blood to maintain their function.

Even if the patient survives the initial stages of sepsis, the later effects can be fatal. Neutrophils stop working properly, so bacteria are not cleared from the blood effectively. In addition, lymphocytes—particularly helper T cells—and dendritic cells undergo apoptosis, affecting the adaptive immune response and leading to immunosuppression that results in increased susceptibility to secondary infection.

Epidemiology

Sepsis is mainly a healthcare-associated disease, reflecting the high incidence of bloodstream infections in hospitalized patients with impaired host defenses. Rates of sepsis have been increasing, likely due to longer life spans, antibiotic suppression of normal microbiota, and immunosuppressive medications. Use of medical equipment where biofilms readily develop—such as respirators, and catheters placed in blood vessels and the urinary system—also increases the risk of sepsis.

Treatment and Prevention

Treatment of sepsis is difficult, because it is a complex syndrome involving several host responses. It is currently treated with antimicrobial medications effective against the causative organism. Bactericidal antibiotics often lyse bacterial cells, however, increasing the release of endotoxin from Gram-negative organisms, worsening the effects of the sepsis. The person is also treated for shock and tissue hypoxia. In addition, fluid replacement and support for organ dysfunction are given.

New treatments for sepsis are being evaluated. Monoclonal antibodies against endotoxin or pro-inflammatory cytokines such as tumor necrosis factor (TNF) can have some benefit when given before shock develops. Another medication being evaluated is drotrecogin alfa (activated human protein C produced by recombinant technology), which opposes the effects of certain pro-inflammatory cytokines. Evaluation of new treatments is difficult because (1) many of the patients have underlying life-threatening illnesses in addition to their sepsis, and (2) animal models for sepsis do not accurately mimic the disease in humans. Other possible targets for new therapies include TLR4 (the TLR that recognizes LPS), complement component C5a, and the autonomic nervous system, which plays a regulatory role in inflammation by controlling release of pro-inflammatory cytokines. Between 30% to 50% of people with sepsis die even with treatment, partly because of their other underlying illnesses. ◀◀ monoclonal antibodies, p. 428

Sepsis can be prevented by quick identification and effective treatment of localized infections, particularly in people with impaired immune defenses. Also, predisposing conditions such as bedsores and pyelonephritis—which commonly lead to sepsis in patients with cancer and diabetes—can usually be prevented. ◀◀ pyelonephritis, p. 613

MicroAssessment 27.2

Bacteria of low virulence can cause serious, even fatal, infections in individuals with a structural abnormality of the heart. A systemic infection represents failure of the body's mechanisms for keeping infections localized to one area. The inflammatory response, although vitally important in localizing infections, can be life-threatening if generalized. People die of Gram-negative sepsis despite antibacterial therapy because of the endotoxin-mediated release of cytokines in the lung and bloodstream.

4. *What is a systemic infection?*
5. *What is the difference between bacteremia and sepsis?*
6. *Why might clots on the heart valves make microorganisms there inaccessible to phagocytic killing?* ■

27.3 ■ Bacterial Diseases of the Lymph Nodes and Spleen

Learning Outcome

4. *Compare and contrast tularemia, brucellosis, and plague.*

Enlargement of the lymph nodes and spleen is a prominent feature of diseases that involve the mononuclear phagocyte system. The examples discussed in this section are now uncommon human diseases in the United States, but they represent a constant threat because they are widespread in animals. ◀◀ mononuclear phagocyte system, p. 340

Tularemia ("Rabbit Fever" or "Deer Fly Fever")

Tularemia is widespread in the United States and is found in wild animals such as rabbits, muskrats, and bobcats. The disease was identified in Tulare County, California, following an outbreak among ground squirrels in the area in 1911. Although the disease affects humans only occasionally in the United States today, it is potentially a bioterrorism threat.

Signs and Symptoms

Signs and symptoms of tularemia depend on how the causative organism enters the body—through mucous membranes or minor cuts and scratches in the skin, or by inhalation. Typically, signs and symptoms appear 2 to 5 days after a person is bitten by an infected tick or insect or handles an infected wild animal. The characteristic sign is an ulcer that develops at the site where the organism enters the body (**figure 27.4**). Regional lymph nodes enlarge, and fever, chills, and achiness occur. Signs and symptoms usually clear in 1 to 4 weeks, but sometimes they last for months.

If the organism is inhaled or infects the lung from the bloodstream, pneumonia may occur. Signs and symptoms include a dry cough and pain beneath the sternum (breastbone), a result of enlarged lymph nodes. Tularemic pneumonia, although rare, is more serious and has a mortality rate as high as 30% if untreated.

Causative Agent

Tularemia is caused by *Francisella tularensis*, a non-motile, aerobic, Gram-negative rod. It was named for Edward Francis, an

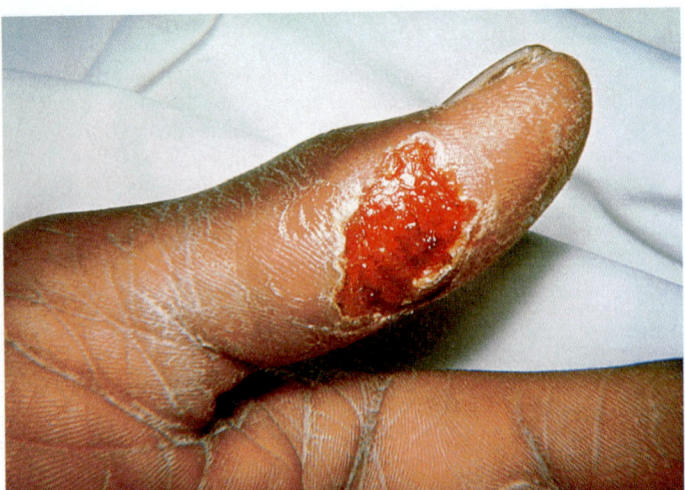

FIGURE 27.4 Tularemic Ulcer of the Thumb This kind of lesion is generally acquired through skinning rabbits or other wild animals.

❓ *In what other ways can tularemia be transmitted?*

American physician who studied tularemia in the early 1900s, and Tulare County, where it was first identified. The organism is unusual because it requires a special medium enriched with the amino acid cysteine in order to grow.

Pathogenesis

Francisella tularensis enters through breaks in the skin or mucous membranes and is carried to the regional lymph nodes, which become large and tender. The nodes may become filled with pus and drain spontaneously. The organism spreads to other parts of the body via the lymphatics and blood vessels. *F. tularensis* is ingested by phagocytic cells and grows within them. This may explain why the disease persists in some people despite the high antibody titers in their blood. Cell-mediated immunity effectively rids the host of this infection. Even without treatment, over 90% of infected people survive. ◀◀ cell-mediated immunity, p. 356

Epidemiology

Tularemia occurs among wild animals in many areas of the Northern Hemisphere, including all the states of the United States except Hawaii. In the eastern United States, summertime human infections commonly result from tick bites. However, cases often occur in the winter months as well, in people skinning rabbits (explaining the common name, "rabbit fever"). People also contract the disease from muskrats, beavers, squirrels, deer, and other wild animals. The animals are generally free of illness but carry the causative organism. In the western United States, infections mostly result from the bites of infected ticks and deer flies (explaining the other common name, "deer fly fever") and usually occur during summer. Generally, 150 to 250 cases of tularemia are reported each year from counties across the United States. Epidemics of inhalation tularemia have occurred from dust arising from mowing lawns or from rodent-infested buildings. Tularemia can also be contracted by ingesting contaminated material. In the United States, *F. tularensis* is listed among the category A (highest-risk) agents of biological terrorism (see Future Challenges 19.1).

Treatment and Prevention

Most cases of tularemia are effectively treated with streptomycin. Ciprofloxacin or gentamicin can also be used.

Tularemia can be prevented by using rubber gloves and goggles or face shields when skinning or handling wild animals. Insect repellants and protective clothing help protect against insect and tick vectors. The disease can also be avoided by removing ticks found after outdoor activity. Meat from wild animals should be cooked thoroughly. The main features of this disease are summarized in **table 27.2.**

Brucellosis ("Undulant Fever" or "Bang's Disease")

Brucellosis is often called "undulant fever," or "Bang's disease," after Frederik Bang (1848–1932), a Danish veterinary professor who discovered the cause of cattle brucellosis. Only about 150 cases of human brucellosis are reported each year in the United States. Many more cases, however, go unreported annually.

Signs and Symptoms

Brucellosis usually starts gradually, and the signs and symptoms are vague. Typically, patients complain of mild fever, sweating, weakness, aches and pains, enlarged lymph nodes, depression and weight loss. The recurrence of fevers over weeks or months in some cases gave rise to the alternative name, "undulant fever." Most people recover within 2 months even without treatment, and only 15% are symptomatic for more than 3 months.

Causative Agent

Brucella sp. are small, aerobic, nonmotile, Gram-negative rods with complex nutritional requirements. The causative agent was discovered by Dr. David Bruce, after whom it was named. Four varieties of

TABLE 27.2	Tularemia ("Rabbit Fever" or "Deer Fly Fever")
Signs and symptoms	Ulcer at site of entry, enlarged lymph nodes in area, fever, chills, achiness
Incubation period	1 to 10 days; usually 2 to 5 days
Causative agent	*Francisella tularensis*, an aerobic, Gram-negative rod
Pathogenesis	Organisms are ingested by phagocytic cells, grow within these cells, and then spread throughout body.
Epidemiology	Present among wildlife in most states of the United States. Risk mainly to hunters, game wardens, and others who handle wildlife. Acquired when the organism penetrates a mucous membrane or enters broken skin, as may occur when skinning rabbits, for example; bite of infected insect or tick.
Treatment and prevention	Treated with gentamicin or ciprofloxacin. Prevention: avoiding bites of insects and ticks; wearing rubber gloves, goggles, when skinning rabbits; taking safety precautions when working with organisms in laboratory.

the genus *Brucella* cause brucellosis in humans. DNA studies show that all members of the genus fall into a single species, *Brucella melitensis,* but traditionally, the different varieties were assigned species names depending largely on their preferred host: *B. abortus* invades cattle; *B. canis,* dogs; *B. melitensis,* goats; and *B. suis,* pigs. The distinctions between the various strains are mainly useful epidemiologically and generally have little pathogenic significance.

Pathogenesis

Brucella organisms penetrate mucous membranes or breaks in the skin and are disseminated via the lymphatic and blood vessels to the heart, kidneys, and other parts of the body. The infection causes spleen enlargement. Like *Francisella tularensis, Brucella* sp. are not only resistant to phagocytic killing but also can grow intracellularly in phagocytes, where they are inaccessible to antibodies and some antibiotics. The mortality rate is low and death is generally due to endocarditis. A serious but rare complication is bone infection (osteomyelitis).

Epidemiology

Brucellosis is typically a chronic infection of domestic animals. It affects the mammary glands and the uterus, contaminating milk and causing abortions in the affected animals (abortion is not a feature of human disease). Worldwide, brucellosis is a major problem in animals used for food, causing yearly losses of many millions of dollars.

Most brucellosis in humans is associated with occupational exposure—workers in the meat packing industry, veterinarians, and slaughterhouse workers can be exposed to the causative agents. People drinking unpasteurized milk or eating soft cheeses made from infected milk can also contract the disease. In the United States, hunters have acquired infections from elk, moose, bison, caribou, and reindeer. *B. melitensis* is listed as a category B (intermediate-risk) bioterrorism agent.

MicroByte

About 20% of the Yellowstone Park bison carry brucellosis, and over 1,000 have been killed when they wandered from the park, in order to prevent transmission of the disease to cattle.

Treatment and Prevention

Brucellosis can usually be treated using doxycycline with rifampin or streptomycin for 6 weeks. In some chronic cases, extended treatment (sometimes up to 6 or more months) may be needed.

The most important control measures for brucellosis are pasteurization of dairy products and inspection of domestic animals for evidence of the disease. Veterinarians, butchers, and slaughterhouse workers should use protective gear such as goggles or face shield and rubber gloves. An attenuated vaccine effectively controls the disease in domestic animals. **Table 27.3** gives the main features of brucellosis.

TABLE 27.3 | Brucellosis ("Undulant Fever" or "Bang's Disease")

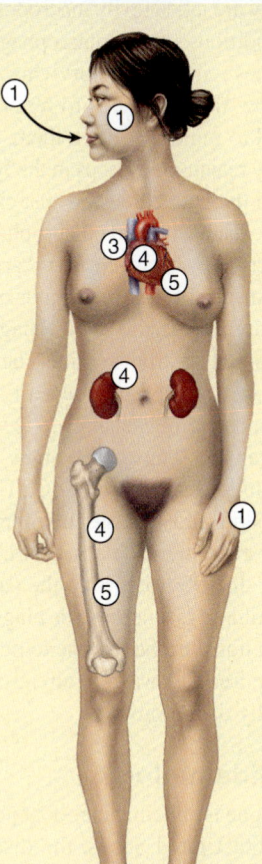

① *Brucella melitensis* enters the body through mucous membranes, skin abrasions, or ingestion of unpasteurized milk.

② The bacteria are taken up by phagocytes but resist digestion and grow within cells.

③ The bacteria enter the lymphatics and bloodstream and are carried throughout the body.

④ Infection is established in other body tissues, such as the heart valves, kidneys, and bones.

⑤ Osteomyelitis is the most common serious complication. Most deaths are due to endocarditis.

Signs and symptoms	Fever, body aches, weight loss, enlargement of lymph nodes; symptoms may subside without treatment but then recur
Incubation period	Usually 5 to 21 days
Causative agent	*Brucella melitensis,* a small, aerobic, Gram-negative rod
Pathogenesis	Organisms penetrate mucous membranes and are carried to heart, kidneys, and other parts of the body via the blood and lymphatic system; they are resistant to phagocytic killing and grow within these cells.
Epidemiology	Main sources of human infections: domestic animals. Disease also occurs in wild animals such as moose, caribou, bison; butchers, farmers, veterinarians, those who drink unpasteurized milk are at risk.
Treatment and prevention	Treatment: doxycycline with rifampin or streptomycin. Prevention: vaccination of domestic animals; pasteurization of milk and milk products; gloves and face protection.

Plague ("Black Death")

Plague, once known as the "black death," was responsible for the death of approximately a quarter of the population of Europe between 1346 and 1350. Crowded conditions in the cities and a large rat population played major roles in the spread of the disease. Plague is a potential bioterrorism disease, listed as category A.

Signs and Symptoms

The signs and symptoms of plague depend on how the disease was acquired. Most commonly, a person is bitten by an infected flea, and symptoms develop 2 to 6 days later. The person characteristically develops markedly enlarged and tender lymph nodes called **buboes**—hence the name **bubonic plague**—in the region that receives lymph drainage from the area of the flea bite. High fever, shock, delirium, and patchy bleeding under the skin quickly develop.

If a person inhales respiratory droplets from an infected patient or animal, they may develop **pneumonic plague** with signs and symptoms of cough and bloody sputum. Pneumonic symptoms arise more quickly than bubonic symptoms, typically in 1 to 3 days.

If the causative organisms spread via the bloodstream, the person may develop **septicemic plague.** Endotoxin released by the causative agent results in shock and disseminated intravascular coagulation (DIC), which causes bleeding into the skin and organs, leading to a red or black patchy rash. ◄◄ DIC, p. 674

Causative Agent

Plague is caused by *Yersinia pestis,* which, like other members of the *Enterobacteriaceae*, is a facultatively anaerobic, Gram-negative rod. It is non-motile and grows best at 28°C. The organism resembles a safety pin when stained with certain dyes because the ends of the bacterium stain more intensely than the middle (**figure 27.5**).

Pathogenesis

Yersinia pestis forms biofilms in the digestive tract of infected fleas, often blocking the tract (**figure 27.6**). This both starves the flea—thereby increasing the likelihood that it will try to feed again—and causes bacteria to be regurgitated into the bite wound as the flea attempts to feed. Even fleas that do not have an obstructed digestive tract can transmit *Y. pestis* because they

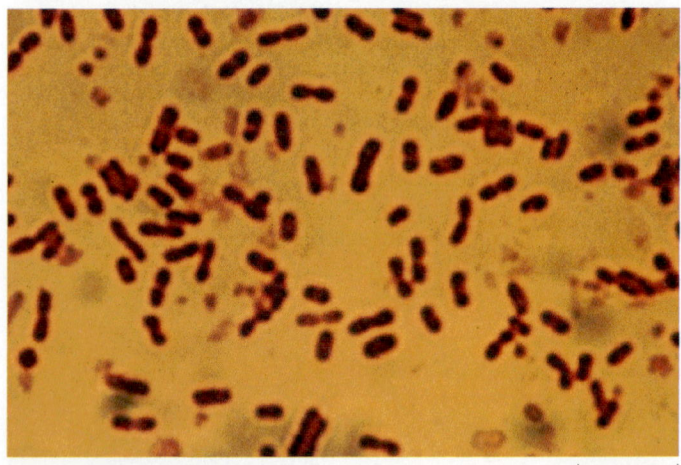

FIGURE 27.5 *Yersinia pestis*

❓ *What is the function of the Yops proteins made by this organism?*

FIGURE 27.6 Fleas Following a Blood Meal (a) Healthy flea with fresh blood in gut. (b) Flea with obstruction due to *Yersinia pestis* infection.

❓ *What is the risk when a flea has a digestive system obstruction?*

(a)

(b)

discharge bacterial cells in their saliva, and these can then be introduced into human tissue when a person scratches the flea bite.

Y. pestis has multiple virulence factors, making it extremely well equipped for avoiding mammalian host defenses. Some virulence factors are encoded on the bacterial chromosome and others are encoded by various plasmids. The characteristics of the most important virulence factors are summarized in **table 27.4.**

Once *Y. pestis* enters via a flea bite, its protease (Pla) clears the lymphatics and capillaries of clots and inactivates certain complement system components, allowing the organisms to spread. The lymphatic vessels carry the bacterial cells to the regional lymph nodes, where they are taken up by macrophages. The bacteria sense the intracellular conditions of the macrophages and respond by turning on various genes required for surviving in the mammalian host. Some of these allow the organisms to resist the killing effects of the macrophages and to multiply within them. After several days, an acute inflammatory reaction develops in the lymph nodes, producing the enlargement and marked tenderness that characterizes bubonic plague.

The infected macrophages eventually die and release the bacteria, which are now "armed" to withstand host defenses. For example, the *Yersinia* outer membrane proteins (YOPs) and capsule allow *Y. pestis* to avoid phagocytosis, and iron acquisition proteins allow it to trap hemin. The lymph nodes may become necrotic, allowing large numbers of the bacterial cells to spread into the bloodstream, causing septicemic plague. Endotoxin released by the organisms causes the signs and symptoms of septicemic plague, including shock and DIC. The dark hemorrhages and dusky color of skin and mucous membranes from DIC probably inspired the common name "black death."

In 10% to 20% of the cases, the infection spreads to the lungs, resulting in pneumonic plague. A person with pneumonic plague can transmit the disease to others in respiratory droplets. Organisms acquired this way are already fully virulent and, therefore, are especially dangerous.

Epidemiology

Plague is endemic in rodent populations worldwide except Australia. In the United States, the disease mostly occurs in about 15 states in the western half of the country. The main reservoirs are prairie dogs and rock squirrels, as well as their fleas. However, rats, rabbits, dogs, and cats can also be hosts. Hundreds of species of

TABLE 27.4	Virulence Factors of *Yersinia pestis*	
Factor	**Coded by**	**Action**
Pla (protease)	pPCP1 plasmid	Activates plasminogen; destroys C3b, C5a and clots
YOPs (proteins)	pCD1 plasmid	Interfere with phagocytosis and the immune response
V antigen	pCD1 plasmid	Controls type III secretion system that delivers YOPs proteins
F1	pMT1 plasmid	Forms antiphagocytic capsule at 37°C
PsaA (adhesin)	Chromosome	Role in attachment to host cells
Complement resistance	Chromosome	Protects against lysis by activated complement
Iron acquisition	Chromosome	Traps hemin and other iron-containing substances; stores iron compounds intracellularly

fleas are able to transmit plague, and the fleas can remain infectious for a year or more. Epidemics of pneumonic plague in humans are caused by respiratory droplets produced by coughing pneumonic plague patients. The number of reported cases of plague is gradually increasing as cities spread into surrounding areas where the main reservoirs live. African countries have the highest number of cases worldwide, but the disease is endemic in 10 Chinese provinces, where earthquakes create conditions for pneumonic plague. ◀◀ endemic disease, p. 438 | ◀◀ reservoir, p. 438

Treatment and Prevention

Plague can be effectively treated with antimicrobial medications including gentamicin, ciprofloxacin, or doxycycline, especially if given within 24 hours of the onset of signs and symptoms. Between 50% and 80% of people with bubonic plague die if not treated. Mortality rate for untreated pneumonic plague is almost 100%.

Plague epidemics can be prevented by rat control measures such as proper garbage disposal, constructing rat-proof buildings, and rat extermination programs. These programs must be combined with the use of insecticides to prevent the escape of infected fleas from dead rats. There is no current vaccine for plague, but new genetically engineered vaccines are being evaluated under a joint agreement among the United States, Canada, and the United Kingdom. The antibiotic tetracycline can be given as a preventive for someone exposed to plague and is useful in controlling epidemics because of its immediate effect. The main features of plague are presented in **table 27.5.**

TABLE 27.5	Plague ("Black Death")

1. *Yersinia pestis* is acquired from the bite of an infected flea or scratching skin contaminated by the flea's feces.
2. The bacteria are carried to regional lymph nodes.
3. Phagocytes ingest the bacteria but the intracellular conditions activate capsule production and other genes responsible for virulence.
4. Fully virulent bacteria break out of the phagocytes, infect the nodes, producing the buboes that characterize bubonic plague.
5. The bacteria may be carried into the bloodstream, causing septicemic plague.
6. The lungs can become infected, producing the highly contagious and lethal pneumonic plague.
7. Bacteria exit with coughing.

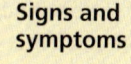

Signs and symptoms	Sudden onset of high fever, large lymph nodes called buboes, skin hemorrhages; sometimes bloody sputum
Incubation period	Usually 1 to 6 days
Causative agent	*Yersinia pestis*, a Gram-negative rod; a member of the *Enterobacteriaceae* with multiple chromosome- and plasmid-coded virulence factors
Pathogenesis	Enters the body with bite of infected flea. Bacteria taken up by macrophages. Intracellular environment causes them to transform into encapsulated organisms capable of producing multiple virulence factors that allow attachment to host cells, and provide defense against the immune system.
Epidemiology	Endemic in rodents and other wild animals, and their fleas, particularly in the western states of the United States. Bubonic plague is transmitted by fleas; pneumonic plague can be transmitted person to person in respiratory droplets. Pneumonic plague is the most dangerous because *Y. pestis* is fully virulent at the time of transmission.
Treatment and prevention	Treatment: prompt diagnosis and antibacterial treatment necessary to prevent high mortality. Prevention: currently no vaccine available; new vaccines are under evaluation. Avoid contact with wild rodents and their burrows. Insecticides and rat control.

Generally, tularemia is a bacteremic disease of wild animals transmitted to humans by exposure to their blood or tissues, or by biting ticks and insects. Brucellosis is most commonly a disease of domestic animals transmitted to humans when individuals handle the animals' flesh or drink unpasteurized milk. Plague is endemic in rodents in the western United States and is transmitted to humans by flea bites. Unlike tularemia and brucellosis, plague has the potential to spread rapidly from human to human by coughing, with a high fatality rate.

7. *Workers in what industry are especially likely to contract brucellosis?*
8. *How would growth within phagocytes protect* Francisella tularensis *from destruction by cell-mediated immunity?*
9. *How would crowded conditions in cities favor spread of plague?* ✚

27.4 ◼ Viral Diseases of the Lymphatic and Blood Vascular Systems

Learning Outcomes

5. *Describe the characteristics of infectious mononucleosis.*
6. *Compare and contrast yellow fever, dengue fever, and viral hemorrhagic fevers.*

A number of viral illnesses affect the lymphatic system and blood vessels. This section discusses infectious mononucleosis, yellow fever, dengue fever, and viral hemorrhagic fevers, which prominently involve the circulatory system. Other examples of viral diseases that affect the lymphoid and circulatory systems are the human immunodeficiency virus (HIV), the cause of AIDS, and cytomegalovirus, which are covered in chapter 28. ▶▶ **human immunodeficiency virus disease, p. 696** ▶▶ **cytomegalovirus disease, p. 711**

Infectious Mononucleosis ("Mono" or "Kissing Disease")

Infectious mononucleosis ("mono") is a disease familiar to many students because of its high incidence among young people. The term *mononucleosis* refers to the fact that people with this disease have an increased number of mononuclear leukocytes in their blood. ◀◀ **leukocytes, p. 339**

Signs and Symptoms

Typically, signs and symptoms of infectious mononucleosis appear after a long incubation period, usually 30 to 60 days. They include fatigue, fever, a sore throat covered with pus, and enlargement of the spleen and lymph nodes. In most cases, the fever and sore throat are gone in about 2 weeks and the enlarged lymph nodes in 3 weeks. Generally, people recover fully within 4 weeks, although some suffer severe exhaustion and difficulty concentrating that affects them for months.

Causative Agent

Infectious mononucleosis is caused by the Epstein-Barr virus (EBV), named after its discoverers, M. A. Epstein and Y. M. Barr.

It is a double-stranded DNA virus of the herpesvirus family, and although identical in appearance to the other known herpesviruses that cause human disease, it is not closely related to any of them.

Pathogenesis

EBV initially infects the mouth and throat and then becomes latent in another cell type (**figure 27.7**). After replicating in the mouth epithelium, salivary glands, and throat, the virus is carried to the lymph nodes. There, it infects B lymphocytes. The B-cell infection can be (1) productive, in which the virus replicates and kills the B cell; or (2) non-productive, in which the virus establishes a latent infection, maintained as either plasmid or provirus (integrated into the host cell chromosome). For most B cells, the infection is non-productive. The virus activates these B cells, causing them to proliferate and produce immunoglobulins. The proliferating B cells are responsible for the large numbers of mononuclear cells that give the disease its name. Their numbers and appearance sometimes cause the incorrect diagnosis of leukemia—this is disproved when the patient spontaneously recovers.

A common consequence of B-cell infection is the appearance of a specific type of antibody useful for diagnosing infectious mononucleosis; it has no pathological significance. This antibody type, called a heterophile antibody, will react with an antigen on the red blood cells of certain animal species, including sheep, horses, and oxen.

The proliferation of the lymphocytes causes lymph node and spleen enlargement. Occasionally, the spleen ruptures—this is the main cause of the rare deaths caused by infectious mononucleosis and is most likely to occur within 3 to 4 weeks of the start of illness.

T cells respond actively to the infection and destroy B cells with productive infections; they do this because these B cells display viral antigens on their surfaces. The abnormal-appearing lymphocytes seen in smears of the patient's blood are effector cytotoxic T cells responding to the infected B cells (**figure 27.8**).

The possibility that EBV plays a role in causing certain malignant tumors (cancers) has been intensely investigated. The EBV genome is detectable in almost all cases of Burkitt's lymphoma and nasopharyngeal carcinoma, but evidence suggests that other factors are also involved in these cancers. EBV may also be a factor in some malignancies in patients with immunodeficiency from AIDS or organ transplantation.

Epidemiology

EBV is distributed worldwide. It infects individuals in crowded, economically disadvantaged groups at an early age without causing significant illness or symptoms of infectious mononucleosis. In more affluent populations, fewer people get infected with EBV, so most lack immunity to it. Adolescents and adults who lack antibody to the virus may get infectious mononucleosis when they are exposed to it later in life, such as when they enter college. Even then, in the age group of 15 to

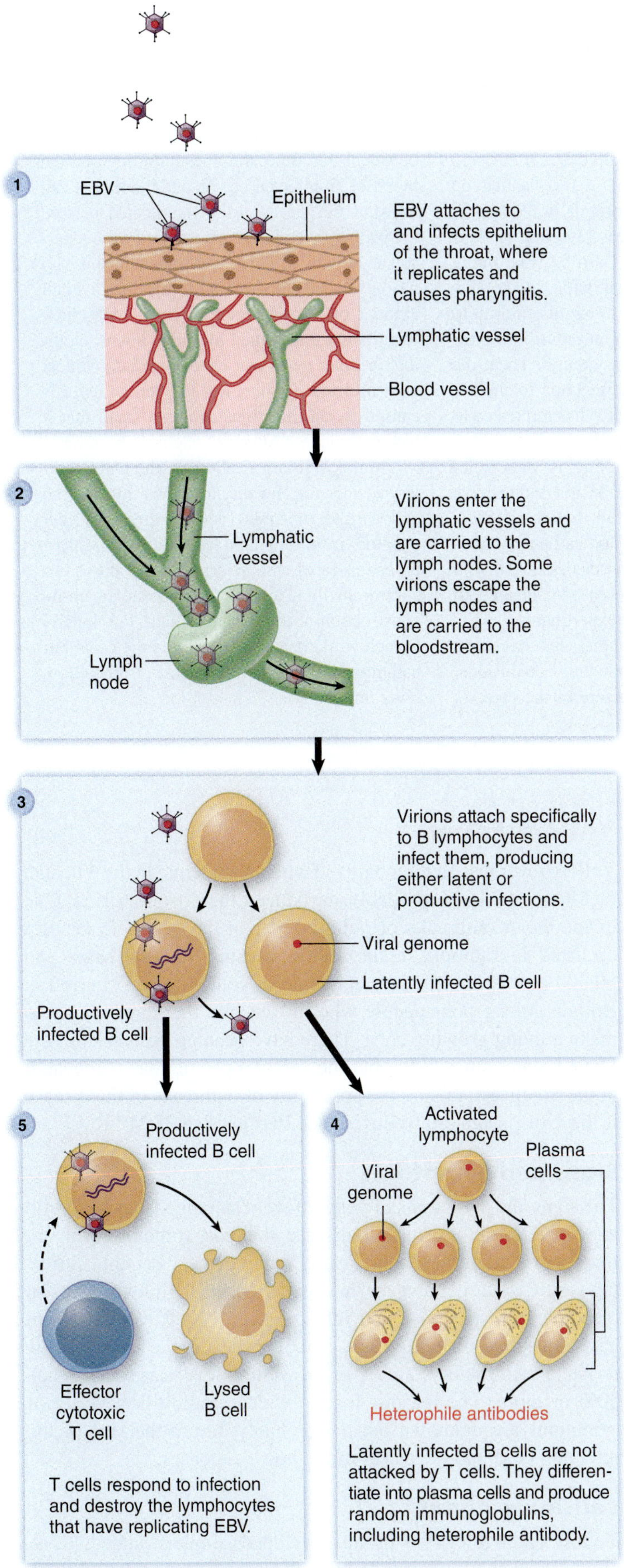

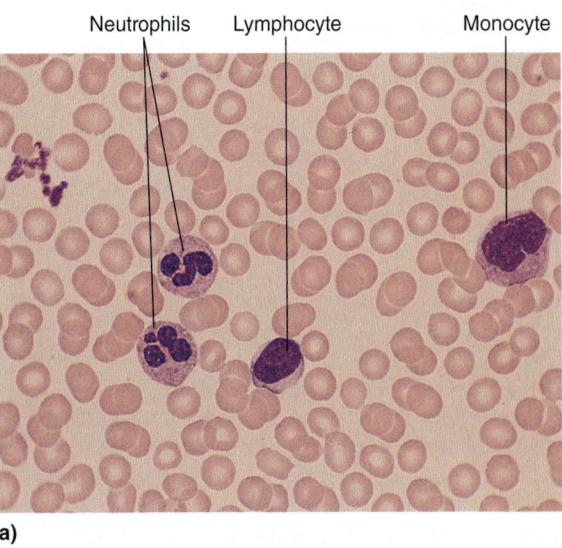

1
EBV
Epithelium
EBV attaches to and infects epithelium of the throat, where it replicates and causes pharyngitis.
Lymphatic vessel
Blood vessel
2
Lymphatic vessel
Lymph node
Virions enter the lymphatic vessels and are carried to the lymph nodes. Some virions escape the lymph nodes and are carried to the bloodstream.
3
Virions attach specifically to B lymphocytes and infect them, producing either latent or productive infections.
Viral genome
Latently infected B cell
Productively infected B cell
5
Productively infected B cell
Effector cytotoxic T cell
Lysed B cell
T cells respond to infection and destroy the lymphocytes that have replicating EBV.
4
Activated lymphocyte
Plasma cells
Viral genome
Heterophile antibodies
Latently infected B cells are not attacked by T cells. They differentiate into plasma cells and produce random immunoglobulins, including heterophile antibody.

FIGURE 27.7 Pathogenesis of Infectious Mononucleosis

? *Why is infectious mononucleosis also called "kissing disease"?*

Neutrophils
Lymphocyte
Monocyte
(a)
Abnormal lymphocyte
(b)

FIGURE 27.8 Normal and Infectious Mononucleosis Blood Smears Pictures of **(a)** a normal blood smear and **(b)** an infectious mononucleosis blood smear showing an abnormal lymphocyte.

? *Why is this disease sometimes misdiagnosed as leukemia?*

24 years, only about half of the EBV infections are symptomatic; the rest have few or no signs or symptoms.

EBV is present in the saliva for up to 18 months after infectious mononucleosis and it occurs intermittently for life. Continuous shedding of virus in saliva is common in people with AIDS or other immunodeficiencies. Mouth-to-mouth kissing is an important mode of transmission in young adults, leading to the name "kissing disease." The donor of the virus is usually asymptomatic and may have been infected in the past without developing symptoms of infectious mononucleosis. By middle age, most people have antibody to the virus, indicating past infection. There is no animal reservoir.

Treatment and Prevention

Antiviral medications such as acyclovir and famciclovir inhibit productive infection by the virus and are helpful in rare serious

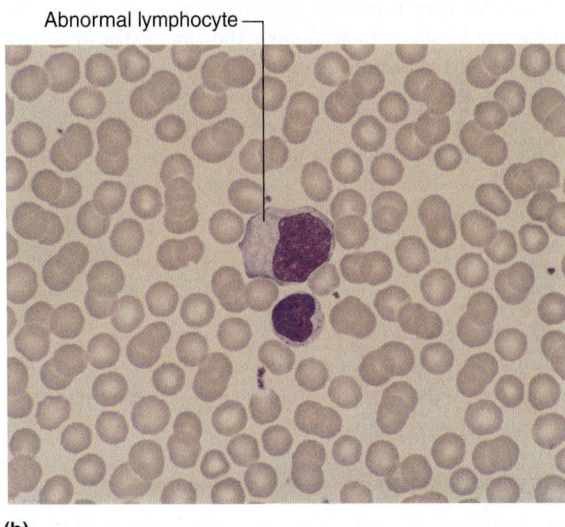

PERSPECTIVE 27.2

Walter Reed and Yellow Fever

Walter Reed was born in Virginia in 1851. After receiving two medical degrees, he entered the Army Medical Corps. He was appointed professor of bacteriology at the Army Medical College in 1893, and in 1900, Reed was named president of the Yellow Fever Commission of the U.S. Army. At that time, the incidence of yellow fever was high and its cause and transmission were unknown.

Reed suspected that an insect was the vector of yellow fever. As early as 1881, Dr. Carlos Finlay of Havana, Cuba, had suggested that a mosquito might be the carrier of this disease, but he was unable to prove his hypothesis. Reed and others in his commission proceeded with a series of experiments to try and prove mosquito transmission. They let laboratory-raised mosquitoes feed on patients with yellow fever and then on members of the commission. Initially, their experiment seemed to have failed, because no one got ill. However, 3 days after an experimental mosquito bite, one member of the commission,

Dr. James Carroll, developed classic yellow fever. Dr. Jesse Lazear, another commission member, noted that the yellow fever patient on which the mosquito had originally fed was in his second day of the disease and that 12 days had elapsed before it bit Carroll. This timing was critical for the transmission of the disease because, as we now know, the mosquito can contract the infection only during a brief period when the patient is viremic. Then, the mosquito can transmit the infection to another individual only after the virus has replicated to a high level in the mosquito.

Lazear was later working in a hospital yellow fever ward where he got bitten by a mosquito, developed yellow fever and died. The yellow fever commission had a mosquito-proof testing facility constructed, called Camp Lazear in honor of their deceased colleague, to house volunteers for their experiments.

As a result of their studies, Dr. Reed and his colleagues made the following conclusions: (1) mosquitoes are the vectors of the disease;

(2) an interval of about 12 days must elapse between the time the mosquito ingests the blood of an infected person and the time it can transmit the disease to an uninfected person; (3) yellow fever can be transmitted from a person acutely ill with the disease to a person who has never had the disease by injecting a small amount of the ill person's blood; (4) yellow fever is not transmitted by soiled linens, clothing, or other items that have come into contact with infected persons; and (5) yellow fever is caused by an infectious agent so small that it passes through a filter that excludes bacteria.

Major William C. Gorgas, the chief sanitary officer for Havana, used this information and started mosquito control measures that resulted in a dramatic reduction in yellow fever cases. Later, Gorgas used mosquito control in the Panama Canal Zone to allow construction of the Panama Canal. Previously, French workers had tried to build the canal but had failed because of heavy losses from yellow fever and malaria.

---

cases; however, they have no activity against the latent infection. Cortisone-like medication is sometimes useful in relieving airway obstruction from swollen tissue.

Infectious mononucleosis can be prevented by avoiding the saliva of another person. This includes avoiding their toothbrushes, drinking glasses, and any other object that may be contaminated with saliva. There is no vaccine for this disease. **Table 27.6** gives the main features of infectious mononucleosis.

TABLE 27.6	Infectious Mononucleosis ("Mono" or "Kissing Disease")
Signs and symptoms	Fatigue, fever, sore throat, and enlargement of lymph nodes
Incubation period	Usually 1 to 2 months
Causative agent	Epstein-Barr virus (EBV), a DNA virus of the herpesvirus family
Pathogenesis	Productive infection of epithelial cells of throat and salivary ducts; latent infection of B lymphocytes; hemorrhage from enlarged spleen is a rare but serious complication.
Epidemiology	Spread by saliva; lifelong recurrent shedding of virus into saliva of asymptomatic, latently infected individuals.

Yellow Fever

Yellow fever was first recognized with an epidemic in the Yucatan of Mexico in 1648, probably introduced there from Africa. One of the worst outbreaks of yellow fever in the twentieth century occurred in Ethiopia in the 1960s, causing 100,000 cases and 30,000 deaths. In 1989, an epidemic of yellow fever occurred in Bolivia among poor people who moved into the jungle to try to make a living growing coca. There have been no outbreaks in the United States since 1905, but the vector mosquito has reappeared in the Southeast, raising the possibility of outbreaks of the disease if the virus is again introduced (see **Perspective 27.2**).

Signs and Symptoms

The signs and symptoms of yellow fever can range from very mild to severe. Symptoms of mild disease, the most common form, are fever and a slight headache lasting a day or two. Patients suffering severe disease, however, may experience a high fever, nausea, bleeding from the nose and into the skin, "black vomit" (from gastrointestinal bleeding), and jaundice (hence, the name *yellow fever*). The mortality rate of severe yellow fever cases can reach 50% or more. The reasons for the wide variation in severity of symptoms are unknown but are probably due to the size of the infecting dose and the status of the host's defenses.

Causative Agent

Yellow fever is caused by an enveloped, single-stranded, RNA arbovirus of the flavivirus family. The virus is transmitted by the bite of infected *Aedes aegypti* mosquitoes, its biological vector.

◀◀ arbovirus, p. 641

Pathogenesis

The yellow fever virus is transmitted by the bite of an infected mosquito. It multiplies in local host cells, enters the bloodstream, and is carried to the liver and other parts of the body. Liver damage results in jaundice and decreased production of clotting proteins. Injury to small blood vessels produces **petechiae**—tiny hemorrhages—throughout the body. Viral replication—as well as the immune response to the infection—damages the circulatory system, leading to bleeding in various tissues, and causing disseminated intravascular coagulation (DIC). Kidney failure is a common consequence of loss of circulating blood and low blood pressure. ◀◀ biological vector, p. 442 ◀◀ DIC, p. 674

Epidemiology

The reservoir of the yellow fever virus is mainly infected mosquitoes and primates living in the tropical jungles of Central and South America and in Africa. Periodically, the disease spreads from the jungle reservoir to urban areas, where it is transmitted to humans by *Aedes* mosquitoes.

Treatment and Prevention

There is no proven antiviral treatment for yellow fever.

Yellow fever can be prevented in urban areas by spraying insecticides and eliminating the breeding sites of its main vector, *Aedes aegypti*. In the jungle, yellow fever control is almost impossible because the mosquito vectors live in the forest canopy and transmit the disease among monkeys. A highly effective attenuated vaccine is available to immunize people who might become exposed, including travelers to endemic areas. The main features of yellow fever are presented in **table 27.7**. ◀◀ attenuated vaccine, p. 423

TABLE 27.7	Yellow Fever
Signs and symptoms	Often only headache and fever. Severe cases characterized by high fever, jaundice, black vomit, and hemorrhages into the skin.
Incubation period	Usually 3 to 6 days
Causative agent	Yellow fever virus, an enveloped, single-stranded RNA virus of the flavivirus family
Pathogenesis	Virus multiplies locally at site of introduction by an infected mosquito; spreads to the liver and throughout the body by the bloodstream. Virus destroys liver cells, causing jaundice and decreased production of blood-clotting proteins. Hemorrhages and decreased strength of the heart result in circulatory failure and kidney failure.
Epidemiology	Virus persists in forest primates and the mosquitoes that feed on them, in Africa and Central and South America; human epidemics occur when the virus infects household mosquitoes that feed on humans.
Treatment and prevention	No proven antiviral therapy is available. There is a highly effective attenuated viral vaccine.

Dengue Fever

Dengue fever is a mosquito-borne viral disease similar to but milder than yellow fever. The disease is also known as "breakbone fever" because it causes severe joint and muscle pain. The disease is found in many tropical areas and is now endemic in more than 100 countries worldwide. It is the most common vector-borne viral disease in the world—the World Health Organization estimates that there are 50 million new cases every year and that 2.5 billion people are at risk of contracting this disease.

MicroByte

Recently, 28 cases of dengue fever were reported in the Florida Keys, a subtropical area of the United States.

Signs and Symptoms

Many cases of dengue fever are asymptomatic or subclinical. If signs and symptoms occur, they begin with a sudden onset of fever that lasts for 2 to 7 days but may recur 12 to 24 hours later (biphasic fever). Other signs and symptoms include headache, joint and muscle pain, pain behind the eyes and a maculopapular rash that may later become hemorrhagic, with development of petechiae. In mild cases, the rash does not develop, often resulting in a misdiagnosis of influenza or other viral disease. Mild mucosal bleeding may occur, typically involving the nose or gums. Patients may also complain of sore throat. In some cases, abdominal pain, vomiting, and diarrhea occur. ◀◀ maculopapular rash, p. 674

Dengue fever is generally self-limiting and is rarely fatal. However, a severe form of the disease—dengue hemorrhagic fever (DHF)—can occur in patients who experience a second dengue infection. DHF may lead to dengue shock syndrome (DSS), which causes a weak, rapid pulse, moist skin, restlessness, and low blood pressure. Petechiae, easy bruising and sometimes massive gastrointestinal and vaginal bleeding occur. Later, widespread blood clotting and DIC may develop. DHF is mostly found in children under 15 years old and can be fatal in this population. ◀◀ DIC, p. 674

Causative Agent

Dengue fever is caused by a single-stranded RNA virus of the flavivirus family. There are four closely related serotypes of the virus (DENV1, DENV2, DENV3, and DENV4), which are transmitted by the bite of an infected mosquito of the *Aedes* family. The most common vector is *Aedes aegypti*, the same species that carries yellow fever. Another important vector of dengue viruses is *Aedes albopictus*, the Asian tiger mosquito, now spreading worldwide. This vector also transmits the denguelike viral disease chikungunya (discussed later). Both of these mosquito species feed only during the day.

Pathogenesis

When an infected mosquito feeds on a human host, dengue viruses are injected into the bite wound, where they infect keratinocytes and epidermal dendritic cells. Some of these cells migrate to the lymph nodes, where monocytes and macrophages are recruited and also become infected. Dengue viruses multiply in the macrophages and are then disseminated systemically in these cells.

The patients were two American boys, 7 and 9 years old, living in Thailand, who developed irritated eyes and slightly runny noses, progressing to fever, headache, and severe muscle pain. It hurt them to move their eyes, and they refused to walk because of pain in their legs. The symptoms subsided after a couple of days, only to recur at lesser intensity. No treatment was given, and they were completely back to normal within a week. Their illness was diagnosed as **dengue** (pronounced DEN-gay), also known as "breakbone fever."

1. What causes dengue and where does it occur?
2. Why is dengue important?
3. What is known about the pathogenesis of dengue?
4. What can be done to prevent dengue?

Discussion

1. Dengue is caused by any of four closely related flaviviruses—DENV1, DENV2, DENV3, and DENV4. The disease occurs in large areas of the tropics and subtropics around the world. In the Americas, the disease is transmitted mainly by *Aedes aegypti* mosquitoes, which are reappearing, after being almost eradicated in the 1960s. This vector now occurs year-round along the Gulf of Mexico. Serious dengue epidemics occurred in Brazil and Cuba in 2002.
2. The two individuals presented here had a mild form of the disease that characteristically involves newcomers to an area where the disease commonly occurs. In endemic areas, the disease is characterized by fever, headache, muscle aches, rash, nausea, and vomiting. In a small percentage of cases, however, the effects of dengue infection are much more serious, resulting in dengue hemorrhagic fever. This form of the disease is characterized by bleeding and leakage of fluid from the capillaries. An important result is that the blood pressure drops and the blood thickens. With expert treatment, the mortality is about 1% to 2%. Dengue shock syndrome is another potential development. It is characterized by profound shock and disseminated intravascular coagulation (DIC) and has a mortality above 40%.
3. About 90% of the cases of hemorrhagic fever or shock occur in people who have previously been infected by a dengue virus and have antibody to it. The remaining cases occur largely in infants who still have transplacentally acquired maternal antibody against dengue viruses. These antibodies, whether from earlier infection or from the mother, attach to the infecting dengue virus strain and promote its uptake by macrophages. The virus, instead of being killed, reproduces in the macrophage. This results in the death of the macrophage and release of chemicals that cause leaky capillaries and shock. T lymphocytes may also play a role in this process in older children and adults. Evidence also exists that the virus causes a depression of the bone marrow, accounting for the very low white blood cell and platelet counts seen in dengue.
4. Scientists at Mahidol University in Bangkok and other research centers around the world are working on vaccines designed to bring dengue under control. At present, control efforts are largely directed at killing mosquitoes and their larvae and eliminating water containers around houses where the mosquitoes breed. Scientists at Colorado State University have successfully genetically engineered mosquitoes to make their cells incapable of reproducing dengue virus. This was done by using another virus to introduce an antisense segment of the dengue genome into the mosquito cells. This was a dramatic accomplishment, but years of work remain before it can play a role in the control of dengue.

Humoral and cell-mediated immune responses clear the viruses. The details of dengue fever pathogenesis are not well understood because there are no animal models for the disease.

The pathogenesis of DHF and DSS is better understood. DHF and DSS generally occur in people who experience a second dengue infection with a different strain of the virus from that of their first infection. **Antibody-dependent enhancement (ADE)** of infection has been suggested to explain severe dengue disease. In this model, preexisting dengue antibodies from a primary dengue infection recognize the viruses in the second infection and bind to them, forming immune complexes. However, because the preformed antibodies are against a different serotype of the virus, they do not neutralize the viruses. Instead, they facilitate viral entry into cells that express Fc receptors, especially macrophages, where the viruses then freely replicate. ADE thus leads to increased numbers of infected cells and a high viral load.

The infected macrophages produce pro-inflammatory cytokines, resulting in a cytokine storm. This, along with activated complement and cytokines released by memory T cells, causes changes in vascular permeability, leading to plasma leakage, that may result in respiratory distress and dehydration. Capillaries become fragile, resulting in the observed hemorrhages (petechiae, easy bruising, and gastrointestinal and vaginal bleeding). Infected macrophages die and release toxic products that cause blood clotting and DIC. Leukopenia (decrease in white blood cell number) occurs. Blood pressure drops, sometimes to life-threatening levels.

Epidemiology

Dengue is an emerging disease and the fastest spreading mosquito-borne viral disease in the world. It is found in mostly tropical and subtropical regions that have high rainfall and plenty of fresh water, conditions that aid in disease development because the mosquito vector breeds in standing water. These regions include Southeast Asia, western Pacific, eastern Mediterranean, the Americas, and rural areas in parts of Africa and the Caribbean. Geographical spread of the disease is linked to international travel, breakdown of vector control measures, and increase in mosquito-breeding habitats—such as used tires and water containers, in which water accumulates.

Treatment and Prevention

There is no specific treatment for dengue fever. Patients are given analgesics (pain relievers). They should avoid aspirin and nonsteroidal anti-inflammatory drugs such as ibuprofen because these medications can worsen bleeding associated with dengue. Aspirin may also be associated with Reye's syndrome. Oral rehydration therapy is given to replace fluid lost from fever and vomiting. Platelet or blood transfusions are given for DHF if bleeding and plasma leakage have occurred. Efforts are being made to develop antiviral medications for dengue. ◄◄ **Reye's syndrome, p. 535** ◄◄ **oral rehydration therapy, p. 571**

Prevention and control of dengue and DHF have become more important with their increasing geographic distribution and incidence. Measures include controlling or eradicating the mosquito population by destroying their breeding sites (standing water). Environmental insecticide use is also important, although increasing insecticide resistance is becoming a problem. Biological control of the vectors (for example, using fish which eat the mosquito larvae), has been successful in some countries. Other prevention measures include educating people in the use of insecticide-treated mosquito bed nets and insect repellents.

Significant effort has been made in developing a dengue vaccine—several live attenuated vaccines are in advanced stages of development. Vaccine development is hampered because creating a vaccine against all four viral serotypes (tetravalent vaccine) is difficult, and there is no animal model for the disease. The main characteristics of dengue fever are presented in **table 27.8.**

Chikungunya

Chikungunya, commonly known as CHIK, got its name from an African word meaning "that which bends up" because people with the disease show bent posture due to severe joint pain. Although CHIK is not a new disease, there have been several recent large outbreaks of it, making it an emerging disease.

The disease is caused by Chikungunya virus, an alphavirus of the *Togaviridae* family. It is transmitted by *Aedes* mosquito species, mostly *A. aegypti* and *A. albopictus*. The signs and symptoms are similar to those of dengue and include fever that typically lasts 2 to 5 days, followed by severe joint pain (especially in the extremities—fingers, toes, wrists, and ankles) that can persist for weeks or months. Sometimes patients develop chronic joint problems—this persistent joint pain differentiates CHIK from dengue fever. Patients often develop a macular or petechial rash. Nonspecific symptoms include headache, conjunctivitis and photophobia, back pain, nausea, and general malaise. The disease is seldom fatal.

CHIK occurs mainly in Africa, Asia, and Southeast Asia, but recent epidemics have been reported in several other areas, associated with returning travelers who inadvertently introduce *A. albopictus* to the affected regions. This vector is now present in Brazil, Central America, the United States, and many European countries, where it has adapted to favorable environmental conditions. In 2007, the first European endemic outbreak of CHIK occurred—Chikungunya virus was transmitted to *A. albopictus* in a northeastern region of Italy by a man returning from India, leading to an epidemic that affected over 200 people.

There is no specific treatment for CHIK. Analgesics and non-steroidal anti-inflammatory drugs such as ibuprofen are used to reduce the joint pain. Fluids are given to reduce dehydration from fever. Chloroquine is increasingly being used successfully to treat chronic cases of the disease. As with most mosquito-borne diseases, CHIK can be prevented by effective vector control—destroying the vector and protecting the population by use of insect repellents and insecticide-impregnated mosquito netting. A vaccine for CHIK is currently in clinical trials.

Emerging Hemorrhagic Fevers

Several viral hemorrhagic fevers are considered emerging. These include Ebola and Marburg hemorrhagic fevers, among others.

Ebola Hemorrhagic Fever

Ebola hemorrhagic fever (EHF) is caused by the Ebola virus, named after the Ebola River Valley in the Democratic Republic of the Congo (in Africa). There are five different species of Ebola virus, four of which cause disease in humans. Ebola virus is one of two virus genera belonging to the *Filoviridae* family.

The Ebola virus causes severe, often fatal disease. Signs and symptoms include rapid onset of fever, headache, abdominal pain, joint and muscle pain, sore throat, macular rash, and bleeding. The virus causes capillaries to become leaky and disrupts clotting—patients bleed from mucous membranes and body orifices; bloody vomit and diarrhea are common. Bleeding is extensive and occurs both internally and externally. The massive bleeding is fatal in 50% to 80% of cases; death is by multiorgan failure or shock.

EHF is found in various African countries. Social and economic conditions often favor the development of epidemics in these countries. Ebola virus has been reported in the United States and England, associated with import of monkeys carrying the virus. No fatal cases have occurred in these countries.

TABLE 27.8	Dengue Fever
Signs and symptoms	Often asymptomatic. Fever, headache, maculopapular rash and severe joint pain. In dengue hemorrhagic fever (DHF), bleeding and shock can occur, as well as disseminated intravascular coagulation (DIC).
Incubation period	Usually 4 to 7 days
Causative agent	Dengue viruses DENV1, DENV2, DENV3, and DENV4—RNA viruses of the flavivirus family
Pathogenesis	Dengue viruses multiply in macrophages and are disseminated throughout the body. Infected macrophages produce pro-inflammatory cytokines, leading to leaky blood vessels and hemorrhaging. In DHF, antibody-dependent enhancement caused by reinfection is thought to enhance viral entry into macrophages, increasing viral load. Macrophages die, causing blood clotting and DIC, which may be fatal.
Epidemiology	Viruses transmitted by mosquitoes of the *Aedes* family, most commonly *Aedes aegypti* but also *A. albopictus*. Found predominantly in tropical and subtropical regions with high rainfall but increasing in geographical spread. DHF usually occurs in children under 15 years old.
Treatment and prevention	No specific treatment. Analgesics for pain; oral rehydration therapy and blood or platelet transfusions if bleeding occurs. Prevention: controlling the vector population. Vaccine development is underway.

There is no standard treatment for EHF. Patients are kept hydrated, and blood and O$_2$ levels maintained. Plasma containing coagulation factors is given to control bleeding. Prevention of EHF is difficult because the natural reservoir is unknown. It is suspected that fruit bats carry the virus. Ebola virus is transmitted to humans from bats and other humans by contact with body fluids. It has been found in other animal species, including gorillas and chimpanzees, but the high mortality in these infected animals makes it unlikely that they are reservoirs.

Once EHF occurs in a community, quick diagnosis and rapid response by healthcare workers is essential to prevent spread to other individuals. Healthcare workers treating infected people should be stringent in their use of protective gear and infection-control measures. A vaccine is in clinical trials.

Marburg Hemorrhagic Fever

Marburg hemorrhagic fever (MHF) is caused by the Marburg virus, the second genus belonging to the family *Filoviridae*. The virus was named for the German town of Marburg when several people died from MHF after handling infected African green monkeys. The signs and symptoms of MHF are similar to those of EHF and include fever, rash, abdominal pain, and nausea. However, bleeding from the orifices is rare. Jaundice and liver failure may occur. Patients may be misdiagnosed because the disease symptoms resemble those of diseases such as malaria or typhoid fever, as well as EHF. MHF has a fatality rate of up to 25%.

MHF is found in Africa. Cases in Europe and the United States have been associated with either handling infected animals or international travel. Treatment of MHF is comparable to that of EHF. Patients receive fluids and supportive therapy, as well as plasma replacement and O$_2$. Prevention is difficult because little is known about the natural reservoir and transmission, but it follows the guidelines given for EHF. Like EHF, the natural reservoir is thought to be fruit bats. Transmission among humans is by body fluids and possibly aerosol droplet. Both Ebola virus and Marburg virus are considered bioterrorism threats, and work is being carried out to develop vaccines against them.

MicroAssessment 27.4

Infectious mononucleosis occurs worldwide and is transmitted from person to person by saliva. Yellow fever virus is transmitted to humans by the mosquitoes of the *Aedes* genus. Dengue fever (breakbone fever) is the most common vector-borne disease in the world. It is also transmitted to humans by *Aedes* mosquitoes and is similar to but milder than yellow fever. A severe form of the disease, dengue hemorrhagic fever, can be fatal. Chikungunya is a newly emerging disease, recently reported in Europe. It resembles dengue in symptoms—like dengue, it causes severe joint pain. Ebola and Marburg hemorrhagic fevers are emerging viral diseases. EHF is a severe, often fatal disease that causes death by massive bleeding and multiorgan failure. MHF is similar to but less serious than EHF.

10. *What characteristic changes occur in the blood of patients with infectious mononucleosis?*

11. *Compare Ebola, Marburg, and Chikungunya hemorrhagic fevers in terms of symptoms, transmission, and prevention.*

12. *Why does it take more than a week before a mosquito just infected with yellow fever virus can transmit the disease?*

27.5 ■ Protozoan Diseases

Learning Outcomes

7. *Outline the* Plasmodium vivax *life cycle.*

8. *Discuss why malaria has been so difficult to control.*

Protozoa infect the blood vascular and lymphatic systems of millions of people globally. Malaria is a widespread protozoan disease that is a leading cause of morbidity (illness) and mortality (death) worldwide. Other causative agents of cardiovascular and lymphatic protozoan diseases include *Trypanosoma brucei,* a bloodstream parasite that causes African sleeping sickness in humans, and *T. cruzi,* which causes Chagas' disease (American trypanosomiasis). The latter often manifests as a chronic heart infection. Protozoans of the genus *Leishmania* cause visceral leishmaniasis, with enormous splenic enlargement. There is more information available about leishmaniasis at the text website (**www.mhhe.com/nester7**). ◀◀ protozoa, p. 291

Malaria

Malaria is an ancient disease, as shown by early Chinese and Hindu writings. During the fourth century B.C., the Greeks noticed the association of malaria with swamps and began drainage projects to control the disease. The Italians gave the disease its name, *malaria,* which means "bad air," in the seventeenth century. In 1902, Ronald Ross received a Nobel Prize for demonstrating the life cycle of the protozoan cause of malaria.

Malaria is the most common serious infectious disease worldwide. In 1955, the World Health Organization (WHO) began a program for the worldwide elimination of the disease using insecticides such as DDT to kill the mosquito vector, and diagnosing and treating infected patients. The program was initially successful—52 nations participated, and by 1960, 10 of them had eradicated malaria. Unfortunately, the mosquito vectors began to develop resistance to insecticides, and malaria began to reappear. In 1976, the WHO acknowledged that the eradication program was a failure. Today, over 300 million people are infected with the malarial parasite annually worldwide, with more than 1 million deaths per year. More people are dying of the disease than when the eradication programs first began.

MicroByte

An estimated 300 to 500 million people contract malaria each year, and a child dies of the disease every 40 seconds.

Signs and Symptoms

The first signs and symptoms of malaria are flulike, with fever, headache, and pain in the joints and muscles. These generally begin about 2 weeks after being bitten by an infected mosquito, but in some cases they start many weeks later. After 2 or 3 weeks, the pattern changes and symptoms fall into three phases: (1) cold phase—the patient feels cold and develops shaking chills that can last for as much as an hour; (2) hot phase—the temperature then begins to rise sharply, often reaching 40°C (104°F) or more; and (3) wet phase—the patient's temperature falls, and drenching sweating occurs. This cycle of intense symptoms—chills, fevers, and sweats—is called a **paroxysm.** Except for fatigue, the patient feels well until 24 or 48 hours later, when the next paroxysm occurs.

Causative Agent

Human malaria is caused by protozoa of the genus *Plasmodium,* which are transmitted by infected female mosquito species of the genus *Anopheles.* Five species of *Plasmodium* are involved— *P. vivax, P. falciparum, P. malariae, P. ovale,* and *P. knowlesi* (this last is a simian species that is more frequently being identified as a cause of human malaria now that molecular testing is available). These species differ in microscopic appearance and sometimes in life cycle. They also cause various types of disease, which differ in severity and treatment.

The *Plasmodium* life cycle is complex, consisting of a liver stage (exoerythrocytic stage; *exo-* means "outside of," *-erythrocytic* refers to red blood cells) and a red blood cell (erythrocytic) stage. ① The infectious form of the protozoan injected into the human host is called a **sporozoite (figure 27.9)**. This form is carried by the bloodstream to the liver,

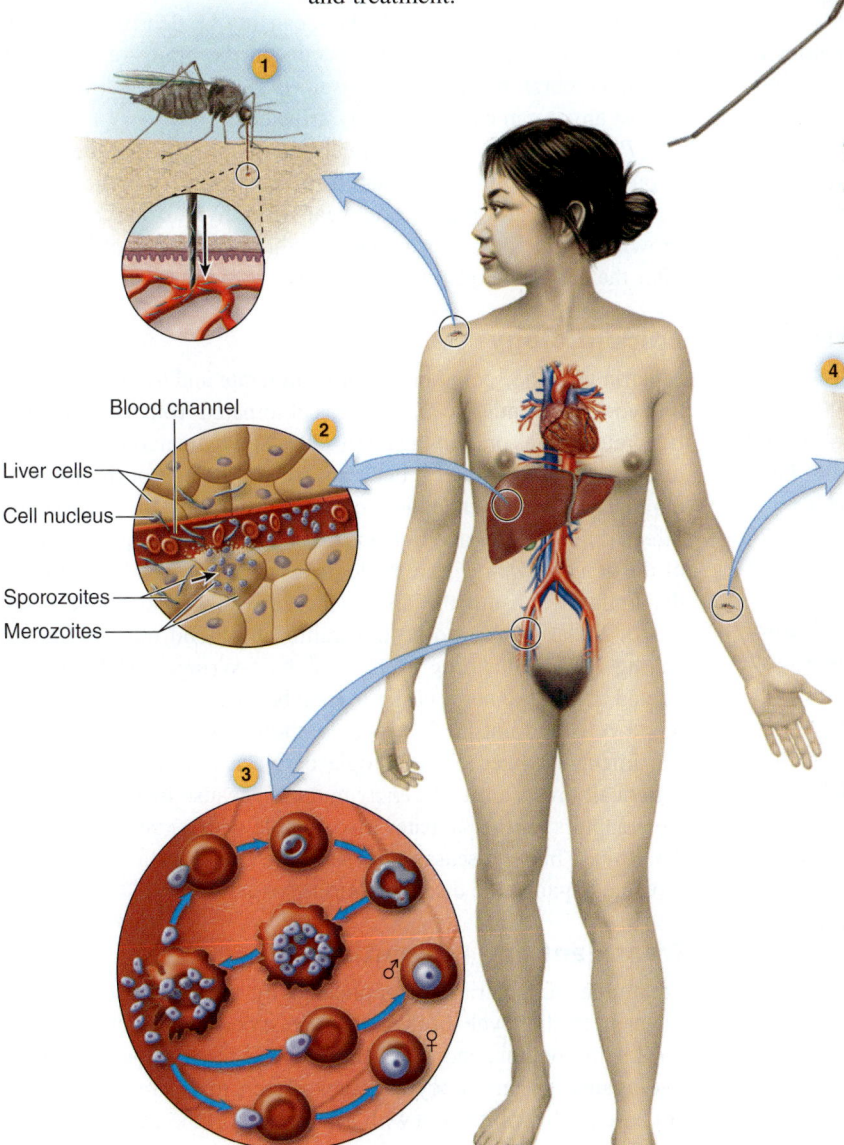

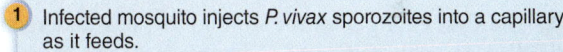

Blood channel

Liver cells

Cell nucleus

Sporozoites

Merozoites

Zygote

Oocyst

① Infected mosquito injects *P. vivax* sporozoites into a capillary as it feeds.

② Sporozoites are carried to the liver, where they multiply in liver cells to form merozoites. The liver cells burst, releasing merozoites into liver blood channels.

③ The merozoites infect and differentiate in red blood cells (RBCs), becoming a ring form, then a trophozoite, then a schizont. The infected RBC breaks open, releasing merozoites. Some merozoites infect new RBCs, repeating this cycle. Others infect new RBCs then differentiate in them, forming male or female gametocytes.

④ Another feeding mosquito ingests RBCs with gametocytes.

⑤ The gametocytes are released as the RBCs are digested.

⑥ The gametocytes become gametes, and fertilization occurs, forming a zygote.

⑦ The zygote becomes motile and penetrates the gut wall.

⑧ In the gut wall, the zygote forms an oocyst and multiplies asexually.

⑨ The oocyst releases sporozoites that infect the mosquito salivary glands.

FIGURE 27.9 Life Cycle of *Plasmodium vivax*

❓ *Why do only female* Anopheles *mosquitoes transmit malaria?*

where it infects hepatocytes (liver cells). **2** There, each parasite enlarges and divides asexually, producing thousands of **merozoites,** which are then released into the bloodstream. Some species of *Plasmodium* are able to form hypnozoites, which can live in the hepatocytes for years before reproducing to form merozoites. The merozoites released from liver cells infect red blood cells (RBCs), beginning the red blood cell stage.

3 In an RBC, the merozoite develops first into a ring form, then into a larger motile **trophozoite** (feeding stage) and then a **schizont** (reproducing stage). The schizont gives rise to merozoites that are released when the RBC ruptures. The merozoites then enter new RBCs and multiply, repeating the cycle.

Some merozoites that enter RBCs develop into **gametocytes** (specialized sexual forms), rather than becoming schizonts. Gametocytes differ from the other forms in both their appearance and susceptibility to antimalarial medications. These sexual forms do not rupture the RBCs. They cannot develop further in the human host and are not important in causing the symptoms of malaria. They are, however, infectious for certain species of *Anopheles* mosquitoes and are thus ultimately responsible for transmitting malaria from one person to another.

4 When a mosquito feeds on an infected person's blood, it ingests the infected RBCs. **5** It digests the RBCs, releasing the gametocytes. Shortly after entering the intestine of the mosquito, and stimulated by the drop in temperature, the male and female gametocytes change in form to become gametes. **6** The male gametocyte transforms into about six tiny, whiplike gametes that swim about until they unite with the female gamete (in much the same way as the sperm and ovum unite in animals), forming a **zygote.**

7 The zygote transforms into a motile form that burrows into the wall of the midgut of the mosquito and forms a cyst. **8** The cyst enlarges as the diploid nucleus undergoes meiosis, dividing asexually into numerous offspring. **9** The cyst ruptures into the body cavity of the mosquito, releasing sporozoites that then find their way to the mosquito's salivary glands and saliva, from which they may be injected into a new human host.

Pathogenesis

The characteristic feature of malaria—recurrent paroxysms followed by feeling healthy again—results from the cycle of growth and release of merozoites from RBCs. Interestingly, the infections in all the millions of RBCs become nearly synchronous. Thus, cell rupture and release of daughter protozoa occur at about the same time for all infected cells, and each release causes a fever.

The spleen characteristically enlarges in malaria to cope with the high levels of foreign material and abnormal RBCs that it removes from the circulation. The spleen may rupture, which can occur with or without trauma. The parasites also cause anemia by destroying RBCs and converting the iron in hemoglobin to a form not readily recycled by the body. The large amount of foreign material in the bloodstream strongly stimulates the immune system. In some cases, the overworked immune system fails and immunodeficiency results.

Generally, people who live in areas where malaria is endemic develop immunity from repeated infections, giving them some protection from the disease. This immunity crosses the placenta and gives partial protection to the newborn. The greatest risk of death from malaria is to children over 6 months of age as this passive immunity decreases.

P. falciparum infections are often very severe, probably because this species can infect all RBCs (leading to high levels of parasitemia), whereas other *Plasmodium* species infect only either young or old RBCs. The *P. falciparum*–infected cells become rigid and stick to each other and to the walls of capillaries. These tiny blood vessels become blocked, causing O_2 deprivation in the tissues they supply. When this occurs in the brain—a condition called cerebral malaria—the result is particularly serious, but almost any organ can be severely affected.

P. vivax and *P. ovale* malaria often relapse because they form hypnozoites (treatment-resistant forms) that survive in the liver in a dormant state. Months or even years later, hypnozoites can begin growing in the liver, starting new erythrocytic cycles of infection after the earlier bloodstream infection has been cured.

Epidemiology

Malaria was once common in both temperate and tropical areas of the world, and endemic malaria was eliminated from the continental United States only in the late 1940s. Today, malaria is mostly a disease of warm climates, but even so, almost half of the world's population lives in endemic areas. The reservoirs for malaria are infected mosquitoes and humans. Human-biting mosquito species of the genus *Anopheles* are biological vectors of malaria, but only the females transmit the disease because males do not feed on blood. Malaria can also be transmitted by blood transfusions or among intravenous drug users who share syringes. Malaria contracted in this manner is easier to treat because it involves only red blood cells and not the liver—only sporozoites from mosquitoes can infect the liver. Some people of black African heritage are genetically resistant to *P. vivax* malaria because their RBCs lack the receptors for the parasite. Also, people with certain genetically determined blood diseases such as sickle cell anemia are partially protected against the disease. ◀◀ biological vector, p. 442

Treatment and Prevention

Treatment of malaria is complicated by the fact that different stages in the life cycle of the parasite respond to different medications. Chloroquine, and chemically related mefloquine, are effective against the erythrocytic stages of sensitive strains, but will not cure the liver infection with *P. vivax* or kill the gametocytes of *P. falciparum*. Primaquine or a newer derivative, tafenoquine, are generally effective against the hypnozoites and the *P. falciparum* gametocytes. These same drugs can be used to prevent malaria.

Malaria prevention has become a global focus. In 1998, a new disease-fighting initiative called Roll Back Malaria was started (involving the WHO, the United Nations Children's Fund [UNICEF], the United Nations Development Program, and the World Bank). The goal was to reduce malaria deaths by 75% by 2015. The program included organizing and funding a sustained

TABLE 27.9 | Malaria

Signs and symptoms	Recurrent cycles of violent chills and fever alternating with feeling healthy
Incubation period	Varies with species; 6 to 37 days
Causative agent	Five species of protozoa of the genus *Plasmodium*
Pathogenesis	Cell burst and release of protozoa causes fever; spleen enlarges in response to removing large amount of foreign material and many abnormal blood cells from the circulation; infected red blood cells stick to each other and to walls of capillaries; vessels block, depriving tissue of O_2.
Epidemiology	Transmitted from person to person by bite of infected *Anopheles* mosquito. Some individuals genetically resistant to infection.
Treatment and prevention	Treatment: usually ACTs; other medicines if sensitivity known. Prevention: chloroquine if in chloroquine-sensitive malarial areas; doxycycline, mefloquine, or other alternative if in chloroquine-resistant areas; after leaving area, primaquine is given for liver stage; ACTs or other medicines for resistant strains; eradication of mosquito vectors; mosquito netting impregnated with insecticide; vaccines under development.

effort to improve access to medical care, strengthening local health facilities, and supplying medications and long-lasting insecticide-treated mosquito nets. Unfortunately, the rate of malaria has continued to increase despite this initiative. This is mostly because chloroquine, the main antimalarial medication used for many years, is no longer effective in many areas of the world (**figure 27.10**). Resistance has also quickly developed to newer medications such as mefloquine.

By 2004, it was generally agreed that for malaria to be controlled, more effective—although much more costly—medications are needed. Derivatives of artemisinin, the active ingredient of an ancient Chinese herbal medicine, are now used in combination with another medication to minimize the risk of developing resistance. With better funding, these combinations, called ACTs (artemisinin-based combination therapies), are now widely available and are helping to reduce the incidence of malaria. A number of potential vaccines are also being tested. The most promising vaccine so far gives only 30% protection to children. Indoor spraying of the insecticide DDT, the use of insecticide-impregnated bed nets, and elimination of mosquito breeding areas are also being emphasized. Adequate funding of malaria control is a continuing problem. **Table 27.9** gives the main features of malaria.

The key features of the diseases covered in this chapter are highlighted in the **Diseases in Review 27.1** table.

MicroAssessment 27.5

There are a number of different stages in the life cycles of the protozoa that cause malaria, and they have differing microscopic appearance, antigenicity, and susceptibility to antimalarial medications. Malaria is a leading serious, infectious disease worldwide, and sustained, well-funded control measures are needed to control it.

13. *Why is malaria contracted from a blood transfusion easier to treat than that contracted from a mosquito?*

14. *What causes the recurrent paroxysms that characterize malaria?*

15. *Are sporozoites diploid or haploid?* ➕

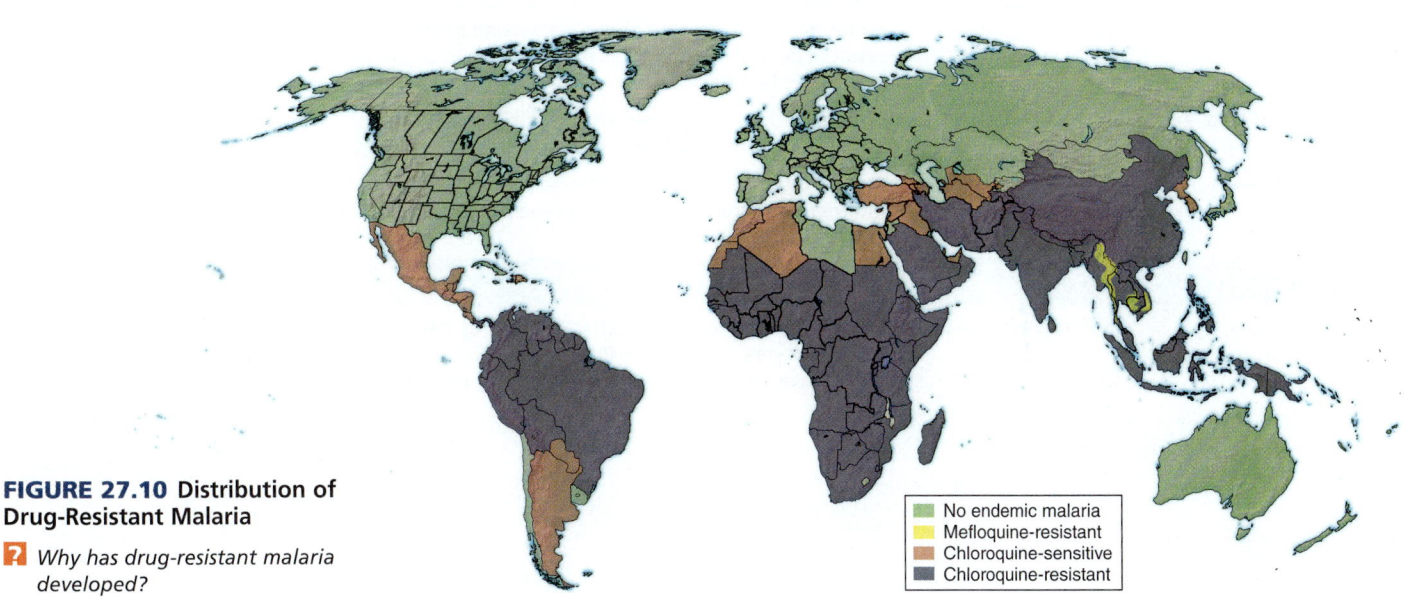

FIGURE 27.10 Distribution of Drug-Resistant Malaria

❓ *Why has drug-resistant malaria developed?*

- ◼ No endemic malaria
- ◼ Mefloquine-resistant
- ◼ Chloroquine-sensitive
- ◼ Chloroquine-resistant

Diseases in Review 27.1

Blood Infections

Disease	Causative Agent	Comment	Summary Table
BACTERIAL INFECTIONS			
Subacute bacterial endocarditis (SBE)	Skin or mouth normal microbiota	Results from organisms colonizing blood clots on abnormal heart valves, causing damage; characterized by progressing fatigue and slight fever; can lead to a stroke.	Table 27.1, p. 673
Sepsis and septic shock	Often *E. coli*, *Bacteroides*, or *Pseudomonas* species	Characterized by fever, shaking chills, and rapid pulse and respiration; due to widespread release of cytokines in response to systemic infection, particularly involving Gram-negative bacteria.	
Tularemia ("rabbit fever")	*Francisella tularensis*	Category A bioterrorism agent. Can be acquired in multiple ways, often by direct contact with an infected animal. Characterized by an ulcer at entry site, fever, and enlarged lymph nodes.	Table 27.2, p. 676
Brucellosis ("undulant fever")	*Brucella melitensis*	Characterized by recurrent fever, weakness, large lymph nodes, and weight loss; acquired via contact with an infected animal or foodborne; causative agent enters via mucous membranes or breaks in the skin.	Table 27.3, p. 677
Plague ("black death")	*Yersinia pestis*	Category A bioterrorism agent; spread by flea bites (bubonic plague) and aerosols (pneumonic plague); endemic in rodent populations.	Table 27.5, p. 679
VIRAL INFECTIONS			
Infectious mononucleosis ("mono")	Epstein-Barr virus (EBV)	Long incubation period, followed by fatigue that can last for months, fever, sore throat, and enlarged lymph nodes; transmitted in saliva; infects B cells, causing both productive and latent infections.	Table 27.6, p. 682
Yellow fever	Yellow fever virus	Transmitted by mosquitoes; disease can be mild or severe; severe form characterized by fever, bleeding, black vomit, and jaundice, and is often fatal. Vaccine available for prevention.	Table 27.7, p. 683
Dengue fever	Dengue virus	Transmitted by mosquitoes; generally mild, although symptoms include severe joint pain; second infection can result in dengue hemorrhagic fever (usually in children), which may be fatal.	Table 27.8, p. 685
Chikungunya (CHIK)	Chikungunya virus	Transmitted by mosquitoes; characterized by fever and severe joint pain, which may become chronic.	
Hemorrhagic fevers	Ebola viruses and Marburg virus	Severe, often fatal diseases characterized by fever and internal and external bleeding; spread person to person by body fluids.	
PROTOZOAN INFECTIONS			
Malaria	*Plasmodium* species	Transmitted by mosquitoes; characterized by paroxysms of chills, fevers, and sweats; *P. falciparum* infections are life-threatening, even in otherwise healthy people.	Table 27.9, p. 689

Rethinking Malaria Control

Previous attempts at malaria control have relied heavily on the use of DDT, a non-biodegradable insecticide. This chemical was banned in the United States in 1972 because of its accumulation in the environment and its damaging effects on birds such as the peregrine falcon and the bald eagle. Many other concerns have been raised about its possible carcinogenicity and potential effects on human fetuses. Although DDT is not generally used in agriculture, more than 20 countries still use it for malaria control, and more would use it if it were affordable. Biodegradable insecticides are available, but they are more expensive than DDT, and are also often toxic. The options for preventing malaria all seem bad, but a choice has to be made of which ones are the least harmful, because malaria is such a problem worldwide. Insecticides may be necessary short term, but other options, including education, economic development, vaccines, biological control of the *Anopheles* vectors, and mosquito habitat alteration, may prove more important in achieving sustained relief from the disease.

The challenge for the future is to better understand the ecology of malaria as it applies in each location, with the aim of minimizing insecticide use and discovering new options for maintaining long-term control.

Summary

27.1 ■ Anatomy, and Physiology, and Ecology (figure 27.1)

The Heart
The left side of the heart receives oxygenated blood from the lungs and pumps it into thick-walled arteries; the right side of the heart receives blood depleted of O_2 from the veins and pumps it through the lungs.

Blood Vessels
Blood vessels include arteries, veins, and capillaries. Arteries have thick, muscular walls. The blood in veins is a dark color because it is depleted of O_2.

Lymphatics (Lymphatic Vessels)
Lymphatics—lymph vessels—take up fluid that leaks from capillaries. They also take up bacteria, which are normally trapped by lymph nodes distributed along the course of the lymphatics.

Spleen
The spleen filters unwanted material such as bacteria and damaged RBCs from the arterial blood. It becomes enlarged in diseases such as infectious mononucleosis and malaria.

27.2 ■ Bacterial Diseases of the Blood Vascular System
Bacteria circulating in the bloodstream can colonize the inside of the heart. These infections can kill a person quickly, as in **acute bacterial endocarditis,** or they can cause a gradual decline, as in **subacute bacterial endocarditis.**

Subacute Bacterial Endocarditis (SBE) (table 27.1)
Subacute bacterial endocarditis (SBE) is commonly caused by members of the normal microbiota, including oral streptococci and *Staphylococcus epidermidis.* Infection usually begins on structural abnormalities of the heart. Prolonged treatment is usually required.

Sepsis and Septic Shock (figure 27.3)
Sepsis, an infection-induced systemic inflammatory response, is commonly a healthcare-associated illness; many afflicted individuals have serious underlying illnesses such as cancer and diabetes. If uncontrolled, it can progress to septic shock. Most fatal cases involve Gram-negative bacteria.

27.3 ■ Bacterial Diseases of the Lymph Nodes and Spleen
Enlargement of the lymph nodes and spleen is a prominent feature of diseases that involve the mononuclear-phagocyte system.

Tularemia ("Rabbit Fever" or "Deer Fly Fever") (table 27.2, figure 27.4)
Tularemia is usually transmitted from wild animals to humans by exposure to the animal's blood or via insects and ticks. The cause is the Gram-negative aerobe *Francisella tularensis,* which is found throughout the United States except Hawaii.

Brucellosis ("Undulant Fever" or "Bang's Disease") (table 27.3)
Brucellosis, caused by *Brucella melitensis,* is usually acquired from cattle or other domestic animals, sometimes from wild animals. Hunters, butchers, and those who drink unpasteurized milk or milk products are at increased risk for the disease. The organisms can infect via mucous membranes and minor skin injuries.

Plague ("Black Death") (table 27.5)
Plague is caused by *Yersinia pestis,* a member of the *Enterobacteriaceae* with many virulence factors that interfere with phagocytosis and immunity (table 27.4, figure 27.5). **Bubonic plague** is transmitted to humans by fleas (figure 27.6). **Pneumonic plague** is transmitted from person to person by aerosols. If the bacteria enter the bloodstream, **septicemic plague** may occur. Untreated, bubonic plague mortality is 50% to 80%; pneumonic plague mortality is almost 100%.

27.4 ■ Viral Diseases of the Lymphatic and Blood Vascular Systems

Infectious Mononucleosis ("Mono" or "Kissing Disease") (table 27.6)
Infectious mononucleosis is caused by the Epstein-Barr virus (EBV), which establishes a lifelong latent infection of B lymphocytes. The disease has a high incidence in young people and can cause exhaustion that lasts for months (figures 27.7, 27.8).

Yellow Fever (table 27.7)
Yellow fever is a zoonosis of mosquitoes and monkeys that exists mainly in tropical jungles; it can become epidemic in humans where a suitable *Aedes* mosquito vector is present. The disease involves the heart and blood vessels throughout the body and is characterized by fever, jaundice, and hemorrhaging. A highly effective attenuated vaccine is available for preventing the disease.

Dengue Fever (table 27.8)
Dengue fever is caused by different viruses that are transmitted by *Aedes* mosquito species. The disease is similar to but milder than yellow fever

and is characterized by joint pain, giving the common name "breakbone fever." A more severe form of the disease, dengue hemorrhagic fever (DHF), may occur in people reinfected with dengue and can be fatal.

Chikungunya

Chikungunya (CHIK) is a newly emerging disease, recently reported in Europe. Like dengue, it causes severe joint pain, which may become chronic.

Emerging Hemorrhagic Fevers

Ebola hemorrhagic fever (EHF) and Marburg hemorrhagic fever (MHF) are emerging viral diseases. EHF is a severe, often fatal disease that causes death by massive bleeding and multiorgan failure. MHF is similar to but less serious than EHF.

27.5 ■ Protozoan Diseases

Malaria is the most widespread of the protozoan blood and lymphatic diseases, and is the most widespread of all serious infectious diseases.

Malaria (table 27.9)

Malaria is caused by five species of *Plasmodium* and is transmitted from person to person by the bite of the females of certain species of *Anopheles* mosquitoes, its biological vector. Now found mainly in warm regions of the world, malaria survives despite massive eradication programs. The life cycle is complex (figure 27.9); different forms of the organism invade different body cells and have different susceptibility to antimalarial medication (figure 27.10). Replication of the organism inside RBCs results in the rupture of all infected cells almost simultaneously and release of the progeny protozoa to infect new red cells.

Review Questions

Short Answer

1. What is the significance of immune complex formation in SBE?
2. What is disseminated intravascular coagulation (DIC)?
3. What activities of humans are likely to expose them to tularemia?
4. Why is brucellosis a threat to big-game hunters?
5. Why might the *Yersinia pestis* from a patient with pneumonic plague be more dangerous than the same organism from fleas?
6. Why might rodent burrows be a source of plague months after they are abandoned?
7. What type of leukocytes does EBV infect?
8. Travelers to and from which areas of the world should have certificates of yellow fever vaccination?
9. Why is a second infection with dengue virus more serious than the first?
10. Which *Plasmodium* species causes the most dangerous form of malaria?

Multiple Choice

1. Which of the following infection fighters are found in lymph?
   a) Leukocytes   b) Antibodies
   c) Complement   d) Interferon
   e) All of the above

2. Which of the following statements about the spleen is *false*?
   a) It is located low on the right side of the abdomen.
   b) It cleanses the blood of foreign material and damaged cells.
   c) It provides an immune response to circulating pathogens.
   d) It can help produce new blood cells.
   e) It enlarges in a number of infectious diseases.

3. Which one of the following statements about SBE is *false*?
   a) It is generally a chronic illness characterized by fatigue and slight fever.
   b) It is usually caused by normal microbiota of the mouth or skin.
   c) Infection typically occurs on the left side of the heart.
   d) Injected-drug abuse can be a risk factor in developing the disease.
   e) It can lead to a stroke.

4. Choose the one *true* statement about sepsis.
   a) It is a rare healthcare-associated disease.
   b) The output of urine increases if shock develops.
   c) It can be caused only by anaerobic bacteria.
   d) An antibiotic that kills the causative organism can be depended on to cure the disease.
   e) Lung damage is an important cause of death.

5. Which of these statements about tularemia is *false*?
   a) It can be contracted from muskrats and bobcats.
   b) Biting insects and ticks can transmit the disease.
   c) The causative organism is closely related to *E. coli.*
   d) A steep-walled ulcer at the site of entry of the bacteria and enlargement of nearby lymph nodes is characteristic.
   e) Without treatment, 9 out of 10 people can be expected to survive.

6. Which of the following statements about brucellosis is *false*?
   a) fevers that come and go over a long period of time gave it the name "undulant fever."
   b) the causative agent can infect via mucous membranes.
   c) the causative agent is readily killed by phagocytes.
   d) the disease in cattle is characterized by chronic infection of the mammary glands and uterus.
   e) butchers are advised to wear goggles or a face shield to help protect against the disease.

7. Which statement about *Yersinia pestis* is *false*?
   a) Growth conditions inside human phagocytes activate virulence genes.
   b) The bacterium can form biofilms in the flea digestive system.
   c) Yops protein increases phagocytosis.
   d) The organism resembles a safety pin in certain stained preparations.
   e) It was responsible for the "black death" in Europe during the 1300s.

8. Which of the following statements about yellow fever is *false*?
   a) There is no animal reservoir.
   b) The name "yellow" comes from the fact that many victims have jaundice.
   c) Certain mosquitoes are biological hosts for the causative agent.
   d) Outbreaks of the disease could occur in the United States because a suitable vector is present.
   e) An attenuated vaccine is widely used to prevent the disease.

9. The malarial form infectious for mosquitoes is called a
   a) gametocyte. b) trophozoite.
   c) sporozoite. d) schizont.
   e) merozoite.

10. Which of the following statements about malaria is *true*?
    a) Transmission cannot occur in temperate climates.
    b) Transmission usually occurs with the bite of a male *Anopheles* mosquito.
    c) The disease is currently well controlled in tropical Africa.
    d) *P. falciparum* infects only old RBCs and therefore causes milder disease than other *Plasmodium* species.
    e) The characteristic recurrent fevers are associated with release of merozoites from RBCs.

Applications

1. Some years ago, dentists and doctors began noticing an association between subacute bacterial endocarditis and prior dental work, and they began advising that an antibiotic be administered at the time of dental procedures to those with known or suspected heart defects. What was the rationale for this advice?

2. A healthcare worker in Honduras is concerned about a potential outbreak of yellow fever in his town. A laborer from a jungle area known to be endemic for the disease had come to the town 2 weeks earlier to work and subsequently developed yellow fever. Several coworkers reported getting mosquito bites while working with him. Why is it important that the healthcare worker determine how long it is since the workers were bitten by the mosquitoes?

Critical Thinking

1. The finding that there is an association between *Chlamydophila pneumoniae* infection and arteriosclerotic lesions raised hopes that new methods to combat arteriosclerosis could be developed. An investigator reviewing this research, however, stated that even a perfect correlation between infection and lesion formation would not prove that infection causes arteriosclerosis. Moreover, even showing that therapeutic antibiotics could prevent infection and lesion formation would not be definitive proof. Is the investigator justified in making this argument? Why or why not?

2. Even though genetically engineered mosquitoes might be developed that do not allow the reproduction of malaria protozoa, these mosquitoes would have little, if any, immediate effect on the spread of the disease. Why should this be so? What would have to happen for these mosquitoes to significantly affect the spread of malaria?

28 HIV Disease and Complications of Immunodeficiency

Color-enhanced TEM of human immunodeficiency virus (HIV).

KEY TERMS

Acute Retroviral Syndrome (ARS) Stage of HIV disease following the incubation stage; often manifests as flulike symptoms.

AIDS (Acquired Immunodeficiency Syndrome) End stage of HIV disease, manifest as severe immunodeficiency.

AIDS-Related Complex (ARC) Group of symptoms (fever, fatigue, diarrhea, and weight loss) that indicate the onset of AIDS.

HAART Highly active antiretroviral therapy; cocktail of medications that interfere with HIV replication.

HIV (Human Immunodeficiency Virus) The causative agent of HIV disease and AIDS.

HIV Set Point Viral load in a person with HIV disease, after the immune system begins to respond to the virus and viral numbers stabilize.

Kaposi's Sarcoma Tumor arising from blood or lymphatic vessels.

***Mycobacterium avium* Complex (MAC)** Group of genetically related bacteria belonging to the genus *Mycobacterium* that causes disease in people with AIDS.

Viral Load Measure of severity of a virus infection; calculated by estimating the concentration of virus particles in involved body fluid.

A Glimpse of History

In 1981, reports were made of an illness characterized by unusual infections, certain malignant tumors, and immunodeficiency in previously healthy, young, homosexual men. This illness came to be known as AIDS (*a*cquired *i*mmuno*d*eficiency *s*yndrome). By 1982, the CDC had convincing evidence that AIDS was caused by a new infectious agent, and scientists around the world rushed to identify it.

To identify the cause of AIDS, the researchers relied on some important scientific advances of the 1960s and 1970s, including the discovery of lymphocyte subsets, the functions of T cells, and the role of cytokines. Another extremely important advance was the discovery of reverse transcriptase by Doctors Howard Temin, David Baltimore, and Renato Dulbecco, who received the Nobel Prize in 1975 for their work. This was followed in early 1978 by the identification of the first human retrovirus, human T-lymphotrophic virus (HTLV), in Dr. Robert Gallo's laboratory at the National Cancer Institute. Dr. Gallo went on to develop a method for cultivating the virus in lymphocytes. ◀◀ lymphocytes, p. 340 ◀◀ cytokines, p. 341 ◀◀ reverse transcriptase, p. 320 ◀◀ retroviruses, p. 320

In January 1983, an important breakthrough occurred in the laboratory of Dr. Luc Montagnier at the Pasteur Institute in Paris. Using the retrovirus culture techniques developed earlier by Gallo, Montagnier and his colleague Françoise Barré-Sinoussi isolated a new virus from a patient with lymphadenopathy syndrome, an AIDS-associated illness. They named the virus LAV, an acronym for *l*ymph*a*denopathy *v*irus. Montagnier could obtain only a small amount of the virus but was able to use it to develop a blood test that showed that all AIDS patients were infected with LAV. He sent a sample of this virus to Gallo. He also applied for a patent on his blood test.

Meanwhile, Gallo's laboratory also recovered a virus from AIDS patients. Gallo was able to obtain large amounts of the virus, which he named HTLV-III because it resembled *h*uman *T-l*ymphotrophic *vi*ruses discovered earlier. Using HTLV-III, Gallo perfected a blood test for diagnosing AIDS and, like Montagnier, applied for a patent. These conflicting patent claims caused a bitter scientific and legal battle that went on for years. A settlement was negotiated in 1987 in which Gallo and Montagnier were named codiscoverers of the test, and both the French and the Americans received royalties from it.

The settlement did not end the conflict. Genetic analysis showed that Gallo's HTLV-III and Montagnier's LAV were actually the same virus. It was eventually discovered that patient samples sent from Montagnier's laboratory contained two different viruses, one of which was the LAV and Gallo's cultures had been accidently contaminated with this virus. In 1994, the earlier settlement was modified.

The American share of the blood test royalties has generated many millions of dollars for the NIH over the years, but the value of the test is greater than any monetary figure. It helped prove that HTLV-III/LAV caused AIDS. The virus was later renamed HIV for *h*uman *i*mmunodeficiency *v*irus. It also showed that many infected people were asymptomatic, that they could transmit the virus, and that the epidemic was far more extensive than previously suspected. These findings helped generate support for controlling the disease.

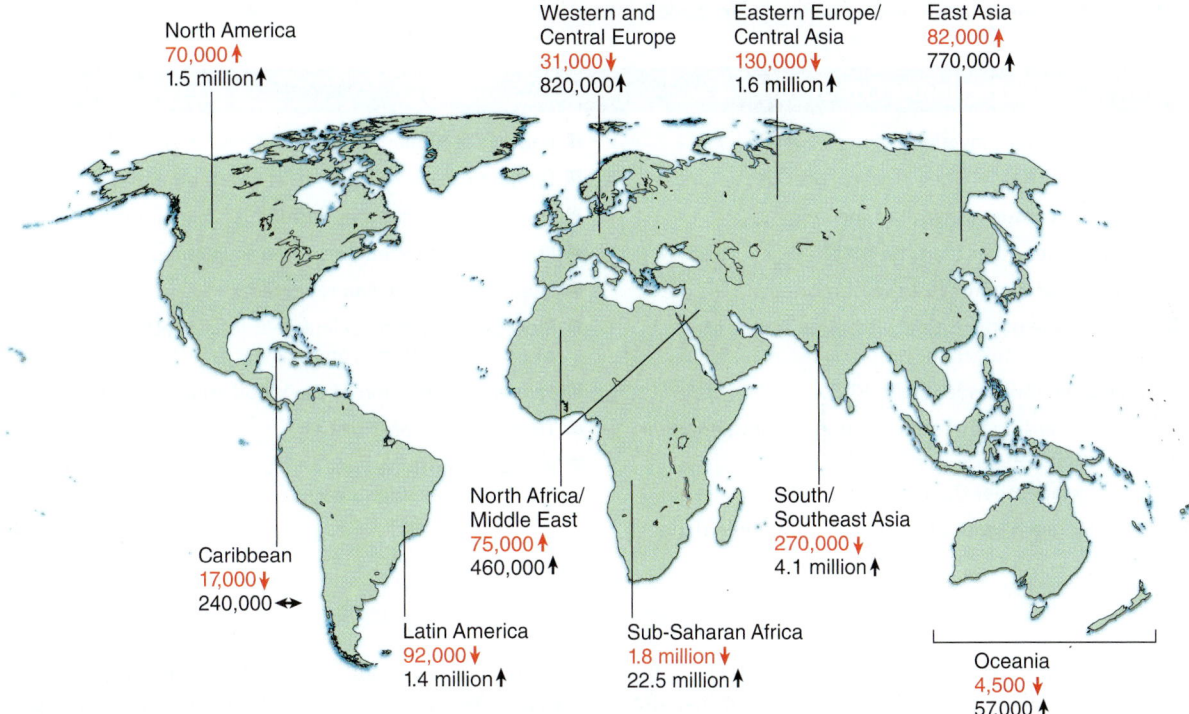

FIGURE 28.1 The Global HIV/AIDS Epidemic at the End of 2009 An estimated 2.6 million people were newly infected during the year (numbers in red). Altogether, 33.3 million people (numbers in black) were living with HIV/AIDS. The arrows indicate an increase (↑) or decrease (↓) in the numbers of cases as compared with 2001. ↔ Shows stable numbers over time period examined.

? *What is one factor that has lead to the decrease in new HIV infection rates in many countries?*

AIDS

In the 20 years after HIV was identified, an enormous number of Americans died from AIDS. By 1994, it had become the leading cause of death among those 25 to 44 years old in the United States. In addition, nearly a million people had contracted HIV, and more than 40,000 were becoming infected every year.

Worldwide in this time period, an estimated 40 million people were living with HIV infection, and more people had died of AIDS than died of the "black death" of Europe during the Middle Ages. It became the primary cause of death in sub-Saharan Africa, and spread rapidly into India and China. Billions of dollars have been spent trying to control the AIDS pandemic, with the result that new HIV infections have declined in many areas (**figure 28.1**). Nevertheless, AIDS is far from conquered. ◄◄ black death, p. 678

28.1 ■ Human Immunodeficiency Virus (HIV) Infection and AIDS

Learning Outcomes

1. *Describe HIV structure and its replication strategy.*
2. *Outline the pathogenesis of HIV disease.*
3. *Discuss the three main ways that HIV is transmitted.*
4. *Describe the treatment and prevention of HIV.*
5. *Describe the prospects for an HIV vaccine.*

AIDS (acquired immunodeficiency syndrome) was first recognized in 1981 after an unusual number of *Pneumocystis* pneumonia cases were identified in previously healthy homosexual and bisexual men in the United States. This type of pneumonia is caused by a fungus now called *Pneumocystis jiroveci,* which infects many people but is of such low virulence that it seldom causes disease in healthy individuals. In 1982, as other groups of people developed unusual opportunistic infections for no apparent reason, the CDC began using the acronym AIDS to describe the unexplained increased susceptibility indicative of underlying immunosuppression. The various opportunistic infections became known as "AIDS-defining conditions"—diseases that indicate a person has AIDS. These were useful in studying the AIDS epidemic before its cause was known, and they still help physicians by alerting them to the possibility that a patient has AIDS (**table 28.1**). ◄◄ immunodeficiency disorders, p. 414

AIDS is caused by the **human immunodeficiency virus (HIV)**. The terms HIV and AIDS are not synonymous, and the distinctions are very important. A person can be infected with HIV but may not be ill. AIDS is only the end stage of a complex disease that has many signs and symptoms that precede immunodeficiency. Three distinct terms are important to keep in mind as you read this chapter:

- **HIV infection** indicates that the virus has entered the body and is replicating, whether or not signs and symptoms are present.

- **HIV disease** indicates that the infection is causing signs and symptoms.

TABLE 28.1	**"AIDS-Defining Conditions"**
■ Cancer of the uterine cervix, invasive	■ Kaposi's sarcoma
■ Candidiasis involving the esophagus, trachea, bronchi, or lungs	■ Lymphomas, such as Burkitt's, or arising in the brain
■ Coccidioidomycosis, of tissues other than the lung	■ Mycobacterial diseases, including tuberculosis
■ Cryptococcosis, of tissues other than the lung	■ Pneumocystis (pneumonia due to *Pneumocystis jiroveci*)
■ Cryptosporidiosis of duration greater than 1 month	■ Pneumonias occurring repeatedly
■ Cytomegalovirus disease of the retina with vision loss or other involvement outside liver, spleen, or lymph nodes	■ Progressive multifocal leukoencephalopathy (a brain disease caused by the JC polyomavirus)
■ Encephalopathy (brain involvement with HIV)	■ *Salmonella* infection of the bloodstream, recurrent
■ Herpes simplex virus causing ulcerations lasting a month or longer or involving the esophagus, bronchi, or lungs	■ Toxoplasmosis of the brain
■ Histoplasmosis of tissues other than the lung	■ Wasting syndrome (weight loss of more than 10% due to HIV); also known as slim disease
■ Isosporiasis (a protozoan disease of the intestine) of more than 1 month's duration	

■ **AIDS** is the last stage of HIV disease, defined by CD4+ lymphocyte (helper T cell) levels and the presence of specific opportunistic infections.

HIV Disease

Almost everyone who becomes infected with HIV develops HIV disease, which slowly destroys the person's immune system and eventually ends in AIDS unless the individual is successfully treated with antiretroviral drugs.

Signs and Symptoms

The first signs and symptoms of HIV disease appear 6 days to 6 weeks after initial infection and are similar to many viral infections. They are referred to as "flulike" and include fever, headache, sore throat, muscle aches, and enlarged lymph nodes. Some people develop a generalized rash and central nervous system symptoms, characterized by moodiness and sometimes seizures and paralysis. Signs and symptoms of this early stage of HIV disease, called the **acute retroviral syndrome (ARS)**, coincide with high levels of virus replication, and typically subside within 4 to 6 weeks. Some infected people either never develop ARS or do not notice the signs and symptoms because they are so mild.

After the early stage of infection, virus levels drop and the disease enters an asymptomatic period that typically lasts for years. This is often referred to as **clinical latency**. HIV continues to replicate during this time period, however, so the virus still causes cell damage and can be transmitted to others. Clinical latency ends as the infected person becomes increasingly immunodeficient due to declining helper T cell numbers. With severe immunodeficiency, the individual contracts various opportunistic bacterial, viral, protozoan, and fungal infections that characterize AIDS. As one might expect, the signs and symptoms at this stage of the disease vary widely according to the kinds of infection. For example, a common sign is a fuzzy white patch on the tongue (hairy leukoplakia) which is a result of latent Epstein-Barr virus (EBV) reactivation (**figure 28.2**). Most people with AIDS suffer fever, weight loss, fatigue, and diarrhea, referred to as the **AIDS-related complex (ARC).** The person may also suffer persistent enlargement of the

lymph nodes, a condition known as **lymphadenopathy syndrome (LAS)**. Without treatment, AIDS is fatal. ◄◄ **helper T cell, p. 356**
◄◄ **opportunistic infections, p. 382** ◄◄ **Epstein-Barr virus, p. 680**

Causative Agent

In the United States, and in most other parts of the world, AIDS is typically caused by human immunodeficiency virus type 1 (HIV-1), a retrovirus belonging to the lentivirus subgroup of the retrovirus family (see **Perspective 28.1**).

HIV-1 variants can be classified into subtypes based on their genomes. Most belong to the group M (for "major") group. Other groups include O ("outlier"), N ("non-M and non-O") and P ("pending identification of further human cases"). The M group, which is responsible for the current AIDS pandemic, is subdivided into genetically related clades (subtypes) designated A, B, C, D, F, G, H, J, K, and CRFs (circulating recombinant forms—hybrids that arise during simultaneous infection with different clades). It is important to be able to recognize the different HIV-1 strains for epidemiological studies.

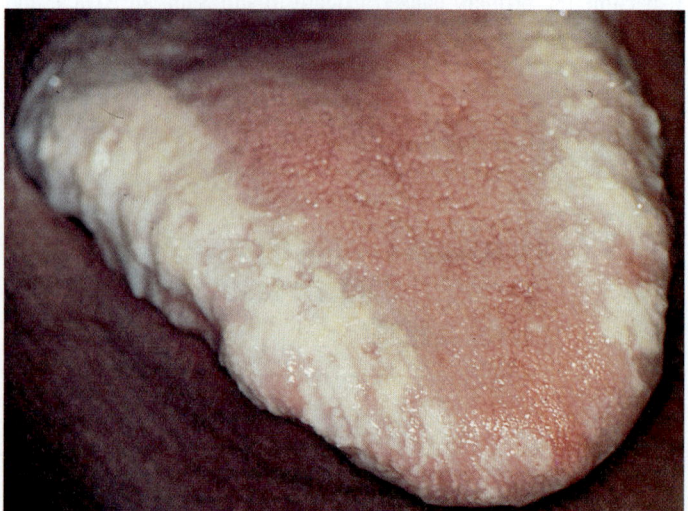

FIGURE 28.2 Hairy Leukoplakia

❓ *What causes this AIDS-related condition?*

Human immunodeficiency virus type 2 (HIV-2) is similar in structure to HIV-1, but differs genetically from HIV-1 by more than 55%. Both HIV-1 and HIV-2 are thought to have arisen by cross-species transmission from different non-human primates. HIV-2 is the main cause of HIV disease and AIDS in parts of West Africa and India, although it has also appeared in the United States and other countries. HIV-2 transmission is generally less efficient than that of HIV-1, and disease progression is slower. Otherwise, the biology of HIV-2 is quite similar to that of HIV-1. For the remainder of this chapter, we will use the term HIV to indicate HIV-1.

HIV Structure HIV is an enveloped virus that has two copies of single-stranded

RNA surrounded by a protein capsid (**figure 28.3a**). The viral envelope originates from the host cell membrane and encloses the viral core. Various proteins make up the viral structure, and these are referred to by two different names: (1) an abbreviation of the functional role or location (for example, CA for capsid protein); and (2) a term that indicates the protein size in kDa (for example, CA is also called p24, being a protein that is 24 kDa in size) (**table 28.2**).

Projecting from the envelope are proteins called Env, which have a complex structure made up of two types of subunits: (1) a transmembrane glycoprotein called TM, or gp41, which anchors Env in the viral envelope; and (2) a surface glycoprotein called SU, or gp120, which rests on the TM protein like a cap and is the part

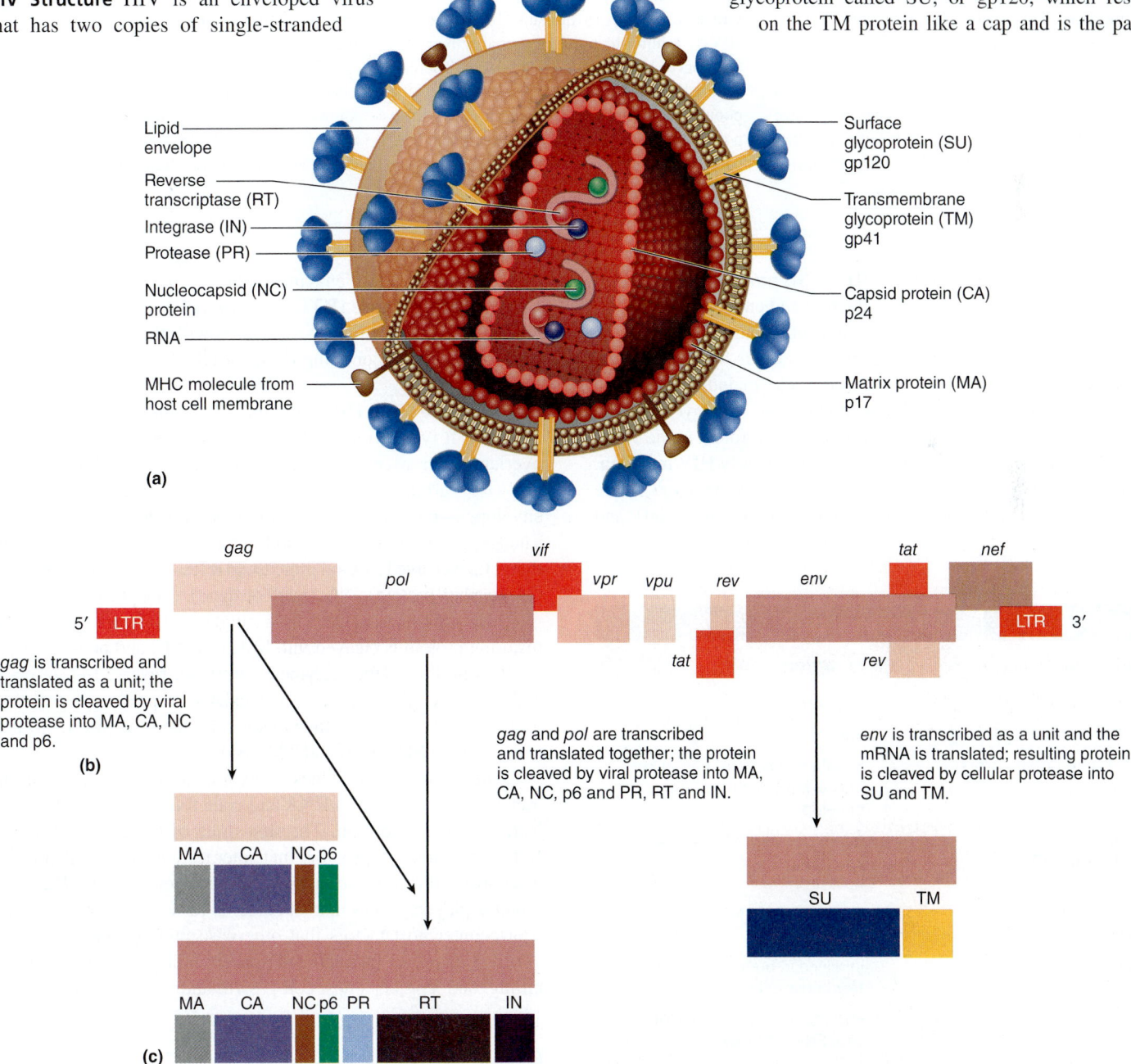

FIGURE 28.3 Human Immunodeficiency Virus Type 1 (HIV-1) **(a)** Diagrammatic representation of the virus showing important antigens. **(b)** Map of the HIV-1 genome. **(c)** HIV-1 gene products. The *gag* and *pol* products must be cleaved by viral protease, and the *env* products must be cleaved by host cell protease.

❓ *What is the function of reverse transcriptase?*

PERSPECTIVE 28.1

Origin of AIDS-Causing Viruses

Where did the AIDS-causing viruses come from? Genetic evidence indicates that the HIV-1 virus mutated to its present form fairly recently, between 50 and 150 years ago. Although it first appeared in the United States in the 1970s, serological evidence indicates that it was present in a few people in Africa in the 1950s.

Viruses similar to HIV exist in a number of wild and domestic animals, including cats, dairy cattle, and monkeys. A virus closely related to HIV-1 has been found in a single West African subspecies of chimpanzees. The virus does not appear to harm the chimpanzees, suggesting that they have been living together for a long time. Another virus, closely related to HIV-2, is present in a species of large monkey, the sooty mangabey. A likely hypothesis is that the AIDS-causing viruses "jumped" to humans from these simian relatives, probably through contact with their blood. Indeed, in Africa, chimps and mangabeys are often killed for food, exposing humans to their blood.

Genetic comparisons of a large number of simian lentiviruses and AIDS-causing viruses from humans support the idea that the simian viruses can jump to humans. In the case of HIV-1, a jump to humans probably occurred only once, between 1910 and 1950. There is no credible evidence to support some of the popular ideas on how HIV got into the human population; for example, it is unlikely that HIV was transferred to humans by inadequately sterilized Salk polio vaccine grown in simian kidney cell cultures and there is no evidence for the suggestion that the viruses resulted from biological warfare experiments by Russia or the United States. It is possible that AIDS-causing viruses have existed for many years in people living in isolated African villages, perhaps even for centuries. According to this idea, population increases and migration to big, crowded cities allowed the viruses to spread rapidly, becoming more virulent in the process.

The answer to the question about the origins of AIDS-causing viruses might never be known precisely, but the question is intriguing and may lead to better understanding of the emergence of new infectious diseases.

---

that initially contacts a host cell. These SU proteins are important because they allow the virus to attach to and enter host cells.

The viral core has several components. Matrix protein (MA), or gp17, lies inside the envelope and has various functions during different stages of the HIV replication cycle, including providing structural support during virion assembly. The next layer in is the protein capsid CA, or p24, the most abundant protein in the virus, which can be measured in the serum to detect early HIV infection. The virion core contains nucleocapsid (NC) protein (p7), core protein p6 (involved in budding of HIV from the host cell), and

TABLE 28.2	HIV Components
Protein	**Function**
Capsid protein (CA or p24)	Coat protein
Matrix protein (MA or p17)	Stabilizes virion
Transmembrane protein (TM or gp41)	Stalklike portion of Env spikes, anchoring Env to the viral envelope
Surface glycoprotein (SU or gp120)	Caplike portion of Env spikes, with which virus attaches to host receptors
Nucleocapsid protein (NC; p7)	Attaches viral RNA to capsid
Core protein (p6)	Involved in budding of HIV from host cell
Reverse transcriptase (RT)	Enzyme that makes copy of viral RNA into DNA
Protease (PR)	Enzyme involved in viral protein processing
Integrase (IN)	Enzyme that inserts viral cDNA into host genome

three important viral enzymes involved in HIV replication—**reverse transcriptase (RT), protease (PR),** and **integrase (IN).** RT, NC, and IN are tightly associated with the RNA. Table 28.2 summarizes the major components of HIV.

HIV Genome The HIV genome is simple yet functionally complex. It typically has only nine genes, but there are multiple overlapping transcription options (figure 28.3b). Three genes—*gag* (for "group antigen"), *pol* (for polymerase), and *env* (for envelope)—code for the structural components just described. The *gag* gene is transcribed and translated into a large polyprotein that is then cleaved by viral protease to release MA (matrix protein), CA (capsid protein), NC (nucleocapsid), and p6. Occasionally, *gag* and *pol* are translated as a single unit (figure 28.3c). The resulting protein is cleaved into MA, CA, NC, and p6 proteins, as well as the three viral enzymes—protease, reverse transcriptase, and integrase. The *env* gene is translated to produce a precursor protein—gp160—that is then cleaved by the host (not viral) protease into the SU/gp120 and TM/gp41 glycoproteins.

The six remaining genes, known as accessory genes, are all translated from spliced mRNA and code for the proteins Tat, Rev, Nef, Vif, Vpr, and Vpu. The first three of these, Tat, Rev, and Nef, are regulatory proteins that interact with host cell proteins to control HIV replication and release from the cell. The other proteins play key roles in host defense; for example, both Vif and Vpu counter host factors that otherwise inhibit HIV replication. Vif also increases infected cell susceptibility to cytotoxic T-cell killing, and Vpr stops host cell growth and induces apoptosis.

Pathogenesis

The ability of HIV to cause a long-term and progressive infection is related to both its replication cycle and the cell types it targets. Understanding key aspects of these has allowed scientists to develop several medications that specifically target HIV.

Attachment and Entry Once HIV enters the body, the SU/gp120 portion of Env attaches to CD4⁺ cells. Recall from chapter 15 that helper T cells are CD4⁺, meaning they have CD4 protein on their surface. These are the main targets of HIV infection. Monocytes, including macrophages and dendritic cells, are also CD4⁺ and they are thus targets of infection as well. Some studies suggest that cells that do not have CD4 can also be infected, but whether this contributes to HIV infection in the host is unclear, and the mechanisms of viral entry are not well understood. ◀◀ **CD4, p. 369**

In addition to needing CD4, the HIV attachment process also requires a co-receptor, usually one of the chemokine receptors CXCR4 or CCR5 (**figure 28.4**). Most strains of HIV use CCR5 as the co-receptor, but some use CXCR4 (particularly strains present at later stages of infection), and some strains can use both. The importance of CCR5 is illustrated by the observation that people who do not make normal amounts of CCR5 are less susceptible to HIV infection. ◀◀ **chemokine, p. 341**

When HIV encounters a CD4 molecule, it binds to it, inducing a conformational change in SU. This change exposes sites in the SU molecule that bind the co-receptor. SU then undergoes another conformational change, leading to fusion of the host and viral membranes, a process facilitated by the TM (gp41). Following fusion, the virus enters the host cell.

Reverse Transcription and Genome Integration After entering the host cell, the HIV-encoded reverse transcriptase makes a DNA copy of the viral RNA genome (**figure 28.5**). The RNA template is degraded and a complementary DNA (cDNA) strand is made, resulting in a double-stranded DNA copy of the original viral RNA. This DNA is circularized and moves into the host nucleus. Once in the nucleus, the virally encoded enzyme integrase inserts the DNA in a linear form into the host cell chromosome as a provirus. No specific nucleotide sequence in the host genome is required for the insertion. Once a DNA copy of the HIV genome is integrated, the provirus is a permanent part of that cell's genome.

Reverse transcriptase makes frequent mistakes as it copies the RNA genome into DNA. Because the enzyme also lacks proofreading ability, these mistakes are left uncorrected. The overall result is that the nucleotide sequence of the integrated DNA copy of the HIV genome is generally slightly different from the parent molecule. These changes allow HIV to evolve quickly, altering its antigens and thereby avoiding the immune response.

Replication and Its Consequences Once the HIV DNA copy is integrated into the host cell genome, it can be transcribed and translated to produce new viral particles. The developing virions bud from the host cell, gaining their envelope as they do so. Some host cells die as a result of the infection, but others survive. Infected helper T cells typically die—the loss of these cells is very important because of their central role in the body's adaptive immune response. Recall that their cytokines normally regulate the action of cytotoxic T cells, antibody production by B cells, and chemotaxis of macrophages. In contrast, infected macrophages usually survive (perhaps because the level of virus replication in them is more moderate), and they release new virions over long periods of time. These cells, along with a particular subset of T cells called resting T cells, become

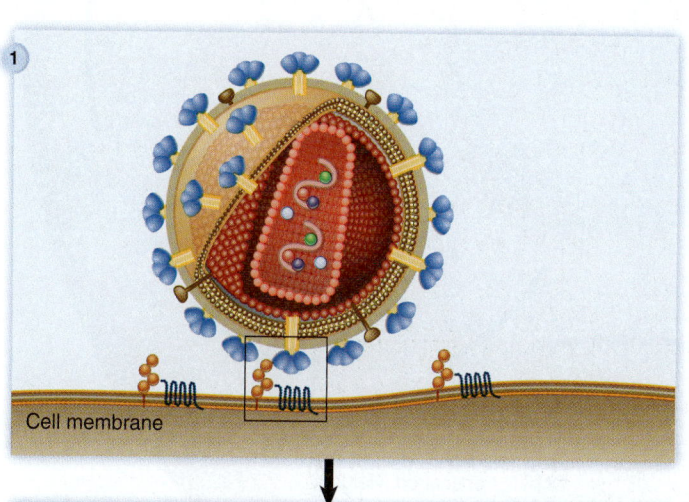

Attachment.
Initial contact occurs when HIV gp120 (SU) binds to the host cell CD4 receptor.

Cell membrane

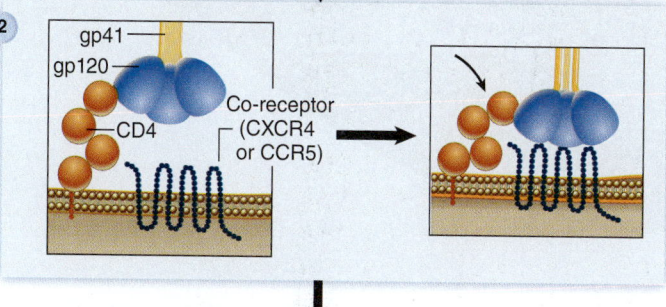

gp41
gp120
CD4
Co-receptor (CXCR4 or CCR5)

Co-receptor binding.
Binding with a co-receptor follows. This causes a conformational change in gp120 that facilitates gp41 (TM) insertion into the host cell membrane.

Fusion.
gp41 mediates host-virus membrane fusion and viral entry into the host cell.

FIGURE 28.4 Attachment and Entry of HIV into a Host Cell **Step 1:** Initial contact occurs between gp120 (SU) and host cell CD4 receptor. **Step 2:** Attachment to a chemokine receptor such as CCR5 then follows. **Step 3:** Membrane fusion, mediated by gp41 (TM).

❓ *How do entry inhibitor medications work?*

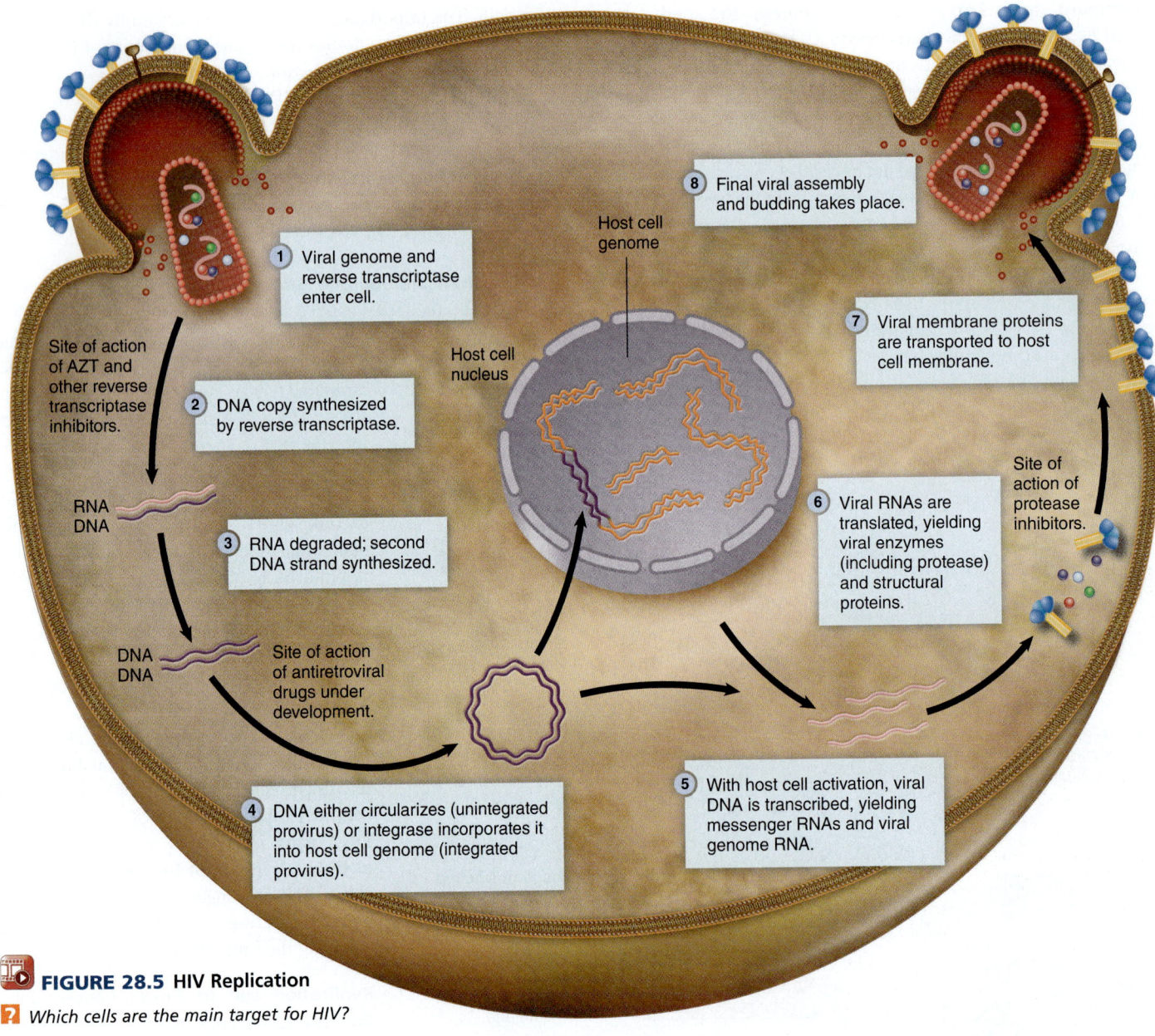

① Viral genome and reverse transcriptase enter cell.

Site of action of AZT and other reverse transcriptase inhibitors.

② DNA copy synthesized by reverse transcriptase.

RNA
DNA

③ RNA degraded; second DNA strand synthesized.

DNA
DNA

Site of action of antiretroviral drugs under development.

④ DNA either circularizes (unintegrated provirus) or integrase incorporates it into host cell genome (integrated provirus).

Host cell genome

Host cell nucleus

⑤ With host cell activation, viral DNA is transcribed, yielding messenger RNAs and viral genome RNA.

⑥ Viral RNAs are translated, yielding viral enzymes (including protease) and structural proteins.

⑧ Final viral assembly and budding takes place.

⑦ Viral membrane proteins are transported to host cell membrane.

Site of action of protease inhibitors.

FIGURE 28.5 HIV Replication

Which cells are the main target for HIV?

reservoirs for replicating HIV, protecting it from the rest of the immune system and providing a continuing source of infectious virus. Although infected macrophages do survive, they show impaired chemotaxis, phagocytosis, and antigen presentation, which further weaken the immune response.

The signs and symptoms of HIV disease and AIDS occur as the number of helper T cells declines, in turn affecting the function of the immune system as a whole. Destruction of helper T cells can occur via multiple mechanisms, including lysis following virus replication, attack by cytotoxic T cells and natural killer cells, and formation of syncytia (fusion of infected cells with large numbers of healthy cells), which are then destroyed. Humoral antibody may play a role in this through antibody-dependent cell cytotoxicity. Apoptosis (programmed cell death) is also accelerated in some HIV-infected cells; this occurs by a number of mechanisms, including induction by viral Tat and production of certain cytokines. Accumulation of viral products

such as viral RNA and unintegrated viral DNA inside the cytoplasm of the infected cell, can also result in its death. ◀◀ **cytotoxic T cells, p. 356** ◀◀ **natural killer cells, p. 374** ◀◀ **antibody-dependent cellular cytotoxicity, p. 361** ◀◀ **apoptosis, p. 350**

Disease Progression The number of HIV particles in the blood, measured by **viral load** (RNA copy number), changes during the course of HIV disease. During the acute retroviral syndrome (ARS), the viral load rises to very high levels. This occurs as progeny virus particles are released from the billions of helper T cells that become infected in the first few weeks of disease (**figure 28.6**). Initially, the person does not have any detectable antibodies against the virus. However, as the viral load increases, the person develops detectable amounts of HIV-specific antibodies, the process of **seroconversion.** The viral load then decreases as the immune response destroys infected cells. The number of helper T cells rapidly decreases as the infected cells are destroyed,

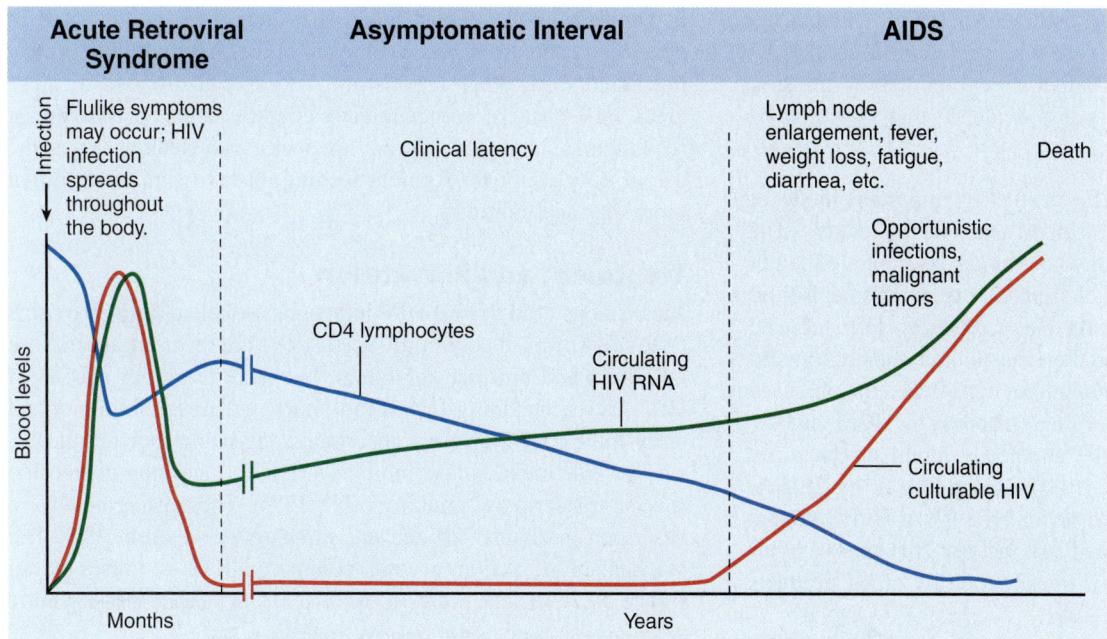

Acute Retroviral Syndrome	Asymptomatic Interval	AIDS

FIGURE 28.6 HIV Disease Progression Levels of virus are very high during the acute retroviral syndrome and at the end of the disease when AIDS occurs. Antibody tests for diagnosing the disease are often negative in the early stage of the disease even though infected people are highly infectious. The disease steadily progresses in the absence of symptoms, as shown by the rising levels of plasma viral RNA and falling CD4+ cell count.

? *What are some of the opportunistic infections and malignant tumors that people with AIDS contract?*

but increases slowly again as the viral load goes down and the T cells are replenished. However, the immune response fails to eliminate the virus completely, and the helper T-cell count typically does not reach the preinfection level. ◄◄ seroconversion, p. 426

During the clinically latent period, which can last for many years, HIV continues to replicate in infected cells. The concentration of virions in the bloodstream during this time remains relatively stable and is referred to as the **HIV (viral) set point.** This set point, which is measured as HIV RNA in plasma, is important because it serves as a predictor of disease progression—the higher the viral set point in a person, the more quickly they are likely to develop AIDS. The observation that the viral set point predicts disease progression has influenced HIV treatment, with a major focus now on keeping the viral set point as low as possible. This affects the infectiousness of the person as well, because HIV levels also correlate with transmission.

The continuous viral replication gradually reduces the helper T-cell population. In most cases, as seen in approximately 80% of HIV-infected people, the peripheral blood helper T-cell count (normally about 1,000 cells per microliter) steadily falls at a rate of roughly 50 cells per microliter per year. Even though the body can normally replace over a billion CD4+ cells per day, the number of new cells made is not enough to replace the number dying.

AIDS Signs and symptoms of AIDS usually appear when helper T-cell counts fall below 200 cells per microliter. As these counts decrease and the immune system stops functioning properly, the viral load rises dramatically. The patient now begins experiencing a variety of opportunistic infections, including malignant tumors caused by infectious agents.

Half of untreated patients progress to AIDS within 9 to 10 years. Rapid progressors (about 10% of infected individuals, who have high viral set points) develop AIDS within a few years of initial infection. At the other end of the spectrum are long-term non-progressors (about 5% to 10% of infected individuals, with low viral set points), who show no decrease in CD4+ cells and

maintain high levels of anti-HIV antibody and HIV-specific CD8+ cytotoxic T cells over many years. At the extreme end is a group called elite controllers, who have very low viral set-point levels, below the detection limit of conventional clinical assays.

MicroByte

Tuberculosis is a leading cause of death among those with HIV disease—in some countries, as many as 80% of people with HIV also have TB.

Epidemiology

HIV is transmitted by sexual contact, through blood and blood product transfusion, and vertically from mother to fetus or infant. It is not spread by insect bites or casual contact such as kissing and sharing food, and there is no evidence that HIV is transmitted by sweat, urine, saliva (spitting), or tears, although it can be found in these fluids in infected people.

Sexual Contact Sex without condom protection is the primary cause of HIV spread. The virus is present in semen, as well as cervical and vaginal secretions. It is thought that HIV enters the vaginal, rectal, penile, or urethral mucosa through tiny lesions that occur during sexual intercourse. It can be transmitted by either male-to-male or male-to-female sexual contact (female-to-female transmission does not appear to occur, or it is very rare). The risk for transmitting HIV through unprotected anal sex is greater than the risk from vaginal sex, probably because the thin rectal mucosa is easily damaged. Although the risk of contracting the virus is greater for the receptive sexual partner, it can also be contracted by the insertive partner. The risk of sexual HIV transmission increases in people who have sex with multiple partners or with a person who has had multiple partners, in people who are the receptive partner (especially receptive anal sex), and in people who experience traumatic sex. It is also higher in those who have any genital irritation or lesion; other sexually transmitted infections can increase the risk of contracting HIV if genital ulcers or lesions are present, as is the case with syphilis, genital herpes, and chlamydia, among others. Saliva is very unlikely to transmit the disease, although there is some evidence that HIV can be transmitted through oral sex.

Blood and Blood Products The next most important mode of transmission of HIV is through blood and blood products. The virus can be transmitted through whole blood, concentrated red or white blood cells, concentrated clotting factors, or plasma. Before tests were available for screening blood products, HIV-infected people unknowingly transmitted the virus to thousands of transfusion recipients, including hemophiliacs who receive transfused clotting factor VIII to treat bleeding episodes. In 1985, an HIV screening test became available, and today blood products are routinely screened for antibodies to HIV-1 and HIV-2 (by ELISA) and for p24 (the capsid protein) or for HIV RNA. Blood donors are also screened for self-reported risk factors. This approach has significantly reduced the risk of transmission by blood products and organ transplants. ◀◀ ELISA, p. 432

People who abuse injected drugs often share needles, so HIV transmission by blood is still a major factor in the disease pandemic. Whenever two people share a hypodermic needle, there is a risk of spreading blood-borne pathogens such as HIV and hepatitis B. People who get tattoos and body piercings are also at risk from this route of transmission, unless universal precautions are followed (using sterile techniques, new needles, new inks, and so on). HIV can be transmitted to healthcare providers by a stick from a needle used on a HIV-positive person, but the risk is very low—0.3% (compared with a risk of nearly 30% with a virus such as hepatitis B).

Vertical Transmission The third most important mode of HIV spread is from mother to infant. HIV can be transmitted to the fetus *in utero* via the placenta, at childbirth, or by breast feeding. Without antiretroviral treatment of the mother, approximately one-third of newborns will contract HIV. However, if the mother does not breast feed and receives antiretroviral treatment, and the infant is delivered by cesarean section—all common practices among HIV-positive women in the United States—this risk drops considerably (to about 1%).

Breast feeding carries a significant risk of mother-infant transmission, contributing about 15% to 20% of infant infections in populations that breast feed. The situation is complicated in developing countries, because the antibodies in breast milk protect the infant from other serious and widespread diseases in those areas. HIV-positive women in these countries are advised to breast feed their infants if they do not have access to clean water, as the use of dirty water for formula feeding leads to significant infant morbidity and mortality.

Treatment and Prevention

Medications used to treat HIV infections are called **antiretrovirals (ARVs).** The most commonly used ARVs are reverse transcriptase inhibitors and protease inhibitors, but there are other classes of HIV ARVs, including fusion inhibitors, integrase inhibitors, and entry inhibitors. Reverse transcriptase inhibitors include nucleoside reverse transcriptase inhibitors (NRTIs) and non-nucleoside reverse transcriptase inhibitors (NNRTIs). These categories were described in chapter 20 and are summarized in **table 28.3.** The mechanism of action of one common NRTI is illustrated in **figure 28.7.** ◀◀ NRTIs, p. 476 ◀◀ NNRTIs, p. 476 ◀◀ protease inhibitors, p. 476 ◀◀ integrase inhibitors, p. 476 ◀◀ entry inhibitors, p. 474

Antiretrovirals are usually given in combinations, an approach referred to as **highly active antiretroviral therapy (HAART).** HAART is effective because each medication contributes to suppressing the virus. Also, when the drugs are used in combination, resistant mutants are less likely to develop, despite the high mutation rate of HIV. This is because it is unlikely that any one virion will simultaneously develop resistance to all the medications at the same time, especially when replication levels are suppressed. Research indicates that early (rather than late) treatment of HIV disease can significantly reduce the risk of death from it.

Most current HAART treatments consist of three drugs: two NRTIs and either a protease inhibitor or an NNRTI. Efforts are being made to simplify the treatment by developing fixed-dose combinations that contain more than one medication. By reducing the number of pills a person needs to take, patient compliance increases. This, in turn, helps prevent resistance from developing and makes the treatment more effective.

HAART does not cure HIV disease. It is used to improve the quality of the person's life, reduce the risk of other complicating diseases, and lower the HIV set point. The latter is particularly important because lowering the number of replicating viral particles decreases the chance that one of them will mutate to drug

TABLE 28.3	HAART Medications		
Medication	**Actions**	**Examples**
Entry inhibitors	Interfere with viral entry into host cell	Fuzeon, maraviroc
Integrase inhibitors	Prevent integration of HIV cDNA into host genome	Raltegravir
Maturation inhibitors	Inhibit *gag* gene processing and virion production	Bevirimat (in clinical trials)
NRTIs (nucleoside reverse transcriptase inhibitors)	Chain terminators that stop DNA replication	AZT, D4T, 3TC
NNRTIs (non-nucleoside reverse transcriptase inhibitors)	Inhibit reverse transcriptase	Nevirapine, delavirdine
Protease inhibitors	Prevent virions from maturing	Saquinavir

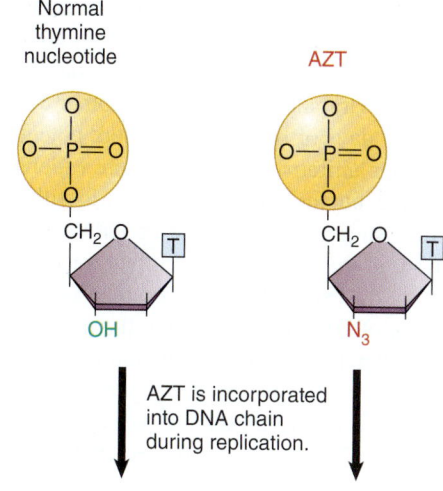

Normal thymine nucleotide

AZT

AZT is incorporated into DNA chain during replication.

The normal nucleotide in (a) ends in 3'OH, to which another nucleotide can be added for elongation of the DNA strand. The AZT in (b) ends in 3'N$_3$, preventing elongation.

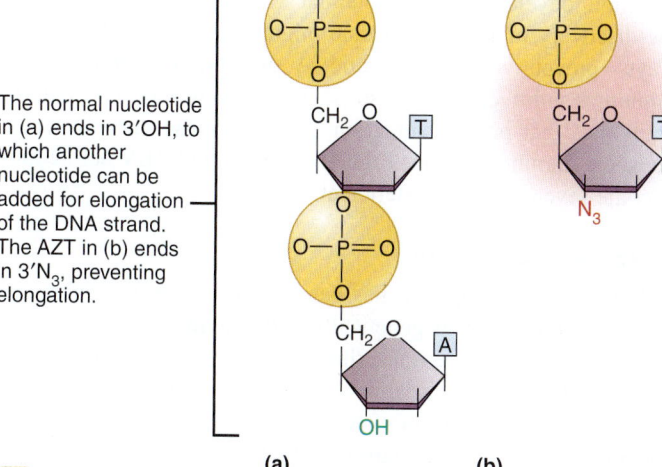

(a)   (b)

FIGURE 28.7 Mode of Action of Zidovudine (AZT)
(a) Normal elongation process of DNA in which reverse transcriptase catalyzes the reaction between the OH group on the chain with the phosphate group of the nucleotide being added. **(b)** Reverse transcriptase catalyzes the reaction of AZT with the growing DNA chain. Since AZT lacks the reactive OH group, no further additions to the chain can occur, and DNA synthesis is stopped. AZT is a nucleoside. It becomes phosphorylated after entry into the cell.

? *Why is AZT given to HIV-positive pregnant women and to their newborns?*

resistance. It is important to remember, however, that HAART does not eliminate HIV provirus hidden in host cell genomes. If medications are stopped, the viral set point rebounds to pretreatment levels.

Although HAART can be very successful in lowering the viral set point, there are problems associated with this treatment. Many HIV strains are now resistant to the medications, and these are spreading. The medications can also have several toxic effects. For example, zidovudine (azidothymidine, or AZT) can cause anemia, low white blood cell count, vomiting, fatigue, headache, and muscle and liver damage. Other NRTIs cause

painful peripheral nerve injury; inflammation of the pancreas; rash, mouth, and esophagus ulceration; and fever. Some of the protease inhibitors cause kidney stones, nausea, diarrhea, and sometimes diabetes.

Another major factor limiting the use of anti-HIV medications is their cost, which can make anti-HIV medications unaffordable in many parts of the world. Negotiated price decreases, the approval of three generic drugs by the U.S. Food and Drug Administration in 2005, and the provision of drugs by the U.S. government to several highly affected countries, have made treatment available to many more HIV patients worldwide.

Preventing Infection There is no approved vaccine that prevents HIV infection, although clinical trials continue. Therefore, prevention efforts must be aimed at avoiding transmission.

Many people with HIV disease do not know they are infected, and therefore all blood and blood products should be considered as potentially containing the virus. Infectious HIV persists in samples of blood plasma for at least a week after they are taken from AIDS patients. HIV on objects and surfaces contaminated by body fluids is easily inactivated by high-level disinfectants and heat at 56°C or more for 30 minutes. Freshly opened household sodium hypochlorite 5.25% bleach, diluted 1:10, is a cheap and effective disinfectant for general use. ◀◀ **high-level disinfectant, p. 116**

Educating people about HIV and how it is transmitted is a powerful weapon against the AIDS epidemic. The virus is not highly contagious, and the risk of contracting and spreading it can be eliminated or significantly reduced by avoiding activities that might transmit it (**table 28.4**). People who are unsure of their HIV status, and especially those at increased risk of contracting

TABLE 28.4	**Behaviors That Help Control the AIDS Epidemic**

1. Abstinence.

2. Staying in a mutually monogamous relationship.

3. Avoiding sex with people at risk for HIV infection (see table 28.5).

4. Avoiding genital and rectal trauma. Small breaks in the skin and mucous membranes allow HIV to infect.

5. Avoiding sex when sores from herpes simplex or other causes are present. These are sites where HIV can infect.

6. Not engaging in unprotected anal intercourse. Receptive anal intercourse carries a high risk of HIV transmission.

7. Using latex condoms from beginning to end of sex. Polyurethane condoms are a reasonable alternative for those allergic to latex. Condoms made from other materials are not reliable for disease prevention, nor are those marketed in many countries outside the United States. Oil-based lubricants can not be used with latex condoms. Condoms for women are available.

8. Postponing pregnancy indefinitely if you are a woman infected with HIV. If you are not sure of your HIV status, have blood tests to rule out HIV disease before considering pregnancy.

9. Using extreme care to avoid needles, razors, toothbrushes, etc., that could be contaminated with someone else's blood.

TABLE 28.5	People at Increased Risk for HIV Disease

1. Injected-drug abusers who have shared needles.

2. People who received blood transfusions or blood products between 1978 and 1985.

3. Sexually promiscuous men and women, especially sex workers, drug abusers, and men who have sex with men.

4. People with history of hepatitis B, syphilis, gonorrhea, or other sexually transmitted infections that may be markers for unprotected sex with multiple partners.

5. People who have had blood or sexual exposure to any of the people listed.

the virus, should get tested—HIV-positive individuals who know their status can ensure that they do not spread the virus to others; they can also receive antiretroviral treatment that lowers the viral set point and helps reduce transmissibility (**table 28.5**).

As mentioned earlier, most HIV transmission occurs between people having unprotected sex. Use of latex condoms during sex significantly reduces HIV transmission. Both male and female condoms are available, although male condoms are less expensive and more readily available. Studies also show that circumcision lowers the risk of HIV transmission to heterosexual men by up to 60%. Several countries that have high HIV infection rates are engaged in developing programs promoting circumcision as a preventive measure. Condom use is still a cheaper way of preventing spread of the disease, however, providing that reliable condoms are available. Further, there is some concern that circumcised men may not realize that they are still at risk of HIV or other sexually transmitted infections if they have unprotected sex.

An approach called pre-exposure prophylaxis, or PrEP, has recently shown great promise in preventing the spread of HIV. In one study, women using a newly developed vaginal microbicide gel that included an active ARV to block HIV infection had significantly reduced HIV infection rates. Another recent study showed that providing two oral ARVs to high-risk, uninfected individuals protected them against HIV infection, reducing their risk of contracting the virus by almost half.

Since 1985, transmission by blood and blood products has been lowered dramatically by screening potential donors for HIV risk factors and testing their blood for HIV and antibodies to HIV. The risk is now estimated to be less than 1 in 1.5 million transfusions. Screening tests have also markedly reduced the risk of HIV transmission from artificial insemination and organ transplantation. Also, medications are effective for preventing AIDS acquired from HIV-contaminated instruments. Hundreds of needle and syringe exchange programs help prevent the spread of HIV and other blood-borne diseases among injected-drug abusers. In these programs, people are given sterile syringes and needles in exchange for used ones. Programs also provide drug rehabilitation efforts and education about condom use and other safer sex practices.

HIV transmission from infected mother to newborn can be prevented in two-thirds of cases by giving AZT to the mother during pregnancy and to the newborn infant for 12 weeks, or by providing nevirapine (NVP) to the mother at delivery and the infant within 3 days of birth. Newer, more potent combinations of AZT and other antiretroviral therapies are widely used and may be more effective, but their long-term safety is still being evaluated. Recent studies suggest that the use of ARVs during the breast feeding period, in the mother or the infant, or both, can reduce transmission by this route. Elective cesarean section significantly reduces the risk of HIV transmission to the newborn baby.

Better treatment and prevention of opportunistic infections and improved antiviral therapy against HIV have significantly lengthened the asymptomatic stage of the disease and prolonged life once AIDS develops (**figure 28.8**). The main features of HIV disease are presented in **table 28.6**.

HIV Vaccine Prospects

There are no approved vaccines for preventing HIV disease, although development of potential vaccines began soon after the discovery of HIV. A vaccine might be used in either of two ways—preventive or therapeutic. A **preventive vaccine** would immunize uninfected individuals against the disease by inducing antibodies that prevent infection. A **therapeutic vaccine** would boost the immunity of those already infected with HIV before they become severely immunodeficient.

Vaccine Challenges

HIV vaccine development has been hampered by several major problems. First, HIV is genetically very variable, due to the high error rate when reverse transcriptase copies the viral genome into DNA. This antigenic variability of HIV enormously complicates the task of developing an effective vaccine against the virus—a vaccine that is effective against one antigenic variant may not be effective against another.

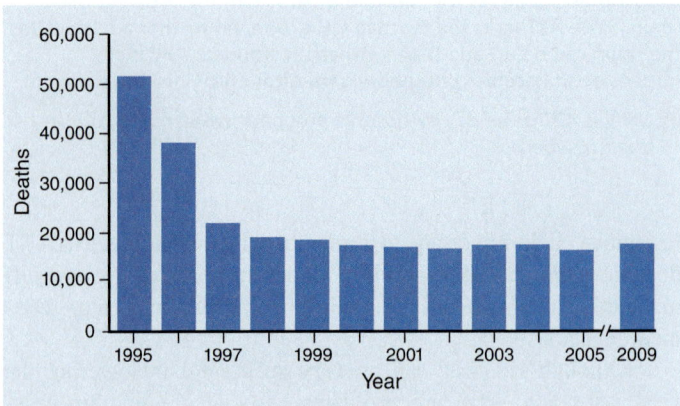

FIGURE 28.8 Deaths Due to AIDS, United States, 1995–2009 Between 17,000 and 18,000 people still die from AIDS every year in this country.

❓ *What caused the marked drop in HIV-related deaths after 1995?*

TABLE 28.6	HIV Disease
Signs and symptoms	Over half develop fever, sore throat, head and muscle aches, rash, and enlarged lymph nodes early in the infection. After an asymptomatic period, symptoms result from immunodeficiency and include unusual malignant tumors, pneumonia, meningoencephalitis, diarrhea, etc.
Incubation period	Usually 6 days to 6 weeks for acute symptoms; immunodeficiency symptoms within 10 years in half the infections (10% within 5 years and 90% within 17 years).
Causative agent	Human immunodeficiency virus type 1 (HIV-1), many subtypes and strains. HIV-2 mainly in West Africa.
Pathogenesis	HIV infects various body cells, notably those vital to specific immunity, CD4$^+$ T lymphocytes and antigen-presenting cells. T cells killed, numbers slowly decline until the immune system can no longer resist infections or development of tumors.
Epidemiology	Three main routes of transmission of HIV: intimate sexual contact, via transfer of blood or blood products, and from mother to child around the time of childbirth. Transmission can also occur by breast milk, and possibly by oral-genital contact.
Treatment and prevention	Treatment: HAART therapy, consisting of combinations of several anti-HIV medications, effective for many AIDS patients, and delays progression of HIV disease to AIDS. Not a cure, and too expensive for most of the world's HIV disease patients. No vaccine yet available. Medications and vaccines can prevent many of the infections that can complicate HIV disease. Anti-HIV medications and cesarean section decrease mother-to-newborn transmission. Effective preventive measures: sex education of schoolchildren, needle exchange programs for drug addicts, use of condoms.

The lack of a good animal model further challenges vaccine development. Until recently, the only species that could be infected with HIV were the great apes (such as chimpanzees); although these animals generally do not develop AIDS when infected with HIV, their antibody response can be monitored. Development of a vaccine would require a better understanding of the immunopathogenesis of HIV disease, which an animal model could provide. Different chimeric mouse models that can be used to study HIV infection and pathogenesis have been developed. These mice have been "humanized," meaning that a variety of relevant human cell types (such as helper T cells and macrophages) have been introduced into them. They can be infected with HIV and they develop some symptoms of HIV disease, thus showing potential as tools for clinical evaluations of HIV disease and to assess candidate HIV vaccines.

A third challenge to HIV vaccine development is the fact that HIV can avoid humoral and cell-mediated immune responses. It does this by becoming latent (persisting as a provirus), and also by causing the formation of syncytia. The syncytia allow virus particles to pass from cell to cell without contacting antibodies, cytotoxic T cells, and other immune components carried by the bloodstream. ◄◄ syncytia, p. 396

Further complicating matters is the fact that infected individuals do indeed have an immune response against the virus, but it is not sufficient to control the infection. Nonetheless, a vaccine is considered the most important way to control the HIV/AIDS pandemic.

Vaccine Research Because HIV does not cause AIDS in species other than humans, vaccine research relies heavily on the use of a model system involving related viruses in non-human primates. This includes SIV (simian immunodeficiency virus) as well as chimeric viruses between HIV and SIV called SHIVs. One of the approaches to producing an effective vaccine was to develop an attenuated HIV strain. Unfortunately, although effective in non-human primates, this vaccine was shown to be too risky for use because HIV's high mutation rate allows it to revert to a pathogenic state.

There are several new approaches to vaccine development currently underway. Candidate vaccines undergo an extensive evaluation for safety and immunogenicity in experimental animals before they are tested in humans. At least 10 experimental vaccines have been developed and tested in humans, with mostly negative results. Vaccine research efforts that have been or are being evaluated include:

■ **Recombinant vector vaccines.** These are viruses that have been engineered to carry HIV genes; for example, vaccinia virus containing HIV envelope genes. Recently, a vaccine using replication-deficient adenovirus containing synthetic HIV-1 *gag*, *pol*, and *nef* genes went to clinical trials (called the STEP study). The trial was stopped in 2007 because it was found that the vaccine did not protect recipients from acquiring HIV, and more importantly, in some cases it actually increased the person's risk of becoming infected. Research is being continued with adenovirus serotypes that are safer for human use. A more promising trial, the RV144 trial in Thailand, tested a canarypox virus also containing HIV genes *env*, *gag*, and *pol*. ◄◄ poxviruses, p. 308

■ **Recombinant HIV-1 envelope glycoprotein vaccines.** These are made by splicing genes for Env glycoproteins into tumor cell lines so that these cells produce large quantities of the glycoproteins that can be used as a vaccine (in a similar way to which the hepatitis B vaccine is made). However, these vaccines protect only against the exact antigenic variant used to make them, and they do not cause a cytotoxic T-cell response.

■ **DNA vaccines.** Plasmid DNA engineered to encode HIV-1 antigens that will be expressed in tissues is a promising vaccine strategy. However, this approach requires multiple injections of high doses of DNA vaccine for an effective immune response. Research is focused on improving these

CASE PRESENTATION   28.1

The patient was a young woman from West Africa showing generalized enlargement of her lymph nodes. She had recently moved to the United States. She had no history of drug abuse or of receiving blood transfusions. She had had three sex partners in her life. Her single pregnancy the year before her arrival was delivered by emergency cesarean section. She and the baby's father had routine tests for HIV at that time, and both tests were negative. The baby and the father remained well. Ten years before her HIV evaluation, the woman had been treated for a fever by scarification (deliberately making superficial cuts in the skin) performed by a native healer; the scarification was done with a razor blade, the sterility of which is unknown to the patient, and had been repeated 4 years before her evaluation.

Her initial test results included non-diagnostic lymph node biopsies and a negative HIV test (ELISA). A repeat HIV test some months later was weakly positive by ELISA, and the confirmatory Western blot showed only questionable reactions of the patient's serum with gp41 and two other HIV antigens. A test for HIV-2 was negative. A CD4+ T lymphocyte cell count was very low.

1. Could this patient have HIV disease? If so, how could the negative tests be explained?
2. Does the history give any possible ways in which she could have contracted HIV disease?
3. How could the diagnosis be established?

4. Could her baby and the baby's father have the disease?
5. Does this case suggest that any changes should be made in the way HIV disease is diagnosed?

Discussion

1. This patient could well have had HIV disease/AIDS because of her persistent, generalized lymphadenopathy and very low CD4+ cell count. Her HIV strain might have differed enough from the pandemic group M HIV-1 strains, so that the usual laboratory tests did not detect antibody to it.
2. The woman could have contracted an AIDS-causing virus during sexual intercourse, or from a blood-contaminated razor blade used for scarification. Contracting such a virus from unsterile instruments during her emergency cesarean is a possibility, but this seems less likely because of the short time span before presenting with severe immunodeficiency.
3. The CDC has a global surveillance system designed to detect cases like this because they raise the possibility of new or rare AIDS-causing viruses being introduced into a population. In the present case, samples of the patient's blood were examined by the CDC and she was shown to be infected with a group O HIV-1 virus. The patient's serum reacted with

peptides specific to group O strains, and the virus was isolated from her blood. Nucleic acid sequences of the *env, gag,* and *pol* genes matched those of previously isolated group O strains. Group O strains of HIV-1 were first found in Cameroon, a country on the West African coast, where they account for 6% of HIV infections.

4. Despite the absence of symptoms, both her baby and the baby's father could have the disease. The baby's risk of infection was reduced but not eliminated by the emergency cesarean section, and infection could have occurred subsequently if it were breast-fed. The mother's very low CD4+ cell count suggests a high level of viremia and, therefore, increased risk of transmitting the disease. She could have infected the baby's father during sexual intercourse, or he could have infected her.

5. This patient was the first case of group O HIV disease identified in the United States. Studies indicate that previous introductions, if any, were not accompanied by spread of the virus. Nevertheless, federal agencies worked with manufacturers of HIV tests to increase sensitivity of the tests to group O strains.

Source: Centers for Disease Control and Prevention. 1996. *Morbidity and Mortality Weekly Report* 45(6): 122.

---

vaccines. Currently a new approach, known as prime-boost, is being evaluated—in this method, a DNA vaccine primes the body for a response, and a second vaccine then boosts this response. One candidate uses a DNA vaccine encoding HIV Gag protein, followed by recombinant vector vaccine.

◀◀ DNA vaccines, p. 425

■ **Peptide vaccines.** The potential of synthetic peptides is also being explored. The amino acid sequence of the surface proteins of HIV is known; thus, peptides that exactly mimic immunologically important segments of these proteins can be synthesized and injected to induce an immune response.

◀◀ peptide vaccines, p. 425

Despite the difficulties of developing an effective HIV vaccine, a vaccine continues to be the best hope for eventually controlling the HIV/AIDS pandemic. Findings that have encouraged vaccine researchers include the observation that some sex-workers in Africa have not become infected, despite

being repeatedly exposed to HIV. Potent and broadly effective antibodies produced against HIV by some of these people are now an important focus of vaccine researchers looking to identify better antigens for vaccines.

MicroAssessment 28.1

The signs and symptoms of people with AIDS are mainly due to the opportunistic infections and tumors that complicate HIV disease. HIV is not highly contagious, and the AIDS pandemic could be stopped by changes in human behavior. Highly active antiretroviral therapy (HAART) has given many patients with AIDS miraculous improvement.

1. *What is the difference between HIV, HIV disease, and AIDS?*
2. *Why is it important for epidemiologists to be able to identify different HIV-1 strains?*
3. *If AIDS was present in Africa in the 1950s, why did it not appear in the United States until the 1970s?* ✚

COMPLICATIONS OF ACQUIRED IMMUNODEFICIENCIES

28.2 ■ Malignant Tumors

Learning Outcomes

6. *Explain how HIV disease might increase the risk of developing malignant tumors.*

7. *Outline the relationship between Kaposi's sarcoma and KSHV (human herpesvirus-8).*

Certain malignant (cancerous) tumors are associated with HIV disease and other acquired immunodeficiencies. Most of these malignancies fall into one of three types: Kaposi's sarcoma, lymphomas, and carcinomas in the anal or cervical epithelium. They tend to metastasize (meaning spread to different areas of the body) and are difficult to treat. Evidence indicates that viruses are involved in their development—it is thought that certain viral antigens, along with cytokines, cause a host cell to multiply rapidly. These cells then mutate (possibly the result of insertion of viral DNA into their genome) and become malignant. The malignant cell escapes detection and destruction by immune surveillance because cell-mediated immunity is defective, so the cell multiplies unchecked. ◀◀ viruses and human tumors, p. 323

Kaposi's Sarcoma

Kaposi's sarcoma is an unusual tumor arising from blood or lymphatic vessels. It is characterized by papular nodules that are red, brown, purple, or black in color (**figure 28.9**). The nodules are typically found on the skin but can develop in the mouth and respiratory and gastrointestinal tracts.

Kaposi's sarcoma (KS) occurs in four epidemiological forms. It was first described as a rare disease in older men of Mediterranean and Eastern European origin—this became known as classic KS. Later, it was described among all age groups in certain regions of sub-Saharan Africa (endemic KS), where it is now the most commonly occurring cancer. Neither classic nor endemic KS is associated with immunodeficiency. The third form of KS

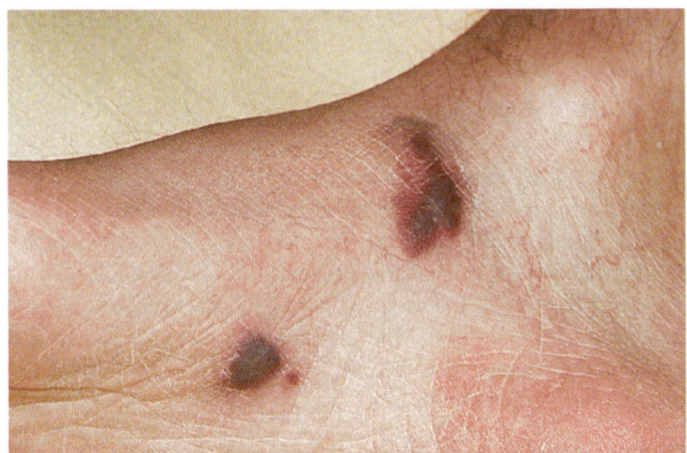

FIGURE 28.9 Kaposi's Sarcoma Lesions

❓ *What virus is detected in all cases of KS?*

(epidemic KS) is associated with HIV disease, and its incidence rose dramatically with the onset of HIV disease and AIDS. It was so common among early AIDS patients that it became an AIDS-defining condition, even though signs and symptoms usually appear before the person becomes severely immunodeficient. AIDS-associated KS is a more aggressive tumor than classic KS and can involve the organs, often becoming fatal. The fourth form of KS (iatrogenic or transplant-related KS) occurs in organ transplant patients who receive immunosuppressant medications.

In 1994, a previously unknown herpesvirus, human herpesvirus-8 (HHV-8), was identified in Kaposi's sarcomas. The virus can be detected in all cases of KS, whether associated with immunodeficiency or not, and it is also referred to as Kaposi's sarcoma–associated herpesvirus (KSHV). KSHV is also associated with other malignant tumors, including multiple myeloma, and can be detected in the saliva of patients with HIV disease. It is present in about 25% of healthy U.S. adults.

In Kaposi's sarcoma, KSHV infects the endothelial cells of blood and lymphatic vessels and persists in the nucleus as a latent episome (not integrated). Although no new virus particles are made, the infected cells express a few virally encoded proteins. One of these, latency-associated nuclear antigen-1 (LANA-1), suppresses genes needed for full viral production. Another causes the infected cells to become spindle-shaped. A few other viral proteins are also made. These cause the host cell to release pro-inflammatory cytokines that induce inflammation, and growth factors that stimulate cell proliferation and survival, and lead to formation of new blood vessels (angiogenesis). Although typically latent, KSHV can become lytic, killing the infected cell. Unexpectedly, cytokines released during the lytic cycle also promote tumor formation and angiogenesis.

KSHV does not always cause Kaposi's sarcoma—in immunocompetent people, infection with this virus is typically asymptomatic. However, almost 50% of people with HIV infection develop KS after acquiring KSHV, indicating that HIV significantly increases the risk of developing this malignancy. This suggests that HIV itself may play a part in developing KS, and research has indicated that the HIV Tat protein is important in activating the lytic life cycle of KSHV. Other studies have started to examine whether host genetic variants, environment, and routes of infection also play a role.

MicroByte ⎯⎯⎯⎯⎯⎯⎯

Kaposi's sarcoma was originally described by Hungarian dermatologist Moritz Kaposi in 1872, but it became an AIDS-defining illness in the 1980s.

B Lymphocyte Tumors

Lymphomas are a group of malignant tumors that arise from lymphocytes. Most lymphomas affect B cells, although T-cell lymphomas also occur. AIDS patients get B-cell lymphomas at a significantly higher rate than that of the general public. Epstein-Barr virus (EBV) plays a role in some types of the AIDS-associated B-cell lymphomas. This virus probably plays an indirect role in lymphoma formation rather than being the direct cause—HIV infection causes activation of latent EBV infection, with release of the virus to infect new B cells. This then causes B-cell proliferation (leading to lymph node enlargement). Malignant B-cell clones are thought to arise from this population of rapidly dividing cells. ◄◄ Epstein-Barr virus, p. 680

EBV-associated lymphomas include cerebral (brain) lymphoma and immunodeficiency-associated Burkitt's lymphoma—both types are significantly more common among AIDS patients than in the general public. Burkitt's lymphoma also occurs in two other forms: endemic and sporadic. The endemic variant is found in Africa, primarily in children who have chronic malaria, which is thought to reduce their resistance to EBV. Sporadic Burkitt's lymphoma resembles the other types and is also thought to be associated with an impaired immune system.

Cervical and Anal Carcinoma

Carcinomas (cancers) of the cervix and anus are strongly associated with human papillomaviruses (HPVs) types 16 and 18. The cells involved in these cancers are epithelial cells and therefore differ from those in Kaposi's sarcomas and lymphomas. HPV is transmitted during sexual activity and infects the cervical and anal epithelium, appearing to cause increased replication of the cells by blocking expression of a cellular gene responsible for controlling cell growth. HPV replication increases as the host's cellular immunity declines with AIDS, organ transplantation, or other immunodeficient conditions. ◄◄ human papillomavirus, p. 631

MicroAssessment 28.2

Certain DNA viruses are strongly associated with development of malignant tumors in patients with HIV disease. These viruses do not cause malignancy by themselves, but require the presence of other conditions. The tumors all arise in a setting of increased cell proliferation caused in part by the viruses.

4. *In what epidemiological forms does Kaposi's sarcoma occur?*
5. *What member of the herpesvirus family is associated with almost all of the B-cell lymphomas of the brain in AIDS patients?*
6. *Would you expect mutations to arise in a population of rapidly dividing cells?* ➕

28.3 ■ Infectious Diseases

Learning Outcomes

8. *Describe the significance of* Pneumocystis jiroveci *infection in people with HIV disease.*
9. *Outline the epidemiology of toxoplasmosis.*
10. *Discuss the relationship between CMV disease and HIV disease.*
11. *Outline the pathogenesis of* Mycobacterium avium *complex infection in immunodeficient people.*

Immunodeficient people are susceptible to the same infections as other people, but the disease is likely to be more severe and may even be fatal. Bacteria, viruses, fungi, protozoa, and even parasitic worms can cause life-threatening illness in those with AIDS. Preventing infections in the immunocompromised is therefore very important. Vaccines should be given to immunocompetent

TABLE 28.7	Examples of Immunizations for Immunocompetent People with HIV Disease
Recommended for All	**Recommended for High-Risk People**
Hepatitis B	Hepatitis A
Influenza	*Haemophilus influenzae* type b
Pneumococcal polysaccharide	Human papillomavirus (HPV)
Diphtheria, tetanus, pertussis (DTaP)	Meningococcal
Measles-mumps-rubella (MMR)*	
Varicella zoster*	

*Only recommended if the person has a CD4+ count above 200 μl/ml.

people with HIV disease as early in the disease process as possible (**table 28.7**). This section presents some examples of infectious diseases common in patients with immunodeficiency.

Pneumocystis pneumonia (PCP)

Pneumocystis pneumonia (PCP), a severe, infectious lung disease, was recognized just after World War II in Europe when it killed malnourished, premature infants in hospitals. Subsequently, occasional cases were identified among immunodeficient patients. With the start of the AIDS epidemic, the incidence of PCP increased rapidly, and it is still a common opportunistic infection in AIDS patients who are not receiving preventive care.

Signs and Symptoms

Many people are infected with the fungus that causes PCP, but the infection is latent because it is kept under control by the immune system. In immunocompromised individuals, however, the organism causes disease. The signs and symptoms of *Pneumocystis* pneumonia typically begin slowly, with gradually increasing shortness of breath and rapid breathing. Fever is usually slight or absent, and only about half of the patients have a cough, which is non-productive. As the disease progresses, a dusky coloration of the skin and mucous membranes appears and gradually worsens—this is caused by poor oxygenation of the blood, and can become fatal.

Causative Agent

Pneumocystis pneumonia is caused by *Pneumocystis jiroveci*, a tiny yeastlike fungus (**figure 28.10**). The organism was formerly classified as *P. carinii* and it is still widely known by that name (explaining the acronym PCP for this type of pneumonia). *P. jiroveci* differs from many fungi in the chemical makeup of its cell wall, which makes it resistant to medications often used against fungal pathogens. The organism forms cysts that have a characteristic appearance, helping in its identification. It has not reliably been cultivated *in vitro*.

Pathogenesis

P. jiroveci spores are easily inhaled into lung tissue. In experimental infections, the spores attach to the alveolar walls, and the

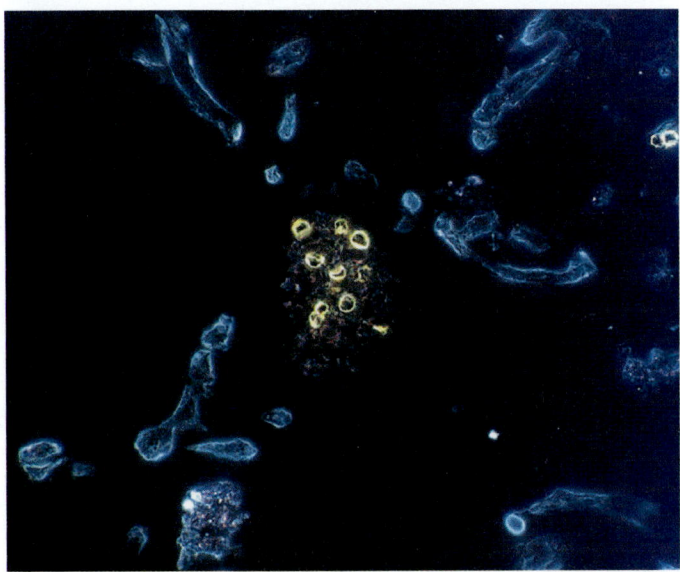

30 μm

FIGURE 28.10 Fluorescent Antibody Stain of *Pneumocystis jiroveci* The yellow circles are *P. jiroveci* cysts.

? *Why is* Pneumocystis *pneumonia known by the acronym PCP?*

TABLE 28.8	*Pneumocystis* Pneumonia
Signs and symptoms	Gradual onset, shortness of breath, rapid breathing, non-productive cough, slight or absent fever, and dusky color of skin and mucous membranes
Incubation period	4 to 8 weeks
Causative agent	*Pneumocystis jiroveci*, a tiny fungus (previously known as *P. carinii*)
Pathogenesis	Pneumocystis can result from reactivation of latent infection or be newly acquired. Spores of *P. jiroveci* enter the body when inhaled, attach to alveolar walls, and multiply. Alveoli fill with fluid, macrophages, and *P. jiroveci*. The walls thicken, impairing O_2 exchange.
Epidemiology	*P. jiroveci* widespread in domestic and wild animals as a latent lung infection, but the source of animal and human infections is unknown. Most humans become infected in early childhood. Disease arises in individuals with immunodeficiency; epidemics can occur in hospitalized premature infants and elderly nursing home residents.
Treatment and prevention	Formerly leading cause of death in those with AIDS, now usually prevented by medication (e.g., trimethoprim-sulfamethoxazole) as soon as the CD4+ lymphocyte count drops to 200 cells/μl. Same medication is used for treatment; alternatives available. Medication is continued for life, or until the CD4+ cell count rises and remains above 200 cells/μl as a result of HAART or other treatment of the underlying immunodeficiency.

alveoli fill with fluid, mononuclear cells, and masses of *P. jiroveci* cells in various stages of development. Later, the alveolar walls become thickened and scarred, preventing the free passage of O_2.

Epidemiology

Pneumocystis jiroveci is distributed worldwide. Most children are infected with *P. jiroveci* at an early age. The infection is asymptomatic and is generally eliminated within a year. The source and transmission of human infections are unknown. Most cases of PCP occur in people with immunodeficiency, but it is uncertain whether their disease is caused by reactivation of a latent infection, or new infection from inhalation of spores. Epidemics among hospitalized malnourished infants and elderly nursing home residents suggest airborne spread, and *P. jiroveci* has been detected in indoor and outdoor air by using polymerase chain reaction (PCR).
◀◀ PCR, p. 227

Treatment and Prevention

PCP is most often treated with trimethoprim-sulfamethoxazole. Alternative medications are given to people who cannot tolerate this medication because of its side effects—mainly rash, nausea, and fever. For unknown reasons, people with HIV disease are more likely to develop these side effects than others. Oxygen therapy and steroids (to reduce inflammation that worsens symptoms) are also given and can significantly decrease mortality. After treatment for PCP, individuals with HIV disease must receive preventive medication indefinitely, or until they have a sustained rise in CD4+ T-cell count to above 200 cells per microliter.

To prevent PCP, HIV-infected people are typically started on medication as soon as they become immunodeficient, as indicated by a CD4+ T-cell count below 200 per microliter, or development of characteristic opportunistic diseases. The main features of *Pneumocystis* pneumonia are presented in **table 28.8**.

Toxoplasmosis

Toxoplasmosis is a protozoan disease that rarely develops among healthy people but can be a serious problem for those with malignant tumors, organ transplant recipients, fetuses, and people with HIV disease.

Signs and Symptoms

Toxoplasmosis causes different signs and symptoms in immunologically healthy people, fetuses, and those with immunodeficiency. Most infections of immunocompetent people are asymptomatic, but 10% to 20% develop signs and symptoms similar to those of infectious mononucleosis. They usually consist of sore throat, fever, enlarged lymph nodes and spleen, and sometimes a rash. These signs and symptoms subside over weeks or months and do not require treatment. Occasionally, life-threatening illness develops, affecting the heart or central nervous system. ◀◀ infectious mononucleosis, p. 680

Fetal toxoplasmosis can be acquired transplacentally if a woman contracts the disease for the first time during her pregnancy. Development of the disease during the first trimester of pregnancy is the least common but most severe, often resulting in miscarriage or stillbirth. Babies born live may have serious birth defects including small or enlarged heads, and lung and liver damage.

Later, these babies can develop seizures or exhibit intellectual disability. However, almost two-thirds of fetal toxoplasmosis cases occur during the last trimester of pregnancy, and the effects on the fetus are usually less severe. Most of these infants appear normal at birth, although later in life, retinitis—infection of the retina, the light-sensitive part of the eye—may occur. This manifests itself as recurrent episodes of pain, sensitivity to light, and blurred vision, usually involving only one eye. Intellectual disability and epilepsy may develop. Congenital toxoplasmosis does not occur in fetuses of women who have had previous exposure to *T. gondii,* most likely because of immune protection.

Toxoplasmosis in immunodeficient people is often life-threatening. It commonly occurs as a reactivation of a latent *T. gondii* infection, and it is associated with various signs and symptoms including those similar to infectious mononucleosis. In more than half the cases, the infected person gets encephalitis (inflammation of the brain meninges), manifested by confusion, weakness, impaired coordination, seizures, stiff neck, paralysis, and coma. They can also develop brain masses much like tumors, which cause headaches and other neurological signs and symptoms such as unstable walking gait, seizures, sensory loss and weakness. Infection of the retina is also common.

Causative Agent

Toxoplasmosis is caused by *Toxoplasma gondii,* a tiny, banana-shaped protozoan that infects many animals, including household pets, pigs, sheep, cows, rodents, and birds (**figure 28.11a**). The life cycle of *T. gondii* is shown in figure 28.11b. The definitive host, in which the organism reproduces sexually, is the domestic cat or other felines (such as bobcats). The protozoan reproduces in the cat's intestinal lining, resulting in many progeny. Some of these spread throughout the body, while others differentiate into male and female gametes (sexual forms). Gametes fuse, forming thick-walled oocysts that are shed in the cat's feces. Millions of the oocysts are released daily, generally over a period of 1 to 3 weeks. In soil, the oocysts undergo further development over 1 to 5 days into an infectious form containing two sporocysts, each with four sporozoites. These can remain viable for up to a year, contaminating soil and water and, secondarily, hands and food. The oocysts are infectious for cats and other animals, including humans. In general, the cats recover from the acute infection and do not shed oocysts again.

When non-feline animals ingest the oocysts, the sporozoites emerge from the oocysts and invade the cells of the small intestine. They do not undergo a sexual cycle. The intestinal infection spreads by the lymphatics and blood vessels throughout the tissues of the host, infecting cells of the heart, brain, and muscles. As host immunity develops, multiplication slows, and a tough, fibrous capsule forms, surrounding large numbers of a smaller form of *T. gondii.* These capsules are called cysts (figure 28.11c), and they remain viable for months or years. The life cycle is completed when a cat becomes infected by eating an animal with cysts in its tissues.

Pathogenesis

The organisms enter the body by ingestion of oocysts or undercooked meat containing tissue cysts. *T. gondii* infects any kind

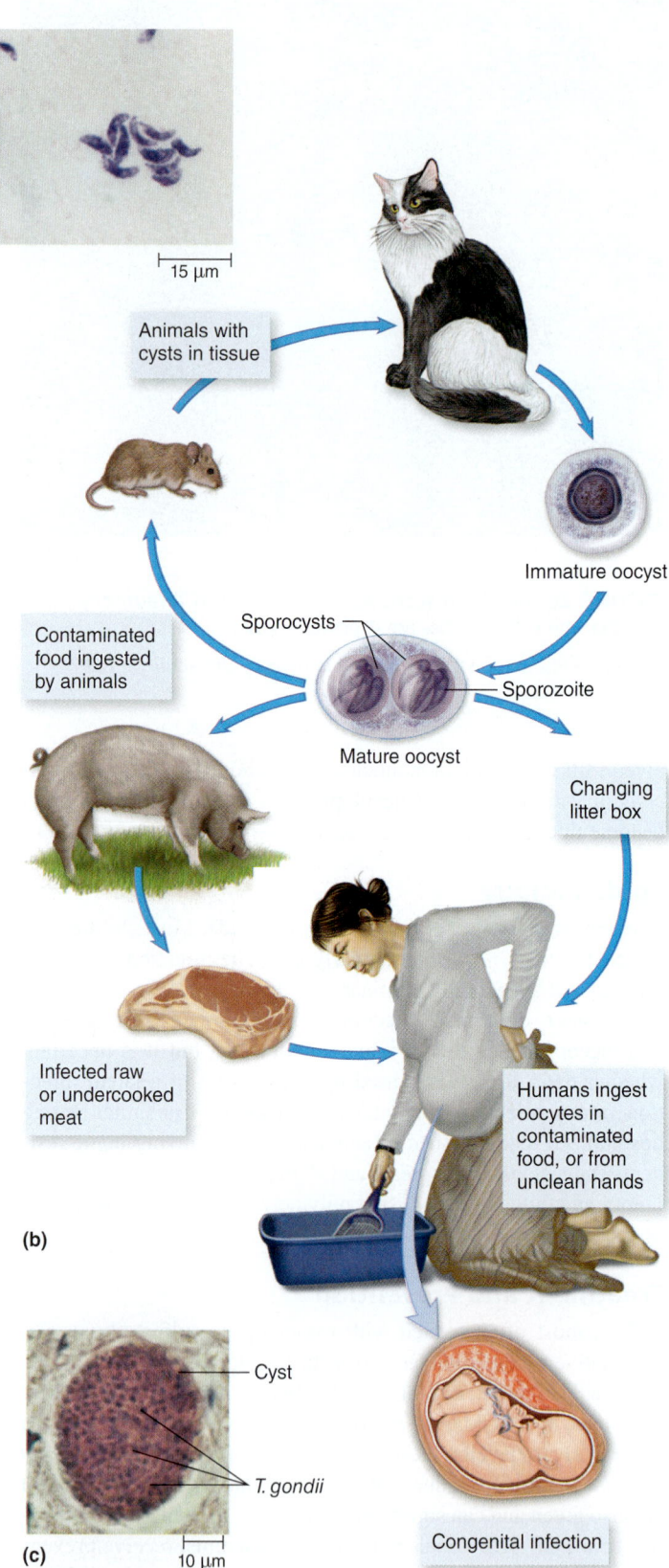

(a)   15 µm

Animals with cysts in tissue

Immature oocyst

Contaminated food ingested by animals

Sporocysts

Sporozoite

Mature oocyst

Changing litter box

Infected raw or undercooked meat

(b)

Humans ingest oocytes in contaminated food, or from unclean hands

Cyst

T. gondii

(c)   10 µm

Congenital infection

FIGURE 28.11 *Toxoplasma gondii* **(a)** Invasive forms. **(b)** Life cycle. Oocysts from cat feces and cysts from raw or inadequately cooked meat can infect humans and many other animals. **(c)** Cyst in tissue.

❓ *Why does eating undercooked meat increase the risk of contracting toxoplasmosis?*

of cell except red blood cells. The organism produces an enzyme that alters the host cell membrane and aids entry into the cell. The infected cells are destroyed by the proliferating *T. gondii*. This process is normally brought under control by the immune response of the host—tissue cysts develop, and infected humans usually show few, if any, signs and symptoms. In patients with immunodeficiency, however, infection can be widespread and uncontrolled, producing many areas of tissue necrosis. The disease process can result from a newly acquired infection, or from reactivation of latent infection that occurs with declining immunity.

Epidemiology

Toxoplasma gondii is distributed worldwide, but it is less common in cold or hot, dry climates. Human infection is widespread. Serological surveys show the infection rate increases with age. Most infections are acquired when a person eats or drinks something contaminated with oocysts, or inhales contaminated dust. Gardening in areas where young, stray cats that eat rodents and birds live can result in *T. gondii* contamination of hands or vegetables. Eating rare meat is risky because it can contain tissue cysts.

Treatment and Prevention

Toxoplasmosis is treated with drugs such as pyrimethamine with sulfadiazine, which are related to trimethoprim-sulfamethoxazole (used to treat PCP). Alternatives are available. All HIV-infected patients and those about to receive immunosuppressant medications are tested for antibody to *T. gondii*. A positive test indicates they have latent infection, and those with positive tests are given prophylactic trimethoprim-sulfamethoxazole if they have fewer than 100 CD4$^+$ T cells per microliter. ◀◀ **PCP, p. 708**

    T. gondii infection can be avoided by washing hands after touching raw meat, soil, or cat litter. Meat, especially lamb, pork, and venison (deer meat), should be cooked thoroughly. Fruits and vegetables should be washed before eating. Litter boxes should be cleaned regularly, before oocysts have a chance to mature. Cats should not be allowed to hunt birds and rodents, nor should they be fed undercooked or raw meat. These measures are especially important for pregnant women and immunodeficient people. The main features of toxoplasmosis are presented in **table 28.9.**

Cytomegalovirus Disease

Cytomegalovirus (CMV) is a member of the herpesvirus family. Like other herpesviruses, CMV is commonly acquired early in life and then remains latent. With immunodeficiency, the infection reactivates and can cause severe illness.

Signs and Symptoms

Signs and symptoms of cytomegalovirus disease follow a pattern similar to that of toxoplasmosis. Acute infections in immunologically healthy people are usually asymptomatic, but adolescents and young adults sometimes develop illness that resembles infectious mononucleosis—fever, fatigue, and enlarged lymph nodes and spleen that can persist for weeks or months.

    If a woman develops an acute CMV infection during pregnancy, her fetus can be severely damaged. This condition is known as congenital cytomegalic inclusion disease and is

TABLE 28.9	Toxoplasmosis
Signs and symptoms	In healthy individuals: sore throat, fever, enlarged lymph nodes, rash; with fetal infections: miscarriage, stillbirth, birth defects, epilepsy, intellectual disability, retinitis; in immunodeficient individuals: confusion, poor coordination, weakness, paralysis, seizures, and coma
Incubation period	Usually indeterminate
Causative agent	*Toxoplasma gondii*, a protozoan infectious for most warm-blooded animals. Sexual reproduction occurs in the intestinal epithelium of cats, the definitive hosts. Infected cats discharge oocysts with their feces. Ingested organisms are released from the oocysts, multiply rapidly, spread throughout the body. As immunity develops, infected cells become filled with the organisms, resulting in tissue cysts, which remain viable and infectious for the lifetime of the animal.
Pathogenesis	Organisms penetrate host cells causing necrosis. With development of immunity cell destruction stops, tissue cysts develop. Most healthy individuals have few or no symptoms unless the number of ingested organisms is very large. Organisms released from tissue cysts if immunity becomes impaired.
Epidemiology	Occurs worldwide, less common in cold or dry locations. Infection acquired by ingesting oocysts from cat feces, or eating inadequately cooked meat.
Treatment and prevention	Treatment: pyrimethamine with sulfadiazine, or alternative medication. Prevention: avoiding foods potentially contaminated with oocysts from cat feces, and not eating inadequately cooked meat. Trimethoprim-sulfamethoxazole is given to immunodeficient persons with CD4$^+$ T lymphocyte counts below 100 cells/µl if they have antibodies to *T. gondii* indicating latent infection.

characterized by jaundice, enlarged liver, anemia, eye inflammation, and birth defects. The vast majority of infected infants appear healthy at birth, but up to a quarter of them develop hearing loss, mental handicap, or other abnormalities later in life.

    In immunodeficient people, latent CMV infection can reactivate, causing a variety of signs and symptoms. Retinitis, leading to blindness, is one of the most common complications of reactivated CMV disease. Other signs and symptoms include fever, loss of appetite, painful joints and muscles, rapid and difficult breathing, ulcerations of the gastrointestinal tract with bleeding, lethargy, paralysis, dementia, coma, and death.

Causative Agent

Human cytomegalovirus is an enveloped, double-stranded DNA virus that looks like other herpesviruses on electron micrographs but has a larger genome. The virus name (*cyto* for "cell" and *megalo* for "large") derives from the fact that cells infected with it are two or more times the size of uninfected cells. Infected cells show a large, intranuclear inclusion body surrounded by a clear halo, inspiring its description as an "owl's eye" (**figure 28.12**). There are many different strains of CMV, and like other herpesviruses, CMV can lyse infected cells or can become latent with possible later reactivation.

Pathogenesis

Cytomegalovirus disease affects a wide variety of tissues, including the eye, central nervous system, lung, and liver (**figure 28.13**). Once the virus enters the host cell, the viral genome can exist in a latent, non-infectious form or a slowly replicating form, or it can direct a productive infection. ◀◀ productive infection, p. 310

Control of viral gene expression depends partly on the type of cell infected. In monocytes, low numbers of infectious viral particles are produced. In T and B lymphocytes, some CMV genes are expressed, but the viral genome is not replicated nor are new viral particles produced. Integration of viral DNA into the host cell genome probably does not occur. If CMV-infected T cells are also infected with HIV, however, productive CMV infection occurs. This involves a wide variety of tissues, causing necrosis.

CMV can be transmitted by transplanted organs and blood transfusions. This indicates that virus-producing cells are present or that non-producing cells in the blood or organ start producing infectious CMV under conditions of immune suppression. Cell-mediated immunity probably plays a role in suppressing production of infectious virus as well as in lysis of infected cells. In cells infected with CMV, however, transfer of MHC molecules to the cell surface is impaired and, therefore, CMV antigens are not recognized as being "foreign." CMV infection is associated with an increase in CD8+ cells and a decrease in CD4+ cells, thus enhancing the effect of HIV infection. Latent infections activate with

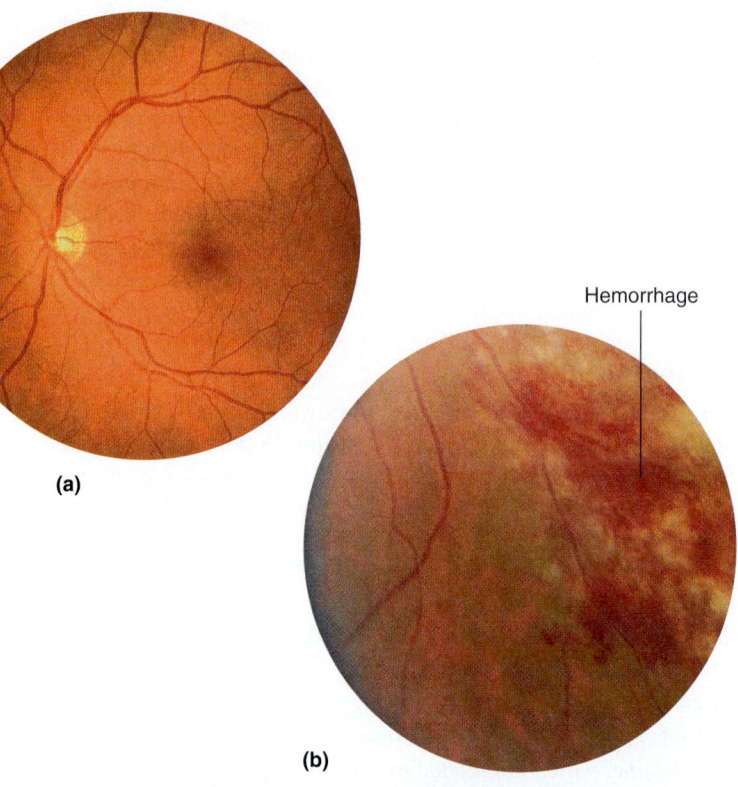

(a)

Hemorrhage

(b)

FIGURE 28.13 Cytomegalovirus (CMV) Retinitis Photograph of a normal retina **(a)** compared with a CMV-infected retina **(b)**.

❓ *What does CMV retinitis commonly lead to in people with AIDS?*

AIDS, organ transplants, and other immunodeficient states. Also, immunodeficient people are highly susceptible to newly acquired infection. ◀◀ MHC molecules, p. 369

Epidemiology

CMV is found worldwide. One U.S. study found that more than 50% of adults, ages 18 to 25 years, and more than 80% of people over age 35 had been infected. Infection is lifelong. Infants born with CMV infection and those who acquire it shortly after birth excrete the virus in their saliva and urine for months or years. Virus is found in saliva, semen, and cervical secretions. Unprotected sex is a common mode of transmission in young adults. Up to 15% of pregnant women secrete the virus, and 1% of newborn infants have CMV in their urine. Breast milk, blood, and tissue transplants may contain CMV and can be responsible for transmission.

Treatment and Prevention

The severity of CMV disease can be reduced by taking the antiviral medications ganciclovir and foscarnet in combination. These medications inhibit DNA polymerase, thus affecting replication of the virus. Both medications have serious side effects, however, mainly bone marrow suppression and kidney damage.

There is no approved vaccine for preventing cytomegalovirus disease. The risk of sexual transmission is effectively reduced by using condoms. Tissue and blood donors can be screened for antibody to CMV; those who have anti-CMV antibody are assumed to be infected and should not donate to those lacking CMV antibody.

Infected cells

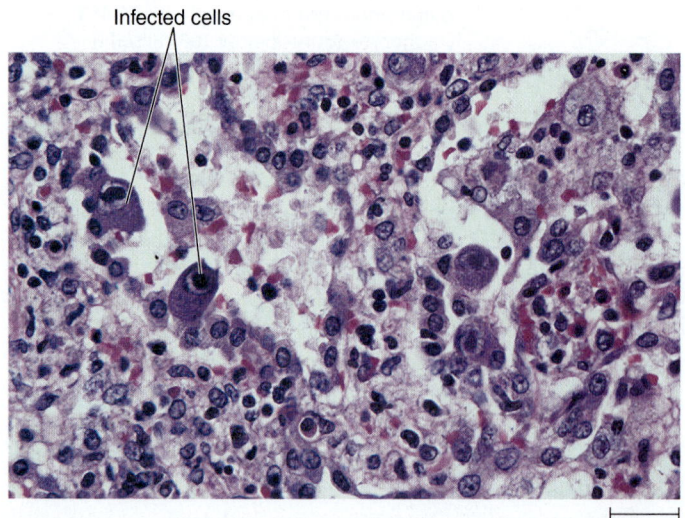

20 μm

FIGURE 28.12 Cytomegalovirus Infection Infected cells are two to four times normal size.

❓ *Why are CMV-infected cells described as having an "owl's-eye" appearance?*

The incidence of CMV retinitis in people with HIV disease can be reduced by half if ganciclovir is given. A ganciclovir implant designed to release the medication into the eye over a long period also delays the progression to blindness. The main features of cytomegalovirus disease are presented in **table 28.10**.

Mycobacterial Diseases

In most immunocompetent people, initial exposure to *Mycobacterium tuberculosis* (the cause of tuberculosis) results in an asymptomatic infection that is controlled by the immune system and becomes latent. However, defects in cell-mediated immunity caused by AIDS and other immunodeficiencies often result in reactivation of latent tuberculosis, leading to active tuberculosis disease, which may be fatal. ◀◀ tuberculosis, p. 502

M. tuberculosis is not the only mycobacterium that causes disease in immunodeficient people, although it is the most common. This section discusses disease caused by organisms of **Mycobacterium avium complex (MAC),** mycobacterial opportunists that complicate immunodeficiency diseases.

TABLE 28.10	Cytomegalovirus Disease
Signs and symptoms	Signs and symptoms rare in immunocompetent individuals, but may resemble infectious mononucleosis if they develop. Infection of the mother during pregnancy can result in disease of the newborn. Immunocompromised individuals may experience blindness, lethargy, dementia, coma, and brain damage.
Incubation period	In immunocompetent adults, 20 to 60 days
Causative agent	Cytomegalovirus, a member of the herpesvirus family.
Pathogenesis	Many tissues susceptible to infection, damage, especially eyes, brain, and liver. CMV latent infection can reactivate, produce infectious virions, tissue necrosis. CD4⁺ T lymphocyte count is depressed and thus can enhance HIV disease.
Epidemiology	Common worldwide; lifelong infection. More than 50% of 18- to 25-year-olds are infected. Infants with congenital infections and those infected shortly after birth shed the virus for months or years. Body fluids, including breast milk, blood, urine, semen, and vaginal secretions, can transmit the disease.
Treatment and prevention	Treatment: ganciclovir plus foscarnet. Prevention: No vaccine available. Condoms decrease transmission. CMV-negative immunodeficient people advised to wash their hands following contact with bodily fluids or feces. Blood and tissue transplants are tested for CMV before being given to CMV-negative individuals. The antiviral medication ganciclovir is considered for immunodeficient individuals who have antibody to CMV and whose CD4⁺ T lymphocyte count falls below 50 cells/μl.

Signs and Symptoms

The vast majority of MAC infections in immunologically healthy people are asymptomatic. However, elderly people, especially those with underlying lung disease from smoking and those with alcoholism, develop a chronic, productive cough. They may also develop lung lesions resembling those of tuberculosis. Children sometimes develop chronic lymph node enlargement on one side of their neck, easily treated by surgical removal of the affected nodes. Patients with AIDS and other severe immunodeficiencies have slowly progressing signs and symptoms ranging from chronic productive cough to fever, drenching sweats, marked weight loss, abdominal pain, and diarrhea.

Causative Agents

MAC consists of groups of more than 24 strains of two closely related acid-fast species—*M. avium* and *M. intracellulare*. The strains are distinguishable by serological tests, optimum growth temperature, and host range. Their growth rate is almost as slow as that of *M. tuberculosis,* but they are easily distinguished from this organism by using biochemical tests and nucleic acid probes.

Pathogenesis

MAC organisms enter the body via the lungs and the gastrointestinal tract. They are phagocytized by macrophages, but resist destruction because they inhibit acid production in the phagosome. Surviving organisms multiply within phagocytes and are carried by the bloodstream to all parts of the body. When cell-mediated immunity is intact, most organisms are destroyed, and disease is localized. With immunodeficiency, the disease spreads throughout the body. In these patients, there is persistent bacteremia, occasionally with counts as high as 1 million bacteria per milliliter of blood. The small intestine contains macrophages packed with high numbers of the bacteria (**figure 28.14**). Despite the presence of enormous numbers of the bacteria, there is little or no inflammatory reaction, and the clinical effect is a slow decline in the patient's well-being rather than a quickly lethal effect. One cause of deterioration may be activation of HIV replication when HIV-infected macrophages are infected with MAC.

MAC bacteria

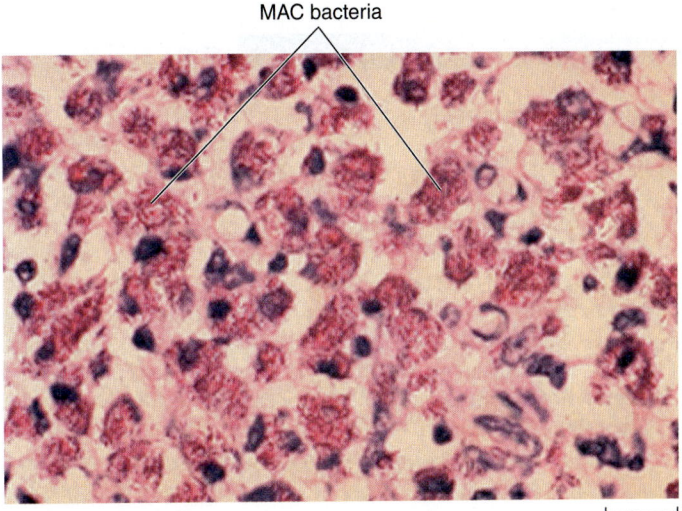

10 μm

FIGURE 28.14 MAC Infection Acid-fast stain showing massive numbers of MAC bacteria infecting cells.

❓ *What species are involved in MAC infection?*

TABLE 28.11	MAC Disease
Signs and symptoms	Usually asymptomatic in healthy people. Children can develop chronic localized enlargement of lymph nodes. Can cause chronic cough similar to tuberculosis in the elderly. Cough, fever, sweating, marked weight loss, abdominal pain, and diarrhea in those with severe immunodeficiencies.
Incubation period	Usually indeterminate
Causative agents	A group of mycobacterial strains in the species *M. avium* and *M. intracellulare*
Pathogenesis	MAC bacteria enter the body via lungs or gastrointestinal tract, are taken up by macrophages. In immunocompetent individuals, most are destroyed and infection is controlled by cell-mediated immunity. In severe immunodeficiency the bacteria resist phagocytosis, are carried by macrophages throughout the body, and grow to enormous numbers in tissues but cause little or no inflammatory reaction.
Epidemiology	MAC organisms are widespread in food, water, soil, and dust. MAC disease of immunodeficient persons could result from environmental sources, or activation of latent infection.
Treatment and prevention	No proven measures to prevent exposure to MAC bacteria. The antibiotic clarithromycin is used to prevent MAC disease in severe immunodeficiency. If MAC bacteremia develops, two or more medications, such as clarithromycin plus ethambutol, are effective.

MicroByte

MAC-infected people may contain 100 billion of the bacteria per gram of tissue.

Epidemiology

MAC organisms are widespread and have been found in food, water, soil, and dust. Some strains are important pathogens of chickens and pigs. In AIDS patients, they are the most common bacterial cause of generalized infection. It is not known whether infection in AIDS patients is mostly newly acquired or a reactivation of latent disease. Most infections are from environmental sources rather than from person-to-person spread.

Treatment and Prevention

If MAC bacteremia develops, patients must be given two or more medications together, such as clarithromycin with ethambutol. There are no effective measures to prevent exposure to MAC organisms. For AIDS patients whose CD4+ count is below 50 cells per microliter, the antibacterial medication clarithromycin is recommended to help prevent MAC disease. The main features of MAC disease are presented in **table 28.11.**

The key features of the diseases covered in this chapter are highlighted in the **Diseases in Review 28.1** table.

MicroAssessment 28.3

Infectious diseases are a major threat to individuals with immunodeficiency. They get the same infectious diseases that afflict healthy people, but these can be life-threatening. In addition, immunocompromised people are much more susceptible to reactivation of latent infections and organisms of low virulence that are generally harmless to others. Prevention of infectious diseases in AIDS patients is important for improving the quality and duration of their lives.

7. *What is the most common opportunistic infection of immunodeficient people?*

8. *Which organisms are involved in MAC bacteremia?*

FUTURE CHALLENGES 28.1

AIDS and Poverty

In the United States and other economically advanced countries, antiviral medications have contributed significantly to reducing mother-child transmission of HIV, to delaying progression of HIV disease to AIDS, and to improving the quality and duration of life of AIDS sufferers. Newer drugs have also decreased transmission of HIV somewhat, thereby slowing the progress of the epidemic. The cost in the United States, while high, amounts to less than 1% of the country's total healthcare expenditure.

Antiviral medications, however, have provided little benefit to most of the world's AIDS sufferers because they cannot afford them. Only about 5.2 million of the 15 million people needing anti-HIV medications are receiving them. Although HIV disease is spreading rapidly in India and China—two countries that together contain more than one-third of the world's population—and in the Caribbean, Southeast Asia, and Eastern Europe, the situation is the most desperate in African countries south of the Sahara Desert. Only 10% of the world's population lives in this area, but two-thirds of the people are infected with HIV. About 75% of the region's young people (ages 15 to 24) infected with HIV are girls or young women.

The humanitarian, social, and economic implications of uncontrolled HIV disease are enormous. Unimaginable suffering; large numbers of starving, uneducated street children; loss of productive work years; spread of other infectious diseases; breakdown of public health; and setbacks for economic development are possible consequences that can affect neighboring countries and the rest of the world.

The challenge is urgent: to find effective HIV control measures, especially for the world's poor.

Diseases in Review 28.1

HIV Disease and Complications Caused by Malignant Tumors and Infectious Diseases

Disease	Causative Agent	Comment	Summary Table
HUMAN IMMUNODEFICIENCY VIRUS AND AIDS			
HIV disease	Human immunodeficiency virus (HIV)	Acute retroviral syndrome marks the early stage; period of clinical latency can last for many years; AIDS begins as the person becomes increasingly immunodeficient; HAART lowers viral set point, extending clinical latency; most antiretroviral drugs target reverse transcriptase or protease; vaccine research has been hampered by the lack of good animal model, HIV's high mutation rate, and HIV's ability to avoid humoral and cellular immune responses.	Table 28.6, p. 705
MALIGNANT TUMORS THAT COMPLICATE ACQUIRED IMMUNODEFICIENCIES			
Kaposi's sarcoma	Human herpesvirus-8 (HHV-8, also called KSHV)	These tumors arising from blood or lymphatic vessels are characterized by red, brown, purple, or black papular nodules on the skin or other body sites; in immunocompetent people, the virus typically causes an asymptomatic infection that becomes latent.	
B lymphocyte tumors	Epstein-Barr virus (EBV)	Latent EBV reactivates as a result of HIV infection; the role of EBV in tumor formation is not well understood.	
Cervical and anal carcinoma	Certain strains of human papillomaviruses (HPVs)	HPV replication increases with decreasing cell-mediated immunity; HPV blocks normal cellular gene that controls cell growth.	
INFECTIOUS DISEASES THAT COMPLICATE ACQUIRED IMMUNODEFICIENCIES			
PROTOZOAN DISEASES			
Pneumocystis pneumonia (PCP)	*Pneumocystis jiroveci*	The most common opportunistic infection in AIDS patients; in immunocompetent people, the fungus typically causes an asymptomatic infection that becomes latent.	Table 28.8, p. 709
Toxoplasmosis	*Toxoplasma gondii*	Latent *T. gondii* reactivates as a result of AIDS, often resulting in encephalitis; acute infections in pregnant women can damage the fetus, resulting in miscarriages or severe birth defects.	Table 28.9, p. 711
VIRAL DISEASE			
Cytomegalovirus disease	Cytomegalovirus (CMV)	Latent CMV reactivates as a result of AIDS, often resulting in retinitis, which causes blindness; infection causes a variety of other signs and symptoms; acute infections in pregnant women can damage the fetus.	Table 28.10, p. 713
BACTERIAL DISEASES			
Mycobacterial diseases	*Mycobacterium* species	Tuberculosis is a significant problem in AIDS patients; a group of opportunists called *Mycobacterium avium* complex (MAC) can cause similar signs and symptoms in AIDS patients.	Table 28.11, p. 714

Summary

AIDS

Since AIDS was first recognized in 1981, it has claimed millions of lives in the United States, and is a leading cause of death in people 25 to 44 years of age. Worldwide, over 34 million people are infected with an AIDS-causing virus **(figure 28.1)**. In sub-Saharan Africa, AIDS is now the number one cause of death, surpassing malaria; it is spreading rapidly in India and into China.

28.1 ■ Human Immunodeficiency Virus (HIV) Infection and AIDS

"AIDS-defining conditions," including serious infections by agents that normally have little virulence, usually reflect immunodeficiency and were especially useful "markers" in early studies of the AIDS epidemic **(table 28.1)**.

HIV Disease (tables 28.2–28.7, figures 28.2–28.7)

Acquired immunodeficiency syndrome (AIDS) is a late manifestation of **human immunodeficiency virus (HIV) disease.** "Flulike" or infectious mononucleosis-like signs and symptoms representing the **acute retroviral syndrome (ARS)** often occur 6 days to 6 weeks after infection by HIV; these subside without treatment. HIV disease progresses asymptomatically for years, although HIV can still be transmitted to others during this time. The asymptomatic period ends with the onset of immunodeficiency (AIDS), marked by opportunistic or reactivated infections, and certain tumors caused by infectious agents. HIV disease and AIDS are caused by human immunodeficiency virus, a single-stranded RNA virus belonging to retrovirus family **(table 28.2, figure 28.3)**. HIV disease is spread mainly by sexual intercourse, by blood or blood products, and from mother to fetus or newborn **(tables 28.4, 28.5)**. Different kinds of cells can be infected by HIV, notably two kinds vital to the immune system—helper T lymphocytes and macrophages **(figure 28.4)**. Inside the host cell, a DNA copy of the viral genome is made by reverse transcriptase, a complementary DNA strand is made, and the resultant double-stranded DNA is inserted into the host genome as a provirus by viral integrase **(figures 28.5, 28.6)**. Antimicrobial medications are used to prevent and treat opportunistic infections, and **highly active antiretroviral therapy (HAART)** is used to maintain a low **HIV set point** and manage HIV disease **(tables 28.3, 28.7; figure 28.7)**.

HIV Vaccine Prospects

There are no approved vaccines for HIV at this time.

COMPLICATIONS OF ACQUIRED IMMUNODEFICIENCIES

28.2 ■ Malignant Tumors

Kaposi's Sarcoma (figure 28.9)

Kaposi's sarcoma is a tumor that arises from blood or lymphatic vessels. The incidence of the tumor is markedly increased among immunodeficient individuals. Infection by human herpesvirus-8 (KSHV) appears to be involved in tumor development.

B Lymphocyte Tumors

Lymphomas are malignant tumors that arise from lymphocytes. Both B and T lymphocytes can give rise to lymphomas in immunodeficient people, but B-cell lymphomas are more common, and these often arise in the brain. Strong evidence suggests that Epstein-Barr virus plays a causative role in these tumors.

Cervical and Anal Carcinoma

There is an increased rate of anal, genital, and cervical carcinoma in people with HIV disease. These cancers are strongly associated with human papillomaviruses (HPVs), transmitted by sexual intercourse.

28.3 ■ Infectious Diseases

Infections that occur in healthy individuals also occur and produce more severe disease in those with immunodeficiency.

Pneumocystis pneumonia (PCP) (table 28.8, figure 28.10)

Before effective preventive regimens were developed, PCP was one of the most common causes of death among AIDS patients. Signs and symptoms develop slowly, with gradually increasing shortness of breath and rapid breathing; patients can die from lack of O_2. The causative agent, *Pneumocystis jiroveci*, a tiny fungus, is widespread in humans, generally causing asymptomatic infections that become latent in immunocompetent people.

Toxoplasmosis (table 28.9)

Toxoplasmosis is caused by *Toxoplasma gondii*, a tiny, protozoan present worldwide. This organism reproduces in the intestinal epithelium of cats but can infect humans and other vertebrates. People contract toxoplasmosis by ingesting oocytes, which are discharged in the feces of acutely infected cats, become infectious in soil, and contaminate food, water, and fingers **(figure 28.11)**. Toxoplasmosis is rare among healthy people, and if it does occur is usually asymptomatic. Signs and symptoms that may develop mimic those of infectious mononucleosis and usually resolve without treatment. This disease, however, can be a serious problem for people with AIDS, who frequently develop encephalitis and brain masses that may be fatal.

Cytomegalovirus Disease (table 28.10)

Cytomegalovirus (CMV) is a common cause of impaired vision in people with AIDS **(figure 28.13)**. It can also cause fever, gastrointestinal bleeding, mental dullness, and blindness in these individuals. CMV is an enveloped, double-stranded DNA virus; infected cells are enlarged and have an "owl's-eye" appearance **(figure 28.12)**. The virus occurs in breast milk, semen, and cervical secretions, and in saliva and urine of infected infants. No vaccine is available. Condoms decrease the risk of sexual transmission. The antiviral medication ganciclovir can be given to prevent CMV retinitis.

Mycobacterial Diseases (table 28.11)

Latent *Mycobacterium tuberculosis* may reactivate and cause active tuberculosis disease in immunocompromised people. *M. avium* complex (MAC) organisms also commonly cause lung infections. Immunodeficient patients may have fever, drenching sweats, severe weight loss, diarrhea, and abdominal pain. MAC organisms resist phagocytic destruction and are carried to all parts of the body in macrophages, multiplying without restriction, and producing massive numbers of the organisms in blood, intestinal epithelium, and other tissues **(figure 28.14)**. No effective preventive measures are available. Prophylactic medication is advised for severely immunodeficient patients, but it can fail to prevent infection.

Review Questions

Short Answer

1. What is the main symptom of patients with lymphadenopathy syndrome (LAS)?

2. Which cells of the immune system are prime targets of HIV?

3. What role do asymptomatic people with HIV disease play in the epidemiology of AIDS?

4. Why might the infant son of a hemophiliac man develop AIDS when the son's parents were strictly monogamous non-abusers of drugs?

5. Give two reasons it is a good idea to know whether you are infected with HIV.

6. What are the three main types of malignant tumors that complicate HIV disease?

7. How do physicians prevent pneumocystis in AIDS patients?

8. In AIDS patients with toxoplasmosis, which part of the body is affected in more than half the cases?

9. Name a feared complication of cytomegalovirus infection in AIDS patients.

10. Where in an AIDS patient's surroundings might MAC organisms be found?

Multiple Choice

1. HIV can be spread by all of the following *except*
   a) blood products. b) hypodermic syringes. c) insect bites.
   d) sexual intercourse. e) organ transplants.

2. All of the following signs and symptoms are characteristic of the AIDS-related complex (ARC) *except*
   a) fever. b) fatigue. c) diarrhea.
   d) blindness. e) weight loss.

3. Which one of the following is *true* of Kaposi's sarcoma?
   a) KSHV is necessary for development of the tumor.
   b) HIV-1 is necessary for development of the tumor.
   c) Both KSHV and HIV-1 are necessary for development of the tumor.
   d) KSHV alone is sufficient for development of the tumor.
   e) Both KSHV and HIV-1 together are sufficient for the tumor to develop.

4. All of the following are HIV accessory genes *except*
   a) *tat.* b) *env.* c) *vpr.* d) *rev.* e) *vpu.*

5. When was AIDS first recognized as representing a new disease?
   a) 1973 b) 1959 c) 1981 d) 1989 e) 1999

6. All of the following are AIDS-defining conditions *except*
   a) influenza.
   b) herpes simplex of the esophagus.
   c) *Pneumocystis jiroveci* pneumonia.
   d) invasive cancer of the uterine cervix.
   e) Kaposi's sarcoma.

7. Which of the following types of cells can be infected by HIV?
   a) Helper T cells
   b) Cytotoxic T cells
   c) B lymphocytes
   d) CD 8$^+$ cells
   e) All of the above

8. All of the following are HIV antigens *except*
   a) CD4. b) TM. c) RT. d) MA. e) CA.

9. Which of the following is a cause of helper T-cell death in HIV disease?
   a) Replication of HIV lyses the cell.
   b) Infected cells are destroyed by cytotoxic T cells (T$_C$).
   c) Infected cells are attacked by natural killer cells.
   d) Cells are killed by fusion and syncytium formation.
   e) All of the above

10. Highly active antiretroviral therapy (HAART) is less than ideal because
    a) it does not eliminate latent HIV infection.
    b) its cost is too great for the majority of AIDS sufferers.
    c) it often has severe side effects.
    d) some HIV strains are resistant to it.
    e) All of the above

Applications

1. An epidemiologist from the CDC was presenting a report on the status of AIDS to a congressional committee. In concluding her remarks, she noted that from an epidemiological perspective it was more important to focus on HIV infection than on AIDS, and urged that the Congress consider redirecting funding of AIDS research to reflect this fact. What was the rationale for her request?

2. A historian researching the influence of society on the spread of communicable disease began to speculate on what it would be like if AIDS had appeared at a different time. What differences might one expect, for example, if AIDS had appeared in 1928 instead of 1978?

Critical Thinking ✚

1. Vaccines have effectively prevented many viral diseases. Attempts over many years to develop an effective vaccine against HIV disease and AIDS, however, have so far met with little success. Why is this so?

2. Why is reverse transcriptase needed in order for HIV to become a provirus?

29 Microbial Ecology

Farming relies on the activities of microorganisms.

A Glimpse of History

For centuries, farmers have understood that growing the same crop on the same piece of land year after year reduces the crop yield. Allowing a field to lie unplanted for one or more seasons lets wild plants grow, and these appear to rejuvenate the soil. It was not until the late nineteenth century that scientists began to discover why this was so. They isolated soil microorganisms associated with certain plants that could take nitrogen from the air and transform it into forms other organisms could use. This process is called nitrogen fixation, and the nitrogen is said to be "fixed."

Although many scientists have worked for years to understand how microorganisms fix nitrogen, one scientist from the Netherlands, Martinus Beijerinck, stands out as an early contributor in these studies. In the late 1880s, he isolated a bacterium from inside the nodules that form on the roots of legumes (plants that bear seeds in pods). Russian microbiologist Sergei Winogradsky then showed that the bacterium forms a symbiotic relationship with the legumes. We now know that the bacterial cells in the nodules fix nitrogen, and they are in a group called rhizobia. ◄◄ rhizobia, p. 269

Beijerinck made major contributions to several areas of microbiology. He worked on yeasts, plant viruses such as tobacco mosaic virus, and plant galls (tumors). He was described as a "keen observer" who was able "to fuse results of remarkable observations with a profound and extensive knowledge of biology and the underlying sciences." This ability was undoubtedly partly responsible for the great success of his work.

Microbes cycle nutrients, maintain fertile soil, and decompose wastes and other pollutants. Without microbial activities, life on earth could not survive. People would quickly become buried by the tons of wastes we generate, and nutrients would be depleted, halting growth and reproduction. In view of the crucial functions microorganisms perform, it seems we should know a great deal about the diverse microbial species that inhabit our surroundings. Quite the opposite is true, however, as less than a mere 1% have been successfully grown in culture.

Even if all microorganisms could be cultivated in the laboratory, the information gained might not accurately reflect their role in the environment. In the laboratory, organisms are grown as pure cultures under controlled conditions that ensure optimal growth. In nature, however, organisms generally grow as members of mixed communities in poorly defined and often changing conditions. Nutrients are normally in short supply, limiting growth. Thus, with respect to environmental microbiology, results obtained in the artificial setting of the laboratory, although useful, must be interpreted with caution. ◄◄ bacterial growth in nature, p. 84

Chapters 4 and 11 discussed prokaryotic growth in nature, and microbial diversity. This chapter will expand on some of those concepts, focusing on activities of microorganisms that make them essential to life.

29.1 ■ Principles of Microbial Ecology

Learning Outcomes

1. *Describe the roles of primary producers, consumers, and decomposers.*
2. *Describe how some microbes are able to grow in low-nutrient environments.*
3. *Compare and contrast microbial competition and antagonism.*
4. *Describe how environmental changes can result in alterations in a microbial community.*
5. *Describe the structural organization of a microbial mat.*
6. *Describe three methods researchers use to better understand complex microbial communities.*

Ecology is the study of the relationships of organisms—plant and animal—to each other and to their environment. Likewise, **microbial ecology** is the study of the relationships of microorganisms to each other and to their environment. Living organisms interact with one another in symbiotic relationships (commensalism, mutualism, and parasitism), described in chapter 16. ◄◄ symbiosis, p. 381

Organisms in a given area, the **community,** interact with each other and the non-living environment, forming an ecological system, or **ecosystem.** Major ecosystems include the oceans, rivers and lakes, deserts, marshes, grasslands, forests, and tundra. Each ecosystem possesses a certain spectrum of organisms and characteristic physical conditions. The region of the earth inhabited by living organisms is called the **biosphere.** Within the biosphere, ecosystems vary both in **biodiversity** (number and variety of species present and their evenness of distribution) and **biomass** (the weight of all organisms present). Microorganisms play a major role in most ecosystems, and many ecosystems host microbes unique to themselves. The role an organism plays in a particular ecosystem is called its **ecological niche.**

The environment immediately surrounding an individual microbe—the **microenvironment**—is most relevant to that cell, but because microorganisms are so small, the microenvironment is difficult to identify and measure. The more readily measured gross environment—the **macroenvironment**—may be very different from the microenvironment. Consider a bacterial cell living within a biofilm (see figure 4.3); growth of aerobic organisms in the biofilm can deplete O_2, creating microzones where obligate anaerobes can grow. Fermenters can produce organic acids that may then be metabolized by other organisms in the film, and various growth factors can be transferred between organisms as well. Thus, certain microorganisms that might seem unexpected in a given macroenvironment actually thrive there because of microenvironments. ◄◄ biofilm, p. 84 ◄◄ growth factor, p. 93

Nutrient Acquisition

Organisms are categorized according to their trophic level (source of food), which is intimately related to the cycling of nutrients. There are three general trophic levels (**figure 29.1**):

■ **Primary producers.** These are autotrophs; they convert CO_2 into organic materials. Producers include both photoautotrophs (use sunlight for energy) and chemolithoautotrophs

FIGURE 29.1 Trophic Levels in an Ecosystem

❓ *Which organisms in the diagram are autotrophs?*

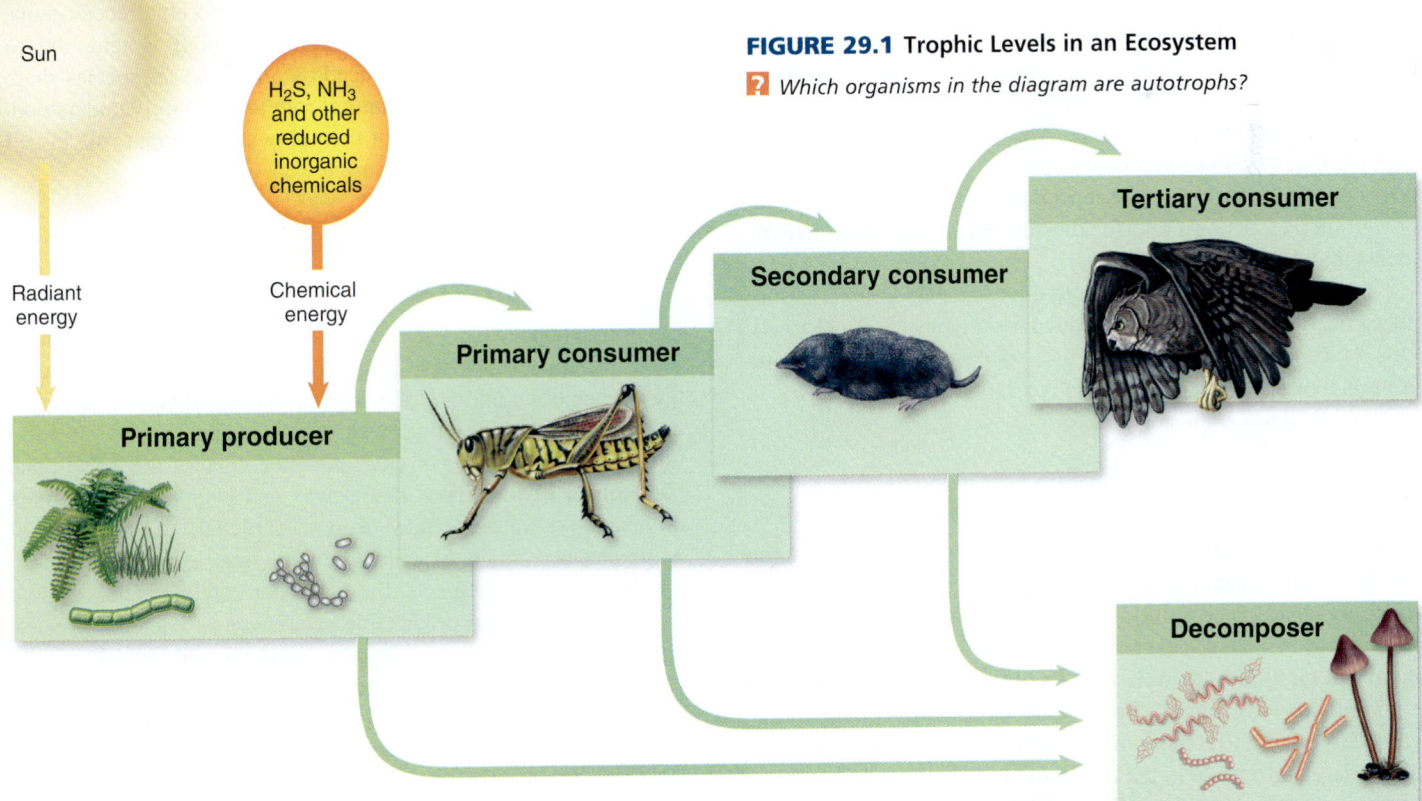

(oxidize inorganic chemicals for energy). Primary producers serve as a food source for consumers and decomposers. ◀◀ **photoautotroph, p. 93** ◀◀ **chemolithoautotroph, p. 93**

- **Consumers.** These are heterotrophs that eat primary producers or other consumers. Herbivores, which eat plants or algae, are primary consumers. Carnivores that eat herbivores are secondary consumers; carnivores that eat other carnivores are tertiary consumers. A chain of consumption is a food chain; interacting food chains are a food web.

- **Decomposers.** These are heterotrophs that digest the remains of primary producers and consumers. The fresh or partially decomposed organic matter used as a food source—including carcasses, excreta, and plant litter—is called detritus. Decomposers specialize in digesting complex materials such as cellulose, converting them into small molecules that can more easily be used by other organisms. The complete breakdown of organic molecules into inorganic molecules such as ammonia, sulfates, phosphates, and carbon dioxide is called mineralization. Microorganisms, particularly bacteria and fungi, play a major role in decomposition because of their ubiquity and unique metabolic capabilities.

Microbes in Low-Nutrient Environments

Low-nutrient environments such as lakes, rivers, and streams are common in nature, so microorganisms that can grow in dilute aqueous solutions are widespread. Most microbial growth in these settings is in biofilms, and the cells are shed from the film into the aqueous solution.

In addition to natural low-nutrient environments, microorganisms even grow in distilled-water reservoirs such as those found in research laboratories and pulmonary mist therapy units used in hospitals. The microbes extract the trace amounts of nutrients absorbed by the water from the air or adsorbed onto the biofilm. Although the organisms grow slowly, they can reach concentrations as high as 10^7 per milliliter. This cell concentration is not high enough to result in a cloudy solution, so the growth usually goes unnoticed. This can have serious consequences for the health of hospitalized patients and for the success of laboratory experiments that depend on water purity.

Organisms that grow in dilute environments contain highly efficient transport systems for moving nutrients inside the cell. Other mechanisms that bacteria use to thrive in dilute aquatic environments are described in chapter 11. ◀◀ **transport systems, p. 55** ◀◀ **thriving in aquatic environments, p. 269**

Microbial Competition and Antagonism

Perhaps nowhere in the living world is competition more fierce and the results of competition more quickly evident than among microorganisms. The ability of an organism to compete successfully for a habitat is generally related to the rate at which the organism multiplies, as well as to its ability to withstand adverse environmental conditions. Because bacteria multiply logarithmically, any small differences in their generation times will result in a very large difference in the total number of cells of each species after a relatively short time (**figure 29.2**).

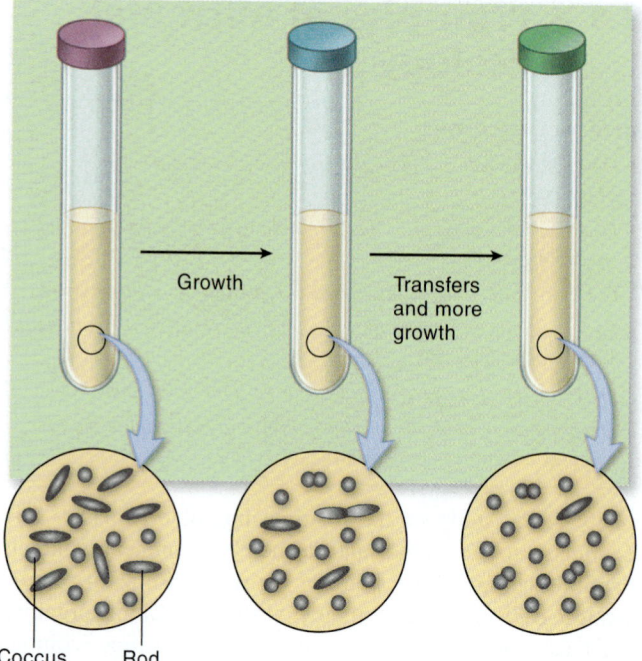

FIGURE 29.2 Competition The bacterium that multiplies faster yields the larger population.

Coccus   Rod

? *Why does transferring some of the culture make it easier to see the effects of competition?*

Antagonism among groups of organisms also helps determine the makeup of a community. In the soil, for example, some microbes resort to a type of chemical warfare, producing antimicrobial compounds. **Bacteriocins,** proteins produced by bacteria that kill closely related strains, are an example of antagonistic chemicals that play an important role in microbial ecosystems, promoting biodiversity through competition. It is tempting to speculate that antibiotics produced by *Streptomyces* species share a similar function, but their natural role is still poorly understood. Recent evidence suggests that they play an important role in cell to cell signaling. ◀◀ **the genus *Streptomyces*, p. 268**

Microorganisms and Environmental Changes

Environmental changes often result in alterations in a community. Those organisms that have adapted to live several inches beneath the surface of an untilled field will probably not be well suited to growth in that field if it is plowed, fertilized, and irrigated. In addition to external sources of environmental change, the growth and metabolism of organisms themselves can alter the environment dramatically. Nutrients may become depleted, and a variety of waste products, many of which are toxic, may accumulate.

In some environments, the changing conditions bring about a highly ordered and predictable succession of bacterial species. An example of this occurs in unpasteurized milk, which usually contains various species of microbes derived mainly from the immediate environment around the cow. Initially, the dominant bacterium is *Lactococcus lactis,* which breaks down the milk sugar lactose, forming lactic acid as a fermentation end product (**figure 29.3**). This sours the milk and also denatures the milk

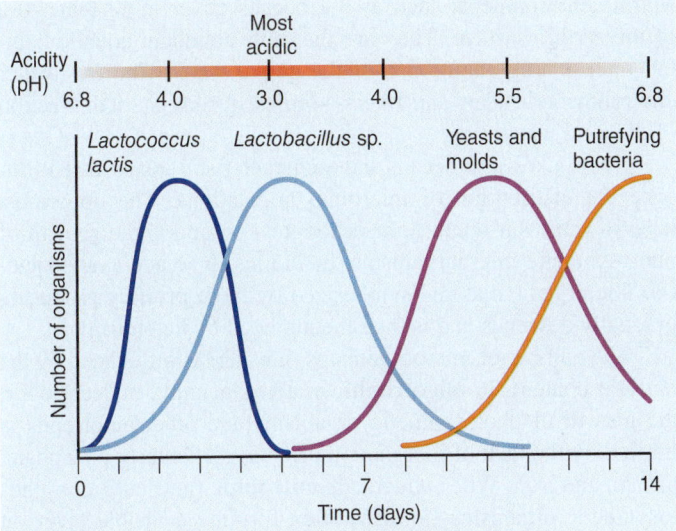

FIGURE 29.3 Growth of Microbial Populations in Unpasteurized Raw Milk at Room Temperature Production of acid causes souring and encourages growth of fungi. Eventually, bacteria digest the proteins, causing putrefaction.

❓ *Why does the pH of the milk first decrease and then increase?*

proteins, causing the milk to curdle. The acid inhibits most other organisms in the milk, and eventually enough acid is produced to inhibit *L. lactis. Lactobacillus* species can multiply in this highly acidic environment, however, and these bacteria metabolize any remaining sugar, forming more acid until their growth is also inhibited. Yeasts and molds, which tolerate even more acid conditions, then become the dominant group. They oxidize the lactic acid, which raises the pH. Most of the sugar has already been used at this point, so the streptococci and lactobacilli cannot resume multiplication. Milk protein (casein) is still available, however, and can be used by protease-producing members of the endospore-forming genus *Bacillus*. This breakdown of protein, known as putrefaction, yields a completely clear and foul-smelling product. The milk thus goes through a succession of changes with time, first souring and finally putrefying. ◀◀ **proteases, p. 150**

Microbial Communities

Microorganisms most often grow as biofilms attached to solid surfaces or at air-water interfaces. General aspects of biofilms were described in detail in chapter 4. In this section, we focus on a specific type of biofilm—a microbial mat. ◀◀ **biofilms, p. 84**

A **microbial mat** is a thick, dense, highly organized structure composed of distinct layers. Frequently they are green, reddish-pink, and black, which indicate the growth of different microbial groups (**figure 29.4**). The top green layer is typically composed of various species of cyanobacteria, and the color is due to their photosynthetic pigments. Directly below the green layer is a reddish-pink layer consisting of purple sulfur bacteria. The light-harvesting pigments of these anoxygenic phototrophs can use wavelengths of light not collected by the cyanobacteria. At the bottom is a black layer, resulting from iron molecules reacting with hydrogen sulfide produced by a group of bacteria called sulfate-reducers. These obligate anaerobes oxidize the organic compounds

FIGURE 29.4 A Microbial Mat A microbial mat is a thick, dense, highly organized structure composed of distinct layers of different groups of microorganisms.

❓ *Why are the middle layers of the mat reddish-pink? Why are the lower layers black?*

produced by the photosynthetic bacteria growing in the mat's upper layers, using sulfate as a terminal electron acceptor. ◀◀ **cyanobacteria, p. 261** ◀◀ **photosynthetic pigments, p. 152** ◀◀ **purple sulfur bacteria, p. 259** ◀◀ **sulfate-reducers, p. 258** ◀◀ **terminal electron acceptor, p. 130**

Although microbial mats can be found in a variety of areas, those near hot springs in Yellowstone National Park are some of the most intensively studied. The mats in these extreme areas are undisturbed by grazing eukaryotic organisms and, consequently, provide an important model for studying microbial interactions.

Studying Microbial Ecology

Because so few microorganisms can be successfully cultivated in the laboratory, studying only those that have been isolated often does not give an accurate picture of what actually occurs in nature. Molecular techniques are now complementing the traditional methods such as culture and microscopy, allowing researchers to better understand complex microbial communities.

Microscopic methods can now be used to examine the composition of microbial populations. For example, certain dyes are made fluorescent by metabolic activities carried out only in living cells, and therefore can be used to selectively observe viable cells (see figure 3.19a). A different technique, fluorescence *in situ* hybridization (FISH), uses nucleic acid probes labeled with a fluorescent molecule to observe only cells that contain specific nucleotide sequences (see figure 9.20). Scanning laser microscopes allow researchers to observe sectional views of a three-dimensional specimen such as a biofilm (see figure 3.8). ◀◀ **fluorescent dyes, p. 49** ◀◀ **FISH, p. 232** ◀◀ **scanning laser microscopes, p. 44**

Polymerase chain reaction (PCR) can be used to detect certain organisms and assess population characteristics. To detect a specific organism, primers are selected that amplify DNA unique to that organism (see figure 9.14). To study the composition of a population, total 16S rRNA gene segments can be amplified. Individual fragments can then be cloned and studied. Alternatively, the set of amplified sequences can be separated and examined using a technique called denaturing gradient gel electrophoresis (DGGE). This procedure gradually denatures double-stranded nucleic acid during gel electrophoresis and, as a consequence, separates fragments of similar size according to their melting point, which is related to the nucleotide sequence. Using DGGE, a mixture of 16S rRNA fragments with different sequences will resolve into a distinct pattern of bands. PCR and DDGE studies have confirmed that standard culture techniques can be poor indicators of the composition of natural microbial populations. The molecular techniques, which show the relative abundance of specific nucleotide sequences, confirm that species predominating in laboratory culture often represent only a fraction of the total population. ◀◀ polymerase chain reaction, p. 227 ◀◀ gel electrophoresis, p. 217

Genomics is also advancing the study of microbial ecology because sequence information gleaned from one species can be applied to others. For example, researchers found that variations of a gene coding for bacterial rhodopsin, a light-sensitive pigment that provides a mechanism for harvesting the energy of sunlight, are widespread in marine bacteria. This gene provides bacteria with a mechanism for phototrophy that does not require chlorophyll and might be an important mechanism for energy accumulation in ocean environments. ◀◀ genomics, p. 184 ◀◀ bacterial rhodopsin, p. 278

Scientists are also using metagenomics—the study of total genomes in a sample—to study microbial populations in their natural environment. ◀◀ metagenomics, p. 184

MicroAssessment 29.1

Microorganisms play a major role in most ecosystems. Organisms are categorized as primary producers, consumers, or decomposers. Competition among microorganisms in a habitat can be intense. Microbial mats have distinct layers. Molecular techniques are allowing researchers to better understand natural microbial communities.

1. *What are the roles of primary producers, consumers, and decomposers?*

2. *Why are molecular techniques important in studying microbial ecology?*

3. *How could FISH (fluorescence* in situ *hybridization) be used to determine the relative proportions of archaea and bacteria in a population?* ✚

29.2 ■ Aquatic Habitats

Learning Outcome

7. *Compare and contrast the habitats provided by marine, freshwater, and specialized aquatic environments.*

Marine environments such as the oceans cover more than 70% of the earth's surface. They are the most abundant aquatic habitat, representing about 95% of the global water. The freshwater environments—lakes and rivers—represent only a small fraction of the total water.

Deep lakes and oceans have characteristic zones that influence the distribution of microbial populations. The uppermost layer—where sufficient light penetrates—supports the growth of photosynthetic microorganisms, including algae and cyanobacteria. The organic material synthesized by these primary producers gradually descends and is then metabolized by heterotrophs.

The number of microorganisms in waters is influenced by the nutrient content. In **oligotrophic** waters, meaning nutrient poor, the growth of photosynthetic organisms and other autotrophs is limited by the lack of inorganic nutrients, particularly phosphate, nitrate, and iron. When waters are **eutrophic** (nutrient rich), photosynthetic organisms flourish, often forming a visible layer on the surface (**figure 29.5**). In turn, photosynthesis produces organic compounds that foster the growth of heterotrophs in lower layers. The heterotrophs consume dissolved O_2 as they metabolize the organic material. Because O_2 consumption can outpace the slow rate of diffusion of atmospheric O_2 into the waters, the environment can become **hypoxic** (very low in dissolved O_2). Insufficient O_2 leads to the death of resident fish and other aquatic animals.

Marine Environments

Marine environments range from the deep sea, where nutrients are scarce, to the shallower coastal regions, where nutrients may

FIGURE 29.5 Eutrophication in a Polluted Stream Photosynthetic organisms flourish in the nutrient-rich water. The organic compounds they produce permit abundant growth of heterotrophs in lower layers.

❓ *How would the cows in the photo contribute to eutrophication?*

be abundant due to runoff from the land. Seawater contains about 3.5% salt, compared with about 0.05% for fresh water. Consequently, it supports the growth of halophilic organisms, which prefer or require high salt concentrations, and halotolerant ones. Temperatures often vary widely at the surface, but decrease with depth until reaching about 2°C in the deeper waters; an exception is the areas around hydrothermal vents, which will be described later. ◄◄ halophilic, p. 92 ►► hydrothermal vents, p. 729

Ocean waters are typically oligotrophic, limiting the growth of microorganisms. The small amounts of organic material produced by photosynthetic organisms is quickly consumed as it descends, so few nutrients reach the sediments below. Even in the deep sea, marine water is O_2-saturated due to mixing associated with tides, currents, and wind action.

The ecology of inshore areas is not as stable as the deep sea and can be dramatically affected by nutrient-rich runoff. An unfortunate example is the **dead zone**—a region devoid of fish and other marine life—that forms in the Gulf of Mexico, as well as other areas (**figure 29.6**). The Mississippi River, carrying nutrients accumulated as it runs through agricultural, industrial, and urbanized regions, feeds into the Gulf. As a consequence of this nutrient-enrichment, populations of algae and cyanobacteria flourish in the spring and summer when sunlight is also plentiful. Heterotrophic microbes then metabolize the organic compounds synthesized by these primary producers, consuming dissolved O_2

FIGURE 29.6 Dead Zone Formation

Algae and cyanobacteria flourish, using photosynthesis to produce organic compounds.

Nutrient-rich water

Organic compounds

Aerobic decomposition of organic compounds by heterotrophic microbes depletes O_2.

Animals flee hypoxic environment or die.

❓ *Why does decomposition by microbes deplete O_2?*

in the process. This causes a large region in the Gulf—sometimes in excess of 7,000 square miles—to become hypoxic. Animals in the area either flee or die. Nutrient enrichment of coastal waters also contributes to blooms of toxin-producing algae. ◄◄ medical importance of algae, p. 291

Freshwater Environments

As with marine environments, the types and relative numbers of microbes inhabiting fresh waters depend on multiple factors including light, concentration of dissolved O_2 and nutrients, and temperature.

Oligotrophic lakes in temperate climates may have anaerobic layers due to thermal stratification resulting from seasonal temperature changes. During the summer months, the surface water warms. This decreases the density of the water, causing it to form a distinct layer that does not mix with the cooler, denser water below. The upper layer, called the epilimnion, is generally O_2 rich due to the activities of photosynthetic organisms. In contrast, the lower layer, the hypolimnion, may be anaerobic due to the consumption of O_2 by heterotrophs. Separating these two layers is the thermocline, a zone of rapid temperature change. As the weather cools, the waters mix, providing O_2 to the deep water.

Rapidly moving waters, such as rivers and streams, are very different from lakes. They are usually shallow and turbulent, facilitating O_2 circulation, so they are generally aerobic. Light may penetrate to their bottom, making photosynthesis possible. Sheathed bacteria such as *Sphaerotilus* sp. and *Leptothrix* sp. commonly adhere to rocks and other solid structures, where they then use nutrients that flow by. ◄◄ sheathed bacteria, p. 269

Specialized Aquatic Environments

Specialized aquatic environments include salt lakes, such as the Great Salt Lake in Utah, which have no outlets. As water in these lakes evaporates, the salt concentrations become much higher than that in seawater. Extreme halophiles thrive in this environment. ◄◄ extreme halophiles, p. 92

Other specialized habitats include iron springs that contain large quantities of ferrous ions; these springs are habitats for species of *Gallionella* and *Sphaerotilus*. Sulfur springs support the growth of both photosynthetic and non-photosynthetic sulfur bacteria. Other aquatic environments include groundwater, stagnant ponds, swimming pools, and drainage ditches, each offering its own opportunity for bacterial growth.

MicroAssessment 29.2

Aquatic habitats include marine, freshwater, and specialized environments. Nutrient-rich waters can become hypoxic. Lakes often exhibit thermal stratification during summer months.

4. *Compare the salt content of seawater with that of fresh water and salt lakes.*

5. *Explain how nutrient-rich runoff can cause waters to become hypoxic.*

6. *Why would the nutrient content of a body of water be more homogeneous than that of a terrestrial environment?* ➕

29.3 ■ Terrestrial Habitats

Learning Outcome

8. *Describe soil as a microbial habitat.*

Although microbes can adhere to and grow on a variety of objects on land, the focus in this section is soil—a critical component of terrestrial ecosystems. Extreme terrestrial habitats, such as volcanic vents and fissures, and some of the extremophiles that inhabit them are described in chapter 11. ◀◀ **archaea that thrive in extreme conditions, p. 278**

One reason scientists are interested in soil is that the microbes there synthesize a variety of useful chemicals. For example, the various species of *Streptomyces* produce over 500 different antibiotic substances, at least 50 of which have useful applications in medicine, agriculture, and industry. The pharmaceutical industry has tested many thousands of soil microorganisms in search of those that produce useful antibiotics. In addition, soil microbes are being investigated for their ability to degrade toxic chemicals, an application of microbiology called bioremediation, which will be discussed in chapter 30. Probably in no other habitat can one find a greater range of biosynthetic and biodegradative capabilities than are represented in the soil. ▶▶ **bioremediation, p. 744**

Characteristics of Soil

Soil is composed of pulverized rock, decaying organic material, air, and water. It teems with life, including bacteria, fungi, algae, protozoa, worms, insects, and plant roots. Soil communities may contain more than 4,000 different species per gram of soil. The top 6 inches of fertile soil may harbor more than 2 tons of bacteria and fungi per acre! Soil represents an environment that can change abruptly and dramatically. Heavy rains, for example, can cause a soil to rapidly become waterlogged. Trees dropping their leaves can suddenly enrich the soil with organic nutrients. Farmers and gardeners rapidly change the nutrient mix by applying fertilizers.

Soil forms as rock weathers. Water, temperature changes, windblown particles, and other physical forces gradually cause the rock to crack and break. Photosynthetic organisms including algae, mosses, and lichens growing on the surfaces of rocks synthesize organic compounds. Various bacteria and fungi then use these compounds as carbon and energy sources, producing acids and other chemicals that gradually decompose the rocks. As soil slowly forms, some plants begin to grow. When these die and decay, the residual organic material functions as a sponge, retaining water and thus allowing more plants to grow. Over time, more organic compounds accumulate, forming a slowly degrading complex polymeric substance called humus. ◀◀ **lichens, p. 286**

The texture of the soil influences its porosity, which in turn impacts the amount of air exchange and how much water can flow through. Finely textured soils, such as clay soils, are more likely to become waterlogged and anaerobic. In contrast, sandy soils that dry quickly allow water to pass through and are generally aerobic.

Microorganisms in Soil

The density and composition of the soil microbiota are dramatically affected by environmental conditions. Wet soils, for example, are unfavorable for aerobic microbes because the spaces in the soil fill up with water, diminishing the amount of air in the soil. When the water content of soil drops to a very low level, as during a drought or in a desert environment, the metabolic activity and number of soil microorganisms decrease. Many species of soil organisms produce survival forms such as endospores and cysts that are resistant to drying. Other environmental influences that affect soil microbes include acidity, temperature, and nutrient supply. For example, acidity suppresses bacterial growth, allowing fungi to thrive with less competition for nutrients. This is why mushrooms often appear in a lawn fertilized with an acid-producing fertilizer such as ammonium chloride.

Prokaryotes are the most numerous soil inhabitants. Their physiological diversity allows them to colonize all types of soil. In general, Gram-positive bacteria are more abundant in soils than Gram-negative bacteria. Among the most common Gram-positive bacteria are members of the genus *Bacillus*. These form endospores, allowing them to survive long periods of adverse conditions such as drought or extreme heat. *Streptomyces* species produce conidia, which are desiccation-resistant structures. They also produce metabolites called geosmins, which give soil its characteristic musty odor. As discussed earlier, *Streptomyces* species produce many medically useful antibiotics. Other bacteria adapted to thrive in terrestrial environments—including myxobacteria and species of *Clostridium, Azotobacter, Agrobacterium,* and *Rhizobium*—were discussed in chapter 11. ◀◀ **endospore-formers, p. 67** ◀◀ **the genus *Streptomyces*, p. 268** ◀◀ **thriving in terrestrial environments, p. 266**

Although prokaryotes are the most numerous soil microbes, the biomass of fungi is much greater. Most fungi are aerobes, so they usually grow in the top 10 cm of soil. The soil fungi degrade complex macromolecules such as lignin (the major component of cell walls of woody plants) and cellulose. Some soil fungi are free-living, and others live in symbiotic relationships. The latter include mycorrhizas—fungi growing in a symbiotic relationship with certain plant roots. ◀◀ **mycorrhizas, p. 286**

In addition to bacteria and fungi, various algae and protozoa are found in most soils. Algae depend on sunlight for energy, so they mostly live on or near the soil surface. Most protozoa require O_2, so they too are found near the surface, typically where microbes on which they feed are plentiful.

The Rhizosphere

The concentration of microbes, particularly Gram-negative bacteria, is generally much greater in the **rhizosphere** (the zone of soil that adheres to plant roots) than in surrounding soil. This is because the root cells secrete organic molecules, enriching the region. Particular bacterial species appear to preferentially interact with certain plants. For example, the rhizosphere of certain grasses can have high concentrations of *Azospirillum* species, which fix nitrogen. ▶▶ **nitrogen fixation, p. 726**

The density and composition of the soil are dramatically affected by environmental conditions. The concentration of microbes in the rhizosphere is generally much higher than that of the surrounding soil.

7. *Why are wet soils unfavorable for aerobic organisms?*

8. *What is the significance of the rhizosphere?*

9. *How can the biomass of fungi in soil be greater considering that bacteria are more numerous?* ✚

29.4 ■ Biogeochemical Cycling and Energy Flow

Learning Outcomes

9. *Diagram the carbon, nitrogen, sulfur, and phosphorus cycles, and describe some of the important microbial contributors.*

10. *Compare and contrast energy cycling in environments with sunlight versus those far removed from sunlight.*

Biogeochemical cycles are the cyclical paths that elements take as they flow through living (biotic) and non-living (abiotic) components of ecosystems. These cycles are important because a fixed and limited amount of the elements that make up living cells exists on the earth and in the atmosphere. Thus, in order for an ecosystem to sustain its characteristic life-forms, elements must continuously be recycled. For example, the organic carbon that animals use as an energy source is then exhaled as carbon dioxide (CO_2); if this inorganic form of carbon were not eventually converted back to an organic form, we would run out of the organic carbon needed for growth. The carbon and nitrogen cycles are particularly important because they involve stable gaseous forms (carbon dioxide and nitrogen gas), which enter the atmosphere and thus have global impacts.

Although elements continually cycle in an ecosystem, energy does not. Instead, energy must be continually added to an ecosystem, fueling the activities required for life.

Understanding the cycling of nutrients and the flow of energy is particularly important as human activities affect the environment in a major way. For example, industrial processes that convert nitrogen gas (N_2) into ammonia-containing fertilizers have boosted food production substantially, but they also alter the nitrogen cycle by increasing the amount of fixed nitrogen available. When these nitrogen sources pollute lakes and coastal areas, eutrophication can result. As a consequence, the dissolved O_2 is depleted, leading to the death of aquatic animals. The extra nutrients also lead to decreased biodiversity in terrestrial ecosystems. Excavation and burning of coal, oil, and other carbon-rich fossil fuels provides energy for our daily activities, but releases additional CO_2 and other carbon-containing gases into the atmosphere. Fossil fuels, the ancient remains of partially decomposed plants and animals, are nutrient reservoirs unavailable without human intervention, and therefore would not normally participate in biogeochemical cycles. The increase of carbon-containing gases in the atmosphere raises global temperatures because the gases absorb infrared radiation and reflect it back to earth.

When studying biogeochemical cycles, it is helpful to bear in mind the role of a given element in a particular organism's metabolism. Elements are used for three general purposes:

■ **Biosynthesis (biomass production).** All organisms require elements for biosynthesis. As an example, nitrogen is required to produce amino acids. Plants and many prokaryotes assimilate nitrogen by incorporating ammonia (NH_3) to synthesize the amino acid glutamate (see figure 6.29a). Some prepare for this step by converting nitrate (NO_3^-) to ammonia. Once glutamate has been synthesized, the amino group can then be transferred to other carbon compounds to produce the necessary amino acids. Animals cannot incorporate ammonia and instead require amino acids in their diet. Some prokaryotes can reduce atmospheric nitrogen to form ammonia—the process of nitrogen fixation. The ammonia can then be incorporated into cellular material. ◀◀ amino acid synthesis, p. 156

■ **Energy source.** Reduced carbon compounds such as sugars, lipids, and amino acids are used as energy sources by chemoorganotrophs. Chemolithotrophs can use reduced inorganic molecules such as hydrogen sulfide (H_2S), ammonia (NH_3), and hydrogen gas (H_2) (see table 4.5). ◀◀ energy source, p. 130

■ **Terminal electron acceptor.** In aerobic conditions, O_2 is used as a terminal electron acceptor. In anaerobic conditions, some prokaryotes can use nitrate (NO_3^-), nitrite (NO_2^-), sulfate (SO_4^-), or carbon dioxide (CO_2) as a terminal electron acceptor. ◀◀ terminal electron acceptor, p. 130

Carbon Cycle

All organisms are composed of organic molecules such as proteins, lipids, and carbohydrates. The carbon travels through the food chain as primary producers are eaten by primary consumers, which are then eaten by secondary consumers. Decomposers then use the remains of primary producers and consumers.

Carbon Fixation

A fundamental aspect of the carbon cycle is carbon fixation, the defining characteristic of primary producers (**figure 29.7**). These organisms all convert CO_2 into an organic form, using mechanisms described in chapter 6. Without primary producers, no other organisms, including humans, could exist. We depend on them to generate the organic carbon compounds we use as an energy source and for biosynthesis. ◀◀ carbon fixation, p. 154

Respiration and Fermentation

When heterotrophs consume organic material, they break it down using respiration and/or fermentation to release the energy, which is captured to make ATP. The processes usually make CO_2.

The type of organic material helps dictate which species degrade it. A wide variety of organisms use sugars, amino acids, and proteins as energy sources, but rapidly multiplying bacteria often

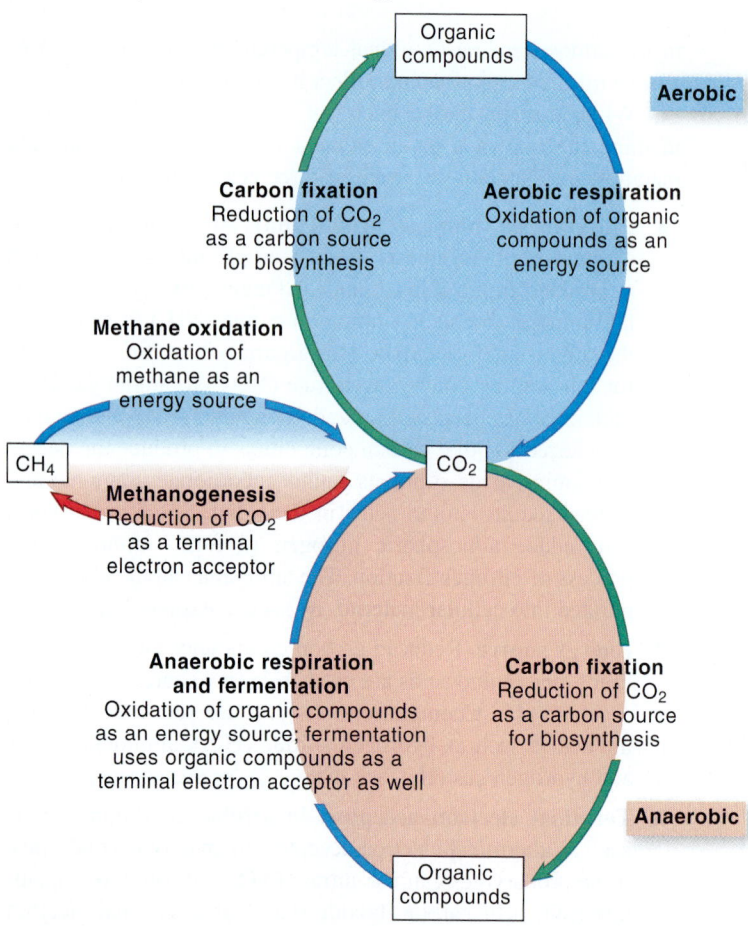

FIGURE 29.7 Carbon Cycle Blue arrows represent steps where carbon compounds are used as energy sources; red arrows represent steps where carbon compounds are used as terminal electron acceptors, and green arrows represent steps where inorganic carbon is used in biosynthesis.

❓ *What is the role of autotrophs in the carbon cycle?*

play the dominant role in decomposing these substances. In contrast, only certain fungi can break down lignin, a major component of wood (**figure 29.8**). Aerobic conditions are required for this degradation, so wood at the bottom of marshes resists decay.

FIGURE 29.8 Wood-Degrading Fungus Growing on a Dead Tree These fungi thrive in wet conditions and digest lignin, the major cell wall component of woody plants.

❓ *Dead trees submerged in marshes resist decay. Why?*

The O_2 supply has a strong influence on the carbon cycle. Not only does O_2 allow degradation of certain compounds such as lignin, it also helps determine the types of carbon-containing gases produced. When organic matter is degraded aerobically, a great deal of CO_2 is produced. When the O_2 level is low, however, as is the case in marshes, swamps, and manure piles, the degradation is incomplete, generating some CO_2 and a variety of other products.

Methanogenesis and Methane Oxidation

In anaerobic environments, CO_2 is used by methanogens. These archaea obtain energy by oxidizing hydrogen gas, using CO_2 as a terminal electron acceptor, generating methane (CH_4). Methane that enters the atmosphere is oxidized by ultraviolet light and chemical ions, forming carbon monoxide (CO) and CO_2. A group of microorganisms called methylotrophs can use methane as an energy source, oxidizing it to produce CO_2. ◄◄ methanogens, p. 256

Nitrogen Cycle

Nitrogen is a component of proteins and nucleic acids. As consumers ingest plants and animals for carbon and energy, they also obtain their required nitrogen. Prokaryotes, as a group, are far more diverse in their use of nitrogen-containing compounds. Some use oxidized nitrogen compounds such as nitrate (NO_3^-) and nitrite (NO_2^-) as terminal electron acceptors; others use reduced nitrogen compounds such as ammonium (NH_4^+) as energy sources. These metabolic activities represent essential steps in the nitrogen cycle (**figure 29.9**).

Nitrogen Fixation

Nitrogen fixation is the process in which nitrogen gas (N_2) is reduced to form ammonia (NH_3), which can then be incorporated into cellular material. The process, catalyzed by the enzyme complex nitrogenase, requires a tremendous amount of energy because N_2 has a very stable triple bond.

Although the atmosphere consists of approximately 80% N_2, relatively few organisms—all of which are prokaryotes—can reduce this gaseous form of the element. Thus, just as humans and other animals depend on primary producers to fix carbon, they rely on prokaryotes to convert atmospheric nitrogen to a form they can assimilate to create biomass.

Some nitrogen-fixing prokaryotes, or **diazotrophs,** are free-living, whereas others form symbiotic associations with higher organisms, particularly certain plants. Among the free-living examples are members of the genus *Azotobacter.* These heterotrophic, aerobic, Gram-negative rods may be the chief suppliers of fixed nitrogen in ecosystems such as grasslands that lack plants with nitrogen-fixing symbionts. The dominant free-living, anaerobic, soil diazotrophs are certain members of the genus *Clostridium.* Certain cyanobacteria are diazotrophs; these photosynthetic bacteria use both nitrogen and carbon from the atmosphere. Symbiotic diazotrophs will be discussed later in the chapter. ►► symbiotic nitrogen-fixers and plants, p. 730 ◄◄ the genus *Azotobacter*, p. 267 ◄◄ the genus *Clostridium*, p. 258 ◄◄ nitrogen-fixing cyanobacteria, p. 262

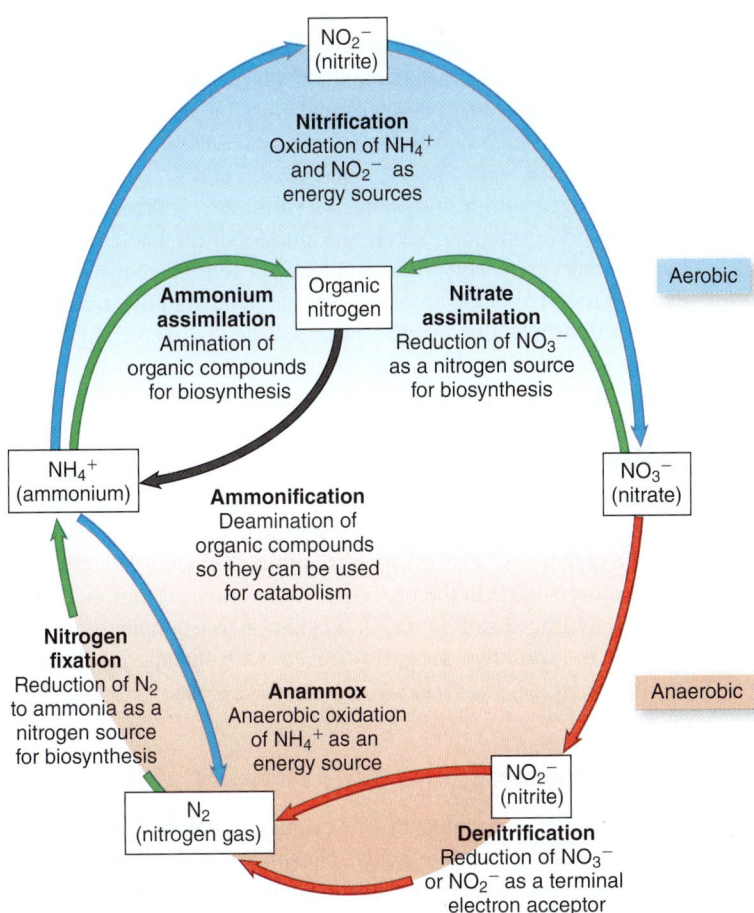

FIGURE 29.9 Nitrogen Cycle Blue arrows represent steps where nitrogen compounds are used as energy sources; red arrows represent steps where nitrogen compounds are used as terminal electron acceptors, and green arrows represent steps where inorganic nitrogen is used in biosynthesis. Not all parts of the cycle are shown.

❓ *Why are prokaryotes critical in the nitrogen cycle?*

Energy-expensive chemical processes can fix nitrogen to make fertilizers, and these are playing an increasingly larger role in the nitrogen cycle. In fact, fixed nitrogen sources associated with human intervention, including fertilizer production and planting crops that foster the growth of symbiotic nitrogen-fixers, now appear to surpass natural biological nitrogen fixation.

MicroByte
Nitrogenase uses approximately 16 molecules of ATP for every molecule of nitrogen fixed.

Ammonification

Ammonification is the decomposition process that converts organic nitrogen into ammonia (NH_3). In alkaline environments, such as heavily limed soil, the gaseous ammonia may enter the atmosphere. In neutral environments, ammonium (NH_4^+) is formed. This positively charged ion adheres to negatively charged particles.

Proteins, which are among the most common nitrogen-containing organic compounds, can be degraded by a wide variety of microbes. The microbes secrete proteolytic enzymes that break down proteins into short peptides or amino acids. These products

are then transported into the cell, and the amino groups removed, releasing ammonium. The decomposer will assimilate much of this to create biomass. Some will be released into the environment, however, where it can be assimilated by plants and other organisms. ◀◀ **deamination, p. 150**

Nitrification

Nitrification is the process that oxidizes ammonium (NH_4^+) to nitrate (NO_3^-). A group of bacteria known collectively as **nitrifiers** do this in a cooperative two-step process, using ammonium and an intermediate, nitrite (NO_2^-), as energy sources. Nitrifiers are obligate aerobes, using O_2 as a terminal electron acceptor. Consequently, nitrification does not occur in waterlogged soils or in anaerobic regions of aquatic environments. ◀◀ **nitrifiers, p. 727**

Nitrification has some important consequences with respect to agricultural practices and pollution. Farmers often apply ammonium-containing compounds to soils as a source of nitrogen for plants. The ammonium is retained by soils, because its positive charge causes it to adhere to negatively charged soil particles. Nitrification converts the ammonium to nitrate, a form of nitrogen more readily used by plants, but rapidly leached from soil by rainwater. To slow nitrification, certain chemicals can be added to ammonium-fertilized soils.

Leaching of nitrate and nitrite from soil is a health concern if the compounds contaminate drinking water. Nitrite is toxic because it can combine with hemoglobin of the blood, reducing blood's O_2-carrying capacity. Even nitrate, which in itself is not very toxic, can be dangerous if high levels are ingested because some intestinal bacteria can use it as a terminal electron acceptor, converting it to nitrite.

Denitrification

Denitrification is the process that reduces nitrate (NO_3^-), converting it to gaseous forms such as nitrous oxide (N_2O) and molecular nitrogen (N_2). This happens when prokaryotes anaerobically respire using nitrate as a terminal electron acceptor. ◀◀ **anaerobic respiration, p. 134**

Denitrification can have negative environmental and economic consequences. Under anaerobic conditions in wet soils, for example, denitrifying bacteria will reduce the oxidized nitrogen compounds of fertilizers, releasing gaseous nitrogen to the atmosphere. In some areas, this process may represent 80% of nitrogen lost from fertilized soil, a considerable economic loss to the farmer. In addition, nitrous oxide contributes to global warming.

Denitrification is not always undesirable. The process can be actively fostered in certain steps of wastewater treatment as a means to remove nitrate. This compound could otherwise act as a fertilizer in the waters to which the wastewater is discharged, promoting algal growth. ▶▶ **microbiology of wastewater treatment, p. 736**

Anammox

Certain bacteria oxidize ammonium under anaerobic conditions, using nitrite as a terminal electron acceptor. This reaction, called **anammox** (for anoxic ammonia oxidation), forms N_2 and might provide an economical means of removing nitrogen compounds during wastewater treatment.

Sulfur Cycle

Sulfur is found in all living matter, chiefly as a component of the amino acids methionine and cysteine. Like the nitrogen cycle, key steps of the sulfur cycle depend on the activities of prokaryotes (**figure 29.10**).

Sulfur Assimilation and Decomposition

Most plants and microorganisms assimilate sulfur as sulfate (SO_4^{2-}), reducing it to form biomass. Like nitrogen, organic sulfur is present chiefly as a part of the amino acids that make up proteins. Decomposition of the sulfur-containing amino acids releases hydrogen sulfide (H_2S), a gas.

Sulfur Oxidation

Hydrogen sulfide (H_2S) and elemental sulfur (S^0) can both serve as an energy source for certain chemolithotrophs. Sulfur-oxidizing prokaryotes, including species of *Beggiatoa*, *Thiothrix*, and *Thiobacillus*, oxidize these molecules to sulfate (SO_4^{2-}). Certain prokaryotes in anaerobic marine environments can oxidize elemental sulfur, using nitrate as a terminal electron acceptor. As discussed in chapter 11, these organisms, including *Thioploca*

species and the largest known bacterium, *Thiomargarita namibiensis*, have unusual mechanisms to cope with the fact that their energy source and terminal electron acceptor are found in two different environments. ◀◀ **sulfur-oxidizing bacteria, p. 262** ◀◀ **sulfur-oxidizing, nitrate-reducing marine bacteria, p. 273**

Hydrogen sulfide and elemental sulfur are oxidized anaerobically by photosynthetic green and purple sulfur bacteria. These bacteria harvest energy from sunlight but require reduced molecules as a source of electrons to generate reducing power. Like the chemolithotrophs that use hydrogen sulfide and elemental sulfur, the photosynthetic sulfur oxidizers produce sulfate. ◀◀ **green sulfur bacteria, p. 260** ◀◀ **purple sulfur bacteria, p. 259**

Sulfur Reduction

Under anaerobic conditions, sulfate generated by the sulfur-oxidizers can then be used as a terminal electron acceptor by certain organisms. The sulfur- and sulfate-reducing bacteria and archaea use sulfate in the process of anaerobic respiration, reducing it to hydrogen sulfide (H_2S). In addition to its unpleasant odor, the H_2S is a problem because it reacts with metals, resulting in corrosion. ◀◀ **sulfur- and sulfate-reducing bacteria, p. 258**

Phosphorus Cycle and Other Cycles

Phosphorus is a component of several critical biological compounds including nucleic acids, phospholipids, and ATP. Most plants and microorganisms can take up phosphorus as orthophosphate (PO_4^{3-}), incorporating it into biomass. From there, the phosphorus is passed along the food web. When plants and animals die, decomposers convert organic phosphate back to the inorganic form.

In many aquatic habitats, growth of algae and cyanobacteria—the primary producers—is limited by low concentrations of phosphorus. When phosphates are added from sources such as agricultural runoff, phosphate-containing detergents, and wastewater, eutrophication can result.

Other important elements, including iron, calcium, zinc, manganese, cobalt, and mercury, are also recycled by microorganisms. Many prokaryotes contain plasmids coding for enzymes that carry out oxidation of metallic ions.

Energy Sources for Ecosystems

All chemotrophs harvest the energy trapped in chemical bonds to generate ATP. This energy cannot be totally recycled, however, because a portion is always lost as heat when bonds are broken. Thus, energy is continually lost from biological systems. To compensate, energy must be added to ecosystems.

Photosynthesis, carried out by chlorophyll-containing plants and microorganisms, converts radiant energy (sunlight) to chemical energy in the form of organic compounds, which can be used by chemoorganotrophs. The requirement for radiant energy has traditionally been used to explain why life is not equally abundant everywhere. However, the discovery of different types of communities far removed from sunlight, including near hydrothermal vents and within rocks, has dramatically altered this idea. These

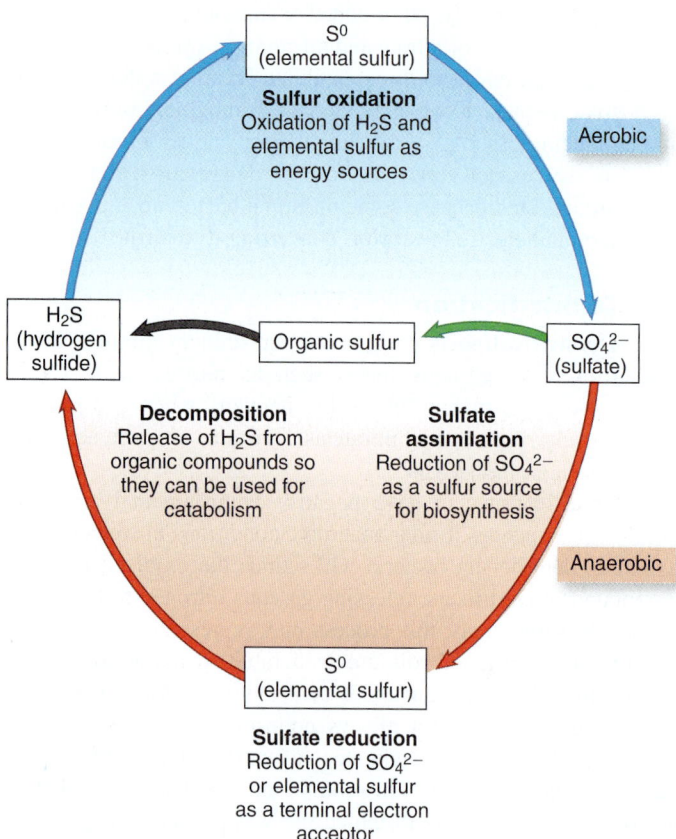

FIGURE 29.10 Sulfur Cycle Blue arrows represent steps where sulfur compounds are used as energy sources; red arrows represent steps where sulfur compounds are used as terminal electron acceptors, and green arrows represent steps where inorganic sulfur is used in biosynthesis.

? *Organisms that oxidize sulfur use it for what purpose in their metabolism?*

communities rely on chemolithoautotrophs, which harvest the energy of reduced inorganic compounds and use it to form organic compounds. ◀◀ **chemolithoautotroph, p. 93**

A number of **hydrothermal vents** have been discovered, some thousands of meters below the ocean surface. These vents form when water seeps into cracks in the ocean floor and becomes heated by the molten rock, finally spewing out in the form of mineral-laden undersea geysers. The hydrogen sulfide discharged supports thriving deep-sea communities, oases in the otherwise desolate ocean floor (**figure 29.11**). Large numbers of sulfur-oxidizing chemolithoautotrophs (bacteria and archaea) are found in and around the vents. Many are free-living but some live in symbiotic association with the large tube worms and clams that inhabit the areas. The chemolithoautotrophs obtain energy by oxidizing hydrogen sulfide, and they fix CO_2, providing the animals with both a carbon and energy source.

Microbial populations have been found almost 3 km under the ground and in iron-rich volcanic rocks from nearly a thousand meters below the surface of the Columbia River (see **Perspective 4.1**). These organisms gain energy from hydrogen (H_2) produced in the subsurface. It has been estimated that if (and it is a big "if" at this point) most similar rocks contain microbes, there could be as much as 2×10^{14} tons of underground microorganisms—equivalent to a layer 1.5 meters thick over the entire land surface of the earth!

MicroAssessment 29.4

Elements are recycled as organisms incorporate them to produce biomass, oxidize reduced forms as energy sources, and reduce oxidized forms as terminal electron acceptors. Carbon fixation uses atmospheric CO_2 to produce organic material; the CO_2 is regenerated during respiration and some fermentations. Prokaryotes are essential for several steps of the nitrogen cycle including nitrogen fixation, nitrification, denitrification, and anammox. Prokaryotes are also essential for several steps of the sulfur cycle, including sulfur reduction and sulfate oxidation.

10. *What are the three general roles of carbon-containing compounds in metabolism?*
11. *Why do farmers try to prevent nitrification?*
12. *Although chemoautotrophs serve as the primary producers near hydrothermal vents, animals there still ultimately depend on the photosynthetic activities of plants and cyanobacteria. Why?* ✛

29.5 ■ Mutualistic Relationships Between Microorganisms and Eukaryotes

Learning Outcome

11. *Describe the mutualistic relationships between fungi and plant roots, symbiotic nitrogen-fixers and plants, and microorganisms and herbivores.*

FIGURE 29.11 Hydrothermal Vent Community (a) This diverse community is supported by the metabolic activities of chemolithoautotrophs. (b) Water escaping from the vent is rich in reduced compounds, including hydrogen sulfide, which can serve as energy sources.

❓ *Photosynthetic organisms are normally considered the most important primary producers; which organisms are primary producers in hydrothermal vent communities?*

(a)

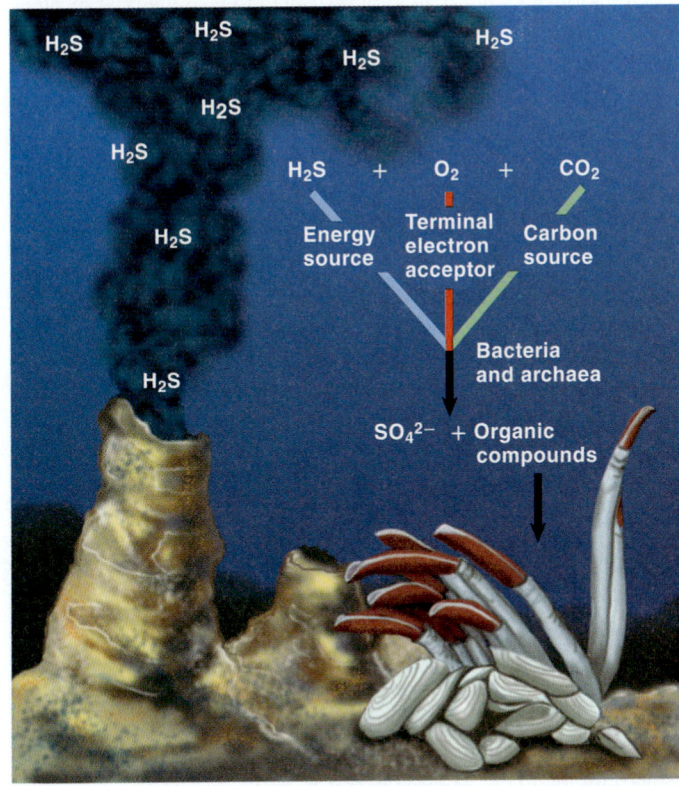

(b)

As described in chapter 16, mutualism is a symbiotic association in which both partners benefit. A variety of other ecologically important symbiotic relationships exist, but mutualistic relationships highlight the vital role of microorganisms to life on this planet. ◀◀ **symbiosis, p. 381**

Mycorrhizas

Mycorrhizas are fungi growing in symbiotic relationships with plant roots (**figure 29.12**). They enhance the competitiveness of plants by helping them take up phosphorus and other substances from the soil. In turn, the fungi gain nutrients for their own growth from root secretions. It is estimated that over 85% of vascular plants (plants with specialized water and food conducting tissues) have mycorrhizas.

There are two common types of mycorrhizal relationships:

■ **Endomycorrhizas.** The fungi penetrate root cells, growing as coils or tight, bushlike masses within the cells. These are by far the most common mycorrhizal relationships and are found in association with most herbaceous plants. Relatively few species of fungi are involved, perhaps only 100 or so, and most appear to be obligate symbionts. The relationship for some plants is also obligate; for example, most orchid seeds will not germinate without the activities of a fungal partner.

■ **Ectomycorrhizas.** The fungi grow around the plant cells, forming a sheath around the root. These fungi mainly associate with certain trees, including conifers, beeches, and oaks. Over 5,000 species of fungi are involved in ectomycorrhizal relationships but many are restricted to a single type of plant.

MicroByte ─────────────
Chanterelles and truffles are examples of commercially valuable ectomycorrhizal fungi.

Symbiotic Nitrogen-Fixers and Plants

Although some free-living bacteria can add fixed nitrogen to the soil, symbiotic nitrogen-fixing organisms are far more significant in benefiting plant growth and crop production. They are important in both terrestrial and aquatic habitats.

Rhizobia

Members of a diverse group of genera, including *Rhizobium, Bradyrhizobium, Sinorhizobium,* and *Azorhizobium*—collectively referred to as **rhizobia**—are the most agriculturally important symbiotic nitrogen-fixing bacteria. These grow as endosymbionts within specialized organs called **nodules** on the roots of legumes (plants that bear seeds in pods) including alfalfa, clover, peas, beans, and peanuts (**figure 29.13a**). The input of soil nitrogen from rhizobia may be about 10 times the annual rate of nitrogen fixation by non-symbiotic organisms. To foster plant growth, farmers often add the appropriate symbionts to the seeds of certain legumes. ◄◄ endosymbiont, p. 76

The plant cells foster the nitrogen-fixation of the endosymbionts by synthesizing a protein called leghemoglobin. This binds to O_2 and regulates its concentration in the nodule, protecting the O_2-sensitive nitrogenase. The plant also provides various nutrients to the endosymbionts. Meanwhile, the endosymbionts provide fixed nitrogen to the plant.

Nodule formation involves extensive chemical communication between the rhizobia and legume partners (figure 29.13b). First, plant root secretions attract the appropriate rhizobial species, which then colonizes the roots. In the most well-characterized

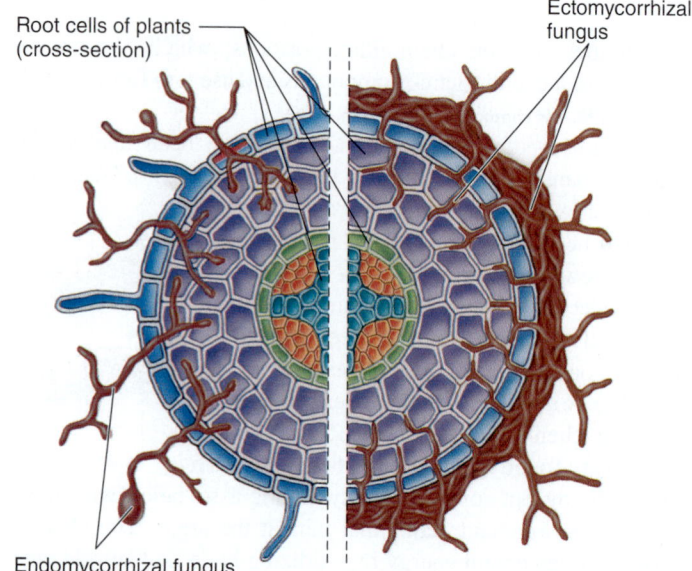

Root cells of plants (cross-section)

Ectomycorrhizal fungus

Endomycorrhizal fungus

(a) Diagram of Mycorrhizas

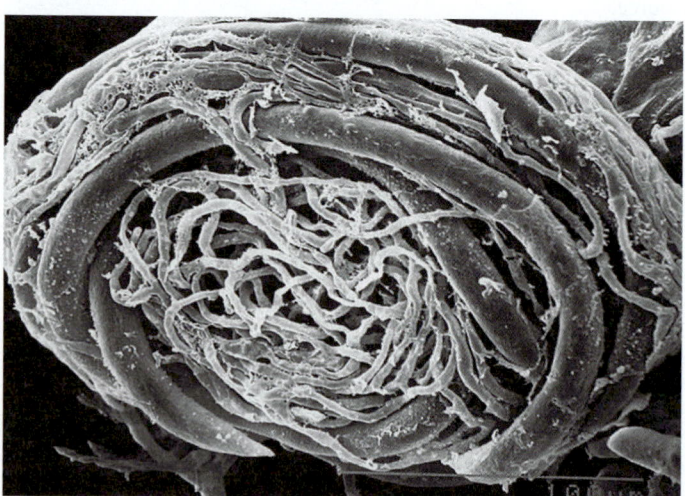

(b) Endomycorrhiza

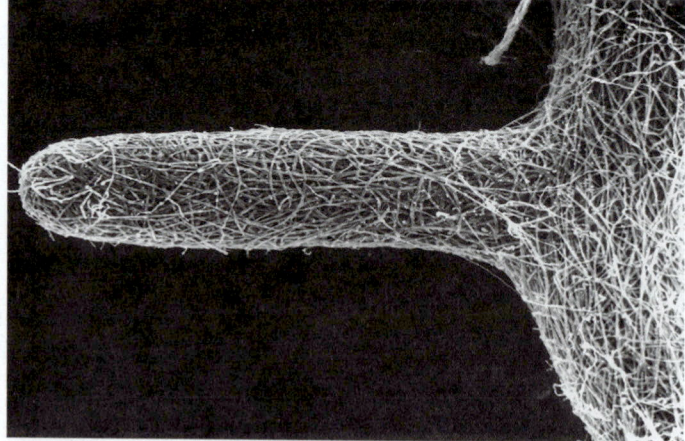

(c) Ectomycorrhiza

FIGURE 29.12 Mycorrhizas (a) Diagram illustrating the two types of mycorrhizas. **(b)** In an endomycorrhizal relationship, the fungi penetrate root cells growing within the cells. **(c)** In an ectomycorrhizal relationship, the fungi grow around the plant root cells, forming a fungal sheath around the root.

❓ *Which type of mycorrhizal relationship is common in herbaceous plants?*

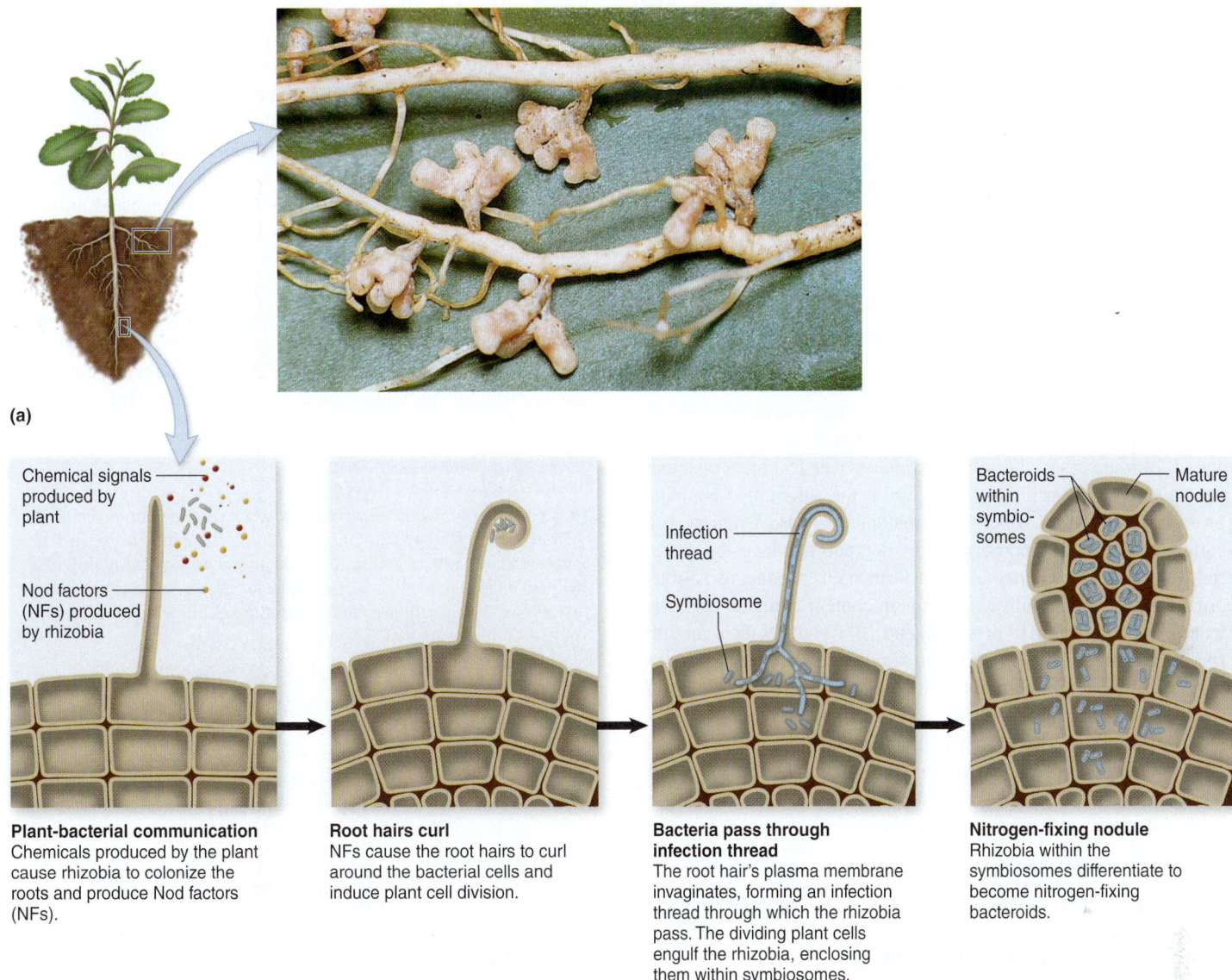

(a)

(b)

Plant-bacterial communication
Chemicals produced by the plant cause rhizobia to colonize the roots and produce Nod factors (NFs).

Root hairs curl
NFs cause the root hairs to curl around the bacterial cells and induce plant cell division.

Bacteria pass through infection thread
The root hair's plasma membrane invaginates, forming an infection thread through which the rhizobia pass. The dividing plant cells engulf the rhizobia, enclosing them within symbiosomes.

Nitrogen-fixing nodule
Rhizobia within the symbiosomes differentiate to become nitrogen-fixing bacteroids.

FIGURE 29.13 Root Nodules (a) Appearance. **(b)** The major steps leading to nodule formation by *Rhizobium* species.

❓ *The pinkish color of the nodules is due to leghemoglobin. What is the role of this protein?*

nodulation systems, some of the chemicals cause the bacteria to produce Nod factors (NFs), which induce a series of events that cause the root hair to curl, trapping the bacterial cells. The NFs also cause underlying plant cells to divide. The plasma membrane of the curled root hair then invaginates, forming a tubelike structure called an infection thread. The rhizobia cells travel through this structure to underlying plant tissues, and are then engulfed by the dividing plant cells, forming membrane-bound structures called symbiosomes. Within a symbiosome, a rhizobial cell changes in shape and function to become a specialized nitrogen-fixing cell called a bacteroid.

Although the relationship between the plant and bacterium is not obligate, it offers a competitive advantage to both partners. The rhizobia do not fix nitrogen in soils lacking legumes, and they compete poorly with other microbes, slowly disappearing from soils in which legumes are not grown. Likewise, legumes compete poorly against other plants in heavily fertilized soils.

Other Nitrogen-Fixing Symbionts

Several genera of non-leguminous trees, including alder and gingko, have nitrogen-fixing root nodules at some stages of their life cycle. The bacteria involved in the symbiosis are members of the genus *Frankia*.

In aquatic environments, the most significant nitrogen-fixers are cyanobacteria. They are especially important in flooded soils such as rice paddies. In fact, rice has been cultivated successfully for centuries without the addition of nitrogen-containing fertilizer because of the symbiotic relationship between the cyanobacterium *Anabaena azollae* and the aquatic fern *Azolla*. The bacterium grows in specialized sacs in the leaves of the fern, providing nitrogen to the fern. Before planting rice, the farmer allows the flooded rice paddy to overgrow with *Azolla* ferns. Then, as the rice grows, it eventually crowds out the ferns. As the ferns die and decompose, their nitrogen is released into the water.

Microorganisms and Herbivores

Another mutualistic relationship occurs between microbes and certain herbivores. In order to subsist on grass and other plant material, herbivores such as cattle and horses rely on a microbial community that inhabits a specialized digestive compartment. The microbes digest cellulose and hemicellulose, two of the major components of plant material, releasing compounds that can then serve as a nutrient source for the animal. The specialized digestive compartment in ruminants such as cattle, sheep, and deer is called a rumen, and it precedes the true stomach. Non-ruminant herbivores such as horses and rabbits have a compartment called a cecum, which serves a similar purpose; it lies between the small intestine and the large intestine. ◄◄ cellulose, p. 32

The rumen functions as an anaerobic fermentation vessel to which nutrients in the form of plant materials are intermittently added. In some cases, the animal produces over 150 liters of saliva per day. In addition to providing water, the saliva also contains bicarbonate, a buffer, which helps maintain the pH. A remarkably complex variety of microorganisms degrade the ingested material, releasing sugars that are then fermented, producing various organic acids. Each milliliter of rumen content contains approximately 10^{10} bacteria, 10^6 protozoa, and 10^3 fungi. Of the over 200 species identified in the rumen, no single one accounts for more than 3% of the total microbiota. The organic acids released during fermentation are absorbed by the cells that line the rumen, providing the animal with an energy and nutrient source. Large quantities of gas—produced as a result of fermentation—are discharged when the animal belches.

After being digested in the rumen, the food mass enters another compartment (the omasum), eventually reaching the acidic true stomach (abomasum). There, more organic acids are absorbed. In addition, lysozyme is secreted, allowing the animal to lyse and then digest members of the microbial population, providing even more nutrients. A critical feature of ruminants is that the microbial population gets the first opportunity to use the ingested nutrients. The animal then uses the metabolic end products as well as the microbial cells themselves.

The cecum of non-ruminant herbivores serves a function similar to a rumen. Microbes in the cecum cannot be used as a food source, however, because of the cecum's location relative to the stomach. The benefit of the location is that the animal can digest and absorb readily available nutrients without competition from microbes.

MicroAssessment 29.5

Mycorrhizal fungi gain nutrients from plant root secretions while helping plants take up substances from soil. Symbiotic nitrogen-fixers provide plants with a source of usable nitrogen while being provided with an exclusive habitat. Microorganisms in the rumen and cecum of herbivores digest cellulose and hemicellulose, allowing the animal to subsist on plant material.

13. *Describe the differences between an endomycorrhiza and ectomycorrhiza.*

14. *Describe the differences between a rumen and a cecum.*

15. *Gardeners sometimes plant clover between productive growing seasons. Why would this practice be beneficial?* ✚

Summary

29.1 ■ Principles of Microbial Ecology

Ecosystems vary in their biodiversity and biomass. The microenvironment is most relevant to a microorganism's survival and growth.

Nutrient Acquisition

Primary producers convert CO_2 into organic material; **consumers** use the organic materials produced by plants; **decomposers** digest the remains of primary producers and consumers (figure 29.1).

Microbes in Low-Nutrient Environments

Microorganisms capable of growing in dilute aqueous solutions are common in nature; often they grow in biofilms.

Microbial Competition and Antagonism

Microorganisms in the environment vie for the same limited pool of nutrients (figure 29.2). A species can competitively exclude others, or produce compounds that inhibit others.

Microorganisms and Environmental Changes

Environmental changes are common, and often result in different species becoming dominant (figure 29.3).

Microbial Communities

A **microbial mat** is a thick, dense, highly organized biofilm composed of distinct colored layers of different groups of microbes (figure 29.4).

Studying Microbial Ecology

Microbial ecology has been difficult to study because so few environmental prokaryotes can be successfully grown in the laboratory. Molecular techniques, including fluorescence *in situ* hybridization (FISH), polymerase chain reaction (PCR), denaturing gradient gel electrophoresis (DGGE), and DNA sequencing are allowing researchers to better understand complex microbial communities. Soil and water metagenomes are being studied.

29.2 ■ Aquatic Habitats

Oligotrophic waters are nutrient poor; **eutrophic** waters are nutrient rich (figure 29.5). Excessive growth of aerobic heterotrophs may cause an aquatic environment to become **hypoxic,** resulting in the death of fish and other aquatic animals.

Marine Environments

Ocean waters are generally oligotrophic and aerobic, but inshore areas can be dramatically affected by nutrient-rich runoff (figure 29.6).

Freshwater Environments

Oligotrophic lakes may have anaerobic layers due to thermal stratification. Shallow, turbulent streams are generally aerobic.

Specialized Aquatic Environments

Salt lakes and mineral-rich springs support the growth of microbes specifically adapted to thrive in these specialized environments.

29.3 ▪ Terrestrial Habitats

Characteristics of Soil

Soil represents an environment that can change abruptly and dramatically.

Microorganisms in Soil

The environmental conditions affect the density and composition of the soil microbiota.

The Rhizosphere

The concentration of microbes in the **rhizosphere** is generally much higher than that of the surrounding soil.

29.4 ▪ Biogeochemical Cycling and Energy Flow

Organisms use elements in biosynthesis to produce biomass, as sources of energy, and as terminal electron acceptors.

Carbon Cycle   (figure 29.7)

One of the fundamental aspects of the carbon cycle is carbon fixation. As consumers and decomposers degrade organic material, respiration and some fermentations release CO_2.

Nitrogen Cycle   (figure 29.9)

The steps of the nitrogen cycle include **nitrogen fixation, ammonification, nitrification, denitrification,** and **anammox.**

Sulfur Cycle   (figure 29.10)

Certain steps of the sulfur cycle—sulfur reduction and sulfate oxidation—depend on the activities of prokaryotes.

Phosphorus Cycle and Other Cycles

Most plants and microorganisms take up orthophosphate, incorporating it into biomass. Iron, calcium, zinc, manganese, cobalt, and mercury are recycled by microorganisms.

Energy Sources for Ecosystems

Photosynthetic organisms convert radiant energy to chemical bond energy in the form of organic compounds. Chemolithoautotrophs harvest energy from reduced inorganic chemicals (figure 29.11).

29.5 ▪ Mutualistic Relationships Between Microorganisms and Eukaryotes

Mycorrhizas

Mycorrhizas help plants take up phosphorus and other substances from soil; in turn the fungal partners gain nutrients for their own growth (figure 29.12). Endomycorrhizal fungi penetrate root cells; ectomycorrhizal fungi grow around root cells.

Symbiotic Nitrogen-Fixers and Plants

Rhizobia reside as nitrogen-fixing endosymbionts in **nodules** on legume roots (figure 29.13). *Frankia* species fix nitrogen in nodules of alder and gingko. A species of cyanobacteria fixes nitrogen in specialized sacs in the leaves of the *Azolla* fern.

Microorganisms and Herbivores

In order to subsist on grass and other plant material, herbivores rely on a community of microbes that inhabit a specialized digestive compartment, either a rumen or a cecum.

Review Questions

Short Answer

1. Describe why a microbial mat has green, reddish-pink, and black layers.
2. Why do lakes in temperate regions stratify during the summer months?
3. Why is there a high concentration of microbes in the rhizosphere?
4. What dictates whether a form of an element is suitable for use as an energy source versus a terminal electron acceptor?
5. Why does wood resting at the bottom of a bog resist decay?
6. What is the importance of nitrogen fixation?
7. Describe the relationship between ammonia oxidizers and nitrite oxidizers.
8. How do hydrothermal vents support thriving communities of microbes, clams, and tube worms?
9. Give examples of free-living and symbiotic nitrogen-fixing microorganisms. Are these prokaryotic or eukaryotic?
10. Describe the steps that lead to the formation of the symbiotic relationship between rhizobia and legumes.

Multiple Choice

1. Cyanobacteria are
   a) primary producers.
   b) consumers.
   c) herbivores.
   d) decomposers.
   e) more than one of the above.

2. Which of the following is *false*?
   a) Culture techniques are an accurate way of determining which members in a microbial community are most common.
   b) Fluorescence *in situ* hybridization (FISH) can be used to distinguish subsets of prokaryotes that contain a specific nucleotide sequence.
   c) Polymerase chain reaction (PCR) can be used to distinguish subsets of prokaryotes based on their 16S rRNA sequences.
   d) Denaturing gradient gel electrophoresis (DGGE) can be used to separate PCR products.
   e) Studying the genome of one organism can give insights into the characteristics of another.

3. Which of the following pairs that relate to aquatic environments does not match?
   a) Oligotrophic—nutrient poor
   b) Hypoxic—oxygen poor
   c) Epilimnion—O_2 poor
   d) Hypolimnion—lower layer
   e) Eutrophic—nutrient rich.

4. Adding high levels of nutrients to a lake or inshore area would have all of the following effects in that environment *except*
   a) death of clams and crabs.
   b) increased growth of heterotrophic microbes.
   c) increased growth of photosynthetic organisms.
   d) increased levels of dissolved O_2.

5. Which of the following pairs that relate to terrestrial environments does not match?

   a) Soil—minimal biodiversity

   b) *Bacillus*—endospores

   c) *Streptomyces*—geosmin production

   d) Fungi—lignin degradation

   e) Rhizosphere—soil that adheres to plant root

6. Atmospheric nitrogen can be used

   a) directly by all living organisms.

   b) only by aerobic bacteria.

   c) only by anaerobic bacteria.

   d) in symbiotic relationships between rhizobia and plants.

   e) in photosynthesis.

7. Which process converts ammonium (NH_4^+) into nitrate (NO_3^-)?

   a) Nitrogen fixation   b) Ammonification   c) Nitrification

   d) Denitrification   e) Anammox

8. Energy for ecosystems can come from

   a) sunlight via photosynthesis.

   b) oxidation of reduced inorganic chemicals by chemoautotrophs.

   c) both a and b.

9. Mycorrhizas represent associations between plant roots and microorganisms that

   a) are antagonistic.

   b) help plants take up phosphorus and other nutrients from soil.

   c) involve algae in the association with plant roots.

   d) form nodules on the plant's leaves.

   e) lead to the production of antibiotics.

10. In symbiotic nitrogen fixation by rhizobia and legumes

    a) the amount of nitrogen fixed is much greater than by nonsymbiotic organisms.

    b) neither the bacteria nor the legume can exist independently.

    c) the bacteria enter the leaves of the legume.

    d) the bacteria operate independently of the legume.

Applications

1. A farmer who was growing soybeans, a type of legume, saw an Internet site advertising an agricultural product for safely killing soil bacteria. The ad claimed that soil bacteria were responsible for most crop losses. The farmer called the agricultural extension office at a local university for advice. Explain what the extension office adviser most likely told the farmer about the usefulness of the product.

2. Recent reports suggest that human activities, such as the generous use of nitrogen fertilizers, have doubled the rate at which elemental nitrogen is fixed, raising concerns of environmental overload of nitrogen. What problems could arise from too much fixed nitrogen, and what could be done about this situation?

Critical Thinking ✚

1. Each colony growing on an agar plate arises from a single cell (see photo). Colonies growing close together are much smaller than those that are well separated. Why would this be so?

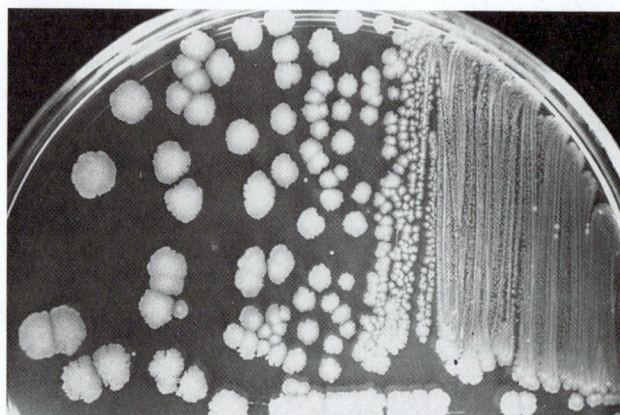

2. An entrepreneur found an economically feasible way of collecting large amounts of sulfur from underwater hot vents in the Pacific Ocean. The sulfur will be harvested from the microorganisms found in the vent areas. A group of ecologists argued that the project would destroy the fragile ecosystem by depleting it of usable sulfur. The entrepreneur argued that the environment would not be harmed because the vents produce more than enough sulfur for the clams and tube worms in the area. Explain who is correct.

30 Environmental Microbiology:
Treatment of Water, Wastes, and Polluted Habitats

A pristine mountain lake.

A Glimpse of History

Delivering fresh water to urban areas and removing human wastes have been practiced at least since Roman times. The ruins of aqueducts that delivered fresh water long distances can be seen today in many parts of Europe. Removing human wastes from cities has been more difficult, however, and the sewers that were used until the mid-nineteenth century were not much more than large, open cesspools.

Long before the discovery of the microbial world, people recognized that some diseases are associated with water supplies. As early as 330 B.C., Alexander the Great had his armies boil their drinking water, a habit that probably contributed to his huge successes. Certainly, many battles have been lost over the years as a result of waterborne diseases that sickened or killed the soldiers. Years before *Vibrio cholerae* was identified as the cause of cholera, people recognized that cholera epidemics were associated with drinking water. The desire for clean, clear water led to sand filtration systems being used in London and elsewhere in the early nineteenth century. Late in that century, Robert Koch showed that this kind of filtration yielded clear water and also removed nearly all bacteria from the water.

As early as the 1840s, Edwin Chadwick, an English activist, championed a new idea on how wastes could be removed. His idea was to construct a system of narrow, smooth ceramic pipes through which water under pressure could be flushed along with solid waste materials. This system would carry the wastes away from the inhabited part of the city to a distant collection site. There, he hoped to collect them and produce fertilizer to sell to farmers.

In 1848, concerned about cholera, the Board of Health in England started widespread reforms and began installing a sewage system along the lines envisioned by Chadwick. New York City did the same in 1866, again in response to a threatened cholera epidemic. By the end of the nineteenth century, most large European and U.S. cities had sewer systems to remove and treat waste materials as well as water systems to deliver safe drinking water. Cholera in the industrialized nations of Europe and North America virtually disappeared.

Most people living in developed countries take for granted that their tap water is safe to drink, their wastes will reliably disappear into sewers or landfills for proper disposal, and pollutants (substances that are harmful or injurious) will not accumulate in the environment. They seldom consider the role that microbes play in these essential aspects of modern life.

Microorganisms are important in the treatment processes described in this chapter for two very distinct reasons. First, we benefit from the fact that microbes are the ultimate recyclers, playing an essential role in the decomposition of our wastes. At the same time, pathogens must be eliminated from sewage before it is discharged and from drinking water for it to be potable (safe for human consumption). Recreational waters such as swimming pools, water parks, lakes, rivers, and shorelines are also monitored to ensure they do not have harmful levels of certain pathogens.

Treatment of water, waste, and polluted habitats is a significant challenge, particularly in densely populated areas. Consider that every day the average American uses about 150 gallons of water, and generates 120 gallons of wastewater and 5 pounds of trash. This means that a city with only 1 million inhabitants is faced with the disposal of approximately 44 billion gallons of wastewater and a million tons of trash each year!

30.1 ■ Microbiology of Wastewater Treatment

Learning Outcomes

1. *Describe the concept of BOD.*
2. *Compare and contrast primary treatment, secondary treatment, advanced treatment, and anaerobic digestion.*
3. *Describe how septic systems function.*

Wastewater, or sewage, is composed of all the material that flows from household plumbing systems, including washing and bathing water and toilet wastes. Municipal wastewater also includes business and industrial wastes. In many cities, storm water runoff that flows into street drains enters the system as well.

The most obvious reason that wastewater must be treated before discharge is that pathogenic microbes—including those that cause diarrheal diseases and hepatitis—can be transmitted in feces. If untreated sewage is released into a river or lake that is then used as a source of drinking water, disease can easily spread. If marine waters become contaminated in a similar manner, eating the local shellfish can result in disease. Shellfish are filter feeders and they concentrate microbes from the waters in which they live.

The high nutrient content of wastewater can also be damaging to the receiving water. When any nutrient-rich substance is added to an aqueous environment, microorganisms quickly use the compounds as energy sources, breaking them down in metabolic pathways such as glycolysis and the TCA cycle (see figure 6.10). As a result, microbes that aerobically respire consume available O_2 in the water, using it as a terminal electron acceptor (see figure 6.20). The amount of dissolved O_2 in lakes and rivers is limited and can easily be depleted during the microbial breakdown of nutrients. Fish and other aquatic animals in the environment die because they require O_2 for respiration (see figure 29.6). Thus, effective wastewater treatment must decrease the level of organic compounds substantially, in addition to eliminating pathogens and pollutants. ◄◄ aerobic respiration, p. 134

Biochemical Oxygen Demand (BOD)

An important goal of wastewater treatment is to decrease the environmental impact by reducing the **biochemical oxygen demand (BOD)**—the amount of O_2 required for the microbial decomposition of organic matter in a given sample. High BOD values indicate that large amounts of degradable materials are present, resulting in correspondingly large amounts of O_2 being consumed during its biological degradation. The BOD of raw sewage is approximately 300 to 400 mg/liter, which could easily deplete the dissolved O_2 in the receiving water. The dissolved O_2 content of natural waters is generally 5 to 10 mg/liter.

To determine the BOD, the O_2 level in a well-aerated sample of microbe-containing test water is first measured. The sample is then incubated in a sealed container in the dark under standard conditions of time and temperature, usually 5 days at 20°C. The O_2 level is then determined again. The difference between the dissolved O_2 at the beginning of the test and at the end reflects the BOD of the sample. In many cases, the sample must be diluted first in order to accurately determine the BOD.

Municipal Wastewater Treatment Methods

Large-scale wastewater treatment plants in the United States use a series of two processes—primary and secondary treatment—as required by the 1972 Federal Water Pollution Control Act, now known as the Clean Water Act. Once treated, the **effluent** (the liquid portion) can be discharged into the receiving water. Additional steps are used to treat **sludge** (the solid portion).

Primary Treatment

Primary treatment is a physical process designed to remove materials that will settle out, removing approximately 50% of the solids and 25% of the BOD. Raw sewage is first passed through a series of screens to remove large objects such as sticks, rags, and trash (**figure 30.1**). Skimmers then remove scum and other floating materials. The sewage then sits in a sedimentation tank, allowing solids to settle out. Once the settling period is complete, the sludge is removed and the primary treated wastewater that remains is sent for secondary treatment.

Secondary Treatment

Secondary treatment is chiefly a biological process that converts most of the suspended solids to inorganic compounds and cell mass that can then be removed, eliminating as much as 95% of the BOD. Microbial growth is actively encouraged during secondary treatment, allowing aerobic organisms to oxidize the biologically degradable organic material to CO_2 and H_2O. Because secondary treatment relies on the metabolic activities of microorganisms, the processes could be devastated if too much toxic industrial waste or hazardous household materials were dumped into wastewater systems, killing the microbial population.

Methods used for secondary treatment include:

■ **Activated sludge process.** This common system relies on mixed populations of aerobic microbes that grow as flocs (suspended biofilms). Although the organisms are often naturally present in wastewater, large numbers are added by introducing a small portion of leftover sludge from the previous load of treated wastes. Plenty of O_2 is supplied

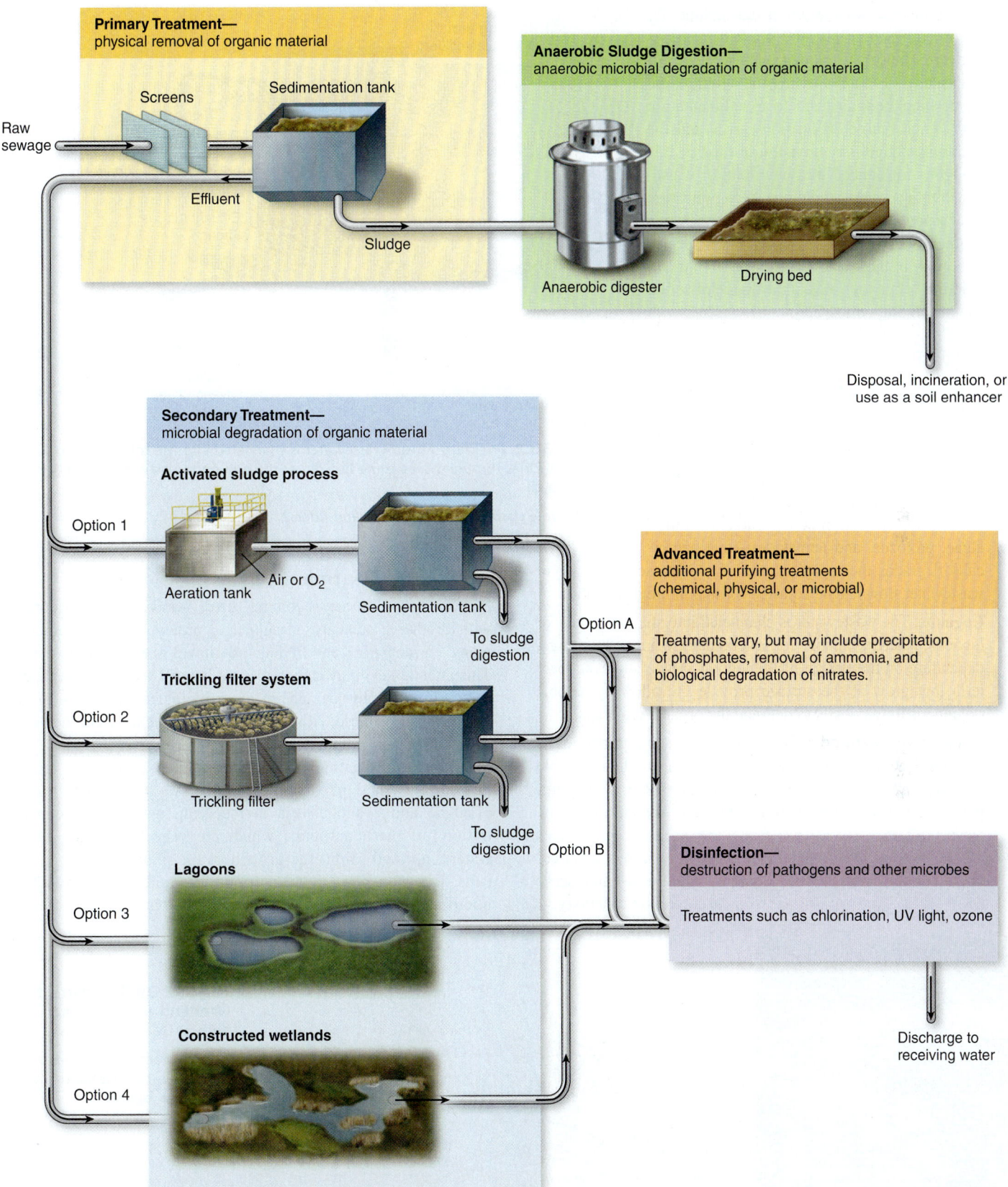

FIGURE 30.1 Municipal Wastewater Treatment The processes consist of primary treatment, secondary treatment, advanced treatment (optional), disinfection, and anaerobic sludge digestion.

❓ *What is the purpose of secondary treatment?*

by mixing the wastewater in an aerator. As the microbes multiply, the organic matter is converted into both biomass and waste products such as CO_2. Following the aeration, the wastewater is again sent to a sedimentation tank. There, most of the flocs settle and the resulting sludge is removed; a portion of this sludge is introduced to a new load of wastewater to act as an inoculum. A complication of the activated sludge process occurs when filamentous bacteria such as *Thiothrix* species overgrow in the wastewater during treatment, creating a buoyant mass that does not settle. This problem, called bulking, interferes with the separation of the solid sludge from the liquid effluent. ◀◀ biofilm, p. 84 ◀◀ *Thiothrix*, p. 262

- **Trickling filter (TF) system.** This method is frequently used in smaller wastewater treatment plants. The TF has a rotating arm that distributes the effluent over a bed of plastic pieces or coarse gravel and rocks (**figure 30.2**). The surfaces of these materials become coated with a biofilm—a mix of bacteria, fungi, algae, protozoa, and nematodes—that aerobically degrades the organic material as it passes. The rate of wastewater flow can be adjusted for maximum degradation.

- **Lagoons.** The wastewater is channeled into shallow ponds, or lagoons, where it remains for several days to a month or more, depending on the design of the lagoon. Algae and cyanobacteria that grow at the surface provide O_2, allowing aerobic organisms in the ponds to degrade the organic materials.

- **Constructed wetlands.** These follow the same principles as lagoons, but their more advanced designs make them suitable habitats for birds and other wildlife (**figure 30.3**). For example, the wastewater treatment processes in Arcata, California, use a series of marshes that now attract a variety of shorebirds and serve as a wildlife sanctuary.

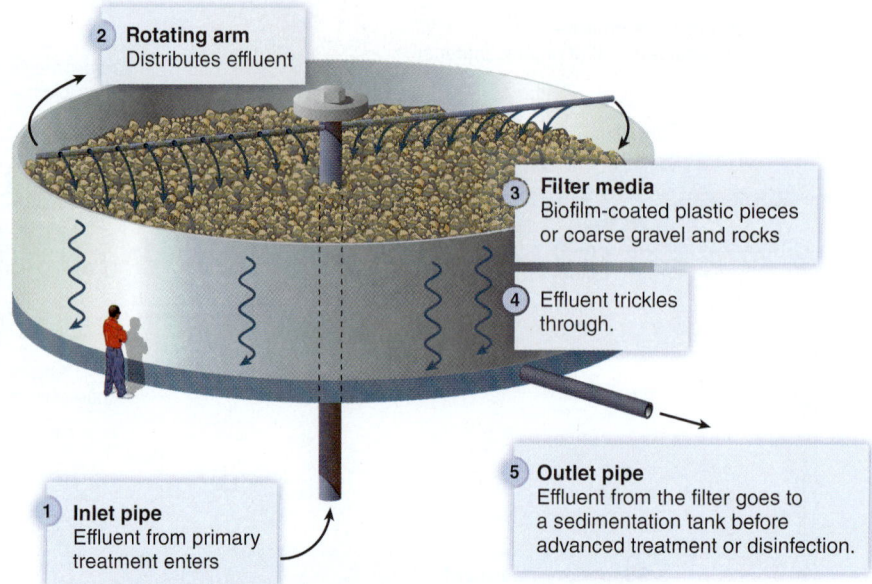

FIGURE 30.2 Trickling Filter Wastewater is channeled into the rotating arm, and then trickles through holes in the bottom of the arm onto plastic or a gravel and rock bed.

? *What role do biofilms play in trickling filters?*

2 **Rotating arm** Distributes effluent

3 **Filter media** Biofilm-coated plastic pieces or coarse gravel and rocks

4 **Effluent trickles through.**

5 **Outlet pipe** Effluent from the filter goes to a sedimentation tank before advanced treatment or disinfection.

1 **Inlet pipe** Effluent from primary treatment enters

Advanced Treatment

As water supplies are becoming scarce—and in order to comply with discharge standards designed to protect the environment—many communities are finding **advanced treatment** necessary. This includes any purification process beyond secondary treatment; it may involve physical, chemical, or biological processes, or any combination of these. Advanced treatment is expensive, however, and has not been common in the past.

Advanced treatment is often designed to remove ammonia, nitrates, and phosphates—compounds that foster growth of algae and cyanobacteria in receiving waters. The concentrations of these nutrients, which are very low in unpolluted waters, normally limit the growth of the photosynthetic organisms. Consequently, if the nutrients are added, photosynthetic microbes proliferate, often leading to buoyant masses of cells

Aquatic plants

Effluent inlet

Effluent outlet

Gradual slope to outlet

Soil or gravel   Watertight membrane

FIGURE 30.3 Constructed Wetland (a) Photograph. **(b)** General components.

? *What advantage does a constructed wetland have over a lagoon?*

(a)   (b)

PERSPECTIVE   30.1

What a Gas!

In rural areas of China, fuel for cooking and heating is often in short supply; as a consequence, the hay that should go for animal feed must be used for fuel. To help solve this problem, many of the farmers build methane-producing tanks on their farms (**figure 1**). Near the family house is an underground cement tank connected to the latrine and pigpen. Human and animal wastes along with water and some other organic materials such as straw are added to the tanks. As the natural process of decomposition occurs, methane gas (CH_4) is produced. This gas rises to the top of the tank and is connected to the house with a hose. Enough gas is produced to provide lights and cooking fuel for the farm family. By producing its own gas, a family also saves money because it does not need to buy coal for cooking. The effluent slurry that accumulates as the wastes decompose can be removed and used as fertilizer.

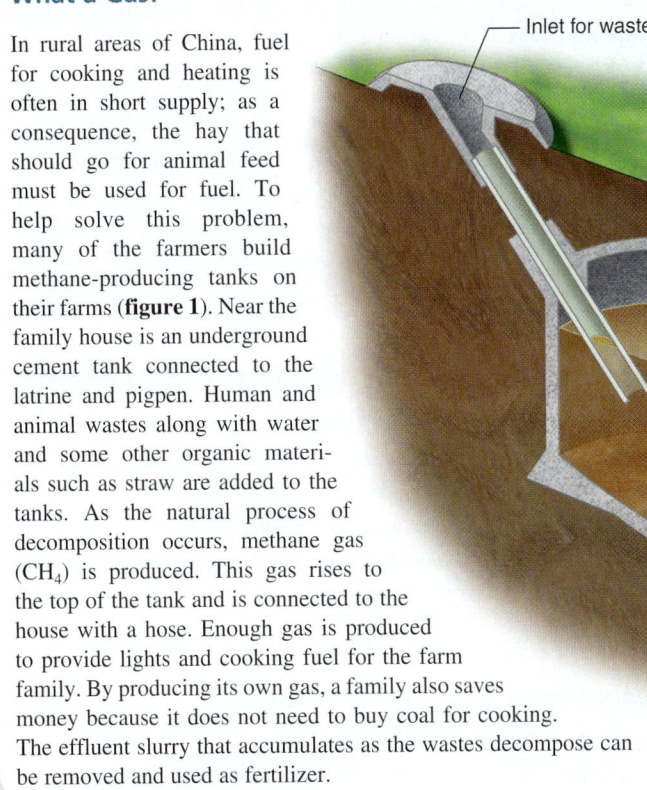

FIGURE 1 Methane-Producing Tank

---

that create a surface scum (see figure 11.7). Accumulations of photosynthetic organisms also provide a source of carbon for other microbes, increasing the BOD and, consequently, threatening other forms of aquatic life.

Ammonia can be removed by a process called ammonia stripping, which liberates gaseous ammonia from the water. Nitrates can be removed using denitrifying bacteria. These organisms use nitrate as a terminal electron acceptor during anaerobic respiration, forming nitrogen gas. This gas is inert, non-toxic, and easily removed. The discovery of anammox bacteria, which use ammonia as an energy source and nitrite as a terminal electron acceptor, provides an alternative means of removing inorganic nitrogen-containing compounds. Phosphates are eliminated using chemicals that combine with phosphates, causing them to precipitate.
◀◀ denitrification, p. 727 ◀◀ anammox, p. 727

Disinfection

Before discharge into the receiving water, the effluent is disinfected with chlorine, ozone, or UV light to decrease the numbers of microorganisms and viruses. If chlorine is used, the disinfected water can then be dechlorinated to avoid releasing excessive amounts of the toxic chemical into the environment. ◀◀ chlorine, p. 118 ◀◀ ozone, p. 120 ◀◀ UV light, p. 115

Anaerobic Digestion

Sludge removed during the sedimentation steps of primary and secondary treatment is transferred to a tank for **anaerobic digestion.** There, various anaerobic microbes act sequentially, ultimately converting much of the organic material to methane:

- Organic compounds $\longrightarrow$ organic acids, CO_2, H_2
- Organic acids $\longrightarrow$ acetate, CO_2, H_2
- Acetate, CO_2, H_2 $\longrightarrow$ methane (CH_4)

Many wastewater treatment plants are equipped to use the methane generated, thereby avoiding the cost of other sources of energy to run their equipment (see **Perspective 30.1**).

After anaerobic digestion, water is removed from the remaining sludge, generating a nutrient-rich product called stabilized sludge. This can be incinerated or disposed of in landfills but may also be used to improve soils and promote plant growth. The sludge generated by the city of Milwaukee, Wisconsin, is used to produce Milorganite, a fertilizer for lawns, gardens, golf courses, and playfields. An increasing number of wastewater treatment facilities are finding similar ways to recycle their treated sludge. Concerns exist, however, about heavy metals and other pollutants that can sometimes be concentrated in the product.

Individual Wastewater Treatment Systems

Rural dwellings customarily rely on **septic systems** for waste-water treatment (**figure 30.4**). Household wastewater is first directed to a large tank, where much of the solid material settles and is degraded by anaerobic microorganisms. The effluent—which has a high BOD—then flows to a series of perforated pipes, where it percolates (slowly passes) through a gravel-containing drain field. The drain field is designed to allow aerobic microorganisms to oxidize the organic material in the same manner described for the trickling filter.

Septic systems must be properly designed and monitored to ensure they work adequately. Circumstances that prevent adequate drainage—such as a clay soil under a drain field—allow anaerobic conditions to develop, preventing the organic material from being oxidized adequately. In addition, toxic materials can inhibit microbial activity in the system. Drainage from a septic tank may contain pathogens; therefore, the tank must never be allowed to drain where it can contaminate water supplies.

MicroAssessment 30.1

Primary treatment is a physical process that removes material that will settle out. Secondary treatment is chiefly a biological process that converts suspended material into inorganic compounds and microbial biomass. Advanced treatment is often designed to remove ammonia, phosphates, and nitrates. Anaerobic digestion converts much of the organic matter to methane. Septic systems are individual wastewater treatment systems.

1. *What does the term BOD mean? What is its significance?*
2. *What is the advantage of removing phosphates and nitrates from wastewater?*
3. *Why is denitrification unlikely to occur during secondary treatment?* ✚

30.2 ■ Drinking Water Treatment and Testing

Learning Outcomes

4. *Describe how drinking water is typically treated.*
5. *Describe why and how drinking water is tested for total coliforms.*

FIGURE 30.4 Septic System (a) The components of a septic system. **(b)** Septic tank, where anaerobic degradation takes place.

❓ *Why is it important that the drainage fields are aerobic?*

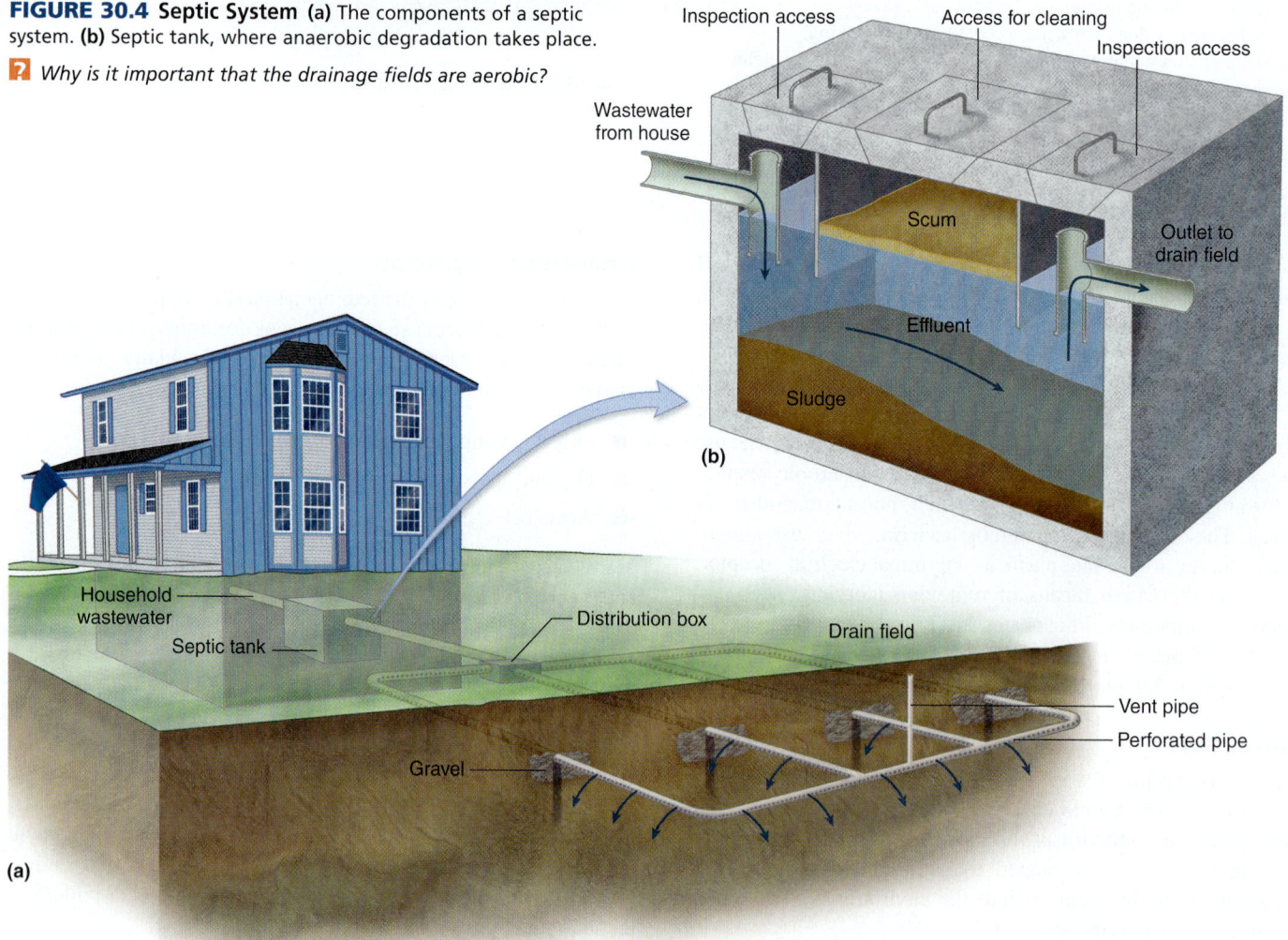

Inspection access   Access for cleaning   Inspection access

Wastewater from house

Scum

Outlet to drain field

Effluent

Sludge

(b)

Household wastewater

Distribution box

Drain field

Septic tank

Vent pipe

Perforated pipe

Gravel

(a)

Large cities generally obtain their drinking water from surface waters such as lakes or rivers. Because surface water may serve as the receiving water for another city's wastewater effluent, drinking water treatment is intimately connected to wastewater treatment. The quality of the surface water is also affected by the characteristics of the watershed (the land over which water flows into the river or lake). Even pristine rivers are likely contaminated with feces of animals that inhabit the watershed.

Smaller communities often use groundwater—pumped from a well—as a source of drinking water. Groundwater is in aquifers (water-containing layers of rock, sand, and gravel) and is replenished as water from rain and other sources seeps through the soil. Because aquifers are not directly exposed to animals and the atmosphere, they are somewhat protected from contamination. However, poorly located or maintained septic systems and sewer lines, as well as sludge or other fertilizers, can lead to groundwater contamination (**figure 30.5**).

Public water systems in the United States are regulated under the Safe Drinking Water Act of 1974, amended in 1986 and 1996. This gives the Environmental Protection Agency (EPA) the authority to set standards in order to control the level of contaminants in drinking water. Standards are modified in response to new concerns; for example, regulations now govern the maximum levels of *Cryptosporidium* oocysts, *Giardia* cysts, and enteric viruses in drinking water. ◀◀ *Cryptosporidium*, p. 604 ◀◀ *Giardia*, p. 602

Water Treatment Processes

Community drinking water treatment is designed to eliminate pathogenic microbes as well as harmful chemicals. First, water flows into a reservoir and is allowed to stand long enough for the particulate matter to settle out (**figure 30.6**). The water is then transferred to a tank where it is mixed with a chemical that causes suspended materials to coagulate (flocculate); an example of a coagulant is alum (aluminum sulfate). The mixture then flows to a sedimentation tank, where the coagulated materials are allowed to slowly sink to the bottom. As they settle, some microbes and other substances are trapped and thereby removed as well.

Following the removal of the coagulated materials, the water is filtered—often through a thick bed of sand and gravel—to remove various microbes including bacteria and protozoan cysts and oocysts. Organic chemicals that may be harmful or give undesirable tastes and odors can be removed by additional filtration. This is done using an activated charcoal filter, which adsorbs dissolved chemicals. An added benefit of filtration is that microorganisms growing in biofilms on the filter materials use carbon from the water as it passes. This lowers the organic carbon content of the water, resulting in less microbial growth in pipes delivering the water. ◀◀ filtration, p. 114

Finally, the water is treated with chlorine or other disinfectants to kill or inactivate harmful bacteria, protozoa, and viruses

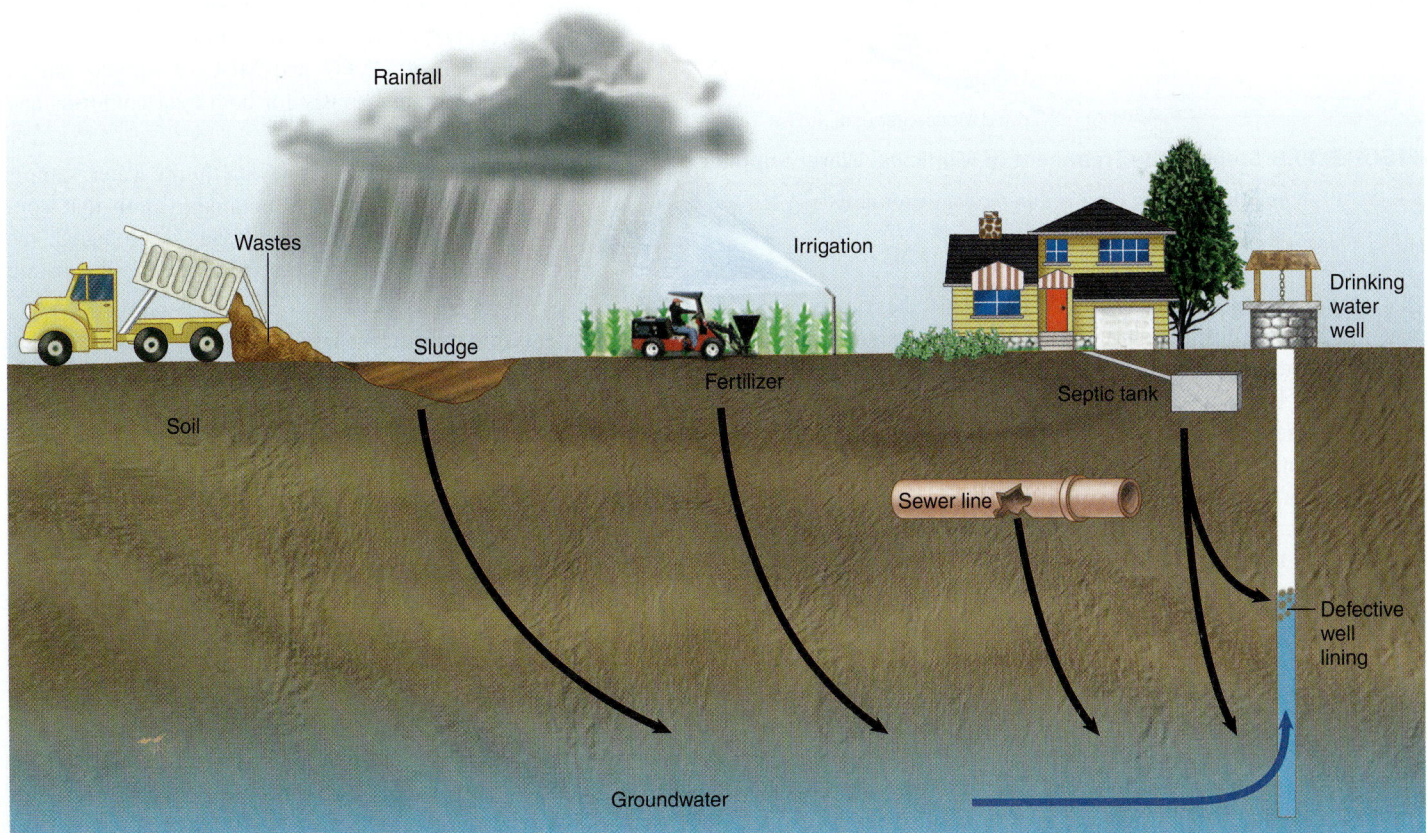

FIGURE 30.5 Groundwater Contamination Poorly located or maintained septic tanks and sewer lines, as well as sludge or other fertilizers, can lead to groundwater contamination.

❓ *How is water in the aquifer replenished?*

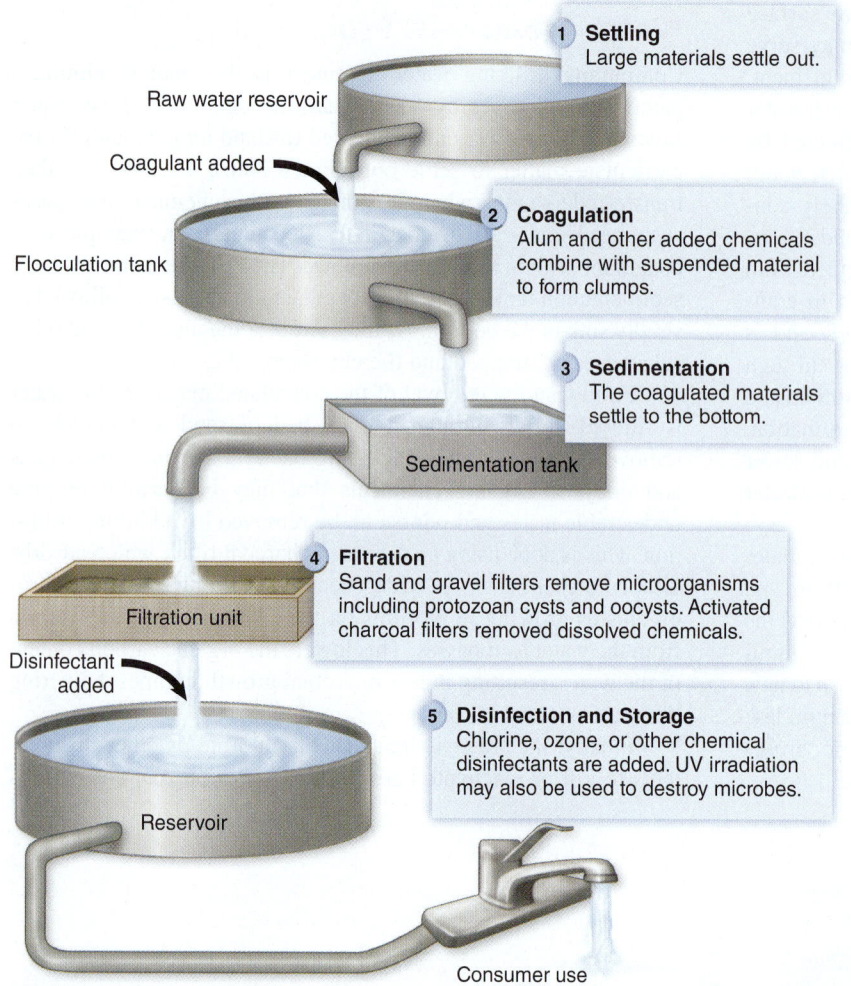

1. **Settling**
Large materials settle out.

Raw water reservoir

Coagulant added

2. **Coagulation**
Alum and other added chemicals combine with suspended material to form clumps.

Flocculation tank

3. **Sedimentation**
The coagulated materials settle to the bottom.

Sedimentation tank

4. **Filtration**
Sand and gravel filters remove microorganisms including protozoan cysts and oocysts. Activated charcoal filters removed dissolved chemicals.

Filtration unit

Disinfectant added

5. **Disinfection and Storage**
Chlorine, ozone, or other chemical disinfectants are added. UV irradiation may also be used to destroy microbes.

Reservoir

Consumer use

FIGURE 30.6 Steps in the Treatment of Municipal Water Supplies

❓ *What methods are used to disinfect municipal water supplies?*

that might remain. A concern with using chlorine, however, is that some of the disinfection by-products might be carcinogenic. In response to this concern, ultraviolet irradiation and ozone are increasingly being used as alternatives, but a small amount of chlorine must still be added to prevent problems associated with post-treatment contamination. Note that disinfection of waters with a high organic content requires more chlorine because organic compounds consume free chlorine. ◀◀ **chlorine, p. 118** ◀◀ **disinfection by-products, p. 110** ◀◀ **ultraviolet irradiation, p. 116** ◀◀ **ozone, p. 120**

Water Testing

A primary concern regarding the safety of drinking water is the possibility that it might be contaminated with any of a wide variety of intestinal pathogens, such as those discussed in chapter 24. It is not feasible to test for all of the pathogens, however, so **indicator organisms** function as surrogates. These microbes are routinely found in feces, survive longer than intestinal pathogens, and are relatively easy to detect. The most common group of bacteria used as indicator organisms in the United States is **total**

coliforms—lactose-fermenting members of the family *Enterobacteriaceae,* including *Escherichia coli.* The group is functionally defined as facultatively anaerobic, Gram-negative, rod-shaped, non-spore-forming bacteria that ferment lactose, forming acid and gas within 48 hours at 35°C.

Although total coliforms are routinely present in the intestinal contents of warm-blooded animals, certain species can also thrive in soils and on plant material. Because of this, their presence does not necessarily imply fecal pollution. To compensate for this shortcoming, **fecal coliforms,** a subset of total coliforms more likely to be of intestinal origin, are also used as indicator organisms. The most common fecal coliform is *E. coli.* Note that although some *E. coli* strains can cause intestinal disease, the species is used in water testing merely to indicate fecal pollution.

Methods used to detect total coliforms in a water sample include:

- **ONPG/MUG test.** A water sample is added to a medium containing ONPG (*o*-nitrophenyl-β-D-galactopyranoside) and MUG (4-methyl-umbelliferyl-β-D-glucuronide). Lactose-fermenting bacteria hydrolyze ONPG, generating a yellow compound; thus, all coliforms turn the medium yellow (**figure 30.7**). *E. coli* produces an enzyme that hydrolyzes MUG, making a fluorescent compound. Because of ONPG and MUG, a sample can be tested simultaneously for both total coliforms and *E. coli.*

- **Presence/absence test.** A 100-ml water sample is added to a lactose-containing broth that contains a vial to trap gas. If gas is produced, the broth is then tested to confirm that coliforms are present.

- **Most probable number (MPN) method.** This statistical assay of cell numbers uses successive dilutions to determine the most probable number of bacteria in a sample (see figure 4.20). To test drinking water, the broth used is similar to that in the presence/absence test. Positive tubes are further tested to confirm that they contain coliforms.

- **Membrane filtration.** A water sample is passed through a filter that retains bacteria (see figure 4.19), concentrating the bacteria from a known volume of water. The filter is then placed on a lactose-containing selective and differential agar medium. ◀◀ **selective media, p. 95** ◀◀ **differential media, p. 95**

The total coliform rule establishes a maximum number of positive samples (100 ml) permitted in water samples. That limit depends on how many samples are routinely collected for monitoring, which relates to the size of the population served by the water system. When at least 40 samples per month are collected, the system is in violation if more than 5% test positive for total coliforms in a month. If fewer samples are collected, the system

FIGURE 30.7 ONPG/MUG Test Coliforms hydrolyze ONPG, yielding a yellow-colored compound. *E. coli* hydrolyzes MUG, generating a blue fluorescent compound.

❓ *What is the significance of finding coliforms in a drinking water supply?*

is in violation if more than one sample tests positive per month. If a sample tests positive, repeat samples within 24 hours are mandated. Samples positive for total coliforms are also tested for either fecal coliforms or *E. coli*. If a water system exceeds the monthly total coliform limit, the state and the public must be notified. Notification is also required if either of two sequential samples that test positive for total coliforms also test positive for fecal coliforms or *E. coli*.

Because total coliform and fecal coliform assays cannot always predict contamination with protozoan cysts and oocysts, alternatives are being explored. Other microbes that can be used as indicators of fecal pollution include enterococci, some *Clostridium* species, and certain types of bacteriophages.

MicroAssessment 30.2

Drinking water may be obtained from surface water or groundwater. Treatment of drinking water is designed to eliminate pathogens and harmful chemicals. In the United States, total and fecal coliforms are the most commonly used indicator organisms.

4. *What is the purpose of coagulation in drinking water treatment?*

5. *Describe two methods of water testing.*

6. *Which would be more likely to cause illness and why—a water sample that tested positive for fecal coliforms or one that tested positive for* E. coli *O157:H7?* ➕

30.3 ■ Microbiology of Solid Waste Treatment

Learning Outcome

6. *Compare and contrast sanitary landfills and composting programs.*

In addition to ridding our environment of wastes in water, we must dispose of the solid wastes (garbage) generated each day. Eliminating these has become an increasingly complex problem.

Sanitary Landfills for Solid Waste Disposal

Sanitary landfills are widely used to dispose of non-hazardous solid wastes in a manner that minimizes damage to human health and the environment. Before sanitary landfills were developed, solid wastes were often piled up on the ground in open-burning dumps, attracting insects and rodents and causing aesthetic and public health problems.

Federal standards dictate that sanitary landfills must be located away from wetlands, earthquake-prone faults, flood plains, or other sensitive areas. The excavated site is lined with plastic sheets or a special membrane on top of a thick layer of clay to prevent contaminates from seeping out into the surrounding environment. A layer of sand with drainage pipes is placed on top of this. When wastes are added to the site, they are compacted and covered with a layer of soil every day. Once a landfill is full, it is covered with soil and plants and can be used for recreation and eventually as a site for construction. Methane and other gases are vented, and the methane is burned or recovered for use.

Sanitary landfills have several disadvantages with respect to waste management. For one thing, only a limited number of sites are available for use near urban and suburban areas. In addition, the methane gas must be removed as the organic waste material anaerobically decomposes, a process that can take more than 50 years. If buildings are constructed before the methane is removed, disastrous gas explosions can occur. Pollutants such as heavy metals and pesticides can leak from landfill sites into the underground aquifers. It is very difficult to purify these aquifers once they have become contaminated.

Sanitary landfills have traditionally been a low-cost method of handling large quantities of solid waste. Because of increased costs and decreased availability of land, however, many cities are looking for ways to decrease the amount of solid waste dumped in landfills. In some cities, the fees charged to people for garbage collection are based on the size of the container collected. The smaller the can, the lower the cost. This is intended to raise people's awareness of how much solid waste they are generating as well as offer an incentive to recycle. Programs to recycle paper, plastics, glass, and metal are being implemented in many cities and counties with great success. Through these programs, landfill areas will be available for a longer period of time.

Municipal and Backyard Composting— Alternative to Landfills

Composting is the natural decomposition of organic solid material. Municipal and home composting programs are becoming popular in many areas and are succeeding in reducing the amount of organic wastes added to landfills. As an added benefit, the black organic material generated is commercially valuable and can be used to improve garden soils.

Backyard composting usually starts with a supply of organic material such as leaves, grass clippings, and kitchen wastes (**figure 30.8**). Composting of meats and fats is generally not recommended because they attract rodents and other pests. Often, some soil and water are added to facilitate the process. As a result of microbial metabolism, the inside of the pile heats up. At 55°C to 66°C, pathogens are killed but thermophilic organisms are not affected. If the pile is frequently aerated, which can be done by physically stirring and turning it, and it is kept moist, the composting can be completed in as little as 6 weeks.

◄◄ thermophile, p. 90

Composting on a large scale offers cities a way to reduce the amount of garbage sent to their landfills. In some cities, yard wastes are collected separately from the main garbage. These are then processed using machinery such as grinders that help make the composting more efficient (**figure 30.9**).

FIGURE 30.9 Municipal Composting

❓ *What is the purpose of grinding the yard waste before putting it in piles?*

(a)

(b)

FIGURE 30.8 Backyard Composting **(a)** A home compost bin. **(b)** Garden debris and many kitchen organic wastes can be composted.

❓ *Why is it important that the compost heap be turned frequently?*

MicroAssessment 30.3

Sanitary landfills are used to dispose of non-hazardous solid wastes. Composting, the natural decomposition of organic solid material, dramatically reduces the need for large landfills.

7. *What are the advantages and disadvantages of landfills?*
8. *List the steps in successful composting.*
9. *Why would soil and water be added to a compost pile?* ➕

30.4 ■ Microbiology of Bioremediation

Learning Outcomes

7. *Describe why pollutants are a problem, and why some persist in the environment.*
8. *Describe how two bioremediation strategies remove pollutants.*

Bioremediation is the use of microorganisms to degrade or detoxify pollutants in a given environment. It generally takes advantage of organisms already present in the polluted environment, but in some cases specific organisms are added to the environment.

Pollutants

Most naturally occurring organic compounds are biodegradable, meaning they can be degraded by one or more species of microorganisms. As oil spills dramatically demonstrate, however, some natural materials can cause devastating effects before they are degraded.

Synthetic compounds are more likely to be biodegradable if their chemical composition is similar to that of naturally occurring substances. In contrast, xenobiotics (synthetic compounds quite different from any in nature) are more likely to persist in the environment because microorganisms are unlikely to have suitable enzymes to break them down. Such enzymes would not give a microbe a competitive advantage in a natural situation.

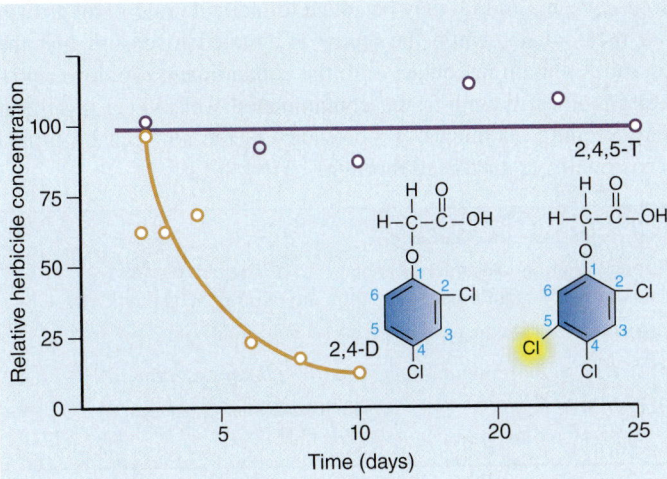

FIGURE 30.10 Comparison of the Rates of Disappearance of Two Structurally Related Herbicides, 2,4-D and 2,4,5-T

❓ *Why might the additional chlorine atom in 2,4,5-T molecule result in slower degradation of the molecule?*

Relatively slight molecular changes significantly alter the biodegradability of a compound. Perhaps the best-studied example involves the herbicides 2,4-dichlorophenoxyacetic acid (2,4-D) and 2,4,5-trichlorophenoxyacetic acid (2,4,5-T). The only difference between these two compounds is the additional chlorine atom on the latter. When 2,4-D is applied to the soil, it disappears within a period of several weeks, as a result of its degradation by microbes in the soil. When 2,4,5-T is applied, however, it is often still present more than a year later (**figure 30.10**). The additional chlorine atom of 2,4,5-T blocks the enzyme that makes the initial attack on 2,4-D.

Most herbicides and insecticides not only are toxic to their target, but also have deleterious effects on fish, birds, and other animals. For example, the pesticide DDT accumulates in the fat of predatory birds through biological magnification (**table 30.1**). Small amounts of the pollutant that contaminate water are concentrated in minute plankton, which are eaten by minnows, accumulating even more in the fish. When large birds eat the fish, the amount of chemical in tissues is tremendously magnified. The continuing ingestion of DDT, which accumulates in fat, results in an ever-greater concentration of the DDT as it passes upward through the food chain. DDT interferes with the reproductive process of birds, leading to the production of fragile eggs, which break before the young can hatch. Although banned in the United States, DDT is still used in other countries, particularly to control mosquitoes that transmit malaria.

TABLE 30.1	Biological Magnification of DDT
Parts per Million DDT	**Source**
0.00005	Water
0.04	Plankton
0.23	Minnow
3.57	Heron
22.8	Merganser (fish-eating duck)

Means of Bioremediation

Many factors influence the degradation rate of pollutants. As a general rule, any practice that favors multiplication of microorganisms will increase the rate of degradation. Thus, providing adequate nutrients, maintaining the pH near neutrality, raising the temperature, and providing an optimal amount of moisture are all likely to promote pollutant degradation. ◀◀ **environmental factors that influence microbial growth, p. 89** ◀◀ **nutritional factors that influence microbial growth, p. 92**

There are two general bioremediation strategies—biostimulation and bioaugmentation. Biostimulation enhances growth of local microbes in a contaminated site by providing additional nutrients. Petroleum-degrading bacteria are naturally present in seawater, but they degrade oil at a very slow rate because the low levels of certain nutrients, including nitrogen and phosphorus, limit their growth. To enhance bioremediation of oil spills, a fertilizer containing these nutrients—and which adheres to oil—was developed. When this fertilizer is applied to an oil spill, microbial growth is stimulated, leading to at least a threefold increase in the speed of degradation (**figure 30.11**). Bioaugmentation relies on activities of microorganisms added to the contaminated material, complementing the resident population. The activated sludge process used during secondary treatment of wastewater is a form of bioaugmentation. A great deal of research is underway to develop microbial strains suited for bioaugmentation. One example, the bacterium *Burkholderia (Pseudomonas) cepacia*, is capable of growth on 2,4,5-T and has been used successfully to remove this chemical from soil samples in the laboratory. However, microbes that thrive under laboratory conditions may not compete well in natural habitats. ◀◀ **activated sludge process, p. 736**

Successful bioremediation may also involve controlling metabolic processes by manipulating the availability of O_2 and specific growth substrates. For example, anaerobic degradation of trichloroethylene (TCE), a solvent used to clean metal parts, results in the

FIGURE 30.11 Oil Spill Bioremediation Bioremediation was used to clean the shoreline contaminated by oil spilled from the *Exxon Valdez*. The growth of local oil-degrading prokaryotes was stimulated by addition of nutrients. Comparing the uncleaned rocks on the left to the rocks cleaned by bioremediation on the right shows the dramatic efficiency of bioremediation.

❓ *Which kind of bioremediation is this—biostimulation or bioaugmentation?*

accumulation of vinyl chloride, a compound more toxic than TCE. Aerobic conditions are important to prevent this buildup. Some pollutants are degraded only when specific substrates are made available to the microbes. This phenomenon, called co-metabolism, occurs because the enzyme produced by the microbe to degrade the additional substrate degrades the pollutant as well. As an example, the enzymes produced by some microbes to degrade methane also degrade TCE. In this case, adding methane enhances the degradation of TCE.

Bioremediation may be done either *in situ* ("in place") or off-site. *In situ* bioremediation generally relies on biostimulation and is less disruptive. Oxygen (O_2) can be added to contaminated groundwater and soil either by injecting hydrogen peroxide, which rapidly decomposes to liberate O_2 and water, or pumping air into soil. Off-site processes may be performed using a bioreactor, a large tank designed to accelerate microbial processes.

Both nutrients and O_2 may be added to facilitate microbial growth and metabolism, while the slurry is agitated to ensure that the microbes remain in contact with the contaminants. A slower process involves mounding the contaminated soil over a layer that traps seeping chemicals. To provide O_2, the soil can be turned occasionally or air forced through.

MicroAssessment 30.4

Bioremediation uses microorganisms to degrade pollutants. Biostimulation and bioaugmentation are two methods employed.

10. *Why do xenobiotics often persist in the environment?*

11. *How is biostimulation different from bioaugmentation?*

12. *In treating an oil spill, why might biostimulation be preferred over bioaugmentation?* ✚

FUTURE CHALLENGES   30.1

Better Identification of Pathogens in Water and Wastes

One of the most important challenges in the field of water and waste treatment is the development of new and better methods to detect waterborne contaminants in both drinking water and environmental water samples. This would make it possible to follow the occurrence and persistence of pathogens in water supplies with greater accuracy, and aid in better reporting of waterborne illnesses. Methods being developed, but not yet perfected for use in this field, include the polymerase chain reaction (PCR) and a variety of fluorescence techniques and radioactivity labeling methods.

Cysts of *Giardia* and oocysts of *Cryptosporidium* have been detected by amplifying specific regions of their DNA by PCR. In fact, using this method, it is possible to detect a single cyst of *Giardia* and to distinguish between species that are pathogenic for humans and those that are not. But problems arise in using these techniques with environmental samples that contain substances that inhibit the reaction. In addition, PCR detects DNA from dead organisms as well as living. Studies are needed to make these techniques feasible for use in water testing.

Viruses can also be detected by PCR. The viruses are concentrated by filtration onto membranes. Many different viruses can be detected simultaneously by combining gene probes from various groups of viruses. A problem is that viruses inactivated by disinfection procedures are still detected by PCR. To overcome this, viruses can be put into cell cultures to allow them to replicate, indicating that they are not inactivated, and the PCR is then performed on the infected cell cultures.

Summary

30.1 ■ Microbiology of Wastewater Treatment

Biochemical Oxygen Demand (BOD)

An important goal of **wastewater** treatment is the reduction of the **biochemical oxygen demand (BOD)**.

Municipal Wastewater Treatment Methods (figures 30.1–30.3)

Primary treatment is a physical process designed to remove materials that sediment out. **Secondary treatment** is chiefly a biological process designed to convert most of the suspended solids to inorganic compounds and microbial biomass, removing most of the BOD. **Advanced treatment** is often designed to remove ammonia, nitrates, and phosphates. Biosolids that result from **anaerobic digestion** of **sludge** can be used to improve soils and promote plant growth.

Individual Wastewater Treatment Systems

Rural dwellings often use **septic systems** for wastewater treatment (figure 30.4).

30.2 ■ Drinking Water Treatment and Testing

Water Treatment Processes

Community drinking water is treated to remove particulate and suspended matter, various microorganisms, and organic chemicals (figure 30.6). Chlorine or other disinfectants are then used to destroy harmful microbes.

Water Testing

Total coliforms and **fecal coliforms** are used as **indicator organisms**; their presence suggests the possible presence of pathogens.

30.3 ■ Microbiology of Solid Waste Treatment

Sanitary Landfills for Solid Waste Disposal

Landfills are used to dispose of solid wastes near towns and cities.

Municipal and Backyard Composting—Alternative to Landfills

Composting reduces the amount of garbage sent to landfills (figures 30.8, 30.9).

30.4 ■ Microbiology of Bioremediation

Pollutants

Synthetic compounds are more likely to be biodegradable if they have a chemical composition similar to that of naturally occurring compounds (figure 30.10).

Means of Bioremediation

Biostimulation can be used to increase the effectiveness of oil degradation by naturally occurring prokaryotes (figure 30.11).

Review Questions

Short Answer

1. Describe how the BOD of a water sample is determined.

2. Which step of wastewater treatment removes most of the BOD?

3. Compare and contrast the activated sludge process and the trickling filter system used in secondary treatment of wastewater.

4. Why is it beneficial to remove nitrates and phosphates in wastewater?

5. How does a septic system work?

6. What is an aquifer?

7. Why do water-testing procedures look for coliforms rather than pathogens?

8. How does the ONPG/MUG test allow a sample to be assayed simultaneously for the presence of both total coliforms and *E. coli*?

9. What aspect of 2,4,5-T makes it more likely to persist in the environment than 2,4-D?

10. Describe the use of bioremediation in the cleanup of oil spills.

Multiple Choice

1. A marked decrease in BOD during secondary treatment indicates
   a) lack of oxidation during treatment.
   b) effective aerobic decomposition during treatment.
   c) effective anaerobic decomposition during treatment.
   d) removal of all pathogenic bacteria.
   e) removal of all toxic chemicals.

2. Advanced treatment is often designed to remove
   a) BOD. b) nitrates and phosphates. c) bacteria.
   d) protozoa. e) methane.

3. Which of the following is not a matching pair?
   a) Potable water—presence of pathogens
   b) High BOD—high organic content
   c) Stabilized sludge—fertilizer
   d) Primary treatment—removal of material that settles
   e) Bulking—growth of filamentous bacteria

4. Which of the following is *false*?
   a) Bulking interferes with trickling filter systems.
   b) Artificial wetlands provide a habitat for wildlife.
   c) Removal of nitrates by microorganisms requires anaerobic conditions.
   d) Methane is a by-product of anaerobic digestion.

5. Which of the following is not a matching pair?
   a) Surface water—watershed
   b) Groundwater—aquifer
   c) Sand and gravel filters—removes organic chemicals
   d) Alum—causes suspended material to coagulate
   e) Disinfection—chlorine, ozone, or ultraviolet light

6. Septic tanks should be placed
   a) as close to the well as possible.
   b) at least 500 feet from the house.
   c) under the house.
   d) in deep clay soil.
   e) where the outflow cannot contaminate any water supply.

7. Which of the following about coliform testing methods is *true*?
   a) All determine the number of *E. coli* present in a sample.
   b) The MPN procedure precisely indicates the concentration of coliforms.
   c) The media used test for the ability to ferment lactose.
   d) A positive test indicates that pathogens are definitely present in the sample.
   e) All coliforms hydrolyze ONPG and MUG.

8. Landfills are often used to dispose of
   a) household wastewater. b) commercial wastewater.
   c) solid wastes. d) petroleum wastes.
   e) wastewater effluent.

9. Backyard composting is an excellent way to dispose of
   a) cooking fats. b) garden debris. c) spoiled meats.
   d) insecticides. e) cleaning supplies.

10. Synthetic compounds are most likely to be biodegradable if they
    a) are totally different from anything found in nature.
    b) have three chlorine atoms per molecule.
    c) are plastics.
    d) are present in very large amounts.
    e) are chemically similar to naturally occurring substances.

Applications

1. A developer is interested in building vacation homes on 150 acres of oceanfront property. A priority is to retain as much natural beauty of the area as possible. Safe and effective wastewater treatment must be part of the plan. What advantages and disadvantages of each of the following options must the developer consider before selecting one?
   a) Individual septic systems for each home
   b) Trickling filter system
   c) Constructed wetlands

2. A public health official is investigating waterborne diseases in Illinois. She notes that over half of the cases of waterborne diseases originating from drinking water were caused by *Giardia lamblia*. Other data showed that most cases of gastroenteritis attributed to exposure to recreational waters were caused by *Cryptosporidium parvum*. What does this suggest about controlling waterborne diseases?

Critical Thinking ✚

1. Why is oil not degraded when in a natural habitat underground yet is susceptible to bioremediation in an oil spill?

2. The accompanying figure shows the effects of different treatments of drinking water on the incidence of typhoid fever in Philadelphia, 1890–1935. If filtration of drinking water caused such a dramatic decrease in the disease incidence, was it necessary to introduce chlorination a few years later? Why or why not?

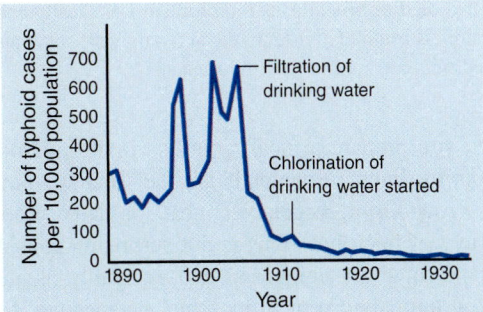

Food Microbiology

Refrigeration is a method of food preservation.

A Glimpse of History

Alice Catherine Evans, the first female president of the Society of American Bacteriology (now the American Society for Microbiology), helped show that unpasteurized milk could be a source of disease in humans. A graduate of both Cornell University and the University of Wisconsin, Evans worked for the U.S. Department of Agriculture, seeking out the sources of microbial contamination of dairy products. In 1917, Evans reported that cases of human brucellosis were related to finding *Brucella abortus* in cows' milk. ◄◄ brucellosis, p. 676

Evans's conclusion that *B. abortus* could be transmitted from cows to humans through milk conflicted with the common view of a number of prominent scientists, including Robert Koch. In 1900, Koch had declared that bovine tuberculosis and brucellosis could not be transmitted to humans. As a result, many scientists and dairy workers would not accept the increasing evidence that diseases were being transmitted from cows to humans through milk. In the late 1930s, after a number of children of dairy workers had died of brucellosis, the problem was finally acknowledged. Today, milk is routinely pasteurized, and only very small amounts of unpasteurized milk are sold in the United States.

When you prepare a meal, you also invite microorganisms to dinner. Practically all the food we purchase or grow—fruit, vegetable, meat, or dairy product—harbors a variety of microbes. This is not surprising considering that bacteria and fungi are ubiquitous and especially plentiful in soil and around animals. From a microbial perspective, food is

an ecosystem. Microbes compete for available nutrients, and the most successful ones predominate.

Microorganisms on foods are not necessarily undesirable. Sometimes, their growth results in pleasant flavors or textures. Foods intentionally altered during production by carefully controlling the activity of bacteria, yeasts, or molds are called **fermented foods** (**figure 31.1**). For example, food manufacturers purposely encourage certain microorganisms to grow in milk in order to produce foods such as sour cream and cheese. Alcoholic beverages, such as beer and wine, and many Asian food condiments, such as soy sauce and miso, also rely on microbial metabolism for their production. Strictly speaking, the term *fermentation* refers to only those metabolic activities that use pyruvate or another organic compound as an electron acceptor, with the result that alcohols and acids are produced. Food scientists, however, use the term more generally, to encompass any desirable change that a microorganism imparts to food. ◄◄ fermentation, p. 147

Undesirable biochemical changes in foods are called spoilage (**figure 31.2**). The processes that cause spoilage are often the same ones involved in fermentation of foods. In fact, a food product considered by one cultural population to be fermented may be considered spoiled by another. Sour milk and moldy bread are examples of foods considered spoiled as a result of microbial growth. The souring of milk, however, involves the same microbial processes that cause the agreeable acidic flavor of sour cream, and the mold growing on bread may be related to the one that causes the blue veining in Gorgonzola cheese.

Growth of pathogens in food can result in foodborne illness, but generally does not cause noticeable changes in food quality. Depending on the type of pathogen, the illness may result from consuming either the living organisms or the toxins they have produced during growth.

31.1 ■ Factors Influencing the Growth of Microorganisms in Foods

Learning Outcome

1. *Describe five intrinsic factors and two extrinsic factors that influence the growth of microorganisms in foods.*

Understanding the factors that influence microbial growth is essential to maintaining food quality when producing fermented foods or prolonging the shelf life of perishable foods. As a general rule, bacteria will predominate in fresh meats and other moist, pH-neutral, nutrient-rich foods. Yeasts and molds can also grow in these foods, but the more rapid increase of bacteria overwhelms the competitors. When conditions such as lack of moisture or high acidity restrict the growth of bacteria, fungi predominate despite their relatively slow growth.

Intrinsic Factors

The inherent conditions in the food—such as water availability, acidity, and nutrient level—are called **intrinsic factors.** They affect which microbes predominate on the product.

Water Availability

Foods vary in terms of how much water is accessible to microorganisms. Fresh meats and milk, for example, have plenty of water to support the growth of many microbes. Bread, nuts, and dried foods, on the other hand, are relatively dry. Jams, jellies, and some other sugar-rich foods are seemingly moist, but most of that water is chemically interacting with the sugar, making it unavailable for use by microbes. Highly salted foods, for similar reasons, have little available moisture.

The term **water activity (a_w)** is used to indicate the amount of water available in foods. By definition, pure water has an a_w of 1.0. Most fresh foods have an a_w above 0.98, whereas ham has an a_w of 0.91, jam has an a_w of 0.85, and some cakes have an a_w of 0.70. Most bacteria require an a_w above 0.90 for growth, which explains why fresh, moist foods spoil more quickly than dried, sugary, or salted foods. Fungi can grow at an a_w as low as 0.80, so

FIGURE 31.1 Fermented Foods

❓ *What makes a food "fermented"?*

Suppressing or limiting microbial growth can preserve the quality of foods and prevent foodborne illnesses. Foods can be canned, pasteurized, or irradiated to eliminate or decrease the numbers of microorganisms. Alternatively, microbial growth can be inhibited by storing food at cold temperatures, or by adding preservatives (growth-inhibiting ingredients). The end products of some fermentation processes can preserve food by preventing the growth of many undesirable microorganisms.

FIGURE 31.2 Examples of Spoiled Food
(a) Rotten apples. (b) Moldy bread. (c) Ear of corn and a lemon after several weeks in the refrigerator.

❓ *How is a spoiled food different from a fermented food?*

(a) (b) (c)

forgotten bread, cheese, jam, and dried foods often become moldy. *Staphylococcus* sp., which are adapted to grow on the dry, salty surfaces of skin, can grow at an a_w of 0.86, lower than the minimum required by most common spoilage bacteria. *Staphylococcus* sp. normally do not compete well with other bacteria, but on salty products such as ham and other cured meats, they can multiply with little competition. Ham is a common vehicle for *S. aureus* food poisoning. ◄◄ *Staphylococcus aureus* **foodborne illness, p. 393**

pH

Many bacterial species, including most pathogens, are inhibited by acidic conditions and cannot grow at a pH below 4.5. An exception is the lactic acid bacteria, which can grow at a pH as low as 3.5. These bacteria produce lactic acid as a result of fermentative metabolism and are used to make yogurt, sauerkraut, and some other fermented foods. They are also prime causes of spoilage of unpasteurized milk and other foods. ◄◄ **lactic acid bacteria, p. 258**

Fungi can grow at a lower pH than most spoilage bacteria, so some acidic foods eventually become moldy. For example, the pH of lemons is approximately 2.2, which inhibits the growth of bacteria, including the lactic acid group. However, some fungi can grow at this low pH.

The pH of a food product can also determine whether toxins can be produced. *Clostridium botulinum* (the bacterium that causes botulism) does not grow or produce toxin below pH 4.5, so it is not considered a danger in highly acidic foods. This is why the canning process for acidic fruits and pickles is less stringent than that for foods with a higher pH. Some newer varieties of tomatoes are less acidic than older types, requiring that acid be added if they are to be safely canned using the less stringent procedures.

Nutrients

An organism requiring a particular vitamin cannot grow in a food lacking that vitamin. A microbe capable of synthesizing that vitamin, however, can grow if other conditions are favorable. Members of the genus *Pseudomonas* often spoil foods because they can synthesize essential nutrients and can multiply in various environments, including refrigeration.

Biological Barriers

Rinds, shells, and other coverings help protect foods from invasion by microorganisms. Eggs, for example, retain their quality much longer with intact shells. Whole lemons keep longer than slices. Even so, microorganisms will eventually break down these coverings and cause spoilage.

Antimicrobial Chemicals

Some foods contain natural antimicrobial chemicals that help prevent spoilage. Egg white, for instance, is rich in lysozyme. If lysozyme-susceptible bacteria breach the protective shell of an egg, they are destroyed by lysozyme before they can cause spoilage. Other examples of naturally occurring antimicrobial chemicals are benzoic acid in cranberries and allicin in garlic. ◄◄ **lysozyme, p. 62**

Extrinsic Factors

Environmental conditions, such as the storage temperature and atmosphere, are called **extrinsic factors.** The extent of microbial growth varies greatly, depending on the conditions under which a food is stored. Microorganisms multiply rapidly in warm, O_2-rich environments such as meats stored at room temperature.

Storage Temperature

The storage temperature affects the growth rate of microorganisms. At low temperatures above freezing, many enzymatic reactions are either very slow or non-existent, with the result that microorganisms multiply slowly, if at all. Microorganisms that grow on refrigerated foods are most likely psychrophiles or psychrotrophs, such as some members of the genus *Pseudomonas*. At freezing temperatures, water becomes crystalline and inaccessible, stopping microbial growth. ◄◄ **psychrophiles, p. 89** ◄◄ **psychrotroph, p. 89**

Atmosphere

The presence or absence of O_2 affects the type of microbial population able to grow in food. Obligate aerobes cannot grow in foods stored under conditions that exclude their required O_2. Keeping O_2 out of food, however, may permit other bacteria to grow, including the obligate anaerobe *Clostridium botulinum*. A case of botulism was traced to the consumption of a thick, homemade stew that had been slowly cooked and then left at room temperature overnight. The cooking process did not destroy the endospores of *C. botulinum* and had driven off the O_2, thereby creating anaerobic conditions in which the organism germinated, multiplied, and produced toxin. ◄◄ **oxygen requirements, p. 90**

MicroAssessment 31.1

Intrinsic factors (such as available moisture, pH, and the presence of antimicrobial chemicals) and extrinsic factors (including storage temperature and atmosphere) influence the type of microorganisms that grow and predominate in a food product.

1. *Why is* Staphylococcus aureus *more likely to be found in high numbers on ham than on fresh meat?*
2. *Which is important to refrigerate: fresh stew or bread? Why?*
3. *Why would the cooking process create anaerobic conditions?* ✚

31.2 ■ Microorganisms in Food and Beverage Production

Learning Outcomes

2. *Explain why lactic acid bacteria are important in food and beverage production.*
3. *Compare and contrast the production of cheese, yogurt, and acidophilus milk.*
4. *Describe the production of pickled vegetables and fermented meat products.*
5. *Compare and contrast the production of wine, beer, distilled spirits, and vinegar.*
6. *Describe the production of soy sauce.*

Fermented foods, such as yogurt, cheese, and pickled vegetables, are perceived as pleasant tasting. In addition, their acids inhibit the growth of many spoilage organisms as well as foodborne pathogens. Thus, fermentation historically has been, and continues to be today, an important method of food preservation, particularly when modern conveniences such as refrigeration are lacking.

Lactic Acid Fermentations by the Lactic Acid Bacteria

The tart taste of yogurt, pickles, sharp cheeses, and some sausages is due to the production of lactic acid by one or more members of a group of bacteria known as the lactic acid bacteria (**table 31.1**). These bacteria—including species of *Lactobacillus, Lactococcus, Streptococcus, Leuconostoc,* and *Pediococcus*—are obligate fermenters that characteristically produce lactic acid as an end product of their metabolism. Some also produce flavorful and aromatic compounds that contribute to the overall quality of fermented foods. ◄◄ **lactic acid bacteria, p. 258** ◄◄ **obligate fermenters, p. 90**

TABLE 31.1	Foods Produced Using Lactic Acid Bacteria	
Food	**Characteristic**
Milk Products	
Cheese (unripened)	Uses a starter culture usually containing *Lactococcus cremoris* and *L. lactis*
Cheese (ripened)	Uses rennin and a starter culture containing *Lactococcus cremoris* and *L. lactis;* ripened for weeks to years; other bacteria and/or fungi may be added to enhance flavor development
Yogurt	Uses a starter culture containing *Streptococcus thermophilus* and *Lactobacillus delbrueckii* subspecies *bulgaricus*
Sweet acidophilus milk	*Lactobacillus acidophilus* added for possible health benefits
Vegetables	
Sauerkraut	Cabbage; succession of naturally occurring bacteria including *Leuconostoc mesenteroides, Lactobacillus brevis,* and *Lactobacillus plantarum*
Pickles	Cucumbers; naturally occurring bacteria
Poi	Taro root; naturally occurring bacteria; Hawaii
Olives	Green olives
Kimchee	Cabbage and other vegetables; Korea
Meats	
Dry and semidry sausages	Uses a starter culture containing species of *Lactobacillus* and *Pediococcus;* meat is stuffed into casings, incubated, heated, and then dried

Cheese, Yogurt, and Other Fermented Milk Products

Milk is sterile in a cow's udder, but rapidly becomes contaminated with a variety of microorganisms during milking and handling. Various species of lactic acid bacteria are inevitably introduced because they commonly reside on the udder. If the milk is not refrigerated, these bacteria grow and ferment lactose—the main sugar in milk—producing lactic acid. The combined effect of removing the primary carbohydrate as a nutrient source and accumulating lactic acid inhibits the growth of many other microbes. Aesthetic features of the milk change as well because lactic acid lowers the pH, which in turn causes the milk proteins to coagulate or curdle, and sours the flavor. ◄◄ **milk spoilage, p. 721**

Today, with high quality control standards and the use of pasteurized milk, the commercial production of fermented milk products does not rely on naturally present lactic acid bacteria. Instead, **starter cultures** containing one or more strains of lactic acid bacteria are added to the milk. These strains are carefully selected to produce the most desirable flavors and textures. Precious starter cultures must be carefully maintained and protected against contamination, particularly by bacteriophages, which can damage or destroy them. ◄◄ **bacteriophage, p. 304**

Cheese Cottage cheese is one of the simplest cheeses to make. Pasteurized milk is inoculated with a starter culture, usually containing *Lactococcus cremoris* and *L. lactis,* and then incubated until fermentation products cause the proteins in milk to coagulate. The coagulated proteins, or **curd,** are heated and cut into small pieces to make it easier to drain the liquid waste portion. Unlike most cheeses that undergo further microbial processes called ripening or curing, cottage cheese is unripened.

The initial steps of ripened cheese production are the same as those of cottage cheese, except the enzyme rennin is added to the fermenting milk to speed protein coagulation (**figure 31.3**). After the proteins coagulate, the whey (the liquid waste portion) is removed. The curds are then salted, pressed, and shaped into the traditional forms, usually bricks or wheels. The cheese is then ripened, resulting in characteristic textural and flavor changes due to the metabolic activities of naturally occurring or starter lactic acid bacteria. Depending on the type of cheese, ripening can take from several weeks to years. Longer ripening creates more acidic, sharper cheeses.

Some cheeses are inoculated with other bacteria or fungi that give characteristics particular to the kind of cheese. For example, the bacterium *Propionibacterium shermanii* ripens Swiss cheese and gives it the characteristic holes and a nutty flavor. This bacterium ferments organic compounds to produce propionic acid and CO_2. The CO_2 gas causes the holes in the cheese, while the propionic acid gives the typical flavor. Propionic acid also inhibits spoilage organisms. Roquefort, Gorgonzola, and Stilton cheeses are ripened by the fungus *Penicillium roquefortii.* Growth of the fungus along cracks in the cheese gives these cheeses the distinctive bluish-green veins. Brie and Camembert are ripened by a white fungus such as *P. candidum* or *P. camemberti* inoculated on the surface of the cheese. As the fungal cells grow into the cheese, they produce enzymes that alter texture and flavor. Limburger cheese is made in a similar manner, but with the bacterium *Brevibacterium linens.*

Coagulation— Lactic acid production and rennin activity cause the milk proteins to coagulate. The coagulated mixture is then cut so that the liquid whey will start separating from the solid curd.

Separation of curds from whey— The curd is heated and cut into small pieces. The liquid whey is removed by draining.

Aging— Curds are salted and pressed into blocks or wheels for aging.

FIGURE 31.3 Commercial Production of Cheese

❓ *Why does a longer ripening process give rise to a sharper cheese?*

Yogurt   To produce yogurt, pasteurized milk is concentrated slightly by evaporation and then inoculated with a starter culture containing *Streptococcus thermophilus* and *Lactobacillus delbrueckii* subspecies *bulgaricus*. The mixture is incubated at 40°C to 45°C for several hours, during which time these thermophilic bacteria grow rapidly. They produce lactic acid and other end products that contribute to the flavor. Carefully controlled incubation conditions favoring the balanced growth of the two species ensure the proper levels of acid and flavor compounds.

Acidophilus Milk   Traditional acidophilus milk is the product of fermentation by *Lactobacillus acidophilus*. The more readily available sweet acidophilus milk retains the flavor of fresh milk because it is not fermented. Instead, a culture of *L. acidophilus*

is added immediately before packaging. The bacteria are simply included for their possible health benefits. Some evidence suggests they may aid in the digestion of lactose as well as prevent and reduce the severity of some diarrheal illnesses, but the role they play in the complex interactions of the human intestinal tract is not clear. Unlike most lactic acid bacteria used as starter cultures, *L. acidophilus* can potentially colonize the intestinal tract.

MicroByte
> The enzyme rennin is found in calves' stomachs, where it helps digest the mother's milk. Today, genetically engineered microbes make it.

Pickled Vegetables

Another fermentation process known as pickling originated as a way to preserve vegetables such as cucumbers and cabbage. Today, pickled products such as sauerkraut (cabbage), pickles (cucumbers), and olives are valued for their flavor. Fermentation of most vegetables uses naturally occurring lactic acid bacteria from the vegetables rather than starter cultures.

One of the most well-studied natural fermentations is the production of sauerkraut. The cabbage is first shredded and layered with salt. The layers are firmly packed to provide an anaerobic environment. The salt draws water and nutrients from the cabbage, creating a brine that inhibits most microbes other than the lactic acid bacteria. Under the correct conditions, natural successions of lactic acid bacteria grow. These bacteria—*Leuconostoc mesenteroides, Lactobacillus brevis,* and *Lactobacillus plantarum*—produce lactic acid, which lowers the pH, further inhibiting undesired microbes. The lactic acid and other fermentation end products give sauerkraut its characteristic tangy taste. When the desired flavor has developed, usually after 2 to 4 weeks at room temperature, the sauerkraut is often canned. Similar processes are used to make some pickles, olives, and other vegetable products.

Fermented Meat Products

Fermented meat products—such as salami, pepperoni, and summer sausage—were traditionally produced by allowing the small numbers of lactic acid bacteria naturally present to multiply to the point of dominance. Relying on the natural fermentation of meat is inherently risky, however, because the incubation conditions can potentially support the growth and toxin production of pathogens such as *Staphylococcus aureus* and *Clostridium botulinum*. Starter cultures are now used because they ensure that lactic acid is rapidly produced, thereby inhibiting the growth of pathogens and improving flavor development. Starter cultures used by U.S. sausage-makers typically contain *Lactobacillus* and/or *Pediococcus* species.

To make fermented sausages, meat is ground and combined with a starter culture and other ingredients including sugar, salt, and nitrite. The sugar serves as a substrate for fermentation, because meat does not naturally contain enough fermentable carbohydrate to produce adequate lactic acid. Salt and nitrite contribute to the flavor and also inhibit the growth of spoilage microorganisms. More importantly, they inhibit *C. botulinum*. After thorough blending, the mixture is stuffed into a casing and incubated from one to several days. The product can then be smoked or otherwise heated to kill bacteria. Finally, it is dried.

Alcoholic Fermentations by Yeast

Some yeasts, such as members of the genus *Saccharomyces*, ferment simple sugars to produce ethanol and CO_2. They are used to make alcoholic beverages as well as vinegar and bread (**table 31.2**).

Wine

Wine is the product of the alcoholic fermentation of naturally occurring sugars in the juices of grapes or other fruits. One of the most important variables in wine is the variety and quality of fruit used. The growing conditions and ripeness as well as other factors affect the content of sugar, acids, and various organic compounds, which in turn critically influences the final product.

The commercial production of wine begins by crushing grapes in a machine that removes the stems and collects the resulting solids and juices, or must (**figure 31.4**). If white wine is to be made, only the clear juices are fermented. For red wine, the entire must of red grapes is put into the fermentation vat. The color and complex flavors of these wines are derived from components of the grape skin and seeds. The solids are removed during fermentation once the desired amount of color and flavor compounds have been extracted. Rose wines obtain their light pink color from the entire crushed red grape fermented for about 1 day, after which the juice is removed and fermented alone.

Fermentation starts when a specially selected strain of *Saccharomyces cerevisiae* is inoculated. Sulfur dioxide (SO_2) is generally added to inhibit the growth of the natural microbial population of the grape, especially acetic acid bacteria. These bacteria convert alcohol to acetic acid (vinegar) and are a common cause of wine spoilage. The *S. cerevisiae* strains used to make wine are

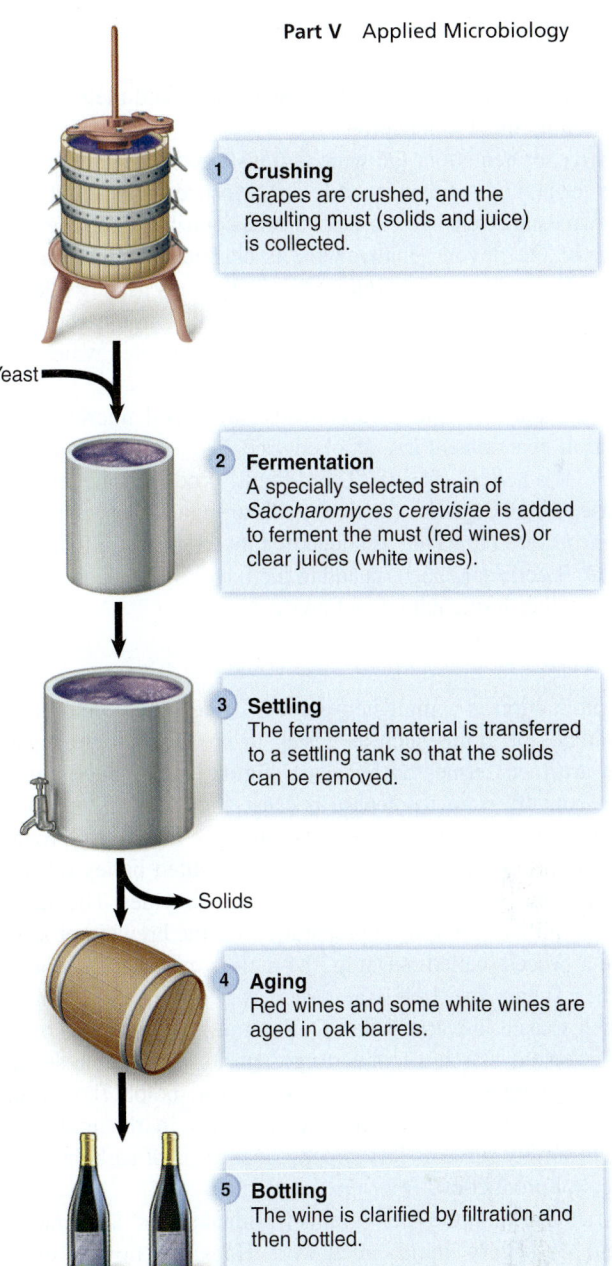

1 **Crushing**
Grapes are crushed, and the resulting must (solids and juice) is collected.

Yeast

2 **Fermentation**
A specially selected strain of *Saccharomyces cerevisiae* is added to ferment the must (red wines) or clear juices (white wines).

3 **Settling**
The fermented material is transferred to a settling tank so that the solids can be removed.

Solids

4 **Aging**
Red wines and some white wines are aged in oak barrels.

5 **Bottling**
The wine is clarified by filtration and then bottled.

FIGURE 31.4 Commercial Production of Wine

❓ *How is the initial step of making white wines different from that of making red wines?*

Product	Characteristic
TABLE 31.2	**Foods and Beverages Produced Using Alcoholic Fermentation by Yeast**
Alcoholic Beverages	
Wine	Sugars in grape juice are fermented by *Saccharomyces cerevisiae*.
Sake	Amylase from mold (*Aspergillus oryzae*) converts the starch in rice to sugar, which is then fermented by *S. cerevisiae*.
Beer	Enzymes in germinated barley convert starches of barley and other grains to sugar, which is then fermented by *S. cerevisiae*.
Distilled spirits	Sugars, or starches that are converted to sugars, are fermented by *S. cerevisiae*; distillation purifies the alcohol.
Vinegar	Alcohol produced by fermentation is oxidized to acetic acid by species of *Gluconobacter* or *Acetobacter*.
Breads	*S. cerevisiae* ferments sugar; expansion of CO_2 causes the bread to rise; alcohol evaporates during baking.

more resistant to the antimicrobial action of SO_2 and produce a higher alcohol content than naturally occurring yeasts.

Fermentation is carried out at a carefully controlled temperature, which varies with the type of wine, for a period ranging from a few days to several weeks. During fermentation most of the sugar is converted to ethanol and CO_2, generally resulting in a final alcohol content of less than 14%. Dry wines result from the complete fermentation of the sugar, whereas sweet wines contain residual sugar.

In addition to the alcoholic fermentation of the grape sugars, a distinctly different type of fermentation, called malolactic fermentation, can occur during wine production. Lactic acid bacteria, primarily species of *Leuconostoc*, convert malic acid to the less acidic lactic acid. Red wines made from grapes grown in cool regions tend to have high levels of malic acid, and their flavor is mellowed by this fermentation.

After fermentation, the wine is transferred to a settling tank, where fermented solids can settle out. Most red wines and some white wines are then aged in oak barrels, contributing to the complexity of the flavor. Finally, wine is clarified by filtration and bottled. The CO_2 produced during fermentation is usually released before the wine is bottled, resulting in a "still" (non-carbonated) wine. Other processes are used to prepare carbonated wines such as champagne.

The Japanese wine sake depends on several microbial fermentation reactions. First, cooked rice is inoculated with the fungus *Aspergillus oryzae*. The fungus produces the enzyme amylase, which degrades the rice starch to sugar. Then, a strain of *Saccharomyces cerevisiae* is added to convert the sugar to alcohol and CO_2. Lactic acid bacteria add to the flavor by producing lactic acid and other fermentation end products.

Beer

Beer production is a multistep process designed to break down the starches of grains such as barley to produce simple sugars, which are then fermented by yeast (**figure 31.5**). Yeasts alone cannot convert grain to alcohol because they lack the enzymes that degrade starch, the primary carbohydrate of grain. Sprouted or germinated barley, however, known as malted barley or malt, naturally contains these and other important enzymes. The malt is dried and milled (ground) in preparation for the brewing process.

In a process called mashing, the malt is mixed with adjuncts (starches, sugars, or whole grains such as rice, corn, or sorghum), and then soaked in warm water. During mashing, enzymes of the malt act on the starches, converting them to fermentable sugars. The final characteristics of the beer, such as color, flavor, and foam, are derived entirely from compounds in the malt. The adjuncts simply serve as less expensive sources of carbohydrates for alcohol production.

After mashing, the spent grains (residual solids) are removed to yield the sugary liquid called wort. Hops, the flowers of the vinelike hop plant, are added to the wort to impart a desirable bitter flavor to the beer and contribute antibacterial substances. The mixture is boiled to extract the flavor components of hops, concentrate the wort, inactivate enzymes, kill most microbes, and precipitate proteins, facilitating their removal. The wort is then centrifuged to remove the solids and cooled before being transferred to the fermentation tank.

Special strains of **brewer's yeasts** (as opposed to baker's yeasts) are commonly used in beer-making. A group of brewer's yeasts, called "bottom yeasts"—which includes a *Saccharomyces cerevisiae* strain referred to as *S. carlsbergensis*—tend to form clumps that sink to the bottom of the fermentation vat. These yeasts ferment best at temperatures between 6°C and 12°C and usually take 8 to 14 days to complete fermentation. In contrast, "top-fermenting yeasts"—other strains of *S. cerevisiae*—are distributed throughout the wort but are carried to the top of the vat by the rising CO_2. They ferment at higher temperatures (14°C to 23°C) and over a shorter period (5 to 7 days). Most American beers are lagers produced by the bottom yeasts. Ales, porters, and stouts are made using top-fermenting yeasts.

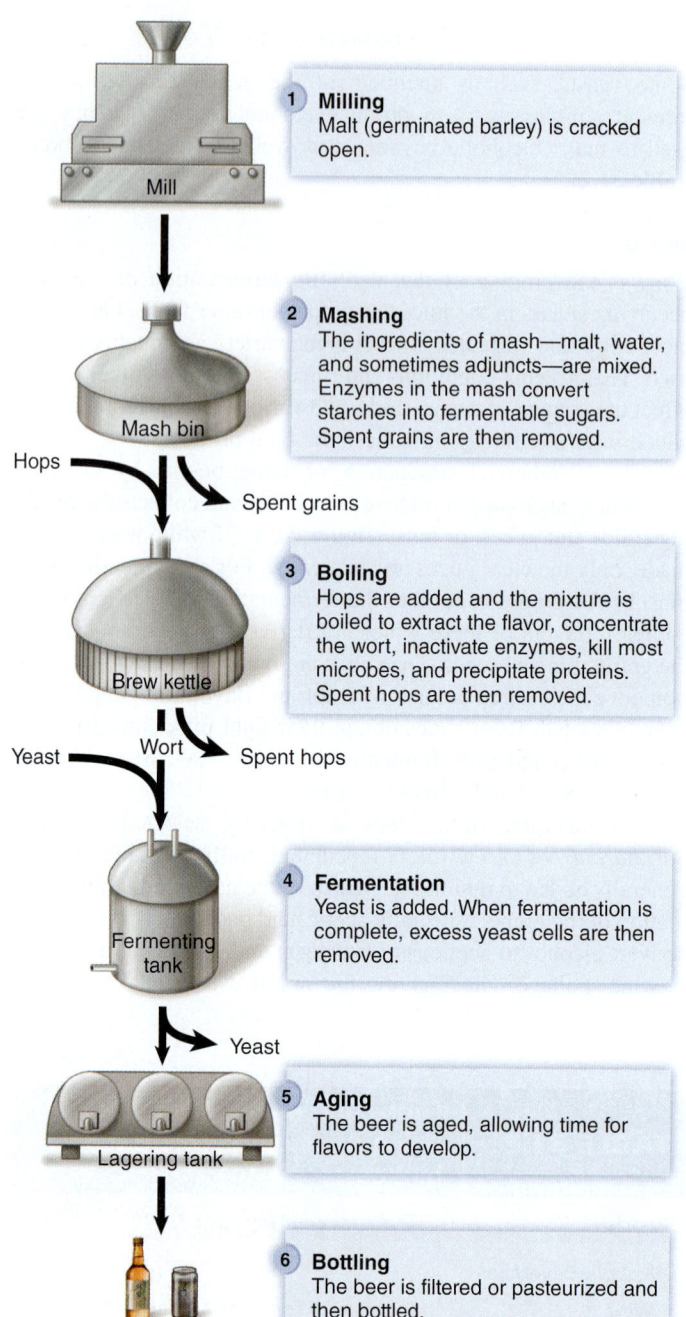

1. **Milling**
Malt (germinated barley) is cracked open.

Mill

2. **Mashing**
The ingredients of mash—malt, water, and sometimes adjuncts—are mixed. Enzymes in the mash convert starches into fermentable sugars. Spent grains are then removed.

Mash bin

Hops

Spent grains

3. **Boiling**
Hops are added and the mixture is boiled to extract the flavor, concentrate the wort, inactivate enzymes, kill most microbes, and precipitate proteins. Spent hops are then removed.

Brew kettle

Yeast Wort Spent hops

4. **Fermentation**
Yeast is added. When fermentation is complete, excess yeast cells are then removed.

Fermenting tank

Yeast

5. **Aging**
The beer is aged, allowing time for flavors to develop.

Lagering tank

6. **Bottling**
The beer is filtered or pasteurized and then bottled.

FIGURE 31.5 Commercial Production of Beer

❓ *How are the yeasts used to make ales different from the ones used to make lagers?*

The fermentation process generates beer with an alcohol content ranging from 3.4% to 6%. Most of the yeasts settle out following fermentation and are removed. These yeasts can be sold as flavor and dietary supplements. The beer is then aged, during which time residual, unwanted flavor compounds are metabolized by remaining yeast cells or settle out. Cask-conditioned beer undergoes a second fermentation, which generates CO_2 in the cask. Other beers must be carbonated to replace the CO_2 that escapes during fermentation. After aging, beer is clarified by filtration, microorganisms are removed or killed using membrane filtration or pasteurization, and the product is packaged.

Distilled Spirits

The manufacturing of distilled spirits such as scotch, whiskey, and gin is initially similar to that of beer, except the wort is not boiled. Consequently, enzymes in the wort continue breaking down the starches during fermentation. When fermentation is complete, the ethanol is collected by distillation.

Different types of spirits are made in different ways. For example, rum is made by fermenting sugar cane or molasses. Malt scotch whiskey is the product of the fermentation of barley that is then aged for several years in oak sherry casks. The wood and the residual sherry contribute both flavor and color to the whiskey as it ages. Lactic acid bacteria are used to produce lactic acid in grain mash for making sour-mash whiskey. The yeast *S. cerevisiae* subsequently ferments the sour mash to form alcohol. The distilled spirit tequila is traditionally made from the fermentation of juices from the agave plant using the bacterium *Zymomonas mobilis*. This bacterium ferments sugars to ethanol and CO_2 via a pathway similar to the yeast alcoholic fermentation pathway.

Vinegar

Vinegar, an aqueous solution of at least 4% acetic acid, is the product of the oxidation of ethanol by the acetic acid bacteria—*Acetobacter* and *Gluconobacter* species. Acetic acid bacteria are strictly aerobic, Gram-negative rods, characterized by their ability to carry out a number of oxidations. They can tolerate high concentrations of acid as they oxidize alcohol to acetic acid.

Alcohol is commercially converted to vinegar using processes that provide O_2 to speed the oxidation reaction. One method uses a vinegar generator, which sprays alcohol onto loosely packed wood shavings that harbor a biofilm of acetic acid bacteria. As the alcohol trickles through the bacteria-coated shavings, it is oxidized to acetic acid. In principle, the vinegar generator operates much like the trickling filter used in waste-water treatment by providing a large surface area for aerobic metabolism. Another method uses a submerged culture reactor, an enclosed system that continuously pumps small air bubbles into alcohol that has been inoculated with acetic acid bacteria.

◄◄ trickling filter, p. 738   ◄◄ biofilm, p. 84

Bread

Yeast bread rises through the action of **baker's yeast**—strains of *Saccharomyces cerevisiae* carefully selected for the commercial baking industry. The CO_2 produced during fermentation causes the bread to rise, producing the spongy texture characteristic of yeast breads (**figure 31.6**). The alcohol evaporates during baking.

Yeast bread is made from a mixture of flour, sugar, salt, milk or water, yeast, and sometimes butter or oil. Packaged baker's yeast that can be reconstituted in warm water is readily available as pressed cakes or dried granules. An excess of yeast is added to allow adequate production of CO_2 in a time period too short to permit multiplication of spoilage bacteria.

Sourdough bread is made with a combination of yeast and lactic acid bacteria. Lactic acid is produced, as well as alcohol and CO_2, giving the bread its sour flavor.

FIGURE 31.6 Bread

❓ *What creates the air pockets in bread?*

Changes Due to Mold Growth

Molds contribute to the flavor and texture of some cheeses, as already discussed. In addition, many traditional dishes and condiments used throughout the world are produced by encouraging the growth of molds on food (**table 31.3**). Successions of naturally occurring microorganisms are often involved. The microbiological and chemical aspects of many of these foods have not been extensively studied.

Soy Sauce

Soy sauce is made by inoculating equal parts of cooked soybeans and roasted cracked wheat with a culture of either *Aspergillus oryzae* or *A. sojae*. The mixture, called koji, is allowed to stand for several days, during which time carbohydrates and proteins in the

TABLE 31.3	Foods Produced Using Molds
Food	**Characteristic**
Soy sauce	Koji is produced by inoculating soybeans and cracked wheat with a starter culture of *Aspergillus oryzae* or *A. sojae*; the mixture is then added to a brine and incubated for many months.
Tempeh	Soybeans are fermented by lactic acid bacteria and then inoculated with a species of the mold *Rhizopus*; Indonesia.
Miso	Rice, soybeans, or barley are inoculated with *Aspergillus oryzae*; Asia.
Cheeses	
Roquefort, Gorgonzola, and Stilton	Curd is inoculated with *Penicillium roquefortii*.
Brie and Camembert	Wheels of cheese are inoculated with selected species of *Penicillium*.

soybeans are broken down, producing a yellow-green liquid containing fermentable sugars, peptides, and amino acids. After this, the mixture is put into a large container with an 18% NaCl solution (brine). Salt-tolerant microorganisms then grow and produce flavor changes over an extended period. Microorganisms involved in this stage of fermentation include lactobacilli, pediococci, and yeasts. After the brine mixture is allowed to ferment for 8 to 12 months, the liquid soy sauce is removed. The residual solids are used as animal feed.

MicroAssessment 31.2

Lactic acid bacteria are used to produce a variety of foods including cheese, yogurt, sauerkraut, and some sausages. Yeasts are used to produce alcoholic beverages and breads. Some cheeses and many traditional foods owe their characteristics to changes caused by molds.

4. *Describe how the metabolism of lactic acid bacteria differs from that of most other microorganisms that can grow aerobically.*

5. *How does the use of starter cultures improve the safety of fermented meat products?*

6. *How could cottage cheese be produced without bacteria?* ✚

31.3 ■ Food Spoilage

Learning Outcome

7. *Distinguish between fermented foods and spoiled foods.*

Food spoilage encompasses any undesirable changes in food. Spoilage microorganisms produce metabolites that have undesirable tastes and odors. Although these are aesthetically disagreeable, they are generally not harmful. This is not surprising when microbial growth requirements are considered. Most human pathogens grow best at temperatures near 37°C, whereas most foods are usually stored at temperatures well below the normal body temperature. Similarly, the nutrients available in fruits, vegetables, and other foods are generally not suitable for the optimum growth of human pathogens. As a result, the non-pathogens can easily outgrow the pathogens when competing for the same nutrients. Spoiled foods are considered unsafe to eat, however, because high numbers of spoilage organisms indicate that foodborne pathogens may be present as well.

Common Spoilage Bacteria

Numerous types of bacteria are important in food spoilage. *Pseudomonas* species can degrade a wide variety of compounds, and grow on and spoil many different kinds of foods, including meats and vegetables. Psychrotrophic species are notorious for spoiling refrigerated foods. Members of the genus *Erwinia* produce enzymes that degrade pectin, and so they commonly cause soft rot of fruits and vegetables. *Acetobacter* species transform ethanol to acetic acid, the principal acid of vinegar. Although this is very beneficial to commercial producers of vinegar, it presents a great problem to wine producers. Milk products are sometimes spoiled by *Alcaligenes* sp. that form a glycocalyx, causing strings of slime, or "ropiness," in raw milk. The lactic acid bacteria, including species of *Streptococcus, Leuconostoc,*

and *Lactobacillus,* all produce lactic acid. Anyone who has unexpectedly consumed sour milk knows that this can be disagreeable. *Bacillus* and *Clostridium* sp. are particularly troublesome causes of food spoilage because their heat-resistant endospores survive cooking and, in some cases, canning. *B. coagulans* and *B. stearothermophilus* spoil some canned foods. ◀◀ **glycocalyx, p. 62**

Common Spoilage Fungi

A wide variety of fungi, including species of *Rhizopus, Alternaria, Penicillium, Aspergillus,* and *Botrytis,* spoil foods. Because fungi grow readily in acidic as well as low-moisture environments, fruits and breads are more likely to be spoiled by fungi than by bacteria. *Aspergillus flavus* grows on peanuts and other grains, producing **aflatoxin,** a potent carcinogen monitored by the Food and Drug Administration.

MicroAssessment 31.3

The metabolites of microbes can spoil foods by imparting undesirable flavors, odors, and textures.

7. *What characteristics of* Pseudomonas *species allow them to spoil such a wide variety of foods?*

8. *Why do fungi most commonly spoil breads and fruits?* ✚

31.4 ■ Foodborne Illness

Learning Outcome

8. *Distinguish between foodborne intoxication and foodborne infection, and give two examples of each.*

Foodborne illness, commonly referred to as food poisoning, occurs when a pathogen, or a toxin it produced, is consumed in a food product. Food production is carefully regulated in the United States to prevent foodborne illness. Federal, state, and local agencies cooperate in inspections to help enforce protective laws. In spite of strict controls, millions of cases of food poisoning are estimated to occur each year from foods prepared either commercially, at home, or in institutions such as hospitals and schools. The vast majority of these cases could have been prevented with proper storage, sanitation, and preparation. **Table 31.4** lists some bacteria that cause foodborne illness in the United States.

To more accurately determine the burden of foodborne illness in the United States, a government program called **FoodNet** (Foodborne Disease Active Surveillance Network) now collects data on laboratory-confirmed cases of diarrheal illness in 10 states, covering approximately 15% of the population (www.cdc.gov/foodnet). By gaining a better understanding of the epidemiology of foodborne diseases, these diseases can hopefully be prevented more easily.

Foodborne Intoxication

Foodborne intoxication is an illness that results from consuming an exotoxin produced by a microorganism growing in a food product. It is the toxin that causes illness, not the living organisms (**figure 31.7**). *Staphylococcus aureus* and *Clostridium botulinum*

TABLE 31.4	Common Foodborne Illnesses	
Organism	**Symptoms**	**Foods Commonly Implicated**
Intoxication		
Clostridium botulinum	Weakness; double vision; progressive inability to speak, swallow, and breathe	Low-acid canned foods such as vegetables and meats
Staphylococcus aureus	Nausea, vomiting, abdominal cramping	Cured meats, creamy salads, cream-filled pastries
Infection		
Campylobacter species	Diarrhea, fever, abdominal pain, nausea, headache	Poultry, raw milk
Clostridium perfringens	Intense abdominal cramps, watery diarrhea	Meats, meat products
Escherichia coli O157:H7	Severe abdominal pain, bloody diarrhea; sometimes hemolytic uremic syndrome	Ground beef, raw vegetables, unpasteurized juices
Listeria monocytogenes	Influenza-like symptoms, fever, may progress to sepsis, meningitis	Raw milk, cheese, meats, raw vegetables
Salmonella species	Nausea, vomiting, abdominal cramps, diarrhea, fever	Poultry, eggs, milk, meat
Shigella species	Abdominal cramps; diarrhea with blood, pus, or mucus; fever; vomiting	Salads, raw vegetables
Vibrio parahaemolyticus	Diarrhea, abdominal cramps, nausea, vomiting, headache, fever, chills	Fish and shellfish

are two examples of bacteria that cause foodborne intoxication. ◀◀ exotoxin, p. 391

Staphylococcus aureus

Many strains of *Staphylococcus aureus* produce a toxin that causes nausea and vomiting when ingested. *S. aureus* does not compete well with most spoilage organisms, but it thrives in moist, rich foods in which other organisms have been killed or their growth inhibited. For example, it can grow with little competition on unrefrigerated salty products such as ham (a_w of 0.91). Creamy pastries and starchy salads stored at room temperature also offer ideal conditions for the growth of this pathogen, because most competing organisms were killed as the ingredients were cooked.

The source of *S. aureus* is usually a human carrier who has not followed adequate hygiene procedures—such as handwashing—before preparing the food. If the organism is inoculated into a food that supports its growth, and the food is left at room temperature for several hours, *S. aureus* can grow and produce the toxin. Unlike most exotoxins, *S. aureus* toxin is heat-stable, so cooking the food will not destroy it. ◀◀ *Staphylococcus aureus*, p. 524

Botulism

Botulism is a deadly paralytic disease caused by ingesting a neurotoxin produced by the anaerobic, spore-forming, Gram-positive rod *Clostridium botulinum* (see **Perspective 31.1**). Unfortunately, growth of the organism and toxin production may not result in any noticeable changes in the taste or appearance of the food. ◀◀ botulism, p. 652 ◀◀ neurotoxin, p. 392

Canning processes for low-acid foods are specifically designed to destroy the endospores of *C. botulinum*. If the endospores are not destroyed or are introduced postprocessing, they can germinate in these

Staphylococcus aureus

- Most bacteria that normally compete with *Staphylococcus aureus* are either killed by cooking or inhibited by high salt conditions.
- A food handler inadvertently transfers *S. aureus* onto food.
- *S. aureus* grows and produces toxin when food is allowed to slowly cool or is stored at room temperature.
- A person ingests the toxin-containing food. Symptoms of staph food poisoning—nausea, abdominal cramping, and vomiting—begin after 4 to 6 hours.

Clostridium botulinum

- *Clostridium botulinum* endospores, common in soil and marine sediments, contaminate many different foods.
- Endospores survive inadequate canning processes. Canned foods are anaerobic.
- Surviving *C. botulinum* endospores germinate, grow, and produce toxin in low-acid canned foods.
- A person ingests the toxin-containing food. Symptoms of botulism, including weakness, double vision, and progressive inability to speak, swallow, and breathe, begin in 12 to 36 hours.

FIGURE 31.7 Typical Events Leading to Foodborne Intoxications by *Staphylococcus aureus* and *Clostridium botulinum*

? *Why is botulism primarily a problem in canned foods rather than fresh ones?*

Botox for Beauty and Pain Relief

The exotoxin produced by *Clostridium botulinum* is one of the most powerful poisons known. These anaerobic soil bacteria are endosporeformers found naturally on many foods. They survive usual cooking methods and inadequate canning procedures. Should conditions become anaerobic in the food, toxin can be produced, and if ingested it causes botulism. A few milligrams of this exotoxin is sufficient to kill the entire population of a large city. The botulinum toxin blocks transmission of acetylcholine nerve signals to the muscles, resulting in paralysis and often in death. This is a very dangerous toxin. Yet surprisingly, in recent years the toxin, known as Botox, has become useful in treating various conditions.

Botox is often used as a cosmetic treatment to remove facial lines, such as frown lines. Extremely dilute Botox is injected directly into the area, paralyzing the muscles that are causing the frown or other lines. Although the lines are erased, the effect is temporary, and the treatment must be repeated after several months.

Botox is also used to relieve a number of very painful and disabling conditions involving muscle contractions such as dystonia (severe muscle cramping). For example, cervical dystonia is a painful disease in which muscles in the neck and shoulders contract involuntarily, causing jerky movements, muscle pain, and tremors. Injections of Botox directly into the affected areas give relief for 3 to 4 months, after which the treatment can be repeated. Another example is Parkinson's disease, a condition in which certain nerve cells are lost, resulting in tremor, impaired movement, and in some cases, dystonia. Botox injected into the affected muscles can give dramatic, although temporary, relief. So, in spite of its powerful and dangerous properties, botulinum toxin, when used with great care, can be a useful therapeutic agent.

---

foods. The resulting vegetative cells can then grow and produce toxin. Manufacturing errors are rare in commercially canned foods, so most cases of botulism are due to improperly processed home-canned foods. As an added safety measure, low-acid, home-canned foods should be boiled for at least 15 minutes immediately before serving it. The toxin is heat-labile (sensitive), so the heat treatment will destroy any toxin that may have been produced.

Cans of foods that are damaged should be discarded, because they might have small holes through which *C. botulinum* could enter. Likewise, bulging cans indicate gas production, which could indicate microbial growth, and should be discarded.

Foodborne Infection

Unlike foodborne intoxication, **foodborne infection** requires the consumption of living organisms. The symptoms of the illness, which usually do not appear for at least 1 day after eating the contaminated food, usually include diarrhea. Thorough cooking of food immediately before consuming it will kill the organisms, thereby preventing infection. *Escherichia coli* O157:H7, *Salmonella* sp., and *Campylobacter* sp. are examples of organisms that cause foodborne infection (**figure 31.8**).

Salmonella and Campylobacter

Salmonella and *Campylobacter* are two genera commonly associated with poultry products such as chicken, turkey, and eggs. Inadequate cooking of these products can result in foodborne infection. **Cross-contamination** of other foods can result in the transfer of pathogens to those foods. For example, if a cutting board on which raw chicken was cut is then immediately used to cut up vegetables for a salad, the salad can become contaminated with *Salmonella* or *Campylobacter* species. ◀◀ *Salmonella*, p. 592 ◀◀ *Campylobacter*, p. 593

Escherichia coli O157:H7

Escherichia coli O157:H7, a common serotype of Shiga toxin–producing *E. coli* (STEC), causes bloody diarrhea. Infection sometimes results in hemolytic uremic syndrome (HUS), a life-threatening condition. ◀◀ hemolytic uremic syndrome, p. 598 ◀◀ STEC, p. 590

Salmonella, Campylobacter

- Incomplete cooking fails to kill all pathogens. Surviving *Salmonella* and/or *Campylobacter* can multiply as food is cooled slowly or stored at room temperature.
- Live organisms are ingested. They multiply in the intestinal tract and cause disease. Symptoms include diarrhea, abdominal pain, and nausea.

E. coli O157:H7

- Incomplete cooking fails to kill all pathogens. Even low numbers of surviving *E. coli* O157:H7 can cause illness.
- Live organisms are ingested. They multiply in the intestinal tract and cause disease. Symptoms include severe abdominal pain and bloody diarrhea.

FIGURE 31.8 Typical Events Leading to Foodborne Infections by *Salmonella, Campylobacter*, and *E. coli* O157:H7

❓ *Why are ground meats more common than steaks as a source of* E. coli *O157:H7 infection?*

The bacterium often colonizes the intestinal tract of healthy cattle and other livestock and is then shed in their feces. Because of this, meats can easily become contaminated. Although the initial contamination normally occurs on the surface, and the bacterial cells are easily destroyed by searing the exterior of meats such as steaks, grinding the meat to create beef patties distributes the pathogen throughout the product. Prevention then involves more thorough cooking so that enough heat reaches the center to kill all the *E. coli* cells, which is why hamburgers are a particularly troublesome source of infection. Outbreaks have also been linked to unpasteurized milk and various kinds of produce that were contaminated with animal manure.

MicroByte

21.7 million pounds of frozen ground beef patties were recalled in 2007 after they were linked to a multistate outbreak of *E. coli* O157:H7.

MicroAssessment 31.4

Foodborne intoxication results from consuming toxins produced by microbes growing in a food. Foodborne infection results from consumption of living organisms.

9. *How does boiling a home-canned food immediately prior to serving it prevent botulism?*

10. *Which foodborne pathogen can cause hemolytic uremic syndrome?*

11. *Why would a large number of competing microorganisms in a food sample result in lack of sensitivity of culture methods for detecting pathogens?* ✚

31.5 ■ Food Preservation

Learning Outcome

9. *Describe the methods used to preserve foods.*

Preventing the growth and metabolic activities of microorganisms that cause spoilage and foodborne illness preserves the quality of food. Some methods of **food preservation,** such as drying and salting, have been known throughout the ages, whereas others have been discovered or developed more recently. The major methods of preserving foods—high-temperature treatment, low-temperature storage, addition of antimicrobial chemicals and irradiation—are briefly summarized here and described in more detail in chapter 5.

- **Canning.** The canning process destroys all spoilage and pathogenic organisms capable of growth at normal storage temperatures. Low-acid foods are processed using steam under pressure (autoclaving) in order to reach temperatures high enough to destroy the endospores of *Clostridium botulinum*. Acidic foods do not require such high heat because *C. botulinum* cannot grow and produce toxin in those foods. ◄◄ canning, p. 113

- **Pasteurization.** Heating foods under controlled conditions at high temperatures for short periods of time destroys non-spore-forming pathogens and reduces the numbers of spoilage organisms without significantly altering the flavor of food. ◄◄ pasteurization, p. 112

- **Cooking.** Cooking, like pasteurization, can destroy non-spore-forming organisms. Cooking obviously alters the characteristics of food, however. Heat distribution may be uneven, resulting in survival of organisms in inadequately heated regions.

- **Refrigeration.** Refrigeration preserves food by slowing the growth rate of microbes. Many organisms, including most pathogens, are unable to multiply at low temperatures. ◄◄ temperature requirements, p. 89

- **Freezing.** Freezing stops microbial growth because water in the form of ice is unavailable for biological reactions. Some of the microbial cells will be killed by damage caused by ice crystals, but those remaining can grow and spoil food once it is thawed.

- **Drying/reducing the a_w.** Drying foods or adding high concentrations of sugars or salts inhibits microbial growth by decreasing the available moisture. Eventually, however, molds may grow. ◄◄ drying food, p. 122

- **Lowering the pH.** Lowering the pH, either by adding acids or encouraging fermentation by lactic acid bacteria, inhibits a wide range of spoilage organisms and pathogens.

- **Adding antimicrobial chemicals.** Organic acids such as propionic acid, benzoic acid, and sorbic acid are naturally occurring antimicrobial chemicals that are added to a variety of foods to inhibit fungal growth. Nitrates are added to cured meats to inhibit growth of *Clostridium botulinum* and other organisms. Wine, fruit juices, and other products are preserved by the addition of sulfur dioxide. ◄◄ chemical preservatives, p. 121

- **Irradiation.** Gamma radiation destroys microorganisms without significantly altering the flavor of foods such as spices and meats. ◄◄ irradiation, p. 115

MicroAssessment 31.5

Food spoilage can be eliminated or delayed by destroying microorganisms or altering conditions to inhibit their growth.

12. *Why are the process temperatures for canning low-acid foods higher then ones for acidic foods?*

13. *Why are nitrates added to cured meats?*

14. *Microorganisms are often grouped according to their optimum growth temperatures. Which groups are most likely to spoil refrigerated foods?* ✚

Using Microorganisms to Nourish the World

To feed the world's steadily increasing population, our limited natural resources must be used efficiently and new protein supplies must be developed. One potential solution that addresses both these needs is to cultivate microorganisms as a protein source, using industrial by-products currently considered wastes as the growth medium.

The term single-cell protein (SCP) was coined in the 1960s to describe the use of unicellular organisms such as yeasts and bacteria as a protein source. Today, the term generally encompasses the use of multicellular microorganisms as well, and might more accurately be called microbial biomass. Although whole cells can be consumed, a more palatable alternative is to extract proteins from those cells and use them to form textured protein products.

Yeasts are considered the most promising, large-scale source of single-cell protein. They multiply rapidly, are larger than bacteria, and are more readily acceptable as a potential food. The most suitable type of yeast depends on the growth medium employed, because different genera use differing carbohydrate sources and conditions for growth. For example *Kluyveromyces marxianus* can be grown on whey, a by-product of cheese-making; *Saccharomyces cerevisiae* can grow on molasses, a by-product of the sugar industry; and *Candida utilis* can multiply on cellulose-containing by-products of the pulp and paper industry. Most of these wastes must be supplemented with a nitrogen source, as well as various vitamins and minerals, to support microbial growth.

To a lesser extent, the use of bacteria as SCP is also being explored. The cyanobacterium *Spirulina maxima* can be cultivated in alkaline lakes and then harvested and dried. Unfortunately, this requires adequate sunlight and warmth, making large-scale, year-round production impractical in most parts of the world.

One of the chief concerns regarding the consumption of microorganisms as a protein source is their high concentration of nucleic acid, primarily RNA. Yeast and bacteria contain as much as eight times more RNA per gram than does meat. High levels of nucleic acid in the diet cause an increase in uric acid in the blood, which can lead to gout and kidney stones. Chemical and enzymatic methods are being developed to decrease the nucleic acid in SCP without altering the nutritional value of the protein.

Summary

31.1 ■ Factors Influencing the Growth of Microorganisms in Foods

Intrinsic Factors

Bacteria require a high a_w. Fungi often grow when the a_w is too low to support bacterial growth. Many bacterial species, including most pathogens, are inhibited by acidic conditions. The nutritional content of a food determines the kinds of organisms that can grow in it. Rinds, shells, and other coverings aid in protecting some foods from invasion by microorganisms. Some foods contain natural antimicrobial chemicals that help prevent spoilage.

Extrinsic Factors

Low temperatures halt or inhibit the growth of most foodborne microorganisms. Psychrophiles and psychrotrophs, however, grow at refrigeration temperatures. The presence or absence of O_2 affects the type of microbial population able to grow in a food.

31.2 ■ Microorganisms in Food and Beverage Production

Not only are fermented foods perceived as pleasant tasting, the acids inhibit the growth of many spoilage organisms and foodborne pathogens.

Lactic Acid Fermentations by the Lactic Acid Bacteria (table 31.1)

The tart taste of yogurt, pickles, sharp cheese, and some sausages is due to the metabolic products of the lactic acid bacteria. **Starter cultures,** and sometimes rennin, are added to pasteurized milk to make cheese. Other bacteria or fungi are sometimes added to cheese to give characteristic flavors or textures (**figure 31.3**). Pickling relies on naturally occurring lactic acid bacteria. Commercial sausage production uses starter cultures to rapidly decrease the pH and prevent the growth of pathogens.

Alcoholic Fermentations by Yeast (table 31.2)

Wine is the product of the fermentation of sugars in fruit juices by specially selected strains of *Saccharomyces cerevisiae* (**figure 31.4**). Beer production is a multistep process designed to break down the starches of grains such as barley to produce simple sugars, which can then serve as a substrate for alcoholic fermentation by yeast (**figure 31.5**). Distilled spirits are produced using distillation to collect the alcohol generated during fermentation. Vinegar is the product of the oxidation of alcohol by the acetic acid bacteria. In bread-making, the CO_2 produced by yeast causes bread to rise; the alcohol is lost to evaporation (**figure 31.6**).

Changes Due to Mold Growth (table 31.3)

Some cheeses and other foods are produced by encouraging the growth of molds on foods. Soy sauce is made by allowing an *Aspergillus* species to degrade a mixture of soybeans and wheat, which is then fermented in brine.

31.3 ■ Food Spoilage

Food spoilage is most often due to the metabolic activities of microorganisms as they grow and use the nutrients in the food.

Common Spoilage Bacteria

Pseudomonas, Erwinia, Acetobacter, Alcaligenes, lactic acid bacteria, and bacteria that form endospores are important causes of food spoilage.

Common Spoilage Fungi

Fungi grow readily in acidic as well as low-moisture environments; therefore, fruits and breads are more likely to be spoiled by fungi than by bacteria.

31.4 ■ Foodborne Illness (table 31.4)

Foodborne Intoxication

Foodborne intoxication results from consuming a toxin produced by a microorganism growing in a food product (**figure 31.7**). Many strains of

Staphylococcus aureus produce a toxin that, when ingested, causes nausea and vomiting. Botulism is caused by ingestion of a neurotoxin produced by the anaerobic, spore-forming, Gram-positive rod *Clostridium botulinum*. As an added safety measure, low-acid home-canned foods should be boiled at least 15 minutes immediately before serving it to destroy any botulinum toxin that could be present.

Foodborne Infection

Foodborne infection requires the consumption of living organisms (figure 31.8). Thorough cooking of food immediately before eating it will kill bacteria, thereby preventing infection. *Salmonella*

and *Campylobacter* species are commonly associated with poultry products. Some outbreaks of *E. coli* O157:H7 have been traced to undercooked contaminated hamburger patties and produce contaminated with manure.

31.5 ■ Food Preservation

Food spoilage can be eliminated or delayed by destroying microorganisms or altering conditions to inhibit their growth. Methods used to preserve foods include canning, pasteurization, cooking, refrigeration, freezing, reducing the a_w, lowering the pH, adding antimicrobial chemicals, and irradiation.

Review Questions

Short Answer

1. What is the purpose of rennin in cheese-making?
2. What causes the bluish-green veins to form in blue cheese?
3. What causes the holes to form in Swiss cheese?
4. What is the difference between traditional acidophilus milk and sweet acidophilus milk?
5. What is the purpose of the mashing step in beer-making?
6. Explain how *Alcaligenes* species cause "ropiness" in raw milk.
7. Explain the significance of *Aspergillus flavus* in grain products.
8. Explain the typical sequence of events that lead to botulism.
9. Explain the typical sequence of events that lead to staphylococcal food poisoning.
10. How does canning differ from pasteurization?

Multiple Choice

1. The a_w of a food product reflects which of the following?
   a) Acidity of the food
   b) Presence of antimicrobial constituents such as lysozyme
   c) Amount of water available
   d) Storage atmosphere
   e) Nutrient content
2. Most spoilage bacteria cannot grow below an a_w of
   a) 0.3.　　b) 0.5.　　c) 0.7.　　d) 0.9.　　e) 1.0.
3. What is a generally minimum pH for growth and toxin production by *Clostridium botulinum* and other foodborne pathogens?
   a) 8.5　　b) 7.0　　c) 6.5　　d) 4.5　　e) 2.0
4. Benzoic acid is an antimicrobial chemical naturally found in which of the following foods?
   a) Apples　　b) Cranberries　　c) Eggs
   d) Milk　　e) Yogurt
5. Which of the following is often added to wine to inhibit growth of the natural microbial population of grapes?
   a) Benzoic acid　　b) Lactic acid　　c) Carbon dioxide
   d) Sulfur dioxide　　e) Oxygen
6. In the brewing process, the sugar and nutrient extract obtained by soaking germinated grain in warm water is called
   a) baker's yeast.　　b) hops.　　c) malt.
   d) must.　　e) wort.
7. Which of the following genera is used in bread, wine, and beer production?
   a) *Lactobacillus*　　b) *Pseudomonas*　　c) *Saccharomyces*
   d) *Streptococcus*　　e) *Staphylococcus*

8. Which group of organisms most commonly spoils breads, fruits, and dried foods?
   a) *Acetobacter*　　b) Fungi
   c) Lactic acid bacteria　　d) *Pseudomonas*
   e) *Saccharomyces*
9. Which of the following organisms cause foodborne intoxication?
   a) *E. coli* O157:H7　　b) *Campylobacter* species
   c) *Lactobacillus* species　　d) *Salmonella* species
   e) *Staphylococcus aureus*
10. Canned pickles require less stringent heat processing than canned beans, because pickles
    a) contain fewer nutrients.
    b) are more acidic.
    c) have a lower a_w.
    d) contain antimicrobial chemicals.
    e) are less likely to be contaminated with endospores.

Applications

1. A small cheese-manufacturing company in Wisconsin is looking for ways to reduce the costs of disposing of whey, a cheese by-product. As a food microbiologist, what would you suggest that the company do with the thousands of liters of whey being produced per month so the company can actually profit from it?
2. A microbiologist is troubleshooting a batch of home-brewed ale that did not ferment properly. She noticed that the alcohol content was only 2%, well below the desired level. Microscopic examination showed numerous yeast cells. Chemical analysis indicated low levels of sugar, high levels of CO_2, and large amounts of protein in the liquid. What did the microbiologist conclude as the probable cause of the beer not coming out properly?

Critical Thinking ✚

1. It has been argued that the nature of the growth of fungi in Roquefort cheese, indicated by the appearance of bluish-green veins, is evidence that these fungi require O_2 for growth. How does this evidence lead to the conclusion?
2. In the production of sauerkraut, a natural succession of lactic acid bacteria is observed growing in the product. What causes the succession? What does this tell you about the optimal growth conditions of the different species of lactic acid bacteria?

Appendix I
Microbial Mathematics

Because prokaryotes are very tiny and can multiply to very large numbers of cells in short time periods, convenient and simple ways are used to indicate their numbers without resorting to many zeros before or after the number. This is one reason why it is important to understand the metric system, which is used in scientific measurements.

The basic unit of measure is the meter, which is equal to about 39 inches. All other units are fractions of a meter:

1 decimeter is one tenth = 0.1 meter

1 centimeter is one hundredth = 0.01 meter

1 millimeter is one thousandth = 0.001 meter

Because prokaryotes are much smaller than a millimeter, even smaller units of measure are used. A millionth of a meter is a micrometer = 0.000001 meter, and is abbreviated μm. This is the most frequently used size measurement in microbiology, because bacteria are in this size range. For comparison, a human hair is about 75 μm wide.

Because it is inconvenient to write so many zeros in front of the 1, an easier way of indicating the same number is through the use of superscript, or exponential, numbers (exponents). One hundred dollars can be written 10^2 dollars. The 10 is called the base number and the 2 is the exponent. Conversely, one hundredth of a dollar is 10^{-2} dollars; thus the exponent is negative. The base most commonly used in biology is 10 (which is designated as $\log_{10}$). The above information can be summarized as follows:

1 millimeter = 1 mm = 0.001 meter = 10^{-3} meter

1 micrometer = 1 μm = 0.000001 meter = 10^{-6} meter

1 nanometer = 1 nm = 0.000000001 meter = 10^{-9} meter

The same prefix designations can be used for weights. The basic unit of weight is the gram, abbreviated g. Approximately 450 grams are in a pound.

1 milligram = 1 mg = 0.001 gram = 10^{-3} g

1 microgram = 1 μg = 0.000001 gram = 10^{-6} g

1 nanogram = 1 ng = 0.000000001 gram = 10^{-9} g

1 picogram = 1 pg = 0.000000000001 gram = 10^{-12} g

Note that the number of zeros before the 1 is one less than the exponent.

The value of the number is obtained by multiplying the base by itself the number of times indicated by the exponent.

Thus, $10^1 = 10 \times 1 = 10$

$10^2 = 10 \times 10 = 100$

$10^3 = 10 \times 10 \times 10 = 1,000$

When the exponent is negative, the base and exponent are divided into 1.

For example, $10^{-2} = 1/10 \times 1/10 = 1/100 = 0.01$

When multiplying numbers having exponents to the same base, the exponents are added.

For example, $10^3 \times 10^2 = 10^5$ (not 10^6)

When dividing numbers having exponents to the same base, the exponents are subtracted.

For example, $10^5 \div 10^2 = 10^3$

In both cases, only if the bases are the same can the exponents be added or subtracted.

Appendix II
Pronunciation Key for Bacterial, Fungal, Protozoan, and Viral Names

A

Acetobacter (a-see′-toe-back-ter)
Achromobacter (a-krome′-oh-back-ter)
Acinetobacter (a-sin-et′-oh-back-ter)
Actinomyces israelii (ak-tin-oh-my′-seez iz-ray′-lee-ee)
Actinomycetes (ak-tin-oh-my′-seats)
Adenovirus (ad′-eh-no-vi-rus)
Agrobacterium tumefaciens (ag-rho-bak-teer′-ee-um too-meh-faysh′-ee-enz)
Alcaligenes (al-ka-li′-jen-ease)
Amoeba (ah-mee′-bah)
Arbovirus (are′-bow-vi-rus)
Aspergillus niger (ass-per-jill′-us nye′-jer)
Aspergillus oryzae (ass-per-jill′-us or-eye′-zee)
Azolla (aye-zol′-lah)
Azotobacter (ay-zoh′-toe-back-ter)

B

Bacillus anthracis (bah-sill′-us an-thra′-siss)
Bacillus cereus (bah-sill′-us seer′-ee-us)
Bacillus coagulans (bah-sill′-us coh-ag′-you-lans)
Bacillus fastidiosus (bah-sill′-us fas-tid-ee-oh′-sus)
Bacillus subtilis (bah-sill′-us sut′-ill-us)
Bacillus thuringiensis (bah-sill′-us thur′-in-jee-en-sis)
Bacteroides (back′-ter-oid′-eez)
Baculovirus (back′-you-low-vi-rus)
Bdellovibrio (del′-o-vib′-re-oh)
Beggiatoa (beg-gee-ah-toe′-ah)
Beijerinckia (by-yer-ink′-ee-ah)
Bordetella pertussis (bor-deh-tell′-ah per-tuss′-iss)
Borrelia burgdorferi (bor-real′-ee-ah berg-dor′-fir-ee)
Bradyrhizobium (bray-dee-rye-zoe′-bee-um)
Brucella abortus (bru-sell′-ah ah-bore′-tus)

C

Campylobacter jejuni (kam′-peh-low-back-ter je-june′-ee)
Candida albicans (kan′-did-ah al′-bi-kanz)
Caulobacter (caw′-loh-back-ter)
Ceratocystis ulmi (see′-rah-toe-sis-tis ul′-mee)
Chlamydia trachomatis (klah-mid′-ee-ah trah-ko-ma′-tiss)
Chlamydophila pneumoniae (klah-mid′-o-fil-ah new-moan′-ee-ee)
Claviceps purpurea (kla′-vi-seps purr-purr′-ee-ah)
Clostridium acetobutylicum (kloss-trid′-ee-um a-seat-tow-bu-till′-i-kum)
Clostridium botulinum (kloss-trid′-ee-um bot-you-line′-um)
Clostridium difficile (kloss-trid′-ee-um dif′-fi-seal)
Clostridium perfringens (kloss-trid′-ee-um per-frin′-gens)
Clostridium tetani (kloss-trid′-ee-um tet′-an-ee)
Coccidioides immitis (cock-sid-ee-oid′-eez im′-mi-tiss)
Coronavirus (kor-oh′-nah-vi-rus)

Corynebacterium diphtheriae (koh-ryne′-nee-bak-teer-ee-um dif-theer′-ee-ee)
Coxsackievirus (cock-sack-ee′-vi-rus)
Cryptococcus neoformans (krip-toe-cock′-us knee-oh-for′-manz)
Cytophaga (sigh-taw′-fa-ga)

D

Desulfovibrio (dee-sul-foh-vib′-ree-oh)

E

Eikenella corrodens (eye-keh-nell′-ah kor-roh′-denz)
Entamoeba histolytica (en-ta-mee′-bah his-toh-lit′-ik-ah)
Enterobacter (en′-ter-oh-back-ter)
Enterococcus faecalis (en′-ter-oh-kock′-us fee-ka′-liss)
Enterovirus (en′-ter-oh-vi-rus)
Epidermophyton (eh-pee-der′-moh-fy-ton)
Epulopiscium (ep′-you-low-pis-se-um)
Escherichia coli (esh-er-ee′-she-ah koh′-lee)

F

Flavivirus (flay′-vih-vi-rus)
Flavobacterium (flay-vo-back-teer′-ee-um)
Francisella tularensis (fran-siss-sell′-ah tu-lah-ren′-siss)
Frankia (frank′-ee-ah)
Fusobacterium (fu′-zoh-back-teer-ee-um)

G

Gallionella (gal-ee-oh-nell′-ah)
Gardnerella vaginalis (gard-nee-rel′-lah va-jin-al′-is)
Geobacillus stearothermophilus (gee′-oh-bah-sill′-us steer-oh-ther-maw′-fill-us)
Giardia lamblia (jee-are′-dee-ah lamb′-lee-ah)
Gluconobacter (glue-kon-oh-back′-ter)
Gonyaulax (gon-ee-ow′-lax)
Gymnodinium breve (jim-no-din′-i-um brev-eh)

H

Haemophilus influenzae (hee-moff′-ill-us in-flew-en′-zee)
Helicobacter pylori (he′-lih-koh-back-ter pie-lore′-ee)
Hepadnavirus (hep-ad′-nah-vi-rus)
Hepatitis virus (hep-ah-ti′-tis vi-rus)
Herpes simplex (her′-peas sim′-plex)
Herpes zoster (her′-peas zoh′-ster)
Histoplasma capsulatum (his-toh-plaz′-mah cap-su-lah′-tum)
Hyphomicrobium (high-foh-my-krow′-bee-um)

I

Influenza virus (in-flew-en′-za vi-rus)

K

Klebsiella pneumoniae (kleb-see-ell′-ah new-moan′-ee-ee)

L

Lactobacillus brevis (lack-toe-ba-sil′-lus bre′-vis)
Lactobacillus bulgaricus (lack-toe-ba-sil′-lus bull-gair′-i-kus)
Lactobacillus casei (lack-toe-ba-sil′-us kay′-see-ee)
Lactobacillus plantarum (lack-toe-ba-sil′-us plan-tar′-um)
Lactobacillus thermophilus (lack-toe-ba-sil′-us ther-mo′-fil-us)
Lactococcus lactis (lack-toe-kock′-us lak′-tiss)
Legionella pneumophila (lee-jon-ell′-ah new-moh′-fill-ah)
Leptospira interrogans (lep-toe-spire′-ah in-ter-roh′-ganz)
Leuconostoc citrovorum (lew-kow-nos′-tok sit-ro-vor′-um)
Listeria monocytogenes (lis-tear′-ee-ah mon′-oh-sigh-to- jen′-eze)

M

Malassezia (mal-as-seez′-e-ah)
Methanobacterium (me-than′-oh-bak-teer-ee-um)
Methanococcus (me-than-oh-ko′-kus)
Microsporum (my-kroh-spore′-um)
Mobiluncus (moh-bi-lun′-kus)
Moraxella catarrhalis (more-ax-ell′-ah kah-tah-rah′-liss)
Moraxella lacunata (more-ax-ell′-ah lak-u-nah′-tah)
Mucor (mu′-kor)
Mycobacterium leprae (my-koh-bak-teer′-ee-um lep-ree)
Mycobacterium tuberculosis (my-koh-bak-teer′-ee-um too-ber-kew-loh′-siss)
Mycoplasma pneumoniae (my-koh-plaz′-mah new-moan′-ee-ee)

N

Neisseria gonorrhoeae (nye-seer′-ee-ah gahn-oh-ree′-ee)
Neisseria meningitidis (nye-seer′-ee-ah men-in-jit′-id-iss)
Neurospora sitophila (new-rah′-spor-ah sit-oh-phil′-ah)

O

Orthomyxovirus (or-thoe-mix′-oh-vi-rus)
Oscillatoria (os-sil-la-tor′-ee-ah)

P

Papillomavirus (pap-il-oh′-ma-vi-rus)
Parainfluenza virus (par-ah-in-flew-en′-zah vi-rus)
Paramecium (pair′-ah-mee-see-um)
Paramyxovirus (par-ah-mix′-oh-vi-rus)
Parvovirus (par′-vo-vi-rus)
Pasteurella multocida (pass-ture-ell′-ah mul-toe-sid′-ah)
Pediococcus soyae (ped-ih-oh-ko′-kus soy′-ee)
Penicillium camemberti (pen-eh-sill′-ee-um cam-em-bare′-tee)
Penicillium roqueforti (pen-eh-sill′-ee-um rok-e-for′-tee)
Peptostreptococcus (pep′-to-strep-to-ko-kus)
Phytophythora infestans (fy′-toe-fy-thor-ah in-fes′-tanz)
Picornavirus (pi-kor′-na-vi-rus)
Plasmodium falciparum (plaz-moh′-dee-um fall-sip′-air-um)
Plasmodium malariae (plaz-moh′-dee-um ma-lair′-ee-ee)
Plasmodium ovale (plaz-moh′-dee-um oh-vah′-lee)
Plasmodium vivax (plaz-moh′-dee-um vye′-vax)
Pneumocystis jiroveci (new-mo-sis′-tis yee-row′-vet-ze)
Poliovirus (poe′-lee-oh-vi-rus)
Polyomavirus (po-lee-oh′-mah vi-rus)
Propionibacterium acnes (proh-pee-ah-nee-bak-teer′-ee-um ak′-neez)

Propionibacterium shermanii (proh-pee-ah-nee-bak-teer′-ee-um sher-man′-ee-ee)
Proteus mirabilis (proh′-tee-us mee-rab′-il-us)
Pseudomonas aeruginosa (sue-dough-moan′-ass aye-rue-gin-o′-sa)

R

Rabies virus (ray′-bees vi-rus)
Retrovirus (re′-trow-vi-rus)
Rhabdovirus (rab′-doh-vi-rus)
Rhinovirus (rye′-no-vi-rus)
Rhizobium (rye-zoh′-bee-um)
Rhizopus nigricans (rise′-oh-pus nye′-gri-kanz)
Rhizopus stolon (rise′-oh-pus stoh′-lon)
Rhodococcus (roh-doh-koh′-kus)
Rickettsia rickettsii (rik-kett′-see-ah rik-kett′-see-ee)
Rotavirus (row′-tah-vi-rus)
Rubella virus (rue-bell′-ah vi-rus)
Rubeola virus (rue-bee-oh′-la vi-rus)

S

Saccharomyces carlsbergensis (sack-ah-row-my′-sees karls-berg-en′-siss)
Saccharomyces cerevisiae (sack-ah-row-my′-sees sara-vis′-ee-ee)
Salmonella enterica (sall-moh-nell′-ah en-ter-i-kah)
Serratia marcescens (ser-ray′-sha mar-sess-sens)
Shigella dysenteriae (shig-ell′-ah diss-en-tair′-ee-ee)
Spirillum volutans (spy-rill′-um vol-u-tanz)
Sporothrix schenckii (spore′-oh-thrix shenk-ee-ee)
Staphylococcus aureus (staff-ill-oh-kok′-us aw′-ree-us)
Staphylococcus epidermidis (staff-ill-oh-kok′-us epi-der′-mid-iss)
Streptobacillus moniliformis (strep-tow-bah-sill′-us mon-ill-i-form′-is)
Streptococcus agalactiae (strep-toe-kock′-us a-ga-lac′-tee-ee)
Streptococcus cremoris (strep-toe-kock′-us kre-more′-iss)
Streptococcus mutans (strep-toe-kock′-us mew′-tanz)
Streptococcus pneumoniae (strep-toe-kock′-us new-moan′-ee-ee)
Streptococcus pyogenes (strep-toe-kock′-us pie-ah-gen-ease)
Streptococcus salivarius (strep-toe-kock′-us sal-ih-vair′-ee-us)
Streptococcus thermophilis (strep-toe-kock′-us ther-moh′-fill-us)
Streptomyces griseus (strep-toe-my′-seez gree′-see-us)

T

Thiobacillus (thigh-oh-bah-sill′-us)
Torulopsis (tore-you-lop′-siss)
Treponema pallidum (tre-poh-nee′-mah pal′-ih-dum)
Trichomonas vaginalis (trick-oh-moan′-as vag-in-al′-iss)
Trichophyton (trick-oh-phye′-ton)
Trypanosoma brucei (tri-pan′-oh-soh-mah bru′-see-ee)

V

Varicella-zoster virus (var-ih-sell′-ah zoh′-ster vi-rus)
Vibrio cholerae (vib′-ree-oh kahl′-er-ee)
Vibrio fischeri (vib′-ree-oh fish′-er-i)

W

Wolbachia (wol-bach′-ee-ah)

Y

Yersinia enterocolitica (yer-sin′-ee-ah en-ter-oh-koh-lih′-tih-kah)
Yersinia pestis (yer-sin′-ee-ah pess′-tiss)

FIGURE III.1 The Entner-Doudoroff Pathway

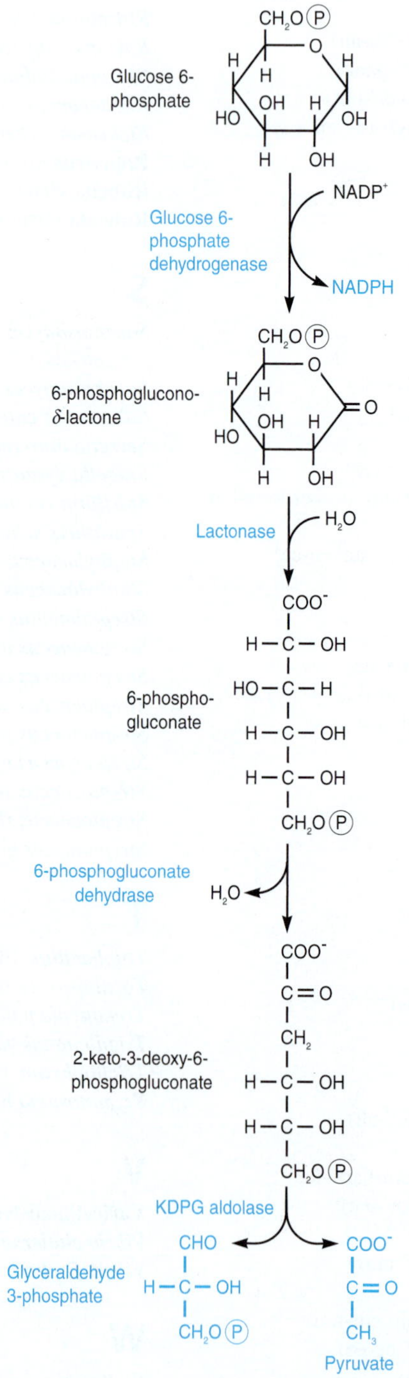

Glucose 6-phosphate

CH_2O (P)

Glucose 6-phosphate dehydrogenase

$NADP^+$

NADPH

6-phosphoglucono-δ-lactone

CH_2O (P)

Lactonase

H_2O

6-phospho-gluconate

COO^-
$H-C-OH$
$HO-C-H$
$H-C-OH$
$H-C-OH$
CH_2O (P)

6-phosphogluconate dehydrase

H_2O

2-keto-3-deoxy-6-phosphogluconate

COO^-
$C=O$
CH_2
$H-C-OH$
$H-C-OH$
CH_2O (P)

KDPG aldolase

Glyceraldehyde 3-phosphate

CHO
$H-C-OH$
CH_2O (P)

COO^-
$C=O$
CH_3

Pyruvate

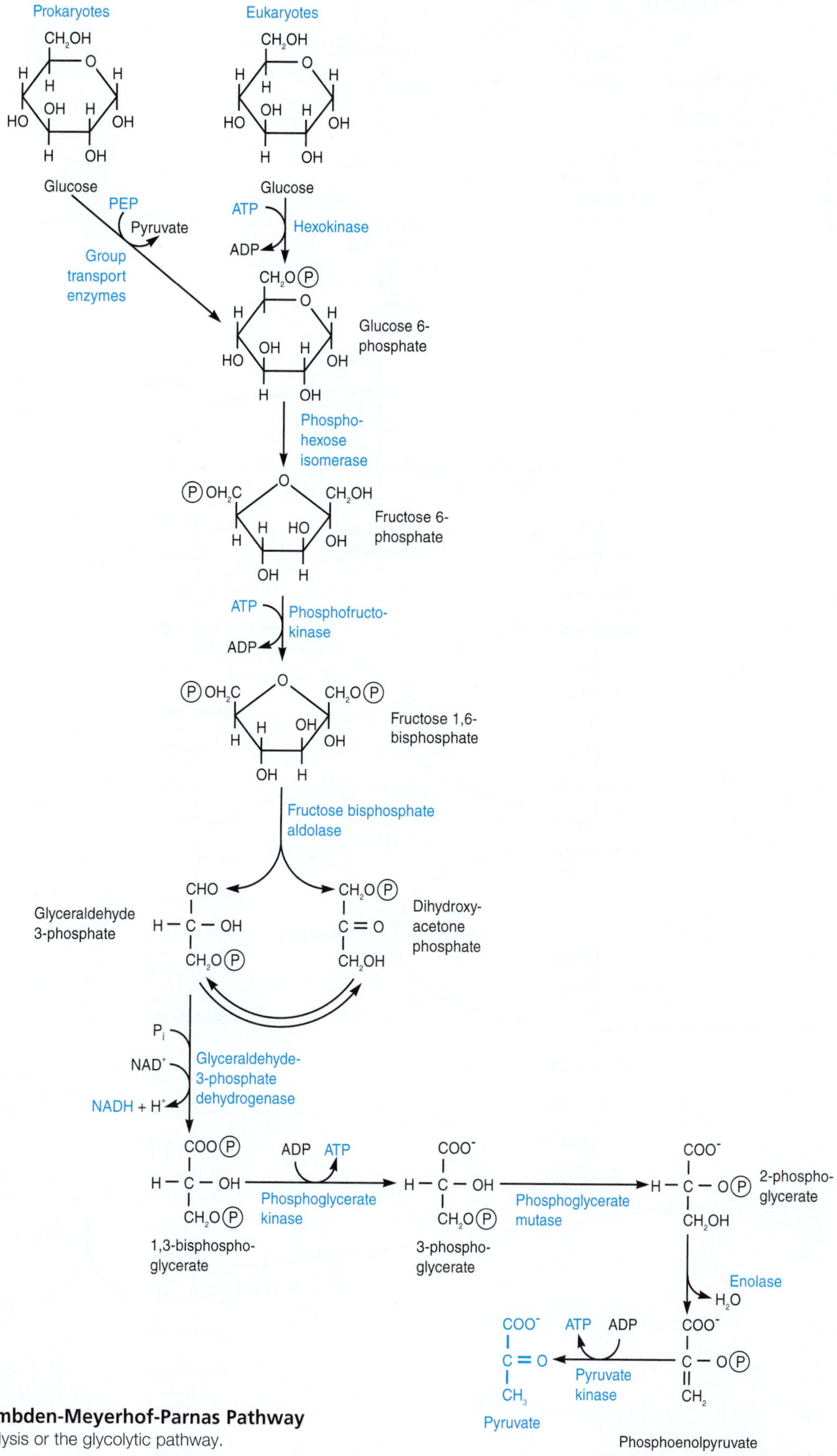

FIGURE III.2 The Embden-Meyerhof-Parnas Pathway
Commonly called glycolysis or the glycolytic pathway.

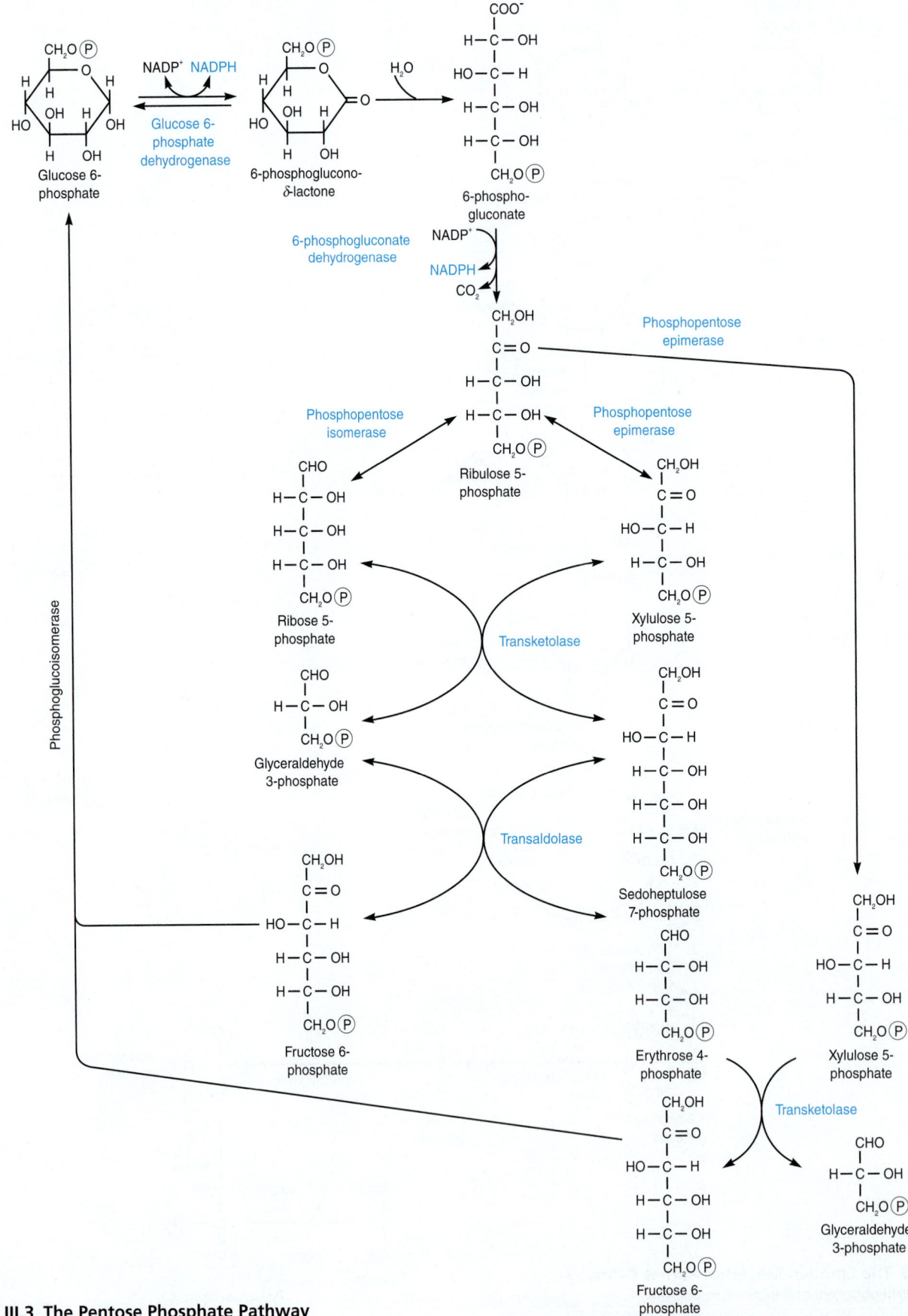

FIGURE III.3 The Pentose Phosphate Pathway

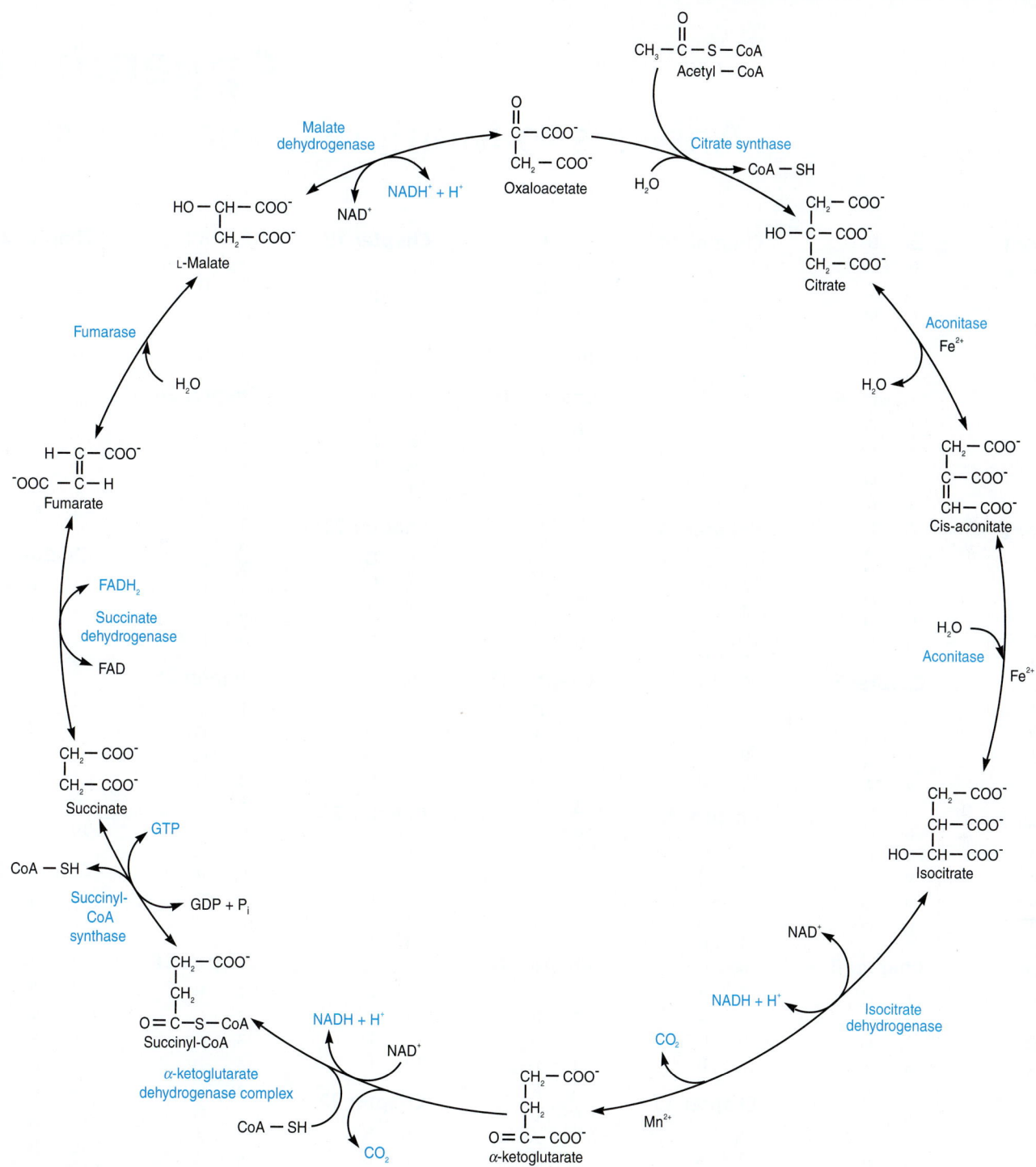

FIGURE III.4 The Tricarboxylic Acid Cycle (TCA Cycle)
Also referred to as the Krebs cycle or the citric acid cycle.

Appendix IV
Answers to Multiple Choice Questions

Chapter 1
1. C
2. C
3. E
4. B
5. D
6. A
7. C
8. A
9. A
10. D

Chapter 2
1. C
2. A
3. E
4. A
5. B
6. C
7. B
8. D
9. E
10. A

Chapter 3
1. A
2. A
3. E
4. B
5. E
6. D
7. A
8. C
9. D
10. A

Chapter 4
1. D
2. A
3. B
4. B
5. B
6. E
7. C
8. A
9. C
10. B

Chapter 5
1. C
2. D
3. A
4. B
5. B
6. C
7. E
8. D
9. C
10. A

Chapter 6
1. D
2. C
3. D
4. E
5. B
6. A
7. A
8. D
9. D
10. E

Chapter 7
1. B
2. A
3. A
4. A
5. C
6. C
7. C
8. B
9. D
10. D

Chapter 8
1. C
2. A
3. D
4. A
5. B
6. D
7. A
8. B
9. A
10. B

Chapter 9
1. C
2. C
3. E
4. B
5. B
6. C
7. B
8. B
9. B
10. D

Chapter 10
1. E
2. B
3. E
4. B
5. D
6. B
7. C
8. D
9. B
10. E

Chapter 11
1. E
2. B
3. E
4. B
5. D
6. A
7. D
8. B
9. C
10. D

Chapter 12
1. C
2. A
3. C
4. A
5. B
6. C
7. A
8. A
9. C
10. B

Chapter 13
1. C
2. A
3. C
4. D
5. E
6. C
7. D
8. A
9. A
10. D

Chapter 14
1. B
2. E
3. A
4. B

Chapter 15
1. E
2. E
3. D
4. D
5. B
6. D
7. D
8. A
9. B
10. D

Chapter 16
1. C
2. A
3. D
4. D
5. A
6. C
7. C
8. D
9. D
10. A

Chapter 17
1. A
2. E
3. A
4. D
5. E
6. C
7. A
8. C
9. A
10. A

Chapter 18
1. C
2. D
3. D
4. E
5. D
6. A
7. C
8. D
9. A
10. C

Chapter 19
1. A
2. B
3. A
4. B
5. E
6. B
7. B
8. E
9. C
10. E

Chapter 20
1. D
2. C
3. A
4. B
5. A
6. B
7. D
8. D
9. A
10. C

Chapter 21
1. E
2. C
3. E
4. A
5. B
6. A
7. A
8. C
9. D
10. D

Chapter 22
1. E
2. E
3. A
4. E
5. E
6. A
7. C
8. A
9. D
10. D

Chapter 23
1. B
2. E
3. C
4. A

Chapter 14 (col top)
5. C
6. A
7. E
8. A
9. C
10. D

Chapter 24
1. C
2. A
3. B
4. D
5. A
6. D
7. C
8. D
9. C
10. E

Chapter 25
1. D
2. D
3. C
4. B
5. C
6. B
7. A
8. B
9. C
10. C

Chapter 26
1. B
2. C
3. C
4. D
5. C
6. E
7. A
8. A
9. B
10. E

Chapter 27
1. E
2. A
3. C
4. E
5. C
6. C
7. C
8. A
9. A
10. E

Chapter 19 (col top)
5. A
6. C
7. D
8. C
9. E
10. B

Chapter 28
1. C
2. D
3. A
4. B
5. C
6. A
7. A
8. A
9. E
10. E

Chapter 29
1. A
2. A
3. C
4. D
5. A
6. D
7. C
8. C
9. B
10. A

Chapter 30
1. B
2. B
3. A
4. A
5. C
6. E
7. C
8. C
9. B
10. E

Chapter 31
1. C
2. D
3. D
4. B
5. D
6. E
7. C
8. B
9. E
10. B

Glossary

abscess A localized collection of pus within a tissue.

A-B toxin Exotoxin composed of an active subunit (A subunit) and a binding subunit (B subunit).

accessory pigments Photosynthetic pigments such as carotenoids that capture light energy not absorbed by chlorophylls.

acid-fast An organism that retains the primary stain in the acid-fast staining procedure.

acid-fast staining A procedure used to stain certain microorganisms, particularly *Mycobacterium* species, that do not readily take up dyes.

acidophiles Organisms that grow optimally at a pH below 5.5.

acquired resistance Development of antimicrobial resistance through spontaneous mutation or acquisition of new genetic information.

actin filaments Cytoskeletal structures of eukaryotic cells that allow movement within the cell.

actinomycetes Filamentous bacteria; many are valuable because they produce antibiotics.

activated macrophages Macrophages stimulated by cytokines to enlarge and become metabolically active, with greatly increased capability to kill and degrade intracellular organisms and materials.

activated sludge method A method of sewage treatment in which wastes are degraded by complex populations of aerobic microorganisms.

activation energy Initial energy required to break a chemical bond.

activator In gene regulation, a protein that enhances the ability of RNA polymerase to initiate transcription.

activator-binding site Nucleotide sequence that precedes an ineffective promoter.

active immunity Protective immunity produced by an individual in response to an antigenic stimulus.

active site Site on an enzyme to which the substrate binds; also known as the catalytic site.

active transport Energy-consuming process by which cells move molecules across a membrane and against a concentration gradient.

active tuberculosis disease (ATBD) A chronic illness caused by *Mycobacterium tuberculosis* and characterized by slight fever, progressive weight loss, night sweating, and persistent cough, often producing blood-streaked sputum.

acute bacterial endocarditis Acute infection of the internal surfaces of the heart.

acute infection Illness characterized by signs and symptoms that develop quickly but last a relatively short time.

acute inflammation Short-term inflammatory response, marked by a prevalence of neutrophils.

acute retroviral syndrome (ARS) Stage of HIV disease following the incubation stage; often includes flulike symptoms.

adaptive immunity Protection provided by host defenses that develop throughout life; involves B cells and T cells.

ADCC (antibody-dependent cellular cytotoxicity) Killing of antibody-coated target cells by natural killer cells, granulocytes, or macrophages.

adenosine diphosphate (ADP) The acceptor of free energy in cells; that energy is used to add an inorganic phosphate (P_i) to ADP, generating ATP.

adenosine triphosphate (ATP) The energy currency of cells. Hydrolysis of its unstable phosphate bonds can be used to power endergonic (energy-consuming) reactions.

adhesin Component of a microorganism that is used to bind to surfaces.

adhesion molecule Molecule on the surface of a cell that allows that cell to adhere to other cells.

adjuvant Substance that increases the immune response to antigen.

ADP Abbreviation for adenosine diphosphate.

advanced treatment Any physical, chemical, or biological purification process beyond secondary treatment of wastewater.

aerobic respiration Metabolic process in which electrons are transferred from the electron transport chain to molecular oxygen (O_2).

aerosol Material dispersed into the air as a fine mist.

aerotaxis Movement toward or away from O_2.

aerotolerant anaerobe Organism that can grow in the presence of O_2 but never uses it as a terminal electron acceptor; an obligate fermenter.

affinity maturation The "fine-tuning" of the fit of an antibody molecule for an antigen; is due to mutations that occur as activated B cells multiply.

aflatoxin Potent toxin made by *Aspergillus flavus*.

agar Polysaccharide extracted from marine algae that is used to solidify microbiological media.

agar plate A Petri dish that contains a solidified culture medium.

agarose Highly purified form of agar used in gel electrophoresis.

agglutination reaction In immunological testing, clumping together of cells or particles by antibody molecules.

AIDS Acquired immunodeficiency syndrome.

AIDS-related complex (ARC) A group of symptoms—fever, fatigue, diarrhea, and weight loss—that mark the onset of AIDS.

alga (plural: **algae**) A primitive photosynthetic eukaryotic organism.

alkalophiles Organisms that grow optimally at a pH above 8.5.

allele One form of a gene.

allergen Antigen that causes an allergy.

allergy Hypersensitivity, especially of the IgE-mediated type.

allograft Organ or tissue transplanted between genetically nonidentical members of the same species.

allosteric Refers to an enzyme or other protein that contains a site to which a small molecule can bind and change the protein's activity.

alpha (α) hemolysis Type of hemolysis characterized by a zone of greenish clearing around colonies grown on blood agar.

alternative pathway Pathway of complement activation initiated by the binding of a complement protein (C3b) to cell surfaces.

alternative sigma factor A sigma factor that recognizes promoters controlling genes needed only in non-routine situations.

Ames test A test that screens for potential carcinogens by measuring the ability of a substance to increase the mutation frequency in a bacterial strain.

amino acids Subunits of a protein molecule.

ammonification The decomposition process that converts organic nitrogen into ammonia (NH_3).

amphibolic pathways Metabolic pathways that play roles in both catabolism and anabolism.

amylases Enzymes that digest starches.

anabolism Cellular processes that synthesize and assemble the subunits of macromolecules, using the energy of ATP; biosynthesis.

anaerobe container A specialized container that can maintain anaerobic conditions; a chemical reaction in a packet generates those conditions.

anaerobic Without molecular oxygen (O_2).

anaerobic chamber An enclosed compartment maintained in an anaerobic environment; a special port is used to add or remove items.

anaerobic digestion Process that uses anaerobic microbes to degrade the sludge obtained during wastewater treatment.

anaerobic respiration Metabolic process in which electrons are transferred from the electron transport chain to a terminal electron acceptor other than O_2.

analytical study In epidemiology, a study done to identify risk factors associated with developing a certain disease.

anammox Anoxic ammonium oxidation.

anaphylaxis (See *systemic anaphylaxis*.)

anion Negatively charged ion.

anneal To form a double-stranded duplex from two complementary strands of nucleic acid.

anoxic Lacking O_2.

anoxygenic phototrophs Photosynthetic bacteria that use H_2S or organic compounds rather than water as a source of electrons for reducing power.

antagonistic In antimicrobial therapy, a combination of antimicrobial medications in which the action of one interferes with the action of the other.

antenna pigments Pigments of photosynthetic organisms that make up a complex that acts as a funnel, capturing light energy and transferring it to reaction-center chlorophyll.

antibacterial drug Chemical used to treat bacterial infections.

antibiogram Antibiotic susceptibility pattern; used to distinguish between different bacterial strains.

The Glossary contains definitions for the majority of bold terms in the text.

G–1

antibiotic Chemical produced by certain molds and bacteria that kills or inhibits the growth of other microorganisms.

antibiotic-associated diarrhea Diarrhea that occurs as a complication of taking antimicrobial medications.

antibody Immunoglobulin protein produced by the body in response to a substance; it reacts specifically with that substance.

anticodon Sequence of three nucleotides in a tRNA molecule that is complementary to a codon in mRNA.

antigen Molecule that reacts specifically with an antibody or lymphocyte.

antigen-antibody complex Antibodies bound to antigen; also called an immune complex.

antigen presentation Process in which animal cells display antigen on MHC molecules for T cells to inspect.

antigen-presenting cells (APCs) Cells such as B cells, macrophages, and dendritic cells that can present exogenous antigen to helper T cells.

antigenic Induces an immune response.

antigenic drift Slight changes in a viral surface antigen that render antibodies made against the previous version only partially protective.

antigenic shift Major change in a viral surface antigen that render antibodies made against the previous version ineffective.

antigenic variation Process by which routine changes occur in a microbial surface antigen.

anti-human IgG antibodies Antibodies that bind the constant region of human IgG molecules.

antimicrobial drug Chemical used to treat microbial infections; also called an antimicrobial.

antiparallel Describes opposing orientations of the two strands of DNA in the double helix.

antiretrovirals (ARVs) Medications used to treat HIV infections.

antiseptic A disinfectant that is non-toxic enough to be used on skin.

antiserum A preparation of serum containing protective antibodies.

antitoxin An antibody preparation that protects against a given toxin.

antiviral drug A medication that interferes with the infection cycle of a virus.

apoptosis Programmed cell death.

apicomplexans A group of protozoa that penetrate host cells by means of a structure called an apical complex.

aquaporins Pore-forming membrane proteins that specifically allow water to pass through.

arbovirus Arthropod-borne virus. One of a large group of RNA viruses carried by insects and mites that act as biological vectors.

Archaea One of the two domains of prokaryotes; many archaea grow in extreme environments.

arthropod Classification grouping of invertebrate animals that includes insects, ticks, lice, and mites.

Arthus reaction Hypersensitivity reaction caused by immune complexes and neutrophils.

artificially acquired immunity Immunity acquired through artificial means such as vaccination or administration of immune globulin.

aseptic technique Use of specific methods and sterile materials to exclude contaminating microbes from an environment.

asexual Reproduction not preceded by the union of cells or exchange of DNA.

asthma Immediate respiratory allergy resulting from the release of mediators from mast cells in the lower airways.

astrobiology Study of life in the universe.

atom The basic unit of all matter.

atomic force microscope Type of scanning probe microscope that has a tip mounted so it can bend in response to the slightest force between the tip and the sample.

ATP Abbreviation for adenosine triphosphate.

ATP synthase Protein complex that harvests the energy of a proton motive force to synthesize ATP.

attaching and effacing (A/E) lesions Intestinal damage characterized by pedestals that form under bacterial cells as a result of induced actin rearrangement in the intestinal cell.

attack rate The percentage of individuals developing illness in a population exposed to an infectious agent.

attenuated vaccine Vaccine composed of a weakened form of a pathogen.

autoantibodies Antibodies that bind to "self" molecules.

autoclave Device that uses steam under pressure to sterilize materials.

autoimmune disease Disease produced as a result of an immune reaction against one's own tissues.

autoradiography Method that uses film to detect a radioactive molecule.

autotroph Organism that uses CO_2 as its main carbon source.

auxotroph A microorganism that requires an organic growth factor.

a_w Abbreviation for water activity.

axon The long thin extension of a nerve cell.

---

bacillus (plural: **bacilli**) Cylindrical-shaped bacterium; also referred to as a rod.

bacteremia Bacterial cells circulating in the bloodstream.

Bacteria One of the two domains of prokaryotes; all medically important prokaryotes are in the domain *Bacteria*.

bactericidal Kills bacteria.

bacteriochlorophyll Type of chlorophyll used by purple and green bacteria; absorbs wavelengths of light that penetrate to greater depths and are not used by other photosynthetic organisms.

bacteriocins Proteins made by bacteria that kill certain other bacteria.

bacteriophage A virus that infects bacteria; often abbreviated to phage.

bacteriostatic Inhibits the growth of bacteria.

baker's yeast Selected strains of *Saccharomyces cerevisiae* that are used to make yeast bread.

balanced pathogenicity Host-parasite relationship in which the parasite persists in the host while causing minimal harm.

base (See *nucleobase*.)

base analog Compound that structurally resembles a nucleobase closely enough to be incorporated into a nucleotide in place of the natural nucleobase.

base-pairing The hydrogen bonding of adenine (A) to thymine (T) and cytosine (C) to guanine (G).

base substitution A mutation in which the wrong nucleotide has been incorporated.

basement membrane Thin layer of fibrous material that underlies epithelial cells.

basophil Leukocyte with large dark-staining granules that contain histamine and other inflammatory mediators; receptors on its surface bind IgE.

batch culture A culture system such as a tube, flask, or agar plate in which nutrients are not replenished and wastes are not removed.

B-cell receptor (BCR) Membrane-bound derivative of the antibody that a B cell is programmed to make; it allows the B cell to recognize a specific epitope.

B cells Lymphocytes programmed to produce antibody molecules.

B lymphocytes (See *B cells*.)

beta-(β) hemolysis Type of hemolysis characterized by a clear zone around a colony grown on blood agar.

beta-(β) lactam drugs Group of antimicrobial medications that inhibit peptidoglycan synthesis and have a shared chemical structure called a β-lactam ring.

beta-(β) lactamase Enzyme that breaks the β-lactam ring of a β-lactam drug, thereby inactivating the medication.

bile Yellow-colored fluid produced by the liver that aids in the absorption of nutrients from the intestine.

binary fission Asexual process of reproduction in which one cell divides to form two daughter cells.

binomial system System of naming an organism with two Latin words that indicate the genus and species.

biochemical oxygen demand (BOD) Measure of the amount of biologically degradable organic material in water.

biodiversity Diversity in the number of species inhabiting an ecosystem and their evenness of distribution.

biofilm Polymer-encased microbial community.

bioinformatics Developing and using computer technology to store, retrieve, and analyze nucleotide sequence data.

biological vector Organism that transmits a pathogen and within which the pathogen can multiply to high numbers.

bioluminescence Biological production of light.

biomass Total weight of all organisms in any particular environment.

bioremediation Process that uses microorganisms to degrade harmful chemicals.

biosphere The sum of all the regions of the earth where life exists.

biosynthesis Cellular processes that synthesize and assemble the subunits of macromolecules, using the energy of ATP.

biotechnology The use of microbiological and biochemical techniques to solve practical problems and produce valuable products.

bioterrorism The intentional use of microbes or their toxins to cause harm.

biotype A group of strains that have a characteristic biochemical pattern different from other strains; also called a biovar.

biovar (See *biotype*.)

blood agar Type of nutrient-rich agar medium that contains red blood cells and can be used to detect hemolysis.

blood-brain barrier Property of the central nervous system blood vessels that restricts passage of infectious agents and certain molecules (such as medications) into the brain and spinal cord.

BOD Abbreviation for biochemical oxygen demand.

boil Painful localized collection of pus within the skin and subcutaneous tissue; a furuncle.

bone marrow Soft material that fills bone cavities and contains hematopoietic stem cells.

botulinum toxin Toxin produced by *Clostridium botulinum* that can cause a fatal paralysis.

brewer's yeast Specially selected strains of *Saccharomyces cerevisiae* that are used in beer-making.

bright-field microscope Type of light microscope that illuminates the field of view evenly.

broad host range plasmid A plasmid that can replicate in a wide variety of unrelated bacteria.

broad-spectrum antimicrobials Antimicrobials that inhibit or kill a wide range of microorganisms, often including both Gram-positive and Gram-negative bacteria.

Bt-toxin Protein crystal naturally produced by *Bacillus thuringiensis* as it forms endospores; toxic to insect larvae that consume it.

bubo Enlarged, tender lymph node characteristic of plague and some sexually transmitted infections.

bubonic plague Form of plague that typically develops when *Yersinia pestis* is injected during the bite of an infected flea.

budding (1) Asexual reproductive technique that involves a pushing out of a part of the parent cell that eventually gives rise to a new daughter cell. (2) A process by which some viruses are released from host cells.

buffer Substance in a solution that acts to prevent changes in pH.

bulking Overgrowth of filamentous microorganisms in sewage at treatment facilities; interferes with the separation of the solid sludge from the liquid effluent.

burst size Number of newly formed virus particles released when the infected host cell lyses

Calvin cycle Metabolic pathway used by many autotrophs to incorporate CO_2 into an organic form; also called the Calvin-Benson cycle.

cAMP Abbreviation for cyclic AMP.

cancer Abnormally growing cells that can spread from their site of origin; malignant tumors.

candidiasis Fungal diseases caused by *Candida albicans.*

CAP Abbreviation for catabolite-activating protein, an activator involved in carbon catabolite repression.

capnophiles Organisms that require increased concentrations of CO_2 (5% to 10%) and approximately 15% O_2.

capped In eukaryotic gene expression, adding a methylated guanine derivative to the 5′ end of the mRNA.

capsid Protein coat that surrounds the nucleic acid of a virus.

capsule A distinct thick gelatinous material that surrounds some types of microorganisms; sometimes required for an organism to cause disease.

capsule stain A staining method used to observe capsules.

carbapenems Group of antimicrobial medications that interferes with peptidoglycan synthesis; very resistant to inactivation by β-lactamases.

carbapenem-resistant *Enterobacteriaceae* (CRE) Members of the *Enterobacteriaceae* that are resistant to carbapenems, as well as nearly all other antimicrobial drugs.

carbohydrate Compounds containing principally carbon, hydrogen, and oxygen atoms in a ratio of 1:2:1.

carbon catabolite repression A regulatory mechanism that allows cells to prioritize their use of carbon/energy sources.

carbon fixation Process of converting inorganic carbon (CO_2) to an organic form.

carbuncle Painful infection of the skin and subcutaneous tissues; manifests as a cluster of boils.

carcinogen A chemical or radiation that causes cancer.

cariogenic Causing dental caries, tooth decay.

carotenoids Type of accessory pigment that increases the efficiency of light capture by absorbing wavelengths of light not absorbed by chlorophylls.

carrier (1) Type of protein found in cell membranes that transports certain compounds across the membrane; may also be called a permease or transporter protein. (2) A human or other animal that harbors a pathogen without noticeable ill effects.

case-control study In epidemiology, a study that compares activities of people who developed a disease with those of people who did not.

case-fatality rate Percentage of people dying of a specific disease within a specific time period.

caseous necrosis Type of localized tissue death resulting in a cheeselike consistency, characteristic of tuberculosis and certain other chronic infectious diseases.

catabolism Cellular processes that harvest the energy released during the breakdown of compounds such as glucose, using it to synthesize ATP.

catabolite Product of catabolism.

catalase Enzyme that breaks down hydrogen peroxide (H_2O_2) to produce water (H_2O) and oxygen gas (O_2).

catalyst Substance that speeds up the rate of a chemical reaction without being altered or depleted in the process.

catheter A flexible plastic or rubber tube inserted into the bladder or other body space, in order to drain it or deliver medication.

cations Positively charged ions.

CD markers Abbreviation for cluster of differentiation markers.

CD4 lymphocytes T lymphocytes bearing the CD4 markers; helper T cells are CD4 cells.

CD8 lymphocytes T lymphocytes bearing the CD8 markers; cytotoxic T cells are CD8 cells.

cDNA DNA obtained by using reverse transcriptase to synthesize DNA from an RNA template *in vitro;* lacks introns that characterize eukaryotic DNA.

cell culture (or **tissue culture**) Cultivation of animal or plant cells in the laboratory.

cell envelope The layers surrounding the contents of the cell; includes the cytoplasmic membrane, cell wall, and capsule (if present).

cell wall Strong barrier that surrounds a cell, keeping the contents from bursting out; in prokaryotes, peptidoglycan provides strength to the cell wall.

cell-mediated immunity (CMI) Immunity due a T-cell response; also called cellular immunity.

cellular immunity (See *cell-mediated immunity.*)

cellulose Polymer of glucose subunits; principal structural component of plant cell walls.

Centers for Disease Control and Prevention (CDC) The U.S. government agency charged with the task of controlling and preventing diseases and injuries.

central metabolic pathways Glycolysis, the TCA cycle, and the pentose phosphate pathway; the transition step is often considered a part of the TCA cycle.

central nervous system (CNS) Brain and spinal cord.

cephalosporins Group of antimicrobial medications that interfere with peptidoglycan synthesis.

cerebrospinal fluid (CSF) Fluid produced within the brain that surrounds the central nervous system.

cestode Tapeworm.

chain terminator A dideoxynucleotide; when this molecule is incorporated into a growing strand of DNA, no additional nucleotides can be added, so elongation of the strand stops.

challenge In immunology, to give an antigen to provoke an immunologic response in a subject previously sensitized to the antigen.

chancre Sore resulting from an ulcerating infection; the "hard chancre" of primary syphilis is typically firm and painless.

channel Pore-forming membrane protein that allows specific ions to diffuse into and out of a cell.

chaperone Protein that helps other proteins fold properly.

chemical bond Force that holds atoms together to form molecules.

chemically defined media A culture medium composed of exact quantities of pure chemicals; generally used for specific experiments when nutrients must be precisely controlled.

chemiosmotic gradient Accumulation of protons immediately outside the cell or mitochondrial matrix due to ejection of protons by the electron transport chain; also called the proton motive force.

chemiosmotic theory The theory that a proton gradient is formed by the electron transport chain and then used to power ATP synthesis.

chemoautotrophs Organisms that use chemicals as a source of energy and CO_2 as the major source of carbon.

chemoheterotrophs Organisms that use chemicals as a source of energy, and organic compounds as a source of carbon.

chemokine Cytokine important in chemotaxis of cells of the immune system.

chemolithoautotrophs Organisms that obtain energy by oxidizing reduced inorganic compounds such as hydrogen gas (H_2), and use CO_2 as a source of carbon.

chemolithotrophs Organisms that obtain energy by oxidizing reduced inorganic chemicals such as hydrogen gas (H_2); in general, chemolithotrophs are chemolithoautotrophs.

chemoorganoheterotrophs Organisms that obtain both energy and carbon from organic compounds.

chemoorganotrophs Organisms that obtain energy by oxidizing organic compounds such as glucose; in general, chemoorganotrophs are chemoorganoheterotrophs.

chemostat Device used to grow bacteria in the laboratory that allows nutrients to be added and waste products to be removed continuously.

chemotaxis Directed movement of an organism in response to a certain chemical in the environment.

chemotherapeutic agent Chemical used as a therapeutic medication to treat a disease.

chemotrophs Organisms that obtain energy by oxidizing chemical compounds.

chickenpox Disease characterized by widespread itchy fluid-filled blisters; caused by varicella-zoster virus (VZV), a herpesvirus.

chlorophylls The primary light-absorbing pigments used in photosynthesis.

chloroplasts Organelles in photosynthetic eukaryotic cells that harvest the energy of sunlight and use it to synthesize ATP, which is then used to fuel the synthesis of organic compounds.

chocolate agar Type of agar medium that contains lysed red blood cells; used to culture fastidious bacteria.

cholesterol Sterol found in animal cell membranes; provides rigidity to eukaryotic membranes.

chromosome Structure that carries an organism's hereditary information.

chronic infections Infections that develop slowly and persist for months or years.

chronic inflammation Long-term inflammatory response, marked by the prevalence of macrophages, giant cells, and granulomas.

chronic periodontitis A destructive inflammatory response that progressively damages the structures that support the teeth.

cilium (plural: **cilia**) Short, projecting hairlike structure of locomotion in some eukaryotic cells, similar in function to a flagellum.

cirrhosis Scarring of the liver that interferes with normal liver function.

clade Subtype of a virus such as HIV, distinguished by similar amino acid sequences of envelope proteins.

class In classification, a collection of similar orders; a collection of several classes makes up a phylum.

class switching The process that allows a B cell to change the antibody class it is programmed to make.

classical pathway Mechanism by which immune complexes activate the complement system.

classification The arrangement of organisms into similar or related groups, primarily to provide easy identification and study.

clinical latency In HIV disease, the time period during which the virus actively replicates without causing obvious signs or symptoms of disease.

clonal selection Process in which a lymphocyte's antigen receptor binds to an antigen, potentially allowing the lymphocyte to proliferate and then differentiate into an effector cell.

clone Group of cells derived from a single cell.

closed system A system such as a tube, flask, or agar plate in which nutrients are not replenished and wastes are not removed as microorganisms grow.

clumping factor Fibrinogen-binding virulence factor of *Staphylococcus aureus* that serves as an identifying characteristic.

clusters of differentiation (CD) markers Surface molecules that allow scientists to distinguish subsets of T cells and other white blood cells.

CMI Abbreviation for cell-mediated immunity.

CO₂ fixation Process of converting inorganic carbon (CO_2) to an organic form.

coagulase Plasma-clotting virulence factor of *Staphylococcus aureus* that serves as an identifying characteristic.

coccobacillus Rod that is so short it can be mistaken for a coccus.

coccus (plural: cocci) Spherical-shaped bacterial cell.

codon Set of three adjacent nucleotides that encode either an amino acid or the termination of the polypeptide.

coenzyme Non-protein organic compound that assists some enzymes, acting as a loosely bound carrier of small molecules or electrons.

cofactor Non-protein component required for the activity of some enzymes.

cohesive ends Single-stranded overhangs generated when DNA is digested with a restriction enzyme that cuts asymmetrically within the recognition sequence; sticky ends.

cohort group In epidemiology, population with a known exposure to a specific risk factor that is followed over time in a prospective study.

collagen Fibrous support protein found in skin, tendons, scars, and other tissues.

coliforms (See *total coliforms*.)

colonization Multiplication of microorganisms on a material, animal, or person without necessarily resulting in tissue invasion or damage.

colony A distinct mass of cells arising from a single cell.

colony blotting Technique that uses a probe to detect a given DNA sequence in colonies growing on an agar plate.

colony-forming unit A single cell or multiple cells attached to one another that give rise to a single colony.

colony-stimulating factors A group of cytokines that direct development of various types of blood cells from hematopoietic stem cells.

combination therapy Administration of two or more antimicrobial medications simultaneously to prevent growth of mutants that might be resistant to one of the antimicrobials.

commensalism Relationship between two organisms in which one partner benefits and the other is unaffected.

commercially sterile Free of all microorganisms capable of growing under normal storage conditions; the endospores of some thermophiles may remain.

common-source epidemic Outbreak of disease due to contaminated food, water, or other single source of infectious agent.

communicable diseases Diseases that are spread from an infected animal or person to another animal or person.

community All of the living organisms in a given area.

competent In horizontal gene transfer, physiological condition in which a bacterial cell is capable of taking up DNA.

competitive inhibition Type of enzyme inhibition that occurs when the inhibitor competes with the normal substrate for binding to the active site.

complement system Series of serum proteins involved with innate immunity; they can be rapidly activated, contributing to protective outcomes including inflammation, lysis of foreign cells, and opsonization.

complementary In DNA structure, the nucleobases that characteristically hydrogen bond to one another; A (adenine) is complementary to T (thymine), and G (guanine) is complementary to C (cytosine).

complex medium Medium for growing bacteria that has some ingredients of variable chemical composition.

compound lipid Molecules that contain fatty acid and glycerol in addition to an element other than carbon, hydrogen, or oxygen.

compound microscope Microscope that uses multiple magnifying lenses, thereby visually enlarging an object by a factor equal to the product of the magnification of each lens.

condenser lens Lens used to focus the illumination of a microscope; it does not affect the magnification.

confocal microscope Type of scanning laser microscope.

congenital rubella syndrome A characteristic pattern of fetal abnormalities that occur when rubella virus infects a developing fetus.

congenital syphilis Outcome of fetal infection by *Treponema pallidum*.

conidia Asexual spores borne on hyphae; produced by both fungi and bacteria of the genus *Streptomyces*.

conjunctivitis Inflammation of the outer surface of the eye.

conjugate vaccine A vaccine composed of a polysaccharide antigen covalently attached to a large protein molecule, thereby converting what would be a T-independent antigen into a T-dependent antigen.

conjugation In bacteria, a mechanism of horizontal gene transfer that involves cell-to-cell contact.

conjugative plasmid Plasmid that carries the genes for sex pili and can transfer copies of itself to other bacteria during conjugation.

constant region The part of an antibody molecule that does not vary in amino acid sequences among molecules of the same immunoglobulin class.

consumers Organisms that eat primary producers or other consumers.

constitutive enzyme An enzyme that is constantly being synthesized by a cell.

constructed wetland Method of wastewater treatment in which water is channeled into ponds where aerobic and anaerobic degradation occurs.

contact dermatitis A T-cell-mediated inflammation of the skin occurring in sensitized individuals as a result of contact with the particular antigen; a form of delayed-type hypersensitivity.

contagious diseases Diseases that are spread from one host to another very readily.

continuous culture Method used to maintain cells in a state of uninterrupted growth by continuously adding nutrients and removing waste products; a type of open system.

contrast In microscopy, the number of different visible shades in a specimen.

convalescence Period of recuperation and recovery from an illness.

convergent evolution Process of evolution when two genetically different organisms independently develop similar structures.

core genome DNA sequences found in all strains of a particular species.

corepressor Molecule that binds to an inactive repressor, thereby allowing it to function as a repressor.

coryneforms Gram-positive cells that are club-shaped and arranged to form V shapes and palisades; resembles the typical microscopic morphology of *Corynebacterium* species.

co-stimulatory molecules Surface proteins expressed by antigen-presenting cells (APCs) when the cell senses molecules that signify an invading microbe or tissue damage; they help dendritic cells activate naive T cells.

Coulter counter An instrument that counts cells in a suspension as they pass through a narrow channel.

counterstain In a differential staining procedure, the stain applied to impart a contrasting color to bacteria that do not retain the primary stain.

covalent bond Strong chemical bond formed by the sharing of electrons between atoms.

CRISPR system Mechanism by which bacterial cells maintain a historical record of phage infections, and thereby become immune to subsequent infections by the same phages; the system also protects against other types of foreign DNA.

critical instruments Medical instruments such as needles and scalpels that come into direct contact with body tissue.

cross-contamination Transfer of pathogens from one item to another.

cross-sectional study Study that surveys a range of people to determine the prevalence of characteristics including disease, risk factors associated with disease, or previous exposure to a disease-causing agent.

croup Acute obstruction of the larynx occurring mainly in infants and young children, often resulting from respiratory syncytial or other viral infections.

crown gall tumor A plant tumor caused by *Agrobacterium tumefaciens*.

cryo-electron microscopy (cryo-EM) A method of electron microscopy in which specimens are rapidly frozen, lessening damage to the sample.

CSF Abbreviation for colony-stimulating factor.

curd Coagulated milk proteins, produced during cheese-making.

cyanobacteria Group of Gram-negative oxygenic phototrophs genetically related to chloroplasts.

cyclic AMP-activating protein (CAP) Protein that binds to cAMP to promote gene transcription.

cyclic photophosphorylation Type of photophosphorylation in which electrons are returned directly to the chlorophyll; used to synthesize ATP without generating reducing power.

cyst Dormant resting protozoan cell characterized by a thickened cell wall.

cysticercus (plural: cysticerci) Cystlike larval form of tapeworms.

cystitis Inflammation of the urinary bladder.

cytochromes Proteins that carry electrons, usually as members of electron transport chains.

cytokine receptor Type of surface receptor that binds a cytokine.

cytokines Small regulatory proteins that cells produce to affect the behavior of other cells.

cytopathic effect Observable change in a cell *in vitro* caused by a viral action such as cell lysis.

cytoplasm Contents of a cell, excluding the nucleus.

cytoplasmic membrane Thin, lipid bilayer that surrounds the cytoplasm and defines the cell boundary.

cytoskeleton The dynamic filamentous network that provides structure and shape to eukaryotic cells.

cytotoxic T cells Type of lymphocyte programmed to destroy infected or cancerous "self" cells.

cytotoxin Toxin that damages a variety of different cell types.

---

danger-associated molecular pattern (DAMP) Molecule that characterizes damaged tissues.

dark-field microscopy Type of microscope that directs light toward the specimen at an angle, so that only light scattered by the specimen enters the objective lens.

dark reactions Process of carbon fixation in photosynthetic organisms; also called light-independent reactions.

dark repair Enzymes of DNA repair that do not depend on visible light.

dead-end host A host from which a parasite typically is not passed to another host, and therefore the parasite cannot complete its life cycle.

dead zone A region lacking fish and other aquatic life, usually due to insufficient dissolved O_2.

death phase In a bacterial growth curve, stage in which the number of viable cells decreases at an exponential rate.

decarboxylation Removal of carbon dioxide from a molecule.

decimal reduction time (D value) Time required for 90% of the organisms to be killed under specific conditions.

decolorizing agent In a differential staining procedure, the chemical used to remove the primary stain.

decomposers Organisms that digest the remains of primary producers and consumers.

decontamination Treatment to reduce the number of pathogens to a level considered safe.

definitive host The host in which a parasite sexually reproduces.

defensins Short antimicrobial peptides produced naturally by a variety of eukaryotic cells to fight infections.

degranulation Release of mediators from a cell's granules, such as histamine is released from mast cells.

dehydration synthesis Chemical reaction in which H_2O is removed with the result that two molecules are joined together.

dehydrogenation Oxidation reaction in which both an electron and an accompanying proton are removed.

delayed-type hypersensitivity Hypersensitivity caused by cytokines released from sensitized T lymphocytes.

denaturation (1) Disruption of the three-dimensional structure of a protein molecule. (2) The separation of the complementary strands of DNA.

dendritic cells Antigen-presenting cells that play an essential role in the activation of naive T cells.

denitrification Bacterial conversion of nitrate to gaseous nitrogen by anaerobic respiration.

dental caries Tooth decay.

dental plaque A biofilm on teeth.

deoxyribonucleic acid (DNA) Macromolecule in the cell that carries genetic information.

deoxyribose A 5-carbon sugar molecule found in DNA.

depth filter Filter with complex passages that allow fluid to pass through while retaining microbes.

dermatophytes Fungi that live on the skin and can be responsible for disease of the hair, nails, and skin.

dermis The layer of skin that underlies the epidermis.

descriptive study Study that characterizes a disease outbreak by determining the attributes of the persons affected as well as the place and time.

desensitization Form of therapy for certain IgE-mediated allergies that involves injecting extremely small but increasing amounts of antigen regularly over a period of months, directing the response from IgE to IgG.

detritus Fresh or partially degraded organic matter used as a food source by decomposers.

diapedesis Movement of leukocytes from blood vessels into tissues in response to a chemotactic stimulus during inflammation.

diatomaceous earth Sedimentary soil composed largely of the remains of diatoms; contains large amounts of silicon.

diauxic growth Two-step growth frequency observed when bacteria are growing in media containing two carbon sources.

diazotroph Organism that can fix nitrogen.

dichotomous key Flowchart of tests used for identifying an organism; each test gives either a positive or negative result.

dideoxy chain termination method A technique used to determine the nucleotide sequence of a strand of DNA; in the procedure, a small amount of chain terminator (a dideoxynucleotide) is added to an *in vitro* synthesis reaction.

dideoxynucleotide (ddNTP) Nucleotide that lacks the 3′ OH group, the portion required for the addition of subsequent nucleotides during DNA synthesis.

differential interference contrast (DIC) microscope Type of microscope that has a device for separating light into two beams that pass through the specimen and then recombine, resulting in the three-dimensional appearance of material in the specimen.

differential media Culture media that contain certain ingredients such as sugars in combination with pH indicators; used to distinguish among organisms based on their metabolic traits.

differential staining Type of staining procedure used to distinguish one group of bacteria from another by taking advantage of the fact that certain bacteria have distinctly different components.

differentiate In cell development, a change in a cell associated with the acquisition of distinct morphological and functional properties.

diffusion Movement of substances from a region of high concentration to a region of low concentration.

digesting Treating with an enzyme to generate degradation products.

diluent Sterile solution used to make dilutions.

dimorphic Able to assume two forms, as the yeast and mold forms of pathogenic fungi.

diphtheroids Club-shaped Gram-positive cells that resemble the typical microscopic morphology of *Corynebacterium diphtheriae*.

diplococci Cocci that typically occur in pairs.

diplomonads A group of flagellated protozoa that lack both mitochondria and hydrogenosomes.

direct contact In epidemiology, physical contact that can transfer microbes.

direct microscopic count Method of determining the number of microbial cells in a measured volume of liquid by counting them microscopically using a special glass slide.

direct selection Technique of selecting mutants by plating organisms on a medium on which the desired mutants but not the parent will grow.

directly observed therapy short-course (DOTS) Method used to ensure that patients comply with their antimicrobial therapy; healthcare workers visit patients and watch them take their medications.

disaccharide Carbohydrate molecule consisting of two monosaccharide molecules.

disease Abnormal condition that results in altered structure or function noticeable by physical examination or laboratory tests.

disinfectant A chemical used to destroy many microorganisms and viruses.

disinfection Process of reducing or eliminating pathogenic microorganisms or viruses in or on a material so that they are no longer a hazard.

disinfection by-products (DBPs) Compounds formed when chlorine or other disinfectants react with naturally occurring chemicals in water.

disseminated gonococcal infection *Neisseria gonorrhoeae* infections that spread through the bloodstream to other parts of the body.

disseminated intravascular coagulation Devastating condition in which clots form in small blood vessels, leading to failure of vital organs.

division Taxonomic rank that groups similar classes; also called a phylum. A collection of similar divisions makes up a kingdom.

DNA Abbreviation for deoxyribonucleic acid.

DNA-based vaccine Vaccine composed of segments of naked DNA that can be introduced directly into muscle tissue; the host tissue expresses the DNA for a short time, producing the antigens encoded by the DNA.

DNA cloning Procedure by which DNA is inserted into a vector to form a recombinant molecule that can then replicate when introduced into cells.

DNA fingerprinting (See *DNA typing*.)

DNA gyrase Enzyme that helps relieve the tension in DNA caused by the unwinding of the two strands of the DNA helix during DNA replication.

DNA library Collection of cloned molecules that together encompass the entire genome of an organism of interest; each clone can be viewed as one "book" of the total genetic information of the organism of interest.

DNA ligase Enzyme that forms covalent bonds between adjacent fragments of DNA.

DNA-mediated transformation Process of horizontal gene transfer in which DNA is transferred as a "naked" molecule.

DNA microarray A solid support that contains a fixed pattern of numerous different single-stranded nucleic acid fragments of known sequences.

DNA polymerases Enzymes that synthesize DNA; they use one strand as a template to make the complementary strand.

DNA probe A piece of DNA, labeled in some manner, that can hybridize to a certain nucleotide sequence as a means to detect that sequence.

DNA replication Duplication of a DNA molecule.

DNA sequencing Determining the sequence (order) of nucleotide bases in a strand of DNA.

DNA typing The use of characteristic patterns in the nucleotide sequence of DNA to match a specimen to a probable source.

domain (1) Level of taxonomic classification above the kingdom level; there are three domains—*Bacteria*, *Archaea*, and *Eucarya*. (2) Distinct parts of proteins associated with certain functions.

donor In horizontal gene transfer, the cell from which the DNA originates.

DOTS Abbreviation for directly observed therapy short-course.

double-blind Type of study where neither the physicians nor the patients know who is receiving the actual treatment.

doubling time Time it takes for the number of cells in a population to double; the generation time.

droplet nuclei Small airborne particles that can remain suspended indefinitely in the presence of even slight air currents.

droplet transmission Transmission of infectious agents through inhalation of respiratory droplets.

Durham tube Small inverted tube placed in a broth of sugar-containing media that is used to detect gas production by a microorganism.

D value Abbreviation for the decimal reduction time.

dysentery Condition characterized by crampy abdominal pain and bloody diarrhea.

ecological niche The role that an organism plays in a particular ecosystem.

ecophysiology The study of the physiological mechanisms organisms use to thrive in a given environment.

ecosystem An environment and the organisms that inhabit it.

ectopic pregnancy Pregnancy in which the embryo develops outside the uterus, usually in a fallopian tube.

eczema Condition characterized by a blistery skin rash, with weeping of fluid and formation of crusts, usually due to an allergy.

edema Swelling of tissues caused by accumulation of fluid.

effector Any molecule that causes an effect.

effector T cell A descendent of an activated T cell that has become armed with specific protective traits.

effluent The liquid portion of treated wastewater.

efflux pump A transporter that moves molecules out of the cell.

electrochemical gradient A separation of charged ions across the membrane.

electromagnetic radiation Form of energy that travels in waves and has no mass; examples include visible and UV light rays and X rays.

electron Negatively charged component of an atom.

electron carriers Molecules cells use to shuttle electrons; they readily accept and then donate electrons.

electron microscope Microscope that uses electrons instead of light and can magnify images in excess of $100,000\times$.

electron transport chain Group of membrane-embedded electron carriers that pass electrons from one to another, and, in the process, move protons across the membrane to create a proton motive force.

electrophoresis Technique that uses an electric current to separate either DNA fragments or proteins.

element A substance that is composed of a single type of atom.

elementary body Small dense-appearing infectious form of *Chlamydia* species that is released upon death and rupture of the host cell.

ELISA Abbreviation for enzyme-linked immunosorbent assay.

emerging disease A diseases that is increasing in incidence or geographic range.

encephalitis Inflammation of the brain.

endemic Constantly present in a population.

endergonic Chemical reaction that requires a net input of energy because the products have more free energy than the starting compounds.

endocarditis Inflammation of the heart valves or lining of the heart chambers.

endocytosis Process through which cells take up particles by enclosing them in a vesicle pinched off from the cell membrane.

endoflagella Characteristic structures of motility in spirochetes.

endogenous antigen An antigen produced within a given host cell.

endogenous pyrogen Fever-inducing substance (such as cytokines) made by the body.

endolysosome Membrane-bound vacuole formed when an endosome fuses with a lysosome.

endonuclease An enzyme that cleaves bonds internally in the backbone of DNA.

endoplasmic reticulum (ER) Organelle of eukaryotes where macromolecules destined for the external environment of other organelles are synthesized.

endosome Vesicle formed when a cell takes up material from the surrounding environment using the process of endocytosis.

endospore A kind of resting cell, characteristic of species in a limited number of bacterial genera; highly resistant to heat, radiation, and disinfectants.

endospore stain A staining method used to observe endospores.

endosymbiont Microorganism that resides within another cell, providing a benefit to the host cell.

endosymbiont theory Theory that the ancestors of mitochondria and chloroplasts were bacteria that had been residing within other cells in a mutually beneficial partnership.

endothelial cell Cell type that lines the blood and lymph vessels.

endotoxic shock Septic shock that occurs as a result of endotoxin (lipopolysaccharide) circulating in the bloodstream.

endotoxin Lipopolysaccharide, a component of the outer membrane of Gram-negative cells that can cause symptoms such as fever and shock; lipid A is responsible for the effects of endotoxin.

end product inhibition Inhibition of gene activity by the end product of a biosynthetic pathway.

energy The capacity to do work.

energy source Compound that a cell oxidizes to harvest energy; also called an electron donor.

enrichment culture Culture method that provides conditions to enhance the growth of one particular species in a mixed population.

enteric fever Typhoid fever or other systemic disease caused by related organisms.

enterics A common name for members of the family *Enterobacteriaceae*.

enterobacteria (See *enterics*.)

enterotoxin Poisonous substance, usually of bacterial origin, that causes diarrhea and vomiting.

enveloped viruses Viruses that have a lipid bilayer surrounding their nucleocapsid.

enzyme A protein that functions as a catalyst, speeding up a biological reaction.

enzyme-linked immunosorbent assay (ELISA) Technique used for detecting and quantifying specific antigens or antibodies by using an antibody labeled with an enzyme.

enzyme-substrate complex Transient form of an enzyme bound to its substrate as the enzyme converts the substrate into a product.

eosinophil A type of white blood cell; thought to be primarily important in expelling parasitic worms from the body.

epidemic A disease or other occurrence for which the incidence is higher than expected within a region or population.

epidemiology The study of factors influencing the frequency and distribution of diseases.

epidermis The outermost layer of skin.

epithelial cell Cell type that lines the surfaces of the body.

epitope Region of an antigen recognized by antibodies and antigen receptors on lymphocytes.

ergosterol Sterol found in fungal cell membranes; the target of many antifungal drugs.

ergot Poisonous substance produced by the fungus that causes rye smut.

erythrocytes Red blood cells.

ethambutol Antimycobacterial drug that inhibits enzymes required for synthesis of mycobacterial cell wall components.

ethidium bromide Mutagenic dye that binds to nucleic acid by intercalating between the nucleobases; ethidium bromide-stained DNA is fluorescent when viewed with UV light.

Eucarya Name of the domain comprising eukaryotic organisms.

eukaryote Organism composed of one or more eukaryotic cells.

eukaryotic cell Complex cell type differing from a prokaryotic cell mainly in having a nuclear membrane surrounding its chromosomes.

eutrophic A nutrient-rich environment supporting the excessive growth of algae and other autotrophs.

evolutionary chronometer A molecule such as rRNA or DNA that can be used to measure the time elapsed since two organisms diverged from a common ancestor.

exanthem A skin rash.

excision repair Mechanism of DNA repair in which a fragment of single-stranded DNA that contains mismatched bases is removed and replaced.

exergonic Describes a chemical reaction that releases energy because the starting compounds have more free energy than the products.

exfoliatin A bacterial toxin that causes sloughing of the outer epidermis.

exocytosis Process by which eukaryotic cells expel material; membrane-bound vesicles inside the cell fuse with the plasma membrane, releasing their contents to the external medium.

exoenzyme Enzyme that acts outside the cell that produces it.

exoerythrocytic Occurring outside the red blood cells, as the developmental cycle in malaria that occurs in the liver.

exogenous antigen An antigen that originated from outside a given host cell.

exogenous pyrogen Fever-inducing substance (such as bacterial endotoxin) from an external source.

exons Portions of eukaryotic genes that will be transcribed and then translated into proteins; interrupted by introns.

exotoxin Soluble poisonous protein substance secreted by a microorganism or released upon lysis.

experimental study Type of study done to assess the effectiveness of measures to prevent or treat disease.

exponential phase In a bacterial growth curve, stage in which cells multiply exponentially; log phase.

extensively drug-resistant *M. tuberculosis* (XDR-TB) Strains of *M. tuberculosis* resistant to isoniazid and rifampin, plus three or more second-line antimicrobials.

extrachromosomal DNA in a cell that is not part of the chromosome.

extremophiles Organisms that live under extremes of temperature, pH, or other environmental conditions.

extrinsic factors In food microbiology, environmental conditions that influence the rate of microbial growth.

F⁻ cell Recipient bacterial cell in conjugation.

F⁺ cell Donor bacterial cell in conjugation, transfers the F plasmid.

F pilus (sex pilus) A protein appendage required for DNA transfer in the process of conjugation.

F plasmid Plasmid found in donor cells of *Escherichia coli* that codes for the F or sex pilus and makes the cell F⁺.

F′ plasmid An F plasmid that carries chromosomal genes.

Fab (fragment antigen-binding) region Portion of an antibody molecule that binds to the antigen.

facilitated diffusion Transport process that moves compounds across a membrane down a concentration gradient; does not require energy.

facultative Flexible with respect to growth conditions; for example, able to live with or without O_2.

facultative anaerobe Organism that grows best in the presence of oxygen (O_2), but can grow in its absence.

FAD Abbreviation for flavin adenine dinucleotide, an electron carrier.

fallopian tubes The tubes that convey ova from the ovaries to the uterus.

family Taxonomic group between order and genus.

fasciitis Inflammation of the fascia, bands of fibrous tissue that underlie the skin and surround muscle and body organs.

fastidious Exacting; refers to organisms that require growth factors.

fatty acid A molecule consisting of long chains of carbon atoms bonded to hydrogen atoms with an acidic group (—COOH) at one end.

Fc receptor Molecule that binds the Fc region of an antibody.

Fc region Stem portion of an antibody molecule.

fecal coliforms Thermotolerant coliform bacteria.

fecal-oral transmission Transmission of organisms that colonize the intestine by ingestion of fecally contaminated material.

feedback inhibition (See *end product inhibition.*)

fermentation Metabolic process that stops short of oxidizing glucose or other organic compounds completely, using an organic intermediate such as pyruvate or a derivative as a terminal electron acceptor.

fermented food Food intentionally altered during production by encouraging the activity of bacteria, yeast, or mold.

fever An increase in internal body temperature to 37.8°C or higher.

fibrin A fibrous protein involved in blood clotting.

fibrinogen A blood protein that is converted to fibrin at a wound site.

fibronectin Glycoprotein occurring on the surface of cells and also in a circulating form that adheres tightly to medical devices; certain pathogens attach to it to initiate colonization.

fimbria (plural: **fimbriae**) Type of pilus that enables cells to attach to a specific surface.

first-line defenses The barriers that separate and shield the interior of the body from the surrounding environment.

first-line drugs Medications preferred because they are most effective as well as least toxic.

5′ end The end of a nucleotide strand that has a phosphate group attached to the number 5 carbon of the sugar.

flaccid paralysis Weakness or loss of muscle tone.

flagella stain A staining method used to observe flagella.

flagellum (plural: **flagella**) (1) In prokaryotic cells, a long protein appendage composed of subunits of flagellin that provides a mechanism of motility. (2) In eukaryotic cells, a long whiplike structure composed of microtubules in a 9+2 arrangement that provides a mechanism of locomotion.

flavin adenine dinucleotide (FAD) An electron carrier derived from the vitamin riboflavin.

flavoprotein Protein to which a flavin such as FAD is attached.

flow cytometer Instrument that counts cells in a suspension by measuring the scattering of light by individual cells as they pass by a laser.

fluid mosaic model Model that describes the dynamic nature of the cytoplasmic membrane.

fluke Short, non-segmented, bilaterally symmetrical flatworm.

fluorescence-activated cell sorter (FACS) Machine that sorts fluorescently labeled cells in a mixture by passing a stream of cells past photodetectors.

fluorescence *in situ* hybridization (FISH) A procedure that uses a fluorescently labeled probe to detect specific nucleotide sequences within intact cells attached to a microscope slide.

fluorescence microscope Special type of microscope used to observe cells that have been stained or tagged with fluorescent dyes.

fluorescent antibody (FA) test Technique that uses fluorescently labeled antibodies to detect specific antigens on cells attached to a microscope slide.

fluoroquinolones Group of antimicrobial drugs that interferes with nucleic acid synthesis.

folliculitis Inflammation of a hair follicle.

fomites Inanimate objects such as books, tools, or towels that can act as transmitters of pathogenic microorganisms or viruses.

food preservation Treating food to prolong the shelf life.

food spoilage Undesirable biochemical changes in foods.

foodborne infection Disease resulting from ingestion of a pathogen-contaminated food.

foodborne intoxication Disease resulting from ingestion of food that contains a toxin produced by a microorganism.

FoodNet Program that collects data on diarrheal illnesses to determine the burden of foodborne illness in the United States.

frameshift mutation Mutation resulting from the addition or deletion of a number of nucleotides not divisible by three.

free energy Amount of energy that can be gained by breaking the bonds of a compound; does not include the energy that is always lost as heat.

fruiting body (1) In myxobacteria, a complex aggregate of cells, visible to the naked eye, produced when nutrients or water are depleted. (2) In fungi, a specialized spore-producing structure.

fungemia Fungi circulating in the bloodstream.

fungicide Kills fungi; used to describe the effects of some antimicrobial chemicals.

fungistatic Able to inhibit the growth of fungi.

fungus (plural: **fungi**) A non-photosynthetic eukaryotic heterotroph.

furuncle A boil; a localized skin infection that penetrates into the subcutaneous tissue, usually caused by *Staphylococcus aureus*.

GALT Abbreviation for gut-associated lymphoid tissue.

gametes Haploid cells that fuse with other gametes to form the diploid zygote in sexual reproduction.

gametocyte Haploid cell that can be used for sexual reproduction.

ganglion (plural: **ganglia**) A tissue mass near the spinal column composed of clusters of sensory neuron cell bodies.

gas chromatography Technique of separating and identifying gaseous components of a substance.

gastroenteritis Acute inflammation of the stomach and intestines; often applied to the syndrome of nausea, vomiting, diarrhea, and abdominal pain.

gas vesicles Small rigid compartments produced by some aquatic bacteria that provide buoyancy to the cell.

gel electrophoresis Technique that uses an electric current to separate either DNA fragments or proteins according to size by drawing them through a slab of gel.

gene The functional unit of a genome.

gene cloning Procedure by which a gene is inserted into a vector to form a recombinant molecule that can then replicate when introduced into cells.

gene expression Transcribing and then translating the information in DNA to produce the encoded protein.

gene fusion The joining of two genes.

generalized transducing particle Bacteriophage progeny that contains part of a bacterial genome instead of phage DNA due to an error during packaging.

generalized transduction Type of horizontal gene transfer that can occur when a phage carries a random piece of bacterial DNA; the phage acquires that DNA when a packaging error occurs during the assembly of phage particles.

generation time Time it takes for the number of cells in a population to double; doubling time.

genetic code Code the correlates a codon (set of three nucleotides) to one amino acid.

genetic engineering Process of deliberately altering an organism's genetic information by changing its nucleic acid sequences.

genetic recombination The joining together of genes from different organisms.

genetics The study of the function and transfer of genes.

genome Complete set of genetic information in a cell.

genomic island Large segment of DNA acquired from another species through horizontal gene transfer; an example is a pathogenicity island.

genomics Study and analysis of the nucleotide sequence of DNA.

genotype The sequence of nucleotides in the DNA of an organism.

genus (plural: **genera**) Taxonomic category of related organisms, usually containing several species. The first name of an organism in the Binomial System of Nomenclature.

germicide Agent that kills microorganisms and inactivates viruses.

germinate The process by which an endospore or other resting cell becomes a vegetative cell.

giant cell Very large cell with many nuclei, formed by the fusion of many macrophages during a chronic inflammatory response; found in granulomas.

gingivitis Inflammation of the gums.

global control The simultaneous regulation of numerous unrelated genes.

glucans Polysaccharides composed of repeating subunits of glucose; involved in the formation of dental plaque.

glucose-salts Type of chemically defined medium that contains only glucose and certain inorganic salts; supports the growth of wild-type *Escherichia coli*.

glycocalyx Polysaccharide layer that surrounds some cells and generally functions as a mechanism of either protection or attachment.

glycogen Polysaccharide composed of glucose molecules.

glycolipids Lipids that have various sugars attached.

glycolysis Metabolic pathway that oxidizes glucose to pyruvate, generating ATP and reducing power.

glycoproteins Proteins with covalently bonded sugar molecules.

glycosylase An enzyme that removes oxidized guanine from DNA by breaking a bond between deoxyribose and the oxidized nucleobase.

goblet cells Mucus-secreting epithelial cells.

Golgi apparatus Series of membrane-bound flattened sacs within eukaryotic cells that serve as the site where macromolecules synthesized in the endoplasmic reticulum are modified before they are transported to other destinations.

G + C content Percentage of guanine plus cytosine in double-stranded DNA; also called the GC content

Gram-negative bacteria Bacteria that lose the crystal violet in the Gram stain procedure and therefore stain pink; their cell wall is composed of a thin layer of peptidoglycan surrounded by an outer membrane.

Gram-positive bacteria Bacteria that retain the crystal violet stain in the Gram stain procedure and therefore stain purple; their cell wall is composed of a thick layer of peptidoglycan.

Gram stain Staining technique that divides bacteria into one of two groups, Gram-positive or Gram-negative, on the basis of color after staining.

granulation tissue New tissue formed during healing of an injury, consisting of small, red, translucent nodules containing abundant blood vessels.

granulocytes White blood cells characterized by the presence of prominent granules; includes basophils, eosinophils, and neutrophils.

granuloma Collections of lymphocytes and macrophages that accumulate in certain chronic infections; an attempt by the body to wall off and contain persistent organisms and antigens.

green bacteria A group of anoxygenic phototrophs that are green or brownish in color.

griseofulvin Antifungal medication that interferes with the action of tubulin, a necessary factor in nuclear division.

group translocation Type of transport process that chemically alters a molecule during its passage through the cytoplasmic membrane.

growth curve Growth pattern observed when cells are grown in a closed system; consists of five stages—lag phase, log phase (or exponential phase), stationary phase, death phase, and the phase of prolonged decline.

growth factors Compounds that a particular bacterium requires in a growth medium because it cannot synthesize them.

gumma Localized area of chronic inflammation and necrosis in tertiary syphilis.

HAART Abbreviation for highly active antiretroviral therapy.

hairy leukoplakia Whitish patch, usually appearing on the tongue of individuals with severe immunodeficiency, thought to be caused by reactivation of latent Epstein-Barr virus (EBV).

half-life Time it takes for one-half of the original number of molecules of a given compound to be eliminated or degraded.

halophile Organism that prefers or requires a high salt (NaCl) medium to grow.

halotolerant Organism that can grow in relatively high salt concentrations, up to 10% NaCl.

haploid Containing only a single set of genes.

hapten Substance to which antibodies bind but that does not elicit the production of those antibodies unless attached to a large carrier molecule.

haustoria Specialized hypha of parasitic fungi that can penetrate plant or animal cell walls.

hay fever Allergic rhinitis; sneezing, runny nose, teary eyes resulting from exposure of a sensitized person to inhaled antigen; an IgE-mediated allergic reaction.

healthcare-associated infections (HAIs) Infections acquired while receiving treatment in a hospital or other healthcare facility.

heavy chain The two higher molecular weight polypeptide chains that make up an antibody molecule; the type of heavy chain determines the class of antibody molecule.

helicase Enzyme that unwinds the DNA helix ahead of the replication fork.

helminth A parasitic worm.

helper T cells Type of lymphocyte programmed to activate B cells and macrophages and also assist other components of adaptive immunity.

hemagglutination Clumping of red blood cells.

hemagglutination inhibition Immunological test used to detect antibodies against certain viruses that naturally cause red blood cells to agglutinate; antibodies that bind the virus inhibit the agglutination.

hemagglutinin A protein important in the virulence of the influenza virus.

hematopoietic stem cells Bone marrow cells that give rise to all blood cells.

hematopoiesis The formation and development of blood cells.

hemolysin Substance that lyses red blood cells.

hemolytic disease of the newborn Disease of the fetus or newborn caused by transplacental passage of maternal antibodies against the baby's red blood cells, resulting in red cell destruction; usually anti-Rhesus (Rh) antibodies are involved and the disease is called Rh disease; also called erythroblastosis fetalis.

hemolytic uremic syndrome (HUS) Serious condition characterized by red cell breakdown and kidney failure; a sequel to infection by certain Shiga toxin–producing strains of *Shigella dysenteriae* and *Escherichia coli.*

hepatitis Inflammation of the liver; various causes, but most commonly the result of viral infection, particularly by the hepatitis viruses.

hepatitis B virus An enveloped DNA virus with an unusual mode of replication involving reverse transcriptase; cause of hepatitis B.

herd immunity Phenomenon that occurs when a critical concentration of immune hosts prevents the spread of an infectious agent.

herpes zoster Another name for shingles; a disease that results from reactivation of the herpesvirus that causes chickenpox.

heterocyst Specialized non-photosynthetic cells of cyanobacteria within which nitrogen fixation occurs.

heteroloboseans A group of protozoa that have amoeboid body forms but also form flagellated cells.

heterophile antibody Antibody that reacts with the red blood cells of another animal.

heterotroph Organism that obtains carbon from an organic compound such as glucose.

Hfr cells (high frequency of recombination cells) Cells that have the F plasmid integrated into their chromosome, allowing them to begin transferring the chromosome by conjugation.

high-efficiency particulate air (HEPA) filters Special filters that remove from air nearly all particles, including microorganisms, that have a diameter 0.3 μm or larger.

high-energy phosphate bond Bond that joins a phosphate group to a compound and is easily broken to release energy that can drive cellular reactions; indicated by the symbol ~.

high-level disinfectant Chemical used to destroy all viruses and vegetative cells, but not endospores.

high-temperature-short-time (HTST) method Most common pasteurization protocol; using this method, milk is pasteurized by holding it at 72°C for 15 seconds.

highly active antiretroviral therapy (HAART) A cocktail of medications that act at different sites during replication of human immunodeficiency virus.

histamine A substance found in basophil and mast cell granules that upon release can cause dilation and increased permeability of blood vessel walls and other effects; a mediator of inflammation.

HIV Abbreviation for human immunodeficiency virus.

HIV disease The illness caused by the human immunodeficiency virus, marked by gradual impairment of the immune system, ending in AIDS.

HIV setpoint Viral load in a person with HIV disease, after the immune system begins to respond to the virus and viral numbers stabilize.

hives Urticaria; an allergic skin reaction characterized by the formation of itchy red swellings.

HLA Abbreviation for human leukocyte antigen.

homologous With respect to DNA, stretches that have similar or identical nucleotide sequences and probably encode similar characteristics.

homologous recombination Genetic recombination between stretches of similar or identical nucleotide sequences.

hops Flowers of the vinelike hop plant; they are added to wort to impart a desirable bitter flavor to beer and contribute antibacterial substances.

horizontal gene transfer Transmission of DNA from one bacterium to another through conjugation, DNA-mediated transformation, or transduction; also called lateral gene transfer.

horizontal transmission Transfer of a pathogen from one person to another through contact, ingestion of food or water, or via a living agent such as an insect.

host Organism in which smaller organisms or viruses live, feed, and reproduce.

host cell In immunology, one of the body's own cells.

host range The range of animals or cell types that a pathogen can infect.

HSV-1 (herpes simplex virus-1) Member of the herpes family of viruses that causes cold sores and other types of infection.

HSV-2 (herpes simplex virus-2) Member of the herpes family of viruses; principal cause of genital herpes.

HTST Abbreviation for high-temperature-short-time pasteurization.

Human Genome Project The now-completed undertaking to sequence the human genome.

Human Microbiome Project Undertaking that uses genomics to study the normal microbiota of the human body.

human leukocyte antigen (HLA) Human MHC molecules.

human immunodeficiency virus (HIV) The causative agent of HIV disease and AIDS.

humoral immunity Immunity due to B cells and an antibody response.

hybridization The annealing of two complementary strands of DNA from different sources to create a hybrid double-stranded molecule.

hybridoma Cell made by fusing a lymphocyte, such as an antibody-producing B cell, with a cancer cell.

hydrogenation Reduction reaction in which an electron and an accompanying proton is added to a molecule.

hydrogen bond Weak attraction between a positively charged hydrogen atom of one compound and a negatively charged atom of another compound; the charges of the two atoms are due to polar covalent bonds.

hydrolysis Chemical reaction in which a molecule is broken down as H_2O is added.

hydrophilic Water loving; soluble in water.

hydrophobic Water fearing; insoluble in water.

hydrothermal vent Undersea geysers that spew out mineral-laden hot water.

hygiene hypothesis The proposal that insufficient exposure to microbes can lead to allergies.

hyperimmune globulin Immunoglobulin prepared from the sera of donors with large amounts of antibodies to certain diseases, such as tetanus; used to prevent or treat the disease.

hypersensitivity Also termed allergy; heightened immune response to antigen.

hyperthermophiles Organisms that have an optimum growth temperature between 70°C and 110°C.

hypha (plural: **hyphae**) Threadlike structure that characterizes the growth of most fungi and some bacterial species.

hypoxic Deficient in O_2.

ID_{50} The number of organisms that, when administered, will cause infection in approximately 50% of hosts.

identification In taxonomy, characterizing an isolate in order to determine the group (taxon) to which it belongs.

IFN Abbreviation for interferon.

IgA proteases Enzymes that degrade IgA; they may have other roles as well.

illness Period of time during which symptoms and signs of disease occur.

immune complex Antibodies bound to antigen.

immune globulin Immunoglobulin G portion of pooled plasma from many donors, containing a wide variety of antibodies; used to provide passive immunity.

immunity Protection against infectious agents and other substances.

immunization Procedure to induce immunity.

immunoassay Tests using immunological reagents such as antigens and antibodies.

immunocompromised A host with weaknesses or defects in the innate or adaptive defenses.

immunodeficiency A state in which the body is unable to produce a normal immune response to antigen.

immunodiffusion tests Precipitation reactions carried out in agarose or other gels.

immunofluorescence Technique used to identify particular antigens microscopically in cells by the binding of a fluorescent antibody to the antigen.

immunogen Antigen that induces an immune response.

immunoglobulin Any of a group of glycoprotein molecules that react specifically with the substance that induced their formation; antibodies.

immunological tolerance (See *tolerance*.)

immunology The study of immunity.

immunosuppression Non-specific suppression of adaptive immune responses.

immunotherapy Techniques used to modify the immune system action for a favorable effect.

impetigo A superficial skin infection characterized by thin-walled vesicles, oozing blisters, and yellow crusts; caused by *Staphylococcus aureus* and *Streptococcus pyogenes*.

inactivated vaccine Vaccine composed of killed bacteria, inactivated virus, or fractions of the agent; the agent in the vaccine is unable to replicate.

inapparent (or **subclinical**) **infections** Infections in which symptoms do not occur or are mild enough to go unnoticed.

incidence rate Number of new cases of a disease within a specific time period in a given population.

inclusion body Microscopically visible structure within a virally infected cell, representing the site at which the virus replicates.

incubation period Interval between entrance of a pathogen into a susceptible host and the onset of illness caused by that pathogen.

index case First identified case of a disease in an epidemic.

indicator organism Microbes commonly found in the intestinal tract whose presence in other environments suggests fecal contamination.

indirect contact Means of transmitting infectious disease via fomites.

indirect selection In microbial genetics, a technique for isolating mutants and identifying organisms unable to grow on a medium on which the parents do grow; often involves replica plating.

induced mutation Mutation that results from the organism being treated with an agent that alters its DNA.

inducer Substance that activates transcription of certain genes.

inducer exclusion A type of carbon catabolite repression that prevents a carbon/energy source from being transported into the cell.

inducible enzyme Enzyme synthesized only under certain environmental conditions.

infection Growth and multiplication of a parasitic organism or virus in or on the body of the host with or without the production of disease.

infectious disease Disease caused by a microbial or viral infection.

infectious dose (ID) Number of microorganisms or viruses sufficient to establish an infection; often expressed as ID_{50} in which 50% of the hosts are infected.

inflammasome Protein complex in macrophages that activates a potent proinflammatory cytokine, thereby initiating an inflammatory response.

inflammation Set of signs and symptoms (swelling, heat, redness, and pain) that characterizes an innate immune response to infection or injury.

inflammatory mediators Chemicals such as histamine that trigger an inflammatory response when released from cells.

inflammatory response Coordinated innate response that leads to inflammation, with the purpose of containing a site of damage, localizing the response, eliminating the invader, and restoring tissue function.

innate immunity Host defenses involving anatomical barriers, sensor systems that recognize patterns associated with microbes or tissue damage, phagocytic cells, and the inflammatory responses.

inorganic molecule A chemical that contains no C—H bonds.

insertion sequence (IS) Short piece of DNA that has the ability to move from one site on a DNA molecule to another; simplest type of transposable element.

insulin-dependent diabetes mellitus (IDDM) Diabetes caused by autoimmune destruction of pancreatic cells by cytotoxic T cells.

integrase Enzyme encoded by HIV or another retrovirus that inserts the DNA copy of the viral genome into a host cell chromosome.

intercalating agents Agents that insert between base pairs in a DNA double helix.

interferons (IFNs) Cytokines that induce cells to resist viral replication.

interleukins (ILs) Cytokines produced by leukocytes.

intermediate filament Component of the eukaryotic cell cytoskeleton.

intermediate-level disinfectant Type of chemical used to destroy all vegetative bacteria including mycobacteria, fungi, and most, but not all, viruses.

intranuclear inclusion body Structure found within the nucleus of cells infected with certain viruses such as cytomegalovirus.

intrinsic factors In food microbiology, the natural characteristics of a food that influence the rate of microbial growth.

intrinsic (innate) resistance Resistance of an organism to an antimicrobial medication due to the inherent characteristics of that type organism.

intron Part of the eukaryotic chromosome that does not code for a protein; removed from the RNA transcript before the mRNA is translated.

in vitro In a test tube or other container as opposed to inside a living plant or animal.

in vivo Inside a living plant or animal as opposed to a test tube or other container.

ion Positively or negatively charged atom or molecule.

ionic bond A chemical bond resulting from the attraction of positively and negatively charged ions.

IS Abbreviation for insertion sequence.

isomer Molecule with the same number and types of atoms as another but differing in its structure.

isotopes Forms of the same chemical element that differ in their number of neutrons.

---

jaundice Yellow color of the skin and eyes caused by buildup of bilirubin in the blood.

Kaposi's sarcoma Tumor arising from blood or lymphatic vessels.

keratin A water-repelling protein found in hair, nails, and the outermost cells of the epidermis.

kinetic energy Energy of motion.

kinetoplastids A group of flagellated protozoa that have a distinct mass in their large, single mitochondrion.

kingdom Taxonomic rank that groups several phyla or divisions; a collection of similar kingdoms makes up a domain.

Kirby-Bauer disc diffusion test Procedure used to determine if a bacterium is susceptible to concentrations of an antimicrobial medication usually present in the bloodstream of an individual receiving the drug.

Koch's postulates The criteria used to determine the cause of an infectious disease by culturing the agent and reproducing the disease.

Koplik spots Lesions of the oral cavity caused by measles virus that resemble a grain of salt on a red base.

---

labeled Tagged with a detectable marker such as a radioactive isotope, a fluorescent dye, or an enzyme.

lac operon Operon that encodes the proteins required for the degradation of lactose; it serves as one of the most important models for studying gene regulation.

lactic acid bacteria Group of Gram-positive bacteria that generates lactic acid as the major end product of their fermentative metabolism.

lactoferrin Iron-binding protein found in leukocytes, saliva, mucus, milk, and other substances; helps defend the body by depriving microorganisms of iron.

lactose Disaccharide consisting of one molecule of glucose and one of galactose.

lacZ' gene Gene used to visually determine whether or not a vector contains a fragment inserted into a multiple cloning site.

lagging strand In DNA replication, the strand that is synthesized as a series of fragments.

lag phase In a bacterial growth curve, stage characterized by extensive macromolecule and ATP synthesis but no increase in cell number.

latent infection Infection in which the infectious agent is present but not active.

latent state The state of a phage when its DNA is integrated into the genome of the host.

latent tuberculosis infection (LTBI) An asymptomatic *Mycobacterium tuberculosis* infection in which the bacterium is held in check by the immune system.

latent viral infection Viral infection in which the virus is silent within the host cell.

LD_{50} The dose that kills approximately 50% of hosts.

leading strand In DNA replication, the DNA strand that is synthesized as a continuous fragment.

leaky Refers to a mutation in which the mutant gene codes for a protein that is partially functional.

lecithin Component of mammalian cell membranes; attacked by the α-toxin of *Clostridium perfringens* and other lecithinases.

lectin pathway Pathway of complement system activation initiated by binding of mannan-binding lectins to microbial cell walls.

leghemoglobin Protein synthesized by leguminous plants that maintains a low O_2 level within a *Rhizobium*-harboring nodule.

lepromatous leprosy Severe form of Hansen's disease that develops when cell-mediated immunity fails to control the growth of *Mycobacterium leprae*.

lethal dose (LD) The dose that causes death; often expressed as LD_{50}, the dose that kills 50% of the test subjects.

leukocidins Substances that kill white blood cells.

leukocytes White blood cells.

L-forms Bacterial variants that have lost the ability to synthesize the peptidoglycan portion of their cell wall.

lichen Symbiotic association of a fungus and either a green alga or a cyanobacterium.

ligand A molecule that specifically binds to a given receptor.

light chains In antibody structure, the two lighter molecular weight polypeptide chains.

light-dependent reactions Processes used by phototrophs to harvest energy from sunlight; the energy-gathering component of photosynthesis.

light-independent reactions Stage of photosynthesis in which the ATP generated in the light-dependent reactions is used to fix carbon; also called dark reactions.

light microscope Microscope that uses visible light to observe objects.

light reactions (See *light-dependent reactions*.)

light repair Process by which bacteria use the energy of visible light to repair UV damage to their DNA.

limiting nutrient Nutrient that limits growth because it is present at the lowest concentration relative to need.

lipid One of a diverse group of organic substances all of which are relatively insoluble in water, but soluble in alcohol, ether, chloroform, or other fat solvents.

lipid A Portion of lipopolysaccharide (LPS) that anchors the molecule in the lipid bilayer of the outer membrane of Gram-negative cells; it plays an important role in the body's ability to recognize the presence of invading bacteria, but is also responsible for the toxic effects of LPS.

lipopolysaccharide (LPS) Molecule formed by bonding of lipid to polysaccharide; a part of the outer membrane of Gram-negative bacteria.

lipoprotein Macromolecule formed by the covalent bonding of lipid to protein.

lipoteichoic acids Component of the Gram-positive cell wall that is linked to the cytoplasmic membrane.

loboseans A group of protozoa that have amoeboid body forms and no flagella.

localized infection Infection limited to one site in or on the body, as a furuncle.

locus Specific location of a gene or nucleotide sequence on a chromosome.

log phase In a bacterial growth curve, stage in which the cells are multiplying exponentially.

low-level disinfectants Type of chemical used to destroy fungi, enveloped viruses, and vegetative bacteria except mycobacteria.

LPS Abbreviation for lipopolysaccharide.

luciferase Enzyme that catalyzes the chemical reactions that produce bioluminescence.

lumen The interior of an organ or organelle.

lymph Clear yellow liquid that contains leukocytes and flows within lymphatic vessels.

lymphadenopathy syndrome (LAS) Enlargement of lymph nodes that often occurs at the end of the period of clinical well-being in HIV disease.

lymphangitis Inflammation of lymphatic vessels.

lymphatic system Collection tissues and organs that bring the population of B cells and T cells into contact with antigens.

lymphatic (lymph) vessels Vessels that carry lymph, which is collected from the fluid that bathes the body's tissues; also called the lymphatics.

lymphatics (See *lymphatic vessels*.)

lymphocyte Leukocyte involved in adaptive immunity; includes B and T cells.

lymphokines Cytokines secreted by lymphocytes.

lyse To burst.

lysogenic conversion Modification of the properties of a cell resulting from expression of phage DNA integrated into a bacterial chromosome.

lysogenic infection Infection by a temperate phage that results in a prophage residing in the host.

lysogens Bacteria that carry a prophage integrated into their chromosome.

lysosome Membrane-bound structure in eukaryotic cells that contains powerful degradative enzymes.

lysozyme Enzyme that degrades the peptidoglycan layer of the bacterial cell wall.

lytic infection Viral infection that causes the host cell to lyse.

lytic phages Bacteriophages that lyse their host.

macroenvironment Overall environment in which an organism lives; opposed to microenvironment.

macrolide A type of antimicrobial medication that interferes with protein synthesis.

macromolecule Very large molecule usually composed of repeating subunits.

macrophages Differentiated phagocytes of the mononuclear phagocyte system that can engulf and destroy microorganisms and other materials.

magnetotaxis Movement by bacterial cells containing magnetite crystals in response to a magnetic field.

major element Chemical elements that make up cell constituents.

major histocompatibility complex (MHC) Cluster of genes coding for key cell surface proteins important in antigen presentation.

malaise Vague feeling of uneasiness or discomfort.

malignant tumor Mass of abnormal cells growing without control, potentially spreading to other parts of the body.

MALT Abbreviation for mucosa-associated lymphoid tissue.

mannan-binding lectins (MBLs) Pattern recognition molecules the body uses to detect polymers of mannose, which are typically found on microbial but not mammalian cells.

mast cells Tissue cells similar in appearance and function to basophils of the blood, with receptors for the Fc portion of IgE; important in the inflammatory response and immediate allergic reactions.

MBC Abbreviation for minimum bactericidal concentration.

M cells Specialized epithelial cells lying over Peyer's patches that collect material in the intestine and transfer it to the lymphoid tissues beneath.

MDR-TB Abbreviation for multidrug-resistant *Mycobacterium tuberculosis*.

mechanical vector Organism such as a fly that physically moves contaminated material from one location to another.

mechanisms of pathogenicity Methods that pathogens use to evade host defenses and cause damage to the host.

medium (plural: **media**) Any material used for growing organisms.

meiosis Process in eukaryotic cells by which the chromosome number is reduced from diploid ($2n$) to haploid ($1n$).

melting Denaturating of double-stranded DNA.

membrane attack complex (MAC) Complex of certain components of the complement system that forms pores in cell membranes, resulting in lysis of the cells.

membrane-damaging toxin Toxin that disrupts plasma membranes of eukaryotic cells.

membrane filter Type of thin microfilter.

membrane filtration Technique used to determine the number of bacterial cells in a liquid sample that has a relatively low number of organisms; concentrates bacteria by filtration before they are plated.

membrane ruffling Characteristic mechanism of engulfment that some bacteria induce by triggering rearrangements of a cell's actin.

memory cells Lymphocytes that persist in the body after an immune response to an antigen.

memory lymphocytes (See *memory cells*.)

memory response (See *secondary response*.)

meninges Membranes covering the brain and spinal cord.

meningitis Inflammation of the meninges.

merozoite Stage in the life cycle of certain protozoa, such as the malaria-causing *Plasmodium* species.

mesophiles Bacteria that grow most rapidly at temperatures between 20°C and 45°C.

messenger RNA (mRNA) Single-stranded RNA that is translated to make protein.

metabolic pathway Series of sequential chemical reactions.

metabolism Sum total of all the enzymatic chemical reactions in a cell.

metabolite Any product of metabolism.

metachromatic granules Polyphosphate granules found in the cytoplasm of some bacteria that appear as different colors when stained with a basic dye.

metagenomics The analysis of the total microbial genomes in a sample.

methanogen Any of a group of *Archaea* that obtains energy by oxidizing hydrogen gas, using CO_2 as a terminal electron acceptor, thereby generating methane (CH_4).

methicillin-resistant *Staphylococcus aureus* (MRSA) Strains of *Staphylococcus aureus* that are resistant to methicillin.

MHC Abbreviation for major histocompatibility complex.

MHC class I molecules Molecules that cells use to present antigen to cytotoxic T cells.

MHC class II molecules Molecules on the surface of antigen-presenting cells (APCs) that present antigen to helper T cells.

MIC Abbreviation for minimum inhibitory concentration.

microaerophiles Organisms that require small amounts of O_2 (2% to 10%) for growth, and are inhibited by higher concentrations.

microarray A solid support that contains a fixed pattern of numerous different single-stranded nucleic acid fragments of known sequences.

microbe General term encompassing microorganisms and viruses.

microbial ecology The study of the relationships of microorganisms to each other and to their environment.

microbial mat A type of microbial community characterized by distinct layers of different groups of microorganisms that together make up a thick, dense, highly organized structure.

microenvironment Environment immediately surrounding an individual microorganism.

microfilter Filter that has a pore size small enough to trap microorganisms but still allow liquids to pass through.

microtiter plate A small tray containing numerous wells, usually 96.

microtubules Cytoskeleton structures of a eukaryotic cell that form mitotic spindles, cilia, and flagella; long hollow cylinders composed of tubulin.

microvillus (plural: microvilli) Tiny cylindrical projections from luminal surfaces of cells such as those lining the intestine; increases surface area of the cell.

minimum bactericidal concentration (MBC) Lowest concentration of a specific antimicrobial medication that kills 99.9% of cells in a culture of a given bacterial strain *in vitro*.

minimum inhibitory concentration (MIC) Lowest concentration of a specific antimicrobial medication that prevents the growth of a given microbial strain *in vitro*.

minus (–) strand (1) The DNA strand used as a template for RNA synthesis. (2) The complement to the plus (or sense) strand of RNA. Also called the negative strand or antisense strand.

missense mutation A mutation that changes the amino acid encoded by DNA.

mismatch repair Repair mechanism in which an enzyme cuts the DNA near a mismatched nucleobase, resulting in the removal and replacement of a short stretch of nucleotides.

mitochondrion Organelle in eukaryotic cells in which the majority of ATP synthesis occurs.

mitogen Substance that induces mitosis; causes proliferation of cells.

mitosis Nuclear division process in eukaryotic cells that ensures the daughter cells receive the same number of chromosomes as the original parent.

MMWR Abbreviation for *Morbidity and Mortality Weekly Report*, published by the Centers for Disease Control and Prevention (CDC).

mobile gene pool DNA sequences found in only some strains of a given species; includes plasmids, transposons, phages, and genomic islands.

molarity The number of moles of a solute in 1 liter of solution.

mold A filamentous fungus.

mole Amount of a chemical in grams that contains 6.022×10^{23} molecules.

molecular Koch's postulates The criteria established to study a microbe's virulence factors using genetic and molecular techniques.

molecule Chemical consisting of two or more atoms held together by chemical bonds.

monocistronic RNA transcript that carries one gene.

monoclonal antibodies Antibody molecules with a single specificity produced *in vitro* by lymphocytes that have been fused with a type of malignant myeloma cell.

monocytes Mononuclear phagocytes of the blood; part of the mononuclear phagocyte system of professional phagocytes.

monomer Subunit of a polymer.

mononuclear phagocytes Monocytes and macrophages.

monosaccharide A sugar; a simple carbohydrate generally having the formula $C_nH_{2n}O_n$, where n can vary in number from three to eight.

morbidity Illness; most often expressed as the rate of illness in a given population at risk.

mordant Substance that increases the affinity of cellular components for a dye.

morphology Form or shape of a particular organism or structure.

mortality Death; most often expressed as a rate of death in a given population at risk.

most probable number (MPN) method Statistical estimate of cell numbers; a sample is successively diluted to determine the point at which subsequent dilutions receive no cells.

M protein A protein in the cell walls of Group A streptococci that is associated with virulence.

mRNA Messenger RNA.

MRSA Abbreviation for methicillin-resistant *Staphylococcus aureus*.

mucociliary escalator Moving layer of mucus and cilia lining the respiratory tract that traps bacteria and other particles and moves them into the throat.

mucosa (See *mucous membrane*.)

mucosa-associated lymphoid tissue (MALT) Lymphoid tissue present in the mucosa of the respiratory, gastrointestinal, and genitourinary tracts.

mucosal immunity Immune response that protects the mucous membranes; typically involves secretory IgA.

mucous membrane Epithelial barrier that is coated with mucus.

multidrug-resistant *Mycobacterium tuberculosis* (MDR-TB) Strains of *Mycobacterium tuberculosis* that are resistant to isoniazid and rifampin.

multilocus sequence typing A method of distinguishing strains that relies on determining the nucleotide sequence of select DNA regions.

multiphoton microscopy A type of scanning laser microscopy that is less damaging to cells than confocal microscopy because it uses a lower- energy light.

multiple-cloning site Short stretch of DNA that contains several unique restriction enzyme recognition sites into which foreign DNA can be cloned.

mushroom Filamentous multicelled fungus with macroscopic fruiting bodies.

mutagen Any agent that increases the frequency at which DNA is altered (mutated).

mutant Organism that has a changed nucleotide sequence in its DNA.

mutation A change in the nucleotide sequence of a cell's DNA that is then passed on to daughter cells.

mutualism A symbiotic association in which both partners benefit.

myasthenia gravis An autoimmune disease characterized by muscle weakness, caused by autoantibodies.

mycelium (plural: mycelia) Tangled, matlike mass of fungal hyphae.

***Mycobacterium avium* complex (MAC)** Group of genetically related bacteria belonging to the genus *Mycobacterium* that causes disease in people with AIDS.

mycology The study of fungi.

mycorrhiza Symbiotic relationship between certain fungi and the roots of plants.

mycosis (plural: mycoses) Disease caused by a fungus.

myxobacteria Group of Gram-negative bacteria that congregate to form complex structures called fruiting bodies.

NAD/NADH Abbreviations for the oxidized/reduced forms of nicotinamide adenine dinucleotide, an electron carrier.

NADP/NADPH Abbreviations for the oxidized/ reduced forms of nicotinamide adenine dinucleotide phosphate, an electron carrier

naked virus Type of virus that does not have a lipid envelope.

narrow host range plasmid Plasmid that only replicates in one or a few closely related species of bacteria.

narrow-spectrum antimicrobials Antimicrobial medications that inhibit or kill a limited range of bacteria.

National Molecular Subtyping Network for Foodborne Disease Surveillance (See *PulseNet*.)

natural killer cell (See *NK cell*.)

natural selection Selection by the environment of those cells best able to grow in that environment.

naturally acquired immunity Immunity acquired through natural means such as exposure to a disease-causing agent, breastfeeding, or transfer of IgG to a fetus *in utero*.

necrotic Dead; refers to dead cells or tissues in contact with living cells, as necrotic tissue in wounds.

necrotizing fasciitis Severe soft-tissue infection that destroys subcutaneous fatty tissues, fascia, and sometimes muscle; commonly called "flesh-eating disease."

negative staining Staining technique that uses an acidic dye to stain the background against which colorless cells can be seen.

negative (–) strand (See *minus [–] strand*.)

Negri body Viral inclusion body characteristic of rabies.

nematodes Roundworms.

neuron Nerve cell.

neurotoxin Toxin that damages the nervous system.

neurotransmitter Any of a group of substances released from the ends of nerve cells when they are stimulated; they cross to the adjacent cell and cause it to be excited or inhibited.

neutron Uncharged component of an atom found in the nucleus.

neutrophiles Organisms that can live and multiply within the range of pH 5 (acidic) to pH 8 (basic) and have a pH optimum near neutral (pH 7).

neutrophils (See *polymorphonuclear neutrophils*.)

nitrification Conversion of ammonia (NH_3) to nitrate (NO_3^-).

nitrifiers Group of Gram-negative bacteria that obtain energy by oxidizing inorganic nitrogen compounds such as ammonia or nitrate.

nitrogen fixation Conversion of nitrogen gas to ammonia.

NOD-like receptors A group of pattern recognition receptors located within cells.

nodules Specialized organs on legume roots within which rhizobia fix nitrogen.

NK cell Type of lymphocyte that induces apoptosis in cells to which antibody has bound or that lack MHC class I molecules on the surface and are stressed.

nomenclature System of assigning names to organisms; a component of taxonomy.

non-communicable diseases Disease that typically cannot be transmitted from one individual to another.

non-competitive inhibition Type of enzyme inhibition that results from a molecule binding to the enzyme at a site other than the active site.

non-critical instruments Medical instruments and surfaces such as stethoscopes and countertops that only come into contact with unbroken skin.

non-cyclic photophosphorylation Type of photophosphorylation in which high-energy electrons are drawn off to generate reducing power; electrons must still be returned to chlorophyll, but they come from a source such as water.

non-homologous recombination The breaking and rejoining of two DNA molecules that do not share similar nucleotide sequences at the site of recombination.

non-polar covalent bond Bond formed by sharing electrons between atoms that have equal attraction for the electrons.

nonsense mutation A mutation that generates a stop codon, resulting in a shortened protein.

normal microbiota (or **normal flora**) That group of microorganisms that colonizes the body surfaces but does not usually cause disease.

nosocomial infection Infection acquired during hospitalization.

notifiable diseases Group of diseases that are reported to the CDC by individual states; typically these diseases are of relatively high incidence or otherwise a potential danger to public health.

nuclear envelope Double membrane that separates the nucleus from the cytoplasm in eukaryotic cells.

nucleic acid amplification techniques (NAATs) *In vitro* methods that increase the amount of a particular region of DNA; PCR is an example.

nucleic acid hybridization (See *hybridization*.)

nucleic acids Ribonucleic acid (RNA) and deoxyribonucleic acid (DNA).

nucleobase The purine or pyrimidine ring structure found in nucleotides; also called a base.

nucleocapsid Viral nucleic acid and its protein coat.

nucleoid Region of a prokaryotic cell containing the DNA.

nucleolus Region within the nucleus where ribosomal RNAs are synthesized.

nucleotides Basic subunits of RNA or DNA consisting of a purine or pyrimidine covalently bonded to ribose or deoxyribose, which is covalently bound to a phosphate molecule.

nucleus (1) Membrane-bound organelle in a eukaryotic cell that contains chromosomes and the nucleolus. (2) The portion of an atom where neutrons and protons are located.

O antigen Antigenic polysaccharide portion of lipopolysaccharide, the molecule that makes up the outer layer of the outer membrane of Gram-negative bacteria.

objective lens The lens or series of lenses of a compound microscope that is closest to the specimen.

obligate aerobes Organisms that require O_2 for growth.

obligate anaerobes Organisms that cannot multiply if O_2 is present; they are often killed by traces of O_2 because of its toxic derivatives.

obligate fermenters Organisms that can grow in the presence of O_2 but never use it as a terminal electron acceptor; also called aerotolerant anaerobes.

obligate intracellular parasites Organisms that grow only inside living cells.

occlusion bodies Masses of viruses inside or outside cells.

ocular lens Lens of a compound microscope that is closest to the eye.

oil A liquid fat.

Okazaki fragment Nucleic acid fragment synthesized as a result of the discontinuous replication of the lagging strand of DNA.

oligonucleotide Short chain of nucleotides.

oligosaccharide Short chain of monosaccharide subunits joined together by covalent bonds; shorter than a polysaccharide.

oligotrophic Nutrient-poor.

oligotrophs Organisms that can grow in a nutrient-poor environment.

oncogene Gene whose activity is involved in turning a normal cell into a cancer cell.

open reading frame (ORF) Stretch of DNA, generally longer than 300 base pairs, that has a reading frame beginning with a start codon and ending with a stop codon; the region likely encodes a protein.

open system Method used to maintain cells in a state of constant growth by continuously adding nutrients and removing waste products; also called a continuous culture.

operator Region located immediately downstream of a promoter to which a repressor can bind; binding of the repressor to the operator prevents RNA polymerase from progressing past that region, thereby blocking transcription.

operon Group of linked genes whose expression is controlled as a single unit.

ophthalmia neonatorum Eye infection of newborns usually caused by *Neisseria gonorrhoeae* or *Chlamydia trachomatis*, acquired from infected mothers during the birth process.

opine Unusual amino acid derivative encoded by a portion of the Ti plasmid of *Agrobacterium tumefaciens*.

opportunistic pathogen Organism that causes disease only in hosts with impaired defense mechanisms or when introduced into an unusual location; also called an opportunist.

opsonin Molecule such as the complement system component C3b and certain antibody classes that binds to invading particles, making it easier for phagocytes to engulf them.

opsonization Enhanced phagocytosis caused by coating of a particle with an opsonin.

optical isomer (or **stereoisomer**) Mirror image of a compound.

optimal proportion Relative proportions of antigen and antibody at which both are fully incorporated in a precipitate.

optimum growth temperature Temperature at which a microorganism multiplies most rapidly.

oral tolerance Decreased reactivity of immune cells resulting from feeding an antigen.

oral rehydration therapy A treatment used to replace fluid and electrolytes lost due to diarrheal disease.

order Taxonomic classification between class and family.

organ A structure composed of different tissues coordinated to perform a specific function.

organelle A cell structure that performs a specific function.

organic compound A compound in which a carbon atom is covalently bonded to a hydrogen atom.

origin of replication Distinct region of a DNA molecule at which replication is initiated.

origin of transfer Short stretch of nucleotides, a part of which is transferred first when a plasmid is transferred to a recipient cell; necessary for plasmid transfer.

osmosis Movement of water across a membrane from a dilute solution to a more concentrated solution.

otitis media Inflammation of the middle ear due to infection.

outbreak Cluster of cases occurring during a brief time interval and affecting a specific population; may herald the onset of an epidemic.

outer membrane (1) In prokaryotic cells, the unique lipid bilayer of Gram-negative cells that surrounds the peptidoglycan layer. (2) In eukaryotic cells, the membrane on the cytoplasmic side of organelles that have double membranes.

oxazolidinones Group of antimicrobial drugs that interferes with protein synthesis.

oxidase test Rapid biochemical test used to detect the activity of cytochrome *c* oxidase.

oxidation-reduction reactions Chemical reactions in which one or more electrons is transferred from one molecule to another.

oxidative phosphorylation Synthesis of ATP using the energy of a proton motive force created by harvesting chemical energy.

oxidized Refers to loss of electrons.

oxygenic phototrophs Phototrophic organisms that produce O_2.

pandemic A worldwide epidemic.

pandemic influenza Widespread influenza epidemic due to antigenic shift in the virus.

papilloma A benign skin tumor; wart.

para-aminobenzoic acid (PABA) Intermediate in the pathway for folate synthesis in bacteria; sulfa drugs have a similar structure to PABA.

parasite An organism or virus that benefits at the expense of a host.

parasitism Living at the expense of a host.

parabasalids A group of flagellated protozoa that lack mitochondria but have hydrogenosomes.

parent strain Refers to the original strain of a bacterium used in an experiment; term is often used in place of wild-type strain.

paroxysm A sudden increase in symptoms.

passive diffusion Process in which molecules flow freely into and out of a cell.

passive immunity Protective immunity resulting from the transfer of antibody-containing serum produced by other individuals or animals.

pasteurization Process of heating food or other substances under controlled conditions to kill pathogens and reduce the total number of microorganisms without damaging the substance.

pathogen Disease-causing organism or virus.

pathogen-associated molecular pattern (PAMP) Molecule that characterizes invading microbes.

pathogenesis Process by which disease develops.

pathogenicity islands Stretches of DNA in bacteria that code for virulence factors and appear to have been acquired from other bacteria.

pattern recognition Method by which the innate immune system recognizes invading microbes; the system uses receptors and other molecules that bind pathogen-associated molecular patterns.

pattern recognition receptors Receptors that bind lipopolysaccharide, peptidoglycan, and other molecular patterns associated with microbes.

PCR product The target fragment amplified exponentially in a PCR reaction.

pediculosis Infestation with lice.

peliosis hepatis Serious condition characterized by formation of blood-filled cysts in the liver, caused by *Bartonella henselae*; usually a complication of AIDS or other severe immunodeficiency.

pelvic inflammatory disease Infection of the fallopian tubes, uterus, or ovaries.

penicillin Antibiotic that interferes with the synthesis of the peptidoglycan portion of bacterial cell walls.

penicillin-binding proteins (PBPs) Target of β-lactam antimicrobial drugs; their role in bacteria is peptidoglycan synthesis.

penicillin enrichment Method for increasing the relative proportion of auxotrophic mutants in a population by using penicillin to kill growing prototrophic cells.

pentose phosphate pathway Metabolic pathway that starts the degradation of glucose, generating reducing power in the form of NADPH, and two precursor metabolites.

peptide bond Covalent bond formed between the —COOH group of one amino acid and the —NH_2 group of another amino acid; characteristic of bonds in proteins.

peptide vaccine Vaccine composed of key antigenic peptides from disease-causing microbes. Vaccines of this type are still in developmental stages.

peptidoglycan Macromolecule found only in bacteria that provides strength to the bacterial cell wall. The basic structure is a chain of two alternating subunits, *N*-acetylmuramic acid (NAM) and *N*-acetylglucosamine (NAG), cross-linked by peptide bridges.

peptone Common component of bacteriological media; consists of proteins originating from any of a variety of sources that have been hydrolyzed to amino acids and short peptides by treatment with enzymes, acids, or alkali.

perforin Molecule produced by T_C cells and NK cells to destroy target cells.

peripheral nervous system Division of the nervous system that carries information to and from the central nervous system (CNS).

periplasm Gel that fills the region between the outer membrane and the cytoplasmic membrane in Gram-negative bacteria.

peristalsis The rhythmic contractions of the intestinal tract that propel food and liquid.

peritrichous flagella Distribution of flagella all around a cell.

peroxidase enzymes Enzymes found in neutrophil granules, saliva, and milk that together with hydrogen peroxide and halide ions make up an effective antimicrobial system.

peroxisome Organelle that uses hydrogen peroxide and superoxide to degrade substances.

persistent infection Infection in which the causative agent remains in the body for long periods of time, often without causing symptoms of disease.

petechia (plural: **petechiae**) Small purplish spot on the skin or mucous membrane caused by hemorrhage.

Petri dish Two-part dish of glass or plastic often used to contain medium solidified with agar, on which bacteria are grown.

Peyer's patches Collections of lymphoid cells in the gastrointestinal tract; part of the mucosa-associated lymphoid tissue (MALT).

pH Scale of 0 to 14 that expresses the acidity or alkalinity of a solution.

phage Shortened term for bacteriophage.

phage induction Process by which phage DNA is excised from bacterial chromosomal DNA.

phagocytes Cells that specialize in engulfing and digesting microbes and cell debris.

phagocytosis (v. **phagocytize**) The process by which certain cells ingest particulate matter by surrounding and enveloping those materials, bringing them into the cell in a phagosome.

phagolysosome Membrane-bound vacuole generated when a phagosome fuses with lysosomes.

phagosome Membrane-bound vacuole that contains the material engulfed by a phagocyte.

pharyngitis Inflammation of the throat.

phase-contrast microscope Type of light microscope that uses special optical devices to amplify the difference in the refractive index of a cell and the surrounding medium, increasing the contrast of the image.

phase of prolonged decline Final stage of the growth curve; most cells die during this phase, but a few are able to grow.

phase variation The reversible and random alteration of expression of certain bacterial structures such as fimbriae by switching on and off the genes that encode those structures.

phenotype The properties of a cell determined by the expression of the genotype.

phospholipase Membrane-damaging toxin that enzymatically removes the polar head group on phospholipids.

phospholipid Lipid that has a phosphate molecule as part of its structure.

photoautotrophs Organisms that use light as the energy source and CO_2 as the major carbon source.

photoheterotrophs Organisms that use light as the energy source and organic compounds as the carbon source.

photooxidation Chemical reaction occurring as a result of absorption of light energy in the presence of O_2.

photophosphorylation Synthesis of ATP using the energy of a proton motive force created by harvesting radiant energy.

photoreactivation (or **light repair**) Using the energy of light to break the covalent bonds joining thymine dimers, thereby restoring the DNA to its original state.

photosynthetic Pertaining to photosynthesis.

photosynthesis Reactions used to harvest the energy of light to synthesize ATP, which is then used to power carbon fixation.

photosystems Protein complexes within which chlorophyll and other light-gathering pigments are organized; located in special photosynthetic membranes.

phototaxis Directed movement of cells in response to variations in light.

phototrophs Organisms that use light as a source of energy.

phycobiliproteins Light-harvesting pigments of cyanobacteria; they absorb energy from wavelengths of light that are not well absorbed by chlorophyll.

phylogenetic tree Type of diagram that depicts the evolutionary heritage of organisms.

phylogeny Evolutionary relatedness of organisms.

phylum (plural: **phyla**) Collection of similar classes; a collection of similar phyla makes up a kingdom; a phylum may also be called a division.

phytoplankton Floating and swimming algae and photosynthetic prokaryotic organisms of lakes and oceans.

pilus (plural: **pili**) Hairlike appendages on many Gram-negative bacteria that function in conjugation and for attachment.

pinocytosis Process by which eukaryotic cells take in liquid and small particles from the surrounding environment by internalizing and pinching off small pieces of their own membrane, bringing along a small volume of liquid and any material attached to the membrane.

placebo A mock drug—it looks and tastes like the experimental drug but has no medicinal value.

plankton Primarily microscopic organisms floating freely in most waters.

plaque (1) Clear area in a monolayer of cells. (2) In dentistry, a polysaccharide-encased community of bacteria (a biofilm) that adheres to a tooth surface.

plaque assay A method used to determine the number of viral particles in a suspension.

plasma Fluid portion of non-clotted blood.

plasma cell Effector B cell, fully differentiated to produce and secrete large amounts of antibody.

plasma membrane Selectively permeable membrane that surrounds the cytoplasm in a cell; also called the cytoplasmic membrane.

plasmid Small extrachromosomal circular DNA molecule that replicates independently of the chromosome; often codes for antibiotic resistance.

plasmodium (1) A multinucleated form of a slime mold. (2) When italicized and the first letter capitalized, the genus name of the protozoan parasite that causes malaria.

plasmolysis Dehydration and shrinkage of cytoplasm from the cell wall as a result of water diffusing out of a cell.

plate count Method used to determine the number of viable cells in a specimen by determining the number of colonies that arise when the specimen is added to an agar medium.

pleomorphic Varying in shape.

pleurisy Inflammation of the pleura, which are membranes that line the lung and chest cavity; the

condition is often marked by a sharp pain associated with breathing.

plus (+) strand (1) The DNA strand complementary to the strand used as a template for RNA synthesis. (2) An mRNA molecule that can be translated to make a protein. Also called the positive strand or sense strand.

PMN Abbreviation for polymorphonuclear neutrophil.

pneumonia Inflammation of the lungs accompanied by filling of the air sacs with fluids such as pus and blood.

pneumonic plague Disease that develops when *Yersinia pestis* infects the lungs.

point mutation Mutation in which only a single base pair is involved.

polar covalent bond Bond formed by sharing electrons between atoms that have unequal attraction for the electrons.

polar flagellum Single flagellum at one end of a cell.

polarity (1) In a chemical, slight positive or negative charges on atoms in a molecule. (2) The 5′ to 3′ directionality of a nucleic acid fragment.

polyadenylation In eukaryotic gene expression, adding a series of adenine derivatives to the 3′ end of an mRNA transcript.

polycistronic An mRNA molecule that carries the information for more than one gene.

polyclonal antibodies A preparation of antibodies that together recognize more than one epitope.

polygenic More than one gene.

polymer Large molecules formed by the joining together of repeating small molecules (subunits).

polymerase chain reaction (PCR) Method used to create millions of copies of a given region of DNA in only a matter of hours.

polymorphic Having different distinct forms.

polymorphonuclear neutrophils (PMNs) Type of phagocytic cell; the nuclei of these cells are segmented and composed of several lobes.

polypeptide Chain of amino acids joined by peptide bonds; also called a protein.

polyribosome Assembly of multiple ribosomes attached to a single mRNA molecule; also called a polysome.

polysaccharide Long chain of monosaccharide subunits.

polysaccharide vaccine Vaccine composed of polysaccharides, which make up the capsule of certain organisms.

polyunsaturated fatty acid Fatty acid that contains numerous double bonds.

porins Proteins in the outer membrane of Gram-negative bacteria that form channels through which small molecules can pass.

portal of entry Place of entry of microorganisms into the host.

portal of exit Place where infectious agents leave the host to find a new host.

positive (+) strand (See *plus [+] strand*.)

potential energy Stored energy; it can exist in a variety of forms including chemical bonds, a rock on the top of a hill, and water behind a dam.

pour-plate method Method of inoculating an agar medium with bacterial cells; the inoculum and melted agar are added to a Petri dish, where the agar hardens; the colonies grow both on the surface and within the medium.

precipitation reaction Reaction of antibody with soluble antigen to form an insoluble substance.

precursor metabolites Metabolic intermediates that can be either used to make the subunits of macromolecules or oxidized to generate ATP.

pre-mRNA A eukaryotic transcript that has not yet had the introns removed.

preservation The process of inhibiting the growth of microorganisms in products to delay spoilage.

prevalence Total number cases, both old and new, in a given population at risk at a point in time.

preventive vaccine Vaccine that prevents a disease.

primary culture Cells taken and grown directly from the tissues of an animal.

primary immune response (See *primary response*.)

primary infection Infection in a previously healthy individual, such as measles in a child who has not had measles before.

primary lymphoid organs Organs in which lymphocytes mature; the thymus and bone marrow.

primary metabolites Compounds synthesized by a cell during the log phase.

primary pathogen Microbe able to cause disease in an otherwise healthy individual.

primary producers Organisms that convert CO_2 into organic compounds; by doing so, they sustain other life forms, including humans.

primary response In immunology, the response that marks the adaptive immune system's first encounter with a particular antigen.

primary stain First dye applied in a multistep differential staining procedures; generally stains all cells.

primary structure Refers to the sequence of amino acids in a protein.

primary treatment In wastewater treatment, a physical process designed to remove materials that will settle out.

primase Enzyme that synthesizes small fragments of RNA to serve as primers for DNA synthesis during DNA replication.

primer RNA molecule that initiates the synthesis of DNA.

prion Infectious protein that causes a neurodegenerative disease.

probe In nucleic acid hybridization, a single-stranded piece of nucleic acid tagged with a detectable marker that is used to detect similar sequences.

probiotics Live, beneficial microorganisms.

prodromal phase Stage consisting of early, vague symptoms indicating the onset of a disease.

productive infection Virus infection in which more virions are produced.

proglottid One of the segments that make up most of the body of a tapeworm.

pro-inflammatory cytokines Any of a group of cytokines that contribute to the inflammatory response.

prokaryote Single-celled organism that does not contain a membrane-bound nucleus.

prokaryotic cell Cell characterized by lack of a nuclear membrane and thus no true nucleus.

promoter Nucleotide sequence to which RNA polymerase binds to initiate transcription.

proofreading The detection and removal by DNA polymerase of an incorrect nucleotide incorporated as DNA is synthesized.

propagated epidemic Outbreak of disease in which the infectious agent is transmitted to others, resulting in steadily increasing numbers of people becoming ill.

prophage Latent form of a temperate phage whose DNA has been inserted into the host's DNA.

prophylaxis Prevention of disease.

prospective study Study that looks ahead to see if the risk factors identified by a retrospective study predict a tendency to develop the disease.

prosthecate bacteria Group of Gram-negative bacteria that have extensions projecting from the cells, thereby increasing their surface area.

protease Enzyme that degrades protein; the protease encoded by HIV is the target of several anti-HIV medications.

protein Macromolecule consisting of amino acid subunits.

protein A Protein produced by *Staphylococcus aureus* that inhibits phagocytosis of the organism by binding to the Fc portion of antibodies.

protein domain A distinct protein part associated with a certain function.

protein subunit vaccine Vaccine composed of key protein antigens or antigenic fragments of an infectious agent, rather than whole cells or viruses.

protist A eukaryotic organism other than a plant, animal, or fungus; may be unicellular or multicellular.

proton Positively charged component of an atom.

proton motive force Form of energy generated as an electron transport chain moves protons across a membrane to create a chemiosmotic gradient.

proton pump Complex of electron carriers in the electron transport chain that ejects protons from the cell.

proto-oncogene A type of gene involved in tumor formation; it codes for a protein that activates transcription.

prototroph Organism that has no organic growth requirements other than a source of carbon and energy.

protozoa Group of single-celled eukaryotic organisms.

provirus Latent form of a virus in which the viral DNA is incorporated into the chromosome of the host.

PrP^C Normal cellular protein that can be converted to an infectious prion protein.

PrP^{SC} Infectious prion protein.

pseudomembrane A tough layer of dead cells and debris accumulated on an epithelial surface.

pseudomembranous colitis Intestinal disease typically caused by toxin-producing strains of *Clostridium difficile;* generally occurs only when a person is taking antimicrobial medications.

pseudopods Transient armlike extensions formed by phagocytes and protozoa; they surround and enclose extracellular material, including bacteria, during the process of phagocytosis.

P-site (or **peptidyl site**) Site on the ribosome where the tRNA that temporarily carries the elongating amino acid chain resides.

psychrophile Microorganism that grows best between –5°C and 15°C.

psychrotroph Organism that has an optimum temperature between 20°C and 30°C.

puerperal fever Childbed fever; infection of the uterus following childbirth, commonly caused by *Streptococcus pyogenes.*

pulsed-field gel electrophoresis Type of gel electrophoresis used to separate very large fragments of DNA.

PulseNet Surveillance network established by the CDC to facilitate the tracking of foodborne disease outbreaks.

pure culture A population of organisms descended from a single cell.

purine A double-ringed nitrogen-containing organic molecule such as the nucleobases adenine and guanine, which are components of nucleotides.

purple bacteria A group of anoxygenic phototrophs that are red, orange, or purple in color.

pus Thick, opaque, often yellowish material that forms at the site of infection, made up of dead neutrophils and tissue debris.

putrefaction Digestion of proteins by enzymes to yield foul-smelling products.

pyelonephritis Infection of the kidneys.

pyoderma Any skin disease characterized by production of pus.

pyogenic Pus-producing.

pyrimidine A single-ringed nitrogen-containing organic molecule such as the nucleobases thymine, cytosine, and uracil, which are components of nucleotides.

pyrogens Fever-inducing substances.

pyroptosis A type of programmed cell death that elicits an inflammatory response.

pyruvate End product of glycolysis; a precursor metabolite used in the synthesis of certain amino acids.

quantal assay In virology, a method that uses several dilutions of a virus culture administered to appropriate hosts to determine virus titer.

quaternary structure Level of structure of a protein molecule consisting of several polypeptide chains.

quinone A lipid-soluble electron carrier that functions in the electron transport chain.

quorum sensing Communication between bacterial cells by means of small molecules, permitting the cells to sense the density of cells.

rDNA DNA that encodes ribosomal RNA (rRNA).

reaction-center pigments Electron donors in the photosynthetic process; an example is chlorophyll *a*.

reactive oxygen species (ROS) Harmful derivatives of O_2 such as superoxide (O_2^-) and hydrogen peroxide (H_2O_2) that are highly toxic to cells.

reading frames Grouping of nucleotides in sequential triplets; an mRNA molecule has three possible reading frames, but only one is typically used in translation.

receptor Type of membrane protein that binds to specific molecules in the environment, providing a mechanism for the cell to sense and adjust to its surroundings.

receptor-mediated endocytosis Type of pinocytosis that allows cells to internalize extracellular ligands that bind to the cell's receptors.

recognition sequence The DNA sequence recognized by a particular restriction enzyme.

recombinant A cell that carries a DNA molecule derived from two different DNA molecules.

recombinant DNA molecule A vector-insert chimera created using recombinant DNA techniques.

recombinant DNA techniques Methods used to join DNA from two different sources *in vitro*.

recombinant vaccines Subunit vaccines produced by genetic engineering.

red tide An algal bloom that discolors water due to the abundant growth of dinoflagellates.

redox reactions Transfer of electrons from one compound to another; one compound becomes reduced and the other becomes oxidized.

reduced Refers to the gain of electrons.

reducing agents Compounds that readily donate electrons to another compound, thereby reducing the other compound.

reducing power Reduced electron carriers such as NADH, NADPH, and FADH$_2$; their bonds contain a form of usable energy.

refractive index A measure of the relative speed of light as it passes through a medium; light rays bend when they pass from a medium with one refractive index to another.

regulatory T cells Type of lymphocyte that helps control the immune response.

regulon Set of related genes transcribed as separate units but controlled by the same regulatory protein.

replica plating Technique for the simultaneous transfer of organisms in separated colonies from one medium to another.

replicase General term for any phage-encoded enzyme that replicates the genome of an RNA phage.

replication fork In DNA synthesis, the site at which the double helix is being unwound to expose the single strands that can function as templates.

replicon Piece of DNA that is capable of replicating; contains an origin of replication.

reporter gene Gene that has a detectable phenotype and can be fused to a gene of interest, providing a mechanism by which to monitor the expression of the gene of interest.

repressible enzyme Enzyme whose synthesis can be turned off by certain conditions.

repressor Protein that binds to the operator site and prevents transcription.

reservoir of infection Source of a pathogen.

resistance plasmid (or R plasmid) Plasmid that carries genetic information for resistance to one or more antimicrobial medications and heavy metals.

resolving power The ability to distinguish two objects that are very close together.

respiration Process that transfers electrons stripped from a chemical energy source to an electron transport chain, generating a proton motive force that is then used to synthesize ATP.

response regulator In a two-component regulatory system, the protein that binds to DNA.

restriction enzyme Type of enzyme that recognizes and cleaves a specific sequence of DNA.

restriction fragment length polymorphism (RFLP) Pattern of fragment sizes obtained by digesting DNA with one or more restriction enzymes.

restriction fragments Fragments generated when DNA is cut with restriction enzymes.

restriction-modification system Bacterial system that uses restriction enzymes to defend against invading foreign DNA; modification enzymes methylate the host DNA to protect it.

reticulate body Replicating, non-infectious intracellular form of *Chlamydia* species.

retrospective study Type of study done following a disease outbreak; compares the actions and events surrounding clinical cases with those of controls.

retroviruses Group of viruses that have a single-stranded RNA genome; their enzyme reverse transcriptase synthesizes a DNA copy that is then integrated into the host cell chromosome.

reverse transcriptase Enzyme that synthesizes double-stranded DNA complementary to an RNA template.

reversion Process by which a second mutation corrects a defect caused by an earlier mutation.

Reye's syndrome Often fatal condition characterized by vomiting, coma, and brain and liver damage, mostly occurring in children treated with aspirin for influenza or chickenpox.

RFLP Abbreviation for restriction fragment length polymorphism.

rheumatic fever A post-streptococcal sequela thought to be due to circulating immune complexes.

rheumatoid arthritis Severe crippling autoimmune disease in which cellular immune responses and antibodies target collagen in connective tissues, most often within joints.

rhizobia A group of Gram-negative nitrogen-fixing bacteria that form symbiotic relationships with leguminous plants such as clover and soybeans.

rhizosphere Zone around plant roots containing organic materials exuded by the roots.

rhuMab (recombinant human monoclonal antibody) Monoclonal antibody derived from a laboratory animal in which part of the molecule has been replaced with the human equivalent.

ribonucleic acid (RNA) Macromolecules in a cell that play a role in converting the information coded by the DNA into amino acid sequences in protein.

ribose A 5-carbon sugar found in RNA.

ribosomal RNA (rRNA) Type of RNA present in ribosomes; the nucleotide sequences of these are increasingly being used to classify and, in some cases, identify microorganisms.

ribosome Structure that facilitates the joining of amino acids during the process of translation; composed of protein and ribosomal RNA.

ribosome-binding site Sequence of nucleotides in bacterial mRNA to which a ribosome binds; the first time the codon for methionine (AUG) appears after that site, translation generally starts.

ribotyping Technique used to distinguish among related strains; detects RFLPs in ribosomal RNA genes.

ribozymes RNA molecules that have a catalytic function.

rifamycins Group of antimicrobial medications that block transcription.

RIG-like receptors A group of pattern recognition receptors within cells; they recognize molecules associated with infecting viruses.

risk factors Specific conditions associated with high frequencies of disease.

RNA Abbreviation for ribonucleic acid.

RNA interference Cellular mechanism that targets specific mRNA molecules for destruction by using small RNA fragments to identify it.

RNA polymerase Enzyme that catalyzes the synthesis of RNA using a DNA template.

RNases Enzymes that degrade RNA.

rod Cylindrical-shaped bacterium; also called a bacillus.

rolling circle replication Mechanism of DNA replication in which a single strand of DNA is synthesized.

rough endoplasmic reticulum Organelle where proteins destined for locations other than the cytoplasm are synthesized.

roundworm A helminth that has a cylindrical, tapered body; a nematode.

R plasmids Plasmids that encode resistance to one or more antimicrobial medications and heavy metals.

rRNA Ribosomal RNA.

SALT Abbreviation for skin-associated lymphoid tissues.

sanitized An item treated to reduce the microbial population to a level that meets accepted health standards.

saprophyte Organism that takes in nutrients from dead and decaying matter.

saturated Refers to a fatty acid that contains no double bonds.

scabies Skin disease caused by a parasitic mite.

scanning electron microscope (SEM) Type of microscope that scans a beam of electrons back and forth over the surface of a specimen.

scanning laser microscope (SLM) Type of microscope that scans laser beans across specimens, thereby allowing a three-dimensional image of a thick structure to be constructed.

scavenger receptor Receptor on phagocytes that facilitates the engulfment of materials that have charged molecules on their surfaces.

schizogony Process of multiple fission in which the nucleus divides a number of times before individual daughter cells are produced.

schizont Multinucleate stage in the development of certain protozoa, such as the ones that cause malaria.

scolex Attachment organ of a tapeworm, the head end.

scrapie Common name for a neurological disease of sheep; it is caused by a prion.

seasonal influenza Yearly influenza outbreaks that result from viral strains undergoing antigenic drift.

sebum Oily secretion of the sebaceous glands of the skin.

secondary infection Infection that occurs along with or immediately following another infection, usually as a result of the first infection.

secondary lymphoid organs Peripheral lymphoid organs where mature lymphocytes function in immune responses; they include the adenoids, tonsils, spleen, appendix, and lymph nodes, among others.

secondary metabolites Metabolic products synthesized during late-log and stationary phase.

secondary response (or memory response) Enhanced immune response that occurs upon second or subsequent exposure to specific antigen, caused by the rapid activation of long-lived memory cells; anamnestic response.

secondary structure Refers to the arrangement of amino acids in a protein; the two major arrangements are helices and sheets.

secondary treatment In treatment of wastewater, a biological process designed to convert most of the suspended solids to microbial mass and inorganic compounds.

second-line drugs Medications that are second choice because they are toxic or less effective.

secretion Releasing a substance from a cell or tissue.

secretory IgA Form of IgA that is transported across mucous membranes, thereby providing mucosal immunity.

segmented virus Virus that has a genome consisting of multiple different nucleic acid fragments.

selectable marker Gene that encodes a selectable phenotype such as antibiotic resistance.

selective enrichment Method of increasing the relative proportion of one particular species in a broth culture by including a selective agent that inhibits the growth of other species.

selective medium Culture medium that inhibits the growth of certain microorganisms and therefore favors the growth of desired microorganisms.

selective toxicity Causing greater harm to a pathogen than to the host.

selectively permeable Material that allows only certain molecules to pass through freely.

"self" cells The body's own cells.

semiconservative replication Nucleic acid replication that results in each of the two double-stranded molecules containing one of the original strands (the template strand) and one newly synthesized strand.

semicritical instruments Medical instruments such as endoscopes that come into contact with mucous membranes, but do not penetrate body tissues.

semipermeable (See *selectively permeable*.)

sensitization Primary exposure to an antigen that can then lead to an allergic reaction when that same antigen is encountered again.

sepsis Acute illness caused by infectious agents or their products circulating in the bloodstream; blood poisoning.

septic shock An array of effects including fever, drop in blood pressure, and disseminated intravascular coagulation, that results from infection of the bloodstream or circulating endotoxin.

septic system An individual wastewater treatment system in which the sludge settles out in a large tank where microorganisms anaerobically degrade it, and the effluent is aerobically degraded as it percolates through a drain field.

septicemic plague Disease that develops when *Yersinia pestis* infects the bloodstream.

serial dilutions Series of dilutions, usually twofold or tenfold, used to determine the titer or concentration of a substance in solution.

seroconversion Change from serum that lacks specific antibodies to serum that has those antibodies.

serology The study of *in vitro* antibody-antigen reactions, particularly those that detect antibodies in serum.

serotype A group of strains that have a characteristic antigenic structure that differs from other strains; also called a serovar.

serovar (See *serotype*.)

serum Fluid portion of blood that remains after blood clots.

serum-resistant Bacteria that have mechanisms to avoid the killing effect of complement system proteins.

serum sickness Systemic immune complex disease that can result from passive immunization using animal serum.

sex pilus Thin protein appendage required for attachment of two bacterial cells prior to DNA transfer by conjugation. The F pilus is an example.

shingles (or **herpes zoster**) Condition resulting from the reactivation of the varicella-zoster virus.

shock Condition with multiple causes characterized by low blood pressure and circulation of the blood inadequate to sustain normal organ function.

siderophore Iron-binding substance produced by bacteria to scavenge iron.

sigma (σ) factor Component of RNA polymerase that recognizes and binds to promoters.

signal sequence Amino acid sequence that directs cellular machinery to secrete the polypeptide.

signal transduction Process that transmits information from outside of a cell to the inside, allowing that cell to respond to changing environmental conditions.

signature sequences Characteristic sequences in the genes encoding ribosomal RNA that can be used to classify or identify certain organisms.

signs Effects of a disease observed by examining the patient.

silent mutation (1) A mutation that does not change the phenotype of the cell. (2) A mutation that does not change the amino acid encoded.

simple diffusion Movement of solutes from a region of high concentration to one of low concentration; does not involve transport proteins.

simple lipid Lipid that contains only carbon, hydrogen, and oxygen.

simple staining Staining technique that uses a basic dye to impart color to cells.

single-cell protein (SCP) Use of microorganisms such as yeast and bacteria as a protein source.

size standard In gel electrophoresis, a series of molecules of known sizes added to a lane of the gel to be used as a basis for later size comparison.

skin-associated lymphoid tissue (SALT) Secondary lymphoid tissue consisting of collections of lymphoid cells under the skin.

slime layer Type of glycocalyx that is diffuse and irregular.

slime mold Terrestrial organism that is similar to fungi but not related genetically.

sludge The solid portion of wastewater that settles to the bottom of sedimentation tanks during primary and secondary treatments.

smear In a staining procedure, the film obtained when a drop of microbe-containing liquid is placed on a microscope slide and allowed to air dry.

smooth endoplasmic reticulum Organelle of eukaryotic cells that is the site of lipid synthesis and degradation and calcium ion storage.

SOS repair Complex, inducible repair process used to repair highly damaged DNA.

specialized transduction Type of horizontal gene transfer that can occur when a temperate phage carries specific bacterial genes; the phage acquires those genes when an error occurs as the prophage excises from a lysogenic cell's chromosome.

species Group of related isolates or strains; the lowest basic unit of taxonomy.

spikes (or **attachment proteins**) Structures on the outside of the virion that bind to host cell receptors.

spirillum (plural: **spirilla**) Curved rod long enough to form spirals.

spirochetes Long helical bacteria that have a flexible cell wall and endoflagella.

splicing Process that removes introns from eukaryotic precursor RNA to generate mRNA.

spontaneous generation Discredited belief that organisms can arise from non-living matter.

spontaneous mutation Mutation that occurs naturally during the course of normal cell processes.

sporadic Occurring irregularly.

spore Type of differentiated, specialized cell formed by certain organisms; includes dormant cells that are resistant to adverse conditions and the reproductive structures formed by fungi.

sporozoite Infectious form of certain protozoa; for example, in malaria, the form entering the body from a mosquito bite, infectious for liver cells.

sporulation In bacteria, the process of producing a resistant spore; eukaryotes can also undergo sporulation.

spread-plate method Technique used to cultivate bacteria by uniformly spreading a suspension of cells onto the surface of an agar plate.

sputum Material coughed from the lungs.

start codon Codon at which translation is initiated; in prokaryotes, typically the first AUG after a ribosome-binding site.

starter cultures Strains of microorganisms added to a food to initiate the fermentation process.

stationary phase In a bacterial growth curve, stage in which the number of viable cells remains constant.

stereoisomer (or **optical isomer**) Mirror image of a compound.

sterilant A chemical used to destroy all microorganisms and viruses in a product, rendering it sterile.

sterile Completely free of all microorganisms and viruses; an absolute term.

sterilization The process of destroying or removing all microorganisms and viruses through physical or chemical means.

steroid Type of lipid with a specific four-membered ring structure.

stock culture Culture stored for use as an inoculum in later procedures.

stop codon Codon that does not code for an amino acid and is not recognized by a tRNA; signals the end of the polypeptide chain.

storage granule An accumulation of a high-molecular-weight polymer synthesized from a nutrient that a cell has in relative excess.

strain Population of cells descended from a single cell.

streak-plate method Simplest and most commonly used technique for isolating bacteria; a series of successive streak patterns is used to sequentially dilute an inoculum on the surface of an agar plate.

streptococcal pyrogenic exotoxins (SPEs) Family of genetically related toxins produced by certain strains of *Streptococcus pyogenes*, responsible for scarlet fever, toxic shock, and "flesh-eating" necrotizing fasciitis.

stromatolite Coral-like mat of filamentous microorganisms.

structural isomers Molecules that contain the same elements but in different arrangements that are not mirror images; structural isomers have different names and properties.

subacute bacterial endocarditis Slowly progressing infection of the internal surfaces of the heart.

subclinical Disease with no apparent symptoms.

substrate (1) Substance on which an enzyme acts to form products. (2) Surface on which an organism will grow.

substrate-level phosphorylation Synthesis of ATP using the energy released in an exergonic (energy-releasing) chemical reaction during the breakdown of the energy source.

sucrose Disaccharide consisting of a molecule of glucose bonded to fructose; common table sugar.

sugar-phosphate backbone Series of alternating sugar and phosphates moieties of a DNA molecule.

sulfa drugs Group of antimicrobial drugs that inhibit folic acid synthesis.

sulfate-reducers Group of obligate anaerobes that use sulfate (SO_4^{2-}) as a terminal electron acceptor, producing hydrogen sulfide as an end product.

sulfur-oxidizing bacteria Group of Gram-negative bacteria that obtain energy by oxidizing elemental sulfur and reduced sulfur compounds, thereby generating sulfuric acid.

S unit Unit of measurement that expresses the sedimentation rate of a compound; reflects the mass and density of the compound; "S" stands for Svedberg.

superantigens Toxins that non-specifically activate many T cells, resulting in excessive cytokine production that leads to severe reactions and sometimes fatal shock.

superficial mycoses Fungal infections that affect the hair, skin, or nails.

superoxide dismutase Enzyme that degrades superoxide to produce hydrogen peroxide.

surface receptors Proteins in the membrane of a cell to which certain signal molecules bind; they allow the cell to sense and respond to external signals.

swarmer cells Motile cells of sheathed bacteria that disperse to new locations.

symbiosis The living together of two dissimilar organisms or symbionts.

symptoms Subjective effects of a disease experienced by the patient, such as pain and nausea.

syncytium (plural: **syncytia**) Multinucleate body formed by the fusion of cells.

synergistic Describes the acting together of agents to produce an effect greater than the sum of the effects of the individual agents.

synergistic infection An infection in which two or more species of pathogens act together to produce an effect greater than the sum of effects if each pathogen were acting alone.

systemic Throughout the body.

systemic anaphylaxis Generalized allergic reaction caused by IgE, resulting in a profound drop in blood pressure.

systemic infection Infection in which the infectious agent spreads throughout the body.

systemic lupus erythematosus (SLE) A systemic autoimmune disease in which antibodies against molecules in the nuclei cause an attack against the body's own cells.

systemic mycoses Fungal infections that affect the tissues deep in the body.

---

tapeworm A helminth that has a segmented, ribbon-shaped body; a cestode.

Taq **polymerase** Heat-stable DNA polymerase of the thermophilic bacterium *Thermus aquaticus*.

target cell In immunology, cell that is the direct recipient of a T cell's effector functions.

taxa Groups into which organisms are classified.

taxonomy The science that studies organisms in order to arrange them into groups; it encompasses identification, classification, and nomenclature.

T_C cell Effector form of a cytotoxic T cell; it induces apoptosis in infected or cancerous "self" cells.

T-cell receptor (TCR) Molecule on a T cell that enables the T cell to recognize a specific antigen.

T cells Lymphocytes that mature in the thymus; they are responsible for cellular immune responses and function as helper cells in the antibody response.

T-dependent antigens Antigens that evoke an antibody response only with the participation of T_H cells.

T-DNA Portion of the Ti plasmid of *Agrobacterium tumefaciens* that is transferred into a plant cell.

T lymphocytes (See *T cells.*)

teichoic acids Gram-positive cell wall component, composed of chains of a common subunit, either ribitol-phosphate or glycerol-phosphate, to which various sugars and D-alanine are usually attached.

temperate phage Bacteriophage that can either become integrated into the host cell DNA as a prophage or direct a productive infection that leads to cell lysis.

template Strand of nucleic acid that a polymerase uses to synthesize a complementary strand.

terminal electron acceptor Chemical that is ultimately reduced as a consequence of fermentation or respiration.

terminator In transcription, a DNA sequence that stops the process.

tertiary structure Level of structure of a protein described by its three-dimensional nature; two major shapes exist, globular and fibrous.

tertiary treatment (See *advanced treatment.*)

tetanospasmin Neurotoxin produced by *Clostridium tetani.*

tetracyclines Group of antimicrobial medications that interfere with protein synthesis.

T_H cell Effector form of a helper T cell; it activates B cells and macrophages, and releases cytokines that stimulate other parts of the immune system.

therapeutic index A measure of the relative toxicity of a medication, defined as the ratio of minimum toxic dose to minimum effective dose.

therapeutic vaccine Type of vaccine used to treat a disease rather than prevent it.

thermophile Organism with an optimum growth temperature between 45°C and 70°C.

thermotaxis Movement in response to temperature.

3′ end The end of a nucleotide strand that has a free hydroxyl group on the number 3 carbon of the sugar.

thrush Infection of the mouth by *Candida albicans.*

thylakoids Membrane-bound disclike structures within the stroma of chloroplasts; they contain chlorophyll.

thymine dimer Two adjacent thymine molecules on the same strand of DNA joined together through covalent bonds.

thymus Primary lymphoid organ, located in the upper chest, in which T lymphocytes mature.

T-independent antigens Antigens that can activate B cells without the assistance of a T_H cell.

Ti plasmid (See *tumor-inducing plasmid.*)

tissue Cooperative associations of cells in a multicellular organism.

tissue culture Laboratory culture of plant or animal cells.

titer Measure of the concentration of a substance in solution.

tolerance Specific unresponsiveness of the adaptive immune system that reflects its ability to ignore any given molecule, such as a normal cellular protein.

toll-like receptors (TLRs) A group of pattern recognition receptors located on the surface of cells and within endosomes.

total coliforms Facultative, non-spore-forming, Gram-negative rods that ferment lactose, producing acid and gas within 48 hours at 35°C; most reside in the intestine, so they are used as indicators of fecal pollution.

toxemia Circulation of toxins in the bloodstream.

toxic shock syndrome Collapse of the blood pressure due to a circulating toxin such as one of the superantigens produced by *Staphylococcus aureus* and *Streptococcus pyogenes.*

toxin Poisonous chemical substance.

toxoid Modified form of a toxin that is no longer toxic but is able to stimulate the production of antibodies that will neutralize the toxin.

trace elements Elements that are required in very minute amounts by all cells; they include cobalt, zinc, copper, molybdenum, and manganese.

trachoma Chronic eye disease caused by certain strains of *Chlamydia trachomatis.*

transamination Transfer of an amino group from an amino acid to a recipient compound, converting the recipient compound to an amino acid.

transcript Fragment of RNA, synthesized using one of the two strands of DNA as a template.

transcription Process of transferring genetic information coded in DNA into messenger RNA (mRNA).

transcytosis Transport of a substance across a border made up of cells.

transduction Mechanism of horizontal gene transfer in which bacterial DNA is transferred inside a phage.

transfer RNA (tRNA) Type of RNA that delivers the appropriate amino acid to the ribosome during translation.

transferrin An iron-binding protein found in blood and tissue fluids.

transformation (DNA-mediated transformation) A mechanism of horizontal gene transfer in which "naked" DNA is transferred.

transfusion reaction Reaction characterized by fever, low blood pressure, pain, nausea, and vomiting, resulting from the transfusion of immunologically incompatible blood.

transgenic Plants and animals into which new DNA has been introduced by genetic engineering.

transition step Step in metabolism that links glycolysis to the TCA cycle; converts pyruvate to acetyl-CoA.

translation Process by which genetic information in the messenger RNA directs the order of amino acids in protein.

translocation Advancement of a ribosome a distance of one codon during translation.

transmissible spongiform encephalopathies (TSEs) Group of fatal neurodegenerative diseases in which brain tissue develops spongelike holes.

transmission electron microscope (TEM) Type of microscope that directs a beam of electrons at a specimen; used to observe details of cell structure.

transport systems Mechanisms cells use to transport molecules across the cytoplasmic membrane.

transposon (or transposable element) A piece of DNA that can move from one site in DNA to another, either in the same molecule or to another molecule in the same cell.

transposition Movement of a piece of DNA from one DNA site to another in the same cell.

trematodes Flatworms known as flukes.

tricarboxylic acid (TCA) cycle Cyclic metabolic pathway that incorporates acetyl-CoA, ultimately generating ATP (or GTP), CO_2, and reducing power, also called the Krebs cycle and the citric acid cycle.

trichomes Filamentous multicellular associations of cyanobacteria that may or may not be enclosed within a sheath.

trickling filter method Wastewater treatment method in which a rotating arm sprays wastewater onto a bed of rocks coated with a biofilm of organisms that aerobically degrades the wastes.

triglyceride Molecule consisting of three molecules of the same or different fatty acids bonded to glycerol.

trimethoprim Antimicrobial medication that interferes with folate synthesis.

tRNA Transfer RNA.

trophozoite Vegetative feeding form of some protozoa.

tubercle Granuloma formed in tuberculosis.

tuberculoid leprosy Form of Hansen's disease in which cell-mediated immunity limits the disease progression.

tumor-inducing (Ti) plasmid Plasmid of *Agrobacterium tumefaciens* that allows the organism to cause tumors in plants; a derivative is used as a vector to introduce DNA into plants by genetic engineering.

tumor necrosis factor (TNF) A cytokine that plays an important role in the inflammatory response and other aspects of immunity.

turbidity Cloudiness in a liquid.

two-component regulatory system Mechanism of gene regulation that uses a sensor and a response regulator.

type 1 diabetes mellitus Organ-specific autoimmune disease caused when cytotoxic T cells destroy a certain type of pancreatic cell.

type III secretions system Mechanism that allows bacteria to transfer gene products directly into host cells.

ubiquity Refers to widespread occurrence.

ultra-high-temperature (UHT) method A method that uses heat to render a product free of all microorganisms that can grow under normal storage conditions.

ultraviolet (UV) light Electromagnetic radiation with wavelengths between 175 and 350 nm.

uncoating In virology, the separation of the protein coat from the nucleic acid of the virion.

unsaturated Refers to a fatty acid with one or more double bonds.

urea A waste product of protein catabolism by the body's cells; present in various body fluids, notably urine.

urinary tract infection Infection of the urinary tract, usually the bladder.

UV Abbreviation for ultraviolet light.

vaccine Preparation of attenuated or inactivated microorganisms or viruses or their components used to immunize a person or animal against a particular disease.

vancomycin Antimicrobial medication that interferes with peptidoglycan synthesis.

vancomycin-intermediate *Staphylococcus aureus* (VISA) Strains of *Staphylococcus aureus* that are intermediate in their susceptibility to vancomycin.

vancomycin-resistant enterococci (VRE) Strains of enterococci that are resistant to vancomycin.

vancomycin-resistant *Staphylococcus aureus* (VRSA) Strains of *Staphylococcus aureus* that are resistant to vancomycin.

variable region The portion of an antibody molecule that contains the antigen-binding sites; tremendous variation exists between the amino acid sequences of variable regions in different antibody molecules.

vector (1) In molecular biology, a piece of DNA that acts as a carrier of a cloned fragment of DNA. (2) In epidemiology, any living organism that can carry a disease-causing microbe; most commonly arthropods such as mosquitoes and ticks.

vegetative cell Typical, actively multiplying cell.

vertical transmission Transfer of a pathogen from a pregnant woman to the fetus, or from a mother to her infant during childbirth or breast feeding.

vesicle A membrane-bound sac used to transport material within a cell.

vibrio (plural: **vibrios**) Short, curved rod-shaped bacterial cell.

villus (plural: **villi**) Fingerlike protrusion from a membrane such as the intestinal lining.

viral load Measure of severity of a virus infection; calculated by estimating the concentration of virus particles in involved body fluid.

viremia Virus particles circulating in the bloodstream.

virion Viral particle in its inert extracellular form.

viroid Piece of RNA that does not have a protein coat but does replicate within living cells.

virulence Relative ability of a pathogen to overcome body defenses and cause disease.

virulence factors Arsenal of mechanisms of pathogenicity of a given microbe.

virulent phages Bacteriophages that routinely lyse their host; they cannot lysogenize the host.

virus Acellular or non-living agent composed of nucleic acid surrounded by a protein coat.

virus-like particle (VLP) vaccine A type of vaccine composed of empty viral capsids.

VISA Abbreviation for vancomycin-intermediate *Staphylococcus aureus.*

vitamin One of a group of organic compounds found in small quantities in natural foodstuffs that are necessary for the growth and reproduction of an organism; usually converted into coenzymes.

VRE Abbreviation for vancomycin-resistant enterococci.

VRSA Abbreviation for vancomycin-resistant *Staphylococcus aureus.*

wastewater Material that flows from household plumbing systems; municipal wastewater also includes business and industrial wastes and storm water runoff.

water activity (a_w) Quantitative measure of the relative amount of water available for microbial growth.

water molds Non-photosynthetic members of the heterokons; similar to the fungi but not related genetically.

Western blotting Procedure that uses labeled antibody molecules to detect specific proteins that have been separated by gel electrophoresis.

wet mount Method of observing a living organisms in a drop of liquid using a microscope.

whey Liquid portion that remains after milk proteins coagulate during cheese-making.

wide host range plasmid Plasmid that can replicate in unrelated bacteria.

wild-type Describes an organism that has the typical characteristics of the species isolated from nature.

World Health Organization (WHO) International agency devoted to improving the health for all peoples.

yeast Unicellular fungus.

zone of inhibition Region around a disc where bacteria are unable to grow due to adverse effects of the compound in the disc.

zoonosis (plural: **zoonoses**) Disease of animals that can be transmitted to humans.

zooplankton Floating and swimming small animals and protozoa found in marine environments, usually in association with the phytoplankton.

zygote Diploid cell formed by the sexual fusion of two haploid cells.

Credits

Photo Credits

Chapter 1
Opener: © Science VU/Visuals Unlimited; **1.3:** © Bettmann/Corbis; **1.5:** © Kathy Talaro/Visuals Unlimited; **1.6:** © Carolina Biological Supply Co./Visuals Unlimited; **1.7a:** © CDC/Janice Haney Carr; **1.7b:** © Dr. Richard Kessel & Dr. Gene Shih/Visuals Unlimited; **1.8:** © Manfred Kage/Peter Arnold; **1.9a:** K.G. Murti/Visuals Unlimited; **1.9b:** © Thomas Broker/Phototake; **1.9c:** © K.G. Murti/Visuals Unlimited; **1.10:** U.S. Department of Agriculture/Dr. Diemer; **1.11:** © Stanley B. Prusiner/Visuals Unlimited; **Perspective 1.1, Fig.1:** Courtesy of Esther R. Angert; **Perspective 1.1, Fig. 2:** Courtesy of Dr. Heide N. Schulz/Max Planck Institute for Marine Microbiology; **Perspective 1.1, Fig. 3:** Courtesy of Reinhard Rachel and Harald Huber, University of Regensburg, Germany.

Chapter 2
Opener: © Scott Camazine/Photo Researchers, Inc.; **Perspective 2.1, Fig. 1a:** © Mark Antman/The Image Works; **Perspective 2.1, Fig. 1b:** © SIU/Peter Arnold.

Chapter 3
Opener: CDC/Janice Haney Carr; **Table 3.1 (Bright-field):** From S. T. Williams, M.E. Sharpe and J.G. Holt (Eds.), Bergey's Manual of Systematic Bacteriology, Vol. 4. Figure 29.3, p. 2454 © 1989 Williams and Wilkins Co., Baltimore. Micrograph from T. Cross, University of Bradford, Bradford, U.K.; **Table 3.1 (Dark-field):** © T. E. Adams/Visuals Unlimited; **Table 3.1 (Phase-contrast):** © Wim van Egmond/Visuals Unlimited; **Table 3.1 (Differential interference contrast):** © Dennis Kunkel Microscopy, Inc./Visuals Unlimited, Inc.; **Table 3.1 (Fluorescence):** © Evans Roberts; **Table 3.1 (Scanning laser):** Courtesy of Michael W. Davidson/National High Magnetic Field Lab; **Table 3.1 (Transmission):** © R. Kessel & C. Shih/Visuals Unlimited; **Table 3.1 (Scanning):** © David M. Phillips/Visuals Unlimited; **Table 3.1 (Atomic Force):** © Torunn Berge/Photo Researchers, Inc.; **3.1:** Courtesy of Leica, Inc., Deerfield, FL; **3.2 (both):** © W. A. Jensen; **3.3a:** © Richard Megna/Fundamental Photographs; **3.4:** © T. E. Adams/Visuals Unlimited; **3.5:** © Wim van Egmond/Visuals Unlimited; **3.6:** © Dennis Kunkel Microscopy, Inc./Visuals Unlimited, Inc.; **3.7:** © Evans Roberts; **3.8 (both):** Courtesy of Michael W. Davidson/National High Magnetic Field Lab; **3.10a:** © R. Kessel & C. Shih/Visuals Unlimited; **3.10b:** © H. Aldrich/Visuals Unlimited; **3.11:** © David M. Phillips/Visuals Unlimited; **3.12:** © Torunn Berge/Photo Researchers, Inc.; **3.14b:** © Leon J. Le Beau/Biological Photo Service; **3.15:** CDC/Dr. George P. Kubica; **3.16:** © Dr Gladden Willis/Visuals Unlimited/Getty; **3.17:** © Jack M. Bostrack/Visuals Unlimited; **3.18:** © E. Chan/Visuals Unlimited; **3.19a:** Courtesy of Molecular Probes, Eugene, OR; **3.19b:** © Richard L. Moore/Biological Photo Service; **3.19c:** © Evans Roberts; **3.20a:** © SciMAT/Photo Researchers, Inc.; **3.20b, 3.20c, 3.20d, 3.20e:** © Dennis Kunkel Microscopy Inc.; **3.21a:** Courtesy of Walther Stoeckenius; **3.21b:** Courtesy of James T. Staley; **3.22a (top):** © George Musil/Visuals Unlimited; **3.22a (bottom):** © David M. Phillips/Visuals Unlimited; **3.22b:** © R. Kessel & C. Shih/Visuals Unlimited; **3.22c:** © Oliver Mecks/Photo Researchers, Inc; **3.23b:** Courtesy of L. Santo, H. Hohl, and H. Frank, "Ultrastructure of Putrefactive Anaerobe 3679h During Sporulation," Journal of Bacteriology 99:824, 1969. American Society for Microbiology; **3.32c and 3.33d:** © Terry Beveridge, University of Guelph; **3.34:** Courtesy of Dr. Edwin S. Boatman; **3.35a:** Courtesy of K.J. Cheng and J.W. Costerton; **3.35b:** Courtesy of A. Progulske and S.C. Holt, Journal of Bacteriology, 143:1003–1018, 1980; **3.36a:** © Fred Hossler/Visuals Unlimited; **3.36b:** © Science VU/Visuals Unlimited; **3.39:** © D. Blackwill and D. Maratea/Visuals Unlimited; **3.40a:** Courtesy of Dr. Charles Brinton, Jr.; **3.40b:** U.S. Department of Agriculture/Harley W. Moon; **3.41a:** © CNRI/SPL/Photo Researchers, Inc.; **3.41b:** © Dr. Gopal Murti/SPL/Photo Researchers; **3.43:** Courtesy of Dr. Edwin S. Boatman; **3.44:** Courtesy of J.F.M. Hoeniger, "Cytology of Spore Germination in Clostridium pectinovorum" Journal of Bacteriology 96:1835, 1968. American Society for Microbiology; **3.46c:** Courtesy of Thomas Fritsche; **3.51b:** © Garry T. Cole/Biological Photo Service; **3.52b:** © Keith Porter/Photo Researchers, Inc.; **3.53:** © George Chapman/Visuals Unlimited/Getty; **3.54:** © R. Bolendar & D. Fawcett/Visuals Unlimited; **3.55:** Courtesy of Charles J. Flickinger.

Chapter 4
Opener: © Science Photo Library RF/Getty Images; **4.2:** © Dr. Dennis Kunkel/Visuals Unlimited; **4.4:** © Michael Gabridge/Visuals Unlimited; **4.5:** © Fred Hossler/Visuals Unlimited; **4.10a:** © Christine Case/Visuals Unlimited; **4.10b:** © L. M. Pope and D. R. Grote/Biological Photo Service; **4.11:** © Dr. Elmer Koneman/Visuals Unlimited; **4.12:** © Becton, Dickson and Company; **4.13:** Courtesy of Thermoforma, Forma Scientific Division; **4.19 (left):** © Dennis Kunkel/Phototake; **4.19 (right):** © Kathy Talaro/Visuals Unlimited; **4.21a:** © Richard Megna/Fundamental Photographs.

Chapter 5
Opener: © Creatas/PunchStock; **5.1a:** © Digital Vision Ltd.; **5.1b:** © Don Tremain/Getty Images; **5.1c:** © Ryan McVay/Getty Images; **5.1d:** © Bob Daemmrich/The Image Works; **5.1e:** © Royalty Free/Corbis; **5.4 (both):** © Evans Roberts; **5.5:** © Pall/Visuals Unlimited; **5.7:** © The McGraw-Hill Companies, Inc./Jill Braaten, photographer; **5.8:** © BananaStock/PunchStock.

Chapter 6
Opener: © John Foxx/Imagestate Media/Imagestate; **6.2:** © Farrell Grehan/Photo Researchers, Inc.; **6.3 (top):** © Photodisc Vol. Series 74, photo by Robert Glusie; **6.3 (bottom):** © Digital Vision/PunchStock; **6.11b, 6.11c:** From Voet: Biochemistry, 1/e, 0471617695, 1990, John Wiley & Sons; **6.23 (yogurt, dairy, pickle), 6.23b (wine, beer), 6.23 (acetone):** © Brian Moeskau/McGraw-Hill; **6.23 (cheese):** © Photodisc/McGraw-Hill; **6.23 (Voges-Proskauer Test), 6.23 (Methyl-Red Test):** © The McGraw-Hill Companies, Inc./Auburn University Photographic Services.

Chapter 7
Opener: © Radius Images/Alamy; **7.5:** From J. Cairns, "The Chromosome of E. coli" in Cold Spring Harbor Symposia on Quantitative Biology, 77, Fig. 2, Pg. 44. © 1963 by Cold Spring Harbor Laboratory Press.

Chapter 8
Opener: © Dr. Gopal Murti/SPL/Custom Medical Stock Photo; **8.6:** © Matt Meadows/Peter Arnold, Inc; **8.21:** Courtesy of C. Brinton, Jr. and J. Carnahan.

Chapter 9
Opener: © Photographer's Choice/Getty Images; **9.2c:** © Richard T. Nowitz/Photo Researchers, Inc.; **9.5:** Courtesy of Pamela Silver and Jason Karana, Harvard Medical School, Dana Farber Cancer Institute. Photo provided by Howell Martin; **9.9:** Reprinted with permission from Edvotek, Inc.; **9.13:** © Josh Westrich/zefa/Corbis; **9.20:** Dr. R. Rachel, and Prof. Dr. K.O. Stetter, University of Regensburg, Lehrstuhl fuer Mikrobiologie, Regensburg, Germany; **9.21:** Courtesy of Rosetta Inpharmics © 2002 Rosetta Inpharmics, Inc. All rights reserved.

Chapter 10
Opener: CDC; **10.2a:** © George Wilder/Visuals Unlimited; **10.2b:** Courtesy of Dr. Thomas R. Fritsche, M.D., Ph.D., Clinical Microbiology Division, University of Washington, Seattle; **10.3a:** © E. Koneman/Visuals Unlimited; **10.3b:** © Biophoto Associates/Photo Researchers, Inc.; **10.4a:** © Denise Anderson; **10.4b, 10.4c:** © Dennis Strete/Fundamental Photographs; **10.6a:** Courtesy of bioMerieux, Inc.; **10.6b:** Courtesy of Becton, Dickinson and Company; **10.10b:** Courtesy of Patricia Exekiel Moor; **10.11, 10.12:** © Evans Roberts.

Chapter 11
Opener: © Science Photo Library RF/Getty Images; **11.1a:** © F. Widdel/Visuals Unlimited; **11.1b:** © Ralph Robinson/Visuals Unlimited; **11.2a:** © Thomas Tottleben/Tottleben Scientific Company; **11.2b:** © David M. Phillips/Visuals Unlimited; **11.3:** © John Walsh/Photo Researchers, Inc.; **11.4a:** Courtesy of Dr. Heinrich Kaltwasser. From ASM News 53(2): Cover, 187; **11.4b:** From M. P. Starr et al (Eds.), The Prokaryotes. Springer-Verlag; **11.5:** From J. G. Holt (Ed.), The Shorter Bergey's Manual of Determinative Bacteriology, 8/e, 1977. Williams and Wilkins Co., Baltimore; **11.6a:** Courtesy of Isao Inouye, University

Index

Information in figures and tables is indicated by *f* and *t* respectively.

A

A-B toxins, 392, 392*f*
ABO blood types, 405
abomasum, 732
abrasions, 549
abscess
  in acne vulgaris, 524
  antimicrobial medications and, 550
  definition of, 521, 548
  formation, 525*f*, 550*f*
  wound, 550, 550*f*
accessory pigments, 152
Accutane, 524
Acetobacter, 755
acid production, in cell product detection, 98*t*, 103
acid-fast, 264
acid-fast stain, 47*t*, 48, 48*f*
Acidithiobacillus, 151, 257*t*, 263, 263*f*
acidophile
  characteristics of, 89*t*
  definition of, 91
  thermophilic extreme, 279
acidophilus milk, 752
Acinetobacter baumanii, 450*t*
acne vulgaris, 523–524, 545*t*
  causative agent, 523
  epidemiology of, 524
  pathogenesis of, 524
  prevention of, 524
  signs and symptoms, 523
  treatment of, 524
acquired immunodeficiency
    syndrome (AIDS),
    414*t*, 415, 540.
    See also human
    immunodeficiency
    virus (HIV)
  anal cancer in, 708
  B lymphocyte tumors in,
    707–708
  cervical cancer in, 708
  complications of, 707–714
  cryptococcal meningoen-
    cephalitis and,
    660, 661
  cytomegalovirus disease in,
    711–713
  deaths from, in U.S., 704*f*
  defining conditions, 696, 696*t*
  definition of, 694
  history of, 694, 696
  infectious diseases in, 708–714

Kaposi's sarcoma in, 707
lymphoma in, 707–708
malignancies in, 707–708
Mycobacterium avium complex
    in, 694, 713–714
poverty and, 714
progression to, 701
related complex (ARC),
    694, 696
toxoplasmosis in, 709–711
acquired resistance, 457, 460,
    471–472
Acremonium cephalosporium, 465
actin, 594
actin filament, 69*f*, 73, 73*f*
Actinomyces, 381*f*
Actinomyces israelii, 560, 561
actinomycosis
  causative agent, 560
  epidemiology of, 561
  pathogenesis of, 560–561
  prevention of, 561
  signs and symptoms, 559
  treatment of, 561
activated lymphocytes, 363, 364*f*
activated macrophages, 348
activated sludge process, 736–738,
    737*f*
activator, 180, 181*f*
activator-binding site, 180, 181*f*
active immunity
  definition of, 419
  in immunization, 420
active transport, 56, 56*t*
active tuberculosis disease
    (ATBD), 503, 504
acute infections, 383
  definition of, 380
  viral, 322, 322*f*
acute inflammation, 350
acute necrotizing ulcerative
    gingivitis (ANUG),
    577, 579*t*
acute post-streptococcal
    glomerulonephritis,
    490, 490*f*
acute retroviral syndrome (ARS),
    694, 696, 701*f*
acute rheumatic fever, 489–490
acyclovir, 475, 475*t*, 533, 537,
    582, 631, 681–682
adaptive immunity
  antibodies in
    adherence prevention by,
    360

avoiding, 390, 396
  cross-linking by, 361
  cross-reactive, 395
  definition of, 354
  discovery of, 354
  in humoral immunity,
    355, 356*f*
  immobilization by, 360–361
  nature of, 359–363
  neutralization by, 360
  opsonization by, 360
  properties of, 360
  structure of, 360, 360*f*
antigens in
  definition of, 354
  discovery of, 354
  nature of, 359
  presentation of, 365
  T lymphocytes and, 368–374
  T-dependent, 359
  T-independent, 359
B lymphocytes in, 365–368
cell-mediated immunity in, 354,
    355, 356–357, 356*f*
clonal selection in, 354,
    363, 364*f*
complexity of, 355
damage associated with, 395
definition of, 354, 355
diversity in, 375–377
history of, research on, 354
humoral immunity in, 354,
    355–356, 356*f*
immunoglobulins in, 361–363,
    362*t*
lymphatic system in, 357–359
primary response in, 355
secondary response in, 355
strategy of, 355–357
tolerance in, 355
addition mutation, 191, 191*f*
ADE. *See* antibody-dependent
    enhancement (ADE)
adenine, 32, 32*f*
adenoids, 485*f*
adenosine diphosphate (ADP),
    24, 128
adenosine triphosphate (ATP), 126
  aerobic respiration and,
    146–147, 146*f*
  in cells, 24, 24*f*
  fermentation and, 134*t*
  generation of, by chemo-
    organoheterotrophs,
    134*t*

glycolysis and, 139
  as measurement of microbial
    growth method,
    98*t*, 103
  in metabolic pathways,
    128–130, 130*f*
  mitochondria and, 74–75
  oxidative phosphorylation and,
    145–146
  respiration and, 134*t*, 146–147,
    146*f*
  synthase, 145–146
  transport systems using, 56
adenoviral respiratory infections,
    495–496, 517*t*
adenovirus, 308*t*
adenylate cyclase, 587
adherence, in infection, 386, 386*f*
adhesins, 386
adhesion molecules, 342
adjuvant
  definition of, 419
  in inactivated vaccines, 424
ADP. *See* adenosine diphosphate
    (ADP)
advanced treatment, 735, 737*f*,
    738–739
A/E. *See* attaching and effacing
    (A/E) lesions
Aedes aegypti, 299*t*
aerobe, obligate, 257*t*, 264–265,
    264*f*, 265*f*
  characteristics of, 89*t*
  definition of, 90
  incubation of, 96
  oxygen requirements of, 91*t*
aerobic chemolithotrophs, 257*t*,
    262–264, 263*f*
aerobic chemoorganotrophs, 257*t*,
    264–265, 264*f*, 265*f*
aerobic respiration, 134, 144, 145*f*,
    146–147, 146*f*
aerotolerant anaerobe
  catalase and, 90–91
  characteristics of, 89*t*
  definition of, 90
  oxygen requirements of, 91*t*
affinity maturation, 366
aflatoxins, 288, 756
African sleeping sickness
  arthropods and, 299*t*
  causative agent, 294*t*, 662
  epidemiology of, 662
  Gambian form of, 662
  pathogenesis of, 662

Streptococcus pneumoniae, 63, 275*t*, 473, 493, 497, 644, 645, 672
Streptococcus pyogenes, 487*f*
  as aerotolerant, 90
  exotoxins produced by, 392*t*
  as lactic acid bacteria, 258
  lysogenic conversion and, 313*t*
  as medically important bacteria, 275*t*
  in otitis media, 493
  in sinusitis, 493
  in strep throat, 487–489
  in wound infections, 551*t*, 553–554
streptogramins, 461*f*, 462*t*, 465*f*, 466
streptokinase, 218*t*
streptolysin O, 487*t*
streptolysin S, 487*t*
Streptomyces, 267*t*, 268, 268*f*, 458, 724
Streptomyces griseus, 458
streptomycin, 458, 460, 462*t*, 466, 677
stroke, 671
Strongyloides stercoralis, 297*t*
strongyloidiasis, 297*t*, 479*t*
STRs. *See* short tandem repeats (STRs)
structure, protein, 27–28, 28*f*
studies
  analytical, 445
  case-control, 445
  cross-sectional, 445
  descriptive, 444–445, 445*f*
  epidemiological, 444–446
  experimental, 445
  prospective, 445
  retrospective, 445
subacute bacterial endocarditis (SBE)
  causative agent, 673
  epidemiology of, 673
  pathogenesis of, 673
  prevention of, 673
  signs and symptoms, 673
  treatment of, 673
subacute sclerosing panencephalitis (SSPE), 538
subclinical infection, 382
subcutaneous layer, of skin, 522, 522*f*
substituted proteins, 29
substrate-level phosphorylation
  definition of, 126
  in metabolic pathways, 130
subunit vaccines, 422
sucker, 298*f*
sucrose, 30*t*
sugar
  fermentation, 243*t*
  preservation and, 121–122
sulfadiazine, 711
sulfamethoxazole, 709

sulfanilamide, 458
sulfate, 728
sulfate-reducing bacteria, 258
sulfhydryl, 24*t*
Sulfolobus, 279, 279*f*
sulfonamides, 461*f*, 462*t*, 467
sulfur assimilation, 728
sulfur bacteria, 150*t*, 257*t*, 258
sulfur cycle, 728
sulfur decomposition, 728
sulfur, functions of, 92, 92*t*
sulfur oxidation, 728
sulfur reduction, 728
sulfur springs, 723
sulfur-oxidizing bacteria, 255, 262–263, 263*f*, 267*t*, 273, 279, 279*f*
sulfur-reducing hyperthermophiles, 278, 279*f*
superantigens, 392*t*, 393, 393*f*, 548, 552
superficial cutaneous mycoses, 543–544, 545*t*
superoxide, 90
superoxide dismutase, 90
suramin, 662
surgery, 107
surveillance, infectious disease, 446–448
swan-necked flask experiment, 2, 2*f*
swarmer cells, 270
sweat, 523
sweat gland, 522*f*, 523
swimmer's itch, 297*t*
swine flu, 508, 511
Swiss cheese, 751
symbiosis, 381
symbiotic relationships, of fungi, 286, 286*f*
Symmetrel, 511
symptomatic infection, 439
symptoms
  definition of, 382
  duration of, 383–384
synapse, 642, 642*f*
syncytia, 511, 705
Synechococcus, 257*t*
Synercid, 466
synergistic infection
  definition of, 548
  in human bite wounds, 562
syphilis
  causative agent, 275*t*, 627–628, 627*f*
  chancre in, 626, 626*f*
  congenital, 627, 629
  end of, 628
  epidemiology of, 626, 628–629
  history of, 457, 611, 626
  HIV infection and, 626, 627
  latent, 627
  microscopic identification of, 43
  pathogenesis of, 628

prevention of, 629
  primary, 626, 628, 629
  secondary, 626, 628, 629
  signs and symptoms, 626–627, 626*f*
  tertiary, 627, 628
  treatment of, 457, 629
  in Tuskegee Experiment, 628
systemic anaphylaxis
  definition of, 401
  in type I hypersensitivity, 404
systemic infection, 384, 670
systemic lupus erythematosus (SLE), 412*t*, 413

T

T cells, 340, 354, 356, 368–374, 377
Taenia saginata, 297*t*
Taenia solium, 297*t*, 479*t*
tafenoquine, 478*t*
tampons, toxic shock syndrome and, 619, 619*f*, 620
Tannerella forsythia, 577
tapeworms, 295, 297*t*, 298
  beef, 297*t*
  fish, 297*t*
  pork, 297*t*, 298*f*
Taq polymerase, 228
tartar, 573, 577
tartaric acid, 17
Tatum, Edward, 5*f*, 161
taxonomy
  definition of, 237
  hierarchies in, 238–239, 238*t*
  numerical, 252
  principles of, 238–241
  viral, 308–310
T$_C$ cell, 354, 357
TCA. *See* tricarboxylic acid (TCA)
TCE. *See* trichloroethylene (TCE)
T-cell receptor (TCR), 356, 368
TCR. *See* T-cell receptor (TCR)
T-dependent antigens, 359, 365
technology, disease emergence and, 448–449
teeth, Hutchinson, 627, 627*f*
teichoic acids, 58
TEM. *See* transmission electron microscope (TEM)
Temin, Howard, 694
tempeh, 755*t*
temperate phages, 312–313, 312*f*
temperature
  in control of microbial growth, 111–116
  disease and, 90
  food preservation and, 90
  fungal growth and, 285
  microbial growth and, 89–90, 89*f*, 89*t*
  preservation and, 121
terbinafine, 477, 477*t*
terminal electron acceptor, 130, 725

terminator, 168, 168*f*
terrestrial environment, 266–269, 724
terrorism, 3, 5*f*, 453, 497, 536
tertiary consumer, 719*f*
tertiary structure, of protein, 27, 28*f*
tertiary syphilis, 627, 628
testicles
  in male genital system, 612, 613*f*
  in mumps, 583
tetanospasmin, 556
tetanus
  age and, 443, 443*f*
  causative agent, 275*t*, 555–556
  epidemiology of, 556
  history of, 548
  neonatal, 557–558
  pathogenesis of, 555–556
  prevention of, 557
  signs and symptoms, 555
  treatment of, 556–557
tetanus immune globulin (TIG), 555
tetanus vaccine, 422*t*, 425*t*, 443, 556, 557*t*
tetracyclines, 461*f*, 462*t*, 465*f*, 466, 499*t*, 500, 561, 679
tetrahydropyrimidines, 479*t*
textiles, healthcare-associated infections and, 452
textured protein products, 760
TF. *See* trickling filter (TF)
T$_H$ cell, 354, 357
Thayer-Martin medium, 94*t*, 95
therapeutic index, 459
therapeutic vaccine, for HIV, 704
thermophile
  characteristics of, 89*t*
  definition of, 90
  extreme, 278–279, 279*f*
Thermoplasma, 279
Thermoplasma acidophilum, 279
Thermotoga maritima, 250
Thermus, 257*t*, 265
Thermus aquaticus, 228
thiabendazole, 479*t*
thiamine, 136*t*, 574
thin-sectioning, 45
Thiobacillus, 257*t*, 728
Thiodictyon, 257*t*, 260
Thiomargarita, 267*t*
Thiomargarita namibiensis, 14, 14*f*, 273, 728
Thioploca, 267*t*, 728
Thiospirillum, 257*t*, 260
Thiothrix, 257*t*, 262, 263, 263*f*, 728
threadworm, 297*t*
three-domain system, 239, 239*f*
3′ end, of DNA, 163, 163*f*
3TC. *See* lamivudine (3TC)
throat
  normal microbiota in, 381*f*
  in respiratory system, 484
Thucydides, 380